(Continued on inside back cover)

Encyclopedia of Plant and Crop Science

edited by

Robert M. Goodman
University of Wisconsin—Madison
Madison, Wisconsin, U.S.A.

MARCEL DEKKER, INC. NEW YORK • BASEL

ISBN: Print: 0-8247-0944-6
ISBN: Online: 0-8247-0943-8
ISBN: Print/Online: 0-8247-4268-0

Library of Congress Cataloging-in-Publication Data
A catalog record of this book is available from the Library of Congress

This book is printed on acid-free paper.

Headquarters
Marcel Dekker, Inc.
270 Madison Avenue, New York, NY 10016
tel: 212-696-9000; fax: 212-685-4540

Eastern Hemisphere Distribution
Marcel Dekker AG
Hutgasse 4, Postfach 812, CH-4001 Basel, Switzerland
tel: 41-61-260-6300; fax: 41-61-260-6333

World Wide Web
http://www.dekker.com

The publisher offers discount on this book when ordered in bulk quantities. For more information, write to Special Sales/ Professional Marketing

Current printing (last digit):
10 9 8 7 6 5 4 3 2

PRINTED IN THE UNITED STATES OF AMERICA

Robert M. Goodman, Editor
University of Wisconsin—Madison
Madison, Wisconsin, U.S.A.

Topical Editor List

Atmosphere and Pollution
Jürg Fuhrer — *Research Leader, Swiss Federal Research Station for Agroecology and Agriculture, Zurich, Switzerland*

Bacterial Diseases
Timothy P. Denny — *Professor of Plant Pathology, University of Georgia, Athens, Georgia, U.S.A.*

Ronald Walcott — *Assistant Professor of Plant Pathology, University of Georgia, Athens, Georgia, U.S.A.*

Biochemistry and Primary Metabolism
Florencio E. Podestá — *Professor and Independent Researcher, Facultad de Ciencias Bioquímicas y Farmacéuticas, Universidad Nacional de Rosario and Centro de Estudios Fotosintéticos y Bioquímicos (CONICET), Rosario, Argentina*

Breeding Methods
William F. Tracy — *Professor of Agronomy, University of Wisconsin—Madison, Madison, Wisconsin, U.S.A.*

Crop Evolution and Domestication
Paul Gepts — *Professor of Agronomy in the Department of Agronomy and Range Science, University of California, Davis, California, U.S.A.*

Cropping Systems
Kenneth G. Cassman — *Professor and Department Head of the Agronomy and Horticulture Department, University of Nebraska, Lincoln, Nebraska, U.S.A.*

Crops and Human History
Garrison Wilkes — *Professor of Biology, University of Massachusetts, Boston, Massachusetts, U.S.A.*

Crops and Industrial Uses
Ken Hunt — *Chief Executive Officer, Anawah Inc., Seattle, Washington, U.S.A.*

Cytogenetics
Bernd Friebe — *Research Professor of Plant Pathology, Kansas State University, Manhattan, Kansas, U.S.A.*

Bikram S. Gill — *Distinguished Professor of Plant Pathology and Director of Wheat Genetics Resource Center, Kansas State University, Manhattan, Kansas, U.S.A.*

Fungal and Oomycete Diseases
Anne E. Dorrance — *Associate Professor of Plant Pathology, Ohio State University, Ohio Agricultural Research and Development Center, Wooster, Ohio, U.S.A.*

Sophien Kamoun — *Associate Professor of Plant Pathology, Ohio State University, Ohio Agricultural Research and Development Center, Wooster, Ohio, U.S.A.*

Genetic Engineering and Biosafety
Emily E. Pullins — *Research Associate for the Institute for Social, Economic, and Ecological Sustainability, University of Minnesota, St. Paul, Minnesota, U.S.A.*

Genetic Engineering and Transgenic Crops
Qifa Zhang — *Professor, Huazhong Agricultural University and National Key Laboratory of Crop Genetic Improvement, Wuhan, P.R. China*

Genetic Resource Conservation
Adi B. Damania — *Data Analyst for the Genetic Resources Conservation Program, University of California, Davis, California, U.S.A.*

Calvin O. Qualset — *Professor and Director Emeritus of the Genetic Resources Conservation Program, University of California, Davis, California, U.S.A.*

Genetic Resources and Germplasm
Wanda W. Collins — *Director of the Plant Sciences Institute, USDA—ARS, Beltsville, Maryland, U.S.A.*

Herbivores and Insects
Marcos Kogan — *Professor Emeritus of Entomology and Director Emeritus of the Integrated Plant Protection Center, Oregon State University, Corvallis, Oregon, U.S.A.*

Sujaya Rao — *Assistant Professor of Entomology, Oregon State University, Corvallis, Oregon, U.S.A.*

Leaves and Phyllosphere
George W. Sundin — *Assistant Professor of Plant Pathology, Michigan State University, East Lansing, Michigan, U.S.A.*

Light and Photoperiod
Emmanuel Liscum — *Associate Professor of Biological Sciences, University of Missouri, Columbia, Missouri, U.S.A.*

Nematodes
Thomas J. Baum — *Associate Professor of Plant Pathology, Iowa State University, Ames, Iowa, U.S.A.*

Photosynthesis
J. Kenneth Hoober — *Professor of Plant Biology at the School of Life Sciences, Arizona State University, Tempe, Arizona, U.S.A.*

Phytochemistry
John Markwell — *Professor of Biochemistry and Agronomy and Horticulture, University of Nebraska, Lincoln, Nebraska, U.S.A.*

Plant Ecology
Fernando Valladares — *Researcher, Centro de Ciencias Medioambientales, Consejo Superior de Investigaciones Cientificas, Madrid, Spain*

Plant Evolution
Michael T. Clegg — *Distinguished Professor of Genetics at the Department of Botany and Plant Sciences, University of California, Riverside, California, U.S.A.*

Plant Nutrition and Nutrient Deficiencies
David A. Dalton — *Professor of Biology, Reed College, Portland, Oregon, U.S.A.*

Population and Quantitative Genetics
Shawn M. Kaeppler — *Professor of Agronomy, University of Wisconsin, Madison, Wisconsin, U.S.A.*

Ronald L. Phillips — *Regents' Professor of Agronomy and Plant Genetics, University of Minnesota, St. Paul, Minnesota, U.S.A.*

Reproductive Biology
Imran Siddiqi — *Scientist, Center for Cellular and Molecular Biology, Hyderabad, India*

Roots and Rhizosphere
Roberto Pinton — *Professor of Plant Nutrition, University of Udine, Udine, Italy*

Seeds
Monica Mezzalama — *Plant Pathologist, CIMMYT Seed Health Unit, Mexico City, Mexico*

Matthew Reynolds — *Head of Wheat Physiology, CIMMYT, Mexico City, Mexico*

Somatic Genetics and Asexual Reproduction

Prakash Lakshmanan — *Senior Scientist at David North Plant Research Centre, Bureau of Sugar Experiment Stations Limited, Brisbane, Australia*

Acram M. Taji — *Professor and Chair of Horticultural Science and Plant Biotechnology, University of New England, Armidale, New South Wales, Australia*

Stress Responses

H. R. Lerner — *Professor of Botany, Institute of Life Sciences at the Hebrew University, Jerusalem, Israel*

Structural and Functional Genomics

Michael Fromm — *Director of the Center for Biotechnology, University of Nebraska, Lincoln, Nebraska, U.S.A.*

Sustainable Agriculture

Frederick Kirschenmann — *Director of the Leopold Center for Sustainable Agriculture, Iowa State University, Ames, Iowa, U.S.A.*

Transgenic Crops for Human Health

Lokesh Joshi — *Associate Professor at Harrington Department of Bioengineering and Arizona Biodesign Institute, Arizona State University, Tempe, Arizona, U.S.A.*

Viral Diseases

Steven A. Lommel — *Assistant Vice-Chancellor for Research and Professor of Plant Pathology, North Carolina State University, Raleigh, North Carolina, U.S.A.*

Water Relations

Neil C. Turner — *Chief Research Scientist at the Centre for Environment and Life Sciences, Commonwealth Scientific and Industrial Research Organization, Wembley (Perth), Australia*

Weeds

Lynn Fandrich — *Department of Crop and Soil Science, Oregon State University, Corvallis, Oregon, U.S.A.*

Carol Mallory-Smith — *Professor of Weed Science, Oregon State University, Corvallis, Oregon, U.S.A.*

List of Contributors

William W. Adams III / *University of Colorado, Boulder, Colorado, U.S.A.*
George C. Allen / *North Carolina State University, Raleigh, North Carolina, U.S.A.*
Leon Hartwell Allen, Jr. / *USDA, Gainesville, Florida, U.S.A.*
G. Nissim Amzallag / *Judean Regional Center for Biological Research, Carmel, Israel*
Radhamani Anandalakshmi / *Yale University, New Haven, Connecticut, U.S.A.*
M. Kat Anderson / *National Resources Conservation Service, University of California, Davis, California, U.S.A.*
Kim A. Anderson / *Oregon State University, Corvallis, Oregon, U.S.A.*
John H. Andrews / *University of Wisconsin, Madison, Wisconsin, U.S.A.*
Murthi Anishetty / *Food and Agriculture Organization of the United Nations, Rome, Italy*
P. Anthony / *University of Nottingham, Loughborough, U.K.*
David L. Asch / *New York State Museum, Albany, New York, U.S.A.*
Hiroshi Ashihara / *Ochanomizu University, Tokyo, Japan*
Mirna Atallah / *Leiden University, Wassenaarseweg, Leiden, The Netherlands*
Joseph E. Atchison / *Atchison Consultants, Inc., Sarasota, Florida, U.S.A.*
T. H. Attridge / *University of East London, Stratford, U.K.*
Daphné Autran / *CINVESTAV—Unidad Irapuato, Irapuato, Mexico*
Adelbert Bacher / *Technische Universität München, Garching, Germany*
P. Stephen Baenziger / *University of Nebraska–Lincoln, Lincoln, Nebraska, U.S.A.*
Luis Balaguer / *Universidad Complutense de Madrid, Madrid, Spain*
James G. Baldwin / *University of California, Riverside, California, U.S.A.*
Carlos L. Ballaré / *Universidad de Buenos Aires, Buenos Aires, Argentina*
Ian Bancroft / *John Innes Centre, Norwich, U.K.*
Ofer Bar-Yosef / *Harvard University, Cambridge, Massachusetts, U.S.A.*
K. R. Barker / *North Carolina State University, Raleigh, North Carolina, U.S.A.*
Jan Bartoš / *Institute of Experimental Botany, Olomouc, Czech Republic*
Gwyn A. Beattie / *Iowa State University, Ames, Iowa, U.S.A.*
D. K. Becker / *Queensland University of Technology, Brisbane, Queensland, Australia*
Philip W. Becraft / *Iowa State University, Ames, Iowa, U.S.A.*
Carl F. Behr / *Wyffels Hybrids Inc., Geneseo, Illinois, U.S.A.*
Robin R. Bellinder / *Cornell University, Ithaca, New York, U.S.A.*
James N. BeMiller / *Purdue University, West Lafayette, Indiana, U.S.A.*
Michael D. Bennett / *Royal Botanic Gardens Kew, Surrey, U.K.*
Tanya Z. Berardini / *Carnegie Institution of Washington, Stanford, California, U.S.A.*
Sant S. Bhojwani / *University of Delhi, Delhi, India*
Paul R. J. Birch / *Scottish Crop Research Institute, Dundee, U.K.*
James A. Birchler / *University of Missouri, Columbia, Missouri, U.S.A.*

M. Deane Bowers / *University of Colorado, Boulder, Colorado, U.S.A.*
Joseph J. Bozell / *National Renewable Energy Laboratory, Golden, Colorado, U.S.A.*
D. S. Brar / *International Rice Research Institute, Los Baños, Laguna, Philippines*
Donald P. Briskin / *University of Illinois, Urbana, Illinois, U.S.A.*
Floor Brouwer / *Agricultural Economics Research Institute, Wageningen, The Hague, The Netherlands*
Patrick H. Brown / *University of California, Davis, California, U.S.A.*
Hikmet Budak / *University of Nebraska—Lincoln, Lincoln, Nebraska, U.S.A.*
Thomas J. Burr / *Cornell University, Geneva, New York, U.S.A.*
Jennifer Lee Busto / *University of Hawaii at Manoa, Honolulu, Hawaii, U.S.A.*
Frederick H. Buttel / *University of Wisconsin, Madison, Wisconsin, U.S.A.*
J. Caballero / *UNAM, Morelia, Michoacán, Mexico*
B. Todd Campbell / *California Cooperative Rice Research Foundation, Biggs, California, U.S.A.*
W. Zacheus Cande / *University of California, Berkeley, California, U.S.A.*
Curgonio Cappelli / *Dipartimento di Arboricoltura e Protezione delle Piante, Perugia, Italy*
Isabelle A. Carré / *University of Warwick, Coventry, U.K.*
Shelly A. Carter / *Samuel Roberts Noble Foundation, Inc., Ardmore, Oklahoma, U.S.A.*
Alexandra M. Casa / *Cornell University, Ithaca, New York, U.S.A.*
A. Casas / *UNAM, Morelia, Michoacán, Mexico*
Kenneth G. Cassman / *University of Nebraska, Lincoln, Nebraska, U.S.A.*
Heriberto Cerutti / *University of Nebraska, Lincoln, Nebraska, U.S.A.*
P. K. Chakrabarty / *University of Florida, Gainesville, Florida, U.S.A.*
A. Chatterjee / *University of Missouri, Columbia, Missouri, U.S.A.*
A. K. Chatterjee / *University of Missouri, Columbia, Missouri, U.S.A.*
David J. Chitwood / *USDA—ARS, Beltsville, Maryland, U.S.A.*
Michael J. Christoffers / *North Dakota State University, Fargo, North Dakota, U.S.A.*
Pier Giorgio Cionini / *Università di Perugia, Perugia, Italy*
Michael T. Clegg / *University of California, Riverside, California, U.S.A.*
David A. Cleveland / *University of California, Santa Barbara, California, U.S.A. and Center for People, Food and Environment, Santa Barbara, California, U.S.A.*
Michael B. Cohen / *University of Alberta, Edmonton, Alberta, Canada*
Joseph Colasanti / *University of Guelph, Guelph, Ontario, Canada*
Anthony G. Condon / *CSIRO Plant Industry, Canberra, Australia*
Patricia L. Conklin / *State University of New York, Cortland, New York, U.S.A.*
David J. Connor / *University of Melbourne, Melbourne, Victoria, Australia*
R. James Cook / *Washington State University, Pullman, Washington, U.S.A.*
David E. L. Cooke / *Scottish Crop Research Institute, Dundee, U.K.*
N. M. Cooley / *CSIRO (Plant Industry), Victoria, Australia*
James E. Cooper / *Queen's University, Belfast, U.K.*
James G. Coors / *University of Wisconsin, Madison, Wisconsin, U.S.A.*
Bart Cottyn / *International Rice Research Institute, Los Baños, Philippines*
Max Crandall / *ESRI, Redlands, California, U.S.A.*
Ian Crute / *Institute of Arable Crops Research, Rothamsted, Harpenden, U.K.*
Christopher A. Cullis / *Case Western Reserve University, Cleveland, Ohio, U.S.A.*
James N. Culver / *University of Maryland Biotechnology Institute, College Park, Maryland, U.S.A.*
Diane A. Cuppels / *Southern Crop Protection and Food Research Centre, London, Canada*
John C. Cushman / *University of Nevada, Reno, Nevada, U.S.A.*

David A. Dalton / *Reed College, Portland, Oregon, U.S.A.*

Henry Daniell / *University of Central Florida, Orlando, Florida, U.S.A.*

Ana C. R. da Silva / *Alellyx Applied Genomics, Campinas, São Paulo, Brazil*

Karabi Datta / *International Rice Research Institute, Metro Manila, Philippines*

Swapan K. Datta / *International Rice Research Institute, Metro Manila, Philippines*

M. R. Davey / *University of Nottingham, Loughborough, U.K.*

Kevin M. Davies / *New Zealand Institute for Crop and Food Research Limited, Palmerston North, New Zealand*

Eric L. Davis / *North Carolina State University, Raleigh, North Carolina, U.S.A.*

Peter de Haan / *Phytovation B.V., Leiden, The Netherlands*

Wil de Jong / *Center for International Forestry Research, Bogor, Indonesia*

Hans de Jong / *Wageningen University, Wageningen, The Netherlands*

A. Delgado-Salinas / *UNAM, Morelia, Michoacán, Mexico*

Barbara Demmig-Adams / *University of Colorado, Boulder, Colorado, U.S.A.*

Barbara Despres / *University of Lausanne, Lausanne, Switzerland*

David A. Dierig / *USDA—ARS, Phoenix, Arizona, U.S.A.*

Brian W. Diers / *University of Illinois, Urbana, Illinois, U.S.A.*

S. P. Dinesh-Kumar / *Yale University, New Haven, Connecticut, U.S.A.*

Xin Shun Ding / *Samuel Roberts Noble Foundation, Inc., Ardmore, Oklahoma, U.S.A.*

Achim Dobermann / *University of Nebraska, Lincoln, Nebraska, U.S.A.*

John Dodds / *Dodds & Associates, Washington, D.C., U.S.A.*

Jaroslav Doležel / *Institute of Experimental Botany, Olomouc, Czech Republic*

K. Dong / *California Department of Food and Agriculture, Sacramento, California, U.S.A.*

María Fabiana Drincovich / *Rosario National University, Rosario, Argentina*

Natalia Dudareva / *Purdue University, West Lafayette, Indiana, U.S.A.*

Dénes Dudits / *Hungarian Academy of Sciences, Szeged, Hungary*

John W. Dudley / *University of Illinois, Urbana, Illinois, U.S.A.*

Robert J. Dufault / *Clemson University, Charleston, South Carolina, U.S.A.*

Stephen O. Duke / *USDA, University, Mississippi, U.S.A.*

Larry Duncan / *University of Florida—IFAS, Lake Alfred, Florida, U.S.A.*

Jim M. Dunwell / *University of Reading, Reading, U.K.*

Lindsey J. du Toit / *Washington State University, Mount Vernon, Washington, U.S.A.*

Jan Dvorak / *University of California, Davis, California, U.S.A.*

Ismail M. Dweikat / *University of Nebraska, Lincoln, Nebraska, U.S.A.*

Volker Ebbert / *University of Colorado, Boulder, Colorado, U.S.A.*

Jode W. Edwards / *University of Alabama, Birmingham, Alabama, U.S.A.*

Sanford D. Eigenbrode / *University of Idaho, Moscow, Idaho, U.S.A.*

Wolfgang Eisenreich / *Technische Universität München, Garching, Germany*

Wolter Elbersen / *Agrotechnological Research Institute, Bornsesteeg, Wageningen, The Netherlands*

Neil J. Emans / *Institute for Molecular Biotechnology, Aachen, Germany*

Takashi R. Endo / *Kyoto University, Kyoto, Japan*

Florent Engelmann / *Centre de Coopération Internationale en Recherche Agronomique pour le Développement, Montpellier, France*

Mustafa Erayman / *Washington State University, Pullman, Washington, U.S.A.*

Ed Etxeberria / *University of Florida, Lake Alfred, Florida, U.S.A.*

Bryce W. Falk / *University of California, Davis, California, U.S.A.*

Lynn Fandrich / *Oregon State University, Corvallis, Oregon, U.S.A.*

Andreas Fangmeier / *University of Hohenheim, Stuttgart, Germany*
Attila Fehér / *Hungarian Academy of Sciences, Szeged, Hungary*
Howard Ferris / *University of California, Davis, California, U.S.A.*
Rainer Fischer / *Fraunhofer Institute for Molecular Biology and Applied Ecology (IME), Aachen, Germany*
C. Robert Flynn / *Arizona Biodesign Institute, Arizona State University, Tempe, Arizona, U.S.A.*
Britta Förster / *Australian National University, Canberra, Australian Capital Territory, Australia*
Michael A. Foster / *Texas A&M University, Pecos, Texas, U.S.A.*
Paul Fransz / *University of Amsterdam, Amsterdam, The Netherlands*
Wayne Frasch / *Arizona State University, Tempe, Arizona, U.S.A.*
Michael E. Fromm / *University of Nebraska, Lincoln, Nebraska, U.S.A.*
David R. Gang / *University of Arizona, Tucson, Arizona, U.S.A.*
Amanda J. Garris / *Cornell University, Ithaca, New York, U.S.A.*
W. Gassmann / *University of Missouri, Columbia, Missouri, U.S.A.*
Paul Gepts / *University of California, Davis, California, U.S.A.*
Godelieve Gheysen / *Ghent University, Ghent, Belgium*
Alfons Gierl / *Technische Universität München, Garching, Germany*
Mark Gijzen / *Agriculture and Agri-Food Canada, London, Ontario, Canada*
Robert L. Gilbertson / *University of California, Davis, California, U.S.A.*
Ken Giller / *Wageningen University, Wageningen, The Netherlands*
Cedric Gillott / *University of Saskatchewan, Saskatoon, Canada*
I. L. Goldman / *University of Wisconsin, Madison, Wisconsin, U.S.A.*
Pierre Goloubinoff / *University of Lausanne, Lausanne, Switzerland*
M. C. Gopinathan / *E.I.D. Parry Ltd., Bangalore, India*
Janet R. Gorst / *Benson Micropropagation, Brisbane, Queensland, Australia*
Richard H. Grant / *Purdue University, West Lafayette, Indiana, U.S.A.*
Michael W. Gray / *Dalhousie University, Halifax, Canada*
Thusitha S. Gunasekera / *Michigan State University, East Lansing, Michigan, U.S.A.*
P. K. Gupta / *Ch. Charan Singh University, Meerut, India*
Vincent P. Gutschick / *New Mexico State University, Las Cruces, New Mexico, U.S.A.*
Arlene Haffa / *Arizona State University, Tempe, Arizona, U.S.A.*
Matthias Hahn / *University of Kaiserslautern, Kaiserslautern, Germany*
Arnel R. Hallauer / *Iowa State University, Ames, Iowa, U.S.A.*
Karen J. Halliday / *University of Bristol, Bristol, U.K.*
Adrienne R. Hardham / *Australian National University, Canberra, Australian Capital Territory, Australia*
Lisa C. Harper / *University of California, Berkeley, California, U.S.A.*
David R. Harris / *University College London, London, U.K.*
John P. Hart / *New York State Museum, Albany, New York, U.S.A.*
John S. Hartung / *USDA—ARS, Beltsville, Maryland, U.S.A.*
Wolfram Hartung / *Universität Würzburg, Würzburg, Germany*
Jacqueline Heard / *Mendel Biotechnology, Hayward, California, U.S.A.*
Murray J. Hill / *New Zealand Seed Technology Institute, Lincoln University, Canterbury, New Zealand*
Philippe Hinsinger / *INRA–ENSA. M, Montpellier, France*
Ann M. Hirsch / *University of California, Los Angeles, California, U.S.A.*
Simon J. Hiscock / *University of Bristol, Bristol, U.K.*
Peter R. Hobbs / *Cornell University, Ithaca, New York, U.S.A.*
Toby Hodgkin / *International Plant Genetic Resource Institute, Rome, Italy*

James B. Holland / *USDA—ARS and North Carolina State University, Raleigh, North Carolina, U.S.A.*
Irene Holst / *Smithsonian Tropical Research Institute, Balboa, Republic of Panama*
David Honys / *Academy of Sciences of the Czech Republic, Praha, Czech Republic*
J. Kenneth Hoober / *Arizona State University, Tempe, Arizona, U.S.A.*
Andreas Houben / *Institute for Plant Genetics and Crop Plant Research, Gatersleben, Germany*
R. A. Hufbauer / *Colorado State University, Fort Collins, Colorado, U.S.A.*
Roger Hull / *John Innes Centre, Colney, Norwich, U.K.*
Richard S. Hussey / *University of Georgia, Athens, Georgia, U.S.A.*
Alberto A. Iglesias / *Instituto Tecnológico de Chascomús (IIB-INTECH), Buenos Aires, Argentina*
Debra Ann Inglis / *Washington State University, Mount Vernon, Washington, U.S.A.*
Terry A. Isbell / *USDA—ARS, Peoria, Illinois, U.S.A.*
Thomas Jack / *Dartmouth College, Hanover, New Hampshire, U.S.A.*
Jaap D. Janse / *Plant Protection Service, Wageningen, The Netherlands*
Johnie N. Jenkins / *USDA—ARS, Mississippi State, Mississippi, U.S.A.*
R. N. Jones / *University of Wales, Aberystwyth, Wales, U.K.*
Stephen S. Jones / *Washington State University, Pullman, Washington, U.S.A.*
Brian R. Jordan / *Lincoln University, Canterbury, New Zealand*
Lokesh Joshi / *Arizona Biodesign Institute, Arizona State University, Tempe, Arizona, U.S.A.*
Brian J. Just / *University of Wisconsin, Madison, Wisconsin, U.S.A.*
Jean-Claude Kader / *Université Pierre et Marie Curie, Paris, France*
Clarence I. Kado / *University of California, Davis, California, U.S.A.*
Nicholas Kalaitzandonakes / *University of Missouri, Columbia, Missouri, U.S.A.*
Faina D. Kamilova / *Leiden University, Institute of Biology, Leiden, The Netherlands*
Manjit S. Kang / *Louisiana State University, Baton Rouge, Louisiana, U.S.A.*
Lawrence Kaplan / *University of Massachusetts, Boston, Massachusetts, U.S.A.*
Anne R. Kapuscinski / *University of Minnesota, St. Paul, Minnesota, U.S.A.*
Douglas L. Karlen / *USDA—ARS, Ames, Iowa, U.S.A.*
Akio Kato / *University of Missouri, Columbia, Missouri, U.S.A.*
James D. Kaufman / *University of Missouri, Columbia, Missouri, U.S.A.*
Jagreet Kaur / *Centre for Cellular and Molecular Biology, Hyderabad, India*
Lou Ellen Kay / *New Mexico State University, Las Cruces, New Mexico, U.S.A.*
Nancy Keller / *University of Wisconsin, Madison, Wisconsin, U.S.A.*
Brian R. Kerry / *Rothamsted Research, Herts, U.K.*
Terrie Klinger / *University of Washington, Seattle, Washington, U.S.A.*
Herbert D. Knudsen / *Natural Fibers Corporation, Ogallala, Nebraska, U.S.A.*
Yasunori Koda / *Hokkaido University, Sapporo, Japan*
Marcos Kogan / *Oregon State University, Corvallis, Oregon, U.S.A.*
Roger T. Koide / *Pennsylvania State University, University Park, Pennsylvania, U.S.A.*
Stephen Kresovich / *Cornell University, Ithaca, New York, U.S.A.*
Jürgen Kreuzwieser / *University of Freiburg, Freiburg, Germany*
Lynn A. Kszos / *Oak Ridge National Laboratory, Oak Ridge, Tennessee, U.S.A.*
Marie Kubaláková / *Institute of Experimental Botany, Olomouc, Czech Republic*
Monto H. Kumagai / *University of Hawaii at Manoa, Honolulu, Hawaii, U.S.A.*
Monlin Kuo / *Iowa State University, Ames, Iowa, U.S.A.*
Ralf G. Kynast / *University of Minnesota, St. Paul, Minnesota, U.S.A.*
Gideon Ladizinsky / *Hebrew University, Rehovot, Israel*

Prakash Lakshmanan / *David North Plant Research Centre, Brisbane, Queensland, Australia*
Kendall R. Lamkey / *Iowa State University, Ames, Iowa, U.S.A.*
Kathryn D. Lardizabal / *Monsanto, Calgene Campus, Davis, California, U.S.A.*
Philip Larkin / *CSIRO Plant Industry, Canberra, Australia*
Brian M. Lawrence / *B.M. Lawrence Consultant Services, Winston-Salem, North Carolina, U.S.A.*
Peter J. Lea / *Lancaster University, Lancaster, U.K.*
James M. Lee / *Washington State University, Pullman, Washington, U.S.A.*
H. R. Lerner / *Hebrew University, Jerusalem, Israel*
Johan H. J. Leveau / *Netherlands Institute of Ecology (NIOO-KNAW), Heteren, The Netherlands*
Eli Levine / *Illinois Natural History Survey, Champaign, Illinois, U.S.A.*
Alex Levine / *The Hebrew University of Jerusalem, Jerusalem, Israel*
Stephen A. Lewis / *Clemson University, Clemson, South Carolina, U.S.A.*
W. Joe Lewis / *USDA—ARS, Tifton, Georgia, U.S.A.*
Chentao Lin / *University of California, Los Angeles, California, U.S.A.*
Su Lin / *Arizona State University, Tempe, Arizona, U.S.A.*
Steven Lindow / *University of California, Berkeley, California, U.S.A.*
Emmanuel Liscum / *University of Missouri, Columbia, Missouri, U.S.A.*
Horst Lörz / *Universität Hamburg, Hamburg, Germany*
J. E. Losey / *Cornell University, Ithaca, New York, U.S.A.*
K. C. Lowe / *University of Nottingham, Nottingham, U.K.*
Tracey L.-D. Lu / *Chinese University of Hong Kong, Hong Kong, P.R. China*
Ben J. J. Lugtenberg / *Leiden University, Institute of Biology, Leiden, The Netherlands*
Adam J. Lukaszewski / *University of California, Riverside, California, U.S.A.*
Michelle R. Lum / *University of California, Los Angeles, California, U.S.A.*
James M. Lynch / *Forest Research, Farnham, Surrey, U.K.*
Carol Mallory-Smith / *Oregon State University, Corvallis, Oregon, U.S.A.*
H. K. Manandhar / *Nepal Agricultural Research Council, Khumaltar, Lalitpur, Nepal*
Rajendra Marathe / *Yale University, New Haven, Connecticut, U.S.A.*
Clarissa J. Maroon-Lango / *USDA—ARS, Beltsville, Maryland, U.S.A.*
Hugh S. Mason / *Arizona State University, Tempe, Arizona, U.S.A.*
S. B. Mathur / *Danish Government Institute of Seed Pathology for Developing Countries, Copenhagen, Denmark*
C. Robertson McClung / *Dartmouth College, Hanover, New Hampshire, U.S.A.*
Timothy R. McDermott / *Montana State University, Bozeman, Montana, U.S.A.*
Bruce A. McDonald / *Swiss Federal Institute of Technology (ETH), Zürich, Switzerland*
Edward C. McGawley / *Louisiana State University, Baton Rouge, Louisiana, U.S.A.*
Samuel B. McLaughlin / *Oak Ridge National Laboratory, Oak Ridge, Tennessee, U.S.A.*
Patricia S. McManus / *University of Wisconsin, Madison, Wisconsin, U.S.A.*
Robert McSorley / *University of Florida, Gainesville, Florida, U.S.A.*
Johan Memelink / *Leiden University, Wassenaarseweg, Leiden, The Netherlands*
Tom Mew / *International Rice Research Institute, Los Baños, Philippines*
Jeffrey C. Miller / *Oregon State University, Corvallis, Oregon, U.S.A.*
Russell K. Monson / *University of Colorado, Boulder, Colorado, U.S.A.*
Tsafrir S. Mor / *Arizona State University, Tempe, Arizona, U.S.A.*
Mario R. Morales / *Purdue University, West Lafayette, Indiana, U.S.A.*
Don W. Morishita / *University of Idaho, Twin Falls, Idaho, U.S.A.*

Peter L. Morrell / *University of California, Riverside, California, U.S.A.*
J. Bradley Morris / *USDA—ARS, PGRCU, University of Georgia, Griffin, Georgia, U.S.A.*
Cindy E. Morris / *Institut National de la Recherche Agronomique, Montfavet, France*
Yasuhiko Mukai / *Osaka Kyoiku University, Osaka, Japan*
Teun Munnik / *University of Amsterdam, Amsterdam, The Netherlands*
Alistair J. Murdoch / *The University of Reading, Reading, U.K.*
Denis J. Murphy / *University of Glamorgan, Cardiff, U.K.*
Deland J. Myers, Sr. / *Iowa State University, Ames, Iowa, U.S.A.*
Paolo Nannipieri / *Università degli Studi di Firenze, Florence, Italy*
Tomás Naranjo / *Universidad Complutense, Madrid, Spain*
Eric B. Nelson / *Cornell University, Ithaca, New York, U.S.A.*
Richard S. Nelson / *Samuel Roberts Noble Foundation, Inc., Ardmore, Oklahoma, U.S.A.*
Lisa Nodzon / *University of Florida, Gainesville, Florida, U.S.A.*
John L. Norelli / *USDA—ARS, Appalachian Fruit Research Station, Kearneysville, West Virginia, U.S.A.*
J. J. Obrycki / *Iowa State University, Ames, Iowa, U.S.A.*
Ron J. Okagaki / *University of Minnesota, St. Paul, Minnesota, U.S.A.*
Jørgen E. Olesen / *Danish Institute of Agricultural Sciences, Tjele, Denmark*
Dawn M. Olson / *USDA—ARS, Tifton, Georgia, U.S.A.*
Rodomiro Ortiz / *International Institute of Tropical Agriculture, Ibadan, Nigeria*
Cheryl Palm / *Columbia University, Palisades, New York, U.S.A.*
Peter Palukaitis / *Scottish Crop Research Institute, Invergowrie, U.K.*
Alyssa Panitch / *Arizona Biodesign Institute, Arizona State University, Tempe, Arizona, U.S.A.*
Roberto Papa / *Università Politecnica delle Marche, Ancona, Italy*
Gyungsoon Park / *Purdue University, West Lafayette, Indiana, U.S.A.*
Rémy S. Pasquet / *IRD-ICIPE, Nairobi, Kenya*
Wojciech P. Pawlowski / *University of California, Berkeley, California, U.S.A.*
Victor Paz / *University of the Philippines, Diliman, Quezon City, Philippines*
Gregory J. Peel / *Purdue University, West Lafayette, Indiana, U.S.A.*
Alessandro Pellegrineschi / *Applied Biotechnology Center and CRC for Molecular Plant Breeding, CIMMYT, Mexico City, México*
P. R. Phifer / *U.S. Fish and Wildlife Service, Portland, Oregon, U.S.A.*
Ronald L. Phillips / *University of Minnesota, St. Paul, Minnesota, U.S.A.*
Gregory C. Phillips / *Arkansas State University, State University, Arkansas, U.S.A.*
Francis Pierce / *Washington State University, Prosser, Washington, U.S.A.*
David Pimentel / *Cornell University, Ithaca, New York, U.S.A.*
Roberto Pinton / *Università di Udine, Udine, Italy*
Jon K. Piper / *Bethel College, North Newton, Kansas, U.S.A.*
Dolores R. Piperno / *Smithsonian Tropical Research Institute, Balboa, Republic of Panama*
William C. Plaxton / *Queen's University, Kingston, Ontario, Canada*
Håkan Pleijel / *Göteborg University, Göteborg, Sweden*
Florencio E. Podestá / *Universidad Nacional de Rosario, Rosario, Argentina*
Barry J. Pogson / *Australian National University, Canberra, Australian Capital Territory, Australia*
Linda Pollak / *Iowa State University, Ames, Iowa, U.S.A.*
Valérie Poncet / *IRD, Montpellier, France*
Archie R. Portis / *USDA and University of Illinois, Urbana, Illinois, U.S.A.*
Alison A. Powell / *University of Aberdeen, Aberdeen, U.K.*

J. B. Power / *University of Nottingham, Loughborough, U.K.*
P. V. Vara Prasad / *University of Florida, Gainesville, Florida, U.S.A.*
Jules Pretty / *University of Essex, Colchester, U.K.*
Stephen A. Prior / *USDA—ARS National Soil Dynamics Laboratory, Auburn, Alabama, U.S.A.*
Seth G. Pritchard / *Belmont University, Nashville, Tennessee, U.S.A.*
Ian J. Puddephat / *Syngenta, Bracknell, Berkshire, U.K.*
Emily E. Pullins / *University of Minnesota, St. Paul, Minnesota, U.S.A.*
Alexander H. Purcell / *University of California, Berkeley, California, U.S.A.*
Nageswararao C. Rachaputi / *Queensland Department of Primary Industries, Kingaroy, Australia*
Agepati S. Raghavendra / *University of Hyderabad, Hyderabad, India*
Glen C. Rains / *University of Georgia, Tifton, Georgia, U.S.A.*
Sujaya Rao / *Oregon State University, Corvallis, Oregon, U.S.A.*
A. L. N. Rao / *University of California, Riverside, California, U.S.A.*
Dennis T. Ray / *University of Arizona, Tucson, Arizona, U.S.A.*
John P. Reganold / *Washington State University, Pullman, Washington, U.S.A.*
M. Ruiz Rejón / *Universidad de Granada, Granada, Spain*
Heinz Rennenberg / *University of Freiburg, Freiburg, Germany*
Seung Y. Rhee / *Carnegie Institution of Washington, Stanford, California, U.S.A.*
David Rhodes / *Purdue University, West Lafayette, Indiana, U.S.A.*
José Luis Riechmann / *California Institute of Technology, Pasadena, California, U.S.A.*
Christine M. H. Riffaud / *Institut National de la Recherche Agronomique, Montfavet, France*
Howard W. Rines / *University of Minnesota, St. Paul, Minnesota, U.S.A.*
Eevi Rintamäki / *University of Turku, Turku, Finland*
Thierry Robert / *Université Paris-Sud, Orsay, France*
A. V. Roberts / *University of East London, London, U.K.*
Philip A. Roberts / *University of California, Riverside, California, U.S.A.*
Jonathan Robinson / *Food and Agriculture Organization of the United Nations, Rome, Italy*
Raoul A. Robinson (Retired) / *Food and Agriculture Organization of the United Nations, Fergus, Ontario, Canada*
Thomas Roitsch / *Universität Würzburg, Würzburg, Germany*
Maria R. Rojas / *University of California, Davis, California, U.S.A.*
Martin Romantschuk / *University of Helsinki, Lahti, Finland*
Ray J. Rose / *The University of Newcastle, Callaghan, Australia*
Todd N. Rosenstiel / *University of Colorado, Boulder, Colorado, U.S.A.*
G. Brett Runion / *USDA—ARS National Soil Dynamics Laboratory, Auburn, Alabama, U.S.A.*
Darren O. Sage / *Horticulture Research International, Wellesbourne, Warwick, U.K.*
V. Sarasan / *Royal Botanic Gardens Kew, Surrey, U.K.*
Aboubakry Sarr / *Université Paris-Sud, Orsay, France*
Takuji Sasaki / *National Institute of Agrobiological Sciences, Tsukuba, Ibaraki, Japan*
Oliver Schabenberger / *SAS Institute Inc., Cary, North Carolina, U.S.A.*
Jennifer L. Schaeffer / *Oregon State University, Corvallis, Oregon, U.S.A.*
Stefan Schillberg / *Fraunhofer Institute for Molecular Biology and Applied Ecology (IME), Aachen, Germany*
Wolfgang Schmidt / *Humboldt University, Berlin, Germany*
Cor D. Schoen / *Plant Research International, Wageningen, The Netherlands*
Ingo Schubert / *Institute for Plant Genetics and Crop Plant Research, Gatersleben, Germany*
Mark Schwarzlaender / *University of Idaho, Moscow, Idaho, U.S.A.*
Liliya D. Serazetdinova / *Universität Hamburg, Hamburg, Germany*

João C. Setubal / *Alellyx Applied Genomics, Campinas, São Paulo, Brazil*
Miti M. Shah / *Arizona Biodesign Institute, Arizona State University, Tempe, Arizona, U.S.A.*
Henry L. Shands / *USDA–ARS National Center for Genetic Resources Preservation, Fort Collins, Colorado, U.S.A.*
John L. Sherwood / *University of Georgia, Athens, Georgia, U.S.A.*
Kazuo Shinozaki / *RIKEN Tsukuba Institute, Koyadai, Tsukuba, Japan*
Imran Siddiqi / *Centre for Cellular and Molecular Biology, Hyderabad, India*
Kadambot H. M. Siddique / *University of Western Australia, Crawley, Australia*
Philipp W. Simon / *USDA—ARS, Madison, Wisconsin, U.S.A.*
Sjef C. M. Smeekens / *University of Utrecht, Utrecht, The Netherlands*
Steven E. Smith / *University of Arizona, Tucson, Arizona, U.S.A.*
M. K. Smith / *Queensland Department of Primary Industries, Queensland, Australia*
C. Michael Smith / *Kansas State University, Manhattan, Kansas, U.S.A.*
Daniela Soleri / *University of California, Santa Barbara, California, U.S.A. and Center for People, Food and Environment, Santa Barbara, California, U.S.A.*
Susanne Someralo / *Dodds & Associates, Washington, D.C., U.S.A.*
Wen-Yuan Song / *University of Florida, Gainesville, Florida, U.S.A.*
Hermona Soreq / *The Hebrew University of Jerusalem, Jerusalem, Israel*
Hubert J. Spiertz / *Wageningen University, Wageningen, The Netherlands*
David M. Spooner / *University of Wisconsin, Madison, Wisconsin, U.S.A.*
David M. Stelly / *Texas A&M University, College Station, Texas, U.S.A.*
W. P. Stephen / *Oregon State University, Corvallis, Oregon, U.S.A.*
John C. Stier / *University of Wisconsin, Madison, Wisconsin, U.S.A.*
Deborah Stinner / *Ohio State University, Wooster, Ohio, U.S.A.*
Duška Stojšin / *Monsanto Company, St. Louis, Missouri, U.S.A.*
Bethany B. Stone / *University of Missouri, Columbia, Missouri, U.S.A.*
G. Stotzky / *New York University, New York, New York, U.S.A.*
Charles W. Stuber / *North Carolina State University, Raleigh, North Carolina, U.S.A.*
George W. Sundin / *Michigan State University, East Lansing, Michigan, U.S.A.*
Katerina P. Svoboda / *Scottish Agricultural College Auchincruive, Ayr, Scotland, U.K.*
Acram M. Taji / *University of New England, Armidale, New South Wales, Australia*
Christa Testerink / *University of Amsterdam, Amsterdam, The Netherlands*
K. Thirumala-Devi / *University of Wisconsin, Madison, Wisconsin, U.S.A.*
Paul B. Thompson / *Michigan State University, East Lansing, Michigan, U.S.A.*
William F. Tracy / *University of Wisconsin, Madison, Wisconsin, U.S.A.*
Patrick J. Tranel / *University of Illinois, Urbana, Illinois, U.S.A.*
A. Forrest Troyer / *University of Illinois, DeKalb, Illinois, U.S.A.*
Federico Trucco / *University of Illinois, Urbana, Illinois, U.S.A.*
Neil C. Turner / *CSIRO Plant Industry, Wembley, Australia*
David Twell / *University of Leicester, Leicester, U.K.*
Richard M. Twyman / *University of York, York, U.K.*
Gregory L. Tylka / *Iowa State University, Ames, Iowa, U.S.A.*
Fernando Valladares / *Centro de Ciencias Medioambientales, Madrid, Spain*
Estela M. Valle / *Facultad de Ciencias Bioquímicas y Farmacéuticas, Rosario, Argentina*
Carroll P. Vance / *USDA—ARS Plant Science Research Unit and University of Minnesota, St. Paul, Minnesota, U.S.A.*
Jan E. G. van Dam / *Agrotechnological Research Institute, Bornsesteeg, Wageningen, The Netherlands*
Jan M. van der Wolf / *Plant Research International, Wageningen, The Netherlands*
Karin van Dijk / *University of Nebraska, Lincoln, Nebraska, U.S.A.*

Zeno Varanini / *University of Udine, Udine, Italy*
Juan M. Vega / *University of Missouri, Columbia, Missouri, U.S.A.*
Richard E. Veilleux / *Virginia Polytechnic Institute & State University, Blacksburg, Virginia, U.S.A.*
Roberto Fontes Vieira / *Embrapa Recursos Genéticos e Biotecnologia, Brasilia, Brazil*
Jean-Philippe Vielle-Calzada / *CINVESTAV—Unidad Irapuato, Irapuato, Mexico*
Alan Vivian / *The University of the West of England, Bristol, U.K.*
Andreas von Tiedemann / *University of Göttingen, Göttingen, Germany*
Jan Vrána / *Institute of Experimental Botany, Olomouc, Czech Republic*
Linda L. Walling / *University of California, Riverside, California, U.S.A.*
Hans-Joachim Weigel / *Institute of Agroecology, Bundesallee, Braunschweig, Germany*
Jim Weller / *University of Tasmania, Hobart, Tasmania, Australia*
James H. Westwood / *Virginia Tech, Blacksburg, Virginia, U.S.A.*
George A. White (Retired) / *USDA—ARS, Beltsville, Maryland, U.S.A.*
Garry C. Whitelam / *University of Leicester, Leicester, U.K.*
Steven A. Whitham / *Iowa State University, Ames, Iowa, U.S.A.*
David M. Wilkinson / *Liverpool John Moores University, Liverpool, U.K.*
Jennifer S. Williams / *University of Queensland, St. Lucia, Queensland, Australia*
Karen A. Williams / *USDA, Beltsville, Maryland, U.S.A.*
David E. Williams / *International Plant Genetic Resources Institute, Cali, Colombia*
Robert D. Willows / *Macquarie University, Sydney, Australia*
Linda M. Wilson / *University of Idaho, Moscow, Idaho, U.S.A.*
Michael Wink / *Universität Heidelberg, Heidelberg, Germany*
J. R. Witcombe / *University of Wales, Bangor, Gwynedd, U.K.*
Barry Wobbes / *University of Utrecht, Utrecht, The Netherlands*
L. L. Wolfenbarger / *University of Nebraska, Omaha, Nebraska, U.S.A.*
John W. Worley / *University of Georgia, Athens, Georgia, U.S.A.*
Graeme C. Wright / *Queensland Department of Primary Industries, Kingaroy, Australia*
C. W. Wrigley / *Food Science and Wheat Australia CRC, North Ryde, Sydney, Australia*
Jin-Rong Xu / *Purdue University, West Lafayette, Indiana, U.S.A.*
Kazuko Yamaguchi-Shinozaki / *Japan International Research Center for Agricultural Sciences, Ohwashi, Tsukuba, Japan*
Charles F. Yocum / *University of Michigan, Ann Arbor, Michigan, U.S.A.*
John M. Young / *Landcare Research, Auckland, New Zealand*
Lucy Zakharova / *University of Missouri, Columbia, Missouri, U.S.A.*
Adel Zayed / *Paradigm Genetics, Inc., RTP, North Carolina, U.S.A.*
Qifa Zhang / *Huazhong Agricultural University, Wuhan, P.R. China*
Daniel Zohary / *Hebrew University of Jerusalem, Jerusalem, Israel*

Contents

Preface

Plants dominate today's biosphere. Because of their numbers, diversity, and ubiquity (on land and in water), plants are by far the most important primary producers on Earth. Through photosynthesis, plants control our atmosphere and capture the incident energy from the sun, which provides nearly all of the chemical energy required to support the entire food chain that makes up the vast biodiversity of life on Earth.

Plants not only support the food and planetary conditions required for life, but also provide a staggering array of natural products that are the basis of many of our medicinals, dyes, spices, plastics, and fine chemicals. The products of photosynthesis are also harvested to provide the fibers with which we make clothing, paper, and lumber. Photosynthesis by plants provides sources of fuel, both fossil (such as coal) and so-called biofuels (such as ethanol).

Agriculture is fundamental to human life on Earth. Crop production supports all aspects of agriculture, including production of livestock. There are an estimated 50,000 edible plant species, but fewer than 300 have been domesticated as crops. Industrial agriculture, with its dependence on machinery, chemicals, and monocropping, is widely practiced today and 90% of the world's land used for crop production is planted with just 15 crop species (the top 5 are rice, wheat, maize, soybeans, and cotton). Yet many other crops play essential roles in food security, nutrition, cultural tradition, and dietary diversity of the world's people.

An encyclopedia that focuses on both plants in general and crop plants is unique and necessary. While academically distinct, the fields of plant biology and crop sciences are interdependent, and so deserve mutual treatment in a single volume. One of the major developments in plant sciences in recent decades has been the discovery of common biological properties, from genomes to development and from metabolism to reproduction, that all plants share. Fundamental studies on model plants, such as *Arabidopsis thaliana*, are contributing to our practical understanding of *all* plants, as have studies on model crops such as rice and tomato. Lessons from plant evolution and ecology have found application in agriculture, and vice versa. In other words, although the subjects in this single reference are located in different academic departments, a goal of this encyclopedia is to place these ideas next to each other. Increasingly, scientists who study plants as botanists and those who study plants as crop scientists have found common ground and common interests, which has strengthened communication across formerly distant disciplines and the disciplines themselves.

The Encyclopedia of Plant and Crop Sciences represents the concerted effort of hundreds of scientists from around the world to present current knowledge about plant life on Earth. The overview format is intended to be accessible to a range of scientists, agriculturalists, policy makers, science writers, students, and the public. Our aim has been to provide authoritative, yet accessible, articles that canvass the major topics from evolution to molecular biology, from ecology to morphology, from physical systems to cropping systems. Selected references accompany every article, intended to help the interested reader delve further into each topic.

Preparing this volume has been a real team effort. I am especially grateful to Oona Schmid, Encyclopedias Editor and Supervisor, along with the able assistance of Gretchen Goode at Marcel Dekker, Inc., who have been delightful to work with and incredibly effective in every way. And I extend my sincere thanks to a superb team of topical editors, authors, and members of the editorial board for hanging in there and making this project a success.

To our readers, we offer you this reference. Should you find it useful, we'd like to hear why and how (EPCS@dekker.com). Should you find it wanting, we also, and especially, want to hear how you believe we can make it better. The publisher as well as the topical editors are committed to continual updating and refinement of what we believe is a valuable resource for transmitting knowledge and understanding about the plants that make life on Earth possible.

Robert M. Goodman
Professor, University of Wisconsin—Madison
August 2003
rgoodman@wisc.edu

Agriculture and Biodiversity

Deborah Stinner
Ohio State University, Wooster, Ohio, U.S.A.

A

INTRODUCTION

No other human activity has greater impact on the earth's biodiversity than agriculture. From its origins some 12,000 years ago, the goal of agriculturists has been to enhance production of desired species over competing species. Expansion of human agricultural activity around the globe historically has resulted in significant impacts on global biodiversity in four major ways: 1) loss of wild biodiversity and species shifts resulting from conversion of native ecosystems by agroecosystems; 2) influence of agroecosystems' structure and function on agrobiodiversity; 3) off-site impacts of agricultural practices; and 4) loss of genetic diversity among and within agricultural species.[1] Although agriculture and biodiversity often are inversely related, biodiversity enhancement can be a key organizing principle in sustainable agroecosystems.[2–4]

AGRICULTURE'S IMPACT ON WILD BIODIVERSITY GLOBALLY

Historically, the earliest subsistence farmers and pastoralists had low population densities and limited technology and their small-scale patchworks of fields, pastures, and home gardens had little net effect on global biodiversity.[2] In some ecosystems, agricultural activity may have actually increased biodiversity because more diverse habitats and ecotones were created—a pattern that may still exist in some areas.[2] However, as surplus agricultural production allowed human populations to increase and with the development of civilizations, the impacts of agriculture on wild biodiversity increased, even to the point that biodiversity loss may have contributed to the decline of some ancient civilizations.[2] Since 1650, there has been at least a 600% increase in the worldwide deforestation of native ecosystems for agriculture and wood extraction that have resulted in radical changes to wild biodiversity globally.[2]

Wild biodiversity is more threatened now than at any time since the extinction of the dinosaurs, with nearly 24% of all mammals, 12% of birds, and almost 14% of plants threatened with extinction.[2] If current trends continue, it is estimated that at least 25% of the earth's species could become extinct or drastically reduced by the middle of this century.[2] Conversion of natural ecosystems to agroecosystems is a primary cause of these alarming trends.[2,4,5] At least 28% of the earth's land area currently is devoted to agriculture to some degree.[1] Intensive agriculture dominates 10% of the earth's total land area and is part of the landscape mosaic on another 17%, while extensive grazing covers an additional 10%–20%.[2] Nearly half of the global temperate broadleaf and mixed-forest and tropical and subtropical dry and monsoon broadleaf forest ecosystems are converted to agricultural use (45.8% and 43.4%, respectively).[1,2] However, agriculture's greatest impact has been on grassland ecosystems, including temperate grasslands, savannas, and shrublands (34.2%); flooded grasslands and savannas (20.2%); and montane grasslands and shrublands (9.8%).[1] Combined, 64.2% of the earth's grassland ecosystems have been converted to agriculture, primarily for production of cereal grasses—maize, rice, and wheat.[1,2] In the past 20 years, net expansion of agricultural land has claimed approximately 130,000 km^2/yr globally, mostly at the expense of forest and grassland ecosystems, but also from wetlands and deserts.[1]

The native ecosystems that agriculture has replaced typically had high biodiversity. A hectare of tropical rain forest may contain over 100 species of trees and at least 10 to 30 animal species for every plant species, leading to estimates of 200,000 or more total species.[6] In contrast, the world's agroecosystems are dominated by some 12 species of grains, 23 vegetable crops, and about 35 fruit and nut crop species.[1] Furthermore, conversion of native ecosystems to agriculture causes dramatic shifts in ecosystem structure and function that affect ecosystem processes above and below ground including energy flow, nutrient cycling, water cycling, food web dynamics, and biodiversity at all trophic levels.[7] The amount of wild biodiversity loss depends on the degree of fragmentation of the native landscape. Whereas some species require vast continuous areas of native habitat, many can survive as long as the appropriate size and number of patches with connecting corridors of native habitat are left intact and provided that barriers to species movement—such as road and irrigation networks—are limited. However, when conversion leads to critical levels of native landscape fragmentation, chain reactions of biodiversity loss have been observed as interdependent species loose the resources they need to survive.[2] Loss of wild biodiversity

Encyclopedia of Plant and Crop Science
DOI: 10.1081/E-EPCS 120010502

at this level leads to loss of numerous ecosystem benefits that are essential to agriculture, e.g., 1) drought and flood mitigation; 2) soil erosion control and soil quality regeneration; 3) pollination of crops and natural vegetation; 4) nutrient cycling; and 5) control of most agricultural pests.[5]

STRUCTURE AND FUNCTION OF AGROECOSYSTEMS AND BIODIVERISTY

The structure and function of agroecosystems are largely determined by local context, including interaction of ecological conditions (including bio-, geo-, and chemical) with social factors, including farmers' economic needs, cultural and spiritual values, and social structure and technology. Two types of agrobiodiversity have been defined:[8] Planned biodiversity is the specific crops and/or livestock that are planted and managed; associated biodiversity is nonagricultural species that find the environment created by the production system compatible (e.g., weeds, insect and disease pests, predators and parasites of pest organisms, and symbiotic and mutualistic species).[8] Planned and associated biodiversity can enhance stability and predictability of agroecosystems.[5] Traditional forms of agriculture—such as home gardens and shade coffee farms in the New and Old World tropics[4,6,8] and traditional Amish dairy farms in North America[9]—have a complex and diverse spatial and vertical structure and high planned and associated biodiversity. For example, traditional neotropical agroforestry systems commonly contain over 100 annual and perennial plant species per field.[4] Traditional agroecosystems create landscape patterns of small-scale diverse patches with many edges, habitat patches, and corridors for wild biodiversity.

In contrast to traditional agroecosystems, the vertical and horizontal structure of modern industrial agroecosystems is simplified into monocultures on a large scale that create landscape patterns of widespread extreme genetic uniformity with few edges, habitat patches, and corridors for dispersal.[2] For example, in the United States 60–70% of the total soybean area is planted with 2–3 varieties, 72% of the potato area with four varieties, and 53% of the cotton area with three varieties.[6] The structure and function of industrial livestock agriculture impose similar negative impacts on biodiversity worldwide.[10] Livestock operations for all major species—particularly swine, poultry, beef, and dairy—are becoming increasingly concentrated, with feed produced in monocultures and brought to the animals in feedlots. Even in more extensive grazing operations, although good management can increase plant biodiversity,[3,10] these systems replace native forests and/or grasslands that once supported highly diverse complexes of coadapted plants and migratory grazing and browsing ungulates and their predators.[10]

OFFSITE IMPACTS OF AGRICULTURAL ACTIVITIES

The third major way that agriculture impacts global biodiversity is through the direct and indirect off-site effects of the various managements used to maintain their structure and function. Growing annual species in large monocultures goes against ecological forces of plant community succession; therefore, a great deal of intervention is required to maintain high levels of production. Fertilizers applied to maximize production of crop plants create favorable habitat for other plant species that are adapted to nutrient enriched conditions, including alien invasive species. Tillage, herbicides, and genetic engineering may prevent competition between crop plants and annual and perennial weeds. Widespread monocultures of nutrient enriched plants create an easily exploited resource for insect pests and disease organisms. Insecticides, fungicides, and genetic engineering may protect crops from these competitors. Furthermore, conventional cropping agroecosystems are notoriously leaky (i.e., the sheer volume of external inputs being applied in combination with soil disturbance and decreased soil quality often exceeds the capacity of the agroecosystem to absorb and process the inputs).[1] Concentrated livestock agriculture also can be a major source of chemical and biological pollution. As a result of these many factors, sediment, excess fertilizer, manure, and pesticides run off into streams and down into groundwater. Hydrological alterations to land and natural streams in combination with chemical and biological pollution cause considerable reductions in aquatic biodiversity that can extend throughout whole watershed systems.[2] The Hypoxia in the Gulf of Mexico is a dead zone that covers 18,000 km^2 where aquatic biodiversity has been drastically reduced by impacts of agriculture in the Mississippi River watershed.[2] A new concern regarding off-site impacts of modern agriculture on biodiversity is the genetic pollution that can result as genetically modified crops expand worldwide.[1] Possible transfer of genes for resistance to weeds, insects, fungi, and viruses could overwhelm wild populations and communities.[2]

LOSS OF DIVERSITY WITHIN AGRICULTURAL SPECIES

Of 7000 crop species, less than 2% are currently important, only 30 of which provide an estimated 90% of the world's calorie intake—with wheat, rice, and maize alone providing more than half of plant-derived calories.[1]

Some 30–40 animal species have been used extensively for agriculture worldwide, but fewer than 14 account for over 90% of global livestock production, whereas some 30% of international domesticated breeds are threatened with extinction.[1] There are additional trends of decreased varietal and landrace diversity within crop species as more farmers adopt modern high-yielding varieties.[1,4] These alarming trends have prompted government policy recommendations whose purpose is to: 1) ensure that current agricultural genetic diversity in plants is preserved in seed banks and plant and germplasm collections (ex situ) or as growing crops (in situ), particularly wild relatives of major crops and livestock breeds in their centers of origin; and 2) ensure that wild crop and livestock relatives are conserved in carefully identified natural systems.[5]

CONCLUSION

Biodiversity as a Principle of Agroecosystem Management

Although industrial agriculture is generally inversely related to biodiversity, there are promising examples of alternative agroecosystems that protect and enhance biodiversity and are also highly productive.[2–4] Some of these include: 1) organic agriculture; 2) sustainable agriculture; 3) permaculture; 4) natural system agriculture; 5) holistic management; and 6) ecoagriculture.[2,3] These models are based on ecological principles and the assumption that biodiversity can contribute significantly to sustainable agricultural production. Within ecoagriculture, the following strategies are proposed to protect and enhance wild biodiversity: 1) create biodiversity reserves that also benefit local farming communities; 2) develop habitat networks in nonfarmed areas; 3) reduce (or reverse) conversion of wild lands to agriculture by increasing farm productivity; 4) minimize agricultural pollution; 5) modify management of soil, water, and vegetation resources; and 6) modify farming systems to mimic natural ecosystems.[2] Examples of specific management practices that sustain or enhance biodiversity include: 1) hedgerows; 2) dykes with wild herbage; 3) polyculture; 4) agroforestry; 5) rotation with legumes; 6) dead and living mulches; 7) strip crops, ribbon cropping, and alley cropping; 8) minimum tillage, no-tillage, and ridge tillage; 9) mosaic landscape porosity; 10) organic farming; 11) biological pest control and integrated pest management; 12) plant resistance; and 13) germplasm diversity.[11] New research, particularly if conducted in a participatory mode with farmers, should lead to many more ways to protect and enhance biodiversity, including: rotational grazing of high-diversity grasslands for dairy and beef cattle production, timber and pulp production systems that use perennial plants, high-diversity mixtures of single annual crops and/or rotational diversity, and precision agriculture that closely matches small-scale soil conditions with optimal crop genotypes.[5] Although specific management practices are helpful, also needed are whole-farm planning approaches and decision-making processes that encompass farmers' values and economic needs, in addition to environmental concerns for biodiversity[2] (e.g., holistic management).[3] More

Fig. 1 Ecoagricultural strategies in practice. (*View this art in color at www.dekker.com.*)

research and education and policies that encourage farmers and consumers to appreciate the ecological, economic, and quality-of-life values of biodiversity are needed to thwart current threats to global biodiversity and agrobiodiversity, and to address the many challenges and opportunities of global sustainable food security.[4]

ARTICLES OF FURTHER INTEREST

Crop Domestication: Fate of Genetic Diversity, p. 333
Crop Improvement: Broadening the Genetic Base for, p. 343
Ecology and Agricultural Sciences, p. 401
Ecology: Functions, Patterns, and Evolution, p. 404
Environmental Concerns and Agricultural Policy, p. 418
Genetic Resource Conservation of Seeds, p. 499

REFERENCES

1. Wood, S.; Sebastian, K.; Scherr, S.J. *Pilot Analysis of Global Ecosystems: Agroecosystems*; International Food Policy Research Institute and World Resources Institute: Washington, DC, 2000.
2. McNeely, J.A.; Scherr, S.J. *Ecoagriculture: Strategies to Feed the World and Save Wild Biodiversity*; Island Press: Washington, 2003.
3. Stinner, D.H.; Stinner, B.R.; Martsolf, E. Biodiversity as an organizing principle in agroecosystem management: Case studies of holistic resource management practitioners in the USA. Agric. Ecosys. Environ. **1997**, *62*, 199–213.
4. http://www.wri.org/wri/sustag/lab-01.html (accessed February 2003).
5. http://www.cast-science.org/bid.pdf (assessed December 2002).
6. Altieri, M.A. The ecological role of biodiversity in agroecosystems. Agric. Ecosys. Environ. **1999**, *74*, 19–31.
7. Swift, M.J.; Anderson, J.M. Biodiversity and Ecosystem Function in Agricultural Systems. In *Biodiversity and Ecosystem Function*; Schuleze, E.D., Mooney, H.A., Eds.; Springer-Verlag: New York, 1993; 15–41.
8. Vandermeer, J. Biodiversity Loss in and Around Agroecosystems. In *Biodiversity and Human Health*; Grifo, F., Rosenthal, J., Eds.; Island Press: Washington, DC, 1997; 111–127.
9. Moore, R.H.; Stinner, D.H.; Kline, D.; Kline, E. Honoring Creation and Tending the Garden: Amish Views of Biodiversity. In *Cultural and Spiritual Values of Biodiversity*; Posey, D.A., Ed.; United Nations Environment Programme, Intermediate Technology Publications: London, 1999; 305–309.
10. Blackburn, H.W.; de Haan, C. Livestock and Biodiversity. In *Biodiversity in Agroecosystems*; Collins, W.W., Qualset, C.O., Eds.; CRC: New York, 1999; 85–99.
11. Paoletti, M.G.; Pimental, D.; Stinner, B.R.; Stinner, D. Agroecosystem Biodiversity: Matching Production and Conservation Biology. In *Biotic Diversity in Agroecosystems*; Paoletti, M.G., Pimental, D., Eds.; Elsevier: New York, 1992; 3–23.

Agriculture: Why and How Did It Begin?

A

David R. Harris
University College London, London, U.K.

INTRODUCTION

Today the world's population exceeds 6000 million people, almost all of whom depend on agriculture for their survival, and yet growing crops and raising domestic animals is a very recent development in the history of humanity. Anatomically modern humans—*Homo sapiens*—began to colonize the continents as foraging hunter-fisher-gatherers some 100,000 years ago, but it was not until about 12,000 years ago that farming began to replace foraging as the main mode of human subsistence. It did so first, and very gradually, in the so-called Fertile Crescent of Southwest Asia. In other regions of the world, such as central China, northern tropical Africa, and Mesoamerica, primary (independent) transitions from foraging to farming also occurred, even later than in Southwest Asia, but by 1500 A.D., when Europeans were beginning to expand overseas, most of the world's population (estimated at 350 million) had become dependent on agriculture.

From today's perspective, this late emergence and very gradual development of agriculture may seem surprising. It prompts the question, why did it not occur much earlier, or, conversely, why did humans remain dependent on hunting, fishing, and gathering for so long? This response stems from a deeply embedded and still prevalent assumption that the transition to agriculture was an inevitable stage in human progress. However, in the long perspective of humanity's foraging past the question that demands an answer is not why did agriculture not develop sooner, but why did it develop at all?

Before pursuing that question, we need to consider what precisely is meant by "agriculture," because failure to define it, and other related terms such as cultivation and domestication, has led to confusion in attempts to explain why and how agriculture arose. Here, cultivation is defined as the sowing or planting, tending, and harvesting of useful domesticated *or* wild plants, which may or may not involve tilling the soil. Domestication is defined as the genetic, physiological, and/or morphological alteration of wild plants that results from deliberate or inadvertent cultural selection and leads to the plants' dependence on humans for their long-term survival. Agriculture is defined as the growing of domesticated crops by methods of cultivation that usually but not always involve systematic tillage of the soil. The distinction between cultivation and agriculture is particularly important because it enables us to differentiate between systems of crop production practiced by farmers and systems of wild-plant production practiced by foragers. Having clarified this distinction, we need next to consider how foragers have cultivated wild plants to enhance their productivity.

PLANT CULTIVATION BY FORAGERS

Many historical and ethnographic accounts of "hunter-gatherers" show that they not only gathered wild plants but often increased the productivity of selected taxa by such methods as controlled burning; vegetation clearance and weeding; harvesting, storing, sowing, and planting seeds, tubers, cuttings, and other propagules, and tilling, draining, and irrigating the soil.[1] Such practices can be regarded as forms of cultivation, but, although they are sometimes described as "protoagricultural,"[2] they do not amount to agriculture (as here defined) because they rarely include fully domesticated crops. This distinction is not just semantic, because the ethnographic, historical, and more limited archaeological evidence we have indicates that cultivation by foragers was usually only a minor activity in their hunting-fishing-gathering systems of subsistence.

Although many forager groups engaged in small-scale cultivation, it did not normally, and certainly not inevitably, lead to full plant domestication and the development of agriculture. However, that this did occasionally occur is undeniable, for if it had not, agriculture would never have arisen. So we must next ask what factors may have caused particular forager groups to invest more time and effort in cultivation and to become progressively more dependent for food and other products on suites of plants that underwent domestication and were transformed into agricultural crops, thus initiating primary transitions from foraging to farming.

FACTORS THAT MAY HAVE PROMOTED PRIMARY TRANSITIONS TO AGRICULTURE

Many factors, singly or in combination, have been proposed as causal agents that could have promoted crop

Encyclopedia of Plant and Crop Science
DOI: 10.1081/E-EPCS 120017093

domestication and the rise of systems of agricultural production. They range from factors external to the human groups concerned, such as natural climatic and vegetational changes, to social-behavioral ones such as reductions in seasonal mobility associated with year-round, long-term occupation of settlements (sedentism) and the elaboration of storage techniques; population pressure; competition for scarce resources; differential access to food and other products associated with the development of social ranking; exchange and trade; and technological innovation.

Underlying the discussion of the relative importance of such factors is the more general question of whether the earliest foragers to develop agriculture were pressured into doing so by factors that induced subsistence stress or whether the process was a more random one by which some groups ''drifted'' voluntarily into progressively greater dependence on a narrower range of plants, some of which were domesticated. It is impossible to resolve such a general question conclusively by appeal to direct archaeological evidence of forager and early agricultural subsistence (which is in any case very meager), but the ethnographic and historical record strongly suggests that agriculture is more demanding of time and energy, and generally more risky, than foraging, even when the latter includes an element of cultivation. It therefore seems unlikely that foragers would have voluntarily and progressively become dependent on agriculture for their main food supply unless they were subjected to some form(s) of subsistence stress.

There is ethnohistorical evidence that some foragers did select and sufficiently modify particular plants that became largely dependent on their cultivation for survival, and can be said to have been at least semi-domesticated,[a] but what most foragers did not do was focus their energy on the cultivation, and domestication, of selected crops to such an extent that they became farmers. This evolutionary pathway appears to have been followed only by a few forager groups in the past whose livelihood came under sustained stress. Such stress could have been generated by several of the factors already mentioned, such as climatic and vegetation changes, population growth, and competition between groups for scarce resources. Very probably, it was when several factors combined in particular situations to exert sustained stress on the subsistence practices of forager groups that transitions to agriculture took place. So we need next to ask in what contexts is this likely to have occurred.

THE CONTEXTS OF TRANSITIONS TO AGRICULTURE

There is little direct archaeological evidence that allows us to trace primary transitions from foraging to farming. Although it is possible to distinguish from their wild progenitors the remains of many domestic plants and animals recovered from sites of early agriculture, there are very few known sites that span and reveal such transitions. One that does is the Levantine site of Tell Abu Hureyra on the Euphrates River in Syria, where large assemblages of charred plant remains of Late Palaeolithic and Neolithic age have been recovered and analyzed. Changes in the composition of the assemblages indicate a transition from the exploitation of the seeds of a wide range of wild plants, including grasses and herbaceous legumes, during the second half of the Late Palaeolithic (Epipalaeolithic) occupation c. 13,000–10,000 years ago, to the cultivation, by the Pre-Pottery Neolithic period at the site c. 9400–7300 years ago, of a small number of domesticated crops, including barley, wheat (einkorn, emmer, and bread wheat), lentil, pea, and faba bean.[4,5]

The evidence suggests that by the Epipalaeolithic period Abu Hureyra was occupied year round and that the inhabitants regularly harvested the seeds of the wild cereals and legumes. They probably began to cultivate them in response to an abrupt change to colder and drier conditions that began about 11,000 years ago and lasted until about 10,000 years ago (the Younger Dryas climatic interval) and progressively reduced stands of the wild plants from the least to the most drought-tolerant species. Toward the end of the Younger Dryas, there is archaeobotanical evidence of increases in weeds typical of dryland cultivation, and through the succeeding Pre-Pottery Neolithic period (c. 10,300–7500 years ago) in the Levant as a whole the number and size of settlements increased in response to population growth, and the remains of the ''founder crops'' of Southwest Asian agriculture[6] appear at an increasing number of sites.[7] By the end of the period, grain farming (and the herding of domestic goats and sheep) had become the mainstay of the human population of the Fertile Crescent and the new agro-pastoral, village-based way of life had begun to spread outward, toward Europe, North Africa, and Central and South Asia.[8,9]

The Southwest Asian Fertile Crescent currently provides a uniquely detailed archaeological record of a primary transition from foraging to farming. It highlights

---

[a]For evidence of semi-domestication of plants by foragers see Ref. [3].

the importance of the interaction of several factors in the process: climatic and vegetational change, sedentism and associated population growth, increased competition for declining food resources, and technological innovation (e.g., in the manufacture of sickles and grindstones). We do not have equivalently comprehensive evidence for primary transitions to agriculture—as opposed to the spread of already established agricultural systems—in other regions of the world where distinctive combinations of plants (and in some cases also animals) were domesticated, such as China, southern India, New Guinea, northern tropical Africa, Mesoamerica, the Andean highlands, Amazonia, and eastern North America. (For more detailed discussion of crop domestication in most of these regions, and in Southwest Asia, see the articles in this volume by Bar-Yosef, Delgado-Salinas et al., Asch and Hart, Lu, Pasquet, and Paz). It is not yet possible to determine with confidence what factors may have interacted to induce foragers to become farmers in these regions, but there is some tentative evidence from central China that may implicate the Younger Dryas climatic interval in the transition from the harvesting of wild rice to its cultivation and domestication in the Late Palaeolithic and the widespread establishment of rice agriculture in the Neolithic.[10,11]

In Mexico and Central and South America, however, there is little evidence of the Younger Dryas. Factors such as the establishment of sedentary settlements associated with population growth and social differentiation, resulting in more intensive exploitation of local wild foods, may have led to greater dependence on the cultivation of particularly productive plants, such as squash, maize, and beans, that ultimately became staple crops. In northern tropical Africa and eastern North America on the other hand, it is possible that short-term changes of climate (more recent than the Younger Dryas) may have stimulated the transitions to agriculture that occurred there later in the Early Holocene.[12,13]

CONCLUSION

It appears that climatic change (to colder and drier conditions) was a key factor in the earliest known transition to agriculture in Southwest Asia. Such changes, in combination with increased sedentism, population growth, and other social factors, may have played an important part in transitions elsewhere. Progress toward a more complete understanding of the process requires the recovery, identification, and accurate dating of plant (and animal) remains from archaeological sites that span periods of transition from foraging to farming. Very few such sites are known and even fewer have been investigated using modern techniques of excavation, dating, and analysis. Until more such research is accomplished, conclusive answers to the questions of why and how agriculture began will remain elusive.

ARTICLES OF FURTHER INTEREST

Crop Domestication in Africa, p. 304
Crop Domestication in China, p. 307
Crop Domestication in Mesoamerica, p. 310
Crop Domestication in Prehistoric Eastern North America, p. 314
Crop Domestication in Southeast Asia, p. 320
Crop Domestication in Southwest Asia, p. 323
Crop Domestication in the American Tropics: Starch Grain Analyses, p. 330
Crop Domestication: Founder Crops, p. 336
Pre-agricultural Plant Gathering and Management, p. 1055

REFERENCES

1. Harris, D.R. Ethnohistorical Evidence for the Exploitation of Wild Grasses and Forbs: Its Scope and Archaeological Implications. In *Plants and Ancient Man: Studies in Palaeoethnobotany*; Van Zeist, W., Casparie, W.A., Eds.; Balkema: Rotterdam, 1984; 63–69.
2. Keeley, L.H. Protoagricultural Practices Among Hunter-Gatherers: A Cross-Cultural Survey. In *Last Hunters—First Farmers: New Perspectives on the Prehistoric Transition to Agriculture*; Price, T.D., Gebauer, A.B., Eds.; School of American Research Press: Santa Fe, NM, 1995; 243–272. See also M.K. Anderson's article in this encyclopedia.
3. Shipek, F.C. An Example of Intensive Plant Husbandry: The Kumeyaay of Southern California. In *Foraging and Farming: The Evolution of Plant Exploitation*; Harris, D.R., Hillman, G.C., Eds.; Unwin Hyman: London, 1989; 159–170.
4. Hillman, G.C. Abu Hureyra 1: The Epipalaeolithic, Overview: The Plant-Based Components of Subsistence at Abu Hureyra 1 and 2. In *Village on the Euphrates: From Foraging to Farming at Abu Hureyra*; Moore, A.M.T., Hillman, G.C., Legge, A.J., Oxford University Press: New York, 2000; 327–399; 416–422.
5. De Moulins, D. Abu Hureyra 2: Plant Remains From the Neolithic. In *Village on the Euphrates: From Foraging to Farming at Abu Hureyra*; Moore, A.M.T., Hillman, G.C., Legge, A.J., Oxford University Press: New York, 2000; 399–416.

6. Zohary, D. The Mode of Domestication of the Founder Crops of Southwest Asian Agriculture. In *The Origins and Spread of Agriculture and Pastoralism in Eurasia*; Harris, D.R., Ed.; UCL Press: London and Smithsonian Institution Press: Washington, DC, 1996; 142–158. See also Zohary's contribution to this encyclopedia.
7. Garrard, A. Charting the emergence of cereal and pulse domestication in South-West Asia. Env. Archaeol. **1999**, *4*, 67–86.
8. Harris, D.R. The origins of agriculture in Southwest Asia. Rev. Archaeol. **1998**, *19* (2), 5–11.
9. Bar-Yosef, O.; Meadow, R.H. The Origins of Agriculture in the Near East. In *Last Hunters—First Farmers: New Perspectives on the Prehistoric Transition to Agriculture*; Price, T.D., Gebauer, A.B., Eds.; School of American Research Press: Santa Fe, NM, 1995; 39–94.
10. Zhao, Z. The Middle Yangtse region in China is one place where rice was domesticated: Phytolith evidence from the Diaotonghuan Cave, northern Jiangxi. Antiquity **1998**, *72* (278), 885–897.
11. Harris, D.R. Climatic Change and the Beginnings of Agriculture: The Case of the Younger Dryas. In *Evolution on Planet Earth: The Impact of the Physical Environment*; Rothschild, L., Lister, A., Eds.; Academic Press: London, 2003; 379–394.
12. Hassan, F.A. Holocene palaeoclimates of Africa. Afr. Archaeol. Rev. **1997**, *14* (4), 213–229.
13. Smith, B.D. *The Emergence of Agriculture,* 2nd Ed.; Scientific American Library: New York, 1998; 210.

Air Pollutant Interactions with Pests and Pathogens

Andreas von Tiedemann
University of Göttingen, Göttingen, Germany

INTRODUCTION

The exponential economic growth in developed countries since the mid-nineteenth century has been accompanied by an increase in the concentrations of various tropospheric trace gases. Among the gaseous pollutants most studied during the past decades are carbon dioxide (CO_2) with a global distribution, ozone and peroxyacetylnitrate (PAN) with a more regional dimension, and others such as sulfur dioxide (SO_2), nitrogen oxide (NO_2), and hydrogen fluoride (HF), which are associated with emissions from local sources. Besides their potential to influence the global climate and their direct effects on plant growth and yield, their effects on plant-pathogen or plant–insect relations may already be, or could become, an important factor affecting plant health. However, evaluation of these effects is difficult because of the diverse temporal/spatial distribution and differences in the chemical behavior and because of the complex interactions between the effects of the trace gases and agronomic factors relevant for plant resistance to diseases and insect pests, including fertilizer and pesticide use, crop variety, soil management, water supply, etc. Also, direct effects of gaseous pollutants on the pathogen or insect cannot easily be separated from indirect effects on the host plants through changes in physiological processes. Experiments under controlled conditions can only reveal a limited picture of the multitude of possible effects that may occur in the field. Despite these limitations, a number of studies mainly carried out between 1970 and 2000 have resulted in a significant amount of data describing specific effects of the main trace gases on the incidence of plant diseases and insect pests. These can be compiled and used to formulate some generalizations.

EFFECTS OF AIR POLLUTANTS AND ELEVATED CO_2 ON PLANT DISEASES

Among the photooxidants, ozone has been the prevalent compound studied for its effects on both plants and plant diseases. Because its photochemical production is favored under conditions of high irradiance, direct effects of ozone on fungal or bacterial pathogens are less likely, as these only grow actively on plant surfaces during wet and cloudy periods. The same applies to soilborne pathogens. Therefore, the main pathway for ozone effects on diseases occurs indirectly via changes in the physiology of the plant. A number of physiological changes induced by ozone are important with respect to plant disease resistance. Numerous physiological effects of ozone may impair the conditions for growth of pathogens, particularly of biotrophs, such as accelerated ageing/premature senescence, degradation of membrane lipids accompanied with increased cellular leakage, reduced net photosynthesis, increased protein degradation, or enhanced ethylene production, in combination with changes in factors directly affecting resistance. These include elevated antioxidant levels, reinforcement of cell walls (lignin, callose, extensins), induction of phytoalexins, or expression of PR-proteins.[1] Although these effects may occur in most plant species, they can lead to contrasting effects on the pathogens, according to their parasitic nature. On one hand, ozone stress can cause a reduction in the growth of bacteria and biotrophic fungi such as powdery mildew and rusts, while on the other hand increased infection by necrotrophic parasites can result from ozone predisposing the plants.[2] It has often been observed that the effect of ozone on a particular pathogen is similar to the effect of ageing and senescence. The profile of ozone exposure and the dose of ozone are important because the exposure may or may not cause visible injury prior to infection. For instance, grey mold induced by *Botrytis cinerea* was enhanced on ozone-injured leaves through the delivery of entry ports, while the disease was restricted on uninjured leaves after exposure to chronic ozone doses. In the latter case, triggering of plant resistance factors may have been involved in increased resistance against fungal invasion. As visible injury is less frequent in the field, the latter situation may be of practical importance.

A few studies have addressed interactions of virus diseases with ozone. These studies consistently looked at changes in ozone sensitivity in virus-infected plants, and they showed that virus infections can protect plants from ozone injury.[2,3]

Most soil fungi can tolerate more than a 10- to 100-fold increase in atmospheric carbon dioxide concentrations. Some pathogenic aerial or soilborne fungi and bacteria were found to be inhibited only at CO_2 concentrations exceeding 3–5%, and others were unaffected or even

Encyclopedia of Plant and Crop Science
DOI: 10.1081/E-EPCS 120005569

stimulated in growth and/or sporulation under these conditions. This suggests that an increase in atmospheric CO_2 from 0.03–0.07% over the next 50 years will probably not have a direct effect on fungal and bacterial plant pathogens but that it may act indirectly via alterations in plant growth, physiology, and metabolic state. Elevated CO_2 can have profound effects on plants, ranging from increased photosynthesis rates to enhanced growth, elevated leaf carbohydrate contents, altered stomata regulation, etc. These changes may favor the growth of biotrophic pathogens, while increased plant growth leading to a denser canopy structure favors foliar pathogens because of more humid microclimatic conditions favoring infection. Increased biomass production will lead to larger amounts of plant litter, which, in turn, has the potential to favor the survival of necrotrophic pathogens during periods with adverse conditions.

Co-occurrence of ozone stress and elevated CO_2 may partly counteract inhibitory or stimulating effects of the two trace gases on plants. This has been demonstrated in studies of combined effects of the two gases on photosynthesis, growth, and yield. However, while CO_2 may offset deleterious effects of ozone on plant growth, the impact of ozone on plant resistance to pathogens seems to be less affected by elevated CO_2.[2]

With the advancement of clean-air technologies in many of the developed countries, the importance of the point-source-related pollutant gases such as sulfur dioxide and hydrogen fluoride has decreased, but in less developed regions, negative effects may still be of great importance. Effects of SO_2 were studied intensively in relation to the occurrence of pathogens associated with forest trees and agricultural crops. In some cases, trees weakened by this pollutant were found to be predisposed to infection, whereas in other cases they were not. Conflicting observations were also made with respect to SO_2 effects on cereal pathogens. The inconsistencies may be related to whether or not injury is caused by the pollutant prior to infection. Exposure to atmospheric hydrogen fluoride (HF) leads to an accumulation of fluoride in plant foliage to levels much higher than those present in the atmosphere. Similarly, fluoride accumulation may occur in plants growing on contaminated soils. At high concentrations, fluoride impairs the growth of some representative plant pathogens, both in vitro and in vivo.[3,4]

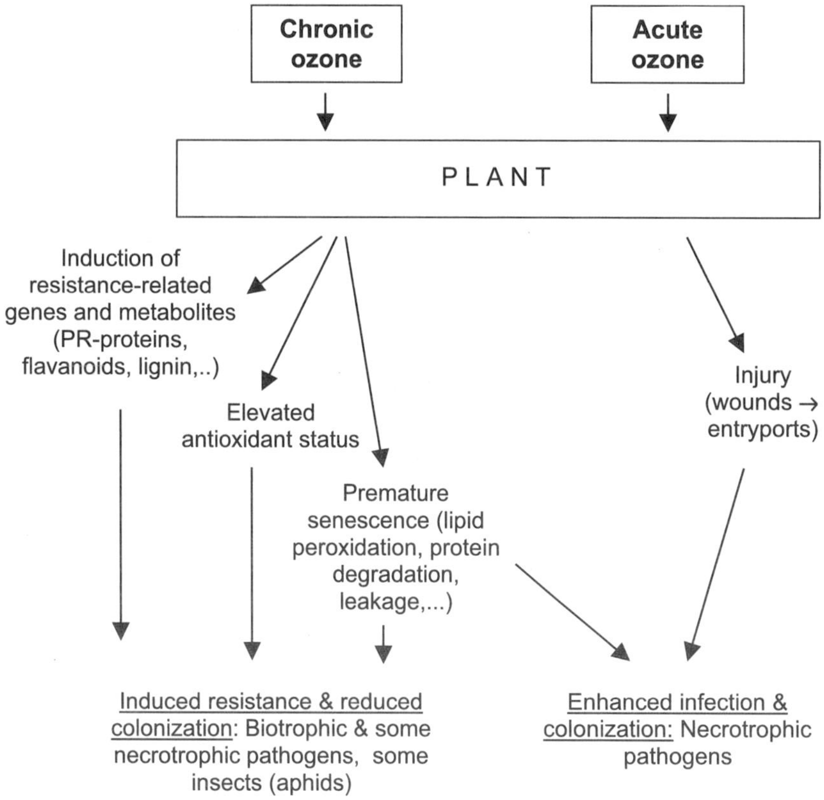

Fig. 1 Summary of the main effects of ozone stress on diseases and pests, mediated by the plant.

EFFECTS OF AIR POLLUTANTS AND ELEVATED CO_2 ON INSECT PESTS

Among the herbivorous insects, aphids attracted the most research interest, much more than, for instance, beetles, moths, and butterflies. Earlier field observations along pollution gradients in the vicinity of industrial plants, urban areas, or motorways and more recent experimental investigations using insect suction traps, filtration systems in urban air, open-top chambers, or field fumigation systems have consistently demonstrated that the growth of herbivorous insect populations in terms of their mean relative growth rate is favored by moderate levels of pollution but inhibited in more polluted environments. This general finding mainly refers to effects of SO_2 and NO_2 in the vicinity of point sources or alongside motorways. In particular, various important aphid species on crop plants such as wheat, barley, broad bean, pea, lupin, and brussels sprout or on tree species including apple, beech, pines, and spruces have consistently performed better in atmospheres with moderate levels of the two pollutants,[3,5–7] both in the field and in closed fumigation chambers. It is generally believed that an increase in available amino acids in polluted plants leads to the stimulation of the growth of aphid populations.

Direct effects of SO_2, NO_2, or O_3 on insects have been less studied, but the available evidence suggests that insects can largely tolerate these air pollutants at realistic levels when exposed either during feeding on artificial diets or on plants. This has been demonstrated when the host plants were exposed prior to or after the transfer of insects. Consequently, by far the most important effects of air pollutants on herbivorous insect populations are mediated through changes in the host plant.

Effects of ozone have been mostly studied in closed chambers, and results have been far more complex than those of SO_2 or NO_2. While many experiments have demonstrated increased growth of aphids under ozone stress, this effect was reversed or abolished either at higher temperatures or when applying ozone continuously instead of episodically. Since, in ambient air, peaks in ozone concentrations typically occur episodically, it remains unclear how both the presence of higher temperatures and a diurnal cycle acting in combination would modulate ozone effects on insect pests in the field. Ozone effects on free amino acid content in plants have been

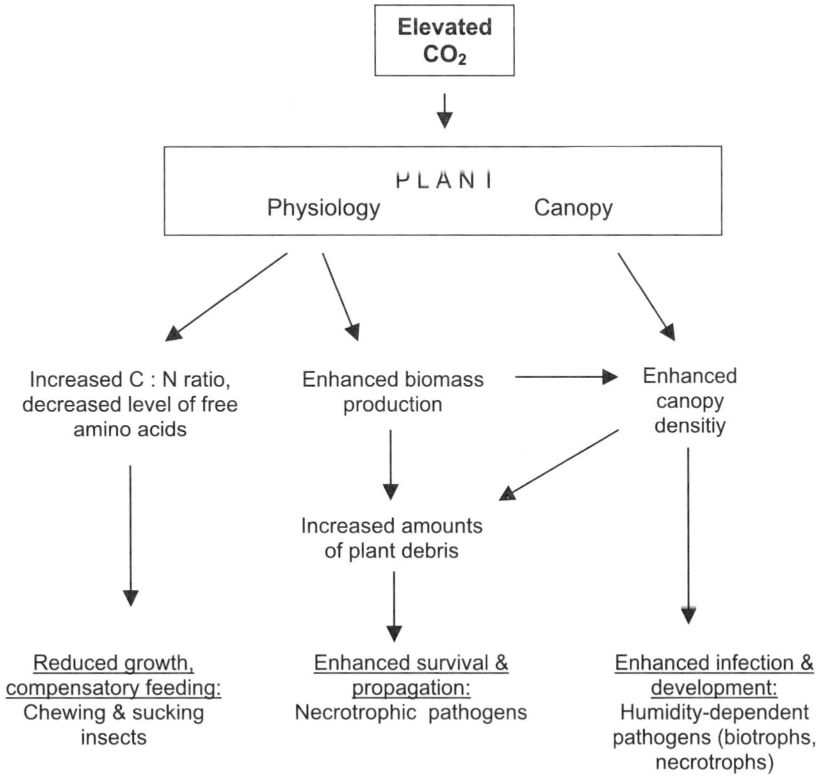

Fig. 2 Summary of important effects of elevated carbon dioxide on diseases and pests, mediated by the altered plant physiology and canopy structure.

found to vary among plant species, pollutant doses, or environmental conditions. Thus, factors other than nutritive traits may be involved in the mediation of ozone effects—for instance, changes in the feeding stimuli on the plant surface.

Limited data on effects of elevated CO_2 demonstrate that insect herbivores, in particular chewing and sucking insects, grow and develop more slowly but consume more plant material. This effect can be attributed to the increased C/N ratio and the reduction in the content of free amino acids in plants grown at elevated CO_2.[7,8] While CO_2 effects on population growth may be negative or not detectable, enhanced compensatory feeding of chewing insects appears to be a consistent phenomenon and may aggravate the damage to plants. Elevated CO_2, a significant growth factor, affects not only the chemical plant composition but also the physical structure and density of the plant canopy, leading to altered microclimatic conditions in the field. Therefore, the combination effects of CO_2, temperature, humidity, nutrient supply, and interactions with other pollutants such as ozone determine the outcome of plant-insect relationships, but present knowledge of these complex interactions is limited.

GENERAL RISK CONSIDERATIONS

Figs. 1 and 2 summarize important effects of ozone and elevated CO_2 on plant diseases and insect pests. However, a general evaluation of the risk for enhanced plant diseases caused by air pollution is difficult because effects vary among different plant-pathogen relationships. Moreover, ozone in combination with elevated CO_2 is less effective as compared to its effects as a single gas. It seems certain, though, that effects of pollutants on diseases and insect pests occur preferentially through alterations in the host plants. However, plants can acclimate to rising CO_2, leading to long-term effects that are different from those observed in controlled short-term experiments. Interacting effects from antagonists, predators, parasitoids, and environmental factors further complicate the evaluation of risks under field conditions. Therefore, a general risk assessment of enhanced diseases and pests caused by ozone stress and/or elevated CO_2 is not possible. Specific risks may exist for some insect infestations in the presence of episodic ozone stress and moderate temperatures. Importantly, SO_2 and NO_2 play a major role in regions of the developing world that, at the same time, suffer from insufficient agricultural production, and an elevated risk caused by insect pests under the prevailing climatic conditions seems likely. Hence, these regions may be most threatened by air pollutant effects on insect pests in the future.

ARTICLES OF FURTHER INTEREST

Air Pollutants: Effects of Ozone on Crop Yield and Quality, p. 13
Air Pollutants: Interactions with Elevated Carbon Dioxide, p. 17
Air Pollutants: Responses of Plant Communities, p. 20
Crop Responses to Elevated Carbon Dioxide, p. 346
Crops and Environmental Change, p. 370
Leaves and the Effects of Elevated Carbon Dioxide Levels, p. 648
Plant Responses to Stress: Nematode Infection, p. 1014
Plant Responses to Stress: Ultraviolet-B Light, p. 1019

REFERENCES

1. Sandermann, H. Ozone and plant health. Annu. Rev. Phytopathol. **1996**, *34*, 347–366.
2. Manning, W.J.; Tiedemann, A.V. Climate change: Potential effects of increased atmospheric carbon dioxide (CO_2), ozone (O_3), and ultraviolet-B (UV-B) radiation on plant diseases. Environ. Pollut. **1995**, *88*, 219–245.
3. Manning, W.J.; Keane, K.D. Effects of Air Pollutants on Interactions Between Plants, Insects, and Pathogens. In *Assessment of Crop Loss From Air Pollutants*; Heck, W.W., Taylor, O.C., Tingey, D.T., Eds.; Elsevier Appl. Science: London, 1988; 365–386.
4. Treshow, M.; Anderson, F.K. *Plant Stress from Air Pollution*; John Wiley & Sons: Chichester, 1989.
5. Brown, V.C. Insect Herbivores and Gaseous Air Pollutants—Current Knowledge and Predictions. In *Insects in a Changing Environment*; Harrington, R., Stork, N.E., Eds.; Academic Press: London, 1995; 219–249.
6. Dohmen, G.P.; McNeill, S.; Bell, J.N.B. Air pollution increases *Aphis fabae* pest potential. Nature **1984**, *307*, 52–53.
7. Whittaker, J.B. Insects in a changing atmosphere. J. Ecol. **2001**, *89*, 507–518.
8. Watt, A.D.; Whittaker, J.B.; Docherty, M.; Brooks, G.; Lindsay, E.; Salt, D.T. The Impact of Elevated Atmospheric CO_2 on Insect Herbivores. In *Insects in a Changing Environment*; Harrington, R., Stork, N.E., Eds.; Academic Press: London, 1995; 197–217.

Air Pollutants: Effects of Ozone on Crop Yield and Quality

Håkan Pleijel
Göteborg University, Göteborg, Sweden

INTRODUCTION

Air pollutants have been known for more than one hundred years to affect both yield and quality of agricultural crops. This is true for a wide range of pollutants including sulfur dioxide, fluorides, and nitrogen-containing pollutants such as nitrogen oxides (NO, NO_2) and ammonia (NH_3). Today, the main focus is on effects of the regional occurrence of elevated tropospheric ozone (O_3). In North America and Europe, where emissions of sulfur dioxide and fluorides have declined, the problem of ozone pollution has received most attention. In contrast, in many developing countries emissions of a number of pollutants are increasing and there are strong indications of adverse effects on crops. However, these effects have been studied to a much lesser extent than in Europe, North America, Japan, and Australia.

DISCOVERY OF OZONE EFFECTS ON CROPS

Systematic studies of ozone effects on crop yield and quality started around 1950 in California, U.S.A.,[1] where the problems caused by ozone and other photochemical oxidants were first discovered. At that time, the studies were mainly of an observational nature, and they were based on visible injury appearing on the plant leaves. Controlled experiments in the laboratory began later; however, initially these yielded largely qualitative data which were not suitable as a basis to estimate the magnitude of potential effects in the field.

But, it should be noted here that in certain crops, such as spinach and tobacco, leaf injury is of large economic importance, apart from potential reductions in absolute yield. In some crops visible leaf injury, expressed as necrotic spots on the leaves, is the most pronounced effect by ozone and other pollutants, while in others a reduction in the leaf life span, the so-called premature senescence, is most important, as it reduces the length of the growth period. The latter occurs in wheat, in which characteristic leaf injury at moderate ozone exposure is generally lacking, while early senescence in response to ozone is pronounced.[2]

OPEN-TOP CHAMBERS

With the introduction of the open-top chamber (OTC) as an exposure system,[3] which permits semicontrolled exposure of plants to gaseous pollutants under ecologically realistic conditions in the field, an important step was taken towards a quantitative understanding of impacts of ozone and other pollutants. An OTC is a transparent plastic cylinder through which air (ambient air, filtered air, or air enriched with various levels of pollutant gases) is ventilated using a fan. OTCs can be mounted in plots of a field-grown crop and kept in place throughout the growing season. Although the system alters the microclimate of the plants to a certain extent,[4] the ecological realism is much larger than in a closed chamber system in the laboratory.

CROP LOSS ASSESSMENT PROGRAMS

In the U.S.A., a large experimental program, the NCLAN (National Crop Loss Assessment Network), involving the use of OTCs of crops, was executed during the 1980s. The NCLAN results showed that ambient levels of ozone had the potential to reduce the yield of a number of crops, including soy bean, wheat, and alfalfa.[5] In addition, economic estimates of crop losses for the U.S.A. indicated an annual loss for the farmers in the range of a few billion U.S. dollars.

A similar program, the European Open-Top Chamber Programme (EOTC), was organized in Western Europe beginning in the second half of the 1980s until the early 1990s. Negative effects of ozone levels typical of wide areas of Europe were observed in beans and spring wheat.[6] In some, but not all, experiments with pastures the main effect of ozone was on the species composition, with clover being replaced by grasses. This effect acts to reduce the protein content and thus the fodder quality of the yield. Towards the end of the 1990s a European research program, Changing Climate and Potential Impacts on Potato Yield and Quality (CHIP), studied the effects of ozone on potato. It was concluded that current ozone levels in Europe can cause significant effects on potato, but that the sensitivity is less than in wheat.

Encyclopedia of Plant and Crop Science
DOI: 10.1081/E-EPCS 120005563

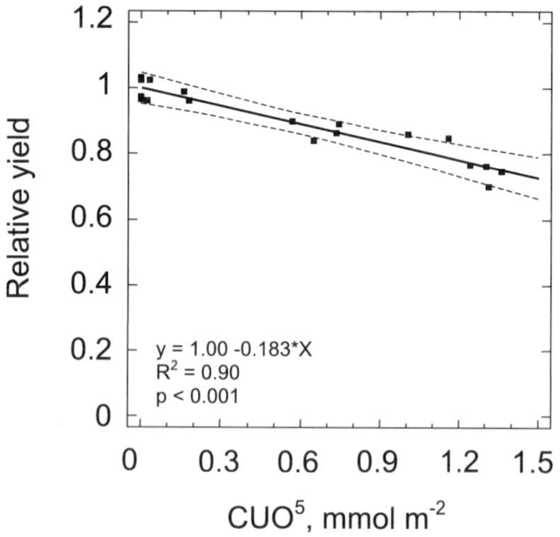

Fig. 1 Relationship between the relative yield of field-grown spring wheat and the cumulative ozone uptake (CUO) by the flag leaves based on stomatal conductance modeling. An ozone uptake rate threshold of 5 nmol $m^{-2}s^{-1}$ was used, since this threshold resulted in the best correlation between relative yield and ozone uptake. Five open-top chamber experiments performed in Sweden were included. (From Ref. 8.)

Based on the experiences of both research programs in North America and Europe, it can be concluded that soybean, wheat, tomato, pulses, and watermelon can be classified as highly sensitive to ozone. Barley and certain fruits are rather insensitive, while potato, rapeseed, sugar beet, maize, and rice are intermediate with respect to ozone sensitivity. There exists a certain degree of intraspecific variation in ozone sensitivity, such as in different bean varieties, but a recent European study resulted in a rather uniform ozone response pattern for different cultivars of wheat.

"CRITICAL LEVELS" OF OZONE

The relationship between grain yield and ozone exposure was most consistent in spring wheat across cultivars and experimental sites, and, considering the pronounced sensitivity observed for this crop, the combined data were used as a basis to derive critical levels for ozone effects under the Convention on Long-Range Transboundary Air Pollution of the United Nations Economic Commission

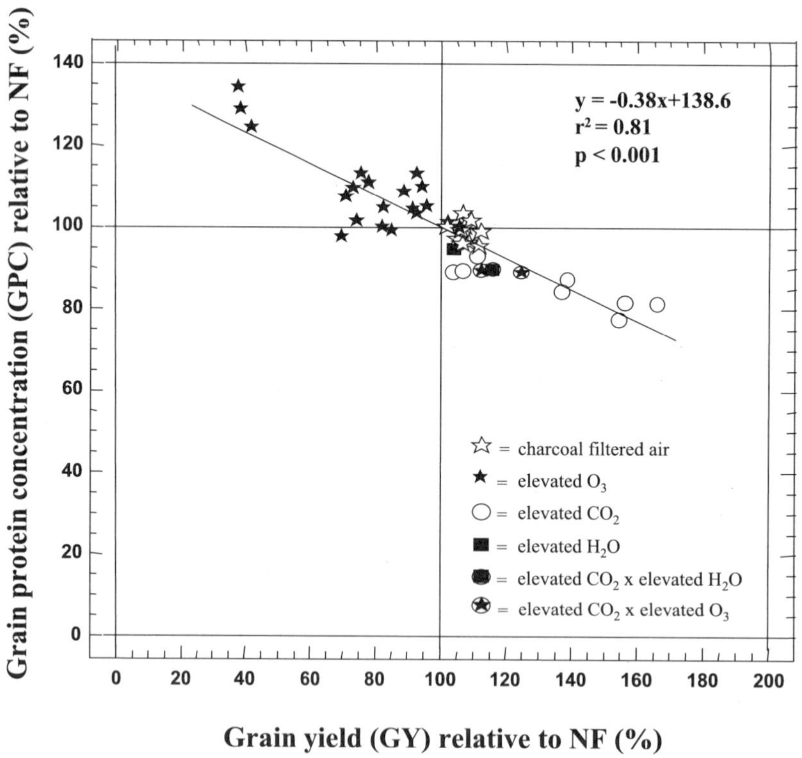

Fig. 2 Relationship between grain protein concentration and grain yield in field-grown spring wheat based on data from open-top chamber experiments performed in Sweden, Finland, Denmark, and Switzerland. The experiments included ozone treatments, carbon dioxide treatments, and a few irrigation treatments. Relative scales were used for both axes: The effects were related to the open-top chamber treatment with non-filtered air (NF) without extra ozone or carbon dioxide fumigation in each of the sixteen experiments included. (From Ref. 9.)

for Europe (UNECE). For this purpose, the ozone exposure index AOT40 (the accumulated exposure over a concentration threshold of 40 nmol mol^{-1} ozone based on hourly averages) was used.[7]

OZONE UPTAKE

An important recent development in the field of crop loss assessments is the consideration of ozone uptake by the plant leaves, in place of a statistical index to characterize the concentration of ozone in the air surrounding the plants, such as the AOT40 exposure index. Key to this method is the quantification of ozone diffusion through the stomata of leaves.

The stomatal conductance of the plant leaves varies with a number of factors, such as soil and air moisture contents, temperature, solar radiation, phenology, and the levels of other pollutants, including carbon dioxide. Stomatal conductance models can now be used to establish relationships between yield loss and ozone uptake. An example is given in Fig. 1 showing the relationship between relative yield in two Swedish wheat cultivars and the calculated, cumulated uptake of ozone by flag leaves, taking into account an uptake rate threshold of 5 nmol m^{-2} s^{-1} (CUO^5).[8] This threshold provided the best correlation between effect and exposure, and it may represent the biochemical defense capacity of the plants. However, it remains to be shown that the two thresholds are directly related to each other.

EFFECTS OF OZONE ON PROTEIN CONTENTS

Much less attention has been paid to the influences of ozone on crop quality as compared to efforts made to understand the relationship with yield. The most discussed quality aspect has been the protein concentration or content, for instance, of wheat grains. The combined results of sixteen open-top chamber experiments performed in four different European countries (Finland, Sweden, Denmark, and Switzerland) were used to show the relationship between grain yield and grain quality (Fig. 2), and grain yield and the off-take of grain protein per unit ground area (Fig. 3), at different ozone or carbon

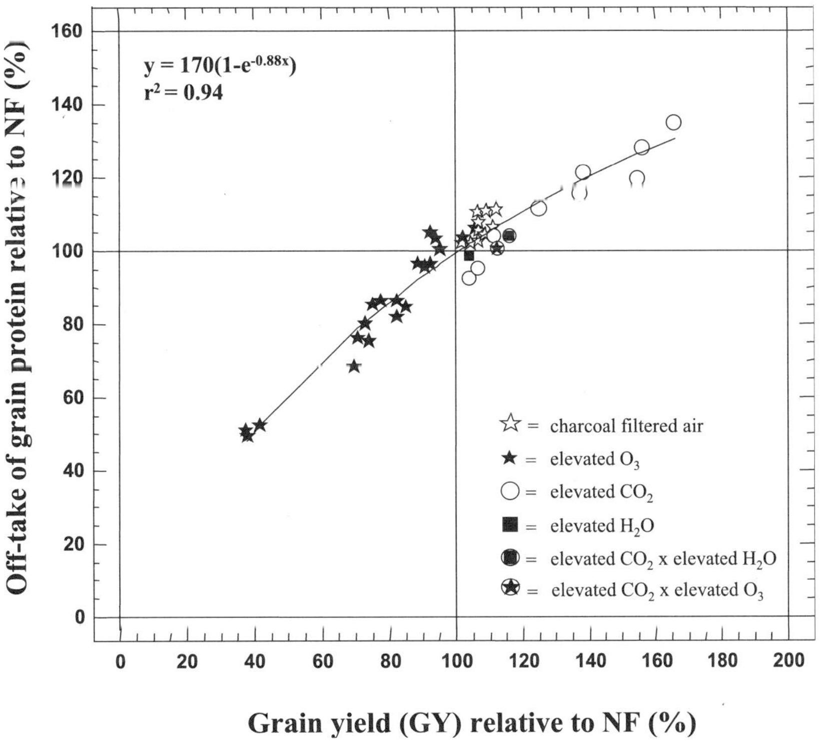

Fig. 3 Relationship between grain protein off-take per unit ground area and grain yield in field-grown spring wheat based on data from open-top chamber experiments performed in Sweden, Finland, Denmark, and Switzerland. The experiments included ozone treatments, carbon dioxide treatments, and a few irrigation treatments. Relative scales were used for both axes: The effects were related to the open-top chamber treatment with non-filtered air (NF) without ozone or carbon dioxide fumigation in each of the sixteen experiments included. (From Ref. 9.)

dioxide levels.[9] From Fig. 2, it can be inferred that, relative to control (always the open-top chamber treatment with non-filtered air), the grain concentration of protein was higher and yield was lower in elevated ozone, which is the opposite result of situations with elevated carbon dioxide or in chambers where the ozone had been reduced by using charcoal filters. This situation could be attributed to a so-called growth dilution effect by elevated carbon dioxide and air filtration stimulating the carbohydrate accumulation relatively more than the uptake of nitrogen at a given nitrogen fertilizer application rate, while in situations of elevated ozone, carbohydrate accumulation was negatively affected more strongly than protein accumulation.

However, Fig. 3 shows that the protein yield per unit ground area was negatively affected by ozone and enhanced by elevated carbon dioxide, although the latter effect tended to saturate at a level determined by the availability of nitrogen in the soil.

CONCLUSION

There is strong evidence that current ozone concentrations over large areas in the industrialized world are high enough to cause yield loss in several important agricultural crops, but it remains a challenge to quantify these effects exactly. New approaches based on the uptake of the pollutant by plants, rather than the concentration in the air surrounding the plant, are promising.

The protein concentration of the yield tends to increase with increasing ozone in some crops in connection with declining yield. On the other hand, the protein yield per unit ground area tends to decrease. In future research, the understanding of additional quality effects of air pollutant exposure should be given more attention.

In the developing world, such as many countries in Asia, there is a great risk for crop losses due to a number of air pollutants for which emission rates are increasing dramatically.[10] A loss in food production in these countries represents a much larger problem than in the industrialized world. More research should be devoted to this problem in the decades to come.

ARTICLES OF FURTHER INTEREST

Air Pollutant Interactions with Pests and Pathogens, p. 9

Air Pollutants: Interactions with Elevated Carbon Dioxide, p. 17

Air Pollutants: Responses of Plant Communities, p. 20

Crop Responses to Elevated Carbon Dioxide, p. 346

Crops and Environmental Change, p. 370

Exchange of Trace Gases Between Crops and the Atmosphere, p. 425

Photosynthesis and Stress, p. 901

Plant Response to Stress: Source-Sink Regulation by Stress, p. 1010

Plant Responses to Stress: Ultraviolet-B Light, p. 1019

REFERENCES

1. Middleton, J.T.; Kendrick, J.B.; Schalm, H.W. Injury to herbaceous plants by smog or air pollution. Plant Dis. Rep. **1950**, *34*, 245–252.
2. Pleijel, H.; Ojanperä, K.; Danielsson, H.; Sild, E.; Gelang, J.; Wallin, G.; Skärby, L.; Selldén, G. Effects of ozone on leaf senescence in spring wheat—Possible consequences for grain yield. Phyton (Austria) **1997**, *37*, 227–232.
3. Heagle, A.S.; Body, D.; Heck, W.W. An open-top filed chamber to assess the impact of air pollution on plants. J. Environ. Qual. **1973**, *2*, 365–368.
4. Unsworth, M.H. Air pollution and vegetation, hypothesis, field exposure, and experiment. Proc. R. Soc. Edinb. **1991**, *97B*, 139–153.
5. Heck, W.W.; Taylor, O.C.; Tingey, D.T. *Assessment of Crop Loss from Air Pollutants*; Elsevier Applied Science: London, 1988.
6. Jäger, H.J.; Unsworth, M.H.; De Temmerman, L.; Mathy, P. *Effects of Air Pollution on Agricultural Crops in Europe. Results of the European Open-Top Chamber Project*; Air Pollution Research Report Series of the Environmental Research Programme of the Commission of the European Communities, Directorate-General for Science, Research and Development, 1992; 46.
7. Fuhrer, J.; Skärby, L.; Ashmore, M. Critical levels for ozone effects on vegetation in Europe. Environ. Pollut. **1997**, *97*, 91–106.
8. Danielsson, H.; Pihl Karlsson, G.; Karlsson, P.E.; Pleijel, H. Ozone uptake modeling and flux-response relationships—Assessments of ozone-induced yield loss in spring wheat. Atmos. Environ. **2003**, *37*, 475–485.
9. Pleijel, H.; Mortensen, L.; Fuhrer, J.; Ojanperä, K.; Danielsson, H. Grain protein accumulation in relation to grain yield of spring wheat (*Triticum aestivum* L.) grown in open-top chambers with different concentrations of ozone, carbon dioxide and water availability. Agric. Ecosyst. Environ. **1998**, *72*, 265–270.
10. Emberson, L.D.; Ashmore, M.R.; Murray, F.; Kuylenstierna, J.C.I.; Percy, K.E.; Izuta, T.; Zheng, Y.; Shimizu, H.; Sheu, B.H.; Liu, C.P.; Agrawal, M.; Wahid, A.; Abdel-Latif, N.M.; van Tienhoven, M.; de Bauer, L.I.; Domingos, M. Impacts of air pollutants on vegetation in developing countries. Water Air Soil Pollut. **2001**, *130*, 107–118.

Air Pollutants: Interactions with Elevated Carbon Dioxide

Hans-Joachim Weigel
Institute of Agroecology, Bundesallee, Braunschweig, Germany

INTRODUCTION

The concentrations of various compounds in the atmosphere have undergone significant changes during the last century, and they continue to change. Many of these compounds interact with the terrestrial biosphere. Depending on the concentration in the atmosphere, gases such as SO_2 or NO_2 may be beneficial to natural and agricultural ecosystems or they may act as air pollutants affecting these systems in a negative or adverse manner. In contrast, carbon dioxide (CO_2) is the basic plant nutrient and has positive and growth stimulating effects on vegetation. Because of their co-occurrence, the assessments of potential effects of atmospheric changes on vegetation have to consider these contrasting impacts and the interactions between effects of air pollutants and elevated CO_2.

ATMOSPHERIC CHANGE: CONCENTRATIONS AND TRENDS

Anthropogenic activities have changed the concentrations of a wide variety of gaseous and particulate compounds in the atmosphere, including carbon dioxide (CO_2), nitrogen monoxide (NO) and nitrogen dioxide (NO_2), sulphur dioxide (SO_2), ozone (O_3), ammonia (NH_3), heavy metals, and volatile organic compounds (VOC).[1] Prominent examples of global importance are CO_2 and tropospheric O_3. Since the beginning of the 19th century, the concentration of CO_2 [CO_2] has increased globally to current values of about 360 ppmv (parts-per-million by volume). It is expected that [CO_2] will continue to increase even more rapidly and may reach about 550–650 ppmv between 2050 and 2100.[2] As CO_2 is the substrate for plant photosynthesis, this increase in [CO_2] will have far reaching consequences for most types of vegetation.[3,4]

Parallel to the increase in [CO_2], ground-level O_3 concentrations ([O_3]) in most industrialized countries have nearly doubled during the last 100 years. Current mean [O_3] in nonurban areas is between 40–75 ppbv (parts-per-billion by volume) during the growing season and 20–35 ppbv as an annual mean.[1] Today, O_3 pollution has also become a major environmental problem in many developing countries. Predictions for the future development of [O_3] are uncertain; in the case that the emission of the precursor compounds nitrogen oxides and volatile organic compounds remain high or continue to increase, [O_3] will follow these emission trends. [O_3] varies considerably in time and space and shows annual and diurnal patterns. In contrast to CO_2, elevated [O_3] is phytotoxic and affects plants negatively.[4,5] However, quantification of the effects of O_3 is difficult due to the large variability of exposure concentrations.

EFFECTS OF CO_2 AND O_3 ALONE

As current atmospheric [CO_2] limits photosynthesis in most C_3 plants, any increase in [CO_2] results in a stimulation of plant physiological and growth processes.[3] The most frequently observed effects of elevated [CO_2] include a stimulation of photosynthesis, enhanced concentrations of soluble carbohydrates, an increase in growth rates and leaf area, and stimulated biomass production and yield. Transpiration rate (per unit leaf area) and stomatal conductance, as well as tissue element concentrations (particularly nitrogen), usually decline. Yield enhancements of up to 25–35% as compared to ambient [CO_2] have been observed when plants were exposed to 550–750 ppmv CO_2.[3,6] The initial stimulation of photosynthesis often decreases under long-term exposure to elevated [CO_2], leading to smaller growth and yield enhancements than expected from the short-term photosynthetic responses. Plant species differ widely in their response to high [CO_2].[6]

By contrast, O_3 is currently regarded as the most important phytotoxic pollutant in the atmosphere.[5] Primary O_3 effects include subtle biochemical and ultrastructural changes, which may result in impaired photosynthesis, alterations of carbon allocation patterns, symptoms of visible injury, enhanced senescence, reduced growth and economic yield, altered resistance to other abiotic and biotic stresses, reduced flowering and seed production, loss of competitive abilities of plant species in communities, and shifts in biodiversity. Current ambient [O_3] in many industrialized areas are high enough to suppress crop yields of sensitive species and to retard growth and development of trees and other plant species of the nonwoody (semi-)natural vegetation. As with CO_2, there is large inter- and intraspecific variability in the O_3 susceptibility of plants.[4,5]

DOI: 10.1081/E-EPCS 120005568

INTERACTIVE EFFECTS OF CO_2 AND AIR POLLUTANTS

While single exposures to elevated [CO_2] or air pollutants may have contrasting effects on plant performance, it is of particular interest to understand how the individual changes in atmospheric constituents may interfere with each other in order to predict the likelihood of combined effects of atmospheric changes on terrestrial ecosystems.[4,7–9]

CO_2 and O_3

A great number of studies on the combined effects of the two gases have shown that high [CO_2] either partially or totally compensate for adverse O_3 effects.[4,7] This has been demonstrated, for example, for some crop species including soybean, wheat, and corn. However, summarized over the total available database with different plant species and cultivars (Table 1), the information is not entirely consistent, as several studies revealed that elevated [CO_2] may not always protect plants from the adverse effects of O_3.[7]

The proposed mechanisms to explain the protective effect of elevated [CO_2] against the phytotoxic effects of O_3 include 1) reduced uptake or flux of O_3 through the stomata due to a CO_2-induced stomatal closure, 2) improved supply of carbon skeletons supporting the synthesis of antioxidants involved in the destruction of O_3 and its toxic products, 3) protection of the Rubisco protein from O_3-induced degradation and 4) CO_2-induced changes in the cell surface/volume ratio.[4,7,8] However, it has been shown that in spite of decreased stomatal conductance under elevated [CO_2], adverse effects of O_3 may still occur. As CO_2 effects on stomatal conductance may be species-specific, it is not yet possible to support a general concept of a CO_2-induced reduction in the flux of O_3 into the plant. Moreover, in a given plant species, protection by high [CO_2] from a particular adverse effect is not necessarily associated with the protection against another adverse effect. For instance, in wheat plants an elevated [CO_2] provided full protection from effects of O_3

Table 1 Selected examples of the effects of elevated [O_3] and [CO_2], singly or in combination, on plant metabolic and growth responses (examples with significant adverse effects of O_3 only)

Species	O_3 Effect	CO_2 Effect	O_3/CO_2 Effect
Potato (cv. Bintje)	Decreased chlorophyll content; visible foliar leaf injury	n.e.	Adverse effect of O_3 on chlorophyll content unchanged; reduced degree of visible O_3-induced leaf injury
Wheat	Visible leaf injury (30%)	n.e.	Reduced degree of visible O_3-induced leaf injury (5%)
Wheat (cv. Minaret)	Loss of Rubisco protein; decline in flag leaf CO_2 assimilation rate	Loss of Rubisco protein; increase in CO_2 assimilation rate	Amelioration of O_3 effects
Wheat (cv. Cocker 9904)	reduced seed yield	Slightly increased seed yield	Amelioration of O_3-induced yield loss
Wheat (cv. Minaret)	Reduced flag leaf photosynthesis	Increased flag leaf photosynthesis	Amelioration of negative O_3 effects
Wheat (cv. Hanno)	Reduced plant relative growth rate; reduced plant biomass	Slight increase in relative growth rate; increased plant biomass	Amelioration of negative O_3 effects
Soybean (cv. Essex)	Reduced seed yield (40%)	Insignificant increase of seed yield	Amelioration of yield suppression
Cotton (cv. Deltapine 51)	Reduced leaf area per mass; reduced starch contents	Increased leaf area per mass and starch contents	Prevention of adverse effects of O_3 by CO_2
Norway spruce	visible leaf injury (chlorotic mottling)	n.e.	No effect of CO_2 on the degree of O_3-induced leaf injury
Trembling aspen (different O_3-sensitive and -tolerant clones)	Reduced tree growth parameters (height, diameter, volume)	Enhancement of growth parameters	No effect of CO_2 on the degree of O_3-induced growth reductions

Compiled from Refs. 4,7–9; n.e. = no effect.

on total plant biomass, but not on grain yield. Hence, the available data are still too limited to allow for a unified view of how these two gases might interact.[7]

CO_2 and Other Air Pollutants

Hardly any studies have addressed the combined effects of elevated [CO_2] and of other air pollutants. Studies of the combined effects of elevated [CO_2] and nitrogen oxides (NO, NO_2) are confined to commercial greenhouses under conditions of horticultural crop production and are not considered here. SO_2 has been found to adversely affect agricultural crops and forest plants in a large number of studies.[8,9] Reduced photosynthesis, altered water relations, growth retardations, yield losses and altered susceptibilities to other stresses are common plant responses observed under SO_2 stress.

It was shown for a range of plant species under various exposure conditions that elevated [CO_2] reduced the sensitivity of the plants to SO_2 injury or protected them from negative effects of SO_2 on growth and yield.[9] With the combined exposure of crop species to both gases, the yield increments were sometimes even larger when compared to the stimulation observed with exposure to elevated [CO_2] alone, suggesting that the plants were able to use the airborne sulphur more effectively under the conditions of enhanced carbon availability. It must be kept in mind that low to moderate SO_2 concentrations may confer a nutritional benefit to plants, particularly under conditions of low sulphur availability in the soil.

CONCLUSION

Rising atmospheric concentrations of CO_2 and air pollutants can have interactive effects on agricultural and wild plant species. Existing evidence on potential interactions is almost exclusively restricted to CO_2 and tropospheric O_3, the concentrations of which are increasing globally. While high CO_2 levels are beneficial to plants, current ambient [O_3] is high enough to impair plants in many regions of the world. There is ambiguous information in the literature concerning the protective effect of elevated [CO_2] from adverse effects of O_3, but it has been demonstrated to occur in many experimental studies. The mechanisms by which elevated [CO_2] and O_3 interact at the physiological and metabolic level remain uncertain. There is also some evidence that rising [CO_2] may protect plants against phytotoxic SO_2 concentrations. Overall the existing database on air pollutant/CO_2 interactions lacks a general conceptual model of the potential modes of interactions. Additional long-term field experiments will be necessary, combined with a better understanding of how other plant and environmental variables, such as plant genotype, soil water deficit, nutrient availability, or temperature, may modify the interaction.

ARTICLES OF FURTHER INTEREST

Air Pollutant Interactions with Pests and Pathogens, p. 9
Air Pollutants: Effects of Ozone on Crop Yield and Quality, p. 13
Air Pollutants: Responses of Plant Communities, p. 20
Breeding: Genotype-by-Environment Interaction, p. 218
Crop Responses to Elevated Carbon Dioxide, p. 346
Crops and Environmental Change, p. 370
Leaves and the Effects of Elevated Carbon Dioxide Levels, p. 648

REFERENCES

1. Dämmgen, U.; Weigel, H.J. Trends in Atmospheric Composition (Nutrients and Pollutants) and Their Interaction with Agroecosystems. In *Sustainable Agriculture for Food, Energy and Industry*; El Bassam, N., Behl, R.K., Prochnow, B., Eds.; James & James (Science Publishers) Ltd.: London, 1998; 85–93.
2. Houghton, J.T.; Ding, Y.; Griggs, D.G.; Noguer, M.; van der Linden, P.J.; Xiaosu, D. *Climate Change 2001: The Scientific Basis*; Cambridge University Press: Cambridge, United Kingdom, 2001.
3. Allen, L.H. Plant responses to rising CO_2 and potential interaction with air pollutants. J. Environ. Qual. **1990**, *19*, 15–34.
4. Olszyk, D.M.; Tingey, D.T.; Watrud, R.; Seidler, R.; Andersen, C. Interactive Effects of O_3 and CO_2: Implications for Terrestrial Ecosystems. In *Trace Gas Emissions and Plants*; Singh, S.N., Ed.; Kluwer Academic Publishers: Netherlands, 2000; 97–136.
5. Lefohn, A.S. *Surface Level Ozone Exposures and Their Effects on Vegetation*; Lewis Publishers, Inc.: Chelsea, 1992.
6. Reddy, K.R.; Hodges, H.F. *Climate Change and Global Crop Productivity*; CABI Publishing: Wallingford, 2000.
7. Polle, A.; Pell, E.J. The Role of Carbon Dioxide in Modifying the Plant Response to Ozone. In *Carbon Dioxide and Environmental Stress*; Luo, Y., Mooney, H.A., Eds.; Academic Press: San Diego, 1999; 193–213.
8. Runeckles, V. Air Pollution and Climate Change. In *Air Pollution and Plant Life*; Bell, J.N.B., Treshow, M., Eds.; John Wiley & Sons Ltd.: Chichester, 2002; 431–454.
9. Groth, J.V.; Krupa, S.V. Crop Ecosystem Responses to Climatic Change: Interactive Effects of Ozone, Ultraviolet-B Radiation, Sulphur Dioxide and Carbon Dioxide on Crops. In *Climate Change and Global Crop Productivity*; Reddy, K.R., Hodges, H.F., Eds.; CABI Publishing: Wallingford, 2000; 387–405.

Air Pollutants: Responses of Plant Communities

Andreas Fangmeier
University of Hohenheim, Stuttgart, Germany

INTRODUCTION

Responses of plant communities to air pollutants have been investigated much less than those of individual plants or single plant species. From studies comparing individuals of different species, it is clear that the sensitivity varies considerably among species. However, because of interactions between species, such as competition and facilitation, individual species responses do not represent the responses of the same individuals growing in communities. From the existing knowledge, it is difficult to draw a generalized picture of community responses. Rather, different communities tend to show individualistic responses.

AIR POLLUTANTS: NUTRIENTS AND/OR TOXIC AGENTS

Atmospheric pollutants can roughly be divided into three groups: 1) phytotoxic compounds that cause only adverse or toxic effects, such as ozone (O_3); 2) essential macro- and micronutrients that can act as fertilizers at low deposition rates but may have adverse effects at high deposition rates,[1] such as sulphur dioxide (SO_2) or nitrogen compounds (NO_x, NH_y); 3) pollutants that in most cases represent an essential resource, such as carbon dioxide (CO_2). In the case of pollutants causing adverse or toxic effects, it can be expected that a species that exhibits high sensitivity to the air pollutants when grown in monoculture will suffer even more when grown in competition with a less sensitive species. In the case of air pollutants that act as potential macro- and micronutrients, the situation is more complex because both the beneficial effects, including growth promotion due to the additional nutrient supply, and the adverse effects, such as lowered stress resistance,[2] must be taken into account on top of any plant–plant interactions.

SCALING FROM SINGLE SPECIES TO PLANT COMMUNITIES

The current knowledge of the responses of plant communities to changes in atmospheric quality is based on a much smaller data base than knowledge of the responses of single species grown in isolation. Roughly less than 1% of the experimental studies with air pollutants, such as O_3, SO_2, NO_x, NH_y, or elevated CO_2, have involved plant communities. The vast majority of experiments have used single plants or monocultures of a species grown under conditions that are not representative of their natural environment. This is mainly because field experimentation with plant communities is demanding and requires exposure experiments with large plot sizes, replicate numbers, and long durations in order to account for the biological variation found in natural habitats.

Sensitivity and responsiveness of plants to air pollutants differ considerably among species, and a sensitive species may be even more affected when grown together with a less responsive species because of the competitive advantages of the latter. Besides competition, other forms of interactions exist between different species, leading to a disadvantage, an advantage (e.g., facilitation), or no effect on either of the partners. Any alteration of the environmental conditions, such as a change in atmospheric quality, may affect these interactions and, hence, the specific responses to the change. Therefore, community reactions cannot easily be predicted from results of exposure experiments with single species. Some studies with communities have shown surprising results, e.g., the success of one particular species, which could never be expected from exposures with the same species grown in isolation. Hence, pollutant effects on plants are modified by the environment and by the presence of other species (Fig. 1). Nevertheless, some basic response patterns have emerged, and examples of illustrative studies are given below.

CASE STUDIES

Tropospheric Ozone—Phytotoxicity Overestimated?

Tropospheric ozone is regarded as the most important and most widespread gaseous air pollutant in many industrialized regions of the world. Its background concentrations have at least doubled over the past century.[3] Recent approaches to define thresholds above which exposures do

Encyclopedia of Plant and Crop Science
DOI: 10.1081/E-EPCS 120005570

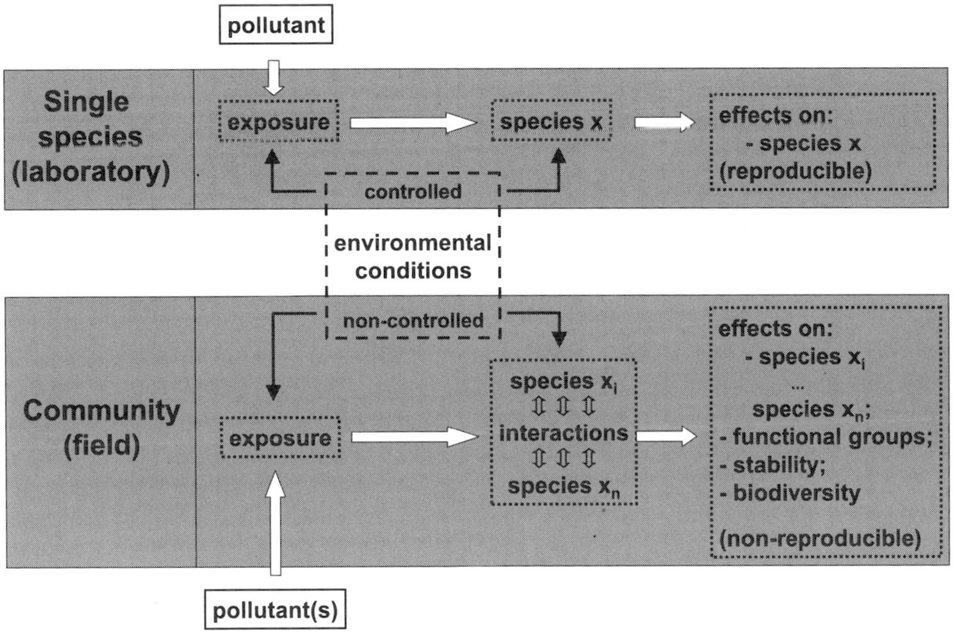

Fig. 1 Scheme of the differences in pollution effects on single species and on plant communities due to species interactions and the degree of variability in environmental conditions. For a better illustration of the problem, two extreme cases are compared: single species exposure under controlled laboratory condition and multispecies exposure in the field.

not harm sensitive vegetation have been based on the Critical Levels concept, which uses the exposure index AOT 40, i.e., the sum of all hourly ozone concentrations exceeding a baseline of 40 ppb during hours when global radiation levels exceed 50 W m^{-2}. For instance, an AOT 40 of 3000 ppb·h calculated over a three-month growing period is currently assumed to protect agricultural crops from negative effects of O_3 on yield[4] and to prevent damage to natural vegetation.

However, from the few experimental field studies with ozone involving plant communities, such thresholds cannot be confirmed. Significant effects on plant species diversity due to ozone in an early successional community were observed when twice ambient ozone concentrations were applied with an AOT 40 reaching 38,000 and 48,000 ppb·h in two consecutive experimental seasons, respectively, but not at lower exposure levels.[5] In an ongoing long-term study with a species-rich meadow, minimal effects of elevated O_3 on the plant community composition could be observed, although the AOT 40 values largely exceeded 3000 ppb·h in each year (P. Bungener, pers. comm.) Thus, it is not clear at present whether O_3 affects the composition of established plant communities, as it was expected from short-term exposure studies with monocultures. Further studies involving herbaceous seminatural ecosystems may give additional insight into the responses of plant communities to tropospheric O_3 by involving modeling based on either mechanistic or plant functional-type principles.[6]

Changes in Plant Communities Exposed to Sulphur Dioxide

Responses of plant communities to SO_2 are known either from field studies around point sources of emission or from fumigation experiments. The general picture emerging from these studies is that of a strong impact of SO_2 leading to different zones of vegetation corresponding to the severity of the pollutant exposure. Acute concentrations may inhibit plant growth completely and even lead to soil erosion because of the high load of acidity and the complete absence of vegetation. An example from a temperate forest region[7] showed that further away from the emission source a zone existed with a low-density cover formed by resistant grasses, herbs, and dwarf shrubs. With increasing distance and decreasing exposure, the herb cover increased and tree species of poor habitats could survive, thus providing the picture of a tree-line "ecotone" (the so-called "Kampfzone"). Symptoms of forest decline could be observed over longer distances from the emission source, and only at exposure levels close to the background concentrations did the upper tree canopy, which is most prone to adverse SO_2 effects, regain its natural shape and performance.

Nitrogen Deposition—An Unwanted Fertilizer

Nitrogen (N), like sulphur (S), is a macronutrient. However, the demand of vegetation for nitrogen is much higher than for sulphur. Typically, N:S ratios in plant tissues approximate 10:1 to 15:1. Therefore, sulphur becomes toxic at much lower rates of deposition compared with nitrogen, and current loads of nitrogen often lead to enhanced growth and productivity rather than to injury. However, eutrophication of ecosystems due to excess nitrogen inputs may also have adverse effects: Nutrient-poor stands that carry some of the most diverse plant communities may be lost, and ecosystems may become more susceptible to different forms of additional stresses.

Most plant species from natural habitats are adapted to nutrient-poor conditions and, because they have a low competitive ability against nitrophilous species, they can only compete successfully in systems with low nitrogen input. Consequently, changes in species composition caused by high nitrogen loads, often associated with a loss of biodiversity, have been observed in several ecosystems. Over the past decades, the forest floor vegetation at many locations in western, central, and northern Europe has seen an increase in nitrophilous species. In heathlands, which represent seminatural ecosystems in most of their area of distribution in western Europe, a transition to grasslands has occurred. Apparently, the equilibrium between nutrient output by periodic removal due to grazing and sod removal and nutrient input by mineralization in these heathlands has been disturbed. This is due to not only the absence of management (e.g., less or no sheep grazing, abandonment of sod removal), but also excess nitrogen deposition from the atmosphere.[8] Somewhat similar effects have been observed in nutrient-poor grasslands, in particular in calcareous grasslands. Because of the soil conditions or as a result of management leading to removal of nutrients by grazing or hay making, these grasslands remain nutrient-poor and carry a high species diversity with many endangered plant and animal species present, and, therefore, they have been set aside as nature reserves. Excess nitrogen deposition to calcareous grasslands strongly stimulates the growth of *Brachypodium pinnatum*, leading to the formation of a dense cover that reduces the light quantity and quality in the lower parts of the canopy, which, in turn, causes a drastic reduction in species diversity.[9]

Besides these effects, nitrogen deposition has been shown to reduce stress resistance. Coniferous trees and heathland shrubs such as *Calluna vulgaris* exhibited lower tolerance to drought and frost after moderate exposure to ammonia (NH_3), and the attack on *C. vulgaris* by an insect herbivore, the heather beetle (*Lochmaea suturalis*), was more severe under conditions of excess nitrogen.[8]

CO_2 Enrichment: The Planet Might Get Greener, But Is This All?

In general, CO_2 enrichment promotes vegetation growth and productivity and makes plants more efficient in their use of resources. Irrespective of the type of the CO_2 fixation pathway (C_3, C_4, or CAM) of a particular species, higher plants profit from CO_2 enrichment by increased water-use efficiency. In addition, plants with C_3-photosynthesis show higher nitrogen-use efficiency when grown at elevated CO_2 levels.

At present, the ambient atmospheric CO_2 concentration limits plant growth in species involving the C_3 fixation pathway. In contrast, C_4 plants possess an efficient CO_2 pre-fixation mechanism, and their photosynthesis is CO_2-saturated at the present atmospheric CO_2 concentration. Therefore, C_3 plants are expected to profit more from CO_2 enrichment than C_4 plants in terms of growth and productivity. This expectation has been confirmed in some, but not all, experimental field studies with plant communities. In a salt marsh under long-term CO_2 enrichment, the C_3 sedge gained competitive advantages over the C_4 grasses.[10] In a tallgrass prairie ecosystem composed of tall warm-season perennial C_4 grasses and smaller cool-season perennial C_3 grasses plus C_3 forbs and C_3 *Cyperaceae* members, eight years of exposure to CO_2 enrichment had little effect on the C_4 grasses but caused a decline in cool-season C_3 grasses and an increase in C_3 forbs and C_3 *Cyperaceae*.[11] In a calcareous grassland composed of C_3 species belonging to different functional groups (graminoids, nonleguminous forbs, and legumes), the legumes were most responsive to CO_2 enrichment in one year, and the forbs in the following year.[12] At the species level, the biomass response ranged from a decrease in several *Trifolium* species to an increase by 271% in *Lotus corniculatus* and by 249% in *Carex flacca*. These examples illustrate the current difficulty in predicting any general responses to CO_2 enrichment across different plant communities.

CONCLUSION

Responses of plant communities to air pollutants may be much more subtle and therefore more difficult to detect than effects on single species or individual plants. Unfortunately, knowledge is restricted to a limited number of experimental exposures with artificial mixtures of species and to an even more limited number of field exposures involving natural plant communities. Thus, at this stage it is difficult to draw general conclusions from limited data. Some general patterns emerge from studies either of the effects of SO_2 pollution, with climax tree species being most susceptible, or of excess nitrogen deposition, e.g., loss of species adapted to nitrogen-poor habitats, whereas no general picture is available concerning effects of O_3 or elevated CO_2. This is due to the lack of data and the fact that individual communities show specific responses. Furthermore, hardly any knowledge exists about the resilience of plant communities in response to air pollutants. From long-term studies on SO_2 effects, it can be concluded that communities and ecosystems are able to regain their structure and function upon termination of the air pollution stress. It may well be possible that complex plant communities can react flexibly to air pollutants, and, consequently, effects are less than expected. Conversely, it seems possible that any pollution level above the natural background could have subtle adverse effects—e.g., in terms of losses in biodiversity, stability, or other ecosystem services and thus should not be tolerated. Furthermore, it should be noted that effects of combinations of air pollutants on plant communities have rarely been investigated. Therefore, the current approach to protect vegetation and ecosystems from adverse effects of pollutants by means of the Critical Levels and Critical Loads concept must be reevaluated regularly and critically as new information on air pollution effects at the community level becomes available.

ARTICLES OF FURTHER INTEREST

Air Pollutant Interactions with Pests and Pathogens, p. 9
Air Pollutants: Effects of Ozone on Crop Yield and Quality, p. 13
Air Pollutants: Interactions with Elevated Carbon Dioxide, p. 17
Crop Responses to Elevated Carbon Dioxide, p. 346
Crops and Environmental Change, p. 370
Ecology: Functions, Patterns, and Evolution, p. 404
Exchange of Trace Gases Between Crops and the Atmosphere, p. 425
Nitrogen, p. 822

REFERENCES

1. Fangmeier, A.; Bender, J.; Weigel, H.J.; Jäger, H.-J. Effects of Pollutant Mixtures. In *Air Pollution and Plant Life*; Bell, J.N.B., Treshow, M., Eds.; Wiley-VCH: Weinheim, 2002; 251–272.
2. Fangmeier, A.; Hadwiger-Fangmeier, A.; Van der Eerden, L.J.M.; Jäger, H.-J. Effects of atmospheric ammonia on vegetation—A review. Environ. Pollut. **1994**, *86*, 43–82.
3. Volz, A.; Kley, D. Evaluation of the Montsouris series of ozone measurements made in the nineteenth century. Nature **1988**, *332*, 240–242.
4. Fuhrer, J.; Skärby, L.; Ashmore, M.R. Critical levels for ozone effects on vegetation in Europe. Environ. Pollut. **1997**, *97*, 91–106.
5. Barbo, D.N.; Chappelka, A.H.; Somers, G.L.; Miller-Goodman, M.S.; Stolte, K. Diversity of an early successional plant community as influenced by ozone. New Phytol. **1998**, *138*, 653–662.
6. BIOSTRESS project website: http://www.uni-hohenheim.de/biostress.
7. Guderian, R. *Air Pollution: Phytotoxicity of Acidic Gases and Its Significance in Air Pollution Control*; Springer: Berlin, 1977; 127 pp.
8. Berdowski, J.J.M. The Effect of External Stress and Disturbance Factors on *Calluna*-Dominated Heathland Vegetation. In *Heathland: Patterns and Processes in a Changing Environment*; Aerts, R., Heil, G.W., Eds.; Kluwer: Dordrecht, 1993; 85–124.
9. Bobbink, R. Effects of nutrient enrichment in Dutch chalk grassland. J. Appl. Ecol. **1991**, *28*, 28–41.
10. Drake, B.G.; Peresta, G.; Beugeling, E.; Matamala, R. Long-Term Elevated CO_2 Exposure in a Chesapeake Bay Wetland: Ecosystem Gas Exchange, Primary Production and Tissue Nitrogen. In *Carbon Dioxide and Terrestrial Ecosystems*; Koch, G.W., Mooney, H.A., Eds.; Academic Press: San Diego, 1996; 197–214.
11. Owensby, C.E.; Ham, J.M.; Knapp, A.K.; Auen, L.M. Biomass production and species composition change in a tallgrass prairie ecosystem after long-term exposure to elevated atmospheric CO_2. Glob. Chang. Biol. **1999**, *5*, 497–506.
12. Leadley, P.W.; Niklaus, P.A.; Stocker, R.; Körner, C. A field study of the effects of elevated CO_2 on plant biomass and community structure in a calcareous grassland. Oecologia **1999**, *118*, 39–49.

Alternative Feedstocks for Bioprocessing

Joseph J. Bozell
National Renewable Energy Laboratory, Golden, Colorado, U.S.A.

INTRODUCTION

Bioprocessing is an inexact term often incorrectly limited to the single context of combining a fermentable sugar with an organism for the purpose of making a chemical. This is a common misconception, and one that must be dispelled, because bioprocessing includes all aspects of the use of renewable and sustainable materials for the production of chemical products, fuels, and power. Alternative feedstocks is a term more easily understood, and for the purposes of this discussion, will mean biomass-derived raw materials and building blocks, i.e., renewables. Bioprocessing deals with three primary issues: supply, separation, and conversion.

DISCUSSION

Supply issues are primarily associated with the growing and collection of the biomass feedstock. Bioprocessing deals with the source of the starting feedstock, its availability, the best geographic location for production, its supply and sustainability, collection, densification, and processing.

In a direct analogy to crude oil, biomass feedstocks are complex mixtures of different materials that must be separated to obtain the primary raw materials for conversion into other products. Bioprocessing involves developing the best methods for separation and isolation of these renewable building blocks, such as lignocellulosic components (cellulose, hemicellulose, lignin), oils (soy oil, canola, etc.), protein, extractives, and higher-value chemicals.

Once isolated, renewable building blocks must be converted into chemical intermediates and final products. Bioprocessing investigates development of the best technology to carry out these transformations. The very limited view of bioprocessing being only a sugar interacting with an organism more correctly fits as a subset of conversion. It is important to realize that bioprocessing does not *require* biotechnology. Bioprocessing also includes the use of conventional chemical technology to carry out transformations on renewable building blocks, as well as hybrid transformations such as combining nonrenewable building blocks with biochemical technology. Lichtenthaler has written extensively about the use of conventional chemical technology to convert carbohydrates to new products.[1]

A comparison between a bioprocessing industry (the "biorefinery") and the existing petrochemical industry is appropriate because each faces the three primary issues. For the most part, issues of supply and separation are generally understood for both bioprocessing and petrochemical refining. However, conversion processes are not as well understood for bioprocessing as they are for petrochemical processing. The primary difference between the two industries is *technology development*. The petrochemical industry has gained amazing transformational control over the behavior of their many crude oil derived building blocks.[2] In contrast, the analogous use of renewables suffers from a much narrower range of discrete building blocks, fewer methods to convert those building blocks to other materials, and a lack of information about the properties and performance available from those products. We are faced with the puzzle of possessing an almost limitless source of raw material in the United States, while being unable to effectively convert it to a wide range of useful products.

This discussion cannot cover all the specifics surrounding these three very broad areas. Accordingly, this brief overview will describe only the major concepts involving bioprocessing and alternative feedstocks for the production of new chemical intermediates and products.

THE CASE FOR RENEWABLES

Well into the 20th century, bioprocessing of renewable feedstocks supplied a significant portion of the United States' chemical needs. The chemurgy movement of the 1930s, led by such notables as William Hale and Henry Ford, promoted the use of farm products as a source of chemicals, with the belief that "anything that can be made from a hydrocarbon could be made from a carbohydrate."[3,4][a] It is only in the period of time

[a]Interestingly, the chemurgical movement was not initiated as a result of a shortage of other feedstocks, but rather as a way to use farm surpluses.

Encyclopedia of Plant and Crop Science
DOI: 10.1081/E-EPCS 120010437

between 1920 and 1950 that we have witnessed the transition to a nonrenewables-based economy.[5]

A vast amount of renewable carbon is produced in the biosphere. About 77×10^9 tons is fixed annually, an amount that could supply almost all domestic organic chemical needs, currently about 7–8% of our total nonrenewables consumption.[6–8] When measured in energy terms, the amount of carbon synthesized is equivalent to about ten times the world consumption.[9] Cellulose, the most abundant organic chemical on Earth, has an annual production of about 100×10^9 tons.

Yet our chemical feedstock supply is overwhelmingly dominated by nonrenewable carbon. Only about 2% comes from renewable sources,[10] thus, relatively few examples of large scale industrial bioprocessing exist. Two notable exceptions are the pulp and paper and the corn wet milling industries. Both convert huge amounts of renewable feedstocks into market products. The corn industry alone produces $8–10 \times 10^9$ bushels of corn/yr. Each bushel contains 33 lb of renewable carbon as glucose, and a corn harvest of 10×10^9 bushels is equivalent to 500×10^6 barrels of crude oil.[11] Over 1×10^9 bushels of this supply is converted to ethanol (EtOH) and high fructose corn syrup. The pulp and paper industry consumes over 100×10^6 metric tons/yr of wood.[12] Specialty dissolving-grade celluloses are used for the production of over 1.4×10^6 metric tons/yr of cellulose esters, ethers, and related materials. Lignin production by the pulp and paper industry is $30–50 \times 10^6$ metric tons/yr.[13] The experience of these industries would seem to indicate that renewables hold considerable promise as feedstocks, complementary to those used by the chemical industry. However, the pulp and paper industry devotes only a small part of its production to chemicals, while the corn wet milling industry is focused largely on starch and its commercial derivatives, ethanol and corn syrup.

Several advantages are frequently associated with bioprocessing of alternative feedstocks:

- The use of biomass has been suggested as a way to mitigate the buildup of greenhouse CO_2 in the atmosphere.[14] Because biomass uses CO_2 for growth through photosynthesis, the use of biomass as a feedstock results in no net increase in atmospheric CO_2 content when the products break down in the environment.[15]b
- It is generally acknowledged that increased use of biomass would extend the lifetime of the available crude oil supplies.[16,17]c
- A chemical industry incorporating a significant percentage of renewable materials is more secure because the feedstock supplies are domestic, leading to a lessened dependence on international "hot spots."
- Biomass is a more flexible feedstock than is crude oil. For example, the advent of genetic engineering has allowed the tailoring of certain plants to produce high levels of specific chemicals.

Moreover, increased use of renewable feedstocks could address broader issues:

- **Global Feedstock Needs:** Recent work has attempted to model when world oil production will peak and has concluded that a decline will begin sometime in the next 5–10 years.[18] Demand will not decrease in line with production. United States energy consumption has increased by more than 28% (about 21 quadrillion btu) during the last 25 years, but more than half of the overall energy growth of the last 25 years (about 11 quadrillion btu) has occurred during the last 6 years.[19]
- **Domestic Energy Consumption:** The United States annually consumes about 94 quads of energy.[19] Of this, 35 quads are used by industry in general, almost 8 quads of which are used in the production of chemicals and paper. This is a significant energy target, and one that could be addressed by a greater use of renewables.

ECONOMICS OF BIOPROCESSING

Economic issues do not appear to present a major hurdle to the use of bioprocessing. The costs of a number of polymeric and monomeric materials compete favorably with nonrenewables (Table 1).[20–22] Other evaluations show that two of the most basic renewable feedstocks, corn and cellulose, are competitive with several fossil feedstocks on both a mass and an energy basis (Table 2).[23] The breakeven price for oil when compared to cellulosic

---

[b]This statement presumes that the CO_2 contained in standing trees, debris, and soil is not released and that consumed biomass is replaced by fresh plantings, i.e., the system must be sustainable.

[c]The concept of "diminishing supplies" is subjective. During the oil crises of the 1970s, projections concluded that oil prices would be about $100/barrel by the late 20th century. In contrast to expectations, crude oil is still relatively inexpensive. A projection by the International Energy Authority predicts that the price of oil (in 1993 dollars) will rise only to $28/barrel by 2050.

Table 1 Costs of some selected renewable feedstocks

Material	US$/kg	US$/lb	Cost type	Source
Polymers				
Cellulose	0.44–1.10	0.20–0.50	Production	a
Lignin	0.07–0.13	0.03–0.06	Production	Fuel value
Carbohydrates				
Glucose	0.60–1.10	0.27–0.50	Sales	a
	0.13–0.26	0.06–0.12	Production	c
Xylose/arabinose	0.07–0.13	0.03–0.06	Production	c
Sucrose	0.40	0.18	Sales	b
Lactose	0.65	0.30	Sales	b
	0.50–1.50	0.23–0.68	Sales	a
Fructose	0.90	0.41	Sales	a
Sorbitol	1.60	0.73	Sales	a
Other				
Levulinic acid	0.18–0.26	0.08–0.12	Production	c

[a]From Ref. 20.
[b]From Ref. 21.
[c]Range of estimates from discussions with various industrial sources.

biomass at $40/ton is $12.7/barrel on an energy basis and $6/barrel on a mass basis.

TECHNOLOGY FOR BIOPROCESSING

Primary examples of bioprocessing in the chemical industry are represented in two major areas:

- **Conventional Processing:** Renewable feedstocks can be converted to marketplace chemicals using conventional chemical transformations. This approach was frequently used in the early part of the 20th century. It offers the advantage of using the vast knowledge possessed by today's chemical industry for the conversion of a new feedstock source to products. However, the technology development issues described earlier have hindered greater use of these kinds of approaches.
- **Biocatalysis:** Biocatalysis offers the benefits of proceeding under very mild conditions, frequently in aqueous reaction media, and the use of readily available renewable feedstocks as starting materials. However, issues of separation, expense, productivity, maintenance of organisms, and new capital investment have so far limited the use of bioprocesses in the chemical industry, except where no other alternatives are available.[24]

Significantly absent from the standard tools available for bioprocessing is the use of nonbiological catalysis. Catalysis is proving to be a powerful technology for the petrochemical-based chemical industry, with 80–90% of all chemical processes involving at least one catalytic

Table 2 Comparative raw material costs[a]

	Oil	Natural gas	Coal	Corn	Lignocellulosics
$/dry tonne	129 ($17.5/barrel) 94 ($12.7/barrel) 44 ($6/barrel)	122 ($2.50/1000 scf)	33	98 (kernels, $2.50/bushel) 19 (stover)	44 (poplar, switchgrass)
$/GJ	3.1 ($17.5/barrel) 2.3 ($12.7/barrel) 1.2 ($6/barrel)	2.3	1.0	5.0 (kernels, $2.50/bushel) 1.0 (stover)	2.3 (poplar, switchgrass)

[a]Dollars are US$.

step.[25] Catalysis will also see increasing use as a tool for bioprocessing.[26–29]

EXAMPLES OF BIOPROCESSING IN THE PRODUCTION OF CHEMICALS

Conventional Processing

Conventional transformation of renewable feedstocks to products is not common in the chemical industry. In many cases, the products are those that can be isolated from existing renewable resources without further structural transformation. Examples include extractives from the pulp and paper industry used in the production of turpentine, tall oils, and rosins,[30] the production of oils from corn or other oil crops,[31] or starch-based polymers.[32] Other products are made by simple derivatization of the materials found naturally occurring in biomass. The pulp and paper industry produces a number of chemicals, including a wide range of cellulose derivatives such as cellulose esters and ethers, rayons, cellophane, etc. A few well-known routes exist for the conversion of renewables to low-molecular-weight monomeric products. Glucose is converted to sorbitol by catalytic hydrogenation, and to gluconic acid by oxidation. Furfural is manufactured by the acidic dehydration of corn byproducts.[33] Xylitol is also produced from xylose by hydrogenation.[34] Vanillin[35] and DMSO have been commercially produced from lignin.

Some renewables-based materials under development have promise, but their production is not yet commercialized. These products include levulinic acid[36] and its derivatives; methyltetrahydrofuran, an automobile fuel extender;[37,38] δ-aminolevulinic acid, a broad spectrum biodegradable herbicide, insecticide, and cancer treatment;[39–43] and diphenolic acid, a material for the production of polymers and other materials.[44] Levoglucosan and levoglucosenone are products of sugar pyrolysis. Hydroxymethylfurfural is made by acid treatment of sugars.[45–47]

Lignin has also been investigated as a chemical feedstock (for example, in the production of quinones.[48]) and has been widely suggested as a component in graft copolymers or polymer blends.[49]

Biocatalysis

The most successful route so far for introducing renewables to the chemical industry has been through biotechnology, although its use for the production of large-volume chemicals is only starting to be realized. Most examples of the use of organisms or enzymatic steps in the production of chemicals have been limited to low-volume, high-value fine chemicals and pharmaceuticals such as oligosaccharides, amino acids, purines, vitamins, nicotine, or indigo.[50–53] This is a sensible first application, given the strict structural requirements of many of these specialty materials. Biocatalysts are generally unchallenged in their ability to provide the stereo-, regio-, and enantioselectivity required by these specialty products. The development of more robust biological systems that can operate in extreme conditions (temperature, low water levels, in organic solvents, under high hydrostatic pressure) will also broaden their applicability.[54–57] An important new example of industrial biotechnology is the Mitsubishi Rayon process for acrylamide, currently operating on a 3×10^{4} metric ton/yr basis by the treatment of nonrenewable acrylonitrile with nitrile hydratase.[58,59]

Table 3 Commercial uses of biotechnology

Compound	1000 tons/yr	US$/ton
Glucose	15,000	600
EtOH	13,000	400
Fructose	1,000	800
Citric acid	800	1,700
Monosodium glutamate	800	1,900
L-lysine	350	2,200
L-lactic acid	70	2,100
L-ascorbic acid	60	1,000
Gluconic acid	40	1,700
Xanthan	30	8,000
Penicillin G	25	20,000
Aspartame	15	40,000

Several examples of large-scale biotechnological processes are known (Table 3).[60] Some operations have been used for many years because there is no equivalent nonbiological route. Ethanol and lactic acid are of particular interest, because they represent chemicals whose original nonbiological production has been almost totally replaced by biochemical manufacture. In addition, the pulp and paper industry has started to incorporate enzyme treatments into their pulping and bleaching sequences,[61] and low-lactose milk (up to 250,000 liters daily) is produced by treating milk with β-galactosidase.[62]

CONCLUSION

The United States possesses sufficient renewable resources to supply all domestic organic chemical needs without sacrificing traditional applications of renewables in the production of food, feed, and fiber.[63] Bioprocessing of renewables will play an important role in the future evolution of the chemical industry. Progress in the use of conventional chemical processing and catalysis for the conversion of renewables to products will see significant growth as nonrenewable crude oil feedstocks diminish and the world turns its attention to new carbon sources.

ARTICLES OF FURTHER INTEREST

Agriculture and Biodiversity, p. 1
Bioinformatics, p. 125
Genetically Modified Oil Crops, p. 509
Molecular Farming in Plants: Technology Plant Forms, p. 753
New Industrial Crops in Europe, p. 813
Sustainable Agriculture: Philosophical Framework, p. 1198
Soy Wood Adhesives for Agro-Based Composites, p. 1169
Starch, p. 1175
Sucrose, p. 1179
Sustainable Agriculture: Definition and Goals, p. 1187
Sweet Sorghum: Applications in Ethanol Production, p. 1201
Switchgrass As a Bioenergy Crop, p. 1207
Wax Esters from Transgenic Plants, p. 1292

REFERENCES

1. Lichtenthaler, F.W. Towards improving the utility of ketoses as organic raw materials. Carbohydr. Res. **1998**, *313*, 69–89.
2. Weissermel, K.; Arpe, H.-J. *Industrial Organic Chemistry*, 3rd Ed.; VCH: Weinheim, 1997.
3. Clark, J.P. Chemurgy. In *Kirk–Othmer Encyclopedia of Chemical Technology*, 4th Ed.; Kroschwitz, J.I., Howe-Grant, M., Eds.; John Wiley & Sons: New York, 1993; Vol. 5, 902.
4. Hale, W.J. *The Farm Chemurgic*; The Stratford Co.: Boston, MA, 1934.
5. Morris, D.; Ahmed, I. *The Carbohydrate Economy: Making Chemicals and Industrial Materials from Plant Matter*; Institute for Local Self Reliance: Washington, D.C., 1992.
6. Donaldson, T.L.; Culberson, O.L. An industry model of commodity chemicals from renewable resources. Energy **1984**, *9*, 693–707.
7. Lipinsky, E.S. Chemicals from biomass: Petrochemical substitution options. Science **1981**, *212*, 1465–1471.
8. Hanselmann, K.W. Lignochemicals. Experientia **1982**, *38*, 176–188.
9. Indergaard, M.; Johansson, A.; Crawford, B., Jr. Biomass Technologies. Chimia **1989**, *43*, 230–232.
10. McLaren, J.S. Future renewable resource needs: Will genomics help? J. Chem. Technol. Biotechnol. **2000**, *75*, 927–932.
11. Varadarajan, S.; Miller, D.J. Catalytic upgrading of fermentation-derived organic acids. Biotechnol. Prog. **1999**, *15*, 845–854.
12. *North American Pulp and Paper Factbook*; Miller Freeman: San Francisco, 2001.
13. Kuhad, R.C.; Singh, A. Lignocellulose biotechnology: Current and future prospects. Crit. Rev. Biotechnol. **1993**, *13*, 151–172.
14. Hall, D.O.; House, J.I. Biomass: A modern and environmentally acceptable fuel. Sol. Energy Mater. Sol. Cells **1995**, *38*, 521–542.
15. Harmon, M.E.; Ferrell, W.K.; Franklin, J.F. Effects on carbon storage of conversion of old-growth forests to young forests. Science **1990**, *247*, 699–702.
16. *The Evolution of the World's Energy System*; Group External Affairs, Shell International Limited, SIL Shell Centre: London, 1995, SE1 7NA.
17. Romm, J.J.; Curtis, C.R. Mideast oil forever? Atl. Mon. **1996**, *277* (4), 57–74.
18. Kerr, R.A. The next oil crisis looms large- and perhaps close. Science **1998**, *281*, 1128–1131.
19. *Energy Information Administration, Annual Energy Review 1997*; DOE/EIA 0384(97): Washington, DC, July 1998, information also available at http://www.eia.doe. gov.
20. Bols, M. *Carbohydrate Building Blocks*; Wiley-Interscience: New York, 1996.
21. Lichtenthaler, F.; Mondel, S. Perspectives in the use of low molecular weight carbohydrates as organic raw materials. Pure Appl. Chem. **1997**, *68*, 1853.
22. Chemical Market Reporter; National Corn Growers Association, weekly publication.
23. Lynd, L.R.; Wyman, C.E.; Gerngross, T.U. Biocommodity engineering. Biotechnol. Prog. **1998**, *15*, 777–793.
24. Rozzell, J.D. Commercial scale biocatalysis: Myths and realities. Bioorg. Med. Chem. **1999**, *7*, 2253–2261.
25. Holderich, W.F.; Roseler, J.; Heitmann, G.; Liebens, A.T. Use of zeolites in the synthesis of fine and intermediate chemicals. Catal. Today **1997**, *37*, 353–366.
26. Bozell, J.J.; Hoberg, J.O.; Claffey, D.; Hames, B.R.; Dimmel, D.R. *Green Chemistry: Frontiers in Benign Chemical Synthesis and Processing*; Anastas, P.T., Williamson, T.C., Eds.; University Press: Oxford, 1998, Chapter 2.
27. Bozell, J.J.; Hoberg, J.O.; Claffey, D.; Hames, B.R. In *New Methodology for the Use of Renewable Materials as Chemical Feedstocks*, Gordon Conference on Green Chemistry, July 1996.
28. Hoberg, J.O.; Bozell, J.J. Cyclopropanation and ring-expansion of unsaturated sugars. Tetrahedron Lett. **1995**, *36*, 6831–6834.
29. Bozell, J.J.; Hoberg, J.O. In *Organometallic Carbohydrate Chemistry. The Reaction of Transition Metal Nucleophiles with Carbohydrate Derivatives, Abstracts of Papers*, American Chemical Society National Meeting, March 1993.
30. Fengel, D.; Wegener, G. *Wood: Chemistry, Ultrastructure and Reactions*; Walter DeGruyter: New York, 1984.

31. Murphy, D.J. *Designer Oil Crops. Breeding, Processing and Biotechnology*; VCH: Weinheim, 1994.
32. Bastioli, C. *Carbohydrates as Organic Raw Material IV*; Praznik, W., Huber, A., Eds.; WUV: Vienna, 1998; 218.
33. *Alternative Feedstocks Program Technical and Economic Assessment: Thermal/Chemical and Bioprocessing Components*; Bozell, J.J., Landucci, R., Eds.; U.S. Department of Energy Office of Industrial Technologies: Washington, D.C., 1993.
34. Melaja, A. Hemalainen L. U.S. Patent 4008285, 1977.
35. Clark, G.S. Vanillin. Perfum. Flavor **1990**, *15*, 45–46.
36. Bozell, J.J.; Moens, L.; Elliott, D.C.; Wang, Y.; Neuenscwander, G.G.; Fitzpatrick, S.W.; Bilski, R.J.; Jarnefeld, J.L. Production of levulinic acid and use as a platform chemical for derived products. Resour. Conserv. Recycl. **2000**, *28*, 227–239.
37. Thomas, J.J.; Barile, R.G. Conversion of Cellulose Hydrolysis Products to Fuels and Chemical Feedstocks. Symposium papers from Energy from Biomass and Wastes VIII, Lake Buena Vista, Florida, January 30–February 3, 1984, p. 1461–1494.
38. Elliott, D.C.; Frye, J.G., Jr. U.S. Patent 5883266 to Battelle Memorial Institute, 1999.
39. Moens, L. U.S. Patent 5907058, to Midwest Research Institute, 1999.
40. Rebeiz, C.A.; Montazer-Zouhoor, A.; Hopen, H.J.; Wu, S.M. Photodynamic herbicides: 1. Concept and phenomenology. Enzyme Microb. Technol. **1984**, *6*, 390–401.
41. Rebeiz, C.A.; Amindari, S.; Reddy, K.N.; Nandihalli, U.B.; Moubarak, M.B.; Velu, J.A. *Porphyric Pesticides: Chemistry, Toxicology, and Pharmaceutical Applications*; ACS Symposium Series, American Chemical Society, Washington, D.C., 1994; Vol. 559.
42. Rebeiz, C.A.; Juvik, J.A.; Rebeiz, C.C. Porphyric insecticides. 1. Concept and phenomenology. Pestic. Biochem. Physiol. **1988**, *30*, 11–27.
43. Rebeiz, N.; Arkins, S.; Rebeiz, C.A.; Simon, J.; Zachary, J.F.; Kelley, K.W. Induction of tumor necrosis by delta-aminolevulinic-acid and 1,10-phenanthroline photodynamic therapy. Cancer Res. **1996**, *56*, 339–344.
44. Isoda, Y.; Azuma, M. Japanese Patent 08053390 to Honshu Chemical Ind., 1996.
45. *Levoglucosenone and Levoglucosans. Chemistry and Applications*; Witczak, Z.J., Ed.; ATL Press: Mount Prospect, IL, 1994.
46. Schiweck, H.; Munir, M.; Rapp, K.M.; Schneider, B.; Vogel, M. *Carbohydrates as Organic Raw Materials*; Lichtenthaler, F.W., Ed.; VCH: Weinheim, 1991; Ch. 3.
47. So, K.S.; Brown, R.C. Economic analysis of selected lignocellulose-to-ethanol conversion technologies. Appl. Biochem. Biotechnol. **1999**, *77–79*, 633–640.
48. Bozell, J.J.; Hames, B.R.; Dimmel, D.R. Cobalt-Schiff base complex catalyzed oxidation of para-substituted phenolics. Preparation of benzoquinones. J. Org. Chem. **1995**, *60*, 2398–2404.
49. Mai, C.; Majcherczyk, A.; Huttermann, A. Chemo-enzymatic synthesis and characterization of graft copolymers from lignin and acrylic compounds. Enzyme Microb. Technol. **2000**, *27*, 167–175.
50. Bommarius, A.S.; Schwarm, M.; Drauz, K. Biocatalysis to amino acid-based chiral pharmaceuticals—Examples and perspectives. J. Mol. Catal., B **1998**, *5*, 1–11.
51. Wandrey, C.; Liese, A.; Kihumbu, D. Industrial biocatalysis: Past, present, and future. Org. Process Res. Dev. **2000**, *4*, 286–290.
52. Schulze, B.; Wubbolts, M.G. Biocatalysis for industrial production of fine chemicals. Curr. Opin. Biotechnol. **1999**, *10*, 609–615.
53. Petersen, M.; Kiener, A. Biocatalysis: Preparation and functionalization of N-heterocycles. Green Chem. **1999**, *1*, 99–106.
54. Sellek, G.A.; Chaudhuri, J.B. Biocatalysis in organic media using enzymes from extremophiles. Enzyme Microb. Technol. **1999**, *25*, 471–482.
55. Adams, M.W.W.; Perler, F.B.; Kelly, R.M. Extremozymes: Expanding the limits of biocatalysis. Bio/Technology **1995**, *13*, 662–668.
56. Niehaus, F.; Bertoldo, C.; Kahler, M.; Antranikian, G. Extremophiles as a source of novel enzymes for industrial application. Appl. Microbiol. Biotechnol. **1999**, *51*, 711–729.
57. Gerday, C.; Aittaleb, M.; Bentahir, M.; Chessa, J.-P.; Claverie, P.; Collins, T.; D'Amico, S.; Dumont, J.; Garsoux, G.; Georlette, D.; Hoyoux, A.; Lonhienne, T.; Meuwis, M.-A.; Feller, G. Cold adapted enzymes: From fundamentals to biotechnology. Trends Biotechnol. **2000**, *18*, 103–107.
58. Ogawa, J.; Shimizu, S. Microbial enzymes: New industrial applications from traditional screening methods. Trends Biotechnol. **1999**, *17*, 13–21.
59. Shimizu, S.; Ogawa, J.; Kataoka, M.; Kobayashi, M. *New Enzymes for Organic Synthesis*; Scheper, T., Ed.; Springer: New York, 1997.
60. Wilke, D. Chemicals from biotechnology: Molecular plant genetics will challenge the chemical and the fermentation industry. Appl. Microbiol. Biotechnol. **1999**, *52*, 135–145.
61. Bajpai, P. Application of enzymes in the pulp and paper industry. Biotechnol. Prog. **1999**, *15*, 147–157.
62. Liese, A.; Filho, M.V. Production of fine chemicals using biocatalysis. Curr. Opin. Biotechnol. **1999**, *10*, 595–603.
63. Okkerse, C.; van Bekkum, H. From fossil to green. Green Chem. **1999**, *1*, 107–114.

Amino Acid and Protein Metabolism

Estela M. Valle
Facultad de Ciencias Bioquímicas y Farmacéuticas, Rosario, Argentina

INTRODUCTION

Amino acid metabolism is one of the most important biochemical processes in plants; similar to other topics in biochemistry, it has been affected by the tremendous developments in science. Following are four actively studied aspects in the field of amino acid metabolism: 1) the identification of new transporters of amino acids and other N-forms like nitrate and ammonium; 2) the characterization of factors controlling the metabolism in situ in distinct cells or subcellular compartments; 3) the regulation of the multiple isoenzymes of amino acid metabolism in the context of a single plant; and 4) the role of amino acids as signaling molecules.

Studies on protein metabolism, on the other hand, have focused on the processes of protein synthesis. The complex regulation of protein degradation is today attracting more attention, particularly because proteolysis is involved in cellular processes such as programmed cell death, circadian rhythm, and the defense response in plants. This chapter summarizes the latest insights in the studies of amino acid metabolism and protein degradation.

The focus of this review is to examine the regulation of amino acid metabolism and protein processing in the context of a single plant. Particular emphasis is given to the enzymes involved in NH_4^+ assimilation, which are often oligomers located in different subcellular compartments. The mechanism operating in oxidative protein cleavage is also discussed.

AMINO ACID METABOLISM

Ammonium is the inorganic N-form to be incorporated into carbon skeletons for the production of amino acids. Several transporters have been identified for NO_3^-, NH_4^+, and amino acids that contribute to a wide array of physiological activities.[1] N-forms assimilated into glutamine or glutamate disseminate into plant metabolism because they are N-donors to other amino acids, nucleotides, chlorophylls, polyamines, and alkaloids.[2] Amino acids are well known as building blocks of proteins, and they are essential in both primary and secondary plant metabolism. Various amino acids also perform other important roles, as signalling molecules or precursors of stress-related compounds under adverse environmental conditions,[3] in particular glutamate, which is found in the intersection of several metabolic pathways. In the chloroplast, one net glutamate molecule is produced by the concerted action of glutamine synthetase (GS) and ferredoxin-dependent glutamate synthase (Fd-GOGAT), which form the GS-GOGAT cycle, responsible for the prevalent NH_4^+ assimilation in plants.[3] Additionally, two molecules of glutamate are generated as the end product of lysine catabolism by the saccharopine pathway in seeds of cereals and dicots.[4] This pathway may be involved in the transient synthesis of glutamate, which then functions as messenger between cells during organ development or in response to environmental changes.[4] In fact, glutamate could be converted into γ-aminobutyric acid, a stress-related signalling molecule; proline, an osmolyte providing drought-tolerance under water stress; and arginine, a precursor of polyamines and nitric oxide generated during stress.[3,4]

In the era of functional genomics and metabolic engineering, new experimental approaches are appearing that help us understand the regulation of amino acid metabolism. Among the analytical methods, nuclear magnetic resonance (NMR) spectroscopy is a promising technique yield insight into the integration and regulation of plant metabolism through nondestructive and noninvasive measurements.[5] Through in vivo NMR methods and under certain experimental conditions, it was possible to differentiate amino acid pools from cytosolic or vacuolar compartments. Certainly, by inducing alkalization in sycamore cells and leaves of Kerguelen cabbage, it was demonstrated that the concentration of amino acids in the cytosol was much higher than in the vacuole, and that proline accumulated to a concentration 2–3 times greater in the cytosol than in the vacuole.[5]

The enzymes involved in NH_4^+ assimilation are generally isoenzymes of different oligomeric arrangements, which are often located in particular subcellular compartments or within different organs and tissues.[3] Whether these enzymes play overlapping (redundant) or distinct (nonredundant) roles, the factors controlling this process during plant growth and development are still a matter of discussion. The strategies used to study these topics are the production of either mutant plants defective in a particular isoenzyme or transgenic

Encyclopedia of Plant and Crop Science
DOI: 10.1081/E-EPCS 120010398

plants overexpressing one gene member of a small gene family regulated differentially during the life span of the cells. To illustrate the application of these technologies, studies were selected showing the role of enzymes involved in the glutamate metabolism.

All plants contained two types of GOGAT, an NADH-dependent enzyme and a Fd-GOGAT, unique to photosynthetic organisms.[3] Thus, while Fd-GOGAT accounts for 96% of the total GOGAT activity in leaves, NADH-GOGAT constitutes the predominant isoenzyme in roots. To assess the in vivo role of GOGAT in primary nitrogen assimilation and in photorespiration, an Arabidopsis mutant deficient in Fd-GOGAT was studied.[3] Gene expression combined with Fd-GOGAT-deficient mutant analyses demonstrated that Arabidopsis contains two expressed genes (*GLU1* and *GLU2*) encoding two distinct Fd-GOGAT isoforms. *GLU1* gene product plays a major role in photorespiration as well as in primary nitrogen assimilation in leaves. The Fd-GOGAT isoenzyme encoded by *GLU2* is proposed to be involved in primary nitrogen assimilation in roots.[3] These contrasting patterns of gene expression suggest nonoverlapping roles for *GLU1* and *GLU2*.

As an example of the second approach is the constitutive overexpression of the cytosolic GS in alfalfa.[6] GS isoenzymes can be localized in the chloroplast (GS_2) or in the cytosol (GS_1), and they have distinct in vivo functions.[7] Plants appear to possess a single nuclear gene encoding GS_2 and multiple GS_1 genes, which are members of small gene families and differentially regulated.[7] In this study, a GS_1 gene was constitutively expressed in all cell types of alfalfa driven by the cauliflower mosaic virus promoter to bypass the transcriptional regulation component.[6] The GS_1 gene was transcribed in these transgenic plants, but GS_1 was unstable and did not accumulate. The results suggested that GS is regulated at multiple steps, besides being regulated at the transcriptional level. One step of regulation is at mRNA stability and may be controlled by the glutamine/glutamate ratio, the ATP/ADP ratio, or the redox balance. Another level of regulation would be protein turnover, and would involve the inactivation of GS by oxygen radicals generated by redox reaction.[6,8]

PROTEIN DEGRADATION

Protein degradation is an important aspect of the cell cycle that occurs in the normal life of the plant. A process of protein selection occurs to specifically degrade proteins; some proteins are degraded when they become damaged. The proteases found in plants can be identified as matrix metalloproteases, processing proteases, and proteases involved in mobilization of storage-protein reserves. Proteolysis also takes place during photoinhibition in the chloroplast, programmed cell death, and photomorphogenesis in the developing seedling involving several subcellular compartments.[9] In the chloroplast, for example, and due to its endosymbiotic origin, each protease is related to a bacterial counterpart. The FtsH1 protein, which is involved in the D1 protein degradation, is related to *Escherichia coli* FtsH1 protein, a metalloprotease and chaperone. Another example is the ClpP, which is responsible for the regulated degradation of the cytochrome b_6f complex.[9] Among the mechanisms operating in protein degradation, the conjugation of proteins by ubiquitin has been implicated. A protein with a chain of at least four ubiquitin subunits is recognized by the proteasome and degraded. The ubiquitin subunits are removed from the substrate by an ubiquitin-specific protease and recycled.[9]

Protein structures are also affected by the state of oxidation of some amino acid residues, which may be the principal parameter affecting the in vivo and in vitro stability of proteins.[10] Transition metals, ozone, nitric oxides, and metal ions are involved in these oxidative modifications of proteins and free amino acids. Protein oxidation was rigorously tested and verified in *E. coli* GS.[10] In plants, GS_2 is extremely prone to oxidative cleavage, and reduced transition metals—presumably resulting from the destruction of iron–sulfur clusters during photoinhibition—play a crucial role in the degradation process.[8] In this process, proteolytic enzymes may not be involved because protease inhibitors provide little or no protection to the light-induced GS degradation.[8]

CONCLUSION

In the past few years the study of amino acid and protein metabolism has diverged enormously, due to the advances of emerging tools in genomics. Understanding this complex regulated network requires the identification of the factors controlling gene expression in response to changes in N status. Data are now emerging to provide evidence that NO_3^-, NH_4^+, or amino acid may also serve as signaling molecules in plants, aside from their role as building blocks. There is also increasing evidence that glutamate may function in a manner that is analogous to its signaling function in the animal nervous system. Several N-metabolizing enzymes have been at least partially characterized and much more information is emerging regarding the organization and control of the genes encoding these enzymes, as well as the modulation of the activity of the enzymes themselves once synthesized. These studies will bring new perspectives to improve the level of essential amino acids in plant seeds.

Regarding protein metabolism, our focus will change in the future to the study of the complex regulation of protein degradation, which still needs more attention.

ARTICLES OF FURTHER INTEREST

Metabolism, Primary: Engineering Pathways, p. 714
Nitrogen, p. 822
Symbiotic Nitrogen Fixation, p. 1218

REFERENCES

1. Williams, L.E.; Miller, A.J. Transporters responsible for the uptake and partitioning of nitrogenous solutes. Annu. Rev. Plant Physiol. Plant Mol. Biol. **2001**, *52*, 659–688.
2. Lea, P.J.; Ireland, R.J. Nitrogen Metabolism in Higher Plants. In *Plant Amino Acids: Biochemistry and Biotechnology*; Singh, B.K., Ed.; Marcel Dekker: New York, 1999; 1–47.
3. Coruzzi, G.; Last, R. Amino Acids. In *Biochemistry & Molecular Biology of Plants*; Buchanan, B., Gruissem, W., Jones, R., Eds.; ASPP: 2000; 358–410.
4. Arruda, P.; Kemper, E.L.; Papes, F.; Leite, A. Regulation of lysine catabolism in higher plants. Trends Plant Sci. **2000**, *5*, 324–330.
5. Ratcliffe, R.G.; Shachar-Hill, Y. Probing plant metabolism with NMR. Annu. Rev. Plant Physiol. Plant Mol. Biol. **2001**, *52*, 499–526.
6. Ortega, J.L.; Temple, S.J.; Sengupta-Gopalan, C. Constitutive overexpression of cytosolic glutamine synthetase (GS_1) gene in transgenic alfalfa demonstrates that GS_1 may be regulated at the level of RNA stability and protein turnover. Plant Cell **2001**, *126*, 109–121.
7. Cren, M.; Hirel, B. Glutamine synthetase in higher plants: Regulation of gene and protein expression from the organ to the cell. Plant Cell Physiol. **1999**, *40*, 1187–1193.
8. Palatnik, J.F.; Carrillo, N.; Valle, E.M. The role of photosynthetic electron transport in the oxidative degradation of chloroplastic glutamine synthetase. Plant Physiol. **1999**, *121*, 471–478.
9. Estelle, M. Proteases and cellular regulation in plants. Curr. Opin. Plant Biol. **2001**, *4*, 254–260.
10. Berlett, B.S.; Stadtman, E.R. Protein oxidation in aging, disease, and oxidative stress. J. Biol. Chem. **1997**, *272*, 20313–20316.

Aneuploid Mapping in Diploids

P. K. Gupta
Ch. Charan Singh University, Meerut, India

INTRODUCTION

Genetic mapping in crop plants (both diploids and polyploids) initially involved the use of recombination frequencies that were treated to be in proportion with genetic distances and, therefore, could be converted into centiMorgan units with the help of a mapping function (Kosambi's mapping function has generally been used). These genetic maps each had a number of linkage groups that generally equaled the haploid chromosome number of the organism concerned. Later, the individual linkage groups could be assigned to specific chromosomes using aneuploids, which are organisms that have a somatic chromosome number that is not an exact multiple of a basic chromosome number. The aneuploids that have been used for mapping differed in diploid and polyploid crops. While trisomics have been used in a large number of diploids such as barley and tomato, monosomics have been more frequently used in polyploids such as bread wheat, cotton, and tobacco. In the case of maize, which was treated earlier as a diploid and is now known to be an archeo-tetraploid, both trisomics and monosomics have been used. Such genetic mapping of chromosomes makes use of abnormal segregation ratios that are obtained in aneuploids, relative to normal diploids, if the gene of interest is located on the chromosome that is involved in aneuploidy. A variety of these aneuploids have been utilized for mapping in diploids. Because much literature is available on the subject, only a summary will be presented in this section.

Only a brief account of the different types of trisomics, their methods of production, and their use in chromosome mapping (both genetic and physical) is given here. Detailed information is available elsewhere.[1,2]

TYPES OF TRISOMICS AND THE TERMINOLOGY

The term trisomic originally referred to a condition in an organism where a particular chromosome is present in three doses, in contrast to each of the other chromosomes being present as a pair.[3] These trisomics were later described as simple primary trisomics to distinguish them from complex primary trisomics (where the extra chromosome is normal, but the remaining constitution is not), and also from the type of trisomics where the extra chromosome is not normal (e.g., secondary, tertiary, telo-, and compensating trisomics). A number of other trisomic types have also been produced and used for mapping. For instance, a complete set of telotrisomics in rice, a variety of compensating trisomics and haplo-triplo disomics in tomato, and a number of acrotrisomics and metatrisomics in barley have been produced and used (discussed later). Balanced tertiary trisomics have also been produced in crops such as barley and utilized for hybrid seed production,[4] but they have rarely been used for mapping.[2]

PRODUCTION OF TRISOMICS IN DIPLOIDS

Methods for the production of trisomics differ depending upon the type of trisomics desired. However, in most diploids, the simple primary trisomics have generally been produced using triploids, often derived from a tetraploid × diploid cross. These triploids, when selfed or crossed (as female parent) with diploids, yield a fairly large number of trisomic plants that can be assembled into a complete set of trisomics and are always equal to the haploid number of chromosomes in the species concerned. Other types of trisomics, including secondary trisomics, telotrisomics and compensating trisomics, often have been obtained in the progeny of simple primary trisomics. Tertiary trisomics, however, have been obtained in the progeny of interchange heterozygotes, due to rare 3:1 meiotic disjunction of interchange quadrivalent, which is characteristic of a translocation heterozygote.

CHARACTERIZATION AND IDENTIFICATION OF TRISOMICS

Simple primary trisomics and other trisomic types have been characterized using one or more of the following criteria: 1) morphological deviations from the normal diploids; 2) karyotype alterations; 3) crosses with known interchange testers; and 4) crosses with the genetic stocks carrying known markers (including molecular markers). However, the most important criterion has been morphological deviation. Meiotic behavior also has been used to

Encyclopedia of Plant and Crop Science
DOI: 10.1081/E-EPCS 120005590

Table 1 Expected gametes and their ratios for a gene present in trisomic condition[a]

	Gametes		Gametic ratio (A:a)
Genotype	(*x*+1)	(*x*)	(AA+Aa+A):(aa+a)
AAa	1AA:2Aa	2A:1a	5A:1a
Aaa	2Aa:1aa	1A:2a	1A:1a

[a]Assuming 1:1 ratio of the two types of gametes.

distinguish the different types of trisomics (e.g., primary, secondary, tertiary, and compensating trisomics).

USE OF TRISOMICS FOR CHROMOSOME MAPPING

Due to the presence of three homologous chromosomes instead of two, a trisomic can have two heterozygous genotypes (AAa, Aaa), which give segregation ratios (described as trisomic ratios) that are different from those obtained in normal diploids (disomic ratios) and thus facilitate chromosome mapping.

Trisomic Genotypes and Segregation Ratios

If the gene to be mapped occurs on the chromosome that is present in an extra dose in a trisomic, and a cross is made between this trisomic and the normal diploid, the resulting F_1 trisomic can be duplex (AAa) or simplex (Aaa) depending upon whether the trisomic carries a dominant or a recessive allele. The segregation ratio for this gene in F_2 deviates from the expected Mendelian ratio in a normal diploid. The ratios that are expected in two types of trisomics have been established (Table 1; Fig. 1). These ratio deviations have been used for assigning genes to specific chromosomes.

Assigning Linkage Groups to Specific Chromosomes

Linkage groups already established in a species through conventional linkage analysis can be assigned to specific chromosomes when at least one marker from each linkage group gives abnormal segregation in one of the trisomics in the F_2 generation. Multiple genetic marker stocks (having markers from more than one linkage group) could be used and would reduce the work involved. This method has been used successfully in a number of crops, including maize, tomato, and barley.

Assigning Genes to Specific Arms of Chromosomes

Once a gene is already assigned to a specific chromosome, secondary, tertiary, and telotrisomics can be used to map this gene to either the short or the long arm of this chromosome. If the dominant allele is present on the arm involved in the extra chromosome of a secondary, tertiary, or telotrisomic, all trisomic progeny in an F_2 population derived from a cross of this trisomic with a recessive genetic stock will exhibit the dominant character (A:a = all:0), thus indicating that the gene is present on this arm. Segregation for this character will suggest that the gene is present on the other arm, which can be confirmed by using telotrisomic for that chromosome arm.

Location of Centromere and Orientation of Linkage Groups

If the linear order of three or more closely linked genes located on two different arms of a chromosome is known, the position of the centromere and the orientation of the chromosome map can be determined. For example, if out of three linked genes a-b-c, a is on the short arm and b

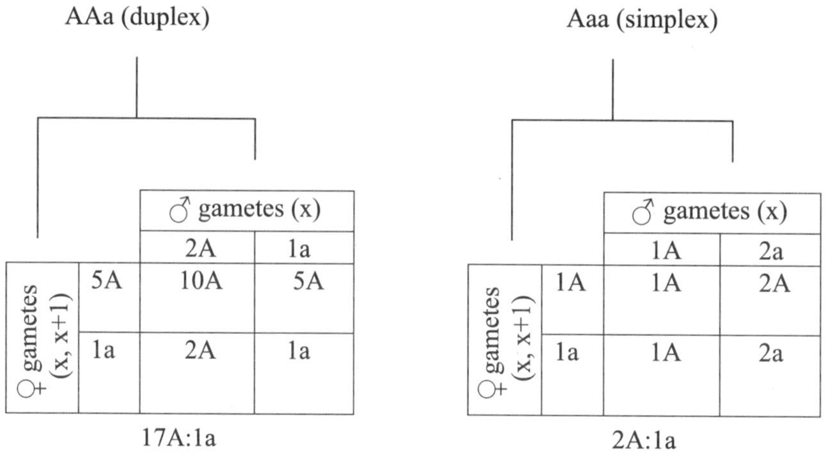

Fig. 1 Formation of gametes and derivation of F_2 phenotypic ratios in a trisomic with AAa (duplex) and with Aaa (simplex) genotype.

and c are on the long arm, the orientation would be a-b-c. In the reverse situation (if a is on long arm and b and c are on the short arm), the orientation would be c-b-a, the centromere being between a and b in both the cases (the genes are read from the end of the short arm to the end of the long arm).

Physical Mapping Using Acro- or Metatrisomics

Physical mapping of genes on chromosomes in diploid crops such as maize and tomato has been accomplished by studying structural changes in chromosomes at pachytene. However, in barley, acrotrisomics (the extra chromosome has one complete arm and a small segment of the other arm) and metatrisomics (having terminal deficiencies in both arms of the extra chromosome) have been produced and used successfully to map individual genes physically on chromosomes. When genes are already assigned to a specific chromosome arm, and acrotrisomics and metatrisomics with segmental deficiencies for this arm are used, they can be mapped physically due to the absence of a trisomic ratio if the gene is present in the deficient region of the extra chromosome.[5]

ALIEN ADDITION LINES FOR MAPPING IN DIPLOIDS

Individual alien chromosome-addition lines have been produced for a number of diploid crops by adding an individual chromosome from the diploid crop to a related polyploid that could tolerate the addition of an extra chromosome. In some cases, telocentric additions have also been successfully produced.[6] Chromosomes of a number of diploid species including rye and barley have been added to tetraploid and hexaploid wheat, and individual maize chromosomes have been added to the oat genome for the purpose of mapping.[7] Alien chromosome additions have also been obtained in rice, Brassica, sugar beet, and cotton. Addition lines have been utilized for mapping molecular markers using barley and rye additions to wheat and maize additions to oat. Deletions produced in the alien chromosomes in these addition lines can also be used for physical mapping.

PRODUCTION AND USE OF MONOSOMICS FOR MAPPING

Monosmics in diploid crops have been produced only sparingly, because loss of a chromosome has a more drastic effect in a diploid than in a polyploid. However, monosomics for some chromosomes in tomato and for all the 10 chromosomes in maize have been successfully produced and used for mapping. For instance, at least three primary monosomics and 25 tertiary monosomics were produced in tomato, and some of them were used for mapping.[8] Similarly, in maize, a complete set of monosomics has been produced using either the B-chromosome system to cause elimination of chromosomes[9] or an *r-xi* deficiency. These monosomics have been used to map genes and molecular markers.[10]

TRANSLOCATION STOCKS FOR MAPPING

Translocation stocks in maize and barley have also been used for mapping. In maize, the linkage between a gene and semisterility due to translocation was observed and facilitated mapping of genes with respect to the translocation breakpoint. B–A translocations (translocation between B-chromosomes and autosomes) were also used for mapping.[11] More recently, translocation stocks were used for physical mapping of restriction fragment length polymorphism (RFLP) markers in relation to the translocation breakpoints in barley. Individual microdissected translocation chromosomes were used for (PCR) driven by primers designed from RFLP markers, so that the positions of translocation breakpoints could be located by identification of closely linked markers (already mapped genetically) that were found to be present on two different chromosomes involved in a translocation.[12]

CONCLUSION

In diploid plants, hyperaneuploids such as trisomics have generally been used for the preparation of genetic and physical maps of chromosomes, although in tomato and maize, monosomics have also been produced and successfully used for mapping. The presence of genes on chromosomes or chromosome arms that are in three doses in a trisomic and a single dose in a monosomic, in contrast to a double dose in a normal diploid, leads to segregation ratios that deviate from normal Mendelian ratios, thus facilitating the assignment of genes to specific chromosomes or chromosome arms.

ARTICLES OF FURTHER INTEREST

Aneuploid Mapping in Polyploids, p. 37
Chromosome Banding, p. 263
Chromosome Manipulation and Crop Improvement, p. 266
Chromosome Rearrangements, p. 270
Chromosome Structure and Evolution, p. 273
Flow Cytogenetics, p. 460
Fluorescence In Situ Hybridization, p. 468

Functional Genomics and Gene Action Knockouts, p. 476
Gene Flow Between Crops and Their Wild Progenitors, p. 488
Genome Rearrangements and Survival of Plant Populations to Changes in Environmental Conditions, p. 513
Molecular Evolution, p. 748

REFERENCES

1. Khush, G.S. *Cytogenetics of Aneuploids*; Academic Press: New York, 1973.
2. Gupta, P.K. *Cytogenetics*; Rastogi Publications: Meerut, India, 1995.
3. Blakeslee, A.F. Types of mutations and their possible significance in evolution. Am. Nat. **1921**, *55*, 254–267.
4. Ramage, R.T. Trisomics from interchange heterozygotes in barley. Agron. J. **1960**, *52*, 156–159.
5. Tsuchiya, T. Chromosome Mapping by Means of Aneuploid Analysis in Barley. In *Chromosome Engineering in Plants: Genetics, Breeding, Evolution, Part A*; Gupta, P.K., Tsuchiya, T., Eds.; Elsevier: Amsterdam, 1991; 361–384.
6. Islam, A.K.M.R. Identification of wheat barley addition lines with N-banding of chromosomes. Chromosoma **1980**, *76*, 365–373.
7. Okagaki, R.J.; Kynast, R.G.; Odland, W.E.; Russel, C.D.; Livingston, S.M.; Rines, H.W.; Phillips, R.L. Mapping maize chromosomes using oat–maize radiation hybrid lines. Crop Sci. Soc. Am. Abstr., Division C-7 **2000**, 188.
8. Khush, G.S.; Rick, C.M. The origin, identification and cytogenetic behaviour of tomato monosomics. Chromosoma **1966**, *18*, 407–420.
9. Rhoades, M.M.; Dempsey, E.; Ghidoni, A. Chromosome elimination in maize induced by B supernumerary chromosomes. Proc. Natl. Acad. Sci. U. S. A. **1967**, *57*, 1626–1632.
10. Weber, D.F. Monosomic Analysis in Maize and Other Diploid Crop Plants. In *Chromosome Engineering in Plants: Genetics, Breeding, Evolution, Part A*; Gupta, P.K., Tsuchiya, T., Eds.; Elsevier: Amsterdam, 1991; 361–384.
11. Beckett, J.B. Cytogenetic, Genetic and Plant Breeding Applications of B–A Translocations in Maize. In *Chromosome Engineering in Plants: Genetics, Breeding, Evolution, Part A*; Gupta, P.K., Tsuchiya, T., Eds.; Elsevier: Amsterdam, 1991; 493–529.
12. Künzel, G.; Korzun, L.; Meister, A. Cytologically integrated physical restriction fragment length polymorphism maps for the barley genome based on translocation breakpoints. Genetics **2000**, *154*, 397–412.

Aneuploid Mapping in Polyploids

David M. Stelly
Texas A&M University, College Station, Texas, U.S.A.

INTRODUCTION

Aneuploidy denotes the condition of having extra or missing chromosomes. For specific aneuploid states that are not overly debilitating, the altered genetic constitution enables gross genome mapping through the localization of genes and sequences to specific genomic, subgenomic, chromosomal, and subchromosomal regions. The information from these analyses lends itself to divide-and-conquer strategies that expedite mapping and other forms of genome analysis that impact efficacy of research, cloning, genetic engineering, interspecific germplasm introgression, marker-assisted selection, and other aspects of breeding.

Plant genomes range about 2500-fold in size, from about 50 Mbp to 125,000 Mbp, with chromosome numbers ranging about 300-fold, from 2 to close to 600. The process of mapping eukaryotic genomes is biologically and technically complex. All mapping methods suffer from technical, statistical, and human limitations. To create robust maps, several orthogonal mapping methods must be extensively integrated, e.g., segregation analysis, aneuploid analysis, molecular cytogenetics, radiation hybrids, contig assembly, and sequencing. Only then is it possible to harness the synergistic benefits from their complementary strengths and weaknesses.

The integration of aneuploid-based mapping with linkage mapping and other orthogonal approaches is especially beneficial and applicable to disomic polyploid plant species, genomes of which are larger and involve many more gene and sequence duplications than diploid related taxa. Aneuploid-based mapping provides a means to establish a sound biological footing to the maps and reduce complexity of the target. Polyploid genomes are more tolerant of the various genic imbalances associated with aneuploidy, and therefore more amenable to aneuploid-based mapping.

ANEUPLOIDY

The most common types of aneuploids are presented diagrammatically in Fig. 1. They can be broadly categorized by whether they have increased or decreased chromosomal (genetic) content, though some involve both. Hypoaneuploids have chromosomal deficiencies, e.g., $2x - 1$, whereas hyperaneuploids have chromosomal excesses, e.g., $2x + 1$. Hypoaneuploids (deficient for chromosomes or chromosome segments) are generally preferred to hyperaneuploids, because they offer greater efficacy for mapping, as well as for chromosome substitution-mediated germplasm introgression.

Irrespective of ploidy, individuals with chromosomally imbalanced sets are called aneuploids. This imbalance may arise from deviations that affect whole chromosomes, chromosome arms, and/or chromosome segments.[2] Certain terms denote changes in chromosome number by indicating the abnormal content remaining, e.g., monosomy for $2n - 1$ and trisomy $2n + 1$. Modifiers indicate content of the affected chromosome, e.g., primary for an intact chromosome and tertiary for a translocated chromosome.[3] Segmental aneuploids may or may not have a normal chromosome number, but all are genetically imbalanced because of abnormal dosage of one or more specific segment(s). Telosomes lack essentially an entire chromosome arm. Other segmental aneuploids have an excess/deficiency of specific chromosome segment(s), due directly to deletion or duplication, or derived indirectly from ancestral heterozygosity for translocations or other rearrangements.

The size and type of each chromosomal abnormality is important to mapping because it determines not only the physical scope of localization, but also what kinds of genes or markers are most amenable to analysis, the method of analysis, and the difficulty and time-requirements of the analysis. Some give immediate results, while others require follow-up analysis, e.g., progeny testing and/or segregation analysis. Where segregation data are gathered, their statistical efficacy can also differ markedly according to the type of analysis and map distances involved.

COMMON ANEUPLOID-BASED MAPPING APPLICATIONS

Aneuploids are commonly applied to five kinds of mapping objectives:

1. Localization of individual mutants or small numbers of loci, e.g., a specific marker, or simple or oligogenic trait.

DOI: 10.1081/E-EPCS 120005595

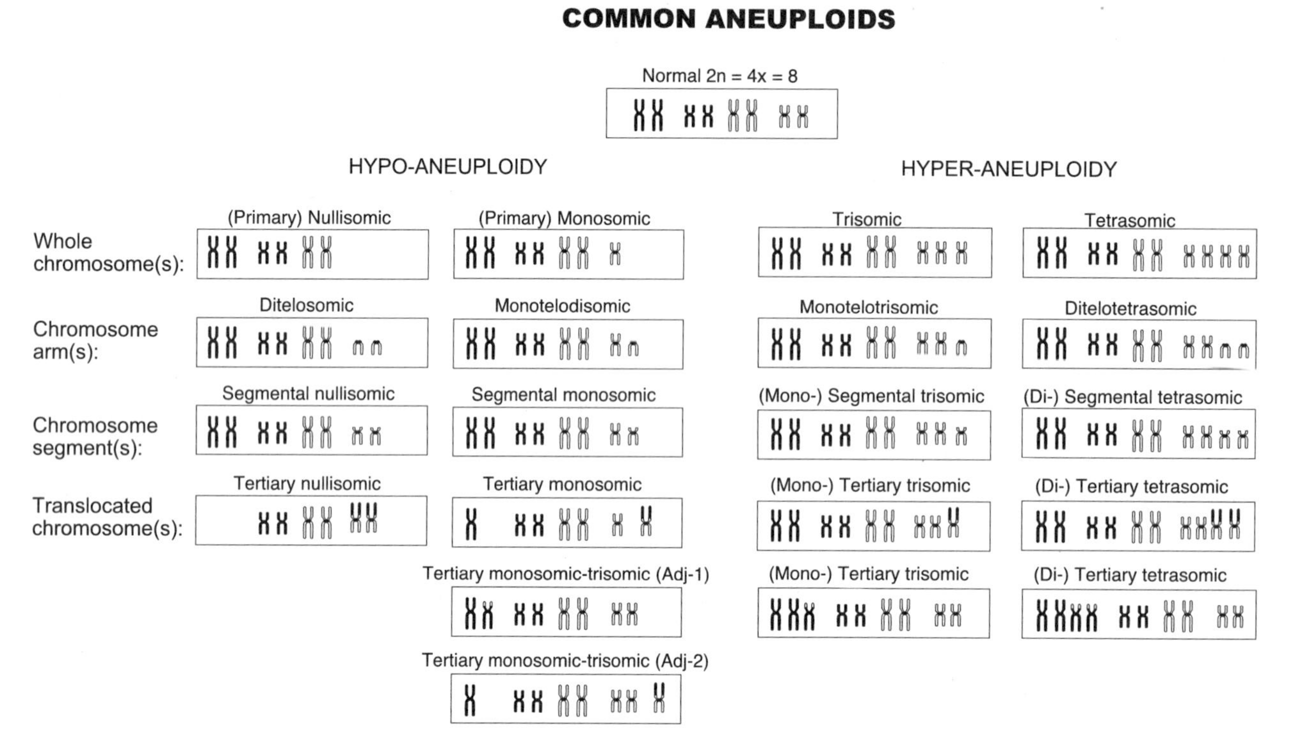

Fig. 1 Common types of aneuploids are depicted for a hypothetical disomic tetraploid species ($2n = 4x = 8$), illustrating differences between hypo- versus hyperaneuploidy, as well as differences with respect to coverage at the chromosome level. The types shown are not exhaustive but illustrate most of the types commonly used for mapping.

2. Genomics and other large-scale genetic endeavors to localize many loci and/or gene products.
3. Centromere mapping.
4. Broad surveys for major chromosome- or segment-specific effects on one or more complex traits, through development of chromosome substitutions or chromosome additions.
5. Development of chromosome-specific mapping populations for genetic dissection of alien germplasm for effects on complex traits, through development of phenotypically characterized mapping populations from chromosome substitutions.

The various types of aneuploids, e.g., whole-chromosome versus telosome aneuploids, differ in their effectiveness for specific mapping goals (Fig. 2).[4,5] The localization of molecular markers and linkage groups to specific chromosomes establishes a biological, cytological, and macromolecular foundation, and enables a logical basis for development of a common nomenclature for linkage groups among research laboratories. Intrachromosomal analyses with telosomes enable placement of centromeres on linkage maps. Telosomes and other segmental aneuploids can be used to establish the orientations of linkage groups with respect to the two arms of each chromosome. Through the accumulation of many segmental aneuploids, the mapping process can be extended to small subchromosomal regions, as in wheat.[6]

PRINCIPLES OF ANEUPLOID-BASED MAPPING

The methods of aneuploid mapping rely on cytogenetic types that are deficient for one or both copies of a locus (hypoaneuploid), or have extra copies (hyperaneuploid). The procedures used for aneuploid-based mapping are determined partly by the type of gene action or marker detection and by the type of aneuploid (Fig. 3). Deletion mapping entails the removal of a locus or allele, the use of which is most efficient when the subject allele or marker is dominant or co-dominant, and not obscured by allele(s) at other loci. Under these circumstances, both the nulli- and the hemizygous states are directly informative about the absence or presence of the particular DNA sequence.

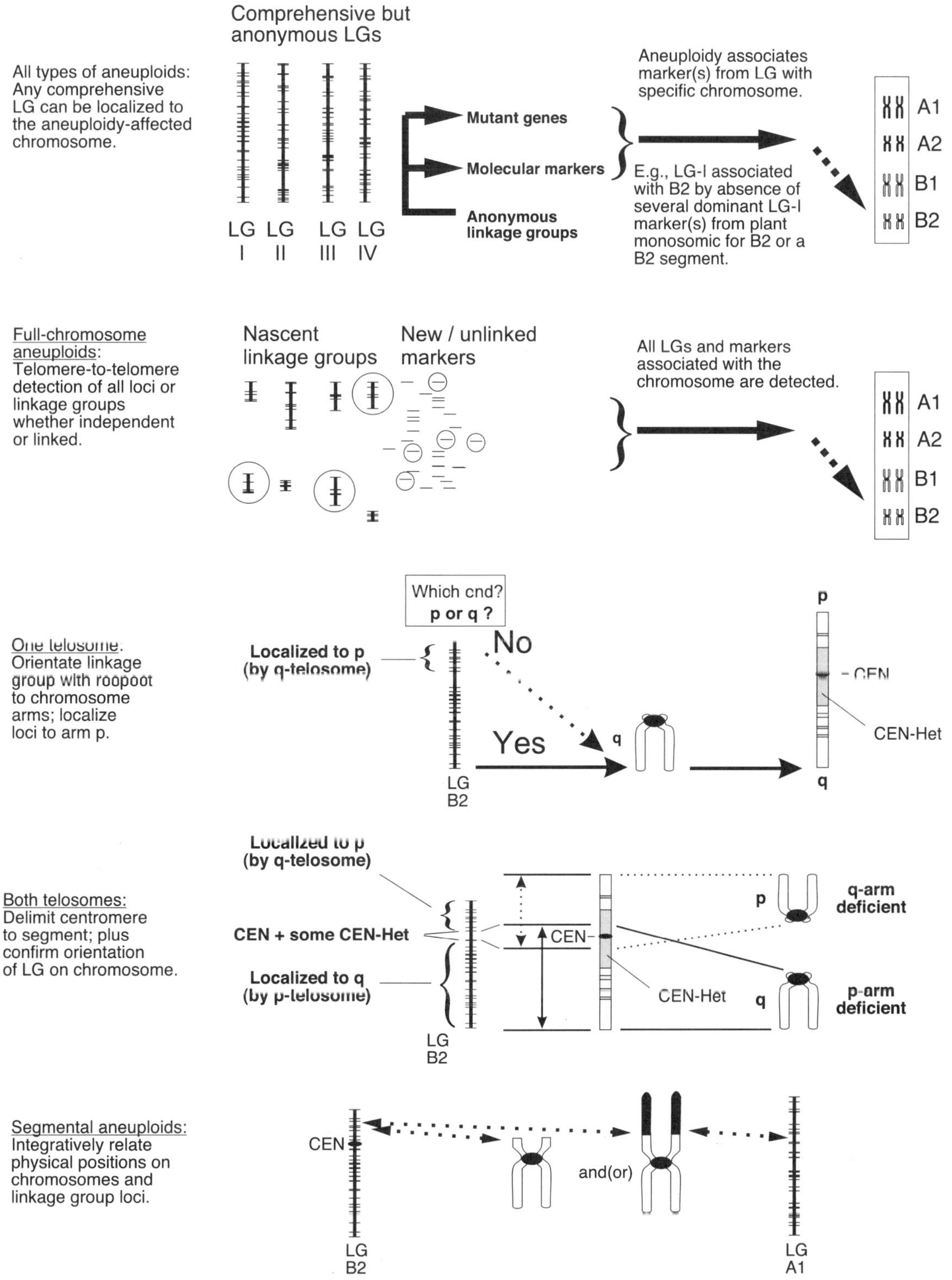

Fig. 2 Common strategies and goals used in aneuploid-based mapping. All sorts of aneuploids enable localization of a comprehensive linkage groups to specific chromosomes. Full-chromosome aneuploids enable detection of all loci and linkage groups. Telosomes are often used to orient linkage groups and delimit the location of the centromeres where the latter activity is to be accomplished with opposing telosomes. Segmental aneuploids enable similar localization and mapping at subchromosomal levels. Tertiary chromosomes enable mapping to segments of two chromosomes at once.

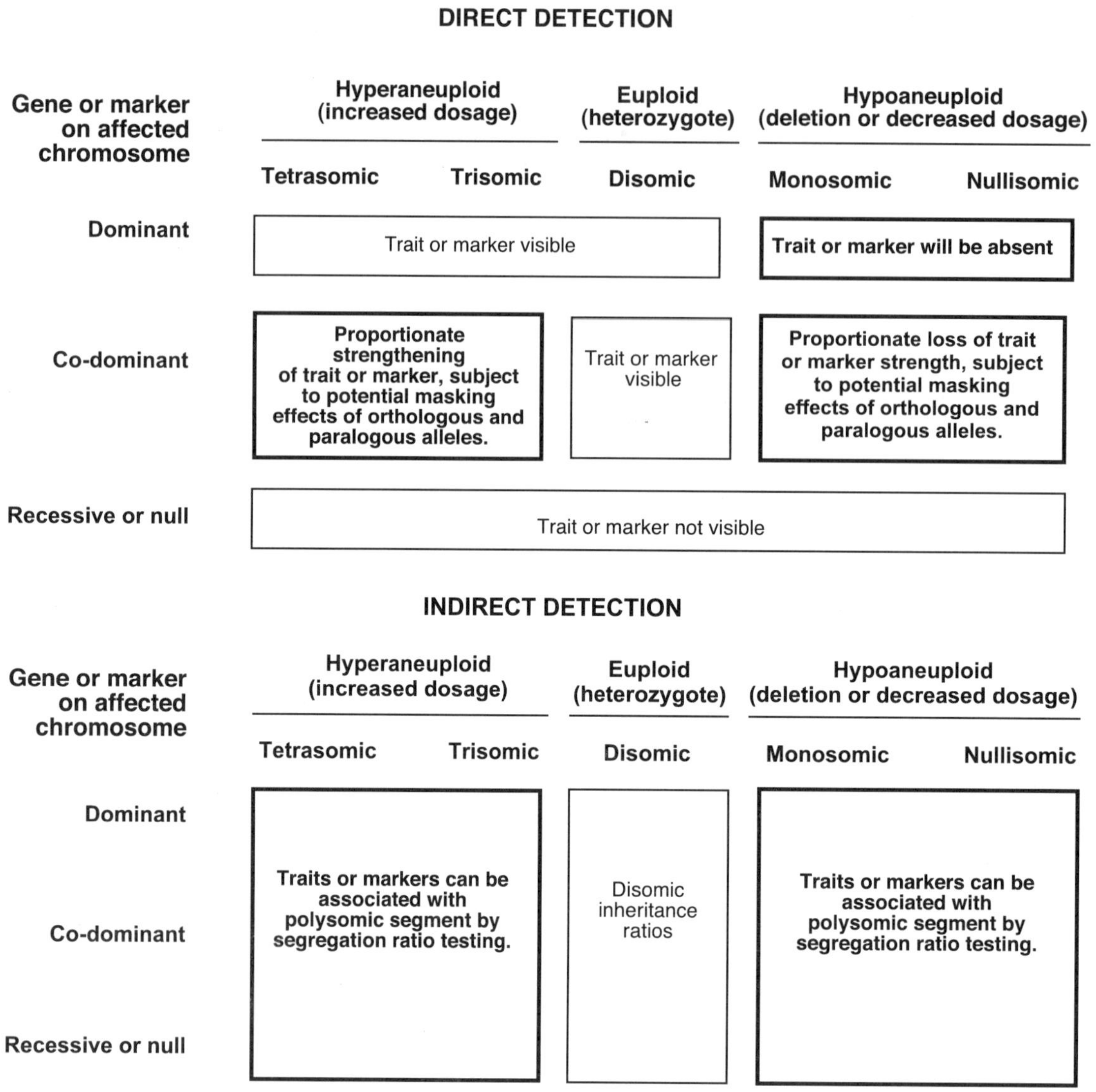

Fig. 3 Effectiveness of direct versus indirect detection systems for mapping by hyper- and hypoaneuploid conditions, according to type of aneuploidy and type of gene or locus ''action.'' Hypoaneuploids that are lacking or deficient for a segment offer the most widely effective means of mapping; elimination allows for direct manifestation of dominant and co-dominant alleles, but not recessive alleles (progeny testing required). In general, hyperaneuploids do not allow for direct mapping in most instances, and must be augmented by progeny testing for polysomic versus disomic inheritance to be effective. Co-dominant alleles and markers are relatively amenable to analysis with both classes of aneuploids.

When the deleted allele or marker is recessive or null, hypoaneuploid deficiency mapping requires progeny testing, because there is no change in phenotype or marker status in the hypoaneuploid. Progeny testing and segregation analysis are generally required for hyperaneuploids, unless 1) new alleles are introduced on the extra chromatin, e.g., as for alien chromosomes or segments from a related species, or 2) sufficiently accurate dosage analysis is possible. Progeny tests can also be used to confirm deletion mapping results from dominant and co-dominant alleles and, thus, bolster reliability of results. In rare instances, recessive traits must be homozygous to be expressed, i.e., they will not be expressed when hemizygous, as when the dominant allele is deleted through monosomy. In these cases, too, deletion mapping alone is insufficient, and progeny testing is required. When a marker is obscured by other similar alleles at the same or different loci, further analysis is often difficult. However, if the marker is readily quantified and present at levels proportional to dosage, then it may be possible to localize the marker by quantitative analysis.

Hypoaneuploids: Nullisomics and Other Deficiency Homozygotes

Where available, nullisomics, ditelosomics, and homozygous deletions are often the tools of choice, because they eliminate a locus or gene. Thus, all molecular markers unique to a nullisomic region are completely absent. For dominant and co-dominant genes and markers, the absence is directly detectable. Deletions of null mutants and recessive alleles are not directly detectable, but can be ascertained by progeny testing. Although nullisomics can be derived in bread wheat, a disomic hexaploid species, they are not available for most plants due to severe effects on viability and fertility. Maintenance across sexual generations requires viability and function of nullihaploid micro- and mega-gametophytes.

Hypoaneuploids: Monosomics and Other Deficiency Hemizygotes

Less severe effects result from monosomy, monotelodisomy, and heterozygosity for segmental deletions. The affected loci are hemizygous, not heterozygous or homozygous. For large-scale mapping applications, the most useful approach has often been to screen for hemizygosity among hypoaneuploids recovered after wide-cross hybridization. For example, a set of different monosomic plants would be mated as female with a divergent genotype, e.g., a different biotype, race, or closely related species. Each hypoaneuploid F_1 hybrid would be highly heterozygous, except for the respective hypoaneuploid chromosome, arm, or segment(s), which must be hemizygous for all loci in it. Differential absence of maternal dominant and co-dominant markers and alleles from a hypoaneuploid but not euploid F_1 hybrids indicates that the locus is associated with the respective chromosome. Progeny tests can be used to confirm results for dominant and co-dominant markers and to detect deletion of cryptic alleles, e.g., null alleles. Highly distorted self and testcross progeny ratios often result due to reduction or absence of sexual transmission, especially through pollen.

Hyperaneuploids

For crops in which hypoaneuploids are not available, hyperaneuploids can usually be obtained. Their use typically entails progeny-segregation analysis to discern between disomic and polysomic inheritance. For example, consider tests of recessive allele "a" across a series of trisomics; if closely linked with the centromere, segregation from the respective A-bearing trisomic will be about 8:1 (A-phenotype : a-phenotype), versus just 3:1 for the euploid and the other trisomics.

Hyperaneuploids may also result from interspecific introgression. If the alien genome does not recombine with the genome of the recurrent parent, monosomic addition lines may be obtainable by iterative backcrossing, and, if transmissible through both gametophytes, disomic additions may be derived from selfing the monosomic additions. These aneuploids and similar segmental derivatives offer high rates of polymorphism relative to the host genome, and thus present an effective tool for direct analysis of the alien genome, and by way of comparative mapping, of the host genome, too.

SOURCES OF ANEUPLOIDS

Because the occurrence of specific aneuploids is typically an unusual or rare event, collections of aneuploid stocks have been developed systematically for a number of organisms, e.g., arabidopsis, corn, cotton, potato, soybean, rice, tomato, and wheat. In some cases, these have been maintained either independently or as part of public genetic or germplasm collections for a number of organisms. Collections may be maintained in plant, seed, vegetative, and/or DNA form. Otherwise, the aneuploids must be generated anew, distinguished, and identified. Typical sources include 1) crosses with an odd-ploid genotype, 2) mutations affecting chromosome or chromatid disjunction in meiosis or mitosis, and treatments with 3) physical mutagens (e.g., X-rays or gamma-rays) or 4) chemicals, e.g., certain mutagens or spindle inhibitors.

SEARCH ENGINE

A panel of aneuploids is useful as a search engine, e.g., to localize grossly one gene, hundreds of genes, or thousands of genes to specific chromosomes or regions. If the panel includes all chromosomes, the search will be comprehensive. Because genomes of most plants contain 5–26 chromosomes per genome, each chromosome represents a significant proportion of a genome, e.g., from 2–20% for most chromosomes. Given a tentative estimate that the basic angiosperm genome contains approximately 30,000 genes, a typical angiosperm chromosome must contain 600–6,000 genes. Thus, aneuploid-based methods are very efficient for large-scale mapping of genes, and yield a "big picture," whereas complementary methods are typically used to provide higher resolution and detail.

ARTICLES OF FURTHER INTEREST

Aneuploid Mapping in Diploids, p. 33
Breeding: Incorporation of Exotic Germplasm, p. 222
Chromosome Manipulation and Crop Improvement, p. 266
Chromosome Rearrangements, p. 270
Genome Size, p. 516
Polyploidy, p. 1038

REFERENCES

1. Bennett, M.D.; Leitch, I.J. *Plant DNA C-Values Database*; (release 2.0, Jan. 2003). http://www.rbgkew.org.uk/cval/homepage.html.
2. Burnham, C.R. *Discussions in Cytogenetics*; Burgess Publishing Co.: Minneapolis, 1962. http://www.agron.missouri.edu/Burnham/contents.html.
3. Khush, G.S. *Cytogenetics of Aneuploids*; Academic Press: New York, 1973; 301.
4. Endrizzi, J.E.; Turcotte, E.L.; Kohel, R.J. Genetics, cytology and evolution of *Gossypium*. Adv. Genet. **1985**, *23*, 271–375.
5. Appels, R.; Morris, R.; Gill, B.S.; May, C.E. *Chromosome Biology*; Kluwer Academic Publishers, 1998.
6. Gill, K.S.; Gill, B.S. Mapping in the realm of polyploidy: The wheat model. BioEssays **1994**, *16*, 841–846.

Anther and Microspore Culture and Production of Haploid Plants

A

Liliya D. Serazetdinova
Horst Lörz
Universität Hamburg, Hamburg, Germany

INTRODUCTION

Haploid and double haploid plants attract the interest of geneticists, plant embryologists, physiologists, and breeders. Their genetic characteristics make them an elegant experimental system for genetic studies as well as an integral part of breeding programs, especially in generating pure lines. Haploid plants can be induced from male as well as female gametophytes. This article focuses on haploid plants derived from the male gametophyte, i.e., anther and microspore cultures, concisely discussing the origin of haploid plants and techniques applied to their production, after a historical overview.

HAPLOIDS

The development and viability of pollen play a key role in the fertility of plants. Besides its importance in sexual reproduction, pollen can be used for haploid plant production. Haploid plants are genetically characterized as plants containing only one set of chromosomes. The haploid state occurs due to the reduction of zygotic (diploid) chromosome number to gametic (haploid) number during meiosis. In nature haploid plants appear via abnormal fertilization, i.e., chromosome elimination or mispairing during the *crossing-over*. Haploid plants are sterile and therefore doubling of the chromosome set is required to produce fertile plants, which are called double haploids (DHs) or homozygous diploids. Two basic genetic features make DHs distinct for genetic studies and breeding.[1] They are

1. The full complement of haploid genome is expressed in the phenotype. For example,

   A. Recessive characters are not supressed by dominant ones.
   B. Lethal mutations or any gene defects lead to the elimination of undesirable genotypes.

2. The production of homozygous diploids is possible in one generation via chromosome doubling of haploid plants.

The best known application of haploids is the F1 hybrid system for the production of homogeneous hybrid varieties. The DH lines are also used for targeted genetic manipulation, mutant breeding and selection, which considerably reduces the time required for the production of new cultivars.

Historical Overview

In 1922, Blakeslee and co-workers first discovered the appearance of natural haploid embryos and plants, which were derived from gametophytic cells of *Datura stramonium*.[2] To date, naturally occurring haploid plants are described in about 100 species of angiosperms.[3] In 1954, Tulecke, for the first time, observed that mature pollen grains of a gymnosperm *Ginkgo biloba* can be induced to proliferate in culture to form haploid callus,[4] but direct formation of embryolike haploid structures from anther culture of *Datura innoxia* was first reported by Guha and Maheswari in 1964. Their experiments clearly demonstrated the feasibility of induction of haploid structures from anther tissues. In 1967, Bourgin and Nitsch succeeded in producing the first haploid plants from cultured anthers of *Nicotiana sylvestris* and *Nicotiana tabacum*. Later in 1974, the first description of microspore culture was also made by Nitsch.[4] Since then, the techniques of microspore and anther culture were optimized for a wide range of economically important dicotyledonous and monocotyledonous plants.[4]

Origin of Haploids: Androgenesis

Two basic strategies are applied to the induction of haploids from higher plants: in vivo and in vitro induction by various physical, chemical, or biological stimulants. The first method for haploid production, developed by Kasha and Kao in 1970, is based on chromosome elimination in hybrid embryos.[4] This methodology exploits the fact that when two unrelated plant species are crossed, the chromosome sets of both parents fail to pair during the crossover stage of meiosis. For example, with crosses between common barley (*Hordeum vulgare*) and its wild ancestor (*Hordeum bulbosum*), the chromosomes of *H. bulbosum* are eliminated with the embryo possessing only

Encyclopedia of Plant and Crop Science
DOI: 10.1081/E-EPCS 120010555

one set of *H. vulgare* chromosomes. The developing haploid embryo is then cultivated on nutrition medium and gives rise to a haploid plant. This technique is known as the *bulbosum* technique and is restricted to a limited number of genotypes.

Anther and pollen culture represent the major techniques for in vitro induction of haploid plants. The development of haploid plants can be induced from pollen via embryogenesis. The formation of embryos from the androgenic (male) tissues is called androgenesis. In this case, the microspores are switched from their normal gametophytic fate to sporophytic development.[5] Different physical and chemical stimuli have been studied for the induction of androgenesis. The most efficient and widely applied techniques include 1) cold pretreatment of spikes (4°C, 4 weeks); 2) starvation—a cultivation of dissected anthers in media without carbon source (20–25°C, 1–4 days); and 3) incubation under higher temperatures (32–36°C, 1–3 days).

Genotype, physiological state of donor plants, stage of microspore development, culture media, and culture conditions are also important determinants of androgenesis.[1,4,6] The anthers containing microspores in mid- or late-uninucleate stage are most suitable for the induction of androgenesis.[2,6] Of the different media components, the carbon source, its concentration, and the ratio of nitrate and ammonium ions (NH^{4+}) are important in achieving embryogenesis and the development of green plants from microspores.[7,8] Based on these findings several basal media were developed for androgenic cultures. Phytohormones, especially the content of cytokinins and auxin; aeration and permanent supply of fresh, well-buffered medium; increased osmotic pressure; and temperature are all critical for successful establishment of androgenic cultures. Regeneration of haploid plantlets from androgenic cultures can be achieved by direct embryogenesis from microspores or via organogenesis.[4]

The doubling of chromosomes in androgenic cultures has been reported to occur spontaneously in culture via endomitosis.[4] It happens at high frequency in wheat (20–50%), rice (72%), and barley (87%).[6] Chromosome doubling can be induced using chemicals such as colchicine.[4] In many plant species, the frequency of spontaneous chromosome doubling in in vitro culture is high enough to produce a significant number of haploid plants.

ANTHER CULTURE

The technique of anther culture is relatively simple and efficient and requires minimal facilities[4,5] (Fig. 1). The androgenesis can be induced by pretreating whole inflorescences (cold pretreatment) or by pretreating dissected anthers (high temperature, starvation). Anthers are cultured on a solid or in a liquid medium on a rotary shaker at 50–60 rpm. The cultures are kept at 24–27°C. First, anthers are cultured on the callus induction medium for about 2 weeks in darkness and subsequently transferred to a regeneration medium containing phytohormones and organic substances at 16-hour photoperiods (2000–8000 lux) for shoot regeneration. Developing plantlets are then transferred to a rooting medium containing lower concentration of carbohydrates and other nutrients. The regeneration frequency of androgenic cultures is usually very high. In barley it ranges from 4.8 to 50 green plants per single anther.[6]

Plant breeding companies routinely use anther culture for the production of haploid plants. The only disadvantage of this technique is the regeneration of plants with different ploidy due to the presence of both gametophytic and sporophytic cells in the culture.

MICROSPORE CULTURE

The technique of microspore (pollen) culture was developed more recently than anther culture. In this technique, pollen grains are separated from the anther tissues and cultured in a liquid medium (Fig. 1). Microspores provide haploid single cells that can be utilized for various biological studies.[2]

Different techniques are applied for the isolation of microspores.[6] The most efficient is the technique of microblending, in which small pieces of inflorescence are put in a blender and quickly cut to release microspores into the isolation solution. The crude preparation is filtered through a sieve and the microspore suspension centrifuged to separate microspores. The plating density (the number of viable microspores per volume of medium) is an important factor in the induction of androgenesis. The optimal population density depends on the genotype, the quality of donor material, and the isolation technique. The induction of androgenesis occurs either while they are still inside the spikes (cold pretreatment) or directly after the isolation (high temperatures, starvation). Microspores are cultured in a liquid induction medium on a rotary shaker. They are kept in the dark at 24–27°C for 3–4 weeks. The emerging calli of visible size are transferred to a solidified medium. Further cultivation of microspore-derived cultures is similar to those derived from the dissected anthers.

The isolated microspore culture offers the possibility of combining selection procedures with the advantages of a haploid system. The nutritional requirements of the isolated microspores are much more complex than those of dissected anthers. The use of isolated microspore culture finds wide application in different fundamental studies and provides greater opportunities for cell

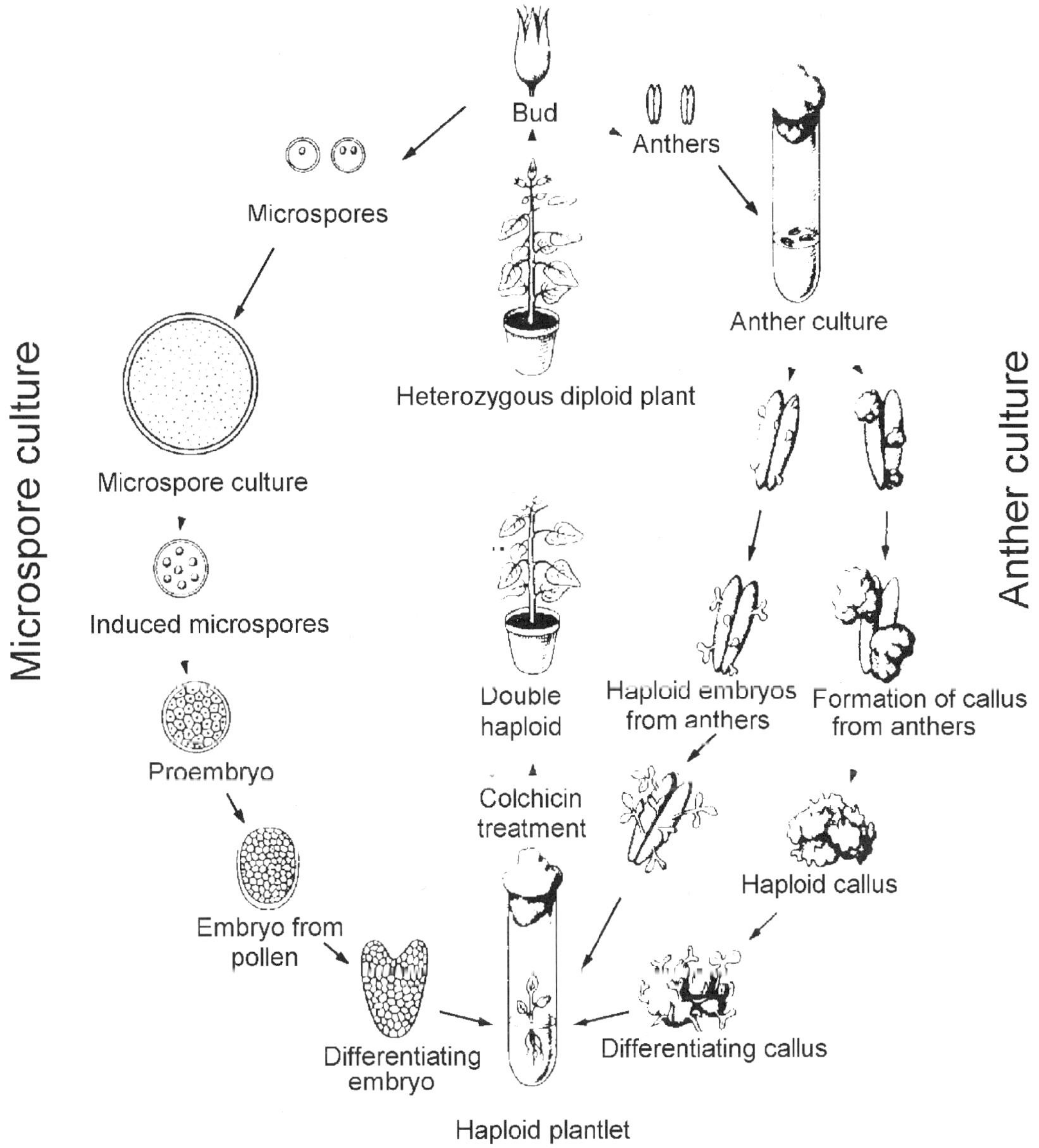

Fig. 1 Methods of haploid production: microspore and anther culture. (Modified from Ref. 9.)

manipulation than does anther culture.[6] Because of their unicellular nature, isolated microspores represent a very promising target for genetic manipulation. Haploid and dihaploid plants with resistance to environmental stress or pathogens can be produced by in vitro cell selection procedures or by transfer of foreign genes within one generation.

In spite of the great potential for haploid plant production, anther and pollen culture methods have two major limitations. Both methods are strongly genotype-dependent and produce a high percentage of albino plants with no practical utility.[6] The percentage of albino plants could reach up to 80% in cultures of monocotyledonous plants.[6]

CONCLUSION

Pollen (microspore) and anther culture can be used for the production of haploid and dihaploid plants. Haploid plants are valuable material for breeding new varieties and biotechnological applications. Fertile dihaploid plants represent an essential source for producing pure inbred lines, which are homogeneous and show no segregation. Haploid and dihaploid plants considerably accelerate and simplify breeding and selection processes. Using the method of haploidization it is possible to obtain a homozygosity for genes in cases where this is normally difficult to achieve, for example for self-incompatible alleles. Haploid cell cultures are also useful material for mutation analysis and cell modification. Microspores as single cells represent an ideal system for in vitro cell selection and genetic manipulation. Microspores can be used for microinjection, electroporation, particle bombardment and cocultivation with *Agrobacterium tumefaciens* resulting in transgenic haploid and homozygous plants.

ARTICLES OF FURTHER INTEREST

Male Gametogenesis, p. 663
Meiosis, p. 711
Plant Cell Culture and Its Applications, p. 931
Plant Cell Tissue and Organ Culture: Concepts and Methodologies, p. 934
Pollen-stigma Interactions, p. 1035
Somaclonal Variation: Origins and Causes, p. 1158
Transformation Methods and Impact, p. 1233

REFERENCES

1. Kasha, K.J. Haploids from Somatic Cells. In *Haploids in Higher Plants: Advances and Potential*; Univ. Press: Guelph, 1974; 67–87.
2. Hu, H. In vitro Induced Haploids in Wheat. In *In vitro Haploid Production of Higher Plants*; Mohan Jain, S., Sopory, S.K., Veilleux, R.E., Eds.; Current Plant Science And Biotechnology in Agriculture, Kluwer Academic Publishers: Dordrecht, 1996; Vol. 4, 73–97.
3. Vasil, I.K. Haploid Production in Higher Plants. In *In vitro Haploid Production of Higher Plants*; Mohan Jain, S., Sopory, S.K., Veilleux, R.E., Eds.; Current Plant Science and Biotechnology in Agriculture, Kluwer Academic Publishers: Dordrecht, 1996; Vol. 4, vii–viii.
4. Bajaj, Y.P.S. Haploids in Crop Improvement. In *Biotechnology in Agriculture and Forestry*; Bajaj, Y.P.S., Ed.; Springer-Verlag: Berlin, 1990; 1–44.
5. Touraev, A.; Vicente, O.; Heberle-Bors, E. Initiation of microspore embryogenesis by stress. Trends Plant Sci. **1997**, *2*, 297–302.
6. Jähne, A.; Lörz, H. Cereal microspore culture. Plant Sci. **1995**, *109*, 1–12.
7. Clement, C.; Audran, J.C. Anther Carbohydrates During in vivo and in vitro Pollen Development. In *Anther and Pollen. From Biology to Biotechnology*; Clement, C., Pacini, E., Audran, J.-C., Eds.; Springer-Verlag: Berlin, 1999; 69–90.
8. Olsen, F.L. Induction of microspore embryogenesis in cultured anthers of *Hordeum vulgare*. The effects of ammonium nitrate, glutamine and asparagines as nitrogen sources. Carlsberg Res. Commun. **1988**, *52*, 393–404.
9. Hess, D. Haploide in der Pflanzenzüchtung. In *Biotechnologie Der Pflanzen*; Hess, D., Ed.; Verlag Eugen Ulmer: Stuttgart, 1992; 142–157.

Arabidopsis thaliana: Characteristics and Annotation of a Model Genome

Tanya Z. Berardini
Seung Y. Rhee
Carnegie Institution of Washington, Stanford, California, U.S.A.

INTRODUCTION

Arabidopsis thaliana is an annual plant of the Brassicaceae family and is commonly found in temperate regions of the world. Its suitability for molecular and genetic experiments has made it one of the most widely studied plants today. *Arabidopsis* is the first plant to be completely sequenced and remains the most completely sequenced eukaryotic genome to date. Approximately 11,000 researchers around the world are currently engaged in unraveling the functions of this genome. Lessons learned from this reference plant will facilitate more systematic and targeted approaches for manipulating and managing plants that impact humans and the environment.

GENOME COMPOSITION AND ORGANIZATION

The genome of *Arabidopsis*, one of the smallest among angiosperms, has been estimated at approximately 146 Mb[1] and is highly dense with genes. To date (January, 2003), 117.3 Mb of nonredundant sequence have been completely sequenced. The remaining gaps are in the centromeres and other highly repetitive regions (The *Arabidopsis* Information Resource (TAIR), http://www.arabidopsis.org/info/agi.html). Approximately 15–20% of each chromosome is composed of heterochromatin around the centromere and, additionally, in two heterochromatic knobs of chromosomes 4 and 5.[1] In the euchromatic regions, the average gene density is 5 kb per gene with 50% of the euchromatic sequence allotted to genes.

Most characteristics such as gene density, distribution of repetitive DNA, and guanine/cytosine (GC) content are constant within and among all 10 euchromatin chromosome arms.[1] This organization is quite different from the organization of most crop plant genomes, where most of the gene-rich tracts are clustered and separated by huge stretches of repetitive DNA.[2] The average GC content over the five chromosomes is 34.9%[1] with about 4–6% of the cytosine residues in the *Arabidopsis* genome being methylated, compared to 30–33% and 22% cytosine methylation in tobacco and wheat, respectively.[3] The repetitive fraction of the genome is more highly methylated than single and low-copy genes.

The most recent genome reannotation by The Institute for Genome Research (TIGR) includes approximately 29,000 genes (http://www.tigr.org/tdb/e2k1/ath1/) containing an average of five exons per gene with a mean unprocessed transcript length of 2085 bp (1584 bp mode) and a mean protein length of 425 amino acid residues (221 mode). Both size distributions are extremely right-skewed with over 90% of the genes being smaller than 4 kb. Approximately a third of the genes identified by computer prediction have been verified with full-length cDNA information and about 64% of the genes have full or partial length cDNA sequences associated with them (TAIR, http://www.arabidopsis.org/servlets/sv). Preliminary results from analyzing hypothetical genes (i.e., genes with no transcript sequence or sequence similarity information) indicate that approximately 80% of them have detectable transcripts.[4] This suggests that approximately 95% of the identified genes (approximately 27,500 transcripts) are part of the transcriptome. In addition, comparison of 5000 full-length cDNAs to the genome reveals that about 2% of the genes are alternatively transcribed, suggesting that the transcriptome size is likely to increase as more experimental data are generated.[5] Taking into account post-translational modifications like phosphorylation and glycosylation, the *Arabidopsis* proteome is likely to be even larger than its transcriptome.

Unlike the human and many other plant genomes, only about 10% of the *Arabidopsis* genome is composed of repetitive DNA. It consists largely of 5S rRNA arrays, 18S-5.8S-25S rRNA arrays, centromere-associated repeat sequences, nucleolar organizers, telomeres, and transposons. *Arabidopsis* contains a rich diversity of most known transposons, as well as some that are structurally unique.

Most repetitive sequences are found in the centromeres and telomeres. Genetically defined centromeres contain a central region composed of 180 bp repeat microsatellites and Athila transposable elements, flanked by sequences containing a number of additional microsatellites, transposable elements, 5s rDNA, and unique sequences containing expressed genes.[6] The unique sequences in the five centromeres are not similar to each other. The

Encyclopedia of Plant and Crop Science
DOI: 10.1081/E-EPCS 120010626

only conserved elements appear to be those in the central domain, suggesting that these structural aspects may be sufficient for centromere function.[1] Telomeres consist of tandemly repeated blocks of CCCTAAA, similar to the DNA patterns found in lower eukaryotes.[7] Unlike animals, *Arabidopsis* can tolerate a severe reduction of telomeric DNA for up to ten generations.[8] It is unknown how this tolerance is achieved, what additional factors

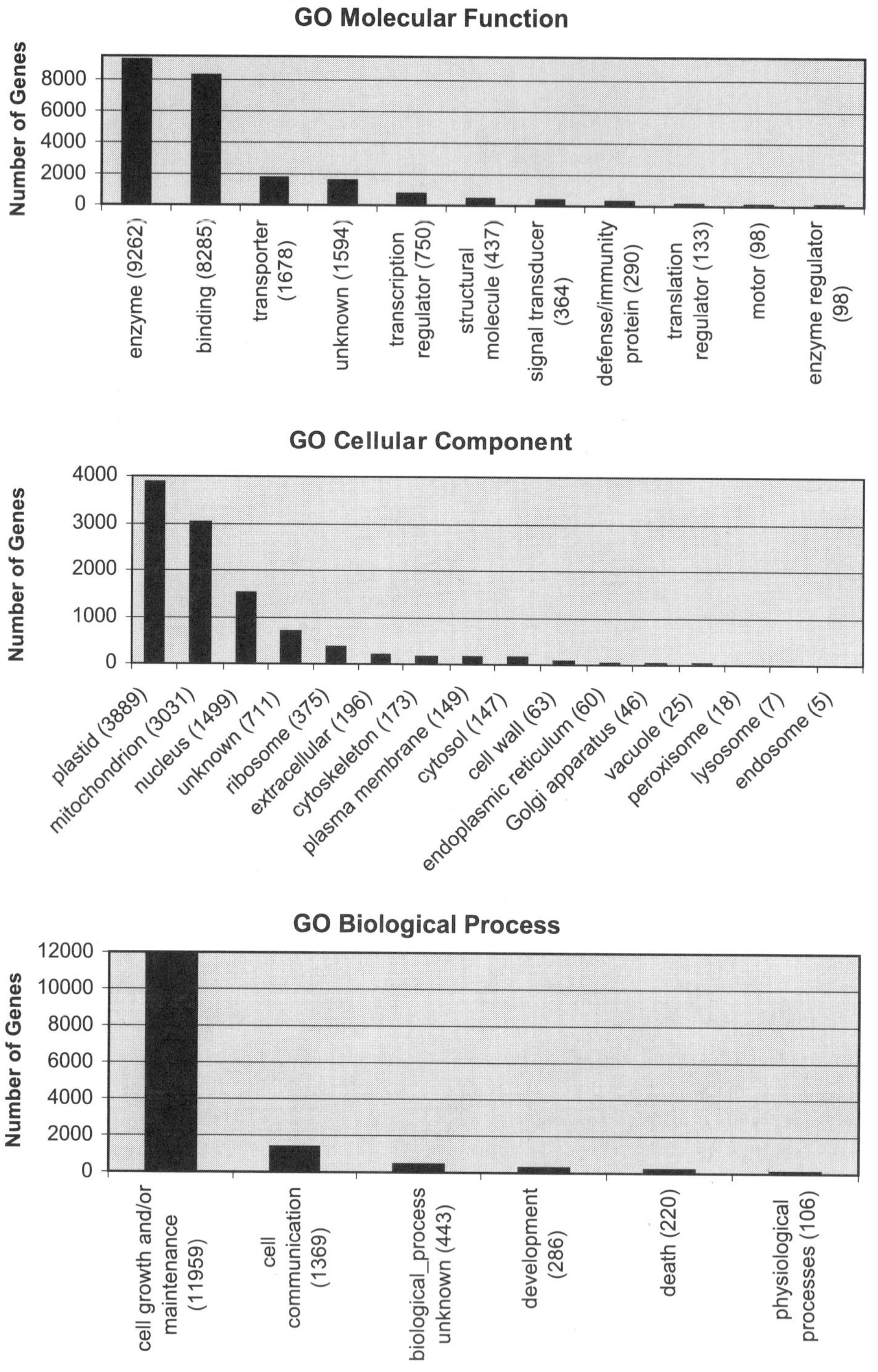

Fig. 1 Genes annotated to the gene ontology (GO) vocabularies as of December 2002. Numbers of genes per term are in parentheses. A. 21,425 genes (74% of all predicted genes) annotated to molecular function terms. B. 16,921 genes (58% of all predicted genes) annotated to cellular component terms. C. 13,816 genes (48% of all predicted genes) annotated to biological process terms.

contribute to the maintenance of chromosomal integrity, or whether this phenomenon occurs in other plants.

Sequence analysis of the *Arabidopsis* genome has revealed a history of genome rearrangements, most notably duplications followed by gene loss.[1] Approximately 60% of the genome has been duplicated, with at least two rounds of large segmental duplications occurring approximately 112 million years ago.[1,9] The estimated timing of the duplication events suggests that the ancient duplications occurred before the divergence of the tomato and *Arabidopsis* lineages.[9] In addition to these large segmental duplications, the genome has undergone many smaller gene duplication events. Approximately 17% of the genes are tandemly repeated and approximately 4000 genes belong to about 1500 gene families with more than five members each.[1]

A very recent rearrangement in the *Arabidopsis* genome involves one of its organellar genomes. Approximately 620 kb of mitochondrial DNA has inserted near the centromeric region of chromosome 2 in the Columbia ecotype.[10] This mitochondrial DNA is not found in other ecotypes such as *Landsberg erecta*, indicating that the rearrangement occurred very recently.

Functional Composition

The *Arabidopsis* genome is being functionally annotated based on data from the literature and from sequence comparisons by TAIR and TIGR. Using the controlled vocabularies developed by the Gene Ontology consortium (http://www.geneontology.org), approximately 20,100 genes have been annotated to terms describing a gene product's molecular function, biological process, and subcellular localization. Fig. 1 illustrates the current distribution of annotations of these genes. About 2020 of the 29,000 genes have been described in the literature (TAIR analysis). Many of these genes are involved in agronomically important processes such as responses to drought, cold, light, and disease as well as processes not found in animal systems such as secondary metabolism.

The challenge of characterizing the remaining 26,980 genes is being addressed using multifaceted approaches to predict gene function followed by in planta assays. Based on conserved domains, there are approximately 11,000 gene families, similar to the number found in other sequenced multicellular eukaryotes such as *Drosophila* and *Caenorhabditis elegans*.[1] An ongoing analysis of the *Arabidopsis* proteome with respect to known metabolic pathways (AraCyc, http://arabidopsis.org/tools/aracyc/) has so far resulted in the assignment of about 1000 genes to 174 pathways, including plant-specific pathways like those involved in secondary metabolism.

Characterization of the *Arabidopsis* transcriptome is facilitated by the extensive use of microarrays to analyze gene expression patterns. Available technologies include targeted cDNA arrays as well as high-density oligonucleotide arrays representing 25,000 genes. Results from over 570 *Arabidopsis* cDNA arrays and multiple high-density oligonucleotide arrays are available to the research community (http://www.arabidopsis.org/info/expression/index.html). Cluster analysis of these experimental results can reveal patterns of possible co-regulation between genes as yet unknown and experimentally characterized genes, and can shed light on the composition and characteristics of the *Arabidopsis* transcriptome.

Finally, functions of genes are being elucidated using forward and reverse genetic approaches. Functional analysis of specific genes is facilitated by the availability of a suite of insertion lines (http://arabidopsis.org/links/insertion.html). The ability to obtain insertions in specific genes allows construction of appropriate double or triple mutant lines to uncover potentially redundant functions among members of gene families. Functions of redundant genes may also be inferred from the phenotypes of engineered dominant mutations created by activation tagging and by the characterization and eventual cloning of loci defined by mutation and quantitative trait loci assessed in natural populations.

CONCLUSION

The sequencing of the *Arabidopsis* genome opened unprecedented opportunities for plant biology and agronomy. It confirmed previous hypotheses about the genome and simultaneously raises many new questions. Understanding the organization and evolutionary history of the *Arabidopsis* genome is important in making an accurate assessment of its relatedness to other plant species and in leveraging knowledge gained from studying this plant to other plants. Using the sequenced genome and a set of rapidly developing genomic analysis tools, researchers are now able to address biological questions in ways that were impossible just a few years ago. These new investigative methods and the results they generate will undoubtedly change the way research is conducted and published. Furthermore, advances made in this small weed serve as both a reference for understanding processes shared with agronomically important plants, and as a tool for understanding the genetic basis for diversity.

ARTICLES OF FURTHER INTEREST

Arabidopsis Transcription Factors: Genome-Wide Comparative Analysis Among Eukaryotes, p. 51
Evolution of Plant Genome Microstructure, p. 421

Genome Structure and Gene Content in Plant and Protist Mitochondrial DNAs, p. 520

REFERENCES

1. Arabidopsis Genome Initiative. Analysis of the genome sequence of the flowering plant *Arabidopsis thaliana*. Nature **2000**, *408* (6814), 796–815.
2. Barakat, A.; Matassi, G.; Bernardi, G. Distribution of genes in the genome of *Arabidopsis thaliana* and its implications for the genome organization of plants. Proc. Natl. Acad. Sci. U.S.A. **1998**, *95* (17), 10044–10049.
3. Fulnecek, J.; Matyasek, R.; Kovarik, A. Distribution of 5-methylcytosine residues in 5S rRNA genes in *Arabidopsis thaliana* and *Secale cereale*. Mol. Genet. Genomics **2002**, *268* (4), 510–517.
4. Xiao, Y.L.; Malik, M.; Whitelaw, C.A.; Town, C.D. Cloning and sequencing of cDNAs for hypothetical genes from chromosome 2 of *Arabidopsis*. Plant Physiol. **2002**, *130* (4), 2118–2128.
5. Haas, B.J.; Volfovsky, N.; Town, C.D.; Troukhan, M.; Alexandrov, N.; Feldmann, K.A.; Flavell, R.B.; White, O.; Salzberg, S.L. Full-length messenger RNA sequences greatly improve genome annotation. Genome Biol. **2002**, *3* (6), RESEARCH0029.
6. Kumekawa, N.; Hosouchi, T.; Tsuruoka, H.; Kotani, H. The size and sequence organization of the centromeric region of *Arabidopsis thaliana* chromosome 4. DNA Res. **2001**, *8* (6), 285–290.
7. Richards, E.J.; Ausubel, F.M. Isolation of a higher eukaryotic telomere from *Arabidopsis thaliana*. Cell **1988**, *53* (1), 127–136.
8. Riha, K.; McKnight, T.D.; Griffing, L.R.; Shippen, D.E. Living with genome instability: Plant responses to telomere dysfunction. Science **2001**, *291* (5509), 1797–1800.
9. Vision, T.J.; Brown, D.G.; Tanksley, S.D. The origins of genomic duplications in *Arabidopsis*. Science **2000**, *290* (5499), 2114–2117.
10. Stupar, R.M.; Lilly, J.W.; Town, C.D.; Cheng, Z.; Kaul, S.; Buell, C.R.; Jiang, J. Complex mtDNA constitutes an approximate 620-kb insertion on *Arabidopsis thaliana* chromosome 2: Implication of potential sequencing errors caused by large-unit repeats. Proc. Natl. Acad. Sci. U. S. A. **2001**, *98* (9), 5099–5103.

Arabidopsis Transcription Factors: Genome-Wide Comparative Analysis Among Eukaryotes

José Luis Riechmann
California Institute of Technology, Pasadena, California, U.S.A.

INTRODUCTION

Transcription factor coding–genes are abundant in the genomes of eukaryotic organisms. In the dicotyledonous plant *Arabidopsis thaliana*, which has a genome of approximately 125 megabase pairs (Mbp) of DNA that contains over 25,000 genes,[1] more than 1500 genes of such type have been identified.[2] The rice (*Oryza sativa*) complement of transcription factors is similar in size and composition to that of Arabidopsis.[3] In addition, different cereals are known to have a similar repertoire and arrangement of genes in their genomes. Thus, the research model plant Arabidopsis has a set of transcriptional regulators similar to those of the main staple crops rice, maize, and wheat. Moreover, many Arabidopsis transcription factor genes have been shown to retain their native functions when introduced as transgenes into other plant species.

TRANSCRIPTION FACTORS: WHAT THEY ARE AND WHAT THEY DO

Gene transcription (the synthesis of RNA molecules from the genomic DNA) is carried out by a multitude of proteins of different biochemical activities that act in concert. These proteins can be classified into different functional groups: the basic transcription apparatus, large multisubunit coactivators, chromatin-related proteins, and transcription factors, which comprise the most numerous of all these groups of proteins.[4] Transcription factors are proteins that show sequence-specific DNA binding and are capable of activating and/or repressing transcription. They are responsible for the selectivity in gene regulation, and are often themselves expressed in a tissue, cell-type, temporal, or stimulus-dependent–specific manner. Transcription factors are modular proteins, and can be grouped into families according to their DNA binding domain.[4]

Many of the biological processes in eukaryotic organisms are controlled at the level of gene expression, primarily through regulation of transcription. In plants, these processes include development, adaptation to the environment, the defense response against pathogens, and metabolic pathways. Moreover, it is now known that morphological changes that occurred during plant domestication and crop improvement were due to mutations in transcription factors, alterations in their expression, or changes in the expression of other types of regulatory proteins,[4] underscoring the importance of this class of genes for plant and crop biotechnology.

TRANSCRIPTION FACTOR GENE CONTENT OF THE ARABIDOPSIS GENOME

The *Arabidopsis thaliana* genome is the first from a higher plant to be sequenced.[1] It comprises approximately 125 Mbp of DNA, and shows a compact organization of high gene density. On average, there is one gene per 4.5 kilobase (kb) of DNA: approximately 2 kb correspond to exons and introns, and approximately 2.5 kb correspond to intergenic regions, which include regulatory sequences such as the promoter and enhancers.[1] Other plants—maize, for example—have genomes that are much larger than that of Arabidopsis, but have similarly organized coding and regulatory sequences. In monocots, active genes are usually distributed in compact gene-rich islands, where much of the genomic DNA corresponds to repetitive sequences. Despite its simplicity, the Arabidopsis genome bears extensive duplications, including many tandem gene duplications and large-scale duplications between different chromosomes,[1] which might affect 40% of its total genes. Duplications can be an obstacle to gene functional analysis, because they often result in functional redundancy or overlap between the duplicated genes.[2,4,5]

The Arabidopsis complement of transcription factor coding–genes has been described and reviewed in detail elsewhere.[2,4] In brief, the Arabidopsis genome codes for at least 1572 transcription factors (or approximately 6% of its approximately 26,000 total genes), which can be grouped into more than 45 different gene families. Such global content of transcriptional regulators is comparable to those of other eukaryotic organisms (for instance,

Encyclopedia of Plant and Crop Science
DOI: 10.1081/E-EPCS 120010627

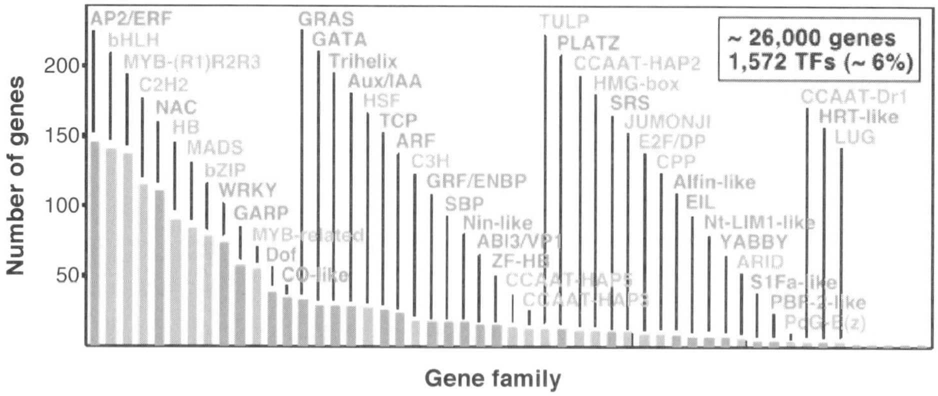

Fig. 1 Content and distribution of transcriptional regulator (TR) coding genes in the Arabidopsis genome. The different families of transcription factors are ordered according to the number of members they contain. Families that are specific to plants are indicated in green (or dark gray), and those that are also present in other eukaryotic kingdoms in blue (or light gray). The data represented in this figure, including the gene family names, are from Refs. 2 and 4. Single-copy transcription factor genes that do not belong to families are depicted, but not listed; these are *LFY*, *NZZ*, *SAP*, *FIE*, and *EYA-like*. (From Ref. 4.) (*View this art in color at www.dekker.com.*)

approximately 4.6% of the genes in the fruit fly *Drosophila melanogaster* code for transcription factors[2]). However, it is well known that many transcription factor gene families exhibit great disparities in abundance among the different eukaryotic kingdoms, and that some families are kingdom-specific. Approximately 45% of the Arabidopsis transcription factors belong to plant-specific gene families, and approximately 53% belong to families found in plants, animals, and fungi. Some of the plant-specific transcription factor families are large, such as AP2/ERF, NAC, WRKY, ARF/IAA, and Dof (Fig. 1). Some other groups, such as the MYB, MADS, and bZIP, which are not particularly numerous in animals or yeast, have been significantly amplified in the plant lineage. This points to the large degree of diversity in transcriptional regulators present in the different eukaryotic kingdoms.[2,4]

In general, it appears that most of the transcription factor families in Arabidopsis are involved in a variety of different biological functions, and vice versa (i.e., in a given function, genes from several families can be involved[2,4]). There are, however, some exceptions; for instance, MADS box genes are most frequently involved in developmental processes.[6,7]

HOW SIMILAR ARE ARABIDOPSIS TRANSCRIPTION FACTORS TO THOSE FROM OTHER PLANTS?

The determination of the sequence of the rice genome, and the large collection of cDNA sequences from other plants available in databases, answer this question. Despite the very different appearance and lifestyle of Arabidopsis and rice, and the fact that the rice genome contains a higher total number of genes, their respective complements of transcription factor genes are similar.[3] The largest transcription factor families in Arabidopsis also appear to be the most prevalent ones in monocotyledonous plants (Fig. 2). In addition, many examples of orthology can be identified among Arabidopsis transcription factor genes and those from rice or maize (for example, in the MADS gene family[6,7]). Putative orthologous MADS-box genes have regularly maintained conserved functions, even after substantial sequence divergence.[7] Moreover, Arabidopsis transcription factors from several different families have been shown to retain their function when introduced into a heterologous species, and vice versa. For example, *LEAFY*, a meristem identity gene that controls the reproductive switch in Arabidopsis, also triggers flowering when introduced as a transgene in aspen or citrus.[8]

In summary, the complement of transcription factors appears to be, in its general characteristics, very similar among monocots and dicots; and individual genes can conserve their native function across species. It is also clear, however, that differences exist. For instance, whereas most of the amplification of the MYB-(R1)R2R3 gene family occurred prior to the separation into monocots and dicots, several subgroups in maize appear to have originated recently or undergone duplication.[9] These recent expansions could have allowed a functional diversification that might not be present in Arabidopsis. Conversely, there are also gene families that are larger in Arabidopsis than in rice (Fig. 2).

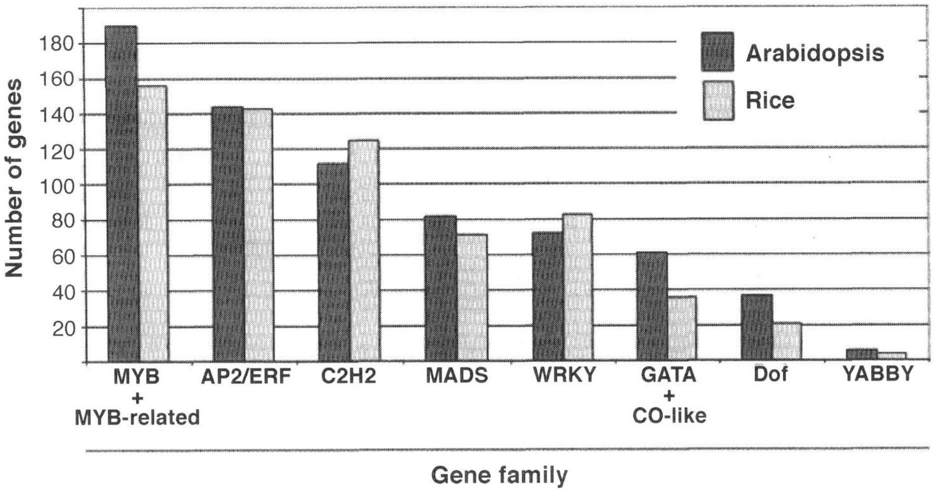

Fig. 2 Comparison of Arabidopsis and rice transcription factor gene families. The number of members of a subset of the transcription factor gene families of rice and Arabidopsis is represented. Arabidopsis data are from Refs. 2 and 4; rice data are from Ref. 3.

FUTURE PROSPECTS

The vast majority of Arabidopsis transcription factors have not been genetically and functionally characterized yet. For those that have been, characterization is usually limited to the description of phenotypic differences between mutant and wild-type plants, and to the determination of their expression patterns. However, there is still very little knowledge of the genes that each of the transcription factors regulates. Thus, the function of the Arabidopsis complement of transcription factors, considered as a whole, and the dynamic relationship between the genome and the transcriptional regulators remain largely unexplored.[4] These areas of research can now be pursued with a variety of reverse genetic methods and functional genomic technologies.[5] In addition to helping elucidate the complex logic of transcription at a genome-wide level in multicellular eukaryotes, such research will have a profound impact on plant biotechnology and agriculture.[4]

ACKNOWLEDGMENTS

I wish to acknowledge my former colleagues at Mendel Biotechnology for their input and work in our transcription factor genomics research program.

ARTICLES OF FURTHER INTEREST

Arabidopsis Thaliana: *Characteristics and Annotation of a Model Genome*, p. 47
Bioinformatics, p. 125
Cross-species and Cross-genera Comparisons in the Grasses, p. 374
Evolution of Plant Genome Microstructure, p. 421
Floral Induction, p. 452
Flower Development, p. 464
Genome Size, p. 516
Gene Expression Modulation in Plants by Sugars in Response to Environmental Changes, p. 484
Molecular Evolution, p. 748
Photoperiodism and the Regulation of Flowering, p. 877
Plant Response to Stress: Regulation of Plant Gene Expression to Drought, p. 999
Rice, p. 1102

REFERENCES

1. Arabidopsis Genome Initiative. Analysis of the genome sequence of the flowering plant *Arabidopsis thaliana*. Nature **2000**, *408*, 796–815.
2. Riechmann, J.L.; Heard, J.; Martin, G.; Reuber, L.; Jiang, C.; Keddie, J.; Adam, L.; Pineda, O.; Ratcliffe, O.J.; Samaha, R.R.; Creelman, R.; Pilgrim, M.; Broun, P.; Zhang, J.Z.; Ghandehari, D.; Sherman, B.K.; Yu, G. Arabidopsis transcription factors: Genome-wide comparative analysis among eukaryotes. Science **2000**, *290*, 2105–2110.
3. Goff, S.A.; Ricke, D.; Lan, T.H.; Presting, G.; Wang, R.; Dunn, M.; Glazebrook, J.; Sessions, A.; Oeller, P.; Varma, H.; Hadley, D.; Hutchison, D.; Martin, C.; Katagiri, F.; Lange, B.M.; Moughamer, T.; Xia, Y.; Budworth, P.; Zhong, J.; Miguel, T.; Paszkowski, U.; Zhang, S.; Colbert, M.; Sun, W.L.; Chen, L.; Cooper, B.; Park, S.; Wood, T.C.; Mao, L.; Quail, P.; Wing, R.; Dean, R.; Yu, Y.; Zharkikh, A.; Shen, R.; Sahasrabudhe, S.; Thomas, A.; Cannings, R.;

Gutin, A.; Pruss, D.; Reid, J.; Tavtigian, S.; Mitchell, J.; Eldredge, G.; Scholl, T.; Miller, R.M.; Bhatnagar, S.; Adey, N.; Rubano, T.; Tusneem, N.; Robinson, R.; Feldhaus, J.; Macalma, T.; Oliphant, A.; Briggs, S. A draft sequence of the rice genome (Oryza sativa L. ssp. japonica). Science **2002**, *296*, 92–100.

4. Riechmann, J.L. Transcriptional Regulation: A Genomic Overview. In *The Arabidopsis Book*; Somerville, C.R., Meyerowitz, E.M., Eds.; American Society of Plant Biologists: Rockville, MD, 2002. DOI: 10.1199/tab0085.
5. Riechmann, J.L.; Ratcliffe, O.J. A genomic perspective on plant transcription factors. Curr. Opin. Plant Biol. **2000**, *3*, 423–434.
6. Ng, M.; Yanofsky, M.F. Function and evolution of the plant MADS-box gene family. Nat. Rev., Genet. **2001**, *2*, 186–195.
7. Theißen, G.; Becker, A.; Di Rosa, A.; Kanno, A.; Kim, J.T.; Munster, T.; Winter, K.U.; Saedler, H. A short history of MADS-box genes in plants. Plant Mol. Biol. **2000**, *42*, 115–149.
8. Martin-Trillo, M.; Martinez-Zapater, J.M. Growing up fast: Manipulating the generation time of trees. Curr. Opin. Biotechnol. **2002**, *13*, 151–155.
9. Rabinowicz, P.D.; Braun, E.L.; Wolfe, A.D.; Bowen, B.; Grotewold, E. Maize R2R3 Myb genes: Sequence analysis reveals amplification in the higher plants. Genetics **1999**, *153*, 427–444.

Archaeobotanical Remains: New Dating Methods

Lawrence Kaplan
University of Massachusetts, Boston, Massachusetts, U.S.A.

A

INTRODUCTION

Since the mid-nineteenth century the disciplines of archaeology and botany have collaborated to form an important tool in the study of crop plant origins. The importance of this collaboration has always been enhanced by some means of dating the plant remains that were recovered. In regions such as Italy and Greece or the Middle East, classical texts and historic reconstruction aided in dating. In the American Southwest, dating of ruins by means of tree ring counts served a similar function. Today, radiocarbon dating is the most important method of determining the age of organic materials.

RADIOCARBON DATING

In 1949 W. F. Libby and associates at the University of Chicago showed that organic remains of known age from ancient sites in Egypt, Syria, and the American Southwest could be accurately dated by assessing their content of radioactive carbon.[1] Since that time radiocarbon dating has become a foundation technique in archaeology and an important tool in the study of ancient plant remains.

Radiocarbon dating is based on the ability to measure the proportion of the radioactive isotope ^{14}C (carbon-14) to the stable (nonradioactive) isotopes ^{12}C (carbon-12) and ^{13}C (carbon-13) in a sample of organic matter. The rate of decline of any radioactive isotope is stated by its "half life." The radioactive isotopes of other elements decay more rapidly or less rapidly than carbon-14, whose half life is about 5730 years,[2] but the half life of each type of radioactive element whether rapid or slow is always the same. Because of this it is possible to determine the age of ancient plant remains and other organic materials.

TRADITIONAL AND ACCELERATOR DATING

Traditional radiocarbon dating, as developed by Libby, measured the emission of beta particles from wood charcoal, or some other organic material and is now sometimes called "indirect dating." Because the amount of carbon-14 in such material is very low to begin with, and continually grows smaller, a fairly large sample is required and the detectable elapsed time from the death of the organism to the present is limited to about 50,000 years. The degree of error in the measurement increases with age.

Another method of carbon-14 dating makes use of instruments (a particle accelerator and a mass spectrometer) combined in a system developed by physical scientists in the late 1970s.[2] The system is called accelerator mass spectrometry (AMS), or sometimes "direct dating" and has the ability to detect and count very small amounts of radiocarbon in a specimen. Both methods are destructive, that is, the sample to be dated must be destroyed in the process of dating. Early in the development of radiocarbon dating it was assumed that the proportion of carbon-14 to other forms of the element had remained constant[2] up until the time of extensive atomic bomb testing. The year 1950 was accepted as the end of constant proportions of carbon-14 in the atmosphere. It soon became apparent, however, that there were differences, or an offset, between age in "radiocarbon years" and age in actual years due to the lack of constancy of carbon-14 levels in the atmosphere.

CALIBRATION

In order to determine the magnitude of the offset, growth ring samples of the remarkably long lived bristlecone pine and the giant sequoia were radiocarbon dated after an actual date had been determined by counting the annual rings of the trees. In order to provide regional standards, the procedure was repeated with other trees and other methods of dating that were applied to diverse organisms including corals. These tests showed that in the past there were variations in the amount of carbon-14 in the atmosphere, probably due to changes in the earth's magnetic field. As noted previously, more recent changes in the proportion of atmospheric carbon-14 have resulted from the testing of atom bombs.

Knowing the magnitude of these variations in atmospheric radiocarbon allowed the age in radiocarbon years to be corrected to give the actual or calibrated age of a specimen. The result of carbon-14 determinations of the age of archaeological plant remains are reported in

Encyclopedia of Plant and Crop Science
DOI: 10.1081/E-EPCS 120017084

uncorrected or uncalibrated age as radiocarbon years before present time (b.p.), where present time is set at 1950. The calibrated age is reported in calendar years; hence, one half of a seed of *Phaseolus acutifolius* (tepary bean) from a cave in the Tehuacán Valley of Mexico dates by AMS to 2300 ± 50 b.p. in radiocarbon years and 400–210 B.C. in dendrocalibrated calendar years.[3]

ACCELERATOR MASS SPECTROMETRY DATES: ADVANTAGES

The great advantage of AMS, over traditional radiocarbon dating, is the ability of AMS to give accurate dates on very small samples, for example, on one-half or less of a single maize kernel. The method is also much more rapid than traditional radiocarbon dating. Prior to the development of AMS the amount of early crop plant material retrieved from a particular level in an archaeological site was seldom sufficient for radiocarbon dating and this was later found to lead to errors. For example, when a piece of wood charcoal excavated in the late 1940s (prior to the development of AMS dating) from Bat Cave in New Mexico was radiocarbon dated to about 6000 b.p. by the traditional method, that date was assigned to other materials associated with it.[4] Among these other materials to receive the date of 6000 b.p. were small maize ears (not sufficient material to be dated by the traditional or indirect C-14 method). These small ears, because of their age, size, and other characteristics, resembled a hypothetical wild maize, which was presumed to be extinct. Years later small samples of the Bat Cave maize were dated by AMS and found to be not older than 3120 b.p. This brought the Bat Cave maize more into line with the earliest dates for prehistoric maize in other Southwestern sites, but it also demonstrated that traditional radiocarbon dates based on associated material could lead to erroneous conclusions. The disturbance of Bat Cave remains by burrowing rats meant that different objects in the same stratum may vary widely in age. Furthermore, because ''smaller'' and ''older'' did not necessarily go together, the smaller ears were not necessarily more primitive.[5]

EARLY AGRICULTURE IN MEXICO

The pursuit of archaeological evidence for the origin of maize brought famed archaeologist R. S. MacNeish to Mexico, where he led major excavation projects, especially in the Tehuacán Valley south of Puebla during the early 1960s. It was in Mexico that the wild growing nearest relatives of maize had been found by botanists. MacNeish and his colleagues were successful in recovering important collections of maize, beans, squashes, and other cultivated and wild plant remains. The strata of prehistoric cave deposits were dated by traditional radiocarbon methods and the results of the excavations were widely recognized as the most important to have been made for the understanding of the origin of major American crop plants. Excavation levels dated within a cultural period extending from 7000 to 5350 b.p. contained small maize ears, which in the opinion then prevalent among maize experts matched the hypothetical wild form. Published in 1967[6] the radiocarbon dates from the time of these earliest maize ears to the most recent, just prior to the Spanish conquest, were widely cited as the most authoritative archaeological timeline for the domestication of maize and the development of agriculture and human society in the high cultures of Mexico. When AMS dating became available, tiny samples of the ancient maize ears themselves were dated and were found to be from 4700 b.p. or about 1500 years younger than the radiocarbon dates previously obtained by the indirect method.[7] These more recent dates are especially significant for understanding the beginnings of Mesoamerican agriculture because they appear to reduce the perplexingly long time period between the first presence of maize and the period in which it became the primary component of the human diet. Accelerator dates on beans recovered from the same sites in the Tehuacán Valley were, like the maize dates, much younger than the dates on associated remains had indicated. In fact, the oldest accelerator dates for beans at Tehuacán were about 2300 b.p.,[3] which shows that, contrary to the historic unity of maize and beans in the Indian diet, there was a substantial period in which maize became a mainstay of the diet without the presence of beans.

PREMAIZE CULTIVATION IN EASTERN NORTH AMERICA

The value and reliability of AMS dates on small samples of seeds from sites where disturbance caused critically important specimens to be mixed with earlier or later charcoal or other organic remains is widely recognized. One of the most significant applications of AMS dating is found in east central North America where seeds and fruit parts of a suite of four indigenous plants [a cucurbita gourd, sunflower, marsh elder, and chenopod (*Chenopodium berlandieri*)] showing characteristics of domesticates were AMS dated to the period of 2000–1000 B.C. The structural characteristics developed during this time period are regarded by most archaeologists specializing in eastern North America as evidence for an agricultural system predating the adoption of maize as the primary crop in this region.[8] The AMS dates of prehistoric maize from this region showed it to be later than previously thought, which fell in nicely with the proposal of a premaize eastern North American cultivation system based on indigenous plant species.[4]

AMS APPLIED TO AFRICAN AND ASIAN CROP PLANTS

The complex and still controversial evolution of domesticated maize contrasts with the evolution of African and Asian cereal grains where the wild ancestors of the domesticates are known and there has been less structural transformation in the evolution of the domesticated crop. Nevertheless, AMS dating now plays an important role in assessing the process of domestication in Africa (especially sorghum) and Asia (especially rice). Carbonized rice grains embedded in ceramic potsherds from east central China[9] were AMS dated to 6400–5800 b.p. In this instance, the antiquity of the pottery, a non-organic ceramic material, as well as the rice, was determined.

CAUTIONS

Radiocarbon dates obtained by AMS and by traditional radiocarbon dating do not always disagree. Nor, when they disagree, are the AMS dates always more recent.

CONCLUSION

The application of structural and taxonomic botanical methods to the analysis of archaeological plant remains furthered the understanding of crop plant origins from the mid-nineteenth century until the present time. When botanical analysis was combined with radiocarbon dating methods, the study of crop plant origins took a great leap forward.

ARTICLE OF FURTHER INTEREST

Crop Domestication in Prehistoric Eastern North America, p. 314

REFERENCES

1. Goldstein, L. Editor's corner. Am. Antiq. **1999**, *64* (1), 5–7.
2. Taylor, R.E. Fifty years of radiocarbon dating. Am. Sci. **2000**, *88* (1), 60–67.
3. Kaplan, L.; Lynch, T.F. *Phaseolus* (Fabaceae) in archaeology: AMS radiocarbon dates and their significance for pre-Columbian agriculture. Econ. Bot. **1999**, *53* (3), 261–272.
4. Meltzer, D.J. North America's vast legacy. Archaeology **1999**, *52* (1), 51–59.
5. Wills, W.H. *Early Prehistoric Agriculture in the American Southwest*; School of American Research Press: Santa Fe, NM, 1988; 126–127.
6. Mangelsdorf, P.C.; MacNeish, R.S.; Galinat, W.C. Prehistoric Wild and Cultivated Maize. In *The Prehistory of the Tehuacan Valley. Vol. 1. Environment and Subsistence*; Byers, D.S., Ed.; University of Texas Press: Austin, TX, 1967; 178–200.
7. Long, A.; Benz, B.F.; Donahue, D.J.; Jull, A.J.; Toolin, L.J. First direct AMS dates on early maize from Tehuacán, Mexico. Radiocarbon **1989**, *31* (3), 1035–1040.
8. Smith, B.D. *The Emergence of Agriculture*; Scientific American Library, W.H. Freeman: New York, 1992; 10–13.
9. Zohary, D.; Hopf, M. *Domestication of Plants in the Old World*, 3rd Ed.; Oxford University Press: New York, 2000; 88–91.

Aromatic Plants for the Flavor and Fragrance Industries

Brian M. Lawrence
B.M. Lawrence Consultant Services, Winston-Salem, North Carolina, U.S.A.

INTRODUCTION

Essential oils—also known as volatile or ethereal oils or essences—are the odoriferous principles found in the aromatic plants that are the raw materials of the flavor and fragrance industries. The term essential, which has alchemical roots, results from the coined phrase ''quinta essentia,'' or ''quintessence,'' the fifth element (the other four being land, fire, wind, and water). Alchemists believed that within the plant kingdom there was a single extractive principle that could prolong life indefinitely, hence the importance of quinta essentia. The term ''oil'' probably originated from the observation that certain plants contain glands or intercellular spaces filled with oily droplets found to be nonmiscible with water. Unlike fatty or fixed oils, which are lipids, essential oils are complex mixtures of volatile compounds that are biosynthesized by living organisms.

For an essential oil to be genuine, it must be isolated by physical means only from a whole plant or plant part of known taxonomic origin. The physical methods used to isolate an essential oil are water, water/steam or steam distillation, or expression (also known as cold pressing, a process unique to the production of citrus peel oils). A limited number of plants (onion, garlic, wintergreen, sweet birch, and bitter almond) yield their oils during processing. In this process, warm water is mixed with the macerated plant material, causing the release of the enzyme-bound volatiles into the water from which the oil is subsequently removed by distillation.

Although the practice of distillation was first discovered in the Indus Valley in Pakistan around 3000 B.C., it was not until around 1300 A.D. that distilled aromatic waters were extensively used as medicaments. Essential oils did not appear as items of commerce until sometime in the 1400s with cedarwood oil, cinnamon oil, rose oil, sage oil, etc. among the earliest.

It is estimated that of the 17,500 aromatic plant species found in the vegetable kingdom, approximately 270 are used to produce essential oils of commerce. Of these 270 oils, ca. 40% are produced from cultivated plants; the other 60% are produced from by-products of a primary industry or from readily accessible wild growing plants.

The plant families that possess species that yield a majority of the most economically important essential oils are Apiaceae or Umbelliferae (fennel, coriander, and other aromatic seed/root oils), Asteraceae or Compositae (chamomile, *Artemisia* sp. oils, etc.), Cupressaceae (cedarwood, cedar leaf, juniper oils, etc.), Geraniaceae (geranium oil), Illiciaceae (star anise oil), Lamiaceae or Labiatae (mint, patchouli, Lavandula sp., and many herb oils), Lauraceae (litsea, camphor, cinnamon, sassafras oils, etc.), Oleaceae (jasmine oil), Pinaceae (pine and fir oils, etc.), Poaceae or Graminae (vetiver and aromatic grass oils), Rosaceae (rose oil), and Santalaceae (sandalwood oil).

WHERE ARE ESSENTIAL OILS FOUND?

Essential oils are contained in special glands or secretory tissues,[1] the type being important characters of the plant family within which the aromatic plant is found. In the Asteraceae (Compositae), schizogenous oil ducts occur in the leaves and stems frequently associated with vascular bundles. In the Lamiaceae (Labiatae), there are characteristic multicellular-headed glandular trichomes present on the leaves and calyx of the plant. The secretion produced by a multicellular head accumulates under a common cuticle that is raised like a blister. In the Lauraceae, there are large oil cells in the stem barks, whereas in the leaf, oil cells tend to be mesophyllic (i.e., found in the interior of the leaf). In the Myrtaceae and Rutaceae, the secretory tissues arise from a special mother cell. This cell divides and the daughter cells separate from one another schizogenously (i.e., they split apart from each other) to leave a central cavity. The cells surrounding the cavity produce an essential oil and the cavity continues to enlarge as the lining walls undergo lysis. Thus, large rounded oil cavities or glands are produced schizolysigenously and are visible to the naked eye (as seen in the oil glands on the peel of an orange, e.g.). Essential oils in the Apiaceae (Umbelliferae) are present in long secretory ducts called vittae that arise schizogenously on the fruits and roots. In the Zingiberaceae, a continuous layer of large rectangular oil cells occurs in the testa (the seed coat); in fact, the oil cells are just within the thin aril and epidermis of the seed and on the rhizome. In addition, isolated oil cells are scattered throughout a scraped rhizome.

Encyclopedia of Plant and Crop Science
DOI: 10.1081/E-EPCS 120010436

Table 1 Oil yield of selected aromatic plants

Species	Common name	Plant part	Oil yield (%)
Syzygium aromaticum (L.) Merr. et L. M. Perry	Clove	Dried buds	16–20%
Myristica fragrans Houtt.	Nutmeg	Kernel	8–16%
Foeniculum vulgare Mill.	Fennel	Dried fruit (seed)	1.5–10.0%
Santalum album L.	Sandalwood	Heartwood	4.5–6.5%
Cinnamomum zeylanicum Blume	Cinnamon	Innerbark	0.8–1.5%
Myrtus communis L.	Myrtle	Twigs and leaves	0.25–1.15%
Ocimum basilicum L.	Basil	Above ground plants	0.01–0.30%

ESSENTIAL OIL EXAMPLES

Examples of plants cultivated for oil production are basil, cinnamon (leaf and bark), dill (weed and seed), mint (cornmint, Native and Scotch spearmint, and peppermint), patchouli, ylang ylang, etc. Examples of by-product oils are citrus peel oils such as orange, (a by-product of the orange juice industry) and eucalyptus leaf and conifer needle oils (by-products of the lumber industry). In contrast, armoise, bois de-rose, galbanum, juniperberry, star anise, Peru balsam, etc. exemplify those oils produced from plants or exudates collected in the wild.

Essential oils are generally found to predominate in one particular organ of a plant such as leaves/calyces (sage), buds (clove), flowers (rose), fruits (lemon), seeds (cardamom), inner bark (cinnamon), heartwood (cedarwood), roots/rhizomes (ginger), resin (mastic), and exudates (Peru balsam). The essential oil content of aromatic plants varies according to the plant and the plant part from which it is obtained (Table 1). In addition to the oils produced from different parts of aromatic plants, some oils are produced from balsams, gums, or exudates. Generally, an incision or hole is made in the stem/trunk of a tree or shrub, from which the balsam or gum exudes. Steam distillation of this exudate produces an oil whose yield varies according to the exudate. A selected list of exudates and their oil yields is shown in Table 2.

From a chemical composition standpoint, an essential oil is a complex mixture of secondary metabolic compounds composed mainly of monoterpenes, sesquiterpenes, and aliphatic and aromatic compounds that can exist as hydrocarbons, ethers/oxides, alcohols, esters, acids, ketones, aldehydes, lactones, phenols, phenol ethers, etc. An essential oil contains between 50 and 300 components present in amounts greater than 1 ppm (0.0001%).

Some aromatic plants yield oils of different composition depending on which plant part is distilled. Cinnamon (*Cinnamomum zeylanicum* Blume) is an example of such a plant, as the leaf oil is rich in eugenol,[2] the flower oil is rich in cinnamyl acetate,[3] the fruit oil is rich in sesquiterpenoid compounds,[4] the inner bark oil is rich in cinnamaldehyde,[2] and the root oil is rich in camphor.[2]

During the nineteenth century, Europe—in particular, France—was the major region for essential oil production. Currently, France is a major producer only of lavandin (*Lavandula xintermedia* Emeric ex. Loisel), and produces only limited quantities of other major volume oils. At present, Asia (China, India, and Indonesia) is by far the largest oil producing region except for the production of citrus oils. For over 100 years the United States has been a major producer of mint, cedarwood (Texas—*Juniperus ashei* Bucholz; Virginia—*J. virginiana* L.), and citrus oils [orange—*Citrus sinensis* (L.), Osbeck, and lemon—*C. limon* (L.) N. L. Burm.]. However, over the past 50 years Central and South America have become major citrus producing regions, surpassing the quantities produced in the United States, e.g., Brazil (orange—*C. sinensis*), Argentina (lemon—*C. limon*), Mexico and Peru (lime—*C.*

Table 2 Oil yield of selected balsams and gums

Species	Common name	Oil yield (%)
Myroxylon balsamum (L.) Harms var. *pereirae* (Royle) Harms	Peru balsam	60–65%
Copaifera reticulate Ducke and other *Copaifera* species	Copaiba balsam	35–60%
Canarium luzonicum Miq.	Elemi	25–30%
Ferula gummosa Boiss.	Galbanum	10–25%
Commiphora myrrha (Nees) Engl.	Myrrh	10–15%
Pistacia lentiscus L.	Mastic gum	1–3%
Cistus ladanifer L.	Labdanum	0.1–0.2%

Table 3 Top twenty essential oils: approximate volumes in 2000

Essential oil	Species	Volume (tonnes)
1. Sweet orange	*Citrus sinensis* (L.) Osbeck	55,000
2. Cornmint	*Mentha arvensis* L.f. *piperascens* Malinv. ex Holmes	17,500
3. Lemon	*Citrus limon* (L.) N. L. Burm.	5600
4. Eucalyptus (cineole-type)	*Eucalyptus globulus* Labill., *E. polybractea* R. T. Baker, and other *Eucalyptus* species	3800
5. Peppermint	*Mentha xpiperita* L.	3500
6. Citronella	*Cymbopogon winterianus* Jowitt and *C. nardus* (L.) Rendle	3000
7. Eucalyptus (citronellal-type)	*Corymbia citriodora* (Hook.) K. D. Hill et L.A.S. Johnson (syn. *Eucalyptus citriodora Hook.*)	2500
8. Clove leaf	*Syzygium aromaticum* (L.) Merr. et L. M. Perry	2000
9. Cedarwood	*Juniperus ashei* Buchholz and *J. virginiana* L.	1300
10. Litsea cubeba	*Litsea cubeba* (Lour.) Pers.	1100
11. Lavandin	*Lavandula xintermedia* Emeric ex Loisel	1100
12. Lime—distilled	*Citrus aurantifolia* (Chrism. et Panzer) Swingle	1000
13. Sassafras—Chinese	*Cinnamomum micranthum* (hayata) Hayata, *C. camphora* L., *C. porrectum* (Roxb.) Kostm, *C. rigidissimum* H. T. Chang and *C. inunctum* Meissn.	1000
14. Native spearmint	*Mentha spicata* L.	980
15. Sassasfras—Brazil	*Ocotea pretiosa* (Nees) Benth.	900
16. Cedarwood—Chinese	*Chamaecyparis funebris* (Endl.) Franco	800
17. Ho or shiu	*Cinnamomum camphora* L.	800
18. Scotch spearmint	*Mentha gracilis* Sole	760
19. Grapefruit	*Citrus paradisi* Macfady	700
20. Patchouli	*Pogostemon cablin* (Blanco) Benth.	650

aurantifolia (Chrism. et Panzer) Swingle. The approximate volumes of the top twenty essential oils are shown in Table 3. A brief description of five major oils follows.

ORANGE OIL

Orange oil (also known as sweet orange oil to differentiate it from bitter orange oil, which is obtained from *Citrus aurantium* L.) is produced as a by-product of the orange juice industry because the juice is the most valuable end product.

Four major processes are used to isolate orange and other citrus oils that are located irregularly in oblate-spherical oil glands found at different depths in the colored portion of the peel (flavedo). These processes are Pellatrice, and Sfumatrice (Italian processes) and Brown Peel Shaver and the Food Machinery Corporation (FMC) process (U.S. processes). FMC, a whole fruit isolation process, has become the predominant method of oil isolation because of its efficiency. In this process, the fruit is positioned in a lower cup made of metal fingers above which is a second cup. Through the converging action of the upper and lower cups, the fruit is peeled and the peel crushed, causing the oil glands to burst. The oil is flushed away with a fine spray of water that removes it from the peel. Simultaneously, a plug is cut from the bottom of the fruit, so that as the two cups come together, the juice, peel remnants, seeds, and pulp are removed into separate streams. The oil is recovered from the oil/water emulsion by filtration followed by a three-stage centrifugation.[5]

Although numerous cultivars of orange are grown, those principally used for oil production are Washington Navel, Parson Brown, Hamlin, Pineapple, Midsweet, and Valencia (Florida, California, Arizona, Texas); Peralina, Valencia, Pera, Westin Natal, and Bahianintia (Sao Paulo state in Brazil); Castellana, Navelina, Newhall, Verna, and Salustiana (Spain); and Biondo Commune, Navelina, Ovale, and Belladonna (Reggio Calabria and Sicily, Italy).[6]

Oranges are sold in box quantities. In Florida, for example, 1000 boxes of Valencia oranges weigh 90,000 lbs. Processing this quantity of oranges yields 39,750 lbs. of peel, pulp, and seeds; 50,000 lbs. of orange juice; and 250 lbs. of cold-pressed orange oil.[7]

For the 1996/1997 season in the United States, the number of boxes of oranges processed were 215,427, Florida (90 lb/box); 12,000, California (75 lb/box); 565, Texas (85 lb/box); and 63, Arizona (75 lb/box).[8] As

Table 4 Regional orange oil production

Country	Quantity (tonnes)
Brazil	28,600
United States	19,250
Italy	1375
Spain	880
Australia	825
Greece	523
Israel	501
Others[a]	3046

[a]Includes Algeria, Argentina, Belize, China, Costa Rica, Cyprus, Guinea, Ivory Coast, Mexico, Morocco, South Africa, Tunisia.

noted in Table 3, ca. 55,000 tonnes of orange oil were produced in 2000. A more specific regional breakdown for this oil production[9] is shown in Table 4.

PEPPERMINT AND NATIVE AND SCOTCH SPEARMINT OILS

As noted in Table 3, approximately 3500 tonnes of peppermint oil, 980 tonnes of Native spearmint oil, and 760 tonnes of Scotch spearmint oil are produced annually. The largest share (more than 90%) of the peppermint oil production is in the United States. A breakdown of the regions of production is shown in Table 5.[10]

Of ca. 980 tonnes of Native spearmint oil produced, 55% was produced in the United States, with 52.5% produced in the West (Washington, Montana, Idaho, Oregon) and 2.5% in the Midwest (Michigan, Indiana, Wisconsin). China and India produced 23% and 19%, respectively.[6] In contrast, of 760 tonnes of Scotch spearmint oil produced worldwide, the United States produced 72% (59% in the West and 13% in the Midwest). The other principal countries of production were Canada (17%) and India (10%).[10]

Although peppermint and both spearmints are produced in the U.S. Midwest and West, the yield of oil is much higher in the West because of its longer, hotter growing season. All U.S. mints are produced clonally from stolons. Because of the widespread incidence of a peppermint wilt fungus (*Verticillium dahliae*), disease-free rootstock is used to replenish or commence a new area of planting. Similarly, both spearmints are susceptible to mint rust (*Puccinia menthae*), so disease-free stolons are used to also commence or replenish new spearmint plantings.[11]

Unlike the spearmints, peppermint is considered a long-day plant because it needs a minimum midsummer day length of 15 hours for flowering and for production of oil of commercially acceptable aroma and quality. This day length can be achieved north of the 40th parallel. To yield commercially acceptable oil quality, the plants are harvested just as the first flowers start to form. This is done to ensure that the undesirable oil component menthofuran[12] is minimized. Day length and flowering have little effect on the oil composition of either Native and Scotch spearmint. In fact, to produce the highest oil yield, both spearmints are harvested in full flower. Even though Scotch and Native spearmints are produced from different *Mentha* species, they are both rich in L-carvone; however, the subtle differences in oil composition make them discernibly different. This in turn affects their end use. Unlike peppermint oil production in the United States, a marketing order is in place for both the Scotch and Native spearmint oils covering an area of the Pacific Northwest due north of the west Wyoming border to the Canadian border, south to the southern border of Utah, and west from the Utah border to the Pacific coast. The administration of the marketing order controls the quantity of oil produced by each grower annually and sets a price for the oil.[13]

ROSE OIL

Although only a small quantity of rose oil is produced annually, it is an extremely high-priced oil, selling for as much as $3500 per kg. The main countries where damask rose (*Rosa damascena* Mill.) is grown are Bulgaria (around Kazanlik), Turkey (in Isparta and Budur provinces), Ukraine, Morocco, and Moldova, whereas *R. centifolia* L. is grown mainly in France and Morocco. The world production of both rose oils is estimated at ca. 5 tonnes (4 tonnes damask and 1 tonne centifolia).

In Turkey, shallow holes 50 cm × 50 cm are prepared in which bare root cuttings are planted and trimmed so that the cutting just shows above the soil level. The cuttings are watered regularly so that 3 years after planting, the

Table 5 Regional quantities of U.S. peppermint oil production in 2000

Region	Quantity (tonnes)
Midwest (Michigan, Indiana, Wisconsin)	390
Willamette Valley (W. Oregon)	360
Madras Valley (E. Oregon)	200
Lagrande (N.E. Oregon)	280
Idaho/Oregon border	540
W. Montana	70
Yakima Valley (Washington)	1330

starts have reached maturity and the roses are ready for harvest. A mature planting yields ca. 5 tonnes of flowers per hectare and can continue producing flowers for 20–30 years.[14] It takes ca. 3.5–4.0 tonnes of flowers to yield 1 kg of oil.[15] The flowers are harvested early in the morning, the time when their oil yield is maximized.

At one time, all rose oil was produced by steam and water distillation in which rose flowers were placed on a grid above the water in the bottom of a still. The water was boiled by direct fire; the steam gencrated released the oil that was condensed with a standard condenser. Today, more than 95% of exported oil is produced with steam distillation in which the steam is generated in a satellite boiler and passed through a still containing evenly spaced, stacked, multiple grids on which the flowers are placed. This allows the oil to be readily removed without the flowers becoming a waterlogged mass, as happens in steam and water stills.

More rose flowers are used to produce a rose concrete and rose absolute than are used to produce rose oil. A concrete is a hydrocarbon extract of the fresh flowers generally produced by percolation of the flowers with hexane, with the spent flowers and residual hexane removed. The concrete is rich in hydrocarbon-soluble materials and devoid of water soluble materials. It is generally a waxy, semisolid, dark-colored mass.[16] An absolute is a highly concentrated alcoholic extract of the concrete. It is prepared from the concrete by dissolving the concrete in hot alcohol and then rapidly cooling the mixture to precipitate the waxes that are removed, generally by filtration or by a cooled surface scraping of the slurry. The alcohol is then removed by distillation under high vacuum. The resultant absolute contains the concentrated aromatic portion of the flowers.[16] The yield of oil from rose flowers is ca. 0.02%; the yield of concrete is 0.22–0.25%, whereas the yield of absolute from the

Table 6 New essential oils introduced over past 25 years

Country of production	Common name	Taxonomic origin
Australia	Lemon-scented tea tree or citratum oil	*Leptospermum petersonnii* F.M. Baill.
	Tantoon oil	*Leptospermum flavescens* Smith
	Lemon myrtle oil	*Backhousia citriodora* F. Muell.
	White cypress oil	*Callitris glaucophylla* J. Thompson et L. A. S. Johnson
	Blue cypress oil	*Callitris interatropica* Baker et Smith
	Rosalina oil	*Melaleuca ericifolia* Smith
Brazil	Lantana oil	*Lantana camara* L.
Canada	Candadian goldenrod oil	*Solidago canadensis* L. var. *canadensis*
	Labrador tea oil	*Ledum groenlandicum* Oeder
Egypt	Black cumin seed oil	*Nigella sativa* L.
France (Reunion)	Santolina oil	*Santolina chamaecyparisus* L.
	Combava leaf and peel oils	*Citrus hystrix* DC
India	Betel leaf oil	*Piper betle* L.
	Jammu lemongrass oil	*Cymbopogon pendulus* (Nees ex Steud.) Wats.
	Jamrosa oil	*Cymbopogon nardus* (L.) Rendle) var. *confertiflorus* (Steud.) Bor. × *C. jwarancusa* (Jones) Schult.
	Clocimum oil	*Ocimum gratissimum* L.
	Curry leaf oil	*Murraya koenigii* (L.) Spreng.
Madagascar	Ravensara oil	*Ravensara aromatica* Sonnerat
Morocco	Blue tansy oil	*Tanacetum annum* L.
	Amni visnaga oil	*Amni visnaga* (L.) Lam.
	Artemisis arborescens oil	*Artemisia arborescens* L.
Nepal	Large or Nepal cardamom oil	*Amomum subulatum* Roxb.
New Zealand	Manuka oil	*Leptospermum scoparium* (J. R. Forst et G. Forst)
	Kanuka oil	*Kunzea ericoides* (A. Rich.)
South Africa	Eriocephalus oil	*Eriocephalus punctulatus* DC.
	Lanyana oil	*Artemisia afra* L.
	Pteronia oil	*Pteronia incana* L.

concrete is 50–60%. It is estimated that ca. 15.0 tonnes of rose concrete are produced annually, from which ca. 8.0 tonnes of rose absolute are produced.

QUALITY OF ESSENTIAL OILS

Commercial essential oils are sold in compliance with standards of physico-chemical characteristics such as odor and color, specific gravity, refractive index, and optical rotation, and chemical characteristics such as gas chromatographic analysis. Standards for oil acceptance can be of national or international origin. Standards are set out as monographs on each oil, such as International Standards Organization (ISO), French Standards (AFNOR), German Standards (DIN), and Food Chemical Codex Standards (FCC).[17]

USES AND TRENDS OF ESSENTIAL OILS

In 1990, it was estimated that 50% of all essential oils produced were used in flavors.[18] This can be readily understood because in the United States the per capita consumption of soft drinks is 166 L annually, whereas the annual per capita consumption for the rest of the world is ca. 6 L. As a result, if soft drink consumption increased 3% worldwide, it would result in a 40% increase in essential oils needs.[18] Over the past 25 years the production levels of some oils have changed for a number of reasons: 1) synthetics have partially replaced their use (spike lavender oil, camphor oil); 2) replacement oils are richer in desirable components than original oil (litsea cubeba oil has partially replaced lemongrass oil because it is a better source of citral, lavandin has replaced lavender because it is a cheaper source of a similar odor-character); 3) wild collection of plant material has been reduced because of the scarcity of raw material (amyris oil, cascarilla oil); and 4) wild collection has been restricted by legislation, by either regional or federal mandate (sassafras in Brazil).

Over this same period, the use of oils such as orange, lemon, and lime has increased. Although citrus flavored soft drinks have mainly driven the increase in use of citrus oils, their use in such top-selling fragrances as K One, Cool Water, Drakkar Noir, Escape (male), Eternity (for men), Hugo, Polo Sport (male and female), and Tommy has also helped maintain this increase.[19]

The production of cornmint oil has also increased, not for the value of the oil but for the natural isolate L-menthol obtained from the oil by freeze crystallization. The pharmaceutical, oral hygiene, cosmetic, tobacco, and confectionery industries are responsible for the use of an estimated 23,000 tonnes of L-menthol (both natural and synthetic).[20] There is a particularly increased use of menthol in India (in chewing tobacco); the use of oral care products in China is also on the increase.

The production of tea tree oil [*Melaleuca alternifolia* (Maiden et Betche Cheel)] for use as a natural additive and home care treatment for burns and wounds has also increased in recent years. This oil is produced in commercial quantities only in Australia, where most households are thought to keep a vial of it in the medicine cabinet.[21] Most other oils have relatively stable levels of production, although their production level in one country may be negatively affected by their production in other countries. Since the mid-1980s there has been increased awareness of essential oils due to the popularization of aromatherapy. Although the use of essential oils in aromatherapy does not affect world production by more than 1.0%, it has led to an increase in the number of oils new to commerce. A list of these oils and their countries of origin is shown in Table 6.

ESSENTIAL OIL MARKET

Although it is difficult to determine the exact size of the essential oil market, it has recently been estimated at $310 million with a projected annual increase of 4%.[22] The $310 million approximation is probably an underestimate, however, because the value of the top twenty oils listed in Table 3 totals ca. $325 million. Consequently, the total worldwide market size for all essential oils is probably closer to $400 million.

CONCLUSION

Aromatic plants are of value because of their aromatic principles, namely their essential oils. The type and methods of essential oil production have not changed much over the centuries. However, the volumes and origins have changed. Over the past century, there has been a shift in essential oil production from Europe and the Americas to Asia primarily because of labor and fuel costs and the urbanization of rural areas. Currently, there is a worldwide reduction in essential oil development funding and the number of scientists familiar with these specialty crops is diminishing. Nevertheless, essential oils continue to be important ingredients of food, cosmetic, and pharmaceutical products. Over the next century, the need for cooperation between the consumer and producer will become mandatory even though there may be further global changes in areas of production.

REFERENCES

1. Fahn, A. *Secretory Tissues in Plants*; Academic Press: New York, 1979.
2. Wijesekera, R.O.; Jayawardene, A.L.; Rajapske, L.S. Volatile constituents of leaf, stem and root oils of cinnamon (Cinnamomum zeylanicum). J. Sci. Food Agric. **1974**, *25*, 1211–1220.
3. Mallavarapu, G.R.; Ramesh, S. Essential oil of the fruits of Cinnamomum zeylanicum Blume. J. Essent. Oil Res. **2000**, *12*, 628–630.
4. Jayaprakasha, G.K.; Mohan Rao, L.J.; Sakariah, K.K. Chemical composition of the flower oil of Cinnamomum zeylanicum Blume. J. Agric. Food Chem. **2000**, *48*, 4294–4295.
5. Flores, J.H.; Segredo, G.T. Citrus oil recovery during juice extraction. Perfum. Flavor. **1996**, *21* (3), 13–15.
6. Saunt, J. *Citrus Varieties of the World*; Sinclair Intl. Ltd.: Norwich, UK, 1990; 128.
7. Kesterson, J.W.; Braddock, R.J. *By-Products and Specialty Products of Florida Citrus*; Tech. Bull., Agricultural Science Univ. Florida: Gainesville, 1976; Vol. 784, 8.
8. USDA. *Foreign Agriculture Circular, Horticultural Products*; US Dept. of Agriculture: Washington, DC, 1999.
9. Adrian, J. La production mondiale d'huile essentielles d'orange. Riv. Ital. EPPOS **1997**, (Numero Speciale), 15.
10. Lawrence, B.M. In *The Commercially Important Mint Oils*, Presented at ISEO 2000, Hamburg, Sept. 10–13, 2000.
11. Lawrence, B.M. The Spearmint and Peppermint Industry of North America. In *3rd International Conference on Aromatic and Medicinal Plants*; Verlet, N., Ed.; Centre Formation Professionelle Promotion Agricole (C.F.P. P.A.): Nyons, France, 1992; 59–90.
12. Cash, D.B.; Hrutfiord, B.F.; McKean, W.T. Effect of individual components on peppermint oil flavor. Food Technol. **1971**, *25*, 53–58.
13. Christensen, R. The Establishment and Functionality of the Farwest Spearmint Marketing Order. In *4th International Conference on Aromatic and Medicinal Plants*; Verlet, N., Ed.; Centre Formation Professionelle Promotion Agricole (C.F.P.P.A.): Nyons, France, 1994; 180–197.
14. Baser, K.H.C. Turkish rose oil. Perfum. Flavor. **1992**, *17* (3), 45–52.
15. Ohloff, G. *Scent and Fragrances*; Pickenhagen, W., Lawrence, B.M., Eds.; Springer Verlag: Berlin, 1994; 154. Translation.
16. Lawrence, B.M. The Isolation of Aromatic Materials from Natural Plant Products. In *A Manual on the Essential Oil Industry*; DeSilva, K.T., Ed.; U.N. Industrial Develop. Organization: Vienna, 1995; 57–154.
17. Baser, K.H.C. Analysis and Quality Assessment of Essential Oils. In *A Manual on the Essential Oil Industry*; DeSilva, K.T., Ed.; U.N. Industrial Develop. Organization: Vienna, 1995; 57–154.
18. Buchel, J.A. In *Flavoring with Essential Oils: State-of-the-Art*, Proceedings of International Conference on Essential Oils Flavours, Fragrances and Cosmetics, Beijing, Oct. 9–13, 1988; International Federation of Essential Oil and Aroma Trades: London, UK, 1990; 32–48.
19. Buccellato, F. Citrus oils in perfumery and cosmetic products. Perfum. Flavor. **2000**, *25* (2), 58–63.
20. Clark, C.S. An aroma profile. Menthol. Perfum. Flavor. **1998**, *23* (5), 33–46.
21. Priest, D. Tea Tree Oil in Cosmeceuticals from Head to Toe. In *Tea Tree, The Genus Melaleuca*; Southwell, I., Lowe, R., Eds.; Harwood Academic Publ.: Amsterdam, 1999; 203–206.
22. Anon. Flavor and fragrance, state of the industry. Perfum. Flavor. **2001**, *26* (4), 29–33.
23. Sterret, F.S. The nature of essential oils. I. The production of essential oils. J. Chem. Educ. **1962**, *39*, 203–210.
24. Rovesti, P. Distillation is 5000 years old. Dragoco Rep. **1977**, (3), 59–62.
25. Gildemeister, E.; Hoffman, Fr. History of Volatile Oils. In *The Volatile Oils*; Kremers, E., Ed.; Pharmaceutical Review Publ. Co.: Milwaukee, 1900; 13–50. Translation.
26. Von Rechenberg, C. *Theorie der Gewinnung und Treenung der ätherischen Öle durch Destillation*; Selbsterverlag von Schimmel & Co.: Miltitz bei Leipzig, Germany, 1910.

Ascorbic Acid: An Essential Micronutrient Provided by Plants

A

Patricia L. Conklin
State University of New York, Cortland, New York, U.S.A.

INTRODUCTION

In 1753 James Lind published his *Treatise on the Scurvy*, in which he described his research on the curative effects of citrus fruits on scurvy, the devastating disease that plagued sailors of the era. However, it was not until 1932 that L-ascorbic acid, the agent responsible for the prevention of scurvy, was purified and chemically synthesized. Today, ascorbic acid is a well-known antioxidant and enzyme cofactor with many roles in human health. Humans, unable to synthesize this micronutrient, depend on obtaining the majority via a diet that includes plants. Despite this, it was only recently that the plant ascorbic acid biosynthetic pathway was unraveled.

AFR AND DHA

Chemically, ascorbic acid can be oxidized to the relatively unreactive ascorbate free radical (AFR). Loss of a second electron produces dehydroscorbate (DHA). Ascorbate free radical can also disproportionate to form DHA and ascorbic acid. Both AFR and DHA can be reduced to ascorbic acid in mammalian cells by a number of systems, including glutaredoxin, thioredoxin reductase, AFR reductase, NADPH-dependent DHA reductase and NADH:ascorbate radical oxidoreductase.[1] Extracellular AFR can be reduced back to ascorbic acid with the use of intracellular ascorbic acid.[2] These chemical properties of ascorbic acid make it an ideal antioxidant. By donation of an electron, it can reduce (and therefore detoxify) highly reactive oxygen intermediates such as singlet oxygen, superoxide, and hydroxyl radicals, forming instead fairly non-reactive AFR or DHA, both of which can be enzymatically recycled back to the fully reduced ascorbic acid. In addition, ascorbic acid can reduce oxidized forms of α-tocopherol (vitamin E), maintaining this membrane antioxidant in its active state[3] (Fig. 1).

ASCORBIC ACID AND HUMAN HEALTH

Oxidative damage is thought to be one of the leading factors contributing to the degenerative processes that result in conditions such as aging cardiovascular disease and cancer. However, in the prevention of cardiovascular disease, the role of ascorbic acid is difficult to differentiate from the overall role of a healthy diet and lifestyle.[4] Ascorbic acid has been shown to decrease oxidative DNA damage. People with low ascorbic acid diets risk elevated DNA damage via oxidation. In contrast, supplementation of healthy subjects with additional ascorbic acid results in no change in the level of oxidized DNA. It is thought that the current US RDA for ascorbic acid may be the level at which the maximum benefit for protection against DNA damage is achieved (reviewed in Ref. 5). However, those who either do not meet the U.S. RDA or have lifestyles that are known to decrease serum ascorbic acid levels (such as smoking) would most likely realize decreased damage to their DNA via supplementation. Protection of DNA against oxidative damage has led to the suggestion that ascorbic acid is involved in cancer prevention. However, the main role of ascorbic acid in prevention of gastric cancer may be due to the vitamin's inhibition of nitrosoamine production[6] rather than via detoxification of reactive oxygen species.

Ascorbic acid is also a cofactor of many dioxygenases and it is a deficiency in this activity that leads to scurvy. Ascorbic acid reduces prosthetic metal ions and also keeps other cofactors such as tetrahydrobiopterin in a reduced state. In addition to its well-known role in collagen biosynthesis, ascorbic acid also acts as a cofactor for enzymes involved in carnitine, progesterone, oxytoxin, catecholamine, and nitric oxide synthesis and has been shown to improve vasodilation by enhancing the synthesis of NO in endothelial cells.[7]

The current U.S. RDA for ascorbic acid is 90 mg/day. In a 1994–95 USDA-sponsored survey, only 63.5% of Americans were meeting 100% of the U.S. RDA for ascorbic acid (two-day average). However, supplementation with single high daily doses of ascorbic acid may not be effective at increasing the intracellular pool of ascorbic acid as expression of SVCT1 (the N_{a2}^{+}-dependent transporter that facilitates uptake of dietary ascorbic acid in epithelial cells) decreases substantially when cells are exposed to high levels of ascorbic acid.[8] This finding may help explain the lack of consensus regarding the role of single daily supplemental doses of ascorbic acid in prevention of degenerative disease.

Encyclopedia of Plant and Crop Science
DOI: 10.1081/E-EPCS 120010380

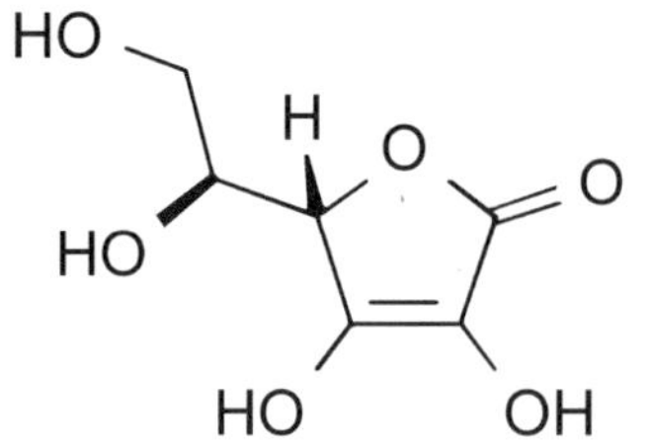

Fig. 1 The chemical structure of L-ascorbic acid.

ASCORBIC ACID PATHWAYS

Despite the dependence of humans on plants for meeting ascorbic acid requirements, the plant ascorbic acid biosynthetic pathway was only recently determined. In 1998, Wheeler and Smirnoff presented evidence for a plant pathway via intermediates that included D-mannose and L-galactose.[9] Prior to this seminal paper, two different plant ascorbic acid pathways were proposed, one analogous to the animal pathway, and the other quite different, involving the osones glucosone and sorbosone. Evidence for the pathway similar to animals rested primarily on the fact that plants harbor a mitochondrial-localized L-galactono-1,4-lactone dehydrogenase with similarity to a L-gulono-γ-lactone oxidase in the animal pathway. This plant enzyme can convert exogenously supplied L-galactono-1,4-lactone to ascorbic acid (reviewed in Ref. 10). The second pathway took into account evidence that (unlike the animal pathway) inversion of the carbon skeleton in the final product (ascorbic acid) relative to that in the primary substrate, D-glucose, does not occur in plants However, there is little evidence for the proposed enzymatic activities necessary for conversion of these osone intermediates into ascorbic acid (reviewed in Ref. 10).

Wheeler and Smirnoff presented two key findings that reconcile the existence of the L-galactono-1,4-lactone dehydrogenase (similar to the animal pathway) and the non-inversion of the glucose carbon skeleton in plants. The first was the demonstration that L-galactono-1,4-lactone is produced in plants from L-galactose by an L-galactose dehydrogenase. Secondly, they found that plants synthesize L-galactose very efficiently from D-mannose, most likely via a GDP-D-mannose-3,5-epimerase. The ascorbic acid biosynthetic pathway constructed from this data is shown in Fig. 2.[9] Early supportive evidence for this biosynthetic pathway came from analysis of the *Arabidopsis* ascorbic acid-deficient mutant, *vtc1*. The *Vtc1* gene was found to encode a GDP-mannose pyrophosphorylase,[11] the activity that catalyzes the generation of GDP-mannose substrate for the aforementioned epimerase. Potato lines expressing an antisense copy of this pyrophosphorylase gene have diminished ascorbic acid levels, independent confirmation of the role of this enzyme in ascorbic acid biosynthesis.[12]

In addition to the GDP-mannose pyrophosphorylase gene (*Vtc1*), several additional genes involved in plant ascorbic acid biosynthesis have been identified. The L-galactono-1,4-lactone dehydrogenase gene has been cloned from several plant species including cauliflower, sweet potato, and tobacco (reviewed in Ref. 10). In addition, annotated sequences with high similarity to known L-galactono-1,4-lactone dehydrogenase sequences are found in the *Arabidopsis thaliana* genomic database. The peptide sequence of a purified GDP-mannose-3,5-epimerase from *Arabidopsis* led to the identification of the epimerase gene.[13] Using a similar strategy, the Smirnoff lab has cloned the L-galactose dehydrogenase gene from *Arabidopsis*.[14] In addition to the ascorbic acid-deficient *Arabidopsis* mutant *vtc1*, three other *VTC* alleles have been identified by virtue of mutant alleles negatively affecting ascorbic acid synthesis.[15] To date, one of these (*Vtc2*) has been cloned[16] although the role of the *Vtc2* gene product in ascorbic acid biosynthesis is yet to be determined.

In the future it will be theoretically possible to engineer plants to produce elevated levels of ascorbic acid as the plant ascorbic acid biosynthetic pathway is better understood, and genes encoding key rate-limiting enzymes in the pathway have been identified. In fact, transgenic tobacco have been described that overexpress the rat L-gulono-1,4-lactone oxidase enzyme (which may have the same activity as plant L-galactono-1,4-lactone dehydrogenase) and accumulate somewhat elevated levels of ascorbic acid.[17]

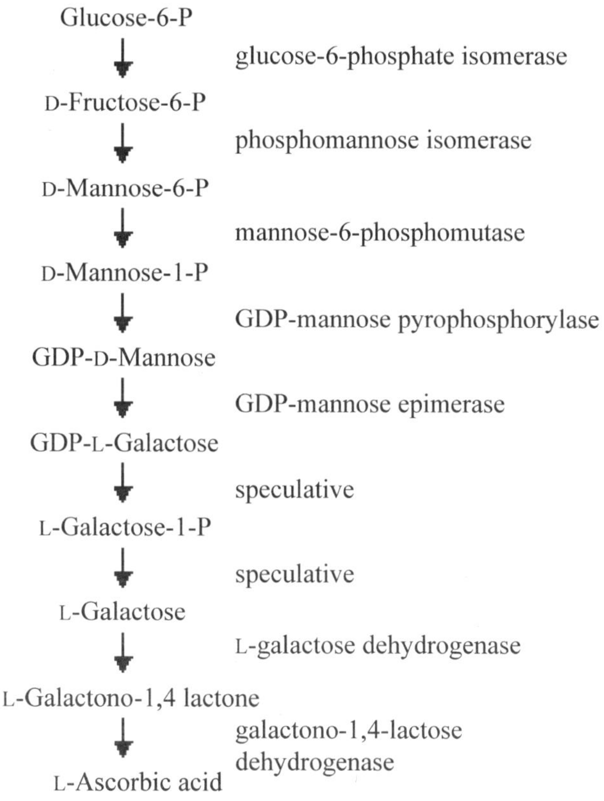

Fig. 2 The biosynthetic pathway of ascorbic acid synthesis in higher plants. (Revised from Ref. 6.)

A

CONCLUSION

Although scurvy in the modern world is quite rare and occurs at <10 mg/day ascorbic acid,[18] it is clear that ascorbic acid has many roles beyond that of a cofactor in collagen biosynthesis, and obtaining at least the U.S. RDA for ascorbic acid is beneficial. Given the fact that many do not meet this minimum RDA, engineering edible plants to produce more of this essential vitamin may lead to improvements in health. Consuming elevated levels of a vitamin such as ascorbic acid via ''functional foods'' would also result in consumption of other beneficial phytochemicals, a benefit that would not be achieved via a vitamin supplement. The introduction of such plants may also reduce the commercial dependence on sulfites as antibrowning agents. Alternatively, crop plants engineered to have increased levels of ascorbic acid may be indirectly beneficial to human health in that such crops may be resistant to environmental stresses that generate ROS and thus produce higher yields, especially in inhospitable environments. As mentioned above, several of the genes in the plant ascorbic acid biosynthetic pathway have been cloned, and research on engineering plants with elevated levels of ascorbic acid is currently underway.

ACKNOWLEDGMENTS

The author would like to thank Carina Barth for her careful review of this manuscript prior to submission. Research in the author's laboratory is supported by grant 98-35100-7000 from the Plant Responses to the Environment Program of the National Research Initiative Competitive Grants Program, U.S. Department of Agriculture.

ARTICLES OF FURTHER INTEREST

Arabidopsis thaliana: *Characteristics and Annotation of a Model Genome,* p. 47
Breeding for Nutritional Quality, p. 182
Metabolism, Primary: Engineering Pathways of, p. 714
Oxidative Stress and DNA Modification in Plants, p. 854
Photosynthesis and Stress, p. 901

REFERENCES

1. Halliwell, B.; Gutteridge, J.M.C. *Free Radicals in Biology and Medicine,* 3rd Ed.; Oxford University Press: Oxford, UK, 1999.
2. VanDuijn, M.M.; Van der Zee, J.; Van den Broek, P.J.A. The ascorbate-driven reduction of extracellular ascorbate free radical by the erythrocyte is an electrogenic process. FEBS Lett. **2001**, *491*, 67–70.
3. Padh, H. Cellular functions of ascorbic acid. Biochem. Cell. Biol. **1990**, *68*, 1166–1173.
4. Asplund, K. Antioxidant vitamins in the prevention of cardiovascular disease: A systematic review. J. Intern. Med. **2002**, *251*, 372–392.
5. Halliwell, B. Vitamin C and genomic stability. Mutat. Res. **2001**, *475*, 29–35.
6. Mirvish, S.S. Effects of vitamins C and E on n-nitroso compound formation, carcinogenesis, and cancer. Cancer **1986**, *58*, 1842–1850.
7. May, J.M. How does ascorbic acid prevent endothelial dysfunction? Free Radic. Biol. Med. **2000**, *28*, 1421–1429.
8. MacDonald, L.; Thumser, A.E.; Sharp, P. Decreased expression of the vitamin C transporter SVCT1 by ascorbic acid in a human intestinal epithelial cell line. Br. J. Nutr. **2002**, *87*, 97–100.
9. Wheeler, G.L.; Jones, M.A.; Smirnoff, N. The biosynthetic pathway of vitamin C in higher plants. Nature **1998**, *393*, 365–369.
10. Smirnoff, N.; Conklin, P.L.; Loewus, F.A. Biosynthesis of ascorbic acid in plants: A renaissance. Annu. Rev. Plant Physiol. Plant Mol. Biol. **2001**, *52*, 437–467.
11. Conklin, P.L.; Norris, S.N.; Wheeler, G.L.; Smirnoff, N.; Williams, E.H.; Last, R.L. Genetic evidence for the role of GDP-mannose in plant ascorbic acid (vitamin C) biosynthesis. Proc. Natl. Acad. Sci. U.S.A. **1999**, *96*, 4198–4203.
12. Keller, R.; Springer, F.; Renz, A.; Kossmann, J. Antisense inhibition of the GDP-mannose pyrophosphorylase reduces the ascorbate content in transgenic plants leading to developmental changes during senescence. Plant J. **1999**, *19*, 131–141.
13. Wolucka, B.A.; Persiau, G.; Doorsselaere, J.V.; Davey, M.W.; Demol, H.; Vandekerckhove, J.; Van Montagu, M.; Zabeau, M.; Boerjan, W. Partial purification and identification of GDP-mannose 3″,5″-epimerase of *Arabidopsis thaliana*, a key enzyme of the plant vitamin C pathway. Proc. Natl. Acad. Sci. U.S.A. **2001**, *98*, 14843–14848.
14. Gatzek, S.; Wheeler, G.L.; Smirnoff, N. Antisense suppression of L-galactose dehydrogenase in *arabidopsis thaliana* provides evidence for its role in ascorbate synthesis and reveals light modulated L-galactose synthesis. Plant J. **2002**, *30*, 541–553.
15. Conklin, P.L.; Saracco, S.A.; Norris, S.R.; Last, R.L. Identification of vitamin C-deficient *Arabidopsis thaliana* mutants. Genetics **2000**, *154*, 847–856.
16. Yander, G. Arabidopsis map-based cloning in the post-genomic era. Plant Physiol. **2002**, *129*, 440–450.
17. Jain, A.K.; Nessler, C.L. Metabolic engineering of an alternative pathway for ascorbic acid biosynthesis in plants. Mol Breeding. **2000**, *6*, 73–78.
18. Levine, M.; Rumsey, S.C.; Daruwala, S.; Park, J.B.; Wang, Y. Criteria and recommendations for vitamin C intake. JAMA **1999**, *281*, 1415–1423.

ATP and NADPH

Wayne Frasch
Arizona State University, Tempe, Arizona, U.S.A.

INTRODUCTION

The energy to drive the metabolic processes of all life-forms on earth is ultimately derived from photosynthetic processes. The principal means used by a living organism to store this energy is through the production of adenosine 5′-triphosphate (ATP) or the reduced form of nicotinamide adenine dinucleotide (NADH) and its phosphorylated variant nicotinamide adenine dinucleotide phosphate (NADPH) that is used in photosynthetic processes. Due to the wide variety of metabolic processes that use ATP as an energy source, this molecule is often referred to as the energy currency of a cell. Energy stored in the high-energy phosphoryl bond of ATP is released when this bond is cleaved by an ATPase reaction to form ADP and inorganic phosphate (Pi). The energy stored in reduced NADPH derives from its negative redox potential relative to other cellular components.

During photosynthesis, ATP is synthesized from ADP and Pi by the chloroplast F_oF_1 ATP synthase. This enzyme is composed of the thylakoid membrane-embedded F_o protein complex, and the attached F_1 protein complex that protrudes into the aqueous stroma of the chloroplast. The F_o is composed of subunits a, b, b′, and c (also known as subunits IV, I, II, and III, respectively), while F_1 contains subunits α, β, γ, δ, and ε. The reduction of $NADP^+$ to NADPH is catalyzed by ferredoxin $NADP^+$ reductase (FNR), a peripheral membrane protein on the stromal side of the thylakoid that is associated with an intrinsic membrane protein in the thylakoid.

INVOLVEMENT OF PHOTOSYNTHETIC LIGHT REACTIONS

The energy used by F_oF_1 to drive the synthesis of ATP from ADP and Pi is derived from the capture of light by light-harvesting and reaction center protein complexes in thylakoid membranes.[1] Once absorbed, light energy is transferred to Photosystem I (PSI) or Photosystem II (PSII) where it induces oxidation of the reaction center chlorophyll and reduction of an electron acceptor that is far more electronegative than the ground state chlorophyll. Since molecules with a more electronegative redox potential store more energy upon reduction, the reduction of the initial electron acceptors in PSI and PSII can be considered to be the step in which light energy is first captured as chemical energy during photosynthesis. Due to the instability of this initial product, a series of sequential redox reactions occurs to prevent the recombination of charges in the reaction center chlorophyll and the initial electron acceptor. These reactions result in the reduction of a molecule that is less electronegative than the initial acceptor but still substantially more electronegative than the reaction center chlorophyll. The oxidized PSII reaction center abstracts electrons from water to release molecular oxygen and protons into the thylakoid lumen (Figs. 1 and 2).

Reduction of plastoquinone by PSII consumes two protons from the stroma. The reduced, lipid-soluble quinol is then oxidized by the cytochrome b_6/f protein complex in the thylakoid, which results in the deposition of the protons in the thylakoid lumen. The reducing equivalents then transfer to oxidized PSI reaction center chlorophyll. The reducing equivalents generated by PSI reduce the iron-sulfur cluster of ferredoxin, a water-soluble protein in the stroma. Ferredoxin in turn reduces $NADP^+$ via the enzyme ferredoxin $NADP^+$ reductase (FNR), a peripheral membrane protein on the stromal side of the thylakoid that is associated with an intrinsic protein in the thylakoid. The net result of the photosynthetic electron transfer reactions is that electrons from water with a redox potential of $+0.81$ V are used to reduce $NADP^+$ that has a redox potential of -0.32 V that accompanies vectorial movement of protons from the stroma to the thylakoid lumen.[1]

For $NADP^+$ to be reduced, it must accept two electrons from a donor. However, the redox active iron-sulfur complex of ferredoxin is capable of transferring only one electron at a time. FNR mediates these one- and two-electron reactions by using the bound coenzyme flavine adenine dinucleotide (FAD) that forms a semiquinone upon accepting an electron from the first ferredoxin. After FAD becomes completely reduced to a hydroquinone by a second ferredoxin, $NADP^+$ is reduced to NADPH. The two-electron reduction of $NADP^+$ is accompanied by one proton and is therefore a hydride (H^-) transfer. FNR consists of two domains, one that binds FAD and the other that contains the $NADP^+$ binding site. Ferredoxin binds in the cleft between the domains in an orientation that positions its iron-sulfur cluster close to FAD.[2]

Encyclopedia of Plant and Crop Science
DOI: 10.1081/E-EPCS 120019401

CHEMIOSMOTIC ENERGY COUPLING

Some of the energy initially captured by the photosynthetic light reactions is used to power the vectorial movement of protons from the stroma to the thylakoid lumen, thereby forming a transmembrane proton gradient.[3] This proton translocation causes a small pH increase of the stroma and a much larger pH decrease in the lumen due to its relatively small volume. This leads to the nonequilibrium condition of a pH difference of 3 to 4 units across the thylakoid membrane during the photosynthetic light reactions. The rate that protons escape from the lumen to the stroma is relatively fast ($t_{1/2}$ of decay of the transmembrane proton gradient is approximately 1 sec). Consequently, the instability of the proton gradient makes it unsuitable for long-term energy storage. Instead, the proton gradient is used by the F_oF_1 ATP synthase as the energy source to drive ATP synthesis. In this manner, the photosynthetic light reactions are coupled to the synthesis of ATP via the proton gradient. Reagents like ammonia that can transport protons across the membrane more rapidly than the F_oF_1 ATP synthase will inhibit ATP synthesis because they collapse the proton gradient. These reagents that uncouple ATP synthesis from electron transfer reactions, called uncouplers, typically increase the rate of the latter reactions by relieving back-pressure from the proton gradient.

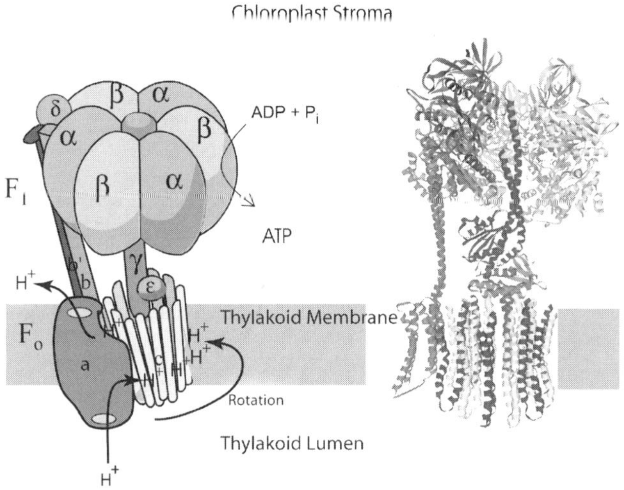

Fig. 1 Subunit composition of the chloroplast F_oF_1 ATP synthase produced from a composite of Protein Data Bank files 1H8E, 1C17, 1L2P, 1ABV, and 1B9U for the analogous subunits of the enzyme from bovine mitochondria and *Escherichia coli* using Web Lab Viewer from Molecular Simulations, Inc. The location of a catalytic site is indicated by Mg^{2+}-ADP, shown in red. An α subunit has been removed to reveal the coiled coil of the γ subunit. (*View this art in color at www.dekker.com.*)

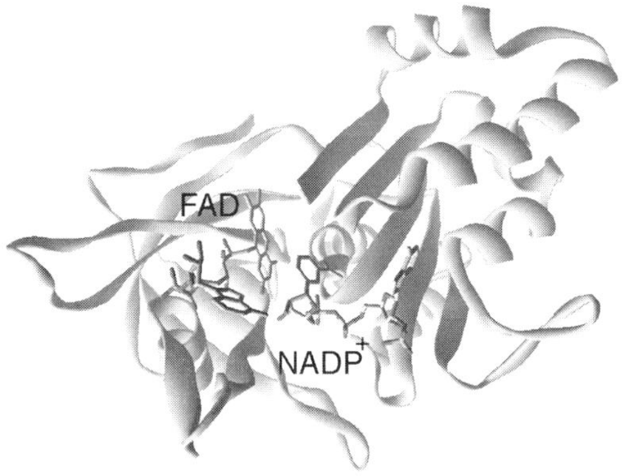

Fig. 2 Spinach chloroplast ferredoxin $NADP^+$ reductase structure produced from Protein Data Bank file 1GJR using Web Lab Viewer from Molecular Simulations, Inc. The domains that bind FAD and $NADP^+$ are shown in green and blue, respectively, with the bound coenzyme. (*View this art in color at www.dekker.com.*)

Energy coupling occurs because F_o serves as an efficient conduit to move protons across the thylakoid membrane and back toward equilibrium with the stroma. Protons move to the stroma in response to the energy gradient that is derived from the concentration difference across the membrane (ΔpH). Because each proton also carries a positive charge, the charge difference across the membrane (ΔΨ) also contributes to this proton energy gradient. A proton concentration difference of about 1000-fold across the membrane (ΔpH = 3) provides sufficient energy for F_oF_1 to drive ATP synthesis. Even though F_oF_1 transports only protons across the membrane, nonequilibrium concentration gradients of other ions such as K^+ can contribute to the energy of the proton gradient by changing the ΔΨ if the ion is permeable to the membrane.

In this chemiosmotic coupling process, the magnitude of the energy gradient, designated the proton-motive force (pmf or μ_{H+}), is related to the transmembrane concentration and charge differences in mV at 30°C by Eq. 1. The proton-motive force translates into more conventional energy terms because it is the sum of the free energy derived from the transmembrane concentration difference (Eq. 2) and the electrical potential gradient generated by the transmembrane concentration gradient of charged species (Eq. 3), where n is the charge on the ion (+1 for protons). This relationship simplifies to Eq. 4.

$$\Delta\mu_{H+} = \Delta\Psi - 60\Delta pH \quad (1)$$

$$\Delta G = 2.3RT\log[H^+_{lumen}]/[H^+_{stroma}] = 2.3RT\Delta pH \quad (2)$$

$$\Delta G = -n\mathrm{F}\Delta E = -n\mathrm{F}\Delta\Psi \quad (3)$$

$$\Delta G = -\mathrm{F}\Delta\mu_{H+} \quad (4)$$

ALTERNATING SITE MECHANISM OF THE F_OF_1 BIOMOLECULAR MOTOR

The F_oF_1 ATP synthase operates as a molecular rotary motor.[4] The protein subunits that compose the rotor and stator of this motor are γ, ε, c and α, β, δ, a, b, b′, respectively. The c subunits assemble into a ring that is currently estimated to contain 14 subunits. Each transmembrane helical c subunit has an aspartic acid that is protonated by the protons of the lumen via a channel located on subunit a. An arginine at a separate location on subunit a displaces the proton from subunit c to release it to the stroma. Each successive proton displacement induces the stepwise rotation of the c subunit ring, driven by the transmembrane proton gradient.

The γ and ε subunits, docked to the c subunit ring, also rotate in response to the proton gradient–driven stepwise rotation and compose the rotor of the biomolecular motor. The N and C termini of the γ subunit form a coiled coil that protrudes through a ring composed of three α and β subunit heterodimers that compose each catalytic site. The ring of αβ heterodimers along with subunits a, b, b′, and δ compose the stator of the motor.

Each catalytic site adopts a conformation in response to the rotational position of the γ subunit such that the three catalytic sites are in different conformations with one site empty. Completion of a catalytic cycle at any one site requires 360° rotation of the γ subunit. In what is known as the binding-change or alternating-site mechanism, the binding of a Mg^{2+}-ADP complex and Pi to the empty site triggers a 120° step rotation of the γ subunit, driven by the proton gradient, that releases ATP from a different catalytic site.[5] The energy from the flux of three protons through the membrane is minimally required to drive a 120° rotation and ATP production, although the measured ratio of protons/ATP is approximately 4.

The F_oF_1 ATP synthase uses the nonequilibrium proton gradient to drive the reaction, ADP + Pi = ATP + H_2O, far beyond the point of equilibrium, in favor of the products. Most enzymes that use ATP hydrolysis as a source of energy derive the energy from the reaction by returning the concentration ratio of ATP/ADP+Pi toward equilibrium. Sequential changes in catalytic site conformation maintain this ratio away from equilibrium. In the initial conformation, the binding of ADP and Pi is preferred to ATP. The first conformational change increases the affinity of the catalytic site for substrates and products and has a low activation energy barrier for ATP synthesis that allows rapid interconversion of substrates and products. The second conformational change converts the site to one that favors ATP over ADP and Pi, while the third decreases the affinity for ATP so that, even when the ATP/ADP + Pi ratio is high, ATP hydrolysis is minimized and ATP dissociates.

REGULATION OF ATP SYNTHESIS IN CHLOROPLASTS

After chloroplasts have reached steady state in the light, the onset of darkness leads to a condition in which the proton gradient rapidly dissipates, while the ATP/ADP + Pi ratio remains high. Under these conditions, F_oF_1 initially catalyzes ATPase-dependent proton pumping back to the lumen, the reverse of synthase activity. This wasteful consumption of the energy captured in light is rapidly halted by formation of a disulfide bond in the γ subunit that stops rotation and ATP hydrolysis.[6] Upon illumination, the first reducing equivalents from PSI that reduce ferredoxin are diverted from FNR to thioredoxin reductase that catalyzes the reduction of thioredoxin. Thioredoxin in turn reduces the disulfide of the γ subunit to activate F_oF_1.

ACKNOWLEDGMENTS

Supported by National Institutes of Health grant GM50202.

ARTICLES OF FURTHER INTEREST

Chlorophylls, p. 258
Exciton Theory, p. 429
Oxygen Production, p. 857
Photosystems: Electron Flow Through, p. 906

REFERENCES

1. Blankenship, R.E. *Molecular Mechanisms of Photosynthesis*; Blackwell Science: Oxford, 2002.
2. Deng, Z.; Alverti, A.; Zanetti, G.; Arakaki, A.K.; Ottado, J.; Orellano, E.G.; Calceterra, N.B.; Ceccarelli, E.A.; Carrillo, N.; Karplus, P.A. A productive $NADP^+$ binding mode of ferredoxin-$NADP^+$ reductase revealed by protein engineering and crystallographic studies. Nat. Struct. Biol. **1999**, *6*, 847–853.
3. Nicholls, D.G.; Ferguson, S. *Bioenergetics 2*; Academic Press: London, 1992.
4. *The Mechanism of F_1F_o-ATPase*; Walker, J.E., Ed.; Biochim. Biophys. Acta; Elsevier Science: Amsterdam, 2000; Vol. 1458, 221–514.
5. Boyer, P.D. ATP synthase past and future. Biochim. Biophys. Acta **1998**, *1365*, 3–9.
6. Frasch, W.D. The F-Type ATPase in Cyanobacteria: Pivotal Point in the Evolution of a Universal Enzyme. In *The Molecular Biology of Cyanobacteria*; Bryant, D.A., Ed.; Kluwer Academic Publ.: Dordrecht, 1994; 361–380.

B Chromosomes

R. N. Jones
University of Wales, Aberystwyth, Wales, U.K.

INTRODUCTION

The earliest record of B chromosomes in plants comes from Anne Lutz in 1916 who found them in Oenothera, and referred to them as "diminutive chromosomes." Later, they also became known as accessories or extra fragment chromosomes. The term B chromosomes is credited to Randolph, who used this name in 1928 to describe certain additional chromosomes found in some plants of maize, and to distinguish them from those of the regular chromosome complement ($2n = 2x = 20$), which are the A chromosomes. B chromosomes is now the universally accepted term, and can be conveniently shortened to B or Bs.

DISCUSSION

The special feature of Bs, which makes them so fascinating and enigmatic, is that they are only found in some individuals of a species and are completely absent from others: In other words, they are dispensable. We may look upon them as optional extras, which is a puzzling idea. It would not be puzzling were it not for the fact that Bs are not rarities at all, but are known in more than one thousand plants and several hundred animals. Bs are part of the genome in those species that carry them, but not an obligatory part like the basic set of A chromosomes. The standard reference that covers all aspects of Bs through the 1980s is by Jones and Rees; two other recent reviews update the story.

The photos of Bs in rye (*Secale cereale*) in Fig. 1 are representative of how we see them in many plant species.

In general terms we can profile them as follows:

- At mitosis, Bs are morphologically distinct from the As in size (usually smaller), centromere position, and status of chromatin (often more heavily heterochromatic), which is how we first recognize them and distinguish them from extra copies of the As.
- Their diagnostic feature is that they show no homology with any of the As and never pair or recombine with them—they follow their own evolutionary pathway.
- They display non-Mendelian modes of inheritance due to their presence in variable numbers in different individuals and their special property of "selfishness" in terms of their numerical increase over generations.
- In high numbers they reduce the vigor and fitness of plants.
- Their phenotypic effects are of a quantitative nature, and they lack genes with major effects.

OCCURRENCE

The latest estimate[1] gives the number of flowering plants with Bs as 1372, of which 12 are conifers and 1360 angiosperms. A few examples are known in ferns and fungi. Among flowering plants they occur in 738 monocots and 622 dicots, and they are found in polyploids as well as diploids. Their presence among families varies enormously, but this variation cannot be interpreted to fit any special pattern. The only thing we can say with any certainty is that they are most often seen in species favored for chromosome studies, such as the Liliaceae and Gramineae.

INHERITANCE

The non-Mendelian mode of transmission of Bs occurs because their number is variable, and meiosis is irregular due to complexities in pairing. In addition, there are systems of mitotic drive in some species (especially Gramineae) based on nondisjunction in gametophytes. In rye there is directed nondisjunction at the first pollen grain (Fig. 1) and first egg cell mitosis; this leads to an unreduced number of Bs being directed into the gametes. In maize the nondisjunction takes place at the second pollen mitosis, followed by preferential fertilization of the egg by the B-carrying sperm, and likewise constitutes a selfish drive mechanism that causes the Bs to spread in natural populations.[2] Equilibrium frequencies of Bs are reached when the forces of accumulation are balanced by those of meiotic loss and reduced fitness.

PHENOTYPIC EFFECTS

In general the phenotypic effects of Bs are either neutral (when they are present in low numbers) or harmful (when

Encyclopedia of Plant and Crop Science
DOI: 10.1081/E-EPCS 120005598

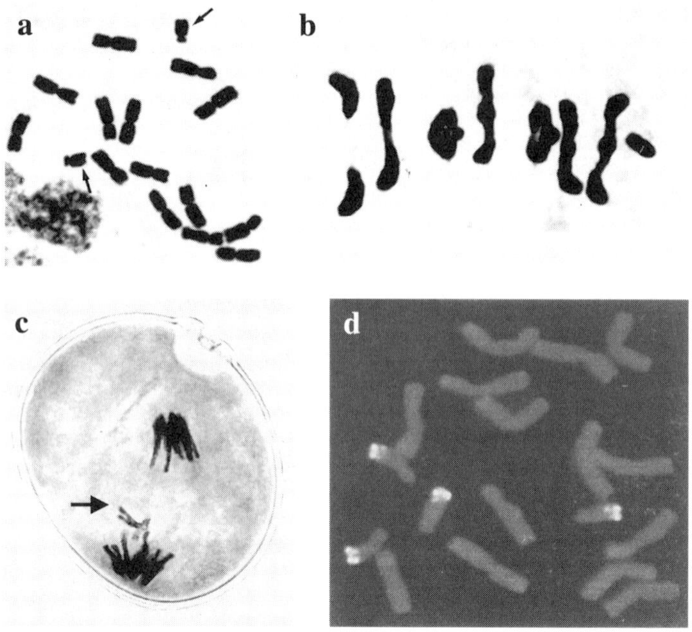

Fig. 1 B chromosomes of rye ($2n = 2x = 14+Bs$) at mitosis and meiosis. (a) c-metaphase in a root meristem cell in a plant with 2 Bs (arrowed). (b) Metaphase I of meiosis showing 7 A-chromosome bivalents and a single unpaired B. The single B will divide at anaphase I and undergo loss at anaphase II. When 2 Bs are present they form a bivalent in most cells, but with higher numbers multivalents are formed and meiosis is irregular. (c) First mitosis of the pollen grain, showing a single B (arrowed) undergoing directed nondisjunction to the generative nucleus. The two chromatids of the B remain joined at sensitive sticking sites on each side of the centromere, and these receptors receive a signal from a genetic element near the end of the long arm of the B. When this element is deleted the Bs disjoin regularly. The spindle at first pollen grain mitosis is asymmetrical, with the equator nearest to the generative pole; it is thought that the sticking of the B chromatids delays their separation long enough for them to be passively included in the generative nucleus as the nuclear membrane encloses them at telophase. (d) c-metaphase in a root meristem cell of a plant with 2 Bs after fluorescent in situ hybridization (FISH) with probes made from the D1100 and E3900 B-specific sequences. (*View this art in color at www.dekker.com.*)

they are found in high numbers). These effects are manifested in all characters in a quantitative manner, but with some intriguing and unexplained differences depending on their occurrence in odd- or even-numbered combinations.[3] Fertility is especially affected in a negative way by high B-numbers. There are some special diploidizing effects on meiosis in hybrid polyploid grass species,[4] which has led to some optimism that Bs may have practical applications.[5]

STRUCTURE AND ORGANIZATION

There are no known cases where the size of the Bs exceeds that of the As: They are not only smaller but often have a number of different morphological forms present in the same individual, as in the case of Aster species where up to 29 polymorphic forms have been found. More important is the question of their genes—or to be precise, their lack of genes. Bs are not only optional for a species, but as far as we know they are a silent part of the genome, with the only exception being some rRNA genes in some species, such as the Australian daisy *Brachcome dichromosomatica*.

Brachycome ($2n = 2x = 4 + 0$–$3Bs$) is interesting in another aspect: It is one of the few species, together with rye and maize, wherein some detailed DNA sequence data are available for the Bs.[6] It seems that up to 10% of the larger of the two kinds of Bs found in *Brachycome* is made of a B-specific family of a 176 base pairs (bp) tandem repetitive DNA sequence, with up to 1.8×10^5 copies of the 3.36×10^8 bp of DNA in each B. The rRNA genes are not transcribed, and there is evidence from chromatin remodeling studies, using immunolabeling with antibodies for differently acetylated forms of histone H4, to suggest that transcriptional activity of the Bs may be silenced in a way similar to that involved in X-chromosome inactivation in mammals.

Molecular studies in rye[7] also point to the absence of genes, and in addition, to two families of sequences, D1100 and E3900, that are B-specific and located at the end of the long arm (Fig. 1). These B-specific families are composed of fragments of a number of unrelated elements, originating from parts of the A chromosomes, which have undergone amplification and simplification. E3900 derives from a Ty3-gypsy retrotransposon.

In maize there is a B-specific repeat sequence (pZmBs) that makes up most of the B centromere and is also found at the telomeres.[8] This sequence is thought to play a key role in the meiotic function of the maize B.

NATURAL POPULATIONS

Large issues in terms of effects are to what extent B chromosomes are of adaptive significance and to what extent their natural polypmorphisms are selfishly generated by mitotic drive, where this exists.[9] There are two schools of thought, according to the system being investigated.

In rye—one of the most intensely studied of all plant species—the situation is clear enough and was resolved by a computer simulation model. This model demonstrated that the drive process based on nondisjunction in the gametophytes is strong enough to overcome all negative effects of the Bs on vigor and reproductive fitness, and that the only factor that can negate the Bs is their own failure to pair and to segregate well at meiosis.[3] The argument then becomes more esoteric and involves the antagonism and coevolutionary dynamics of Bs and anti-B genes in A chromosomes. We enter the dimension of host–parasite interactions, where we find the plants as hosts and the Bs as genome parasites.[10] The B-drive in maize is less strong than in rye, but the outcome is essentially the same and resolves the issue of how Bs come to exist so extensively in natural or seminatural populations.[10] In maize, as in rye, Bs are not found in modern cultivars of these crops, and this is because of heavy selection for high fertility that quickly eliminates the B chromosomes.

Chives (*Allium schoenoprasum* $2n = 2x = 16 + 0$–20 Bs) have Bs in natural populations, and tell a different story. There is no obvious method of accumulation of the Bs in this species. In fact they are transmitted at a mean of 0.39 in some populations, which is lower than the Mendelian rate of 0.5, and there is a net loss in progenies compared with their parents. Furthermore, their presence in high numbers reduces fitness of the carriers, and this prompts the question of how they are maintained in nature. The hypothesis that is proposed, based on greenhouse experiments, is that there is differential selection operating at the seedling stage (based on drought tolerance) and that Bs confer some selective advantage in this respect.

CONCLUSION

In the final analysis Bs remain enigmatic and intriguing. There is a wealth of data about their occurrence in populations, but a serious deficiency of understanding about the dynamics of their natural world. The glimpses we have been able to glean suggest a rich harvest of population genetics still awaits to be gathered before we even begin to close the book on this story. To complete the narrative we will also need more sequence data and a view of their mode of origin, which is missing at the present time.

ARTICLES OF FURTHER INTEREST

Agriculture and Biodiversity, p. 1
Breeding Hybrids, p. 186
Breeding: Incorporation of Exotic Germplasm, p. 222
Chromosome Manipulation and Crop Improvement, p. 266
Chromosome Rearrangements, p. 270
Chromosome Structure and Evolution, p. 273
Crop Improvement: Broadening the Genetic Base for, p. 343
Female Gametogenesis, p. 439
Fluorescence In Situ Hybridization, p. 468
Genetic Resource Conservation of Seeds, p. 499
Genome Rearrangements and Survival of Plant Populations to Changes in Environmental Conditions, p. 513
Genome Size, p. 516
Interphase Nucleus, the, p. 568
Interspecific Hybridization, p. 619
Male Gametogenesis, p. 663
Meiosis, p. 711
Mitosis, p. 734
Molecular Analysis of Chromosome Landmarks, p. 740
Molecular Evolution, p. 748
Polyploidy, p. 1038
Population Genetics, p. 1042

REFERENCES

1. Jones, R.N. B chromosomes in plants. New Phytol. **1995**, *131* (4), 411–434.

2. Jones, R.N. B-chromosome drive. Am. Natur. **1991**, *137* (3), 430–442.
3. Jones, R.N.; Rees, H. *B Chromosomes*; Academic Press: London, 1982.
4. Evans, G.M.; Macefield, A.J. The effect of B chromosomes on homoeologous pairing in species hybrids. I. *Lolium temulentum x Lolium perenne*. Chromosoma **1973**, *41* (1), 63–73.
5. Jones, R.N. Cytogenetics of B-chromosomes in Crops. In *Chromosome Engineering in Plants: Genetics, Breeding, Evolution. Part A*; Gupta, P.K., Tsuchiya, T., Eds.; Elsevier: Oxford, 1991; 141–157.
6. Houben, A.; Belyaev, N.D.; Leach, C.R.; Timmis, J.N. Differences in histone acetylation and replication timing between A and B chromosomes of *Brachycome dichromosomatica*. Chromosom. Res. **1997**, *5* (4), 233–237.
7. Langdon, T.; Seago, C.; Jones, R.N.; Ougham, H.; Thomas, H.; Forster, J.W.; Jenkins, G. De novo evolution of satellite DNA on the rye B chromosome. Genetics **2000**, *154* (2), 869–884.
8. Kaszás, E.; Birchler, J.A. Meiotic transmission rates correlate with physical features of rearranged centromeres in maize. Genetics **1998**, *50* (4), 1683–1692.
9. Bougourd, S.M.; Jones, R.N. B chromosomes: A physiological enigma. New Phytol. **1997**, *137* (1), 43–54.
10. Puertas, M.J. Nature and evolution of B chromosomes in plant: A non-coding but information-rich part of plant genomes. Cytogenet. Genome Res. **2002**, *96* (1–4), 198–205.

Bacterial Attachment to Leaves

Martin Romantschuk
University of Helsinki, Lahti, Finland

B

INTRODUCTION

Attachment of bacterial cells to colonizable surfaces is a common event that in many environments absolutely determines the fate of the bacterium. In other environments and for other bacteria, the requirement for attachment is not absolute, but the adherence may still give a selective advantage, and therefore capacity to adhere is a fitness factor. In this context, the term ''surface'' should be understood broadly, to include not only colonizable tissues or inanimate surfaces but also the surface of other microbes and bacterial cells in, for example, biofilms.

In the case of plant leaf–associated bacteria, both plant pathogenic and saprophytic bacteria harbor a capacity for more or less specific adherence, and they often do carry the genes for expressing attachment-associated surface structures such as pili or fimbriae. But apparently leaf surface attachment is in most, if not all, cases not an absolutely required trait—at best merely a fitness factor improving the chances for successful colonization in competition with other microbes. This may, however, be decisive in natural environmental conditions where any contribution to fitness determines the outcome of the bacterial competition of the limited resources of the leaf surface or the limited access to the inner leaf tissue.

BACTERIAL STRUCTURES INVOLVED IN ATTACHMENT

When spreading through rainsplash, etc., the initial interaction of a bacterial cell with a plant leaf surface is a random event. The distribution of bacteria deposited on a leaf surface after the droplet has dried is, however, far from random (Fig. 1). The distribution—high relative bacterial cell densities on trichomes, leaf veins, and stomates—is likely to reflect nonspecific hydrophobic interaction in combination with degree of access to leaf surface structures, rather than specific interaction of a bacterial adhesin with a receptor, although this type of specificity has not been ruled out. At any rate, the presence or absence of bacterial appendages, such as pili, fimbriae, flagella, and other bacterial cell surface structures, does influence the ability to attach.

Pili and Fimbriae

The main, but not exclusive, focus will be on pili and fimbriae produced by various plant-associated, and in most cases plant-pathogenic, bacteria (Table 1). In contrast to polysaccharides and flagella, most pili or fimbriae have one primary function, and that is attachment. The direct effect of the presence of these surface appendages is adhesion of the carrier bacterial cell to surfaces, solid particles, etc. The secondary events following attachment may be very diverse, including successful colonization of a surface, twitching motility, biofilm formation, aggregation of bacterial cells, which helps in moisture retention and UV tolerance, etc. On the other hand, neither attachment nor pili are indispensable in most cases, but they are apparent epiphytic fitness factors, although very few studies have been performed to confirm this.

Type IV Pili

Type IV pili that also go by several alternative names (type 4 fimbriae, common pili, bundle-forming pili, etc.) are produced by a great number of gram-negative bacteria, including, as it seems, most gram-negative plant pathogens (Table 1). The genes for biogenesis of type IV pili show homology to those of type II secretion (general secretory pathway), and although many bacteria carry the genes required for pilus biogenesis, pilus production has not always been observed. The fact that pili have not been observed in the case of a certain strain does not signify lack of pili, since a clear phenotype is not always associated with the presence of pili. The pili are often expressed in very low numbers and have in some cases been shown to be inducible in certain specific conditions. In the case of *Pseudomonas syringae* this type of pili was observed at an early stage as the receptors for phage $\phi6$, but only later[1] their identity was confirmed as type IV pili. At that point, it was already known that the pili also have a function in promoting attachment of bacteria to plant leaf surfaces,[2] in aggregating bacteria to form pellicle on the surface of stationary cultures[3] and clusters, which influences the UV resistance of the bacteria,[4] and possibly also in formation of biofilms on plant surfaces. No plant surface receptor for the type IV

Encyclopedia of Plant and Crop Science
DOI: 10.1081/E-EPCS 120010618

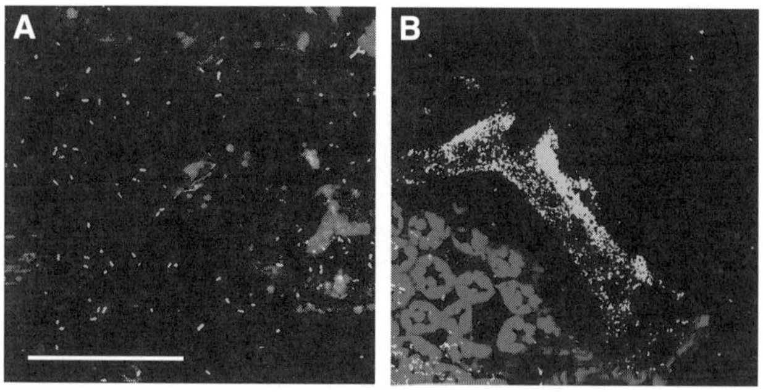

Fig. 1 Confocal laser scanning micrograph of bacteria inoculated by spraying onto plant leaves followed by live/dead staining of the bacterial cells. The bar denotes 50 μm. Living bacteria appear green against a background of dark red fluorescing leaf tissue. Brighter red spots indicate the presence of chloroplasts. A: *Pseudomonas syringae* pv. *tomato* DC3000 scattered over the lower surface of a tomato leaf. The guard cells and opening of a stomate is seen in the middle of the picture. Plant cell outlines cannot be distinguished, but in equivalent scanning electron micrographs (not shown) bacteria can be seen in crevices between neighboring cells and occasionally clustering at and on top of stomata as well as attached to trichomes. B: *Erwinia carotovora* subsp. *carotovora* cells aggregating onto the trichome of an *Arabidopsis* leaf. (Pictures provided by Tristan Boureau.) (*View this art in color at www.dekker.com.*)

pilus has been isolated, and it is possible that the attachment is merely a result of hydrophobic interaction between the pilus and the leaf cuticle. These pili are not required for pathogenesis as such, since nonpiliated mutant strains still cause HR in nonsusceptible plants and disease symptoms in susceptible plants when infiltrated into the plant tissue. The pili are, however, apparently epiphytic fitness factors, the effect of which is seen in a situation of competition. Nonpiliated bacterial mutants initiate colonization less efficiently than the parental wild-type strain, and this difference is maintained in a situation of competition. In the absence of competition, even the nonpiliated strain reaches high population densities but is more sensitive to dislocation by flushing water.[5]

Among the plant pathogens, type IV pili have been observed in at least *Xanthomonas hyacinthi*,[6] *Xanthomonas campestris* pv. *vesicatoria*,[7] many pathovars of *P. syringae*,[1,8] and *Ralstonia solanacearum*,[9] and the list is likely to grow as the genomes of more bacteria become characterized. Indeed, in the case of Pierce's disease of grape and similar diseases of other plants caused by *Xylella fastidiosa*, the presence of pili was suspected when it was found that the genome contained the type IV pilus genes. The bacterium is spread by insects and colonizes the vascular tissue of the plant with wilting as the result. The bacterium forms colonylike structures in the plant tissue, and it is conceivable that type IV pili are required to cement the colonies and to resist the flushing by water in the xylem. Table 1 lists plant-associated bacteria where pili have been observed and partially characterized. Not all of those bacteria are leaf-associated; some are specialized root colonizers (*Klebsiella* spp., *Azoarcus*, *Pseudomonas fluorescens*) or pathogens (*Ralstonia solanacearum*),

Table 1 Pili/fimbriae of some plant-pathogenic and saprophytic bacteria

	Type of pilus/fimbria					
	1	**3**	**IV**	**Hrp**	**T/F**	**N-acetyl-lactosamine**
Species						
Agrobacterium tumefaciens					+	
Azoarcus			+			
Erwinia carotovora	+			+		
Erwinia chrysanthemi				+		
Erwinia rhapontici						+
Klebsiella aerogenes		+				
Klebsiella pneumoniae	+	+				
Pantoea stewartii				+		
Pseudomonas syringae pvs			+	+		
Pseudomonas fluorescens			+	+		
Ralstonia solanacearum			+	+		
Xanthomonas campestris			+	+		
Xanthomonas hyacinthi			+			
Xylella fastidiosa			+?			
Function	Surface adhesion	Surface adhesion	Adhesion, aggregation, twitching	Transfer path determinants	DNA transfer	Surface adhesion

whereas others are found both in the rhizosphere and in the phyllosphere (*Agrobacterium*, *Erwinia* spp.).

Fimbriae: Type 1, Type 3, etc.

Many bacteria of the Enterobacteriaceae family form type 1 and type 3 fimbriae, which have been shown to mediate attachment of bacterial cells to target tissues in the case of animal diseases. Also, in some plant-pathogenic and other plant-associated bacteria, these fimbriae have been observed[10] (Table 1), but a direct correlation to pathogenicity or virulence is in most cases missing. *Erwinia rhapotici* produces fimbriae with a specific adherence to N-acetyllactoseamine.[11] The bacterium infects rhubarb leaves but also infects wheat, causing pink grains to be formed. Receptor analogues, particularly N-acetyllactoseamine, were able to inhibit bacterial attachment and grain coloration. The true pathogenic and ecological role of the Enterobacterial fimbriae/pili is, however, unclear.

Attachment Inside the Leaf Tissue: The Hrp Pilus

The Hrp pathogenicity island (PAI) is essential for pathogenesis of hypersensitivity reaction in the case of many biotrophic plant pathogenic bacteria species such as *P. syringae*. Recently Hrp genes were also shown to play a role in the interaction of the necrotrophic pathogen *Erwinia carotovora* ssp. *carotovora* with its host,[12] and the Hrp-PAI has been observed in many additional bacterial species (see Table 1 for partial listing). The Hrp pathogenicity island contains a type III secretion system with homology to the gene cluster required for flagellar synthesis. Type III protein secretion is widespread among gram-negative pathogens of both animals and plants. Apparently, the products of so-called avirulence genes and some pathogenicity determinants are translocated from the bacterial cell, possibly directly inside the plant cell cytoplasm. An idea of how this might take place emerged when the Hrp pilus was discovered.[13] Recently the pilus was shown to grow by addition of pilin (HrpA) monomers at the pilus tip[14] and to function as a fountain secreting harpin (HrpZ) at the growing tip.[14,15] The pilus grows through the plant cell wall, until it reaches and possibly penetrates the plasma membrane, enabling pathogenicity determinants to enter the plant cell.[15,16] At this point, the bacteria are attached and directly linked to the plant cell, and although the Hrp pilus–mediated attachment is not necessarily physically strong, it is required for proper functioning of the interaction. In *R. solanacearum* bacterial cells were observed to attach in a polar manner to plant cells, but this form of attachment was not dependent on Hrp pili. Instead, a different pilus/fimbria might be involved.[17] Whether foliar pathogens depend on a primary or additional adhesin for successful Hrp pilus–mediated interaction is not known.

The results obtained with the Hrp pilus show that proteins can be transported in the lumen of the apparently hollow pilus. It is likely that in order to fit through the tube that the pilus constitutes, the protein has to be unfolded. The pilin itself, HrpA, possibly folds as it emerges at the pilus tip and attaches to the previous HrpA residue. Whether a specific link to a plant receptor is formed when the pilus tip reaches the plant cell membrane is presently not known, but other bacterial proteins, such as the harpin, may also play a role here.

Attachment of *Agrobacterium*

Attachment to plant cells is a necessary and early step in disease of dicot plants caused by *Agrobacterium*. The bacterium colonizes the rhizosphere but attaches also to wounded aerial plant tissues. Chemotactic motility towards wounds is followed by attachment, the first step of which apparently is dependent on one or more bacterial cell surface protein(s). Several types of mutants with impaired binding have been observed, but it appears that in some cases the effect is pleiotrophic and can be overcome by changing the incubation conditions, such as temperature or osmolarity.[18] After initial binding, cellulose fibrils produced by the bacterium anchor the cells to the wounded plant tissue.[19] Early binding steps are followed by specific cell-to-cell interactions characteristic for only *Agrobacterium*. The bacterial cell transfers genetic material directly from the bacterial cell to the cytoplasm of the plant cell through a direct link formed by the bacterium. The molecular and structural details of how the transfer takes place is still partly obscure; whether transfer requires a direct contact between participating cells, which are brought together by retraction of the *Agrobacterium* T-pilus, or whether the translocation of the protein-DNA complex is mediated by the elongated pilus is still a matter of controversy. The *Agrobacterium* type IV secretion system, showing homology to the F-conjugation system, shares no homology to the Hrp system (type III secretion) of *P. syringae* and other pathogens, but these two secretion systems do show mechanistic similarities.

CONCLUSION

Bacterial attachment to leaf surfaces is not an absolutely required trait, shown by the fact that plant-associated bacteria can colonize and plant-pathogenic bacteria can cause disease even if genes normally associated with attachment are mutated. The testing scenarios in these cases are, however, never natural. The bacteria are applied alone and usually in large quantities, whereas in natural conditions where any inoculum is likely to be mixed, chance plays a great part, in combination with the fact that there is competition for space and resources and a battle against harsh environmental conditions, such as high UV irradiation or alternation between flushing water and drought or between high and low temperatures. Here, even small differences in fitness translate into successful colonization or failure. This should be kept in mind when considering the whole ecology of a plant-associated bacterium. Such matters are of particular significance when attempting to achieve biocontrol of plant disease or when trying to understand the epidemiology of bacterial diseases the causative agent of which is a more or less opportunistic pathogen. Sequencing of a growing number of bacteria is likely to reveal that they all carry genes associated with attachment and that these genes are there not as a genetic load but for a specific reason—optimal fitness.

ARTICLES OF FURTHER INTEREST

Bacterial Pathogens: Early Interactions with Host Plants, p. 89
Bacterial Products Important for Plant Disease: Cell-Surface Components, p. 92
Bacterial Survival and Dissemination in Natural Environments, p. 108
Bacterial Survival and Dissemination in Seeds and Planting Material, p. 111
Bacterial Survival Strategies, p. 115
Bacteria-Plant Host Specificity Determinants, p. 119
Management of Bacterial Diseases of Plants: Regulatory Aspects, p. 669

REFERENCES

1. Roine, E.; Nunn, D.; Paulin, L.; Romantschuk, M. Characterization of *pilP*, a gene required for pilus expression in *Pseudomonas syringae* pahtovar *phaseolicola*. J. Bacteriol. **1996**, *178*, 410–417.
2. Romantschuk, M.; Bamford, D.H. The causal agent of halo blight in bean, *Pseudomonas syringae* pv. *phaseolicola*, attaches to stomata via its pili. Microb. Pathog. **1986**, *1*, 139–148.
3. Romantschuk, M.; Roine, E.; Björklöf, K.; Ojanen, T.; Nurmiaho-Lassila, E.L.; Haahtela, K. Microbial Attachment to Plant Aerial Surfaces. In *Microbiology of Aerial Plant Surfaces*; Morris, C.E., Nicot, P., Nguyen-The, C., Eds.; Plenum Publishing Corporation: New York, 1996; 43–57.
4. Roine, E.; Raineri, D.M.; Romantshuk, M.; Wilson, M.; Nunn, D.N. Characterization of type IV pilus genes in *Pseudomonas syringae* pathovar *tomato* D3000. Mol. Plant-Microb. Interact. **1998**, *11*, 1048–1056.
5. Suoniemi, A.; Björklöf, K.; Haahtela, K.; Romantschuk, M. Pili of *Pseudomonas syringae* pv. syringae enhance initiation of bacterial epiphytic colonization of bean. Microbiology **1995**, *141*, 497–503.
6. Van Doorn, J.; Boonekamp, P.M.; Oudega, B. Partial characterization of fimbriae of Xanthomonas campestris pv. hyacinthi. Mol. Plant-Microb. Interact. **1994**, *7*, 334–344.
7. Ojanen-Reuhs, T.; Kalkkinen, N.; Westerlund-Wikström, B.; van Doorn, J.; Haahtela, K.; Nurmiaho-Lassila, E.L.; Wengelnik, K.; Bonas, U.; Korhonen, T. Characterization of the *fimA* gene encoding bundle-forming fimbriae of the plant pathogen *Xanthomonas campestris* pv. *vesicatoria*. J. Bacteriol. **1997**, *179*, 1280–1290.
8. Romantschuk, M.; Nurmiaho-Lassila; Suoniemi, A.; Roine, E. Pilus-mediated adhesion of *Pseudomonas syringae* to the surface of host and non-host plants. J. Gen. Microbiol. **1993**, *139*, 2251–2260.
9. Liu, H.; Kang, Y.; Genin, S.; Schell, M.A.; Denny, T.P. Twiching motility of Ralstonia solanacearum requires a type IV pilus system. Microbiology **2001**, *147*, 3215–3229.
10. Romantschuk, M. Attachment of plant pathogenic bacteria to the surface of plants. Annu. Rev. Phytopathol. **1992**, *30*, 225–243.
11. Korhonen, T.K.; Haahtela, K.; Pirkola, A.; Parkkinen, J. A N-acetyllactoseamine-specific cell-binding activity in a plant pathogen, Erwinia rhapontici. FEBS Lett. **1988**, *236*, 163–166.
12. Rantakari, A.; Virtaharju, O.; Vähämiko, M.; Taira, S.; Palva, E.T.; Saarilahti, H.; Romantschuk, M. Cloning and partial characterization of the *hrp*-gene cluster of *Erwinia carotovora* subsp. *carotovora*. Mol. Plant-Microb. Interact. **2001**, *14*, 962–968.
13. Roine, E.; Wei, W.; Yuan, J.; Nurmiaho-Lassila, E.L.; Kalkkinen, N.; Romantschuk, M.; He, S.Y. Hrp pilus: A novel hrp-dependent bacterial surface appendage produced by *Pseudomonas syringae*. Proc. Natl. Acad. Sci. U.S.A. **1997**, *94*, 3459–3464.
14. Li, C.-M.; Brown, I.; Boureau, T.; Mansfield, J.; Romantschuk, M.; Taira, S. Bacterial virulence proteins are injected through the type III secretion system's molecular needle. EMBO J. **2002**, *21*, 1909–1915.
15. Brown, I.R.; Mansfield, J.W.; Taira, S.; Roine, E.; Romantschuk, M. Immunocytochemical localization of HrpA and HrpZ supports a role for the Hrp pilus in the transfer of effector proteisn from *Pseudomonas syringae* pv. *tomato* across the host plant cell wall. Mol. Plant Microb. Interact. **2001**, *14*, 394–404.
16. Jin, Q.; He, S.Y. Role of the Hrp pilus in type III protein secterion in *Pseudomonas syringae*. Science **2001**, *294*, 2556–2558.
17. Van Gijsegem, F.; Vasse, J.; Camus, J.C.; Marenda, M.; Boucher, C. Ralstonia solanacearum produces Hrp-dependent pili that are required for PopA secretion but not for attachment of bacteria to plant cells. Mol. Microbiol. **2000**, *36*, 249–260.
18. Bash, R.; Matthysse, A.G. Attachment to roots and virulence of chvB mutant of Agrobacterium tumefaciens are temterature sensitive. Mol. Plant-Microb. Interact. **2002**, *15*, 160–163.
19. Matthysse, A.G.; White, S.; Lightfoot, R. Genes required for cellulose synthesis in *Agrobacterium tumefaciens*. J. Bacteriol. **1995**, *177*, 1069–1075.

Bacterial Blight of Rice

Bart Cottyn
Tom Mew
International Rice Research Institute, Los Baños, Philippines

B

INTRODUCTION

Bacterial blight caused by *Xanthomonas oryzae* pv. *oryzae* is one of the most serious diseases of rice worldwide. The disease was first reported in 1884 from Japan. In the 1960s, bacterial blight epidemics occurred in other Asian regions as a result of the introduction of modern, high yielding but susceptible rice cultivars such as TN1 and IR8. Subsequently, the disease was also reported in Latin America in the 1970s, and in Africa and North America in the 1980s.

Over the past 20 years, mainly due to the spectacular developments of new molecular techniques, advances have been made in understanding the epidemiology, population biology, and host–pathogen interaction of bacterial blight of rice. Substantial research has been done on the population genetics of *X. oryzae* pv. *oryzae* to understand the adaptation of the pathogen population on deployed resistance in the field. The cloning and characterization of pathogen avirulence genes and plant resistance genes has provided new insights into the molecular mechanisms of pathogenesis. The practical approaches of applying knowledge of *X. oryzae* pv. *oryzae* population genetics in disease management strategies have made the rice-bacterial blight system one of the most widely used models for studying host–pathogen interaction.

SYMPTOMS

Bacterial blight is prevalent in the tropics during the rainy season in irrigated or rain-fed lowland and deepwater rice production systems. Bacterial blight symptoms usually develop in the field at the tillering stage, and the disease incidence increases with plant growth, peaking at the flowering stage. Lesions on the leaf blade are initially water-soaked and typically associated with the leaf tips and edges. Lesions gradually enlarge, turn yellow, and may coalesce to cover the entire leaf blade. Older lesions appear as bleached, white to straw-colored necrotic areas, and severely infected leaves wither quickly (Fig. 1). Young plants are the most susceptible, and a severe form of the disease termed ''kresek'' may develop if roots or leaves are damaged and infected during transplanting. Such early infection usually results in seedling death within one to six weeks.

THE PATHOGEN

Taxonomy

The present taxonomic status of *Xanthomonas oryzae* pv. *oryzae*[1] is the result of integrated phenotypic and genotypic analyses. Earlier classifications were *Pseudomonas oryzae* (by Uyeda and Ishiyama in 1926) *Xanthomonas oryzae* (by Dowson in 1943), and *Xanthomonas campestris* pv. *oryzae* (by Dye in 1978). The species *X. oryzae* includes the two pathovars *oryzae* and *oryzicola*, the causal organism of bacterial leaf streak of rice.[1] The two pathovars share a genomic relatedness of over 85% DNA-DNA homology and occupy a distinct position in the genus *Xanthomonas*, which supported the creation of a new species, *X. oryzae*. The different symptoms produced and the distinct phenotypes of the two rice pathogens supported their classification as separate pathovars within the species *X. oryzae*. The two pathogens can also be differentiated by whole-cell fatty acid and protein profiles, reaction to monoclonal antibodies, and DNA fingerprinting.

Population Structure

Traditionally, *X. oryzae* pv. *oryzae* populations have been characterized by virulence typing on a set of differential cultivars carrying different resistance genes, thus establishing races or pathotypes.[2] In the Philippines, ten races of *X. oryzae* pv. *oryzae* have been defined based on a set of near-isogenic lines (IRBB lines) carrying 12 individual bacterial blight resistance genes in a common genetic background (Table 1). Analysis of genomic variation of the pathogen was initiated with the development of DNA fingerprinting techniques in the 1980s and the discovery of repetitive DNA elements in *X. oryzae* pv. *oryzae*.[3] Several repetitive elements such as insertion sequences (IS*1112* and IS*1113*), transposable elements (TNX6 and TNX7), and avirulence genes, have been identified

Encyclopedia of Plant and Crop Science
DOI: 10.1081/E-EPCS 120010586

Fig. 1 Rice field severely infected by bacterial blight in Yunnan province, China. Inserts illustrate bacterial leaf blight symptoms at various stages of infection. (Photographs by I. Oña and compiled by E. Panisales, IRRI.)

in copy numbers varying from 3 to 80 in the genome of *X. oryzae* pv. *oryzae*. A high genomic diversity within *X. oryzae* pv. *oryzae* was detected by DNA fingerprinting using the repetitive elements and avirulence genes as markers (Fig. 2), which, combined with virulence typing, allowed the description of *X. oryzae* pv. *oryzae* populations across major rice growing regions[4] and the monitoring of changes in the pathogen population structure.

THE DISEASE CYCLE

X. oryzae pv. *oryzae* is a vascular pathogen and infects the plant through hydathodes or wounds. Upon entering the leaf through hydathodes, the pathogen multiplies in the intercellular spaces of the mesophyll and gains access to the xylem vessels. Its multiplication and spread within leaf-lesions is more extensive in compatible (susceptible) than incompatible (resistant) interactions. Histological studies of diseased leaf tissue indicated that in resistant interactions the pathogen is enveloped by plant polymers resulting in densely packed bacterial cells that are unable to colonize the intercellular spaces.[5] More recently, genes encoding peroxidases that are predominantly expressed in rice leaves during resistant interactions have been identified, and increased peroxidase has been correlated with accumulation of lignin-like phenolic compounds. During the development of the resistant and the susceptible reaction, almost identical biochemical changes occur in the rice leaves, but their induction is delayed in the susceptible reaction. The pathogen multiplies rapidly during the susceptible interaction and drops of bacterial ooze, which are produced on the surface of leaf-lesions, can be passed directly from plant to plant by contact, or indirectly via irrigation water. Secondary spread of *X. oryzae* pv. *oryzae* in the tropics is associated strongly with occurrence of typhoons, which not only disperse inoculum but also cause wounds that promote infection.

X. oryzae pv. *oryzae* survives in rice stubble and on numerous weed hosts, which are considered as primary sources of initial inoculum for bacterial blight epidemics. The pathogen does not survive well outside of its host and saprophytic survival in soil is short-lived. *X. oryzae* pv. *oryzae* can be seedborne, but the epidemiological significance of its seedborne phase remains to be determined. The *X. oryzae* pv. *oryzae* strains from North America and Africa, although previously assumed to be introduced via seed from Asia, are clearly different from Asian strains based on phenotypic and genotypic characteristics. Isolation of *X. oryzae* pv. *oryzae* from seed is difficult due to its low population size, its slow growth, and its poor competitive ability relative to the large background seed microflora. *X. oryzae* pv. *oryzae* detection methods that require cultivation are inadequate and lack sensitivity for routine seed testing. Moreover, it has become clear in the past 10 years that saprophytic xanthomonads similar in appearance can commonly be isolated from rice seed,[6] and may have contributed to conflicting reports regarding the seed transmission of bacterial blight. The presence of nonpathogenic xanthomonads may also present a potential identification problem for seed certification agencies. A polymerase chain reaction (PCR)-based detection method that provides the required sensitivity for seed assays and can differentiate nonpathogenic xanthomonads from *X. oryzae* pv. *oryzae* has been developed. A disadvantage of the PCR method, however, is that it cannot ascertain viability or genotype of the detected cells.

DISEASE MANAGEMENT

Chemical control of bacterial blight by mercuric and copper compounds, and antibiotics such as streptomycin and its derivatives, has not proven effective. Cultural practices such as fallow and drying the field after harvest, removing of diseased rice straw, and eradication of possible alternate host plants around paddy fields are useful but often difficult to implement.

The deployment of host plant resistance has been the most efficient control strategy for bacterial blight. Over 25 genes that confer resistance to specific races or clusters of races of *X. oryzae* pv. *oryzae* have been identified from cultivated varieties, wild rices, and induced mutation. Two resistance genes, *Xa21* and *Xa1*, have been cloned and molecularly characterized.[7,8] So far, *Xa4* has been

Table 1 Reactions[a] to Philippines races of *Xanthomonas oryzae* pv. *oryzae* of IRBB near-isogenic lines (NILs) carrying different bacterial blight resistance (*Xa*) genes

NILs	*Xa*-gene(s)	Race 1 (PXO61)[b]	Race 2 (PXO86)	Race 3B[c] (PXO79)	Race 3C (PXO340)	Race 4 (PXO71)	Race 5 (PXO112)	Race 6 (PXO99)	Race 7 (PXO145)	Race 8 (PXO280)	Race 9a (PXO339)	Race 10 (PXO341)
IRBB1	*Xa1*	S	S	S	S	S	S	S	S	S	S	S
IRBB2	*Xa2*	S	S	S	S	S	S	S	S	S	S	S
IRBB3	*Xa3*	S	S	S	S	S	S	S	S	S	S	S
IRBB4	*Xa4*	R	S	S	S	MR-MS	R	S	R	R	S	R
IRBB5	*xa5*	R	R	R	R	S	R	S	R	R	R	R
IRBB7	*Xa7*	MS	R	R	R	S	R	S	R	R	S	R
IRBB8	*xa8*	S	S	S	S	S	MR	S	S	S	S	S
IRBB10	*Xa10*	S	R	S	S	S	R	S	R	S	S	S
IRBB11	*Xa11*	S	S	S	S	S	S	S	S	S	S	S
IRBB13	*xa13*	S	S	S	S	S	S	R	S	S	S	S
IRBB14	*Xa14*	S	S	S	S	S	R	S	S	R	S	S
IRBB21	*Xa21*	R	R	R	MR	R	R	MR	MR	MR	MR	S
IRBB50	*Xa4+xa5*	R	R	R	R	R	R	S	R	R	R	R
IRBB51	*Xa4+xa13*	R	S	S	S	MS	R	R	R	MR	S	R
IRBB52	*Xa4+Xa21*	R	R	R	MR	MR	R	MR	R	R	MR	R
IRBB54	*xa5+Xa21*	R	R	R	R	R	R	R	R	R	R	R
IRBB55	*xa13+Xa21*	R	R	R	MR	R	R	R	MR	MR	MR	S
IRBB56	*Xa4+xa5+xa13*	R	R	R	R	R	R	MR	R	R	R	R
IRBB57	*Xa4+xa5+Xa21*	R	R	R	R	R	R	R	R	R	R	R
IRBB59	*xa5+xa13+Xa21*	R	R	R	R	R	R	R	R	R	R	R
IRBB60	*Xa4+xa5+ xa13+Xa21*	R	R	R	R	R	R	R	R	R	R	R
IRBB61	*Xa4+xa5+Xa7*	R	R	R	R	MR	R	S	R	R	R	R
IRBB62	*Xa4+Xa7+Xa21*	R	R	R	R	R	R	MR	R	R	MS	R
IRBB63	*xa5+Xa7+xa13*	R	R	R	R	R	R	R	R	R	R	R
IRBB64	*Xa4+xa5+ Xa7+Xa21*	R	R	R	R	R	R	R	R	R	R	R
IRBB65	*Xa4+Xa7+ xa13+Xa21*	R	R	R	R	R	R	R	R	R	MR	R
IRBB66	*Xa4+xa5+Xa7+ xa13+Xa21*	R	R	R	R	R	R	R	R	R	R	R

[a]Resistance or susceptibility of rice plants to *X. oryzae* pv. *oryzae* is expressed in lesion lengths measured at 14 days after inoculation. Resistant (R): <5 cm; moderately resistant (MR): 5 to 10 cm; moderately susceptible (MS): 10 to 15 cm; susceptible (S): >15 cm. All NILs were inoculated at 40 to 45 days after sowing.

[b]A representative strain of *X. oryzae* pv. *oryzae* for each of the defined Philippine races is given in parentheses.

[c]Race 3B belongs to race 3, lineage B; race 3C belongs to race 3, lineage C; race 9a differs from race 9b and 9c in the absence of *avrXa7*, the avirulence gene that corresponds to *Xa7*.

(Data provided by C.M. Vera Cruz, IRRI.)

B

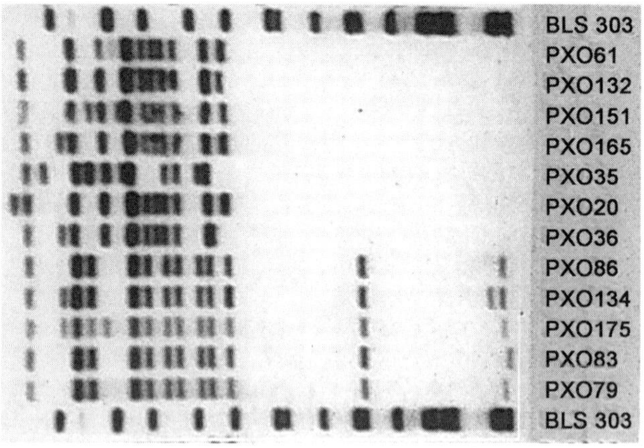

Fig. 2 Restriction fragment length polymorphism (RFLP) analysis of strains of *Xanthomonas oryzae* pv. *oryzae* using a cloned avirulence gene *avrXa10* as hybridization probe. (Data provided by C. M. Vera Cruz, IRRI.)

the most commonly deployed gene in many modern semi-dwarf cultivars grown in the tropics. However, the widespread deployment in monoculture of single-gene resistant cultivars often led to the breakdown of plant resistance as a result of rapid changes in the pathogen population and the emergence of new virulent races.[5] The three avirulence genes (*avrxa5*, *avrXa7*, and *avrXa10*) isolated from *X. oryzae* pv. *oryzae* belong to the *avrBs3* gene family, a common type of avirulence gene found in different species and pathovars of *Xanthomonas*. Mutagenesis analysis of the avirulence genes[9] indicated that assessing the fitness penalty imposed on the pathogen by the loss of both avirulence and aggressiveness functions of an avirulence gene could be useful in predicting the durability of a corresponding plant resistance gene.[10] Current breeding programs are being optimized by incorporating molecular marker techniques, and aim to transfer resistance genes identified as being effective against the predominant population of the bacterial blight pathogen. Recently, near-isogenic lines with pyramids of the resistance genes have been developed for use as resistance donors in breeding programs (Table 1). Strategic deployment of individual race-specific resistance genes or pyramiding of genes that have complementary resistance against multiple pathogen races may limit the build up of particular races.

CONCLUSION

The population biology of the bacterial blight pathogen has been intensively studied to understand the population structure changes in response to planting resistant cultivars. The integration of knowledge of *X. oryzae* pv. *oryzae* population genetics into breeding and deployment strategies for resistant cultivars may provide a basis to design gene-based disease management strategies. The principle behind such a population approach is that use of host plant resistance should be guided by knowledge of temporal and spatial changes in the pathogen population structure. Detailed knowledge of the pathogen population diversity in the field can also be useful in clarifying questions of the initial source of inoculum for bacterial blight epidemics. The information that will become available from the large-scale genome sequencing efforts will provide new opportunities to investigate genetic mechanisms of host–pathogen interactions and to analyze factors important for durable resistance.

ARTICLES OF FURTHER INTEREST

Bacterial Pathogens: Detection and Identification Methods, p. 84
Bacterial Pathogens: Early Interactions with Host Plants, p. 89
Bacterial Products Important for Plant Disease: Diffusible Metabolites As Regulatory Signals, p. 95
Bacterial Products Important for Plant Disease: Extracellular Enzymes, p. 98
Bacterial Survival and Dissemination in Natural Environments, p. 108
Bacterial Survival and Dissemination in Seeds and Planting Material, p. 111
Bacteria-Plant Host Specificity Determinants, p. 119
Breeding for Durable Resistance, p. 179
Breeding Self-Pollinated Crops Through Marker-Assisted Selection, p. 202
Classification and Nomenclature of Plant Pathogenic Bacteria, p. 286
Genomic Approaches to Understanding Plant Pathogenic Bacteria, p. 524
Management of Bacterial Diseases of Plants: Regulatory Aspects, p. 669
Plant Diseases Caused by Bacteria, p. 947
Population Genetics, p. 1042
Rice, p. 1102

REFERENCES

1. Swings, J.; Van der Mooter, M.; Vauterin, L.; Hoste, B.; Gillis, M.; Mew, T.W.; Kersters, K. Reclassification of the causal agents of bacterial blight (*Xanthomonas campestris* pv. *oryzae*) and bacterial leaf streak (*Xanthomonas campestris* pv. *oryzicola*) of rice as pathovars of *Xanthomonas oryzae* (ex Ishiyama 1922) sp. nov., nom. ret. Int. J. Syst. Bacteriol. **1990**, *40* (3), 309–311.
2. Mew, T.W. Current status and future prospects on research

on bacterial blight of rice. Annu. Rev. Phytopathol. **1987**, *25*, 359–382.
3. Leach, J.E.; White, F.F.; Rhoads, M.L.; Leung, H. A repetitive DNA sequence differentiates *Xanthomonas campestris* pv. *oryzae* from other pathovars of *Xanthomonas campestris*. Mol. Plant-Microb. Interact. **1990**, *3* (4), 238–246.
4. Adhikari, T.B.; Vera Cruz, C.M.; Zhang, Q.; Nelson, R.J.; Skinner, D.Z.; Mew, T.W.; Leach, J.E. Genetic diversity of *Xanthomonas oryzae* pv. *oryzae* in Asia. Appl. Environ. Microbiol. **1995**, *61* (3), 966–971.
5. Mew, T.W.; Alvarez, A.M.; Leach, J.E.; Swings, J. Focus on bacterial blight of rice. Plant Dis. **1993**, *77* (1), 5–12.
6. Cottyn, B.; Regalado, E.; Lannoot, B.; De Cleene, M.; Mew, T.W.; Swings, J. Bacterial populations associated with rice seed in the tropical environment. Phytopathology **2001**, *91* (3), 282–292.
7. Song, W.Y.; Wang, G.-L.; Chen, L.-L.; Kim, H.-S.; Pi, L.-Y.; Holsten, T.; Gardner, J.; Wang, B.; Zhai, W.-X.; Zhu, L.-H.; Fauquet, C.; Ronald, P. A receptor kinase-like protein encoded by the rice disease resistance gene, *Xa21*. Science **1995**, *270* (5243), 1804–1806.
8. Yoshimura, S.; Yamanouchi, U.; Katayose, Y.; Toki, S.; Wang, Z.X.; Kono, I.; Kurata, N.; Yano, M.; Iwata, N.; Sasaki, T. Expression of *Xa1*, a bacterial blight-resistance gene in rice, is induced by bacterial inoculation. Proc. Natl. Acad. Sci. U. S. A. **1998**, *95* (4), 1663–1668.
9. Bai, J.; Choi, S.-H.; Ponciano, G.; Leung, H.; Leach, J.E. *Xanthomonas oryzae* pv. *oryzae* avirulence genes contribute differently and specifically to pathogen aggressiveness. Mol. Plant-Microb. Interact. **2000**, *13* (12), 1322–1329.
10. Leach, J.E.; Vera Cruz, C.M.; Bai, J.; Leung, H. Pathogen fitness penalty as a predictor of durability of disease resistance genes. Annu. Rev. Phytopathol. **2001**, *39*, 187–224.

B

Bacterial Pathogens: Detection and Identification Methods

Jan M. van der Wolf
Cor D. Schoen
Plant Research International, Wageningen, The Netherlands

INTRODUCTION

Detection and identification of plant pathogenic bacteria is a prerequisite for implementing effective disease management strategies. In the past, bacterial diseases were detected by field inspections and accompanied by lengthy isolation and identification procedures, including laborious biochemical reactions and pathogenicity tests. These procedures are often impractical for routine detection. Consequently, there is a continuing search for rapid and reliable methods that can replace those used traditionally. This chapter discusses the current techniques used in the detection of phytopathogenic bacteria as well as newer technology that promises to improve the efficiency of identification.

SEROLOGICAL ASSAYS

The introduction of serological assays in the 1960s signaled an important advance in the identification of phytopathogenic bacteria.[1] In particular, the enzyme-linked immunosorbent assay (ELISA) and immunofluorescence (IF) antibody staining have been widely adopted in testing programs. For example, in the Netherlands, 60,000 potato samples are tested annually with IF to detect *Ralstonia solanacearum*, the causal organism of brown rot. For epidemiological studies, immunofluorescence colony-staining (IFC), in which bacterial colonies grown in agar pour plates are stained with fluorophore-tagged antibodies and detected by fluorescence microscopy, provides a sensitive and quantitative method for detecting culturable plant pathogenic bacteria in complex substrates.[1]

Initially, much of the serology-based assays relied on polyclonal antibodies that, in some cases, lacked the desired specificity and were difficult to reproduce. However, the introduction of hybridoma and phage display technologies significantly improved the selection of consistent and specific monoclonal antibodies.[2]

The latest development in serology-based techniques is the application of flow cytometry (FCM), which allows the rapid and quantitative detection of IF-stained bacterial cells[3] (Fig. 1). In FCM, large numbers of individual bacterial cells are counted as they flow through a capillary tube and pass a laser beam. Multiple cellular parameters are determined simultaneously based on the cell's fluorescence and its ability to scatter light. A flow cytometer is expensive, but it may replace labor intensive IF microscopy for analysis of antibody stained cells. Sample preparation for FCM analysis is simple and comprises filtration of the cells and suspension and incubation with fluorochrome-labeled antibodies. It also allows combining antibody staining and the use of fluorescent markers for viability (vital stains), such as propidium and hexidium iodide for red fluorescent staining of dead and carboxy fluorescein diacetate and calcein AM for green fluorescent staining of viable cells.

NUCLEIC ACID–BASED DETECTION AND IDENTIFICATION METHODS

A second major breakthrough in routine detection of phytopathogenic bacteria was the introduction of techniques targeting nucleic acid sequences. In particular, techniques that rely on specific amplification of nucleic acid sequences, such as the polymerase chain reaction (PCR), are widely accepted.[4] These techniques are rapid, specific, and allow detection of genes associated with pathogenicity and virulence. PCR-based procedures have been developed for many economically important plant pathogenic bacteria. Under optimal conditions, PCR allows detection of nucleic acids from 1 target cell per reaction. However, inefficient DNA extraction and/or inhibition of amplification by sample contaminants frequently result in reduced sensitivity. Inclusion of sample controls, which are co-extracted and co-amplified, can indicate false-negative results due to poor extraction and inhibition. False-positive results, caused by cross-contamination, can be circumvented by using real-time PCR amplification procedures in sealed tubes.[4] During real-time PCR amplification, amplicon production is coupled with the light emission generated by the excitation of

Encyclopedia of Plant and Crop Science
DOI: 10.1081/E-EPCS 120010572

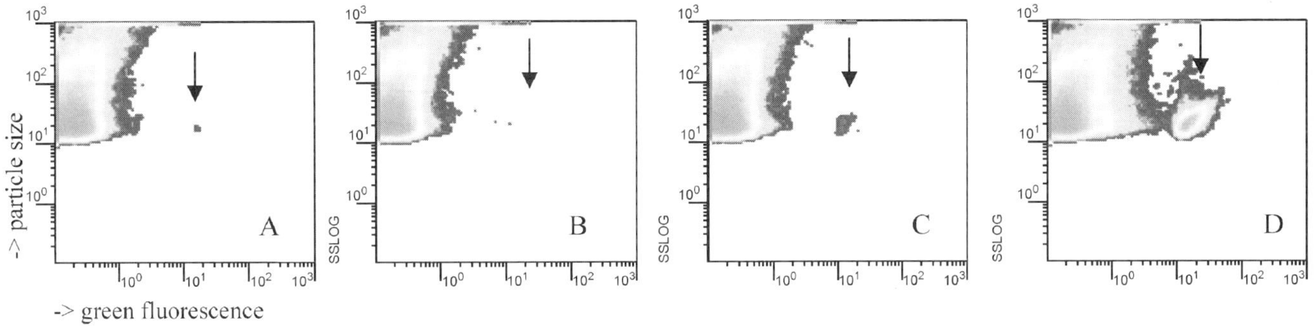

Fig. 1 Flow cytometry (FCM) for detection of *Xanthomonas campestris* pv. *campestris* (*Xcc*), stained with fluoroscein isothiocyanate (FITC)–labelled monoclonal antibodies, in a cabbage seed extract. Analysis was done on the basis of particle size (Y-axis) and green fluorescence intensity (X-axis). On the left side of each graph a large group of background bacteria with low green fluorescence intensity is visible. The *Xcc* population is indicated with an arrow. A. no *Xcc* added, B. 10^3*Xcc* cells/ml added, C. 10^5*Xcc* cells/ml added, D. 10^7*Xcc* cells/ml added. (*View this art in color at www.dekker.com.*)

fluorophores. The increasing fluorescent signal can be automatically monitored by a fluorometer indirectly indicating increases in amplicon concentration. The found cycler threshold values are related to initial concentrations of template DNA; hence, real-time PCR can allow initial populations of target bacteria to be estimated. Additionally, real-time PCR is rapid and eliminates the need for electrophoretic analysis of amplicons.

Enrichment of target bacteria by plating samples on selective growth media prior to PCR can result in improved detection sensitivity. Moreover, tedious DNA extraction methods can be avoided, because PCR inhibitory substances are absorbed into the agar medium and bacterial cells harvested from plates can be detected via PCR without DNA extraction.

Recently, nucleic acid sequence-based amplification (NASBA) has been introduced for detecting plant pathogenic bacteria. NASBA results in the exponential amplification of single-strand RNA molecules, is performed at a constant 41°C, and is based on concurrent activity of three enzymes, reverse transcriptase, RNaseH, and T7 RNA polymerase.[5] Amplicons can be detected by Northern blotting, and in real-time by using molecular beacons (single-stranded oligonucleotides with a stem-loop structure that specifically anneal with the amplicons, generating a fluorescent signal). This homogeneous assay, called AmpliDet RNA, allowed detection of *R. solanacearum* and *Clavibacter michiganensis* subsp. *sepedonicus* rRNA in potato peel extracts, with a detection threshold of 1000 colony-forming-units (CFUs)/mL.[5] One drawback of standard PCR, targeting DNA, is the inability to disinguish between viable and nonviable cells. Most RNA molecules rapidly degrade after a cell dies and its presence can indicate cell viability. Therefore, AmpliDet RNA can be an important epidemiological tool.

DNA amplification procedures revolutionized not only detection but also identification of plant pathogenic bacteria. Large portions of a bacterial genome can be analyzed with genetic fingerprint techniques. Using small oligonucleotides (approximately 10 nucleotides (nt)) as primers in random amplified polymorphic DNA (RAPD)-PCR, multiple fragments from complex bacterial genomes are generated. The use, however, of small oligonucleotides can lead to inconsistent results. Robustness is higher in rep-PCR techniques that employ larger oligonucleotides (approximately 20 nt) to amplify fragments flanked by conserved repetitive DNA sequences that are randomly dispersed throughout the bacterial genome. The availability of representative fingerprint libraries allows rep-PCR techniques to be employed for identification of unknown isolates within the group of *Clavibacter michiganensis*, phytopathogenic xanthomonads, and pseudomonads.[6] Most other genotypic fingerprinting methods lack these libraries and are used only to study genomic diversity and population structures. Bacteria can also be identified and classified by the RiboPrinter system based upon an automated restriction fragment length polymorphism analysis of phylogenetically conserved rRNA sequences. The system includes DNA extraction from bacteria, followed by enzymatic restriction digestion and probing with a hypervariable region of the rRNA operon. Restriction patterns are compared with those in a database, which comprises patterns of over 600 strains, although most of these are medical strains.

Nearly all nucleic acid–based detection methods involve an amplification reaction. One exception is the fluorescence in situ hybridization (FISH) technique, by which DNA in permeabilized bacterial cells is stained with an oligonucleotide labeled with a fluorescent dye. Bacteria can subsequently be visualized by UV microscopy.[7] Cell permeabilization procedures are often time consuming, and bacterial cells stained by FISH are often not as bright as those stained by IF. Specificity of detection can be improved by combining FISH with IF.

FATTY ACID METHYL ESTER ANALYSIS

Fatty acid profiling is the technique by which the composition of (methylated) fatty acids in bacteria is analyzed by gas chromatography. The equipment is expensive, but when bacteria are cultivated under standardized conditions, the technique has proven to be highly reliable for identification of many groups of bacteria, even at the subspecies level.[1] In general, the automated Microbial Identification System from Microbial ID (Delaware, U.S.A.) is used, which includes an extended database for plant pathogenic bacteria.

SUBSTRATE UTILIZATION ASSAYS

The MicroLog^{TM} system (Biolog, Hayward, CA, U.S.A.) allows identification of bacteria based on their ability to utilize a combination of 95 different substrates in a convenient microplate format. Conversion of an indicator (tetrazolium violet, a redox dye) by metabolically active bacteria results in a color change, which can be observed visually or measured with a microplate reader. Substrate utilization profiles are compared with those in an extended library that comprises over 1900 bacteria. Unfortunately, the range of phytopathogenic bacteria included in the Biolog database is limited.[8]

INTEGRATED DETECTION AND IDENTIFICATION PROCEDURES

Micro-array technology represents the next generation of DNA diagnostic tools for the simultaneous identification of multiple plant pathogenic organisms. Micro-arrays have many oligonucleotides (>10,000) spotted or synthesized in approximately 1 cm^2 area on a solid surface to serve as hybridization targets. Single-strand (ss) DNA isolated from a bacterial sample is then hybridized to immobilized complementary DNA targets, generating double strand DNA. Fluorescent dyes, most commonly used to detect the capture of target oligonucleotides are excited by laser energy, and the emission is analyzed with confocal microscopy or charge-coupled device technologies (Fig. 2).[9] Alternatively, the difference in electrical conductivity between dsDNA and ssDNA can be used to indicate whether hybridization has occurred, thus avoiding use of fluorophore-tagged DNA probes.[10] The presence of many different targets will improve specificity, allow detection of variants of the pathogens in question, and avoid laborious confirmation procedures.

Microfabrication processes are also being explored for manufacturing silicon, glass, or plastic arrays with diverse analytical functions. Analyzers using these arrays hold great promise in simplifying the processing of crude biological samples, running biochemical reactions, and detecting the results. Systems with these characteristics

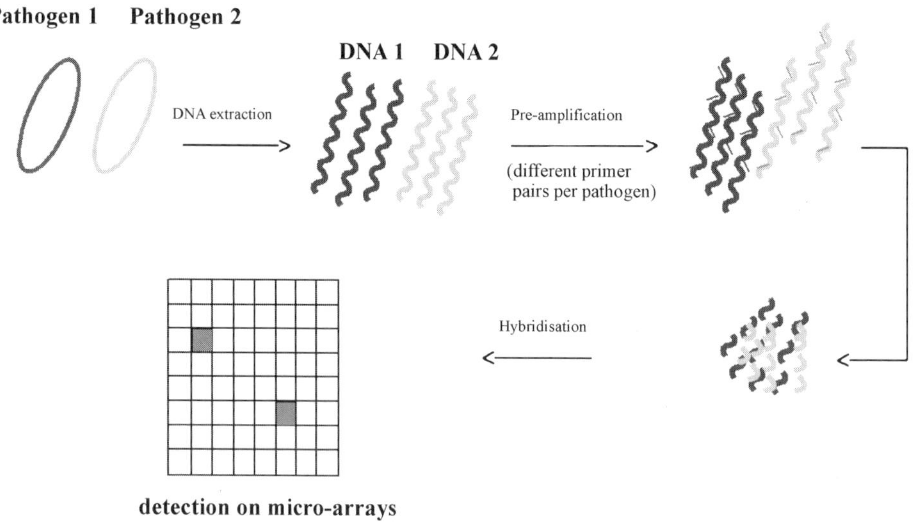

Fig. 2 Schematic diagram of multiplex micro-array detection and identification. This technique can be developed by using multiple primers and probes for each pathogen. (*View this art in color at www.dekker.com.*)

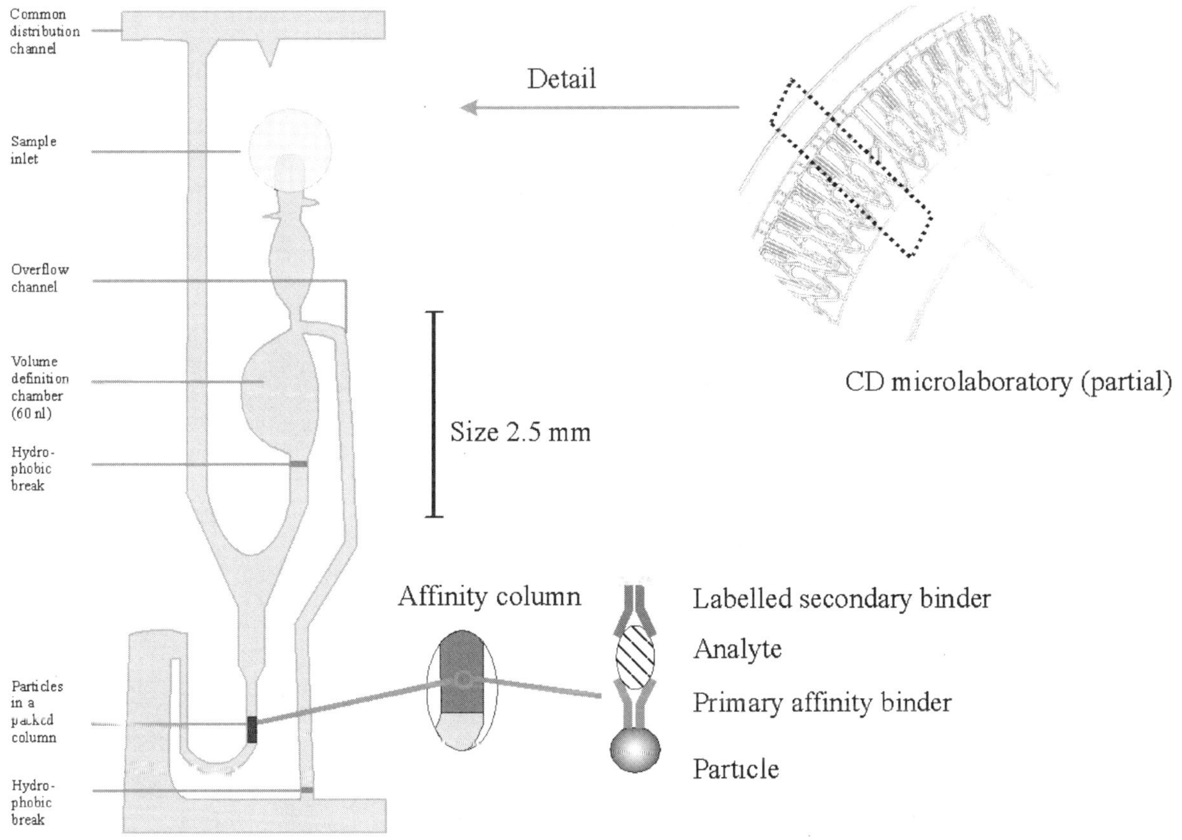

Fig. 3 Schematic diagram of application-specific microstructures replicated onto a CD micro-laboratory (400 in total). Incorporation of a column, functionalized with a primary affinity binder, into each microstructure determines assay specificity. (Modified from Gyros.) (*View this art in color at www.dekker.com.*)

have been termed ''laboratory-on-a-chip,'' which implies the ability to process crude biological samples to isolate target molecules of interest.[11] An integrated computer-controlled nanoliter device for DNA analysis that automatically extracts and amplifies target DNA and analyzes the resulting amplicons using electrophoresis[12] has been recently described (Fig. 3). All the system components are contained on single glass and silicon substrates. This technology provides a glimpse of the future techniques for detection and identification of phytopathogenic bacteria.[13]

ARTICLES OF FURTHER INTEREST

Bacterial Attachment to Leaves, p. 75
Bacterial Blight of Rice, p. 79
Bacterial Pathogens: Early Interactions with Host Plants, p. 89
Bacterial Products Important for Plant Disease: Cell-Surface Components, p. 92
Bacterial Products Important for Plant Disease: Diffusible Metabolites As Regulatory Signals, p. 95
Bacterial Products Important for Plant Disease: Extracellular Enzymes, p. 98
Bacterial Products Important for Plant Disease: Toxins and Growth Factors, p. 101
Bacterial Survival and Dissemination in Insects, p. 105
Bacterial Survival and Dissemination in Natural Environments, p. 108
Bacterial Survival and Dissemination in Seeds and Planting Material, p. 111
Bacterial Survival Strategies, p. 115
Bacteria-Plant Host Specificity Determinants, p. 119
Classification and Nomenclature of Plant Pathogenic Bacteria, p. 286
Crown Gall, p. 379
Fire Blight, p. 443
Fluorescence In Situ Hybridization, p. 468
Genomic Approaches to Understanding Plant Pathogenic Bacteria, p. 524
Management of Bacterial Diseases of Plants: Regulatory Aspects, p. 669

Plant Diseases Caused by Bacteria, p. 947
Virus Assays: Detection and Diagnosis, p. 1273

REFERENCES

1. Klement, Z.; Rudolph, K.; Sands, D.C. *Methods in Phytobacteriology*; Akademiai Kiado: Budapest, Hungary, 1990.
2. Griep, R.A.; Van Twisk, C.; Van Beckhoven, J.R.C.M.; Van der Wolf, J.M.; Schots, A. Development of specific recombinant monoclonal antibodies against the lipopolysaccharides of *Ralstonia solanacearum* race 3. Phytopathology **1998**, *88*, 795–803.
3. Chitarra, L.G.; Langerak, C.J.; Bergervoet, J.H.W.; Van den Bulk, R.W. Detection of the plant pathogenic bacterium *Xanthomonas campestris* pv. *campestris* in seed extracts of *Brassica* sp. applying fluorescent antibodies and flow cytometry. Cytometry **2002**, *47*, 118–126.
4. Schaad, N.W.; Jones, J.B.; Chun, W. *Laboratory Guide for Identification of Plant Pathogenic Bacteria,* 3rd Ed.; APS Press: St. Paul, Minnesota, USA, 2002.
5. Szemes, M.; Schoen, C.D.; Van der Wolf, J.M. Molecular Beacons for Homogeneous Real-Time Montoring of Amplification Products. In *Molecular Microbial Ecology Manual*; Akkermans, A.D.L., Van Elsas, J.D., De Bruijn, F.J., Eds.; Kluwer Academic Publishers: Dordrecht, 2001; 2.3.2/1–2.3.2/17.
6. Rademaker, J.L.W.; Hoste, B.; Louws, F.J.; Kersters, K.; Swings, J.; Vauterin, L.; Vauterin, P.; De Bruijn, F.J. Comparison of AFLP and rep-PCR genomic fingerprinting with DNA–DNA homology studies: *Xanthomonas* as a model system. Int. J. Syst. Bacteriol. **2000**, *50*, 665–677.
7. Li, X.; De Boer, S.H.; Ward, L.J. Improved microscopic identification of *Clavibacter michiganensis* subsp. *sepedonicus* cells by combining in situ hybridization with immunofluorescence. Lett. Appl. Microbiol. **1997**, *24*, 431–434.
8. Jones, J.B.; Chase, A.R.; Harris, G.K. Evaluation of the Biolog GN MicroPlate system for identification of some plant pathogenic bacteria. Plant Dis. **1993**, *77*, 553–558.
9. O'Donnell, M.J.; Smith, C.L.; Cantor, C.R. The development of microfabricated arrays for DNA sequencing and analysis. Trends Biotechnol. **1996**, *14*, 401–407.
10. Lewis, F.D.; Wu, T.; Zhang, Y.; Letsinger, R.L.; Greenfield, S.R.; Wasielewski, M.R. Distance-dependent electrontransfer in DNA hairpins. Science **1998**, *277*, 673–676.
11. Cheng, J.; Sheldon, E.L.; Wu, L.; Heller, M.J.; O'Connel, J.P. Isolation of cultured cervical carcinoma cells mixed with peripheral blood cells on a bioelectronic chip. Anal. Chem. **1998**, *70*, 2321–2326.
12. Cheng, J.; Sheldon, E.L.; Wu, L.; Uribe, A.; Gerrue, L.O.; Carrino, J.; Heller, M.J.; O'Connel, J.P. Preparation and hybridization analysis of DNA\RNA *from E. Coli* on microfabricated bioelectronic chips. Nat. Biotechnol. **1998**, *16*, 541–546.
13. Anderson, R.C.; Bogdan, G.J.; Barniv, Z.; Dawes, T.D.; Winkler, J.; Roy, K. Microfluidic biochemical analysis systems. Transducers **1997**, *97*, 477–479.

Bacterial Pathogens: Early Interactions with Host Plants

B

Gwyn A. Beattie
Iowa State University, Ames, Iowa, U.S.A.

INTRODUCTION

From the perspective of inducing plant diseases, the importance of the early interactions of pathogenic bacteria with plant hosts is usually in the build-up of pathogen populations. The ability of most pathogenic bacteria to develop large populations can be attributed to their invasion and multiplication within healthy plant tissue, traits that are rare among the diverse bacteria that associate with plants. The ability to establish large populations often involves bacterial modification of the inhabited microsites—such as by inducing the secretion of plant nutrients—and may be enhanced by cooperative interactions among the bacteria. Although large pathogen populations are usually required for disease induction, this alone is not sufficient, as indicated by the fact that large populations of foliar pathogens are often present on asymptomatic leaves.[1] In addition to large pathogen populations, disease induction requires a susceptible host and environmental conditions conducive to disease.

COLONIZATION OF PLANT SURFACES

Bacterial pathogens may immigrate directly into the plant tissue, for example via an insect vector, but most immigrate onto the surfaces of leaves, blossoms, fruits, and roots. The presence of large phytopathogen populations on aerial plant surfaces suggests that bacteria can colonize these surface sites. This is supported by micrographs showing their abundance on plants, by their recovery in large numbers in plant washings, and by their ability to be killed by treating plants with surface sterilants.[1,2] Furthermore, pathogen inoculation onto plants often results in large pathogen populations on the surface before disease symptoms appear. Several pathogens have been demonstrated to multiply on plant surfaces. The foliar pathogen *Rhodococcus fascians* multiplies on leaf surfaces before entering the plant, although this pathogen is unusual in that it also induces symptoms in the absence of large internal (endophytic) populations.[3] The pathogen *Erwinia amylovora* multiplies on the stigma surface in pear and apple blossoms, a surface that is particularly moist due to hygroscopic stigmatic secretions. Instead of multiplication, large surface populations may reflect efficient movement to the surface by bacteria that multiplied within the plant tissue. For example, endophytic populations of many *Pseudomonas syringae* and *Xanthomonas campestris* pathovars extrude through stomata and onto the leaf surface.[2]

The bacterial traits required for colonizing plant surfaces are probably distinct from those required for colonizing internal plant tissues.[2] For example, bacterial colonization of leaf surfaces may require the ability to tolerate large and rapid fluctuations in environmental conditions. In contrast, bacteria that colonize internal leaf tissues may need to overcome the defense responses induced in the plant during microbial invasion. If the maintenance of traits that promote multiplication on plant surfaces imposes a fitness cost on a pathogen, then those traits are likely to be lost. The extent to which most pathogenic bacteria maintain traits for surface colonization is currently unknown.

MOVEMENT INTO THE PLANT TISSUE

Entry of bacterial pathogens into plants occurs through natural openings such as stomata and hydathodes on leaves, modified stomata known as nectarthodes on blossoms, lenticels on woody stems and tubers, and at the sites of lateral root emergence. Bacteria also enter wounds, including those resulting from broken trichomes, damage by hail and wind-blown sand, and various cultivation practices. Bacterial pathogens may migrate to these entry sites on the plant surface. For example, *Agrobacterium tumefaciens* and *P. syringae* pv. *phaseolicola* exhibit directed movement, or chemotaxis, toward wounds.[4] Although other pathogens exhibit chemotaxis toward plant extracts in vitro, most pathogens have not been evaluated for their ability to move toward entry sites on plants, nor for a role for chemotaxis in promoting bacteria migration into the plant.

For many pathogens, bacterial motility enhances bacterial movement to, or into, the entry sites on the plant. As evidence of this, nonmotile mutants of several pathogens were significantly reduced in virulence in tests that required bacteria to enter through natural entry sites.[5,6] Interestingly, these nonmotile mutants were fully

Encyclopedia of Plant and Crop Science
DOI: 10.1081/E-EPCS 120010569

virulent when introduced directly into the plant tissue, indicating that bacterial motility may not be important for movement after entry into the plant.

MULTIPLICATION IN INTERCELLULAR SPACES

Suppression of Plant Defense Responses

While bacteria on the plant surface are separated from the epidermal cells by the waxy plant cuticle, bacteria within the plant are in close contact with plant cells. This contact is believed to trigger plant defense responses. Pathogenic bacteria may exhibit a higher level of tolerance to these defense responses than nonpathogenic bacteria. Furthermore, at least one bacterial pathogen, *P. syringae* pv. *phaseolicola*, has been demonstrated to suppress the plant defense responses in plant tissue.[7] Most pathogenic bacteria have a type III protein secretion system that transfers bacterial proteins into plant cells.[8] This secretion system and many of the secreted proteins contribute to the establishment of large endophytic populations. Although the functions of the secreted proteins are mostly unknown, one or more of them may suppress plant defense responses that are normally induced in the presence of bacterial invaders.

Modification of Environment in Plant Tissue

The internal tissue of healthy plants protects bacteria from large environmental fluctuations, and it may provide sufficient resources to support limited bacterial growth. Active modification of this habitat to increase nutrient concentrations and water availability is probably critical to the establishment of large endophytic populations. Exposure of plant cells to pathogens typically increases the permeability of the plant cell membrane. Some pathogens also exhibit specific traits that may increase this permeability, including the production of particular phytotoxins and plant growth hormones.[2] All bacterial pathogens that exhibit extensive multiplication within plants have the type III secretion system described above. One or more of the secreted proteins may function to alter the plant physiology and make the plant tissue more favorable for bacterial growth, such as by inducing the release of nutrients.

Another bacterial trait that could alter the local environment of a pathogen in plant tissues is the production of a hygroscopic matrix, composed of one or more types of exopolysaccharide (EPS), around the cells. This matrix probably increases the amount of water available to the bacteria. The introduction of purified EPS from some foliar phytopathogens into leaves induces persistent water-soaking (the release of cellular water into intercellular spaces), which has been shown to promote bacterial growth in these spaces. EPS can affect bacteria–plant interactions in many ways, including enhancing bacterial colonization and survival within plant tissues as well as contributing to wilting and other disease symptoms.[9]

Cooperation Among Bacteria

Extensive habitat modification may be augmented by, or depend on, cooperative interactions among bacteria. A major form of cooperation is the density-dependent induction of specific bacterial traits. This density dependence is mediated by the production of extracellular diffusible signal molecules, which upon achieving a threshold concentration induce the expression of various target genes. Density-dependent signal molecules in pathogenic bacteria include *N*-acyl homoserine lactones (HSL) and a volatile fatty acid. Pathogenic bacteria have been demonstrated to regulate a diversity of traits in this manner,[10] including the production of EPS, exoenzymes, antibiotics, pigments, and conjugal plasmid transfer. Several of these traits can strongly influence the interactions between the plant and the pathogen. For example, the production of exoenzymes by *Erwinia carotovora* subsp. *carotovora* (*Ecc*) is controlled by an HSL derivative. These enzymes—including pectinases, polygalacturonases, and cellulases—are particularly good elicitors of plant defense responses. At low cell densities, *Ecc* can multiply in the plant without producing exoenzymes, and thus can avoid inducing a major plant defense response. When the *Ecc* cells attain a high density, however, they cooperate to produce sufficiently large amounts of exoenzymes to degrade the plant tissue without being hindered by a plant defense response. This plant tissue degradation greatly increases the nutrients available to the pathogen.

MOVEMENT WITHIN THE PLANT

Some pathogens exhibit little migration from the initial site of infection, whereas others exhibit extensive movement within the plant. For example, some vascular pathogens that enter the stomata or root cracks must spread through the intercellular spaces of the inner cortex until they reach the vascular parenchyma. This movement may involve flagellar motility, twitching motility, or spreading due to an increased mass of bacterial cells and EPS.

CONCLUSION

From the perspective of bacteria, the plant is merely a habitat for growth and survival. Following their immigration to a plant, bacteria may colonize the sites at which they arrive, move to sites that are more favorable for multiplication, and/or modify the environment in their immediate habitat to enhance their growth. Populations that develop on the plant surface result, at least in part, from bacterial movement from the internal tissues, and these surface populations serve as an important inoculum reservoir for further infections. However, with the exception of only a few bacterial pathogens—including the tumor-inducing pathogens *Agrobacterium tumefaciens* and *Rhodococcus fascians*—disease induction requires that large populations develop within the plant tissues. Traits that are currently believed to favor this population development include motility and/or chemotaxis toward entry sites, tolerance and/or suppression of plant defenses, communication within bacterial populations, and active habitat modification. Further efforts to better define the bacterial traits contributing to endophytic colonization are key to improving our understanding of the early interactions of pathogenic bacteria with plants.

ARTICLES OF FURTHER INTEREST

Bacterial Attachment to Leaves, p. 75
Bacterial Products Important for Plant Disease: Cell Surface Components, p. 92
Bacterial Products Important for Plant Disease: Diffusible Metabolites As Regulatory Signals, p. 95
Bacterial Products Important for Plant Disease: Extracellular Enzymes, p. 98
Bacterial Products Important for Plant Disease: Toxins and Growth Factors, p. 101
Bacterial Survival and Dissemination in Natural Environments, p. 108
Bacterial Survival Strategies, p. 115
Bacteria-Plant Host Specificity Determinants, p. 119
Biological Control in the Phyllosphere, p. 130
Classification and Nomenclature of Plant Pathogenic Bacteria, p. 286
Fire Blight, p. 443
Leaf Cuticle, p. 635
Leaf Structure, p. 638
Leaf Surface Sugars, p. 642
Leaves and Canopies: Physical Environment, p. 646
Plant Diseases Caused by Bacteria, p. 947

REFERENCES

1. Hirano, S.S.; Upper, C.D. Bacteria in the leaf ecosystem with emphasis on *Pseudomonas syringae*—A pathogen, ice nucleus, and epiphyte. Microbiol. Mol. Biol. Rev. **2000**, *64* (3), 624–653.
2. Beattie, G.A.; Lindow, S.E. Bacterial colonization of leaves: A spectrum of strategies. Phytopathology **1999**, *89* (5), 353–359.
3. Cornelis, K.; Ritsema, T.; Nijsse, J.; Holsters, M.; Goethals, K.; Jaziri, M. The plant pathogen *Rhodococcus fascians* colonizes the exterior and interior of the aerial parts of plants. Mol. Plant-Microb. Interact. **2001**, *14* (5), 599–608.
4. Vande Broek, A.; Vanderleyden, J. The role of bacterial motility, chemotaxis, and attachment in bacteria–plant interactions. Mol. Plant-Microb. Interact. **1995**, *8* (6), 800–810.
5. Hatterman, D.R.; Ries, S.M. Motility of *Pseudomonas syringae* pv. *glycinea* and its role in infection. Phytopathology **1989**, *79*, 284–289.
6. Tans-Kersten, J.; Huang, H.; Allen, C. *Ralstonia solanacearum* needs motility for invasive virulence on tomato. J. Bacteriol. **2001**, *183* (12), 3597–3605.
7. Jakobek, J.L.; Smith, J.A.; Lindgren, P.B. Suppression of bean defense responses by *Pseudomonas syringae*. Plant Cell **1993**, *5*, 57–63.
8. Galán, J.E.; Collmer, A. Type III secretion machines: Bacterial devices for protein delivery into host cells. Science **1999**, *284*, 1322–1328.
9. Denny, T.P. Involvement of bacterial polysaccharides in plant pathogens. Annu. Rev. Phytopathol. **1995**, *33*, 173–197.
10. Pierson, L.S.III; Wood, D.W.; Pierson, E.A. Homoserine lactone-mediated gene regulation in plant-associated bacteria. Annu. Rev. Phytopathol. **1998**, *36*, 207–225.

Bacterial Products Important for Plant Disease: Cell-Surface Components

P. K. Chakrabarty
University of Florida, Gainesville, Florida, U.S.A.

A. Chatterjee
W. Gassmann
A. K. Chatterjee
University of Missouri, Columbia, Missouri, U.S.A.

INTRODUCTION

Cell surface components play important roles in pathogenicity and virulence of plant pathogenic bacteria. Besides maintaining the cellular integrity, these components play major roles in mobility, communication, survival, and interaction with host cells. They consist of three main components: cell envelope, cell appendages, and extracellular polysaccharides (EPSs).

CELL ENVELOPE

The cell envelope of gram-negative bacteria consists of an inner cytoplasmic membrane, a periplasmic space, a thin layer of peptidoglycan, and an outer lipoprotein membrane. Gram-positive bacteria have much thicker cell walls relative to gram-negative bacteria and lack outer lipoprotein membranes. The lipopolysaccharide (LPS) molecules on the outer surface of lipoprotein membranes of gram-negative bacteria have a major role in determining the surface characteristics of bacterial cells and are also largely responsible for their antigenicity. LPSs are structurally diverse, ubiquitous, indispensable components that apparently have diverse roles in bacterial plant pathogenesis.[1] The polysaccharide moiety of LPSs (the O-antigen) is presumed to contribute to the exclusion of plant-derived antimicrobial compounds and thereby facilitates plant infection by the pathogen. The O-antigen layer is also important in the initial interaction between bacterium and plant cells in determining compatibility or incompatibility.[2]

The presence of membrane-bound receptors is thought to be important for bacterial functions both outside and within plants.[3] A few examples are the outer membrane receptor of siderophore complexes in *Pseudomonas syringae* pv. *syringae* and *Erwinia chrysanthemi* and the VirA inner membrane chemoreceptor for acetosyringone in *Agrobacterium tumefaciens*. Mutants of *P. syringae* pv. *syringae* deficient in iron-pyoverdin receptors fail to take up chelated iron, while the tumor formation by *A. tumefaciens* is abolished with mutation of *virA*.

CELL SURFACE APPENDAGES

Flagella

The mobility of bacteria is an important aspect of infection and is mediated by helical rotation of flagella, long and delicate threadlike proteinaceous appendages that extend beyond the outer lipoprotein membrane of bacterial cells. Flagella are generally thin (diameters of approximately 20 nm) and long, with some having a length 10 times the diameter of the cell. Bacteria may be monotrichous with a single polar flagellum (as in *Xanthomonas*), lophotrichous with a tuft of polar flagella (as in *Pseudomonas*), or peritrichous with flagella arranged around the entire cell (as in *Erwinia*). *A. tumefaciens* swims with its flagella after sensing a chemical signal emanating from the host. Nonflagellated mutants of this pathogen are defective in virulence and induce fewer and smaller tumors on its host.[4] Similarly, *Escherichia carotovora* ssp. *atroseptica* mutants defective in flagellin biosynthesis and motility are less virulent compared to the wild-type strain. $HexA^-$ (hyperproduction of exoenzymes) mutants of the same species, which overexpress flagellar biosynthesis genes and are hypermotile, are also hypervirulent.

Fimbriae or Pili

In addition to flagella, many bacteria possess appendages referred to synonymously as fimbriae or pili. Pili are typically thin proteinaceous filaments radiating from bacterial cells and are smaller and more numerous compared to flagella. The presence of certain pili has been demonstrated

Encyclopedia of Plant and Crop Science
DOI: 10.1081/E-EPCS 120021338

in several plant pathogenic bacteria, including *Ralstonia solanacearum*, *P. syringae*, and *A. tumefaciens*.[5] There are multiple types of pili, but they may be divided into two groups: adhesive pili and pili required for transfer of DNA and protein between cells.[6]

Adhesive pili have been found to play an important role in pathogenesis by enabling bacterial attachment to host surfaces. In the case of *P. syringae* pv. *phaseolicola*, a direct correlation exists between the number of pili per bacterial cell and bacterial attachment to the plant surface. Cells devoid of pili do not show any increase in attachment above a low basal level, while wild-type and superpiliated cells show a progressive increase in association over time. When directly infiltrated into leaf tissue, both piliated and nonpiliated strains of this bacterium cause disease, but only piliated strains cause infection when sprayed onto the outer surface of the leaf. This suggests that pili-mediated attachment to leaf surfaces may be important in early stages of the infection process. Pili in *R. solanacearum* are shown to be important for virulence, autoaggregation, biofilm formation, and polar attachment of the bacterium.[7] Adhesive pili are also involved in twitching motility, bacteriophage adsorption, and UV tolerance based on cell aggregation on leaf surfaces.[6,8]

The second group of pili includes the *Escherichia coli* sex pilus (F pilus), the *A. tumefaciens vir* gene–encoded T pilus, and the *P. syringae hrp* gene–encoded Hrp pilus. The F pilus is required for DNA transfer between bacteria, whereas the T pilus functions in the transfer of T-DNA and proteins from *Agrobacterium* to plant cells. Finally, the Hrp pilus plays a distinct role in virulence protein secretion via the type III secretion system in *Pseudomonas*. They have been the subject of active research[9] and several recent reviews.[10,11] *R. solanacearum* mutants defective in Hrp pilus production are impaired in interactions with plants and in secretion of the PopA protein involved in eliciting the host defense response. However, they are not impaired in their ability to attach to plant cells.[12]

EXTRACELLULAR POLYSACCHARIDES

Most plant pathogenic bacteria are enveloped in a viscous material mainly composed of polysaccharides, but polypeptides and glycopeptides may also occur. When thin and diffuse, this material is called a slime layer, whereas it is called a capsule when the layer is thick and spatially well-defined. EPSs play important roles in the pathogenicity and virulence of several bacterial pathogens, including species of *Erwinia, Clavibacter, Pantoea, Pseudomonas, Ralstonia*, and *Xanthomonas*.[13] Specifically, EPSs are thought to function in pathogenicity through induction and maintenance of water soaking, protection of bacteria from desiccation, prevention of detection by the host, and occlusion of xylem vessels of infected plants causing restriction of water movement.

Erwinia amylovora, the fire blight pathogen of apple and pear, produces amylovoran, which is required for its pathogenicity.[14] Amylovoran aids in bacterial movement through the cortex of host plants, clogs the vascular system, and causes wilting. Similarly, *Pantoea stewartii* ssp. *stewartii* causes vascular wilt and leaf blight of corn by producing EPSs, which are implicated in pathogenicity by causing water-soaked lesions and systemic wilting. At least five genes govern EPS production in this pathogen, all of which are required for wilting. *R. solanacearum*, the causal agent of vascular wilt of diverse plants, produces large amounts of EPSs, which are present as a soluble slime rather than as an insoluble capsule. Mutants of *R. solanacearum* specifically blocked in EPS synthesis rarely wilt or kill plants, even when large numbers of cells are directly injected into the stem.[15] At least two gene clusters important for EPS biosynthesis have been identified in this pathogen. *Xanthomonas campestris* pv. *campestris*, the causal agent of black rot of crucifers, produces the EPS xanthan, which is implicated in its virulence and pathogenicity.[16] EPS has also been shown to be important in the pathogenicity of *X. axonopodis* pv. *malvacearum*, which causes angular leaf spot and blight of cotton.[17] In contrast, EPSs appear to play an ancillary role in *E. chrysanthemi* pathogenesis, as EPS-deficient strains are still capable of causing disease, but disease symptoms are less severe as compared to those caused by wild-type strains.

CONCLUSION

Cell surface components of plant pathogenic bacteria play a major role in bacterial survival and pathogenicity. They not only provide rigidity and structural integrity but also include external appendages that function in mobility and conjugation. Cell membranes are the active sites for signal receptors and transport of components associated with pathogenicity. Capsules or slime layers form outer envelopes that prevent desiccation of bacteria and are directly or indirectly involved as virulence factors. EPSs that form the bulk of capsule or slime layers are necessary for several pathogens to cause disease symptoms such as water soaking and wilting. Far from being static, the formation of cell surface components such as pili and EPSs is often induced by physiological conditions within the host or in a cell density–dependent manner. They are thus part of the dynamic response of bacterial pathogens to overcome preformed and induced plant defense responses.

ARTICLES OF FURTHER INTEREST

Bacterial Attachment to Leaves, p. 75
Bacterial Survival and Dissemination in Natural Environments, p. 108
Bacterial Survival Strategies, p. 115
Genomic Approaches to Understanding Plant Pathogenic Bacteria, p. 524

REFERENCES

1. Dow, M.; Newman, M.A.; von Roepenack, E. The induction and modulation of plant defense responses by bacterial lipopolysaccharides. Annu. Rev. Phytopathol. **2000**, *38*, 241–261.
2. Newman, M.A.; Dow, J.M.; Daniels, M.J. Bacterial lipopolysaccharides and plant-pathogen interactions. Eur. J. Plant Pathol. **2001**, *107*, 95–102.
3. Sigee, D.C. Bacterial Virulence in Plant Disease. In *Bacterial Plant Pathology: Cell and Molecular Aspects*; Cambridge University Press: Cambridge, 1993; 172–211.
4. Dumenyo, C.K.; Chatterjee, A.; Chatterjee, A.K. Phytobacteriology. In *Encyclopedia of Plant Pathology*; Maloy, O.C., Murray, T.D., Eds.; John Wiley & Sons: California, 2001; Vol. 2, 762–769.
5. Romantschuk, M. Attachment of plant pathogenic bacteria to plant surfaces. Annu. Rev. Phytopathol. **1992**, *30*, 225–243.
6. He, S.Y. Pili. In *Encyclopedia of Plant Pathology*; Maloy, O.C., Murray, T.D., Eds.; John Wiley & Sons: California, 2001; Vol. 2, 772–774.
7. Kang, Y.W.; Liu, H.L.; Genin, S.; Schell, M.A.; Denny, T.P. *Ralstonia solanacearum* requires type 4 pili to adhere to multiple surfaces, and for natural transformation and virulence. Mol. Microbiol. **2002**, *46*, 427–437.
8. Liu, H.L.; Kang, Y.W.; Genin, S.; Schell, M.A.; Denny, T.P. Twitching motility of *Ralstonia solanacearum* requires a type IV pilus system. Microbiology **2001**, *147*, 3215–3229.
9. Jin, Q.; He, S.Y. Role of the Hrp Pilus in Type III protein secretion in *Pseudomonas syringae*. Science **2001**, *294*, 2556–2558.
10. Lindgren, P.B. The role of *hrp* genes during plant-bacterial interactions. Annu. Rev. Phytopathol. **1997**, *35*, 129–152.
11. He, S.Y. Type III protein secretion systems in plant and animal pathogenic bacteria. Annu. Rev. Phytopathol. **1998**, *36*, 363–392.
12. Van Gijsegem, F.; Vasse, J.; Camus, J.C.; Marenda, M.; Boucher, C. *Ralstonia solanacearum* produces hrp-dependent pili that are required for PopA secretion but not for attachment of bacteria to plant cells. Mol. Microbiol. **2000**, *36*, 249–260.
13. Denny, T.P. Involvement of bacterial polysaccharides in plant pathogens. Annu. Rev. Phytopathol. **1995**, *33*, 173–197.
14. Chatterjee, A.K.; Dumenyo, C.K.; Liu, Y.; Chatterjee, A. *Erwinia*: Genetics of Pathogenicity Factors. In *Encyclopedia of Microbiology*; Lederberg, J., Ed.; Academic Press: California, 2000; Vol. 2, 106–129.
15. Schell, M.A. Control of virulence and pathogenicity genes of *Ralstonia solanacearum* by an elaborate sensory network. Annu. Rev. Phytopathol. **2000**, *38*, 263–292.
16. Katzen, F.; Ferreiro, D.U.; Oddo, C.G.; Ielmini, M.V.; Becker, A.; Puhler, A.; Ielpi, L. *Xanthomonas campestris* pv. *campestris* gum mutants: Effects on xanthan biosynthesis and plant virulence. J. Bacteriol. **1998**, *180*, 1607–1617.
17. Chakrabarty, P.K.; Mahadevan, A.; Raj, S.; Meshram, M.K.; Gabriel, D.W. Plasmid-borne determinants of pigmentation, exopolysaccharide production and virulence in *Xanthomonas campestris* pv. *malvacearum*. Can. J. Microbiol. **1995**, *41*, 740–745.

Bacterial Products Important for Plant Disease: Diffusible Metabolites As Regulatory Signals

P. K. Chakrabarty
University of Florida, Gainesville, Florida, U.S.A.

A. Chatterjee
W. Gassmann
A. K. Chatterjee
University of Missouri, Columbia, Missouri, U.S.A.

INTRODUCTION

Many plant pathogenic bacteria produce diffusible secondary metabolites that act as quorum-sensing signals. These metabolites accumulate in bacterial cultures and regulate a diverse array of genes, including those involved in virulence and pathogenicity. Quorum-sensing derives its name from the observation that these diffusible signals reach a threshold level based on bacterial cell density (a "quorum") and thus regulate expression of target genes within a population in a cell density–dependent manner. This phenomenon was originally observed with bioluminescence induction in the marine bacterium *Vibrio fischeri* and was termed autoinduction because induction occurred in sealed cultures without external stimuli.

AHL-DEPENDENT QUORUM SENSING

The best-studied quorum-sensing signaling system, which is conserved across many bacterial genera, involves signaling molecules that are classified as *N*-acyl homoserine lactones (AHL). The basic structure of AHLs consists of a homoserine lactone ring connected to a fatty acyl side chain (Fig. 1). AHLs vary with respect to the length of their *N*-acyl chain, which may range from 4 to 14 carbon atoms, be saturated or unsaturated, and have a hydroxy or an oxo group at the third carbon.[1,2]

AHL-mediated gene regulation generally involves a transcriptional activator (a homologue of LuxR, which controls bioluminescence in *V. fischeri*) and an AHL synthase (encoded by a *luxI* homologue of *V. fischeri*). In general, LuxR homologues bind AHL, and the complex activates the expression of AHL synthase and other genes regulated by quorum-sensing. Thus, homologues of *luxR* and *luxI* typically form a regulatory pair controlling many phenotypes. AHL production has been reported to regulate pathogenicity and virulence factors in several genera of plant pathogenic bacteria including *Agrobacterium*, *Erwinia*, and *Pantoea*.[3–6] In plant-associated pseudomonads, multiple traits are regulated by more than one quorum-sensing system. For example, in the nonpathogen *Pseudomonas aureofaciens*, a potential biocontrol agent of several fungal pathogens, two separate quorum-sensing systems control cell surface features, exoprotease activity, and antibiotic and secondary metabolite production. All these traits are important for survival of bacteria on the plant surface.[7]

Among plant pathogens, the most studied quorum-sensing systems are those operating in *Erwinia carotovora* ssp. *carotovora*, *Erwinia chrysanthemi*, and *Pantoea (Erwinia) stewartii* ssp. *stewartii*. In the soft-rot bacterium *E. carotovora* ssp. *carotovora*, *N*-3-oxo-hexanoyl-L-HSL (OHL) regulates production of extracellular enzymes, harpins, and antibiotics.[8] The acyl-HSL synthases involved in the production of OHL have been identified in three strains of *E. carotovora* ssp. *carotovora*: ExpI in strain SCC3193,[9] CarI in strain GS101,[10] and HslI in strain 71.[11] In strain SCC3193, the *expI* (*luxI* homologue) and *expR* (*luxR* homologue) loci are linked. ExpI and ExpR are presumed to constitute a quorum-sensing regulatory pair coordinating the cell density–dependent expression of the genes encoding extracellular enzymes. In *E. carotovora* ssp. *carotovora* strain GS101, CarI coordinates the synthesis of the antibiotic carbapenem along with the expression of macerating enzymes. OHL-regulated expression of carbapenem biosynthetic enzymes is dependent on the LuxR homologue CarR. Interestingly, CarR does not activate expression of macerating enzymes, indicating that the CarI-synthesized OHL may bind to two distinct transcriptional factors. Mutants of *E. carotovora* ssp. *carotovora* strain 71 deficient in HslI are devoid of AHL activity, indicating that OHL probably is the only AHL analogue produced by this bacterium.[12] OHL production by HslI appears to be constitutive and not inducible via a cognate LuxR homologue.

In contrast to the single quorum-sensing systems in *E. carotovora* ssp. *carotovora* strains, multiple AHLs

DOI: 10.1081/E-EPCS 120010570

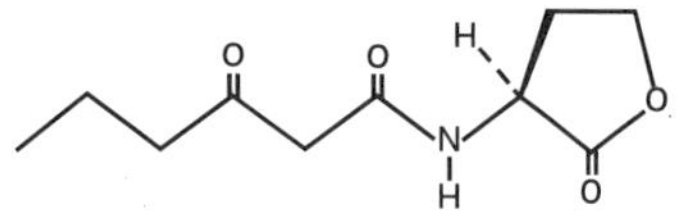

Fig. 1 Structure of the AHL *N*-3-oxo-hexanoyl-L-homoserine lactone (OHL).

are produced in *E. chrysanthemi* strain 16. Similarly, *E. chrysanthemi* strain 3937 produces three AHLs, of which two OHLs are synthesized by *expI* and a third compound, *N*-(decanoyl)-L-homoserine lactone, is synthesized by an unidentified gene. ExpR, a LuxR homologue of *E. chrysanthemi* strain 3937, binds and activates pectate lyase gene promoters in the presence of AHL but binds its own promoter in its absence, presumably to repress its expression.

In *Pantoea stewartii* ssp. *stewartii*, the causal agent of Stewart's wilt of corn, stewartan is an exopolysaccharide that functions as a major virulence factor. Growth-phase-dependent production of stewartan in strain DC283 is regulated by OHL.[13] The *esaI* gene encoding OHL synthase and the *esaR* gene encoding the cognate transcription factor are tightly linked. In contrast to typical LuxR-type activators of other bacteria, EsaR in strain DC283 functions as a repressor of stewartan production in the absence of OHL. $EsaR^-$ mutants overproduce stewartan and are less virulent than the wild-type strain, indicating that premature or excessive production of stewartan interferes with steps in the infection cycle of *P. stewartii* ssp. *stewartii*.

In *Agrobacterium*, two bioactive and related AHLs (3-oxo-C8-HSL and 3-oxo-C6-HSL) are involved in tumor induction in host plants and in conjugation.[6,14] Conjugal transfer of tumor-inducing Ti plasmids between *Agrobacterium* cells is stimulated by high cell densities and requires opines produced by host tumor tissue. These AHLs were therefore originally called conjugation factors (CF). The *Agrobacterium* LuxR homologue, TraR, is the receptor of the AHLs and induces the expression of conjugal transfer (*tra*) genes only in the presence of AHLs. Expression of TraR, in turn, is regulated by opines. The additional requirement for an external factor in regulation of *tra* gene expression is distinct from the original autoinduction model formulated for *V. fischeri*.

In *Pseudomonas syringae* pv.*syringae*, the synthesis of AHL has not been linked to the production of any pathogenicity factor. AHL mutants produced extracellular protease and phytotoxins and elicited the hypersensitive response in tobacco leaves. Possibly, the AHL in *P. syringae* pv. *syringae* strains plays a role in determining fitness on plant surfaces.[15] Likewise, a canonical quorum-sensing system dependent on the *luxI/luxR* homologues *solI/solR* has been described in *Ralstonia solanacearum*, but the function of this system in regulating expression of virulence factors appears limited.[16]

NON-AHL CELL DENSITY–SENSING SYSTEMS

Pathogenicity of some plant-associated bacteria is regulated through non-AHL cell density–sensing signals.[16] In *R. solanacearum*, the Phc (phenotypic conversion) bacterial sensor kinase/response regulator two-component system controls production of exopolysaccharides (EPS) and other pathogenicity factors. The activating signal of this regulatory system is the volatile 3-hydroxy palmitic acid methyl ester (3-OH PAME).[17] The target of the 3-OH PAME–responsive two-component system is PhcA, which, although showing homology to LysR, is a global transcriptional regulator of many genes. Accumulation of 3-OH PAME results in increased functionality of PhcA and expression of virulence factors. Mutants in PhcA display severely reduced virulence. Unlike the canonical quorum-sensing systems, the Phc regulatory system is not exclusively cell density–dependent, since exogenous addition of 3-OH PAME to cultures at low cell density does not activate gene expression.

In *Xanthomonas campestris* pv. *campestris* strain 8004, production of extracellular enzymes important for pathogenicity is regulated by a diffusible signal factor (DSF) of unknown structure.[18] No AHL has been identified in *X. campestris* pv. *campestris*, and the genes involved in regulating expression of these extracellular enzymes do not show homology to *luxI/luxR* genes. *X. campestris* pv. *campestris* strain B24 produces another non-AHL diffusible signal called diffusible factor (DF), which regulates production of the membrane-bound pigment xanthomonadin and EPS. DF may be a butyrolactone derivative and is active at very low concentrations.[19] Available evidence suggests that DSF and DF are independent molecules regulating extracellular enzyme and xanthomonadin production, respectively, but both also regulate EPS production in the respective *X. campestris* pv. *campestris* strains.

CONCLUSION

It is clear that several plant pathogenic bacteria utilize diffusible metabolites to regulate the expression of various genes that encode the products for interacting with their hosts (e.g., pathogenicity/virulence factors) or with other bacteria (e.g., antibiotics or bacteriocins). In many cases, the quorum-sensing system of bacterial pathogens is interlinked with other regulatory systems of gene expression. Therefore, the exact roles of diffusible signals in regulating bacterial pathogenicity are not always clearly understood. One plausible role is delaying the expression of pathogenicity/virulence factors, which may prevent elicitation of host defense responses until the bacterial population has reached a quorum (threshold) sufficient to overcome these responses.

ARTICLES OF FURTHER INTEREST

Bacterial Products Important for Plant Disease: Cell-Surface Components, p. 92

Bacterial Products Important for Plant Disease: Extracellular Enzymes, p. 98

Bacterial Products Important for Plant Disease: Toxins and Growth Factors, p. 101

Bacterial Survival Strategies, p. 115

Fire Blight, p. 443

Genomic Approaches to Understanding Plant Pathogenic Bacteria, p. 524

Plant Diseases Caused by Bacteria, p. 947

REFERENCES

1. Fuqua, C.; Winans, S.C.; Greenberg, E.P. Census and consensus in bacterial ecosystems: The LuxR-LuxI family of quorum-sensing transcriptional regulators. Annu. Rev. Microbiol. **1996**, *50*, 727–751.
2. Pierson, L.S., III; Wood, D.W.; Pierson, E.A. Homoserine lactone-mediated gene regulation in plant associated bacteria. Annu. Rev. Phytopathol. **1998**, *36*, 207–225.
3. Loh, J.; Pierson, E.A.; Pierson, L.S., III; Stacey, G.; Chatterjee, A. Quorum sensing in plant associated bacteria. Curr. Opin. Plant Biol. **2002**, *5*, 285–290.
4. Whitehead, N.A.; Barnard, A.M.; Slater, H.; Simpson, N.J.; Galmond, G.P. Quorum-sensing in Gram-negative bacteria. FEMS Microbiol. Rev. **2001**, *25*, 365–404.
5. Miller, M.B.; Bassler, B.L. Quorum sensing in bacteria. Annu. Rev. Microbiol. **2001**, *55*, 165–199.
6. Pierson, L.S., III; Wood, D.W.; Beck von Bodman, S. Quorum Sensing in Plant-Associated Bacteria. In *Cell-Cell Signaling*; Dunney, G.M., Winans, S.C., Eds.; American Society for Microbiology Press: Washington, D.C., 1999; 101–116.
7. Zhang, Z.; Pierson, L.S., III A second quorum sensing system regulates cell surface properties but not phenazine antibiotic production in *Pseudomonas aureofaciens*. Appl. Environ. Microbiol. **2001**, *67*, 4305–4315.
8. Mukherjee, A.; Cui, Y.; Liu, Y.; Chatterjee, A.K. Molecular characterization and expression of the *Erwinia carotovora* $hrpN_{Ecc}$ gene, which encodes an elicitor of the hypersensitive reaction. Mol. Plant-Microbe Interact. **1997**, *10*, 462–471.
9. Pirhonen, M.; Flego, D.; Heikinheimo, R.; Palva, T.E. A small diffusible signal molecule is responsible for the global control of virulence and exoenzyme production in the plant pathogen Erwinia carotovora. EMBO J. **1993**, *12*, 2467–2476.
10. Swift, S.; Throup, J.P.; Williams, P.; Salmond, G.P.C.; Stewart, G.S.A.B. Quorum sensing: A population-density component in the determination of bacterial phenotype. Trends Biochem. Sci. **1996**, *21*, 214–219.
11. Chatterjee, A.; Cui, Y.; Liu, Y.; Dumenyo, C.K.; Chatterjee, A.K. Inactivation of *rsmA* leads to overproduction of extracellular pectinases, cellulases, and proteases in *Erwinia carotovora* subsp. *carotovora* in the absence of the starvation/ cell density-sensing signal, *N*-(3-oxohexanoyl)-L-homoserine lactone. Appl. Environ. Microbiol. **1995**, *61*, 1959–1967.
12. Chatterjee, A.K.; Dumenyo, C.K.; Liu, Y.; Chatterjee, A. Erwinia: Genetics of Pathogenicity Factors. In *Encyclopedia of Microbiology*; Lederberg, J., Ed.; Academic Press: California, 2000; Vol. 2, 106–129.
13. Beck von Bodman, S.; Majerczak, D.R.; Coplin, D.L. A negative regulator mediates quorum-sensing control of exopolysaccharide production in *Pantoea stewartii* subsp. *Stewartii*. Proc. Natl. Acad. Sci. U. S. A. **1998**, *95*, 7687–7692.
14. Zhang, L.; Murphy, P.J.; Kerr, A.; Tate, M.E. Agrobacterium conjugation and gene regulation by N-acyl homoserine lactones. Nature **1993**, *362*, 446–448.
15. Dumenyo, C.K.; Mukherjee, A.; Chum, W.; Chatterjee, A.K. Genetic and physiological evidence for the production of *N*-acyl homoserine lactones by *Pseudomonas syringae* pv. *syringae* and other fluorescent Pseudomonas species. Eur. J. Plant Pathol. **1998**, *104*, 569–582.
16. Denny, T.P. Autoregulator-dependent control of extracellular polysaccharide production in phytopathogenic bacteria. Eur. J. Plant Pathol. **1999**, *105*, 417–430.
17. Flavier, A.B.; Clough, S.J.; Schell, M.A.; Denny, T.P. Identification of 3-hydroxypalmitic acid methyl ester as a novel autoregulator controlling virulence in *Ralstonia solanacearum*. Mol. Microbiol. **1997**, *26*, 251–259.
18. Barber, C.E.; Tang, J.L.; Feng, J.X.; Pan, M.Q.; Wilson, T.J.G.; Slater, H.; Dow, J.M.; Williams, P.; Daniels, M.J. A novel regulatory system required for pathogenicity of *Xanthomonas campestris* is mediated by a small diffusible signal molecule. Mol. Microbiol. **1997**, *24*, 555–566.
19. Chun, W.; Cui, J.; Poplawski, A. Purification, characterization and biological role of a pheromone produced by *Xanthomonas campestris* pv. *campestris*. Physiol. Mol. Plant Pathol. **1997**, *51*, 1–14.

Bacterial Products Important for Plant Disease: Extracellular Enzymes

P. K. Chakrabarty
University of Florida, Gainesville, Florida, U.S.A.

A. Chatterjee
W. Gassmann
A. K. Chatterjee
University of Missouri, Columbia, Missouri, U.S.A.

INTRODUCTION

The degradation of plant cell walls is an important feature of soft-rot bacteria and bacteria that cause tissue necrosis and vascular wilt diseases. Once inside the host, these pathogens secrete a wide array of macerating enzymes to facilitate degradation of host cell wall components that in turn can be absorbed and assimilated by the invading pathogens. This chapter will focus on the extracellular enzymes that degrade three main constituents of plant cell walls (cellulose, pectin, and proteins) and the secretion pathways used by these enzymes.

CELLULASES

Cellulose is an unbranched polymer of β-1,4-linked D-glucose and is the most abundant plant polysaccharide, accounting for 15 to 30% of the dry mass of all primary cell walls. The cellulose polymers in plant cell walls form microfibrils, which are paracrystalline assemblies of parallel β-1,4-glycan chains hydrogen-bonded to one another.[1] Microfibrils in turn are linked to each other by hydrogen bonding with another class of polysaccharide, the cross-linking glycans (formerly called hemicelluloses). Cross-linking glycans are linear chains with a β-1,4-glucan backbone and relatively short side chains. Unlike cellulose, cross-linking glycans do not hydrogen-bond with each other and as such do not aggregate to form microfibrils.

Enzymes that hydrolyze cellulose are commonly produced by soft-rot pathogens and are generally called cellulases.[2] The cellulase activity of enzymes is mostly determined by their ability to hydrolyze substrates such as carboxymethyl cellulose (CMC) or fibrous cellulose, which are different from paracrystalline cellulose itself. Accordingly, these enzymes are sometimes also specified as carboxymethyl cellulases. However, since the term cellulase is generally accepted in the literature for enzymes that hydrolyze both paracrystalline cellulose or constituents thereof, we will follow this convention for simplicity unless stated otherwise.

The breakdown of cellulose by cellulases results in the production of glucose by a series of separate enzymatic reactions. Thus, some cellulases attack cellulose by cleaving cross-linking glycans, and others break cellulose chains into shorter fragments. The hydrolyzed products are then acted upon by a third group of cellulases, called 1,4-endoglucanases, which produce cellobiose (β-1,4-linked disaccharide of glucose). Cellobiose is the substrate for β-glucosidase, which catalyzes its hydrolysis into glucose molecules.

Two cellulase (*cel*) genes each have been cloned from the two phytopathogenic bacteria *Erwinia chrysanthemi* and *Erwinia carotovorum* ssp. *carotovorum*.[3,4] In *E. carotovorum* ssp. *carotovorum* strain LY34, two cellulases, referred to as CelA and CelB, have been characterized based on their hydrolysis of carboxymethyl cellulose. Five cellulase enzymes have been detected in *E. chrysanthemi* by gel activity staining on CMC-SDS-PAGE (carboxymethyl cellulose–sodium dodecyl sulfate–polyacrylamide gel electrophoresis), although only two cellulase genes, *cel*5Z and *cel*8Y, have been cloned and characterized.[4] Strains in which the genes for major cellulases have been deleted display reduced virulence. *Ralstonia solanacearum* possibly produces two extracellular glucanases based on sequence homology, and the hydrolytic activity on β-1,4 glycosidic linkages has been demonstrated for one of them.[5] Cellulase-deficient *R. solanacearum* mutants are also significantly less virulent on tomato plants.

PECTINASES

Pectic acid is a polymer composed chiefly of galacturonic acid units linked by α-1,4-glycosidic bonds. Pectin is

Encyclopedia of Plant and Crop Science
DOI: 10.1081/E-EPCS 120021340

formed by methyl esterification of the carboxyl groups in the pectic acid chains. Pectic substances are a major component of the middle lamella and matrix of primary cell walls and thus play an important role in tissue cohesion. The enzymes that degrade pectic substances are known as pectinases or pectolytic enzymes and are of three main types: hydrolases such as polygalacturonase (Peh), lyases such as pectate lyase (Pel) and pectin lyase (Pnl), and esterases such as pectin methyl esterase (Pem). Peh hydrolyzes α-1,4-glycosidic linkages between two galacturonic acid units. Pel and Pnl break the α-1,4-glycosidic bond by elimination, creating a double bond between C4 and C5 in the galacturonosyl residue. Peh, Pel, and Pnl enzymes are further subdivided into endopectinases and exopectinases. Endopectinases act at random sites within pectin chains, whereas exopectinases act only on terminal linkages. Lastly, Pem hydrolyzes the methyl ester bonds in pectin to yield pectic acid and methanol. Pem enzymes do not affect overall chain length but alter the solubility of pectic substances and the rate at which they are degraded by the chain-splitting enzymes.

The role of pectinases in bacterial pathogenicity has been well documented for the soft-rot bacteria *E. chrysanthemi* and *E. carotovorum* ssp. *carotovorum*, and in the vascular wilt pathogen *R. solanacearum*.[3,5,6] *E. chrysanthemi* strain EC16 produces four Pel enzymes encoded by four independently regulated genes.[7] *E. chrysanthemi* strain 3937 secretes an arsenal of at least eight endo-Pels, two exo-Pels, and four Pehs, including the most recently characterized polygalacturonase N.[8] Multiplicity of pectolytic enzymes found in soft-rot bacteria may reflect the availability of different substrates occurring in different hosts. In addition, there may be a requirement for cooperation between different enzymes for optimal maceration of invaded tissues. For example, Peh differs from Pel in being inhibited by higher calcium ions and pH. At the beginning of tissue maceration, Peh may play a major role in cell wall degradation, but increasing levels of calcium and pH would result in increased Pel activity.

Deletion of all Pel genes in *E. carotovora* ssp. *carotovora* results in 98% reduction in maceration of potato tuber tissue.[9] A single Pel gene is necessary for pathogenicity of *Pseudomonas viridiflava*, and mutation of the single Peh gene of *Agrobacterium vitis* decreased the virulence of this pathogen on grape. *R. solanacearum* produces one Pem and three Pehs (Peh-A, B, C) but no Pel enzyme. A PehA-PehB–deficient double mutant of *R. solanacearum* invades and colonizes stems more slowly than wild type strains. However, deletion of individual pectinases in these pathogens does not always affect pathogenicity. For example, endo-Peh defective mutants of *E. carotovora* with unaltered Pel activity still retain pathogenicity on several hosts.[9]

PROTEASES

The primary cell wall matrix contains several classes of glycoproteins in addition to polysaccharides. Secretion of proteases (Prt) appears widespread among phytopathogenic bacteria.[7] *Xanthomonas campestris* pv. *campestris* produces four proteases. At high inoculum levels, protease-deficient mutants of *X. campestris* pv. *campestris* show wild-type virulence on turnip leaves. However, a pronounced reduction in virulence is observed at low inoculum levels.[10] Most soft-rotting *Erwinia* produce extracellular proteases, many of which are metalloproteases.[3] Strain SCC3193 of *E. carotvora* ssp. *carotovora* produces an extracellular metalloprotease designated PrtW,[11] which is distinct from previously described proteases from the same pathogen. Mutants deficient in PrtW produced normal levels of Peh, Pel, and Cel, but were considerably reduced in their virulence. *E. chrysanthemi* also produces four protease isozymes organized in two clusters.[12] Among soft-rot pseudomonads, protease production rather than pectolytic enzyme production was more strongly correlated with the ability to macerate plant tissues.[6]

SECRETION SYSTEMS FOR EXTRACELLULAR ENZYMES

Extracellular enzymes must be secreted out of bacterial cells into host intercellular spaces to function in pathogenicity. In plant-associated gram-negative bacteria, there are four distinct pathways for secretion of extracellular pathogenicity factors, referred to as type I to type IV secretion systems.[13] The exoenzymes discussed in this chapter are substrates for type I or type II secretion systems. Type III and type IV secretion systems are associated with export of Hrp/Avr pathogenicity factors and *Agrobacterium* T-DNA, respectively (see articles by Vivian and by Burr).

The type I secretion system, exemplified by the secretion of metalloproteases of *E. chrysanthemi*, targets substrates directly through bacterial inner and outer membranes to the exterior. Translocation across the cytoplasmic membrane is performed by PrtD, an ATP-binding cassette (ABC) transporter, which recognizes a C-terminally located secretion signal in the secreted protein.[13] The one-step secretion of proteases in *E. chrysanthemi* requires two additional proteins: PrtE, which belongs to the membrane fusion protein family and spans the inner and outer membrane, and PrtF, an outer membrane protein. Besides extracellular proteases, such systems export siderophores for metal uptake, toxins, and antimicrobial peptides in several phytopathogenic bacteria.[14]

The general or type II secretion system has two parts: the export pathway for transport across the inner membrane and the so-called main terminal branch pathway for secretion through the outer membrane.[13] Secretion via the type II secretion system requires an N-terminal signal peptide that is cleaved off after the protein is translocated across the inner membrane into the periplasm, where it is folded into its mature form prior to transport across the outer membrane. Many bacteria utilize this system, e.g., *Erwinia* for secretion of pectinases and cellulases, *X. campestris* for secretion of pectinases, cellulases, and proteases, and *R. solanacearum* for secretion of Pehs, Pem, and endoglucanase.

CONCLUSION

Bacterial plant pathogens produce an arsenal of extracellular enzymes capable of degrading plant cell wall components composed chiefly of pectin, cellulose, and glycoproteins. These host components are liable to be degraded by bacterial pectinases, cellulases, and proteases, respectively. These enzymes cause softening and disintegration of cell wall materials and facilitate penetration and spread of the pathogen in the host. In the case of vascular diseases, the liberation of large oligomers into transpiration channels may interfere with normal movement of water. The extracellular enzymes of bacterial plant pathogens constitute pathogenicity/virulence factors and are secreted out of bacterial cells through specialized protein secretion systems. Model representatives of type I and type II secretion systems have been well studied in terms of membrane topology, as well as the structure and biochemical functions of the constituent proteins. A better understanding of the mechanisms and regulation of extracellular enzyme secretion in phytopathogenic bacteria may allow improved control of bacterial diseases in plants.

ARTICLES OF FURTHER INTEREST

Bacterial Products Important for Plant Disease: Cell-Surface Components, p. 92

Bacterial Products Important for Plant Disease: Diffusible Metabolites As Regulatory Signals, p. 95

Bacterial Products Important for Plant Disease: Toxins and Growth Factors, p. 101

Bacteria-Plant Host Specificity Determinants, p. 119

Fire Blight, p. 443

Genomic Approaches to Understanding Plant Pathogenic Bacteria, p. 524

Management of Bacterial Diseases of Plants: Regulatory Aspects, p. 669

Management of Fungal and Oomycete Vegetable Diseases: Crops, p. 681

Plant Diseases Caused by Bacteria, p. 947

Symbiotic Nitrogen Fixation, p. 1218

REFERENCES

1. Carpita, N.; McCann, M. The Cell Wall. In *Biochemistry and Molecular Biology of Plants*; Buchanan, B.B., Gruissem, W., Jones, R.L., Eds.; American Society of Plant Physiologists: Rockville, MD, 2000; 52–108.
2. Barras, F.; van Gijsegem, F.; Chatterjee, A.K. Extracellular enzymes and pathogenesis of soft-rot Erwinia. Annu. Rev. Phytopathol. **1994**, *32*, 201–234.
3. Chatterjee, A.K.; Dumenyo, C.K.; Liu, Y.; Chatterjee, A. Erwinia: Genetics of Pathogenicity Factors. In *Encyclopedia of Microbiology*; Lederberg, J., Ed.; Academic Press: California, 2000; Vol. 2, 106–129.
4. Cho, S.J.; Park, S.R.; Kim, M.K.; Lim, W.J.; Lim, W.J.; Ryu, S.K.; An, C.L.; Hong, S.Y.; Kim, H.; Cho, Y.U.; Yun, H.D. Cloning of the *cel*8Y gene from *Pectobacterium chrysanthemi* PY35 and its comparison to *cel* genes of soft-rot *Pectobacterium*. Mol. Cells **2002**, *13*, 28–34.
5. Schell, M.A. Control of virulence and pathogenicity genes of *Ralstonia solanacearum* by an elaborate sensory network. Annu. Rev. Phytopathol. **2000**, *38*, 263–292.
6. Collmer, A.; Keen, N.T. The role of pectic enzymes in plant pathogenesis. Annu. Rev. Phytopathol. **1986**, *24*, 383–409.
7. Sigee, D.C. Bacterial Virulence in Plant Disease. In *Bacterial Plant Pathology: Cell and Molecular Aspects*; Cambridge University Press: Cambridge, 1993; 172–211.
8. Hugouvieux-Cotte-Pattat, N.; Schevchik, V.E.; Nasser, W. PehN, a polygalacturonase homologue with a low hydrolase activity, is coregulated with the other *Erwinia chrysanthemi* polygalacturonases. J. Bacteriol. **2002**, *184*, 2664–2673.
9. Freeman, S. Extracellular Enzymes, Pectin. In *Encyclopedia of Plant Pathology*; Maloy, O.C., Murray, T.D., Eds.; John Wiley & Sons: New York, 2001; Vol. 1, 438–440.
10. Dow, J.M.; Davies, H.A.; Daniels, M.J. A metalloprotease from *Xanthomonas campestris* that specifically degrades proline/hydroxyproline-rich glycoproteins of the plant extra-cellular matrix. Mol. Plant-Microbe Int. **1998**, *11*, 1085–1093.
11. Marits, R.; Koiv, V.; Laasik, E.; Mae, A. Isolation of an extracellular protease gene of *Erwinia carotovora* subsp. *carotovora* strain SCC3193 by transposon mutagenesis and the role of protease in phytopathogenicity. Microbiology **1999**, *145*, 1959–1966.
12. Wandersman, C.; Delepelaire, P.; Letoffe, S.; Schwartz, M. Characterisation of *Erwinia chrysanthemi* extracellular proteases: Cloning and expression of the protease genes in *Escherichia coli*. J. Bacteriol. **1987**, *169*, 5046–5053.
13. Salmond, G.P.C. Secretion of extracellular virulence factors by plant pathogenic bacteria. Annu. Rev. Phytopathol. **1994**, *32*, 181–200.
14. Dumenyo, C.K.; Chatterjee, A.; Chatterjee, A.K. Extracellular Proteins: Secretion. In *Encyclopedia of Plant Pathology*; Maloy, O.C., Murray, T.D., Eds.; John Wiley & Sons: New York, 2001; Vol. 1, 440–443.

Bacterial Products Important for Plant Disease: Toxins and Growth Factors

P. K. Chakrabarty
University of Florida, Gainesville, Florida, U.S.A.

W. Gassmann
A. Chatterjee
A. K. Chatterjee
University of Missouri, Columbia, Missouri, U.S.A.

INTRODUCTION

Toxins are low molecular weight, non-enzymatic virulence factors produced by many phytopathogenic bacteria. They have a wide range of physiological and biochemical effects on the host plant and produce diverse symptoms, such as chlorosis, water soaking, necrosis, growth abnormalities, and wilting. They cause chemical injury to the host either by affecting the permeability of cell membranes or by inactivating or inhibiting host enzymes. Unlike many fungal toxins, bacterial toxins are not host-selective and therefore do not determine host range. These toxins, generally considered as secondary metabolites, are secreted by bacteria growing both in vitro and in planta.

Some pathogens also produce naturally occurring plant growth regulators such as auxins, cytokinins, gibberellins, and ethylene in their hosts during infection. This causes an imbalance in the host's hormonal system, thereby resulting in abnormal growth and function. The role of microbial growth factors in plant pathogenesis has been extensively studied with reference to gall- and tumor-forming bacterial pathogens.

TOXINS

Some well-studied toxins secreted by plant pathogenic bacteria are listed in Table 1. Most of these are produced by pseudomonads. Coronatine, syringomycin, syringopeptin, tabtoxin, and phaseolotoxin are the most intensively studied phytotoxins of *Pseudomonas syringae*, and each contributes significantly to bacterial virulence.[1] Their modes of action, regulation, and biosynthesis have been reviewed recently by Bender et al.[2]

Coronatine

P. syringae pvs. *atropurpurea* (a pathogen of rye grass), *glycinea* (soybean), *maculicola* (bean), *morsprunorum* (*Prunus*), and *tomato* (tomato) produce the toxin coronatine. This toxin consists of a polyketide structure linked to a cyclopropane component. Although rare outside *P. syringae* pathovars, coronatine has been reported to be produced by *Xanthomonas campestris* pv. *phormiicola*, a pathogen of flax.[3] Coronatine functions partly as a mimic of methyl jasmonate, a hormone synthesized by plants undergoing biological stress, and induces symptoms of chlorosis, stunting, and hypertrophy. Coronatine has been shown to play a distinct role in virulence based on studies of coronatine biosynthesis mutants of *P. syringae* pathovars.[2]

Syringomycin and Related Toxins

Syringomycin is representative of cyclic lipodepsinonapeptide phytotoxins produced by most strains of *P. syringae* pv. *syringae* from diverse plant hosts.[2] It is composed of a polar peptide head of nine amino acids and a hydrophobic 3-hydroxy fatty acid tail (Fig. 1). An amide bond attaches the 3-hydroxy fatty acid to an N-terminal serine residue, which in turn is linked to 4-chlorothreonine at the C terminus by an ester linkage to form a macrocyclic lactone ring. Other distinctive structural features are three uncommon amino acids (2,3-dehydroxyaminobutyric acid, 3-hydroxyaspartic acid, and 4-chlorothreonine) at the C terminus and the presence of D-isomers of serine and 2,4-diaminobutyric acid. Syringopeptins represent another class of lipodepsipeptide phytotoxins produced by strains of this pathogen.[2] In contrast to lipodepsinonapeptides, syringopeptins contain either 22 or 25 amino acids depending on the specific bacterial strain (Fig. 2). An ester bond between allo-threonine and the C-terminal tyrosine residue forms a lactone ring. Both syringomycin and syringopeptin are necrosis-inducing toxins; however, the two related toxins differ in their biological properties and antimicrobial specificity against different groups of microbes.[4] These toxins form pores in plasma membranes that lead to electrolyte leakage. In addition to being phytotoxic, they

DOI: 10.1081/E-EPCS 120021339

Table 1 Toxins in plant pathogenesis

Toxins	Bacterium	Host	Target
Coronatine	*Pseudomonas syringae*		Not known
	pv. *atropurpurea*	Rye	
	pv. *glycinea*	Soybean	
	pv. *tomato*	Tomato	
	pv. *morsprunorum*	Prunes	
	pv. *maculicola*	Beans	
Tagetitoxin	*P. syringae* pv. *tagetis*	Tagetus	Chloroplast RNA polymerase
Phaseolotoxin	pv. *phaseolicola*	Beans	Ornithine carbamoyl transferase
Syringomycin Syringopeptin	pv. *syringae*	Stone fruits, pome fruits, and grasses	Plasma membrane
Tolaasin	pv. *tolaasii*	Mushroom	Plasma membrane
Tabtoxin	pv. *tabaci*	Tobacco	Glutamate synthetase
	pv. *atropurpurea*	Rye	
	pv. *coronafaciens*	Oats	
	pv. *garcea*		
Rhizobitoxine	*Pseudomonas andropogonis*	Corn, sorghum, sudan grass	β-cystathionase
	Rhizobium sp.	Legumes	
Carboxylic acid	*Xanthomonas oryzae*	Rice	Not known
	Xanthomonas pv. *manihot*,		
	Xanthomonas campestris pv. *campestris*		

exhibit broad antibiotic activity against prokaryotes and eukaryotes. Toxin-deficient mutants of *P. syringae* pathovars are either reduced in their virulence or are non-pathogenic.

Tabtoxin

Tabtoxin is a monocyclic β-lactam produced by *P. syringae* pvs. *tabaci*, *coronafaciens*, and *garcae*.[1] It is a dipeptide composed of threonine linked by a peptide bond to an uncommon amino acid, tabtoxinine-β-lactam (Fig. 3). Tabtoxin itself is not toxic but is hydrolyzed by aminopeptidases of plant or bacterial origin in the plant's intercellular spaces. This releases the biologically active compound tabtoxinine-β-lactam, which is actively taken up by the plant amino acid transport system. The toxin inactivates the enzyme glutamine synthetase that is involved in assimilation of ammonia within chloroplasts, and consequently leads to an accumulation of ammonia to toxic levels. The latter uncouples photophosphorylation and destroys the thylakoid membrane of the chloroplasts, causing chlorosis and eventually necrosis of the infected tissues. Tabtoxin is associated with the symptoms of wildfire disease of tobacco and halo blight of oats but is considered to be a virulence factor rather than an essential component of these diseases. This is because the non-toxigenic strains of these pathogens can still induce necrosis without producing the characteristic yellow halos.

Phaseolotoxin

Phaseolotoxin is a modified ornithine-alanine-arginine tripeptide carrying a sulfodiaminophosphinyl group (Fig. 4). In plants, the tripeptide is cleaved by peptidases

3— hydroxy fatty acyl — L-Ser — D-Ser — D-Dab — L-Dab — L-Arg — L-Phe — Z-Dhb — L-Asp (3-OH) — L-Thr (4-Chl)

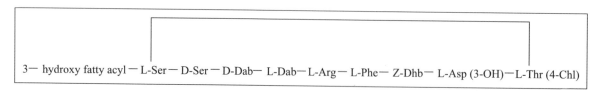

Fig. 1 Structure of Syringomycin. Abbreviations of nonstandard amino acids: Asp(3-OH), 3-hydroxyaspartic acid; Dab, 2-4-diaminobutyric acid; Dhb, 2,3-dehydroaminobutyric acid; Thr(4-Chl), 4-chlorothreonine.

$CH_3(CH_2)_{6\text{-}8}CH(OH)CH_2CO$-Dhb-Pro-Val-Val-Ala-Ala-Val-Val-Dhb-Ala-Val-Ala-Ala-Dhb-*a*Thr-Ser-Ala-Dhb

Tyr-Dab-Dab-Ala

Fig. 2 Structure of Syringopeptin form SP22. Abbreviations of nonstandard amino acids: Dab, 2-4-diaminobutyric acid; Dhb, 2,3-dehydroaminobutyric acid; *a*Thr, *allo*threonine.

to release alanine, arginine, and the biologically active moiety, sulfodiaminophosphinyl ornithine. The toxin inactivates ornithine carbamoyl transferase (OCT), an enzyme of the urea cycle, which normally converts ornithine to citrulline, a precursor of arginine. Inhibition of OCT leads to accumulation of ornithine and depletion of citrulline and arginine. Phaseolotoxin also seems to inhibit biosynthesis of pyrimidine nucleotide, reduce the activity of ribosomes, interfere with lipid synthesis, change the permeability of plant cell membranes, and result in accumulation of large starch grains in the chloroplasts. A direct correlation exists between the amount of phaseolotoxin produced by strains of *P. syringae* pv. *phaseolicola* and the extent of chlorotic haloes on infected plants. Toxin-deficient mutants survive and multiply at the site of inoculation but fail to move systemically inside host plants.

Other Toxins

Tagetitoxin is a cyclic compound[1] produced solely by *P. syringae* pv. *tagetis*, a pathogen of several members of the Compositae family. Unlike other toxins, tagetitoxin is mainly produced in planta, where it induces symptoms of apical chlorosis and yellow haloes on leaves. The toxin mainly affects chloroplasts by inhibiting chloroplast RNA polymerase.[5] Tolaasin is a lipodepsipeptide composed of 18 amino acid residues with a β-octanoic acid at the N terminus. This toxin is produced by *P. syringae* pv. *tolaasii*, a pathogen of mushroom, and it causes disruption of the host plasma membrane.[6] Rhizobitoxine secreted by *Rhizobium* and the unrelated pathogen *Pseudomonas andropogonis* is a vinylglycine compound that causes inhibition of homocysteine synthesis by β-cystathionase inactivation.[7] Carboxylic acid toxins are produced by several pathovars of *Xanthomonas campestris* in culture and may be involved in disease symptom induction.[5]

In addition to producing toxins directed against the host, many plant pathogenic bacteria produce secondary metabolites to compete with other bacteria. Several strains of *Erwinia carotovora* ssp. *carotovora* and *Pantoea agglomerans* produce antibiotics and bacteriocins.[8,9] Bacteriocins are mostly proteinaceous antibiotics with very narrow specificity against strains of the same or closely related species.

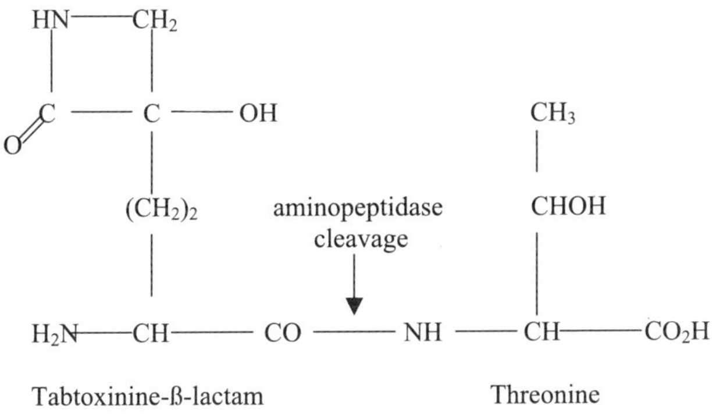

Fig. 3 Structure of tabtoxin, which consists of the toxic moiety tabtoxinine-β-lactam linked to threonine. The arrow shows the site of aminopeptidase cleavage, which releases tabtoxinine-β-lactam.

GROWTH FACTORS

Growth hormones are important in the pathogenicity of gall-forming phytopathogenic bacteria such as *Agrobacterium*, *Pseudomonas* and *Erwinia*.[10] They initiate high rates of cell division (hyperplasia) and cause extensive cell enlargement (hypertrophy) in infected tissue. This proliferation of host tissue results in a range of distinct symptoms characteristic of the causative pathogen. These include crown gall and tumors of pome, stone, and other fruits by *Agrobacterium tumefaciens*, hairy root of apple caused by *A. rhizogenes*, galls of olive and oleander by *P. savastanoi* pv. *savastanoi*, and galls of *Gypsophila paniculata* by *Erwinia herbicola* pv. *gypsophilae*. Two major phytohormones that are associated with these diseases are auxins and cytokinins. In the case of

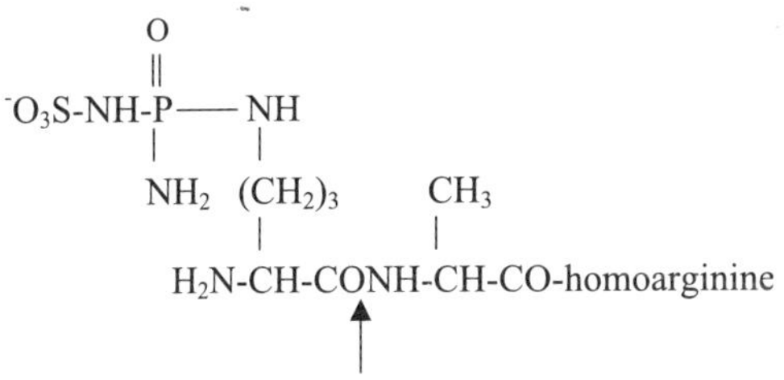

Fig. 4 Structure of phaseolotoxin. Plant peptidases cleave phaseolotoxin (arrow) to release alanine and arginine resulting in the formation of (N′ sulfodiaminophosphonyl)-L-ornithine.

P. savastanoi pv. *savastanoi*, the roles of auxins and cytokinins in disease development are not well understood, although auxin production has been demonstrated to increase pathogen fitness. In the case of *A. tumefaciens*, the production of auxins and cytokinins plays a direct role in pathogenicity by increasing the cell mass of transformed host cells that produce opines, which in turn serve as a nutrient source only for the pathogen (see the related article "Crown Gall"). In *E. herbicola* pv. *gypsophilae*, although gall initiation is triggered by other virulence factors, the biosynthesis of auxin and cytokinin enhances the gall formation in *G. paniculata*.[11]

Genes involved in the synthesis of auxins and cytokinins have been cloned and characterized in *P. savastanoi* pv. *savastanoi*, *A. tumefaciens* and *E. herbicola* pv. *gypsophilae*. A high degree of homology exists between the auxin and cytokinin genes of these pathogens.

CONCLUSION

Many bacterial pathogens produce an array of toxins that are important virulence factors. In recent years, some studies have used molecular approaches to construct bacteria deficient in individual toxins and have clearly shown that toxins can contribute quantitatively to pathogen virulence and disease development. Several of these toxins affect the metabolic processes of the host, thereby facilitating the multiplication and spread of the pathogen. Toxins can also alter permeability of the plasma membrane and produce visible chlorosis and necrosis of the host. Plant pathogen-derived phytohormones produce deleterious effects on the host's growth regulatory system in infected plants. While bacterial toxins appear to be indiscriminate weapons against the host, the deployment of growth regulators reveal an intricate and well-adapted interplay between host and pathogen.

ARTICLES OF FURTHER INTEREST

Bacterial Products Important for Plant Disease: Cell-Surface Components, p. 92
Bacterial Products Important for Plant Disease: Diffusible Metabolites As Regulatory Signals, p. 95
Bacterial Products Important for Plant Disease: Extracellular Enzymes, p. 98
Crown Gall, p. 379
Plant Diseases Caused by Bacteria, p. 947

REFERENCES

1. Gross, D.C. Molecular and genetic analysis of toxin production by pathovars of *Pseudomonas syringae*. Annu. Rev. Phytopathol. **1991**, *29*, 247–248.
2. Bender, C.L.; Alarcon-Chaidez, F.; Gross, D.C. *Pseudomonas syringae* phytotoxins: Mode of action, regulation and biosynthesis by peptide and polyketide synthetases. Microbiol. Mol. Biol. Rev. **1999**, *63*, 266–292.
3. Tamura, K.; Takikawa, Y.; Tsuyumu, S.; Goto, M.; Watanabe, M. Coronatine production by *Xanthomonas campestris* pv. *phormiicola*. Ann. Phytopathol. Soc. Jpn. **1992**, *58*, 276–281.
4. Lavermicocca, P.; Iacobellis, N.S.; Simmaco, M.; Graniti, A. Biological properties and spectrum of activity of *Pseudomonas syringae* pv. *syringae* toxins. Physiol. Mol. Plant Pathol. **1997**, *50*, 129–140.
5. Sigee, D.C. Bacterial Virulence in Plant Disease. In *Bacterial Plant Pathology: Cell and Molecular Aspects*; Cambridge University Press: Cambridge, 1993; 172–211.
6. Rainey, P.B.; Brodey, C.L.; Johnstone, K. Biological properties and spectrum of activity of tolaasin, a lipodepsipeptide toxin produced by the mushroom pathogen *Pseudomonas tolaasii*. Physiol. Mol. Plant Pathol. **1991**, *39*, 57–70.
7. Mitchell, R.E.; Frey, E.J. Rhizobitoxine and hydroxythreonine production by *Pseudomonas andropogonis* strains, and the implications to plant disease. Physiol. Mol. Plant Pathol. **1988**, *33*, 335–341.
8. Chatterjee, A.K.; Dumenyo, C.K.; Liu, Y.; Chatterjee, A. Erwinia: Genetics of Pathogenicity Factors. In *Encyclopedia of Microbiology*; Lederberg, J., Ed.; Academic Press: California, 2000; Vol. 2, 106–129.
9. Vidaver, A.K. Bacteriocin. In *Encyclopedia of Plant Pathology*; Maloy, O.C., Murray, T.D., Eds.; John Wiley & Sons: California, 2001; Vol. 1, 95–96.
10. Costacurta, A.; Vanderleyden, J. Synthesis of phytohormones by plant-associated bacteria. Crit. Rev. Microbiol. **1995**, *21*, 1–18.
11. Manulis, S.; Haviv-Chesner, A.; Brandle, M.T.; Lindow, S.E.; Barash, I. Differential involvement of indole-3-acetic acid biosynthetic pathways in pathogenicity and epiphytic fitness of *Erwinia herbicola* pv. *gypsophilae*. Mol. Plant-Microbe Interact. **1998**, *11*, 634–642.

Bacterial Survival and Dissemination in Insects

Alexander H. Purcell
University of California, Berkeley, California, U.S.A.

B

INTRODUCTION

Most bacterial plant pathogens are spread from one plant to another by wind, rain, or human activities, but many of them infect plants only through the intervention of insect vectors. Non-vector insects may pick up microbial pathogens by feeding on infected plants and can retain large numbers of the microbes within their bodies, but only those insects that can transmit a pathogen or parasite to a host are vectors. The transmission process can be a simple mechanical transfer of a pathogen by an insect to a susceptible site on a plant. In these cases, insects are usually not the only means by which the pathogen can disperse and infect plants. At the other extreme, transmission can require a complex sequence of steps for the pathogen to circulate and multiply within specific locations before the vector can transmit the pathogen to a plant. Complex transmission processes confer considerable specificity as to which insects can serve as vectors.

PATHOGENS WITH NON-OBLIGATE VECTORS

Fireblight of apple, pear, and other pome fruits is an example of a disease where many kinds of insects can increase the spread of the causal bacterium (*Erwinia amylovora*) from plant to plant, but insects are not essential. For example, flower-visiting insects can transport *E. amylovora* to flowers and create tiny wounds that the bacterium can invade, but pruning tools and splashing rain are more important ways by which the bacterium spreads.[1] Likewise, insects may be important in moving *E. carotovora* subsp. *carotovora* from potato cull piles into potato production fields, increasing the incidence of soft rot in the crop.[2] Removal or burial of cull piles from field areas to deny insect access is an important disease control measure.

PATHOGENS WITH OBLIGATE VECTORS

Chewing Insects as Vectors

Bacterial wilt of cucurbits (caused by *Erwinia tracheiphila*) and Stewarts wilt of maize (caused by *Pantoea* [*Erwinia*] *stewartii* subsp. *stewartii*) require chewing insect vectors to spread the causal bacteria from plant to plant. Cucumber beetles (family Chrysomelidae) that feed frequently on cucurbits are the most important vectors of *E. tracheiphila*. Flea beetles (Chrysomelidae, subfamily Alticinae) are the key vectors of the Stewarts wilt bacterium.[3] Controlling the insect vectors is important for the management of both diseases. The beetle vectors of each of these pathogens harbor the bacteria within their guts during the winter and then establish new infections in plants the following growing season. The bacteria do not otherwise survive the winter in soil or plant debris. A winter cold severity index can predict the severity of Stewarts wilt, because cold temperatures lessen the survival of flea beetle vectors and disease severity depends on the number of independent bacterial infections per leaf, which originate in the margins of the small holes chewed by the flea beetle vectors. Not many details are known about how the beetle vectors transmit these two pathogens or why other, similar chewing insects cannot serve as vectors.

Sucking Insects as Vectors

Most insect vectors of plant pathogens are sucking insects that feed on plant sap or cell contents. Bacterial pathogens that are strict parasites of plant vascular systems (phloem or xylem) usually have specific sucking insect vectors. Pathogens with a high degree of vector specificity have only a single or few (but usually related) insects as vectors, although other insects may acquire but not transmit these pathogens from infected plants by feeding on infected tissues.

A major disadvantage of xylem or phloem sap as a primary food source is that it has low or unbalanced concentrations of nutrients.[4,5] Consequently, all insects that feed on plant sap probably harbor one or more bacterial symbionts, housed inside the cells of specialized organs, that provide essential nutrients required for the host insect to develop and reproduce.[4,5] The symbionts invade the developing eggs within the mother insect to ensure that all offspring have these essential bacteria.[5] These codependent partnerships between insect and bacteria are ancient and may predispose plant sap–feeding

Encyclopedia of Plant and Crop Science
DOI: 10.1081/E-EPCS 120010567

insects to be capable of supporting the multiplication and internal circulation of other bacteria within the host insect.

Noncirculative Transmission of Bacteria

The best known of the bacterial parasites of plants that are strictly limited to xylem is *Xylella fastidiosa*. This bacterium has a broad range of vectors, but they are all sucking insects that specialize on feeding on xylem sap.[6] Vectors include spittlebugs (family Cercopidae) and leafhoppers (family Cicadellidae) in the subfamily Cicadellinae, which are commonly called sharpshooters. Clues as to how vectors transmit the bacteria are 1) that there is no delay between vector acquisition of the bacteria from infected plants and the vectors transmitting the bacterium to plants (called the latent period) and 2) that vectors stop transmitting after shedding their outer skin. These properties suggest that the vector transmits to plants bacteria that are attached to the lining of the mouth (the foregut, which is shed with molting of the outer skin), but the exact location in the foregut is still unclear.[7] Bacteria can be seen attached to various parts of the vector foregut, where they experience flow velocities of more than 5 to 50 centimeters per second.

Circulative Transmission of Bacteria

Some insect vectors transmit bacterial parasites that are restricted to the plant's phloem cells. In contrast to the nutrient-dilute xylem sap, the food-conducting phloem contains sap that is high in sugars and other solutes. The lack of a rigid cell wall may be an advantage for phloem-inhabiting bacteria to counter the osmotic stresses of the high sugar environment of phloem sap, and the most commonly recognized bacterial parasites of phloem are Gram-positive bacteria (class Mollicutes) that lack a cell wall. The absence of a rigid cell wall also allows the small bacteria (0.2 to 0.5 microns wide, lengths variable to several microns) to deform and squeeze through the pores connecting phloem elements.

Two types of mollicutes are plant pathogens. Spiroplasmas are helical in shape, and phytoplasmas are spherical to filamentous in form (pleomorphic). To date, these mollicutes cannot be mechanically inoculated into plants, so vectors are essential for infection. Citrus stubborn disease and corn stunt are two important plant diseases caused by leafhopper-transmitted spiroplasmas. The number and diversity of phytoplasma diseases (at least 300) is much greater than that caused by spiroplasmas.

The mollicutes have a high degree of vector specificity because of the complex route within the vector that must be completed for transmission to occur. This requires vector uptake of the mollicute by feeding in the phloem, passage of the mollicute through the vectors gut and its surrounding basement membrane, and multiplication in one or more sites or organs within the vectors body cavity (hemocoel), followed by movement to the interior of certain salivary gland cells and ejection into plant phloem in such a way as to establish a new mollicute population in the plant. All of these consecutive steps result in a latent period of many weeks until the vectors are able to first transmit. The principal mollicute vectors are leafhoppers, planthoppers, and psyllids. Spiroplasmas occur in great diversity in insects or nectar, but only a relative few are plant pathogens.

Grapevine yellows is the name of a complex of diseases with the same or very similar symptoms, but caused by a variety of phytoplasmas, each with different vectors.[8] This complex provides an example of how different mollicute pathogens with specific vectors vary in how they spread and thus how they can be controlled. The flavescence dorée (FD) type of grapevine yellows is transmitted by a single leafhopper species (*Scaphoideus titanus*); the bois noir type is transmitted by a planthopper (*Hyalesthes obsoletus*). At least two other types of grapevine yellows caused by phytoplasmas are genetically unique from FD or bois noir and are presumed to be closely related to leafhopper-transmitted phytoplasmas. The FD phytoplasma can be transmitted to plant species other than grape, but the strong specificity of its vector for feeding and reproducing on grape limits the occurrence of FD in the field to grape. This vector feeding preference thus clearly target vineyards and wild grapes for controlling the vector with insecticides. In addition, eliminating FD-grapevines is an important part of control measures to reduce the percentage of remaining vectors that carry the FD phytoplasma. In contrast, the bois noir phytoplasma and its vector can be found in numerous weed and crop species in or near vineyards, complicating the control of the vector and making the elimination of plants harboring the phytoplasma very difficult.[9]

The best known and most important example of a disease caused by phloem-limited bacteria that has a typical rigid cell wall is citrus greening disease.[10] The milder, African version of greening disease is presumed to be caused by *Candidatus* Liberobacter africanum and the more severe Asian form (Huanglongbin) by *Candidatus* Liberobacter asiaticum. Each type of greening disease has it own species of psyllid (superfamily Psylloidea) vectors. Psyllid vectors begin to transmit the bacteria to plants after a latent period of many days to weeks, depending on temperature. This is presumably because the bacteria must pass through the vectors gut wall, multiply within the insects body cavity (hemocoel), and enter the salivary glands before they can be ejected out of the insect's body into plants with saliva secreted by the insect during feeding. The occurrence of the psyllid vectors only on citrus may limit the greening bacterium to citrus.

VECTOR MOVEMENTS

The transmission efficiency of any vector species is only one factor in determining the effectiveness of that species in the spread of pathogens. Of equal importance are the vectors feeding preferences, population sizes, and movements. Vector species vary enormously in their capacity or inclination to disperse. Examples of long distance migrants are the beet leafhopper vector of *S. citri* and the aster leafhopper vector of aster yellows phytoplasma.[11] Each of these leafhoppers can migrate many hundreds of kilometers during spring months to establish new vector populations that are more localized during summer. In contrast, the principal vectors of *X. fastidiosa* in California mainly disperse into vineyards that are only about 100 meters from their breeding habitats during the spring months that are critical for establishing chronic (permanent) infections that cause Pierces disease of grapes.[7,12]

CONCLUSION

The kinds of relationships that have evolved between insects, plant pathogenic bacteria, and plants are almost as diverse, but fortunately not as common, as insect relationships with nonpathogenic bacteria such as symbionts or gut-associated bacteria. All of the bacterial plant pathogens transmitted by obligate vectors persist in the vector, meaning that they can be transmitted for the lifetime of the vector (in the case of *X. fastidiosa*, for the length of a developmental stage). This increases the difficulty of using vector control to reduce the spread of these plant pathogens because infective vectors that survive long enough to disperse over long distances are still infective when they enter a crop field. In addition, the wide plant host ranges of many vector-borne bacterial pathogens means that sanitation by removing disease sources may have little effect on pathogen spread within a crop. As illustrated by two different pathogens of the grape yellows complex of phytoplasma diseases, what is effective for control of one disease may be ineffective for another. The specifics of pathogen, vector, and crop biology must be understood to devise effective control strategies on a case-by-case basis.

ARTICLE OF FURTHER INTEREST

Pierce's Disease and Others Caused by Xylella fastidiosa, p. 928

REFERENCES

1. Van Der Zwet, T. *Fire Blight—Its Nature, Prevention, and Control: A Practical Guide to Integrated Disease Management*; Cornell Univ. and U.S. Dept. of Agriculture Superintendent of Documents: Washington, DC, 1995.
2. Burgess, P.J.; Blakeman, J.P.; Perombelon, M.C.M. Contamination and subsequent multiplication of soft rot erwinias on healthy potato leaves and debris after haulm destruction. Plant Pathol. (Oxford) **1994**, *43* (2), 286–299.
3. Pepper, E.H. *Stewart's Bacterial Wilt of Corn*; Monograph no. 4, American Phytopathological Society: St. Paul, MN, 1967.
4. Douglas, A.E. Nutritional interactions in insect microbial symbioses: Aphids and their symbiotic bacteria *Buchnera*. Annu. Rev. Entomol. **1998**, *43*, 17–37.
5. Buchner, P. *Endosymbiosis of Animals with Plant Microorganisms*; John Wiley & Sons: New York, 1965.
6. Purcell, A.H. Homopteran Transmission of Xylem-Inhabiting Bacteria. In *Current Topics in Vector Research*; Harris, K.F., Ed.; Springer-Verlag: New York, 1989; 6, 243–266.
7. Purcell, A.H.; Hopkins, D.L. Fastidious xylem limited bacterial plant pathogens. Annu. Rev. Phytopathol. **1996**, *34*, 131–151.
8. Daire, X.; Clair, D.; Larrue, J.; Boudon-Padieu, E.; Alma, A.; Arzone, A.; Carraro, L.; Osler, R.; Refatti, E.; Granata, G.; Credi, R.; Tanne, E.; Pearson, R.; Caudwell, A. Occurrence of diverse MLOs in tissues of grapevine affected by grapevine yellows in different countries. Vitis **1993**, *32*, 247–248.
9. Cousin, M.-T.; Boudon-Padieu, E. Phytoplasmes et phytoplasmoses: Caracteristiques, symptomes et diagnostic. Cah. Agric. **2002**, *11* (2), 115–126.
10. Garnier, M.; Danel, N.; Bové, J.M. Aetiology of citrus greening disease. Ann. Microbiol. **1984**, *135A*, 169–179.
11. Carter, W. *Insects in Relation to Plant Disease,* 2nd Ed.; John Wiley & Sons: New York, 1973.
12. Andersen, P.C.; Brodbeck, B.V.; Mizell, R.F. Feeding by the leafhopper *Homaladisca coagulata* in relation to xylem fluid chemistry and tension. J. Insect Physiol. **1992**, *38*, 611–612.

Bacterial Survival and Dissemination in Natural Environments

Steven Lindow
University of California, Berkeley, California, U.S.A.

INTRODUCTION

A variety of different bacterial species attack all possible plant parts. The inoculum of such pathogens can survive in different ways, depending on the habitat in which infection occurs. Very often such inoculum comes from other plants, and can be removed from former hosts and arrive at new hosts in a variety of ways. The various survival methods exploited by plant pathogenic bacteria and the methods of inoculum dispersal will be explored, with an emphasis on those pathogens that attack above-ground plant parts.

SURVIVAL OF PLANT PATHOGENIC BACTERIA

Just as there are many different kinds of plant pathogenic bacteria attacking different plant parts, there are many different strategies for their survival during periods when either host plants are unavailable or environmental conditions required for infection are not conducive. Especially in the case of annual agricultural crop plants, pathogenic bacteria survive from one season to another on or within seed or propagative plant parts, such as tubers. Some pathogens also survive in or on living crop plants or weeds. For example, the fire blight pathogen *Erwinia amylovora* survives the winter in cankers in infected branches of pear and apple trees. Since survival in seed and planting material will be addressed in detail in another article, it will not be discussed here. Almost all bacterial plant pathogens can also survive in debris of infected plants,[1] but because they are generally very poor saprophytes, this phase is unimportant in either maintaining or increasing pathogen abundance.[1,2]

Many plant pathogens that attack foliar plant parts may exist in a ''resident phase'' on healthy shoots, buds, or roots.[1–3] Termed epiphytes, such bacteria can colonize both susceptible host plants as well as nonhost plants. The resident phase of a crop pathogen on a weed species can be of consequence in the epidemiology of crop plants nearby. The significance of such epiphytic inoculum was first described for *Pseudomonas syringae* pv. *syringae*, which colonizes hairy vetch plants that survive winters under snow cover in the northern United States and can subsequently be transferred to nearby bean crops in the spring.[4] The epiphytic phase of many foliar plant pathogens has since been described on nonhost plants.[1–6]

Because plant disease occurs only when suitable environmental conditions coincide with sufficient pathogen populations on a susceptible plant, the epiphytic phase is important in allowing the persistence of the pathogen on host plants during periods that are not conducive to infection. For example, inoculum of *P. syringae* pv. *syringae* on bean plants emerging from infested seed develops epiphytic populations on leaves.[3,6] The epiphytic population can be redistributed across the susceptible plant parts, and more importantly, can persist for extended periods of time in the absence of disease. The population sizes of epiphytic bacteria can be very dynamic, with death of the epiphytes commonly occurring during periods of dry weather.[5–7] The survival of epiphytes on leaves may be facilitated by the occurrence of at least some of the epiphytic cells in ''protected locations'' within leaves.[5] Plant pathogens, unlike nonpathogens, may thus have some ability to escape stresses on plants. Such adaptations might, however, apply only to their colonization of susceptible plants, and may account for the fact that populations of plant pathogens are usually higher on host plants than on nonhost plants.[5]

Additionally, physical stresses on the leaf may reduce the culturablity of epiphytes without actually reducing their viability. Cells that enter a viable but nonculturable (VBNC) state are indistinguishable from dead cells by standard plate-count methods of estimating bacterial populations, but could still be available as a source of inoculum should conditions in which the cells are ''revived'' be encountered.[8] The conditions, if any, under which epiphytes may become VBNC must be understood to better assess epiphytic populations as a means of cell survival on plants.

DISSEMINATION OF PLANT PATHOGENS FROM PLANTS

Bacterial cells can be removed from plants in at least two different ways. The most attention has been placed on rain as a vector to remove cells from leaves. Rain can wash off a substantial proportion (up to 50%) of bacteria from

Encyclopedia of Plant and Crop Science
DOI: 10.1081/E-EPCS 120010566

leaves.[9] However, a large variation in bacterial population sizes (although a little smaller overall) will remain among these leaves.[6] In addition, the reductions in bacterial population that occur during some rain events[4] are sometimes small compared to the magnitude of increases that occur following rain.[6] The relatively small reductions in bacterial populations during rain may be due in part to their having become attached to the leaf surface.[5] In addition, epiphytes that are dislodged during rain may reattach to other leaves before they are lost from the plant canopy. Some of the dislodged bacteria may instead become incorporated into small droplets of water that can be dispersed away from the original plant.[10] Adjacent plants can recapture some of these aerosol droplets by impaction and sedimentation, thereby minimizing the loss of bacteria from plant canopies during rain.[9–11]

Substantial numbers of epiphytic bacteria also leave the surface of plants in dry aerosol particles. A net upward flux of bacteria can occur above plant canopies, especially during mid-day when plants are dry and winds are at their maximum velocity.[9,11] In fact, the upward flux of bacteria from dry plants is much greater than that from either wet plants or from bare soil.[9,11] A net daily introduction to the atmosphere of about 5×10^6 particles bearing viable bacteria per square meter of plant canopy having large bacterial population sizes (approximately 10^6 cells/cm^2 of leaf) has been observed.[11] This would represent an emigration of only about 0.0001% of the total epiphytic population each day.[2] Only if a very high fraction of the emigrant cells lost viability or entered a VBNC state soon after removal from a plant canopy (and hence would not be counted) could the number of total emigrant cells approach even a few percent of the total bacterial population. Thus, emigration away from a plant appears to contribute little to the reduction of epiphytic bacterial populations on a plant. Other modes of removing of bacteria from plant surfaces, such as by insect vectoring or the physical abrasion of adjacent leaves, are probably inconsequential, but have not been quantified.

CAPTURE OF BACTERIA BY PLANTS

Immigration of bacteria to a leaf is coupled strongly with their emigration from another leaf, since most dispersed bacteria have foliar plant parts as their primary habitat. Immigration to a leaf may occur via several modes of transportation.

1. Many bacteria can be transported to a leaf via rain splash. Although rain deposits a large percentage of the bacteria released from plants onto the soil,[4,9,11] substantial lateral movement of bacteria can occur during rain.[10] In general, however, more bacteria are removed from a plant with an established epiphytic microflora than are deposited from adjacent plants by rain.[6,9]
2. A number of phytopathogenic bacteria can be transferred from infected plants to healthy plants by insect vectors. There are some bacterial pathogens that are disseminated due to intimate associations with an insect vector,[10,12] but most bacteria are transmitted via insects that are contaminated during their foraging or nectar collecting activities.[12] This latter phenomenon is most well studied in the case of the vectoring of *E. amylovora* from cankers or infected flowers of pear and apple trees to newly opened flowers where infection can occur.[12] Although such transmission has been demonstrated, the number of cells that are transferred is unknown, but probably very small.
3. Plant pathogenic bacteria might be disseminated with infected leaves that become airborne. Since bacteria can survive for long periods of time on dead and/or dry infected leaves,[1,7] it is likely that as leaf fragments are dispersed in the wind, cells of phytopathogens could be transferred to healthy leaves.[10] Again, the prevalence of such a phenomenon and the number of bacteria that might potentially be transferred to new leaves by this process are unknown.
4. Many bacterial plant pathogens produce exopolysaccharides (slime). This slime can be quite substantial in the case of certain pathogens such as *E. amylovora*, whose slimes can produce strands that are as much as 10 cm long in infected tissues. It has been speculated that these strands could fragment as they dry and disperse as small particles to other plants.[2]
5. The deposition of dry aerosol particles has been rather well studied. On average, about 10^3 particles containing viable bacteria were deposited in an area the size of a bean leaf (approximately 100 cm^2) each day.[2,6,9,11] Although many aerosol droplets are scrubbed from the air by other raindrops, some can disperse beyond the immediate site of release and potentially can be deposited onto other plants.[2,6,9,11] The contribution of immigrant wet aerosol particles to the size of bacterial populations has not been well studied, but is probably less than that of dry particles.

CONCLUSION

Since the survival and dispersal of bacterial plant pathogens often involve relatively small numbers of cells, studying these processes has been hampered over the years due to inadequate methods to quantify and differentiate the cells. New molecular tools can detect very small numbers of cells, as well as cells that are not culturable, and should prove very useful in epidemiological studies of plant pathogens. As in the case of human bacterial pathogens that commonly exist in a VBNC state but are still infectious, there may be many plant pathogens that survive undetected in an unculturable state. It will be important to examine this possibility with the new tools available. Although the epiphytic phase has become well accepted as a reservoir for plant pathogens, their endophytic phase in host and nonhost plants is largely unexplored and may also constitute a sizable reservoir of inoculum.

ARTICLES OF FURTHER INTEREST

Bacterial Attachment to Leaves, p. 75
Bacterial Blight of Rice, p. 79
Bacterial Pathogens: Detection and Identification Methods, p. 84
Bacterial Pathogens: Early Interactions with Host Plants, p. 89
Bacterial Survival and Dissemination in Insects, p. 105
Bacterial Survival and Dissemination in Seeds and Planting Material, p. 111
Bacterial Survival Strategies, p. 115
Fire Blight, p. 443
Leaf Cuticle, p. 635
Leaf Structure, p. 638
Leaves and Canopies: Physical Environment, p. 646
Plant Diseases Caused by Bacteria, p. 947

REFERENCES

1. Leben, C. How plant-pathogenic bacteria survive. Plant Dis. **1981**, *65* (8), 633–637.
2. Lindow, S.E. Role of Immigration and Other Processes in Determining Epiphytic Bacterial Populations: Implications for Disease Management. In *Aerial Plant Surface Microbiology*; Morris, C.E., Nicot, P.C., Nguyen-the, C., Eds.; Plenum Publishing Co.: New York, 1996; 155–168.
3. Hirano, S.S.; Upper, C.D. Bacteria in the leaf ecosystem with emphasis on *Pseudomonas syringae*: A pathogen, ice nucleus, and epiphyte. Microbiol. Mol. Biol. Rev. **2000**, *64* (3), 624–653.
4. Ercolani, G.L.; Hagedorn, D.J.; Kelman, A.; Rand, R.E. Epiphytic survival of *Pseudomonas syringae* on hairy vetch in relation to epidemiology of bacterial brown spot of bean in Wisconsin. Phytopathology **1974**, *64*, 1330–1339.
5. Beattie, G.A.; Lindow, S.E. The secret life of bacterial colonists of leaf surfaces. Annu. Rev. Phytopathol. **1995**, *33*, 145–172.
6. Hirano, S.S.; Rouse, D.I.; Clayton, M.K.; Upper, C.D. *Pseudomonas syringae* pv. *syringae* and bacterial brown spot of bean: A study of epiphytic phytopathogenic bacteria and associated disease. Plant Dis. **1995**, *79* (11), 1085–1093.
7. Henis, Y.; Bashan, Y. Epiphytic Survival of Bacterial Leaf Pathogens. In *Microbiology of the Phyllosphere*; Fokkema, N.J., van den Heuvel, J., Eds.; Cambridge University Press: New York, 1986; 252–268.
8. Wilson, M.; Lindow, S.E. Viable but Nonculturable Cells in Plant-Associated Bacterial Populations. In *Nonculturable Microorganisms in the Environment*; Colwell, R.R., Grimes, D.J., Eds.; ASM Press: Washington, DC, 1999; 229–241.
9. Upper, C.D.; Hirano, S.S. Aerial Dispersal of Bacteria. In *Assessing Ecological Risks of Biotechnology*; Ginzburg, L.R., Ed.; Butterworth-Heinemann: Stoneham, MA, 1991; 75–94.
10. Venette, J.R. How Bacteria Find Their Hosts. In *Phytopathogenic Prokaryotes*; Mount, M.S., Lacy, G.H., Eds.; Academic Press: New York, 1982; Vol. 2, 75–94.
11. Lindemann, J.; Upper, C.D. Aerial dispersal of epiphytic bacteria over bean plants. Appl. Environ. Microbiol. **1985**, *50* (5), 1229–1232.
12. Harrison, M.D.; Brewer, J.W.; Merrill, L.D. Insect Transmission of Bacterial Plant Pathogens. In *Vectors of Plant Pathogens*; Harris, K.F., Maramorosch, K., Eds.; Academic Press: New York, 1980; 201–292.

Bacterial Survival and Dissemination in Seeds and Planting Material

B

Diane A. Cuppels
Southern Crop Protection and Food Research Centre, London, Canada

INTRODUCTION

Many phytopathogenic bacteria can survive in or on the seed or other propagules of their host plant. Contaminated or infected planting material provides a major inoculum source for many of the economically important diseases of plants caused by bacteria. Often there are no visible symptoms or signs of infection on the seed. Provided that the association is not completely superficial, most bacteria survive as long as their host seeds. Seed and vegetatively propagated planting material frequently are produced at locations distant from their intended market; thus a pathogen may be introduced to regions where it had not previously existed. Because of this practice, we now have a worldwide distribution of diseases such as fire blight of apple and pear, bacterial wilt of potato and tomato, and bacterial leaf blight of rice. Also, diseases such as bacterial spot of tomato, whose causal agent does not overwinter in northern climates, may be reintroduced each year on seed or transplants that were produced in warmer regions. Preventing the spread of disease through seed and other propagating material requires the institution of quarantines and routine inspections, the use of disease-free production areas, and the eradication or reduction of seedborne inoculum.

BACTERIAL INVASION AND SURVIVAL ON PROPAGATIVE MATERIAL

Seed may carry phytopathogenic bacteria internally or externally. If the bacteria establish themselves within the seed, then it is considered an infection; if they are found adhering to the seed surface or mixed with seed, then it is considered an infestation or contamination. Bacteria usually are located in the seed coat, and are rarely seen in the generative tissues. Successful establishment of the pathogen in or on seed depends upon a number of factors, including the host and pathogen genotypes and environmental conditions. Invasion may occur by a number of different routes and a pathogen may not be limited to just one mode of entry.[1] Only a few bacterial pathogens actually are known to be transmitted to the developing seed through the tissues of the mother plant. Generally, these bacteria are vascular pathogens such as the cabbage black rot pathogen *Xanthomonas campestris* pv. *campestris*. Systemic infection of seed occurs through flower or fruit stalks (pedicels) or through ovule (seed) stalks (funiculi). If the black rot pathogen is present at flowering, then it may enter the xylem of the pedicels and travel through the suture vein of pods to the funiculi. The tomato pathogen *Clavibacter michiganensis* subsp. *michiganensis* invades the placenta via the calyx and then progresses to the funiculi. Some bacteria gain entry through lesions on fruit or pods (a dry and dehiscent form of fruit). *Pseudomonas syringae* pv. *phaseolicola* penetrates the pod wall of bean directly, whereas *Xanthomonas axonopodis* pv. *phaseoli* enters via the dorsal suture of pea pods. The hilum—which is the scar produced when the seed detaches from the funiculus—is a natural entry point for bacteria. Bacteria such as *X. axonopodis* pv. *phaseoli* also can penetrate seed through the micropyle (an integumentary opening of the ovule). Bacteria also may reside on the seed surface. Seed coats may be covered with cracks that provide an ideal sanctuary for bacteria, as do the substomatal chambers of the seed coat. Infected plant parts or small clumps of infested soil may become mixed with the seed during the harvesting, extraction, or cleaning of seed. Bacteria can be found loosely associated with individual seeds, bonded to the seed coat or lodged among the seed hairs. The rough and textured surface characteristic of some seed coats facilitates adherence by bacteria (Fig. 1). Although invaded seed typically do not show signs of infection, symptoms do occasionally develop and include watersoaking around the hilum, yellow or brown discolorations, wrinkling, and buttery deposits.

As mentioned, most phytopathogenic bacteria will remain viable as long as their host seed.[2] The seed moisture content typically is low; thus the bacteria are in a state of dormancy. The bacterial speck pathogen *Pseudomonas syringae* pv. *tomato* has been recovered from twenty-year-old tomato seed, whereas *X. axonopodis* pv. *phaseoli* can survive on bean seed for at least 15 years. However, studies with the bean pathogens *P. syringae* pv. *phaseolicola* and *X. axonopodis* pv. *phaseoli* showed that bacterial populations will decline as their time in storage

Encyclopedia of Plant and Crop Science
DOI: 10.1081/E-EPCS 120010568

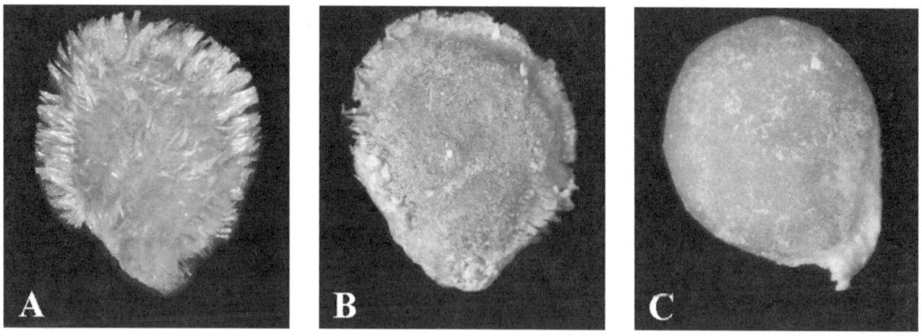

Fig. 1 The effect of seed extraction and treatment on the seed coat surface of tomato seeds. (A) Fuzzy, untreated seed; (B) hydrochloric acid-treated seed; and (C) chlorine-treated seed. (*View this art in color at www.dekker.com.*)

increases. If the moisture content and temperature of the host seed are lowered, bacterial longevity will increase. Bacteria that are more deeply situated in seed tissue are better protected than those on the surface and thus have a better chance of survival. Many plant-associated bacteria produce exopolysaccharides that may shield them from desiccation and other harsh environmental conditions. The persistence of phytopathogenic bacteria on seed also is strongly influenced by the presence or absence of other microorganisms.

Phytopathogenic bacteria can be carried on vegetatively propagated planting material such as cuttings, grafts, transplants, tubers, rhizomes, corms, and bulbs. Frequently, these infections are latent with the host showing no visible signs of disease. The bacterial population associated with a latent infection is usually small and thus can be difficult to detect. The situation is particularly serious in nurseries of perennial crops such as grapevines and trees. If the nursery soil becomes infested with the crown gall pathogen *Agrobacterium tumefaciens* (see the article titled ''Crown Gall'' by T. Burr), it will very quickly spread among the young plants, which will often show no signs of disease until after they have been planted in destination vineyards or orchards. Likewise, symptomless greenhouse-grown tomato transplants often can be an inoculum source for diseases such as bacterial speck and bacterial canker on field tomatoes.[3]

DISEASE TRANSMISSION THROUGH PROPAGATIVE MATERIAL

Although many phytopathogenic bacteria can infect or infest seed, systemic infection of young seedlings through infected seed does not occur frequently; nonsystemic transmission resulting in postemergence symptoms is more common.[1] Again, vascular pathogens such as *X. campestris* pv. *campestris* are most likely to be transmitted systemically. Many host and environmental factors influence pathogen movement from seed. High humidity and free moisture are required for primary cotyledon infection. Symptoms include leaf and stem spots, wilting, root rots, and blights. Initially, diseases caused by seedborne pathogens may develop slowly. However, once established, pathogens can spread very quickly, particularly if the seedlings are grown in high density in transplant greenhouses or nurseries (Fig. 2). The clipping practices sometimes used to prepare a uniform and vigorous shipment of transplants will enhance pathogen transmission throughout the nursery. One contaminated seed in 10,000 may be enough to cause an epidemic and economic loss, depending upon the pathogen, host, and environmental conditions.

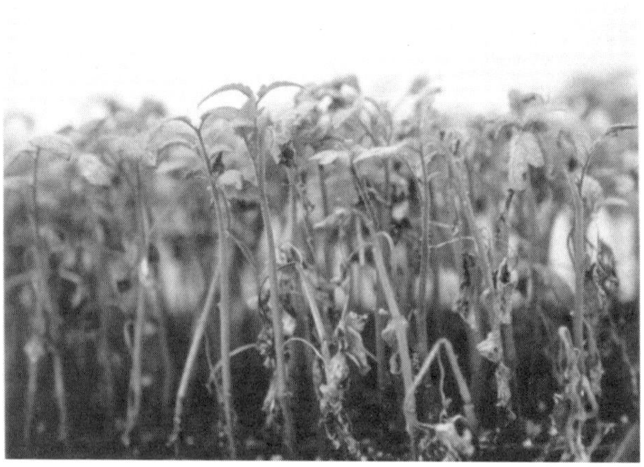

Fig. 2 Six-week-old tomato seedlings heavily infected with the bacterial spot pathogen *Xanthomonas vesicatoria*. These seedlings were grown in high density (288-well plug trays) in a tomato and pepper transplant greenhouse. (*View this art in color at www.dekker.com.*)

PREVENTION OF BACTERIAL DISSEMINATION ON PROPAGATIVE MATERIAL

Bacterial diseases carried on seed or vegetatively-propagated planting material are best controlled by an integrated approach, employing exclusion through quarantine measures, disease-free areas for production, resistant cultivars (if available), treatment to reduce bacterial contamination and routine seed health testing or certification. Most countries have plant quarantine laws and regulations based on the International Plant Protection Convention of the Food and Agriculture Organization (FAO) of the United Nations. The European and Mediterranean Plant Protection Organization (EPPO) has published two quarantine lists: A1 gives the quarantine pests not present in the area (includes 13 bacteria) and A2 records the quarantine pests present but officially controlled and not widely distributed (includes 21 bacteria).[4] A list of regulated plant pests of concern in the United States is maintained by the U.S. Department of Agriculture Animal and Plant Health Inspection Service.[5] In addition, the North American Plant Protection Organization (NAPPO) operates the Phytosanitary Alert System,[6] which provides up to date information on pests of importance in North America.

Commonly used seed treatments to reduce or eliminate pathogenic bacteria include thermotherapy (hot water, dry heat, or aerated steam), acid dips, antibiotic soaks, and chlorination.[2] Although there are several biological control based products that target bacterial pathogens, most are not intended for seed application. The antibiotic most commonly used on seed is streptomycin (see the article titled "Chemical Methods" by D. Hopkins); unfortunately, it may have a phytotoxic effect depending upon the dosage and duration of treatment and the sensitivity of the host. Phytotoxicity can be a problem with other treatments as well; the dose that is effective at eradicating the pathogen may also reduce seed viability. Many of the sites on the seed-coat surface where bacteria might adhere are eliminated by acid or chlorine treatment (Fig. 1). For transplants or other vegetatively-propagated planting material, copper-based bactericide sprays are often used to reduce pathogen populations. Soil in tree nurseries may be disinfested using steam or solarization. Several biopesticides employing *Agrobacterium radiobacter* strains antagonistic to the crown gall pathogen are now commercially available for ornamental, fruit, and nut stock protection.[7]

For seed health tests to be an effective tool in certification and quarantine programs, they must reliably detect and estimate the pathogen population present on seed.[8] The inoculum threshold for a pathogen, defined as the amount of seed infection/infestation that causes disease under field conditions and results in economic loss, is based upon such a test; unfortunately, inoculum thresholds have not been adequately established for many of the bacterial pathogens. Determining inoculum thresholds is a difficult process that is affected by numerous environmental and cultural practices. Because they can spread very quickly through a production field, bacteria usually have low inoculum thresholds.[9] For crops started from transplants grown in high density, seed often is given a zero-tolerance level for bacterial contamination. One uncomplicated but relatively insensitive assay for bacteria on seed is the growing-on test, in which a large sample of seed is planted and the resulting seedlings are scored for the presence of disease. Other simple detection methods include direct planting on semi-selective media and host plant inoculation.[2] Serological and nucleic acid-based methods have been developed for the identification of a number of bacterial plant pathogens.[10] Although more expensive, these methods offer greater sensitivity and a shorter response time than conventional assays.

CONCLUSION

Although the International Seed Testing Association (ISTA) provides a standardized set of seed health testing methods,[11] they are not universally accepted by the various agencies and laboratories responsible for screening seed. The situation is further complicated by the emergence of several new molecular detection methods. These new methods have great potential but they must first be evaluated and validated before they can be incorporated into any standardized set of recommendations. Such tests should be helpful not only in assessing the emerging new seed treatment technologies but also in determining inoculum thresholds and realistic tolerance levels for seed-borne diseases—one of the greatest challenges facing the seed industry today.

ACKNOWLEDGMENTS

For the Department of Agriculture and Agri-Food, Government of Canada, © Minister of Public Works and Government Services Canada, 2002.

ARTICLES OF FURTHER INTEREST

Bacterial Pathogens: Detection and Identification Methods, p. 84

Bacterial Survival Strategies, p. 115
Crown Gall, p. 379
Management of Bacterial Diseases of Plants: Regulatory Aspects, p. 669

REFERENCES

1. Agarwal, V.K.; Sinclair, J.B. *Principles of Seed Pathology*, 2nd Ed.; CRC Press: Boca Raton, FL, 1996.
2. Maude, R.B. *Seedborne Diseases and their Control*; CAB International: Wallingford, United Kingdom, 1996.
3. Cuppels, D.A.; Elmhirst, J. Disease development and changes in the natural *Pseudomonas syringae* pv. *tomato* populations on field tomato plants. Plant Dis. **1999**, *83*, 759–764.
4. http://www.eppo.org/QUARANTINE/quarantine.html (accessed May 2002).
5. http://www.invasivespecies.org (accessed May 2002).
6. http://www.pestalert.org/introduction.html (accessed May 2002).
7. http://www.apsnet.org/online/feature/biocontrol/ (accessed May 2002).
8. Reeves, J.C. Nucleic Acid Techniques in Testing for Seedborne Diseases. In *New Diagnostics in Crop Sciences*; Skerritt, J.H., Appels, R., Eds.; CAB International: Oxford, United Kingdom, 1995; 127–151.
9. Schaad, N. Inoculum thresholds of seedborne pathogens: Bacteria. Phytopathology **1988**, *78*, 872–875.
10. DeBoer, S.; Cuppels, D.A.; Gitaitis, R.D. Detecting Latent Bacterial Infections. In *Advances in Botanical Research*; DeBoer, S., Andrews, J., Tommerup, I.C., Eds.; Academic Press: London, 1996; Vol. 23, 27–57.
11. http://www.seedhealth.org/ (accessed May 2002).

Bacterial Survival Strategies

Cindy E. Morris
Christine M. H. Riffaud
Institut National de la Recherche Agronomique, Montfavet, France

INTRODUCTION

Bacteria in the phyllosphere have significant roles in plant health and general plant biology, in the hygienic and market qualities of food products derived from plants, in global climatological processes, and in the recycling of elements. The phyllosphere is a heterogeneous environment that can engender stress due to UV radiation, temperature fluctuation, dehydration, osmotic and pH conditions, oxidizing agents and other antimicrobial compounds, starvation, predation, and parasites. Examples are given of the two types of strategies—stress avoidance and resistance to unavoidable stress—exploited by phyllosphere bacteria to survive in this habitat.

IMPORTANCE OF THE PHYLLOSPHERE AS A HABITAT FOR BACTERIA

Plants inhabit over 90% of the 1.5×10^8 km^2 of terrestrial surface of the Earth, providing over 10^8 km^2 of potential habitat for microorganisms in the form of leaf surfaces.[1] Bacteria are generally the most abundant of the microbes associated with leaves. In temperate climates, bacteria attain population densities of at least 10^4 cells cm^{-2}. It has been estimated that the global population size of bacteria in the phyllosphere is roughly 10^{24} to 10^{26}.[1] Many of the bacteria in the phyllosphere have well-described roles in environmental processes or in the biology of the host plant. These include plant pathogens; animal pathogens; ice nucleation-active bacteria; bacteria that produce active phytohormones, allergens, toxins, and aromatic compounds; and antagonists of plant pathogens. Phyllosphere bacteria probably also participate in global cycling of carbon and nitrogen and in the fossilization of plant material. Hence, leaves harbor bacteria that have significant roles in plant health and general plant biology, in the hygienic and market qualities of food products derived from plants, in global climatological processes, and in the recycling of elements. The importance and diversity of the roles of bacteria associated with leaves has prompted interest in understanding how bacteria persist and proliferate in this habitat.

STRATEGIES FOR ADAPTING TO STRESS

In light of the physical, chemical, and biological nature of the phyllosphere habitat,[2] bacteria are likely to experience stresses due to UV radiation, temperature, dehydration, osmotic and pH conditions, oxidizing agents and other antimicrobial compounds, starvation, predation, and parasites. These stresses are accentuated by the ever-changing nature of the phyllosphere as leaves senesce, die, and fall to the ground thereby delivering the residing bacteria to a markedly different habitat. Furthermore, because leaves are generally in open environments, phyllosphere bacteria can be taken up by insects, become airborne or waterborne, be transported to inert surfaces, etc. The diversity of situations in which phyllosphere bacteria may be found suggests that they are capable of surviving an even broader range of conditions than those encountered in the phyllosphere.

For bacteria in general, much is known about the physiological alterations induced by the different sources of stress indicated above. Oxidizing agents, for example, damage membranes. Other antimicrobial compounds may cause denaturation of proteins. Temperature extremes can cause membranes to move out of the liquid crystalline phase essential for them to function correctly. UV radiation leads to DNA damage. Dehydration also can lead to DNA damage, protein denaturation, and an increase in the melting point of cell membranes, causing transition to the gel phase. However, bacteria in the phyllosphere as well as in other environments experience multiple forms of stress simultaneously. For example, drying of water on leaf surfaces would be accompanied by changes in osmotic conditions, pH, and availability of nutrient sources as solubilized nutrients become more and more concentrated and eventually crystallize. Furthermore, bacteria can turn on a common regulatory network in response to many different environmental stresses. Hence, in summarizing the mechanisms exploited by bacteria to survive stress, it is difficult to distinguish mechanisms that are specific to a given stress.

Mechanisms for survival exploited by bacteria in general can be grouped into categories describing the processes by which these mechanisms are expressed as listed in Table 1. These processes reflect two overall

Encyclopedia of Plant and Crop Science
DOI: 10.1081/E-EPCS 120010611

Table 1 Bacterial survival strategies

Type of stress response	Examples for bacteria in general	Examples for phyllosphere bacteria
Stress avoidance		
Environmental modification by bacteria.	Production of wetting agents, of mucus, and of enzymes and toxins causing tissue leakage or degradation of the colonized substrate leading to increased availability or improved sequestering of water and nutrients; induction of root nodules by rhizobacteria; induction of opine synthesis by genetic transformation of plant hosts by *Agrobacterium* spp.; production of exopolymeric-enrobed biofilms.	Production of biosurfactants, exopolysaccharides, pectolytic enzymes, toxins, phytohormones, and biofilms by a wide range of phyllosphere bacteria.
Dispersal into heterogeneous niches.	Coexistence of planktonic and attached components of bacterial populations in aquatic and other liquid-saturated systems.	Dissemination of bacteria by air, water, and insects; colonization of seeds; colonization of a range of leaf features (trichomes, grooves over veins, substomatal cavities, etc.); formation of biofilms.
Coexistence with other microbial species having active stress-resistance mechanisms leading to protection of coinhabitants.	Coexistence with bacteria producing extracellular enzymes capable of degrading or inactivating antibiotics or liberating nutrient sources.	Not yet described for phyllosphere bacteria.
Responses to unavoidable stress		
Constitutive factors.	Resistance to certain antimicrobial compounds and to certain parameters of the physical-chemical environment (temperature, salt, etc.).	Expression of the *uvrB* gene, involved in UV resistance in *Xanthomonas campestris.*
Factors induced by physical–chemical environmental conditions.	Enhanced resistance to acids and active oxygen species in response to low doses of these compounds; production of small cells with reduced metabolic activity in response to starvation; generalized stress resistance induced during starvation; changes in lipid content of membranes during nutrient stress leading to reduced permeability to antimicrobials; changes in cell hydrophobicity during nutritional stress leading to changes in capacity to adhere to surfaces; viable-but-not-culturable states; formation of biofilms.	Resistance to certain heavy metals that are constituents of pesticides (Cu^{2+}, etc.); osmoadaptation in *Erwinia carotovora*; production of siderophores (an iron chelator) induced under conditions of iron limitation in *Pseudomonas* spp. and *Erwinia* spp.; induction of a viable-but-not-culturable state.
Population density–dependent factors.	Factors triggered by the process of quorum sensing.	Antibiosis against congenerics induced at high population densities in *Erwinia carotovora.*
Life cycle–dependent processes.	Formation of spores; production of dense cell masses and microcysts during the life cycle of fruiting myxobacteria; production of flagellated swarming cells during the life cycle of sheathed bacteria; expression of metabolic processes specifically during stationary phase.	Formation of spores by *Bacillus* spp.
Rapid generation of genetic diversity.	Phase variation; appearance of hypermutators; horizontal gene transfer.	Development of antibiotic-resistant populations (*Erwinia* spp., *Pseudomonas* spp., *Xanthomonas* spp.) due to horizontal transfer of antibiotic resistance genes associated with conjugative plasmids and transposable elements.

strategies: 1) stress avoidance and 2) resistance of unavoidable stresses. Stress avoidance includes the following: 1) modification of environmental conditions by bacteria to ensure availability of water and nutrients; 2) dispersal of cells (passive or active) into a wide range of environmental conditions; and 3) coexistence with other microorganisms who ensure protection of coinhabitants by producing extracellular enzymes that inactivate antibiotics or liberate nutrient sources, for example. The following are responses to unavoidable stress: 1) constitutive expression of genes involved in resistance to antimicrobial compounds, to salt, to extremes of temperature, etc.; 2) induction of a wide range of traits by conditions of the physical or chemical environment; 3) expression of traits triggered by the population density–dependent process of quorum sensing; 4) development of cell types and cell states characteristic of specific phases of the life cycle (spores, swarming cells, etc.); and 5) rapid generation of genetic diversity. These processes are not exclusive. For example, environmental conditions may be involved in induction of different life-cycle phases such as spore production or swarming. Furthermore, population density may foster the environmental conditions leading to induction of these life-cycle phases via depletion of nutrients or accumulation of metabolites. Ultimately, the specific molecular signals triggering the processes listed above may be very limited in number and may explain why certain regulatory pathways are common to a wide range of stress responses.[3]

Phyllosphere bacteria express many of the processes listed in Table 1. In some cases, the contribution of these processes to enhanced survival in the face of stress has been demonstrated, and in other cases data are not yet available to support their importance in survival. The production of toxins, phytohormones, and pectolytic enzymes by phyllosphere bacteria contributes to modification of local environmental conditions by causing the release of nutrients for bacterial growth.[4] Bacterial-produced biosurfactants can enhance availability of free water on leaf surfaces. Numerous phyllosphere bacteria produce exopolysaccharides that protect them from desiccation. These exopolysaccharides might also contribute to the formation of microbial biofilms, ubiquitous on leaf surfaces.[5] Biofilms, in general, foster the establishment of environmental conditions protecting bacteria from the harshness of the external environment. Although this may also be true for biofilms on leaf surfaces, it is not yet clear which, if any, phyllosphere bacteria gain significant survival advantages by residing in biofilms. Biofilms are also one of the diverse niches inhabited by phyllosphere bacteria. Other niches include the varied features of the leaf surface such as trichomes, grooves over leaf veins, hydathodes, and substomatal cavities. Viable phyllosphere bacteria have also been found in irrigation waters, in lakes and rivers, and in air and have been associated with debris in soil. Widespread dissemination into a diversity of niches may lead bacteria to new, favorable sites for colonization, thereby maximizing survival for the population as a whole. It has been suggested that the ice nuclei produced by *Pseudomonas* spp., *Xanthomonas* spp., and *Erwinia* spp. play this very role. In the form of aerosols, these bacteria are transported from leaf surfaces into the stratosphere. By initiating ice formation and subsequent rainfall while in the stratosphere, ice nucleation–active bacteria ensure dissemination to new plants.[6]

Few of the bacterial traits associated with survival on leaf surfaces are clearly expressed constitutively. Expression of the *uvrB* gene, involved in UV resistance in *Xanthomonas campestris*, is one example.[7] On the other hand, the conditions of the physical-chemical environment can induce expression of a wide range of traits. Osmoadaptation, the production of iron scavengers (siderophores) and resistance to certain heavy metals that are constituents of pesticides can be induced by conditions of the chemical environment and are probably important in survival on leaf surfaces. Production of small cells manifesting generalized stress resistance is a common response of bacteria to oligotrophic conditions. Reduction in cell size has been observed for bacteria growing on leaf surfaces, but cell size per se is not clearly related to enhanced forms of resistance in these bacteria. Temperature influences the expression of numerous genes of certain plant pathogenic bacteria.[8] To date, the genes described are involved in pathogenicity to plants. Although the expression of these genes confer enhanced fitness, the significance of temperature induction is not clear. In other bacteria, adaptation to cold or heat shock involves modification of membrane fluidity, nucleic acid conformation, protein flexibility, and repair of misfolded proteins.[3] The temperature adaptations of plant pathogenic bacteria described above may be involved in reducing the cost of pathogenicity-related processes for the bacterial population.

Population density–dependent responses to stress are generally mediated by quorum sensing, enabling bacteria to mount a unified response that is advantageous to the population as a whole in the face of stress. Quorum-sensing operons appear to be typically found in bacterial species from fluctuating, heterogeneous environments and absent in bacteria from stable environments.[9] Numerous species of phyllosphere bacteria produce pheromones that, upon accumulation, induce expression of, for example, pectolytic activity or production of bacteriocins active against congenerics. Dense cell masses such as those found in biofilms or in substomatal cavities would be ideal sites for accumulation of bacterial pheromones. However, expression of quorum-sensing operons on leaves has not yet been demonstrated.

Little is known about the role of life cycle–dependent processes in the survival of bacteria on leaf surfaces and in particular those that occur during stationary phase. Another important survival strategy for a broad spectrum of bacteria involves increasing the genetic diversity of populations. One form of this strategy, referred to as phase variation, results in reversible, high-frequency genetic variation of specific contingency loci. Another common process of generating genetic diversity involves hypermutation. Among phyllosphere bacteria, hypermutation has not been described. Phase variation, observed in terms of changes in lipopolysaccharides or in pathogenicity to plants, is well-known for certain phyllosphere species, but its contribution to survival has not been elucidated. On the other hand, phyllosphere bacteria are known to benefit from gene acquisition via horizontal transfer. The development of antibiotic resistant populations of *Erwinia* spp., *Pseudomonas* spp., and *Xanthomonas* spp. occurs due to transfer of antibiotic resistance genes associated with conjugative plasmids and transposable elements.

CONCLUSION

Bacteria associated with leaves exploit numerous processes to survive the environmental stresses experienced in this habitat. The diversity of these processes reflects the complex ecology of bacteria associated with living tissue in open, heterogeneous environments. However, it is likely that these processes are aimed at a few primordial goals: protecting membranes, protein, and DNA from irreversible damage and preventing cells from being killed by predators and phages. The next frontier in the study of survival strategies of phyllosphere bacteria will be the search for the molecular mechanisms involved in protecting the vital functions of bacterial cells common to these different processes.

ARTICLES OF FURTHER INTEREST

Bacteria Survival and Dissemination in Natural Environments, p. 108

Bacterial Survival and Dissemination in Seeds and Planting Material, p. 111

Leaf Cuticle, p. 635

Leaf Surface Sugars, p. 642

Trichomes, p. 1254

UV Radiation Effects on Phyllosphere Microbes, p. 1258

REFERENCES

1. Morris, C.E.; Kinkel, L.L. Fifty Years of Phyllosphere Microbiology: Significant Contributions to Research in Related Fields. In *Phyllosphere Microbiology*; Lindow, S.E., Poinar, E., Elliot, V., Eds.; APS Press: Minneapolis, 2002; 353–363.
2. Morris, C.E. Phyllosphere. In *Encyclopedia of Life Sciences*; Nature Publishing Group: London, 2001. http://www.els.net (accessed October 2001).
3. Ramos, J.L.; Gallegos, M.-T.; Marqués, S.; Ramos-González, M.-I.; Espinosa-Urgel, M.; Segura, A. Responses of Gram-negative bacteria to certain environmental stressors. Curr. Opin. Microbiol. **2001**, *4* (2), 166–171.
4. Beattie, G.A.; Lindow, S.E. Bacterial colonization of leaves: A spectrum of strategies. Phytopathology **1999**, *89* (5), 353–359.
5. Morris, C.E.; Barnes, M.B.; McLean, R.J.C. Biofilms on Leaf Surfaces: Implications for the Biology, Ecology and Management of Populations of Epiphytic Bacteria. In *Phyllosphere Microbiology*; Lindow, S.E., Poinar, E., Elliot, V., Eds.; APS Press: Minneapolis, 2002; 138–154.
6. Wolber, P.K. Bacterial ice nucleation. Adv. Microb. Physiol. **1993**, *32*, 203–237.
7. Lee, T.C.; Lee, M.C.; Hung, C.H.; Weng, S.F.; Tseng, Y.H. Sequence, transcriptional analysis and chromosomal location of the *Xanthomonas campestris* pv. *campestris uvrB* gene. J. Mol. Microbiol. Biotechnol. **2001**, *3* (4), 519–528.
8. Ullrich, M.S.; Schergaut, M.; Boch, J.; Ullrich, B. Temperature-responsive genetic loci in the plant pathogen *Pseudomonas syringae* pv. *glycinea*. Microbiology **2000**, *146* (10), 2457–2468.
9. Swift, S.; Downie, J.A.; Whitehead, N.A.; Barnard, A.M.; Salamond, G.P.; Williams, P. Quorum-sensing as a population-density-dependent determinant of bacterial physiology. Adv. Microb. Physiol. **2001**, *45*, 199–270.

Bacteria-Plant Host Specificity Determinants

B

Alan Vivian
The University of the West of England, Bristol, U.K.

INTRODUCTION

Our understanding of the bacteria-plant interaction has grown rapidly since the pioneering work of H. H. Flor set the theoretical basis for its investigation. It seems that bacterial pathogens possess a battery of virulence (*vir*) determinants, some of which the plant has learned to recognize and respond strongly to (these are the products of avirulence (*avr*) genes)—a fact that has enabled their isolation. These virulence proteins are delivered inside plant cells by the type III secretion system—some likely to the nucleus and others to the cytosol—where they function as effectors to promote disease. If specific molecular recognition occurs, then programmed plant cell death—called the hypersensitive response (HR)—results in a localized necrotic lesion around the point of bacterial invasion that limits and curtails the infection.

PATHOGEN RACE DEFINITION

Races have been most clearly identified among pathogenic varieties (pathovars, pv.) of *Pseudomonas syringae* and *Xanthomonas campestris* by their interaction with host plants that have been bred to produce cultivated varieties (cultivar, cv.) differing in their resistance to the pathogen. Races are defined by their differential interactions with a set of cultivars for a given pathovar-host system. The most clearly defined race structures among pathogenic bacteria are those for *P. syringae* pv. *pisi* and pea (*Pisum sativum*), *P. syringae* pv. *phaseolicola* and bean (*Phaseolus* spp.), and for *P. syringae* pv. *glycinea* and soybean (*Glycine max*).[1] The basis of this race structure is matching resistance (*R*) genes in the host and *avr* genes in the bacterium, represented in the classical quadratic check (Fig. 1). The *R* gene content of cultivars has been determined by genetic crossing and testing of the reactions of the progeny to bacterial attack, and many *avr* genes have been cloned and their specificity confirmed. An example (Table 1) postulates a system based on five established gene pairs, together with a possible sixth pair.[1] Races may not be genetically homogeneous: In *P. syringae* pv. *pisi*, races 3 and 4 include strains from two distinct genomic groups that appear to have evolved separately to parasitize pea.[4]

RESISTANCE GENES AND THE HYPERSENSITIVE RESPONSE

R genes have been isolated from a number of plant species, including tomato (*Lycopersicon esculentum*) and *Arabidopsis thaliana*. They appear to encode protein receptors for signal transduction pathways that potentiate plant responses to bacterial attack. Six major classes of R proteins have been identified.[5] These proteins often possess a leucine-rich repeat (LRR) domain, potentially involved in recognition, and domains for signal relay. A useful concept has been the guard hypothesis,[6] which now appears to provide a molecular basis to account for how at least three pairs of Avr/R proteins interact.[7] It proposes that an R protein acts as a guard to monitor the status of a go-between protein that interacts directly with both the R and Avr proteins. When the Avr protein binds to or modifies (for example, by peptide cleavage) the go-between protein, this somehow activates the R protein to initiate the phenotypic expression of a HR. The release of putative antimicrobial factors, which are thought to include active oxygen species and phytoalexins, limits the spread of the pathogen.

BACTERIAL TYPE III SECRETION OF EFFECTOR PROTEINS

Some bacterial pathogens of both animal and plant cells possess a specialized protein secretion system, called the type III system.[8] In plant pathogens, the genes that are conserved from animal type III systems are designated *hrc*, and the remaining genes *hrp* (HR and pathogenicity). The system was originally recognized through the creation of bacterial mutations that resulted in the concomitant loss of the ability to cause disease in the usual host and induce a HR in a nonhost, such as tobacco (*Nicotiana tabacum*).[9] Although direct observation of the delivery of proteins into plant cells via a type III system has not been reported, there is considerable circumstantial evidence to support this conclusion.[10] Secretion of effector proteins is dependent on a functional type III system, and expression of the *hrc*/*hrp* genes is regulated coordinately with the secretion system. Some effector genes have been shown to have multiple functions. An example

Encyclopedia of Plant and Crop Science
DOI: 10.1081/E-EPCS 120010571

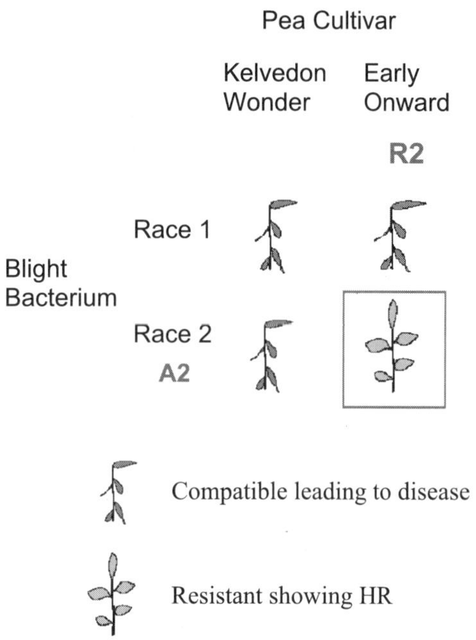

Fig. 1 Gene-for-gene: The quadratic check. A simplified comparison of two races of *Pseudomonas syringae* pv. *pisi* inoculated on two pea cultivars. Specificity resides in the matching of the avirulence gene (here designated A2) with the resistance gene R2, resulting in resistance to disease. All other combinations of races and cultivars result in susceptibility and a disease outcome. (*View this art in color at www.dekker.com.*)

is *avrPphF*, a plasmidborne gene from the bean pathogen *P. syringae* pv. *phaseolicola*, which confers cultivar-specificity through its avirulence activity in bean cultivars carrying the *R1* resistance gene. In native races a second effector gene, *avrPphC*, blocks the virulence function of *avrPphF*. The presence of *avrPphF* in a nonpathogenic, plasmid-cured derivative of the bean pathogen lacking a number of effector genes, including *avrPphC*, confers an enhanced HR in cv. Canadian Wonder and virulence in cv. Tendergreen. This example and the failure to detect loss of virulence when individual effector genes are inactivated suggest that virulence is redundantly encoded and requires the concerted actions of a number of effector proteins.[5]

Structurally, effector genes are of two kinds: those that resemble *avrBs3* (see Fig. 2) and those that do not. The *avrBs3*-like gene products appear to be targeted to the plant cell nucleus, where it is suggested they act (directly or indirectly) as transcription factors modulating host gene expression. Thus far, the *avrBs3*-like genes have only been found in *Xanthomonas* spp. and *Ralstonia solanacearum*. The specificity of these effector proteins appears to lie in centrally located repeats of 34 amino acids, which vary in number and precise sequence.[10] The remaining group of *avr* genes (which are predominantly from *P. syringae*) are often clustered on the bacterial genome, tend to have low G+C content of their coding sequences, and appear to code for peptides that are hydrophilic and range

Table 1 Gene-for-gene relationships between pea cultivars and races of *Pseudomonas syringae* pv. *pisi*

+ = susceptible response
− = resistance response
? = gene probably present
. = gene absent

							Races/avirulence genes						
							1	2	3[a]	4[a]	5	6	7
							1	.	.	.	.	.	.
							.	2	.	.	2	.	2
							3	.	3	.	.	.	3
							4	.	.	4	4	.	4
							.	.	.	.	5	.	.
							6?	.	.	.	6?	.	.
Cultivar	Resistance genes												
Kelvedon Wonder	.	.	.	.	.	.	+	+	+	+	+	+	+
Early Onward	.	2	.	.	.	.	+	−	+	+	−	+	−
Belinda	.	.	3	.	.	.	−	+	−	+	+	+	−
Hurst Greenshaft	.	.	.	4	.	6?	−	+	+	−	−	+	−
Partridge	.	.	3	4	.	.	−	+	−	−	−	+	−
Sleaford Triumph	.	2	.	4	.	.	−	−	+	−	−	+	−
Vinco	1	2	3	.	5	.	−	−	−	+	−	+	−
Fortune	.	2	3	4	.	.	−	−	−	−	−	+	−

[a]Races 3 and 4 are further subdivided into genomic groups that do not differ in their respective host specificity. Thus races 3A and 4A are in genomic group II and races 3B and 4B are in genomic group I (Ref. 4).
(Based on Ref. 1.)

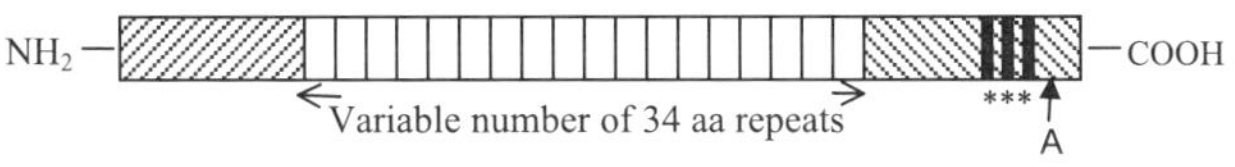

Fig. 2 Structural features of the AvrBs3 family of effector proteins. A diagrammatic representation of the protein whose central region contains between 17.5 and 25 repeats of a 34-amino sequence. Asterisks = nuclear localization signals; A = acidic transcriptional activation domain.

in size from 18 to 100 kiloDaltons (kDa). All known *P. syringae* effector genes are associated with Hrp box promoters, whose consensus sequence is 5′-GGAACC-NA-N_{13-14}-CCACNNA-3′ and which lies upstream of the coding region of the gene. This appears to ensure that their regulation is coordinated with the expression of the type III secretion system. Other systems of regulation are found in other plant pathogenic bacteria.[5,10]

CONCLUSION

Disease and host specificity are intimately linked and involve an interaction between two living organisms, in this case a bacterium and a plant host. A feature of such encounters is the concept of matching genes, the products of which interact in a very precise way. Pathogenic bacteria deliver effector proteins—the products of *avr* and *vir* genes—into plant cells via a specialized type III protein secretion system (in phytopathogens, called Hrp). Resistance genes in plants specify a surveillance system (which detects avirulence gene products), and the phenotypic outcome is seen as a HR in the plant. Disease occurs in the presence of appropriate virulence gene products, delivered by the Hrp machinery but conditioned on the absence of a functional avirulence gene product or resistance protein that can recognize the virulence protein.[11,12]

The cloning of *R* genes has opened the potential for engineering disease resistance in susceptible plants. But the redundancy seen in bacterial effectors suggests that reliance on individual engineered R genes will be as unreliable as their introduction through classical breeding.[12] Thus, it is vitally important to study genomes of both bacterium and plant to set all aspects of the bacterium–plant interaction in their true context. This in turn should enable us to gain insights into the global control mechanisms that drive pathogen evolution. The phenomenon of PAIs, which contain many effector genes, should be investigated together with their accompanying bacteriophage and insertion sequence neighbors to discover how their horizontal spread may be mediated and controlled.

ARTICLES OF FURTHER INTEREST

Bacterial Pathogens: Early Interactions with Host Plants, p. 89
Bacterial Products Important for Plant Disease: Cell Surface Components, p. 92
Bacterial Products Important for Plant Disease: Diffusible Metabolites As Regulatory Signals, p. 95
Bacterial Products Important for Plant Disease: Extracellular Enzymes, p. 98
Bacterial Products Important for Plant Disease: Toxins and Growth Factors, p. 101
Fire Blight, p. 443
Plant Diseases Caused by Bacteria, p. 947
Plant–Pathogen Interactions: Evolution, p. 965

REFERENCES

1. Bevan, J.R.; Taylor, J.D.; Crute, I.R.; Hunter, P.J.; Vivian, A. Genetic analysis of resistance in *Pisum sativum* cultivars to specific races of *Pseudomonas syringae* pathovar *pisi*. Plant Pathol. **1995**, *44* (1), 98–108.
2. Galan, J.E.; Collmer, A. Type III secretion machines: Bacterial devices for protein delivery into host cells. Science **1999**, *284*, 1322–1328. May 21.
3. Ellingboe, A.H. Genetics of Host–Parasite Interactions. In *The Encyclopedia of Plant Physiology*; Heitefuss, Williams, Eds.; New Series Physiological Plant Pathology, Springer-Verlag: New York, 1976; Vol. 4, 761–778.
4. Vivian, A.; Arnold, D.L. Bacterial effector genes and their role in host–pathogen interactions. J. Plant Pathol. **2000**, *82* (3), 163–178.
5. Jones, J.D.G. Putting knowledge of plant disease resistance genes to work. Curr. Opin. Plant Biol. **2001**, *4* (4), 281–287.
6. Van der Biezen, E.A.; Jones, J.D.G. Plant disease-resistance proteins and the gene-for-gene concept. Trends Plant Sci. **1998**, *23* (12), 454–456.
7. Schneider, D.S. Plant immunity and film noir: What gumshoe detectives can teach us about plant–pathogen interactions. Cell **2002**, *109*, 537–540.
8. Cornelis, G.R.; Van Gijsegem, F. Assembly and function of type III secretory systems. Annu. Rev. Microbiol. **2000**, *54*, 735–774.
9. Lindgren, P.B. The role of *hrp* genes during plant–bacterial interactions. Annu. Rev. Phytopathol. **1997**, *35*, 129–152.
10. Lahaye, T.; Bonas, U. Molecular secrets of bacterial type III effector proteins. Trends Plant Sci. **2001**, *6* (10), 479–485.
11. Crute, I.R. Gene-for-gene recognition in plant–pathogen interactions. Philos. Trans. R. Soc. Lond., B **1994**, *346*, 345–349.
12. *The Gene-for-Gene Relationship in Plant–Parasite Interactions*; Crute, I.R., Holub, E.B., Burdon, J.J., Eds.; CAB International: Wallingford, UK, 1997.

Biennial Crops

I. L. Goldman
University of Wisconsin, Madison, Wisconsin, U.S.A.

INTRODUCTION

The common environmental signals that regulate the processes of vegetative and reproductive growth cycles in plants are temperature and photoperiod. Mechanisms have evolved in many plants to synchronize reproductive development with a particular environmental cue or cues, thereby leading to improved reproductive success and, therefore, improved fitness. One such adaptation is spring flowering preceded by a season of vegetative growth. Plants with this particular adaptation, or those that require two seasons in order to complete reproductive growth, may be considered biennials.

WHAT ARE BIENNIALS?

Biennials typically consist of a root system, a compressed stem, and a rosette of leaves close to the soil surface following the first season of growth. Biennials generally do not become woody during the second season of growth; however, there are many examples in which root or hypocotyl tissues undergo secondary growth.[1] In such cases, the swollen tissues resulting from this may be used as a crop, such as the hypocotyl and root that are present in the carrot or beet. In biennial plants, food reserves that accumulate in root tissues during the first season of growth are used to produce reproductive structures during the second season of growth. In addition, hormones such as gibberellins and cytokinins are produced in root meristems and transported via the xylem to the shoot, where cell growth and development will occur during this second season of growth.[1]

A number of important crop plants are considered biennial, including cabbage and related *Brassica* crops, sugar beet, Swiss chard, table beet and related *Chenopodiaceae* crops, carrot and related *Apiaceae* crops, onion and related *Alliaceae* crops, and a wide variety of ornamental species. Although many of these plants are considered biennial from a horticultural standpoint, few of them are "true" biennials and instead represent genetic modifications for maximization of vegetative growth prior to reproduction. Certain grains that are sown in the fall and flower in the spring, such as winter wheat and rye, may be considered winter annuals. These plants possess a facultative vernalization requirement. Therefore, they will flower more quickly with a cold treatment, but the cold treatment is not required for flowering. Biennials, by contrast, are obligate from a vernalization point of view.

Many cultivated vegetable crops are considered biennials. They are typically consumed after the first season of growth, and thus we usually consider them to be biennials that are cultivated as annuals. This article is largely focused on biennial vegetable crops. Cultivars of most winter annual grain crops are pure lines, and methods of breeding pure lines are discussed in another section. The term "vegetable" is problematic from a scientific point of view because it lacks biological meaning and instead refers only to a cultural phenomenon. Still, there are two important criteria for vegetables that allow us to place these crops into a particular category for classification: They are immature plant parts that are of high moisture content. For those vegetable crops that are biennials, the immature plant parts that form during the first season of growth are those that are desired. In fact, it is likely that these organs are the products of artificial selection for enhanced biomass in annual plants.

VEGETABLE CROPS

In the case of many vegetable crops, the large vegetative structures desired by human cultures necessitated longer periods of vegetative growth. Selection under the process of domestication likely modified the life cycle of such crops from annuals to biennials, thereby increasing the vegetative growth period and providing a substantial reserve for subsequent reproductive growth. Three common examples of this are cabbage, carrot, and beet, all of which were likely domesticated from annual ancestors and modified by European agriculturists to adhere to a biennial life cycle.

Cabbage (*Brassica oleracea*) was selected from leafy forms of *Brassica* into a headed form in order to allow for a storage form of this edible plant. Likely, this transition took place as the leafy forms moved from a warm to a cool climate during cabbage domestication. As selection for a headed form was carried out, the life cycle was lengthened to allow for maximal biomass production during the first season of growth. Thus, an annual ancestor was converted

Encyclopedia of Plant and Crop Science
DOI: 10.1081/E-EPCS 120016291

into a biennial derivative by enhancing vegetative growth at the expense of reproductive growth. The late-summer flowering associated with the annual forms became early-spring flowering in the biennial forms, thereby allowing for only vegetative tissues to be produced during the first season of growth.

A similar evolutionary history took place for the biennial vegetable crops carrot (*Daucus carota*) and table beet (*Beta vulgaris*).[2,3] Annual forms with small root systems were gradually converted to biennial forms as selection was practiced for enhanced root and hypocotyl growth. In the climates where this was practiced, the swollen-rooted forms could be stored for a longer period of time, thereby providing a substantial food source during the winter months.

Table beet has been cultivated for millennia as both a root and a leafy vegetable crop. Its origins trace back to the development of a leafy vegetable by the Romans from wild species of *Beta* growing in the Mediterranean region (reviewed in Ref. 3). As the crop moved into Northern Europe, the growing season was shorter and the winter was longer. This may have caused selection pressure to favor a transition toward a biennial life cycle by selecting a swollen hypocotyl or root as an overwintering propagule. Some have suggested that swollen roots may have been selected from leafy beets cultivated in Assyrian, Greek, and Roman gardens.[4,5]

BIOLOGY OF BIENNIALISM

The term ''vernalization'' comes from the Latin ''vernus,'' which means spring. Vernalization is an adaptation to environments in which it is advantageous to flower in the spring, following a season of vegetative growth. Vernalization accelerates the ability of a plant to flower but is not responsible for the formation of flower primordia or for the breaking of dormancy.[6] Vernalization can be obligate or facultative in plants; however, in biennial plants the vernalization requirement is obligate.[6]

Typically, temperatures of 1–7°C are required for adequate vernalization, although many exceptions exist.[6] The duration of the vernalization requirement varies; however, two to three months is average. For many crops, flowering may occur with a minimal period of vernalization, but maximal flowering may require additional time. Some species are sensitive to fluctuations in temperature during vernalization, while others are not.[7] A vernalization treatment can be rendered ineffective with a heat treatment, and this process is known as devernalization. Although seed can be vernalized in some species, most plants must reach a more advanced developmental stage in order to be vernalized. The shoot apex is the most likely spot for the perception of vernalization.[6,8] It is also possible for vegetative tissues to be vernalized and ultimately regenerated into whole plants.

The hormone gibberellic acid (GA) is often implicated in regulating the flowering response in plants. Applications of GA to biennial plants can cause flowering in the absence of vernalization, and this may allow for seed production in a biennial plant during a single growing season. Michaels and Amasino[9] found that the late-flowering vernalization-responsive *Arabidopsis* mutants respond normally to cold treatment in the presence of the *gal-3* allele, which is a deletion in kaurene synthase, a gene involved in GA biosynthesis. This finding suggests GA may not be involved directly in the vernalization pathway. Interestingly, the vernalization requirement of a biennial plant can be eliminated by grafting. Non–cold-treated plants can be grafted to cold-treated biennial plants and induced to flower.[8] Lang suggested the possible presence of a vernalization hormone (vernalin) that might be produced constitutively in plants that do not require vernalization as well as in those that do; however, such a hormone has never been identified.

Bolting, or the appearance of a flower stalk during the vegetative growth stage, is detrimental to crop production. Bolting is a fairly common occurrence in early-planted biennial crops such as carrot and table beet, particularly in temperate environments. Significant yield losses can be expected when bolting has occurred, and selection against bolting is routinely practiced. Jaggard et al.[10] reported that 50% of field-grown sugar beet plants bolted when temperatures were less than 12°C for 60 days during vegetative growth. Vernalization typically takes place for 12 weeks during the standard breeding cycle, during which time temperatures are maintained at approximately 2–5°C.

Although seemingly complex from a physiological point of view, biennialism is often controlled by relatively few loci. Both dominant and recessive alleles have been identified that control the biennial versus annual habit. *Hyraceum niger* has a single dominant allele conditioning biennialism, while sugar beet has a recessive allele, *b*, conditioning the biennial habit.[11] Recent work by Michaels and Amasino[12] points toward major regulatory genes controlling the response to vernalization and the transition from vegetative to reproductive growth in biennial plants, which is consistent with the finding of a relatively simple genetic control of biennialism.

CONCLUSION

Much of their recent molecular work has been conducted with *Arabidopsis*, a model plant that has shed much light on the biology of biennialism. *Arabidopsis* plants flower in response to long days and vernalization. Many researchers use a rapid-cycling summer-annual ecotype of *Arabidopsis* in their genetic studies. However, many ecotypes of *Arabidopsis* are extremely late flowering unless vernalized, and thus they behave as winter annuals. Two loci, *frigida* and *flowering locus C* (FLC), are responsible for the vernalization-responsive late-flowering habit of these winter-annual ecotypes. The vernalization requirement for late-flowering ecotypes is created when the floral inhibitor FLC is up-regulated. After a cold treatment, FLC transcripts are down-regulated and remain low for the remainder of the plant's life cycle, and it is during this phase that flowering occurs.[12] Interaction and expression of these genes may be responsible for conditioning the biennial habit, and it will be interesting to determine whether biennialism is conditioned by these genes in other species.

ARTICLE OF FURTHER INTEREST

Breeding Biennial Crops, p. 171

REFERENCES

1. Raven, P.; Evert, R.F.; Eichorn, S. *Biology of Plants*, 4th Ed.; Worth Publishers: New York, 1986; 775 pp.
2. Simon, P. Domestication, historical development, and modern breeding of carrot. Plant Breed. Rev. **2000**, *19*, 157–190.
3. Goldman, I.L.; Navazio, J. History and breeding of table beet in the United States. Plant Breeding Rev. **2003**, *22*, 357–388.
4. Ford-Lloyd, B.V.; Williams, J.T. A revision of *Beta* section *Vulgares* (Chenopodiaceae), with new light on the origin of cultivated beets. Bot. J. Linn. Soc. **1975**, *71*, 89–102.
5. Williams, J.T.; Ford-Lloyd, B.V. The systematics of the chenopodiaceae. Taxon **1974**, *23*, 353–354.
6. Michaels, S.D.; Amasino, R.M. Memories of winter: Vernalization and the competence to flower. Plant Cell Environ. **2000**, *23*, 1145 1153.
7. Thompson, H.C. Further studies on effect of temperature on initiation of flowering in celery. Proc. Am. Soc. Hortic. Sci. **1944**, *35*, 425–430.
8. Lang, A. Physiology of Flower Initiation. In *Encyclopedia of Plant Physiology*; Rushland, W., Ed.; Springer Verlag: Berlin, 1965; 1371–1536.
9. Michaels, S.D.; Amasino, R.M. The gibberellic acid biosynthesis mutant *gal-3* of *Arabidopsis thaliana* is responsive to vernalization. Dev. Genet. **1999a**, *25*, 194–198.
10. Jaggard, J.W.; Wickens, R.; Webb, D.J.; Scott, R.K. Effects of sowing date on plant establishment and bolting and the influence of these factors on yields of sugar beet. J. Agric. Sci. Cambridge **1983**, *101*, 147–161.
11. Abegg, F.A. A genetic factor for the annual habit in beets and linkage relationship. J. Agric. Res. **1936**, *53*, 493–511.
12. Michaels, S.D.; Amasino, R.M. Flowering locus C encodes a novel MADS domain protein that acts as a repressor of flowering. Plant Cell **1999b**, *11*, 949–956.

Bioinformatics

B

Michael T. Clegg
Peter L. Morrell
University of California, Riverside, California, U.S.A.

INTRODUCTION: MINING MOLECULAR SEQUENCE DATA

DNA sequences diverge over time because of the gradual accumulation of mutational differences. Mutations erode DNA coding information, and if sufficient time has elapsed since the separation of lineages, it may be difficult to discern any historical similarity between two DNA sequences that once shared a common ancestor. The bioinformatician is presented with DNA sequences, protein sequences, or derivative data such as DNA hybridization signals associated with microarrays. A major challenge is to detect regions of similarity between different DNA sequences (or surrogate measures) to test the inferences of a common history. A common origin implies a shared function, or a related function, and this provides important clues in the assignment of a preliminary function to raw sequence data from a new source. A second major challenge is the management and curation of the vast stores of new data originating from the genomics enterprise. A variety of computational tools are available to organize and retrieve useful results from these vast databases. Genomics data are maintained in a number of publicly available databases. Perhaps the most important of these is the NCBI database, which can be accessed via the World Wide Web (http:www.ncbi.nlm.nih.gov) and provides a number of data analysis tools.

Bioinformatics combines computer science, genetics, and statistics to meet the challenge of mining useful information from the vast stores of new molecular sequence data. In this article we will touch on a few of the issues that confront bioinformations with a specific focus on DNA sequence data.

Several computer scripting or programming languages are used to handle DNA and protein sequence data. Perl and Python scripting are both commonly used; example code from Bioperl is available at http://bioperl.org/ and for Biopython at http://biopython.org/. Many tools that simulate DNA sequence evolution are written in C or C++. Dr. Richard Hudson and his colleagues have made source code to simulation-related tools available at http://home.uchicago.edu/~rhudson1/source.html and http://molpopgen.org/.

ANALYTICAL METHODS: THE CHALLENGE OF SEQUENCE ALIGNMENT

The proper alignment of molecular sequence data is fundamental to many aspects of bioinformatics. The raw data are usually strings of DNA, RNA, or protein sequences for a particular gene or protein drawn from a set of organisms. These data may also take the form of a position of a fragment on a gel (e.g., RFLPs or microsatellites) where identity in the location in the gel is assumed to imply identity in the underlying DNA or protein state. In what follows, we will restrict our discussion to DNA sequence data for simplicity. We use the term "string of nucleotides" to refer to the DNA sequence for a gene obtained from a particular organism (or a particular copy of a multigene family from within a genome). Thus, S_1 refers to the string from source 1, and S_n refers to the string from source n. The first analytical task is to align the strings of nucleotides (S_1, S_2, ..., S_n) to minimize the number of nucleotide differences across the set {S}. This involves finding the minimum of a weighted function of the number of indel (insertion/deletion) events and nucleotide site differences over the set. The weights are assigned based on some prior assumptions about the likelihood of indels versus nucleotide site differences (Fig. 1). The final alignment is assumed to be the "best" representation of the number of mutational events that occurred over the evolutionary time spanned by the set {S}. We denote the aligned set as {A}. More-sophisticated alignment algorithms use a tree-fitting iteration to generate a best alignment by simultaneously estimating the phylogeny and alignment (e.g., Clustal W). Various alignment algorithms are readily available from a wide variety of DNA sequence analysis packages.

When the sequences in {S} are so diverged that it is difficult to obtain an alignment from DNA sequences, a translation into the derived amino acid sequence is useful. Protein change occurs at a slower rate than DNA sequence change, and alignment of a protein sequence may be relatively straightforward even when the underlying DNA sequences are highly diverged.

It should be clear from this limited discussion that the alignment step is a crucial operation, and all subsequent

Encyclopedia of Plant and Crop Science
DOI: 10.1081/E-EPCS 120006085

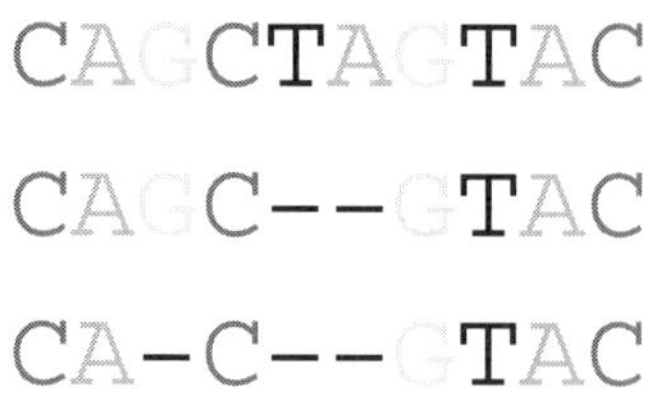

Fig. 1 A nucleotide sequence alignment with two indels. Alternative alignments could include a single indel but would require a nucleotide substitution.

calculations (e.g., distance measures, see below) depend on the accuracy of this operation. An operation that assumes a very high penalty (weight) for indels will force a larger number of nucleotide site differences on the alignment; conversely, assuming a very low penalty for indels will force a large number of small indels at the expense of nucleotide site differences (Fig. 1). These distortions will affect all subsequent inferences based on the sequence data. Indels are much more common in noncoding regions. When they occur in coding sequence, they often involve nucleotide triplets and preserve the gene's reading frame (Fig. 2). As indels become superimposed on one another during the course of evolution (this is especially common in noncoding sequence), it is increasingly difficult to determine the boundaries of individual indel events. As a consequence, it is usually not possible to count the number of indels that separate two sequences, and because the process of insertion/deletion is not stochastically regular, it is also not feasible to construct mathematical models of the process to provide a basis for estimation of the number of events. One practical way out of this dilemma is to select a set {S} where the time of separation between each S_i ($i = 1, 2, \ldots, n$) is sufficiently small so that at most one indel event will have occurred in a region. It is then possible to count the number of events across the set {S}.

ESTIMATION OF DISTANCE METRICS

Once an alignment is obtained, various calculations can be made on the aligned set {A}. One of the most common and useful calculations is to estimate the evolutionary distance between a pair of sequences (see Ref. 1 for a detailed discussion of distance metrics). The simplest distance measure is the percent divergence between two sequences, where divergence is measured as nucleotide site differences and indels are omitted from the calculation. This is a satisfactory measure for sequences that have been separated for a ''sufficiently brief'' period of evolutionary time. In operational terms, ''sufficiently brief'' means that the likelihood of two mutations hitting the same site is small enough to be neglected. Over longer periods of time, the likelihood of two or more hits at a site cannot be neglected, so a model of the substitution process must be introduced. Mathematical models permit estimation of the total number of events both observed and unobserved.

A number of mathematical models of nucleotide substitution have been introduced over the years. They all have the following fundamental assumptions in common: 1) The probability of a substitution event per site is assumed to be small so that multiple events per site have close to a zero probability over small intervals of time; 2) statistical independence of mutational events over time and over sites is assumed; and 3) an assumption must be made about the equilibrium frequency of nucleotides in the sequence (usually a uniform frequency of 25% per

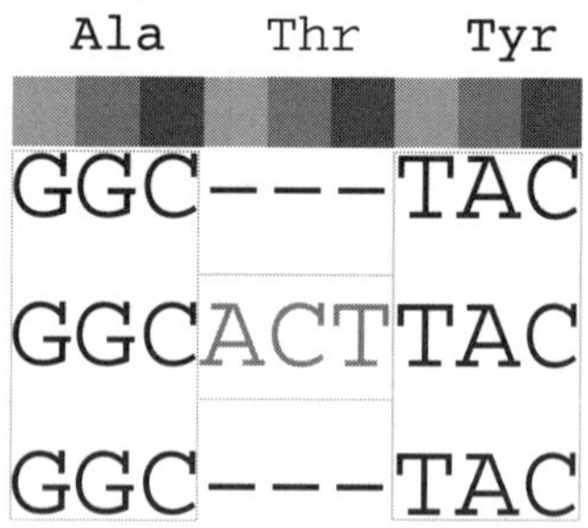

Fig. 2 In this alignment, first-, second-, and third-position nucleotides are indicated above each base. A three-base-pair indel results in an amino acid sequence with threonine inserted between the alanine and tyrosine in the first and third sequence.

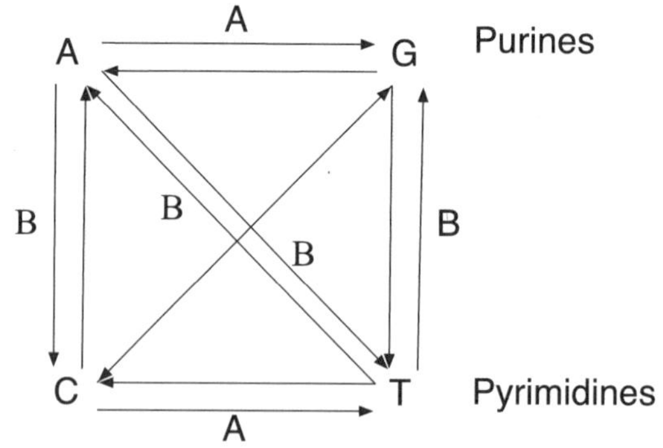

Fig. 3 Under the simplest (one-parameter) model of nucleotide substitution, the rate of change between any pair of nucleotides is equivalent. Under the two-parameter model (pictured here), the rate of transitions (A) may differ from the rate of transversions (B).

nucleotide type is assumed [A = 0.25, T = 0.25, G = 0.25, C = 0.25]). The simplest model of nucleotide change assumes that all possible nucleotide interchanges occur with an identical probability,[2] which requires that only a single mutation parameter be estimated.

Other progressively more complex models are also available. For example, a slightly more complex model posits two mutation parameters—one for the probability of transition mutations and one for the probability of transversion mutations (a transition is an interchange of a purine for another purine or a prymidine for another prymidine and a transversion is a purine to prymidine interchange or the reverse) (Fig. 3). There are now computer programs available that allow researchers to choose the model of DNA substitution that best fits their data.[3,4] A number of programs implement various distance calculations from aligned sequence sets; among the best known is the PHYLIP package,[5] at http://evolution.genetics.washington.edu/phylip.html. The PHYLIP Web page provides links to many other computer programs useful for the study of sequence evolution.

At its simplest, the distance between two sequences (1 and 2), denoted $D_{1,2}$, is a function of the mutation rate μ and the time ($2t$) since the two sequences separated from a common ancestor (the factor 2 occurs because both sequences trace back t time units to their common ancestor). It follows that $D_{1,2}$ is approximately equal to $2\mu t$. This simple relation is very useful in many rough calculations such as obtaining time estimates based on a molecular clock assumption.

Molecular Clock Analyses: As noted previously, the time between separation of lineages can be estimated if mutation rates are constant per site per unit time and if no other forces intervene to alter the rate of accumulation of mutational change between molecules that trace back to a single common ancestor. The most likely force to affect the accumulation of nucleotide change is natural selection. Natural selection acts as a kind of editor of the sequence messages transmitted through time (see article on population genetics). Mutations that arise and are deleterious to function are edited out by selection because their carriers are less likely to survive and reproduce. Hence, it is important to attempt to use sequence regions that are less likely to be perceived by selection for clock calculations (see forthcoming section on partitions of DNA sequence data).

Another likely cause of rate variation is variation in generation time. Recall that we assumed that μ is constant per unit time. Our conventional measurement of time is in years (celestial time), but the more natural measure of time among organisms is in terms of generations. It seems likely that μ is closely correlated with replication cycles that in turn depend on generation time. We must therefore regard molecular clock calculations as rough approximations.

INFERRING HISTORY: THE PROBLEM OF ESTIMATING PHYLOGENETIC TREES

The sequence alignment also provides the starting point for estimating a phylogeny. We will not delve into the technical details of phylogeny estimation, but rather give a brief outline of the major methods in common use (the interested reader is referred to Ref. 1 or Ref. 6 and the PHYLIP documentation (cited previously) for a detailed discussion of this topic). All methods assume that the elements of {A} are connected through a dichotomous branching tree. The problem is to deduce the branching network from {A} subject to some constraint or optimality criterion. There are three different approaches to tree estimation in common use: 1) parsimony; 2) maximum likelihood; and 3) distance-based methods.

Parsimony attempts to minimize the number of site changes over the tree (''character state changes'' in the jargon of systematics). Thus, the parsimony optimality criterion is minimum evolution or minimum number of character state changes. Well-developed algorithms exist for parsimony calculations, and tree estimation is relatively fast for large data sets. This method tends to be the choice of workers in systematics because it conforms to a philosophy that emphasizes the importance of calculating phylogenetic trees from ancestrally derived characters. Parsimony does have at least two drawbacks. First, it does not provide a unique tree. Frequently a number of trees will give the same minimum number of character state changes (i.e., a number of trees may satisfy the optimality criterion). This is not a particularly important drawback as it is a simple consequence of the fact that the parsimony optimality criterion is based on a discrete measure (number of site or character state changes) rather than a continuous measure (as is the case with maximum likelihood and distance-based approaches discussed below). Second, parsimony can be biased when evolutionary rates differ substantially over the tree. This bias is manifested by a tendency for long edges (rapidly evolving branches) to be joined together when in fact they should join other branches on the tree. This is a more significant concern in view of the fact that rate variation appears to be common over long evolutionary distances.

Maximum likelihood is a statistically derived method that depends on a mathematical model of the nucleotide substitution process (analogous to the distance estimation models introduced above). This method has the virtue that the machinery of statistical hypothesis testing can be implemented to discriminate among candidate trees. It also provides a direct statement about the uncertainty associated with any tree in terms of the variances of branch length estimates. Finally, assuming that the model is appropriate, maximum likelihood takes proper account of rate variation. It is possible to test the fit of various models to data (as discussed above); so, model validation is a feature of the method. Maximum likelihood has the serious drawback that it requires large amounts of computer time for moderate data sets (20–30 sequences), and larger data sets cannot be analyzed in reasonable lengths of time. Several maximum likelihood programs are available over the World Wide Web for tree estimation, and faster versions are under development (see the PHYLIP Web site).

Distance-based methods, as the name implies, begin by estimating a matrix of distances from {A}(denoted $D_{i,j}$) using the estimation methods introduced in the section on evolutionary distances. The optimality criterion seeks to minimize the total distance over the tree. A number of different algorithms have been introduced to estimate a tree from the $[D_{i,j}]$ or from some transformation of the $[D_{i,j}]$ that attempts to adjust for variation in evolutionary rates. Distance-based methods are computationally fast and can be very useful for data exploration. In the case of both parsimony and distance methods, it is easy to use resampling methods (such as the bootstrap) to evaluate the statistical support for a particular tree. Because the various estimation methods differ in assumptions and methodologies, many workers view consistency among methods as an indication that a particular phylogeny is robust to these variations.

A phylogenetic tree is the fundamental starting point for much analytical work in genomics because it provides our best estimate of the pattern of historical relationships among a set of sequences and hence among the set of organisms that donated those sequences. A wide variety of questions are addressed by using the phylogeny as an organizing framework. To illustrate this point, we mention a few important problems that rest on a phylogenetic analysis.

1. For many purposes, it is useful to deduce the pattern of duplication among members of a multigene family. This is accomplished by estimating a gene family phylogeny.
2. An important question is whether the pattern of transposon evolution is consistent with vertical transmission (and hence consistent with a phylogenetic tree).
3. Phylogenetic analysis allows us to ask which partitions of sequence along a chromosome are allelic and share a common history of transmission (inconsistent combinations of mutations that distinguish alleles might indicate intralocus recombination in the gene phylogeny).
4. It is often desirable to map other molecular or phenotypic changes on a tree to determine their time of origin and the temporal order in which major evolutionary events occurred.

ANALYSIS OF SEQUENCE DATA: PARTITIONS OF DATA SETS

It is often informative to divide {S} into various subsets that reflect natural partitions of the data. A natural partition might be untranslated regions and exons; a further partition could be 5′ untranslated regions, introns, exons, and 3′ untranslated regions. In addition, exon regions can be further partitioned into codons, and codons can be partitioned into synonymous versus replacement sites. Finally, synonymous sites can be partitioned into twofold, fourfold, and sixfold degenerate sites based on codon degeneracy patterns. All the calculations introduced above can be performed on various partitions of the data.

One partition that is especially informative is the partition of exon regions into replacement sites (sites where a substitution induces an amino acid change) and synonymous sites (sites where a substitution does not induce an amino acid change). Because synonymous changes do not change the protein, they are thought to be approximately neutral to natural selection. Let us denote D_s as the distance estimated from the synonymous partition of {A} and D_n as the distance estimated from the replacement partition of {A}. Then, the ratio D_s / D_n measures the average strength of selective constraint on amino acid change as compared to synonymous change. The ratio is one when replacement sites change with the same frequency as synonymous sites. Typically this ratio is observed to be approximately 10, indicating a tenfold selective retardation in protein change relative to the assumed mutational input. (This follows because if synonymous changes are neutral, then D_s approximates $2t\mu$ and $2t$ cancels in the ratio.) It is possible to plot this ratio on branches of a phylogenetic tree to search for indications of acceleration or retardation of protein evolution on particular branches of the tree. A plot of this kind allows the detection of regions of the tree where evolutionary patterns have shifted. A ratio of one or greater indicates an acceleration of protein evolution, which may be an indicator of an adaptive shift. For example, Jia et al.[7] have used this approach to identify regions of accelerated protein evolution for the Myb family of plant transcription factors. A D_s / D_n ratio of one or greater in population data can also indicate a balanced polymorphism as is evidently the case with the major histocompatibility polymorphism of humans.

A different partition is a moving average of the D_s / D_n ratio across the sequence graphed as a function of a sliding window. For example, a window of 100 nucleotides may be chosen, and the averages of D_n and D_s are calculated for this window beginning with nucleotide 1 to 100 in the sequence. The window is then moved one nucleotide to the right (2 to 101), and the averages are recalculated and so forth until the end of the sequence is reached. Large changes in the plots may indicate regions of the sequence that are subject to differing evolutionary forces.

Bioinformatics is a relatively new and rapidly evolving area of research, but there are several excellent books that provide either a detailed introduction to the field[8–11] or a more in-depth exploration of particular problems.[12]

CONCLUSIONS

The mining and analysis of genomics data is still in an early phase. Only a few of the large number of potential applications have been touched on in this article. During the 1990s the technology for producing molecular sequence data grew faster than our ability to extract all useful knowledge from these data sets. Despite this lag, there has been a fortuitous correspondence between the growth of computational power and the growth of genomics databases. There is every reason to expect that the combination of increasing computational power, new analytical techniques, and the expansion of databases will lead to a continuing stream of new discoveries that will have a revolutionary impact on biology, medicine, agriculture, and environmental management during the next quarter century.

ARTICLES OF FURTHER INTEREST

Agriculture and Biodiversity, p. 1
Arabidopsis thaliana: *Characteristics and Annotation of a Model Genome*, p. 47
Arabidopsis Transcription Factors: Genome-Wide Comparative Analysis Among Eukaryotes, p. 51
Chromosome Structure and Evolution, p. 273
Crop Domestication: Fate of Genetic Diversity, p. 333
Crop Improvement: Broadening the Genetic Base for, p. 343
Genetic Diversity Among Weeds, p. 496
Genetic Resource Conservation of Seeds, p. 499
Genome Rearrangements and Survival of Plant Populations to Changes in Environmental Conditions, p. 513
Genome Size, p. 516
Molecular Analysis of Chromosome Landmarks, p. 740
Molecular Evolution, p. 748
Mutational Processes, p. 760
Polyploidy, p. 1038
Population Genetics, p. 1042

REFERENCES

1. Nei, M.; Kumar, S. *Molecular Evolution and Phylogenetics*; Oxford University Press: New York, 2000.
2. Jukes, T.H.; Cantor, C.R. Evolution of Protein Molecules. In *Mammalian Protein Metabolism*; Munro, H.N., Ed.; Academic Press: New York, 1969; 21–132.
3. Posada, D.; Crandall, K.A. MODELTEST: Testing the model of DNA substitution. Bioinformatics **1998**, *14* (9), 817–818.
4. Muse, S.; Kosakovsky Pond, S. *HyPhy. Hypothesis Testing Using Phylogenies*; 2001.
5. Felsenstein, J. *Inferring Phylogenies*; Sinauer: Massachusetts, 2003.
6. Nei, M. *Molecular Population Genetics and Evolution*; Neuberger, A., Tatum, E.I., Eds.; North-Holland: Amsterdam, 1975; Vol. 40. http://www.bio.psu.edu/People/Faculty/Nei/Lab/publications.htm.
7. Jia, L.; Clegg, M.T.; Jiang, T. Excess nonsynonymous substitutions suggest that positive selection episodes operated in the DNA-binding domain evolution of *Arabidopsis* R2R3-MYB genes. Plant Mol. Biol. **2003**, *52* (3), 627–642.
8. Baxevanis, A.D.; Ouellette Francis, B.F. *Bioinformatics: A Practical Guide to the Analysis of Genes and Proteins*, 2nd Ed.; Wiley-Interscience: New York, 2001.
9. Krane, D.E.; Raymer, M.L. *Fundamental Concepts of Bioinformatics*; Benjamin Cummings: San Francisco, 2002.
10. Meidanis, J.; Setubal, J.C. *Introduction to Computational Molecular Biology*; PWS Publishing Company: Boston, MA, 1997.
11. Mount, D.W. *Bioinformatics: Sequence and Genome Analysis*; Cold Springs Harbor Press: Cold Springs Harbor, 2001.
12. Gusfield, D. *Algorithms on Strings, Trees, and Sequences: Computer Science and Computational Biology*; Cambridge University Press: Cambridge, 1997.

Biological Control in the Phyllosphere

John H. Andrews
University of Wisconsin, Madison, Wisconsin, U.S.A.

INTRODUCTION

This article addresses biological control of plant disease in the aerial environment—construed broadly to mean the use or facilitation of organisms, genes, or gene products to disfavor a pathogen. Thus, this discussion may be taken to include manipulation of a host plant genetically or in other ways, but the emphasis is on microorganisms (antagonists) that are introduced and interfere directly or indirectly with pathogens. This article considers 1) the status of biological control efforts in the phyllosphere; 2) key aspects of microbial and phyllosphere ecology that influence the likelihood of success; and 3) future prospects and directions.

THE STATUS OF PHYLLOSPHERE BIOCONTROL

Since the first attempts to control disease by applying antagonists to forest soil in the 1920s, there have been hundreds if not thousands of reports on microbial antagonism and biocontrol. A tangible product of this wealth of information is eight biocontrol products currently registered in the United States for application in the phyllosphere (Table 1), and approximately twice this number for use in the rhizosphere or elsewhere.[1] These are underestimates because they exclude agents not registered in the United States, or that are under development, or that are developed but not yet registered. Of the eight formulations, one product is based on the fungal hyperparasite *Ampelomyces quisqualis* (AQ-10 Biofungicide™); five are based on bacteria (three *Pseudomonas* spp. and two *Bacillus subtilus* strains); and two are chemicals (a protein and a benzothiadiazole) that activate host defenses. In addition to these commercial products, one benefit of decades of research on biocontrol has been the significantly increased understanding of microbial ecology and plant-microbe interactions.

PHYLLOSPHERE ECOLOGY

The main science underpinning biological control is applied ecology pertaining to the antagonist, the pathogen, the plant, and the abiotic and biotic environment in which the antagonist must function. It remains debatable whether the optimal approach is to use relevant ecological information in advance to screen candidates and then introduce what is supposedly the best among them to a particular crop ecosystem, or to operate empirically by using field releases at the outset to select the best among unscreened candidates. Ultimately, the success or failure of an introduced agent will depend on the existing ecological milieu.

Occasionally a biocontrol agent functions relatively inertly—essentially as a pesticide—in which case it is typically released in repeated, inundative fashion. The best example is *Bacillus thuringiensis* (BT) for insect control, although there are also some cases in plant pathology.[2] Much more commonly, the antagonist must colonize the habitat to be effective, i.e., it must be ecologically competent. Colonization of an introduced biological control agent is analogous to successful invasion of a foreign species, and similar ecological concepts should apply.[3–5] However, because invasiveness (or competitiveness) is a complex phenotype, it is likely that multiple genes and interactions will almost always be involved. This will make genetic analysis—a key step in one line of efforts to understand and improve biocontrol—complicated and refractory.

What are the determinants of competency in the phyllosphere and success as a control agent? Important intrinsic (biological) factors include the following: 1) pathogen vulnerability to attack at a weak link in its life cycle, typically at the preinfectional stage; 2) a mode of antagonism effective against, and preferably specific to, the pathogen in question; and 3) antagonist growth dynamics that are compatible with constraints imposed by the phyllosphere environment. Important extrinsic (commercial) factors include the type of crop and the economics of producing it; the economic significance of the pathogen; and the feasibility of mass culturing, formulating, and delivering the biocontrol agent to the site where it is needed. That there are not more registered biocontrol agents is testimony to the difficulty in meeting the dual stringent demands of biology and commerce. The need for biocontrol to be comparable to pesticides in being cheap, effective, reliable, and adaptable will be mitigated to the extent to which agriculture moves away from the profitability paradigm toward one based on sustainability.

Much has been written about the phyllosphere and, more broadly, the aerial environment of plants,[6,7] including

Encyclopedia of Plant and Crop Science
DOI: 10.1081/E-EPCS 120010612

Table 1 Commercial biocontrol products for phyllosphere diseases in the United States[a,b]

Organism or agent	Category	Brand name	Pathogens or diseases controlled	Host plants
Ampelomyces quisqualis M-10	Fungus	AQ-10 Biofungicide (Ecogen, Inc.)	Powdery mildews	Fruits, vegetables, and ornamentals
Pseudomonas aureofaciens TX-1	Bacterium	Bioject Spot-less (Eco Soil Systems, Inc.)	Anthracnose and soilborne pathogens	Turf
Pseudomonas fluorescens A506	Bacterium	BlightBan 506 (NuFarm, Inc.)	Frost damage, fireblight, fruit russetting	Tree fruits, small fruits, vegetables
Pseudomonas chlororaphis	Bacterium	Cedomon (BioAgri AB)	*Drechslera* spp, *Bipolaris sorokiniana*, *Ustilago hordei*	Barley, oats
Bacillus subtilis QST713	Bacterium	Serenade (AgraQuest, Inc.)	Powdery and downy mildews; leaf spots and blights	Diverse crops
B. subtilis MBI600	Bacterium	MBI600 Foliar (Becker Underwood, Inc.)	*Botrytis*, powdery mildew, leaf spot, eyespot, blotch	Cereals, fruits, vegetables, ornamentals
Hrp N harpin protein from *Erwinia amylovora*	Plant defense activator	Messenger (EDEN Bioscience Corp.)	Diverse fungal, viral, and bacterial pathogens	Ornamentals, fruits, vegetables, field crops
Acibenzolar-S-methyl	Plant defense activator	Actigard 50WP (Syngenta Crop Protection)	Downy mildew, tobacco blue mold, bacterial leaf spots	Leafy vegetables, tobacco, tomato

Based on Ref. 1 and manufacturers' Websites.

[a]See also EPA Website www.epa.gov/pesticides/biopesticides.

[b]Certain of these products may also be used to control soilborne diseases, or may be considered primarily as chemicals rather than biologicals (plant defense activators).

the proceedings of seven international phyllosphere microbiology conferences since 1970. Microbial population levels in the phyllosphere can reach an average of about 10^6–10^7 cells per cm^2, but densities range over several orders of magnitude among leaves as well as over time. Spatial variation in populations on the surface of a given leaf is likewise high, with some features—such as veins—densely colonized while the majority of the leaf remains relatively unoccupied. Epiphytic bacterial and yeast populations are highly responsive to environmental changes. Many (>100) species of filamentous fungi have been isolated from the phylloplane, but most of these are represented by quiescent propagules deposited from the air spora rather than as actively growing members of the community. Leaf surfaces are a unique and highly variable habitat that differ, for example, from root surfaces in their longevity, topography, aeration, nutrient flux, continuity of moisture, and microclimate. These characteristics influence the microbial inhabitants, including any natural or introduced biocontrol agents. Moreover, unlike roots, leaves do not reside in a medium that can be managed over time to promote antagonism. This implies that the potential for biocontrol will be different in the two habitats.

FUTURE PROSPECTS AND DIRECTIONS

Modification of plants, by conventional breeding or through genetic engineering, to resist pathogens or to favor beneficial microbes is a major future direction. A recent notable example is the transgenic resistance conferred by virus coat protein expressed in papaya to *Papaya ringspot virus*.[8] Transgenic papaya varieties have been established in Hawaii in commercial quantity since 1998; the fruit is widely marketed. As another example, transgenic sugarcane expressing the albicidin detoxifying gene from the leaf scald pathogen *Xanthomonas albilineans* was resistant to the disease and to multiplication of *Xanthomonas*.[9] This suggests that, through plant gene expression approaches founded on knowledge of molecular genetics and developmental biology, it may be possible to disrupt pathogenesis factors or key stages in the pathogen's life cycle (e.g., sclerotia production). However, to date the release of genetically modified organisms and the sale of such foods have generally been met with consumer skepticism and opposition. The prospects of this strategy for the immediate future are thus clouded and reinforce the earlier comment that an intrinsically effective biocontrol agent alone can not ensure commercial success.

Efforts to discover natural products from microorganisms, and other sources are intensifying.[10] Such compounds may have activity against plant pathogens or serve as models for laboratory synthesis of related chemicals of higher potency. For example, the fungicide fludioxonil, which is structurally related to the natural antifungal metabolite pyrrolnitrin from *Pseudomonas* spp, has been recently introduced.[10] Other compounds that show potential in biocontrol are the cercropins, small antimicrobial peptides with lytic activity.[11] An antifungal cercropin has been expressed in *Saccharomyces cerevisiae*; the transformed yeast cells, when coinoculated into wounded tomato fruit with spores of *Colletotrichum coccodes*, inhibited growth of the pathogen and decay.[11] In principle, a similar method could be used in the phyllosphere, where the goal would be to transform a competitive yeast colonist.

Two recently introduced products are at the vanguard of a new era in control strategy based on activation of the plant's resistance response. The active ingredient of one such product, Messenger™, is the harpin protein produced by *Erwinia amylovora* (Table 1). Harpin induces host gene expression and the generalized response is effective against several categories of pathogens, evidently with minimal impact on non-target organisms. Actigard™, based on the chemical inducer benzothiadiazole, acts similarly to trigger a systemic acquired resistance-like response (Table 1) and there are other examples.[12] Preliminary evidence from many laboratories is emerging that the silencing of specific genes at the messenger RNA (mRNA) level by the phenomenon known as RNA interference (RNAi) may be in part a plant defense mechanism. Knowledge of the mechanism may lead to novel biological means to control disease.

One final route to improved biocontrol is through a more insightful strategy based on microbial antagonists. Countless failed experiments on field release of putative biocontrol agents testify starkly to the need for refinements and a more judicious approach. There are some circumstances and environments (evidently including greenhouses)[13] wherein antagonists can perform effectively and reproducibly. However, the hallmark of microbial biocontrol is its variability,[14] although this may be offset by combinations of agents[15] and likely by more attention to matching the agent to the local environment in which it must function. An encouraging note is that, given a mode of antagonism amenable to improvement by targeted genetic modification, substantial improvement in control can result.[10]

CONCLUSION

Several commercial products with diverse modes of action are now available for biocontrol of foliar diseases. In general, these hold relatively small, "niche" markets, but nevertheless demonstrate that biocontrol in the phyllosphere is a commercial reality in some situations. Future research in microbial ecology, molecular biology, and plant physiology will provide new insights, and ultimately new products, for biocontrol. It is likely that the real and lasting contribution of biocontrol in the phyllosphere, as is the case in soil systems, will be as one component in integrated programs that are a part of good crop production practice.

ACKNOWLEDGMENTS

I thank G. Boland for comments on the manuscript and the U.S. Department of Agriculture (Hatch project no. WIS04446) and National Science Foundation (project no. DEB-0075358) for research support.

ARTICLES OF FURTHER INTEREST

Bacterial Pathogens: Detection and Identification Methods, p. 84
Bacterial Pathogens: Early Interactions with Host Plants, p. 89
Bacterial Products Important for Plant Disease: Cell-Surface Components, p. 92
Bacterial Products Important for Plant Disease: Diffusible Metabolites As Regulatory Signals, p. 95
Bacterial Products Important for Plant Disease: Extracellular Enzymes, p. 98
Bacterial Products Important for Plant Disease: Toxins and Growth Factors, p. 101
Integrated Pest Management, p. 612
Management of Bacterial Diseases of Plants: Regulatory Aspects, p. 669
Sustainable Agriculture: Ecological Indicators, p. 1191

REFERENCES

1. McSpadden Gardner, B.B.; Fravel, D.R. *Biological Control of Plant Pathogens: Research, Commercialization, and Application in the U.S.A. Plant Health Progr*; **2002**. doi:10.1094/PHP-2002-0510-01-RV.
2. Hjeljord, L.G.; Stensvand, A.; Tronsmo, A. Antagonism of nutrient-activated conidia of *Trichoderma harzianum* (*atroviride*) P1 against *Botrytis cinerea*. Phytopathology **2001**, *91*, 1172–1180.
3. Andrews, J.H. Biological control in the phyllosphere: Realistic goal or false hope? Can. J. Plant Pathol. **1990**, *12*, 300–307.
4. Andrews, J.H. Biological control in the phyllosphere. Annu. Rev. Phytopathol. **1992**, *30*, 603–635.
5. Marois, J.J.; Coleman, P.M. Ecological succession and biological control in the phyllosphere. Can. J. Bot. **1995**, *73* (suppl.), S76–S82.
6. Andrews, J.H.; Harris, R.F. The ecology and biogeography of microorganisms on plant surfaces. Annu. Rev. Phytopathol. **2000**, *38*, 145–180.
7. Lindow, S.E. *Phyllosphere Microbiology*; Hecht-Poinar, E.I., Elliott, V.J., Eds.; APS Press: St. Paul, MN, 2002.
8. Ferreira, S.A.; Pitz, K.Y.; Manshardt, R.; Zee, F.; Fitch, M.; Gonsalves, D. Virus coat protein transgenic papaya provides practical control of *Papaya ringspot virus* in Hawaii. Plant Dis. **2002**, *86*, 101–105.
9. Zhang, L.; Xu, J.; Birch, R.G. Engineered detoxification confers resistance against a pathogenic bacterium. Nature Biotechnol. **1999**, *17*, 1021–1024.
10. Ligon, J.M.; Hill, D.S.; Hammer, P.E.; Torkewitz, N.R.; Hofmann, D.; Kempf, H.-J.; van Pée, K.-H. Natural products with antifungal activity from *Pseudomonas* biocontrol bacteria. Pest Manag. Sci. **2000**, *56*, 688–695.
11. Jones, R.W.; Prusky, D. Expression of an antifungal peptide in *Saccharomyces*: A new approach to biological control of the postharvest disease caused by *Colletotrichum gloeosporioides*. Phytopathology **2002**, *92*, 33–37.
12. Cohen, Y.R. β-aminobutyric acid-induced resistance against plant pathogens. Plant Dis. **2002**, *86*, 448–457.
13. Paulitz, T.C.; Bélanger, R.R. Biological control in greenhouse systems. Annu. Rev. Phytopathol. **2001**, *39*, 103–133.
14. Boland, G.J. Stability analysis for evaluating the influence of environment on chemical and biological control of white mold (*Sclerotinia sclerotiorum*) of bean. Biol. Control **1997**, *9*, 7–14.
15. Guetsky, R.; Shtienberg, D.; Elad, Y.; Dinoor, A. Combining biocontrol agents to reduce the variability of biological control. Phytopathology **2001**, *91*, 621–627.

Biological Control of Nematodes

Brian R. Kerry
Rothamsted Research, Herts, U.K.

INTRODUCTION

Nematode populations are regulated by a wide range of natural enemies, including predators, antagonists, parasites, and pathogens.[1] Although predatory protozoa, nematodes, turbellarians, tardigrades, mites, and insects are common, especially in uncultivated soils, little is known of their impact on nematode communities, and most research has concentrated on microbial agents that attack specific nematode parasites. In some soils, nematophagous fungi and bacteria have been reported to control the multiplication of pest species on nematode-susceptible crops (Table 1), but despite the commercialization of a few organisms, none is in widespread use. Continued economic, environmental, and health concerns over the use of nematicides have resulted in the development of control strategies that depend less on the use of these chemicals. However, biocontrol agents are yet to play a significant role in the integrated pest management of nematodes and they have proved difficult to exploit.

MICROBIAL NATURAL ENEMIES

Antagonists

Several bacteria, actinomycetes, and fungi produce bioactive compounds that are toxic or affect the behavior of nematodes without infecting them.[2] Bacteria such as *Pseudomonas fluorescens* may reduce nematode hatch and invasion of roots by affecting their exudates, whereas *Agrobacterium radiobacter* and *Bacillus sphaericus* induce plant resistance.[3] Endophytic fungi may compete for resources in roots or may produce systemic effects that reduce nematode development.

The toxin produced by the actinomycete *Streptomyces avermitilis* and the metabolites of the fungus *Myrothecium verrucaria* have been used in commercial anthelmintics and a nematicide, respectively. Antagonists may be a valuable source of biorational nematicidal compounds, but the production of toxins on nutrient media in vitro is no indication of an organism's ability to produce such compounds in situ. Some leguminous crops encourage nematode-antagonistic bacterial communities in their rhizospheres, which could be used in pest control strategies.[4]

Parasites and Pathogens

Most research on the biocontrol of nematodes has been done on parasitic or pathogenic microorganisms. Those bacteria and fungi that parasitize active stages of nematodes produce adhesive spores or specialist traps. The Gram-negative bacteria *Pasteuria* spp., have a wide host range among plant parasitic nematodes, but individual isolates may be highly specific.[5] Host recognition appears to reside in spore hydrophobin interactions with protein receptors on the nematode cuticle. Spores of *Pasteuria* spp. that parasitize cyst and root-knot nematodes usually do not germinate until the infective second-stage juveniles have invaded the roots. The bacteria spread throughout the body and significantly reduce or prevent egg production, completing their life cycle in 20–30 days at 30°C. *Pasteuria penetrans* and *P. nishizawae* have been associated with the natural control of root-knot and cyst nematodes, respectively. However, problems in the in vitro production of these bacteria have limited their exploitation as biocontrol agents.

Nematophagous species occur in most fungal groups. The much-studied trapping fungi produce adhesive or constricting traps that ensnare their hosts. In *Arthrobotrys oligospora*, host recognition is thought to depend on lectin carbohydrate interactions. Although host–parasite interactions influence trap production and rate of capture, nematode trapping fungi do not depend on nematode prey for their nutrition. The extent of the saprophytic phase in soil is unclear and differs between species. *Duddingtonia flagrans*, which produces adhesive trap networks, is being evaluated for the biocontrol of nematode parasites of domestic animals, but the use of trapping fungi for the control of plant parasitic nematodes has had limited success.

Some fungi, such as *Catenaria anguillulae,* infect their nematode hosts with zoospores that aggregate around natural openings, encyst on the cuticle, and produce a germ tube that penetrates the host, whereas both *Hirsutella rhossiliensis* and *Drechmeria coniospora* produce adhesive spores on conidiophores that develop from infected cadavers. Although *H. rhossiliensis* is associated with the decline of *Criconomella xenoplax* populations on peaches, it requires large numbers of spores in soil to maintain small host populations, and this has limited its potential as a biocontrol agent.

Encyclopedia of Plant and Crop Science
DOI: 10.1081/E-EPCS 120010396

B

Table 1 Some examples of nematode-suppressive soils and their general characteristics

Nematode pest	Host crop	Country	Causal agent(s) of suppression
Examples			
Heterodera avenae	Temperate cereal monocultures	N. Europe	*Nematophthora gyrophila* *Pochonia chlamydosporia*
Meloidogyne spp.	Peaches	California	*Dactylella oviparasitica*
Criconemella xenoplax	Peaches	S. Carolina	*Hirsutella rhossiliensis* *Pseudomonas aureofaciens*
M. arenaria	Peanuts	Florida	*Pasteuria penetrans*
Meloidogyne spp.	Vegetables	W. Africa	*P. penetrans*
Meloidogyne spp.	Vines	Australia	*P. penetrans*

Characteristics

- Suppression of nematode populations occurs only after 4–5 years of susceptible crops and may be specific to a single pest species.
- Monocultures of nematode-susceptible crops or perennial crops are usually needed to support the development of a nematophagous microflora.
- Nematode damage may be severe in the early stages of the cropping cycle, and control strategies that reduce nematode populations may delay the onset of suppression.
- Soil amendments and crops may be used to manipulate the natural enemy community in the rhizosphere and increase biocontrol.
- Sustainable management usually depends on a diverse (intra- or interspecific) microbial community.
- Suppressiveness can be removed by biological treatments.

(From Ref. 2.)

Fungi that attack the sedentary stages (eggs and saccate females) of nematodes are generally rhizosphere colonizers that do not produce special infection structures other than an appressorium. More than 250 species of fungi have been isolated from the females and eggs of cyst and root-knot nematodes; *Paecilomyces lilacinus* and *Pochonia chlamydosporia* have been most studied. Both are facultative parasites that colonize the rhizosphere. Infection of eggs is associated with the production of an alkaline serine protease in both species, which may be a virulence and host-range determinant. These fungi are readily cultured in vitro and are being evaluated as potential commercial agents. *Pochonia chlamydosporia* is a key species in soils suppressive to the cereal cyst nematode in many soils in northern Europe.

DYNAMICS OF MICROBIAL AGENTS IN SOIL

There is much intraspecific variation among species of fungi and bacteria in suppressive soils, but the importance of such variation in the regulation of host populations is unknown. Those bacteria and fungi that occur in the rhizosphere are much affected by interactions between the plant and the nematode host.[6] Density-dependent interactions with hosts have been reported for both obligate and facultative parasites, but the dearth of quantitative methods to estimate parasite/pathogen populations has made observations beyond microcosms in controlled conditions difficult. However, it is clear that the dynamics of individual microorganisms differ greatly and have a profound influence on the potential of such agents for biocontrol.[7,8]

BIOCONTROL APPROACHES

Relatively few natural enemies have been evaluated as biocontrol agents for nematodes, but of those tested, none is likely to provide rapid and effective levels of pest control if applied alone. Biocontrol strategies have been extensively reviewed.[1,2] Although suppressive soils have provided the most sustainable method of nematode management for a few pest species in some soils, applying selected agents must be integrated with other control measures. Application of microbial agents following solarization or the addition of soil amendments, or in combination with nematicides or resistant hosts, has improved their performance. However, work on optimizing production and formulation methods remains to be done for most microbial natural enemies that have potential for biocontrol.

CONCLUSION

Nematode pests are obligate parasites on plants and spend different amounts of time in the rhizosphere, depending on their mode of parasitism. A detailed understanding of interactions between nematodes and their natural enemies in the rhizosphere is essential for the development of rational biocontrol strategies. Molecular biological and immunological techniques that discriminate and quantify individual isolates of microbial agents are needed to help understand the host–parasite dynamics in soil. Also, in combination with biochemical methods they would provide powerful tools for studies of the multitrophic interactions that occur in the rhizosphere, especially the factors that affect the switch from the saprophytic to the parasitic state in facultative parasitic fungi. Natural enemies may also provide novel bioactive compounds that could be developed as new nematicides or be incorporated in a transgenic approach to nematode management.

ARTICLES OF FURTHER INTEREST

Agriculture and Biodiversity, p. 1
Bacterial Pathogens: Detection and Identification Methods, p. 84
Bacterial Products Important for Plant Disease: Extracellular Enzymes, p. 98
Bacterial Survival and Dissemination in Natural Environments, p. 108
Bacterial Survival Strategies, p. 115
Biological Control of Oomycetes and Fungal Pathogens, p. 137
Biological Control of Weeds, p. 141
Integrated Pest Management, p. 612
Management of Nematode Diseases: Options, p. 684
Nematode Biology, Morphology, and Physiology, p. 781
Nematode Infestations: Assessment, p. 788
Nematode Population Dynamics, p. 797
Nematode Problems: Most Prevalent, p. 800
Nematodes and Host Resistance, p. 805
Organic Agriculture As a Form of Sustainable Agriculture, p. 846
Rhizosphere: An Overview, p. 1084
Rhizosphere: Microbial Populations, p. 1090
Rhizosphere Management: Microbial Manipulation for Biocontrol, p. 1098
Sustainable Agriculture: Definition and Goals of, p. 1187

REFERENCES

1. Stirling, G.R. *Biological Control of Plant Parasitic Nematodes: Progress, Problems and Prospects*; CAB International: UK, 1991.
2. Kerry, B.R.; Hominick, W.M. Biological Control. In *The Biology of Nematodes*; Lee, D.L., Ed.; Taylor and Francis: London, 2001; 483–509.
3. Sikora, R.A. Interrelationship between plant health promoting rhizobacteria, plant parasitic nematodes and soil microorganisms. Mededelingen Faculteit Landbouwwetenschappen Rijksuniversiteit Gent **1988**, *53*, 867–878.
4. Kloepper, J.W.; Rodriguez-Kabana, R.; McInroy, J.A.; Collins, D.J. Analysis of populations and physiological characterization of microorganisms in rhizospheres of plants with antagonistic properties to phytopathogenic nematodes. Plant Soil **1991**, *136*, 777–783.
5. Chen, Z.X.; Dickson, D.W. Review of *Pasteuria penetrans*: Biology, ecology and biological control potential. J. Nematol. **1998**, *30*, 313–340.
6. Kerry, B.R. Rhizosphere interactions and the exploitation of microbial agents for the biological control of nematodes. Annu. Rev. Phytopathol. **2000**, *38*, 423–441.
7. Atkinson, H.J.; Dürschner-Pelz, U. Spore transmission and epidemiology of *Verticillium balanoides*, an endozoic fungal parasite of nematodes in soil. J. Invertebr. Pathol. **1995**, *65*, 237–242.
8. Jaffee, B.A.; Phillips, R.; Muldoon, A. Density-dependent host-pathogen dynamics in soil microcosms. Ecology **1992**, *73*, 495–506.

Biological Control of Oomycetes and Fungal Pathogens

B

Eric B. Nelson
Cornell University, Ithaca, New York, U.S.A.

INTRODUCTION

Oomycetes and fungi are economically damaging plant pathogens whose presence and activities invoke the use of repeated fungicide applications to minimize losses in plant yield, quality, or aesthetics. Increasing environmental and human health concerns associated with widespread fungicide use has prompted scientists and plant producers to explore biological methods of disease control.

Biological control strategies make use of microorganisms to mitigate disease losses. Such disease suppressive microorganisms are commonly found in many different habitats. Biological control strategies attempt to enhance the activities of these disease-suppressive microorganisms either by introducing high populations of specific microorganisms or by enhancing the conditions that enable microorganisms in their natural habitats to suppress diseases. Common strategies for manipulation of biological control microbes will be discussed along with the commercialization potential of biological disease control strategies in agriculture.

MICROBIAL INOCULANTS

Microbial inoculation strategies attempt to increase soil or foliar populations of specific disease-suppressive microbes temporarily and dramatically. Microorganisms commonly studied and deployed for control of oomycete and fungal diseases include species of bacteria in the genera *Bacillus*, *Pseudomonas*, and *Streptomyces* and fungi in the genera *Coniothyrium*, *Gliocladium*, and *Trichoderma*.[1] Several organisms have served as model systems for generating many of our concepts of biological disease control. These include specific strains of *Bacillus cereus* and *B. subtilus*, *Enterobacter cloacae*, *Pseudomonas fluorescens*, and *Trichoderma harzianum*. Often specific strains of bacteria and fungi are deployed for biological control. Many times such strains, operating under rather narrow modes of action, have relatively narrow ranges of pathogen species for which they are effective.

The processes and traits expressed by biocontrol microbes result in reduced disease development by either directly disrupting stages of the pathogen's life cycle or by indirectly altering the biochemistry of the leaf surface, spermosphere, or rhizosphere such that pathogenesis is disrupted. This can be accomplished by producing antibiotic compounds active against the target pathogen, by competing with the target pathogen for specific resources such as iron or carbon, or by parasitizing hyphae or reproductive structures of the target pathogen. Both fungal and bacterial biocontrol microbes can also induce the expression of resistance responses in the host plant, making the host less susceptible to infection or disease development.

Various formulations of *Trichoderma harzianum* have been the most widely commercialized and efficacious inoculants used for the control of oomycete and fungal diseases. Of particular importance is strain T-22 of *T. harzianum* sold under the trade names RootShield™, PlantShield™, and TurfShield™ (Table 1). This inoculant is sold in the U.S. to the greenhouse, row crop, and turf industries. During 1999, retail sales of T-22 products totaled around $3 million[2] and were expected to increase in the next years.

COMMERCIALIZATION HURDLES

Despite decades of research on the biological control of oomycete and fungal plant diseases, there remain few widely adopted and commercially successful microbial inoculants. Fewer than 25 microbial species have been commercialized worldwide (Table 1). There are a number of reasons for the lack of development and grower adoption. Among the more important are problems in formulation and delivery, variability in performance, and problems with poor efficacy under optimum conditions for disease development. There are countless examples of biological control organisms that perform effectively under defined laboratory conditions but fail when introduced on different crops under varying field conditions. Still others might perform effectively in the field, but exhibit strong year-to-year or site-to-site variability. Additionally, the economics and level of biological knowledge necessary for growers to implement biological control strategies is still not favorable for many cropping systems, making adoption levels low.

Encyclopedia of Plant and Crop Science
DOI: 10.1081/E-EPCS 120019935

Table 1 Commercial microbial inoculants used worldwide for control of oomycete and fungal plant diseases

Microbial inoculant	Target disease(s)	Countries registered	Product names
Fungi and oomycetes			
Ampelomyces quisqualis	Powdery mildews	U.S.A., South Africa	AQ-10™
Candida oleophila	Post-harvest diseases[c]	U.S.A., Israel	Aspire™
Chaetomium cupreum/C. globosum	Diseases caused by *Phytophthora* and other root rot fungi	China, Philippines, Russia, Thailand, and Vietnam	Ketomium (R)™
Coniothyrium minitans	Diseases caused by *Sclerotinia*	U.S.A., Austria, France, Italy, Luxembourg, Germany, Mexico, Poland	Contans WG™, Intercept WG™
Cryptococcus albidus	Postharvest fruit diseases	South Africa	YieldPlus™
Fusarium oxysporum	Fusarium wilt diseases	Italy, France	Biofox C™, Fusaclean™
Gliocladium catenulatum	Fungal diseases of greenhouse crops	U.S.A.	Primastop™
Gliocladium virens	Seed, seedling, and root rots[b]	U.S.A., Finland	Soil Guard™, Gliomix™
Phlebiopsis gigantea	*Heterobasidion annosum* on pine and spruce trees	Britain, Sweden, Norway, Switzerland, and Finland	Rotstop™, PG Suspension
Pseudozyma flocculosa	Powdery mildew diseases	U.S.A., Canada	Sporodex L™
Pythium oligandrum	Damping-off of sugar beets	Slovak Republic	Polygandron™
Reynoutria sachalinensis	Powdery mildew, gray mold	U.S.A.	Milsana bioprotectant
Trichoderma harzianum	Seed and root rotting diseases[e]	U.S.A., Canada, Europe, Israel, Australia, South Africa	RootShield™, PlantShield™, TurfShield™, BioTrek 22G™, Supresivit™, T-22G™, T-22HB™, Trichodex™, EcoT™, Harzan 1™
Trichoderma polysporum	Tree-wound pathogens	U.S.A., Britain, Sweden, Denmark, Chile, Germany	Binab™
Trichoderma viride	*Armillaria*, *Botryosphaeria*, and others	New Zealand	Trichopel™, Trichojet™, Trichodowels™, Trichoseal™
Bacteria			
Bacillus licheniformis	Turfgrass diseases[a]	U.S.A.	EcoGuard™
Bacillus pumilus	Soybean seed and root rots[b]	U.S.A.	GB34™
Bacillus subtilis	Various foliar and root diseases[a]	U.S.A., Britain, Japan, South Africa, New Zealand	Kodiak™, Epic™, Concentrate™, Kodiak™, HB™, Quantum 4000, HB™, System 3™, Taegro™, Botkiller™, Serenade™, Subtilex™
Burkholderia cepacia	Seed, seedling, and root rots[b]	U.S.A.	Avogreen™, Deny™, Blue Circle™, Intercept™
Pseudomonas aureofaciens	Turfgrass diseases	U.S.A.	SpotLess™
Pseudomonas chlororaphis	Seed, seedling, and root rots[b]	U.S.A., Sweden	AtEze™, Cedomon™
Pseudomonas syringae	Postharvest fruit diseases	U.S.A.	Bio-Save 10, 11, 100, 110, 1000™
Streptomyces griseoviridis	Wilts, seed, and root rots[d]	U.S.A., Canada, Finland, Netherlands	Mycostop™
Streptomyces lydicus	Wilts, seed, and root rots[d]	U.S.A.	Actinovate™
Streptomyces hygrospinosis var. beijingensis	Various fungal diseases	China	AB 120™

[a]Diseases cause by species of *Rhizoctonia*, *Fusarium*, *Alternaria*, *Aspergillus*, and other fungi and Oomycetes powdery mildew, downy mildew, early leaf spot, early blight, and late blight diseases.
[b]Diseases caused by species of *Rhizoctonia*, *Fusarium*, and *Pythium*.
[c]Diseases caused by species of *Botrytis* and *Penicillium*.
[d]Diseases caused by species of *Fusarium*, *Alternaria*, *Rhizoctonia*, *Phomopsis*, *Pythium*, *Phytophthora*, and *Botrytis*.
[e]Diseases caused by species of *Pythium*, *Rhizoctonia*, *Verticillium*, *Sclerotium*, *Botrytis*, and others.
(Largely from the EPA Biopesticides website http://www.epa.gov/pesticides/biopesticides/; the PAN Pesticides Database http://www.pesticideinfo.org/index.html; and ATTRA Microbial Pesticides, Manufacturers & Suppliers Resource List http://attra.ncat.org/attra-pub/microbials.htm.)

Table 2 Examples of soils suppressive to oomycete and fungal plant diseases

Pathogen	Disease	Microorganisms responsible
Aphanomyces euteiches	Root rot of pea	Unknown
Armillaria mellea	Root rot of conifers	Unknown
Cephalosporium graminearum	Stripe of wheat	Unknown
Didymella lycopercisi	Stem rot of tomato	Unknown
Fusarium oxysporum	Wilts of various crops	Nonpathogenic *Fusarium* spp.; *Pseudomonas* spp.
Fusarium solani	Root rot of bean	Unknown
Fusarium culmorum	Foot rot of barley	Unknown
Gaeumannomyces graminis	Take-all of cereals	*Pseudomonas* spp.
Phytophthora capsici	Damping-off of tomato	Unknown
Phytophthora cinnamomi	Root rots of various crops	Various bacteria
Phytophthora sojae	Root rot of soybean	Unknown
Poria weirii	Root rot of conifers	Unknown
Pseudocercosporella herpotrichoides	Root rot of cereals	Unknown
Pythium aphanidermatum	Root rot of many crops	Unknown
Pythium splendens	Damping-off of cucumber	Unknown
Pythium ultimum	Damping-off of cotton	Seed-colonizing bacteria
Rhizoctonia solani	Root rot of many crops	Unknown
Sclerotium rolfsii	Rot of tomato	Unknown
Thielaviopsis basicola	Root rot of tobacco	Unknown
Verticillium albo-atrum	Wilt of potato	Unknown

(From Ref. 8.)

SUPPRESSIVE SOILS

Disease and pathogen suppressive soils have been known for over a century and identified from many parts of the world (Table 2).[3] They occur naturally or may be induced either through continuous monoculture or through the addition of organic amendments. Among the best-known are those suppressive to diseases caused by oomycetes and fungi such as *Pythium*, *Phytophthora*, *Fusarium*, *Rhizoctonia*, and *Gaeumannomyces*. The level of disease control is related predominantly and directly to unique microbiological properties associated with the soils themselves or, more importantly, with the spermosphere or rhizosphere of plants grown in these soils.[4] Although the specific microorganisms providing the disease control are generally not known, suppressive soils provide some of the best examples of effective biological control where they can serve as models for understanding how microorganisms in their natural habitats might be manipulated to reduce plant disease losses.

Among the best examples of naturally suppressive soils are those suppressive to Fusarium wilt diseases of various crops.[5] These soils are characterized by elevated populations of nonpathogenic *Fusarium* and fluorescent *Pseudomonas* species. These organisms compete with pathogenic species of *Fusarium* for carbon and iron, resulting in reduced plant infection.

Suppressiveness can be induced in soils either through the introduction of organic amendments or, in some cases, from crop monoculture. Organic soil amendments, particularly composts, have been studied extensively for their ability to induce suppressiveness to many oomycete and fungal diseases.[6] These studies confirm the involvement of compost-associated and soil-enhanced microbial communities in the suppressive properties of amended soils. The identities of the specific microorganisms contributing to this suppressiveness remain elusive.

The most widely studied example of induced disease suppressiveness in soils is the take-all decline phenomenon following cereal monoculture. Continuous cereal monoculture has been observed worldwide to result in the gradual decline in the severity of take-all disease caused by *Gaeumannomyces graminis* var. *tritici* (*Ggt*). The natural selection provided by monoculture results in the buildup of specific antibiotic producing *Pseudomonas* species on and in *Ggt* lesions on cereal roots that suppress root infection by Ggt.[7] If this cropping strategy is interrupted, disease suppression is lost.

CONCLUSION

The unpredictable nature of biological control systems has plagued research in this field for over 80 years and to date, no clear means of overcoming this variability has been discovered. Insights into how individual or communities of biological control organisms succeed and how they fail can come only from a detailed understanding of how these organisms function during their interaction with the pathogen, the plant, and the environment in which they are placed.

Knowledge of specific microbial traits essential for biological control activity and the important regulatory role played by the plant, whether it be in eliciting pre-infection developmental responses of pathogens, or in regulating the pathogen-suppressive behavior of introduced biological control microorganisms, is crucial to the ultimate success of biological control strategies. By understanding how the host, pathogen, and associated microbes contribute to biological control processes, we will be able to predict better and manipulate microbial behavior with the goal of enhancing biological disease control.

ARTICLES OF FURTHER INTEREST

Biological Control in the Phyllosphere, p. 130
Fungal and Oomycete Plant Pathogens: Cell Biology, p. 480
Management of Fungal and Oomycete Diseases: Fruit Crops, p. 678
Management of Fungal and Oomycete Diseases: Vegetable Crops, p. 681
Mechanisms of Infection: Oomycetes, p. 697
Oomycete-Plant Interactions: Current Issues in, p. 843
Population Genetics of Plant Pathogenic Fungi, p. 1046
Rhizosphere Management: Microbial Manipulation for Biocontrol, p. 1098

REFERENCES

1. Cook, R.J.; Baker, K.F. *The Nature and Practice of Biological Control*; The American Phytopathological Society: St Paul, 1983.
2. Harman, G.E. Myths and dogmas of biocontrol—Changes in perceptions derived from research on *Trichoderma harzianum T-22*. Plant Dis. **2000**, *84* (4), 377–393.
3. Hornby, D. Suppressive soils. Annu. Rev. Phytopathol. **1983**, *21*, 65–85.
4. Whipps, J.M. Microbial interactions and biocontrol in the rhizosphere. J. Exp. Bot. **2001**, *52*, 487–511.
5. Alabouvette, C. Fusarium wilt suppressive soils: An example of disease-suppressive soils. Aust. Plant Pathol. **1999**, *28* (1), 57–64.
6. Hoitink, H.A.J.; Boehm, M.J. Biocontrol within the context of soil microbial communities: A substrate-dependent phenomenon. Annu. Rev. Phytopathol. **1999**, *37*, 427–446.
7. Weller, D.M.; Raaijmakers, J.M.; Gardener, B.B.M.; Thomashow, L.S. Microbial populations responsible for specific soil suppressiveness to plant pathogens. Annu. Rev. Phytopathol. **2002**, *40*, 309–348.
8. Schneider, R.W. *Suppressive Soils and Plant Disease*; The American Phytopathological Society: St. Paul, 1982.

Biological Control of Weeds

Linda M. Wilson
Mark Schwarzlaender
University of Idaho, Moscow, Idaho, U.S.A.

B

INTRODUCTION

Biological control of weeds is the deliberate introduction and release of a weed's natural enemies with the goal of reducing the weed infestation. To date, biocontrol of weeds has been limited to nonindigenous, exotic weeds. Insects are the most widely used organisms in weed biological control programs worldwide. Other natural enemies include mites, nematodes, and plant pathogens. Biological control of terrestrial and aquatic weeds using insects is described, and important approaches, strategies, and considerations in biological weed control are discussed.

DEVELOPING A BIOLOGICAL CONTROL OF WEEDS PROGRAM

Through global commerce and tourism, humans have traveled the world, consciously or unconsciously accompanied by plants. Non-indigenous plant species have been intentionally or accidentally introduced into many, if not most, countries around the world. Examples of introductions are purple loosestrife (*Lythrum salicaria*) and toadflax (*Linaria* spp.); both are valued as ornamental plants in Europe, but are major weeds in North America. Exotic plant species become invasive in the new location because they encounter favorable environments, adequate disturbance, and reduced competition, and they also lack their specialized natural enemies. In their native plant communities, these plant species do not attain pest densities; their natural enemies, a system of insects, mites, nematodes, and pathogens, regulate their population density.[1]

Nearly all of the terrestrial and aquatic weeds targeted for biological control worldwide are exotic, nonindigenous, perennial plants.[2] Uncontrolled expansion of invasive plants into large, monospecific stands of weeds, on relatively low-value land, rendered conventional weed control methods impractical; it is in these situations that biological weed control has been most effective. One approach, classical biological control, is the introduction and release of one or more natural enemies that are expected to establish, multiply, disperse, and damage their host weed with little or no manipulation following introduction and release into the environment.

Since the late 18th century, biological control has become a useful and important tool in pest management. The first record of an intentional biological control of weeds program was the introduction of a cochineal scale insect (*Dactylopius ceylonicus*) into India in 1795 from Brazil to control an introduced cactus (*Opuntia vulgaris*).[3] Successful, large-scale biological control of weeds began in the 1920s with the importation of the moth *Cactoblastis cactorum* from Argentina for the control of prickly pear cactus (*Opuntia* spp.) in Australia. The first use of biological weed control in the United States was in 1903 with the importation of natural enemies into Hawaii for the control of lantana (*Lantana camara*), followed in 1939 with the importation of two leaf-feeding beetles (*Chrysolina* spp.) for the control of St. Johnswort (*Hypericum perforatum*).[4] Between 1950 and 1998, programs were initiated for over 130 additional terrestrial and aquatic weeds in 57 countries worldwide, involving 490 species of natural enemies.[4] The major regions in which biological control is used include North America, South America, Australia and New Zealand, Central and Southern Africa, and Indonesia and the Pacific Islands.[5]

When successful, biological control is cost-effective, environmentally safe, and sustainable. Limitations to biological control include the time required to develop programs, and the fact that effects are not immediate and not always adequate. Biological control has a good record of environmental safety, but it is not without risks. These risks include threats to non–target plant species and resulting indirect ecological effects to the environment.[6] Careful selection, screening, and monitoring of potential biological control agents is imperative to minimize risks to native and other non–target plants. To emphasize safety, foreign exploration, quarantine, rearing, release, and host specificity testing all follow a comprehensive set of protocols. In considering a biological control program for any given weed, researchers evaluate the extent of the weed problem and its suitability for biological control, the potential natural enemies, the feeding range of the potential organisms, and their general suitability as biological control agents.

Encyclopedia of Plant and Crop Science
DOI: 10.1081/E-EPCS 120020283

Table 1 Selected terrestrial and aquatic weeds, their most commonly used biological control agents, and location

Target weed	Natural enemy	Type	Location
Alligatorweed	*Agasicles hygrophila*	flea beetle	Australia, China,
Alternanthera philoxeroides	*Vogtia malloi*	stem borer	New Zealand, U.S.A.
Giant salvinia	*Cyrtobagous salviniae*	root weevil	Australia, Southern Africa, Ghana,
Salvinia molesta			South Pacific Islands, India
Gorse	*Apion ulicis*	seed weevil	Australia, New Zealand, U.S.A.
Ulex europaeus	*Agonopterix ulicit*	soft shoot moth	
	Tetranychus lintearius	spider mite	
	Sericothrips staphylinus	thrips	
	Agonopterix nervosa	tip moth	
Hawkweeds	*Aulacidea subterminalis*	gall wasp	New Zealand
Hieracium spp.	*Microlabis pilosellae*	gall midge	
	Chelosia urbana	root-feeding hover fly	
Houndstongue	*Mogulones cruciger*	root weevil	Canada
Cynoglossum officinale	*Longitarsus quadriguttatus*	root-feeding flea beetle	
Knapweeds	*Agapeta zoegana*	root moth	Canada, U.S.A.
Centaurea spp.	*Cyphocleonus achates*	root weevil	
	Larinus minutus, Larinus obtusus	seed head weevil	
	Metzneria paucipunctella	seed head moth	
	Sphenoptera jugoslavica	root gall beetle	
	Urophora affinis,	seed head gall fly	
	Urophora quadrifasciata		
	Chaetorellia acrolophi	seed head fly	
	Bangasternus fausti	seed head weevil	
	Pterolonche inspersa	root moth	
Leafy spurge	*Aphthona* spp. (6 species)	root beetle	Canada, U.S.A.
Euphorbia esula	*Chamaesphecia hungarica*	root moth	
	Hyles euphorbiae	defoliating moth	
	Oberea erythrocephala	root beetle	
	Spurgia esulae	bud gall midge	
	Pegomya curticornis	root fly	
Mediterranean sage	*Phrydiuchus tau*	root weevil	U.S.A.
Salvia aethiopis			
Parthenium weed	*Zygogramma bicolorata*	leaf beetle	Australia, India
Parthenium hysterophorus	*Epiblema strenuana*	stem-galling moth	
	Bucculatrix parthenica	leaf-mining moth	
	Listronotus setosipennis	stem-boring weevil	
Paterson's Curse	*Longitarsus echii*	flea beetle	Australia, South Africa
Echium plantagineum	*Meligethes planiusculus*	pollen beetle	
	Dialectica scalariella	leaf-mining moth	
	Mogolunes larvatus	crown weevil	
	Mogolunes geographicus	root weevil	
Prickly pear and other cacti	*Cactoblastis* spp.	stem-boring moth	Australia, India, Indonesia,
Opuntia spp.	*Dactylopius* spp.	scale insect	New Zealand, U.S.A.
	Chelinidea tabulata	leaf-sucking bug	
Purple loosestrife	*Galerucella calmariensis,*	defoliating weevil	U.S.A.
	Galerucella pusilla		
Lythrum salicaria	*Hylobius transversovittatus*	root weevil	
	Nanophyes marmoratus	root weevil	
Rush skeletonweed	*Bradyrrhoa gilveolella*	root moth	Australia, U.S.A.
Chondrilla juncea	*Cystiphora schmidti*	gall midge	
	Eriophyes chondrillae	gall mite	
	Puccinia chrondrillina	rust fungus	
Scotch broom	*Apion fuscirostre*	seed weevil	U.S.A., New Zealand, Australia

(*Continued*)

Table 1 Selected terrestrial and aquatic weeds, their most commonly used biological control agents, and location (*Continued*)

Target weed	Natural enemy	Type	Location
Cytisus scoparius	*Bruciodius villosus*	seed beetle	
	Leucoptera spartifoliella		
	Arytainilla spartiophila		
St. Johnswort	*Chrysolina hyperici*, *Chrysolina quadrigemina*	leaf beetle	Canada, Chile, South Africa, U.S.A.
Hypericum perforatum	*Agrilus hyperici*	root-boring beetle	
Tansy ragwort	*Pegohylemyia seneciella*	seed head fly	U.S.A., Canada
Senecio jacobaea	*Tyria jacobaeae*	defoliating moth	
	Longitarsus jacobaeae, *Longitarsus flavicornis*	flea beetle	
	Cochylis atricapitana	crown-boring moth	
Thistles	*Rhinocyllus conicus*	seed weevil	U.S.A., Canada, Australia, New Zealand, South Africa
Carduus, Cirsium,	*Urophora stylata*	gall fly	
Onopordum spp.	*Trichosirocalus horridus*	rosette weevil	
	Larinus latus	seed weevil	
Toadflax	*Gymnetron linariae*	root-galling weevil	U.S.A., Canada
Linaria spp.	*Eteobalea intermediella*	root moth	
	Mecinus janthinus	stem-boring weevil	
	Calophasia lunula	defoliating moth	
	Brachypterolus pulicarius	seed weevil	
Waterhyacinth	*Neochetina eichhorniae*, *Neochetina bruchi*	leaf beetle	U.S.A., Australia, Africa, Malaysia
Eichhornia crassipes	*Sameodes albiguttalis*	petiole moth	
	Niphograpta albiguttalis	stem-boring moth	
	Thripiticus spp.	stem fly	
Yellow starthistle	*Bangasternus orientalis*	seed head weevil	U.S.A., Canada
Centaurea solstitialis	*Chaetorellia australis*, *Chaetorellia succinea*	seed head fly	
	Eustenopus villosus	seed head weevil	
	Larinus curtus	seed head weevil	
	Urophora sirunaseva	seed head gall fly	

BIOLOGICAL WEED CONTROL AGENTS

Phytophagous arthropods (insects and mites) comprise over 70% of the species imported for biological control of terrestrial and aquatic weeds worldwide.[5] Insects are often selected as biological control agents because they are relatively easy to study, and many species exhibit a high degree of host specificity, feeding on only a few plant species closely related to the target weed. Host range is the set of plant species on which a natural enemy can feed, develop, and complete its life cycle. Those selected for biological control must have a very narrow host range; ideally, they survive only on the target weed. Insects that feed inside the plant are more likely to be host-specific and have a more highly evolved association with the target weed than do externally feeding insects.[1] In addition, a number of insect species may occur on a single plant or within the same part of the plant. For example, spotted knapweed (*Centaurea stoebe*) root feeders include three moth and two beetle species.[7] Table 1 provides a partial list of the major groups of biological control agents established on weeds worldwide.

Insects damage weeds either directly or indirectly. Direct feeding can severely injure or kill the plant. For example, insects suck out plant fluids, defoliate, bore into roots, shoots, and stems, and eat flowers or seeds. Indirect damage, resulting from secondary pathogen infections, can also weaken or kill the plant.

The insect orders most commonly used in weed biological control are beetles (Coleoptera), flies (Diptera), and moths (Lepidoptera) and to a lesser degree wasps (Hymenoptera), true bugs (Hemiptera), and thrips (Thysanoptera). Beetles have been the most effective and successful insects used in terrestrial and aquatic weed biocontrol.[5] Families of beetles used in biological control are weevils (Curculionidae), leaf beetles (Chrysomelidae), long-horn beetles (Cerambycidae), wood-boring beetles (Buprestidae), and seed beetles (Bruchidae). Beetles may feed on seed heads, bore in stems and roots, or defoliate plants.

Flies such as gall midges (Cecidomyiidae) or fruit flies (Tephritidae) are used in many terrestrial weed biocontrol programs. Examples of gall midges include the rush skeletonweed stem and leaf gall midge (*Cystiphora schmidti*) and the leafy spurge shoot tip gall midge (*Spurgia esulae*). Examples of seed-feeding fruit flies are the knapweed gall flies (*Urophora* spp. and *Terellia virens*) and the yellow starthistle seed flies (*Chaetorellia* spp.). Moths are also widely used in weed biocontrol programs; they are typically seed-feeders or root-borers. Examples of moths include the leafy spurge root moth (*Chamesphecia* spp.), the parthenium weed stem boring moth (*Epiblema strenuana*), and the ragwort defoliating moth (*Tyria jacobaeae*). Examples of less commonly used types of insects in weed biological control include the hawkweed gall wasp (*Aulacidea subterminalis*), the prickly pear defoliating bug (*Dactylopius opuntiae*), and the gorse shoot thrips (*Seriothrips staphylinus*).

Although not insects, mites, with their two body regions, four pairs of legs, and microscopic size, are also used in weed biocontrol programs. Both gall mites and spider mites damage the plant by sucking valuable fluids from the stems and leaves; gall mites also cause plants to develop galls. Examples are the rush skeletonweed gall mite (*Eriophyes chondrillae*) and the gorse spider mite (*Tetranychus linearius*).

When initiating a weed biological control program, researchers survey natural enemies associated with the target weed in its native range. Candidate insect species are selected depending on their degree of specificity with the host, their ease of collecting and rearing, and their impact on the weed.[8] Extensive host specificity studies are conducted over several years to determine that the insects attack only the target weed and a few closely related species, and do not pose a threat to species unrelated to the target plant, commercial or horticultural species, or threatened and endangered species located in the area where the insect is intended for introduction. Accurate identification of the selected insects and the host plant is vitally important in biological control of weeds.

The success of biological control of weeds programs historically has been based on the degree to which the weed population declined following the introduction of the biological control agent(s). Harris[9] proposed a sequential, four-step process for evaluating the success of a biological control release, which includes establishment success, biological success, host impact, and control success. Cullen[10] further proposed that the ultimate success of an insect in weed biological control is a result of a combination of three major factors: 1) the damage an individual insect can do to a plant, 2) the ecology of the insect in determining its density, and therefore, the total damage produced, and 3) the ecology of the weed in determining if the damage is significant in reducing its population. These categories of success provide evaluation of biological weed control within a practical context. Objective assessment in weed biological control programs using manipulative experiments[10] and economic cost-benefit analyses[11] are also being applied to evaluate weed biocontrol programs.

CONCLUSION

The aim of biological control of weeds is not to eradicate weed, but rather to suppress the weed to a more acceptable density level. Biological control insects can affect weeds in plant communities by killing them or suppressing their competitive ability in the community. Land managers should note that successful biological suppression of a weed is not strictly the weed management end point. Without further management and intervention, the successional vegetation may be as undesirable. Very likely the opening created by suppression of one weed will be filled with an unrelated, potentially more invasive weed that, after gaining dominance, may eventually become a future target for biological control. Care must be taken to avoid this ''biological control treadmill''[12] and to prevent the need for and expense of introducing new organisms into the environment. In the most successful integrated weed management programs, biological weed control has combined together with competition from perennial grasses, prescribed grazing, and herbicides to achieve successful weed suppression and restoration of desirable, perennial plant communities.

ARTICLES OF FURTHER INTEREST

Chemical Weed Control, p. 255
Insect Host Plant Resistance in Crops, p. 605
Insect–Plant Interactions, p. 609
Integrated Pest Management, p. 612
Weed Management in Less Developed Countries, p. 1295

REFERENCES

1. Strong, D.R.; Lawton, J.H.; Southwood, R. *Insects on Plants: Community Patterns and Mechanisms*; Harvard Univ. Press: Cambridge, MA, 1984.
2. Julien, M.H. Biological control of weeds worldwide: Trends, rates of success and the future. Biocontrol News Inf. **1989**, *10*, 299–306.
3. Moran, V.C.; Zimmerman, H.G. The biological control of cactus weeds: Achievements and prospects. Biocontrol News Inf. **1984**, *3*, 18–26.

4. Goeden, R.D. A capsule history of biological control of weeds. Biocontrol News Inf. **1988**, *9,* 55–61.
5. Julien, M.H.; Griffiths, M.W. *Biological Control of Weeds: A World Catalogue of Agents and Their Target Weeds,* 4th Ed.; CABI Publishing: Wallingford, U.K., 1998.
6. McFadyen, R.E.C. Biological control of weeds. Annu. Rev. Entomol. **1998**, *43,* 369–393.
7. Wilson, L.M.; Randall, C.B. *Biology and Biological Control of Knapweeds*, FHTET-2001-07; USDA Forest Service: Washington, D.C., 2002.
8. Harris, P. The selection of effective agents for the biological control of weeds. Can. Entomol. **1973**, *105,* 1495–1503.
9. Cullen, J.M. In *Predicting Effectiveness: Fact and Fantasy,* Proc. VIII Int. Symp. Biol. Contr. Weeds, Lincoln Univ., Canterbury, New Zealand, February 2–7, 1992; Delfosse, E.S., Scott, R.R., Eds.; DSIR/CSIRO: Melbourne, 1995; 103–109.
10. McEvoy, P.B.; Rudd, N.T.; Cox, C.S.; Huso, M. Disturbance, competition, and herbivory effects on ragwort *Senecio jacobaea* populations. Ecol. Monogr. **1993**, *63,* 55–75.
11. Tisdell, C.A.; Auld, B.A. In *Evaluation of Biological Control Projects*, Proc, VII Int. Symp. Biol. Contr. Weeds, Rome, Italy, March 6–11, 1988; Delfosse, E.S., Ed.; Ist. Sper. Patol. Veg. (MAF): Rome, 1989; 93–100.
12. McCaffrey, J.P.; Wilson, L.M. In *Assessment of Biological Control of Exotic Broadleaf Weeds in Intermountain Rangelands*, Proc, Symp. on Ecology and Management of Annual Rangelands, Boise, ID, May 18–22, 1992; Monsen, S.B., Kitchen, S.G., Eds.; USDA-ARS: Logan, UT, 1994; 101–102. INT-GTR-313.

B

Biosafety Applications in the Plant Sciences: Expanding Notions About

Emily E. Pullins
University of Minnesota, St. Paul, Minnesota, U.S.A.

INTRODUCTION

The expanding and evolving definition of biosafety has important implications for efforts to manage and safely use plants that are modified using modern molecular methods (biotechnology). The concept of biosafety has evolved over forty years, beginning with the recognition of hazards derived from biological science activities. Biohazards were defined as early as 1965 as a ''risk to mankind or the environment, especially. one arising out of biological or medical work.''[1] This definition included hazardous living microorganisms such as bacteria, nonliving hazardous biological matter such as viruses or prions, as well as human error and material waste byproducts that could facilitate the agency of these hazards, such as the improper disposal of used syringes.

WHAT IS BIOSAFETY?

The term biosafety, coined in 1977, was defined as processes that assured the safe use of potential biohazards through the systematic and system-wide implementation of safety programs, most often within the context of a laboratory or clinical setting.[2] With the advent of modern molecular methods, some scientists raised concerns about novel hazards in organisms altered with these methods, such as recombinant DNA technology.[3] Thirty years of public debate about the likelihood and assessment of these organisms as potential hazards have entered the plant sciences as plants have been altered through these same techniques. Some microbiologists take issue with the inclusion of plants and animals derived using biotechnology in the formal biosafety definition, because this may cause the public to unnecessarily doubt the safety of these products, and may distract from occupational or public safety concerns about traditionally recognized biohazards.[4]

Debates among plant scientists have pivoted on what constitutes an appropriate biosafety evaluation and management process, particularly focusing on the extent to which gaps in information and unanticipated effects of these organisms should govern their use.[5] Narrower definitions of biosafety science tend to focus primarily on risk assessment and management tasks related to a single trait within a single species derived using biotechnological methods; these may be limited further by the temporal and spatial aspects of assessment. For example, risk assessments may occur prior to commercialization, focused on risk issues related to approved and expected use of the trait, and within a spatial range smaller than that after commercial release of the product. Some biologists argue that, based on existing scientific evidence from these risk assessments, there are no ''immediate or significant risks'' to environmental and food biosafety in the case of the array of plants most recently modified by biotechnological methods.[6]

Broader definitions of biosafety consider additional risk issues, such as when genetic material, related traits, living organisms, or their nonliving products developed using biotechnological methods transcend anticipated physical or biological boundaries, temporal or spatial scales, or particular political jurisdictions.[7] The settings in which safety evaluation and management of plants derived from modern molecular methods as potential hazards has been expanded by some to include both inside and outside controlled laboratory settings; in the trade of products across international borders; and in public, occupational, and environmental spheres. Broader biosafety concepts may also include a proactive safety program approach, in which risk assessment and management are tools used within a comprehensive safety prioritization framework;[8] additional activities may include the formation of safety standards to govern the development, use, testing, and monitoring of these products.

These various notions of biosafety have been used for over a decade in vernacular and professional settings and are prominent in international and national policy documents. There is now political support for the notion that ''[e]very country that wants to use GM [genetic modification] technology also needs to establish a strong biosafety system to ensure that any risks that these new crops might bring can be handled safely.''[9] As a result, considerable efforts are being made to develop and use new scientific, technological, and political safety evaluation and management methods to address potentially hazardous aspects of these organisms in use.[10–12] Recent critiques of some national policies related to the safety of

Encyclopedia of Plant and Crop Science
DOI: 10.1081/E-EPCS 120010601

transgenic plants reveal that governance of these organisms tends to consist of a patchwork of risk-focused government regulations that, in some instances, do not comprehensively address present or future safety concerns.[13–15] Several organizations have initiated institutional-capacity building activities in developing countries to facilitate biosafety research and governance programs;[16,17] concerns about funding for national and international biosafety programs is a consistent theme in reports on these activities. The U.N.'s Convention on Biological Diversity is responding to such critiques by moving toward the establishment and support of formal biosafety programs based on the broader definition of biosafety, which include setting standards for acceptable risks, validation of scientific information related to risks such as post-commercialization testing and monitoring of products for safety, and training for safe management of biotechnologies.[18,19] Most notable among the international policy activities addressed by this convention was the development of the Cartagena Protocol on Biosafety, which sought to "contribute to ensuring an adequate level of protection in the field of the safe transfer, handling, and use of living modified organisms resulting from modern biotechnology that may have adverse effects on the conservation and sustainable use of biological diversity, taking also into account risks to human health, and specifically focusing on transboundary movements."[18] This protocol was adopted by consensus at the Extraordinary Conference of the Parties to the Convention on Biological Diversity, January 29, 2000, and has been ratified by 23 countries.[20] Substantive support for recommendations made in the protocol has come from The Global Environment Facility through the United Nations Environment Program, which has contributed over 3.6 million dollars to pilot national biosafety frameworks and to sponsor related activities worldwide.[21,22]

LOOKING AHEAD

Biosafety presents new challenges to thinking about safe practices in the relatively new agricultural biotechnology industry, not the least of which are the myriad safety evaluations necessary for use of or exposure to such products in multiple spheres, including occupational and worker safety, environmental toxicology and chemical safety, environmental ecological safety, food safety, consumer safety, and medical or pharmaceutical safety. Assuring biosafety across these spheres may require more than expanded definitions and policy change. Effective and efficient safety evaluations for products that are derived from new biological technologies may require both a coordination of existing, diverse safety expertise, and the construction of new formal safety programs that prioritize novel safety issues related to these products. This approach puts safety as a singular priority in the process of engineering—thinking systematically about the conception, implementation, and use of an engineered design in order to prevent harm; this includes, and goes beyond, risk assessment and management activities.[23] The rich history of formal safety program advocacy and development in the previous century, which was in part motivated by intense social and philosophical debates, resembles current controversies that have fueled the expansion of the biosafety concept.[24] Therefore, future activities that may provide governance and direction to biosafety programs in the agricultural biotechnology industry may closely resemble the formation of safety programs in other industries. For the agricultural biotechnology industry, the adoption of the expanded notion of biosafety principles and practices would represent an historical opportunity to contribute to new developments and discoveries in the safety engineering sciences.

Formal safety engineering programs in other industries that serve as models are those that work to establish safety standards and a framework for an industry-wide safety program for products using a process that utilizes the elements of criteria setting, verification, follow-up, and safety leadership.[a] In criteria setting, safety criteria are developed to systematically analyze the possible harm of a product. This involves the rigorous identification of hazards, the assessment of risk, and planning to reduce risk. Establishing a complete and scientifically reliable set of safety design criteria rests on two requirements: establishing rigorous criteria at the outset of development of a new product and independently validating those criteria before they are used. Verification is the design of rigorous tests to fully challenge the product and credibly demonstrate that it meets the pre-set and government-approved safety criteria. Designing these tests requires the application of the best available scientific methodologies and information from all relevant fields. Follow up efforts require the setting of criteria and conducting of tests that verify the product continues to meet its safety criteria in commercial use. Open-minded and scrupulous monitoring of the product in all its uses is also required for effective follow-up; the discovery of problems needs to be followed up with meaningful and timely corrective action. Safety leadership assures the consistent and proper implementation of the criteria setting, verification, and follow-up procedures.

Safety leadership needs to occur in three areas. The first area is the establishment of rigorously trained and independently certified safety engineers who are valued employees of industry firms. The second area is the

[a]From Ref. 23.

encouragement of a company management style that fosters broad thinking, application of the best scientific methodologies and information, self-imposed responsibility to make safe products, responsiveness to evidence of real hazards and problems, and independent review of all aspects of the product safety program. The third area is the creation of a framework for managing the application of industry-wide safety standards, including an independent audit function. Assuming that the trend toward safety-oriented management programs for plants derived using new biological technologies continues, the life science industries may adopt these elements of safety programs to form a more extensive framework for biosafety. By wedding such programs to a transparent, public deliberation process, we may better address the fundamental fears of consumers and achieve successful negotiations about when products are safe enough.

CONCLUSION

This section is an introductory review of the science and policy efforts that inform biosafety science and policies for plant crops derived from agricultural biotechnology, and includes speculation regarding future directions for this trend. Topics covered in this section include an overview of scientific research on risk issues specific to transgenic or genetically engineered crop plants, such as gene flow between such plants, the persistence of pesticides in soils that contain transgenic crop plant residue, management for insect resistance to pest-protected crop plants, and considerations for management of disease-resistant transgenic crop plants. In addition, a review of adaptive biosafety assessment and management principles and practices provides a model for biosafety management of transgenic crop plants that incorporates lessons learned from a wide array of scientific and political specializations.

ARTICLES OF FURTHER INTEREST

Biosafety Approaches to Transgenic Crop Plant Gene Flow, p. 150
Biosafety Considerations for Transgenic Insecticidal Plants: Non-Target Herbivores, Detritivores, and Pollinators, p. 153
Biosafety Considerations for Transgenic Insecticidal Plants: Non-Target Predators and Parasitoids, p. 156
Biosafety Programs for Genetically Engineered Plants: An Adaptive Approach, p. 160
Biosafety Science: Overview of Plant Risk Issues, p. 164
Transgenes (GM) Sampling and Detection Methods in Seeds, p. 1238
Transgenetic Plants: Breeding Programs for Sustainable Insect Resistance, p. 1245
Transgenic Crop Plants in the Environment, p. 1248
Transgenic Crops: Regulatory Standards and Procedures of Research and Commercialization, p. 1251

REFERENCES

1. *Oxford English Dictionary Online,* 2nd Ed.; Simpson, J., Ed.; Oxford University Press: Oxford, England, 2002.
2. Gilpin, R.W. Elements of a Biosafety Program. In *Biological Safety Principles and Practices,* 3rd Ed.; Fleming, D.O., Hunt, D.L., Eds.; Amer. Soc. for Microbiology: Washington, D.C., 2000; 443.
3. Wade, N. Genetics: Conference sets strict controls to replace moratorium. Science **1975**, *187*, 931–935.
4. Doblhoff-Dier, O.; Collins, C.H. Biosafety: Future priorities for research in health care. J. Biotech. **2001**, *5*, 227–239.
5. Gupta, A. *Framing "Biosafety" in an international context. ENRP Discussion Paper E-99-10*; Kennedy School of Government, Harvard University, 1999. http://environment.Harvard.edu/gea.
6. Stewart, N.S., Jr.; Richards, H.A., IV; Halfhill, M.D. Transgenic plants and biosafety: Science, misconceptions, and public perceptions. Biotechniques **2000**, *29*, 832–843.
7. Kapuscinski, A.R.; Nega, T.; Hallerman, E. In *Adaptive Biosafety Assessment and Management Regimes for Aquatic Genetically Modified Organisms in the Environment*, Towards Policies for Conservation and Sustainable Use of Aquatic Genetic Resources: A Think Tank, ICLARM Conference Proceedings 59, Bellagio, Como, Italy, April 14–18, 1998; Pullin, R.S.V., Bartley, D.M., Kooiman, J., Eds.; ICLARM: Penang, Malaysia, 1999; 225–251. Described the case of aquatic genetically modified organisms.
8. Making Safety First A Reality: Final Report of the March 2–3, 2001 Workshop; Safety First: Active Governance of Genetic Engineering for Human and Environment Health Worldwide Workshop, Institute for Social, Economic and Ecological Sustainability, University of Minnesota, March 2–3, 2001; Kapuscinski, A.R., Jacobs, L.R., Pullins, E.E., Eds.; ISEES: Minneapolis, Minnesota, August, 2001. http://www.fw.umn.edu/isees/biotech/sfinit.htm.
9. Raeves, J. Are first world fears causing the third world to go hungry? Time.com **2001, 9 July**. http://www.time.com/time/world/printout/0,8816,166925,00.html. Quoting K. Raworth, UNDP Economist.
10. Report on the Third Session, Joint FAO/WHO Food Standard Programme Codex Alimentarius Commission Meeting, Yokohama, Japan, March 4–8, 2002; FAO/WHO: Rome, Italy, 47–73. In *Codex Ad Hoc Intergovernmental Task Force on Food Derived from Biotechnology. Appendix II. Draft Principles for the Risk Analysis of*

Foods Derived from Modern Biotechnology. ftp://ftp.fao.org/ codex/alinorm03/Al03_34e.pdf.

11. Scientists' Working Group on Biosafety. In *Manual for Assessing Ecological and Human Health Effects of Genetically Engineered Organisms. Part One: Introductory Text and Supporting Text for Flowcharts. Part Two: Flowcharts and Worksheets*; The Edmonds Institute: Edmonds, Washington, 1998. http://www.edmonds-institute.org/manual.html.
12. *Genetically Engineered Organisms: Assessing Environmental and Human Health Effects*; Letourneau, D.K., Burrows, B.E., Eds.; CRC Press: New York, 2001.
13. Committee on Environmental Impacts Associated with Commercialization of Transgenic Crops. *Environmental Effects of Transgenic Crops: The Scope and Adequacy of Regulation*; National Academy Press: Washington, DC, 2002.
14. Committee on Genetically Modified Pest-Protected Plants. *Genetically Modified Pest-Protected Plants: Science and Regulation*; National Academy Press: Washington, DC, 2000.
15. Expert Panel on the Future of Food Biotechnology. *Executive Committee on the Future of Food Biotechnology Report*; Royal Soc. Can.: Ottawa, Ontario, Canada, 2001. http://www.rsc.ca/foodbiotechnology/GMreportEN.pdf.
16. United States Agency for International Development. Appendix 1—Agricultural Biotechnology Support Project Evaluation. In *Request for Applications; Program for Biosafety Systems*; USAID: Washington, DC, 2002; 18–65.
17. James, C.; Krattiger, A.F. The ISAAA Biosafety Initiative: Institutional Capacity Building Through Technology Transfer. In *Biosafety for Sustainable Agriculture: Sharing Biotechnology Regulatory Experiences of the Western Hemisphere*; Krattiger, A.F., Rosemarin, A., Eds.; ISAAA: Ithaca, NY, 1994; 225–237. http://www.isaaa.org/publications/biosafety/Biosafety_bk.pdf.
18. Secretariat of the Convention on Biological Diversity. *Cartagena Protocol on Biosafety to the Convention on Biological Diversity*; UNEP: Montreal, 2000. http://www.biodiv.org/doc/legal/cartagena-protocol-en.pdf.
19. Secretariat of the Convention on Biological Diversity. *CBD Biosafety Capacity Building Projects and Other Initiatives Database*; CBD: Washington, DC, 2000. http://bch.biodiv.org/pilot/CapacityBuildingStart.asp.
20. Secretariat of the Convention on Biological Diversity. *Cartagena Protocol on Biosafety Signatures and Ratifications*; CBD: Washington, DC, 2002. http://www.biodiv.org/biosafety/signinglist.asp?order=date.
21. Secretariat of the Global Environment Facility. *GEF Action on Biodiversity*; GEF: Washington, DC, 2000. http://www.gefweb.org/Projects/Focal_Areas/BiodiversityBooklet.pdf.
22. Secretariat of the Global Environment Facility, GEF Council. *Approves 20 Projects Worth \610.9 Million*; GEF: Washington, DC, 17 May, 2002. http://www.gefweb.org/Outreach/Media/Press_Releases/May_2002_work_program_press_release.pdf.
23. Final Report of the April 22, 2002 Meeting of the Executive Advisory Board and Steering Committee of the Safety First Initiative, Institute for Social, Economic and Ecological Sustainability, University of Minnesota, April 22, 2002; In *Safety First: Making It a Reality for Biotechnology Products*, Kapuscinski, A.R., Hann, S.D., Jacobs, L.R., Pullins, E.E., Stenquist, B.C.K., Eds.; ISEES: Minneapolis, Minnesota, August, 2002. http://www.fw.umn.edu/isees/biotech/sfreport.pdf.
24. Aldrich, M. *Safety First: Technology, Labor and Business in the Building of American Worker Safety 1870–1939*; Johns Hopkins University Press: Baltimore, 1997.

Biosafety Approaches to Transgenic Crop Plant Gene Flow

Terrie Klinger
University of Washington, Seattle, Washington, U.S.A.

INTRODUCTION

Gene flow from transgenic (or GMO) crops to wild relatives will introduce transgenes into wild plant populations at rates in excess of those in the absence of the commercial release of GMO crops. Unintended introduction of transgenes confers potential hazard to agricultural ecosystems and to wild populations and the ecosystems in which they exist. Hazards associated with transgene introduction by gene flow can be eliminated only when the likelihood of crop-to-wild hybridization is zero. When the likelihood of hybridization is greater than zero, precautionary management approaches require the implementation of risk-averse strategies that appropriately minimize potential negative impacts to populations and the environment. Effective management strategies will take into account socially acceptable levels of risk, will incorporate uncertainty due to spatial and temporal variability in gene flow and to error associated with the estimation of gene flow, and will become increasingly precautionary as levels of uncertainty increase.

HYBRIDIZATION AND INTROGRESSION

Conventional crop plants are known to hybridize with their wild relatives where they occur within mating distance.[1,2] That some transgenic (or GMO) crops are or will be capable of effecting similar crop-to-wild matings is undisputed. Only where hybridization is impossible (e.g., where crops and wild relatives do not co-occur) can the risks of hybridization and introgression be assumed to be nil. In all other cases, development of risk-averse strategies for the management of hybridization and its associated impacts is appropriate. It is important to note in this context that the apparent inability to hybridize (e.g., due to apparent incompatibility or sterility) does not necessarily confer an absolute or permanent inability to hybridize.

Hybridization between crops and wild relatives can lead to gene introgression and spread through wild populations. Introgression and spread are most likely to occur when the introduced traits confer a selective advantage, for example by increasing fitness due to higher rates of reproduction or survival. Introgression can alter genetic structure within recipient populations, and in extreme cases can increase the risk of extinction in small populations.[3] Where GMO crops are capable of hybridizing with weedy or wild plants, their introduction can lead to the introgression and spread of transgenes through non-target populations. Some traits introduced by transgenes are likely to be undesirable in wild populations. For example, the generation of herbicide-resistant weeds could negatively impact both wild and agricultural populations by altering competitive interactions in favor of hybrids. Other environmentally relevant traits could be conferred to wild plants through hybridization with GMOs, including traits such as increased drought or salt tolerance, or changes in mating success, life history traits, or competitive abilities. The accelerating use of GMO crops and the growing number of genetic constructs borne by individual GMOs increases the likelihood that multiple engineered genes will be introduced simultaneously or sequentially into wild populations. The occurrence of multiple introductions will substantially reduce our ability to predict the outcome of such introductions, because the complexity of the problem grows with the number of new introductions, and because interactions between multiple engineered genes in wild-type backgrounds are difficult or impossible to anticipate.

While hybridization does not always constitute a limiting step in gene introgression among plants, it does constitute an important first step in the introgression and spread of transgenes, and therefore is an important factor to consider in the development of risk-averse management strategies that seek to minimize the negative impacts of transgenes released into the environment. One essential element of such management strategies is the incorporation of uncertainty into the estimation of gene flow between transgenic crops and wild relatives.

FACTORS AFFECTING RATES OF HYBRIDIZATION

Numerous factors are known to affect rates of hybridization in plants. Genetic factors (genotype, mating system,

Encyclopedia of Plant and Crop Science
DOI: 10.1081/E-EPCS 120010604

and compatibility between mates) and biological and physical aspects of populations and the local environment (size, shape, number, density and distance between donor and recipient populations, pollinator behavior and abundance, and spatial and temporal variability in physical factors such as temperature and rainfall) all can affect hybridization rates.[4] Consequently, spatial and temporal variability in rates of hybridization can be quite high, hindering confident estimation of distance-dependent gene flow.

Distance-dependent hybridization rates typically are reported as estimated average values.[5] However, high levels of variability in rates of gene flow render average measures inappropriate for management purposes, because average measures can dramatically underestimate maximum rates of hybridization,[4] leading to substantial underestimation of risk and of potential adverse effects.

Several studies have described "hot spots" of hybridization among crop plants or between crops and wild relatives that could not have been predicted from average rates of hybridization (e.g., Refs. 4 and 6). Significantly, discrepancies between average and maximum rates of hybridization have been found to increase with distance between donor and recipient populations. For one case in which hybridization between cultivated and wild radish (*Raphanus sativus*) was studied in an agricultural field setting, observed maximum rates of crop-to-wild hybridization at 1000 m distance from the crop exceeded average hybridization rates by a factor of about 30.[4] The same study found that the observed number of hybrid progeny at 1000 m distance from the crop exceeded the expected number of hybrids by a factor of about 315. These findings imply that actual rates of hybridization can exceed average rates by a substantial amount. Therefore, the use of average rates of hybridization in risk assessment is decidedly not risk-averse.

USING MODELS TO DESCRIBE HYBRIDIZATION

Numerical models have been used to describe the expected decay of hybridization rates with distance.[7,8] For example, maximum likelihood techniques have been applied to problems of distance-dependent dispersal of pollen from source populations.[9,10] However, these models require large amounts of case-specific data for adequate parameterization, and such data are unlikely to be available in most instances. Numerical models have yet to be successfully applied to situations in which multiple source populations contribute transgenes to one or more recipient populations, nor are they adequate to describe low-frequency but high-magnitude hybridization events that can lead to hot-spots of hybridization. Models might better be used to help design monitoring programs than to predict hybridization rates; however, the choice of inappropriate models could bias the distribution of monitoring effort, leading to undetected impacts due to sampling error. Adequate management strategies require that policy decisions be made regarding levels of risk that are acceptable to society. Models used to develop management strategies must incorporate these policy decisions in addition to incorporating uncertainty in all model parameters.

RISK-AVERSE STRATEGIES

It must be assumed that crop-to-wild hybridization will occur wherever crops and wild relatives grow within mating distance. Hybridization between transgenic crops and wild populations will impose some degree of hazard, for example, through the displacement or extirpation of native species, or the alteration or loss of ecosystem function. Depending on their individual probabilities of occurrence, these potential hazards will confer risk to wild populations and the environment. Consequently, the most risk-averse strategy is to avoid planting transgenic crops in regions where they co-occur with wild relatives. Where this is impractical or undesirable for economic or social reasons, risk-averse strategies must be developed to account for uncertainty in rates of hybridization and introgression. In particular, it is important that management strategies become increasingly conservative as the amount of uncertainty increases, as the hazard posed by individual or aggregate engineered traits increases, and as the number of engineered traits or donor populations increases. Examples of risk-averse strategies include changes in cropping practices to decrease the potential for hybridization and changes in genetic engineering techniques to decrease the likelihood of gene transfer by pollen.

CONCLUSION

In a world where the human population is expected to increase by 50% over the next 75 years, the expanded use of GMOs to enhance crop production is extremely likely. Changes in human land use practices combined with the effects of global change will place additional and increasing stress on already burdened agricultural and wild ecosystems. These ecosystems will be highly susceptible to further disruption caused by the introduction of transgenic constructs into wild populations, and effects may intensify as the number and magnitude of interacting stressors grows.

Under these conditions, the management of risks associated with transgenic crops and other agricultural products will become extremely important in protecting ecosystem function. A precautionary approach to management requires that management practices become adequately risk-averse *before* substantial negative effects are manifest in agricultural or wild ecosystems. Risk-averse management requires the development of policies that take into account socially-acceptable levels of risk concerning the function of agricultural and wild ecosystems. In addition, risk-averse management strategies must incorporate uncertainty in all parameters used to predict the likelihood of adverse effects of the use of GMOs in crop production.

ARTICLES OF FURTHER INTEREST

Agriculture and Biodiversity, p. 1
Biosafety Applications in the Plant Sciences: Expanding Notions About, p. 146
Biosafety Considerations for Transgenic Insecticidal Plants: Non-Target Herbivores, Detritivores, and Pollinators, p. 153
Biosafety Considerations for Transgenic Insecticidal Plants: Non-Target Predators and Parasitoids, p. 156
Biosafety Science: Overview of Plant Risk Issues, p. 164
Chromosome Manipulation and Crop Improvement, p. 266
Crop Domestication: Fate of Genetic Diversity, p. 333
Gene Flow Between Crops and Their Wild Progenitors, p. 488
Genetically Engineered Crops with Resistance Against Insects, p. 506
Genetically Modified Oil Crops, p. 509
Herbicide-Resistant Weeds, p. 551
Transformation Methods and Impact, p. 1233
Transgenes (GM) Sampling and Detection Methods in Seeds, p. 1238
Transgenes: Expression and Silencing of, p. 1242
Transgenic Crops: Regulatory Standards and Procedures of Research and Commercialization, p. 1251

REFERENCES

1. Ellstrand, N.C.; Prentice, H.C.; Hancock, J.F. Gene flow and introgression from domesticated plants into their wild relatives. Ann. Rev. Ecolog. Syst. **1999**, *30*, 539–563.
2. Jørgensen, R.B.; Andersen, B.; Landbo, L.; Mikkelsen, T. Spontaneous hybridization between oilseed rape (*Brassica napus*) and weedy relatives. Acta Hortic. **1996**, *407*, 193–200.
3. Ellstrand, N.C. Gene flow by pollen: Implications for plant conservation genetics. Oikos **1992**, *63*, 77–86.
4. Klinger, T. Variability and Uncertainty in Crop to Wild Hybridization. In *Genetically Engineered Organisms: Assessing Environmental and Human Health Effects*; Letourneau, D.K., Burrows, B.E., Eds.; CRC Press: Boca Raton, 2002; 1–15.
5. Klinger, T.; Arriola, P.E.; Ellstrand, N.C. Crop-weed hybridization in radish (*Raphanus sativus*): Effects of distance and population size. Am. J. Bot. **1992**, *79* (12), 1431–1435.
6. Hokanson, S.C.; Grumet, R.; Hancock, J.F. Effect of border rows and trap/donor ratios on pollen-mediated gene movement. Ecol. Appl. **1997**, *7* (3), 1075–1081.
7. Kareiva, P.; Morris, W.; Jacobi, C.M. Studying and managing the risk of cross-fertilization between transgenic crops and wild relatives. Mol. Ecol. **1994**, *3*, 15–21.
8. Kareiva, P.; Parker, I.M.; Pascual, M. Can we use experiments and models in predicting the invasiveness of genetically engineered organisms? Ecology **1996**, *77* (6), 1670–1675.
9. Nurminiemi, M.; Tufto, J.; Nilsson, N.-O.; Rognli, O.-A. Spatial models of pollen dispersal in the forage grass meadow fescue. Evol. Ecol. **1997**, *12*, 487–502.
10. Tufto, J.; Engen, S.; Hindar, K. Stochastic dispersal processes in plant populations. Theor. Popul. Biol. **1997**, *52*, 16–26.

Biosafety Considerations for Transgenic Insecticidal Plants: Non-Target Herbivores, Detritivores, and Pollinators

J. E. Losey
Cornell University, Ithaca, New York, U.S.A.

J. J. Obrycki
Iowa State University, Ames, Iowa, U.S.A.

R. A. Hufbauer
Colorado State University, Fort Collins, Colorado, U.S.A.

INTRODUCTION

Transgenic insecticidal plants produce proteins that are toxic to particular groups of insects. In addition to killing pest insects feeding directly on the crop, they may affect non-target organisms that feed on litter from the crop or on nectar or pollen that expresses the toxins. Detrimental effects on non-target species have been documented, but no immediate catastrophic impacts have been identified. Although protocols have been developed to quantify impacts on non-target biodiversity, complete assessment of non-target effects will necessitate determination of changes in the ability of non-target organisms to perform ecological functions such as weed suppression, decomposition, and pollination. Losey et al. (this volume) focus on predators or parasitoids, and in this entry we discuss effects on direct consumers.

BACKGROUND

A wide variety of crops have been modified to produce insecticidal proteins derived from genes transferred from the bacterium *Bacillus thuringiensis* (Bt). Transformed corn, known as ''Bt corn,'' is the most widely planted transgenic insecticidal crop in the world. Genetic material from different strains of *B. thuringiensis* produces toxins effective against different groups of insects. Currently, the only commercially available hybrids are derived from the Bt *kurtstakii* strain (Bt*k* corn) and were developed for selected lepidopteran species that feed on aboveground portions of the corn plant.[1] By 1999 over 6 million hectares of Bt corn was planted, and adoption reached at least this level in 2000 and 2001.[2] Transgenic Bt corn is now the most common management tactic for the European corn borer, *Ostrinia nubilalis*, throughout the United States. Corn hybrids transformed with genetic material from the Bt *tenebrionis* strain, which is active against coleoptera (beetles), are in the final stages of registration. If the new hybrids are approved, they appear destined for widespread use because corn rootworms (*Diabrotica* spp.) cause more damage than European corn borer and are the target of considerably more total kilograms of insecticide.

The potential benefits of transgenic insecticidal corn include reduction of resources devoted to scouting for pests, reduction in broad-spectrum insecticide applications, increased or protected yields due to season-long control of *O. nubilalis*,[3] protection of stored corn from lepidopteran pests,[4] and lower mycotoxin levels due to a reduction in fungal plant pathogens associated with *O. nubilalis* feeding.[5] The varying magnitude of these benefits is discussed in.[6]

Balanced against these potential benefits are possible negative aspects of growing these crops.[6] In general, negative effects of genetically modified crops could include selection for resistance among populations of the target pest, exchange of genetic material between the transgenic crop and related plant species, and impact on non-target species. The negative impact on non-target species can be separated into direct effects on organisms that feed on living or dead corn tissue (e.g., herbivores, pollenivores, detritivores) and indirect effects on organisms that primarily prey upon those direct consumers (predators). The existence of these four functional groups illustrates the often underestimated complexity of the many agroecosystems and the multiple mechanisms for potential impact (Fig. 1). It is important to note that direct consumers provide invaluable ecological services including weed suppression, pollination, and decomposition, while indirect feeders contribute greatly to suppression of insect pests. Interference with these processes could lead to

Encyclopedia of Plant and Crop Science
DOI: 10.1081/E-EPCS 120010607

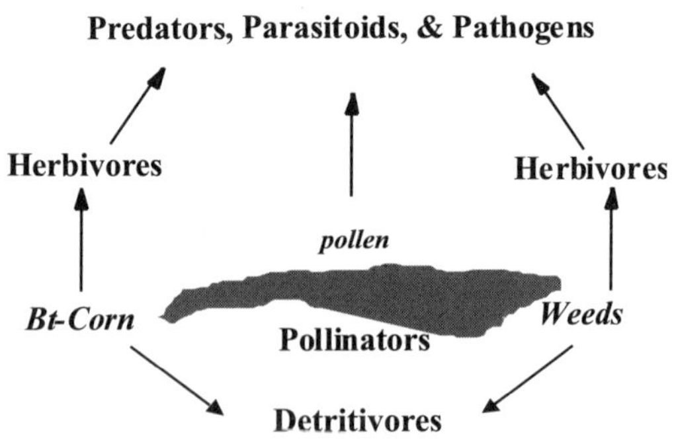

Fig. 1 Functional groups to be considered for assessment of risk from Bt corn and their relationship to Bt corn. (Adapted from Ref. 6.)

increased competition from weeds or delayed breakdown in plant material, both of which could lead to lower yields.

INSECT HERBIVORES

Currently commercialized hybrids of Bt corn and cotton express toxins that are active against lepidoptera, and many herbivorous non-target species are likely to be directly susceptible. There are two major crops that express a beetle toxin: Bt*t* corn and Bt*t* potatoes. Since Bt*t* corn is not commercially available and Bt*t* potatoes have not been widely adopted as of this writing, we will focus primarily on effects on lepidopteran non-targets. However, it is important to note that many families of beetles contain species that may directly consume living or dead corn tissue. Many of the same protocols used to identify which species of butterflies and moths are most at risk may be effective for beetles as well.

Since lepidopteran herbivores that feed on corn plant tissue within the cornfield are considered ''target pests,'' we will consider non-target herbivores to be those species that may contact corn pollen on weedy plant species within fields or on plants outside of fields. Since cotton is not wind-pollinated, the small amount and low mobility of the pollen produced minimizes the potential of impact through pollen drift. It is important to note that different events of Bt corn produce variable amounts of toxin in their pollen, with some events expressing very little. The lepidopteran species most likely affected by Bt corn pollen can be determined by examining their distribution and phenology.[7] Factors that will determine which lepidopteran species are most likely to be affected by pollen from Bt*k* corn include the following: 1) which plant species grow in and around corn; 2) which lepidopteran species feed on those plant species; 3) temporal overlap of corn pollen shed and larval feeding by the non-target lepidoptera; and 4) susceptibility of potentially affected species to the Bt toxin.[7] Integrating distribution, phenology, and susceptibility can allow a ranking of the risk to specific lepidopteran species. Species that may be at particularly high risk could then be identified for further testing.

The monarch butterfly, *Danaus plexippus*, provides one example of how a species might be evaluated after it has been determined to be potentially at risk from Bt*k* corn pollen. Observations that the monarchs' host plant, milkweed, was common in cornfields and often exposed to pollen led to initial studies that confirmed the toxicity of the pollen to monarch larvae.[8,9] Further studies determined that the risk of a short-term catastrophic impact on monarch populations was negligible based on the level of pollen on milkweed leaves and the temporal overlap of pollen shed and larval feeding.[10] However, this study also points out the importance of assessing longer-term, more subtle impacts on monarch populations.[10] Using a ''coarse filter'' such as the four-point one we propose above, other species that warrant more detailed ecological studies like those done for the monarch butterfly could be identified.

POLLINATORS

An assessment of the impact of each Bt corn hybrid on pollinators is required for EPA registration.[1] Although the toxins expressed in Bt corn pollen are specific for lepidoptera, several studies raise questions about its effects on pollinators, i.e., domesticated and wild bees. In the registration documentation, pollen from Bt corn is reported to have no effect on survival of either larval or adult domesticated bees.[1] However, unexpected effects of transgenic plants on domesticated bees have been reported. For example, a preparation of Bt*t*, reported to be specific for coleoptera, caused significant mortality in domesticated bees.[11] Proteins, other than Bt, produced in transgenic rapeseed pollen, targeted for coleoptera and lepidoptera, interfered with learning by domesticated bees.[12] When toxins are not expressed in the pollen, the process of transforming a plant may reduce pollen output, lowering availability of an important food source to pollinators.[13] These studies raise concerns about the precision of genetic transformations and unintended side effects of genetic transfers. In addition, although wild bees provide a substantial amount of the pollination in many systems, they apparently were not tested for Bt corn registration, and we are not aware of any studies that examined the impact of Bt pollen on wild bees.

DECOMPOSERS

The insecticidal toxin (CryIA(b)) found in one type of transgenic corn (event 176) caused significant mortality and reduced reproduction in the soil-dwelling collembolan, *Folsomia candida*.[1] A previous study had shown no effects of feeding on transgenic cotton leaves by *F. candida*.[14] The same insecticidal protein was present in both transgenic crops (cotton and corn); however, the higher dose in the corn appears to have caused the adverse effects. Even though the EPA reports this adverse non-target effect, they conclude that there is a 200-fold safety factor in the levels of toxin that would occur in the field.[1] In addition, because no buildup of corn stalk residues have been observed following use of soil insecticides in cornfields, which presumably would have a negative effect on collembola, ''an observable deleterious effect on the soil ecosystem is not expected to result from the growing of CryIA(b)-endotoxin-containing corn plants.''[1] This conclusion may need to be reconsidered, because there are potentially important differences in the seasonal occurrence of soil insecticides (at planting) to that of transgenic Bt toxins (from roots,[15] pollen deposition, and stalk residues at harvest) relative to the seasonal life cycles of collembola.

CONCLUSIONS

The results of studies on the direct effects of Bt crops on organisms that feed on crop tissues (e.g., living tissue, litter, pollen) are mixed. While no short-term catastrophic impacts have been identified, several impacts that warrant further study have been documented. Research up to this point has focused solely on relatively simple measures of the biodiversity of non-target organisms. The rapid adoption of Bt*k* corn and the predicted equally rapid adoption rate for Bt*t* corn as well as other transgenic plants make a complete assessment of both positive and negative impacts on non-target organisms imperative. A complete assessment of non-target impacts needs to include measures of how ecological functions (e.g., weed suppression, pollination, decomposition) are impacted by transgenic crops in comparison to how they are impacted by conventional pest management tactics.

REFERENCES

1. Anonymous. *EPA Office of Pesticide Programs. Biopesticide Fact Sheet.* http://www.epa.gov/pesticides/biopesticides/pips/bt_brad.htm (accessed 2003).
2. Benbrook, C.M. http://www.gefoodalert.org/library/admin/uploadedfiles/When_Does_It_Pay_To_Plant_Bt_Corn.pdf (accessed 2003).
3. Rice, M.E.; Pilcher, C.D. Potential benefits and limitations of transgenic Bt corn for management of the European corn borer (Lepidoptera: Crambidae). Am. Entomol. **1988**, *44*, 75–78.
4. Giles, K.L.; Hellmich, R.L.; Iverson, C.T.; Lewis, L.C. Effects of transgenic *Bacillus thuringiensis* maize grain on *B. thuringiensis*-susceptible *Plodia interpunctella* (Lepidoptera: Pyralidae). J. Econ. Entomol. **2000**, *93*, 1011–1016.
5. Munkvold, G.P.; Hellmich, R.L.; Rice, L.G. Comparison of fumonisin concentrations in kernels of transgenic Bt maize hybrids and nontransgenic hybrids. Plant Dis. **1999**, *83*, 130–138.
6. Obrycki, J.J.; Losey, J.E.; Taylor, O.; Hansen, L.C. Transgenic insecticidal corn: Beyond insecticidal toxicity to ecological complexity. BioScience **2001**, *51*, 353–361.
7. Losey, J.E.; Obrycki, J.J.; Hufbauer, R.A. Impacts of Genetically-Engineered Crops on Non-Target Herbivores: Bt-Corn and Monarch Butterflies as a Case Study. In *Genetically Engineered Organisms: Assessing Environmental and Human Health Effects*; Letourneau, D.K., Burrows, B.E., Eds.; CRC Press, Inc.: Boca Raton, FL, 2001.
8. Losey, J.E.; Rayor, L.S.; Carter, M.E. Transgenic pollen harms monarch larvae. Nature **1999**, *399*, 214.
9. Jesse, L.C.H.; Obrycki, J.J. Field deposition of Bt transgenic corn pollen: Lethal effects on the monarch butterfly. Oceologia **2000**, *125*, 241–248.
10. Stanley-Horn, D.E.; Dively, G.P.; Hellmich, R.L.; Mattila, H.R.; Sears, M.K.; Rose, R.; Jesse, L.C.H.; Losey, J.E.; Obrycki, J.J.; Lewis, L. Assessing the impact of Cry1Ab-expressing corn pollen on monarch butterfly larvae in field studies. Proc. Natl. Acad. Sci. **2001**, *98*, 11931–11936.
11. Vandenberg, J.D. Safety of four entomopathogens for caged adult honey bees (Hymenoptera: Apidae). J. Econ. Entomol. **1990**, *83*, 755–759.
12. Picard-Nioi, A.L.; Grison, R.; Olsen, L.; Arnold, C.P.G.; Pham-Delegue, M.H. Impact of proteins used in plant genetic engineering: Toxicity and behavioral study in the honeybee. J. Econ. Entomol. **1997**, *90*, 1710–1716.
13. Malone, L.A.; Pham-Delègue, M. Effects of transgene products on honey bees (*Apis mellifera*) and bumblebees (*Bombus* sp.). Apidologie **2001**, *32*, 287–304.
14. Yu, L.; Berry, R.E.; Croft, B.A. Effects of *Bacillus thuringiensis* toxins in transgenic cotton and potato on *Folsomia candida* (Collembola: Isotomidae) and *Oppia nitens* (Acari: Oribatidae). J. Econ. Entomol. **1997**, *90*, 113–118.
15. Stotzky, G. Release, Persistence, and Biological Activity n Soil of Insecticidal Proteins from *Bacillus thuringiensis*. In *Genetically Engineered Organisms: Assessing Environmental and Human Health Effects*; Letourneau, D.K., Burrows, B.E., Eds.; CRC Press, Inc.: Boca Raton, FL, 2002.

Biosafety Considerations for Transgenic Insecticidal Plants: Non-Target Predators and Parasitoids

J. E. Losey
Cornell University, Ithaca, New York, U.S.A.

J. J. Obrycki
Iowa State Univeristy, Ames, Iowa, U.S.A.

R. A. Hufbauer
Colorado State University, Fort Collins, Colorado, U.S.A.

INTRODUCTION

The range of transgenic insecticidal crops grown in the United States and the potential benefits associated with those crops have been outlined in Losey et al. (this volume). As noted by Losey et al., the negative impact on non-target species can be separated into direct effects on insects that feed on living or dead corn tissue (e.g., herbivores, pollenivores, detritivores) and indirect effects on organisms that are primarily predaceous or parasitic on those direct consumers. Losey et al. focus on direct consumers, and in this entry we focus on predators and parasitoids.

BACKGROUND

Transgenic insecticidal plants produce proteins that are toxic to particular groups of insects. In addition to their toxicity to herbivorous insects feeding directly on plant tissues, the engineered plants may affect predator and parasitoid species. The proteins may be directly toxic to predators and parasitoids that supplement their diet with pollen, nectar, or tissues such as roots. Simply feeding upon prey or hosts that have consumed insecticidal plant tissue also may be detrimental to both predators and parasitoids. Finally, reductions in predator and parasitoid populations may be linked to reductions in prey and host populations associated with insecticidal plants. The few empirical studies to date bearing on these issues have reported mixed results.

Previous research showing some negative effects of microbial insecticide formulations of *Bacillus thuringiensis* on natural enemy species[1] indicates a need to assess the impact of Bt corn on populations of insect predators and parasitoids in the corn ecosystem.[2] Numerous species of insect natural enemies attack both the European corn borer, *Ostrinia nubilalis*, and the corn rootworm, *Diabrotica* spp., in North America. *O. nubilalis* natural enemies include several predatory species that tend to have relatively broad host ranges and several specific insect parasitoids.[3] Although there are no significant parasitoids of corn rootworms, there are several important soil- and litter-dwelling predators that may be affected by transgenic Bt corn including carabid and staphylinid beetles, ants, spiders, and mites.[3] In addition to preying upon corn rootworm, carabid and staphylinid beetles are also known to consume the lepidopteran corn pests: black cutworm, armyworm, fall armyworm, common stalk borer, and European corn borer.[3] Transgenic corn may affect natural enemies via three modes: 1) direct feeding on corn tissues, e.g., pollen, roots; 2) feeding on hosts that have fed on corn; and 3) through reductions in host populations.[4] Data submitted for governmental registration of transgenic crops appear to focus primarily on the first mode.[5]

INSECT PREDATORS

Several species of insect predators that attack the corn borer also feed on corn pollen (Table 1). Direct consumption of transgenic corn pollen by immature stages of three predatory species commonly found in cornfields did not affect development or survival.[6] However, increased mortality of lacewing (*Chrysoperlacarnea*) larvae, which may consume pollen, was observed when *C. carnea* larvae fed on artificial diet containing Bt toxin.[7] Predator species may consume other corn tissues as well. Many carabid species which are primarily predators, such as *Stenolophus comma* and *Clivinia impressefrons*, also feed directly on plant roots and tissues.[a] However, there

[a]Schmaedick, M., Cornell University, personal communication.

Encyclopedia of Plant and Crop Science
DOI: 10.1081/E-EPCS 120015445

Table 1 Interactions between natural enemies and transgenic insecticidal Bt Corn

Species	Location of Study: L = Lab; F = Field	Effect: − = negative; + = positive; 0 = no effect	Results of comparison between transgenic and non-transgenic corn	Reference
		Insect Predators		
Neuroptera: Chrysopidae				
Chrysoperla carnea	F	0	Number of adults in transgenic fields	10
	L	0	Larval development and survival on transgenic pollen	6
	L	−	Decreased larval survival on transgenic pollen or prey exposed to Bt toxins	7
	L	0	Larval development and survival on aphid prey reared on Bt corn	15
Coleoptera: Coccinellidae				
Coleomegilla maculata	F	0	Number of adults and larvae	9
	F	0, +	Increased number of adults	10
	L	0	Larval survival and development on pollen	6
Cycloneda munda	F	0	Number of adults	10
Hippodamia convergens	F	0	Number of adults	10
Hemiptera: Anthocoridae				
Orius insidiosus	F	0	Number of adults and nymphs	9
	F	−, +	Number of adults	10
	L	0	Nymphal survival and development on pollen	6
Orius majusculus	L	0	Nymphal survival and development on thrips prey reared on Bt corn	8
		Insect Parasitoids		
Hymenoptera: Braconidae				
Macrocentris cingulum (formerly *M. grandii*)	F	−	30 to 60% reduction in adults in transgenic fields	10
	F	0	Parasitism of larval hosts on non-transgenic plants within transgenic plots	9
Hymenoptera: Ichneumonidae				
Eriborus terebrans	F	0	Parasitism of larval hosts on non-transgenic plants within transgenic plots	10

(Adapted from Ref. 2)

are no data bearing on the effects of Bt corn tissues on these organisms.

There is also little data on indirect consumption of Bt toxins through prey or hosts that have fed on Bt corn. Lacewing larvae that preyed upon corn borers or other lepidopteran larvae that had fed on transgenic corn show increased mortality,[7] but similar developmental times and survival rates were observed when the predator *Oriusmajusculus* was fed a thrips species that had been reared on either Bt or non-Bt corn.[8]

Negative effects on invertebrate predators have not been documented in the field (Table 1); sampling from transgenic and non-transgenic cornfields has detected no differences in predator abundance.[9,10] In one field study, higher numbers of predators were observed in Bt cornfields (Table 1). However, in a two-year field study, abundance of the parasitoid species *Macrocentriscingulum* (previously *Macrocentrisgrandii*), specific to corn borer larvae, was lower in Bt cornfields in Iowa[10] (Table 1). This reduction is expected because of significant reductions of larval hosts in Bt corn. The abundance of a second parasitoid species, *Erioborusterrebrans*, may also decline in transgenic fields due to the lack of corn borer hosts, although a field study reported no effects of transgenic corn on *E. terebrans* parasitism.[9] In Orr & Landis,[9] relatively small non-transgenic plots, were planted within larger transgenic plots, and *O. nubilalis* larval hosts were parasitized in these non-transgenic plants. Effects of Bt corn on *E. terebrans* parasitism may only be detectable in field studies conducted on a larger scale.

The potential trophic-level effects of Bt corn on vertebrate predators also should be considered in an ecological assessment of this biotechnology because bats and birds are known to prey on larvae and adults of several lepidopteran corn pests. Feeding Bt toxin directly to bobwhite quail for 14 days showed no effect on the quail.[5] We are not aware of any studies that have considered the indirect effects on bird populations resulting from declines in *O. nubilalis* densities following use of transgenic corn. However, if lepidoptera and their predators and parasitoids are significantly reduced in Bt cornfields and adjacent margins, we might expect the insect prey available for birds, rodents, and amphibians to decrease (see Ref. 11 for a simulation of the potential effects of herbicide-tolerant crops on seed-eating birds).

Long-term field studies are needed to determine if the widespread planting of transgenic corn creates an ''ecological desert'' with relatively few hosts for natural enemies. This type of ecological pattern has been observed following the overuse of insecticides or regional planting of highly resistant crop varieties.[12] The interactions among natural enemy and pest populations will likely occur within a mosaic of transgenic and non-transgenic cornfields, due to the current requirement for non-transgenic corn refugia to maintain susceptible corn borer populations. If corn borer densities are significantly suppressed by the use of transgenic corn, it might follow that significant reductions in natural enemy densities will occur, which may influence the rate of development of resistant pest populations.[13] Natural enemies currently cause substantial levels of mortality of the corn borer.[14] If this level of mortality were reduced and corn borer populations developed resistance, the result could be higher densities of the corn borer. Thus, negative impacts on natural enemies raise the possibility that overuse of transgenic corn could lead to the types of resurgence and secondary pest outbreaks that are associated with misuse of synthetic broad-spectrum insecticides.

CONCLUSIONS AND RECOMMENDATIONS—CONSIDERATIONS OF THE RISKS AND BENEFITS OF Bt CORN

Clearly more data are needed on the potential impact of Bt*k* and Bt*t* corn on predators and parasitoids. Studies on direct consumption of corn tissues are the simplest to carry out, but impacts mediated through the consumption of toxic prey or reductions in prey densities may have equal or greater importance in the field. Predicting the impact of future transgenic crops is difficult based on current studies because virtually all available data regarding impacts on predators come from Bt*k* corn, which targets butterflies and moths (lepidoptera), essentially none of which are predaceous. Impacts seem more likely in Bt*t* corn since the toxin affects beetles, the largest and most important group of predators. The greater risk to predators associated with Bt*t* corn highlights the need for a functional approach to assessment of non-target impacts. Important aspects of a functional assessment of predator impact would include field-level measures of predation rate and pest population suppression. Understanding the interaction between biological control and biotechnology will greatly facilitate the integration of these two important pest management strategies and increase the probability of avoiding the problems associated with the rapid adoption of pesticides.

REFERENCES

1. Glare, T.R.; O'Callaghan, M. *Bacillus Thuringiensis Biology, Ecology and Safety*; John Wiley & Sons: New York, 2000; 350.
2. Obrycki, J.J.; Losey, J.E.; Taylor, O.; Hansen, L.C. Transgenic insecticidal corn: Beyond insecticidal toxicity to ecological complexity. BioScience **2001**, *51*, 353–361.
3. Steffey, K.L.; Rice, M.E.; All, J.; Andow, D.A.; Gray, M.E.; Van Duyn, J.W. *Handbook of Corn Insects*; Entomol. Soc. Amer.: Lanham, MD, 1999; 164.
4. Hoy, C.W.; Feldman, J.; Gould, F.; Kennedy, G.G.; Reed, G.; Wyman, J.A. *Naturally Occurring Biological Controls in Genetically Engineered Crops*; Barbosa, P., Ed.; Conservation Biological Control Academic Press: New York, 1998; 185–205.
5. Anonymous; EPA Office of Pesticide Programs *Biopesticide Fact Sheet*; 1999. http://www.epa.gov/pesticides/biopesticides/pips/bt-brad.htm.
6. Pilcher, C.D.; Obrycki, J.J.; Rice, M.E.; Lewis, L.C. Preimaginal development, survival, and field abundance of insect predators on transgenic *Bacillus thuringiensis* corn. Environ. Entomol. **1997**, *26*, 446–454.

7. Hilbeck, A.; Moar, W.J.; Pusztai-Carey, M.; Filippini, A.; Bigler, F. Prey-mediated effects of Cry1Ab toxin and protoxin and Cry2A protoxin on the predator *Chrysoperla carnea*. Entomol. Exp. Appl. **1999**, *91*, 305–316.
8. Zwahlen, C.; Nentwig, W.; Bigler, F.; Hilbeck, A. Tritrophic interactions of transgenic *Bacillus thuringiensis* corn, *Anaphothrips obscurus* (Thysanoptera: Thripidae), and the predator *Orius majusculus* (Heteroptera: Anthocoridae). Environ. Entomol. **2000**, *29*, 846–850.
9. Orr, D.B.; Landis, D.A. Oviposition of European corn borer (Lepidoptera: Pyralidae) and impact of natural enemy populations in transgenic versus isogenic corn. J. Econ. Entomol. **1997**, *90*, 905–909.
10. Pilcher, C.D. Phenological, Physiological, and Ecological Influences of Transgenic Bt Corn on European Corn Borer Management. In PhD Thesis; Iowa State University, 1999; 204.
11. Watkinson, A.R.; Freckleton, R.P.; Robinson, R.A.; Sutherland, W.J. Predictions of biodiversity response to genetically modified herbicide-tolerant crops. Science **2000**, *289*, 1554–1557.
12. Gould, F. The evolutionary potential of crop pests. Am. Sci. **1991**, *79*, 496–507.
13. Gould, F.; Kennedy, G.G.; Johnson, M.T. Effects of natural enemies on the rate of herbivore adaptation to resistant host plants. Entomol. Exp. Appl. **1991**, *58*, 1–14.
14. Phoofolo, M.W.; Obrycki, J.J.; Lewis, L.C. Quantitative assessment of biotic mortality factors of the European corn borer, *Ostinia nubilalis* (Lepidoptera: Crambidae) in field corn. J. Econ. Entomol. **2001**, *94*, 617–622.
15. Lozzia, G.C.; Furlanis, C.; Manachini, B.; Rigamonti, I.E. Effects of Bt corn on *Rhopalosiphum padi* L. (Rhynchota Aphididae) and on its predator *Chrysoperla carnea* (Neuroptera Chrysopidae). Boll. Zool. Agrar. Bachic. Ser II. **1998**, *30*, 153-1.

B

Biosafety Programs for Genetically Engineered Plants: An Adaptive Approach

Anne R. Kapuscinski
Emily E. Pullins
University of Minnesota, St. Paul, Minnesota, U.S.A.

INTRODUCTION

Biotechnologists are rapidly diversifying the kinds of genetically engineered plants under development and production, thus offering a range of potential benefits to farmers and consumers. Approaches to the characterization, analysis, and deliberation of risks associated with genetically engineered plants have steadily improved with the development of the technologies and products themselves. This chapter reviews how lessons learned in environmental management, as well as in recent deployments of genetically engineered crops, are motivating a shift from single-plant risk assessment and management to more comprehensive and adaptive biosafety programs, within which traditional risk assessment and management tools are embedded.

ADAPTIVE BIOSAFETY PROGRAMS

Biosafety programs for genetically engineered plants should employ recent advances in environmental management of living organisms, such as grasslands, forests, and fisheries. According to recent reviews,[1,2] environmental management programs that have failed to meet their economic and environmental objectives have made four major errors. They: 1) largely ignored critical ecological interactions, 2) assumed that surrounding environmental conditions would remain fairly constant, 3) did not undertake long-term monitoring, and 4) considered only very narrow temporal and spatial scales. Biosafety management programs for genetically engineered plants can avoid these errors, many of which were recognized in a report on the environmental effects of transgenic plants.[3]

The carefully examined cases of failure in environmental management of living organisms show that the responsible institutions and the users were sometimes blind-sided by surprising social and ecological feedbacks in the system or, to use the terminology of Senge,[4] ''fixes that backfire.'' Analyses of such failures have led to a major rethinking of the problems that environmental policies attempt to solve[5] and have contributed to the shaping of adaptive management or adaptive environmental assessment and management approaches promulgated by Holling[6] and developed by Lee,[7] Walters,[8] and others.

These approaches are ''adaptive'' because they are designed to increase the chances that users and managers of a resource will consider all the relevant knowledge and experience before approving a proposed action and will detect and learn from expected and unexpected effects, which could be beneficial or harmful effects, as rapidly as possible even when there is a time lag in manifestation of the effects. These approaches are ''adaptive'' also because they include mechanisms to increase the ability of the responsible institutions and users to revise decisions in light of what they learn. In our case, the resource under management would be a particular genetically engineered plant produced in one or more kinds of agro-ecosystems, followed by the harvested whole plant, plant parts, or processed products being distributed through various public and marketing channels. The responsible institutions and users would be groups such as biotechnology companies, seed companies, farmers, food processors and retailers, government regulators, consumer groups, and environmental and other public interest groups. Implementation of an adaptive biosafety program could confirm the predicted environmental safety of producing the genetically engineered plant in a particular agro-ecosystem. This suggests it may be possible to reduce safety management controls over its production. Alternatively, the adaptive biosafety program might detect an unexpected environmental harm, triggering the need to revise controls over its production or, in the few cases where the harm may be severe and irreversible, to halt its production.

STEPS OF ADAPTIVE BIOSAFETY ASSESSMENT

Figure 1 represents the basic steps involved in adaptive biosafety assessment and management. These are summarized below and more fully elaborated in a proposed framework of adaptive biosafety assessment and management by Kapuscinski et al.[2] This framework builds on

Encyclopedia of Plant and Crop Science
DOI: 10.1081/E-EPCS 120010602

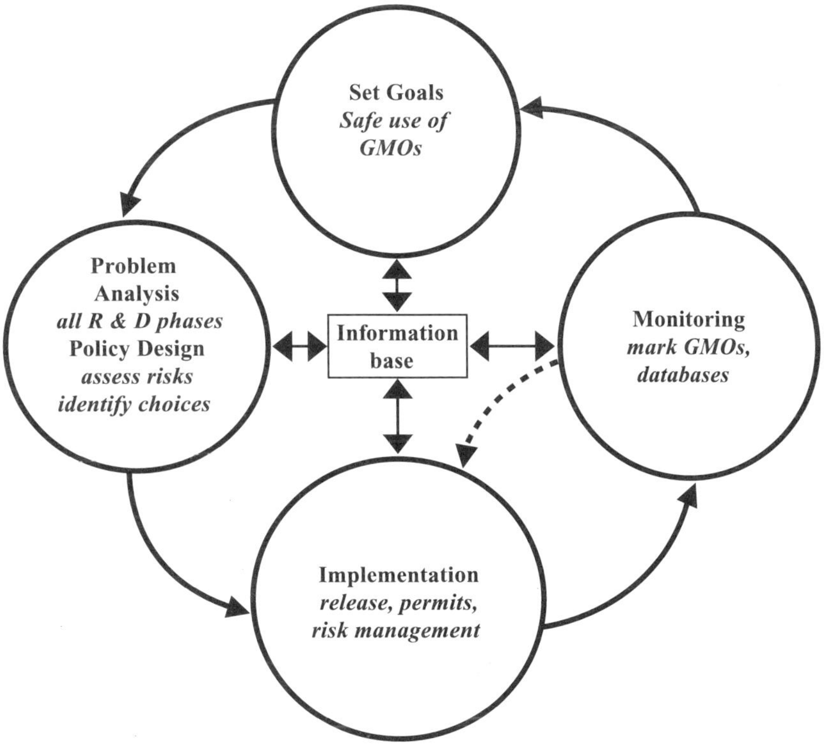

Fig. 1 The interconnected phases of adaptive biosafety assessment and management for uses of genetically engineered organisms (GEOs) in diverse societal-environmental systems. These phases should be applied iteratively and across multiple spatial scales from the local to global, with adequate provisions for information exchange among people implementing biosafety policies at different scales. (From Genetically Engineered Organisms: Assessing Environmental and Human Health Effects, CRC Press, New York, 2002.)

the essential, systematic steps of risk assessment and risk management that are part of biosafety assessment of genetically engineered organisms (GEOs).[9]

Goals

Adaptive biosafety assessment and management begins with clear identification and statement of biosafety goals. At a minimum, goals should aim to anticipate/determine ecological risks posed by a given GEO and discern whether the GEO will yield the benefits it was designed to provide.

Problem Analysis and Policy Design

This stage of the adaptive biosafety framework involves clear definition of the ecological and social problems associated with attempts to meet stated goals. It involves bringing together disciplinary specialists and theorists and resource users, with knowledge about the social and environmental contexts of the problems. This group considers a range of objectives, from unchecked release to a complete ban on uses of specific GEOs. Though these extremes may be unrealistic, they nonetheless provide the limits within which any realistic objective would fall; thus, they help the group to choose among policy options for testing.

Implementation

The biosafety policy action chosen for a particular kind of plant GEO should be implemented at the spatial and political scales likely to be impacted by the GEOs at issue. Local, national, and regional laws and regulations and international agreements, such as the Cartagena Protocol on Biosafety, may come into play to govern release and trade of specific GEOs. A comprehensive set of biosafety policies with coordination between the international and national scale would include measures for management of risks that cross spatial and political boundaries (following assessment of these risks at the problem analysis phase), biosafety capacity building programs, national permitting of trade and uses of GEOs (based on assessment and management of risks that might occur at the national or subnational scale), advanced informed agreements on transnational trade of GEOs, and an international system of liability and compensation for addressing cases in which harms occur.

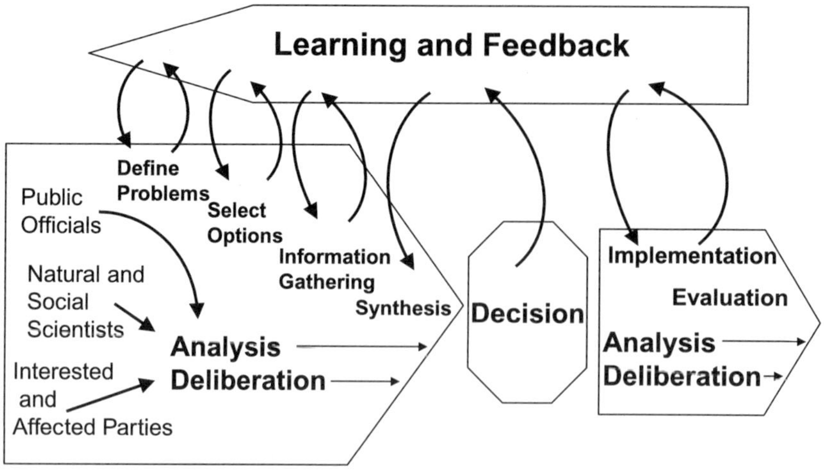

Fig. 2 The risk decision process. (From Ref. 15.)

Monitoring

An adaptive approach requires that, once a GEO is approved for large-scale use, there is a systematic, technically and financially feasible plan in place for postrelease monitoring. The plan would stipulate indicators and variables to be monitored, as well as agreed upon threshold limits of ecological change, derived from the biosafety objectives for case-by-case combinations of the genetically engineered plant and agro-ecosystems where it may end up.[4] Planning should include the selection of sampling designs for the monitoring of GEOs that reduce statistical errors that over- or underestimate the risks of a GEO, including estimating the costs of such improved sampling designs, and the selection of identification technologies for GEOs and their products, potentially at the molecular level (e.g., Ref. 10). A project to measure the effects on nontarget grassland bird species of changes in patterns of herbicide applications in fields planted with herbicide-tolerant soybeans exemplifies monitoring for indirect ecological effects of genetically engineered plants.[11]

As long as monitoring data indicate ''normal'' system behavior, the adaptive cycle runs between implementation and monitoring (depicted by the dotted line in Fig. 1). When monitoring indicates ''abnormal'' system behavior, the resulting data are used to review and, if deemed necessary, redefine goals, then to reanalyze problems, next to review and revise implementation, and finally to revise monitoring accordingly.

Information Base

The information base serves as a ''source and sink'' of the best available knowledge for each phase of the adaptive biosafety cycle. Continual updating and consultation of the information base also helps to bypass or shorten the time spent at each phase by ensuring that lessons learned from past experiences are incorporated into revised policies and by highlighting what remains unknown. Through continual interaction with the information base, adaptive biosafety programs can continually unravel uncertainties and adaptively reformulate policies. This kind of actively adaptive biosafety assessment and management provides the most effective way for humans to address the uncertainty inherent to the interconnected natural and social systems.[1,7] A recent example is the inclusion of new information in the assessment and management of risks of particular Bt corn varieties in interaction with Monarch butterflies; innovative research methods and varieties with altered trait expression resulted in the identification of commercial varieties that have improved environmental biosafety.[a]

CONCLUSION

The above five steps serve to strengthen the scientific reliability, technical efficacy, and institutional responsiveness of a biosafety program. But the success of biosafety programs depends equally on winning and sustaining public trust in decisions made under them. Thus, adaptive biosafety programs need to utilize transparent decision-making processes and involve deliberation among potentially affected and interested parties at critical points in decision-making (Fig. 2). Implementation of the various stages of the adaptive cycle should involve political and regulatory decision makers, appropriate disciplinary specialists, methodologists, people with experiential knowledge about the social and environmental conditions, and all affected parties (e.g., Ref. 13). This kind of civic engagement in the development of the

[a]See five articles in Special Edition Section of Ref. 12.

knowledge base and implementation of biosafety policies can ensure that all phases of adaptive biosafety are broadly understood, built upon all relevant knowledge, and responsive to the concerns of the affected parties.[14] Although even the best designed analytic-deliberative process cannot eliminate all controversy, this kind of transparency has been shown to increase the public's trust in diverse cases of risk decision-making.[15]

ACKNOWLEDGMENTS

Support for the authors while writing this article came, in part, from the Pew Marine Fellows Program; Minnesota Sea Grant College Program supported by the NOAA Office of Sea Grant, U.S. Department of Commerce, under grant No. NA16RG1046; the Pew Initiative on Food and Biotechnology; and the University of Minnesota's Interdisciplinary Center for the Study of Global Change. The U.S. Government is authorized to reproduce and distribute reprints for government purposes, not withstanding any copyright notation that may appear hereon.

ARTICLES OF FURTHER INTEREST

Biosafety Applications in the Plant Sciences: Expanding Notions About, p. 146
Biosafety Approaches to Transgenic Crop Plant Gene Flow, p. 150
Biosafety Considerations for Transgenic Insecticidal Plants: Non-Target Herbivores, Detritivores, and Pollinators, p. 153
Biosafety Considerations for Transgenic Insecticidal Plants: Non-Target Predators and Parasitoids, p. 156
Biosafety Science: Overview of Plant Risk Issues, p. 164
Transgenes (GM) Sampling and Detection Methods in Seeds, p. 1238
Transgenetic Plants: Breeding Programs for Sustainable Insect Resistance, p. 1245
Transgenic Crop Plants in the Environment, p. 1248
Transgenic Crops: Regulatory Standards and Procedures of Research and Commercialization, p. 1251

REFERENCES

1. *Barriers and Bridges to the Renewal of Ecosystems and Institutions*; Gunderson, L.H., Holling, C.S., Light, S.S., Eds.; Columbia University Press: New York, 1995.
2. Kapuscinski, A.R.; Nega, T.; Hallerman, E.M. Adaptive Biosafety Assessment and Management Regimes for Aquatic Genetically Modified Organisms in the Environment. In *Towards Policies for Conservation and Sustainable Use of Aquatic Genetic Resources*; Pullin, R.S.V., Bartley, D.M., Koooiman, J., Eds.; International Center for Living Aquatic Resources Management, Conf. Proc., 1999; 225.
3. Committee on Environmental Impacts Associated with Commercialization of Transgenic Crops. *Environmental Effects of Transgenic Crops: The Scope and Adequacy of Regulation*; National Academy Press: Washington, DC, 2002.
4. Senge, P.M. *The Fifth Discipline: The Art and Practice of the Learning Organization*; Currency Doubleday: New York, 1994.
5. Holling, C.S. What Barriers? What Bridges? In *Barriers and Bridges to the Renewal of Ecosystems and Institutions*; Gunderson, L.H., Holling, C.S., Light, S.S., Eds.; Columbia University Press: New York, 1995; 3.
6. Holling, C.S. *Adaptive Environmental Assessment and Management*; John Wiley and Sons: Chichester, U.K., 1978.
7. Lee, K.N. *Compass and Gyroscope: Integrating Science and Politics for the Environment*; Island Press: Washington, DC, 1993.
8. Walters, C.J. *Adaptive Management of Renewable Resources*; MacMillan: New York, 1986.
9. Kapuscinski, A.R. Controversies in Designing Useful Ecological Assessments of Organisms. In *Genetically Engineered Organisms: Assessing Environmental and Human Health Effects*; Letourneau, D.K., Burrows, B.E., Eds.; CRC Press: Washington, DC, 2002.
10. Miller, L.; Kapuscinski, A.R. Historical analysis of genetic variation reveals low effective population size in a northern pike, *Esox lucius*, population. Genetics **1997**, *147*, 1249–1258.
11. Wolfenbarger, L.L.; McCarty, J.P. *Large Scale Ecological Effects of Herbicide Tolerant Crops on Avian Communities and Reproduction. Biotechnology Risk Assessment Research Grants Program*; U.S. Department of Agriculture, August 2002.
12. Proceedings of the U.S. National Academy of Sciences. Proc. Natl. Acad. Sci. **2001**, *98*, 11908–11942.
13. Renn, O.; Webler, T.; Rakel, H.; Dienel, P.; Johnson, B. Public participation in decision making: A three-step procedure. Policy Sci. **1993**, *26*, 189.
14. Kapuscinski, A.R.; Goodman, R.M.; Hann, S.D.; Jacobs, L.R.; Pullins, E.E.; Johnson, C.S.; Kinsey, J.D.; Krall, R.L.; La Viña, A.G.M.; Mellon, M.G.; Ruttan, V.W. Making 'safety first' a reality for biotechnology products. Nat. Biotechnol. **2003**, *21* (6), 599–601.
15. Committee on Risk Characterization. In *Understanding Risk: Informing Decisions in a Democratic Society*; Stern, P.C., Fineberg, H.V., Eds.; National Academy Press: Washington, DC, 1996.

Biosafety Science: Overview of Plant Risk Issues

L. L. Wolfenbarger
University of Nebraska, Omaha, Nebraska, U.S.A.

P. R. Phifer
U.S. Fish and Wildlife Service, Portland, Oregon, U.S.A.

INTRODUCTION

Biosafety science endeavors to assess the human health and environmental risks associated with the creation and release of a genetically engineered organism into the environment.[1] Plant risk issues associated with the introduction of genetically engineered crops have been discussed extensively,[2–4] and these generally focus on how engineered crops may impact biodiversity and biotic interactions in ecosystems as well as genetic diversity within populations and species. There are three broad categories of risks that might cause adverse changes in biodiversity and in biotic interactions in ecosystems: potential invasiveness, non-target effects, and new viral diseases. The likelihood of these risks will vary according to the trait engineered, the crop variety, the environmental conditions, and the surrounding ecosystem.

POTENTIAL INVASIVENESS

Humans have a long history of introducing plants for horticultural or agricultural purposes. Sometimes these practices include introducing species into non-native habitats or introducing new varieties or cultivars into an existing range. The vast majority of the nonindigenous species and novel cultivars are benign or beneficial when introduced. A small percentage of these, however, known as invasive species, spread widely in their non-native ecosystems and cause substantial environmental harm.[5] Invasive species can degrade natural ecosystem functions and structure and are categorized as one of three most pressing environmental problems, along with global climate change and habitat loss,[6] that cost an estimated $137 billion annually in the United States alone.[7]

A novel trait introduced into plants through artificial selection or genetic engineering has the potential to produce changes that enhance an organism's ability to become an invasive species. Ultimately, it is the individual plant's novel phenotype(s) and its interactions with the surrounding biotic and abiotic environment that determine whether it will become invasive. These interactions involve existing opportunities within the environment for unintended establishment (e.g., a currently stressed ecosystem may be more susceptible to invasion), and the persistence and gene flow of the introduced organism. Each interaction, in turn, depends on various components of survival and reproduction of the organism or its hybrids. Given the multiple elements involved in the establishment of a new invasive species, it is difficult to predict with significant confidence whether an introduced organism will become invasive.

NON-TARGET EFFECTS

Direct Effects on Survivorship and Reproduction

Like other potential stressors introduced into the environment, genetically engineered crops have the potential to affect the survival and reproduction of non-target species. For example, transgenic crops with insecticidal properties may have lethal or sublethal effects on non-target insects. Peer-reviewed studies have focused on the effects of Bt corn on a small number of insect species,[8] particularly two species of butterflies: the monarch (*Danaus plexippus*) and the black swallowtail (*Papilio polyxenes*). The work on butterflies indicated that effects on survivorship and larval growth vary according to the transformation event (type of Bt corn) and species.[9,10] Some genetically engineered crops and microorganisms have been shown to affect soil ecosystems.[11,12] No published studies have examined the significance (if any) of these differences or transient changes with an experimental design that mimics standard practices used with a transgenic organism compared to nontransgenic alternatives. However, lethal and even sublethal effects at the individual level can destabilize the population dynamics of species and the function and structure of ecosystems.[13]

Non-target effects have traditionally been identified using laboratory tests on representative organisms to identify standard measures of toxicity (e.g., LD50, the dose at which 50% of subject organisms die). However, these laboratory tests may not always accurately represent

Encyclopedia of Plant and Crop Science
DOI: 10.1081/E-EPCS 120010603

complex ecological systems, creating the need for extensive field studies.[8]

Indirect Effects Caused by Bioaccumulation

Genetically engineered crops have the potential to indirectly impact populations of species whose survival or reproduction depends on the pests or weeds controlled. For example, population models suggest that more effective control of weeds by using herbicide-tolerant crops could lead to lower food availability for birds specializing on seeds.[14]

Large-scale use of insecticidal pesticides is known to disrupt ecological interactions within naturally occurring biological control systems[15] and a similar risk exists for genetically engineered crops that produce proteins with insecticidal properties. Specifically, genetically engineered crops with insecticidal properties may cause indirect non-target effects if bioaccumulation of pesticidal proteins can occur as predators consume prey items that contain these proteins. Such negative tri-trophic-level effects in the laboratory have been reported for the green lacewing (*Chrysoperla carnea*) reared on prey that have fed on Bt corn,[16] but field studies have not been conducted to quantify what typical exposures to such prey items may be. Knowing the effects of pesticidal proteins on the survivorship and reproduction of organisms at multiple trophic levels of the food chain as well as understanding the fate and transport of these proteins are key components to understanding the potential for non-target effects.

While the detailed discussion here has focused on transgenic plants with insecticidal properties, any biologically active transgenic product will have the potential to have lethal or sublethal, direct or indirect effects on non-target organisms. In the future these could include crops that produce pharmaceuticals, vitamins, nutrients or micronutients, or compounds that are novel for plants used for industrial purposes (e.g., polymer or oil production).

NEW VIRAL DISEASES

Viruses with new biological characteristics could potentially arise in transgenic virus-resistant plants through recombination and heteroencapsidation. Recombination dominates viral evolution[17] and has been observed in a wide range of viruses, including regions encoding the capsid protein,[18] which is a commonly engineered product for transgenic plants. The presence of viral coat protein genes in transgenic plants offers the opportunity for recombination between incoming viruses and RNA transcribed by viral coat protein genes. Recombination between the transgene's mRNAs and RNA of ''challenge viruses'' can drive the development of chimeric genomes that contain segments of two distinct RNA species. Such events may generate a virus with properties that differ from either progenitor virus.

Laboratory studies show that recombination does occur between a viral genome and RNA from viral sequences present in transgenic plants.[18] Yet, researchers have noted the uncertainty in predicting the effects of recombination on symptomatology and on alterations of host ranges,[19] as well as the need for broad field experimentation to provide realistic assessments about the types of recombinants that might emerge, the frequency of emergence of novel viruses, and what steps can be taken to minimize the appearance of virulent recombinant.[20]

Heteroencapsidation of viral nucleic acids by transgenic coat proteins could also create new viral strains, but they would not propagate because the genome would not code for the transgenic coat protein.

The actual ecological significance of any viral change depends on what impacts it causes in natural habitats and communities. Impacts could occur directly if new ecological niches for viruses change plant communities in a negative way, or indirectly if these viruses create epidemics that alter agricultural practices that increase negative ecological effects.

CONCLUSION

The risks associated with genetically engineered plants will vary spatially, temporally, and according to species and cultivar and to the ecosystem encountered. These risks include risks from potential invasiveness, non-target effects, and new viral diseases. As more organisms, including genetically engineered organisms, are introduced into non-native environments around the globe, the possibility increases that these organisms will come in contact with sensitive or susceptible plant species or habitats, thereby, elevating risk. Our ability to determine the risks plant species and communities face from the products of genetic engineering, however, is similar to our ability to predict the risks associated with the introduction of a novel, conventionally produced organism. This ability is imprecise, given the complexity of the organism and the environment into which it is introduced, and further efforts to improve these risk assessment abilities are needed.

ARTICLES OF FURTHER INTEREST

Agriculture and Biodiversity, p. 1
Biosafety Applications in the Plant Sciences: Expanding Notions About, p. 146

REFERENCES

1. Scientists' Working Group on Biosafety. *Manual for Assessing Ecological and Human Health Effects of Genetically Engineered Organisms*; Edmonds Institute: Seattle, WA, 1998. http://www.edmonds-institute.org/manual.html.
2. Dale, P.J.; Clarke, B.; Fontes, E.M.G. Potential for the environmental impact of transgenic crops. Nat. Biotechnol. **2002**, *20* (6), 567–574.
3. Marvier, M. Ecology of transgenic crops. Am. Sci. **2001**, *89* (2), 160–167.
4. Wolfenbarger, L.L.; Phifer, P.R. The ecological risks and benefits of genetically engineered plants. Science **2000**, *290* (5499), 2088–2093.
5. Mack, R.N.; Simberloff, D.; Lonsdale, W.M.; Evans, H.; Clout, M.; Bazzaz, F. Biotic invasions: Causes, epidemiology, global consequences and control. Issues Ecol. **2000**, *5*, 1–20.
6. Sala, O.E.; Chapin, F.S.; Armesto, J.J.; Berlow, E.; Bloomfield, J.; Dirzo, R.; Huber-Sanwald, E.; Huenneke, L.F.; Jackson, R.B.; Kinzig, A.; Leemans, R.; Lodge, D.M.; Mooney, H.A.; Oesterheld, M.; Poff, N.L.; Sykes, M.T.; Walker, B.H.; Walker, M.; Wall, D.H. Global biodiversity scenarios for the year 2100. Science **2000**, *287* (5459), 1770–1774.
7. Pimentel, D.; Lach, L.; Zuniga, R.; Morrison, D. Environmental and economic costs of nonindigenous species in the United States. Bioscience **2000**, *50* (1), 53–65.
8. Obrycki, J.J.; Losey, J.E.; Taylor, O.R.; Jesse, L.C.H. Transgenic insecticidal corn: Beyond insecticidal toxicity to ecological complexity. Bioscience **2001**, *51* (5), 353–361.
9. Losey, J.E.; Obrycki, J.J.; Hufbauer, R.A. Impacts of Genetically Engineered Crops on Non-Target Herbivores: *Bt* Corn and Monarch Butterflies as Case Studies. In *Genetically Engineered Organisms: Assessing Environmental and Human Health Effects*; Letourneau, D.K., Burrows, B.E., Eds.; CRC Press: Boca Raton, FL, 2001; 143–166.
10. Zangerl, A.R.; McKenna, D.; Wraight, C.L.; Carroll, M.; Ficarello, P.; Warner, R.; Berenbaum, M.R. Effects of exposure to event 176 *Bacillus thuringiensis* corn pollen on monarch and black swallowtail caterpillars under field conditions. Proc. Natl. Acad. Sci. U. S. A. **2001**, *98* (21), 11908–11912.
11. Dunfield, K.E.; Germida, J.J. Diversity of bacterial communities in the rhizosphere and root interior of field-grown genetically modified *Brassica napus*. FEMS Microbiol. Ecol. **2001**, *38* (1), 1–9.
12. Watrud, L.S.; Seidler, R.J. Nontarget Ecological Effects of Plant, Microbial and Chemical Introductions to Terrestrial Systems. In *Soil Chemistry and Ecosystem Health Special Publication No. 52*; Soil Science Society of America: Madison, WI, 1998.
13. van der Heijden, M.G.A.; Klironomos, J.N.; Ursic, M.; Moutoglis, P.; Streitwolf-Engel, R.; Boller, T.; Wiemken, A.; Sanders, I.R. Mycorrhizal fungal diversity determines plant biodiversity, ecosystem variability and productivity. Nature **1998**, *396* (6706), 69–72.
14. Watkinson, A.R.; Freckleton, R.P.; Robinson, R.A.; Sutherland, W.J. Predictions of biodiversity response to genetically modified herbicide-tolerant crops. Science **2000**, *289* (5484), 1554–1557.
15. Croft, B.A. *Arthropod Biological Control Agents and Pesticides*; John Wiley and Sons: New York, 1990.
16. Hilbeck, A. Transgenic Host Plant Resistance and Non-Target Effects. In *Genetically Engineered Organisms: Assessing Environmental and Human Health Effects*; Letourneau, D.K., Burrows, B.E., Eds.; CRC Press: Boca Raton, FL, 2001; 167–186.
17. Gibbs, M.J.; Armstrong, J.; Weiller, G.F.; Gibbs, A.J. Virus Evolution: The Past, a Window on the Future. In *Virus Resistant Transgenic Plants: Potential Ecological Impact*; Tepfer, M., Balazs, E., Eds.; Springer-Verlag: Berlin, 1997; 1–19.
18. Kiraly, L.; Bourque, J.E.; Schoelz, J.E. Temporal and spatial appearances of recombinant viruses formed between cauliflower mosaic virus (CaMV) and CaMV sequences present in trangenic *Nicotiana bigelovii*. Mol. Plant-Microb. Interact. **1998**, *11* (4), 309–316.
19. Schoelz, J.E.; Wintermantel, W.M. Expansion of viral host range through complementation and recombination in transgenic plants. Plant Cell **1993**, *5*, 1669–1679.
20. Borja, M.; Rubio, T.; Scholthof, H.B.; Jackson, A.O. Recombination with host transgenes and effects on virus evolution: An overview and opinion. Mol. Plant-Microb. Interact. **1999**, *12* (2), 87–92.

Boron

Patrick H. Brown
University of California, Davis, California, U.S.A.

INTRODUCTION

Though boron (B) has been recognized as an essential plant element for 80 years, it was only in the late 1990s that significant advances were made in understanding the metabolism of B in plants. Primary among these advances has been the determination of the chemical form and function of B in plant cell walls, the identification of the role of polyhydric alcohols in transport of B in the phloem, and the characterization of the processes of transport of B across membranes. Boron is now known to be essential for cell wall structure and function, probably through its role as a stabilizer of the cell wall pectic network and subsequent regulation of cell wall pore size. Boron uptake and transport is unique among the essential plant nutrients. At adequate levels in the soil B passively permeates the plasma membrane through aquaporin type channels and direct membrane diffusion. At low levels of B supply, carrier mediated transport may occur. The transport of B within a plant is unique in that species differ markedly in mobility of B in the phloem. This phenomenon is now well understood and has resulted in novel molecular approaches to enhance plant tolerance of environments that are deficient in B. The discovery that B is essential for animals (which lack a cellulose rich cell wall) suggests that B has additional, undetermined, functions in biology. Careful consideration of the physical and chemical properties of B in biological systems, and of the experimental data from both plants and animals suggests that B plays a critical role in membrane structure and hence function.

UPTAKE

Boron is essential for plant growth. Even brief reductions in availability of B to roots, particularly during reproductive growth, have a profound effect on plant growth and productivity. Boron is also present at toxic levels in many arid soils, and many agricultural species are sensitive to high levels of B in soils and waters. Understanding the biology of B uptake by plants is therefore critical to the management of B in natural and agricultural systems. Boron is present in the environment and acquired by plants, as the uncharged molecule H_3BO_3, for which there is no useful radioisotope. B also rapidly forms variably stable complexes with a wide range of biological molecules. These characteristics have long hampered research into the mechanism of B uptake. It was only in 2000 that new methodology and experimentation provided the first direct measurement of membrane permeability of B and demonstrated that B permeates passively through aquaporin-type channels and directly through the plasma membrane. Theoretical calculations suggest that passive B permeation can satisfy plant B needs at adequate levels of B supply but would be inadequate at conditions of marginal B supply. The hypothesis that an active, carrier-mediated process is involved in B uptake at low B supply is supported by considerable experimentation; however, no specific carrier proteins have been characterized and no genetic mutants lacking in B uptake have been identified.

A model of B acquisition by plants is provided in Fig. 1. At adequate to excessive levels of B supply, both passive transmembrane permeation and channel proteins provide B uptake. Small differences in membrane permeability coefficients probably contribute to species variation in B uptake. At lower levels of B supply (<1.0 μM), passive B permeation appears to be inadequate to satisfy B demand, and uptake may be facilitated by an inducible, active carrier system.

PHLOEM TRANSLOCATION

Boron is unique in that species differ dramatically in their capacity for transport in the phloem. In the majority of plants, B has restricted phloem mobility, while in species that produce polyols (sorbitol, mannitol, dulcitol, etc.), B is readily translocated in the phloem to satisfy the demands for B in growth.[1] Due to these differences in phloem transport, B toxicity symptoms are also fundamentally different among these plants. Non–polyol-producing plants show burning of the tip and margin of old leaves under B toxicity, while in the polyol-producing species these symptoms are absent and toxicity is expressed as meristematic dieback.[2] The ability to remobilize B from old to young leaves in polyol-producing species reduces the occurrence of transient B deficiencies and provides remarkable tolerance of short term B deprivation. Species that do not transport B in the phloem are very sensitive to short term B deprivation. Tobacco plants genetically

Encyclopedia of Plant and Crop Science
DOI: 10.1081/E-EPCS 120010507

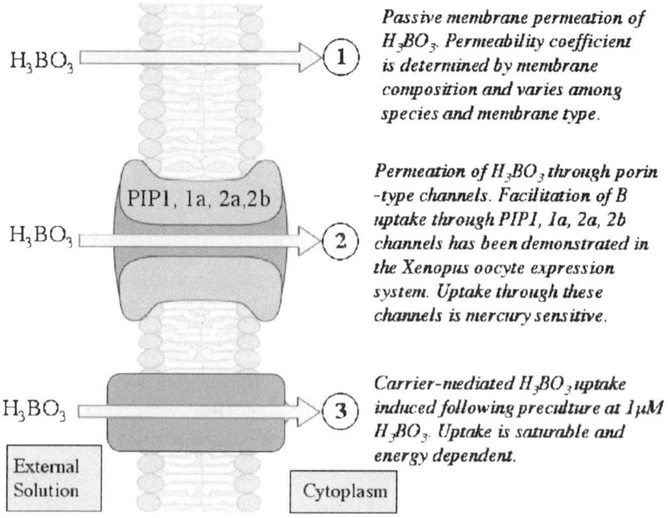

Fig. 1 Model of H_3BO_3 across plant membranes. 1. Passive B transport by transmembrane diffusion. Undissociated boric acid diffuses directly through the plasma membrane and is influenced by the permeability coefficient, membrane source (plasma membrane, tonoplast membrane) and the concentration gradient. 2. Channel-mediated B transport through porin-type channels. Evidence for B transport through channel proteins was derived from experiments with channel blocking mercurials and from experiments in which the porin-type channel proteins (PIP 1, 1a, 2a, 2b) expressed in the *Xenopus* oocyte system resulted in a 30–40% stimulation in B uptake. 3. Proposed inducible carrier-mediated transport. Preculture in the presence of 1 μM H_3BO_3 stimulates synthesis of carrier-mediated H_3BO_3 transport. Km values for H_3BO_3 transport of 2 μM (*Chara nitella*) and 15 μM (*Helianthus anuus*) have been determined. (*View this art in color at www.dekker.com.*)

engineered to produce sorbitol exhibit a marked increase in within-plant B mobility and a resultant tolerance of B deficient conditions. The development of enhanced tolerance of B deficiency in transgenic tobacco was the first effective manipulation of a plant nutrition character.[3]

BORON FUNCTION AND DEFICIENCY SYMPTOMS

Boron deficiencies occur widely and have a significant agronomic impact throughout the world. Deficiency symptoms include the inhibition of apical and extension growth, necrosis of terminal buds, cracking and breaking of stems and petioles, abortion of flower initials, and shedding of fruits. Boron deficiency also causes many physiological and biochemical changes including altered cell wall structure, altered membrane integrity and function, changes in enzyme activity, and altered production of a wide range of plant metabolites. The essential function of B in plants was not established until 2001, when it was demonstrated that B is critical for the formation of the plant cell wall and hence essential for plant growth.[4] Though B's role in plant cell walls is inadequate to explain all of the observed effects of B deficiency seen in plants, there is currently no definitive evidence to suggest that B plays a specific role in any other metabolic function in plants.

A role for B in the cell wall of plants has long been predicted on the basis of several historical observations and broad interpretations of anatomical observations under B deficiency. It is well known that B is essential for organisms with carbohydrate-rich cell walls and symptoms of B deficiency suggest that B plays an important role in the cell wall. A B-polysaccharide complex was isolated from radish roots in 1993 and subsequently identified as rhamnogalacturonan-II (RG-II). In plant cell walls, RG-II is present predominantly as a dimeric molecule (dB-RG-II), in which two chains of monomeric RG-II (mRG-II) are cross-linked by a 1:2 borate ester with two of the four apiosyl residues of RG-II side chains. There is no compelling evidence to suggest that B is directly involved in the synthesis of cell wall material because no decrease in the production of pectic substances or cell wall precursors is observed under B deficient conditions. Experimental evidence does suggest, however, that B influences the incorporation of proteins, pectins, and/or precursors into the existing and extending cell wall.

A hypothetical scheme for the role of B in cell walls is provided in Fig. 2. In this hypothesis the formation of dB-RG-II influences plant growth and metabolism through its effect on wall pore size. Under adequate B conditions, the formation of dB-RG-II is critical for optimal pore formation. The pore size in the cell wall affects the passage of large molecules such as proteins (including wall-modifying enzymes), influences the transport of precursors to sites of cell wall deposition, and optimizes the environment that is probably critical for normal processes of wall development and cell growth. In the absence of adequate B, pore size is increased and the processes of cell wall deposition and growth are disrupted.

A function for B in cell walls is inadequate to explain all observed effects of B deficiency. Evidence now suggests that B probably plays a role in membrane structure and function. This hypothesis is supported by the observation that B deficiency disrupts membrane transport processes and alters the composition of the cell membrane. Boron deficiency in animals disrupts processes that are highly membrane-specific or that require synthesis of new membranes. It has been proposed that B plays a specific function in cellular membranes through its role in the formation and function of membrane rafts, as illustrated in Fig. 3.

Plant membranes contain a variety of hydroxyl- and amine-containing constituents with the potential to form complexes with borate. These include sugars (e.g., galactose and mannose) and amino acids (e.g., serine and

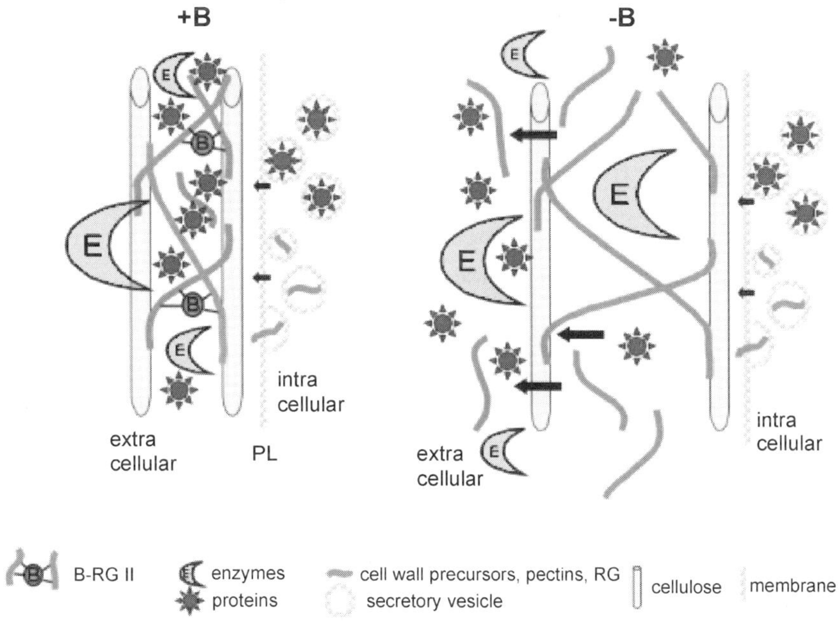

Fig. 2 Hypothetical scheme illustrating the role of B on cell wall pore size and function. Under adequate B conditions (+B in diagram), the formation of dB-RG-II is critical for optimal pore size formation and as a consequence affects the normal cell wall passage of large molecules including proteins (wall-modifying enzymes), influences the transport of cell wall precursors to sites of cell wall deposition, and optimizes the cell wall environment critical for normal processes of wall development and cellular growth. In the absence of adequate B (−B in diagram), pore size is increased and the normal processes of cell wall deposition and growth are disrupted. (*View this art in color at www.dekker.com.*)

tyrosine). These potential B-binding molecules are particularly abundant in membrane rafts, which are functionally discrete and physiologically active membrane subdomains. Membrane rafts are thought to have a specific physiological role in membrane signal transduction and to serve as the sites for glucosylphosphatidyl-inositol (GPI) protein association. Boron deficiency disrupts the incorporation of GPI proteins into the cell membrane. Boron may therefore have a specific function in cellular membranes through its role in the formation, stability, and function of membrane rafts.[6] Perhaps the most significant finding of recent years does not relate to the role of B

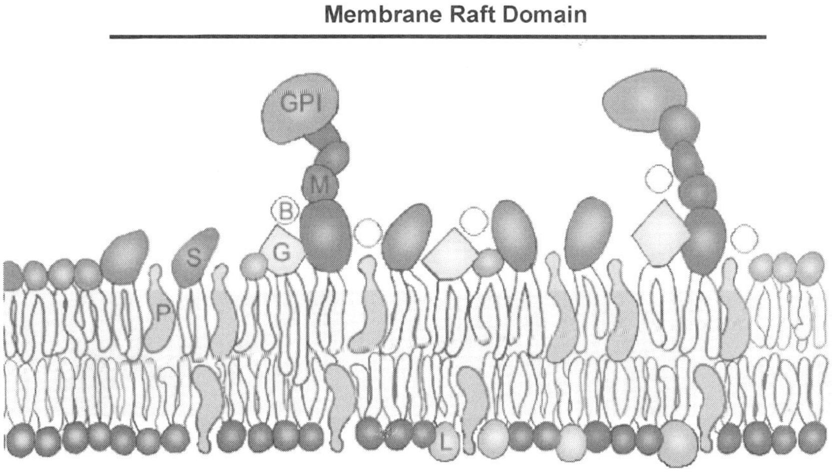

Fig. 3 Model indicating proposed function of boron (B) in cell membrane structure and function. The liquid ordered phase of a membrane raft is delineated by the black line. The exoplasmic leaflet is enriched in glycosphingolipids (G), sphingomyelin (S), and glucosylphosphatidyl-inositol (GPI) anchored proteins. The inner leaflet is shown to contain glycerolipids (L). It is hypothesized that B may attach to mannose-rich domains (M) of GPI proteins or to sugars in the glycolipid headgroups, thereby influencing chemical or physical membrane charactersitics. (Adapted from Ref. 5.) (*View this art in color at www.dekker.com.*)

in plants but rather the observation that B is essential for animals. This discovery clearly implies a broader role for B in biology that has yet to be determined.

CONCLUSION

Though our understanding of the biology of B has developed dramatically, much still remains uncertain. Of all the effects of B on plant biology, none is more significant, or more puzzling, than the unique sensitivity of all species to B deficiency during reproductive growth. This is particularly significant since 80% of all agricultural production is based on reproductive structures. Recent advances in our understanding of B uptake by roots, B function in cell walls, and B transport within the plant provide a framework within which we can now attempt to discern the role of B in reproductive growth. Advances in quantification of B and in the use of stable isotopes will facilitate this research, as will breakthroughs in the molecular regulation of flowering. These approaches provide researchers with a means to induce flowering in a predictable pattern and to dissect the possible role of B through careful utilization of the host of available flowering mutants and molecular techniques.

ARTICLES OF FURTHER INTEREST

Cropping Systems: Irrigated Continuous Rice Systems of Tropical and Subtropical Asia, p. 349
Flower Development, p. 464
In Vitro Pollination and Fertilization, p. 587
Minor Nutrients, p. 726
Nitrogen, p. 822
Phosphorus, p. 872
Photosynthate Partitioning and Transport, p. 897
Potassium and Other Macronutrients, p. 1049
Root Membrane Activities Relevant to Plant-Soil Interactions, p. 1110

REFERENCES

1. Brown, P.H.; Shelp, B.J. Boron mobility in plants. Plant Soil **1997**, *193* (1–2), 85–101.
2. Brown, P.H.; Hu, H.; Roberts, W.G. Occurrence of sugar alcohols determines boron toxicity symptoms of ornamental species. J. Am. Soc. Hortic. Sci. **1999**, *124* (4), 347–352.
3. Brown, P.H.; Bellaloui, N.; Hu, H.; Dandekar, A. Transgenically enhanced sorbitol synthesis facilitates phloem boron transport and increases tolerance of tobacco to boron deficiency. Plant Physiol. **1999**, *119* (1), 17–20.
4. O'Neill, M.A.; Eberhard, S.; Albersheim, P.; Darvill, A.G. Requirement of borate cross-linking of cell wall rhamnogalacturonan II for *Arabidopsis* growth. Science **2001**, *294* (5543), 846–849.
5. Simons, K.; Ikonen, E. Functional rafts in cell membranes. Nature **1997**, *387*, 569–572.
6. Brown, P.H.; Bellaloui, N.; Wimmer, M.A.; Bassil, E.S.; Ruiz, J.; Hu, H.; Pfeffer, H.; Dannel, F.; Romheld, V. Boron in plant biology. Plant Biol. **2002**, *4* (2), 205–223.

Breeding Biennial Crops

I. L. Goldman
University of Wisconsin, Madison, Wisconsin, U.S.A.

INTRODUCTION

Biennial plants require two seasons of growth to complete their life cycle. The breeding of biennial crops typically involves crop or propagule production during the first season of growth, followed by a vernalization period in controlled environment and a subsequent flowering and hybridization period in either controlled environment or the open field. Thus, breeders of biennial crops have at least two opportunities to evaluate the quality of the crop: at the end of the first season's growth and after the vernalization period. In this way, biennial crop breeding allows breeders to evaluate crop quality and productivity in the field and to further refine their selections after evaluating storage traits during the vernalization phase. Biennial crop breeding may be compressed into an annual cycle if controlled environments are used; however, in many cases two full seasons are required for a single breeding cycle, making it a more time-consuming process than the breeding of annual crops.

BREEDING METHODS FOR BIENNIALS

Perhaps the most important innovation for breeding biennial crops was the introduction of controlled environments. The use of greenhouses and growth chambers, beginning in the early part of the 20th century, allowed plant breeders to compress the two-season life cycle of a biennial plant into a calendar year. This, in turn, greatly facilitated breeding procedures and allowed for greater gains from selection as well as enhanced precision in hybridization. One of the first uses of this procedure was in the cabbage breeding program of J.C. Walker at the University of Wisconsin. Dr. Walker made use of greenhouses during the winter months to allow for the flowering and hybridization of cabbage plants during the development of yellows-resistant cabbage.[1] The use of greenhouses for breeding allowed Walker to reduce the amount of time it took to develop improved cabbage germplasm by one-half. Furthermore, the use of a controlled environment such as a greenhouse allowed for precise pollen control and improved seed set, which are crucial to the success of any breeding program.

Breeding programs for biennial crops, such as those for vegetable crops, use the first season of growth to produce the crop or propagule. In this discussion, propagule is synonymous with crop because the crop itself is later used as the propagule for seed production. For example, in the case of a crop such as carrot, the crop and the propagule are one and the same. In temperate regions, crop production may take place during the spring and summer months, so that by the end of summer or the beginning of autumn the crop is harvested and ready to be vernalized. Vernalization then takes place under controlled environment for a specified period, often coinciding with autumn. Following a suitable period of vernalization, plant propagules are brought into growth chambers or greenhouses where they are placed under long days and allowed to flower. Hybridization and seed production take place during this period, usually coinciding with winter and early spring. Thus, by the end of the pollination and seed production season, the biennial life cycle has been compressed into a single calendar year for the purpose of efficiency.

This scheme requires a vernalization chamber or access to a climate in which propagules can successfully overwinter outside, as well as access to greenhouses or growth chambers for reproduction. Therefore, while it is very efficient from a calendar point of view, the scheme requires large inputs of energy and resources in order to accomplish the compression of the life cycle. This program has been highly successful with many biennial crops, although not all biennial crops can be bred in such a way. Onion (*Allium cepa*), for example, requires a long vegetative period and long vernalization period, which typically do not allow its life cycle to be compressed into a single calendar year. In such cases, the vernalized propagules are held in cold storage for a longer period of time, perhaps as long as five months, during which time it is possible that they will have become de-vernalized. These propagules are planted as quickly as possible in the spring, where elongation of the flower stalk and flowering commences immediately.

Unlike the breeding of annual crops, the breeding of biennial crops allows for at least two primary opportunities to evaluate crop performance and quality. This is true whether the biennial cycle is compressed into a calendar year or whether two full seasons are used to

DOI: 10.1081/E-EPCS 120010529

complete the life cycle. The reason for these two opportunities is that the propagules harvested for vernalization must be handled at first to prepare them for vernalization and then again after vernalization and prior to planting for reproductive purposes. For example, the breeding of table beet follows an annual cycle with two opportunities for selection.[2] Seed is sown in May in the field. After approximately 70–90 days, plants are harvested and roots are separated from leaves with a scissors. Roots are selected, washed, trimmed, and placed in paper bags containing wood shavings. These paper bags are placed in plastic bags that are then sealed. The microclimate inside the paper bag where the propagules are placed is humid enough to prevent desiccation but dry enough to reduce damage caused by plant pathogens. Following a period of vernalization that typically lasts 10–12 weeks, roots are removed from the vernalization chamber and reselected.

Performance in storage is as crucial a trait for a biennial crop as yield performance during the first season of growth. The reason for this is that the biennial crop must be vernalized for an extended period of time, and therefore traits associated with high quality during this period are of great value in a crop cultivar. In addition, breeders of biennial crops realize that many of these same traits are those preferred by consumers for these same crops sold out of storage. For example, carrot germplasm that has been selected for superior traits during vernalization (retention of color and flavor, resistance to storage pathogens, inhibition of sprouting) will be more valuable in a breeding program as well as in a supermarket. Thus, selection for traits associated with quality factors during vernalization will serve the dual purpose of being valuable at a commercial level once a cultivar is produced.

In this way, biennial crops are often selected twice: once in the field following the harvest of the first season's growth and again following the vernalization period. Both of these opportunities for selection allow the breeder a chance to improve the crop for characteristics such as field performance and storability, which are both of great value in determining the success of a cultivar.

Following this second selection, propagules may be planted under controlled environment conditions, such as a greenhouse, where they will bolt and begin to flower. Typically, greenhouse conditions for hybridization and seed production are under long days, which will promote flowering in biennial plants.

Breeders of biennial crops have made use of genes that alter crop life cycles and thereby enhance the efficiency of breeding. One such example is the use of the *B* allele, which conditions an annual habit in *Beta vulgaris*. Plants carrying *bb* are biennial; however, a single *B* allele will result in an annual growth habit. W.H. Gabelman obtained the *B* allele conditioning annual flowering habit from sugar beet breeding material from Dr. V.F. Savitsky.[2,3] In general, the *B* allele allows for efficient development of sterile inbreds since spring-sown plants carrying *Bb* flower in the midwestern United States by mid-August. A cross of the constitution $Bb \times bb$ will give rise to 50% annual (*Bb*) progeny, which, because they are flowering, can be classified for sterility and other floral traits in the field. These annual sterile plants can then be decapitated, vernalized, and reflowered in winter in the greenhouse nursery, ensuring continuous inbreeding of the sterile line with its maintainer line.[2]

When biennial plants carrying sterile cytoplasm are desired, such as during the latter stages of an inbred development program, the remaining 50% of the segregating progeny from the above-described cross that were not flowering can be chosen for appropriate test crosses or commercial use.[3] These are of the desired genotype *bb*. In practice, use of the *B* allele in table beet breeding allows for greater flexibility and precision in inbred development because one can choose annual or biennial (or both) plants in the field and more accurately choose and plan the crosses to be made during winter months.

CONCLUSION

Biennial crop breeding, like much of modern scientific plant breeding in the United States during the 20th century, has followed a path toward the inbred-hybrid method. The inbred-hybrid method of breeding, which was developed in maize during the early decades of the 20th century, set the pattern for breeding techniques in many crops.[4] The inbred-hybrid method allowed for the development of F_1 hybrids, which offered superiority in terms of early season vigor, productivity, and uniformity. Procedures similar to those used in annual crops have been used to apply the inbred-hybrid method to the breeding of biennial crops. The primary difference found in biennial crops bred using this method is the increased length of time required for each cycle of breeding, although as discussed above this can be greatly shortened with controlled environment nurseries.

ARTICLES OF FURTHER INTEREST

Biennial Crops, p. 122
Breeding Clones, p. 174
Breeding Hybrids, p. 186
Breeding Plants and Heterosis, p. 189
Breeding Plants with Transgenes, p. 193
Breeding Pure Line Cultivars, p. 196
Breeding Self-Pollinated Crops Through Marker-Assisted Selection, p. 202
Breeding Synthetic Cultivars, p. 205

REFERENCES

1. Walker, J.C. *Plant Pathology*; McGraw-Hill: New York, 1950; 699 pp.
2. Goldman, I.L.; Navazio, J. History and breeding of table beet in the United States. Plant Breeding Rev. **2003**, *22*, 357–388.
3. Bosemark, N.O. Genetics and Breeding. In *The Sugar Beet Crop: Science into Practice*; Cooke, D.A., Scott, R.K., Eds.; Chapman and Hall: London, 1993; 67–119.
4. Goldman, I.L. Prediction in plant breeding. Plant Breeding Rev. **2000**, *19*, 15–40.

B

Breeding Clones

Rodomiro Ortiz
International Institute of Tropical Agriculture, Ibadan, Nigeria

INTRODUCTION

Clones are propagules arising from asexual (or vegetative) propagation. The most important vegetatively propagated food crops are potato, cassava, sweet potato, yam, plantain/banana, sugar cane, and fruit trees. Other crops with asexual propagations are some ornamentals, grasses, and forages. Among the most common planting materials are tubers (e.g., potatoes and yams), vines (sweet potatoes), stem cuttings (cassava), and suckers (plantains and bananas). In vegetatively propagated crops the common origin of planting materials is crucial to having uniform trials. Tissue culture-derived plantlets are also promising planting materials to achieve propagule uniformity in some of these food crops.

CROSSBREEDING

Crossbreeding methods for vegetatively propagated crops rely on sexual hybridization, i.e., seeds are needed for producing new genotypes after crossing selected parents. Hence, special protocols are used to maximize flowering in some vegetatively propagated crops. Time (i.e., photoperiod) and intensity of light are among the most important factors affecting flowering in these crops.

The main goal of breeding clones will be to obtain genotypes that are phenotypically uniform (homogeneous) but often highly heterozygous, particularly if nonadditive gene action controls the commercial trait(s) of interest. Nonadditive gene action may arise from intra- or inter-allelic (epistasis) interactions. The conventional plan for breeding clones consists of: 1) selecting appropriate parents for crossing schemes; 2) early or late selection in clonal generations, which will be determined by the heritability of the targeted trait(s); and 3) adequate environmental sampling (i.e., number of locations and years) for testing advanced breeding materials leading to cultivar development. The steps in the most common breeding scheme are given in Fig. 1.

ANALYTICAL BREEDING

Genetic manipulations of complete chromosome sets are called ploidy manipulations: scaling up and down chromosome numbers of a species within a polyploid series. The most important vegetatively propagated food crops (potato, sweetpotato, yam, plantain/banana, and some fruit trees) possess well-endowed genetic resources from their wild species, which are often of lower ploidy.[1] Chromosome sets are manipulated with haploids, $2n$ gametes, and through interspecific–interploidy crosses. Analytical breeding schemes rely mainly on ploidy manipulations to ''capture'' diversity from exotic (wild or nonadapted germplasm) and use $2n$ gametes to incorporate this genetic diversity through unilateral (USP; $n \times 2n$ or $2n \times n$) or bilateral (BSP; $2n \times 2n$) polyploidization.[2] Haploids are propagules with the gametophytic chromosome number (n) and $2n$ gametes possess the sporophytic chromosome number of the parental source. The most interesting examples of analytical breeding are in vegetatively propagated species such as potato,[3] sweet potato,[4] and cassava[5] among roots and tubers, and plantain/banana[6] among fruit crops. This breeding approach appears promising in sugar cane,[7] blackberry,[8] blueberry,[9] strawberry,[10] and other fruit crops.[11]

Potato may be regarded as the model crop either for breeding clones by conventional methods[12] or for broadening the genetic base of crop production, particularly through analytical breeding.[13] In potato ploidy manipulations, chromosome sets are easily managed with wild species, maternal haploids obtained through parthenogenesis, $2n$ gametes arising from meitoic mutants, and the endosperm balance number (EBN). This endosperm dosage system, also common to other angiosperm genera, requires a 2:1 ratio of maternal to paternal contributions to achieve normal seed development after hybridization.[14] The wild species (mostly diploids) bring new genetic variation to the breeding pool, whereas haploids ''capture'' this genetic diversity by crossing them with diploid wild species. The resulting haploid-species hybrids—producing $2n$ gametes—and the EBN are the means for broadening the genetic base of the cultivated potato through USP or BSP (Fig. 2), which recent analysis with genetic markers has confirmed.[15] Furthermore, such

Encyclopedia of Plant and Crop Science
DOI: 10.1081/E-EPCS 120010527

Source population (5,000–100,000 seedlings) after crossing selected parents

↓ Defect elimination or mild-selection for specific attributes

Single plots of (100–3,000 selected clones) for clonal evaluation

↓ Screening for specific attributes as per breeding plan

Preliminary yield trial (25–100 clones) with 2 replications

↓ Screening to confirm attributes and early yield assessment

Advanced yield trial (10–25 clones) with 3 to 4 replications in at least 3 locations

↓ Further yield assessment

Uniform yield trial (5–15 best clones) with 4 replications in many locations

↓ Yield assessment and testing stability across location range

On-farm participatory testing of elite materials (2–5 clones)

↓ Farmer (and sometimes end-user) testing

Multiplication of selected clone(s) and cultivar release

(through appropriate national committee)

Fig. 1 Common breeding scheme for vegetatively propagated tropical crops at the International Institute of Tropical Agriculture. Propagule numbers are crop-dependent.

analysis suggests that the need for broadening the genetic base in potato may be met by specific chromosomes or regions within chromosomes.

EVOLUTIONARY APPROACH

Genetic bottlenecks could happen during the evolution of vegetatively propagated crops because breeders of these crops (farmers in the early days, but nowadays mostly trained professionals) may select a few sports (or mutants) with the desired characteristic, which could replace old cultivars in a large scale area. Triploid plantains provide an interesting example, in which most of the variation observed in approximately 120 cultivars (or landraces) known world-wide resulted from mutations accumulated throughout the history of cultivation of this crop and from farmer selection of a few strains.[16] In this triploid crop, gene flow through pollen was prevented due to the low male fertility of the crop.[17] Diploid banana species and plantain producing $2n$ eggs were the

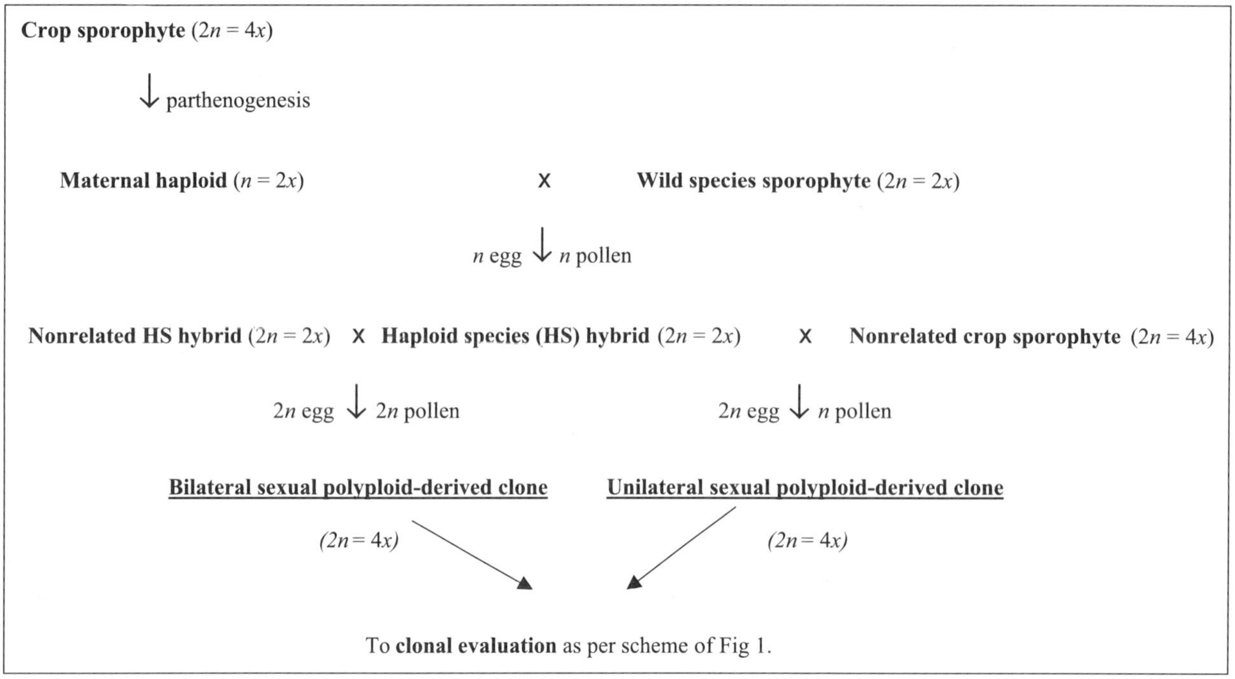

Fig. 2 Ploidy manipulations (or scaling up and down chromosome numbers) for breeding clones with haploids and $2n$ gametes in a crop within a polyploid series. The example refers to a tetraploid ($2n = 4x$) crop with diploid wild species.

tools for broadening the genetic base in this important tropical starchy crop. Tetraploid hybrids may be obtained by hybridizing $2n$ eggs from plantains with n pollen from diploid accessions. Plantain-derived diploids also result from such crosses.

In plantains, heterozygous triploid clones (which are farmers' selections that have been around for a long time) are the sources of allelic diversity. These alleles are released to the tetraploid hybrids through $2n$ eggs and are further broadened by the alleles provided by the diploid bananas. Advanced ploidy manipulations may lead to secondary triploids resulting from crosses between selected tetraploids and elite diploid stocks, both producing n gametes.[18] Triploid *Musa* hybrids may also occur as a result of to USP among selected diploid stocks, because one of the parents produces $2n$ gametes. Such breeding methods for plantain should be regarded as part of an evolutionary improvement approach, because conventional breeding will be enhanced by innovative knowledge-led methods, as described above, for introducing additional genetic variation.[19] Genetic markers may assist in the process of recurrent selection of plantain germplasm, and combining ability tests will assist the selection of elite parents at any ploidy level.[20,21] Recent results from this approach show that prospects for plantain (and banana) breeding "are unlimited, and increased efforts will at once initiate a new phase of *Musa* evolution."[22]

SOMACLONAL VARIATION

Irrespective of the advantages of tissue culture for vegetatively propagated crops, somaclonal variation may affect the true-to-type during micropropagation. Somaclonal variation refers to genetic variation arising from tissue culture regeneration among plants from the same original genotype. Careful plant breeders minimize somaclonal variation by: 1) selecting deliberately stable sources of materials for primary explants; 2) limited subculturing and multiplication (less cycles, short time for subculturing, and a few hundred plants per primary explant); and 3) nursery screening to detect and rogue off-types. Somaclonal variation may provide a potential source for genetic improvement of some vegetatively propagated crops. However, in most crops, the range of somaclonal variants recovered through shoot-tip culture seems to be narrow, mimics naturally occurring variation, or produces defective genotypes.

FUTURE PROSPECTS

Hybrid clones may result from artificial hand-pollination or through poly-crosses among parents that are selected for their specific combining ability. Poly-crossing refers to a pollination system based on natural random mating among selected genotypes grown together in isolated plots. The seeds from the multiple hybrid mixtures may be also regarded as synthetic cultivars because they are derived from selected genotypes that combine well among them. A recent report on potato suggests that it will be feasible to obtain the same or better performance with a minimum number of parents in the source synthetic population and by allowing it to open-pollinate in an isolation plot.[23]

Hybrid seed from poly-crosses are obtained from isolated plots to avoid contamination with pollen from other nonselected clones. Synthetic populations derived from these poly-crosses are tested in other locations to identify promising offspring for selection and cultivar development. Local selections follow a dynamic conservation approach to genetic resources because target farmers preserve distinct, locally adapted and improved genotypes across locations. Ploidy manipulations coupled with this breeding approach broaden the genetic base of vegetatively propagated crops, thereby enhancing crop adaptation and sustaining genetic gains in respective breeding pools. As pointed out, early locally adapted germplasm with enhanced adaptation to stress-prone environments or resistance to pest and diseases will allow the sustainable and environment-friendly production of vegetatively propagated crops, which often are affected by many biotic or abiotic stresses.

ACKNOWLEDGMENTS

Some of these ideas for crop breeding are the result of some stimulating discussions with my academic mentors: Prof. Francisco Delgado de la Flor, supervisor for the MSc degree at Univ. Nacional Agraria–La Molina, Perú; Em. Campbell-Bascom Prof. Stanley J. Peloquin, advisor for the Ph.D. degree at the Univ. of Wisconsin–Madison, USA); and Dr. Masaru Iwanaga (with whom I started my international professional career at the Centro Internacional de la Papa, Perú). These initial ideas were further developed in my early professional years at the International Institute of Tropical Agriculture (Nigeria) with the late Dirk R. Vuylsteke, a meticulous *Musa* scientist and humanitarian, who dedicated his short but outstanding professional life to improving African agriculture.

ARTICLES OF FURTHER INTEREST

Breeding: Choice of Parents, p. 215
Breeding: Genotype-by-Environment Interaction, p. 218
Breeding: Incorporation of Exotic Germplasm, p. 222
Breeding: Mating Designs, p. 225
Breeding: Participatory Approaches, p. 229
Breeding: Recurrent Selection and Gain from Selection, p. 232
Breeding: The Backcross Method, p. 237
Breeding for Durable Resistance, p. 179
Breeding Hybrids, p. 186
Breeding Plants and Heterosis, p. 189
Breeding Synthetic Cultivars, p. 205
Chromosome Manipulation and Crop Improvement, p. 266
Crop Domestication: Fate of Genetic Diversity, p. 333
Crop Improvement: Broadening the Genetic Base for, p. 343
Farmer Selection and Conservation of Crop Varieties, p. 433
Gene Flow Between Crops and Their Wild Progenitors, p. 488
Polyploidy, p. 1038
Quantitative Trait Locus Analyses of the Domestication Syndrome and Domestication Process, p. 1069

REFERENCES

1. Ortiz, R. Germplasm Enhancement to Sustain Genetic Gains in Crop Improvement. In *Managing Plant Genetic Diversity*; Engels, J.M.M., Brown, A.H.D., Jackson, M., Eds.; 2002; 275–290. IPGRI: Rome, Italy; CAB International: Wallingford.
2. Peloquin, S.J.; Ortiz, R. Techniques for Introgressing Unadapted Germplasm to Breeding Populations. In *Plant Breeding in the 1990s*; Stalker, H.T., Murphy, J.P., Eds.; CAB International: Wallingford, 1992; 485–507.
3. Ortiz, R. Potato breeding via ploidy manipulations. Plant Breed. Rev. **1998**, *16*, 15–86.
4. Iwanaga, M.; Freyre, R.; Orjeda, G. Use of *Ipomoea trifida* (HBK) G. Don germplasm for sweet potato improvement. 1. Development of synthetic hexaploids of *I. trifida* by ploidy manipulations. Genome **1991**, *24*, 201–208.
5. Hahn, S.K.; Bai, K.V.; Asiedu, R. Tetraploids, triploids and 2n pollen from diploid interspecific crosses with cassava. Theor. Appl. Genet. **1990**, *79*, 433–439.
6. Vuylsteke, D.; Ortiz, R.; Ferris, R.S.B.; Crouch, J.H. Plantain improvement. Plant Breed. Rev. **1998**, *16*, 267–320.
7. Bremer, G. Problems in breeding and cytology of sugar cane. Euphytica **1961**, *10*, 59–78.
8. Hall, H.K. Blackberry breeding. Plant Breed. Rev. **1990**, *8*, 244–312.
9. Ortiz, R.; Bruederle, L.P.; Vorsa, N.; Laverty, T. The origin of polysomic polyploids via 2n pollen in *Vaccinium* section *Cyanococcus*. Euphytica **1992**, *61*, 241–246.
10. Bringhurst, R.S.; Voth, V. Breeding octoploid strawberries. Iowa State Univ. J. Res. **1984**, *58*, 371–382.
11. Sanford, J.C. Ploidy Manipulations. In *Methods in Fruit Breeding*; Moore, J.N., Janick, J., Eds.; Purdue Univ. Press: West Lafayette, IN, 1983; 100–123.

12. Tarn, T.R.; Tai, G.C.C.; de Jong, H. Breeding potatoes for long-day, temperate climates. Plant Breed. Rev. **1992**, *9*, 217–332.
13. Ortiz, R. The State of Use of Potato Genetic Diversity. In *Broadening the Genetic Bases of Crop Production*; Cooper, H.D., Spillane, C., Hodgkin, T., Eds.; 2001; 181–200. Food and Agriculture Organization of the United Nations (FAO): Rome, Italy; International Plant Genetic Resources Institute: Rome, Italy; CAB International: Wallingford.
14. Ehlenfeldt, M.K.; Ortiz, R. On the origins of endosperm dosage requirements in *Solanum* and other angiosperma genera. Sex. Plant Reprod. **1995**, *8*, 189–196.
15. Ortiz, R.; Huamán, Z. Allozyme polymorphism in tetraploid potato gene pools and the effect of human selection. Theor. Appl. Genet. **2001**, *103*, 792–796.
16. Crouch, H.K.; Crouch, J.H.; Madsen, S.; Vuylsteke, D.; Ortiz, R. Comparative analysis of phenotypic and genotypic diversity among plantain landraces (*Musa* spp., AAB group). Theor. Appl. Genet. **2000**, *101*, 1056– 1065.
17. Ortiz, R.; Madsen, S.; Vuylsteke, D. Classification of African plantain landraces and banana cultivars using a phenotypic distance index of quantitative descriptors. Theor. Appl. Genet. **1998**, *96*, 94–101.
18. Ortiz, R. Secondary polyploids, heterosis and evolutionary crop breeding for further improvement of the plantain and banana genome. Theor. Appl. Genet. **1997**, *94*, 1113–1120.
19. Ortiz, R.; Vuylsteke, D.R. *Musa* scientist and humanitarian. Plant Breed. Rev. **2001**, *21*, 1–25.
20. Ortiz, R.; Craenen, K.; Vuylsteke, D. Ploidy manipulations and genetic markers as tools for analysis of quantitative trait variation in progeny derived from triploid plantains. Hereditas **1997**, *126*, 255–259.
21. Tenkouano, A.; Crouch, J.H.; Crouch, H.K.; Ortiz, R. Genetic diversity, hybrid performance and combining ability for yield in *Musa* germplasm. Euphytica **1998**, *102*, 281–288.
22. Vuylsteke, D. Breeding bananas and plantains: From intractability to feasibility. Acta Hortic. **2000**, *540*, 149–156.
23. Golmirzaie, A.M.; Ortiz, R. Inbreeding and true seed in tetrasomic potato. IV. Synthetic cultivars. Theor. Appl. Genet. **2002**, *104*, 161–164.

Breeding for Durable Resistance

Raoul A. Robinson (Retired)
Food and Agriculture Organization of the United Nations, Fergus, Ontario, Canada

B

INTRODUCTION

Since the recognition of Mendel's laws in 1900, plant breeders have shown a marked preference for single-gene resistances to crop pests and diseases. This preference has been greatly strengthened by the techniques of molecular modification, which, of necessity, also involve single-gene resistances. These single-gene, qualitative resistances have many advantages but, unfortunately, they are almost invariably *within* the capacity for genetic change of the crop parasite, and they endure only until the parasite produces a new strain that is unaffected by the resistance in question. A parallel phenomenon is seen with many synthetic insecticides and fungicides. Durable resistance, often called horizontal resistance, on the other hand, is normally controlled by polygenes. Such resistance endures in the same way, and for the same reason, that the effectiveness of copper and dithiocarbamate fungicides, or rotenone insecticides, endures. These protection mechanisms are *beyond* the capacity for genetic change of the crop parasite in question.

BRIEF LITERATURE REVIEW

The most comprehensive review of breeding for durable resistance is that of Simmonds.[1] He gives examples of durable resistance in twenty-one species of crop, functioning variously against airborne and soil-borne fungal pathogens, bacteria, viruses, insects, and nematodes.

In 1992, Stoner[2] reviewed 705 papers on plant host resistance to insects and mites in vegetables, and she also quotes reviews of this topic in grain crops, alfalfa, and cotton. She comments that, in most studies, the resistance is a quantitative trait, but she adds that there has been little plant breeding for resistance to insects.

Robinson[3] has described the techniques of breeding for durable resistance, and the present article is a brief summary of those techniques. Robinson[4] has also published free downloads of his books on the Internet.

BREEDING FOR QUANTITATIVE VARIABLES

Because durable resistance is almost invariably controlled by polygenes, it must be treated as a quantitative variable with a maximum and a minimum. As a rule, the maximum level of quantitative resistance is represented by a negligible loss of crop in the absence of crop protection chemicals. The minimum level of quantitative resistance is represented by a total loss of crop in the absence of crop protection chemicals. Many modern cultivars have levels of quantitative resistance to some of their parasites that come close to this minimum.

Breeding for quantitative resistance is thus a progressive and cumulative process, in the sense that a good cultivar need never be replaced, except with a better cultivar. Even modest increases in the level of quantitative resistance will reduce the need for crop protection chemicals.

Breeding for quantitative variables requires recurrent mass selection, in which the best individuals in a large population become the parents of the next generation. The resistance accumulates mainly by transgressive segregation, and the entire program is likely to require some 10–15 breeding cycles. This approach is clearly a major departure from the classic breeding of autogamous crops such as wheat, beans, and rice. The recurrent mass selection must be conducted in accordance with the following rules.

Original Parents

Some 10–20 modern cultivars with high yield and high quality of crop product should be chosen as the original parents of the recurrent mass selection. While it does no harm to choose parents with a reasonably high quantitative resistance, this is not essential because transgressive segregation will accumulate adequate resistance from susceptible parents, provided the genetic base is reasonably wide. Robinson[3] has described the remarkable effect of this process in maize in tropical Africa.

Cross-Pollination

The original parents must be cross-pollinated in a half-diallel cross (i.e., each parent is crossed with every other parent, without any self-pollinations, and without any distinction between male and female parents). Similar cross-pollination must occur with the selected individuals of each screening generation that become the parents of the next breeding cycle. The techniques of pollination depend on the nature of the crop in question. Beek[5] used a

Encyclopedia of Plant and Crop Science
DOI: 10.1081/E-EPCS 120010539

male gametocide to achieve several million crosses in a rather small population of wheat, and it appears that this technique can be used with other cereals. A marker gene can be used in crops such as beans, in which the levels of natural cross-pollination are low. Hand-pollination, following hand-emasculation, is useful when a single pollination produces many seeds, as with crops in the *Solanaceae* and *Cucurbitaceae*. In only a few crops (e.g., *Cicer arietinum*) is a labor-intensive hand-pollination inevitable.

On-Site Screening

On-site screening means that the screening population is screened 1) in the area of future cultivation; 2) in the time of year of future cultivation; and 3) according to the farming system of future cultivation (e.g., organic versus artificial fertilizers, irrigated versus rain-fed). On-site screening is necessary because the epidemiological competence of crop parasites varies markedly and differentially between agro-ecosystems. A cultivar that has all its quantitative resistances in balance with one agro-ecosystem, is likely to have too much resistance to some parasites, and too little to others, when taken to a different agro-ecosystem. However, most agro-ecosystems are large enough to justify a breeding program for each of their important species of crop.

Screen for Yield and Quality

The screening population should be screened primarily for high yield, on the grounds that susceptible individuals cannot yield well in the presence of parasites. As the original parents were high quality cultivars, it is necessary to ensure that the level of this quality does not decline and that individuals with a reduced quality should not become parents in the next breeding cycle.

Inoculate to Prevent Chance Escape

It is imperative that that every individual in the screening population is parasitized with every locally important parasite. This is because the level of quantitative resistance can be assessed only in terms of the level of parasitism. A chance escape from parasitism will provide an entirely false indication of resistance, and using such an individual as a parent would seriously reduce the genetic advance.

Ensure That No Single-Gene Resistances Are Functioning

The level of quantitative resistance can be assessed only by the level of parasitism. Consequently, it is essential that no single-gene resistances are functioning in the screening population. If these resistances cannot be eliminated genetically, they can be inactivated, usually by the "one-pathotype technique" described elsewhere by Robinson.[3]

Avoid the Vertifolia Effect

Vanderplank[6] first recognized a phenomenon that he called the "vertifolia effect." This is the loss of quantitative resistance that occurs during breeding for single-gene resistance. The vertifolia effect also occurs if crop protection chemicals are used to protect the screening population. In breeding potatoes, the vertifolia effect has probably been occurring with quantitative resistance to blight (*Phytophthora infestans*) since the discovery of Bordeaux mixture in 1882, and subsequently during many years of breeding for single-gene resistances. In cotton, a vertifolia effect in insect resistance was probably initiated with the discovery of DDT.

Selection Pressures

During the early screening generations, the selection pressures for quantitative resistance are likely to be too high. There is then a danger that the screening population, which has a very low level of resistance, will be destroyed entirely by its parasites. Should this appear likely, crop protection chemicals should be used, as late as safety permits, in order to preserve the minority of the least susceptible individuals. Conversely, as quantitative resistance accumulates in the later screening generations, the selection pressures for resistance decline. Inoculation of the screening population will then be necessary to determine the levels of resistance.

TWO SOURCES OF MAJOR ERROR

Robinson[3] has discussed two sources of major error that have apparently been responsible for the general neglect of quantitative resistance during the past century. These sources of error have so denigrated the appearance of quantitative resistance as to make it appear valueless or even nonexistent. And they indicate that a considerably lower level of resistance than we may believe will achieve an adequate control of crop parasites. They are biological anarchy and parasite interference.

Biological Anarchy

Robinson[3] defined biological anarchy as the loss of biological control that occurs when the various organisms contributing to biological control are reduced or eliminated by crop protection chemicals. The crop parasites then behave with a savagery that would be impossible if their numbers had been kept down by their natural enemies. The importance of biological anarchy is that it obscures levels of quantitative resistance that might provide a complete control if the natural biological controls are functioning. In other words, we probably need considerably lower levels of quantitative resistance than we may currently believe. The importance of biological anarchy can be assessed by the general usefulness of the entomological techniques of integrated pest management (IPM). The best way to restore biological control is to use horizontal resistance. And the best way to enhance the effects of horizontal resistance is to restore biological controls. The two effects are mutually reinforcing.

Parasite Interference

Parasite interference was first recognised by Vanderplank[6] and it occurs because crop parasites can move from one field plot to another. James et al.[7] assessed its importance in potatoes. Parasite interference can easily increase the levels of parasitism in test plots by several orders of magnitude. Perhaps the most dramatic example of parasite interference is seen in the small plots used by small cereal breeders working with single-gene resistances. These family-selection (i.e., ear-to-row) plots exhibit the hypersensitive flecks that result from non-matching infections, but these flecks may occur in millions. There can be so many of them that the resistant cereal appears to be severely diseased, and the breeders warn that this phenomenon must not be mistaken for true disease.

Parasite interference also occurs during recurrent mass selection because an individual resistant plant is likely to be surrounded by more susceptible individuals. That resistant individual will then have a level of parasitism far higher than it would in a farmer's field where there is no parasite interference. For this reason, all assessments of the level of parasitism during recurrent mass selection must be relative measurements. Only the least parasitized (or highest yielding) individuals are selected, however severely parasitized (or low yielding) they may be in more general terms.

PLANT BREEDING CLUBS

Robinson[3] has argued that breeding for durable resistance is so easy that it can be undertaken by amateurs. The proof of this comes from the remarkably successful, but illegal, amateur breeding of marijuana. Amateur breeding would be at its most successful when amateurs organized themselves into a plant breeding club. He suggests that the most useful and productive clubs would be student clubs in colleges and universities.

ARTICLES OF FURTHER INTEREST

Breeding: Participatory Approaches, p. 229
Breeding: Recurrent Selection and Gain from Selection, p. 232
Farmer Selection and Conservation of Crop Varieties, p. 433
Integrated Pest Management, p. 612

REFERENCES

1. Simmonds, N.W. Genetics of horizontal resistance to diseases of crops. Biol. Rev. **1991**, 189–241.
2. Stoner, K.A. Bibliography of plant resistance to arthropods in vegetables, 1977–1991. Phytoparasitica **1992**, *20* (2), 125–180.
3. Robinson, R.A. *Return to Resistance; Breeding Plants to Reduce Pesticide Dependence*; agAccess: Davis, CA, 1996.
4. Robinson, R.A. Free downloads of books at www.sharebooks.ca.
5. Beek, M.A. *Selection Procedures for Durable Resistance in Wheat*; Agric. University: Wageningen, 1988. Papers 88-2.
6. Vanderplank, J.E. *Disease Resistance in Plants*; Academic Press: New York, 1968.
7. James, W.C.; Shih, C.S.; Callbeck, L.C.; Hodgson, W.A. Interplot interference in field experiments with late blight of potato (*Phytophthora infestans*). Phytopathology **1973**, *63*, 1269–1275.

B

Breeding for Nutritional Quality

Brian J. Just
University of Wisconsin, Madison, Wisconsin, U.S.A.

Philipp W. Simon
USDA—ARS, Madison, Wisconsin, U.S.A.

INTRODUCTION

Breeding plants for improved nutritional quality is an old idea. One can imagine early gatherers and farmers avoiding eating certain plants that made them ill. In more modern times, selection for sugar content of sugarbeet has been ongoing for over 250 years, whereas selection for oil and protein content in maize and soybean is approaching 100 years in practice. As research in the nutritional sciences continues to demonstrate the importance to our health and well-being of plant-derived nutrients in foods, more breeding is underway to increase the type and amount of nutrients in cereals, legumes, vegetables, and fruits. The human diet consists of a diverse array of crops. Because specific nutrient levels in many of these crops are often under genetic control, the prospects for improving the nutritional quality of food through plant breeding are promising.

TARGETS FOR IMPROVEMENT

Traditionally, nutrients in the human diet have been defined as components required for the proper growth and function of the body. These essential nutrients are usually categorized into two classes: the macronutrients and the micronutrients. Macronutrients are compounds that are required in large amounts and include carbohydrates, proteins, and lipids. Over 90% of the U.S. dietary carbohydrates come from plant sources (Table 1). Micronutrients are needed in smaller amounts but are nonetheless critical for general maintenance of the body. The micronutrients include minerals and vitamins. Over 90% of the vitamin C; 70% of the thiamin, folate, Fe, and Cu; and 50% of vitamin E, niacin, Mg, and K come from plant sources (Table 1). In less developed countries, plants contribute much higher proportions of these nutrients to the diet.

Beyond essential nutrients, phytonutrients are another class of compounds derived from plants that have also been recently studied. Phytonutrients (also called phytochemicals) are not specifically required for bodily function, but have been documented or implicated as having a positive impact on human health. Many phytonutrients act as antioxidants and help prevent diseases such as cancer, heart disease, and strokes. There is a wide diversity of phytonutrients and they include pigments (e.g., carotenoids and anthocyanins), glucosinolates, terpenoids, saponins, and a host of other organic molecules. In the future, many more phytonutrients may be discovered.

In addition to nutrients, antinutrients and allergens are known to exist in some crops. Antinutrients are substances that have a toxic effect or inhibit the uptake of nutrients. Examples of antinutrients are glycoalkaloids in some potatoes and phytate in some legumes. Compounds from each of these nutrient and antinutrient classes have been the targets of, or have been suggested as future targets for, plant breeding efforts.

BREEDING METHODS AND EXAMPLES OF BREEDING FOR IMPROVED NUTRITIONAL QUALITY

Breeding for nutrient composition of a crop species is not fundamentally different from breeding for other quality or yield traits. Choice of method depends on limitations imposed by the biology of the species, number of genes controlling the trait, level of variation in source material utilized in the breeding program, ease of measuring the level of the nutrient, and extent to which environment affects nutrient levels.

Breeding for any trait begins with a survey of the variability within a crop species of the trait of interest. Once useful variability for a nutritional quality trait is identified, the breeder usually elucidates the mode of inheritance. Nutritional quality traits can be either monogenic (qualitative), oligogenic, or polygenic (quantitative). Within the germplasm to be utilized, a heritability estimate is obtained that gives the breeder an idea of how much of the observed variation is genetic in nature and how easily that genetic variation can be exploited by various breeding methods.

Encyclopedia of Plant and Crop Science
DOI: 10.1081/E-EPCS 120010538

Table 1 Plant sources of macronutrients, vitamins, and minerals in the U.S. food supply: 1990

		Breeding effort		
Nutrient	**Plant sources—total%; Major sources**	**Developing countries, for food**	**Developed countries, for food**	**Developed countries, for extraction/feed**
Macronutrients				
Carbohydrates	95; Cereals, potatoes, vegetables, sugar	X	X	X
Proteins	30; Cereals, legumes	X		X
Fats	25; Vegetable oils			X
Micronutrients: Vitamins				
A	40; Vegetables	X	X	
Thiamine (B_1)	70; Cereals, potatoes, vegetables			
Riboflavin (B_2)	70; Cereals, vegetables, fruits			
Niacin (B_3)	55; Cereals, potatoes			
B_6	45; Potatoes, cereals, vegetables			
Folate	75; Vegetables, cereals, fruits		X	
C	95; Fruits, vegetables, potatoes		X	
E	50; Vegetable oils			
Micronutrients: Minerals				
Ca	20; Vegetables, legumes, cereals, fruits			
P	35; Cereals, legumes, vegetables, potatoes			
Mg	55; Cereals, legumes, vegetables, potatoes			
Fe	70; Cereals, legumes, vegetables, potatoes	X	X	
Zn	30; Cereals, legumes, vegetable, potatoes			
Cu	70; Cereals, legumes, vegetables, potatoes	X	X	
K	55; Vegetables, fruits, and potatoes			

(From Ref. 1.)

If a nutritional trait is governed by a single gene, the usual approach is to introgress the allele causing the desired phenotype into an appropriate background. This can be achieved by backcross breeding. Numerous single gene mutations have been identified in many crop species that affect the level of a nutritionally important compound; examples are listed in Table 2. After introgressing the allele of choice into commercially acceptable backgrounds, quantitative variation for expression of the trait may be observed and exploited for further improvement. However, introgressed genes can also negatively affect other agronomically important traits. These undesirable effects can be because of pleiotropy or because the gene of interest is tightly linked to genes with undesirable effects. Further breeding may be necessary to eliminate or reduce the negative attributes caused by the introgressed gene or genes linked to it. For example, the *opaque-2* mutation in corn, which confers the desirable high-lysine phenotype, is associated with a soft endosperm that is commercially unacceptable. Recurrent selection has recently been used to increase endosperm hardness in high-lysine maize synthetics.[6]

If a trait exhibits quantitative variation, which is characteristic of traits under polygenic control, recurrent

Table 2 Examples of loci with large effect on production of nutrient or anti nutrient compounds

Locus	Crop	Nutrient class	Phenotype	Reference
opaque-2	Corn	Macronutrient	Alters protein profile to include more lysine, an essential amino acid	2
hp	Tomato	Micronutrient	Increases provitamin A and vitamin C content in fruits	3
P_1	Carrot	Phytochemical	Induces anthocyanin production in the root	4
Bi	Squash	Anti-nutrient	Controls production of cucurbitacin E, a toxin in squash fruits	5

(From Refs. 2–5.)

selection methods can be used to increase the mean level of the nutrient in a population over several cycles of selection. An example of this approach is illustrated by the breeding of the high-carotene mass (HCM) carrot population. Over eleven cycles of mass selection, the provitamin A carotenoid content was increased to over four times the level found in typical carrots.[7] The longest continuous plant breeding effort has been the Illinois project in which selection for oil and protein levels in corn for 100 generations continues to be effective.[8]

In some cases, the necessary variability of the nutritional trait may not be available in cultivated germplasm. It may then be necessary to use exotic germplasm such as wild relatives of the crop plant or unadapted landraces to improve nutritional quality. Species related to the cultivated tomato (*Lycopersicon pimpinellifolium*) were used to more than double vitamin C content of tomatoes. However, use of this vitamin C–enhanced germplasm has been limited because tomato breeders have not yet been able to overcome negative effects on yield associated with the nutrient enhancing genes.[9]

Recent developments in genetic engineering offer new ways to improve nutrient content in plants. This approach allows geneticists to introduce genetic variation into a species that is unavailable in its germplasm pool. Genetic engineering of plants can either be used to introduce a completely novel trait or to manipulate a known biochemical pathway. This metabolic engineering approach was used to induce rice plants to accumulate β-carotene (provitamin A) in the endosperm of the grain, albeit at low levels.[10] It may be possible to further increase the β-carotene levels in this transgenic rice (known as Golden Rice) by recurrent selection or by backcrossing into different backgrounds.

OTHER ASPECTS OF NUTRITIONAL QUALITY IMPROVEMENT

The success of a breeding program to improve nutritional quality is largely dependent on the ability of the breeder to accurately assess and improve the phenotype. For some nutrients, such as those involving beneficial plant pigments, selection by visual examination may be successful. However, more sophisticated chemical analysis is required to assess levels of most plant nutrient compounds. A wide variety of analytical chemistry methods exist to assay the level of nutrients in plants. If a particular target nutrient is difficult, time-consuming, or expensive to assay, it may be worthwhile to develop molecular markers linked to the genes that affect nutrient level. These markers can then be used to assist the breeder in selection for the trait.

Another important factor in the breeding of plants with increased nutritional quality is environmental effects on expression. For example, crops bred for increased uptake and storage of a mineral nutrient must have that nutrient available in the soil. Similarly, environmental effects such as temperature and disease pressure can alter the chemistry of the plant, which may in turn affect the nutrient composition. This variation due to nongenetic factors accounts for genotype by environment interactions. A genotype with increased levels of a nutrient in one environment may have lower levels of the same nutrient in a different environment. Hence, multilocation testing is essential for the development of nutritionally enhanced cultivars.

Breeding plants for increased levels of a nutrient per se may not be effective in increasing the level of that nutrient in the human diet. The term ''bioavailability'' refers to the ease with which a nutrient can be assimilated and utilized by the consumer. For example, bioavailability of carotenoids is influenced by the matrix in which these pigments exist in the plant, the postharvest handling of the crop, the method of food preparation, and the nutritional status of the consumer.[11] Very little is known about the factors that contribute to the bioavailability of many micronutrients. Future research will be needed to determine the factors in the plant that contribute to nutrient bioavailability and whether those factors can be manipulated through plant breeding.

CONCLUSION

The end use for plant varieties with increased nutrient levels may be direct consumption or for extraction of the nutrient for use in supplements. Breeding for extractable nutrients such as plant oils, carbohydrates, protein, pigments, and vitamins has long been successfully pursued to yield the ingredients for home and processed food industry use. With these ingredients, food products can be enriched or fortified through processing. However, breeding for increased nutrients in whole foods for consumption is important because it provides a more direct way to deliver nutrients to consumers, especially in developing countries where the enhancement of diet is most critical. Furthermore, nutritionally improved cultivars can provide a locally sustainable nutrient source and even an additional source of farm income in developing countries. As our knowledge of plant genetics and human nutrition grows, breeding for nutritional quality will no doubt become a more important target for plant improvement.

ARTICLES OF FURTHER INTEREST

Ascorbic Acid: An Essential Micronutrient Provided by Plants, p. 65
Carotenoids in Photosynthesis, p. 245
Legumes: Nutraceutical and Pharmaceutical Uses, p. 651
Metabolism, Primary: Engineering Pathways of, p. 714
Metabolism, Secondary: Engineering Pathways of, p. 720
Minor Nutrients, p. 726
Nutraceuticals and Functional Foods: Market Innovation, p. 839
Phytochemical Diversity of Secondary Metabolites, p. 915
Secondary Metabolites As Phytomedicines, p. 1120

B

REFERENCES

1. Gerrior, S.A.; Zizza, C. *Nutrient Content of the U.S. Food Supply, 1909–1990*; Home Economic Research Report No. 52, U.S. Department of Agriculture, Agricultural Research Service: Washington, DC, 1994; 1–114.
2. Mertz, E.T.; Bates, L.S.; Nelson, O.E. Mutant gene that changes protein composition and increases lysine content of maize endosperm. Science **1964**, *145* (3629), 279–280.
3. Jarret, R.L.; Sayama, H.; Tigchelaar, E.C. Pleiotropic effects associated with the chlorophyll intensifier mutations high pigment and dark green in tomato. J. Am. Soc. Hortic. Sci. **1984**, *109* (6), 873–878.
4. Simon, P.W. Inheritance and expression of purple and yellow storage root color in carrot. J. Heredity **1996**, *87* (1), 63–66.
5. Gorski, P.M.; Jaworski, A.; Shannon, S.; Robinson, R.W. Rapid TLC and HPLC test for cucurbitacins. Cucurbit Gen. Coop. **1985**, *8*, 69–70.
6. Lambert, R.J.; Chung, L.C. Phenotypic reccurrent selection for increased endosperm hardness in two high lysine maize synthetics. Crop. Sci. **1995**, *35* (2), 451–456.
7. Simon, P.W.; Wolff, X.Y.; Peterson, C.E.; Kammerlohr, D.S.; Rubatzky, V.E.; Strandberg, J.O.; Bassett, M.J.; White, J.M. High carotene mass carrot population. HortScience **1989**, *24* (1), 174–175.
8. Dudley, J.W.; Lamert, R.J. Ninety generations of selection for oil and protein in maize. Maydica **1992**, *37* (1), 81–87.
9. Lincoln, R.E.; Kohler, G.W.; Silver, W.; Porter, J.W. Breeding for increased ascorbic acid content in tomatoes. Bot. Gaz. **1950**, *111* (3), 343–353.
10. Ye, X.; Al-Babili, S.; Klöti, A.; Zhang, J.; Lucca, P.; Beyer, P.; Potrykus, I. Engineering the provitamin A (β-carotene) biosynthetic pathway into (carotenoid-free) rice endosperm. Science **2000**, *287* (5451), 303–305.
11. Takyi, E.E.K. Bioavailability of Carotenoids from Vegetables Versus Supplements. In *Vegetables, Fruits, and Herbs in Health Promotion*; Watson, R.R., Ed.; CRC Press: Boca Raton, 2001; 19–34.

Breeding Hybrids

Arnel R. Hallauer
Iowa State University, Ames, Iowa, U.S.A.

INTRODUCTION

Hybrids are crosses between two parents that either occur naturally in plant populations or occur when plant breeders and geneticists manually produce crosses between selected parents.[1,2] The relative importance of hybrids varies among plant species, but in nearly all instances hybrids have a role in plant improvement. In allogamous (cross-fertilizing) plant species each plant is theoretically a hybrid because of the union of gametes between the male and female parents. Except for specific instances, the frequency of hybrids would be less in autogamous (self-fertilizing) plant species. In most plant breeding programs, hybrids are produced between elite parents to develop F_2 populations by either selfing or sibbing (plant-to-plant crosses) the hybrid (F_1) plants. Selfing and selection is initiated within the F_2 populations to develop pure lines (or inbreds) that may be used either as pure-line cultivars for autogamous plant species or as parents to produce hybrids for allogamous crop species. Hybrids for autogamous crop species are vehicles to develop segregating populations in which selection is practiced to develop pure-line cultivars for use by the producers. In allogamous plant species hybrids are often the starting point to develop pure lines that are ultimately used as parents to produce hybrids for use by the producers. Generally hybrids are associated with greater productivity and vigor, but this varies among and within autogamous and allogamous crop species and for different plant traits.

Hybrids have become the predominant form of cultivar in many crops. In industrialized countries, many allogamous crops including maize, sunflower, brassicas, cucurbits, carrots, beets, and onions are predominately produced as hybrid cultivars. Hybrid cultivars are also common in certain autogamous crops, including sorghum, tomato, and peppers. In nonindustrialized countries use of hybrid cultivars in allogamous species is increasing and hybrid rice has become important in China.

HISTORY OF HYBRIDS

The occurrence and interest in hybrids were studied during the 18th and 19th centuries.[1,2] Mendel and Darwin also produced hybrids and reported that the hybrids tended to have greater vigor and growth than the parents of the hybrids. Beal studied hybrids of open-pollinated varieties (variety crosses) of maize (*Zea mays* L.) in 1880 and reported that the yields of the hybrids exceeded yields of the parent varieties.[3] Further crosses among maize varieties were produced, but the hybrid yields were not consistently greater than the parents of the hybrids. Data for 244 variety hybrids of maize produced during the first two decades of the 20th century were summarized and the chances were about equal that yield of the variety hybrids was or was not better than the greater yielding parent.[4,5] Because the hybrids were often produced in a haphazard manner and yields of the hybrids were not consistently greater than the parental varieties, hybrids of open-pollinated maize varieties were not widely accepted. Choice of parents, production methods, and genetic variability of the parent varieties were contributing factors for the inconsistent performance of the maize variety hybrids.

Early studies of hybrids were of interest, but the researchers did not recognize the commercial possibilities of hybrids. They were examining potential of the uniqueness and benefits of hybrids for the gardeners and farmers. Success in development of maize hybrids stimulated interest in other crop species, and success was attained in most instances.[6] Greater interests and efforts were emphasized in those plant species that had potential to be exploited commercially. Ownership of the parents of successful hybrids permitted control by the originator of the parents of the hybrids.

The modern concept of hybrids was developed from inbreeding and crossing studies of maize conducted 1908 to 1910.[7–9] Methods were explicitly described for development of pure lines (inbreds), crossing of the inbreds to produce single-cross hybrids ($A \times B$ and $C \times D$), and testing of single-cross hybrids to determine the best hybrid to provide to the producers. The concepts were not readily accepted because of the difficulties of producing adequate quantities of hybrid seed on the weak, poor-yielding inbreds. The cost limitation was reduced when it was suggested that single-cross hybrids ($A \times B$ and $C \times D$) be used to produce double-cross hybrids $(A \times B)(C \times D)$. The four inbreds (A, B, C, and D) were not closely unrelated, and the yield of the double-cross hybrids was similar to

Encyclopedia of Plant and Crop Science
DOI: 10.1081/E-EPCS 120010530

the yield of the single-cross hybrids. The suggestion for use of double-cross hybrids stimulated greater interest in hybrids, and maize research programs were initiated during the 1920s to test the proposed concepts.[7,8]

HYBRID MAIZE

The genetic studies conducted by G. H. Shull were the basis for developing maize hybrids.[7] The concept of and research on the commercialization of maize hybrids have been the model used to develop hybrids in other plant species.[6] Development of superior maize hybrids includes isolation of pure lines (or inbreds), usually by self-pollination, and the evaluation of inbreds in hybrids to determine which hybrids have superior performance. The original suggestion was for single-cross hybrids [crosses of two inbreds (A × B)]. Extensive research on inbreds and double-cross hybrids was conducted during the 1920s for inbred development, relation between inbred traits and their hybrid traits, initial evaluation of inbreds for combining ability in testcrosses, prediction of performance of inbreds in double-cross hybrids, and evaluation of predicted double crosses. Initially, yields of the first double-cross hybrids were not significantly greater than the landrace cultivars from which the inbreds were derived. During the 1930s, double-cross hybrids that were superior to the landrace cultivars became available. Entrepreneurs recognized the potential commercial value of hybrids, and organizations were developed to produce and sell consistently high-quality seed to the producers. Inbreds used to produce the double-cross hybrids were developed by public sector breeding programs.[3]

Acceptance of maize hybrids was rapid and nearly 100% of the U.S. Corn Belt area was planted with double-cross hybrids by 1950. Consequently, commercial organizations rapidly expanded and developed their own breeding programs. Commercial breeding programs developed improved inbreds and hybrids by pedigree selection within elite-line crosses. Vigor and seed yields of the recycled inbreds were better than the original (first-cycle) inbreds developed from the landrace varieties. Because breeding, selection, and testing methods were simpler for single-cross hybrids, the improved inbreds could be used to produce single-cross hybrids at costs acceptable to the producers. Starting in the 1960s, single-cross hybrids were produced and sold to the producers. Presently, nearly 100% of the U.S. maize area is planted with single-cross hybrids, which was the original type of hybrid described in 1910.[7]

The development of the inbred-hybrid concept has had tremendous impact on U.S. maize production. Average U.S. maize yields seldom exceeded 1.88 tons per hectare from 1866 to 1935. From 1935 to 1965, average U.S. maize yields doubled and from 1965 to 1995 average U.S. maize yields again doubled. By 2000, U.S. producers averaged more than 8.75 tons per hectare. Since 1932, average U.S. maize yields have increased more than fivefold and total production of more than 5 billion metric tons are produced on nearly 30% less area. The inbred-hybrid concept can rightfully claim to be one of greatest plant breeding achievements of the 20th century.

GENETIC BASIS OF HYBRID VIGOR

Hybrids are defined as being superior to either the mean of their parents or to the high parent. For practical purposes, hybrids greater than the better parents are desired. Although crosses between parents have generally been superior to their parents, the genetic basis of hybrid superiority (hybrid vigor or heterosis) has not been resolved.[6,10] Different genetic hypotheses have been proposed, but definitive evidence in support of one specific hypothesis has not been adequate for explaining the phenomenon of hybrid vigor. Genetic explanations of hybrid vigor have included additivity of favorable alleles with partial to complete dominance effects and nonadditive effects due to overdominance and epistatic effects. Although research has focused on whether hybrid vigor is due to additive gene effects or nonadditive gene effects, in reality, however, all types of genetic effects are probably involved in the expression of hybrid vigor. The exact genetic basis of hybrid vigor may never be determined because the genetic effects will vary among crosses within a plant species and among plant species. Hybrid vigor is dependent on two factors: differences in gene frequencies between the two parents and some level of dominance. Hybrids among different inbreds identify some crosses that have greater hybrid vigor than other crosses. Origin of parents may suggest certain patterns where greater hybrid vigor may be expected. Hybrids have been tested extensively in maize breeding programs, and sources of parents for hybrids have been identified that have greater hybrid superiority than other sources. For example, inbreds that trace their origin to Reid Yellow Dent and Lancaster Sure Crop open-pollinated maize species are designated as heterotic groups because hybrids that include one parent from Reid Yellow Dent and one parent from Lancaster Sure Crop have greater hybrid vigor than hybrids whose parents are both from either Reid Yellow Dent or Lancaster Sure Crop. Inbreds of Reid Yellow Dent origin are usually the female parents to produce the single-cross seed. Similar heterotic groups have been identified in other areas of the world for maize. Similar patterns of heterotic groups have not been identified for other plant species, probably because research has not been as extensive as in maize.

ARTICLES OF FURTHER INTEREST

Breeding Plants and Heterosis, p. 189
Breeding Widely Adapted Cultivars: Examples from Maize, p. 211
Breeding: Choice of Parents, p. 215

REFERENCES

1. Burton, G.W. Utilization of Hybrid Vigor. In *Crop Breeding*; Wood, D.R., Rawal, K.M., Wood, M.N., Eds.; Crop Science Society of America: Madison, WI, 1983; 89–107.
2. Stoskopf, N.C.; Tomes, D.T.; Christie, B.R. *Plant Breeding: Theory and Practice*; Westview Press, Inc.: Boulder, CO, 1993; 287–308.
3. Crabb, R. *The Hybrid Corn Makers*; West Chicago Publishing Company: West Chicago, 1993; 1–331.
4. Richey, F.D. The experimental basis for the present status of corn breeding. J. Amer. Soc. Agron. **1922**, *14*, 1–17.
5. Hallauer, A.R.; Miranda, F.J.B. *Quantitative Genetics in Maize Breeding*; Iowa State University Press: Ames, IA, 1988; 337–373.
6. Coors, J.G.; Pandey, S. *Proceedings International Symposium on the Genetics and Exploitation of Heterosis in Crops*; Crop Science Society of America: Madison, WI, 1999; 1–524.
7. Shull, G.H. Development of Heterosis Concept. In *Heterosis*; Gowen, J.W., Ed.; Iowa State University Press: Ames, IA, 1952; 14–65.
8. Gowen, J.W. *Heterosis*; Iowa State University Press: Ames, IA, 1952; 1–552.
9. Hallauer, A.R.; Russell, W.A.; Lamkey, K.R. Corn Breeding. In *Corn and Corn Improvement*; Sprague, G.F., Dudley, J.W., Eds.; Crop Science Society of America: Madison, WI, 1988; 463–564.
10. Sprague, G.F. The experimental basis for hybrid maize. Biol. Rev. **1946**, *21*, 101–120.

Breeding Plants and Heterosis

B

Kendall R. Lamkey
Iowa State University, Ames, Iowa, U.S.A.

Jode W. Edwards
University of Alabama, Birmingham, Alabama, U.S.A.

INTRODUCTION

Heterosis may be one of the most important biological concepts to emerge from the 20th century because of its fundamental role in the hybrid corn industry. Crabb considers the development of hybrid corn to be one of the seminal achievements of the 20th century.[1] Corn yields were stagnant during the period when open-pollinated varieties were predominantly grown (pre-1930s), and since the discovery and implementation of the inbred-hybrid concept of breeding corn, yields have continued to increase and there are no signs of a plateau in yield.

DEFINITION

The term heterosis was coined by G. H. Shull in a lecture he gave in Berlin, Germany in 1912.[2] Shull coined the word heterosis to avoid the implications that hybrid vigor was Mendelian in origin.[3] Heterosis and hybrid vigor are now used as synonyms and refer to the increased vigor of a cross or hybrid over the average of its parents. Shull defined heterosis as "... the interpretation of increased vigor, size, fruitfulness, speed of development, resistance to disease and to insect pests, or to climatic rigors of any kind, manifested by crossbred organisms as compared with corresponding inbreds, as the specific results of unlikeness in the constitutions of the uniting parental gametes."[2]

Shull's definition of the heterosis concept is a phenomenological definition and suggests no hypothesis about the underlying genetic basis of heterosis. The genetic basis of heterosis and the exploitation of heterosis in crop plants have been major foci of research since Shull's[4,5] seminal papers on inbreeding and outcrossing. Two major volumes[6,7] and several minor volumes and reviews have been devoted to the historic, scientific, and breeding aspects of heterosis.

SHULL'S COMPOSITION OF A FIELD OF MAIZE

The concept of heterosis can best be appreciated by reviewing Shull's description of the composition of an open-pollinated field of maize.[8] After observing the effects of inbreeding and cross-breeding in maize, Shull observed "The obvious conclusion to be reached is that an ordinary corn field is a series of very complex hybrids produced by the combination of numerous elementary species. Self fertilization soon eliminates the hybrid elements and reduces the strain to its elementary components." What Shull was saying was that every plant in an open-pollinated field of corn is a hybrid resulting from the cross of two random (i.e., different) gametes. Selfing the individual plants of the open-pollinated variety to homozygosity would result in the fixation of the individual gametes in the population in inbred lines. Crossing these inbred lines together would "re-create" the population and restore it to its original state of being a collection of hybrids.

Because an individual plant (which is a hybrid) of an open-pollinated field of corn cannot be reproduced identically by sexual or asexual methods, corn breeders must reduce segregating populations of corn to inbred lines to identify the superior gametes to cross for producing hybrids. Inbred lines are homozygous and homogeneous and can be maintained either by allowing them to open-pollinate in isolation or by self-pollinating. The reproducibility of inbreds means that hybrids can be reproduced indefinitely by making crosses between the inbred lines. Inbred lines, however, generally yield 40 to 70% less than the hybrid plant from which they were developed. This decrease in performance is known as inbreeding depression and is the converse of heterosis. Thus, to have heterosis of economic importance, the cross between two inbred lines must recover the performance lost from inbreeding depression and give an additional increase in performance above and beyond that lost by

Encyclopedia of Plant and Crop Science
DOI: 10.1081/E-EPCS 120010524

inbreeding depression.[9] Many believe that the key to understanding the genetic basis of heterosis lies with understanding inbreeding depression. Edwards and Lamkey have provided evidence that inbreeding depression is a heritable trait and that at least part of the observed inbreeding depression is due to a few genes with a large effect.[10]

Thus, with the publication of Shull's paper,[4] the hybrid corn industry was born. The quantitative genetics of agronomic traits and understanding the genetic mechanisms underlying heterosis became the focus of many research programs. Shull's concept that an open-pollinated population is a collection of hybrids was important conceptually to understanding the quantitative genetic basis of heterosis.

THE QUANTITATIVE GENETICS OF HETEROSIS

Studies in the genetics of heterosis have been mainly focused on issues of gene action, in particular, whether dominant or overdominant gene action is the basis of heterosis. There is little empirical evidence, from any species, to support overdominant gene action as the major cause of heterosis.[11] The issue of overdominance versus dominance primarily affects choice of selection methodology in crop plants. Selection methods used in hybrid crops such as maize are optimal for both dominant and overdominant types of gene action. The debate over the type of gene action controlling heterosis was significant in that it served as the stimulus for much of the quantitative genetics research that has been conducted since the 1940s. This research has been reviewed extensively in the literature.[11]

The terms ''hybrids'' and ''heterosis'' are not synonymous, as is often assumed. It is possible to produce hybrids that exhibit no heterosis, but it is not possible to have heterosis without producing a hybrid. The performance of a hybrid and the heterosis that a hybrid exhibits are two different functions of phenotype. Heterosis is most commonly defined as the difference between the performance of the hybrid and the average of its parents, which is usually called midparent heterosis, i.e., $\acute{H} = \bar{F}_1 - \overline{MP}$, where $\acute{H}$ = heterosis, $\bar{F}_1$ = the hybrid performance, and $\overline{MP}$ = the average of the two parents. Heterosis is often expressed as a percentage of the midparent, but this measure is difficult to interpret genetically and statistical tests are nearly impossible.[9]

With this equation, heterosis can be calculated for any hybrid as long as phenotypic measurements exist for the hybrid and the two parents. The genetic expectation of this quantity is, however, a more difficult matter. For example, it would not be valid to calculate heterosis from the cross of two inbred lines (F (inbreeding coefficient) = 1) and compare that value with the heterosis obtained from crossing two random mating and non-inbred populations ($F = 0$). The genetic expectation of the resulting estimate of heterosis is different for the two cases for two reasons.[9] First, the inbreeding levels of the parents are different for the two cases and second, the reference populations are different. Hybrids made from more inbred parents will necessarily exhibit higher heterosis. Because heterosis is reference-population dependent, quantitative genetic interpretation of comparisons of heterosis among hybrids requires that the two estimates share a common reference population and have parents with equivalent inbreeding levels. For example, if the difference in heterosis between two single-cross hybrids is to be genetically interpreted, the parents of the two single-cross hybrids must be derived from the same reference population and have equivalent inbreeding coefficients.

THE PRACTICAL APPLICATION OF HETEROSIS

Despite the scientific interest in understanding the genetic basis of heterosis, hybrid crop breeders do not select hybrids with high heterosis and farmers do not plant hybrids because they have high heterosis. Both farmers and breeders want hybrids with the highest hybrid (F_1) yields. There appears to be no relationship between the increase in hybrid yields and an increase in heterosis. Yields in corn have gone up over time, but the experimental data suggest that heterosis has not increased over time.[12] This would suggest that the benefits of heterosis have already been realized and that future gains in crop improvement in hybrid crops will not come from increasing heterosis (Fig. 1).

There are many advantages for farmers who grow F_1 hybrids. Chief among them is uniformity. Another advantage from a breeding perspective is that hybrids always outyield open-pollinated varieties because hybrid breeders are working in the tails of the distribution of phenotypes of an open-pollinated variety. For example, if we assume that an open-pollinated variety is a collection of hybrids that are normally distributed with some mean and variance (Fig. 2), farmers growing this open-pollinated variety will report yields near the mean of the distribution of hybrids. Hybrid breeders, however, will be working to identify, extract, and reproduce the best hybrid from the upper tail of the distribution. Even if open-pollinated breeders improve the population's performance, hybrid breeders can still take advantage of this shift in the mean of the distribution and extract F_1 hybrids even better than the ones from the unimproved population (Fig. 2). This shifting of the mean of the distribution while

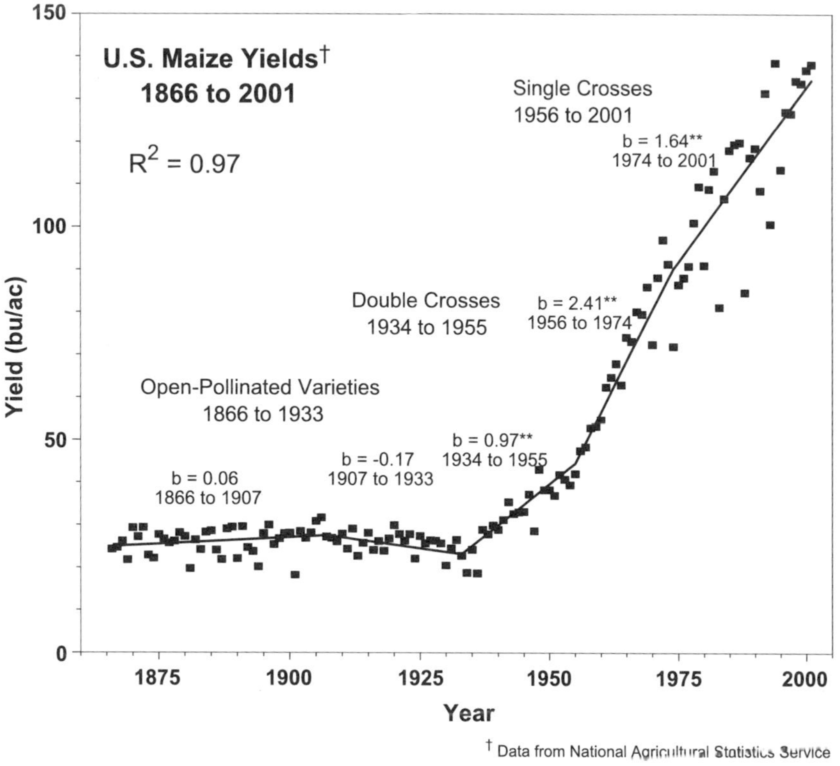

Fig. 1 Maize yields in the United States from 1866 to 2001. The regression coefficients (b) for each of the periods of time represent the increase in grain yield (bu/ac) per year. (Data from the USDA-National Agricultural Statistics Service.)

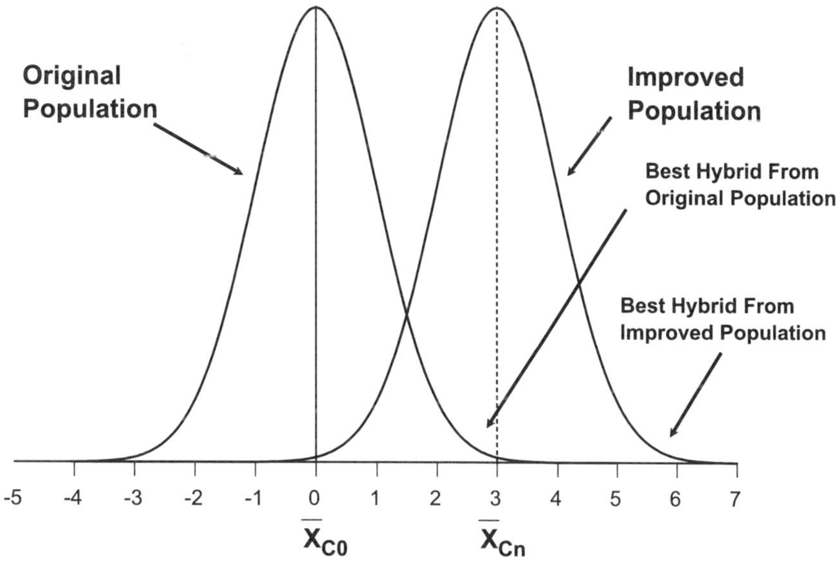

Fig. 2 A comparison of the phenotypic distributions of two open-pollinated varieties before (with mean $\bar{X}_{C0}$) and after (with mean $\bar{X}_{Cn}$) selection showing that even if open-pollinated variety breeders are successful in improving their varieties, the best hybrids will still come from the tails of that distribution and be 2 to 4 standard deviations better than the mean of the variety. (Adapted from Ref. [13].)

maintaining the variance of the distribution is one of the fundamental objectives of recurrent selection programs and provides the long-term foundation for genetic variation that hybrid breeders need to continue making genetic gains.[14]

ARTICLES OF FURTHER INTEREST

Breeding Hybrids, p. 186
Breeding: Recurrent Selection and Gain from Selection, p. 232

REFERENCES

1. Crabb, R. *The Hybrid Corn-Makers*; West Chicago Publishing Co.: Illinois, 1993.
2. Shull, G.H. Beginnings of the Heterosis Concept. *Heterosis*, Proceedings of the Conference, Ames, IA, July 15–20, 1950; Gowen, J.W., Ed.; Iowa State College Press: Ames, 1952; 14–48.
3. Sprague, G.F. *Heterosis in Maize: Theory and Practice*. In *Monographs on Theoretical and Applied Genetics Vol. 6 Heterosis*; Frankel, R., Ed.; Springer-Verlag: Berlin Heidelberg, 1983; 47–70.
4. Shull, G.H. A pure line method of corn breeding. Am. Breed. Assoc. Rep. **1909**, *5*, 51–59.
5. Shull, G.H. Hybridization methods in corn breeding. Am. Breed. Mag. **1910**, *1*, 98–107.
6. In *Heterosis*, Proceedings of the Conference, Ames, IA, July 15–20, 1950; Gowen, J.W., Ed.; Iowa State College Press: Ames, 1952.
7. In *The Genetics and Exploitation of Heterosis in Crops*, Proceedings of the International Symposium, CIMMYT, Mexico City, Mexico, Aug. 17–22, 1997; Coors, J.G., Pandey, S., Eds.; ASA, CSSA, and SSSA: Madison, WI, 1999.
8. Shull, G.H. The composition of a field of maize. Am. Breed. Assoc. Rep. **1908**, *4*, 296–301.
9. Lamkey, K.R.; Edwards, J.W. Quantitative Genetics of Heterosis. In *The Genetics and Exploitation of Heterosis in Crops*, Proceedings of the International Symposium, CIMMYT, Mexico City, Mexico, Aug. 17–22, 1997; Coors, J.G., Pandey, S., Eds.; ASA, CSSA, and SSSA: Madison, WI, 1999; 31–48.
10. Edwards, J.E.; Lamkey, K.R. Quantitative genetics of inbreeding in a synthetic maize population. Crop Sci. **2002**, *42* (4), 1094–1104.
11. Crow, J.F. Dominance and Overdominance. In *The Genetics and Exploitation of Heterosis in Crops*, Proceedings of the International Symposium, CIMMYT, Mexico City, Mexico, Aug. 17–22, 1997; Coors, J.G., Pandey, S., Eds.; ASA, CSSA, and SSSA: Madison, WI, 1999; 49–58.
12. Duvick, D.N. Heterosis: Feeding People and Protecting Natural Resources. In *The Genetics and Exploitation of Heterosis in Crops*, Proceedings of the International Symposium, CIMMYT, Mexico City, Mexico, Aug. 17–22, 1997; Coors, J.G., Pandey, S., Eds.; ASA, CSSA, and SSSA: Madison, WI, 1999; 19–29.
13. Eberhart, S.A. Factors affecting efficiencies of breeding methods. African Soils **1970**, *15*, 669–680.
14. Coors, J.G. Selection Methodologies and Heterosis. In *The Genetics and Exploitation of Heterosis in Crops,* Proceedings of the International Symposium, CIMMYT, Mexico City, Mexico, Aug. 17–22, 1997; Coors, J.G., Pandey, S., Eds.; ASA, CSSA, and SSSA: Madison, WI, 1999; 225–245.

Breeding Plants with Transgenes

B

Duška Stojšin
Monsanto Company, St. Louis, Missouri, U.S.A.

Carl F. Behr
Wyffels Hybrids Inc., Geneseo, Illinois, U.S.A.

INTRODUCTION

Transgenic crop breeding and evaluation strategies are based on the same general principles as those practiced by breeders working with nontransgenic crops. There are, however, specific considerations associated with evaluation of transgenic crops due to inheritance differences, evaluation criteria, and regulatory requirements.

Generally there are two distinct breeding stages associated with the commercial development of a transgene. The first stage involves evaluating and selecting a transgenic event, while the second phase involves integrating the selected event into commercial germplasm.

STAGE 1: CONSTRUCT AND EVENT SELECTION

A construct contains an engineered DNA sequence that is inserted into the plant genome via a transformation method. Numerous transformants are usually generated for a construct, each resulting in unique insertion site(s). These independent transformation events need to be evaluated in order to select those with superior performance. The construct and event evaluation process is based on five major criteria:[1] 1) transgene expression/efficacy, 2) molecular characterization of the insert, 3) segregation of the trait of interest, 4) agronomics of the developed lines, and 5) stability of the transgene expression. Selection criteria usually stay the same throughout the evaluation process, but stringency of testing generally increases for each criterion in more advanced generations. As with traditional breeding, large population of independent transformation events and more thorough evaluation result in the greater chance of success.[2]

1) Expression of a transgene is the level of protein or protein product controlled by the inserted DNA construct. Efficacy is the phenotypic expression of that gene of interest (GOI). Factors that can affect the level of expression and efficacy of a given event are insertion site, transgene copy number, intactness of the transgene, zygosity of the GOI, level of inbreeding associated with a genotype, the genetic background, and environmental conditions.

2) Molecular characterization aids selection for transgenes with single, intact inserts. Molecular characterization requires a variety of assays to determine presence of GOI, copy number, insert number, insert complexity, presence of the vector backbone, and development of event-specific assays. Molecular characterization is a tool that assists in selection of events that exhibit simple inheritance and stable expression level of a transgene. It also provides tools for determining event purity and identity, and is a required component of various regulatory submissions.

3) Segregation of the trait of interest is tested in order to select transgenic events that follow a single-locus segregation pattern. A direct approach is to evaluate the segregation of the trait of interest; an indirect approach is to assess the selectable marker segregation (if one is associated with the transgene). Regardless of the approach, this criterion becomes more important once breeders can evaluate a large number of plants, family segregation, and the self-pollinated progeny of a given event.

4) Agronomic characteristics may vary among events due to somaclonal variation, insertional effects, homozygosity of a transgene, level of inbreeding of the plant genome, and genetic background. In advanced generations, agronomic performance should be evaluated in several genetic backgrounds in replicated trials. For hybrid crops, agronomic trials should be conducted in both inbred and hybrid background.

5) Stability of transgene expression needs to be tested in different generations, environments, and in different genetic backgrounds. Event instability over generations is often caused by transgene inactivation due to multiple transgene copies, zygosity level, highly methylated insertion site, or level of stress.[3] Such events should be discarded even when gene silencing is reversible.

STAGE 2: COMMERCIAL PLANT BREEDING INVOLVING TRANSGENES

Transgenes used in commercial breeding programs generally have been selected for the criteria discussed above

Encyclopedia of Plant and Crop Science
DOI: 10.1081/E-EPCS 120010534

before being transferred to commercial plant breeding programs. Generally, events with a single intact insert will be inherited as a single dominant gene and follow Mendelian segregation ratios. Commercial breeding strategies include 1) backcrossing and 2) forward breeding.

1) Backcrossing is an efficient breeding method when the goal is to recover the genotype of an elite inbred with the addition of a transgene. In each backcross generation, plants that contain the transgene are identified and crossed to the recurrent parent. Several backcross generations with selection for recurrent parent phenotype are generally used by commercial breeders to recover the genotype of the elite parent. During backcrossing the transgene is kept in a hemizygous state; therefore, at the end of backcrossing it is necessary to self- or sib-pollinate plants containing the transgene to fix the transgene in a homozygous state. The number of backcross generations can be reduced by molecular assisted backcrossing (MABC). The MABC method uses molecular markers to identify plants that are most similar to the recurrent parent in each backcross generation. With the use of MABC and appropriate population size, it is possible to identify plants that have recovered over 98% of the recurrent parent genome after only two or three backcross generations. By eliminating several generations of backcrossing, it is often possible to bring a commercial transgenic product to market one year earlier than a product produced by conventional backcrossing. Backcrossing and MABC are routinely used in the United States to develop transgenic crop products.

2) Forward breeding is any breeding method that has the goal of developing a transgenic variety, inbred line, or hybrid that is genotypically different, and superior, to the parents used to develop the improved genotype. When forward breeding a transgenic crop, selection pressure for the efficacy of the transgene is usually applied during each generation of the breeding program. Additionally, it is usually advantageous to fix the transgene in a homozygous state during the breeding process as soon as possible to uncover potential agronomic problems caused by unfavorable transgene × genotype interactions. Forward breeding is used extensively in the United States to develop Roundup Ready® varieties of soybean [*Glycine max* (L.) Merr.] and cotton (*Gossypium hirsutum* L.), as well as Bollgard™ cotton.

After integrating the transgene into commercial germplasm, the final product should be tested in multiple locations. Testing typically includes yield trials in trait neutral environments (e.g., insect free environments for insect-protected plants), as well as typical environments in the target market. If the new transgenic product has been derived from backcrossing, it is usually tested for equivalency by comparing it to the nontransgenic version in all environments.

QUALITY ASSURANCE AND QUALITY CONTROL

Because transgenic events are regulated by government agencies, it is important that commercial transgenic crops do not contain unintended transgenic events. Breeders working with transgenes must have appropriate quality assurance (QA) and quality control (QC) programs in place to avoid and/or detect errors that may result during the event evaluation and selection process.

1) Quality assurance is the set of best practices used to ensure the overall quality of the final product from the standpoint of transgene purity.[4] The setup of a transgenic breeding nursery should be designed to minimize the possibility of contamination between nontransgenic and transgenic material or among different transgenic events. Nursery designs should use spatial and/or temporal isolation between nontransgenic and transgenic plants, as well as among different transgenic events. Other important QA methods include utilizing a consistent and concise pedigree nomenclature that clearly identifies the germplasm and transgene being integrated, checking phenotypic segregation ratios for the gene(s) of interest in each generation, and utilizing differences in germplasm to maximize the identification of off-types due to outcrossing. Additional QA measures are often practiced in cross-pollinated crops due to the increased risk of outcrossing. Some of these include restricting movement of a pollinating crew from one nursery to another, or increasing the spatial isolation between different transgenic events.

2) Quality control is the testing conducted to verify that the desired quality is actually achieved.[4] Quality control is done with the objective to test the seed purity and integrity for the presence of an intended event. In addition, it is important to test for the absence of unintended transgenes, regardless of whether they control the same trait as the event of interest or different traits. Stringent QC standards ensure that the event purity and integrity is maintained in the evaluation, breeding, and seed-increase portion of transgene development.

CONCLUSION

The product of transgenic crop breeding is a new cultivar containing a specific transgene at a specific locus that controls or modifies expression of a specific trait. During the development and selection, events with single intact inserts and high and stable transgene expression should be selected. Efficacy and agronomic performance of the transgene needs to be demonstrated in different genetic backgrounds and in different environments. After integrating the transgene into commercial germplasm, the final product should also be tested in multiple locations. Regulated handling of transgenic seed adds to the complexity of developing transgenic crops. However, social, economic, and environmental benefits associated with these products make these efforts worthwhile.

ARTICLES OF FURTHER INTEREST

Biosafety Considerations for Transgenic Insecticidal Plants: Non-Target Herbivores, Detritivores, and Pollinators, p. 153
Biosafety Considerations for Transgenic Insecticidal Plants: Non-Target Predators and Parasitoids, p. 156
Gene Silencing: A Defense Mechanism Against Alien Genetic Information, p. 492
Genetically Engineered Crops with Resistance Against Insects, p. 506
Genetically Modified Oil Crops, p. 509
Herbicide-Resistant Weeds, p. 551
Medical Molecular Pharming: Therapeutic Recombinant Antibodies, Biopharmaceuticals and Edible Vaccines in Transgenic Plants Engineered via the Chloroplast Genome, p. 705
Transformation Methods and Impact, p. 1233
Transgenes: Expression and Silencing of, p. 1242
Transgenes (GM) Sampling and Detection Methods in Seeds, p. 1238
Transgenic Crop Plants in the Environment, p. 1248
Transgenic Crops: Regulatory Standards and Procedures of Research and Commercialization, p. 1251
Transgenetic Plants: Breeding Programs for Sustainable Insect Resistance, p. 1245

REFERENCES

1. Stojšin, D.; Behr, C.F.; Heredia, O.; Stojšin, R. Evaluation and breeding of transgenic corn. Genetika **2000**, *32* (3), 419–430.
2. Conner, A.J.; Christey, M.C. Plant breeding and seed marketing options for the introduction of transgenic insect-resistant crops. Biocontrol Sci. Technol. **1994**, *4*, 463–473.
3. Zhong, G.Y. Transgene gentics and breeding: What do we know? Proc. 53rd Annu. Corn Sorghum Res. Conf. **1998**, *53*, 250–269.
4. Hall, M.; Moeghji, M.; Parker, G.; Peterman, C.; Yates, D. Strategies to maximize quality assurance (QA) when integrating transgenes in breeding programs. Ill. Corn Breed. Sch. **2002**, *38*.

Breeding Pure Line Cultivars

P. Stephen Baenziger
University of Nebraska-Lincoln, Lincoln, Nebraska, U.S.A.

Mustafa Erayman
Washington State University, Pullman, Washington, U.S.A.

Hikmet Budak
University of Nebraska-Lincoln, Lincoln, Nebraska, U.S.A.

B. Todd Campbell
California Cooperative Rice Research Foundation, Biggs, California, U.S.A.

INTRODUCTION

Pure line cultivars are defined as cultivars that are "genetically pure or homozygous plants resulting from continued inbreeding or self-fertilization." Plant breeding methods are usually determined by the pollination or reproductive biology of the plant that the breeder is trying to improve. Plants can be generally divided into three classes: self-pollinated [e.g., barley (*Hordeum vulgare* L.), rice (*Oryza sativa* L.), soybeans (*Glycine max* (l.) Merr.), wheat (*Triticum aestivum* L.)]; cross-pollinated [e.g., alfalfa (*Medicago sativa* L.), maize (*Zea mays* L.), sunflowers (*Helianthus annuus* L.)]; or apomictic [e.g., buffelgrass (*Cenchrus ciliaris* L.)]. Apomictic plants form seed from an asexual process that occurs in the ovary without the fusion of a male and a female gamete. Self-fertilization (selfing) occurs naturally in self-pollinated crops and can be done by hand pollinations in cross-pollinated crops. Hence, breeding pure lines is normally associated with self-pollinated crops, but is also an important part of breeding inbred lines that are used to make hybrids in self- or cross-pollinated crops. The inbred parents of hybrid crops are generally not available to farmers or producers, so they will not be discussed in detail here. However, their development is the same as a pure line that is intended to be released as a cultivar to farmers or producers; hence, the techniques described below are totally applicable.

THE PHASES OF PURE LINE PLANT BREEDING

There are three phases of pure line plant breeding, specifically: 1) the introduction of genetic variation; 2) inbreeding and selection; and 3) evaluation of the selected lines.[1,2] The first phase is to create variation for selection. The second phase uses inbreeding to separate the genetic variation into individual lines that can be selected. The third phase is to evaluate the selected variants in numerous environments to determine where the selected lines can be successfully grown and where they should not be grown.

Introduction of Genetic Variation

For self- or cross-pollinated plants, the introduction of variation is usually done by making a cross between two or more parents (a sexual hybrid). Variation can also be introduced by mutagens and genetic transformation (genetic engineering), but crossing is by far the most common method. When pure lines are used as parents, the crosses are often Parent 1 × Parent 2; (Parent 1 × Parent 2) × Parent 3; or (Parent 1 × Parent 2) × (Parent 3 × Parent 4). The crosses are known as single cross, three-way cross, or four-way (syn. double) cross. The progeny of these crosses are referred to as F_1 seed, where F is for filial (literally defined as "pertaining to a son or daughter")[3] and $_1$ is the first generation of the cross. Succeeding generations are numbered in sequence, e.g., F_2, F_3, F_4, . . ., F_n. The F_1 seed form the initial population for the inbreeding and selection phase. Random mating populations (generally available only in cross-pollinated plants) can also be used as the initial populations for the inbreeding and selection phase. In most breeding programs, numerous populations are created each year.

Inbreeding and Selection

To understand inbreeding and selection in pure line breeding, the important concepts are: 1) the nature of inbreeding (usually selfing, but could be intermating

Encyclopedia of Plant and Crop Science
DOI: 10.1081/E-EPCS 120010528

closely related individuals); 2) the approach to homozygosity; and 3) the type of selection involved (artificial or natural). Selfing has the quickest approach to homozygosity, and in self-pollinated crops it is the standard form of inbreeding. For a locus where the two parents have different alleles (e.g., AA vs. aa), the F_1 will be Aa (heterozygous). A heterozygous locus in the F_1 will segregate in later generations, so it is called a segregating locus. The F_2 will segregate $^1/_4$ AA: $^1/_2$ Aa: $^1/_4$ aa (half of the progeny will be heterozygous and half of the progeny will be homozygous). In each successive generation without selection, heterozygosity will decrease by $^1/_2$ and the level of homozygosity will be 1 minus the level of heterozygosity (Table 1). A locus can be either heterozygous or homozygous, so the sum of those progeny that are heterozygous and those progeny that are homozygous must equal 1. Plant genomes have tens of thousands of genes,[4] so more than one important locus is heterozygous in the F_1. The formulae in Table 1 can be generalized, when m loci are heterozygous, to: 1) the proportion of progeny that are heterozygous at every segregating locus in F_n generation is $(1/2^n)^m$; and 2) the proportion of plants that are homozygous at every locus is $(1-(1/2^n))^m$. The sum of the progeny that are heterozygous at every segregating locus and the progeny that are homozygous at every segregating locus does not sum to one because many individuals can be heterozygous at some loci and homozygous at other loci. The importance of knowing how many progeny are homozygous at every locus is that a pure line is genetically homozygous and should not segregate for important traits. The greater the number of segregating loci, the greater the number of selfing generations that are needed to create a pure line (Table 1).

With selection, the approach to homozygosity is improved and the number of lines with favorable alleles is increased because unfavorable genotypes are eliminated. If aa is undesirable, for example, it can be removed at each generation, leaving only AA and Aa. Selection can be either artificial selection (selection done by the breeder) or natural selection (selection done by nature). Examples of artificial selection include the breeder selecting for earliness, plant type, or end-use quality. Some of these characteristics may be important in the marketplace, but do not affect plant performance. Examples of natural selection include: 1) disease or insect resistant plants producing more seed during infections/infestations than susceptible plants; and 2) fall-sown cereals surviving the winter to produce grain in the spring or summer. Plants that are diseased, eaten by insects, or winterkilled are removed from the segregating population by nature and not by the breeder (Fig. 1), hence the term "natural selection." Every time a plant is grown in the field, it is affected by natural selection.

The common inbreeding methods are pedigree, bulk, and single-seed descent.[1,5,6] Briefly, in the pedigree system plants are selected in the early generations and their progeny are planted in a row (called a family). Early generation selection is for characters that are easily measured and highly heritable (e.g., the progeny will look like the parents). Often, early generation selection is based on culling, removing the plants with obviously poor or unwanted characters (e.g., disease susceptibility). Based on the phenotype of the family-row, individual plants are selected within each row that have the desirable characteristics for advancing to the next generation. No plants are selected in the families that do not have most or all of the desirable characteristics. As the family rows approach uniformity, whole families are selected, rather than individual plants within the rows (Fig. 2). Pedigree breeding uses both artificial selection and natural selection. In

Table 1 The consequences of inbreeding at a single locus and at multiple segregating loci on the proportion of lines homozygous at every locus[a]

		Single segregating locus		**5 segregating loci**	**10 segregating loci**
Generation		**Proportion heterozygous**	**Proportion homozygous**	**Proportion homozygous**	**Proportion homozygous**
F1		1.0000	0.0000	0.0000	0.0000
F2	1	0.5000	0.5000	0.0313	0.0010
F3	2	0.2500	0.7500	0.2373	0.0563
F4	3	0.1250	0.8750	0.5129	0.2631
F5	4	0.0625	0.9375	0.7242	0.5245
F6	5	0.0313	0.9688	0.8532	0.7280
F7	6	0.0156	0.9844	0.9243	0.8543
F8	7	0.0078	0.9922	0.9615	0.9246
F*n*		(1/2)^*n*-1	1 – (1/2)^*n*-1		

[a]Ideally, a pure line should be completely homozygous. The more inbreeding, the greater the proportion or chance of selecting a homozygous line. Note that a homozygous line can have either the favorable or the unfavorable allele at the locus, in the absense of selection.

Fig. 1 Winter wheat F_2 populations growing at Mead, Nebraska in 2001. Surviving populations have lush, green plants; winterkilled populations are the bare spots in the field. The surviving and winterkilled populations are an example of natural selection, in that cold temperatures during the winter selected winterhardy plants and killed wintertender plants. No selection for winterhardiness by the breeder was needed.

the pedigree system, selection is done in every generation, and record keeping can become quite complex.

In the bulk breeding system, the progeny of a cross (or occasionally a number of crosses) are planted using commercial seeding rates and harvested in bulk in the early generations. The bulk breeding philosophy is that it is important to inbreed to a level of homozygosity before doing artificial selection. In later generations, lines are selected from the bulk and planted as families, similar to the pedigree system. If the plants within a family are uniform, selection is based on the family. If the plants within a family are not uniform, then selection within the family is similar to that in the early generations of the pedigree system. In the bulk breeding system, natural selection occurs in every generation, but little artificial selection is done. Competition within a bulk population

Fig. 2 Experimental lines in the F_5 generation grown at Lincoln, Nebraska in 2002. Each line is grown in a four-row plot. The plots marked with red paint have been selected for harvest; the unmarked plots will be discarded. This is an example of family selection in which the families are sufficiently uniform that it is unnecessary to select plants within the family to further inbreed for uniformity.

consisting of different plant types is a concern, because competition may cause some desirable plant types that would be valuable as pure lines to be lost from the bulk population.[7] Because the populations are kept as bulks, record keeping is minimal and very little time per bulk is needed. The simplicity of the bulk breeding method allows many populations to be in a breeding program concurrently.

In the single-seed descent breeding method, a large F_2 population is grown from the F_1 plants. A single seed from each F_2 plant is advanced to the next generation and the process of selfing and advancing a single seed is repeated for a number of generations, hence the name ''single-seed descent.'' After the appropriate number of generations, individual plants are harvested and their progeny are grown as families for selection. In single-seed descent breeding, the progeny of every plant is advanced, so there is no artificial or natural selection. Because there is no selection, the plants can be grown in environments that do not represent normal growing conditions. For example, multiple generations are often grown in greenhouses or off-season nurseries under conditions that are completely unrepresentative of normal field conditions. The philosophy of single-seed descent breeding is that early-generation selection is not as valuable as inbreeding—often very rapid inbreeding.

Each of these three methods has its advantages and disadvantages, and often they are not used in their ''pure'' form, but rather are modified as needed. Pedigree breeding is often the most labor-intensive, because every plant in every generation is recorded and measured until it is discarded from the program. In crops such as maize, where selfing is done by hand and each plant is individually cared for, it is the only method that works.

Pedigree breeding, if resources were not limiting, would be the method of choice for most pure line breeders because it provides more information about each line than the other breeding procedures. The advantage of the bulk breeding method is that numerous populations can be easily manipulated concurrently. Bulk breeding is widely used in self-pollinated cereal crops. It is not unusual for a moderate-size wheat breeding program to create 500 to 700 new populations each year and to have over a million plants in each bulk generation. The disadvantage of bulk breeding is that less is known about a line and there is less artificial selection when compared to the pedigree breeding method. Single-seed descent breeding has the advantage that generations can be quickly advanced, often three generations or more per year. Hence, it is the quickest method to reach homozygosity, or uniformity, and it is commonly used in breeding soybeans. Its greatest disadvantage is that little selection is done during the selfing generations, so many undesirable lines are advanced. The lack of selection can be an advantage if artificial or natural selection may select the wrong lines. For example, single-seed descent can be used in Hawaii or Puerto Rico, due to their climate, for crops that will never be grown there. Because there is no selection, types that are adapted to other environments are maintained during selfing and their progeny can be selected in the appropriate environment.

An example of a breeding procedure using a mixture of these methods would be to use: 1) the bulk breeding method for the first two generations, to allow inbreeding and natural selection to reduce unwanted types; 2) the pedigree method in the middle generations, to select better plant types; and 3) single-seed descent for two or more rapid generations in the greenhouse or Hawaii, to increase inbreeding and line uniformity of the now highly selected lines.

Evaluation of Selected Lines

In this phase, the goal is to carefully evaluate the selected lines to determine which truly have potential for being produced as cultivars, to determine their area of adaptation, and to provide useful information for those growing the lines in the future. Regardless of the parents of the population or the breeding method used, the evaluation method will be the same, in that it involves replicated testing of advanced experimental lines in numerous environments. Replicated trials are expensive and labor intensive, so only the best lines are evaluated. Though there may be millions of plants in the early generations, less than 1% of the progeny reach the advanced experimental stage and are evaluated each year in replicated trials. Usually, fewer than five—and often none—are released as cultivars in a year.

CONCLUSION

With the advances in plant genomics and transformation, sources of genetic variation are greatly expanding. Inbreeding and selection will continue to be very important in breeding pure lines and will be augmented by marker-assisted selection using molecular markers. The near-finished products (the advanced experimental lines) will continue to be evaluated, using replicated field trials, to ensure they have the attributes that make them worthy of production.

ARTICLES OF FURTHER INTEREST

Breeding Biennial Crops, p. 171
Breeding Clones, p. 174
Breeding Hybrids, p. 186
Breeding Plants and Heterosis, p. 189
Breeding Self-Pollinated Crops Through Marker-Assisted Selection, p. 202
Breeding Using Doubled Haploids, p. 207
Breeding: Choice of Parents, p. 215
Breeding: Genotype-by-Environment Interaction, p. 218
Breeding: Incorporation of Exotic Germplasm, p. 222
Breeding: Mating Designs, p. 225
Breeding: The Backcross Method, p. 237
Crop Improvement: Broadening the Genetic Base for, p. 343
Genetic Resources Conservation of Seeds, p. 499

REFERENCES

1. Stroskopf, N.C.; Tomes, D.T.; Christie, B.R. *Plant Breeding: Theory and Practice*; Westview Press, Inc.: Bolder, CO, 1993; 531.
2. Fehr, W.R.; Hadley, H.H. *Hybridization of Crop Plants*; American Society of Agronomy and Crop Science Society of America: Madison, WI, 1980; 765.
3. Morris, W. *The American Heritage Dictionary of the English Language*; American Heritage Publishing Co., Inc. and Houghton Mifling Company: Boston, MA, 1969; 1505.
4. Yu, J.; Hu, S.; Wang, J.; Wong, G.K.-S.; Li, S.; Liu, B.; Deng, Y.; Dai, L.; Zhou, Y.; Zhang, X.; Cao, M.; Liu, J.; Sun, J.; Tang, J.; Chen, Y.; Huang, X.; Lin, W.; Ye, C.; Tong, W.; Cong, L.; Geng, J.; Han, Y.; Li, L.; Li, W.; Hu, G.; Huang, X.; Li, W.; Li, J.; Li, L.; Liu, J.; Qi, Q.; Liu, J.; Liu, J.; Li, L.; Li, L.; Wang, X.; Lu, H.; Wu, T.; Zhu, M.; Ni, P.; Han, H.; Dong, W.; Ren, X.; Feng, X.; Cui, P.; Li, X.; Wang, H.; Xu, X.; Zhai, W.; Xu, Z.; Zhang, J.; He, S.;

Zhang, J.; Xu, J.; Zhang, K.; Zen, X.; Dong, J.; Zeng, W.; Tao, L.; Ye, J.; Tan, J.; Ren, X.; Chenc, X.; He, J.; Liu, D.; Tian, W.; Tian, C.; Xia, H.; Bao, Q.; Li, G.; Gao, H.; Cao, T.; Wang, J.; Zhao, W.; Li, P.; Chen, W.; Wang, X.; Zhang, Y.; Hu, J.; Wang, J.; Liu, S.; Yang, J.; Zhang, G.; Xiong, Y.; Li, Z.; Mao, L.; Zhou, C.; Zhu, Z.; Chen, R.; Hao, B.; Zheng, W.; Chen, S.; Guo, W.; Li, G.; Liu, S.; Tao, M.; Wang, J.; Zhu, L.; Yuan, L.; Yang, H. A draft sequence of the rice genome (*Oryza sativa* L. ssp. *indica*). Science **2002**, *296*, 79–92.

5. Goff, S.A.; Ricke, D.; Lan, T.-H.; Presting, G.; Wang, R.; Dunn, M.; Glazebrook, J.; Sessions, A.; Oeller, P.; Varma, H.; Hadley, D.; Hutchinson, D.; Marting, C.; Katagiri, F.; Lange, B.M.; Moughamer, T.; Xia, Y.; Budworth, P.; Zhong, J.; Miguel, T.; Paskowski, U.; Zhang, S.; Colbert, M.; Sun, W.L.; Chen, L.; Cooper, B.; Park, S.; Wood, T.C.; Mao, L.; Quail, P.; Wing, R.; Dean, R.; Yu, Y.; Zharkikh, A.; Shen, R.; Sahasrabudhe, S.; Thomas, A.; Cannings, R.; Gutin, D.; Pruss, J.; Reid, S.; Tavtigan, J.; Mitchell, G.; Eldredge, T.; Scholl, R.M.; Miller, S.; Bhatnagar, N.; Adey, T.; Rubano, N.; Tusneem, R.; Robinson, J.; Feldhaus, T.; Macalma, A.; Oliphant, S. A draft sequence of the rice genome (*Oryza sativa* L. ssp. *japonica*). Science **2002**, *296*, 92–100.
6. Fehr, W.R. *Principles of Cultivar Development*; Theory and Technique, Macmillan Publishing Co.: USA1987; Vol. 1; 536.
7. Khalifa, M.A.; Qualset, C.O. Intergenotypic competition between tall and dwarf wheats. I. In mechanical mixture. Crop Sci. **2002**, *14*, 795–799.

B

Breeding Self-Pollinated Crops Through Marker-Assisted Selection

Brian W. Diers
University of Illinois, Urbana, Illinois, U.S.A.

INTRODUCTION

Plant breeders have made great strides in improving crop plants. These improvements have been made primarily through traditional breeding methods. These methods typically include the development of segregating germplasm and the selection of those genotypes with the best performance for traits of interest through direct trait evaluation. Marker-assisted selection (MAS) can be used by breeders to improve the efficiency of selection. With MAS, breeders select for traits of interest through genetic markers linked to genes controlling the trait or through backcrossing for the recurrent parent genome. For some traits, MAS could potentially increase the efficiency of selection compared to phenotypic selection.

The discussion in this article will be limited to MAS in self-pollinated crop plants that are grown commercially as inbred cultivars. Marker-assisted selection in hybrid crops will be covered in another article. Inbred cultivars are commonly grown for small grain and legume crops.

GENETIC MARKERS

The genetic markers typically applied in MAS are DNA polymorphisms. These polymorphisms are abundant in genomes and can be revealed through a number of methods. A widely used system to reveal polymorphisms is simple sequence repeat (SSR) markers. These markers are widely used because they are highly polymorphic, inexpensive, and relatively easy to use. These markers are being employed by breeders to map traits and conduct MAS. Additional genotyping methods such as allele-specific hybridizations, and Taqman™ probes have also been developed for automated assays of specific genetic regions.

USES OF MARKER-ASSISTED SELECTION

Before MAS can be done, markers must be identified that are tightly linked to the genes being selected. These linkage associations are typically identified in genetic populations that are phenotyped for the trait of interest and screened with a battery of genetic markers. These data are analyzed and the positions of the genes relative to markers are determined.

A primary use of marker-assisted selection is in the identification of plants or experimental lines within segregating populations that carry favorable alleles. In the development of inbred cultivars, breeders typically cross parents to develop segregating populations.[1] Plants in the populations would be selfed until an acceptable level of homozygosity is reached and inbred lines would be derived from the plants. Selection may occur during inbreeding, or single-seed descent[2] may be done without selection. Marker-assisted selection could be employed any time during this process to assist in the identification of desired individuals.

With MAS, breeders would use markers linked to genes of interest to select plants or lines that have a high probability of carrying the desired genes. Compared to traditional phenotypic selection, MAS could potentially reduce costs, increase the total number of plants or lines available for selection, increase the effectiveness of selection, and reduce the time required for a breeding cycle. The efficiency of MAS for quantitative traits has been evaluated using modeling and simulation methods.[3,4] The overall conclusions from these studies are that MAS could be more efficient than phenotypic selection when populations are large and traits have a low heritability. An additional consideration is the cost of phenotypic selection. Even when traits are highly heritable, if the cost of phenotypic selection is high, MAS would be cost effective if markers linked closely to the genes of interest are available.

An example of a trait that is effectively selected with markers is resistance to soybean cyst nematode (SCN, *Heterodera glycines* Inchinohe) in soybean [*Glycine max* (L.) Merr.]. Because of the importance of SCN resistance in soybean cultivars, this is a major breeding objective worldwide. Phenotypic selection for SCN resistance is done by counting the number of female nematodes produced on the roots of inoculated plants, which is tedious and expensive. Although SCN resistance in populations is quantitative, a few major quantitative trait loci (QTL) control a large proportion of the resistance and these

Encyclopedia of Plant and Crop Science
DOI: 10.1081/E-EPCS 120010523

QTL have been targets of selection. Mudge et al.[5] showed that by selecting two SSR markers that flank the major SCN resistance gene *rhg1*, they could identify 98% of the lines that had a 70% reduction in female nematodes on the roots compared to the susceptible check. Concibido et al.[6] estimated that the cost of phenotypic screening for SCN resistance was $1.70 per data point, whereas PCR-based MAS cost $1.50 per data point. Since the writing of Concibido et al.[6] in 1996, the cost of phenotypic screening has increased and the cost of MAS has been reduced. Both private and public breeders have incorporated MAS for SCN resistance into their breeding programs.

In contrast to disease resistance, MAS in segregating populations for other traits such as yield will likely be difficult in elite germplasm. This difficulty exists because it is often not obvious which parent of an elite population carries each positive allele for QTL until mapping is done within the population. This is inefficient because it requires the collection of phenotypic data prior to MAS. In addition, there is likely insufficient linkage disequilibrium present in crop species to make predictions of what QTL alleles are present in parents based on all but very closely linked markers.[7]

Selection during backcrossing is another important use of genetic markers in the development of inbred varieties. Backcrossing is a breeding method in which one or more genes is transferred from a donor parent into a recurrent parent through a series of crosses back to the recurrent parent. The recurrent parent typically has high performance, but is deficient for one or more traits carried by the donor parent. For example, this method is often employed to add new disease resistance genes or transgenic traits into cultivars that are used as recurrent parents. Markers can increase the speed of backcrossing through selection both for genes being backcrossed and for the recurrent parent genome.

Markers are especially useful in selecting the gene being backcrossed when the phenotype of the gene can not be easily assayed. This includes cases in which the gene is recessive, a destructive test is required for phenotyping, the phenotype is only expressed after pollination, or the gene contributes to a quantitative trait that can be evaluated only in replicated tests. Selections can be made, based on markers linked to the gene being backcrossed, that alleviate the need for progeny testing between cycles of backcrossing. It is especially important to have markers tightly linked to the gene or there is a significant risk that recombination between markers and the gene may occur. Such an event would result in the loss of the gene during backcrossing.

The recovery speed of the recurrent parent can be increased using markers in the selection of backcross individuals that carry the greatest proportion of the recurrent parent genome. For example, during each generation of backcrossing, F_1 individuals can be tested with a set of markers spaced throughout the genome. Those individuals fixed for the greatest proportion of recurrent parent alleles are selected. Using modeling experiments, Hospital et al.[8] determined that if fewer than 10% of the backcross F_1 individuals were selected based on markers each generation, the amount of recurrent parent genome in individuals after three generations of backcrossing is about equal to or greater than expected without selection at the fifth backcross generation. A difficulty that many breeders working with self-pollinated crops face in selecting for the recurrent parent genome is in generating sufficient backcross F_1 seed to select among. This is especially true in legume species such as soybean, where the production of large numbers of hybrid seed is time consuming and expensive.

An example of MAS during backcrossing was provided by Zhou et al.[9] They improved grain quality through MAS by incorporating the *Waxy* gene into the cultivar Zhenshan 97 through three backcrosses and one selfing. During backcrossing, MAS was done to select for the *Waxy* gene, to identify recombinants close to the gene, and to recover the recurrent parent genome. The incorporation of the gene improved grain quality of both the line by itself and the line in a hybrid combination. The agronomic performance of both the backcross line and the hybrid were essentially identical to the recurrent parent with the exception of reduced grain weight.

CONCLUSION

It is expected that the cost of marker technology will continue to decrease. In addition, more genes controlling quantitative and qualitative traits will be mapped and cloned. These factors are expected to result in more widespread use of MAS in the future.

ARTICLES OF FURTHER INTEREST

Breeding Plants with Transgenes, p. 193
Breeding Pure Line Cultivars, p. 196
Breeding: Choice of Parents, p. 215
Breeding: Incorporation of Exotic Germplasm, p. 222
Breeding: Recurrent Selection and Gain from Selection, p. 232
Breeding: The Backcross Method, p. 237

B

Management of Diseases in Seed Crops, p. 675
Management of Nematode Diseases: Options, p. 684

REFERENCES

1. Fehr, W.R. *Principles of Cultivar Development*; Macmillan: New York, 1987.
2. Brim, C.A. A modified pedigree method of selection in soybeans. Crop Sci. **1966**, *6*, 220.
3. Lande, R.; Thompson, R. Efficiency of marker-assisted selection in the improvement of quantitative traits. Genetics **1990**, *124*, 743–756.
4. Wittaker, J.C.; Curnow, R.N.; Haley, C.S.; Thompson, R. Using marker-maps in marker-assisted selection. Genet. Res. **1995**, *66*, 255–265.
5. Mudge, J.; Cregan, P.B.; Kenworthy, J.P.; Kenworthy, W.J.; Orf, J.H.; Young, N.D. Two microsatellite markers that flank the major soybean cyst nematode resistance locus. Crop Sci. **1997**, *37*, 1611–1615.
6. Concibido, V.C.; Denny, R.L.; Lange, D.A.; Orf, J.H.; Young, N.D. RFLP mapping and marker-assisted selection of soybean cyst nematode resistance in PI 209332. Crop Sci. **1996**, *36*, 1643–1650.
7. Rafalski, J.A. Novel genetic mapping tools in plants: SNPs and LD-based approaches. Plant Sci. **2002**, *162*, 329–333.
8. Hospital, F.; Chevalet, C.; Mulsant, P. Using markers in gene introgression breeding programs. Genetics **1992**, *132*, 1199–1210.
9. Zhou, P.H.; Tan, T.F.; He, Y.Q.; Xu, C.G.; Zhang, Q. Simultaneous improvement for four quality traits of Zhenshan 97, an elite parent of hybrid rice, by molecular marker-assisted selection. Theor. Appl. Genet. **2003**, *106*, 326–331.

Breeding Synthetic Cultivars

Steven E. Smith
University of Arizona, Tucson, Arizona, U.S.A.

INTRODUCTION

Synthetics and the specialized populations derived from them—known as synthetic cultivars (also commonly referred to as synthetic varieties,[1] which are considered completely equivalent to synthetic cultivars here)—are common products of plant breeding activities in a wide array of cross-pollinated species. Various definitions have been applied to these populations and some plant breeders have considered them to be equivalent, although this can lead to confusion. Following Lonnquist,[2] a synthetic is an open-pollinated population maintained in isolated plantings that is derived from the random mating of selfed plants or lines or other genotypes (parents) produced from mass selection. As such, a synthetic is simply the bulked seed resulting from one or more cycles of population improvement that involve artificial selection.

WHAT ARE SYNTHETIC CULTIVARS?

Synthetic cultivars have generally come to represent a specific type of synthetic that is intended for commercial (on-farm) use.[3] As such, the parents of synthetic cultivars are also preserved for future synthesis of the cultivar and may be inbred or sibbed lines, clones, F_1 hybrids, or populations.[4] When open-pollinated populations are intermated, the resulting population is sometimes referred to as a composite or composite variety, in contrast to synthetics or synthetic cultivars.[5] The original concept behind the production of synthetic cultivars is attributed to Hayes and Garber[6] and their work with maize. They described the "synthetic production of a variety" as involving hybridization among several inbred lines, with selection among F_1 progenies and advanced generations to produce an improved open-pollinated population. In early formal definitions of synthetic cultivars, the selection of parents was necessarily based on some test of their combining ability, which could be used to differentiate synthetic cultivars from synthetics or typical open-pollinated populations. However, some plant breeders have broadened the use of the term "synthetic cultivar" to include any open-pollinated population produced in plant breeding that is intended for direct commercial use.[5,7]

Specialized abbreviations are used to describe the generations represented by individual synthetics or synthetic cultivars.[2] Most commonly, genotypes initially intermated to produce a synthetic (or synthetic cultivar) represent the Syn-0 generation. Likewise, the Syn-1, Syn-2, etc. generations represent the seed produced by intermating progenies produced by Syn-0 and Syn-1 plants, respectively.

PARENTAL PERFORMANCE

Parental performance due to additive gene action is preserved within synthetic cultivars. The use of synthetic cultivars also allows for the controlled exploitation of heterosis. This is most important in cases where the production of hybrid varieties is not possible because it is not economical to control pollination adequately for the production of hybrid seed. With completely random mating, the Syn-1 generation will result from all $n(n-1)/2$ possible crosses between n parents, and is assumed to contain equal numbers of progenies from each of these crosses. The performance of advanced generations in synthetics depends on the number of parents (n), the mean performance of the parents themselves ($\bar{P}$), the mean performance of all possible hybrid combinations among the parents ($\bar{F}_1$), (which is equivalent to general combining ability), and the amount of self fertilization that occurs. If only a few parents are included, the average performance of Syn-1 offspring would be expected to be higher, but this would also be associated with a higher coefficient of inbreeding in later generations. A simple relationship, now commonly known as Wright's formula, has been developed to estimate the performance of the Syn-2 generation (denoted by $\hat{F}_2$) where parents are in Hardy-Weinberg equilibrium:[5]

$$\hat{F}_2 = \bar{F}_1 - \frac{(\bar{F}_1 - \bar{P})}{n}$$

The rationale behind this relationship is based on the value $\bar{F}_1 - \bar{P}$, representing performance attributable to heterosis and the theoretical expectation that $1/n$ of the heterosis in the F_1 (Syn-1) will be lost in the F_2 (Syn-2) or, alternatively, $(n-1)/n$ of this heterosis will be retained.[8] Assuming random mating and no selection, no loss of heterosis is expected in later generations in diploid orga-

Encyclopedia of Plant and Crop Science
DOI: 10.1081/E-EPCS 120010531

nisms. As the number of parents in a synthetic increases, the performance of the synthetic will approach that of the source population. While there remains much disagreement, the optimum number of parents for a synthetic cultivar may be as few as four, although in practice larger numbers of parents are common.[9] Very large numbers of parents may be used in cases where stability of performance is considered more important than absolute performance. Extensive description of the theory related to the prediction of synthetic cultivar performance and gene action responsible for this has been presented.[9,10]

CONCLUSION

Synthetics are a common component of population improvement programs in most cross-pollinated crop species, although the term may not be routinely applied by plant breeders. Cultivars in many perennial forage crops are regularly referred to as synthetic varieties.[3,11] In these species the broadest definition of the synthetic cultivar is generally adopted and parents are usually highly heterozygous, are typically not selected for combining ability, and are most often preserved for resynthesis as vegetative propagules. Natural intermating and successive generations of seed increase are important elements of the synthetic cultivar concept in these species because commercial quantities of seed may not be available until Syn-3 or Syn-4 generations.[3] Other than in these perennial forage species, synthetic cultivars are most common in maize, where parents are often inbred lines. Such synthetic cultivars are generally intended for use in environments where stability of performance may be paramount and the infrastructure necessary for the production of hybrid varieties does not exist.[12] Limited efforts have also been directed toward the development of synthetic cultivars in some partially self-pollinated crop species.[13]

ARTICLES OF FURTHER INTEREST

Breeding: Choice of Parents, p. 215
Breeding: Recurrent Selection and Gain from Selection, p. 232
Breeding Hybrids, p. 186
Breeding Plants and Heterosis, p. 189
Breeding Widely Adapted Cultivars: Examples from Maize, p. 211

REFERENCES

1. Fehr, W.R. *Principles of Cultivar Development. Theory and Technique*; Macmillan Publishing: New York, 1987; Vol. 1, 1–536.
2. Lonnquist, J.H. Progress from recurrent selection procedures for the improvement of corn populations. Nebr. Agric. Exp. Stn. Res. Bull. **1961**, *197*, 1–33.
3. Rumbaugh, M.D.; Caddel, J.L.; Rowe, D.E. Breeding and Quantitative Genetics. In *Alfalfa and Alfalfa Improvement*; Hanson, A.A., Barnes, D.K., Hill, R.R., Jr., Eds.; Am. Soc. Agronomy: Madison, WI, 1988; 777–808.
4. Simmonds, N.W. *Principles of Crop Improvement*; Longman Group Limited: London, 1981; 1–408.
5. Hallauer, A.R.; Miranda, J.B.FO. *Quantitative Genetics in Maize Breeding*; Iowa State Univ. Press: Ames, 1988; 1–468.
6. Hayes, H.K.; Garber, R.J. Synthetic production of high-protein corn in relation to breeding. J. Am. Soc. Agron. **1919**, *11* (8), 309–318.
7. Tysdal, H.M.; Crandall, B.H. The polycross progeny performance as an index of the combining ability of alfalfa clones. J. Am. Soc. Agron. **1948**, *40* (4), 293–306.
8. Busbice, T.H. Predicting yield of synthetic varieties. Crop Sci. **1970**, *10* (3), 265–269.
9. Wricke, G.; Weber, W.E. *Quantitative Genetics and Selection in Plant Breeding*; Walter de Gruyter: Berlin, 1986; 1–406.
10. Gallais, A. Why develop synthetic varieties? Agronomie **1992**, *12* (8), 601–609.
11. Taylor, N.L. Forage Legumes. In *Principles of Cultivar Development*; Fehr, W.A., Ed.; Macmillian Publishing: New York, 1987; Vol. 2, 209–248.
12. Paliwal, R.L.; Smith, M.E. Tropical Maize: Innovative Approaches for Sustainable Productivity and Production Increases. In *Crop Improvement. Challenges in the Twenty-First Century*; Kang, M.S., Ed.; Food Products Press: New York, 2002; 43–73.
13. Maalouf, F.S.; Suso, M.J.; Moreno, M.T. Choice of methods and indices for identifying the best parentals for synthetic varieties in faba bean. Agronomie **1999**, *19* (8), 705–712.

Breeding Using Doubled Haploids

B

Richard E. Veilleux
Virginia Polytechnic Institute & State University, Blacksburg, Virginia, U.S.A.

INTRODUCTION

Haploids are organisms with the gametic chromosome number (i.e., half the number of somatic cells). In seed plants, haploids can be derived from the tissue culture of gametic cells (microspores, megaspores) occurring in the anthers or ovules of flowers or by wide crosses. Doubling the chromosome number of haploids results in completely homozygous plants useful in various plant breeding applications. In self-pollinating crops (e.g., barley, tomato, wheat), doubled haploids (DH) have been used to construct linkage maps of molecular markers and to estimate the number of genes controlling complex traits and even directly released as cultivars. In cross-pollinating crops (e.g., maize, potato, cucumber), DH have been used to examine the effects of inbreeding and to derive inbred lines. Such lines can be used for hybrid cultivar production or for studying the inheritance of useful traits. Molecular maps based on DH of cross-pollinated crops have many skewed loci due to selection against deleterious alleles during haploid derivation. A single cycle of haploidization and chromosome doubling can dramatically reduce the time required to develop homozygous lines, compared to the usual method of inbreeding for many generations.

OCCURRENCE OF HAPLOIDS

In nature, haploids occur sporadically as weak, sterile plants with half the usual chromosome number for a given species. Charles Rick, an eminent plant geneticist, scoured the tomato fields of central California, selecting unfruitful plants. By cytogenetic analysis, he reported that two of 66 unfruitful plants were haploid ($2n = 1x = 12$), occurring at a frequency of 0.0036%.[1] Spontaneous haploids often occur in twin seedlings, where two plants emerge from the same seed: one from a normal zygotic embryo after fertilization of an egg cell by a sperm, and the other through adventive embryogenesis of a haploid cell within the embryo sac, likely a synergid. The frequency of spontaneous haploids varies with species and cultivar. The general sterility of haploids precludes their direct utilization in breeding programs, so for many years they were regarded as botanical curiosities; it seemed surprising that higher plants could even survive with so few chromosomes.

ANTHER/MICROSPORE CULTURE

Sipra Guha, a graduate student in botany studying plant developmental biology under Satish C. Maheshwari at the University of Delhi in the 1960s, cultured floral organs of the jimson weed (*Datura stramonium*) to observe their development in plant tissue culture. The appearance of embryos in cultured anthers after several weeks was unexpected; that these embryos had the haploid chromosome number was even more surprising.[2] The process whereby the gametophytic development of immature pollen is redirected to sporophytic development of haploid embryos and plants is called androgenesis. The anther culture technique is now available for many species. Isolated microspore culture—wherein immature microspores are extracted from anthers, purified, and cultured on a synthetic growth medium—has also been used, albeit with less success. Anther culture is a reasonably simple technique: Immature buds are selected with microspores at the late uninucleate stage (just at the first pollen mitosis that ordinarily results in the formation of vegetative and generative nuclei); after surface sterilization, the buds are dissected in a laminar-flow hood and the detached anthers placed in either liquid or solid sterile culture medium that may be amended with plant growth regulators or other substances; after 4–6 weeks, embryos or plantlets may emerge from the cultured anthers. The anther culture response is variable, the frequency of embryos from cultured anthers is often low. Embryos must be transferred to a different medium—usually with a high cytokinin/auxin ratio—to encourage regeneration (Fig. 1). Plantlets can be transferred directly to basal medium. Haploids can be verified by chromosome counts, flow cytometry of nuclei stained with a DNA stain, or chloroplast counts in guard cell pairs. If anther culture is unsuccessful, haploids may be obtained by ovule culture, used successfully for beet and onion.

WIDE CROSSES

Haploid plants can also be obtained through interspecific hybridization, e.g., in barley [*Hordeum vulgare* × *H. bulbosum*[3]], wheat [*Triticum aestivum* × *Zea mays*[4]], potato [*Solanum tuberosum* × *S. phureja*[5]], and tobacco [*Nicotiana tabacum* × *N. africanum*[6]]. In such crosses the

DOI: 10.1081/E-EPCS 120010535

Fig. 1 Regenerating barley embryos from anther culture (left) and doubled haploid plantlets after subculture (right). Photos supplied by the Department of Primary Industries Barley Breeding program, Hermitage Research Station, Warwick, Queensland. (*View this art in color at www.dekker.com.*)

endosperm may develop normally but the unfertilized egg grows into a haploid embryo. Otherwise, the embryo resulting from a fertilized egg may undergo elimination of paternal chromosomes. Elimination may be incomplete, resulting in the transmission of fragments of the paternal genome to predominantly maternal haploids. Embryo rescue—the excision and culture of immature embryos—may be required to prevent premature embryo abortion. In some cases (e.g., in apple and potato), a dominant homozygous marker has been linked to the locus that facilitates haploid induction in the pollinator. With this refinement, true hybrids carry the easily scored marker and can be discarded; individuals without the marker are putative haploids.

CHROMOSOME DOUBLING

Colchicine—an extract of the autumn crocus, *Colchicum autumnale*—has been used to double the chromosome number of plants,[7] thus providing a means to convert haploids to homozygous lines. Colchicine may be incorporated into the medium used for anther or microspore culture to increase the frequency of DH directly. Colchicine can otherwise be applied in a separate treatment to previously identified haploid plants. Alternatively, DH can be spontaneously regenerated by tissue culture of nonmeristematic explants of haploids (Fig. 2). Thus, androgenic or gynogenic DH populations can be obtained for many crops.

DOUBLED HAPLOIDS

The genetic architecture of haploid and doubled haploid populations differs with the breeding structure of the species. Self-pollinated crops tolerate inbreeding. They harbor no onerous genetic load (i.e., the recessive deleterious and lethal alleles found in cross-pollinated species). Therefore, a DH population derived from a hybrid of two inbred lines of a self-pollinated species is expected to represent a random gametic array of genotypes. However, with cross-pollinated species, exposure of deleterious and lethal alleles on haploidization, when there is no alternative viable allele to compensate, would result in the death of haploid embryos or plants bearing such lethals. Hence, the rigorous selection imposed by haploidization may result in a population of plants that differs dramatically from products of a random gametic array.

BREEDING WITH DOUBLED HAPLOIDS IN SELF-POLLINATING CROPS

In self-pollinating crops with true-breeding cultivars (i.e., wherein the progeny resemble the parents except for infrequent mutations), the general breeding strategy is to cross parents with contrasting desirable traits. The progeny are then selected over several generations of self-pollination until complementary traits are fixed in an inbred line that, after several years of evaluation, becomes a new cultivar. An obvious advantage of employing DH in such a breeding scheme is to fix complementary traits in inbred lines by extraction of haploids from the F_1 hybrid and doubling the chromosome number. Depending on the life cycle of the species, this can reduce by several years the time for developing a new cultivar. Anther-derived DH of wheat released as cultivars in France and Hungary required approximately half the usual time to develop.[8]

Because this process requires fewer generations than does inbreeding, it involves fewer cycles of genetic recombination. For DH derived from an F_1 hybrid, only a single recombination occurs before homozygosity is fixed; for inbreeding by self-pollination in a diploid

Fig. 2 Monoploid potato at left with 12 chromosomes and its doubled monoploid counterpart at right with 24 chromosomes. (*View this art in color at www.dekker.com.*)

species, six cycles are required to reach 98% homozygosity, with each cycle affording another opportunity for genetic recombination. Actually, fewer cycles did not appreciably affect the quality of barley inbreds.[9] Although there were slight agronomic advantages among plants that had undergone a second cycle of recombination (e.g., taller plants, higher yield, more kernels per spike), the marginal advantage did not warrant postponing DH production.

Considerable effort has been focused in recent years on constructing genetic linkage maps of crop plants. The complete sequence of the model plant *Arabidopsis thaliana* has been determined.[10] Its comparatively small genome size has assisted this effort. DH populations derived from an F_1 between parents that differ for many traits have been used to construct linkage maps in barley,[11,12] rice,[13] and canola (*Brassica napus*),[14] among others. Parallel linkage maps have been constructed from rice hybrids using both a DH and a recombinant inbred line (RIL) population of common genetic origin.[15] RIL were derived by self-pollinating and selecting single seedlings over nine generations. The major difference between the two populations was expected to reflect the number of opportunities for genetic recombination. With only a single exception, the markers retained the same relative positions on the two maps, but were further apart in the DH population. The RIL map was 70% of the length of the DH map.

The method of derivation of DH can also affect the length of a genetic map. Because of a generally higher recombination rate during microsporogenesis (male gamete formation) compared with megasporogenesis (female gamete formation) in plants,[16] maps based on anther-derived DH are predicted to be shorter than those derived from gynogenic DH, as observed with barley.[11]

BREEDING WITH DOUBLED HAPLOIDS IN CROSS-POLLINATING CROPS

In cross-pollinating crops, hybrid cultivars that result from crossing selected inbred lines dominate the market. Such cultivars may have a significant yield advantage over other types of cultivars. In addition, seed companies retain proprietary rights to hybrid cultivars. Because hybrid cultivars do not breed true, growers must purchase fresh seed each year rather than save seed from the previous year's harvest. Therefore, development of superior inbred lines and their test-crossing to evaluate hybrid combinations are important for breeding cross-pollinating crops. DH are an important source of homozygous lines. Asparagus is a unique situation because it has both male and female plants (dioecy) with a sex determination system similar to that in humans (i.e., XX females and XY males). The male plants are considered more desirable because of their more tender spears. Through the use of DH, both XX and YY plants can be produced, although the latter (supermales) would not naturally occur. By crossing XX and YY DH, uniform all-male hybrid cultivars are possible.

In many cross-pollinated crops, self-pollination reveals inbreeding depression, whereby lethal and deleterious recessive genes—usually masked by heterozygosity—are expressed. Eventually, the plants become too weak to be of any breeding value. The same process occurs on haploidization except that the weakest plants perish and the survivors represent unique genetic arrays with adequate vigor. These DH may not be fertile because sexual fertility is not prerequisite to survival, as for self-pollination. Haploids and their corresponding DH may be the only homozygous germplasm available in many cross-pollinating crops. However, their utilization as parents of hybrid cultivars is likely to require additional breeding. The first maize inbreds developed in the early 20th century were too weak to generate vigorous F_1 hybrids directly; therefore, hybrid seed produced by double crosses involving four different inbreds were used. Ultimately, more vigorous inbreds were developed such that true F_1 hybrids now dominate production fields. We could expect similar results with initial homozygotes of other highly heterozygous cross-pollinated species.

When DH are easily obtained, as for tobacco or canola, a DH population from an F_1 hybrid can be used to reduce genetic complexity of segregation patterns. As demonstrated in Fig. 3, for a dihybrid cross without linkage, only 25% of the F_2 are homozygous. However, with DH extraction from the same F_1 plant, the resulting population consists of only four classes of homozygous individuals, thus simplifying segregation ratios by eliminating heterozygous genotypes. The frequency of homozygous recessive

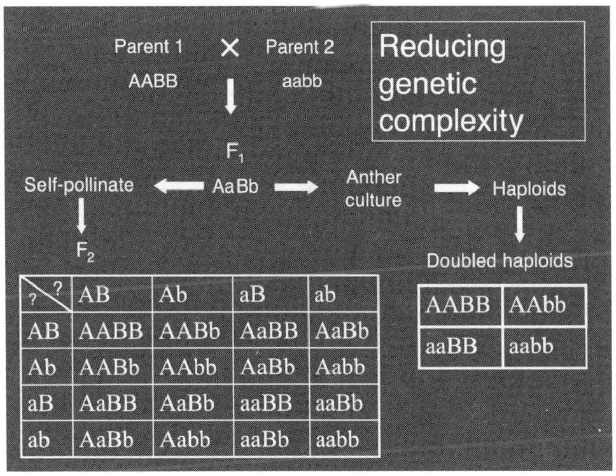

Fig. 3 Genotypic segregation in an F_2 vs. a doubled haploid population resulting from a dihybrid cross (AABB × aabb). (*View this art in color at www.dekker.com.*)

genotypes in an F_2 population is calculated as $(1/2)^{2n}$, whereas that of the DH population is $(1/2)^n$ where n = the number of independent genes that differ among parents. As the number of genes differentiating parents increases, the difference in genetic segregation between the two types of populations becomes more dramatic.

CONCLUSION

DH are only recently available for many crops. Yet they have already been exploited in plant breeding and genetic research. Cultivars have been released for several crops using DH parents, and many genetic maps have been constructed using DH populations. With self-pollinated crops, DH can easily be integrated into cultivar development schemes. However, with cross-pollinated crops, DH represent genetic equivalents to primitive inbred lines and will need considerable improvement before they have a significant impact on breeding.

ARTICLES OF FURTHER INTEREST

Anther and Microspore Culture and Production of Haploid Plants, p. 43
Breeding Hybrids, p. 186
Breeding Plants and Heterosis, p. 189
Breeding Pure Line Cultivars, p. 196
Breeding Self-Pollinated Crops Through Marker-Assisted Selection, p. 202
Female Gametogenesis, p. 439
In Vitro Chromosome Doubling, p. 572
In Vitro Morphogenesis in Plants—Recent Advances, p. 579
In Vitro Plant Regeneration by Organogenesis, p. 584
Male Gametogenesis, p. 663
Meiosis, p. 711
Plant Cell Culture and Its Applications, p. 931
Plant Cell Tissue and Organ Culture: Concepts and Methodologies, p. 934
Polyploidy, p. 1038
Population Genetics, p. 1042

REFERENCES

1. Rick, C.M. A survey of cytogenetic causes of unfruitfulness in the tomato. Genetics **1945**, *30*, 347–362. (July).
2. Guha, S.; Maheshwari, S.C. Cell division and differentiation of embryos in the pollen grains of *Datura* in vitro. Nature **1966**, *212* (5057), 97–98.
3. Kasha, K.J.; Kao, K.N. High frequency haploid production in barley (*Hordeum vulgare* L.). Nature **1970**, *225* (5235), 874–876.
4. Laurie, D.A.; Bennett, M.D. The production of haploid wheat plants from wheat × maize crosses. Theor. Appl. Genet. **1988**, *76* (3), 393–397.
5. Hougas, R.W.; Peloquin, S.J.; Ross, R.W. Haploids of the common potato. J. Heredity **1958**, *47* (3), 103–107.
6. Burk, L.G.; Gerstel, D.U.; Wernsman, E.A. Maternal haploids of *Nicotiana tabacum* L. tobacco from seed. Science **1979**, *206* (4418), 585.
7. Blakeslee, A.F.; Avery, A.G. Methods of inducing doubling chromosomes in plants. J. Heredity **1937**, *28*, 393–411. (December).
8. Pauk, J.; Kertesz, Z.; Beke, B.; Bona, L.; Csosz, M.; Matuz, J. New winter wheat variety—GK Delibab developed via combining conventional breeding and in vitro androgenesis. Cereal Res. Commun. **1995**, *23* (3), 251–256.
9. Iyamabo, O.E.; Hayes, P.M. Effects of selection and opportunities for recombination in doubled-haploid populations of barley (*Hordeum vulgare* L.). Plant Breed. **1995**, *114* (2), 131–136.
10. The Arabidopsis Initiative. Analysis of the genome sequence of the flowering plant *Arabidopsis thaliana*. Nature **2000**, *408* (6814), 796–815.
11. Devaux, P.; Kilian, A.; Kleinhofs, A. Comparative mapping of the barley genome with male and female recombination-derived, doubled haploid populations. Mol. Gen. Genet. **1995**, *249* (6), 600–608.
12. Marquez-Cedillo, L.A.; Hayes, P.M.; Kleinhofs, A.; Legge, W.G.; Rossnagel, B.G.; Sato, K.; Ullrich, S.E.; Wesenberg, D.M. QTL analysis of agronomic traits in barley based on the doubled haploid progeny of two elite North American varieties representing different germplasm groups. Theor. Appl. Genet. **2001**, *103* (4), 625–637.
13. Shen, L.S.; He, P.; Xu, Y.B.; Tan, Z.B.; Lu, C.F.X.; Zhu, L.H. Genetic molecular linkage map construction and genome analysis of rice doubled haploid population. Acta Bot. Sin. **1998**, *40* (12), 1115–1122.
14. Ferreira, M.E.; Williams, P.H.; Osborn, T.C. RFLP mapping of *Brassica napus* using doubled haploid lines. Theor. Appl. Genet. **1994**, *89* (5), 615–621.
15. He, P.; Li, J.Z.; Zheng, X.W.; Shen, L.S.; Lu, C.F.; Chen, Y.; Zhu, L.H. Comparison of molecular linkage maps and agronomic trait loci between DH and RIL populations derived from the same rice cross. Crop Sci. **2001**, *41* (4), 1240–1246.
16. Kearsey, M.J.; Ramsay, L.D.; Jennings, D.E.; Lydiate, D.J.; Bohuon, E.J.R.; Marshall, D.F. Higher recombination frequencies in female compared to male meioses in *Brassica oleracea*. Theor. Appl. Genet. **1996**, *92* (3–4), 363–367.

Breeding Widely Adapted Cultivars: Examples from Maize

A. Forrest Troyer
University of Illinois, DeKalb, Illinois, U.S.A.

INTRODUCTION

Adaptation is the driving force of evolution. Local adaptation was important for open-pollinated varieties of maize (*Zea mays* L.). The number of varieties increased from about 250 to more than 1000 as the United States grew westward during the 1800s. Some varieties—Reid Yellow Dent, Minnesota 13, Lancaster Sure Crop, Northwestern Dent, and Leaming Corn—were more widely adapted and thus more widely grown. They persisted and then prevailed in the background of today's hybrids because they had more genes for adaptation to the U.S. Corn Belt. Hybrid maize provided wider adaptation. Aggressive, research-oriented maize seed companies grew at the expense of smaller, local companies. The number of maize seed companies decreased from thousands in 1930 to fewer than 300 today. Widely adapted hybrids do better in variable weather conditions. The farmer worries less about the weather.

CREATING VARIABILITY FOR ADAPTATION

Darwin's first topic in *Origin of Species* is causes of variability which he attributes to changed conditions of life.[1] Flint maize variability increased during domestication in tropical southern Mexico, during travel northwest to what is now the southwest United States, and then during travel northeast across the Great Plains to New England. Dent maize variability increased as Spanish Conquistadors moved it to Florida from Mexico via Cuba. American colonists and settlers selected in flint by dent crosses to obtain larger ears with straighter rows of kernels and greater numbers of rows. Corn Belt dent variability increased as maize moved with American westward expansion. Resulting open-pollinated varieties contain enormous variability caused by past adaptation to many changes in conditions of life over 6500 years of travel.

NATURAL ENVIRONMENTS AND ADAPTATION

Stevens[2] states temperate climates have fewer but more widely adapted species than tropical climates. Temperate climates are more variable and stressful. Temperate growing seasons have longer and more variable day length, cooler average and minimum temperatures, less frequent and less total rainfall, and shorter seasons. Flint maize has been in what is now the United States 2500 years longer than dent maize. Northern flints are typically earlier maturing and lower yielding than southern dents. Flint by dent crosses with subsequent natural and human (artificial) selection provided better adapted, higher yielding, open-pollinated varieties for all environmental niches in the United States.

Adapted maize flowers late enough in the growing season to provide adequate plant size, yet early enough to complete or nearly complete grain filling in an average length season.[3] Maize maturity zones are based on accumulated heat units during the frost free period. They match season warmth and length with plant flowering time and grain harvest moisture. Full-season maize usually yields more where season length is limiting. Short-day photoperiod reaction limits north–south adaptation.

The U.S. corn belt is a large area with several adaptive zones. Maize grown west of the Mississippi River requires more heat and drought tolerance, and less ear and stalk rot disease tolerance due to hotter, drier summers and drier autumns. Stronger ear retention is required because of greater European corn borer (*Ostrinia nubilalis*) pressure and windier autumns. East of the Mississippi River, the climate is more humid and maize requires more tolerance to fungal diseases and better root strength. Virus tolerance (MDMV and MCDV) may be needed for the Ohio River Valley, Tennessee, and wherever Johnson grass (*Sorghum halepense* L.) is present.

ADAPTATION AND OPEN-POLLINATED VARIETY DEVELOPMENT

Human (artificial) selection emphasized good ears because more good ears were desired. Successful seed maize companies spanned human generations to modify varieties and establish reputations. More drought tolerance and earlier maturity were needed for westward expansion. A few varieties were more widely adapted and more popular than others: Reid Yellow Dent, Leaming Corn, and Lancaster Sure Crop varieties were selected under more

DOI: 10.1081/E-EPCS 120010526

favorable natural conditions; Minnesota 13 and Northwestern Dent were selected under lower-rainfall, cooler-temperature, stressful natural conditions.[4]

Darwin, in *Origin of Species*, notes the association of wider adaptation with greater taxonomic subdivision because larger areas contain more environmental niches where natural selection causes subdivisions.[1] Adding human selection for cultural practices and different uses resulted in further subdivision and more open-pollinated varieties. Montgomery estimated 250 varieties existed before America's westward expansion and 1000 varieties afterward.[5] Newer varieties were better adapted farther north and west as the southern corn belt of the 1830s moved to the Midwest by the 1880s. Adaptation for narrower environmental niches occurred just as Darwin predicted.[1]

ADAPTATION AND HYBRID DEVELOPMENT

Hybrid maize began to replace open-pollinated varieties in the 1930s. Hybrid maize no longer relied on local selection and production to become better adapted. Aggressive, research-oriented, hybrid maize seed companies grew at the expense of smaller companies. The number of maize seed companies decreased from thousands in 1930 to fewer than 300 today. Hybrids tested at several locations for a few years replaced varieties developed with many years of natural and human selection at one location. The five major, widely adapted varieties were outnumbered (91 to 5) by more locally adapted varieties in the initial development of inbreds for hybrid maize.[6] But, the widely adapted germplasm persisted in hybrid breeding programs, and now these five varieties make up 87% of the known background of today's hybrids because they contained more genes for adaptation to the U.S. corn belt environment.[4,7]

In 1921 G.S. Carter produced the first commercial maize hybrid (Burr-Leaming double cross) near Clinton, Connecticut. In 1924 H.A. Wallace's Copper Cross hybrid was sold in Iowa. In 1933 hybrid corn grew on 54,675 hectares (ha) (135,000 acres or about 0.1% of U.S. acres). In 1945 Iowa farmers grew 99% hybrid maize; 39,000 mechanical pickers harvested 75% of the Iowa crop.

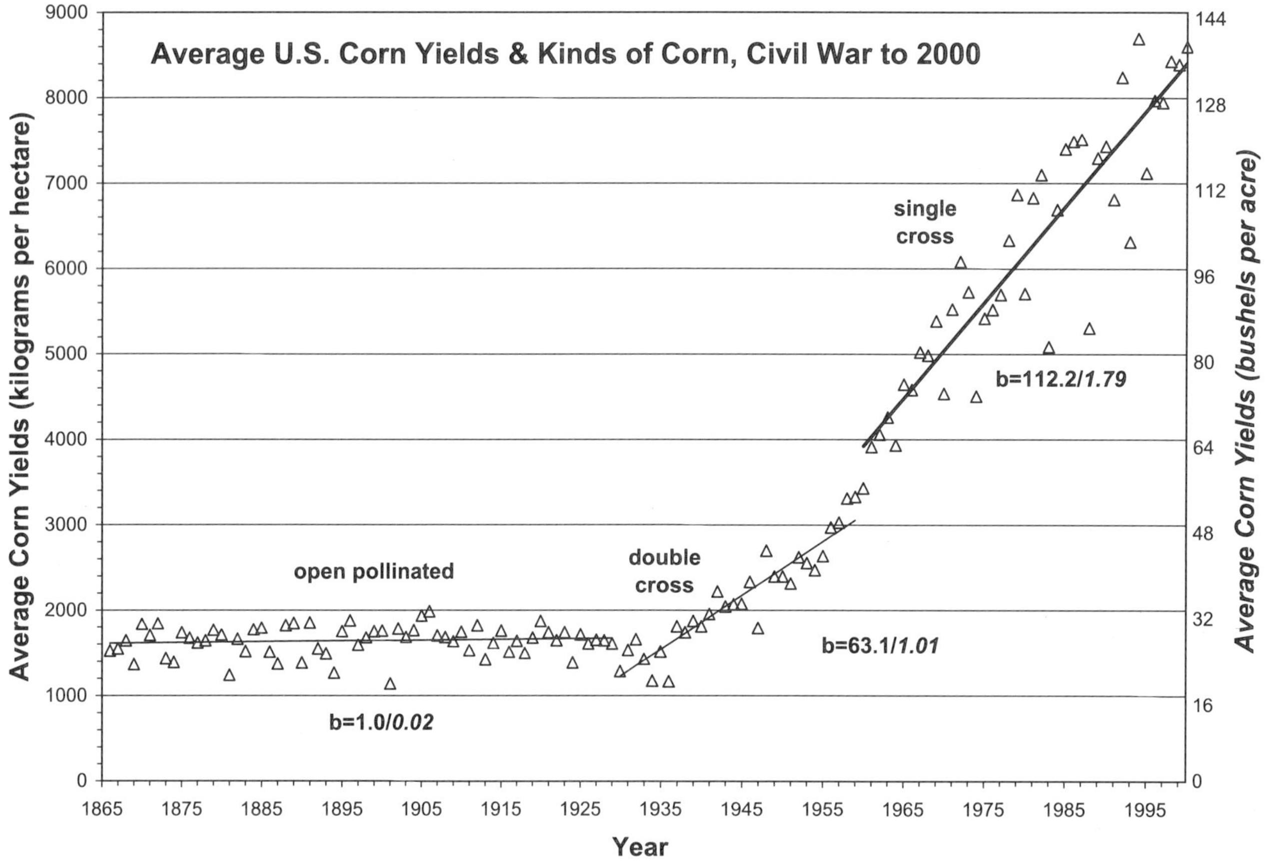

Fig. 1 Average U.S. corn yields and kinds of corn, Civil War to 2000; periods dominated by open-pollinated, by four-parent crosses, and by two-parent crosses are shown; ''b'' values (regressions) indicate average gain per year. (Data compiled by the USDA.)

In 1950 the Corn Belt grew 99% hybrid maize. From 1960 to 1980, single-cross hybrids replaced double-cross hybrids (Fig. 1).

ARTIFICIAL ENVIRONMENTS (CULTURAL PRACTICES) AND ADAPTATION

Dobzhansky states in *Mankind Evolving* that ethnic culture modifies environments to help people adapt.[8] Agronomic cultural practices modify environments to help maize adapt. Successful maize breeders selected for better adaptation to high intensity cultural practices[9–11] (Figs. 1 and 2). Inorganic fertilizers proved useful in the early 1900s. Plant population densities in the Corn Belt tripled in 100 years and doubled in the last 40 (Fig. 2).[10] From 1946 to 1950, Dr. Scarseth of Purdue conducted higher fertility, continuous corn trials.[12] In 1954 2,4-D herbicide was first commercially used. In the early 1960s, agricultural nitrogen became much cheaper. Farmers averaged 50 kg/ha (45#/A.) in 1960 and 146 kg/ha (130#/A.) by 1980 (Fig. 2).[11] In 1965 Atrazine® herbicide was first commercially used. In the late 1960s maize row width narrowed from 102 to 76 cm (40 to 30 in.). In the 1970s combines (field shelling) provided harvest-time yield results (faster feedback). Tax credits encouraged the purchase of larger machinery for more timely operations. In the 1980s nitrogen rates leveled off but plant densities continued to increase.

GENOTYPE-BY-ENVIRONMENT INTERACTION AND ADAPTATION

Genetic diversity (multiple genotypes) is an adaptive response to different environments. A more homogenous environment allows fewer genotypes to prevail. A completely homogenous environment allows a single genotype to always be superior. More plants per acre, earlier planting, more nitrogen, narrower rows, and better pest control provide a more homogenous environment that requires fewer genotypes. Such an environment particularly favors those genotypes surviving natural and artificial selection for adaptation to these resources and practices.[9,13]

Ceccarelli emphasizes that plant characteristics for maximum yield are different under optimum conditions than under stressful conditions.[14] He generates significant variety crossover interactions with widely different stability regression coefficients (0.80 vs. 1.17) extended to very low 1820 kg/ha (29 bu/acre) yield levels. He advocates breeding for these lower yield levels. Stability coefficients less than 1.0 are conservative (for stressed, lower-yield environments) and regressions greater than 1.0 are aggressive (for higher-yield environments). Farmers choose aggressive and conservative hybrids based on risk tolerance. Stability regression coefficients for popular hybrids in the U.S. corn belt usually range from 0.95 to 1.10. United States maize farmers average 9400 kg/ha (140 bu/ac) and expect more than 1820 kg/ha (29 bu/acre)

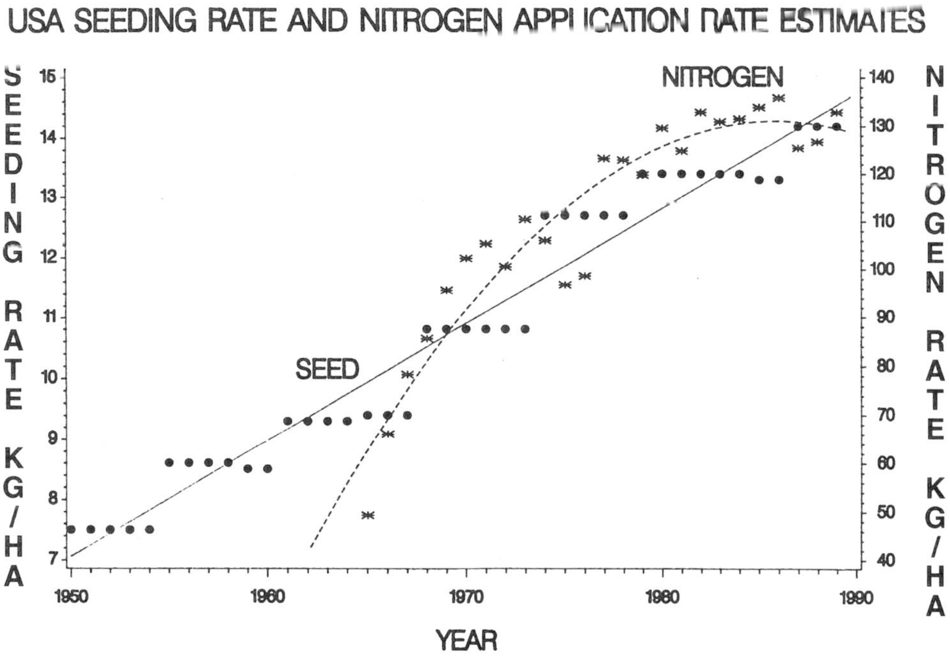

Fig. 2 Corn plant density (determined every five to seven years) and nitrogen fertilizer increase over time (1950–1990) in the U.S. corn belt. (From Refs. 9 and 10.)

at planting (hybrid choice) time—no market exists for special low-yielding maize hybrids.

Lower-yielding environments are important when breeding for wide adaptation. Calculating performance as the percent of location mean equalizes high- and low-yielding environments. Thus each environment will have equal weight when selecting top performers. Farmers produce maize and exchange it for money to pay bills. Yield results in local market dollars per unit area on date of harvest compensate lower-yielding seasons and earlier-maturity hybrids via higher prices. This identifies, over time, the higher-value hybrids for an area. Hybrid by year (growing season) is the most important interaction. Years (growing seasons) may be warm, cold, long, short, wet, or dry. Hybrid by location interaction in the U.S. midwest is usually caused by differing moisture availability at flowering time.

WIDELY ADAPTED HYBRIDS

Widely adapted hybrids perform well under variable weather conditions and are more likely to produce higher yields. The farmer worries less and profits more.[13] Fewer hybrids are necessary for all concerned. Regional hybrids have deficiencies that limit wider adaptation. Multiple-year results measure the consistency of hybrid performance and the value of the company's hybrid graduation system. Weather and cultural practices greatly affect maize performance. Many testing sites are necessary to adequately sample the total environment. More testing sites identify more genes for adaptation. Modern testing programs eliminate artifacts, use modern information management, and provide pertinent hybrid comparisons. Strip tests (large plots) mimic farmers' fields with less border effect, no alleys, and farmer management. Adaptation is critical; quality of germplasm is more important than quantity (number of chances) or diversity of germplasm. Successful U.S. corn belt maize breeding programs improve the adaptation of a tropical crop to a temperate environment.[4,7]

High plant-density stress is the ultimate stress for maize. Selection against silk delay and for good ear development at higher plant densities is survival of the fittest in its purest form.[9] It develops tougher inbreds and hybrids that are more widely adapted.[13] Pioneer brand hybrids 3780 and 3732 were the world's two most popular maize hybrids. Three of their four parent inbreds were developed and the two hybrids were evaluated under high plant-density stress. We applied 368 kg/ha (300#/acre) of nitrogen to nurseries and to yield test fields.[13] Darwin states that excess of nutriment is perhaps the single most efficient exciting cause of variability.[1]

ARTICLES OF FURTHER INTEREST

Agriculture and Biodiversity, p. 1
Breeding: Genotype-by-Environment Interaction, p. 218
Breeding Hybrids, p. 186
Competition: Responses to Shade by Neighbors, p. 300
Crop Domestication: Fate of Genetic Diversity, p. 333
Drought and Drought Resistance, p. 386
Plant Response to Stress: Critical Periods in Plant Development, p. 981

REFERENCES

1. Darwin, C.R. *On the Origin of Species*; John Murray: London, 1859; 1–479.
2. Stevens, G. The latitudinal gradient in geographical range: How so many species coexist in the tropics. Am. Nat. **1989**, *133* (2), 240–242.
3. Troyer, A.F.; Brown, W.L. Selection for early flowering in corn. Crop Sci. **1972**, *12*, 301–304.
4. Troyer, A.F. Background of U.S. hybrid corn. Crop Sci. **1999**, *39*, 601–626.
5. Montgomery, E.G. *The Corn Crops*; McMillan: New York, 1916; 1–347.
6. Jenkins, M.T. Corn Improvement. In *Yearbook of Agriculture*; Bressman, E.S., Ed.; GPO, 1936; 455–522.
7. Troyer, A.F. Background of U.S. hybrid corn II: Breeding, climate, and food. Crop Sci. **2004**, *44* (2).
8. Dobzhansky, T. *Mankind Evolving*; Yale University Press: New Haven, CT, 1962; 1–381.
9. Troyer, A.F.; Rosenbrook, R.W. Utility of higher plant densities for corn performance testing. Crop Sci. **1983**, *23*, 863–867.
10. Anonymous. *USDA Agricultural Statistics*; Agricultural Statistical Service: Washington, DC, 1989.
11. Anonymous. *Fertilizer Use with Price Statistics*; Stat. Bul., USDA-ERS: P.O. Box 1608, Rockville, MD, 1989; Vol. 780.
12. Scarseth, G.D. *Man and His Earth*; Iowa State University Press: Ames, IA, 1962; 1–199.
13. Troyer, A.F. Breeding widely adapted, popular maize hybrids. Euphytica **1996**, *92*, 163–174.
14. Ceccarelli, S. Wide adaptation: How wide? Euphytica **1989**, *40*, 197–205.

Breeding: Choice of Parents

John W. Dudley
University of Illinois, Urbana, Illinois, U.S.A.

B

INTRODUCTION

The choice of parents for use in a plant breeding program is one of the most important decisions a breeder makes. It will determine the ultimate success, or lack thereof, in a breeding program. This critical choice should be based on the objectives of the breeding program; the germplasm, technology, and resources available to the breeder; the ease of manipulation of reproduction in the species; and some basic scientific principles. The purpose of this article is to outline the basic principles involved in choice of parents and to provide some basic guidelines useful to a breeder.

BREEDING STRATEGIES

A clear statement of the objectives of a particular breeding program is a necessary first step in determining choice of parents. For example, if the objective is to add a single gene for disease resistance to a cultivar with otherwise acceptable agronomic and quality characteristics, one of the parents must have the disease resistance gene, regardless of its other characteristics. On the other hand, if the objective is to develop a higher yielding cultivar of wheat (*Triticum aestivum*, L.) by crossing two good lines, then the parents should both be high-yielding to maximize the opportunity for producing a new line with higher yields than either of the parents' yields.

The types of breeding strategies for which parents are chosen can be grouped into three major categories, regardless of mode of reproduction: 1) Backcross programs in which the objective is to add a single gene or small number of genes to an existing elite cultivar. The usual objective is to add a gene for disease resistance or a quality factor without disturbing the unique combination of genes that make the cultivar elite; 2) crossing of elite lines to produce a segregating generation from which new lines with characteristics superior to either of the parents' characteristics will be selected. This second category is usually used for quantitative traits such as grain yield; and 3) development of segregating populations such as synthetic varieties, either for direct use in production or as a pool for recurrent selection.

BACKCROSSING

In the past, the choice of parents for backcrossing programs was limited by the availability of the single genes of interest in species that could be crossed with the cultivar to be improved. However, with the introduction of transformation technology, the source of a gene for improvement of a specific species can be any species. A prime example is the development of corn resistant to the European corn borer (*Ostrinia nubilalis*) using a gene from the bacterium *Bacillus thuringensis*. The gene from the bacterium is introduced into a particular corn inbred by transformation. Once the gene is stabilized in the new inbred, the breeding procedure is the same as if a single gene had been found in a corn (*Zea mays*, L.) inbred. This topic is developed further in the articles on Backcross Breeding (Tracy, this volume) and Breeding with transgenes.

SELECTION WITHIN AN F_2

The basic principles for selection within a segregating population (such as within an F_2,) are the same whether the species is self-pollinated (as, e.g., wheat) or is cross-pollinated and the inbred–hybrid system is used as in corn. The objective is to obtain a segregating population with a high mean and adequate genetic variance. Such a population would satisfy the usefulness criterion of Schnell.[1] This criterion is expressed mathematically as $U_{(\alpha)}=Y\pm\Delta G_{(\alpha)}$, where $U_{(\alpha)}$ is the usefulness of a population, Y is the mean of the unselected population, and $\Delta G_{(\alpha)}$ is the expected gain from selection. $\Delta G_{(\alpha)}$ is a function of the genetic variance in the population, the selection intensity, and a mating factor based on whether selection can be done prior to or after pollination. Thus, all other things being equal, a population with both a high mean and a high genetic variance would be more useful. This principle holds whether the population being developed is an F_2 from a cross between two pure lines or a synthetic from intercrossing several lines.

The identification of parents with potential to produce segregating populations with a high mean and a high genetic variance has been the subject of several studies.

Encyclopedia of Plant and Crop Science
DOI: 10.1081/E-EPCS 120010520

Prediction of the mean of a cross is often based on the midparent value of the lines to be crossed. This prediction can be enhanced by use of the Best Linear Unbiased Prediction (BLUP) technique,[2] which takes advantage of relationships among lines being considered. Prediction of genetic variance in a particular cross has been attempted by using genetic distance between parents based on either known pedigree relationships or distance calculated from molecular marker data.[3] A general recommendation is to select as parents the pairs of lines with the highest predicted midparent values and within that set, select the pair or pairs that are most genetically diverse.

The usual axiom for plant breeders is to cross good by good in order to obtain something better. This basic principle for choosing parents of a cross between two inbreds is supported by theory based on the concept of classes of loci.[4] For any pair of homozygous lines, there are four classes of loci that can be designated as *i*, *j*, *k*, and *l*. Class *i* loci are ++ for both parents, class *j* are ++ for P_1 and −− for P_2, class *k* are −− for P_1 and ++ for P_2, whereas class *l* loci are −− for both P_1 and P_2. Thus the only classes of loci useful for improvement of the population are classes *j* and *k*. The maximum probability of isolating a new line, homozygous for more favorable alleles than either P_1 or P_2, is obtained when the number of class *i* loci is equal to the number of class *j* loci.[5] When one parent has several more loci with favorable alleles in classes *i* and *j* than the other, backcrossing to the parent with the largest number of loci with favorable alleles enhances the probability of obtaining a new line with more favorable alleles than either of the parents' alleles.[4,6]

The concept of classes of loci can be extended to the choice of parents for improving a single-cross hybrid. In this case, three lines are involved: P_1, P_2, and a donor line (P_w). As shown by Dudley,[7] eight classes of loci exist for three lines. The objective is to incorporate alleles from P_w into either P_1 or P_2 to improve the performance of the hybrid between them. Four basic questions need to be answered in choosing a particular P_w:[7] 1) Which hybrid should be improved? The usual answer is the best hybrid for the target set of environments; 2) which parent line should be improved? In hybrid breeding programs based on heterotic patterns, the parent to be improved is the one to which the donor line is most closely related. This choice maintains the heterotic pattern, thus maximizing heterosis in the new hybrid; 3) which donor line should be selected? The short answer is: the line with the maximum number of loci with favorable alleles at loci where neither parent has favorable alleles and a minimum at loci where the donor has unfavorable alleles, the parent to be improved has favorable alleles, and the other parent has unfavorable alleles; 4) should selfing begin in the F_2 or should backcrossing be used? The answer to this question is based on the principle that if the donor and the line to be improved have similar numbers of loci homozygous for favorable alleles at loci where they differ, then the maximum probability for identifying a line with more loci with favorable alleles than either parent is obtained. If the parent to be improved has many more loci homozygous favorable than the donor has, then backcrossing is indicated. These questions and their answers can be applied to use of a donor population as well as to a donor inbred.[7]

SELECTION OF PARENTS FOR A SYNTHETIC

Again, the principle of usefulness applies. However, the prediction of mean performance of a potential synthetic is a function of the mean of all possible crosses among the parents and inbreeding depression.[8] The equation $Y_2 = Y_1 - (Y_1 - Y_0)/n$ (where Y_2 is the mean of a synthetic obtained by intercrossing all possible single crosses among a set of *n* lines, Y_1 is the average performance of all possible single crosses, and Y_0 is the mean performance of the *n* parental inbreds) predicts the synthetic mean. This prediction says nothing about the variation in the population.

CONCLUSION

The basic principle of crossing good by good is not likely to change in plant breeding programs in the future. Current techniques for identifying parents are useful and provide an advantage over breeder's intuition. However, the availability of genomic technology may lead to refined methods of identifying parents having complementary sets of genes that will provide useful genetic variability for selection. This is particularly likely to be true if an adequate understanding of the genetic control of metabolic pathways involved in stress resistance and yield is obtained.

ARTICLES OF FURTHER INTEREST

Breeding Biennial Crops, p. 171
Breeding Hybrids, p. 186
Breeding Synthetic Cultivars, p. 205
Breeding: Genotype-by-Environment Interaction, p. 218
Breeding: Incorporation of Exotic Germplasm, p. 222
Breeding: Mating Designs, p. 225
Breeding Plants and Heterosis, p. 189
Breeding Pure Line Cultivars, p. 196

Breeding: Recurrent Selection and Gain from Selection, p. 232

Crop Improvement: Broadening the Genetic Base for, p. 343

REFERENCES

1. Lamkey, K.R.; Schnicker, B.J.; Melchinger, A.E. Epistasis in an elite maize hybrid and choice of generation for inbred line development. Crop Sci. **1995**, *35*, 1272–1281.
2. Panter, E.M.; Allen, F.L. Using best linear unbiased predictions to enhance breeding value for yield in soybean. I. Choosing parents. Crop Sci. **1995**, *35*, 397–405.
3. Burkhamer, R.L.; Lanning, R.J.; Martens, J.M.; Talbert, L.E. Predicting progeny variance from parental divergence in hard red spring wheat. Crop Sci. **1998**, *38*, 243–248.
4. Dudley, J.W. Theory for transfer of alleles. Crop Sci. **1982**, *22*, 631–637.
5. Bailey, T.B., Jr.; Comstock, R.E. Linkage and the synthesis of better genotypes in self-fertilizing species. Crop Sci. **1976**, *16*, 363–370.
6. Bailey, T.B., Jr. In *Selection Limits in Selfing Populations*, International Conference on Quantitative Genetics, Aug. 16–21, 1976; Pollak, E., Kempthorne, O., Bailey, T.B., Jr., Eds.; Iowa State Univ. Press: Ames, IA, 1977; 399–412.
7. Dudley, J.W. In *Theory for Identification of Lines or Populations Useful for Improvement of Elite Single Crosses*, Proc. 2nd Intern. Conf. on Quantitative Genetics, Raleigh, NC, June 1–5, 1987; Weir, B., Eisen, E., Goodman, M., Namkoong, G., Eds.; Sinauer Assocs.: Sunderland, MA, 1988; 451–461.
8. Hallauer, A.R.; Mirando, FO, J.B. *Quantitative Genetics in Maize Breeding*; Iowa State University Press: Ames, IA, 1988.

B

Breeding: Genotype-by-Environment Interaction

Manjit S. Kang
Louisiana State University, Baton Rouge, Louisiana, U.S.A.

INTRODUCTION

Genotype-by-environment interaction (GEI)—a universal issue—relates to all living organisms, from bacteria to plants to humans. The subject is important in agricultural, genetic, evolutionary, and statistical research. Genotype-by-environment interaction refers to differential responses of different genotypes across a range of environments. Gene expression is dependent upon environmental factors and may be modified, enhanced, silenced, and/or timed by the regulatory mechanisms of the cell in response to internal and external forces. A range of phenotypes can result from a genotype in response to different environments; the phenomenon is called norms of reaction, or phenotypic plasticity. Norms of reaction represent the expression of phenotypic variability in individuals of a single genotype. The lack of phenotypic plasticity is called canalization.

In this article, discussion of GEI is limited to plant breeding—the art and science of improving plants. Plant breeding helps increase crop productivity by overcoming challenges posed by changes in soil and climatic factors, altered spectrum of pests, consumer demand, and economic policies.

GENE EXPRESSION AND ENVIRONMENT

Gene expression is the process whereby a gene produces a chemical product (protein) that carries out its designated function. The genotype (genetic make up—deoxyribose nucleic acid, DNA, or gene) of an individual and environmental conditions determine the physical appearance, or phenotype, of an individual. Genes carry the blueprint of an organism, which is translated into a phenotype under proper environmental conditions. Transcription of a gene into messenger ribose nucleic acid (mRNA) and its translation into a polypeptide (protein) represent the central dogma of genetics, i.e., DNA $\Rightarrow$ mRNA $\Rightarrow$ protein. The current, modified version of the central dogma is DNA $\Longleftrightarrow$ mRNA $\Rightarrow$ protein, because reverse transcription does occur in certain organisms (retroviruses).

To understand GEI, it is important to know how genes express in different environments and how qualitative and quantitative characters are influenced by environmental factors.[1] Qualitative traits (controlled by one or two genes) are, in general, highly heritable, i.e., they are not affected by environmental factors. They exhibit a discontinuous variation in segregating generations, e.g., the second filial generation (F_2). For example, crossing of round (*RR*) and wrinkled (*rr*) seeded garden pea plants always yields the first filial generation (F_1) with round seeds, and F_2 plants segregate in a discrete, 3 round (1 *RR* + 2 *Rr*) to 1 wrinkled (*rr*) ratio, no matter where the cross is grown in the permissive range of environments. Certain qualitative traits, however, are influenced rather drastically by environmental factors. For example, the sun-red gene in maize (*Zea mays* L.) produces red kernels if kernels are exposed to direct sunlight, but in the absence of sunlight, the kernels remain white.[2]

Quantitative traits, on the contrary, are controlled by several genes and invariably exhibit low heritability. They are, in general, highly influenced by environmental factors and display continuous variation.[3,4] Most economically important plant traits, e.g., harvestable grain or forage yield, can be classified as quantitative or multigenic.[5] Plant adaptation relative to quantitative traits is intimately connected to environment.[6]

WHAT IS GENOTYPE-BY-ENVIRONMENT INTERACTION?

Crop performance—the observed phenotype—is a function of genotype (variety or cultivar), environment, and GEI. For interaction to be detected, there must be at least two distinct genotypes evaluated in at least two different environments. From a statistical standpoint

$$y_{ij} = \mu + g_i + e_j + (ge)_{ij}$$

where y_{ij} represents phenotype (e.g., yield) for *i*th genotype in *j*th environment; μ, overall mean performance; g_i, genotype effect; e_j, environment effect; and $(ge)_{ij}$, interaction between g_i and e_j. Thus, for a given genotype there can be many phenotypes, depending upon environmental conditions and the extent of GEI.

Fig. 1a represents absence of interaction, where responses of the two genotypes A and B to environments are similar, i.e., parallel responses. The norms of reaction for

Encyclopedia of Plant and Crop Science
DOI: 10.1081/E-EPCS 120010525

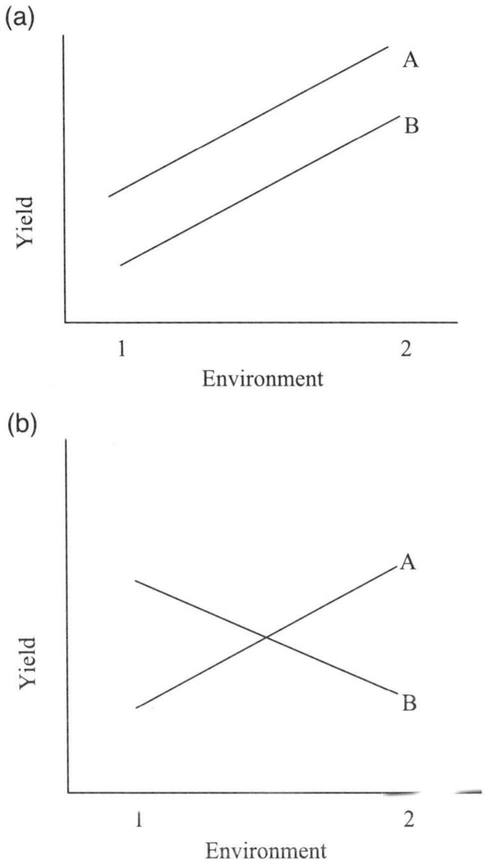

Fig. 1 (a) This graph shows no genotype-by-environment interaction. Genotypes A and B respond to the two environments in a similar manner. (b) This graph depicts a crossover type of interaction. Genotypes A and B switch ranks between the two environments.

the two genotypes are additive: A and B increase with change in environment.[4]

Genotype-by-environment interaction represents nonparallel or nonadditive norms of reaction. Fig. 1b illustrates a crossover GEI (genotypes switch ranks between environments). The environmental modification of the two genotypes is opposite in direction: A increases but B decreases.

IMPORTANCE OF GENOTYPE-BY-ENVIRONMENT INTERACTION

With the expected doubling of world population from the current 6 billion to 10–12 billion by the mid-21st century, GEI and stability of crop performance across environments should become more relevant issues for breeders. Land and water resources being limited, a greater emphasis will need to be placed on sustainable agricultural systems and on proper exploitation and use of GEI.

Information on GEI can help determine whether or not cultivars should be developed for an entire region in a single breeding program. The presence of a crossover GEI necessitates partitioning of a breeding program to develop cultivars for specific environments. In the complete absence of GEI, a single variety of any crop would produce its maximum yield the world over, and also variety trials would need to be conducted at only one location to provide universal results—but this is an unrealistic scenario.[7] Genotype-by-environment interaction is routinely detected in multienvironment crop yield trials.

IMPLICATIONS OF GENOTYPE-BY-ENVIRONMENT INTERACTION IN PLANT BREEDING

Genotype-by-environment interaction offers both challenges and opportunities for breeders. Abiotic and biotic stresses are generally causes of GEI. Suboptimal levels of water (drought stress) can help identify water-use efficient genotypes. At superoptimal levels (flooding), flood-tolerant genotypes can be identified.

The larger the GEI component, the lower the heritability. The lower the heritability of a trait, the greater the difficulty in improving that trait. Other implications of GEI are:

1. Problem in identifying superior cultivars: As the magnitude of interaction between G and E increases, the usefulness and reliability of their main effects decrease. Crossover GEI increases the difficulty in identifying truly superior genotypes across environments.
2. Increased evaluation cost: A large GEI necessitates use of additional environments to obtain reliable results, which increases the cost of evaluating breeding lines. If weather patterns and/or management practices differ in target areas, testing must be done at several sites representative of the target areas.

Multienvironment testing allows identification of cultivars that perform consistently from year to year (low temporal variability) and those that perform consistently from location to location (low spatial variability). Temporal stability benefits growers, whereas spatial stability benefits seed companies and breeders. Analyses of multienvironment trial data allow identification of genotypes of broad adaptation as well as those specifically adapted to a particular environment. Broad adaptation, or stability of performance (reliability) across environments, helps conserve limited breeding resources. The desirability of stability of performance across environments depends on

Table 1 Mean yield of five barley cultivars (genotypes) from six environments

	Environments→					
Genotypes↓	**E1**	**E2**	**E3**	**E4**	**E5**	**E6**
G1	26.95	41.17	30.90	36.45	27.55	25.77
G2	31.28	42.92	30.40	30.55	23.15	23.97
G3	33.35	43.82	32.48	36.70	27.63	24.38
G4	32.82	56.53	45.20	44.38	25.20	32.27
G5	30.42	42.30	36.53	33.42	30.73	31.68

(Adapted from Yates, F. and Cochran, W.G. The analysis of groups of experiments. Journal of Agricultural Science **1938**, 28, 556–580.)

whether or not environmental differences are predictable. Breeders seek performance stability (a lack of GEI) against uncontrollable factors. If GEI is attributable to unpredictable environmental factors, e.g., weather variables, varieties with stable performance across a range of conditions should be selected or developed.[8] If GEI is caused by variations in predictable factors, e.g., soil type and cultural practices, varieties specifically adapted to an environment should be developed.[8] For certain genotypes to do well, additional agronomic inputs may be necessary, which tends to create a genotype-by-environment correlation.[1]

CAUSES OF GENOTYPE-BY-ENVIRONMENT INTERACTION

Biotic stresses (insects, diseases, and weeds) are major constraints to crop productivity. Differences in insect and disease resistance among genotypes can be associated with stable or unstable performance across environments.[1] Knowledge of the cause(s) of a significant interaction helps make accurate predictions of genotype performance in diverse environments. Understanding genotypic responses to individual factors aids in interpreting and exploiting GEI.

EXPLOITATION OF GENOTYPE-BY-ENVIRONMENT INTERACTION

There are numerous methods of analyzing GEI.[8–10] Methods that integrate performance and stability of performance across environments are useful breeding tools.[1,8] Only a brief reference will be made to some of them here. The Additive Main effects and Multiplicative Interaction (AMMI) method is useful for understanding complex GEI.[7] The Shifted Multiplicative Model (SHMM) is helpful in identifying subsets of genotypes or environments with negligible rank changes.[10]

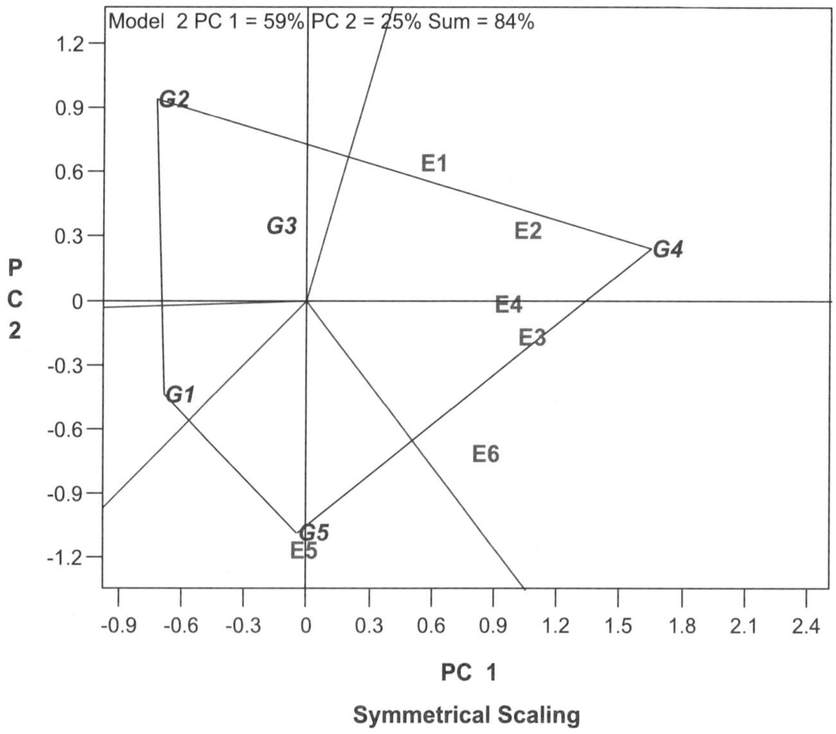

Fig. 2 This polygon shows performances of five genotypes in six environments. Genotype G4 had the highest performance in all environments except E5. Genotype G5 had the best performance in E5. This graph was generated via the GGEbiplot software. (*View this art in color at www.dekker.com.*)

A recent method called GGE Biplot is a powerful tool for visualizing and interpreting two-way GEI data.[8] Yield data shown in Table 1 were analyzed via GGEbiplot. In all environments except E5, G4 is the highest yielding cultivar (Fig. 2). Cultivar G5 yielded the most in E5. The other cultivars were not the highest yielding in any environment. The sites Regression Model (SREG) is similar to the GGE biplot technique, which helps in assessing both general and specific adaptation of genotypes.[10]

CONCLUSION

Methods such as GGEbiplot, AMMI, and SHMM will play an increasingly greater role in interpreting and exploiting GEI in plant breeding in the 21st century. DNA-based markers (tags) should enhance our understanding and use of GEI. Proper understanding, use, and exploitation of GEI can help ease the food problem in the world.

ARTICLES OF FURTHER INTEREST

Breeding Widely Adapted Cultivars: Examples from Maize, p. 211

Crops and Environmental Change, p. 370

REFERENCES

1. Kang, M.S. Using genotype-by-environment interaction for crop cultivar development. Adv. Agron. **1998**, *62*, 199–252.
2. Rédei, G.P. *Genetics Manual: Current Theory, Concepts, Terms*; World Scientific: River Edge, NJ, 1998.
3. Bradshaw, A.D. Evolutionary significance of phenotypic plasticity in plants. Adv. Gen. **1965**, *13*, 115–155.
4. Schlichting, C.D.; Pigliucci, M. *Phenotypic Evolution: A Reaction Norm Perspective*; Sinauer Associates: Sunderland, MA, 1998.
5. Stuber, C.W.; Polacco, M.; Senior, M.L. Synergy of empirical breeding, marker-assisted selection, and genomics to increase crop yield potential. Crop Sci. **1999**, *39*, 1571–1583.
6. *Plant Adaptation and Crop Improvement*; Cooper, M., Hammer, G.L., Eds.; CAB International, ICRISAT, IRRI: Wallingford, U.K., 1996.
7. Gauch, H.G., Jr.; Zobel, R.W. AMMI Analysis of Yield Trials. In *Genotype-by-Environment Interaction*; Kang, M.S., Gauch, H.G., Jr., Eds.; CRC Press: Boca Raton, FL, 1996; 85–122.
8. Yan, W.; Kang, M.S. *GGE Biplot Analysis: A Graphical Tool for Breeders, Geneticists, and Agronomists*; CRC Press: Boca Raton, FL, 2002.
9. *Quantitative Genetics, Genomics and Plant Breeding*; Kang, M.S., Ed.; CABI Publishing: Wallingford, U.K., 2002.
10. Crossa, J.; Cornelius, P.L. Linear-bilinear Models for the Analysis of Genotype-environment Interaction. In *Quantitative Genetics, Genomics and Plant Breeding*; CABI Publishing: Wallingford, U.K., 2002; 305–322.

Breeding: Incorporation of Exotic Germplasm

James B. Holland
USDA—ARS and North Carolina State University, Raleigh, North Carolina, U.S.A.

INTRODUCTION

Exotic germplasm refers to crop varieties unadapted to a breeder's target environment, and is an important resource for crop improvement. Because genetic diversity within elite cultivars of a crop is limited compared to the variability within the species and its relatives worldwide, genes from exotic germplasm can protect the crop against new biotic and abiotic stresses, and may represent unique alleles for productivity that are absent from elite crop gene pools. Introducing substantial amounts of genetic material from exotic sources into elite crop gene pools while maintaining their productivity is difficult, however. Exotic germplasm incorporation programs require long-term commitments and appropriate breeding strategies, and may be assisted by DNA marker technologies.

CROP DOMESTICATION AND GENETIC BOTTLENECKS

Cultivars of major crops in industrialized nations represent only a small sample of the genetic variability available in those species worldwide.[1,2] Plant breeding per se often reduces genetic variation in crop species[1–3] because only superior genotypes are selected, but Darwin[4] suggested that selection by plant and animal breeders was a major cause of *increased* variation within domesticated species. Depending on circumstances, plant breeding may contribute to either increases or decreases in crop genetic variation.

Domestication of many crops involved genetic bottlenecks, initially reducing genetic diversity.[3,5] However, selection for crop adaptation to widely varying agroecological habitats and for diverse uses by farmers resulted in subsequent increases in variability. For example, bread wheat is significantly less variable than its close wild relatives because it underwent a severe genetic bottleneck during domestication.[5] Nevertheless, genetic variability exists among wheat varieties because human selection acted to preserve rare favorable variants in varieties adapted to different habitats and uses.[5] However, modern plant breeding for industrial agriculture generally results in reduced genetic variation, because uniformity of type is demanded by most farmers, commodity handlers and processors, and consumers.[2] Therefore, plant breeding per se does not necessarily reduce genetic variation. This observation provides hope that the genetic bases of modern crops can be enhanced through plant breeding.

The limited genetic variation within modern crops is a concern because it may result in widespread crop-yield and quality losses if new pathogen populations or unusual abiotic stresses occur.[2] Incorporation of new alleles from exotic germplasm that confer pathogen resistance can alleviate this genetic vulnerability.[5,6] Furthermore, improvements in crop productivity may be achieved by incorporating exotic germplasm into elite gene pools since it is highly unlikely that all favorable alleles were sampled in the ancestors of modern cultivars.

CROP ENHANCEMENTS

What is the best way to enhance crops with new alleles from exotic parents and wild species? Simmonds[1] distinguished between two general strategies: introgression and incorporation. Introgression involves backcrossing a few chromosome segments with easily identifiable effects (often disease resistances) into elite cultivars. In contrast, incorporation aims to create populations that are adapted to a breeder's target set of environments but that are also genetically distinct from elite cultivars. Incorporation requires isolating exotic populations from locally adapted populations and conducting many cycles of mild selection for adaptation while maximizing recombination. Only when exotic populations have been improved to a level of reasonably good productivity should they be crossed to elite germplasm. The aim of introgression is to disrupt elite genetic backgrounds as little as possible during the introduction of a relatively small number of exotic alleles. The objective of incorporation is to produce new breeding populations that have very high proportions of unique, exotic-derived alleles in order to broaden substantially the crop's genetic base.

EXAMPLES

Successful germplasm incorporation programs have been conducted in potato and sugarcane. In both cases,

Encyclopedia of Plant and Crop Science
DOI: 10.1081/E-EPCS 120010536

broad-based exotic populations were introduced to target production environments and subjected to mild selection for adaptation, disease resistance, and productivity over many generations of sexual recombination.[1] After about 20 to 30 years of selection in both crops, improved exotic populations had yields similar to local cultivars. Furthermore, crosses between local and exotic populations exhibited substantial high-parent heterosis and formed the basis for new commercial cultivars.[1]

Similarly, Goodman et al.[7] reported successful incorporation of exotic tropical maize into the narrow gene pool of agriculturally elite temperate maize. This program emphasized identifying superior exotic germplasm sources, creating selection populations by intercrossing tropical hybrids, and enhancing recombination during inbred line development by sib-mating rather than self-fertilizing. Although the tropical-derived inbreds themselves had inferior performance compared to adapted inbred lines, hybrids created from crosses between tropical-derived inbreds and temperate germplasm were agronomically competitive with commercial hybrids.

The most successful incorporation programs to date were conducted in clonally-propagated or hybrid crops because the heterosis achieved in the first generation of exotic-by-adapted crosses could be fully captured. However, many open pollinated crops (including most forages) are not easily inbred; rather, they are managed as populations. In these crops, even though substantial heterosis often occurs in the F_1 generation of exotic-by-adapted crosses, pure F_1 hybrids cannot be created on a commercial scale.[8] However, production of semihybrids formed through population crosses could capitalize commercially on a large portion of the heterosis, and should present new opportunities for germplasm incorporation in open-pollinated crops.[8]

EXOTIC GERMPLASM IN ATYPICAL SITUATIONS

Incorporation is more difficult to implement in naturally self-fertilizing crops[9] because of two factors that hinder maintaining exotic germplasm independently from adapted populations. First, the best sources of genetic variation for many self-fertilizing crops are wild species that are entirely unadaptable to agriculture. Second, less recombination occurs during selfing than outcrossing generations. Therefore, selection within semiexotic populations created by crossing exotic and adapted germplasm is an appropriate strategy to introduce unique alleles at many loci into the narrow gene pools of self-pollinated crops, although it is not incorporation in the strict sense.

Frey and colleagues (reviewed in Refs. 9,10) extensively evaluated methods for introducing exotic germplasm into self-pollinated crop gene pools using wild and cultivated oats as a model. Their results indicated that superior transgressive segregants were obtained following a few backcrosses of wild species germplasm into cultivated oats. Another successful approach was to hybridize diverse wild and cultivated oat germplasm to form genetically broad-based populations, and to enforce outcrossing among selected early-generation lines over many cycles of recurrent selection.

Even in cross-pollinated crops, development of semiexotic populations may be required to adapt exotic germplasm sufficiently so that it can contribute to the local gene pool. For example, "conversion" programs to adapt exotic tropical maize and sorghum germplasm rapidly to temperate environments have been implemented to overcome photoperiod-related maturity problems.[6,7] Mild selection only for the most important adaptation characters is critical to avoid massive loss of exotic alleles from semiexotic populations before they have a chance to recombine with alleles carried on chromosomes from adapted parents.

Some attempts to incorporate exotic germplasm into elite gene pools have failed. A major difficulty is the inability to identify superior sources among the overwhelming numbers of samples stored in germplasm banks for many crops. Use of randomly chosen exotic germplasm has been unsuccessful.[7,9] Information on the breeding value of exotic germplasm sources would be most helpful[6] but phenotypic evaluations in local target environments (to which exotic germplasm is unadapted) are often worthless because beneficial alleles may not be expressed or may be masked by other genes that confer maladaptation.[1,3,7]

One approach to selecting superior exotic germplasm is to evaluate first in environments to which exotic materials are adapted. If genotype-by-environment interactions are sufficiently strong, however, there may be no relationship (or worse, a negative relationship) between the performances of exotic germplasm in their regions of origin and as breeding parents with elite germplasm in local environments. This hypothesis is testable, however, and results in maize suggest that performance within environments to which exotics are adapted sufficiently predicts breeding value in temperate environments to be useful as an initial screen for source materials.[7]

CONCLUSION

Incorporation of exotic germplasm is the best means to enhance the genetic base of modern crops substantially, but it is neither easy nor rapid. New directions for incorporation and introgression involve using DNA markers to characterize the value of specific genomic regions in exotic germplasm sources.[3] DNA marker analysis combined with phenotypic evaluations of exotic-by-adapted progenies permits identification of favorable alleles in exotic germplasm regardless of the effects of unfavorable alleles. Substantial gains in agronomic performance from marker-assisted backcrossing of chromosomal segments from wild tomato and rice strains into their cultivated counterparts have been reported.[3] DNA marker-assisted selection could also be used to aid exotic ''conversion'' programs by ensuring introgression of only adaptation alleles from elite into exotic germplasms. However, DNA markers alone cannot solve the primary problem of how to identify superior sources of exotic germplasm. DNA marker-based strategies may complement, but cannot replace, long-term incorporation programs based on phenotypic selection.

ARTICLES OF FURTHER INTEREST

Breeding Clones, p. 174
Breeding Hybrids, p. 186
Breeding Plants and Heterosis, p. 189
Breeding Pure Line Cultivars, p. 196
Breeding Self-Pollinated Crops Through Marker-Assisted Selection, p. 202
Breeding: Choice of Parents, p. 215
Breeding: Genotype-by-Environment Interaction, p. 218
Breeding: Recurrent Selection and Gain from Selection, p. 232
Breeding: The Backcross Method, p. 237
Crop Domestication: Fate of Genetic Diversity, p. 333
Crop Domestication: Role of Unconscious Selection, p. 340
Crop Improvement: Broadening the Genetic Base for, p. 343
Genetic Resource Conservation of Seeds, p. 499
Farmer Selection and Conservation of Crop Varieties, p. 433
Genetic Resources of Medicinal and Aromatic Plants from Brazil, p. 502
Molecular Technologies and Their Role in Maintaining and Utilizing Genetic Resources, p. 757

REFERENCES

1. Simmonds, N.W. Introgression and incorporation. Strategies for the use of crop genetic resources. Biol. Rev. **1993**, *68*, 539–562.
2. Anonymous. *Genetic Vulnerability of Major Crops*; National Academy of Sciences: Washington, DC, 1972.
3. Tanksley, S.D.; McCouch, S.R. Seed banks and molecular maps: Unlocking genetic potential from the wild. Science **1997**, *277* (5329), 1063–1066.
4. Darwin, C. *On the Origin of Species*; John Murray: London, 1859.
5. Cox, T.S.; Wood, D. The Nature and Role of Crop Biodiversity. In *Agrobiodiversity: Characterization, Utilization, and Management*; Wood, D., Lenne, J.M., Eds.; CABI Publishing: Wallingford, UK, 1999; 35–57.
6. Maunder, A.B. Identification of Useful Germplasm for Practical Plant Breeding Programs. In *Plant Breeding in the 1990s*; Stalker, H.T., Murphy, J.P., Eds.; CAB International: Wallingford, U.K., 1992; 147–169.
7. Goodman, M.M.; Moreno, J.; Castillo, F.; Holley, R.N.; Carson, M.L. Using tropical maize germplasm for temperate breeding. Maydica **2000**, *45*, 221–234.
8. Brummer, E.C. Capturing heterosis in forage crop cultivar development. Crop Sci. **1999**, *39* (4), 943–954.
9. Holland, J.B. Oat Improvement. In *Crop Improvement for the 21st Century*; Kang, M.S., Ed.; Research Signpost: Trivandrum, India, 1997; 57–98.
10. Frey, K.J. Genetic Resources and Their Use in Oat Breeding. In *Proceedings of the Second International Oat Conference*; Lawes, D., Thomas, H., Eds.; Martinus Nijhoff Publishers: Dordrecht, 1986; 7–15.

Breeding: Mating Designs

Charles W. Stuber
North Carolina State University, Raleigh, North Carolina, U.S.A.

B

INTRODUCTION

Mating designs used by breeders and geneticists range from simple to complex, including some types of multipurpose designs. Selecting the kinds of mating techniques and arrangements depends upon: 1) predominate type of pollination (self or cross); 2) type of crossing used (artificial or natural); 3) type of pollen dissemination (wind or insect); 4) unique features, such as cytoplasmic or genetic sterility; 5) purpose of project (breeding or genetic); and 6) size of population required. A simple mating design, the top cross, is frequently used to evaluate a series of breeding lines, selections, or clones. Several designs of intermediate complexity are available for either generating synthetic populations or recombining selected entries in recurrent selection programs. Complex, multipurpose designs are available that may be used both to estimate genetic variances, or effects, and to generate families for use in either full-sib or half-sib selection schemes. All designs involve some type of hybrid crosses. Some may also include self-pollinations.

BASIC TYPES OF HYBRID CROSSES

Although the term ''hybrid'' is defined as the cross between two genetically dissimilar parents, it is most commonly used to denote the F_1 from a cross of two inbred lines. Crosses among clones, open-pollinated cultivars, or other genetically dissimilar populations are included in the F_1 hybrid population category. The four types of hybrid crosses most frequently used are the single (two-way) cross, three-way cross, backcross, and double cross. The single cross involves two parents:

$A \times B \rightarrow F_1 = \text{single cross}$

As the name implies, the three-way cross involves three different parents:

$(A \times B \rightarrow F_1) \times C \rightarrow \text{threeway cross}$

The backcross involves the mating of a single cross to one of its parents:

$(A \times B \rightarrow F_1) \times A \rightarrow \text{backcross}$

The double cross involves the mating of two different single crosses as follows:

$(A \times B \rightarrow F_1) \times (C \times D \rightarrow F_1) \rightarrow \text{double cross}$

TOPCROSS MATING SCHEME

The topcross mating scheme involves the crossing of a number of selections, lines, or clones to a common parent (tester), which may be a cultivar, an inbred line, a single cross, etc., where the tester is the same for each mating. In cross-pollinated crops, such as corn, this type of mating is commonly an inbred-cultivar cross.[1] It is frequently used for evaluation of general combining abilities for a group of lines, clones, or selections. The topcross method also has been used for initial evaluation of breeding potentials in exotic corn races.

If an isolated crossing block is available, natural hybridization can be used for topcrossing in many species. The tester is used as the male parent, and the females (lines or clones) to be tested are either male-sterile or self-incompatible or are emasculated before pollen shed. The ratio of number of male rows to female rows varies with species. Alternating rows may be required in crops such as sugar beet, whereas two male rows may be adequate to pollinate 10 to 12 female rows in castor. It may be necessary to delay planting dates if the male and female parents differ in days to flowering. If wind is a factor in pollen dissemination, tester rows should be planted perpendicular to prevailing winds. If isolation is not possible, the tester may be planted in paired rows with the materials to be evaluated. Hand pollinations are made between rows, and the tester may be used as either the male or the female parent.

BACKCROSS BREEDING METHOD

The backcross breeding method is a form of recurrent hybridization that is widely and effectively used to improve lines or cultivars that excel in most attributes but lack one or a few desirable characteristics.[2] Characteristics transferred by backcrossing in traditional breeding projects are usually highly heritable traits, such

Encyclopedia of Plant and Crop Science
DOI: 10.1081/E-EPCS 120010521

as resistance to disease or insect pests. The technique is used in both cross- and self-pollinated species but has been used infrequently in most perennial forage crops. Backcrossing is normally initiated by making an F_1 hybrid between the nonrecurrent (donor) parent and the recurrent parent. This F_1 is then backcrossed to the recurrent parent to produce the BC_1. A number of BC_1 individuals are grown, and those with the desired trait from the donor parent are selected and crossed to the recurrent parent to produce the BC_2. BC_2 individuals selected for the desired trait are crossed to the recurrent parent. This procedure, shown below, is continued until the desired number of backcrosses is completed:

$$(A \times B \rightarrow F_1) \times A \rightarrow BC_1 \times A \rightarrow BC_2 \times A \ldots \rightarrow Bc_n$$

For simply inherited traits, usually five to eight backcrosses are considered sufficient to effectively recover the genotype of the recurrent parent with the incorporated gene from the donor parent. However, with the advent of DNA-based marker technology, the number of required backcrosses can be reduced greatly because both the marker region of the donor gene and the marker complement of the recurrent parent can be monitored and selected during the backcrossing procedure.[3] This technique is being used widely in the transfer of transgenes to elite lines.

POLYCROSS MATING SCHEME

The polycross is a mating arrangement for interpollinating a group of cultivars or clones using natural hybridization in an isolated crossing block.[4] If an isolation block is not available, hand-crossing is required, and the entries must be planted to facilitate the required interpollinations. The polycross is used frequently for forage grasses and legumes, sweet potato, and sugarcane. It is often used for generating synthetic cultivars and may be used for recombining selected entries or families in recurrent selection programs. Progeny from each entry have a common parent in the polycross design. Thus, half-sib families are generated, and these are frequently used for evaluating general combining abilities.

Because the purpose of the polycross is to provide an equal opportunity for each entry to be crossed with every other entry, the field layout is the critical feature of the design. If the number of entries is 10 or less, the Latin square is often used, as this places every entry in every row and column of the layout. If more than 10 cultivars or clones are to be interpollinated, a completely randomized block design, with adequate replications, is normally preferred for the polycross. When the seed is harvested, equal quantities from each replication of each entry are bulked. Although the polycross requires a minimum of effort for intermating a group of entries, deviations from random mating may occur unless entries flower simultaneously. Planting of early entries can be delayed appropriately to avoid this problem in annual crops if days to flowering are known in advance.

DIALLEL AND PARTIAL-DIALLEL DESIGNS

The diallel is a mating design to cross three or more parents in all possible combinations. This design has been used extensively for developing breeding populations for recurrent selection.[5] It probably has been used more frequently than any other design in crop plants to estimate general and specific combining ability effects and variances. Genetic interpretations and analyses are provided in numerous papers.[6,7] The complete diallel includes all possible matings, including selfs and reciprocals, among a set of entries. These entries are usually homozygous inbred lines, pure lines, or clones, although an alternative design has been proposed and the analysis described for cultivar cross diallels.[8] For the complete diallel, there are p^2 crosses among p parents. With no selfs or reciprocals, $p(p-1)/2$ crosses are included in the design. In crops such as corn, the parental lines are planted in paired rows for each mating. This requires $p-1$ rows for each parent and a total of $p(p-1)$ rows. Reciprocals may or may not be made, as desired, with this arrangement. Unless the two rows in a pair flower simultaneously, reciprocal crossings may not be possible. Selfs can be made on individual plants in the crossing block or in a separate selfing block.

A major disadvantage of the diallel is the large number of crosses generated in the mating scheme. However, several types of partial diallels have been evaluated and reviewed.[9] It is suggested that an evaluation of five to seven crosses per parent in a partial diallel should give reliable estimates of general combining ability. Therefore, a large number of parents can be screened for their general combining abilities in a partial-diallel scheme, and the selected parents then might be evaluated in a full-diallel fashion.

MULTIPURPOSE MATING DESIGNS

Three widely used multipurpose mating designs—Design I, Design II, and Design III—have been proposed and described.[10,11] Design I is a nested type of mating design in which each member of a group of parents used as males is mated to a different group of parents used as females. A frequently used arrangement involves the crossing of each male to four female parents, thus generating four full-sib families nested within a half-sib family. No female parent is involved in more than one mating. Males can be interchanged with females in the design if the matings are more convenient. The design is used to estimate additive and dominance variances and can also be used to generate families for evaluation in full-sib or half-sib recurrent selection. Frequently, both functions are performed simultaneously.

Design II, a factorial design, is essentially a modification of Design I.[10,11] It is used to estimate genetic variances and to evaluate inbred lines for combining ability. Each member of a group of parents used as males is mated to each member of another group of parents designated as females. Frequently four males and four females are designated to form each mating group, which will generate a total of 16 full-sib families. This design is well-suited to multiflowered plants because each plant can be used repeatedly as both male and female. For single-eared corn, the design is suitable only if the parents are inbred lines from some specific population. Thus, the female is repeated by using several plants from each female line.

Design III involves backcrossing of F_2 plants to the two parent inbred lines from which the F_2 was derived.[10,11] The F_2 plants are used as male parents; the number of inbred plants crossed to each F_2 should be large enough to ensure sufficient seed for field evaluations. Design III is used infrequently and primarily to estimate the average dominance of genes. Estimates of genetic parameters tend to be more biased from epistasis in this design than in Designs I and II.

RECIPROCAL SELECTION DESIGNS

Design I is often used to generate families for recurrent selection programs and is particularly well-adapted to the reciprocal recurrent selection scheme.[12] When Design I is used for reciprocal recurrent selection, males are selfed as well as crossed to females from the opposite population. After evaluation of half-sib progenies from each male, selfed seed from each selected male is used to produce plants for the recombination cycle, which generates the population for the next selection cycle.

A variation of the above scheme[13] is called reciprocal full-sib selection and allows concurrent population improvement and development of single-cross hybrids. The design requires the ability to produce selfs and crosses on the same plant; thus, multiflowered (multieared for corn) plants are necessary. With the selfing feature, inbred lines are developed concurrently with the selection process, thus providing rapid development of new single crosses.

OTHER INTERMATING DESIGNS

Objectives for intermating include synthesis of populations, recombination of selected plants or families, and maintenance or increase of populations. Matings generally are designed to ensure thorough recombination and an equal genetic contribution of each unit in a random fashion. Although designs such as the polycross and diallel (or partial diallel) are frequently used for this purpose, other arrangements may be more suitable. One procedure involves chain (cyclic) crossing whereby each plant is used once as a male parent and once as a female parent, as follows:

$$A \rightarrow B \rightarrow C \rightarrow D \rightarrow E \rightarrow F \rightarrow G \rightarrow H$$

where the arrows indicate the direction of pollen transport. This procedure ensures that each entry is equally represented, if equal amounts of seed are bulked from each mating. However, the procedure provides only a limited amount of recombination. It is useful, however, as a sibbing technique for population or synthetic cultivar increases or for population maintenance if a large number of plants are sampled (>200).

Another intermating method, called the bulk entry method for population development, provides for thorough recombination when a large number of parents are composited.[14] In this method (designed specifically for corn), replicated plantings of individual entries are made in isolation, and plants are detasseled before pollen shed. Rows of bulked seed of all entries are planted between rows or ranges of the individual entries to provide pollen as in a topcross nursery. Ears of each entry are saved and bulked over replications at harvest to represent that entry in the following season. If desired, selection can be imposed and only ears from the best plants would be bulked. Other procedures for intermating a number of entries for population development may include combinations of diallel, partial-diallel, and chain-crossing designs.

CONCLUSION

Recombination and intermating in many self-pollinated species is difficult because of problems associated with artificial pollination and poor seed set. Genetic male sterility has been used successfully in crops such as barley and soybean to obtain natural crosses for synthesizing and intermating composite populations. Innovative researchers may use various combinations of the designs outlined above to accomplish the matings required in their breeding programs.

ARTICLES OF FURTHER INTEREST

Breeding: Choice of Parents, p. 215
Breeding: Genotype-by-Environment Interaction, p. 218

Breeding: Incorporation of Exotic Germplasm, p. 222
Breeding: Mating Designs, p. 225
Breeding: Participatory Approaches, p. 229
Breeding: Recurrent Selection and Gain from Selection, p. 232
Breeding Biennial Crops, p. 171
Breeding Clones, p. 174
Breeding for Durable Resistance, p. 179
Breeding for Nutritional Quality, p. 182
Breeding Hybrids, p. 186
Breeding Plants and Heterosis, p. 189
Breeding Plants with Transgenes, p. 193
Breeding Pure Line Cultivars, p. 196
Breeding Self-Pollinated Crops Through Marker-Assisted Selection, p. 202
Breeding Synthetic Cultivars, p. 205
Breeding Using Doubled Haploids, p. 207
Breeding Widely Adapted Cultivars: Examples from Maize, p. 211

REFERENCES

1. Jenkins, M.T.; Brunson, A.M. Methods of testing inbred lines of maize in crossbred populations. J. Am. Soc. Agron. **1932**, *24*, 523–530.
2. Briggs, F.N. The use of the backcross in crop improvement. Am. Nat. **1938**, *72*, 285–292.
3. Stuber, C.W. Breeding Multigenic Traits. In *DNA-Based Markers in Plants*; Phillips, R., Vasil, I., Eds.; Kluver Academic Publishers, 2001; 115–137.
4. Tysdal, H.M. History and development of the polycross technique in alfalfa breeding. Alfalfa Improv. Conf. Rep. **1948**, *11*, 36–39.
5. Jensen, N.F. Composite breeding methods and DSM system in cereals. Crop Sci. **1978**, *18*, 622–626.
6. Hayman, B.I. The theory and analysis of diallel crosses. Genetics **1954**, *39*, 789–809.
7. Griffing, B. Concept of general and specific combining ability in relation to diallel crossing systems. Aust. J. Biol. Sci. **1956**, *9*, 463–493.
8. Gardner, C.O.; Eberhart, S.A. Analysis and interpretation of the variety cross diallel and related populations. Biometrics **1966**, *22*, 439–452.
9. Dhillon, B.S. The application of partial-diallel crosses in plant breeding—A review. Crop Improv. **1975**, *2* (1–2), 1–8.
10. Comstock, R.E.; Robinson, H.F. The components of genetic variance in populations of biparental progenies and their use in estimating the average degree of dominance. Biometrics **1948**, *4*, 254–266.
11. Comstock, R.E.; Robinson, H.F. Estimation of Average Dominance of Genes. In *Heterosis*; Iowa State College Press: Ames, IA, 1952; 494–516.
12. Comstock, R.E.; Robinson, H.F.; Harvey, P.H. A breeding procedure designed to maximize use of both general and specific combining ability. Agron. J. **1949**, *41*, 360–367.
13. Hallauer, A.R.; Eberhart, S.A. Reciprocal full-sib selection. Crop Sci. **1970**, *10*, 315–316.
14. Eberhart, S.A.; Harrison, M.N.; Ogada, F. A comprehensive breeding system. Zuchter **1967**, *37*, 169–174.

Breeding: Participatory Approaches

J. R. Witcombe
University of Wales, Bangor, Gwynedd, U.K.

B

INTRODUCTION

An efficient plant breeding system—the breeding of new varieties and their testing, release, and popularization—should result in farmers growing diverse, recently released varieties. Unfortunately, this is the case in few, if any, developing countries. One important reason may be the inadequate participation of farmers in the process, so participatory plant breeding (PPB) methods and participatory varietal selection (PVS) have been advocated in both marginal and favorable environments. PPB actively involves farmers at an early stage, from the setting of goals to selecting among early generation materials, whereas PVS is restricted to selecting among finished varieties that could simply be the products of conventional breeding.

Participatory methods can be used to increase the efficiency of formal breeding programs in producing and popularizing varieties appropriate for resource-poor farmers or to empower farmers, and promote development in farmers' communities.

PARTICIPATORY VARIETAL SELECTION

PVS has met with outstanding success among low-resource farmers for many crops in marginal areas and countries.[1–3] Farmers identify varieties of greater utility than the ones they are growing because they have higher grain yield, earlier maturity, higher fodder yield, improved grain quality, or other favored characteristics. Usually suitable varieties already exist for marginal areas that are better than those currently grown, but farmers have simply never had the opportunity to try them.[3]

PVS in marginal areas overcomes the difficulties of multilocational trials that often fail to represent well the environments in farmers' fields. Genetic improvement should be focused on traits of economic importance, but formal trial systems place an undue emphasis on grain yield, do not measure all economically important traits, and rarely employ a system of trade-offs between the traits that are measured.[4] In PVS, farmers grow a wide choice of varieties and evaluate many traits that they trade off, for example, by accepting lower grain yields for higher quality or earlier maturity. PVS is a simpler, more direct approach than the alternative of using a selection index in multilocational trials. It also accounts for differences in preferences among socio-economic groups and localities and accounts for temporal changes in farmers' preferences with, for example, trends in market price and fodder availability.

PVS is a powerful method of assessing quality traits that are difficult or expensive to evaluate in conventional trials, e.g., milling percentage based on large seed quantities, cooking, and maintaining quality, taste, and market price.

PVS is now broadly accepted and well documented (reviewed in Ref. 5), and several international agricultural research centers have substantial networks for PVS, e.g., CIMMYT for maize[6] and WARDA for rice.[7]

PARTICIPATORY PLANT BREEDING

PVS can be efficiently followed by PPB since farmer-preferred cultivars are the ideal parents for PPB programs. PPB is less well-documented and well-accepted than PVS and, with a few exceptions, results from PPB programs are only now emerging, e.g., in rice,[8] maize,[9] and cassava.[10]

PPB can be consultative and collaborative: Farmers are consulted to set goals and choose parents, for example, and they collaborate by growing and selecting breeding materials in their own fields. The choice of consultative or collaborative methods will depend on the crop and the available resources.

Participation inevitably decentralizes the breeding program to farmers' fields, so the benefits of participation and decentralization are confounded. Selection in the target environment should result in faster genetic progress.[11] There is also a gain in exploiting specific adaptation to a target environment, although this advantage may be outweighed by reduced economies of scale if more specifically adapted varieties have smaller areas of adoption.

FARMERS AND GOAL SETTING

Much conventional breeding produces widely adapted varieties for many farmers over a wide area, and farmers

Encyclopedia of Plant and Crop Science
DOI: 10.1081/E-EPCS 120010457

participate little in setting breeding goals. Participatory goal setting helps in breeding varieties specifically adapted to well-defined physical or socioeconomic environments. It can be combined with PVS to help identify farmers' preferences for traits in diverse genetic material. In many cases, early maturity—perhaps even earlier than that of existing landraces or cultivars—is found to be as important as yield. Farmers sometimes set goals for quality traits not previously envisioned by breeders, such as pericarp color in rice.[8]

Participatory breeding also identifies traits that breeders had not considered important or of which they were previously unaware. Examples are appetite delay in rice (farmers want varieties that satisfy appetite for longer) and strict requirements for ease of threshing in areas where threshing is done manually.

INVOLVING FARMERS SHOULD CHANGE BREEDING METHODS

Participation requires different breeding methods if resources are to be used efficiently. Farmers can easily and cheaply grow large populations, but growing complex nurseries of many entries is more difficult and can be done only with training or assistance from scientists. Hence, in outbreeding crops, selection can be restricted to a single composite, while in inbreeding crops, fewer crosses can reduce the number of entries. With fewer crosses, the population size of each cross can be much larger than is common in conventional plant breeding. Although contrary to common practice, using few crosses with large populations is supported strongly by theory, and choice of crosses is perhaps the most critical factor for success.[12,13] PVS can identify superior parents, making it easier to carefully choose a few crosses, or varieties or landraces already popular with farmers can be used as parents.

Using only a few crosses in one inbreeding crop, rice, has been effective. From one cross, two rice varieties have been released from a PPB program in eastern India[14] and from the same cross, farmers in Nepal are accepting varieties for different ecosystems in both early-season and main-season rice. A few-crosses strategy is effective because PPB is noncompetitive and simply aims to breed a better variety than the one farmers are growing. However, this is not sufficient in commercial breeding, because breeders have to produce a variety that is also superior to those of their competitors. Under these cirumstances, more crosses may be appropriate.[13]

Bulk-population methods have been recommended as more suited to participatory approaches in inbreeding crops.[15] This is supported by theory, despite the reliance of many breeding programs on pedigree breeding.[a] Collaborative plant breeding using bulk breeding[b] has been very effective. From heterogeneous bulks of populations of nearly homozygous lines, farmers have selected varieties that perform well. Another method tests pure lines[c] derived from bulk populations. Here consultative participation—joint evaluation by farmers and breeders in the field—has been effective for selecting the best pure lines. Farmers subsequently evaluate the selected pure lines for grain quality traits.

In outbreeding crops, simple approaches such as mass selection have been effective in PPB for maize in both eastern and western India.[9] It has proved easier to carry out selection on researcher-managed populations, because avoiding unwanted cross-pollination in crops grown in farmers' fields is more difficult.

PVS FOLLOWS PPB

PVS follows seamlessly from PPB. As soon as potential varieties have been produced by PPB, farmers test them using PVS. This is an important advantage, because the results of PPB reach farmers more quickly than the results of conventional breeding, where varieties typically are tested with farmers after a long delay for on-station testing and multiplication.[1] This is an important advantage of PPB, since delay in obtaining benefits reduces the rate of return on the investment in plant breeding.[16]

PPB AND CONVENTIONAL BREEDING ARE COMPLEMENTARY STRATEGIES

Conventional breeding and PPB are complementary. Many PVS and PPB programs use the widely adapted products of conventional breeding either as varieties per se or as parents.[17] PPB uses more specialized parents than conventional breeding and produces different germplasm, so conventional breeding can benefit from this genetic resource. Screening large germplasm collections or strategic breeding that uses wide crosses between cultivated and wild species is beyond the scope of PPB. Partnerships between conventional and participatory

[a]Where individual plants or lines are selected on the basis of their desirability, judged individually and by their pedigree record (ancestry).

[b]Growing the progeny of a cross as a single population (bulk populations). Selection in the bulk can be done by pooling the seed from superior plants.

[c]A pure line results from inbreeding until there is effectively no genetic variation within the line.

approaches are needed, and institutional structures will determine exactly how this is done.

ARTICLES OF FURTHER INTEREST

Breeding Pure Line Cultivars, p. 196
Breeding: Choice of Parents, p. 215
Breeding: Genotype-by-Environment Interaction, p. 218
Farmer Selection and Conservation of Crop Varieties, p. 433
Genetic Resources of Medicinal and Aromatic Plants from Brazil, p. 502

REFERENCES

1. Witcombe, J.R.; Joshi, A.; Joshi, K.D.; Sthapit, B.R. Farmer participatory crop improvement. I. Varietal selection and breeding methods and their impact on biodiversity. Exp. Agric. **1996**, *32*, 445–460.
2. Sperling, L.; Loevinsohn, M.E.; Ntabomvura, B. Rethinking the farmer's role in plant breeding: Local bean experts and on-station selection in Rwanda. Exp. Agric. **1993**, *29*, 509–519.
3. Joshi, A.; Witcombe, J.R. Farmer participatory crop improvement. II. Participatory varietal selection, a case study in India. Exp. Agric. **1996**, *32*, 461–477.
4. Witcombe, J.R.; Virk, D.S.; Raj, A.G.B. Resource Allocation and Efficiency of the Varietal Testing System. In *Seeds of Choice; Making the Most of New Varieties for Small Farmers*; Witcombe, J.R., Virk, D.S., Farrington, J., Eds.; Oxford and IBH Publishing, New Delhi and Intermediate Technology Publishing: London, 1998; 135–142.
5. Weltzein, E.; Smith, M.; Meitzner, L.S.; Sperling, L. *Technical and Institutional Issues in Participatory Plant Breeding from the Perspective of Formal Plant Breeding. A Global Analysis of Issues, Results and Current Experiences*; PPM Monograph No. 1. CGIAR Systemwide Programme on Participatory Research and Gender Analysis for Technology Development and Institutional Innovation Centro Internacional de Agricultura Tropical (CIAT), 2003; 1–229.
6. Bänziger, M.; De Meyer, J. Collaborative Maize Variety Development for Stress-Prone Environments in Southern Africa. In *Farmers, Scientists and Plant Breeding: Integrating Knowledge and Practice*; Cleveland, D.A., Soleri, D., Eds.; CABI Publishing: Wallingford, UK, 2001; 269–296.
7. WARDA (West Africa Rice Development Association). *Participatory Varietal Selection. The Spark that Lit a Flame*; WARDA: Bouaké, Côte d'Ivoire, 1999.
8. Sthapit, B.R.; Joshi, K.D.; Witcombe, J.R. Farmer participatory crop improvement. III. Participatory plant breeding, a case study for rice in Nepal. Exp. Agric. **1996**, *32*, 479–496.
9. Goyal, S.N.; Joshi, A.; Witcombe, J.R. In *Participatory Crop Improvement in Maize in Gujarat, India*, Proceedings of an International Symposium on Participatory Plant Breeding and Participatory Plant Genetic Resource Enhancement: An Exchange of Experience from South and South East Asia, Pokhara, Nepal, May 1–5, 2000; CGIAR System Wide Programme on Participatory Research and Gender Analysis for Technology Development and Institutional Innovation Centro Internacional de Agricultura Tropical (CIAT), 2001; 237–242.
10. Fukuda, W.M.G.; Saad, N. *Participatory Research in Cassava Breeding with Farmers in Northeastern Brazil*; EMBRAPA/CNPMF, Brazil. PRGA Program, Colombia, 2001; http://www.prgaprogram.org (accessed November 2001).
11. Atlin, G.N.; Frey, K.J. Predicting the relative effectiveness of direct versus indirect selection for oat yields in three types of stress environments. Euphytica **1989**, *44*, 137–142.
12. Allard, R.W. *Principles of Plant Breeding,* 2nd Ed.; John Wiley & Sons, Inc.: Toronto, 1999; 91.
13. Witcombe, J.R.; Virk, D.S. Number of crosses and population size for participatory and classical plant breeding. Euphytica **2001**, *122*, 589–597.
14. Kumar, R.; Singh, D.N.; Prasad, S.C.; Gangwar, J.S.; Virk, D.S.; Witcombe, J.R. In *Participatory Plant Breeding in Rice in Eastern India*, Proceedings of an International Symposium on Participatory Plant Breeding and Participatory Plant Genetic Resource Enhancement: An Exchange of Experience from South and South East Asia, Pokhara, Nepal, May 1-5, 2000; CGIAR System Wide Programme on Participatory Research and Gender Analysis for Technology Development and Institutional Innovation Centro Internacional de Agricultura Tropical (CIAT), 2001; 233–236.
15. Witcombe, J.R.; Subedi, M.; Joshi, K.D. In *Towards a Practical Participatory Plant Breeding Strategy in Predominantly Self-Pollinated Crops*, Proceedings of an International Symposium on Participatory Plant Breeding and Participatory Plant Genetic Enhancement: An Exchange of Experience from South and South East Asia, Pokhara, Nepal, May 1–5, 2000; CGIAR System Wide Programme on Participatory Research and Gender Analysis for Technology Development and Institutional Innovation Centro Internacional de Agricultura Tropical (CIAT), 2001; 243–248.
16. Pandey, S.; Rajatasereekul, S. Economics of plant breeding: The value of shorter breeding cycles for rice in Northeast Thailand. Field Crops Res. **1999**, *64*, 187–197.
17. Witcombe, J.R. The Impact of Decentralized and Participatory Plant Breeding on the Genetic Base of Crops. In *Broadening the Genetic Base of Crop Production*; Cooper, H.D., Spillane, C., Hodgkin, T., Eds.; CABI Publishing: Wallingford, UK, 2001; 407–417.

Breeding: Recurrent Selection and Gain from Selection

James G. Coors
University of Wisconsin, Madison, Wisconsin, U.S.A.

INTRODUCTION

There are many ways to conduct selection in plants, and plant breeders usually develop specific breeding procedures that best capture the genetic potential of the germplasm at hand. Breeding systems must accommodate factors such as mating system and progeny size that are unique to particular species. Nonetheless, there are overall patterns that characterize a broad array of breeding techniques. Recurrent selection is one of the most common overlying patterns.

RECURRENT SELECTION

Recurrent selection is a cyclic process of population improvement (Fig. 1). Cycles consist of distinct steps involving evaluation, selection, and recombination to improve breeding populations continuously. Recurrent selection is designed to increase the frequency of favorable alleles for a specific trait in a breeding population over cycles of selection. Breeding populations are sources of potential cultivars such as synthetics, inbreds, or hybrids that may be developed using pedigree or some other selection technique. As long as the frequency of desirable alleles increases in the breeding population, the chance of extracting or developing desirable cultivars from the breeding population will also increase.

Recurrent selection is a long-term endeavor. Gene frequencies change slowly over cycles of selection, and commercial breeders with near-term objectives do not expend much effort on upgrading breeding populations from which adapted cultivars are developed. With the exception of maize (*Zea mays* L.), classical recurrent selection is rarely used for crop species where inbred or hybrid cultivars are common. Recurrent selection techniques are most often used for open-pollinated species where improved breeding populations may be used directly as cultivars. In maize, recurrent selection techniques are highly refined and many have been used extensively for research purposes, as well as to develop widely-used breeding populations, such as the Stiff Stalk Synthetic developed by Iowa State University.[4]

Methods of Recurrent Selection

Methods of recurrent selection vary according to whether breeding populations are being improved for per se performance (intrapopulation improvement) or for improved combining ability with complementary breeding populations or tester lines (interpopulation improvement). Methods are further classified by type of progeny evaluated (''evaluation'' units) and the type of progeny selected and recombined (''recombination'' units).

Intrapopulation recurrent selection

Several intrapopulation recurrent selection methods are used to improve performance of breeding populations (Table 1). They differ based on the nature of the evaluation and recombination units. A simple mass selection program for an open-pollinated species might involve the evaluation of a number of plants for a specific trait, e.g., seed yield, where the seed obtained from the most desirable plants is harvested, composited, and replanted the next season to start the next cycle. In this case, the plants are the evaluation units, and the harvested seeds from the selected plants are the recombination units. Since seed yield is measured after seeds are formed via random pollination from nearby plants, only the maternal contribution (seed-yielding ability of a plant) is subject to selection. This is often referred to as uniparental mass selection. For some traits, e.g., seedling vigor, evaluation and selection can occur prior to fertilization, and both maternal and paternal contributions can be controlled by intermating only selected plants during the recombination phase. In such cases, the recombination units are the individual selected plants, and the selection system is termed biparental mass selection. As might be expected, expected gains from biparental selection are twice those from uniparental selection.

When traits are difficult to measure on individual plants, half-sib, full-sib, or inbred (S_1, S_2, etc.) families can be used as evaluation units. Families can be planted in multiple environments, and traits can be measured on several plants in each family. There are several options for recombination. For example, if half-sib

Encyclopedia of Plant and Crop Science
DOI: 10.1081/E-EPCS 120010522

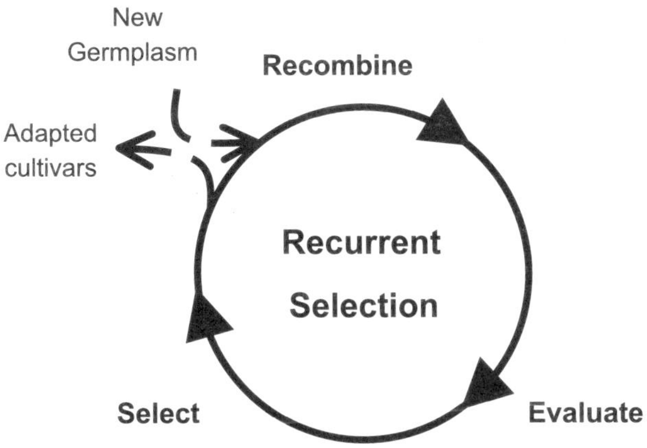

Fig. 1 Recurrent selection is a cyclic process involving evaluation, selection, and recombination to improve continuously breeding populations that are used as sources for cultivar development.

families are evaluated, recombination can be via: 1) compositing of open-pollinated seed harvested from the selected half-sib families, 2) compositing of remnant seed from the selected half-sib families, or 3) intermating the parents that gave rise to the selected half-sib families. On a cycle basis, method 3) is four times as effective as method 1) and twice as effective as method 2), but method 3) requires an additional season to plant and intermate the selected parental plants. The other family-based methods of intrapopulation recurrent selection have similar recombination options. Intrapopulation selection methods can easily be combined when appropriate.

Interpopulation recurrent selection

When breeders want to exploit hybrid vigor, they may choose to improve two breeding populations (Pop. A and Pop. B) simultaneously for combining ability, and the selection procedures are collectively referred to as reciprocal recurrent selection (Table 1). The evaluation units can be half-sib and full-sib families, but these families have a different structure than in intrapopulation recurrent selection. The mating of a single plant in Pop. A with a single plant from Pop. B would produce a full-sib family for reciprocal recurrent selection. A half-sib family might be created by pollinating a single plant from one population by pollen from a number of random plants from the other. In this example, there are two types of half-sib families for evaluation, those derived from individual maternal plants in Pop. A, and those derived from individual maternal plants in Pop. B.

For both full-sib and half-sib reciprocal recurrent selection, recombination occurs within each population in order to keep the breeding populations distinct from one another. Depending on the flowering habit and life cycle, there may be a number of different recombination options. One would be to recombine S_1 families derived by self-pollinating plants that were used to make the selected crosses. Or, if the parental plants can be vegetatively propagated and maintained for several seasons, the parent plants used to make the selected crosses can be directly intermated within each population to create seed for the improved breeding populations for the next cycle of selection.

Plant breeders often want to improve the combining ability of a single breeding population with a tester line, typically an elite inbred line used for hybrid seed

Table 1 Examples of typical intrapopulation and interpopulation recurrent selection programs

Method	Evaluation unit	Recombination unit
Intrapopulation improvement		
Mass, uniparental control	Individual plants	Open-pollinated seeds from selected plants
Mass, biparental control	Individual plants	Individual selected plants
Half-sib family, method 1	Half-sib families	Open-pollinated seed from selected half-sib families
Half-sib family, method 2	Half-sib families	Remnant seed from selected half-sib families
Half-sib family, method 3	Half-sib families	Common parents of selected families
Full-sib family	Full-sib families	Full-sib families
Selfed family	S_1 or S_2 families	S_1 or S_2 families
Interpopulation improvement		
Half-sib reciprocal recurrent	Half-sib families	S_1 or S_2 families
Full-sib reciprocal recurrent	Full-sib families	S_1 or S_2 families
Testcross	Testcross families	S_1 or S_2 families

(Adapted from Refs. 1 and 4.)

production. In this case, a breeder could develop inbred (S_1, S_2, etc.) families that would all be crossed to the tester line. The testcrosses would be evaluated, and remnant seed from the selected inbreds from the breeding population would be used to recombine selected families for the next cycle.

GAINS FROM SELECTION

Expected Gains from Recurrent Selection

Expected gains can be calculated based on the form of the evaluation units, the form of recombination units, selection intensity, genetic and phenotypic variance of the trait, and number of seasons required for each cycle.[6,7] The general form of the expected gains equation is

$$\Delta G = \frac{S}{\sigma^2_{Ph}} \sum_{i=1} COV(p_i, g_i) \frac{\partial \overline{g_i}}{\partial p_i}$$

where S is the selection differential, σ^2_{Ph} is the phenotypic variance, $COV(p_i, g_i)$ is the covariance of allele frequency with genotypic value at locus i, and

$$\frac{\partial \overline{g_i}}{\partial p_i}$$

is the rate of change in mean genotypic value in the population as the allele frequency changes for locus i.

The general expected gains equation is used to develop specific formulae, such as those provide in Table 2, that compare the relative potential of various selection schemes. While the expected gains presented in Table 2 are based on gain per cycle of selection, there are several other ways to express gain, including gain per unit time (e.g., gain per year) and gain per unit input (e.g., gain per dollar invested). From a plant breeder's perspective, gain per unit time and gain per unit input are often the most useful indicators of selection efficiency. These expected gains can be easily derived from gain per cycle once a particular breeding method is devised.

Realized Gains from Recurrent Selection

Measurement of realized gains from recurrent selection requires extensive evaluation over several cycles of selection. Response is highly variable from cycle to cycle due to environmental variation, sampling errors, and genetic drift. One of the most common ways to calculate realized gain is to regress the mean performance of the breeding population at each cycle on the cycle number. The linear regression coefficient then provides an estimate of gain per cycle. However, there are complicating issues that warrant other estimation techniques in particular situations.[8]

Gains from selection are often less than expected when the effective population size is small. Small population size leads to unanticipated variation in response due to genetic drift, and genetic variation in the breeding population may be unnecessarily reduced over cycles if the number of selected plants or families recombined each cycle is small.

Table 2 Expected gains per cycle for several intrapopulation and interpopulation recurrent selection programs

Method	Expected gain	Generations per cycle
Intrapopulation improvement		
Mass selection, uniparental control	$(1/2)i\sigma^2_A/\sigma_P$	1
Mass selection, biparental control	$i\sigma^2_A/\sigma_P$	1
Half-sib selection, method 1	$(1/8)i_1\sigma^2_A/\sigma_{HS}$	1
Half-sib selection, method 2	$(1/4)i\sigma^2_A/\sigma_{HS}$	2
Half-sib selection, method 3	$(1/2)i\sigma^2_A/\sigma_{HS}$	3
Full-sib selection	$(1/2)i\sigma^2_A/\sigma_{FS}$	3
S_1 selection	$i(\sigma^2_A + C_1)/\sigma_{S_1}$	3
S_2 selection	$(3/2)i(\sigma^2_A + C_2)/\sigma_{S_2}$	4
Interpopulation improvement		
Half-sib reciprocal recurrent	$(1/4)i\sigma^2_{A12}/\sigma_{HS12} + (1/4)i\sigma^2_{A21}/\sigma_{HS21}$	3
Full-sib reciprocal recurrent	$(1/2)i\sigma^2_{A(12)}/\sigma_{FS(12)}$	3

i = selection intensity in standardized units; σ^2_A = additive genetic variance; σ^2_{A12} and σ^2_{A21} are homologs of σ^2_A and refer to the additive variance expressed among half sibs developed from each population used in reciprocal recurrent selection; σ_P = phenotypic standard deviation of single-plant measurements; σ_{HS}, σ_{FS}, σ_{S_1}, and σ_{S_2} are the standard deviations of family means; factors C_1 and C_2 for S_1 and S_2 selection depend on the degree of dominance in the population.

(Adapted from Ref. 4.)

The longest-running structured recurrent selection program known in plants is that for oil and protein in maize started at the University of Illinois in 1896 by C.G. Hopkins.[9] More than 100 selection cycles have been completed for high oil and protein concentration. Mass selection was used in the initial cycles, but most of the selection has been by various half-sib family procedures. Even with 100 cycles completed, selection for high oil and protein continues without conspicuous loss of response. Unfortunately, there are few other such long-term recurrent selection studies in plants. Two mass selection programs in maize equal or exceed 30 cycles, one for ear length[10,11] and the other for prolificacy.[12] The longest-running reciprocal recurrent selection study is that at Iowa State University involving 15 cycles with the Iowa Stiff Stalk Synthetic and Iowa Corn Borer Synthetic #1.[13] Most other selection studies in plants have been discontinued after fewer than 10 cycles. Some of the most informative studies regarding the nature of genetic variation and selection response are those from non-plant species with short generation times. For example, one of the longest known selection studies involves a 20,000-generation experiment with *Escherichia coli*.[14]

Much of the research on recurrent selection in plants involves selection for grain yield in maize. In a survey of 133 selection studies in maize, the mean of all types of intrapopulation improvement averaged approximately 2 to 3% gain per cycle or 50 to 90 kg per ha year.[15] The average gain for interpopulation improvement was higher: approximately 3 to 4% per cycle or 80 to 110 kg per ha year.

Intrapopulation recurrent selection methods based solely on additive genetic variance have been successful, and comparisons among selection methods for relative rate of improvement agree with theoretical expectations in most instances. Interpopulation improvement has been especially productive. Breeders have been able to use dominance efficiently to increase annual response rates in those species with heterotic potential. The fundamental message from recurrent selection, which also applies to nearly all breeding systems, is that rapid progression from evaluation to selection and finally recombination is essential for long-term germplasm improvement.

WHY IS RECURRENT SELECTION IMPORTANT?

The are two essential goals for plant breeders: 1) creating commercially viable varieties for the near-term and 2) increasing the genetic potential of germplasm so that breeders will continue to make gains in the future. At times, these goals are in conflict. While most individual breeders' careers stand or fall on the production of commercial varieties, the long-term success of the seed industry depends on continued improvement of productive breeding populations from which commercial varieties will be developed by future breeders. The contrast between cultivar development and recurrent selection reflects, in part, the tradeoff plant breeders must make between short- and long-term goals.

ARTICLES OF FURTHER INTEREST

Breeding Plants with Transgenes, p. 193
Breeding Pure Line Cultivars, p. 196
Breeding Self-Pollinated Crops Through Marker-Assisted Selection, p. 202
Breeding: Choice of Parents, p. 215
Breeding: Incorporation of Exotic Germplasm, p. 222
Breeding: Mating Designs, p. 225
Breeding: The Backcross Method, p. 237
Crop Domestication: Fate of Genetic Diversity, p. 333
Crop Domestication: Founder Crops, p. 336
Crop Domestication: Role of Unconscious Selection, p. 340
Crop Improvement: Broadening the Genetic Base for, p. 343
Evolution of Plant Genome Microstructure, p. 421
Genetic Resource Conservation of Seeds, p. 499
Genetic Resources of Medicinal and Aromatic Plants from Brazil, p. 502
Germplasm Acquisition, p. 537
Germplasm Collections: Regeneration in Maintenance, p. 541
Germplasm Maintenance, p. 544
Germplasm: International and National Centers, p. 531
Molecular Evolution, p. 748
Molecular Technologies and Their Role Maintaining and Utilizing, p. 757
Mutational Processes, p. 760
Population Genetics, p. 1042
Quantitative Trait Location Analyses of Domestication Syndrome and Domestication Process, p. 1069

REFERENCES

1. Bernardo, R. *Breeding for Quantitative Traits in Plants*; Stemma Press: Woodbury, MN, 2002.
2. Briggs, F.N.; Knowles, P.F. *Introduction to Plant Breeding*; Reinhold Publishing Co.: New York, NY, 1967.
3. Fehr, W.R. *Principles of Cultivar Development. Vol. 1:*

Theory and Technique; Macmillan Publishing Co.: New York, NY, 1987.
4. Hallauer, A.R.; Miranda, J.B.FO. *Quantitative Genetics in Maize Breeding*; Iowa State University Press: Ames, IA, 1988.
5. Wricke, G.W.; Weber, E. *Quantitative Genetics and Selection in Plant Breeding*; Walter de Gruyter: New York, NY, 1986.
6. Empig, L.T.; Gardner, C.O.; Compton, W.A. *Theoretical Gains for Different Population Improvement Procedures*; Nebraska Agricultural Extension Bulletin No. M26 (revised), Univ. of Nebraska: Lincoln, NE, 1981.
7. Rowe, D.E.; Hill, R.R., Jr. *Theoretical Improvement of Autotetraploid Crops: Interpopulation and Intra Population Selection*, USDA Tech. Bull. No. 1689; Agricultural Research Service: Washington, DC, 1984.
8. Falconer, D.S.; Mackay; Trudy, F.C. *Introduction to Quantitative Genetics*; Longman: Essex, England, 1996.
9. Dudley, J.W. 100 Generations of Selection for Oil and Protein in Corn: People, Progress, and Promise. In *Plant Breeding Reviews Vol. 24, Part 1, Long Term Selection: Maize*; Coors, J.G., Dentine, M.R., Dudley, J.W., Lamkey, K.R., Eds.; Wiley, 2003. *in press*.
10. Lopez-Reynoso; de Jesus, J.; Hallauer, A.R. Twenty-seven cycles of divergent mass selection for ear length in maize. Crop Sci. **1998**, *38*, 1099–1107.
11. Hallauer, A.R.; Ross, A.J.; Lee, M. Long-Term Divergent Selection for Ear Length in Corn. In *Plant Breeding Reviews Vol. 24, Part 1, Long Term Selection: Maize*; Coors, J.G., Dentine, M.R., Dudley, J.W., Lamkey, K.R., Eds.; Wiley, 2003. *in press*.
12. de Leon, N.; Coors, J.G. Twenty-four cycles of mass selection for prolificacy in the Golden Glow maize population. Crop Sci. **2002**, *42*, 325–333.
13. Labate, J.A.; Lamkey, K.R.; Lee, M.; Woodman, W.L. Temporal changes in allele frequencies in two reciprocally selected maize populations. Theor. Appl. Genet. **1999**, *99*, 1166–1178.
14. Lenski, R. Phenotypic and Genomic Evolution During a 20,000-Generation Experiment with *E. coli*. In *Plant Breeding Reviews Vol. 24, Part 2, Long Term Selection: Crops, Animals, and Bacteria*; Coors, J.G., Dentine, M.R., Dudley, J.W., Lamkey, K.R., Eds.; Wiley, 2003. *in press*.
15. Coors, J.G. Selection Methodologies and Heterosis. In *Genetics and Exploitation of Heterosis in Crops*; Coors, J.G., Pandey, S., Eds.; CIMMYT, ASA, CSSA: Madison, WI, 1999; 225–245.

Breeding: The Backcross Method

William F. Tracy
University of Wisconsin, Madison, Wisconsin, U.S.A.

INTRODUCTION

Backcross breeding has been an important tool for plant breeders for more than 80 years, and with the relatively recent incorporation of transformation as a plant breeding tool the importance of the backcross method has increased. The goal of most backcrossing programs is to improve a particular strain (recurrent parent) for a specific characteristic, usually a single gene, obtained from a donor parent. In most backcross programs the objective is to recover the recurrent parent essentially unchanged except for the introgression of the new characteristic. Backcrossing allows the plant breeder more precise control of allele frequencies than other traditional plant breeding methods. The addition of molecular markers to backcrossing programs allows even greater precision and more rapid incorporation of alleles into cultivars.

GENETIC BASIS OF BACKCROSSING

In a typical backcross program, with the objective of introgressing allele A^2 into the recurrent parent, an F_1 (A^1A^2) is made by crossing the recurrent parent (A^1A^1) and the donor parent (A^2A^2). The following growing season, the F_1 (A^1A^2) is backcrossed to the recurrent parent (A^1A^1). The progeny resulting from this cross is the backcross 1 (BC_1) generation. The backcrossing is repeated for a number of cycles. The number of cycles depends upon the objectives of the breeding program and other factors.[1] In the progeny of the last backcross the donated allele will be heterozygous. For recessive traits one generation of self-pollination followed by progeny testing is needed to isolate and identify plants homozygous for the donated allele. For dominant traits two cycles of self-pollination followed by progeny testing are required to identify homozygous plants. If a codominant molecular marker is tightly linked to the dominant gene of interest it is possible to eliminate the second self-pollination and progeny test.

50% of the alleles in the F_1 individuals are from the recurrent parent and 50% are from the donor parent. In each successive backcross, average percentage of alleles from the recurrent parent increases by 50% (Table 1). Thus the average percent of recurrent alleles in the BC_1 is 75% and in the BC_2 87.5%. The equation to calculate average recovery of recurrent alleles is $1-(1/2)^{n+1}$, where n = the number of backcross cycles.[1] If the recurrent parent is an inbred cultivar, then this equation also gives the average homozygosity for each backcross. Recovery is given as an average because in each backcross there is a range among plants for the number of recurrent alleles. The individual plants in each backcross are a sample of the range of possible allele frequencies.

Selection and genetic linkage alter the average percent recovery of recurrent alleles.[1] If plants backcrossed to the recurrent parent are chosen because they are phenotypically similar to the recurrent parent, then the recovery of recurrent alleles will increase. This is especially true in the early cycles of backcrossing when there is more phenotypic variation within backcross families. The use of genetic markers to determine the most similar plants in each backcross cycle can greatly increase the rate of recovery.[2,3]

Genetic linkage reduces the recovery of those recurrent alleles linked to the locus that is being substituted. As the genetic distance between the desired donated allele, A^2, and an undesirable allele, B^2, decreases the recovery of the desired haplotype, A^2B^1, becomes more difficult. The probability of eliminating allele B^2 is $1-(1-c)^n$, where c is the crossover rate and n is the number of backcrosses.[4] If A^2 and B^2 are unlinked ($c = 0.5$), the probability that B^2 will be eliminated after five backcrosses is 0.969. On the other hand, if A^2 and B^2 are tightly linked, i.e., $c = 0.01$, the probability that B^2 will be eliminated after five backcrosses is 0.049. Backcrossing, however, does increase the likelihood of identifying favorable recombinants relative to selfing.[4] Molecular markers can be used to identify desirable rare recombinants. Markers flanking the location of the donated allele can be used to set precisely the size of the donated section of DNA.[3,5]

Another factor that slows the recovery of recurrent alleles is the effect of modifying genes. Backcrossing is most commonly used with simply inherited traits, but modifying genes or background effects affect the phenotype of many simply inherited traits. For example, the *shrunken2* allele, when backcrossed into corn (*Zea mays* L.) inbreds results in high sucrose concentration in the endosperm, which is useful for commercial sweet corn.[6] However, in many genetic backgrounds the high sugar

Encyclopedia of Plant and Crop Science
DOI: 10.1081/E-EPCS 120010532

Table 1 Average recovery of alleles of the recurrent parent during backcrossing

Generation	% Alleles from recurrent parent
F_1	50
BC_1	75
BC_2	87.500
BC_3	93.750
BC_4	96.875
BC_5	98.4375
BC_6	99.21875

phenotype can result in very poor seed germination.[7] Genes that improve germination are often present in the donor parent and in some cases must be introgressed into the recurrent parent along with the *shrunken2* gene.[8] Each modifying gene will be linked to undesirable alleles from the donor parent and therefore reduce the average recovery of recurrent alleles. Usually the genetic locations of modifying genes are unknown and their effects are small. Thus they are less amenable to marker-assisted backcrossing than major genes. In cases when it is important for modifying genes to be introgressed, the breeder might opt for fewer cycles of backcrossing, followed by self-pollination and selection for types that have the best combination of modifiers and therefore best express the donated trait.

PROGENY TESTING

During backcrossing the presence of the donated allele in the backcross generations must be determined. When derived from two inbred parents the F_1 generation will be heterozygous (A^1A^2). In the BC_1 generation 50% of the plants will be heterozygous at the gene of interest (A^1A^2), and 50% of the plants will be homozygous for the allele from the recurrent parent (A^1A^1). If the donated allele is dominant and expressed prior to pollination, then the heterozygous plants can be identified and used for pollination and no progeny testing is needed. If the donor allele is recessive, then a progeny test must be done to determine which plants are heterozygous for the donated allele. Depending on the reproductive system, plants in the BC_1 generation can be self-pollinated or crossed to a tester to determine if the donated allele is present. For many species the testcross and backcross can be performed in the same generation by using the paired cross or paired progeny selection method.[9] In this method BC_1 plants are numbered, self-pollinated, and backcrossed to the recurrent parent. Progeny derived from the backcross is given the same number as progeny from the self-pollination. Following seed maturation the backcross and the selfed progenies are grown. The selfed progeny is evaluated for the trait of interest. The backcross progeny corresponding to those selfed progeny that have a copy of the donated allele (A^1A^2) are backcrossed again. Those backcross progeny that correspond to selfed progeny that do not have the gene of interest (A^1A^1) are discarded, as are all of the selfed progeny. The paired cross method allows a backcross to be performed each growing season, saving time but using more labor for pollination and space.

Progeny testing need not be done every cycle of backcrossing as long as the number of individuals in each backcross generation is large enough to ensure that heterozygous individuals exist in the population.[1]

PARENTS AND DIRECTION OF CROSS

Backcrossing is usually used to introgress a desirable trait into an otherwise high-performing cultivar. Since backcrossing requires a number of cycles it is important that the recurrent parent still be useful following the backcross program. If the recurrent parent is a marginal cultivar and it is likely that it will be quickly surpassed by newer cultivars even after the donated allele has been introgressed, then the time and dollars devoted to backcrossing will be wasted. Backcrossing a new allele into a well established, widely used cultivar has a high probability of resulting in a cultivar that will be successful. The donor parent should be the best adapted germplasm available that has the allele of interest. Wild or exotic germplasm will have more deleterious traits, and more cycles of backcrossing may be required to remove the deleterious traits.[1]

The recurrent parent should be used as the female in the initial cross with the donor parent. This will preserve the relationship between the nuclear and cytoplasmic genomes of the recurrent parent. In the following backcrosses the direction of the cross is unimportant because both the parents will have the same cytoplasm. The exception is when the goal is to transfer an entire nuclear genome into a different cytoplasm. The nuclear genome would be from the recurrent parent, and the donor parent contributes the cytoplasm. This is commonly done when developing cytoplasmic male sterile lines.[9]

BACKCROSSING OPEN-POLLINATED CULTIVARS

Backcrossing is simplest when the recurrent parent is homozygous and homogeneous, such as an inbred or pure line cultivar; however, it is possible to backcross traits into heterozygous, heterogeneous populations. In this case adequate sampling of the alleles in the population must be done during each cycle of backcrossing. Numerous different individuals from the recurrent parent must be crossed to different individuals from each backcross generation. The number of individuals needed varies depending on the level of variability in the recurrent parent. A minimum of 100 individuals from the recurrent parent should contribute to the next generation. However, on average, only 50% of the backcrosses will be with heterozygous plants, so to obtain 100 successful pollinations more than 200 would need to be done. Fewer plants may be used in early cycles but the number should increase in later cycles. Progeny testing will have to be done for each backcross pollination, and this can require significant space and time.

NUMBER OF CYCLES

The number of cycles of backcrossing depends on the importance of fully recovering the phenotype of the recurrent parent, the amount of selection imposed during backcrossing, the adaptation of the donor parent, linkage between desirable and undesirable genes,[1] and the availability of marker assisted selection systems. If the goal is to recover the recurrent parent essentially unchanged except for the donated allele, then a minimum of five backcrosses is needed. In many cases significant differences may persist beyond five or six backcrosses, and more backcrosses will be required. If the donor parent is unadapted or the donated allele is tightly linked to deleterious alleles, more than five backcrosses will be required. Selection, especially when based on molecular markers, decreases the number of backcrosses required.[2,5] Marker-assisted backcrossing can reduce the number of backcrosses needed by 50% or more. Time is saved but cost per backcross cycle increases due to the cost of collecting marker data.

Many breeders will develop new lines from populations derived from one, two, or three cycles of backcrossing. The advantage of limited backcrossing is the possibility of identifying individuals superior to the recurrent parent. The likelihood of identifying transgressive segregants decreases with each cycle of backcrossing.[1] Limited backcrossing may also be used when, in addition to the donated allele, modifying genes from the donor are important. Increased cycles of backcrossing decreases the probability of fixing multiple loci with small effects.

POLYGENIC TRAITS

Backcrossing can also be used to transfer traits controlled by multiple genes. To accomplish the transfer of polygenic traits, selection for the trait of interest must be done after each cycle of backcrossing. For example, in the growing season following the first backcross, a number of BC_1 plants are self-pollinated. In the following growing season these BC_1F_2 progeny are grown and evaluated for the trait of interest and similarity to the recurrent parent. Selected individuals will then be backcrossed to the recurrent parent. Polygenic traits are usually metric traits and affected by the environment. Therefore, evaluation and selection among the BC_1F_2 families may need to be based on replicated trials and multiple environments. This greatly increases the amount of resources required. Due to dealing with multiple loci spread throughout the genome the recovery of the parental type will generally be less complete than when backcrossing a simply inherited trait.

The advanced backcross system uses molecular markers to increase the precision and speed of incorporating polygenic traits.[10] The advanced backcross method has been applied in numerous crops in crosses with wild germplasm to identify quantitative trait loci that may benefit the domesticated species.[11,12] The linked markers are then used to introgress the alleles into elite germplasm using marker assisted backcrossing.

EVALUATION OF NEW CULTIVAR

The purpose of backcross breeding is to recover the recurrent parent essentially unchanged except for the donated trait. Backcross breeding programs have been very successful in the development of improved cultivars. As in most complex systems, however, unexpected and

Fig. 1 Rust- (*Puccinia sorghi*) susceptible (left) and rust-resistant (right) sweet corn hybrids. Rust resistance is controlled by a single gene, Rp1, which was incorporated by backcrossing. (Photo by Jerald Pataky.) (*View this art in color at www.dekker.com.*)

undesirable outcomes occur. It is imperative that any new cultivar, developed by any method, be extensively evaluated in the intended area of production prior to release. Because the backcross method is very effective at recovering the phenotype of the recurrent parent, breeders have been tempted to reduce the amount of evaluation prior to cultivar release. Rushing a new cultivar to market without adequate testing can lead to crop failure and puts the farmer at economic risk. Such failures are seldom published in scientific literature, but most breeders have numerous anecdotes regarding spectacular failures of new cultivars derived by backcross breeding. In my own breeding program I used the backcross method to introgress an allele (*Rp1d*) for resistance to common rust (*Puccinia sorghi*) into an elite commercial sweet corn inbred. After six backcrosses I had developed a new line that appeared identical to the recurrent parent except for improved rust resistance. Prior to release, to increase seed amounts, the original and new lines were grown in the same environment. The original inbred (recurrent parent) yielded 1000 kg/ha, while the backcross-derived line yielded 225 kg/ha. Clearly this would have been disastrous if released directly to farmers based solely on its phenotypic appearance (Fig. 1).

CONCLUSION

Backcross breeding is a powerful tool for rapidly and precisely introgressing traits of interest into established cultivars. It is easiest and fastest when used with pure lines and traits controlled by a single gene. Backcrossing can be used with open-pollinated cultivars and polygenic traits, but such programs require more dollars, space, and time. Marker technology can increase the precision and speed of backcrossing. Despite the apparent ease in recovering the phenotype of the recurrent parent it is critical to test the performance of the newly derived cultivar carefully prior to release to ensure that the new cultivar performs as expected.

ARTICLES OF FURTHER INTEREST

Breeding for Nutritional Quality, p. 182
Breeding Plants with Transgenes, p. 193
Breeding Pure Line Cultivars, p. 196
Breeding Self-Pollinated Crops and Marker-Assisted Selection, p. 202
Breeding Using Doubled Haploids, p. 207
Breeding: Incorporation of Exotic Germplasm, p. 222
Breeding: Mating Designs, p. 225
Genetically Engineered Crops with Resistance Against Insects, p. 506
Genetically Modified Oil Crops, p. 509

REFERENCES

1. Fehr, W.R. *Principles of Cultivar Development; Vol. 1 Theory and Technique*; Macmillan Pub. Co.: New York, 1987.
2. Frisch, M.; Bohn, M.; Melchinger, A.E. Comparison of selection strategies for marker-assisted backcrossing of a gene. Crop Sci. **1999**, *39*, 1295–1301.
3. Hospital, F.; Decoux, G. Popmin: A program for the numerical optimization of population sizes in marker-assisted backcross programs. J. Heredity **2002**, *93*, 383–384.
4. Allard, R.W. *Principles of Plant Breeding*, 2nd Ed.; John Wiley and Sons: New York, 1999.
5. Hospital, F. Size of donor chromosome segments around introgressed loci and reduction of linkage drag in marker-assisted backcross programs. Genetics **2001**, *158*, 1363–1379.
6. Laughnan, J.R. The effect of sh2 factor on carbohydrate reserves in the mature endosperm of maize. Genetics **1953**, *38*, 485–499.
7. Tracy, W.F. Potential of field corn germplasm for the improvement of sweet corn. Crop Science **1990**, *30*, 1041–1045.
8. Tracy, W.F. History, breeding, and genetics of supersweet corn. Plant Breed. Rev. **1997**, *14*, 189–236.
9. Jones, D.F.; Mangelsdorf, P.C. The production of hybrid seed corn without detasseling. Conn. Agric. Exp. Sta. Bull. **1951**, *550*.
10. Tanksley, S.D.; McCouch, S.R. Seed banks and molecular maps: Unlocking genetic potential from the wild. Science **1997**, *277*, 1063–1066.
11. Ho, J.C.; McCouch, S.R.; Smith, M.E. Improvement of hybrid yield by advanced backcross QTL analysis in elite maize. TAG **2002**, *105*, 440–448.
12. Xiao, J.H.; Li, J.M.; Grandillo, S.; Ahn, S.N.; Yuan, L.P.; Tanksley, S.D.; McCouch, S.R. Identification of trait improving quantitative trait loci alleles from a wild rice relative, *Oryza rufipogon*. Genetics **1998**, *150*, 899–909.

C_3 Photosynthesis to Crassulacean Acid Metabolism Shift in *Mesembryanthenum crystallinum*: A Stress Tolerance Mechanism

John C. Cushman
University of Nevada, Reno, Nevada, U.S.A.

INTRODUCTION

Water is an essential requirement for all plant life. However, more than 35% of the earth's landmass is considered semiarid or arid. Crassulacean acid metabolism (CAM), one of three major modes of photosynthetic carbon fixation, is present in approximately 7% of terrestrial plants distributed across 33 families and 328 genera. CAM is a highly plastic adaptation to water limitation, which results in a five- to tenfold increase in water use efficiency under comparable conditions relative to C_4 and C_3 plants. Studies of the ecological distribution of CAM plants in predominantly arid habitats reinforce the notion that CAM confers a competitive advantage in hot, dry climates. Environmental conditions such as temperature, light intensity, and water availability have a profound effect on the degree to which this alternative photosynthetic carbon fixation pathway is manifested within the context of an evolutionary diverse continuum of intermediate, inducible, and obligate modes of CAM. A number of physiological, biochemical, and molecular genetic factors—such as carbohydrate and metabolite concentrations and a circadian clock—control the genotypic and phenotypic plasticity of CAM.

DEFINING CAM

The basic biochemical reactions of the CAM cycle are confined within single chloroplast-containing cells and constitute a pattern of CO_2 uptake, C_4 acid, and glucan formation that has been described as the diel repetition of four phases.[1] Specifically, CAM involves the nocturnal fixation of atmospheric and respiratory CO_2 by phospho*enol*pyruvate carboxylase (PEPC) resulting in the the formulation of oxaloacetate that is reduced to malate and stored in the vacuole (phase I). Subsequent daytime decarboxylation of these organic acids to release CO_2, which is refixed by ribulose-1,5-bisphosphate carboxylase/oxygenase (Rubisco) in the chloroplast, leads to carbohydrate production (phase III). In fully watered obligate CAM plants this phase is flanked by two transient periods of atmospheric CO_2 uptake, which mark the transition from C_4 to C_3 metabolism in the morning (phase II) and from Rubisco to PEPC activity in the afternoon (phase IV). CO_2 release during the day when stomata remain closed concentrates CO_2 around Rubisco, suppressing its oxygenase activity. The temporal separation of the C_3 and C_4 carboxylase activities and the associated diurnal fluctuation of carbon are supported by a complex and tightly controlled system of enzymatic and metabolite transport activities. These activities are, in turn, initiated by developmental and environmental factors, and maintained by a circadian clock. This complexity is orchestrated, in large part, by an intricate network of control mechanisms that originates with changes in gene expression.

CAM PLASTICITY

The traditional definition of CAM within the earlier discussed four-phase framework was derived primarily from measuring changes in patterns of leaf–gas exchange, acidity, and malate concentrations in fully watered obligate CAM plants. However, developmental, morphological, and environmental conditions—such as temperature, light, and water availability—play important roles in modulating the performance of CAM against the backdrop of a continuum of CAM modes governed by a combination of ontogenetic and evolutionary factors.[2,3] C_3-CAM intermediate species such as *Sedum telephium*, which do not progress beyond CAM-cycling, exhibit negligible nocturnal CO_2 uptake under well-water conditions along with some refixation of respiratory CO_2. However, following drought stress the relative contribution of nocturnal CO_2 uptake can increase in the absence of daytime stomatal closure. In inducible, highly plastic CAM species such as the well studied dicotyledonous tree *Clusia minor*, CAM induction can be extremely rapid (within one day) and is fully reversible.[2] In the halophyte *M. crystallinum*, salinity or water limitation accelerates and magnifies a developmentally programmed C_3-CAM transition over several days and is not fully reversible.[4] However, even well studied obligate CAM species such as *Kalanchoë*

Encyclopedia of Plant and Crop Science
DOI: 10.1081/E-EPCS 120010656

daigremontiana will adjust to water deficit by reducing uptake during phases II and IV.[5] Surveys of CAM plasticity within the Crassulaceae have revealed that the magnitude of nocturnal CO_2 fixation tends to correlate well with a greater degree of succulence in species that predominate in more arid climates.[6] Furthermore, thinner-leafed *Kalanchoë* species are more plastic in photosynthetic expression than thicker-leafed, succulent relatives, which are bound to nocturnal CO_2 fixation, presumably due to the extreme CO_2-diffusion limits within their tissues.[3]

CONTROL OF CAM INDUCTION

Although no unique enzymes are required to perform CAM, large changes in the abundance and regulation of an assortment of enzymes involved in organic acid and carbohydrate formation, turnover, and intracellular transport functions occur during CAM induction to meet the increased flow of carbon through the pathway and to avoid futile cycling of carbon skeletons contained within a single photosynthetic cell. To satisfy the metabolic demands for diel (day-night) carbon flux, CAM-specific members of multigene families appear to have become recruited during evolution.[7] Recruited genes typically display elevated expression patterns, whereas other isoforms that presumably fulfill anapleurotic or tissue-specific functional roles undergo little change in expression and generally are expressed at low levels in CAM-performing tissues. Alternatively, a single gene may fulfill both C_3 photosynthesis-and CAM-specific functions. Nuclear run-on transcription assays and transient assays using promoter–reporter gene fusion studies have confirmed that transcriptional activation is the primary control point responsible for modulating the expression of many CAM-specific genes. PEPC is the most abundant CAM enzyme and plays a major regulatory role in controlling circadian patterns of CO_2 fixation leading to malate synthesis.[8] Flux through PEPC is mediated by reversible protein phosphorylation, catalyzed by a circadianly controlled PEPC kinase that renders the enzyme considerably less sensitive to inhibition by negative effectors (e.g., L-malate) at night, but both more active and more sensitive to activation by positive effectors during the day (e.g., glucose-6-phosphate, triose phosphate). The kinase activity is, in turn, controlled by a circadian clock at the level of gene expression, with PEPC kinase transcripts up to twentyfold more abundant at night. Recent findings provide evidence for the coordinated regulation of Rubisco and PEPC over the diurnal phases of CAM in *K. daigremontiana*. Rubisco activation is exceptionally protracted over the diurnal course. During phase II, PEPC activity dominates, and Rubisco activation increases with elevated CO_2, but does not attain maximal activity until phase IV. Delayed activation is thought to be a result of delayed action of inhibitors during phase II[9] as well as delayed synthesis of Rubisco activase protein,[5] which is in turn regulated at the level of transcript abundance under the control of the circadian clock. Thus, both PEPC and Rubisco appear to be highly regulated in order to prevent competing carboxylations, to minimize futile carbon cycling between the C_3 and C_4 pathways and to optimize CO_2 fixation during both short-term and long-term changes in environmental conditions.[5]

A major metabolic constraint for CAM induction is the ability to distinguish between the large day–night reciprocating pool of carbohydrates required for CAM and those carbon skeletons destined for nitrogen metabolism, respiration, or growth. Acclimation to salinity and induction of CAM in *M. crystallinum* necessitates a reallocation of photosynthetically fixed carbon in order to fuel polyol accumulation and the dark reactions of CAM.[10] This is accomplished by coordinated increases in the activities of a number of glycolytic, gluconeogenic, starch biosynthetic, and degradative enzymes and plastidic transport activities, which acquire pronounced diel changes in activity that are governed by corresponding circadian changes in gene expression. In addition, posttranscriptional and posttranslational processes are likely to be key control points for these activity changes. Elucidating the relative functional contribution of various enzymes and transporters involved in carbohydrate partitioning and turnover will require a comprehensive characterization of the corresponding genes or gene families in a variety of different CAM species exhibiting different carbohydrate storage strategies.

SIGNALING EVENTS IN CAM INDUCTION AND CIRCADIAN CONTROL

Investigation of the signaling mechanisms that regulate CAM induction and the diurnal or circadian regulation of the CAM cycle have focused almost exclusively on *M. crystallinum*.[11] Signal transduction events initiated by water deficit, salinity stress, or abscisic acid (ABA) treatments that trigger expression of CAM genes are mediated by intracellular Ca^{2+}, Ca^{2+}/calmodulin-dependent protein kinases, protein phosphatases 2A and 1, phosphoinositides, and protein synthesis. Phytochrome and UV-A/Blue light have also been demonstrated to mediate CAM induction. Recent inhibitor studies in detached *M. crystallinum* leaves suggest that both ABA-dependent and -independent pathways may mediate CAM induction.[11] In addition to ABA, various plant growth regulators, such as cytokinins and gibberellic acid, have also been implicated as signaling molecules that participate in CAM induction in different ways depending on their mode of application.

Pharmacological treatments of detached leaves of *M. crystallinum* have produced evidence for the participation of phosphoinositide-dependent phospholipase C, inositol 1,4,5 triphosphate (IP_3)-gated tonoplast calcium channels, and a putative Ca^{2+}-calmodulin-dependent protein kinase in the regulation of nocturnal PEPC kinase activity and the phosphorylation state of PEPC.[12] Protein dephosphorylation events have also been implicated in the circadian signaling pathway that regulates PEPC kinase expression. Daytime/early dark phase alkalinization of the cytosol was suggested to be the trigger for initiating PEPC phosphorylation by PEPC kinase. However, in contrast to C_4 plants, elevations in cytosolic pH appear to have little or no influence on PEPC kinase activity in *M. crystallinum* or *K. fedtschenkoi*, respectively. Both RNA and protein synthesis also appear to be essential to the signaling cascade that regulates PEPC kinase activity, an output of the circadian oscillator. PEPC kinase transcript abundance is inversely correlated with cytosolic malate concentrations in plants in which malate concentrations were manipulated by withholding CO_2.[5] This observation suggests that cytosolic malate concentrations exert a negative effect on PEPC kinase gene expression or mRNA stability and override its circadian control. Temperature-gating experiments have provided further evidence that malate modulation of PEPC kinase expression is a secondary effect of the circadian clock, rather than a primary effect.[12] Alternatively, other metabolites such as phospho*enol*pyrovate (PEP), may act as a feed-forward activator of PEPC kinase expression.[3] Cytosolic malate is likely to be controlled by transport of malate across the tonoplast membrane, a view that is well supported by temperature effects on tonoplast function and modeling studies.[12] Thus, response to environmental factors that alter organic acid content or malate partitioning between the vacuole and cytosol may be able to override circadian rhythms of PEPC kinase activity, providing a possible fine-tuning mechanism for the rapid alterations in PEPC activity observed in some CAM species. However, it remains unclear whether the tonoplast functions as a master switch in generating circadian oscillations in CAM, or whether clock-controlled components exert their effect on malate transporters situated within the tonoplast membrane.

CONCLUSION

Although often viewed as an insignificant photosynthetic adaptation expressed in a small number of highly specialized, slow growing, arid land plants, it should be remembered that many horticulturally and agronomically important plants—including orchids (e.g., vanilla), bromeliads (e.g., pineapple), and cacti and agaves (used for tequila production)—perfom CAM.[3] Moreover, it is noteworthy that the photosynthetic plasticity and water-conserving features of CAM do not necessarily compromise the potential for high productivity. Steady advances in our understanding of the ecophysiology, physiology, and biochemical aspects of the induction and circadian regulation of CAM have been made through detailed investigations of a variety of well studied models. In the future, the development of a genetic model could facilitate the rapid exploration and understanding of the molecular mechanisms that underlie CAM induction and the circadian control of this specialized photosynthetic adaptation. Since the serendipitous discovery of the stress-inducible switch from C_3 photosynthesis to CAM in *M. crystallinum*, this species has been extensively studied. The inducible nature of CAM in this species allows a clear distinction to be made between molecular, biochemical, and physiological processes associated with C_3 photosynthesis versus those governing CAM.[2] *M. crystallinum* grows rapidly, produces large numbers of seeds (10,000–15,000) per plant, and is self-fertile; large mutant collections exist for this species.[13] Furthermore, *M. crystallinum* has a relatively small genome (390 Mb) relative to other CAM models and a large expressed sequence tag (EST) database with associated gene index.[14] The availability of such molecular genetic resources will permit large-scale gene expression profiling to elucidate the regulatory control circuits involved in CAM induction and circadian rhythmicity and to provide insights into the functional significance of genes and enzymes that optimize physiological performance in arid environments. Finally, a reliable transformation technology will be required to perform reverse-genetic screens and to enable future studies of gene function by suppression or over-expression of specific genes of interest. In the near future, transcriptome, proteome, and metabolite profiling studies will provide an integrated view of the complex and remarkably plastic responses to environmental, developmental, and circadian cues that dictate the diel patterns of CO_2 fixation characteristic of CAM plants.

ARTICLES OF FURTHER INTEREST

Circadian Clock in Plants, p. 278
Crop Responses to Elevated Carbon Dioxide, p. 346
Crops and Environmental Change, p. 370
Drought and Drought Resistance, p. 386
Drought Resistant Ideotypes: Future Technologies and Approaches for the Development of, p. 391
Ecophysiology, p. 410
Environmental Concerns and Agricultural Policy, p. 418

REFERENCES

1. Osmond, C.B. Crassulacean Acid Metabolism: A Curiosity in Context. In *Annual Reviews of Plant Physiology*; Academic Press: San Diego, 1978; Vol. 29, 379–414.
2. Cushman, J.C.; Borland, A.M. Induction of Crassulacean acid metabolism by water limitation. Plant Cell Environ. **2002**, *25*, 295–310.
3. Dodd, A.N.; Borland, A.M.; Haslam, R.P.; Griffiths, H.; Maxwell, K. Crassulacean acid metabolism: Plastic, fantastic. J. Exp. Bot. **2002**, *53*, 569–580.
4. Adams, P.; Nelson, D.E.; Yamada, S.; Chmara, W.; Jensen, R.G.; Bohnert, H.J.; Griffiths, H. Growth and development of *Mesembryanthemum crystallinum* (Aizoaceae). New Phytol. **1998**, *138*, 171–190.
5. Griffiths, H.; Helliker, B.; Roberts, A.; Haslam, R.P.; Girnus, J.; Robe, W.E.; Borlan, A.M.; Maxwell, K. Regulation of Rubisco activity in crassulacean acid metabolism plants: Better late than never. Funct. Plant Biol. **2002**, *29*, 689–696.
6. Kluge, M.; Razanoelisoa, B.; Brulfert, J. Implications of genotypic diversity and phenotypic plasticity in the ecophysiological success of CAM plants, examined by studies on the vegetation of Madagascar. Plant Biol. **2001**, *3*, 214–222.
7. Cushman, J.C.; Bohnert, H.J. Crassulacean acid metabolism: Molecular genetics. Annu. Rev. Plant Physiol. Plant Mol. Biol. **1999**, *50*, 305–332.
8. Nimmo, H.G.; Fontaine, V.; Hartwell, J.; Jenkins, G.I.; Nimmo, G.A.; Wilkins, M.B. PEP carboxylase kinase is a novel protein kinase controlled at the level of expression. New Phytol. **2001**, *151*, 91–97.
9. Maxwell, K. Resistance is useful: Diurnal patterns of photosynthesis in C_3 and crassulacean acid metabolism epiphytic bromeliads. Funct. Plant Biol. **2002**, *29*, 679–687.
10. Borland, A.M.; Dodd, A.N. Carbohydrate partitioning in crassulacean acid metabolism plants: Reconciling potential conflicts of interest. Funct. Plant Biol. **2002**, *29*, 707–716.
11. Taybi, T.; Cushman, J.C.; Borland, A.M. Enivonmental, hormonal and circadian regulation of crassulacean acid metabolism. Funct. Plant Biol. **2002**, *29*, 669–678.
12. Hartwell, J.; Nimmo, G.A.; Wilkins, M.B.; Jenkins, G.I.; Nimmo, H.G. Probing the circadian control of phosphoenolpyruvate carboxylase kinase expression in *Kalanchoë fedtschenkoi*. Funct. Plant Biol. **2002**, *29*, 663–668.
13. Cushman, J.C. Crassulacean acid metabolism: A plastic photosynthetic adaptation to arid environments. Plant Physiol. **2001**, *127*, 1439–1448.
14. http://www.tigr.org/tdb/tgi/mcgi/ (accessed September 2002).

Carotenoids in Photosynthesis

C

Britta Förster
Barry J. Pogson
Australian National University, Canberra, Australian Capital Territory, Australia

INTRODUCTION

Carotenoids are plant pigments that are essential for the survival and productivity of all photosynthetic organisms. Their function in harvesting light energy complements the role of chlorophylls. The structure and assembly of the photosystems is dependent on the availability of specific carotenoids that assist in correct folding and maintain stability of photosystem proteins. In addition, carotenoids are essential for several of the photoprotective mechanisms employed by plants to effectively dissipate (as heat) excess photon energy absorbed by chlorophyll, thus preventing formation of highly reactive oxygen species. Carotenoids also deactivate singlet oxygen generated in the reaction centers and antennae.

BACKGROUND

Carotenoids are red, orange, and yellow pigments that are found in very diverse organisms, ranging from bacteria to plants and animals that either synthesize the pigments or acquire them through nutrition. There are two main classes—the nonpolar carotenes and the oxygenated xanthophylls. Xanthophylls received their name after the coloration of autumn leaves (xanthos = yellow and phyll = leaf) in 1837.[1] Carotenoids have been intensively researched for over a century. This article focuses on the involvement and physiological relevance of carotenoids in photosystem II (PSII) of higher plants. In particular, roles of carotenoids in PSII assembly, energy transfer to and from chlorophylls, thermal dissipation of excess absorbed light, and minimization of singlet oxygen generation and its deactivation are discussed.[2] Carotenoid chemistry and biosynthesis have been reviewed comprehensively elsewhere.[3]

Light-harvesting and photoprotective mechanisms must be balanced to ensure optimal photoassimilate production for sustained plant growth while preventing lethal free radical damage to the plant. Photon capture and energy transfer from the antenna drive the electron transfer reactions of photosynthesis. Factors that disrupt the balance between these reactions lead to the production of reactive oxygen species, including singlet oxygen ($^1O_2$*), superoxide anion radicals (O_2^-), hydroxyl radicals ($^{\cdot}OH$), and hydrogen peroxide (H_2O_2), which potentially cause oxidative damage to lipids, proteins, and nucleic acids.[4] Therefore, effective acclimatory mechanisms at whole plant, cellular, and molecular levels are essential to accommodate short- and long-term exposure to potentially photodamaging full sunlight and environmental stresses that can cause plant death or greatly reduced crop yields.

Consequently, complementary and redundant protective mechanisms have evolved that include 1) harmless thermal dissipation of excess energy involving xanthophylls; 2) quenching of triplet chlorophylls by carotenoids; 3) free radical detoxification via antioxidants (e.g., ascorbate, tocopherols, and carotenoids) and antioxidant enzymes such as catalase, superoxide dismutase, and ascorbate peroxidase; 4) down-regulation of light absorption via decreasing light-harvesting antenna size; and 5) repair of photodamage. Carotenoids are known to confer protection against photo-oxidative damage by stabilizing the pigment–protein complexes and membranes, quenching of excited-state chlorophylls and reactive oxygen species, and reduction of the reactive $P680^+$ in PSII.[4] The PSII complex is central to understanding these processes, as it is the primary site of stress-induced photoinactivation under most natural conditions.

CAROTENOIDS IN PHOTOSYSTEM ASSEMBLY AND FUNCTION

Six primary photosynthetic pigments are found in the photosystems of higher plants: chlorophyll *a*, chlorophyll *b*, and four almost ubiquitous carotenoids—lutein, β-carotene, violaxanthin, and neoxanthin (Fig. 1). Zeaxanthin and antheraxanthin are two other xanthophylls common in chloroplasts, whereas only a few plant species accumulate additional carotenoids.

The ubiquitous nature of β-carotene in the reaction centers of oxygenic photosynthetic organisms suggests that it fulfills a fundamental, although not yet clearly defined, role. Carotenes are required for quenching chlorophyll triplet states in purple bacterial reaction centers, and their loss results in cell death. In higher plants, β-carotene is unable to effectively quench PSII core chlorophylls, but

DOI: 10.1081/E-EPCS 120019334

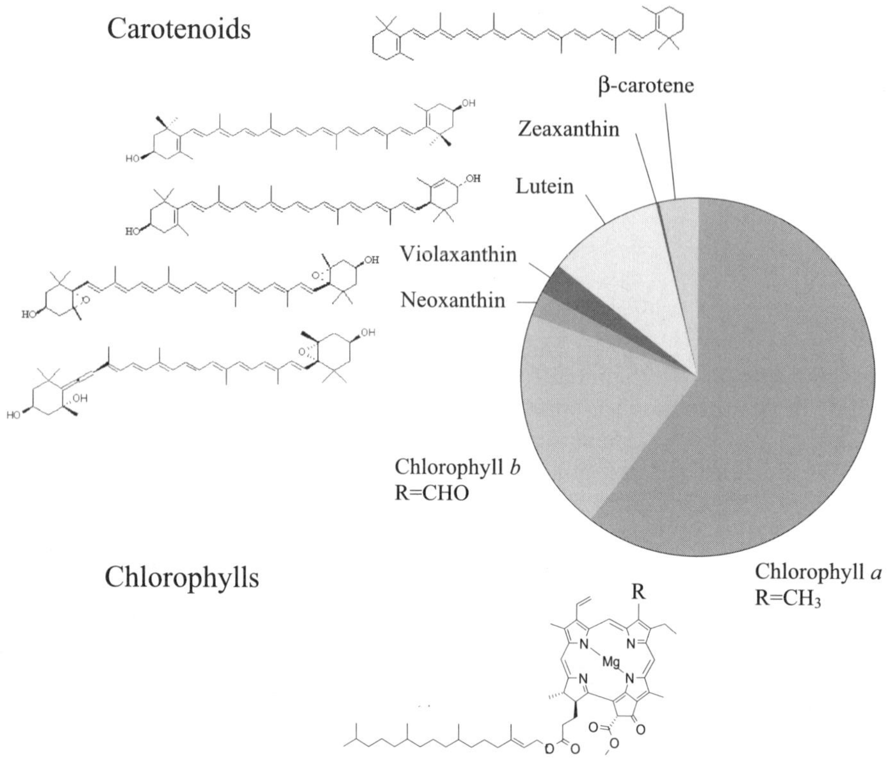

Fig. 1 Pigments of higher plant chloroplasts. The chart represents the typical pigment composition in chloroplasts of *Arabidopsis* (μg pigment/g fresh weight) when plants are exposed to moderate light (Ref. 2). Chemical structures are shown next to each pigment. (*View this art in color at www.dekker.com.*)

it may afford photoprotection by quenching $^1O_2^*$ or by signalling damage of the D1 reaction center protein that triggers repair of photoinactivated PSII complexes.

The xanthophylls are located in the light-harvesting pigment complexes (LHCs) and the core antenna proteins (Fig. 2), where they act as accessory light-harvesting pigments by absorbing those wavelengths in the visible spectrum not absorbed by chlorophyll. The absorbed energy is then transferred to chlorophyll. The carotenoid-binding sites have been determined using a combination of structural studies, in vitro reconstitutions of recombinant proteins, and analyses of pigment-binding to native LHCs.[5] Four distinct carotenoid-binding sites have been identified in LHCs (L1, L2, N1, V1), but only three of these are occupied in plants grown at moderate light intensities.[5] In LHCIIb (Fig. 2), the L1 and L2 sites are occupied by lutein, although either violaxanthin or zeaxanthin can be accommodated, whereas the N1 site is specific for neoxanthin. The V1 site is proposed to function in high light stress responses by loosely binding violaxanthin, which facilitates deepoxidation to the photoprotective zeaxanthin.[5] In the minor LHCs (Fig. 2), the L1 site is occupied by lutein and L2 by either violaxanthin or neoxanthin. *β*-Carotene is typically excluded from LHCs, although it is found in Lhca1, the major LHC of PSI.

Despite the apparent plasticity of carotenoid-binding sites, there are limits to the flexibility of substitutions in vitro and in vivo, which seem to correlate with specific structural properties of the pigments. Lactucaxanthin, a xanthophyll with two hydroxylated ε-rings, cannot support LHC assembly in vitro. Mutations resulting in exclusive accumulation of zeaxanthin, a xanthophyll with planar *β*-rings, are associated with delayed greening in higher plants and decreased photoprotective energy dissipation in algae. These phenotypes probably reflect disrupted LHC assembly due to lack of the nonplanar ε-ring of lutein or the epoxidated *β*-ring of violaxanthin. LHCIIb trimers are absent in lutein-deficient plant and algal mutants, suggesting that lutein is compulsory for LHC trimerization.

CAROTENOIDS IN PHOTOPROTECTION

The intensity of light in the middle of a sunny day may exceed a plant's capacity to use the harvested light for photochemistry, which can result in a decrease in the production of photoassimilate (sugars, etc.) by about 65%.[6] Other environmental factors such as cold reduce this capacity further. The first and critical strategy employed by all plants is to safely utilize and dissipate

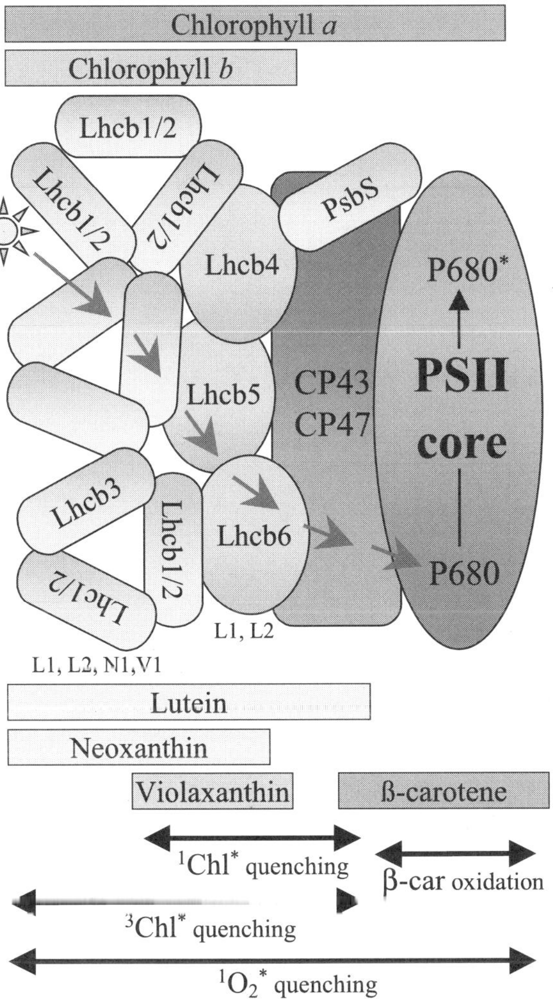

Fig. 2 Model of pigment localization and protein subunits in PSII. The major LHC antennae (LHCIIb or Lhcb1–3) and the minor LHC antennae (Lhcb4–6) and PSII core antenna (CP43, CP47) are shown surrounding the PSII reaction center core. PsbS is shown to be associated with the PSII, although its exact location is subject to debate. Bars indicate the distribution of each of the chlorophylls and carotenoids. Carotenoid-binding sites (Ref. 5) are described in the text. Sites of the various types of photoprotection are indicated by arrows; however, the location of some mechanisms (such as ^{1}Chl* quenching) is not clear. *(View this art in color at www.dekker.com.)*

absorbed energy, thereby preventing the formation of free radicals and oxidizing species. Absorption of light energy results in the excitation of chlorophyll from the ground state to its singlet state (^{1}Chl*) (Fig. 3). Under optimal conditions this energy is transferred to the PSII reaction center to catalyze the photochemical reactions. However, excess light absorption results in a greater proportion of chlorophyll molecules in excited electronic states, which can be deactivated by heat emission, return to ground state by emitting fluorescence, or undergo intersystem crossing (i.e., reverse the spin direction of an excited electron) to form triplet state chlorophyll (^{3}Chl*). The latter process is nearly an order of magnitude faster than fluorescence emission, and because ^{3}Chl* can effectively interact with ground-state molecular oxygen ($^3O_2$) to generate highly reactive and potentially toxic $^1O_2$*, the risk of oxidative damage to nearby organic compounds markedly increases under these conditions.

Carotenoids are able to quench both singlet and triplet excited states of chlorophyll (Fig. 3).[7] Carotenoids participate in ^{1}Chl* deactivation via nonphotochemical quenching (NPQ), a process that thermally dissipates the energy of ^{1}Chl*. The different components and relative contribution of NPQ to energy dissipation are determined from changes in chlorophyll fluorescence in response to saturating pulses of light.[8] The predominant component of NPQ is fast; reversible; and requires acidification of the thylakoid lumen, synthesis of specific xanthophylls, and protonation of the PsbS (CP22) protein of PSII. The synthesis of the specific xanthophylls requires activation of the xanthophyll cycle to convert violaxanthin into zeaxanthin and antheraxanthin. The acidification of the thylakoid lumen activates the cycle and results in protonation of lumenally located residues of LHCs and PsbS. The excited-state energy levels of zeaxanthin, antheraxanthin, and lutein, and their proximity to chlorophylls would allow them to efficiently accept excitation energy from ^{1}Chl* via Förster Dexter electron exchange mechanisms. These mechanistic implications are even more plausible, considering that those carotenoids are clearly necessary for induction and maintenance of NPQ both in vitro and in vivo.[9] However, an appropriate protein environment, such as the aforementioned protonation, is also required for operation of NPQ in vivo.

The *npq1* mutant and transgenic plants deficient in the high light-induced accumulation of antheraxanthin and zeaxanthin are unable to induce more than 14–60% reversible NPQ, which is essentially eliminated when lutein is absent.[9] In fact, irrespective of genetic backgrounds, altering lutein levels has an effect on NPQ consistent with lutein contributing to NPQ, particularly in the early phase.

The necessity for the PsbS protein in NPQ has been a key discovery.[10] Mutations in PsbS that prevent its accumulation or protonation result in almost no NPQ despite accumulation of zeaxanthin.[10] PsbS is capable of binding zeaxanthin in vitro, which correlates with a characteristic NPQ-induced change in absorbance at 535 nm (ΔA_{535}). Furthermore, the level of NPQ is directly proportional to the amount of PsbS, and PsbS increases up to tenfold in overwintering species subject to cold-induced light stress. This suggests that changes in PsbS levels may be employed for modulation of NPQ levels in

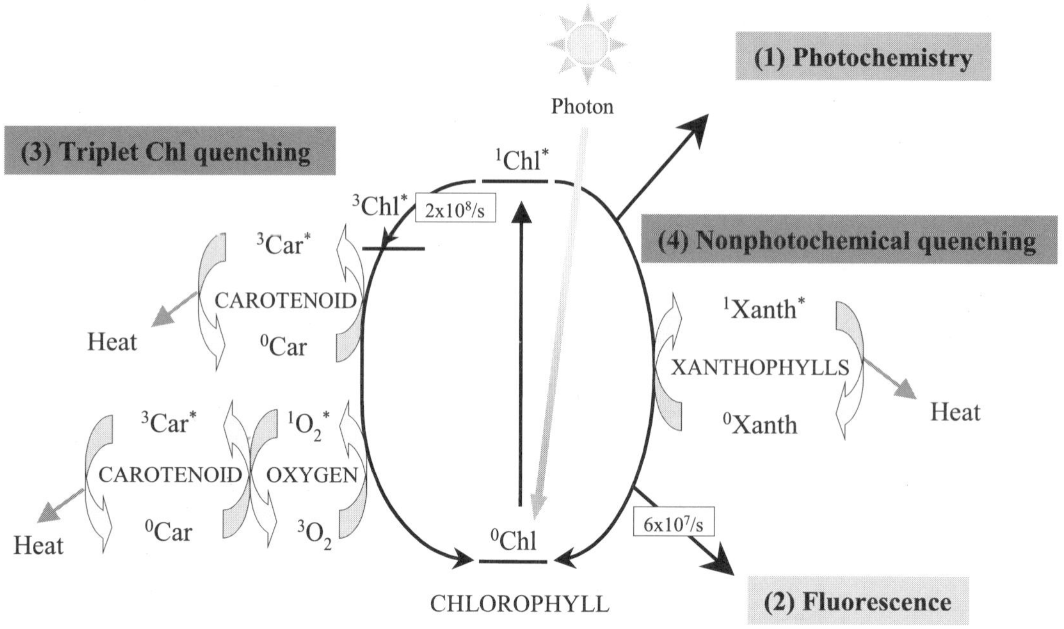

Fig. 3 Photoprotective roles of carotenoids and singlet chlorophyll quenching. Photon absorption raises ground-state chlorophyll (^{0}Chl) to the excited singlet state (^{1}Chl*) energy level. Ideally, this energy is utilized to drive photochemical electron transfer of photosynthesis (Ref. 1). If photon absorption exceeds photosynthetic capacity, ^{1}Chl* in the antennae and reaction center dissipates energy via fluorescence at a rate constant of $6 \times 10^7\ s^{-1}$ corresponding to a lifetime of about 15–20 ps (Ref. 2). Formation of triplet-state chlorophyll (^{3}Chl*) competes with fluorescence at a rate constant of $2 \times 10^8\ s^{-1}$, resulting in a lifetime of about 5 ps. Carotenoids can directly quench ^{3}Chl* (Ref. 3), thus preventing ^{3}Chl* from reacting with ground-state oxygen ($^3O_2$) to form the phototoxic singlet oxygen ($^1O_2$*). Alternatively, $^1O_2$* can be scavenged directly by carotenoids (Ref. 3). Furthermore, ^{1}Chl* energy can be dissipated thermally by certain xanthophylls, which is apparent as nonphotochemical quenching of chlorophyll fluorescence (NPQ) and involves formation of zeaxanthin in the xanthophyll cycle (Ref. 4). (*View this art in color at www.dekker.com.*)

response to variations in environmental conditions during plant development and growth.

CONCLUSION

Although there is a tendency to focus on the relationship of carotenoids and NPQ in photoprotection (as both are readily measured), sufficient consideration must be given to the fact that carotenoids have a range of significant constitutive functions in preventing oxidative damage by quenching excited triplet states of chlorophyll and scavenging free radicals (Fig. 3). Although all carotenoids can quench chlorophyll triplets in vitro, lutein bound to L1 has also been identified as the main site of triplet quenching in reconstituted LHCs.[7] Zeaxanthin-deficient mutants exhibit enhanced lipid peroxidation and leaf necrosis upon exposure to light or chilling stress (which appears to reflect a role for carotenoids in inhibiting lipid peroxidation, consistent with their localization in the chloroplast and thylakoid membranes), in addition to binding directly to LHC apoproteins.

What is the physiological importance of carotenoids? Plants with altered carotenoid compositions show a wide range of phenotypes from full viability and normal greening to high-susceptibility, bleaching, delayed greening, and semilethality. Even apparently healthy and viable mutants lacking zeaxanthin are reduced in seed production by 30–50% under field conditions,[11] which constitutes a selective disadvantage. Clearly, optimal carotenoid composition is required to sustain plant fitness and survival.

ACKNOWLEDGMENTS

Our thanks to the many colleagues, students, and staff who provided many of the results and insights that form the basis of this review.

ARTICLES OF FURTHER INTEREST

Chlorophylls, p. 258
Fluorescence: A Probe of Photosynthetic Apparatus, p. 472

Photosynthesis and Stress, p. 901
Plant Response to Stress: Modification of the Photosynthetic Apparatus, p. 990

REFERENCES

1. Govindjee. On the requirement of minimum number of four versus eight quanta of light for the evolution of one molecule of oxygen in photosynthesis: A historical note. Photosynth. Res. **1999**, *59*, 249–254.
2. Pogson, B.J.; Rissler, H.M.; Frank, H.A. The Roles of Carotenoids in Photosystem II of Higher Plants. In *Photosystem II: The Water/Plastoquinone Oxido-Reductase in Photosynthesis*; Satoh, K., Wydrzynski, T., Eds.; Advances in Photosynthesis and Respiration Series, Kluwer Academic Publishers: Dordrecht, The Netherlands, 2003; Vol. 12, *in press*.
3. Cunningham, F.J.; Gantt, E. Genes and enzymes of carotenoid biosynthesis in plants. Annu. Rev. Plant Physiol. Plant Mol. Biol. **1998**, *49*, 557–583.
4. Niyogi, K.K. Photoprotection revisited: Genetic and molecular approaches. Annu. Rev. Plant Physiol. Plant Mol. Biol. **1999**, *50*, 333–359.
5. Bassi, R.; Caffarri, S. Lhc proteins and the regulation of photosynthetic light harvesting function by xanthophylls. Photosynth. Res. **2000**, *64*, 243–256.
6. Osmond, C.B.; Anderson, J.M.; Ball, M.C.; Egerton, J.J.G. Compromising Efficiency: The Molecular Ecology of Light-Source Utilization in Plants. In *Physiological Plant Ecology*; Press, M.C., Scholes, J.D., Barker, M.G., Barker, M.G., Eds.; Blackwell Science: Oxford, 1998; 1–24.
7. Formaggio, E.; Cinque, G.; Bassi, R. Functional architecture of the major light-harvesting complex from higher plants. J. Mol. Biol. **2001**, *314*, 1157–1166.
8. Krause, G.H.; Weis, E. Chlorophyll fluorescence and photosynthesis: The basics. Annu. Rev. Plant Physiol. Plant Mol. Biol. **1991**, *42*, 313–349.
9. Niyogi, K.K.; Shih, C.; Chow, W.S.; Pogson, B.J.; DellaPenna, D.; Björkman, O. Photoprotection in a zeaxanthin- and lutein-deficient double mutant of Arabidopsis. Photosynth. Res. **2001**, *67*, 139–145.
10. Li, X.-P.; Björkman, O.; Shih, C.; Grossman, A.R.; Rosenquist, M.; Jansson, S.; Niyogi, K.K. A pigment-binding protein essential for regulation of photosynthetic light harvesting. Nature **2000**, *403*, 391–395.
11. Külheim, C.; Agren, J.; Jansson, S. Rapid regulation of light harvesting and plant fitness in the field. Science **2002**, *297*, 91–93.

Cereal Sprouting

C. W. Wrigley
Food Science and Wheat Australia CRC, North Ryde, Sydney, Australia

INTRODUCTION

Rain at harvest time causes several problems for the cereal grain farmer. In addition to difficulty in drying the grain for storage, rain brings the possibility that the grain may start to germinate. This event is likely to make the harvested grain unsuitable for most types of utilization, thereby drastically reducing its market value.

Severe sprouting is visually evident by the presence of roots and shoots emerging from the germ of the cereal grain, but significant damage may be done before there are visual signs of damage, by the development of hydrolytic enzymes, especially alpha-amylase. Testing for the production of these enzymes is thus an important part of evaluating grain quality at harvest. If such damage is detected, the rain-damaged grain must be segregated, so as to maintain the quality of the sound grain with which it might otherwise be mixed.

There are further possibilities for overcoming the problems caused by sprouting. These include the breeding of varieties that have built-in dormancy to provide tolerance to these effects of wet harvest conditions. In addition, there are possibilities of predicting situations where there are increased risks of sprout damage, based on combinations of variety and climate (mainly moisture and temperature).

SEED MORPHOLOGY AND PHYSIOLOGY

Seed or Grain?

A seed is a plant's way of forming another plant. For this purpose, the seed is provided with the following essential anatomical parts (Fig. 1):

- An embryo (germ) from which the new plant will form.
- A store of nutrients (starch and protein in the endosperm of the cereal grain) to nurture the emerging plant until its own photosynthesis can take over.
- An outer layer (the bran of the cereal grain) to protect the seed against damage.

The great advantage of the seed as a means of perpetuating the species is that it can remain dormant (resting unchanged) for long periods of time. In this way, the seed awaits the ''rainy day'' (a suitable combination of moisture and temperature) when the conditions are conducive for it to commence the germination process that will start it on its way to producing another plant. This process starts with the swelling of the germ as it takes up moisture, followed by the splitting of the germ covering to allow the roots and shoots to emerge and elongate, seeking soil and light, respectively. Long ago, mankind found the cereal seed to be an important answer to his need for nutrition, based on the two important attributes, namely, nutrients suited to the diet of humans (and of animals) and storage for long periods of time without loss of nutritional value. The consequent rise in the growing of grain crops changed early man from a hunter-gatherer to an agriculturalist, with the advantages of fixed residence and spare time to develop cultural pursuits. Consequently, we regard the ''seed,'' with its prime function to perpetuate the plant, as a ''grain,'' whose function is the provision of food for man and beast.

The Germination Process

Germination is essential for the role as ''seed,'' but it may not suit the ''grain'' role. This is because important biochemical changes have commenced in the germinating seed, even before the anatomical changes become evident.[1] These biochemical changes include the production of hydrolytic enzymes, the agents that have the physiological function of breaking down the stored nutrients to simpler compounds that can be used to produce new tissues for plant growth, in particular the emerging roots and shoots.

The most significant of these enzymes is alpha-amylase, the primary enzyme involved in reducing starch (a polymer of the monosaccharide glucose) to the component simple sugars for transport to the growing embryo. Figure 2 shows how amylase attack has ''eaten away'' at a starch granule during germination. Protease action is also needed during germination to break down the stored proteins of the endosperm, providing amino acids for protein synthesis in the embryo.

Encyclopedia of Plant and Crop Science
DOI: 10.1081/E-EPCS 120021908

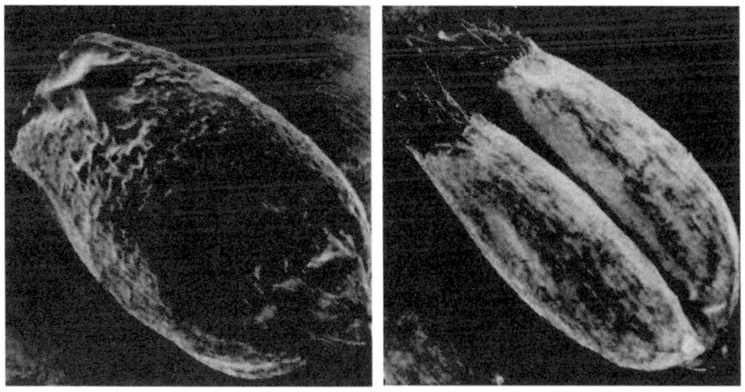

Fig. 1 The anatomy of a wheat grain, as shown in the scanning electron microscope, dorsal and ventral (crease) views. The germ (embryo) is seen as a bulge at the lower right of the dorsal view (at left). (*View this art in color at www.dekker.com.*)

Dormancy

Ideally, there is a significant lag period (dormancy) before germination commences in the newly harvested grain. However, it is not unusual for the dormancy period to be short or absent, with the consequence that germination may commence even before the grain is harvested if it gets wet. This is illustrated in Fig. 3; the grains in the wheat head on the left have sprouted, whereas the variety at right has shown adequate dormancy. Barley grains are shown in Fig. 4 in varying degrees of sprout damage.[2]

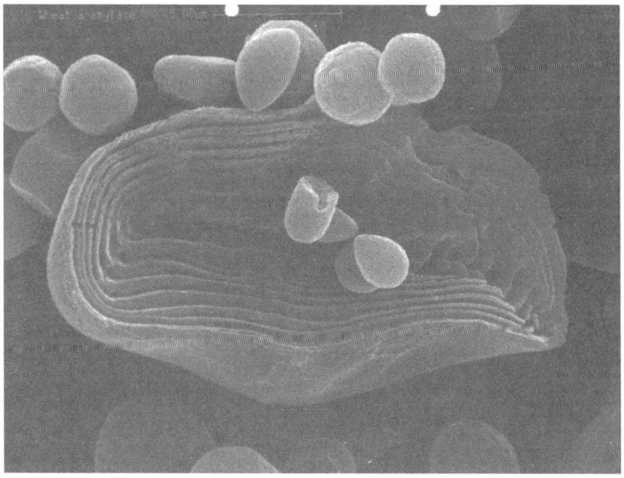

Fig. 2 Scanning electron micrograph of a wheat starch granule that has been attacked by amylase during germination. Small starch granules can also be seen. Amylase action has accentuated the layers of deposition in the eroded starch granule. The eroded starch granule is about 20 micrometers across the wide dimension.

PROCESSING PROBLEMS OF SPROUTING

Sprouting is a major problem for wheat. Even if there are no visible signs of roots and shoots, the production of alpha-amylase due to incipient sprouting causes processing problems for food products such as bread (sticky crumb and difficulty in slicing), noodles (poor color and disintegration of structure), and Chinese steamed breads (poor crumb structure). Rye grain is especially susceptible to sprout damage, causing difficulty in the production of rye breads.

For barley, germinated grain is unsuitable for malting and brewing, the premium use of barley. Uniform germination of all barley grains is needed for the malting process to be successful. A germination percentage of over 95% is generally required for sound malting barley, but this quality requirement may not be met for germinated barley. Even if rain damage has been only minor, the affected barley is likely to have a shortened storage life due to a premature drop in germination percentage. This is illustrated in Fig. 5, which shows that the germination rate falls during storage, depending on the moisture and temperature conditions. Incipient sprouting

Fig. 3 Grains in the wheat head at left have germinated, seen by the protrusion of roots and shoots, following the head being exposed to rain. In contrast, grains in the head on the right have shown adequate dormancy to tolerate the effects of the rain. (*View this art in color at www.dekker.com.*)

Fig. 4 Barley grains showing various degrees of visible sprouting, seen as roots and shoots emerging from the germ end of the grains. (Adapted from Fitzsimmons and Wrigley, Ref. 2.) (*View this art in color at www.dekker.com.*)

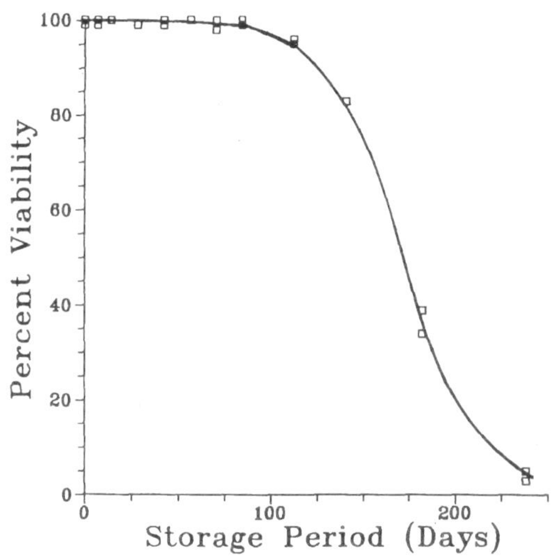

Fig. 5 The rate of germination decreases during the storage of malting quality for barley.[8] The rate of this loss is greater for barley that has undergone incipient sprouting at harvest.

accelerates the rate of this loss of germination, with consequent loss of value because the grain must be used for animal feed instead of for malting.

Premature sprouting also occurs in other cereals, including maize, oats, and rice, although the economic consequences are not so serious as they are for the processing of wheat, rye, and barley. Nevertheless, seed viability during storage is reduced for germinated grain.

TESTING FOR SPROUT DAMAGE

Traditionally, sprout damage has been detected by visual inspection, by the appearance of a swollen or split covering on the germ or in the extreme by the appearance of roots and shoots. Inspection standards have specified minimum proportions of such grains for a grain load to qualify for specific grades. However, this basis of sprout evaluation ignores the basic nature of the problem, namely, the enzymes of germination.

Accordingly, various tests have been developed to determine the activity of alpha-amylase, either on the starch of the sample or on an added substrate. Tests of the former category involve milling a representative sample

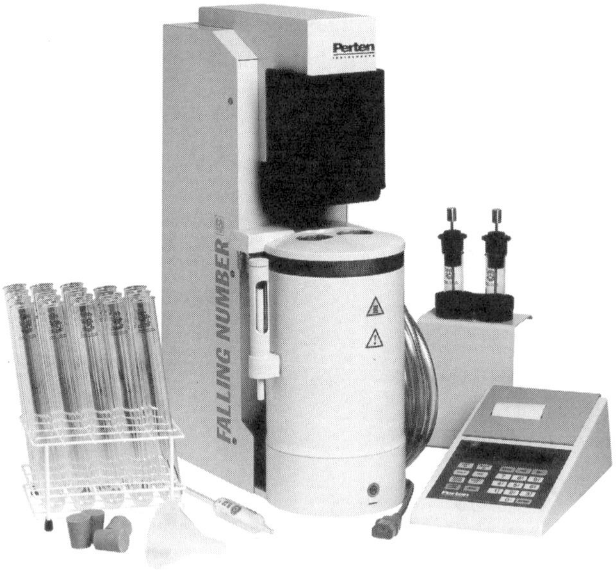

Fig. 6 Falling Number equipment. A suspension of milled grain is mixed with water in a long glass tube immersed in boiling water, using the vertical movement of a plunger. After 60 seconds of mixing, the plunger is raised and allowed to fall under its own weight. Falling time is long for a sound sample but is short if the amylase of sprouted grain has hydrolyzed the starch of the suspended grain particles. (*View this art in color at www.dekker.com.*)

of grain and heating it to accelerate the action of the amylase, followed by some form of measurement of the viscosity of the heated suspension. In the Falling Number test,[3] remaining starch viscosity is determined by timing the fall of a plunger through the suspension of milled grain (Fig. 6). Changes in viscosity are measured by the power required to stir the suspension in the Stirring Number test, which uses the Rapid Visco Analyzer[4] (Fig. 7).

In other test systems, the amylase enzyme is extracted from the milled grain for direct analysis[5] or by immunoassay.[6] This latter method, based on the reaction of antibodies specific to the alpha-amylase of cereal grains (Fig. 8), is rapid (about five minutes) and can be used in the field (literally) without the need for additional equipment.

REMEDIES FOR SPROUT DAMAGE

Breeding for Increased Dormancy

Genetic sources of tolerance to sprout damage are available, as is indicated by the fact that the degree of dormancy varies among genotypes of the various cereals. It has therefore been a priority of plant breeders to incorporate genes for dormancy in varieties by crossbreeding, especially in regions where sprout damage is a frequent occurrence. In red-grained wheats, which have a reputation for greater dormancy than white-grained varieties, dormancy is associated with genes for red coloration of the bran layers, located on the Group 3 chromosomes, together with at least one other gene.[7,8] Genetic research is thus discovering the genetic basis of dormancy in wheat and barley, thereby assisting breeders to select for tolerance to sprout damage more efficiently.

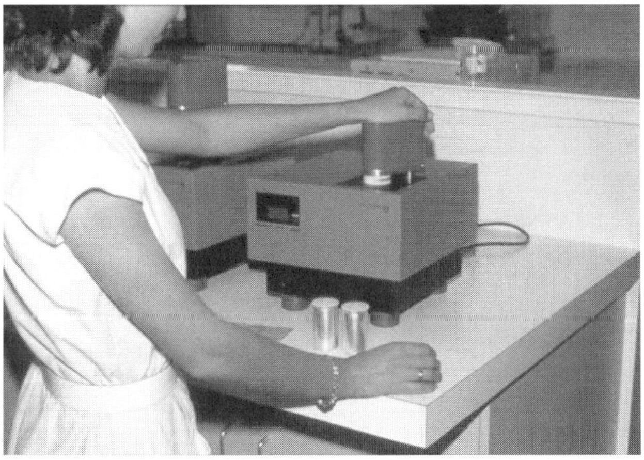

Fig. 7 The Rapid Visco Analyzer, used to determine Stirring Number values. A suspension of milled grain is placed in a disposable cup, clamped in a heating block. The apparent viscosity of the suspension is measured continuously during stirring. A low viscosity after three minutes indicates that the sample is sprouted. (*View this art in color at www.dekker.com.*)

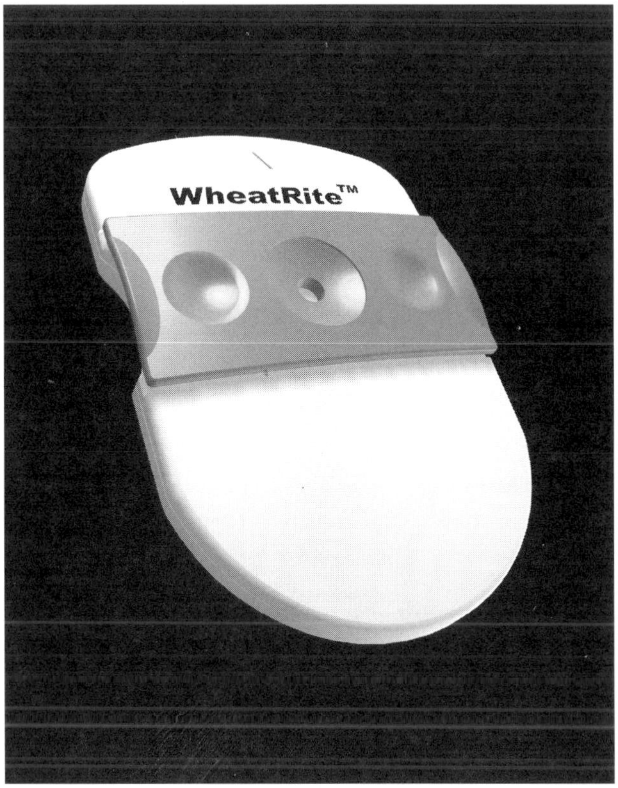

Fig. 8 The WheatRite test cassette for immunoassay of amylase in sprouted grain. An extract of crushed grain is spotted onto the card, which is then closed to permit immunochromatography to take place. The degree of sprout damage is indicated by the intensity of a colored band appearing in the window of the closed card. (*View this art in color at www.dekker.com.*)

Prediction of Problems Due to Sprouting

A knowledge of local weather conditions just before grain delivery is an obvious means for the staff of a mill or grain elevator to be forewarned that grain delivered from some local areas may be sprout damaged. Longer-term prediction involves risk assessment based on predicted climate patterns obtained from historical climate data. In both of these cases, predictions should also take into account the sprouting susceptibility of the varieties involved.

A further aspect of the value of prediction relates to the storage of malting barley. If the rate of fall of viability can be predicted for various consignments, then the grain with the shortest "safe" storage life can be used before the grain that is sound, with consequent longer storage life. This form of prediction involves assessing the degree of sprout damage and using this measure in a mathematical model that reflects the rate of loss of germination rate, depending on the conditions of storage, especially temperature and moisture.[9] Another factor that may be relevant is the variety involved, because barleys may differ in the rate of loss of viability.

CONCLUSION

Sprouting damage has long been a problem, especially in climates where wet harvest conditions are common. For this reason, considerable research effort has been expended on breeding genotypes that have longer dormancy and thus better tolerance to sprout damage. These efforts have already provided valuable results for many varieties, but the breeder must balance this genetic trait against many other breeding objectives. As more efficient procedures become available to select for sprout tolerance, tolerant varieties will become more common. Possibilities for such breeding methods are likely to come from the identification of molecular markers and their use in screening for such benefits.[10] In the meantime, the problem of sprouted grain will become less severe with the implementation of better methods of testing and prediction.

ARTICLES OF FURTHER INTEREST

Breeding: Genotype-by-Environment Interaction, p. 218
Plant Response to Stress: Critical Periods in Plant Development, p. 981
Seed Dormancy, p. 1130

REFERENCES

1. Meredith, P.; Pomeranz, Y. Sprouted grain. Adv. Cereal Sci. Technol. **1985**, *7*, 239–320.
2. Fitzsimmons, R.W.; Wrigley, C.W. *Australian Barleys: Identification of Varieties, Grain Defects and Foreign Seeds,* 2nd Ed.; CSIRO: Melbourne, Australia, 1984.
3. Anon AACC Method Number 56–81B. Determination of Falling Number. Approved November, 1972. In *Approved Methods of the American Association of Cereal Chemists,* 10th Ed.; The Association: St Paul, MN, 2000.
4. Anon AACC Method Number 22–08. Method of Alpha-Amylase Activity with the Rapid Visco Analyser. Approved November, 1995. In *Approved Methods of the American Association of Cereal Chemists,* 10th Ed.; The Association: St Paul, MN, 2000.
5. McCleary, B.V.; Sheehan, H. Measurement of cereal alpha-amylase: A new assay procedure. J. Cereal Sci. **1987**, *6*, 237.
6. Skerritt, J.H.; Heywood, R.H. A five-minute field test for on-farm detection of pre-harvest sprouting in wheat. Crop Sci. **2000**, *40*, 742–756.
7. Mares, D.J. The Seed Coat and Dormancy in Wheat Grains. In *Eighth International Symposium on Preharvest Sprouting in Cereals, 1998*; Weipert, D., Ed.; Association for Cereal Research, Federal Centre for Cereal, Potato and Lipid Research: Detmold, Germany, 1999; 77–81.
8. Flintham, J.; Adlam, R.; Gale, M. Seed Coat and Embryo Dormancy in Wheat. In *Eighth International Symposium on Preharvest Sprouting in Cereals, 1998*; Weipert, D., Ed.; Association for Cereal Research, Federal Centre for Cereal, Potato and Lipid Research: Detmold, Germany, 1999; 67–76.
9. Bason, M.L.; Ronalds, J.A.; Wrigley, C.W. Prediction of 'safe' storage life for sound and weather-damaged malting barley. Cereal Foods World **1993**, *38*, 361–363.
10. Mares, D.J.; Mrva, K. Mapping quantitative trait loci associated with variation in grain dormancy in Australian wheat. Aust. J. Agric. Resour. **2001**, *52*, 1257–1265.

Chemical Weed Control

Don W. Morishita
University of Idaho, Twin Falls, Idaho, U.S.A.

INTRODUCTION

Herbicides are chemicals used for the control of weeds or other unwanted vegetation. Most herbicides in use today are synthetically derived organic compounds. A few herbicides are derived from naturally occurring chemicals found in plants. The Weed Science Society of America recognizes 315 herbicides that have been developed over the years. Herbicides are just one of many types of pesticides and represent about 60% of the pesticides used in the United States. Other pesticides include insecticides, fungicides, bactericides, nematicides, rodenticides, as well as others. The greatest use of chemical weed control is in crop production. In addition, herbicides are used to control weeds growing in turf grass, as in golf courses and parks, ornamental flowerbeds, pastures, rangeland, forests, aquatic areas, and industrial sites. The types of herbicide used in these different areas vary widely, as do the methods of their use.

Herbicides affect plant growth in many ways. Some herbicides are applied only to plant foliage, some only to soil, and some are applied only to water for controlling unwanted aquatic vegetation. Herbicides also can vary in how long they persist in the environment. Herbicides that persist in the soil for long periods are used only in areas where long-term weed control is desired. Herbicides that break down rapidly typically are used in situations where residual activity of the herbicide is undesirable (e.g., a persistent herbicide used to control crop weeds may be undesirable if it takes so long to break down that it could injure the crop subsequently grown in the same field).

There are several advantages and disadvantages to the use of herbicides. From an agricultural food production standpoint, herbicides provide more economical weed control, reduce human labor requirements, allow for increased farming acreage per person, can reduce the number of tillage operations and soil erosion, and permit weed control where cultivation is impossible. They can also be beneficial for increasing aesthetic and land value through removal of unwanted vegetation. Herbicides, like other pesticides, are often criticized for their impact on the environment and potential injury to humans and other animals. When misused, herbicides can contaminate the environment, injure non-target vegetation, and persist for periods longer than desired. When use properly and judiciously, herbicides provide many environmental and economical benefits.

This article discusses the history and use of herbicides, and briefly explains how herbicides affect plant growth and behave in the environment, and how governmental regulation of herbicides ensures their safety.

HISTORY OF HERBICIDES

The use of chemicals for the control of vegetation dates back hundreds of years. Sodium chloride (table salt) is probably the oldest known chemical used as a herbicide.[1] Historical reports of Romans using salt in 146 B.C. to sack Carthage indicate the earliest use of a chemical to kill plants. In this case, the Romans used salt to prevent crop growth. From 1937 to 1950, salt was used for controlling weeds on Kansas highway rights-of-way at rates of up to 20 tons per acre.[2] Other inorganic chemicals were used as herbicides. One was calcium carbonate (lime), recommended for controlling horsetail (*Equisetum* sp.) in Germany as early as 1840. Other inorganic chemicals used as herbicides prior to 1940 included arsenical compounds such as sodium arsenite, sodium chlorate, and sodium borate. From 1927 to 1935, 4 million pounds of sodium chlorate were used in the state of Idaho alone for controlling perennial weeds such as field bindweed (*Convolvulus arvensis*) and Canada thistle (*Cirsium arvense*).[2] Other chemicals used as herbicides prior to the 1940s included sulfuric acid, copper sulfate, carbon bisulfide, salt of dinitrophenol, iron sulfate, ammonium sulfate, and kerosene.

Sinox, a precursor to dinoseb, was the first synthetically produced organic herbicide. A major breakthrough was the production of 2,4-D, discovered in 1941 and originally tested as a fungicide and insecticide.[1] Soon after, it was discovered that 2,4-D could selectively kill broadleaf or dicotyledonous plants without hurting grass or monocotyledonous plants. The discovery of 2,4-D is considered by many to be the beginning of modern herbicides. One of the most widely used herbicides in the world today continues to be 2,4-D.[3]

DOI: 10.1081/E-EPCS 120020268

CLASSIFICATION OF HERBICIDES

The ways herbicides are used and how they kill weeds vary greatly. Consequently, herbicides are classified many different ways.[1,4–7] The two most common classifications are based on the herbicide's chemical structure and its mode of action. Herbicides with similar chemical structures are grouped into a common chemical family. The structure of a herbicide influences how it will act in plants and in the environment. For example, metsulfuron and tribenuron are two sulfonylurea herbicides used for broadleaf weed control in cereal crops[8] (Fig. 1). However, slight differences in their chemical structure alter not only which weed species they control, but also how long they persist and how much they may leach or move in the soil. A sensitive broadleaf crop such as canola can be planted 60 days after tribenuron application. If metsulfuron is used, canola cannot be planted for as long as 22 months after application. Metsulfuron also has a higher potential for moving in the soil compared to tribenuron[9] (Table 1).

Herbicides classified in the same chemical family typically have the same mode of action. The mode of action is defined as the sequence of events beginning with introduction of the herbicide into the environment through the death of the plant.[1] Weed scientists also refer to a mechanism of action, relating to the primary biochemical or biophysical process leading to plant death.[1] The location in the plant where herbicides exert their toxicity at the cellular level is called the site of action.

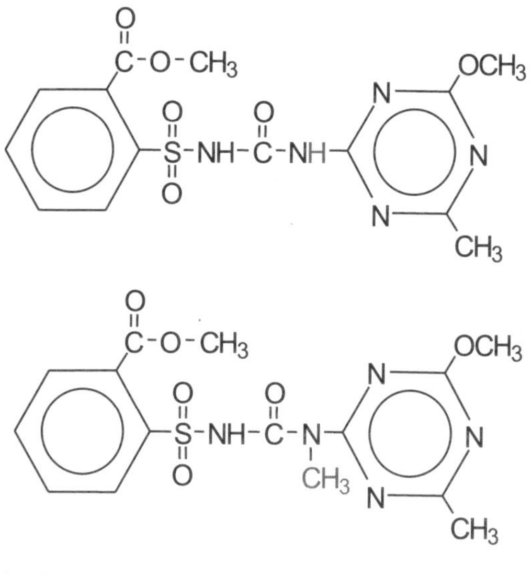

Fig. 1 Structural similarity between metsulfuron (top) and tribenuron (bottom). Although these two herbicides have only slight differences in chemical structure, metsulfuron persists in the soil about three times longer and controls many weeds that tribenuron does not. Tribenuron is used extensively in cereal grain production, whereas metsulfuron is used in cereal grain production, rangeland, and noncrop areas. (*View this art in color at www.dekker.com.*)

Herbicides can also be classified based on whether they selectively kill weeds and whether they translocate in the plant. Some herbicides (like 2,4-D) are classified as selective because they control weeds in cereal crops without injuring the crop. Nonselective herbicides such as glyphosate will injure or kill nearly all plants regardless of species. Herbicide selectivity may also be dependent on factors such as plant health, environmental conditions, and application rate, timing, and method.

Herbicides such as 2,4-D and glyphosate are also classified as systemic herbicides because they are taken up by the plant after they are applied, and move or translocate to a site of action in the plant that causes injury or death to the plant. Nonsystemic herbicides are called contact herbicides. Bromoxynil and paraquat are two examples of contact herbicides. A contact herbicide, once taken up by the plant, is not translocated in order to injure or kill the plant.

HOW HERBICIDES KILL WEEDS

Some herbicides are applied to the foliage to kill weeds, whereas other herbicides control weeds only when they are applied to the soil. Other herbicides control weeds when they are applied to the foliage and/or the soil. Glyphosate and 2,4-D are both foliar-applied herbicides and have little or no effect on plants when applied to the soil. Metolachlor and trifluralin are two examples of soil-applied herbicides, both of which have little or no effect on plants when applied to the foliage. Soil-applied herbicides kill weeds as they germinate and begin to grow in the soil. Picloram and atrazine are two examples of herbicides that have both foliar and soil activity.

As previously mentioned, herbicides kill plants in many different ways. The mode of action of herbicides is sometimes broadly categorized into the following: plant growth regulators, seedling growth inhibitors, photosynthetic inhibitors, amino acid synthesis inhibitors, lipid synthesis inhibitors, and cell membrane disruptors. Some examples of herbicides and their modes of action are listed in Table 1. Once a herbicide is applied to plant foliage or to the soil, it must be absorbed, or taken up, by the plant. If it is a systemic herbicide, it must be translocated to a site of action. If it is a contact herbicide, the site of action will be very close to where the herbicide was absorbed. A plant that is tolerant of the herbicide is usually able to break down or metabolize it into a nontoxic form. Sometimes, repeated use of herbicides with the same mode of action can lead to the development of herbicide-resistant weeds. These weeds are able to tolerate more than normal amounts of herbicide.

C

Table 1 Common herbicides' mode of action, persistence,[a] and relative mobility[b] in soil

Herbicide	Product name(s)	Mode of action	Persistence	Mobility in soil
2,4-D	Many names	Plant growth regulator (synthetic auxin)	Short	Moderate
Atrazine	Aatrex, other names	Photosynthesis inhibitor (PS II inhibition)	Moderate	High
Glyphosate	Roundup, other names	Amino acid synthesis inhibitor (EPSP synthase inhibition)	Moderate	Extremely low
Metsulfuron	Ally, Escort	Amino acid synthesis inhibitor (ALS inhibitor)	Moderate	High
Metolachlor	Dual II, other names	Seedling growth inhibitor (lipid synthesis inhibitor)	Moderate	High
Paraquat	Gramoxone, other names	Photosynthesis inhibitor/cell (membrane disruptor)	Persistent	Very low
Picloram	Tordon	Plant growth regulator (synthetic auxin)	Moderate	Very high
Tribenuron	Express	Amino acid synthesis inhibitor (ALS inhibitor)	Short	Moderate
Trifluralin	Treflan, other names	Seedling growth inhibitor (microtubule assembly inhibition)	Moderate	Very low

[a]Persistence is categorized based on a pesticide's half-life (i.e., the time required for a pesticide to degrade to half its original concentration). Pesticides categorized as Short have less than a 30-day half-life; Moderate indicates a 30- to 100-day half-life; Persistent indicates more than a 100-day half-life.
[b]Pesticide mobility ratings are based on an empirically derived Groundwater Ubiquity Score (GUS). Pesticides with a GUS less than 0.1 are considered to have an extremely low potential to move toward groundwater. Values of 1.0–2.0 are low; 2.0–3.0 are moderate; 3.0–4.0 are high; and values greater than 4.0 have very high potential to move toward groundwater.

HERBICIDE REGULATION

In order for a herbicide or any other pesticide to become available for use in the United States, it must be registered by the Environmental Protection Agency (EPA), which follows criteria established by the Food Quality Protection Act. The EPA requires a set of rigorous tests to ensure that a pesticide will be safe to use and not harm people, animals, or the environment. Manufacturers must prove that a herbicide can be used to perform its intended function without reasonable adverse effects on the environment.[1]

CONCLUSION

Herbicides have been and continue to be an important tool for successful integrated weed management. However, it must be remembered that herbicides are not the only method of controlling weeds. The most successful weed management programs utilize integrated weed management practices—preventive, cultural, mechanical, biological, and chemical control. The use of herbicides for controlling weeds has reduced dependency on hand labor, and allowed land managers to utilize more land. Herbicides will continue to be an important tool for controlling weeds. However, overreliance on chemicals for weed control increases the risk of environmental contamination, selection for herbicide-resistant weeds, and likelihood of failure.

REFERENCES

1. Zimdahl, R.L. Introduction to Chemical Weed Control. In *Fundamentals of Weed Science*; Academic Press, Inc.: San Diego, CA, 1993; 208–224.
2. Timmons, F.L. A history of weed control in the United States and Canada. Weed Sci. **1970**, *18*, 294–307.
3. Donaldson, D.; Kiely, T.; Grube, A. *Pesticide Industry Sales and Usage: 1998 and 1999 Market Estimates*; U.S. Environmental Protection Agency Office of Pesticide Programs: Washington, DC, 2002; 4–14.
4. Monaco, T.J.; Weller, S.C.; Ashton, F.M. Herbicide Registration and Environmental Impact. In *Weed Science: Principles and Practices,* 4th Ed.; John Wiley & Sons: New York, 2002; 84–97.
5. Anderson, W.P. Modes and Sites of Action of Herbicides. In *Weed Science: Principles and Applications,* 3rd Ed.; West Publishing Co.: St. Paul, MN, 1996; 97–107.
6. Aldrich, R.J.; Dremer, R.J. Herbicide Use. In *Principles in Weed Management,* 2nd Ed.; Iowa State University Press: Ames, IA, 1997; 263–272.
7. Ross, M.A.; Lembi, C.A. Plant-Herbicide Interactions. In *Applied Weed Science,* 2nd Ed.; Prentice Hall: Upper Saddle River, NJ, 1999; 97–125.
8. Weed Science Society of America. 2003, Homepage, <http://www.wssa.net/> (April 2003).
9. Vogue, P.A.; Kerle, E.A.; Jenkins, J.J. *OSU Extension Pesticide Properties Database*; 1994, http://ace.orst.edu/info/npic/ppdmove.htm.

Chlorophylls

Robert D. Willows
Macquarie University, Sydney, Australia

INTRODUCTION

Chlorophylls are the pigments that make plants green. Chlorophylls are arguably the most important compounds on earth, because they are required for the harvesting and transduction of light energy in photosynthesis. The structures and spectra of chlorophylls, the details of how chlorophylls are synthesized, the challenges faced by plants in synthesizing chlorophylls, and the degradation of chlorophylls will be discussed.

STRUCTURE OF CHLOROPHYLLS FOUND IN PLANTS

The chlorophylls belong to a class of compounds known as tetrapyrroles, which include other pigment molecules such as heme and vitamin B_{12}. Chlorophylls are distinguished from other tetrapyrroles by the presence of a centrally coordinated magnesium and a fifth isocyclic ring. Chlorophyll *a* and chlorophyll *b* (Fig. 1.) are the major types of chlorophylls found in plants; they have a characteristic green color due to strong absorbance of blue and red light. Chlorophylls also show a characteristic red fluorescence when light is absorbed. The spectral properties are essential for the function of chlorophyll in harvesting light energy and in the transduction of that light energy for photosynthesis.

The ratio of chlorophyll *a* to *b* varies from 2.0–2.8 for shade-adapted plants to 3.5–4.9 for plants adapted to full-sun conditions. The variation in chlorophyll *a*/*b* ratio is due to differences in the ratio of photosystem I (PSI) and photosystem II (PSII) and the size and composition of the light harvesting complexes (LHCs) associated with each photosystem. The photosystems contain chlorophyll *a* but not chlorophyll *b*. In contrast the LHCs contain significant amounts of chlorophyll *b*. Shade-adapted plants have more LHCs and thus have lower chlorophyll *a*/*b* ratios than sun-adapted plants.[1,2]

A number of spectral variants of chlorophyll *a* are detected in vivo. These spectral differences result from chlorophyll molecules in different environments within LHCs and photosystems. Other minor chlorophyll derivatives have also been reported, with the most important ones being chlorophyll *a*′, found in the PSI reaction center, and pheophytin, which is found in the PSII reaction center. Other organisms such as algae, cyanobacteria, and photosynthetic bacteria have other types of chlorophylls or bacteriochlorophylls within their LHCs and photosystems.[3]

BIOSYNTHESIS OF CHLOROPHYLLS

Chlorophyll is synthesized within the chloroplast of plants and algae, and most of the chlorophyll biosynthetic enzymes are encoded by the nuclear genome. Despite the complete sequencing of a number of plant, algal, and photosynthetic bacterial genomes, not all of the genes involved in chlorophyll biosynthesis have been positively identified. Thus, although many of the details of chlorophyll biosynthesis have been elucidated, many gaps in our knowledge remain. Figs. 2–4 show an overview of the chlorophyll biosynthetic pathway. The interesting and novel reactions in each section of the pathway are discussed below.

The tetrapyrrole ring system is synthesized from eight molecules of 5-aminolevulinic acid (ALA). ALA is the universal precursor of all tetrapyrroles and there are two known pathways for the synthesis of ALA, as shown in Fig. 2. The most common pathway for ALA biosynthesis in nature is the C-5 pathway found in plants, archaea, and most bacteria. Glutamyl-tRNA and glutamate-1-semialdehyde are the two intermediates in this pathway. The glutamyl-tRNA is also used for protein biosynthesis. The Shemin pathway, which biosynthesizes ALA from glycine and succinate, was discovered first but is only found in α-proteobacteria and most non-photosynthetic eukaryotes.

Both heme and chlorophyll have a common biosynthetic route from ALA to protoporphyrin IX, as shown in Fig. 3. ALA dehydratase, also known as porphobilinogen synthase, catalyzes the condensation of two molecules of aminolevulinic acid to form the monopyrrole porphobilinogen. Four porphobilinogens are then condensed to form a linear tetrapyrrole called hydroxymethylbilane, a reaction catalyzed by hydroxymethylbilane synthase, which is also known as porphobilinogen deaminase.

Encyclopedia of Plant and Crop Science
DOI: 10.1081/E-EPCS 120019335

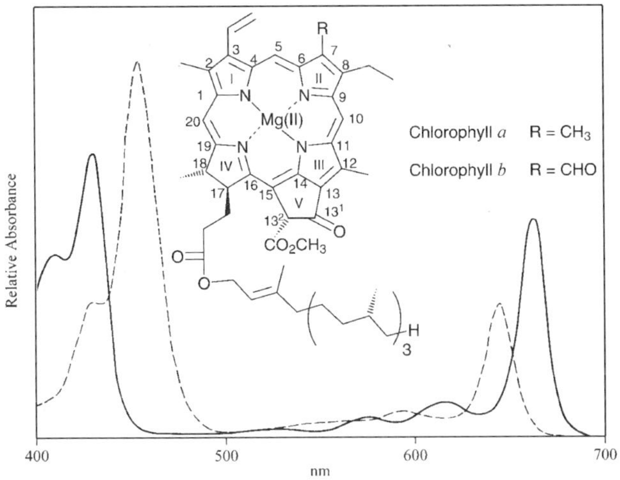

Fig. 1 Absorbance spectra and structures of chlorophyll *a* (solid line) and chlorophyll *b* (dotted line). Spectra are in ether and are normalized based on extinction coefficients given by Ref. 2.

This enzyme has a covalently-attached dipyrrole cofactor. Porphobilinogen is added sequentially to this cofactor so that a linear hexapyrrole forms as an intermediate on the enzyme before release of hydroxymethylbilane and regeneration of the enzyme. Hydroxymethylbilane is unstable and will cyclize non-enzymatically to uroporphyrinogen I, which is non-functional. The enzyme uroporphyrinogen III synthase is required to cyclize hydroxymethylbilane to produce uroporphyrinogen III, in which ring IV is inverted relative to the other pyrrole rings. Porphobilinogen can be readily condensed to form uroporphyrinogen isomers of which 50% are isomer III. Isomer III was probably scavenged by the earliest organisms before they were able to synthesize their own tetrapyrroles. When organisms subsequently evolved to synthesize their own tetrapyrroles, they would have had to evolve a mechanism for synthesizing isomer III. Uroporphyrinogen decarboxylase converts uroporphyrinogen III to coproporphyrinogen III, which is then converted to the last colorless intermediate protoporphyrinogen IX by coproporphyrinogen oxidase.[4]

Protoporphyrinogen oxidase converts protoporphyrinogen to protoporphyrin IX, the first colored intermediate in the pathway. Protoporphyrinogen oxidase is the primary target of the diphenyl ether class of herbicides, which inhibits the protoporphyrinogen oxidase found in both mitochondria and chloroplasts. As a consequence, protoporphyrinogen accumulates and moves into the cytosol where it is oxidized to protoporphyrin IX. Protoporphyrin IX accumulation results in photo-oxidative damage and cell death. The potential to cause photo-oxidative damage by the colored intermediates in chlorophyll biosynthesis highlights the challenges faced by plants in synthesizing chlorophyll from protoporphyrinogen onwards.[5,6]

A metal ion is then inserted into protoporphyrin IX by a metal ion chelatase. When Fe^{2+} is inserted by the enzyme ferrochelatase, heme is the end product. When Mg^{2+} is inserted by magnesium chelatase, chlorophyll will be the end product. Thus, insertion of Mg^{2+} by magnesium chelatase represents a commitment to chlorophyll biosynthesis, as shown in Fig. 4. Magnesium chelatase is a complex enzyme consisting of three subunits and belongs to a class of proteins known as AAA+ molecular machines. This enzyme hydrolyzes ATP and the mechanism appears to have similarities to some molecular chaperons. After Mg^{2+} insertion, an S-adenosylmethionine-dependent O-methyltransferase adds a methyl group to the 13-propionate sidechain. Molecular oxygen is incorporated when ring V is formed by an oxidative cyclase whose genes have yet to be positively identified. The next step in the pathway is the reduction of ring IV, which can be catalyzed by two types of enzyme, both known as protochlorophyllide oxidoreductases (POR). The more ancient of the two types of POR has not been found in flowering plants but occurs in almost all other photosynthetic organisms, including gymnosperms and algae. This POR has structural similarities to nitrogenase, with three different protein subunits encoded by the chloroplast genome. The second type of POR is found in all land plants and algae. It is a single subunit enzyme that requires light and reduced nicotinamide adenine dinucleotide phosphate (NADPH) to catalyze the reduction of ring IV to produce a chlorin. This requirement for light appears to be an important developmental adaptation for angiosperms. The reduction of the vinyl group on ring II occurs at either this stage or at some stage

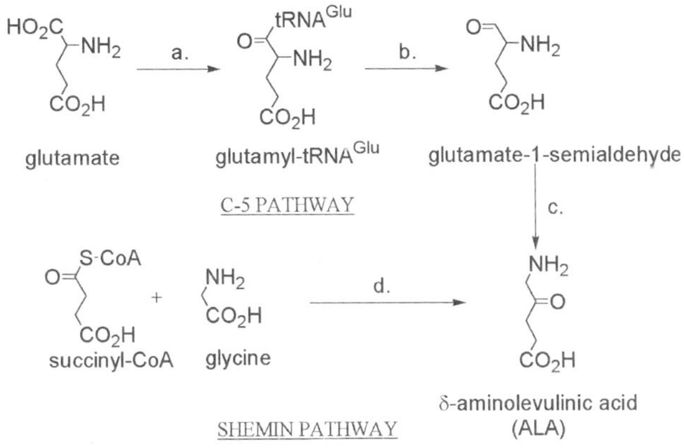

Fig. 2 ALA biosynthetic pathways. a. glutamyl-tRNA synthetase; b. glutamyl-tRNA reductase; c. glutamate-1-semialdehyde aminotransferase; d. ALA synthase.

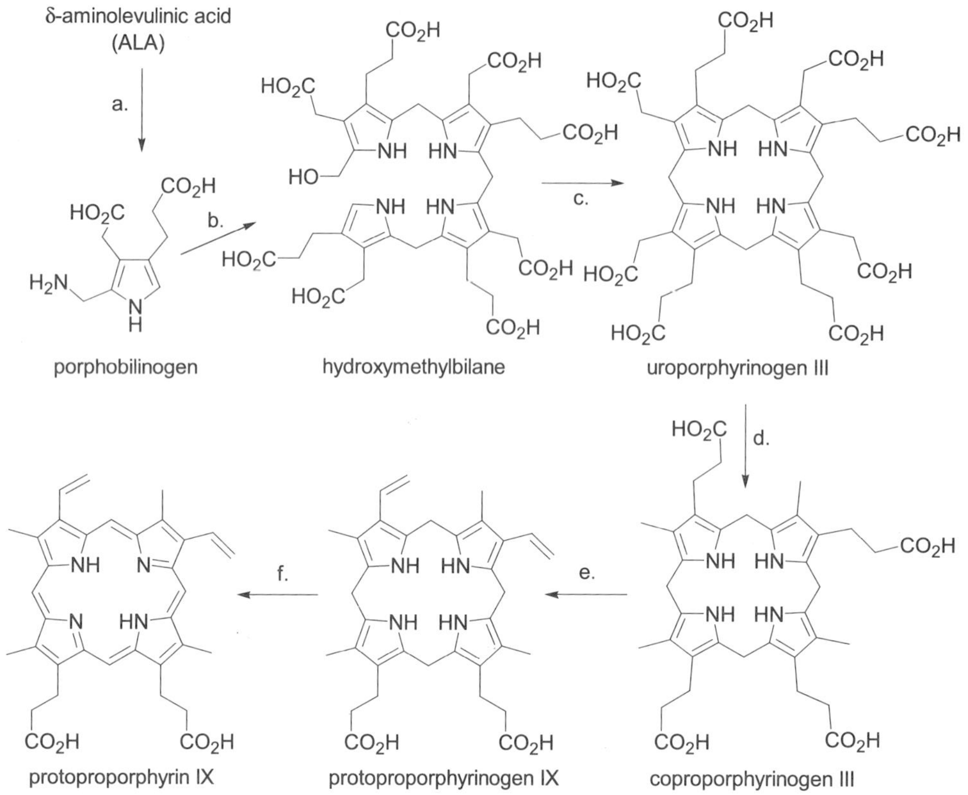

Fig. 3 Biosynthetic pathway from ALA to the first colored intermediate protoporphyrin IX. a. ALA dehydratase; b. porphobilinogen deaminase; c. uroporphyrinogen III synthase; d. uroporphyrinogen decarboxylase; e. coproporphyrinogen oxidase; f. protoporphyrinogen oxidase.

from protoporphyrin IX onwards. The enzyme responsible for this reduction has not been identified. The final step in chlorophyll *a* synthesis is esterification of the alcohol phytol onto the remaining propionate sidechain of chlorophyllide *a* to make chlorophyll *a* by chlorophyll synthase. Chlorophyll *a* can then be converted to chlorophyll *b* by chlorophyll *a* oxygenase. Chlorophyll *b* can be converted back to chlorophyll *a* via chorophyllide *b* and chlorophyllide *a*. However, the enzymes responsible for this conversion have not been identified.[3]

REGULATION OF CHLOROPHYLL BIOSYNTHESIS

Chlorophyll itself, and the intermediates in the pathway from protoporphyrin IX onwards, are phototoxic when not bound to protein in a way that allows absorbed light energy to be dissipated or transduced. When pigments are irradiated in the presence of molecular oxygen, highly reactive singlet oxygen is produced. To limit the formation of singlet oxygen, the biosynthesis of chlorophyll is tightly regulated and is also coordinated with synthesis of LHC and photosystem proteins. This regulation and coordination occurs by both feedback inhibition of enzyme activities as well as regulation of the quantities of key enzymes in the pathway. Some of the key regulatory features of chlorophyll biosynthesis are described below. However, we are still far from understanding the entire regulatory network.[7]

ALA biosynthesis is the primary regulatory point in the pathway, because feeding ALA results in accumulation of intermediates. Evidence exists for feedback inhibition of ALA biosynthesis in plants by heme and the magnesium containing tetrapyrroles. In greening seedlings, light activates production of ALA biosynthetic enzymes as well as other enzymes in the pathway, and in some species, light causes the rapid degradation of one POR isozyme while activating the synthesis of a second POR isozyme. Light also activates the transcription of magnesium chelatase genes and a circadian rhythm controls the transcription of these genes. In addition to these controls,

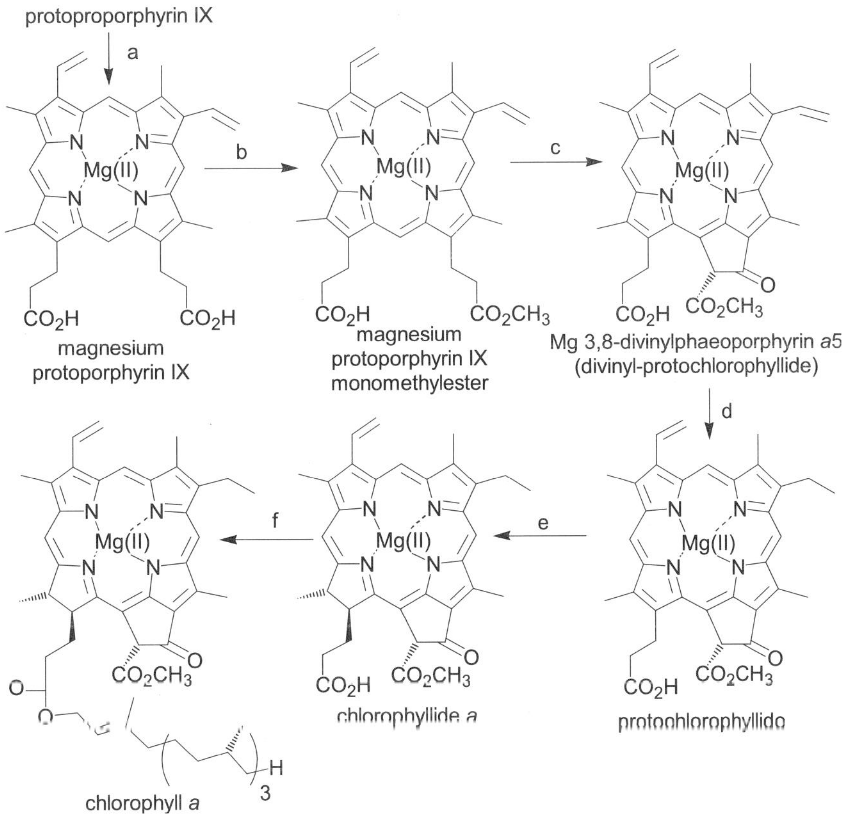

Fig. 4 Chlorophyll biosynthetic pathway from protoporphyrin IX. a. magnesium chelatase; b. S-adenosylmethionine:magnesium protoporphyrin IX O-methyltransferase; c. magnesium protoporphyrin IX monomethylester oxidative cyclase; d. 8-vinyl reductase; e. protochlorophyllide oxidoreductase; f. chlorophyll synthase.

phytochrome and cytokinins have also been implicated in the regulation of the quantity of chlorophyll biosynthetic enzymes or mRNA.[5–7]

The chlorophyll biosynthetic intermediates, Mg-protoporphyrin IX and Mg-protoporphyrin IX monomethyl ester, are directly involved in the transcriptional regulation of LHCs. This finding links chlorophyll biosynthesis to the regulation of pigment-binding protein synthesis and is likely to be important in coordination of these two processes.[8,9]

DEGRADATION

The color change of leaves in autumn is due to degradation of chlorophyll to colorless compounds when leaves senesce. This decolorization is essential so that plants can recover the protein nitrogen from the leaves. Chlorophyll is phototoxic and can produce singlet oxygen when not associated with LHCs or photosystems. Thus, in order to recover the protein from senescing leaves, the chlorophyll must be degraded to colorless compounds so that the protein nitrogen can be efficiently recovered. This degradation is an ordered process involving removal of the centrally coordinated magnesium and opening of the tetrapyrrole ring by a monoxygenase.[10]

CONCLUSION

Chlorophyll biosynthesis and degradation are complex processes that are vital for the growth and development of photosynthetic organisms, although not yet fully understood. The current focus of research into these processes aims to 1) identify of all of the genes/enzymes involved; 2) provide a structural and mechanistic understanding of the enzymes and regulatory proteins; and 3) determine the regulatory mechanisms by which they are coordinated with other cellular and developmental processes.

ARTICLES OF FURTHER INTEREST

Carotenoids in Photosynthesis, p. 245
Photosystems: Electron Flow Through, p. 906

REFERENCES

1. Anderson, J.M. Photoregulation of the composition, function, and structure of thylakoid membranes. Annu. Rev. Plant Physiol. **1986**, *37*, 93–136.
2. Porra, R.J. The checkered history of the development and use of simultaneous equations for the accurate determination of chlorophylls *a* and *b*. Photosynth. Res. **2002**, *73*, 149–156.
3. Willows, R.D. Biosynthesis of chlorophylls from protoporphyrin IX. Nat. Prod. Rep. **2003**, *20*, 1–16.
4. Beale, S.I. Enzymes of chlorophyll biosynthesis. Photosynth. Res. **1999**, *60*, 43–73.
5. Rudiger, W. Chlorophyll metabolism—From outer space down to the molecular level. Phytochemistry **1997**, *46*, 1151–1167.
6. Reinbothe, S.; Reinbothe, C.; Apel, K.; Lebedev, N. Evolution of chlorophyll biosynthesis—The challenge to survive photooxidation. Cell **1996**, *86*, 703–705.
7. Grimm, B. Regulatory Mechanisms for Eukayotic Tetrapyrrole Biosynthesis. In *The Porphyrin Handbook II*; Kadish, K.M., Smith, K., Guilard, R., Eds.; Academic Press: San Diego, 2003; Vol. 12, 1–32.
8. Strand, A.; Asami, T.; Alonso, J.; Ecker, J.R.; Chory, J. Chloroplast to nucleus communication triggered by accumulation of Mg-protoporphyrin IX. Nature **2003**, *421*, 79–83.
9. Larkin, R.M.; Alonso, J.M.; Ecker, J.R.; Chory, J. Gun4, a regulator of chlorophyll synthesis and intracellular signalling. Science **2003**, *299*, 902–906.
10. Matile, P.; Hortensteiner, S.; Thomas, H. Chlorophyll degradation. Annu. Rev. Plant Physiol. Plant Mol. Biol. **1999**, *50*, 67–95.

Chromosome Banding

Takashi R. Endo
Kyoto University, Kyoto, Japan

C

INTRODUCTION

A chromosome is a discrete body carrying many genes and is composed of DNA and protein. Treatment of chromosomes with alkali or phosphate buffer, followed by staining with Giemsa stain, produces a pattern of bands along the chromosomes. The bands are regarded to represent constitutive heterochromatin, which is composed of repetitive DNA and is highly condensed. C-banding and N-banding are the most commonly used methods for the identification of individual chromosomes and chromosomal structural changes in plants. The power of chromosome banding will be enhanced when it is combined with in situ hybridization.

The particular chromosome complement termed ''karyotype'' is specific to each species. A karyotype is defined by the number and morphology of the chromosomes. The morphology of a chromosome is characterized by the absolute size, position of centromere, and the presence or absence of nucleolar and secondary constrictions. However, the karyotype allows only a few chromosomes of unusual size or shape to be identified unequivocally.

BACKGROUND INFORMATION

In the early 1970s, several chromosome-banding techniques were first developed in mammals. The techniques termed G-, C-, N-, R-, and Q-banding involve treatments with trypsin, alkali, or phosphate buffer, followed by staining with Giemsa or a fluorochrome dye to produce a pattern of bands along the chromosomes.[1] The nature of the bands is regarded to be related to constitutive heterochromatin, which is composed of repetitive DNA and is highly condensed. These chromosome-banding techniques are very powerful tools for the identification of chromosomes and chromosomal structural changes. Among different chromosome banding techniques, G-banding is usually the method of choice for the investigation of human chromosomes, because G-banded preparations contain many bands.

Stimulated by the success in chromosome banding for human and animal chromosomes, cytogeneticists started to apply the same banding methods to plant chromosomes in the 1970s. G-banding does not produce bands to a useful extent in plants. C-banding and N-banding produce distinct bands in many plant species, although not as many as G-bands in animals. At present, these have been the most commonly used methods for the identification of individual chromosomes and chromosomal structural changes in plants, especially in species with large chromosomes, such as rye, barley, wheat, onion, broad bean, and anemone.[2]

C-BANDING

C-bands were first described as pericentric bands, corresponding to the satellite DNA regions in mammalian species, after the treatment of preparations for in situ hybridization. The bands were originally termed C-bands because of their correlation with the centromeric heterochromatin. However, C-banding in plants produces bands not only in pericentric regions but also in terminal and interstitial regions of chromosomes (Fig. 1). Several procedures have been developed for staining constitutive heterochromatin in plants. The C-banding procedure includes an alkali treatment, usually in a barium hydroxide solution ($Ba(H)_2$), incubation in 2 × SSC (standard saline citrate) solution, and then staining with Giemsa stain that contains Azur, eosin, and Methylene Blue (Leishman's stain or Wright stain containing only eosin and Methylene Blue is also used). C-banding in plants does not always produce pericentric bands and is not the only technique that differentially stains constitutive heterochromatin. Therefore,C-banding might be best described as a Giemsa banding method that selectively stains most regions of constitutive heterochromatin after treatment with a barium hydroxide solution followed by incubation in 2 × SSC.[2]

C-bands are regarded as corresponding to constitutive heterochromatin regions containing a mass of repetitive DNA sequences of various kinds.[3] The nature of action of the chemicals in C-banding is as follows: The hydrochloric acid (or acetic acid) treatment depurinates the chromosomes halfway without degrading all of the DNA; the alkali treatment denatures the DNA, which aids solubilization; incubation in warm SSC breaks the sugar-phosphate backbone and DNA fragments pass into solution. Giemsa staining looks like a precipitation reaction

Encyclopedia of Plant and Crop Science
DOI: 10.1081/E-EPCS 120005683

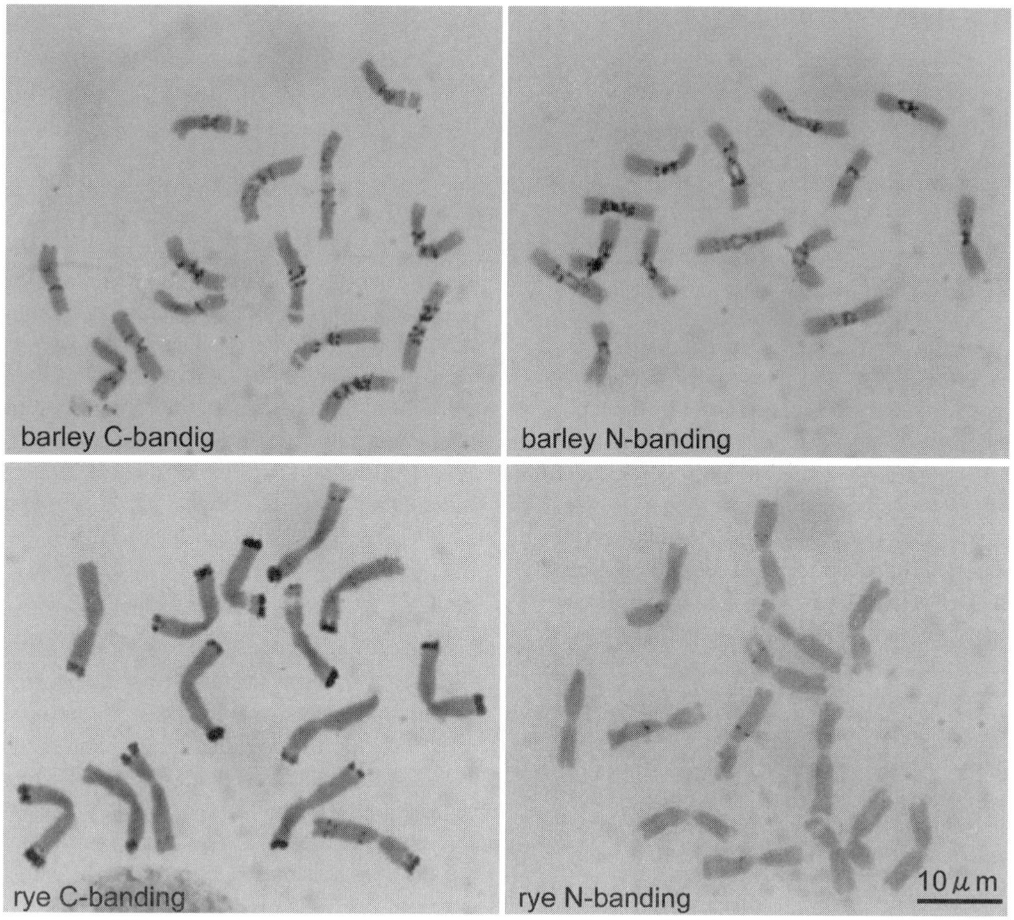

Fig. 1 C-banding and N-banding of barley and rye. Note both methods produce very similar banding patterns in barley, but in rye N-banding produces only a few of the bands produced by C-banding.

on the part of the chromosome that still retains the constitution of the highly compacted heterochromatin.[1] Banding patterns produced by different C-banding methods are basically the same, although some minor bands do not appear or are enhanced depending on the procedures used.[4]

C-banding allows geneticists to construct accurate idiograms for comparing the karyotypes of different plant species and varieties. C-band karyotyping was conducted in the aneuploid series of common wheat (*Triticum aestivum*, $2n = 6x = 42$) to identify individual wheat chromosomes based on their particular banding patterns.[5] The genome evolution of wheat and its related wild species was further analyzed by C-banding beyond the genome analysis based on meiotic chromosome pairing.[6] C-banding further allows the screening for deficient chromosomes in the progeny of a certain line of common wheat in which chromosomal breakage occurs by a genetic mechanism, and a series of deletion stocks have been produced in common wheat.[7]

N-BANDING

N-banding was developed as a simple, one-step banding technique to localize nucleolus organizers in animal and plant chromosomes. However, this method was found to stain differentially the regions of constitutive heterochromatin, not nucleolus organizers, in cereal species. N-banding involves incubation in a hot sodium dihydrogenphosphate (NaH_2PO_4) solution.[2] The N-banding pattern is very similar to that generated by in situ hybridization with a probe of satellite DNA, $(GAA)_m(GAG)_n$.[8] The N- and C-banding patterns look similar in some species but are very different in other species (Fig. 1). All the N-bands appear to be included in the C-bands. Probably, C-banding can differentially stain various types of heterochromatin concurrently.

Chromosome banding is important in crop science because it allows us to identify the chromosomal constitution of interspecies hybrids, alien addition lines, and derivative lines containing alien chromosomal

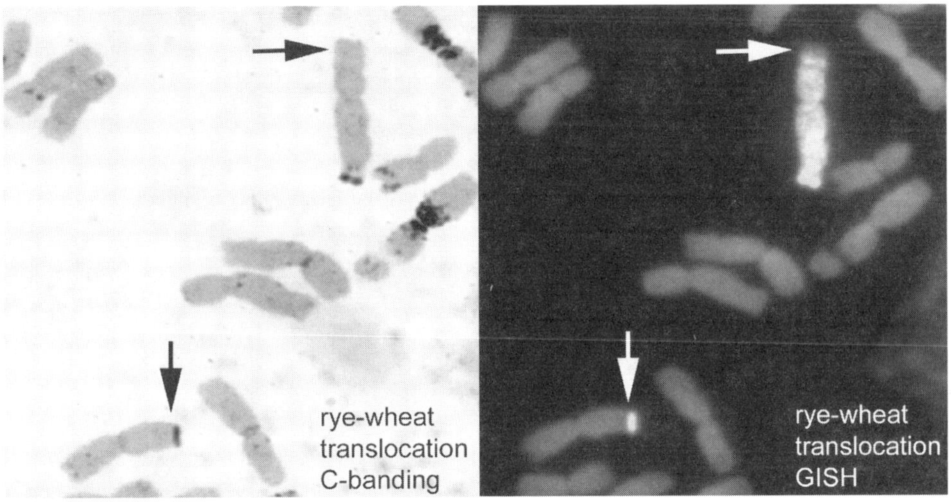

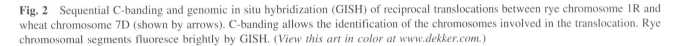

Fig. 2 Sequential C-banding and genomic in situ hybridization (GISH) of reciprocal translocations between rye chromosome 1R and wheat chromosome 7D (shown by arrows). C-banding allows the identification of the chromosomes involved in the translocation. Rye chromosomal segments fluoresce brightly by GISH. (*View this art in color at www.dekker.com.*)

segments. In situ hybridization, the latest technique of molecular cytogenetics, allows the detection of a sequence of interest, such as ribosomal DNA sequences. If highly repetitive DNA sequences are present in large blocks in the genome of an organism, their presence can also be detected as bands by in situ hybridization. In situ hybridization banding is potentially powerful for karyotyping in plants; however, C- and N-banding are still more suitable methods of karyotype analysis in plants because they produce more bands than in situ hybridization.

The power of chromosome banding will be enhanced when it is combined with in situ hybridization. By genomic in situ hybridization chromosomes of one genome can be distinguished from those of another genome, but it is impossible to identify individual chromosomes. For analysis in wheat, when C- or N-banding is followed by in situ hybridization,[9] chromosome banding can identify individual chromosomes and the subsequent in situ hybridization shows the DNA nature of the chromosomes, such as SAT-chromosomes containing a nucleolus organizer region and alien chromosomes containing different types of repetitive DNA sequences. The sequential chromosome banding and in situ hybridization technique is now used to pinpoint the break points of wheat-alien chromosome translocations[10] (Fig. 2).

ARTICLES OF FURTHER INTEREST

Breeding: Incorporation of Exotic Germplasm, p. 222
Chromosome Manipulation and Crop Improvement, p. 266
Chromosome Rearrangements, p. 270
Chromosome Structure and Evolution, p. 273
Fluorescence In Situ Hybridization, p. 468
Mitosis, p. 734

REFERENCES

1. Clark, M.S.; Wall, W.J. Chromosome Identification. In *Chromosomes*, Chapman and Hall: London, 1996; 72–93.
2. Friebe, B.; Endo, T.R.; Gill, B.S. Chromosome Banding Methods. In *Plant Chromosomes*; Fukui, K., Ed.; CRC Press: Boca Raton, 1996; 123–153.
3. Mukai, Y.; Friebe, B.; Gill, B.S. Comparison of C-banding patterns and in situ hybridization sites using highly repetitive and total genomic rye DNA probes of 'Imperial' rye chromosomes added to 'Chinese Spring' wheat. Jpn. J. Genet. **1992**, *67*, 71–83.
4. Gill, B.S.; Kimber, G. Giemsa C-banding and the evolution of wheat. Proc. Natl. Acad. Sci. U. S. A. **1974**, *71*, 4086–4090.
5. Gill, B.S.; Friebe, B.; Endo, T.R. Standard karyotype and nomenclature system for description of chromosome bands and structural aberrations in wheat (*Triticum aestivum*). Genome **1991**, *34*, 830–839.
6. Friebe, B.; Gill, B.S. Chromosome Banding and Genome Analysis in Diploid and Cultivated Polyploid Wheats. In *Methods of Genome Analysis in Plants*; Fukui, P.P., Ed.; CRC Press: Boca Raton, 1996; 39–60.
7. Endo, T.R.; Gill, B.S. The deletion stocks of common wheat. J. Heredity **1996**, *87*, 295–307.
8. Dennis, E.S.; Gerlach, W.L.; Peacock, W.J. Identical polypyrimidine-polypurine satellite DNAs in wheat and barley. Heredity **1980**, *44*, 349.
9. Jiang, J.; Gill, B.S. Sequential chromosome banding and in situ hybridization analysis. Genome **1993**, *36*, 792.
10. Shi, R.; Endo, T.R. Genetic induction of chromosomal rearrangements in barley chromosome 7H added to common wheat. Chromosoma **2000**, *109*, 358–363.

Chromosome Manipulation and Crop Improvement

Adam J. Lukaszewski
University of California, Riverside, California, U.S.A.

INTRODUCTION

Chromosomes carry most of the genetic information of a cell. Any genetic experiment or a breeding effort involves, de facto, a degree of chromosome manipulation. However, in some situations, especially when interspecific or intergeneric hybrids are made for breeding purposes, special steps must be taken to induce chromosomes to perform some unusual feats and generate unusual variants.

In diploids, introgressions from wild relatives are standard, if challenging, breeding strategies. Many desirable characteristics have been introduced in this fashion into such crops as barley, maize, potato, rice, soybeans, tomato, and many others.[5] Some unusual chromosome behavior has been observed along the way, such as apparent absence of chiasma interference in wide hybrids of rice or a change in the distribution of recombination in wide hybrids of grasses, but the procedures are driven strictly by breeding demands. Once the hybrids are made, they do not require any special protocols.

WHAT IS CHROMOSOME ENGINEERING?

An amazing array of techniques and approaches to chromosome manipulation have been devised and tried in research, but they may not lead to the development of successful crops. The criterion for this section, therefore, is successful utilization of the chromosome constructs in commercial agriculture. Conscious chromosome manipulation in crop improvement was termed ''chromosome engineering'' by its pioneer, E. R. Sears.[1] The term refers to a series of actions leading to the introgression into a crop species of a fragment of a chromosome carrying a desirable characteristic, most often a disease or pest resistance locus. Usually, the donors are wild relatives. The practical limits to such introgressions are set by the ability of the donor and recipient species to hybridize and to produce hybrids with at least trace fertility. Somatic cell fusion, while promising to expand the boundaries of wide hybridization and already used successfully in potato breeding,[2] is yet to prove itself in commercial agriculture.

The gene pools of wild relatives of crops are not adapted to agriculture. Therefore, the amount of the introgressed chromatin is of primary concern so that as little as possible of the undomesticated gene pool is transferred along with the targeted locus. Theoretically, single loci can be transferred, but such goals are unrealistic. Often, the process of introgression is frustrated by tight linkages that introduce undesirable characteristics or eliminate desirable loci and may require repeated cycles of backcrosses and selection to remedy.[3] Pleasant surprises also happen, such as the yield-enhancing effects of two alien introgressions into wheat.[4]

WHEAT ENGINEERING

The situation is more complex in wheat. Wheat is an allopolyploid, composed of three closely related genomes. To ensure diploid-like behavior in meiosis, and only bivalent pairing guarantees chromosome stability, wheat evolved a genetic system that controls the stringency of recombination.[6] The stringency criteria are so high that even homologues in intervarietal hybrids may be unable to pair. Under these conditions, there is practically no wheat-alien chromosome pairing. To force it, the pairing control system has to be disabled. This is done by the removal of the main locus, *Ph1*, either by a deletion or nullisomy for chromosome 5B or by making use of its dominant suppressors.

Practical chromosome engineering in wheat is illustrated in Fig. 1. An alien chromosome with the targeted locus is introgressed into its proper (homoeologous) position in one of the genomes by the use of monosomics. This introgression is combined with the *Ph1* mutation or a suppressor, and recombinant wheat-alien chromosomes are recovered in the progeny. Usually, these primary recombinants are single breakpoint translocations. They appear in two configurations: with wheat telomeres and alien centromeres and vice versa. The primary recombinants are tested for the locations of the breakpoints and the presence of the target locus. From each configuration, a chromosome is selected that carries the targeted locus and has the breakpoint as close to it as possible. These two chromosomes are combined in one plant with the *Ph1* locus. This permits only homologous recombination. The only region of homology shared by the two recombinants

Encyclopedia of Plant and Crop Science
DOI: 10.1081/E-EPCS 120005593

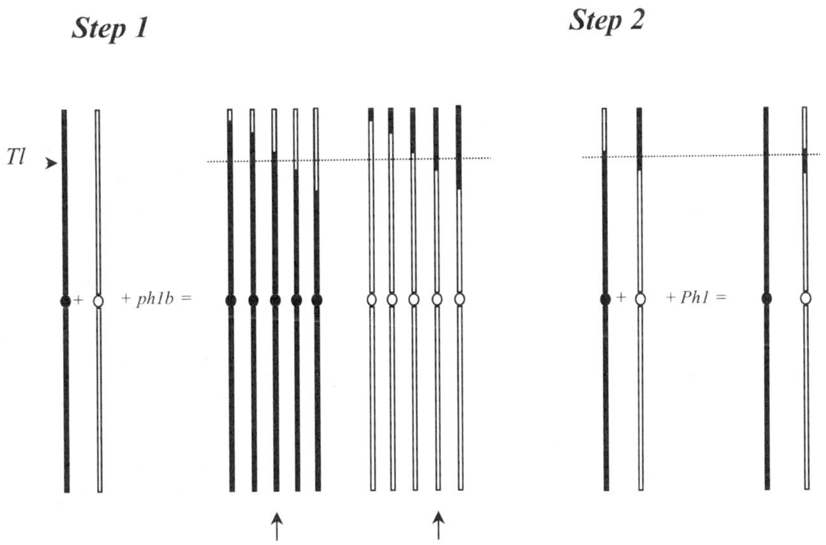

Fig. 1 A two-step approach to engineering intercalary alien transfers into wheat. In Step 1, a donor alien (solid black) and a recipient wheat (open white) chromosome are induced to pair and recombine by the absence of the *Ph1* locus, and two classes of single breakpoint translocations are recovered. From each class, recombinants with the target locus (*Tl*) and the closest possible breakpoint are selected (arrows). In Step 2, the two selected recombinants are combined in one plant. In the presence of wild-type *Ph1*, only homologous recombination is permitted. A crossover in the overlapping region of homology produces a wheat chromosome with an intercalary alien introgression that carries *Tl* and a normal donor chromosome.

is between the two breakpoints: A crossover within that segment produces a secondary recombinant, a chromosome of wheat with an intercalary segment of alien chromatin, and a normal donor chromosome.

PRIMARY RECOMBINANT CHROMOSOMES

The precision of this approach depends on the number of primary recombinant chromosomes available and the precision of breakpoint mapping. Larger populations of primary recombinants offer higher probability of breakpoints close to the targeted locus. The number of available recombinants is a function of the pairing frequency of the donor and recipient chromosomes and the size of screened populations. Because pairing of alien chromosomes with wheat can be very low even in the absence of *Ph1*, the screened populations may have to be very large to eventually generate small intercalary transfers. For example, if the targeted locus in the final chromosome is to be flanked by no more than 1 cM of alien chromatin, and the donor and the recipient pair 1% of the time, 29,956 progeny must be screened for a 95% probability of recovery of the two necessary breakpoints.

Very high numbers of the primary recombinants are not necessarily advantageous. With the current marker technology, high-precision breakpoint mapping is possible, if costly. However, even if the positions of tightly spaced breakpoints can accurately be resolved, and the best two are selected to be combined into an intercalary introgression, the probability of a crossover in their shared segment of homology may be too low to be practically useful. In fact, the contrasting structure of the two primary recombinants may prevent them from forming any chiasmata, even if sufficient homology exists. No practical limits for the chiasma establishment under such conditions have been determined, but the author observed a threefold reduction in the crossover frequencies in pairs of selected primary recombinants, relative to the frequencies inferred on the basis of their breakpoint positions.

Once a series of primary recombinant chromosomes have been produced, any higher-order recombinants can be generated, depending on the practical demands. Four breakpoint translocations were produced in an attempt to remedy the quality defect of the 1RS.1BL wheat-rye translocation in wheat.[7] Primary recombinants can also be used in homoeologous recombination to direct the crossovers into proximal locations that would not be easily accessible in the first round of recombination, thereby extending the range of chromosome locations accessible to manipulation.

The limit of this engineering approach is set by chromosome affinity. If there is no meiotic pairing of the donor with the recipient, the recombination approach will

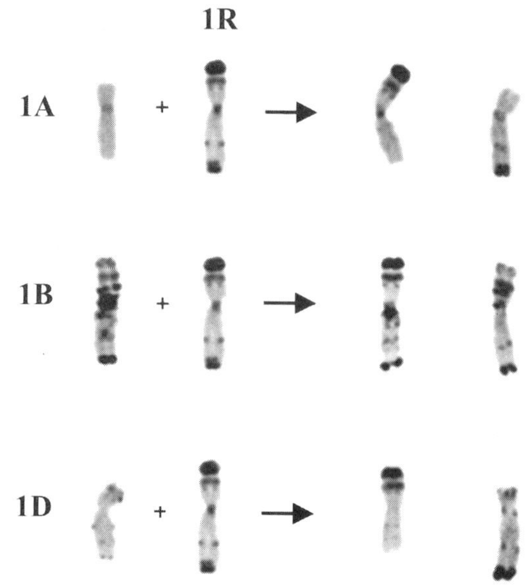

Fig. 2 Whole-arm translocations of the same rye chromosome (1R) with each of the group-1 chromosomes of wheat (1A, 1B, 1D). From all possible combinations of random fusion of centric fission products, only compensating translocations, that is, those involving the short arm of rye (1RS) with the long arms of wheat (1AL, 1BL, and 1DL) and the short arms of wheat (1AS, 1BS, and 1DS) with the long arm of rye (1RL), were selected.

not succeed. Such recalcitrant chromosomes may still be engineered but with considerably less precision. This is done by chromosome fragmentation, either centric fission-fusion where at least the recipient can be selected or by completely random processes of chromosome breakage and reunion of broken fragments.

FISSION-FUSION

The centric fission–fusion approach exploits a tendency of unpaired chromosomes in meiosis to break across the centromeres. From a range of random fission-fusion products of a donor and a recipient, compensating whole-arm translocations can be selected (Fig. 2). One of the most successful alien introgressions in wheat, the 1RS.1BL wheat-rye translocation, must have originated by centric fission-fusion.

Irradiation of plants or pollen offers the means of introgressing chromosome fragments smaller than whole arms. The fragmentation approach uses brute force to break chromosomes and relies on the DNA repair mechanisms to fuse the broken fragments into new chromosomes. No affinity of the chromosomes is required, but the resulting translocation still must compensate the absence of the original recipient chromosome or it will not be accepted in agriculture. Built into this approach is an assumption, or hope, that a large enough scale offers a sensible chance of finding compensating translocations that carry the targeted loci from the donors. Given the astronomical numbers of possible breakpoint positions and fusion combinations, the fragmentation approach does not offer realistic chances for successful transfers. With few exceptions,[8] chromosome translocations produced in this fashion are noncompensating and have not been widely utilized in practical plant breeding.[9] However, the fragmentation approach lures like a lottery: It teases with a large payoff for a low investment. On the other hand, introgressions based on recombination, if done with sufficient care, offer a high probability of success but can be extremely laborious.

Chromosomes of crop species can be engineered to produce duplications of small segments of the genome. This is accomplished by combining certain reciprocal translocations into double translocation homozygotes. In barley, some of such duplications produced up to 35% yield increases.[10] It is doubtful that this approach will be used on a wider scale as very few crops have large enough collections of chromosome translocations to suit this purpose, and no efforts to produce them are apparent.

CONCLUSION

The techniques of molecular biology will likely have a major impact on future chromosome manipulation approaches in plant breeding. Already, genetic maps and sets of DNA markers increase the precision of chromosome engineering. Perhaps the techniques of gene identification, cloning, and transformation will eventually eliminate all need for painstaking assembly of chromosome pieces into viable constructs. However, it is likely that in their search for new sources of variation, breeders will keep probing the fringes of the possible. Such fringes do not offer sufficient economic incentives for routine applications of sophisticated and expensive techniques. Therefore, the quest will probably continue even if the details of the approaches change.

ARTICLES OF FURTHER INTEREST

Breeding: Incorporation of Exotic Germplasm, p. 222
Chromosome Banding, p. 263
Chromosome Rearrangements, p. 270
Fluorescence In Situ Hybridization, p. 468
Germplasm: International and National Centers, p. 531

Interspecific Hybridization, p. 619
Meiosis, p. 711
Polyploidy, p. 1038

REFERENCES

1. Sears, E.R. Transfer of Alien Genetic Material to Wheat. In *Wheat Science: Today and Tomorrow*; Evans, L.T., Peacock, W.J., Eds.; Cambridge Univ. Press: Cambridge, 1981; 75–89.
2. Helgeson, J.P.; Pohlman, J.D.; Austin, S.; Haberlach, G.T.; Wielgus, S.M.; Ronis, D.; Zambolim, L.; Tooley, P.; McGrath, J.M.; James, R.V.; Stevenson, W.R. Somatic hybrids between *Solanum bulbocastanum* and potato: A new source of resistance to late blight. Theor. Appl. Genet. **1998**, *96*, 738–742.
3. Duvick, D.N. The Romance of Plant Breeding and Other Myths. In *Gene Manipulation in Plant Improvement*; Gustafson, J.P., Ed.; Plenum Press: New York, 1990; 39–54.
4. Singh, R.P.; Huerta Espino, J.; Rajaram, S.; Crossa, J. Agronomic effects from chromosome translocations 7DL.7Ag and 1BL.1RS in spring wheat. Crop Sci. **1998**, *38*, 27–33.
5. Gupta, P.K.; Tsuchiya, T. *Chromosome Engineering in Plants: Genetics, Breeding, Evolution*; Elsevier: Amsterdam, 1991.
6. Dubcovsky, J.; Luo, M.-C.; Dvorak, J. Differentiation between homoeologous chromosomes 1A of wheat and $1A^{m}$ of *Triticum monococcum* and its recognition by the wheat *Ph1* locus. Proc. Natl. Acad. Sci. U. S. A. **1995**, *92*, 6645–6649.
7. Lukaszewski, A.J. Manipulation of the 1RS.1BL translocation in wheat by induced homoeologous recombination. Crop Sci. **2000**, *40*, 216–225.
8. Sears, E.R. The transfer of leaf rust resistance from *Ae. umbellulata* to wheat. Brookhaven Symp. Biol. **1956**, *9*, 1–22.
9. Friebe, B.; Jiang, J.; Raupp, W.J.; McIntosh, R.A.; Gill, B.S. Characterization of wheat–alien translocations conferring resistance to diseases and pests: Current status. Euphytica **1996**, *91*, 59–87.
10. Hagberg, A.; Hagberg, P. Production and Analysis of Chromosome Duplications in Barley. In *Chromosome Engineering in Plants: Genetics, Breeding, Evolution*; Gupta, P.K., Tsuchiya, T., Eds.; Elsevier: Amsterdam, 1991; 401–410.

Chromosome Rearrangements

Tomás Naranjo
Universidad Complutense, Madrid, Spain

INTRODUCTION

Chromosome rearrangements and polyploidy have produced a huge variation in the number, size, and morphology of chromosomes among, and often within, plant taxa. Structural chromosome modifications invariably occur in all species. These changes appear in one or a few individuals and occasionally become fixed in a given population. Some chromosome rearrangements may accompany the origin of new species or be shared by related species that inherited them from a common ancestor. Cytogenetic studies that detect and explain the origin and evolutionary implications of chromosome rearrangements are considered in this report, with special emphasis on Triticeae, a tribe of the family *Poaceae* (*Gramineae*) with diploid and polyploid species and chromosome rearrangements at the intraspecific, specific, and supraspecific levels.

IDENTIFICATION OF CHROMOSOMAL ABERRATIONS

Identification of structural changes is a prerequisite for understanding their effect and implication on genome evolution. Multivalent configurations at metaphase I were evidence for the evolution of multiple translocation complexes in the genera *Oenothera* and *Rhoeo*. Translocations accumulated during the evolution of these genera are maintained in heterozygous condition because of the action of lethal genes or mechanisms in the gametophyte that prevent gametes with a particular chromosome set from participating in fertilization. However, chromosome rearrangements are usually fixed in homozygous condition and do not require additional mechanisms to become permanent. Other cytogenetic approaches permit the identification of these structural changes. In some species, the chromosome structure has been studied using chromosome banding techniques.[1] Pairing data between chromosomes from related (homoeologous) genomes at metaphase I in interspecific hybrids (Fig. 1) provides information on chromosome structural differentiation.[2] The fluorescence in situ hybridization (FISH) technique reveals the physical location and organization of DNA sequences on chromosomes. A variant of FISH, genomic in situ hybridization (GISH), permits visualization of entire genomes and allows identification of intergenomic chromosome exchanges (Fig. 2) produced after allopolyploid formation.[3] Finally, linkage and gene order in genomes with variable degrees of relatedness may be compared from the construction of genetic maps.[4]

TRITICEAE AS A MODEL FOR STUDYING CHROMOSOME REARRANGEMENTS

The tribe Triticeae contains about 500 diploid and polyploid species, including the agronomically important crops, wheat, barley, and rye. Diploid species of the tribe diverged from an ancestral species with $2n=14$ chromosomes. Autopolyploids and allopolyploids appeared in a subsequent evolutionary step. Common bread wheat, *Triticum aestivum*, is an allohexaploid species ($2n=6x=42$, genome AABBDD) that arose from the hybridization between tetraploid wheat *T. turgidum* ($2n=4x=28$, AABB) and *Aegilops tauschii* (DD). Hybridization between *T. urartu* (AA) and the donor species of the B genome gave rise to *T. turgidum*. Development of engineered nullisomic–tetrasomic lines of bread wheat as well as wheat-alien addition and substitution lines helped identify homoeologous chromosomes. Comparative genome analysis based on homoeologous pairing revealed that the D genome of wheat and most of the A and B genome chromosomes preserve the ancestral arrangement. Two interchanges involving chromosome arms 5AL/4AL and 4AL/7BS, and a large pericentric inversion of chromosome 4A that occurred during the evolution of wheat were detected.[3] These rearrangements are present in *T. turgidum*, which inherited the 5AL/4AL translocation from *T. urartu*. Comparative mapping demonstrated the conserved synteny of wheat chromosomes only disrupted by the above mentioned rearrangements in addition to a paracentric inversion of 4AL.[4] GISH analysis confirmed the absence of translocations between homoeologous chromosomes.[5]

T. timophevii (A^tA^tGG), another tetraploid species formed from the hybridization between *T. urartu* and *Ae. speltoides*, also inherited the 4AL/5AL translocation from *T. urartu*. Tetraploid wheats share no other evolutionary chromosome rearrangement, although four

Encyclopedia of Plant and Crop Science
DOI: 10.1081/E-EPCS 120005589

species-specific translocations, $6A^tL/1GS$, 1GS/4GS, $4GS/4A^tL$, and $4A^tL/3A^tL$, are known in *T. timopheevii*.[3,6] Structural chromosome differences suggest that *T. turgidum* and *T. timopheevii* originated from two independent hybridization events. Several landraces and wild populations of tetraploid wheats have additional chromosome modifications, whose histories may be traced. For instance, the karyotypic diversity of wild populations of *T. timopheevii* in different geographical regions suggests that the center of origin and the center of primary diversity of this species is in Iraq.[7]

Diploid species show a variable number of chromosome rearrangements relative to wheat.[8] Genomes of *Ae. umbellulata* and *Secale cereale* (rye) have eleven and seven rearrangements, respectively—mainly translocations. It needs to be determined whether two inversions detected in *Hordeum vulgare* (barley) and another two in *H. bulbosum* are species-specific. *Ae. longissima*, *T. urartu*, and *T. monococcum* have one translocation each, and *Ae. speltoides*, *Ae, sharonensis*, and *Ae. tauschii* show no structural modification. Rye and *Ae. umbellulata* have an interchange between chromosomes 4/5 similar to the 4AL/5AL translocation of the genus *Triticum*. However, these translocations arose independently, because the D genome, which is more closely related to the A genome than to the genomes of rye or *Ae. umbellulata*, has no translocations. Dissimilar numbers of rearrangements, either in outbreeders such as rye and *Ae. speltoides*, or in predominantly self-pollinated species such as *Ae. umbellulata* or *Ae. sharonensis*, indicate no relationships between structural chromosome differentiation and the reproductive system. The translocation present in *Ae. longissima* could be the primary cause of the reproductive isolation that led to the formation of this species. The same process could have occurred in the differentiation of the A genome of the genus *Triticum*. Likewise, rearrangements present in rye most likely accumulated during the evolution of the genus *Secale*. However, chromosome differentiation is not a prerequisite for divergence of diploid species among the Triticeae. The genera *Hordeum* and *Secale* and the *Triticum-Aegilops* group are representative members of separate lineages in the tribe Triticeae, and originated in that order and are not parallel to the accumulation of rearrangements.

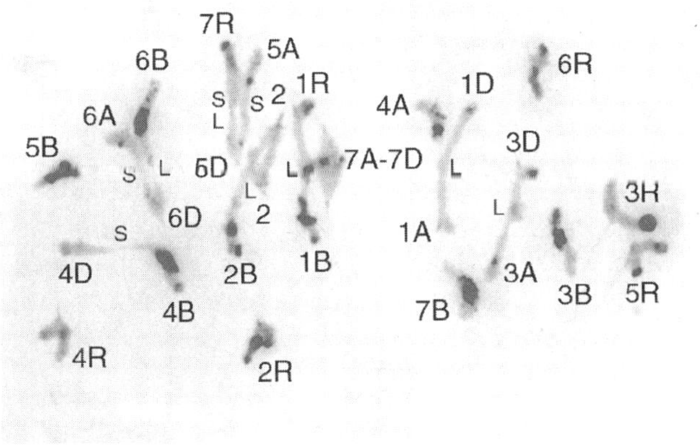

Fig. 1 Homoeologous pairing at metaphase I in a *ph1bT. aestivum* (AABBDD) × *S. cereale* (RR) hybrid. Association 7RS-5DL denotes that chromosome arm 7RS carries a translocated segment from 5RL. L, long arm, S, short arm.

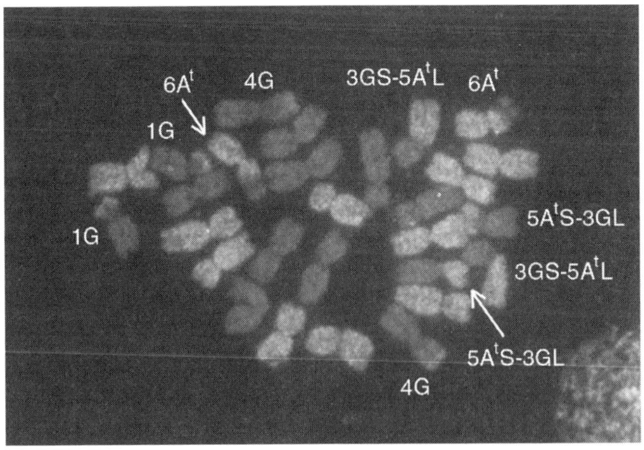

Fig. 2 Intergenomic translocations in *T. timopheevii* chromosomes detected by GISH. (From Ref. 6 by permission of NRC Research Press.) (*View this art in color at www.dekker.com.*)

The highly conserved structure of the diploid progenitors of polyploid wheats contrasts with the number—three and four, respectively—of evolutionary chromosome rearrangements produced in tetraploid wheats. All these structural changes had to arise in the primitive tetraploids, because transient structures of their sequence were not found. The formation of new polyploids is often accompanied by extensive genomic modifications within a short period of time. Evolutionary chromosome rearrangements of tetraploid wheats may represent a fraction of the overall genomic reorganization that occurred immediately after polyploidization. Data from comparative mapping support a high rate of chromosome rearrangements following the whole genome replication that occurred in the ancestors of *Brassica* and maize. Some of these changes could be involved in the subsequent reduction of the basic chromosome number.

ORIGIN OF CHROMOSOME REARRANGEMENTS

Intergenomic translocations also exist in polyploids from other taxa such as *Avena* or *Nicotiana*. Recombination between homoeologous chromosomes in the primitive polyploid has been suggested as a possible origin of such translocations. However, this is not the case in polyploid wheats, where evolutionary intergenomic rearrangements involve nonhomoeologous chromosomes. Both intergenomic and intragenomic rearrangements could have a similar origin. The position of the breakpoints at or near the centromere in many intraspecific chromosome rearrangements of tetraploid wheats[9] contrasts with the intercalary or distal position of evolutionary rearrangements. This nonrandom distribution may reflect different mechanisms generating each type of chromosome modification. Translocations with centromeric breakpoint may arise by rejoining the arms of nonhomologous chromosomes that misdivide simultaneously at anaphase I after failure of synapsis or recombination. Another possibility is that translocations derive from the effect of gametocidal genes, which cause chromosomal breaks in the gametophyte. Most breakpoints are noncentromeric, although centromeric breaks also occur. The broken chromosomes can rejoin to produce translocated chromosomes. Finally, transposable elements may mediate the production of chromosome rearrangements. Transposable elements have been found to be associated with chromosomal rearrangements such as deletions, duplications, inversions, and translocations in different species. Homologous recombination between elements at different locations and alternative transposition have been proposed as two possible mechanisms by which chromosome rearrangements may occur.[10]

FUTURE PROSPECTS

The cytogenetic approaches used in the last 15 years reveal that intraspecific and evolutionary chromosome rearrangements occurred in diploid and polyploid species from different plant taxa. Among diploids, very distant species preserve their ancestral chromosome structure, although others accumulate several chromosome rearrangements. Thus, chromosome rearrangements are not a prerequisite for speciation. By contrast, speciation caused by polyploidy is often accompanied by structural chromosome differentiation.

An immediate aim in the study of chromosome rearrangements is the determination of the DNA sequence at the breakpoints. This may help to explain the origin of rearrangements and phenomena such as their accumulation in some chromosomes, as in wheat chromosome 4A, or their recurrence, as in the translocation between chromosomes 4/5 in Triticeae. On the other hand, structural changes may modify the topological arrangement of chromatin in the nucleus. The consequences that chromosome rearrangements may have for gene function and its evolutionary implications should be the focus of new studies.

ARTICLES OF FURTHER INTEREST

Chromosome Banding, p. 263
Chromosome Manipulation and Crop Improvement, p. 266
Chromosome Structure and Evolution, p. 273
Fluorescence In Situ Hybridization, p. 468
Molecular Evolution, p. 748
Polyploidy, p. 1038

REFERENCES

1. Friebe, B.; Gill, B.S. Chromosome Banding and Genome Analysis in Diploid and Cultivated Polyploid Wheats. In *Methods of Genome Analysis in Plants*; Hauhar, P.P., Ed.; CRC Press: Boca Raton, 1996; 39–60.
2. Naranjo, T.; Roca, A.; Goicoechea, P.G.; Giráldez, R. Arm homoeology of wheat and rye chromosomes. Genome **1987**, *229*, 873–882.
3. Jiang, J.; Gill, B.S. Different species-specific chromosome translocations in *Triticum timopheevii* and *T. turgidum* support the diphyletic origin of polyploid wheats. Chromosome Res. **1994**, *2*, 59–64.
4. Devos, K.M.; Gale, M.D. Comparative genetics in the grasses. Plant Mol. Biol. **1997**, *35*, 3–15.
5. Sánchez-Morán, E.; Benavente, E.; Orellana, J. Analysis of karyotypic stability of homoeologous pairing (*ph*) mutants in allopolyploid wheats. Chromosoma **2001**, *110*, 371–377.
6. Rodríguez, S.; Perera, E.; Maestra, B.; Díez, M.; Naranjo, T. Chromosome structure of *Triticum timopheevii* relative to *T. turgidum*. Genome **2000**, *43*, 923–930.
7. Badaeva, E.D.; Badaev, N.S.; Gill, B.S.; Filatenko, A.A. Intraspecific karyotype divergence in *Triticum araraticum* (*Poaceae*). Plant. Syst. Evol. **1994**, *192*, 117–145.
8. Maestra, B.; Naranjo, T. Genome Evolution in Triticeae. In *Chromosomes Today*; Olmo, E., Redy, C.A., Eds.; Birhauser: Verlag: Basel, 2000; Vol. 13, 115–167.
9. Badaeva, E.D.; Jiang, J.; Gill, B.S. Detection of intergenomic translocations with centromeric and noncentromeric breakpoints in *Triticum araraticum*: Mechanism of origin and adaptive significance. Genome **1995**, *38*, 976–981.
10. Gray, Y.H.M. It takes two transposons to tango: Transposable-element-mediated chromosomal rearrangements. Trends Genet. **2000**, *16*, 461–468.

Chromosome Structure and Evolution

Ingo Schubert
Andreas Houben
Institute for Plant Genetics and Crop Plant Research, Gatersleben, Germany

INTRODUCTION

Plants share with other eukaryotes the basic structural organization of chromosomes and their subdivision into functional domains, such as telomeres, centromeres, nucleolus organizers, euchromatin, and heterochromatin.[1] The same holds true for the primary and secondary events that contribute to evolutionary alteration of number, size, shape, and content of chromosomes.[2] Constraints restricting the variability of chromosome size are indicated.

CHROMOSOME STRUCTURE

Plant nuclear chromosomes, like those of other eukaryotes, are composed of DNA, RNA, and proteins forming the chromatin. They consist of one linear DNA double helix per unreplicated chromosome (or per sister chromatid after reduplication during S-Phase of the cell cycle). The DNA double helix encoding the genetic information is wound approximately 1.75 turns, corresponding to approximately 160–200 base pairs (bp), around octamers of two molecules each of the histones H2A, H2B, H3, and H4—the nucleosomal core—while the linking stretches of DNA are usually associated with histone H1. This ''bead on a string'' structure represents the thin chromatin fibril that in turn is hierarchically structured by coils, loops, and/or spirals eventually resulting in the most dense ''transport'' form that becomes visible microscopically as individual chromosomes during nuclear division. This structuring, for which the precise mechanism is not yet known, yields a condensation by approximately 5 orders of magnitude from naked DNA to mitotic chromosomes (Fig. 1a).

Post-translational modifications of specific amino acids of nucleosomal histones (acetylation, methylation, phosphorylation, and ubiquitination) are involved in regulation of several nuclear processes (''epigenetic histone code.''[3]).

Each diploid species is characterized by a specific set of chromosome pairs that may vary considerably in number ($2n = 4$ to >100), size (<1–>10 μm), and shape (symmetric/symmetric arm length, with/without satellites; see Fig. 1b). The chromosome complement of an organism—the karyotype—may consist of chromosomes that are similar or different in size and/or morphology.

Although the estimated number of genes in higher eukaryotes is similar (between approximately 20,000 to 60,000), the DNA content of (unreplicated) nuclear plant genomes may vary from approximately 1.5×10^2 Megabase pairs (Mbp) in *Arabidopsis* to $>10^5$ Mbp in some *Fritillaria* species. This high variability is due mainly to the varying content of tandem and dispersed repetitive sequences. The latter are represented mainly by redundant, potentially mobile transposon and retrotransposon sequences. These rapidly evolving sequences have a major impact on chromosome evolution, and probably also on speciation and regulation processes. Discussion of the main structurally and functionally distinct domains of chromosomes (Fig. 1c) follows.

Telomeres

The chromosome ends, called telomeres, are responsible for stable maintenance of linear chromosomes.[4] Highly conserved TTTAGGG tandem repeats represent the typical telomeric sequences of all plant phyla. In some Asparagales these repeats are substituted with the vertebrate-specific sequence $TTAGGG_n$ and in Alliaceae they are secondarily lost during evolution, as are the insect repeats $TTAGG_n$ in Diptera. Canonical telomeric repeats form single-stranded 3′-overhangs that pair into proximal double-stranded regions generating a ''T-loop.'' The terminal T-loops are stabilized by telomere-binding proteins, thus preventing exonucleolytic degradation and recombination between DNA ends that occur during repair of internal double-strand breaks. Because all known DNA-polymerases need RNA-primers to add deoxy-nucleotides in the 5′ to 3′ direction, each replication cycle results in the shortening of linear DNA molecules. This shortening is compensated by a reverse transcriptase—the telomerase—that, according to an internal complementary RNA-template, adds telomeric repeats to the 3′ DNA ends (Fig. 1d). In the absence of telomerase, telomeres may, in some cases, become extended by nonreciprocal recombination (conversion). However, this has not yet been reported for plants. Frequently, telomeric sequences are proximally followed by species-specific subtelomeric tandem repeats.

Encyclopedia of Plant and Crop Science
DOI: 10.1081/E-EPCS 120005587

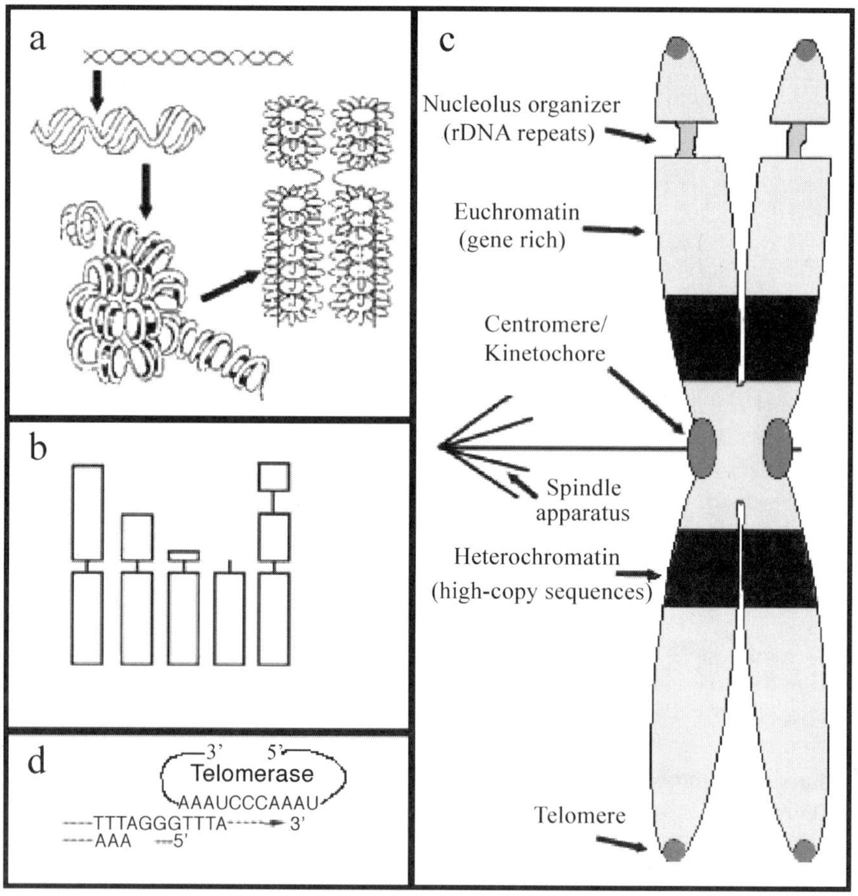

Fig. 1 Structure of eukaryotic chromosomes. a) Structural condensation from naked DNA into metaphase chromosomes. b) Types of chromosome morphology from left to right: metacentric; submetacentric; acrocentric; telocentric; and metacentric chromosome with satellite, separated from its arm by the nucleolus-organizing secondary constriction. c) Schematic presentation of a metaphase chromosome and its domains. d) Elongation of terminal telomeric sequence repeats according to the complementary internal RNA component of the reverse transcriptase ''telomerase.''

Centromeres

Another domain of eukaryotic chromosomes essential for their stable inheritance is the centromere,[5,6] the primary construction of monocentric chromosomes. (Some plant genera, e.g., *Luzula*, possess poly- or holocentric chromosomes, with centromeres not restricted to a single site on the chromosome.) At the centromeres >20 highly conserved proteins are permanently or transiently (during the nuclear divisions) assembled within the kinetochore

Fig. 2 Different types of chromosome rearrangements. a) Alteration of chromosome number by reversible fusion of telocentric into bi-armed metacentric chromosomes without (extensive) loss of sequences (centromeres: black, telomeres: triangles). b) Basic mechanisms of primary structural chromosome rearrangements. c) ''Breakage-fusion-bridge'' cycle of dicentric chromosomes (resulting from asymmetric chromosome translocations) may secondarily yield deletions (top, right) and duplications and inversions (bottom, right) due to random disruption of chromosome bridges during nuclear divisions until the breakage products become stabilized by the gain of telomeric sequences at the broken ends. d) Secondary chromosome rearrangements (bottom, right) arising by crossover (x) between homologous regions of translocation chromosomes within a meiotic multivalent ([]) in individuals heterozygous for two translocations. (As a second recombination product, the ancestral karyotype is reconstituted. Compare top and bottom left.) e) Alteration of diploid chromosome number by meiotic mis-segregation from a meiotic multivalent ([]) of individuals heterozygous for two translocations between one bi-armed and two one-armed chromosomes. When the two bi-armed translocation chromosomes segregate into one gamete and the remaining four one-armed chromosomes segregate into the other, chromosome numbers alter into $n-1$ and $n+1$ simultaneously, accompanied by only minor deletions or duplications (]), respectively.

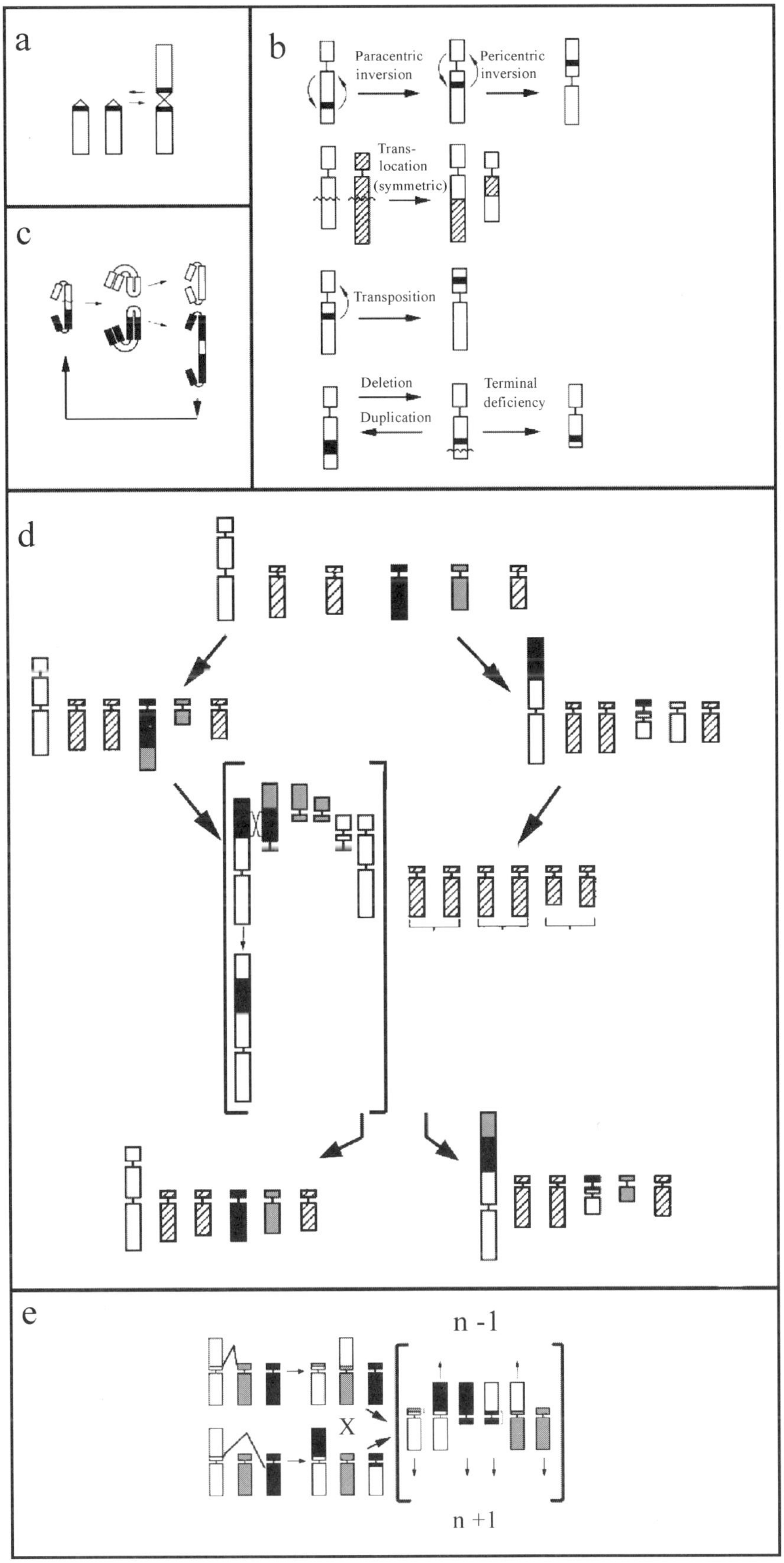
a
b
Paracentric
inversion
Pericentric
inversion
Trans-
location
(symmetric)
Transposition
Deletion
Duplication
Terminal
deficiency
c
d
e
n -1
X
n +1

Fig. 2

complex. Distinct kinetochoric proteins are responsible for the onset of nuclear division (anaphase-promoting complex), for cohesion of sister chromatids at centromeres until anaphase, and for correct segregation of sister chromatids (mitosis, meiosis II) or homologous chromosomes (meiosis I) into the daughter nuclei via interaction with the fibers of the spindle apparatus.

In contrast to telomeres, centromeres are not specified by highly conserved DNA sequences. Tandem repeats (and in cereals, also ''gypsy-type'' retroelements) often occur at centromeres, and bind centromere-specific histone H3 variants. Nevertheless, their functional importance is unclear because they are not present in all centromeres/neocentromeres of rearranged chromosomes. Therefore, an epigenetic mechanism for kinetochore assembly cannot be excluded.

Nucleolus Organizers

Nucleolus organizing regions (NORs) occur on one or several chromosome pairs of a complement and harbor the most redundant genes (up to 10,000 copies per locus). They encode the transcription units for the large ribosomal rRNA fractions (5.8S, 18S, 25S RNA). If not at terminal position or suppressed by ''nucleolar dominance,'' they are visible as achromatic threads—the ''secondary constrictions,'' separating the distal ''satellite'' from the remaining chromosome arm (Fig. 1b,c)—and form a nucleolus, where ribosomes are synthesized, in interphase nuclei. Specific proteins of active NORs can be visualized by silver staining.

Heterochromatin

Heterochromatin forms the most dense structures of interphase nuclei. It represents late replicating, genetically inert, or transcriptionally inactive chromatin. On dividing chromosomes, heterochromatin is made visible by various techniques yielding species-specific band patterns, mostly around centromeres but also at interstitial and/or terminal positions. In addition to constitutive heterochromatin, sex chromosomes or entire parental chromosome sets may appear as ''facultative'' heterochromatin (not yet found in plants). Constitutive heterochromatin of large genomes often consists of highly methylated tandem-repetitive DNA sequences associated with heterochromatin-specific proteins and nucleosomal histones H3 and H4, showing a low level of lysine acetylation. Lysine 4 of H3 is regularly less intensely and lysine 9 mostly more intensely di-methylated in heterochromatin than in euchromatin domains. Within small genomes (e.g., that of Arabidopsis) heterochromatin is restricted mainly to pericentromeric regions, containing many of the potentially mobile sequences (transposons, retrotransposons) in addition to tandem repeats, and inactive rDNA. Heterochromatin is rarely involved in meiotic recombination but it represents ''hot spots'' for structural chromosome aberrations if repair of DNA damage interferes with replication. Positioning of genes into or close to heterochromatin may lead to suppression of their transcription (position effect) via ''heterochromatinization.''

Heterochromatin represents the most variable chromosomal domain with a high potential for evolutionary alteration and epigenetic modification of nuclear processes such as replication, transcription, repair, and recombination.

Euchromatin

Euchromatin is of less dense structure in interphase nuclei, replicates earlier than heterochromatin, and is enriched in genes (approximately 1 gene/5 kilobase pairs (kb) in *Arabidopsis*). It contains higher levels of acetylated histones and H3 di-methylated at lysine 4. Euchromatin of large plant genomes may harbor considerable amounts ($\geq$90%) of diverse repetitive sequences, in particular retroelements, interspersed between genes/gene clusters. Such sequences are inactivated by DNA di-methylation and associated with H3 methylated at lysine 9. As to both features, euchromatin of large plant genomes resembles the pericentromeric heterochromatin of the small Arabidopsis genome.

CHROMOSOME EVOLUTION

During evolution, the number, shape, size, and content of chromosomes may change according to mechanisms similar in principle for all eukaryotes.

Primary Rearrangements

Apart from interspecific hybridization, adding chromosome complements or single chromosomes to that of a host spindle disturbances may result in poly- or aneuploid chromosome numbers.

At the euploid level, chromosome number, size, and shape may change due to fusion of telocentrics resulting in metacentric chromosomes. This is reversible if centromeric and telomeric sequences from both telocentrics persist within the fusion chromosome[7] (Fig. 2a). Otherwise, centric fission may yield stable telocentrics when telomeric sequences are patched to the breakpoints.

Size and/or morphology also may be modified by reciprocal translocations, para-/pericentric inversions, interstitial or terminal deletions (tolerable only when dispensable/multiple sequences are involved), or by sequence insertions into breaks (Fig. 2b). Such ''primary'' structural rearrangements usually reflect misrepair of

DNA damage, particularly during nonhomologous end-joining of double-strand breaks.

Sequence amplification/deletion may also occur via ''replication slippage'' or due to unequal (out-of-frame) recombination (meiotic crossover, somatic sister chromatid exchange) between (partially) homologous sequences. Chromosomes may ''grow'' by active dispersion of retroelements or ''shrink'' by excision of transposons or exonucleolytic degradation of break ends during repair processes.

Secondary Rearrangements

Unstable products of primary rearrangements, e.g., dicentric chromosomes (resulting from asymmetric translocations), or ring chromosomes (resulting from intrachromosomal translocations), may initiate ''breakage-fusion-bridge'' cycles and eventually yield stable chromosomes with further inversions, duplications, or deletions[7] (Fig. 2c).

Even stable primarily reconstructed chromosomes may cause secondary rearrangements by meiotic cross-over between homologous regions of translocation chromosomes when combined in heterozygous condition[7] (Fig. 2d).

Mis-segregation from meiotic multivalents within individuals heterozygous for two whole-arm translocations involving one meta- and two acrocentric chromosomes may alter the euploid chromosome number of a complement simultaneously in both directions[8] (Fig. 2e). In principle, all types of chromosome modification may contribute individually or in combination to chromosome evolution. However, comparative mapping, chromosome banding, and/or painting often revealed preferences for specific types of modifications during the chromosomal evolution of related groups of organisms.

Tolerance Limits for Chromosome Evolution

Finally, there are apparently upper and lower (species-specific) tolerance limits for chromosome size. Chromosome arms longer than half of the spindle axis dimension at telophase yield disturbances during nuclear divisions, caused by incomplete separation of the corresponding sister chromatid arms.[9]

On the other hand, chromosomes falling below a lower size limit (<1% of the genome) frequently do not segregate correctly during meiosis (even if an original centromere is present), and thus are not transmitted to the progeny.[10] Although the reason for this behavior (e.g., lack of homologous pairing or need for a minimal ''lateral support'' of centromeres) is not yet evident, it bears severe consequences for the construction of artificial mini-chromosomes.

ARTICLES OF FURTHER INTEREST

Chromosome Banding, p. 263
Chromosome Rearrangements, p. 270
Genome Size, p. 516
Interphase Nucleus, The, p. 568
Meiosis, p. 711
Mitosis, p. 734
Molecular Evolution, p. 748

REFERENCES

1. Appels, R.; Morris, R.; Gill, B.S.; May, C.E. *Chromosome Biology*; Kluwer Acad. Publ.: Boston, 1998.
2. Stebbins, G.L. *Chromosomal Evolution in Higher Plants*; Arnold: London, 1971.
3. Jenuwein, T.; Allis, C.D. Translating the histone code. Science **2001**, *293* (5532), 1074–1083.
4. Biessmann, H., Mason, J.M. Telomere maintenance without telomerase. Chromosoma **1997**, *106* (2), 63–69.
5. Choo, K.H.A. *The Centromere*; Oxford Univ. Press: Oxford, 1997.
6. Yu, H.-G.; Hiatt, E.N.; Dawe, R.K. The plant kinetochore. Trends Plant Sci. **2000**, *5* (11), 543–547.
7. Schubert, I.; Rieger, R.; Künzel, G. Karyotype Reconstruction in Plants with Special Emphasis on *Vicia faba* L. In *Chromosome Engineering in Plants: Genetics, Breeding, Evolution. (Part A)*; Gupta, P.K., Tsuchiya, T., Eds.; Elsevier: Amsterdam, 1991; 113–140.
8. Schubert, I.; Rieger, R.; Fuchs, J. Alteration of basic chromosome number by fusion–fission cycles. Genome **1995**, *38* (6), 1289–1292.
9. Schubert, I.; Oud, J.L. There is an upper limit of chromosome size for normal development of an organism. Cell **1997**, *88* (4), 515–520.
10. Schubert, I. Alteration of chromosome numbers by generation of minichromosomes—Is there a lower limit of chromosome size for stable segregation? Cytogenet. Cell Genet. **2001**, *93* (3–4), 175–181.

Circadian Clock in Plants

C. Robertson McClung
Dartmouth College, Hanover, New Hampshire, U.S.A.

INTRODUCTION

The circadian clock is an endogenous oscillator that drives rhythms with periods of approximately 24 hours. By definition, these circadian (*circa*, approximately; *dies*, day) rhythms persist in constant conditions and reflect the activity of an endogenous biological clock. Plants are richly rhythmic and the circadian clock regulates a number of key metabolic pathways and stress responses. In addition, the circadian clock plays a critical role in the photoperiodic regulation of the transition to flowering in many species.

AN OVERVIEW OF THE CIRCADIAN SYSTEM

The circadian clock has a number of defining characteristics.[1,2] The period of the rhythm is approximately—but seldom exactly—24 hours. The clock is self-sustaining, meaning that oscillations persist in constant conditions. Environmental time cues entrain the clock, synchronizing it with the local daily cycle. Thus, the circadian system consists of three components: input pathways that entrain the clock, the central oscillator (clock), and output pathways to generate overt rhythms (Fig. 1).

OUTPUTS: RHYTHMIC PROCESSES IN PLANTS

Many plant processes exhibit circadian oscillations. The earliest described rhythms were in leaf movement.[1,2] These are often seen in legumes, in which they are generated by a specialized organ called the pulvinus. Circadian-gated fluxes of ions and water cause cells in the extensor and flexor regions of the pulvinus to swell in antiphase (i.e., 180° out of phase) to drive a circadian oscillation in leaf position. These leaf movements may have adaptive value in regulating perception of photoperiodic light signals.[2] *Arabidopsis thaliana* lacks a pulvinus, but displays rhythms in cell elongation, and thus in growth rate. For example, there is a circadian rhythm in the elongation rate of the abaxial and adaxial cells of the petiole that confers an oscillation in position of cotyledons and leaves. Similarly, there are oscillations in the elongation rates of the hypocotyl and the inflorescence stem. Such rhythms are easily monitored by video imaging.[1,2] Other physiological properties—such as stomatal aperture and conductance, the rate of CO_2 assimilation, and the activities of enzymes of the Calvin cycle—are also regulated by the clock in some species.[2] Examples of rhythmic processes of Arabidopsis are shown in Fig. 2.

The circadian clock controls the expression of many plant genes. Rhythms in mRNA abundance are observed in 5–10% of all genes, and the peaks in mRNA abundance for different genes occur at distinct circadian phases.[3] Because circadian regulation of transcription underlies the rhythmic expression of many genes, it seems likely that the clock directly regulates the expression of transcription factors; the different transcription factors then coordinately confer circadian expression patterns on suites of genes.[2] For example, 23 genes encoding enzymes of phenylpropanoid biosynthesis are coordinately regulated, oscillating with mRNA peaks about 4 hr before subjective dawn.[3] Oscillating together with these genes is PRODUCTION OF ANTHOCYANIN PIGMENT 1 (*PAP1*), which encodes a Myb domain transcription factor shown to regulate the transcription of several genes encoding enzymes of the pathway. This offers the likely scenario that the clock regulates PAP1 expression, which in turn regulates the entire phenylpropanoid biosynthetic pathway.

Although nuclear run-on experiments have been employed to demonstrate circadian regulation of transcription, the current method of choice is the measurement of light production from plant promoter::firefly luciferase gene fusions (Fig. 2C). Using this approach, a number of promoters have been characterized and it is now possible to select a promoter to drive transcription at essentially any circadian phase of choice.[2]

The timing of flowering in many species is photoperiodic, and circadian timekeeping is essential for photoperiodic time measurement. Many mutations that affect circadian rhythms in gene expression and leaf movement also affect flowering timing.[4] CONSTANS (*CO*) encodes a Zn finger transcription factor that plays a key role in the integration of circadian timekeeping and day-length

Encyclopedia of Plant and Crop Science
DOI: 10.1081/E-EPCS 120010674

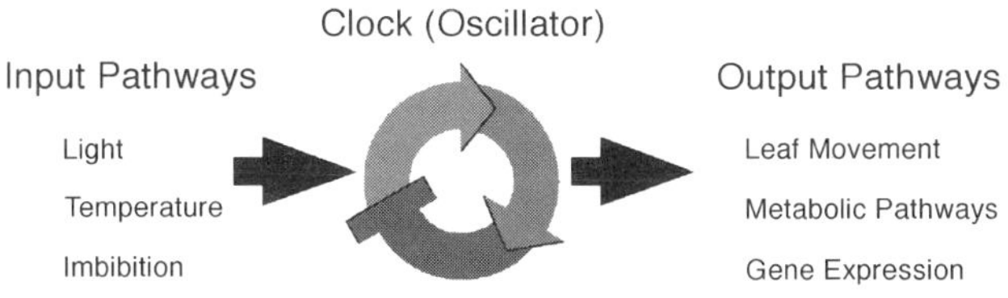

Fig. 1 A simplified scheme illustrating the three components of the circadian system. (*View this art in color at www.dekker.com.*)

perception.[5] *CO* mRNA abundance oscillates and the phase of the peak in *CO* mRNA varies with photoperiod such that only in long days does *CO* mRNA abundance peak in the light. Post-transcriptional regulation of CO activity by light is required for the transcriptional activation of FLOWERING LOCUS T (*FT*), a CO target gene whose expression is sufficient to induce flowering.[5] Thus, in long days CO is activated, *FT* is transcribed, and flowering is promoted.

ENTRAINMENT (INPUT)

Any biological circadian clock must be reset in response to the local daily cycle. For example, imbibition (the hydration of the dry seed) will synchronize the clocks

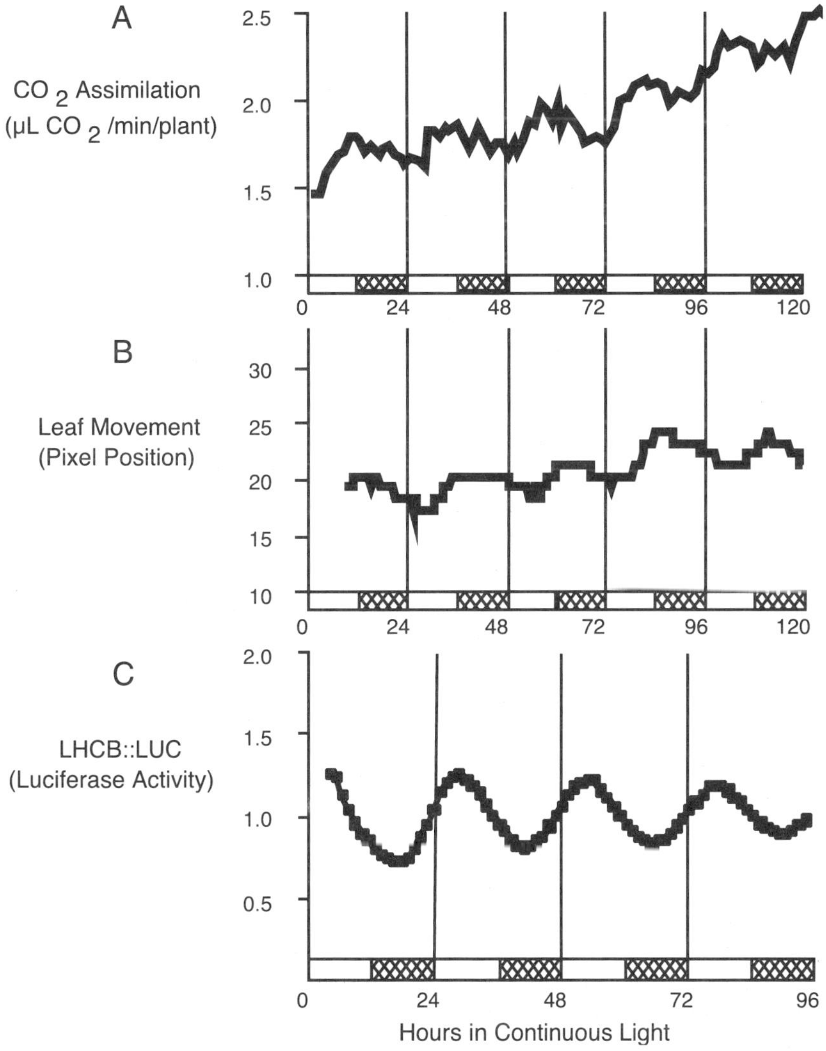

Fig. 2 Typical clock outputs in *Arabidopsis thaliana*. A. CO_2 assimilation; B. Leaf (cotyledon) movement; C. Transcription of the LIGHT-HARVESTING CHLOROPHYLL a/b BINDING PROTEIN gene (*LHCB*) as measured by luciferase (luc) activity in seedlings transformed with *LHCB::LUC* transcriptional gene fusions. For each assay, wild type seedlings (Accession Columbia) were grown in a light–dark (12–12) cycle and then transferred into continuous light at $T = 0$. The entraining light–dark regimen is indicated by the bars underneath each graph, where open bars indicate subjective day and the hatched bars indicate subjective night.

in a population of seedlings.[2] The most common stimuli that entrain the circadian clock are light and temperature. Little is known of the mechanism by which temperature is perceived and the signal is transmitted to the clock. Considerably more is known about photoperception and the clock. Plants have a number of photoreceptors. Both phytochromes (PHY, sensitive to red/far-red light, and also to blue light) and cryptochromes (CRY, sensitive to blue light) provide light signals to establish period length and phase and to entrain the clock.[6] The sensitivity of the clock to entraining stimuli varies at different times of day. Typically, light prior to dawn advances and light after dusk delays the phase of the clock, whereas light at midday has little effect. The abundance of PHY and CRY photoreceptors is itself a clock output[7] and light signaling is modulated (gated) by the clock, providing at least part of the mechanistic explanation of how the clock modulates its own sensitivity to light.

Additional plant photoreceptors are known. The phototropins do not seem to provide input to the circadian clock. However, other potential photoreceptors may provide input to the clock. Three LOV (light, oxygen, voltage) DOMAIN KELCH PROTEINS [LKP, also called ZEITLUPE, (ZTL), FLAVIN-BINDING, KELCH REPEAT, F-BOX (FKF), or ADAGIO (ADO)] contain the LOV domain, which may function in photoreception.[2,6] Misexpression (loss of function or overexpression) of these proteins confers circadian defects. Kelch repeats allow protein–protein interactions, whereas the F-box is a motif that allows the targeting of other specific proteins for ubiquitination and degradation by the proteasome, suggesting a role for these proteins in the light-regulated degradation of circadian oscillators or in input or output pathway components.

Several other components have been implicated in circadian photoperception. Loss of EARLY FLOWERING 3 (*ELF3*) function results in early flowering, hypocotyl elongation, and conditional arrhythmicity in continuous light.[2,6] *ELF3* is clock-regulated; both transcript and protein accumulation peak at dusk. *ELF3* encodes a nuclear protein that interacts with PHYB and is a negative modulator of PHYB signaling to the clock. GIGANTEA (*GI*) is another clock-controlled gene implicated in PHYB-mediated light input. *gi* mutants are altered in the leaf movement and gene expression rhythms of multiple genes.[2,6] The period-shortening effect of *gi-1* on gene expression rhythms is less severe in extended dark than in continuous light. The extension of period length seen in light of decreasing fluence is less pronounced in *gi-1* than in wild type, which indicates that GI acts in light input. *gi* mutants are late-flowering—opposite to the early-flowering phenotype of *elf3* and *phyB* null alleles. Perhaps GI and ELF3 mediate PHYB signaling to the clock, but their effects on flowering time are mediated through another signaling pathway.

THE OSCILLATOR: INTERLOCKED NEGATIVE AND POSITIVE FEEDBACK LOOPS

Circadian oscillators typically are composed of two interconnected feedback loops, one positive and one negative.[8] With the determination of the complete sequence of the Arabidopsis and rice genomes, no obvious orthologs to most known clock proteins can be found, demonstrating that at least part of the plant clock mechanism is novel. Key oscillator components include two single Myb domain transcription factors: CIRCADIAN CLOCK ASSOCIATED 1 (CCA1) and LATE ELONGATED HYPOCOTYL (LHY), and TIMING OF CAB EXPRESSION 1 [also called ARABIDOPSIS PSEUDO-RESPONSE-REGULATOR 1 (TOC1/APRR1)].

mRNA abundances of *CCA1*, *LHY*, and *TOC1* oscillate. If oscillation is necessary for clock function—as would be predicted for a clock component—then clamping mRNA or protein abundance at either high or low levels should eliminate clock function. Indeed, overexpression of any of these three genes results in arrhythmicity of multiple clock outputs.[2,9] Although loss of CCA1 or of LHY function shortens the period of mRNA oscillation in multiple clock-controlled genes, the plants retain rhythmicity. However, the double *cca1 lhy* mutant is arrhythmic in leaf movement, which is consistent with CCA1 and LHY playing necessary, albeit redundant, roles for sustained oscillator function.[10,11] Loss of TOC1 function shortens the period length, suggesting that another component plays a partially overlapping role with TOC1 in clock function. The identification of this component is a key goal. CCA1 overexpression results in reduced accumulation of *TOC1* mRNA, indicating that CCA1 is a negative regulator of TOC1. Loss of TOC1 function reduces accumulation of *CCA1* mRNA, suggesting that TOC1 is a negative regulator of CCA1.[10] Thus CCA1, LHY, and TOC1 compose a negative feedback loop of the sort found in other circadian oscillators (Fig. 3).[2,9,12] TOC1 is a member of a family of five APRR genes. mRNA abundance of the other four members also oscillates. However, whether these genes play roles in input pathways or in the oscillator itself—or simply represent clock outputs—remains to be established.[2]

Another link to other circadian oscillators is the role of phosphorylation by casein kinase 2 (CK2). Binding of CCA1 to its recognition site in clock-regulated

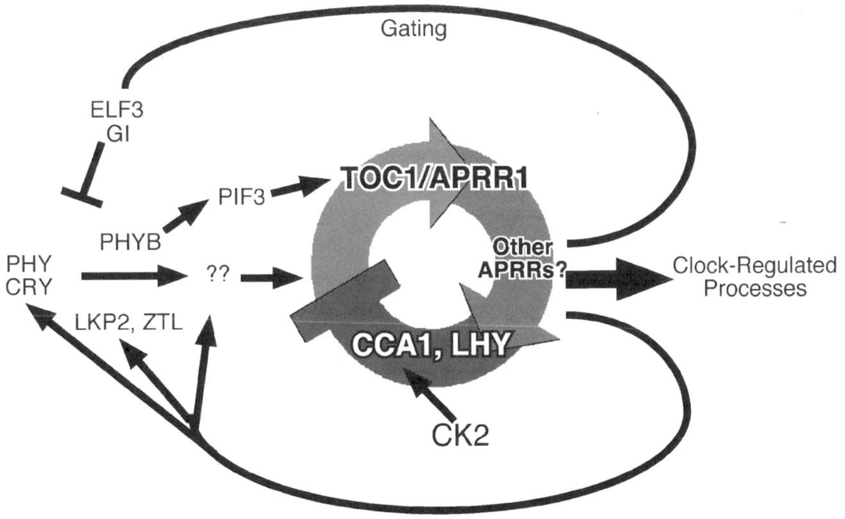

Fig. 3 A model of an Arabidopsis circadian system. Light input to the clock occurs via PHY and CRY, whose expression is clock regulated. In addition, light signaling to the clock is gated by the clock-regulated activity of ELF3 and GI. ZTL binds to PHYB and CRY1, and may mediate their degradation. At least some PHYB signaling occurs via PHYTOCHROME-INTERACTING FACTOR 3 (PIF3)—which binds to CCA1 and LHY promoters and possibly to other targets in the clock—and activates transcription when bound by PHYB. A central oscillator with a number of oscillator components is illustrated. CCA1 and LHY are phosphorylated by CK2, which is required for binding of CCA1 to the *LHCB* promoter. A family of TOC1 related ARABIDOPSIS PSEUDO-RESPONSE REGULATORS (APRRs) may function within the oscillator or as input (or output) pathway components. Output pathways may emanate from any of the putative oscillator components. *(View this art in color at www.dekker.com)*

promoters requires its phosphorylation by CK2.[2] CK2 abundance is not clock-regulated. Nonetheless, misregulation of CK2 disrupts clock function. Proteolytic degradation has been shown to play important roles in other clock systems; phosphorylation is also required for substrate recognition by F-box proteins such as the LKP family. The degradation of clock components in response to environmental time cues may be a critical step in plant oscillator function, and may also contribute to resetting the clock.

CONCLUSION

The plant circadian clock is composed of a negative feedback loop in which one component (TOC1) is a positive regulator of a second component (CCA1 and LHY) which, in turn, negatively regulates the expression of the first component. The clock is entrained to environmental time cues via input pathways that monitor environmental stimuli such as light and temperature. The clock regulates its sensitivity by regulating the expression and activity of input pathway components. Multiple output pathways emanate from oscillator components to drive overtly rhythmic processes.

ACKNOWLEDGMENTS

My work on circadian rhythms is supported by grants IBN-9817603 and MCB-0091008 from The National Science Foundation. I thank Mary Lou Guerinot, Todd Michael, and Patrice Salomé for comments on the manuscript. I apologize to all whose work could not be directly cited due to space limitations.

ARTICLES OF FURTHER INTEREST

Flower Development, p. 464
Photoperiodism and the Regulation of Flowering, p. 877
Photoreceptors and Associated Signaling I: Phytochromes, p. 881
Photoreceptors and Associated Signaling II: Cryptochromes, p. 885
Photoreceptors and Associated Signaling III: Phototropins, p. 889

Photoreceptors and Associated Signaling IV: UV Receptors, p. 893

REFERENCES

1. *Biological Rhythms and Photoperiodism in Plants*; Lumsden, P.J., Millar, A.J., Eds.; Bios Scientific Publishers: Oxford, 1998.
2. McClung, C.R.; Salomé, P.A.; Michael, T.P. The Arabidopsis Circadian System. In *The Arabidopsis Book*; Somerville, C.R., Meyerowitz, E.M., Eds.; American Society of Plant Biologists: Rockville, MD, 2002. DOI 10.1199/tab.0044. http://www.aspb.org/publications/arabidopsis/.
3. Harmer, S.L.; Hogenesch, J.B.; Straume, M.; Chang, H.-S.; Han, B.; Zhu, T.; Wang, X.; Kreps, J.A.; Kay, S.A. Orchestrated transcription of key pathways in Arabidopsis by the circadian clock. Science **2000**, *290*, 2110–2113.
4. Simpson, G.G.; Gendall, A.R.; Dean, C. When to switch to flowering. Annu. Rev. Cell Dev. Biol. **1999**, *15*, 519–550.
5. Suárez-López, P.; Wheatley, K.; Robson, F.; Onouchi, H.; Valverde, F.; Coupland, G. *CONSTANS* mediates between the circadian clock and the control of flowering in *Arabidopsis*. Nature **2001**, *410*, 1116–1120.
6. Devlin, P.F. Signs of the time: Environmental input to the circadian clock. J. Exp. Bot. **2002**, *53*, 1535–1550.
7. Tóth, R.; Kevei, É.; Hall, A.; Millar, A.J.; Nagy, F.; Kozma-Bognár, L. Circadian clock-regulated expression of phytochrome and cryptochrome genes in Arabidopsis. Plant Physiol. **2001**, *127*, 1607–1616.
8. Young, M.W.; Kay, S.A. Time zones: A comparative genetics of circadian clocks. Nat. Rev., Genet. **2001**, *2*, 702–715.
9. Carré, I.A.; Kim, J.-Y. MYB transcription factors in the Arabidopsis circadian clock. J. Exp. Bot. **2002**, *53*, 1551–1557.
10. Alabadí, D.; Yanovsky, M.J.; Más, P.; Harmer, S.L.; Kay, S.A. Critical role for CCA1 and LHY in maintaining circadian rhythmicity in *Arabidopsis*. Curr. Biol. **2002**, *12*, 757–761.
11. Mizoguchi, T.; Wheatley, K.; Hanzawa, Y.; Wright, L.; Mizoguchi, M.; Song, H.-R.; Carré, I.A.; Coupland, G. *LHY* and *CCA*1 are partially redundant genes required to maintain circadian rhythms in Arabidopsis. Dev. Cell **2002**, *2*, 629–641.
12. Alabadí, D.; Oyama, T.; Yanovsky, M.J.; Harmon, F.G.; Más, P.; Kay, S.A. Reciprocal regulation between *TOC1* and *LHY/CCA1* within the *Arabidopsis* circadian clock. Science **2001**, *293*, 880–883.

Classification and Identification of Nematodes

James G. Baldwin
University of California, Riverside, California, U.S.A.

C

INTRODUCTION

Although nematodes (Phylum Nematoda) are the most abundant multicellular organisms on Earth, plant parasites probably include less than 1% of all nematode species, and only a few hundred principal species are responsible for the billions of dollars of crop losses annually. However, a broad perspective of nematode classification is pertinent to agriculture. Many species that are not plant parasites nevertheless impact crops. For example, soil microbivore nematodes are a critical component of the soil ecosystem and play a crucial role in regulating nutrient recycling. Furthermore, some entomo pathogenic nematodes are biocontrol agents for insect pests.

CLASSIFICATION

All plant parasites feed by a protrusible stylet or spear, and this puncturing tool has arisen independently in three highly divergent orders, Tylenchida, Dorylaimida, and Triplonchida (Table 1); all of these orders also include nonplant parasites that use stylet or spear adaptations to feed on fungi, algae, or small soil organisms.[1]

GOALS AND TOOLS OF CLASSIFICATION

A goal of taxonomy is to develop a classification that reflects the evolutionary (phylogenetic) history of nematodes, because such a system is sure to have the greatest predictive value for extending knowledge (including pathogenicity) about well-known model species to newly-discovered or less-studied taxa. A predicitive classification also serves as the basic tenet of science: repeatability. In contrast, past nematode classification systems often have been largely arbitrary or based on convergent ecological behaviors or typological approaches to morphological characters. Often these morphological characters have been too few, and their homology and polarity inadequately understood to infer patterns of evolution. In many cases, these classifications have been fragmented and contradictory across practical fields of plant pathology, parasitology, medicine, entomology, and ecology.

New tools and methodologies, are now testing, integrating, and sometimes challenging extant classification systems using computer-based algorithms for integrating large matrices of phylogenetically informative novel characters from molecules, fine structure, and comparative development.[2] Molecular-based phylogenetic analyses, while substantially congruent with morphological phylogenetic analyses, are the impetus for revised classification systems that realistically place plant parasites within their nonparasitic context. Whereas most have considered nematodes to include two classes, Secernentea and Adenophorea (Table 1), new characters are resurrecting old challenges of Adenophorea as an evolutionary unit, and may thus lead to its elimination from classification systems.[2] New morphological and molecular characters also justify removal of trichodorid plant parasites, formerly in the Dorylaimida, into the Triplonchida, which is closely aligned with enoplids, a group largely comprising aquatic and marine nematodes.[2–4]

Advances in a phylogenetic classification of nematodes require knowledge of new characters and inclusion of far more representative taxa in phylogenetic analyses. This is particularly true for the Tylenchida, the order including the majority of crop parasites (Table 1). Although molecular tools have been applied in tylenchid diagnostics and toward resolving phylogenetic relationships of limited groups of genera or species, large-scale molecular-based phylogenies of the order are still being developed and are presently unlikely to resolve some of the more problematic families for revised classification.[5,6] Monophyly (sharing a unique evolutionary history) of Heteroderinae, including cyst (e.g., *Heterodera, Globodera*) and root knot (*Meloidogyne*) nematodes is challenged, as is the monophyly of Pratylenchidae. An issue of particular significance is whether Tylenchina share a unique evolutionary history with all Aphelenchina, and how these are best represented in classification, since new phylogenetically informative characters have only been considered for a few representatives in the latter.

Encyclopedia of Plant and Crop Science
DOI: 10.1081/E-EPCS 120010385

Table 1 Overview of extant higher classification of some important plant–parasitic nematodes

Class—SECERNENTEA
Order—TYLENCHIDA[a]
Suborder—TYLENCHINA
Superfamily—TYLENCHOIDEA
Family—ANGUINIDAE
Anguina—seed gall
Ditylenchus—stem and bulb
Family—BELONOLAIMIDAE
Tylenchorhynchus—stunt
Belonolaimus—sting
Family—PRATYLENCHIDAE
Pratylenchus—lesion
Radopholus—burrowing
Hirschmanniella
Nacobbus—false root knot
Family—HOPLOLAIMIDAE
Hoplolaimus—lance
Helicotylenchus—spiral
Rotylenchus—spiral
Rotylenchulus—reniform
Family—HETERODERIDAE
Heterodera—cyst
Globodera—cyst
Meloidogyne—root knot
Superfamily—CRICONEMATOIDEA
Family—CRICONEMATIDAE
Criconemoides—ring
Macroposthonia—ring
Hemicycliophora—sheath
Family—TYLENCHULIDAE
Tylenchulus—citrus
Suborder—APHELENCHINA[b]
Superfamily—APHELENCHOIDOIDEA
Family—APHELENCHOIDIDAE
Aphelenchoides—foliar
Family—PARASITAPHELENCHIDAE
Bursaphelenchus—pinewood
Rhadinaphelenchus—red ring
Class—ADENOPHOREA[c]
Order—DORYLAIMIDA
Suborder DORYLAIMINA
Superfamily—DORYLAIMOIDEA
Family—Longidoridae
Longidorus—needle
Xiphinema—dagger
Order TRIPLONCHIDA[c]
Suborder DIPHTHEROPHORINA
Superfamily Trichodoroidea
Family—Trichodoridae
Trichodorus—stubby root

[a]Adapted from Ref. 9. See Ref. 10 for a more recent classification of Tylenchida.
[b]Many consider Aphenchina to be in a distinct order, Aphelenchida. For a detailed classification see Ref. 11.
[c]See Refs. 2–4 for a discussion of Triplonchida relative to Dorylaimida and related issues of monophyly of Adenophorea.

DIAGNOSTICS

Where the goal of classification is to reflect evolutionary history, the assigned level/rank of higher taxonomic categories takes on less importance than the overall relationships reflected by the nesting of those categories. What is not arbitrary, however, is ''species,'' a category widely understood to be the fundamental evolutionary clade but with considerable discussion as to both the ideal of a species as a real entity and practical operational approaches for recognizing species.[7] In practice, most species of nematodes have been delimited by morphological uniqueness. The result often has been confusion in failing to recognize cryptic species (morphologically similar but evolutionarily independent), or to recognize intraspecific polymorphism or frequent hybridization. Yet, meaningful repeatable research with predictive power requires minimizing errors in recognizing and recovering true species.[7]

A wide range of morphological and molecular tools may be useful toward discovering the evolutionary units of species, and once species are recognized with some certainty, practical tools for routine diagnosis often can be developed. Ideally, these tools are morphological characters, molecular markers, or behaviors (i.e., host range) that are easily scored and so perfectly congruent with ''true species'' that they represent a shortcut to identification.

On the most practical level, diagnosis of plant disease caused by soil nematodes requires special consideration, because typical symptoms include general poor health, chlorosis, sensitivity to drought, and increased vulnerability to additional diseases. Such symptoms are often ambiguous and can be attributed to factors other than nematodes. In field crops, patches of distressed plants among a general pattern of otherwise healthy plants may be indicative of parasitic nematodes. In many cases, diagnosis is aided by special knowledge of vulnerability of a particular crop to a certain nematode; such recognition can narrow the diagnostic process to a few simple tests. For example, evaluation of the cause of unthrifty patches of soybean plants might always include the possibility of the soybean cyst nematode. Similarly, heavily galled tomato roots or badly distorted carrot roots are a strong indicator of the presence of one or more species of root-knot nematode. In these cases, diagnostic confirmation involves isolating nematodes from roots and surrounding soil using a combination of washing and sieving. Sometimes the isolation procedure involves centrifugation to separate nematodes from soil particles, and perhaps (depending on the species) it may be practical to dissect individual nematodes directly from roots. Nematodes collected in a small dish typically then are examined with a compound microscope and the diagnosis is made based on specific morphology of the nematode. We have noted that new tools are emerging to complement morphological diagnosis, but important challenges remain.

SHORTCOMINGS OF DIAGNOSTIC TECHNIQUES

Extant diagnostic techniques have important limitations. For example, perineal patterns, host ranges, or esterase electrophoresis patterns have been used to diagnose *Meloidogyne* species; unfortunately, because of intraspecific variability, any one criterion falls short of certainty. Similarly, molecular markers such as ITS region or D2/D3 regions of 28S rDNA are often developed based on limited screening, with the possibly faulty assumption that these characters are congruent with species. The difficulties of recognizing species are illustrated by the long controversy related to the burrowing nematode *Radopholus similis*. This species, as currently defined,[8] includes geographically limited isolates that attack citrus, and widespread isolates that do not. Although there has been a long search for morphological, karyotypical, biochemical and behavioral markers to separate this host polymorphism and define separate species (i.e., *R. similis* and *R. citrophilus*), ITS and D2/D3 examination of globally distributed isolates was used to vindicate their placement, regardless of host range, in a single species. Conversely, one burrowing nematode isolate from Australia, while morphologically similar to *R. similis*, had unique molecular markers (35 autapomorphies) and did not share a clade with *R. similis*.[8]

In agricultural nematology, there is sometimes a conflict of goals between those that would define and diagnose species as an evolutionary unit and those that seek a practical diagnosis to predict host range and pathogenicity for management strategies, including crop rotation, host resistance, or quarantine. In the latter case, the ideal is a marker, not necessarily for species as evolutionary unit, but whatever unit (including sub- or super-specific) is linked to a particular host range or pathogenicity. With advances in genomics, we anticipate that in addition to reliable markers for species diagnosis, it ultimately will be feasible to independently diagnose isolates for particular genetic markers for pathogenicity. Presently, for example, quarantines designed to protect citrus target the species *R. similis*, but we anticipate a future in which quarantines can be focused at those isolates of *R. similis* that are pathogenic to citrus.

Presently, for most taxa, morphological diagnosis of species is the primary option, and for others it remains the benchmark for evaluating the diagnostic effectiveness of molecular and other markers. However, new tools bring promise of classifications that reflect evolution, support research on models to understand pathways for pathogenicity, and more specifically diagnose and manage crop disease agents.

ARTICLES OF FURTHER INTEREST

Agriculture and Biodiversity, p. 1
Bioinformatics, p. 125
Biological Control of Nematodes, p. 134
Management of Nematode Diseases: Options, p. 684
Nematode Biology, Morphology, and Physiology, p. 781
Nematode Feeding Strategies, p. 784
Nematode Infestations: Assessment, p. 788
Nematode Population Dynamics, p. 797
Nematode Problems: Most Prevalent, p. 800
Nematodes and Host Resistance, p. 805
Nematodes: Ecology, p. 809
Plant Responses to Stress: Nematode Infections, p. 1014

REFERENCES

1. Baldwin, J.G.; Nadler, S.A.; Freckman, D.W. Nematodes—Pervading the Earth and Linking All Life. In *Nature and Human Society, the Quest for a Sustainable World*; Raven, P., Williams, T., Eds.; National Academy Press: Washington, DC, **1999**, 176–191. http://books.nap.edu/books/0309065550/html/176.html.
2. De Ley, P.; Blaxter, M. Systematic Position and Phylogeny. In *The Biology of Nematodes*; Lee, D.L., Ed.; Taylor & Francis: London, **2002**, 1–30.
3. Blaxter, M.L.; De Ley, P.; Garey, J.R.; Liu, L.X.; Scheldeman, P.; Vierstraete, A.; Vanfleteren, J.R.; Mackey, L.Y.; Dorris, M.; Frisse, L.M.; Vida, J.T.; Thomas, K., et al. A molecular evolutionary framework for the phylum Nematoda. Nature **1998**, *392*, 71–75.
4. Decraemer, W. *The Family Trichodoridae: Stubby Root and Virus Vector Nematodes*; Kluwer Academic Publishers: Boston, **1995**, 1–360.
5. De Ley; Vanfleteren. Personal communication.
6. Adams; Powers. Personal communication.
7. Adam, B.J. Species concepts and the evolutionary paradigm in modern nematology. J. Nematol. **1998**, *30*, 1–21.
8. Kaplan, D.T.; Thomas, W.K.; Frisse, L.M.; Sarah, J.L.; Stanton, J.M.; Speijer, P.R.; Marin, D.H.; Opperman, C.H. Phylogenetic analysis of geographically diverse *Radopholus similis* via rDNA sequence reveals a monomorphic motif. J. Nematol. **2000**, *32*, 134–142.
9. Maggenti, A.A.; Luc, M.; Raski, D.; Fortuner, R.; Geraert, E. A reappraisal of Tylenchina (Nemata). 2. Classification of the suborder Tylenchina (Nemata: Diplogasteria). Rev. Nematol. **1987**, 132–142.
10. Siddiqi, M.R. *Tylenchida Parasites of Plants and Insects*, 2nd Ed.; CABI Publishing: New York, **2000**, 1–833.
11. Hunt, D.J. *Aphelenchida, Longidoridae and Trichodoridae: Their Systematics and Bionomics*; CAB International: Oxon, UK, **1993**, 1–352.

C

Classification and Nomenclature of Plant Pathogenic Bacteria

John M. Young
Landcare Research, Auckland, New Zealand

INTRODUCTION

Fundamental to bacterial taxonomy is the interdependence of classification, nomenclature (naming), and identification. Without identified strains it is impossible to perform comparative taxonomy to classify and to name taxa. Without the establishment of phenotypic descriptions, bacterial strains cannot be allocated to taxa and novel taxa cannot be recognized.

The Approved Lists of Bacterial Names recognizes only the names of bacterial species for which there are modern descriptions permitting species to be distinguished from one other, and at least one extant strain that can be accepted as the type, or name-bearing, strain. In 2002, 5806 bacterial species in about 1094 genera were recognized. Plant pathogenic bacteria are recorded in 132 species in 29 genera. Within these taxa are several hundred pathogens, many recognized as pathovars (see below) that are, to some degree, specific to their host plant species and genera.

APPROACHES TO CLASSIFICATION

Classification is the ordering of bacteria into natural groups. Different approaches lead to different classifications and hence, to different nomenclature. Phenetic classifications (those based on the overall similarities and differences between bacteria) and phylogenetic classifications (those based on the inferred ancestral relationships of bacteria) of plant pathogenic bacteria have been discussed in detail elsewhere.[1] Polyphasic classification, based on a consensus of data gathered by all available methods that would be consistent with phylogenetic classification, has been suggested to be the best approach to bacterial classification.[2] Where several methods are employed in taxonomic studies, as in the polyphasic approach, there is an assumption that there will be a concurrence of groupings resulting from the different methods of study.[3,4] However, in practice, where comprehensive studies of species have been made, this assumption is not always borne out.[5]

With the refinement of bacterial classification offered by the development of molecular methods and the recognition of previously unacknowledged bacterial diversity, a phylogenetic species concept that emphasizes DNA–DNA reassociation values has become the standard for species definition. In a recent revision, a species concept was proposed that took account of both DNA–DNA reassociation and phenotypic studies.[4] A protocol for generic characterization emphasized the need for clarity of descriptions based on phenotypic properties at the level of genera, the impracticability of defining genera solely on the basis of phylogenetic data, and the priority of phenotypic characterization over phylogenetic inference.[3]

IDENTIFICATION

The rapid and reliable identification of isolates may be the most important task in taxonomy. Strains of taxa (species and genera) that were described using older technologies, and for which determinative keys were developed, can still be readily identified.[6,7] A problem arises for strains of taxa based on genomic and multiple-character methods alone, because they cannot readily be allocated to their genus or species due to a lack of determinative tests. There are now many named species found on plant surfaces that, if isolated from a plant other than their host, would be almost impossible to identify directly according to the most recent nomenclature.

Molecular techniques have greatly improved detection and identification of some organisms, because they provide probes that are specific at generic, specific, and lower levels. Probes are now regularly made that can identify bacterial pathogens as specific pathovars. There is now a choice of many probes with different levels of sensitivity that can be used to establish the presence of specified bacteria as clonal groups, infrasubspecies, and higher taxonomic groups.[7] Most importantly, molecular probes are now available to identify phytoplasmas or mycoplasma-like organisms that cannot yet be cultivated. Probes have been developed as antibodies, DNA oligonucleotides, and polymerase chain reaction (PCR) primers that can be used to amplify specific DNA sequences in target organisms. Their use is restricted to the identification of

Encyclopedia of Plant and Crop Science
DOI: 10.1081/E-EPCS 120010565

the particular organism for which the probe is designed and do not permit the identification of bacterial isolates in general. The search now is for a universal character or small set of characters, revealed in a single analytic step or by a small number of steps, that identifies species and genera that have been comprehensively classified by phenetic or polyphasic methods.

MISUNDERSTANDINGS OF BACTERIAL NOMENCLATURE

Type Strains Are Not Necessarily Typical

For the publication of a valid bacterial species name, *The International Code of Nomenclature of Bacteria* requires proposal of a name, a description of the taxon, and the designation of a nomenclatural type strain. The type strain should not be expected to be representative of the taxon as a whole and it is not necessarily the most typical or representative strain of the taxon. Mistakes can be expected when taxa are being tested for the presence or absence of small numbers of characters, and deductions are made based on the type strain alone.

Names of Taxa Are Not Necessarily Descriptive

For bacteria, there has been a strong tradition of proposing names that are descriptive of a significant character of the taxon. Descriptive terms necessarily refer to one or a few characters, often regulated by small numbers of genes and therefore not present in all members of the taxon (e.g., not all members of *Agrobacterium tumefaciens* cause tumors, and not all members of *Xanthomonas* are yellow). In many cases, revisions of descriptions, especially of species, result in complete or almost complete dissociation between a taxon's description and the descriptive element shown by the name. Indeed, as the extent of bacterial heterogeneity becomes more apparent, the expectation that the members of taxa established using natural phenetic or polyphasic classifications will share more than a small number of common characters diminishes. This caution should also be applied to special purpose names such as those of pathovars. Pathovar names, commonly ascribed on the basis of the host plant from which the pathogen was first isolated, can become misleading as a wider host range of the pathogen may be realized.[1]

Species Can Have Several Valid Names

Formal revisions of bacterial classification regularly lead to changes in the names of bacterial species and genera. Although any taxon that is based on the same type strain can have only one name, subdivisions and amalgamations of taxa mean that there can be several valid names applied to taxa, from which a choice may be made. Revisions can sometimes produce incomplete and unsatisfactory nomenclature. No error is committed provided names are chosen, even those not based on the most recent classification, that are legitimate and validly published.[5,8]

PATHOVARS

Historically, many names of plant pathogens were applied to bacterial populations that could not be classified as species and so could not be included in the *Approved Lists*. After 1980, pathogens were therefore classified in a special purpose classification at the infrasubspecies level, as pathovars. The International Society for Plant Pathology proposed that pathovars be named according to nomenclatural standards[9,10] complementary to *The International Code of Nomenclature of Bacteria*. Pathovar classification and nomenclature is based on the capacity of particular bacterial populations to cause distinctive disease syndromes in plants in nature, or refers to their distinct host ranges. Such populations are characterized by their pathogenicity mechanisms (implying distinct combinations of pathogenicity genes).[1,12]

There is ample evidence to show that many pathovars are specific to relatively small numbers of host species and genera.[13] Some of these are highly virulent, with proved, well-adapted mechanisms of pathogenicity. However, some pathovars, e.g., *Pseudomonas syringae* pv. *syringae*, are represented by diverse populations of strains varying in degrees of virulence, with wide and sometimes overlapping host ranges.[1] Furthermore, it is possible, even likely, that pathovars such as *P. syringae* pv. *lachrymans* and *Xanthomonas campestris* pv. *alfalfae* do not have the pathogenic specificity attributed to them. There may be examples in *Xanthomonas* of weak pathogens with wide host ranges, which has resulted in proposals for new pathovar names that are in fact synonyms.[13]

CONCLUSION

Bacterial classification can be based on a number of taxonomic models, each leading to different interpretations and to nomenclature with differing utility. Phylogenetic, phenetic, and polyphasic classifications vie in claims to being the most natural and therefore offering the best nomenclature. Recently, molecular data, particularly selected DNA sequence data, has led to the subdivision of existing genera. However, such revisions commonly do not take account of conflicts arising with analyses of alternative databases. Although there are examples of more than one complete chromosomal sequence representative of one species or genus, the extent of heterogeneity between such sequences points to a complexity that may not lead to simple bacterial classifications. Furthermore, taxonomic revisions sometimes fail to take account of the need of practising bacteriologists for stable nomenclature, for whom evolutionary classifications give rise to taxa that are difficult to identify, and therefore, to nomenclature that is of marginal utility. The wider use of special purpose nomenclature may be needed increasingly as natural classification becomes divorced from bacterial groups of applied interest to mankind.

ARTICLE OF FURTHER INTEREST

Bacterial Pathogens: Detection and Identification Methods, p. 84

REFERENCES

1. Young, J.M.; Takikawa, Y.; Gardan, L.; Stead, D.E. Changing concepts in the taxonomy of plant pathogenic bacteria. Annu. Rev. Phytopathol. **1992**, *30*, 67–105.
2. Vandamme, P.; Pot, B.; Gillis, M.; De Vos, P.; Kersters, K.; Swings, J. Polyphasic taxonomy, a consensus approach to bacterial systematics. Microbiol. Rev. **1996**, *60*, 407–438.
3. Murray, R.G.E.; Brenner, D.J.; Colwell, R.R.; De Vos, P.; Goodfellow, M.; Grimont, P.A.D.; Pfenning, N.; Stackebrandt, E.; Zavarzin, G.A. Report of the ad hoc committee on approaches to taxonomy within the proteobacteria. Int. J. Syst. Bacteriol. **1990**, *40*, 213–215.
4. Stackebrandt, E.; Frederiksen, W.; Garrity, G.M.; Grimont, P.A.D.; Kampfer, P.; Maiden, M.C.J.; Nesme, X.; Rossello-Mora, R.; Swings, J.; Trüper, H.G.; Vauterin, L.; Ward, A.C.; Whitman, W.B. Report of the ad hoc committee for the re-evaluation of the species definition in bacteriology. Int. J. Syst. Evol. Microbiol. **2002**, *52*, 1043–1047.
5. Young, J.M. Recent Systematics Developments and Implications for Plant Pathogenic Bacteria. In *Applied Microbial Systematics*; Priest, F.G., Goodfellow, M., Eds.; Chapman Hall: London, 2000; 133–160.
6. Bradbury, J.F. *Guide to Plant Pathogenic Bacteria*; CAB International Mycological Institute: Kew, London, England, 1986.
7. *Laboratory Guide for Identification of Plant Pathogenic Bacteria*; Schaad, N.W., Jones, J.B., Chun, W., Eds.; APS Press: St. Paul, MN, 2001.
8. Young, J.M.; Saddler, G.S.; Takikawa, Y.; De Boer, S.H.; Vauterin, L.; Gardan, L.; Gvozdyak, R.I.; Stead, D.E. Names of plant pathogenic bacteria 1864–1995. Rev. Plant. Pathol. **1996**, *75*, 721–763.
9. Dye, D.W.; Bradbury, J.F.; Goto, M.; Hayward, A.C.; Lelliott, R.A.; Schroth, M.N. International standards for naming pathovars of phytopathogenic bacteria and a list of pathovar names and pathotype strains. Rev. Plant. Pathol. **1980**, *59*, 153–168.
10. http://www.isppweb.org/tppbints.htm (accessed December, 2002).
11. http://www.isppweb.org/nppb.htm (accessed December 2002).
12. Hayward, A.C. The Hosts of *Xanthomonas*—Introduction. In *Xanthomonas*; Swings, J., Civerolo, E.L., Eds.; Chapman Hall: London, England, 1993; 1–18.

Coevolution of Insects and Plants

M. Deane Bowers
University of Colorado, Boulder, Colorado, U.S.A.

INTRODUCTION

This article briefly outlines the development of the concept of coevolution since it was first proposed in 1964. It touches on both criticisms of and support for the theory. Several examples of coevolution are given, including both pollination and herbivory. These examples include the parsnip webworms and their parsnip hostplants, flea beetles and their hostplants, yucca moths and yucca, and fig wasps and figs.

WHAT IS COEVOLUTION?

The term "coevolution" was first used in a historic paper on butterflies and their host plants published by Ehrlich and Raven.[1] They defined coevolution as "patterns of interactions between two major groups of organisms with a close and evident ecological relationship." They used data on larval host plants of related butterfly taxa to suggest that chemical compounds in plants (so-called secondary compounds, natural products, secondary metabolites, or allelochemicals) were important in the specialization of butterfly taxa on particular plant taxa, as well as in the evolutionary diversification of both butterflies and angiosperms.

As formulated by Ehrlich and Raven,[1] the process of coevolution involves reciprocal selection by plants on their insect herbivores and by insect herbivores on their host plants. They proposed the following sequence of events: 1) Plants produce novel secondary metabolites by mutation or recombination; 2) these compounds reduce the suitability of the plant as food for insects; 3) as a result, the plants "escape" from these insects and are able to undergo evolutionary radiation into a new adaptive zone; 4) by mutation or recombination, insects evolve resistance to these secondary metabolites; 5) these adapted insects are able to use a host plant unavailable to other herbivores and thus enter a new adaptive zone in which they are able to undergo their own evolutionary radiation. This reciprocal selective process can continue, resulting in the evolutionary diversification of plants, including secondary compounds, and their insect herbivores.

Over the years, authors have argued about the definition of coevolution, whether they believed that the system they were describing was a coevolved one, and even whether coevolution occurred at all.[2] More recently, the term coevolution was more precisely defined as "an evolutionary change in a trait of the individuals in one population in response to a trait of the individuals of a second population followed by an evolutionary response by the second population to the change in the first."[3] This definition more specifically defines the requirements of both specificity and reciprocity: Each trait evolves in response to the other and both traits evolve.

Coevolutionary interactions may be classified as pairwise (=specific) or diffuse. Pairwise interactions are those involving only two species (or populations) or sets of pairs of species (or populations). Such pairwise interactions follow well from Janzen's definition above. However, if the requirement of specificity is less stringently applied, one or more species or populations may evolve in response to a trait or set of traits in several species, resulting in diffuse coevolution. At some point it may be impossible to distinguish diffuse coevolution, as so broadly defined, from evolution; however, the distinction lies in the reciprocity of the interactions.

Coevolution between insects and plants may include parasitic relationships such as herbivory and frugivory, as well as mutualistic relationships such as pollination and seed dispersal. While coevolutionary theory has traditionally been built on the view of insects as herbivores,[1] more recently there is increasing evidence from studies of pollination and other mutualistic interactions.

SOME ARGUMENTS AGAINST COEVOLUTION

Criticisms of coevolutionary theory and its importance in shaping the interactions between insects and plants range from rejection of coevolution[2] to evidence that hostplant chemistry, predators, sequential colonization, or biogeography have played the more significant role.[4,5] One common criticism pertains to an important tenet of the theory, that of reciprocal selection. Reciprocal selection requires that insects act as selective agents on their hostplants. Some authors believe that insect herbivory does not influence plant evolution because it rarely (if

Encyclopedia of Plant and Crop Science
DOI: 10.1081/E-EPCS 120010481

ever) reduces plant fitness.[2] This lack of influence is attributed to several factors, including the rarity of insects relative to their host plants, low mean levels of damage by herbivorous insects (5–15%), and the greater importance of bacterial and fungal pathogens in influencing plant evolution.

SOME EVIDENCE FOR COEVOLUTION

While not all researchers may agree that coevolution is one of the most important processes in generating the diversity of organisms, there are examples of insects and plants that may demonstrate the role of coevolution in producing these interactions. They range from mutualisms such as pollination to predation and parasitism such as herbivory. The possibilities provided by molecular techniques have generated a great deal of interest because they can provide data (DNA sequences) that are not linked to the interaction, as may be the case with some (though not necessarily all) morphological characters.

Several different systems of insects and plants have been used to assess the role of coevolution in the interactions of insects and plants, including herbivores and plants, pollinators and plants, and dispersal agents and plants. In addition, different means of assessing the role of coevolutionary processes in these interactions were used. For example, Berenbaum[6] used a combination of direct empirical evidence coupled with information from the literature to argue that species of the carrot family, the Apiaceae, and certain insect herbivores feeding on these plants provide a premier example of coevolution. This coevolutionary relationship, she suggested, was mediated by a group of secondary compounds called furanocoumarins. Since her first paper in 1983, substantial evidence has been accumulated by Berenbaum and others that supports the ideas she originally developed,[6] and evidence for coevolution has been reported for other relationships among different species, as well.

In an effort to combine these ideas of reciprocal adaptive radiation with the emerging field of phylogenetics, Mitter and Brooks[7] suggested that coevolutionary interactions between groups of organisms might be evident in the existence of parallel phylogenies of the interacting groups (=parallel cladogenesis). That is, if two groups of interacting organisms (such as a certain group of plants and their associated pollinators or herbivores) have reciprocally affected each other's evolution, i.e., coevolved, then this will be reflected in phylogenetic trees that mirror each other. Such methods provide a historical approach to answering the question of whether groups of organisms have coevolved.

Yucca moths (*Tegiticula*, family Prodoxidae) and yucca (*Yucca*, family Agavaceae) have a mutualistic relationship in which the female moths use specialized mouthparts to collect pollen from *Yucca* flowers, which they actively transfer to the stigma of a flower in another inflorescence. Before transfer of the pollen, the females oviposit into the ovary of the flower and the larvae eat some of the resulting seeds. Thus the yuccas need the moths for pollen transfer and moths need the yuccas as food for their progeny. Phylogenetic analyses of both *Yucca* and *Tegiticula*, however, indicate that this relationship arose as the result of yucca moths colonizing a partially diversified set of host taxa rather than parallel cladogenesis.[8]

Figs (*Ficus*, Moraceae) are pollinated and fed upon by members of the wasp family Agaonidae (Chalcidoidea: Hymenoptera). Oviposition in the fig is typically linked with pollination of the fig flowers. Some of the fig flowers are pollinated, producing seeds and food for the next generation of fig wasps, while others serve as larval food. As with yucca moths, there may be ''cheaters'' who take advantage of the system, using the resource for food, but not pollinating the flowers. Females of pollinating species exhibit a set of morphological adaptations for using figs, and these pollinators may be actively pollinating, that is storing pollen in thoracic pockets and depositing it on the stigma, or passively pollinating, that is, there is no active pollination behavior.[9] This system is considered a good example of coevolution because there is evidence for parallel cladogenesis between figs and their pollinators (as well as parasites), as well as for reciprocity in the evolution of interacting traits.[9]

Beetles in the monophyletic leaf-beetle genus *Phyllobrotica* (Galerucinae: Chrysomelidae) feed primarily on a genus of mints, the skull-caps, *Scutellaria* (Lamiaceae). Larvae are root feeders and adults are leaf and flower feeders, and both life stages are very specialized in their feeding habits: One species of beetle feeds on a single host species in both larval and adult stages (with a couple of exceptions).[10] Comparison of the phylogenetic hypotheses generated by use of morphological characters for the beetles and from published data for the host plants indicate almost complete concordance, suggesting that the relationship of these beetles and their hostplants is a premier example of coevolution between insects and their hostplants.[10]

CONCLUSION

While the number of good examples of coevolution is still relatively few, those described thus far have been important in shaping our current thinking about evolution in a broad sense, as have those that have not shown evidence of coevolution. In his book, Thompson[8] argues that coevolution may not be detected in particular populations because there is a geographic mosaic of coevolution in which some populations of interacting species are coevolving, while others are not. Such a structure would make identification of coevolving taxa more challenging. Nonetheless, the identification and study of such taxa may be crucial to our understanding of population and community dynamics, population and evolutionary genetics, as well as the process of coevolution itself.

ACKNOWLEDGMENTS

Thanks to Yan Linhart, Kasey Barton, Eric DeChaine, and an anonymous reviewer for helpful comments.

ARTICLES OF FURTHER INTEREST

Floral Scent, p. 456
Genetically Engineered Crops with Resistance Against Insects, p. 506
Insect/Host Plant Resistance in Crops, p. 605
Insect–Plant Interactions, p. 609
New Secondary Metabolites: Potential Evolution, p. 818
Oomycete–Plant Interactions: Current Issues in, p. 843
Physiology of Herbivorous Insects, p. 910
Phytochemical Diversity of Secondary Metabolites, p. 915
Plant Defenses Against Insects: Constitutive and Induced Chemical Defenses, p. 939
Plant Defenses Against Insects: Physical Defenses, p. 944
Plant–Pathogen Interactions: Evolution of, p. 965
Root-Feeding Insects, p. 1114

REFERENCES

1. Ehrlich, P.R.; Raven, P. Butterflies and plants: A study in coevolution. Evolution **1964**, *18*, 586–608.
2. Jermy, T. Evolution of insect/host plant relationships. Am. Nat. **1984**, *124*, 609–630.
3. Janzen, D. When is it coevolution? Evolution **1980**, *34*, 611–612.
4. Bernays, E.A.; Graham, M. On the evolution of host specificity in phytophagous insects. Ecology **1988**, *69*, 886–892.
5. Becerra, J. Insects on plants: Macroevolutionary trends in host use. Science **1997**, *276*, 253–256.
6. Berenbaum, M. Chemical mediation of coevolution: Phylogenetic evidence for Apiaceae and associates. Ann. Mo. Bot. Gard. **2001**, *88*, 45–59.
7. Mitter, C.; Brooks, D. Phylogenetic Aspects of Coevolution. In *Coevolution*; Futuyma, D.J., Slatkin, M., Eds.; Sinauer: Sunderland, MA, 1983; 65–98.
8. Thompson, J.C. *The Coevolutionary Process*; Univ. of Chicago Press: Chicago, 1994.
9. Wieblen, G.D. How to be a fig wasp. Annu. Rev. Entomol. **2002**, *47*, 299–330.
10. Farrell, B.; Mitter, C. Phylogenesis of insect/plant interactions: Have *Phyllobrotica* leaf beetles (Chrysomelidae) and the Lamiales diversified in parallel? Evolution **1990**, *44*, 1389–1403.

Columbian Exchange: The Role of Analogue Crops in the Adoption and Dissemination of Exotic Cultigens

David E. Williams
International Plant Genetic Resources Institute, Cali, Colombia

INTRODUCTION

The violent collision of cultures that began with Christopher Columbus's landfall in 1492 set off an unprecedented and sustained intercontinental exchange of crops that has enriched and altered agricultural systems and diets worldwide. Native American crops such as the potato, maize, tomato, cacao, cassava, and tobacco have had an immeasurable impact upon the rest of the world, while Old World crops such as coffee, sugar cane, bananas, wheat, and rice now constitute the basis of the agricultural economies of many countries in the Americas. During the initial exchanges that accompanied the European conquest of the Americas, dozens of staple crops from both hemispheres quickly became established on foreign soil where in many cases they soon became more important and more widespread than in their land of origin. Other crops were much slower in gaining acceptance. Crops such as the Andean pseudocereals, quinoa and amaranth, have still not achieved significant acceptance beyond their areas of origin despite concerted efforts to promote their undisputed agronomic and nutritional qualities. Some local cultigens, particularly in the Americas, were so overwhelmed by the advent of the exotic crops and farming systems that their use declined nearly to the point of extinction. How can we explain such variation in the acceptance of crops involved in the Columbian exchange?

ANALOGUE CROPS

Clearly, absence of pests and diseases and climatic suitability were important factors favoring adoption.[1–3] Various biological, agricultural, and socioeconomic factors have been proposed as causes of marginalization of native crops in the Americas following the Columbian exchange.[4,5] However, among the many historical and cultural factors affecting crop and food acceptance, the preexistence of analogue crops in both hemispheres offers a partial explanation, at least in some cases. Analogue crops are pairs of species that were independently domesticated, usually on separate continents, but which share important traits of appearance, management, and/or use and are therefore somewhat "familiar" to farmers and consumers upon introduction. Analogous crop pairs typically pertain to the same botanical family and sometimes the same genus. Some analogue crops are botanically unrelated yet occupy similar niches in the agroecosystems and/or cuisines of their respective homelands (see Table 1).

The similarity of introduced crops to their native analogues was often sufficient incentive for adoption. When maize was first introduced to Africa and Central Asia, it was certainly recognized as a relative of the native sorghum, whose plant it strongly resembles, thereby facilitating its widespread adoption. However, sorghum and the other African and Asian millets did not have the same impact in the Americas where they are cultivated on a limited scale for fodder and birdseed. After maize, the potato and tomato claim the largest tonnage of New World crops produced in the Old World. Once the initial European prejudices against these solanaceous introductions were overcome, the Andean tuber was found to be well adapted to the cool conditions of northern Europe and, as a starchy food, the potato combined well with the fatty meats, butter, and cream of the Western diet. Meanwhile, the sweet but acidy fruit of the tomato became so appreciated that it is now the most widely eaten "vegetable" in the world.

Of the grain legumes, common beans (*Phaseolus*) from the Americas and cowpeas (*Vigna*) from Asia are perfect analogues, so similar in morphology and use that they were all classified by Linnaeus in 1753 within the genus *Phaseolus* before more modern taxonomists determined that they pertain to distinct genera endemic to different hemispheres.[6] Today, cowpeas and common beans have either replaced or are grown alongside one another throughout the world. In contrast, distinct species of grain lupines (*Lupinus*) were independently domesticated in the Old and New Worlds, yet neither found acceptance across the Atlantic.

Another legume, the peanut (*Arachis hypogaea*), experienced seemingly instantaneous acceptance as soon as

Encyclopedia of Plant and Crop Science
DOI: 10.1081/E-EPCS 120010414

it was introduced from South America into West Africa around 1500. West Africa is home to the Bambara groundnut (*Vigna subterranea*), the peanut's analogue. Both crops are geocarpic (i.e., fruits formed underground) and produce high-calorie seeds with similar culinary properties. The larger size and yield of the American peanut caused it to displace its African analogue in many areas. In Asia, the peanut has another analogue, the soybean (*Glycine max*), both species being valued as oilseeds. Following its early introduction, the peanut became fully integrated into traditional Asian farming systems and cuisines, and China and India are currently the world's biggest peanut producers. The soybean, on the other hand, gained widespread acceptance in the New World only in the past few decades but has become one of the most important commodity crops in the United States, Brazil, Paraguay, Bolivia, and elsewhere. However, despite its phenomenal recent expansion, the soybean is grown in the Americas exclusively as a cash crop and has not entered the local cuisines except as a "hidden" ingredient in the form of soy oil, protein meal, or animal feed. Ironically, large-scale expansion for soybean production in southwestern Brazil and adjacent regions of Bolivia and Paraguay is transforming the natural ecosystems that are home to the peanut's closest wild relatives and threatens these unique genetic resources, some still undiscovered, with extinction.[7]

The Columbian exchange of root crops is notable for its relative lack of analogue species. The physical similarities between the plants of the South American potato and the Old World poisonous nightshades evidently delayed the acceptance of the tuber in Europe. While the "Irish" and sweet potatoes (*Solanum tuberosum* and *Ipomoea batatas*) eventually received wide acceptance throughout the world, numerous other domesticated Andean root and tuber crops (e.g., *Oxalis*, *Tropaeolum*, *Ullucus*, *Lepidium*, *Arracacia*, *Canna*, etc.) have not, and their use is currently in decline even in their area of origin.[8] The widespread adoption of cassava (*Manihot*) in Africa and Asia was not replicated in the Americas by the Old World yams (*Dioscorea* spp.).[9,10] In contrast, the domesticated aroids of Asia (*Colocasia*) and America (*Xanthosoma*) are perfect analogues and today are commonly grown together throughout the lowland tropics where, in some places such as West Africa and the Caribbean, the exotic cultigen has assumed greater importance than its native analogue.[6]

The New World squashes, pumpkins, and gourds (*Cucurbita* spp.) turned out to be a fair trade for their Old World analogues, the melons and cucumber (*Citrullus* and *Cucumis*); all these crops are now widely cultivated across the globe.

The quest for spices and condiments was one of the primary objectives of Columbus's first voyage.[11] While unsuccessful in finding a shorter route to the source of black pepper (*Piper nigrum*) as hoped, he did encounter the red peppers (*Capsicum* spp.), which quickly became one of the most widely cultivated crop genera, particularly *C. annuum*.[12] Black pepper, on the other hand, has made limited inroads in tropical America, where it is cultivated commercially only in Brazil. An interesting pair of analogue crops are the Old World coriander (*Coriandrum sativum*) and the New World culantro (*Eryngium foetidum*), both domesticated umbelliferous herbs with nearly identical flavors. In Central America and the Caribbean, the two herbs are often used interchangeably.

Of the perennial fruit crops, grapes were well known to people on both sides of the Atlantic before Columbus's time, but the European wine grape (*Vitis vinifera*) was assiduously planted throughout the New World by early colonists and missionaries. In the late 1800s, when the aphid pest *Phylloxera* began decimating vineyards around the world, American *Vitis* species began traveling abroad as resistant rootstocks and today are used almost exclusively wherever commercial grapes are produced.[10] Walnuts (*Juglans*) have analogue species in both hemispheres, yet commercial groves of English walnut (*Juglans regia*) in North America are grafted onto rootstocks of the hardier American species, *Juglans hindsii*. Roseaceous fruits such as apples and cherries (*Malus* and *Prunus*) have native species on both sides of the Atlantic, but only the Old World apples and cherries successfully made the transoceanic trip. Although no apples were domesticated in the Americas, there was a domesticated applelike hawthorn (*Crataegus pubescens*) in Mexico and a domesticated black cherry (*Prunus serotina*) in Mexico and the Andes, but these crops have mostly remained confined to their areas of origin.[13] Major Old World fruit crops without analogues are the banana and plantain, citrus, and mango, all of which have made impressive inroads in the Americas. New World fruits with worldwide diffusion include the pineapple and, more recently, the avocado, for which there are no Old World analogues.

Different species of cotton (*Gossypium*) were cultivated in each hemisphere for some 5000 years before the advent of the Columbian exchange. After the long-staple cottons from the New World (*Gossypium hirsutum* and *Gossypium barbadense*) were introduced to Asia and Africa, they almost entirely displaced their short-staple analogues (*Gossypium herbaceum* and *Gossypium arboreum*).[7,14] The analogous pair of botanically unrelated stimulating beverage crops, coffee (*Coffea*) and cacao (*Theobroma*), participated reciprocally in the Columbian exchange and eventually established themselves as major

Table 1 Some representative pairs of analogue crops and their respective adoption and diffusion as a result of the Columbian exchange

Old World	New World	Family	Use	Direction and extent of adoption and diffusion[a]	
				OW>NW	NW>OW
Sorghum bicolor	*Zea mays*	Poaceae	Cereal	2	3
Vicia faba	*Lupinus mutabilis*	Fabaceae	Pulse	3	–
Lupinus albus	*Lupinus mutabilis*	Fabaceae	Pulse	–	–
Vigna unguiculata	*Phaseolus vulgaris*	Fabaceae	Pulse	3	3
Vigna subterranea	*Arachis hypogaea*	Fabaceae	Pulse	–	3
Glycine max	*Arachis hypogaea*	Fabaceae	Oilseed	3	3
Cucumis spp.	*Cucurbita* spp.	Cucurbitaceae	Fruit/vegetable	3	3
Colocasia	*Xanthosoma*	Araceae	Root crop	3	3
Piper nigrum	*Capsicum* spp.	Piperaceae/Solanaceae	Spice	1	3
Coriandrum	*Eryngium*	Apiaceae	Spice	3	–
Atropa belladonna	*Solanum tuberosum*	Solanaceae	Poisonous root crop	–	(3)
Atropa belladonna	*Lycopersicon esculentum*	Solanaceae	Poisonous vegetable	–	(3)
Malus	*Crataegus*	Rosaceae	Fruit	3	–
Vitis vinifera	*Vitis* spp.	Vitaceae	Fruit/rootstock	3	(3)
Juglans regia	*Juglans hindsii*	Juglandaceae	Nut/rootstock	3	(2)
Gossypium herbaceum, *G. arboreum*	*Gossypium hirsutum*, *G. barbadense*	Malvaceae	Fiber	–	3
Coffea arabica, *C. rustica*	*Theobroma cacao*	Rubiaceae/Sterculiaceae	Beverage	(3)	(3)
Papaver somniferum	*Erythroxylum coca*	Papaveraceae/ Erythroxylaceae	Narcotic	2	–

[a]Explanation of symbols:

- **NW > OW** = New World crop adopted in the Old World.
- **OW > NW** = Old World crop adopted in the New World.
- Numbers indicate degree of adoption and diffusion of crop in the new hemisphere: 1 = limited, 2 = significant, 3 = widespread, '–' = not adopted.
- Parentheses ( ) around numbers indicate delayed adoption or diffusion.

commodity crops on opposite ends of the earth from their areas of origin.[14] Even illicit crops such as the opium poppy (*Papaver somniferum*) and coca (*Erythroxylum coca*) can be regarded as analogues. However, while the Asian opium poppy is now cultivated illegally in significant quantities in Mexico, Guatemala, and Colombia, it is difficult to explain why coca cultivation has never expanded beyond its traditional area of production in northwestern South America. The high-value narcotics derived from both plants are now consumed worldwide in both licit and illicit forms.

CONCLUSION

The Columbian exchange of crops was influenced by the existence of analogue crops, and an examination of their effect can help us understand the different degrees of adoption and diffusion of those crops. However, as shown by the examples presented above, the presence of native analogues didn't always favor the adoption of introduced crops, and sometimes the opposite was the case where the native analogues served to delay or even prevent the adoption of the exotic species. Nor were analogous crop pairs always reciprocally adopted: Sometimes the exchange was unidirectional when only one of the analogues found acceptance abroad. Moreover, many crops without analogues were readily adopted and prospered in their new environment. The identification of analogue pairs and the determination of their effect on the adoption and dispersal of one another require interpretation on a case-by-case basis. By and large, however, the presence of analogous crop pairs has tended to favor their post-Columbian adoption.

The far-reaching impact of the Columbian exchange has been one of incalculable benefit to world agriculture, diet, and cuisine. The Columbian exchange marked the beginning of an ongoing process of global movement of crop germplasm resulting in a strong interdependence between all nations in terms of the plant genetic resources upon which the world's food security and agricultural sustainability depend. With no more crop species becoming available beyond those that we already know, the need to maintain and increase the international exchange of existing crop genetic resources is now more important than ever for ensuring the future well-being of farmers and consumers everywhere.

ACKNOWLEDGMENTS

The author expresses his thanks to Luigi Guarino, Garrison Wilkes, Sandra Williams, and an anonymous reviewer for their helpful comments on earlier versions of this essay.

ARTICLES OF FURTHER INTEREST

Crop Domestication in Africa, p. 304
Crop Domestication in China, p. 307
Crop Domestication in Mesoamerica, p. 310
Crop Domestication in Prehistoric Eastern North America, p. 314
Crop Domestication in Southeast Asia, p. 320
Crop Domestication in Southwest Asia, p. 323
Crop Domestication: Fate of Genetic Diversity, p. 333
Crop Domestication: Role of Unconscious Selection, p. 340
Crop Improvement: Broadening the Genetic Base for, p. 343
Farmer Selection and Conservation of Crop Varieties, p. 433
Germplasm Acquisition, p. 537

REFERENCES

1. Crosby, A.W., Jr. *The Columbian Exchange: Biological and Cultural Consequences of 1492*; Greenwood Press: Westport, CT, 1972.
2. Sperling, C.R.; Williams, D.E. Horticultural Crop Germplasm: 500 Years of Exchange. In *International Germplasm Transfer: Past and Present: CSSA Special Publication 23*; Crop Science Society of America: Madison, WI, 1995; 47–60.
3. Weatherford, J.W. *Indian Givers: How the Indians of the Americas Transformed the World*; Fawcett Columbine: New York, 1988.
4. Clement, C.R. 1492 and the loss of Amazonian crop genetic resources. I. The relation between domestication and human population decline. Econ. Bot. **1999**, *53* (2), 188–202.
5. Martinez-Alfaro, M.A.; Ortega-Paczka, R.; Cruz-Leon, A. Introduction of Flora from the Old World and Causes of Crop Marginalization. In *Neglected Crops: 1492 from a Different Perspective*; Hernandez-Bermejo, J.E., Leon, J., Eds.; FAO Plant Production and Protection Series, 1994; Vol. 26, 23–33. FAO, Rome.
6. Smartt, J.; Simmonds, N.W. *Evolution of Crop Plants*, 2nd Ed.; Longman Scientific and Technical: Essex, 1995.
7. Jarvis, A.; Guarino, L.; Williams, D.; Williams, K.; Hyman, G. Spatial analysis of wild peanut distributions and the implications for plant genetic resources con-

servation. Plant Genet. Resour. Newsl. **2002**, *131*, 29–35.
8. National Research Council. *Lost Crops of the Incas: Little-Known Plants of the Andes with Promise for Worldwide Cultivation*; National Academy Press: Washington, DC, 1989.
9. Pickersgill, B. Crop Introductions and the Development of Secondary Areas of Diversity. In *Plants for Food and Medicine*; Prendergast, H.D.V., Etkin, N.L., Harris, D.R., Houghton, P.J., Eds.; Royal Botanic Gardens: Kew, 1998; 93–105.
10. *Seeds of Change*; Viola, H.J., Margolis, C., Eds.; Smithsonian Institution Press: Washington, DC, 1991.
11. Columbus, C. *Journals and Other Documents on the Life and Voyages of Christopher Columbus*; Morison, S.E., Ed.; Heritage Press: New York, 1963.
12. Eshbaugh, W.H. Peppers: History and Exploitation of a Serendipitous New Crop Discovery. In *New Crops*; Janick, J., Simon, J.E., Eds.; John Wiley & Sons: New York, 1993; 132–139.
13. *Neglected Crops: 1492 from a Different Perspective*; Hernandez-Bermejo, J.E., Leon, J., Eds.; FAO Plant Production and Protection Series, 1994; Vol. 26. FAO, Rome.
14. Sauer, J.D. *Historical Geography of Crop Plants: A Select Roster*; CRC Press: Boca Raton, FL, 1993.

Commercial Micropropagation

M. C. Gopinathan
E.I.D. Parry Ltd., Bangalore, India

C

INTRODUCTION

Cultivation of plants for ornamental purpose is a wide spread activity throughout the world. Growing and trading flowers and live plants has become a big industry. The world consumption of floriculture products is estimated to be $50 billion. In most of these crops, the availability of planting material is a major limiting factor. Although several of them can be propagated easily, there are others that have no known form of vegetative propagation. Even in cases where traditional methods work, the rate of multiplication is either slow or limited. Further, occurrence of various viral diseases and their transmittance through conventional methods of multiplication make floriculture a high risk industry. Development of plant tissue culture methods such as micropropagation has enabled floriculture industry to tide over these problems associated with plant propagation. The advantages and limitations of commercial micropropagation are briefly discussed in this article.

MICROPROPAGATION

Plant Propagation by tissue culture began with Morel and Martin,[1] who demonstrated the elimination of virus from Dahlia stock through shoot apex culture. Morel's group successfully extended this to other crops, eventually including a Cymbidium mosaic virus–infected Cymbidium orchid.[2] Morel observed the emergence of many plantlets from a single explant in his early experiments. Most orchids reproduce very slowly. Orchid propagators quickly adopted Morel's technique as a commercial practice. Success of rapid propagation of orchids resulted in the application of these techniques to other ornamental crops.

COMMERCIAL MICROPROPAGATION

The ability to control plant propagation and growth provided a big stimulus to entrepreneurs in both horticulture and agriculture. Producers of nursery planting materials saw this as a big business opportunity. This stimulated fast development of commercial in vitro propagation in the early 1970s. At the same time the term "micropropagation" denoting vegetative propagation or cloning of plants in sterile environment, was coined. The 1970s and 1980s were decades of boom for micropropagation industry worldwide. This was facilitated not only by the technological improvements but also by the dedicated work of a large number of scientists, researchers, and technicians. Today, the list of plants that can be propagated by tissue culture has expanded considerably.[3–10]

There has been a steady increase in the number of commercial micropropagation laboratories since the beginning of 1970s. These companies are both small- and medium-size, some of which are privately held and others of which are owned by large corporations.[11] The globalization of micropropagation industry has been active during the 1980s and commercial laboratories were primarily located in the United States, the United Kingdom, Australia, Belgium, Canada, France, Israel, Italy, Japan, The Netherlands, Taiwan, Korea, Malaysia, Thailand, and China. Micropropagation is generally more expensive than conventional propagation because of the high cost of technology and large labor input. The situation has not changed much during the last 30 years, irrespective of the introduction of labor saving and efficiency improvement strategies such as mechanization, modular buildings, computer-based monitoring, and expert management systems. The increase in labor-cost led to the closure of many laboratories in developed countries and shifted the production base to developing countries.

COMMERCIAL MICROPROPAGATION TECHNOLOGY

From a commercial perspective, the application of micropropagation may be best organized into three important areas in the development and marketing of an improved product: product development, product enhancement, and marketability of the product.

Encyclopedia of Plant and Crop Science
DOI: 10.1081/E-EPCS 120010558

Product Development

There are several ways by which micropropagation can shorten product development time and speed the release of improved varieties. These are

Rapid multiplication—production of a large number of plants from a limited stock in a relatively short time.
Product uniformity—since a large portion of production cycle take place under artificial conditions, production with a high degree of uniformity can be achieved.
High volume—allows for the production of large number of plants in a relatively small space.

Product Enhancement

Improved phenotype—for some species micropropagation can effectively enhance the quality and desirable characteristics of a plant and thereby provide a product with enhanced value. For example, micropropagated Syngonium shows a greater degree of basal branching, which is often cited as a desirable characteristic appreciated by commercial growers.

Disease free plants—micropropagation provides the means of eliminating pathogens, especially virus. In general, the aseptic environment of the tissue culture process results in the production of cleaner and healthier plants.

Marketability of the Product

Among the benefits associated with commercial micropropagation are those allowing for ease of product flow and marketing.

Product diversity—the micropropagation system allows for rapid production of a diverse range of plant species and type of propagules, such as plants fully established in soil, unrooted shoots or micro cuttings, and cultured clumps of shoots, for the grower.

Movement of product—the potential to produce material certified free of particular pathogens allows easy exchange of plant material without being subjected to quarantine procedures. An increasing quantity of bare root plants (free of soil) are being shipped in vitro. Plant tissue culture thus allows the utilization of offshore production to capitalize on lower labor costs and expanded markets.

Non-seasonal production—since plants are produced in an artificial, controlled environment, production can be carried out year round, even 24 hours a day. Thus, once micropropagation requirements for a given species are defined, production cycles can be scheduled to meet peak demands.

LIMITATIONS OF MICROPROPAGATION

Although numerous benefits and advantages are associated with micropropagation, there are three major limitations: product line, customer acceptance, and high product cost.

Product Line

Currently, the choice of crops to be produced by micropropagation is limited to the species for which acceptable micropropagation protocols have been defined.[3,6,7,10] Further, the product line is determined by the market demand. Although there are numerous reports in literature of micropropagation systems for a wide range of plant species, they are not often amenable to commercial levels of production. Most of this additional development work is conducted by commercial micropropagation companies as part of their in-house research programs.

Customer Acceptance

Product quality—the ability to deliver a product with consistently high quality is of prime importance and directly impacts customer acceptance. To ensure a high level of quality, it is very important to direct particular effort towards quality control, including rouging of off-types, grading by size, and product trials if possible.

Product delivery—the ability to schedule crops accurately and deliver a defined quantity of product consistently are important factors in customer acceptance. Currently, micropropagation is best suited to provide a steady stream of plant material rather than adapt to a customer's seasonal requirements. Most commercial labs require a purchase commitment from customer and a notice of at least six months prior to a change in requested quantities. In addition, familiarity of the grower with micropropagation procedures is required for successful operation if the plants are to be established in soil.

High Product Cost

Relatively large capital investment is required to establish a commercial micropropagation laboratory with its associated facilities. For many species, micropropagation is currently not competitive compared with other conventional methods such as cuttings and seeds. Thus, micropropagation is profitable only when there is an associated advantage over conventional methods.

PRODUCTION TECHNOLOGY

Micropropagation technology generally follows standard tissue culture protocols.[5,6,9,10,12] For example, a small piece of plant to be multiplied (explant) is isolated from the healthy plant and surface sterilized. The sources of explant material will vary depending on the species, but shoot tips, leaf, stem, and lateral buds have been successfully used. The sterilized explant is rinsed with sterile water and placed in sterile closed containers with specific nutrient media. The explant may produce proliferating shoots or may undergo callus or embryogenic growth in the incubation stage. Media constituents and its combinations mainly determine the nature of growth. Proliferating cultures are transferred periodically to fresh medium to obtain desired number of plants. These plants are rooted individually or in clumps, acclimatized in green or shade houses, packed and shipped to the customers. Although the plants can be multiplied by several techniques, the methods most commonly utilized by commercial micropropagation companies include enhanced axillary shoot branching, adventitious shoot production, and somatic embryogenesis. Depending on the species, a well-defined, species-specific protocol of plant regeneration will be used for plant production.

The adoption of Murashige's[13] Stage I, II, III & IV production scheme greatly simplifies daily operation, accounting, and product cost analysis of a commercial production facility, and allows for greater care in communication with customers and other laboratories.

PRODUCTION FACILITY

The ultimate success of a commercial micropropagation business is largely influenced by the design, location, and cost of the facility. Commercial tissue culture facilities usually require specialized areas and equipment for culture media preparations, sterilization, and storage of nutrient media, aseptic manipulation of plant material, incubation and maintenance of cultures under controlled conditions of temperature, light and humidity, acclimatization in greenhouse, and packing and shipping.

The laboratories can be classified by the product type offered; the range is from Stage II (culture in vessels) through Stage III (bare root plantlets) to finished Stage IV liners. The product type largely reflects whether or not the company is equipped with acclimization and greenhouse facilities. Other potentially available services may include disease indexing and research capability to serve both in-house and external contract requirements.

ARTICLES OF FURTHER INTEREST

Anther and Microspore Culture and Production of Haploid Plants, p. 43
In Vitro Morphogenesis in Plants—Recent Advances, p. 579
In Vitro Plant Regeneration by Organogenesis, p. 584
In Vitro Tuberization, p. 594
Somatic Embryogenesis in Plants, p. 1165

REFERENCES

1. Morel, G.; Martin, C. Guerison de dahlias atteints d'une maladie a virus. C. R. Acad. Sci. **1952**, *235*, 1324–1325.
2. Morel, G. Producing virus-free Cymbidium. Am. Orchid Soc. Bull. **1960**, *29*, 495–497.
3. George, E.F.; Puttock, D.J.M.; George, H.J. *Plant Culture Media—Formulations and Uses*; Exegetics Limited: England, 1987; Vol. 1, 1–493.
4. George, E.F.; Puttock, D.J.M.; George, H.J. *Plant Culture Media—Commentary and Analysis*; Exegetics Limited: England, 1988; Vol. 2, 1–420.
5. George, E.F. *Plant Propagation by Tissue Culture*, 2nd Ed.; The Technology Exegetics Limited: England, 1993; 1–574. Part. 1.
6. George, E.F. *Plant Propagation by Tissue Culture*, 2nd Ed.; Exegetics Limited: England, 1996; 1–799. Part 2, In Practice.
7. *Biotechnology in Agriculture and Forestry*; Bajaj, Y.P.S., Ed.; Springer and Verlag, 1997; Vol. 40, 1–397.
8. *Transplant Production in the 21st Century*; Kubota, C., Chun, C., Eds.; Academic Publishers, 2000; 1–300.
9. Razdan, M.K. *Introduction to Plant Tissue Culture*; Science Publishers: India, 2002; 1–376.
10. Herman, E.B. *Recent Advances in Plant Tissue Culture*; Agritech Publications: New York, USA, 2002; Vol I–VI.
11. http://aggie-horticulture.tamu.edu/tisscult/microprop/microprop.html. (accessed May 2002).
12. Bhojwani, S.S.; Razdan, M.K. *Plant Tissue Culture—Theory and Practice*; Elsevier: New York, 1983; 1–473.
13. Murashige, T. Plant propagation through tissue cultures. Annu. Rev. Plant Physiol. **1974**, *25*, 135–166.

Competition: Responses to Shade by Neighbors

Carlos L. Ballaré
Universidad de Buenos Aires, Buenos Aires, Argentina

INTRODUCTION

In dense plant populations, neighbors are a central determinant of the growth and performance of each individual plant. Neighbors compete for light and soil resources and influence the interactions between plants and their consumers. Not surprisingly, plant species that often occur in dense canopies have evolved exquisite sensing mechanisms to detect the proximity of neighboring plants. Some of these mechanisms are based on the detection of changes in the light environment. This detection is accomplished by means of dedicated photoreceptors, which constantly monitor the light climate and relay molecular signals to mechanisms that control plant physiology and morphogenesis. Most field crops are grown at high density, in order to maximize yield per unit area. Therefore, variations in the way in which individual crop plants "read" their light environment can have important consequences on the growth rate and yield of the whole canopy. In this article I briefly describe light-sensing mechanisms used by plants to detect neighbor proximity and discuss the ecological and agricultural significance of the responses elicited by these photoreceptors. For a comprehensive review, the reader is referred to Refs. 1 and 2.

FORAGING FOR LIGHT

Plants acclimate to shading by other plants with plastic morphological and physiological adjustments. These adjustments often include changes in the stoichiometry and organization of the photosynthetic apparatus and large changes in shoot morphology and biomass allocation patterns. A typical morphological response to shading is the production of thinner leaves, which has the overall effect of increasing photosynthetic area per unit of carbon invested in leaf production. Other responses vary among taxa and species habitats. Typical responses among dicots from open habitats, such as grasslands and arable land, include increased internode and petiole elongation, production of more erect leaves, reduced branching, and acceleration of flowering. Grasses often respond to shading with reduced tillering, increased sheath length, and increased tiller angles. All these morphological changes tend to improve the access of the younger leaves to the upper canopy strata, increasing light exposure. In patchy canopies, where the light environment is also heterogeneous over the horizontal dimension, plants actively project leaf area into canopy gaps. This projection is the result of phototropic bending of the shoots and differential branching and growth responses. The combination of morphological and physiological adjustments that plants generate as they grow in a heterogeneous canopy matrix, which results in concentration of photosynthetic power in well-illuminated spots, is often referred to as "shade-avoidance" or "foraging for light."[1,3] Interestingly, plants of open habitats are able to detect their neighbors remotely, using changes in the spectral composition of reflected sunlight, and forage for light when growth and morphological development are still not limited by severe shading.[2]

HOW DO PLANTS READ THE LIGHT ENVIRONMENT?

Plants detect changes in their light environment using specific photoreceptors. In *Arabidopsis thaliana*, three families of informational photoreceptors have been identified: phytochromes (Phy), cryptochromes (Cry), and phototropins (Phot).[4–6] Phy receptors are maximally sensitive to the red (R, 600–700 nm) and far-red (FR, 700–800 nm) regions of the solar spectrum. There are five Phy receptors in *Arabidopsis* (PhyA–E), which are encoded by a family of five divergent genes (*PHYA–E*). All Phy receptors have the same chromophore, a linear tetrapyrrole. Phy receptors play a central role in regulating nearly all aspects of plant morphogenesis, including seed germination, de-etiolation, stem elongation, leaf morphology and orientation, tropisms, flowering, and organ senescence.[3] Cry receptors[5] are specialized sensors of blue light (B, 400–500 nm) and UV-A radiation (λ max = 370 nm). There are two Cry receptors characterized in *Arabidopsis* (cry1 and cry2). These are encoded by *CRY*1 and *CRY*2, respectively, and they both use flavin as chromophore. Cry proteins bear similarities to microbial photolyases but have no photolyase (DNA repair) activity. These receptors play a role in controlling hypocotyl elongation in seedlings and also in the entertainment of the

Encyclopedia of Plant and Crop Science
DOI: 10.1081/E-EPCS 120010518

circadian clock. Finally, two Phot receptors (Phot1 and Phot2) have been found in *Arabidopsis*.[6] These are also specialized B light receptors that use flavin chromophores and control phototropic responses at several levels, from whole organs and organelles to stomatal opening.

HOW DO PLANTS USE SPECIFIC PHOTORECEPTORS TO FORAGE FOR LIGHT AND RESPOND TO SHADING?

Chlorophyll molecules in photosynthetic organs absorb strongly in the B and R regions of the light spectrum. In contrast, the absorption in the green and FR regions is much lower, and most photons in these wavebands are either transmitted or reflected. Therefore, the proximity of neighboring plants not only reduces the total amount of light energy available for photosynthesis but also has a profound influence on the spectral composition of radiation.[1–3] Of particular importance, given the spectral characteristics of plant photoreceptors, are the depletion of B light, which can be sensed by Phot and Cry receptors, and the reduction in R-to-FR ratio (R:FR), which is detected by members of the Phy family.

For the purpose of illustration, two general, simplified scenarios will be considered. The first scenario is that of a seedling that emerges under an established leaf canopy and has to deal with a light environment already depleted in photosynthetic light availability (e.g., under a forest canopy). The second scenario is that encountered by seedlings that emerge from bare soil, with no overtopping neighbors but surrounded by other seedlings of similar age and size (e.g., weed and crop seedlings emerging after seed bed preparation in arable land, or cohorts of regenerating seedlings emerging in a clear-cut). The mechanisms whereby plants sense and respond to deep overhead shading (first scenario) have been difficult to disentangle, because multiple environmental factors are affected under the shade of a dense canopy.[2] Physiological experiments and experiments with mutants suggest that changes in R:FR, perceived by PhyB, are important. However, other factors of the radiation environment, such as reduced overall fluence rate, reduced ultraviolet radiation, and other microenvironmental changes (reduced wind speed and increased humidity), are likely to play a role in controlling plant morphogenesis under natural shade. Mutants deficient in PhyB sometimes show strong responses to shading, confirming the involvement of other sensing mechanisms.

Foraging for light in populations of plants that are approximately even-height (second scenario) often involves perception of FR radiation reflected by neighboring plants (Fig. 1). Back-reflection of FR photons by neighboring leaves lowers the R:FR ratio in the light that impinges laterally on plant shoots, signaling the proximity of potential competitors before there is significant shading among neighbors.[1,2,7] Most plants from open habitats use the drop in lateral R:FR as an early warning signal of oncoming competition and respond with a rapid increase in stem (or petiole) elongation, production of more erect leaves (particularly rosette plants), and reduced branching. Studies with photoreceptor mutants show that PhyB plays a key role in the early detection of potential competitors.[2] In even-height plant populations, the proximity of neighboring plants can also reduce the light fluence rate received by the stems well before there is a significant

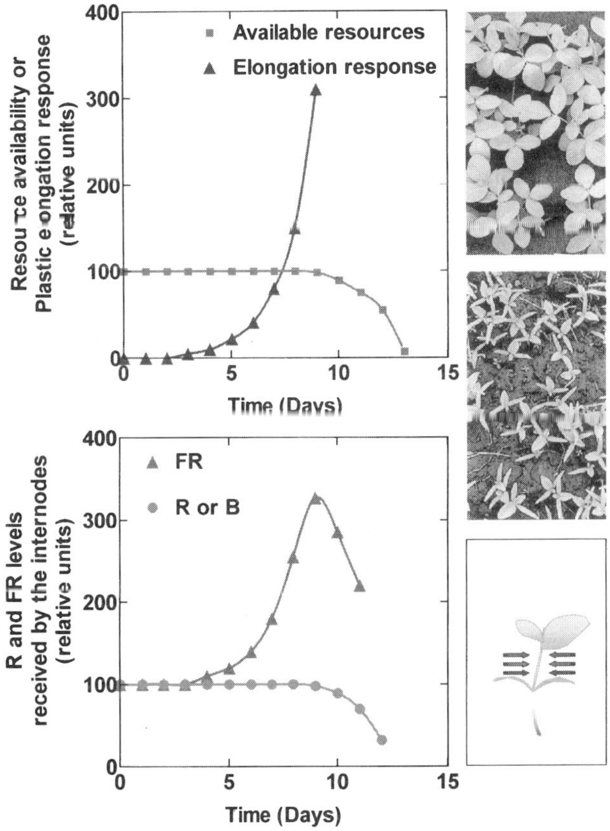

Fig. 1 Reflected FR radiation is an early signal of competition. In even-height canopies, such as those shown on the right-hand side panels (cohorts of soybean seedlings and *Datura ferox*), stem elongation responses to crowding occur well before neighboring plants cause a significant reduction in the availability of light for photosynthesis (upper graph). Early responses such as these are triggered by changes in the spectral composition of the light that impinges laterally on the shoots, which is rapidly enriched in FR as neighboring plants grow and the leaf area index of the canopy increases (lower graph and diagram). (For a review, see Refs. 1 and 2.) *(View this art in color at www.dekkker.com.)*

depletion in light availability at leaf level. This reduction in fluence rate promotes stem elongation, and both B light receptors and the phytochromes are involved in the perception of fluence-rate signals.[2] The generation of rapid elongation responses to early signals of competition increases the likelihood of individual plant survival in a system where access to the light resource rapidly becomes limiting as the plants grow and occupy canopy space.[1,8]

In patchy canopies, reflected FR is also a directional signal. Plants project leaf area away from potential competitors using negative phototropic responses driven by reflected FR and controlled by PhyB. In addition, plants located at or near the edge of a canopy gap are exposed to intense B light gradients, as B levels are very low within the canopy and high in the gap area. Since plants are positively phototropic to B light (via phot1 and phot2 receptors), these gradients also serve as a directional signal that attracts new shoots to canopy gaps, complementing the negative phototropic responses to FR.[2]

ECOLOGICAL SIGNIFICANCE

As indicated in the foregoing, and as supported by experiments, foraging for light in response to proximity signals perceived by dedicated photoreceptors is essential for plant survival in rapidly growing canopies.[1,8] Photomorphogenic signaling among neighbors also has important implications for the functioning of the canopy as a whole.[1,2] It has been shown that in monocultures of mutants with impaired R:FR sensitivity, size-structuring (i.e., the development of size differences among neighbors) is much more pronounced than in populations of photomorphogenically competent plants.[9,10] This is because photomorphogenically normal plants respond to early signals of shading with adaptive morphological changes that improve light interception, and the intensity of this response tends to be greater in small (more shaded) plants than in large plants (at least as long as growth is not severely limited by lack of resources).[9] This morphological adjustment, which is absent in R:FR–blind mutants, attenuates the suppressing effect that large plants have on small (more shaded) plants and thereby retards the development of size hierarchies within the population.

CONCLUSION

Individual crop plants change their shape and allocation patterns to improve light capture in the canopy. What are the implications for crop yield? Can the light-foraging pattern be controlled and improved to increase crop productivity? Researchers have begun to answer these questions using a variety of experimental approaches, which in most cases have involved manipulations of the light environment and morphogenic sensitivity using model plants such as *Arabidopsis* and tobacco. Two basic hypotheses have been developed and partially tested.[1] The first hypothesis is that light foraging and shade-avoidance responses, while essential for individual plant survival in the wild, are deleterious to crop yield. The reasoning behind this idea is that height growth responses utilize assimilates that crop plants could otherwise invest in the generation of leaf area or in the development of reproductive structures and harvestable organs. The second hypothesis conveys exactly the opposite idea, i.e., that foraging for light, driven by photomorphogenic mechanisms, is essential to ensure optimal deployment and redistribution of the crop photosynthetic capacity during canopy development and to buffer the crop population against the development of size inequalities. Therefore, eliminating plant sensitivity to light signals of neighbor proximity would have negative impacts on crop yield. Clearly, these hypotheses have very different practical implications for breeding programs that attempt to increase crop yield by manipulation of light sensitivity. Experimental evidence has accumulated in favor of both hypotheses, suggesting that the impacts of altering photomorphogenic behavior on crop growth and yield may vary with crop species, developmental phase, and overall agronomic scenario.[1] Physiological experiments have shown that directed overexpression of plant photoreceptor genes can be used to molecularly mask specific plant organs to proximity photosignals without altering the light sensitivity of other plant parts. This strategy could be used to target specific plant organs or processes in biotechnological programs aimed at manipulating selected morphogenic responses in cultivated plants.[11] Alternatively, targeted alterations of crop plant photomorphogenesis could be obtained by directing the bioengineering efforts to signaling components that couple the photoreceptors with the molecular controls of a particular physiological function.

ARTICLES OF FURTHER INTEREST

Breeding Plants with Transgenes, p. 193
Ecology: Functions, Patterns, and Evolution, p. 404
Leaves and Canopies: Physical Environment, p. 646
Photoreceptors and Associated Signaling I: Phytochromes, p. 881
Photoreceptors and Associated Signaling II: Cryptochromes, p. 885
Photoreceptors and Associated Signaling III: Phototropins, p. 889

Photoreceptors and Associated Signaling IV: UV Receptors, p. 893
Phytochrome and Cryptochrome Functions in Crop Plants, p. 920
Shade Avoidance Syndrome and Its Impact on Agriculture, p. 1152

REFERENCES

1. Ballaré, C.L.; Scopel, A.L.; Sánchez, R.A. Foraging for light: Photosensory ecology and agricultural implications. Plant Cell Environ. **1997**, *20*, 820–825.
2. Ballaré, C.L. Keeping up with the neighbours: Phytochrome sensing and other signalling mechanisms. Trends Plant Sci. **1999**, *4*, 97–102.
3. Smith, H. Phytochromes and light signal perception by plants—An emerging synthesis. Nature **2000**, *407*, 585–596.
4. Casal, J.J. Phytochromes, cryptochromes, phototropin: Photoreceptor interactions in plants. Photochem. Photobiol. **2000**, *71*, 1 11.
5. Ahmad, M.; Cashmore, A.R. Seeing blue: The discovery of cryptochrome. Plant Mol. Biol. **1996**, *30*, 851–861.
6. Briggs, W.R.; Christie, J.M. Phototropins 1 and 2: Versatile plant blue-light receptors. Trends Plant Sci. **2002**, *7*, 204–210.
7. Ballaré, C.L.; Scopel, A.L.; Sánchez, R.A. Far-red radiation reflected from adjacent leaves: An early signal of competition in plant canopies. Science **1990**, *247*, 329–332.
8. Schmitt, J. Is photomorphogenic shade avoidance adaptive? Perspectives from population biology. Plant Cell Environ. **1997**, *20*, 825–829.
9. Ballaré, C.L.; Scopel, A.L.; Jordan, E.T.; Vierstra, R.D. Signaling among neighboring plants and the development of size inequalities in plant populations. Proc. Natl. Acad. Sci. USA. **1994**, *91*, 10094–10098.
10. Ballaré, C.L.; Scopel, A.L. Phytochrome-signalling in canopies. Testing its population-level implications with photoreceptor mutants of Arabidopsis. Funct. Ecol. **1997**, *11*, 441–450.
11. Rousseaux, M.C.; Ballaré, C.L.; Jordan, E.T.; Vierstra, R.D. Directed overexpression of PHYA locally suppresses stem elongation and leaf senescence responses to far-red radiation. Plant Cell Environ. **1997**, *20*, 1551–1558.

C

Crop Domestication in Africa

Rémy S. Pasquet
IRD-ICIPE, Nairobi, Kenya

INTRODUCTION

Several plants, including sorghum, were domesticated in Africa. These indigenous plants provided an adequate base for the development of sedentary agriculture. Furthermore, the highly developed cultures of Nok, Ife, and Benin, as well as the Sudanic kingdoms, were supported by an indigenous African agriculture.

TIMING OF PLANT DOMESTICATION

Plant domesticates appeared in Egypt around 5000 B.C., but these belonged to the Southwest Asia complex, coming from the Levant where agriculture was established by 6000 B.C., without any southern influence. Among the preserved plant materials found in ancient Egyptian tombs, there is not a single African crop.[1,2] Strikingly, in Africa south of the Sahara, archaeological and linguistic data show that the domestication of animals preceded cereal agriculture by two to four millennia. Cattle were present in Sahara by 5000 B.C., while the earliest dates for crop domesticates are around 1000 B.C. Collection and maybe cultivation of wild grasses was important for early pastoralists, long before plant domestication.[2,3]

THE NORTHERN SAVANNAS

Vavilov[4] identified the Ethiopian highlands as a center of plant domestication, and Harlan[5] considered an area covering all of sub-Saharan Africa north of the equator as a noncenter of plant domestication. Most African crops are savanna species, including the ones cultivated in forest areas. Savannas south of the equator appear not to have been involved in crop domestication. Agriculture spread there with the move of Bantu-speaking farmers. This southward move is well dated archaeologically and started during the first millennium B.C.[6] However, northern savannas are a complex area, from which three language phyla originated. These three phyla are assumed to have originated in the same ecological zone and adjacent to one another: Niger-Congo around the bend of the Niger, Nilo-Saharan east of Lake Chad, and Afro-Asiatic around the upper Nile.[3] Therefore, following Ehret,[7] it should bepossible to consider five domestication centers, all part of Harlan's[5] noncenter.

MANDE CENTER

Several species are definitely West African, and they all can be linked with Mande language speakers. The African rice, *Oryza glaberrima*, is cultivated from Senegal to Chad. The progenitor is an annual, autogamous grass, *Oryza breviligulata*, adapted to savanna water holes that fill up during the rains, from Senegal to Central Africa. There is also a complex of wild, weedy, and domesticated forms. Highest genetic diversity is found in the inland Niger delta on the one hand, in Guinea and southern Senegal on the other hand.[1,3] Two small-seeded millets also belong to that center. As a domesticate, *Brachiaria deflexa* is only known from Guinea, while wild *B. deflexa* is widespread all over northern Africa. Fonio (*Digitaria exilis*) extends from Cape Verde to Nigeria and is more important in the western part of that area, when considering both the number of varieties and the acreage. The wild progenitor could be *Digitaria longiflora*, a widely distributed species. The present distribution and archaeological data from Mauritania suggest that these millets were the first cultivated cereals, progressively replaced by rice and pearl millet.[1,3]

Although the evidence is less clear, pearl millet (*Pennisetum glaucum*), the most important African cereal after sorghum, seems to be of West African origin also. Pearl millet is an allogamous plant cultivated from Senegal to India and southern Africa. The wild forms occur discontinuously in the Sahel and subdesert zone from the Atlantic to the Red Sea, with large hybrid swarms between wild and domesticated types. Regarding the domesticated forms, from a nucleus including late-maturing West African, East African, and Indian cultivars, authors separate early-maturing West African cultivars on the one hand and cultivars from southern Africa on the other hand, an organization reminiscent of that of sorghum and cowpea.[8] The cereal reached India early, like sorghum and finger millet, which would suggest an East African origin.[3] However, it is the first cereal to appear in West African archaeological sites, where it predates sorghum: from 1500 B.C. in Mauritania

Encyclopedia of Plant and Crop Science
DOI: 10.1081/E-EPCS 120017102

to 800 B.C. in the Lake Chad area, while sorghum does not appear before 800 A.D. in the latter area.[2]

BENUE-CONGO CENTER

Yams are the main staple crop in an area from eastern Ivory Coast to Cameroon, and yam names are well correlated with the Kwa and Benue-Congo languages.[1] There are several species, but the main ones belong to the *Dioscorea cayenensis–Dioscorea rotundata* complex. Nevertheless, yams do not seem to be true domesticates in Africa. Yam "domestication" involves vegetative propagation under cultivation conditions adapted for wild material, with wild forms that have been cultivated very recently growing together with older cultivated forms. When the wild *Dioscorea praehensilis* is cultivated, the *D. cayenensis* phenotype is obtained after a few generations, i.e., the amount of fiber in the tuber declines markedly as the starch content increases, there is a drop in the number of thorns, and the number of cataphylls (scale-like modified leaf) borne on the main stem decreases spectacularly. The polyploid complex should be of hybrid origin (*D. cayenensis sensu stricto* and *D. praehensilis–Dioscorea abyssinica*, with other minor species), and there is gene flow between wild and cultivated forms.[8] Two pulses, *Macrotyloma geocarpum* and *Sphenostylis stenocarpa*, belong to that center. *M. geocarpum* is cultivated from Burkina Faso to Nigeria, and its wild progenitor is not really known. Its vocabulary could suggest a link with Gur language groups. *S. stenocarpa* is cultivated from Burkina Faso to Cameroon (mainly for its seeds) and further south in Zaire (mainly for its tuber). Bambara groundnut (*Vigna subterranea*) could also belong to that area. This pulse is widely cultivated in most of Africa, but its wild progenitor is restricted to savannas from Nigeria to Sudan. Its vocabulary suggests a link with Adamawa-Ubangi and Benue-Congo languages. Surprisingly, all the plants from this center are "subterranean."

EAST CENTRAL CENTER

Sorghum (*Sorghum bicolor*) is widely cultivated all over Africa, although it was recently superseded by maize in some areas. Wild-race verticilliflorum is the primary progenitor candidate and is extremely abundant in the eastern half of Africa, from Chad and Sudan to southern Africa. Domesticated sorghums are classified into five races. Race Bicolor, the most primitive race, includes sorghums cultivated for their sweet stems, for beer brewing, and for dye. Race Guinea is primarily West African and is adapted to high rainfall conditions. Race Caudatum is grown in the area from Lake Chad to Ethiopia. Race Kafir is found in southern Africa, and race Durra has its main center of cultivation in India. Race Bicolor was disseminated everywhere, and all the other races evolved from race Bicolor, each in its own geographic area.[3,8] Archaeological data suggest that sorghum was domesticated in the eastern Sahara, where pearl millet is not reported. These earliest finds have been dated to slightly before 1000 B.C. The reliability of claims for domesticated sorghum at earlier sites in Africa, Arabia, and India cannot be substantiated at present,[2] though sorghum, along with finger millet, was brought to India at an early time.[3] Like sorghum, cowpea (*Vigna unguiculata*) is cultivated almost everywhere in Africa. In addition to seed production, cowpea is used as fodder and was an important fiber plant. Organization of the domesticated genepool is reminiscent of sorghum. From a nucleus of primitive cultivars (cv.-gr. Biflora), cv.-gr. Melanophthalmus with thin seed testa was developed in West Africa, while photoperiod-independent cv.-gr. Unguiculata and cv.-gr. Sesquipedalis (the yard-long bean) were developed in southern Africa and Asia, respectively. The wild progenitor is a weed encountered all over African savannas, and introgressions between wild and domesticated types are widespread though not frequent. Vocabulary suggests a link with Chadic and Nilo-Saharan speakers.[8]

EAST CENTER

Finger millet (*Eleusine coracana*) is a small-seeded cereal cultivated from Ethiopia to Cameroon and in southern Africa and India. In the wild, it is a widespread tropical weed, especially abundant in the highlands of East Africa, where there is a crop-weed complex. The crop was introduced early on in India.[3] Noog (*Guizotia abyssinica*) is an oilseed and a major crop in Ethiopia only, where the complete range of wild, weedy, and domesticated races is found in great abundance.[1,3] Tef (*Eragrostis tef*) is grown on a very large scale in Ethiopia, but almost nowhere else. The wild progenitor is *Eragrostis pilosa*, a common and widely distributed grass. There is no crop-weed complex since both wild and domesticated races are cleistogamous.[3]

ENSETE CENTER

Ensete (*Ensete ventricosum*) is widespread as a wild bananalike tree, from Ethiopia to Cameroon and southern Africa. As a starchy crop, it is cultivated only in the Omotic language area of Ethiopia, and like yam, it is not really a true domesticate. However, it has been hypothesized that the distribution of wild ensete and that of the banana-based cultures of central and eastern Africa, whose traditional life focused on the cultivation of banana, are clearly very ancient. This might suggest that the East Asian domesticate (*Musa acuminata*) could be a replacement for an older, similar crop such as ensete.[6,7]

CONCLUSION

Plant domestication in Africa was a late event, which took place during the second millennium B.C. or later. All plants were domesticated within the savanna area, which stretches from the Atlantic to the Red Sea. Due to the ethnic and linguistic diversity of Africa, it is possible to identify five domestication centers within this area.

ARTICLES OF FURTHER INTEREST

Crop Domestication in Southwest Asia, p. 323
Crop Domestication in the American Tropics: Starch Grains Analyses, p. 330
Crop Domestication: Founder Crops, p. 336
Crop Domestication: Role of Unconscious Selection, p. 340
Cropping Systems: Slash-and-Burn Cropping Systems of the Tropics, p. 363
Gene Flow Between Crops and Their Wild Progenitors, p. 488

REFERENCES

1. *Origins of African Plant Domestication*; Harlan, J.R., De Wet, J.M.J., Stemler, A.B.L., Eds.; Mouton Publishers: The Hague, 1976; 1–498.
2. *The Exploitation of Plant Resources in Ancient Africa*; Van der Veen, M., Ed.; Kluwer Academy: New York, 1999; 1–283.
3. *The Archaeology of Africa: Food, Metals and Towns*; Shaw, T., Sinclair, P., Andah, B., Okpoko, A., Eds.; Routledge: London, 1993; 1–857.
4. Vavilov, N.I. Studies on the origin of cultivated plants. Bull. Appl. Bot. Genet. Pl. Breed. **1926**, *16*, 1–245.
5. Harlan, J.R. Agricultural origins: Centers and non centers. Science **1971**, *174*, 468–474.
6. *The Growth and Farming Communities in Africa from the Equator Southward*; Sutton, J.E.G., Ed.; British Institute in Eastern Africa: Nairobi, 1996; 1–338.
7. *From Hunters to Farmers: The Causes and Consequences of Food Production in Africa*; Clark, J.D., Brandt, S.A., Eds.; University of California Press: Berkeley, 1984; 1–443.
8. *Tropical Plant Breeding*; Charrier, A., Jacquot, M., Hamon, S., Nicolas, D., Eds.; Science Publishers: Enfield, USA, 2001; 1–569. CIRAD: Montpellier, France.

Crop Domestication in China

Tracey L.-D. Lu
Chinese University of Hong Kong, Hong Kong, China

INTRODUCTION

Since J.G. Andersson discovered a piece of rice husk on a Neolithic potsherd at the Yangshao village in the Yellow River Valley in 1927, the origin of agriculture in China has been a topic of scholarly interest.[1] N. Vavilov and Y. Ding also investigated this issue in the 1920s and the 1930s.[1] Today, archaeological discoveries from the 1970s, as well as agronomic and genetic studies, have provided information for our understanding of the domestication of several crops and plants in China, but the questions on how, why, and precisely where and when agriculture occurred in this area still remain.

EVIDENCE FOR CROP DOMESTICATION IN CHINA

Breakthrough archaeological evidence for crop domestication in China was first found in the 1970s.[1] In the Hemudu assemblage in the lower Yangzi River Valley, dated between 7000 and 5300 years ago, rice (*Oryza sativa*) grains and straws were discovered. Remains of foxtail and broomcorn millets (*Setaria sativa* and *Panicum miliaceum*, respectively) were found in the Cishan and Peiligang assemblages in the middle Yellow River Valley, both dated to 7800 years ago (Fig. 1).[1]

Additional archaeological remains were found in the Yellow and the Yangzi River Valleys from the late 1980s to the 1990s, among them the Jiahu assemblage located between the Yellow and the Yangzi River Valleys, and the Xianrendong, Yuchanyan, Pengtoushan and Bashidang assemblages in the Yangzi River Valley (Fig. 1).[1] These discoveries and multi-disciplinary analyses of the new data have profoundly enriched our knowledge on the origin of rice farming in China. An initial isotopic analysis on human remains found in Xianrendong suggests that wild rice was gathered and consumed as food by the Xianrendong occupants by 12,000 years ago, probably even earlier.[2] Phytolith analysis further suggests that the Xianrendong occupants might have started rice cultivation between 10,000 and 9000 years ago.[2] Meanwhile, rice husks found in the Yuchanyan cave suggest that rice might have also been gathered at the boundary area between the Yangzi River Valley and South China at the beginning of the Holocene.[1] The rice husk and grains found in Jiahu, Pengtoushan, and Bashidang, associated with house remains, burials, tools, pottery, and even musical instruments (in Jiahu) illustrate that rice was grown in the Huai River Valley and the middle Yangzi River Valley by sedentary and affluent farmers as early as 9000 to 8500 years ago.[1]

In addition, Chinese cabbage (*Brassica pekinensis*) seeds were found in Banpo in the middle Yellow River valley, dated to around 6800 years ago (Fig. 1).[3] Remains of domesticated soybean (*Glycine max*) and peach (*Prunus persica*) were recently discovered in Zaojiaoshu, the middle Yellow River Valley, dated to 3900 to 3600 years ago (Fig. 1).[4] However, no archaeobotanic remains of tea in prehistoric China have been reported so far.

THE PROCESS OF CROP DOMESTICATION

Crop domestication is a process of manipulating wild plants in such a way that selection under cultivation leads to new plant types that meet human beings' needs. In China's context, the major subjects of this process are the annual green foxtail (*S. viridis*) and the perennial wild rice (*O. rufipogon* Griff.). The wild progenitor of broomcorn millet (*P. miliaceum*) has not been identified to date. Based on plant observations and measurements, it is clear that there are salient differences between green foxtail and wild rice and their domesticated counterparts (Table 1).[1]

Genetic study of green foxtail is insufficient at the moment. For rice, however, recent genetic research indicates that many phenotypes of the perennial wild rice, such as seed shattering and longer growth cycle, are controlled by several genes, many of them being dominant.[5,6] In other words, the majority of genes controlling the human-favored phenotypes such as tough rachis, shorter growth cycle, and lower degree of dormancy, etc., are recessive. Humans have made the recessive genes dominant in the domesticated plants through cultivation by consciously and unconsciously selecting plants containing certain recessive genotypes. Archaeological data in China today are still insufficient to fully illustrate this human interference process, but some clues are apparent.

Encyclopedia of Plant and Crop Science
DOI: 10.1081/E-EPCS 120017099

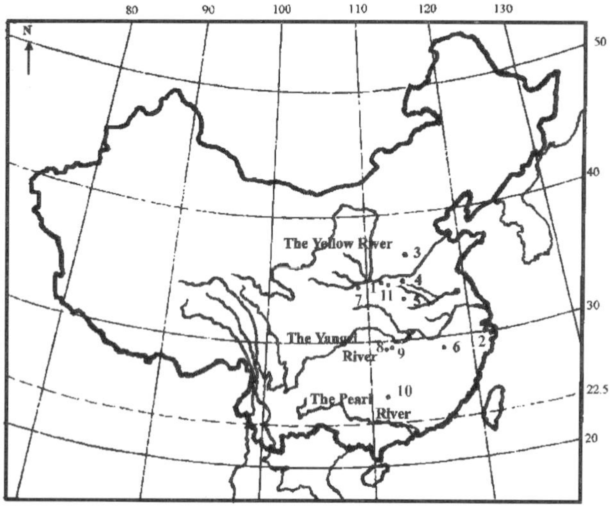

Fig. 1 Archaeological sites mentioned in the text. 1. Yangshao 2. Hemudu 3. Cishan 4. Peiligang 5. Jiahu 6. Xianrendong 7. Banpo 8. Pengtoushan 9. Bashidang 10. Yuchanyan 11. Zaojiaoshu.

As mentioned above, rice gathering might have begun by 10,000 b.p. in the Yangzi River Valley, followed by cultivation at around 9000 years ago. The rice remains found in Jiahu and Bashidang have been identified as a transitional type between wild and domesticated rice, characterized by smaller grains compared to that of current domesticated rice, but displaying various traits of both the *japonica* and *indica* subspecies.[7] As the quantities of rice grains found in these sites are substantial, they represent the early stage of rice domestication by intensive cultivation.

Compared to rice domestication, data on the domestication process of millets are very limited. Although an initial use-wear analysis indicates that wild grasses might have been harvested in the middle Yellow River Valley before 13,000 years ago,[8] there is no solid evidence for wild millet gathering to date. Because millets do not seem to produce diagnostic phytolith cells,[9] the effectiveness of phytolith analysis for millet farming is significantly hindered. This further affects the study of the origin of millet farming, as seeds are not always well preserved in prehistoric deposits. In order to collect data from different aspects, a green foxtail cultivation experiment has been carried out in the Yellow River Valley since 1999.[10] The initial result of this experiment, plus ethnographic data from Taiwan indigenous foxtail millet cultivators[11] and Chinese historic documents, suggests that millets might have been domesticated by rotational cultivation and by consciously selecting plants with larger seeds and fewer panicles.[10,11]

The result of the green foxtail cultivation experiment also indicates that the return of initial millet cultivation is very low; therefore, the early farmers could not have purely relied on farming for their survival.[10] A wild rice cultivation experiment conducted by the author in South China since 2000 also suggests a very low output for wild rice farming. Therefore, the process of crop domestication in China might have been a gradual one, and the early farmers might have remained foragers until the output of farming increased.

Archaeological data to date seem to indicate that domesticated cabbage, soybean, peach, etc., occurred well after the domestication of millets and rice in the Yellow and the Yangzi River Valleys. This may suggest that the development of horticulture and the domestication of other plants occurred after farming based on millet and rice became major subsistence strategies in the above regions.

CONCLUSION

Crop domestication in China occurred at the beginning of the Holocene in temperate zones, when the climate began to warm up after the last glacial episode. Current research indicates that wild rice was gathered and consumed as food by 10,000 years ago in the Yangzi River Valley, followed by rice cultivation in the same region by 9000 years ago.[1] This process from gathering to cultivation is

Table 1 Some major differences between wild and domesticated foxtail millet and rice

Criteria	Wild millets and rice	Domesticated millets and rice
Tillers per plant	More tillers	Fewer tillers
Seed shattering	At a very high rate	Very low rate
Panicle diameter	Usually small	Comparatively bigger
Flowering and ripening	Heterochronous—last for more than one month	Simultaneous
Seeds per panicle	Much fewer seeds	More seeds
Size of seeds	Small	Big
Growth cycle	Longer	Shorter

(From Refs. 1 and 10.)

similar to that in other domestication centers of the world. On the other hand, the process of millet domestication in the Yellow River Valley is much less clear, although it is certain that millets were intensively grown in that region by 7800 years ago.

Archaeological discoveries made in recent years have shed more light on the issue of plant domestication in China, but many questions still remain. First, are the Yellow and the Yangzi Valleys two independent centers for plant domestication, or just one center? Second, what is the impetus to the transition from foraging to farming in China? Third, who are the early millet farmers in the Yellow River Valley, and precisely when did millet cultivation begin in that area? Last, but not the least, what is the taxonomic link between the cultivar groups of domesticated rice (*japonica*, *indica*, and *javanica*)? Which one was domesticated first? Were they the result of separate domestications or of a single domestication followed by divergence under cultivation? Was the *japonica* subspecies domesticated in China and the *indica* domesticated in India? At present archaeologists, agronomists, and geneticists are debating this issue, and no consensus has been reached. All of these questions await further study.

ARTICLES OF FURTHER INTEREST

Agriculture and Biodiversity, p. 1
Agriculture: Why and How Did It Begin?, p. 5
Crop Domestication: Fate of Genetic Diversity, p. 333
Crop Domestication: Role of Unconscious Selection, p. 340
Mutational Processes, p. 760

REFERENCES

1. Lu, T.L.-D. *The Transition from Foraging to Farming and the Origin of Agriculture in China*; British Archaeology Reports International Series, Archaeopress: Oxford, UK, 1999; Vol. 774.
2. *Origins of Rice Agriculture*; MacNeish, R., Libby, J.G., Eds.; The University of Texas Press: El Paso, USA, 1995.
3. Chen, W.H. *Atlas on Archaeological Agriculture in China*; Science and Technology Press: Nancheung, 1994.
4. The Luoyang Archaeology Team. *Luoyang Zaojiaoshu (The Zaojiaoshu Site in Luoyang City)*; Science Press: Beijing, 2002.
5. Xiong, L.Z.; Liu, K.D.; Dai, X.K.; Xu, C.G.; Zhang, Q. Identification of genetic factors controlling domestication-related traits of rice using an F2 population of a cross between *Oryza sativa* and *O. rufipogon*. Theor. Appl. Genet. **1999**, *98*, 243–251.
6. Cai, H.-W.; Morishima, H. Genomic regions affecting seed shattering and seed dormancy in rice. Theor. Appl. Genet. **2000**, *100*, 840–846.
7. Wang, X.K. Major Progress on the Origin of Rice Domestication in China (in Chinese). In *The Origin and Evolution of Rice Domestication in China*; Wang, X.K., Sun, C.Q., Eds.; Chinese Agronomic University Press: Beijing, 1996; 2–7.
8. Lu, T.L.-D. Les outils de récolte de céréales néolithiques de la vallée du Fleuve Jaune. Ann. Fond. Fyssen **2001**, *16*, 103–114.
9. Zhijun, Z. Personal communication.
10. Lu, T.L.-D. A cultivation experiment of green foxtail (*Setaria viridis*) in the middle yellow river valley and some related issues. Asian Perspect. **2002**, *1*, 1–14.
11. Fogg, W.H. Swidden Cultivation of Foxtail Millet by Taiwan Aborigines: A Cultural Analogue of the Domestication of *Setaria italica*. In *The Origins of Chinese Civilization*; Keightley, D.N., Ed.; University of California Press: Berkeley, 1983; 95–115.

Crop Domestication in Mesoamerica

A. Delgado-Salinas
J. Caballero
A. Casas
UNAM, Morelia, Michoacán, Mexico

INTRODUCTION

Mesoamerica is the cultural area that extends from northern Mexico (except for the northwestern states) south to Costa Rica. This region has been the origin of numerous domesticated plants, many of which are important sources of food and other needs for human populations all over the world. With a vast biodiversity established in arid, temperate, and tropical ecosystems, Mesoamerican human populations diversified into a broad array of cultural groups that have explored, experimented, and modified through selection a great number of plant resources. Mesoamerican peoples have depended on the continued output of food, fiber, medicine, and other essentials derived from wild, weedy, and domesticated plants. Their agricultural landscape has been characterized since early days by irregular patterns of multiple cropping agroecosystems, home gardens, and nearby human-modified vegetation areas (transformed habitats that hold mixed stands of prominent species that provide families with food security and nutritional balance).

ARCHAEOBOTANICAL EVIDENCE

Archaeological research in Mesoamerica has been conducted in regions of Mexico and Central America and provides a glimpse of some of the considerable diversity of both wild and domesticated plants throughout human history. In the highland sites of Mexico (Oaxaca, Tamaulipas, and Tehuacán), changes in major and minor crops, as well as wild and manipulated plants, are represented through different chronological ages. Evidence shows plants imported over great distances or in different stages of domestication. The archaeobotanical record from the highland sites includes macrobotanical remains of maize (*Zea mays*), wild and domesticated beans (*Phaseolus acutifolius*, *Phaseolus coccineus*, *Phaseolus vulgaris*), squash and gourds (*Cucurbita argyrosperma*, *Cucurbita moschata*, *Cucurbita pepo*, and *Lagenaria siceraria*), chili peppers (*Capsicum annum*), husk tomato (*Physalis* sp.), foxtail grass (*Setaria* sp.), chenopods and amaranths (*Chenopodium* sp.; *Amaranthus cruentus*, *Amaranthus hypochondriacus*), chayote (*Sechium edule*), cotton (*Gossypium* sp.), avocado (*Persea americana*), zapote blanco (*Casimiroa edulis*), zapote negro (*Diospyros digyna*), and guajes (*Leucaena esculenta* and *Leucaena leucocephala*).[1,2] Evidence from lowland sites (Tabasco, Mexico, Guatemala, El Salvador, Honduras, Costa Rica, and Panama) represents later stages in the archaeobotanical record and partially documents the innovative work of early agriculturalists that locally domesticated cotton (*Gossypium hirsutum*), jícama (*Pachyrhizus erosus*), chayotes (*Sechium edule*), arrowhead (*Maranta arundinacea*), and other crops, in addition to vegetatively propagated vanilla (*Vanilla planifolia*) and henequén (*Agave fourcroydes*). Remains from ca. 6000 b.p. also document the occurrence of maize (pollen), beans, squash, chili peppers, sunflowers, avocado, plums (*Spondias* sp.), and palms (such as fruits of *Acrocomia* sp.).[1,3]

Recently, age dating of some of these plant remains has been modified by the use of accelerator mass spectrometry (AMS), which has provided new and unexpected information on the history of domesticated plants. Archaeological remains from Guilá Naquitz cave in Oaxaca, México, indicate that ancient populations of domesticated squash (*C. pepo*) are represented as far back as 10,000 years b.p.[2] Surprisingly, what had been considered the earliest beans records, from Tehuacán, Mexico, have recently been dated as later than thought, at 2300 years b.p.[4] Furthermore, a domesticated sunflower seed (*Helianthus annuus*) dated ca. 4000 years b.p. has been discovered in Tabasco, Mexico.[3] This controversial discovery challenges the hypothesis that sunflower domestication originated in eastern United States. In addition, chemical evidence based on chromatography and mass spectrometry of Mayan ceramic vessels from northern Belize has moved back the earliest recorded use of chocolate (*Theobroma cacao*) as a beverage to as early as 2600 b.p.[5] (Table 1). Some plants gradually faded from the archaeological record (such as *Canavalia* and *Helianthus* in Mexico, and *Setaria*), possibly because of ecological changes or different agronomic practices. The latter explanation was suggested for *Setaria*, which was replaced by maize.[6]

Encyclopedia of Plant and Crop Science
DOI: 10.1081/E-EPCS 120017097

Table 1 Species diversity and earliest remains of major domesticates and sunflower in Mesoamerica

Genera	No. of species W/D	Archaeological site region	Earliest remains (b.p.)	Plant remains
Capsicum	30/5	Highlands	7,000	seed
Cucurbita	15/5	Highlands	10,000*	seed
Helianthus	70/2	Lowlands	4,000*	seed/achene
Phaseolus	50/5	Highlands	2,285*	pod
Zea	4/1	Lowlands	6,000	pollen

Species diversity (Wild and Domesticated) was obtained from Ref. 12 and for maize from Ref. 8. Remains ages were estimated by indirect radiocarbon method[1,3] and by Accelerator Mass Spectrometry (AMS, **1:4, 8, 15**)* and are given in years before present (b.p.).

SYSTEMATICS AND MOLECULAR EVIDENCE

Within the last 20 years, molecular systematists and geneticists, based on morphological and molecular evidence, have shed light on the systematics, genetic structure, and evolution of several crops.[7] These studies have assessed genetic variation and determined relationships between wild and domesticated species. Research has also determined centers of origin for some species, such as maize.[8] Gene pools in centers of genetic diversity have been studied for various species (e.g., *P. vulgaris*).[9] Also, indirect and direct estimates of gene flow between cultivars and wild counterparts have been obtained for several crops.[10]

Understanding the phylogeny and genomic evolution of crops has also made possible (albeit not without some problems of calibration) the calculation of rates of evolution of crop lineages.

SINGLE AND MULTIPLE DOMESTICATION EVENTS

Early Mesoamerican agriculturalists, and those from other regions, have converged in their selection of the same widespread genera and, consequently, have often domesticated multiple congeneric species. For example, separate domestications have been proposed for different species of *Gossypium* in both Old and New Worlds.[11] Also, among others, different species of *Amaranthus*, *Annona*, *Chenopodium*, and *Pachyrhizus* were domesticated in both the northern and southern hemispheres of the New World.[12] In addition, throughout their distributional ranges, some species have been domesticated more than once. Morphological and molecular evidence has confirmed independent domestication events in *C. moschata*, *Phaseolus lunatus*, and *P. vulgaris* in both Mesoamerican and Andean regions.[9,13] *C. pepo* was domesticated in both southern Mexico and eastern United States.[14] Often these vicarious domestications have accentuated population differences within species, such as in *P. vulgaris*, where Mesoamerican and South American gene pools are molecularly distinguishable.[9] However, many species have been domesticated only once in their history. A single domestication of maize is proposed from southern Mexico ca. 9000 years ago.[8] The wild progenitor of tomato (*Solanum lycopersicum* var. *cerasiforme*) was apparently domesticated once, in Mesoamerica, even though the genus is highly diversified in South America. Molecular evidence has shown that the domestication region of cocoa (*Theobroma cacao*) was in fact the Amazon basin, and not in southern Mesoamerica as was earlier thought.[15]

IN SITU DOMESTICATION BY SELECTIVE TOLERANCE

Manipulation of plant resources by local peoples has influenced plant evolution in various ways and continues to operate in Mesoamerica. At present, indigenous peoples in Mesoamerica utilize more than 6000 plant species for food, medicine, and a great variety of other purposes. Management of plant resources involves different forms of manipulation, which increase the availability of plants in space and time and provide a glimpse of different stages of domestication. Most plant species are harvested from the natural vegetation, but some plant resources, mainly trees and other perennials, are selectively spared when cutting down the forest to establish maize fields or settlements. In some cases, plant individuals or their populations are protected by controlling or eliminating their competitors and predators. The conscious dispersal of sexual and vegetative propagules of plants is another way in which selection may be imposed. Cultivation of wild plants in home gardens and croplands is also a common practice that maximizes reproduction and productivity of plant resources. Many species are managed simultaneously in two or more ways in different regions, and even within a single region or locality, by the same people. Studies on morphological variation of some species of central Mexico show that selective tolerance and other forms of plant management constitute active processes of in situ domestication. It is known that when gathering plants in the wild, people frequently harvest only those individuals that they recognize as more desirable, because they have better flavor, texture, color, shape, disease or insect resistance, etc. For example, the Mixtec and the Nahua of central Mexico recognize differences in the quality of edible fruits such as guayaba (*Psidium guajava*), guamuchil (*Pithecellobium dulce*), nance (*Byrsonima crassifolia*), and ciruela (*Spondias purpurea*). The recognition of variation in wild plants by indigenous peoples may be the first step in the process of conscious selection of the most desired phenotype. In situ domestication by selective tolerance seems to be a common and current process in Mesoamerica. Selective tolerance has been described for *Leucaena esculenta* ssp. *esculenta*, *Opuntia* spp, and several species of columnar cacti such as *Stenocereus stellatus*, *Escontria chiotilla*, *Polaskia chichipe*, and *Polaskia chende*. In all these cases, selection has favored phenotypes with larger, more flavorful seeds, although they are more susceptible to predation, in the case of *L. esculenta*, and fruits that are larger, sweeter, thinner skinned, and less spiny, in the case of cacti. In the case of *P. chichipe*, selection has also favored individuals that self-pollinate and germinate more rapidly. This form of management allows local people to modify the genetic structure of a population by increasing the frequency of the most desired phenotypes while eliminating those that are not desirable. At least 37 other wild species have been reported as selectively tolerated and sometimes promoted by local people. This number could be much higher given the relatively scant attention that ethnobotanists have paid to the study of this kind of plant management. Moreover, in situ domestication has favored the maintenance of vital coevolutionary processes of long-term ecological associations between crop plants and their wild pathogens and symbionts.

CONCLUSIONS

Since most plants subject to in situ domestication are common elements of the landscape of the indigenous regions of Mesoamerica, the crops, pasturelands, and fallow fields of the region can be seen as complex agrosilvicultural systems. All elements of these systems, including the apparently wild plants, have a role in the local economy and are the result of some degree of selection and manipulation by local people. Cutting down the natural vegetation for agricultural purposes involves a significant loss of biodiversity; conversely, domestication processes add an important amount of human-created biodiversity to nature.

The important ecological and evolutionary role that skilled indigenous Mesoamerican peoples have played and continue to play is undeniable. Their activities result in the creation, maintenance, and enhancement of plant genetic resources, which results in worldwide contributions to food support and other human needs.

ACKNOWLEDGMENTS

Thanks to Lawrence M. Kelly, who kindly provided critical and helpful comments.

ARTICLES OF FURTHER INTEREST

Agriculture and Biodiversity, p. 1
Agriculture: Why and How Did It Begin?, p. 5
Archaeobotanical Remains: New Dating Methods, p. 55
Crop Domestication in Africa, p. 304
Crop Domestication in China, p. 307
Crop Domestication in Prehistoric Eastern North America, p. 314
Crop Domestication in Southeast Asia, p. 320
Crop Domestication in Southwest Asia, p. 323
Crop Domestication in the American Tropics: Phytolith Analyses, p. 326
Crop Domestication in the American Tropics: Starch Grains Analyses, p. 330
Crop Domestication: Fate of Genetic Diversity, p. 333
Crop Domestication: Founder Crops, p. 336
Crop Domestication: Role of Unconscious Selection, p. 340
Crop Improvement: Broadening the Genetic Base for, p. 343
Cropping Systems: Slash-and-Burn Cropping Systems of the Tropics, p. 363
Gene Flow Between Crops and Their Wild Progenitors, p. 488
Pre-agricultural Plant Gathering and Management, p. 1055
Quantitative Trait Location Analyses of The Domestication Syndrome and Domestication Process, p. 1069

REFERENCES

1. McClung de Tapia, E. The Origins of Agriculture in Mesoamerica and Central America. In *The Origins of Agriculture*; Cowan, C.W., Watson, P.J., Eds.; Smithsonian Institution Press: Washington, 1992; 143–171.
2. Smith, B.D. The initial domestication of *Cucurbita pepo* in the Americas 10, 000 years ago. Science **1997**, *276*, 932–934.

3. Pope, K.C.; Pohl, M.E.D.; Jones, J.G.; Lentz, D.L.; von Nagy, C.; Vega, F.J.; Quitmyer, I.R. Origin and environmental setting of ancient agriculture in the lowlands of Mesoamerica. Science **2001**, *292*, 1370–1374.
4. Kaplan, L.; Lynch, T.F. *Phaseolus* (Fabaceae) Archeology: AMS radiocarbon dates and their significance for pre-Columbian agriculture. Econ. Bot. **1999**, *53* (3), 261–272.
5. Hurst, J.W.; Tarka, S.M., Jr.; Powis, T.G.; Valdez, F., Jr.; Hester, T.R. Archaeology: Cacao usage by the earliest Maya civilization. Nature **2002**, *418*, 289–290.
6. Flannery, K.V. The origins of agriculture. Annu. Rev. Anthropol. **1973**, *2*, 271–310.
7. Gepts, P. The use of molecular and biochemical markers in crop evolution studies. Evol. Biol. **1993**, *27*, 51–94.
8. Matsuoka, Y.; Vigouroux, Y.; Goodman, M.M.; Sánchez, G.J.; Buckler, E.; Doebley, J. A single domestication of maize shown by multilocus microsatelites genotyping. Proc. Natl. Acad. Sci. U. S. A. **2002**, *99* (9), 6080–6084.
9. Gepts, P. Origin and evolution of common bean: Past events and recent trends. HortScience **1998**, *33*, 1124–1130.
10. Ellstrand, N.C.; Prentice, H.C.; Hancock, J.F. Gene flow and introgression from domesticated plants into their wild relatives. Ann. Rev. Ecolog. Syst. **1999**, *30*, 539–563.
11. Cronn, R.C.; Small, R.L.; Haselkorn, T.; Wendel, J.F. Rapid diversification of the cotton genus (*Gossypium*: Malvaceae) revealed by analysis of sixteen nuclear and chloroplast genes. Am. J. Bot. **2002**, *89* (4), 707–725.
12. *Evolution of Crop Plants,* 2nd Ed.; Smartt, J., Simmonds, N.W., Eds.; Longman: London, United Kingdom, 1995.
13. Doebley, J. Molecular Systematics and Crop Evolution. In *Molecular Systematics of Plants*; Soltis, P.S., Soltis, D.E., Doyle, J.J., Eds.; Chapman and Hall: New York, 1992; 202–222.
14. Decker-Walters, D.S.; Staub, J.E.; Sang-Min, C.; Nakata, E.; Quemada, H.D. Diversity in free-living populations of *Cucurbita pepo* (Cucurbitaceae) as assessed by random amplified polymorphic DNA. Syst. Bot. **2002**, *27* (1), 19–28.
15. De la Cruz, M.; Whitakus, R.; Gómez-Pompa, A.; Mota-Bravo, L. Origins of cacao cultivation. Nature **1995**, *375*, 333–345.

Crop Domestication in Prehistoric Eastern North America

David L. Asch
John P. Hart
New York State Museum, Albany, New York, U.S.A.

INTRODUCTION

At European Contact, eastern North American Indian agriculture featured the New World cosmopolitan "three sisters:" maize, beans, and squash. Maize and beans had diffused from the tropics as domesticates, as did some squashes. The dominance of this triad in temperate eastern North America was recent. Maize became an important crop only about 1000 years ago, and beans entered the region at 850 b.p. But before maize became preeminent—as early as 3500 b.p.—there was an "Eastern Agricultural Complex" (EAC), which consisted of several indigenous crops. EAC was largely an indigeneous development; its origins can be traced back at least 7300 years.

EARLIEST AGRICULTURE

Eastern North American agriculture (Tables 1 and 2) began with the growing of yellow-flowered gourd and bottle gourd cucurbits whose hard-shelled fruits presumably had toxic, bitter flesh. These plants have edible seeds, and the fruits were useful as containers, fishnet floats, and rattles. By c. 7300 b.p., the first was present in a habitation-site midden in Illinois (Koster) and the second with human burials in Florida (Windover) (see Fig. 1 for site locations).[1–4]

A yellow-flowered gourd subspecies presently grows spontaneously and persistently in naturally disturbed floodplains from east Texas to the Ozarks.[3,4] In Florida, seeds of yellow-flowered gourd are found in Late Pleistocene sediments predating human presence.[4] Seeds and fruit rinds from Middle Holocene archaeological sites located far north and east of these regions (Table 2) imply that humans, carrying the plant from site to site, were responsible for a major range extension.[5,11] Although self-seeding would have occurred, occasional planting and care probably were essential for long-term persistence in parts of the archaeological range where today yellow-flowered gourd is evanescent without planting.[3] Separate domestication of a second subspecies occurred in Mexico.[4]

The geographic range of bottle gourd before it entered into a mutualistic relationship with humans is undetermined, but the species probably originated in tropical Africa. It has been recovered at Mesoamerican and Andean archaeological sites of about the same age as in Florida, or older.[1,3] Unless the fruit was transported by ocean currents to Florida, it arrived there by human hands. Its presence in southeast Missouri at c. 5000 b.p.[2] was surely due to human transport. Bottle gourd persists spontaneously in areas of human disturbance, but long-term survival requires occasional planting or care.[3]

Marshelder was the first species to be cultivated (i.e., planted) primarily for food in eastern North America. At Napoleon Hollow, Illinois, the mean size of its achenes at 4000 b.p. is larger than in modern wild populations, implying domesticatory selection.[2] Marshelder may have been cultivated earlier at nearby Koster where it is the most common seed from a 5700–4900 b.p. occupation, but those achenes are indistinguishable from wild plants. Giant ragweed is the second most common seed in that deposit. At Marble Bluff, Arkansas, giant ragweed achenes were preserved in a 2850-b.p. cache with seeds of known crops.[10] This species is much less frequent in later prehistoric seed assemblages. Although evidence is lacking for range extension or morphological changes, the temporal pattern of its archaeological occurrence suggests that it was an early crop that was later abandoned.

EASTERN AGRICULTURAL COMPLEX

Over the millennia, several annual species grown primarily for their edible seeds were added to the agricultural ensemble (Table 2). By 2000 b.p., six were cultivated in the Midwest: oily-seeded marshelder and sunflower, and starchy-seeded goosefoot, knotweed, maygrass, and little barley.[1,5] Biological domestication of marshelder, sunflower, and probably knotweed is attested by gradually increasing achene size.[2,3,13] Goosefoot domestication is evinced by occurrence of two seed forms that have a testa thinner than in wild plants, perhaps resulting from unconscious selection.[1,8] It is likely that thin-testa seeds germinated promptly when sown, which in an optimal garden-bed environment would give the seedlings an advantage in competition for light and nutrients.[1] Morphological changes in maygrass and little barley are

Encyclopedia of Plant and Crop Science
DOI: 10.1081/E-EPCS 120017103

Table 1 Agricultural species of aboriginal eastern North America, 7300 b.p. to European contact

Species	Common name	Family	How propagated	Major uses[a]
Amaranthus hypochondriacus L.	Grain amaranth	Amaranthaceae	Annual from seed	Edible seed
Ambrosia trifida L.[b]	Giant ragweed	Asteraceae	Annual from seed	Edible seed
Chenopodium berlandieri Moq.	Goosefoot	Chenopodiaceae	Annual from seed	Edible seed
Cucurbita argyrosperma Huber	Cushaw squash	Cucurbitaceae	Annual from seed	Fruit with edible flesh
Cucurbita pepo subsp. *ovifera* (L.) Decker	Yellow-flowered gourd; squashes: crookneck, scallop-forms, and other forms	Cucurbitaceae	Annual from seed	Fruits: hard-shelled forms for containers, rattles; some forms with edible flesh; edible seed
Cucurbita pepo L. subsp. *pepo*	Pumpkin	Cucurbitaceae	Annual from seed	Fruit with edible flesh; edible seed
Echinochloa muricata (P. Beauv.) Fernald[b]	Barnyard grass	Poaceae	Annual from seed	Edible seed
Helianthus annuus L.	Common sunflower	Asteraceae	Annual from seed	Edible seed
Helianthus tuberosus L.[c]	Jerusalem artichoke	Asteraceae	Herbaceous perennial from tubers	Edible tubers
Hordeum pusillum Nutt.	Little barley	Poaceae	Annual from seed	Edible seed
Ilex vomitoria Ait.	Yaupon holly, cassine	Aquifoliaceae	Perennial shrub, transplanted	Leaves and twigs brewed to make ceremonial/medicinal "black drink"
Iva annua L.	Marshelder, sumpweed	Asteraceae	Annual from seed	Edible seed
Lagenaria siceraria (Molina) Standl.	Bottle gourd	Cucurbitaceae	Annual from seed	Fruits for containers, rattles; edible seed
Nicotiana quadrivalvis Pursh	Tobacco	Solanaceae	Annual from seed	Leaves smoked for ceremonial or medicinal purposes
Nicotiana rustica L.	Tobacco	Solanaceae	Annual from seed	Leaves smoked for ceremonial or medicinal purposes
Passiflora incarnata L.[c]	Maypops	Passifloraceae	Herbaceous perennial vine from seed or root division	Sweet, slightly acid, edible fruit
Phalaris caroliniana Walt.	Maygrass	Poaceae	Annual from seed	Edible seed
Phaseolus vulgaris L.	Common bean	Fabaceae	Annual from seed	Edible pods (green) and seed
Polygonum erectum L.	Erect knotweed	Polygonaceae	Annual from seed	Edible seed
Zea mays L.	Maize, corn	Poaceae	Annual from seed	Edible seed

[a]The term "seed" is employed in a common-language sense.
[b]Probably planted.
[c]Spontaneous in fields; probably planted as well.
(From Refs. 1–11, which include citations of primary literature.)

absent or less convincingly substantianted, but cultivation can be inferred from archaeological occurrences at locations beyond the extended modern ranges of these weedy species.[6]

For the cucurbit species, there is unambiguous evidence of larger seed and fruit size and thicker fruit rind, with a diversification of forms that increased overall technological utility. At an undetermined time, a yellow-flowered gourd was selected whose fruit had nonbitter edible flesh, which thus can be called a squash.[1,3]

Of the six seed crops, four mature in late summer/autumn, two in late spring. Immature squash could be a summer vegetable. Effective harvests of marshelder, maygrass, little barley, and perhaps other species would

Table 2 Records of eastern North American crops

Dates (b.p.)[a]	Trends, species, evidence	Exemplary sites
7500–5000	**Range extension of gourds:**	
	Cucurbita sp., range expansion from presumed native occurrence of *C. pepo* subsp. *ovifera* along Gulf coastal plain or Ozark drainages. Phenotypic change not evident but would be difficult to recognize in archaeological specimens consisting of small, carbonized fruit-rind fragments.	Koster, IL Napoleon Hollow, IL Anderson, TN Sharrow, ME Memorial Park, PA
	Lagenaria siceraria, small fruits placed with human burials in FL; probably introduced from New World tropics by humans; small fruit size suggests FL plant was similar to wild progenitor of domesticate.	Windover, FL
5000–4000	**Continued use of gourds; initial indigenous seed crops:**	
	Cucurbita pepo subsp. *ovifera*, slight increase of seed size.	Phillips Spring, MO
	Lagenaria siceraria, range extension to continental interior.	Phillips Spring, MO
	Iva annua, wild-size achenes at Koster (5500 b.p.), enlarged achenes in IL by 4000 b.p.	Napoleon Hollow, IL
	Helianthus annuus, range extension from Plains e. to TN (probably not indigenous e. of long. 95° W), increase in achene size.	Hayes, TN
	Phalaris caroliniana, range extension to e. TN	Hayes, TN Bacon Bend, TN
	Chenopodium berlandieri, wild-type seeds more frequent in archaeobotanical assemblages.	Napoleon Hollow, IL
	Ambrosia trifida, in IL, kernels well represented in some assemblages, without evidence of morphological change.	Napoleon Hollow, IL
4000–3000	**Phenotypic changes in indigenous plants resulting from selection within agroecologies:**	
	Chenopodium berlandieri, testa thinner than in wild seeds.	Newt Kash, KY Cloudsplitter, KY
	Phalaris caroliniana, association with known crops, range extension to e. KY.	Newt Kash, KY
	Smoking pipes, first appearance (IL) at 3500 b.p.; *Nicotiana* not demonstrably present.	Robeson Hills, IL
3000–2000	**Continued development of indigenous crop complex; first evidence of maize:**	
	Cucurbita pepo subsp. *ovifera* & *Lagenaria siceraria*, large thick-shelled container forms present; seeds documented in human feces.	Cloudsplitter, KY Salts Cave, KY
	Ambrosia trifida, abundant in cache with other crop plants (AR).	Marble Bluff, AR
	Helianthus annuus, larger achenes.	Higgs, TN Salts Cave, KY
	Iva annuua, larger achenes, range extension to e. KY.	Cloudsplitter, KY Salts Cave, KY
	Phalaris caroliniana, seeds identified in human feces, grown beyond indigenous range without evident phenotypic change.	Salts Cave, KY
	Hordeum pusillum, first appearance in archaeological assemblages (IL) but without evident range extension or phenotypic change.	Ambrose Flick, IL
	Zea mays, first macrobotanical evidence from e. N. Am. at 2100 b.p. (IL).	Holding, IL
2000–1000	**Wider distribution of indigenous crop complex; earliest tobacco; spread of maize:**	
	Iva annua & *Helianthus annuus*, continuing increase in achene size.	Many sites
	Polygonum erectum, common in IL after 2000 b.p., without morphological change.	Smiling Dan, IL
	Hordeum pusillum, range extension.	Several sites
	Nicotiana sp., earliest identified seed in e. N. Am., 1800 b.p. (IL).	Smiling Dan, IL
	Zea mays, more widely distributed, especially after 1500 b.p.; not common until end of millennium.	Edwin Harness, OH Icehouse Bottom, TN Grand Banks, ON Memorial Park, PA
	Indigenous seed complex present on Plains; evident e. of Appalachian Front only with first appearance of maize.	Many sites

(*Continued*)

Table 2 Records of eastern North American crops (*Continued*)

Dates (b.p.)[a]	Trends, species, evidence	Exemplary sites
1000–500	**Maize the most important crop throughout e. N. Am. as maize–beans–squash triad becomes dominant; indigenous crop complex variably declines; other species introductions from Mesoamerica or Southwest:**	
	Polygonum erectum & *Iva annua*, increase in achene size.	Hill Creek, IL
	Helianthus annuus, marked increase in size of disks & achenes, probable presence of monocephalic form.	Ozark rockshelters, AR–MO
	Echinochloa muricata, common at some Prairie Peninsula sites, without evident phenotypic change.[12]	13ML175
	Cucurbita argyrosperma, *Cucurbita pepo* subsp. *pepo*, & *Amaranthus hypochondriacus*, introduced from Mesoamerica or Southwest.	Ozark rockshelters, AR–MO
	Passiflora incarnata, common at Southeastern archaeological sites.	Many sites
	Zea mays, grown to geographic limit of modern corn agriculture; stable carbon isotope analysis of human bone collagen indicates it was widely a major source of dietary protein.	Many sites
	Phaseolus vulgaris, first occurrence at 850 b.p. in w. MO & NE, rapid spread e. of Mississippi R. at 700 b.p.	23PL16
	Indigenous seed crops & maize commonly co-occurring before 750 b.p.; indigenous seeds less important thereafter.	Many sites
500	**Dominance of maize–beans–squash, with tobacco & bottle gourd as nonfood agricultural plants; disappearance of native seed complex, except sunflower; several native crops observed by Europeans that are not archaeologically evident (mostly unlisted); crops obtained from Europeans (unlisted):**	
	Planted by Indians before a.d. 1800, according to European records: *Zea mays*, *Phaseolus vulgaris*, *Cucurbita pepo*, *C. argyrosperma*, *Nicotiana rustica*, *N. quadrivalvis*, *Lagenaria siceraria*, *Helianthus annuus*, *H. tuberosus* (planting inferred from occurrence in e. MA), *Chenopodium* sp. (1 or 2 records), *Passiflora incarnata*, & *Ilex vomitoria*.	

[a]Radiocarbon years b.p., uncorrected for fluctuations in atmospheric ^{14}C.
(From Refs. 1–11, which include citations of primary literature.)

require planting in dense stands. Sunflower and cucurbit plants could be individually tended.

Generally in eastern North America, plant materials decay rapidly and are preserved at archaeological sites only when carbonized, making it difficult to determine dietary importance of EAC crops. Critical evidence on diet comes from a few caves and rockshelters (mostly in the Ozarks and Kentucky) where desiccating environments have preserved uncarbonized perishables.[10,14] Indians mining mirabilite in Salts Cave, Kentucky at 2500 b.p. deposited feces that are preserved intact. The fecal bulk is mostly seed remains. In a sample of 100 paleofeces, Yarnell[14] identified EAC seeds in 98 specimens, with a mean of 3 EAC species per specimen. Starchy and oily seeds co-occurred in most feces. Twenty percent contained both spring- and fall-harvested crops, an indication of seed storage for out-of-season use. On the cave floor were thick-shelled gourds fashioned into containers.

For several millennia, eastern North American agriculture developed in near-total isolation from other agricultural centers. However, tobacco, an exotic, appeared 1800 years ago during EAC ascendancy.[3] The earliest macrobotanical record of maize is 2100 b.p., but maize was uncommon for centuries thereafter.[5,8,9] Possibly there were other extraregional contacts before the late prehistoric period. Little barley and maygrass are indigenous in both the southwestern and southeastern United States, and they occur archaeologically in the southwest.[6] Domesticated common sunflower has been reported at a prehistoric site in Mexico,[15] and domesticated *Chenopodium berlandieri* occurs today in Mexico (as subsp. *nuttalliae* H.D. Wilson & Heiser).

LATE PREHISTORIC AND HISTORIC AGRICULTURE

Introductions of domesticated plants from beyond eastern North America continued in late prehistoric times—including common bean, a squash domesticated from the Mexican lineage of yellow-flowered gourd, a second squash species, and a grain amaranth.[5,8] Several

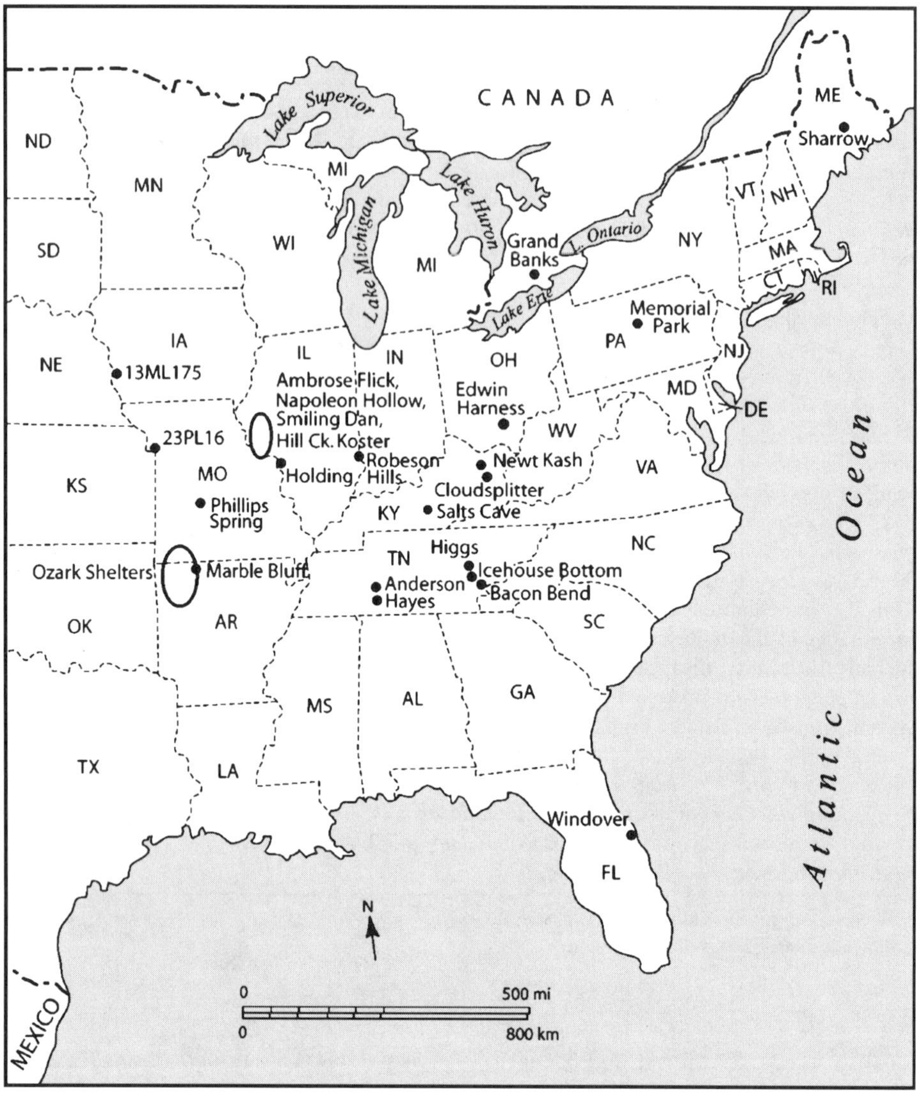

Fig. 1 Archaeological sites listed in Table 2 and mentioned in the text.

domesticates accompanying Europeans in the early Contact period (e.g., watermelon and peaches) were assimilated rapidly.[3]

After maize became the dietary staple around 1000 b.p., EAC crops continued to be grown for a few centuries, at least in the Midwestern core area. Ultimately, the maize–beans–squash triad supplanted EAC crops, except for sunflower. Cultivars of several indigenous species that had evolved during millennia of domestication were lost.[1,5]

Some species grown before European Contact have probably not been recognized on account of their low archaeological visibility or difficulty of distinguishing them from wild plants. For instance, Europeans recorded that Southeastern Indians transplanted the yaupon holly shrub from the seacoast to interior villages. This assured a convenient supply of the caffeine-containing leaves and twigs from which was brewed the ''black drink'' used in social/ceremonial contexts.[3] A survey of historic documents from Contact to1940 found that Indians of Eastern North American grew over 20 ''specialty plants''—i.e., species having nonfood uses or eaten for reasons other than their caloric contribution.[3,8] These were mostly nonweedy perennials, in contrast to the EAC crop progenitors, which were weedy annuals. The specialty plants were seldom grown in fields with annual plants and sometimes were replanted in the habitat where they were collected; Indian plant husbandry included activities that do not resemble agriculture from a European perspective.

PLANT HUSBANDRY

Husbandry or management of ecosystems may have altered availability of so-called wild foods.[5] Historic records document Indian planting of wild rice (*Zizania* spp.) to establish it in northern lakes where it was absent.[3] In the Northeast, groundnut (*Apios americana* Medik.) occurs exclusively as a sterile triploid that reproduces from the tubers (which were an important Indian food); Indians possibly established groundnut in that part of its range.[5] Finally, effective large-scale hickorynut (*Carya* spp.) harvesting, which dates to as early as 7500 b.p.[2,5] would have been possible only with access to the concentrated masts that open-grown trees can produce. Intentional burning, selective girdling of trees, and opening the forests for agriculture likely contributed to this resource's availability.

ARTICLES OF FURTHER INTEREST

Agriculture: Why and How Did It Begin, p. 5
Archaeobotanial Remains: New Dating Methods, p. 55
Crop Domestication in Mesoamerica, p. 310
Crop Domestication: Founder Crops, p. 336
Crop Domestication: Role of Unconscious Selection, p. 340
Pre-Agricultural Plant Gathering and Management, p. 1055

REFERENCES

1. Smith, B.D. *Rivers of Change: Essays on Early Agriculture in Eastern North America*; Smithsonian Institution Press: Washington, 1992.
2. *Prehistoric Food Production in North America*; Ford, R.I., Ed.; Anthropological Papers, University of Michigan, Museum of Anthropology: Ann Arbor, 1985; Vol. 75.
3. Asch, D.L. *Aboriginal Specialty-Plant Propagation: Illinois Prehistory and an Eastern North American Post-Contact Perspective*; Ph.D. Diss.; University of Michigan: Ann Arbor, 1995.
4. J. Ethnobiol. **1993**, *13* (1).
5. *People and Plants in Ancient Eastern North America*; Minnis, P.E., Ed.; Smithsonian Institution Press: Washington, 2003.
6. Asch, D.L.; Asch Sidell, N. Archeobotany. In *Early Woodland Occupations at the Ambrose Flick Site in the Northern Sny Bottom of West-Central Illinois*; Stafford, C.R., Ed.; Research Series, Center for American Archeology: Kampsville, IL, 1992; Vol. 10, 177–293.
7. *Foraging and Farming in the Eastern Woodlands*; Scarry, C.M., Ed.; University Press of Florida: Gainesville, 1993.
8. *Agricultural Origins and Development in the Midcontinent*; Green, W., Ed.; Report, University of Iowa, Office of the State Archaeologist: Iowa City, 1994; Vol. 19.
9. Riley, T.J.; Walz, G.R.; Bareis, C.J.; Fortier, A.C.; Parker, K. Accelerator mass spectrometry (AMS) dates confirm early *Zea mays* in the Mississippi River Valley. Am. Antiq. **1994**, *59*, 490–497.
10. *People, Plants, and Landscapes: Studies in Paleoethnobotany*; Gremillion, K.J., Ed.; University of Alabama Press: Tuscaloosa, 1997.
11. *Current Northeast Paleoethnobotany*; Hart, J.P., Ed.; New York State Museum Bulletin, University of the State of New York: Albany, 1999; Vol. 494.
12. Asch, D.L.; Green, W. Archaeobotanical Analysis. In *Phase III Excavations at 13ML118 and 13ML175, Mills County, Iowa*; Morrow, T.A., Ed.; Contract Completion Report, University of Iowa, Office of the State Archaeologist: Iowa City, 1995; Vol. 469, 59–69, 113–120.
13. Yarnell, R.A. Domestication of Sunflower and Sumpweed in Eastern North America. In *The Nature and Status of Ethnobotany*; Ford, R.I., Ed.; Anthropological Papers, University of Michigan, Museum of Anthropology: Ann Arbor, 1978; Vol. 67, 289–299.
14. *Archeology of the Mammoth Cave Area*; Watson, P.J., Ed.; Academic Press: New York, 1974.
15. Lentz, D.L.; Pohl, M.E.D.; Pope, K.O.; Wyatt, A.R. Prehistoric sunflower (*Helianthus annuus* L.) domestication in Mexico. Econ. Bot. **2001**, *55*, 370–376.

Crop Domestication in Southeast Asia

Victor Paz
University of the Philippines, Diliman, Quezon City, Philippines

INTRODUCTION

The time depth for the origin of Southeast Asian agriculture is not well established. The nature of the early agriculture, though, is clear—it is based on the domestication of root crops and the cultivation of tubers. The characteristics of early agriculture in the region are also tightly intertwined with the practice of arboriculture. Southeast Asia may be geographically divided into two main parts: Mainland and Island. The extent of the Mainland is within the land area of present-day Myanmar, Thailand, Cambodia, Laos, Vietnam, and the Malay Peninsula. Island Southeast Asia is within the bounds of the Philippines, Malaysia, Brunei, and Indonesia. The present nature of agriculture in Southeast Asia may be characterized as dominated by cereal cultivation. This was not the case, however, at the origin of a way of life that depended on the cultivation and domestication of plants. Cereals, such as rice (*Oryza sativa* L.), are the main crop of current agriculture in Southeast Asia.

ORIGINS OF CEREAL AGRICULTURE

With the absence of any well-founded evidence for the earliest cereal agriculture and the lack of any early evidence for early domestication of field-working animals in the region, it is not likely that agriculture based on rice or millet (*Echinochloa frumentacea* Link or *Setaria italica* (L.) P. Beauv.) started out in Southeast Asia earlier than in East Asia, where a whole suite of evidence supports the presence of cereal agriculture not later than 8000 before present (b.p.).[1–3] The question of the origin of agriculture can best be addressed by evidence of human utilization of plants and the presence of cultivars at an established time depth. In both cases, the antiquity of the evidence depends on the nature of the archaeological data.

From well-determined archaeobotanical remains, the beginning of rice and millet cultivation in Mainland Southeast Asia may be roughly placed around 5000 to 3000 b.p., and in Island Southeast Asia at 4000 to 3850 b.p.[4,5] On the other hand, there is a clear picture of tubers and nut utilization in the region as far back as the late Pleistocene and early Holocene, based on plant remains coming from several sites in Island and Mainland Southeast Asia, e.g., Niah West Mouth Cave in Sarawak, Ulu Leang in Sulawesi, and Spirit Cave and Banyan Valley Cave in Thailand (Fig. 1).[5,6] There is also substantial evidence of root/tuber consumption and the exploitation of tree products continuing all throughout the later epochs of the region's archaeological record.[6,7]

NONCEREAL PLANT UTILIZATION

The accumulated current evidence from the archaeology of the region, which is still growing, shows the existence in deep history of a people–plant relationship that may be described as tuber-based agriculture coupled with arboriculture. In this type of relationship, humans depended mainly on the exploitation of food resources coming from the forest, which involved the utilization of a substantial number of plants that eventually led to the cultivation of a considerable number of species and the local domestication of some of these plants. It may also be argued that the start of tuber-based agriculture in Southeast Asia coincided with the development of the greater yam *Dioscorea alata* L. from its two known wild progenitors—*D. hamiltonii* Hook f. and *D. persimilis* Prain and Burkill[8,9]—somewhere in northern Mainland Southeast Asia. As a cultivar, *D. alata* is definitely the consequence of a practice that involved the cultivation of several nondomesticated vegetative plants for generations.

NATIVE AND INTRODUCED PLANT UTILIZATION

The time depth for the dominance of forest plant utilization perhaps began as soon as the appearance of modern humans in the region, with the number of species cultivated increasing through time—even during and after the introduction of cereal crops and other root crops. Although we talk of local plants cultivated or domesticated, it is significant to note that introduced plants were also integrated into the system in later times, as reflected best by the integration into the system of *Ipomoea batatas* (L.) Lam. and *Manihot esculenta* Crantz, both coming from South America.[10]

Encyclopedia of Plant and Crop Science
DOI: 10.1081/E-EPCS 120017100

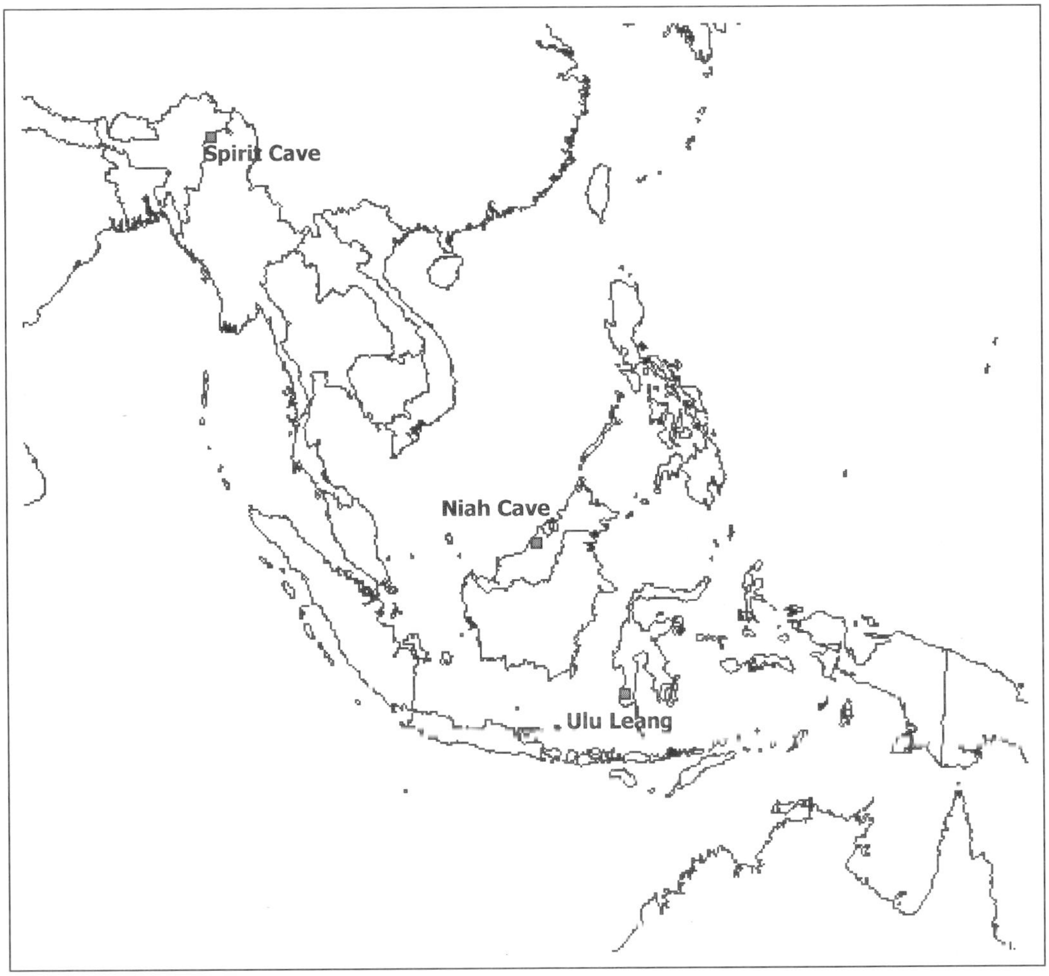

Fig. 1 Locations of plant remains from the late Pleistocene and early Holocene. (*View this art in color at www.dekkker.com.*)

There is a strong possibility that the beginnings of an agricultural way of life in the region revolved around the cultivation of a variety of yams and the pantropical domesticate *Colocasia esculenta* Schott (taro), together with other cultivated ''wild'' species of *Dioscorea* and *Alocasia*. The diet was further supported by *Musa* spp. (plantains) and nut-bearing trees such as *Canarium* spp., mostly cultivated in their natural habitats inside the forest or at the fringes of settlements. The exploitation of cultivated and a few wild tuber-yielding plants inside the forest corresponded to an arboricultural relationship of people with the rest of the forest's plant population.

CONCLUSION

The foregoing description briefly illustrates the nature of the origin of agriculture in Southeast Asia. Tuber-based agriculture and arboriculture may perhaps be the main reasons why early introduction of cereal agriculture had a slower and more fragmented acceptance. Future archaeobotanical work coupled with more palynological sequences from the late Pleistocene to later periods will further improve our knowledge of the nature and time depth of Southeast Asian agriculture.

ARTICLES OF FURTHER INTEREST

Agriculture and Biodiversity, p. 1
Archaeobotanical Remains: New Dating Methods, p. 55
Agriculture: Why and How Did it Begin? p. 5
Crop Domestication in Prehistoric Eastern North America, p. 314
Crop Domestication in Southwest Asia, p. 323
Crop Domestication in the American Tropics: Phytolith Analyses, p. 326
Pre-Agricultural Plant Gathering and Management, p. 1055
Rice, p. 1102

REFERENCES

1. Chang, T.-T. Rice. In *The Cambridge History of Food*; Kiple, K., Ornelas, K.C., Eds.; Cambridge University Press: Cambridge, England, 2000; Vol. 1, 132–148.
2. Barnes, G.L. *The Rise of Civilization in East Asia: The Archaeology of China, Korea and Japan*; Thames & Hudson: London, 1999; 89–97.
3. Lu, T.L.-d. The transition from foraging to farming in China. Indo-Pacific prehistory: The Melaka papers. **1999**, *2*, 77–80.
4. Bellwood, P.; Gillespie, R.; Thompson, G.B.; Vogel, J.S.; Ardika, I.W.; Datan, I. New dates for prehistoric Asian rice. Asian Perspect. **1992**, *31* (2), 161–170.
5. Barker, G.; Badang, D.; Barton, H.; Beavitt, P.; Bird, M.; Daly, P.; Doherty, C.; Gilbertson, D.; Glover, I.; Hunt, C.; Manser, M.; McLaren, S.; Paz, V.; Pyatt, B.; Reynolds, T.; Rose, J.; Rushworth, G.; Stephens, M. The Niah cave project: The second (2001) season of fieldwork. Sarawak Mus. J. **2001**, *LVI* (77), 37–120.
6. Bellwood, P. *Prehistory of the Indo-Malaysian Archipelago*; University of Hawaii Press: Honolulu, 1997; 202–223.
7. Paz, V.J. Archaeobotany and Cultural Transformation: Patterns of Early Plant Utilization in Northern Wallacea. Ph.D. Dissertation; Department of Archaeology, University of Cambridge: Cambridge, U.K., 2001; 181–258.
8. Alexander, J.; Coursey, D.G. The Origins of Yam Cultivation. In *The Domestication and Exploitation of Plants and Animals*; Ucko, P., Dimbleby, G.W., Eds.; Duckworth: London, 1969; 405–425.
9. Burkill, I.H. The rise and decline of the greater yam in the service of man. Adv. Sci. **1951**, *7*, 443–448.
10. Piperno, D.R.; Pearsall, D.M. *The Origins of Agriculture in the Lowland Neotropics*; Academic Press: New York, 1998; 9,312,120.

Crop Domestication in Southwest Asia

Ofer Bar-Yosef
Harvard University, Cambridge, Massachusetts, U.S.A.

C

INTRODUCTION

In Southwest Asia, the transition from foraging to cultivation and systematic animal husbandry that resulted in the establishment of farming communities, domesticated plants, and animals is a process that began some 11,600 years ago and lasted for a few millennia. The archaeological, archaeobotanical, and archaeozoological records indicate that continuous hunting and gathering accompanied the introduction of this new mode of subsistence and the departure from the former lifeways of mobile and semisedentary foragers. This process, which probably emerged as an abrupt change in a small core area, is recorded in the archaeology of the Levant, or the western wing of the Fertile Crescent, with the ensuing diffusion of agricultural products into the Zagros foothills (the eastern wing of the Fertile Crescent) and the colonization of the Anatolian plateau. Incipient seafaring facilitated the inhabitation of Cyprus and possibly the arrival of farmers in the Nile delta.

THE EARLIEST PHASES OF CULTIVATION

Hypotheses regarding the impetus for cultivation and intentional herding of wild goat, sheep, cattle, and pigs vary between ''push'' and ''pull'' models.[1–8] Paleoclimatic information acquired during the last decade reflecting a general agreement between marine and terrestrial pollen and isotope records indicates that the Younger Dryas, a well-known cold and dry period in the Northern Hemisphere (ca. 13,000–11,600 cal. b.p.), resulted in major environmental deterioration in most areas. It seems feasible that these were the conditions that in a well-populated region such as the Levant created the need to modify subsistence strategies and settlement patterns by local foragers. The best data sets are as yet limited to the central-southern Levant where the Early Natufian, a culture of sedentary and semisedentary hunter-gatherers, practiced extensive hunting and plant gathering from ca. 14,500 cal. b.p. The knowledge of cereal and legume exploitation was already established, as evidenced by the well-preserved plant remains from the water-logged site Ohallo II, uncovered in the Sea of Galilee (Lake Kinneret), dated to ca. 23,000 cal. b.p.[9]

Reactions to the worsening conditions varied among different groups of foragers. The Harifian culture in the Negev and northern Sinai were Late Natufian groups who improved their hunting techniques through the invention of the Harif point and increased their mobility. Another example is the sedentism of groups such as those who built the village of Hallan Çemi.[1,5,6,8,10]

The current archaeobotanical evidence indicates that the first Neolithic cultivators sowed wild cereals, whether einkorn, wheat, barley, or rye.[1,4,6,7,11] Concurrently, these farming communities, grouped archaeologically under the chronological term of Pre-Pottery Neolithic A (abbreviated as PPNA), were geographically spread from Çayönü in the Taurus foothills, through Jerf el Ahmar and Mureybet in the Euphrates Valley, through Jericho and Drha in the lower Jordan Valley and the Dead Sea area. Continued hunting, trapping, and gathering wild fruits, seeds, and leaves supplemented the diets of these communities. The full appearance of the domesticated forms in the Levantine Corridor seems to have occurred in the course of the ensuing several hundred years.

The stone tools of the PPNA communities included the first aerodynamically shaped projectiles, knapped and polished axes, sickle blades, special grinding slabs, either flat and rounded or with small cupholes, and handstones. Their dwellings were rounded and oval pit-houses, walls built by plano-convex, loaf-shaped mud bricks. Floors were sometimes covered by mats. Special attention was given to the stone tower in Jericho, built inside the village and protected by walls from alluvial accumulations. It was hypothesized that on top it had a brick-built shrine.

The growth of human populations during the PPNA is documented by an increase in the size of the largest sites from 0.2 hectares during the Natufian period to 2.0 hectares. This could have been the result of increased sedentism, established plant cultivation, the presence of storage facilities, and probably the clustering of human groups as the need to defend exploitable and arable territories evolved. Predictable supplies of weaning foodstuffs ensured the survival of newborns and together with greater reliability of food supplies caused a drop in the age of menarche and facilitated a longer period of fertility. The results led to the restructuring of social institutions. Large villages, with about 300–400 people, became independent, viable biological units. Markers for past

Encyclopedia of Plant and Crop Science
DOI: 10.1081/E-EPCS 120017101

cosmologies are the use of female figurines, animal figurines, and abstract signs on the back of "shaft straighteners" (employed in preparing wooden shafts as part of the archery equipment). The presence of special built-up rounded pit-houses indicates the role of particular members of the community (perhaps the "elders"). Similar information is conveyed by the burial of adults from which skulls were removed and placed in buildings (e.g., in Mureybet). Children's graves were not disturbed, thus reflecting their different social value. The sense of group individuality and ownership reached a level expressed in styles of artifacts, decorations, and building types and in the importation of precious rocks or products made of obsidian and chlorite.[1,2]

ESTABLISHED VILLAGE SOCIETIES

During the time of the PPNB (ca. 9500–8200 cal. b.p.), large and numerous hamlets and villages were established across the Levant, expanding into the Zagros foothills and some intermontane valleys, as well as the Anatolian plateau. Colonizers inhabited Cyprus and by the end of the period the Nile delta, Crete, and the edges of the Balkans and the Greek mainland. Contemporary foragers survived through most of the arid belt stretching from Sinai through the Syro-Arabian desert and east, north, and west of the Taurus-Zagros mountains.

Domestic architecture in PPNB villages was dominated by rectangular plans, where rooms could be added if necessary.[10] In many cases, the floors were plastered. Upper structure was made of mudbricks or wattle and daube. Particular plans are the "grill buildings," the early form in Çayönü, possibly reflecting division between storage and living quarters. "Corridor buildings" in Beidha and 'Ain Ghazal probably had a second floor as evidenced in the village of Basta where all of the walls were built of stones. Storage facilities, garbage pits, kilns for burning limestone, and some walls at the edge of the habitation (Jericho, Beidha) are recorded from various sites.

Within the territories of farming communities, cultic centers emerged (e.g., Göbekli Tepe, Kfar HaHoresh), and in more than one site, shrines or temples were uncovered (e.g., Navali Çori, 'Ain Ghazal). The large and heavy T-shaped pillars carved with animal imagery, large human sculptures, human plaster statues, plastered skulls, stone masks, complex burial features, and other elements reflect a rich and varied ideology but also demonstrate geographic differences. Tool kits, often made on blades, were produced by artisans who practiced a particular technique of core reduction known as "naviform" cores.[2,12] Special workshops, sometimes adjacent to the outcrops such as those of central Anatolian obsidian, shipped their products to other villages. The shaped stone tools include large arrowheads and bifacially knapped and polished celts, and their geographic distribution expresses the westward colonization of Neolithic farmers. The different projectile forms and use of microliths indicate the individuality of the eastern Mesopotamia interaction sphere.[12]

Cultivated plants include einkorn and emmer wheat, barley, pea, lentil, horsebean (or faba bean), chickpea, and flax. The exploitation of wild plants, fruits, and seeds continued. The domestication of goat and sheep, cattle, and pig began by husbandry of these animals in the context of farmer-hunter villages, during the PPNB or even slightly earlier. Goat, sheep, cattle, and pigs most likely first occurred in villages on the hilly flanks of the Taurus/Zagros (papers in Ref. 6) where these animals had been hunted for many millennia and local inhabitants were familiar with their behavior. Moving them with fallow deer to Cyprus reflects the emergence of intentional herding of wild forms, as well as the presence of efficient navigation and marine transportation, both indicating a more intricate social organization.[1,2,13] The expansion of early Neolithic groups in various directions marks population increase and the need for arable land and pastures.

The transmission of the new economy eastward to the Zagros foothills, from Kurdistan in the north to Khuzistan in the south, was probably done by diffusion without major displacements of human communities.[12] This process is reflected in the continuation of the tradition of manufacturing microlithic artifacts by local populations from the Late Palaeolithic through the Neolithic[14] and by the geographic trajectory of the radiocarbon dates for early villages.

Neolithic economy and groups of farmers spread through the Mediterranean basin by coastal navigation and by inland movements. Transporting wild animals to Cyprus means that the process of domestication as expressed in morphological changes took longer than assumed and that wild species such as the deer played a role in the economy and perhaps the ideology of these groups.

THE COLLAPSE OF THE PPNB CIVILIZATION

Stratigraphic unconformities indicating site abandonment were not uncommon during the PPNB period, but the shift to the next period (commonly known as the Pottery Neolithic time zone in the Levant) is well marked in the physical evidence. The establishment of new hamlets and farmsteads in the southern Levant, new villages in western Anatolia, and the evolution of local pottery-making traditions mark the new period. Not many changes in the suite of domesticated plants or animals can be seen, but the shift in the settlement pattern reflects a major societal change. This is also the time to which pastoral stations in semiarid areas are attributed. A decrease in the overall population at the desert margins is recorded and greater attachment to water sources is evidenced. While the change could have been the results of overexploitation of the environment, it seems that the climatic crisis around 8400–8200 cal. b.p, as recorded in the ice, marine, and pollen cores as well as the terrestrial speleothemes, was the culprit. A short cold and dry spell expressed in series of drought years in most of southwestern Asia possibly resulted in lower grain yields, depletion of pastures, and some water shortages. Under the new circumstances, a complex society that subsisted on farming and herding, in which the demands of better-off individuals (or families) drove the flow of foreign commodities, could not continue to accumulate surplus. The shift in the pattern of seasonal precipitation imposed the search for pastures further away. Hence, the economic deterioration resulted in a series of societal changes expressed in the disappearance of previously large villages and the establishment of smaller villages or hamlets. The new conditions probably enhanced the reliance on the more flexible subsistence strategy of pastoral nomads.

REFERENCES

1. Bar-Yosef, O. From Sedentary Foragers to Village Hierarchies: The Emergence of Social Institutions. In *The Origin of Human Social Institutions*; Runciman, G., Ed.; Proceedings of the British Academy, Oxford University Press: Oxford, 2001; Vol. 110, 1–38.
2. Cauvin, J. *The Birth of the Gods and the Origins of Agriculture*; Cambridge University Press: Cambridge, 2000. Translated by T. Watkins.
3. Flannery, K.V. The origins of the village revisited: From nuclear to extended households. Am. Antiq. *67*, 417–433.
4. Garrard, A.N. Charting the emergence of cereal and pulse domestication in South West Asia. Env. Archaeol. **1999**, *4*, 67–86.
5. Goring-Morris, A.N.; Belfer-Cohen, A. The articulation of cultural processes and late quaternary environmental changes in Cisjordan. Paleorient **1997**, *23*, 71–94.
6. *The Origins and Spread of Agriculture and Pastoralism in Eurasia*; Harris, D., Ed.; UCL Press: London, 1996.
7. Hillman, G.C.; Hedges, R.; Moore, A.; Colledge, S.; Pettitt, P. New evidence of Lateglacial cereal cultivation at Abu Hureyra on the Euphrates. Holocene **2001**, *11* (4), 383–393.
8. Smith, B.D. *The Emergence of Agriculture*, 2nd Ed.; Scientific American Library: New York, 1998.
9. Kislev, M.E.; Nadel, D.; Carmi, I. Epi-Palaeolithic (19,000 B.P.) cereal and fruit diet at Ohalo II, Sea of Galilee, Israel. Rev. Palaeobot. Palynol. **1992**, *71*, 161–166.
10. Hole, F. Is Size Important? Function and Hierarchy in Neolithic Settlements. In *Life in Neolithic Farming Communities: Social Organization, Identity, and Differentiation*; Kuijt, I., Ed.; Plenum Press: New York, 2000; 191–209.
11. Hillman, G.C. Abu Hureyra 1: The Epipalaeolithic. In *Village on the Euphrates: From Foraging to Farming at Abu Hureyra*; Oxford University Press: Oxford, 2000; 327–399.
12. Kozlowski, S.K. *The Eastern Wing of the Fertile Crescent: Late Prehistory of Greater Mesopotamian Lithic Industries*; British Archaeological Reports International Series, Archaeopress: Oxford, 1999; Vol. 760.
13. Peltenburg, E.; Croft, P.; Jackson, A.; McCartney, C.; Murray, M.A. Well-Established Colonists: Mylouthkia 1 and the Cypro-Pre-Pottery Neolithic B. In *The Earliest Prehistory of Cyprus: From Colonization to Exploitation*; Swiny, S., Ed.; Cyprus American Archaeological Research Institute Monograph Series, American Schools of Prehistoric Research: Boston, MA, 2001.
14. Hole, F. The Spread of Agriculture to the Eastern Arc of the Fertile Crescent: Food for the Herders. In *The Origins of Agriculture and Crop Domestication*; Damania, A.B., Valkoun, J., Willcox, G., Qualset, C.O., Eds.; ICARDA: Aleppo, Syria, 1998.
15. Zeder, M.A.; Hesse, B. The initial domestication of goats (*Capra hircus*) in the Zagros Mountains 10,000 years ago. Science **2000**, *287*, 2254–2257.

C

Crop Domestication in the American Tropics: Phytolith Analyses

Dolores R. Piperno
Smithsonian Tropical Research Institute, Balboa, Republic of Panama

INTRODUCTION

The lowland American tropical forest was one of the world's primary centers of agricultural origins. Many important seed, root, and tree crops—including manioc (*Manihot esculenta* Crantz) and sweet potato (*Ipomoea batata* L.), at least two species of squashes (*Cucurbita argyrosperma* Huber and *Cucurbita moschata* Duchesne), and perhaps maize (*Zea mays* L.)—were brought under cultivation and domesticated within its floristically diverse borders. The inimical conditions for plant preservation in these humid forests severely hampered the study of their agricultural history. Since 1980, archaeobotanists have developed a technique, phytolith analysis, that provides tangible evidence of early Neotropical plant cultivation and crop dispersals, as well as of the crop production systems that supported nascent agriculture. This article summarizes the results and ramifications of some of these efforts.

PHYTOLITH ANALYSIS

Phytoliths are microscopic bodies of silica (SiO_2 nH_2O) that form in growing plants and are released into the environment after a plant has died and decayed. Phytoliths (literally, plant stones) resist processes that cause the decomposition of other plant materials, and are thus preserved over long periods of time.[1] Formerly thought to be waste products secreted by plants, scientists now identify a number of different roles phytoliths play in plants. Among the most important may be their role in deterring herbivory and pathogenic fungi.

For example, studies of phytolith and lignin formation in wild and domesticated species of *Zea* and *Cucurbita* indicate that phytolith formation is genetically determined, and that the same loci that govern lignin deposition also regulate the production of phytoliths.[2,3] The genetic loci involved are teosinte glume architecture (*tga1*) in Zea and hard rind (*Hr*) in *Cucurbita*, which, as its name implies, makes the fruit exterior hard. Lignin's role in plant defense is unquestioned; fortifying seed bracts or fleshy fruit rinds with two hard and undigestible substances appears to be a better strategy for some plants than using just one. Because silicification sites in many other plants are likely to cause maximal discomfort to overeager herbivores (e.g., leaf hairs, seed and leaf epidermes),[1,3,4] more demonstrations of genetic control over phytolith formation are expected.

CROP PLANT IDENTIFICATION THROUGH PHYTOLITH STUDY

Numerous gymnosperms, monocotyledons, and dicotyledons heavily silicify their vegetative, reproductive, and underground organs, and produce phytoliths of taxonomic utility at levels varying from the family to the species.[1,4] Phytolith morphology is diagnostic in part because phytoliths are often silicified casts of the cells and plant tissues in which they form. Hence, phytoliths reflect the gross morphology of these structures, obviously useful in plant identification.[1,3–5] This aspect becomes particularly important for studying plant domestication when, as has been demonstrated for *Zea* and *Cucurbita*, the genetic loci that control phytolith formation in glumes, cupules, and fruit rinds also accounted for significant phenotypic changes, and thus in their phytoliths during the domestication process.[2,3]

Crop plants that were either originally domesticated or heavily used in the Neotropical forest and that produced diagnostic phytoliths upon multiple independent investigation of different regional flora include maize,[1,5,6] squashes and gourds of *Cucurbita* spp. and bottle gourd (*Lagenaria siceraria* L.),[1,3,5,7,8] and arrowroot (*Maranta arundinacea*).[4,5] Phytoliths from the rachis, seed glumes, and leaves of Balsas teosinte (*Z. mays* ssp. *parviglumis*) differ from those of its domesticated product, maize[1,2,6,9] (Fig. 1a,b). Phytoliths from some species of squashes (e.g., *Cucurbita maxima* ssp. *andreana*; *Cucurbita maxima*, *Cucurbita moschata*) appear to differ on a morphological basis (Fig. 2a–d), and phytolith size also differentiates many domesticated from wild taxa.[3,8]

Encyclopedia of Plant and Crop Science
DOI: 10.1081/E-EPCS 120017085

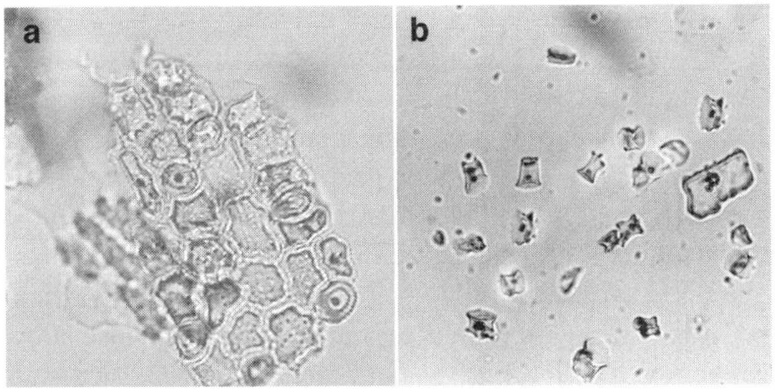

Fig. 1 (a) A group of articulated phytoliths from a fruitcase (lower glume and rachis segment) of Balsas teosinte. (b) Phytoliths from a cob (cupule and glumes) of maize, Race Maiz Ancho. The following differences can be observed: In teosinte, there are many more long-cell phytoliths, some of which are unique, and the spherical short-cell phytoliths (called rondels by phytolith researchers) are more highly decorated than in maize, where undecorated types of rondels dominate the phytolith assemblage. These differences are caused by the locus *tga1*, which regulates which cells become silicified in these structures of wild and domesticated *Zea*. The significant contrasts in ornamentation between wild and domesticated *Zea* phytoliths largely result from the differing degrees of lignification in these taxa, also primarily controlled by *tga1*.

PHYTOLITHS AND NEOTROPICAL CROP PLANT EVOLUTION

The phytolith record from the Neotropical forest, where systematic studies addressing a broad range of economic plant taxa have been carried out longer than in other regions of the world,[4,5] reveals considerable new data on plant use. This section summarizes some of this evidence, with special attention to earlier parts of the archaeological record and two important genera, *Cucurbita* and *Zea*, for which information garnered from previous studies of inadequately preserved macroplant fossils is particularly slim.

Squashes and Gourds of *Cucurbita* spp.

The high oil and protein content of *Cucurbita* spp. seeds made them favorite resources of human populations in the New World. In the lowland tropics, as in highland Mexico (Delgado-Salinas et al., this volume), the genus appears to have been cultivated at an early date. Phytolith analysis of an important preceramic cultural tradition on the coast of Ecuador, called Las Vegas, indicates that a wild species of *Cucurbita* was exploited by 10,000 b.p., and that varieties of squashes and gourds having phytoliths with sizes characteristic of modern, domesticated plants developed between 10,000 and 9500 b.p.[8,10,11] The particular species involved was probably the only free-growing *Cucurbita* known from Ecuador, *C. ecuadorensis*.

Cucurbita phytoliths the size of modern domesticated species also left records dating to about 8000 b.p. in the Colombian Amazon,[11] where no wild species are distributed today. In Panama, phytoliths identifiable on a morphological basis as *C. moschata* (Fig. 2b) occur in contexts that predate 7000 b.p., perhaps by 1000 years (Piperno, unpublished information). Phytoliths from bottle gourd are also present at all of these sites, supporting evidence from highland areas that this plant was dispersed and well used at an early date in the Americas.

Phytoliths and the Origins and Dispersals of Maize

Mexico is the undisputed cradle of origin for maize, but the timing and routes of maize domestication are under active debate. Some investigators,[12,13] relying on the existing macrofossil evidence from a few sites in the dry Mexican highlands, adhere to a mid-Holocene time frame for maize's beginnings (ca. 5500 to 5000 b.p.). Others[11,14] estimate that domestication by ca. 7000 b.p. is more accurate, given the existing microfossil evidence

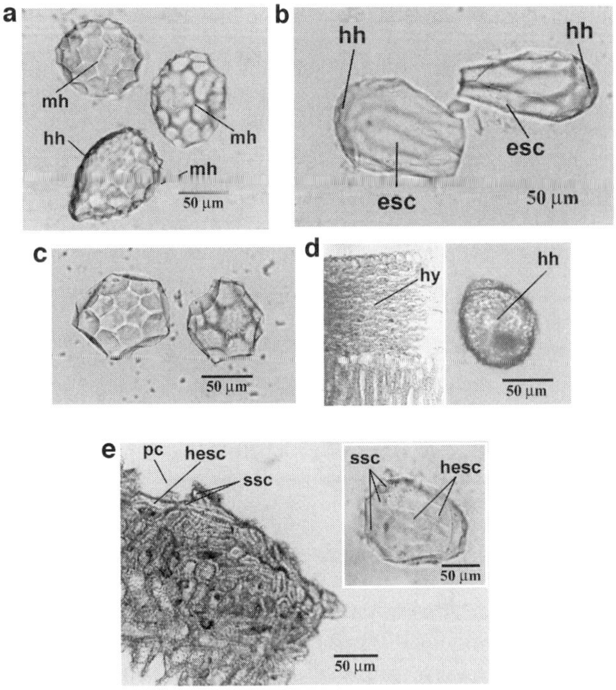

Fig. 2 a–e. Phytoliths from the fruit rinds of various species of *Cucurbita* and bottle gourd. They are formed in specialized spaces at the interface of the hypodermis and upper mesocarp of the rind, and thus acquire the characteristics of these cell arrangements, which are diagnostic of *Cucurbita* and *Lagenaria* and can be species-specific in *Cucurbita*. (a) *Cucurbita moschata*; (b) *C. moschata*; (c) *C. maxima* ssp. *andreana*; (d) *C. maxima*; (e) *Lagenaria siceraria*. (Published with permission from the Proceedings of the National Academy of Sciences.)

from south of Mexico (discussed later). Because the earliest known maize macrofossils date to 5400 b.p. at Guilß Naquitz Cave and are already domesticated, and because no prior evidence for the exploitation of a wild or domesticated *Zea* exists from this site or others located in the dry Mexican highlands,[15] the record is clearly incomplete. Recent pollen evidence indicating that maize spread to the Gulf Coast lowlands of Mexico by ca. 6200 b.p., where it was likely grown under slash-and-burn methods of cultivation, helps to fill this gap.[14]

Regardless of their chronological stripes, many investigators, relying on the molecular evidence obtained from modern maize and its closest living wild relatives, looked to the tropical deciduous forest of the Central Balsas River Valley in southwestern Mexico as a likely region for maize's origins.[4,12,14,16] A recent molecular analysis indicates that the oldest surviving maize races are from the Mexican highlands, although the lower-elevation Central Balsas populations of parviglumis teosinte are still identified as being genetically closest to maize.[17] Whether these findings reflect a different distribution of Balsas teosinte during the early Holocene than seen today, an inadequate representation of prehistoric lowland maize diversity in extant Mexican maize, or simply the nearly total lack of archaeological information from sites in the Balsas River Valley, requires further archaeological and paleoecological research in the Balsas and other lowland regions and more excavations in the highlands.

Of greatest concern is that microfossils of maize—be they pollen or phytoliths—are routinely recovered from archaeological and paleoecological contexts in southern Central America and northern South America that date from ca. 7000 to 5000 b.p.[11] A middle-Holocene time frame for maize domestication does not seem to account for these findings. At the Aguadulce Rock Shelter, Panama, where discrete assemblages of phytoliths from maize cupules and chaff are stratified in deep, securely dated, late preceramic sediments (7000 to 6000 b.p.), residues from plant grinding stones from the same contexts also yield starch grains from maize kernels and maize chaff/cupule phytoliths[11,18] (Piperno and Holst, this volume). Future work in Mexico will reveal the age of *Zea* as a domesticated taxon, but newer research to the south increasingly indicates it was dispersed into the tropical lowlands of Central America and northern South America at an earlier date than once believed. It may have been a valued plant in the ceremonial life of social communities, and consumed as a fermented beverage.

Phytoliths and the Prehistory of Slash-and-Burn Cultivation

Indigenous agriculture in the tropical forest today often involves the practice of swidden, or slash-and-burn, cultivation by people whose settlements are small and impermanent clusters of houses exercising social and political autonomy. Swidden cultivation is land-extensive; that is, one to two hectares of land are typically required to feed a single family, meaning that large areas of terrain are abandoned after a few years of use. Therefore, although early human settlements may be hard to identify archaeologically, past agricultural practices in the forest should leave identifiable signatures in paleoecological records, where a regional portrait of vegetational change reflecting forest clearing, burning, and plant succession may be obtained through pollen and charcoal studies.[4]

Phytolith analysis of lake sediment cores provides comparable signals of these activities.[19] When phytolith, pollen, and charcoal studies are carried out in tandem, more refined pictures of past agricultural practices can be obtained because each technique has strengths that redress the other's shortcomings in terms of the production and taxonomic specificity of major, indicator arboreal and herbaceous taxa.[19] Such multiproxy evidence for the beginnings of slash-and-burn cultivation in the Neotropics has been obtained from lake sediments in Panama and Ecuador, dating to 7000 b.p. and ca. 5300 b.p., respectively.[11] In these sequences, phytoliths and pollen from cultivars, including maize, are embedded within large phytolith and pollen populations of early successional herbaceous and arboreal species, and numerous micro- and macrofragments of charcoal. In Panama, the paleoecological data originate from the same region and time periods providing the archaeological phytolith and starch grain evidence noted in the foregoing. Other paleoecological findings wherein phytolith studies are not yet incorporated are similar, with agricultural systems containing maize and other cultivars (e.g., manioc and tree crops) being evidenced on the Gulf Coast of Mexico at 6200 b.p.,[14] and in the Cauca Valley, Colombia, and the Colombian Amazon by 5200 and 4700 b.p., respectively.[11]

It thus appears that techniques of swidden cultivation have considerable antiquity in the American tropics. Combined archaeobotanical and paleoecological data indicate that earlier (ca. 9000 to 7000 b.p.) food production seems to have been a more simple and inexpensive kind of horticulture, perhaps practiced largely in house gardens, which did not involve significant field preparation and the progressive removal of primary forest trees from large areas.[11]

CONCLUSION

Phytolith studies provide empirical data relating to the early use and spread of some important crop plants in the American tropics. The results, especially when combined with those from palynology and starch grain analysis (see Piperno and Holst, this volume), indicate that the lowland Neotropical forest was an early and important center of plant husbandry in the Americas. Although Neotropical plant husbandry has been typically viewed as root and tree crop based, some seed crops (e.g., *Cucurbita*) were incorporated into early cultivation practices. We can expect continued accumulation of information, as new sites are excavated or existing sediments and other cultural materials are analyzed using phytolith techniques.

ARTICLES OF FURTHER INTEREST

Archeobotanical Remains: New Dating Methods, p. 55
Crop Domestication in Africa, p. 304
Crop Domestication in China, p. 307
Crop Domestication in Mesoamerica, p. 310
Crop Domestication in the American Tropics: Starch Grain Analyses, p. 330
Crop Domestication: Role of Unconscious Selection, p. 340

REFERENCES

1. Piperno, D.R. *Phytolith Analysis: An Archaeological and Geological Perspective*; Academic Press: San Diego, 1988.
2. Dorweiler, J.E.; Doebley, J. Developmental analysis of Teosinte Glume Architecture1: A key locus in the evolution of maize (*Poaceae*). Am. J. Bot. **1997**, *84* (10), 1313–1322.
3. Piperno, D.R.; Holst, I.; Wessel-Beaver, L.; Andres, T.C. Evidence for the control of phytolith formation in Cucurbita fruits by the hard rind (Hr) genetic locus: Archaeological and ecological implications. Proc. Natl. Acad. Sci. U. S. A. **2000**, *99* (16), 10923–10928.
4. Piperno, D.R. Paleoethnobotany in the Neotropics from microfossils: New insights into ancient plant use and agricultural origins in the tropical forest. J. World Prehist. **1998**, *12* (4), 393–449.
5. Pearsall, D.M. *Paleoethnobotany: A Handbook of Techniques*; Academic Press: San Diego, 2000.
6. Bozarth, S.R. Maize (*Zea mays*) cob phytoliths from a Central Kansas Great Bend Aspect archaeological site. Plains Anthropol. **1993**, *38* (146), 279–286.
7. Bozarth, S.R. Diagnostic opal phytoliths from rinds of selected Cucurbita species. Am. Antiq. **1987**, *52* (3), 607–615.
8. Piperno, D.R.; Holst, I.; Andres, T.C.; Stothert, K.E. Phytoliths in Cucurbita and other Neotropical Cucurbitaceae and their occurrence in early archaeological sites from the lowland American tropics. J. Archaeol. Sci. **2000**, *27*, 193–208.
9. Piperno, D.R.; Pearsall, D.M. Phytoliths in the reproductive structures of maize and teosinte: Implications for the study of maize evolution. J. Archaeol. Sci. **1993**, *20*, 337–362.
10. Piperno, D.R.; Stothert, K.E. Phytolith evidence for early Holocene Cucurbita domestication in southwest Ecuador. Science *299*, 1054–1057.
11. Piperno, D.R.; Pearsall, D.M. *The Origins of Agriculture in the Lowland Neotropics*; Academic Press: San Diego, 1998.
12. Smith, B.D. *The Emergence of Agriculture*; Scientific American Library: New York, 1998.
13. Fritz, G. Is New World agriculture getting younger? Curr. Anthropol. **1994**, *35* (3), 305–309.
14. Pope, K.O; Pohl, M.E.D.; Jones, J.G.; Lentz, D.L.; von Nagy, C.; Vega, F.J.; Quitmyer, I.R. Origin and environmental setting of ancient agriculture in the lowlands of Mesoamerica. Science **2001**, *292*, 1370–1373.
15. Piperno, D.R.; Flannery, K.V. The earliest archaeological maize (*Zea mays* L.) from highland Mexico: New accelerator mass spectrometry dates and their implications. Proc. Natl. Acad. Sci. U. S. A. **2001**, *98* (4), 2101–2103.
16. Smith, B. Documenting plant domestication: The consilience of biological and archaeological approaches. Proc. Natl. Acad. Sci. U. S. A. **2001**, *98* (4), 1324–1326.
17. Matsuoka, Y.; Vigouroux, Y.; Goodman, M.M.; Sanchez, J.; Buckler, E.; Doebley, J. A single domestication for maize shown by multilocus microsatellite genotyping. Proc. Natl. Acad. Sci. U. S. A. **2002**, *99*, 6080–6084.
18. Piperno, D.R.; Ranere, A.J.; Holst, I.; Hansell, P. Starch grains reveal early root crop horticulture in the Panamanian tropical forest. Nature **2000**, *407*, 894–897.
19. Piperno, D.R. Phytoliths. In *Tracking Environmental Change Using Lake Sediments. Volume 3: Terrestrial, Algal, and Siliceous Indicators*; Smol, J.P., Birks, H.J.B., Last, W.M., Eds.; Kluwer Academic Publishers: Dordrect, The Netherlands, 2002; 235–251.
20. Harlan, J.R. *Crops and Man*, 2nd Ed.; American Society of Agronomy and Crop Science Society of America: Madison, 1992.
21. Sauer, C.O. Cultivated Plants of South and Central America. In *Handbook of South American Indians*; Steward, J., Ed.; Bureau of American Ethnology Bulletin, No. 143, U.S. Government Printing Office: Washington, DC, 1950; Vol. 6, 487–543.

Crop Domestication in the American Tropics: Starch Grain Analyses

Dolores R. Piperno
Irene Holst
Smithsonian Tropical Research Institute, Balboa, Republic of Panama

INTRODUCTION

A cardinal attribute of New World agriculture is the large number of plants that were taken under cultivation and domesticated for their starch-rich underground organs.[1,2] The lowland tropical forest contributed several of these, including manioc (*Manihot esculenta* Crantz), sweet potato (*Ipomoea batatas* (L.) Lam.), yams (*Dioscorea trifida* L.f.), yautia (*Xanthosoma sagittifolium* (L.) Schott & Endl.), arrowroot (*Maranta arundinacea* L.), and lirén (*Calathea allouia* (Aubl.) Lindl.).[1–3] However, research bearing on when and where they were originally domesticated and dispersed out of their cradles of origin has been hampered by a paucity of data. In all but the most arid climates, soft, starchy plant structures are not preserved in archaeobotanical records, and most root crops similarly contribute sparse pollen records and few or otherwise unidentifiable phytoliths.[4]

Starch grain studies can help us to elucidate these problems. This paper discusses some present applications and future directions of starch grain analysis in tropical archaeology. It describes how starch analysis may document not only past root crop cultivation, but also the early use and spread of important seed plants such as maize (*Zea mays* L.) and beans (*Phaseolus* spp.).

STARCH GRAINS: THEIR PROPERTIES AND ARCHAEOLOGICAL IDENTIFICATION

Starch grains, which are found in large quantities in most higher plants, are the major form in which plants store their carbohydrates, or energy.[5,6] Starch grain molecules are primarily composed of amylose and amylopectin. They can be found in all organs of a higher plant, including roots, rhizomes, tubers, leaves, fruits, and flowers. However, only subterranean organs and seeds commonly possess what is called reserve starch, which is differentiated from another type of starch called chloroplast or transitory starch. The latter is principally formed in leaves and other vegetative structures and can also be found in pollen.[5,6] An important difference between chloroplast and reserve starch relevant to archaeological research is that reserve starches are produced in a highly diverse array of forms that may be genus-and even species-specific, while transitory granules are mostly of the same type and of limited use in identification. Also, transitory starch, as its name suggests, is formed during the day and utilized at night, while reserve starch is stored and utilized later in the cycle of the plant. Therefore, it is the reserve starch, formed in tiny organelles called amyloplasts, that is most useful for archaeobotanical enquiry. Because reserve starch grains are quite different in morphology from transitory starches, one can also identify the source of many reserve starches as seeds or underground plant organs.

There is a large amount of literature on starch grain properties and morphology that researchers interested in archaeological applications can refer to.[5–8] Any number of atlases and keys of starch grains exist, among the most extensive of which are Reichert's[5] and Seidemann's,[9] which contain descriptions and photographs of starches from hundreds of economically important tropical and other plants. A dedicated starch journal, *Die Stärke*, also exists, in which starch grains of various taxa are routinely described and illustrated. It is widely acknowledged by botanists that the morphology of starch granules can be specific to, and diagnostic of, an individual genus or species. The morphological features that allow for identification can be observed with a compound light microscope and include granule shape and size, form and position (centric or eccentric) of the hilum (the botanical center of the granule), presence and types of fissures (natural cracks on the grains at the hila) and lamellae (growth bands), and number and characteristics of pressure facets.[5,10] Studies undertaken in the authors' laboratory of over four hundred species of crop and other plants of economic importance indicate that starch grains from maize, *Phaseolus* and *Canavalia* beans, manioc, sweet potato, yams, arrowroot, yautia, yam beans (*Pachyrrizus* spp.), peanuts (*Arachis hypogaea* L.), squashes (*Cucurbita* spp.), and palms possess the same morphological attributes noted by previous investigations and are distinguishable from each other and those of different genera in this reference collection (Fig. 1a–h).

Encyclopedia of Plant and Crop Science
DOI: 10.1081/E-EPCS 120017086

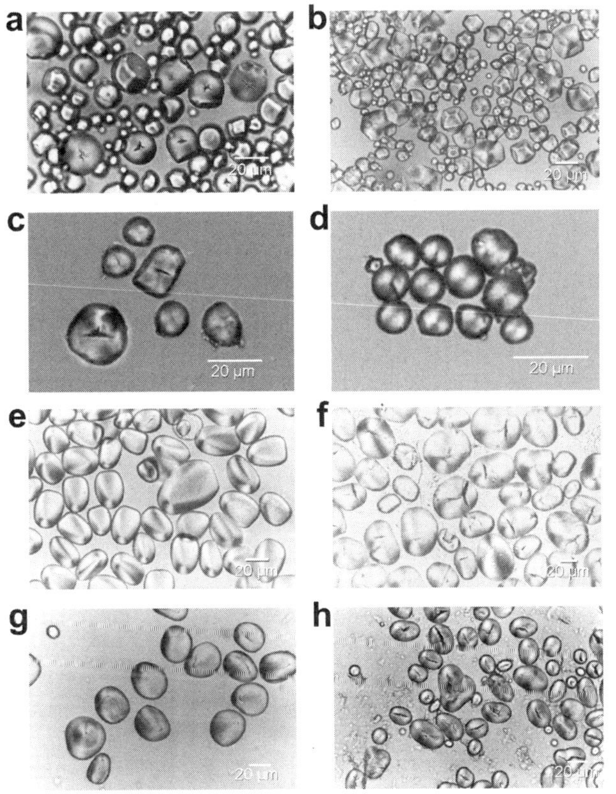

Fig. 1 Starch grains from various crop plants of the Neotropics showing how they are differentiable on the basis of overall shape and characteristics of the hila, fissures, pressure facets, and lamellae. **a** manioc roots; **b** sweet potato roots; **c** maize kernel, Race Jala; **d** maize kernel, Race Harinoso de Ocho (eight-rowed flour corn); **e** yam (*D. trifida*); **f** arrowroot root (*M. arudinacea*); **g** lirén root (*Calathea allouia*); **h** common bean (*Phaseolus vulgaris*). Also notice that starch grains from hard and soft endosperm maize varieties can be distinguished (**c** and **d**) because the manner in which they congregate at formation creates differences in their starch populations (rough, angled, and irregular in hard endosperm vs. circular, smooth, and without fissures in soft endosperm).

The large corpus of literature on starch grain morphology was mainly compiled by researchers focused on the purely botanical aspects and commercial uses of starches. Less attention was understandably given to how grain size and morphology in domesticated crops might differ from those in closely related wild species. This will become an area of intense interest for paleoethnobotanists, as they seek to explore the potential of starch grain studies in investigating the earliest histories of some important roots and tubers. Studies carried out recently, for example, by the authors and others indicate that starch grains from bitter and sweet forms of manioc sampled from Central and South America can be distinguished from those of its wild ancestor, *M. flabellifolia* (Pohl) Ciferri[11] and other wild species of *Manihot* occurring in southern central America on the basis of both morphology and size.[10,12,13] More work of this kind will eventually determine which other domesticated species might be identifiable in their cradles of origin.

ARCHAEOLOGICAL RECOVERY AND APPLICATIONS

As with phytolith studies, starch analysis is predicated on the notion that when the macro-structures of tubers decay, some of the starch grains they contain will survive in a largely unaltered state and be retrievable for study. The properties of starch grains and their sensitivity to degradation under various conditions, which largely determine their capacity for long-term survival, are fairly well understood and can be summarized as follows: Starches are highly sensitive to heat, strong acids and bases, and oxidizing compounds. Many grains start to gelatinize, whereby they melt and lose their diagnostic properties, at temperatures of between 40 and 50°C. These factors probably mean that they may often be more poorly preserved in some contexts associated with ancient human settlements (e.g., leached and acidic soils, alkaline shell middens, locations near the heat of hearths) than are either phytoliths or pollen grains. Similarly, starches may not commonly remain as identifiable residues in archaeological ceramics (unless the pots had a storage function), although this question is in need of study.

The number of archaeological starch grain studies carried out to date in the humid Neotropics is limited, but results have been rewarding. Starch grains from a variety of plants survive well on the used facets of plant grinding stones, where the superficial cracks and crevices on the stones apparently afford them long-term protection from the various processes that degrade them. In contrast, and not surprisingly, associated archaeological sediments examined thus far from the tropical forest contain almost no starch. For example, the authors isolated starch grains from early Holocene-aged (ca. 10,000–9000 b.p.) grinding stones from the Upper Cauca Valley, Colombia.[3] A variety of taxa were represented, including *Maranta* spp. and unidentified legumes and grasses. Grinding stones from late preceramic (7000–5000 b.p.) and early ceramic archaeological contexts (ca. 5000–3000 b.p.) in Central Pacific Panama yielded starch grains from a variety of plants, including manioc, maize, arrowroot, yams, and legumes.[10,12] This study also indicated that (a probably sweet form of) manioc was domesticated and then spread into southern Central America from southern Amazonia, its area of origin,[11] by ca. 7000 b.p. to 6000 b.p., and it supported previously obtained archaeological and paleoecological phytolith and pollen data for an early development of agriculture in central Panama using both root and seed crops.[13]

Another recent archaeological starch grain study involves analyses of small, longitudinal chips of stone, called ''grater chips'', retrieved from archaeological sites in the Middle Orinoco Basin dated to between A.D. 430 and 720.[13,14] On the basis of an ethnographic analogy, these quartz flakes were thought to have been used for processing bitter forms of manioc. The starch studies, however, resulted in the recovery of hundreds of starch grains, none of them from manioc. With the use of a large reference sample of plants from the region and published starch grain atlases, starches from maize, yams, palms, and *Calathea* spp were identified, showing a more diverse assemblage of plants than was expected on the artifacts based on inferences derived from ethnographic analogies.[13,14] The absence of manioc starches also raises new questions concerning the antiquity of bitter manioc-based economies in this region of South America. Finally, maize-like starch grains have been found in Mayan teeth calculi,[15] where they were the most abundant starch remains present. Possible manioc grains were also present on the teeth. Food residue of various types found in tooth calculi, while not commonly studied at present, have considerable potential for the elucidation of dietary trends in the Neotropics.

CONCLUSIONS

Starch grain research in the humid tropics is in its beginning stages, but the considerable potential is obvious. Armed with good modern reference collections, archaeobotanists can begin addressing a number of important questions concerning agricultural history in the lowland Neotropics. With the considerable amount of basic information already available on starch grain morphology, researchers should be able to fairly expeditiously move into questions concerning the discrimination of starches from crop plants and their closest living wild relatives. Archaeological specimens from newly excavated cultural occupations or well-curated artifacts and sediments retrieved from previous research may form the basis for starch grain analysis.

ARTICLES OF FURTHER INTEREST

Archaeobotanial Remains: New Dating Methods, p. 55
Crop Domestication in Africa, p. 304
Crop Domestication in China, p. 307
Crop Domestication in Mesoamerica, p. 310
Crop Domestication: Fate of Genetic Diversity, p. 333
Crop Domestication: Role of Unconscious Selection, p. 340

REFERENCES

1. Harlan, J.R. *Crops and Man,* 2nd Ed.; American Society of Agronomy and Crop Science Society of America: Madison, 1992.
2. Sauer, C.O. Cultivated Plants of South and Central America. In *Handbook of South American Indians*; Steward, J., Ed.; Bureau of American Ethnology Bulletin, No. 143, U.S. Government Printing Office: Washington, DC, 1950; Vol. 6, 487–543.
3. Piperno, D.R.; Pearsall, D.M. *The Origins of Agriculture in the Lowland Neotropics*; Academic Press: San Diego, 1998.
4. Piperno, D.R. Paleoethnobotany in the Neotropics from microfossils: New insights into ancient plant use and agricultural origins in the tropical forest. J. World Prehist. **1998**, *12* (4), 393–449.
5. Reichert, E.T. *The Differentiation and Specificity of Starches in Relation to Genera, Species, Etc.*; Carnegie Institution of Washington: Washington, D.C., 1913.
6. *Starch: Chemistry and Technology*; Whistler, R.L., Bemiller, J.N., Paschall, E.F., Eds.; Academic Press: Orlando, 1984.
7. Loy, T.H. Methods in the Analysis of Starch Residues on Stone Tools. In *Tropical Archaeobotany: Applications and New Developments*; Hather, J.G., Ed.; Routledge: London, 1994; 86–114.
8. Ugent, D.; Pozorski, S.; Pozorski, T. Archaeological potato tuber remains from the Casma Valley of Peru. Econ. Bot. **1982**, *36* (2), 182–192.
9. Seidemann, J. *Stärke-Atlas*; Paul Parey: Berlin, 1966.
10. Piperno, D.R. Starch Grains and Root Crops. In *Documenting Domestication: New Molecular and Archaeological Paradigms*; Zeder, M., Decker-Walters, D., Smith, B., Eds.; Smithsonian Institution Press: Washington, DC, *in press.*
11. Olsen, K.M.; Schaal, B.A. Evidence on the origin of cassava: Phylogeography of *Manihot esculenta*. Proc. Natl. Acad. Sci. U. S. A. **1999**, *96* (10), 5586–5591.
12. Piperno, D.R.; Holst, I.; Ranere, A.J.; Hansell, P. Starch grains reveal early root crop horticulture in the Panamanian tropical forest. Nature **2000**, *407*, 894–897.
13. Perry, L. *Prehispanic Subsistence in the Middle Orinoco Basin: Starch Analyses Yield New Evidence*; Unpublished PhD Dissertation, Southern Illinois University: Carbondale, 2001.
14. Perry, L. Starch analyses reveal multiple functions of quartz ''manioc'' grater flakes from the Orinoco basin, Venezuela. Intercienca **2002**, *27* (11), 635–639.
15. Scott Cummings, L.; Magennis, A. A Phytolith and Starch Record of Food and Grit in Mayan Human Tooth Tartar. In *The State of the Art of Phytoliths in Soils and Plants*; Pinilla, A., Juan-Tresserras, J., Machado, M.J., Eds.; Monografias del Centro de Ciencias Medioambientales: Madrid, 1997; 211–218.

Crop Domestication: Fate of Genetic Diversity

C

Gideon Ladizinsky
Hebrew University, Rehovot, Israel

INTRODUCTION

Nearly all crop plants originated directly from wild ancestors, some of which are still thriving in natural habitats. Populations of wild progenitors of crop plants, and of other wild plant species, usually exhibit various degrees of diversity in morphological and physiological traits as well as in molecular markers. Crop plants usually possess only part of the genetic diversity of their progenitors, but are characterized by traits that never occurred in their wild parents. As crop plants and their wild progenitors are still capable of free gene exchange, what were the processes and forces that caused such a shift in genetic diversity between the two? As is shown in this article, human selection (unconscious and deliberate) and the adaptive value of domesticated characters under this selection played a key role in that shift in genetic diversity.

THE SHAPE OF GENETIC VARIATION IN CROP PLANTS AND THEIR WILD PARENTS

Crop plants and their wild parents differ from each other in a few key characters known as the domestication syndrome. These sharp differences are the main reason that in classical taxonomy the two are still treated as separate species.

Variation in natural populations of wild progenitors is believed to be regulated mainly by natural selection. Yet, the adaptive value of variation in molecular markers is often questioned and has even been regarded as neutral. Evidence of the adaptive nature of molecular variation is not prevalent, as consistent differences in allozyme and microsatellites pattern between wild wheat plants occurring across narrow boundaries of basalt and terra-rossa soil types[1] indicate that these differences are of adaptive nature.

Whereas populations of wild progenitors have perpetuated themselves for millennia in their natural habitats, crop plants cannot survive without human care. This is because of the loss of several traits that are vital for survival in nature but are selected against in cultivated fields. In most seed crops the loss of the seed dispersal mechanism marks the transition from wild to domesticated state. Wild barley, *Hordeum spontaneum*, disperses its seeds shortly after maturity when the spike axis (rachis) disintegrates into individual dispersal units. Each unit contains at the bottom a segment of the spike axis, a single seed of the central spikelet ending in a long awn, accompanied by two sterile lateral spikelets. This arrowhead structure ensures quick burial of wild barley seeds in the ground. In domesticated barley the spike axis remains intact after maturity and occasionally even upon threshing (Fig. 1). Tough spike axis is controlled by each of two recessive genes. Hence, the transition from wild to domesticated state in barley is due to a single mutation. Other components of the domestication syndrome in barley and in other crops that are unique to domesticated forms are also the result of mutation in a single gene or in a small number of genes.[2]

HOW MUCH VARIATION OF THE WILD PROGENITOR IS POSSESSED BY THE CROP?

Barley and many other grain crops are selfers and any individual plant is homozygous for most of its genes. Because the transition from wild to domesticated state in barley and many other crops is due to mutation in a single gene, the entire crop could theoretically have evolved from a single wild plant. If this is true, one would expect that of each polymorphic locus of the wild progenitor, only one or two alleles would be included in the crop. This must also be true for vegetatively propagated crops.

Comparison of DNA variation attempted between domesticated forms and their wild ancestor of many crops has shown that numerous (but not all) alleles of the wild progenitor's polymorphic loci are present in the crop plant. Molecular markers are suitable for such comparisons because, since the beginning of domestication, they have not been subjected to human selection, at least not directly. The results of these comparisons have two main implications for the nature of variation in crop plants: 1) domesticated products have undergone considerable genetic bottleneck; and 2) there must have been circumstances by which more than one allele of the wild progenitor's polymorphic loci could have been introduced into the crop plant.

Encyclopedia of Plant and Crop Science
DOI: 10.1081/E-EPCS 120017090

Fig. 1 Wild and domesticated barley, natural size. At right, dispersal unit of wild barley; at left, postthreshing intact spike axis of two-row domesticated barley.

DOMESTICATION AS A FOUNDER EFFECT

Founder effect is a situation wherein a few individuals, possessing a portion of the variation of their mother polymorphic population, establish a new population in isolation from the mother population, and therefore are much more nonomorphic.[3]

Whenever sufficient evidence is available on the geographic distribution of the wild ancestor of the crop, the geographic distribution of the crop's wild genetic stock, and the locations where the crop was firstly domesticated, it becomes obvious that only a few wild populations could have given rise to the crop. For example, the lentil's wild progenitor, ssp. *orientalis*, is native to the Middle East and Central Asia. All the tested domesticated lentils are interfertile, possess the same chromosomal architecture, and are almost monomorphic for their cytoplasmic genome (cpDNA). All the tested ssp. *orientalis* populations from Israel and Lebanon differ from the cultigen by a single chromosomal rearrangement. Other populations from Turkey differ from the cultigen by other chromosomal aberrations, and others are cross-incompatible with the cultigen. Sub sp. *orientalis* accessions from Central Asia possess the chromosomal architecture of the cultigen but differ in their cpDNA. Only populations from northern Syria and southern Turkey match the chromosomal architecture of the cultigen or share its cytoplasmic genome and crossability potential. Again, hidden traits that could not be selected for or against by humans and that are polymorphic in the wild progenitor but monomorphic in the cultigen, can be utilized for determining the wild genetic stock of the cultigen.[4] It is safe to say that domesticated lentil evolved from wild populations in a rather restricted area of the wild lentil's distribution range; consequently, genetic variation of other populations was excluded upon domestication. Concomitance between domestication and founder effect is not unique to lentil. Of the three chromosomal types of wild pepper *Capsicum annum* var. *minimum*,[5] only one is present in the cultigen. Of several phaseoline seed protein types of wild bean, only one is present in the domesticated bean.[6] Similarly, sunflower is monomorphic for several enzymes and cpDNA, and is identical to one of the wild sunflower subsets.[7] Alleles of enzymic genes occurring in wild soybean,[8] barley,[9] and radish[10] are missing in their domesticated counterparts. These examples and many others indicate a severe bottleneck situation during domestication as a result of founder effect.

POSSIBLE SOURCES OF PARALLEL VARIATION IN THE CROP AND ITS WILD FORM

Crop plants differ from their wild ancestors by a set of characters favored by humans and better adapted to their agricultural technology, but negatively selected in the wild. In other words, they are the outcome of changes in selection pressure exerted by humans. Traits not involved in that adaptation have been affected by natural selection and in the wild. As indicated in the foregoing, some of the molecular alleles of the wild progenitor are missing in the cultigen, although a great many of them are present. With the notion that crops derive from a single mutant plant, it is assumed that the parallel variation of molecular genes in crops and in their wild ancestors is due to postdomestication mutations in the crops, or to gene flow between the crops and their wild form.

It is highly improbable that postdomestication mutation in the crop has created similar allelic frequency of molecular genes in the crop and its wild ancestor, particularly in the relatively short time since domestication began (see Kaplan contribution). Gene flow between the two is also an inadequate explanation, even when they grow side-by-side. When hybridization between the two takes place, the wild plant is usually the pistil parent because the amount of pollen released by the crop is far greater than that produced by the wild form (see Papa and Gepts contribution). Gene flow from the crop to its wild form can be detected by the aid of the crop's traits that never occurred in the wild. Such a character in barley is six-row spike, which resulted from a mutation changing the two lateral sterile spikelets at each spike node into fertile ones. Hybrids between six-rowed barley and *H. spontaneum* have brittle six-rowed spikes known as Agriocriton. Small populations of the Agriocriton type may occur at edges of fields and along roadsides, and some persist for many years. Similarly in maize, gene flow from the cultigen to its wild form, teosinte, created a seed dormancy free teosinte type that became a serious weed in parts of Mexico.[11] Hybridization wherein the crop plant is the pistil parent must occur as well, particularly in cross-pollinating plants and garden crops, where gene flow from the wild parent is prevented by weeding out such hybrids or hybrid derivatives.[11]

Similar allelic frequency of molecular genes in the crop and its wild form cannot be accounted for by gene flow at the transitional period when farmers grew mixtures of domesticated and wild forms.[12]

The conclusion from all these is that many crop plants have evolved not from a single mutant plant but from several independent mutants, each of which contributed to the crop the same or different alleles of molecular genes. The number of domesticated mutants that gave rise to the barley crop has been estimated to be about 100.[13] This number could allow the introduction of esterase alleles into the cultigen, occurring at the rate of 0.01 in wild barley.

CONCLUSION

Founder effect and bottleneck situations during the first stages of domestication have caused only part of the genetic variation of the crop's wild progenitor to be present in the domesticated forms. Domestication occurred in a restricted area of the distribution range of the crop's wild progenitor, and was involved in a relatively small number of individuals. Parallel variation in the crop and its wild form is due mainly to the number of the crop's founder mutants, and only negligibly to postdomestication gene flow. Under domestication, crop plants have acquired new traits that are negatively selected in the wild and have never been recorded in the wild form. Many of these traits are monogenic and highly adapted to agriculture practices and technology. Their establishment could occur automatically or with the aid of active human selection.

Because only part of the wild form's genetic variation is present in the crop, the wild form remains a valuable source of economically important traits for crop improvement. Most notable is pest and disease resistance, but stress-tolerant food quality and other traits may be demanded in the future.

REFERENCES

1. Li, Y.C.; Fahima, T.; Penge, J.H.; Röder, M.S.; Kirzhner, V.M.; Beiles, A.; Korol, A.B.; Nevo, E. Edaphic microsatellite DNA divergence in wild emmer wheat, *Triticum dicoccoides*, at a microsite: Tabigha, Israel. Theor. Appl. Genet. **2000**, *101*, 1029–1038.
2. Ladizinsky, G. *Plant Evolution Under Domestication*; Kluwer: Dordrecht, 1998.
3. Mayr, E. *Animal Species and Evolution*; Harvard Univ. Press: Massachusetts, 1963.
4. Ladizinsky, G. Identification of the lentil's wild genetic stock. Genet. Resour. Crop Evol. **1999**, *46*, 143–147.
5. Pickersgill, B. Relationships between weedy and cultivated forms of some species of chili peppers (genus *Capsicum*). Evolution **1971**, *25*, 683–691.
6. Gepts, P.; Bliss, F.A. Phaseolin variability among wild and cultivated common bean (*Phaseolus vulgaris*) from Columbia. Econ. Bot. **1986**, *40*, 469–478.
7. Riesberg, L.H.; Sieler, G. Molecular evidence and the origin and development of the domesticated sunflower (*Helianthus annus* L.). Econ. Bot. **1990**, *44S*, 79–91.
8. Kiangs, Y.T.; Marshall, M.B. Soybeans. In *Isozymes in Plant Genetics and Breeding*; Tanksley, S.D., Orton, S.D., Eds.; 1983; 295–328. Part B.
9. Kahler, A.L.; Allard, R.W. Worldwide pattern of genetic variation among four esterase loci in barley. (*Hordeum vulgare* L.). Theor. Appl. Genet. **1981**, *59*, 101–111.
10. Ellstrand, N.C.; Marshall, D.L. The impact of domestication on distribution of allozyme variation within and among cultivars of radish, *Raphanus sativus* L. Theo. Appl. Genet. **1984**, *69*, 393–398.
11. Wilkes, H.G. Hybridization of maize and teosinte in Mexico and Guatemala and the improvement of maize. Econ. Bot. **1977**, *31*, 245–293.
12. Ladizinsky, G.; Genizi, A. Could early gene flow have created similar allozyme gene frequency in cultivated and wild barley. Genet. Resour. Crop Evol. **2000**, *48*, 101–104.
13. Ladizinsky, G. How many tough-rachis mutants gave rise to domesticated barley. Genet. Resour. Crop Evol. **1999**, *45*, 411–414.

Crop Domestication: Founder Crops

Daniel Zohary
Hebrew University of Jerusalem, Jerusalem, Israel

INTRODUCTION

At present, the questions of when and where plant domestication started and how agricultural systems evolved are only partly answered. This entry tries to sketch what is already known on 1) the nuclear areas where plant domestication was independently initiated in several parts of the world and 2) the founder crops that, in each of these core areas, triggered the shift from hunting and gathering to food production. The survey focuses on the origin and subsequent development of the region of Mediterranean agriculture (Southwest Asia, the Mediterranean basin, and temperate Europe). This is the most extensively explored agricultural system. It is also the region in which the earliest, definite evidence on plant domestication has been discovered to date.

KINDS OF EVIDENCE FOR ELUCIDATING THE BEGINNING OF AGRICULTURE

The study of the origin of domesticated plants is an interdisciplinary venture based on information obtained from numerous fields. However, the modern synthesis[1–3] leans heavily on two principal sources: 1) the archaeological evidence obtained from examining plant remains uncovered in archaeological excavations and 2) the botanical and genetic evidence extracted from the living plants. Archaeology supplies the fossil evidence and a radiocarbon ^{14}C timetable for the reconstruction of the history of farming. The study of the living plants identifies (by genetic tests) the wild ancestry of the crops, and uses the geographic distribution ranges of the wild progenitors to define the general areas in which domestication could have taken place. In addition, comparisons between the wild forms and their domestic counterparts reveal the changes that the crops underwent under domestication. Molecular surveys of the genetic polymorphism found in the crops and in their wild counterparts provide clues as to how and, more specifically, where these plants could have been domesticated.

In both disciplines, considerable information has already been assembled.[1,2,4] However, this evidence is very uneven. A few parts of the world (southwest Asia, Europe, Meso- and North America) have been extensively studied—both archaeologically and botanically. In these territories, critical information on the start of agriculture is already available, permitting a relatively safe evaluation. The archaeological exploration of several other large landmasses (East Asia, the Indian subcontinent, South America) is much scarcer. Yet the finds do provide some clues about the early history of farming in these territories. In still other large parts of the world (Africa south of the Sahara, most of the tropical belts of America and Asia), the archaeological evidence is still deplorably insufficient. (Yet, in tropical crops analyses of starch grains and phytoliths might soon change this picture.) All together, the information on the beginning and early expansion of agriculture is strongly skewed. At present, what can be reliably evaluated is not the full global picture, but only the relatively better-explored regions.

FOUNDER CROPS AND NUCLEAR AREAS

In spite of the large lacunae that still exist in the archaeobotanical information, the combined evidence from the excavations and from the living plants clearly shows that farming was independently initiated in several parts of the world. In each such separated, relatively small nuclear area, or cradle of agriculture, indigenous wild plants were taken into cultivation. They evolved into the first crops (the founder crops) that initiated food production in these core areas. Consequently the growing of these cultigens triggered the development of distinct, agricultural systems—each with its characteristic and largely unique crops. To date, the following five nuclear areas (Table 1) have been widely accepted by crop plant evolutionists and archaeologists:[2,4] 1) The Fertile Crescent belt in the Near East; 2) The valleys of the Yangtze River and of the Yellow River in China; 3) Southwest Mexico; 4) The Central Andes; 5) and the eastern United States. Additional nuclear areas have been proposed by various authors (for example, the Sahel belt and the Ethiopian highlands in Africa; and the Papuan highlands in New Guinea). However, in the view of the present author, the archaeo-botanical evidence essential for backing these claims is, as yet, far from being sufficient.

It is also worth mentioning that not in every traditional agricultural system was farming started de novo by

Encyclopedia of Plant and Crop Science
DOI: 10.1081/E-EPCS 120017091

Table 1 The five commonly accepted "nuclear areas" of the world (**1–5**), two examples of suspected ones (**6?–7?**), and the main founder crops that were uncovered in each of them[a]

Nuclear area	Characteristic founder crops	Earliest definite signs of farming	Resulting agricultural system
1 The Fertile Crescent belt in the Near East	Emmer wheat, einkorn wheat, barley, pea, lentil, flax	9500 ^{14}C yrs b.p. (=10,500 cal yrs b.p.)	S.W. Asia, the Mediterranean basin, and temperate Europe
2 Valleys of the Yangze River and the Yellow River	Asian rice, foxtail millet	8500 ^{14}C yrs b.p. (=9500 cal yrs b.p.)	East Asian farming
3 Southwest Mexico	Maize, squash, common bean	4700 ^{14}C yrs b.p. (=5300 cal yrs b.p.)	Mesoamerican farming
4 Central Andes	Potato, quinoa, common bean	4500 ^{14}C yrs b.p. (=5100 cal yrs b.p.)	High altitude farming in the Andes
5 Eastern United States	Goosefoot, sunflower, squash	4500 ^{14}C yrs b.p. (=5100 cal yrs b.p.)	Died out
6? The Sahel belt and/or the Ethiopian highlands in Africa	Sorghum, pearl millet, cow pea, African rice	Sites containing early contexts were not uncovered yet	The savanna belt south of the Sahara
7? New Guinea highlands	Bananas, sugar cane, taro	6100 ^{14}C yrs b.p. (=7000 cal yrs b.p.)	Tropical S.E. Asian farming

[a]Also listed are the ages before present (b.p.) of the earliest definite signs of farming in each core area (both radiocarbon ^{14}C age and calibrated, calendar age). For more information about ^{14}C age calibration consult Ref. 5, particularly the "Calibration Table for Radiocarbon Ages" on p. 253.

isolated, independent domestication of native founder crops. The available archaeo-botanical information clearly indicates that in some parts of the world, agriculture was initiated by the arrival, from outside, of fully domestic alien crops, while domestication of a rich variety of indigenous plants came only later. The Indian subcontinent seems to be a relatively well-documented example for this kind of development. There, food production seems to have been started by the arrival (from the West) of the Near Eastern package of crops.[1] Indigenous cultigens appeared in this landmass only later.

THE RISE AND SPREAD OF MEDITERRANEAN AGRICULTURE

In terms of the amount of information assembled on the origin of domesticated plants, the traditional region of Mediterranean agriculture (comprising Southwest Asia, the Mediterranean basin, and temperate Europe) is the best studied one. Its wild flora (including the identification of the ancestors of domesticated plants) is well recorded. Moreover, plant remains that have been expertly identified and radiocarbon dated are now available from hundreds of Neolithic and Bronze Age sites scattered all over this vast region. The earliest, definite signs of agriculture in this region were found in the Fertile Crescent belt of the Near East.[1,2,4] Here, a string of early Neolithic farming villages appeared at the second half of the 10th millennium before present (b.p.) uncalibrated radiocarbon ^{14}C time (=second half of the 11th millennium b.p. calibrated time). (For further information on calibration of radiocarbon ^{14}C ages consult.[5]) Moreover, this colonization intensified in the next thousand years. Remains retrieved from these Pre-Pottery Neolithic B early sites show that eight plants growing wild in the Fertile Crescent, namely emmer-type hard wheat, einkorn wheat, barley, pea, lentil, chickpea, bitter vetch, and flax had already entered domestication.[1] Common (and most revealing) are the remains of the wheat and barley. In contexts starting from 9500 ^{14}C yrs b.p. (=10,500 cal yrs b.p.) onward, forms with nonshattering ears appear. This is a reliable indicator of domestication. It shows that there and then these cereals were already grown as crops. By 8500–8000 ^{14}C yrs b.p. (=9500–8800 cal yrs b.p.) convincing signs of sheep, goat, cattle, and swine domestication appear, as well, and the Near Eastern Neolithic food production "package" was formed.

Once this package was assembled, and the early Neolithic farming villages were established, this new technology started to expand,[1] and it did so explosively. By 8000 ^{14}C yrs b.p. (=8800 cal yrs b.p.) this type of agriculture reached Greece. By 7000 ^{14}C yrs b.p. (=7800 cal yrs b.p.) it had already established itself in southern Italy, Serbia, the Caucasus, and Turkmenistan. Less than 800–1000 years later, grain agriculture (as well as the rearing of sheep, goat, and cattle) was already widely practiced in the loess soil belt of temperate Europe—from the Ukraine to northern France. More or less at this time, this new technology spread to the western parts of the Mediterranean basin, and to the Nile Valley, reaching also the Indus basin. All over these vast territories, agriculture was started by the introduction of the same Near Eastern founder crops. Additional Mediterranean plants, some native to other parts of this huge agriculture region, were incorporated only later. Most of the alien crops, domesticated outside the Mediterranean system, appeared much later.

Some three thousand years after the start of Neolithic grain agriculture, fruit tree cultivation (based on the invention of vegetative propagation) appeared in this region.[1] As with grain crops, the earliest convincing signs of fruit crop horticulture were found in the Near East. Here, the native olive, fig, grape vine, and date palm have been introduced into cultivation at Chalcolithic times, some 5500 ^{14}C yrs b.p. (=6300 cal yrs b.p.). Fruit growing, too, spread quickly. By the Early Bronze Age around 5000 ^{14}C yrs b.p. (=5700 cal yrs b.p.), olive, grape vine, and fig were already principal elements of food production in the Levant countries. Their cultivation was soon extended to the Aegean belt. Date palm groves flourished in Mesopotamia and the warm fringes of the Near East. In Egypt, date cultivation seems to have started somewhat later. Remains of its fruits appear in masses from the Middle Kingdom times onward. More or less at this time the cultivation of this palm extended eastward and reached the Indus Plain.

From the Bronze Age onward there are also sound indications of cultivation of vegetables.[1] Melon, watermelon, onion, garlic, leek, and lettuce were apparently the first vegetable crops grown in Egypt and Mesopotamia. Definite signs of their cultivation appear by 4500–3500 ^{14}C yrs b.p. (=5100–3800 cal yrs b.p.). By 2800–2000 cal yrs b.p. the list of Mediterranean and Southwest Asian vegetable crops had grown considerably, and beet, turnip, cabbage, radish, carrot, parsnip, celery, parsley, and asparagus had also entered cultivation. More or less at the same time (about 2400–2000 cal yrs b.p.), a second group of native fruit trees (those in which cultivation depends on grafting) were also added. Most conspicuous among them are the apple, pear, plum, cherry and carob trees. Contrary to the earlier crops that were almost all introduced into cultivation in the Near East, many of the vegetables and the later fruit trees were probably picked up not in the Near Eastern nuclear area but in other parts of the Mediterranean system of agriculture. Thus, starting in the early Neolithic and ending in classic times, an impressive assemblage of native crops were domesticated, and they diffused all over Southwest Asia, the Mediterranean basin, and temperate Europe. Most of them remained economically important until today.

MODE AND PLACE OF DOMESTICATION

Recently, the comparison of the amount of genetic polymorphism present in domesticated crops with that found in their wild progenitors provided effective tools for answering the following questions: 1) What was the mode of domestication of these founder crops? Were their wild progenitors introduced into cultivation only once, and in a single locality (and therefore had a single or monophyletic origin), or, alternatively, were they taken into cultivation several times and in different places (and therefore had a multiple or polyphyletic origin)? 2) Where (within the geographic distribution range of each of the wild progenitors) could domestication have taken place?

For elucidating the mode of domestication, one tries to assess what part of the genetic polymorphism found in the wild progenitor is also present in the crop.[6] In cases of a single event, only a limited fraction of the total genetic variation present in the wild progenitor should be expected to enter the domesticated gene pool. In contrast, when multiple domestications occur, a much larger fraction of wild genetic variation has a chance to enter the domestic gene pool. Indicative of this is also the nature of the genes controlling key domestication traits (i.e., traits that were automatically and immediately selected for once the wild progenitor was introduced into cultivation). If in all cultivars of the crop a given domestication trait is found to be governed by the same major gene (or the same combination of genes), this uniformity suggests a single origin. In contrast, when in different cultivars (within the crop) such a domestication trait is governed by different nonallelic mutations, one should suspect multiple domestications. Comparisons have already been made in the following Near East founder crops: einkorn wheat, emmer wheat, barley, lentil, and pea.[6–8] In all five founder crops, the available data suggest a monophyletic origin or, at most, very few domestication events.

For pinpointing the place of domestication, it is essential to test representative samples of the wild progenitor obtained from throughout its geographic range and to compare these samples with equally representative sampling of the domestic gene pool. The location where the progenitor's populations are genuinely wild and exhibit the closest genetic similarity to the crop should be suspected to be the place of domestication of the cultigen. In some of the Near Eastern founder crops (particularly in the wheats) extensive comparisons have already been carried out.[7,8] Thus the place of domestication of cultivated einkorn wheat has been pinpointed to a small area in Southeast Turkey. Also in Mesoamerica, maize was similarly surveyed.[2,6] Here populations of teosinte (the wild progenitor of domestic maize) showing full genetic similarity with the crop were found to be confined to a small territory in Southwest Mexico. In contrast, another basic crop of the New World, namely the common bean, apparently had a polyphyletic origin.[10] The available evidence convincingly shows that different wild forms of this bean were taken independently into cultivation, both in Meso- and in South America.

CONCLUSION

As previously noted, our knowledge of where, when, and how agriculture evolved is still fragmented and skewed. Yet the available evidence fully supports the notion that farming was independently initiated in several nuclear areas. In each such cradle of agriculture, indigenous, wild plants were taken into cultivation. Some evolved as successful founder crops, initiated farming, and later frequently triggered the development of a distinct agriculture system. Equally impressive is the fact that many of the earliest crops retained their central role in food production all through the history of agriculture. This is the case with wheat, barley, pea, and lentil, i.e., the main founder crops in the Near East, rice in China, maize and beans in Meso america, and (probably) bananas and taro in New Guinea. All seem to have founded agriculture; all are economically leading crops even today.

ARTICLES OF FURTHER INTEREST

Agriculture: Why and How Did It Begin?, p. 5
Archaeobotanial Remains: New Dating Methods, p. 55
Crop Domestication in Africa, p. 304
Crop Domestication in China, p. 307
Crop Domestication in Mesoamerica, p. 310
Crop Domestication in Prehistoric Eastern North America, p. 314
Crop Domestication in Southeast Asia, p. 320
Crop Domestication in Southwest Asia, p. 323
Crop Domestication in the American Tropics: Phytolith Analyses, p. 326
Crop Domestication: Role of Unconscious Selection, p. 340
Crop Domestication in the American Tropics: Starch Grain Analyses, p. 330

REFERENCES

1. Zohary, D.; Hopf, M. *Domestication of Plants in the Old World,* 3rd Ed.; Oxford Univ. Press: Oxford, United Kingdom, 2000.
2. Smith, B.D. *The Emergence of Agriculture*; Scientific American Library: New York, USA, 1995.
3. Harlan, J.R. *Crops and Man,* 2nd Ed.; American Society of Agronomy: Madison, WI, USA, 1992.
4. Diamond, J. *Guns, Germs and Steel: The Fate of Human Societies*; Jonathan Cape, Random House: United Kingdom, 1997.
5. Roberts, N. *The Holocene: An Environmental History,* 2nd Ed.; Blackwell: Oxford, United Kingdom, 1998.
6. Zohary, D. Monophyletic vs. polyphyletic origin of the crops on which agriculture was founded in the Near East. Genet. Resour. Crop Evol. **1999**, *46*, 133–142.
7. Heun, M.; Schäfer-Pregl, R.; Klawan, D.; Castagna, R.; Accerbi, M.; Borghi, B.; Salamini, F. Site of einkorn wheat domestication identified by DNA fingerprinting. Science **1997**, *278*, 1312–1314.
8. Salamini, F.; Özkan, H.; Brandolini, A.; Schäfer-Pregl, R.; Martin, W. Genetics and geography of wild cereals domestication in the Near East. Nat. Rev. Genet. **2002**, *3*, 429–441.
9. Doebley, J. Molecular evidence and the evolution of maize. Econ. Bot. **1990**, *44*, 6–27, (Suppl.).
10. Gept, P. Biochemical evidence bearing on the domestication of *Phaseolus* (Fabaceae) beans. Econ. Bot. **1990**, *44*, 28–38, (Suppl.).

Crop Domestication: Role of Unconscious Selection

Daniel Zohary
Hebrew University of Jerusalem, Jerusalem, Israel

INTRODUCTION

The following two types of selection are associated with domestication. They operate (and complement each other) when wild plants are introduced into cultivation: 1) conscious selection applied deliberately by the growers for traits of interest to them; 2) unconscious or automatic selection brought about by the fact that the plants concerned were picked from their original wild habitats and placed in a new (and frequently very different) human-made environment. This shift in the ecology led automatically to drastic changes in selection pressures. In response to the introduction of the plants into the anthropogenic environment, numerous adaptations vital for survival in the wild environment lost their fitness and broke down. New traits were automatically selected to fit the new conditions, resulting in the build-up of characteristic domestication syndromes—each fitting the specific agricultural conditions provided by the domesticators.

It is now widely accepted that unconscious and conscious selection are closely intertwined and played an important role in shaping many of the domestication traits that characterize crops and distinguish them from their wild ancestors. This article outlines the role of unconscious selection in crop plant evolution. It evaluates some of the principal changes in the environment that these wild plants were exposed to, once taken from their natural habitats and transferred into cultivation. It traces some of the main changes in selection pressures that could have been caused by this shift in ecology, and it points out some of the morphological, physiological, and chromosomal developments expected to have evolved in response to these environmental changes.

MAINTENANCE PRACTICES AND THEIR IMPACT

Two main practices are used by the cultivator to grow plants: 1) planting of seed and 2) vegetative or clonal propagation. The choice between these two agronomic methods is also the choice between two contrasting patterns of selection and evolution under domestication.

With very few exceptions (of apomictic crops), seed planting can be equated with sexual reproduction. Cultivated plants maintained by seed (the bulk of grain crops, numerous vegetable and truck crops) undergo a recombination-and-selection cycle every sowing. In other words, such crops have been subjected, under domestication, to hundreds (or even thousands) of cycles of selection.[1,2] They have been continually molded either as 1) clusters of inbred lines (in predominantly self-pollinated crops) or as 2) distinct cultivated races (in cross-pollinated plants). In numerous sexually reproducing crops, the results of such repeated selection are indeed impressive. Under domestication, these crops diverged considerably from their wild progenitors. At present, they are distinguished from them by complex syndromes of morphological, developmental, physiological, and biochemical traits.

In contrast, vegetatively propagated crops (most of the fruit crops and the tuber crops, some of the vegetables) have had an entirely different history of selection. Cultivars in these crops are not true genetic races, but just clonal replications of exceptional individuals, excelling in fruit or tuber qualities, which as a rule are also highly heterozygous. They were originally picked up from variable, panmictic, wild populations, and later also selected from among segregating progeny produced by the cultivars. In terms of selection, domestication of clonally propagated crops is largely a single step operation. With the exception of rare somatic mutations, selection is completed once a given clone is picked up. In traditional horticulture, the turnover of clones has been quite low, and appreciated genotypes were frequently maintained for long periods. Thus, clonal crops underwent, under domestication, only few recombination-and-selection cycles. In sharp contrast with sexually reproducing grain crops, their cultivars do not represent true breeding races, but only clones that, as a rule, segregate widely when progeny tested. Significantly, the large majority of such segregating progeny are economically worthless. Moreover, they often regress towards the mean found in wild populations, showing striking resemblance to the wild forms.

Seed and pollen fertility (including stable behavior of chromosomes in meiosis) are additional traits in which one finds wide differences between seed-planted crops and clonally propagated ones. In sexually reproducing populations, fertility is automatically safeguarded each generation by stabilizing selection. Mutations affecting fertility are promptly weeded out. As a rule, sexually

Encyclopedia of Plant and Crop Science
DOI: 10.1081/E-EPCS 120017092

reproducing cultivars are generally fully fertile. In contrast, the shift from sexual reproduction (in the wild) to vegetative propagation (under domestication) brings about drastic relaxation of the stabilizing selection that safeguards fertility.[1,2] Under such maintenance practice, sterile or semi-sterile clones are tolerated, including unbalanced chromosomal situations such as triploid, pentaploid, and aneuploid clones. Indeed, intracrop chromosomal polymorphism is quite common among clonally propagated crops. Besides, it should be pointed out that in clonally propagating crops, fertility is affected not only by the maintenance practice, but also by the choice of the desired parts (see next section).

THE PURPOSE FOR WHICH THE PLANT IS GROWN

Some cultivated plants are grown for their vegetative parts (roots, tubers, leaves, stems, etc.). In others, the reproductive parts (inflorescences, flowers, fruits, seeds) are used. The choice of the desired parts introduces automatically contrasting selection pressures, particularly in traits associated with the reproductive biology of the crops.[2]

As already noted, when crops are grown for their seeds, they are protected (like their wild relatives) by stabilizing selection, which safeguards their fertility. Grain crops have the most rigid protection of this kind. Yields in these sexually reproducing crops depend decisively on normal chromosome behavior in meiosis and on streamlined development of flowers, fruits, and seed. Deviants are promptly weeded out and the reproductive system is kept in balance. It is no wonder that among cultivated plants, grain crops are the most conservative in this regard. In spite of the fact that many of the grain crops have already produced hundreds (or even thousands) of generations under domestication, they show very little intracrop chromosome divergence or chromosome instability. With very few exceptions (such as the formation of hexaploid bread wheat), chromosome sets in grain crops are identical with those found in their wild progenitors, and wild and tamed forms are fully interfertile.

In contrast with the sexually reproducing grain crops, considerable reduction of pollen and seed fertility (as well as chromosome stability) is tolerated by the bulk of the vegetatively propagated crops grown for their fruits. Since in these crops the target of the grower is the fruit, the reproductive parts of the crop (inflorescences, flowers, fruits) are unconsciously kept intact. Yet, in fruit crop culture, the following conflict has to be resolved: The growers consciously select clones producing fleshy, tasty fruits. They are also equally attracted to seedless fruits or to fruits with reduced number of pips or stones. However, in most plants, the development of fruits commences only after fertilization and initiation of seed development. Several solutions for curtailing seed set without harming fruit formation evolved automatically in fruit trees.[2,3] Most conspicuous among them are mutations conferring parthenocarpy, that is, induction of fruit development without fertilization and without seed set (e.g., in bananas, common fig, and some pear cultivars).

Crops maintained by vegetative propagation and grown for their vegetative parts exhibit the most drastic disruption of their flowering and fruiting system, and the most bizarre chromosomal situations. In contrast to the fruit trees, conscious selection in these crops focuses on the increase of vegetative output. Because these crops are clonally propagated, this pressure is rarely counterbalanced by stabilizing selection to retain the reproductive organs. Root and tuber crops are outstanding examples of such evolution. Cultivated clones of cassava, yams, taro, sweet potato, or garlic frequently show drastic reduction of the amount of flowering. Some rarely produce any flowers. When rare flowers do appear they are frequently abnormal and sterile. Many of these crops are also exceptionally chromosomally variable, and frequently contain several ploidy levels and/or aneuploid number of chromosomes. Thus, in the yams, *Dioscorea alata* is known to contain all ploidy levels between $3x$ and $8x$, where as in *D. esculenta* $4x$, $6x$, $9x$ and, $10x$ cultivars are reported.[4] Sugarcane confronts us with an even more complex chromosome picture.[5] Cultivated clones in this crop are all highly polyploid, and frequently also aneuploid. Modern cultivars contain $2n = 100$ to $2n = 125$ chromosomes. Older cultivars vary from $2n = 80$ to $2n = 124$.

THE IMPACT OF SOWING AND REAPING

Traditional grain agriculture is based on sowing the seeds of the crop in a tilled field, harvesting the mature plants, and threshing out the grains. The introduction of grain plants into the system of tilling, sowing, and reaping triggers automatic selection towards the following changes,[1,2,6,7] setting them apart from their wild progenitors:

First, there is an automatic selection for retaining the mature seed on the mother plant, i.e., for the breakdown of the wild mode of seed dispersal. Most conspicuous is the shift from shattering spikes or panicles in the wild cereals to the nonshattering condition in their domesticated counterparts, or the parallel evolvement of nondehiscent pods in the cultivated legumes. The loss of the wild-type devices for seed dissemination is one of the most conspicuous outcomes of the introduction of grain plants into cultivation. It is also the most reliable indicator of domestication in grain crop remains retrieved from archaeological excavations.[3] Moreover, both theoretical considerations and experimental evidence suggest that at least in the wheats and in barley, the establishment of nonshattering mutants under the system of sowing and reaping could have been a fast process. It could have been accomplished in the course of only several scores of generations.[8]

A second major outcome of introducing wild grain plants into a regime of tilling, sowing, and reaping is the breakdown of the wild-type inhibition of germination. Wild plants are adapted to spread the germination of their seed over time. A common, vital adaptation (especially in annuals) is spreading the germination of the seed yield over several years.[3,6,9] Again, under a regime of cultivation, such germination inhibition is automatically selected against. Most grain crops have lost their wild-type regulation of germination. Practically all seed produced by their cultivars germinate immediately and synchronically upon imbibition.

Numerous other traits seem to have been automatically selected for once grain plants were introduced into the regime of tilling, sowing, and reaping.[6,9] Some are noted here for illustration. Under such practice, dense and uniform stands are frequently established in the tilled fields, and plants with erect habit will be favored. In response to the way the crops are harvested, synchronous ripening will be selected for. Because the seeds are stored (and protected) in granaries, thinner shells will evolve. Since seeds are sown deeper in the ground under tilling compared to the situation in the wild environment, increase in seed size would be expected. Since tilling enhances soil fertility, there is unconscious selection for increasing the number of fertile flowers in the inflorescences.

FRUIT TREE CULTIVATION AND ITS CONSEQUENCES

Most fruit trees under domestication are derived from cross-pollinated wild progenitors in which this pollination system is safeguarded either by self-incompatibility or by dioecy (male and female flowers borne on different individuals). Because of this background, the shift from sexual reproduction (in the wild) to the planting of vegetatively propagated clones, introduced serious limitations on fruiting.[10] Planting a single self-incompatible clone, or alternatively a female clone (or clones), would not bring about fruit set. Several horticultural inventions to assure fruit set were made by the growers (e.g., mixed planting of several genotypes in self-incompatible fruit crops, adding male individuals or practicing artificial pollination in dioecious crops). They were accompanied by unconscious selection for several types of mutations that resolved the restrictions set by self-incompatibility and/or sex determination.

In several self-incompatible crops (peach, apricot, sour cherry, as well as in several varieties of almond or olive), mutations appeared that caused the breakdown of self-incompatibility. In several originally dioecious species (such as grape vine and the carob), changes from dioecy to hermaphroditism evolved and rendered cross-pollination unnecessary. Finally, as previously mentioned, in numerous self-incompatible and dioecious fruit crops, pollination has been dispensed altogether by incorporation of mutations conferring parthenocarpy (fruit development without pollination and without seed development).

CONCLUSION

The present survey illustrates how conscious selection and unconscious selection have operated closely and complemented each other in crop plant evolution. Several innovative decisions made by the ancient farmers such as 1) in what ways to reproduce or propagate the crops, 2) how to raise and harvest them, and 3) which parts of the plants would be of use, were critical. They automatically set the chosen plants on different and contrasting courses of evolution under domestication.

ARTICLE OF FURTHER INTEREST

Crop Domestication: Founder Crops, p. 336

REFERENCES

1. Darlington, C.D. *Chromosome Botany and the Origin of Cultivated Plants*, 3rd Ed.; Allen and Unwin: London, UK, 1973.
2. Zohary, D. Modes of Evolution Under Domestication. In *Plant Biosystematics*; Grant, W., Ed.; Academic Press: Toronto, Canada, 1984; 579–596.
3. Zohary, D.; Hopf, M. *Domestication of Plants in the Old World*, 3rd Ed.; Oxford Univ. Press: Oxford, UK, 2000.
4. Hahn, S.K. Yams. *Evolution of Crop Plants*, 2nd Ed.; Smartt, J., Simmonds, N.W., Eds.; Longman: Harlow, UK, 1995; 112–120.
5. Roach, B.T. Sugarcanes. In *Evolution of Crop Plants*, 2nd Ed.; Smartt, J., Simmonds, N.W., Eds.; Longman: Harlow, UK, 1995.
6. Heiser, C.B. Aspects of unconscious selection and the evolution of domesticated plants. Euphytica **1988**, *37*, 77–81.
7. Harlan, J.R. *Crops and Man*, 2nd Ed.; American Society of Agronomy: Madison, WI, USA, 1992.
8. Hillman, G.C.; Davies, M.S. Measured domestication rates in wild wheat and barley under primitive cultivation and their archaeological implications. J. World Prehist. **1990**, *4*, 157–222.
9. Harlan, J.R.; de Wet, J.M.J.S.; Price, G. Comparative evolution of cereals. Evolution **1973**, *27*, 322–325.
10. Zohary, D.; Spiegel-Roy, P. Beginning of fruit growing in the Old World. Science **1975**, *187*, 319–327.

Crop Improvement: Broadening the Genetic Base for

Toby Hodgkin
International Plant Genetic Resource Institute, Rome, Italy

C

INTRODUCTION

Crop improvement occurs through selection operating on genetic variability and depends on the continuing availability and use, by plant breeders, of sufficient desirable genetic variation. The difficulty in many plant breeding programs is to determine what constitutes "sufficient desirable genetic variation." Too much variation can become unmanageable, while too little will result in a failure to obtain any significant advances from selection.

In crops, the amount of variability available for selection is limited by comparison with their wild relatives. The process of domestication has usually involved a dramatic reduction in the genetic diversity of the crop genepool. This limitation in the amount of diversity within the crop has often been followed by further losses of genetic diversity in specific areas or materials—as crops were taken to new geographic areas, as the types desired by farmers became more narrowly determined, as selection to deal with specific disease outbreaks was required, or as plant breeders tended to restrict the materials they used to meet increasingly rigorous standards of performance and uniformity.

In its widest sense, base broadening includes increasing the amount of genetic variation used in cultivar selection programs and increasing the genetic diversity present in production systems (through increasing the range of diversity available to farmers and increasing the numbers of crops or cultivars grown in production systems). In this article, various ways of determining the need for base broadening will be illustrated and some methods for increasing the amounts of diversity in plant breeding programs, will be briefly described. A much fuller treatment of crop improvement issues and aspects of base broadening in production systems has recently been provided by Cooper, Spillane, and Hodgkin.

THE NEED FOR BASE BROADENING

This continuous narrowing of the genetic base of a crop is widely believed to be undesirable and even to have had dramatic consequences on production and human well-being. It has been blamed for the Irish potato famine, caused by late blight, and for the maize southern leaf blight epidemic. This latter event led to a report by the U.S. Academy of Sciences that recommended broadening the genetic base of major staple crops. Despite this recommendation, there appear to have been few substantial efforts to develop and implement base-broadening programs, and modern cultivar production often continues to depend on recycling the variations present in selected elite materials.

Evidence that there is a need (or, perhaps better, an opportunity) for base broadening in a crop can come from the identification of bottlenecks, from the detection of limited amounts of genetic diversity, or from production characteristics suggesting that variation available for improvement has become limiting.

Bottlenecks

The first, and often the most significant bottleneck in crop plants occurs on domestication. In many crops there are very large amounts of genetic diversity present in the nearest related wild taxa.[1] Domestication is usually accompanied by substantially reduced gene flow between the crop and related wild taxa, even where they belong to the same primary genepool. In many crops there are no wild relatives in the primary genepool, and the barrier to gene flow is virtually complete (e.g., *Triticum aestivum*—bread wheat; *Brassica napus*—oilseed rape; *Cocos nucifera*—coconut). Other crops considered to have a narrow genetic base because of domestication bottlenecks include rice, durum wheat, *Phaseolus* beans, tomato, pigeon pea, and chick pea.

Significant bottlenecks have also occurred when crops have been taken to new production locations. These founder effects result in a set of cultivars drawn from a restricted sample of the total diversity of a crop. A classic example of this has been the introduction of potato in W. Europe. Other examples include maize in W. Africa, soybean in the United States, lentil in S. Asia, and tropical commodity crops such as cocoa, coffee, and rubber.

The development of specific crop types to satisfy particular needs has also resulted in a noticeable reduction in diversity in different types and limited gene flow between them. For example, as the different horticultural *Brassica oleracea* crop types became established (e.g., cabbage, Brussels sprout, cauliflower, kohlrabi, calabrese)

DOI: 10.1081/E-EPCS 120010456

they ceased to be intercrossed, and some of the genetic diversity found in one of the different crops no longer occurs in others. This effect becomes more extreme as selection becomes more intense. Highly specialized crop types such as spring-malting barley may show significant loss of diversity, even when compared with the full range of modern barley cultivars—let alone when compared with local cultivars from the crop's center of origin. The narrow genetic base of U.S. maize is a classic example in this respect.

Genetic Diversity

Studies on the amount and distribution of genetic diversity can provide an important way of quantifying the effect of a known or believed bottleneck. Of course, such studies could also indicate when the effect of an identifiable bottleneck has been limited by subsequent gene flow. However, most of the investigations to date, confirm substantial reductions in the diversity of modern cultivar material, as compared with traditional cultivars or wild relatives. Miller and Tanksley[4] found that diversity in RFLPs of tomato cultivars was virtually absent compared with traditional cultivars and with other *Lycopersicon* species. Allard[5] noted that modern California cultivars of barley had half the number of alleles per locus found in M. Eastern traditional cultivars, which themselves had half the number of alleles found in wild *H. spontaneum*. Sonnante et al.[6] detected a similar pattern in the amounts of diversity found in different *Phaseolus* bean germplasm classes.

In the United States, a number of crops including wheat rice have a narrow genetic base, as indicated by their coefficient of parentage.[7] Current varieties derive from a very limited number of parents, and over 70% of the diversity found in U.S. soybean is estimated to be derived from just seven parents. In other cases, such as oats in the United States or soybeans in China, the genetic base is much broader, and in China it is estimated that 70% of the genetic diversity found in current soybean cultivars can be attributed to more than 70 parents. Of course, a broad genetic base in plant breeding, obtained through the use of large numbers of parents, may not of itself lead to diversity in production systems. This requires a breeding program that produces large numbers of genetically distinct varieties, as measured by direct genetic diversity studies.

While it has been suggested that the genetic base will continue to erode, this does not seem always to be the case. In the United Kingdom, the amount of diversity present in nationally listed wheat cultivars appears to have remained largely unchanged over the last 70 years, since the 1930s.[8]

Production Problems

A number of aspects related to production characteristics are considered to indicate that there may be a need for base-broadening actions. The most substantial of these is evidence that a yield plateau has been reached and that little further progress is being made in crop performance. Continued crop failure through vulnerability to specific diseases or pests may also indicate an absence of desired variability in breeding material. In these cases, breeders often tend to look to more exotic material (e.g., wild relatives) to find genes that meet their needs. This may even involve using the tertiary genepool and in vitro culture to obtain desired traits.

Farmers and consumers can themselves provide indicators that base broadening is needed. If there is clear evidence that farmers' cultivar needs are not being met or that the crop is becoming unprofitable for growers who are failing to obtain adequate returns, this may be because necessary characteristics are not present in breeding material.

OPTIONS FOR BROADENING THE BASE

As a result of the international efforts to conserve plant genetic resources, there is available to the plant breeder a very substantial amount of desirable genetic diversity in genebanks throughout the world. There may be problems in accessing the material, and it may be difficult for breeders to know what is the most useful material to include in breeding programs, but for most major crops, obtaining diversity is not the biggest problem.

Often, the plant breeder's preferred approach when producing new cultivars will be introgression of the desired new trait into an elite background, incorporating as little additional genetic material as possible. This is particularly the case when the material with the desired trait belongs to the secondary or tertiary genepool. However, this process does not necessarily result in a significantly broader genetic base for the crop. Indeed, one aim of such a program is usually to limit introgression to the single desired gene, and molecular markers are substantially improving the efficiency of this process. Base-broadening approaches are different, in that they seek to deliberately enhance the amount of variation in the breeding programs. A number of ways to achieve this objective in a more or less systematic way have been described.

Population Management

Populations can be established by crossing a number of genetically different parents, often from different geographic areas so as to maximize their distinctness. These populations are then commonly established at a number of different locations and allowed to adapt to new conditions with minimum artificial selection. Examples of these include the barley composite cross populations[5] and the dynamic wheat populations developed in France.[9]

Incorporation

Simmonds[10] describes procedures that combined repeated cycles of recombination and mild selection in the target environment on sets of extremely diverse material. The objective is the wide-scale incorporation of genetic variability into existing crop material so as to enhance diversity and provide a basis for the identification of new, useful variation. Incorporation programs have been developed for potato and sugar cane, and in the latter have led to the development of much useful germplasm.

Germplasm Enhancement

A number of procedures have been developed which allow breeders to identify and include new material in breeding programs in a systematic way. These include the Genetic Enhancement of Maize (GEM) program and the Hierarchical Open-Ended (HOPE) system for broadening the breeding base, also in maize. Similar approaches have been developed for maize in France and for sugar beet in Europe.

Traditional and Participatory Breeding

Conscious inclusion of traditional cultivars from a wide genetic base, as in barley at ICARDA[12] and by CIMMYT in its wheat breeding work,[12] will materially increase diversity in the breeding program. While there is much that can be done in this way, the base-broadening effect will be rather variable depending on the parents chosen and selection pressure used. The increasing interest in participatory plant breeding is also likely to result in increased variability in production systems and, potentially, a broader base for crop improvement.[13]

REFERENCES

1. Harlan, J.R.; de wet, J.M.J. Toward a rational classification of cultivated plants. Taxon **1970**, (20), 509–517.
2. National Research Council. *Genetic Vulnerability of Major Crops*; National Academy of Sciences: Washington, DC, 1972.
3. Cooper, H.D.; Spillane, C.; Hodgkin, T. *Broadening the Genetic Base of Crop Production*; CAB International: Wallingford, United Kingdom, 2001; pp. 452+xxii.
4. Miller, J.C.; Tanksley, S.D. RFLP analysis of phylogenetic relationships and genetic variation in the genus *Lycopersicon*. Theor. Appl. Genet. **1990**, (80), 437–448.
5. Allard, R.W. Predictive Methods of Germplasm Identification. In *Plant Breeding in the 1990s*; Stalker, H.T., Murphy, J.P., Eds.; CAB International: Wallingford, United Kingdom, 1992; 119–146.
6. Sonnante, G.; Stockton, T.; Nodali, R.O.; Becera Velasquez, V.L.; Gepts, P. Evolution of genetic diversity during the domestication of the common-bean (*Phaseolus vulgaris* L.). Theor. Appl. Genet. **1994**, (89), 629–635.
7. Gizlice, Z.; Carter, T.E., Jr.; Gerig, T.M.; Burton, J.W. Genetic diversity patterns in North American public soybean cultivars based on coefficient of parentage. Crop Sci. **1996**, (36), 753–765.
8. Donini, P.; Law, J.R.; Koebner, R.M.D.; Reeves, J.C.; Cooke, R.J. Temporal trends in the diversity of UK wheat. Theor. Appl. Genet. **2000**, (100), 912–917.
9. Goldringer, I.; Enjalbert, J.; David, J.; Paillard, S.; Pham, J.L.; Brabant, P. Dynamic Management of Genetic Resources: A 13-Year Experiment on Wheat. In *Broadening the Genetic Base of Crop Production*; Cooper, H.D., Spillane, C., Hodgkin, T., Eds.; CAB International: Wallingford, United Kingdom, 2001; 245–260.
10. Simmonds, N.W. Introgression and incorporation. Strategies for the use of crop genetic resources. Biol. Rev. **1993**, (68), 539–562.
11. Ceccarelli, S. Specific adaptation and breeding for marginal conditions. Euphytica **1994**, (77), 205–219.
12. Rejesus, R.M.; Smale, M.; Van Ginkel, M. Wheat breeders' perspectives on genetic diversity and germplasm use: Findings from an international survey. Plant Var. Seeds **1996**, (9), 129–147.
13. Witcombe, J.R. Does Plant Breeding Lead to a Loss of Genetic Diversity? In *Agrobiodiversity—Characterization, Utilization and Management*; Wood, D., Lenne, J., Eds.; CABI International: Wallingford, United Kingdom, 1999; 245–274.

Crop Responses to Elevated Carbon Dioxide

Leon Hartwell Allen, Jr.
USDA, Gainesville, Florida, U.S.A.

P. V. Vara Prasad
University of Florida, Gainesville, Florida, U.S.A.

INTRODUCTION

Atmospheric carbon dioxide (CO_2) concentration has increased from 280 ppm (parts per million, mole fraction basis) in preindustrial times to 370 ppm today. As concentrations of CO_2 and other greenhouse gases rise, global temperature is anticipated to increase.[1] Elevated CO_2 will improve crop yields due to increased photosynthesis. However, at above-optimum temperatures for reproductive growth processes, the benefits of elevated CO_2 could be overwhelmed by negative effects of high temperature, leading to lower seed yield.

The extent of growth and yield responses of plants to elevated CO_2 depends on the photosynthetic pathway. Crops with C_3 photosynthesis will respond markedly to increasing CO_2 concentrations. Common C_3 crops are small grain cereals (wheat, rice, barley, oat, and rye); grain legumes or pulses (soybean, peanut, various beans and peas); root and tuber crops (potato, cassava, sweet potato, sugar beet, yams); most oil, fruit, nut, vegetable, and fiber crops; and temperate-zone (cool-climate) forage and grassland species. In contrast, plants with C_4 photosynthesis will respond little to rising atmospheric CO_2 because a mechanism to increase the concentration of CO_2 in leaves causes CO_2 saturation of photosynthesis at current ambient concentrations. Common C_4 crops are maize (corn), sugarcane, sorghum, millet, and many tropical and subtropical zone (warm-climate) grass species. This article focuses on responses to elevated CO_2 and increased temperature of C_3 crops. Response patterns are similar, but not the same, across a broad range of species and conditions.[2]

EFFECTS OF CO_2 AND TEMPERATURE

Photosynthesis and Respiration

Doubling of CO_2 concentration will increase photosynthesis of C_3 crop species by 30–50%.[2–4] The primary enzyme in leaf photosynthesis of C_3 plants, ribulose 1,5-bisphosphate carboxylase/oxygenase (Rubisco), can bind to either CO_2 or O_2. An increase in the concentration of CO_2 enables this molecule to better compete with dissolved O_2 for binding sites on the Rubisco protein, thus leading to an increase of photosynthesis of C_3 species. The CO_2 concentrating mechanism of C_4 plants is mediated by the enzyme phosphoenolpyruvate carboxylase (PEPcase). The contrasting effect of CO_2 on photosynthesis of C_3 and C_4 plants is illustrated in Fig. 1. Response curves of photosynthesis versus CO_2 are nonlinear, and little benefit will accrue above 700 ppm.

The hypothesis that elevated CO_2 has a direct, immediate effect in decreasing the respiration rate of plants seems to have little basis. However, the indirect, long-term effect of elevated CO_2 can cause an increase in respiration via an increase in the amount of living biomass. Rice plants grown in CO_2 ranging from 160 to 900 ppm had respiration rates directly proportional to the total nitrogen content (protein content) of the plant.[5] However, elevated temperatures can increase plant dark respiration rates regardless of CO_2 concentration. Furthermore, elevated temperature decreases solubility of CO_2 relative to O_2 in the cytosol, thereby decreasing photosynthesis, but this solubility effect on photosynthesis is usually offset more in high CO_2 than in ambient CO_2.

Stomatal Conductance, Transpiration, and Water Use

Increasing CO_2 causes partial closure of stomata, the small pores (formed by slits between two flexible guard cells) on leaves that govern photosynthetic CO_2 uptake and transpiration (water vapor loss). Stomatal conductance for water vapor decreases about 40% for a doubling of CO_2. Decreased stomatal conductance decreases transpiration of leaves, but not in direct proportion to the decrease of stomatal conductance because leaf temperature increases by 1–2°C in doubled CO_2 due to decreased evaporational cooling. In turn, vapor pressure of water inside leaves increases and causes a greater leaf-to-air vapor pressure difference, which is the driving force for transpiration. This effect partially offsets decreased stomatal conductance, and thus whole-crop transpiration

Encyclopedia of Plant and Crop Science
DOI: 10.1081/E-EPCS 120005566

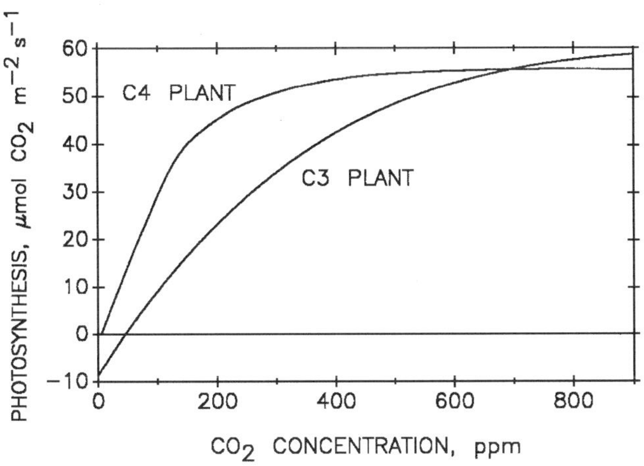

Fig. 1 Typical leaf photosynthetic rate responses of C_3 and C_4 plants to CO_2 concentration when measured in non-limiting (high light) conditions.

is maintained only slightly lower (10%) than would exist at ambient CO_2.[6]

Although crop transpiration might decrease slightly in elevated CO_2, water use will increase if temperatures rise. Fig. 2 shows the increase of average daily transpiration of a soybean crop with increasing temperature at two levels of CO_2. The reduction in water use by doubled CO_2 was about 9% at the mean temperature of 23°C. Crop water use might increase about four-fold over the average daily temperature range of 20–40°C. Therefore, small increases in temperatures would more than offset the water-saving effect of CO_2 via reduced stomatal conductance.[6]

Shoot and Root Growth

Crops exposed to elevated CO_2 generally grow larger.[2] Plants such as soybean have a higher percentage of total biomass in stems to support leaves and seed pods. Crops such as rice and wheat produce a larger number of tillers, which leads to greater yield because of the greater number of seed heads per plant. Leaves may be larger or thicker and accumulate more starch, especially for plants like soybean.

Elevated temperatures may either increase or decrease the vegetative biomass production of crops. Vegetative biomass of warm-climate species or cultivars of forages, sugarcane, soybean, and peanut may increase slightly with temperature increases, whereas vegetative biomass of cool-climate cultivars tends to decrease with increasing temperature.

Elevated CO_2 generally increases biomass, volume, and length of roots, as well as increasing biomass allocation to roots (increased root-shoot ratio). Root and tuber crops tend to have a greater yield response to elevated CO_2 than seed or forage crops. Increased photosynthesis also favors symbiotic nitrogen fixation in legumes. Since legumes can supply nitrogen via symbiotic nitrogen fixation, crop legumes (both seed and forage crops) might respond relatively more to a rise in CO_2 concentration than non-legumes.

Seed Yield and Quality

Seed yields generally increase nonlinearly in response to increasing CO_2, but this increase is not quite as much as the increase in photosynthesis.[2] Part of the additional carbon fixed goes into producing more plant vegetative biomass. Increases in seed yields of many C_3 crops range between 20% and 35%,[3] whereas increases for C_4 crops are only about 10% to 15%. Elevated CO_2 may cause higher carbohydrate and lower nitrogen content of small cereal grains, but no changes tend to occur in grain legumes.[7] Although wheat and barley showed increases in seed numbers (about +15%) in elevated CO_2, seed N concentration was even more strongly reduced (about −20%). Under limiting water or nutrient conditions, relative yield responses to elevated CO_2 may increase, although absolute yields will decrease.

Increasing temperature is detrimental to seed production, as illustrated for tropical lowland rice and kidney bean in Fig. 3.[4,8] The quantitative responses of seed yield reduction to increasing temperature vary among species and crop cultivars, but the pattern is the same. Each crop has an optimum temperature for reproductive growth processes. Seed yields decline about 10% per °C to zero at about 10°C above the optimum temperature. Seed yields decline to zero at about 32°C for a cool-climate cultivar of kidney bean, 36°C for tropical lowland rice, and 40°C for warm-climate cultivars of peanut and soybean. Fig. 3

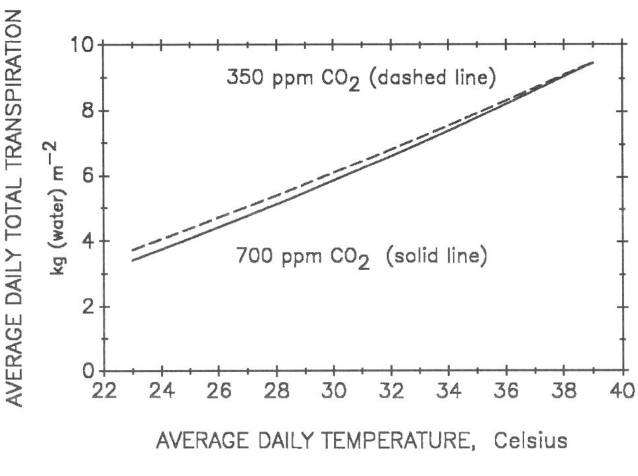

Fig. 2 Typical average daily whole-crop transpiration of C_3 plants when grown at two levels of CO_2 and across a mean daily temperature range of 20–40°C. (Adapted from Ref. 6.)

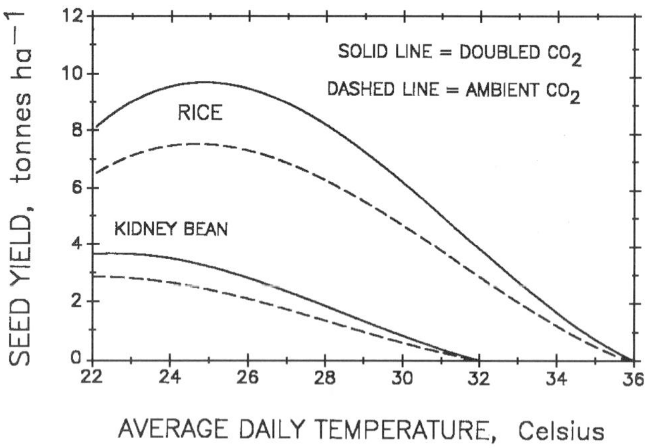

Fig. 3 Typical seed yields of rice and kidney bean at two CO_2 concentrations (ambient and double ambient) and across a range of temperatures. (Adapted from Refs. 4 and 8.)

shows that elevated CO_2 does not offset the decline of seed yield with increasing temperature.

Crops are especially sensitive to elevated temperature from a few days before pollen maturation through fertilization of the ovule.[9] Important processes during this period are viable pollen production, pollen shedding, pollen tube growth, and fertilization. Crops may also be sensitive to temperature during seed-filling processes, the time when the seeds load up with proteins, carbohydrates, oils, and other nutrients.[7] Soybean and kidney bean seeds increasingly fail to fill properly as temperatures increase and form smaller, shriveled seeds with reduced seed germination capability and nutritional quality than under optimum temperatures.

CONCLUSION

Photosynthesis and growth of C_3 crops are increased when grown at high CO_2; however, the extent of stimulation varies with temperature among species and cultivars. The potential decrease in transpiration caused by partial closure of stomata in elevated CO_2 is largely negated by the energy balance between the crop and environment, which results in similar total water use in similar climatic conditions. Seed yields are increased by elevated CO_2 under optimal temperature. However, at supra-optimal temperature, seed yields are decreased under both ambient and elevated CO_2.[2,5,8] If increases in temperature accompany increases in CO_2 concentration, seed yields will decrease in regions where temperatures are at or above optimum. Future research should be directed toward identifying high-temperature tolerant cultivars that can produce more seeds under harsh climatic conditions.

ARTICLES OF FURTHER INTEREST

Air Pollutants: Interactions with Elevated Carbon Dioxide, p. 17
Crops and Environmental Change, p. 370
Drought and Drought Resistance, p. 386
Ecophysiology, p. 410
Leaves and Canopies: Physical Environment, p. 646
Leaves and the Effects of Elevated Carbon Dioxide Levels, p. 648
Osmotic Adjustment and Osmoregulation, p. 850
Photosynthate Partitioning and Transport, p. 897
Photosynthesis and Stress, p. 901
Seed Vigor, p. 1139
Water Use Efficiency Including Carbon Isotope Discrimination, p. 1288

REFERENCES

1. Rosenzweig, C.; Hillel, D. *Climate Change and the Global Harvest*; Oxford University Press: New York, 1998.
2. *Climate Change and Global Crop Productivity*; Reddy, K.R., Hodges, H.F., Eds.; CABI Publishing: Oxon, UK, 2000.
3. Ainsworth, E.A.; Davey, P.A.; Bernacchi, C.J.; Dermody, O.C.; Heaton, E.A.; Moore, D.J.; Morgan, P.B.; Naidu, S.L.; Ra, H.S.Y.; Zhu, X.G.; Curtis, P.S.; Long, S.P. A meta-analysis of elevated CO_2 effects on soybean (*Glycine max*) physiology, growth and yield. Glob. Chang. Biol. **2002**, *8* (8), 695–709.
4. Baker, J.T.; Allen, L.H. Contrasting Crop Species Responses to CO_2 and Temperature: Rice, Soybean and Citrus. In *CO_2 and Biosphere*; Rozema, J., Lambers, H., van de Geijn, S.C., Cambridge, M.L., Eds.; Kluwer Academic Publishers: Dordrecht, 1993; 239–260.
5. Baker, J.T.; Laugel, F.; Boote, K.J.; Allen, L.H. Effects of daytime carbon dioxide concentration on dark respiration of rice. Plant Cell Environ. **1992**, *15* (2), 231–239.
6. Allen, L.H.; Pan, D.; Boote, K.J.; Pickering, N.B.; Jones, J.W. Carbon dioxide and temperature effects on evapotranspiration and water-use efficiency of soybean. Agron. J. **2003**, *95* (4), 1071–1081.
7. Jablonski, L.M.; Wang, X.Z.; Curtis, P.S. Plant reproduction under elevated CO_2 conditions: A meta-analysis of reports on 79 crop and wild species. New Phytol. **2002**, *156* (1), 9–26.
8. Prasad, P.V.V.; Boote, K.J.; Allen, L.H.; Thomas, J.M.G. Effects of elevated temperature and carbon dioxide concentration on seed-set and yield of kidney bean (*Phaseolus vulgaris* L.). Glob. Chang. Biol. **2002**, *8* (8), 710–721.
9. Prasad, P.V.V.; Craufurd, P.Q.; Kakani, V.G.; Wheeler, T.R.; Boote, K.J. Influence of high temperature during pre- and post-anthesis stages of floral development on fruit-set and pollen germination in peanut. Aust. J. Plant Physiol. **2001**, *28* (3), 233–240.

Cropping Systems: Irrigated Continuous Rice Systems of Tropical and Subtropical Asia

Achim Dobermann
Kenneth G. Cassman
University of Nebraska, Lincoln, Nebraska, U.S.A.

INTRODUCTION

Irrigated rice is grown in bunded, puddled fields with assured irrigation for one or more crops a year on alluvial floodplains, terraces, inland valleys, and deltas in the humid and subhumid subtropics and humid tropics of Asia. Favorable climatic conditions and fertile soils in combination with irrigation allow farmers to grow one to three crops per year in submerged soil. The irrigated rice ecosystem accounts for 55% of the global harvested rice area and 75% of the world's annual rice production.

By 2020, average yields of irrigated rice must rise by about 20 to 25% to meet expected demand. However, growth rates of both yield and total irrigated rice production have slowed down in recent years, raising concerns about the sustainability of intensive irrigated rice systems and future rice supply.

Future yield increases will require germplasm with increased yield potential and substantial improvements in soil and crop management—particularly with regard to nutrient, water, and pest management—to lift average farm yields to about 70% of the yield potential. Technological advances to achieve such improvements must be synergistic with dynamic changes in the socioeconomic and biophysical environment in Asia, where competition for natural and human resources continues to intensify. This article characterizes intensive, irrigated rice systems and discusses the critical current and future challenges to crop improvement and management.

IMPORTANCE, GEOGRAPHICAL DISTRIBUTION, AND PRODUCTION TRENDS

Worldwide, about 79 million hectares (ha) of rice (*Oryza sativa* L.) is grown under irrigated conditions (55% of the global harvested area), accounting for about 75% of the annual rice production. Irrigated rice-based cropping systems include single-crop rice, rice–upland crop double cropping, or continuous monoculture with two to three rice crops per year.[1] Double- and triple-crop monoculture systems occupy a land area of about 24 million ha in Asia, allowing 49 million ha of rice to be harvested annually.[2] These systems account for 40% of the global rice supply and feed about 1.8 billion Asians. Rice accounts for 30 to 80% of the calories consumed in most countries of Asia. The ability to produce a rice surplus on good irrigated land has contributed much to the economic development and political stability in that region.

Continuous rice systems are mostly located on flood plains along major rivers, terraces, inland valleys, and coastal plains in the humid and subhumid subtropics and tropics. Rainfall ranges from 1000 to more than 2000 mm per year. The warm climate and access to water allow farmers to grow two to three short-duration rice crops per year (Fig. 1). Rice is grown in dry seasons with low rainfall and high solar radiation and in humid seasons with lower yield potential due to cloudy conditions and high rainfall. Soils vary widely, but are mostly of relatively high quality. Common soil types include Inceptisols, Alfisols, Entisols, Vertisols, Mollisols, and Ultisols. Irrigated rice is also grown on more marginal soils that have various problems, which may cause mineral nutrition deficiencies or toxicities in some cases.[3]

Double cropping became common in China about 1000 years ago, and triple cropping probably started in the 14th century.[4] Intensification since the mid 1960s has involved an increase in the number of crops grown per year and greater yield per crop cycle. Higher yields have resulted from the combination of increased yield potential of modern varieties, improved crop nutrition made possible by fertilizer application, and improved host-plant resistance and pest management.[5] The current average grain yield of irrigated rice is about 5.3 mega grams per hectare (Mg ha^{-1}) per crop. Growth rates of yield and total irrigated rice production have slowed down, but they vary among countries and regions (Fig. 2). From 1967 to 1984, rice production grew at an annual rate of 3.2%, mainly because of yield increases (2.5% yr^{-1}), but declined to 1.5% yr^{-1} (production) and 1.2% yr^{-1} (yield) during the period from 1984 to 1996. This slowdown is partly due to lower rice prices and the slowdown in demand growth, but concern was also raised about resource degradation. Yield declines were observed in

Encyclopedia of Plant and Crop Science
DOI: 10.1081/E-EPCS 120010544

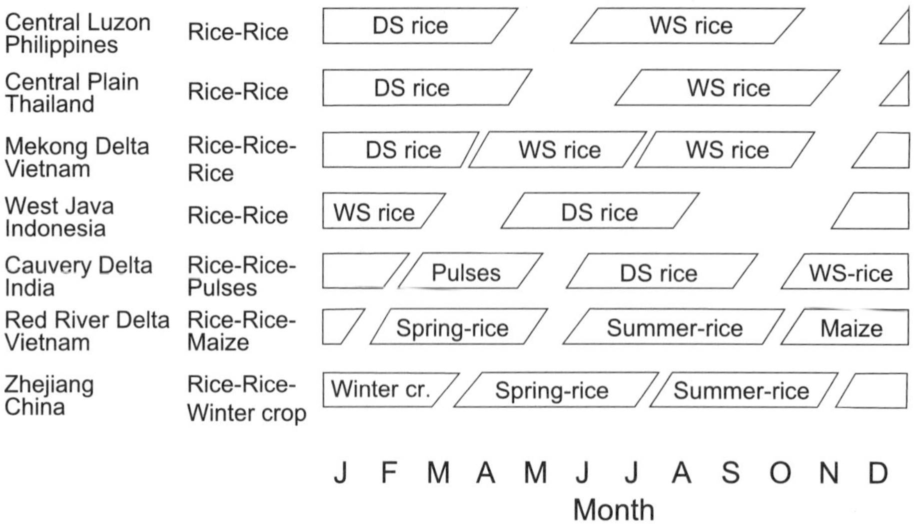

Fig. 1 Cropping systems and the annual cropping calendar in major irrigated rice areas of south, southeast, and east Asia. Each site shown represents a large irrigated area in which intensive rice cropping is the dominant agricultural enterprise. DS = dry season; WS = wet season.

some long-term research experiments, but appear not to be widespread at current production levels.[6] In some countries and large rice production domains where farmers were early adopters of modern irrigated rice production technologies, yields have stagnated since the mid-1980s (Fig. 2).

SOIL AND CROP MANAGEMENT

Germplasm

Nearly all irrigated rice is produced with modern rice varieties. The release of high-yielding rice varieties such as IR8 (1966), IR20 (1969), IR36 (1976), IR64 (1985), and IR72 (1988) helped fuel the Green Revolution in Asia. The new varieties provided a quantum leap in yield potential compared to the traditional land races they replaced because of shorter stature, lodging resistance, greater harvest index, and responsiveness to nitrogen (N). Due to their short growth duration (95 to 115 days) they became a key factor in the expansion of double- and triple-crop rice systems. Rice hybrids with a 10% increase in yield potential above that of the best inbred varieties presently account for about 18 million ha (50%) of the Chinese rice harvest area. Adoption of hybrids is also beginning to occur in other countries, in a total of about 0.5 million ha.

Breeding programs during the past 30 years have focused on incorporating disease and insect resistances into irrigated rice varieties, which has increased yield stability under continuous cropping. Although there has been little increase in the yield potential of inbred rice varieties since IR8 was introduced, efforts are in progress to develop a new plant type with a 25% larger yield potential (12 to 12.5 Mg ha^{-1} in tropical regions) compared to the best inbred indica varieties.[7] Another key challenge is to combine the traits of high yield with high grain quality. Genetic engineering and molecular breeding techniques have been used to improve host-plant resistance to pests, or, more recently, to improve specific grain-quality traits such as vitamin A or iron content.

Nutrient Management

A rice crop yielding 6 Mg ha^{-1} takes up about 105 kg N ha^{-1}, 18 kg P ha^{-1}, 100 kg K ha^{-1}, 11 kg S ha−1, and 0.3 kg Zn ha^{-1}.[3] About 40% of the N, 80–85% of the K, 30–35% of the P, 40–50% of the S, and 60% of the Zn absorbed by rice remains in vegetative plant parts at maturity. For centuries, naturally occurring sedimentation, nutrient inflow by irrigation, organic residues and amendments, biological N_2 fixation, and carbon assimilation by floodwater flora and fauna[8] maintained soil fertility in the relatively low-yielding traditional irrigated rice systems. With intensification, annual crop nutrient removal has increased and mineral fertilizers have become the primary nutrient source. Nitrogen-fixing green manures are not widely used because their main purpose of providing N has been replaced by fertilizer N. Straw is the major organic material available to most rice farmers,

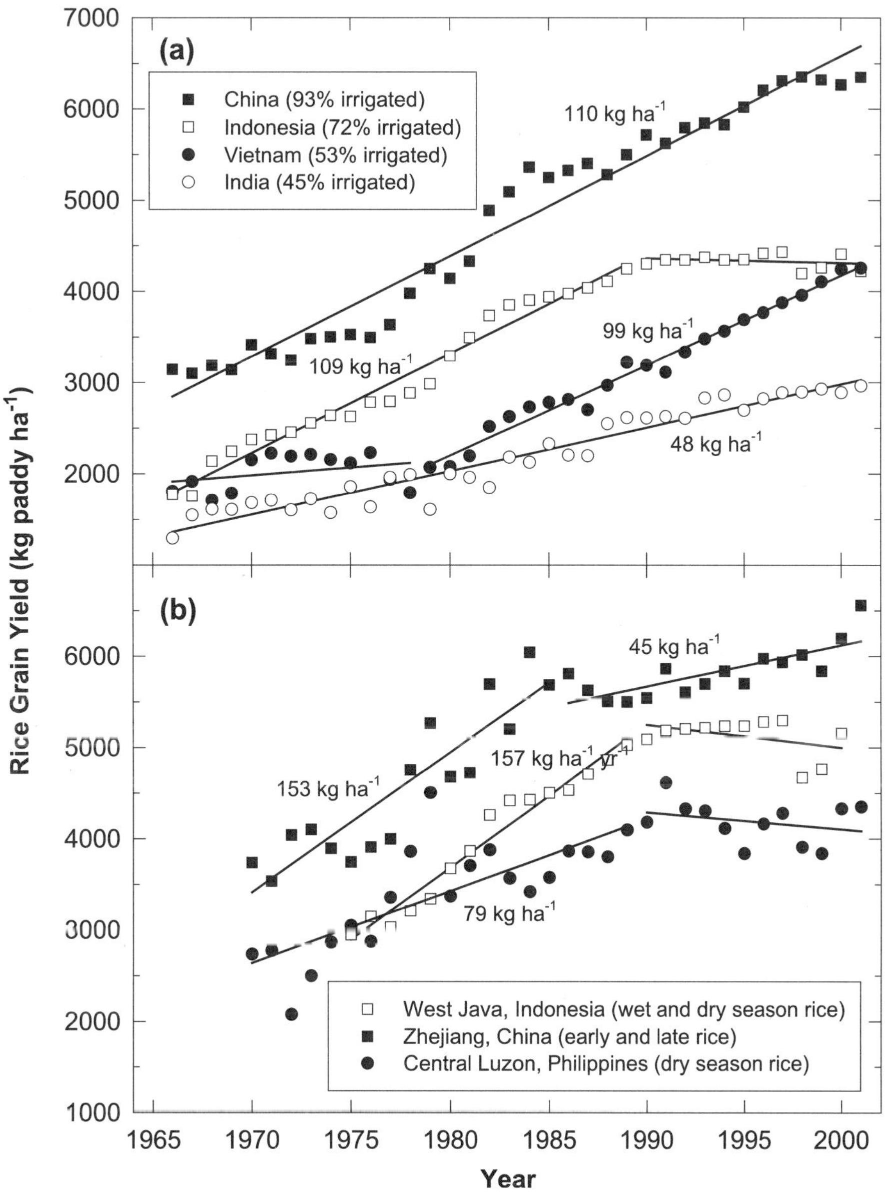

Fig. 2 Trends of national average rice yields in several Asian countries (a) and average irrigated rice yield in three provinces in which farmers were early adopters of modern rice and continuous annual double-crop systems during the late 1960s (b). Annual rates of yield increase (kg ha^{-1} per year, estimated from linear regression) are shown for periods with statistically significant yield changes.

but it is removed from the field, burned in the field, incorporated into the soil, or used as mulch for the following crop.

Fertilizer rates in most irrigated rice farms of Asia typically range from 80 to 140 kg N, 10 to 25 kg P, and 0 to 50 kg K ha^{-1} per crop. Fertilizers are mostly broadcast by hand, and the number of N applications varies from one or two up to five or six in certain regions. Soil testing is not common and most fertilizer recommendations are given as blanket recommendations for larger regions. On average, only 30% of the applied N is taken up by the crop and only 20% of all farmers achieve uptake efficiencies of more than 50%.[9] The main reason for the low N-use efficiency is a lack of congruence between nutrient supply and crop demand, resulting in significant nitrogen losses due to denitrification and ammonia volatilization.[10] Nutrient balance estimates (Table 1) suggest that, at the average farm level, phosphorus (P) applications generally are in balance with crop P removal, although P deficiency occurs in some areas. Due to low fertilizer rates and

Table 1 Estimated average input-output balance of N, P, and K in intensive rice systems of south and southeast Asia with an average yield of 5.2 Mg ha^{-1}

	kg ha^{-1} $crop^{-1}$		
Inputs and outputs	**N**	**P**	**K**
Inputs			
Fertilizer	117	18	17
Farmyard manure	5	2	5
Biological N_2-fixation	50	0	0
Outputs			
Gaseous losses	87	0	0
Net removal with grain	58	12	13
Net removal with straw	20	2	35
Input–output balance	**+7**	**+6**	**−26**

(From Ref. 17.)

significant straw removal, negative potassium (K) input–output balances are common. About 80% of the intensive rice fields in Asia have a negative K balance, but K budgets may vary from −100 to +100 kg K ha^{-1} per crop.[11]

Tillage and Crop Establishment

Wet-soil tillage is predominant in irrigated rice systems of Asia.[12] It involves plowing, puddling, and harrowing passes to create a homogeneous slurry-like surface soil and to reduce water percolation. These operations are accomplished using manual labor, traction animals, and small 2-wheel hand-held tractors. Four-wheel tractors are becoming available in some areas. Puddling helps control weeds and incorporates crop residue and fertilizer. Transplanting 10- to 30-day-old rice seedlings is the predominant crop establishment method, but direct seeding is becoming common in many intensive rice areas. Broadcast seeding of presoaked seed at rates of 100 to 200 kg seed ha^{-1} is used in irrigated rice areas with high labor cost or insufficient labor availability. Other forms of direct seeding include row seeding or dry seeding of rice.

Water Management

More than 80% of the developed freshwater resources in Asia are used for irrigation purposes and more than 90% of the total irrigation water is used for rice production. Traditionally, irrigated transplanted rice has been grown in a permanent floodwater layer of 5 to 20 cm depth throughout the whole growing period. Besides supplying water to the plant, the ponded water layer helps suppress weed growth and increases nutrient availability.[12] Direct-seeded rice is typically grown with a floodwater-free period of two to three weeks after seeding, which is then followed by flooding throughout the remaining growth period.

Increasing competition for water among agriculture, industry, and the rapidly growing urban population will force many rice farmers to use less water and increase their water use efficiency. Water-saving intermittent irrigation techniques have been developed, but they require a high degree of management control at both the farm and irrigation system levels. They are often associated with increased herbicide use and larger N losses. Research is ongoing to understand the complex interactions between water management and other crop management practices.

Pest Management

Hand-weeding is widespread in transplanted rice areas, but is labor-intensive (150 to 200 h ha^{-1} per crop). In recent years, herbicide use for weed control has become a common practice, particularly in areas with direct seeding. Strong host-plant resistance to major disease and insect pests has been the focus of rice breeding programs during the past 30 years and provides the foundation of integrated pest management.[13] Modern approaches to crop protection rely on pest management rather than control or eradication. Predatory and parasitic natural enemies contribute to keeping insect pests in check. However, insecticide use ranges from 0.1 to more than 1.0 kg active ingredient ha^{-1} per crop, and fungicide use ranges from 0 to 0.3 kg active ingredient ha^{-1}. Fungal and bacterial diseases are mostly a problem in high-yielding environments, especially in humid environments. Excessive use of N fertilizer may attract insect pests as well as increase disease incidence, whereas elements such as potassium and silicon improve the resistance to diseases.

SOIL QUALITY AND ENVIRONMENTAL CONCERNS

Several soil characteristics contribute to the unique biophysical sustainability of continuous rice farming systems.[4] Acidification is a relatively minor problem because the physical chemistry of flooded soil systems causes soil pH to stabilize around 6.5 to 7 after flooding. Nutrients tend to be leached into lowland soils rather than out of them because of their landscape position. Erosion rarely occurs because fields are well leveled and surrounded by bunds. Phosphorus is maintained in more readily available forms than in aerated, upland soils. Significant input of C and N is derived from biological activity in the soil-floodwater system. Increases in soil C and N content occur over time, even with complete removal of aboveground plant biomass. Recent concerns have mainly centered on a possible decline in soil nutrient availability. In continuously flooded systems, the accumulating N is likely to be stored in organic matter pools that are less plant-available than in well-aerated soils.[10] Soil K stocks are declining in many intensive rice areas, even on the most fertile lowland rice soils, and K deficiency is becoming more frequent. Constraints such as micronutrient deficiencies, salinization, and iron toxicity are of concern in some irrigated rice areas of Asia.

Gaseous losses of N often exceed 50% of applied fertilizer N, mostly due to ammonia volatilization. Nitrous oxide emissions occur as a result of nitrification–denitrification during periods of alternating soil wetting and drying. In irrigated rice systems with good water control, N_2O emissions are small except when excessively high N fertilizer rates are applied to fertile rice soils. Nitrate leaching losses are usually below 10% of the applied fertilizer N. Irrigated rice monoculture systems in Asia sequester atmospheric CO_2, but they also account for about 2 to 5% of the global methane (CH_4) emissions into the atmosphere.[14] Methane emission can be managed through a variety of means, including organic and inorganic amendments as well as crop management practices that also affect nutrient dynamics. Misuse of pesticides and other agrochemicals may adversely affect human health and water quality in intensive, irrigated rice areas.[15]

FUTURE CHALLENGES

Total rice demand will continue to increase at an annual rate of about 1%.[16] There will be little net increase in rice cropping area or in the amount of irrigated land available for rice production. Therefore, average yields of irrigated rice must reach 6.5 to 7 Mg ha^{-1} to meet expected demand in 2020. Labor costs will continue to rise faster than the cost of energy and fertilizer, favoring the adoption of mechanized technologies. Competition for water resources will intensify. Where feasible, cropping systems will diversify, but rice monoculture in double- or triple-crop systems will remain the preferred choice in lowlands with heavy clay soils and other constraints to the production of upland crops.

Increases in rice yield potential through germplasm improvement will require substantial research investments. Meanwhile, increased production from irrigated rice systems will come largely from improved crop, soil, water, and pest management that slowly closes the existing gap between average and potential yield levels. Increasing nutrient and water use efficiency, improving seed quality and crop establishment, and reducing crop losses due to pests are the key challenges for fine-tuning soil and crop management. Nutrient management will require a more dynamic, site-specific approach to account for spatial and temporal variability in indigenous nutrient supplies and crop nutrient needs.[9] New technologies must not conflict with the need to reduce labor and other production costs or the need to contribute to reducing potential environmental impacts from rice production. The application of biotechnology will be crucial for improving grain quality traits—especially those that can improve nutritional status for low-income people who rely on rice as their primary staple—and for maintaining adequate levels of durable pest resistance.

ARTICLES OF FURTHER INTEREST

Cropping Systems: Irrigated Rice and Wheat of the Indo-gangetic Plains, p. 355
Rice, p. 1102

REFERENCES

1. IRRI. *Rice Almanac*; International Rice Research Institute (IRRI): Los Baños, Philippines, 1997.
2. Huke, R.E.; Huke, E.H. *Rice Area by Type of Culture: South, Southeast, and East Asia. A Revised and Updated Database*; International Rice Research Institute: Los Baños, Philippines, 1997.
3. Dobermann, A.; Fairhurst, T.H. *Rice: Nutrient Disorders and Nutrient Management; Potash and Phosphate Institute*; International Rice Research Institute: Singapore, 2000.
4. Greenland, D.J. *The Sustainability of Rice Farming*; CAB International, International Rice Research Institute: Oxon, 1997.
5. Cassman, K.G.; Pingali, P.L. Intensification of irrigated rice systems: Learning from the past to meet future challenges. GeoJournal **1995**, *35* (3), 299–305.
6. Dawe, D.; Dobermann, A.; Moya, P.; Abdulrachman, S.; Bijay Singh; Lal, P.; Li, S.Y.; Lin, B.; Panaullah, G.; Sariam, O.; Singh, Y.; Swarup, A.; Tan, P.S.; Zhen, Q.X. How widespread are yield declines in long-term rice experiments in Asia? Field Crops Res. **2000**, *66*, 175–193.

7. Peng, S.; Cassman, K.G.; Virmani, S.S.; Sheehy, J.E.; Khush, G.S. Yield potential trends of tropical rice since the release of IR8 and the challenge of increasing rice yield potential. Crop Sci. **1999**, *39*, 1552–1559.
8. Roger, P.A. *Biology and Management of the Floodwater Ecosystem in Ricefields*; International Rice Research Institute, ORSTOM: Los Baños, Philippines, 1996.
9. Dobermann, A.; Witt, C.; Dawe, D.; Gines, G.C.; Nagarajan, R.; Satawathananont, S.; Son, T.T.; Tan, P.S.; Wang, G.H.; Chien, N.V.; Thoa, V.T.K.; Phung, C.V.; Stalin, P.; Muthukrishnan, P.; Ravi, V.; Babu, M.; Chatuporn, S.; Kongchum, M.; Sun, Q.; Fu, R.; Simbahan, G.C.; Adviento, M.A.A. Site-specific nutrient management for intensive rice cropping systems in Asia. Field Crops Res. **2002**, *74*, 37–66.
10. Cassman, K.G.; Peng, S.; Olk, D.C.; Ladha, J.K.; Reichardt, W.; Dobermann, A.; Singh, U. Opportunities for increased nitrogen use efficiency from improved resource management in irrigated rice systems. Field Crops Res. **1998**, *56*, 7–38.
11. Dobermann, A.; Cassman, K.G.; Mamaril, C.P.; Sheehy, J.E. Management of phosphorus, potassium and sulfur in intensive, irrigated lowland rice. Field Crops Res. **1998**, *56*, 113–138.
12. De Datta, S.K. *Principles and Practices of Rice Production*; J.Wiley: New York, 1981.
13. Heong, K.L.; Teng, P.S.; Moody, K. Managing rice pests with less chemicals. GeoJournal **1995**, *35* (3), 337–349.
14. Matthews, R.B.; Wassmann, R.; Knox, J.W.; Buendia, L.V. Using a crop/soil simulation model and GIS techniques to assess methane emissions from rice fields in Asia. IV. Upscaling to national levels. Nutr. Cycl. Agroecosyst. **2000**, *58*, 201–217.
15. Pingali, P.L.; Roger, P.A. *Impact of Pesticides on Farmer Health and the Rice Environment*; International Rice Research Institute: Los Baños, Philippines, 1995.
16. Rosegrant, M.W.; Paisner, M.S.; Meijer, S.; Witcover, J. *Global Food Projections to 2020: Emerging Trends and Alternative Futures*; IFPRI: Washington, DC, 2001.
17. Dobermann, A.; Cassman, K.G. Plant nutrient management for enhanced productivity in intensive grain production systems of the United States and Asia. Plant Soil **2002**, *247*, 153–175.

Cropping Systems: Irrigated Rice and Wheat of the Indo-gangetic Plains

David J. Connor
University of Melbourne, Melbourne, Victoria, Australia

INTRODUCTION

Rice and wheat are the world's two most important cereal crops that have evolved in distinct geographic distributions—rice in the tropics and wheat in temperate regions. Suitable thermal conditions for both crops are found during the annual cycle of the Indo-Gangetic Plains (IGP), where 13.5 Mha (85% of cropped area) are now devoted to rice-wheat (R-W) production sequences. There are an additional 9.5 Mha of R-W sequences in south and south–central China, where climatic conditions and access to irrigation also allow double cropping. The expansion was driven by the increasing demand for food and was made possible by the development of short-duration cultivars of both species, and in some places by subsidies and increased use of irrigation, fertilizers, and pesticides. The rice-wheat system is now the most important production system for the food security of south Asian countries. Demand for high-aggregate yield from the R-W system makes it a case study in contemporary agriculture: the search for high-yielding systems that are also sustainable.

The brief treatment presented here concentrates on the IGP and on the main crops of rice and wheat, dealing with productivity and the key management processes related to soil, nutrients, and water. Readers are referred to recent reviews[1,2] for more detail and to the Web site of the R-W Consortium.[3]

DISTRIBUTION AND SYSTEM CHARACTERISTICS

The R-W areas of the IGP currently cover 10 Mha in India, 2.2 Mha in Pakistan, and 0.5 Mha each in Bangladesh and Nepal. The origin of the system differs from east to west. In the east, wheat was introduced into rice areas as a winter crop (e.g., in Bangladesh), while in the west rice was introduced as a summer crop after wheat (e.g., in the Punjab and Haryana states of India). There are two broad production systems.[4] In the west, as in the Indian and Pakistani Punjab, monsoon rice is followed by wheat. Rice is transplanted from late May to July and harvested from September to late November. Wheat is then grown from October/November to March/April. In the eastern, warmer areas, rice is grown during the summer to early winter, usually from June/July to October/November, and wheat from November/December to March. This allows for the possibility of a third crop, often a legume, after wheat and before rice.[4] A detailed agroclimatic analysis of the R-W zones of India is available for Ref. 5.

Rice–wheat systems are practiced on soil types ranging from sandy loam to heavy clay. The common requirement is sufficient clay content to hold ponded surface water for rice production, but in some areas of coarse-textured soils on which the system has developed (e.g., the Indian Punjab) soils do not hold water, so rice crops are continuously irrigated. Fertilizer use is variable. Many subsistence farmers (e.g., in Bangladesh) still rely on farm-yard manure combined with small amounts of inorganic fertilizer, while up to 300 kg N/ha are applied to rice and wheat crops combined in the Indian Punjab.[6] Rice is mostly transplanted by hand, with mechanical transplanters being adopted in some areas (e.g., the Indian and Pakistani Punjab). Much rice is now direct-seeded, but there is no reliable data on its extent in R-W systems. Machinery is increasingly available for sowing and harvesting, and many small-scale farmers in the eastern IGP (e.g., Bangladesh) have access to small machinery powered by two- and four-wheeled tractors.

PRODUCTIVITY AND NUTRIENT BALANCE

The R-W system is highly extractive at the levels of yield that are being sought. That condition is exacerbated by the frequent harvest or burning of straw and by the small amount of organic matter that can be returned as farm-yard manure because of its competing use as fuel for cooking.

The average annual productivity of the R-W system in IGP is small (5 to 7 Mg/ha) by comparison to currently attainable (8 to 10 Mg/ha) and site-potential (12 to 19 Mg/ha) yields.[6] India is an exception, with current yields of 9 Mg/ha.[5] Although concern has been expressed about yield stagnation of the system, a recent

Encyclopedia of Plant and Crop Science
DOI: 10.1081/E-EPCS 120010543

analyses of long-term experiments (1983 to 1999) at six locations in the IGP[7] does not support that conclusion. The analysis reveals that crop performance depends upon nutrient management. When the results were averaged over locations, rice yields exhibited a declining yield only in those treatments that received no fertilizer. Yield trends were positive with recommended fertilizer application, especially where nutrient demand was met by a combination of inorganic fertilizer and organic manure. In wheat, yield declined without fertilizer but remained constant with applied fertilizer. Consideration of nutrient extraction (in harvest) and input by fertilizer explains these results, revealing that greater yields will require greater and more carefully targeted fertilization. There are two parts to this process.

The first is attention to providing adequate and balanced macronutrients needed to sustain the system.[8] Based on average nutrient concentrations in grain and straw, a yield of 10 Mg/ha equally divided between rice and wheat requires an estimated replacement for continuing productivity of 163, 34, and 47 kg/ha of N, P, and K, respectively, to account for grain harvest. This increases to 243, 45, and 63 kg/ha when stubble is also removed. The second part of the process is to correct micronutrient deficiencies of Zn, Fe, Mn, B, Cu, and Mo that are now identified across the region, depending mostly on soil type but also on previous cropping history.[9]

The inclusion of legumes is seen by many as the way to supply N to the system. Nutrient balances show, however, that legumes make only small contributions to the N requirement, relative to the high yields required of the system.[4] Further, legume N is not a free good but is accumulated at cost to land, time, labor, water, fertilizer, and the opportunity cost of other crops. It is not surprising that in terms of N balance alone, few farmers find legumes to be economic at the current price of N fertilizer. There are, however, other reasons for crop diversification: to assist soil, disease, and weed management; to improve human nutrition; and to increase or stabilize economic returns.

PUDDLING AND THE WET–DRY CYCLE

Preparation of land for rice by puddling, followed by inundation until crop maturity, have significant effects on the physical, chemical, and biological status of soils, influencing growing conditions for all crops in the system.[10,11] Puddling, the repeated tillage of saturated (inundated) soil in preparation for a rice crop, is undertaken because it offers significant advantages to rice production. It softens soil to facilitate transplanting of rice seedlings, promotes root growth, aids weed control, and reduces water and nutrient losses caused by leaching. As an additional benefit, inundation mobilizes phosphorus and holds nitrogen in the ammonium form. Puddling, however, disperses soil aggregates, destroys soil structure, and in fine-textured soils forms a massive topsoil that sets hard and often cracks widely when dry. When repeated from year to year, puddling may form a compacted layer up to 5 to 10 cm thick at 10 to 40 cm depth. In R-W systems, the destruction of surface soil structure and the formation of hardpan as the compacted layers dry can be serious liabilities to the establishment and performance of wheat and other upland crops grown after rice.[10,11]

An important question asks if this wet–dry transition, when managed by puddling, is inimical to the high-aggregate productivity of all crops in the system? If so, under what conditions are alternative management systems required?

WEEDS, PESTS, AND DISEASES

The productivity of R-W systems may be significantly reduced by competition from weeds, insects, rodents, termites, and nematodes, as well as diseases.[4] The wet–dry transition provides a barrier to carryover from crop to crop—for example, by termites, rodents, and some weeds and soil-borne diseases—but significant problems remain. The outstanding weed is phalaris (*Phalaris minor*), which has become widespread in wheat crops. Insect problems include the rice pink stem borer (*Sesamia inferens*) and the shoot fly (*Antherigona oryzae*). Interestingly, both insects originally attacked only rice, but they now infect wheat also and have become a significant general threat. Nematodes are soil-living organisms that feed on root systems, and while some species are specifically associated with either rice or wheat, others (e.g., *Meloidogyne* spp.) are polyphagus, can attack all crops in R-W sequences, and pose a serious threat to the system. There is also a range of soil- and seed-borne diseases that can be transmitted from crop to crop in R-W sequences. Some soil-borne pathogens (e.g., *Fusarium*, *Rhizoctonia*, and *Sclerotinium*) survive the anaerobic conditions during the rice phase and build up in continuous R-W cropping. Leaf blights (*Helminthosporium* spp.) are considered the most pathogenic, and because they can proliferate on crop residues they are set to become a greater problem in developing zero-tilllage systems.

DIRECTIONS FOR THE FUTURE

The R-W system of the IGP has developed quickly from both rice- and wheat-based origins, but it is unlikely that its area can increase significantly. Change will continue to characterize the system as it struggles to intensify productivity within environmental constraints. Productivity relates to the advantages of crop diversification and new management options, while environmental constraints are principally the availability of water and labor and pollution from stubble burning. The challenge can be summarized as the need for greater productivity combined with greater use-efficiencies of water, nutrients, and labor.

Two directions are likely to lead to a marked divergence within the system.[12] First, improvements will be made to the traditional system by gradually improving nutrient management as cultivars are improved; by mechanization that allows more timely sowing and harvesting; and by residue management techniques to improve soil structure. Second, the present early initiatives to modify production toward raised-bed culture of both rice and wheat will offer an immediate increase in yield and profitability associated with greater use-efficiencies of water, nutrients, and labor, and will be associated with greater opportunities for crop diversification.

ACKNOWLEDGMENT

Dr. Jagadish Timsina offered valued comment on the manuscript.

ARTICLES OF FURTHER INTEREST

Cropping Systems: Irrigated Continuous Rice Systems of Tropical and Subtropical Asia, p. 349
Integrated Pest Management, p. 612
Managemenet of Nematode Diseases: Options, p. 684
Phosphorus, p. 872
Potassium and Other Macronutrients, p. 1049
Rice, p. 1102
Sustainable Agriculture and Food Security, p. 1183
Symbiotic Nitrogen Fixation: Special Roles of Micronutrients, p. 1226
Water Deficits: Development, p. 1284

REFERENCES

1. Jiaguo, Z. Rice-Wheat Cropping System in China. In *Soil and Crop Management Practices for Enhanced Productivity of the Rice-Wheat Cropping System in Sichuan Province of China*; Hobbs, P.R., Gupta, R.K., Eds.; Rice-Wheat Consortium Paper Series 9, Rice-Wheat Consortium for the Indo-Gangetic Plains: New Dehli, 2000; 1–10.
2. http://www.rwc-prism.cgiar.org/rwc. (accessed 13 June 2002).
3. *Improving the Productivity and Sustainability of R-W Systems of the Indo-Gangetic Plains: A Synthesis of NARS-IRRI Partnership Research*; Ladha, J.K., Fischer, K.S., Hossain, M., Hobbs, P.R., Hardy, B., Eds.; Discussion Paper No. 40, IRRI: Manila, 2000; 1–31.
4. Timsina, J.; Connor, D.J. Productivity and management of rice–wheat systems: Issues and challenges. Field Crops Res. **2001**, *69* (2), 93–132.
5. Narang, R.S.; Virmani, S.M. *Rice-Wheat Cropping Systems of the Indo-Gangetic Plain of India*; Rice-Wheat Consortium Paper Series 11, Rice-Wheat Consortium for the Indo-Gangetic Plains: New Dehli, 2001; 1–36.
6. Aggarwal, P.K.; Talukdar, K.K.; Mall, R.K. *Potential Yields of Rice-Wheat System in the Indo-Gangetic Plain of India*; Rice-Wheat Consortium Paper Series 10, Rice-Wheat Consortium for the Indo-Gangetic Plains: New Dehli, 2000; 1–12.
7. Yadav, R.L.; Dwivedi, K.; Prasad, K.; Tomar, O.K.; Shurpali, N.J.; Pandey, P.S. Yield trends and changes in soil organic-C and available NPK in a long-term rice-wheat system under integrated use of manures and fertilizers. Field Crops Res. **2000**, *68* (3), 219–246.
8. Yadvinder-Singh, B.-S. Efficient management of primary nutrients in the rice-wheat system. J. Crop Prod. **2001**, *4* (1), 23–86.
9. Nayyar, V.K.; Aroara, C.L.; Kataki, P.K. Management of soil micronutrient deficiencies in the rice-wheat cropping systems. J. Crop Prod. **2001**, *4* (1), 87–132.
10. Aggarwal, G.C.; Sidhu, A.S.; Sekhon, N.K.; Sandhu, K.S.; Sur, H.S. Puddling and N management effects on crop response in a rice-wheat cropping system. Soil Tillage Res. **1995**, *36* (3–4), 129–139.
11. Kirchhof, G.; Priyono, S.; Utomo, W.H.; Adisarwanto, T.; Dacanay, E.V.; So, H.B. The effect of soil puddling on the soil physical properties and the growth of rice and post-rice crops. Soil Tillage Res. **2000**, *56* (1–2), 37–50.
12. Hobbs, P.R. Tillage and crop establishment in south Asian rice-wheat cropping systems: Present practices and future options. J. Crop Prod. **2001**, *4* (1), 1–22.

Cropping Systems: Rain-Fed Maize-Soybean Rotations of North America

Douglas L. Karlen
USDA—ARS, Ames, Iowa, U.S.A.

INTRODUCTION

A two-year maize-soybean rotation became the dominant land use in the midwestern United States during the last half of the 20th century. This occurred primarily through public and private research and development efforts devoted to the genetic improvement of maize and to making soybean a truly "miracle crop," but also coincided with major changes in the livestock industry that decreased demand for oats and alfalfa. Herein, soil, plant, weed, and insect management practices used to enhance the maize-soybean rotations are reviewed. Economic, environmental, and social effects associated with the rapid expansion and dominance of maize-soybean rotations are also discussed.

NORTH AMERICAN MAIZE AND SOYBEAN PRODUCTION

North American maize and soybean production totaled approximately 259 and 75 million metric tons respectively, in 1999–2000,[1] or about 44 and over 50% of global production. Of this total, Mexico and Canada produce approximately 7 and 4%, respectively. In the Unite States, these two crops were grown on approximately 32 and 30 million hectares (ha), generally in a two-year rotation. Assuming the irrigated proportion was similar to 1998,[2] rain-fed production accounted for approximately 85 and 93%, respectively.

WHY ARE MAIZE-SOYBEAN ROTATIONS INCREASING?

Table 1 shows that between 1950 and 2000, soybean increased by 500%[3] while oat and alfalfa declined by 90 and 20%. Domesticated in northeast China around the 11th century B.C., soybeans were first brought to the U.S. in 1804 as ballast in a cargo ship.[4] United States farmers first grew the crop in 1829, primarily for soy sauce or as "coffee berries" when real coffee was scarce. During the late 1880s, farmers began growing the crop for forage or green manure. George Washington Carver began studying soybean at the Tuskegee Institute in 1904 and made several discoveries regarding its use for protein and oil.

World War II cut U.S. imports of edible fats and oil by 40%, forcing processors to soybean oil. Then through public and private investment in plant genetics, processing techniques, and new uses (e.g., soybean meal for livestock and poultry) soybean's popularity quickly increased. The bushy, green plants often grown in rotation with maize flower during the summer and produce 60–80 pods, each holding 2–4 pea-sized beans. Harvested in the fall, each bushel (27.3 kg) of seed yields approximately 22 kg of protein-rich meal and 5 kg of oil. Now grown in more than 30 states, soybean is the second largest cash crop in the United States.

Soybeans contain eight essential amino acids that are not produced naturally in the human body. The oil is used for cooking and numerous food and nonfood products.[4] Lecithin (a mixture of the diglycerides of stearic, palmitic, and oleic acids linked to the choline ester of phosphoric acid) is extracted from soybean oil and used for products ranging from pharmaceuticals to protective coatings. It is a natural emulsifier and lubricant. Soybean hulls provide a natural source of dietary fiber for bran breads, cereal and snacks. After removing the oil, soybean flakes can be processed into various edible soy protein products or used to produce meal for animal feeds.

Maize is a staple food in Mexico and several other Central and South American countries, but in the United States it is used primarily as the energy source in livestock feed and as a sweetener to replace sugar. World trade is also an important factor with 66% of exported maize originating in the United States.[5] Each bushel of maize (25.4 kg) provides 14.3 kg of starch or 15 kg of sweetener or 9.45 L of ethanol fuel, plus 6.1 kg of gluten feed (20% protein), 1.2 kg of gluten meal (60% protein), and 0.7 kg of oil. Annual research and development investments of more than $137 million dollars by private industry focus on improving maize hybrids for farmers.[6] This investment has helped promote maize-soybean rotations by providing improved plant genetics, better production practices, new uses for maize, and a stable market. Maize genetic improvement has generally focused on yield enhancement and stability under drought and disease stress, while

Encyclopedia of Plant and Crop Science
DOI: 120010542/E-EPCS 120010542

Table 1 Estimated U.S. maize, soybean, oat, and alfalfa production for 1950 and 2000[a]

Crop	1950	2000
Maize	33.55	32.20
Oat	18.24	1.81
Alfalfa	30.42	24.23
Soybean	6.09	30.16

[a]In million hectares. To convert hectares to acres, multiply by 2.47.

production issues have focused on tillage, fertilizer, and water use efficiency.

Industrial uses for maize include a wide array of bio-carbon products including starch, plastics, and ethanol. Research on the extraction of polyol derivatives for ethylene glycol (antifreeze) and propylene glycol (food and health products), development of new commercial manufacturing processes to convert maize fiber into value-added chemical feedstocks, development of a microbe that will simultaneously convert the glucose, xylose and arabinose from maize to ethanol, development of technologies to produce 1,3-propanediol for plastics, and studies on the maize genome are all contributing to the support for maize-soybean rotations.

MANAGEMENT CHALLENGES ASSOCIATED WITH MAIZE-SOYBEAN ROTATIONS

Compared to monoculture maize, a two-year maize-soybean rotation will generally increase yields of both crops by 10% or more.[7] However, several factors must be managed well if both crops are to achieve their full genetic potential. As shown for soybean (Fig. 1), multiple factors including herbicide injury, nematodes, diseases, and other stresses[8] interact to reduce potential yield. One advantage of rotation is that some stressors are host-specific (e.g., soybean cyst nematode), but others, such as the northern corn rootworm, are no longer managed by rotation[9] because of natural selection for extended egg diapause. This adaption allows the insect's eggs to survive in the soil through two growing seasons. Similarly, some populations of western corn rootworm are no longer managed by the rotation because they have adapted behaviorally with the females ovipositing in soybean fields during the nonmaize phase of the rotation.

Weed control within a maize-soybean rotation is complementary because maize is a C4 grass crop, whereas soybean is a broadleaf C3 crop. As a result, herbicides applied for weed control are rotated from year to year. Furthermore, through development of herbicide-resistant genetically modified (GMO) crops, weed control is generally not a major production challenge.

Nematodes affecting maize have received less attention than those affecting soybean, but if damaging levels are found, control measures should be used. In infested fields, maize plants may appear to be moisture-stressed, stunted and chlorotic, or exhibit less extreme signs of poor plant growth.[10] The most important maize nematodes in North America and their type of damage are listed in Table 2.

Nutrient management, row and plant spacing, N rates, winter and early-spring weeds, and insects (e.g., borers, root and ear worms, aphids, beetles, grubs, and maggots) can all affect maize within the rotation. Applying starter fertilizer may increase root proliferation and can promote uniform early-season plant growth and development if soil-test phosphorus (P) and, sometimes, if potassium (K) levels are low. More rapid early-season growth and development through hybrid vigor, fertility, and disease and insect resistance may hasten flowering (i.e., tassel and silk development) by 1 to 3 days. More rapid canopy closure can reduce late-season weed development and perhaps reduce the potential for water stress during pollination and grain-fill. Plant spacing influences light

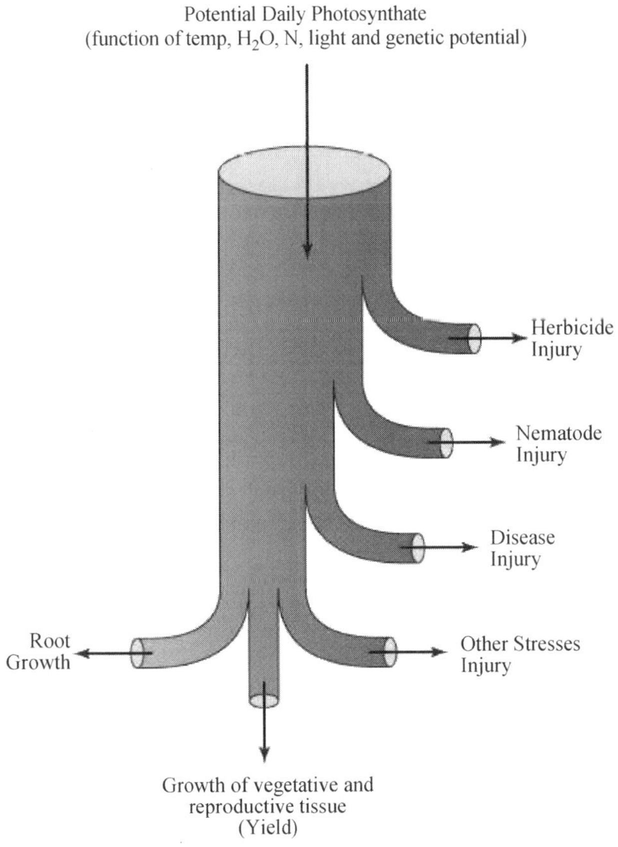

Fig. 1 Factors affecting soybean yield potential. (*View this art in color at www.dekker.com.*)

Table 2 Important North American maize nematodes

Common name	Latin name	Frequency and type of damage
Dagger	*Xiphinema* spp.	Occasional and moderate damage. Severe plant stunting, chlorosis, few fine feeder roots.
Lance	*Hoplolaimus* spp.	Occasional and moderate damage. Reduces root system, darkened discolored roots, moderate stunting, and chlorosis.
Lesion	*Pratylenchus* spp.	Very common and moderate damage. Small root system, darkened and discolored roots, moderate stunting.
Root knot	*Meloidogyne* spp.	Occasional and moderate damage. Swollen roots, moderate stunting, associated with a small root system.
Spiral	*Helicotylenchus* spp.	Occasional and damaging at high populations. Smaller than normal root system, moderate stunting, chlorosis.
Sting	*Belonolaimus* spp.	Rare but very damaging. Severe stunting, chlorosis, small and devitalized root system.
Stubby-root	*Trichodorus* spp.	Rare but very damaging. Stubby lateral roots, coarse roots, excessive upper roots, severe stunting, chlorosis.
Stunt	*Tylenchorhynchus* spp.	Occasional and damaging at high populations. Smaller than normal root system, moderate stunting, chlorosis.

(Adapted from Ref. 9.)

interception (radiant energy-use efficiency), and by encouraging more efficient photosynthate partitioning, may affect stress tolerance. Poorly spaced or missing maize plants generally decrease yield, whereas an occasional extra plant tends to increase it. The use of variable N rates offers potential economic and environmental benefits, but adoption has been slow because of uncertainty associated with predicting available soil N, especially in tile-drained Midwestern soils.[11] Several new systemic insecticide seed treatments and genetic engineering technologies (e.g., development of Bt hybrids) are being used to minimize European corn borer and corn rootworm damage, although each of these increases the overall cost of production.

Planting dates and rates, row spacing, fertilization strategies, and genetics are also important for soybean. Recently, earlier planting dates have been used to increase soybean yield potential, but this exposes the crop to stand establishment, disease, and frost risks. Sudden death syndrome (SDS) is one disease that has spread northward throughout the U.S. SDS is caused by a specific strain of a common soil fungus (*Fusarium solani*) that infects the roots and produces a toxin that damages the soybean leaves. The incidence of SDS is often higher if the crop is exposed to cool, moist soils early in the growing season. Management strategies include selecting tolerant varieties, delaying planting, improving drainage, reducing compaction, and reducing crop stress.

Green stem syndrome (GSS) is another disease that has increased dramatically in North America. Plants exhibiting this disease retain green stems, leaves, and petioles late in the season after normal leaf drop. They are also susceptible to leaf and pod distortion, shattering, and seed coat stains. Multiple factors appear to be contributing to this disease, but due to the similarity of symptoms, GSS is often attributed to virus infection. However, not all plants with GSS are infected with known viruses, and plants may be virus-infected without showing symptoms of GSS. Large increases in bean leaf beetle (*Ceraotoma trifurcata*) and aphid populations are presumably responsible for increased virus infections. The bean leaf beetle is a vector contributing to the spread of the bean pod mottle virus, a disease that can result in multiple symptoms including mottled leaves, stunted plants, reduced top growth and nodule weight, and mottled or stained seeds.

The bean leaf beetle has also been identified as a very important problem for producers of organically grown soybean[12] destined for human consumption in tofu or other products. In Iowa, only 5 percent of the large-seeded, high-protein organic soybean was rejected from the tofu market because of seed staining in 1998. However, in 1999 rejection rates increased dramatically to nearly 50%. Soybean aphid is another pest affecting the crop in northern states. The insect was found in only one Minnesota county in 2000, but within one year, significant damage was reported throughout the state. The rapid increase in these insect and disease problems is definitely one of the management challenges associated with the relatively simple maize and soybean rotation.

ARE MAIZE-SOYBEAN ROTATIONS SUSTAINABLE?

Considering the near ''miracle'' impact that soybean has had during the past 50 years, many agricultural and political leaders may consider this question heresy. Is this denial[13] or simply a failure to consider all aspects of this agricultural system? Without question, maize and soybean have become very important crops and their production in a two-year rotation is very complementary. The machinery (tillage, planters, harvesters, etc.), grain storage, and marketing systems are very compatible, making the maize-soybean system ''easier'' to manage with less expense for overhead or infrastructure. But what impact has this simple rotation had on soil resources, water quality, biodiversity, wildlife corridors, and rural communities?

The question will undoubtedly be debated for some time, because of differences in how agricultural systems are defined. The boundaries could be a plant, field, farm, watershed, region, or nation. Fueling the debate are observations that the two-year maize-soybean rotation is associated with 1) fewer and larger farm operations, 2) fewer livestock farms, 3) more pest problems, 4) increased iron chlorosis, 5) narrow profit margins resulting in greater government loan deficiency payments (LDPs), and 6) increased soil erosion and further degradation of soil quality. Some argue that this is no different than during the 1950s when an average of ten Wisconsin dairy farms went out of business each day.[14] However, others[15] conclude that to help shape a sustainable future, it is important to understand where we are and how we got here.

Economically, the 1999 and 2000 prices for soybean crops averaged \$173 per metric ton,[3] the lowest price since 1972 when gasoline prices were \$0.07 L^{-1} (\$0.25 gal^{-1}). Assuming operating costs (\$196 ha^{-1}) and overhead costs (\$416 ha^{-1}) were similar to those for 1998,[4] soybean farmers lost approximately \$60 ha^{-1} each year. To reduce overhead, producers turn to economies of scale and become more reliant on government subsidies. This results in even larger operations and a further reduction in the number of farms and farm families.

Environmentally, maize-soybean rotations grown using conventional tillage and residue management practices often result in greater soil erosion because prior to mid-July and again in the autumn there is very little crop growth or residue and therefore minimum soil surface protection from intensive rainstorms. The increased erosion and lack of surface residues have a negative effect on soil quality through their effect on soil structure and tilth. Furthermore, the small amounts of water use prior to mid-June for maize or mid-July for soybean can lead to increased drainage and runoff of spring rainfall. This reduces water-use efficiency and can contribute to nitrate leaching or loss of phosphorus and organic matter through runoff.

Ecologically, the loss of crop diversity and dominance of maize-soybean rotations leads to a loss of habitat for insects, birds, and small animals. Greater numbers and diversity of wildlife are generally considered highly favorable in a rural ecosystem and contribute substantially to the rural quality of life.[13] Increased crop diversity also helps sustain a better balance among insects and can often minimize the severity of various pests.

The sociological impact of having maize-soybean rotations dominate the landscape is more subtle and indirect than the other effects, but as the number of farms and farm families decrease there is a measurable decline in the viability of small, rural communities. Student numbers and church memberships decrease, requiring consolidation and merger; the number of businesses declines; and fewer local youth remain after high school graduation. Producers also begin to bypass the local community as inputs are purchased from larger regional outlets where prices are cheaper.[13] Collectively, these trends have contributed to the ''hollowing out'' of rural communities and, if they continue, will threaten the vitality of rural America as we know it.

CONCLUSION

Balancing the benefits of maize-soybean rotations with the ecological, environmental, and social losses will not be easy. However, the process that led to the development of soybean may serve as a model for the future. Public and private investment in research and technology were critical drivers in making soybean a source for literally thousands of bio based products. Therefore, if natural (i.e., soil, water, air, biological, and landscape), human (i.e., intellectual, experiential, values, and leadership), social (e.g., trust, relationships, attitudes), and financial resources were invested in more diverse cropping systems and in managing land and human resources at community or watershed boundaries, the outcome could be much different than if we continue to pursue consolidation and simplification. Increased crop diversity could enable maize and soybean to achieve more of their genetic potential by minimizing stress. Greater crop diversity across rural landscapes would enhance esthetics and quality of life, and could result in significant improvements in soil, water, and air quality. The new vision for North American agriculture would be the production of the right crop in the right place at the right time based upon natural resource, climate, and market conditions.

ARTICLES OF FURTHER INTEREST

Agriculture and Biodiversity, p. 1
Crops and Environmental Change, p. 370
Ecology and Agricultural Sciences, p. 401
Environmental Concerns and Agricultural Policy, p. 418

Herbicide-Resistant Weeds, p. 551
Insect–Plant Interactions, p. 609
Social Aspects of Sustainable Agriculture, p. 1155
Sustainable Agriculture and Food Security, p. 1183
Sustainable Agriculture: Definition and Goals, p. 1187
Sustainable Agriculture: Ecological Indictors, p. 1191

REFERENCES

1. http://www.usda.gov/oce/waob/wasde/latest.pdf. (accessed May 2002).
2. http://www.nass.usda.gov/census/census97/fris/tbl22.txt. (accessed May 2002).
3. http://www.usda.gov/nass/pubs/histdata.htm. (accessed May 2002).
4. http://www.unitedsoybean.org/soystats2000/. (accessed May 2002).
5. http://www.ncga.com/03world/World2002/. (accessed May 2002).
6. Duvick, D.N.; Cassman, K.G. Post-green revolution trends in yield potential of temperate maize in the North-Central United States. Crop Sci. **1999**, *39*, 1622–1630.
7. Karlen, D.L.; Varvel, G.E.; Bullock, D.G.; Cruse, R.M. Crop rotations for the 21st century. Adv. Agron. **1994**, *53*, 1–45.
8. http://www.iasoybeans.com/srdc/srdcjl12.htm. (accessed May 2002).
9. Karlen, D.L.; Buhler, D.D.; Ellsbury, M.M.; Andrews, S.S. Soil, weed, and insect management strategies for sustainable agriculture. OnLine J. Biol. Sci. **2002**, *2* (1), 58–62.
10. Ireland, D. Corn Nematodes. In *Crop Insights*; Pioneer Hi-Bred International, Inc.: Johnston, IA, 2001; Vol. 11 (No. 17).
11. Dinnes, D.L.; Karlen, D.L.; Jaynes, D.B.; Kaspar, T.C.; Hatfield, J.L.; Colvin, T.S.; Cambardella, C.A. Nitrogen management strategies to reduce nitrate leaching in tile-drained Midwestern soils. Agron. J. **2002**, *94*, 153–171.
12. Wood, M. Organic Grows on America. In *Agricultural Research Magazine*; USDA-ARS: Beltsville, MD, Feb., 2002; 4–9.
13. http://sroc.coafes.umn.edu/Soils/Published%20Abstracts/Present-Day%20Agriculture.pdf. (accessed May 2002).
14. Apps, J. Adjustment: 1920–1960. Chapter 5. In *Cheese: The Making of a Wisconsin Tradition*; Amherst Press: Amherst, WI, 1998.
15. Flora, C.B. A vision for rural America. Word World **2000**, *20* (2), 170–178.

Cropping Systems: Slash-and-Burn Cropping Systems of the Tropics

Ken Giller
Wageningen University, Wageningen, The Netherlands

Cheryl Palm
Columbia University, Palisades, New York, U.S.A.

INTRODUCTION

This article describes the key features of slash-and-burn agriculture (otherwise known as shifting cultivation). Such systems are useful in demonstrating the concept of agricultural sustainability in relation to increasing pressure on land due to population growth. The largely closed nutrient cycles under natural vegetation are opened up, resulting in losses that can only be restored through long fallows or through intensification of production.

SLASH-AND-BURN IS THE OLDEST FORM OF AGRICULTURE

In the hierarchy of farming systems, slash-and-burn (or shifting cultivation) is essentially the basic form of agriculture, and remains the perfect example to illustrate the concept of sustainability in agriculture in relation to intensity of land use. Shifting cultivation can be simply defined as the "alternation of cropping periods on cleared plots and lengthy periods when the soil is rested."[1] Although we tend to associate slash-and-burn with typical management systems in tropical rainforests, various types of shifting cultivation occur in areas of both forest and savanna, and were in fact the dominant form of agriculture practiced in the conversion of temperate-zone forests and woodlands. Various forms of shifting cultivation can be identified in pollen records from many parts of the world dating back several thousand years. From the great complexity of causes of tropical deforestation, agricultural clearance by smallholder farmers is but one force—currently a relatively minor one.[2] Currently, major forces driving the clearance of forests are the exploitation of tropical timber and land clearance for commercial agriculture.

Currently about 37 million people practice shifting cultivation on 1035 million hectares of land in the tropics (Table 1). Although this is only about 3% of the agricultural population, it encompasses 22% of the agricultural land area of the tropics. Regionally, similar numbers of people practice shifting cultivation in Africa, Latin America, and Asia. The area in Latin America is two times more than that in Africa and three times that in Asia.

THE FOUR PHASES OF SHIFTING CULTIVATION

Four main phases can typically be identified in a cycle of slash-and-burn agriculture: clearing, burning, cropping, and abandonment. The abandonment phase involves the movement of activity to a new location, sometimes in very extensive systems involving movement of whole settlements. (This is essentially a long fallow phase during which the productivity of the soil (system) recovers.) Crop-fallow systems in temperate regions are generally associated with soil left bare while the land is rested, but in most tropical regions the fallow phase is associated with the rapid regrowth of vegetation.

The clearance phase involves the felling of trees and slashing of the shrub layer (understorey). This phase usually occurs at or before the onset of the dry season so that the slashed vegetation can dry out to allow burning at the end of the dry season or beginning of the rainy season. Fire is the basic tool used to clear away the vegetation. This job would otherwise require significant time and labor. The burn has several positive and negative effects on the systems' productivity. The intensity and effects of burning depend substantially on how the vegetation and fire are managed. Particularly important factors are the length of time the cut vegetation is left to dry, and whether heaps or piles are made before burning. Piling of biomass tends to achieve a more complete burn, and leaves behind little organic matter.

The clearance and burning phases result in the opening up of what are relatively closed nutrient cycles under forest, where the perennial root system ensures efficient capture of nutrients into the vegetation and recycles through litterfall and decomposition.[4] Elements that are readily oxidized to gases (notably C, N, and S) are lost during the burn, although the degree of nutrient loss depends on the intensity of the fire. Other nutrients—notably

DOI: 10.1081/E-EPCS 120010540

Table 1 Land area and population practicing shifting cultivation in the tropics

	Million hectares	Agricultural population (millions of people)
Africa	263	11
Latin America	600	11
South and Southeast Asia	172	15
Tropical total	1035	37

(From Sanchez and Palm, forthcoming; adapted from Ref. 3.)

the basic cations Ca, Mg, and K (and to some extent P)—are returned to the soil in the ash and serve as fertilizer for the subsequent crops. As many of the soils on which shifting cultivation is practiced in the tropics are strongly acidic (often with high saturation of aluminium ions), the addition of substantial amounts of base cations can have an important liming effect and ameliorate conditions for plant growth by decreasing aluminium toxicity. Burning also increases the direct impact of rain hitting the soil. With the loss of plant cover and destruction of the litter layer, erosion can result, causing drastic loss of highly enriched surface soil.

Cropping is the third phase of shifting cultivation and is extremely variable. It can be as short as two cropping seasons on inherently infertile soils, or as long as five cropping seasons on more fertile soils.[4] The first crops are typically fast-growing and nutrient-demanding, such as cereal crops (including maize, upland rice, sorghum, or millet). Subsequent crops tend to be slower-growing and less nutrient-demanding (such as cassava, bananas, or legumes). Characteristics of the cropping phase are a rapid decline in soil organic matter and soil fertility, described in the classic text of Nye and Greenland.[5] This decline is often accompanied by an increase in weed pressure. Eventually the investment of labor in weeding exceeds the return in crop productivity, so that moving and clearing new plots is more favorable than continued cropping. At this point, the vegetation is left to regrow into the fallow phase. The length of fallow required to restore the original productivity of the land depends on many factors, including the length of the preceding cropping phase. Periods of at least 15–20 years appear to be necessary in West Africa.[4,6]

Ruthenberg proposed a useful classification of systems based on the intensity of land use,[1] where a value R is assigned for the proportion of land cultivated annually, or the proportion of time any given piece of land is held under cultivation. If the proportion of land cultivated $R = 0.15$, then the dwellings also move; if $R = 0.30$, then a greater portion of dwellings tend to be static. When R rises above 0.33, such land uses are no longer considered to be shifting cultivation, and instead are classified as fallow systems. Systems with values of R above 0.70 are regarded as permanent farming.

SHIFTING CULTIVATION SYSTEMS ILLUSTRATE SUSTAINABILITY

The concept of sustainability of the natural resource base and agricultural production is illustrated in Fig. 1, based on the early analyses of Guillemin.[7] The first case represents a situation where land is abundant, so that there is ample time for the land's productivity to recover to its original status before cropping (Fig. 1a). Under this scenario there is a period during which the land is rested unnecessarily. Greater productivity can be achieved when land is cropped as soon as soil productivity is restored to its earlier status (Fig. 1b). The time required to recover after each cropping phase is indicated as increasing, presumably as the rate of vegetation recovery decreases with repeated clearance. As population pressure on land increases, the length of the fallow restoration period is shortened, so that the land is cleared and cultivated before it reaches its prior soil fertility status, resulting in a productivity decline to a new equilibrium value (Fig. 1c). Although recovery of soil fertility is often stated as one of the main reasons for the fallow phase, the level of nutrients in the soil often decline's during the fallow phase as nutrients are transferred from the soil to the regrowth vegetation.[8] As such, it is the total nutrient stocks in the soil plus vegetation system that is important to the recovery of fertility.

A fairly well documented example of shifting cultivation is the Chitemene system found in northeast Zambia.[9,10] The whole region is covered by open savanna *miombo* woodland dominated by nonnodulating legume trees belonging to the subfamily *Caesalpinioideae* of the genera *Brachystegia* and *Julbernardia*. Circular areas of land are opened by lopping the high branches from the trees, leaving trunks 2–4 meters in length in the fields. The branches are heaped into the center of the opened circle, typically an area 10 times that of the cultivated area needed to provide sufficient vegetation to obtain adequate crop yields. The predominant soils in northern Zambia are strongly acid Oxisols and the large amount of ash is important in ameliorating aluminium toxicity in the soil. The land is typically sown to the relatively nutrient-demanding crop of finger millet (*Eleusine coracana*) followed by a crop of longer duration of cassava, often intercropped with Bambara groundnut (*Vigna subterranea*). The land is then abandoned and left fallow, with trees fairly rapidly regrowing from the lopped stumps. The population pressure in this region long ago exceeded the carrying capacity of the land,[11] so that although

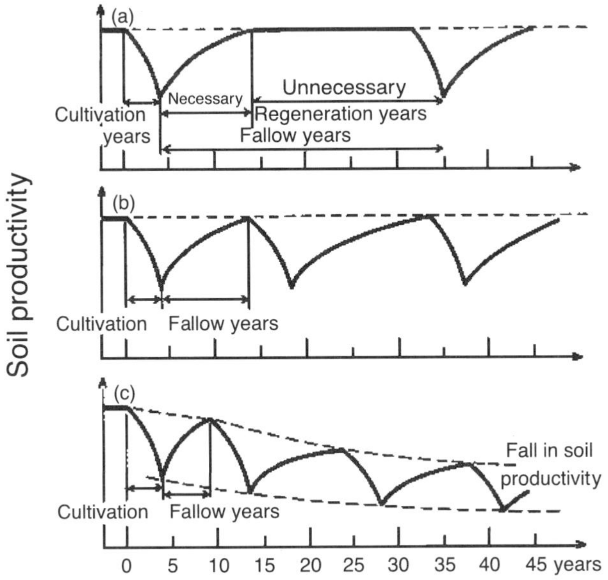

Fig. 1 The theoretical relationship between the length of fallow and soil productivity. (From Refs. 1 and 7.)

Chitemene agriculture is still widely practiced, much of the population depends on permanent farming or on income from relatives living in cities.

UNDERSTANDING SHIFTING CULTIVATION

Shifting cultivation is an extensive form of food production. Various attempts to intensify agriculture in remote areas have tended to fail for reasons clearly summarized by Ruthenberg:[1]

> Fallow systems exist by soil-mining, and efforts to prevent it are usually not economic (....) given the price relations in the location and the preference of the people concerned. The return to soil-preserving measures (green manuring, compost, terracing, etc.) is too low in relation to the disutility of effort, and there is not yet the scarcity of land which would bring about a change in preferences.

As such, there is an almost inevitable degradation in the natural resource base with the intensification of shifting cultivation, until the point that the land becomes scare and the returns in soil-preserving measures become high. This process of land degradation and subsequent investment in and recuperation of land is well described by Boserup.[12] The search for alternative forms of sustainable livelihoods for people living on the forest margins of the tropics has been a major aim of the Alternatives to Slash-and-Burn Program (http://www.asb.cgiar.org/).

CONCLUSION

By combining data into simple mathematical models, substantial insights have been gained in our understanding of shifting cultivation. The first such attempt can be traced to the classical work of Nye and Greenland,[5] who described the basis of declining soil C content, depending on the relative lengths of cropping and fallow phases. Trenbath[13] used a wide variety of data sources from Southeast Asia to demonstrate how the length of the fallow recovery phase increased with longer periods of cropping, to the point that after six crops, vegetation recovery was deflected to anthropic savanna dominated by dense grass cover, typically of *Imperata cylindrica*. This analysis has been further developed by van Noordwijk to explore the relationships between population pressure, productivity of agriculture, and other ecosystem services such as C stocks, biodiversity, and clean water provision.[14,15]

ARTICLES OF FURTHER INTEREST

Organic Agriculture As a Form of Sustainable Agriculture, p. 846
Sustainable Agriculture: Philosophical Framework for, p. 1198
Reconciling Agriculture with the Conservation of Tropical Forests, p. 1078
Sustainable Agriculture and Food Security, p. 1183
Sustainable Agriculture: Definition and Goals, p. 1187

REFERENCES

1. Ruthenberg, H. *Farming Systems in the Tropics*, 3rd Ed.; Clarendon Press: Oxford, 1980; 101.
2. Geist, H.J.; Lambin, E.F. Proximate causes and underlying driving forces of tropical deforestation. Bioscience **2002**, *52*, 143–150.
3. Dixon, J.; Gulliver, A.; Gibbon, D. *Farming Systems and Poverty: Improving Farmers' Livelihoods in a Changing World*; FAO: Rome, 2001.
4. Sanchez, P.A. *Properties and Management of Soils in the Tropics*; John Wiley: New York, 1976; 618.
5. Nye, P.H.; Greenland, D.J. *The Soil Under Shifting Cultivation*, Technical Communication No. 51; Commonwealth Agricultural Bureaux: Harpenden, UK, 1960.
6. Szott, L.T.; Palm, C.A.; Buresh, R.J. Ecosystem fertility and fallow function in the humid and subhumid tropics. Agrofor. Syst. **1999**, *47*, 163–196.
7. Guillemin, R. Evolution de l'agriculture autochthone dans les savannes de l'Oubangui. Agron. Trop., Nogent **1956**, *12*. Nos, 1,2,3.
8. Szott, L.T.; Palm, C.A. Nutrient stocks in managed and natural humid tropical fallows. Plant Soil **1996**, *186*, 293–309.

9. Stromgaard, P. Biomass, growth, and burning of woodland in a shifting cultivation area of south central Africa. For. Ecol. Manag. **1985**, *12*, 163–178.
10. Stromgaard, P. Soil nutrient accumulation under traditional African agriculture in the miombo woodland of Zambia. Trop. Agric. (Trinidad) **1991**, *68*, 74–80.
11. Chidumayo, E.N. A shifting cultivation land use system under population pressure in Zambia. Agrofor. Syst. **1987**, *5*, 15–25.
12. Boserup, E. *The Conditions of Agricultural Growth*; Aldine: New York, 1965.
13. Trenbath, B.R. The Use of Mathematical Models in the Development of Shifting Cultivation Systems. In *Mineral Nutrients in Tropical Forest and Savanna Ecosystems*; Proctor, J., Ed.; Blackwell Scientific Publications: Oxford, 1989; 353–371.
14. van Noordwijk, M. Scale effects in crop-fallow rotations. Agrofor. Syst. **1999**, *47*, 239–251.
15. van Noordwijk, M. Scaling trade-offs between crop productivity, carbon stocks and biodiversity in shifting cultivation landscape mosaics: The FALLOW model. Ecol. Model. **2002**, *149*, 113–126.

Cropping Systems: Yield Improvement of Wheat in Temperate Northwestern Europe

C

Hubert J. Spiertz
Wageningen University, Wageningen, The Netherlands

INTRODUCTION

Cereal production in Western Europe is mainly determined by wheat and barley production, with oats and rye as minor cereal crops. During the period 1996–2001 the harvest area averaged 17.4 and 11.4 Mha for wheat and barley, respectively. France, Germany, and the United Kingdom are the countries in the temperate region with the largest area of wheat production: 5.1, 2.8, and 1.9 Mha, respectively. According to the (FAOSTAT) database, average wheat yields (14% moisture) in Western Europe have increased since 1972 from 3950 to 5860 kg ha^{-1} on average. However, in the better endowed regions of the coastal zones of Northwestern Europe, such as the Netherlands, the wheat yields increased from 4820 to 8200 kg ha^{-1}. The progress in raising yields has been even more successful in winter wheat than in spring wheat because of the much longer growing season, which allows extended duration of tillering and spikelet development as well as postfloral photosynthetic activity and grain filling.

DISCUSSION

Yields of winter wheat in regions with long days and a mild climate, e.g., Northern Germany and Scotland, currently average about 9000 kg ha^{-1}, with top yields up to 11,000 kg ha^{-1} under conditions with optimal fertilization and no occurrence of pests and diseases. With the change in regulations on the use of pesticides and on food safety the occurrence of weeds, pests, and diseases will be less under chemical control. Fig. 1 indicates that winter wheat yields in the Netherlands since 1995 have clearly leveled off from the steady yield advance over several decades. This yield stagnation might result from a decrease in the use of pesticides and nitrogen in response both to lower cereal prices and to societal concerns about the effects of these inputs on environmental quality.

The characteristics of the climate and the soils in Western Europe were presented in *Ecosystems of the World*.[1] Besides climate and soil quality, technological innovations have been and will continue to be an important factor in the development of European cereal production. As a consequence of a change in the European Community's Common Agricultural Policy (CAP) the prices of cereal bulk commodities have decreased considerably during the last decade. As a result, the European policy has made a shift from quantity to quality in baking and brewing, as well as for other food and convenience and feed-related products. The priority for food quality and safety provides an opportunity to reorganize the agrifood chain according to the expectations of consumers and to make the agricultural and food sectors more competitive, while respecting safety and environmental requirements. Besides the shift from a bulk commodity to a high-quality produce used for food and feed, cereals will continue to play an important role in arable cropping systems because of the need to control soil-borne diseases without the use of pesticides.

WHEAT RESEARCH: THE CASE OF CROP PHYSIOLOGY

The yield performance of wheat has been studied on the level of single plants and even organs as well as on the level of the crop and cropping system. Fundamental research on plant development and crop growth during the second half of the 20th century in the Netherlands and the United Kingdom has been of great importance to understanding yield formation in wheat. Yield potential is defined as the yield of a cultivar adapted to a specific environment grown without biotic (pests and diseases) and abiotic (water and nutrients) stress or other yield-limiting factors.[2] The earliest assessments of the potential yield of wheat for a defined environment were made on the basis of leaf photosynthesis measurements and theoretical assumptions about light interception, respiration, and assimilate distribution.[3] The first estimates of potential grain yields in a temperate climate under favorable growing conditions centered around 10,000 kg ha^{-1}. These calculations of grain yield were based on a sound understanding of biophysical and physiological processes at the canopy level. Such theoretical studies and experimental evidence made it very clear that there existed a considerable gap between actual and potential grain yields.

DOI: 10.1081/E-EPCS 120010541

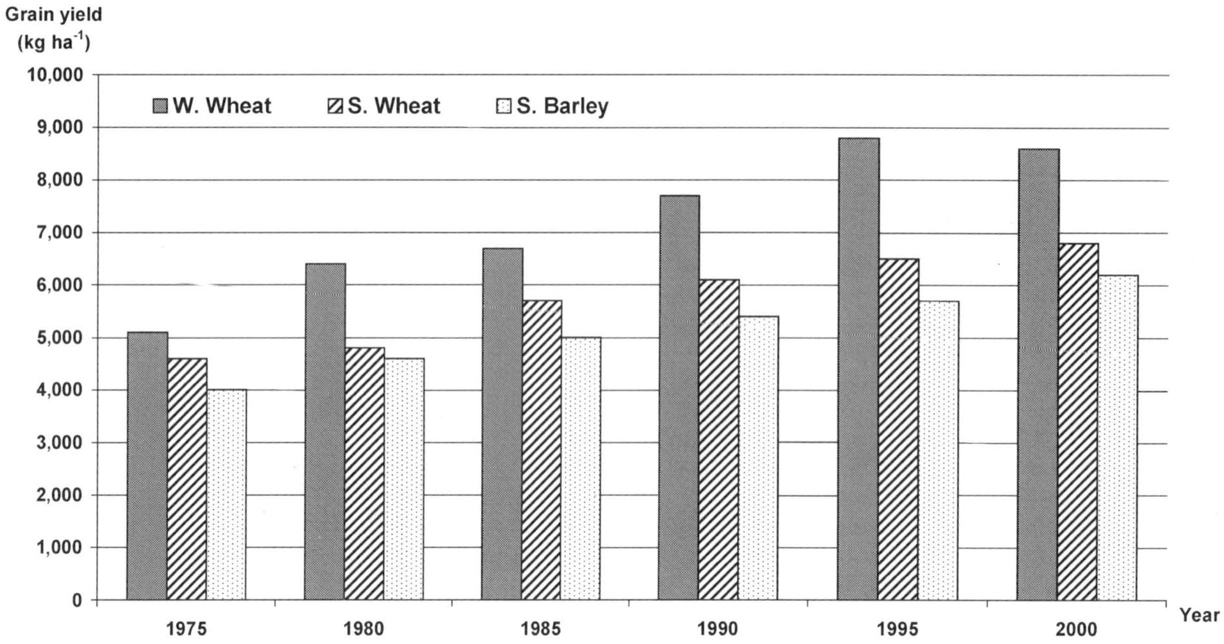

Fig. 1 Trends in grain yields of winter wheat, spring wheat, and spring barley in the Netherlands from 1975 to 2000.

Grain yield under favorable growing conditions is mainly sink-limited, especially for wheat. From the experiments under controlled conditions it may be concluded that a 25 to 30% increase in the potential grain yield level will be possible. Potential grain yields of about 15,000 kg ha^{-1} require a stand of at least 600 heads per m^2 with a minimum of 50 grains per head. Our understanding of processes governing initiation of organ development and related changes in carbon and nitrogen fluxes is much weaker than our knowledge of process controls on light interception and crop photosynthesis.

OPPORTUNITIES FOR GENETIC IMPROVEMENT

It is almost a matter of common sense that yields have limits. Within the crop, there are limits to the rates of processes that produce dry matter, and harvest index cannot be increased beyond certain limits.[4,5] Because current wheat cultivars are already close to this theoretical threshold, the scope for future increases in potential yield by increasing harvest index is very small. Therefore, increasing biomass production must be considered, sooner or later, as the main route toward further raising potential yield.[6] Of the total leaf nitrogen, 50 to 80% is allocated to photosynthetic machinery. Accordingly, short-term regulation and long-term acclimation of photosynthesis, with respect to nitrogen costs, are also major subjects of ecophysiological studies for yield improvement in relation to nitrogen requirements.

The recent advances in functional genomics have allowed us to understand the detailed genetic basis of many complex traits, such as flowering time, culm length, and stay-green characteristics. Within the next few years, it is likely that other important agronomic traits will be linked to a relatively small number of key genes. This will have an impact on breeding perspectives when it is possible not only to quantify the effect of gene activity on physiology and morphology, but also to integrate these effects over the life cycle of the crop.[7]

In the future, more emphasis should be given to defining new ideotypes that are adapted to milder temperatures during winter and spring and to temperature extremes during flowering and grain filling. Concurrently, breeding should be more focussed on improving the composition of the grain in relation to the demands of the food processing industry and consumer demands.[8] The protein demand of such high-quality grains can only partially be met by reallocation of nitrogen from the vegetative parts. A substantial amount of nitrogen uptake must occur during the postflowering period.

CROP AND CANOPY MANAGEMENT

During the first five decades of the last century, yields of cereals stagnated because the risk of lodging and infestations of pests and diseases placed limits on increasing nitrogen supply to cereal crops. A breakthrough came when the yielding ability of new varieties could be expressed under controlled management practices: split-dressings of nitrogen, use of chlormequat to improve lodging resistance, and the use of systemic pesticides and fungicides.[9] In the period from 1970 to 1990, annual increases in actual wheat yields were about 2.5% instead of the historical 0.5 to 1%, which may be considered as the European ''green revolution.'' This achievement was the result of a technology package that increased biomass production by means of enhanced photosynthetic longevity of the canopy of high-yielding cultivars.

The synergism between the genetic improvement of wheat cultivars and improved crop management resulted in higher biomass yield and an increase in the harvest index as well. This additive effect stimulated farmers to boost wheat yields by increasing inputs under the favorable economic conditions for growing cereals in the European Union. It turned out that farmers could considerably reduce inputs, without risk of a yield penalty, by precision management. The strategic research on growth, development, and yield formation of cereals has contributed to research-based crop management with a time- and dose-specific crop protection and nitrogen fertilization approach, thereby securing yield performance and stability in Northwestern Europe.[1] At issue is whether this approach will allow *continued* increases in yields, however, as farmers are motivated by policies and incentives to reduce inputs of fertilizers and pesticides in response to societal interests in the protection of environmental quality and natural resources.

CONCLUSION

In Europe, wheat has been the model crop for studying progress made by breeding, as well as for pioneering studies on actual and potential crop production. In the future, more insight into gene-plant-crop-environment interactions will contribute to a better and earlier assessment of the yield potential and stability of new genotypes. Molecular-based plant breeding has become so powerful in modifying the genetic base of a genotype that plant breeding can no longer rely only on testing of the new material in field plots. New, advanced methods of evaluation under standardized conditions with the use of crop models are needed.

The transformation from high-input cereal production to an ecologically based ''green'' or ''organic'' crop production has had a huge impact on cereal research, breeding, and management practices. It is the start of a new era, in which knowledge of gene expression can be linked to better understanding and monitoring of crop functioning under optimal and stress conditions. Based on real-time monitoring of the crop, fine-tuning of the time, space, and dose management in will become possible. The aim will be to produce healthy food with a controlled use of natural resources and external inputs.

ARTICLES OF FURTHER INTEREST

Breeding for Nutritional Quality, p. 182
Cropping Systems: Irrigated Rice and Wheat of the Indo-gangetic Plains, p. 355
Ecophysiology, p. 410
Nitrogen, p. 822

REFERENCES

1. Spiertz, J.H.J.; van Heemst, H.D.J.; van Keulen, H. Field-Crop Systems in North-Western Europe. In *Ecosystems of the World; Field Crop Ecosystems*; Pearson, C.J., Ed.; Elsevier: Amsterdam, 1992; 357–371.
2. Evans, L.T.; Fischer, R.A. Yield potential: Its definition, measurement and significance. Crop Sci. **1999**, *39*, 1544–1551.
3. de Wit, C.T. Photosynthesis on leaf canopies. Wagening. Agric. Res. Rep. **1965**, *663*, 1–57.
4. Austin, R.B. Yield of wheat in the United Kingdom: Recent advances and prospects. Crop Sci. **1999**, *39*, 1604–1610.
5. Hay, R.K. Harvest index: A review of its use in plant breeding and crop physiology. Ann. Appl. Biol. **1995**, *126*, 197–216.
6. Dreccer, F.M. *Radiation and Nitrogen Use in Wheat and Oilseed Rape Crops*; Thesis Wageningen University, 2000; 133 pp.
7. Spiertz, J.H.J.; Schapendonk, A.H.C.M. Opportunities for Wheat Improvement; The Role of Crop Physiology Revisited. In *Proceedings Warren E. Kronstad Symposium*; Reeves, J., McNab, A., Rajaram, S., Eds.; CIMMYT, Mexico, 2002.
8. Triboi, E.; Triboi-Blondel, A.-M. Productivity and grain or seed composition: A new approach to an old problem. Eur. J. Agron. **2002**, *16* (3), 163–186.
9. Spiertz, J.H.J.; de Vos, N.M. Agronomical and physiological aspects of the role of nitrogen in yield formation of cereals. Plant Soil **1983**, *75*, 379–391.

Crops and Environmental Change

Jørgen E. Olesen
Danish Institute of Agricultural Sciences, Tjele, Denmark

INTRODUCTION

Crops are directly affected by any changes in the external environment, which can influence crop development, growth, and resource use efficiency. These environmental changes include changes in climatic conditions (primarily temperature and rainfall), levels of atmospheric CO_2, SO_2, NO_x, ozone, and ultraviolet-B (UV-B) radiation. Depending on the character of the change and the current climate and soil conditions, these changes may be beneficial, neutral, or detrimental to crop yield and quality. Agricultural crops also contribute to environmental change by solute and gaseous emissions affecting environment at local to global scales. These emissions include nitrate leaching, phosphorus losses by erosion, and gaseous emissions of ammonia, nitrous oxide, and methane.

GLOBAL ENVIRONMENTAL CHANGE

Global change is defined in this article as actual and prospective anthropogenic changes in land use, atmospheric composition, nutrient deposition, climate, and UV-B radiation. The earth is undergoing rapid environmental changes because of human actions. The natural rates of nitrogen addition and phosphorus liberation to terrestrial ecosystems have been doubled, and atmospheric concentrations of CO_2, CH_4, and N_2O have been increased to 40%, 130%, and 17%, respectively, above preindustrial levels. The atmospheric concentration of these greenhouse gases are projected to further increase, leading to a global warming of 1.7 to 5.4°C by 2100.[1] Stratospheric ozone depletion leads to increased (UV-B) radiation. At local to regional scales emissions of NH_3, SO_2, and NO_x may lead to plant damages, soil acidification, and eutrofication. Photochemical reactions with air pollutants lead to the formation of tropospheric ozone. Many of these emissions and effects are related to land use changes.

CLIMATE IMPACTS ON CROPS

Crop production is based primarily on photosynthesis, and is thus dependent on incoming radiation. However, the potential for production set by the radiation is greatly modified by temperature and rainfall. The main effect of temperature is to control the duration of the period when growth is possible in each year. Also, other processes linked with the accumulation of dry matter (leaf area expansion, photosynthesis, respiration, etc.) are directly affected by temperature. Rainfall and soil water availability may affect the duration of growth through effects on leaf area duration and the photosynthetic efficiency through stomatal closure (Table 1).

There is an optimum temperature range for cool climate crops such as wheat, potato, and soybean of 15–20°C and for warm climate crops such as rice and maize of 25–30°C.[2] However, there is considerable variation between crops and cultivars in their critical low and high temperatures. For determinate crop species, which include most annual crops and many of the perennial ones, crop duration is determined by thermal time. For these crops a temperature increase will reduce the duration of crop growth and thus biomass accumulation often leading to lower yields (Fig. 1). For nondeterminant species, which include managed grasslands and biannual crops such as beets, a temperature increase will often increase the duration of crop growth in cool and temperate climates, leading to higher yields (Fig. 1).

Changes in seasonal rainfall distribution and intensities will in most cases be more important for crop production than changes in the annual amount. Drought stress or excessive moisture may lead to crop failure, or more timely rainfall may be beneficial to crops in more arid areas. Crop yield and quality responds in nonlinear ways to changes in temperature and rainfall. Such responses are often most pronounced in marginally suitable crop areas. The result is that increased variability of temperature and rainfall reduces in average yields in addition to increasing yield variability.[3] Overall, climate change is expected to increase yields at high- and mid-latitudes, and lead to decreases at lower latitudes. The global food supply is expected to be relatively unaffected. However, regional and continental effects may be severe.[4]

EFFECT OF OTHER ENVIRONMENTAL CHANGES

Plant photosynthesis responds strongly to CO_2 concentration.[5] However, the response depends on the photosynthetic pathway, and the increased photosynthesis with

Encyclopedia of Plant and Crop Science
DOI: 10.1081/E-EPCS 120010501

Table 1 Sensitivity of cropping systems to changes in selected environmental conditions

Component	Temperature	Rainfall	CO_2	$O_3/SO_2/NO_x$
Plants	Growth duration	Dry matter growth	Dry matter growth Water use	Dry matter growth
Water	Irrigation demand Soil salinization	Soil moisture Water table Workability	Soil moisture	
Soil	SOM decomposition Nutrient cycling	Soil erosion Nutrient leaching	Litter input to soil	Acidification
Pests/diseases	Proliferation of pests	Bacterial and fungal infections	Host biomass quality	
Weeds	Herbicide effectiveness	Herbicide effectiveness	Crop competition	

elevated CO_2 is strongest for C_3 crops (Fig. 1). CO_2 enrichment also reduces stomatal aperture and stomatal density, causing a reduction in crop transpiration. The resulting effects of these responses to higher CO_2 are increasing resource-use efficiencies for radiation, water, and nitrogen.[6] The highest response is seen for water-use efficiency, and this effect is seen in both C_3 and C_4 species. C_3 species capable of N_2-fixation have shown particular high growth stimulation to elevated CO_2.[7]

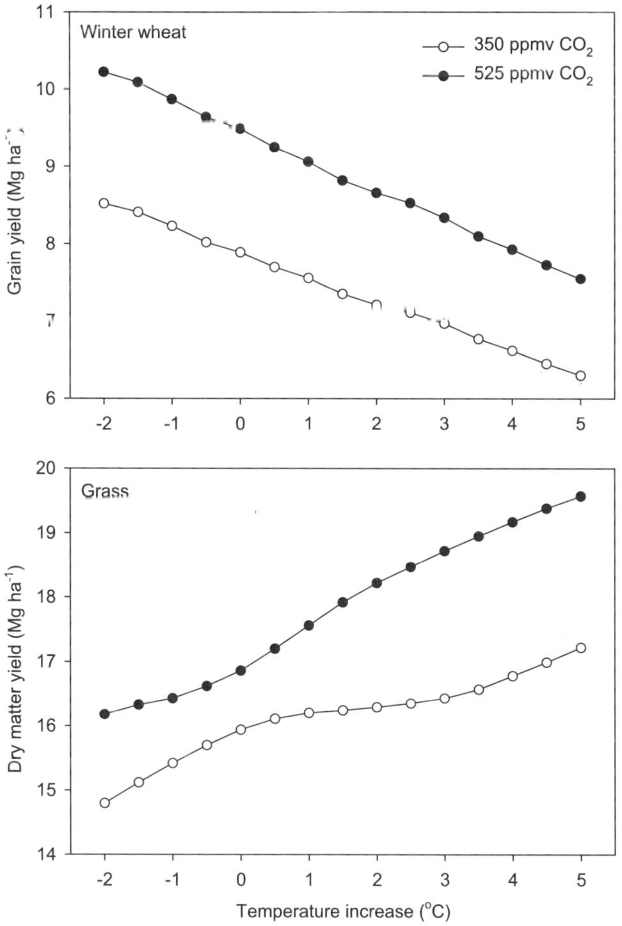

Fig. 1 Simulated average yield of winter wheat and cut grassland with changes in temperature for baseline (350 ppmv) and 50% higher (525 ppmv) atmospheric CO_2 concentration using the CLIMCROP model on loamy sand soil in Denmark.

Plant photosynthesis is sensitive to high UV-B irradiance. However, crops are generally resistant to increased UV-B radiation, and realistic UV-B irradiances in the field do not appear to have any significant effects on crop photosynthesis.[8] Increased tropospheric ozone concentration reduces photosynthesis through damage to the photosynthetic systems. In Europe, a doubling of tropospheric ozone concentration has been shown to reduce wheat yields by 9%.[9] High concentrations of SO_2 and NO_x have also been found to reduce photosynthesis and plant growth.

INTERACTIVE AND INDIRECT EFFECTS

Positive effects of elevated CO_2 in absolute terms are usually enhanced by increasing temperature and light. At higher CO_2 levels, damage to plant growth from air pollutants such as NO_x, SO_2, and O_3 is partly reduced because of reduced stomatal conductance.[5,9]

Crops depend on soils for water and nutrient supply. Soil organic matter plays a key role in building and sustaining soil fertility. Increased temperature will increase the turnover rate of organic matter with the risk of long-term reductions in soil fertility (Table 1). The weather also influences the practicality of managing soils and crops properly. This often determines the range of profitable crops that can be grown.

The majority of pest and disease problems are closely linked with their host crops. This makes major changes in plant protection problems less likely. However, increased temperature is more favourable for the proliferation of insect pests and many diseases in warmer climates. Unlike pests and diseases, weeds are also directly

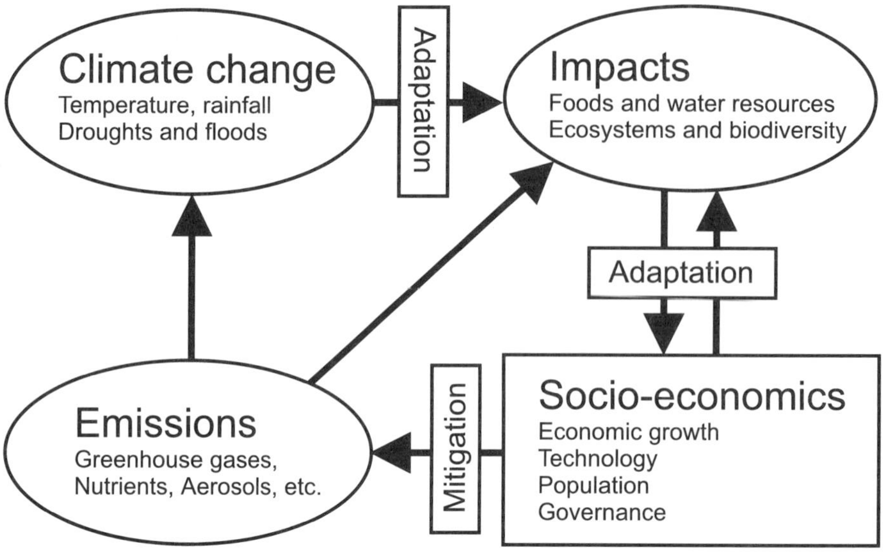

Fig. 2 Adaptation is the influence of emissions, climate change, and other factors on crop production systems and their environment, whereas mitigation is the attempt of society to prevent or reduce such impacts through emissions reductions.

influenced by changes in atmospheric CO_2 concentration. However, the competitive balance between the crops and the weeds depends on the specific interaction. The control of weeds, pests, and diseases is also likely to be affected by these changes.

ENVIRONMENTAL EFFECTS OF CROP PRODUCTION

Crop production is itself a major contributor to gaseous emissions of CH_4, NH_3, N_2O, and NO, which causes environmental problems at local (NO and NH_3) to global (CH_4 and N_2O) scales. Emissions of nitrates by leaching and phosphorus by leaching and surface runoff are also negatively affecting surrounding natural ecosystems through eutrofication. Most of these emissions are closely linked with the intensity of agricultural production.[10,11]

ADAPTATION AND MITIGATION TO CHANGE

The negative impacts of environmental change can be reduced through adaptation of cropping practices and agricultural systems (Fig. 2). Several adaptation strategies have been proposed involving both short-term adjustments of production systems (e.g., sowing date, cultivar choice, fertilizer, and pesticide inputs) and long-term adaptations involving major structural changes (e.g., land use, crop choice, land management, and irrigation systems).

Mitigation involves reducing the environmental emissions, and agriculture has a range of options to do so. Greenhouse gas emissions may be reduced, either directly by reducing energy use and emissions of methane and nitrous oxide or by substitution of fossil energy use and carbon sequestration in soils.[11] Leaching and erosion losses of nitrates and phosphorus can be reduced through improved crop and soil management aimed at closer nutrient cycles.[10]

CONCLUSION

Crop production is affected by environmental changes occurring at local to global scales. The changes may be both beneficial and detrimental, but effects are largest in regions that are currently marginal for a given crop. Adaptations in land use and crop management to the environmental changes will be necessary. As part of these adaptations, measures to reduce the emissions of environmentally harmful gases and solutes should also be considered.

ARTICLES OF FURTHER INTEREST

Air Pollutants: Effects of Ozone on Crop Yield and Quality, p. 13

Air Pollutants: Interactions with Elevated Carbon Dioxide, p. 17

Crop Responses to Elevated Carbon Dioxide, p. 346

Environmental Concerns and Agricultural Policy, p. 418

Exchange of Trace Gases Between Crops and the Atmosphere, p. 425

Sustainable Agriculture: Definition and Goals, p. 1187
Sustainable Agriculture: Economic Indicators of, p. 1195

REFERENCES

1. Houghton, J.T.; Ding, Y.; Griggs, D.J.; Noguer, M.; van der Linden, P.J.; Dai, X.; Maskell, K.; Johnson, C.A. Climate Change 2001: The Scientific Basis. In *Contribution of Working Group I to the Third Assessment Report of the Intergovernmental Panel on Climate Change*; Cambridge University Press: Cambridge, UK, 2001; 1–881.
2. Rötter, R.; van de Geijn, S.C. Climate change effects on plant growth, crop yield and livestock. Clim. Change **1999**, *43*, 651–681.
3. Mearns, L.O.; Rosenzweig, C.; Goldberg, R. Mean and variance change in climate scenarios: Methods, agricultural applications, and measures of uncertainty. Clim. Change **1997**, *35*, 367–396.
4. Parry, M.; Rosenzweig, C.; Iglesias, A.; Fischer, F.; Livermore, M. Climate change and world food security: A new assessment. Global Environ. Change **1999**, *9*, S51–S67.
5. Allen, L.H. Plant responses to rising carbon dioxide and potential interactions with air pollutants. J. Environ. Qual. **1990**, *19*, 15–34.
6. Drake, B.G.; Gonzàlez-Meler, M.A. More efficient plants: A consequence of rising atmospheric CO_2? Ann. Rev. Plant Physiol. Plant Mol. Biol. **1997**, *48*, 609–639.
7. Kimball, B.A.; Kobayashi, K.; Bindi, M. Responses of agricultural crops to free-air CO_2 enrichment. Adv. Agron. **2002**, *77*, 293–368.
8. Allen, D.J.; Nogués, S.; Baker, N.R. Ozone depletion and increased UV-B radiation: Is there a real threat to photosynthesis? J. Exp. Bot. **1998**, *49*, 175–188.
9. van Oijen, M.; Ewert, F. The effects of climatic variation in Europe on the yield response of spring wheat cv. Minaret to elevated CO_2 and O_3: Analysis of open-top chamber experiments by means of two crop growth simulation models. Eur. J. Agron. **1999**, *10*, 249–264.
10. Power, J.F.; Wiese, R.; Flowerday, D. Managing farming systems for nitrate control: A research review from management systems evaluation areas. J. Environ. Qual. **2001**, *30*, 1866–1880.
11. Rosenzweig, C.; Hillel, D. Soils and global climate change: Challenges and opportunities. Soil Sci. **2000**, *165*, 47–56.

Cross-species and Cross-genera Comparisons in the Grasses

Ismail M. Dweikat
University of Nebraska, Lincoln, Nebraska, U.S.A.

INTRODUCTION

The grass family (Poaceae) contains about 10,000 species, 700 genera, and six sub-families. Although other angiosperm families contain more species and more genera, the Poaceae exceeds all other families in ecological dominance. Grasses are believed to have originated in the late Cretaceous period more than 66 million years ago, and they now populate almost every land habitat known in both temperate and tropical regions. Members of this family include some of our most important agricultural food and feed crops, wheat (*Triticum* spp.), maize, (*Zea mays*), rice (*Oryza sativa*), barley (*Hordeum vulgare*), oats (*Avena sativa*), sugarcane (*Saccharum* spp.), rye (*Secale cereale*), pearl millet (*Pennisetum glaucum*), and sorghum (*Sorghum bicolor*).

Genomic comparison studies among grass species have become quite extensive and a remarkable amount of information regarding genomic relationships within the grass family is now available. A graphic summary of comparative mapping data among grasses, greatly simplified, has become known as "The Circle Diagram," because of the method used to associate expressed gene sequence maps of different grass species on one radial axis. The diagram shown includes twelve species from four subfamilies: Pooides (wheat and oat), Panicoids (maize, sorghum, sugarcane, and foxtail millet), Oryzoids (rice), and the Chlorinoids (finger millet). The extraordinary pattern that emerges visually illustrates that all the grasses in the four subfamilies contain genes in the same order despite huge differences in their DNA content and chromosome numbers. In fact, it is possible to establish that a limited number of rice linkage groups is sufficient to summarize the marker arrangement on the 12 rice, seven wheat, and 10 maize chromosomes. However, these types of comparisons also reveal a reasonable frequency of large chromosomal rearrangements often shared by particular lineages of grass species.

Evidence is accumulating that much of the fortyfold variation in genome size among the grasses is due to variation in the prevalence of one specific class of repetitive DNA, retrotransposons, both active and ancient. Retrotransposon elements account for more than 70% of the total maize genome. Despite the huge difference in genome size, chromosome number variation from 2–19, and their ancient origin, comparative mapping reveals a remarkable degree of synteny (conserved clustering of gene/unique sequences) among closely related grass species. The extensive conservation of gene content and order among grass chromosomes has led to the proposal of a single progenitor genome structure for all grasses.

RICE: THE MODEL GRASS SPECIES

In the past, categorization of living organisms was based on morphology and anatomy. However, as genetic knowledge has improved, classifications are increasingly reliant on genomic differences among species. With advances in biotechnology, molecular techniques facilitate the classification more rapidly and precisely. According to the theory of evolution, it is believed that all species have evolved from the same ancestor, which has subsequently mutated to create a specific identity for each species. In the order of evolution, closely related species should have more similar genomes than distant ones, and the ancestor should have a smaller and simpler genome organization than its descendants, in which repetitive DNA has accumulated to result in increased genome size. Deletion of a chromosome fragment can occur, but this generally is disadvantageous to survival. Hence, genome size tends to enlarge rather than lessen over time. Based on these assumptions, rice, which has the smallest genome (430 mbp) of the grass family, is proposed to be a grass ancestor and a model for studying comparative genetics and evolution of the grass family. Figure 1 shows a minimized (four subfamilies) phylogenetic tree and includes most of the cultivated grass species. The family exhibits large variation in DNA content among the diploid grasses, the 1C genome size (DNA content of the unreplicated haploid set of chromosome) varying by fortyfold.

Among rice species exists remarkable genome uniformity. Comparative studies of *O. sativa* and the wild species *Oryza officinalis* shows that marker order in both genetic maps is largely conserved.[1] Likewise, Aggarwal et al.[2] fingerprinted more than 23 species of wild rice using AFLP markers to investigate their phylogenetic

Encyclopedia of Plant and Crop Science
DOI: 10.1081/E-EPCS 120010629

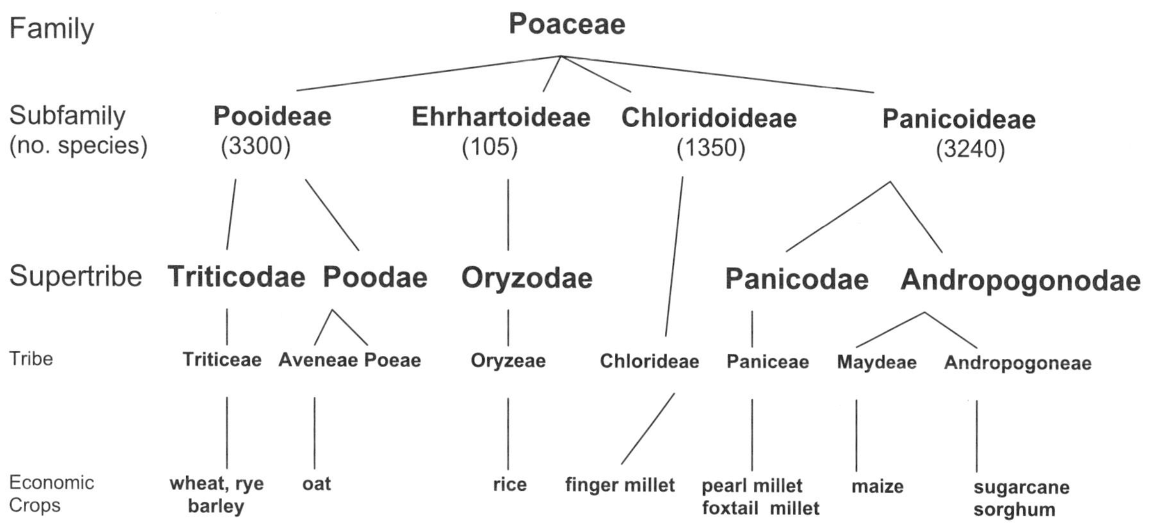

Fig. 1 Simplified taxonomy of the cultivated grass species.

relationships and concluded that all the species originate from one common ancestor.

STRIKING GROSS GENOMIC SIMILARITY IS EVIDENT WITHIN THE POOIDES SUBFAMILY

Genetic linkage maps can be compared among species using a common set of DNA probes. Comparison among the molecular maps of *Triticum* spp. indicates that the orders of molecular markers on the linkage map of these species, detected with the same probes, were essentially homosequential. Consequently, consensus maps that represent the unified linkage maps of each chromosome of these species can be constructed. This RFLP cross-mapping exposed the previously unrecognized generality that genes and their linear arrangement along the chromosomes of different cereal species are remarkably conserved.[3–5] Initially, this finding was restricted to close relatives, e.g., within *Triticeae* cereals, between *Oryza* species, and between maize and sorghum. However, recent evidence suggests that this conservation has been maintained over long evolutionary periods to extend across the entire grass family (Fig. 2).

Wheat, a hexaploid, is postulated to have originated from the combined genomes of three diploid species, represented by present-day *Triticum monococcum*, *Triticum uratu*, and *Triticum tauschii*. These three genomes of wheat were shown to be highly collinear (conserved in order of genes or markers). Apart from a few translocations that occurred early in the evolution of the hexaploid wheat, chromosomes of the A, B, and D genomes are found to be entirely collinear.[6] The A, B, and D genomes within hexaploid wheat have not undergone extensive gene loss or rearrangement since polyploid formation. Likewise, comparison of *Hordeum bulbosum*, a wild relative of barley, and cultivated barley (*H. vulgare*) reveals a high level of conserved synteny between the two species.[7,8]

Advances in genomic analysis of the grasses has encouraged researchers to expand their vision to what might be possible if they examine species further out on the evolutionary tree. The goal of these efforts is to maximize information transfer among species using knowledge of their relationships at different levels. The conserved gene orders and the feasibility of sharing DNA probes and primers across species has greatly extended the power of mapping analysis of corresponding chromosomal regions to more distant species.

In the *Triticeae* tribe, wheat, barley, and rye share a basic chromosome number of 7, with similar DNA content. A consensus map was developed for the *Triticeae* based on a common set of markers mapped onto the respective linkage groups of *Triticum aestivum*, *Triticum monococcum*, *Triticum tauschii*, and *Hordeum* species.[7–9] Significant syntenic relationships and conserved gene order are evident within the tribe. Likewise, only a few well-defined rearrangements distinguish the rye and wheat genomes at this level of analysis.[9] The close relationships that exist among the *Triticeae* species has even allowed the transfer of genomic and cDNA clones from one species to another.

MORE DETAILED GENOMIC ANALYSIS WITHIN THE PANICOID SUBFAMILY

In the Andropogoneae tribe, maize and sorghum both have a chromosome number of 10. In comparisons with an

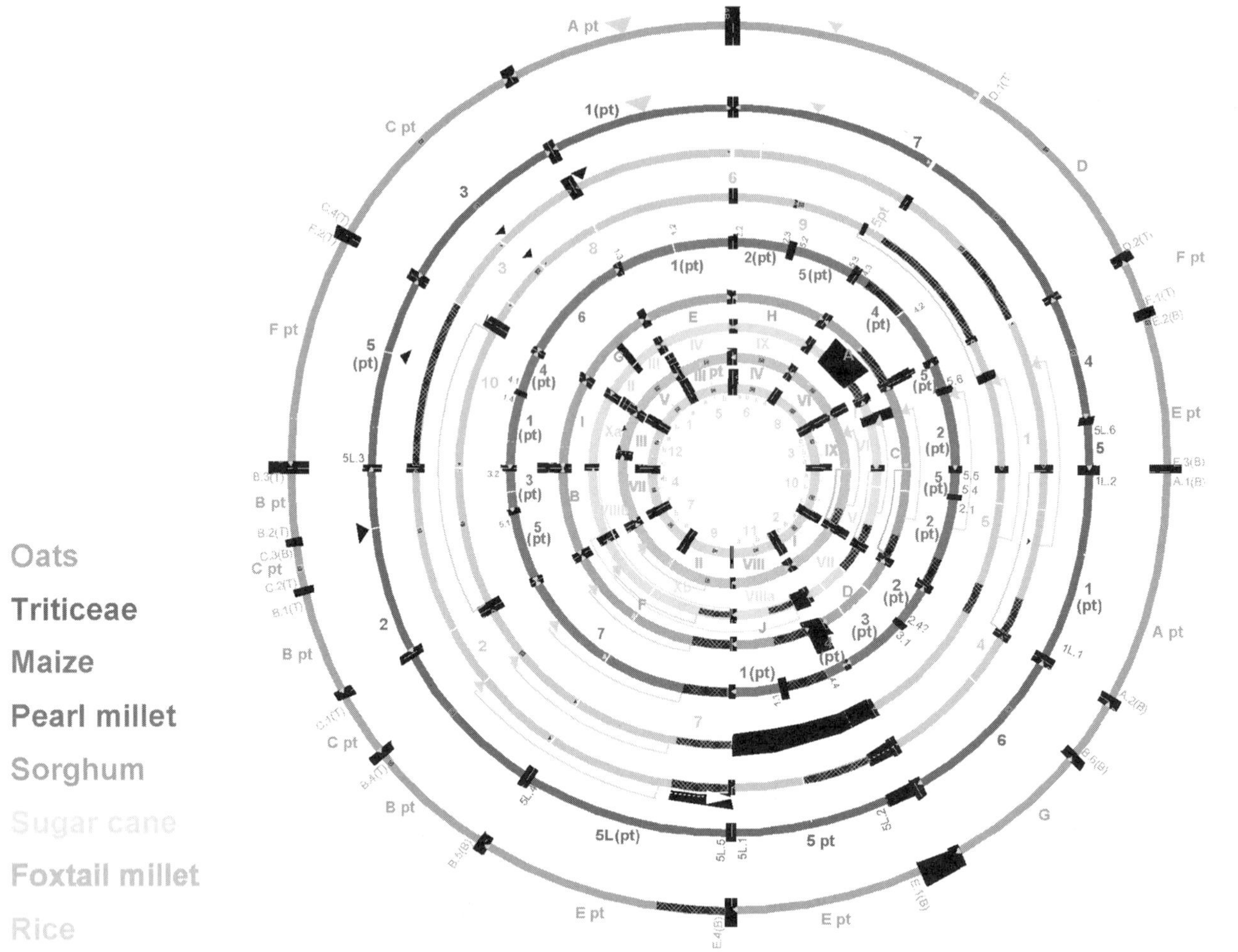

Fig. 2 Genomes of 12 grass species aligned to form one consensus map. Each circle represents the chromosomal complement of a single grass genome. The alignment of seven genomes was arranged relative to rice, the smallest known grass genome. The arrows indicate inversions and translocations, relative to rice, postulated to describe present day chromosomes. Locations of telomeres (■) and centromeres (▲) are shown where known. Hatched areas indicate chromosome regions for which very little comparative data exist. L, long arm; S, short arm; T, top of chromosome; B, bottom of chromosome; and pt, part (3). (*View this art in color at www.dekkker.com.*)

ancient species such as maize, differences between homeologous chromosome sets emerge that involve major genomic rearrangements, although most appear to be associated with small local events.[10] Maize is considered to be an ancient polyploid that has, during its evolution, undergone genomic diploidization.[11] In accordance with the polyploid ancestry of maize, a rice linkage group corresponds to two different maize chromosomes.[12] Comparative mapping between maize and sorghum similarly confirms the duplicated nature of the maize genome. Most sorghum chromosomal segments are found to be collinear to pairs of duplicated regions in maize, with finer chromosomal rearrangement also evident between the two genomes. Sugarcane linkage groups also show syntenic relationships to the duplicated regions of maize. However, the genomes of sugarcane and sorghum appear to be much more closely related to one another, with respect to chromosome organization, than either is to maize.[13]

Sugarcane possesses one of the most complex genome organizations of the grasses. Sugarcane germplasm is composed of polyploid, aneuploid clones derived from interspecific hybridization between two species, *Saccharum officinarum* ($2n = 80$) and *S. spontaneum* ($2n = 40$–128). In spite of this enhanced genomic complexity and ploidy, a high degree of collinearity exists among all *Saccharum* species.[14]

The extensive genomic similarity evident among grass species by DNA mapping studies has led logically to detailed cross-species comparative sequencing efforts that involve orthologous chromosome segments. Several such studies, comparing maize with sorghum and rice, and wheat with barley and rice, have provided important insight to the events that differentiate grass genomes and likely participate in the speciation process. Although at the gross chromosomal level genome organization is remarkably conserved, gene sequences demonstrate over 20% alteration in organization or content, distinguishing even closely related species.[15] These extensive small genomic rearrangements include deletions, duplications, translocations, and inversion events. Consequently, at a finer level of analysis, genomic synteny is surprisingly low.

COMPARATIVE MAPPING AS A TOOL FOR GENE ISOLATION

The evolutionary implications of genome collinearity were elegantly demonstrated in the mapping of major loci involved in the domestication of sorghum, maize, and rice several years ago. Paterson et al.[16] showed the influence of convergent selection for important agronomic traits over the evolution of these crops, with corresponding quantitative trait loci (QTLs) identified to control the same complex traits in species that diverged over 65 million years ago.

With data derived from global comparative mapping endeavors, several groups have now reported the identification of putative orthologous loci in a range of grass genera. Lin et al.[17] reported that QTLs with major effects on height and flowering in sorghum have counterparts in homeologous segments of the rice, wheat, barley, and maize genomes. Pereira and Lee[18] similarly identified genomic regions affecting plant height in sorghum and maize. Quantitative trait loci (QTL) controlling the important agronomic traits such as shattering and dwarfing, seed size, and photoperiod sensitivity genes are also collinear between grass species.[6,18,19] Harrington et al.[20] mapped starch branching enzyme III in rice and used comparative maps to predict the location of this enzyme in maize, oat, and wheat. The maps that provide the basis for combining the genetic information available in these crops are based on probes that hybridize across the cereals.

A map-based cloning approach in rice has been used for the isolation of the wheat *Ph1* gene that controls chromosome pairing.[21] Likewise, Asnaghi et al.[22] utilized the synteny among sorghum, maize, sugarcane, and rice to determine the map location of a sugarcane rust-resistance gene. Synteny in the vicinity of *rpg4*, a barley stem rust-resistance gene, was investigated with rice and barley molecular markers. This strategy was successful in physically and genetically positioning the locus.[23]

With the sequencing of the rice genome, the ability to locate important genes of interest within physically defined intervals will be greatly enhanced. Efforts are now underway to develop collections of rice insertion mutants. Thus, it will soon be feasible to clone loci efficiently in rice of major agronomic importance based on their functional identification. With these genomic tools in place, it will also be possible to assess fully and exploit the genomic synteny that exists among the grasses.

CONCLUSION

Comparative mapping in the grasses has revealed extensive genome collinearity at the gross chromosomal level between the studied species. However, analyses conducted at a finer scale suggest a high degree of repeat expansion, small-scale deletion/insertions, inversions and translocations distinguishing species. These alterations should not, for the most part, impose major limitations on the use of comparative mapping for gene cloning and fine-scale mapping of important traits. Comparative analysis of closely related grass species that differ in genome size has revealed both the sources of genome size differences and the utility of small genome species for cross-species gene localization efforts. The availability of the genome sequence of rice spurs a wealth of comparative and functional studies in all cereal species. Although the conservation of microsynteny between rice and the other cereals is often incomplete, the possibility to use the rice sequence as a source of markers and candidate genes will play an important role in future gene cloning and function assignments.

ACKNOWLEDGMENTS

The author would like to extend his sincere appreciation to Dr. Mike Gale, John Innes Institute, U.K., for providing the consensus map figure.

ARTICLES OF FURTHER INTEREST

REFERENCES

1. Jena, K.K.; Khush, G.S.; Kochert, G. Comparative RFLP mapping of a wild rice, *Oryza officinalis*, and cultivated rice, *O. sativa*. Genome **1994**, *37*, 382–389.

2. Aggarwal, R.; Brar, D.; Nandi, S.; Huang, N.; Khush, G. Phylogenetic relationships among Oryza species revealed by AFLP markers. Theor. Appl. Genet. **1999**, *98*, 1320–1328.
3. Gale, M.D.; Devos, K.M. Plant comparative genetics after 10 years. Science **1998**, *282*, 656–659.
4. San Miguel, P.; Bennetzen, J. Evidence that a recent increase in maize genome size was caused by the massive amplification of intergene retrotransposons. Ann. Bot. **1998**, *82*, 37–44.
5. Bennetzen, J.L.; Freeling, M. Grasses as a single genetic system: Genome composition, collinearity and compatibility. Trends Genet. **1993**, *9*, 259–261.
6. Devos, K.; Gale, M. Genome relationships: The grass model in current research. Plant Cell **2000**, *12*, 637–646.
7. Laurie, D.A.; Pratchett, N.; Bezant, J.H.; Snape, J.W. RFLP mapping of five major genes and eight quantitative trait loci controlling flowering time in winter x spring barley (Hordeum vulgare L.) cross. Genome **1995**, *38*, 575–585.
8. Laurie, D.; Devos, K. Trends in comparative genetics and their potential impacts on wheat and barley research. Plant Mol. Biol. **2002**, *48*, 729–740.
9. Devos, K.M.; Gale, M.D. Comparative genetics in the grasses. Plant Mol. Biol. **1997**, *35*, 3–15.
10. Tikhonov, A.P.; San Miguel, P.J.; Nakajima, Y.; Gorenstein, N.M.; Bennetzen, J.F. Collinearity and its exceptions in orthologous *adh* regions of maize and sorghum. Proc. Natl. Acad. Sci. U. S. A. **1999**, *96*, 7409–7414.
11. Gaut, B.; Doebley, J. DNA sequence evidence for the segmental allotetraploid origin of maize. Proc. Natl. Acad. Sci. U. S. A. **1997**, *88*, 2060–2064.
12. Moore, G.; Devos, K.M.; Wang, Z.; Gale, M.D. Cereal genome evolution—Grasses, line up and form a circle. Curr. Biol. **1995**, *5*, 737–739.
13. Grivet, L.; D'Hont, A.; Dufour, P.; Hamon, P.; Roques, D.; Glazmann, J. Comparative genome mapping of sugarcane with other species within the Andropogoneae Tribe. Heredity **1994**, *73*, 500–508.
14. Paterson, A.; Bowers, J.; Burow, M.; Draye, X.; Elisk, C.; Jiang, C.; Kaster, C.; Lan, T.; Lin, Y.; Ming, R.; Wright, R. Comparative genomics of plant chromosomes. Plant Cell **2000**, *12*, 1523–1539.
15. Bennetzen, J.; Ramakrishna, W. Numerous small rearrangements of gene content, order and orientation differentiate grass genomes. Plant Mol. Biol. **2002**, *48*, 821–82715.
16. Paterson, A.H.; Lin, Y.R.; Li, Z.; Schertz, K.F.; Doebley, J.F.; Pinson, S.R.M.; Liu, S.C.; Stansel, J.W.; Irvine, J.E. Convergent domestication of cereal crops by independent mutations at corresponding genetic loci. Science **1995**, *269*, 1714–1718.
17. Lin, Y.; Schertz, K.; Paterson, A. Comparative analysis of QTLs affecting plant height and maturity across the Poaceae, in reference to an interspecific sorghum population. Genetics **1995**, *141*, 391–411.
18. Pereira, M.G.; Lee, M. Identification of genomic regions affecting plant height in sorghum and maize. Theor. Appl. Genet. **1995**, *90*, 380–388.
19. Gaut, B.S. Evolutionary dynamics of grass genomes. New Phytol. **2002**, *154*, 15–28.
20. Harrington, S.; Bligh, H.; Park, W.; Jones, C.; McCouch, S. Linkage mapping of starch branching enzyme III in rice (Oryza sativa L.) and prediction of location of orthologous genes in other grasses. Theor. Appl. Genet. **1997**, *94*, 564–568.
21. Foote, T.; Roberts, M.; Kurata, N.; Sasaki, T.; Moore, G. Detailed comparative mapping of cereal chromosome region corresponding to the *Ph1* locus in wheat. Genetics **1997**, *147*, 801–807.
22. Asnaghi, C.; Paulet, F.; Kaye, C.; Grivet, L.; Deu, M.; Glaszmann, J.C.; D'Hont, A. Application of synteny across poaceae to determine the map location of a sugarcane rust resistance gene. Theor. Appl. Genet. **2000**, *101*, 962–969.
23. Druka, A.; Kudrna, D.; Han, F.; Kilian, A.; Steffenson, B.; Frisch, D.; Tomkins, J.; Wing, R.; Kleinhofs, A. Physical mapping of the barley stem rust resistance gene *rpg4*. Mol. Gen. Genet. **2000**, *264*, 283–290.

Crown Gall

Thomas J. Burr
Cornell University, Geneva, New York, U.S.A.

C

INTRODUCTION

Crown gall is a plant disease characterized by the development of fleshy or woody galls on plants. The galls, or plant tumors, may develop on roots, at the crown, or on aerial parts of the plant. Generally, they develop on woody plant tissues and not on green growing shoots. Crown gall affects more than 90 families of plants and occurs worldwide. Infected plants may grow poorly and die prematurely. The unsightly appearance of crown gall is especially important in nursery operations where affected plants are usually discarded, resulting in great economic loss. Crown gall is particularly important on fruit and nut crops such as almonds, cherries, peaches, raspberries, and grapes. It is also important on ornamentals such as cypress, euonymus, forsythia, hibiscus, lilac, privet, rose, virburnum, and willow.

Crown gall is caused by bacteria belonging to the genus *Agrobacterium*, a member of the family Rhizobiaceae, and therefore related to *Rhizobium* spp. that are involved with fixation of atmospheric nitrogen in the nodules of legume plants. *Agrobacterium* spp. are gram-negative, aerobic, motile bacteria. It was recently proposed that the genus *Agrobacterium* be eliminated and the present species be placed in *Rhizobium*; however, the acceptance of the new nomenclature by bacteriologists is still being debated. Currently, within *Agrobacterium* there are four recognized plant pathogenic species: *A. tumefaciens*, *A. rhizogenes*, *A. rubi*, and *A. vitis*. The first two species occur on a wide range of hosts, whereas *A. rubi* causes cane gall on *Rubus* spp., and *A. vitis* causes disease only on grape. As scientists continue to isolate and characterize bacterial strains from crown gall-infected plants, it is becoming clear that additional host-associated species of *Agrobacterium* are likely to be discovered.

PATHOGEN BIOLOGY AND THE INFECTION PROCESS

Pathogenic (gall-forming) and nonpathogenic forms of *Agrobacterium* spp. are commonly detected in association with plants and in soils. In some cases, such as grape, the bacterium may be disseminated in propagation material that appears to be healthy. There are selective culture media that facilitate the isolation and identification of the bacteria. Colony growth is characteristic on the media and isolates are examined subsequently for their ability to cause galls on various plants such as sunflower, tobacco, and tomato. Other biochemical and physiological tests are also used to confirm *Agrobacterium* species.[1] Analysis of characteristic DNA sequences in the bacterium by polymerase chain reaction (PCR) is also useful for identifying species and for determining whether strains carry genes that are associated with gall formation (see the following).

The process by which *Agrobacterium* infects plants has been a very active and exciting research area for many years. Following are few of the major discoveries related to crown gall infection: 1) in 1958 it was determined that during infection the bacterium transfers a tumor-inducing factor to the plant, after which, galls can be cultured in the absence of the bacterium and without added hormones;[2] 2) crown gall tumors produce amino acid derivatives called opines;[3] and 3) infections result from transfer of part of the bacterial Ti plasmid (the T-DNA) into the plant where it is expressed.[4] The T-DNA is bordered by specific DNA recognition sequences and is composed of genes that are expressed in the plant, including those that encode the production of the plant hormones auxin and cytokinin. Thus, infected plant cells have increased hormone production resulting in rapid and uncontrolled plant cell growth that leads to gall development (Fig. 1).

Many of the steps involved in transfer of the T-DNA to the plant have been identified, for which process there are excellent reviews.[5,6] A general scheme of the infection process is shown in Fig. 2. In general, galls develop at sites on plants where wounds have occurred. Wounds may play several functions, including the release of chemicals such as sugars and phenolic materials that are sensed by the bacterium. Prior to infection, the bacterium attaches to plant cells. Then, following perception of the chemical signals, pathogenicity-related genes in the bacterium are activated, thereby initiating the process of preparing the T-DNA for its transfer to the plant cell. Steps include extracting the T-DNA from the Ti plasmid in a single-stranded form (the T-strand) and the binding of certain proteins (VirD2 and VirE2) to the T-strand that facilitate its transfer and protect it from enzymatic degradation. Other genes encode proteins (VirB1 to VirB11) that build a pore in the bacterial membrane, through which the T-strand can

Encyclopedia of Plant and Crop Science
DOI: 10.1081/E-EPCS 120010579

Fig. 1 Crown gall disease on trunks of grapevines. (*View this art in color at www.dekker.com.*)

travel on its way to the plant. A pilus (threadlike channel) connecting the bacterium to the plant cell has been visualized through which the T-strand probably travels on its way into the plant cell. Although an impressive number of discoveries have been made concerning *Agrobacterium* infection, additional research is necessary to completely understand the infection process.

Agrobacterium can be considered a natural genetic engineer of plants. It is the only known example of the occurrence of interkingdom transfer of DNA (from a bacterium to plant). This amazing phenomenon has been utilized to genetically engineer plants with genes for crop improvement. In this case, the disease-associated T-DNA genes are removed from the Ti plasmid and substituted with genes to be expressed in plants. Examples of such genes are those for virus resistance and herbicide tolerance, and those that make plants resistant to insect feeding.

Other key discoveries related to *Agrobacterium* infection were that molecules known as opines are produced by infected cells (in the galls) and that the bacterium has the ability to utilize the opines as selective nutrient sources.[3]

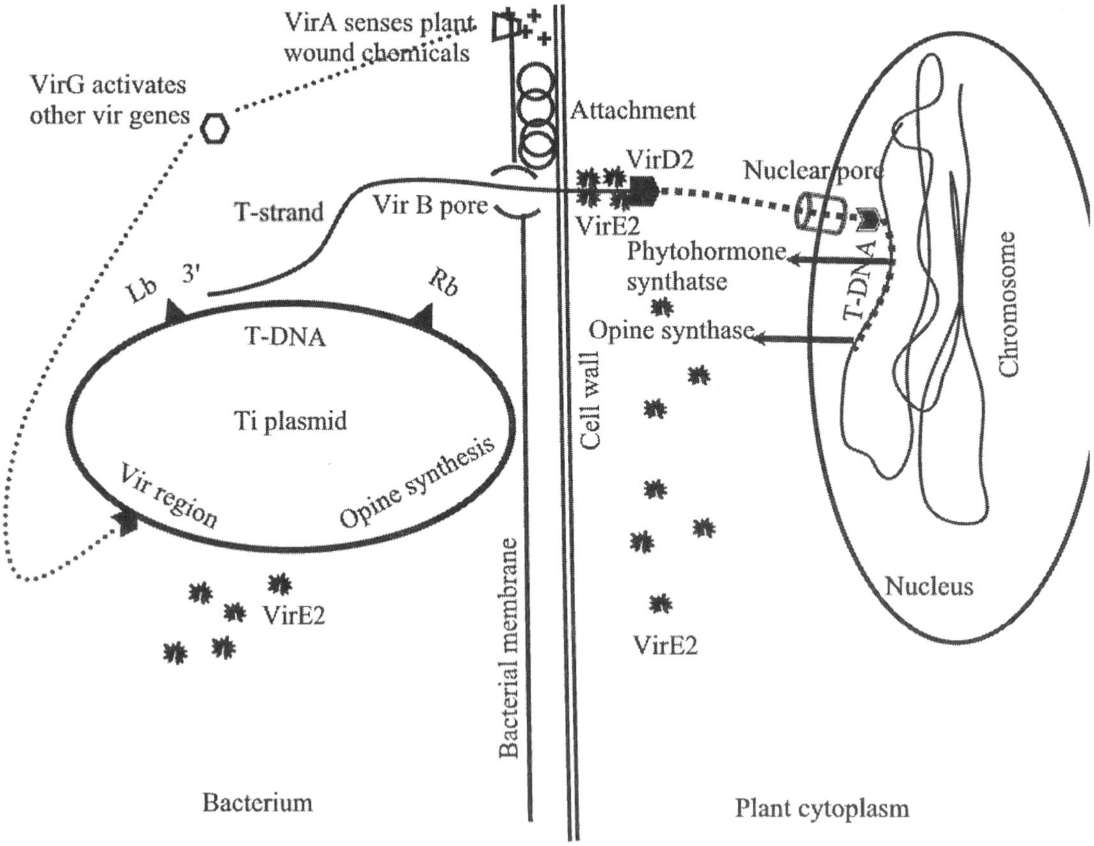

Fig. 2 Crown gall infection process. After the bacterium attaches to the plant cell, the VirA–VirG regulatory proteins sense plant wound signals and induce expression of other *vir* genes. VirD2 attaches to the 5′ end of the T-strand, which migrates though the pore made by multiple VirB proteins. In the plant, the T-strand is coated with VirE2, migrates through the nuclear pore, becomes integrated in the plant chromosome, and is expressed.

Certain opines also function as signals for conjugal transfer of Ti plasmids. Genes for opine synthesis reside on the T-DNA and are therefore transferred on the T-strand to the plant during infection. Genes for catabolism of opines are on the Ti plasmid, outside the T-DNA. Therefore crown galls provide the bacterium a nutritional environment whereby it has a selective niche for survival and growth. Many types of opines associated with Ti plasmids of *Agrobacterium*, including octopine and nopaline, have been identified.

Scientific examination of collections *Agrobacterium* strains make it apparent that there is a significant level of genetic diversity even within a species. Strains may differ in the opine genes they carry. Diversity has also been evaluated by determining the differences in the structures of Ti plasmids and by genetic fingerprinting of certain regions of the bacterial chromosome. Ti plasmids may have multiple T-DNA regions, all carrying different components of genes that are essential for gall formation. Genetic elements such as insertion sequences are also found in Ti plasmids and on the bacterial chromosome. Significant progress has been made in predicting phylogenic relationships within *Agrobacterium*.[7] Recently, the complete genome of *A. tumefaciens* strain C58 was completed and published.[8,9]

DISEASE CYCLE

Agrobacterium survives in soil and in association with plants and plant debris. It is often disseminated along with plants that appear to be healthy. In such cases it may persist in soil clinging to plants, in latent infections, or as an endophyte in the vascular system. In plants such as grape, the bacterium can be isolated from sap that is collected from bleeding plants in the early spring. Crown gall typically develops at wounds on roots or at the crown of plants, but on some hosts, such as grape, it is prevalent on trunks. The specific cells plant that are infected by the bacterium appear to be those involved in wound-healing. Galls are soft and fleshy during their early development and subsequently become necrotic and dry, at which time it is difficult to isolate the bacterium from them.

CONTROL OF CROWN GALL

Plant varieties often differ greatly in their susceptibility to crown gall. For example, wine grapes, *Vitis vinifera,* are generally quite susceptible, whereas *Vitis labrusca* and hybrid varieties are often more resistant. Such differences in varietal susceptibility have been noted for other plant species as well. When considering control strategies, therefore, it is may be possible to select plant varieties that are resistant. Another consideration in disease management is to avoid injuring plants. This may involve selecting proper planting sites where plants will be less prone to freeze injury or using cultivation practices that minimize wounding. Although *Agrobacterium* is sensitive to antibiotics such as streptomycin, treating plants with bactericides is not generally recognized as an effective control. Similarly, fumigation has not been effective for eradicating *Agrobacterium* from soils. One reason for the failure of chemical controls may be that the bacterium resides within plant tissues where it is unlikely to contact chemicals applied on the plant or to soil.

One highly successful way to prevent crown gall is with biological control (see also Management of Bacterial Diseases: Biological Control). In the early 1970s a non-tumorigenic strain of *Agrobacterium* (strain K84) was discovered in Australia.[10] This strain produces at least one antibiotic (agrocin 84) that is lethal to many strains of pathogenic *Agrobacterium*. Roots of plants are soaked in a suspension of K84 cells prior to planting as a means of protecting infection sites. A genetically modified form of K84 (strain K1026) was developed by deleting genes that would allow K84 to transfer genes associated with agrocin production to pathogenic forms of the bacterium (which then become resistant to the biological control). K84 and K1026 are sold commercially in several regions of the world and have been proven effective for controlling crown gall on several types of plants.

CONCLUSION

Crown gall is a disease of worldwide importance on many different plants. Four plant pathogenic species of *Agrobacterium* are recognized, and diverse tumorigenic and nontumorigenic strains exist in association with plants and in soil. Disease control is through selection of resistant varieties, cultural practices, and biological control, which has been quite successful. *A. tumefaciens* causes crown gall by transferring a portion of its Ti plasmid, the T-DNA, to the plant where it is integrated into the plant genome and is expressed. Gene expression leads to increased plant hormone production resulting in gall formation. This natural ability of *Agrobacterium* to function as a gene vector has been harnessed by scientists for the genetic engineering of numerous plant species.

Research on crown gall and its causal bacterium have yielded many exciting discoveries related to mechanisms of disease and plant-microbe interactions. Continued *Agrobacterium* research is focused on how bacteria communicate with each other, what plant genes are associated with infection, and how improvements can be made to the *Agrobacterium* plant transformation system. Results of such research will also likely lead to effective controls for crown gall.

ARTICLES OF FURTHER INTEREST

Bacterial Pathogens: Detection and Identification Methods, p. 84
Bacterial Pathogens: Early Interactions with Host Plants, p. 89
Bacterial Products Important for Plant Disease: Diffusible Metabolites As Regulatory Signals, p. 95
Bacterial Survival and Dissemination in Natural Environments, p. 108
Bacterial Survival and Dissemination in Seeds and Planting Material, p. 111
Classification and Nomenclature of Plant Pathogenic Bacteria, p. 286
Management of Bacterial Diseases of Plants: Regulatory Aspects, p. 669
Plant Diseases Caused by Bacteria, p. 947

REFERENCES

1. Moore, L.W.; Bouzar, H.; Burr, T.J. Agrobacterium. In *Laboratory Guide for Identification of Plant Pathogenic Bacteria*; Schaad, N.W., Jones, J.B., Chun, W., Eds.; American Phytopathological Society Press: St. Paul, MN, 2001; 17–39.
2. Braun, A.C. A physiological basis for autonomous growth of the crown-gall tumor cell. Proc. Natl. Acad. Sci. U. S. A. **1958**, *44*, 344–349.
3. Dessaux, Y.; Petit, A.; Farrand, S.K.; Murphy, P.J. Opines and Opine-Like Molecules Involved in Plant-Rhizobiaceae Interactions. In *The Rhizobiaceae: Molecular Biology of Model Plant-Associated Bacteria*; Spaink, H.P., Kondorosi, A.P., Hooykaas, P.J.J., Eds.; Kluwer Academic Publishers: Dordrecht, The Netherlands, 1998; 173–197.
4. Chilton, M.-D.; Drummond, M.H.; Merlo, D.J.; Sciaky, D.; Montoya, A.L.; Gordon, M.P.; Nester, E.W. Stable incorporation of plasmid DNA into higher plant cells: The molecular basis of crown gall tumorigenesis. Cell **1977**, *11*, 263–271.
5. Zhu, J.; Oger, P.M.; Shrammeijer, B.; Hooykaas, P.J.J.; Farrand, S.K.; Winans, S.C. The basis of crown gall tumorigenesis. J. Bacteriol. **2000**, *182*, 3885–3895.
6. Zupan, J.; Muth, T.R.; Draper, O.; Zambryski, P. The transfer of DNA from *Agrobacterium* tumefaciens into plants: A feast of fundamental insights. Plant J. **2000**, *23*, 11–28.
7. Otten, L.; DeRuffray, P.; Momol, E.A.; Momol, M.T.; Burr, T.J. Phylogenetic relationships between *Agrobacterium vitis* isolates and their Ti plasmids. Mol. Plant-Microb. Interact. **1996**, *9*, 782–786.
8. Wood, D.W.; 50 co-authors. The genome of the natural genetic engineer *Agrobacterium* tumefaciens C58. Science **2001**, *294*, 2317–2323.
9. Goodner, B.; 30 co-authors. Genome sequence of the plant pathogen and biotechnology agent *Agrobacterium* tumefaciens C58. Science **2001**, *294*, 2323–2328.
10. Kerr, A. Biological control of crown gall: Seed inoculation. J. Appl. Bacteriol. **1972**, *35*, 493–497.

Cytogenetics of Apomixis

Hans de Jong
Wageningen University, Wageningen, The Netherlands

INTRODUCTION

Apomixis in plants is a natural alternative to sexual reproduction, in which embryos are formed without paternal contribution and, hence, produce offspring that are genetically identical to the mother. This clonal reproduction through seeds is accomplished by circumventing the reductional part of female meiosis and fertilization. It has been described for more than 400 polyploid species (0.1% of all angiosperms) and is common in the *Rosaceae*, *Poaceae*, and *Asteraceae*. Apomixis is a complex reproduction pathway with three main modes: 1) adventitious embryony, 2) apospory, and 3) diplospory. Adventitious embryony, in which embryos with suspensors are formed directly from somatic tissue, is found in citrus. In aposporous apomicts, somatic ovary cells develop via unreduced embryo sacs into unreduced embryos, whereas in the diplosporous type, meiosis in the megasporocyte is incomplete, giving rise to unreduced gametes that later develop into unreduced embryo sacs. In addition, apomicticlike traits were described for various species, including alfalfa, barley, lily, and potato, which exhibit only components of apomixis such as $2n$ gamete formation, female and male parthenogenesis, and aberrant endosperm formation.

Apomixis is especially significant for its potential to develop a breeding system that can capture and fix heterosis and propagate complex traits in seed crops, including those that determine the health and nutritional value of foodstuffs. However, apomixis is very rare among crop species and so far, all attempts to transfer apomixis to cultivated plants by sexual crosses have resulted only in partially fertile, agronomically unsuitable addition lines with one or more alien chromosomes. Research groups worldwide now join efforts to understand better the genetics, molecular organization, and embryological development of apomixis and to transfer the trait from apomictic model species into crops, e.g., *Pennisetum squamulatum* to pearl millet, *Tripsacum* to maize, *Paspalum* to rice, *Taraxacum* to lettuce, and *Boechera* (*Arabis*) to *Arabidopsis*.

METHODOLOGY

In most natural apomicts, the mode of reproduction is related to its ploidy level, which means that diploids are sexual and polyploids apomictic (Fig. 1).[1,2] Bulked-segregant analysis of populations obtained from a sexual × apomictic cross therefore can detect molecular markers linked to the apomictic genes. Although RAPDs, AFLPs, RFLPs, and SCARs are favorites for linkage studies, SSRs (microsatellites) are superior because multiple molecular variants per locus can distinguish homologous chromosomes in a polyploid cross. RFLPs with cDNA probes previously mapped in maize, rice, and other grass species take advantage of the known genomic synteny observed in grass genomes.

Determining chromosome numbers in polyploid and aneuploid populations requires a time-consuming analysis of chromosome preparations. Now this technique is generally replaced by flow cytometry, which has been shown to be most effective for the screening of $2n$ pollen and seed samples.[3] However, bright field and Nomarski-DIC microscopy remain irreplaceable for more detailed assessment of meiotic chromosome behavior, spindle formation, and cell dynamics in male and female organs, as well as embryo developmental alterations in cleared whole-mount preparations of anthers and ovules (see Fig. 2).

Where genetic and physical mapping have produced bacterial artificial chromosomes or other large insert-vectors containing chromosomal regions with markers tightly linked to the apomictic components, the region can be mapped directly on the chromosomes by Fluorescence in situ Hybridization (FISH). This molecular cytogenetic technique elucidates the position of the apomixis trait with respect to centromere, telomere, and nucleolar organizer regions. The positions of apomictic-linked sequences in heterochromatin are of special interest, as these chromosome regions are rich in repetitive DNA and often prone to gene silencing.

GENETICS OF APOMIXIS

Genetic studies of apomixis are tedious and time consuming due to variation in the expression (penetrance) of the apomixis trait, the complex mode of polyploid inheritance, the need for microscopic assessment of the gamete and embryological phenotype, and the need for extensive progeny tests. Because apomixis occurs in

Encyclopedia of Plant and Crop Science
DOI: 10.1081/E-EPCS 120005602

Fig. 1 The common dandelion (*Taraxacum officinale*) is a well-known apomictic representative of the Compositae family. Triploid and tetraploid apomictic forms are distributed in northern and central Europe, whereas the diploid sexual forms are more frequent in the southern part of Europe. (*View this art in color at www.dekker.com.*)

several plant families, it is tempting to believe that their origin is polyphyletic, each with their own specific features and mechanisms. In pioneering studies, parthenogenesis was found to be strictly associated with apospory and, hence, inherited as a single Mendelian trait. Recent investigations favor a model with a few tightly linked dominant genes responsible for at least unreduced female gamete formation and parthenogenetic development of the embryo in the absence of fertilization.[4] In *Pennisetum squamulatum,* a chromosomal region containing 14 molecular markers linked to apospory and hemizygous in apomictic genotypes was found to be highly conserved with collinearity in the related apomictic buffelgrass species *Cenchrus ciliaris*.[5] In dandelion (*Taraxacum officinale*), sexual diploids pollinated with apomictic triploids revealed apomictic recombinants with diplospory and autonomous endosperm development but lacking parthenogenesis.[6] Similar examples of such apomictic recombinants, *Erigeron annuus* and *Poa pratensis*, suggest that the different elements of apospory were indeed controlled by separate gene systems. Additional genetic analyses in *Taraxacum* demonstrated that a dominant trait linked to an SSR marker and located on one of the NOR chromosomes appears to control diplospory. Because this marker is absent in sexual diploid progeny, it was postulated that the apomixis loci are linked to recessive lethals and cannot be transmitted through haploid pollen.

Comparative mapping with maize or rice markers in *Brachiaria decumbens*, *Tripsacum dactyloides*, and *Paspalum simplex* showed a clear lack of recombination in the region associated with the apospory locus, which in the corresponding sexual diploids cover a region ranging from 15 to 40 cM. In *B. decumbens*, apospory was found to be inherited as a single Mendelian locus that is embedded in a chromosomal region homologous to maize chromosome 5 and rice chromosome 2. *Tripsacum* is an important donor of apomixis to maize, and various intergeneric hybrids between them were produced for transferring the apomictic traits to the crop species. Bulk-segregant analysis revealed few RFLP markers that cosegregated with diplospory from *T. dactyloides*. Their restriction fragments were previously mapped on the distal end of chromosome 6 of maize. In a similar approach, the apospory locus in *B. decumbens* could show synteny with the homologous regions in maize chromosome five and rice chromosome 2. In *P. simplex*, apomixis is inherited as a single genetic unit on a large chromosome region that shows synteny of five markers from the telomeric region of rice chromosome 12 and one marker located on rice chromosome 8.

No natural apomicts have been described for wheat, but a synthetic sexual ''Salmon'' line was produced with the capability of parthenogenesis. In this system, a line with a T1BL·1RS translocation chromosome that was transferred into *Aegilops caudata* or *Ae. kotchyi* cytoplasm displayed autonomous embryo formation due to the synergy of a cytoplasmic factor and two nuclear genes.[7]

Apomixis is very rare in the Brassicacae, but the *Boechera holboellii* (*drumondii*) complex is one important representative that is a close relative of the dicot model plant for molecular biologists, *Arabidopsis thaliana*, which brings a huge treasure of molecular and genetic knowledge directly on female meiosis and the embryological aspects of apomixis. The *B. holboelli* complex is widespread in North America and Greenland

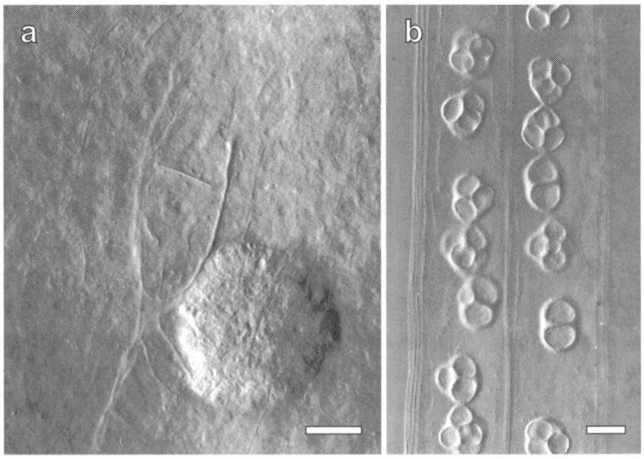

Fig. 2 Normarski-DIC photomicrographs of female and male organs of a triploid apomictic dandelion. **a** Ovary with dyad containing an unreduced megaspore and a degenerating megaspore. **b** Pollen sac with dyads of unreduced gametes and tetrads of reduced microspores. Bar equals 10 μm.

where apomictic forms occur even on the diploid level. Polyploidy and aneuploidy are associated with apomixis, and genetic analysis of independently derived aneuploid apomicts always indicates the presence of the same chromosome, which has undergone morphological changes and no longer pairs at meiosis with its putative progenitors.

SEGREGATION DISTORTION OF APOMIXIS LOCI

Most studies on apomixis demonstrated unequivocally that apomixis and diploidy are mutually exclusive. The general hypothesis assumes that the loci for apospory or diplospory are linked to recessive lethals and so are not transmitted to diploid progeny through haploid gametes. Various transmission studies of markers linked to apospory traits further revealed the following explanations for the absence of apomixis in diploids: 1) selection against haploid gamete formation; 2) meiotic drive, in which the apomictic trait containing chromosome remains as a univalent at metaphase I and always ends up in the diploid gamete; 3) postzygotic lethality.[8]

FUTURE PROSPECTS

Current research on apomixis focuses on 1) the genetics and molecular mechanisms of how apomixis is regulated in naturally apomictic species and 2) the identification of individual components of apomixis (nonreduction of the chromosome number, parthenogenesis, and the control of endosperm development) in well-characterized sexual model systems such as *Arabidopsis*, rice, and maize. If the molecular bases of apomictic components are sufficiently elucidated and its DNA sequence characterized, traits can then be transferred to crop species by genetic transformation, thus avoiding long breeding programs and hybridization barriers that prevent introgression of the desired trait from unrelated species. Further cytogenetic research is also needed to elucidate the role of unsaturated meiotic chromosome pairing in epigenetically changing the chromosome regions for triggering apomictic pathways. When chromosome regions with apomixis-linked markers have single-copy sequences sufficient for FISH studies, cytogenetic mapping of the region on the chromosomes will then be possible. We also need to understand how apomixis traits can be expressed and transmitted and to what extent these processes are sex-specific. Many of the aforementioned apomicts have sexual relatives with recently sequenced genomes (*Paspalum*-rice, *Boechera*-*Arabidopsis*) or for which extensive EST libraries are available (*Taraxacum*-lettuce). The genome collinearity and sequence similarity between these species will further accelerate fine mapping and gene identification efforts.

ACKNOWLEDGMENTS

The author thanks Drs. Peter van Dijk and Peter van Baarlen for providing the figures for this paper and for discussions on the manuscript.

ARTICLES OF FURTHER INTEREST

Aneuploid Mapping in Polyploids, p. 37
B Chromosomes, p. 71
Breeding Clones, p. 174
Breeding Hybrids, p. 186
Genomic Imprinting in Plants, p. 527
Polyploidy, p. 1038

REFERENCES

1. Nogler, G.A. Gametophytic Apomixis. In *Embryology of Angiosperms*; Johri, B.M., Ed.; Springer Verlag: New York, 1984; 475–518.
2. Bretagnolle, F.; Thompson, J.D. Gametes with the somatic chromosome number: Mechanisms of their formation and role in the evolution of autopolyploid plant. New Phytol. **1995**, *129*, 1–22. Tansley Review No. 78.
3. Matzk, F.; Meister, A.; Schubert, I. An efficient screen for reproductive pathways using mature seeds of monocots and dicots. Plant J. **2000**, *21*, 97–108.
4. Grimanelli, D.; Leblanc, O.; Espinosa, E.; Perotti, E.; Gonzalez, D.L.D.; Savidan, Y. Mapping diplosporous apomixis in tetraploid Tripsacum: One gene or several genes? Heredity **1988**, *80*, 33–39.
5. Roche, D.; Hanna, W.W.; Ozias Akins, P. Is supernumerary chromatin involved in gametophytic apomixis of polyploid plants? Sex. Plant Reprod. **2001**, *13*, 343–349.
6. van Dijk, P.J.; Tas, I.C.Q.; Falque, M.; Bakx-Schotman, T. Crosses between sexual and apomictic dandelions (Taraxacum). II. The breakdown of apomixis. Heredity **1999**, *83*, 715–721.
7. Matzk, F. The ‘Salmon system’ of wheat—A suitable model for apomixis research. Hered. Landskrona **1996**, *125*, 299–301.
8. Grossniklaus, U.; Nogler, G.A.; van Dijk, P.J. How to avoid sex: The genetic control of gametophytic apomixis. Plant Cell **2001**, *13*, 1491–1498.

Drought and Drought Resistance

Graeme C. Wright
Nageswararao C. Rachaputi
Queensland Department of Primary Industries, Kingaroy, Australia

INTRODUCTION

Drought can be considered as a set of climate pressures that can result from a combination of heat, aerial, or soil water deficits, as well as salinity. The diversity of drought created from these phenomena has led to the selection of numerous types of resistance mechanisms that operate at different levels of life organization (molecule, cell, organ, plant, and crop). Decades of research have been dedicated to the understanding of these mechanisms, with a premise that the improved understanding would contribute to the long-term improvement of plant and crop production under drought conditions.[1]

DISCUSSION

This article will concentrate on crop production, and in that context drought is a term used to define circumstances in which growth or yield of the crop is reduced because of insufficient water supply to meet the crop's water demand. During the 1960s to 1980s most of the drought research was dedicated to understanding the mechanisms of survival and growth under drought conditions. It is only in fairly recent times that attention has been given to recognizing the complex nature of drought and to separating the *productivity* of crop plants under drought from *survival* mechanisms. Drought resistance in modern agriculture requires sustainable and economically viable crop production, despite stress. However, plant survival can be a critical factor in subsistence agriculture, where the ability of a crop to survive drought and produce some yield is of critical importance.

Hence, in the context of agricultural production, drought resistance in a crop can be best defined in terms of the optimization of crop yield in relation to a limiting water supply.[2] Multitudes of options exist for farmers to alleviate the effects of drought on crop yield, depending on the probability of drought. These can be categorized into management and genetic options, although they can be integrated into a package to manage drought in the target environment. The basis of most management technologies adopted by farmers revolves around optimizing water conservation and its subsequent utilization by the crop. Examples include the use of deep tillage to increase rainfall infiltration, stubble retention to minimize soil evaporation,[3] and intercropping.[4] Genetic options include the use of the best locally adapted varieties or landraces, as well as relaying and intercropping varieties with varying phenology to exploit differences in timing and severity of drought patterns.[5]

Future advances in crop drought resistance and associated improvements in productivity under drought are most likely to come from genetic improvement programs that can apply the wealth of knowledge created over the past century. As Richards[6] states, however, it will never be possible to *overcome* the effects of drought, any progress is likely to be slow, and the gains will only be small. The following sections will therefore concentrate on opportunities and emerging technologies for the improvement of drought resistance in crop plants, using genetic enhancement.

DROUGHT RESISTANCE TRAITS

Levitt[7] has proposed a terminology for drought resistance and its subdivision into different categories based on different mechanisms. These three categories of drought resistance have been widely accepted, and they continue today in a slightly modified form to provide a framework for evaluating potential traits for use in crop breeding programs.[8–10] They are drought escape, dehydration postponement, and dehydration tolerance.

Drought Escape

Matching the phenology to the expected water supply in a given target environment has been an important strategy for improving productivity in water-limited environments.[11] In most crop species there is large genetic variability in phenological traits, and these traits are highly heritable and amenable to selection in large-scale breeding programs.[12] Matching phenology has proven to be a highly successful approach in environments that

Encyclopedia of Plant and Crop Science
DOI: 10.1081/E-EPCS 120010407

have a high probability of end-of-season drought stress pattern.[13]

Dehydration Postponement

Crops use a variety of mechanisms to maintain turgor in leaves and reproductive structures despite declining water availability. They can effectively regulate water loss from leaves via stomatal control, with large varietal differences in stomatal conductance in response to leaf water potential recorded in cereals[14] and grain legumes.[15]

Production of abscisic acid (ABA) has been implicated as a mechanism behind stomatal control.[16] Other benefits of ABA, including maintenance of turgor in wheat spikelets and subsequent grain set, have also been reported.[17] Subsequent stimulation of research activity into the use of the ABA trait in crop breeding programs has followed. However, Blum[18] recently concluded that while ABA is undoubtedly involved in plant response to drought stress and even perhaps in desiccation tolerance, its value in the context of drought resistance breeding is still questionable.

Osmotic adjustment (OA) has been reported as an important drought-adaptive mechanism in crop plants where solutes accumulate in response to increasing water deficits, thereby maintaining tissue turgor despite decreases in plant water potential.[19] OA has been shown to maintain stomatal conductance, photosynthesis, and leaf expansion at low water potential,[20] as well as reducing flower abortion[21] and improving soil-water extraction despite declining water availability.[22,23] More recent research has confirmed that OA is directly associated with grain yield in a number of crops, including wheat,[24,25] sorghum,[26] and chickpea.[27]

Dehydration Tolerance

The ability of cells to continue metabolism at low-leaf water potential is known as dehydration tolerance. Membrane stability and the associated leakage of solutes from the cell[28] provide one measure of the ability of crop/genotypes to withstand dehydration. Sinclair and Ludlow[29] have suggested that the lethal water potential is a key measure of dehydration tolerance, with significant variation among crops and cultivars observed. Some stages of the plant's life cycle are less susceptible to large reductions in water content. For instance, prior to establishment of a large root system, seedlings may often survive large reductions in water content.[30] Protection against lethal damage in seedlings and seeds is correlated with the accumulation of sugars and proteins.[31] Proline has also been implicated in cellular survival of water deficits, and has also been involved in osmotic adjustment. Its role as a selection trait for enhanced drought resistance has been questioned.[32] Although the work on drought resistance mechanisms has produced a few promising leads, their application in practical breeding programs has been limited.

HEIRARCHY OF DROUGHT RESISTANCE TRAITS

Richards[6] suggests there are two major principles to consider when identifying critical traits to use in breeding programs aimed at improving productivity under drought, namely, the influence of the trait in relation to time scale and to level of organization.

Time Scale

Traits that influence drought resistance in crops can span a wide range of time scales. Short-term responses to water deficit, for example, include many of the processes covered earlier (heat-shock proteins, stomatal closure, OA, ABA), which Passioura[2] suggests are often primarily concerned with "metabolic housekeeping." Although these processes are important, they tend to be associated with crop survival rather than with events that influence crop productivity. At the other end of the time scale are longer-acting processes such as control of leaf area development, which can be modulated by the crop to adjust water supply to prevailing demand.[33] It is not always clear which processes are operating to control these balances, but presumably hormonal signals are involved. Passioura[2] argues that researchers need to distinguish between traits linked to short-term responses that might be important for overall drought resistance and those that are unimportant when integrated over longer time scales.

Level of Organization

The capacity of the trait to influence yield is related to the level of organization (molecule–cell–organ–plant–crop) in which the trait is likely to be expressed.[34] Richards[6] cites an example that despite the doubling of crop yields since 1900, the rate of leaf photosynthesis, which is expressed at the cellular-organ level, has remained the same or decreased. Increases in leaf area during this period have been largely responsible for the yield increases. It is concluded that the closer the trait is to the level of organization of the crop the more influence it will have on productivity (Fig. 1).

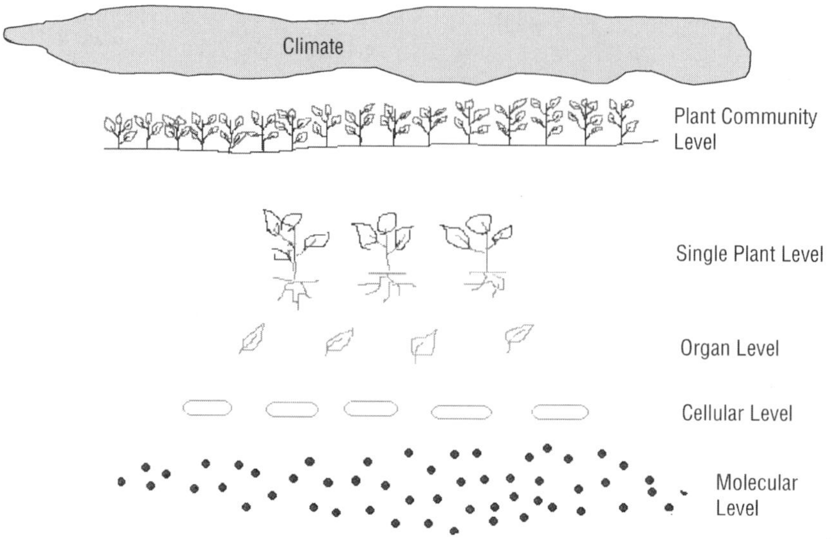

Fig. 1 The hierarchy of processes leading to crop yield.

DROUGHT RESISTANCE TRAITS IN TERMS OF THE PASSIOURA WATER MODEL

Passioura[2,34,35] argues that there are no traits that confer global drought resistance. Also, given the earlier discussion which clearly suggests that short term responses to drought stress operating at the cellular level may have no bearing on the yield of water-limited crops, crop productivity is best analyzed from the top down. Here the thinking has shifted from understanding defense mechanisms for survival to applying this knowledge to optimize economic productivity under a given water-limited condition.

Passioura proposed that when water is the major limit, grain yield (GY) is a function of the amount of water transpired by the crop (T), the efficiency of use of this water in biomass production (WUE), and the proportion of biomass that is partitioned into grain, or harvest index (HI), thus:

$$GY = T \times WUE \times HI \tag{1}$$

It was argued that individual traits could be assessed in terms of their contribution to each of these functional yield components and thereby increase yield under drought. The identity has provided a framework to more critically identify and evaluate important drought resistance traits. The following examples demonstrate the utility of this approach.

In a number of crops, improvement in the WUE trait has been achieved via selection for carbon isotope discrimination,[36] or correlated surrogate measures.[37–39] Readers are referred to recent reviews for more details on this subject.[6,40,41]

In maize, increased partitioning to the grain (or higher HI) has been brought about by using an ideotype selection index that focused on a reduction in anthesis to silking interval.[42] Grain yield of the final selections increased by 108 kg/ha/cycle and came about by an increase in HI, with no change in final biomass relative to the parents.

Wright et al.[43] proposed that estimates of each of the water model components for a peanut crop could be derived from simple and low-cost measurements of total and pod biomass at harvest, from estimates of WUE from correlated measures of specific leaf area, and by reverse-engineering the TDM component of the water model, such that T=TDM/TE. A selection study is being conducted on 4 peanut populations, in which a selection index approach utilising T, WUE, and HI is currently assessing the value of these traits in a large-scale breeding program in India and Australia.[44]

CONCLUSION

Although drought is commonly referred to as ''prolonged water deficit'' periods during a crop's life, it is indeed a complex syndrome with various climate pressures operating together in infinite combinations, resulting in significant reductions in crop performance. Historically, options for managing agricultural drought have revolved around management techniques such as deep tillage, mulching, intercropping, etc. However, future options for drought management will increasingly be based on the genetic improvement of crops for targeted environments. With genetic options, matching phenology to water availability in target environments has proven to be highly a successful approach. Although our understanding of drought resistance mechanisms has improved significantly, it is essential to distinguish the traits linked to short term responses from those which are important when integrated over longer time scales.

ARTICLES OF FURTHER INTEREST

Drought Resistant Ideotypes: Future Technologies and Approaches for the Development of, p. 391
Osmotic Adjustment and Osmoregulation, p. 850
Plant Response to Stress: Regulation of Plant Gene Expression to Drought, p. 999
Water Deficits: Development, p. 1284
Water Use Efficiency Including Carbon Isotope Discrimination, p. 1288

REFERENCES

1. Schulze, E.D. Adaptation Mechanisms of Non-Cultivated Arid-Zone Plants: Useful Lessons for Agriculture. In *Drought Research Priorities for the Dryland Tropics*; Bidinger, F.R., Johansen, C., Eds.; ICIRSAT: Patancheru, A.P. 502324, India, 1988; 159–177.
2. Passioura, J.B. Drought and Drought Tolerance. In *Drought Tolerance in Higher Plants: Genetical, Physiological, and Molecular Biological Analysis*; Belhassen, E., Ed.; Kluwer Academic Publishers: The Netherlands, 1996; 1–6.
3. Unger, W.P.; Jones, O.R.; Steiner, J.L. Principles of Crop and Soil Management Procedures for Maximising Production per Unit Rainfall. In *Drought Research Priorities for the Dry Land Tropics*; Bidinger, F.R., Johansen, C., Eds.; ICIRSAT: Patancheru, A.P. 502324, India, 1988; 97–112.
4. Natarajan, M.; Willey, R.W. The effects of water stress on yield advantages of intercropping systems. Field Crops Res. **1986**, *13*, 117–131.
5. Nageswara Rao, R.C.; Wadia, K.D.R.; Williams, J.H. Intercropping of short and long duration groundnut genotypes to increase productivity in environments prone to end-of-season drought. Exp. Agric. **1990**, *26*, 63–72.
6. Richards, R.A. Defining Selection Criteria to Improve Yield Under Drought. In *Drought Tolerance in Higher Plants: Genetical, Physiological, and Molecular Biological Analysis*; Belhassen, E., Ed.; Kluwer Academic Publishers: The Netherlands, 1996; 71–78.
7. Levitt, J. *Response of Plants to Environmental Stresses*; Academic Press: New York, 1972.
8. Turner, N.C. Crop water deficits: A decade of progress. Adv. Agron. **1986**, *39*, 1–51.
9. Turner, N.C.; Wright, G.C.; Siddique, K.H.M. Adapatation of grain legumes (pulses) to water-limited environments. Adv. Agron. **2001**, *71*, 193–231.
10. Turner, N.C. Drought Resistance: A Comparison of Two Frameworks. In *Management of Agricultural Drought: Agronomic and Genetic Options*; Saxena, N.P., Johansen, C., Chauhan, Y.S., Rao, R.C.N. Eds.; Oxford and IBH: New Delhi, 2000.
11. Subbarao, G.V.; Johansen, C.; Slinkard, A.E.; Rao, R.C.N.; Saxena, N.P.; Chauhan, Y.S. Strategies for improving drought resistance in grain legumes. Crit. Rev. Plant Sci. **1995**, *14*, 469–523.
12. Jackson, P.; Robertson, M.; Cooper, M.; Hammer, G. The role of physiological understanding in plant breeding; From a breeding perspective. Field Crops Res. **1996**, *49*, 11–37.
13. Siddique, K.H.M.; Loss, S.P.; Regan, S.P.; Jettner, R. Adaptation of cool season grain legumes in Mediterranean-type environments of south-western Australia. Aust. J. Agric. Resour. **1999**, *50*, 375–387.
14. Wright, G.C.; Smith, R.C.G.; Morgan, J.M. Differences between two grin sorghum genotypes in adaptation to drought stress. III. Physiological responses. Aust. J. Agric. Resour. **1983**, *34*, 637–651.
15. Flower, D.J.; Ludlow, M.M. Contribution of osmotic adjustment to the dehydration tolerance of water-stressed pigeonpea (*Cajanus cajan* (L.) millsp.) leaves. Plant Cell Environ. **1986**, *9*, 33–40.
16. Davies, W.J.; Tardieu, F.; Trejo, C.L. How do chemical signals work in plants that grow in drying soil? Plant Physiol. **1994**, *104*, 309–314.
17. Morgan, J.M. Possible role of abscisic acid in reducing seed set in water-stressed wheat plants. Nature (Lond.) **1980**, *285*, 655–657.
18. Blum, A. Crop Responses to Drought and Interpretation of Adaptation. In *Drought Tolerance in Higher Plants: Genetical, Physiological, and Molecular Biological Analysis*; Belhassen, E., Ed.; Kluwer Academic Publishers: The Netherlands, 1996; 57–70.
19. Morgan, J.M. Osmoregulation and water stress in higher plants. Annu. Rev. Plant Physiol. **1984**, *35*, 299–319.
20. Jones, M.M.; Rawson, H.M. Influence of the rate of development of leaf water deficits upon photosynthesis, leaf conductance, water use efficiency, and osmotic potential in sorghum. Physiol. Plant. **1979**, *45*, 103–111.
21. Morgan, J.M.; King, R.W. Association between loss of leaf turgor, abscisic acid levels and seed set in two wheat cultivars. Aust. J. Plant Physiol. **1984**, *11*, 143–150.
22. Morgan, J.M.; Condon, A.G. Water use, grain yield and osmoregulation in wheat. Aust. J. Plant Physiol. **1986**, *13*, 523–532.
23. Wright, G.C.; Smith, R.C.G. Differences between two grin sorghum genotypes in adaptation to drought stress. II. Root water uptake and water use. Aust J. Agric. Resour. **1983**, *34*, 627–636.
24. Morgan, J.M. Osmoregulation as a selection criterion for drought tolerance in wheat. Aust. J. Agric. Resour. **1983**, *34*, 607–614.
25. Blum, A.; Mayer, J.; Gozlan, G. Associations between plant production and some physiological components of

drought resistance in wheat. Plant Cell Environ. **1983**, *6*, 219–225.
26. Santamaria, J.M.; Ludlow, M.M.; Fukai, S. Contributions of osmotic adjustment to grain yield in *Sorghum bicolor* (L.) Moench under water-limited conditions. I. Water stress before anthesis. Aust. J. Agric. Resour. **1990**, *41*, 51–65.
27. Morgan, J.M.; Rodriguez-Maribona, B.; Knights, E.J. Adaptation to water-deficit in chickpea breeding lines by osmoregulation: Relationship to grain-yields in the field. Field Crops Res. **1991**, *27*, 61–70.
28. Blum, A. *Plant Breeding for Stress Environments*; CRC Press: Boca Raton, FL, 1988.
29. Sinclair, T.R.; Ludlow, M.M. Influence of soil water supply on the plant water balance of four tropical grain legumes. Aust. J. Plant Physiol. **1986**, *13*, 329–341.
30. Chandler, P.M.; Munns, R.; Robertson, M. Regulation of Dehydrin Expression. In *Plant Responses to Cellular Dehydration During Environmental Stress*; Close, T.J., Bray, E.A., Eds.; American Society of Plant Physiologists: Rockville, MD, 1993; 159–166.
31. Close, T.J.; Fenton, R.D.; Yang, A.; Asghar, R.; DeMason, D.A.; Crone, D.E.; Meyer, N.C.; Moonan, F. Dehydrin: The Protein. In *Plant Responses to Cellular Dehydration During Environmental Stress*; Close, T.J., Bray, E.A., Eds.; Amer. Society of Plant Physiology, 1993; 104–118.
32. Hanson, A.D.; Hitz, W.D. Metabolic responses of mesophytes to plant water deficits. Annu. Rev. Plant Physiol. **1982**, *33*, 163–203.
33. Mathews, R.B.; Harris, D.; Williams, J.H.; Nageswara Rao, R.C. The physiological basis for yield differences between four groundnut genotypes in response to drought. II. Solar radiation interception and leaf movement. Exp. Agric. **1988**, *24*, 203–213.
34. Passioura, J.B. The Interaction Between Physiology and the Breeding of Wheat. In *Wheat Science—Today and Tomorrow*; Evans, L.T., Peacock, W.J., Eds.; Cambridge University Press: Cambridge, 1981; 191–201.
35. Passioura, J.B. Grain yield, harvest index and water use of wheat. J. Aust. Inst. Agric. Sci. **1977**, *43*, 117–120.
36. Farquhar, G.D.; Richards, R.A. Isotopic composition of plant carbon correlates with water-use efficiency in wheat genotypes. Aust. J. Plant Physiol. **1984**, *11*, 539–552.
37. Nageswara Rao, R.C.; Wright, G.C. Stability of the relationship between specific leaf area and carbon isotope discrimination across environments in peanuts. Crop Sci. **1994**, *34*, 98–103.
38. Masle, J.; Farquhar, G.D.; Wong, S.C. Transpiration ratio and plant mineral content are related among genotypes of a range of species. Aust. J. Plant Physiol. **1992**, *19*, 709–721.
39. Clark, D.H.; Johnson, D.A.; Kephart, K.D.; Jackson, N.A. Near infrared reflectance spectroscopy estimation of ^{13}C discrimination in forages. J. Range Manag. **1995**, *48*, 132–136.
40. Subbarao, G.V.; Johansen, C.; Rao, R.C.N.; Wright, G.C. Transpiration Efficiency: Avenues for Genetic Improvement. In *Handbook of Plant and Crop Physiology*; Pessarakli, M., Ed.; Marcel Dekker: New York, NY, 1994; 785–806.
41. Wright, G.C.; Rachaputi, N.C. Transpiration Efficiency. In *Water and Plants (Biology) for the Encyclopedia of Water Science*; Marcel Dekker, Inc., 2003; 982–988.
42. Fischer, K.S.; Johnson, E.C.; Edmeades, G.O. Breeding and Selection for Drought Resistance in Tropical Maize. In *Drought Resistance in Crops with Emphasis on Rice*; International Rice Research Institute: Los Banos, 1983; 377–399.
43. Wright, G.C.; Rao, R.C.N.; Basu, M.S. A Physiological Approach to the Understanding of Genotype by Environment Interactions—A Case Study on Improvement of Drought Adaptation in Groundnut. In *Plant Adaptation and Crop Improvement*; Cooper, M., Hammer, G.L., Eds.; CAB International: Wallingford, 1996; 365–381.
44. Nigam, S.N.; NageswaraRao, N.C.; Wright, G.C. In *Breeding for Increased Water-Use Efficiency in Groundnut*, New Millenium International Groundnut Workshop, Shandong Peanut Research Institute, Qingdao, China, Sept. 4–7, 2001; 1–2.

Drought Resistant Ideotypes: Future Technologies and Approaches for the Development of

D

Nageswararao C. Rachaputi
Graeme C. Wright
Queensland Department of Primary Industries, Kingaroy, Australia

INTRODUCTION

Historically, agriculture has been based on the commodity market. However, in addition to increased production of commodities, future global markets will require ''safe'' food products. Global phenomena such as increasing scarcity of irrigation water, global warming, increasing populations, and increased consumer awareness about food safety will continue to mount pressure on crop breeding programs to improve the productivity as well as the quality of dry land crops. Hence there is an urgent need for crop scientists to develop new and novel approaches to designing and developing suitable crops and varieties that are well adapted to drought environments.

IDEOTYPE

The ideotype concept, which involves pyramiding drought resistance traits to create an ''ideal'' genotype, remains an as yet unrealized aspiration: to improve crop yield by breeding by design.[1] The development of an ideotype for a given drought environment is a complex task requiring integration of knowledge of the environment, the potential capacity of the plant, its constitutive traits, and its adaptive responses to the specific levels and timing of stress—all in the context of final productivity.[2] However, recent progress in information technology, crop simulation modeling, and biotechnological approaches offers some exciting possibilities for better integration of current knowledge of drought resistance to complement conventional breeding approaches. This article describes some of those emerging technologies, which hopefully can improve the success rate of crop breeding programs engaged in the development of drought-resistant ideotypes.

Using Simulation Modeling to Characterize Target Population of Environments (TPEs)

Careful consideration must be given to assessing the importance of a specific trait in each target environment. For example, where deep soil-water reserves are available on soils with high water-holding capacity, one would expect that genotypes with extensive root traits would be better adapted. Characterizing the target environment, therefore, becomes a critical requirement in any successful crop improvement effort. Simple soil–water balance models that require easily gathered inputs such as daily rainfall, pan evaporation, and estimates of soil-water holding capacity and rooting depth, allow ready estimates of crop-available water and/or relative transpiration.[3,4] This historical analysis can be further evaluated using more sophisticated crop modeling, which allows the pattern of seasonal crop water stress to be quantified for specific environments in a probabilistic framework. Differences in the amount and frequency of rainfall, in combination with differences in soil water-holding capacity for a range of seasons and sites, can be easily combined and integrated using these techniques.[5] Wright[6] uses the approach to define the most likely drought stress patterns occurring in the rainfed peanut production areas in Queensland. The relationship between crop thermal time and a crop water stress index (or relative transpiration) derived from the model allowed explicit quantification of the seasonal crop water stress patterns over 85 years of historical climate data for this site. Cluster analysis allows similar stress patterns to be grouped, such that the 85 years of data can be clustered down into 5 groups (Fig. 1). Such an analysis then allows the breeder to gain some insight into the probability of likely stress patterns at specific sites. These analytical techniques also highlight how characterization of a water-limited environment in terms of a probabilistic framework can greatly assist the breeder in defining breeding objectives and the drought characteristics that may be the most appropriate for the target environment.

Analysis of Physiological Traits in an Analytical Model Framework

Crop analytical models described by Wright and Rachaputi (2002) in this encyclopedia enable the ''dissection'' of yield into a small number of independent physiological components that effectively integrate numerous complex processes into fewer ''biologically meaningful''

Encyclopedia of Plant and Crop Science
DOI: 10.1081/E-EPCS 120017851

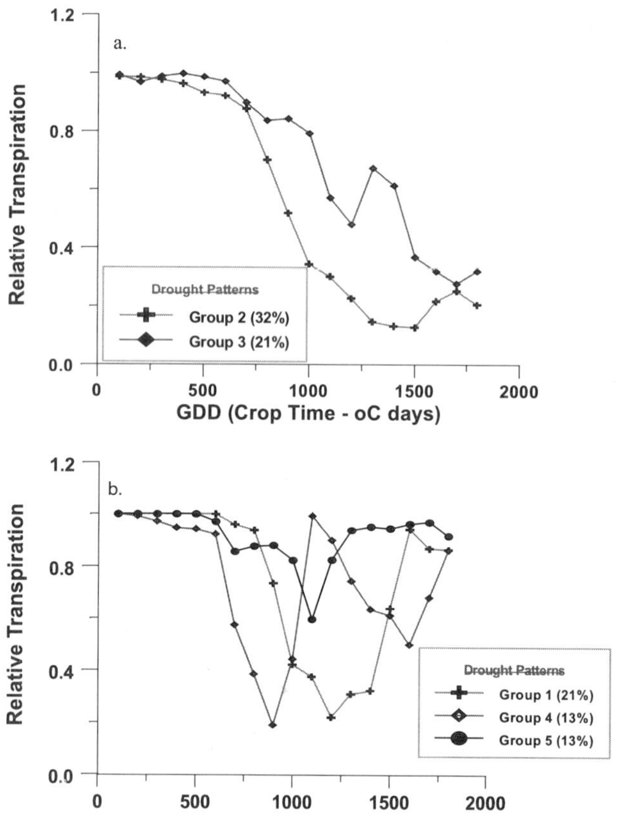

Fig. 1 Changes in calculated relative transpiration with growing degree days from long-term climatic data at Kingaroy, Queensland, Australia for (a) two groups of years showing terminal droughts, and (b) three groups of years showing different intermittent droughts.

parameters. In a practical breeding program the genotype-by-environment interaction (G × E) can potentially be reduced by applying these integrated parameters that contribute to the yield, rather than using yield alone, which is known to have a high G × E interaction.[7] Each subcomponent of these relationships represents an integrated function of a number of developmental, morphological, physiological, and biochemical attributes.[8] Any attribute thought to be beneficial for drought adaptability should then be evaluated in terms of its functional relationship and strength of correlation to one of the yield components.

The key to utilizing these analytical relationships in crop improvement programs is critically dependent on the ability to obtain reliable measurements for each attribute for the large number of genotypes within large-scale breeding programs.[9] Obtaining reliable data for the parameters in the relationships for the large number of plots (genotypes) has been recently addressed, following some novel and pragmatic approaches to quantify the parameters.[7,10] A more thorough coverage of this topic is reported in Turner et al.[11]

Simulating Traits and Breeding Approaches Using Crop Models

Crop models allow the opportunity to quantitatively assess the value of hypothetical genotypes with desired levels of a given trait(s) under different environmental and management conditions. A number of workers have assessed the impact of varying levels of expression of traits in grain sorghum, including osmotic adjustment (OA), greater water extraction, phenology, radiation use efficiency, tiller production, and staygreen (e.g., Refs. 12–14). A limitation of this approach is that traits are considered physiologically independent and do not take into account the genetic correlations among traits that are known to occur. There is a lack of information on the extent of genetic variation, genetic correlations, and physiological mode of action for many putative traits.[9,14] Targeted experimentation is necessary to define these parameters, along with necessary changes in model structure to accommodate known physiological linkages. Boote et al.[15] suggest that simulation analysis may be used in the future to help determine which genetic traits would be most amenable for molecular geneticists to manipulate in order to maximize yield and returns for specific environments.

Gene-to-Phenotype Modeling

Research into gene-to-phenotype modeling in plant breeding has recently commenced in a number of laboratories around the world. While the major focus of much drought research has been to determine the molecular basis of adaptation to drought environments, much of the work is limited to response at cellular level rather than crop adaptation, with often-overrated claims of potential benefits to agriculture. Research on functional controls of crop plants requires an understanding of the gene-to-trait-to-phenotype adaptation. Researchers at the University of Queensland and the Agricultural Production Systems Research Unit in Queensland have recently initiated a program on modeling plant adaptation and plant breeding that integrates world-leading quantitative genetics research (QU-GENE) and crop modeling (APSIM) simulation platforms.[16] The aim is to determine the best breeding methods that will find the optimal combinations of genes and cropping system management to exploit target environments. The major research challenge lies in linking understanding of the plant functional basis of phenotype expression to gene function, given that large numbers of genes and complex gene networks are involved in the genetic architecture of these complex traits.[17]

Experimental data is used in QU-GENE to define the gene–trait and gene–gene interactions and the gene–marker associations for the entire population of possible genotypes. APSIM takes trait values for all combinations and, using weather and soil data for any number of years and locations in the target environment, simulates the entire genotype–environment landscape for expressed target traits and their pleiotropic and epistatic effects on yield. A QU-GENE module samples the population and applies a selection methodology (e.g., mass, recurrent, pedigree, indirect, marker-assisted, or combinations of these). Millions of combinations of different methods and/ or method parameters are evaluated to identify those that are most efficient in accumulating superior gene combinations. This simulation platform is the result of many years of experimental work in crop physiology and plant breeding. This technology will become a major tool in the future to conduct ''virtual'' plant breeding, using hypothetical parents and making selections for a given trait or combination of traits, and conducting virtual multilocation studies. These tools should also give physiologists and breeders the opportunity to cooperatively decide on the optimal selection and breeding strategy for yield (and quality) improvement in a given environment.

Transgenic Technologies

In situations where specific technical challenges are difficult to handle through traditional crop breeding, transgenic approaches may offer exciting new possibilities for developing new ideotypes suitable for hostile environments, including drought, salinity, nutrient toxicities, and heat. The genes that confer resistance to these complex stresses might have to be isolated from other plants or organisms and introduced into the desired crop species or varieties to develop a genetically modified plant (ideotype) with desired resistance qualities.

A genetically modified (GM) plant is one that contains rDNA or has been developed through the use of genetic engineering or rDNA technology. GM plants are also referred to as GE (genetically engineered), transgenic, or rDNA (recombinant DNA) plants. In simple terms, the development of a GM plant involves two steps: 1) isolating the gene that encodes the desired trait, and 2) introducing the gene into the recipient plant. The gene can be a synthetic construct or a sequence taken from a donor organism. An additional ''tracker'' gene is added to the gene for the desired trait to allow the new trait to be tracked throughout the process of producing a GM plant. Two common types of tracker genes are those that confer either antibiotic resistance or herbicide resistance. Once the tracker gene has been added to the gene for the desired trait, the genetic construct must be inserted into the recipient plant cells.

The introduction (or insertion) of the gene into recipient plant cells is commonly done either by involving bacterium-mediated transfer of DNA or by shooting (or projecting) DNA into the cells of the recipient plant.

Plant cells that incorporate the donor DNA into their genome are transformed or genetically modified. Following the gene insertion process, the plant cells are then grown in a medium that contains an agent to help identify only those plant cells that are genetically modified. The cells are induced to form plants using plant growth hormones. Only transformed plants that have incorporated the donor DNA will show the characteristic of the tracker gene; nonmodified plants will not. The plants that have the characteristic of the tracker are then verified as containing the DNA for the desired trait.

Although the technology has been successfully tested to develop GM plants with insect pest and virus resistance, the literature suggests that there has been limited research into development of GM plants with resistance to abiotic stresses such as drought and heat.[18] Unfortunately, most of the traits contributing to drought resistance involve multiple genes and complex interactions among them. It will probably be some time before multigene-governed traits are successfully introduced into cultivated species and dry land farmers benefit from the growing of such transgenic crops. While transgenic technology holds promise for the future, we believe that in the short term successes in improving drought adaptation of cultivated crops may be derived from marker-assisted selection and breeding approaches.

CONCLUSION

This article highlights the need for developing a drought-resistant ideotype approach to designing crop genotypes for water-limited environments. Such an approach would require integration of current knowledge with more recently developed innovative tools such as crop simulation models and gene-to-phenotype modeling to assess the value of individual traits in the target population environments. Future research on gene-to-phenotype modeling will develop software-based tools to construct genotypes with desired traits and develop knowledge on how to use them in ''virtual'' breeding programs to assess the performance of progeny in target environments. Major challenges lie ahead in linking the understanding of the plant functional basis of phenotype expression to gene function, given that there are large numbers of gene networks operating in the plant architecture. In situations where specific technical challenges are difficult to handle through traditional crop breeding, transgenic technology may offer great possibilities for developing plants with tolerance to complex stresses such as salinity, drought, nutrient toxicities, and heat.

ARTICLES OF FURTHER INTEREST

Breeding Self-Pollinated Crops Through Marker-Assisted Selection, p. 202
Breeding: Choice of Parents, p. 215
Breeding: Genotype-by-Environment Interaction, p. 218
Breeding: Incorporation of Exotic Germplasm, p. 222
Drought and Drought Resistance, p. 386
Osmotic Adjustment and Osmoregulation, p. 850
Plant Response to Stress: Regulation of Plant Gene Expression to Drought, p. 999
Water Deficits: Development, p. 1284
Water Use Efficiency Including Carbon Isotope Discrimination, p. 1288

REFERENCES

1. Hamblin, A.P.; Kyneur, G. *Trends in Wheat Yields and Soil Fertility in Australia*; Australian Government Publishing Service: Canberra, 1993.
2. Blum, A. Crop Responses to Drought and Interpretation of Adaptation. In *Drought Tolerance in Higher Plants: Genetical, Physiological, and Molecular Biological Analysis*; Belhassen, E., Ed.; Kluwer Academic Publishers: The Netherlands, 1996; 57–70.
3. Ritchie, J.T. A User-Oriented Model of the Soil Water Balance in Wheat. In *Wheat Growth and Modelling*; Day, W., Atkin, R.K., Eds.; Plenum Press: New York, NY, 1985; 293–305.
4. Robertson, M.J.; Fukai, S. Comparison of water extraction models for grain sorghum under continuous soil drying. Field Crops Res. **1994**, *36*, 145–160.
5. Muchow, R.C.; Cooper, M.; Hammer, G.L. Characterising Environmental Challenges Using Crop Simulation. In *Crop Adaptation and Crop Improvement*; Cooper, M., Hammer, G.L., Eds.; CAB International: Wallingford, 1996; 349–364.
6. Wright, G.C. Management of Drought in Peanuts—Can Crop Modelling Assist in Long-Term Planning Decsisions? Proceedings of the 2nd Australian Peanut Conference, Gold Coast, Queensland, 1997; Qld Department of Primary Industries, pp. 26–29.
7. Williams, J.H. Concepts for the Application of Crop Physiological Models to Crop Breeding. In *Groundnut—A Global Perspective: Proceedings of an International Workshop*; Nigam, S.N., Ed.; International Crops Research Institute for the Semi-Arid Tropics: Patancheru, India, 1992; 345–352.
8. Hardwick, R.C. Critical Physiological Traits in Pulse Crops. In *World Crops: Cool Season Food Legumes*; Summerfield, R.J., Ed.; Kluwer Academic Publishers: Dordrecht, 1988; 885–896.
9. Jackson, P.; Robertson, M.; Cooper, M.; Hammer, G. The role of physiological understanding in plant breeding; from a breeding perspective. Field Crops Res. **1996**, *49*, 11–37.
10. Turner, N.C.; Wright, G.C.; Siddique, K.H.M. Adapatation of grain legumes (pulses) to water-limited environments. Adv. Agron. **2001**, *71*, 193–231.
11. Williams, J.H.; Saxena, N.P. The use of non-destructive measurement and physiological models of yield determination to investigate factors determining differences in seed yield between genotypes of 'desi' chickpeas (*Cicer arietln:um*). Ann. Appl. Biol. **1991**, *119*, 105–112.
12. Jordan, W.R.; Dugas, W.A.; Shouse, P.J. Strategies for crop improvement for drought prone regions. Agric. Water Manag. **1983**, *7*, 281–299.
13. Hammer, G.L.; Vanderlip, R.L. Studies on genotype X environment interaction in grain sorghum. III. Modelling the impact in field environments. Crop Sci. **1989**, *29*, 385–391.
14. Hammer, G.L.; Butler, D.G.; Muchow, R.C.; Meinke, H. Integrating Physiological Understanding and Plant Breeding via Crop Modelling and Optimisation. In *Crop Adaptation and Crop Improvement*; Cooper, M., Hammer, G.L., Eds.; CAB International: Wallingford, 1996; 419–441.
15. Boote, K.J.; Jones, J.W.; Pickering, N.B. Potential uses and limitations of crop models. Agron. J. **1996**, *88*, 704–716.
16. Podlich, D.W.; Cooper, M. QU-GENE: A platform for quantitative analysis of genetic models. Bioinformatics **1998**, *14*, 632–653.
17. Cooper, M.; Chapman, S.C.; Podlich, D.W.; Hammer, G.L. The GP Problem: Quantifying Gene-to-Phenotype Relationships. In Silico Biology 2, 0013 (2002). In *Functional Genomics*; Wingender, E., Ed.; Bioinformation Systems e.V Dagstuhl Seminar, 2002. (http://www.bioinfo.de/isb/2002/02/0013/main.html).
18. Fresco, L.O. In *Genetically Modified Organisms in Food and Agriculture: Where Are We? Where Are We Going?*, Proc. on ''Crop and Forest Biotechnology for the Future,'' September 16–18, 2001; Royal Swedish Academy of Agriculture and Forestry: Falkenberg, Sweden.

Echinacea: Uses As a Medicine

E

Mario R. Morales
Purdue University, West Lafayette, Indiana, U.S.A.

INTRODUCTION

Echinacea was one of the main medicinal plants used by Native Americans before colonization. It was used to treat infections, snakebites, wounds, inflammations, and several other ailments. Later, lay European settlers incorporated echinacea into their list of home remedies. It was not until the 1880s that the plant was prescribed by physicians. Echinacea became quite popular and was subjected to intensive research from 1887 to 1937. The popularity of echinacea declined when sulfa remedies were discovered in the 1930s. Echinacea surged in popularity again in the 1980s, which stimulated intensive investigation.

The chemical composition of echinacea is complex. Several components, especially caffeic acid derivatives, alkylamides, polysaccharides, and glycoproteins, are responsible for its medicinal properties. Immunostimulatory, anti-inflammatory, and antibiotic properties of echinacea have been demonstrated through in vitro and in vivo experiments. Clinical studies have not been conclusive, but they suggest that echinacea may be effective in treating but not preventing upper respiratory tract infections. Plant preparations will be more effective if taken at the onset of the infection.

HISTORY

Native Americans were aware of the medicinal qualities of echinacea before the arrival of the first Europeans. They used the plant to treat a broad spectrum of ailments, especially upper respiratory tract infections (URTI). Upon arrival, colonists readily incorporated echinacea into their list of home remedies. During the 18th and 19th centuries, echinacea was a popular medicinal plant, well accepted by lay people but rejected by medical doctors. Professional interest in the plant started in 1887, when physicians recognized the medicinal qualities of the plant and added it to their list of prescribed plant remedies. Echinacea became quite popular and was intensively studied from 1887 to 1937.[1] With the discovery of sulfas (antibiotics) in the 1930s, the use of echinacea declined dramatically in the United States. Scientific and commercial interest in the plant moved to Germany, where Gerhard Madaus, a pioneer in medicinal plant research, began experimentation and large-scale cultivation of *E. purpurea*.[1] Echinacea became popular again in the United States in the 1980s and 1990s. A 1990 survey showed that 60 million Americans (34% of the population) used at least one of 16 alternative medicines. The Dietary Supplement Health and Education Act of 1994 (DSHEA), which deregulated the dietary and herbal industry, allowed public access to a large diversity of products in several formulations and combinations and helped to increase this figure to 83 million (44%) by 1997. Annual sales of echinacea products increased from $80 million in 1997 to more than $300 million in 1999.

The genus *Echinacea* (*Asteraceae*) has nine species, all native to North America, but only *E. purpurea*, *E. angustifolia*, and *E. pallida* (Fig. 1) are used medicinally. The plant is herbaceous and perennial with large, coarse, and ovate to lanceolate leaves and attractive composite flowers with colorful rays at the top of 0.4–1.2 m-long stems.

PHARMACOLOGY

The chemical composition of echinacea is complex. The most relevant constituents are caffeic acid derivatives, alkylamides, polysaccharides, and glycoproteins.[2] Concentration of these compounds varies according to species, plant part (roots, leaves, stems, flowers), and mode of processing. The curative properties of echinacea may not be due to the effect of a single compound but perhaps to the synergistic effect of several of them combined. These compounds are reported to enhance the immune system and provide echinacea with antiviral, antibacterial, fungicidal, insecticidal, and anti-inflammatory properties. However, the action mechanism, bioavailability, potency, and synergism of these compounds remain unclear and need further investigation.

Caffeic Acid Derivatives

Several caffeic acid derivatives (echinacoside, cichoric or chicoric acid, caftaric acid, chlorogenic acid, cynarin)

Encyclopedia of Plant and Crop Science
DOI: 10.1081/E-EPCS 120010426

Fig. 1 Aerial parts and roots of *E. purpurea* (a) and *E. pallida* (b), and aerial parts of *E. angustifolia* (c) growing in central Indiana. (*View this art in color at www.dekker.com.*)

have been isolated from the echinacea plant. Echinacoside contains two molecules of glucose and one molecule each of rhamnose, caffeic acid, and catecholethanol. It is present in the roots and flowers of *E. angustifolia* and *E. pallida* but absent in *E. purpurea*. Echinacoside has a weak antibiotic activity. Cichoric acid is abundant in roots and flowers of *E. purpurea*, while caftaric acid is present in *E. purpurea* and *E. pallida*. Chlorogenic acid is present in *E. angustifolia* and *E. pallida*. Cynarin, found only in *E. angustifolia* roots, could be used to differentiate *E. angustifolia* from *E. pallida*. The three echinacea species were found to have antioxidant activities and free-radical scavenging capacities, probably due to the content of caffeic acid derivatives.[3] Cichoric acid is a powerful stimulant of the immune system and stimulates phagocytosis in vitro and in vivo[4] and inhibits human immunodeficiency virus type 1 (HIV-1) integrase.[5] Echinacoside has a low antibacterial and antiviral potency but does not have immunostimulatory properties.[6]

Alkamides (Alkylamides)

Echinacea alkamides are mainly isobutylamides of highly unsaturated carboxylic acids with olefinic and/or acetylenic bonds.[7] Fourteen of them were identified in *E. angustifolia*, eleven in *E. purpurea*, and six in *E. pallida* roots.[8] Alkamide levels were much higher in dried roots (2.65–3.28 mg/g) than in dried leaves (0.10–0.18 mg/g) of *E. purpurea*.[9] The isomers dodeca-2E,4E,8Z,10E/Z-tetraenoic acid isobutylamides (tetraene alkamides 8/9) have been found to be the most abundant in roots (43–49% of total alkamides) and the aerial part (55–76% of total alkamides) of *E. purpurea*.[9,10] Since they have shown less bioactivity than the total alkamide extract, it is not clear yet which of the *E. purpurea* alkamides has the most immunostimulatory activity; this alkamide remains to be identified.[9] Alkamide fractions stimulate phagocyte activity in vitro and in vivo[11] and work as anti-inflammatory agents by inhibiting cyclooxygenase and 5-lipooxygenase.[12] It was found recently that 100 mg/ml of several *E. purpurea* alkamides inhibited cyclooxygenase-I and-II in the range of 36–60% and 15–46%, respectively, and that this dose killed 100% of the mosquito *Aedes aegyptii* larvae.[13]

Polysaccharides and Glycoproteins

Two polysaccharides, PS I (4-O-methyl-glucuronoarabinoxylan) and PS II (acidic arabinorhamnogalactan), were isolated by systematic fractionation from the aqueous extract of the aerial parts of *E. purpurea* and were found, by pharmacologic testing, to stimulate phagocytosis in vitro and in vivo and enhance the production of oxygen radicals by macrophages. However, the extent to which these compounds occur in echinacea preparations is not completely known. Methods to routinely detect pharmacologically active polysaccharides in echinacea preparations still need to be developed. Glycoproteins isolated from *E. purpurea* and *E. angustifolia* roots contain proteins (ca. 3%), the sugars arabinose (64–84%) and galactose (2–5%), and glucosamines (6%). *E. angustifolia* and *E. purpurea* roots contain similar amounts of glycoproteins, while *E. pallida* contains less.

IN VITRO AND IN VIVO STUDIES

In vitro and in vivo studies have demonstrated the potential of echinacea compounds as stimulants of the immune system. For instance, the polysaccharide arabinogalactan activated macrophages to cytotoxicity against tumor cells and the microorganism *Leishmania enriettii* and induced the macrophages to produce tumor necrosis factor (TNF-alpha), interleukin-1 (IL-1), and interferon-beta-2 (IFN-beta-2), all potent enhancers of natural-killer (NK) cells.[14] Root extract administered to leukemic mice at the onset of the tumor induced a 2.5-fold increase in the absolute number of splenic NK cells after nine days of treatment and the survival of 1/3 of the mice beyond the three months of the experiment.[15] The stimulatory action of echinacea phytochemicals on NK cells and monocytes, both mediators of nonspecific immunity and well-demonstrated killers of virus-containing cells, was also demonstrated by Sun.[16] Recently, mixtures of purified cichoric acid, polysaccharide, and alkylamide from *E. purpurea* were able to increase the phagocytic activity and phagocytic index of the alveolar macrophages in a dose-dependent fashion in an in vivo study.[17] However, the oral intake of freshly expressed juice of echinacea herbs by healthy humans did not enhance the phagocytic activity on polymorphonuclear leukocytes or monocytes and did not influence the production of TNF-alpha and IL-1 beta by stimulated monocytes,[18] proving that oral intakes are probably not effective.

Tests of the anti-inflammatory and wound-healing properties of pure echinacoside and *E. pallida* and *E. purpurea* root alcohol extracts applied topically on rats found echinacoside and the *E. pallida* extract to be markedly effective at reducing the edematous process after 48 and 72 hours, and while rats in the control and *E. purpurea* groups still showed signs of inflammation and open wounds, rats in the echinacoside and *E. pallida* extract groups had no signs of inflammation and had enhanced wound healing at the end of the experiment.[19] Echinacoside's antihyaluronidase activity and type III collagen protection from damage caused by free radicals had been previously reported.[20]

CLINICAL STUDIES

Clinical studies in humans have been inconclusive regarding the efficacy of echinacea for the prevention and treatment of URTI, colds, and flu mainly because of lack of standardization of the plant preparations and deficiencies in design and/or experimental methodology. Melchart et al.[21] reviewed 26 controlled trials, 18 randomized and 11 double-blind, that were published from 1961 to 1992. Six trials studied echinacea for treatment URTIs and six for prophylaxis of URTIs, with eleven trials concluding that echinacea was superior to the control. Most of these studies, however, were of low methodological quality, and the claim that echinacea was efficient for the treatment or prevention of URTIs or the common cold was unclear in all but one, which administered 900 mg/day of an ethanolic extract of *E. purpurea* root and concluded that the preparation was effective in relieving URTI symptoms.[22]

In 1999, Barrett et al.[23] published a review of 13 new trials on the use of echinacea for treatment or prevention of URTIs; the trials were randomized, double-blind, and placebo-controlled. There were nine treatment trials, and eight of them concluded that echinacea is beneficial against URTIs. There were four prevention trials, but none of them reported significant evidence that echinacea was effective in preventing URTIs. These studies also had methodological deficiencies, but the reviewers concluded that echinacea might be effective in the treatment but not the prevention of acute URTIs. Several clinical studies published later support this conclusion. Three preventive studies, the first evaluating two ethanolic extracts from the roots of *E. purpurea* and *E. angustifolia* on 289 healthy individuals,[24] the second measuring the effects of the expressed juice from the aerial parts of flowering *E. purpurea* plants on 108 participants,[25] and the third assessing the effectiveness of an echinacea preparation with 0.16% cichoric acid content,[26] failed to significantly reduce the frequency, duration, or severity of the common cold symptoms. Two treatment studies, one evaluating the trade drug Echinaforce (6.78 mg of *E. purpurea* extract, 95% herb and 5% root), a concentrated fresh plant extract (48.27 mg of *E. purpurea*, 95% herb and 5% root), and a fresh plant extract (29.60 mg of *E. purpurea* root)[27] and the second the herbal tea preparation Echinacea Plus®, a blend of leaves, flowers, and stems of *E. purpurea* and *E. angustifolia* plus a water-soluble dry extract of *E. purpurea* root,[28] found that Echinaforce, the concentrated *E. purpurea* extract, and Echinacea Plus® were significantly effective at reducing and relieving cold and flu symptoms when given at the early onset of the infection.

PROCESSING

The content of active compounds in echinacea preparations varies considerably because of genetic variation, planting location, growth conditions, maturity at harvest, crop storage conditions, parts of the plant used, processing system and conditions, type of preparation (tablet, capsule, tea, expressed juice, chemical extractions), and dosage. These factors determine the level of potency and efficacy of the echinacea preparations. Cichoric acid is highly susceptible to oxidation and degrades almost completely during the preparation of fresh plant pressed juices. Water extracts from ground echinacea tops and roots lose within seconds more than 50% of the phenolic compounds. Losses are due to the effect of the enzyme polyphenol oxidase (PPO); extraction with 70% ethanol denatures the enzyme.[29] Echinacea juices extracted in the presence of 50–100 mM ascorbic acid and stabilized with 40% ethanol have high contents of cichoric acid and good long-term stability.[30] Most echinacea tinctures have 20–25% alcohol, which may destroy polysaccharides such as arabinogalactan while other active ingredients remain intact and active. It was found recently that the addition of citric acid, malic acid, or hibiscus extract to glycerin extracts of *E. purpurea* greatly improved the stability of the caffeic acid derivatives and that the stability was dependent on the concentration of the antioxidant added.[31] Supercritical fluid extraction shows potential for the recovery of alkylamides from dried echinacea roots.[32] If fresh echinacea is not processed immediately, it should be freeze-dried to avoid enzymatic degradation and destruction of the active compounds.

STANDARDIZATION

Standardization of medicinal plants should guarantee the concentration, therapeutic effectiveness, stability, and consistency of the active compounds in the preparation. On occasion, standardizations have been made based on compounds with no medicinal value. *E. angustifolia* root preparations have been standardized using echinacoside content, but echinacoside does not seem to be a major medicinal component, which renders the standardization useless because it says nothing about the major active medicinal compounds. Many active compounds in most medicinal plants act additively or synergistically, which makes standardization more difficult. The roots of *E. purpurea* have more than 80 compounds, and to measure the intrinsic and additive therapeutic properties of all of them would be almost impossible. Some studies are now concentrating on using arabinogalactan as a marker to standardize echinacea.

CONCLUSION

Echinacea has curative properties that were recognized first by Native Americans and then by American colonists and Europeans. In vitro and in vivo studies have demonstrated that this plant possesses immunostimulant, anti-inflammatory, and antibiotic properties, which are due to the presence of caffeic acid derivatives, alkylamides, polysaccharides, and glycoproteins in roots and aerial parts. Results from clinical studies have been mixed because of deficient experimental design, inappropriate drug processing, and lack of good standards. Scientists agree that echinacea may be effective for the treatment but not the prevention of URTIs and colds. More research is needed on the physical, chemical, and medicinal properties of the active components in order to develop useful standardizing procedures. It is also important to improve the design of the clinical studies to more accurately evaluate the curative properties of echinacea.

ARTICLES OF FURTHER INTEREST

Biennial Crops, p. 122
Genetic Resources of Medicinal and Aromatic Plants from Brazil, p. 502
Medical Molecular Pharming:Therapeutic Recombinant Antibodies, Biopharmaceuticals and Edible Vaccines in Transgenic Plants Engineered via the Chloroplast Genome, p. 705

REFERENCES

1. Hobbs, C. *The Echinacea Handbook*; Eclectic Medical Publications: Portland, OR, 1989.
2. Bauer, R. Chemistry, Pharmacology and Clinical Applications of Echinacea Products. In *Herbs, Botanicals & Teas*; Mazza, G., Oamah, B.D., Eds.; Technomic Publishing Company, Inc.: Lancaster, PA, 2000; 45–73.
3. Hu, C.; Kitts, D.D. Studies on the antioxidant activity of Echinacea root extract. J. Agric. Food Chem. **2000**, *48*, 1466–1472.
4. Bauer, R.; Remiger, P.; Jurcic, K.; Wagner, H. Influence of Echinacea extracts on phagocytotic activity. Z. Phytother. **1989**, *10*, 43–48.
5. Robinson, W.E. L-chicoric acid, an inhibitor of human immunodeficiency virus type 1 (HIV-1) integrase, improves on the in vitro anti-HIV-1 effect of Zidovudine plus protease inhibitor (AG1350). Antivir. Res. **1998**, *39*, 101–111.
6. Cheminat, A.; Zawatzky, R.; Becker, H.; Brouillard, R. Caffeoyl conjugates from Echinacea species: Structures and biological activity. Phytochemistry **1988**, *27*, 2787–2794.
7. Greger, H. Alkamides: Structural relationships, distribution and biological activity. Planta Med. **1984**, *50*, 366–375.
8. Bauer, R.; Remiger, P. TLC and HPLC analysis of alkamides in Echinacea drugs. Planta Med. **1989**, *55*, 367–371.
9. Hyun-Ock, K.; Durance, T.D.; Scaman, C.H.; Kitts, D.D. Retention of alkamides in dried *Echinacea purpurea*. J. Agric. Food Chem. **2000**, *48*, 4187–4192.
10. Willis, R.B.H.; Stuart, D.L. Alkiylamide and cichoric acid levels in *Echinacea purpurea* grown in Australia. Food Chem. **1999**, *67*, 385–388.
11. Bauer, R.; Juric, K.; Puhlmann, J.; Wagner, H. Immunological in vivo and in vitro examinations of Echinacea extracts. Drug Res. **1988**, *38*, 276–281.
12. Muller-Jakic, B.; Breu, W.; Probstle, A.; Redl, K.; Greger, H.; Bauer, R. In vitro inhibition of cyclooxygenase and 5-lipooxygenase by alkamides from Echinacea and Achillea species. Planta Med. **1994**, *60*, 37–40.
13. Clifford, L.J.; Nair, M.G.; Rana, J.; Dewitt, D.L. Bioactivity of alkamides isolated from *Echinacea purpurea* (L.) Moench. Phytomedicine **2002**, *9* (3), 249–253.
14. Luettig, B.; Steinmuller, C.; Gifford, G.E.; Wagner, H.; Lohmann-Matthes, M.-L. Macrophage activation by the polysaccharide arabinogalactan isolated from plant cell cultures of *Echinacea purpurea*. J. Natl. Cancer Inst. **1989**, *81*, 669–675.
15. Currier, N.L.; Miller, S.C. *Echinacea purpurea* and melatonin augment natural-killer cells in leukemic mice and prolong life span. J. Altern. Complement. Med. **2001**, *7*, 241–251.
16. Sun, L.Z.-Y.; Currier, N.L.; Miller, S.C. The American coneflower: A prophylactic role involving nonspecific immunity. J. Altern. Complement. Med. **1999**, *5*, 437–446.
17. Goel, V.; Chang, C.; Slama, J.V.; Barton, R.; Bauer, R.; Gahler, R.; Basu, T.K. Alkylamides of *Echinacea purpurea* stimulate alveolar macrophage function in normal rats. Int. Immunopharmacol. **2002**, *2*, 381–387.
18. Schwarz, E.; Metzler, J.; Diedrich, J.P.; Freudenstein, J.; Bode, C.; Bode, J.C. Oral administration of freshly expressed juice of *Echinacea purpurea* herbs fail to stimulate the nonspecific immune response in healthy young men: Results of a double-blind, placebo-controlled crossover study. J. Immunother. **2002**, *25*, 413–420.
19. Speroni, E.; Govoni, P.; Guizzardi, S.; Renzulli, C.; Guerra, M.C. Anti-inflammatory and cicatrizing activity of *Echinacea pallida* Nutt. root extract. J. Ethnopharmacol. **2002**, *79*, 265–272.
20. Facino, R.M.; Carini, M.; Aldini, G.; Saibene, L.; Pietta, P.; Mauri, P. Echinacoside and caffeoyl conjugates protect collagen from free radical-induced degradation: A potential use of Echinacea extracts in the prevention of skin photodamage. Planta Med. **1995**, *61*, 510–514.
21. Melchart, D.; Linde, K.; Worku, F.; Bauer, R.; Wagner, H. Immunomodulation with Echinacea—A systematic review of controlled clinical trials. Phytomedicine **1994**, *1*, 245–254.
22. Braunig, B.; Dorn, M.; Knick, E. *Echinacea purpureae* radix for the enhancement of the body's own immune defense mechanisms in influenza-like infections. Z. Phytother. **1992**, *13*, 7–13.
23. Barrett, B.; Vohmann, M.; Calabrese, C. Echinacea for upper respiratory infection. J. Fam. Pract. **1999**, *48* (8), 628–635.

E

24. Melchart, D.; Walther, E.; Linde, K.; Brandmaier, R.; Lersch, C. Echinacea root extracts for the prevention of upper respiratory tract infections: A double-blind, placebo-controlled randomized trial. Arch. Fam. Med. **1998**, *7*, 541–545.
25. Grimm, W.; Muller, H.H. A randomized controlled trial of the effect of fluid extract of *Echinacea purpurea* on the frequency and severity of colds and respiratory infections. Am. J. Med. **1999**, *106*, 138–143.
26. Turner, R.B.; Riker, D.K.; Gangemi, J.D. Ineffectiveness of Echinacea for prevention of experimental rhinovirus colds. Antimicrob. Agents Chemother. **2000**, *44* (6), 1708–1709.
27. Brinkeborn, R.M.; Shah, D.V.; Degenring, F.H. Echinaforce and other Echinacea fresh plant prepartions in the treatment of the common cold. A randomized placebo controlled, double-blind clinical trial. Phytomedicine **1999**, *6* (1), 1–5.
28. Lindenmuth, G.F.; Lindenmuth, E.B. The efficacy of Echinacea compound herbal tea preparation on the severity and duration of upper respiratory and flu symptoms: A randomized, double-blind placebo-controlled study. J. Altern. Complement. Med. **2000**, *6* (4), 327–334.
29. Perry, N.B.; Burgess, E.J.; Glennie, V.A. Echinacea standardization: Analytical methods for phenolic compounds and typical levels in medicinal species. J. Agric. Food Chem. **2001**, *49*, 1702–1706.
30. Nuslein, B.; Kurzmann, M.; Bauer, R.; Keis, W. Enzymatic degradation of cichoric acid in *Echinacea purpurea* preparations. J. Nat. Prod. **2000**, *63*, 1615–1618.
31. Bergeron, C.; Gafner, S.; Batcha, L.L.; Angerhofer, C.K. Stabilization of cafeic acid derivatives in *Echinacea purpurea* L. glycerin extract. J. Agric. Food Chem. **2002**, *50*, 3967–3970.
32. Sun, L.; Rezaei, K.A.; Temelli, F.; Ooraikul, B. Supercritical fluid extraction of alkylamides from *Echinacea angustifolia*. J. Agric. Food Chem. **2002**, *50*, 3947–3953.

Ecology and Agricultural Sciences

Jon K. Piper
Bethel College, North Newton, Kansas, U.S.A.

INTRODUCTION

Modern industrialized agriculture has been enormously successful in terms of producing high yields. Over the past several decades, agricultural output has steadily increased, and will need to continue increasing to meet the demands of the growing human population. In recent decades, however, critics have pointed to several unfortunate environmental consequences of modern agriculture. These include soil erosion, contamination of the off-farm environment, and heavy reliance on nonrenewable resources. As far back as the 1930s, proponents of permanent or sustainable agriculture began calling for an imbuing of "ecological principles" into agricultural philosophy and practice. There are several notable examples where this environmental ethic has taken hold in the agricultural establishment, yet many areas still need improvement in order to protect the environment while ensuring a safe and permanent food supply for the indefinite future.

THE PRACTICE OF AGRICULTURE AND THE NATURE OF AGRICULTURAL SCIENCES

Agriculture is one of civilized humankind's oldest activities. Arguably, the practice of agriculture in all its forms has changed the face of the earth more than any other enterprise. As much as 23% of the Earth's land surface is presently devoted to row-crop agriculture, converted pastureland, and urban settlements.[1] In contrast to hunting–gathering systems, in which societies partake of nature's bounty without appreciably changing the structure of wild ecosystems, agriculture manipulates the landscape to maximize production of harvestable products.

In the modern era, the agricultural sciences have included agronomy, horticulture, and—somewhat more holistically—range management, forestry, and soil science. The goals of the agricultural establishment are to ensure a reliable, nutritious, and affordable food supply while providing economic opportunities for farmers by expanding global markets. An additional, implied purpose is to maintain the long-term health and productivity of agricultural, forest, and range lands. This dual purpose may be exemplified by two of the largest agencies within the United States Department of Agriculture. The Agricultural Research Service supports research that seeks to maximize agricultural production; the Natural Resources Conservation Service addresses the protection of such critical agricultural resources as soil and freshwater.

Modern agriculture has certainly benefited most segments of society. For consumers in developed nations, it provides a dependable supply of relatively inexpensive, high-quality food. Although people in less-developed countries have not always partaken of this bounty reliably or equitably, in general the Green Revolution has forestalled widespread starvation within the growing world population. Multinational food corporations, large-scale and corporate farming operations, and the companies that supply them with seed, chemicals, and machinery have all profited particularly well.

The priority for most university researchers within agriculture programs has been to maintain a steady increase in yield via "improvements" in chemical fertilizer technology and delivery, genetics, and machinery. Accessing these improvements, however, often requires large capital investments by farmers.

THE PROBLEM(S) OF AGRICULTURE

Agriculture is by nature extractive, altering ecosystems to emphasize useful products and in the process reducing local biodiversity. The natural world can accommodate this practice where human population density is low, people are nomadic, and technology is primitive. In subsistence forms of agriculture, by which a farmer mainly grows enough food for the immediate family, each person's effect on the landscape is small. There are several examples, however, in which preindustrial societies depleted or despoiled their environmental resource base, leading to their collapse. This has led some writers to characterize the issue as "the problem *of* agriculture" rather than "the problems *within* agriculture."[2] Either way, the process of environmental degradation has only been exacerbated by improving agricultural prowess in machinery, chemicals, irrigation, and the ability to farm fragile and otherwise marginal landscapes.

The various negative consequences of modern agriculture have been well documented.[3] The environmental

Encyclopedia of Plant and Crop Science
DOI: 10.1081/E-EPCS 120010500

failures include soil erosion, loss of native habitat, contamination of surface- and groundwaters, harmful effects of biocides on non-target organisms, salinization, evolution of pest resistance to pesticides, and depletion of nonrenewable resources. Moreover, modern agricultural policy and practice have led to the destruction of rural economies and increasing domination of the global food industry by fewer and fewer corporate entities. For example, from 1940 through the 1980s, the number of farms in the United States dropped from over 6 million to around 2 million, and average farm size increased correspondingly.[4]

In short, in recent decades there has been growing criticism of the agricultural research establishment's apparent disregard for ecological considerations and the health and value of rural culture. Indeed, the focus of much agricultural research tends to be short-term and narrow, placing the natural world lower on its list of considerations or even ignoring it entirely.

THE NATURE OF ECOLOGICAL SCIENCE

Ecology is the scientific study of the relationships between organisms and the biotic and abiotic components of their environment. Traditionally, ecologists have focused on wild organisms in settings relatively unaffected by human activity. Ecology's greatest contribution to modern scientific thought probably is the concept of the ecosystem, the integration of the living and the nonliving worlds via linkages of energy flow and nutrient cycling. From a human society perspective, this concept means that what we do to the natural world we ultimately do to ourselves.

An important subdiscipline within ecology is the field of applied ecology. The purpose of applied ecology is to use the techniques of ecological study to understand and improve resource use in human-managed systems. Such an ecological perspective has greatly benefited such areas as soil conservation, range management, fisheries management, and forestry. The application of ecological principles to the development and management of agricultural systems is called agroecology.

THE "MARRIAGE" OF ECOLOGY AND AGRICULTURE

As long ago as 1938, Herbert Hanson, then president of the Ecological Society of America, called for an "invasion" of ecology into the realms of agriculture and resource conservation.[5] The rationale was that the science of ecology provides the concepts that are needed for achieving harmonious relationships among human beings, their domesticated plants and animals, and the environments that sustain them. Furthermore, natural areas serve as the standards or benchmarks against which such practices as agriculture, grazing, or logging should be measured.

With the rise of the modern environmental movement in the 1960s and 1970s came a blending of ecological holism into agricultural thinking, thus creating a new ethical foundation for farming. This concept of sustainability involves a reorientation of human values, from a short-term technological world view to a longer-term ecological world view. Farming was no longer to be just about producing food, but increasingly about conserving elements of the natural environment. In addition, the sustainable agriculture movement incorporated a social agenda as well as an ecological directive.[6] Hence, the basis of the new farming would be not only technical but also cultural and spiritual, since technical efforts alone had failed to bring about stability in the farming sector.

During the 1980s, agroecology became a legitimate scientific field in its own right.[7,8] Hence, farms are considered as ecosystems that should comply with regional ecological guidelines regarding energy cycles, food webs, biodiversity, and a tendency toward ecological stability with the highest possible yield. Relevant ecological concepts include cropping systems that mimic natural succession, nutrient recycling, rotations that enhance the soil community, and natural methods of fighting pests and diseases.

By the end of the 20th century, several varieties of sustainable agriculture had become commonplace. Each to a large extent incorporates a holistic, more ecological ethic and a sense of respect for nature and rural society. For example, organic farming refers to practices that seek to avoid the use of synthetic herbicides, pesticides, and fertilizers, while at the same time building soil organic matter. Permaculture, a system that originated in Australia, emphasizes small-scale technology and the individual farm as a self-sufficient unit. Permaculture integrates people and the landscape to provide food, energy, and shelter in a systematic way.[9] The term natural systems agriculture has been applied to the concept of agricultural systems that mimic the structure and function of natural ecosystems.[10] The best-known example is the notion of perennial grain polycultures that mimic North American prairie ecosystems. The potential benefits of such systems include protection against soil erosion, efficient nutrient cycling, and biological management of insect pests, pathogens, and weeds.

Nature provides several crucial services for humanity's agriculture—it purifies the air and water, harbors insects and vertebrates that pollinate crops or consume many agricultural pests, and provides soil organisms that convert crop residues to humus and assimilable plant nutrients. Agricultural ecosystems are embedded within larger natural systems with which they exchange organisms and materials. Thus, they can not be divorced from the greater natural systems of which they are a part. The major question is whether our agricultural systems can be redesigned to run harmoniously with—instead of in opposition to—the larger system.

AN AGENDA FOR THE 21ST CENTURY

By the close of the 20th century, proponents of sustainable agriculture had successfully established their ideas as legitimate solutions for many of the problems confronting modern farming and food production. Legislation has helped to reduce soil erosion, decrease water pollution, protect prime farmland and ecologically sensitive areas, and provide additional income to farmers. The agricultural establishment now formally acknowledges that environmental concerns must be balanced with the economic concerns of farmers, agribusinesses, and consumers. However, a great deal more still needs to be accomplished in reforming agriculture along environmental lines.

Industrialized humanity has proceeded as though the raw materials of our economy were discrete entities and not parts of larger, complex, interacting systems. We have kept careful ledgers on the benefits, but generally not on the environmental costs of our activities. We have mined energy and mineral resources, harvested forests, farmed virtually every arable acre, and used bodies of water as though there would be no significant effects downstream. Sustainable agriculture is a system that for generations to come will not only be productive and profitable but must also conserve resources, protect the environment, and enhance the health and safety of the citizenry. Specifically, this means maintaining renewable resources at their current levels, avoiding the production of wastes beyond the environment's capacity to assimilate them, protecting the Earth's biodiversity, and reducing the dependence on nonrenewable resources while maintaining the capacity to ensure an acceptable standard of living long into the future.

ARTICLES OF FURTHER INTEREST

Agriculture and Biodiversity, p. 1
Crops and Environmental Change, p. 370
Ecology: Functions, Patterns and Evolution, p. 404
Environmental Concerns and Agricultural Policy, p. 418
Herbicides in the Environment: Fate of, p. 554
Integrated Pest Management, p. 612
Organic Agriculture As a Form of Sustainable Agriculture, p. 846
Crop Domestication in the American Tropics: Phytolith Analyses, p. 326
Pre-agricultural Plant Gathering and Management, p. 1055
Reconciling Agriculture with the Conservation of Tropical Forests, p. 1078
Social Aspects of Sustainable Agriculture, p. 1152
Sustainable Agriculture and Food Security, p. 1183
Sustainable Agriculture: Definition and Goals, p. 1187
Sustainable Agriculture: Ecological Indicators, p. 1191
Sustainable Agriculture: Economic Indicators, p. 1195

REFERENCES

1. Vitousek, P.M.; Mooney, H.A.; Lubchenco, J.; Melillo, J.M. Human domination of Earth's ecosystems. Science **1997**, *277*, 494–499.
2. Jackson, W.; Piper, J. The necessary marriage between ecology and agriculture. Ecology **1989**, *70*, 1591–1593.
3. Soule, J.D.; Piper, J.K. *Farming in Nature's Image: An Ecological Approach to Agriculture*; Island Press: Washington, DC, 1992; 1–286.
4. Strange, M. *Family Farming: A New Economic Vision*; University of Nebraska Press: Lincoln, NE, 1988; 1–311.
5. Hanson, H.C. Ecology in agriculture. Ecology **1939**, *20*, 111–117.
6. Beeman, R.S.; Pritchard, J.A. *A Green and Permanent Land: Ecology and Agriculture in the Twentieth Century*; University Press of Kansas: Lawrence, KS, 2001; 1–219.
7. Gliessman, S.R. *Agroecology: Ecological Processes in Sustainable Agriculture*; Ann Arbor Press: Ann Arbor, MI, 1998; 1–357.
8. Thomas, V.G.; Kevan, P.G. Basic principles of agroecology and sustainable agriculture. J. Agric. Environ. Ethics **1993**, *3*, 1–19.
9. Mollison, B.C. *Permaculture: A Designer's Manual*; Tagari Publications: Tyalgum, Australia, 1988; 1–576.
10. Piper, J.K. Natural Systems Agriculture. In *Biodiversity in Agroecosystems*; Collins, W.W., Qualset, C.O., Eds.; CRC Press: Boca Raton, 1999; 167–196.

Ecology: Functions, Patterns, and Evolution

Fernando Valladares
Centro de Ciencias Medioambientales, Madrid, Spain

INTRODUCTION

Ecology is the study of living organisms and their environment in an attempt to explain and predict. While natural history involves the accumulation of detailed data with emphasis on the autecology of each species, the objectives of ecology in general and of functional ecology in particular are to develop predictive theories and to assemble the data to develop general models. Functional ecology has three basic components: 1) constructing trait matrices through screening of various plant and animal species, 2) exploring empirical relationships among these traits, and 3) determining the relationships between traits and environments.[1] Studies in functional ecology encompass a wide range of approaches, from individuals to populations; from mechanistically detailed to deliberately simplified, black-box simulations; and from deductive to inductive.[2] Functional ecology is concerned with the links between structure and function, the existence of general patterns among species, and the evolutionary connections among these patterns. And functional ecology is, above all, timely and pertinent, because the environmental degradation associated with human development is rapidly destroying the very systems that ecologists seek to understand. If we are to anticipate the extent and repercussions of global change in natural habitats, we first need to understand how organisms and ecosystems function.

EXPLORING THE MECHANISMS: FROM PHYSIOLOGY TO ECOPHYSIOLOGY AND FUNCTIONAL ECOLOGY

The study of plant functions has largely followed a reductionistic approach aimed at explaining the functions in terms of the principles of physics and chemistry. However, plant physiology, which is focused on molecules, organelles, and cells, has not been able to provide reliable predictions of the responses of vegetation to changes in their environment, due to the multiple interactions involved in the responses of plants and the hierarchical nature of plant organization.[3] The realization of the fact that the responses of higher organizational levels are not predictable from the dynamics of those of smaller scales led to the advent of plant ecophysiology, which is focused on organs—a more relevant scale of organization to address questions regarding plant performance. Ecophysiologists have primarily focused on the structural and functional properties of leaves. Leaves display wide variation in morphology and physiology, including differences in specific mass, carbon and nitrogen investments, stomatal densities, optical properties, and hydraulic and photosynthetic characteristics. The emphasis of ecophysiological studies on leaves is due to the profound implications of the interactions between leaf structure and function for the performance of plants in natural habitats.[4,5] However, the traditional ecophysiological approach proved to be insufficient in predicting plant distribution and responses to changing environments, which led to the development of functional plant ecology.[6] Functional ecology is centered on whole plants as the unit of analysis, encompassing a range of scales of organization from organs to whole organism architecture, and is based on a much broader conception of plant functions than that formulated by the earlier practitioners of ecophysiology.[7]

THE SEARCH OF GENERAL PATTERNS: COMPARATIVE ECOLOGY

Since elucidation of the range of possible functional responses of plants is not possible with the use of model organisms, such as those typically used in plant physiology, functional ecology arises as an essentially comparative science. Ideally, functional ecology deals with traits measured on a large number of species in order to minimize the influence of the peculiarities of the autecology of each species. Two main approaches have been followed to find general ecological patterns in nature: 1) screening, that is, the design of bioassays for a trait or a set of traits measured simultaneously on a large number of species, as in the classic study by Grime and Hunt[8] of the relative growth rate of 132 species of British flora, and 2) empiricism, or the search for quantitative relationships between measurable dependent and independent variables (e.g., correlations among pairs

Encyclopedia of Plant and Crop Science
DOI: 10.1081/E-EPCS 120010498

of traits or traits and environments) producing quantitative models using traits and not species, as in the general revision of leaf traits by Reich et al.[4] Models using traits are more general than those based on species and can be more easily transferred to different floras.[1] The question arises as to which trait must be measured. A possible answer can be obtained by analyzing the basic functions that organisms perform (i.e., resource acquisition, the ability to tolerate environmental extremes, and the ability to compete with neighbors) and then either looking for traits which measure these functions or carrying out direct bioassays of them. The most common limitations of this kind of study are the difficulties in finding unambiguous linkages of a trait to a specific function, and the so-called phylogenetic constraints that are due to the fact that phylogenetic proximity among species can influence their functional similarities.

Individual plants demonstrate important degrees of phenotypic variation, which must be considered in comparative studies. No two individuals of the same species exhibit the same final shape or functional features, regardless of how similar the genotypes of two individuals may be.[9] Part of this variation is due to phenotypic plasticity, that is, the capacity of a given genotype to render different phenotypes under different environmental conditions, and part is due to other reasons (ontogenetic stage, developmental instability). Plants of the same chronological age can be ontogenetically different, so interpretation of differences in phenotypic traits will depend on whether comparisons are made as a function of age, size, or developmental stage.

EXPERIMENTAL ECOLOGY

Broad-scale comparisons must be driven by hypotheses and must be based on robust statistical designs. However, the finding of statistically significant patterns is no guarantee of underlying cause-and-effect relationships, which must be tested experimentally. Gradient analysis, where functional responses are examined along a clearly defined environmental gradient, is a powerful approach to exploring the relationships between plant function and environment, but it is prone to spurious relationships when there is a hidden factor covarying with the factor defining the gradient.[3] Inferences from gradient analysis and broad-scale comparisons are statistical in nature and must be confirmed experimentally.

Experiments are designed to test hypotheses, but sufficient knowledge must be available to specify more than a trivial hypothesis before thorough experimentation can be undertaken.[2] Unfortunately, this is not the case for many natural systems. Although experiments are a common practice in physiology and ecophysiology, where mechanisms can be explored and hypotheses tested under relatively well-controlled conditions, they are less common in functional ecology studies, especially when hypotheses must be tested in natural habitats. Experimental design should meet a number of standards that are not easy—and in certain cases not possible—to meet in field ecology. The controls should be randomly intermixed with the treatments, both in space and time, and both the control and the experiments must be replicated enough. Replicates are not easy to find in natural scenarios, and when they can be found they are frequently not truly independent of one another, leading to a weak experimental design due to pseudoreplication. In addition, multiple causality and indirect effects, which significantly complicate the interpretation of results, are commonplace in ecology. However, experimental ecology is burgeoning despite all this adversity, because experiments are irreplaceable elements in achieving relevant progress in our understanding of ecological processes, as has been revealed by a number of experimental manipulations of natural populations.[10]

EVOLUTIONARY ECOLOGY

Functional ecology eventually leads to evolutionary ecology. Trends in functional traits across species and mechanisms linking cause and effect contribute to our understanding of evolutionary processes, especially when they are considered on a time scale long enough to allow for changes in gene frequencies, the essence of evolution. Whereas functional ecology is interested in the immediate influence of environment on a given trait, evolutionary ecology is aimed at understanding why some individuals have left the most offspring in response to long-term consistent patterns of environmental conditions. Functional ecology, focused on an ecological time scale (*now* time), asks questions of ''how?'' and is concerned with the proximate factors influencing an event. Evolutionary ecology is focused on an evolutionary time scale (geological time), asking questions of ''why?'' and concerned with the ultimate factors influencing an event. Neither is more correct than the other, and they are not mutually exclusive because ecological events can always be profitably considered within an evolutionary framework, and vice versa.[10]

There are five agents of evolution: natural selection (differential reproductive success of individuals within a population); genetic drift (random sampling bias in small populations); gene flow (migration movements of individuals among and between populations with different gene frequencies); meiotic drive (segregation distortion of certain alleles that do not follow the Mendelian lottery of meiosis and recombination); and mutation. Of these agents, only natural selection is directed, resulting in conformity between organisms and their environments.[10] Darwin's theory of natural selection is a fundamental unifying theory, and the many studies that have been carried out over the last century in support of it demonstrate the power of the rigorous application of the genetic theory of natural selection to population biology.

ARTICLES OF FURTHER INTEREST

Breeding: Genotype-by-Environment Interaction, p. 218
Coevolution of Insects and Plants, p. 289
Competition: Responses to Shade by Neighbors, p. 300
Cross-species and Cross-genera Comparisons in the Grasses, p. 374
Ecology and Agricultural Sciences, p. 401
Ecophysiology, p. 410
Photosynthesis and Stress, p. 901
Plant Response to Stress: Mechanisms of Accommodation, p. 987
Population Genetics, p. 1042
Shade Avoidance Syndrome and Its Impact on Agriculture, p. 1152

REFERENCES

1. Keddy, P.A. A pragmatic approach to functional ecology. Funct. Ecol. **1992**, *6*, 621–626.
2. Mentis, M.T. Hypothetico-deductive and inductive approaches in ecology. Funct. Ecol. **1988**, *2*, 5–14.
3. Duarte, C. Methods in Comparative Functional Ecology. In *Handbook of Functional Plant Ecology*; Pugnaire, F.I., Valladares, F., Eds.; Marcel Dekker: New York, 1999; 1–8.
4. Reich, P.B.; Ellsworth, D.S.; Walters, M.B.; Vose, J.M.; Gresham, C.; Volin, J.C.; Bowman, W.D. Generality of leaf trait relationships: A test across six biomes. Ecology **1999**, *80*, 1955–1969.
5. Niinemets, U. Global-scale climatic controls of leaf dry mass per area, density, and thickness in trees and shrubs. Ecology **2001**, *82*, 453–469.
6. Pugnaire, F.I.; Valladares, F. *Handbook of Functional Plant Ecology*; Marcel Dekker: New York, 1999.
7. Valladares, F.; Pearcy, R.W. The functional ecology of shoot architecture in sun and shade plants of *Heteromeles arbutifolia* M. Roem., a Californian chaparral shrub. Oecologia **1998**, *114*, 1–10.
8. Grime, J.P.; Hunt, R. Relative growth rate: Its range and adaptive significance in a local flora. J. Ecol. **1975**, *63*, 393–422.
9. Coleman, J.S.; McConnaughay; Ackerley, D.D. Interpreting phenotypic variation in plants. Trends Ecol. Evol. **1994**, *9*, 187–191.
10. Pianka, E.R. *Evolutionary Ecology*; Benjamin-Cummings: San Francisco, 2000.

Economic Impact of Insects

David Pimentel
Cornell University, Ithaca, New York, U.S.A.

INTRODUCTION

Insects, plant pathogens, and weeds are major pests of crops in the United States and throughout the world. Approximately 70,000 species of pests attack crops, with about 10,000 species being insect pests worldwide. This article focuses on the economic consequences for agriculture, but it is important to establish that there are serious consequences for humans and other animals in the food chain from insects, topics that can't be explored in detail in this entry.

SCOPE OF PROBLEM

There are approximately 2500 insect and mite species attacking U.S. crops. About 1500 of these species are native insect/mite species that moved from feeding on native vegetation to feeding on our introduced crops. Thus, an estimated 60% of the U.S. insect pests are native species and about 40% are introduced invader species.[1,2] Both groups of pests contribute nearly equally to the current annual 13% crop losses to insect pests, despite all the insecticides that are applied plus other types of controls practiced in the United States. Most or about 99% of U.S. crops are introduced species; therefore, many U.S. native insect species find these introduced crops an attractive food source.[3] For example, the Colorado potato beetle did not feed originally on potatoes but fed on a native weed species. Although the center of origin of the potato is the Andean regions of Bolivia and Peru, the plant was introduced into the United States from Ireland, where it has been cultivated since the mid-17th century.[3] Today, the Colorado potato beetle is the most serious pest of potatoes in the United States and elsewhere in the world where both the potato and potato beetle have been introduced.

An estimated 3 billion kg of pesticides are applied annually in the world in an attempt to control world pests.[4] Approximately 40% of the pesticides are insecticides, 40% herbicides, and 20% fungicides. About 1 billion kg of pesticides are applied within the United States; however, only 20% are insecticides.[4] Most or 68% of the pesticides applied in the United States are herbicides.

Despite the application of 3 billion kg of pesticides in the world, more than 40% of all crop production is lost to the pest complex, with crop losses estimated to be 15% for insect pests, 13% weeds, and 12% plant pathogens.[4] In the United States, total crop losses are estimated to be 37% despite the use of pesticides and all other controls, with crop losses estimated to be 13% due to insect pests, 12% weeds, and 12% plant pathogens.

The 3 billion kg of pesticides applied annually in the world costs an estimated $32 billion per year or slightly more than $10 per kg. With about 1.5 billion ha of cropland in the world, it follows that about 2 kg of pesticides is applied per ha, or $20 per ha is invested in pesticide control.

INSECT PESTS

Note that despite a tenfold increase in total quantity (weight) of insecticide used in the United States since about 1945 when synthetic insecticides were first used, crop losses to insect pests have nearly doubled from about 7% in 1945 to 13% today.[1] Actually, the situation is a great deal more serious because during the past 40 years the toxicity of pesticides has increased 100- to 200-fold. For example, in 1945 many of the insecticides were applied at about 1 to 2 kg per ha; however, today many insecticides are applied at dosages of 10 to 20 grams per ha.[1]

The increase in crop losses to insect pests from 7% in 1945 to 13% today, despite a 100–200-fold increase in the toxicity of insecticides applied per ha, is due to changes in agricultural technologies over the past 50 years.[2] These changes in agricultural practices include the following: 1) the planting of some crop varieties that are more susceptible to insect pests than those planted previously; 2) the destruction of natural enemies of some pests by insecticides (e.g., destruction of cotton bollworm and budworm natural enemies), thereby creating the need for additional insecticide applications; 3) insecticide resistance developing in insect pests, thus requiring additional applications of more toxic insecticides; 4) reduction in

Encyclopedia of Plant and Crop Science
DOI: 10.1081/E-EPCS 120010475

crop rotations, which caused further increases in insect pest populations (e.g., corn rootworm complex); 5) lowering of the Food and Drug Administration (FDA) tolerances for insects and insect parts in foods, and enforcement of more stringent "cosmetic standards" for fruits and vegetables by processors and retailers; 6) increased use of aircraft application of insecticides for insect control, with significantly less insecticide reaching the target crop; 7) reduced field sanitation and more crop residues for the harboring of insect pests; 8) reduced tillage that leaves more insect-infested crop residues on the surface of the land; 9) culturing some crops in climatic regions where insect pests are more severe; and 10) the application of some herbicides that improve the nutritional makeup of the crop for insect invasions.[2]

The 13% of potential U.S. crop production represents a crop value lost to insect pests of approximately $35 billion per year. In addition, nearly $2 billion in insecticides and miticides are applied each year for control. Ignoring other control costs, combined crop losses and pesticides thus total about $37 billion per year.[3]

Total crop losses to insects and mites worldwide are estimated to be approximately $400 billion per year. Given that more than 3 billion of the 6.2 billion people on earth are malnourished, this loss of food to insect and mite pests each year, despite the use of 3 billion kg of pesticides, is an enormous loss to society.

INSECTICIDE AND MITICIDE PEST CONTROLS

In general, insecticide and miticide control of insects and mites return approximately $4 for every $1 invested in chemical control.[1] This is an excellent return, but not as high as some of the nonchemical controls. For example, biological pest control has reported earnings of $100 to $800 per $1 invested in pest control. It must be recognized that the development of biological controls, although highly desirable, are not easy to develop and implement.

Both in crops and nature, host plant resistance and natural enemies (parasites and predators) play an important role in pest control. In nature, seldom do insect pests and plant pathogens remove more than 10% of the resources from the host plant. Host plant resistance, consisting of toxic chemicals, hairiness, hardness, and combinations of these, prevents insects from feeding intensely on host plants.[5] Some of the chemicals involved in plants resisting insect attack include cyanide, alkaloids, tannins, and others. Predators and parasites that attack insects play an equally important role in controlling insect attackers on plants in nature and in agroecosystems.

INSECT TRANSMISSION OF PLANT PATHOGENS

Insects with sucking mouthparts, such as aphids and plant bugs, play a major role in the transmission of plant pathogens from plant to plant. It is estimated that about 25% of the plant pathogens are transmitted by insects. The most common pathogens transmitted are viruses. These pathogens include lettuce yellows and pea mosaic virus. Fungal pathogens are also transmitted by insects. For example, Dutch elm disease is transmitted by two bark beetle species that live under the bark of elm trees. In infected trees, the bark beetles become covered with fungal spores. When the beetles disperse and feed on uninfected elm trees, they leave behind fungal spores that in turn infect the healthy elm trees.

ENVIRONMENTAL AND PUBLIC HEALTH IMPACTS OF INSECTS

Some insect species have become environmental and public health pests. For example, the imported red fire ant kills poultry chicks, lizards, snakes, and ground-nesting birds. Investigations suggest that the fire ant has caused a 34% decline in swallow nesting success as well as a decline in the northern bobwhite quail populations in the United States. The ant has been reported to kill infirm people and people who are highly sensitive to the sting of the ant. The estimated damages to wildlife, livestock, and public health in the United States is more than $1 billion per year, with these losses occurring primarily in southern United States.[3]

In another example, the Formosan termite that was introduced into the United States has been reported to cause more than $1 billion per year in property damage, repairs, and controls. As it spreads further in the nation, the damages will increase.[3]

CONCLUSION

Insect and mite pests in the United States and world are causing significant crop, public health, and environmental damages. Just for crop losses in the United States, it is estimated that insect and mite species are causing $37 billion per year, if control costs are included. Worldwide crop losses to insects and mites are estimated to be $400 billion per year. The public health and environmental damages in the United States and world are estimated to be valued at several hundred billion dollars per year.

ARTICLES OF FURTHER INTEREST

Genetically Engineered Crops with Resistance Against Insects, p. 506
Insect Life History Strategies: Development and Growth, p. 598
Insect Life History Strategies: Reproduction and Survival, p. 601
Insect/Host Plant Resistance in Crops, p. 605
Insect–Plant Interactions, p. 609
Integrated Pest Management, p. 612
Mutualisms in Plant–Insect Interactions, p. 763
Physiology of Herbivorous Insects, p. 910
Plant Defenses Against Insects: Constitutive and Induced Chemical Defenses, p. 939
Plant Defenses Against Insects: Physical Defenses, p. 944
Root-Feeding Insects, p. 1114

REFERENCES

1. Pimentel, D. *Handbook on Pest Management in Agriculture, Three Volumes*; CRC Press: Boca Raton, FL, 1991.
2. Pimentel, D.; Lehman, H. *The Pesticide Question: Environment, Economics and Ethics*; Chapman and Hall: New York, 1993.
3. Pimentel, D.; Lach, L.; Zuniga, R.; Morrison, D. Environmental and economic costs of non-indigenous species in the United States. BioScience **2000**, *50* (1), 53–65.
4. Pimentel, D. *Techniques for Reducing Pesticides: Environmental and Economic Benefits*; John Wiley: Chichester, UK, 1997.
5. Pimentel, D. Herbivore population feeding pressure on plant host: Feedback evolution and host conservation. Oikos **1988**, *53*, 289–302.

Ecophysiology

Luis Balaguer
Universidad Complutense de Madrid, Madrid, Spain

INTRODUCTION

Plant ecophysiology is science that shares with other disciplines its focus on plant performance under field conditions, but differs from them in its ecosystem approach. The main challenge in ecophysiology is to scale up, from physiological processes at the organ level to whole plants, canopies, landscapes, and even to a global level. Across these levels, ecophysiological studies deal with the response to stress, disturbance, or interactions among organisms. Progress is directed toward assessing plant response to finer-scale patterns of biotic, spatial, and temporal heterogeneity. Species characterization on the basis of their responses to different factors and at different organization levels enables arranging species into functional groups, each of which represents a type of plant homeostasis in a changing environment. Understanding of these patterns is significantly limited by the difficulty of dealing with the disparities in the spatial and temporal scales used. The significance of physiological processes to explain plant community composition and ecosystem dynamics is constrained by taxon- and site-specific factors. Ecosystem processes are sensitive to initial conditions, interactive effects, and long-term restrictions. Future research on ecophysiology should give special attention to below-ground plant performance, long-term effects of short- to medium-term events, multiple trophic interactions, feed-forward response to environmental stress, and the reliability and accuracy of regional scale predictions.

PLANT ECOPHYSIOLOGY

Plant ecophysiology is a discipline included in the interface of botany, plant physiology, and ecology. This intersection, widely considered, includes the study of the function and performance of plants in their environment. Within such a broad scope, there is a gradient from autecology, at the boundaries of pure physiology, to synecology, at the boundaries of pure ecology. Closer to autecology, environmental physiology has been defined as the physiology of acclimation, and so it involves a consideration of optimal patterns of response to limiting resources, such as light, CO_2, water, or nitrogen. Optimization is thus its central theme. In contrast, the studies in physiological ecology or ecophysiology are based on an ecosystem approach.

Few authors distinguish between physiological ecology, as a discipline centered on the characterization of the plant role in the community, and ecophysiology, devoted to the study of the plant role in the ecosystem.[1] According to this distinction, the former discipline aims to characterize plant niche and the latter to describe the effects of environmental, phenological, morphological, and physiological processes on mass and energy exchange through a plant. Such differentiation is in consonance with the acknowledgement of two types of analytical problems: defining plant tolerances and describing plant performance.[2]

Most researchers, however, consider both terms synonymous, aiming to integrate, rather than segregate, both perspectives. In this conception, ecophysiology endeavors to explain ecological processes in physiological terms.[2] The scientific challenge is to scale upward, from organs to whole plants, canopies, landscapes, and even to a global level. At the ecosystem level, this scaling up constitutes an alternative approach to the conceptual simplification implicit in the big leaf models, which do not consider the internal heterogeneity of the system. On a global scale, ecophysiological scaling finds alternative models to those derived from the climate envelope approach, which assume that the predicted climate changes will only shift the geographical distribution of plant communities considered as homogeneous and invariable units. Progress in ecophysiology is directed toward assessing plant response to finer-scale patterns of biotic, spatial, and temporal heterogeneity across ecosystems. In consonance, there has been a shift from an earlier ecophysiological perspective, focused on the acclimation of plants to stable environmental gradients, to an increasing interest in the analysis of the mechanisms by which plants exploit gaps, patches, favorable microsites, and transient events.[3]

GRIME'S TRIANGLE

Ecophysiological studies are related to at least one of the three axes of Grime's triangle: stress, disturbance, and competition.[4] Stress is defined as the extent to which any combination of environmental conditions reduces plant performance.[5] Among them, those whose effects have

Encyclopedia of Plant and Crop Science
DOI: 10.1081/E-EPCS 120010505

been most commonly studied are drought, salinity, nutrient shortage, extreme irradiance, and temperatures. Disturbance is defined as the extent to which any change in the resource base in a habitat induces a change in plant population response.[6] Disturbances are, among others, clearing of vegetation, changes in land use, pollution, fires, flooding, hurricanes, landslides, trampling, and the action of herbivores and pathogens. Ecophysiological studies deal not only with the thresholds of resistance but also with plant strategies to withstand stress and disturbance. In this context, strategy is understood as the constitutive character syndrome and the complex suite of inducible responses that allow occupation of a particular environment. Plant strategies can be considered the equivalent of behavior in animals, and thus ecophysiology shares with ethology this subject matter and the interest in its ontogeny and phylogeny. The comparative analysis of plant performance in response to stress or disturbances groups plant species in three categories: tolerants, which endure unfavorable environmental conditions and resist their effects; avoiders, which trigger inducible mechanisms that prevent deleterious effects; and evaders, with a life cycle or metabolic activity restricted to favorable periods. Some authors consider this latter category as a particular case of the avoidance strategy.[7]

In relation to the third axis of Grime's triangle, ecophysiology is concerned with any kind of interaction between organisms, including competition. These interactions among individuals change in intensity and even in direction over space and time, ranging from obligate facilitation to exclusion.

The descriptive study of the response to stress, disturbance, and interaction with other organisms enables the ordering of plant species into functional groups, which allows a theoretical reduction of the diversity of species to a more limited number of functions and structures. On this basis, ecophysiological studies attempt to connect biological diversity with ecosystem functioning. Their conclusions are relevant for modeling the dynamics of plant communities in the context of ecological succession, and also for predicting ecosystem drifts triggered by disturbances.[8]

GLOBAL IMPACT

Ecophysiology has a key role in the understanding of the potential effects of global change and contributes new insights to restoration ecology. The value of ecophysiology in this context derives from its potential to characterize homeostatic adjustments at different scales. On a temporal scale shorter than an individual lifetime, homeostasis is limited to the flexibility of plant response to stress, disturbance, or biotic interactions. Homeostasis at the individual level can involve reversible adjustments to rapid environmental fluctuations (environmental tracking), the ability of a single genotype to express different phenotypes in response to different environments (phenotypic plasticity), or the ability to adjust the performance of established modules in response to environmental changes (acclimation).[6] On a temporal scale longer than the lifespan of a single generation, homeostasis involves adaptation, which is a process of genetic change whereby the average state or the range of variation of a feature becomes prevalent because of its selective value.[9] At this level, ecophysiology is concerned with the analysis of the physiological basis of population divergence and ecotypic differentiation.[2]

However, conceptual difficulties arise from the contrasted magnitudes of the spatiotemporal scales that are relevant in the plant ecophysiological approach. The complexity of the scaling up lies not only in the increasing number of units, components, or modules involved, but mainly in the intrinsic control mechanisms that arise at each level. These regulatory mechanisms do not emerge from the assembling of individual components but from the interactions among them. Neglecting this principle constitutes one of the most common sources of error in ecophysiology. An example is to assume an invariable adaptive value of physiological processes from the organ level to the interpretation of plant communities and ecosystems. This assumption emanates from erroneous faith in the power of natural selection as an optimizing agent, without considering the phylogenetic, developmental, and architectural constraints that derived from scaling up to the organism level.[10] When these constraints are acknowledged but dismissed as unimportant or inscrutable, it is often erroneously accepted that every single trait necessarily reports a benefit just limited by the costs imposed by its integration into complex organisms, communities, or ecosystems. This notion of suboptimality is what Gould and Lewontin have termed the Panglossian Paradigm.[10] Physiological factors are not enough to explain how the potential pool of plant species comes down to those that actually form part of a community in a particular site at a given instant of time. Historical filters have to be necessarily considered if the output of ecophysiology is to account in physiological terms for the ways plants can live where they do (Table 1).[5]

CONCLUSION

Future directions in ecophysiological research will address current gaps in knowledge. Some areas have been neglected due to the fact that scientific progress in this field has been severely limited by methodological constraints. The first area that deserves attention is the

Table 1 Taxon- and site-specific constraints that limit the significance of physiological processes to explain plant community composition and ecosystem dynamics

Taxon-specific constraints	Site-specific constraints
Sensibility to initial conditions	
Plant species can be older than the current suite of environmental conditions, and many of their traits have not been selected for their current functions.	The iterative nature of the ecological succession determines the dependence of the composition of every plant community on the local availability of plant species.
Interactive effects	
Plant traits are genetically and physiologically linked. Selective pressures promote character suites or syndromes.	Interaction with other organisms contributes to determine the distribution of plant species. This ecological sorting should not be equated with selection.
Long-term restrictions	
Evolution of individual species is limited by phylogenetic constraints.	Current composition, structure, and function of plant communities are affected by disturbance regimes and irreversible transitions, such as climate changes.

characterization of below-ground plant performance. Secondly, most of the ecophysiological research has been undertaken with seedlings, saplings, or short-lived plant species, and over limited periods of time. A major challenge is to assess the relevance of short- to medium-term events in long-lived woody plants. Exciting new insights are also expected from the characterization of tritrophic plant-insect-herbivore-natural enemy systems, with clear implications for the biological control of pests, as well as from the study of plant response to likely changes, that is plant sensing and reaction to environmental cues in a feed-forward manner. Finally, the ultimate goal of ecophysiology, still far from achieved, is to provide regional-scale predictions accurate and reliable enough to be useful for managers and policy makers.

ACKNOWLEDGMENTS

I am grateful to Professor Hans Lambers for his valuable comments and suggestions.

ARTICLES OF FURTHER INTEREST

Air Pollutants: Interactions with Elevated Carbon Dioxide, p. 17
Air Pollutants: Responses of Plant Communities, p. 20
Biological Control in the Phyllosphere, p. 130
Breeding: Genotype-by-Environment Interaction, p. 218
C_3 Photosynthesis to Crassulacean Acid Metabolism Shift in Mesembryanthenum crystallinum: *A Stress Tolerance*, p. 241
Coevolution of Insects and Plants, p. 289
Competition: Responses to Shade by Neighbors, p. 300
Drought and Drought Resistance, p. 386
Ecology: Functions, Patterns, and Evolution, p. 404
Plant–Pathogen Interaction: Evolution, p. 965
Fluorescence: A Probe of Photosynthetic Apparatus, p. 472
Genome Rearrangements and Survival of Plant Populations to Changes in Environmental Conditions, p. 513
Insect–Plant Interactions, p. 609
Leaves and Canopies: Physical Environment, p. 646
Plant Response to Stress: Biochemical Adaptations to Phosphate Deficiency, p. 976
Plant Response to Stress: Critical Periods in Plant Development, p. 981
Plant Response to Stress: Genome Reorganization in Flax, p. 984
Plant Response to Stress: Mechanisms of Accommodation, p. 987
Plant Response to Stress: Modifications of the Photosynthetic Apparatus, p. 990
Plant Responses to Stress: Nematode Infections, p. 1014
Plant Response to Stress: Phosphatidic Acid As a Secondary Messenger, p. 995
Plant Response to Stress: Regulation of Plant Gene Expression to Drought, p. 999
Plant Response to Stress: Role of the Jasmonate Signal Transduction Pathway, p. 1006
Plant Response to Stress: Role of Molecular Chaperones, p. 1002
Plant Response to Stress: Source-Sink Regulation by Stress, p. 1010
Plant Responses to Stress: Ultraviolet-B Light, p. 1019
Rhizosphere: An Overview, p. 1084
Photosynthesis and Stress, p. 901
UV Radiation Penetration in Plant Canopies, p. 1261
Water Deficits: Development, p. 1284
Water Use Efficiency Including Carbon Isotope Discrimination, p. 1288

REFERENCES

1. Alm, D.M.; Hesketh, J.D.; Capman, W.C.; Begonia, G.B. Plant Ecology. In *Encyclopedia of Agricultural Science*; Arntzen, J., Ritter, E.M., Eds.; Academic Press, Inc.: New York, 1994; Vol. 3, 259–274.
2. Lange, O.L.; Nobel, P.S.; Osmond, C.B.; Ziegler, H. *Physiological Plant Ecology I: Responses to the Physical Environment*; Springer-Verlag, Cop.: Berlin, Germany, 1981.
3. Bell, G.; Lechowicz, M.J. Spatial Heterogeneity at Small Scales and How Plants Respond to It. In *Exploitation of Environmental Heterogeneity by Plants: Ecophysiological Processes Above- and Belowground*; Caldwell, M.M., Pearcy, R.W., Eds.; Academic Press, Inc.: San Diego, CA, 1994; 391–414.
4. Grime, J.P. *Plant Strategies, Vegetation Processes, and Ecosystem Properties*; John Wiley, Cop.: New York, 2001.
5. Lambers, H.; Chapin, F.S., III; Pons, T.L. *Plant Physiological Ecology*; Springer-Verlag New York, Inc.: New York, 1998.
6. Bazzaz, F.A. *Plants in Changing Environments: Linking Physiological, Population, and Community Ecology*; Cambridge Univ. Press: Cambridge, United Kingdom, 1996.
7. Jones, H.G. *Plants and Microclimate: A Quantitative Approach to Environmental Plant Physiology*; Cambridge Univ. Press: Cambridge, United Kingdom, 1992.
8. Smith, T.M.; Shugart, H.H.; Woodward, F.I. *Plant Functional Types: Their Relevance to Ecosystem Properties and Global Change*; Cambridge Univ. Press: Cambridge, United Kingdom, 1998.
9. Futuyma, D.J. *Evolutionary Biology*; Sinaver Associates, Cop.: Sunderland, MA, 1998.
10. Gould, S.J.; Lewontin, R.C. The spandrels of San Marco and the panglossian paradigm: A critique of the adaptationist programme. Proc. R. Soc. Lond., B **1979**, *205* (1161), 581–598.

Endosperm Development

Philip W. Becraft
Iowa State University, Ames, Iowa, U.S.A.

INTRODUCTION

The endosperm is the "other" product of double fertilization; one pollen sperm fertilizes the egg to produce a zygote, while the other sperm fertilizes the central cell to produce the triploid primary endosperm cell. Endosperm development is often overlooked because it represents a dead-end, often transitory, tissue. However, endosperm development is proving to be unexpectedly interesting, entailing many novel and highly specialized processes. The list includes novel cytoskeletal behavior, cytokinesis, cell wall formation, and cell cycle regulation. Questions of cell fate acquisition and signaling, programmed cell death, and morphogenesis also apply. Given the importance of cereal grains to human and livestock nutrition and industrial uses, understanding the fundamental processes of endosperm development has great potential for application to grain improvement.

TYPES OF ENDOSPERM DEVELOPMENT

There are three types of endosperm development: cellular, nuclear, and helobial. In cellular development, mitosis is accompanied by cytokinesis from the outset. In contrast, nuclear development undergoes a period of free nuclear division leading to a multinucleate syncitium; cytokinesis and cellularization ensue later. In helobial development, the first division includes cytokinesis. Subsequently, one of the daughter cells continues the cellular pattern of development, with mitosis accompanied by cytokinesis, while the other daughter undergoes a period of free nuclear division. This review will focus on nuclear development, which is the most common among crops and other well-studied plant systems.

The endosperm can also be either transitory or persistent. The most obvious example of a persistent endosperm is in cereal grains where the endosperm is the major seed storage tissue and composes the bulk of the seed. In other plants, the cotyledons assume the primary storage function, and most of the endosperm degenerates during seed development. Still other plants are intermediate with both endosperm and cotyledons participating in food storage.

EPIGENETIC REGULATION OF ENDOSPERM DEVELOPMENT

Endosperm development is normally triggered by fertilization. The inhibition of endosperm development prior to fertilization appears to require chromatin-level gene regulation because mutations in polycomb group genes, including *MEDEA*, *FIE*, and *FIS2*, allow fertilization-independent seed and endosperm development. These genes are imprinted such that only the maternal copy is transcribed in the fertilized seed.[1] In addition, the normal balance of maternal to paternal chromosomes is important to endosperm development because, in unbalanced situations, several cell types including basal transfer cells and aleurone cells fail to differentiate properly and seeds usually abort.

SYNCITIAL DEVELOPMENT AND CELLULARIZATION

The early syncitial divisions appear ordered because Barbara McClintock showed that clonal sectors form reproducible patterns, dividing the endosperm in halves, quarters, and so forth. Phragmoplasts form between daughter nuclei, but a cell wall is not deposited, indicating that cytokinesis is blocked by inhibiting phragmoplast function rather than formation. As free nuclear divisions progress, the central vacuole enlarges, restricting the nuclei to the periphery of the coenocyte. Radial arrays of microtubules surrounding each nucleus divide the cytoplasm into nucleo-cytoplasmic domains and create a regular spacing pattern (Fig. 1).

At the onset of cellularization, a novel form of cytokinesis occurs whereby cell plates form between non-daughter nuclei in the absence of mitosis. Phragmoplasts develop where the radial microtubule arrays from neighboring nuclei meet, and cell plates are deposited perpendicular to the outer endosperm cell wall. The cell plates coalesce to form tubular cell walls, or alveoli, around each nucleus. The microtubular arrays extend centripetally as the alveolar walls continue to grow inward. The nuclei then undergo periclinal mitotic divisions, each contributing a daughter nucleus toward the interior. This division is

Encyclopedia of Plant and Crop Science
DOI: 10.1081/E-EPCS 120012917

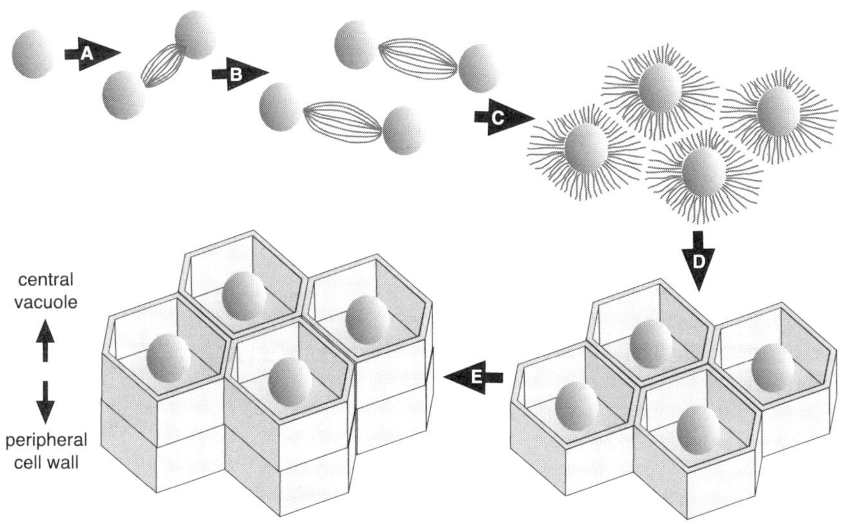

Fig. 1 Cellularization. (A, B) Endosperm nuclei undergo mitotic divisions without cytokinesis. (C) Radial microtubule arrays divide the endosperm into nucleo-cytoplasmic domains. (D) Cell walls form between the nuclei, creating alveoli. (E) Alveolar nuclei undergo a periclinal division accompanied by cytokinesis. The result is a cellular peripheral layer and an alveolar interior layer. *(View this art in color at www.dekker.com.)*

accompanied by cytokinesis, yielding a cellular peripheral layer and an alveolar interior layer (Fig. 1). This process reiterates in the internal layer until the interior is completely filled and cellular.

Subsequent development varies, depending on whether the endosperm is persistent. In *Arabidopsis*, most of the endosperm degenerates as it is displaced by the growing embryo. In cereals, extensive cell division occurs after cellularization. Divisions initially occur throughout the endosperm but then become localized to the periphery as internal cells accumulate storage products and undergo maturation. Thus, there is a general maturation gradient with the most mature cells at the interior and the young, actively dividing cells at the periphery.

CELL FATE SPECIFICATION

The highly specialized endosperm of cereal grains contains three major cell types (Fig. 2). At the base of the endosperm, near the attachment site to the maternal plant, is the basal endosperm transfer layer. These cells are specialized for the uptake of solutes and nutrients from the plant, for the developing grain. The aleurone is the outermost layer of cells surrounding the periphery of the endosperm. While performing some storage functions, the major function is to secrete digestive enzymes that break down storage products in the endosperm for the germinating seedling. The majority of the endosperm consists of starchy endosperm cells, the primary site of starch and storage protein deposition. The starchy endosperm can be subdivided into several cell types with specialized morphological characteristics or gene expression patterns. The subaleurone is the meristematic region at the periphery of the endosperm, just internal to the aleurone layer. A specialized region called the "embryo surrounding region" expresses several genes of unknown function called *ESRs*.

The molecular mechanisms that specify the identities of the various cell types found in the endosperm are not yet understood.[2,3] Positional information is present at very early stages because in barley, maize, and *Arabidopsis* various transcripts or reporters are localized to specific regions of the syncitial endosperm. ZmMRP1 transcript is localized to the basal region of the precellular maize endosperm and codes for a transcription factor that drives expression of *BETL* genes in the basal transfer layer.[4] *ESR* transcripts are also expressed in a localized region of the early cellular maize endosperm, about four days after pollination (Fig. 2).

Not all cell fate decisions are restricted to early stages. Aleurone cell fate remains plastic throughout endosperm development. Positional cues are required to specify and maintain aleurone identity up through late cell divisions. Disruption of the ability to perceive or respond to these cues causes aleurone cells to switch identity to starchy endosperm. Restoration of endosperm cells' ability to perceive the positional cue late in development allows the acquisition of aleurone fate.

Two genes involved in the aleurone cell fate process are known in maize. Mutations in either gene result in partial or complete loss of the aleurone layer. The *crinkly4*

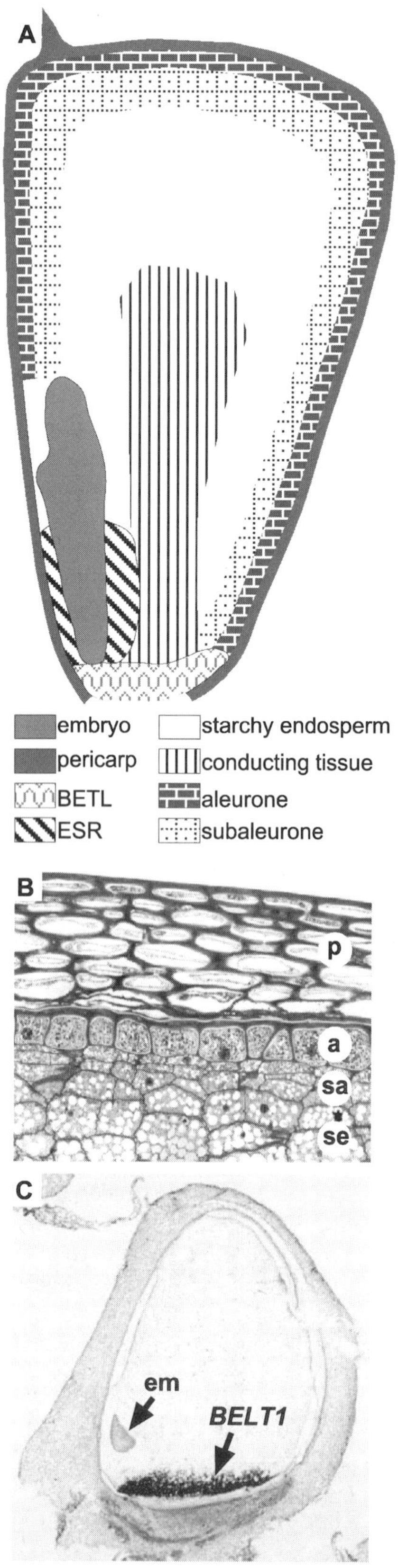

gene encodes a receptor kinase that is hypothesized to function in the perception of the cue that specifies aleurone cell identity. The *dek1* gene encodes a novel membrane-localized calpain protease.[5] Genetic evidence suggests that the *crinkly4* and *dek1* gene products function in related processes, but whether the gene products function in the same signal transduction pathway is not yet clear.[6]

STORAGE PRODUCT ACCUMULATION

As cells of a cereal endosperm mature, they begin to accumulate storage products. The major storage products are starch and protein. Starch accumulates in highly ordered grains located in amyloplasts.[7] The mechanisms of starch synthesis and deposition are surprisingly complex. Starch grains contain two types of starch molecules, unbranched amylose and branched amylopectin. Packing of starch molecules into crystalline grains involves the coordinated activities of starch synthases, starch branching enzymes, and starch debranching enzymes, but how the cooperative activity of these enzymes leads to the observed starch structure is poorly understood.

Maturing cells also accumulate storage proteins in protein bodies.[8] Storage proteins greatly impact the nutritional value of grains, as well as their flour and baking characteristics. Globulins are storage proteins found in all seeds and are soluble in saline buffers. Prolamines are specific to cereals and are only soluble in denaturing solvents due to their hydrophobicity. Several types of protein bodies are found in endosperm, some derived directly from endoplasmic reticulum and others vacuolar. Each contains mixtures of storage proteins whose assembly into protein bodies is accomplished through interactions among the constituent proteins. Storage protein genes show differential expression in various cell types and in different regions of the starchy endosperm, contributing to the different properties observed in various regions of grains.

As cells accumulate storage products, their volume enlarges greatly. Concomitantly, cells undergo chromosomal endoreduplication. This genomic amplification

Fig. 2 Cell types of a mature maize endosperm. (A) Diagrammatic representation of the various tissues of a maize kernel and specialized cell types within the endosperm. (B) A section through the peripheral region of a maize kernel showing starchy endosperm (se), subaleurone (sa), aleurone (a), and pericarp (p). (C) In situ hybridization showing the localized expression of the *BETL1* transcript in the basal endosperm transfer layer. (Courtesy of Richard Thompson.) *(View this art in color at www.dekker.com.)*

might be required to support the large size and high metabolic activity of endosperm cells. Endoreduplication is accomplished by incompletely understood modifications to the cell cycle, leading to an inhibition of M-phase.

MATURATION

In a mature cereal grain, all the endosperm cells die except the aleurone. As starchy endosperm cells complete storage product accumulation, they undergo programmed cell death. This occurs progressively from the interior out toward the periphery, following the general pattern of cell division and storage product accumulation. Programmed cell death is stimulated by the hormone ethylene.

The aleurone survives to serve as a digestive tissue during germination, secreting amylase to hydrolyze the starch stores and make glucose available to the growing seedling. Mature seeds dehydrate to levels that would be lethal to normal plant cells. Maturation includes a specialized genetic and physiological program that confers desiccation tolerance to the aleurone and embryo. A set of proteins called dehydrins accumulate and are thought to act as protectants, stabilizing membranes and maintaining protein structures. Maturation is promoted by the hormone abscisic acid (ABA) while gibberellic acid (GA) inhibits maturation and promotes the germination process.[9] The VIVIPAROUS1 transcription factor is central to the maturation process; *vp1* mutants are ABA insensitive, fail to undergo maturation and desiccation tolerance, and directly enter germination.

CONCLUSION

The deceptively simple endosperm is proving to be as complex and interesting as it is important. Recent technical advances, the availability of genome sequences of *Arabidopsis* and rice, and the coordinated efforts of multiple groups coming to bear on questions of endosperm development are sure to produce exciting progress in the near future.

ARTICLES OF FURTHER INTEREST

Female Gametogenesis, p. 439
Flower Development, p. 464
Genomic Imprinting in Plants, p. 527

REFERENCES

1. Grossniklaus, U.; Spillane, C.; Page, D.R.; Kohler, C. Genomic imprinting and seed development: Endosperm formation with and without sex. Curr. Opin. Plant Biol. **2001**, *4* (1), 21–27.
2. Becraft, P.W. Cell fate specification in the cereal endosperm. Semin. Cell. Dev. Biol. **2001**, *12*, 387–394.
3. Olsen, O.A. Endosperm development: Cellularization and cell fate specification. Annu Rev. Plant Physiol. Plant Mol. Biol. **2001**, *52*, 233–267.
4. Gomez, E.; Royo, J.; Guo, Y.; Thompson, R.; Hueros, G. Establishment of cereal endosperm expression domains: Identification and properties of a maize transfer cell-specific transcription factor, ZmMRP-1. Plant Cell **2002**, *14* (3), 599–610.
5. Lid, S.E.; Gruis, D.; Jung, R.; Lorentzen, J.A.; Ananiev, E.; Chamberlin, M.; Niu, X.; Meeley, R.; Nichols, S.; Olsen, O.-A. The *defective kernel* 1 (*dek1*) gene required for aleurone cell development in the endosperm of maize grains encodes a membrane protein of the calpain gene superfamily. Proc. Natl. Acad. Sci. U. S. A. **2002**, *99*, 5460–5465.
6. Becraft, P.W.; Li, K.; Dey, N.; Asuncion-Crabb, Y.T. The maize *dek1*gene functions in embryonic pattern formation and in cell fate specification. Development **2002**, *129*, 5217–5225.
7. Myers, A.M.; Morell, M.K.; James, M.G.; Ball, S.G. Recent progress toward understanding biosynthesis of the amylopectin crystal. Plant Physiol. **2000**, *122* (4), 989–998.
8. Shewry, P.R.; Halford, N.G. Cereal seed storage proteins: Structures, properties and role in grain utilization. J. Exp. Bot. **2002**, *53* (370), 947–958.
9. Wobus, U.; Weber, H. Seed maturation: Genetic programmes and control signals. Curr. Opin. Plant Biol. **1999**, *2* (1), 33–38.

Environmental Concerns and Agricultural Policy

Floor Brouwer
Agricultural Economics Research Institute, Wageningen UR, The Hague, The Netherlands

INTRODUCTION

Europe and North America are key producers of food in the world market and major users of crop protection products. For a number of commodities (e.g., cereals, oilcrops, potatoes, and citrus fruit) they have a considerable share of global food production. Western Europe and North America also account for more than half of the world market of agrochemicals. The E.U., U.S.A., Japan, and Brazil account for about three-quarters of the global market of agrochemicals. When treated properly, the use of crop protection products extends the lifetime of food products in the agrifood chain. Also, food can be supplied in a more uniform manner and have a better appearance. However, residues of these products may harm the natural environment by contamination of groundwater resources, surface waters, and soils, as well as spray drift affecting air quality and biodiversity. A major challenge to agriculture is to internalize these social costs of production properly in farm management practices. Environmental policy measures are taken across the globe to control such external effects of production on the natural environment.

Agricultural policy is a main area of public intervention in the agricultural sector, and the use of crop protection products is therefore also affected by such policy measures. Market and price support measures have encouraged the use of crop protection products. More recently, however, efforts have been made to internalize environmentally harmful effects of crop protection products into farming practices, and reforms of agricultural policies could stimulate farmers to change their practice.

Agricultural and environmental policy measures could both support the effort to reduce pressure on the natural environment, and their interactions with the use of crop protection products are of major societal interest. This is the subject of the current contribution.

PATTERNS OF USING CROP PROTECTION PRODUCTS

During the 1990s, producers of crop protection products were faced with stagnant or even declining markets (at least in the developed countries) and with more stringent environmental constraints applied to agriculture.[1] Total use of crop protection products in the E.U. amounted to some 325 million kg of active ingredients during the early 1990s, which dropped to slightly below 300 million kg in 1996 (Table 1). The declining trends during the first half of the 1990s seems to have reversed in the years thereafter with patterns of use rising again. Some divergent trends are observed across Member States. The European Commission has developed a strategy with the objective of reducing the impact of crop protection products on human health and the environment and, more generally, more sustainable use of such agrochemicals.[2]

In addition to reforms of agricultural policy (price reduction of cereals, with set-aside requirements and the introduction of direct payments), several factors have also contributed to the downward trend in the E.U.:

- Product innovations that have allowed for the provision of new compounds requiring lower dosages. Several active ingredients have been authorized for agricultural use and entered the market during the 1990s. Such chemical substitution allows for a smaller dosage to suffice for treating plants compared to what was used in the past. This type of innovation has been observed on broader scales, and many developed countries faced a diminishing (e.g., Japan) or slow growing (e.g., U.S.A.) market.
- Improved technologies to target application better. The application of the MLHD-approach (Minimum Lethal Herbicide Dosage), for example, allowed for a considerable reduction in the use of herbicides. Mechanical weed-control methods also replaced agrochemicals. Also, aerial spraying is one of the most restricted practices in developed agriculture that contributed to the improved application.[3]
- Improved farm management practices, including the use of Integrated Crop Management (ICM) practices and Integrated Pest Management (IPM) practices. Environmentally related agricultural programs have been initiated in several countries (e.g., Canada, U.S.A.) to reduce usage of crop protection products through research, education, technology transfer, and the encouragement of more widespread use of IPM and ICM.

Encyclopedia of Plant and Crop Science
DOI: 10.1081/E-EPCS 120010516

Table 1 Sales of crop protection products by member state (in tons of active ingredients)

Country	1991	1996	1999	(kg per ha)
Belgium	9,969	10,403	9,202	6.6
Netherlands	17,306	9,918	10,232	5.2
Portugal	9,355	12,457	15,412	4.0
France	103,434	92,889	114,695	3.8
Italy	58,123	48,050	50,850	3.3
Greece	7,860	9,870	10,153	2.6
United Kingdom	29,022	35,659	35,668	2.2
Germany	36,944	35,085	35,403	2.0
Luxembourg	253	–	421	2.0
Spain	39,147	33,236	33,614	1.2
Denmark	4,628	3,664	2,874	1.1
Austria	4,488	3,566	3,419	1.0
Sweden	1,837	1,528	1,698	0.6
Ireland	2,006	2,568	2,071	0.5
Finland	1,734	933	1,158	0.5
E.U.-15	325,653	299,826	326,870	2.6

(From Eurostat, New Cronos (2002).)

- National mandatory reduction schemes, including specific measures developed in the context of environmental policy, have also contributed to the reduction of usage of agrochemicals.
- Market incentives also stimulate environmental awareness of agricultural producers. Such efforts might respond to government regulation and to consumer preferences of environmentally friendly products.

ENVIRONMENTAL CONCERNS: RISKS AND HAZARDS

Environmental concerns on the agricultural use of crop protection products tend to cause concern in a large number of countries. They may cause contamination of groundwater resources and surface water by leaching, run-off, and spray drift. Aerial spraying of crop protection products and the contamination of surface water is an issue in several countries, as are potential threats to non-target species and to human health (especially from residues in drinking water). In addition, human health concerns are due to residues in food. A comparison of environmental standards among main traders in the world market indicate that crop protection products are a main issue, being a high priority in policy in the E.U. and U.S.A.[4]

Monitoring in the E.U. indicates that water quality standards could be exceeded in between 5 and 25% of the samples taken in regions with intensive arable production and horticulture. Standards for pesticides in drinking water are part of the Drinking Water Directive 80/778/EEC, and a limit of 0.1 μg per litre applies to individual pesticides and 0.5 μg per litre for the sum of all pesticides. Such limits are not based on toxicological criteria, but in fact are a surrogate for zero detection. Human health concerns in the U.S.A. mainly relate to residues in food. Harmful effects on water quality related to the use of crop protection products are a main issue in this country. The issue is identified as a problem and of significant concern in Australia, where air quality problems associated with pesticide drift are an important problem and large-scale aerial spraying is common. In contrast, it is not considered a problem for farming in New Zealand.

AGRI-ENVIRONMENTAL POLICY AND THE USE OF CROP PROTECTION PRODUCTS

Farmers do respond to changes in income and commodity prices, and agricultural policy is also an important area for consideration with the use of crop protection products. Market interventions and subsidies have been key instruments in agricultural policy for most of the developed countries. Major instruments applied are market-price support conferred through border measures (e.g., tariffs, export subsidies, quantitative and qualitative restrictions) and administrative pricing regimes, other production-linked supports such as deficiency payments, and subsidies for intermediate inputs.

It is widely agreed that market interventions and price-support measures have encouraged greater agrochemical use than would otherwise be the case. There is limited evidence of environmentally beneficial effects of reducing price support measures on agrochemical use, because the price elasticity of demand for agrochemicals tends to be low in the short run.

Falconer and Oskam[5] suggest that the combined effects of a reduction in the intervention prices of cereals and the introduction of set-aside requirements in the E.U. reduced the use of pesticides for growing cereals by less than 10% and the use of pesticides overall by 3%. The effects of the agricultural reform measures from the late 1990s on pesticide use are estimated to be very modest. One reason for the limited reduction of pesticide use under this reform is that price decreases and set-aside schemes work in opposite directions. Pesticide use has also been reduced through acreage restrictions in the U.S.A. There, the Conservation Reserve Program (CRP) reduced crop acreage and, hence, pesticide use on land that might otherwise have remained in production. However, agricultural support programs can provide incentives to increase pesticide use on land that is not set aside.[6]

Pesticide use in aggregate tends to be unresponsive to small price changes, though demand for individual products can be very responsive to even small price changes. Most empirical studies offer evidence on price elasticities regarding demand for pesticides of between –0.2 and –0.5.[7] The high price elasticity of herbicides (between –0.7 and –0.9) relates to the options available to control pests and diseases without treatment. Mechanical weed control, for example, can often substitute for herbicides. The price elasticity of demand for pesticides differs for specific crops. For cereals, for example, it could be as high as –1.1.

Codes for Good Agricultural Practices (GAP) are developed in the context of agricultural policy to identify environmental constraints to direct payments provided to farmers. The adoption of such a code through ICM or IPM techniques might be a requirement for farmers to be eligible for full compensation to hectare payments. In such a case, farmers who were not fulfilling a condition to adopt such a Code would face withdrawal of part of their income compensation. A range of environmental constraints are currently put on farmers in the context of cross compliance.[8] Retailers and food processors are demanding better and audited farming systems in response to the changed consumer demands. Therefore, agriculture will increasingly respond to and work with others in the agri-food chain. The adoption of ICM and IPM would qualify for such demands.

CONCLUSION

Contamination of soil, leaching of nutrients and pesticides, water extraction, and drainage pose a widespread threat to the environment. Nitrate and phosphate loading and the run-off of livestock wastes cause significant water pollution problems. Also, the excessive use of crop protection products poses a widespread threat to human health and the environment. High levels of crop protection products are associated mainly with areas of intensive arable crops and horticulture.

Societal debate that started in the late 1980s has given incentives to better control the harmful effects of plant protection products on health and the environment. Since then, the interest moved towards a more targeted and rationalized use of plant protection products. Mandatory requirements on their use increasingly tend to include farm management aspects (rather than measures to reduce total use), focus on specific measures (rather than general measures that apply to the whole agricultural sector), and link environmental quality with food safety aspects.

ARTICLES OF FURTHER INTEREST

Agriculture and Biodiversity, p. 1
Cropping Systems: Yield of Improvement of Wheat in Temperate Northwestern Europe, p. 367
Crops and Environmental Change, p. 370
Economic Impact of Insects, p. 407
Integrated Pest Management, p. 612
Sustainable Agriculture and Food Security, p. 1183
Sustainable Agriculture: Definition and Goals, p. 1187
Sustainable Agriculture: Economic Indicators, p. 1195

REFERENCES

1. Mackenzie, W. *The Current Situation and Future Prospects of the EU Crop Protection Industry*; Wood Mackenzie Consultants Limited: Edinburgh, 1997.
2. Commission of the European Communities (CEC). *Towards a Thematic Strategy on the Sustainable Use of Pesticides*; Communication from the Commission to the Council, the European Parliament and the Economic and Social Committee: Brussels, 2002. COM(2002) 349 final. http://europa.eu.int/eur-lex/en/com/pdf/2002/com2002_0349en01.pdf.
3. Brouwer, F.; Ervin, D.E. Environmental and Human-Health Standards Influencing Competitiveness. In *Public Concerns, Environmental Standards and Agricultural Trade*; Brouwer, F., Ervin, D.E., Eds.; CABI Publishing: Wallingford, 2002; 255–284.
4. Brouwer, F.; Baldock, D.; Carpentier, C.; Dwyer, J.; Ervin, D.; Fox, G.; Meister, A.; Stringer, R. *Comparison of Environmental and Health-Related Standards Influencing the Relative Competitiveness of EU Agriculture vis-à-vis Main Competitors in the World Market*; Agricultural Economics Research Institute (LEI), The Hague, 2000. Report 5.00.07.
5. Falconer, K.; Oskam, A. The Arable Crops Regime and the Use of Pesticides. In *CAP Regimes and the European Countryside: Prospects for Integration Between Agricultural, Regional and Environmental Policies*; Brouwer, F., Lowe, P., Eds.; CABI Publishing: Wallingford, 2000; 87–102.
6. U.S. Department of Agriculture. *Agricultural Resources and Environmental Indicators, 1996–97*; Economic Research Service, U.S. Department of Agriculture: Washington, DC, Agricultural Handbook No. 712.
7. Rayment, M.; Bartram, H.; Curtoys, J. *Pesticide Taxes: A Discussion Paper*; Royal Society for the Protection of Birds, Sandy, UK, 1998.
8. Petersen, J.-E.; Shaw, K. In *Environmental Standards in Agriculture*, Proceedings of a Pan-European Conference, Madrid, London, October 5–7, 2000; Institute for European Environmental Policy.

Evolution of Plant Genome Microstructure

Ian Bancroft
John Innes Centre, Norwich, U.K.

INTRODUCTION

The evolution of gene structure has been studied for many years by the comparison of the nucleotide sequences of related genes. At the other extreme of scale, the gross organization of the genomes of related plant species has been investigated using molecular genetic markers. Such investigations have shown how individual genes evolve by changes in nucleotide sequences and how whole genomes evolve by rearrangement. The recent availability of whole genome sequences and the application of high-throughput physical mapping strategies have enabled us to investigate the intermediate-scaled organization—or microstructure—of plant genomes. Understanding the evolutionary processes involved in shaping genome structure on the scale of gene-by-gene organization is particularly important for the use of comparative positional cloning of genes in crops based on sequence information from model species. Comparative analyses of plant genome microstructure—conducted between both closely and distantly related species, and between duplicated segments within a genome—have allowed us to advance our understanding of the mechanisms involved in genome evolution. Some of the milestone comparative analyses of plant genome microstructure, focusing on the arrangement of genes, will be reviewed here and the results interpreted in relation to mechanisms involved in genome evolution.

APPROACHES TO THE ANALYSIS OF GENOME MICROSTRUCTURE

The resource of preference for the analysis of genome microstructure is complete, high-quality genome sequence. For plants, this is presently available only for *Arabidopsis thaliana*.[1] However, high-quality sequences of large insert genomic clones (usually as Bacterial Artificial Chromosomes, BACs) can also be very valuable for comparative analyses, particularly for comparison to the genome of *A. thaliana*.[2,3] Although the sequencing of BACs is now routine, it is an expensive approach. Where a partial or complete genome sequence is already available from one plant, hybridization-based physical mapping approaches using BAC libraries can be used to investigate the relative conservation of genes in the corresponding segments of other plant genomes.[4]

The evolution of genome structure is usually studied by the comparative analysis of orthologous genome segments in species descended from a common ancestor. However, we can also study the divergence of related segments that arose by duplication events and are retained by an organism. The analysis of structural evolution following genome duplication is of particular importance, as polyploidy (the amplification of whole genomes) is be very common in plants.[5]

COMPARATIVE ANALYSIS OF RELATED GENOME SEGMENTS

The most distantly related plant species used for the comparative analysis of genome microstructure are *A. thaliana* and rice, the lineages of which diverged ca. 200 million years ago (MYA). The first region of the rice genome compared in detail consisted of 340 kilobase pair (kb) of DNA and contained 56 predicted genes.[3] Twenty-two of the genes were found to have homologues in *A. thaliana* indicative of conservation of microstructure in five parallel segments, as shown in Fig. 1. Subsequent analyses confirmed the generality of conservation of genome microstructure in *A. thaliana* and rice, but only 17% of genes conserved were in identified homoeologous segments.[6] The lineages of *A. thaliana* and tomato diverged more recently, ca. 150 MYA. Four parallel segments of the *A. thaliana* genome were found to have conserved microstructure with a sequenced genomic clone of tomato.[2] These findings demonstrate that comparative analysis of genome segments in very distantly related plant species is feasible. Because both analyses identified conservation of microstructure in several segments of the *A. thaliana* genome, those segments themselves must be related, permitting

DOI: 10.1081/E-EPCS 120010631

the comparative analysis of duplicated segments within a genome.

Extensive duplication of the *A. thaliana* genome has been identified by the analysis of the genome sequence. This has been interpreted as the result of a tetraploidy event.[1] that, as estimated by the extent of sequence divergence of the conserved genes, happened ca. 65 MYA.[7] The organization of the segments related by this event differ primarily by the interspersion of conserved genes by nonconserved genes, and by rearrangements and translocations.[1] An example of the gene-by-gene comparison of two of the genome segments from Fig. 1 is shown in Fig. 2. This comparison illustrates another unexpected feature of the *A. thaliana* genome: Tandemly repeated genes are very common, accounting for 17% of all ca. 25,000 genes of *A. thaliana*.[1]

A. thaliana is related to the cultivated *Brassica* crops, the lineages of which diverged ca. 19 MYA. Comparative analyses of the genomes of *A. thaliana* and *Brassica oleracea* using hybridization-based physical mapping approaches identified extensive conservation of genome microstructure. There was, however, an interspersed pattern of conserved and nonconserved genes and evidence for rearrangements.[4] Differences in genome microstructure can also be identified, again by hybridization-based physical mapping approaches, between *Brassica* species (e.g., *B. oleracea* and *B. rapa*) and within species (e.g., *B. rapa* ssp. *trilocularis* and *B. rapa* ssp. *pekinesis*) (T. van den Boogaart, D. Rana, Y.-P. Lim and I. Bancroft, in preparation), despite times since divergence among the lineages of one million years or less. The genomes of the *Brassica* species appear to be derived from a triplication of an *A. thaliana*-like genome; this complexity is thought to be associated with the rapid rate of change of genome microstructure observed in *A. thaliana* and *Brassica* species.[8] Differences in genome microstructure, including relocation of genes in the genome, have also been identified in different ecotypes of *A. thaliana*.[1]

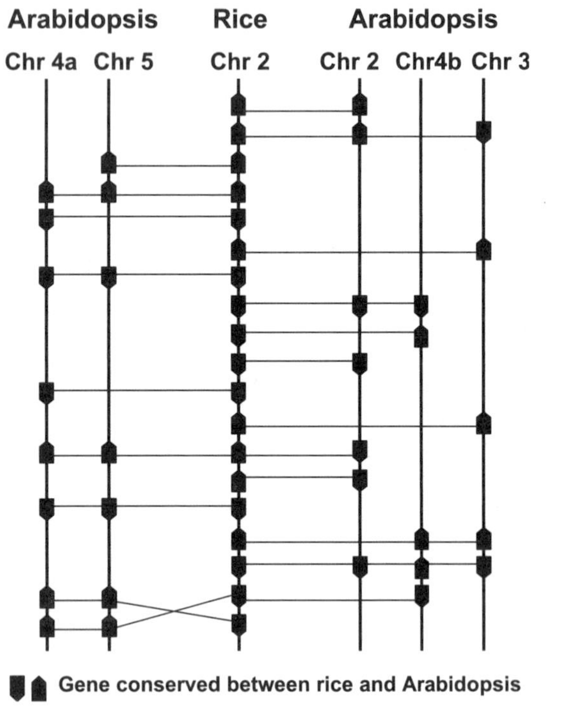

Fig. 1 Comparison of the presence and relative stand orientation of genes conserved in a sequenced region of the genome of rice and five segments of the genome of *A. thaliana*. (From Ref. 3.)

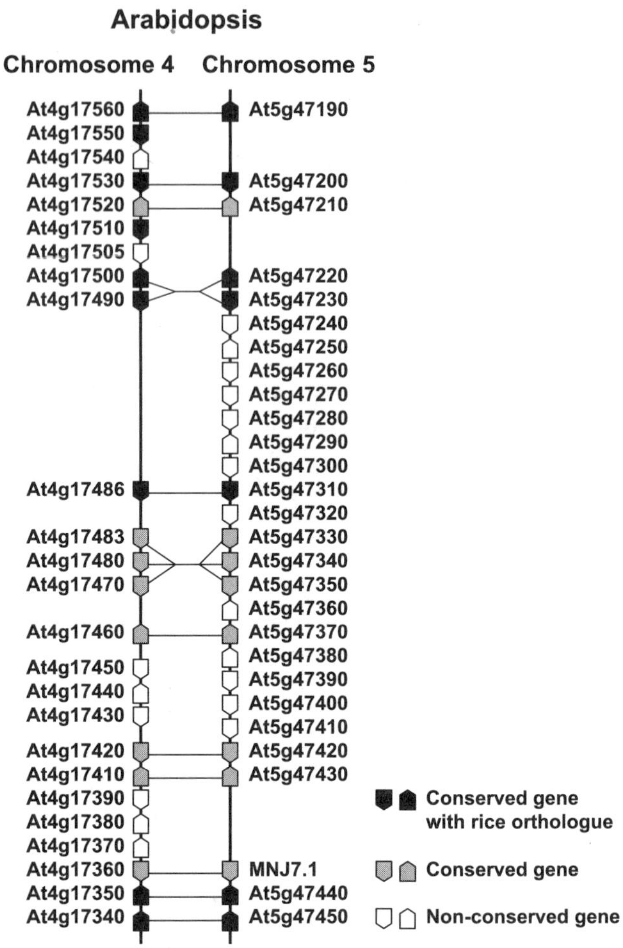

Fig. 2 Comparison of the microstructures of related segments of the *A. thaliana* genome. Conserved genes were identified by BLAST alignment to the complete *A. thaliana* genome sequence, updating a previously reported analysis. (From Ref. 9.)

EVOLUTIONARY MECHANISMS

Related genome segments, with a wide range of times since divergence, can readily be identified in plants. Comparative analyses of the most closely related genomes indicate that the events most frequently affecting genome microstructure are sequence polymorphisms and small insertion/deletion events. Such events are likely to be the result of errors in DNA replication or repair. They also provide evidence for the transposition of gene-sized DNA segments, not recognizable as transposons, to new genomic locations.[1] It is not known what kind of transposition mechanism might bring this about. It may not be via reverse transcription of mRNA and integration into the genome (such as can lead to pseudogenes), because small clusters of genes also appear to be replicatively transposed.[9] The relatively frequent occurrence of transposition of single genes or small gene clusters will result in a background of genes in novel genomic locations, which could confuse comparative approaches to the identification of orthologous genes.

Tandem duplication of genes is very common within plant genome;[1] once duplicated, such tandem genes can persist in the genome. For example, Fig. 2 shows tandemly duplicated and triplicated conserved genes in genome segments that are thought to have diverged from a common ancestor 65 MYA.[7] Presumably the duplicated genes have assumed new and selectively advantageous functions. The interspersed pattern of conserved and nonconserved genes is a striking feature of duplicated plant genome segments and is also shown in Fig. 2. Both tandem duplications of genes and interspersed loss of genes are potential outcomes of unequal crossover, depending on whether recipient or donor chromosome is inherited, respectively. The data from comparisons of plant genome microstructure are consistent with the hypothesis that unequal crossover is a very important mechanism in the evolution of genome microstructure in plants.[8] The loss of genes from chromosome segments by unequal crossover will have no effect on the fitness of a plant if those genes are duplicated elsewhere. Thus gene duplication and polyploidy produce a genome buffered against detrimental effects of this kind of evolutionary mechanism.

Although important, unequal crossover is far from the only mechanism involved in the evolution of plant genome microstructure. Gross-scale genome rearrangement is a notable feature and appears prevalent in species with polyploid ancestry such as the *Brassica* species.[10] Over long periods of time, the extent of reshuffling will erode the conserved blocks to a small size, such as now conserved between *A. thaliana* and rice.[6] Other events—such as inversions and translocations identified in duplicated genome segments—also occur, as does the apparent transposition of individual genes and small groups of genes.[1,9] It is the combination of all of these mechanisms that shapes plant genome microstructure.

CONCLUSION

The vision is emerging of the microstructure of plant genomes evolving by a variety of mechanisms, with detrimental effects buffered by extensive genome duplication. Over time, duplicated segments will tend toward complementary gene content until redundancy is reduced to the extent that loss of genes becomes detrimental to fitness. The size of conserved blocks both within and between genomes will degenerate by rearrangement over time; transpositionlike events will further erode conservation of microstructure. Thus a proportion of genes in related genome segments will be identifiable between or within genomes, although many will not. The conservation of microstructure will often permit the identification of orthologous genes in different plant genomes, and the identification of putatively functionally redundant genes within a genome. However, there will be exceptions, even between closely related species; therefore, functional interpretation and comparative genome structural analyses must be integrated with caution. For very distantly related species, the orthology of genes will rarely be demonstrable due to the variety of mechanisms contributing to the evolution of plant genome microstructure.

ARTICLES OF FURTHER INTEREST

Arabidopsis thaliana: *Characteristics and Annotation of a Model Genome*, p. 47
Chromosome Structure and Evolution, p. 273
Genome Rearrangements and Survival of Plant Populations to Changes in Environmental Conditions, p. 513
Genome Size, p. 516

REFERENCES

1. The Arabidopsis Genome Initiative Analysis of the genome of the flowering plant *Arabidopsis thaliana*. Nature **2000**, *408*, 796–815.
2. Ku, H.-M.; Vision, T.; Liu, J.; Tanksley, S.D. Comparing sequenced segments of the tomato and *Arabidopsis*

genomes: Large-scale duplication followed by selective gene loss creates a network of synteny. Proc. Natl. Acad. Sci. U. S. A. **2000**, *97*, 9121–9126.
3. Mayer, K.; Murphy, G.; Tarchini, R.; Wambutt, R.; Volckaert, G.; Pohl, T.; Düsterhöft, A.; Stiekema, W.; Entian, K.-D.; Terryn, N.; Lemcke, K.; Haase, D.; Hall, C.R.; van Dodeweerd, A.-M.; Tingey, S.V.; Mewes, H.W.; Bevan, M.W.; Bancroft, I. Conservation of microstructure between a sequenced region of the genome of rice and multiple segments of the genome of *Arabidopsis thaliana*. Genome Res. **2001**, *11*, 1167–1174.
4. O'Neill, C.; Bancroft, I. Comparative physical mapping of segments of the genome of *Brassica oleracea* var *alboglabra* that are homoeologous to sequenced regions of the chromosomes 4 and 5 of *Arabidopsis thaliana*. Plant J. **2000**, *23*, 233–243.
5. Wendel, J.F. Genome evolution in polyploids. Plant Mol. Biol. **2000**, *42*, 225–249.
6. Salse, J.; Piegu, B.; Cooke, R.; Delseny, M. Synteny between *Arabidopsis thaliana* and rice at the genome level: A tool to identify conservation in the ongoing rice genome sequencing project. Nucleic Acids Res. **2002**, *30*, 2316–2328.
7. Lynch, M.; Conery, J.S. The evolutionary fate and consequences of duplicate genes. Science **2000**, *290*, 1151–1155.
8. Bancroft, I. Duplicate and diverge: The evolution of plant genome microstructure trends in genetics. **2001**, *17*, 89–93.
9. Bancroft, I. Insights into the structural and functional evolution of plant genomes afforded by the nucleotide sequences of chromosomes 2 and 4 of *Arabidopsis thaliana*. Yeast **2000**, *17*, 1–5.
10. Lagercrantz, U. Comparative mapping between *Arabidopsis thaliana* and *Brassica nigra* indicates that *Brassica* genomes have evolved through extensive genome replication accompanied by chromosome fusions and frequent rearrangements. Genetics **1998**, *150*, 1217–1228.

Exchange of Trace Gases Between Crops and the Atmosphere

Jürgen Kreuzwieser
Heinz Rennenberg
University of Freiburg, Freiburg, Germany

INTRODUCTION

The atmosphere mainly consists of nitrogen (78% by volume) and oxygen (21% by volume). The remaining 1% is known as trace gases, with the noble gas argon the most abundant. Concentrations of other trace gases are typically present in the range of parts per trillion by volume (pptv) to parts per million by volume (ppmv); they include the greenhouse gases carbon dioxide (CO_2), methane (CH_4), nitrous oxide (N_2O), ethane, water vapor, and ozone (O_3); the air pollutants sulfur dioxide (SO_2), ammonia (NH_3), nitric oxide (NO), nitrogen dioxide (NO_2), peroxyacylnitrates (PAN), nitric acid (HNO_3), and carbon monoxide (CO); and a number of volatile organic compounds (VOCs). Among VOCs are isoprenoids (mainly isoprene and monoterpenes) and many oxygenated species such as alcohols, aldehydes, and organic acids which, due to their reactivity, strongly affect the oxidation capacity of the troposphere and influence the concentration and distribution of several other trace gases, including CH_4 or CO. On a regional scale, VOCs significantly contribute to the formation of tropospheric O_3. The main source of VOCs (approximately 90%) is natural emission by vegetation.

IMPACTS OF TRACE GASES ON CROPS

The effects of atmospheric trace gases on crops are quite diverse and depend on the type of gas and its concentration (Table 1), the duration of exposure, and a range of plant internal factors. Direct phytotoxic effects due to exposure to high concentrations of pollutants such as O_3 and SO_2 on crop plants include, among others, visible leaf injury; changes in chloroplast structure and cell membranes; disturbances of stomatal regulation, respiration, and photosynthesis; and reductions in growth and yield.[3] However, because sulfur (S) is an essential nutrient for plants, SO_2 absorbed by foliage may also be used as an additional sulfur source in polluted areas, in addition to sulfate from the soil.[3] The same principle applies to nitrogen (N). Thus, effects of trace gases can be divided into phytotoxic effects caused by protons, organic compounds, SO_2, NO_2, NH_3, and O_3; and nutritional effects caused by S- and N-containing gases, and CO_2.[4]

FACTORS CONTROLLING TRACE GAS EXCHANGE

Trace gases can be exchanged between the atmosphere and aboveground plant parts by: 1) dry deposition as gases or aerosols; 2) wet deposition as dissolved compounds in rainwater or snow; or 3) interception of compounds dissolved in mist or cloud water[5] (Fig. 1). The direction of the exchange (i.e., emission versus deposition) and its velocity are controlled by the physicochemical conditions and by internal plant factors.

The gradient in the gas concentration between substomatal cavities and the atmosphere is the driving force for gas exchange. The gas flux is a diffusive (passive) process and can be described by Fick's law. Accordingly, the net flux of a trace gas is zero when the substomatal concentration is equal to the concentrations in the surrounding ambient air. This concentration is referred to as the compensation point for the particular gas. When trace gas concentrations outside the leaves are higher than those in the substomatal cavities, a net flux into the leaves will take place (deposition), and vice versa. Therefore, crops may act both as a source (if ambient concentrations are lower than substomatal concentrations) or as a sink of a specific gas (Table 2). This dual behavior has been observed for a variety of gases (SO_2, H_2S, NO_2, NH_3, organic acids, and aldehydes). Compensation points for pollutants such as NH_3 range between 0.4 and 15 parts per billion by volume (ppbv).[6] They depend mainly on the plant species or cultivar, development stage, temperature, and status of N nutrition of the plants. Generally, compensation points increase with increasing availability of nutrients in the soil, which suggests that this is one mechanism by which plants cope with excess nutrient supply.[7]

Encyclopedia of Plant and Crop Science
DOI: 10.1081/E-EPCS 120005571

Table 1 Range of ambient concentrations of different trace gases in the atmosphere

Trace gas	Ambient concentrations
Ar	9,340 ppm
CO_2	365 ppm
CH_4	1745 ppb
N_2O	314 ppb
Isoprene	ppt–several ppb
Monoterpenes	ppt–several ppb
Alcohols	1–30 ppb
Carbonyls (aldehydes, ketones)	1–30 ppb
Alkanes	1–3 ppb
Alkenes	1–3 ppb
Esters	ppt–several ppb
Carbonic acids	0.1–16 ppb
SO_2	0.5–50 ppb
H_2S	0–0.2 ppb
NO_2	4–200 ppb
O_3	20–80 ppb

PLANT PHYSIOLOGICAL CONTROLS

The existence of a compensation point depends on the capacity of a plant to produce the trace gas to be exchanged, or to consume it. Therefore, compensation points do not exist for compounds that cannot be produced (e.g., O_3) or consumed (e.g., isoprene). It is evident that for each of the many gases exchanged between crop plants and the atmosphere, specific metabolic pathways exist, not all of which are considered here. As an example, and because of their increasing importance in recent years, some details on the exchange of nitrogen compounds are presented (Fig. 2).

Both NO_2 and NH_3 can be taken up by aboveground parts of plants, mainly via the stomata of leaves.[5] In the aqueous phase of the apoplast, NO_2 is either disproportionated yielding equal amounts of NO_2^- and NO_3^-, or it reacts with apoplastic ascorbate.[8] Because disproportionation of NO_2 in water is slow at atmospheric NO_2 concentrations, the reaction with ascorbate may be of more importance. Upon conversion to NO_3^- or NO_2^- these anions are transported to the cytoplasm, where they are reduced by the assimilatory nitrate reduction pathway yielding NH_4^+ and the amino acid glutamine. Atmospheric NH_3 dissolves in the aqueous phase of the apoplastic space to yield NH_4^+, which is then taken up into the cytoplasm.

Both NH_3 and NO_2 can also be emitted by plants. NH_3 may be released from cellular NH_4^+ pools when plants are exposed to excess nitrogen in the soil. In addition, it can be released from drying water films at the leaf surface. During this process the remaining NH_4^+ concentrations on the surface will increase. In contrast, the chemical source of NO_2 emitted by the leaves is largely unknown; it has been proposed that nitrate reductase may be involved in the reduction of NO_2^- to NO_2.[7]

The rate of trace gas emission does not necessarily depend on the actual rate of production. Some volatile compounds are produced and then stored in particular pools. For example, some plant reservoirs contain high amounts of monoterpenes, which can be emitted throughout the day and night independent of light-dependent biosynthesis.

Through their effect on biochemical pathways, biotic and abiotic factors (e.g., stress factors) and the developmental stage of plants influence the rate of trace gas emission. For example, stress caused by wounding, chilling, iron deficiency, O_2 deficiency, or induction of

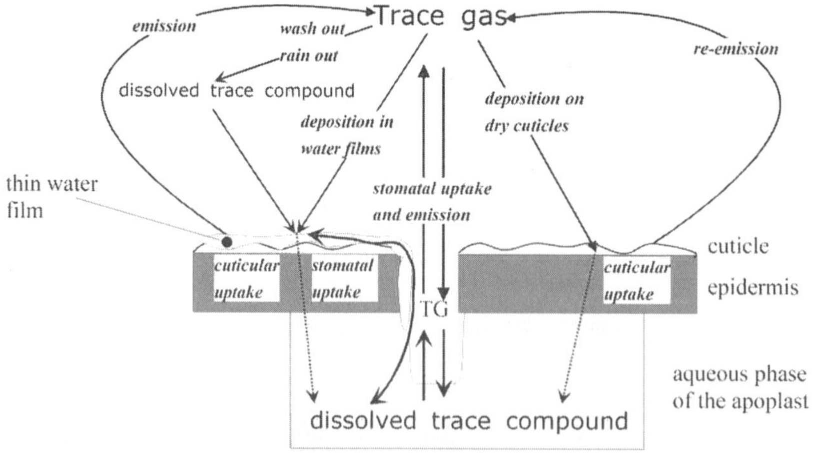

Fig. 1 Main access routes for trace gases to enter a leaf: Uptake via the stomata and cuticular uptake. (Modified according to Ref. 7.)

Table 2 Typical range of rates of exchange for different trace gases between crops and the atmosphere

Trace gas	Range of exchange[a] [µg g^{-1} (leaf d.wt.) h^{-1}]
SO_2	−300 to 0
H_2S	−2 to 2
O_3	0 to −300
NO_2	−3 to 1
NH_3	−3 to 3
Isoprene	0 to 0.1
Monoterpenes	0 to 20
Oxygenated VOCs	−50 to 50

[a]Negative values indicate deposition; positive values indicate emission of respective trace gas.

oxidative stress caused by exposure to O_3 or SO_2 can lead to the production of the VOCs hexenal, hexanal, formaldehyde, formate, ethene, ethane, ethanol, and acetaldehyde.[9]

INTERNAL TRANSFER RESISTANCES

The transfer of gases in and out of plants is often described by a resistance analogy. Along the path from the sites of production (or consumption) to the atmosphere, a series exists of mainly internal plant resistances. Because biosynthesis and consumption of volatile compounds usually take place in the cytoplasm or in other compartments of the cell, the gas must pass across the bordering membranes. The diffusive flux through these membranes is determined by the molecular size and the lipophilic character of the particular compound. Polar molecules such as organic and inorganic acids are not likely to be dissolved in the lipophilic membranes; therefore, the diffusive flux should be slow. Carrier proteins can facilitate the transport of these polar compounds across membranes from the cytoplasm into the apoplastic space, similar to the transport of organic acids.[10]

The compounds are transferred from the liquid phase in the apoplastic space to the gaseous phase in the substomatal cavity, or vice versa. The volatilization into the internal leaf air space depends on 1) chemical properties of both the aqueous phase and the individual compound to be emitted (e.g., its solubility in the apoplastic solution); and 2) physical factors such as the ambient concentration of the gas, its vapor pressure, and temperature.[10] A reduction of the apoplastic pH reduces the resistance for inorganic and organic acids, because these compounds become protonated and thereby more volatile. When present in the gaseous phase of the apoplastic space, trace gases can escape from the leaves through either the cuticle or the stomata. However, because of their polarity, the lipophilic cuticle constitutes a strong barrier, and therefore diffusion through the stomata is the main pathway. However, nonstomatal emission and deposition have also been observed in air pollutants such as NO, NO_2, and SO_2, but at much lower rates than stomatal exchange. For this reason, factors that influence the stomatal aperture exert a strong influence on the rate of gas exchange between plants and the atmosphere. The control by the stomata of emission and deposition of NO_2, SO_2, O_3, PAN, and other trace gases was observed in many studies, including in investigations of crop plants. Thus, concentration and time of exposure are not the only factors determining the effect of an air pollutant on vegetation. Plants usually close their stomata during hot, dry conditions when, for instance, O_3 levels are high. This may provide some protection for the plants from O_3 injury. Alternatively, in northern Europe where O_3 concentrations are lower than in southern and central Europe, the potential O_3 uptake at a given O_3 concentration may be higher because of higher air humidity, leading to high rates of stomatal O_3 uptake.

CONCLUSION

Trace gases influence not only natural ecosystems but also agricultural crops. Future studies should focus on the impact of different combinations of air pollutants (e.g., increased nitrogen input combined with elevated O_3 concentrations) on plants and should include aspects of global climate change (e.g., higher temperatures,

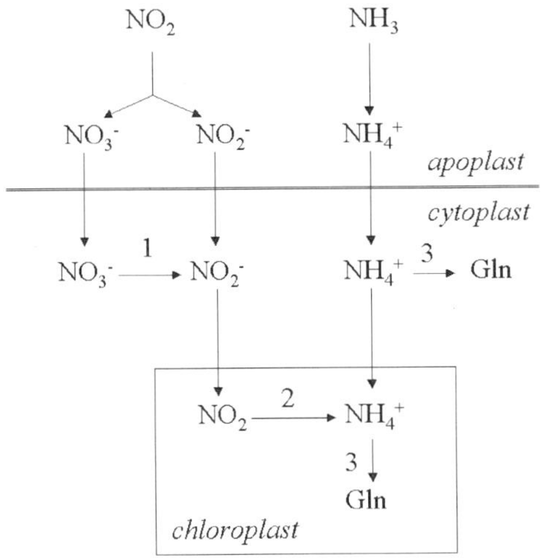

Fig. 2 Main processes involved in the assimilation of atmospheric NH_3 and NO_2 taken up by the leaves. 1: nitrate reductase, 2: nitrite reductase, 3: glutamine synthetase (cytoplasmatic and chloroplastic isoforms). (Modified according to Ref. 7.)

droughts, and increased frequency of heavy rainfalls and droughts in combination with elevated CO_2 concentrations). Future management practices should allow for optimum plant growth, while simultaneously reducing the loss of pollutants from the plant–soil system.[4]

ARTICLES OF FURTHER INTEREST

Air Pollutants: Effects of Ozone on Crop Yield and Quality, p. 13
Air Pollutants: Interactions with Elevated Carbon Dioxide, p. 17
Air Pollutants: Responses of Plant Communities, p. 20
Isoprenoid Metabolism, p. 625
Nitrogen, p. 822
Oxidative Stress and DNA Modification in Plants, p. 854
Plant Response to Stress: Critical Periods in Plant Development, p. 981

REFERENCES

1. Trainer, M.; Williams, E.J.; Parrish, D.D.; Buhr, M.P.; Allwine, E.J.; Westberg, H.H.; Fehsenfeld, F.C.; Liu, S.C. Models and observations of the impact of natural hydrocarbons on rural ozone. Nature **1987**, *329*, 705–707.
2. Guenther, A.B.; Hewitt, C.N.; Erickson, D.; Fall, R.; Geron, C.; Graedel, T.; Harley, P.; Klinger, L.; Lerdau, M.; McKay, W.A.; Pierce, T.; Scholes, B.; Steinbrecher, R.; Tallamraju, R.; Taylor, J.; Zimmerman, P. A global model of natural volatile organic compound emissions. J. Geophys. Res. **1995**, *100*, 8873–8892.
3. Podlesna, A. Air pollution by sulfur dioxide in Poland—Impact on agriculture. Phyton **2002**, *42*, 157–163.
4. Rogasik, J.; Schroetter, S.; Schnug, E. Impact of air pollutants on agriculture. Phyton **2002**, *42*, 171–182.
5. Wellburn, A.R. Why are atmospheric oxides of nitrogen usually phytotoxic and not alternative fertilizers? New Phytol. **1990**, *115*, 395–429.
6. Omasa, K.; Endo, R.; Tobe, K.; Kondo, T. Gas diffusion model analysis of foliar absorption of organic and inorganic air pollutants. Phyton **2002**, *42*, 135–148.
7. Rennenberg, H.; Geßler, A. Consequences of N Deposition to Forest Ecosystems—Recent Results and Future Research Needs. In *Forest Growth Responses to the Pollution Climate of the 21st Century*; Sheppard, L.J., Neil Cape, J., Eds.; Kluwer Academic Publ.: Dordrecht, The Netherlands, 1999; 47–64.
8. Ramge, P.; Badeck, F.W.; Ploechl, M.; Kohlmaier, G.H. Apoplastic antioxidants as decisive elimination factors within the uptake of nitrogen dioxide into leaf tissues. New Phytol. **1993**, *125*, 771–785.
9. Kreuzwieser, J.; Schnitzler, J.-P.; Steinbrecher, R. Biosynthesis of organic compounds emitted by plants. Plant Biol. **1999**, *1*, 149–159.
10. Gabriel, R.; Schäfer, L.; Gerlach, C.; Rausch, T.; Kesselmeier, J. Factors controlling the emissions of volatile organic acids from leaves of *Quercus ilex* L. (Holm oak). Atmos. Environ. **1999**, *33*, 1347–1355.

Exciton Theory

Su Lin
Arizona State University, Tempe, Arizona, U.S.A.

INTRODUCTION

Excitons are electronic excitations capable of motion in a nonmetallic condensed medium. Their generation usually results from absorption of energy as photons. Exciton decay results in products such as light emission, charge-separated state formation, or heat. The theory of excitons is essential to the interpretation of spectroscopic properties and electronic excitation dynamics in complex systems.[1] It is also essential to understanding the initiation of photochemical processes in photosynthesis.

Systems that perform photosynthesis contain large numbers of interacting pigments, mainly chlorophylls and bacteriochlorophylls. Photosynthesis is initiated by photon absorption in the region of 400–1100 nm, which generates excited states of these pigments. This is followed by many energy transfer steps. Exciton theory therefore plays an essential role in photosynthesis studies.[2,3]

CONCEPT AND PROPERTIES

Excitons are quanta of electronic excitation energy traveling in a periodic structure such as a crystal lattice or diffusing in a less well-ordered array of biomolecules. They transfer excitation energy from one place to another without transferring charge. Different types of excitons exist.[4] Only the Frenkel, or small radius, exciton is of importance to photosynthesis. It is an excited state of a condensed system closely resembling a molecular excited state, carrying the excitation energy either by hopping from site to site or by the motion of an exciton wavepacket.

Exciton coherence describes the degree of correlation of the phase of the excitation on different molecules. As the wavelength of the absorbed light is much greater than the dimensions of a typical pigment-protein complex, a certain degree of coherence is always present in excitons at the time of their creation. Exciton scattering by phonons and disorder within a complex leads to the loss of coherence. During the coherence time, a molecular system can be treated as a supercomplex. In the limiting case of a totally incoherent exciton, excitation energy can be transferred between individual molecules by the Förster mechanism, to be discussed shortly.

For a system consisting of excitonically coupled molecules, each having two energy levels, the K-th exciton state is given in terms of wavefunctions

$$\Psi_K = \sum_i c_{Ki}\phi_i^1 \prod_{j \neq i} \phi_j^0 \tag{1}$$

where the coefficients c_{Ki} represent the relative contribution of individually excited molecules to the exciton state, and ϕ_i^0 and ϕ_i^1 are the wavefunctions of the ground and excited states of the i-th molecule, respectively. The energy an exciton carries is the energy difference between the initial and the final states.

To describe the absorption strength of a material, the concept of transition dipole moment is used. Its square is a measure of the probability that the radiation will produce an excited state. For an excitonically coupled molecular system, the transition dipole moment of an excited state is a linear combination of those from the individual molecules:

$$\mu_K = \sum_i c_{Ki}\mu_i \tag{2}$$

where the coefficients c_{Ki} are the same as in Eq. 1. The integrated intensity of an absorption band correlates directly with the square of the transition dipole moment.

Excitonic coupling can be demonstrated in a molecular dimer, in which two identical molecules interact strongly, resulting in their excited states being completely mixed. The wavefunctions of the excited dimer are

$$\begin{aligned}\Psi_+ &= \frac{1}{\sqrt{2}}(\phi_1^1\phi_2^0 + \phi_1^0\phi_2^1)\\ \Psi_- &= \frac{1}{\sqrt{2}}(\phi_1^1\phi_2^0 - \phi_1^0\phi_2^1)\end{aligned} \tag{3}$$

and the corresponding energies are

$$\begin{aligned}E_+ &= E + V_{12}\\ E_- &= E - V_{12}\end{aligned} \tag{4}$$

where E_+ and E_- are the dimer excited state energies, E is the monomer excited state energy, and V_{12} is the coupling energy between the monomer excitations. The energy levels of a dimer are shifted away from the monomeric energy (Fig. 1a), which can be seen in the absorption spectrum (Fig. 1b). The amount of spectral shift and the intensity distribution over the exciton bands are

Encyclopedia of Plant and Crop Science
DOI: 10.1081/E-EPCS 120019313

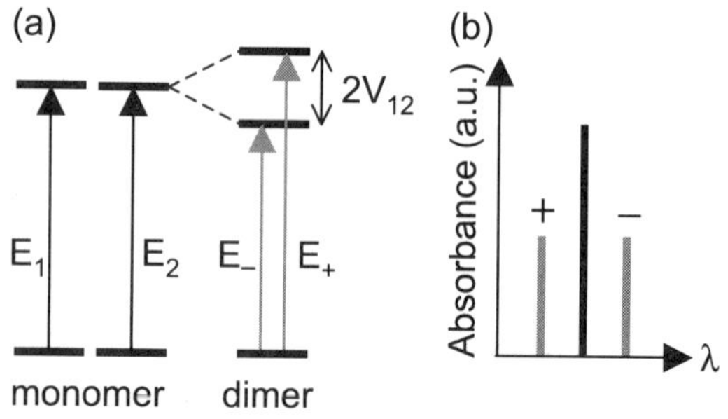

Fig. 1 (a) Energy levels of monomers and dimer (b) their corresponding absorption band positions.

determined by the monomers' geometric arrangement and electronic properties.

When the interactions between molecules are weak, excitations are essentially localized on individual molecules, corresponding to the incoherent exciton case. Förster theory[2,3] is employed to find the energy transfer rate between a donor and acceptor molecule, separated by a distance R:

$$k_{d \to a} = \frac{1}{\tau_0}\left(\frac{R_0}{R}\right)^6 \tag{5}$$

where τ_0 is the excited-state lifetime of the donor in the absence of transfer and R_0 is a characteristic distance related to the overlap between the donor emission spectrum and acceptor absorption spectrum, the relative orientation of molecular transition moments, and the system refractive index. Typical values of R_0 are 20–60 Å. Transfer occurs when $R < R_0$. When $R > R_0$ the donor is more likely to lose energy by other means. The rate of transfer between identical molecules is the same in both directions, but when one (called #2) has an excited state at a higher energy, the back-to-forward rate ratio is the Boltzmann factor,

$$\frac{k_{1 \to 2}}{k_{2 \to 1}} = e^{-(\Delta G/RT)} \tag{6}$$

where $\Delta G = G_2 - G_1$ is the free energy difference of the excited levels. This equation governs the excitation distribution within a heterogeneous system, therefore determining the profiles of fluorescence and transient absorption spectra.

Excitation energy migration, whether coherent or incoherent, involves more than two molecules. In most cases, migration is a sequence of transfers. During the process of exciton transfer and migration, because of the system heterogeneity, the exciton's energy can be transferred to an excited state with an energy lower than that of the exciton. The excitation might thereby lose its mobility, becoming trapped in that particular state. When more than one exciton is present in an excitonically coupled system, collisions between them can take place during their lifetime. Exciton-exciton interaction can result in the formation of a new exciton with twice the energy of one exciton, one exciton and heat, or heat alone. Although annihilation is of little importance under physiological light intensities, it has proved to be a useful effect in experimental work.

THE EXCITON IN PHOTOSYNTHETIC SYSTEMS

Initiating photosynthesis, the exciton transfers absorbed energy among neighboring chromophores (the "antenna") before it is trapped by the reaction center, where the energy is converted and stored as chemical energy.

An example of excitonic coupling effecting spectral changes is the LH2 antenna from purple bacterium *Rhodopseudomonas acidophila*, containing large numbers of interacting pigments. Fig. 2 illustrates that the Q_y transition of *monomeric* bacteriochlorophyll (Bchl)-*a* peaks at 780 nm, while the absorption of Bchl-*a* molecules in LH2 exhibits two bands at 800 and 850 nm. The pigments absorbing at 800 nm are from the B800 ring in which all the Bchl molecules are parallel to the plane of the ring. The pigments are separated approximately 21 Å from each other, resulting in weak interactions between molecules. The arrangement of pigments absorbing at 850 nm is quite different. The quantum mechanical calculation of excitonic properties in dimers can be used here to explain the spectral differences with the structure of LH2. Pairs of Bchl-*a* molecules form closely coupled dimers, and nine such dimers form the B850 ring with all molecules separated at a distance of 9 Å. All the pigments are approximately perpendicular to the plane of the B850 ring. The initial excitation is delocalized over the major

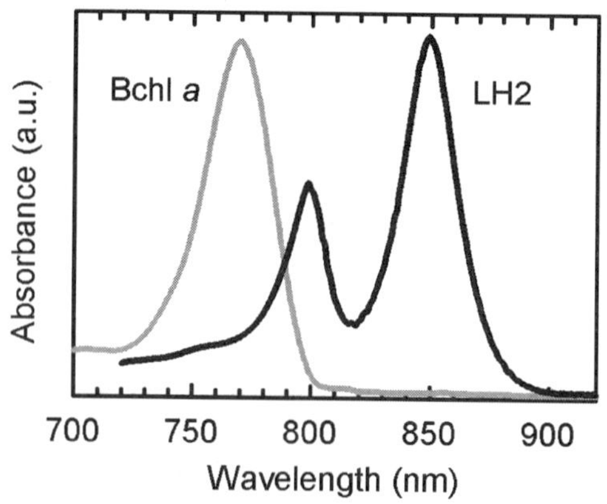

Fig. 2 Absorption spectra of Bchl *a* (gray) and LH2 (black).

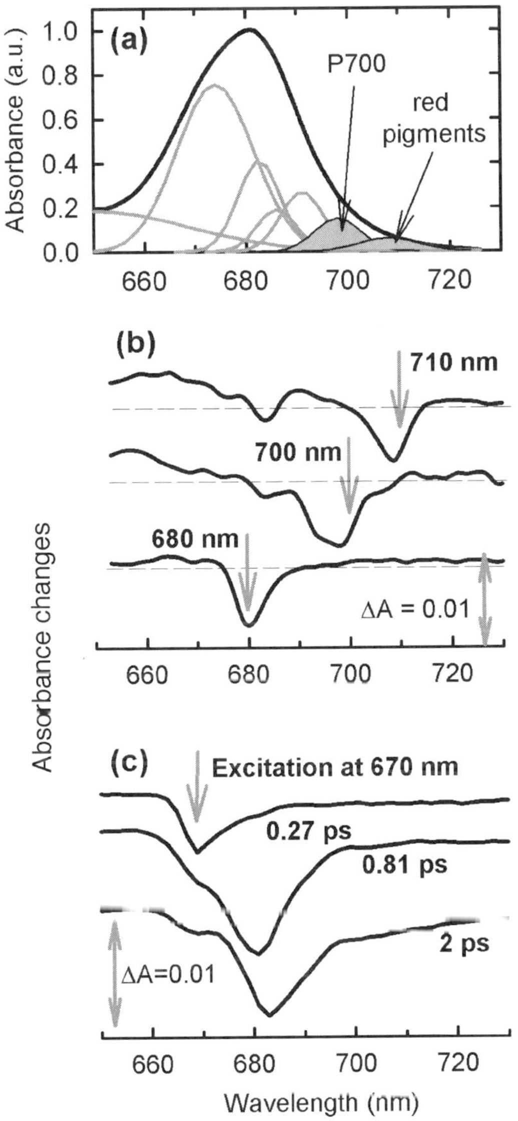

Fig. 3 (a) Room temperature Q_Y absorption band of PSI fitted with Gaussian components. (b) Transient absorption changes at 77 K recorded at 200 fs after laser excitations at marked wavelenghts. (c) Time-resolved absorption spectra at 77 K with 670 nm excitation.

part of the B850 ring when excited directly, while the excitation of B800 creates much less coherence. The B850 coherence is partially lost within a few hundred femtoseconds because of electron-phonon interaction and scattering by disorder.

Another system studied extensively using exciton theory is PSI, one of the essential complexes in green plants and cyanobacteria. X-ray analysis of *Synechococcus elongatus* reveals that each unit of PSI complex contains about 100 chlorophyll (Chl)-*a* molecules. About 75% of the inter-pigment distances fall in the range of 10–20 Å. Except for a moderate spectral shift due to the pigment-protein interaction, the molecules maintain their individual characteristics. Therefore, PSI is usually modeled as a loosely packed molecular crystal. However, the inter-pigment distances are small enough compared with the Chl-*a* R_0, which is at least 50 Å, that the Förster mechanism is believed to govern the energy transfer between pigments. There are several clusters of Chl-*a* molecules showing inter-pigment distances of 6–9 Å, with their chlorin planes almost parallel to each other, including the special pair in the reaction center. Those geometric arrangements create strong excitonic coupling between the molecules in these clusters. Various inter-pigment interactions in PSI are clearly reflected in both steady-state absorption and transient absorbance spectra. The Q_y transition band fitted with Gaussian components (Fig. 3a) shows that two bands are shifted more than 20 nm away from their dominant absorption around 680 nm. The 700-nm band is identified as the absorption of a special pair called P700, formed by two parallel Chl-*a* molecules at a 3.6 Å interplanar distance. The redmost band at 710 nm is believed to originate in a group of strongly coupled molecules. The early-time transient absorption spectra in Fig. 3b show that 680-nm excitation induces a narrow bleaching band originating in the pigments with transition energies close to the 680-nm photon energy. Excitations at 700 or 710 nm induce much broader initial bleachings and spectral changes at the other wavelengths. The exciton delocalization features disappear within a few hundred femtoseconds. Excitation transfer and migration processes within PSI are reflected in its spectral dynamics in Fig. 3c. The initial bleaching at the excitation wavelength 670 nm is broadened and shifted to the red at later times. The dominant contribution to the spectral redistribution is from exciton migration within the complex governed by the Förster mechanism.

CONCLUSION

Exciton theory has been successfully applied to study the spectral dynamic properties of photosynthetic antenna and reaction center systems, thereby broadening and deepening our understanding of their structure-function relationships. Research on initial photochemical processes in photosynthesis not only is greatly benefited by exciton theory but also provides further understanding and tests of exciton theory.

ACKNOWLEDGMENTS

I would like to thank my former advisor Robert S. Knox for his encouragement and support for this writing. I would also like to thank Alexander Melkozernov, Kenneth Hoober, and Robert E. Blankenship for providing data and suggestions.

ARTICLES OF FURTHER INTEREST

Chlorophylls, p. 258
Fluorescence: A Probe of Photosynthetic Apparatus, p. 472
Photosystems: Electron Flow Through, p. 906

REFERENCES

1. Knox, R.S.; Knox, W.H. Excitons *Encyclopedia of Applied Physics*; VCH Publishers, 1991; Vol. 6, 311–325.
2. Van Amerongen, H.; Valkunas, L.; van Grondelle, R. The Exciton Concept. In *Photosynthetic Excitons*; World Scientific Publishing: Singapore, 2000; 47–72.
3. Blankenship, R.E. Antenna Complexes and Energy Transfer Processes. In *Molecular Mechanisms of Photosynthesis*; Blackwell Science Ltd.: Oxford, 2002; 61–94.
4. Dexter, D.L.; Knox, R.S. Formal Exciton Theory. In *Excitons*; Interscience Publishers: New York, 1965; 40–62.

Farmer Selection and Conservation of Crop Varieties

Daniela Soleri
David A. Cleveland
University of California, Santa Barbara, California, U.S.A. and
Center for People, Food and Environment, Santa Barbara, California, U.S.A.

INTRODUCTION

After domestication, plant species were often transported widely, and many genetically distinct farmers' varieties (FVs, crop varieties traditionally maintained and grown by farmers) developed in specific locations.[1] FVs continue to be grown today by many small-scale farmers in traditionally-based agricultural systems (TBAS), fulfilling both local or regional consumption needs, as well as the larger social need for the conservation of genetic diversity.[2]

Crop genetic variation (V_G) is a measure of the number of alleles and degree of difference between them, and their arrangement in plants and populations. A change in V_G over generations is evolution, though one form of this change, microevolution, is reversible. Farmers and the biophysical environment select plants within populations. Farmers also choose between populations or varieties. This phenotypic selection and choice together determine the degree to which varieties change between generations, evolve over generations, or stay the same. Conservation in a narrow sense means the preservation of the V_G present at a given time. However, in situ conservation in farmers' fields is commonly understood to mean that the specific alleles and genetic structures contributing to that V_G may evolve in response to changing local selection pressures, while still maintaining a high level of V_G.[3] In contrast, ex situ conservation in genebanks attempts to conserve genetic diversity present at a given location and moment in time, preserving the same alleles and structures over time. Thus, different forms of conservation include different amounts and forms of change.

Sometimes farmers carry out selection or choice intentionally to change or conserve V_G. However, much of farmer practice is intended to further production and consumption goals and affects crop evolution unintentionally. Therefore, in order to understand farmer selection and conservation, it is important to understand the relationship 1) between production, consumption, selection, and conservation in TBAS, and 2) between farmer knowledge and practice and the basic genetics of crop populations and their interactions with growing environments [genetic variation, environmental variation and genotype-by-environment interaction (G×E), and response to selection] (Table 1).

FARMERS AND FVs IN TRADITIONALLY-BASED AGRICULTURAL SYSTEMS

TBAS are characterized by the integration within the household of production, consumption, selection, and conservation, whereas in industrial agriculture these functions are spatially and structurally separated. Farm households in TBAS typically rely on their own food production for a significant proportion of their consumption and this production is essential for feeding the population in TBAS now and in the future, even with production increases in industrial agriculture[4]—by 2025 three billion people will depend on agricultural production in TBAS.[5]

TBAS are also characterized by marginal growing environments (relatively high stress, high temporal and spatial variability, and low external inputs) and the continued use of FVs, even when modern crop varieties (MVs) are available.[6] FVs include landraces, traditional varieties selected by farmers, MVs adapted to farmers' environments by farmer and natural selection, and progeny from crosses between landraces and MVs (sometimes referred to as "creolized" or "degenerated" MVs). The V_G of farmer-managed FVs is not well documented, but is presumed to support broad resistance to multiple biotic and abiotic stresses, making them valuable not only for farmers because they decrease the production risks in marginal environments, but also for plant breeders and conservationists as the basis for future production in industrial agriculture.[7] Farmers value FVs for agronomic traits, such as drought resistance, pest resistance and photoperiod sensitivity, as well as for traits contributing to storage, food preparation, taste, market value, and appearance (e.g., maize varieties grown for purple husks used in tamale production).

Encyclopedia of Plant and Crop Science
DOI: 10.1081/E-EPCS 120017094

FARMER CHOICE: GENETIC VARIATION, CLASSIFICATION, GENOTYPE × ENVIRONMENT INTERACTION, AND RISK

The way farmers classify and value traits, which can vary between women and men, and between households in a community,[2] affects adoption and abandonment of varieties and populations, farmers' tolerance of intravarietal gene flow, and, thus, intraspecific V_G. Experimental evidence suggests that farmers can choose among large numbers of genotypes—in Syria, farmers were able to identify efficiently high yielding barley populations from among 208 entries, including 100 segregating populations.[7] Farmers' choice of varieties and populations when adopting or abandoning them from their repertoires, saving seed for planting, and procuring seed, does not change the genetic makeup of those units directly, and there is no evidence that farmers have any expectation of changing them. However, farmers' choice of crops, varieties, and populations does affect the total V_G farmers manage and the number of populations within which farmers can select plants.

The FV reproductive system, in combination with farmers' propagation methods, are important determinants of interspecific and intraspecific V_G both directly and indirectly, because resulting differences in the consistency of the V_G present over generations affects farmers' perception and management.[8] V_G in asexually propagated outcrossing crops such as cassava is exactly replicated in amount and structure between generations with discrete, fixed types (clones) or groups of types maintained as distinct varieties,[9,10] that may be either homo- or heterozygous. Intrapopulation V_G, affected by the genetics of the particular trait, becomes more dynamic and less structured with the intentional inclusion by farmers of sexually propagated individuals into clonal populations based on morphological similarity.[10] The same increase in dynamism occurs with increasing rates of outcrossing in

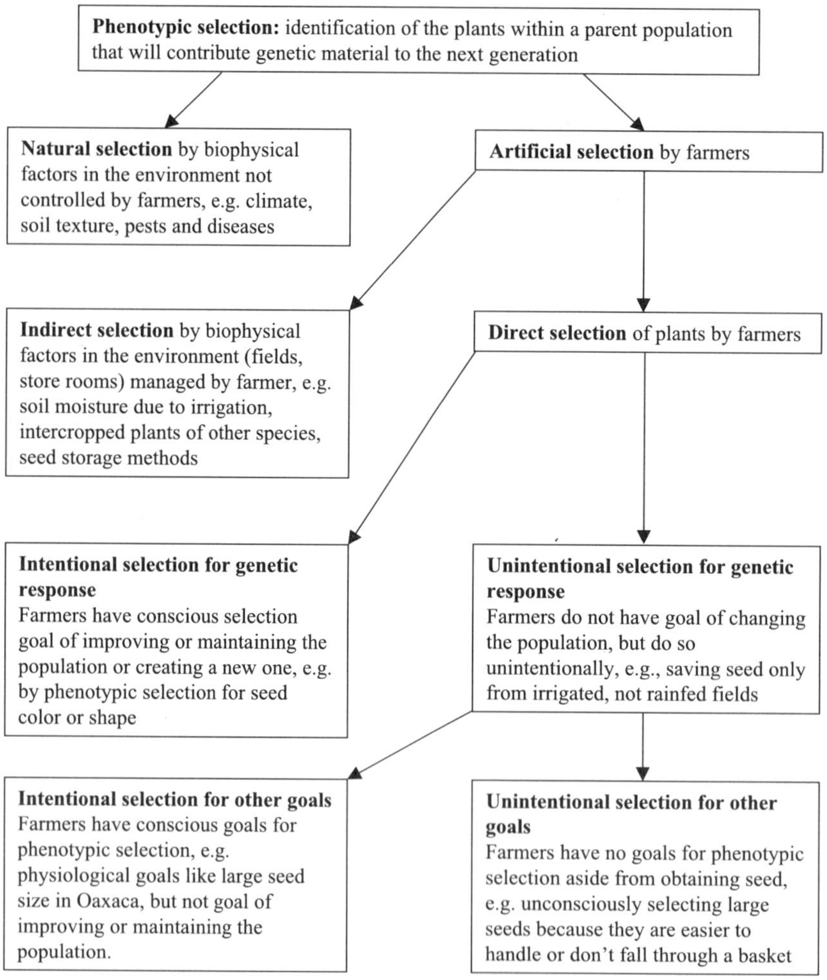

Fig. 1 Classification of farmer selection leading to genetic response, according to the agent of phenotypic selection and intent of farmer as agent.

sexually propagated crops, because variation can be continuous within a population. Moreover, segregation, crossing-over, recombination, and other events during meiosis and fertilization result in strong variability of V_G between generations. In predominantly allogamous crops such as maize, heterozygosity can be high, making it difficult to discern discrete segregation classes particularly in the presence of environmental variation, and retention of distinguishing varietal characteristics requires maintenance selection.[11] Highly autogamous crops such as rice are predominantly homozygous, making exploitation of V_G and retention of varietal distinctions easier, even if varieties are composed of multiple, distinct lines.

Farmers' choices depend in part on the range of spatial, temporal, and management environments present, the V_G available to them, and the extent to which genotypes are widely versus narrowly adapted. In turn, environmental variation in these growing environments interacts with V_G (G×E) to produce variation in yield of grain, straw, roots, tubers, leaves, and other characteristics over space and time. As a result, farmers may have different criteria for different environments, as in Rajasthan, India, where pearl millet farmers realize there is a trade-off between panicle size and tillering ability: Farmers in a less stressful environment prefer varieties producing larger panicles, while those in a more stressful environment prefer varieties with high tillering under their conditions.[12]

Farmers' seed management and choice of growing environments determine the possible extent of pollen flow between populations or varieties. In Jalisco, Mexico, farmers regularly mix maize populations together by classifying seed obtained from diverse sources as the same variety, which, together with planting patterns, leads to a 1–2% level of gene flow between maize plots during one crop cycle as detected by isozyme analysis, affecting genetic composition over several crop cycles.[13] The morphological and genetic continuum across the four major local varieties suggests that traits from a variety introduced 40 years ago have introgressed into the other varieties.

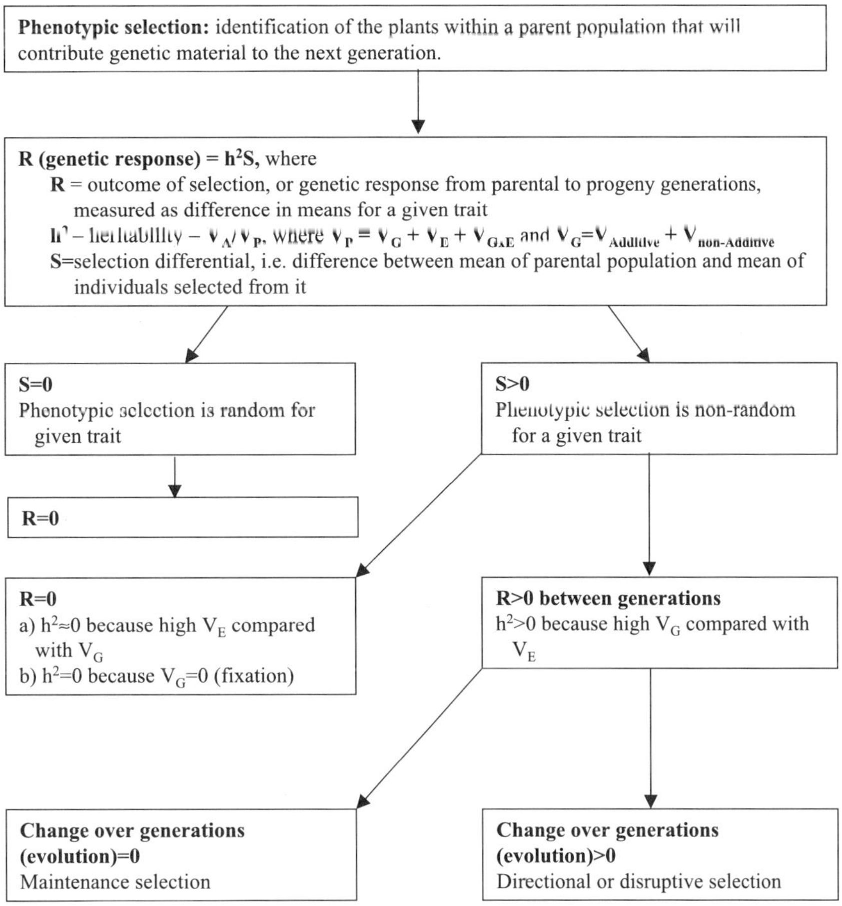

Fig. 2 Classification of farmer selection according to genetic response as an outcome of phenotypic selection. V_A = additive genetic variation, V_P = phenotype variation, V_G = genetic variation, V_E = environmental variation, V_{GXE} = variation in genotype by environmental interaction.

Patterns of variation in yield affect farmers' choice of crop variety via their attitude toward risk. In response to scenarios depicting varietal G×E and temporal variation, farmers from more marginal growing environments were more risk-averse compared to those from more favorable environments, the former preferring a crop variety with low but stable yields across temporal environments, the latter choosing a variety highly responsive to favorable conditions but with poor performance under less favorable conditions.[14]

Table 1 Farmer selection and choice and the change and conservation of crop varieties

Farmer knowledge (including values) on which practice may be based	Farmer practice	Potential effect of farmer practice on selection and conservation of populations/varieties	Example
	Indirect selection/conservation by farmer-managed growing and storage environment		
Understanding of G×E	Allocation of varieties to spatial, temporal, and management environments	Selection pressures in environments result in maintenance of existing or development of new populations/varieties, including evolution of wide or narrow adaptation	*Spatial*: varieties specified for different soil or moisture types; rice, Nepal; pearl millet, India. *Temporal*: varieties with different cycle lengths, maize, Mexico
	Management of growing environments	Changing selection pressures	Changes in fertilizer application, maize, Mexico.
Risk, values, G×E	Choice of environments for testing new populations/varieties	↑ or ↓ V_G	High stress, rice, Nepal; Optimal conditions, barley, Syria
Escape from economic or political pressure; desire for different way of life	Abandonment of fields or farms, reduced field size	↓ V_G within due to reduced area for planting, ↓ effective population size, genetic drift	Pooling of subvarieties, maize, Hopi and Zuni; Reduction in area, potatoes, Peru; maize, Mexico
	Direct selection/conservation, intentional re. population change		
Discount rate (values re. future), altruism (values re. community)	Conservation of varieties for the future, for other farmers	↑ intraspecific V_G	Rice, Thailand; maize, Hopi
Interest and expertise in experimentation	Deliberate crossing	↑ V_G	Maize-teosinte, Mexico; MV-FV pearl millet, India; MV-FV and FV-FV, maize, Mexico.
Understanding of h^2	Selection of individuals (plants, propagules) from within parent population	↑ or ↓ V_G via R	Among seedlings, cassava, Peru; among panicles, pearl millet, India.
	Direct, selection/conservation, unintentional re. population change, but intentional re. other goals, as result of production/consumption practices		
Attitudes towards risk re. yield stability	Adoption and abandonment of FVs, MVs	↑ or ↓ intraspecific diversity	Maize, Hopi Rice, Nepal
	Adoption and abandonment of lines in multiline varieties of self pollinated crops; seedlots in cross-pollinating crops	↑ or ↓ intravarietal diversity	Common bean, East Africa Maize, Mexico
Agronomic, storage, culinary, aesthetic and ritual criteria, implicit and explicit	Selection or choice based on production/consumption criteria	↑ or ↓ intra- and intervarietal diversity	Storage and culinary criteria: maize, Mexico; and ritual criteria, rice, Nepal
Choice criteria	Acquisition of seed, seed lots	Gene flow via seed then pollen flow, hybridization, recombination within varieties	Cycle length, maize, Mexico; cuttings and seedlings, cassava, Guyana

G×E: genotype by environment interaction
V_G: genetic variation
↑ or ↓: increase or decrease
h^2: heritability in the narrow sense
R: response selection
MV: modern crop variety, product of formal breeding system
FV: farmer developed crop variety

SELECTION: HERITABILITY, PHENOTYPIC SELECTION DIFFERENTIAL, AND RESPONSE

Phenotypic selection is the identification of individual plants within a population that will contribute genetic material to the next generation. Phenotypic selection of FVs in TBAS can be classified according to the agent of selection (natural environment, farmer-managed environment, or farmer) and according to farmers' goals for selection (Fig. 1). Farmer selection can also be classified according to the outcome (Fig. 2). Geneticists and plant breeders tend to think of phenotypic selection as seeking to produce genetic change, but farmers often do not. Whether or not farmer selection changes the genetic makeup of the population (i.e., effects genetic response or R) depends on heritability (h^2), or the proportion of phenotypic variation that is genetic and can be inherited; and the selection differential (S), or the difference between the means of the parental population and sample selected from it: $R = h^2S$. The extent to which selection maintains potentially useful V_G is a measure of its contribution to in situ conservation.

Heritability is often understood by farmers who distinguish between high and low heritability traits and consciously select for the former, while often considering it not worth while or even possible to select for the latter, especially in cross-pollinating crops.[14] When farmers' selection criteria centers on low heritability traits such as large ear size in maize, they may achieve high S, and little or no R. However, they persist in selection because they have other goals, such as improving the quality of planting seed, not high R.[11,15]

In terms of seeking genetic response, farmers may practice intentional selection either to create new varieties, best documented in vegetatively propagated and self-pollinating crops,[9] or for varietal maintenance or improvement, although much evidence for the latter is anecdotal. Unintentional selection—that is, not seeking genetic response—, as documented with maize farmers in Mexico, may be undertaken for varietal maintenance and/or to ensure planting seed quality, although this can also result in genetic response.

Quantitative research on the goals and outcomes of farmers' selection is relatively new. Selection exercises in two independent investigations of maize in Mexico found farmers' selections to be significantly different from the original population for a number of ear traits, resulting in high S values.[11,15] However, R values calculated in the Oaxaca study were zero for these as well as other morphophenological traits.[15] Similarly, the Jalisco study found that selection served to diminish the impact of gene flow, but not to change the population being selected on.[11] Indeed, a recent study across four sites each with different crops found that often a majority of farmers in a site did not see their seed selection as a process of cumulative, directional change.[14]

However, intentional phenotypic selection for goals other than genetic response is practiced by nearly all farmers in that study and probably in TBAS, the reasons documented to date being seed quality (germination and early vigor) and purity and because this is "the way we know," that is, because farmers may not want to change (viz. 'improve') a variety, although genetic response may result unintentionally. To understand this from the farmers' perspective, it is necessary to take into account the multiple functions of crop populations in TBAS: production of food and seed, consumption, conservation, and improvement.

CONCLUSIONS

Selection and conservation in TBAS contrast substantially with industrial agricultural systems, and understanding farmers' practices, and the knowledge and goals underlying them, is critical for supporting food production, food consumption, crop improvement, and crop genetic resources conservation for farm communities in TBAS and for long-term global food security. The urgency of understanding farmer selection and conservation will increase in the future with the on going loss of genetic resources, the rapid spread of transgenic crop varieties with limited genetic diversity, the development of a global system of intellectual property rights in crop genetic resources, and the movement to make formal plant breeding more relevant to farmers in TBAS through plant breeding and conservation based on direct farmer and scientist collaboration.

ACKNOWLEDGMENTS

The research on which this article is based was supported in part by NSF grant no. SES-9977996.

ARTICLES OF FURTHER INTEREST

Crop Domestication in Africa, p. 304
Crop Improvement: Broadening the Genetic Base for, p. 343
Crops and Environmental Change, p. 370

Gene Flow Between Crops and Their Wild Progenitors, p. 488
Pre-Agricultural Plant Gathering and Management, p. 1055

REFERENCES

1. Harlan, J.R. *Crops and Man,* 2nd Ed.; American Society of Agronomy, Inc. and Crop Science Society of America, Inc.: Madison, WI, 1992.
2. Smale, M. Economics Perspectives on Collaborative Plant Breeding for Conservation of Genetic Diversity on Farm. In *Farmers, Scientists and Plant Breeding: Integrating Knowledge and Practice*; Cleveland, D.A., Soleri, D., Eds.; CAB International: Oxon, UK, 2002; 83–105.
3. Brown, A.H.D. The Genetic Structure of Crop Landraces and the Challenge to Conserve Them in situ on Farms. In *Genes in the Field: On-Farm Conservation of Crop Diversity*; Brush, S.B., Ed.; Lewis Publishers: Boca Raton, FL, 1999; 29–48.
4. Heisey, P.W.; Edmeades, G.O. Part 1. Maize Production in Drought-Stressed Environments: Technical Options and Research Resource Allocation. In *World Maize Facts and Trends 1997/98*; CIMMYT, Ed.; CIMMYT: Mexico, D.F., 1999; 1–36.
5. Evans, L.T. *Feeding the Ten Billion: Plants and Population Growth*; Cambridge University Press: Cambridge, UK, 1998.
6. Brush, S.B.; Taylor, J.E.; Bellon, M.R. Technology adoption and biological diversity in Andean potato agriculture. J. Dev. Econ. **1992**, *39*, 365.
7. Ceccarelli, S.; Grando, S.; Tutwiler, R.; Bahar, J.; Martini, A.M.; Salahieh, H.; Goodchild, A.; Michael, M. A methodological study on participatory barley breeding I. Selection phase. Euphytica **2000**, *111*, 91.
8. Cleveland, D.A.; Soleri, D.; Smith, S.E. A biological framework for understanding farmers' plant breeding. Econ. Bot. **2000**, *4*, 377.
9. Boster, J.S. Selection for perceptual distinctiveness: Evidence from Aguaruna cultivars of Manihot esculenta. Econ. Bot. **1985**, *39*, 310.
10. Elias, M.; Penet, L.; Vindry, P.; McKey, D.; Panaud, O.; Robert, T. Unmanaged sexual reproduction and the dynamics of genetic diversity of a vegetatively propagated crop plant, cassava (Manihot esculenta Crantz), in a traditional farming system. Mol. Ecol. **2001**, *10*, 1895–1907.
11. Louette, D.; Smale, M. Farmers' seed selection practices and maize variety characteristics in a traditional Mexican community. Euphytica **2000**, *113*, 25.
12. Weltzien, R.E.; Whitaker, M.L.; Rattunde, H.F.W.; Dhamotharan, M.; Anders, M.M. Participatory Approaches in Pearl Millet Breeding. In *Seeds of Choice*; Witcombe, J., Virk, D., Farrington, J., Eds.; Intermediate Technology Publications: London, 1998; 143–170.
13. Louette, D.; Charrier, A.; Berthaud, J.D. In situ conservation of maize in Mexico: Genetic diversity and maize seed management in a traditional community. Econ. Bot. **1997**, *51*, 20.
14. Soleri, D.; Cleveland, D.A.; Smith, S.E.; Ceccarelli, S.; Grando, S.; Rana, R.B.; Rijal, D.; Ríos Labrada, H. Understanding Farmers' Knowledge as the Basis for Collaboration with Plant Breeders: Methodological Development and Examples From Ongoing Research in Mexico, Syria, Cuba, and Nepal. In *Farmers, Scientists and Plant Breeding: Integrating Knowledge and Practice*; Cleveland, D.A., Soleri, D., Eds.; CAB International: Oxon, UK, 2002; 19–60.
15. Soleri, D.; Smith, S.E.; Cleveland, D.A. Evaluating the potential for farmer and plant breeder collaboration: A case study of farmer maize selection in Oaxaca, Mexico. Euphytica **2000**, *116*, 41.

Female Gametogenesis

Jagreet Kaur
Imran Siddiqi
Centre for Cellular and Molecular Biology, Hyderabad, India

F

INTRODUCTION

The female gametophyte or embryo sac in seed plants is a specialized structure containing the egg cell and a small number of associated cells that are essential for fertilization and seed development. In higher plants, the female gametophyte is encased within the ovule and represents an evolutionary reduction of a distinct gametophytic generation that in lower plants exists as separate free living organisms. The key events of fertilization, embryogenesis, and endosperm development leading up to the formation of a mature seed all take place within the embryo sac. We discuss here the stages in development of the embryo sac, its cellular organization and structure in the context of its unique function in plant reproduction, and its role in fertilization and seed development. We have emphasized recent advances that have come from the use of genetic approaches to understand biogenesis of the embryo sac and its control of seed development.

THE FORMATION OF THE EMBRYO SAC

The first step in the pathway leading to female gametophyte development is the specification of the archesporial cell at the tip of the growing ovule. The archesporial cell enlarges and either directly becomes the megaspore mother cell (MMC; also called the megasporocyte) or in many species undergoes an additional division to give rise to the MMC. The development of the embryo sac consists of two stages subsequent to the formation of the MMC: megasporogenesis comprising meiosis and formation of the functional megaspore, and megagametogenesis in which the functional megaspore develops into the female gametophyte (Fig. 1).

Megasporogenesis

The MMC undergoes meiosis to produce a tetrad of four spores. Most commonly in the monosporic type of development, three of the four spores degenerate, and the remaining one becomes the functional megaspore.[1] Prior to meiosis, the MMC undergoes a process of polarization with the movement of the nucleus and organelles (mitochondria and plastids) toward one end of the cell, the end that will generally give rise to the functional megaspore. During meiosis, the cell wall of the MMC forms a layer of callose, a β(1,3)-glucan, which has low permeability and serves to insulate the MMC from the surrounding cells as well as to separate the megaspores after meiosis. Callose is removed from the cell wall of the functional megaspore early in megagametogenesis. Meiosis proceeds contemporaneously with the specification and development of the functional megaspore. Based upon studies in *Arabidopsis*, many of the genes required for meiotic chromosome organization and recombination are conserved between plants and other eukaryotes; however, others appear to be unique to plants.[2]

Megagametogenesis

In the most common type of development, the functional megaspore undergoes three rounds of nuclear division without cytokinesis to produce eight nuclei, accompanied by extensive cell expansion to form a large embryo sac. A prominent vacuole occupies most of the central region of the embryo sac. The nuclei migrate to particular parts of the embryo sac and differentiate into specific cell types: an egg cell, synergids, and antipodal cells. The embryo sac is polarized and consists of the egg apparatus made up of an egg cell and two synergids in most species, at the end close to the micropyle. A cluster of antipodal cells occupy the opposite (chalazal) pole, and a central cell nucleus is found toward the middle. The dihaploid central cell nucleus is formed by fusion of two polar nuclei. The egg cell develops into the embryo and the central cell into the endosperm after double fertilization. In *Arabidopsis* the antipodals degenerate close to the time of fertilization, whereas in maize the antipodals proliferate and form a cluster of cells at the chalazal pole of the embryo sac.

The embryo sac is enclosed by the integuments and develops in very close contact with the sporophyte on which it depends for nutrition as well as developmental cues. The endothelium, which forms the inner layer of the inner integument of the ovule, closely contacts the embryo sac and acts as an interface between the female gametophyte and the maternal sporophyte. Plants are unique in

Encyclopedia of Plant and Crop Science
DOI: 10.1081/E-EPCS 120012913

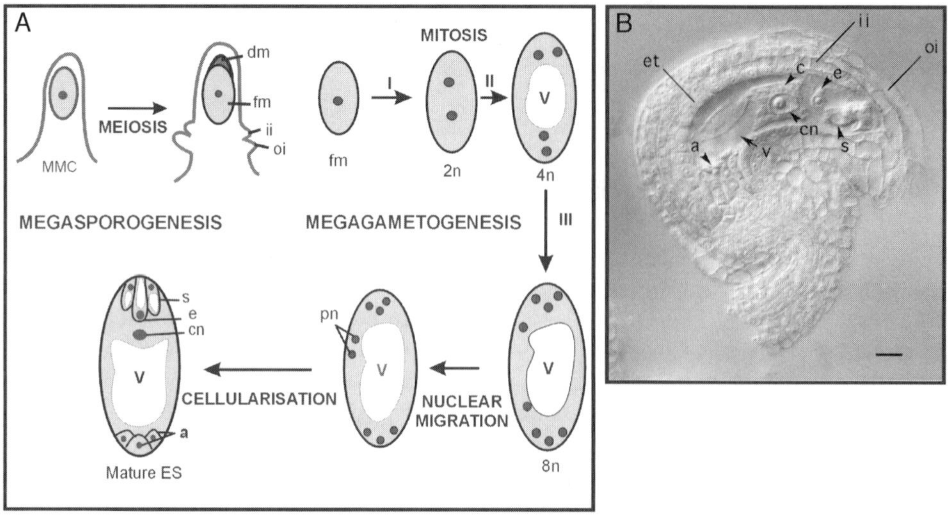

Fig. 1 A) Schematic showing stages of female gametophyte development of the *Polygonum* type. B) Mature ovule of *Arabidopsis thaliana* (reprinted from Development **2000**, *127*, 197–207). a, antipodals; c, central cell; cn, central cell nucleus; dm, degenerating megaspores; e, egg cell; et, endothelium; fm, functional megaspore; ii, inner integument; mmc, megaspore mother cell; oi, outer integument; pn, polar nuclei; s, synergids; v, vacuole. Bar, 10 μm. (*View this art in color at www.dekker.com.*)

undergoing two distinct fertilization events for the formation of the embryo and endosperm, each having a different ploidy and maternal-to-paternal genomic ratio. This requirement is reflected in the development of the female gametophyte where fusion of the two polar nuclei sets up endosperm ploidy. The cytoskeleton plays an important role in the control of nuclear division and nuclear positioning, and mutants in which nuclear division and positioning are altered also show disruption of microtubules and the cytoskeleton in the embryo sac.

EMBRYO SAC FUNCTION

The embryo sac promotes fertilization and development of the embryo and the endosperm. In sexually reproducing species, both embryogenesis and endosperm development are initiated only after double fertilization takes place. However, in apomictic species, activation of one or both of these programs takes place without fertilization. In the *fis* class of mutants of Arabidopsis, endosperm development is initiated in the absence of fertilization.[3] This suggests that the necessary elements for the early part of the program of embryogenesis and endosperm development are already established in the female gametophyte prior to fertilization and that fertilization may act as a trigger to overcome a block in the cell division cycle, as is the case for animal embryogenesis. Maternal control of embryogenesis in plants can be of two kinds: sporophytic control, which is exerted by the ovule surrounding the female gametophyte, and gametophytic control, which comes from the female gametophyte. There is evidence for both types of control in plants.[4]

An important function of the mature female gametophyte is to attract the pollen tube containing the two male sperm cells to enter the ovule through the micropyle. The pollen tube penetrates the wall of the embryo sac and one of the synergid cells, releasing the two sperm cells into the cytoplasm of the cell. The synergids play an accessory role in fertilization. Close to the time of entry of the pollen tube, the synergid cell degenerates and the cytoplasm undergoes reorganization to form a band of F-actin, which is thought to direct movement of the two sperm cells to fuse with the egg and central cells, respectively.[5] The synergid cells also play a direct role in providing the signal to attract the pollen tube into the embryo sac.[6]

DEVELOPMENTAL GENES AND MECHANISMS

In recent years, genetic and molecular approaches have led to significant advances in understanding the different stages in the pathway of female gametogenesis. Properties of the *mac1* mutant of maize suggest that the gene may play a role in a lateral inhibition process that restricts the number of archesporial cells to one per ovule.[7] The *spl*

Table 1 Representative genes acting at different stages of female gametophyte development

Gene name	Species	Stage	Mutant phenotype	Affected sex	Gene product
MAC1	Maize	Sporocyte formation	Multiple megasporocytes; male meiotic arrest	Both	–
SPL	*Arabidopsis*	Sporocyte formation	Lacks sporocyte	Both	Nuclear protein similar to MADS box transcription factors
AM1	Maize	Sporogenesis	Defective synapsis; meiotic arrest	Both	–
SDS	*Arabidopsis*	Sporogenesis	Defective synapsis and bivalent formation	Both	Cyclin-like protein
SYN1/DIF1	*Arabidopsis*	Sporogenesis	Defects in meiotic chromosome condensation and segregation	Both	Cohesin (RAD21-like)
DYAD/SWI1	*Arabidopsis*	Sporogenesis	Defective chromosome organization and meiotic progression	Both	Novel protein
AKV	*Arabidopsis*	Sporogenesis	Megaspores fail to degenerate	Female	–
FEM2	*Arabidopsis*	Gametogenesis	Arrests at one nucleate stage	Female	–
IG	Maize	Gametogenesis	Defects in division and migration of nuclei	Female	–
GFA2	*Arabidopsis*	Gametogenesis	Defects in synergid cell death and fusion of polar nuclei	Female	Mitochondrial DnaJ protein
LO2	Maize	Gametogenesis	Nuclear division defects	Both	–

gene of *Arabidopsis* has been proposed to be a transcriptional regulator of sporocyte development.[8] Genes affecting megagametogenesis can act by primarily affecting the ovule surrounding the embryo sac, or by acting directly on the female gametophyte. Mutant screens for genes that act in the female gametophyte and are required for its function (female gametophytic genes) were first carried out in maize and more recently in *Arabidopsis* using transposon and T-DNA tagging.[9] The strategies have been based on a combination of reduced seed set and segregation distortion of a dominant antibiotic resistance marker present on the DNA used as the insertional mutagen. The rationale is that if a plant is heterozygous for a mutation in a gene that has an essential function in the female gametophyte at the haploid stage of development after meiosis, then half the ovules (those in which the female gametophyte carries the mutant allele) would be sterile and the mutant allele would be transmitted only through the pollen. Further, if the mutation is due to insertion of DNA carrying a dominant antibiotic marker such as Kanamycin resistance, then the seeds from the plant would give a Kanamycin-resistant–Kanamycin-sensitive segregation ratio that departs from the normal 3:1 and is closer to 1:1. The use of differential cDNA screening has also proved to be a successful approach in isolating embryo sac specific genes.[10] These approaches are beginning to reveal the molecular players behind the developmental events at different stages in the pathway of female gametophyte development (Table 1).

CONCLUSION

The pathway of female gametogenesis provides a well-defined system to address basic issues in plant development in the context of a small number of specialized cells that make up the female gametophyte. Questions whose answers are the goals of future research include the following:

1. How are the megasporocyte and, later, the functional megaspore specified?
2. What is the mechanism underlying the control of programmed cell death in the remaining megaspores?
3. What is the basis of polarity and the nature of positional information in the developing embryo sac that leads genetically identical nuclei to adopt distinct cellular identities?
4. What are the cell cycle controls that trigger embryo and endosperm development after fertilization and how are these altered in apomictic plants?

The understanding of apomixis in relation to female gametogenesis is also of considerable practical importance with respect to its potential application in plant breeding.

ARTICLES OF FURTHER INTEREST

Cytogenetics of Apomixis, p. 383
Genomic Imprinting in Plants, p. 527

REFERENCES

1. Willemse, M.T.M.; van Went, J.L. The Female Gametophyte. In *Embryology of Angiosperms*; Johri, B.M., Ed.; Springer-Verlag: New York, 1984; 159–196.
2. Agashe, B.A.; Prasad, C.K.; Siddiqi, I. Identification and analysis of DYAD: A gene required for meiotic chromosome organization and female meiotic progression in *Arabidopsis*. Development **2002**, *129*, 3935–3943.
3. Grossniklaus, U.; Spillane, C.; Page, D.R.; Köhler, C. Genomic Imprinting and seed development: Endopserm formation with and without sex. Curr. Opin. Plant Biol. **2001**, *4*, 21–27.
4. Ray, A. New paradigms in plant embryogenesis: Maternal control comes in different flavors. Trends Plant Sci. **1998**, *3*, 325–327.
5. Huang, B.Q.; Russell, S.D. Fertilization in *Nicotiana tabacum*: Cytoskeletal modifications in the embryo sac during synergid degeneration. Planta **1994**, *194*, 200–214.
6. Higashiyama, T.; Yabe, S.; Sasaki, N.; Nishimura, Y.; Miyagishima, S.; Kuroiwa, H.; Kuroiwa, T. Pollen tube attraction by the synergid cell. Science **2001**, *293* (5534), 1480–1483.
7. Sheridan, W.F.; Avalkina, N.A.; Shamrov, I.I.; Batygina, T.B.; Golubovskaya, I.N. The *mac1* gene: Controlling the commitment to the meiotic pathway in maize. Genetics **1996**, *142*, 1009–1020.
8. Yang, W.C.; Ye, D.; Xu, J.; Sundaresan, V. The *SPOROCYTELESS* gene of Arabidopsis is required for initiation of sporogenesis and encodes a novel nuclear protein. Genes Dev. **1999**, *13*, 2108–2117.
9. Yang, W.-C.; Sundaresan, V. Genetics of gametophyte biogenesis in *Arabidopsis*. Curr. Opin. Plant Biol. **2000**, *3*, 53–57.
10. Cordts, S.; Bantin, J.; Wittich, P.E.; Kranz, E.; Lorz, H.; Dresselhaus, T. ZmES genes encode peptides with structural homology to defensins and are specifically expressed in the female gametophyte of maize. Plant J. **2001**, *25* (1), 103–114.

Fire Blight

John L. Norelli
USDA—ARS, Appalachian Fruit Research Station,
Kearneysville, West Virginia, U.S.A.

INTRODUCTION

Fire blight, caused by the bacterium *Erwinia amylovora* (Burr.) Winslow et al. is a destructive disease of apple, pear, and woody ornamentals. The disease has been reported to occur on approximately 130 species of plants, all of which are in the rose family (*Rosaceae*). The name "fire blight" was derived from the characteristic blackening of vegetative tissue caused by the disease, often making trees appear as if they are burnt.

DISCUSSION

Fire blight is indigenous to North America. It was first reported in 1780 on pear and quince in the Hudson Valley of New York. Research to determine the causal agent of fire blight conducted by Thomas J. Burrill and J.C. Arthur between 1877 and 1886 played an important role in acceptance of bacteria as plant pathogens.[1] The disease currently occurs throughout North America, in most of Europe, in the Middle East, and in New Zealand. Asian pear blight (also known as bacterial shoot blight of pear), caused by the closely related pathogen *E. pyrifoliae*, is a fire blight-like disease that occurs in Korea and Japan.[2]

Since the mid-19th century, fire blight has caused serious economic losses for apple and pear growers. In 1914 in Illinois, fire blight caused losses of apples and pears valued at $1.5 million.[3] From 1900 to 1910 the loss of pear trees resulting from the introduction of the bacterium to California was catastrophic; the U.S. Census indicated a 28% decline in total pear trees in California during this period.[1] In 2000, a fire blight epidemic in Michigan caused the death of approximately 350,000 trees resulting in a total economic loss of $42 million.[4] These examples illustrate why fire blight is one of the plant diseases most feared by fruit growers.

SYMPTOMS

E. amylovora can infect blossoms, stems, immature fruits, woody branches, tree trunks, and root crowns. Infected blossoms and peduncles first appear water-soaked or gray-green, then shrivel and turn brownish to black (Fig. 1A). Droplets of ooze (plant sap and *E. amylovora*) sometimes exude from the blossom peduncle. Similarly on immature fruit, symptomatic tissues first appear water-soaked and later become brown to black, with droplets of ooze frequently present (Fig. 1B). Symptoms on shoots generally appear first in young leaves with a blackening of the petiole and leaf midrib. Blackening will spread through the secondary veins, and the entire leaf blade will become necrotic. Blighted shoots often wilt and form a shepherd's crook (Fig. 1C). Stem tissue will shrivel and blacken. Droplets of ooze sometimes exude from stem or leaf petioles (Fig. 1D). In addition, aerial bacterial strands that can be dispersed by wind can form on stems. Necrosis can spread downward from infected blossoms, shoots, and fruits through the current season's growth and into older wood. Infected wood will appear shriveled or sunken and eventually turn dark brown to black. Abundant ooze can sometimes flow along the bark of infected branches, trunk, or the rootstock shank (Fig. 1E). In general, symptoms on apple appear brown and those on pear appear black.

In addition to Asian pear blight, other disorders with fire blight-like symptoms include pear blast, caused by *Pseudomonas syringae* pv. *syringae*; *Nectria* twig blight; European pear dieback, caused by *Phomopsis tanakae*; and damage resulting from the twig borer beetle (*Polyycaon confertus*).

DISEASE EPIDEMIOLOGY

E. amylovora most commonly overwinters in fire blight cankers (infected woody tissue from the previous season). In spring, bacteria multiply at the margin between infected and healthy tissue. Usually only a small proportion of the cankers in an orchard will produce inoculum. Although the production of ooze is common, *E. amylovora* can be isolated from active cankers in the absence of ooze. In addition to cankers, bacteria have been reported to overwinter in dormant buds and in internal nonsymptomatic plant tissue (endophytic bacteria). Although contaminated planting material is an important factor in the

Encyclopedia of Plant and Crop Science
DOI: 10.1081/E-EPCS 120010580

Fig. 1 Fire blight symptoms on apple and pear. A: infected pear blossoms (photo courtesy of Dr. Tom van der Zwet, USDA, ARS, AFRS, Kearneysville, WV); B: infected immature apple fruits with bacterial ooze (photo courtesy of Dr. Alan Jones, Michigan State Univ., East Lansing); C: blighted pear shoot with typical form of shepherd's crook (photo courtesy of Dr. Ken Hickey, The Pennsylvania State University, Biglerville); D: bacterial ooze exuding from infected apple shoot (photo courtesy of Dr. Alan Jones); and E: infected apple rootstock crown with abundant ooze. (*View this art in color at www.dekker.com.*)

introduction of the pathogen to new areas, infested buds and endophytic bacteria are not considered important sources of primary inoculum in orchards with a prior history of fire blight (Fig. 2).

During bloom, bacteria can be transferred from active cankers to blossoms by water (rain, irrigation, or dilute chemical sprays) or by insects (ants and flies). Bees are important in spreading bacteria among blossoms. In blossoms, *E. amylovora* can multiply on the surface of the stigma without causing disease.[5] Large epiphytic populations can develop when warm temperatures occur during bloom, enhancing the likelihood of blossom infection. Blossom infection usually occurs when bacteria are washed by rain to nectaries located in the hypanthium (floral cup), although other flower parts can also be invaded.[2,5]

Much less is known about the process by which shoots become infected by *E. amylovora*. Shoot infection

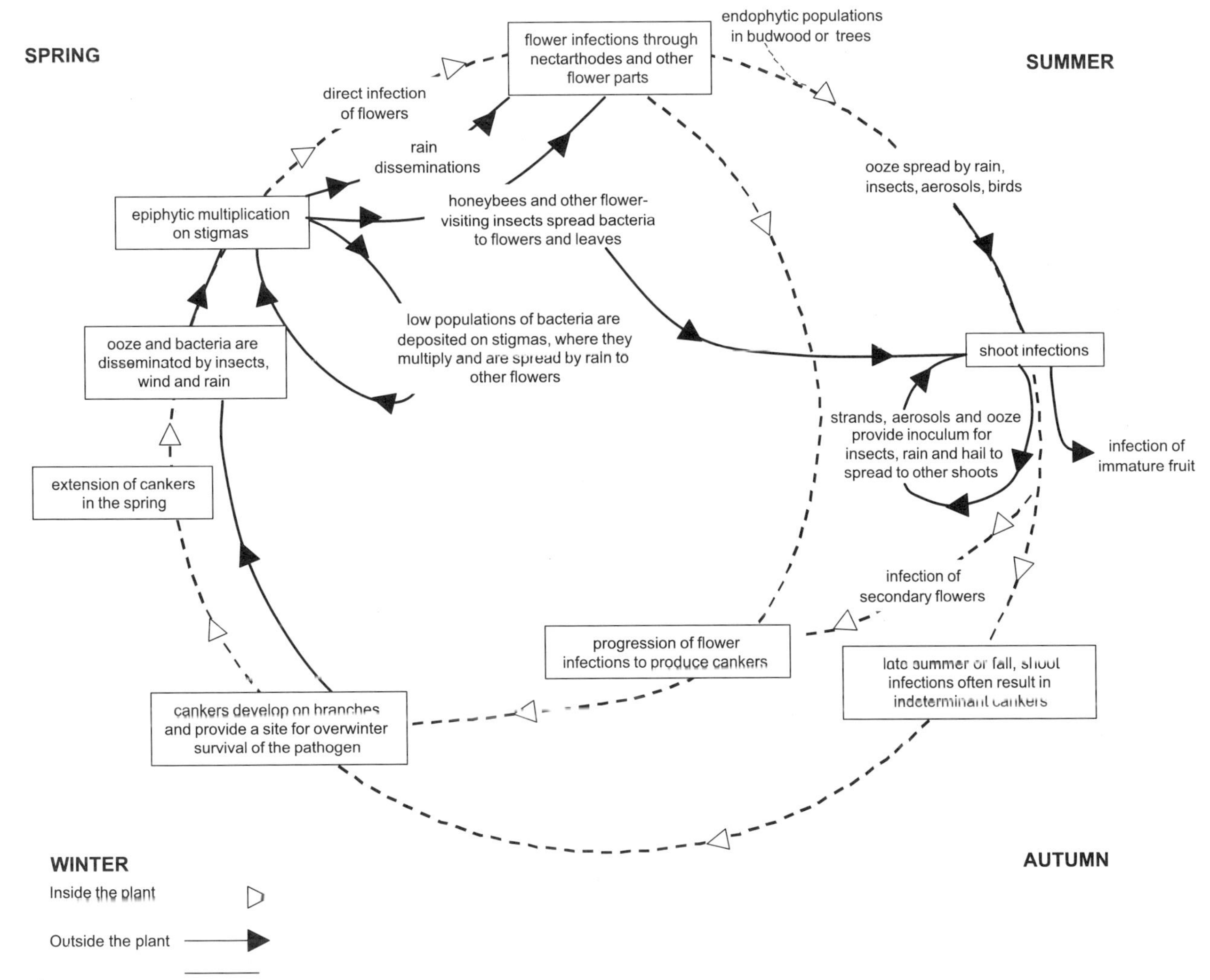

Fig. 2 Disease cycle of fire blight. (From Ref. 5.)

can occur following injury caused by hail or high winds if inoculum is present. *E. amylovora* can also move systemically from blossom infections to proximal shoots to initiate infection. Endophytic *E. amylovora* can also be isolated from more distal symptomless shoots; however, the role of endophytic bacteria in initiating new shoot infections is poorly understood. In addition, it is thought that insect feeding in the presence of inoculum can result in shoot infection.[6] Although almost 100 species of insects have been reported to be associated with the dissemination of fire blight,[2] there is a lack of information demonstrating the importance of specific species in the shoot-blight phase of the disease.

Recently, fire blight of apple rootstocks has become a serious economic problem in high-density orchards. Most apple growing regions have adopted the use of high-density orchard systems that depend on dwarfing rootstocks to control tree size. The commonly used dwarfing rootstocks, Malling (M.) 9 and M.26, are highly susceptible to *E. amylovora,* and infection usually kills trees by girdling the rootstock. Several avenues of rootstock infection have been demonstrated, including infection of rootstock suckers (vegetative shoots developing from the rootstock), internal spread of endophytic bacteria, or direct infection of the rootstock through discontinuities in the bark caused by growth or various injuries.

Several terms refer to the various phases of fire blight. Blossom, shoot and rootstock infections are referred to as "blossom blight," "shoot blight," and "rootstock blight," respectively. "Trauma blight" describes fire blight infections due to wounding by hail or high wind. "Canker blight" refers to the renewed activity of overwintering cankers that can result in oozing from bark and infection of new shoot growth.[7]

DISEASE MANAGEMENT

Despite the destructive potential of fire blight, it is sporadic in its occurrence. The disease can cause little damage in an area for several years, followed by severe epidemics once environmental conditions are favorable. Effective management of fire blight requires the integration of several practices aimed at reducing the amount of initial inoculum, imposing barriers to successful establishment of the pathogen, and reducing host susceptibility to infection.

Removal of cankers during winter pruning is a critical component of fire blight management. Management strategies also include the application of early-season copper sprays to reduce primary inoculum in the orchard and pruning out early season infections after bloom to reduce inoculum available for shoot infection.[6]

Preventing blossom infection is also a critical component of fire blight management. Antibiotics are effective in preventing blossom infection, but must be applied precisely to coincide with fire blight infection periods. Several models including the Thomson mean temperature line, Maryblyt™, Cougar Blight, and the Billing Integrated System can predict the occurrence of blossom infection periods.[8] In general, antibiotic applications are recommended during bloom when weather has been relatively warm and there is a high probability of rain and temperatures higher than 18°C in the next 24 hours. Although blossom blight control has relied heavily on antibiotic sprays, significant progress has been made in the biological control of blossom infection, and commercial products are currently available.[9] In general, biological control agents should be applied during early bloom so they can become established prior to the occurrence of infection periods. Antibiotics are not recommended for the control of shoot infection due to their limited efficacy and to reduce the chance of antibiotic resistance developing within *E. amylovora* populations. However, antibiotic application is recommended following a hailstorm in orchards with fire blight infections.

Management practices that minimize excessive vegetative growth will reduce host susceptibility to *E. amylovora* infection. Gibberellin biosynthesis inhibitors that reduce shoot growth also reduce fire blight intensity.[10] Excessive applications of nitrogen, including barnyard manure, should be avoided. Serious fire blight damage can usually be avoided by planting scion varieties and rootstocks with resistance. Unfortunately, current markets and production economics are encouraging the fruit industry to move in exactly the opposite direction.

CONCLUSION

After more than a century of research, a great deal is known about fire blight and its causal agent, *E. amylovora*, yet the disease remains a serious threat to the pome fruit industry. The ability of *E. amylovora* to infect different host tissues and the limited number of effective management practices make it difficult to stop or slow the progress of epidemics. In addition, the sporadic nature of the disease encourages growers to become lax in implementing costly control practices after long periods without serious fire blight outbreaks.

Due to the widespread use of streptomycin to control fire blight, streptomycin-resistant *E. amylovora* strains are now common in several regions of the United States.[11] Finding alternative control agents to replace (or supplement) streptomycin is an important goal for future research. Currently, no effective cultural practices or chemical treatments are available to control the rootstock phase of fire blight. Development of apple rootstocks that combine desirable pomological characteristics with resistance to *E. amylovora* has the potential to provide practical control for rootstock blight in the future. Chemical treatments that enhance host resistance,[10] biological control,[9] and cultivars with genetically engineered resistance[12] hold great promise for the future control of fire blight.

ARTICLE OF FURTHER INTEREST

Management of Bacterial Diseases of Plants: Regulatory Aspects, p. 669

REFERENCES

1. Kim, W.-S.; Hildebrand, M.; Jock, S.; Geider, K. Molecular comparison of pathogenic bacteria from pear trees in Japan and the fire blight pathogen *Erwinia amylovora*. Microbiology (Reading) **2001**, *147*, 2951–2959.
2. van der Zwet, T.; Keil, H.L. *Fire Blight—A Bacterial Disease of Rosaceous Plants*; Agric. Hdbk., US Dept. Agric.: Washington, DC, 1979; Vol. 510, 1–200.
3. Baker, K.F. Fire blight of pome fruits, the genesis of the concept that bacteria can be pathogenic to plants. Hilgardia **1971**, *40* (18), 603–633.
4. http://www.msue.msu.edu/vanburen/fb2000.htm (accessed July 2002).

5. Thomson, S.V. Epidemiology of Fire Blight. In *Fire Blight, the Disease and Its Causative Agent, Erwinia Amylovora*; Vanneste, J.L., Ed.; CABI Publishing: Wallingford, UK, 2000; 9–36. http://www.cabi-publishing.org/bookshop/readingroom/0851992943/2943ch2.pdf.
6. van der Zwet, T.; Beer, S.V. *Fire Blight—Its Nature, Prevention and Control; A Practical Guide to Integrated Disease Management*; Agric. Information Bulletin, U.S. Dept. Agric.: Washington, DC, 1995; Vol. 631, 1–91.
7. Steiner, P.W. Integrated Orchard and Nursery Management for the Control of Fire Blight. In *Fire Blight, the Disease and Its Causative Agent, Erwinia Amylovora*; Vanneste, J.L., Ed.; CABI Publishing: Wallingford, UK, 2000; 339–358. http://www.cabi-publishing.org/bookshop/readingroom/0851992943/2943ch17.pdf.
8. Billing, E. Fire Blight Risk Assessment Systems and Models. In *Fire Blight, the Disease and Its Causative Agent, Erwinia Amylovora*; Vanneste, J.L., Ed.; CABI Publishing: Wallingford, UK, 2000; 293–318. http://www.cabi-publishing.org/bookshop/readingroom/0851992943/2943ch15.pdf.
9. Johnson, K.B.; Stockwell, V.O. Biological Control of Fire Blight. In *Fire Blight, the Disease and Its Causative Agent, Erwinia Amylovora*; Vanneste, J.L., Ed.; CABI Publishing: Wallingford, UK, 2000; 319–337.
10. Yoder, K.S.; Miller, S.S.; Byers, R.E. Suppression of fireblight in apple shoots by prohexadione-calcium following experimental and natural inoculation. HortScience **1999**, *34* (7), 1202–1204.
11. Jones, A.L.; Schnabel, E.L. The Development of Streptomycin-Resistant Strains of *Erwinia Amylovora*. In *Fire Blight, the Disease and Its Causative Agent, Erwinia Amylovora*; Vanneste, J.L., Ed.; CABI Publishing: Wallingford, UK, 2000; 235–251.
12. Aldwinckle, H.S.; Norelli, J.L. Transgenic Pomaceous Fruit with Fire Blight Resistance. U.S. Patent 6,100,453, August 8, 2000.

F

Fixed-Nitrogen Deficiency: Overcoming by Nodulation

Michelle R. Lum
Ann M. Hirsch
University of California, Los Angeles, California, U.S.A.

INTRODUCTION

One of the most studied symbioses is that between plants and bacteria of the gram-negative Rhizobiaceae (*Rhizobium, Azorhizobium, Sinorhizobium, Bradyrhizobium*, and *Mesorhizobium*). This association is restricted to the family Fabaceae, with the exception of *Parasponia* (Ulmaceae). The association culminates in the formation on plant roots of a novel organ, the nodule, that houses the bacteria in a protected and specialized environment. The bacteria provide fixed nitrogen to the plant in exchange for carbohydrates. Gram-positive, filamentous, actinomycetes (Frankiaceae) also engage in a symbiotic nitrogen association, but with actinorhizal plants, a diverse group of dicotyledons represented by eight different families. Nodules are again the key plant structure formed in response to the symbiosis. The benefits of these nitrogen-fixing symbioses have long been recognized. It is known that the Greeks engaged in agricultural practices utilizing leguminous plants as early as the 5th century B.C.

NITROGEN

Nitrogen is required for amino acid, nucleotide, protein, and nucleic acid production, necessary for the fundamental processes governing plant growth and development. However, soils become readily depleted of available nitrogen, particularly if nitrogen-containing plant parts are not plowed back into the field. Over 70% of the atmosphere is made up of molecular nitrogen (N_2). However, the triple bond holding the two nitrogen atoms together results in a very stable molecule. Only some prokaryotes—those that synthesize nitrogenase—fix N_2 and reduce it to ammonia (NH_3), which is then converted to nitrate (NO_3^-), a form used by plants. Most nitrogen in the fertilizer supplied to crop plants is in the form of NH_3, produced by the energy-consuming Haber-Bosch process. However, the high cost associated with producing nitrogen fertilizer and the problems associated with its application (such as nitrification of the soil and runoff into water supplies) make biological nitrogen fixation significant, particularly in countries where fertilizers are not subsidized.

THE *Rhizobium*-LEGUME SYMBIOSIS

The process leading to the development of the nodule involves constant communication between the partners. Under nitrogen-limiting conditions, leguminous plants secrete flavonoids—phenolic compounds composed of two benzene rings linked by a heterocylic pyran or pyrone ring. There are more than 4000 flavonoids, differentiated by the number of hydroxyl substituents, methyl groups, sugars, and other substituents (Fig. 1A).[1] The release of flavonoids causes *Rhizobium* in the rhizosphere to be attracted to the plant roots. Specific flavonoids induce rhizobial *nod* genes, resulting in the production of Nod factor—monoacylated chains of 3–5 β-1,4 linked, N-acetyl glucosamines with various chemical modifications (Fig. 1B). Nod factor is the key signaling molecule produced by *Rhizobium* because it is absolutely required for progression of the symbiosis.[2] Specific rhizobia attach to the plant root, causing root hair deformation, root hair curling, and 360° curls (shepherd's crooks). In addition, plant early nodulin (*ENOD*) genes are induced.[3] An infection thread, through which the bacteria enter the root hair, is formed from the invagination of the plant plasma membrane accompanied by the deposition of cell wall material. Either inner or outer cortical cells of the root begin to divide and eventually the nodule develops (Fig. 2A, Figs. 3A and B). The infection thread elongates and branches within the newly divided cells, and the bacteria become endocytosed into infection droplets. In the differentiated parts of the nodule, the rhizobia develop into bacteroids and fix atmospheric nitrogen. The peribacteroid membrane (PBM) always separates the bacteria from the cytoplasm of the plant cell and is the location at which nutrient exchange occurs. The bacteroid, the PBM, and the interface between are referred to as the symbiosome (Fig. 2C).

THE *Frankia*-ACTINORHIZAL SYMBIOSIS

In contrast to legume nodules, in which the central infected tissue is surrounded by nodule parenchyma and peripheral vascular bundles (Fig. 2A), actinorhizal nodules resemble lateral roots, having a central vascular bundle. The peripheral infected tissue is surrounded by cortical

Encyclopedia of Plant and Crop Science
DOI: 10.1081/E-EPCS 120010654

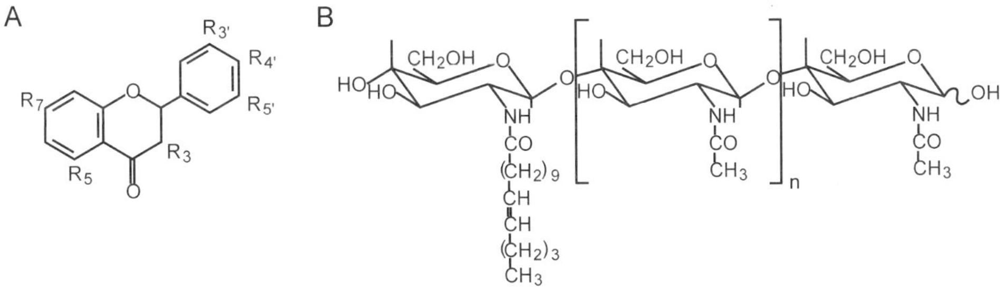

Fig. 1 Signal molecules involved in the *Rhizobium*-legume symbiosis. (A) Generic structure of flavonoids. R refers to the various substituents. (B) Generic structure of Nod factor; n = number of glucosamine residues in the backbone.

nodule parenchyma (Figs. 2B, 3C). Cell divisions that initiate the nodule primordium take place in the pericycle instead of the cortex. Root hair deformation and infection thread formation also characterize the *Frankia*-actinorhizal symbiosis; however, the early signals involved are not well understood, although flavonoids may be involved. In the nitrogen-fixing zone of the nodule, filament branches differentiate into specialized structures known as vesicles that express nitrogenase and fix nitrogen (Fig. 2D).[4] To date, *Frankia* genes involved in establishing the nodule have not been identified.

VARIATIONS IN THE PROGRESSION OF THE SYMBIOSES

Entrance of the bacteria in both symbioses can occur by intercellular infection through gaps in the epidermis where

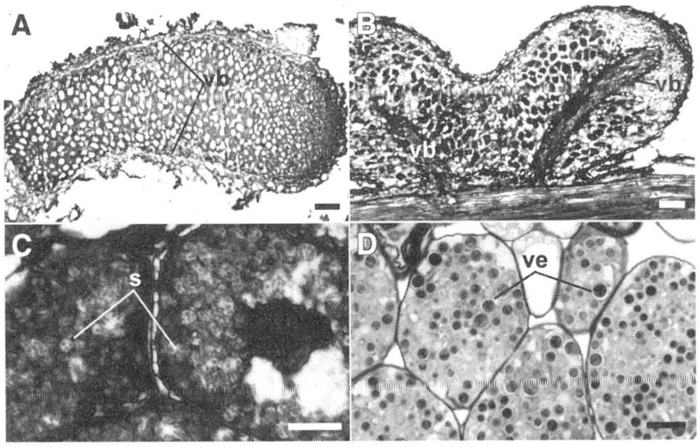

Fig. 2 Symbiotic structures. (A) Section of an indeterminate nodule of *Melilotus alba*; vb = vascular bundle. (B) Section of an actinorhizal nodule of *Alnus*; vb = vascular bundle. (C) Symbiosomes within cells of a *Pisum sativum* nodule; s = symbiosome. (D) Nitrogen-fixing vesicles of *Frankia* bacteria in an *Elaeagnus* nodule; ve = vesicle. Scale bars (A, B) represent 100 μM; scale bars (C, D) represent 10 μM. (*View this art in color at www.dekker.com.*)

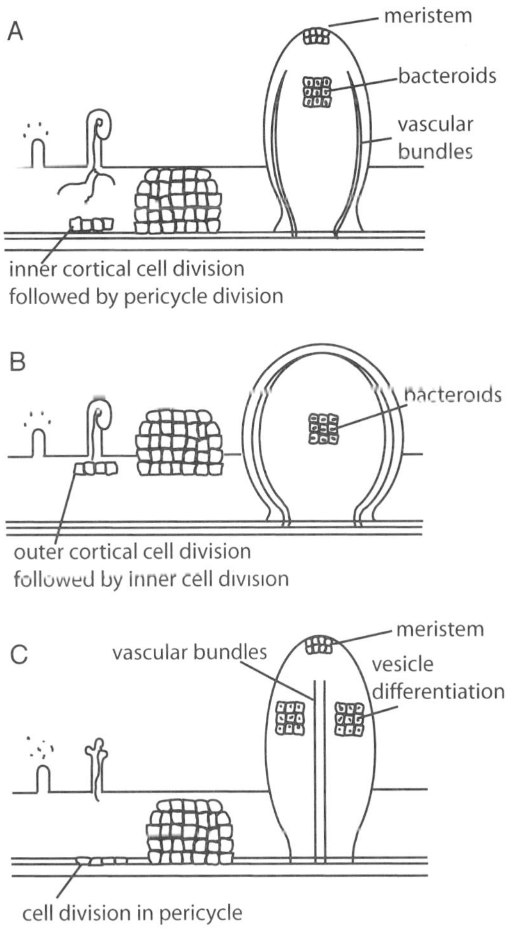

Fig. 3 Schematic of different types of nodules. (A) Indeterminate nodule in which the infected central tissue is surrounded by the nodule parenchyma and peripheral vascular bundles. (B) Determinate nodule in which cell divisions initiate in the outer cortex. (C) Actinorhizal nodule in which the central vascular bundle is flanked by infected cells and cell divisions originate from the pericycle. (Based in part on Ref. 5.)

lateral or adventitious roots emerge from roots or stems, or by penetration of the middle lamella of intact cells. *Parasponia*, the only nonlegume known to associate with *Rhizobium*, develops nodules that resemble modified lateral roots. Nodule morphology depends on the host. For example, indeterminate nodules are produced in *Medicago*, *Melilotus*, *Pisum*, and *Trifolium*, where initial cell divisions in the inner cortex (usually opposite the protoxylem pole) result in the formation of the nodule primordium (Fig. 3A). Because of a persistent nodule meristem, nodules are usually elongated. In contrast, *Phaseolus*, *Glycine*, and *Lotus* produce determinate nodules. Cell divisions initiate in the outer cortex, followed by division of the inner cortical cells. The nodule meristem stops dividing early in nodule development, resulting in a spherical nodule (Fig. 3B).

METABOLISM

Nitrogenase is highly sensitive to oxygen, and the nodule provides a specialized environment for rhizobia. Oxygen is maintained at low levels in the cell, partially due to nodule structure and to leghemoglobin, which acts as an oxygen scavenger. In contrast, *Frankia* bacteria provide oxygen protection for themselves via the lipid envelope that surrounds the nitrogen-fixing vesicles. In addition, the high rate of respiration in infected cells and the action of hemoglobin may limit the presence of oxygen.[6]

The peribacteroid membrane that surrounds the bacteroids is the site of nutrient exchange in the rhizobia-legume symbiosis. Channels have been identified that catalyze the transport of fixed nitrogen to the plant, which is then fed into the glutamine/glutamate synthase cycle. Nitrogen is subsequently transferred to the rest of the plant through the vascular system. Carbohydrates are imported to bacteroids as dicarboxylic acids by active transporters found in the bacteroid membrane.[7]

The details of nitrogen and carbon metabolism and transport in the actinorhizal symbiosis are not as well defined. However, in both symbioses, carbon is critical for energy and reducing power. Globally, the nitrogen contribution by the *Rhizobium*-legume symbiosis is an estimated 24–584 kg N ha^{-1} y^{-1}, whereas the *Frankia*-actinorhizal symbiosis contributes 2–362 kg N ha^{-1} y^{-1}.[8]

EVOLUTIONARY ASPECTS

All the plant species that engage in nitrogen-fixing nodule symbioses are members of the Rosid I clade. Thus evolutionarily, there appears to be a single origin for the plant's susceptibility to entrance by bacteria.[9] However, nodulation may have multiple origins of evolution, as indicated by varying nodule morphology and infection methods used by different hosts. Within the Rosid I clade, there are four major lineages of nodulating plants:[10] One encompasses legumes and the others consist of nonlegumes, including the actinorhizal families Casuarinaceae and Myricaceae, as well as the rhizobia-nodulated *Parasponia*. Within the three subfamilies of the Fabaceae—the Papilionoideae, Mimosoideae, and Caesalpinioideae—over 80% of the first two groups nodulate, whereas less that 25% nodulate in the more basal Caesalpinioideae. Based on phylogenetic analysis of *rbcL*, nodulation may have had three evolutionary origins: at the base of the papilionoids, prior to the evolution of the mimosoids, and in the branch leading to *Chamaecrista*, one of the few caesalpinioid genera that nodulates.[9]

CONCLUSION

Nitrogen is critical for plant growth and development, and some plants have the advantage of engaging in a symbiotic association with nitrogen-fixing bacteria. Interaction with rhizobia is primarily restricted to legumes, but *Frankia* bacteria associate with different dicotyledonous families. Both types of symbioses culminate in the formation of a novel plant organ, the nodule. Although nodule structure and the form in which the bacteria fix nitrogen varies between the symbioses, an evolutionary predisposition for nodulation seems likely. Advances made in determining the common mechanisms involved in both associations will further our understanding of biological nitrogen fixation.

ARTICLES OF FURTHER INTEREST

Bacteria-Plant Host Specificity Determinants, p. 119
Cropping Systems: Rain-Fed Maize-Soybean Rotations of North America, p. 358
Mycorrhizal Evolution, p. 767
Mycorrhizal Symbioses, p. 770
Nitrogen, p. 822
Phosphorus, p. 872
Rhizosphere: An Overview, p. 1084
Rhizosphere: Microbial Populations, p. 1090
Rhizosphere: Nutrient Movement and Availability, p. 1094
Symbioses with Rhizobia and Mycorrhizal Fungi: Microbe/Plant Interactions and Signal Exchange, p. 1213
Symbiotic Nitrogen Fixation, p. 1218
Symbiotic Nitrogen Fixation: Plant Nutrition, p. 1222

Symbiotic Nitrogen Fixation: Special Roles of Micronutrients, p. 1226

REFERENCES

1. Harborne, J.B.; Williams, C.A. Advances in flavonoid research since 1992. Phytochemistry **2000**, *55*, 481–504.
2. Long, S.R. *Rhizobium* symbiosis: Nod factors in perspective. Plant Cell **1996**, *8*, 1885–1898.
3. Crespi, M.; Galvez, S. Molecular mechanisms in root nodule development. J. Plant Growth Regul. **2000**, *19*, 155–166.
4. Wall, L.G. The actinorhizal symbiosis. J. Plant Growth Regul. **2000**, *19*, 167–182.
5. Akkermans, A.D.I.; Hirsch, A.M. A reconsideration of terminology in *Frankia* research: A need for congruence. Physiol. Plant. **1997**, *99*, 574–578.
6. Huss-Danell, K. Actinorhizal symbioses and their N_2 fixation. New Phytol. **1997**, *136*, 375–405.
7. Luyten, E.; Vanderleyden, J. Survey of genes identified in *Sinorhizobium meliloti* spp., necessary for the development of an efficient symbiosis. Eur. J. Soil Biol. **2000**, *36*, 1–26.
8. Shantharam, S.; Mattoo, A.K. Enhancing biological nitrogen fixation: An appraisal of current and alternative technologies for N input into plants. Plant Soil **1997**, *194*, 205–216.
9. Doyle, J.J. Phylogenetic perspectives on nodulation: Evolving views of plants and symbiotic bacteria. Trends Plant Sci. **1998**, *3*, 473–478.
10. Soltis, D.E.; Soltis, P.S.; Morgan, D.R.; Swensen, S.M.; Mullin, B.C.; Dowd, J.M.; Martin, P.G. Chloroplast gene sequence data suggest a single origin of the predisposition for symbiotic nitrogen-fixation in angiosperms. Proc. Natl. Acad. Sci. U.S.A. **1995**, *92*, 2647–2651.

F

Floral Induction

Joseph Colasanti
University of Guelph, Guelph, Ontario, Canada

INTRODUCTION

The switch from vegetative to reproductive growth signifies an important change in the life cycle of higher plants. In terms of agricultural production, coordinated timing of the transition to flowering is essential for optimal fruit set and seed production. Numerous physiological studies based on specific floral inductive treatments have defined key features of the transition event. Yet, despite the importance of understanding flowering time and the large amount of research devoted to its study, there is much to be learned about the genetic and biochemical signals that underlie the transition to reproductive growth. Recent molecular genetic approaches are yielding insights about the genes that control floral induction.

THE NATURE OF THE FLORAL INDUCTION EVENT

By observing plants growing throughout a season cycle it seems obvious that flowering time is synchronized with environmental cues. Most plants flower at a particular time in the season, often in the spring or summer in temperate climates, in order to produce the next generation of progeny before harsh winter conditions ensue. It was assumed that the reason for this seasonal timing of flower formation was due to an increase in photoassimilate production resulting from longer days. However, in the early 1920s Garner and Allard first reported a genetic connection between flower induction and environmental cues, and suggested the causes of flowering were more complex than the buildup of resources to a critical level.[1] They discovered a mutant of tobacco, Maryland Mammoth, that flowers only when plants are exposed to short days. The discovery of this phenomenon, which they called photoperiodism, prompted physiologists to classify plants as either short-day plants (SDPs), long-day plants (LDPs), or day-neutral plants (DNPs). SDPs and LDPs require a critical length of night or day, respectively, to induce flowering, whereas DNPs flower regardless of the photoperiod to which they are exposed. Plants that do not flower unless they receive an inductive photoperiod are termed obligate SDPs or LDPs. Most plants are facultative SDPs or LDPs; that is, flowering time is accelerated by inductive photoperiods, but even in the absence of such treatment they will flower eventually.

The identification of photoperiod-responsive plants provided physiologists with a tool that allowed them to apply precise inductive stimuli and then evaluate the flowering response in a quantitative way. Numerous experiments based on this premise were carried out with both SDPs and LDPs in an attempt to determine the nature of the floral stimulus. The ability to graft parts of induced plants to uninduced plants, and to apply specific photoperiod treatments to individual leaves, was particularly useful in dissecting the mechanisms of the flowering process (Fig. 1). Based on these compelling studies, several important features of floral induction were elucidated. First, it was confirmed that the shoot apex is the target of the floral stimulus and, for floral evocation to occur, the shoot apex must become competent to receive the floral stimulus. Another significant finding is that the floral stimulus originates in leaves and is transported to the shoot apex. Grafting experiments indicated that the floral stimulus had properties of a substance with a measurable mobility that is transported only through living tissue. Therefore the phloem—in addition to its role of transporting photoassimilates from source to sink tissues—also acts as a conduit for flower-inducing signals. Given these characteristics, and in light of the discovery that plant-growth regulators such as gibberellins and auxin are relatively simple chemical compounds, attempts were made to isolate this putative flower-inducing compound, sometimes referred to as "florigen." Although these studies generated a considerable body of data supporting the mechanism of action of the floral induction process, the biochemical nature of the floral stimulus remained elusive.[2]

MUTATIONS THAT AFFECT FLORAL INDUCTION

While the search for the floral stimulus continued, studies with garden pea investigated the genetic basis of floral induction. The ability to graft peas, and the availability of a variety of mutants defective in various aspects of the transition to flowering, reinforced the findings of

Encyclopedia of Plant and Crop Science
DOI: 10.1081/E-EPCS 120012909

physiological studies that a leaf-generated floral stimulus moved to the shoot apex.[3] Studies of pea mutants with altered flowering times laid the groundwork for molecular genetic studies in other species. In particular, many mutations that affect flowering time have been characterized in the small crucifer *Arabidopsis thaliana*. Genetic analysis of these mutants has allowed the construction of a model of floral induction that incorporates previous findings of physiologists. The Arabidopsis model divides the floral induction process into two basic components: the endogenous pathway and the environmental pathway (Fig. 2). Endogenous signals include properties intrinsic to plant growth such as size, leaf number, and age. DNPs probably utilize these factors when assessing the appropriate time to flower. Environmental signals such as photoperiod or temperature play a critical role in informing the plant that the correct season for flowering has arrived. Genetic studies of Arabidopsis have identified genes that act in the endogenous pathway, the environmental pathway, or both (discussed later).

As with past physiological studies, much of the recent work with Arabidopsis has focused on the genes that act in the environmental pathway. Genes that regulate photoperiod, light perception, and circadian rhythms have been studied extensively.[4] Significant progress has been made in establishing that photoreceptors such as phytochromes and cryptochromes are important links between the photoperiod signals and the ultimate floral inductive response. Temperature perception is another facet of the environmental pathway, and a web of interacting genes that coordinate regulation of the vernalization response is being revealed. Vernalization plays a critical role in signaling the correct season to flower in important crops such as wheat.

Genes that regulate gibberellic acid (GA) biosynthesis are essential components of the endogenous pathway in Arabidopsis and other plants. However, the role GA plays in controlling flowering in many other plant species is uncertain. Because it is not a universal inducer of flowering, and because it is not expressed exclusively in leaves in response to inductive external signals, GA does not possess the traits required of the presumed mobile floral stimulus. More likely, GA acts as an accessory to the floral inductive response in certain species. Overall, the genetic control of the endogenous signaling pathway is less well understood than the environmental pathway.

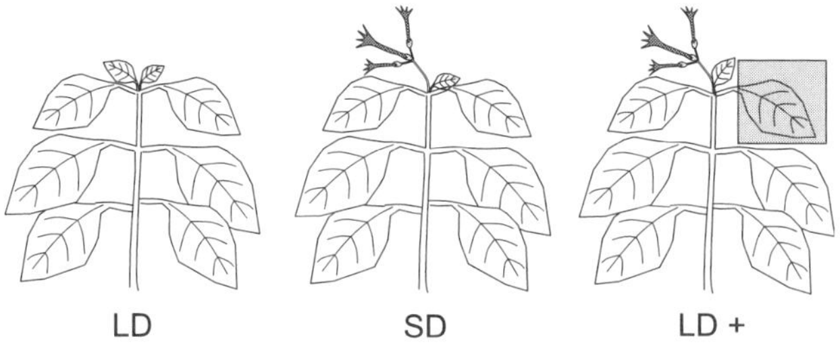

Fig. 1 Floral induction by exposing single leaf to short days. Left: a short-day (SD) plant grown under long-day (LD) conditions continues to grow vegetatively. Center: if given short-day inductive treatments (SD) the plant will flower. Right: an SD plant grown under long days but with one leaf (shaded box) exposed to inductive short days will flower (LD+). (*View this art in color at www.dekker.com.*)

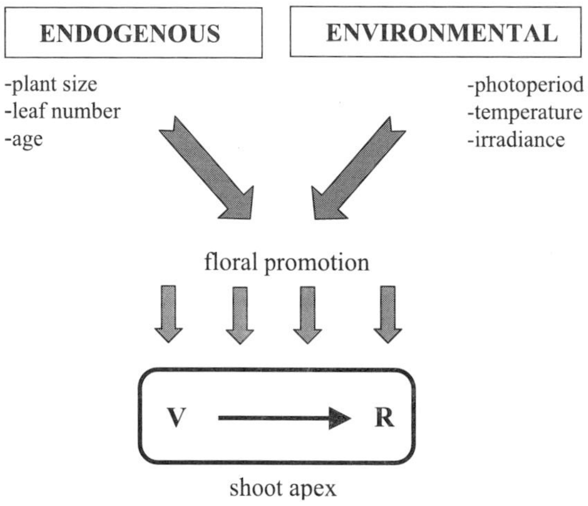

Fig. 2 Floral induction model based on physiological and genetic studies. The boxed area at bottom represents the transition from vegetative (V) to reproductive (R) growth at the shoot apex. (*View this art in color at www.dekker.com.*)

INSIGHTS ABOUT THE MOLECULAR NATURE OF THE FLORAL STIMULUS

Arabidopsis is the first plant to have its genome sequenced and it has become a valuable model for the molecular genetic dissection of the floral induction process.[5,6] Many of the genes that control flowering in Arabidopsis have been isolated and their gene products identified. In some cases the functional equivalents of Arabidopsis flowering-time genes have been found in other plants, including important crop plants such as rice[7] (Table 1). Several Arabidopsis flowering-time genes were found to encode regulatory proteins; that is, they function by controlling the expression of other genes. Some of the first genes isolated in Arabidopsis were found to encode proteins that appear to function as regulators of other genes. For example, *CONSTANS* (*CO*), one of the most intensely studied Arabidopsis genes, encodes a transcription factor that has been found to directly activate the expression of genes involved in the transition to flowering.[6] The *CO* gene plays an important role in transducing

Table 1 Genes from different species that control flowering time

Gene[a]	Species	Flowering phenotype[b]	Possible role of gene product	Pathway	Mechanism of action
CO	Arabidopsis	Late	Transcription	Photoperiod	Perceives circadian rhythms, activates genes
LD	Arabidopsis	Late	Transcription	Endogenous	Activates genes
CRY2	Arabidopsis	Late	Photoreceptor	Photoperiod	Perceives blue light
EMF1	Arabidopsis	Early	Transcription	Endogenous	Unknown
VRN2	Arabidopsis	Late	Transcription/chromatin struct	Vernalization	Perceives prolonged cold periods, activates genes
Hd1	Rice	Early	Transcription	Photoperiod	Perceives photoperiod
Hd6	Rice	Early	Kinase	Photoperiod	Photoperiod perception
id1	Maize	Late	Transcription	Endogenous	Regulates leaf derived mobile floral stimulus

[a]*CO, CONSTANS*; *LD, LUMINEDEPENDENS*; *CRY2, CRYPTOCHROME 2*; *EMF1, EMBRYONIC FLOWER 1*; *VRN2, VERNALIZATION 2*; *Hd1* and *Hd6*, *Heading Date 1* and *6*, respectively; *id1*, *indeterminate 1*.
[b]Phenotypes are for loss of function alleles.
(From Refs. 5 and 10.)

photoperiod signals; thus, loss of *CO* gene function results in mutants that do not flower early under the long day conditions that accelerate flowering in wild type Arabidopsis. The importance of the *CO* gene in controlling flowering may extend to other species as well. Recently a gene in rice that controls flowering time (or *heading date*), *Hd1*, was found to encode a protein with similarity to *CO*. Like CO, *Hd1* acts in a pathway that senses day length, although the exact mechanism of its action is still being determined.

Other genes that may act as gene regulators were found in the vernalization pathway as well. Most recently it was discovered that the Arabidopsis *VRN2* gene encodes proteins that affect gene expression by controlling chromatin structure.[5] Given the widespread changes associated with the transition to flowering, it might be expected that mutations in important regulatory genes, such as transcription factors, might have the greatest effect on the transition to flowering.[8] Discovering the genes controlled by these regulators is a crucial first step in revealing the metabolic and biochemical changes associated with the transition to reproductive growth.

Analysis of Arabidopsis flowering time mutants has generated a complex model that includes both

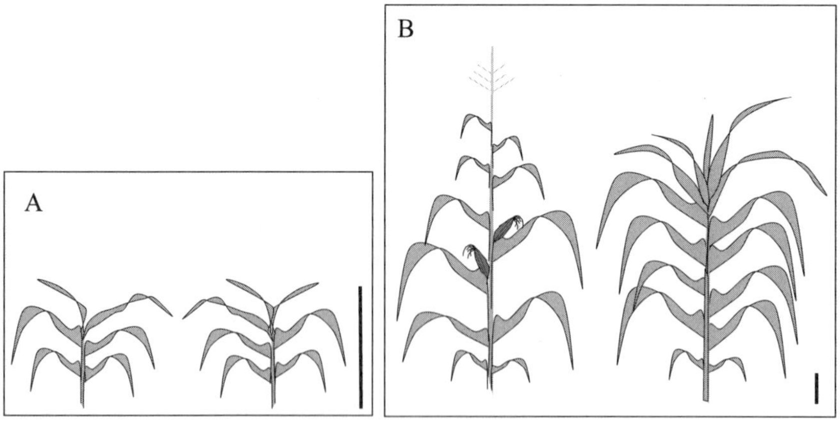

Fig. 3 Comparison of normal and *id1* mutant maize. A. Drawing of maize seedlings at seedling stage (about 3 weeks). Normal plant on the left, *id1* mutant plant on the right. Mutant and normal plants are indistinguishable at this early seedling vegetative stage of growth. B. Mature maize plants (about 12 weeks). A normal plant (left) has made a tassel and ears (male and female inflorescences, respectively) and fertilization has occurred. An *id1* mutant (right) continues to grow vegetatively. (Scale: bar = approx. 30 cm.) (*View this art in color at www.dekker.com.*)

environmental and endogenous elements in the control of floral induction.[5,6] However, so far this model provides no explicit clues about the nature of the floral stimulus. Analysis of a flowering time mutant in maize, *indeterminate1* (*id1*), may lead to a better understanding of the floral stimulus. Maize plants that have lost *id1* gene function remain in a prolonged state of vegetative growth and flower extremely late (Fig. 3). As was found with many Arabidopsis flowering time genes, the *id1* gene encodes a transcriptional regulator.[9] However, unlike the Arabidopsis regulatory genes that control flowering time, the *id1* gene is unique in that it is expressed only in maize leaves, and not at the shoot apex. Therefore the ID1 protein may regulate the expression of genes that mediate the production and/or transmission of a leaf-derived floral stimulus.[10] The maize *id1* gene is also unique in that it does not appear to have a functional equivalent in Arabidopsis. Therefore *id1* may be a flowering time regulator unique to maize, or perhaps grasses in general.

CONCLUSION

Physiological studies have established a solid framework for understanding the floral induction process, and genetic analysis reveals many loci with specific roles in regulating the transition to flowering. Significant progress has been made in identifying genes that are important regulators of the endogenous and environmental signals that bring about the onset of flowering. Further progress will rely on comparisons of normal and nonflowering plants in order to establish the biochemical and metabolic profiles associated with the flowering transition. Understanding how floral inductive signals are produced, perceived, and transmitted could eventually reveal intricate connections between basic plant processes such as general metabolism, nutrient flow, and source/sink relations in plants. The ability to control time to flowering, or to prevent it completely, could have significant agricultural applications.

ARTICLES OF FURTHER INTEREST

Arabidopsis thaliana: Characteristics and Annotation of a Model Genome, p. 47
Arabidopsis Transcription Factors: Genome-Wide Comparative Analysis Among Eukaryotes, p. 51
Biennial Crops, p. 122
Circadian Clock in Plants, p. 278
Crop Domestication in Prehistoric Eastern North America, p. 314
Flower Development, p. 464
In Vitro Flowering, p. 576
Leaf Structure, p. 638
Photoperiodism and the Regulation of Flowering, p. 877
Photoreceptors and Associated Signaling I: Phytochromes, p. 881
Photoreceptors and Associated Signaling II: Cryptochromes, p. 885
Photosynthate Partitioning and Transport, p. 897
Phytochrome and Cryptochrome Functions in Crop Plants, p. 920

REFERENCES

1. Garner, W.W.; Allard, H.A. Effect of the relative effect of day and night and other factors of the environment on growth and reproduction in plants. J. Agric. Res. **1920**, *18*, 553–606.
2. Hillman, W.S. *The Physiology of Flowering*; Biology Studies, Holt, Rinehart and Winston: New York, 1962; 164.
3. Weller, J.L.; Reid, J.B.; Taylor, S.A.; Murfet, I.C. The genetic control of flowering in pea. Trends Plant Sci. **1997**, *2* (11), 412–418.
4. Samach, A.; Coupland, G. Time measurement and the control of flowering in plants. Bioessays **2000**, *22* (1), 38–47.
5. Simpson, G.G.; Dean, C. Flowering—Arabidopsis, the rosetta stone of flowering time? Science **2002**, *296* (5566), 285–289.
6. Mouradov, A.; Cremer, F.; Coupland, G. Control of flowering time: Interacting pathways as a basis for diversity. Plant Cell **2002**, *14*, S111–S130.
7. Yano, M.; Katayose, Y.; Ashikari, M.; Yamanouchi, U.; Monna, L.; Fuse, T.; Baba, T.; Yamamoto, K.; Umehara, Y.; Nagamura, Y.; Sasaki, T. Hd1, a major photoperiod sensitivity quantitative trait locus in rice, is closely related to the Arabidopsis flowering time gene CONSTANS. Plant Cell **2000**, *12* (12), 2473–2483.
8. Doebley, J.; Lukens, L. Transcriptional regulators and the evolution of plant form. Plant Cell **1998**, *10* (7), 1075–1082.
9. Colasanti, J.; Yuan, Z.; Sundaresan, V. The *indeterminate* gene encodes a zinc finger protein and regulates a leaf-generated signal required for the transition to flowering in maize. Cell **1998**, *93* (4), 593–603.
10. Colasanti, J.; Sundaresan, V. 'Florigen' enters the molecular age: Long-distance signals that cause plants to flower. Trends Biochem. Sci. **2000**, *25* (5), 236–240.

Floral Scent

Natalia Dudareva
Purdue University, West Lafayette, Indiana, U.S.A.

INTRODUCTION

Flowers of many plants attract pollinators by producing and emitting volatile compounds, which belong to a broad category of secondary metabolites. Floral scent is a key modulating factor in plant–insect interactions and plays a central role in successful pollination, and thus in fruit development, of many crop species. The composition and total output of floral scent changes during the life span of the flower in relation to flower age, pollination status, environmental conditions, and diurnal endogenous rhythms. Biochemistry, physiology, and molecular biology of floral scent, as well as molecular mechanisms controlling its accumulation and release, are discussed in this article with regard to implications for future directions in this field.

FLORAL SCENT AND ITS IMPORTANCE

Floral scent is typically a complex mixture of low molecular weight compounds [100–200 (Daltons)] emitted by flowers into the atmosphere. The relative abundances of and interactions between these compounds give the flower its unique, characteristic fragrance. Flowers of many plant species produce floral scent to attract pollinators. Volatiles emitted from flowers function as both long- and short-distance attractants and play a prominent role in the location and selection of flowers by insects, especially in moth-pollinated flowers, which are detected and visited at night.[1] To date, little is known about how insects respond to individual components found within floral scents, but it is clear that they are capable of distinguishing complex scent mixtures. In addition to attracting insects to flowers and guiding them to food resources within the flower, floral volatiles are essential in allowing insects to discriminate among plant species and even among individual flowers of a single species. Closely related plant species that rely on different types of insects for pollination produce different odors, reflecting the olfactory sensitivities or preferences of the pollinators. By providing species-specific signals, flower fragrances facilitate an insect's ability to learn particular food sources, thereby increasing its foraging efficiency. At the same time, successful pollen transfer and thus, sexual reproduction—which are beneficial to plants—are ensured. About 73% of cultivated plant species is pollinated at least partially by a variety of bees; many others are dependent on other pollinators such as flies, beetles, butterflies, moths, or other animals.[2] Moreover, one-third of our total diet depends, directly or indirectly, on insect-pollinated plants including most fruit trees, berries, nuts, oilseeds, and vegetables. Pollination not only affects crop yield, but also the quality and efficiency of crop production. Many crops require most, if not all, ovules to be fertilized for optimum fruit size and shape (e.g., apple, berries, and watermelon). A decrease in fragrance emission reduces the ability of flowers to attract pollinators, whereas intensifying the flower odor can increase the pollinator's recruitment rate. In addition, floral scent plays a significant role in the food, flavor, floriculture, cosmetic, and fragrance industries.

Some volatile compounds found in floral scent also have important functions in vegetative processes.[3] They may attract natural predators to herbivore-damaged plants. As a consequence, some plants reduce the number of herbivores more than 90% by releasing volatiles.[4] They may also function as repellents against herbivores or as airborne signals that activate disease resistance via the expression of defense-related genes in neighboring plants and in the healthy tissues of infected plants.

BIOCHEMISTRY OF FLORAL SCENT

The investigation of the chemical composition of floral scents has been conducted by headspace analysis.[5] In this procedure, a flower that is still connected to the rest of the plant is placed inside a small glass or plastic chamber and its emitted volatiles are collected by continually purging the air inside the chamber through a cartridge packed with polymer that binds these volatiles. After a fixed period of time, trapped volatiles are extracted from the cartridge with an organic solvent (a variation of this procedure—the highly sensitive solid phase microextraction method, or SPME—allows for ''instant'' sampling of headspace volatiles).[6] The eluted solution is injected into a gas chromatograph that separates the different volatiles, and then each volatile is identified by mass spectrometry.

Encyclopedia of Plant and Crop Science
DOI: 10.1081/E-EPCS 120005951

Headspace analysis in combination with gas chromatography and mass spectrometry have revealed over 700 compounds from 441 taxa in 174 genera in 60 plant families.[7] Floral fragrances belong to a broad category of secondary metabolites and are dominated by terpenoids (monoterpenes and sesquiterpenes), phenylpropanoid, and benzenoid compounds. Fatty acid derivatives and a range of other chemicals, especially those containing nitrogen or sulfur, are also sometimes present. Each of these classes is, in turn, represented by compounds having different functional groups, e.g., hydrocarbons, alcohols, aldehydes, ketones, acids, ethers, and esters.[7] Terpenoids are synthesized from isopentenyl diphosphate by different mono- and sesquiterpene synthases.[8] Phenylpropanoids, including benzenoids, derive from L-phenylalanine through the action of the pivotal enzyme phenylalanine ammonium lyase. Although the biosynthetic pathways to these volatile compounds have not yet been completely characterized, common modifications such as hydroxylation, acetylation, and methylation reactions have been described. Other fragrance compounds, such as short-chain alcohols and aldehydes, are formed by metabolic conversion or degradation of phospholipids and fatty acids through the concerted action of lipoxygenases, hydroperoxide lyases, isomerases, and dehydrogenases.

The chemical composition of flower fragrances varies widely among species in terms of the number, identity, and relative amounts of constituent volatile compounds[7] and is correlated with the type of visiting pollinators. Species pollinated by bees and flies have sweet scents, whereas those pollinated by beetles have strong musty, spicy, or fruity odors.[1] The most common constituents of floral scents (Fig. 1) are monoterpenes (linalool, limonene, myrcene, trans-β-ocimene, and geraniol); some sesquiterpenes (farnesene, nerolidol, caryophyllene, and germacrene); and benzenoid compounds (benzylacetate, eugenol, (iso)methyleugenol, methylsalicylate, and methylbenzoate).

PHYSIOLOGY OF FLORAL SCENT

In the flowers of many plant species the petals are the principal emitters of volatiles, although various other parts of the flower may also participate in fragrance emission.[1,9] Whereas the same floral scent components are often emitted from all parts of the flower (although not

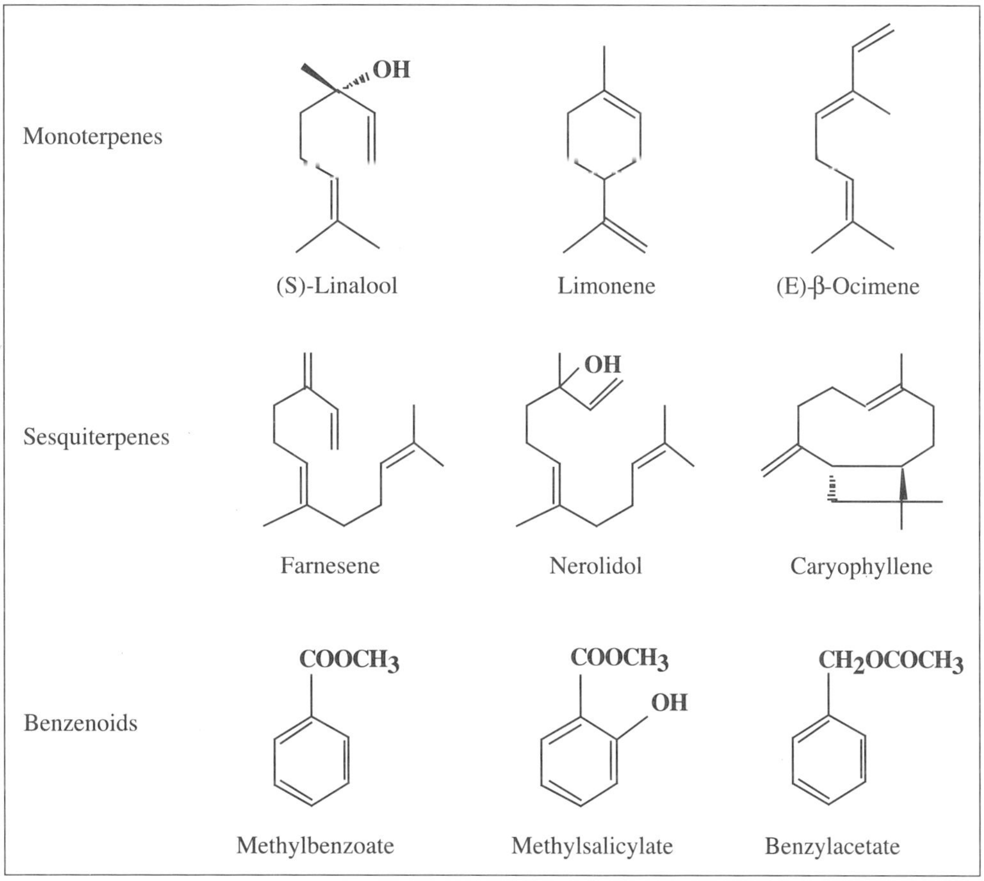

Fig. 1 Structures of representative floral scent volatile compounds.

necessarily the same amount or at the same rate), sometimes specific compounds may be emitted from only a subset of the floral organs. Some plants—orchids for example—have developed highly specialized anatomical structures, called scent glands (i.e., osmophores), for fragrance production; in other plants, the nonspecialized floral epidermal cells are recruited for scent production and emission.[3]

The composition of floral scent—as well as total output—changes during the life span of the flower. These changes occur in relation to flower age, pollination, environmental conditions, and diurnal endogenous rhythms.[9] Plants tend to have their scent output at maximal levels only when the flowers are ready for pollination and concomitantly when their potential pollinators are active. During flower development newly opened and young flowers, which are not ready to function as pollen donors because their anthers have not yet dehisced, produce fewer odors and are less attractive to pollinators than older flowers.[9] Once a flower has been sufficiently pollinated, quantitative and/or qualitative changes of floral bouquets lead to lower attractiveness of these flowers and help to direct pollinators to the unpollinated flowers, thereby maximizing the reproductive success of the plant. The rhythmic release of flower scent during the daily light/dark cycle generally coincides with the foraging activity of potential pollinators. Plants that are pollinated by insects with maximum activity during daytime (e.g., bees) show a diurnal rhythmicity, whereas flowers pollinated by nocturnal insects such as moths tend to have maximal scent output during the night. While the circadian clock controls the nocturnal emission of volatiles, daytime emission in most cases is noncircadian and controlled by irradiation levels. However, circadian control of fragrance production has been shown for rose and snapdragon, both diurnally emitting plants.

MOLECULAR BIOLOGY OF FLORAL SCENT

Whereas the chemistry of plant volatiles is well understood, less is known about the biosynthesis of this diverse group of compounds. The initial breakthrough began in 1994 when (S)–linalool synthase (LIS), an enzyme responsible for the formation of the acyclic monoterpene linalool, was purified from *Clarkia breweri* flowers. This first enzyme of floral volatile formation provided information on an amino acid sequence that facilitated the isolation of the corresponding gene. A similar protein-based cloning strategy was used for isolation of four additional genes (Table 1): S-adenosyl-L-methionine (SAM):(iso) eugenol O-methyl transferase (IEMT), acetyl-coenzyme A:benzyl alcohol acetyltransferase (BEAT), S-adenosyl-L-methionine:salicylic acid carboxyl methyl transferase (SAMT), and S-adenosyl-L-methionine:benzoic acid carboxyl methyl transferase (BAMT), which encode the enzymes responsible for the formation of methyl(iso)eugenol, benzylacetate, methylsalicylate, and methylbenzoate, respectively.[3] All of these genes were isolated from *C. breweri*, except for BAMT, which was cloned from flowers of snapdragon in which methylbenzoate is one of the major fragrance compounds. The isolation of genes responsible for the formation of

Table 1 Current isolated and characterized genes involved in floral scent production[a]

Gene	Product	Species	Approach used
Linalool synthase (LIS)	(S)-Linalool	*Clarkia breweri*	Protein-based cloning
S-adenosyl-L-methionine: (iso) eugenol O-methyl transferase (IEMT)	Methy(iso)eugenol	*Clarkia breweri*	Protein-based cloning
acetyl-coenzyme A:benzyl alcohol acetyltransferase (BEAT)	Benzylacetate	*Clarkia breweri*	Protein-based cloning
S-adenosyl-L-methionine:salicylic acid carboxyl methyl transferase (SAMT)	Methylsalicylate	*Clarkia breweri*	Protein-based cloning
S-adenosyl-L-methionine:benzoic acid carboxyl methyl transferase (BAMT)	Methylbenzoate	*Antirrhinum majus*	Protein-based cloning
Benzoyl-coenzyme A:benzyl alcohol benzoyl transferase (BEBT)	Benzylbenzoate	*Clarkia breweri*	Functional genomic
S-adenosyl-L-methionine:salicylic acid carboxyl methyl transferase (SAMT)	Methylsalicylate	*Antirrhinum majus*	Functional genomic
Orcinol O-methyl transferase (OOMT1 and OOMT2)	Orcinol dimethyl ether (3,5-dimethoxytoluene)	*Rosa hybrida*	Functional genomic
Germacrene D synthase	Germacrene D	*Rosa hybrida*	Functional genomic

[a]Function of the gene was confirmed by overexpression in *Escherichia coli*.

floral scent volatiles allowed investigation of the regulation of scent biosynthesis. It has been shown that the flowers synthesize scent compounds de novo in the tissues from which these volatiles are emitted. The emission levels, corresponding enzyme activities, and the expression of genes encoding scent biosynthetic enzymes are all temporally and spatially regulated during flower development. Although highest in petals, expression of scent genes is relatively uniform and restricted to surfaces of floral tissues (epidermal cells), from which volatile compounds can easily escape into the atmosphere after being synthesized. Transcriptional regulation of the expression of these genes at the site of emission and the level of supplied substrates for the reactions were found to be the major factors controlling scent production, and indirectly, scent emission during flower development. When regulation of rhythmic emission of methylbenzoate was investigated, it was found that the level of substrate (benzoic acid) also plays a major role in the regulation of circadian emission of methylbenzoate in diurnally (snapdragon) and nocturnally (*Petunia* cv Mitchell and *Nicotiana suaveolens*) emitting flowers.

FUTURE DIRECTIONS IN FLORAL SCENT

Research into floral scent genes using high throughput technologies has recently been initiated in *C. breweri*, snapdragon, and rose.[10] These projects combine detailed fragrance analyses with the creation of petal expressed sequence tag (EST) databases. In addition, the *Escherichia coli* expression system allows rapid functional characterization of fragrance genes. The integration of a functional genomic approach with floral scent research will lead to high-throughput identification of novel scent genes. Indeed, this approach has already successfully accelerated discovery and characterization of new scent genes in *Clarkia*, roses, and snapdragon (Table 1). The identification of genes for the production of floral scent opens up new opportunities for its manipulation, which would have great economic impact on ornamentals that have reduced floral scent due to decades of classical breeding.

ARTICLES OF FURTHER INTEREST

Aromatic Plants for the Flavor and Fragrance Industry, p. 58
Circadian Clock in Plants, p. 278
Insect–Plant Interactions, p. 609
In Vitro Flowering, p. 576
In Vitro Pollination and Fertilization, p. 587
Metabolism, Secondary: Engineering Pathways of, p. 720
New Secondary Metabolites: Potential Evolution, p. 818
Phenylpropanoids, p. 868
Phytochemical Diversity of Secondary Metabolites, p. 915

REFERENCES

1. Dobson, H.E.M. Floral volatiles in insect biology. In *Insect-plant interactions*; Bernays, E., Ed.; CRC Press: Boca Raton, FL, 1994; Vol. V, 47–81.
2. Buchmann, S.L.; Nabhan, G.P. *The Forgotten Pollinators*; Island Press: Washington, D.C., 1996.
3. Dudareva, N.; Pichersky, E. Biochemical and molecular genetic aspects of floral scents. Plant Physiol. **2000**, *122*, 627–633.
4. Kessler, A.; Baldwin, I.T. Defensive function of herbivore-induced plant volatile emissions in nature. Science **2001**, *291*, 2142–2143.
5. Raguso, R.A.; Pellmyr, O. Dynamic headspace analysis of floral volatiles: A comparison of methods. OIKOS **1998**, *81*, 238–254.
6. Matich, A.J.; Rowan, D.D.; Banks, N.H. Solid phase micro-extraction for quantitative headspace sampling of apple volatiles. Anal. Chem. **1996**, *68*, 4114–4118.
7. Knudsen, J.T.; Tollsten, L.; Bergstrom, G. Floral scents—A checklist of volatile compounds isolated by head-space techniques. Phytochemistry **1993**, *33*, 253–280.
8. Trapp, S.C.; Croteau, R.B. Genomic organization of plant terpene synthases and molecular evolutionary implications. Genetics **2001**, *158* (2), 811–832.
9. Dudareva, N.; Piechulla, B.; Pichersky, E. Biogenesis of floral scent. Hortic. Rev. **1999**, *24*, 31–54.
10. Vainstein, A.; Lewinsohn, E.; Pichersky, E.; Weiss, D. Floral fragrance. New inroads into an old commodity. Plant Physiol. **2001**, *127*, 1383–1389.

Flow Cytogenetics

Jaroslav Doležel
Marie Kubaláková
Jan Vrána
Jan Bartoš
Institute of Experimental Botany, Olomouc, Czech Republic

INTRODUCTION

Flow cytogenetics uses flow cytometry and sorting to classify and purify isolated mitotic chromosomes. The chromosomes are analyzed while moving at high speed in a liquid stream. This gives high resolution and statistical precision, because large populations of chromosomes can be analyzed in a short time. Flow sorting allows purification of large quantities of specific chromosomes according to their fluorescence, which can be used for physical genome mapping, production of recombinant DNA libraries, and protein analysis. Chromosome classification by flow cytometry (flow karyotyping) may be used to detect structural and numerical chromosome changes.

PRINCIPLES

Flow Cytometry

Flow cytometry involves the optical analysis of microscopic particles, constrained to flow in single file within a fluid stream (sheath fluid) through a narrow beam of excitation light (Fig. 1). During the short time each particle is in the light beam, the light is scattered and fluorescent molecules bound to the particle are excited. The scattered light and the fluorescence are collected, spectrally filtered through dichroic mirrors and filters, and converted to electrical pulses by photodetectors. The rate of analysis ranges from hundreds to thousands of particles per second. Flow sorters can, in addition, remove any selected particle from the suspension. The liquid stream emerging from the flow chamber into air is broken into droplets by the action of a vibrating transducer. Droplets containing particles of interest are electrically charged as they separate from the stream and are deflected by passage through an electric field. Charges can be negative or positive, so that two categories of particles can be sorted simultaneously.

Preparation of Chromosome Suspensions

Preparation of an aqueous suspension of intact, dispersed chromosomes consists of induction of cell-cycle synchrony, accumulation of cells in metaphase, and release of chromosomes into isolation buffer.[1] Chromosomes have been isolated from suspension-cultured cells, leaf mesophyll protoplast cultures, and root-tip meristems.[2] Because root-tip cells are karyologically stable and roots can be easily obtained from seeds,[3] these have been used most frequently. Cell-cycle synchrony is induced by treatment with a DNA synthesis inhibitor that accumulates cycling cells at the G_1/S interface. After its removal, the cells transit S and G_2 phases and enter mitosis synchronously, where they are accumulated at metaphase by the action of a mitotic spindle inhibitor. Intact chromosomes are released from metaphase cells either by hypotonic lysis of protoplasts prepared by enzymatic removal of cell walls, or by mechanical tissue homogenization.[1,3] The latter is more popular, because it is rapid and suitable for root tips.

Flow Cytometric Analysis

Chromosomes in suspension are stained with a DNA-binding fluorochrome and classified according to their relative DNA content.[2] The result is displayed as a histogram of relative fluorescence intensity (flow karyotype). Ideally, each chromosome is represented by a single well-discriminated peak on the flow karyotype. Chromosomes can also be stained by a pair of fluorochromes that bind preferentially to A + T-rich and G + C-rich DNA, and classified by DNA base composition.[2] Bivariate flow karyotypes thus obtained may be displayed as contour plots, dot plots, or isometric plots. However, due to negligible differences in DNA base content between chromosomes, bivariate flow karyotyping has not often been used in plants. Another possibility is to classify chromosomes according to the number of copies of DNA repeats after fluorescent labeling.

Chromosome Sorting

To sort a specific chromosome, its peak is selected on a flow karyotype and marked as a sort window. In most plant species, only a few chromosome types can be resolved; the others form overlapping composite peaks. Translocation lines, deletion lines, and chromosome addition lines can be

Encyclopedia of Plant and Crop Science
DOI: 10.1081/E-EPCS 120005592

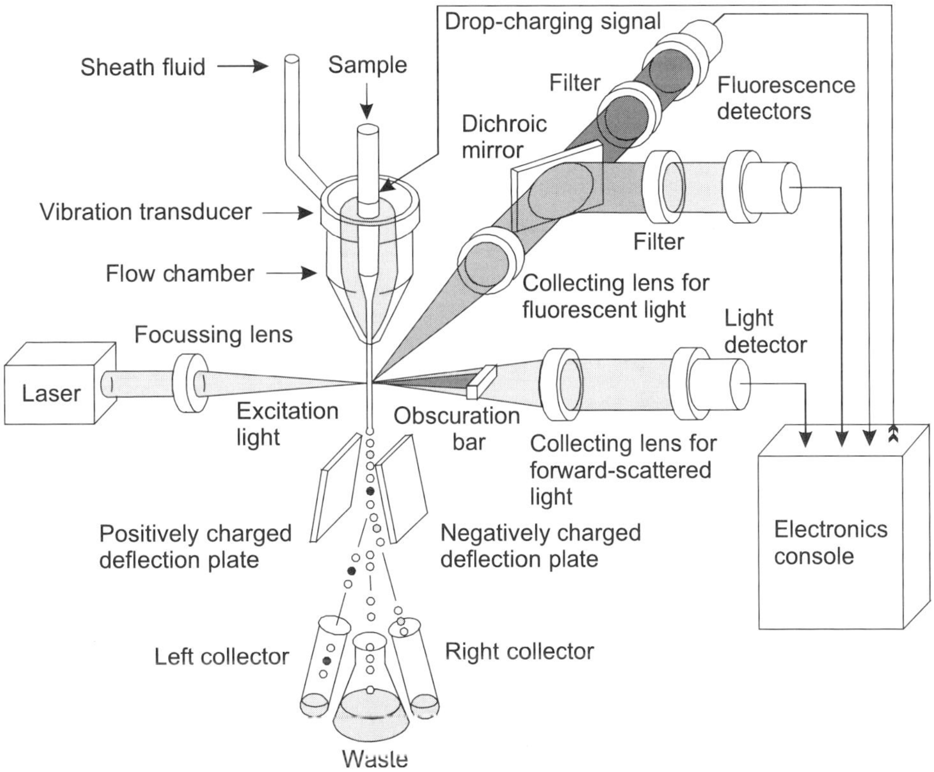

Fig. 1 Schematic view of a flow cytometer and electrostatic droplet sorter capable of simultaneous analysis of two fluorescence signals and forward-scattered light. Two categories of particles may be sorted simultaneously.

used to overcome this problem.[4] The usefulness of a sorted chromosome fraction depends on its purity, which is primarily determined by the resolution of the chromosome peak (the degree of overlap with other peaks), and by the presence of clumps and fragments in the chromosome suspension. Simultaneous analysis of light scatter and chromosome length helps to discriminate such particles. Purities >95% have been achieved. The rate at which a chromosome can be sorted depends on its frequency of occurrence and the resolution of its peak; it is possible to purify specific plant chromosomes at rates up to 20 per second. The identity and purity of sorted chromosomes are best determined microscopically after labeling the DNA repeats, which allows unambiguous chromosome identification (Fig. 2).[5] Microscopic analysis is usually complemented by polymerase chain reaction (PCR) with primers for chromosome-specific markers.

APPLICATIONS

Flow Karyotyping

Flow karyotyping was first developed for *Haplopappus gracilis*[6] and subsequently for a number of species, including economically important legumes and cereals.[4] Flow karyotyping provides unbiased, representative data on the frequency of occurrence of individual chromosome types. As the peak position on the abscissa of the flow karyotype reflects chromosome DNA content, and peak area reflects the frequency of occurrence of that chromosome, both structural and numerical chromosome aberrations can be detected. Successful applications include quantitative analysis of the frequency of occurrence of B chromosomes, and identification and transmission stability of translocation chromosomes, deletion chromosomes, and alien chromosomes in chromosome addition lines (Fig. 2). Flow karyotyping can identify structural chromosome changes that result in detectable changes in DNA content. The method is sensitive enough to detect chromosome polymorphism in wheat.[7]

Chromosome Sorting

Chromosome sorting has proven useful for assigning peaks on the flow karyotype to specific chromosomes, and for preparation of purified chromosome fractions. Because most agricultural plant species have complex genomes, chromosome sorting facilitates their analysis. Flow-sorted chromosomes have been used for physical mapping of

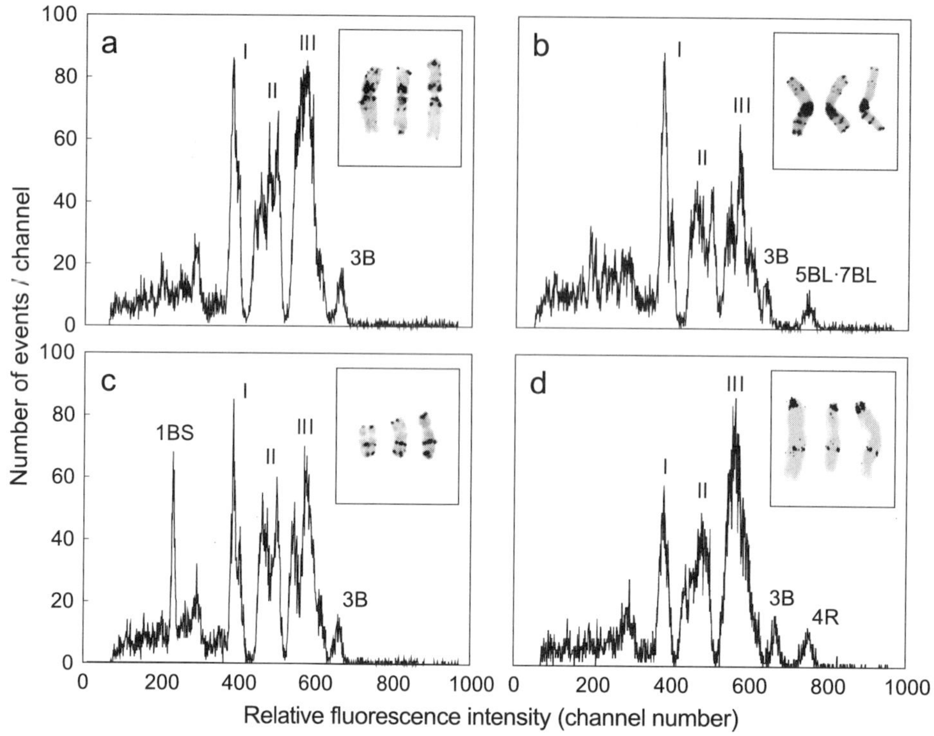

Fig. 2 Flow karyotypes obtained after analysis of 4′, 6-diamino-2-phenylindole (DAPI)-stained chromosome suspensions from three hexaploid ($2n=6x=42$) wheat lines (a–c) and from a wheat–rye chromosome addition line (d). (a) The ''Chinese Spring'' karyotype contains three composite peaks (I–III) representing groups of chromosomes that cannot be discriminated individually, and a peak representing chromosome 3B. (*Insert:* chromosomes 3B after GAA banding); (b) ''Cappelle Desprez'' has three composite peaks (I–III) and well-discriminated peaks representing 3B and translocation chromosome 5BL·7BL. (*Insert:* chromosomes 5BL·7BL after GAA banding); (c) ''Pavon;'' ditelosomic for 1BS, shows three composite peaks (I–III), the peak of 3B, and a clearly discriminated peak of 1BS. (*Insert:* GAA-banded 1BS); (d) The peak of rye chromosome 4R is well-discriminated on the flow karyotype of the ''Chinese Spring''/''Imperial'' addition line (insert: chromosome 4R after FISH with a probe for pSc119.2 repeat.)

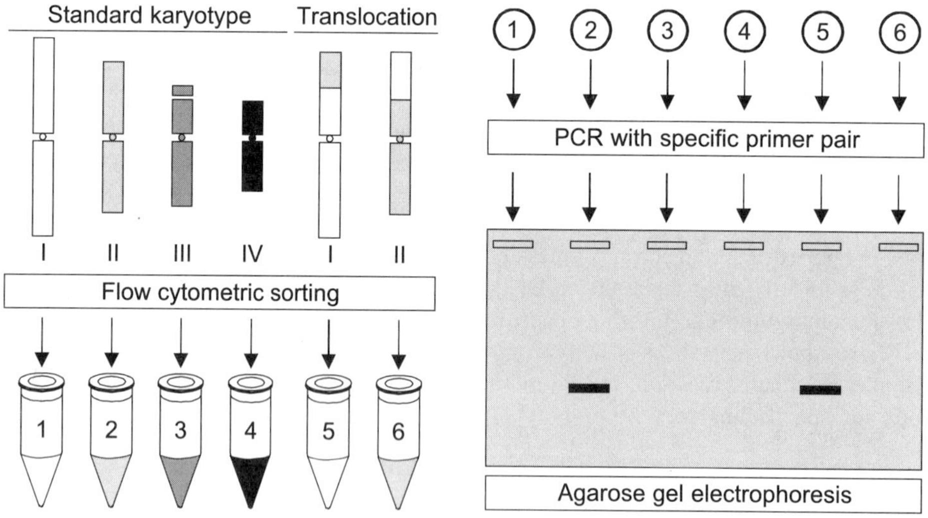

Fig. 3 Schematic representation of gene mapping using flow-sorted chromosomes and PCR. All chromosome types of a standard karyotype (I–IV) are sorted and subjected to PCR with a pair of specific primers. After ethidium bromide-stained agarose gel electrophoresis of PCR products, the chromosome fraction containing the target sequence shows a bright band of the appropriate size. All other chromosome fractions are negative. Availability of translocation chromosomes permits localization of the DNA sequence to a specific chromosome region.

DNA sequences using dot-blot hybridization or PCR.[8] The use of translocation chromosomes facilitates mapping to subchromosomal regions (Fig. 3). Targeted isolation of molecular markers for saturation of genetic linkage maps and preparation of chromosome-specific DNA libraries are other important uses.[9,10] Flow cytometry is the only method that can purify chromosomes in microgram quantities, which are needed for direct cloning and construction of large-insert DNA libraries. Sorted chromosomes are also used for physical mapping of DNA sequences using fluorescence in situ hybridization (FISH) and primed in situ DNA labeling (PRINS), preparation of linearly stretched chromosomes for high-resolution physical mapping, and localization of chromosomal proteins using immunofluorescence.

CONCLUSION

Flow cytogenetics has been developed for a number of species, including economically important legumes and cereals, and is becoming increasingly useful for plant genome analysis. Flow karyotyping facilitates analysis of the frequency of occurrence and transmission stability of specific chromosome types, including translocation and deletion chromosomes. Chromosome sorting simplifies the analysis of complex plant genomes, and the use of sorted chromosomes for physical mapping of DNA sequences and targeted isolation of molecular markers is now routine. The biggest potential is in the construction of large-insert chromosome-specific DNA libraries cloned in bacterial artificial chromosome (BAC) vectors. Such libraries would be extremely helpful for construction of physical maps and for map-based cloning in crops with large genomes.

ACKNOWLEDGMENTS

We apologize to those colleagues whose work could not be cited due to space limitations.

ARTICLES OF FURTHER INTEREST

Aneuploid Mapping in Diploids, p. 33
Aneuploid Mapping in Polyploids, p. 37
B Chromosomes, p. 71
Chromosome Rearrangements, p. 270

REFERENCES

1. Doležel, J.; Macas, J.; Lucretti, S. Flow Analysis and Sorting of Plant Chromosomes. In *Current Protocols in Cytometry*; Robinson, J.P., Darzynkiewicz, Z., Dean, P.N., Dressler, L.G., Orfao, A., Rabinovitch, P.S., Stewart, C.C., Tanke, H.J., Wheeless, L.L., Eds.; John Wiley & Sons, Inc.: New York, 1999; 5.3.1–5.3.33.
2. Doležel, J.; Lucretti, S.; Schubert, I. Plant chromosome analysis and sorting by flow cytometry. Crit. Rev. Plant Sci. **1994**, *13* (3), 275–309.
3. Doležel, J.; Číhalíková, J.; Lucretti, S. A high-yield procedure for isolation of metaphase chromosomes from root tips of *Vicia faba* L. Planta **1992**, *188* (1), 93–98.
4. Doležel, J.; Lysák, M.A.; Kubaláková, M.; Šimková, H.; Macas, J.; Lucretti, S. Sorting of Plant Chromosomes. In *Methods in Cell Biology*, 3rd Ed.; Darzynkiewicz, Z., Crissman, H.A., Robinson, J.P., Eds.; Academic Press: San Diego, 2001; Vol. 64, Part B, 3–31.
5. Kubaláková, M.; Lysák, M.A.; Vrána, J.; Šimková, H.; Číhalíková, J.; Doležel, J. Rapid identification and determination of purity of flow-sorted plant chromosomes using C-PRINS. Cytometry **2000**, *41* (2), 102–108.
6. De Laat, A.M.M.; Blaas, J. Flow-cytometric characterization and sorting of plant chromosomes. Theor. Appl. Genet. **1984**, *67* (3), 463–467.
7. Kubaláková, M.; Vrána, J.; Číhalíková, J.; Šimková, H.; Doležel, J. Flow karyotyping and chromosome sorting in bread wheat (*Triticum aestivum* L.). Theor. Appl. Genet. **2002**, *104* (8), 1362–1372.
8. Neumann, P.; Požárková, D.; Vrána, J.; Doležel, J.; Macas, J. Chromosome sorting and PCR-based physical mapping in pea (*Pisum sativum* L.). Chrom. Res. **2002**, *10* (1), 63–71.
9. Macas, J.; Gualberti, G.; Nouzová, M.; Samec, P.; Lucretti, S.; Doležel, J. Construction of chromosome-specific DNA libraries covering the whole genome of field bean (*Vicia faba* L.). Chrom. Res. **1996**, *4* (7), 531–539.
10. Požárková, D.; Koblížková, D.; Román, B.; Torres, A.M.; Lucretti, S.; Lysák, M.; Doležel, J.; Macas, J. Development and characterization of microsatellite markers from chromosome 1-specific DNA libraries of *Vicia faba*. Biol. Plant. **2002**, *45* (3), 337–345.

Flower Development

Thomas Jack
Dartmouth College, Hanover, New Hampshire, U.S.A.

INTRODUCTION

Experiments over the last 15 years in genetically tractable plants, primarily *Antirrhinum* and *Arabidopsis*, have elucidated the general mechanisms of floral induction, floral meristem identity, and floral organ identity. In *Arabidopsis*, four inputs stimulate the transition from vegetative growth to reproductive growth: long-day photoperiod, gibberellins, vernalization, and the age of the plant. Pathways mediating these four inputs have been genetically dissected. These pathways feed into three key floral integrators: *FT*, *SOC1/AGL20*, and *LFY*. The identity of the floral meristem itself is determined primarily by expression of *LFY* together with a second gene, *AP1*. In addition to playing a key role in integrating floral inductive signals, *LFY* also plays an important role in activating the floral organ identity genes *AG* and *AP3*.

SHIFT FROM VEGETATIVE TO REPRODUCTIVE GROWTH

For the majority of the 20th century, plant biologists were interested in trying to isolate and characterize substances that could induce plants to shift from vegetative growth to reproductive growth. Experiments in the 1930s demonstrated that a transmissible substance referred to as ''florigen'' was capable of inducing flowering. Efforts to purify florigen in subsequent decades were unsuccessful. Obviously, the identification of floral-promoting substances has important agricultural implications for controlling the timing of flowering of crop plants.

Molecular genetic experiments in *Arabidopsis* have led to the identification of a number of key genes whose activity is sufficient to accelerate flowering (Fig. 1).[1,2] *Arabidopsis* is a facultative long-day plant in that long days accelerate flowering, but even under noninductive short days, Arabidopsis does flower, though flowering is delayed. Expression of any of five key genes—*LEAFY* (*LFY*), *APETALA1* (*AP1*), *FLOWERING LOCUS T* (*FT*), *CONSTANS* (*CO*), and *SUPPRESSOR OF OVEREXPRESSOR OF CO* (*SOC1*) (also referred to as *AGL20*)—is sufficient to induce flowering in vegetative *Arabidopsis* (Table 1). Collectively, these five genes are responsive to four flower-promoting pathways: a long-day photoperiod promotion pathway, a gibberellin responsive pathway, a vernalization promotion pathway, and an autonomous pathway.

The components of the day-length promotion pathway were identified based on mutants that flower late under long-day photoperiods but exhibit a flowering time similar to wild type under short-day photoperiods. Day length is sensed by *Arabidopsis* by a combination of photoreceptors such as phytochromes and cryptochromes. Mutations in *PHYTOCHROME A (PHYA)* and *CRYPTOCHROME 2 (CRY2)* result in late flowering in long days. Light input feeds into a circadian oscillator that regulates the daily rhythm of a large number of genes. Mutations in circadian clock components *GIGANTEA* (*GI*) and *LHY* result in late flowering in long-day photoperiods. One gene regulated by the circadian clock is *CONSTANS* (*CO*). Mutations in *CO* are late flowering in long days. The circadian peak in *CO* RNA expression occurs in the evening. When plants are grown in short-day photoperiods, the circadian peak of *CO* RNA expression occurs in the dark. By contrast, under long-day photoperiods, the peak of *CO* expression occurs in the light.[3] In addition to controlling the circadian rhythm of *CO*, light also directly activates *CO* (the molecular basis of this activation is unclear at present), but direct activation by light occurs only under long-day growth conditions. Thus, *CO* is active only when the circadian clock is functional and when the photoperiod exceeds a certain minimum length.

CO directly activates the floral activator/integrator *FT*. To a lesser degree, *CO* also functions to activate a second floral activator/integrator, *SOC1/AGL20*.

In *Arabidopsis*, gibberellins (GA) are required for flowering under short-day photoperiods but play only a minor role in stimulating flowering in long days. Mutants that are unable to synthesize active GA (e.g., *ga1*) do not flower when grown under short-day photoperiods. The promotion of flowering by GA occurs, at least in part, by activating *LEAFY* (*LFY*), a third floral activator/integrator. A *cis*-acting element in the *LFY* promoter is responsive to GA. *LFY* is also upregulated by long-day photoperiods, but the *cis*-acting promoter element that responds to long-day photoperiods is separable from that which is responsive to GA.[4]

Even under noninductive photoperiods, *Arabidopsis* does flower, suggesting that endogenous factors such as

Encyclopedia of Plant and Crop Science
DOI: 10.1081/E-EPCS 120012910

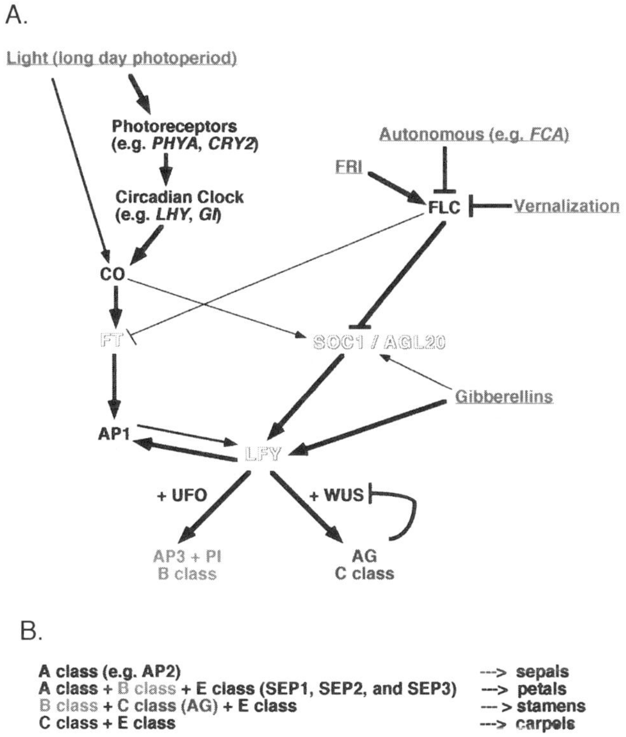

Fig. 1 A) The major pathways mediating floral induction are indicated. The four major input pathways are shown in red (at www.dekker.com) or gray with underline (long-day promotion, autonomous, vernalization, and GA). These pathways are integrated primarily at three floral activators/integrators (*FT*, *LFY*, and *SOC/AGL20*) shown in green or outline type. *LFY* is upregulated in the floral meristem after the floral transition and functions to activate floral organ identity genes such as *AP3* (pink/light gray) and *AG* (blue/dark gray). The thickness of the lines indicates the relative strength of the regulatory interaction; in general terms, thicker lines predominate over thinner lines. B) Combinations of A, B, C, and E activities specify organ identity in the flower. (*View this art in color at www.dekker.com.*)

age of the plant are important for the stimulation of flowering. Mutants in the autonomous pathway flower later than wild type under both long-and short-day conditions (i.e., these mutants are still responsive to photoperiod). The autonomous pathway (or pathways) functions by downregulating the *FLOWERING LOCUS C* (*FLC*) floral repressor. In autonomous pathway mutants such as *fca*, the levels of *FLC* RNA and protein are high, which represses flowering resulting in a late-flowering phenotype.[5] A second input to *FLC* is *FRIGIDA* (*FRI*), a positive activator of *FLC*. There are many ecotypes of Arabidopsis that represent the wide geographical distribution of this species. Some ecotypes are early flowering; for example, the common lab strains Landsberg and Columbia flower after about 20–25 days when grown in long-day photoperiods. By contrast, other ecotypes flower much later, at 50–60 days under inductive long-day conditions. Late-flowering ecotypes of *Arabidopsis* contain wild-type copies of both *FLC* and *FRI* and are very late flowering, while early-flowering ecotypes have naturally occurring mutations in *FRI* and/or *FLC*.[6]

The vernalization response also acts primarily via *FLC*. Vernalization accelerates flowering after an extended cold period (e.g., six weeks at 4°C). The extended cold period mimics overwintering and results in rapid flowering in the spring when the plant encounters inductive photoperiods. Vernalization results in a quantitative and cumulative reduction in *FLC* RNA. The reduction in *FLC* RNA occurs epigenetically. The present model is that exposure to cold demethylates the *FLC* gene, which alters chromatin structure and leads to transcriptional inactivation of *FLC*. The inactivation of *FLC* is maintained throughout the life of the plant, but *FLC* is reactivated in progeny plants. *FLC* functions primarily by repressing the floral activator/integrator *SOC1/AGL20*. Although the primary inputs to *SOC1* are via the autonomous/vernalization pathways, *SOC1* is activated secondarily by both gibberellins and the photoperiod promotion pathway acting via *CO*. Similarly, although *FT* is activated primarily by the long-day promotion pathway, *FT* also receives inputs via *FLC*. One clear message that is emerging is that there is considerable cross talk among the four major floral inductive pathways.

Table 1 Critical genes in flower development in *Arabidopsis*

Gene	Gene product
PHYA	Phytochrome red/far-red photoreceptor
CRY2	Cryptochrome blue-light photoreceptor
LHY	single myb domain protein, Sequence-specific DNA-binding protein
GI	Putative transmembrane protein
CO	Zn finger, nuclear protein
FT	Similarity to kinase inhibitor proteins
FCA	RNA-binding protein
FLC	MADS transcription factor
SOC1/AGL20	MADS transcription factor
FRI	No similarity to proteins in databases
LFY	Sequence-specific DNA-binding protein
AP1	MADS transcription factor
TFL1	Similarity to kinase inhibitor proteins
CAL	MADS transcription factor
WUS	Homeodomain transcription factor
UFO	F-box protein
AP2	AP2 domain transcription factor
AP3	MADS transcription factor
PI	MADS transcription factor
AG	MADS transcription factor
SEP1,SEP2, SEP3	MADS transcription factor

SPECIFICATION OF THE FLORAL MERISTEM

FT and *SOC1* are broadly expressed in the plant. By contrast, RNA and protein expression of *LFY* and *AP1* is highest in the floral meristem primordia. *LFY* is dramatically upregulated in the floral meristem primordia prior to the morphological differentiation of the floral primordia from the shoot apical meristem. Presumably, this spatial upregulation of *LFY* is dependent on broadly expressed activation signals dependent on GA, *FT*, and *SOC1* combined with repression of *LFY* activation in the shoot apex, which likely is mediated by genes such as *TERMINAL FLOWER 1* (*TFL1*). In the floral meristem anlagen, *LFY* activates a second gene, *AP1*. Like *LFY*, *AP1* is necessary for proper execution of the floral development program. Although *LFY* activates *AP1*, *AP1* also responds to floral inductive pathways independently of *LFY*. Consistent with this idea, both *lfy* and *ap1* mutants exhibit a partial conversion of flowers to shoots. In *lfy* mutants, the basal positions on the inflorescence are converted to shoots, while in more apical positions, abnormal flowers develop. The flower-to-shoot transformation observed in *ap1* mutants is different from *lfy* in that the flowers themselves exhibit indeterminate growth characteristics; specifically, secondary flowers develop in the axils of the first whorl organs. *AP1* functions redundantly with a closely related gene called *CAULIFLOWER* (*CAL*). *cal* single mutants exhibit a wild-type phenotype, but *cal* dramatically enhances the *ap1* phenotype resulting in the replacement of flowers with proliferating masses of meristem tissue that resemble the vegetable cauliflower. Similarly, *lfy ap1* double mutants also exhibit an enhanced phenotype; all positions on the inflorescence normally occupied by flowers instead develop as shoots that exhibit only slight floral character. This residual floral character is removed in the *lfy ap1 cal* triple mutant that exhibits vegetative, but lacks floral, characteristics.

SPECIFICATION OF FLORAL ORGAN IDENTITY

As described above, activation of *LFY* is dependent on floral inductive signals. In turn, *LFY* functions as a key activator of floral organ identity genes.[7] The identity of the sepals, petals, stamens, and carpels is specified by four broad classes of organ identity genes referred to as A-, B-, C-, and E-class genes (Fig. 2). A-class genes specify the identity of sepals and petals; in A-class mutants such as *apetala2* (*ap2*), sepals and petals fail to develop with the correct identity. B-class genes specify the identity of petals and stamens; in *apetala3* (*ap3*) and *pistillata* (*pi*) mutants, petals develop as sepals, and stamens develop as

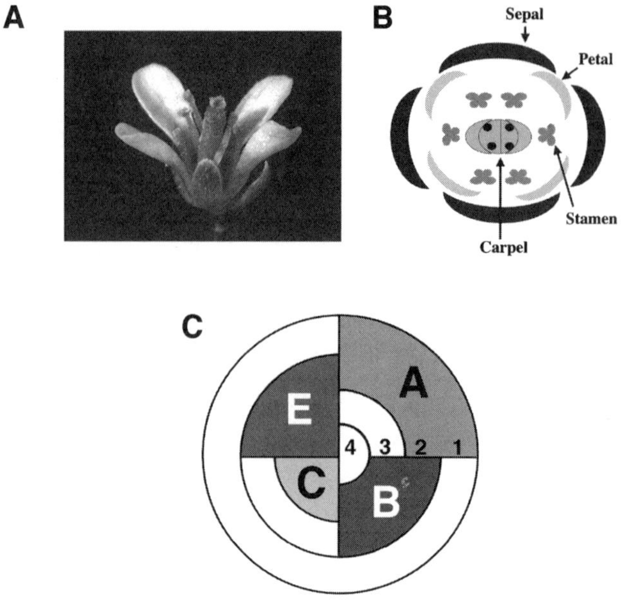

Fig. 2 A) Photograph of an *Arabidopsis* flower. B) Floral diagram of an *Arabidopsis* flower indicating the positions of the sepals, petals, stamens, and carpels that develop in whorls 1, 2, 3, and 4, respectively. C) The ABC model of flower development, as well as the revised ABCE model, postulates that combinations of A, B, C, and E activities specify organ identity in the four whorls of the flower. (*View this art in color at www.dekker.com.*)

carpels. The C-class gene *AGAMOUS* (*AG*) specifies the identity of stamens and carpels. In *ag* mutants, stamens develop as petals and the carpels develop as a new flower; thus, *ag* mutants are indeterminate. Unlike mutants in the A-, B-, and C-class genes, single mutants in any of the three E-class genes, *SEPALLATA1* (*SEP1*), *SEP2*, or *SEP3*, fail to exhibit a dramatic floral phenotype. However, when mutations in all three *SEP* genes are combined together in a *sep1 sep2 sep3* triple mutant, the petals, stamens, and carpels develop as sepal-like organs and the flowers are indeterminate. This demonstrates that *SEP1*, *SEP2*, and *SEP3* redundantly specify petals, stamens, carpels, and floral meristem determinacy.[8]

The floral organ identity genes are expressed in region-specific patterns in the developing floral primordia. For example, *AG* RNA is expressed in the floral meristem in the precursor cells of the stamens and carpels while *AP3* and *PI* are expressed in the precursor cells of the petals and stamens. In *lfy* mutants, *AP3* and *PI* expression is dramatically reduced and petals and stamens fail to develop, suggesting that *LFY* is a critical positive regulator of *AP3* and *PI*. *LFY*, which is expressed throughout the floral meristem, leads to the activation of *AP3* and *PI* in a subset of floral meristem cells destined to develop into petals and stamens. The present model is that *LFY* functions together with the *UNUSUAL FLORAL ORGANS* (*UFO*) coregulator. *UFO* encodes an F-box protein and is postulated to function by targeting a repressor of *AP3* for ubiquitin-mediated protein degradation.

LFY also plays a role in the direct activation of *AG*. *LFY* activates *AG* together with *WUSCHEL* (*WUS*), a gene that is necessary for maintaining meristem cells in an undifferentiated state. Organ identity in the flower is specified from outside to inside, i.e., the sepals are specified first, followed by petals, stamens, and carpels. *WUS* is expressed in the floral meristem prior to floral organ differentiation. The combination of *LFY* plus *WUS* in the stamen and carpel primordia directly activates *AG*. *AG* leads to the specification of stamen and carpel identity, and one of the functions of *AG* is to downregulate *WUS*, which leads to elimination of the stem cell identity in the flower.

Once activated, the ABCE genes are postulated to function combinatorially to specify floral organ identity. The original version of this model involved only the A-, B-, and C-class genes and was called the ABC model.[9] With the discovery of the important role of the SEP genes, the ABC model has been expanded to include the SEP genes as an E class. According to the revisionist ABCE model,[10] sepals are specified by A alone, petals by A + B + E, stamens by B + C + E, and carpels by C + E.

CONCLUSION

Until recently, the molecular basis for the combinatorial nature of the ABCE model was unclear. AG, AP3, PI, and the three SEP proteins are all members of the MADS transcription factor family. Although MADS proteins are capable of binding to DNA as dimers, several lines of evidence support a model that postulates that MADS proteins are capable of functioning in higher-order complexes, e.g., trimers or tetramers. This raises the possibility that different tetramers of MADS proteins might be specifying organ identity; for example, stamens might be specified by AP3+PI+AG+SEP while carpels are specified by AG+AG+SEP+SEP. Future work should tell us whether such complexes exist and are functional in plants.

ARTICLES OF FURTHER INTEREST

Arabidopsis Transcription Factors: Genome-Wide Comparative Analysis Among Eukaryotes, p. 51
Circadian Clock in Plants, p. 278
Floral Induction, p. 452
Male Gametogenesis, p. 663
Photoperiodism and the Regulation of Flowering, p. 877
Photoreceptors and Associated Signaling I: Phytochromes, p. 881
Photoreceptors and Associated Signaling II: Cryptochromes, p. 885

REFERENCES

1. Simpson, G.G.; Dean, C. Arabidopsis, the Rosetta stone of flowering time. Science **2002**, *296*, 285–289.
2. Mouradov, A.; Cremer, F.; Coupland, G. Control of flowering time: Interacting pathways as a basis for diversity. Plant Cell **2002**, *14*, S111–S130.
3. Suarez-Lopez, P.; Wheatley, K.; Robson, F.; Onouchi, H.; Valverde, F.; Coupland, G. *CONSTANS* mediates between the circadian clock and the control of flowering in Arabidopsis. Nature **2001**, *410*, 1116–1120.
4. Blazquez, M.A.; Weigel, D. Integration of floral inductive signals in Arabidopsis. Nature **2000**, *404*, 889–892.
5. Rouse, D.T.; Sheldon, C.C.; Bagnall, D.J.; Peacock, W.J.; Dennis, E.S. FLC, a repressor of flowering, is regulated by genes in different inductive pathways. Plant J. **2002**, *29*, 183–191.
6. Johanson, U.; Johanson, U.; West, J.; Lister, C.; Michaels, S.; Amasino, R.; Dean, C. Molecular analysis of *FRIGIDA*, a major determinant of natural variation in Arabidopsis flowering time. Science **2000**, *290*, 344–347.
7. Lohmann, J.U.; Weigel, D. Building beauty: The genetic control of floral patterning. Dev. Cell **2002**, *2*, 135–142.
8. Pelaz, S.; Ditta, G.S.; Baumann, E.; Wisman, E.; Yanofsky, M.F. B and C floral organ identity functions require *SEPALLATA* MADS-box genes. Nature **2000**, *405*, 200–203.
9. Coen, E.S.; Meyerowitz, E.M. The war of the whorls: Genetic interactions controlling flower development. Nature **1991**, *353*, 31–37.
10. Theissen, G. Development of floral organ identity; Stories from the MADS house. Curr. Opin. Plant Biol. **2001**, *4*, 75–85.

Fluorescence In Situ Hybridization

Yasuhiko Mukai
Osaka Kyoiku University, Osaka, Japan

INTRODUCTION

Fluorescence in situ hybridization (FISH) was first applied to visualize DNA sequences on plant chromosomes more than one decade ago.[1] In that time, the technique has become very important in genome analysis of plants to assess the homology between genomes and localize genes and DNA sequences on individual chromosomes or extended DNA fibers. Genomic in situ hybridization (GISH), an offshoot of the FISH technique, is a powerful tool where total genomic DNA is used as a probe for visualizing genome homology between polyploid species and their progenitors and to supplement information on the genomic origin of polyploids. With improvements in detection sensitivity and resolution of FISH technique, it is now possible to detect even unique sequences. Mapping of genes or DNA sequences on extended DNA fibers and intact cloned DNA has further increased the resolution power of the FISH technique. FISH was successful in showing the position of agronomically important genes on the chromosomes of crop plants. The technique has also revealed gene synteny in different crops.

PRINCIPLE AND METHODS OF FISH

In situ hybridization (ISH), pioneered by Gall and Pardue,[2] provides a sensitive method for the detection of DNA or RNA sequences in cytological preparations. In this technique, radioactive probes were initially used for localizing genes or specific DNA sequences on metaphase chromosomes. The chromosomal DNA and probes are denatured, and then complementary sequences in the probe and target DNA are allowed to reanneal. After washing and autoradiographic detection, signals are visible at the site of probe hybridization. In the 1980s, nonradioactive probes labeled with biotin, digoxigenin, and fluorochrome were developed for ISH. The advantages of using nonradioactive probes over hybridization with isotope-based probes include longer probe stability, speed, higher sensitivity, spatial resolution, and simultaneous detection of more than one probe. In detection by FISH, fluorescently labeled affinity reagents are used. FISH is rapidly replacing standard ISH procedures and becoming a conventional cytogenetic tool.

Mitotic cells from root-tip meristems or meiotic cells from pollen mother cells are used for cytological preparations. Slides are incubated at 70°C for 2 min in a solution containing 70% formamide/2xSSC to melt the chromosomal DNA into single strands. For probes, cloned DNA from plasmid, lambda phage, phase cosmid, BAC libraries, amplified DNA by PCR, and genomic DNA are used. The two types of ribosomal RNA genes (rDNA), 18S-5.8S-26S rDNA and 5S rDNA, have been used extensively as probes for physical mapping in higher plants because these genes are arranged in tandem arrays clustered at some sites per haploid genome (Fig. 1). Parts of these sequences are highly conserved beyond families. Probes are labeled with reporter molecules (biotin-16-dUTP or digoxigenin-11-dUTP) by means of nick-translation or random primer method. Probe mixtures are applied to slide, and the slide is placed in a humid chamber at 37°C for 6 h or longer to facilitate probe hybridization in situ. After hybridization, probes are detected with fluorochrome-conjugated avidin or antibody (Table 1). The hybridization signals are observed under a fluorescence microscope.

APPLICATIONS OF FISH

Multicolor FISH

Currently, a large number of probe-labeling and fluorescent reagents are available in different colors for simultaneous detection of multiple-target sequences (Table 1). Two or more sequences can be detected in the same cell by using fluorochromes of different colors (Fig. 1). Reid et al.[3] succeeded in simultaneously visualizing different DNA probes on human chromosomes by FISH using a combination of fluorescence and digital-imaging microscopy. Using three haptens (single, double, and triple labeling and three fluorochromes), a total of seven probes can be resolved ($2^3 - 1 = 7$). Increasing the number of hapten-fluorochromes to four would allow 15 different probes to be detected ($2^4 - 1 = 15$; five hapten-fluorochromes would detect 31 probes).

Encyclopedia of Plant and Crop Science
DOI: 10.1081/E-EPCS 120005681

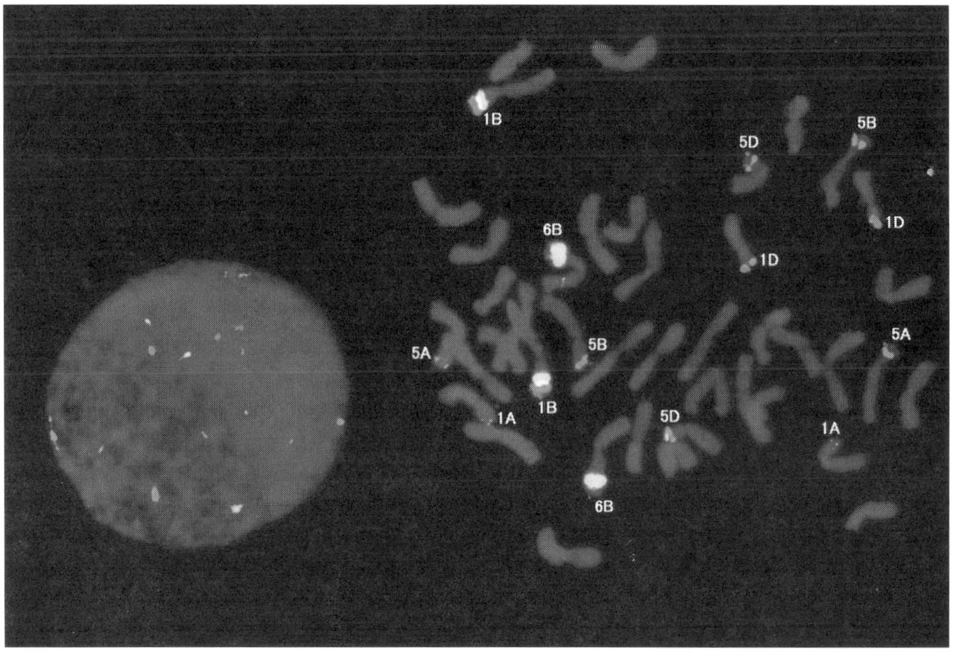

Fig. 1 Simultaneous mapping of 18S-5.8S-26S rRNA (green fluorescence, light grey) and 5S rRNA genes (red fluorescence, dark grey) on mitotic chromosomes and an interphase nucleus of hexaploid wheat using multicolor FISH. Chromosomes were counterstained by DAPI. (*View this art in color at www.dekker.com.*)

Multiple visualization of targets in a single hybridization experiment is of major importance for the physical mapping of genes along plant chromosomes. Using repeated sequences or chromosome-specific marker DNA for chromosome identification, another DNA probe labeled with a different reporter molecule can be allocated onto a single metaphase plate, dispensing with the need for further chromosome identification by banding analysis.[4]

Table 1 Labeling and detection reagents in FISH

Reporter molecules	
Indirect labeling	
Biotin-16-dUTP	
Digoxigenin-11-dUTP	
Direct labeling	
Fluorescein-12-dUTP	
Rhodamine-5-dUTP	
Cy3-dUTP	
Cy5-dUTP	
Fluorochrome	
Blue	AMCA
Green	FITC, Cy2, Alexa 488
Red	Rhodamine, Cy3, Alexa 594
Far red	Cy5, Cy7
Counterstain	
Blue	DAPI
Green	YOYO1
Red	PI

FISH Using Total Genomic DNA

Genomic in situ hybridization (GISH) has been an excellent technology for visualizing whole genomes, because the total genomic DNA used as a probe reflects the overall molecular composition of the genome.[5] The classical methods of genome analysis, involving cumbersome analysis of meiotic pairing behavior and pollen fertility of F_1 hybrids have various limitations. The GISH technique has overcome all these problems and has helped in the unequivocal discrimination of genomes of many polyploid species including wheat, oat, rice, finger millet, cotton, peanut, tobacco, coffee, and banana. Furthermore, because GISH can be successfully applied at somatic cell level to discriminate between genomes, it can be used as a promising tool in the genome analysis of woody species, which otherwise require a long time to reach sexual reproductive phase in order to observe chromosome association at meiosis. We can obtain more information about the molecular characterization of chromosomal abnormalities by combining the total genomic DNA probe as a genome marker with a cloned DNA probe, over chromosome banding. Translocations and insertions between different genomes are detected mostly in interspecific hybrids or alien chromosome-transfer lines.[6]

In Situ PCR

The in situ polymerase chain reaction (in situ PCR) is a newly developed method that combines the extreme sensitivity of PCR with the cytological localization of DNA or RNA sequences through in situ hybridization. This technique was first used to map plant genes by Mukai and Appels,[7] where the in situ location of rye-specific spacer region on metaphase chromosomes was determined using two pairs of primers designed for rye Nor-R1 and rye 5S-RNA-R1 sequences with the size of amplificants being 386 bp and 107 bp, respectively. The in situ PCR is a useful method for amplifying the small regions of DNA sequences of specific plant chromosomes and for mapping low-copy genes of interest.

FISH Using BAC Clones (BAC FISH)

The construction of large-insert genomic DNA libraries is essential for analyzing complex plant genomes. Bacterial artificial chromosome (BAC) libraries are used most widely at present because of several advantages over yeast artificial chromosomes (YACs). In FISH-mapping studies of plants, probes containing less than several kb of single-copy DNA do not reliably produce detectable signals using current FISH techniques in spite of its amenability of coupling with CCD camera and digital-imaging analysis. Alternative methods for detecting unique sequences involve the use of large genomic clones such as lambda phages, cosmids, BACs, and YACs. The signals from repeated sequences are suppressed by competitive hybridization with unlabeled total genomic or Cot1 DNA. Physical mapping by FISH using BAC clones as probes has been successful in plant species with relatively small genomes, such as sorghum, cotton, rice, *Arabidopsis*, tomato, and potato. In crops with large genomes such as wheat and onion, however, physical mapping of BAC clones is not successful even if the signals from repeated sequences are suppressed by competitive hybridization with unlabeled Cot1 DNA.

FISH on Extended DNA Fibers

In physical-mapping studies, FISH on extended DNA fibers from interphase nuclei is a useful tool for determining the sizes of target DNA sequences, the order of genes or clones, and their distances in a large chromosome region.[8] Currently, fiber FISH has the potential to trace the target sequences with lengths of up to 2.0 Mb on single, extended DNA fibers, a spatial resolution of 1 kb between adjacent targets, and a target detection sensitivity of as little as 700 bp in plants (Table 2). In transgenic plants, fiber FISH can physically map the transgenes directly on extended DNA fibers. This method is complementary to PCR, Southern blot, and sequence analyses. The fiber FISH technique contributes to the construction of BAC contigs, in chromosome walking, and map-based cloning. Molecular combing, a recently evolved method, has enabled direct mapping of purified BAC DNA molecules.[9]

Table 2 Detection sensitivity and mapping resolution in FISH in plants

	Detection sensitivity	Mapping resolution
Mitotic chromosomes	10 kb	5.0 Mb
Pachytene chromosomes	2.0 kb	1.0 Mb
Extended DNA fibers	700 bp	1.0 kb

Limitation of FISH Mapping

At present, target sequences less than 10 kb on a chromosome cannot reliably be detected through conventional FISH methods (Table 2). However, low-copy sequences, such as agronomically important genes encoding starch synthetic enzymes, seed storage proteins, and grain hardness, have been routinely mapped in wheat and rye using lambda phage clones containing inserts of 11 to 20 kb of genomic DNA sequence.

FUTURE PROSPECTS

FISH is now the technique of choice to physically visualize genomes and chromosomes and the order of chromosome segments, genes, and DNA sequences. Many applications and refinements in the technology now are available for microscopic visualization of DNA manifestation in situ, previously confined to gel-blot hybridization. With the various modifications and refinements for higher-resolution FISH now available, identifying genomes, chromosomes, and genes through sequence localization and orientation is suited to specific experimental objectives, including identifying agronomically useful genes and their localization, integration sites, and copy number in transgenics.

Through FISH analysis using human chromosome-specific painting probes, chromosome synteny between human and other mammalian species has been studied. This comparative chromosome painting was termed zoo-FISH. Lately, chromosome painting has been applied successfully in *Arabidopsis* using chromosome-specific BAC clones and multicolor FISH.[10] This method could help establish chromosomal homologies between *Arabidopsis* chromosomes and other Brassicaceae species. FISH using total repetitive DNA of a reference species as a probe, such as rice/*Arabidopsis* DNA, also could be used to identify repetitive-DNA and bar-coding markers for chromosome identification and comparative mapping between species.[11] This would have tremendous implications in the microidentification of cryptic structural changes and chromosome diversity over populations. Pooled BAC DNA clones derived from individual chromosomes of an anchor species are used as probes for FISH on chromosomes of other related species. This innovative FISH technology visualizes cytogenetic homologies that refine the comparative maps constructed by molecular gene mapping of individual loci and opens new avenues for genomics by facilitating the extrapolation of results from the genome projects.

REFERENCES

1. Yamamoto, M.; Mukai, Y. Application of fluorescence in situ hybridization to molecular cytogenetics of wheat. Wheat Inf. Serv. **1989**, *69*, 30–32.
2. Gall, J.G.; Pardue, M.L. Formation and detection of RNA-DNA hybrid molecules in cytological preparations. Proc. Natl. Acad. Sci. U. S. A. **1969**, *63*, 378–383.
3. Reid, T.; Baldini, A.; Rand, T.C.; Ward, D.C. Simultaneous visualization of seven different DNA probes by in situ hybridization using combinatorial fluorescence and digital imaging microscopy. Proc. Natl. Acad. Sci. U. S. A. **1992**, *89*, 1388–1392.
4. Mukai, Y. Multicolor Fluorescence in situ Hybridization: A New Tool for Genome Analysis. In *Methods of Genome Analysis in Plants*; Jauhar, P.P., Ed.; CRC Press: Boca Raton, FL, 1996; 181–192.
5. Mukai, Y. Molecular-cytogenetic Analysis of Plant Chromosomes by in situ Hybridization. In *Plant Genome and Plastome: Their Structure and Evolution in Commemoration of Professor Hitoshi Kihara's Centennial*; Tsunewaki, K., Ed.; Kodansha Scientific, Ltd.: Tokyo, 1995; 45–51.
6. Mukai, Y.; Friebe, B.; Hatchett, J.H.; Yamamoto, M.; Gill, B.S. Molecular cytogenetic analysis of radiation-induced wheat-rye terminal and intercalary chromosomal translocations and the detection of rye chromatin specifying resistance to Hessian fly. Chromosoma **1993**, *102*, 88–95.
7. Mukai, Y.; Appels, R. Direct chromosome mapping of plant genes by in situ polymerase chain reaction (in situ PCR). Chromosome Res. **1996**, *4*, 401–404.
8. de Jong, J.H.; Fransz, P.; Zabel, P. High resolution FISH in plants-techniques and applications. Trends Plant Sci. **1999**, *4*, 258–263.
9. Jackson, S.A.; Dong, F.; Jiang, J. Digital mapping of bacterial artificial chromosomes by fluorescence in situ hybridization. Plant J. **1999**, *17*, 581–587.
10. Lysak, M.A.; Pecinka, A.; Schubert, I. Recent progress in chromosome painting of *Arabidopsis* and related species. Chromosome Res. **2003**, *11*, 195–204.
11. Lavania, U.C. Chromosome diversity in population: Defining conservation units and their microidentification through genomic in situ painting. Curr. Sci. **2002**, *83*, 124–127.

Fluorescence: A Probe of Photosynthetic Apparatus

J. Kenneth Hoober
Arizona State University, Tempe, Arizona, U.S.A.

INTRODUCTION

Absorption of light by chlorophyll is the cardinal event in photosynthesis. The absorption of a photon, with the exact amount of energy required to raise an electron to a higher orbital, generates an excited state from which chlorophyll molecules readily donate an electron to another molecule. The process of transferring an electron to an acceptor, a photochemical oxidation–reduction reaction, results in productive initiation of the process of photosynthesis. This reaction, in which the energy of the photon is trapped by a chemical reaction, is often referred to as photochemical quenching of the excited state. However, when an electron acceptor is not available, the excited state of the chlorophyll molecule decays, in part by release of heat and in part by emission of a photon of lower energy—the process of fluorescence.

FLUORESCENCE TRANSIENTS IN GREEN CELLS

Chlorophyll *a* is a highly fluorescent molecule (Fig. 1). In organic solvents, an environment that minimizes quenching by the solvent, the quantum yield (the number of photons emitted per number of photons absorbed) is in the range of 0.31 to 0.35. Chlorophyll is relatively insoluble in water, and thus it functions within the nonpolar environment provided by thylakoid membranes. The interior of the membrane is rich in the hydrocarbon portions of the membrane lipids, and thus provides an organic phase within the chloroplast. Although free chlorophyll in a membrane can retain much of its inherent fluorescence, most if not all chlorophyll is attached to proteins that position the molecules in such a fashion that absorbed energy is readily transferred to other molecules and thus eventually trapped. More than 90% of absorbed light quanta are productively converted to electrons in photosynthesis. Therefore, chlorophyll in thylakoid membranes in vivo exhibits a low level of fluorescence, with a maximal quantum yield of only 0.03 to 0.05.[1] This level is reached when the photosynthetic apparatus is saturated, a physiological state that does not allow further trapping of energy.

Changes in the low level of fluorescence of chlorophyll in chloroplasts reflect its functional state, and thus measurement of fluorescence under a variety of conditions is a sensitive means to monitor photosynthetic activity. The fluorescent yield depends upon the ability of the remainder of the photosynthetic system to trap absorbed energy. Fig. 2 illustrates measurements made on a short time scale that describe the status of the reaction centers of photosystem II. In this experiment, cells of the green alga *Chlamydomonas reinhardtii* were dark-adapted for several minutes and then exposed to a modulated red light source (wavelength of maximal intensity, 650 nm) at a low intensity of 2.5 μmol photons $m^{-2}s^{-1}$.[2] (Light of this wavelength is absorbed primarily by chlorophyll *b* in the light-harvesting antenna. Because of the close association of chlorophyll *b* with chlorophyll *a* within light-harvesting complexes of the antenna, energy is rapidly transferred to chlorophyll *a* on a time scale of several hundred femtoseconds to a few picoseconds.) A low, intrinsic level of fluorescence, designated F_o, is released by the antenna when the reaction centers and electron acceptors are fully oxidized. Upon exposure to a higher intensity (80 μmol photons $m^{-2}s^{-1}$) actinic red light, a rapid increase in fluorescence was observed on the time scale of several hundred milliseconds.

The first group of antenna complexes to become fully fluorescent were those not functionally connected to reaction centers, indicated by F_{pl}, a plateau level that varies with conditions; these unconnected complexes were relatively highly fluorescent. The rise in fluorescence to the peak value, F_p, is a measure of the rate at which a plastoquinone molecule designated Q_A, the primary acceptor of photosystem II, is reduced. When the rate of reduction of Q_A is greater than the rate at which electrons can be removed from the reaction center by subsequent electron carriers, the reaction center becomes saturated and any additional energy absorbed by the antenna, which cannot be trapped, is then released by heat and/or fluorescence. The rate at which fluorescence rises from F_{pl} to F_p is a function of the intensity of actinic light as well as the size of the antenna that absorbs the light.[3] The subsequent fall from the peak level of fluorescence occurs when the pathway for carbon fixation is activated by light and electrons are transported through photosystem I to $NADP^+$. Opening the gate at the end of the photochemical

Encyclopedia of Plant and Crop Science
DOI: 10.1081/E-EPCS 120019318

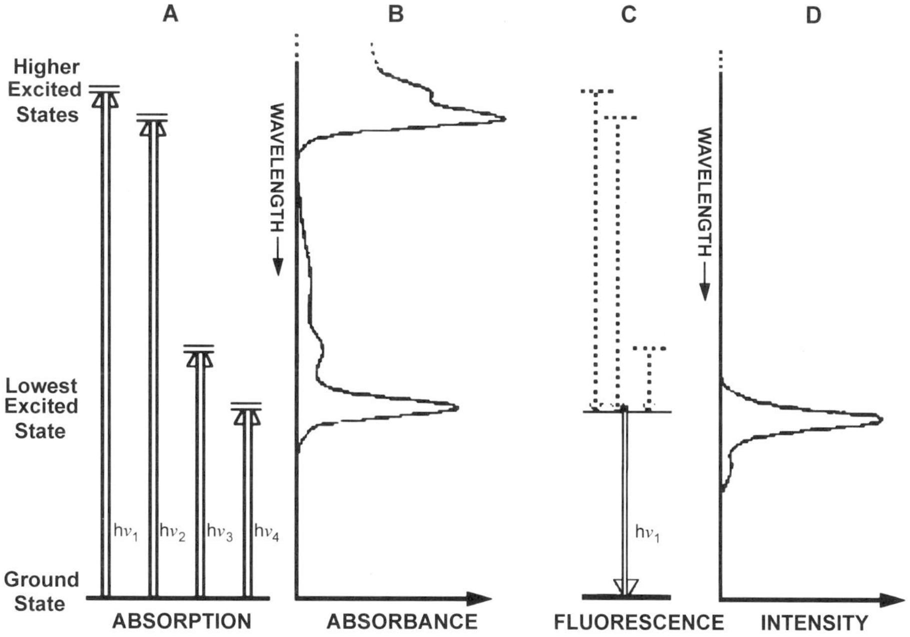

Fig. 1 Absorption and fluorescence of chlorophyll *a*. A. Energy levels within the molecule allow absorption of photons, which are described by the absorption spectrum shown in B. Increases in energy states by $h\nu_1$ to $h\nu_4$ describe absorption bands referred to as the B_y and B_x Soret and the Q_x and Q_y peaks, respectively. C. Return to the ground state from higher excited states initially occurs by internal, radiationless decay to the lowest excited state, from which the molecule achieves the ground state by emission of energy as fluorescence, whose spectrum is shown in D. (Adapted from Ref. 9.)

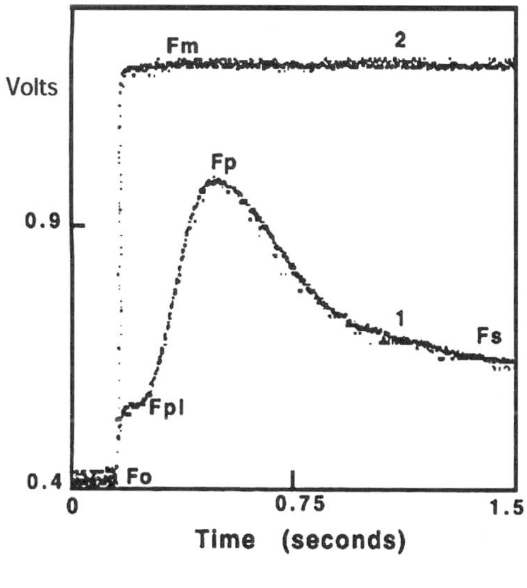

Fig. 2 Fluorescence induction kinetics displayed by green cells of the alga *Chlamydomonas reinhardtill.* Fluorescence was measured at room temperature in the absence (curve 1) or presence (curve 2) of 10 μM DCMU, a herbicide that blocks exit of electrons from photosystem II. (From Ref. 2.)

pathway results in re-oxidation of the plastoquinone pool and thus makes more electron acceptors available for photosystem II. The final steady-state level of fluorescence (F_s) is an indication of how well the two photosystems are connected, or how readily electrons can be transferred through the complete system to the carbon fixing pathway.

Curve 1 in Fig. 2 describes the typical fluorescence transient for healthy chloroplasts. Deviation from this pattern is diagnostic of a change in the physiological status of photosystem II. For example, some herbicides act by blocking the transfer of electrons from Q_A of photosystem II to the next electron carrier. Curve 2 in Fig. 2 shows the change in fluorescence upon exposure to actinic light when such an inhibitor was added. The electron acceptor Q_A no longer donated electrons to Q_B and thus rapidly became reduced. Energy flow through the reaction center from the antenna was blocked, and all the energy absorbed by the antenna had to reemitted, some of it as fluorescence. This condition, therefore, yields the maximal amount of fluorescence that can be generated in the assay, which is termed F_m. Maximal fluorescence of the system can also be obtained by exposure to an intense burst of white light, which floods the reaction center with excitons at a higher rate than Q_A can release electrons.[4]

The maximal quantum yield of photosystem II is expressed by a simple relationship:

$$\Phi_F = F_m - F_o \text{ (variable fluorescence}, F_v) / F_m \text{ (maximal fluorescence)}$$

This value is between 0.7 and 0.8 for healthy leaves. When the reaction center is damaged, one result is an increase in F_o, which reduces the value of F_v. The decrease in this ratio indicates a reduced yield of photosynthesis.

State 1 is the condition in which the protein–chlorophyll complexes of the major light-harvesting antenna (LHCII) are connected to photosystem II. When the intensity of incident light is very high and Q_A exists in a relatively reduced state, LHCII dissociates from photosystem II and transfers more energy into photosystem I. This arrangement is designated state 2. Photosystem I is kinetically a faster complex than photosystem II,[5] and thus trapping of energy absorbed by LHCII is more efficient in state 2. The maximal level of fluorescence of the antenna that can be achieved with an intense flash of light is thereby reduced. Because photosystem II is located within the granal stacks of thylakoid membranes, whereas photosystem I is more peripherally distributed,[6] the term 'state transition' implies a reorganization of membrane components.

A reduction in the maximal emission intensity, F_m, is also achieved by the process of nonphotochemical quenching, by which absorbed light energy is dissipated as heat. This mechanism requires the LHCII-associated protein PsbS, the xanthophyll zeaxanthin, and a relatively low pH within the thylakoid lumen. The extent of nonphotochemical quenching is determined from the attenuation of F_m to lower values, termed F_m', during a series of light flashes.

FLUORESCENCE SIGNATURE OF DEVELOPING THYLAKOID MEMBRANES

The change in pattern of the fluorescence transient provides important information on the status of photosystems during chloroplast development. During the early phase, thylakoid membranes initially form as small vesicles, which expand into lamellar structures that eventually adhere to form grana.[7] Segregation of photosystems I and II does not occur until granal stacks are formed. Thus, in the separate, initially formed vesicles, reaction centers and antenna complexes should be distributed randomly within the membrane, with the light-harvesting complexes feeding exciton energy to photosystem I as well as photosystem II. Electrons produced by photosystem II are then rapidly pulled through the electron transport chain connecting the photosystems, and Q_A remains oxidized. This arrangement is characteristic of a state 2 condition. As shown in Fig. 3A, at the beginning of chloroplast development in *yellow* mutants of *Chlamydomonas reinhardtii*, reaction centers were not available for the small number of antenna complexes present, and addition of DCMU, an inhibitor of electron transfer from Q_A to Q_B, had no effect. After a short period of chloroplast development, actinic light caused a rise to the same plateau level of fluorescence (Fig. 3B). The typical rise in fluorescence to F_p was not observed, even though a large increase to F_m occurred upon addition of DCMU, which demonstrated that functional photosystems had formed and some of the energy absorbed by LHCII was efficiently trapped by photosystem II. In these experiments, the typical fluorescence transient shown in Fig. 2 emerged gradually as the growing thylakoid membranes adhered to form grana.[2]

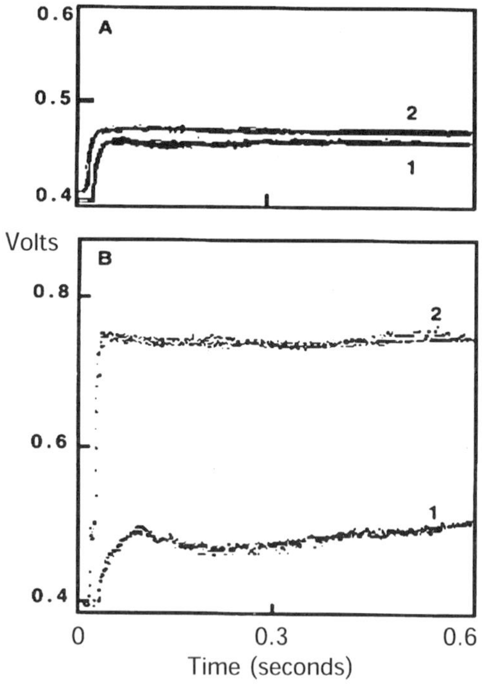

Fig. 3 Fluorescence induction displayed by yellow cells of the alga *Chlamydomonas reinhardtii* (A) at the beginning of chloroplast development and (B) after one hour of exposure to light. The cellular content of chlorophyll increased approximateldy four-fold during this period. Curve 1 in each panel was obtained with cells in the absence of DCMU, whereas curve 2 in each panel was obtained with cells treated with 10 μM DCMU. (From Ref. 2.)

CONCLUSION

Short-term fluorescence transients provide a sensitive and detailed probe with which to monitor the functional state of photosystem II. These measurements can also be used to monitor changes over extended periods of illumination. The general concept of chlorophyll fluorescence as a functional probe of the physiological state of photosystem II can be used to monitor a number of different parameters in addition to those shown here.[8]

ARTICLES OF FURTHER INTEREST

Chlorophylls, p. 258
Exciton Theory, p. 429
Photosystems: Electron Flow Through, p. 906

REFERENCES

1. Krause, G.H.; Weis, E. Chlorophyll fluorescence and photosynthesis: The basics. Annu. Rev. Plant Physiol. Plant Mol. Biol. **1991**, *42*, 313–349.
2. White, R.A.; Hoober, J.K. Biogenesis of thylakoid membranes in *Chlamydomonas reinhardtii y*1. Plant Physiol. **1994**, *106*, 583–590.
3. Strasser, R.J.; Srivastava, A.; Govindjee. Polyphasic chlorophyll *a* fluorescence transient in plants and cyanobacteria. Photochem. Photobiol. **1995**, *61*, 32–42.
4. Schreiber, U.; Bilger, W.; Hormann, H.; Neubauer, C. Chlorophyll Fluorescence as a Diagnostic Tool: Basics and Some Aspects of Practical Relevance. In *Photosynthesis: A Comprehensive Treatise*; Raghavendra, A.S., Ed.; Cambridge Univ. Press: Cambridge, United Kingdom, 1998; 320–336.
5. Trissl, H.W.; Wilhelm, C. Why do thylakoid membranes from higher plants form grana stacks? Trends Biochem. Sci. **1993**, *18*, 415–419.
6. Albertsson, P.-Å. The structure and function of the chloroplast photosynthetic membrane—A model for the domain organization. Photosynth. Res. **1995**, *46*, 141–149.
7. Hoober, J.K.; Boyd, C.O.; Paavola, L.G. Origin of thylakoid membranes in *Chlamydomonas reinhardtii y*-1 at 38°C. Plant Physiol. **1991**, *96*, 1321–1328.
8. Fracheboud, Y. http://www.ab.ipw.agrl.ethz.ch (accessed March 2003).
9. Sauer, K. Primary Events and the Trapping of Energy. In *Bioenergetics of Photosynthesis*; Govindjee, Ed.; Academic Press: New York, 1975; 115–181.

Functional Genomics and Gene Action Knockouts

Jacqueline Heard
Mendel Biotechnology, Hayward, California, U.S.A.

INTRODUCTION

Functional genomics describes an organism-wide approach to the experimental characterization of gene function that capitalizes on information gathered as a consequence of genome sequencing. An era of intense genomic analysis of plant species has just begun. In 2000, the first complete genome sequence of a model plant, *Arabidopsis thaliana* (Arabidopsis), was released.[1] Very recently, a publicly available draft sequence of the rice genome was also released.[2] Arabidopsis and rice represent model species for the study of dicots and monocots, respectively—the two most important lineages in the plant kingdom with respect to the commercial application of genomic research. These two genomes provide the first view into which genes and proteins shape the growth and development of plants. Functional genomics seeks to uncover gene function using tools on a genome-wide scale, placing the transcripts, proteins, and metabolites into complex control networks with the goal of dissecting and understanding mechanisms that control key plant traits.

GENOMICS, STRUCTURE TO FUNCTION, AND THE CONVERGENCE OF THEORETICAL AND EXPERIMENTAL BIOLOGY

Structural and comparative genomics provide a foundation for the organism-wide analysis of gene function, and are vital to our ability to leverage the information we learn from model species to other genomes of commercial value. It is generally anticipated that by examining structural components of genomes—understanding the degree of synteny or colinearity between model species and crop species—the functional analysis of a protein in a model system will lead to the subsequent identification of the functional ortholog in crops. Certainly, the shortcomings of any model system will become especially apparent as the evolutionary distance between organisms increases. As a result, several plant species are being developed for genomic-scale analysis across major crop taxa, including the Brassicaceae, Poaceae, Fabaceae, and Solanaceae (Table 1).

Particularly vital to the functional characterization of genomes is the determination of the complement of coding regions—one of the first steps to the structural characterization of a genome. Gene structure predictions based on current gene-finding algorithms are correct only about half the time.[3] It is clear that at this time, predictive algorithms must be supplemented by various experimental approaches, such as the use of microarray technology and the collection of full-length cDNAs in order for the genome annotation to be accurate and comprehensive.[3]

Once the complement of genes that make up a genome has been predicted, sophisticated methodologies that have come to ''define'' functional genomics research facilitate the characterization of expressed genes (transcriptome), proteins (proteome), and metabolites (metabolome) of a model organism in the context of their cellular and subcellular location, a given environment, and genotypic background. The uses of microarray analysis, proteomics, and metabolomics for genome-wide analyses in plants have been reviewed recently.[4–7] These approaches, combined with the phenotypic characterization of gene action knockouts and transgenic lines ectopically expressing a gene of interest, will provide the experimental basis for the assignment of gene function.

Massive data collection efforts and computational modeling of pathways will be needed to finally place the transcripts, proteins, and metabolites into complex control networks. It is likely that functional genomics research will increasingly challenge the way we process data, requiring data integration across several diverse disciplines, and will require an unprecedented level of convergence between theoretical and experimental approaches.

ASSIGNING FUNCTIONS TO GENES: INSERTIONAL MUTAGENESIS AND GENE ACTION KNOCKOUTS

An estimated 30–40% of the genes in a given genome encode proteins of unknown function, those for which no functional information is available, and those for which a putative function cannot be predicted based on

Encyclopedia of Plant and Crop Science
DOI: 10.1081/E-EPCS 120010632

Table 1 Examples of model plants across major crop taxa: EST count was accessed June 2002 from GenBank at http://www.ncbi.nlm.nih.gov/dbEST/dbEST_summary.html

Model plant	Crop taxa	Genome size	EST # genbank	Mutagenesis stategies	Resource links
Arabidopsis	Brassicaceae	~125 Mb	~175 K	T-DNA *Ac/Ds* *En/Spm*	http://arabidopsis.org/info/agi.html
Tomato	Solanaceae	~900 Mb	~160 K	*Ac/Ds* Fast neutron EMS	http://www.sgn.cornell.edu/; http://www.tigr.org/tdb/tgi/lgi/;
Lotus	Fabaceae	~400 Mb	~32 K	EMS Fast neutron T-DNA *Ac/Ds*	http://www.kazusa.or.jp/en/plant/lotus/EST
Medicago	Fabaceae	~500 Mb	~160 K	T-DNA EMS	http://www.noble.org/medicago/index.html; http://medicago.org/; http://medicago.toulouse.inra.fr/EU/mtindex.htm
Rice	Poaceae	~140 Mb	~105 K	*Tos17*	http://btn.genomics.org.cn/rice; http://genome.sinica.edu.tw/; http://www.rice-research.org/; http://www.tmri.org; http://www.genomics.org.cn:7001/index.jsp; http://bioserver.myongji.ac.kr/ricemac.html; http://www.gramene.org/
Maize	Poaceae	~2500 Mb	~160 K	*AC/DS* *MuDR/Mu*	http://www.maizemap.org/index.htm; http://www.stanford.edu/~walbot/walbot_genedisc.html; http://www.cbc.umn.edu/ResearchProjects/Maize/AZ_info.html; http://mtm.cshl.org/; http://www.zmdb.iastate.edu/zmdb/sitemap.html; http://www.agron.missouri.edu/

biochemical properties. For the remaining roughly 50–60% of the proteins—those with a putative function to which no true functional assignment has been made—a fairly simple database search of a model organism would mine the genome for a biochemical functionality of interest. However, simply knowing that a gene encodes a MADS-box type transcription factor, for example, does not inform us about the function it actually performs in the plant. Transcript profiling and proteomics will allow predictions to be made regarding gene function, but in the end the hypotheses generated will need to be validated *in planta*. Gene function must ultimately be assigned on the basis of phenotypic evidence, and is most easily done in model organisms. Insertion mutagenesis is probably the most technically feasible approach for the systematic analysis of gene function in the context of the whole organism. A variety of mutagenesis strategies have been devised to create and screen for tagged knockouts in genes of interest using transfer-DNA (T-DNA) and transposons.[8] For over a decade, Arabidopsis has been used intensively as a model plant species for the determination of gene function. Forward genetics has generated functional information for less than 5% of the Arabidopsis genes to date.[9] A reverse-genetics approach, where a collection of knockouts in all genes in a genome can be screened for a mutation in a gene of interest, allows a large-scale systematic assignment of function to genes in the genome—an organism-through-phenotype characterization.

Once function has been assigned to a protein in a model organism such as Arabidopsis, examining the function of the putative ortholog in a crop species will almost certainly follow. Knockout collections have been and are being generated that allow the retrieval of a mutation in a gene of interest through PCR screening in a variety of plant species including Arabidopsis, maize, petunia, Lotus, alfalfa, tomato, snapdragon, rice, and *Brachypodium*. Table 1 summarizes information regarding a number of the species listed above. In the case of Arabidopsis, a variety of screenable collections exist and more recently, a catalog of sequenced insertion events from individual lines is being created where a scientist simply requests seed for a particular mutant of interest.[10] This resource is being created as part of the ambitious multinational project, the Arabidopsis 2010 project funded by The National Science Foundation. The goal of the 10-year endeavor is to complete the functional analysis of the Arabidopsis genome by the year 2010.[9] This academic project is unconventional by design, in that it plans for ''technology centers'' to provide resources such as insertional mutants and microarray data.[9] The participants recognize the efficiency gains that can be made by centralizing work that is best suited for efficient high-throughput methods.[9] The project also plans for the development of databases and bioinformatics tools that will facilitate public access to the data. The information and tools generated will be used for computer modeling in an effort to move toward the creation of a ''virtual plant.''[9]

OTHER MUTAGENESIS STRATEGIES FOR REVERSE GENETICS

The limitations to creating and using a collection of insertional mutants in any organism are genome size, transformability, the degree of redundancy, and ease of growth. When creating a saturating population of insertional mutants is problematic due to genome size and the lack of efficient gene transfer methods, other mutagenesis strategies such as high-density fast neutron[11] and chemical mutagenesis can be used. An elegant technique called Targeting Induced Local Lesions in Genomes (TILLING) has recently been developed wherein a point mutation in any gene of interest in theoretically any plant of interest can be detected by heteroduplex analysis, provided some gene sequence knowledge is available.[12]

Another common characteristic of plant genomes that limits the power of insertional mutagenesis is the degree of duplication. This problem can be quite substantial in plants in which many proteins have co-expressed isovariants, making it difficult to identify function through a single insertion event in any one of the isovariants.[13] Having the complete genome sequence available allows for predictions of functional redundancy wherein gene knockouts would be needed in several genes in order to reveal function; however, many genes for isovariants will be closely linked and therefore difficult to make double insertion mutant knockouts. Where ease of gene transfer is not an issue, gene-silencing approaches such as post-transcriptional gene silencing (PTGS) or RNA interference (RNAi) could be used to knock out the activity of several genes simultaneously.[14]

The reader is referred to other articles in the section on Structural and Functional Genomics for a detailed description of the mutagenesis strategies discussed here.

CONCLUSION

The goal of functional genomics is not only to improve our basic knowledge of plants, but also to provide plant biotechnology with the information and technology needed to create improved crop cultivars. Clearly, one of the steps needed to facilitate the application of our knowledge is the development of new strategies to manipulate plant genomes and to reduce the time and expense for commercial development of new traits. Controlling the integration of transgenes through recombinase-directed plant transformation offers a way to reduce the unpredictability in transgene expression levels, and allows for the removal of unwanted selectable markers.[15] Gene-targeting approaches offer similar advantages; however, homologous recombination occurs at extremely low frequency in higher plants and is not a feasible strategy at this point for either gene discovery or commercial applications.[16] Targeted mutations can also be created using RNA/DNA oligonucleotide hybrids[17] and with efficiency improvements, the technique offers another potentially promising tool to facilitate the application of the knowledge gained through functional genomics research to the improvement of crop species.

ARTICLES OF FURTHER INTEREST

Arabidopsis thaliana: Characteristics and Annotation of a Model Genome, p. 47
Genome Size, p. 516
RNA-Mediated silencing, p. 1106
Transformation Methods and Impact, p. 1233
Transgenes: Expression and Silencing of, p. 1242

REFERENCES

1. The Arabidopsis Genome Initiative. Analysis of the genome sequence of the flowering plant *Arabidopsis thaliana*. Nature **2000**, *408* (6814), 796–815.
2. http://btn.genomics.org.cn/rice (accessed June 2002).
3. Pertea, M.; Salzberg, S.L. Computational gene finding in plants. Plant Mol. Biol. **2002**, *48* (1), 39–48.
4. Aharoni, A.; Vorst, O. DNA microarrays for functional plant genomics. Plant Mol. Biol. **2002**, *48* (1–2), 99–118.

5. Kersten, B.; Burkle, L.; Kuhn, E.J.; Giavalisco, P.; Konthur, Z.; Lueking, A.; Walter, G.; Eickhoff, H.; Schneider, U. Large-scale plant proteomics. Plant Mol. Biol. **2002**, *48* (1–2), 133–141.
6. Roberts, J.K.M. Proteomics and a future generation of plant molecular biologists. Plant Mol. Biol. **2002**, *48* (1–2), 143–154.
7. Fiehn, O. Metabolomics—The link between genotypes and phenotypes. Plant Mol. Biol. **2002**, *48* (1–2), 143–154.
8. Maes, T.; De Keukeleire, P.; Gerats, T. Plant tagnology. Trends Plant Sci. **1999**, *4* (3), 90–96.
9. Somerville, C.; Dangl, J. Genomics. Plant Biology 2010. Science **2000**, *290* (5499), 2077–2078.
10. http://signal.salk.edu/tabout.html (accessed June 2002).
11. Li, X.; Song, Y.; Century, K.; Straight, S.; Ronald, P.; Dong, X.; Lassner, M.; Zhang, Y. A fast neutron deletion mutagenesis-based reverse genetics system for plants. Plant J. **2001**, *27* (3), 235–242.
12. McCallum, C.M.; Comai, L.; Greene, E.A.; Henikoff, S. Targeted induced local lesions in genomes (TILLING) for plant functional genomics. Plant Physiol. **2000**, *123* (2), 439–442.
13. Vision, T.J.; Brown, D.G.; Tanksley, S.D. The origins of genomic duplications in Arabidopsis. Science **2000**, *290* (5499), 2114–2117.
14. Wang, M.B.; Waterhouse, P.M. Application of gene silencing in plants. Curr. Opin. Plant Biol. **2002**, *5* (2), 146–150.
15. Ow, D.W. Recombinase-directed plant transformation for the post-genomic era. Plant Mol. Biol. **2002**, *48* (1–2), 183–200.
16. Puchta, H. Gene replacement by homologous recombination in plants. Plant Mol. Biol. **2002**, *48* (1–2), 173–182.

F

Fungal and Oomycete Plant Pathogens: Cell Biology

Adrienne R. Hardham
Australian National University, Canberra, Australian Capital Territory, Australia

INTRODUCTION

Plants have developed many strategies to stop potential pathogens from being able to infect and establish disease. These strategies include preformed and induced physical barriers (such as strong cell walls) and chemical inhibitors. However, there are many thousands of species of eukaryotic microbes, such as fungi and funguslike organisms, that have evolved mechanisms to overcome these defenses. They are able to colonize and reproduce in selected host plants, causing huge economic losses in important food and commodity crops as well as environmental damage in natural ecosystems.

Traditionally, the fungal kingdom included six major groups of organisms: Basidiomycetes, Ascomycetes, Zygomycetes, Deuteromycetes, Chytridiomycetes, and Oomycetes. All form hyphae that grow by extension at the hyphal apex, and all have an absorptive mode of nutrition. In all but the Oomycetes, chitin is the main microfibrillar component of the cell wall, and all but the Oomycetes are haploid during a major part of their life cycle. Recent molecular sequence analysis has confirmed that the structural, biochemical, and genetic differences between the Oomycetes and the other five groups are indicative of the distinct phylogeny of the Oomycetes, which are more closely related to brown algae than they are to true fungi. Nevertheless, Oomycete plant pathogens—and there are many that cause devastating plant diseases—must overcome the same range of plant defenses encountered by the true fungi. As details of the cellular and molecular basis of the infection process unfold, the similarities and differences in the infection strategies employed by these two groups of plant pathogens will become increasingly clear.

THE INITIATION OF PLANT INFECTION

Spore Adhesion

Fungal and Oomycete pathogens usually reach a potential host plant as spores that are dispersed in wind or water. Having made initial contact, it is important that the spores quickly become attached to the plant surface so that they are not blown or washed away before establishing infection. In some cases, including the highly successful and destructive pathogens *Magnaporthe grisea* (causing rice blast) and *Phytophthora infestans* (causing late blight of potato), the spores release adhesive material within minutes of making contact with the plant. *M. grisea* conidia store preformed adhesive in the periplasm at the spore tip[1] and release it upon hydration. *Phytophthora* species store preformed adhesive in small vesicles under the ventral surface of motile spores that are able to position themselves with precision on the plant surface before secreting the adhesive (Fig. 1: 1–4).[2] In other cases, the adhesive material is released more slowly and is apparently synthesized after contact with the potential host. Pathogen spores also secrete other compounds onto the host surface, including mucilaginous material that may prevent desiccation (Fig. 1: 4) and enzymes (such as cutinases) that may degrade the plant surface.[3]

Spore Germination and Hyphal Tip Growth

Once attached to the plant surface, spores germinate and seek suitable sites from which to attempt to penetrate the plant's outer surface. In many cases, the site of spore germination is determined by physical and chemical factors in the spore's environment, but in other cases, the site of spore germination is predetermined. The spores of some rust fungi, for example, have preformed germination pores; in *Phytophthora* spores, the germ tube emerges from the center of the ventral surface, which has been oriented toward the plant surface.[2] Germ tubes grow by extension at the hyphal tip, achieved by the localized fusion of numerous small apical vesicles carrying wall materials and enzymes. Many pathogens orient their hyphal growth according to the topography of the underlying plant surface. Hyphae may grow along (Plate II (5)) or perpendicular to the grooves formed by the anticlinal walls of the epidermal cells.[4] This thigmotropic growth is believed to enhance the pathogen's chances of finding a suitable infection site.

Encyclopedia of Plant and Crop Science
DOI: 10.1081/E-EPCS 120019925

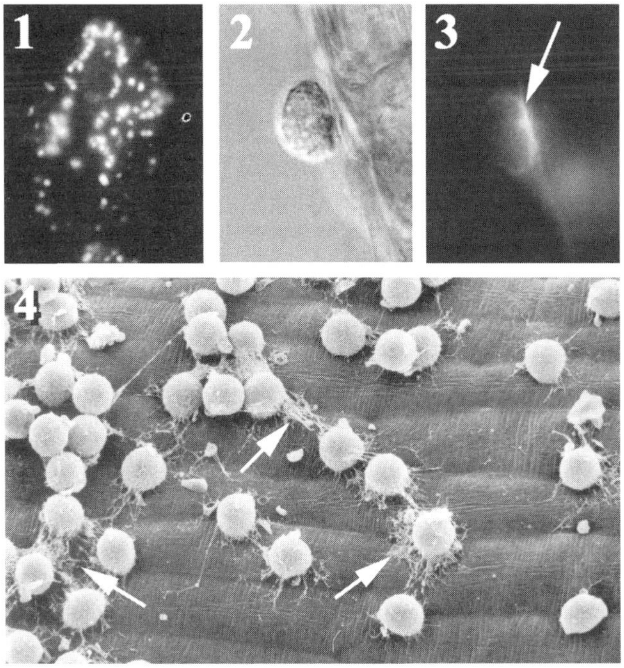

Fig. 1 (**1**) Zoospore of *Phytophthora cinnamomi* labeled with an antibody (Vsv1) that reacts with adhesive proteins stored in small vesicles under the ventral surface of the spore. Immunofluorescence micrograph (×700). (**2**) Encysted spore of *P. cinnamomi* attached to the surface of an onion root. Differential interference contrast micrograph (×700). (**3**) The same spore as shown in (2). Adhesive material secreted by the spore is labeled by an antibody (Vsv1) that reacts with the adhesive protein (arrow). Immunofluorescence micrograph (×700). (**4**) Spores of *P. cinnamomi* that have preferentially settled over the grooves formed by the anticlinal walls of onion root epidermal cells. Remnants of mucilaginous material can be seen surrounding the spores (arrows). (Scanning electron micrograph (×500) courtesy of Division of Entomology, CSIRO.)

HOST COLONIZATION

Host Penetration

Many fungal and Oomycete species that are foliar pathogens enter the leaf through the stomatal pore between two guard cells (Fig. 2: 5 and 6). Other species invade the plant surface, be it root or shoot, by growing between the epidermal cells along the anticlinal wall (Fig. 2: 7) or by penetraing directly through the outer epidermal cell wall. In each of these situations, the hyphal apex often differentiates into an appressorium (Fig. 2: 5 and 6; Fig. 3: 7 and 8), a cell specialized to overcome the formidable barrier presented by the outer epidermal cell wall. In rust fungi, appressorium differentiation is triggered by contact with the ridges formed by the stomatal guard cells (Fig. 2: 5 and 6), a response that can also be induced by ridges of similar dimensions on inert substrata,

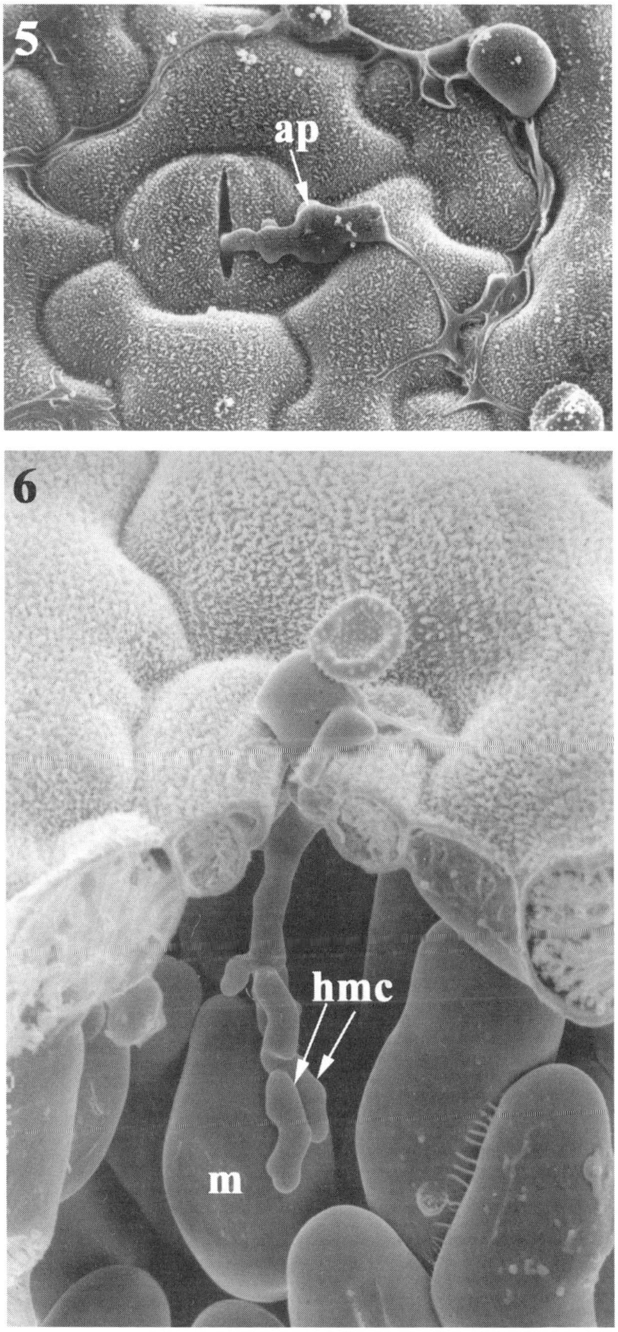

Fig. 2 (**5**) Urediniospores of the flax rust fungus, *Melampsora lini*, germinated on the surface of a flax leaf. The germ tubes have grown along the grooves formed by the anticlinal walls of the epidermal cells. Contact with a guard cell has triggered the formation of an appressorium (ap) from which the fungus has grown into the stomatal pore. (Cryo-scanning electron micrograph (×600) courtesy of Dr. I. Kobayashi and the Division of Plant Industry, CSIRO.) (**6**) A *M. lini* appressorium has produced a germ tube that has grown through a stomatal pore into the substomatal cavity in a flax leaf. Infection hyphae have contacted a mesophyll cell (m) and have differentiated into haustorial mother cells (hmc). (Cryo-scanning electron micrograph (×860) courtesy of Dr. I. Kobayashi and the Division of Plant Industry, CSIRO.)

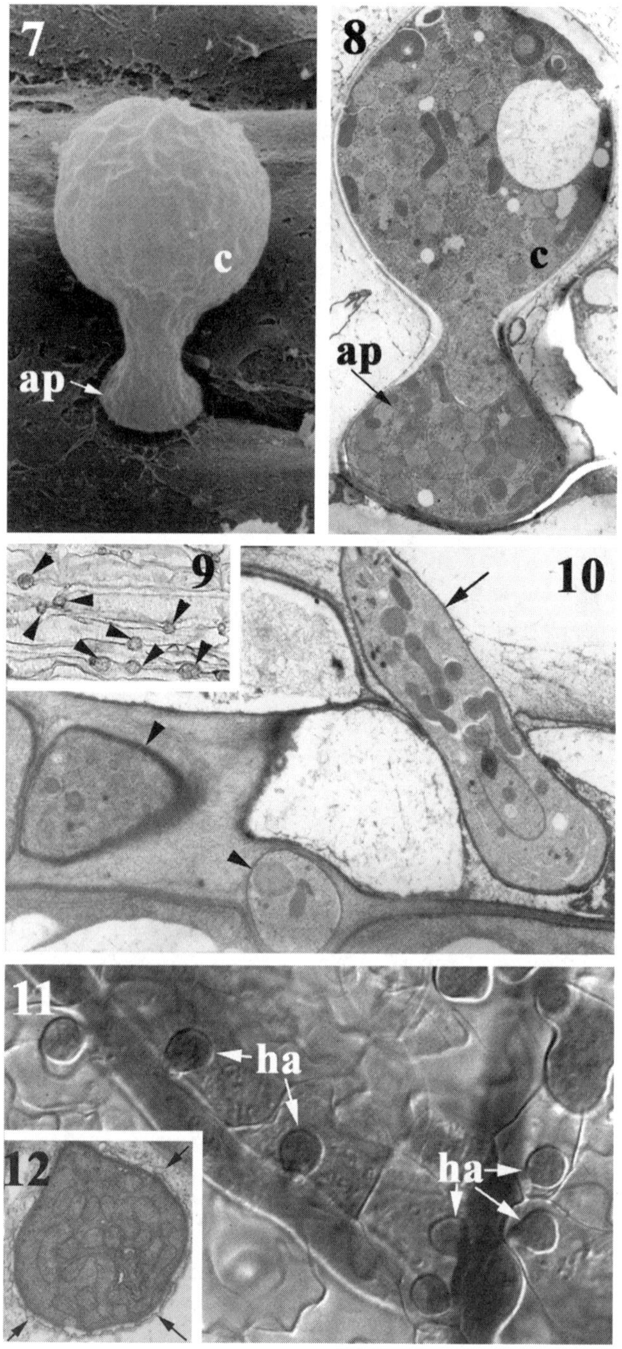

indicating a purely physical signaling mechanism.[4] In the rice blast fungus, *M. grisea*, contact with a hydrophobic surface can trigger appressorium formation, and it is in this organism that appressorium development and function have been studied in greatest detail.[1] In *M. grisea*, the wall of the appressorium becomes melanized, a feature that makes the wall impermeable to virtually all molecules except water. Within the cytosol, the appressorium accumulates high concentrations of glycerol (up to about 3 M), which leads to the buildup of extraordinarily high hydrostatic (turgor) pressures. In *M. grisea*, a pressure as high as 8 MPa has been measured. The base of the appressorium is attached tightly to the plant surface, and at its center a fine penetration peg pierces the underlying plant cell wall. Experimental studies have shown that the penetration pegs of *M. grisea* appressoria are capable of puncturing inert surfaces (such as mylar sheets) of hardness similar to that of a rice leaf surface.[1] However, it has also been found that a rice leaf is penetrated more quickly than a mylar sheet of similar hardness, indicating that enzymatic weakening of the plant cell wall accompanies the mechanical puncturing of the plant surface. Once inside the plant tissues, the pathogen hyphae may grow intercellularly (Fig. 3: 9 and 10) or intracellularly (Fig. 3: 10–12).

Haustoria and Nutrient Acquisition

Necrotrophic pathogens obtain the nutrients they need from dead cells, and thus necrotrophic infections are characterized by widespread damage to plant cells and large expanding necrotic lesions. Biotrophic pathogens, on the other hand, establish a stable relationship with living plant cells in order to obtain the nutrients they require for growth and development. In foliar biotrophs such as the rust fungi, an infection hypha that has developed from the substomatal vesicle makes contact with the surface of a mesophyll cell and differentiates to form a haustorial mother cell (Fig. 2: 6). The haustorial mother cell produces a fine penetration peg that grows through the mesophyll

Fig. 3 (**7**) *P. cinnamomi* cyst (c) that has germinated on the surface of an onion root and formed an appressoriumlike structure (ap) before penetrating the root surface along the anticlinal wall between two epidermal cells. Scanning electron micrograph (×3000). (**8**) *P. cinnamomi* cyst (c) and appressoriumlike swelling (ap) on the surface of a *Eucalyptus sieberi* root. Transmission electron micrograph of material prepared by high-pressure freezing and freeze-substitution (×3500). (**9**) Hyphae of *P. nicotianae* (arrowheads) growing intercellularly between the cells of a tobacco root. Light micrograph of methacylate-embedded material (×400). (**10**) Hyphae of *P. cinnamomi* growing intercellularly (arrowheads) and intracellularly (arrow) within a root of *E. sieberi*. Transmission electron micrograph of material prepared by high-pressure freezing and freeze-substitution (×4000). (**11**) Haustoria (ha) formed by a virulent isolate of *Peronospora parasitica* within epidermal cells of an *Arabidopsis* cotyledon. (Light micrograph (×500) of cleared material stained with trypan blue, courtesy of Dr. D. Takemoto.) (**12**) Haustorium of *M. lini* within a mesophyll cell of a flax leaf. Arrows indicate the extrahaustorial membrane. (Transmission electron micrograph (×13,500) of chemically fixed material, courtesy of Dr. L. Murdoch.)

cell wall but does not penetrate the plant cell plasma membrane. Lack of mechanical disruption in the plant cell wall suggests that the penetration peg secretes enzymes that locally degrade the cell wall. The penetrating hypha then evaginates to form a haustorium, a cell specialized for nutrient uptake from the host plant cell. Because the plant plasma membrane has not been breached, the fungal haustorium remains surrounded by the host plasma membrane, which becomes specialized and is termed the extrahaustorial membrane (Fig. 3: 12). The extrahaustorial matrix lies between the extrahaustorial membrane and the wall of the haustorium. The pathogen orchestrates structural and biochemical changes in the infected plant cell in order to enhance the synthesis of nutrients (especially sugars and amino acids) by the host cell and facilitate their secretion into the extrahaustorial matrix, from where they can be absorbed by the haustorium. Molecular evidence of this process has been recently obtained following the cloning of genes from the rust fungus, *Uromyces fabae*, that encode a proton-pumping ATPase, and amino acid and sugar transporters.[5] These proteins have been shown to reside in the haustorial plasma membrane. It is believed that the H^+-ATPase powers the uptake of amino acids and sugars through the membrane-bound transporters. Colonization of host tissues and acquisition of nutrients allows the pathogen to sporulate and initiate a new cycle of plant infection.

CONCLUSION

Fungal and Oomycete plant pathogens are able to detect and respond to both chemical and physical signals from potential host plants and to regulate the polarity of their growth and the development of specialized infection structures accordingly. Modern approaches to studies of pathogenesis in these organisms are rapidly increasing our understanding of the molecules involved in signal exchange between host and pathogen, and in the induction, differentiation, and function of the specialized infection structures that form a vital part of the strategies to overcome plant defenses and establish disease in susceptible plants.

ACKNOWLEDGMENTS

I thank Professor K. Mendgen (University of Konstanz) for assistance with the high-pressure freezing of material shown in Fig. 3: 8 and 10.

ARTICLES OF FURTHER INTEREST

Biological Control of Oomycetes and Fungal Pathogens, p. 137
Management of Fungal and Oomycete Diseases: Fruit Crops, p. 678
Management of Fungal and Oomycete Diseases: Vegetable Crops, p. 681
Oomycete–Plant Interactions: Current Issues in, p. 843

REFERENCES

1. Howard, R.J.; Valent, B. Breaking and entering: Host penetration by the fungal rice blast pathogen *Magnaporthe grisea*. Annu. Rev. Microbiol. **1996**, *50*, 491–512.
2. Hardham, A.R. Cell Biology of Fungal Infection of Plants. In *The Mycota Vol. VII; Biology of the Fungal Cell*; Howard, R.J., Gow, N.A.R., Eds.; Springer-Verlag: Heidelberg, 2001; 91–123.
3. Mendgen, K.; Hahn, M.; Deising, H. Morphogenesis and mechanisms of penetration by plant pathogenic fungi. Annu. Rev. Phytopathol. **1996**, *34*, 367–386.
4. Staples, R.C.; Hoch, H.C. Physical and Chemical Cues for Spore Germination and Appressorium Formation by Fungal Pathogens. In *The Mycota V Part A. Plant Relationships*; Carroll, G., Tudzynski, P., Eds.; Springer-Verlag: Berlin, 1997; 27–40.
5. Mendgen, K.; Struck, C.; Voegele, R.T.; Hahn, M. Biotrophy and rust haustoria. Physiol. Mol. Plant Pathol. **2000**, *56*, 141–145.

Gene Expression Modulation in Plants by Sugars in Response to Environmental Changes

Barry Wobbes
Sjef C. M. Smeekens
University of Utrecht, Utrecht, The Netherlands

INTRODUCTION

Sugars play pivotal roles in the life cycle of plants. Sugars provide energy for metabolic processess and serve as building blocks for the structural carbohydrates necessary for plant growth. A second important function of sugars is as signaling compounds that can modulate gene expression. Over the past decade it has become apparent that sugars are involved as signaling molecules in many processess important for different stages of the life cycle of the plant. The sugar-induced feedback inhibition of photosynthesis is one example of a sugar-regulated process. When carbohydrates accumulate in mature source leaves, repression of the genes involved in photosynthesis is observed, and as a consequence photosynthesis is reduced. This affects source/sink relationships on the whole plant level because sugar abundance induces storage and utilization programs, and sugar depletion results in upregulation of mobilization and export processess. But sugars are also important for regulation of genes involved in pathogen defense, storage protein accumulation, secondary metabolism, cell cycle, and germination. The type of sugar—mono- or disaccharide—can have its own specific effects. Sugars exert these effects at different levels. Next to transcription, effects on messenger RNA (mRNA) stability and posttranslational modifications have also been described. Over the past few years it has become clear that sugar signals interact with many other signaling pathways, such as light perception and signaling, N assimilation, and plant hormone-signaling cascades. This illustrates the versatile and important roles that sugars play in plants, and the complexity of the underlying signaling cascades.

HEXOSE AND DISACCHARIDE SENSING AND SIGNALING

Glucose Sensing and Signaling

The fact that sugars are able to modulate gene expression[1] indicates the existence of specific sugar-sensing mechanisms in plants. Since the 1980s it has been known that the baker's yeast *Saccharomyces cerevisiae* is able to sense glucose and repress genes involved in the metabolism of other carbon sources—a system known as catabolite repression.[2] During later years, it became apparent that yeast possesses multiple glucose-sensing systems, each responsible for initiating specific signaling cascades. The two glucose transporterlike proteins, RGT2 and SNF3, sense extracellular glucose concentrations and are responsible for the subsequent induction of hexose transporter (HXT) proteins in yeast. Low amounts of glucose are sensed by SNF3 and lead to induction of high-affinity HXT, whereas RGT2 senses high glucose levels and initiates signaling, leading to induction of low-affinity HXTs. Furthermore, a G-protein–coupled receptor responsible for sensing high glucose levels was identified that operates in the cyclic Adenosine Mono Phosphate (cAMP) pathway. Both types of sensors bind glucose extracellularly and, via a conformational change, transduce the signal over the plasma membrane. In contrast, hexokinase (HXK) functions as an enzymatic and signaling factor in the carbon catabolite-repression pathway. Evidence for this dual function of hexokinase has mounted; feeding experiments with glucose analogues that are substrates for HXK and not further metabolized are able to induce the carbon catabolite-repressed state.[2] Furthermore, *hxk2* alleles have been identified that showed uncoupling of enzymatic and signaling functions.[2] The molecular details as to how HXK is involved in glucose repression remain to be resolved.

The evidence for the involvement of HXK in plants as sugar sensors stems from experiments with nonmetabolizable glucose analogues and transgenic plants with altered HXK levels. Overexpression of AtHXK1 leads to a glucose hypersensitivity phenotype, whereas lower levels of the main signaling hexose kinase lead to hyposensitivity.[3] Similarly, glucose analogues that are substrates for HXK can mimic the glucose repression of photosynthesis[4] and glyoxylate cycle genes.[5] These results indicate the possible role HXK could fullfill as a glucose sensor. Plants possess at least three separate glucose signaling systems, each responsible for regulating a specific set of

Encyclopedia of Plant and Crop Science
DOI: 10.1081/E-EPCS 120010663

genes.[6] Two HXK-dependent systems are present: a system for which glucose phosphorylation by itself is sufficient for its activation, and another system that relies on further glucose metabolism. The pathogenesis-related PR1 and PR5 genes require further glucose metabolism for induction,[7] whereas regulation of several genes involved in photosynthesis and nitrogen metabolism depends solely on glucose phosphorylation. Furthermore, a HXK-independent glucose-sensing/-signaling pathway is present that regulated transcription of a number of genes involved in a variety of processes such as carbon and nitrogen metabolism and UV-protection.

Work with yeast on the metabolism of trehalose, a disaccharide consisting of two glucose moieties, has revealed that trehalose-6-phosphate (T6P) is a negative regulator of HXK activity and thereby interacts with glucose signaling. In plants, the trehalose-metabolizing enzymes trehalose phosphate synthase (TPS) and trehalose phosphate phosphatase (TPP) have been expressed and shown to enhance and decrease photosynthesis, respectively.[8] This led to the hypothesis that T6P inhibits HXK activity in plants also. However, evidence for such regulation lacks experimental substantiation. Nevertheless, in plants the trehalose system serves a signaling function.

Sucrose-Specific Sensing

Specific sucrose sensors have not been identified from yeast, plants, or any other organism. Recent experimental data, however, indicate that plants possess such sensors, because expression of the basic leucine-zipper type transcription factor *ATB2* is repressed in a sucrose-specific manner.[9] Repression operates posttranscriptionally and occurs when seedlings are exposed to elevated but physiologically relevant sucrose concentrations. Application of similar concentrations of glucose and fructose—the two products of sucrose hydrolysis—did not cause repression. Deletion of the 5′UTR resulted in loss of the sucrose response, indicating the involvement of *cis*-acting elements in transducing the signal. Feeding sucrose to detached sugar beet leaves also revealed the potential of sucrose to negatively regulate sucrose symporter activity by decreasing messenger levels.[9] Addition of glucose or the HXK inhibitor mannoheptulose did not result in decreased transport activity, indicating that HXK-dependent signaling pathways are not involved. The fact that glucose could not substitute for sucrose indicates sensing of sucrose occurs before it is hydrolyzed. This conclusion was confirmed by results from a study on the repressive effects of (nonmetabolizable) disaccharides on GA-induced α-amylase expression in barley. Repression by sucrose and glucose was shown to be mediated by destabilizing the transcript,[10] whereas nonmetabolizable sucrose analogues affected α-amylase expression via repression of transcriptional induction. These results indicate the existence of different systems to sense and signal the presence of glucose and sucrose in plants.

SUGAR-SIGNALING INTERMEDIATES

The SNF1 kinase complex is an important component of sugar-signaling cascades in yeast. Activation of this protein kinase occurs under low-glucose conditions and leads to derepression of glucose-repressed genes, enabling yeast to grow in the presence of carbon sources other than glucose. SNF1 kinase homologues from a large number of plant species (SnRKs) have been isolated and shown to complement the yeast SNF1 mutation. This indicates that the complex has been evolutionarily conserved and could function in sugar sensing/signaling in plants as well. The yeast SNF1 kinase complex and plant SnRKs are related to the mammalian AMP-activated protein kinase (AMPK), suggesting they could be activated as a consequence of intracellular changes in AMP levels. Plant SnRKs have been shown to regulate expression and activity of key enzymes in carbon and nitrogen metabolism, such as sucrose phosphate synthase (SPS), sucrose synthase and nitrate reductase (NR).[11] Interestingly, activities of phosphorylated SPS and NR are inhibited only by the binding of 14-3-3 proteins adding another layer of regulation. The activity of the SnRK complex itself is also regulated. Mutations in PRL1, a SnRK-interacting protein, results in derepression of glucose-repressed genes and hypersensitivity to glucose and sucrose.[12] In the *prl* background, activity of SnRK1 is approximately 50% higher indicating PRL1 is a negative regulator of the SnRK complex in plants.[13] The *prl1* phenotype is highly pleiotropic, showing augmentation of sensitivity to cytokinin, ethylene, abscisic acid (ABA), and auxin. Furthermore, *prl1* accumulates sugars and starch in leaves and shows inhibition of cell elongation and root growth. The pleiotropic phenotype of *prl1* and the variety of SnRK targets indicate this complex could function as an important cross point between hormone, nitrogen, and sugar signaling.

INTERACTION OF SUGAR WITH HORMONE SIGNALS

Over the last few years, extensive cross talk between sugar- and phytohormone-signaling cascades has been discovered, as described for the *PRL1* locus. But also more specific interaction between sugar- and hormone-signaling cascades has also become clear;[14,15] the interaction of ABA and ethylene with sugar signaling has been characterized in some detail. ABA and ethylene are both involved in a plethora of physiological processess in plants. ABA is best known as a stress hormone and its biosynthesis is induced (or sensitivity to ABA is enhanced under adverse growth conditions). The classical ethylene-regulated processes include fruit ripening and the triple-response. The three-way interaction of ABA, ethylene- and sugar- signaling pathways was revealed in mutational and epistatic analyses.[16] The *abi4* locus was originally identified as an ABA-signaling mutant, but in recent years it was also isolated from various sugar-related mutant screens. In one such screen for mutants lacking the ability to repress PC expression by sucrose, *sun6* (sucrose uncoupled) was isolated and shown to be allelic to *abi4*.[9] From another sucrose-response screen, *abi4* was isolated as the *isi3* (impaired sucrose induction) mutant and shown to lack the ability to induce the *ApL3* gene by sucrose.[14] ABA itself is not able to induce *ApL3* but enhances the sensitivity of tissues to sugars. Furthermore, the *isi4* mutant was shown to be allelic to the ABA-deficient mutant *aba2*, confirming the ABA-sugar interaction.[14] Ethylene signaling also interacts with sugar responses as indicated by the facts that ethylene overproduction or constitutive signaling (*eto* and *ctr1*) causes insensitivity to glucose and that ethylene-signaling mutants (*etr1* and *ein2*) are glucose-hypersensitive.[16] Application of ethylene to wild-type seedlings phenocopies the glucose-insensitive phenotype; together with the behavior of the ethylene mutants, these results show that ethylene desensitizes seedlings for sugars. Because ABA itself is negatively affected by ethylene, the ethylene effect on sugar sensitivity may depend on ABA interaction with the sugar-signaling systems. It has thus become clear that extensive cross talk between ABA, ethylene signalling, and sugar signaling can occur (Fig. 1). The question remains as to the physiological relevance of these interactions and to what extent stress-signaling is involved under the high-sugar conditions used.

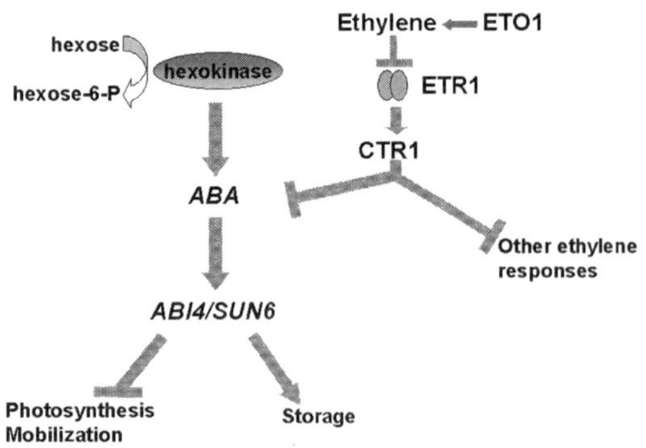

Fig. 1 Schematic presentation of interactions between hexose-, ABA-, and ethylene signaling pathways. (*View this art in color at www.dekker.com.*)

INTERACTION OF CARBON AND NITROGEN SIGNALS

In addition to mechanisms that perceive and signal the presence of sugars, plants also sense nitrogen compounds and enhance or limit nitrogen uptake and assimilation as needed. Coordination of sugar production via photosynthesis, nitrogen uptake, and assimilation allows plants to efficiently use these compounds to optimize growth. For example, genes involved in nitrate transport and reduction are induced by sugars. On the contrary, low carbon levels have inhibiting effects on N assimilation.[17] Furthermore, the reciprocal induction of photosynthetic gene expression and photosynthesis by high nitrate further indicates the existance of C:N balancing mechanisms. Nitrate also can repress starch production by decreasing *AGPS*, a key regulator of starch biosynthesis. These interactions result in the enhanced availability of carbohydrates to sustain production and transport of amino acids when N availability is high. The molecular details of regulation by C:N ratios are largely unknown, but the ability of SnRKs to phosphorylate such diverse targets as NR and SPS, and the subsequent inactivation by 14-3-3 protein binding show the ability of this complex to play a role in controling internal C:N balance.

CONCLUSION

Sugar signaling is of crucial importance in many if not all stages of the plant life cycle. It is now clear that sugar-signaling pathways are tightly interwoven with other signaling systems, such as those for hormones and nitrate. The challenge remains to further explore these interactions and understand their physiological significance.

ARTICLES OF FURTHER INTEREST

Nitrogen, p. 822
Photosynthate Partitioning and Transport, p. 897
Plant Response to Stress: Abscisic Acid Fluxes, p. 973
Plant Response to Stress: Source-Sink Regulation by Stress, p. 1010

Starch, p. 1175
Sucrose, p. 1179

REFERENCES

1. Koch, K.E. Carbohydrate-modulated genes in plants. Annu. Rev. Plant Physiol. Plant Mol. Biol. **1996**, *47*, 509–540.
2. Rolland, F.; Winderickx, J.; Thevelein, J.M. Glucose-sensing mechanisms in eukaryotic cells. Trends Biochem. Sci. **2001**, *26* (5), 310–317.
3. Jang, J.-C.; León, P.; Sheen, J. Hexokinase as a sugar sensor in higher plants. Plant Cell **1997**, *9*, 5–19.
4. Jang, J.-C.; Sheen, J. Sugar sensing in higher plants. Plant Cell **1994**, *6*, 1665–1679.
5. Graham, I.A.; Denby, K.J.; Leaver, C.J. Carbon catabolite repression regulates glyoxylate cylce gene expression in cucumber. Plant Cell **1994**, *6*, 761–772.
6. Sheen, J.; Zhou, L.; Jang, J.-C. Sugars as signaling molecules. Curr. Opin. Plant Biol. **1999**, *2*, 410–418.
7. Xiao, W.; Sheen, J.; Jang, J.-C. The role of hexokinase in plant sugar signal transduction and growth and development. Plant Mol. Biol. **2000**, *44*, 451–461.
8. Paul, M.; Pellny, T.; Goddijn, O. Enhancing photosynthesis with sugar signals. Trends Plant Sci. **2001**, *6*, 197–200.
9. Smeekens, S. Sugar sensing and signaling. Annu. Rev. Plant Physiol. Plant Mol. Biol. **2000**, *51*, 49–81.
10. Loreti, E.; Alpi, A.; Perata, P. Glucose and disaccharide-sensing mechanisms modulate the expression of α-amylase in barley embryos. Plant Physiol. **2000**, *123*, 939–948.
11. Halford, N.G.; Hardie, D.G. SNF1-related protein kinases: Global regulators of carbon metabolism in plants? Plant Mol. Biol. **1998**, *37*, 735–748.
12. Németh, K.; Salchert, K.; Putnoky, P.; Bhalerao, R.; Koncz-Kálmán, Z.; Stankovic-Stangeland, B.; Bakó, L.; Mathur, J.; Ökrész, L.; Stabel, S.; Geigenberger, P.; Stitt, M.; Rédei, G.P.; Schell, J.; Koncz, C. Pleitropic control of glucose and hormone responses by PRL1, a nuclear WD protein, in Arabidopsis. Genes Dev. **1998**, *12*, 3059–3073.
13. Bhalerao, R.P.; Salchert, K.; Bakó, L.; Ökrész, L.; Szabados, L.; Muranaka, T.; Machida, Y.; Schell, J.; Koncz, C. Regulatory interaction of PRL1 WD protein with *Arabidopsis* SNF1-like protein kinases. Proc. Natl. Acad. Sci. U. S. A. **1999**, *96*, 5322–5327.
14. Rook, F.; Corke, F.; Card, R.; Munz, G.; Smith, C.; Bevan, M.W. Impaired sucrose-induction mutants reveal the modulation of sugar-induced starch biosynthetic gene expression by abscisic acid signalling. Plant J. **2001**, *26*, 421–433.
15. Gibson, S.I.; Laby, R.J.; Kim, L. The *sugar-insensitive1* (*isi*1) mutant of *Arabidopsis* is allelic to *ctr*1. Biochem. Biophys. Res. Comm. **2001**, *280*, 196–203.
16. Gazzarrini, S.; McCourt, P. Genetic interactions between ABA, ethylene and sugar signaling pathways. Curr. Opin. Plant Biol. **2001**, *4*, 387–391.
17. Coruzzi, G.M.; Zhou, L. Carbon and nitrogen sensing and signaling in plants: Emerging 'matrix effects.' Curr. Opin. Plant Biol. **2001**, *4*, 247–253.

Gene Flow Between Crops and Their Wild Progenitors

Roberto Papa
Università Politecnica delle Marche, Ancona, Italy

Paul Gepts
University of California, Davis, California, U.S.A.

INTRODUCTION

Gene flow occurs when there is migration of individuals (e.g., seeds) or gametes (e.g., pollen) between populations. Along with drift, selection, and mutation, it represents one of the main evolutionary forces causing changes in gene frequencies. The main effect of gene flow is the reduction of differentiation between populations accompanied by a parallel increase in differences between individuals within a population. The life history and demographic factors are also important in the determination of the effect of gene flow on the structure of genetic diversity (e.g., domestication bottleneck). Here we will briefly illustrate the current knowledge relative to the level, causes, and consequences of gene flow in the specific context of crops and their wild progenitors.

IMPORTANCE OF GENE FLOW

Wild-to-domesticated gene flow has important implications in relation to the evolution of crop plants. After domestication, gene flow in the centers of origin can partially restore the low genetic diversity included in these first domesticated populations. For instance, the differentiation of domesticated barley from the Himalayas and India compared to the Near East germplasm is probably due to introgression from Asian populations of wild barley (*Hordeum vulgare* ssp. *spontaneum*) after domestication.[1] Gene flow still plays an important role as a source of new alleles for domesticated crops where traditional farming continues to be practiced.[2]

Besides this natural gene flow, we should also consider the human-driven gene flow from wild to domesticated populations due to modern plant breeding, whereby useful wild alleles are integrated into the domesticated crop species. Many sources of resistance to pathogens and parasites have been introduced from wild germplasm by breeders and, more recently, genes relating to quantitative traits, such as fruit size in tomato and grain yield in rice.[3] Gene flow from domesticated to wild populations is also an important issue relative to the release of transgenic varieties because of the potential effect on the genetic diversity of the wild relatives and the possible production of new aggressive weeds.

FACTORS AFFECTING GENE FLOW

Table 1 summarizes the various factors affecting gene flow. In order to exchange genes, individuals need first to be sexually compatible; in other words, gene flow occurs between populations of the same biological species, or between populations of closely related species (introgressive hybridization), among which hybrids are partially fertile and can yield fertile progeny. In most cases, crops and their wild progenitors belong to the same biological species.[4] However, in some cases domesticated crops and their progenitors belong to different biological species, in particular when domestication has involved polyploidization and/or interspecific hybridization, such as is seen for bread wheat (*Triticum aestivum* L.), a hexaploid originated by hybridization between a domesticated tetraploid and the wild diploid *Aegilops tauschii*. In these cases, even with the reproductive barrier due to the different ploidy level between the domesticated and wild progenitors, hybridization can still occur and can produce fertile progeny.

Gene flow between wild and domesticated populations is also limited by their phenology, geographic distribution, and spatial arrangement. Gene flow can occur if plant populations have overlapping flowering periods and are at a suitable distance, depending on the seed and pollen dispersal ability and the environmental factors such as wind, humidity, and biotic factors (pollinators and other animals favoring seed dispersal). Pollen and seed dispersal rates are strictly correlated with distance, such that dispersal rapidly decreases with distance to a very low value (e.g., within 50–200 meters), although a relatively low level of dispersal may occur even over very great distances (e.g., several kilometers). Evidence of gene flow between wild progenitors and domesticated crops has been documented for almost all crop species, including allogamous, autogamous, and vegetatively propagated

Encyclopedia of Plant and Crop Science
DOI: 10.1081/E-EPCS 120017095

Table 1 Factors related to gene flow between wild progenitors and domesticated crop populations

Factors	Key aspect(s)	General consequences
Prezygotic		
Reproductive barrier I	Pollen competition, sexual compatibility	Partial incompatibility will reduce gene flow. Pollen competition may reduce gene flow or increase it just in one direction.
Geographical distance	Pollen/seed dispersal ability	Gene flow occurring only within the center of origin of a crop or within the (sympatric) areas of distribution of wild populations
Phenology	Genotype x environment interaction	Gene flow occurring only if flowering period is overlapping in the areas of sympatry
Dispersal ability	Biotic and abiotic factors affecting dispersal/breeding and propagation system	Rate of gene flow higher in allogamous than autogamous or vegetative propagating species
Population size	Relative size of wild and domesticated populations	Different sizes of domesticated and wild populations may lead to one-way gene flow.
Weedy populations	Presence of uncultivated fields and disturbed areas. Weed control.	Gene flow will be limited by weed control and favored by the presence of disturbed or uncultivated fields.
Postzygotic		
Reproductive barrier II (Hybrid fertility)	Different ploidy, chromosomic mutation	Partial sterility will reduce gene flow.
Selection in wild populations	Level of differentiation between wild and domesticated/genetic control of the domestication syndrome	Selection will act only in the segregant progeny with a lower fitness of progeny homozygous for domesticated alleles (when wild alleles of the domestication syndrome are dominant).
Selection in domesticated populations	Farmers' conscious selection (e.g., seed)/agronomic practices/genetic control of the domestication syndrome	Selection will limit gene flow as a post-zygotic reproductive barrier by reducing the fitness of first generation hybrids (when domesticated alleles of the domestication syndrome are dominant).
Breeding system	Effective recombination (out-crossing and heterozygosity)	In allogamous species, introgression will be limited only for selected loci; in autogamous species, introgression will also be limited for loci linked to selected loci (hitchhiking).
Demography	Population size (bottleneck)	Reduced population size will increase hitchhiking and will extend the effect of selection following gene flow events.

species. One of the very few exceptions is the strict selfing species *Arachis hypogea*.[2,5] According to a theoretical model, the amount of gene flow needed to prevent the genetic isolation and differentiation of populations and their independent evolution has been shown to be relatively low, about one migrant per generation,[6] suggesting that even if the rate of dispersal and out-crossing is very variable among and even within species, gene flow is expected to be an important evolutionary force for most species.

Nevertheless, the reproductive and the propagation systems of domesticated crops represent crucial factors affecting the rate of gene flow between wild and domesticated populations. Clearly in an allogamous species the higher out-crossing rate results in a much higher frequency of domesticated-to-wild hybridizations than in an autogamous species. For vegetatively propagated crops, such as fruit trees, gene flow from wild to domesticated populations will not occur unless farmers/breeders use sexual reproduction to obtain new cultivars. In contrast, gene flow will be very intensive in the opposite direction (domesticated to wild) leading to unilateral gene flow (one-way migration). One-way migration may also occur in sexually propagated species. For instance, in the common bean (*Phaseolus vulgaris* L.), the gene flow from the domesticated to the wild populations has been found to be about 3–4-times higher than that in the opposite direction.[7] In this case, one-way migration can be promoted by the presence of differences in population sizes between the wild and domesticated populations because the domesticated genes will gradually increase in the wild populations, while in the domesticated populations, the few immigrant genes will be diluted and hence have a very low effect on gene frequencies.

Farmers' fields usually contain a large number of individuals, while wild populations close to such fields will often be made up of a very small number of individuals. Consequently, hybridization events will be more frequent when wild plants are the maternal parent than when they are the paternal parent.

Farmers themselves can also affect the rate of gene flow from wild to domesticated crops. Indeed, in several cases traits related to the domestication syndrome are recessive, and, hence, first generation hybrids are usually more similar to the wild plants, and are thus easily detectable. For this reason farmers can discard first generation hybrids by choosing the seeds for the planting of the next generation, producing an effect analogous to that of a post-zygotic reproductive barrier. Gene flow is also limited by weed control and intensive cultivation (e.g., absence of uncultivated areas), which eliminate wild plants growing within and around the crop. It is likely that in the early ages of agriculture the opposite situation would have occurred, with the predominant direction of gene flow being from wild to domesticated populations because of the smaller crop populations and the lower differentiation between wild and domesticated forms.

In some cases, hybridization between wild and domesticated populations leads to the development of weedy populations[1,7] that can be found in farmers' fields or that can colonize other disturbed environments (i.e., field borders, abandoned fields, roadsides) and present intermediate characteristics between wild and domesticated forms.[4] These weedy forms can also originate as ''escapes'' from cultivation. In either case, their presence facilitates the exchange of genes between domesticated and wild populations.

GENE FLOW AND SELECTION

In the previous paragraphs we have demonstrated that gene flow between wild progenitors and domesticated populations is a significant phenomenon in almost all crop species. However, wild progenitors and domesticated crops maintain their distinct phenotypes even in sympatry. This suggests that selection has a prominent role in limiting the introgression between the wild and domesticated forms. However, even if there is little direct evidence, selection is likely to vary greatly between the wild and the domesticated environments, as among the different genes involved in the control of the domestication syndrome, and among different crop species[8] and different agronomic systems. As previously indicated, for several key traits of the domestication syndrome, the wild alleles are dominant (i.e., shattering, dormancy, growth habit, photoperiodic sensitivity), and the first generation hybrids are more similar to the wild than to the domesticated forms. Considering also that hybrids may show heterosis, the first generation hybrids will have a much higher reproductive success in the wild environment than in the domesticated environment, where farmers may easily eliminate them by selecting the seeds or because the progeny will not be included in the next harvest due to shattering or dormancy. Consequently, in the wild environment, selection against domesticated alleles will mainly occur in the segregating progeny following the F_1 generation, thus favoring the introgression of genes from domesticated populations because of recombination. Both selection and asymmetric gene flow will favor the introgression from domesticated to wild populations rather than in the opposite direction; this may explain why in domesticated populations low levels of introgression from the wild progenitors are often seen,[1,7] even for allogamous species such as maize.[8]

Asymmetric gene flow and different types of selection can be considered as possible causes of the displacement of genetic diversity in the wild progenitor populations, as has been observed in cotton and rice,[5] and to a lesser extent in the common bean.[7] In addition to the target loci (i.e., genes for domestication traits), selection may affect the surrounding chromosome regions because of linkage (hitchhiking). Indeed, in allogamous species, which present a high level of heterozygosity, selection will affect (by elimination of the migrant alleles) only loci under selection because of recombination. In contrast, in autogamous species, selection will indirectly reduce the introgression for neutral loci linked to those under selection (hitchhiking). The extent of hitchhiking can be very low (a few hundred base pairs) in allogamous species such as maize,[9] but can also become very large (several cM) in autogamous species, because the reduction of the out-crossing rate drastically reduces the level of effective recombination. In addition, the extent of hitchhiking also varies according to the level of recombination in different parts of the genome, the demography of the population (e.g., the existence of bottlenecks in the evolution of the population versus the species as a whole), and other evolutionary factors such as selection.[10–12]

CONCLUSION

Gene flow and introgression between domesticated crops and their wild progenitors occurs in most cases, although its intensity and effects are very variable in relation to any given species (of crop and its wild progenitor), its life history, the environment (space and time), genome location (in relation to domestication syndrome genes), the agro-ecosystem, and human activities. Knowledge of this subject is currently growing rapidly because of the interest in evaluating the potential effects of transgene release into the environment and the role of wild progenitor genetic diversity in conservation and breeding.

ARTICLES OF FURTHER INTEREST

Agriculture and Biodiversity, p. 1
Biosafety Approaches to Transgenic Crop Plant Gene Flow, p. 150
Biosafety Science: Overview of Plant Risk Issues, p. 164
Breeding: Incorporation of Exotic Germplasm, p. 222
Crop Domestication: Fate of Genetic Diversity, p. 333
Crop Improvement: Broadening the Genetic Base for, p. 343
Farmer Selection and Conservation of Crop Varieties, p. 433
Genetic Resource Conservation of Seeds, p. 499
Interspecific Hybridization, p. 619
Molecular Evolution, p. 748
Molecular Technologies and Their Role Maintaining and Utilizing Genetic Resources, p. 757
Population Genetics, p. 1042
Quantitative Trait Locus Analyses of the Domestication Syndrome and Domestication Process, p. 1069
Transgenic Crop Plants in the Environment, p. 1248

REFERENCES

1. Salamini, F.; Ozkan, H.; Brandolini, A.; Schafer-Pregl, R.; Martin, W. Genetics and geography of wild cereal domestication in the Near East. Nat. Rev., Genet. **2002**, *3*, 429–441.
2. Jarvis, D.I.; Hodgkin, T. Wild relatives and crop cultivars: Detecting natural introgression and farmer selection of new genetic combinations in agroecosystems. Mol. Ecol. **1999**, *8*, S159–S173.
3. Tanskley, S.D.; McCouch, S.R. Seed banks and molecular maps: Unlocking genetic potential from the wild. Science **1997**, *277*, 1063–1066.
4. Harlan, J.R. *Crops and Man*, 2nd Ed.; American Society of Agronomy: Madison, WI, 1992.
5. Ellstrand, N.; Prentice, H.; Hancock, J. Gene flow and introgression from domesticated plants into their wild relatives. Ann. Rev. Ecolog. Syst. **1999**, *30*, 539–563.
6. Hartl, D.; Clark, A.G. *Principles of Population Genetics*, 3rd Ed.; Sinauer Associates: Sunderland, MA, 1997.
7. Papa, R.; Gepts, P. Asymmetry of gene flow and differential geographical structure of molecular diversity in wild and domesticated common bean (*Phaseolus vulgaris* L.) from Mesoamerica. Theor. Appl. Genet. **2002**, *106*, 239–250.
8. Matsuoka, Y.; Vigouroux, Y.; Goodman, M.M.; Sanchez, G.J.; Buckler, E.; Doebley, J. A single domestication for maize shown by multilocus microsatellite genotyping. Proc. Natl. Acad. Sci. U. S. A. **2002**, *30*, 6080–6084.
9. Wang, R.-L.; Stec, A.; Hey, J.; Lukens, L.; Doebley, J. The limits of selection during maize domestication. Nature **1999**, *398*, 236–239.
10. Nordborg, M.; Borevitz, J.; Bergelson, J.; Berry, C.; Chory, J.; Hagenblad, J.; Kreitman, M.; Maloof, J.; Noyes, T.; Oefner, P.J.; Stahl, E.A.; Weigel, D. The extent of linkage disequilibrium in *Arabidopsis thaliana*. Nat. Genet. **2002**, *30*, 190–193.
11. Ching, A.; Caldwell, K.; Jung, M.; Dolan, M.; Smith, O.; Tingey, S.; Morgante, M.; Rafalski, A. SNP frequency, haplotype structure and linkage disequilibrium in elite maize inbred lines. BMC Genet. **2002**, *3*, 19. http://biomedcentral.com/1471-2156/3/19.
12. Tian, D.; Araki, T.; Stahl, E.A.; Bergelson, J.; Kreitman, M. Signature of balancing selection in Arabidopsis. Proc. Natl. Acad. Sci. U. S. A. **2002**, *99*, 11525–11530.

Gene Silencing: A Defense Mechanism Against Alien Genetic Information

Peter de Haan
Phytovation B.V., Leiden, The Netherlands

INTRODUCTION

Gene expression in eukaryotic cells is tightly controlled by regulatory mechanisms acting at the transcriptional level in the nucleus or at the posttranscriptional level in the cytoplasm. The last few years it has become clear that besides proteins, noncoding RNA molecules play an important role in gene regulation. Several independent lines of research relating control of gene expression led to the discovery of a novel posttranscriptional regulatory mechanism referred to as RNA silencing. RNA silencing uses RNA instead of proteins as the signaling and target molecules.

In plants, RNA silencing appears to play an important role in morphogenesis. In addition, this cytoplasmic RNA surveillance mechanism is involved in protection against intracellular molecular parasites such as the transposable elements, viroids, and viruses.

RNA SILENCING TO CONTROL GENE EXPRESSION

RNA molecules play prominent roles as signals and targets in gene regulation. This implies that the central dogma in molecular biology (where DNA is transcribed into RNA, which is translated into proteins, which play major roles in gene regulation and development) clearly needs to be updated.

A number of observations originally made about plants—such as transgene-induced silencing of genes (e.g., flower pigmentation), RNA-mediated virus-resistance, and virus-induced gene silencing—and later on about other organisms—such as quelling in fungi and RNA interference in nematodes, insects, and mammalian cells—turned out to rely on a similar molecular process. This process, which can generally be termed RNA silencing, is induced by overexpressed and double-stranded RNA (dsRNA) molecules, and involves sequence-specific RNA degradation in the cytoplasm of cells from higher eukaryotes.[1]

The key step in the induction of RNA silencing is the formation of dsRNA by host RNA-dependent RNA polymerases (RdRp), which is recognized and cleaved by a dsRNA-specific RNAse III-type nuclease (denoted DICER) to yield small (21–25 nucleotides long), short, interfering RNAs (siRNAs). The siRNAs have the capacity to specifically bind to complementary mRNA molecules and thereby enable the host-encoded RdRps to produce a second generation of dsRNA molecules, which are again cleaved by DICER to yield secondary siRNAs. Genetic studies revealed additional genes and proteins (such as helicases) active in the RNA silencing pathway in plants and animals. All these proteins are associated with cytoplasmic nuclease complexes—denoted RNA-induced silencing complexes (RISC)—approximately 500 kD in size.[2]

Besides siRNAs, other classes of noncoding transcripts have been found that play important roles in gene regulation. One particular class of noncoding transcripts—denoted micro-RNAs (miRNAs)—has been identified in nematodes, mammals, and recently plants, that was hitherto overlooked using standard methods for identifying genes. In plants, there are indications that miRNAs are involved in morphogenesis and in stress responses. The miRNA genes are clustered on the chromosomal DNA, suggesting they are processed from long precursors. The miRNAs have extensive duplex secondary structures that are cleaved by DICER to yield short temporal RNAs (stRNAs) resembling siRNAs. These molecules are targeted to the 3′ ends of cognate mRNAs and thereby inhibit their translation. Because this process requires (some of) the same components as the RNA silencing pathway and also takes place in cytoplasmic complexes with size similar to that of RISC, it is tempting to assume that miRNAs initiate RNA silencing in RISC. It has remained unclear which host genes are precisely targeted by miRNAs and how their expression patterns are modified. It is possible that transcriptional control of certain genes by their own promoters is not stringent enough; miRNAs here serve as an extra lock to prevent accumulation of cognate proteins in specific tissues at specific points in time.[3]

Silenced host genes are kept in a silent state either by continuous supply of stRNAs and/or siRNAs, or by transcriptional gene silencing. A number of reports, mainly dealing with transgene-induced gene silencing, show that siRNAs are able to induce methylation of homologous sequences on the chromosomal DNA. This coincides with alterations in the local chromatin structure,

Encyclopedia of Plant and Crop Science
DOI: 10.1081/E-EPCS 120010661

rendering euchromatin into heterochromatin, which finally results in inhibition of transcription of the affected gene. A number of genes have been identified in *A. thaliana*, including one encoding a methyltransferase (MET1) and one encoding a protein involved in chromatin remodelling (DDM1), which play roles in transcriptional silencing. Hence, it appears that RNA silencing is not an exclusive posttranscriptional process but that it also has components acting at the transcriptional level.[4]

The main question remains, how does RISC discriminate between RNAs to be degraded and RNAs to be retained? Most structural RNAs, e.g., transfer RNAs (tRNAs) and ribosomal RNAs (rRNAs), have extensive double-stranded structures, but they do not initiate RNA silencing. It is also known that these structural RNAs are closely associated with proteins to form ribosomes, and may therefore not be accessible to the RISC complex. Protein encoding mRNA molecules are covered with ribosomes in polysome complexes, immediately after translocation to the cytoplasm. Hence, all structural and coding RNAs are covered with proteins or imbedded in higher-order structures and very seldom occur as naked RNA. RISC was discovered by copurification with polysomes; it could therefore very well be that in vivo RISC is associated with ribosomes. Remarkably, one of the RISC components ARGONAUT shows sequence homology to translation initiation factor elF2C, which might be another indication of the tight linkage between translation and RNA silencing.

The covered or imbedded RNAs are hidden from the surveillance activity of RISC and protected from being degraded. The most plausible explanation for the selectivity of RISC is therefore that exclusively RNA molecules, which accumulate free in the cytoplasm such as noncoding (ds)RNAs, miRNAs, and overabundant mRNAs, are targeted by RISC and enter the degradation pathway. Resulting siRNAs reside in RISC and target complementary RNAs or induce transcriptional silencing (Fig. 1). In this process the dsRNA and miRNA species are direct prey for DICER, whereas the overexpressed RNAs are first copied into dsRNA by a host RdRp, most likely in a primer-independent fashion.[5] Alternatively, it may be that RNA synthesis is randomly primed by tRNAs.

RNA SILENCING TO CONTROL INTRACELLULAR PARASITES

All living organisms are hosts for parasites. A number of them are intracellular parasites that use host components for their replication. Plants suffer from three classes of intracellular parasites: transposable DNA elements, viroids, and viruses.

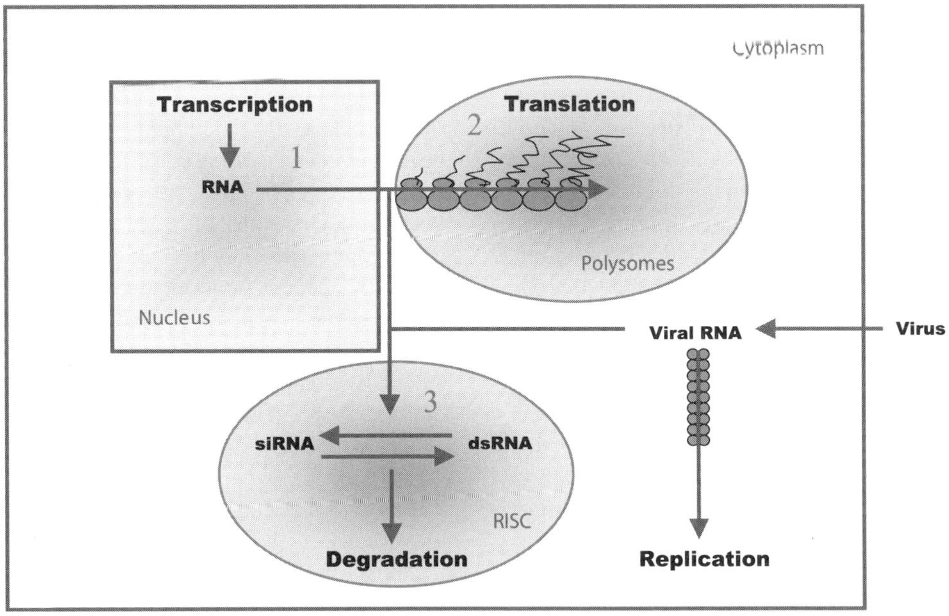

Fig. 1 Schematic representation of gene regulation in eukaryotes. 1: In the nucleus, genes are transcribed to yield RNA molecules. The RNAs play regulatory roles in RNA processing in the nucleus, or they are transported to the cytoplasm. 2: The structural RNAs are imbedded in polysomes, which are responsible for translation of host and viral mRNAs. 3: RISC performs a surveillance function and targets unprotected RNAs for degradation. The resulting siRNAs in RISC bind to complementary RNAs and induce their degradation. Viral replication intermediates are always wrapped with CP or NP (or with ribosomes in the case of plus-strand RNA viruses). (*View this art in color at www.dekker.com.*)

Transposable elements such as retroelements and transposons are integrated in host chromosomal DNA and have the capacity to jump to other positions. They were first described in maize in the early 1950s, and turned out to be highly abundant in eukaryotes, including most crop species. Because of their mobilizing capacities, transposable elements represent mutagens and thus have the potential to destroy the genetic makeup of individuals.[6] It has recently been shown that eukaryotes use RNA silencing to protect their genomes from jumping DNA elements. Inserted transposable elements are usually methylated and transcriptionally silent, whereas specific siRNAs can be found in the cytoplasm. These elements are most likely kept immobilized by transcriptional silencing induced either by overproduced transcripts or by dsRNAs, which are produced from multiple tightly linked copies of a particular element.[7]

In tissues such as meristems, RNA silencing appears not to be active, so that one could imagine that during active cell division (both in mitosis and meiosis), RNA silencing is inactive and DNA methylation patterns are reset. As a result, transposable elements can become active and jump to other loci. The elements are stabilized again in the differentiated cells. Besides having negative effects, transposable elements are also beneficial to their hosts. Because mobilization creates additional variability in progeny populations, conferring a selective advantage to the host, it may be in the host's interest to retain transposable elements that are active during cell division but that are controlled by RNA silencing in vegetative tissues.

Another class of intracellular parasites are viroids, which are pathogenic small circular noncoding single-stranded RNAs folded into dumbbell structures. They replicate in the nucleus of infected cells and from there migrate to the cytoplasm.[8] As noncoding and almost perfect dsRNA entities, they seem ideal prey for RISC. However, in the cytoplasm viroids most likely mimic small ribosomal RNAs, hide in ribosomes, and thereby escape from being massively degraded.[5]

Most crops are hosts for viruses, the most complex intracellular parasites. Viral infections cause symptoms varying from unnoticeable to severe necrosis, and may lead to dramatic yield losses. Most of the plant-infecting viruses have RNA genomes and exclusively replicate in the cytoplasm of infected cells. Relatively few plant viruses have DNA genomes that replicate in the nucleus. However, their transcripts are normally directed to the cytoplasm where the viral proteins are produced. Also, virus replication in plants is controlled by RNA silencing.[9] It has been suggested that double-stranded replication intermediates induce silencing, but this seems unlikely. DNA viruses also hide their replication intermediates in the nucleus and are susceptible to silencing. Moreover, if RNA viruses would expose their dsRNA replication intermediates as naked RNA in the cytoplasm, this would induce an immediate and massive RNA-silencing response. The genomic sense and antisense RNA molecules are therefore always carefully wrapped with coat proteins (CP) or ribosomes in case of plus-strand RNA viruses or nucleocapsid proteins (NP) in case of negative-strand RNA viruses. The amount of CP or NP accumulating in the cytoplasm determines whether transcription or replication takes place by the viral RdRp. In the beginning of the infection process, with low amounts of CP/NP, transcription or translation occurs, yielding mRNAs encoding viral proteins. At a later stage, when high amounts of CP/NP accumulate, RNA replication takes place, yielding genomic RNAs assembled into progeny virus particles. Virus replication is therefore controlled by simultaneous activity of two types of RdRps: viral and host RdRps. It seems likely that threshold levels of viral mRNAs in the cytoplasm instead of double-stranded replication intermediates trigger RNA silencing (Fig. 1). In this process, a host RdRp copies the viral mRNAs, most likely in a primer-independent or tRNA-primed fashion.

One should bear in mind that RNA silencing is not capable of creating virus resistance in many if not all cases. True resistance is mediated by virus resistance genes, such as the N gene from tobacco or the Tm-2a gene from tomato, both conferring immunity against tobamovirus infections in solanaceous crops. RNA silencing is merely involved in keeping viral titres low, thereby minimizing the detrimental effects of the virus infection on the plant.

It is not surprising that as part of the ongoing battle between parasite and host, many viruses carry functions to suppress this intracellular defense mechanism. A number of plant viral RNA-silencing suppressors have recently been identified. Mutations in the viral RNA-silencing suppressor genes frequently lead to a dramatic drop in virulence. Therefore, the viral RNA-silencing suppressors can be regarded as the major virulence factors. The suppressors identified and studied so far are highly heterogeneous and appear to inhibit different components of the RNA silencing pathway. This indicates that silencing suppressors are independently acquired by viruses, and that this hence has been a relatively recent event.

One of the best studied suppressors is cucumber mosaic virus (CMV) P2b.[10] Next to RNA silencing suppression, P2b is also indicated as being involved in long-distance movement of the virus. The protein is targeted to the nucleus and prevents silencing in newly emerging tissues. It is not able to reverse silencing, once established. Compared to other plant viral silencing suppressors, P2b is relatively weak; this might explain why CMV infections generally cause mild mosaic symptoms.

Another well studied suppressor is potato virus Y (PVY) HC-Pro. This suppressor prevents degradation of dsRNAs into siRNAs by DICER, and is able to reverse RNA silencing. The NSs gene from *Tomato spotted wilt virus* (TSWV)—a negative-strand RNA virus—seems to inhibit the same step in the RNA silencing pathway as PVY HC-Pro. Both silencing suppressors have strong activities, which may explain why potyviruses and tospoviruses rank among the most economically detrimental groups of plant viruses, generally causing severe disease symptoms.

In contrast to HC-pro or NSs, the P19-silencing suppression protein of tombusviruses does not prevent the formation of siRNAs and directly interacts with the siRNAs or with RISC and inhibits their action. The 2b protein of CMV and the p25 protein of PVX seem to interfere with processes even further downstream. The C2 protein of tomato yellow leaf curl virus (TYLCV) is the first silencing suppressor identified from a plant DNA virus. All plant viruses most likely carry RNA-silencing suppressor genes, which are important if not essential in overcoming this general intracellular defense response.

CONCLUSION

RNA silencing is a newly discovered gene regulation mechanism, involving cytoplasmic sequence-specific RNA degradation. RNA-silencing components are clustered into a riboprotein complex called RISC that seems associated with polysomes. RISC has a surveillance function and targets free uncovered RNAs, such as over-expressed RNA and dsRNAs for degradation. The central molecules are the degradation products of the DICER enzyme—siRNAs—that serve as a sequence-specific memory for the recognition of complementary RNAs to be degraded, and as a signal for transcriptional gene silencing in the nucleus. RNA silencing plays a dual function in eukaryotic organisms. First, it is involved in morphogenesis and possibly other developmental processes. Second, it counteracts alien intracellular genetic information: It is responsible for keeping transposable elements latent and for suppression of virus disease development. RNA silencing has been analyzed in detail in plants.Although many question await an answer, this led to an enormous body of information on this newly discovered regulatory mechanism. This knowledge has already led to a number of applications. Engineered virus-resistant crops based on RNA-silencing principles have recently reached the market. Gene silencing is employed to produce crops resistant to other pathogens. This is achieved by blocking the expression of recessive dispensable genes such as powdery mildew resistance in cereals. Plant viral RNA-silencing suppressors can be used as gene expression enhancers to produce recombinant proteins in plants.

Current genomics efforts will reveal additional information on RNA-silencing machinery, especially on the biological functions of noncoding RNAs. With detailed knowledge of this intriguing process, many more applications will undoubtedly become available in the near future.

ACKNOWLEDGMENTS

I would like to thank Dr. M. W. Prins for critically reading the manuscript. I apologize to a crowd of researchers who have contributed to the increase in knowledge of RNA silencing, whose work I was unable to mention here.

ARTICLES OF FURTHER INTEREST

Genomic Imprinting in Plants, p. 527
Plant Diseases Caused by Subviral Agents, p. 956
Plant DNA Virus Diseases, p. 960
Plant RNA Virus Diseases, p. 1023
Plant Viral Synergisms, p. 1026
Plant Virus: Structure and Assembly, p. 1029
Plant Viruses: Initiation of Infection, p. 1032
RNA-mediated Silencing, p. 1106
Transgenes: Expression and Silencing of, p. 1242
Virus-Induced Gene Silencing, p. 1276

REFERENCES

1. Lindbo, J.A.; Silva-Rosales, L.; Proebsting, W.M.; Dougherty, W.G. Induction of a highly specific antiviral state in transgenic plants: Implications for regulation of gene expression and virus-resistance. Plant Cell **1993**, *5*, 1749–1759.
2. Hutvagner, G.; Zamore, P.D. RNAi: Nature abhors a double-strand. Curr. Opin. Genet. Dev. **2002**, *12*, 225–232.
3. Storz, G. An expanding universe of noncoding RNAs. Science **2002**, *296*, 1260–1263.
4. Vaucheret, H.; Fagard, M. Transcriptional gene silencing in plants: Targets, inducers and regulators. Trends Genet. **2001**, *17*, 29–35.
5. Schiebel, W.; Pelissier, T.; Riedel, L.; Thalmeir, S.; Kempe, D.; Lottspeich, F.; Sanger, H.L.; Wassenegger, M. Isolation of an RNA-directed RNA polymerase-specific cDNA clone from tomato. Plant Cell **1998**, *10*, 2087–2101.
6. Bennetzen, J.L. Transposable element contributions to plant gene and genome evolution. Plant Mol. Biol. **2000**, *42*, 251–269.
7. Ketting, R.F.; Haverkamp, T.H.; Van Luenen, H.G.; Plasterk, R.H. Mut-7 of *C. elegans*, required for transposon silencing and RNA interference, is a homolog of the Werner syndrome helicase and RNAseD. Cell **1999**, *99*, 133–141.
8. Diener, T.O. The viroid: Biological oddity or evolutionary fossil? Adv. Virus Res. **2001**, *57*, 137–184.
9. Ahlquist, P. RNA-dependent RNA polymerases, viruses and RNA silencing. Science **2002**, *296*, 1270–1273.
10. Ding, S.W. RNA silencing. Curr. Opin. Biotechnol. **2000**, *11*, 152–156.

Genetic Diversity Among Weeds

Michael J. Christoffers
North Dakota State University, Fargo, North Dakota, U.S.A.

INTRODUCTION

Weeds are plants that grow where they are not desired, often in areas disturbed by humans. Genetic diversity is considered an important component of plant adaptability and is therefore important to the ability of weeds to establish and proliferate in areas and/or environments to which they are not native. Tools for genetic diversity research are expanding from heritable phenotypic traits to DNA-based molecular markers. Genetic diversity among weeds is influenced by breeding system, population history, gene flow, selection, and genetic features such as ploidy. Weed genetic diversity can be used to investigate the history of weedy populations and may also help predict the likelihood of survival and spread of weed populations.

RESEARCH TOOLS

Phenotype is the expression of genotype as influenced by environment. However, phenotype is often a poor indicator of genotype because of environmental effects. Weeds often have high phenotypic plasticity that masks underlying genetic components. Selection also may encourage divergence among genetically related weeds or select for similarities among genetically unrelated weeds. Thus, the study of weed genetic diversity using genetic markers that are selectively neutral and stable in dynamic environments is desirable.

Isozymes

Isozymes are enzyme variants that are separable on electrophoretic gels. Isozymes may be encoded by genes at different loci or by alleles of a single locus. The latter are often termed allozymes. Isozyme markers are advantageous for genetic diversity studies because of their ease of use and low cost. Allelic isozymes are codominant, which allows heterozygotes and homozygotes to be distinguished and facilitates estimates of allele frequency. However, the number of isozymes available for use is limited, so genetic diversity is sometimes underestimated. Also, isozymes are expressed proteins, so their selective neutrality cannot always be ensured.

Restriction Fragment Length Polymorphisms (RFLPs) and Amplified Fragment Length Polymorphisms (AFLPs)

Tools to directly assess variation at the molecular DNA level have been successfully used to study weed genetic diversity.[1] One technique is based on the presence/absence or distance between DNA sequences, typically four to six base pairs, recognized and cut by restriction enzymes. Variations are observed as fragment length polymorphisms and are usually considered selectively neutral. Nuclear RFLPs involve detection of specific fragments through hybridization to labeled probes and are usually codominant. RFLPs are highly reliable but require relatively large DNA samples and the availability of probes with homology to the sequences being investigated. Analyses involving large numbers of RFLPs can be expensive and time-consuming.

RFLPs are also commonly used to assess organellar genome variation, especially using chloroplast DNA (cpDNA). The relative abundance of cpDNA decreases the amount of plant tissue needed for RFLP analysis and may also eliminate the need for probes. Organellar genomes are typically but not always maternally inherited.

AFLPs involve the attachment of oligonucleotide adapters to fragments generated by restriction enzyme digestion followed by fragment amplification using the polymerase chain reaction (PCR). AFLPs require little prior sequence knowledge and more easily detect genome-wide diversity compared with RFLPs. AFLPs can also be used with small samples and, like RFLPs, are highly reproducible. However, AFLPs are dominant markers and do not allow classification of individuals as heterozygous or homozygous.

Randomly Amplified Polymorphic DNAs (RAPDs)

RAPDs are one of the most popular DNA markers for weed genetic diversity studies. Markers for RAPD analyses are generated using single, typically 10-base-pair PCR primers of arbitrary sequence to randomly amplify sequences in a genome. RAPDs require little prior sequence knowledge and do not require large DNA samples. A single RAPD primer may generate several bands, but

Encyclopedia of Plant and Crop Science
DOI: 10.1081/E-EPCS 120020234

RAPDs are dominant and provide less information than markers that distinguish between homozygotes and heterozygotes. RAPDs are considered selectively neutral but are subject to concerns about reproducibility.

Simple Sequence Repeats (SSRs) and Inter-Simple Sequence Repeats (ISSRs)

A selectively neutral PCR-based system that can generate codominant marker data involves the analysis of SSRs, a.k.a., microsatellites. SSRs most useful for genetic diversity studies are those with tandem repetition of two to four base pairs. Repeat numbers are often highly variable among allelic SSRs, providing ample variability for genetic diversity studies. SSR analyses involve PCR primers anchored to SSR flanks, allowing amplification of intervening repeats, and discrimination of repeat number using high-resolution electrophoresis. Since sequences surrounding SSRs often vary among species that are not closely related, development of SSR primers may be necessary prior to new weed diversity studies.

A second tool for weed genetic diversity studies based on SSRs is the PCR amplification of ISSRs. In this technique, a single primer based on SSR sequence is used to amplify regions between adjacent SSR loci. ISSRs are considered relatively easy to use, reliable, readily adaptable to divergent weed species, and selectively neutral. ISSR markers are usually scored as dominant without identification of heterozygotes, but some ISSRs are codominant.

DNA Sequencing

DNA sequence data provides the most definitive and accurate analysis of genetic variation. However, sequencing is expensive and labor-intensive and is not practical for most weed diversity studies. The most popular sequencing application for weed research has been characterization of genes known to be important for weedy traits, such as those that confer herbicide resistance. However, the diversity among such genes is often not selectively neutral.

APPLICATIONS

Weed genetic diversity studies are used to address many aspects of weed ecology and biology. Studies may involve the analysis of individual, population, or species relatedness to trace historical influences on weed evolution. Molecular data derived from cpDNA RFLPs has been used to investigate relatedness among populations of leafy spurge, revealing similarities between several North American populations and those from Russia.[2] Isozyme diversity among tall and common waterhemp has been used to support grouping the two species into one,[3] while RAPD markers have supported the possibility of different barnyardgrass species in Arkansas.[4]

The origin and spread of herbicide resistance is an important area of weed research and has been studied in wild oat using ISSR and RAPD markers.[5] Gene flow among weed species[5] and relationships among wild, weedy, and cultivated forms of plant species[6] have also been investigated using molecular genetic markers.

Genetic diversity is considered critical to the adaptability of species.[7] Weed adaptability facilitates the establishment of populations in new areas and survival of weeds in dynamic environments. Comparisons of inter- and intra-population genetic diversity can help predict weed population adaptability. For example, high intra-population genetic diversity based on SSRs suggests that barren brome on English farms may also be diverse for traits such as herbicide resistance or seed dormancy.[8]

FACTORS THAT INFLUENCE GENETIC DIVERSITY

Weed genetic diversity is important for population establishment and survival, but diversity is also influenced by varied factors. Weed biology components and environment are major factors influencing genetic diversity among weeds.

Breeding Systems and Gene Flow

Self-pollination and/or asexual reproduction are considered advantageous for establishment of new weed populations due to reproductive assurance. Lack of sexual recombination among individuals also helps maintain adapted genotypes. As a general rule, plant species that tend toward self-pollination or asexual reproduction have lower intra-population but higher inter-population genetic diversity compared with predominantly outcrossing species.[9] High inter-population diversity is due to minimal pollen exchange leading to increased genetic drift among populations of self-pollinating or asexual species. However, gene flow among populations via seed or other propagules can maintain intra-population diversity without sexual recombination. Many weeds are especially adapted to dispersal through movement of seed or whole plants via wind or animals. Agricultural activity also promotes exchange of weed propagules among populations through movement by farm implements, crop seed, or irrigation water.

Ploidy

Genetic diversity of individuals is also important to weed adaptability. Polyploids are found with increased frequency among weeds compared with other plant species.[10] Polyploids, especially allopolyploids, often display high genetic diversity due to fixed heterozygosity among their genomic copies. Maintenance of this diversity in self-pollinating species minimizes potential inbreeding depression. The potential diversity carried by an individual weed, however, is limited. Large genomes are correlated with characteristics detrimental to weeds including relatively slow growth and development.[10]

Environment

Environment may select for specific genotypes within populations and reduce overall genetic diversity. However, weeds with newly selected traits such as herbicide resistance often have higher-than-expected intra-population diversity. Genetic diversity may be maintained by hybridization, multiple genes or multiple gene origins for traits, or phenotypic plasticity. Dynamic environments may also favor different genotypes at different times, which maintains overall genetic diversity.

CONCLUSION

Genetic diversity is considered critical for weed population adaptability. Weeds have diverse means of maintaining genetic diversity based on their biology and responses to environment. Selectively neutral genetic markers have been successfully used to investigate relationships among weed populations, but their neutrality may make them poor predictors of weed adaptability. Continued research of genetic diversity underlying selectable weed traits is necessary to optimize weed management.

ACKNOWLEDGMENTS

This material includes work supported by the Cooperative State Research, Education, and Extension Service, U.S. Department of Agriculture, under Agreement No. 2002-34361-11781.

ARTICLES OF FURTHER INTEREST

Herbicide-Resistant Weeds, p. 551
Molecular Biology Applied to Weed Science, p. 745
Molecular Evolution, p. 748
Polyploidy, p. 1038
Population Genetics, p. 1042

REFERENCES

1. Jasieniuk, M.; Maxwell, B.D. Plant diversity: New insights from molecular biology and genomics technologies. Weed Sci. **2001**, *49* (2), 257–265.
2. Nissen, S.J.; Masters, R.A.; Lee, D.J.; Rowe, M.L. DNA-based marker systems to determine genetic diversity of weedy species and their application to biocontrol. Weed Sci. **1995**, *43* (3), 504–513.
3. Pratt, D.B.; Clark, L.G. *Amaranthus rudis* and *A. tuberculatus*—One species or two? J. Torrey Bot. Soc. **2001**, *128* (3), 282–296.
4. Rutledge, J.; Talbert, R.E.; Sneller, C.H. RAPD analysis of genetic variation among propanil-resistant and -susceptible *Echinochloa crus-galli* populations in Arkansas. Weed Sci. **2000**, *48* (6), 669–674.
5. Cavan, G.; Biss, P.; Moss, S.R. Herbicide resistance and gene flow in wild-oats (*Avena fatua* and *Avena sterilis* ssp. *ludoviciana*). Ann. Appl. Biol. **1998**, *133* (2), 207–217.
6. Crouch, J.H.; Lewis, B.G.; Lydiate, D.J.; Mithen, R. Genetic diversity of wild, weedy and cultivated forms of *Brassica rapa*. Heredity **1995**, *74* (5), 491–496.
7. Fisher, R.A. *The Genetical Theory of Natural Selection*; Clarendon Press: Oxford, 1930.
8. Green, J.M.; Barker, J.H.A.; Marshall, E.J.P.; Froud-Williams, R.J.; Peters, N.C.B.; Arnold, G.M.; Dawson, K.; Karp, A. Microsatellite analysis of the inbreeding grass weed barren brome (*Anisantha sterilis*) reveals genetic diversity at the within- and between-farm scales. Mol. Ecol. **2001**, *10* (4), 1035–1045.
9. Hamrick, J.L.; Godt, M.J.W. Allozyme Diversity in Plant Species. In *Plant Population Genetics, Breeding, and Genetic Resources*; Brown, A.H.D., Clegg, M.T., Kahler, A.L., Weir, B.S., Eds.; Sinauer Associates: Sunderland, MA, 1990; 43–63.
10. Bennett, M.D.; Leitch, I.J.; Hanson, L. DNA amounts in two samples of angiosperm weeds. Ann. Bot. **1998**, *82* (Supplement A), 121–134.

Genetic Resource Conservation of Seeds

Florent Engelmann
Centre de Coopération Internationale en Recherche Agronomique pour le Développement, Montpellier, France

INTRODUCTION

In the 1950s and 1960s, major advances in plant breeding brought about the "green revolution," which resulted in wide-scale adoption of high-yielding varieties and genetically uniform cultivars of staple crops, particularly wheat and rice. Consequently, global concern about the loss of genetic diversity in these crops increased, as farmers abandoned their locally adapted landraces and traditional varieties, replacing them with improved, yet genetically uniform modern ones. The International Agricultural Research Centers (IARC) of the Consultative Group on International Agricultural Research (CGIAR) started to assemble germplasm collections of the major crop species within their respective mandates. The International Board for Plant Genetic Resources (IBPGR) was established in 1974 in this context to coordinate the global effort to systematically collect and conserve the world's threatened plant genetic diversity. Today, as a result of this effort, over 1300 genebanks and germplasm collections exist around the world, maintaining approximately 6,100,000 accessions, largely of major food crops including cereals and some legumes, i.e., species that can be conserved easily as seed.

This article reviews the current storage technologies and management procedures developed for seeds, describes the problems and achievements with seed storage, and identifies priorities for improving the efficiency of seed conservation.

CONSERVATION OF ORTHODOX SEEDS

Many of the world's major food plants produce so-called orthodox seeds, which tolerate extensive desiccation and can be stored dry at low temperature. Storage of orthodox seeds is the most widely practiced method of ex situ conservation of plant genetic resources (FAO 1996), since 90% of the accessions stored in genebanks are maintained as seed. Following drying to low moisture content (3–7% fresh weight basis, depending on the species), such seeds can be conserved in hermetically sealed containers at low temperature, preferably at –18°C or cooler, for several decades.[1] All relevant techniques are well established, and practical documents covering the main aspects of seed conservation, are available, including design of seed storage facilities for genetic conservation, principles of seed testing for monitoring viability of seed accessions maintained in genebanks, methods for removing dormancy and germinating seeds, and suitable methods for processing and handling seeds in genebanks.[2]

In addition to being the most convenient material for genetic resource conservation, seeds are also a convenient form for distributing germplasm to farmers, breeders, scientists, and other users. Moreover, since seeds are less likely to carry diseases than other plant material, their use for exchange of plant germplasm can facilitate quarantine procedures.

CONSERVATION OF NONORTHODOX SEEDS

In contrast to orthodox seeds, a considerable number of species, predominantly from tropical or subtropical origin, such as coconut, cacao, and many forest and fruit tree species, produce so-called nonorthodox seeds, which are unable to withstand much desiccation and are often sensitive to chilling. Nonorthodox seeds have been further subdivided in recalcitrant and intermediate seeds based on their desiccation sensitivity, which is high for the former and lower for the latter group. Nonorthodox seeds cannot be maintained under the storage conditions described above, i.e., low moisture content and temperature, and have to be kept in moist, relatively warm conditions to maintain viability. Even when stored in optimal conditions, their lifespan is limited to weeks, occasionally months. Of over 7000 species for which published information on seed storage behavior exists,[3] approximately 10% are recorded as nonorthodox or possibly nonorthodox.

Genetic resources of nonorthodox species are traditionally conserved as whole plants in field collections. This mode of conservation is faced with various problems and limitations, and cryopreservation (liquid nitrogen, –196°C) currently offers the only safe and cost-effective option for the long-term conservation of genetic resources of problem species. However, cryopreservation research is still at a preliminary stage for nonorthodox species.[4]

DOI: 10.1081/E-EPCS 120020301

MANAGEMENT OF SEED COLLECTIONS

Only a limited number of genebanks operate at very high standards, whereas many others are facing difficulties due to inadequate infrastructures, lack of adequate seed processing and storage equipment, unreliable electricity supply, funding and staffing constraints, and inadequate management practices. Therefore, seeds are often stored under suboptimal conditions and require more frequent regeneration, thus bearing additional costs on often already insufficient operating budgets of genebanks. Great difficulties are faced in particular by many countries with regeneration of seed collections.[5] Seed storage technologies are relatively easy to apply. The problems relate more to resource constraints that impact the performance of essential operations. Efficient and cost-effective genebank management procedures have now become key elements for long-term ex situ conservation of plant genetic resources.

Other important aspects of genebank operations concern germplasm characterization and documentation. The extent to which germplasm collections are characterized varies widely between genebanks and species[5] but is far from complete in many instances. Concerning documentation, the situation is highly contrasted.[5] Some, mainly developed, countries have fully computerized documentation systems and relatively complete accession data, while many others lack information on the accessions in their collections, including the so-called passport data.

MAIN CHALLENGES AND PRIORITIES FOR IMPROVING SEED CONSERVATION

Various priority areas have been identified for orthodox and nonorthodox seed conservation research and for improving seed genebank management procedures.

Research Priorities for Orthodox Seed Species

Determining critical seed moisture content

The preferred conditions recommended for long-term seed storage are 3–7% moisture content, depending on the species, at −18°C or lower.[1] However, it has been shown that drying seeds beyond a critical moisture content provides no additional benefit to longevity and may even accelerate seed aging rates, and that interactions exist between the critical relative humidity and storage temperature. Research should be pursued on this topic of high importance.

Developing low-input storage techniques

Various research projects have focused on the development of the ultra-dry seed technology, which allows storing seeds desiccated to very low moisture contents at room temperature, thereby suppressing the need for refrigeration equipment. Although drying seed to very low moisture prior to storage seems to have fewer advantages than was initially expected, ultra-dry storage is still considered to be a useful, practical, low-cost technique in those circumstances where no adequate refrigeration can be provided.[6] Research on various aspects of the ultra-dry seed storage technology and on its applicability to a broader number of species should therefore be continued.

Improving and monitoring viability

The viability of conserved accessions depends on their initial quality and how they have been processed for storage, as well as on the actual storage conditions. There is evidence that a very small decrease in initial seed viability can result in substantial reduction in storage life. This needs further investigation, and research should be performed notably on the effect of germplasm handling in the field during regeneration and during subsequent processing stages prior to its arrival at the genebank, as well as on growing conditions, disease status, and time of harvest of the plants.

Research Priorities for Nonorthodox Seed Species

Understanding seed recalcitrance

A number of physical and metabolic processes or mechanisms have been suggested to confer, or contribute to, desiccation tolerance.[7] Different processes may confer protection against the consequences of loss of water at different hydration levels, and the absence, or ineffective expression, of one or more of these could determine the relative degree of desiccation sensitivity of seeds of individual species. Additional research is needed to improve our understanding of the mechanisms involved in seed recalcitrance.

Developing improved conservation techniques

Various technical options exist for improving storage of nonorthodox species. Especially with species for which no or only little information is available, it is advisable, before undertaking any ''high-tech'' research, to examine the development pattern of seeds and to run preliminary experiments to determine their desiccation sensitivity as well as to define germination and storage conditions.

IPGRI, in collaboration with numerous institutions worldwide, has developed a protocol for screening tropical forest tree seeds for their desiccation sensitivity and storage behavior,[8] which might be applicable to seeds of other species after required modification and adaptation. For long-term storage of nonorthodox species, cryopreservation represents the only option. Numerous technical approaches exist, including freezing of seeds, embryonic axes, shoot apices sampled from embryos, adventitious buds, or somatic embryos, depending on the sensitivity of the species studied.[4]

Germplasm Management Procedures

The objective of any genebank management procedure is to maintain genetic integrity and viability of accessions during conservation and to ensure their accessibility for use in adequate quantity and quality at the lowest possible cost. As mentioned previously, many genebanks face financial constraints that hamper their efficient operation. It is therefore very important to improve genebank management procedures to make them more efficient and cost-effective. In this aim, the use of molecular markers should be increased to improve characterization and evaluation, improved seed regeneration and accession management procedures should be established, the use of collections should be enhanced through the development of core collections, germplasm health aspects including new biotechnological tools for detection, indexing, and eradication of pathogens should be better integrated in routine genebank operations, and improved documentation tools should be developed. Special attention should be given to developing appropriate techniques for genebanks in developing countries, where specialized equipment is frequently lacking and resources are usually limited.

CONCLUSION

The improvements in the seed storage techniques and genebank management procedures resulting from the research performed on the priority areas identified above will further increase the key role of seeds in the ex situ conservation of genetic resources of many species. However, it is now well recognized that an appropriate conservation strategy for a particular plant genepool requires a holistic approach, combining the different ex situ and in situ conservation techniques available in a complementary manner.[9] Selection of the appropriate methods should be based on a range of criteria, including the biological nature of the species in question, practicality and feasibility of the particular methods chosen (which depends on the availability of the necessary infrastructures), their efficiency, and the cost-effectiveness and security afforded by their application. An important area in this is the linkage between in situ and ex situ components of the strategy, especially with respect to the dynamic nature of the former and the static, but potentially more secure, approach of the latter.

ARTICLES OF FURTHER INTEREST

Agriculture and Biodiversity, p. 1
Crop Improvement: Broadening the Genetic Base for, p. 343
Germplasm Acquisition, p. 537
Germplasm Collections: Regeneration in Maintenance, p. 541
Germplasm Maintenance, p. 544
Germplasm: International and National Centers, p. 531
Seed Vigor, p. 1139
Seeds: Pathogen Transmission Through, p. 1142

REFERENCES

1. FAO/IPGRI. *Genebank Standards*; Food and Agriculture Organization of the United Nations: Rome and International Plant Genetic Resources Institute: Rome, 1994.
2. http://www.ipgri.cgiar.org/system/page.asp?theme=7 (accessed March 2003).
3. Hong, T.D.; Linington, S.; Ellis, R.H. *Seed Storage Behaviour: A Compendium*; Handbooks for Genebanks, International Plant Genetic Resources Institute: Rome, 1996; Vol. 4.
4. Engelmann, F.; Takagi, H. *Cryopreservation of Tropical Plant Germplasm—Current Research Progress and Applications*; 2000. Japan International Center for Agricultural Sciences: Tsukuba, Japan and International Plant Genetic Resources Institute: Rome.
5. FAO. *Report on the State of the World's Plant Genetic Resources for Food and Agriculture*; Food and Agriculture Organization of the United Nations: Rome, 1996.
6. Walters, C. Ultra-dry seed storage. Seed Sci. Res. **1998**, *8*. Suppl. No 1.
7. Pammenter, N.W.; Berjak, P. A review of recalcitrant seed physiology in relation to desiccation-tolerance mechanisms. Seed Sci. Res. **1999**, *9*, 13–37.
8. IPGRI/DFSC. Desiccation and Storage Protocol—March 1999. In *The Project on Handling and Storage of Recalcitrant and Intermediate Tropical Forest Tree Seeds, Newsletter*; April 5, 1999; 23–39.
9. Maxted, N.; Ford-Lloyd, B.V.; Hawkes, J.G. Complementary Conservation Strategies. In *Plant Genetic Resources Conservation*; Maxted, N., Ford-Lloyd, B.V., Hawkes, J.G., Eds.; Chapman & Hall: London, 1997; 15–39.

Genetic Resources of Medicinal and Aromatic Plants from Brazil

Roberto Fontes Vieira
Embrapa Recursos Genéticos e Biotecnologia, Brasilia, Brazil

INTRODUCTION

Approximately two-thirds of the biological diversity of the world is found in tropical zones, mainly in developing countries. Brazil is considered the country with the greatest biodiversity on the planet, with nearly 55,000 native species distributed over six major biomes: Amazon Forest, Cerrado, Caatinga, Atlantic Forest, Pantanal, and Meridional Forest and Grassland.

Embrapa (Brazilian Agricultural Research Corporation), through its Genetic Resources and Biotechnology Research Center and in collaboration with several universities and other Embrapa research centers, has been developing efforts to establish germplasm banks for medicinal and aromatic species. Twenty germplasm collections of medicinal and aromatic plants have been established in Brazil, and 70 species recently were defined as priority for germplasm conservation. The main challenge now is to develop strategies for the conservation, cultivation, and sustainable management of each of these species.

GERMPLASM CONSERVATION

The Brazilian Amazon Forest covers nearly 40% of all national territory. This ecosystem is rather fragile, and its productivity and stability depend on the recycling of nutrients whose efficiency is directly related to the biological diversity and the structural complexity of the forest.[1]

The Cerrado is the second largest ecological dominion of Brazil, covering approximately 23% of Brazilian territory, where a continuous herbaceous stratum is joined to an arboreal stratum, with variable density of woody species. The Caatinga extends over areas of the states of the Brazilian Northeast and is characterized by the xerophitic vegetation typical of a semiarid climate. The soils that are fertile, due to the nature of their original materials and the low level of rainfall, experience minor runoff.[1]

The Atlantic Forest extends over nearly the whole Brazilian coastline and is one of the most endangered ecosystems of the world, with less than 10% of the original vegetation remaining. The climate is predominantly hot and tropical, and annual precipitation ranges from 1000 to 1750 mm. The territory of the Meridional Forests and Grasslands includes the mesophytic tropical forests, the subtropical forests, and the meridional grasslands of the states of southern Brazil. The climate is humid, tropical and subtropical, with some areas of temperate climate. Pantanal is a geologically lowered area filled with sediments that have settled in the basin of the Paraguay River. Pantanal flora is formed by species from both Cerrado and Amazon vegetation.[1]

Ex situ conservation of threatened germplasm includes seed banks, field preservation, tissue culture, and cryopreservation. Seed storage is considered the ideal method. Seeds considered orthodox can be dried and preserved at subzero temperatures (−20°C), whereas recalcitrant seeds, including most tropical species, lose their seed viability when subjected to the same conditions.[2] Seeds of maize (*Zea mays* L.) have been maintained at Embrapa for more than 12 years in a long-term cold chamber at −20°C. Maintenance of the germplasm in field collections is costly, requires large areas, and can be affected by adverse environmental conditions. Tissue culture or cryopreservation techniques also can be considered in some cases. In an ex situ procedure, the germplasm is collected from fields, markets, small farms, and other sites in the form of seeds, cuttings, underground systems, and sprouts. The collected samples should represent the original population with passport data and herbarium vouchers.

The following species have been recognized as priority for germplasm conservation.[3]

Psychotria ipecacuanha (Brot.) Stokes, Rubiaceae (Ipecac)

Ipecac (*P. ipecacuanha*) (Fig. 1A) is a shrub whose medicinal value relates to the production of emetine in the roots. Ipecac is found in the humid forests of the southern part of the Amazon Forest in the states of Rondônia and Mato Grosso, and in the Atlantic Forest in the states of Bahia, Espírito Santo, Minas Gerais, and Rio de Janeiro.[4]

Ipecac is a powerful emetic used to treat gastrointestinal diseases, diarrhea, and intermitent fevers. It is employed as an expectorant in bronchitis, bronchopneumonia, asthma, and mumps, and also as a vasoconstrictor.

Encyclopedia of Plant and Crop Science
DOI: 10.1081/E-EPCS 120010421

Fig. 1 Medicinal plants from Brazil: (A) *Psychotria ipecacuanha*, (B) *Pilocarpus microphyllus*, (C) *Maytenus ilicifolia*, and (D) *Pfaffia glomerata*. (*View this art in color at www.dekker.com.*)

In 1988, Embrapa began a project for the recollecting and conservation of Ipecac genetic variability. Five collecting expeditions covering seven states were undertaken, collecting a total of 86 accessions[4] that are now maintained in field germplasm banks at Embrapa Ocidental Amazon, PA. Recently, more germplasm accessions were collected by the North Fluminense University, including 10 accessions that originated in the Atlantic Forest.

Pilocarpus microphyllus Stapf., Rutaceae (Jaborandi)

Jaborandi (*ia-mbor-end*) (Fig. 1B) is an indigenous name of the species *P. microphyllus*, which contains the highest pilocarpine content in the leaves. This plant is an understory species of the Amazonian rainforest that reaches six to eight meters in height.

Pilocarpine is an imidazolic alkaloid that stimulates secretions of the respiratory tract and the salivary, lachrymal, gastric, and other glands.[5] In the treatment of glaucoma, the alkaloid pilocarpine acts directly on cholinergic receptor sites, thus mimicking the action of acetylcholine. Intraocular pressure is thereby reduced, and despite its short-term action, pilocarpine is the standard drug used for initial and maintenance therapy in certain types of primary glaucoma.[6]

In 1991, Embrapa initiated a project for recollecting and conservation of the genetic variability of *P. microphyllus* and its closely related species. From 1991 to 1993, two collection expeditions were undertaken, collecting a total of 27 accessions in the form of seeds and seedlings.

A Jaborandi germplasm bank was established at Maranhão State University, São Luis, MA and at Embrapa Ocidental Amazon, Belém, PA.

Pilocarpus microphyllus seeds are considered orthodox. Seeds can be dried down to 6–8% moisture content and can be conserved for a long period at −18°C and 5% relative humidy.[7] The wild harvest of leaves from wild *P. microphyllus* has been carried out to such an extent that it has significantly reduced the natural populations. Currently, this species can be found only in indigenous areas and on private lands, and is inluded in the official list of endangered plants from Brazilian flora.

Phyllanthus niruri L., Euphorbiaceae (Quebra Pedra)

Quebra Pedra is a small, erect annual herb growing up to 30–40 cm in height. Although several species are recognized by this common name, *P. niruri* and *P. sellovianus* are the most scientifically studied. The antispasmodic activity of alkaloids in *P. sellovianus* explains the popular use of the plant to treat kidney and bladder stones. The alkaloid extract demonstrates smooth muscle relaxation specific to the urinary and biliary tracts, which facilitates the expulsion of kidney or bladder calculi.[8,9]

Quebra Pedra has gained worldwide attention due to its effects against Hepatitis B.[10] There have been no side effects or toxicity reported in any of the clinical studies or in its many years of reported use in herbal medicine.

Several species are called Quebra Pedra and contain the same or similar active compounds. A germplasm collection to study the genetic and chemical variation of this species, as well as its seed physiology, is necessary and warranted.

Maytenus ilicifolia Martius *ex.* Reiss., Celastraceae (Espinheira Santa)

Espinheira Santa (Fig. 1C) is a small, shrublike evergreen tree reaching up to 5 m in height. It is native to many parts of southern Brazil, mainly in the Paraná and Santa Catarina states.

Leaves of *Maytenus* species are used in the popular medicine of Brazil for their reported antiacid and antiulcerogenic activity. Attempts to detect general depressant, hypnotic, anticonvulsant, and analgesic effects were reported by Oliveira et al.[11] The potent antiulcerogenic effect of Espinheira Santa leaves was demonstrated to be effective compared to two leading antiulcer drugs, Ranitidine and Cimetidine.[12] Toxicological studies have demonstrated the plant's safety.

Seeds of *Maytenus ilicifolia* can be classified as orthodox and can be stored in long-term cold chambers at −20°C. In 1995, the Forestry Department of the University of Paraná began a project to study the genetic variability of natural populations of *M. ilicifolia*, and 78 accessions were collected in the states of Parana, Santa Catarina, and Rio Grande do Sul. Field collections are maintained at an Embrapa center in Ponta Grossa, PR.

Pfaffia glomerata (Spreng.) Pedersen, Amaranthaceae (Brazilian Ginseng)

Pfaffia glomerata (Fig. 1D) is a large, shrubby ground vine, which has a deep root system. It grows mainly in the borders of the Paraná River, and predatory collection has greatly reduced its natural populations. *Pfaffia* is known as "Brazilian ginseng" because it is widely used as an adaptogen for many ailments and to overcome weakness, much like American and Asian ginseng (*Panax* spp.).

This action is attributed to the anabolic agent—ecdysterone and three novel ecdysteroid glycosides, which are found in high amounts in *Pfaffia* roots. The root of *Pfaffia* also contains about 11% saponins. These saponins include a group of novel chemicals called pfaffosides as well as pfaffic acids, glycosides, and nortriperpenes. These saponins have clinically demonstrated the ability to inhibit cultured tumor cell melanomas and to help regulate blood sugar levels.[13]

In 2001, Embrapa Genetic Resources, in collaboration with Paraná State Rural Assistance Corporation and São Paulo State University (Unesp), Botucatu, launched a field expedition that collected 15 accessions, including more than 200 hundred individuals from all along the borders of the Paraná River. This material has been evaluated for chemical and molecular markers and has been deposited in both filed and in vitro collections.

ARTICLES OF FURTHER INTEREST

Agriculture and Biodiversity, p. 1
Echinacea: Uses as a Medicine, p. 395
Genetic Resource Conservation of Seeds, p. 499
Genetic Resources of Medicinal and Aromatic Plants from Brazil, p. 502
Germplasm Acquisition, p. 537
Germplasm Collections: Regeneration in Maintenance, p. 541
Germplasm Maintenance, p. 544
Germplasm: International and National Centers, p. 531
Herbs, Spices, and Condiments, p. 559
New Industrial Crops in Europe, p. 813

Phytochemical Diversity of Secondary Metabolites, p. 915
Secondary Metabolites as Phytomedicines, p. 1120

REFERENCES

1. Anon. *FAO International Conference and Programme for Plant Genetic Resources (ICPPGR). Country Report*; FAO, 1995. http://www.fao.org/waicent/faoinfo/agricult/agp/agps/pgrfa/pdf/brazil.pdf. (accessed July 2001).
2. Faiad, M.G.R.; Goedert, C.O.; Wetzel, M.M.V.S.; Silva, D.B.; Pereira Neto, L.G. *Banco de germoplasma de Sementes da Embrapa*; Série Documentos, Embrapa: Brasília, 2001; Vol. 71, p. 31.
3. Vieira, R.F.; Silva, S.R.; Alves, R.B.N.; Silva, D.B.; Dias, T.A.B.; Wetzel, M.M.; Udry, M.C.; Martins, R.C. *Estratégias para Conservação e Manejo de Recursos Genéticos de Plantas Medicinais e Aromáticas. Resultados de 1ª Reunião Técnica*; Embrapa/Ibama/CNPq: Brasília, 2002; p. 184.
4. Skorupa, L.A.; Assis, M.C. Collecting and conserving Ipecac (*Psychotria ipecacuanha*, Rubiaceae) germplasm in Brazil. Econ. Bot. **1998**, *52*, 209–210.
5. Morton, J.F. *Major Medicinal Plants*; Charles C. Thomas: Illinois, 1977; 187–189.
6. Lewis, W.H.; Elvin-Lewis, P.F. *Medical Botany*; Wiley: New York, 1977.
7. Eira, M.T.S.; Vieira, R.F.; Mello, C.M.C.; Freitas, R.W.A. Conservação de sementes de Jaborandi (*Pilocarpus microphyllus* Stapf.). Revista Brasileira de Sementes **1992**, *14* (1), 37–39.
8. Calixto, J.B. Antispasmodic effects of an alkaloid extracted from *Phyllanthus sellowianus*: Comparative study with papaverine. Braz. J. Med. Biol. Res. **1984**, *17* (3–4), 313–321.
9. Santos, A.R. Analgesic effects of *callus* culture extracts from selected species of *Phyllanthus* in mice. J. Pharm. Pharmacol. **1994**, *46*, 755–759.
10. Wang, M. Herbs of the genus *Phyllanthus* in the treatment of chronic hepatitis B: Observations with three preparations from different geographic sites. J. Lab. Clin. Med. **1995**, *126*, 350–352.
11. Oliveira, M.G.M.; Monteiro, M.G.; Macaubas, C.; Barbosa, V.P.; Carlini, E.A. Pharmacologic and toxicologic effects of two *Maytenus* species in laboratory animals. J. Ethnopharmacol. **1991**, *34*, 29–41.
12. Souza-Formigoni, M.L.O.; Oliveira, M.G.M.; Monteiro, M.G.; Silveira-Filho, N.G.; Braz, S.; Carlini, E.A. Antiulcerogenic effects of two *Maytenus* species in laboratory animals. J. Ethnopharmacol. **1991**, *34* (1), 21–27.
13. Takemoto, T.; Nishimoto, N.; Nakai, S.; Takagi, N.; Hayashi, S.; Odashima, S.; Wada, Y. Pfaffic acid, a novel nortriterpene from *Pfaffia paniculata* Kuntze, Tetrahedron Lett. **1983**, *24*, 1057–1060.

Genetically Engineered Crops with Resistance Against Insects

Johnie N. Jenkins
USDA—ARS, Mississippi State, Mississippi, U.S.A.

INTRODUCTION

Many insect pests reduce yield and quality of agricultural crops, and a large worldwide market exists for insecticides for their control. Even though insecticides are used for their control, insects in the order Lepidoptera and Coleoptera remain major worldwide pests of important food and fiber crops. The common soil bacterium, *Bacillus thuringiensis* (Berliner), produces a crystalline protein that is toxic when consumed by certain larvae in those orders. Insecticides based on crystalline proteins from *B. thuringiensis* have been useful in the control of Lepidopterous caterpillars. These *B. thuringiensis* pesticides have been used successfully and safely for over 50 years. They are not toxic to humans, to other mammals, or to other higher animals.

Host plant resistance to insects has been developed for several crops using conventional breeding of genes within the plant species.[1] Genetic transformation technology offers a means of moving genes across the species barrier. With the advent of this technology in the early 1980s, scientists began attempts to transform plants using genes from *B. thuringiensis*, thereby providing control of selected insects using a bacterial gene genetically engineered into a plant. The first reported use of a toxin gene from *B. thuringiensis* expressed in plants occurred in 1987 when tobacco plants, *Nicotianatabacum* L., were developed that produced enough of the toxin to kill first instar larvae of tobacco hornworm, *Manduca sexta* L., placed on leaves of transformed plants.[2–4]

GENETICALLY ENGINEERED CROPS

Transformation of plants by the bacterium *Agrobacterum tumifacens* became routinely possible in the early 1980s. In 1987, cotton plants, *Gossypium hirsutum* L., were first regenerated from cells into complete plants via tissue culture. Scientists had previously attempted to transform cotton plants with DNA sequences of the delta endotoxin gene from *B. thurgiensis* var kurstaki. Scientists at Monsanto Co. first transformed tomato, *Lycopersicon esculentum*, to express the Bt toxin gene and field tested these plants.[5] Scientists at Agracetus were the first to produce cotton plants for field testing that expressed the *B. thuringiensis* gene.[6] However, when Agracetus cotton plants were grown in field plots they did not express the toxin at a level sufficient to control targeted insects.[7] Monsanto was the first company to produce transformed cotton, corn, *Zea maize*, and potato, *Solanum tuberosum,* plants that expressed toxin(s) from *B. thuringiensis* at levels needed for commercial control of targeted insect pests.[8,9] This increased toxin expression in transformed plants resulted from modifications by Monsanto scientists[8] to the DNA sequence of a specific toxin gene of *B. thuringiensis.*

All the transgenic, insect-resistant crop species that have been commercialized to date express one of the various crystalline proteins from *B. thuringiensis*. Three transgenic crop species are currently approved by the U.S. Environmental Protection Agency (EPA) for growing in the United States.[10] The Cry1Ac protein in cotton was registered by Monsanto as Bollgard® and targeted to tobacco budworm, *Heliothis virescens* (Fab.), bollworm, *Helicoverpa zea* (Boddie), and pink bollworm, *Pectinophora gossypiella* (Saunders). The Cry3A protein in potato was registered by Monsanto as NewLeaf® and targeted against the Colorado potato beetle, *Leptinotarsa decemlineata* Say. The Cry1Ab protein in corn was registered by Monsanto as YieldGard® for full commercial use in field corn. The Cry1Ab protein in corn was registered by Syngenta Seeds, Inc. as YieldGard® and Attribute™ for full commercial use in field and sweet corn, respectively.

Transgenic Bollgard cotton was first grown commercially in the United States in 1996. In 2000, cotton with a gene from *B. thuringiensis* was grown on 1.5 million hectares (ha) and cotton with this gene plus a herbicide resistance gene was grown on 1.7 million ha in the world, primarily in the United States, China, Australia, Argentina, Mexico, and South Africa.[11] The United States is the largest user of Bollgard cotton with 39% of the U.S. crop in 2000 planted to Bollgard or Bollgard plus herbicide resistant cotton.[11] In China, cotton with a *B. thuringiensis* gene was grown for the first time in 1997 on 63,000 ha and in 2000 it was grown on 500,000 ha.[11] In 2000, growers of Bollgard cotton accrued $168 million in economic benefit from lower production costs and increased yield.[12] They applied 1.04 million fewer pounds

Encyclopedia of Plant and Crop Science
DOI: 10.1081/E-EPCS 120010493

of insecticide. Additional benefits were reduced pesticide exposure risks, improved preservation of beneficial insects, and increased wildlife benefits.[12] In China, the Chinese Academy of Agricultural Sciences (CAAS) and Monsanto/DeltaPine each have versions of Bt cotton grown by an estimated 3 million small farmers in 2000.[11] Transgenic Bt cotton offers the small farmers of China a significant increase in income and a significant decrease in pesticide usage and associated pesticide problems.

Corn is another major crop with a significant area planted to transgenic hybrids. Corn with a gene from *B. thuringiensis* was first planted in the United States in 1996. In 2000, 6.8 million ha were planted on a global basis, with 92% planted in the United States.[11]

HOW PLANTS WITH GENES FROM *B. thurgiensis* KILL INSECTS

The toxin produced by transgenic plants transformed to express the toxin gene from *B. thuringiensis* affects susceptible insects as the protein[13] is solubilized and proteolytically processed into polypeptide molecules that bind with specific affinity to receptor sites in the insect's midgut epithelial cells. This causes pores or ion channels to develop in the cell membranes, thus disturbing cellular osmotic balance and causing cells to swell and lyse and causing paralysis in the insect's midgut and mandibles. Death then occurs through a combination of starvation and septicemia.

MANAGING OR DELAYING DEVELOPMENT OF INSECT RESISTANCE TO Bt IN GM CROPS

Management of the development of resistance to *B. thuringiensis* insecticidal toxins in transgenic plants is of great concern to many people.[13] The U.S. EPA requires a unique resistance management plan as part of its approval of transgenic Bt cotton and corn. This resistance management plan requires delayed development of populations of pest insects that are resistant to the toxin. The plan was developed around a high-toxin dosage and a concept of refuge in which susceptible insects are produced in sufficient numbers to mate with the low numbers of resistant insects that may survive on the transgenic crops. Because no actual data had ever been developed to test this strategy, insect population models were developed and used to define refuge sizes and a strategy for use. General conclusions from population models were that 1) the level of toxic protein should be very high relative to the target pests LD_{90}; 2) these crops should be planted in a manner that allows high levels of mating among susceptible and resistant pest insects; and 3) movement of larvae between transgenic and nontransgenic plants must be avoided.[14]

CURRENT STATUS OF *B. thuringiensis* GENES IN PLANTS

Approximately 40 different insect resistance genes have been introduced into crop plants.[15] At least 10 Bt genes encoding different Bt toxins (cry1Aa, cry1Ab, cry1Ac, cry1Ba, cry1Ca, cry1H, cry2Aa, cry3A, cry6A, and cry9c) have been genetically engineered into 26 species of plants; however, codon-optimized genes have been transformed into only cotton, corn, potato, broccoli, *Brassica oleracea italica*, cabbage, *B. oleracea capitata*, and alfalfa, *Medicago* sp.[15]

CONCLUSION

In the future many types of genes will be used to genetically engineer insect resistant plants. Some of the promising non-cry genes from *B. thuringiensis* and *B. cereus* are the vegetative phase insecticidal proteins VIP 1, VIP 2, and VIP 3A.[15] Candidate genes, other than those from *B. thurgiensis*, are plant protease inhibitors, plant amylase inhibitors, plant lectins, and chitinases. From animals, serine protease inhibitors and chitinase genes are being investigated. At present, only plants with *B. thuringiensis* cry genes have been commercialized.[15] Over 240 insecticidal cry proteins produced by the various strains of *B. thuringiensis* have been classified.[16] In addition to these other genes, there seems to be ample genetic variability among *B. thuringiensis* strains to allow scientists to use these cry genes to develop many different versions of crop resistance.

ARTICLES OF FURTHER INTEREST

Breeding for Durable Resistance, p. 179
Breeding Plants with Transgenes, p. 193
Breeding: Incorporation of Exotic Germplasm, p. 222

REFERENCES

1. Lynch, R.E.; Baozhu, G.; Timper, P.; Wilson, J. Improving host–plant resistance to pests. Pest Manag. Sci. **2003**, *59*, 718–727.
2. Adang, M.J.; Firoozabady, E.; Klein, J.; DeBoer, D.; Sekar, V.; Kempl, J.D.; Murray, E.; Rocheleau, T.A.; Rashka, K.; Staffeld, G.; Stock, C.; Sutton, D.; Merlo, D.J. Expression of

a *Bacillus thuringiensis* insecticidal crystal protein gene in tobacco leaves. Mol. Strategy Crop Prot. **1987**, 345–353.
3. Barton, K.A.; Whiteley, H.R.; Yang, N.S. *Bacillus thuringiensis* delta endotoxin expressed in transgenic *Nicotiana tabacum* provides resistance to Lepidopteran insects. Plant Physiol. **1987**, *85*, 1103–1109.
4. Vaeck, M.; Reynaerts, A.; Hofte, H.; Jansens, S.; DeBeukeleer, M.; Dean, C.; Zabeau, M.; Van Montagu, M.; Leemans, J. Transgenic plants protected from insect attack. Nature **1987**, *328*, 33–37.
5. Fischoff, D.A.; Bowdish, K.S.; Perlak, F.J.; Marrone, P.G.; McCormick, S.M.; Niedermeyer, J.G.; Dean, D.A.; Rusano-Kretzmer, K.; Mayer, E.J.; Rochester, D.E.; Rogers, S.G.; Fraley, R.T. Insect tolerant transgenic tomato plants. Bio/Technology **1987**, *5*, 807–813.
6. Umbeck, P.; Johnson, G.; Barton, K.; Swain, W. Genetically transformed cotton (*Gossypium hirsutum* L.) plants. Biotechnology **1987**, *5*, 263–266.
7. Jenkins, J.N.; Parrott, W.L.; McCarty, J.C.; Barton, K.A.; Umbeck, P.F. *Field Test of Transgenic Cottons Containing a Bacillus thuringiensis Gene*; Mississippi Agricultural and Forestry Experiment Station: Mississippi State, MS, 1991; 1–6.
8. Perlak, F.J.; Deaton, R.W.; Armstrong, T.A.; Fuchs, R.L.; Sims, S.R.; Greenplate, J.T.; Fishoff, D.A. Insect resistant cotton plants. Biotechnology **1990**, *8*, 939–943.
9. McIntosh, S.C.; Stone, T.B.; Sims, S.R.; Hunter, P.L.; Greenplate, J.T.; Marrone, P.G.; Perlak, F.J.; Fischoff, D.A.; Fuchs, R.L. Specificity and efficacy of purified *Bacilllus thuringiensis* proteins against agronomically important insects. J. Invertebr. Pathol. **1990**, *56*, 258–266.
10. EPA Biopesticide registration document: *Bacillus thuringiensis* plant incorporated protectants. http://www.epa.gov/pesticides/biopesticides/reds/brad/_bt_pip.htm (accessed April 2, 2002).
11. James, C. *Global Review of Commercialized Transgenic Crops: 2000*; International Service for the Acquisition of Agri-Biotech Applications (ISAAA Briefs), 2000; 110.
12. Leonard, R.; Smith, R. IPM and environmental impacts of Bt cotton, a new era of crop protection and consumer benefits. Monsanto Company Brochure **2002**.
13. Ferre, J.; Van Rie, J. Biochemistry and genetics of insect resistance to *Bacillus thuringiensis*. Annu. Rev. Entomol. **2002**, *47*, 501–533.
14. Gould, F. Sustainability of transgenic insecticidal cultivars: Integrating pest genetics and ecology. Annu. Rev. Entomol. **1998**, *43*, 701–726.
15. Schuler, T.H.; Poppy, G.M.; Kerry, B.R.; Denholm, I. Insect-resistant transgenic plants. Btibtech **1998**, *16*, 168–174.
16. http://www.biols.susx.ac.uk/home/Neil_Crickmore/BT/index.html. (accessed April 3, 2002).

Genetically Modified Oil Crops

Denis J. Murphy
University of Glamorgan, Cardiff, U.K.

G

INTRODUCTION

For over five thousand years, oil crops have been sources of many useful nonfood products ranging from lubricants to lamp fuels. Since the late 19th century, such nonfood uses of plant oils have declined due to the twin pressures of relatively cheap petroleum-based alternatives and the need to feed the ever-increasing human population of our planet. By the late 20th century, less than 15% of plant-derived oils were used for industrial purposes. The vast bulk of vegetable oils are currently traded as commodities destined for human consumption in such products as margarines, cooking oils, and processed foods. More recently, however, interest has revived in the possible exploitation of oil crops for a wide spectrum of nonedible products including cosmetics, biodegradable plastics, and even high-value pharmaceuticals.

MARKETS AND CROPS

Global vegetable oil markets are dominated by the ''big four'' crops—soybean, oil palm, rapeseed, and sunflower—which together make up over 86% of the total global traded production of almost 90 million tonnes. The major aim of oil crop engineering has been to alter the fatty acid profiles of the major oil crops. As shown in Table 1, these crops have relatively narrow fatty acid profiles dominated by C16 and C18 groups that are not optimal for many industrial uses. On the other hand, there are many examples of minor crops and noncrop plants that accumulate very high levels of a diverse range of novel fatty acids with chain lengths from C8 to C24, and with useful chemical functionalities such as hydroxyl, epoxy, and acetylenic groups. Such oils can be used for the manufacture of products such as adhesives, paints, detergents, lubricants, and nylons, to name but a few. In Table 2, some examples of oil-bearing seeds that already produce some of these novel and potentially useful fatty acids are shown. Over the past decade, many of these plants have been used as sources of genes encoding fatty acid biosynthetic enzymes for transfer into mainstream oil crops in the hope that those crops would then accumulate the novel oils on a scale of up to millions of tonnes per year. This concept has been termed ''designer oil crops.''[1]

Rapeseed was successfully transformed as early as 1984, followed by sunflower and soybean in the 1990s. More recently, marker transgenes have been inserted into oil palm[2] and lipid biosynthetic genes will doubtless soon follow, although it will probably be another decade before alteration of the oil composition of this important tree crop has been achieved. Most of the genes related to fatty acid biosynthesis were isolated during the 1990s, either from the various exotic oilseeds or from model plants such as *Arabidopsis thaliana*, which as well as being an important research plant is also an oilseed that accumulates storage lipid as almost half of its seed weight. Therefore, in principle, it should now be possible to engineer transgenic oil crops to produce the desired range of novel fatty acids for a variety of industrial applications.

PROBLEMS AND PROSPECTS

Unfortunately, it appears that the accumulation of high levels of a single desired fatty acid in the storage oil, although fairly common in nature, is not readily achievable by simply inserting a few lipid biosynthetic genes into a given transgenic plant. There is increasing evidence that fatty acid modifications may behave as quantitative traits that are controlled to a greater or lesser extent by numerous genes.[3] It may be the case that in order to achieve levels of 90% of a particular fatty acid in a crop plant, the insertion of at least four and as many as ten transgenes may be required. This will add considerably to the cost and timescale of such manipulations.

Another problem in oil crops such as rapeseed is that exotic fatty acids designed for sequestration in the storage oils may sometimes also accumulate in membrane lipids, with possible deleterious consequences. Plants that naturally accumulate exotic fatty acids such as lauric acid (C12), which is a powerful membrane-destabilizing detergent, have evolved mechanisms to ''channel'' these deleterious fatty acids into the storage lipid pool. This mechanism, which probably involves specific phospholipases and acyltransferases, is found in lauric-accumulating seeds such as *Cuphea spp* and *Umbelluria californica*,

Encyclopedia of Plant and Crop Science
DOI: 10.1081/E-EPCS 120010439

Table 1 Percentage fatty acid composition of the "big four" oil crops

Fatty acid[a]	Soybean	Oil palm[b]	Rapeseed	Sunflower
16:0	11	45	5	6
18:0	4	5	1	5
18:1	22	38	61	20
18:2	53	11	22	69
18:3	8	0.2	10	0.1

[a]Fatty acids are denoted by their carbon chain length followed by the number of double bonds.
[b]Mesocarp.

but does not appear to be particularly active in rapeseed.[4] A further potential technical challenge to the engineering of designer oil crops is the finding that some of the gene promoters most commonly used to achieve the seed-specific expression of transgenes may also direct gene expression in other tissues, including roots.[5,6] While it is possible that in many cases the accumulation of exotic fatty acids in nonseed tissues may not be problematic, this may not always be true, and it underlines the need for the thorough metabolic profiling of all transgenic varieties before their general release.

The consequence of these and other complexities of plant molecular genetics and metabolism is that, despite many impressive achievements in isolating oil-related genes and producing transgenic plants with modified seed oil compositions, it has not been possible yet to achieve the kind of high levels (80–90%) of novel fatty acids that will make possible their widespread commercial exploitation. In the case of the high-lauric transgenic rapeseed (canola) crop, difficulties in its commercialization are also due to the existence of a competing source of lauric oil, namely palm kernel oil from the Far East. Palm oil is both cheaper to produce than rapeseed oil and in much more plentiful supply. The lauric-oil variety of rapeseed was improved from 40% to 60% lauric content by the insertion of several additional transgenes,[3] but it still remains far from being a commercial success and is no longer under development as a crop variety in the United States.

The availability of many genes involved in fatty acid modification and the good progress in transforming the main oil crop species will doubtless encourage further efforts to resolve the challenge of low levels of novel fatty acid production. But even if such efforts are successful, the commercial success of transgenic oil crops will remain problematic. It will be necessary to identify and develop robust markets for transgenic oil products; simply substituting for low-cost petroleum-derived products is unlikely to be economical for many decades. The additional costs of identity preservation will probably preclude the use of such transgenic oils as large-scale, low-value commodities in competition with conventional plant oils, even for industrial applications. In summary, transgenic oil crops producing novel fatty acids may have promise for the long-term future, but their commercial prospects over the next few years remain decidedly uncertain.

An attractive alternative to novel fatty acid production in oil crops is to engineer them to accumulate biopolymers instead. Virtually all of our conventional plastics are made from nonrenewable petroleum-derived products

Table 2 Accumulation of novel fatty acids by some oil-producing plants

Fatty acid[a]	Amount[b]	Plant species	Uses
8:0	94%	*Cuphea avigera*	Fuel, food
10:0	95%	*Cuphea koehneana*	Detergents, food
12:0	94%	*Litsea stocksii*	Detergents, food
14:0	92%	*Knema globularia*	Soaps, cosmetics
16:0	92%	*Myrica cerifera*	Food, soaps
18:0	65%	*Garcinia cornea*	Food, confectionery
20:0	33%	*Nephelium lappaceum*	Lubricants
22:0	48%	*Brassica tournefortii*	Lubricants
24:0	19%	*Adenanthera pavonina*	Lubricants
$18{:}1_{\Delta 6}$	76%	*Coriandrum sativum*	Nylons, detergents
$18{:}1_{\Delta 9}$	78%	*Olea europaea*	Food, lubricants
$22{:}1_{\Delta 13}$	58%	*Crambe abyssinica*	Plasticizers, nylons
$18{:}2_{\Delta 9,12}$	75%	*Helianthus annuus*	Food, coatings
$\alpha 18{:}3_{\Delta 9,12,15}$	60%	*Linum usitatissimum*	Paints, varnishes
$\gamma 18{:}3_{\Delta 6,9,12}$	25%	*Borago officinalis*	Therapeutic products
18:1–hydroxy	90%	*Ricinus communis*	Plasticizers, cosmetics
18:2–epoxy	60%	*Crepis palestina*	Resins, coatings
18:2–triple	70%	*Crepis alpina*	Coatings, lubricants
18:3–oxo	78%	*Oiticica*	Paints, inks
18:3–conj	70%	*Tung*	Enamels, varnishes
20:1/22;1wax	95%	*Simmondsia chinensis*	Cosmetics, lubricants

[a]Fatty acids are denoted by their carbon chain length followed by the number of double bonds or the nature of other functionalities.
[b]Percentage of total fatty acids; data are taken from Ref. 10.

such as adipic acid and vinyl chloride. Some soil bacteria such as *Ralstonia eutrophus* are able to accumulate up to 80% of their mass in the form of nontoxic biodegradable polymers called polyhydroxyalkanoates (PHAs). The PHAs are made up of β-hydroxyalkanoate subunits that are synthesized from acetyl-CoA via a relatively short pathway involving as few as three enzymes for the most common PHA, polyhydroxybutyrate.[7] The cost of PHAs could be considerably reduced if they were produced on an agricultural scale in transgenic crops. This prospect has led several companies, including Monsanto and Metabolix, to attempt to develop transgenic rapeseed plants containing the bacterial genes responsible for PHA biosynthesis. Provided the PHAs accumulate in the plastids, and not in the cytosol, it is possible to obtain modest yields of the polymer from either leaves or seeds.[8] A major and as yet unresolved technical hurdle is how to extract biopolymers from plant tissues in an efficient and cost-effective manner. Another complexity is that polyhydroxybutyrate, which is the most widespread PHA, is a rather brittle plastic and is not suitable for most applications. The best-performing plastics are copolymers of polyhydroxybutyrate with other PHAs, such as polyhydroxyvalerate. Although the production of such copolymers in transgenic plants is considerably more difficult than the production of single-subunit polymers, progress has recently been made in this area.[9] While there are several companies and academic labs attempting to make commercially extractable PHAs in plants (including one in oil palm), it seems unlikely that these environmentally friendly products will be commercially available for quite a few years to come.

Apart from these scientific and technical challenges, engineered oil crops also face considerable challenges regarding their management and economics. The major managerial problem concerns the need to segregate a transgenic crop variety producing a novel product from nontransgenic commodity crops and from other transgenic varieties of the same species that accumulate different products. This is a formidable task, given the intricacy of the supply chain from breeder to grower to crusher to processor and so on, all the way to the retailer and, ultimately, to the consumer. The difficulties in ensuring strict segregation of otherwise indistinguishable transgenic crops have consistently been underestimated by many in the industry. Several well-publicized failures in the segregation of transgenic rapeseed and maize crops in recent years (e.g., the STARLink affair, although note that this did not involve genes modified for oil composition)[10] have thrown this issue into much sharper focus. The contamination of a batch of seeds containing, for example, a hydroxy oil designed for industrial use with another batch of seeds containing a high-oleic oil for edible consumption (or vice versa) would result in a mixture that would be useless for both purposes. Efficient segregation is likely to be both difficult to control and expensive to implement. This may limit the cultivation of transgenic crops producing novel oils to geographically remote areas and/or to relatively high-value niche markets, where the additional costs of identity preservation can be met by the added value of the product.

CONCLUSION

The conclusion that engineered oil crops may be best suited to relatively low-volume, high-value markets allows the possible expansion of the target oilseed species beyond the ''big four'' oil crops to include minor oil crops like safflower or linseed, or even noncrop species like *Cuphea*. Use of such oil crops to produce novel products would have the advantage that segregation from sexually compatible food crop varieties would be fairly straightforward. Another innovative development of transgenic oil crops is the use of the oil as a carrier for recombinant high-value proteins such as pharmaceutical peptides and industrial enzymes.[11] Therefore, the prospects for oil crop biotechnology are now significantly different from the 1990s vision of large-scale ''designer oil crops,'' but the prospects remain positive, albeit rather more long-term than was originally envisaged. For this new vision to be realized, more investment in research must be coupled with a better appreciation of the economic, managerial, and public acceptability challenges that will confront the new crops.

ARTICLES OF FURTHER INTEREST

Genetically Engineered Crops with Resistance Against Insects, p. 506
Lesquerella Potential for Commercialization, p. 656
Lipid Metabolism, p. 659
Metabolism, Primary: Engineering Pathways of, p. 714
Natural Rubber, p. 778
New Industrial Crops in Europe, p. 813
Transformation Methods and Impact, p. 1233

REFERENCES

1. Murphy, D.J. *Designer Oil Crops*; VCH Press: Weinheim, Germany, 1994.
2. Parveez, G.K.A.; Masri, M.M.; Zainal, A.; Majid, N.A.; Yunus, A.M.M.; Fadilah, H.H.; Parid, O.; Cheah, S.C. Transgenic oil palm: Production and projection. Biochem. Soc. Trans. **2000**, *28*, 969–971.

G

3. Voelker, T.A.; Hayes, T.R.; Cranmer, A.M.; Turner, J.C.; Davies, H.M. Genetic engineering of a quantitative trait: Metabolic and genetic parameters influencing the accumulation of laurate in rapeseed. Plant J. **1996**, *9*, 229–241.
4. Wiberg, E.; Banas, A.; Stymne, S. Fatty acid distribution and lipid metabolism in developing seeds of laurate-producing rape (*Brassica napus* L.). Planta **1997**, *203*, 341–348.
5. Baumlein, H.; Boerjan, W.; Nagy, I.; Bassuner, R.; van Montagu, M.; Inze, D.; Wobus, U. A novel seed protein from *Vicia faba* is developmentally regulated in transgenic tobacco and *Arabidopsis* plants. Mol. Gen. Genet. **1991**, *225*, 459–467.
6. Murphy, D.J.; Hernandez-Pinzon, I.; Patel, K. Roles of lipid bodies and lipid-body proteins in seeds and other tissues. J. Plant Physiol. **2001**, *158*, 471–478.
7. Steinbüchel, A.; Fuchtenbusch, B. Bacteria and other biological systems for polyester production. Trends Biotechnol. **1998**, *16*, 419–427.
8. Snell, K.D.; Peoples, O.P. Polyhydroxyalkanoate polymers and their production in transgenic plants. Metab. Eng. **2002**, *4*, 29–40.
9. Slater, S.; Mitsky, T.A.; Houmiel, K.L.; Hao, M.; Reiser, S.E.; Taylor, N.B.; Tran, M.; Valentin, H.E.; Rodriguez, D.J.; Stone, D.A.; Padgette, S.R.; Kishore, G.; Gruys, K.J. Metabolic engineering of Arabidopsis and Brassica for poly(3-hydroxybutyrate-co-3-hydroxyvalerate) copolymer production. Nat. Biotechnol. **1999**, *10*, 960–961.
10. Murphy, D.J. Biotechnology, Its Impact and Future Prospects. In *Molecular to Global Photosynthesis*; Archer, M.A., Barber, J., Eds.; Imperial College Press: London, 2003, *in press*.
11. van Rooijen, G.J.H.; Moloney, M.M. Plant seed oil bodies as carriers for foreign proteins. Bio/Technology **1995**, *13*, 72–77.

Genome Rearrangements and Survival of Plant Populations to Changes in Environmental Conditions

G

Pier Giorgio Cionini
Università di Perugia, Perugia, Italy

INTRODUCTION

The capability of plants to develop plastic response reactions to adapt to changes in environmental conditions is unique in the biological world. However, the idea that the effect of the environment on an organism is, as a rule, physiological—with the genome remaining unchanged—has been a corollary of the biological tenet that a genome must be constant—apart from changes in chromosome number and type—on account of the absolute primacy of DNA over all biological phenomena and of its function as the basis of heredity. The concept that the nuclear genome may be intrinsically plastic due to its content of independent replicative units still meets with incredulity.[1] However, there is a growing consensus of opinion at the present time that changes in genome size and organization may be not restricted to species divergence, and that fluid domains do exist in the nuclear DNA in addition to more stable portions, particularly in plants.

Genomic rearrangements occurring in plant populations have been revealed by changes of genome size, confirmed and specified by molecular investigation, and shown to be related to alterations of plant developmental dynamics and phenotypic characteristics. The following sections show that changes in the nuclear DNA, which as a rule consist of redundancy modulations of repeated DNA sequences and which hence alter the organization of the genome: 1) may differentiate both populations and plants within one and the same population; 2) may have a role in overcoming both changes and changeability of the environment; and 3) may represent either the results of selection processes or direct responses of plant genomes to environmental stimuli.

GENOMIC VARIATION BETWEEN POPULATIONS

Broad bean (*Vicia faba*) is a good example of genomic rearrangements which may differentiate plant populations. Local populations of this species scattered all along the Mediterranean Basin differ 1.35-fold in genome size. Molecular analyses show that the copy number of both tandemly arranged and interspersed DNA repeats differs between populations and correlates significantly with the genome size.[2] Similar observations have been made in other species. A highly significant, positive correlation was found to exist between the mean genome size of *Dasypyrum villosum* populations and the altitudes of the stations where they grow.[3] The amounts of DNA and heterochromatin vary according to latitude and altitude in North American lines of maize (*Zea mays*).[4] The genome sizes of populations of wild barley (*Hordeum spontaneum*) differ significantly and correlate positively with January temperature at the stations, and *BARE*-1 retrotransposons have been shown to vary in redundancy in the nuclear DNA of wild barley plants from stations having microclimates sharply differing in solar irradiation and aridity.[5] The amount of nuclear DNA may vary up to 1.32-fold in Italian populations of fescue (*Festuca arundinacea*) due to variations in the amount of heterochromatin and to changes in the redundancy of DNA sequences which belong largely to a particular fraction of repetitive DNA and are, at least in part, transposable elements or their remnants. The genome sizes of the fescue populations correlate positively with the mean temperature for the year and for the coldest month at the stations and correlate negatively with their latitudes.[6]

The adaptive significance of these genomic changes—their role in buffering environmental pressure and allowing a species to grow in a range of ecogeographical conditions and climates—is clearly shown by their effects on the developmental dynamics of the plants. Cell proliferation and enlargement are affected by the genomic alterations occurring in broad bean populations, and the interactions between these two developmental factors cause the germination power of the seeds and the growth rate of plant organs to differ significantly between populations (compare Ref. 2). In *F. arundinacea*, the germination power of the seeds and the growth rates of roots and leaves, which are developmental events taking place during winter and early spring, are correlated negatively with the genome size, whereas the height of culms and other quantitative characteristics of plant organs developing later in the warmer season are correlated positively.[7]

Encyclopedia of Plant and Crop Science
DOI: 10.1081/E-EPCS 120010650

GENOMIC VARIATION WITHIN POPULATIONS

Changes in the redundancy of repeated DNA sequences, which result in genome size alterations, may also differentiate plants within populations. An intriguing example of intrapopulation variability of genome size and organization is found in sunflower (*Helianthus annuus*). In this species, genomic changes are continuously produced during reproduction, even with selfing and homozygosity. Embryos developing in different portions of the flowering head acquire different genome sizes, with a gradient, based on the genome size of the mother plant, that decreases from the pheriphery to the middle of the head. These genomic rearrangements are due to complex, specific changes in the redundancy and methylation of DNA repeats belonging to different sequence families.[8] Another example is *D. villosum*. Up to a 1.66-fold difference was shown to occur between the genome size of individual plants belonging to given populations of this species, and variability of DNA contents was found to be greater, as a rule, in populations from mountain sites, where the environment is expected to be particularly limiting and/or variable.[3]

It was shown, in both *D. villosum* and sunflower, that plants differing in genome size also differ in given aspects of their development and, remarkably, in their flowering interval. It is demonstrated in the following section that genomic rearrangements can represent direct responses of the plant genome to environmental stimuli. Therefore, changes of genome size and organization between individual plants of a population may be due to the diversity of environmental (or microenvironmental in the case of sunflower) conditions occurring during the reproduction and /or development of each of them. Intrapopulation variability of genome size and organization may be seen as an evolutionary factor allowing plant populations to withstand the variability of conditions in a given environment.[3,9]

DIRECT RESPONSES OF THE PLANT GENOME TO ENVIRONMENTAL STIMULI

Genomic rearrangements in plant populations may be the result of selection processes after the occurrence of uncontrolled events in the genome. However, adaptive changes in nuclear DNA also may represent direct responses of the plant genome to environmental variation. The case with flax (*Linum usitatissimum*) is now classic. When plants carrying given genotypes are germinated and grown under different temperatures and/or in unbalanced mineral nutrients, small and large plants are obtained (genotrophs). These phenotypic characters remain heritable, and genotrophs differ 1.16-fold in genome size. The changes occurring in the nuclear DNA were extensively characterized at the molecular level and shown to be due to copy number variations in given repeated sequence families.[10]

Redundancy modulation of DNA repeats may be specifically controlled. The copy number of certain interspersed repeats increases in plantlets of *F. arundinacea* obtained by germinating seeds at 30°C, while that of different repeats increases at 10°C.[6] In *H. annuus*, several repetitive sequence families are more represented in large-size genomes, but there are DNA repeats whose redundancy is greater in the relatively small-size genomes.[8] This specificity suggests that given DNA domains are plastic, not simply instable, with changes occurring under plant-level control and not independently of the genome as a whole.

CONCLUSION

Genomic rearrangements may be the paramount factors, together with the processes of development and reproduction, that allow the adaptive responses of plants to environmental pressure made particularly stringent by immobility. Nucleotypic effects—which can modulate cell proliferation and enlargement and, consequently, developmental dynamics—and/or more specific controls of the nuclear activity may represent the way(s) by which genome rearrangements, determined by quantitative changes of repeated sequences having different origin and chromosomal organization, achieve environmental adaptation These rearrangements may help to explain the extraordinary plasticity at the morphological and physiological level that is a common feature of plants and may constitute an answer to the debated question about the functional role(s) of repeated sequences in eukaryotic genomes, where they represent the DNA fraction that is by far the largest, especially in plants.

ARTICLES OF FURTHER INTEREST

B Chromosomes, p. 71
Chromosome Rearrangements, p. 270
Crops and Environmental Change, p. 370
Genetic Diversity Among Weeds, p. 496
Genome Size, p. 516
Molecular Evolution, p. 748
Oxidative Stress and DNA Modification in Plants, p. 854
Plant Response to Stress: Genome Reorganization in Flax, p. 984

Plant Response to Stress: Mechanisms of Accomodation, p. 987
Polyploidy, p. 1038
Somaclonal Variation: Origins and Causes, p. 1158

REFERENCES

1. Greilhuber, J. Intraspecific variation in genome size: A critical reassessment. Ann. Bot. **1998**, *82* (Supplement A), 27–35.
2. Frediani, M.; Gelati, M.T.; Maggini, F.; Galasso, I.; Minelli, S.; Ceccarelli, M.; Cionini, P.G. A family of dispersed repeats in the genome of *Vicia faba*: Structure, chromosomal organization, redundancy modulation, and evolution. Chromosoma **1999**, *108*, 317–324.
3. Caceres, M.E.; De Pace, C.; Scarascia Mugnozza, G.T.; Kotsonis, P.; Ceccarelli, M.; Cionini, P.G. Genome size variation within *Dasypyrum villosum*: Correlations with chromosomal traits, environmental factors and plant phenotypic characteristics and behaviour in reproduction. Theor. Appl. Genet. **1998**, *96*, 559–567.
4. Rayburn, A.L.; Auger, J.A. Genome size variation in *Zea mays* ssp. *mays* adapted to different altitudes. Theor. Appl. Genet. **1990**, *79*, 470–474.
5. Kalendar, R.; Tanskanen, J.; Immonen, S.; Nevo, E.; Schulman, H.H. Genome evolution of wild barley (*Hordeum spontaneum*) by BARE-1 retrotransposon dynamics in response to sharp microclimatic divergence. Proc. Natl. Acad. Sci. U.S.A. **2000**, *97*, 6603–6607.
6. Ceccarelli, M.; Esposto, M.C.; Roscini, C.; Sarri, V.; Frediani, M.; Gelati, M.T.; Cavallini, A.; Giordani, T.; Pellegrino, R.M.; Cionini, P.G. Genome plasticity in *Festuca arundinacea*: Direct response to temperature changes by redundancy modulation of interspersed DNA repeats. Theor. Appl. Genet. **2002**, *104*, 901–907.
7. Ceccarelli, M.; Minelli, S.; Falcinelli, M.; Cionini, P.G. Genome size and plant development in hexaploid *Festuca arundinacea*. Heredity **1993**, *71*, 555–560.
8. Natali, L.; Cavallini, A.; Cionini, G.; Sassoli, O.; Cionini, P.G.; Durante, M. Nuclear DNA chnages within *Helianthus annuus* L.: Changes within single progenies and their relationships with plant development. Theor. Appl. Genet. **1993**, *85*, 506–512.
9. Cavallini, A.; Natali, L.; Giordani, T.; Durante, M.; Cionini, P.G. Nuclear DNA changes within *Helianthus annuus* L.: Variations in the amount and methylation of repetitive DNA within homozygous progenies. Theor. Appl. Genet. **1996**, *92*, 285–291.
10. Cullis, C.A. The Environment as an Active Generator of Adaptive Genomic Variation. In *Plant Responses to Environmental Stresses*; Lerner, H.R., Ed.; Marcel Dekker, Inc.: New York, 1999; 149–160.

Genome Size

Michael D. Bennett
Royal Botanic Gardens Kew, Surrey, U.K.

INTRODUCTION

The DNA amount in the unreplicated gametic nucleus is refered to as its C-value. Estimates for nearly 4000 species are accessible in the Plant DNA C-values database. C-values differ more than 1000-fold for plants of similar organismic complexity reflecting differences in ploidy level or amounts of repeated DNA sequences. Genome size is evolutionarily labile and molecular mechanisms responsible for DNA gain or loss are beginning to be understood. C-value and genome size are key diversity characters with important consequences and uses.

DEFINING GENOME SIZE AND DNA C-VALUE

Plants contain DNA in nuclear and organellar (e.g., chloroplast) genomes. This article concerns only the former. Today the term genome may refer to either all the nuclear DNA or only some of it. Thus, a paper on hexaploid *Triticum aestivum* can say that ''loci are triplicated in the wheat genome'' but also talk of ''the three genomes of hexaploid bread wheat.'' Genome originally defined a basic (monoploid) chromosome set (x), and the term is restricted to this cytogenetic definition below.

To avoid confusion with chromosome number, Swift introduced the term ''C-value'' (''C'' standing for ''Constant'') in 1951. 1C refers to the DNA amount in an unreplicated gametic chromosome set (n) of an organism. This usage applies irrespective of ploidy level, so it is important to understand that whereas genome size equals the 1C DNA amount in a diploid species, genome size is always less than the 1C DNA amount in polyploid species (Table 1). DNA amounts are usually expressed in picograms (pg) or in megabase pairs of nucleotides (Mb) (NB 1 pg = 10^{-12} g; 1 Mb = 10^6 nucleotide base pairs; 1 pg = 980 Mb).

METHODS USED TO ESTIMATE DNA AMOUNTS

Since 1950 DNA amounts have been measured by various techniques. Chemical methods using extracted DNA were little used after the 1960s, while reassociation kinetics was hardly used after the 1970s as faster and simpler techniques were developed that estimate the amount of DNA in single cells or nuclei using microdensitometry, flow cytometry, or video-based image analysis. C-values obtained by these methods are estimates due to inherent technical errors, but complete genome sequencing offers the chance of measuring genome size more exactly.

WHOLE GENOME SEQUENCING

In 2000 the Arabidopsis Genome Initiative published the first ''complete genome sequence'' in a flowering plant (*Arabidopsis thaliana*). Its genome size was given as 125 Mb, comprising 115.4 Mb in sequenced regions plus a rough estimate of 10 Mb in unsequenced regions.[1] This is an underestimate, as the estimate for unsequenced gaps was too low. 125 Mb disagrees with many estimates made by other methods that placed 1C in *Arabidopsis* in the range 147–172 Mb. Moreover, comparison with *Caenorhabditis elegans* (whose 1C-value is known from complete genome sequencing to be 100.25 Mb) gave a 1C DNA amount of approximately 157 Mb for *Arabidopsis*.

In 2002 draft sequences for the first crop species were published. The 1C-value for indian rice *Oryza sativa* ssp. *indica* was given as 466 Mb[2] with a total scaffold length of 362 Mb plus 104 Mb of masked reads. This value agrees closely with estimates made previously using nonmolecular methods (Table 2).

PLANT DNA C-VALUE REFERENCE SOURCES

DNA C-values for plants come from more than 500 widely scattered original sources. Estimates for angiosperms have been listed for easy reference since 1976[3] and combined in an electronic database since 1997. C-values for angiosperms, gymnosperms, pteridophytes, and bryophytes were pooled in the Plant DNA C-values database,[4] and in 2002, nuclear DNA amounts for almost 4000 plant species (excluding algae) were listed.

Encyclopedia of Plant and Crop Science
DOI: 10.1081/E-EPCS 120005684

Table 1 DNA C-value and genome size in a hexaploid triticale and its parents

Species	Ploidy level	Chromosome number (2n)	Genomic constitution	1C DNA amount (pg)	Genome size (pg)
Secale cereale	2x	14	RR	8.28	8.28
Triticum turgidum	4x	28	AABB	12.28[a]	6.14[a]
Triticale	6x	42	AABBRR	19.80[a]	6.60[a]

[a]Mean for 2 or 3 genomes = 1C DNA amount/half ploidal level.

INTERSPECIFIC VARIATION IN DNA AMOUNT

Comparison shows large differences in numbers of estimates and representation between major plant groups (Table 3), and striking variation in C-value within some but not all groups. C-values differ more than 1000-fold in angiosperms (Table 4, Fig. 1), and 450-fold from 0.16 pg in *Selaginella kraussiana* to 72.7 pg in *Psilotum nudum* in pteridophytes. However, C-values vary only 12-fold, from 0.17 pg in *Holomitium arboretum* to 2.05 pg in *Mnium marginatum* in bryophytes, and only 14-fold in gymnosperms from 2.25 pg in *Gnetum ula* to 32.2 pg in *Pinus nelsonii*.

INTRASPECIFIC VARIATION

DNA amount is relatively constant within species and characteristic of taxa, but limited intraspecific variation is known. Such variation can reflect differences in both type and number of chromosomes (e.g., sex and B-chromosomes) and in repeated DNA sequences. A well studied example concerns corn (*Zea mays*), where C-values for different lines can vary by up to 35%. Much of this variation correlates with the presence and size of heterochromatic "knobs" in the genome. Knob number is also related to geographical distribution and may reflect environmental adaptation.[5] However, such examples are few, and recent comparisons of onion (*Allium cepa*) cultivars from widely different regions detected no differences in genome size.

Table 2 1C-values (Mb) in *Arabidopsis thaliana* cv. Columbia and *Oryza sativa* ssp. *indica* line 93-11 based on complete genome sequencing or non-molecular methods

Method	Arabidopsis	Oryza
Whole genome sequencing	125 (3)	466 (4)
Feulgen microdensitometry	167 (2)	490 (2)
Flow cytometry	147–172 (2)	441–468 (2)

The relative constancy of genome size within species is controversial. Reinvestigation of several claimed examples shows them to be technical artifacts.[6] Variation at the DNA-sequence level may be triggered by events such as wide hybridization. Yet the relative constancy seen in many species suggests that DNA amount may normally be subject to innate controls by counting mechanisms that detect and regulate genome size characters in tightly defined limits. If so, the mechanisms responsible are unknown.

GENOME SIZE EVOLUTION—MECHANISMS OF DNA GAIN AND LOSS

The existence of major differences in C-value raises questions concerning its origin and significance. Huge variation in genome size unrelated to organismal complexity was epitomised as "the C-value paradox" by Thomas in 1971. Such differences are now largely attributed to changes in the proportion of noncoding, repetitive DNA sequences (e.g., transposable elements) and the extent of genome duplication. The puzzle now is to understand the molecular mechanisms and evolutionary pressures that determine the amounts of repetitive DNA in species genomes and thus their genome sizes.[7]

Evolutionary studies were long flawed by the lack of a rigorous phylogenetic framework, but new phylogenetic data recently facilitated meaningful comparisons. Superimposing genome size data onto a robust phylogenetic tree[8] showed that ancestral angiosperms had small genomes (i.e., 1C $\leq$ 1.4 pg) that are retained in most extant taxa. Possession of large ($\geq$14 pg) and very large ($\geq$35 pg) genomes is a derived condition that arose independently at least six times during angiosperm evolution. A similar analysis for grasses suggested that plants may have a "one way ticket to genome obesity,"[9] as only mechanisms capable of generating rapid genome expansion (such as transposable element amplification) were known. Perhaps plants with small genomes have effective mechanisms to suppress retrotransposition activity, whereas large genomes arose from the release of retrotransposition suppression. More recent studies have

Table 3 Minimum, maximum, and range (max./min.) of 1C DNA estimates in major groups of land plants and their representation in the Plant DNA C-values database

Group	Minimum (pg)	Maximum (pg)	Range	No. of species with C-values	No. of species recognized	Representation (%)
Bryophytes	0.17	2.05	12.1	171	c. 18,000	1.0
Pteridophytes	0.16	72.68	450	82	c. 9000	0.9
Gymnosperms	2.25	32.20	14.3	181	730	24.8
Angiosperms	c. 0.1	127.40	1274	3493	250,000	1.4

(From Ref. 2.)

begun to shed light on several mechanisms involving recombination that can bring about a decrease in genome size.[10] These are poorly understood, but mounting evidence suggests that genome contraction is a widespread phenomenon operating more extensively than once recognised. Plant genome size is evolutionarily labile and the particular genome size of a plant species reflects the dynamic balance between the opposing evolutionary forces of expansion and contraction.[11]

USES OF GENOME SIZE DATA

DNA C-values have many uses. C-value reference lists provided data for comparative studies at levels ranging from the biosphere to genome organization, and in diverse disciplines including phylogeny, ecology, genomics, cell biology, conservation, physiology, and development.[12]

NUCLEOTYPIC EFFECTS

DNA affects the phenotype in two ways: first by its genic content, and second by the physical consequences of its

Table 4 The range of 1C DNA estimates in angiosperms

Species	Chromosome number (2n)	Ploidy level (x)	1C DNA Amount (pg)	(Mb)
Aesculus hippocastanum	40	2	0.13	123
Oryza sativa	24	2	0.50	490
Zea mays	20	2	2.73	2670
Hordeum vulgare	14	2	5.55	5400
Secale cereale	14	2	8.28	8110
Vicia faba	12	2	13.33	13,060
Allium cepa	14	2	16.75	16,415
Lilium longiflorum	24	2	35.20	34,500
Fritillaria assyriaca	48	4	127.40	124,850

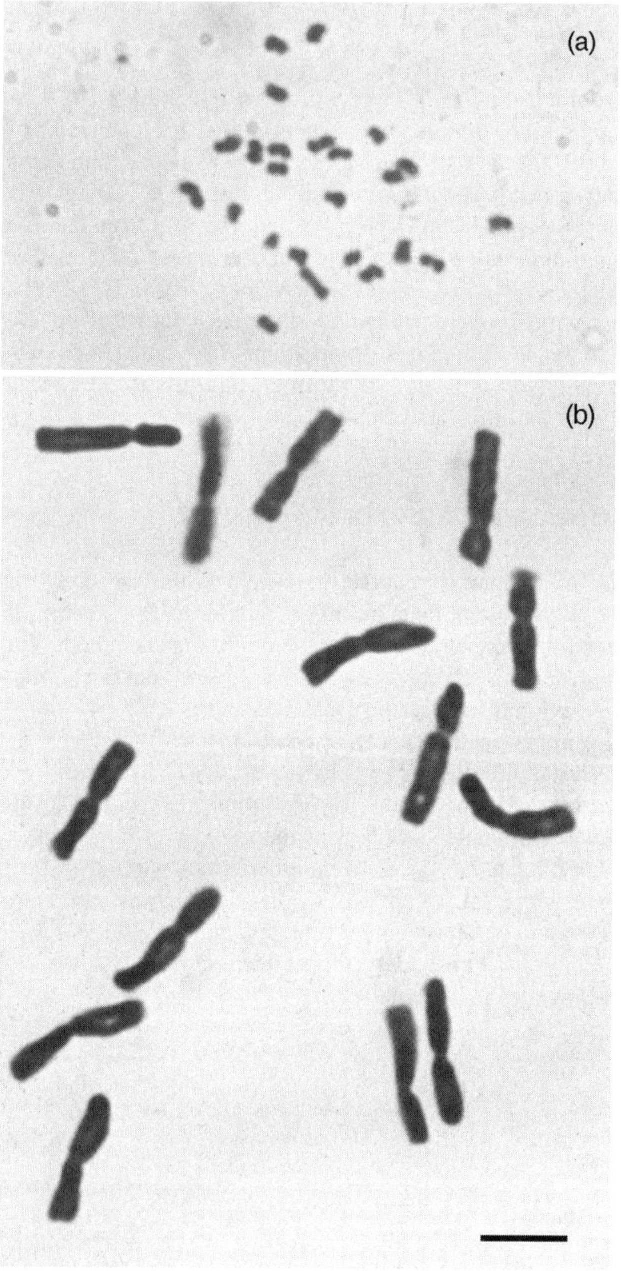

Fig. 1 Metaphase chromosomes of two cereal crops with widely different 4C DNA amounts: (a) *Oryza sativa*—2 pg, (b) *Secale cereale*—33.16 pg. Scale bar = 5 microns.

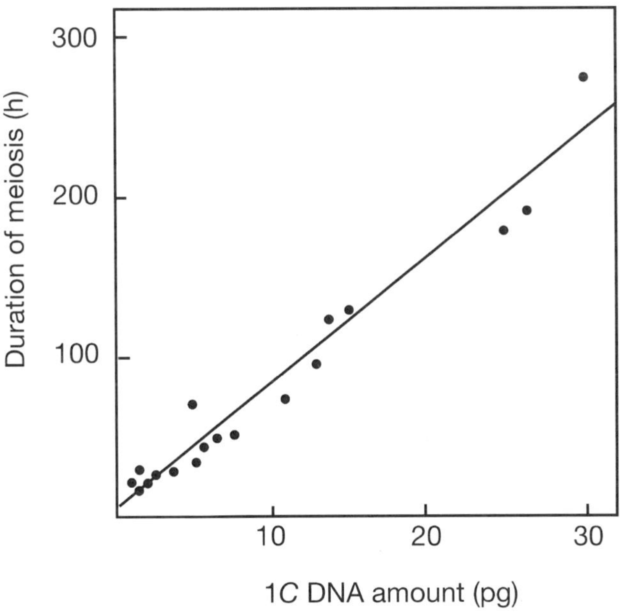

Fig. 2 Close nucleotypic relationship between 1C DNA amount and the duration of meiosis in 18 diploid angiosperm species grown at 20°C. (Redrawn from Ref. 13.)

mass and volume, independently of its encoded information (i.e., nucleotypic effects). Nuclear DNA amount shows strikingly close nucleotypic correlations with many widely different phenotypic and phenological characters at cell, tissue, and organismic levels, e.g., chromosome size, chloroplast number, pollen volume, duration of meiosis (Fig. 2), seed weight, and minimum generation time. Thus, C-value is a key factor in scaling the size and rate of development of living systems.

The closer the correlation, the more useful C-value data are as predictors. Interest in C-values as predictors extends to a broad range of ecological and environmental factors, including invasiveness, crop distribution and biomass, and predicting responses of vegetation to global warming or a nuclear winter.[12]

CONCLUSION

Future work on plant DNA amounts offers exciting prospects. New C-value estimates for plant species will soon give complete representation for angiosperm families and for several island floras, and meaningful representation for African, Asian and South-American floras. The availability of genome sequences for several species and of databases linking species C-values with environmental information will allow tests at sequence level of theories about molecular mechanisms responsible for DNA gain and loss, and of possible links between plant ecology, extinction rates, and genome size.

ARTICLES OF FURTHER INTEREST

Arabidopsis Thaliana: Characteristics and Annotation of a Model Genome, p. 47
Flow Cytogenetics, p. 460
Polyploidy, p. 1038

REFERENCES

1. Arabidopsis Genome Initiative. Analysis of the genome sequence of the flowering plant *Arabidopsis thaliana*. Nature (Lond.) **2000**, *408* (6814), 796–815.
2. Yu, J.; and 99 others. A draft sequence of the rice genome (*Oryza sativa* L. ssp. *indica*). Science **2002**, *296* (5565), 79–92.
3. Bennett, M.D.; Smith, J.B. Nuclear DNA amounts in angiosperms. Philos. Trans. R. Soc. Lond., B **1976**, *274* (933), 227–274.
4. Bennett, M.D.; Leitch, I.J. Plant DNA C-values database (release 2.0, Dec. 2002). http://www.rbgkew.org.uk/ (accessed December 2002).
5. Rayburn, A.L.; Price, H.J.; Smith, J.D.; Gold, J.R. C-Band heterochromatin and DNA content in *Zea mays* (L.). Am. J. Bot. **1985**, *72* (10), 1610–1617.
6. Greilhuber, J. Intraspecific variation in genome size: A critical reassessment. Ann. Bot. **1998**, *82* (Supplement A), 27–35.
7. Gregory, T.R. Coincidence, coevolution, or causation? DNA content, cell size, and the C-value enigma. Biol. Rev. **2001**, *76* (1), 65–101.
8. Leitch, I.J.; Chase, M.W.; Bennett, M.D. Phylogenetic analysis of DNA C-values provides evidence for a small ancestral genome size in flowering plants. Ann. Bot. **1998**, *82* (Supplement A), 85–94.
9. Bennetzen, J.L.; Kellogg, E.A. Do plants have a one-way ticket to genomic obesity? Plant Cell **1997**, *9* (9), 1509–1514.
10. Wendel, J.F.; Cronn, R.C.; Johnston, J.S.; Price, H.J. Feast and famine in plant genomes. Genetica **2002**, *115* (1), 37–47.
11. Petrov, D.A. Mutational equilibrium model of genome size evolution. Theor. Popul. Biol. **2002**, *61* (4), 531–544.
12. Bennett, M.D.; Bhandol, P.; Leitch, I.J. Nuclear DNA amounts in angiosperms and their modern uses—807 New estimates. Ann. Bot. **2000**, *86* (4), 859–909.
13. Bennett, M.D. The time and duration of meiosis. Philos. Trans. R. Soc. Lond., B **1977**, *277* (955), 201–277.

Genome Structure and Gene Content in Plant and Protist Mitochondrial DNAs

Michael W. Gray
Dalhousie University, Halifax, Canada

INTRODUCTION

Complete sequencing of mitochondrial DNA (mtDNA) has provided much information about the size, gene content, and architecture of this organellar genome. One of the striking features of mtDNA is its exceptional physical and organizational plasticity, which stands in marked contrast to its conserved genetic function (i.e., specification of a limited number of components of the mitochondrial electron transport and translation systems). This structural variability is especially pronounced within the plant and protist lineages of eukaryotes; at the same time, these two groups are distinguished in having a generally similar—and larger—gene content than either animal or fungal mtDNAs. Although we are beginning to unravel the pathways of mitochondrial genome restructuring, as yet we have gained only limited understanding of how (and particularly why) the mitochondrial genome has evolved so differently in the various eukaryotic lineages.

MITOCHONDRIAL GENOMES IN PROTISTS

Protists (protozoa and algae) are defined on the basis of exclusion: They are not animals, plants, or fungi. Most protists are single-celled, but multicellular groups do exist (e.g., among the red, brown, and green algae). To date, complete mitochondrial genome sequences have been published for more than 30 different protists (with at least 20 additional sequences in progress), covering a phylogenetically broad range.[1] Protist mtDNAs include the most gene-poor (*Plasmodium falciparum*) and most gene-rich (*Reclinomonas americana*) mitochondrial genomes known (Table 1). Most protist mtDNA sequences assemble as single circular molecules, but linear protist mtDNAs are also found (e.g., in the ciliate protozoon, *Tetrahymena pyriformis*, and the green alga, *Chlamydomonas reinhardtii* (Table 1)).

Protist mitochondrial genomes divide roughly into two basic groups, which have been termed "ancestral" and "reduced derived"[2] Ancestral mtDNAs are those that most closely resemble, in gene content and organization, the α-Proteobacteria-like genome from which contemporary mitochondrial genomes originated.[3] Ancestral mitochondrial genomes are tightly packed, comprising mostly coding sequence. Compared to animal and fungal mtDNAs, they contain extra genes, especially ones specifying ribosomal proteins; they encode strikingly eubacterialike large subunit (LSU, 23S-like) and small subunit (SSU, 16S-like) rRNAs, as well as 5S rRNA; they carry a complete or almost complete set of tRNA genes whose tRNA products have conventional, bacterialike secondary structures; with few exceptions, they are mostly intron-poor; they often display eubacterialike gene clusters (e.g., *rps* and *rpl* genes); and they use the standard genetic code for translation.

Ancestral mtDNAs are exemplified in Table 1 by *R. americana* and *Prototheca wickerhamii*, with the former protist (a jakobid flagellate) having the most eubacteria-like mitochondrial genome yet described.[4] While not the largest mtDNA in size (those of land plants are considerably larger; see below), *R. americana* mtDNA contains a greater number of genes of known function than any other mitochondrial genome, including some 18 protein-coding genes not previously found in any other characterized mtDNA.

Reduced derived mitochondrial genomes (such as those of *P. falciparum* and *C. reinhardtii*; Table 1) are characterized by extensive loss of genes (tRNA as well as protein-coding); loss of the 5S rRNA gene; aberrant rRNA and tRNA secondary structures; radical divergence in rDNA structure (in both *P. falciparum* and *C. reinhardtii*, the mitochondrial rRNA genes are fragmented into modules that are rearranged and dispersed throughout thegenome); an accelerated rate of sequence divergence (in both protein-coding and rRNA genes); a highly biased codon usage, including complete absence of certain codons; and (often) a nonstandard genetic code. Mitochondrial genome reduction has proceeded to an extreme in *P. falciparum*. Not only is its mtDNA the smallest known (a mere 6000 bp), it encodes the fewest genes: only three protein-coding (*cob*, *cox1*, and *cox3*) and two rRNA (*rnl* and *rns*). In fact, with the exception of *cox3*, these genes are

Encyclopedia of Plant and Crop Science
DOI: 10.1081/E-EPCS 120010633

Table 1 Genome size and form and gene content in protist and plant mitochondrial DNAs

	Form	Size (bp)	rRNA Genes[a]			Protein-coding genes			tRNA genes	Introns[b]
			LSU	SSU	5S	EC/OP[c]	Ribosomal[d]	Other[e]		
Plasmodium falciparum (apicomplexan)	Linear[f]	5 966[g]	+	+	–	3	0	0	0	0
Chlamydomonas reinhardtii (green alga)	Linear	15 758[h]	+	+	–	7	0	1	3	0
Chondrus crispus (red alga)	Circular	25 836	+	+	+	17	5	4	23	1 (II)
Prototheca wickerhamii (green alga)	Circular	55 328	+	+	+	17	13	6	26	5 (I)
Acanthamoeba castellanii (amoeboid protozoon)	Circular	41 591	+	+	+[i]	18[j]	16	7	15[k]	3 (I)
Reclinomonas americana[l] (jakobid flagellate)	Circular	69 034	+	+	+	24	27	16	26	1 (II)
Marchantia polymorpha	?[m]	186 609	+	+	+	18[n]	16	89	27	7 (I), 25 (II)

[a]LSU, large subunit; SSU, small subunit.
[b]Roman numerals in parentheses specify intron type (group I and/or group II).
[c]EC/OP, electron transport/oxidative phosphorylation (*nad*, NADH dehydrogenase, Complex I; *sdh*, succinate dehydrogenase, Complex II; *cob*, apocytochrome *b*, Complex III; *cox*, cytochrome *c* oxidase, Complex IV; *atp*, ATP synthase, Complex V).
[d]*rps*, small subunit; *rpl*, large subunit.
[e]Conserved open reading frames (ORFs) of unknown function; unique ORFs; intronic ORFs.
[f]Head-to-tail tandem repeat of 6 kb unit.
[g]Length of repeat unit.
[h]Includes 492-bp subterminal inverted repeats and terminal 40-nucleotide 3′ single-strand extensions.
[i]C. Bullerwell and M. W. Gray, Discovery and characterization of *Acanthamoeba castellanii* michondrial 5S rRNA. RNA **2003**, *9* (3), 287–292.
[j]A single ORF encodes both subunits 1 and 2 (*cox*1 and *cox*2) of cytochrome *c* oxidase.
[k]Transcripts of most *A. castellanii* mitochondrial tRNA genes (12 of 15) undergo substitutional editing at one or more of the first three positions of the acceptor stem.
[l]Genome also encodes the RNA component (*rnpB*) of a mitochondrial RNase P.
[m]Circular-mapping genome, but biochemical data indicate a linear genome (see Ref. [6]).
[n]Genome also contains a *nad7* pseudogene.

the only ones that are universally contained in all characterized mtDNAs. In a number of cases, tRNA genes are completely (e.g., *P. falciparum*) or almost completely (e.g., *C. reinhardtii*) absent from reduced derived mtDNAs, with nucleus-encoded tRNA species being imported into mitochondria to make up the deficit of tRNAs required for mitochondrial translation.

The division into ancestral and reduced derived groupings serves to emphasize that reduced derived genomes are (in an evolutionary sense) a degenerate form of mtDNA. However, the division into these two categories is not absolute. For example, the *A. castellanii* mitochondrial genome (Table 1), which has many of the characteristics of an ancestral mtDNA, also displays a few ''degenerate'' features, such as fusion of the *cox1* and *cox2* genes into a single ORF; use of TGA to specify an amino acid (tryptophan) rather than translation termination; and loss of many tRNA genes, with transcripts of most of the remaining ones undergoing a novel form of RNA editing. Likewise, the *Chondrus crispus* (a red algal) mitochondrial genome is mostly ancestral in character but has lost almost all of the ribosomal protein genes that are typically found in ancestral protist mtDNAs (Table 1).

MITOCHONDRIAL GENOMES IN PLANTS

Land plant mtDNA exemplifies a third type of mitochondrial genome organization, designated expanded ancestral.[2] In this type, gene content approximates that of ancestral protist mtDNAs, a 5S rRNA gene is present, rRNA and tRNA secondary structures are conventional, and the standard genetic code is employed during mitochondrial translation (Table 1). However, the genome is large (180 to 2400 kb in different plants) and spacious, consisting mostly of noncoding sequence. Spacers are long, numerous repeats are present, and the genome contains many introns and intron ORFs.[5]

Because the plant mitochondrial genome is so large, few complete sequences have been determined, although these do span the range from primitive bryophytes (e.g., *Marchantia polymorpha*) to advanced angiosperms (the dicots, *Arabidopsis thaliana* and *Beta vulgaris*, and the monocot, rice (*Oryza sativa*)).[1] As well, partial sequences have been published for many other land plant mtDNAs. These data indicate that the expanded ancestral pattern must have emerged in the earliest stages of land plant evolution. Mapped and/or sequenced plant mtDNAs can be assembled into a ''master circle'' that contains all of the identified genes as well as various repeated sequences.[5] Whether the master circle, a conceptual model, is an accurate description of the physical state of the plant mitochondrial genome has been questioned by studies designed to examine the actual form of isolated plant mtDNA.[6]

A hallmark of plant mtDNA is the presence of numerous repeated sequences that are recombinationally active; when such repeats are directly oriented in a circular chromosome, they promote the formation of subcircular products, each of which contains a portion of the master circle.[5] At the sequence level, plant mtDNA is one of the most slowly evolving genomes known; in contrast, on an evolutionary timescale, it undergoes extraordinarily rapid rearrangement, undoubtedly facilitated by the genome's high proportion of noncoding sequence.

Because it is mostly noncoding, plant mtDNA is likely predisposed to accept foreign DNA. Indeed, sequences evidently derived from both nuclear and chloroplast DNA have been found as part of the mitochondrial genome of many angiosperms;[5] in some cases, the incorporated chloroplast DNA contributes functional tRNAs that supplement a limited set of mtDNA-encoded (''native'') tRNA species.[7] Excluding mobile introns, sequences from nuclear or chloroplast genomes have not been found in the mitochondrial genome outside of the land plant lineage.

CONCLUSIONS

We may expect that further insights into the organization, function, and evolution of plant and protist mtDNAs will come from continued sequencing and comparative analysis of mitochondrial genomes from these two major eukaryotic groups. Particularly interesting is the question of how reduced derived and expanded ancestral forms of mtDNA arose from ancestral forms, and when these transitions occurred in various lineages. A case in point concerns streptophytes (land plants and charophyte algae) and chlorophytes (all other green algae), two monophyletic lineages of green plants that share a common ancestor to the exclusion of other eukaryotes. Although, as noted above, land plants exhibit the expanded ancestral type of mtDNA, primitive chlorophytes (prasinophytes) contain the ancestral mitochondrial genome type typical of many nongreen protists. At what point the evolutionary transition from compact to expanded genome occurred is a question that could readily be explored by judicious sequencing of mtDNA within additional primitive prasinophytes and within the charophyte algae. Comparative mitochondrial genome analysis will not only establish pathways but may also suggest mechanisms by which mtDNA evolves.[8,9]

ARTICLES OF FURTHER INTEREST

Arabidopsis thaliana: Characteristics and Annotation of a Model Genome, p. 47
Evolution of Plant Genome Microstructure, p. 421
Genome Size, p. 516
Mitochondrial Respiration, p. 729
Molecular Evolution, p. 748
Mutational Processes, p. 760
Nucleic Acid Metabolism, p. 833

REFERENCES

1. http://megasun.bch.umontreal.ca/ogmp/projects/other/mtcomp.html (accessed August 2002).
2. Gray, M.W.; Lemieux, C.; Burger, G.; Lang, B.F.; Otis, C.; Plante, I.; Turmel, M. Mitochondrial Genome Organization and Evolution within the Green Algae and Land Plants. In *Plant Mitochondria: From Gene to Function*; Møller, I.M., Gardeström, P., Glimelius, K., Glaser, E., Eds.; Backhuys Publishers: Leiden, The Netherlands, 1998; 1–8.
3. Gray, M.W.; Burger, G.; Lang, B.F. Mitochondrial evolution. Science **1999**, *283* (5407), 1476–1481.
4. Lang, B.F.; Burger, G.; O'Kelly, C.J.; Cedergren, R.; Golding, G.B.; Lemieux, C.; Sankoff, D.; Turmel, M.; Gray, M.W. An ancestral mitochondrial DNA resembling a eubacterial genome in miniature. Nature **1997**, *387* (6632), 493–497.
5. Hanson, M.R.; Folkerts, O. Structure and function of the higher plant mitochondrial genome. Int. Rev. Cytol. **1992**, *141*, 129–172.
6. Oldenburg, D.J.; Bendich, A.J. Mitochondrial DNA from the liverwort *Marchantia polymorpha*: Circularly permuted linear molecules, head-to-tail concatemers, and a 5′ protein. J. Mol. Biol. **2001**, *310* (3), 549–562.
7. Maréchal-Drouard, L.; Weil, J.H.; Dietrich, A. Transfer RNAs and transfer RNA genes in plants. Annu. Rev. Plant Physiol. Plant Mol. Biol. **1993**, *44*, 13–32.
8. Gray, M.W.; Lang, B.F.; Cedergren, R.; Golding, G.B.; Lemieux, C.; Sankoff, D.; Turmel, M.; Brossard, N.; Delage, E.; Littlejohn, T.G.; Plante, I.; Rioux, P.; Saint-Louis, D.; Zhu, Y.; Burger, G. Genome structure and gene content in protist mitochondrial DNAs. Nucleic Acids Res. **1998**, *26* (4), 865–878.
9. Lang, B.F.; Gray, M.W.; Burger, G. Mitochondrial genome evolution and the origin of eukaryotes. Annu. Rev. Genet. **1999**, *33*, 351–397.

Genomic Approaches to Understanding Plant Pathogenic Bacteria

João C. Setubal
Ana C. R. da Silva
Alellyx Applied Genomics, Campinas, São Paulo, Brazil

INTRODUCTION

Nowadays, a genomic approach to understanding any bacterium usually requires a detailed description of the organism's complete genome, which means the entire nucleotide sequence and a list of all potential genes and their known or predicted functions. In the case of pathogenic bacteria, we can then concentrate further study on the genes directly involved in or related to pathogenicity. This approach tries to relate the genome information to what is known about disease symptoms, ideally also including genomic data from host plants. The ultimate goal is to understand the molecular mechanisms of plant-pathogen interactions and their evolutionary history.

BACTERIAL GENOME SEQUENCING

In the whole genome approach, the aim is to determine the DNA sequence of all chromosomes and plasmids present in the organism. The process starts with purification of DNA from a pure culture of bacteria. Using the shotgun technique, bacterial DNA is randomly cut into small (between 1000 and 5000 base pairs (bp) long) pieces that are then ligated into plasmid vectors forming a genomic library of inserts. The sequence of both ends of each insert is determined (each end resulting in a read) by machines called automatic DNA sequencers, each of which is currently capable of delivering about 700 reads per day. Reads must then be assembled into a consensus sequence by computer programs. Because DNA fragmentation is never entirely random, and because of experimental errors as well as the presence of repetitive sequences in the genome, it is necessary to sequence as many as 10 times more base pairs in reads than are in the actual genome. This multiplicative factor is known as the genome coverage. Assembly is not yet an entirely automated process. A few genomic regions will require special care, and some gaps take much longer to close than others. Special software is used to determine the error probability, which can also be expressed as a quality value of each assembled base. The desired result is a gapless sequence with an estimated error of less than one base per 10,000 bp. The process of gap closure and improvement of consensus base qualities is called finishing. Bacterial genomes vary from about one million to 10 million bp. With today's technology, and using just one sequencing machine, a five-million-bp genome can be completely sequenced, assembled, and finished in about six months by a team of about 10 people.

GENOME ANNOTATION

Once the sequence is available, it should be annotated. This means, at a minimum, finding the potential genes and assigning functions to them when possible. Most, if not all, of this is done in silico, that is, with the aid of computer programs (studied and developed in the field of bioinformatics[1]). Function assignment relies extensively on sequence similarity. This means that we assign a function to a gene in a newly sequenced genome based on its similarity to already available gene sequences in sequence databases such as GenBank (http://www.ncbi.nlm.nih.gov). The most widely used similarity-detection tool is the BLAST program,[2] although many others exist. After being given a new sequence (which can be in nucleotides or already translated into amino acids), BLAST will search a sequence database and possibly return ''hits,'' that is, database sequences that have similarity to the given sequence above a statistical threshold. If the significant hits found by BLAST come from sequences for which a function has been assigned, then this function can be assigned to the new gene. Genes must then be classified according to their assigned function. Each genome project usually develops its own classification scheme, but most are based on one originally developed for *Escherichia coli* by M. Riley.[3] For plant pathogenic bacteria, genome annotation requires special attention to genes associated with pathogenicity, and the classification scheme should include categories such as avirulence, hypersensitive response, exopolysaccharides, surface proteins, toxin production and detoxification, and host cell wall degradation. Genes related to flagellar and chemotaxis systems are also important for pathogenicity.[4]

After function assignment, genome annotation can proceed to more specialized analyses. One of these is pathway

Encyclopedia of Plant and Crop Science
DOI: 10.1081/E-EPCS 120010578

analysis, where an attempt is made to map the genes onto known gene networks. The best-studied of these networks are the metabolic pathways. This mapping can show that one or more pathways are missing or that certain pathways are almost complete, suggesting that the role of missing enzymes is played by genes not recognized as having that function. This can happen because up to 40% of the potential genes in most bacterial genomes do not resemble known genes and thus are not assigned a likely function.

More-sophisticated analyses are required to find genes that are associated with pathogenicity but that are not similar to known genes. One approach makes use of regulatory sequences that are present in the DNA region upstream of genes and that play a part in their transcription and regulation. It is known that in certain cases genes that play similar roles in an organism will have similar regulatory sequences even if the genes themselves are not similar to each other (but similarity of regulatory sequence does not always imply similar function). If the regulatory sequence, *R*, of certain pathogenicity genes is known, then one can search for additional co-regulated genes by searching for a sequence similar to *R*. The similarity may be subtle and may therefore require the use of sophisticated techniques such as hidden Markov models. On the other hand, the conservation of *R* may happen even across species, such that an *R* sequence identified in one organism can be used to help identify genes in the same family in another species.[5]

Another approach to gain a deeper understanding of pathogenicity genes comes from careful and detailed comparison of the genomes of two or more related species. If both species are pathogenic, one may try to understand the similarities or differences they have in terms of symptoms or host range, or both, through the study of their shared or dissimilar genes. Such studies show the importance of having the complete genome of each organism so that conclusions about the absence of genes can be drawn more reliably. Complete genomes also make it possible to build gene/protein sequence databases. When including several organisms, such databases allow multiorganism comparisons that can yield important insights. For example, van Sluys et al.[9] used this approach to obtain a list of genes shared by eight plant-associated bacteria; such genes are good candidates to be fundamental in plant-bacteria interactions. The inclusion of genomic information from plants (such as *Arabidopsis thaliana* and rice, whose genomes have been sequenced) in these databases will turn them into even more powerful resources.

The topic of horizontally transferred genes is also of particular relevance for pathogenicity studies.[7] Genes are horizontally (or laterally) transferred when they enter the organism not through inheritance but through invasion by phages or mobile genetic elements. If such genes do not kill the organism but are instead kept in its genome, this means that they confer some selective advantage. Lateral transfer between bacteria is well documented in the literature. One example is from Da Silva et al.,[12] who present strong evidence that many pathogenicity genes found in *Xanthomonas axonopodis* pv. *citri* have been laterally transferred. Conversely, sometimes genes are lost because they no longer are important for the organism's survival, and their presence in a related species may suggest their continued importance in the environment of the other organism.

FUNCTIONAL STUDIES

Genome annotation forms the foundation for follow-up functional analyses in the laboratory, which will confirm or refute the hypothesis raised in silico. One technique is targeted gene disruption or knock out. This is commonly done using a special plasmid vector that does not replicate inside the bacteria under study and which has been engineered to carry part of the gene to be knocked out. After insertion of the vector, which also carries an antibiotic resistance gene, into the bacteria, it will either insert into the genome by homologous, site-specific recombination or be lost. Selection of bacteria for those expressing the vector-encoded antibiotic resistance yields cells with the vector inserted into the target gene.

Another laboratory technique is the gene expression microarray. A DNA microarray (a gene chip) is made by depositing short DNA sequences of many or all genes from a genome on a glass slide in high densities (more than 2000 spots in 2 cm^2). Similar to the traditional Northern blot technique, microarrays are probed with labeled mRNA from cells so that it hybridizes to immobilized, homologous target DNAs. If the RNA was labeled with a fluorescent tag, then the intensity of fluorescence in each spot is proportional to the amount of RNA retained by the target DNA. When RNA has been isolated from bacteria exposed to two different environmental or cultural conditions, the preparations can be labeled with different fluorescent tags. Hybridization of these probes to identical microarrays, followed by suitable image and data processing, reveals genes that are differentially expressed under those conditions.

CONCLUSION

As of this writing, five plant-pathogen bacterial genomes have been completed and published:

- *Xylella fastidiosa*, strain 9a5c[10]
- *Agrobacterium tumefaciens* C58[6,11]
- *Ralstonia solanacearum*[8]

- *Xanthomonas axonopodis* pv. *citri*[12]
- *Xanthomonas campestris* pv. *campestris*[12]

These projects have yielded a wealth of genomic information and have placed phytopathology at a new level. For a detailed review of these projects, see Van Sluys et al.[9] No concrete and effective new ways to deal with the diseases caused by the plant pathogens have yet resulted from these projects. However, most researchers have no doubt that these genomic approaches will eventually yield disease management benefits. This may still take a great deal of time and effort; genomics is only the first step.

ARTICLES OF FURTHER INTEREST

Arabidopsis thaliana: Characteristics and Annotation of a Model Genome, p. 47
Arabidopsis Transcription Factors: Genome-Wide Comparative Analysis Among Eukaryotes, p. 51
Bioinformatics, p. 125
Functional Genomics and Gene Action Knockouts, p. 476

REFERENCES

1. Mount, D. *Bioinformatics: Sequence and Genome Analysis*; Cold Spring Harbor Laboratory Press: New York, 2001.
2. Alstchul, S.F.; Madden, T.L.; Schaffer, A.A.; Zhang, J.; Zhang, Z.; Miller, W.; Lipman, D.J. Gapped BLAST and PSI-BLAST: A new generation of protein database search programs. Nucleic Acids Res. **1997**, *25*, 3389–3402.
3. Riley, M. Functions of the gene products of in *Escherichia coli*. Microbiol. Rev. **1993**, *57*, 862–952.
4. Alfano, J.R.; Collmer, A. Mechanisms of Bacterial Pathogenesis in Plants: Familiar Foes in a Foreign Kingdom. In *Principles of Bacterial Pathogenesis*; Groisman, E.A., Ed.; Academic Press: San Diego, 2001; 179–226.
5. Terai, G.; Takagi, T.; Nakai, K. Prediction of co-regulated genes in *Bacillus subtilis* on the basis of upstream elements conserved across three closely related species. Genome Biol. **2001**, *2* (11), 0048.1–0048.12.
6. Goodner, B., et al. Genome sequence of the plant pathogen and biotechnology agent *Agrobacterium tumefaciens* C58. Science **2001**, *294*, 2323–2328.
7. Bushman, F. Lateral DNA Transfer. Cold Spring Harbor Laboratory Press, 2002.
8. Salanoubat, M., et al. Genome sequence of the plant pathogen *Ralstonia solanacearum*. Nature **2002**, *415*, 497–502.
9. van Sluys, M.A.; Monteiro-Vitorello, C.B.; Camargo, L.E.A.; Menck, C.F.M.; da Silva, A.C.R.; Ferro, J.A.; Oliveira, M.C.; Setubal, J.C.; Kitajima, J.P.; Simpson, A.J. Comparative genomic analysis of plant-associated bacteria. Annu. Rev. Phytopathol. **2002**, *40*, 169–189.
10. Simpson, A.J.G., et al. The genome sequence of the plant pathogen *Xylella fastidiosa*. Nature **2000**, *406*, 151–157.
11. Wood, D.W., et al. The genome of the natural genetic engineer *Agrobacterium tumefaciens* C58. Science **2001**, *294*, 2317–2323.
12. da Silva, A.C.R., et al. Comparison of the genomes of two *Xanthomonas* pathogens with differing host specificities. Nature **2002**, *417*, 459–463.

Genomic Imprinting in Plants

G

Daphné Autran
Jean-Philippe Vielle-Calzada
CINVESTAV—Unidad Irapuato, Irapuato, Mexico

INTRODUCTION

The differentiation and fusion of male and female gametes to initiate the formation of a new organism is a common feature in the life cycle of higher eukaryotes. In the ovule of flowering plants, the embryo develops following fertilization of the haploid egg cell by one of the sperm cells delivered by the pollen tube, giving rise to a diploid zygote. A second sperm cell fuses with the binucleated (homodiploid) central cell to give rise to the triploid endosperm having a 2:1 maternal to paternal genomic ratio. The embryo and the endosperm, each with different ploidy and parental gene dosage, coordinately develop inside diploid maternal tissues.

Plant gametes carry distinct haploid genetic constitutions that are usually considered to be functionally equivalent, both contributing active alleles that will influence the phenotype of the newly formed individual. For some traits, the activity of an allele is influenced by the parent from which it came. Parent-of-origin effects can be the consequence of differences in transcription between paternally and maternally inherited genes in the embryo and/or the endosperm. Genomic imprinting refers to a specific type of genetic regulation resulting from a mitotically stable epigenetic modification that consistently inactivates one of the parental alleles. Loci that are regulated by genomic imprinting will be differentially expressed in a parent-of-origin dependent manner. If a gene is controlled by genomic imprinting, transcription occurs exclusively from one of the two parental gene copies. Hence, imprinted genes are inherited in a silent state from one of the two parents, and in a fully active form from the other. Genomic imprinting results in a functional nonequivalence of parental genomes during embryogenesis and/or endosperm development. Due to particular features of the plant life cycle, it is often difficult to distinguish between the effects of maternal factors and the consequences of genomic imprinting on the regulation of a specific locus. A maternal effect refers strictly to a genetic definition: Any reciprocal cross resulting in differential phenotypes that are exclusively determined by the genotype of the female parent (or the megagametophyte) is the consequence of a maternal effect. This broad type of parent-of-origin effect can result from the action of imprinted genes, but also from: 1) gene products that are stored in the cytoplasm of the egg cell; 2) dosage-sensitive genes acting in the embryo and/or the endosperm; or 3) products encoded by the genome of maternally inherited organelles. Maternal effects caused by cytoplasmic protein storage and genomic imprinting are clearly distinguishable if transcripts are present only prior to fertilization but not after (perdurance, implying a maternal effect sensu stricto) or if transcripts are present only after fertilization but not before (imprinting).

SEED DEVELOPMENT AND GENOMIC IMPRINTING

Whereas genomic imprinting has been extensively studied in mammals and insects, the first demonstration of a gene regulated by genomic imprinting was obtained for the *r1* locus in maize.[1] Subsequently, pioneering studies by Lin[2] and Kermicle used exceptional mutants in maize to manipulate genome dosage in the endosperm. In the *indeterminate gametophyte* (*ig*) mutant, several haploid nuclei fuse in the central cell, generating an endosperm of variable ploidy. By conducting reciprocal crosses involving genotypes with different ploidy levels, they showed that deviations from the normal 2:1 maternal to paternal genome ratio in the fertilized central cell resulted in an aborted endosperm. They inferred that the parental origin of each of the genomic contributions is crucial for normal endosperm formation. Similar conclusions were obtained from lines carrying chromosomal translocations in maize: Deletions of particular regions of a specific chromosome that were paternally inherited lead to a significant reduction in endosperm growth. The defects are not rescued by providing an additional maternal copy of the translocated region.[3] These important results allowed dosage-dependent effects to be distinguished from the occurrence of genomic imprinting during endosperm development.

Encyclopedia of Plant and Crop Science
DOI: 10.1081/E-EPCS 120012918

The FIS Class of Genes Is Regulated by Genomic Imprinting in *Arabidopsis*

More recently, direct evidence of genomic imprinting has been obtained for a handful of genes.[4] Whereas in maize several genes are differentially expressed in a parent-of-origin–dependent manner, only members of the *FERTILIZATION-INDPENDENDENT SEED* class of genes (*FIS*) have been shown to be regulated by genomic imprinting in *Arabidopsis*. These three *Arabidopsis* genes were identified on the basis of their gametophytic maternal effect mutant phenotype, which causes seed abortion, and by their ability to initiate endosperm formation in the absence of fertilization.[5] Female gametophytes of *medea* (*mea*), *fertilization independent endosperm* (*fie*), and *fis2* give rise to abnormal seeds irrespective of the paternal genomic contribution. The embryo and/or endosperm of *fis* mutants show an increase in cell number, suggesting that the primary role of the *FIS* genes is to restrict cell proliferation. Consistent with this role, *MEA* and *FIE* encode Polycomb group proteins (PcG). In *Drosophila*, these proteins are known to form repressor complexes that regulate higher-order chromatin structure during animal development. The phenotypes shown by *mea* and *fie* suggest that their function is conserved in flowering plants. The *Arabidopsis FIS2* gene encodes a Zn-finger protein implicated in the formation of the PcG complex.[6] All *FIS* genes are expressed during female gametophyte development prior to fertilization; *MEA* and *FIE* are coexpressed in the embryo and the chalazal domain of the free nuclear endosperm, whereas expression of *FIS2* could only be detected during the early stages of endosperm development. A combination of in situ hybridization and reverse-transcriptase polymerase chain reaction (RT-PCR) experiments provided the evidence that following fertilization *MEA* is only transcribed from maternally inherited copies in both the developing embryo and the endosperm, demonstrating that the gene is regulated by genomic imprinting.[7] Similar approaches were used to show that *fie* and *fis2* are also imprinted.[8]

Genomewide Mechanism of Paternal Imprinting in *Arabidopsis*

Because plant embryos can develop from somatic cells or microspores, it was generally considered that the maternal contribution is not crucial for embryogenesis. Recent studies in *Arabidopsis* suggest that early embryo development is mainly under maternal control due to a combination of maternal products deposited in the female gametophyte and a genomewide mechanism of paternal imprinting.[9] Using reciprocal crosses to wild-type, more than 20 genes detected by gene trap approaches were shown to be expressed only when maternally inherited

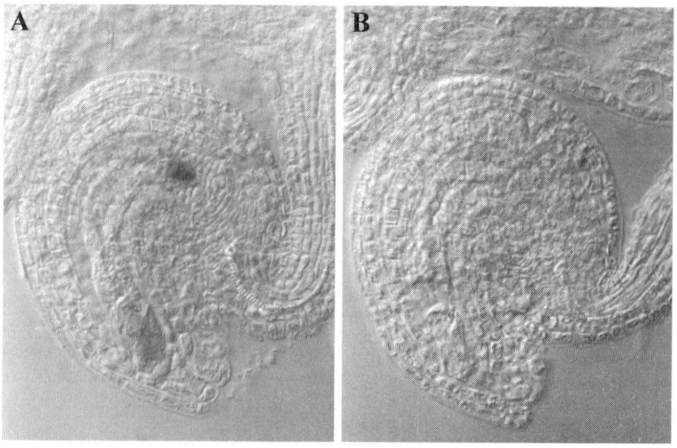

Fig. 1 Silencing of paternally inherited gene activity during seed development in *Arabidopsis*. (A) When an enhancer detector is used as a female parent in crosses to wild-type, reporter gene expression can be detected in the developing embryo and endosperm of F1 seeds 50 hours after pollination (HAP). (B) When the same enhancer detector is used as the male parent in crosses to wild-type, reporter gene expression is absent from the embryo and the endosperm 50 HAP. (Figure courtesy of Gerardo Acosta-García.) (*View this art in color at www.dekker.com.*)

(Fig. 1). Analysis of gene expression using allele-specific RT-PCR demonstrated that the absence of transcription resulted from the silent state of the endogenous genes and not from the inactivation of the reporter gene present in the gene trap transgenic construct. All of these genes are randomly distributed across the *Arabidopsis* genome and encode a wide range of proteins, showing that the observed paternal silencing is not limited to a specific genomic region or a particular developmental mechanism. For some of these genes the activation of the paternal genome occurred 96 hours following fertilization, i.e., at the midglobular stage of embryo development. The correlation between allele-specific silencing and genetic activity was demonstrated by analyzing early embryonic defects in *emb30*, a recessive embryo-lethal mutant defective in the first zygotic division of *Arabidopsis*. Heterozygous *emb30/EMB30* embryos that inherited a wild-type allele from the male parent did show a mutant phenotype, confirming that early defects cannot be rescued by an active paternal *EMB30* allele. The degree of inactivation and the timing of initiation of paternal transcription may vary from gene to gene.[10] Additional experiments have suggested that some specific genes or their regulatory elements escape this genomewide mechanism of paternal silencing and are active early following fertilization. Baroux et al.[11] showed that embryo abortion can be induced as early as the 2- to 4-cell stage using a transactivation system that depends on the transmission of the elicitor through male gametes. These results suggest that either some regulatory elements can escape paternal imprinting (in this case, the cyclin B1 promoter and the heterologous pOp promoter), or that the activity of the paternal genome is not completely abolished but rather attenuated by a very low level of transcription during early seed development.

THE REGULATION OF GENOMIC IMPRINTING

Based on the existing evidence, it currently appears that two distinct classes of imprinting mechanisms regulate seed development in flowering plants.[12] A delayed activation of the paternal genome indicates that a wide mechanism of imprinting ensures that both parental genomes are not functionally equivalent following fertilization. In addition, specific loci such as members of the *FIS* class of genes could be imprinted until later stages of seed development, as has been shown for *MEA*. In both cases the evidence implies that a mechanism of allele-specific inactivation must be established during gametogenesis. Maintenance of the resulting imprint must be ensured through successive haploid mitotic cycles and finally lost in diploid cells at some stage of sporophytic development before being reestablished in the germline.

Despite extensive research in mammals, the regulation of genomic imprinting is far from understood. To date, the most common mechanism associated with the regulation of genomic imprinting in plants is DNA methylation. In animals, there is a correlation between methylation (epigenetic modification of CpG to 5-methyl-CpG) and the imprinted status of most of the genes described. In *Arabidopsis*, reciprocal crosses of transgenic *methyltransferase 1* antisense (*met1*a/s) to wild-type diploids produced seeds with defects similar to those observed in interploidy crosses. For example, hypomethylated females crossed to wild-type males gave rise to paternalized seeds containing a large endosperm and embryo. The reciprocal cross leads to the opposite effect. Moreover, mutations in the *FIS* genes can be rescued by reducing global methylation levels by either *met1*a/s[6] or *ddm1* pollen.[7] DDM1 is a member of the SWI/SNF family of ATP-dependent chromatin remodeling enzymes, and *ddm1* mutations reduce genomic methylation by 70%.[13] In most cases, rescue of the embryo or the endosperm occurs even if the paternal genome carries a *fis* mutant allele, suggesting that unknown genes can substitute for *FIS* function in early seed development, and that *FIS* paternal alleles remain silent even in a hypomethylated context, implying that their regulation most likely involves additional mechanisms yet to be discovered. Additional epigenetic mechanisms involving chromatin remodelling, histone modification, or specific methylation could ensure the imprinted memory necessary for the ocurrence of mitotically stable reversible changes in gene expression.[5] The fact that *FIS* genes encode chromatin remodelling factors and are regulated by genomic imprinting suggest that their target genes, involved in the regulation of seed development, are probably also imprinted. Interestingly, DEMETER (DME)—a DNA glycosylase with nuclear localization domains—has been shown to be required for the expression of the *MEA* maternal transcript in the developing endosperm.[14] DME is preferentially expressed in the developing endosperm, and homozygous *dme* plants show a maternal effect, suggesting that DEMETER could also be imprinted.

CONCLUSION

How did genomic imprinting evolve? Haig and Westoby[15] proposed that parent-of-origin specific effects evolved as a consequence of an intragenomic conflict over the allocation of nutrients from the mother to its offspring. This theory predicts that paternal interests favor selfish fitness for their own offspring, and not the fitness of full siblings from a distinct father. Therefore, paternally expressed genes would tend to promote the growth of the embryo. In contrast, maternal interests favor survival of all siblings irrespective of their paternal origin. Maternally expressed genes would tend to optimize overall survival by reducing the size of the embryo. The role of genomic imprinting in regulating parental interests may have been conserved in organisms that have acquired placental habits. Supporting evidence has been provided by studies of imprinted genes in mice and humans, but also in plants. Most imprinted genes identified in mammals fit the parental conflict theory. In flowering plants, the manipulation of gene dosage in the endosperm and the embryo of maize and *Arabidopsis* has shown that increasing paternal dosage promotes growth of the endosperm, whereas an increase in the maternal dosage reduces its size. In *Arabidopsis*, disruption of *MEA* leads to overproliferation of embryonic cells, albeit embryos arrested at an earlier stage than wild-type. In constrast to mammals, genes exclusively transcribed from the paternally inherited copy have yet to be discovered in flowering plants.

Although parental conflicts at the gene level are just starting to be investigated in flowering plants, the elucidation of the genetic basis and molecular mechanisms regulating a genomewide paternal silencing during early seed development has yet to be initiated. The synchronous evolution of parent-of-origin effects in mammals and flowering plants could be closely related to their sexual habits. In that regard, the investigation of imprinting in apomictic plants, in combination with the fast emergence of genomic technology, should provide important clues to determine the number of genes that are regulated by genomic imprinting and their overall impact on plant growth and development.

ACKNOWLEDGMENTS

We thank Gerardo Acosta-García for providing the micrographs shown in Fig. 1. Research in our laboratory is supported by CONACyT (grants Z029 and B34324) and Biotec Internacional S.A. de C.V (Monterrey NL) and the Howard Hughes Medical Institute (HHMI). Jean-Philippe Vielle-Calzada is an International Scholar of HHMI.

ARTICLES OF FURTHER INTEREST

Cytogenetics of Apomixis, p. 383
Endosperm Development, p. 414
Female Gametogenesis, p. 439
Male Gametogenesis, p. 663
Meiosis, p. 711
Polyploidy, p. 1038
Sex Chromosomes in Plants, p. 1148
Somaclonal Variation: Origins and Causes, p. 1158

REFERENCES

1. Kermicle, J.L. Dependance of the *R-mottled* aleurone phenotype in maize on the mode of sexual transmission. Genetics **1970**, *66*, 69–85.
2. Lin, B.Y. Ploidy barrier to endosperm development in maize. Genetics **1984**, *107*, 103–115.
3. Kermicle, J.L.; Alleman, M. Genomic imprinting in maize in relation to the angiopserm life cycle. Dev. Supp. **1990**, *1*, 9–14.
4. Evans, M.M.S.; Kermicle, J.L. Interactions between maternal effect and zygotic effect mutations during maize seed development. Genetics **2001**, *159*, 303–315.
5. Grossniklaus, U.; Spillane, C.; Page, D.R.; Köhler, C. Genomic imprinting and seed development: Endopserm formation with and without sex. Curr. Opin. Plant Biol. **2001**, *4*, 21–27.
6. Luo, M.; Bilodeau, P.; Dennis, E.S.; Peacock, J.; Chaudbury, A. Expression and parent-of-origin effects for FIS2, MEA and FIE in the endosperm of developing *Arabidopsis* seeds. Proc. Natl. Acad. Sci. U. S. A. **2000**, *96*, 296–301.
7. Vielle-Calzada, J.P.; Thomas, J.; Spillane, C.; Coluccio, A.; Hoeppner, M.A.; Grossniklaus, U. Maintenance of genomic imprinting at the *Arabidopsis medea* locus requires zygotic *DDM1* activity. Genes Dev. **1999**, *13*, 2971–2982.
8. Kinoshita, T.; Yadegari, R.; Harada, J.H.; Goldberg, R.B.; Fisher, R.L. Imprinting of the *MEDEA* polycomb gene in the *Arabidopsis* endopserm. Plant Cell **1999**, *11*, 1945–1952.
9. Vielle-Calzada, J.P.; Baskar, R.; Grossniklaus, U. Delayed activation of the paternal genome during seed development. Nature **2000**, *404*, 91–94.
10. Weijers, D.; Geldner, N.; Offringa, R.; Jürgens, G. Early paternal gene activation in *Arabidopsis*. Nature **2001**, *414*, 709–710.
11. Baroux, C.; Blainvillain, R.; Gallois, P. Paternally inherited transgenes are down-regulated but retain low activity during early embryogenesis in *Arabidopsis*. FEBS Lett. **2001**, *509*, 11–16.
12. Baroux, C.; Spillane, C.; Grossniklaus, U. Genomic imprinting during seed development. Adv. Genet. **2002**, *46*, 165–214.
13. Jeddeloh, J.A.; Stokes, T.L.; Richards, E.J. Maintenance of genomic methylation requires a SWI2/SNF2-like protein. Nat. Genet. **1999**, *22*, 94–97.
14. Choi, Y.; Gerhing, M.; Johnson, L.; Hannon, M.; Harada, J.J.; Goldberg, R.B.; Jacobsen, S.E.; Fisher, R.L. DEMETER, a DNA glycosylase domain protein, is required for endosperm gene imprinting and seed viability in *Arabidopsis*. Cell **2002**, *110*, 33–42.
15. Haig, D.; Westoby, M. Genomic imprinting in endosperm: Its effect on seed development in crosses between species, and betweeen different ploidies of the same species, and its implication for the evolution of apomixis. Philos. Trans. R. Soc. Lond., B **1991**, *333*, 1–13.

Germplasm: International and National Centers

Jonathan Robinson
Murthi Anishetty
Food and Agriculture Organization of the United Nations, Rome, Italy

INTRODUCTION

It is estimated that ex situ plant germplasm collections comprise six million accessions, over half of which represent base collections, kept solely for conservation. The remainder are active, working collections available for distribution and use by plant breeders and researchers. One third of the total number of accessions is thought to be unique. The 11 genebanks of the Consultative Group on International Agricultural Research (CGIAR) and International Agricultural Research Centers (IARCs) together house major collections of accessions (ca. 600,000). The rest are conserved in about 1300 regional and national genebanks. The CGIAR collections have been built up over 25 years and, following agreements signed in 1994, are held in trust for the world community under the intergovernmental authority of the Food and Agriculture Organization of the United Nations (FAO).

DISCUSSION

Most plant germplasm is conserved as seeds, but approximately 500,000 accessions, many of which represent fruit trees, are maintained in field genebanks. An additional 40,000 accessions are conserved in vitro through cryopreservation or tissue culture. Botanical gardens, arboreta, and herbaria are also important centers for plant germplasm conservation. There are approximately 1500 such centers spread around the globe. The majority are located in Europe, the Commonwealth of Independent States, and the United States. A little over 10% are privately owned. Germplasm seedbanks are also maintained by 150 botanical gardens. About half the botanical gardens conserve germplasm of ornamental species, crop relatives, medicinal plants, and forest trees.

International and regional centers conserve broad ranges of plant germplasm derived from diverse locations. Many national germplasm centers conserve mainly indigenous material, whereas others, including those in Australia, the United Kingdom, and United States for example, conserve mainly imported germplasm. National genebanks such as the N.I. Vavilov All-Russian Research Institute for Plant Industry (VIR) in Saint Petersburg, Russia, maintain accessions collected from around the world and more resemble international centers than national ones. The Institute of Plant Genetics and Crop Plant Research (IPK) at Gatersleben, Germany, is another such institute that houses large and diverse collections of crop germplasm. Whereas the international and regional centers for plant germplasm are generally able to conserve collections under optimal conditions, the state of national collections varies enormously. The number of unique accessions is often difficult to determine, as duplication—deliberate as well as unintentional—is often not easy to gauge. Developed countries invariably have seed and field genebanks and botanical gardens, but many developing nations do not. However, substantial germplasm collections do exist in some developing countries, including Brazil, China, Ethiopia, and India. India, Indonesia, and Sri Lanka, moreover, have among the best maintained tropical botanical gardens in the world, at Calcutta, Bogor, and Kandy, respectively.

Useful information on genebanks and their operations is available in several general texts.[1–4]

INTERNATIONAL CENTERS

The CGIAR centers have mandates to breed and research a limited number of crop species, including most of the major staples of the developing world; their genebanks consequently house germplasm geared toward those ends. Several of the collections—including CIMMYT's wheat (*Triticum* spp.), IRRI's rice (*Oryza* spp.), CIAT's bean (*Phaseolus* spp.) and cassava (*Manihot* spp.), and ICRISAT's chickpea (*Cicer arietinum*) and pigeon pea (*Cajanus cajan*)—represent the world's largest ex situ collections of those crops. Detailed information on international agricultural research center holdings is available from the System-wide Information Network for Genetic Resources (SINGER) database.[5] These data are summarized in Table 1.

Encyclopedia of Plant and Crop Science
DOI: 10.1081/E-EPCS 120010461

Table 1 In-trust germplasm collections of the International Agricultural Research Centers of the Consultative Group on International Agricultural Research

Center[a]	Crop	Genus	No. of accessions
CIAT	Cassava	*Manihot*	> 5,700
	Forages		> 16,300
	Beans	*Phaseolus*	> 28,700
CIMMYT	Maize	*Zea*	> 19,500
	Wheat	*Triticum*	>79,900
CIP	Potato	*Solanum*	>5,000
	Sweet potato	*Ipomoea*	>6,400
	Andean roots & tubers		>1,100
ICARDA	Barley	*Hordeum*	>24,200
	Chickpea	*Cicer*	>9,100
	Faba bean	*Vicia*	>9,000
	Forages		>24,500
	Lentil	*Lens*	>7,800
	Wheat	*Triticum*	>30,200
ICRAF		*Sesbania*	25
ICRISAT	Chickpea	*Cicer*	>16,900
	Groundnut	*Arachis*	>14,300
	Pearl millet (spp.)	*Pennisetum*	>21,200
	Minor millets	*Setaria* etc.	>9,000
	Pigeon pea	*Cajanus*	>12,600
	Sorghum	*Sorghum*	>35,700
IITA	Bambara groundnut	*Voandzeia*	>2,000
	Cassava	*Manihot*	>2,000
	Cowpea	*Vigna*	>16,600
	Soybean	*Glycine*	>1,900
	Yams	*Dioscorea*	>2,800
ILRI	Forages		>11,500
IPGRI/INIBAP	Banana/plantain	*Musa*	>900
IRRI	Rice	*Oryza*	>80,600
WARDA	Rice	*Oryza*	>14,900
TOTAL			>513,700

[a]CIAT—Centro Internaciónal de Agricultura Tropical; CIMMYT—Centro Internaciónal de Mejoramiento del Maíz y Trigo; CIP—Centro Internaciónal de la Papa; ICARDA—International Center for Agricultural Research in the Dry Areas; ICRAF—International Center for Research in Agroforestry; ICRISAT—International Crop Research Institute for the Semi-Arid Tropics; IITA—International Institute for Tropical Agriculture; ILRI—International Livestock Research Institute; IPGRI—International Plant Genetic Resources Institute; INIBAP—International Institute for Bananas and Plantains; IRRI—International Rice Research Institute; WARDA—West African Rice Development Association. (From Ref. 6.)

REGIONAL CENTERS

There are five principal regional centers for plant germplasm. The Asian Vegetable Research and Development Center (AVRDC) is located in Taiwan and holds over 47,000 accessions, concentrating on tomato (*Lycopersicon esculentum*), pepper (*Capsicum* spp.), soybean (*Glycine* spp.), and mung bean (*Vigna* spp.) germplasm. It is one of the world's largest genebanks for vegetable species. The Centro Agronómico Tropical de Investigación y Enseñanza (CATIE) in Costa Rica conserves more than 35,000 accessions of curcurbits, peppers, beans, coffee (*Coffea* spp.), and cocoa (*Theobroma* spp.) in seedbanks and in decentralized field genebanks. Large ranges of temperate fruit and berry crops are maintained in the field genebanks of the Nordic Gene Bank (NGB), headquartered in Sweden. In total it manages more than 27,000 accessions, including vegetable, root, oil, and pulse crops. The Southern African Development Community (SADC) has a plant genetic resources center in Zambia that contains approximately 5000 germplasm accessions, representing collections and duplicate collections of southern African national collections. Last, the Arab Center for the Studies of Arid Zones and Dry Lands (ACSAD) in Syria has a sizeable collection of fruit trees from West Asia and North Africa. Other regional genebanks are being developed. For instance, a taro (*Colocasia* spp.) genebank is being established in Fiji that will serve the 22 member states of the South Pacific Commission. There is also a multiple-site International Coconut (*Cocus nucifera*) Genebank hosted by Brazil, Côte d'Ivoire, India, Indonesia, and Papua New Guinea.

NATIONAL CENTERS

National genebanks comprise a range of institutions administered by governments, universities, nongovernment organizations (NGOs), and the private sector. Most countries in Europe have established genebanks. Many of those genebanks contain working collections, but university genebanks, for example, often hold important collections of genetic stocks. The Royal Botanic Gardens at Kew in England have substantial living plant collections—including native and exotic species—at two sites. There is also a national reference collection of over seven million plant specimens (including 250,000 type specimens) and 20,000 seed accessions in the herbarium carpological collections. The Millennium Seed Bank is the largest seed bank in the world for wild plants and holds seed of more than 5000 species from 600 genera. The Russian national genebank, VIR, has vast germplasm collections of a range of crop species, including large collections of maize, *Zea mays* (18,000 accessions) and wheat (36,000) and the world's largest potato (*Solanum tuberosum*) collection (8500 accessions). Some major genebank holdings for a range of crops are given in Table 2.

Africa has relatively few centers for plant germplasm that function effectively. Facilities are most advanced in Côte d'Ivoire, Ethiopia, Kenya, and Nigeria. However, more than 80% of oil palm (*Elaeis guineensis*) accessions are maintained in the Democratic Republic of the Congo, and Uganda has an important collection of banana (*Musa* spp.) germplasm (250 accessions). In the Near East, Turkey has one of the few facilities with long-term storage capacity, but genebanks in Iran, Israel, Pakistan, and Turkey operate at a high level. The Hebrew University of Jerusalem has an important sesame (*Sesamum indicum*) collection. Field genebanks in the Near East are quite numerous. Turkey has several field genebanks devoted to crops including fruit trees, olives (*Olea europea*), and garlic (*Allium sativum*). In South Asia, the genebank facilities in India are the most advanced. The Indian Agricultural Research Institute, for example, has the world's largest collection of maize germplasm (25,000 accessions) and the National Bureau of Plant Genetic Resources has the world's second largest chickpea collection (ca. 15,000 accessions).

There are several major national collections of germplasm in the East Asia region, especially in China, Japan, and the Republic of Korea, as indicated in Table 2. Those collections include soybean, citrus, and sesame. In terms of total numbers of genebank accessions, China ranks second in the world (350,000) after the United States (555,000) and ahead of India (342,000). In Southeast Asia and the Pacific, Indonesia, Malaysia, the Philippines, and Thailand have important centers for germplasm conservation, including many important field genebank collections. Only two genebanks holding substantial numbers of accessions of banana and plantain germplasm are in the center of origin of the crop. These collections are in Papua New Guinea (The National Agricultural Research Institute), which has the highest number of unique banana/plantain accessions, and the Philippines (Davoa National Crop Research and Development Center). The National Plant Genetic Resource Center of the Philippines holds the world base collections of several minor crops and duplicate collections of some Asian vegetables. Kebun Raya, the Indonesian botanical gardens at Bogor, were established in 1817 and now cover four sites and contain over 15,000 species of trees and plants. Oil palm was introduced into Southeast Asia in 1859 from plants grown there.

In the Americas, the United States and Canada have numerous large seed genebank and field genebank collections. Argentina, Brazil, Chile, Mexico, and Venezuela also have good long-term storage facilities. The two Brazilian agencies, Centro Nacional de Recursos Genéticos e Biotecnologia (CENARGEN) and Empresa Brasileira de Pesquisa Agropecuária (EMBRAPA), manage large collections of a range of crops including beans, cassava, cocoa, citrus, rice, and soybean. There are several major bean germplasm collections in Mexico. Honduras (Fundación Hondureña de Investigación Agricola, FHIA) has a large and very important collection of 430 banana and plantain accessions. Cuba is relatively alone in having adequate facilities in the Caribbean, but Trinidad and Tobago holds a major cocoa germplasm collection (2300 accessions). The United States Department of Agriculture (USDA) Agricultural Research Service (ARS) has national plant germplasm collections at about 30 sites. The National Small Grains Collection (NSGC) at the University of Idaho, Aberdeen, has sizeable collections of wheat (>46,500); barley (ca. 27,000); oats, *Avena sativa* (21,000); rice (>17,000); rye, *Secale cereale* (>2000); triticale, X *Triticosecale* (ca. 2,000); and *Aegilops* (>2000). The USDA, moreover, has a very important potato germplasm collection at Sturgeon Bay and a substantial fruit, nut, and horticultural crop germplasm collection in Corvallis, at Oregon State University (11,430 accessions representing 54 genera and 745 species). The National Seed Storage Laboratory is located in Fort Collins, Colorado, and is a base collection repository for 23,827 accessions representing 207 genera and 514 species. There are numerous other germplasm centers at the many agricultural universities in the United States, including a particularly important

Table 2 Major national germplasm holding for a range of crops

Crop	Institute	No. accessions	Total no. recorded accessions[a,b]
Banana	CIRAD, Guadeloupe Research Station, France	>400	13,125[a]
	Fundación Hondureña de Investigación Agricola (FHIA), Honduras	>400	
	Centre de Researche Régionales sur les Bananiers et Plantains (CRBP), Cameroon	>300	
Beans	USDA-ARS-WRPIS, USA	>14,000	268,400[b]
	Centro Nacional de Pesquisa de Arroz e Feijao (CNPAF) EMBRAPA, Brazil	ca. 10,700	
	Instituto Nacional de Investigaciones Forestales, Agrícolas y Pecuarias (INIFAP), Mexico	ca. 10,500	
Cocoa	Cocoa Research Unit, Trinidad and Tobago	>2,000	12,750[a]
	CENARGEN, Brazil	ca. 2,300	
	Centro Nacional de Investigaciónes Agropecuárias (FONAIAP), Venezuela	ca. 1,000	
Cassava	Centro Nacional de Pesquisa de Mandioca e Fruticultura (CNPMF), EMBRAPA, Brazil	>2,600	27,900[b]
	Central Tuber Crops Research Institute, India	>1,600	
	National Root Crops Research Institute, Nigeria	>1,100	
Chickpea	National Bureau of Plant Genetic Resources, India	>14,000	70,000[b]
	Seed Plant Improvement Institute, Iran	ca. 5000	
	USDA, USA	ca. 4,600	
Citrus	National Institute of Agrobiological Resources, Japan	>2,000	6,170[b]
	Instituto Agronomico de Campinas, Brazil	ca. 1,700	
	SRA INRA-CIRAD, France	ca. 1,100	
Coconut	Estación Local Irapa, FONAIAP, Venezuela	ca. 1,000	1,352[a]
	Philippine Coconut Authority, Samboanga Research Centre, Philippines	>140	
	Central Plantation Crops Research Institute (CPCRI), India	>100	
Maize	Directorate of Maize Research, India	ca. 25,000	328,000[a]
	Institute of Crop Germplasm Resources, CAAS, China	>15,000	
	Instituto Nacional de Investigaciones Forestales, Agrícolas y Pecuarias (INIFAP), Mexico	>15,000	

Pigeon pea	National Bureau of Plant Genetic Resources, India	>4,000	ca. 25,000[b]
	Malawi PGR Centre, Chitedze Agricultural Research Station, Malawi	ca. 500	
	National Plant Genetic Resources Laboratory, IPB/UPLB, Philippines	ca. 400	
Potato	N.I. Vavilov All-Russian Research Institute of Plant Industry (VIR), Russia	ca. 8,500	ca. 30,000[b]
	USDA Potato Genebank, U.S.A.	>5,500	
	IPK-Gatersleben, Germany	>5,000	
Rice	Institute of Crop Germplasm Resources, CAAS, China	>64,000	>420,000[b]
	National Bureau of Plant Genetic Resources, India	>53,000	
	National Institute of Agrobiological Resources, Japan	>36,000	
Sesame	Genetic Resources Division, NSMO, RDA, Korea	>8,000	>31,000[a]
	Oil Crops Research Institute, CAAS, China	>4,000	
	National Bureau of Plant Genetic Resources, India	>2,800	
Soybean	Institute of Crop Germplasm Resources, CAAS, China	>30,000	>184,000[a]
	USDA-ARS, U.S.A.	ca. 18,000	
	Genetic Resources Division, NSMO, Korea	>17,000	
Taro	National Agricultural Research Institute, Papua New Guinea	>850	ca. 3,600[c]
	National Bureau of Plant Genetic Resources, India	ca. 470	
	Plant Genetic Resources Centre, VASI, Vietnam	ca. 400	
Tomato	N.I. Vavilov All-Russian Research Institute of Plant Industry (VIR), Russia	>7,200	ca. 86,000[a]
	University of California, U.S.A.	ca. 5,800	
	Dept. Horticultural Sciences, NY State Agricultural Research Center, U.S.A.	ca. 4,000	
Wheat	USDA National Small Grains Collection, U.S.A.	>46,000	>970,000[a]
	Germplasm Institute of the Italian National Research Council, Italy	ca. 41,000	
	Institute of Crop Germplasm Resources, CAAS, China	ca. 40,000	

[a]Total accessions recorded from the IPGRI database.
[b]Total accessions listed in the FAO State of the World's Plant Genetic Resources for Food and Agriculture report.[7]
[c]Total accessions recorded from the FAO WIEWS database.
(From Ref. 8.)

tomato germplasm collection maintained at the University of California, comprising nearly 6000 accessions.

CONCLUSION

The recently signed International Treaty on Plant Genetic Resources for Food and Agriculture, adopted by the thirty-first session of the FAO Conference (November, 2001) represents a legally binding international commitment to the improvement of the world's key food and feed crops. The Treaty is based on a multilateral system of facilitated access and benefit sharing. This multilateral system covers plant genetic resources for food and agriculture, including a list of crops established according to the criteria of food security and interdependence. The list of crops covers species from 54 major crop genera, but excludes some genera of major importance, including *Glycine*, *Arachis*, and *Elaeis*. Also included are over 80 species of forages from 29 genera. It is anticipated that if there is unrestricted access to the listed genetic resources they will be better conserved for the future.

Although the germplasm in the genebanks of the developed world is generally well conserved, that in many of the developing countries is in a precarious position. Moreover, some germplasm collections—particularly field genebanks—are naturally vulnerable. Banana and taro collections are two examples prone to hurricane damage and disease, respectively. In addition, potentially important sources of wild relative germplasm grow in vulnerable ecosystems and environments, which are at risk of being lost. Through an initiative begun by the System-wide Genetic Resources Program of the CGIAR, termed the Global Conservation Trust, it is hoped to ensure that some of the most important and vulnerable germplasm collections in the world can be conserved for the well-being of humankind.

ARTICLES OF FURTHER INTEREST

Genetic Resource Conservation of Seeds, p. 499
Germplasm Collections: Regeneration in Maintenance, p. 541

REFERENCES

1. *Biodiversity in Trust. Conservation and Use of Plant Genetic Resources in CGIAR Centres*; Fuccillo, D., Sears, L., Stapleton, P., Eds.; Cambridge University Press, 1997.
2. Managing Global Genetic Resources. *Agricultural Crop Issues and Policies*; Board on Agriculture National Research Council, National Academy Press: Washington, D.C., 1993; 449 pp.
3. Plucknett, D.L.; Smith, N.J.H.; Williams, J.T.; Anishetty, N.M. *Gene Banks and the World's Poor*; Princeton University Press: Princeton, NJ, USA, 1987.
4. WIEWS. http://apps3.fao.org/wiews/wiews/jsp.
5. http://singer.cgiar.org/.
6. The System-wide Genetic Resources Programme, **2000**. http://www.sgrp.cgiar.org/cgiar_geneticresources-germpla.html.
7. FAO. *The State of the World's Plant Genetic Resources for Food and Agriculture*; FAO: Rome, Italy, 1998; 510 pp.
8. IPGRI GRST. *Development of a Scientifically Sound and Financially Sustainable Global Genebank System. An SGRP-Supported Collaborative Activity. Final Report on the Technical Research Phase*; IPGRI: Rome, September 2001.

Germplasm Acquisition

David M. Spooner
University of Wisconsin, Madison, Wisconsin, U.S.A.

Karen A. Williams
USDA, Beltsville, Maryland, U.S.A.

G

INTRODUCTION

Plant genetic resources—varying from wild relatives of crops to farmers' varieties—form the raw materials of crop improvement programs. They have proven economic value to improve a wide range of traits, ranging from disease and pest resistances to improved yields and improved agronomic traits, and are critical to global food security. Their importance has long been recognized, and many national and international institutes have been set up to acquire, increase, store, distribute, and use them for crop improvement. This article briefly reviews the steps involved in the acquisition of germplasm, from the planning and conduct through the follow-up stages of germplasm collecting expeditions.

WHY COLLECT GERMPLASM?

Genetic resources, including landraces (farmer varieties) and wild relatives of our crops, are crucial to global food security. These resources are distributed worldwide, but with a concentration of diversity south of the tropic of Cancer (Fig. 1). The size and distribution of economic gains from yield increases in the major U.S. crops attributable to genetic improvements is impressive. Farmers have benefited from a 1% yield increase per year, half of this due to genetic improvements. Based on these assumptions, a one-time permanent estimation of the worldwide value to consumers of germplasm, as assessed by reduced food prices, is between \$8.1 billion and \$15.4 billion.[1] Although the United States enjoys 50–60% of these benefits, consumers in developing and transitional economies enjoy between \$6.1 and \$11.6 billion of the benefits. Whereas the ultimate goal of germplasm acquisition is crop improvement, a major subsidiary benefit is the availability of these collections for characterization studies (e.g., taxonomic studies) to aid breeders.

Germplasm may be acquired either by exploration or from existing germplasm collections. Exploration is a means of obtaining germplasm that does not exist in ex situ collections. For the major crops, relatively good coverage exists in some national programs and in the International Agricultural Research Centers (IARCs) of the Consultative Group on International Research (CGIAR). Examples are the International Rice Research Institute (IRRI) in the Philippines and the International Maize and Wheat Improvement Center (CIMMYT) in Mexico.[2,3] However, collecting needs remain, even in the major crops; and many minor orphan crops have major collecting needs.

HISTORY OF PLANT COLLECTING

Germplasm acquisition has a long history, with records of exchanges between cultures dating back thousands of years. International germplasm collecting expeditions in the 16th and 17th centuries were largely focused on exotic foods and ornamentals for botanic and university gardens. The Royal Botanic Gardens at Kew, England, established a widespread network of botanic gardens, resulting in the movement of enormous numbers of samples worldwide. Botanic gardens published new collections in seed lists that facilitated free worldwide exchanges of germplasm. The number of introductions brought into cultivation by botanic gardens exceeded 80,000.[4]

Private industry and gardening societies also collected, maintained, and exchanged germplasm, as did state and federal agencies. The early 19th century saw the development of major national collections, such as the All-Union Institute of Plant Introduction in St. Petersburg, Russia, later renamed the N.I. Vavilov All-Union Scientific Research Institute of Plant Industry (VIR). This institute, initiated by Nicolay Vavilov, sponsored collections worldwide and established an extensive system of national institutes to maintain, characterize, and use them. The U.S. Department of Agriculture (USDA) plant exploration program began formally in 1898 with the creation of the Section of Seed and Plant Introduction that evolved into an organized U.S. National Plant Germplasm System (NPGS). From 1898 to 2001, the USDA conducted 540 explorations, 80% to foreign countries.

Encyclopedia of Plant and Crop Science
DOI: 10.1081/E-EPCS 120010453

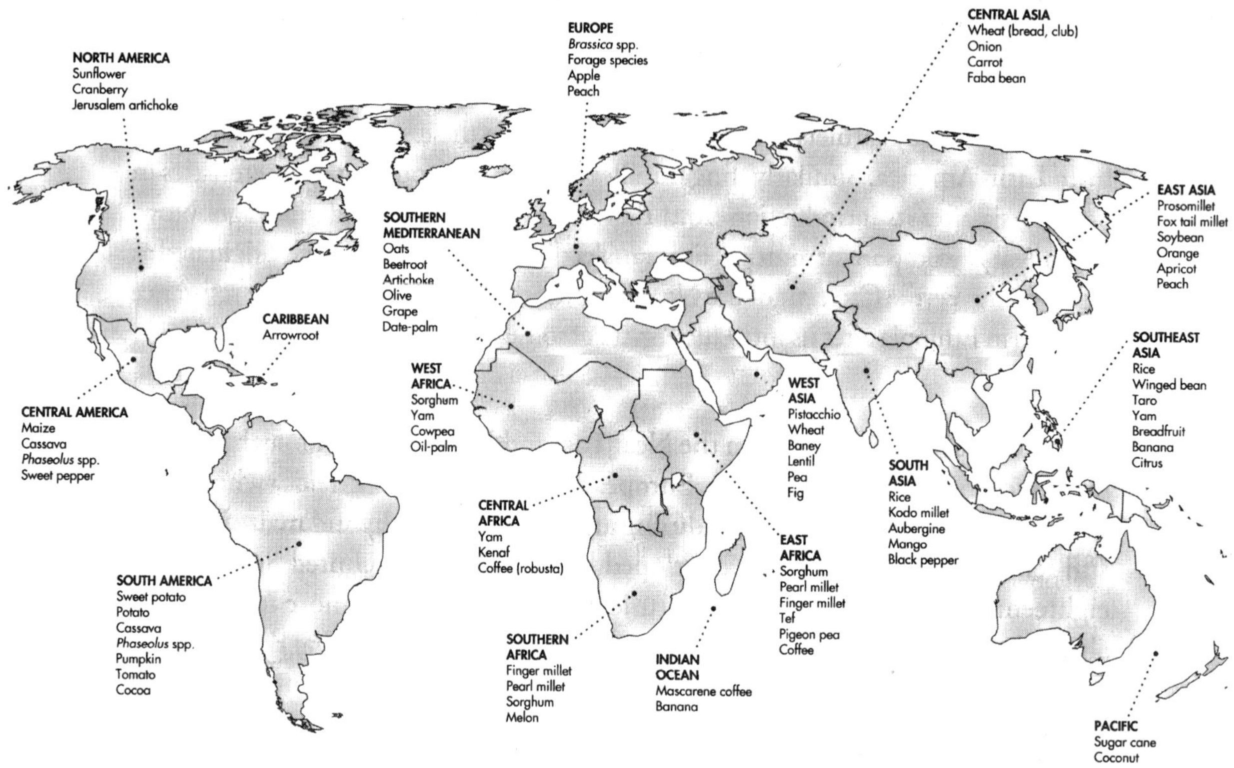

Fig. 1 Regions of diversity of major cultivated plants. (Figure from Ref. 2. Used with permission of the United Nations Food and Agriculture Organization.)

Explorations were made in every year except the war years, 1942–1945. Today the NPGS is the largest national germplasm system in the world, with 450,000 samples covering more than 10,000 species.[5] At least a quarter of the yearly distributions made by the NPGS are to scientists outside the U.S. National genetic resources programs are established in many other countries, and over six million collections are estimated to occur in ex situ genebanks worldwide.[2] In the future, the International Treaty on Plant Genetic Resources for Food and Agriculture (IT) will govern access to germplasm for Plant Genetic Resources for Food and Agriculture, which has been agreed to by country members of FAO and will enter into force upon ratification by the required number of countries. The implementation of the Treaty will determine the future ease of access.

ACCESS

In the initial stage of germplasm collecting, and continuing through the 1960s and 1970s, most germplasm was generally freely available. Exchange of plant genetic resources is now the subject of the IT, which was adopted by the FAO Commission on Plant Genetic Resources in November, 2001, and will become international law once it is ratified by 40 countries. The treaty establishes a multilateral system for the exchange of key crops and forage species. National laws of countries that are parties to the IT must be adjusted to conform. In some countries, national genetic resources programs are not yet established or their regulations have not yet been implemented. In others, access (including that required for plant exploration) has already been subject to new regulations. In order to address developing countries' concerns regarding benefit sharing, which is required by the new treaty, some developed nations are exploring the association of increased nonmonetary benefits, such as student training; the transfer of equipment, information; and technology; and other collaborative exchanges, with agreements on access.[6]

PLANNING AN EXPEDITION

Extensive background work is critical to the success of germplasm explorations, beginning with the definition of goals. Most explorations are initially intended to fill in gaps in current collections, as determined by taxonomic, ecological, or geographic criteria. Thus, an exhaustive search of available collections (national genebanks, international genebanks, and individual research collections) should be made before an exploration is contemplated. Some apparently available germplasm is not accessible because it is not increased, is diseased, or is otherwise restricted. Sometimes, discordant taxonomies of the same group make the task of comparing all sources of information difficult because of competing names and classifications (see an example in wheat at http://wheat.pw.usda.gov/ggpages/DEM/9IWGS/taxonomy.html and potato).[3] Searching for germplasm currently available in genebanks is simplified by the on-line availability of the germplasm in the genebanks of the CGIAR http://www.singer.cgiar.org/) and the U.S. NPGS http://www.ars-grin.gov/npgs/), but germplasm holdings of many smaller genebanks require direct correspondence with managers. The determination of all germplasm collections available is facilitated by indices of common collections held in common among genebanks,[7] although such indices are rare. Monographs (treatment of a particular group) and floras (treatment of all plants in a particular region) provide locality data, as well as identify regional herbaria. (An index of herbaria, *Index Herbariorum*, is available at http://www.nybg.org/bsci/ih/ih.html.)

Once the goals of an exploration have been identified, successful planning involves attention to access agreements, identifying and establishing contacts with knowledgeable in-country field collaborators, timing, logistics, and field equipment.[4,8,9] The new political climate can make access agreements difficult to obtain, and planning must begin well in advance. Agreements with host countries should clearly state details of the collection, export, increase, intended use, and distribution of germplasm. An experienced collaborator—ideally one who knows the logistics, culture, and crop—is essential. Timing of the expedition is crucial, especially for highly seasonable species for which differences in rainfall the prior or current year can drastically affect the growth of plants in a given year and the availability of germplasm. Germplasm collecting expeditions are expensive and time-consuming. Logistical challenges of some areas are often daunting, and investments in reliable equipment, especially vehicles, pay off. Field equipment needs vary according to the crop and terrain, but a geographic positioning system (GPS—determines latitude and longitude), altimeter (in mountainous areas), and maps are essential for reliable locality data. Some maps can be obtained only in country. Time must be allotted at the beginning of the exploration to purchase maps, visit herbaria, and meet local officials.

CONDUCTING THE EXPEDITION

Important considerations in the field include number of sites to visit, number of plants to sample, sampling techniques, and the number and type of propagules to sample from each plant.[4] These differ among species with different breeding systems and dispersal mechanisms. The number of collecting sites must be planned to maximize the amount of genetic variation sampled, within the constraints of time and funding. Marshall and Brown and others[4] proposed mathematical formulas for these that are useful when germplasm is common, as when collecting landraces. For many crops, populations are often so scarce that these theoretical calculations give way to the practicality of collecting sufficient germplasm of all populations encountered to ensure a successful germplasm increase, taking care to maintain populations intact in the wild. Collecting methodologies depend on a number of factors, including the biology of the targeted taxon and the objectives of the expedition. Different types of propagules, whether seed or vegetative, require different sampling and handling techniques in the field.

Local markets provide a relatively easy means of collecting landraces for many crops, and are important sources of information on the diversity of a crop in a given area. However, visits to local farmers provide better information on plant characteristics, uses, and cultural methods; higher quality germplasm; and varieties grown only for home consumption. Visiting farmers of different ethnicity is as important as visiting different ecogeographic regions.

Minimal data to be recorded when collecting germplasm should describe country, lower-level administrative units, locality, latitude, longitude, altitude, collector number, type of material (seed, vegetative, pollen, in vitro material), improvement status (wild, weedy, landrace), abundance, morphological description of the accession, and habitat. Additional data that may be collected include slope, aspect, landform, and various descriptors for soil, drainage, and vegetation. Such basic data collected in the field have been referred to as passport data (as used elsewhere in this volume). The more data collected the better, but the need for data must be balanced with time constraints and the need to visit additional sites. Some descriptors applying only to landraces include farmer name, cultural methods used by farmers, length and time of growing season, history of landrace, and traits perceived by the farmer and other users. Herbarium vouchers are also needed to document the collection, with enough duplicates to ensure that sets are left in the country and others left in recognized institutions upon return home. Collections ultimately are studied by a wide array of broadly defined prebreeding studies to guide the breeders in their use, as taxonomy, diversity screening, or resistance evaluations. Such studies vastly increase the value of collections, and can be greatly aided by the careful collection of the field data previously described.

CONCLUSION

Most agreements stipulate the deposition of duplicate germplasm and herbarium samples in the country, although others may state otherwise (e.g., collectors may be allowed to leave with all germplasm if the host country is provided with a sample of the first germplasm increase). Successful passage of samples through quarantine to the germplasm station is greatly facilitated by cleaning and processing samples properly. For example, herbarium specimens and fruits often harbor adult insects or larvae, so proper fumigation and seed extraction are crucial to successful introductions. Notes taken in the field are often scanty; it is best to write a complete report soon after the expedition, before facts are forgotten. Well-planned and conducted expeditions often merit publication in peer-reviewed crop-specific journals or journals devoted to germplasm collections and evaluations, such as *Genetic Resources Newsletter*, FAO. Such publication makes results of an expedition accessible to the world germplasm community. When published with cooperators, it generates goodwill and opens doors for continued collaboration.[10]

ARTICLES OF FURTHER INTEREST

Crop Domestication in Mesoamerica, p. 310
Crop Domestication: Fate of Genetic Diversity, p. 333
Crop Improvement: Broadening the Genetic Base for, p. 343
Genetic Resource Conservation of Seeds, p. 499
Genetic Resources of Medicinal and Aromatic Plants from Brazil, p. 502
Germplasm Collections: Regeneration in Maintenance, p. 541

REFERENCES

1. Frisvold, G.; Sullivan, J.; Raneses, A. Who gains from genetic improvements in U.S. crops? AgBioForum **1999**, *2* (3&4).
2. FAO. *The State of the World's Plant Genetic Resources for Food and Agriculture*; FAO: Romc, Italy, 1987.
3. Spooner, D.M. Exploring Options for the List Approach, Plant Genetic Resources for Food and Agriculture in situ and ex situ: Where are the Genes of Food Security Likely to Come From? In *Proceedings of an International Workshop, Interdependence and Food Security: Which List of PGRFA for the Future Multilateral System?*; Instituto Agronomico per L'Oltremare: Florence, Italy, 1998; 133–164.
4. *Collecting Plant Genetic Diversity. Technical Guidelines*; Guarino, L., Ramanatha Rao, V., Reid, R., Eds.; CAB International: Wallingford, U.K., 1995.
5. Board on Agriculture; National Research Council; National Academy of Sciences. *Managing Global Genetic Resources, the U.S. National Plant Germplasm System*; National Academy Press: Washington, DC, 1991.
6. Williams, K.A. Sharing the Benefits of Germplasm Exploration. In *People and Plants Handbook*; Martin, G.J., Barrow, S., Eyzaguirre, P.B., Eds.; Issue 7: Growing Diversity, 2001; 34–35.
7. Huamán, Z.; Hoekstra, R.; Bamberg, J. The intergenebank potato database and the dimensions of available wild potato germplasm. Am. J. Potato Res. **2000**, *77* (6), 353–362.
8. *Crop Genetic Resources for Today and Tomorrow*; Frankel, O.H., Hawkes, J.G., Eds.; Cambridge University Press: Cambridge, 1975.
9. Plucknett, D.L.; Smith, N.J.H.; Williams, J.T.; Anishetty, N.M. *Gene Banks and the World's Food*; Princeton Univ. Press: Princeton, NJ.
10. Kloppenburg, J.R. *First the Seed: The Political Economy of Plant Biotechnology*; Cambridge Univ. Press: Cambridge, 1988.

Germplasm Collections: Regeneration in Maintenance

Linda Pollak
Iowa State University, Ames, Iowa, U.S.A.

INTRODUCTION

When seed samples are collected from populations and saved in a germplasm bank, they are maintained as separate accessions (entries) along with the passport data gathered during collection. When the seed viability decreases or supply falls short due to inability to meet user demand, regeneration is necessary. Vegetatively propagated plants have special problems of regeneration. The goal of regeneration is to preserve the genetic variability that existed in the population when it was collected.

QUALITY CONTROL MAINTENANCE OF SEED-PROPAGATED COLLECTIONS

Allard[1] gives four primary reasons why genetic variability changes during maintenance or storage: differential survival rate during maintenance, selection during regeneration, outcrossing with other accessions, and genetic drift. There are substantial differences in the ability of different genotypes within a species to survive in storage, drastically altering the genetic composition of an entry.

During regeneration, selection can take place even under conditions designed to maximize survival and minimize selection, and can bring about substantial change in genetic composition.[1] Over several cycles of regeneration, the entry could have little resemblance to the parent collected in nature. Many of the reasons why unintended selection takes place are environmental and are a result of regenerating accessions in an environment to which they are not adapted.

In outcrossing species, outcrossing with other entries can be a problem when pollination is not carefully controlled.[1] Outcrossing can also be a problem with inbreeding species (self-pollinators) where pollination is not normally controlled, because a small percentage of outcrossing can still occur in self-pollinators. This can be a problem when regenerating a very large collection, because the percentage of outcrossing can vary among accessions. It is not normally feasible to control pollination in self-pollinators because it is usually difficult and time consuming, and therefore expensive.

When sample size of an entry is small (i.e., few seeds), there can be drift toward fixation of alleles (loss of variability) during regeneration in both outcrossing and inbreeding species.[1] Fixation is largely independent of selective value, so many entries become fixed for deleterious alleles and are difficult to maintain. If the collection is large, however, genetic variability can be preserved because genes will become fixed randomly in entries.

UNINTENDED SELECTION WHILE REGENERATING UNADAPTED ACCESSIONS

One needs to be aware of the breeding system of the species being regenerated regarding overdominance, additive effects of blocks of genes, level of ploidy and genome structure (auto or allopolyploidy), vegetative behavior (annual or perennial), and competitive ability at various stages of development.[2] These characteristics are interactive so that the population can flexibly react to selective pressure in its natural environment. Therefore, all elements of this system have to be taken into account carefully when we preserve and regenerate the population in order to avoid inadvertent changes in the genetic makeup, which would result in permanent loss of some variability.

Transfer of ecotypes from original conditions to a collection center disturbs normal selection pressures.[2] When artificial cross-pollinating is necessary, plant density needs to be optimal, and limiting factors of the environment not present in the area from which the population derives must be controlled (including abnormal frost, drought, heat, and pathogenic attack).

The primary environmental factors influencing flowering are day length (photoperiod) and temperature.[3] In photoperiod-sensitive species, such as maize, photoperiod determines the location where successful regeneration of short-day accessions can take place. Response to day length is also influenced by temperature. Most plants have critical minimum and maximum temperatures, and temperature's effect on floral initiation and development is highly variable, even within species. Many species exhibit some form of dormancy.[4] For example, some winter annuals and biennials require exposure to low temperatures before reproductive development can be initiated. Multiple planting dates while regenerating may restrict subpopulation segregation by day length or maturity, and thus maintain population outcrossing.

Encyclopedia of Plant and Crop Science
DOI: 10.1081/E-EPCS 120010455

For wind-pollinated species problems can arise from a change in weather conditions.[2] Pollen shedding is dependent on temperature, humidity, day length, etc. Under other climatic environments it could be reduced, and self-pollination would increase abnormally.

Adequate soil moisture is necessary for growing vigorous plants with minimal flower and seed abortion.[3] Moisture stress can reduce seed production if the stress occurs during flowering and early seed filling. Moisture stress can also influence seed set of species that develop their fruit below ground. Maturity can be hastened by reducing available water toward the end of the growing season. Relative humidity is important at pollinating, but there is an upper limit above which pollen clumping and disease infections can occur. Successful flowering and seed set occur when soil fertility is maintained at a level that will produce healthy green plants.[3] However, some plants have special nutrient requirements for seed development.

The following examples illustrate what can happen when accessions are regenerated in an environment without the natural pressures that occur in their environment of adaptation.[2] In one example, insect pollinators have specific habits that allow pollination of plants normally pollinated by the insects. When transferring these plant populations we also have to consider transferring their original population of pollinators.[2] There can be drastic consequences on the plant progeny when, after transfer to new conditions, the plant population does not have the proper pollinator. The problem can be alleviated by enclosing the regenerated population in screened cages with the appropriate insect, if it is available.[4] Another example occurs in tropical grain crops or forages, where there are species complexes consisting of wild and cultivated varieties that undergo coevolution by permanent exchange of genic flow.[2] Weed forms can give variability and fitness to cultivated forms, or compete in eliminating weaker phenotypes. Weeds contribute as pollinators but escape harvest, so crop collections do not include the specific weed population. After regeneration of these accessions, a bias occurs in the reproduced cultivar if the weed pollinator is not available.

Populations also coexist with an agricultural system of specific biological, chemical, and physical environments.[2] There is a negative correlation between the sophisticated control system in the most intensive agricultural systems and the genetic diversity of the plant. When regenerating populations under management techniques that optimize crop production from an economic point of view, there may be effects causing reduced diversity. Thus the agricultural system must also travel with the population.

Changes in Allele Frequency

Allelic frequency is the percentage of a particular allele at a given locus in a population gene pool, considering all alleles at that locus. While regenerating an accession, it is very important to maintain the frequency of alleles so that rare alleles are not lost. It is less important to preserve a representative sample of the target species than to preserve at least one copy of each of the different alleles.[5] The common alleles are of far greater interest to plant breeders, but rare alleles may be important for special uses. Losses of samples and alleles can be controlled by judicious management of the collection and reduced to acceptable levels.

Allelic diversity in accessions decays during regeneration, based on three variables: the size of the population grown for regeneration; the mating system; and the variation in the number of gametes per plant.[6] Procedures for maintaining accessions should be chosen based on optimizing these variables within the limited amount of resources available. Population size is the most important variable, but if populations of the needed size are not practical due to a large number of accessions that need to be maintained, the other two variables become more important.

Yonezawa et al.[6] found that a single-seed-type regeneration (each plant leaves one progeny) gave the largest effective population size in regenerating seed of moderately or highly selfing species, whereas a biparental regeneration (plants are pollinated in pairs with one offspring left from each of the paired plants) gave the largest effective size for outcrossing species. However, differences between the systems were not appreciably large unless the accessions were regenerated over 10 or more cycles with 50 or fewer plants. They suggest that a single-seed system combined with selfing may be the most effective procedure for regenerating the seed of outcrossing species if the species is self-compatible and sufficiently tolerant to inbreeding depression.

Whenever possible, regeneration procedures should control the number of pollen parents through controlled hand pollination and control the number of female parent gametes by harvesting equal numbers of kernels from each seed plant.[7] When pollen and seed parents are controlled during regeneration the effective population size is twice the size of the original population. Crossa et al.[7] recommended a sample size of 150–350 maize plants to capture alleles at frequencies of 0.03–0.05 or higher in each of 150 loci, with a 90–95% probability.

Some studies show that genetic drift is not a large problem. In comparing three methods of pollination (plant to plant, chain crossing, and mixing pollen) there were no significant differences among frequencies observed using seeds of maize accessions with floury and flint endosperm characteristics in frequencies of 0.97 and 0.03, respectively.[8] Wheat microsatellite markers were used to analyze eight bulks of seeds stored more than 50 years in a seed reference collection at room temperature and regenerated up to 24 times.[9] No contamination due to foreign pollen or incorrect handling was discovered. In one accession genetic drift was observed, and in another heterogeneity for two markers was maintained. This study showed microsatellites to be a simple and reliable marker system for the verification of the integrity and genetic stability of gene bank accessions. Biotechnology methods have primarily been of benefit for crops difficult to maintain, however, such as those requiring vegetative propagation and those with recalcitrant seeds that cannot be dried to low levels for optimum storage.[10]

Vegetatively Propagated Plants

Vegetatively propagated germplasm is the most difficult and expensive to regenerate, and generally requires considerable space.[4] Although some of these crops are sexually fertile, it is often not convenient to propagate them commercially from seed because of high levels of heterozygosity, and because breeders require uniform clones. Many are sterile or polyploids, or have reduced fertility.

Vegetative organs are short lived and deteriorate rapidly after harvest, unless stored in ideal storage conditions.[11] Annual regeneration is costly, and there is danger of disease infection. It is extremely difficult to keep vegetatively propagated plants free from viruses, which leads to degeneration of clonal stocks. Because of these special problems, tissue culture or in vitro techniques are used.[11] Keeping the cultures in ultra-low temperatures opens the possibility of storing the germplasm indefinitely. Tissue culture has a major advantage, in that a large number of genotypes can be stored in a relatively small area at a fraction of the cost of growing the material in the field. The two major problems with using tissue culture are high levels of somaclonal variation, which occurs when tissue is regenerated into seedlings, and the limited length of storage time before regeneration is required. Improved tissue culture techniques are leading to lower somaclonal variation levels, and work is proceeding on cryopreservation that will allow tissue to be preserved for long periods.[10]

Cost-Benefit Ratios of Regeneration

The labor requirement of regeneration is reduced as storage conditions improve.[1] If materials are extremely similar they can be combined as more is learned about them. We should aim to conserve sufficient stocks of each species to saturate the breeder's capacity to evaluate and utilize the conserved germplasm, both now and in the future.[5] Thus, evaluation and utilization should be the major limiting factors in germplasm conservation.

G

ARTICLES OF FURTHER INTEREST

Agriculture and Biodiversity, p. 1
Breeding: Incorporation of Exotic Germplasm, p. 222
Crop Domestication: Fate of Genetic Diversity, p. 333
Genetic Resource Conservation of Seeds, p. 499
Germplasm Acquisition, p. 537
Germplasm Maintenance, p. 544
Germplasm: International and National Centers, p. 531

REFERENCES

1. Allard, R.W. Problems of Maintenance. In *Genetic Resources in Plants – Their Exploration and Conservation, IBP Handbook no. 11. 554 pp*; Frankel, O.H., Bennett, E., Eds.; Blackwell Scientific Publications: Oxford, 1970; 491–494.
2. Demarly, Y. *Seed Regeneration in Cross-Pollinated Species*; Porceddu, E., Jenkins, G., Eds.; A.A. Balkema: Rotterdam, 1982; 7–31.
3. Major, D.J. Environmental Effects on Flowering. In *Hybridization of Crop Plants*; Fehr, W.R., Hadley, H.H., Eds.; ASA-CSSA: Madison, WI, 1980; 1–15.
4. Forsberg, R.A.; Smith, R.R. Sources, Maintenance, and Utilization of Parental Material. In *Hybridization of Crop Plants*; Fehr, W.R., Hadley, H.H., Eds.; ASA-CSSA: Madison, WI, 1980; 65–81.
5. Marshall, D.R. *Crop Genetic Resources for Today and Tomorrow, Int. Biological Programme 2*; Frankel, O.H., Hawkes, J.G., Eds.; Cambridge University Press: Cambridge, 1975; 53–80.
6. Yonezawa, K.; Ishii, T.; Nomura, T.; Morishima, H. Effectiveness of some management procedures for seed regeneration of plant genetic resources accessions. Genetica **1996**, *43*, 517–524.
7. Crossa, J.; Taba, S.; Eberhart, S.A.; Bretting, P.; Vencovsky, R. Practical considerations for maintaining germplasm in maize. Theor. Appl. Genet. **1994**, *89*, 89–95.
8. Pantuso, F.; Ferrer, M.; Eyérabide, G.; Suárez, E. *Evaluation of Regeneration Methods for Maize Conservation in Germplasm Banks*; Proc. VII Congreso Nacional de Maiz: Pergamino, Argentina, 2001.
9. Bomer, A.; Chebotar, S.; Korzun, V. Molecular characterization of the genetic integrity of wheat (Triticum aestivum L.) germplasm after long-term maintenance. Theor. Appl. Genet. **2000**, *100*, 494–497.
10. Rao, V.R.; Riley, K.W. The use of biotechnology for conservation and utilization of plant genetic resources. Plant Genet. Resour. Newsl. **1994**, *97*, 3–20.
11. Ford-Lloyd, B.; Jackson, M. *Plant Genetic Resources: An Introduction to their Conservation and Use*; Edward Arnold Ltd: London, 1986; 61–68.

Germplasm Maintenance

Henry L. Shands
USDA–ARS National Center for Genetic Resources Preservation, Fort Collins, Colorado, U.S.A.

INTRODUCTION

The apparently mundane task of maintaining germplasm[a] for a genebank[b] is anything but mundane. Huge responsibilities are incumbent upon all those involved in the processing of germplasm, from acquisition to distribution. Scientists expect a high-quality and adequately documented product. To ensure quality in germination, freedom from disease, and adequate documentation, managers must examine all aspects of management, from initial receipt of seed to storage conditions.[1]

SOURCE OF GERMPLASM

Sites receiving germplasm must review the process through which they received it because authenticity is essential. Plant collecting is most easily documented through direct collectors[2] using Global Positioning System (GPS) locators and herbarium specimens. When it is received from another genebank's collection, there may be gaps in the existing information, especially if the donor genebank is not the original source. If there are many points of handling between field source and genebank, there are more chances for errors in the documentation and integrity of the material. This is especially true if the sample has had multiple growouts since being acquired. End users must assess the source and determine their risk in data and germplasm accuracy. Maintenance of a nonoriginal source of germplasm is only as good as the many intervening waypoints.

Resource Documentation

If the material is not collected first-hand, documentation is often absent, inaccurate, or deficient. Most samples acquired through exchanges include little information provided with the sample.[3] For some end users, this is a minor omission but for others, it is of serious concern. Breeders screening for a specific trait will evaluate the sample for that trait while assessing other virtues. To the extent possible, those data will be put into the database. Curators will assess all regular passport data characters and qualities and document them under their growout conditions to confirm and compare with any original passport data. The Food and Agriculture Organization of the United Nations (FAO) International Treaty on Plant Genetic Resources[4] requires that data be supplied to the requester. If data are not available, however, the requirement is hollow.

Field Growouts of Seed Crops

For the curator to properly maintain germplasm, sufficient critical data must be provided relative to the initial sample. Wild relatives of crop plants may be uniquely restricted to a specific point of origin but collectors try to collect many different samples throughout an area. A landrace would likely be more variable than a bred variety.

If a landrace acquisition is not the original sample, the curator must assess what population size and environmental biases have already occurred or been introduced. A recommended population size is dependent on numerous factors, including pollination method (selfing or outcrossing), heterogeneity of population, and goals of the curator to purify the sample or retain any heterogeneity or heterozygosity.[5]

If the sample has been regenerated under curator conditions (in contrast to in a farmer's field), how many plants were grown and how were they pollinated? Unknown contamination from foreign pollen of like or compatible species is a concern to curators. To maintain the genetic integrity of the sample, the curator must take measures such as maintaining appropriate isolation distances or bagging to prevent contamination. The isolation distance is dependent on method of pollination (whether windblown or insect commuted), flowering type (whether open or closed, self or cross-pollinated), and, for some species, whether self-compatible or not. If pollination is carried out by hand, isolation distance is immaterial but some of the other factors remain critical.

If the material is from a tropical or subtropical environment, what biases have been or are being introduced

[a]Germplasm—a whole organism or any propagule of genetic resources that can be stored in ex-situ or in-situ conditions and used to identify, multiply or preserve the original individual's genetic identity.
[b]Genebank—a repository for germplasm.

Encyclopedia of Plant and Crop Science
DOI: 10.1081/E-EPCS 120010454

by growing it in a temperate zone? When it is grown outside its normal eco-zone the population will likely be biased due to a truncated flowering period. Photoperiod-induced late flowering and subsequent seed set and seed development may be impaired and not represented in the harvest sample. The flowering date of the next population may thus be biased to be earlier, and the population may lose genes and traits physically associated with linkage groups containing flowering factors.

Some concern is raised about maintaining base collection samples free from gene constructs of genetically modified organisms (GMOs). Where regeneration occurs in areas where commercial GMO crops are grown, adventitious GMOs may result from cross-pollination.[6] When base collection supplies are replaced by regenerated samples that possibly contain GMOs, notation to database records should be made. As with any contamination of the base sample, future regenerations will likely further deviate from the original description. The deviation will reflect any selective advantage of the GMO even if the regeneration site is removed from areas of GMO crops. Field collections of non-GMO clonally propagated crops would not normally be affected by the presence of new fields of GMO cultivars of the same species, or be affected by the same issues as seed crops as long as their reproduction is through the somatic buds of the original clone or its sports. However, seed sources would likely be affected.

Field, Laboratory, and Greenhouse Maintenance

Regeneration is one of the most costly operations for a genebank. This cost is not only in time and money but also in the risks involved. In all aspects of genebank management, clean seed/propagule practices must be followed expressly in seed room and laboratory preparation, in hand or mechanical planting, in hand bagging and pollination, and in harvest and seed processing techniques. Precision of accession and location information can be assisted with bar coding, but labels must be carefully applied. Mixtures from improper human procedure are as problematical as those from weather-related events. Torrential rains—whether alone, of associated with tropical storms, or coupled with tornadoes' destructive winds—can move or destroy plants in the field, greenhouse, or laboratory. Maps of exact plantings should be safely stored for reference. Maps of clonal orchards are critical to reconstructing locations of individual plants or trees. Whatever the causal event for the mechanical mixture of seed, propagules, pollen, or plants, the result is the same—an inferior product to deliver to future generations. Preventive measures should be taken to minimize the risk of mixture and loss of genetic purity at all steps.

GENETIC STOCK MAINTENANCE

Not all germplasm is varietal material. Genetic stocks as a group include various gene mutations, chromosomal aberrations (such as translocations, additions/deletions, and inversions), and even increases or decreases in levels of ploidy of one or more of the chromosomes or whole chromosome sets.[7] For most of these stocks, some genetic or cytogenetic analysis must be made of the resultant progeny to confirm the changes contained therein. A growout test is usually necessary to confirm the presence of a lethal mutant that must be maintained in a heterozygous condition with its allele. In all cases, the subsequent generation is analyzed to validate the genetic constitution of the progeny of the original regeneration. One must be careful not to expend valuable seeds in needless growouts. Also, a competent analysis is often required by a skilled scientist. Seed distribution is often limited to a few seeds due to their unique nature and cost. Genetic stocks are thus found mostly at special stock centers where the necessary expertise is available to maintain them. The same stock centers may also maintain the related wild relatives used in improving the crop. Scientists will often conduct interspecific and intergeneric crosses to move potentially useful alien genes and chromosomes into the cultivated form. This differs from the GMO issue discussed earlier because the generations subsequent to the crosses are subject to natural chromosome rejection, and few plants survive or lead to useful crop plants. The process is quite random—more in the realm of research than in the realm of contamination, as in the case of GMO field crops.

Storage of Germplasm

Storage is often the most cost-effective process in a genebank's operation.[8] Proper storage helps abate the high cost of regeneration and risk of loss of genetic integrity. Seed storage protocols for any seed crop begin at the time of pollination when procedures are put in place to assure accuracy of bagging notation. Insecticides are often dusted in the bags to reduce loss to predator insects that find the warm, moist environment enticing for feeding and reproduction. Seed maturity and drydown are important, so harvest must be made timely and seeds must be dried to remove excess moisture.[9–11] Overheating in dryers and stacking in greenhouses without aeration or elsewhere where rodents and other animals can do damage is obviously poor management.

Once at the genebank, the seed material must be dehydrated through a combination of heat and dehumidification to reach optimum seed storage moisture.[11,12] The moisture level is critical prior to placing seed in sealed foil bags or in liquid nitrogen vapor (cryopreservation). Data are evolving to suggest that long-term storage in liquid nitrogen vapor has merit over the conventional 20°C in extending seed life (Walters, personal communication). In some cases too little moisture will result in damage in storage.[12] Data are available for many seed crops showing that lipids are a critical determinant of optimum seed storage moisture[10,12] and that lipids influence aging kinetics[10,13] and cell damage during cryopreservation.[14]

Clonally propagated material in tissue culture must be carefully managed to reduce transfers and opportunity for somatic mutation.[15] Cool storage and experimentation with culture media to extend storage life before transfer are essential. Cryopreservation for vegetative tissue of a limited number of species has only begun, and new protocols are evolving, giving hope that life expectancy can be improved and samples stored for long periods of time.[15,16]

Although DNA storage has been given considerable discussion, there is little evidence that it will play a large role in genebank storage. It obviously will play a role in helping to validate genetic change and to elucidate individual's genetic makeup for many purposes in the future.

CONCLUSION

Germplasm maintenance requires implementation of many pieces of information, careful techniques of propagation, and rigidly employed laboratory-type sanitation for stocks and their environment. Anything less subjects the germplasm to loss of genetic integrity and truism, and incorporates imperfections to pass on to future generations.

ACKNOWLEDGMENTS

The author wishes to acknowledge and thank Dr. Christina Walters[10] for her suggestions and comments on this article.

REFERENCES

1. Sackville Hamilton, N.R.; Chorlton, K.H. *Regeneration of Accessions in Seed Collections: A Decision Guide*; Hankbook for Genebnaks, International Plant Genetic Resources Institute: Rome, 1997; Vol. 5.
2. Purdue, R.E.; Christenson, G.M. Plant Exploration. In *Plant Breeding Reviews*; The National Plant Germplasm System, Timber Press: Portland, 1989; Vol. 7, 67–94.
3. Roath, W.W. Evaluation and Enhancement. In *Plant Breeding Reviews*; The National Plant Germplasm System, Timber Press: Portland, 1989; Vol. 7, 183–211.
4. FAO. *International Treaty on Plant Genetic Resources for Food and Agriculture*; Food and Agriculture Organization of the United Nations: Rome, 2001. (ftp://ext-ftp.fao.org/waicent/pub/cgrfa8/iu/ITPGRe.pdf Article 12.3 (c).
5. Clark, R.L. Seed Maintenance and Storage. In *Plant Breeding Reviews*; The National Plant Germplasm System, Timber Press: Portland, 1989; Vol. 7, 95–128.
6. Eastham, K.; Sweet, J. *Genetically Modified Organisms (GMOs): The Significance of Gene Flow Through Pollen Transfer*; European Environment Agency: Copenhagen, 2002. (http://www.eea.eu.int).
7. Shands, H. Plant genetic resources conservation: The role of the gene bank in delivering useful genetic materials to the research scientist. J. Heredity **1990**, *81*, 7–10.
8. Eberhart, S.A.; Roos, E.E.; Towill, L.E. Strategies for Long-Term Management of Germplasm Collections. In *Genetics and Conservation of Rare Plants*; Falk, D.A., Holsinger, K.E., Eds.; Oxford University Press, 1991; 135–148.
9. Justice, O.L.; Bass, L.N. *Principles and Practices of Seed Storage*; USDA Ag Handbook, US Government Printing Office: Washington, 1978; Vol. 506, 1–289.
10. Walters, C. Understanding the mechanisms and kinetics of seed aging. Seed Sci. Res. **1998a**, *8*, 223–244.
11. Walters, C.; Towill, L.E. Seeds and Pollen. In *The Commercial Storage of Fruits, Vegetables, and Florist and Nursery Stocks*; Gross, K., Ed.; USDA Ag Handbook, US Government Printing Office: Washington, 2002; Vol. 66.
12. Walters, C. Ultra-dry technology: Perspective from the National Seed Storage Laboratory U.S.A. Seed Sci. Res. **1998b**, *8* (Suppl. 1), 11–14.
13. Priestley, D.A. *Seed Aging, Implications for Seed Storage and Persistence in the Soil*; Cornell University Press: Ithaca, 1986. ISBN 0-8014-1865-8.
14. Vertucci, C. Effects of cooling rate on seeds exposed to liquid nitrogen temperatures. Plant Physiol. **1989**, *90*, 1478–1485.
15. Withers, L.A.; Engleman, F. In Vitro Conservation of Plant Genetic Resources. In *Biotechnology in Agriculture*; Altman, A., Ed.; Marcel Dekker Inc.: New York, 1997.
16. Engelmann, F. In Vitro Conservation Methods. In *Biotechnology and Plant Genetic Resources: Conservation and Use*; Ford-Lloyd, B.V., Newbury, H.J., Callow, J.A., Eds.; CABI: Wallingford, UK, 1997; 119–162.

Glycolysis

Florencio E. Podestá
Universidad Nacional de Rosario, Rosario, Argentina

INTRODUCTION

Glycolysis is a universal metabolic route present in all living cells, at least in part. It is commonly defined as the catabolic pathway leading to the conversion of sugars into pyruvate.

$$\text{glucose} + 2\ \text{ADP} + 2\ \text{Pi} + 2\ \text{NAD} \rightarrow 2\ \text{pyruvate} + 2\ \text{ATP} + 2\ \text{NADH}$$

As presented, this equation, which applies well to animal or yeast glycolysis, implies a simple, straight path of degradation from glucose to pyruvate. This assertion has become increasingly outdated, when applied to plants, by the vast amount of new information gathered within the past three decades of research.

DISTINCTIVE FEATURES OF PLANT GLYCOLYSIS

Research in the carbohydrate metabolism area has provided unequivocal information showing that plant glycolysis has distinctive features that encompass the existence of unique enzymes, the use of pyrophosphate instead of ATP as phosphate donor, and a different mode of regulation.[1–3] In plants, carbon may enter glycolysis from two pools: the hexose phosphate pool and the triose phosphate pool.[2] The first is fed by sucrose or starch degradation. The second pool is the result of hexose-phosphate degradation, photosynthetic carbon fixation, or the action of the oxidative pentose phosphate pathway.[2] Additionally, the presence of enzymes specific to plant glycolysis can change the final product and/or the ATP yield.[1] Having these facts in mind, it is easy to visualize that the glycolytic metabolism in plants is not a mere linear sequence of reactions leading from one substrate to a final product; on the contrary, the actual substrate and products will depend on the metabolic status of the plant tissue at that particular time.

Plant Glycolysis Is Compartmentalized

Unlike heterotrophic organisms, plants are able to synthesize hexoses through photosynthesis and thus do not depend on an external carbon source. In chloroplasts, carbon is fixed and transiently stored as starch. In nonphotosynthetic tissue, plastids also serve as carbon stores, using imported carbohydrate that comes in the form of sucrose. Thus, at least part of the glycolytic machinery has to be used in plastid carbon metabolism to degrade starch products and provide ATP, NADH, pyruvate for fatty acid synthesis, triose phosphates that are exported to the cytosol, and intermediates for secondary metabolism.[1]

Meanwhile, cytosolic glycolysis is the source of pyruvate for the tricarboxylic acid cycle, with the concomitant production of ATP and NADH. It also generates the carbon skeletons needed in N assimilation through associated anapleurotic reactions and the PEP used in primary carbon fixation in crassulacean acid metabolism and C_4 plants.[4] The plastid and cytosolic hexose-phosphate and triose-phosphate pools are connected through the Pi translocator of the inner plastid membrane.

It is of paramount importance to realize that glycolysis in plants goes far beyond being a merely catabolic metabolism designed to obtain energy and reductive power from hexoses. Its products serve as precursors for several anabolic pathways leading to the synthesis of amino acids, lipids, and other cell constituents. The separation of roles and operational timeframe among the cytosol and plastid carbon metabolism calls for different regulatory needs, and compartmentation provides a means of integrating different metabolic pathways with diverse control requirements.

THE OPERATION OF PLANT GLYCOLYSIS

Individual Reactions

Beyond a description of each of the reactions involved in glycolysis, this section presents a summary of the main features that characterize plant glycolysis and differentiate it from animal and yeast counterparts, offering an account of the recent knowledge regarding its particular chemistry and mode of regulation.

The first step of glycolysis consists of the phosphorylation of hexoses that will then enter the hexose phosphate pool (Fig. 1). Hexose phosphates are produced

Encyclopedia of Plant and Crop Science
DOI: 10.1081/E-EPCS 120010403

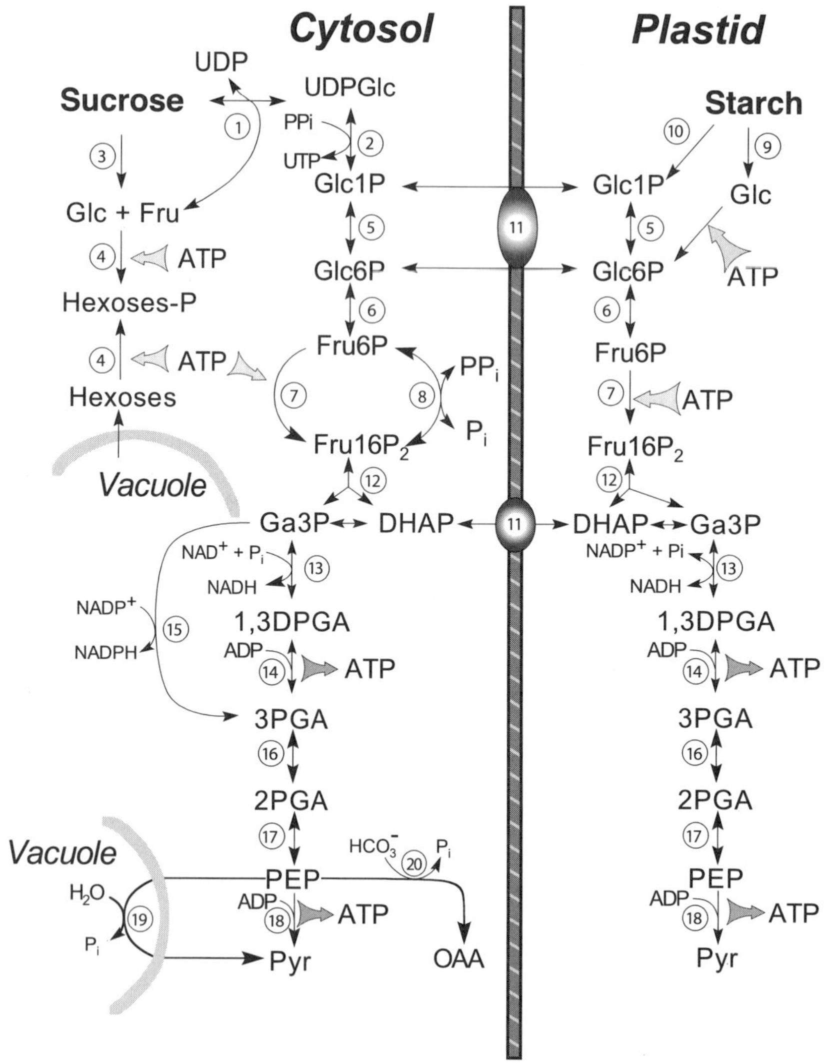

Fig. 1 Glycolysis in plant cells. Individual reactions are catalyzed by the following enzymes or transporters: (1) sucrose synthase; (2) UDPglucose pyrophosphorylase; (3) invertase; (4) hexokinase; (5) phosphoglucomutase; (6) phosphoglucoisomerase; (7) PFK; (8) PFP; (9) amylase; (10) starch phosphorylase; (11) phosphate translocator; (12) aldolase; (13) glyceraldehyde-3-phosphate dehydrogenase; (14) 3-phosphoglycerate kinase; (15) non-phosphorylating glyceraldehyde 3-phosphate dehydrogenase; (16) phosphoglyceromutase; (17) enolase; (18) pyruvate kinase; (19) PEP phosphatase; (20) PEP carboxylase. (*View this art in color at www.dekker.com.*)

in the cytosol during degradation of sucrose by sucrose synthase and UDP glucose pyrophosphorylase. The action of invertase gives rise to hexoses, which are phosphorylated by a hexokinase. As in animal cells, plant hexokinase has also been reported to be bound to mitochondria.[2] Cytosolic hexokinase has also been implicated in the sugar-sensing pathway that governs specific carbohydrate metabolism gene expression.[5] Plastid hexokinase is used to phosphorylate glucose produced by amylolytic starch degradation, although its presence is not universal in all plants.

The hexose phosphate pool exists in a thermodynamic equilibrium maintained through phosphoglucomutase and phosphoglucoisomerase, which catalyze reversible reactions. The next step is where one of the crucial differences between animal and plant glycolysis surfaces. The phosphorylation of Fru-6-P in the cytosol can be achieved by two different enzymes: the classical PFK reaction using ATP, or the alternative reaction that uses PP_i as phosphoryl donor, catalyzed by the pyrophosphate-dependent PFK or PFP (Fig. 1).

The latter reaction is exclusive to the plant cytosol, and although it is also present in some bacteria and unicellular parasites it has not been found in animals or yeast.[1] PFP, unlike PFK, catalyzes a reversible reaction, which is remarkable considering that the Fru-6-P/Fru-1,6-P_2

interconversion is a key regulatory step in glycolysis. The presence of PFP and the fact that PP_i concentrations remain stable in the plant cytosol under a variety of conditions make this reaction a bypass to that catalyzed by PFK (and FBPase in the opposite direction), eliminating the use of one ATP molecule. Although its precise role in plant metabolism is not yet clearly defined, it could fulfill a role in the glycolytic breakdown of sugars under nutrient stress conditions.[6] In certain CAM plants, where a massive flux of carbon through the glycolytic pathway is required, PFP activity can exceed PFK by an order of magnitude.[7] Recent studies also hint at this enzyme being a potentially important target for glycolysis regulation.[8]

Fru-1,6-P_2 is converted to triose phosphates by aldolase. Plant aldolases, both cytosolic and plastidic, are structurally related class I aldolases. Although not a regulatory enzyme, aldolase plays an important role in photosynthesis control.[9]

Upon interconversion of DHAP in Ga3P, the ATP-yielding phase of glycolysis starts (Fig. 1). Ga3P is oxidized to 1,3DPGA by GaPDH, which conserves the energy obtained from oxidation by the formation of a high-energy mixed anhydride between the phosphate and a carboxyl group. In this reaction, reductant is formed as NADH in the cytosol or NADPH in plastids. This energy-conserving step may be bypassed by an enzyme unique to the plant cytosol, the nonphosphorylating GaPDH. It uses $NADP^+$ but does not incorporate a phosphate group, rendering 3PGA that cannot be used to produce ATP.[6]

The following reactions do not offer substantial differences from classical glycolysis, and they lead to the formation of PEP through the combined action of 3PGA kinase, phosphoglyceromutase, and enolase.

The metabolism of PEP is a metabolic branchpoint, both in plastids and in the cytosol. In plastids, PEP is used by pyruvate kinase to obtain pyruvate and ATP, or it may enter the aromatic amino acid and secondary metabolites biosynthetic pathway.[1]

In the cytosol, the situation is more complex. A cytosolic pyruvate kinase may also catalyze the energy-conserving reaction that yields ATP and pyruvate, but PEP can also be used by the ubiquitous cytosolic enzyme PEP carboxylase, which carboxylates PEP to oxaloacetate. PEP carboxylase exists in various isoforms. One is a housekeeping enzyme that has an anapleurotic role and is implicated in providing building blocks for amino acid synthesis; another is a C_4 or CAM plants-specific isoform, present at high levels and implicated in the auxiliary CO_2 fixation mechanisms specific to these plants.[10] The PEP carboxylase and pyruvate kinase reactions are strictly coordinated through different regulatory mechanisms. Additionally, a third enzyme catalyzes PEP hydrolysis. This enzyme, PEP phosphatase,[11] has no certain function, but it has been implicated in a bypass reaction of carbon metabolism under Pi stress, as mentioned above for PFP, in which the nonphosphorylating GaPDH[6] also intervenes.

Regulation

Regulation of plant glycolysis follows a notably different strategy from that observed in other organisms. One of the main differences is the role of the signal metabolite Fru-2,6-P_2. In animals, this compund activates PFK and inhibits FBPase. In plants, Fru-2,6-P_2 does not affect cytosolic PFK, whereas it strongly activates PFP and inhibits FBPase.[1,2] Since PFP is saturated by cytosolic Fru-2,6-P_2 levels, its participation in regulating carbon metabolism in vivo has been argued. Recent work shows that physiological concentrations of metabolites—in particular Pi—reduce the affinity of PFP for Fru-1,6-P_2 to the extent that physiological variations of this metabolite could affect glycolysis in vivo.[8] Although plant PFKs are not responsive to Fru-2,6-P_2, both plastid and cytosolic isozymes show strong inhibition by PEP, which is relieved by P_i.[1,2] Thus, activity of PFK depends on the activation state of the enzymes involved in PEP metabolism. Within this regulatory scheme, an increase in pyruvate kinase (not sensitive to FBP activation in plants) or PEP carboxylase activity, for instance, would lower PEP levels, thus activating PFK. This bottom-up regulation model clearly differs from the classical top-down scheme proposed for animal glycolysis, in which flow through the lower part of glycolysis depends on the activation state of PFK.

On the other hand, cytosolic pyruvate kinase and PEP carboxylase are subject to a variety of regulatory mechanisms. Depending on the tissue, cytosolic pyruvate kinase is inhibited in a pH-dependent fashion by several tricarboxylic acid cycle intermediates, ATP and notably the amino acids Asp and Glu, linking glycolysis to N metabolism. Accordingly, plant PEP carboxylases are strongly feedback-inhibited by malate and activated by glucose-6-phosphate. This regulation is, in turn, modulated by a sophisticated signal transduction mechanism that affects the phosphorylation state of the enzyme, which is linked to the illumination state of the leaf and other signals.[10] The reduction state of disulfide groups in glycolytic enzymes has also been implicated in regulation, as is the case for plastid and cytosolic GaPDH or PFP, which are activated by reduction.

GLYCOLYSIS AND CROP PRODUCTIVITY

Given the widespread impact of the glycolytic pathway in the general metabolism of plant cells, it is not surprising that an important effort is being conducted to fully understand its operation. The beneficial aspects of the potential manipulation of plant glycolysis are not restricted to higher crop yields in terms of total starch or sucrose produced; a change in the nutritional content of crop plants is also possible as a result of the cross talk with nitrogen and lipid metabolisms.

ABBREVIATIONS

ATP-dependent phosphofructokinase	PFK
PPi-dependent phosphofructokinase	PFP
dihydrohyacetone phosphate	DHAP
glyceraldehyde 3-phosphate	Ga3P
3 phosphoglycerate	3PGA
1,3 diphosphoglycerate	1,3DPGA
phosphoenolpyruvate	PEP
fructose-1,6-bisphosphate	Fru-1,6-P_2
fructose-2,6-bisphopshate	Fru-2,6-P_2
fructose 1,6-bisphosphate	FBPase
glyceraldehyde-3-phosphate dehydrogenase	GaPDH
Crassulacean Acid Metabolism	CAM

REFERENCES

1. Plaxton, W.C. The organization and regulation of plant glycolysis. Annu. Rev. Plant Physiol. Plant Mol. Biol. **1996**, *47*, 185–214.
2. Dennis, D.T.; Blakeley, S.D. Carbohydrate Metabolism. In *Biochemistry and Molecular Biology of Plants*; Buchanan, B.B., Gruissem, W., Jones, R.L., Eds.; American Society of Plant Physiologists: Rockville, MD, USA, 2000; 630–675.
3. Givan, C.V. Evolving concepts in plant glycolysis: Two centuries of progress. Biol. Rev. **1999**, *74*, 277–309.
4. Iglesias, A.A.; Podestá, F.E.; Andreo, C.S. Structural and Regulatory Properties of the Enzymes Involved in C_3, C_4 and CAM Pathways for Photosynthetic Carbon Assimilation. In *Handbook of Photosynthesis*; Pessarakli, M., Ed.; Marcel Dekker: New York, 1997; 481–504.
5. Jang, J.-C.; Sheen, J. Sugar sensing in plants. Trends Plant Sci. **1997**, *2*, 208–214.
6. Duff, S.M.G.; Moorhead, G.B.G.; Lefebvre, D.D.; Plaxton, W.C. Phosphate starvation inducible 'bypass' of adenylate and phosphate dependent glycolytic enzymes in *Brassica nigra* suspension cells. Plant Physiol. **1989**, *90*, 1275–1278.
7. Trípodi, K.E.J.; Podestá, F.E. Purification and structural and kinetic characterization of the pyrophosphate: Fructose-6-phosphate 1-phosphotransferase from the Crassulacean Acid metabolism plant, pineapple. Plant Physiol. **1997**, *113*, 779–786.
8. Theodorou, M.E.; Kruger, N.J. Physiological relevance of fructose 2,6-bisphosphate in the regulation of spinach leaf pyrophosphate: Fructose 6-phosphate 1-phosphotransferase. Planta **2001**, *213* (1), 147–157.
9. Haake, V.; Zrenner, R.; Sonnewald, U.; Stitt, M. A moderate decrease of plastid aldolase activity inhibits photosynthesis, alters the levels of sugars and starch, and inhibits growth of potato tubers. Plant J. **1999**, *14*, 147–157.
10. Chollet, R.; Vidal, J.; O'Leary, M.H. Phosphoenolpyruvate carboxylase: A ubiquitous, highly regulated enzyme in plants. Annu. Rev. Plant Physiol. Plant Mol. Biol. **1996**, *47*, 273–298.
11. Duff, S.M.G.; Lefebvre, D.D.; Plaxton, W.C. Purification and characterization of a phosphoenolpyruvate phosphatase from *Brassica nigra* suspension cells. Plant Physiol. **1989**, *90*, 734–741.

Herbicide-Resistant Weeds

Carol Mallory-Smith
Oregon State University, Corvallis, Oregon, U.S.A.

INTRODUCTION

The introduction of synthetic herbicides revolutionized weed control. However, the repeated use of herbicides led to the selection of herbicide-resistant weeds. The selection of resistant weeds reduces control options and may increase control costs. In many cases, the most effective and economical herbicide is lost because of resistance. Herbicide-resistant weeds are one of the major challenges in weed management. Although resistance presents a challenge, herbicide-resistant weeds can be controlled with herbicides with different sites of action or other methods of weed control.

BACKGROUND INFORMATION

Herbicide resistance is the inherited ability of a biotype to survive and reproduce following exposure to a dose of a herbicide that is normally lethal to the wild type.[1] Herbicide resistance is an evolved response to selection pressure by a herbicide.

Herbicide-resistant biotypes may be present in a weed population in very small numbers. The repeated use of one herbicide or herbicides with the same site of action allows these resistant plants to survive and reproduce. The number of resistant plants increases until the herbicide is no longer effective. There is no evidence that the herbicide causes the mutations that lead to resistance.

Cross-resistance is the expression of one mechanism that provides plants with the ability to withstand herbicides from different chemical classes.[2] For example, a single point mutation in the enzyme acetolactate synthase (ALS) may provide resistance to five different chemical classes including the widely used sulfonylurea and imidazolinone herbicides.[3] However, cross-resistance at the whole-plant level is difficult to predict because a different point mutation in the ALS enzyme may provide resistance to one chemical class and not others. Cross-resistance also can result from increased metabolic activity that leads to detoxification of herbicides from different chemical classes.

Multiple-resistance is the expression of more than one mechanism that provides plants with the ability to withstand herbicides from different chemical classes.[2] Weed populations may have simultaneous resistance to many herbicides. For example, a common waterhemp (*Amaranthus rudis*) population in Illinois is resistant to triazine and ALS-inhibiting herbicide classes. However, resistance to these two different classes of herbicides is endowed by two different mechanisms within the same plant.[4] The weed species has two target site mutations, one for each herbicide class. An annual ryegrass (*Lolium rigidum* Gaud.) population in Australia is resistant to at least nine different herbicide classes.[5] In this case, herbicide options may become very limited. As with cross-resistance, multiple-resistance is difficult to predict; therefore, management of weeds with these types of resistance is complicated.

HISTORY OF HERBICIDE RESISTANCE

It was not long after the commercial use of herbicides that Able[6] and Harper[7] discussed the potential for weeds to evolve resistance. However, the first well-documented example for the selection of a herbicide-resistant weed was triazine-resistant common groundsel (*Senecio vulgaris* L.), which was identified in 1968 in Washington State.[8] The resistant biotype was found in a nursery that had been treated once or twice annually for 10 years with triazine herbicides. There were earlier reports of differential responses within weed species, but these variable responses to herbicides were not necessarily attributed to resistance.[8]

To date, 281 herbicide-resistant biotypes from 168 species have been identified.[9] The reason that the biotype number and the species number are different is that the same species has been identified with resistance to different herbicides in different locations. Of the 168 species, 100 are dicots and 68 monocots. Resistance has occurred to most herbicide chemical families.

MECHANISMS RESPONSIBLE FOR RESISTANCE

Several mechanisms theoretically could be responsible for herbicide resistance. Those mechanisms include reduced herbicide uptake, reduced herbicide translocation,

Encyclopedia of Plant and Crop Science
DOI: 10.1081/E-EPCS 120020280

herbicide sequestration, herbicide target-site mutation, and herbicide detoxification. In the cases where the resistance mechanism has been determined, the mechanism responsible in most instances has been either target-site mutations or detoxification by metabolism.[10]

Target-site mutations have been identified in weeds resistant to herbicides that inhibit photosynthesis, microtubule assembly, or amino acid production. Most often there is a point mutation, a single nucleotide change, which results in an amino acid change and is responsible for the resistance.[10] The shape of the herbicide binding site is modified and the herbicide can no longer bind.

Metabolism-based resistance does not involve the binding site of the herbicide, but instead the herbicide is broken down by biochemical processes that make it less toxic to the plant. Several groups of enzymes are involved in the process. Enzymes that are thought to be most important in herbicide metabolism are glutathione *S*-transferases and cytochrome P450 monooxygenases.[11,12]

FACTORS THAT INFLUENCE THE SELECTION OF RESISTANT BIOTYPES

Resistance usually occurs when a herbicide has been used repeatedly, either year after year or multiple times during a year, and the herbicide is highly effective, killing more than 90% of the treated weeds. The more effective a herbicide, the higher the selection pressure for resistance. Therefore, all herbicides do not exert the same selection pressure.

If a herbicide has only one site of action, it is easier to select a resistant biotype because only one mutation is needed. A herbicide that has soil residual activity will provide more selection pressure because the herbicide remains in the environment and any new seedlings will be exposed to the herbicide. If a herbicide has a very short residual, repeated applications during one growing season can have the same effect.

Herbicide factors influence the selection of herbicide-resistant weeds, but agronomic factors can also be important. Many resistant weed species have been selected in monoculture production systems. These systems result in the repeated use of the same herbicide. The increased reliance on herbicides for weed control along with a concurrent decrease in other weed management tactics further increases selection of resistant biotypes. The introduction of herbicide-resistant crops also increased the use of a single herbicide with decreased alternative controls.

Some weed species seem to be more prone to herbicide resistance than others. Some species have increased mutation rates or increased genetic variability. Increased variability is usually found in cross-pollinating species. Selection of resistant biotypes varies depending upon how likely it is for the resistance mutation to be lethal.

Weeds that produce more than one generation per year may be exposed to herbicides with the same site of action more often than those that produce only one generation per year. Because the selection of a resistant biotype is dependent on the number of individuals that are exposed to the herbicide, those weed species that produce more seeds may also produce a resistant individual more quickly.

Breeding systems and inheritance of the trait will influence how fast resistance spreads once it occurs. If the trait is controlled by one recessive gene, the heterozygote and the homozygote dominant plants will be susceptible, so only $^1/_4$ of the population will survive herbicide treatment. If the trait is dominant, the heterozygote and the homozygote dominant plants will be resistant, and the population will build quickly because $^3/_4$ of the plants will survive herbicide treatment. The trait will be readily moved within and between populations if the weed species outcrosses. In a selfing population, the trait will have reduced movement with pollen. Maternal inheritance will prevent herbicide resistance from moving with the pollen. An example of maternal inheritance is triazine resistance.[13]

Fitness is the reproductive ability of an individual, and competitive ability is the capacity of a plant to acquire resources. Initially, researchers assumed that herbicide-resistant weed species would have reduced fitness and competitive ability. Indeed, triazine-resistant weed species do have reduced growth and competitive ability.[14–16] However, weeds resistant to ALS-inhibiting herbicides have not consistently had reduced fitness or competitive ability.[17,18]

Refuges are a common tactic for the management of insect and pathogen resistance. Susceptible populations are maintained in surrounding areas and can be used to swamp resistance alleles. Generally, pollen and seed are not sufficiently mobile to swamp resistant weed populations. Migration is probably more important in the movement of resistance to a susceptible population than vice versa. This is particularly true with the tumbleweeds. Long-distance movement of a herbicide resistance gene out of an area is most likely to occur through seed movement.

PREVENTION AND MANAGEMENT OF HERBICIDE-RESISTANT WEEDS

Any management strategy that reduces the selection pressure from a herbicide will reduce the selection for a herbicide-resistant weed in the system. Recommendations for the prevention or management of herbicide-resistant weeds are often the same.[1] The recommendations from the herbicide industry and university personnel include many common factors. Common recommendations are to rotate herbicides with different sites of action to reduce selection pressure, to use short-residual herbicides so that selection pressure during a cropping season is reduced, to use multiple weed-control methods in conjunction with herbicides, and to plant certified crop seed so that herbicide-resistant weed seeds are not introduced into a field. Growers need to keep accurate records of herbicides that have been used on a field so that they can adopt a weed management plan that reduces the selection of herbicide-resistant weeds. The integration of these recommendations will reduce selection pressure for herbicide-resistant weeds.

CONCLUSION

Herbicide-resistant weeds will continue to be an issue for weed management as long as herbicides are used for weed control. Herbicide-resistant weeds can be managed, and no herbicides have been removed from the marketplace because of resistance. When growers use multiple weed-management techniques, the selection of herbicide-resistant weeds is reduced. An integrated approach using crop rotation, herbicides with different sites of action, and alternative weed control such as physical and mechanical weed control is useful in both the prevention of herbicide-resistant weeds and the management of herbicide-resistant weeds if they do occur.

ARTICLES OF FURTHER INTEREST

Chemical Weed Control, p. 255
Genetic Diversity Among Weeds, p. 496
Molecular Biology Applied to Weed Science, p. 745
Mutational Processes, p. 760
Photosystems: Electron Flow Through, p. 906
Population Genetics, p. 1042

REFERENCES

1. Retzinger, E.J.; Mallory-Smith, C. Classification of herbicides by site of action for weed resistance management strategies. Weed Sci. **1997**, *11*, 384–393.
2. Hall, L.M.; Holtum, J.A.M.; Powles, S.B. Mechanisms Responsible for Cross Resistance and Multiple Resistance. In *Herbicide Resistance in Plants*; CRC Press: Boca Raton, FL, 1994; 243–261.
3. Tranel, P.J.; Wright, T.R. Resistance of weeds to ALS-inhibiting herbicides: What have we learned? Weed Sci. **2002**, *50*, 700–712.
4. Foes, M.J.; Liu, L. A biotype of common waterhemp (*Amaranthus rudis*) resistant to triazine and ALS herbicides. Weed Sci. **1998**, *46*, 514–520.
5. Burnet, M.W.M.; Hart, Q.; Holtum, J.A.M.; Powles, S.B. Resistance to nine herbicide classes in a population of rigid ryegrass (*Lolium rigidum*). Weed Sci. **1994**, *42*, 369–377.
6. Abel, A.L. The rotation of weedkillers. Proc. Brit. Weed Control Conf. **1954**, *2*, 249–255.
7. Harper, J.C. The evolution of weeds in relation to resistance to herbicides. Proc. Brit. Weed Control Conf. **1956**, *3*, 179–188.
8. Ryan, G.F. Resistance of common groundsel to simazine and atrazine. Weed Sci. **1970**, *18*, 614–616.
9. Heap, I. *The International Survey of Herbicide Resistant Weeds*; www.weedscience.com (accessed August 2003).
10. Preston, C.; Mallory-Smith, C.A. Biochemical Mechanisms, Inheritance, and Molecular Genetics of Herbicide Resistance in Weeds. In *Herbicide Resistant Weed Management in World Grain Crops*; CRC Press: Boco Raton, FL, 2001; 23–60.
11. Barrett, M. The Role of Cytochrome P450 Enzymes in Herbicide Metabolism. In *Herbicides and Their Mechanisms of Action*; CRC Press: Boca Raton, FL, 2000; 25–37.
12. Edwards, R.; Dixon, D.P. The Role of Glutathione in Herbicide Metabolism. In *Herbicides and their Mechanisms of Action*; CRC Press: Boca Raton, FL, 2000; 38–71.
13. Souza-Machado, V.; Bandeen, J.D.; Stephenson, G.R.; Lavigne, P. Uniparental inheritance of chloroplast atrazine tolerance in *Brassica campestris*. Can. J. Plant Sci. **1978**, *58*, 977–981.
14. Conard, S.G.; Radosevich, S.R. Ecological fitness of *Senecio vulgaris* and *Amaranthus retroflexus* biotypes susceptible and resistant to atrazine. J. Appl. Ecol. **1979**, *16*, 171–177.
15. Marriage, P.B.; Warwick, S.I. Differential growth and response to atrazine between and within susceptible and resistant biotypes of *Chenopodium album*. L Weed Res. **1980**, *20*, 9–15.
16. Williams, M.M.I.I.; Jordan, N. The fitness cost of triazine resistance in jimsonweed (*Datura stramonium* L.). Am. Midl. Nat. **1995**, *133*, 131–137.
17. Alcocer-Rutherling, M.; Thill, D.C.; Shafii, B. Differential competitiveness of sulfonylurea resistant and susceptible prickly lettuce (*Lactuca serriola*). Weed Tech. **1992**, *6*, 303–309.
18. Thompson, C.R.; Thill, D.C.; Shafii, B. Growth and competitiveness of sulfonylurea-resistant and –susceptible kochia (*Kochia scoparia*). Weed Sci. **1994**, *42*, 172–179.

Herbicides in the Environment: Fate of

Kim A. Anderson
Jennifer L. Schaeffer
Oregon State University, Corvallis, Oregon, U.S.A.

INTRODUCTION

Herbicides are an integral part of our society; they are used by the general public, governments, institutions, foresters, and farmers. The benefits of herbicides need to be delivered without posing unacceptable risk to non-target sites. Therefore, understanding (and predicting) the environmental fate of herbicides is critically important. Environmental fate is determined by the individual fate processes of transformation and transport. Chemical and physical processes are primary determinants of transformation and transport of herbicides applied to soils and plants. Transformation determines what herbicides are degraded to in the environment and how quickly, while transport determines where herbicides move in the environment and how quickly. Jointly, these processes affect how much of a pesticide and its metabolites (degradation products) are present in the environment, where, and for how long. Herbicide fate also varies in response to changes in environmental conditions and application management practices (e.g., spray drift, volatilization), so it is important to understand these variables as well.

Major environmental compartments for herbicides can be considered as surface waters, the subsurface (soil and groundwater), and the atmosphere. Each medium has its own unique characteristics; however, there are many similarities when considering herbicide movement. Herbicides are rarely restricted to only one medium; therefore, chemical exchange among the compartments must be considered.

Wind erosion, volatilization, photo-degradation, runoff, plant uptake, sorption to soil, microbial or chemical degradation and leaching are potential pathways for loss from an application site. Chemical and microbial degradation are critically important factors in the fate of herbicides. Chemical conditions in soil are important secondary determinants of herbicide transport and fate. The importance of interactions between herbicides and solid phases of soils, soil water, and air within and above soil depends on a variety of chemical factors. Adsorption of herbicides from soil water to soil particle is one of the most important chemical determinants that limit mobility in soils. Environmental fate of herbicides depends on the chemical transformations, degradation, and transport in each environmental compartment.

TRANSFORMATION

Transformations determine how long a herbicide will stay in the environment. Molecular interactions of herbicides are based in part on the herbicide's chemical nature and are predicated on the physical-chemical properties and reactivities of the herbicide. Several generalized exchange processes between compartments are shown in Fig. 1. In order to understand (and therefore predict) environmental fate, the physical-chemical properties of herbicides must be known. Physical-chemical properties such as molecular formula, molecular weight, boiling point, melting point, decomposition point, water solubility, organic solubility, vapor pressure (V_p), Henry's law constant (K_H), octanol/water partitioning (K_{ow}), acidity constant (pKa), and soil sorption (K_d, K_{oc}) are important in predicting transformations. How some of these chemical-physical properties affect herbicide fate is briefly discussed (Fig. 2). Table 1,[1–4] demonstrates the wide variation of chemical-physical properties of a selected group of herbicides. Water solubility, the solubility of a herbicide in water at a specific temperature, determines the affinity for aqueous media and affects movement between air, soil, and water compartments. Vapor pressure, the pressure of the vapor of a herbicide at equilibrium with its pure condensed phase, measures a herbicide's tendency to transfer to and from gaseous environmental phases. Vapor pressure is critical for predicting either the equilibrium distribution or the rate of exchange to and from natural waters. Henry's law constant (K_H), the air-water distribution ratio for neutral compounds in dilute solutions in pure water, determines how a chemical will distribute between the gas and aqueous phase at equilibrium. Henry's law constant, therefore, only approximates the air-water partition in natural

Encyclopedia of Plant and Crop Science
DOI: 10.1081/E-EPCS 120020276

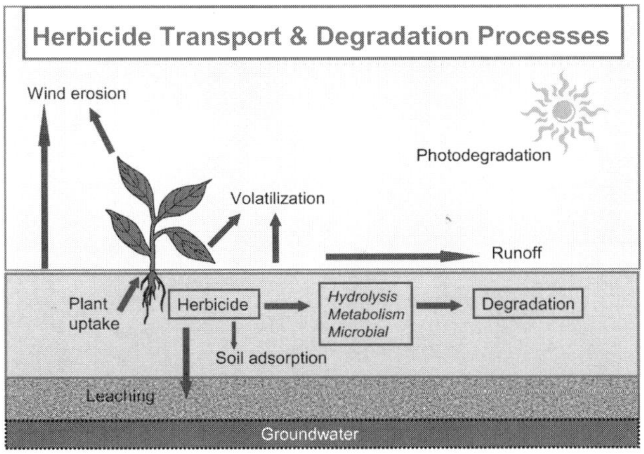

Fig. 1 Illustrated are pathways for loss of herbicide from a herbicide application site. Each pathway (arrow) illustrates a particular loss process occurring at some rate constant (k). The importance of each pathway and magnitude of each rate constant will vary substantially between herbicides and environmental conditions. (*View this art in color at www.dekker.com.*)

waters. Octanol-water partition coefficient (K_{ow}), the partition of organic compounds between water and octanol, is used to estimate the equilibrium partitioning of nonpolar organic compounds between water and organisms. The octanol-water partition coefficient is also proportional to partitioning of organic compounds (herbicides) into soil humus and other naturally occurring organic phases.

In the environment, many herbicides are present in a charged state (not neutral). Charged species have different properties and reactivities as compared to their neutral counterparts. Therefore, the extent to which a compound forms ions in environmental ecosystems is important. The pKa is a measure of the strength of an acid relative to water. Strong organic acids (pKa $\cong$ 0–3) in ambient natural waters (pH 4–9) will be present predominantly as anions. Conversely, very weak acids (pKa $\geq$ 12) in ambient natural waters will be present in their associated form (neutral). In an analogous fashion, strong bases (pKa $\geq$ 11) will be present as ions.[5] Examples of weak acids are 2,4-D and triclopyr, and examples of weak bases are atrazine, dicamba, and simazine (Table 1).

Sorption of a herbicide is when the herbicide binds to the soil or sediment particles (K_d). Soil sorption is an important process since it can dramatically affect herbicide fate. Figure 2 illustrates the relationship between herbicides sorbed to soil particles versus those dissolved in soil water. Soil sorption is dependent on soil type and influenced heavily by the organic matter content. Typically, soils high in clay and organic matter have a higher sorption capacity. To account for organic matter content, sorption is sometimes normalized for % organic matter, referred to as K_{oc}. K_d, K_H, and V_p describe the potential for exchange of herbicide compounds between soil, water, and air over short distances

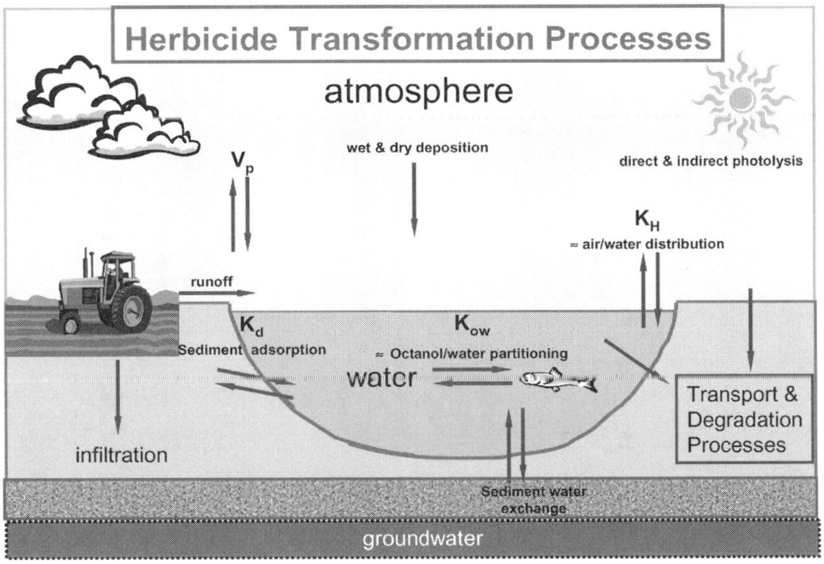

Fig. 2 Illustrated are pathways for transformation processes for loss of herbicide from a herbicide application site. Each pathway (arrow) illustrates a particular loss process occurring at some rate constant (k). The importance of each pathway and magnitude of each rate constant will vary substantially between herbicides and environmental conditions. (*View this art in color at www.dekker.com.*)

Table 1 Physiochemical properties of selected herbicides[a]

Chemical name	CAS number	Chemical class	Solubility (mg/L)	Vapor pressure (mm Hg)	K_d soil sorption K_{oc} normalized for organic intent	Log K_{ow} octanol/water partition constant	K_H Henry's constant (atm-m^3/mole)	pKa acidity constant
2,4-D	94-75-7	Chlorphenoxy acid	677	8.25×10^{-5} at 20°°C	0.08–0.94	2.81	3.54×10^{-8}	2.73
Atrazine	1912-24-9	Triazine	34.7	2.89×10^{-7}	K_{oc} 100	2.61	2.36×10^{-9}	1.7
Bromoxynil	1689-84-5	Hydroxybenzonitrile	130	4.8×10^{-6}	K_{oc} 2.48	2	1.4×10^{-6}	4.06
Carbaryl	63-25-2	Carbamate	110 at 22°°C	1.36×10^{-6}	2.45–4.69	2.36	3.27×10^{-9}	NA
Chlorsulfuron	64902-72-3	Sulfonylurea	2.8	2.3×10^{-11}	K_{oc} 1.02	2	3.9×10^{-15}	3.6
Dicamba	1918-00-9	Benzoic acid	8310	3.38×10^{-5}	0.07–0.53	2.21	2.18×10^{-9}	1.97
Diuron	330-54-1	Substituted urea	42	6.9×10^{-8}	2.9–14	2.68	5.04×10^{-10}	−1
Metolachlor	051218-45-2	Chloroacetanilide	530 at 20°°C	3.14×10^{-5}	1.5–10	3.13	9×10^{-9} at 20°°C	NA
Metsulfuron	74223-64-6	Sulfonylurea	9500	2.5×10^{-12}	0.36–1.40	2.2	1.32×10^{-16}	3.64
Pendimethenalin	40487-42-1	2,4-Dinitroanaline	0.3 at 20°°C	3×10^{-5}	30–380	5.18	8.56×10^{-7}	NA
Pentachlorophenol	87-86-5	Chlorinated phenol	14	1.1×10^{-4}	K_{oc} 262–38905	5.12	2.45×10^{-8} at 22°°C	4.7
Prometon	1610-18-0	Triazine	750	2.3×10^{-6} at 20°°C	0.373–2.61	2.99	9.09×10^{-10} at 20°°C	4.3
Simazine	122-34-9	Triazine	6.2 at 22°°C	2.21×10^{-8}	0.48–4.31	2.18	9.42×10^{-10}	1.62
Thifensulfuron	79277-27-3	Sulfonylurea	230	1.28×10^{-10}	0.08–1.38	1.56	4.08×10^{-14}	4
Triasulfuron	82097-50-5	Sulfonylurea	815	1.5×10^{-8}	K_{oc} 73.4–190.6	0.58	9.9×10^{-7} at 20°°C	4.64
Tribenuron	101200-48-0	Sulfonylurea	2040	3.97×10^{-10}	0.19–2.0	1.17	1.01×10^{-8}	5
Trifluralin	1582-09-8	Dinitroanaline	0.184	4.58×10^{-5}	18.6–54.8	5.34	1.03×10^{-4} at 20°°C	NA

[a]Data at 25°C unless otherwise noted.

by diffusion. Transport over longer distances involves mass transfer (advection-dispersion).

TRANSPORT

Transport of herbicides from runoff from soil is an important route of entry into surface waters. Herbicide runoff can be classified into two categories: herbicides that are either dissolved or suspended in runoff waters. Dissolved herbicides are characterized by low adsorption and high water solubility, while suspended herbicides are characterized by high soil sorption. Leaching depends strongly on local environmental conditions, such as percolation rates through local soils. Adsorption decreases herbicide leaching mobility by reducing the amount of herbicide available to the percolating soil water. However, herbicide runoff and leaching also are controlled by the amount of herbicide degradation.

DEGRADATION

Herbicide degradation ultimately ends with the formation of simple stable compounds, such as carbon dioxide; however, there are intermediates of varying stability on the way to complete mineralization (e.g., H_2O and CO_2). The rate of degradation of a particular herbicide can vary widely. The chemical nature of the herbicide is important, as discussed previously, but the degradation rate also depends on the availability of other reactants, as well as environmental factors. There are three major degradation pathways for herbicides: photo-degradation, chemical degradation, and microbial degradation.

Photo-degradation is the breakdown of a herbicide by sunlight at the plant, soil, or water surface. Direct photolysis occurs when the herbicide absorbs light (some portion of the solar spectrum), and this leads to dissociation of some kind (e.g., bonds break). Indirect photolysis occurs when a sensitizer molecule is radiatively excited and is sufficiently long-lived to transfer energy, such as an electron, a hydrogen atom, or a proton to another "receptor" molecule. The receptor molecule (herbicide), without directly absorbing radiation, can be activated via the "sensitizer" molecule to undergo dissociation or other kinds of chemical reactions leading to photo-degradation.

Chemical degradation is the breakdown of herbicides by processes not involving living organisms (abiotic). Hydrolysis may be one of the more important mechanisms of degradation for herbicides. Hydrolysis is the reaction of a herbicide where water interacts with the herbicide, replacing a portion of the molecule with OH. The following functional groups and chemical classes are known to be susceptible to hydrolysis: ethers, amides, phenylurea compounds, nitrile, carbamates, thiocarbamates, and triazines. Hydrolysis is influenced by environmental conditions including pH, water hardness, dissolved organic matter, dissolved metals, and temperature. Half life ($t_{1/2}$) is the amount of time it takes the parent compound (herbicide) to decay to half its original concentration. The half life for chemical degradation via the hydrolysis pathway is strongly dependent on environmental conditions. For example, the $t_{1/2}$ for 2,4-D at pH 6 is four years, while the $t_{1/2}$ at pH 9 is 37 hours.[6]

Microbial degradation is the breakdown of compounds (herbicides) by microorganisms. Bacteria and fungi are the primary microorganisms responsible for biotransformations (biotic reactions). Rates of microbial degradation are largely determined by environmental conditions such as temperature, pH, reduction-oxidation conditions, moisture, oxygen, organic matter, and food. Complete biodegradation, mineralization, yields carbon dioxide, water, and minerals.

CONCLUSION

The perfect herbicide would be one that controls weeds as necessary and then quickly breaks down and never moves off-site. The development of new herbicides over the last decade has focused in part on minimizing unacceptable risk to non-target sites. This has resulted in the development of more biologically active herbicides that greatly reduce application rates. Other research developments include minimizing the leaching of herbicides. When herbicides are applied in the environment, many transport and transformation processes are involved in their dissipation. Predicting herbicide environmental fate and behavior is complicated and is determined by the chemical processes for each environmental compartment.

In addition to understanding the fate of herbicides in the environment, emerging research also endeavors to look at the risk to non-target organisms. The risk of herbicides to organisms is relative to the bioavailability of the herbicide. Bioavailability may be approximately inversely proportional to K_d. Herbicides with larger K_d's tend to have lower bioavailability. Along with the determination of transport and transformation chemical processes, understanding the complete risk to non-target sites, therefore, should include the determination of the bioavailability of the herbicide.

ACKNOWLEDGMENTS

The authors thank Dr. Jeffery Jenkins, Oregon State University, for helpful discussions.

ARTICLES OF FURTHER INTEREST

Chemical Weed Control, p. 255
Environmental Concerns and Agricultural Policy, p. 418
Sustainable Agriculture and Food Security, p. 1183
Sustainable Agriculture: Definition and Goals of, p. 1187
Sustainable Agriculture: Ecological Indicators, p. 1191
Sustainable Agriculture: Economic Indicators, p. 1195

REFERENCES

1. University of Guelph Ontario Agricultural College. http://www.uoguelph.ca/OAC/env/bio/PROFILES.pdf, March 2003.
2. The Extension Toxicology Network. http://ace.orst.edu/info/extoxnet/pips/ghindex.html, March 2003.
3. United States Department of Agriculture, Agricultural Research Services. http://wizard.arsusda.gov/acsl/ppdb3.html, March 2003.
4. Environmental Science Syracuse Research Corporation. http://esc.syrres.com/interkow/physdemo.htm, March 2003.
5. Montgomery, J.H. *Agrochemicals Desk Reference: Environmental Data*; Lewis Publishers: Ann Arbor, 1993.
6. Kamrin, M.A. *Pesticide Profiles: Toxicity, Environmental Impact, and Fate*; CRC Press LCC: New York, 1997.

Herbs, Spices, and Condiments

Katerina P. Svoboda
Scottish Agricultural College Auchincruive, Ayr, Scotland, U.K.

INTRODUCTION

Herbs and spices include plants whose leaves, stems, flowering tops, fruits, and roots are used—either fresh, dried, or frozen—in seasoning of foods; in the beverage industry; as dietary supplements; in pharmaceutical, perfumery, and cosmetic products; and for numerous other purposes, including dyeing, potpourris, insect deterrents, and aromatherapy. The distinction between culinary herbs and spices is an imprecise one, and recently the most common term used for both groups is ''aromatic and medicinal plants.'' Scientific and regulatory information is systematically presented and published world-wide at a high rate.

CULINARY HERBS

Culinary herbs are usually defined as edible plants, consumed in small quantities and providing flavor and aroma to foods and beverages.[1,2] They are predominantly grown in temperate and Mediterranean regions, and can be consumed as vegetables or as condiments. Extracts of these plants, in the form of essential oils and oleoresins, are also increasingly employed in various industries.[3]

Herbs have been used traditionally for centuries to add flavor to monotonous diets. Onion, garlic, leek, parsley, sage, celery, mustard, dill, and lovage (Fig. 1) can serve as examples. Each country and locality, however, offers the local population a collection of potential herbs, shrubs, and woody plants that have been employed in culinary traditions among ethnic groups. Among culinary herbs, 85 genera are described comprehensively by Small,[4] who gives examples of traditional herbs, edible sprouts, and edible flowers. Many herbs can contribute significantly to nutrient food value, specifically for their content of vitamins and minerals, and for their antioxidant and/or anticancer properties. Over 100 species of herbs and spices are used to brew tea, but the safety of some of these is open to question. It is important to be sure about high standards for the labeling, safety, and efficacy of any herb obtained on the market.

There are few true varieties of even the most familiar culinary herbs that would pass the rigorous testing applied to major crops (distinctness, uniformity, and stability). Most named varieties arise from simple selection, and they can display large variation in morphology, physiology, and the chemistry of both primary and secondary metabolites. Many varieties are selected on the basis of their suitability for a particular region, and the behavior of such cultivars in another environment may be quite different. Herbs can be subjected to various treatments such as extraction, distillation, expression, fractionation, purification, concentration, or fermentation (Fig. 2). The extraction process has a big influence on the composition of the resulting extract. It is important to control every part of the process in order to obtain reproducible and commercially marketable extracts.

The commercial growing of a herbal crop requires a stable supply of uniform seeds or propagules. The composition and quantity of flavor and fragrance chemicals in plants changes during the growing season; consequently, harvest time is very important. The effects of fertilizers, herbicides, pesticides, and heavy metals have to be considered for each species. Sun radiation, rain, altitude, and day-length also play an important role in the production of these chemicals in plants, and, finally, post-harvest processing and storage can markedly influence the quality of the final product.

MEDICINAL HERBS

Knowledge of medicinal properties of herbs has been handed down from generation to generation for thousands of years. Ancient herbal records from China, India, and Egypt dating back to 3000–1000 B.C. belong to the oldest archaeological finds. Plants mentioned in Western herbals, mainly by Greek physicians (for example, *De Materia Medica* by Dioscorides, from the 1st century A.D.), followed by herbals from Middle Ages Europe (Gerard, Culpeper), and many new herbals from the 18th and 19th centuries were used continuously until the onset of synthetic medicinal products in the 20th century. These plants are potentially a vast source of new drugs. Thousands of compounds have been isolated and tested for their pharmacological activities: antibacterial, antiviral, antitumor, antiinflammatory, antioxidant, hypo-

Encyclopedia of Plant and Crop Science
DOI: 10.1081/E-EPCS 120010435

Fig. 1 *Levisticum officinale* (lovage) grown for seed. (*View this art in color at www.dekker.com.*)

glycemic, and antiulcer.[5] Many other biologically active compounds can be identified, with hallucinogenic, cardiovasular, muscle relaxant, and antileukemic properties. Some 80% of the world's population still relies on traditional, plant-based medicine, despite major advances in developing synthetic drugs. The World Health Organization has compiled a list of 20,000 medicinal plants in use.

Medicinal plants possess therapeutic properties, and many plant species are not just medicinal, but can be used also as herbs or as an edible vegetable. Certain species can be highly poisonous, such as *Hyoscyamus niger* (henbane), *Datura stramonium* (Jimson weed), *Veratrum album* (white hellebore), and *Digitalis lanata*. A large group of herbs is not poisonous, however, and these are widely used for their medicinal properties. Their pharmacological activities depend on plant metabolism products: carbohydrates, lipids, proteins, phenols, coumarins, flavonoids, anthocyanins, tannins, quinones, terpenoids, steroids, and alkaloids.[6] These plant products form starting materials for the synthesis of drugs, food additives, fragrances and flavors, and insecticides and pesticides.

Claims for the effectiveness and safety of herbal remedies are generally based on historical experience, but therapeutic effects have been recently demonstrated in scientific investigations, and this validation process continues. The world-wide herbal products industry is worth an estimated £11 billion per annum, and this is increasing at 5–10% per annum. The increased demand for herbal medicines and alternative remedies is driven by consumers for a number of reasons, including an aging population, increased recognition of the importance of diet to long-term health, and an increase in the availability of pharmaceuticals and natural remedies. Significant numbers of herbs (70–90% of 21,000 species used all over the world) are obtained through commercial collection from their natural habitat. The undergound parts include the root (*radix*), rhizome (*rhizoma*), tuber (*tuber*), and bulb (*bulbus*); the aerial parts are the leaf (*folium*), herbage (*herba*), flower (*flos*), fruit (*fructus*), seed (*semen*), and bark (*cortex*). There are particular problems in collecting from the wild, including a mistaken identification of the plants leading to poisoning, and the localized extinction of plants leading to reduced biodiversity. At least 150 medicinal and aromatic plant species are threatened as a result of overcollection and destructive harvesting techniques, as well as habitat loss and changes in their area of distribution.

Qualified medical advice should be sought before using the information provided by various books and leaflets. Herbal remedies are very powerful and are not safe simply because they are natural. Standardization is done in order to obtain herbs and their extracts with reproducible therapeutic activity from batch to batch. The herbs and their extracts must not be toxic, must not interact with other medicines that the consumer may be taking, and

Fig. 2 Field-scale mobile distillation equipment for chamomile harvest. Norfolk, UK, summer 2001. (*View this art in color at www.dekker.com.*)

must not be contaminated with pesticide residues, heavy metals, or microbes. The main concerns involving yeasts and filamentous fungi are caused by species of *Aspergillus*, *Penicillium*, *Mucor*, *Rhisopus*, *Absidia*, *Alternaria*, *Cladosporium*, and *Trichoderma*. Both fresh and dried herbal material can also support heavy bacterial loads. Imported plant material can be of poor quality due to the use of irrigation water contaminated with sewage sludges. The guidelines for good agricultural practice (GAP) for aromatic and medicinal plants in the European Union are intended to be applied to the growing and primary processing of all these plants used in the food, feed, medicinal, flavoring, cosmetic, and perfume industries.

SPICES

Spices are usually of tropical or subtropical origin, highly aromatic, and with a high percentage of volatile oils.[7] Spices from plant sources have been used throughout history for their presumed preservative properties, and for perfumery and flavor purposes. The economies of many countries were based on trade in spices, which brought them wealth and power. At present, the bulk of the world spice trade is concentrated in black and white pepper, clove, nutmeg, cardamom, cinnamon, ginger, mace, and allspice. Accurate statistics are difficult to obtain, because the figures for individual species are not recorded separately. The American Spice Trade Association provides up-to-date information on importing, market trends, quality standards, production, and processing.

Spices can be used either whole or ground, single or blended, or in dried or frozen form. They can be further processed, encapsulated, dispersed, solubilized, and sterilized. Detailed specification standards concerning the amount of volatile oils, total ash, fiber, moisture, specific gravity, and other characteristics are increasingly more strict.

ESSENTIAL OILS

Essential oils are isolated from various parts of the plant: leaves (mint, savory, oregano, lavandula), bark and wood (cinnamon, frankincense, juniperus), root (vetiver), rhizome (ginger), flower bud (cloves), seed (nutmeg, caraway, dill), or flowers (jasmine, roses). Over 3000 essential oils have been identified from a vast number of plants belonging to some 87 families. The knowledge of botany, taxonomy, chemotaxonomy, and biochemistry greatly added to the production and utilization of these oils in the food, drink, pharmaceutical, and cosmetics industries.

Accumulation and secretion of volatile oils in plants is generally associated with the presence of glandular structures, either external or internal, such as oil cells, glands, glandular hairs, resin ducts, canals, and laticifiers.[8] It is assumed that the volatile oils are produced within the secretory cells of these structures, which are characteristic for each plant species (Fig. 3). Extraction techniques (using olive oil, for example) and basic distillation have been known since pre-Christian times. Considerable improvement and development occurred from the 14th century, when various pharmacies started to prepare plant-based oil remedies. In the first half of the 19th century the production of essential oils was industrialized and was followed by isolation of single organic compounds from the oil (e.g., cinnamaldhyde from cinnamon oil). Later, the first synthetic aroma molecules were synthesized by the chemical industry. At present, thousands of flavor and fragrance molecules have been isolated, and their structures have been elucidated. Consequently, these molecules can be synthesized. Natural products are obtained directly from plant sources. Nature-

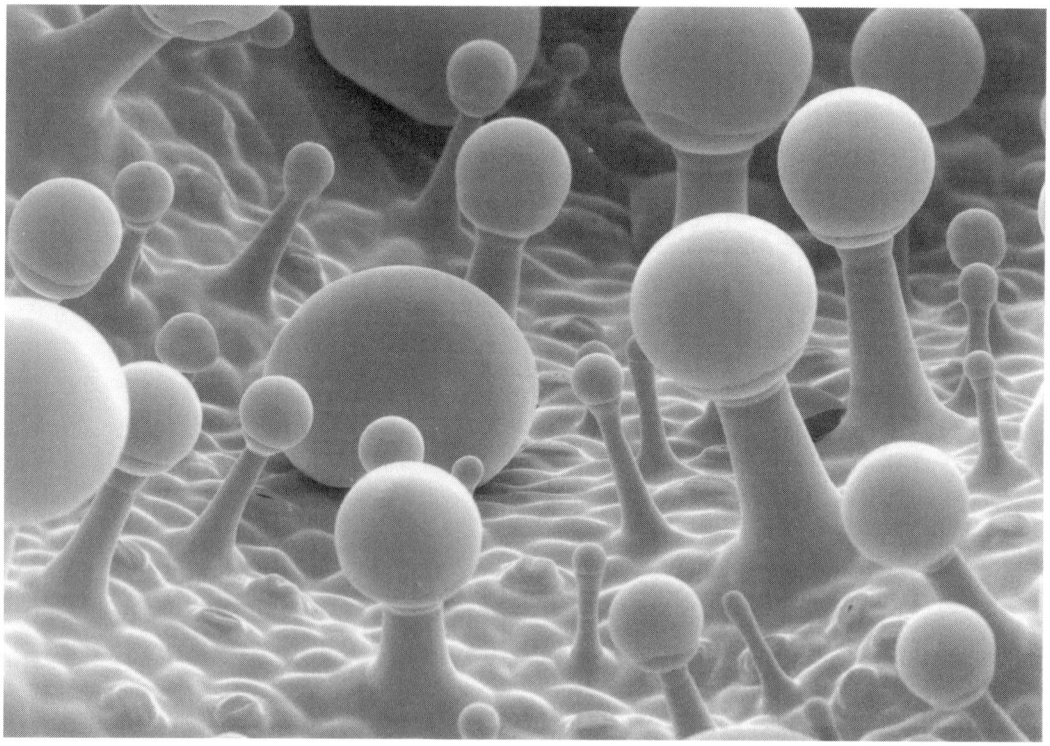

Fig. 3 Stalked and sessile secretory cells (×437) on the flower surface of *Salvia sclarea* (clary sage).

identical compounds are produced synthetically, and they are chemically identical to their natural counterparts. Artificial flavor substances have not been yet identified in plant or animal products, but they are produced artificially.[9] Lists of approved flavoring substances (GRAS—generally recognized as safe) are published in the CFR (Code of Federal Regulations of the U.S.A.); by the FEMA (Flavor and Extract Manufacturers Association of the United States); and by EC Regulation No. 2232/96. Hundreds of monographs on fragrance material and essential oils have been published by the Research Instititute for Fragrance Materials in Food and Chemical Toxicology.

CONDIMENTS

The word condiment is derived from the Latin *condimentum*, which means seasoning, and from *condire*, which means to pickle. Salt, pepper, mustard, various sauces derived from fruit, vegetables, herbs and spices (soya, horseradish, wasabi, mooli, onion, garlic, chilli, mint, cranberry, dill, watercress, curry, basil), chutneys, relishes, and preserves belong to this group. Practically any type of herb or spice can be used for flavor, preserves, spiced sauces, and drinks.[10]

REFERENCES

1. Bown, D. *The Royal Horticultural Society: Encyclopedia of Herbs and Their Uses*; Dorling Kindersley Limited: London, UK, 1995; 1–424.
2. *CAB Abstracts*; CABI Publishing, CABInternational: Wallingford, Oxon OX10 8DE, UK, 1995–2002.
3. Bauer, K.; Gerbe, D.; Suburg, H. *Common Fragrance and Flavour Materials. Preparation, Properties and Uses*; Wiley-VCH Verlag GmbH: Weinheim, Germany, 1997; 1–278.
4. Small, E. *Culinary Herbs*; National Research Council of Canada: Otawa, Ontario, Canada, 1997; Vol. 4, 1–710.
5. Cowan, M.M. Plant products as antimicrobial agents. Clin. Microbiol. Rev. **1999**, 564–582.
6. Dewick, P.M. *Medicinal Natural Products: A Biosynthetic Approach*; John Wiley and Sons Inc.: New York, USA, 1998; 1–466.
7. Burdock, G.A. *Fenaroli's Handbook of Flavour Ingredients*; CRC Press: London, UK, 1995; 1–327.
8. Svoboda, K.P.; Svoboda, T.G. *Secretory Structures of Aromatic and Medicinal Plants—A Review and Atlas of Micrographs*; Miscroscopic Publications: Powys, UK, 2000; 1–60.
9. *Chemicals from Plants*; Walton, N.J., Brown, D.E., Eds.; Imperial College Press, 1999; 1–425.
10. Ortiz, E.L. *The Encyclopedia of Herbs, Spices and Flavourings*; Dorling Kindersley Ltd.: London, UK, 1992; 1–288.

Human Cholinesterases from Plants for Detoxification

Tsafrir S. Mor
Arizona State University, Tempe, Arizona, U.S.A.

Hermona Soreq
The Hebrew University of Jerusalem, Jerusalem, Israel

INTRODUCTION

Acetylcholinesterase (AChE) is a major component of central and peripheral neurotransmission, terminating signals at cholinergic synapses by hydrolyzing the neurotransmitter acetylcholine. During the past decade, nonenzymatic roles for AChE and the homologous enzyme butyrylcholinesterase (BChE) were recognized. Cholinesterases in the bloodstream may play a protective role by metabolic inactivation of many nerve agents, including natural and synthetic anticholinesterases (e.g., organophosphate nerve gases and pesticides), street drugs (e.g., cocaine and heroin), and muscle relaxants (e.g., succinylcholine). The use of exogenously applied cholinesterase in treatment and prevention of intoxication by these toxicants is being evaluated. For therapy based on cholinesterases to be feasible, large amounts of properly folded and stable enzymatic preparations that are free of mammalian pathogens are needed. Transgenic plants are emerging as one of the more promising alternative production systems for pharmaceutically important proteins; however, very little is known about the rules controlling posttranslational processing of such proteins, including protein folding, glycosylation, etc. Human AChE in tomato plants has recently been expressed; this system is being evaluated as an expression system for other variants of cholinesterases. Beyond the importance of these cholinesterases as scavengers of a wide scope of toxicants, these enzymes are an excellent model for studying posttranslational modifications of recombinant proteins produced in plants.

HUMAN CHOLINESTERASES

Cholinesterases (ChEs) can be distinguished by their substrate and inhibitor-binding specificities.[1–3] The hydrolytic activity of acetylcholinesterase (AChE) is limited to the neurotransmitter acetylcholine (ACh) which selectively interacts with a narrow group of inhibitors.[1,2] Mammals possess a homologous serum enzyme—butyrylcholinesterase (BChE)—that has a broader catalytic spectrum with preference for longer-chain substrates[3] and noncholine substrates such as cocaine.[4,5]

In humans, AChE and BChE are encoded by single genes, with the primary transcript of the *ACHE* gene subjected to alternative splicing, yielding three mRNA variants resulting in three protein isoforms with distinct C-termini with clearly defined spatiotemporal expression patterns, subunit structures, subcellular localization, and subtly different enzymatic properties. (Fig. 1).[6] The catalytic core of human AChE, common to all variants, is sufficient for ACh hydrolysis. Posttranslational modifications of ChEs include glycosylation oligomerization and membrane anchoring via protein-protein interactions (AChE-S) or a glycolipid (AChE-E).[7]

Exposure to anti-AChE agents leads to overstimulation of cholinergic pathways and potentially death.[2] Whereas some naturally occurring AChE inhibitors are very potent, clinically relevant human exposure to them is rare. However, synthetic anti-AChE compounds—especially organophosphates (OPs)—are widely used as pesticides and pose a substantial occupational and environmental risk. Even more ominous is the fear of deliberate use of OPs as chemical warfare agents. Current medical intervention in the case of acute exposure to anticholinesterase agents includes use of the muscarinic ACh-receptor antagonist, atropine, and oximes to reactivate the OP-modified AChE. The reversible carbamate pyridostigmine bromide is also used for prophylaxis.[2] However, these conventional treatments have limited effectiveness and may have serious short- and long-term side effects, including significant performance deficits and even permanent brain damage.

A different approach in treatment and prevention of anti-AChE toxicity seeks to mimic one of the physiological lines of defense. In humans, ChEs can be found in the bloodstream, either in the serum (BChE and AChE-R) or anchored to erythrocytes (AChE-E). Evidence suggests that these enzymes act as scavengers of potentially harmful substances, preventing them from accumulating and reaching their targets. For example, these circulating ChEs can bind and inactivate anticholinesterases, thereby protecting AChE-S, the vital synaptic isoform of AChE.

Encyclopedia of Plant and Crop Science
DOI: 10.1081/E-EPCS 120010382

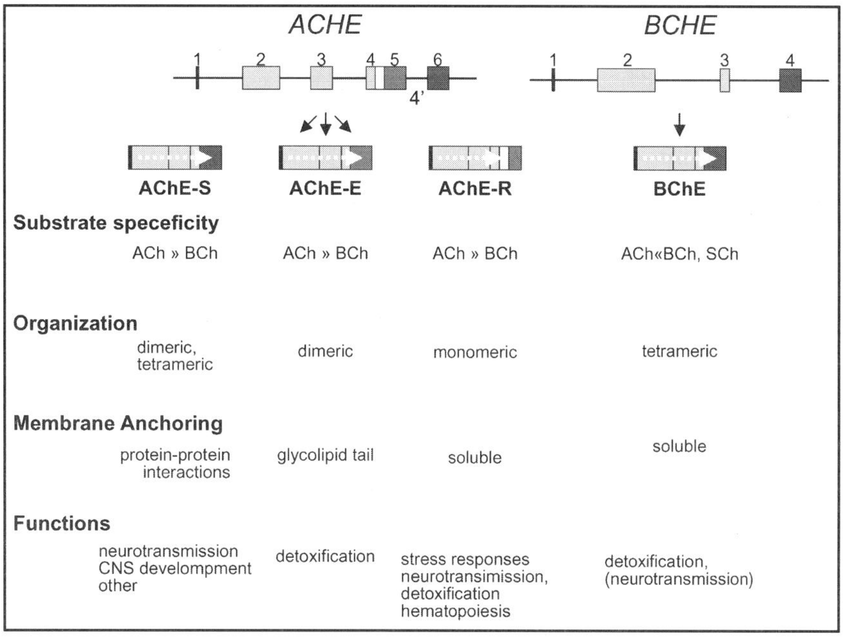

Fig. 1 Molecular biology of human ChEs. 3′-alternative splicing events of the primary transcripts of the *ACHE* gene yield three mRNA species, which give rise to three AChE variants with distinct C-termini. In contrast, transcription and splicing of the *BCHE* gene product results in only one mRNA, giving rise to a single BChE protein. These ChEs have distinct biochemical properties and expression patterns. CNS = central nervous system; NMJ = neuromuscular junction.

Administration of exogenous ChEs to boost the potential of the native circulating enzymes to scavenge anticholinergic agents is an efficacious and safe approach for the treatment of anti-AChE toxicity. The efficacy of this therapeutic paradigm to protect against a challenge of organophosphates was proven in a variety of animal models.[8,9] Similarly, administered BChE is effective in the metabolic inactivation of cocaine (in case of overdose) and muscle relaxants (e.g., to treat succinylcholine post-anesthesia apnea that occurs in patients with genetic or acquired low-plasma BChE activities).[4,5]

HUMAN AChE PRODUCTION IN PLANTS

To be used as detoxifiers, ChEs can be purified from human or animal blood, but these sources are supply-limited and have inherent problems (e.g., potential contamination with human pathogens and prions). Various cell cultures have been explored as systems for the production of recombinant ChEs but suffer from disadvantages. Recombinant ChEs expressed in *Escherichia coli* must be denatured and renatured, and even then have only partial activity. In addition, they are extremely labile as compared to the native enzymes.[10] Production of ChEs by yeast cultures[10] or in insect cells by the baculovirus vector system[11] is possible but remained small scale, perhaps reflecting yield problems. Production in mammalian cell cultures is possible and was extensively studied;[12] however, the costs of production (and scaling up) as well as the safety of this system are major shortcomings.

We recently introduced transgenic plants as a novel production system for human AChE, when we successfully expressed the catalytic domain of the enzyme (common to all variants) in tomato plants and reported on the initial characterization of this recombinant protein (Fig. 2).[13] This was the first demonstration of expression in plants of a key protein component of the human nervous system. Some of the transgenic tomato lines express high levels of AChE activity, comparable to those reported for production of the enzyme in yeast. The activity is typical of authentic human AChE activity as judged by its substrate and inhibitor interactions. The plant-derived enzyme is relatively stable in the crude plant extract.[13] More recently, we introduced the same construct to *Arabidopsis thaliana* and demonstrated the presence of AChE in such putative transformants by activity assays. Indeed, the AChE activities measured in leaf extracts of these plants were higher than previously seen in our best-expressing tomato lines (Fig. 2), probably reflecting the presence of a naturally occurring ChE inhibitor in the tomato extracts (Mor et al., manuscript submitted).

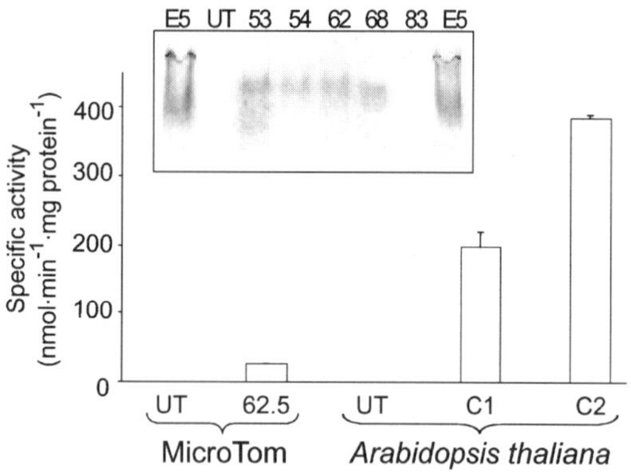

Fig. 2 Expression of core domain of AChE in plants. Leaf extracts of two *Arabidopsis thaliana* AChE transformants, an AChE-expressing tomato plant, and their untransformed (UT) counterparts were assayed for AChE activity. Insert: Plant-derived AChE migrates as a discrete band in nondenaturing gel electrophoresis. Protein samples from the indicated transgenic tomato plant lines, untransformed plant (UT), and a commercially available preparation of AChE-E were resolved on a nondenaturing gel, which was then stained for AChE activity. (*View this art in color at www.dekker.com.*)

CONCLUSION

Anticipating a growing need for large amounts of ChE that will be properly folded, stable, enzymatically active, and free of mammalian pathogens, we advocate transgenic plants as a suitable system for production of these human enzymes. Our previous results demonstrate that as an expression system for ChEs, the potential of transgenic plants is still far from being fully realized. To make this prospect a reality the system must be optimized to maximize expression of the recombinant protein. For example, we have redesigned the coding regions of the DNA constructs encoding the various AChE isoforms to conform their codon usage to that which prevails in plants and to eliminate deleterious sequences that might hamper their expression in plants. As this encyclopedia goes to print, we are analyzing transgenic plants that harbor these plant-optimized synthetic genes, that show a dramatic increase in their expression levels. We are also exploring various options for targeting the transgenic enzyme to subcellular compartments that would allow maximal accumulation of the recombinant proteins, and for screening for the best host plant species for the purpose. Further challenges include establishing a purification scheme for the recombinant plant-produced AChE (made easier with increased expression levels) and subsequent testing of the enzyme's ability to protect animals from OP intoxication.

In more general terms, whereas the basic requirements for expressing mammalian proteins in plants are quite well established, little is known of the rules controlling the posttranslational processing events (including protein folding, glycosylation, etc.) that such recombinant proteins undergo in plant cells. ChEs are appropriate models for studying these topics because in mammals they exist as different, wellstudied variants with distinct posttranslational modifications that give them their unique biochemical properties. By biochemical characterization of the plant-produced ChEs and comparison to their native counterparts, important insights can be gained to establish the universal utility of plants for the production of pharmaceutically valuable proteins.

ACKNOWLEDGMENTS

The authors thank their respective lab members. This work was supported by a research contract from the Defense Advanced Research Project Agency (#N66001-01-C-8015).

ARTICLE OF FURTHER INTEREST

Vaccines Produced in Transgenic Plants, p. 1265

REFERENCES

1. Soreq, H.; Seidman, S. Acetylcholinesterase—New roles for an old actor. Nat. Rev. Neurosci. **2001**, *2* (4), 294–302.
2. Glick, D.; Shapira, M.; Soreq, H. Molecular Nueorotoxicolocgy Implications of Acetylcholinesterase Inhibition. In *Site-Selective Neurotoxicity,* 1st Ed.; Talyor and Francis: London, 2002; 116–133.
3. Small, D.H.; Michaelson, S.; Sberna, G. Non-classical actions of cholinesterases: Role in cellular differentiation, tumorigenesis and Alzheimer's disease. Neurochem. Int. **1996**, *28* (5–6), 453–483.
4. Carmona, G.N.; Jufer, R.A.; Goldberg, S.R.; Gorelick, D.A.; Greig, N.H.; Yu, Q.S.; Cone, E.J.; Schindler, C.W. Butyrylcholinesterase accelerates cocaine metabolism: In vitro and in vivo effects in nonhuman primates and humans. Drug Metab. Dispos. **2000**, *28* (3), 367–371.
5. Ashani, Y. Prospective of human butyrylcholinesterase as a detoxifying antidote and potential regulator of controlled-release drugs. Drug Dev. Res. **2000**, *50* (3–4), 298–308.
6. http://www.ensam.inra.fr/cgi-bin/ace/index (accessed September 2002).
7. Nachon, F.; Nicolet, Y.; Viguie, N.; Masson, P.; Fontecilla-Camps, J.C.; Lockridge, O. Engineering of a monomeric and low-glycosylated form of human butyrylcholinesterase: Expression, purification, characterization

and crystallization. Eur. J. Biochem. **2002**, *269* (2), 630–637.
8. Maxwell, D.M.; Brecht, K.M.; Doctor, B.P.; Wolfe, A.D. Comparison of antidote protection against soman by pyridostigmine, HI-6 and acetylcholinesterase. J. Pharmacol. Exp. Ther. **1993**, *264* (3), 1085–1089.
9. Allon, N.; Raveh, L.; Gilat, E.; Cohen, E.; Grunwald, J.; Ashani, Y. Prophylaxis against soman inhalation toxicity in guinea pigs by pretreatment alone with human serum butyrylcholinesterase. Toxicol. Sci. **1998**, *43* (2), 121–128.
10. Heim, J.; Schmidt-Dannert, C.; Atomi, H.; Schmid, R.D. Functional expression of a mammalian acetylcholinesterase in *Pichia pastoris*: Comparison to acetylcholinesterase, expressed and reconstituted from *E. coli*. Biochim. Biophys. Acta **1998**, *1396* (3), 306–319.
11. Radic, Z.; Gibney, G.; Kawamoto, S.; MacPhee-Quigley, K.; Bongiorno, C.; Taylor, P. Expression of recombinant acetylcholinesterase in a baculovirus system: Kinetic properties of glutamate 199 mutants. Biochemistry **1992**, *31* (40), 9760–9767.
12. Kronman, C.; Chitlaru, T.; Elhanany, E.; Velan, B.; Shafferman, A. Hierarchy of post-translational modifications involved in the circulatory longevity of glycoproteins. Demonstration of concerted contributions of glycan sialylation and subunit assembly to the pharmacokinetic behavior of bovine acetylcholinesterase. J. Biol. Chem. **2000**, *275* (38), 29488–29502.
13. Mor, T.S.; Sternfeld, M.; Soreq, H.; Arntzen, C.J.; Mason, H.S. Expression of recombinant human acetylcholinesterase in transgenic tomato plants. Biotechnol. Bioeng. **2001**, *75* (3), 259–266.

Interphase Nucleus, The

Paul Fransz
University of Amsterdam, Amsterdam, The Netherlands

INTRODUCTION

Most cells of a multicellular organism contain the same genetic information, yet their gene expression pattern, morphology, and function may differ considerably. Cell identity is determined by the nuclear program and established by a complex molecular interplay between DNA sequence and proteins involved in higher order chromatin structure. The organization of the nucleus, therefore, plays a crucial role in cell type and stage-specific processes. The nucleus is a complex organelle showing a high degree of compartmentalization. The linear organization of several eukaryotic genomes is completed, but we know very little about the higher order chromatin structure. Our knowledge of the functional organization of the interphase nucleus is also still very limited. Most of what we know about nuclear architecture comes from microscopic studies with animal cells. Apart from chromatin, a number of subnuclear structures have been characterized, including the nucleolus, splicing-factor compartments, and many different small nuclear foci. The molecular mechanism underlying the nuclear architecture is currently being unravelled in animals, plants, and yeast. Here, the organization of the interphase nucleus in plants is discussed with respect to chromosome structure and function.

CHROMOSOMES OCCUPY NUCLEAR TERRITORIES

The interphase nucleus is the organelle that accommodates the chromosomes and, thus, the majority of the genetic information that makes up a eukaryote organism.[1] In addition, nuclear processes involve maintaining and passing the genetic information by repairing, replicating, and transcribing the DNA sequence. The haploid genome of higher eukaryotes varies between 10^8 (*Caenorhabditis elegans* and *Arabidopsis thaliana*) and 10^{11} (*Amphiuma* and *Fritillaria*) base pairs. One of the main questions in studies of interphase nuclei is how the eukaryote genome, spanning up to several meters of DNA, fits into a nucleus with a diameter of only a couple of microns. In order to understand nuclear organization it is essential to study individual chromosomes, which occupy the major part of the nucleus as discrete domains. In 1888, Boveri was the first to suggest that interphase chromosomes occupy nuclear territories. One century later, this theory was demonstrated in human cells by chromosome painting using fluorescence in situ hybridization (FISH) with chromosome-specific DNA probes.[2] Chromosome painting appeared impossible in plants with large genomes such as wheat, *Vicia faba* or *Pisum sativum*, probably because the complex molecular organization of their genomes includes abundant dispersed repetitive sequences. However, alien chromosomes present in monosomic or disomic addition lines can be successfully visualized by FISH using the alien genomic DNA as probe. This technique is also known as genomic in situ hybridization (GISH).[3] In interphase nuclei of cereals with large genomes, GISH reveals elongated chromosome territories. Plants with small genomes, such as *Arabidopsis* contain less repeats and have a higher gene density. The low repeat content allows the use of large genomic DNA probes to visualize chromosome-specific regions up to an entire chromosome territory.[4] Differential chromosome painting can monitor position and size of the homologous chromosome territories and subchromosomal regions at subsequent cell cycle stages (Fig. 1).

Although chromosomal territories are discrete nuclear regions, close association of different chromosomes may occur. Long before the FISH technique was developed, fusion of interphase chromosomes was suggested by cytogeneticists who analysed plant species such as the crucifers, which contain nuclei with discrete, heterochromatic chromocenters. The FISH technology identified the subchromosomal regions involved in homolog association. In polyploid wheat, but not in its diploid progenitors, homologous centromeres associate in the xylem cells of developing roots.[5] Similarly, in parenchyma cells of *Arabidopsis*, chromocenters with homologous rDNA repeats associate nonrandomly. Between 10% (chromosome 5) and 75% (chromosome 2) of the homologs associate at their chromocenters.[6] Moreover, correct alignment of homologous regions occurs in 6% ($n = 114$) of the nuclei (Fig. 2A). In comparison association of homologous chromosome territories in mammalian cells was found for gene-dense, small chromosomes. These observations suggest a physical interaction between homologous chromosome territories in somatic cells.

Encyclopedia of Plant and Crop Science
DOI: 10.1081/E-EPCS 120005594

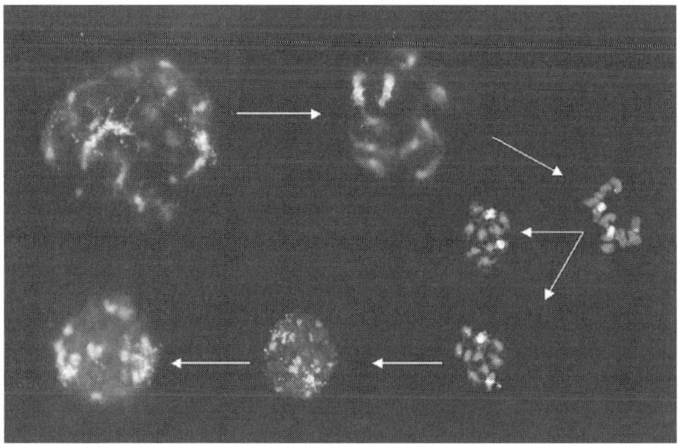

Fig. 1 Different cell cycle stages of *Arabidopsis* parenchyma cells after in situ hybridization with 118 individually labeled BAC DNA clones covering the long arm. (From Ref. 4.) (*View this art in color at www.dekker.com.*)

genome size, such as position or size of the nucleolus and chromosome size. In *Arabidopsis*, for example, the telomeres are clustered around the nucleolus (Fig. 3C). Furthermore, the genome of *Crepis capillaris* (1C = 2000 Mb), which shows a Rabl configuration, is smaller than 3000 Mb, but the chromosomes ($1n = 3$) are longer than in maize (2540 Mb, $1n = 10$), which shows a non-Rabl pattern. The Rabl orientation in wheat (1C = 16000 Mb) and rye (1C = 7500 Mb) corresponds with the elongated chromosome territories in these species. Human chromosomes (1C = 3000 Mb) generally do not show a Rabl orientation, but a polar orientation of chromosome domains with the major heterochromatin segments towards the nuclear periphery has been suggested based on 1) the observation of late-replicating (G band) DNA near the nuclear periphery and 2) electronmicroscopical studies showing a tendency for perinuclear positioning of heterochromatin domains.

CHROMOSOME ARRANGEMENT

Each chromosome consists of several distinct domains such as the centromere region, telomere, heterochromatin, and euchromatin. The chromosome territory is likely organized in such a way that each chromosomal subregion occupies a specific position corresponding to its function. The pericentromeric heterochromatic regions are generally positioned towards the periphery of the nucleus. The location of telomeres, however, appears variable. Organisms with large genomes (haploid genome size >3000 Mb) generally exhibit a Rabl orientation with telomeres and centromeres at opposite nuclear poles, whereas organisms with relatively small genomes show non-Rabl patterns (Fig. 3).[7] The orientation of chromosomes during interphase may depend on factors other than

CHROMOSOME ORGANIZATION

Several models of higher-order chromatin structures have been proposed based on microscopic investigation of human interphase chromosomes. The basic unit of chromatin is the nucleosome, which consists of the core histone proteins H2A, H2B, H3, and H4, around which approximately 150 bp of DNA is wrapped. The nucleosomes form a 10 nm fiber, which is visible under the electron microscope as a beads on the string arrangement. Higher-order structures of chromatin are formed in the presence of histone H1. A chromatin structure, termed the solenoid, has been proposed, which consists of a 30 nm fiber, in which nucleosomes are packaged into a helical coil with six nucleosomes per turn and histone H1 positioned along the central axis of the coil.

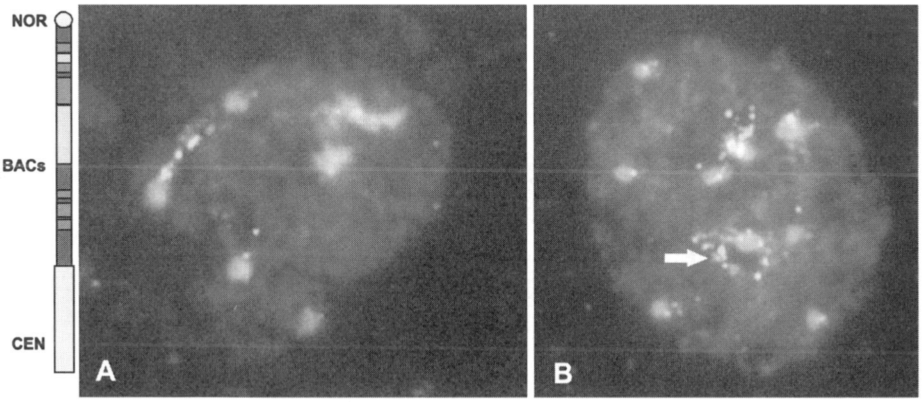

Fig. 2 FISH images of *Arabidopsis* nuclei hybridized with 18 BACs from chromosome arm 4S (see diagram on the left) show aligned chromosome arms (**A**) and a megabase-sized loop (arrow) covering the entire arm 4S in one of the homologs (**B**). (*View this art in color at www.dekker.com.*)

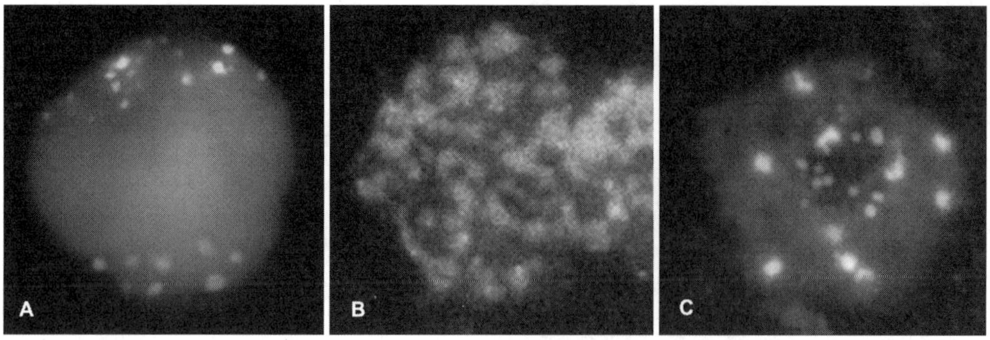

Fig. 3 Telomere position (red) in barley (**A**), tomato (**B**), and *Arabidopsis* (**C**), which have an average chromosome size of 4900 Mb, 950 Mb, and 130 Mb, respectively. The centromere in barley is shown in green opposite to the telomeres, while in tomato and *Arabidopsis* the centromeres are in the heterochromatin islands. (Fig. 3A is kindly provided by Zuzana Jasencakova.) (*View this art in color at www.dekker.com.*)

Further winding of nucleosomes into higher organization levels of chromatin generates irregular folding patterns that are difficult to interpret. Structures such as the scaffold or the matrix have been proposed that may form a supportive skeleton keeping the higher-order chromatin loops together. However, the existence of these supportive structures is a matter of debate, because they have been demonstrated only with severe extraction techniques.

In *Arabidopsis*, nuclei chromosomes have a relatively simple organization. Each chromosome consists of a heterochromatic chromocenter from which euchromatic loops emanate.[6] Chromocenters contain all major tandem repeats and the majority of the dispersed repeats. In contrast, euchromatin loops are gene-rich and range in size from 200 kb up to an entire chromosome arm (Fig. 2B). Together, chromocenter and loops form a chromosome territory (Fig. 4). This chromocenter-loop organization suggests a clear functional differentiation within a chromosome territory. Most of the DNA in chromocenters is heavily methylated, while the tails of histone H3 are methylated at lysine position 9. In contrast, the euchromatin loops contain less methylated DNA and are rich in histone H4 acetylated at lysine positions 5 and 8, and histone H3 methylated at lysine position 4. These modifications of DNA and histones imply that chromocenters represent transcriptionally silent domains, whereas euchromatic loops contain potentially active gene regions.

The model reflects the simple linear organization of the *Arabidopsis* chromosomes, which contain only 15% heterochromatin, confined to the pericentromeric and the subtelomeric 45S ribosomal regions. The majority of the chromosome consists of gene-rich euchromatin. In comparison, wheat and rye contain more than 95% heterochromatin. Their chromosomes are not likely to have a similar

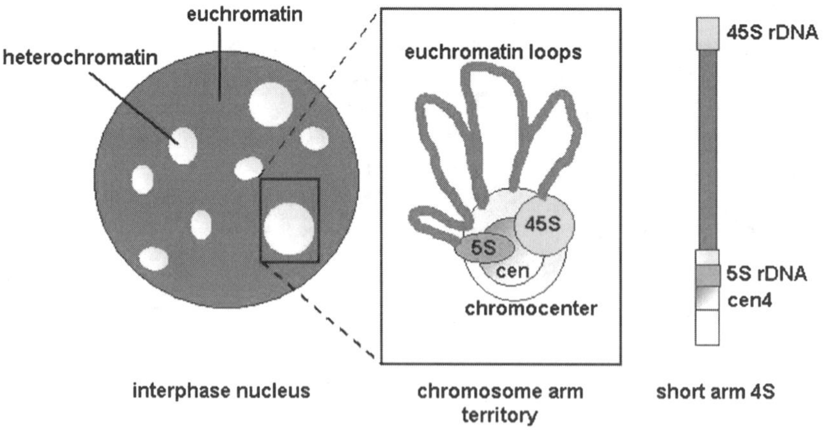

Fig. 4 The chromocenter-loop model for the organization of *Arabidopsis* chromosomes. Each chromosome consists of a heterochromatic chromocenter from which euchromatin loops emanate. The loops may span 0.2 kb up to an entire arm of 3 Mb. (*View this art in color at www.dekker.com.*)

simple chromocenter-loop organization. Based on the genomic sequence, the content of the average chromosome territory of *Arabidopsis* is estimated at 25 Mbp of DNA with 5200 genes. This is in contrast with the average human chromosome territory, which contains five times more DNA (130 Mbp) but only approximately 1700 genes. Megabase-sized loops have been demonstrated indirectly for the histocompatibility complex locus on human chromosome 6. Whether these loops are equivalent to the loops found in *Arabidopsis* and whether the chromocenter-loop model also accounts for other eukaryotes remains to be investigated.

CONCLUSION

The description of territories, euchromatin loops, and heterochromatin domains may suggest a rigid, non-flexible nuclear architecture. However, the interphase nucleus and the chromatin structure appear to be highly dynamic. Advanced 3D-fluorescence microscopy has recently provided new tools for studying the dynamics of chromatin. By tagging chromatin proteins with GFP, it is possible to track specific chromosome sites in real time and measure the kinetic properties of protein complexes.[8] Nuclear components are shown to be transient complexes of nucleic acids and proteins that associate and dissociate dependent on internal or external stimuli. We soon will be able to visualize in detail chromatin remodelling in differentiating cells. More and more we tend to consider the nucleus as a membrane-bound bag filled with DNA, RNA, and protein complexes and lacking a solid skeleton. The flexible, higher-order structure of chromatin is controlled by a complex interplay between DNA and protein complexes. Modulation of chromatin structure and function takes place via covalent modification of histone tails by acetylation, methylation, phosphorylation, or ubiquitination.

These modifications affect the composition of chromatin-associated protein complexes and alter the activity of the gene regions.[9] Several chromatin modifiers have been identified and their effect on gene expression demonstrated in different species. For example, methylation of histone H3 at lysine position K9 by histone methyltransferase SUV39 forms a stable mark that is recognized by heterochromatin protein (HP1). This results in binding of HP1 and subsequent spreading of heterochromatin into native chromatin. The same basic molecular mechanism has been reported for yeast, animals, and plants. The characterization of proteins and protein complexes is currently being established, which will eventually lead to the identification of the factors that are responsible for the higher order organization of chromatin in interphase nuclei.

ARTICLES OF FURTHER INTEREST

B Chromosomes, p. 71
Chromosome Banding, p. 263
Chromosome Manipulation and Crop Improvement, p. 266
Chromosome Rearrangements, p. 270
Chromosome Structure and Evolution, p. 273
Cytogenetics of Apomixis, p. 383
Fluorescence In Situ Hybridization, p. 468
Genome Size, p. 516

REFERENCES

1. Cremer, T.; Cremer, C. Chromosome territories, nuclear architecture and gene regulation in mammalian cells. Nat. Rev. Genet. **2001**, *2* (4), 292–301.
2. Lichter, P.; Cremer, T.; Borden, J.; Manuelidis, L.; Ward, D.C. Delineation of individual human chromosomes in metaphase and interphase cells by in situ suppression hybridization using recombinant DNA libraries. Human Genet. **1988**, *80*, 224–234.
3. Schwarzacher, T.; Leitch, A.R.; Bennett, M.D.; Heslop-Harrison, J.S. In situ localization of parental genomes in a wide hybrid. Ann. Bot. **1989**, *64*, 315–324.
4. Lysak, M.A.; Fransz, P.F.; Ali, H.B.M.; Schubert, I. Chromosome painting in *Arabidopsis thaliana*. Plant J. **2001**, *28*, 689–697.
5. Martinez-Perez, E.; Shaw, P.; Moore, G. The Ph1 locus is needed to ensure specific somatic and meiotic centromere association. Nature **2001**, *411*, 204–207.
6. Fransz, P.; De Jong, J.H.; Lysak, M.; Ruffini Castiglione, M.; Schubert, I. Interphase chromosomes in *Arabidopsis* are organised as well-defined chromocenters from which euchromatin loops emanate. Proc. Natl. Acad. Sci. **2002**, *99*, 14584–14589.
7. Dong, F.; Jiang, J. Non-Rabl patterns of centromere and telomere distribution in the interphase nuclei of plant cells. Chromosome Res. **1998**, *6* (7), 551–558.
8. Misteli, T. Protein dynamics: Implications for nuclear architecture and gene expression. Science **2001**, *291*, 843–847.
9. Fransz, P.F.; De Jong, J.H. Chromatin dynamics in plants. Curr. Opin. Plant Biol. **2002**, *5*, 560–567.

In Vitro Chromosome Doubling

V. Sarasan
Royal Botanic Gardens, Kew, Surrey, U.K.

A. V. Roberts
University of East London, London, U.K.

INTRODUCTION

Chromosome doubling occurs when a cell fails to divide into two daughter cells at mitosis. As a result, a single restitution nucleus is formed that contains the chromosomes that would normally have segregated to two daughter nuclei. A polyploid plant may emerge if the cell with doubled chromosomes develops into a complete plant. In nature, polyploid plants may be created by the spontaneous doubling of chromosomes in somatic cells or by the fusion of unreduced gametes. In the laboratory, it can be induced by the application of antimitotic agents to sporophytic or gametophytic tissues. Chromosome doubling also occurs when protoplasts of somatic cells are fused. This review focuses on the objectives of chromosome doubling, the use of antimitotic agents, and methods of detecting and isolating polyploidy variants.

POLYPLOIDY AND ITS ADVANTAGES

The chromosomes of a gamete are denoted by the letter n, and the chromosomes in each genome of a gamete by x. Thus, a diploid cell can be represented by the formula $2n = 2x$ and a tetraploid plant obtained by chromosome doubling by $2n = 4x$. Doubling the number of chromosomes of a tetraploid results in an octoploid ($2n = 8x$). Hybrids between diploid and tetraploid plants are triploids of usually low fertility. Doubling the chromosome number of a triploid results in a hexaploid ($2n = 6x$). A distinction is drawn between autopolyploids and allopolyploids on the basis of their origin and fertility.[1] An autopolyploid is derived from a fertile plant, such as a typical representative of a species, whereas an allopolyploid is of interspecific origin. An autopolyploid is typically less fertile than the original undoubled plant because, as each chromosome has more than one homologous partner, irregular pairing occurs at meiosis. An allopolyploid is usually fertile because chromosome doubling provides each chromosome with just one homologous partner and meiosis is regular. Allopolyploids benefit from a type of hybrid vigor, which results from the combination of the genes of two or more species. A well known example is evolution, in cultivation, of the bread wheat *Triticum aestivum*. This allohexaploid originated in two stages, each of which involved interspecific hybridization associated with chromosome doubling. First there was the formation of an allotetraploid from two diploid species and then an allohexaploid from the allotetraploid and another diploid species. There was an increase in vigor at each stage.

Although chromosome doubling of somatic cells and fusion of unreduced gametes are events that occur at low frequency in nature, as many as 80% of angiosperm species may be of polyploid origin.[1] The commercial importance of polyploids in agriculture and horticulture has led to the use of antimitotic agents that induce chromosome doubling in somatic cells by interfering with spindle formation. One common reason for inducing polyploidy is to obtain a fertile allopolyploid from a sterile hybrid. Another is to obtain a plant with the larger leaves, flowers, or fruits often associated with both auto- and allopolyploidy.

ANTIMITOTIC AGENTS

During cell division, chromosomes are segregated to opposite poles of a cell by spindle fibers. The spindle fibers are composed of microtubules that are formed rapidly at the onset of cell division by a process that involves the polymerization of α- and β-tubulin polypeptides.[2] Antimitotic agents such as colchicine typically destabilize the microtubules and prevent new ones from forming. Consequently, the daughter chromosomes cannot segregate to opposite poles of the cell at anaphase. The two chromatids of each chromosome are initially held together at the centromere but eventually the association lapses, creating separate chromosomes. Cytokinesis fails to occur, leading to the formation of a restitution nucleus and a polyploid cell.

Encyclopedia of Plant and Crop Science
DOI: 10.1081/E-EPCS 120016562

Colchicine was first used for chromosome doubling by Levan[3] and has been widely used for that purpose to date. However, in many species it is reported to cause undesirable effects such as sterility and abnormal growth.[4] More recently, chromosome doubling has been achieved with antimitotic herbicides, which prevent the synthesis or polymerization of the tubulin molecules into the microtubules of the spindle. Of these, the dinitroaniline herbicides oryzalin and trifluralin, the phosphoric amide herbicide amiprophos methyl (APM), and pronamide have been found to be effective in chromosome doubling.[5] Colchicine has a higher affinity for animal than plant microtubules but the reverse applies to the herbicides.

DETECTION OF POLYPLOIDS

The traditional method for assessing ploidy is by counting chromosomes in root meristematic cells. The recent method of flow cytometry[6] enables ploidy to be determined more rapidly and, as several thousands of cells can be assessed in a single leaf sample, cytochimersim can be detected more readily. Some morphological characteristics such as stomatal length, the number of chloroplasts in guard cells, and pollen grain diameter (which are often greater in tetraploids than diploids) can sometimes be used as indicators of ploidy. Other features that have been commonly noted following chromosome doubling include shorter stems, thicker and darker green leaf laminas with more serrated margins, thicker petioles, thicker stems with shorter internodes, and wider crotch angles (see references listed in Table 1).

STRATEGIES AND OBJECTIVES IN IN VITRO CHROMOSOME DOUBLING

Treatment of Sporophytic Tissues

Antimitotic agents can be applied to somatic tissues in vitro or in vivo. The treatment of seeds may be satisfactorily achieved in vivo by soaking germinating seeds in solutions of an antimitotic agent. There is a good chance the antimitotic agent will be imbibed when the seed swells and that it will reach the target tissue at a suitable concentration. However, the delivery of the antimitotic agent to the meristems of an established plant can be more difficult. In this event, in vitro culture is a widely used approach (Table 1). Delivery of the antimitotic agent may be easier through cut surfaces of an explant than through the protective cuticle of an in vivo plant. In some species, plants can be regenerated through adventitious shoots or somatic embryos after treatment, which greatly reduces the occurrence of cytochimerism.

If all cells in a meristem divided at the same rate, the optimum period of exposure to a spindle inhibitor would correspond to the cell cycle time. However, there are slow cycling cells that may remain undoubled and, therefore the duration of antimitotic agent application required for polyploidization must be determined empirically for each species. Compared to colchicine, oryzalin was found to be, for example, more phytotoxic to the leaves of *Actinidia deliciosa*[7] and more efficient in inducing polyploidy in *Rhododendron*,[8] but reports on the relative merits of oryzalin and colchicine vary. Recent examples of chromosome doubling in sporophytic tissues are listed in Table 1.

Table 1 Recent examples of chromosome doubling of sporophytic and gametophytic cells in vitro

Species	Tissue	Method of induction	Ref.
Sporophytic			
Actinidia deliciosa	Shoots and leaves	COL, ORY	7
Humulus lupulus	Apical shoot buds	COL	13
Rhododendron hybrids	Shoots	COL, ORY	8
Solanum tuberosum	Tuber discs	SPO	14
Trifolium nigrescens x	Axillary meristems	COL	15
Gametophytic			
Allium cepa	Ovule-derived shoots	COL, ORY	9
Beta vulgaris	Ovules	APM, ORY, PRO, TRI	16
Gerbera jamesonii	Ovule-derived shoots	COL, ORY	12
Helianthus annuus[a]	Haploid embryos	SPO, COL	11
Triticum aestivum	Microspores	SPO	17

Abbreviations: APM: amiprophos methyl, COL: colchicine, ORY: oryzalin, PRO: pronamide, SPO: spontaneous, TRI: trifluralin.
[a]After pollination with irradiated pollen.
(Data from Refs. 7–9 and 11–17.)

Treatment of Gametophytic Tissues

The major objective of chromosome doubling of gametophytic tissues is to produce homozygous dihybrids. These can be used as pure breeding parents and two homozygous lines can be hybridized to produce genetically uniform F_1 hybrid seed. The more conventional method of producing pure breeding parents involves the slow procedure of selfing over about seven generations. Culture of haploids can be achieved through androgenic or gynogenic culture (Table 1). This can be combined with spontaneous or induced chromosome doubling to give dihaploids. The suitability of androgenic or gynogenic culture differs according to the species involved. In some species, regeneration is difficult to achieve in the presence of a chromosome-doubling agent. It may then be more efficient to first establish haploid shoot cultures to which antimitotic agents can subsequently be applied, as in gynogenetic cultures of *Allium cepa*.[9]

Androgenic culture can be achieved by anther or microspore culture. Chromosome doubling may occur spontaneously or as a result of treatment with antimitotic agents. Adjustment of in vitro culture conditions may increase the frequency of spontaneous chromosome doubling. Gynogenic cultures may be obtained from unfertilized ovules. Haploids and dihaploid embryos may also be obtained by pollination with pollen of the same species that has been exposed to ionizing radiation, which disrupts normal chromosome function but permits pollen germination. This method was applied by Todorova et al.[10] to sunflower. In gerbera, 76–100% of plants regenerated from unpollinated ovule cultures were haploid but treatment with colchicine resulted in 34% dihaploid plants. Oryzalin was found to be less phytotoxic than colchicine in doubling the chromosome number of gynogenic haploids of *Gerbera*.[11]

CONCLUSION

The use of chromosome doubling for crop improvement has increased in the last decade because two crucially important techniques—plant tissue culture and flow cytometry—are now more widely available. It can be expected that this trend will continue as in vitro techniques, including those for adventitious regeneration, are successfully applied to recalcitrant species. As a method for crop improvement, interspecific hybridization linked with chromosome doubling has some advantages over the development of transgenic crops. It is a relatively low-cost technology, and can be used by small or large commercial enterprises with no public resistance. It is also likely that new applications of this technology will emerge in the future. Conservation of endangered species with reduced reproductive potential is an example. In *Ramosmania rodriguesii*, an endangered species,[12] stylar incompatibilities prevent fertilization and seed setting. Chromosome doubling of sporophytic tissues may in itself result in self-compatible plants. Alternatively, doubling of gametophytic tissues may result in dihaploids with the essential diversity of genotypes required for sexual compatibility.

ARTICLES OF FURTHER INTEREST

Anther and Microspore Culture and Production of Haploid Plants, p. 43
Breeding for Durable Resistance, p. 179
Breeding Hybrids, p. 186
Breeding Using Doubled Haploids, p. 207
Breeding: Choice of Parents, p. 215
Commercial Micropropagation, p. 297
Flow Cytogenetics, p. 460
In Vitro Morphogenesis in Plant—Recent Advances, p. 579
In Vitro Production of Triploid Plants, p. 590
Plant Cell Culture and Its Applications, p. 931
Plant Cell Tissue and Organ Culture: Concept And Methodologies, p. 934
Polyploidy, p. 1038
Somatic Embryogenesis in Plants, p. 1165

REFERENCES

1. Leitch, I.J.; Bennett, M.D. Polyploidy in angiosperms. Trends Plant Sci. **1997**, *2*, 470–476.
2. Downing, K.H. Structural basis for the interaction of tubulin with proteins and drugs that affect microtubule dynamics. Annu. Rev. Cell Dev. Biol. **2000**, *16*, 89–111.
3. Levan, A. The effect of colchicine on meiosis in *Allium*. Hereditas **1939**, *25*, 9–26.
4. Wan, Y.; Petolini, J.F.; Widholm, J.M. Efficient production of doubled haploid plants through colchicine treatment of anther-derived maize callus. Theor. Appl. Genet. **1989**, *77*, 889–892.
5. Vaughan, K.C.; Lehnen, L., Jr. Mitotic disrupter herbicides. Weed Sci. **1991**, *39*, 450–457.
6. Mottley, J.; Yokoya, K.; Roberts, A.V. D-flowering, the Flow Cytometry of Plant DNA. In *Living Color: Protocols in Flow Cytometry and Cell Sorting*; Diamond, R., DeMaggio, S., Eds.; Springer-Verlag: Berlin, 2000.
7. Chalak, L.; Legave, J.M. Oryzalin combined with adventitious regeneration for an efficient chromosome

doubling of trihaploid kiwifruit. Plant Cell Rep. **1996**, *16*, 97–100.
8. Vainola, A. Polyploidization and early screening of Rhododendron hybrids. Euphytica **2000**, *112*, 239–244.
9. Geoffriau, E.; Kahane, R.; Bellamy, C.; Rancillac, M. Ploidy stability and in vitro chromosome doubling in gynogenic clones of onoin (*Allium cepa* L.). Plant Sci. **1997**, *122*, 201–208.
10. Todorova, M.; Ivanov, P.; Shindrova, P.; Christov, M.; Ivanova, I. Doubled haploid production of sunflower (*Helianthus annuus* L.) through irradiated pollen-induced parthenogenesis. Euphytica **1997**, *97*, 249–254.
11. Tosca, A.; Pandolfi, R.; Citterio, S.; Fasoli, A.; Sgorbati, S. Determination by flow-cytometry of the chromosome doubling capacity of colchicine and oryzalin in gynogenetic haploids of *Gerbera*. Plant Cell Rep. **1995**, *14*, 455–458.
12. Owens, S.J.; Jackson, A.; Maunder, M.; Rudall, P.; Mat, J. The breeding systems of *Ramosmania heterophylla*—Dioecy or heterostly. Bot. J. Linn. Soc. **1993**, *113*, 77–86.
13. Roy, A.T.; Legget, G.; Koutoulis, A. In vitro induction and generation of tetraploids from mixoploids in hop (*Humulus lupulus* L.). Plant Cell Rep. **2001**, *20*, 489–495.
14. Imai, T.; Aida, R.; Ishige, T. High-frequency tetraploidy in *Agrobacterium*-mediated transformants regenerated from tuber discs of diploid potato lines. Plant Cell Rep. **1993**, *12*, 299–302.
15. Hussain, S.W.; Williams, W.M.; Mercer, C.F.; White, D.W.R. Transfer of clover cyst nematode resistance from *Trifolium nigrescens* Viv to *T. repens* L by interspecific hybridisation. Theor. Appl. Genet. **1997**, *95*, 1274–1281.
16. Hansen, A.L.; Gertz, A.; Joersobo, M.; Andersen, S.B. Antimicrobial herbicides for in vitro chromosome doubling in *Beta vulgaris* L. ovule culture. Euphytica **1998**, *101*, 231–237.
17. Liu, W.; Zheng, M.Y.; Konzak, C.F. Improved green plant production via isolated microspore culture in bread wheat (*Triticum aestivum* L.). Plant Cell Rep. **2002**, *20*, 821–824.

In Vitro Flowering

Prakash Lakshmanan
David North Plant Research Centre, Brisbane, Queensland, Australia

Acram M. Taji
The University of New England, Armidale, New South Wales, Australia

INTRODUCTION

Flowering—the transition from the vegetative to the reproductive phase in plants—is an important developmental process with considerable biological and economic significance. Despite decades of research and rapid advances in technology, our understanding of this important developmental process is still fragmentary. From the results of previous research, it is evident that the majority of plants use environmental cues to regulate flowering. Environmental variables with regular seasonal patterns such as temperature, photoperiod, and irradiance are the key signals for floral transition. These factors are perceived by different plant parts, and strong and diverse interactions between the environmental variables are required for floral induction to occur in many species.

Classical physiological, genetic, and grafting experiments, although invaluable in deciphering various aspects of flowering, have failed to unravel the true nature of the flowering stimulus or the mechanism(s) by which various environmental cues induce flowering. Novel approaches involving in vitro and molecular techniques offer unique opportunities to investigate the flowering process from new perspectives. This article summarizes the major achievements of in vitro techniques in understanding flowering, and highlights the merits and future prospects of such approaches in resolving the questions surrounding floral transition.

ADVANTAGES OF IN VITRO FLOWERING

As compared to whole plant-based, in vivo experimental conditions, in vitro methods present several advantages to investigate flowering. They allow for much greater control of experimental conditions under investigation. For instance, precise and efficient application of different treatments and elimination of interfering factors can be achieved relatively easily under in vitro conditions. This ensures more reliable analysis of the role of growth regulators, nutrients, and environmental variables on flowering. Under tissue culture conditions, whole plants and individual plant parts (including isolated meristems) can be used for floral induction. In vitro cultures are more amenable to bioassays and model systems development than in vivo methods because growth conditions can be precisely controlled and monitored in culture. In vitro systems have been successfully used for the determination of molecular elements involved in flowering and reproductive development.[2] They can be exploited for in vitro breeding, which may reduce the breeding cycle considerably in perennial crops.

FACTORS INFLUENCING IN VITRO FLOWERING

Successful induction of flowering in vitro has been reported for numerous species from both monocotyledonous and dicotyledonous families. Considering the complexity of the regulation of flowering, it is not surprising that no common set of growth conditions exist that could be applied to in vitro floral induction in diverse species. From the available evidence, the key regulatory factors that induce flowering in vitro include photoperiod, irradiance, temperature, growth stage, and phytohormones. Besides these controlling factors, in vitro studies also showed that various nutrients, particularly nitrogen and carbon, and many nonhormonal bioactive compounds such as polyamines, jasmonates, and benzoic acid derivatives play a role in the regulation of vegetative to reproductive phase transition.

Phytohormones and In Vitro Flowering

Whereas there is no conclusive evidence to establish a particular compound as the master signal evoking floral transition under in vivo conditions, hormones in general are emerging as important factors regulating floral induction in in vitro cultures in a large number of species.[3] As with any other developmental process, flowering in vitro is promoted by some hormones while inhibited by others. More intriguingly, in some species, the same hormone

Encyclopedia of Plant and Crop Science
DOI: 10.1081/E-EPCS 120010559

elicits contrasting flowering responses under different culture conditions.

Auxins in general do not promote flowering in vitro. However, there are examples of auxin-induced flowering in culture. For instance, *Pharbitis nils* seedlings grown in vitro could be induced to flower in noninductive conditions by the application of α-naphthaleneacetic acid (NAA). Auxin also promoted flowering in *Perilla fructescence*, *Streptocarpus nobilis*, *Phlox drummondi*, and *Torenia fournieri*.[3] Nonetheless, more evidence points to an inhibitory role for auxin in flower induction. Indeed, experiments with day-neutral tobacco cultures indicate that auxin has the potential to even revert the plants that are already in the reproductive phase to vegetative state. Working with *Torenia*—the most extensively investigated plant in relation to auxin involvement in flowering—Tanimoto and Harada[4] concluded that auxin is unlikely to be a component of floral stimulus, although it promotes flower bud differentiation.

Among the different phytohormones, cytokinins are the most effective inducers of flowering in vitro. In most of the species tested, at least one cytokinin was required for flower induction and normal development of floral parts.[3] It is important to note that in many species, for example, *Plumbago indica*, *P. fructescence*, and *Browallia demissa*, cytokinins were able to induce flowering in explants obtained from strictly vegetative plants maintained under noninductive conditions. Floral induction occurred only in a narrow range of cytokinin concentrations, and supraoptimal levels of cytokinin in the medium always resulted in vegetative bud formation in all the species investigated. Although the vast majority of the reports indicate a promotive role for cytokinin, it inhibited flowering in cultures of *Scrofularia arguta* and different culivars of *Kalanchoe* and tobacco.

Different plant species require different types of cytokinins to induce flowering in culture. As an example, 6-benzyladenine is the most effective cytokinin for flower production in many orchids, bamboos, and passion fruit, whereas kinetin was more favorable than others for floral induction in *Arabidopsis* stem explants. In many species, however, a combination of different cytokinins or of cytokinin and auxin was needed for in vitro flower induction.

Gibberellins, the most potent florigenic compounds in vernalization-requiring plants, received much less attention in in vitro studies compared with cytokinins and auxins. Both promotive and inhibitory roles have been suggested for gibberellins with regard to in vitro flowering. Unlike cytokinins, there is little conclusive evidence to suggest that gibberellins evoke floral induction in vitro, as the explants used for floral induction studies involving gibberellin were derived from already flowering or florally-determined plants. Considering its importance in the development of reproductive structures, there is an obvious void of information about the inductive role of gibberellin in flowering of tissue-cultured plants.

Despite the fact that inhibitors have been implicated in the regulation of reproductive development, little work has been done on the action of abscisic acid (ABA) and ethylene on in vitro flowering. ABA induced flowering in in vitro cultures of *T. fournieri* and *P. nils*, and markedly enhanced flower production in *P. indica* and *P. fructescence*. A close parallel between the endogenous ABA level and the ability to flower in vitro was noticed in *Torenia*. In this species, applied ABA inhibited flower bud formation in tissues taken from old plants, whereas a promotive effect on flowering was noticed in explants from younger plants.

Nonhormonal Compounds and In Vitro Flowering

Besides the known phytohormones, many bioactive compounds were found to have florigenic activity in vitro. The most significant ones are nicotinic acid, salicylic acid, benzoic acid, and coumaric acid.

Knowledge about the inductive role of mineral nutrients in in vitro flowering is limited. Investigations on *Torenia* found that low salt concentration and the absence of NH_4NO_3 were supportive of flowering in vitro. Nitrate appears to be the most florigenic of all the forms of nitrogen, and a high carbon:nitrogen ratio brings about flowering in some plants. In *Sinapsis alba* and *Cuscuta reflexa*, sucrose could induce or eliminate the requirement of high light intensity for flower induction, whereas even a brief removal of tobacco explants from glucose delayed flower formation.

CONTRIBUTION OF IN VITRO STUDIES ON FLOWERING

From the foregoing discussion it is evident that considerable effort has been expended to develop in vitro flowering systems for a large number of species. Many of these culture systems could be used to develop reproducible and sensitive bioassay systems to verify the role of chemical, biological, and environmental factors in controlling flowering. In addition, such systems are often the only ways to gain more insight into certain aspects of flowering, which are otherwise impossible to achieve in an in vivo environment. For example, experiments with florally determined tobacco and orchid plants have shown that determination to flower can be carried through from the flowering parent plant to the explant, and that it can persist for long periods in culture, often for several generations in orchids.

In vitro approaches provide unique opportunities to test and interpret the theories of the multicomponent model of flowering. To cite an example, it is almost impossible to verify the antagonistic effects exerted by the root on flowering without resorting to tissue culture systems. With appropriate experimental strategies, this aspect of flowering has been elegantly demonstrated in *Kalanchoe, Cichorium*, and sunflower.

The value of having a precisely defined and reproducible in vitro system to further our knowledge of flowering is best exemplified in the recent research by Yu and Goh on orchids.[2,5] They exploited a cytokinin-induced flowering system of *Dendrobium* orchid to investigate the profile of gene expression during flowering,[2] and identified three MADS-box genes of the AP1/AGL9 subfamily expressed specifically at the time of floral transition.[5] In addition, the *Dendrobium* in vitro flowering system allowed them to identify and characterize *DOH1*, a gene required for maintenance of basic plant architecture and floral induction.[6]

IN VITRO FLOWERING AND PLANT BREEDING

An in vitro system producing fertile flowers would be an invaluable breeding tool, especially for difficult-to-flower plants such as trees, bamboos, etc. In bamboo, which usually has 12 to 24 years of juvenile growth and is gregarious (i.e., all the bamboos in a local population flower simultaneously), the development of an efficient in vitro flowering system capable of producing viable seeds will be an important advancement in breeding this species.[7] A similar tissue culture system producing fertile flowers and viable seeds has been described for various orchids.

CONCLUSION

From the preceding discussion, it is apparent that in vitro flowering is not only an efficient approach to understanding various facets of flowering, but could also be exploited for practical applications. Considering the success obtained with numerous plant species, the development of an in vitro flowering system for most of the angiosperms is now within reach. The physiological and developmental status of the donor plant, composition and concentration of mineral salts in the medium, and presence of cytokinin are emerging as key factors controlling floral transition in culture. Clearly, in vitro systems are presenting unique opportunities to unravel the mysteries of flowering. With advances in molecular biology, remarkable progress in the molecular characterization of this developmental process can be expected in the near future.

ARTICLES OF FURTHER INTEREST

Anther and Microspore Culture and Production of Haploid Plants, p. 43
Floral Induction, p. 452
In Vitro Morphogenesis in Plants—Recent Advances, p. 579
In Vitro Plant Regeneration by Organogenesis, p. 584
In Vitro Pollination and Fertilization, p. 587
In Vitro Production of Triploid Plants, p. 590
Plant Cell Tissue and Organ Culture: Concepts and Methodologies, p. 934

REFERENCES

1. Hempel, F.D.; Welch, D.R.; Feldman, L.J. Floral induction and determination: Where is flowering controlled. Trends Plant Sci. **2000**, *5*, 17–21.
2. Yu, H.; Goh, C.J. Differential gene expression during floral transition in an orchid hybrid *Dendrobium* Madame Thong-In. Plant Cell Rep. **2000**, *19*, 926–931.
3. Van Staden, J.; Dickens, C.W.S. In Vitro Induction of Flowering and Its Relevance to Micropropagation. In *Biotechnology in Agriculture and Forestry, High-Tech and Micropropagation 1*; Bajaj, Y.P.S., Ed.; Springer-Verlag: Berlin, 1991; Vol. 17, 85–115.
4. Tanimoto, S.; Harada, H. Effect of IAA, ammonium nitrate and sucrose on the initiation and development of floral buds in *Torenia* stem segments cultured in vitro. Plant Cell Physiol. **1981**, *22*, 1553–1560.
5. Yu, H.; Goh, C.J. Identification and characterisation of three orchid MADS-box genes of the AP1/AGL9 subfamily during floral transition. Plant Physiol. **2000**, *123*, 1325–1336.
6. Yu, H.; Yang, S.H.; Goh, C.J. DOH1, a class 1 *knox* gene, is required for maintenance of the basic plant architecture and floral transition in orchid. Plant Cell **2000**, *12*, 2143–2160.
7. Nadgauda, R.S.; Parsharami, V.A.; Mascarenhas, A.F. Precocious flowering and seeding behaviour in tissue cultured bamboos. Nature **1900**, *344*, 335–336.

In Vitro Morphogenesis in Plants—Recent Advances

Gregory C. Phillips
Arkansas State University, State University, Arkansas, U.S.A.

INTRODUCTION

The capacity of cultured plant tissues and cells to undergo morphogenesis, resulting in the formation of discrete organs or whole plants, has provided opportunities for numerous applications of in vitro plant biology in studies of basic botany, biochemistry, propagation, breeding, and development of transgenic crops. Whereas the fundamental techniques to achieve in vitro plant morphogenesis have been well established for a number of years, innovations in particular aspects of the technology continue to be made. Tremendous progress has been made in recent years regarding the genetic bases underlying both in vitro and in situ plant morphogenesis, stimulated by progress in functional genomics research. Advances in the identification of specific genes that are involved in plant morphogenesis in vitro, as well as some selected technical innovations, will be discussed.

FUNDAMENTAL ASPECTS OF IN VITRO MORPHOGENESIS

The two primary morphogenic pathways leading to whole plant regeneration—which is a prerequisite for most plant breeding and genetic and transgenic applications of in vitro biology—involve either somatic embryogenesis, or shoot organogenesis followed by root organogenesis. Both developmental pathways can occur either directly without a callus intermediate stage, termed adventitious; or indirectly following an unorganized callus stage, termed de novo.[1] Few plant species have been shown to regenerate by both organogenic and somatic embryogenic pathways, but many plant species can regenerate by one or the other of these pathways.

Somatic embryogenesis may be the best example of totipotency expressed among a large number of plants.[2] Various culture treatments can be manipulated to optimize the frequency and morphological quality of somatic embryos, which are bipolar structures containing both shoot and root apices and developing in a manner parallel to that of zygotic embryos. Typical treatment factors include the plant growth regulator sources and concentrations (especially the auxin), choice of explant nutrient medium composition (especially inorganic vs. organic nitrogen sources and carbohydrate sources and concentrations), culture environment (including the physical form of the medium, e.g., liquid or semisolid; pH; humidity; light quality and quantity or absence of light; temperature; gaseous environment), and osmotic potential. Many of these factors must be adjusted (e.g., carbohydrates, nitrogen sources) or completely changed (e.g., withdrawal or reduction in auxin signal; perhaps an increase in other plant growth regulators such as abscisic acid; osmotic potential change to encourage desiccation) during maturation of somatic embryos, during which time they become competent for conversion into plantlets.[2]

Many of the same culture factors described above for somatic embryogenesis are also manipulated to induce and optimize organogenesis, but often these factors are manipulated in different ways.[3] For example, a high auxin signal (often specifically using 2,4-dichlorphenoxyacetic acid) is usually important to induce somatic embryogenesis, whereas a high cytokinin-to-auxin ratio (or high cytokinin with no auxin) is typically required to induce shoot organogenesis. Root initiation also typically requires a moderate to high auxin signal—but rarely use of 2,4-dichlorophenoxyacetic acid. Rather, a more ''natural'' source of auxin is used.[1] Because regenerated organs are unipolar, two distinct organogenic induction signals—one to induce shoots and the other to induce roots—are required to regenerate a whole plant. In contrast, bipolar somatic embryos are induced by a single induction signal.

GENETIC COMPONENTS OF MORPHOGENESIS

One of the most exciting advances in recent years is the discovery of specific genes involved in plant regeneration in vitro. Such genes are being explored in order to increase transformation efficiency and to develop marker-free transgenic plants.[4] Because a primary factor in optimizing somatic embryogenesis and organogenesis is phytohormone models, it is of interest that receptors for each of the major phytohormone classes have now been identified and many of the corresponding genes have been cloned.[4,5] Examples of specific genes involved in the major plant morphogenesis pathways are summarized in Table 1.

Encyclopedia of Plant and Crop Science
DOI: 10.1081/E-EPCS 120010554

Table 1 Examples of genes involved in various plant morphogenesis pathways

Gene	Putative function	Reference(s)
Somatic embryogenesis		
LEC2	Initiates ectopic somatic embryogenesis	4
WUS (*PGA6*), *SERK*, *LEC1*	Involved in the vegetative–to–embryogenic transition	4,6,7
SHR	Establishes ground tissue via asymmetric cell division	8
CLV, *WUS*	Regulate stem cell fate	8,9
CLV1, *CLV3*, *STM*	Regulate shoot apical meristem development	8,9
LEC1, *FUS3*, *ABI3*	Regulate embryo maturation	8
Shoot organogenesis		
CYCD3	Involved in acquisition of competence for organogenesis	9,10
SRD3	Competence for shoot organogenesis	10,11
SRD1, *SRD2*	Competence for redifferentiation of shoots	10,11
ESR1	Enhances shoot regeneration, vegetative–to–organogenic transition	4
CRE1	Cytokinin receptor	4
CKI1	Cytokinin perception	4,9
CLV, *WUS*	Preserve stem cell identity in shoot apical meristem	9
KN1, *STM*	Initiate ectopic shoot meristems, shoot apical meristem function	9
SHO, *MGO*	Modifiers of the shoot apical meristem involved in leaf founder cell recruitment, lateral organ primordia	9
Root organogenesis		
SRD2	Competence for root organogenesis	10,11
PKL	Transition of embryonic root cells to grow vegetatively	7
RML	Root apical meristem function	11,12
CYCD4;1	Involved in lateral root formation	13
RAC	Involved in adventitious root formation and auxin transduction	11
Floral organogenesis		
LFY	Switch to reproductive development, floral meristem identity	9,14
AP1	A-class gene involved in establishing the first floral whorl: petals	14
UFO	Interacts with *LFY* by providing regional specificity within floral meristems and to control B-class signals that establish the second floral whorl: sepals	14
WUS	Interacts with *LFY* to control C-class genes	14
AG	C-class gene typifying the class; interacts with B-class signals to produce the third floral whorl: stamens; C-class genes acting alone produce the fourth floral whorl: carpels	14
SEP	Cofactors for A-, B-, and C-class genes to convert vegetative leaves into floral organs	14

Genetic Aspects of Somatic Embryogenesis

Transgenic expression of the *LEC2* (leafy cotyledon) gene (Table 1) is sufficient to initiate somatic embryogenesis with high viability but some abnormalities persist in morphology.[4] Several genes appear to be involved in the vegetative-to-embryogenic transition, such as *WUS* (wuschel, or *PGA6*, plant growth activator), *LEC1*,[6] *SERK* (somatic embryogenesis receptor kinase),[4] and *PT1* (primordial timing).[7] *SHR* (short root) establishes the ground tissue through the first asymmetric cell division,[8] *CLV* (clavata) and *WUS* interact to determine stem cell fate, and *CLV* and *STM* (shoot meristemless) regulate development of the shoot apical meristem.[8,9] *LEC1*, *ABI3* (abscisic acid-insensitive), and *FUS3* (fusca) are involved in somatic embryo maturation.[8]

Genetic Aspects of Shoot Organogenesis

CYCD3 is involved in the acquisition of competence for shoot regeneration,[9,10] as is *SRD3* (shoot redifferentiation)[10,11] (Table 1). Shoot redifferentiation also involves *SRD1* and *SRD2*.[10,11] The vegetative-to-shoot organogenesis transition is promoted by *ESR1* (enhancer of shoot regeneration).[4] Two genes representing potentially independent pathways involved early in shoot organogenesis signal transduction include *CRE1* (cytokinin receptor) and *CKI1* (cytokinin independent).[4,11] The shoot apical meristem stem cell identity is regulated by *CLV* and *WUS*,[9] parallel to that observed in the shoot apical meristem of somatic embryos (described earlier). *STM* and *KN1* (knotted) are involved in the function of the shoot apical meristem, and overexpression leads to the formation of ectopic shoot meristems.[9,10] Other regulators involved in shoot apical meristem organization and lateral shoot formation include multiple *SHO* (shoot organization) and *MGO* (mgoun) genes.[9]

Genetic Aspects of Root Organogenesis

Competence to regenerate root organs is affected by *SRD2*[10,11] (Table 1). The transition of embryonic root cells to initiate vegetative growth is controlled by *PKL* (pickle).[7] *RML1* (root meristemless) and *RML2* play specific roles in the root apical meristem,[11] and interact with components of the apical dominance system.[12] The *RAC* (rooting auxin cascade) gene is involved in an early stage of auxin perception specific to the formation of adventitious roots,[11] and CYCD4;1 is directly involved in lateral root primordia formation.[13]

Genetic Aspects of Floral Organogenesis

Floral organs arise as determinate structures out of the indeterminate shoot apical meristem.[9] The concept of floral organs being specified by the A-, B-, and C-class genes is well established.[14] *LFY* (leafy) is a key gene involved in the switch to reproductive growth and in establishing floral meristem identity[9,14] (Table 1). *LFY* activates the key A-class gene *AP1* (apetala), establishing the petals or outermost whorl of the floral organ.[14] *LFY* and *UFO* (unusual floral organs) interact to control the B-class genes, with *UFO* providing regional specificity within meristems and thereby establishing the sepal whorl. *LFY* and *WUS* interact to control the C-class genes typified by *AG* (agamous), and the C-class genes interact with B-class genes to establish the stamens in the third whorl. C-class genes also act alone to establish the fourth or innermost whorl composed of carpels (because *AG* suppresses the action of *WUS*, thereby resulting in a suppression of the B-class components). Three MADS-Box *SEP* (sepellata) genes act as cofactors with the A-, B-, and C-class genes to convert vegetative leaves into floral organs.

TECHNICAL INNOVATIONS

In the past decade, many of the technical improvements resulting in improved in vitro plant regeneration systems have been related to manipulation of the gaseous and/or physical environment of the cultures. Falling outside this category are a few other noteworthy innovations pertaining to thin cell layer techniques and synthetic seeds.

Manipulation of the Gaseous and/or Physical Environment

Cultured plant tissues are known to interact with the culture medium and gaseous environment. Forced ventilation and use of ventilated culture vessels, for example, have facilitated optimization of in vitro morphogenesis systems, and high CO_2 treatments have permitted establishment of photoautotrophic cultures.[15] Control of the amount of ethylene released by the cultured tissues into the head space of the culture vessel—or alternatively, inhibition of ethylene synthesis or action—have led to improved morphogenic responses.[16]

Efforts to improve bioreactor designs to facilitate economical large-scale production of plants or plant products have continued. Key issues that must be addressed with bioreactor designs for plant cell and tissue growth include aeration and minimization of shear damage. Advances in automation and computer controls have rendered bioreactor performance more reliable.[17] One of the most exciting developments in bioreactor design has been the temporary immersion system, which alternates immersion of the plant tissues in the liquid culture medium with exposure to the air space at timed intervals.[18] Temporary immersion bioreactors have been demonstrated to improve yields of shoot proliferation cultures, microtubers, and somatic embryos, as well as improve the quality and vigor of the propagules with reduced frequencies of abnormalities and hyperhydricity.

Another interesting development is the use of perfluorochemicals and commercially stabilized bovine hemoglobin as gas carriers to enhance cell performance in liquid culture systems such as bioreactors. Perfluorochemicals are recyclable (can be used to deliver gases, then be recovered from the culture and recharged), and emulsion with the surfactant Pluronic F-68® appears to synergistically enhance effectiveness. These gas carriers have been shown to improve cell division rates, stimulate biomass production, improve yields of cellular products, and enhance morphogenic totipotency.[19] A technical innovation with a more physical impact on the culture environment is the use of semipermeable cellulose acetate membranes to enhance citrus somatic embryogenesis and particularly to normalize somatic embryo development.[20]

Applications of Thin Cell Layer and Synthetic Seed Techniques

Thin cell layer culture, an approach involving mainly the manipulation of explant size to induce and optimize regeneration, has been used for many years with dicotyledonous species to study in vitro morphogenesis. Thin cell layer cultures can be manipulated for rigorously controlled programming of different morphogenic responses: callus formation, shoot organogenesis, root organogenesis, floral organogenesis, or somatic embryogenesis.[21] In recent years the thin cell layer technique has been extended to a variety of species formerly considered to be recalcitrant to in vitro morphogenesis. Evidence is also gathering that thin cell layer techniques can be useful for recovering transgenic plants from species heretofore considered recalcitrant to genetic transformation.

There continues to be interest in developing synthetic seed technology based on artificial encapsulation of somatic embryos suitable for direct field sowing with reliable conversion into viable plants. The most important technical advances in this area involve the use of automated bioreactors to improve yields, combined with the use of computer imaging to sort out the somatic embryos possessing sufficient quality for encapsulation and subsequent conversion.[22] Even more exciting are the advances in use of nonembryogenic (unipolar) structures for encapsulation as synthetic seed.[23] There seems to be a lower risk of somaclonal variation using unipolar structures such as microbulbs; microtubers; rhizomes; corms; shoots; or nodes containing either apical or axillary buds, meristemoids, and bud primordia for encapsulation. Synthetic seed technology can be extended to a wider variety of genotypes.

CONCLUSION

Basic research has begun to dissect the complex genetic pathways involved in various aspects of plant morphogenesis, including all of the major pathways leading to in vitro plant regeneration. A number of candidate genes are being identified that can be expressed transgenically to enhance or even to initiate plant regeneration from cultured cells and tissues. Such genes are being explored for potential use in developing marker-free transgenic systems as well as to potentially enhance the frequencies of transgenic plant recovery.[4] These advances, as well as advances in specific culture systems such as thin cell layers,[21] offer the prospects of extending more efficient in vitro plant regeneration techniques to previously recalcitrant crops, and of developing more efficient genetic transformation methods. Such advances in controlling in vitro morphogenesis should play important roles at the applied level in developing new crop cultivars and reducing the cost of micropropagation, and in furthering basic research in the area of functional genomics by testing of transgenes in a wider array of plant species.

ACKNOWLEDGMENTS

The author thanks Dr. Oluf Gamborg, Dr. Trevor Thorpe, and Dr. Prakash Lakshmanan for helpful comments regarding topics appropriate to this review.

ARTICLES OF FURTHER INTEREST

Anther and Microspore Culture and Production of Haploid Plants, p. 43
Flower Development: The Pathway to, p. 464
In Vitro Flowering, p. 576
In Vitro Plant Regeneration by Organogenesis, p. 584
In Vitro Production of Triploid Plants, p. 590
In Vitro Tuberization, p. 594
Plant Cell Culture and Its Applications, p. 931
Plant Cell Tissue and Organ Culture: Concepts and Methodologies, p. 934
Regeneration from Guard Cells of Crop and Other Plant Species, p. 1081
Somatic Embryogenesis in Plants, p. 1165

REFERENCES

1. *Plant Cell, Tissue and Organ Culture—Fundamental Methods*; Gamborg, O.L., Phillips, G.C., Eds.; Springer-Verlag: Heidelberg, NY, 1995.
2. Thorpe, T.A. Somatic embryogenesis: Morphogenesis, physiology, biochemistry and molecular biology. Korean J. Plant Tissue Cult. **2000**, *27* (4), 245–258.
3. Joy, R.W., IV; Thorpe, T.A. Shoot Morphogenesis: Structure, Physiology, Biochemistry and Molecular Biology. In *Morphogenesis in Plant Tissue Cultures*; Soh, W.-Y., Bhojwani, S.S., Eds.; Kluwer Academic Publishers: Dordrecht, 1999; 171–214.
4. Zuo, J.; Niu, Q.-W.; Ikeda, Y.; Chua, N.-H. Marker-free transformation: Increasing transformation frequency by the use of regeneration-promoting genes. Curr. Opin. Biotechnol. **2002**, *13*, 173–180.
5. Møller, S.G.; Chua, N.-H. Interactions and intersections of plant signaling pathways. J. Mol. Biol. **1999**, *293*, 219–234.
6. Zuo, J.; Niu, Q.-W.; Frugis, G.; Chua, N.-H. The *WUSCHEL* gene promotes vegetative-to-embryonic transition in *Arabidopsis*. The Plant J. **2002**, *30* (3), 349–359.
7. Harada, J.J. Signaling in plant embryogenesis. Curr. Opin. Plant Biol. **1999**, *2*, 23–27.
8. von Arnold, S.; Sabala, I.; Bozhkov, P.; Dyachok, J.; Filonova, L. Developmental pathways of somatic embryogenesis. Plant Cell Tissue Organ Cult. **2002**, *69* (3), 233–249.
9. Fletcher, J.C. Coordination of cell proliferation and cell

fate decisions in the angiosperm shoot apical meristem. BioEssays **2002**, *24* (1), 27–37.
10. Sugiyama, M. Organogenesis in vitro. Curr. Opin. Plant Biol. **1999**, *2*, 61–64.
11. Sugiyama, M. Genetic analysis of plant morphogenesis in vitro. Int. Rev. Cytol. **2000**, *196*, 67–84.
12. Anderson, J.V.; Chao, W.S.; Horvath, D.P. A current review on the regulation of dormancy in vegetative buds. Weed Sci. **2001**, *49*, 581–589.
13. De Veylder, L.; de Almeida Engler, J.; Burssens, S.; Manevski, A.; Lescure, B.; van Montagu, M.; Engler, G.; Inzé, D. A new D-type cyclin of *Arabidopsis thaliana* expressed during lateral root primordia formation. Planta **1999**, *208*, 453–462.
14. Lohmann, J.U.; Weigel, D. Building beauty: The genetic control of floral patterning. Dev. Cell **2002**, *2*, 135–142.
15. Buddendorf-Joosten, J.M.C.; Woltering, E.J. Components of the gaseous environment and their effects on plant growth and development in vitro. Plant Growth Regul. **1994**, *15*, 1–16.
16. Kumar, P.P.; Lakshmanan, P.; Thorpe, T.A. Regulation of morphogenesis in plant tissue culture by ethylene. In Vitro Cell. Dev. Biol., Plant **1998**, *34* (2), 94–103.
17. Paek, K.-Y.; Hahn, E.-J.; Son, S.-H. Application of bioreactors for large-scale micropropagation systems of plants. In Vitro Cell. Dev. Biol., Plant **2001**, *37* (2), 149–157.
18. Etienne, H.; Berthouly, M. Temporary immersion systems in plant micropropagation. Plant Cell, Tissue Organ Cult. **2002**, *69* (3), 215–231.
19. Lowe, K.C.; Davey, M.R.; Power, J.B. Perfluorochemicals: Their applications and benefits to cell culture. Trends Biotechnol. **1998**, *16*, 272–277.
20. Niedz, R.P.; Hyndman, S.E.; Wynn, E.T.; Bausher, M.G. Normalizing sweet orange [C. *sinensis* (L.) Osbeck] somatic embryogenesis with semi-permeable membranes. In Vitro Cell. Dev. Biol., Plant **2002**, *38* (6), 552–557.
21. Nhut, D.T.; Teixeira da Silva, J.A.; Aswath, C.R. The importance of the explant on regeneration in thin cell layer technology. In Vitro Cell. Dev. Biol., Plant **2003**, *39*, 266–276.
22. Ibaraki, Y.; Kurata, K. Automation of somatic embryo production. Plant Cell Tissue Organ Cult. **2001**, *65* (3), 179–199.
23. Standardi, A.; Piccioni, E. Recent perspectives on synthetic seed technology using nonembryogenic in vitro-derived explants. Int. J. Plant Sci. **1998**, *159* (6), 968–978.

In Vitro Plant Regeneration by Organogenesis

Janet R. Gorst
Benson Micropropagation, Brisbane, Queensland, Australia

INTRODUCTION

Efficient regeneration of plants from cells and tissues through organogenesis is an important prerequisite for the successful application of biotechnology to crop improvement. In vitro regenerability is a highly variable genetic trait that can be introgressed into nonregenerating (recalcitrant) lines by conventional breeding. The availability of mutants with distinctive regenerative characteristics and the rapid developments in molecular biology have considerably advanced our understanding of the phenomenon of in vitro organogenesis in the recent past. While the last 10 years have seen remarkable progress in defining molecular mechanisms underlying plant processes, the specific area of organogenesis in vitro has not yielded many molecular secrets.

TOTIPOTENCY

The concept of totipotency is central to understanding in vitro regeneration. The term is used in the context of differentiation not being an irreversible process as a cell undergoes maturation, i.e., a living plant cell with overt functional and structural specialization still carries all the information necessary to divide and undergo a morphogenetic process in the form of either organogenesis [which can be either rhizogenesis (root formation), caulogenesis (shoot formation) or, occasionally, flower formation] or embryogenesis, or to develop directly into a specialized cell type (e.g., as seen in xylogenesis). It is clear, however, from observations of regeneration in even highly regenerative explants that not all living differentiated cells of an explant participate in the regeneration process. This may be due to: 1) an inability to achieve in vitro the necessary conditions for totipotent expression; 2) genetic (physical changes to chromosomes, e.g., loss of DNA or nucleotide substitution) or epigenetic (changes in DNA pexpression as a consequence of development, e.g., DNA methylation or the isolation of DNA into heterochromatin) blocks that interfere with the expression of totipotency; 3) the fact that not all cells are totipotent, i.e., although all cells may appear to be the same in a particular tissue, only some possess special characteristics that enable them to regenerate plants when isolated and cultured under inductive conditions.

The first step in the expression of totipotency, where it occurs, is for mature cells to reenter the cell cycle and resume cell division (a process known as dedifferentiation). The next step is redifferentiation, either through direct formation of organized structures (direct regeneration) or by the formation of an intervening callus stage from which organized structures may later be induced (indirect regeneration). An early appreciation of the mechanisms underlying regeneration of whole plants, or parts of plants, from cells came with the classic observations of Skoog and Miller[1] that the direction of differentiation could be influenced by the ratio of the exogenously supplied growth regulators auxin and cytokinin. They observed in tobacco stem pith cultures that a high ratio of auxin to cytokinin led to initiation of roots, whereas a low ratio led to development of shoots. Although there are many species for which this simple manipulation will not work, in principle, this is the basis for regeneration in plant tissue culture systems. The two groups of growth regulators play a pivotal role in unlocking and realizing totipotent expression by influencing both dedifferentiation and redifferentiation. Note, however, that other medium conditions such as nitrogen, carbon source, and pH are also extremely important.

The process whereby differentiated cells respond to inductive phenomena leading to organogenesis involves two major phases—competence and determination. These phases reflect the two-stage practice of exposing cultures first to an ''induction'' medium and then to a ''regeneration'' medium during the regeneration process,[2] although there are cultures for which both phases will occur on the same medium, particularly in the case of direct regeneration.

COMPETENCE

This is a transient state in which cells can be induced to follow an organogenic pathway[2,3] and mechanical wounding is the most effective biological trigger for shifting cells into the competent state. Competence can be thought of as having two distinct components, one for cell division and the other for organogenesis.

Encyclopedia of Plant and Crop Science
DOI: 10.1081/E-EPCS 120010547

Competence for Cell Division

In order to sustain cell division following wounding, exogenous auxin (+/− cytokinin) is usually required. Progress in understanding the molecular basis of the action of auxins and cytokinins in the initiation and maintenance of cell proliferation has been slow, and the complex interaction between exogenously applied growth regulators—overlayed with the unknown of endogenous synthesis—makes it difficult to differentiate the individual roles of auxin and cytokinin.

Proteins known as cyclin-dependent kinases (Cdks) govern the onset of S-phase and mitosis in all eukaryotic cells and require an activating cyclin subunit, which leads to the formation of Cyclin/Cdk complexes. In particular, homologues of a 34 kDa protein kinase known as p34 (coded for by the Cdk CDC) have been found in all eukaryotes that have been investigated. Induction of the competence of transformed protoplasts to divide in the presence of auxin and, to a lesser degree, cytokinin, was shown to be accompanied by expression of the *Arabidopsis CDC2a* gene, even if there was no subsequent cell division,[4] and led to the proposal of a linkage between the expression of *CDC2a* and competence for proliferation. A role for the *Arabidopsis* gene *SRD2* in conferring competence for cell division has been hypothesized,[5] and the expression of *CDC2a* in the *srd2* mutant is being undertaken. At another level of complexity, the gene *AINTEGUMENTA* (*ANT*) has also been implicated in cell cycle progression, but in addition it influences organ growth.[6] Ectopic expression of the gene gives rise to transformed plants that show spontaneous callus formation and regeneration (of roots, leaves, or shoots) at wound or senescence sites. Seen as a gene that maintains meristematic competence in cells, *ANT* could give a molecular basis to the frequent observation in tissue culture that explants derived from immature tissue (such as from embryos) are much more likely to succeed in producing regenerable cultures than those obtained from mature tissue. In other words, competence in vitro may be correlated with continuing meristematic activity in vivo.

Competence for Organogenesis

A gene in *Arabidopsis* (*IRE1*) that acts very early in dedifferentiation confers the ability of cells to respond later to specific regeneration stimuli such as auxin and cytokinin.[7] The work of Ozawa et al.[5] with *Arabidopsis* mutants *srd1*, *srd2*, and *srd3*, which are defective in their ability to regenerate, identified three sequentially acquired states associated with organogenic competence. Initially there is IC (incompetent with respect to both cell proliferation and organogenesis), which requires the gene *SRD2* in order to progress to CR (competent with respect to rhizogenesis). Finally, *SRD3* is involved in the progress from CR to CSR (competent with respect to shoot and root organogenesis).

DETERMINATION

This is a process in which cells follow a specific developmental pathway. The distinction between determination and competence can be illustrated by the work of Christianson and Warnick.[2] They found that callus produced on *Convolvulus* explants was initially developmentally interchangeable, i.e., it was competent to follow two developmental pathways—root formation and shoot formation. Once induction of shoots began, the cells involved in shoot formation became determined, and transfer to a root-inducing medium did not affect the formation of shoots. In other words, as determination proceeds, cells become more and more committed, and the developmental potential becomes restricted unless there is a catastrophic event—such as wounding—that cuts across the determined state. The realizing of commitment is considered to be a third phase in the process of organogenesis.[2]

In the Skoog and Miller model,[1] caulogenesis is stimulated by exogenous cytokinin. The work of Ozawa et al.[5] indicated that the genes *SRD1* and *SRD2* play essential roles in the caulogenesis induced by culturing competent explants on a medium containing cytokinin. Three genes, *CKI1* isolated from *Arabidopsis*,[8] *ESR1* isolated from *Arabidopsis*,[9] and *PkMADS1* isolated from *Paulownia*,[10] have been identified as regulators of shoot regeneration. CKI1 is thought to function as a cytokinin receptor in the process of cytokinin induction of shoot organogenesis. *ESR1* expression is induced by cytokinins, but transcripts of the gene accumulate only after acquisition of organogenic competence. *PkMADS1* is hypothesised to be a shoot meristem identity gene whose expression is necessary in activating the developmental pathway leading to direct regeneration of shoots. Cytokinin also induces another shoot meristem gene, *PASTICCINO*, but the outcome is an inhibitory one that prevents excessive cell proliferation,[11] thus eliminating abnormal shoot development. The expression of homeobox genes in plants can be induced by cytokinin, and two such genes—*KNOTTED1* and *STM (SHOOT MERISTEMLESS)*—have been implicated in shoot formation.[12]

The most frequent type of regeneration occurring in cultured cells is root formation, and this is stimulated by auxin.[1] However, there is some complexity in the action of auxin, because although it stimulates root initiation, its continued presence in the culture medium can inhibit the outgrowth of roots. Experimental systems looking at the molecular basis of rhizogenesis are few and deal almost exclusively with lateral root formation and the development of adventitious roots on stem cuttings. Lund et al.,[13] working with a tobacco root mutant (*rac*) that fails to initiate adventitious roots in response to exogenous auxin, concluded that the *RAC* gene is involved in an auxin signal transduction pathway acting prior to the first organized divisions that lead to the formation of root meristems. During the determination phase of rhizogenesis the *LRP1* (*LATERAL ROOT PRIMORDIUM1*) gene is expressed,[14] and during both determination and commitment certain S-adenosylmethionine synthetase-encoding genes (*SAMS*) are also up-regulated.[15] The *Agrobacterium rhizogenes* infection system is a potentially useful tool for characterizing events in rhizogenesis. Cells infected with the bacterium show an increased sensitivity to auxin, consistent with the ability of such cells to undergo root meristem neoformation and proliferation.

CONCLUSION

Regeneration through organogenesis represents an amazing developmental plasticity that sets plant cells apart from most animal cells. It is an extraordinarily complex phenomenon influenced by an array of internal and external factors. The molecular work to date suggests that there is certainly no single ''totipotency'' gene, and the existence of a conserved suite of genes that defines a group of cells as organogenic or recalcitrant in vitro is not very evident either.

ARTICLES OF FURTHER INTEREST

Anther and Microspore Culture and Production of Haploid Plants, p. 43
In Vitro Flowering, p. 576
In Vitro Morphogenesis in Plants: Recent Advances, p. 579
Plant Cell Culture and Its Applications, p. 931
Plant Cell Tissue and Organ Culture: Concepts and Methodologies, p. 934
Protoplast Culture and Regeneration, p. 1065
Regeneration from Guard Cells of Crop and Other Plant Species, p. 1081
Somatic Embryogenesis in Plants, p. 1165

REFERENCES

1. Skoog, F.; Miller, C.O. Chemical regulation of growth and organ formation in plant tissues cultivated in vitro. Symp. Soc. Exp. Biol. **1957**, *11*, 118–131.
2. Christianson, M.L.; Warnick, D.A. Competence and determination in the process of in vitro shoot organogenesis. Dev. Biol. **1983**, *95*, 288–293.
3. Sugiyama, M. Organogenesis in vitro. Curr. Opin. Plant Biol. **1999**, *2*, 61–64.
4. Hemerly, A.S.; Ferreira, P.; de Almeda, J.; Van Montagu, M.; Engler, G.; Inzé, D. *cdc2a* expression in arabidopsis is linked with competence for cell division. Plant Cell **1993**, *5*, 1711–1723.
5. Ozawa, S.; Yasutani, I.; Fukuda, H.; Komamine, A.; Sugiyama, M. Organogenic responses in tissue culture of srd mutants of *Arabidopsis thaliana*. Development **1998**, *125*, 135–142.
6. Mizukami, Y.; Fischer, R.L. Plant organ size control: *AINTEGUMENTA* regulates growth and cell numbers during organogenesis. Proc. Natl. Acad. Sci. **2000**, *97*, 942–947.
7. Cary, A.C.; Uttamchandani, S.J.; Smets, R.; Van Onckelen, H.A.; Howell, S.H. *Arabidopsis* mutants with increased organ regeneration in tissue culture are more competent to respond to hormonal signals. Planta **2001**, *213*, 700–707.
8. Kakimoto, T. CKI1, a histidine kinase homolog implicated in cytokinin signal transduction. Science **1996**, *274*, 982–985.
9. Banno, H.; Ikeda, Y.; Niu, Q.-W.; Chua, N.-H. Overexpression of Arabidopsis *ESR1* induces initiation of shoot regeneration. Plant Cell **2001**, *13*, 2609–2618.
10. Prakash, A.P.; Kumar, P.P. *PkMADS1* is a novel MADS box gene regulating adventitious shoot induction and vegetative shoot development in *Paulownia kawakamii*. Plant J. **2002**, *29*, 141–151.
11. Faure, J.D.; Vittorioso, P.; Santoni, V.; Fraiaier, V.; Prinsen, E.; Barlier, I.; Van Onckelen, H.; Caboche, M.; Bellini, C. The *PASTICCINO* genes of *Arabidopsis thaliana* are involved in the control of cell division and differentiation. Development **1998**, *125*, 909–918.
12. Rupp, H.-M.; Frank, M.; Werner, T.; Strnad, M.; Schmülling, T. Increased steady-state mRNA levels of the *STM* and *KNAT1* homeobox genes in cytokinin-overproducing *Arabidopsis thaliana* indicate a role for cytokinins in the shoot apical meristem. Plant J. **1999**, *18*, 557–563.
13. Lund, S.T.; Smith, A.G.; Hackett, W.P. Differential gene expression in response to auxin treatment in the wild type and *rac*, an adventitious rooting-incompetent mutant of tobacco. Plant Physiol. **1997**, *114*, 1197–1206.
14. Ermel, F.E.; Vizoso, S.; Charpentier, J.-P.; Jay-Allemand, C.; Catesson, A.-M.; Couée. Mechanisms of primordium formation during adventitious root development from walnut cotyledon explants. Planta **2000**, *211*, 563–574.
15. Lindroth, A.M.; Saarikoski, P.; Flygh, G.; Clapham, D.; Grönroos, R.; Thelander, M.; Ronne, H.; von Arnold, S. Two S-adenosylmethionine synthetase-encoding genes differentially expressed during adventitious root development in *Pinus contorta*. Plant Mol. Biol. **2001**, *46*, 335–346.

In Vitro Pollination and Fertilization

Acram M. Taji
University of New England, Armidale, New South Wales, Australia

Jennifer S. Williams
University of Queensland, St. Lucia, Queensland, Australia

INTRODUCTION

Pollination followed by fertilization normally leads to the production of an embryo that, in the intact plant, is linked with normal seed development. Most angiosperms are outbreeders, as such self-pollination (selfing) is limited. Furthermore, hybridization between species and/or genera is also rare in nature. In plant breeding, however, selfing and hybridization are methods commonly used to obtain desirable crosses. If no fertilization takes place after self- or cross-pollination, plant breeders resort to special procedures such as in vitro pollination to bring about fertilization.

In vitro pollination and fertilization is a technique wherein male and female gametes are isolated and introduced to each other under conditions suitable for zygote formation. It involves pollen tube penetration of the embryo sac via manipulation of maternal tissue by methods other than the normal in situ process. Initially developed to bypass prezygotic incompatibility barriers, this technique has been used for the production of hybrids, the induction of haploid plants, overcoming sexual self-incompatibility, and in the study of reproductive processes and pollen physiology. The diversity of applications is mirrored by the diversity of species to which this technique has been applied.

IN VITRO POLLINATION AND FERTILIZATION

Indian scientists Kanta, Rangaswamy, and Maheshwari[2] led the first successful forays into the concept of in vitro pollination and fertilization. By 1962 they established methods for the in vitro pollination and fertilization of *Papaver somniferum* (opium plant). Successful in vitro pollination and fertilization is dependent on several basic elements,[3] including: 1) use of pollen grain and ovules at the required developmental phase; 2) sufficient pollen germination; 3) appropriate growth of pollen tubes and correct gametogenesis (sperm development); 4) pollen tube penetration into ovules; 5) successful fertilization; and 6) appropriate nutrient media (to support pollen germination, pollen tube growth, fertilization, and embryo development).

In vitro pollination and fertilization has been most successful in species such as those belonging to the *Papaveraceae*, *Caryophyllaceae*, and *Solanaceae*, whose ovaries are large, contain many ovules, and are situated on nutrient-rich placenta.[4] Compared to in vivo methods, in vitro pollination and fertilization offers increased control of environmental conditions, and thus enhances the accuracy and repeatability of experiments.

Types of In Vitro Pollination and Fertilization

There are three main methods of in vitro pollination and fertilization, each suitable for slightly different purposes. Successful application of a method to one species or family does not guarantee its successful use in another species or family.

In vitro stigmatic pollination and fertilization

In this method, pollen from an externally sterilized ripe anther is placed on the stigma of an emasculated flower. This technique is suitable for plants with premature ovary loss. *Nicotiana rustica*, *Nicotiana tabacum*, *Pisum sativum*, and *Zea mays* are examples of species in which this technique has been successful.[5]

In vitro placental pollination and fertilization

In this technique, placenta from explants with unfertilized ovules are dissected from an externally sterilized flower, onto which pollen grains are placed. Because this method has worked well with certain species of *Brassica*, it is thought that in vitro cross-pollination could be successfully applied to all members of the Brassicaceae to avoid pollen/pistil incompatibility.[4] This method has also been used successfully on members of the Caryophyllaceae and *Zea mays*. Most important, in vitro placental pollination and fertilization can sometimes be achieved with plants

Encyclopedia of Plant and Crop Science
DOI: 10.1081/E-EPCS 120010561

(such as *Petunia axillaris*) that are completely self-incompatible in vivo.[6]

In vitro ovular pollination and fertilization

Both cross-pollination and self-pollination can be achieved using this technique, in which pollen is applied to excised ovules from an externally sterilized flower. Selected ovules must be suitable for pollination by their own or foreign pollen. Success with this technique has been seen in *Cichorium intybus* L.[7] and various species of Brassicaceae.[8]

PRACTICAL APPLICATIONS

There are many practical applications of in vitro pollination and fertilization. These techniques often allow barriers to pollination and fertilization to be overcome, achieving cross-pollination/fertilization and self-pollination/fertilization where it is impossible to realize these objectives in vivo. This method has the potential to aid in the rescue of endangered species by overcoming incompatibilities, and in the creation of higher-quality commercial products through development of hybrids. It also enables further study into the process of fertilization, and the physiology of pollen.

OVERCOMING INCOMPATIBILITY BARRIERS (SELF-INCOMPATIBILITY AND INTERSPECIFIC INCOMPATIBILITY)

Incompatibility barriers can be classed as prezygotic (occurring prior to fertilization) and postzygotic (occurring after fertilization). Prezygotic incompatibility factors include those that impede effective fertilization (e.g., failure of pollen to germinate and penetrate the stigma, disintegration of the pollen tubes in the style, inadequate length of pollen tubes to effect fertilization). Circumventing prezygotic barriers by using in vitro techniques does not rule out the existence of postzygotic barriers to success in achieving the desired cross.

Postzygotic barriers occur during or immediately following syngamy (union of gametes during fertilization), and result in embryo starvation and abortion due largely to abnormal endosperm development.[9] When used in tandem with embryo culture techniques, in vitro pollination and fertilization can overcome sufficient incompatibility barriers to produce viable seeds and whole plants. Overcoming incompatibility barriers will potentially lead to the production of better commercial plant products, and could assist in saving endangered species.

Self-incompatibility or interspecific incompatibility is often manifested in pollen/stigma or pollen/style rejection responses, which can be bypassed through use of in vitro pollination and fertilization techniques. In vitro placental pollination and fertilization enabled Zenkteler et al.[8] to bypass pollen/stigma self-incompatibility in *Brassica campestris*. In vitro ovular pollination and fertilization was the technique used by Rangaswamy and Shivanna[6] to overcome self-incompatibility in *P. axillaris*.

PRODUCTION OF HYBRIDS

In vitro pollination and fertilization, particularly when used in conjunction with embryo culture techniques, enables incompatibilities to be overcome, which allows interspecific, interfamiliar, and intergeneric hybridization where conventional crossing techniques fail. Combining in vitro pollination and fertilization with new pollination methods, such as in vitro cut-style pollination and in vitro grafted-style techniques,[10] has the potential to overcome even greater incompatibilities than those overcome by the use of in vitro pollination and fertilization alone. The combination of in vitro pollination and the grafted-style method (GSM) has successfully overcome hybridization barriers between several incongruent lily species.[10]

Stylar barriers preventing hybridization (in particular, *Nicotiana* interspecific combinations) have been overcome by use of in vitro pollination and fertilization.[9] In vitro placental pollination and fertilization enabled Zenkteler et al.[8] to obtain hybrids between *Brassica napus* and *B. campestris*. Hybrids between *Brassica chinensis* and *Brassica pekinensis* were obtained using in vitro ovular pollination and fertilization.[11]

INDUCTION OF HAPLOID PLANTS

Haploid parthenogenesis can be induced by pollination with pollen of unrelated species or with inactivated pollen from the same species.[7] *Melandrium album* (used as the female plant) can be pollinated with pollen from species of *Nicotiana* to obtain 1% haploid plants.[4]

STUDYING REPRODUCTIVE PROCESSES AND POLLEN PHYSIOLOGY

In vitro methods offer new ways to study plant reproductive processes and pollen physiology. The development of an in vitro fertilization method without the application of electrical pulses for higher plant gametes could be used for studies of cell-cell interaction (adhesion, fusion, recognition).[12] Haldrup and Bruun[13] showed that in vitro pollination could be employed to study the time course of, and processes related to, pollen-tube development. Rougier et al.[14] predicted that in vitro fertilization systems (readily available for studying the processes of zygote formation in flowering plants) may provide new tools for molecular analysis of fertilization and early embryo development. In vitro pollination has been used as a model for the study of fertilization in *Z. mays*.[7]

CONCLUSION

The basic concept of in vitro pollination and fertilization has evolved to be applied to various components of the pistil, and to be adapted to an enormously diverse range of plant species.

In vitro pollination and fertilization is a technique that has been used for the production of hybrids, the induction of haploid plants, overcoming sexual self-incompatibility, and the study of reproductive processes and pollen physiology. Although its potential has not yet been fully explored, in vitro pollination and fertilization has already assisted in the conservation of endangered species[15] and in the creation of new hybrids. In vitro pollination and fertilization will almost certainly result in the development of stable novel genotypes of economically superior plants, the most important of which are edible crop species. Greater genotypic variation could mean decreased disease susceptibility and increased crop survival in inhospitable climates—very desirable traits for crop improvement.

In this era of genetic engineering, in vitro pollination and fertilization represents an essential tool with which to gain a deeper understanding of the mechanisms of fertilization in plants and the physiology of pollen. In vitro pollination and fertilization undoubtedly has a role to play in the search for genes affecting pollination and fertilization in *Arabidopsis*, the recently sequenced model plant.

In the future, in vitro pollination and fertilization will be used in conjunction with myriad techniques of cell engineering to aid researchers in their quest to better understand the mysteries of plants.

REFERENCES

1. Stewart, J.M.D. In vitro fertilization and embryo rescue. Environ. Exp. Bot. **1981**, *21*, 301–315.
2. Kanta, K.; Rangaswamy, N.S.; Maheshwari, P. Test-tube fertilization in a flowering plant. Nature **1962**, *194*, 1214–1217.
3. Taji, A.M.; Kumar, P.; Lakshmanan, P. In Vitro Pollination and Fertilization. In *In Vitro Plant Breeding*; Food Products Press: New York, 2002; 57–67.
4. Zenkteler, M. In-vitro fertilization of ovules of some species of *Brassicaceae*. Plant Breed. **1990**, *105*, 221–228.
5. Pierik, R.L.M. Test Tube Fertilization. In *In Vitro Culture of Higher Plants*; Martinus Hijhoff Publishers: Dordrecht, 1987; 239–242.
6. Rangaswamy, N.S.; Shivanna, K.R. Induction of gamete compatibility and seed formation in axenic cultures of a diploid self-incompatible species of *Petunia*. Nature **1967**, *216*, 937–939.
7. Castano, C.I.; De Proft, M.P. In vitro pollination of isolated ovules of *Cichorium intybus* L.. Plant Cell Rep. **2000**, *19*, 616–621.
8. Zenkteler, M. In vitro placental pollination in *Brassica campestris* and *B. napus*. J. Plant Physiol. **1987**, *128*, 245–250.
9. DeVerna, J.W.; Myers, J.R.; Collins, G.B. Bypassing prefertilization barriers to hybridization in *Nicotiana* using in vitro pollination and fertilization. Theor. Appl. Genet. **1987**, *73*, 665–671.
10. Van Tuyl, J.M.; Van Dien, M.P.; Van Creij, M.G.M.; Van Kleinwee, T.C.M.; Franken, J.; Bino, R.J. Application of in vitro pollination, ovary culture, ovule culture and embryo rescue for overcoming incongruity barriers in interspecific *Lilium* crosses. Plant Sci. **1991**, *74*, 115–126.
11. Kameya, T.; Hinata, K. Test-tube fertilization of excised ovules in *Brassica*. Jpn. J. Breed. **1970**, *20*, 253–260.
12. Kranz, E.; Lorz, H. In vitro fertilization with isolate, single gametes results in zygotic embryogenesis and fertile maize plants. Plant Cell **1993**, *5*, 739–746.
13. Haldrup, A.; Bruun, L. Self-incompatibility reactions and compatible pollen-tube growth are retained with in vitro pollinations of sugar beet, *Beta vulgaris* L. Sex. Plant Reprod. **1993**, *6*, 46–51.
14. Rougier, M.; Antoine, A.F.; Aldon, D.; Dumas, C. New lights in early steps of in vitro fertilization in plants. Sex. Plant Reprod. **1996**, *9*, 324–329.
15. Taji, A.M.; Williams, R.R. Perpetuation of the self-incompatible rare species *Swainsona laxa* R. BR by pollination in vitro and in situ. Plant Sci. **1987**, *48*, 137–140.

In Vitro Production of Triploid Plants

Sant S. Bhojwani
University of Delhi, Delhi, India

INTRODUCTION

Most of the flowering plants are diploid with two sets of chromosomes. The plants with a higher number of chromosomes are called polyploids. Many spontaneous and induced polyploid varieties of crop plants are under cultivation because of their better vigor, improved fruit size/quality, and/or attractive flowers or foliage. The triploid plants, with three sets of chromosomes, are seed-sterile due to disturbance in gamete formation. Therefore, triploids hold great commercial potential where seedlessness is desirable or is of no serious consequence. Triploid varieties of banana, citrus, hops, and watermelon are already under commercial cultivation for their seedless fruits. Some natural triploids of tomato bear larger and tastier fruits than their diploid counterpart. Triploidy is of special significance for crops, which are grown for their vegetative parts or flowers. Some examples where triploids have proved superior to their diploid or tetraploid counterparts for a specific economic trait are high-yielding cassava (IPGRI Newsletter for Asia, the Pacific and Oceania, May–August, 2001), disease-resistant mulberry, more vigorous and ornamental petunia, and poplar with more desirable pulpwood.

The natural occurrence of triploids is very rare. On the other hand, the traditional method of artificial production of triploids is tedious and lengthy. It involves treating the diploids with colchicine to raise tetraploids, with four sets of chromosomes, followed by backcrossing the superior tetraploids with the diploids. In many cases, this is not possible due to sexual sterility of the tetraploid plants or failure of the cross. In contrast, in vitro regeneration of plants from endosperm, a natural triploid tissue, offers a direct, single-step approach to raising triploids. The sexually sterile triploids can be multiplied by micropropagation.

ENDOSPERM DEVELOPMENT

The angiosperms are characterized by double fertilization. During sexual reproduction, the male gametophyte delivers two sperms in the female gametophyte. Of these, one sperm fuses with the egg, which forms the well-organized diploid embryo, whereas the other sperm fuses with two nuclei in the central cell, which gives rise to an unorganized endosperm tissue.[2] Since the endosperm is derived from the fusion product of three haploid nuclei, it is triploid. It is the main source of nutrition for the developing embryo. In some plants (pea, orange, and watermelon) the endosperm is completely consumed before seed maturation, whereas in others (castor bean, maize, and coconut) it persists in mature seeds as a massive tissue and supports the growth of the embryo during seed germination.

ENDOSPERM CULTURE

Following the discovery of double fertilization around 1898, the embryologists regarded endosperm as a second embryo modified to serve as a nutritive tissue for the zygotic embryo. Sargant[3] remarked that the triploid nature of the endosperm might be responsible for its formless structure. However, the in vitro studies have clearly demonstrated the ability of the endosperm tissue for unlimited growth and to regenerate full plants.[4]

The work on endosperm culture was initiated in 1933, but the establishment of continuously growing tissue cultures of maize endosperm was first achieved in 1949, by LaRue.[5] Although maize endosperm persists in mature seeds, callus cultures could be established only from immature endosperm; 8–11 days after pollination (DAP) was the best stage. This is also true for other cereals, where the bulk of the mature endosperm is dead. Thus, the age of the endosperm at the time of culture could be a critical factor in raising tissue cultures of some species.

In 1963, Mohan Ram and Satsangi[6] demonstrated that divisions could be induced in mature endosperm cells of castor bean by treating the seeds with 2,4-dichlorophenoxyacetic acid (2,4-D) before germination. In the same year, establishment of continuously growing callus cultures from the mature endosperm of sandalwood was reported by Rangaswamy and Rao.[7] Since then, it has been possible to initiate tissue cultures from mature and immature endosperm of many plants. A critical factor in the induction of cell divisions in mature endosperm is its initial association with the embryo. There is ample evidence to suggest that the germinating embryo contributes some factor, which is essential to trigger cell divisions in the mature endosperm. The mature endosperm from dried

Encyclopedia of Plant and Crop Science
DOI: 10.1081/E-EPCS 120010563

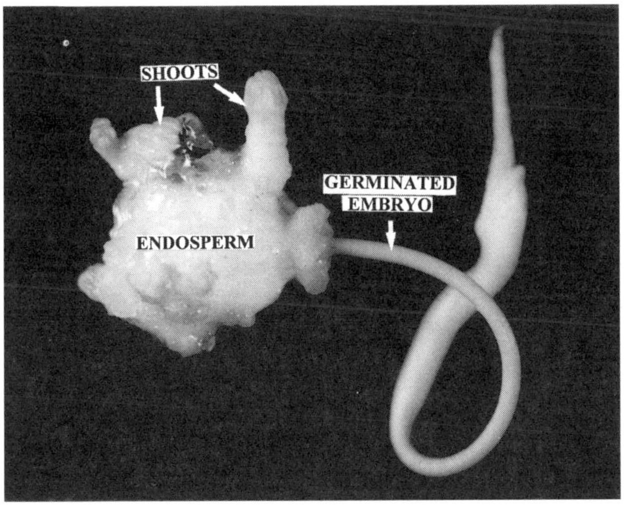

Fig. 1 A five-week-old culture of decoated seed of *Exocarpus cupressiformis*. The embryo has germinated and multiple shoot buds have differentiated from the surface of the endosperm.

seeds of castor bean did not callus, but if the seeds were soaked in $Ca(OCl)_2$ solution 22 hours prior to the excision of the endosperm, it exhibited callusing. The frequency of callusing increased if the seeds were allowed to germinate, and the number of cultures showing proliferation was directly related to the germination period before the excision of the endosperm. For tomato, soaking the seeds for three days was optimum in terms of the number of cultures showing endosperm callusing. Therefore, to initiate callus cultures from mature endosperm, it is cultured along with the embryo intact, and once the endosperm callusing has initiated (after 7–8 days) the embryo is removed to ensure that it does not callus and contaminate the endosperm callus. In some plants, the embryo factor could be substituted by gibberellic acid (GA_3). Although it is now possible to induce divisions in mature endosperm of many species (castor bean, croton, jatropha, putranjiva, neem, and sandalwood), immature endosperm continues to be the explant of choice in cereals and in those species where endosperm is consumed before seed maturation (acacia, citrus, cucumber, and mulberry).

The maize endosperm callus grew best on a medium containing yeast extract. However, in most other systems an auxin (generally 2,4-D), a cytokinin, and a rich source of organic nitrogen (yeast extract or casein hydrolysate) are required for optimal growth. The callus can be multiplied for unlimited period by periodic subcultures.

PLANT REGENERATION

By 1964 the cellular totipotency of plant cells had been well established, but the endosperm cells were still

Table 1 Plant species from which shoots, embryos, or full plants have been regenerated from cultured endosperm

Species	Reference
Acacia nilotica (Acacia)	9
Actinidia chinensis (Kiwifruit)	10
Actinidia deliciosa (Kiwifruit)	11
Citrus grandis (Orange)	12
Citrus sinensis (Orange)	13
Codiaeum variegatum (Croton)	14
Coffea sp. (Coffee)	15
Dendrophthoe falcata (Mistletoe)	16
Exocarpus cupressiformis (Cherry ballart)	8
Jatropha panduraefolia (Jatropha)	17
Juglans regia (Walnut)	18
Leptomeria acida (Native current)	16
Mallotus philippensis (Kamala tree)	19
Malus pumila (Apple)	20
Morus alba (Mulberry)	21
Oryza sativa (Rice)	22
Petroselinum hortense (Parsley)	23
Prunus persica (Peach)	24
Putranjiva roxburghii (Putranjiva)	25
Pyrus malus (Apple)	26
Santalum album (Sandalwood)	27
Scurrula pulverulenta (Leafy mistletoe)	28
Taxillus vestitus (Leafy mistletoe)	16

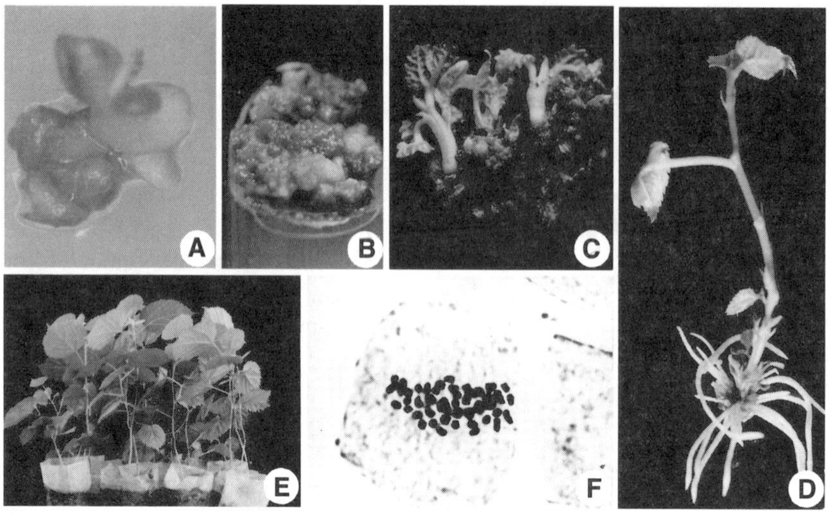

Fig. 2 Plant regeneration from mulberry endosperm. *A*. A seven-day-old culture of immature endosperm with the embryo intact. The endosperm has started to callus and the embryo has turned green and germinated. *B*. A well-established endosperm callus. *C*. The endosperm callus has differentiated multiple shoots. *D*. A plantlet of endosperm origin. *E*. A root tip cell of a plant regenerated from endosperm callus, with triploid number of chromosomes (42). (Based on Ref. 21.) (*View this art in color at www.dekker.com.*)

regarded as recalcitrant for plant regeneration. In 1965 Johri and Bhojwani,[8] working with *Exocarpus cupressiformis*, first reported the differentiation of well-organized shoot buds from the mature endosperm. Histological studies confirmed the origin of the shoots from the peripheral cells of the endosperm (Fig. 1). As expected, the shoots were triploid. Since then, shoot/embryo/plant regeneration from the endosperm tissue has been reported for at least 24 species (Table 1).

In the cultures of mature endosperm, shoot differentiation may occur directly from the endosperm or after callusing. However, in immature endosperm cultures, it is always preceded by a callus phase. Plant regeneration from the callused endosperm of acacia, citrus, and sandalwood occurred via somatic embryogenesis.

For shoot bud differentiation from endosperm tissue, a cytokinin is always essential. Whereas *Scurrula pulverulenta* and *Taxillus vestitus* differentiated shoots in the presence of a cytokinin alone, *Dendrophthoe falcata* and *Leptomeria acida* required an auxin in addition to a cytokinin. The immature endosperm of mulberry (17–20 DAP) exhibited good callusing on a medium containing 5 μM 6-benzylaminopurine (BAP) and 1 μM α-naphthaleneacetic acid (NAA). Further addition of coconut milk and yeast extract improved the callusing response. The initial association of the embryo was promotive for endosperm callusing. The callus differentiated multiple shoots when transferred to a medium containing BAP or thidiazuron. However, maximum regeneration occurred when BAP was added in conjunction with NAA[21] The endosperm callus of citrus initiated on a medium containing 2,4-D (2 mg hr l^{-1}), BAP (5 mg hr l^{-1}), and casein hydrolysate (1000 mg hr l^{-1}) differentiated globular embryos upon transfer to a medium with GA_3 (1 mg hr l^{-1}) as the sole hormone.[13] Further development of the embryos and plant regeneration required doubling the salt concentration and raising GA_3 concentration to 15 mg hr l^{-1}. In *Taxillus vestitus*, injury to the endosperm enhanced shoot bud differentiation (Fig. 2).

Most of the organs and plants regenerated from endosperm are triploid. Full triploid plants of endosperm origin have been developed for apple, kiwifruit, mulberry, neem, and sandalwood. The triploids of mulberry and sandalwood have been established in soil.

CONCLUSION

There is enough experimental evidence to suggest that endosperm cells are totipotent, and the technique of endosperm culture holds great potential in raising triploids of crop plants. Endosperm culture from crosses of selected diploid parents is likely to give better triploids. Triploid plants of *Petunia axillaris* derived from cultured microspores were much more ornamental than their parental anther-donor diploids.[29]

ARTICLES OF FURTHER INTEREST

Crop Improvement: Broadening the Genetic Base for, p. 343

Endosperm Development, p. 414
In Vitro Morphogenesis in Plants—Recent Advances, p. 579
In Vitro Plant Regeneration by Organogenesis, p. 584
Plant Cell Culture and Its Applications, p. 931
Plant Cell Tissue and Organ Culture: Concepts and Methodologies, p. 934
Polyploidy, p. 1038
Somatic Embryogenesis in Plants, p. 1165

REFERENCES

1. Kagn-zur; Mills, D.; Mizrahi, Y. Callus formation from tomato endosperm. Acta Hortic. **1990**, *280*, 139–142.
2. Bhojwani, S.S.; Bhatnagar, S.P. *The Embryology of Angiosperms*; Vikas Publishing House: New Delhi, 1999; 1–357.
3. Sargant, E. Recent work on the results of fertilization in angiosperms. Ann. Bot. **1900**, *14* (LVI), 689–712.
4. Bhojwani, S.S.; Razdan, M.K. *Plant Tissue Culture: Theory and Practice*; Elsevier: Amsterdam, 1996; 1–767.
5. LaRue.
6. Mohan Ram; Satsangi.
7. Rangaswamy; Rao.
8. Johri; Bhojwani.
9. Garg; et al.
10. Mu; et al.
11. Machno, D.; Przywara, L. Endosperm culture of *Actinidia* spp. Acta Biol. Crac., Ser. Bot. **1997**, *39*, 55–61.
12. Wang; Chang.
13. Gmitter; et al.
14. Chikannaiah; Gayatri.
15. Raghuramalu, Y. Anther and endosperm culture of coffee. J. Coffee Res. **1989**, *19*, 71–81.
16. Nag; Johri.
17. Srivastava.
18. Tulecke, W.; Mc Ganaham, G.; Ahmadi, H. Regeneration by somatic embryogenesis of triploid plants from endosperm of walnut, *Juglans regia* L. cv. Manregian. Plant Cell Rep. **1988**, *7* (5), 301–304.
19. Sehgal, C.B.; Abbas, N.S. Induction of triploid plantlets from the endosperm culture of *Mallotus philippensis* Mull. Phytomorphology **1996**, *46* (3), 283–289.
20. Mu; Liu.
21. Thomas, T.D.; Bhatnagar, A.K.; Bhojwani, S.S. Production of triploid plants of mulberry (*Morus alba* L.) by endosperm culture. Plant Cell Rep. **2000**, *19* (4), 395–403.
22. Nakano; et al.
23. Masuda; et al.
24. Liu; Liu.
25. Srivastava.
26. Mu; et al.
27. Lakshmi Sita; et al.
28. Bhojwani; Johri.
29. Gupta, P.P. Genesis of microspore-derived triploid petunias. Theor. Appl. Genet. **1982**, *61* (2), 327–331.

In Vitro Tuberization

Yasunori Koda
Hokkaido University, Sapporo, Japan

INTRODUCTION

Various plants such as potato, yam, and Jerusalem artichoke develop tubers as organs for reproduction. When nodal stem segments of these plants are cultured aseptically in vitro under appropriate conditions, in vitro tubers are formed synchronously on lateral shoots that emerge from the nodes. Tubers formed in vitro are usually called microtubers. The culture system has provided a useful mechanism for various basic studies related to storage organ formation. In this article, in vitro potato tuberization is mainly described, with regard to the factors affecting the process, the mechanism, and a simple method of induction.

TUBERIZATION IN THE FIELD

After the first report by Barker on potato tuberization in vitro,[1] the culture system has been used in various basic studies. The nature of tuberization in the field is described first, since there are several analogies between tuberization in vitro and in the field. Tuberization in field-grown plants is controlled mainly by photoperiod. Short days stimulate the process, whereas long days prevent or retard it. The response to photoperiod interacts with many other factors. Both high temperature and high levels of reduced nitrogen in the soil inhibit tuberization. Plants from physiologically old mother tubers form tubers irrespective of photoperiod, so aging makes the photoperiod requirement for tuberization redundant. As to the genotype, late-maturing cultivars have a shorter critical photoperiod than the early-maturing cultivars. In other words, tuberization occurs earlier in the early-maturing cultivars under the natural field conditions from summer to autumn. The differences in their tuberizing ability seems to be due to the balance between two endogenous factors, one being inhibitive and the other promotive. The inhibitive factor is gibberellin (GA) and promotive one seems to be jasmonates (jasmonic acid, JA, and related compounds).[2] GA that is necessary for formation and elongation of stolon strongly inhibits tuberization, and reduction of GA level is a prerequisite to tuberization.

Potato tuberization is initiated by expansion of cells at the subapical region of a stolon, i.e., a kind of lateral shoot. The swelling at this early stage is mainly caused by radial expansion of cells. Subsequently, vigorous thickening growth due to expansion and division of cells occurs, and starch accumulates. The initial cell expansion is attributed to both an increase in osmotic pressure in the cells due to the accumulation of sucrose and changes in cell wall architecture that increase extensibility of the wall. The direction of cell growth (elongation or expansion) is controlled by the orientation of cortical microtubules (MTs). Characteristic reorientation of MTs, from perpendicular to parallel with respect to the axis of the stolon, is found when a slight swelling of the stolon occurs. Jasmonic acid (JA) is capable of inducing both reorientation of MTs and expansion of cells.[2] After the commencement of the swelling, cytokinins are accumulated in the expanded region, leading to cell division.

FACTORS AFFECTING IN VITRO TUBERIZATION

Various factors affecting in vitro tuberization are summarized in Table 1. Each factor exhibits the opposite effect on shoot growth. Longer duration of culture period tends to stimulate tuberization. The low tuberizing ability of late-maturing cultivars seems to be due partly to higher endogenous GA.[3] Application of an inhibitor for GA biosynthesis, such as ancymidol, uniconazol, or paclobutrazol at concentrations below 10 μM, stimulates in vitro tuberization in late-maturing cultivars. JA is also capable of inducing microtubers in these cultivars, but it induces only smaller tubers.

Sucrose at a high concentration (more than 6%) is indispensable for in vitro tuberization. Although the supplied sucrose is rapidly hydrolyzed into glucose and fructose, these monosaccharides cannot replace sucrose,[4] indicating that sucrose is not a mere regulator of osmotic pressure. Cytokinins (benzyladenine, kinetin, and zeatin, etc.) at a concentration of 10 μM can promote tuberization.

Culture conditions, such as vessel size, headspace gas exchange, and culture density, affect in vitro tuberization and yield of microtubers. Although a large-scale bioreactor with forced aeration is favorable for good yield, the

Encyclopedia of Plant and Crop Science
DOI: 10.1081/E-EPCS 120010556

Table 1 Various factors affecting in vitro tuberization and shoot growth

Factors		Tuberization	Shoot growth
Environmental			
Photoperiods	Short days	Promotive	Inhibitive
	Long days	Inhibitive	Promotive
Temperature	Cool (20–22°C)	Promotive	Inhibitive
	Warm (25°C)	Inhibitive	Promotive
Nutritional			
Sucrose	High (8%)	Promotive	Inhibitive
	Low (3%)	Inhibitive	Promotive
Reduced nitrogen		Inhibitive	Promotive
Genotypic			
Early-maturing cultivars		Promotive	Inhibitive
Late-maturing cultivars		Inhibitive	Promotive

system itself affects the cost of seed tuber production. A vessel that contains a smaller amount of medium (e.g., a 500 ml bottle containing 50 ml medium) can produce enough microtubers. When the bottle is sealed tightly with aluminum foil, many lenticels are formed on the surface of the microtubers, reducing their storability. Lenticle formation seems to be mediated by ethylene production, because an application of silver thiosulfate (ethylene antagonist) to the medium or sealing the culture vessels with air-permeable membrane can reduce lenticel formation. The density of shoots in the culture also affects microtuber production; usually, the larger the density, the higher the rate of production.

MICROTUBER DORMANCY

Prolonged or highly variable dormancy is a significant obstacle to the efficient use of microtubers as seed tubers. The dormancy seems to be induced by endogenous abscisic acid (ABA) and/or ethylene.[5] Microtubers that developed in the presence of fluridone (an inhibitor for synthesis of carotenoid including ABA) or silver nitrate (ethylene antagonist) exhibit a loss of dormancy and sprout on the mother shoots (precocious sprouting). On the other hand, GA can release tubers from dormancy. Varying photoperiodic treatments and light intensities during the microtuber induction affect the duration of dormancy. The lower the light intensity, the longer the dormant period. An 8-hour photoperiod instead of continuous darkness reduces dormancy.[6] Although several exogenous chemicals could break dormancy, chemical treatments sometimes affect the growth of plants developed from microtubers. Rindite and bromoethane have a practical use as dormancy-releasing regents.[6] The duration of dormancy can be manipulated by various methods, but development of cultivar-specific protocols is necessary to control dormancy.

A SIMPLE METHOD FOR IN VITRO TUBERIZATION

An in vitro tuberization system can be established easily from tubers using the procedure shown in Fig. 1. This

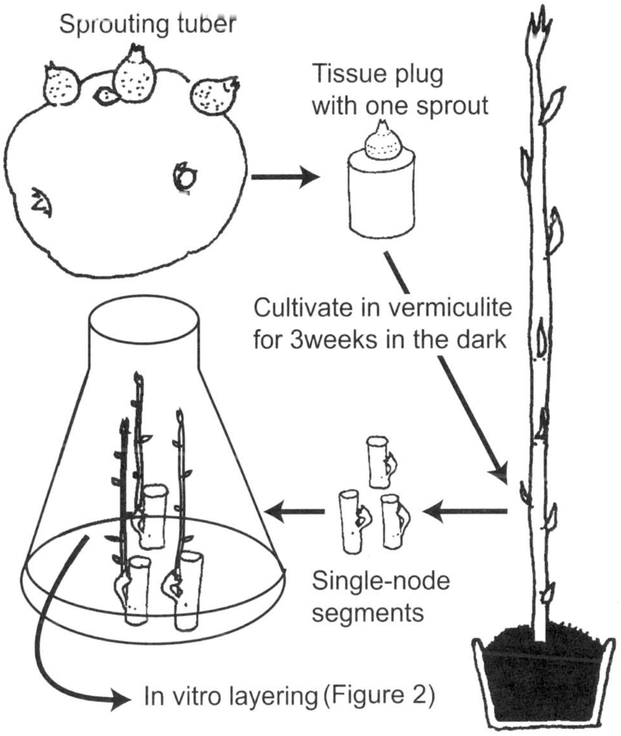

Fig. 1 Schematic diagram showing the procedure to establish an in vitro tuberization system from potato tubers.

method enables one tuber to produce many nodal segments as materials for culture, and is also applicable to bioassay for tuber-inducing activity. Surface-sterilization of tubers and tissue plugs is necessary to avoid microbial contamination, and relative humidity in the dark room should stay below 30% in order to raise healthy etiolated shoots. After surface-sterilization, nodal segments of the shoots are planted on agar-solidified Murashige–Skoog (MS) medium (3% sucrose) and cultured at 25°C for 2 to 3 weeks to obtain lateral shoots. If some tuber-inducing compound such as JA is applied to the medium, tubers are formed on the laterals. To produce microtubers routinely, two culture steps are usually employed (Fig. 2). The first step is called in vitro layering, which involves shoot propagation by recycling segments of shoots back to the shoot culture. Vigorously growing shoots are selected and subcultured every four to six weeks in liquid medium, under long-day conditions (16 h photoperiod), at 25°C. Viruses can be eliminated by meristem culture at any time during this step. To induce in vitro tuberization, the propagated shoots are transferred to fresh liquid medium that contains 8% sucrose and are cultured in the dark, or under short-day conditions, at 20–22°C. Many microtubers are formed on the lateral shoots within a week, and the tubers continue to grow for another seven to eight weeks. The final fresh weight of the microtubers is usually between 0.2–1.0 g. Recycling smaller microtubers back into propagation of shoots is a preferable option for mass production.[7] Since somaclonal mutation scarcely occurs during these steps, this method enables long-term conservation of disease-free germplasm.[8] The in vitro layering system is also applicable to the conservation of potato viruses. For in vitro tuberization in other plants such as yam (*Dioscorea* spp) and Jerusalem artichoke (*Helianthus tuberosus*), a similar method can be employed.

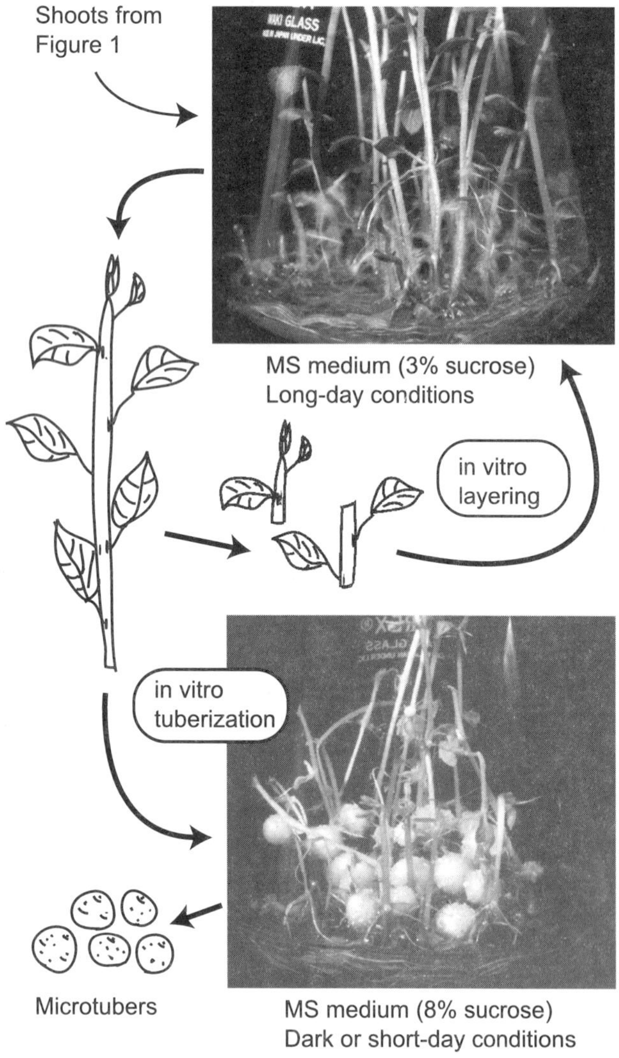

Fig. 2 Schematic diagram showing the procedure for successive microtuber production.

MICROTUBERS AS SEED TUBERS

The degree of periderm (skin) development affects the storability of microtubers. Greening of microtubers by light irradiation improves the storability. Microtubers retain their viability for three years under low temperature and high relative humidity storage conditions. Each microtuber produces a single shoot when planted in the field. Early growth of the plant is much slower than that from true tubers, but the plant produces many branches and grows vigorously thereafter. Interestingly, the plant's final yield of tubers is comparable to that of the plant grown from a true tuber.

CONCLUSION

Besides practical use for production of disease-free seed tubers, the in vitro tuberization system is applicable to experimental tools for various basic studies, such as studies on the mechanism of tuberization, carbohydrate metabolism, starch synthesis, and the transformation of plastids (from amyloplasts to chloroplasts). Furthermore, *Agrobacterium*-mediated gene transformation is easily carried out using this system.[9] Several attempts have been made to improve protein quality and to introduce foreign carbohydrate in tubers. Perhaps in the near future, this method will be applied to produce various useful seed tubers for molecular farming.

ARTICLES OF FURTHER INTEREST

Breeding Plants with Transgenes, p. 193
Commercial Micropropagation, p. 297
Germplasm Maintenance, p. 544
In Vitro Morphogenesis in Plants—Recent Advances, p. 579
Plant Cell Tissue and Organ Culture: Concepts and Methodologies, p. 939
Vaccines Produced in Transgenic Plants, p. 1265

REFERENCES

1. Barker, W.G. A method for the in vitro culturing of potato tubers. Science **1953**, *118*, 384–385.
2. Koda, Y. Possible involvement of jasmonate in various morphogenic events. Physiol. Plant. **1997**, *100*, 639–646.
3. Koda, Y.; Kikuta, Y. Effects of jasmonates on in vitro tuberization in several potato cultivars that differ greatly in maturity. Plant Prod. Sci. **2001**, *4*, 66–70.
4. Yu, W.C.; Joyce, P.J.; Cameron, D.C.; McCown, B.H. Sucrose utilization during potato microtuber growth in bioreactors. Plant Cell Rep. **2000**, *19*, 407–413.
5. Suttle, J.C. Involvement of ethylene in potato mictoruber doirmancy. Plant Physiol. **1998**, *118*, 843–848.
6. Coleman, W.K.; Coleman, S.E. Modification of potato micotuber dormancy during induction and growth in vitro or ex virto. Am. J. Potato Res. **2000**, *77*, 103–110.
7. Khuri, S.; Moorby, J. Nodal segments or microtubers as explants for in vitro microtuber production of potato. Plant Cell, Tissue Organ Cult. **1996**, *45*, 215–222.
8. Wang, P.-J.; Hu, C.-Y. In vitro mass tuberization and virus-free seed potato production in Taiwan. Am. Potato J. **1982**, *59*, 33–37.
9. Synder, G.W.; Belknap, W.R. A modified method for routine *Agrobacterium* mediated transformation of in vitro grown potato microtubers. Plant Cell Rep. **1993**, *12*, 324–327.

Insect Life History Strategies: Development and Growth

Jeffrey C. Miller
Oregon State University, Corvallis, Oregon, U.S.A.

INTRODUCTION

The growth and survival of individuals, family units, and populations perpetuate the species. Adaptations to specific modes of growth and survival lead to the evolution of life history traits. Furthermore, combinations of life history traits involving physiological, morphological, behavioral, and ecological characteristics result in suites of adaptive tactics providing for the expression of certain features that may occur in coincidence or opposition with other features. Combinations and tradeoffs in tactics used to achieve growth and survival are repeated to some degree among species that capitalize on essential resources exhibiting similar attributes regarding spatial and temporal characteristics. For instance, many species capable of rapid colonization of ephemeral habitats exhibit short generation times and relatively high reproductive potential. Such traits exhibited in combination with other traits, such as weak interspecific competitive abilities, provide an adaptive suite of traits generally referred to as life history strategies. The life history characteristics that in sum create the suite of traits involved in life history strategies can be considered in at least eleven component areas: type of metamorphosis, location of life stages, acquisition of food resources, developmental rates, phenology, reproductive potential, dispersal and colonization, competitive behavior, defensive mechanisms, living in extreme environments, and group living and domiciles. Each of these components of life history strategies will be discussed in the context of adaptive suites and tradeoffs in risks and benefits relating to growth and survival.

METAMORPHOSIS

Three principle processes occur in the developmental pathway from the egg, to the immature stage, to adulthood. The most primitive process is ametabolous development. This process involves three life stages: egg, immature, and adult, where sequential molts of immature individuals (instars) do not result in an appearance or ecology that differs to any significant degree from that of the adult, and adults that are reproductively mature may continue to molt. The primitively wingless insects, such as, silverfish, firebrats, and jumping bristletails are some of the more notable groups characterized by ametabolous development. In contrast, the hemimetabolous (including paurometablous) insects also develop through three life stages, but the immature stages (nymphs or naiads) are distinguished by external wing buds, no pupal stage, and adults that are reproductively mature and do not continue to molt. Fully developed wings are present following the molt from the final instar to the adult. Typically, the ecology of immatures and adults that exhibit hemimetablous development is very similar. Hemimetabolous development is characteristic of many insect groups exhibiting a diverse array of habits. For instance, aquatic insects such as dragonflies, damselflies, mayflies, and stoneflies. Also, numerous terrestrial plant-chewing and plant-sucking insects exhibit hemimetabolous development, such as grasshoppers, earwigs, walking sticks, true bugs, bark lice, aphids, whiteflies, mealybugs, and scales. The holometabolous insects develop through four distinct life stages: egg, larva, pupa, and adult. The immature stages, unlike in ametabolous or hemimetabolous development, exhibit a morphology and ecology typically very different from the adult stage. The transition from immature to adult occurs most dramatically in the pupal stage when the larva with internal wingbuds transforms into the reproductively mature adult with fully developed external wings. Holometabolous development is characteristic of the groups comprising a majority of insect biodiversity, such as beetles, butterflies, moths, lacewings, fleas, caddisflies, true flies, sawflies, wasps, bees, and ants.

Regardless of which developmental process is followed by an insect species, the major function in the immature stage is acquisition of food for growth, while the major function of the adult stage is reproduction and dispersal. Uncommonly, the immature stages are the primary dispersal agents and select life history tactics are associated with this condition, as discussed under colonization. The fact that most insect species exhibit holometabolous development suggests that holometaboly facilitates resource partitioning between immatures and adults, which reduces competition between the life stages and in turn enhances the exploitation of specialized niches, resulting in higher levels of speciation.

Encyclopedia of Plant and Crop Science
DOI: 10.1081/E-EPCS 120010478

LOCATION OF LIFE STAGES

The sites for development of eggs, immatures, the pupa, and adults may involve many types of environments and microhabitats. For instance, herbivorous insects may occur on the surfaces of plants or within plant organs. The location of the various life stages is relevant to life history strategies because specialized adaptations are necessary to mediate physical and biological conditions encountered in the variety of microhabitats throughout the environment. Adaptations may be needed for defense against soil microbes, extracting nitrogen from a low nitrogen source like plant sap, or tolerance of temperature extremes. The life history of a typical geometer moth illustrates the complexity of a holometabolous life cycle where each of the four life stages occurs in a different microenvironment: Eggs are placed on the plant; caterpillars feed on leaves; pupation occurs in the soil; and the adult is free-flying and feeds on nectar or does not feed at all. In general, immatures of herbivorous insects function in the role of seed-, root-, crown-, or stem-borers, leaf-miners, leaf-chewers, or phloem- or xylem-feeders. The type of food consumed may be a specific resource or a generalized set of resources. An insect herbivore that is a specialist feeder may be either a taxonomic specialist, which means the insect feeds on only one plant species or a few species within the same genus, or a plant part specialist, which means the insect feeds only on seeds or roots, etc. In extreme cases, a generalist insect herbivore might feed on a hundred or more plant species across several plant families. The acquisition of suitable food resources and the condition of being a food specialist or generalist can involve many types of life history traits regarding morphology, physiology, and behavior (Fig. 1).

Fig. 1 Caterpillars represent a major group of herbivorous insects; they are the larval life stage of Lepidoptera and exhibit holometabolous development. The caterpillar will molt into the pupal stage, from which the adult butterfly or moth will emerge. (Photo by Jeffrey C. Miller.) (*View this art in color at www.dekker.com.*)

FOOD RESOURCES

Numerous specialized morphological adaptations and numerous life history traits are associated with the behavioral and physiological adaptations involved in avoiding starvation, locating food, and food acceptability. Among the tactics employed to mitigate starvation are cannibalism, oosorption (females are able to reabsorb nutrients from eggs still in the reproductive tract), protein reclamation from muscle degeneration, and development through an extra (supernumary) molt. One of the dominant themes regarding food resources is the condition of specialist and generalist feeding habits and vegetation as apparent and unapparent plants.[8] Specialist feeders have a low probability of randomly finding food; therefore, they must possess traits that place them in the proper habitat at the proper time for exploitation of their respective food resources. For specialists the proper habitat can be discovered by detecting certain cues such as unique form and color, or unique and often toxic chemicals characteristic of apparent plants. In addition to serving as cues in the process of host plant location, certain otherwise toxic plant chemicals may serve as phagostimulants and as a protein source for nutrition and defense mechanisms. On the other hand, generalist feeders may have a higher probability of encountering a seemingly random but acceptable food resource, but do not derive notable chemically based defensive properties from the unapparent plants, because the foliage is devoid of acutely toxic compounds. The timing of life stages to coincide with the availability of required food resources is an issue of phenology and is discussed after developmental rates.

DEVELOPMENTAL RATES

The rate of development depends on three main factors: temperature, nutrition, and genetics. Temperature is critical to insect development and is the focus of mathematical models capable of predicting insect developmental rates. The temperature-dependent growth models require information regarding two key variables in insect life history traits: 1) temperature thresholds (upper and lower) for development and 2) heat-unit requirements above the lower developmental threshold needed to complete development of each life stage.[9] Typically, insects inhabiting temperate latitudes exhibit a lower developmental threshold in the range of 5–15°C. One of the tradeoff considerations regarding temperature-dependent growth and life history traits is that a low lower developmental threshold might allow for a rapid accumulation of heat-units and, therefore, rapid growth rates. However, a low lower developmental threshold may also compromise tolerance regarding the upper developmental threshold, which, if exceeded, will result in death due to an intolerance of heat. Likewise, a higher value for the lower developmental threshold may confer a cold intolerance and a slower rate of heat-unit accumulation that could result in a slower growth rate. As with other life history traits, the genetically based developmental thresholds are, to one degree or another, characteristic of family units, populations, and species. The ability of an insect to thermoregulate, increase body temperature through rapid muscle contraction or basking in the sun, involves life history traits that allow certain species to defy environmental temperatures that otherwise would preclude an activity such as feeding or flight. Because temperature is a dominant factor in determining growth rates, it is also the case that temperature is influential in determining the seasonal occurrence of life stages and therefore the phenology of various events that must occur in synchrony, such as the timing of egg hatch with the occurrence of bud break and leaf maturation.

PHENOLOGY

Three aspects of phenology are particularly relevant to life history strategies: 1) diapause, 2) overwintering, and 3) voltinism. The onset of diapause and overwintering are often coincidental in nature.[10] That is, particularly among temperate zone insects, overwintering life stages will be in a state of diapause that was triggered in part by declining photoperiod in the autumn. The gradual conclusion of the diapause condition is triggered by a cold period followed by warming above the lower developmental threshold. A combination of factors, including developmental threshold values, heat-unit requirements, diapause conditions, and the temperatures characteristic of the habitat, will determine how many generations an insect population may exhibit within one year. Univoltine species cycle through only one generation per year and typically exhibit relatively narrow variance in the seasonal occurrence of the various life stages, whereas, multivoltine populations may cycle through many generations per year and exhibit a wide variance in the presence of various life stages and have overlapping generations. Among other considerations, a consequence of voltinism in combination with the length of time necessary to complete a generation is the effect on potential reproductive potential expressed by the population or species.

ARTICLES OF FURTHER INTEREST

Agriculture and Biodiversity, p. 1
Coevolution of Insects and Plants, p. 289
Ecology and Agricultural Sciences, p. 401
Ecology: Functions, Patterns, and Evolution, p. 404
Insect Life History Strategies: Reproduction and Survival, p. 601
Insect–Plant Interactions, p. 609
Physiology of Herbivorous Insects, p. 910
Plant Defenses Against Insects: Physical Defenses, p. 944

REFERENCES

1. *Insect Life History Patterns*; Denno, R.F., Dingle, H., Eds.; Springer-Verlag: New York, 1981.
2. *Evolution and Genetics of Life Histories*; Dingle, H., Hegman, J.P., Eds.; Springer-Verlag: New York, 1982.
3. Fox, C.W.; Czesak, M.E. Evolutionary ecology of progeny size in arthropods. Annu. Rev. Entomol. **2000**, *45*, 341–369. Annual Reviews: Palo Alto, CA.
4. Price, P.W. *Insect Ecology*; John Wiley and Sons: New York, 1984.
5. Roff, D.A. *The Evolution of Life Histories*; Chapman and Hall: New York, 1992.
6. *The Evolution of Insect Life Cycles*; Taylor, F., Karban, R., Eds.; Springer-Verlag: New York, 1986.
7. Zera, A.J. The physiology of life history trade-offs in animals. Annu. Rev. Ecolog. Syst. **2001**, *32*, 95–126. Annual Reviews: Palo Alto, CA.
8. Rhoades, D.F.; Cates, R.G. Toward a General Theory of Plant Antiherbivore Chemistry. In *Biochemical Interactions Between Plants and Insects*; Wallace, J.W., Mansell, R.L., Eds.; Recent Advances in Phytochemistry. Plenum Pr.: New York, 1976; Vol. 10, 168–213.
9. Heinrich, B. *The Hot-Blooded Insects*; Harvard University Press: Cambridge, MA, 1993.
10. Tauber, M.J.; Tauber, C.A.; Masaki, S. *Seasonal Adaptations of Insects*; Oxford University Press: New York, 1986.

Insect Life History Strategies: Reproduction and Survival

Jeffrey C. Miller
Oregon State University, Corvallis, Oregon, U.S.A.

INTRODUCTION

Insect herbivore life history traits related to development and growth are integral to understanding the life history traits regarding reproduction and survival. In fact, the relationships cannot fully be appreciated unless the adaptive nature of each life history trait is considered in the context of natural selection and its affects and effects concerning variation across temporal and spatial scales, predictable and unpredictable aspects in environmental constancy and heterogeneity, variation in genetics, the fitness value of a given combination of alleles, and the high diversity of combinations in life history traits. This article on insect life history strategies concentrates on adaptations regarding reproductive potential, dispersal and colonization, competitive behavior, defensive mechanisms, living in extreme environments, and group living and domiciles.

REPRODUCTIVE POTENTIAL

The total number of offspring, the timing of delivery of offspring, and sex ratio are central themes in life history strategy analysis.[1–3] In fact, the concept of r- and K-selection[4] is predicated on the dynamics of birth and death rates that on balance represent the reproductive capacity of a given population or, in a very general application, the species. Species that exhibit a high fecundity (total egg production per female) may also exhibit short generation times and may be multivoltine. The annualized reproductive capacity of such a species is extremely high. On the other hand, the rates of mortality in such a species may also be extremely high, resulting in highly volatile population densities. Other life history traits that may be associated with highly fecund species are rapid colonizing ability (discussed shortly), generalist food requirements, intense intraspecific competition for limited resources, and inferior interspecific competitive ability.[5,6] Conversely, species exhibiting lower levels of reproductive potential may be slower to colonize newly available habitats, but once they do they may retain their presence by being more specialized in food requirements and superior interspecific competitors (Fig. 1).

Many insect species may control the sex ratio of their offspring through various mechanisms where the female can regulate the fertilization of her eggs with sperm stored in a special pouch called the spermatheca. One sex determination mechanism is arrhenotoky, also known as haplo-diploidy, where haploid offspring are males and diploid (fertilized eggs) offspring are females. Some species of weevils and all of the Hymenoptera are arrhenotokous. The ratio of female to male offspring has the direct consequence of altering the dynamics of population growth rates. Some species do not possess males during any part of their life cycle or only during certain generations in a multivoltine life cycle. For instance, aphids develop through a sexual generation from the fall through winter and into the spring; however, in late spring and during the summer, many species will develop through numerous parthenogenetic generations that lack males. In the case of aphids, other life history traits are associated with the alternation of sexual generations, such as shifts in specific host plant affinities between generations. During the summer, asexually produced females are wingless, the parthenogenetic females give birth to live young, and summer populations typically occur in large aggregations on plants. Each of the individual features illustrated by the complex aphid life cycle contributes to an extremely high reproductive potential due to a mix of life traits that in combination contribute to rapid population growth. On the other hand, aphid life history strategies are associated with considerable risk regarding mortality due to predation. A diverse complex of predators, parasitoids, and diseases, each with their own life history strategies, may at times counterbalance the high reproductive potential inherent in aphid populations and thereby regulate aphid numbers to very low population densities. An effective defense against natural enemies is dispersal, an action allowing escape and affecting survival in other ways as well.

DISPERSAL AND COLONIZATION

The movement of individuals is critical for the exchange of genes among and between populations, which in turn is essential for adaptation to changing environments and for placing an individual in a certain life stage into a suitable

Encyclopedia of Plant and Crop Science
DOI: 10.1081/E-EPCS 120024250

Fig. 1 Ants are excellent examples of how insects exhibit an array of life history strategies involving defensive behavior, domiciles, herbivory by farming, dispersal, competition, and surviving in extreme environments. (Photo by Jeffrey C. Miller.) (*View this art in color at www.dekker.com.*)

habitat at an appropriate time of the year. Long distance movement, including migration, in insects is a mode of active dispersal generally controlled by flight. Migrations are not a prevalent feature in life history strategies of herbivorous insects but are well known in a few species, such as the monarch butterfly, the milkweed bug, and some species of noctuid moths.[7] However, aerial passive dispersal over distances measured in kilometers is a relatively common behavior. Aerial passive dispersal is strongly influenced by wind currents, thus the insect must be buoyant and may be incapable of flight. For instance, certain Lepidoptera disperse in the larval stage. The caterpillars typically are hairy or spin silk for buoyancy and have generalist feeding requirements, and the females are flightless but highly fecund, whereas the male is winged and capable of flight.[8] Flightlessness is an obvious condition of adult insects exhibiting aptery and brachyptery with the activity of dispersal dependent on a passive mode; however, the feature of flightlessness is not necessarily a fixed trait. Species may possess various states of wingedness that are expressed through a suite of polymorphic conditions that in some cases are sexually determined. The expression of sexually dimorphic traits, polymorphic features, differences in colonizing abilities, and reproductive potential also interrelate with considerations of competitive behavior and defensive mechanisms.

COMPETITIVE BEHAVIOR

Insect herbivores compete within and among species for limited, essential resources. Therefore, life history traits exist that either enhance the competitive edge of one species over another or provide a means for one species to minimize competition for limited, essential resources such as mates, oviposition sites, food, and shelter. Two of the primary tactics employed by insects to gain a competitive advantage are 1) the use of chemical markers and 2) the expression of aggressive behavior. Numerous insect species mark parts of the habitat with pheromones. An example of habitat marking is the placement of a pheromone with a newly deposited egg, which has the effect of deterring oviposition by other females of the same species at or near the same place. Pheromones are also prevalent in mating behavior, trail maintenance, colony integrity, and warning signals. Interestingly, the pheromones that may alleviate intraspecific or interspecific competition for food among herbivorous species may serve as a cue in aiding natural enemies of the herbivores in prey location. The expression of aggressive behavior is manifested in life history traits such as cannibalism and various degrees of combat, from pushing and shoving to biting and stinging to the death of one or both of the combatants. Many herbivorous insects are cannibalistic, that is, they will kill and in some cases consume individuals of their own species. Cannibalism is most commonly encountered in species exhibiting a solitary lifestyle. For instance, caterpillars are immature Lepidoptera, a group that is virtually herbivorous in nature, and yet individual caterpillars will fight with and consume other caterpillars of the same species, or different species, when they encounter one another on their host plant. Aggressive behavior also provides an important component of defensive mechanisms.

DEFENSIVE MECHANISMS

Numerous life history traits are involved in defense mechanisms, such as, as previously mentioned, flight for escape, mimicry for appearing to be another species that possesses a threatening or toxic defense, aposematic coloration for warning predators of toxic compounds, crypsis for camouflage and for hiding out in the open, startle behavior for scaring off a potential attacker, feigning for appearing to be dead, expulsion of liquids for repelling an attacker, spines and poison glands, alarm pheromones, domicile construction, and guarding behaviors for protection of progeny or a colony. Each of these features, and many more, may be found occurring in combination with one another in an insect species.[9] In general, the adaptive nature of any particular suite of defensive mechanisms is dependent upon the intensity of the selective pressure exerted by a given mortality factor. For instance, a species of beetle, such as the Colorado potato beetle, whose larva is a specialist feeder on the foliage of potato, which is a toxic plant, may gain protection from enemies by sequestering the plant allelochemicals that otherwise confer protection to the plant against generalist herbivores. The ability to sequester toxic compounds will have limits and may be genetically determined and, thus, subject to variability within the population, as in monarch butterflies where certain individuals lack the cardiac glycosides responsible for the aposematic coloration of the adult. If the ability to sequester the toxic compound has an energetic cost, perhaps measured relative to reproductive potential, then individuals that are not capable of sequestering the toxic compounds are at risk, but they might gain a benefit if predation is not a primary mortality factor. Other defensive mechanisms involve mimicry and crypsis, which are expressed through morphological characters and coloration, traits that can vary based on selective pressures and genetically based polymorphisms. As mortality due to predation changes in time and space within the environment so will the gene frequency regulating the expression of each particular state of the polymorphism. Each genotype within the polymorphic suite of traits can exhibit a different fitness value depending on the presence and absence of predators. Thus, the tradeoffs in benefits and risks can have the consequence of changing the trend of directional selection. The principle of tradeoffs in benefits and risks in life history strategies is also well illustrated by considering adaptations to extreme environments.

EXTREME ENVIRONMENTS

Specialized morphological features and character reduction are two common types of adaptations to extreme environments. Some examples of these are a lack of color in certain species living in habitats devoid of light, such as caves; a lack of sight, as in certain soil-dwelling blind ground beetles; extremely dorso-ventrally flattened bodies typical of many leaf-mining caterpillars; and exaggerated features, such as the enlarged antennae associated with cave-dwelling crickets or the fossorial legs present in the soil-dwelling nymphs of cicadas. Similarly, insects in arid environments will exhibit specialized traits such as an extremely waxy cuticle, high numbers of Malphigian tubules, or physiological means for excreting uric acid instead of ammonia. Each one of these traits provides a selective advantage due to benefits regarding water conservation. Overall, growth and survival in extreme environments will depend on a combination of life history traits involving other physiological mechanisms and behavioral activities regarding the timing of dispersal to avoid harsh environmental conditions, phenology of life stages and diapause or aestivation, reproductive potential to compensate for high mortality, food resource requirements, and developmental biology, which determines growth rates and voltinism. One additional feature in life history strategies relates to the mitigation of adaptations to extreme environments through the behavior of group living and the activity of domicile construction.

GROUP LIVING AND DOMICILES

Insects may occur as solitary individuals, in groups consisting of a few to hundreds of individuals, or in a family unit consisting of thousands of individuals. Some of the benefits of group living are protection from natural enemies, cooperation for brood care, and in some cases construction of nests, as occurs with social insects such as termites, bees, and ants.[10] However, one of the risks of group living, whether it is as a simple aggregation of aphids on the exterior of a leaf or as a social unit inside a nest, is the supply of adequate food reserves. Among the social insects, the means for gathering and storing large quantities of food involves specialized life history traits including morphological features such as enlarged foreguts or hindguts, pollen baskets, physiological processes resulting in caste differentiation, behavioral mechanisms such as feeding via trophollaxis, and sanitation. Also, social insects are noted for their communal construction of domiciles; however, living within a domicile is not restricted to the social insects. Numerous species of herbivorous, nonsocial insects live within structures of some sort. For instance, gall-making insects, notably certain beetles, flies, wasps, and aphids, may stimulate their host plant to create a gall that includes a supply of nutritive tissue as well as a place to house the immatures and provide a barrier to guard against the outside environment. Although the benefits of a gall are obvious, one of the risks is that a predator capable of invading the gall space can readily consume its otherwise defenseless prey.

CONCLUSION

The subject of insect life history strategies is a well-studied topic on the basis of empirical data, concepts, and models. Basically, the topic focuses on considerations of variability in the environment, variability in genetics of the organisms responding to the variable environments, and the relative advantages and disadvantages of various combinations of life history traits across spatial and temporal components of the environment.

ARTICLES OF FURTHER INTEREST

Agriculture and Biodiversity, p. 1
Coevolution of Insects and Plants, p. 289

Ecology and Agricultural Sciences, p. 401
Ecology: Functions, Patterns, and Evolution, p. 404
Insect Life History Strategies: Development and Growth, p. 598
Insect–Plant Interactions, p. 609
Physiology of Herbivorous Insects, p. 910
Plant Defenses Against Insects: Physical Defenses, p. 944

REFERENCES

1. Cole, L.C. The population consequences of life history phenomena. Q. Rev. Biol. **1954**, *29*, 103–137.
2. Stearns, S.C. *The Evolution of Life Histories*; Oxford Press: Oxford, UK, 1992.
3. Southwood, T.R.E. Bionomic Strategies and Population Parameters. In *Theoretical Ecology*; May, R.M., Ed.; W.B. Saunders: Philadelphia, PA, 1976; 26–48.
4. MacArthur, R.H.; Wilson, E.O. *The Theory of Island Biogeography*; Monographs in Population Biology, Princeton University Press: Princeton, NJ, 1967; Vol. 1.
5. Hassell, M.P. *The Dynamics of Arthropod Predator–Prey Systems*; Monographs in Population Biology, Princeton University Press: Princeton, NJ, 1978; Vol. 13.
6. Miller, J.C. Niche relationships among parasitic insects occurring in a temporary habitat. Ecology **1980**, *61*, 270–275.
7. *Evolution of Insect Migration and Diapause*; Dingle, H., Ed.; Springer-Verlag: New York, 1986.
8. Barbosa, P.; Krishik, V.; Lance, D. Life-history traits of forest—Inhabiting flightless *Lepidoptera*. Am. Midl. Nat. **1989**, *122*, 262–274.
9. Turpin, F.T. *Insect Appreciation*; Kendall Hunt: Dubuque, IA, 2000.
10. Wilson, E.O. *The Insect Societies*; Harvard University Press: Cambridge, MA, 1974.

Insect/Host Plant Resistance in Crops

I

C. Michael Smith
Kansas State University, Manhattan, Kansas, U.S.A.

INTRODUCTION

The development of plants with resistance to insects is a proven method for managing insect pest populations. Many insect-resistant crop varieties have played vital, integral roles in sustainable systems of agricultural production. Resistant varieties are nonpolluting; ecologically, biologically, and socially acceptable; and economically feasible as a means of pest control. In the eighteenth and nineteenth centuries, insect resistant varieties of cereals and fruits were first developed and cultivated in Europe and North America. When the grape phylloxera, *Phylloxera vittifolae*, destroyed the French wine industry during the late 1880s, resistant rootstocks from the United States were grafted to French scions, and the industry recovered.

TERMINOLOGY

Plant resistance to insects is the sum of the inheritable genetic qualities that result in a plant of one variety or species being less damaged than a susceptible plant lacking these qualities under a similar pressure of herbivory.[1] Resistance is measured on a relative scale, and the degree of resistance is based on comparison to susceptible plants that lack resistance in the same experiment.[2] Variation in insects, test plants, and the environment may affect the expression of resistance, often resulting in pseudoresistance in normally susceptible plants. The effect of each variable should be determined before concluding that a plant is resistant.[3]

Associational plant resistance to insects occurs when mixtures of plant species slow pest insect population development on normally susceptible plants growing in association with resistant plants[4] or plants infected with fungal endophytes that produce alkaloids that kill or delay the development of pest insects.[5] Searches for resistance can be directed at selection for either allopatric resistance (plants evolving in the absence of insect pressure—no previous evolutionary contact) or sympatric resistance (plants evolving in the presence of insect pressure).[6]

FUNCTIONAL CATEGORIES OF RESISTANCE

Painter[7] defined the negative effects on insect biology of chemical and morphological plant defenses in resistant plants as antibiosis. These defenses may be overcome by the expression of virulence genes in an insect pest population after prolonged exposure to high levels of antibiosis. Populations expressing virulence are known as insect biotypes. Many biotypes are parthenogenic (reproduce asexually) aphid species, or Dipteran and Heteropteran species with high reproductive potentials that infest large cereal (rice and wheat) monocultures. Biotypes have been identified by phenotypic plant reaction for many years, but DNA marker techniques are beginning to be employed for more accurate identification of biotypes.

Resistance is also manifested as antixenosis, where chemical or morphological plant factors result in pest insect selection of alternate host plants.[8] The resistant plant may also be able to withstand or recover from insect damage and yield as much or more biomass than a protected susceptible plant. In this case, resistance is expressed as tolerance to the pest insect. An advantage of tolerance is that although the pest population level is not reduced, an ample food source for pest predators and parasites is provided, and synergizes the beneficial effects of these biological control agents. Frequently, resistant plants exhibit multiple resistance categories.

In addition to enhancing biological control, even moderate levels of resistance combined with insecticides can greatly reduce insecticide use costs and the amount of insecticide residues. This effect has been demonstrated in many different agricultural systems.[9] In some cases, insect-resistant varieties offer producers genetically incorporated insect control for the cost of the seed alone. Insect-resistant varieties provide a substantially greater rate of return on investment than do insecticides. The current global economic value of all insect-resistant crop plants, based on increased crop yields and reduced insecticide use, is approximately $2.7 billion per year[9] (Table 1).

Encyclopedia of Plant and Crop Science
DOI: 10.1081/E-EPCS 120010492

Table 1 Approximate annual value of resistance genes deployed in all arthropod-resistant crops produced globally

Crop plant	Pest arthropod	Geographic location(s)	Value (US$ × 1,000)
Cotton (*Bt*)	*Helicoverpa zea*	Southeast U.S.	1,350
	Heliothis virescens		
Rice	*Nilaparvata lugens*	South/Southeast Asia	1,125
	Nephotettix cincticeps		
Sorghum	*Schizaphis graminum*	Midwest/Southwest U.S.	85
	Blissus leucopterous		
Wheat	*Aceria toschillea*	North America	170
	Mayetiola destructor	Southeast U.S.	19
	Diuraphis noxia	Western U.S./South Africa	14
	Cephus cinctus	North America	17

INSECT RESISTANCE TRANSGENES

Two changes occurred in the development of insect-resistant varieties near the end of the twentieth century. First, *Agrobacteirum* transformation systems and biol-oistic projectile devices were used to transfer genes encoding insecticidal crystal (cry) toxins from the soil bacterium *Bacillus thuringensis* (Bt) into crop plants. After ingestion, crystals are solubilized in the alkaline environment of the insect gut, where active toxic frag-ment(s) are released by insect digestion, bind to specific receptors on the midgut cells of susceptible larvae, and then cause osmotic lysis of those cells, and insect death. Several crop plants have also been transformed with transgenes expressing insect digestive enzyme inhibitors of both plant and insect origin.

Before the development of Bt transformants, numerous studies determined that insects became virulent to the cry toxin gene after prolonged exposure to high doses of Bt, similar to how high doses of conventional pesticide or high levels of conventional gene expression lead to virulence in target insects. One early research study indicated that Bt corn was toxic to larvae of the monarch butterfly, *Danaus plexippus*. Subsequently, the U.S. National Academy of Sciences reviewed a series of several other experiments and concluded that the risk to monarchs from Bt corn is not significant. One Bt corn event grown on a very small U.S. hectarage was shown to be toxic to monarch larvae and is being eliminated from production.

To obtain their maximum longevity, Bt insect-resistant transgenes in corn and cotton are deployed with non-Bt plant refuges that enable the survival of pest moths from susceptible larvae to mate with moths produced from larvae virulent to Bt. Shifting the mortality of larvae heterozygous for virulence from 50% to 95% provides a tenfold delay in time before the development of virulence.[10]

Although numerous transgenic crop plants have been developed, only the production and marketing of transgenic corn, cotton, and potato varieties have proceeded in North America and Asia. However, because these are essentially insecticidal plants, their deployment strategy has proven complex, and their production and use have met with strong opposition by environmentalists, primarily in Europe. Estimates are that recommended insecticide use against one insect pest of corn, *Ostrinia nubilalis*, dropped approximately 30% after the commercialization of Bt corn in North America.

DNA RESISTANCE MARKERS

The second major shift in the development of insect-resistant plants is the adaptation of genotypic screening techniques to identify insect resistance genes. This change involves isolating DNA from resistant and susceptible parents, amplifying complementary DNA with molecular markers from known chromosome locations, separating the DNA amplification products by electrophoresis, and identifying primers that differentially amplify the DNA of resistant and susceptible parents and their progeny in a polymorphic (informative) pattern. DNA markers have several advantages over phenotypic markers. They can be codominant and detect heterozygous patterns of inheritance, where morphological markers behave in a dominant/recessive manner and do not. The allelic variation of DNA markers is greater than for morphological markers. Finally, molecular markers are unaffected by the environmental fluctuations that have been shown to have a significant effect on morphological markers. DNA markers can be used to map resistance genes inherited as dominant traits at single loci, or as quantitative traits, located at multiple loci and linked to several minor genes contributing to insect resistance.

The chromosome location of many constitutive insect resistance genes has been identified in a wide number of crop plants using molecular markers.[11] However, the only insect resistance gene cloned to date is Meu1 from wild tomato, *Lycopersicon peruvianum*, which confers resistance to the potato aphid, *Macrosiphum euphorbiae*, and the root knot nematode, *Meloidogyne incognita*.[12] Meu1 is a member of the nucleotide-binding site and leucine-rich region (NBS–LRR) family of disease and nematode resistance genes.

EXPRESSED RESISTANCE GENES

While research on constitutive herbivore resistance genes is only beginning, progress is also being made in the identification of unique mRNA transcripts expressing induced insect resistance genes that are produced by several species of plants. In essence, tissue damage caused by components of insect saliva contacting plant tissues activates an octadecanoid signaling cascade that leads to biosynthesis of signals such as jasmonic acid, salicylic acid, ethylene, and abscissic acid that in turn activate the production of many different types of insect antifeedant compounds. Chewing insects elicit plant responses different from those induced in response to feeding by piercing/sucking insects.[13] In addition, the end result of hypersensitive responses to insect feeding is often the production of enzymes that lead to cell wall thickening in resistant varieties, hindering insect feeding and digestion. In cereal plants, induced disease and nematode resistance gene loci occur in clusters[14] and there is evidence that insect resistance genes occur in a similar manner.[15]

CONCLUSION

Insect-resistant crops play an important role in world sustainable agricultural systems, and will become more prominent as world food needs increase. Resistant varieties have proven to be ecologically and socially acceptable to consumers, and economically feasible for producers. Although there have been many successes in deploying insect-resistant genes in improved varieties, it is not completely clear how plants recognize the attack and feeding by different insects. However, chemical and mechanical stimuli perceived by plants during insect feeding contribute to plant recognition of feeding. Very limited information exists about the molecular aspects of plant response to insects, but a partial understanding is emerging. The continued use and refinement of molecular marker techniques to genotype constitutive and induced insect resistance genes in plants should continue to provide the necessary information to determine the molecular bases of plant genetic defenses against insects.

The use of functional genomics will enable the sequencing of additional insect resistance genes in many plants, and will help develop refined genetic maps of these genes. A more complete understanding of pest resistance genes will reduce the possible development of virulent, resistance-breaking insect biotypes and lead to more rapid development of crop varieties with greater and more diverse levels of insect resistance.

ACKNOWLEDGMENTS

I appreciate the assistance of Dr. Elena Boyko, Dr. Randall Higgins, and Dr. Xuming Liu, Department of Entomology, Kanasas State University, and Dr. Phil Sloderbeck, Kansas Cooperative Extension Service, for suggestions and comments that helped guide the development of the manuscript.

ARTICLES OF FURTHER INTEREST

Genetically Engineered Crops with Resistance Against Insects, p. 506
Plant Defenses Against Insects: Constitutive and Induced Chemical Defenses, p. 939
Plant Defenses Against Insects: Physical Defenses, p. 944
Transgenetic Plants: Breeding Programs for Insect Resistance, p. 1245

REFERENCES

1. Smith, C.M. *Plant Resistance to Insects—A Fundamental Approach*; John Wiley & Sons: New York, 1989.
2. *Economic, Environmental, and Social Benefits of Resistance in Field Crops*; Wiseman, B.R., Webster, J.A., Eds.; Thomas Say Publication in Entomology, Entomological Society of America: Lanham, 1999.
3. Chesnokov, P.G. *Methods of Investigating Plant Resistance to Pests*; National Science Foundation: Washington, DC, 1953. (Israel Program for Scientific Translation).
4. Letourneau, D.K. Associational resistance in squash monocultures and polycultures in tropical Mexico. Environ. Entomol. **1986**, *15*, 285–292.
5. Clement, S.L.; Kaiser, W.J.; Eischenseer, H. *Acremonium* Endophytes in Germplasms of Major Grasses and Their Utilization for Insect Resistance. In *Biotechnology of Endophytic Fungi of Grasses*; Bacon, C.W., White, J.F., Eds.; CRC: Boca Raton, 1994; 185–199.
6. Harris, M.K. Allopatric resistance: Searching for sources of insect resistance for use in agriculture. Environ. Entomol. **1975**, *4*, 661–669.

7. Painter, R.H. *Insect Resistance in Crop Plants*; University of Kansas Press: Lawrence, 1951.
8. Kogan, M.; Ortman, E.E. Antixenosis-a new term proposed to replace Painter's ''Non-preference'' modality of resistance. Bull. Entomol. Soc. Am. **1978**, *24*, 175–176.
9. Smith, C.M. Plant Resistance to Insects. In *Biological and Biotechnological Control of Insects*; Rechcigl, J., Rechcigl, N., Eds.; Lewis Publishers: Boca Raton, 1999; 171–208.
10. Gould, F. Sustainability of transgenic insecticidal cultivars: Integrating pest genetics and ecology. Ann. Rev. Entomol. **1998**, *43*, 701–726.
11. Yencho, G.C.; Cohen, M.B.; Byrne, P.F. Applications of tagging and mapping insect resistance gene loci in plants. Ann. Rev. Entomol. **2000**, *45*, 393–422.
12. Vos, P.; Simons, G.; Jesse, T.; Wijbrandi, J.; Heinen, L.; Hogers, R.; Frijters, A.; Groenendijk, J.; Diergaarde, P.; Reijans, M.; Fierens-Onstenk, J.; de Both, M.; Peleman, J.; Liharska, T.; Hontelez, J.; Zabeau, M. The tomato *Mi*-1 gene confers resistance to both root-knot nematodes and potato aphids. Nat. Biotechnol. **1998**, *16*, 315–1316.
13. Stout, M.J.; Fidantsef, A.L.; Duffey, S.S.; Bostock, R.M. Signal interactions in pathogen and insect attack: Systemic plant-mediated interactions between pathogens and herbivores of the tomato, *Lycopsericon esculentum*. Physiol. Mol. Plant Pathol. **1999**, *54*, 115–130.
14. Boyko, E.V.; Kalendar, R.; Korzun, V.; Korol, A.; Schulman, A.; Gill, B.S. A high density genetic map of *Aegilops tauschii* incorporating retrotransposons and defense-related genes: Insights into cereal chromosome structure and function. Plant Mol. Biol. **2002**, *48*, 767–790.
15. Liu, X.M.; Smith, C.M.; Gill, B.S.; Tolmay, V. Microsatellitemarkers linked to six Russian wheat aphid resistance genes in wheat. Theor. Appl. Genet. **2001**, *102*, 504–510.

Insect–Plant Interactions

Sujaya Rao
Oregon State University, Corvallis, Oregon, U.S.A.

INTRODUCTION

Insects and plants have evolved together for millions of years, and their relationships are both antagonistic and mutualistic. Antagonistic associations arise because the energy trapped in plants during photosynthesis is used by approximately half of the species of insects. Insects of the six largest orders (Orthoptera, Hemiptera, Lepidoptera, Coleoptera, Diptera, and Hymenoptera) derive most or all of their food directly from plants.[1] However, despite extensive herbivory by insects, plants continue to flourish, in part because of defenses that they have evolved against insect attack. As a result, no insect species feeds on all plants, and no plant species is eaten by all insect herbivores. Certain insect species have, however, evolved effective means to overcome plant defenses, and in turn new plant defenses have evolved. Coevolution has also resulted in the development of mutualistic interactions that benefit both plants and insects in diverse ways. This review will outline plant–insect interactions that are beneficial to insects, those that are beneficial to plants, and certain mutualistic interactions.

INTERACTIONS OF BENEFIT TO INSECTS

Feeding

Insects profit by using plants as sources of food. All parts of a plant may be eaten by phytophagous insects: roots, sap, stems, leaves, buds, pollen, nectar, fruits, seeds, and decaying plant tissue. Monophagous insects feed on one or a few plant species, whereas polyphagous insects include a wide range of plants in their diet. Plants provide insect herbivores with nutrients necessary for growth and development, such as water, nitrogen, carbohydrates, amino acids, lipids, minerals, and vitamins.[2] Plants also serve as sources of sterols required for production of hormones.

Plant growth is affected negatively when photosynthetic tissue is eaten, phloem sap is drained, or uptake of water and minerals is affected as a result of feeding by chewing and sucking insects. Chewing insects feed either externally or internally, as borers or miners. Sucking insects imbibe juices and cause weakening and yellowing of plants. These influences result in stunted growth, lowered production of seeds, weakened competitive abilities, or plant death. In addition, damage to flowers, fruits, and seeds directly affects population growth of plants. Further, some insects inject toxins or disease organisms such as viruses during feeding, which can have additional detrimental impacts on host plants. Certain insect species such as leafcutter bees transport pieces of leaves to line their nests, whereas leafcutter ants use parts of leaves as a substrate for growing fungi in their nests.

Several species of phytophagous insects induce the production of abnormal growths called galls in their host plants. Galls are found in buds, leaves, stems, flowers, or roots, and their shapes and locations are often characteristic of the plant and insect species concerned. They provide insects with a rich source of food and protection from predators and the elements.[3]

Oviposition

Plants provide sites for oviposition that enable immatures to commence feeding soon after egg hatch. Egg laying is a critical step in the life cycle of insects, since survival of progeny is highly dependent on recognition of suitable host plants by gravid females. Newly emerged larvae are often limited in their dispersal abilities, and hence host selection is a function of the ovipositing females. Eggs are often attached to plants and may be laid singly or in groups on any plant part. In certain insect orders such as Hemiptera, several species conceal their eggs by insertion into plant tissues.

Additional Benefits

Not all interactions beneficial to insects are detrimental to plants. Plants provide insect herbivores with shelter from natural enemies and the elements, rain and sunlight. Plants can also act as sites for mating.

INTERACTIONS OF BENEFIT TO PLANTS

Despite the great number of phytophagous insect species that exist, plants continue to dominate the landscape. This suggests that plants are a formidable evolutionary barrier

Encyclopedia of Plant and Crop Science
DOI: 10.1081/E-EPCS 120010482

that many insects are not able to overcome.[4] Plants possess morphological or chemical defenses (discussed below) that enable them to escape insect attack.

Carnivorous Plants

Certain plant species use insects to supplement their nutrition. These carnivorous plants are found in nitrogen-poor soils, particularly in acid bogs and heavy volcanic clays, and their root systems are not extensive. Carnivorous plants trap and digest insects. Trapping mechanisms include sticky exudates that entangle insects, such as in sundews; structural modifications, such as the pitcher on pitcher plants; or modified leaves such as those in the Venus flytrap.[5] Trapped insects are digested by enzymes secreted from various glands, and amino acids—the nitrogenous compounds in insects that are beneficial to the plant—are absorbed by special tissues in the plant.

MUTUALISTIC INTERACTIONS

Different species or guilds of species that have coevolved to the extent that they benefit each other are considered mutualistic. Mutualistic interactions between insects and plants have evolved on several occasions. Many angiosperms, or flowering plants, are dependent on insects for pollination. More specialized mutualistic interactions have evolved such as those between neotropical species of acacia trees that produce secretions attractive to an aggressive ant species, which in return protects the trees from insect herbivores.[6]

A lesser-known mutualistic interaction that has gained attention in recent years exists between plants and the natural enemies of herbivorous insects. Natural enemies, parasitoids and predators of insect herbivores, cue in on stimuli associated with plants in habitats where their hosts are likely to occur. Predators and parasitoids benefit by reduction of the time spent in searching, whereas plants that provide appropriate cues for attracting natural enemies benefit by the reduction in herbivory.[7] Although undamaged plants attract natural enemies, the influence is enhanced when plants are damaged by herbivores.

FACTORS AFFECTING INSECT–PLANT INTERACTIONS

Insect species use a restricted range of taxonomically related plant species. Distinct food preferences are to a large extent attributable to an insect's ability to identify acceptable host plants on the basis of plant-related traits, amid a multitude of nonhosts. Host location by herbivorous insects is a complex process involving a hierarchy of cues, some acting from a distance and some at close range.

Host plant selection in insects includes host habitat finding, host finding, host recognition and acceptance, and host suitability. Plant-related factors can influence each of these steps. Differential colonization of plant varieties by the same insect species may be due to differences in morphological or chemical traits. Plant traits such as architecture, thickened cell walls, or trichomes (small hairs on plant surfaces) impede surface penetration or herbivore movement, and thereby influence insect herbivores directly. Such traits can benefit herbivorous insects indirectly when they exert negative impacts on predators and parasitoids, thereby providing enemy-free space. As a result, parasitoids and predators are more successful in locating and accessing their hosts or prey on certain plant species than on others. Chemical traits in plants that have negative impacts on insect herbivores include secondary plant metabolites produced by plants.[8] These include alkaloids such as caffeine and nicotine, pyrethrum, tannins, and diverse other compounds that are difficult to handle and hence deter herbivory. Plants also produce compounds that function as protein inhibitors or insect growth regulators, and thereby affect metabolism and development.

Insects and plants have evolved chemical messenger systems involving diverse allelochemicals (Fig. 1). An allelochemical is a chemical involved in an interaction between organisms belonging to different species. When plant compounds evoke a reaction in a receiving herbivorous insect that is favorable to the plant but not the insect, the compound is called an allomone. Allomones include toxic plant compounds mentioned above that serve as defenses against herbivory. In chemically-mediated

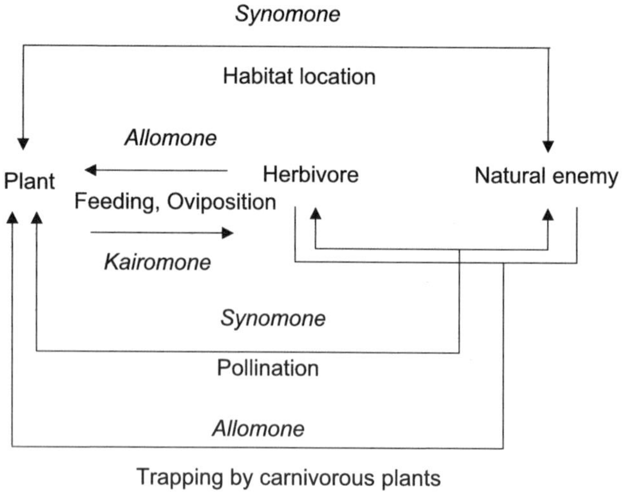

Fig. 1 Allelochemicals in plant–insect interactions (the direction of arrows indicates beneficiaries in the interactions).

interactions where the receiver benefits but not the emitter, the chemical involved in the relationship functions as a kairomone. In plant–insect interactions, examples of kairomones include phytochemicals that attract herbivorous insects to their host plants. When plants produce chemical cues that draw natural enemies to habitats of their hosts, the chemicals involved in the interactions function as synomones, because the emitter and receiver of the chemical both benefit from the interaction.[9]

Plant-related allelochemicals are used extensively by insects for location of appropriate host plants for feeding or oviposition purposes. Typically volatile compounds from plants serve to draw insects from a distance, whereas more stable compounds serve as short range cues for arrestment and ultimate acceptance of the plant. Attraction of an insect to a plant may be due to the presence of specific compounds such as allyl isothiocyanate in cruciferous plants, the presence of specific blends of compounds, or the absence of deterrent compounds.

Some insects are able to overcome the detrimental effects of chemical defenses in plants by detoxification, sequestration, or excretion. In addition, certain species that sequester toxic phytochemicals use these compounds to their advantage in their relationships with conspecifics or members of other species. For example, moths of *Utetheisa ornatrix* sequester pyrrolizidine alkaloids acquired from their food plants during larval development. The alkaloids are involved in mate selection and are also transferred to eggs, providing protection against predation.[10]

FUTURE CHALLENGES

The examples presented here provide some insights into the diversity of interactions between plants and insects. Coevolutionary processes in nature continue to modify these relationships. In addition, human practices bring about further changes in associations between plants and insects. For decades, plant–insect interactions have been exploited in conventional plant breeding programs for development of cultivars that are resistant to attack by insect pests, but little attention has been directed toward impacts on natural enemies. With biotechnology advances, evolutionary changes in plant–insect interactions are likely to be expedited. Evaluation in risk assessment programs of the impacts of genetically modified plants on herbivorous insects and their natural enemies is critical, and it presents a tremendous challenge due to the complexity and diversity of plant–insect interactions.

ARTICLES OF FURTHER INTEREST

Mutualisms in Plant–Insect Interactions, p. 763
Plant Defenses Against Insects: Constitutive and Induced Chemical Defenses, p. 939
Plant Defenses Against Insects: Physical Defenses, p. 944

REFERENCES

1. Daly, H.V.; Doyen, J.T.; Purcell, A.H. *Introduction to Insect Biology and Diversity*; Oxford University Press: New York, 1998.
2. Dadd, R.H. Insect nutrition: Current developments and metabolic implications. Annu. Rev. Entomol. **1973**, *18*, 381–420.
3. Williams, M.A.J. Plant Galls: A Perspective. In *Plant Galls: Organisms, Interactions, Populations*; Williams, M.A.J., Ed.; Clarendon Press: Oxford, 1994; 1–7.
4. Bernays, E.A.; Chapman, R.E. *Host–Plant Selection by Phytophagous Insects*; Chapman and Hall: New York, 1994.
5. Juniper, B.E.; Robins, R.J.; Joel, D.M. *The Carnivorous Plants*; Academic Press: London, 1989.
6. Janzen, D.H. Coevolution of mutualism between ants and acacias in Central America. Evolution **1966**, *20*, 249–275.
7. Price, P.W.; Bouton, C.E.; Gross, P.; McPheron, B.A.; Thompson, J.N.; Weis, A.E. Interactions among three trophic levels: Influences of plants on interactions between insect herbivores and natural enemies. Ann. Rev. Ecolog. Syst. **1980**, *11*, 41–65.
8. Fraenkel, G.S. The raison d'etre of secondary plant substances. Science **1959**, *129*, 1466–1470.
9. Nordlund, D.A. Semiochemicals: A Review of the Terminology. In *Semiochemicals: Their Role in Pest Control*; Nordlund, D.A., Jones, R.L., Lewis, W.J., Eds.; John Wiley and Sons: New York, 1981; 13–28.
10. Eisner, T.; Eisner, M.; Rossini, C.; Iyengar, V.; Roach, B.L.; Benedikt, E.; Meinwald, J. Chemical defense against predation in an insect egg. Proc. Natl. Acad. Sci. **2000**, *97*, 1634–1639.

Integrated Pest Management

Marcos Kogan
Oregon State University, Corvallis, Oregon, U.S.A.

INTRODUCTION

In the context of integrated pest management (IPM), a pest is any living organism whose life system conflicts with human interests, economy, health, or comfort. Pests belong to three main groups: 1) invertebrate and vertebrate animals; 2) disease-causing microbial pathogens; and 3) weeds or undesirable plants growing in desirable sites.

The concept of pest is entirely anthropocentric. There are no pests in nature, in the absence of humans. An organism becomes a pest only if it causes injury to crop plants in the field or to their products in storage or, outside an agricultural setting, if it affects structures built to serve human needs. This article focuses on the impact of pests on plants.

BACKGROUND

Since the beginnings of agriculture some 10,000 years ago, humans competed with other organisms for the same food resources. Primitive tools to fight pests were probably limited to handpicking and methods such as flooding or fire, although at times of despair agriculturists resorted to magic incantations or religious proclamations.[1] But as early as 2500 B.C., Sumerians used sulfur to control insects and by 1000 B.C., Egyptians and Chinese were using plant-derived insecticides to protect stored grain.[2] Real progress in the fight against agricultural pests occurred upon advances in biological sciences, resulting in a better understanding of arthropod pests and the relationships between microbial pathogens and plant diseases. Until the middle of the twentieth century, insect pest control ingeniously combined knowledge of pest biology and cultural and biological control methods. Trichlorodiphenyltrichloroethane (DDT) in the early 1940s ushered in an era of organo-synthetic insecticides. Initial results were spectacular and many entomologists foresaw the day when all insect pests would be eradicated. This optimistic view was premature, as insects quickly evolved mechanisms to resist the new chemicals, requiring more frequent applications, higher dosages, or the replacement of old insecticides with more powerful ones. This cycle has become known as the insecticide treadmill.[1] At the end of the 1950s, scientists around the world became aware of the risks associated with the application of powerful insecticides, as resistance rendered one chemical after the other useless and as beneficial organisms were indiscriminately killed along with target pest species. Such killing of insect parasitoids and predators caused the resurgence of pests at even higher levels. In addition, populations of other plant-eating species, usually balanced by the pests' natural enemies, exploded in outbreak proportions, causing upsurges of secondary pests and pest replacements. As a result of mounting concern about resistance, resurgence, and upsurges of secondary pests, the concept of integration emerged—an attempt to reconcile pesticides with the preservation of natural enemies.[3]

INTEGRATION

The next phase in the evolution of IPM was the expansion of the term "integrated" to include all available methods (biological, cultural, mechanical, physical, legislative, as well as chemical) and to span all pest categories of importance in the cropping system (invertebrates, vertebrates, microbial pathogens, and weeds). Because of the multiple meanings, integration in IPM was conceived at three levels of increasing complexity. Level I integration refers to multiple control tactics against single pests or pest complexes; level II is integration of all pest categories and the methods for their control; level III integration applies the steps in level II to the entire cropping system. At level III, the principles of IPM and sustainable agriculture converge.[4]

THE ECOLOGICAL BASES OF IPM

Ecology offered the tools to describe how populations expand and contract under natural conditions; how biotic communities are organized and how their component organisms interact; how abiotic (nonliving) factors (climate, soil, topography, hydrology) shape biotic communities; how biodiversity stabilizes ecological systems; and, perhaps most significantly, how to understand the effects of

Encyclopedia of Plant and Crop Science
DOI: 10.1081/E-EPCS 120010494

disturbances in the dynamics of biotic communities.[5,6] Thus, IPM relied upon a solid ecological foundation.[4,5,7]

THE TOOLS OF PEST MANAGEMENT

Management of pests in an IPM system is accomplished through selection of control tactics singly or carefully integrated into a management system. These tactics usually fall into the following main classes: 1) chemical; 2) biological; 3) cultural/mechanical; and 4) physical control methods (these four methods apply to all pests); 5) host plant resistance (applicable to invertebrate and microbial pests); 6) behavioral; and 7) sterile insect technique, a genetic control method (relevant only to insects).

Chemical Control

Chemical pesticides are used to kill or interfere with the development of pest organisms. Pesticides are called insecticides, acaricides, nematicides, fungicides, or herbicides if the target pests are insects, mites, nematodes, fungi, or weeds. Pesticides are useful tools in IPM if used selectively and if the pest population threatens to reach the economic injury level (described later). Much care is necessary to handle pesticides and avoid their potentially negative effect on the environment.[8]

Biological Control

Biological control refers to the action of parasitoids, predators, and pathogens in maintaining another organism's population density at a lower average than would occur in their absence.[9] In IPM practice, biological control includes intentional introduction of natural enemies to reduce pest population levels (classical biological control), the mass release of biocontrol organisms (inundative biocontrol), or the timely inoculation of natural enemies (augmentative biocontrol).[10]

Cultural Control/Mechanical

Cultural practices become cultural controls when adopted for intentional effect on pests. Thus, tillage and other operations for soil preparation have a direct impact on weeds and on soilborne insect pests. The manipulation of the crop environment to favor natural enemies is called habitat management, and is of growing interest as a form of conservation biological control.[10,11]

Physical Control

Fire, water, electricity, and radiation are some of the main physical forces used in IPM. Flooding in paddy rice cultivation is a major adjuvant method in many parts of the world.

Host Plant Resistance

This means of control is based on breeding crop plants to incorporate genes whose products cause plants to be unsuitable for insects feeding on them (antibiosis) or impair insects' feeding behavior (antixenosis).[12] Resistance is the only effective method for control of plant viral diseases. Techniques of genetic engineering have opened new opportunities to expand plant resistance in IPM, but the approach is not without risks and should be adopted cautiously, within a strict IPM framework.[13]

Behavioral Control

Behavioral control uses chemicals (allelochemicals) to interfere with normal patterns of mainly sexual (mating) and feeding behaviors of arthropods . Sex pheromones are used for mating disruption.[14,15] Feeding excitants or deterrents are used to disrupt normal feeding behavior or to attract and kill insects.[16]

Sterile Insect Technique

This genetic control method disrupts normal progeny production of the target species by the mass release of sterile insects.[17]

THE SCALE OF IPM SYSTEMS

Most IPM programs focus on single fields because of local variability in physical, crop, and pest conditions. Some pests, however, are highly mobile and require a regional approach for their control. Furthermore, certain control tactics are effective only if deployed over large areas. Mating disruption for control of the codling moth in apple and pear orchards in the western United States used an areawide approach that required a minimum operational unit of about 160 ha. Advancement of IPM to higher levels of integration will require planning and implementation at the landscape or even ecoregional levels. Advanced technologies of geographic information systems (GIS) and remote sensing are essential for development of such programs. Such planning is still at its infancy in IPM.

DECISION SUPPORT SYSTEMS FOR IPM IMPLEMENTATION

IPM uses objective criteria for making decisions about the need for a control action and selection of the most

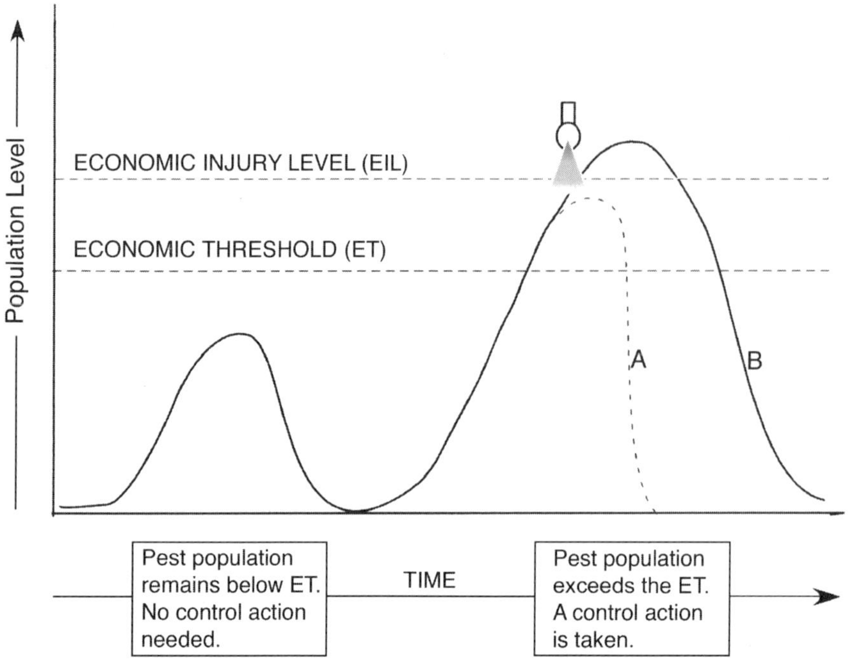

Fig. 1 Fundamental decision-making concepts for insect pest management: economic threshold (ET) and economic injury level (EIL). Illustrated is an insect pest population with two peaks in one year. The first peak remains below ET and no treatment is needed; the second peak exceeds ET. If a treatment is applied, the population crashes (curve A); if a treatment is not applied, the population exceeds EIL and an economic crop loss occurs (curve B). (*View this art in color at www.dekker.com.*)

appropriate tactics. In arthropod pest control the criteria are based on assessments of the extant pest population in the field, its damage potential (if not controlled), and the relative costs of treatment and crop value. The main parameters are the economic injury level (EIL) and economic threshold (ET). ET is the population level that will trigger a control action. Fig. 1 describes these parameters. To use ETs it is necessary to monitor pest populations in the field using well-defined sampling methods, which vary with the pest species. Although useful, the concept of EIL is less effective as a decision-making support tool for weed management and is partially applicable for microbial disease management.[18]

CONCLUSION

It is difficult to assess the global impact of IMP. It is safe, however, to state that the concept drastically changed the approach to pest control among progressive growers, the chemical industry, research establishments, and policy makers. One of the few economic analyses of the impact of IPM was done in Brazil on soybean production costs. The study suggested that technologies available in 1980 (date of the study), if adopted throughout the entire production area (8.5 million ha), would have resulted in savings of US$216 million annually. Most major agricultural crops have benefited from the adoption of IPM, although IPM is in its infancy and major advances are still to be expected as modern technology comes to be incorporated into the management of all pests.

ARTICLES OF FURTHER INTEREST

Agriculture and Biodiversity, p. 1
Biological Control in the Phyllosphere, p. 130
Biological Control of Nematodes, p. 134
Biological Control of Oomycetes and Fungal Pathogens, p. 137
Biosafety Considerations for Transgenic Insecticidal Plants: Non-Target Herbivores, Detritivores, and Pollinators, p. 153
Biosafety Considerations for Transgenic Insecticidal Plants: Non-Target Predators and Parasitoids, p. 156
Breeding for Durable Resistance, p. 179
Ecology and Agricultural Sciences, p. 401
Economic Impact of Insects, p. 407
Genetically Engineered Crops with Resistance Against Insects, p. 506

Insect/Host Plant Resistance in Crops, p. 605
Management of Bacterial Diseases of Plants: Regulatory Aspects, p. 669
Management of Fungal and Oomycete Diseases: Fruit Crops, p. 678
Management of Fungal and Oomycete Diseases: Vegetable Crops, p. 681
Management of Nematode Diseases: Options, p. 684
Nematodes and Host Resistance, p. 805
Nematodes: Ecology, p. 809
Rhizosphere Management: Microbial Manipulation for Biocontrol, p. 1098
Sustainable Agriculture: Definition and Goals, p. 1187
Weed Management in Less Developed Countries, p. 1295

REFERENCES

1. Van den Bosch, R. *The Pesticide Conspiracy*; Doubleday: Garden City, NY, 1978; pp. viii, 226.
2. Norris, R.F.; Caswell-Chen, E.P.; Kogan, M. *Concepts in Integrated Pest Management*; Prentice Hall: Upper Saddle River, NJ, 2003; 586.
3. Kogan, M. Integrated pest management: Historical perspectives and contemporary developments. Annu. Rev. Entomol. **1998**, *43*, 243–277.
4. Kogan, M. Integrated pest management theory and practice. Entomol. Exp. Appl. **1988**, *49*, 59–70.
5. Levins, R.L.; Wilson, M. Ecological theory and pest management. Annu. Rev. Entomol. **1980**, *25*, 287–308.
6. Kogan, M. *Ecological Theory and Integrated Pest Management Practice*; Environmental Science and Technology, Wiley: New York, 1986; pp. xvii, 362.
7. Huffaker, C.B.; Gutierrez, A.P. *Ecological Entomology*; Wiley: New York, 1999; pp. xix, 756.
8. Wheeler, W.B. *Pesticides in Agriculture and the Environment*; Marcel Dekker: New York, 2002; pp. x, 330.
9. Bellows, T.S.; Fisher, T.W. *Handbook of Biological Control: Principles and Aplications of Biological Control*; Academic Press: San Diego, 1999; pp. xxiii, 1046.
10. Gurr, G.M.; Wratten, S.D. Integrated biological control: A proposal for enhancing success in biological control. Int. J. Pest Manag. **1999**, *45*, 81–84.
11. Landis, D.A.; Wratten, S.D.; Gurr, G.M. Habitat management to conserve natural enemies of arthropod pests in agriculture. Annu. Rev. Entomol. **2000**, *45*, 175–201.
12. Smith, C.M. *Plant Resistance to Insects: A Fundamental Approach*; Wiley: New York, 1989; pp. viii, 286.
13. Atherton, K.T. *Genetically Modified Crops: Assessing Safety*; Taylor & Francis: London, UK, 2002; 256.
14. Carde, R.T.; Bell, W.J. *Chemical Ecology of Insects 2*; Chapman & Hall: New York, 1995; pp. viii, 433.
15. Carde, R.T.; Minks, A.K. *Insect Pheromone Research: New Directions*; Chapman & Hall: New York, 1997; pp. xxii, 684.
16. Metcalf, R.L.; Deem-Dickson, L.; Lampman, R.L. Attracticides for the Control of Diabroticite Rootworms. In *Pest Management: Biologically Based Technology*; Vaughn, J.L., Ed.; American Chemical Society: Beltsville, MD, 1993; 258–264.
17. Calkins, C.O.; Klassen, W.; Liedo, P. *Fruit Flies and the Sterile Insect Technique*; CRC Press: Boca Raton, FL, 1994; 258.
18. *Economic Thresholds for Integrated Pest Management*; Higley, L.G., Pedigo, L.P., Eds., The University of Nebraska Press: Lincoln, NE, 1996; 327.

Intellectual Property and Plant Science

Susanne Someralo
John Dodds
Dodds & Associates, Washington, D.C., U.S.A.

INTRODUCTION

Intellectual property (IP) refers to creations of the human mind. Intellectual property may be inventions, artistic works, designs, names, images, and so on. The basic means to protect intellectual property are patents, copyrights, trademarks, and trade secrets. Plant Breeder's Right is a specific type of protection for new sexually propagated plant varieties.

Even if the intellectual property rights are largely national rights, there are several international treaties, the most important being the TRIPS agreement (Trade Related Aspects of Intellectual Property Rights), regulating the contents of the national intellectual property laws. In addition to international treaties, there are international organizations such as WIPO (World Intellectual Property Organization) and UPOV (International Union for the Protection of New Varieties of Plants) administering certain intellectual property practices.

In the following article we will briefly go through various means for protection relevant to plant sciences. Thereafter, we will deal with the current issues of intellectual property and modern plant sciences, and lastly we will briefly discuss the development of intellectual property legislation and international treaties affecting it.

MEANS TO PROTECT INTELLECTUAL PROPERTY

The basic means to protect intellectual property are patents, copyrights, trademarks, trade secrets, and Plant Breeder's Rights. There are variations in the national laws, and some countries provide broader protection than others. For example, a concept of plant patent is specific for the protection system of the United States. A basic understanding of these mechanisms is essential for anyone whose research may lead to an invention.

Patents

Historically, a patent was a grant made by a sovereign that would allow for the monopoly of a particular industry, service, or goods. Over time, the concept has been refined from a public policy perspective and has evolved to an agreement between the government and the inventor/creator.

In return for the right to exclude others from the practice of the invention, the government requests the inventor to fully disclose the enablement of the invention; furthermore, the monopoly is now limited by time and clearly is only applicable in the territory under the jurisdiction of the government.

A patent, as stated above, is an agreement between the government and the inventor, and in the United States a fundamental right is provided in Article I, Section 8, of the Constitution. Congress is empowered to "...promote the progress of science and useful arts by securing for limited times to authors and inventors the exclusive right to their respective writings and discoveries."

In exchange for a limited-term right (usually 20 years) to exclude others from making, using, or selling the invention, the inventor must provide a complete and accurate public description of the invention and the best mode of "practicing" it. This provides others with the ability to use that information to invent further, thus promoting technology development for the benefit of society.

This right to exclude means that a patent is a "negative right" since a patent holder may only exclude others from using, manufacturing, copying, or selling his or her invention.

Copyrights

A copyright is a type of intellectual property protection for "authors" of original works. Basically, a copyright protects an original work and allows the author an exclusive right to reproduce the work, prepare derivatives, distribute copies of the work, and perform the copyrighted work publicly.

Historically, copyrights have been important in protecting the rights of artists and authors. Today, copyrights are becoming more and more important in protecting the rights of database creators. Copyrights may be relevant means of protecting, for example, GIS databases or databases containing gene sequences.

Encyclopedia of Plant and Crop Science
DOI: 10.1081/E-EPCS 120024249

Trademarks

A trademark is a word, phrase, symbol, or design, or a combination of those, that distinguishes the source of one's goods or services from those of others. A trademark can be valid only when it is used in connection with the goods or services in commerce. Trademarks are important means for distinguishing a product or a technology; for example, plant seeds may be trademarked.

Geographical Indications

Geographical indications are defined in the TRIPS agreement as a type of intellectual property. A geographical indication is a sign used on goods that have a specific geographical origin and possess qualities or a reputation that are due to that place of origin. In the United States, geographical indications are treated as trademarks.

Most commonly, a geographical indication consists of the name of the place of origin of the goods. Agricultural products typically have qualities that derive from their place of production and are influenced by specific local factors, such as climate and soil.

Trade Secrets

Trade secrets are the cheapest way to protect one's intellectual property: Having a trade secret simply requires that the intellectual property is kept secret. A trade secret could, for example, be composition of a culture medium or a method to transform a plant species.

The positive aspect of trade secrets in addition to their cheapness is that there is no expiration date. However, the negative side is that once the secret is out the protection is gone.

Plant Breeder's Rights

The TRIPS agreement obliges the member countries to provide protection for plant varieties either through patents or by an effective sui generis system or a combination of them. The most common sui generis system is Plant Breeder's Rights (PBR). PBR allows the developers of new plant varieties to control multiplication and sale of the reproductive material of a new variety.

Several countries have joined UPOV, established in 1961 and last revised in 1991. UPOV sets forth the minimum requirements that the national protection should grant.

In the United States the right is called Plant Variety Protection (PVP), and it is granted for sexually reproduced or tuber-propagated plants. The term of protection is 20 years for most crops and 25 years for trees, shrubs, and vines.

Plant Patent

Plant patent is a specific type of patent granted in the United States to an inventor who has invented or discovered and asexually reproduced a distinct and new variety of plant species. The provision excludes tuber-propagated plants or plants found in an uncultivated state. Plant patents are issued for 20 years.

CURRENT ISSUES

Patenting Life

The intellectual property system is a dynamic system that needs to be reevaluated in the course of developing technologies and trade practices. The emergence of modern biotechnology more than 20 years ago prompted the U.S. Supreme Court to reevaluate the concept of patentability. In 1980 the Court gave the famous decision[1] stating that a genetically modified bacteria qualifies as patentable subject matter and that "everything under the sun made by man is patentable." Five years later, the Board of Appeals and Interferences of the U.S. Patent and Trademark Office held that genetically modified plants are patentable.[2]

However, as recently as 2001, the U.S. Supreme Court still discussed whether utility patents may be issued for plants.[3] The ruling was that newly developed plant breeds fall within the terms of utility patent. Therefore, a sexually reproduced plant variety may receive a double protection by PVP and by utility patent.

The issue of patentability of transgenic plants is not resolved in Europe yet. The European Patent Office (EPO) acts under European Patent Convention (EPC). EPC prohibits patenting of plant varieties. In 1995 the EPO Board of Appeal held that a transgenic plant was a new plant variety and therefore not patentable. However, cells from the transgenic plant were held to be "microbiological products" and therefore patentable.[4]

The European Biotechnology Directive entered into force on July 30th, 2000. The Directive confirms that biological material per se is not unpatentable. Inventions having the required novelty, inventive step, and industrial applicability shall be patentable even if they comprise biological material or utilize a process by means of which biological material is produced, processed, or used. Therefore, an isolated gene or gene sequence may be patentable provided that its function is known and a suitable industrial application is derived from that product.

According to the Directive, plant and animal varieties are not patentable, but the definition of plant variety is so narrow that most transgenic plants do not fit into it. The European Commission is soon going to release legislation for so-called Community Patent that would cover all the member countries. The Community Patent would follow the standard of the Biotechnology Directive and therefore basically allow patenting of transgenic plants.

Traditional Knowledge

Traditional knowledge refers to the knowledge, innovations, and practices of indigenous and local communities around the world. Traditional knowledge has been transmitted orally from generation to generation and tends to be collectively owned. Among many other things, agricultural practices, including the development of plant species and animal breeds, may be traditional knowledge. This kind of traditional knowledge is based on genetic resources. The relationship between genetic resources, traditional knowledge, and intellectual property rights is one of the most important issues on the agenda of several international organizations.

The Convention of Biological Diversity (CBD) constitutes the central instrument concerning biodiversity at the international level. CBD affirms the sovereign rights of states to exploit their own resources pursuant to their own environmental policies.

The Convention also provides a broad framework for member states' policies concerning access, development, and transfer of technologies and acknowledges the necessity for all parties to recognize and protect intellectual property rights in this field. It points to the need for equitable sharing of benefits arising from the use of traditional knowledge, innovations, and practices relevant to the conservation of biodiversity and the sustainable use of its components.

The Food and Agricultural Organization (FAO) adopted the International Treaty on Plant Genetic Resources (ITPGR) in 2001. The objectives of the Treaty are the conservation and sustainable use of plant genetic resources for food and agriculture and the fair and equitable sharing of the benefits arising out of their use, in harmony with the Convention of Biological Diversity, for sustainable agriculture and food security. The contracting parties agree to establish a multilateral system to facilitate access to plant genetic resources for food and agriculture and to share the benefits arising from the utilization of these resources in a fair and equitable way.

The issues related to traditional knowledge are also closely related to IP protection in the developing world. People in many developing countries are asset-poor but knowledge-rich, and poor communities seldom gain from that knowledge. As it might be difficult for indigenous people to invest in conventional IP protection, let alone litigations to protect their IP, new IP protection and contract models may be needed to protect such knowledge and to bring the benefit to the poorest.

CONCLUSION

Even if the concept and means for intellectual property are old, the issues emerging with the rapid development of life sciences together with ethical concerns have made it necessary to reevaluate the concept. Presently there are several international organizations trying to create harmonized guidelines for protection of intellectual property in industrialized countries as well as in developing countries. A detailed overview of the current issue of intellectual property rights in agricultural biotechnology is given in recent review.[5] Especially interesting issues are those related to the protection of traditional knowledge.

REFERENCES

1. Diamond v. Chacrabarty 447 U.S. 303, 65 (1980).
2. In Re. Hibbert 227 U.S.P.Q 443, 444 (BNA, 1985).
3. J.E.M. Ag. Supply, Inc. v. Pioneer Hi-Bred Int'l, Inc., 534 US 124 (2001).
4. Decision T356/93 Technical Board of Appeal 3.3.4. *Plant Cells/Plant Genetic Systems (Feb. 1995)*; O.J. EPO 1995; 545.
5. *Intellectual Property Rights in Agricultural Biotechnology*, 2nd Ed.; Erbish, F.H., Maredia, K.M., Eds.; Michigan State University: East Lansing, USA, 2003.

Interspecific Hybridization

D. S. Brar
International Rice Research Institute, Los Banos, Laguna, Philippines

INTRODUCTION

Genetic variability for agronomic traits is a prerequisite for any plant breeding program. To introduce such variability, plant breeders usually employ intervarietal hybridization. However, in several cases, useful variability for certain traits is either limited or lacking in the cultivated species. In these situations, breeders look for diverse sources and resort to interspecific hybridization. The wild relatives of crop plants are an important reservoir of useful genes and offer enormous potential to introduce such genes into cultivated species. Interspecific hybridization has been one of the important plant breeding methods to broaden the gene pool of crops. Numerous examples are available of the production of interspecific hybrids and their utilization in genetics and breeding research. Several pre- or postzygotic barriers hinder the production of interspecific hybrids and the transfer of genes into crops through interspecific hybridization. However, several techniques of chromosome manipulation, tissue culture, (embryo rescue, anther culture, somatic cell hybridization), molecular markers, and molecular cytogenetics can be employed to overcome these barriers.

Interspecific hybridization has provided a wealth of information on genomic relationships among species, nucleo-cytoplasmic interactions, synteny among distant genomes, the nature of reproductive barriers, mechanisms controlling chromosome pairing, chromosome elimination, nuclear restitution, and alien-gene introgression into the genomes of cultivated species. Numerous examples are available on the transfer of useful genes into crop plants through interspecific hybridization. Notable examples include the transfer of resistance to diseases and pests; leaf and stem rust, Hessian fly, green bug in wheat; crown rust and mildew in oat; mosaic virus and blank shank in tobacco; grassy stunt virus in rice; powdery mildew in barley; black arm in cotton; aphid in lettuce; mosaic virus in okra; bacterial wilt and root knot nematode in tomato; late blight and virus A, X, Y in potato; nematode resistance in sugarbeet; wider adaption to warmer and drier region in peas; hardiness in grapes; cytoplasmic male sterility in wheat and cotton; high protein content in wheat; and highly soluble solids and increased salt tolerance in tomato.

CYTOGENETICS OF INTERSPECIFIC HYBRIDS AND THEIR DERIVATIVES

Interspecific hybrids and their derivatives exhibit peculiar features such as hybrid inviability, hybrid sterility, hybrid breakdown, limited recombination, and linkage drag.[1,2] Factors such as genomic disharmony, nuclear instability, unfavorable nuclear-cytoplasmic interactions, the presence of deleterious genes, chromosome elimination, allocycly, and centromeric affinity account for limited progress in introgression of useful genes through interspecific hybridization[3] (Table 1). As an example, Fig. 1 shows production of interspecific hybrids through embryo rescue between distantly related species of *Oryza*.

Hybrid Sterility

Hybrid sterility may be due to differences in structure and number of chromosomes, lack of chromosome homoeology resulting in a variable number of univalents, and production of unbalanced gametes. There are numerous reports of sterile interspecific hybrids and their derivatives. This kind of sterility could be due to gene substitutions and/or cryptic structural differences. Crosses between species with similar genomic constitutions (*Oryza sativa* × *Oryza glaberrima*; *Oryza sativa* × *Oryza longistaminata*) are highly sterile.

Allocycly or Genomic Disharmony

The parental genomes in interspecific crosses generally show disharmony resulting in meiotic instability, aneuploidy, and unbalanced gametes, and thus to increased sterility and reduced grain yield. Differences in the cell-cycle rhythm of parents, genomic ratio, presence of telomeric heterochromatin, and the cytoplasm donor may contribute singly or collectively to disharmony of the two genomes.

Nuclear Instability

Nuclear instability refers to any deviant nuclear behavior producing a nucleus (or nuclei) of abnormal structure,

Encyclopedia of Plant and Crop Science
DOI: 10.1081/E-EPCS 120005682

Table 1 Techniques for overcoming barriers to interspecific hybridization

Barrier	Technique for overcoming barrier
Prefertilization	
Failure of pollen germination	• Mechanical removal of pistil followed by pollination of the exposed end of the style
	• Use of recognition pollen
Slow pollen tube growth	• Use of recognition pollen, growth hormones
	• In vitro fertilization
Pollen tube unable to reach the style	• Shortening the style
Arresting of pollen tube in the style, ovary and ovule	• In vitro fertilization
Failure to obtain sexual hybrids	• Protoplast fusion
Differences in ploidy level	• Chromosome doubling of species or species hybrid before hybridization with the recipient species
	• Bridging species
	• Reducing chromosome number of polyploid species before hybridization
Postfertilization	
Hybrid inviability and weakness	• Embryo rescue
Embryo abortion	• In vivo/vitro embryo rescue/embryo implantation
Embryo abortion at very young stages	• Ovary, ovule culture
	• In vitro fertilization
Lethality of F_1 hybrids	• Reciprocal crosses
	• Grafting of hybrids
	• Regenerating plants from callus
Chromosome elimination	• Altering genomic ratios of two species
	• Inducing chromosomal exchanges before onset of elimination
Hybrid sterility	• Chromosome doubling (amphiploid production)
	• Backcrossing with the recurrent parent
Hybrid breakdown	• Growing larger F_2 populations
	• Inducing premeiotic chromosomal exchanges in hybrids through mutagenesis and/or tissue culture
Limited alien recombination	• Inducing homoeologous recombination through genetic manipulation of chromosome pairing system
	• Inducing chromosomal exchanges through tissue culture and/or irradiation
Low recovery of alien recombinants	• Selection of recombinants with alien chromosome segments in early generations using molecular markers and in-situ hybridization techniques
	• Selection of lines with smaller alien segment substitutions for enhanced transmission to the progenies

karyotype, or behavior. Nuclear instability and mitotic abnormalities in coenocytic endosperm affect seed fertility and endosperm development in triticale. Nuclear instability in triticale endosperm often begins with the formation of a chromatin bridge at anaphase. Restitution nuclei result in degenerated endosperm. The probable cause of nuclear instability may be the higher DNA content in rye. Unlike wheat chromosomes, rye chromosomes have large telomeric segments of heterochromatin. Late-replicating DNA (mostly telomeric heterochromatin) in rye chromosome may cause bridge formation at anaphase. Such bridges may cause the production of abnormally polyploid endosperm nuclei leading to sterility. First-division restitution (FDR), second-division restitution (SDR) in potato, and indeterminate meiotic restitution (IMR) in lily are common features of interspecific crosses.

Hybrid Breakdown and Reversion to Parental Types

The phenomenon of hybrid breakdown refers to situations where the F_1 is fertile but segregants in the F_2 or later generations are weak or sterile. The phenomenon of hybrid breakdown is not well understood; probable causes include centromeric affinity, cryptic structural hybridity, gene substitution, and unfavorable nuclear-cytoplasm interactions.

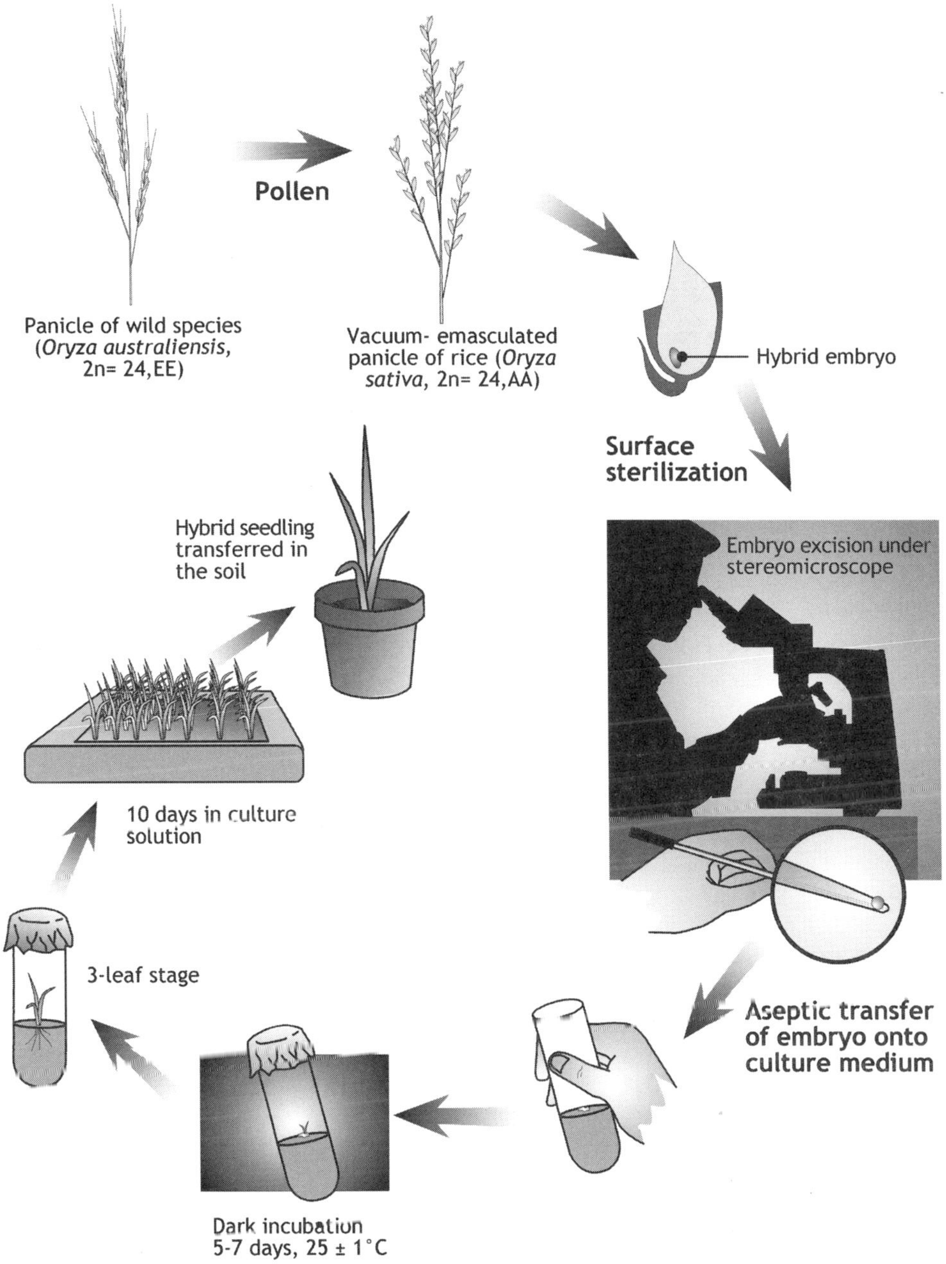

Fig. 1 Production of interspecific hybrids through embryo rescue between distantly related species of *Oryza*. *(View this art in color at www.dekker.com.)*

Limited Recombination

Reduced chromosome pairing and lack of recombination between genomes of alien and cultivated species are key factors limiting alien gene transfer. The extent of recombination depends on the genetic homoeology of two genomes, which may be different for different chromosomes or chromosome segments. Until precise methods for inducing or enhancing homoeologous recombination and allosyndetic pairing become available, it will be difficult to transfer alien genes into commercial varieties.

Presence of Deleterious Genes and Linkage Drag

Breeders in general are reluctant to use wild species for the improvement of cultivated species because of linkage drag, wherein some undesirable characters of wild parents are transferred along with useful genes. Some interspecific derivatives (such as Transec) have not been used extensively in wheat breeding because they are associated with lowered yield. Similarly, the Agatha strain of wheat (possessing rust resistance from *Agropyron*) shows undesirable association with yellow flour color. A number of available wheat lines with substituted chromosomes of *Aegilops*, *Agropyron*, and *Secale* have disease and pest resistance or high protein. However, only a few of them are widely used in cultivar improvement because of linkage drag.

ALIEN GENE TRANSFER

Major advances have been made in understanding genomic relationships, genetic control of chromosome pairing, and characterization of alien chromatin in interspecific derivatives. Methods have been developed for incorporating complete alien genomes (amphiploids), single chromosomes (alien addition/substitution), small chromosome segments (alien translocations), or a few alien genes from wild into cultivated species. Numerous examples are available of alien gene transfer into crop plants.[2,4,5]

Synthetic Amphiploids

A majority of the interspecific hybrids show sterility to varying degrees. Fertility can be restored by doubling the chromosome number of F_1 hybrids (amphiploid production). The other option is to backcross with the recurrent parent until fertile progenies (introgression lines) with stable chromosome number become available. Several amphiploids of hybrids involving cultivated species and wild relatives have been produced during the last 45 years, especially since the discovery of colchicine. With the exception of triticale, however, none has become a new crop.

In amphiploids such lack of genomic harmony causes meiotic instability and several other undesirable features resulting from nuclear-cytoplasmic interactions. However, they are important for extracting desirable derivatives involving crosses of natural × synthetic allopolyploids.

Extracted Allopolyploids

Several amphiploids in wheat, cotton, tobacco, and *Brassica* have been used to extract stable allopolyploids through intercrosses of synthetic × natural allopolyploids. Triticale is the classic example for demonstrating the usefulness of extracted polyploids. The hexaploid triticales extracted from the progenies of 8*x* triticale × 6*x* triticale and 6*x* wheat × 6*x* triticale are cytologically more stable. They are superior to the primary triticales in meiotic stability, grain yield, and grain quality.

The production of amphiploids, alien chromosome addition and substitution lines, and backcross (introgression) progenies are important genetic stocks for transferring alien genes to crop plants. Once these stocks are available, they can be further manipulated by directed chromosome engineering.

Characterization of Alien Chromatin in Interspecific Derivatives through C-Banding and In Situ Hybridization

Several methods are available to characterize alien chromatin in interspecific hybrids and their derivatives. Morphological traits, isozyme, molecular markers, C-banding, fluorescence in situ hybridization (FISH), and genomic in situ hybridization (GISH) have been employed to characterize parental genomes, extra alien chromosome, translocated chromosome segments, and introgressed genes. Fiber FISH can further enhance the precision for characterization of homoeologous pairing in interspecific hybrids and their derivatives. C-banding has been employed to characterize changes in karyotype and alien chromosome translocations.[6] Several kinds of molecular markers, BAC and YAC, including repetitive DNA; total genomic DNA; and species-specific markers, have been used in characterizing alien chromatin in a great number of interspecific derivatives.

RESEARCH PRIORITIES AND OUTLOOK ON INTERSPECIFIC HYBRIDIZATION

With advances in tissue culture, genomics, and molecular cytogenetics the outlook for interspecific hybridization in crop improvement seems more promising. Following are some priorities for future research on interspecific hybridization.

Map-Based Cloning of Gene(s) Controlling Chromosome Pairing

Isolate *Ph* gene(s) and introduce into other crops that lack a similar system to promote homoeologous recombination. As in the case of wheat, there is a need to identify pairing controlling genes in crops such as *Brassica*, cotton, rice, soybean, legumes, and millets to enhance alien gene transfer from different species.

Identifying and Introgressing Chromosome Regions in Wild Species Carrying Novel Alleles for Enhanced Yield Potential

Molecular analyses of mapping populations derived from interspecific crosses have revealed novel QTLs-yield–enhancing loci. Molecular-marker technology should be used to identify and introgress such novel alleles from wild species into cultivated species.

Physical Mapping Using Alien Chromosome Addition Lines and Radiation Hybrids

Partial hybrids or alien addition lines have been produced through chromosome elimination from oat × maize crosses. Such cytogenetic stocks and radiation hybrids should be developed in other systems. Species-specific DNA sequences, retroelements, chromosome-specific markers, and GISH could be used in the physical mapping of genomes of crop plants.

Identifying Agronomically Important Gene-Rich Chromosome Regions

Identify gene-rich regions in different species. Physical mapping of such genes would provide useful information on the possibility for map-based cloning and introgression into genomes of cultivated plants. Genes in the proximal region of chromosomes may be difficult to introgress because of limited recombination. New methods of tissue culture or transposons should be used to transfer such genes into elite germplasm.

Enhancing Chromosomal Exchanges through Tissue Culture among Genomes of Cultivated and Distantly Related Species

Induce chromosomal exchanges between genomes of species that otherwise lack chromosome pairing. Tissue culture of interspecific (sexual and somatic) hybrids and alien addition lines is suggested to achieve this.

FISH-Assisted Selection for Alien Recombinants

Advances in GISH/FISH have made it possible to characterize lines carrying alien chromatin. Emphasis should be given to develop protocols to select interspecific derivatives in early generations based on GISH, important for the recovery of recombinants and to enhance the selection efficiency of alien gene introgression.

Characterizing Genomic Relationships and Microsynteny among Species

Conventional cytogenetic techniques have provided a wealth of information on the genomic homoeology and evolutionary relationships of different species. Molecular cytogenetic techniques involving molecular markers, GISH, FISH, and fiber-FISH should be used to understand precisely the nature and extent of cryptic structural differences, microsynteny, and homoeologous pairing during early meiotic divisions among distant genomes to enhance the efficiency of alien gene introgression.

Producing Haploids from Interspecific Crosses through Chromosome Elimination

Chromosome elimination through species crosses has proven to be a valuable mechanism to produce haploids in wheat, barley, and oat. Haploid-inducing systems need to be identified in other crops such as *Brassica*, cotton, rice, and legumes. This would be an important complementary approach to produce haploids where anther culture response is limited to only specific genotypes.

Broadening Gene Pool of Crops for Tolerance to Major Biotic and Abiotic Stresses

An integrated approach using tissue culture, molecular markers, and molecular-cytogenetic techniques should be used to monitor alien chromatin and introgress useful genetic variability from wild species into crop plants through interspecific hybridization.

ARTICLES OF FURTHER INTEREST

Chromosome Banding, p. 263
Chromosome Manipulation and Crop Improvement, p. 266
Crop Improvement: Broadening the Genetic Base for, p. 343
Fluorescence In Situ Hybridization, p. 468
In Vitro Pollination and Fertilization, p. 587
Molecular Analysis of Chromosome Landmarks, p. 740
Polyploidy, p. 1038

Radiation Hybrid Mapping, p. 1074
Somaclonal Variation: Origins and Causes, p. 1158

REFERENCES

1. Stebbins, G.L. The inviability, weakness and sterility of interspecific hybrids. Adv. Genet. **1958**, *9*, 147–215.
2. Brar, D.S.; Khush, G.S. Wide Hybridization and Chromosome Manipulation in Cereals. In *Handbook of Plant Cell Culture Vol. 4; Techniques and Applications*; Evans, D.H., Sharp, W.R., Ammirato, P.V., Eds.; MacMillan Publ.: New York, 1986; 221–263.
3. Khush, G.S.; Brar, D.S. Overcoming the Barriers in Hybridization. In *Distant Hybridization of Crop Plants*; Kalloo, G., Chowdhury, J.B., Eds.; Monograph on Theoretical and Applied Genetics, Number 16 supplementary, Springer-Verlag, 1992; 47–61.
4. Brar, D.S.; Khush, G.S. Transferring Genes From Wild Species into Rice. In *Quantitative Genetics, Genomics and Plant Breeding*; Kang, M.S., Ed.; CAB International: Wallingford, UK, 2002; 197–217.
5. In *Interspecific Hybridization in Plant Breeding*, Proceedings of the Eight Congress of EUCARPIA, Madrid, Spain, Sanchez-Monge, E., Garcia-Olmedo, F., Eds.; 1977; 1–407.
6. Friebe, B.; Jiang, J.; Raupp, W.J.; McIntosh, R.A.; Gill, B.S. Characterization of wheat-alien translocations conferring resistance to diseases and pests: Current status. Euphytica **1996**, *91*, 59–87.

Isoprenoid Metabolism

Russell K. Monson
Todd N. Rosenstiel
University of Colorado, Boulder, Colorado, U.S.A.

INTRODUCTION

Plants synthesize a wide range of molecules with complex structural and functional attributes from the branched, five-carbon monomer precursor known as the isoprene unit. Isoprenoid molecules range in size from those with 5 carbon atoms (the hemiterpenes) to those with more than 40 carbon atoms (the polyterpenes). The various isoprenoid groups have a broad range of functions in plants, including those associated with herbivore deterrence, metabolic electron transport, and hormonal activity. Isoprenoid compounds are used in numerous ways to support human activities, including their use as food flavoring agents (e.g., mint essential oils), perfumes (e.g., monoterpenes), pharmaceuticals (e.g., the antitumor compound, taxol), and insecticides (e.g., pyrethrins). Isoprenoid compounds also have important roles in ecological interactions, including their role in plant–insect interactions (e.g., as antiherbivore agents and as attractants for pollinators and parasites alike) and as allelopathic agents. Recently, it has become clear that the loss of volatile isoprenoid molecules from plants has important ramifications for atmospheric processes. These molecules cause much of the active photochemistry that occurs in the lower atmosphere, including those reactions that lead to tropospheric ozone production and the oxidation of methane, a potent greenhouse gas.[1,2]

BIOSYNTHESIS OF ISOPRENOID PRECURSORS

All isoprenoids are biosynthesized from just two 5-carbon (C_5) metabolic precursors: isopentenyl diphosphate (IPP) and its allylic isomer, dimethylallyl diphosphate (DMAPP) (Table 1). Until recently, the biosynthesis of these precursors in plant cells was thought to occur exclusively via the well known mevalonic acid pathway, as occurs in yeast, fungi, and animal cells. However, several results inconsistent with the mevalonate pathway led to the discovery of an additional non mevalonate pathway of isoprenoid biosynthesis operating in certain bacteria, green algae, and plants.[3] The discovery and partial elucidation of the deoxyxlulose phosphate (DOXP) pathway represents one of the most exciting advances in basic plant physiology to occur near the closure of the 20th century.

In plant cells, IPP synthesis not only results from two independent biochemical pathways, but these pathways occur in distinct cellular compartments (Fig. 1). The mevalonate route to isoprenoids appears to reside solely in the cytosol-endoplasmic reticulum (ER) compartment where sterols, sesquiterpenes (C_{15}), triterpenes (C_{30}), and polyterpenes are synthesized. Hemiterpenes—such as isoprene (C_5), monoterpenes (C_{10}), diterpenes (C_{20}), and tetraterpenes (C_{40}) as well as the side chains of plastoquinone and α-tocopherol—are synthesized in plastids from IPP formed predominately via the DOXP pathway. A third cellular compartment, the mitochondrion, also participates in isoprenoid biosynthesis by using IPP from the cytosol-ER pathway to make the prenyl side chain of ubiquinone, a component of this organelle's electron transport chain. This extensive subcellular partitioning of biochemical pathways and products highlights the significant role of compartmentation in plant isoprenoid metabolism. However, the compartmental separation of the two different IPP biosynthetic pathways is not absolute because plastids may supply IPP to the cytosol for use in biosynthesis and vice versa. For example, the first two C_5 units of the sesquiterpenes of chamomile are formed via the DOXP pathway, whereas the third unit is derived from both biosynthetic routes.[4] Although crosstalk between the two pathways appears to be small (<1%), elucidation of the nature of the metabolites exchanged between the compartments and regulation of this process remains a significant challenge to plant biochemists.

The biochemical steps involved in the formation of IPP via the mevolonic acid pathway have been firmly established and are known to result from the stepwise condensation of three molecules of acetyl coenzyme A. As in the situation of animals, numerous studies indicate that 3-hydroxy-3-methylglutaryl-CoA reductase (HMGR), a key enzyme in this pathway, also catalyzes a rate-determining step in sterol formation in plants. Although accumulated evidence suggests that the activity of the plant HMGR is highly regulated, it remains unclear how differential regulation of HGMR activity in the cytosol facilitates the production of distinct isoprenoid families.

Encyclopedia of Plant and Crop Science
DOI: 10.1081/E-EPCS 120005950

Table 1 Primary groups of isoprenoid compounds and examples of functional roles in plants

Type of compound	Number of C atoms	Role in plants
Hemiterpenes (isoprene)	5	Possible thermal protectant
Monoterpenes	10	Antiherbivore; pollinator attractant
Sesquiterpenes	15	Antiherbivore; antimicrobial; abscisic acid
Diterpenes	20	Phytol side-chain of chlorophyll; gibberellins; phytoalexins
Triterpenes	30	Phytosterols of membranes; antiherbivore
Tetraterpenes	40	Carotenoids
Polyterpenes	>40	Plastoquinone and ubiquinone electron carriers

The DOXP pathway of IPP formation in plastids begins with pyruvate and glyceraldehyde 3-phosphate. In this pathway, pyruvate reacts with a thiamine pyrophosphate (TPP) cofactor to yield a two-carbon molecule, hydroxyethyl-TPP, which condenses with glyceraldehyde 3-phosphate. TPP is released to form a five-carbon intermediate, 1-deoxy-D-xylulose 5-phosphate (for which the pathway receives its name), which is rearranged and reduced to form 2-C-methyl-D-erythritol 4-phosphate and subsequently is transformed to yield IPP. The enzymes and corresponding gene sequences of the first five steps in the new pathway are now fairly well documented.[5] However, the genes, enzymes and reactions involved in the final steps leading to IPP and DMAPP formation remain unknown.

The identification of the regulatory steps of the DOXP pathway is an issue of major importance for both theoretical and applied reasons. So far, efforts have focused on the first two enzymes of the pathway (1-deoxy-D-xylulose 5-phosphate synthase, DXS; and 1-deoxy-D-xylulose 5-phosphate reductoisomerase, DXR) and preliminary results indicate both enzymes may contribute to regulating

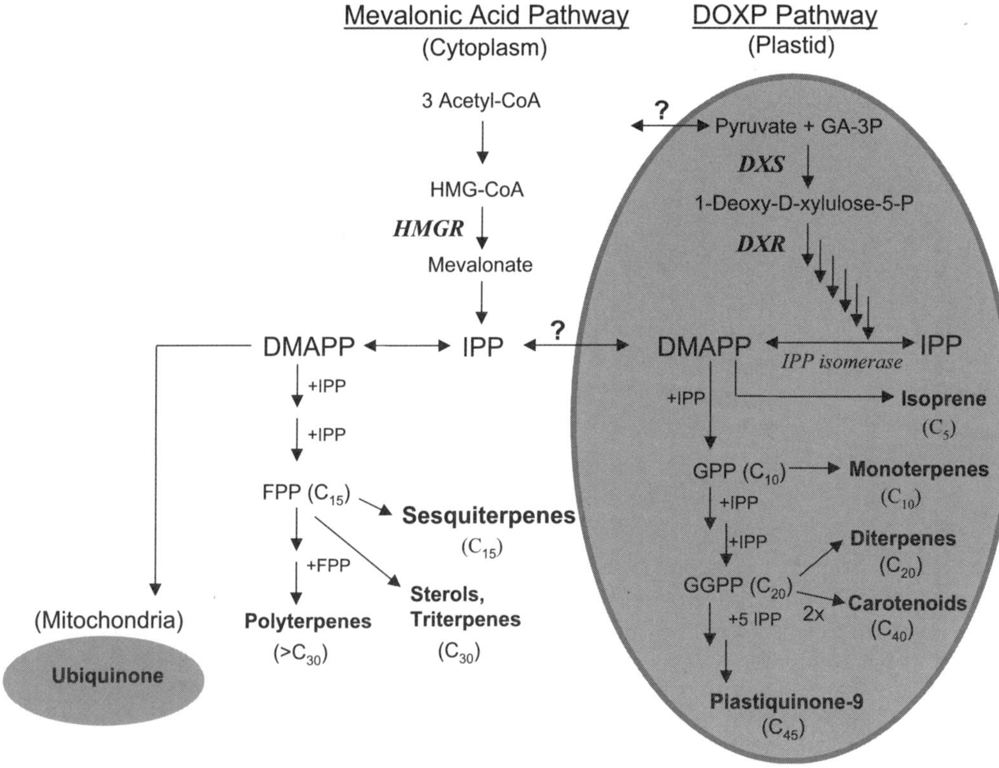

Fig. 1 Compartmentation of isoprenoid biosynthesis in higher plants between the cytosol (mevalonic acid pathway) and plastids (DOXP pathway). The role of cytosolic IPP in the formation of mitochondrial ubiquinones is also indicated. Abbreviations used: DMAPP, dimethylallyl diphosphate; IPP, isopentenyl diphosphate; GPP, geranyl diphosphate; FPP, farnesyl diphosphate; GGPP, geranylgeranyl diphosphate; HMGR, HMG-CoA reductase, DXS, deoxyxylulose phosphate synthase; DXR, deoxyxylulose phosphate reductoisomerase. (*View this art in color at www.dekker.com.*)

plant isoprenoid production. Transgenic manipulation of the DOXP pathway holds much promise for the production of commercially important isoprenoids. In one unique example, overexpression of DXR in peppermint led to a dramatic increase in total essential oil accumulation.[6] Similar successes in the genetic engineering of isoprenoid metabolism have been achieved in the area of isoprenoid vitamins, including α-tocopherol (vitamin E) production in oilseeds and enhanced β-carotene in rice and rapeseed. Because the DOXP pathway is uniquely restricted to plants and certain bacteria (particularly human pathogens), selective inhibition of the DOXP pathway might be the basis for the development of new herbicides as well as novel antibiotics and antimalarial agents.[7] Despite rapid advances in our understanding of the novel DOXP pathway, significant uncertainties remain. Because constraints on precursor flow will ultimately limit the effectiveness of transgenes for isoprenoid synthesis, information about the flux controls on IPP and DMAPP formation as well as the source of carbon precursors supplying the DOXP pathway is sorely needed.

THE ISOPRENE RULE AND FORMATION OF ALLYLIC DIPHOSPHATE ESTERS

In 1910, the German chemist Otto Wallach was awarded the Nobel Prize in Chemistry for elucidation of the isoprene rule. Simply stated, the isoprene rule revealed that isoprenoid molecules in plants are synthesized by repetitive head-to-tail condensations of a five-carbon, alkene monomer. This rule was modified in the 1930s by Leopold Ruzicka to give the biogenetic isoprene rule, which stated that an isoprenoid molecule is one that is derived from an isoprenoid precursor, even if it undergoes subsequent chemical modification that results in a product with obscure connections to its isoprene origins. Ruzicka was awarded the Nobel Prize in Chemistry in 1939. Thus, the history of isoprenoid chemistry in plants has a rich award-winning legacy.

The head-to-tail condensation of IPP by prenyltransferase enzymes forms three types of allylic diphosphate ester compounds—geranyl diphosphate (GPP), farnesyl diphosphate (FPP) and geranylgeranyl diphosphate (GGPP)—which are subsequently converted into a broad range of terpenes by various terpene synthase enzymes. In the first step of allylic diphosphate ester compounds, IPP is converted to dimethylallyl diphosphate (DMAPP) by IPP isomerase. In the presence of a divalent metal cation, the allylic double bond of the DMAPP molecule forms a resonance-stabilized carbocation that can subsequently react with a second IPP molecule to form GPP (C_{10}) (Fig. 2). Further addition(s) of IPP forms FPP (one additional IPP) or GGPP (two additional IPPs).

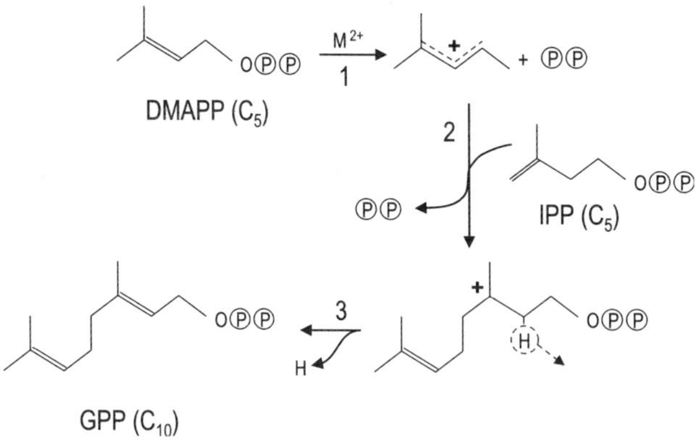

Fig. 2 Condensation reaction of prenyltransferase to join together two hemiterpenes (C_5) to form geranyl diphosphate (GPP). In Step 1, divalent metal cofactors facilitate delocalization of a carbocation, allowing DMAPP to join with IPP through a condensation reaction (Step 2). In Step 3, a proton is extracted from the enzyme-bound intermediate, producing GPP.

TERPENE SYNTHASE ENZYMES

The production of terpenoid compounds from GPP, FPP or GGPP involves an initial ionization to promote cyclization, the primary catalyst being terpene synthase (or terpene cyclase) enzymes. (Acyclic terpenes are known to exist, but most are cyclic.) Many terpene synthases catalyze the synthesis of multiple terpene products; this is permitted because of the multiple fates that are possible for the carbocation intermediate that is formed from the cyclization catalyzed by terpene synthases.[8] Over thirty different terpene synthases from angiosperms and gymnosperms, with roles in both primary and secondary metabolic pathways, have been identified on the basis of cDNA sequences. Sequence analysis has revealed that the evolutionary diversification of terpene synthases has occurred through gene duplication and divergence of enzymes; the primary direction of diversification appears to be from ancestral genes involved in primary metabolism toward derived genes involved in secondary metabolism.[9] Given the ubiquitous distribution of terpene synthase genes in plants, it is likely that future phylogenetic studies of divergent plant groups will benefit from studies of sequence homology among terpene synthase genes.

THE ISOPRENOID CONNECTION TO ATMOSPHERIC CHEMISTRY

In 1960, Fritz Went recognized the potential for plants to add large amounts of volatile isoprenoid molecules to the atmosphere; using the concentration of terpene-containing leaf oils in sagebrush, he estimated global terpene emissions to represent 175 Tg C y^{-1} (1 Tg$=10^{12}$ g).[10]

More recent estimates suggest that as much as 500–1000 Tg C y^{-1} may enter the atmosphere as volatile isoprenoid compounds.[11] Although estimates of the global magnitude of these emissions contain many uncertainties, it is clear that the emissions have a profound effect on photochemical reactions in the lower atmosphere. Isoprene, which is formed in the leaves of many tree species, has a boiling point of 39°C, causing it to volatilize from leaves at high rates immediately after being formed. Once in the atmosphere, isoprene has an average lifetime of only 0.2 hours before its double bonds are oxidized by reaction with hydroxyl radicals. A similar fate exists for monoterpene molecules that escape their storage reservoirs in the leaves and enter the atmosphere. The oxidation of isoprenoid molecules by OH causes an increase in the lifetime of CH_4, a compound with strong potential to cause climate warming, because the primary path for the oxidative breakdown of CH_4 also requires OH molecules. The atmospheric oxidation of isoprenoid molecules forms a series of secondary compounds, which participate in even further chemical oxidation. The net result of these oxidations is the formation of organic peroxy compounds, which can react with NO to form NO_2 and ultimately promote the reaction of NO_2 with O_2 to form tropospheric O_3, an important pollutant and radiatively active greenhouse gas. Globally, the emission of isoprenoid molecules from plants is estimated to promote a 30% increase in tropospheric O_3 and to increase the lifetime of CH_4 by 15%.[12]

CONCLUSION

The formation of isoprenoid molecules in plants has important ramifications for issues associated with agriculture and human nutrition, plant adaptation to stress, and global ecology. Studies of the metabolic pathways leading to isoprenoid molecules have revealed several fundamental principles of plant biochemistry, and have provided the foundation for highly recognized scientific research. These fundamental discoveries will allow future research on this topic to focus on the genes that control the expression of isoprenoid biochemistry and permit transgenic approaches to be developed for the manipulation of isoprenoids in various plant tissues.

ARTICLES OF FURTHER INTEREST

Carotenoids in Photosynthesis, p. 245
Chlorophylls, p. 258
Ecophysiology, p. 410
Exchange of Trace Gases Between Crops and the Atmosphere, p. 425
Metabolism, Secondary: Engineering Pathways of, p. 720
Natural Rubber, p. 778
Plant Defenses Against Insects: Constitutive and Induced Chemical Defenses, p. 939

REFERENCES

1. Monson, R.K. Volatile organic compound emissions from terrestrial ecosystems: A primary biological control over atmospheric chemistry. Israel J. Chem. **2002**, *42*, 29–42.
2. Monson, R.K.; Holland, E.A. Biospheric trace gas fluxes and their control over tropospheric chemistry. Annu. Rev. Ecol. Syst. **2001**, *32*, 547–576.
3. Lichtenthaler, H.K. The 1-deoxy-D-xyulose-5-phosphate pathway of isoprenoid biosynthesis in plants. Annu. Rev. Plant Physiol. Plant Mol. Biol. **1999**, *50*, 47–65.
4. Adam, K.P.; Zapp, J. Biosynthesis of the isoprene units of chamomile sesquiterpenes. Phytochemistry **1998**, *48*, 953–959.
5. Eisenreich, W.; Rohdich, F.; Bacher, A. Deoxyxylulose phosphate pathway to terpenoids. Trends Plant Sci. **2001**, *6*, 1360–1385.
6. Mahmoud, S.S.; Croteau, R.B. Metabolic engineering of essential oil yield and composition in mint by altering expression of deoxyxylulose phosphate reductoisomerase and menthofuran synthase. Proc. Natl. Acad. Sci., U. S. A. **2001**, *98*, 8915–8920.
7. Jomma, H.; Wiesner, J.; Sanderbrand, S.; Altincicek, B.; Weidemeyer, C.; Inhibitors of the nonmevalonate pathway of isoprenoid biosynthesis as antimalarial drugs. Science **1999**, *285*, 1573–1576.
8. Savage, T.J.; Hatch, M.W.; Croteau, R. Monoterpene synthases of Pinus contorta and related conifers. A new class of terpenoid cyclase. J. Biol. Chem. **1994**, *269* (6), 4012–4020.
9. Trapp, S.C.; Croteau, R.B. Genomic organization of plant terpene synthases and molecular evolutionary implications. Genetics **2001**, *158* (2), 811–832.
10. Went, F.W. Blue hazes in the atmosphere. Nature **1960**, *187*, 641–643.
11. Guenther, A.B.; Hewitt, C.N.; Erickson, D.; Fall, R.; Geron, C.; A global model of natural organic compound emissions. J. Geophys. Res. **1995**, *100*, 8873–8892.
12. Poisson, N.; Kanakidou, M.; Crutzen, P.J. Impact of non-methane hydrocarbons on tropospheric chemistry and the oxidizing power of the global troposphere: three-dimensional modeling results. J. Atmos. Chem. **2000**, *36*, 157–230.

Larvicidal Proteins from *Bacillus thuringiensis* in Soil: Release, Persistence, and Effects

G. Stotzky
New York University, New York, New York, U.S.A.

INTRODUCTION

Bacillus thuringiensis (*Bt*) produces a parasporal, proteinaceous, crystalline inclusion during sporulation. This inclusion is solubilized and hydrolyzed in the midgut of larvae of susceptible insects, releasing polypeptide toxins that cause death.[1–3] Distinct insecticidal crystal protein (IPC) *cry* genes code for larvicidal Cry proteins: Cry1 and Cry2B proteins are specifically toxic to Lepidoptera, Cry2A proteins to Lepidoptera and Diptera, Cry3 proteins to Coleoptera, and four Cry4 proteins to Diptera. Two genes (*cytA*, *cytB*) that code for cytolytic proteins (CytA, CytB) are present with the Cry4 proteins. Some ICPs exhibit activity against Homoptera, Hymenoptera, Orthoptera, Mallophaga, nematodes, mites, *Collembola*, protozoa, and other organisms.[1–3]

DISCUSSION

Sprays of *Bt* containing cells, spores, and ICPs have been used as insecticides for more than 30 years with generally no unexpected toxicities, probably because *Bt* does not survive or grow well in soil, and its spores are rapidly inactivated by ultraviolet (UV) light.[1,2,4,5] Hence, there is little production of the proteins in soil, and persistence of introduced toxins is a function primarily of the 1) amount added, 2) rate of consumption and inactivation by insect larvae, 3) rate of degradation by microorganisms, and 4) rate of abiotic inactivation. However, when genes that code for these proteins are genetically engineered into plants, the proteins continue to be synthesized during growth. If production exceeds consumption, inactivation, and degradation, toxins could accumulate and enhance control of target pests; constitute a hazard to nontarget organisms, such as the soil microbiota, beneficial insects, and other animals; and result in the selection and enrichment of toxin-resistant target insects.[1,4–6] Persistence is enhanced when the proteins are bound on surface-active particles (e.g., clays and humic substances) in soil and, thereby rendered less accessible for microbial degradation, although toxicity is retained.[4,5]

Potential hazards and benefits are affected by modifications (e.g., truncation, rearrangement of codons)[2] of introduced *cry* genes to code only for synthesis of "active" toxins, or a portion of the toxins, rather than code for synthesis of nontoxic ICPs. Consequently, it will not be necessary for an organism to have a high midgut pH (ca. 10.5) for solubilization of ICPs and specific proteolytic enzymes to cleave them into toxic subunits. Nontarget insects and organisms in higher and lower trophic levels could, therefore, be susceptible, leaving only the third barrier apparently responsible for host specificity; i.e., specific receptors on midgut epithelium that are often, but not always, present in larger numbers in susceptible larvae.[1,2,5]

Binding of toxins from *B. thuringiensis* subsp. *kurstaki* (*Btk*, 66 kDa, active against Lepidoptera), subsp. *tenebrionis* (*Btt*, 68 kDa, active against Coleoptera), and subsp. *israelensis* (*Bti*, 25–130 kDa, active against Diptera) on montmorillonite (M) and kaolinite (K); clay-, silt-, and sand-size fractions of soil; humic acids from different soils, and complexes of clay-humic acid-Al hydroxypolymers have been studied.[4,5] M and K are predominant clay minerals in many soils and differ in structure, physicochemical characteristics, and effects on biological activity.[7] The purpose of these in vitro studies, the results of which are summarized in Table 1, was to determine whether toxins in transgenic *Bt* plants or ICPs in sprays are bound on surface-active soil particles and whether binding results in the proteins becoming resistant to biodegradation and to persist while retaining larvicidal activity.[4,5]

INTERACTIONS IN SOIL OF LARVICIDAL PROTEINS IN TRANSGENIC *Bt* PLANTS

Biodegradation of Biomass of *Bt* Plants

Addition of biomass from transgenic corn (*Zea mays* L.) expressing Cry1Ab protein resulted in a significantly lower gross metabolic activity (as measured by CO_2 evolution) of soil than addition of isogenic nontransgenic biomass (Flores, Saxena, and Stotzky, unpublished.[4,5]

DOI: 10.1081/E-EPCS 120010605

Table 1 Summary of interactions of purified *Bt* toxins with surface-active particles: Effects on persistence and larvicidal activity

- Larvicidal proteins from *Bacillus thuringiensis* subspp. *kurstaki* (*Btk*; antilepidopteran), *tenebrionis* (*Btt*; anticoleopteran), and *israelensis* (*Bti*; antidipteran) bound rapidly and tightly on clays, humic acids, and complexes of clay-humic acid-Al hydroxypolymers; binding was pH dependent and greatest near the isoelectric point (pI) of the proteins; binding of the toxin from *Btk* was greater than binding of the toxin from *Btt*, even though the M_r of both was similar (66 and 68 kDa, respectively).
- Bound toxins retained their structure, antigenicity, and insecticidal activity.
- Intercalation of clays by the toxins was minimal.
- Biodegradation of the toxins was reduced when bound; microbial utilization of the toxins as a source of carbon was reduced significantly more than utilization as a source of nitrogen.
- Larvicidal activity of bound toxins was retained.
- Larvicidal activity of the toxin from *Btk* was detected 234 days after addition to nonsterile soils (longest time studied).
- Persistence of larvicidal activity was greater in acidic soils, in part probably because microbial activity was lower than in less acidic soils; persistence was reduced when the pH of acidic soils was raised to ca. 7.0 with $CaCO_3$.
- Persistence was similar under aerobic and anaerobic conditions and when soil was alternately wetted and dried or frozen and thawed, which indicated tight binding.
- Persistence in soil was demonstrated by dot-blot enzyme-linked immunosorbent assay (ELISA), flow cytometry, Western blots, and larvicidal assays.
- Toxins from *Btk, Btt*, and *Bti* had no microbiostatic or microbicidal effect against a spectrum of bacteria (gram positive and negative), fungi (filamentous and yeast), and algae, neither in pure nor in mixed cultures.

Soils amended with biomass of *Bt* corn were lethal to larvae of the tobacco hornworm (*Menduca sexta*), whereas there was essentially no mortality with soils amended with biomass of non-*Bt* corn or unamended. Similar results were obtained with biomass of *Bt* rice, cotton, canola, tobacco, and potato.

The lower biodegradation of *Bt* corn biomass was not the result of differences in the C:N ratio of the biomasses, as altering the ratio did not significantly alter the relative differences in biodegradation. It was apparently not the result of inhibition of the soil microbiota, as numbers of culturable bacteria and fungi and the activity of enzymes representative of those involved in the degradation of plant biomass were not significantly different in soil amended with biomass of *Bt* as opposed to non-*Bt* corn. These results confirmed in vitro observations that the Cry proteins from *Btk*, *Btt*, and *Bti* were not toxic to pure and mixed cultures of microbes and in situ observations of no consistent and lasting effects of biomass of transgenic *Bt* plants on the soil microbiota.[4,5]

Lignin Content of *Bt* Plants

The lignin content of 10 hybrids of *Bt* corn was 33–97% higher than that of their respective non-*Bt* near-isolines.[8] It was significantly higher in plants transformed by event Bt11 than by event MON810; the lignin content of the only available hybrid transformed by event 176 was lowest. There were no significant differences among isogenic non-*Bt* hybrids. However, the lignin content of the biomass of the other plant species, which was considerably lower than that of corn, was not significantly different in *Bt* and non-*Bt* plants. Hence, the higher lignin content of *Bt* corn may not be an important factor in its lower biodegradation. Nevertheless, modifications in lignin content could have ecological implications, e.g., higher lignin content in *Bt* corn may provide greater resistance to attack by second-generation European corn borer; reduce susceptibility to molds; and retard litter degradation and decomposition, which may be beneficial, as organic matter from *Bt* corn may persist and accumulate longer and at higher levels, thereby improving soil structure and reducing erosion.[8] By contrast, longer persistence may extend the time the toxin is present in soil and enhance the hazard to nontarget organisms and selection of toxin-resistant target insects.

Release of Cry Proteins in Root Exudates

Cry1Ab protein was released in root exudates from 13 hybrids of *Bt* corn (transformation events Bt11, MON810, and 176).[9] Presence of the toxin in sterile hydroponic culture was indicated by a major band migrating on sodium dodecyl sulfate polyacrylamide gel electrophoresis to a position corresponding to a molecular mass (M_r) of 66 kDa, the same as that of Cry1Ab protein, and confirmed by immunological and larvicidal assays. After 25 days, when the hydroponic culture was no longer sterile, the 66-kDa band was not detected (there were several new protein bands of smaller M_r) and immunological and larvicidal assays were negative, indicating that microbial proteases had hydrolyzed the protein. By contrast, toxin was detected after 180 days—the longest time studied—from *Bt* corn grown in nonsterile soil in a plant-growth room, confirming that toxin was bound on surface-active soil particles, which protected it from hydrolysis.

To estimate the importance of clays and other physicochemical characteristics—which influence activity and ecology of microbes in soil[5,7,10] —to persistence of toxin released in root exudates, studies were done in soil amended to 3, 6, 9, or 12% with M or K. Forty days after germination, all samples of rhizosphere soil from *Bt* corn were immunologically positive for the Cry1Ab protein and toxic to larvae of *M. sexta*, whereas there was no significant mortality with soil from non-*Bt* corn or without plants.[9] The weight of surviving larvae exposed to soils from *Bt* corn was 50–92% lower than the weight of larvae exposed to soils from non-*Bt* corn or without plants.

Larvicidal activity was higher initially in soil amended with M than with K, probably because M—a swelling 2:1, Si:Al, clay with a significantly higher cation-exchange capacity and specific surface area than K, a nonswelling 1:1, Si:Al, clay—bound more toxin in root exudates, as also observed with pure toxin.[4,5] However, mortality was essentially the same after 40 days, indicating that over a longer time, persistence of larvicidal activity appeared to be independent of clay mineralogy and other physicochemical characteristics of soils, and that toxin was concentrated when adsorbed on surface-active soil components.

Immunological and larvicidal assays of rhizosphere soil from *Bt* corn grown in the field were also positive, even in soil collected after frost from plants that had been dead for several months.[5,9] There were no consistent differences in exudation of toxin between hybrids derived from different transformation events, neither in the plant-growth room nor in the field.

In addition to toxin introduced to soil in plant biomass after harvest and in pollen during tasseling,[11] toxin will be released to soil from roots during growth of *Bt* corn, which could improve control of insect pests, enhance selection of toxin-resistant target insects, and/or constitute a hazard to nontarget organisms. Because *Bt* corn contains truncated genes that encode toxin rather than nontoxic ICP, potential hazards are exacerbated. This is because it is not necessary for an organism to have a high gut pH and specific proteases, and receptors for toxin are present in target and nontarget insects.[1] Consequently, nontarget insects and other organisms in higher and lower trophic levels could be susceptible.[5]

CrylAc protein was not detected in hydroponic culture or nonsterile soil in root exudates of *Bt* cotton, canola, or tobacco, whereas Cry1Ab protein was detected in root exudates of *Bt* rice. Cry3A protein was detected in root exudates of *Bt* potato immunologically and by their toxicity to larvae of the Colorado potato beetle (*Leptinotarsa decemlineata*) (Saxena and Stotzky, unpublished). It is not clear how such large proteins (e.g., the 66-kDa Cry1Ab protein) are released intact from roots, because the release of molecules with high M_r usually requires the presence of a "signal peptide."[12] The endoplasmic reticulum is apparently close to or associated with the plasma membrane in roots of corn and, perhaps, of rice and potato but apparently not of cotton, canola, or tobacco. Although some toxin from corn, rice, and potato was probably derived from sloughed and damaged root cells, the major portion was from root exudates, as there was no discernable root debris in sterile hydroponic cultures. Moreover, no toxin was detected from cotton, canola, or tobacco in soil, again indicating that any damage to roots was a minor source of the toxins.

L

Effects of Toxins on Worms, Nematodes, and Microbes

There were no significant differences in mortality and weight of earthworms (*Lumbricus terrestris*) after 40 days in soil planted with *Bt* or non-*Bt* corn or not planted and after 45 days in soil amended with biomass of *Bt* or non-*Bt* corn or not amended.[13] Toxin was present in casts and guts of worms grown in *Bt*-contaminated soil, indicating again that toxin had bound on surface-active particles in soil, which protected it from biodegradation. When worms were transferred from *Bt*-containing soil to fresh soil, toxin was cleared from guts in 1–2 days. There were also no statistically significant differences in numbers of nematodes and culturable protozoa, bacteria (including actinomycetes), and fungi between rhizosphere soil of *Bt* and non-*Bt* corn or between soil amended with *Bt* or non-*Bt* biomass. Soil with biomass of *Bt* corn and from the rhizosphere of *Bt* corn was immunologically positive for toxin and lethal to larvae of *M. sexta* after 45 and 40 days, respectively. Although these results suggested that toxin released in root exudates of *Bt* corn or from degradation of biomass of *Bt* corn is not toxic to a variety of organisms in soil, only one species of earthworms and only culturable microorganisms and nematodes were evaluated. More detailed studies, including techniques of molecular biology, are necessary to confirm the absence of effects of Cry1Ab and other Cry proteins on biodiversity in soil.

Uptake by Plants of Cry Proteins from Soil

When non-*Bt* corn, carrot, radish, and turnip were grown in nonsterile soil amended with Cry1Ab protein or biomass of *Bt* corn or in soil in which *Bt* corn had been grown, the toxin was not detected in any plants, whereas it was present in soil even after 180 days, the longest time evaluated.[14,15] No Cry1Ab protein was detected in non-*Bt* corn grown aseptically in hydroponic culture in which *Bt* corn had been grown aseptically, whereas it was easily

Table 2 Summary of fate and effects of *Bt* toxins in root exudates and biomass of transgenic plants

- Biodegradation of biomass of transgenic *Bt* corn, measured by CO_2 evolution, was significantly lower than that of near-isogenic non-*Bt* corn.
- No consistent statistically significant differences in the numbers of culturable bacteria, fungi, and the activity of representative enzymes between soil amended with *Bt* or non-*Bt* corn or not amended.
- Reduced metabolic activity of soil amended with *Bt* corn may have been result of significantly higher lignin content in *Bt* than in non-*Bt* corn.
- Biodegradation of biomass of *Bt* rice, cotton, canola, tobacco, and potato was also significantly lower than that of biomass of near-isogenic non-*Bt* plants, but lignin content of these plant species, which was significantly lower than that of corn, was not significantly different in *Bt* and non-*Bt* biomass.
- CryIAb protein was released in root exudates of *Bt* corn (13 hybrids representing three transformation events) and persisted in rhizosphere soil in vitro and in situ; protein accumulated more in soil amended (3 to 12%) with montmorillonite than with kaolinite.
- CryIAb protein released in root exudates or from biomass of *Bt* corn appeared to have no effect on numbers of earthworms, nematodes, protozoa, bacteria, and fungi in soil.
- CryIAc protein was not released in root exudates of *Bt* canola, tobacco, or cotton. Cry1Ab protein was released in root exudates of rice, and Cry3A protein was released in root exudates of *Bt* potato.
- CryIAb protein released in root exudates and from biomass of *Bt* corn was not taken up from nonsterile soil or sterile hydroponic culture by non-*Bt* corn, carrot, radish, or turnip, even though the toxin persisted for at least 180 days in soil (the longest time studied).
- CryIAb protein—purified, in root exudates, and from biomass of *Bt* corn—moved through soil during leaching with water; movement was less in soils amended with montmorillonite than with kaolinite and decreased as the concentration of added clays increased.
- Toxins from *Bt* could persist, accumulate, and remain insecticidal in soil as the result of binding on clays and humic substances and, therefore, pose a hazard to nontarget organisms, enhance selection of toxin-resistant target species, or enhance control of insect pests.

detected in the solution. Hence, the apparent lack of uptake of toxin from soil was not the result of binding on surface-active particles, as no such particles were in the solution.

Movement of Cry Proteins in Soil

CryIAb protein added to the top of columns of soil was detected in leachates 1 and 3 hours after addition, with ca. 75% detected from soil not amended with clay and ca. 16% detected from soil amended to 12% with M or K; intermediate amounts were leached from soils amended to 3, 6, or 9% with the clays.[16] Larvicidal activity was higher with leachates from soil not amended or amended to 3 or 6% than from soil amended to 9 or 12%. After 12 and 24 h, no protein was detected, indicating that it bound on the soils and desorption was reduced. Vertical distribution, in the columns, of protein not recovered in leachates confirmed that it moved less through soil amended with higher amounts of clay. Larvicidal activity of soil generally decreased with depth as clay concentration increased.

CryIAb protein was present in leachates from soil columns in which *Bt* corn hybrids of the three transformation events were grown, indicating some vertical movement from the rhizosphere. The protein was also present in leachates from soil amended three years earlier with *Bt* corn biomass, indicating that as the biomass degraded, toxin was released, and some bound on soil particles and some dissolved in and moved down with soil water.

Movement of CryIAb protein through soil was influenced by its tendency to stick on soil particles.[4,5,7,10] The protein exhibited stronger binding and higher persistence in soils with higher clay concentrations—especially of M—and it remained near the soil surface, increasing the probability of it being transported to surface waters via erosion and runoff. In contrast, the protein was leached more through soils with lower clay concentrations, and may contaminate groundwater. Contamination of surface or groundwater, which depends greatly on the desorption of the protein and on the amount of water impacting soil as rain, irrigation, snow melts, etc., may pose a hazard to nontarget aquatic Lepidoptera, which are more plentiful in water than in soil.[17] Without sufficient water, the protein will remain within the biologically-active root zone, where some protein, especially that not bound on particles, will be mineralized.

CONCLUSION

The interaction of larvicidal proteins with surface-active particles that differ greatly in composition and structure demonstrates the importance of such particles to the biology of natural habitats. These studies (summarized in Tables 1 and 2) also confirm and extend previous observations on the influence of surface-active particles on the activity, ecology, and population dynamics of microbes and viruses, as well as on the transfer of genetic information among bacteria in soil and other habitats.[7,10,18]

Persistence of bound toxins from *Bt* could pose a hazard to nontarget organisms and result in selection of toxin-resistant target insects and thereby negate benefits of using a biological insecticide. However, persistence of bound toxins could also enhance control of target pests, particularly soil-borne ones. These aspects require more study, especially a case-by-case evaluation of each toxin.

These studies also indicate that caution must be exercised before the release to the environment of transgenic plants and animals that are genetically modified to function as ''factories'' (''biopharms'') for production of vaccines, hormones, antibodies, blood substitutes, toxins, and other pharmaceuticals. As with *Bt* plants, biomass of these plant factories will be incorporated into soil. Feces, urine, and even carcasses containing bioactive compounds from transgenic animals will also reach soil and other habitats. If these compounds (including prions from contaminated animal carcasses) bind on clays and humic substances, they may persist, and if they retain their bioactivity, they may affect the biology of these habitats. Consequently, before extensive use of such plant and animal factories, their persistence and the potential effects of their products on inhabitants of soil and other habitats must be evaluated.

ACKNOWLEDGMENTS

Most of the studies from the Laboratory of Microbial Ecology discussed in this article were supported, in part, by grants from the U.S. Environmental Protection Agency (most recently, R826107-01), U.S. National Science Foundation, and NYU Research Challenge Fund. Opinions expressed herein are not necessarily those of the Agency, Foundation, or Fund. Sincere appreciation is expressed to colleagues who contributed to these studies.

ARTICLES OF FURTHER INTEREST

Bacterial Products Important for Plant Disease: Toxins and Growth Factors, p. 101
Biological Control in the Phyllosphere, p. 130
Biological Control of Nematodes, p. 134
Biosafety Applications in the Plant Sciences: Expanding Notions About, p. 146
Biosafety Programs for Genetically Engineered Plants: An Adaptive Approach, p. 160
Biosafety Science: Overview of Plant Risk Issues, p. 164
Breeding Plants with Transgenes, p. 193
Genetically Engineered Crops with Resistance Against Insects, p. 506
Medical Molecular Pharming: Therapeutic Recombinant Antibodies, Biopharmaceuticals and Edible Vaccines in Transgenic Plants Engineered via the Chloroplast Genome, p. 705
Mycotoxins Produced by Plant Pathogenic Fungi, p. 773
Plant Defenses Against Insects: Constitutive and Induced Chemical Defenses, p. 939
Rhizosphere: An Overview, p. 1084
Rhizosphere: Biochemical Reactions, p. 1087
Rhizosphere: Microbial Populations, p. 1090
Rhizosphere Management: Microbial Manipulation for Biocontrol, p. 1098
Root Membrane Activities Relevant to Plant-Soil Interactions, p. 1110
Transgenetic Plants: Breeding Programs for Sustainable Insect Resistance, p. 1245
Transgenic Crop Plants in the Environment, p. 1248
Transgenic Crops: Regulatory Standards and Procedures of Research and Commercialization, p. 1251
Vaccines Produced in Transgenic Plants, p. 1265

REFERENCES

1. Höfte, H.; Whiteley, H.R. Insecticidal crystal proteins of *Bacillus thuringiensis*. Microbiol. Rev. **1989**, *53*, 242–255.
2. Schnepf, E.; Crickmore, N.; Van Rie, J.; Lereclus, D.; Baum, J.; Feitelson, J.; Zeigler, D.R.; Dean, D.H. *Bacillus thuringiensis* and its pesticidal crystal proteins. Microbiol. Mol. Biol. Rev. **1998**, *62*, 775–806.
3. Crickmore, N.; Zeigler, D.R.; Feitelson, J.; Schnepf, E.; Van Rie, J.; Lereclus, D.; Baum, J.; Dean, D.H. Revision of the nomenclature for the *Bacillus thuringiensis* pesticidal crystal proteins. Microbiol. Mol. Biol. Rev. **1998**, *62*, 807–813.
4. Stotzky, G. Persistence and biological activity in soil of insecticidal proteins from *Bacillus thuringiensis* and of bacterial DNA bound on clays and humic acids. J. Environ. Qual. **2000**, *29*, 691–705.
5. Stotzky, G. Release, Persistence, and Biological Activity in Soil of Insecticidal Proteins from *Bacillus thuringiensis*. In *Genetically Engineered Organisms: Assessing Environmental and Human Health Effects*; Letourneau, D., Burrows, B., Eds.; CRC Press: Boca Raton, 2001; 187–222.
6. *Genetically Engineered Organisms: Assessing Environmental and Human Health Effects*; Letourneau, D., Burrows, B., Eds.; CRC Press: Boca Raton, 2001.
7. Stotzky, G. Influence of Soil Mineral Colloids on Metabolic Processes, Growth, Adhesion, and Ecology of Microbes and Viruses. In *Interactions of Soil Minerals with Natural Organics and Microbes*; Huang, P.M., Schnitzer, M., Eds.; Soil Sci. Soc. Am.: Madison, 1986; 305–428.
8. Saxena, D.; Stotzky, G. *Bt* corn has a higher lignin content than non-*Bt* corn. Am. J. Bot. **2001**, *88*, 1704–1706.
9. Saxena, D.; Flores, S.; Stotzky, G. *Bt* toxin is released in root exudates from 12 transgenic corn hybrids representing

three transformation events. Soil Biol. Biochem. **2002**, *34*, 133–137.

10. Stotzky, G. Soil as an Environment for Microbial Life. In *Modern Soil Microbiology*; van Elsas, J.D., Wellington, E.M.H., Trevors, J.T., Eds.; Dekker: New York, 1997; 1–20.
11. Obrycki, J.J.; Losey, J.E.; Taylor, O.R.; Jesse, L.C.H. Transgenic insecticidal corn: Beyond insecticidal toxicity to ecological complexity. Bioscience **2001**, *51*, 353–361.
12. Borisjuk, N.V.; Borisjuk, L.G.; Logendra, S.; Peterson, F.; Gleba, Y.; Raskin, I. Production of recombinant proteins in plant root exudates. Nature Biotech. **1999**, *17*, 466–469.
13. Saxena, D.; Stotzky, G. *Bacillus thuringiensis* (*Bt*) toxin released from root exudates and biomass of *Bt* corn has no apparent effect on earthworms, nematodes, protozoa, bacteria, and fungi in soil. Soil Biol. Biochem. **2001**, *33*, 1225–1230.
14. Saxena, D.; Stotzky, G. *Bt* toxin is not taken up from soil by plants. Nature Biotech. **2001**, *19*, 199.
15. Saxena, D.; Stotzky, G. *Bt* toxin is not taken up from soil or hydroponic culture by corn, carrot, radish, or turnip. Plant Soil **2002**, *239*, 165–172.
16. Saxena, D.; Flores, S.; Stotzky, G. Vertical movement in soil of insecticidal Cry1Ab protein from *Bacillus thuringiensis*. Soil Biol. Biochem. **2002**, *34*, 111–120.
17. Williams, D.D.; Feltmate, B.W. *Aquatic Insects*; CAB International: Oxford, 1994; 358.
18. Yin, X.; Stotzky, G. Gene transfer among bacteria in natural environments. Adv. Appl. Microbiol. **1997**, *45*, 153–212.

Leaf Cuticle

Gwyn A. Beattie
Iowa State University, Ames, Iowa, U.S.A.

INTRODUCTION

The leaf cuticle is the waxy layer that covers the surface of leaves. Its main function is to protect the plant against water loss, and it is therefore critical to life in an aerial environment. It offers protection against environmental stresses as well as microbial pathogens and herbivorous insects, and contributes to the ability of many plant species to keep their leaves clean. These functions are related to the structure and composition of the cuticle, and particularly its main functional component, the cuticular and epicuticular waxes.

THE STRUCTURE AND FUNCTION OF LEAF CUTICLES

The thin continuous layer known as the cuticle is present on the epidermal cell surfaces of leaves, including on the trichomes and guard cells, as well as on the cells lining the substomatal cavities. This nearly impermeable membrane was likely a key development during the evolution of plants from aquatic to terrestrial habitats.[1] Plant cuticles consist of a network of insoluble structural polymers, typically cutin, that is infused with a complex mixture of lipids, or cuticular waxes.[2] These waxes are usually very long-chain fatty acids and their oxygenated derivatives, including alcohols, aldehydes, ketones, and esters. In contrast to cellular lipids, which are usually 15 to 18 carbons in length, cuticular waxes are usually much longer, often 20 to 40 carbons in length. Some of the waxes are extruded onto the surface of the cutin layer (Fig. 1). These epicuticular waxes form a relatively smooth layer on some plant surfaces, and on others they form crystals (Fig. 2). Epicuticular wax crystals vary greatly across plant species in their morphology and distribution.[1,3]

A major function of the leaf cuticle is to serve as a diffusion barrier; it minimizes the loss of water and solutes while also limiting the inward movement of aqueous pollutants. The leaf cuticle can reflect light, particularly when crystalline epicuticular waxes are present, and this provides protection against both damaging UV radiation and light-driven temperature increases and their consequent reductions in transpiration, thus improving water conservation.[3] The presence of crystalline epicuticular waxes can confer water repellency. Water applied to the surface of water-repellent leaves forms droplets that roll off, and these droplets collect nonadhering contaminating particles such as dust and microbes, thus effectively cleaning the leaf surface.[4] Water repellency is particularly common to the adaxial surfaces of many plant leaves, where it may help prevent the formation of water films over the stomata in humid environments. Low leaf wettability resulting from the hydrophobic waxes—particularly crystalline waxes—contributes to a small contact area between aqueous solutions and the leaf surface. This is likely of great importance to photosynthetic gas exchange and thus plant growth, to the ecology of the leaf-associated microflora (including the potential infectivity of pathogens), and to the retention and uptake of aqueous pollutants and agricultural sprays such as pesticides, herbicides, and plant growth regulators.[3,5] Last, the leaf cuticle contributes to protection from herbivorous insects and phytopathogenic microbes by a variety of mechanisms.

FACTORS INFLUENCING LEAF CUTICLES

The quantity, composition, and structure of the cuticular waxes are specific to the plant species, the plant age, and even the adaxial versus abaxial leaf surface. For example, as leaves age, crystalline waxes appear to be degraded and this changes the surface characteristics and topography of the leaf. Environmental conditions that affect the waxes include light intensity, UV-B exposure, water stress and temperature.[3] For example, plants grown under elevated light intensity or drought stress often produce more, or different, cuticular waxes than nonstressed plants. Wind can disrupt or remove the waxes by rubbing, flexing, or particle impaction, and also has been suggested to melt the waxes due to frictional heat. Rain can remove epicuticular waxes from the leaves (as illustrated by the presence of crystalline waxes in rain drops after they impact the leaf surface) and can alter wax composition by selectively removing certain classes of crystals. Air pollutants and acid rain can cause structural degradation of the waxes, stomatal occlusions, and shifts in wax composition, as has been well documented with conifer needles. In fact, wax

Encyclopedia of Plant and Crop Science
DOI: 10.1081/E-EPCS 120010615

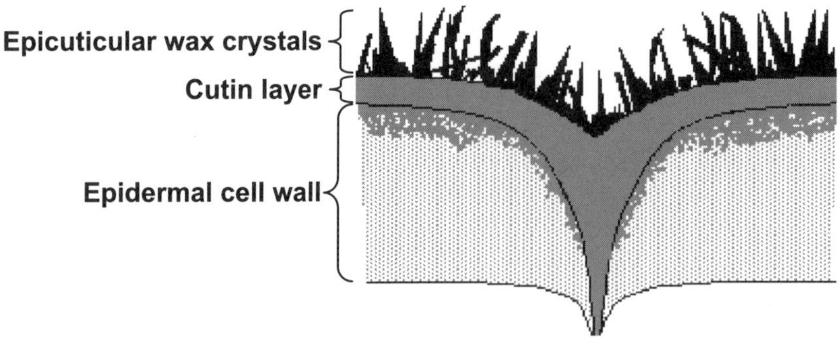

Fig. 1 Generalized structure of a leaf cuticle. The epicuticular waxes may be crystalline or amorphous. Additional cuticular waxes are infused in the cutin layer and are often impregnated in the upper epidermal cell wall.

degradation and compositional shifts have been used as bioindicators of air pollution in forests.[3]

LEAF CUTICLES AS PLAYING FIELDS FOR INSECT HERBIVORES AND THEIR PREDATORS

Leaf–insect interactions are dominated by the cuticle as an obstacle to insect attachment and movement.[6] Cuticular waxes interfere with attachment and movement of insects primarily by exfoliation of the wax crystals. In fact, carnivorous pitcher plants have capitalized on wax exfoliation. (The pitcher plant captures insects by producing wax plates that are readily released upon contact by an insect.) This exfoliation can deter insect herbivores, thus providing defense against herbivory, but may also deter the predators of these herbivores. Among the many diverse surfaces that support insects, leaves with crystalline waxes are unusual in the effectiveness with which they prevent insect adherence.

The chemical and physical properties of leaf cuticles influence host plant selection by herbivorous insects.[6] For example, the crystalline waxes can dictate the appearance of leaves, often giving them a whitish color, which affects their attractiveness to some herbivorous insects. Similarly, specific waxes can influence the acceptance or rejection of a potential plant host for feeding or for the deposition of eggs, which often requires selection of a host that is suitable for larval development. Insect behavior may be affected by the ability of cuticular waxes to serve as solvents for insect pheromones or other volatile compounds from the environment; however, this area has been relatively understudied considering its potential ecological significance.

INTERACTIONS BETWEEN LEAF CUTICLES AND THE LEAF MICROFLORA

The leaf cuticle influences the immigration of waterborne and airborne microbes by affecting water repellency and the formation of a still-air boundary layer surrounding the leaf. The cuticle also affects the attachment of some bacteria, yeasts, and fungi to leaves, as illustrated by the finding that leaf surface hydrophobicity

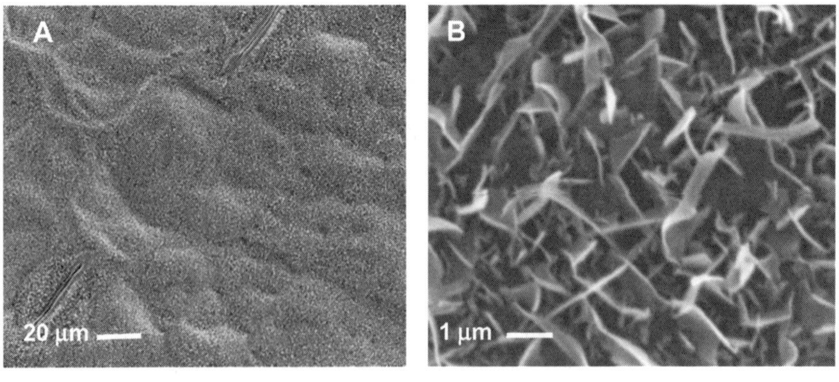

Fig. 2 Epicuticular crystalline waxes on the surface of maize (*Zea mays*) leaves under (A) low (500X) and (B) high (2000X) magnification.

promotes attachment by many fungal spores. Similarly, although not well characterized, the cuticle can influence the size of the resident bacterial and yeast populations, an influence that may be mediated, to a large degree, by leaf cuticle–water relations.[5] A few recent studies suggest that bacteria may have mechanisms for penetrating plant cuticles;[5] however, bacteria are generally regarded as unable to penetrate intact cuticles. This is supported by the finding that bacterial production of known cuticle-degrading enzymes—namely cutinases—is a rare trait, and by knowledge that phytopathogenic bacteria invade leaves via natural openings, wounds, and insect damage, but not via direct penetration. In contrast, phytopathogenic fungi have evolved multiple mechanisms for penetrating plant cuticles, and clearly respond to distinct cuticle-dependent signals.[7] Spores and germlings of phytopathogenic fungi are often associated with visible erosion of the cuticle, which likely results from enzymatic degradation. Cutin-degrading enzymes have been characterized in many fungal species, although wax-degrading enzymes have not been identified. Some fungal species penetrate leaves without producing highly developed infection structures, whereas others produce appressoria (structures that aid penetration). These structures often employ turgor pressure as a mechanism of force. Simple contact with a surface—particularly a hydrophobic surface—induces appressorium formation as well as adhesion and germination in some fungi. For example, upon contact with the leaf cuticle, *Erysiphe graminis* secretes a liquid that aids adherence and creates a localized hydrophilic region, presumably by degrading cutin, and the resulting hydrophilicity stimulates germination. Similar to their effect on insects, specific cuticular waxes can act as stimulants or deterrents of specific fungal behaviors, including germination and appressorium formation.

CONCLUSION

Current research interests in leaf cuticles—particularly in cuticular waxes—include their diversity and biosynthesis; their role in plant evolution and development; their role in plant responses to insects and microbes, as discussed here, as well as to drought, freezing temperatures, solar radiation, and air pollution; their role in the retention and penetration of agricultural sprays; their use as industrial plant waxes; their use in breeding of ornamental and agronomic plants with improved performance; and their contribution to atmosphere aerosols.[8] Mutants that are altered in loci involved in the biosynthesis, secretion, and regulation of cuticular wax production have been identified for several plant species. The cloning and characterization of these and other loci involved in cuticle synthesis and regulation, along with current tools in plant molecular biology, should benefit these research areas by increasing our fundamental understanding of cuticle formation and regulation and by providing information useful for cuticle modification, such as for crop improvement.

ARTICLES OF FURTHER INTEREST

Bacterial Survival and Dissemination in Natural Environments, p. 108
Coevolution of Insects and Plants, p. 289
Drought and Drought Resistance, p. 386
Insect–Plant Interactions, p. 609
Leaf Structure, p. 638
Leaves and Canopies: Physical Environment of, p. 646
Lipid Metabolism, p. 659
Plant Defenses Against Insects: Constitutive and Induced Chemical Defenses, p. 939
Plant Defenses Against Insects: Physical Defenses, p. 944
Plant Diseases Caused by Bacteria, p. 947
Water Use Efficiency Including Carbon Isotope Discrimination, p. 1288

REFERENCES

1. Gülz, P.G. Epicuticular leaf waxes in the evolution of the plant kingdom. J. Plant Physiol. **1994**, *143*, 453–464.
2. Jeffree, C.E. Structure and Ontogeny of Plant Cuticles. In *Plant Cuticles: An Integrated Functional Approach*; Kerstiens, G., Ed.; BIOS Scientific Publishers: Oxford, 1996; 33–82.
3. Jenks, M.A.; Ashworth, E.N. Plant epicuticular waxes: Function, production, and genetics. Hort. Rev. **1999**, *23*, 1–68.
4. Barthlott, W.; Neinhuis, C. Purity of the sacred lotus, or escape from contamination in biological surfaces. Planta **1997**, *202*, 1–8.
5. Beattie, G.A. Leaf Surface Waxes and the Process of Leaf Colonization by Microorganisms. In *Phyllosphere Microbiology*; Lindow, S.E., Hecht-Poinar, E.I., Elliott, V.J., Eds.; APS Press: Minneapolis, MN, 2002; 3–26.
6. Eigenbrode, S.D. Plant Surface Waxes and Insect Behaviour. In *Plant Cuticles: An Integrated Functional Approach*; Kerstiens, G., Ed.; BIOS Scientific Publishers: Oxford, 1996; 201–221.
7. Mendgen, K.; Hahn, M.; Deising, H. Morphogenesis and mechanisms of penetration by plant pathogenic fungi. Annu. Rev. Phytopathol. **1996**, *34*, 367–386.
8. Simoneit, B.R.T.; Mazurek, M.A. Organic matter of the troposphere. II. Natural background of biogenic lipid matter in aerosols over the rural western United States. Atmos. Environ. **1982**, *16* (9), 2139–2159.

Leaf Structure

Vincent P. Gutschick
Lou Ellen Kay
New Mexico State University, Las Cruces, New Mexico, U.S.A.

INTRODUCTION

Leaves are constructed mechanically and biochemically primarily for their function in photosynthesis. They must be constructed at low metabolic cost and protect themselves against mechanical stresses, climatic extremes, excess light, and damage by herbivores and pathogens. Concurrently, they must balance water use efficiently. Some leaves are specialized for additional functions such as support of the whole plant or storage. The leaf surface, parenchyma, and vascular system show unique contributions to each of these functions. The overall size and shape of leaves are also relevant for construction cost, energy transfer, light interception, and mechanical properties. The development of each leaf and its functions, from inception to death, is subject to a number of controls. These respond to the environment (nutrition, water and solute status, etc.) in ways that are often manifestly adaptive. Leaf traits exist in very diverse combinations in response to natural selection on the net benefit of a leaf in all its functions. Artificial selection imposes different weightings of traits; the optimal combinations in agriculture are in some cases quite divergent from wild types, as in leaf area per mass and geometry of leaf display.

PERFORMING PHOTOSYNTHESIS

The primary function of a leaf is photosynthesis. Leaves are typically thin, providing maximum surface area for light interception at the lowest cost in metabolites and nutrients. Nonetheless, leaves are diverse in thickness, size, shape, reflectivity and other optical properties, and presence of trichomes. The diversity reflects demands to 1) use all resources—not just light—efficiently (particularly water and nitrogen and sometimes other nutrients), and not just for photosynthesis but also for defense;[1] and 2) avoid damage, both biotic (from herbivores, fungi, etc.) and abiotic (from wind, extreme temperatures, ultraviolet radiation, acute shortage of water, etc.).

STRUCTURE OF THE LEAF

The leaf surface is covered by a cuticle (Fig. 1). This hydrophobic layer limits losses of water and ions, offers the first defense against pathogens, and conditions the optical properties of the leaf. When thick and scalelike, it reduces light interception as a protection against photoinhibition and thermal loading. The outermost layer of cells is the epidermis. This layer provides some structural strength and significant ultraviolet screening via absorptive chemicals. A small percentage of epidermal cells is specialized as guard cells, forming the stomata that regulate the exchange of gases, primarily CO_2 and water vapor but also leaf volatiles such as terpenes and a portion of the O_2. Other specialized cells (trichomes) may be present, presenting long hairs that have many functions—defense, radiative balance, control of surface wetting, etc.

The largest volume of cells is parenchyma cells specialized for photosynthesis. Palisade cells are columnar and strongly photosynthetic. Spongy mesophyll cells are also photosynthetic, while offering air passages for gas exchange. Complete photosynthesis (both the capture of light to produce reductant and carbon fixation) is present in all these cells in most plants. In plants with the C_4 pathway, the parenchyma cells other than the bundle sheath capture CO_2 in a transportable form (such as malate) and generate reductant. Final sugar generation occurs inside the bundle sheath, which is poorly gas-permeable, allowing high partial pressures of CO_2 to build up.

Photosynthetic cells typically have thin cell walls for high conductivity to dissolved gases. Stress-tolerant plants—the sclerophylls—toughen their leaves with secondary cell wall growth, plus dense packing; consequently, they have low photosynthetic rates. Other structural features of the leaf interior include sclereids (as in sclerophylls) and, in grasses, silica bodies (whose evolution may even have caused some herbivore extinction[2]). Some leaves contain laticifers, fused cells that transport quantities of insect-deterring latex.

Veins are a major leaf structure. The xylem imports water, of which usually more than 99% is used over the life of the plant for transpiration as opposed to for new

Encyclopedia of Plant and Crop Science
DOI: 10.1081/E-EPCS 120010616

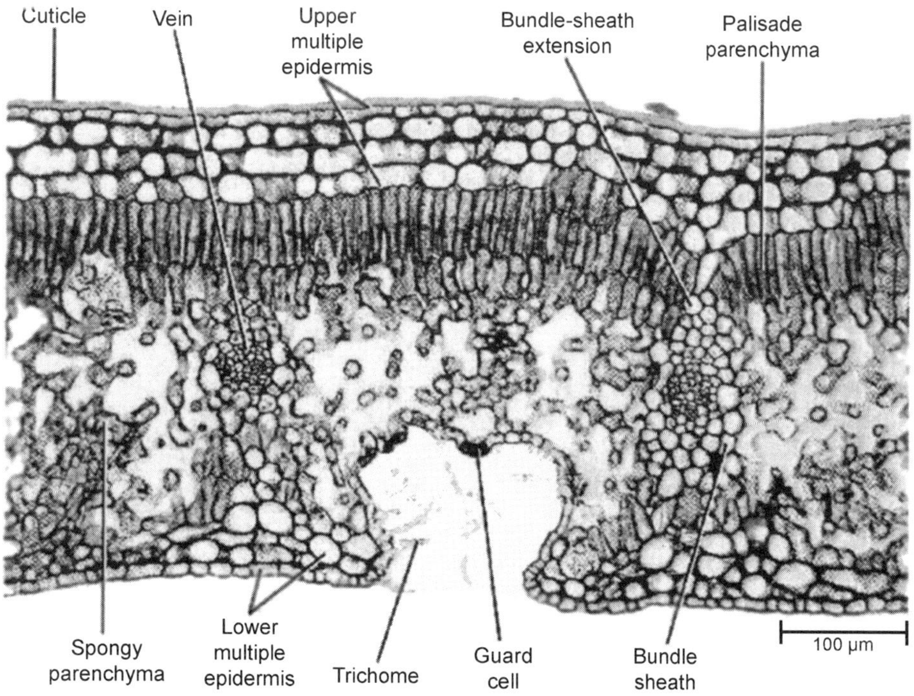

Fig. 1 Transverse section through the leaf of oleander (*Nerium oleander*), a xerophyte with thick cuticle, multilayered epidermis, and recessed stomata. (From *Biology of Plants*, 6th Ed., P. H. Raven et al.; 1999; W. H. Freeman; New York; used with permission.)

growth. The xylem also imports some nutrients. It may be strongly lignified; this condition, plus the vascular geometry as branched cylinders, is the dominant contribution to biomechanical integrity. Leaves must be strong (resisting high forces of gravitational loads, winds, and herbivore action), stiff (resisting bending, to maintain light capture), tough (requiring much work to break them), and resistant to crack propagation.[3] Phloem is the other major vascular component. It transports photosynthesized sugars out of mature leaves. It also transports hormones, some photosynthetically reduced nutrients (N and S compounds), and nutrients that recycle as signals to regulate root activity. The phloem possesses some parenchyma cells, some fibers, and sclereids as lesser contributors to structure and defense. The vascular bundles of xylem and phloem occur commonly in nets or, in monocots and ginkgos, as parallel strands. The pattern is in part adaptive (e.g., grasses withstand grazing by regrowth from the base, easiest biomechanically with parallel veins) and is in part an evolutionary constraint. In C_4 plants, mesophyll and bundle-sheath cells cluster tightly around xylem and phloem in the Kranz anatomy, which enables concentration of CO_2 for efficient photosynthesis.

A number of variations occur in internal structure. Thick mesophyll of high water content stores water in succulent species. In those species performing CAM photosynthesis, this also serves as inexpensive, copious storage for malate, reducing the limit on growth rate.

LEAF MORPHOLOGY

The leaf proper commonly has a petiole, supporting it away from the stem and other leaves. The petiole, although minimally photosynthetic has a significant fraction of total leaf mass, minimization of this mass is often important in carbon gain per mass and growth rate.

Overall shapes of leaves and their arrangement on the stem are usually diagnostic of a species or an ecotype; occasionally the shape is plastic, as in sassafras (lobing) and aquatic plants (long, thin water leaves contrast with aerial leaves). Leaf types divide fundamentally into simple and compound (Fig. 2). Compound leaves appear to cost less energy to construct but are less durable.[4] Other geometric features include entirety (no reentrant curves) vs. dissection into discrete lobes. Dissection increases the boundary layer conductance for gases and heat;[5] dissected leaves remain closer to air temperature than do entire leaves. Another marked division is between flat leaves and the needle-shaped leaves of conifers (most rushes and sedges have tiny leaves but photosynthetic stems). Needle leaves have low photosynthetic area per unit mass but possess strength and damage-resistance.

Leaf size also varies among species, and is more plastic than is geometric shape within a species. Thick leaves have low area per mass (i.e., per specific leaf area). All else being equal, thickness confers high photosynthetic capacity per area but correlates with smaller leaf area. The

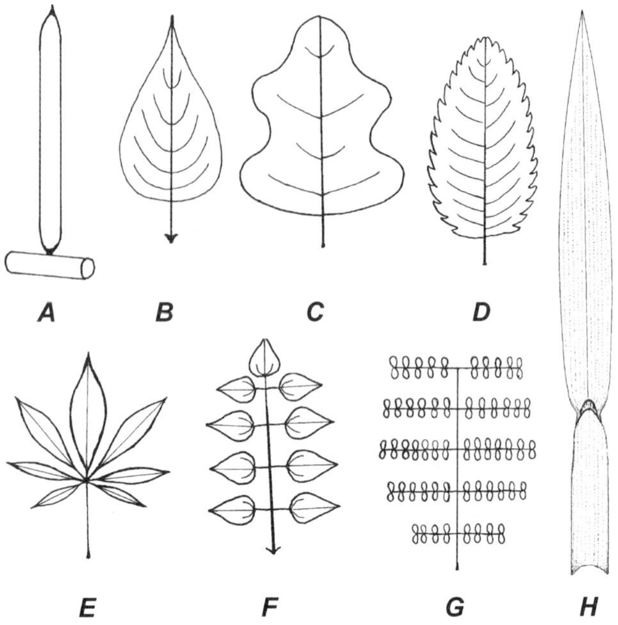

Fig. 2 A selection of leaf shapes and their terminology: (A) needle leaf; (B) entire leaf (ovate shape); (C) lobed leaf; (D) dissected leaf; (E) palmately compound leaf with many leaflets; (F) pinnately compound leaf; (G) doubly (bipinnately) compound leaf; (H) grass leaf (a residual meristem at the base of the blade (at the narrow point in drawing) supports continuing leaf growth). (Artwork by L. E. Kay.)

net effect on photosynthesis per mass—the most important correlate of relative growth rate—is nearly neutral.[6] Thinness, however, is correlated with higher N content that raises photosynthesis per unit mass, and offers greater shading of competitors. The lower structural content makes such leaves more damage-prone; they are typically short-lived and never evergreen. Leaves of an individual plant can acclimate in thickness. Leaves receiving more sunlight develop higher photosynthetic capacity, commonly from larger cells and more layers. Higher photosynthetic capacity relative to stomatal conductance can also confer higher water-use efficiency by virtue of a lower operating point in leaf-internal CO_2.

CONTROLS OVER DEVELOPMENT AND DEATH

Leaves vary widely in development rate and lifespan. Initiation has numerous controls, both intrinsic and environmental. Leaf expansion (initiation time and duration) is paced to thermal time in degree-days, which has been shown to have a molecular basis in cyclin proteins.[7] Cell division and cell enlargement are restricted to rather fixed intervals; adverse conditions during early expansion (low light, dryness of air or soil) permanently eliminate some growth potential, which shows up later as smaller leaves.[7] Leaf thickness and attendant photosynthetic capacity responds as well to temperature and irradiance during expansion. Patterns are apparently adaptive, as in making leaves of greater protein content at low temperature, largely compensating for lower activity per unit mass of protein.

Leaves lose function gradually from causes such as UV damage and abruptly from causes such as salt accumulation.[8] Ultimately, even in the absence of lethal insults such as pathogens or frost, a leaf's death is programmed by internal and environmental signals (e.g., by the photoperiod as a reliable signal of the end of frost-free or drought-free days). Leaves thus can be evergreen or deciduous in response to drought or photoperiod. Leaf death is an orderly process. An abscission zone forms at the base, and typically half of the nutrients are scavenged, to be stored or used in newer leaves. Limits to the fraction scavenged are set by the need to maintain export functions.

ADDITIONAL MODES OF STRESS PROTECTION AND SPECIALIZATION

Leaves withstand wind stresses by drag reduction as well as by simple strength and toughness. The lamina must bend and fold properly, requiring specialized shapes.[3] Excess light or heat loading induces in selected species (e.g., grasses, legumes) alterations of leaf shape (e.g., rolling) and gross orientation (heliotropism). These changes complement the protection offered by reflective cuticles and pubescence. Tolerance of trampling in grasses is conferred in part by flexible and readily repaired vasculature. Leaves have other structures ancillary to photosynthesis but important nonetheless. These include salt glands for ion balance, trichomes, or hairs for physical and chemical defense.

Pathogens are deterred by many defenses, including cuticular integrity, chemical composition, and stomatal design. Drip tips in the wet tropics reduce wetness duration and thus fungal challenges. Herbivory by insects and vertebrates is deterred chemically and structurally, as noted earlier. Leaves of sclerophylls are tough and thick primarily for such defense during their long lifetimes; inelasticity is no particular advantage for handling turgor pressure changes in water stress.

Some leaves are modified for functions remote from photosynthesis. Some tendrils—which provide support for the whole plant—are modified leaves, as in the garden pea. Bulbs incorporate thick leaves modified for food storage; the petioles of celery function similarly. Leaves may be modified into spines for protection. Leaves may trap and digest insects, as in pitcher plants, sundews, and flytraps. These modifications are almost always an adaptation to low nutrient availability in soil.

TRAIT SELECTION AND THE NET VALUE OF A LEAF

What do these diverse leaf designs achieve under the diversity of growing conditions? One can estimate costs of constructing and maintaining a leaf in operation,[6,9] as well as current and lifelong-integrated benefits. A sine qua non for survival—much less for productivity—is that a leaf must pay back its own construction costs in photosynthate. This constrains the design lifetime, the nutrient content, and other features. Leaves must also be efficient in using scarce resources such as N and water. Conifers and sclerophylls appear to be selected for this. A conifer may have a low photosynthetic rate (1/4 that of a forb), but may maintain this rate over longer times (threefold) with less N (half as much). The conifer's N-use efficiency is (3/4)/(1/2) = 1.5-fold greater than the forb's.

A great many traits—structural, biochemical, and phenological—are thus under selection. Crop plants retain many traits and trait combinations selected under predomestication natural conditions. Artificial selection goals now conflict with fitness criteria of old; this conflict is important to understand in breeding improved crops.[6] An example is breeding for plants with erect leaves that share light interception well with other plants of the same stand; this design consciously forgoes competitor shading. In this task, it is well to consider that selection is diluted among many traits,[10] such that many tradeoffs in structure (as in specific leaf area) currently operate near neutrality; performance is nearly insensitive to variations near the current mean. Extreme events may be most important in selection. These are rarely monitored or quantified in importance in current breeding or crop ecology.

ARTICLES OF FURTHER INTEREST

Insect–Plant Interactions, p. 609
Leaf Cuticle, p. 635
Plant Defenses Against Insects: Constitutive and Induced Chemical Defenses, p. 939
Plant Defenses Against Insects: Physical Defenses, p. 944
Plant Response to Stress: Ultraviolet-B Light, p. 1019

REFERENCES

1. Hartley, S.L.; Jones, C.G. Plant Chemistry and Herbivory, or Why the World is Green. In *Plant Ecology*, 4th Ed.; Crawley, M.J., Ed.; Blackwell Science: Oxford, 1997; 284–324.
2. Kaiser, J. Tracking vanishing mammals and elusive nitrogen. Science **1998**, *281*, 1274–1275.
3. Vogel, S. *Cats' Paws and Catapults*; W.W. Norton: New York, 1998; 82–127.
4. Givnish, T.J. On the Adaptive Significance of Compound Leaves, in Particular Reference to Tropical Trees. In *Tropical Trees as Living Systems*; Tomlinson, P.B., Zimmermann, M.H., Eds.; Cambridge University: Cambridge, 1978; 351–380.
5. Gurevitch, J.; Scheupp, P.H. Boundary layer properties of highly dissected leaves: An investigation using an electrochemical fluid tunnel. Plant Cell Environ. **1990**, *13*, 783–792.
6. Gutschick, V.P. *A Functional Biology of Crop Plants*; Croom Helm: London, 1987.
7. Tardieu, F.; Granier, C. Quantitative analysis of cell division in leaves: Methods, developmental patterns and effects of environmental conditions. Plant Mol. Biol. **2000**, *43*, 555–567.
8. Munns, R. Physiological processes limiting plant growth in saline soils: Some dogmas and hypotheses. Plant Cell Environ. **1993**, *16*, 15–24.
9. Harper, J.L. The value of a leaf. Oecologia **1989**, *80*, 53–58.
10. Gutschick, V.P.; BassiriRad, H. Extreme events as shaping physiology, ecology, and evolution of plants: Toward a unified definition and evaluation of their consequences. New Phytol. **2003**, *160*, 21–42.

Leaf Surface Sugars

Johan H. J. Leveau
Netherlands Institute of Ecology (NIOO-KNAW), Heteren, The Netherlands

INTRODUCTION

Plant leaves attract a wide variety of organisms. For most of them, the appeal lies in the fact that the leaf is the primary site of photosynthesis, and thus represents an abundant source of carbohydrates such as glucose, fructose, and sucrose, as well as oligosaccharides (e.g., raffinose, stachyose, and verbascose) and sugar alcohols (e.g., mannitol and sorbitol). The waxy cuticle that covers the leaf effectively confines these compounds to the leaf's interior, but it cannot entirely prevent small amounts from finding their way to the outside. Furthermore, it offers only minimal protection against the attacks of herbivorous insects, by which substantial amounts of leaf sap can reach the leaf surface. Some plants actually invite the visitation of insects and other organisms by the release of highly localized puddles of sugar on their leaf surfaces. In this regard, the leaf surface, or phyllosphere, functions as an interface where sugars serve different purposes to the plant and to its visitors.

The first part of this article lists the shapes and abundances in which sugars can be found on the surfaces of plant leaves. In the second part, the fates and relevance of leaf surface sugars are discussed.

SOURCES OF LEAF SURFACE SUGARS

Leachates

Leaching is defined as the process by which organic and inorganic metabolites are lost from above-ground plant parts by leakage from the leaf interior. It is considered a passive and reversible process that is driven by the difference in metabolite concentrations across the leaf cuticle, i.e., between the inside of the leaf and the leaf surface. When there is water present on the leaf, for example, as drops of rain or dew, leaching to the surface is greatly stimulated, probably because the water acts as an effective sink for the leachates.

The most common method for assessing the occurrence and abundance of leachates is by repetitious rinsing of the leaf surface, usually with water. Such leaf washings contain an impressive collection of inorganic and organic substances.[1] The bulk of leachates consist of sugars, with fructose, sucrose, and glucose being most predominant. Usually, leachates are not found in very great abundance. For example, only about 2.5 μg of fructose, glucose, and sucrose can be washed off a greenhouse-grown bean plant leaf.[2] A leaf like this contains between 1 and 5 mg sugars on the inside, so that losses due to leaching account for only 0.05 to 0.25%. While such amounts may appear small on a per-leaf basis, they are quite impressive on a larger scale. One report details the loss of 800 pounds of carbohydrates per year per acre from an apple tree orchard as a result of leaching.[1]

There are numerous factors that influence the rate by which sugars are leached to the leaf surface; Tukey provides a comprehensive list.[1] One is the wettability of the leaf, which is often directly related to its waxiness. Leaves of pea, orange, and sugar beet are wetted with more difficulty than those of bean, squash, banana, cacao, and coffee, and hence are less susceptible to leaching. Leaf surface structures such as hairs and veins can hold water by capillary action and may in that way enhance leaching. Other important factors are the time, intensity, and amount of rain, mist, or dew that a plant is exposed to. In rainwater, for example, sugar concentrations have been measured in the range of tens of milligrams per liter, whereas in dew, which is more stagnant, they were 10 times higher. A continuous light drizzle increases leaching more than the same amount of water in a short, heavy rain. This may be explained partly by the increased runoff from leaves during heavy rain.

Extrafloral Nectaries

Extrafloral nectaries are secretory glands on leaves and stems of plants. They produce a liquid with nectarlike qualities, and they have been found on plants from over 1000 species. An extrafloral nectary can appear nonstructural, as a modified stipule, or as modified trichomes on the margins or surfaces of a leaf. Extrafloral nectaries secrete the sugars glucose, fructose, and sucrose, along with smaller amounts of protein and inorganics. The source of the sugars is the phloem, which often—but not always—ends near the nectary. Extrafloral nectar tends to be sweeter than phloem sap, in part because it is exposed to the air, and therefore more prone to evaporation. Extrafloral nectaries of broad bean (*Vicia faba*) secrete a nectar that can contain

Encyclopedia of Plant and Crop Science
DOI: 10.1081/E-EPCS 120010617

several hundreds of micrograms of sugar per µl. Leaves of the castor bean (*Ricinus communis*) produce an average of about 2 microliters of nectar per day, consisting of 75% sugars (sucrose, glucose, and fructose).

Guttation Fluid

Guttation is caused when the transpiration from a leaf is so low that the hydrostatic pressure in the xylem is able to force apoplastic fluid out to the leaf surface. This occurs in the early morning or late evening when humidity is high, temperature is low, and soil moisture is abundant. Guttation fluid follows the path of least resistance, and usually comes out of modified stomata called hydathodes. Depending on the plant species, hydathodes are located at the tips or margins of leaves, or on the leaf surface. Guttation fluids contain relatively low concentrations of sugar, usually anywhere from less than 1 µg per liter up to nearly 40 µg per liter. Guttation fluids from rye, wheat, and barley seedlings contain mostly glucose and galactose. A secondary effect of guttation fluids is leaf injury caused by an accumulation of salts when guttation droplets on the leaf surface evaporate.

Manna

Manna is a sugary substance that is excreted from leaves and stems of certain trees and eucalypts (*Eucalyptus* and *Angophora* species). It exudes after injury from incisions or from feeding by insects such as the gum tree bug (*Amorbus obscuricornus*). The exudate crystalizes and forms white beads that oftentimes fall to the ground. Manna generally consists of 60% sugars, most of which is raffinose, a trisaccharide of fructose, glucose, and galactose, with smaller amounts of other sugars such as melibiose (a disaccharide of glucose and galactose), sucrose, glucose, fructose, and stachyose (a tetrasaccharide of fructose, glucose, and two units of galactose). A major component of the manna from the ash tree (*Fraxinus ornus*) is the sugar alcohol mannitol, also called mannite or manna sugar. Manna may resemble phloem sap in sugar composition, but the proportional abundances of individual sugars can be markedly different, which in part may be due to the presence of sugar-specific enzymes such as sucrases and oxidases in the saliva of the feeding insect. Manna is oftentimes confused with other insect-mediated sources of leaf surface sugar. For instance, one hypothesis for the biblical reference to manna points to honeydew (see below).

Honeydew

Honeydew is the sugary excretion of nymphal stages of aphids, coccids, and insects belonging to the family *Psyllidae* (plant lice). These phloem feeders insert their stylets into the plant leaf and take advantage of the positive pressure of the phloem for the collection of leaf sap. Because phloem is high in sugars but low in nitrogen, the insect must ingest large amounts of phloem to meet its nitrogen demands. The excess phloem sugars pass through the gut of the insect and are excreted as waste. This honeydew is ejected with force from the end of the insect's abdomen, which prevents the insects from sticking to one another and to the leaf surface. Honeydew contains about one-third sugars, most of which are small polysaccharides with additional glucose, fructose, and sucrose. Other sugars such as melezitose, erlose, raffinose, and trehalose have been found in the honeydew of some sapfeeders. These sugars are synthesized from plant-derived sugars during passage through the insect gut, and they are thought to help the sapfeeder keep its osmotic balance while feeding.

Lerp

Lerp is a waxy material secreted as a protective scale by psyllids such as the lerp psyllid (*Glycaspis brimblecombei*). A lerp (the Australian Aborigine word for house) is a small conical structure under which the larva feeds and grows to adulthood. It is made by the larva itself from excreted honeydew, and is composed for 90% of glucose polymers.

Pollen and Fungal Spores

Pollen consists of up to 10% carbohydrates (dry weight), including starch and soluble sugars (fructose, glucose, sucrose). It is also a good source of protein and lipids. The diameter of a pollen grain is about 15–50 µm, which makes it highly motile in the wind. High concentrations of up to thousands of grains per square centimeter can easily be found deposited on leaf surfaces. The pollen can also be readily washed from leaves again, for example, by rain. Wind and rain also help disperse fungal spores that can be covered with a gelatinous coat containing considerable amounts of sugar.

FATES AND RELEVANCE OF LEAF SURFACE SUGARS

Food

Many organisms rely on plant leaf sugars for nutrition. For some, it is their main source of carbohydrates; for others, an occasional delicacy. Native Americans reportedly collected honeydew and ate it as candy. In Turkey, Iran, and Iraq honeydew is harvested from branches infested by aphids by shaking or beating the honeydew off. It is then dissolved in water, strained, mixed with eggs, almonds and spices, and sold as a popular sweet. Aboriginal Australians ate lerp as a staple food, and they prepared manna from the leaves of the eucalyptus tree.

Manna, honeydew, and lerp are also common food sources for a variety of bird species.[3] Their high carbohydrate content makes them very attractive sources of energy. Many of these birds have to supplement this diet by feeding on insects to meet their protein requirements. Many species of ants are readily attracted to the sugar in extrafloral nectaries, honeydew, and manna as sources of carbohydrates.[4,5] Forest- or leaf-honey is produced by bees from the honeydew of certain aphids and various other insects.

Microbes that visit or inhabit leaf surfaces have quantity requirements for sugars that are very different from those of humans, birds, or insects. For example, while a human may crave 300 grams of carbohydrates per day just for maintenance, a bacterium is able to replicate itself with 10^{15} times less, or a mere 0.3 picograms of glucose.[6] Obviously then, the tiny amounts of sugars that are found in leachates and guttation fluids are most relevant to such bacteria because they are sufficient to allow the generation of a substantial population size on plant leaves. For example, 3 micrograms of leached sugar could and often does produce 10 million bacterial cells on a single bean leaf under environmentally favorable conditions.

For many bacteria that have made the plant leaf surface their preferred habitat, sugars appear to be the limiting factor for growth. Not unexpectedly, many of these bacteria have adopted ways to gain access to more of the sweet stuff. One is the ability to lower the surface tension of water, thereby, for example, making rain drops on the leaf surface flatter so that the efficiency of leaching is increased as a result of the larger surface area.[7] Another bacterial adaptation is the ability to synthesize the plant hormone auxin, which is thought to trick the plant into the release of nutrients, including sugars. Because sugars are often a limiting factor in bacterial growth, they can evoke fierce competition.[8] Such competition is the basis for many of the strategies for the biological control of microbial plant pests.

Redistribution and Removal of Sugars from the Leaf Surface

Water that moves freely across the leaf surface is able to redistribute sugars and wash them off the leaf surface. These washed-off nutrients can land on lower leaves, where they can be taken up again by the plant, or on the ground below, where they can be used by the root system or other organisms. Such leachates may contain considerable amounts of nutrients; in one experiment, bean plants were grown through one complete generation on the leachates from squash leaves.[1]

Carriers of Information

Some organisms rely on leaf surface sugars as carriers of information during their interaction with the plant and with other organisms. For example, when choosing a spot to lay her eggs, a female of the european corn borer butterfly (*Ostrinia nubilalis*) relies on the surface sugars fructose, sucrose, and glucose to assess the carbohydrate content of the leaf.[9] Extrafloral nectaries attract ants, which in turn protect the plant by removing herbivorous insects. Parasitic wasps and flies use these extrafloral nectaries as meeting places for mates. Some plants have ways to control the amount, composition, and viscosity of this nectar and are so able to selectively attract or deter protectors or herbivores. The mutualistic interaction between aphids and ants is based on the attraction of ants to the honeydew.[4] For the aphid, removal of the sugary compound prevents itself and the leaf surface from becoming too sticky. Aphids also can choose their mutualists and antagonists by making the honeydew selectively attractive through the synthesis of honeydew-specific sugars, such as melezitose or erlose.

CONCLUSION

Because of its exposed character, the leaf is subject to many physical, chemical and biological influences that greatly affect the form and quantities of sugars found on its surface. This makes the leaf a highly dynamic and heterogeneous environment, for which the ecological role of sugars and changes in their abundance is not always easy to interpret or predict. This holds true in particular for microbial visitors to the leaf surface. For the most part this is due to the wholly different scale at which these microbes experience the leaf surface and its sugars. Because their carbohydrate requirements are so small, even small changes in sugar availability can have tremendous consequences. Furthermore, spatial variation in sugar availability is much more pronounced on the micrometer scale at which these microbes live. Only recently a successful attempt was made to follow the sugar consumption of individual bacteria in the phyllosphere.[6] From this study and more to come, an understanding will emerge that will enhance our appreciation for the role of sugars in microbial colonization of the leaf surface, and which could well help to improve current methods for biological control of plant pests.[10]

ARTICLES OF FURTHER INTEREST

REFERENCES

1. Tukey, H.B., Jr. Leaching of metabolites from aboveground plant parts and its implications. Bull. Torrey Bot. Club **1966**, *93* (6), 385–401.
2. Mercier, J.; Lindow, S.E. Role of leaf surface sugars in colonization of plants by bacterial epiphytes. Appl. Environ. Microbiol. **2000**, *66* (1), 369–374.
3. Gartrell, B.D. The nutritional, morphologic, and physiologic bases of nectarivory in australian birds. J. Avian Med. Surg. **2000**, *14* (2), 85–94.
4. Wäckers, F. It pays to be sweet: Sugars in mutualistic interactions. Biologist **2002**, *49* (4), 1–5.
5. Steinbauer, M.J. A note on manna feeding by ants (Hymenoptera: Formicidae). J. Nat. Hist. **1996**, *30* (8), 1185–1192.
6. Leveau, J.H.J.; Lindow, S.E. Apetite of an epiphyte: Quantitative monitoring of sugar consumption in the phyllosphere. Proc. Natl. Acad. Sci. U. S. A. **2001**, *98* (6), 3446–3453.
7. Knoll, D.; Schreiber, L. Plant–microbe interactions: Wetting of ivy (*Hedera helix* L.) leaf surfaces in relation to colonization by epiphytic microorganisms. Microb. Ecol. **2000**, *40* (1), 33–42.
8. Lindow, S.E.; Leveau, J.H.J. Phyllosphere microbiology. Curr. Opin. Biotechnol. **2002**, *13* (3), 238–243.
9. Derridj, S.; Boutin, J.P.; Fiala, V.; Soldaat, L.L. Primary metabolites composition of the leek leaf surface: Comparative study, impact on the host–plant selection by an ovipositing insect. Acta Bot. Gall. **1996**, *143* (2–3), 125–130.
10. Kinkel, L.L. This volume.

Leaves and Canopies: Physical Environment

George W. Sundin
Michigan State University, East Lansing, Michigan, U.S.A.

Cindy E. Morris
Institut National de la Recherche Agronomique, Montfavet, France

INTRODUCTION

The phyllosphere supports the growth of a diverse flora of bacteria and fungi; however, as a habitat for microbes, the phyllosphere presents a stressful environment that is strongly influenced by physical factors. Phyllosphere microbial residents grow through the utilization of the limited resources available in this habitat; survival is also predicated upon the ability of organisms to cope with the varied environmental stress conditions including fluctuating water availability, heat, osmotic stress, and exposure to solar UV radiation (UVR).

PHYSICAL NATURE OF THE LEAF SURFACE

Leaves are surrounded by a boundary layer of air that can influence temperature, relative humidity, and concentrations of gases and volatile metabolites relative to that of the ambient air. Within the leaf boundary layer, temperature can be as much as 4–7° different (generally greater) from that of the air[1] and RH as much as 20% above ambient conditions.[2] Humidity effects, especially the increased RH within thick closed canopies, tend to favor the epidemic development of some plant pathogens. Modification of the canopy environment of a crop, if possible, can affect disease incidence. For example, closed-canopy cultivars of bean are more susceptible to white mold than open-canopy cultivars.[3] The epidermal cells of leaves are covered by a cuticle. Cuticles are significant barriers to the diffusion of water-soluble compounds from the leaf apoplast toward the surface of the leaf and thereby probably foster oligotrophic conditions on leaf surfaces. Cuticles also influence the hydrophobicity of leaves, the reflectance of light, and leaf cooling. Leaf surfaces have topographical features such as leaf hairs, glandular trichomes, stomata, junctions of epidermal cells, hydathodes, and grooves over leaf veins, some of which are associated with thinner cuticles and boundary layers or may facilitate local accumulation of water. Terrestrial plants experience incessant fluctuations in the quantity of free water present on their leaves. Condensation and impaction of water droplets on leaf surfaces alternate with evaporation and runoff of water in cycles that are generally diurnal. The ebb and flow of free water in the phyllosphere lead to variations in local pH, as substances such as ammonia, sulfur dioxide, and calcium carbonate that may be present on leaf surfaces are dissolved, and also lead to variation in local concentrations of leaf surface nutrients.[4]

Rapid fluctuations in the physical environment of the phyllosphere are normal; those parameters of most importance to microbes include ultraviolet radiation (UVR), water and nutrient availability, temperature, and osmotic stress. The geography of leaf surfaces and the physiology of leaves of different ages co-occurring on the same plant also influence microbial growth in the phyllosphere habitat. The ecological view of leaves as islands and of each leaf as an individual habitat is instructive with particular reference to plant pathogen interactions and the potential for biological control.

SOLAR ULTRAVIOLET RADIATION

Leaf size, angle, and the density of plant canopies can markedly reduce UVR penetration to lower leaves; conversely, the relative humidity and free surface moisture within lower areas of dense canopies is typically maintained at higher levels. The abaxial (upper) surface of completely exposed leaves is one of the most light-intense of the terrestrial microbial habitats. Several reports have shown that microbial populations tend to be larger on adaxial surfaces,[5] although some fungal species can survive equally well on abaxial surfaces.[6] UVR filtering by leaves higher in the canopy can modulate microbial populations on lower leaves. The UVR-sensitive fungal plant pathogen *Exobasidium vexans* colonizes and incites disease symptoms on lower leaves of its tea host, and the fungus is otherwise highly sensitive to solar UVB radiation.[7] Solar UVR affects microbial populations on a daily basis;[8] however, long-term seasonal effects, although evident,[9] are more difficult to discern because of the confounding effects of temperature and desiccation.

Encyclopedia of Plant and Crop Science
DOI: 10.1081/E-EPCS 120010619

PLANT CANOPY EFFECTS

Plant canopies affect plant disease epidemiology, particularly in the rain-splash dispersal of fungal spores. Understanding the physics of rain splash and impacting droplets has been important to modeling studies of spore dislodging and movement.[10] Drop size plays a major role in the overall splash process.[11] Canopy effects include secondary movement of raindrops; "drip" drops, or secondary drops falling from leaves, travel at reduced speeds and play a reduced role in spore dispersal. Thus, thicker canopies tend to reduce spore dispersal possibly due to decrease in spore removal by drip drops or because fewer raindrops reach impaction sites, and by direct interception of spore-carrying droplets by intervening plant surfaces prior to spore "escape."[12] However, it should be noted that thicker canopies can have a positive effect on spore germination by lengthening the duration of leaf wetness. The momentum of rain is important in stimulating the growth of the plant-pathogenic bacterium *Pseudomonas syringae* in the bean phyllosphere.[13] Although the effect of plant canopy or canopy position is not known for this example, it is intriguing that the ecology of both plant pathogenic bacteria and fungi can be strongly influenced by rain, the effect of which is markedly conditioned by the canopy micro- and macroenvironment.

Leaves are also exposed to exogenous compounds that may be deleterious to bacteria. These include pesticides, the major air pollutants originating from the combustion of fossil fuels and from intensive agricultural practices, and heavy metals from smelting factories.[10] Exposure of leaves to agricultural chemicals utilized for pest or pathogen control is another area influenced by plant canopies. In these cases, uniform pesticide distribution is the desired outcome. However, this goal is typically achieved only in situations of low density of crop canopies.[14]

CONCLUSION

The physical environment of the phyllosphere and the role of environmental stress in modulating microbial phyllosphere populations remains a field of investigation still relatively early in its development. While environmental parameters have been examined singly, the integration of multiple environmental effects has been experimentally difficult and, for the most part, not undertaken. Because of the direct effect of the environment on phyllosphere microbes, and the influence of some of these microbes on plant health, an expanded understanding of the phyllosphere physical environment is a goal for a more integrated understanding of plant disease management.

REFERENCES

1. Grace, J. *Plant Response to Wind*; Academic Press: London, 1977; 204 pp.
2. Boulard, T.; Mermier, M.; Fargues, J.; Smits, N.; Rougier, M.; Roy, J.C. Tomato leaf boundary layer climate: Implications for microbiological whitefly control in greenhouses. Agric. For. Meteorol. **2002**, *110*, 159–176.
3. Steadman, J.R. White mould—A serious yield limiting disease of bean. Plant Dis. **1983**, *67*, 346–350.
4. Morris, C.E. Phyllosphere. In *Encyclopedia of Life Sciences*; Nature Publishing Group: London; http://www.els.net (accessed October 2001).
5. Beattie, G.A.; Lindow, S.E. The secret life of foliar bacterial pathogens on leaves. Annu. Rev. Phytopathol. **1995**, *33*, 145–172.
6. Moody, S.A.; Newsham, K.K.; Ayres, P.G.; Paul, N.D. Variation in the responses of litter and phylloplane fungi to UV–B radiation (290–315 nm). Mycol. Res. **1999**, *103*, 1469–1477.
7. Gunasekera, T.S.; Paul, N.D.; Ayres, P.G. The effects of ultraviolet B (UV–B: 290–320 nm) radiation on blister blight disease of tea (*Camellia sinensis*). Plant Pathol. **1997**, *46*, 179–185.
8. Sundin, G.W.; Jacobs, J.L. Ultraviolet radiation (UVR) sensitivity analysis and UVR survival strategies of a bacterial community from the phyllosphere of field-grown peanut (*Arachis hypogeae* L.). Microb. Ecol. **1999**, *38*, 27–38.
9. Jacobs, J.L.; Sundin, G.W. Effect of solar UV–B radiation on a phyllosphere bacterial community. Appl. Environ. Microbiol. **2001**, *67*, 5488–5496.
10. Madden, L.V. Rainfall and the dispersal of fungal spores. Adv. Plant Pathol. **1992**, *8*, 39–79.
11. Fitt, B.D.L.; McCartney, H.A.; Walklate, P.J. The role of rain in dispersal of pathogen inoculum. Annu. Rev. Phytopathol. **1989**, *27*, 241–270.
12. Boudreau, M.A.; Madden, L.V. Effects of strawberry density on dispersal of *Colletotrichum acutatum* by simulated rain. Phytopathology **1994**, *84*, 934–941.
13. Hirano, S.S.; Baker, L.S.; Upper, C.D. Raindrop momentum triggers growth of leaf-associated populations of *Pseudomonas syringae* on field-grown snap bean plants. Appl. Environ. Microbiol. **1996**, *62*, 2560–2566.
14. Travis, J.W.; Skroch, W.A.; Sutton, T.B. Effect of canopy density on pesticide deposition and distribution in apple trees. Plant Dis. **1987**, *71*, 613–615.

Leaves and the Effects of Elevated Carbon Dioxide Levels

Stephen A. Prior
USDA—ARS National Soil Dynamics Laboratory, Auburn, Alabama, U.S.A.

Seth G. Pritchard
Belmont University, Nashville, Tennessee, U.S.A.

G. Brett Runion
USDA—ARS National Soil Dynamics Laboratory, Auburn, Alabama, U.S.A.

INTRODUCTION

Since the onset of the Industrial Revolution, the use of fossil fuel and the clearing of land have led to a dramatic rise in the concentration of carbon dioxide (CO_2) in the atmosphere. This increase has significant implications for green plants because CO_2 is the prime input to photosynthesis. Leaves are the first point of contact for the majority of carbon transfer from atmosphere to biosphere. This capture of carbon by leaves enables the growth of plants, which play a vital role in maintaining the environment and form the base of agriculture. Humans and other organisms depend on this thin layer of leaves for food. Leaves collect highly dilute CO_2 from the atmosphere and concentrate and transform it into organic compounds. These compounds ultimately combine to form useful products (food and fiber). Leaf litter decomposition (by microbes) is a key process in agroecosystems through its role in carbon and nutrient cycling and storage. Leaves and their functions are pivotal in agroecosystems (and are sensitive to environmental change, especially rising CO_2); this article provides a short overview.

PHYSIOLOGICAL PROCESSES IN LEAVES

Changes in atmospheric CO_2 level have been shown to impact two major functions of leaves—photosynthesis and transpiration.[1–3] Crops grown under CO_2 enrichment usually exhibit increased mass, which is attributed to more photosynthetic capacity and enhanced water use efficiency (ratio of CO_2 assimilated to water transpired). There are, however, differences among plants. For example, plants with a C_3 photosynthetic pathway often exhibit a greater response to high CO_2. The CO_2-concentrating mechanism in C_4 species limits the response, but C_4 plants can nevertheless exhibit growth stimulation because elevated CO_2 can increase water use efficiency. The C_3 plant generally benefits from both increased photosynthesis and water use efficiency. Increases in water use efficiency are notable, although the combined effects of CO_2 on decreasing stomatal conductance and increasing leaf area and leaf temperature have resulted in modest reductions in estimated whole plant water use.[3]

An often reported consequence of enhanced photosynthesis is the accumulation of nonstructural carbohydrates.[1,2] Photosynthetic acclimation (a decline in photosynthesis over time) can be related to the buildup of nonstructural carbohydrates in leaves. The coupling of carbohydrate accumulation in leaves and acclimation is probably related to source-sink imbalances, with sink strength an important controlling factor. Some evidence shows a correlation between rooting volume and photosynthetic capacity as well as limitations in resource availability (e.g., soil nitrogen) influencing source-sink activity.[1,2] Increases in soluble carbohydrate concentration may repress photosynthetic gene expression, leading to reductions in photosynthetic pigments (e.g., chlorophyll and carotenoids) and important soluble proteins, including rubisco and antioxidant enzymes.[4] The photosynthetic apparatus may be more susceptible to photodamage due to decreased carotenoid content. There are some indications that increases in carbohydrates and decreases in protein are associated with a slight suppression of growth respiration and construction costs of leaves.[5] Rubisco represents a sizable portion of leaf nitrogen and plays a significant role in the reduction of foliar nitrogen concentration commonly found under high CO_2 conditions.[1,2] Reductions in antioxidant enzyme activities observed under high CO_2 may be reflective of less oxidative stress resulting from growth in CO_2-enriched atmospheres.[4] Furthermore, antioxidant systems can differ according to plant genotype, suggesting differential resistance to biotic and abiotic stresses.

LEAF STRUCTURE AND DEVELOPMENT

Changes in the internal structure of leaves can occur under elevated CO_2 conditions.[6] Consideration of leaf

Encyclopedia of Plant and Crop Science
DOI: 10.1081/E-EPCS 120010613

structural changes may help interpret divergent findings with regard to photosynthetic acclimation. Fixation of CO_2 occurs at specific sites within the chloroplast and, under high CO_2, starch accumulation may change chloroplast structure and function (e.g., altered chloroplast integrity may limit carbohydrate transport to other organs). Both photosynthetic and assimilate transport capacity may also be altered by CO_2-induced shifts in mesophyll and vascular tissue. Leaves grown under elevated CO_2 have exhibited an extra layer (a third layer) of palisade cells as well as increased total mesophyll cross-sectional area and vascular tissue area. Stomatal density has been shown to decrease with elevated CO_2 in some, but not all cases. A further alteration is changes in epicuticular wax on high CO_2-grown leaves.[7,8] Little is known about how CO_2 will affect trichomes. Changes in the leaf surface could affect water relations, susceptibility to pests and diseases, and surface properties that are important in chemical protectant application.

Plants grown under high CO_2 often exhibit increased area per leaf, greater leaf thickness, more leaves per plant, and higher total leaf area per plant,[6] but effects on leaf area index are variable.[2,3] Exposure to elevated CO_2 is thought to have little impact on rates of leaf initiation. Reported increases in leaf thickness and decreases in specific leaf area (leaf area/total leaf dry weight) are often the result of altered anatomy or increased starch accumulation. Although highly variable, increases in cell expansion may contribute to larger leaf size more than increased cell division. Increased expansion appears to result from greater cell wall relaxation and/or greater cell turgor. There is little information on how high CO_2 will affect leaf shape or duration.

ENVIRONMENTAL CONSIDERATIONS

CO_2-induced increases in leaf production and changes in leaf composition can significantly affect other trophic level organisms that utilize leaf tissue as a food substrate. The nitrogen concentration of green leaf material is often lower under high CO_2, whereas leaf carbon:nitrogen ratio, nonstructural carbohydrates, and secondary defense compounds can be increased. Effects on senescent leaf quality are more variable, but tend to be lower in magnitude.[1,3,5] Changes in green leaves are relevant to phyllosphere organisms (fungi and bacteria living on leaf surfaces) and grazers (insects and livestock). The reproductive success of insects may be altered by diets of high CO_2-leaf tissue. However, the sparseness of information prevents generalization concerning the life cycle strategies of these organisms in a high CO_2 world. Information on this topic is critical for effective pest management (rates, timing, and number of pesticide applications may change under higher CO_2 environments). Trace gas emission (e.g., methane) associated with cattle production is related to low forage quality, thus any CO_2-induced downward shift in leaf nitrogen concentration may be significant.

Senesced leaves are a food source for soil organisms and a major component of carbon inputs for agroecosystems.[9] Leaf litter input can be increased by high CO_2, but decomposition rates will vary by crop species. Some decomposition work suggests that CO_2-enriched cropping systems may store more carbon, decomposition may be limited by nitrogen, and nitrogen release from litter may be slowed. A more thorough understanding of carbon and nitrogen cycling is needed to predict the potential for soil carbon storage in agroecosystems.[1,9] Green leaf material can also contribute to litter production in systems using cover crops. Plant growth stage and leaf tissue nitrogen need to be considered for cover crop termination to optimize resource availability (nitrogen and/or water) to the following crop. There are some indications that CO_2 can alter crop phenology,[3] which may alter kill time for cover crops and subsequent timing of farm operations, including planting. Another complication is that herbicide efficacy can vary by weed species under elevated CO_2.[10] This has important implications for future weed management (frequency of spraying/rate adjustments), including control of invasive species, and also raises questions about herbicide efficacy on cover crops.

CONCLUSION

Leaf responses to elevated atmospheric CO_2 will need to be further elucidated if we are to accurately understand and predict the development and growth of crops under future environmental conditions. Future research will need to address:

1. How alterations in leaf structure are related to both cellular and higher-level growth processes;
2. Whether increased cell division is driven by greater rates of cell expansion or by molecular cues;
3. The role of ultrastructural, anatomical, and morphological leaf adjustments in photosynthetic acclimation;
4. How CO_2-induced changes in leaf structure and leaf function will alter the degree of impact of multiple crop stresses (abiotic and biotic);
5. How CO_2-induced shifts in leaf quality and/or changes in leaf surface features (e.g., epicuticular waxes, trichomes) will alter herbivore and pathogen attacks;
6. Mechanisms for herbicide tolerance of plants grown under high CO_2;
7. How CO_2 alters leaf nutrient composition of cover crops over time (growth stage and phenology); and

8. With regard to leaf residue (green and senesced) supplying carbon and nitrogen to the soil, how elevated CO_2 will affect available crop nitrogen and the residence time of leaf-derived carbon in the soil.

ARTICLES OF FURTHER INTEREST

Air Pollutants: Interactions with Elevated Carbon Dioxide, p. 17
Crop Responses to Elevated Carbon Dioxide, p. 346
Crops and Environmental Change, p. 370
Ecophysiology, p. 410
Environmental Concerns and Agricultural Policy, p. 418
Leaf Cuticle, p. 635
Leaf Structure, p. 638
Leaves and Canopies: Physical Environment, p. 646
Photosynthate Partitioning and Transport, p. 897
Starch, p. 1175
Sucrose, p. 1179
Trichomes, p. 1254
Water Deficits: Development, p. 1284
Water Use Efficiency Including Carbon Isotope Discrimination, p. 1288

REFERENCES

1. Amthor, J.S. Terrestrial higher-plant response to increasing atmospheric [CO_2] in relation to the global carbon cycle. Glob. Chang. Biol. **1995**, *1*, 243–274.
2. Drake, B.G.; Gonzalez-Meler, M.A. More efficient plants: A consequence of rising atmospheric CO_2? Annu. Rev. Plant Physiol. Plant Mol. Biol. **1997**, *48*, 609–639.
3. Kimball, B.A.; Kobayashi, K.; Bindi, M. Responses of agricultural crops to free-air CO_2 enrichment. Adv. Agron. **2002**, *77*, 293–368.
4. Pritchard, S.G.; Ju, Z.; Van Santen, E.; Qui, J.; Weaver, D.B.; Prior, S.A.; Rogers, H.H. The influence of elevated CO_2 on the activities of antioxidative enzymes in two soybean genotypes. Aust. J. Plant Physiol. **2000**, *27*, 1061–1068.
5. Poorter, H.; Van Berkel, Y.; Baxter, R.; Den Hertog, J.; Dijkstra, P.; Gifford, R.M.; Griffin, K.L.; Roumet, C.; Roy, J.; Wong, S.C. The effect of elevated CO_2 on the chemical composition and construction costs of leaves of 27 C_3 species. Plant Cell Environ. **1997**, *20*, 472–482.
6. Pritchard, S.G.; Rogers, H.H.; Prior, S.A.; Peterson, C.M. Elevated CO_2 and plant structure: A review. Glob. Chang. Biol. **1999**, *5*, 807–837.
7. Prior, S.A.; Pritchard, S.G.; Runion, G.B.; Rogers, H.H.; Mitchell, R.J. Influence of atmospheric CO_2 enrichment, soil N, and water stress on needle surface wax formation in *Pinus palustris* (Pinaceae). Am. J. Bot. **1997**, *84*, 1070–1077.
8. Thomas, J.F.; Harvey, C.N. Leaf anatomy of four species grown under continuous CO_2 enrichment. Bot. Gaz. **1983**, *144*, 303–309.
9. Torbert, H.A.; Prior, S.A.; Rogers, H.H.; Wood, C.W. Elevated atmospheric CO_2 effects on agro-ecosystems: Residue decomposition processes and soil C storage. Plant Soil **2000**, *224*, 59–73.
10. Ziska, L.H.; Teasdale, J.R.; Bunce, J.A. Future atmospheric carbon dioxide may increase tolerance to glyphosate. Weed Sci. **1999**, *47*, 608–615.

Legumes: Nutraceutical and Pharmaceutical Uses

J. Bradley Morris
USDA—ARS, PGRCU, University of Georgia, Griffin, Georgia, U.S.A.

INTRODUCTION

Legumes are one of the three largest families of flowering plants, with approximately 690 genera and about 18,000 species. Many legumes are currently used for nutraceutical and pharmaceutical purposes, whereas others contain medically important phytochemicals with potential for further use.

The legumes currently recognized by the Pharmacist's letter/Prescriber's letter natural medicines comprehensive database for nutraceutical effectiveness include guar [*Cyamopsis tetragonloba* (L.) Taub.], soybean [*Glycine max* (L.) Merr.], winged bean [*Psophocarpus tetragonolobus* (L.) DC.], and fenugreek [*Trigonella foenum-graecum* L.]. Several other legumes are currently recognized for pharmaceutical effectiveness.

NUTRACEUTICALS

Many other legumes are or contain phytochemicals that are possibly effective or have been shown to have potential therapeutic effects. One of these is field beans (*Phaseolus* spp.), which provide one of the most important sources of dietary protein for human consumption. These legumes are rich in the amino acids lysine and tryptophane. Dry adzuki beans [*P. angularis* (Willd.) W. Wight] contain about 22% protein. Mung bean (*P. aureus* Roxb.) seed contains about 23.6% protein and 3.3% fiber. Rice bean (*P. calcaratus* Roxb.) pulse contains about 21.7% protein and 5.2% fiber. Lima (also known as butter bean) (*P. lunatus* L.) pulse contains 20.7% protein and 4.3% fiber. The most popular and widely used bean is the common bean known as French bean, kidney bean, runner bean, snap bean, and string bean. These bean pods have been shown to be possibly effective when used orally as a supportive treatment for the inability to urinate.[1,2] In addition, *P. vulgaris* L., winged bean, jicama [*Pachyrrhi zus erosus* (L.) Urban], *Lablab purpureus* (L.) Sweet, soybean, and kudzu [*Pueraria montana* var. *lobata* (Willd.) Maesen & S. Almeida] guar [Cyamopsis tetragonoloba (L.) Taub] are excellent sources of fiber (Fig. 1).[3] Winged bean leaves, flowers, shoots, immature pods, immature dry seeds, and tubers are edible and highly nutritious. The mature dry seeds are especially nutritious because of their high protein content (30–42%). However, the seeds can be soaked for 10 hours and then boiled for 30 minutes to rid them of any antinutritional components. Winged bean contains oil consisting of tocopherols that are antioxidants. These antioxidants improve the utilization of vitamin A in the human body.

Peanut (*Arachis hypogaea* L.) seed is rich in oil, containing about 53% oleic acid and 25% linoleic acid with about 30% protein. Peanut contains two important antioxidants: protocatechuic acid and lecithin.[3] Peanut oil is potentially useful for lowering cholesterol and preventing heart disease; however, allergic reactions are fairly common among the human population.[1] Aesculetin found in the peanut plant has been shown to be anti-inflammatory and a cancer preventive.[3] Lecithin exists in the peanut seed and has been found to range from 5000 to 7000 ppm. Lecithin has been shown to have antioxidant activity and some activity related to antialzheimeran.[3] It also appears to be possibly effective for treating gall bladder disease and hypercholesterolemia.[1]

Kudzu is used in traditional Chinese medicine for managing alcoholism, fever, cold, flu, and neck stiffness.[1] Kudzu contains other important phytochemicals–namely, isoflavones including daidzin, daidzein, and puerarin. Both daidzein and daidzin decrease alcohol consumption in rats.[4] A few products containing kudzu extracts are on the market; however, studies are needed to investigate the effects of kudzu in humans. The product known as Estroven, marketed by Amerifit, contains kudzu root isoflavones; a nutraceutical known as kudzu root is marketed by General Nutrition Center (GNC).

Velvetbean [*Mucuna pruriens* (L.) DC.] seed extract contains L-dopa, which is used as an antiparkinsonian.[3] Sabinsa Corporation markets velvetbean seed extract in powder form for use as a nutraceutical (Fig. 2).

Astragalus [*Astragalus membranaceus* (Fisch.) ex Link] has shown some preliminary evidence that it can reduce the chance of developing the common cold. The

Encyclopedia of Plant and Crop Science
DOI: 10.1081/E-EPCS 120010422

Fig. 1 Functional foods containing guar gum. (*View this art in color at www.dekker.com.*)

active components appear to be in the root, consisting of saponins, flavonoids, polysaccharides, and coumarins.[5] It is an antioxidant and improves the immune response by potentiating the effects of interferon.[1] Several *Vigna* spp. including [*V. aconitifolia* (Jacq.) Marechal], *V. angularis* (Willd.) Ohwi & H. Ohashi, *V. mungo* (L.) Hepper, *V. radiata* (L.) R. Wilczek, and *V. subterranean* (L.) Verdc. contain fiber from seed ranging from 9000 to 79,591 ppm. Plant fiber ranges from 31,700 to 286,000 ppm, whereas fruit fiber ranges from 9000 to 299,000 ppm in *V. aconitifolia*, *V. mungo*, and *V. subterranean*. Sprout seedling fiber ranges from 7200 to 94,500 ppm in *V. radiata*.[3] See Table 1 for therapeutically effective nutraceuticals.

PHARMACEUTICALS

See Table 2 for currently recognized legumes with pharmaceutical effectiveness. The following discussion will include additional effective qualities for the phytopharmaceuticals listed in Table 2.

Beta-sitosterol found in jackbean [*Canavalia ensiformis* (L.) DC.] is a plant sterol similar to cholesterol; however, artherogenesis does not occur with beta-sitosterol because less than 5% is actually absorbed. Beta-sitosterol significantly improves urinary symptoms, increases maximum urinary flow, and decreases postvoid residual urine volume. It also significantly reduces total and low-density lipoprotein (LDL) cholesterol levels, but has little or no effect on high-density lipoprotein (HDL) cholesterol levels.[1] Pectin is possibly effective for lowering cholesterol.[1] Currently most of the therapeutic qualities from pectin have been discovered in citrus and apples. However, a substantial portion has been identified in the sunn hemp [*Crotalaria juncea* (L.)] stem.[3] Guar gum (Fig. 3) is possibly effective for reducing triglycerides.[1] Abbott Laboratories markets Ensure Glucerna, in which the key ingredient is guar gum for helping people with diabetes (Fig. 4). Several legumes have or contain potentially useful phytochemicals with pharmaceutical properties. Luteolin found in [*Chamaecrista mimosoides* (L.) E. Greene] and kaempferol from [*Sesbania sesban* (L.) Merr.] have anti-inflammatory, antioxidant, and cancer-preventive potential. Pinitol from [*Macrotyloma uniflorum* (Lam.) Verdc.] has antidiabetic and expectorant possibilities. Lecithin found in velvet bean (Fig. 5) and other legumes is likely effective when used for reducing hepatic steatosis in long-term parenteral nutrition patients and when used topically for dermatitis and dry skin.[1] Some noteworthy discoveries for cancer treatments include plant-derived isoflavones. Genistein has thus far been shown to be more abundant in subclover (*Trifolium subterraneum* L.) than soybean or red clover (*Trifolium pratense* L.).[6] Clinical trials have revealed that genistein inhibits the growth of both MCF-7 breast cancer cells and human mammary epithelial cells in vitro, whereas biochanin, another isoflavone found in subclover, is just as effective against these tumor cells but is five times less potent in the inhibition of normal cells.[7] Furthermore, genistein has been shown to inhibit bladder cancer.[8] Mimosine found in Leucaena species (*Leucaena* spp.) blocks cell cycle progression in MDA-MB-453 human breast cancer cells.[9]

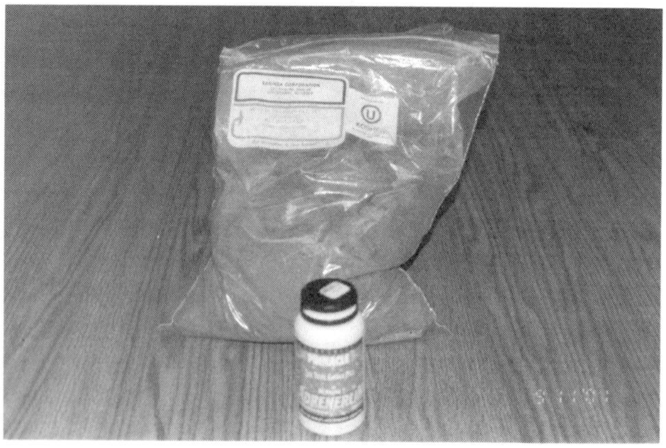

Fig. 2 Nutraceutical and pharmaceutical products from velvetbean. (*View this art in color at www.dekker.com.*)

Table 1 Legumes with therapeutically effective nutraceuticals

Taxon	Common name	Nutraceutical	Use	Product[a]/Company
Cyamopsis tetragonoloba	Guar	Guar gum	Cholesterol reduction, fiber source, antidiabetes	Multifiber complex/Natrol Guar gum/Atrium, Inc. Centrum Kids/Whitehall-Robbins Healthcare Acutrim Natural P.M./Heritage Consumer Products Acutrim Natural A.M./Heritage Consumer Products Fibersol/TwinLab
Glycine max	Soybean	Soy	Cholesterol reduction	One-A-Day Menopause Health/Bayer One-A-Day Cholesterol Health/Bayer One-A-Day Bone Strength/Bayer
Glycine soja	Soybean	Soybean oil	Parenteral nutrient	Coenzyme Q10/Leiner Health Products
Psophocarpus tetrgonolobus	Winged bean	Fiber		
Trigonella foenum-graecum	Fenugreek		Lowers blood sugar in diabetics	

[a]Many of these products contain the mentioned nutraceutical as an ingredient.
(From Refs. 1 and 3.)

Table 2 Legumes with therapeutically effective pharmaceuticals

Taxon	Common name	Phytochemical	Use	Product[a]/Company
Canavalia ensiformis	Jack bean	Beta sitosterol[a]	Prostatic hyperplasia, lowers cholesterol	Aspen-Maximal Sterol Complex/Aspen Group, Inc. Phytosterol Complex/ Progressive Labs
		Rutin[a]	Treating osteoarthritis	C-Complex 1000/ Puritan's Pride
Crotalaria juncea	Sunn hemp	Pectin	Treat diarrhea	
Cyamopsis tetragonolobus	Guar	Guar gum	Laxative	
Glycine max	Soybean	Lecithin	Reduces hepatic steatosis	Glycobar: Peanut butter and jelly/Pharmanex Lecithin-E-Nutrilite/Nutrilite Daily Essentials for Men/ Health Smart Vitamin
Medicago sativa	Alfalfa	Saponins	Lowers cholesterol	
Melilotus officinalis	Sweet clover	Coumarin	Decreases pain	
Senna septemtrionalis	Senna	Quercetin	Treats chronic, nonbacterial prostatitis	
Trifolium pratense	Red clover	Isoflavones	Prevent osteoporosis	Promensil/Novogen
Trigonella foenum-graecum	Fenugreek		Lowers blood sugar in diabetics	

[a]Found in jackbean; however, companies develop beta-sitosterol and rutin from other plant species.
(From Ref. [1].)

Fig. 3 Productive guar growing in a field (Griffin, GA). (*View this art in color at www.dekker.com.*)

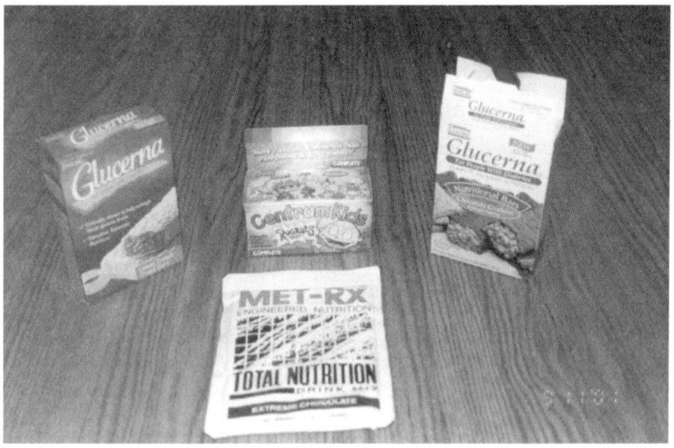

Fig. 4 Nutraceuticals containing guar gum. (*View this art in color at www.dekker.com.*)

Fig. 5 Productive velvetbean growing in a field (Griffin, GA). (*View this art in color at www.dekker.com.*)

CONCLUSION

Both nutraceutical and pharmaceutical products have been made from legumes. Many legumes are currently known to be therapeutically effective, whereas others are undergoing research to ascertain whether they are effective in benefiting health-related problems. The use of such natural products has stimulated the interest of the general public because of the health-promoting properties of legumes. These possibilities make legumes very important for the future—and not just a hill of beans.

ARTICLES OF FURTHER INTEREST

Agriculture and Biodiversity, p. 1
Echinacea: Uses as a Medicine, p. 395
Genetic Resources of Medicinal and Aromatic Plants from Brazil, p. 502
New Secondary Metabolites: Potential Evolution, p. 818
Phytochemical Diversity of Secondary Metabolites, p. 915
Secondary Metabolites as Phytomedicines, p. 1120

REFERENCES

1. *Pharmacist's Letter/Prescriber's Letter Natural Medicines Comprehensive Database,* 3rd Ed.; Jellin, J.M., Gregory, P., Batz, F., Hitchens, K., Eds.; Therapeutic Research Faculty: Stockton, CA, 2000; 1–1529.

2. *The Complete German Commission E Monographs: Therapeutic Guide to Herbal Medicines*; Blumenthal, M., Ed.; American Botanical Council: Boston, MA, 1998. Trans. S. Klein.
3. Beckstrom-Sternberg, S.M.; Duke, J.A. *The Phytochemical Database*; http://ars-genome.cornell.edu/cgi-bin/WebAce/webace?db=phytochemdb. (Data version July 1994.)
4. Lin, R.C.; et al. Effects of isoflavones on *alcohol* pharmacokinetics and alcohol-drinking behavior in rats. Am. J. Clin. Nutr. **1998**, *68*, 1512S–1515S. (6 Suppl.)
5. Leung, A.Y.; Foster, S. *Encyclopedia of Common Natural Ingredients Used in Food, Drugs and Cosmetics,* 2nd Ed.; John Wiley and Sons: New York, NY, 1996.
6. Duke, J.A. A Mini Course in Medical Botany. In *Syllabus. Module 2: Phytochemicals (Mineral, Phytamins, and Vitamins).* http://www.ars-grin.gov/duke/syllabus/module2.htm.
7. Peterson, T.G.; Coward, L.; Kirk, M.; Falany, C.N.; Barnes, S. The role of metabolism in mammary epithelial cell growth inhibition by the isoflavones genistein and biochanin A. Carcinogenesis **1996**, *17*, 1861–1869.
8. Theodorescu, D.; Laderoute, K.R.; Calaoagan, J.M.; Guilding, K.M. Inhibition of human bladder cancer cell motility by genistein is dependent on epidermal growth factor receptor but not p21ras gene expression. Int. J. Cancer **1998**, *78*, 775–782.
9. Kulp, K.S.; Vulliet, P.R. Mimosine blocks cell cycle progression by chelating iron in asynchronous human breast cancer cells. Toxicol. Appl. Pharmacol. **1996**, *139*, 356–364.

Lesquerella Potential for Commercialization

David A. Dierig
USDA—ARS, Phoenix, Arizona, U.S.A.

Michael A. Foster
Texas A & M University, Pecos, Texas, U.S.A.

Terry A. Isbell
USDA—ARS, Peoria, Illinois, U.S.A.

Dennis T. Ray
University of Arizona, Tucson, Arizona, U.S.A.

INTRODUCTION

Lesquerella produces a triglyceride oil that is two-thirds or more lesquerolic acid, a hydroxy fatty acid (HFA). Hydroxy fatty acids are now used in nylon-11 and nylon-6,10, lithium greases, coatings, sulfated and sulfonated oil, sebacic acid, ethoxylated oil, food grade lubricants, polyurethanes, and cosmetics. The current source of HFAs is ricinoleic acid from castor (*Ricinus communis* L.), which was formerly cultivated in the United States. Approximately 41,000 million tons are now imported, primarily from Brazil and India, at a value exceeding $100 million per year. Lesquerella is native to the United States. and could be established as a domestic HFA source and reduce reliance on castor oil importation.

Lesquerella has several novel properties that set it apart from castor and other oilseeds. One property, is its oil functionality, including difunctional hydroxy moities (in contrast to other seed oils such as castor that are trifunctional). Lesquerella has been reported to contain natural estolides (secondary esters derived from the addition of a fatty acid moiety to the hydroxyl functionality of the hydroxy triglyceride) that improve performance of the oil. Estolides have been encountered in only a few other seed oils, and are not naturally found in castor oil (although they can be synthetically fabricated). Estolides have been shown to improve vegetable oil performance in motor oils by improving pour points (temperature at which the oil no longer pours), and estolides have also been used as viscosity modifiers in lubricating oils. The second of lesquerella's novel properties is its antioxidants, converted from the seed meal fraction that contain unique glucosinolates and have superior oxidative stability properties. The seed meal also has a binder application, unique for seed meals. The third novel property is the seed coat of lesquerella, which contains a unique gum that has rheological properties equivalent to guar or xanthan that can be useful in coatings and food thickeners.

MARKET POTENTIAL

Large markets exist for hydroxylated oils as feedstock for lithium greases, polymers in paints and coatings, basestocks in lubricants, and in applications in the personal care industry.[1] Lesquerella could be established as a reliable domestic supply of oilseed for all these markets and simultaneously provide an alternative crop for farmers that would indirectly increase local profits. The unique chemical structure of lesquerella oil offers distinct advantages for the development of new applications.[2]

Increasing attention to environmental issues drives the lubricant industry to increase the ecological friendliness of its products.[3] The industry has, for the last decade, been trying to formulate biodegradable lubricants with technical characteristics superior to those based on mineral oil (petroleum). Volumes of lubricants, especially engine oils and hydraulic fluid, are relatively large and most are petroleum based. Hydraulic fluids are consumed at 5 million metric tons (MT)/year in the United States, and have the highest need for biodegradable lubricants.[3]

Until now, petroleum-based lubricants have been superior to biodegradable vegetable oils because of their lower pour points and higher oxidative stability.[3] Some vegetable oils (such as soybean) require up to 30% synthetic additives to improve their performance. However, the triglyceride estolides that are found in species of the genus *Lesquerella* overcome this vegetable oil deficiency by blocking unsaturation and providing branches that lead to lower pour points.[4] In fact, vegetable-based lubricants with synthetic additives still do not approach the properties of estolides. The commercialization of lesquerella

Encyclopedia of Plant and Crop Science
DOI: 10.1081/E-EPCS 120010432

oil, with its natural estolides, could lead to the development of a lubricant derived directly from the seed, significantly lowering processing costs. The lower pour point of estolides could also allow soybean oil lubricants to be formulated without synthetic additives. Tests indicate a pour point between −15 and −20°C and a cloud point of −1°C for commercial products made from soybean oil. The potential exists that estolide additives from lesquerella (pour point −33°C, cloud point −37°C) could be used to lower the pour and cloud points of soybean-based lubricants that do not presently match the performance of petroleum-based products (pour points of −21 to −36°C) currently on the market.

Lesquerella glucosinolate, 3-(methylsulfinyl)propyl glucosinolate, is unique in that it breaks down in the meal enzymatically when the seed meal is mixed with water to form 1-isothiocyanato-3-(methylsulfinyl) propane almost exclusively. The isothiocyanate is the precursor to thiourea antioxidants made with natural amines. The isothiocyanate also is a by-product of lesquerella oil deodorization. In both cases the recovered isothiocyanate can be easily converted to an amine and reacted back with the isothiocyanate to form unique thiourea antioxidants.

POTENTIAL FOR YIELD INCREASES

A wealth of genetic variability has been provided for a lesquerella-plant-breeding program that allows for improvements in both plant growth characteristics and oil quantity and quality. Improved germplasm has been developed in the past few years and publically released through the Agricultural Research Service, U.S. Water Conservation Laboratory (USWCL).[5] Another accomplishment has been the development of hybrids among *L. fendleri* and related species. *L. fendleri* has a limited amount of variation for seed oil lesquerolic acid content (HFA). Other species have much higher HFA content seed oil (89% compared to 54%), but lower seed yields. This trait was found to be maternally inherited in hybrids; a series of backcrosses have allowed the transfer of this trait into *L. fendleri*.

CURRENT PRODUCTION TECHNIQUES

Lesquerella is a low-growing herbaceous shrub with much less biomass for disposal than other oilseed crops. Standard farm equipment with minor modifications can be used for all routine farming practices. Lesquerella should be planted in the fall with planting dates ranging from August 15 to September 1 in Texas and New Mexico, and October 1 to 15 in Arizona. Harvests will occur 8 to 9 months later in the spring.

Lesquerella requires 25 to 26 inches of irrigation water for maximum seed yield, as compared to 30 to 40 inches for cotton.[6] Nitrogen/acre amounts of 50 to 100 lb. increased lesquerella biomass and seed yields.[7] Lesquerella can be efficiently harvested by direct combining. Maximum seed yields have been obtained with a planting density of about 400,000 plants/acre (approximately 5.0 to 6.0 lb. of seed/ac).[8] Crop rotations have been studied and there have been no reports of allelopathy.

Because lesquerella will be grown in irrigated arid and semiarid regions of the southwest United States, a salt-tolerant line has been developed[9] that is able to grow in these areas where drainage effluents are reused and where the crop could benefit marginal agricultural lands. In addition, lesquerella has been shown to be a potential phytoremediator of selenium (Se)-contaminated soils, because the plant can accumulate this harmful trace element in its leaves and stems. High levels of Se found in areas such as the San Joaquin Valley of California pose health threats to humans, livestock, and wildlife.[10] Following harvest, the seed oil would be utilized for the proposed industrial purposes described earlier, and the bioaccumulated plant parts that remain could be removed from the area.

In the production areas where lesquerella is likely to be grown, control of weeds is extremely important. In fact, weed control could be a major factor limiting establishment of commercially-viable lesquerella production in the United States Lesquerella is slow to establish, making the crop less competitive with grass and broadleaf weeds that emerge in crop rows and cannot be removed mechanically. For lesquerella to be produced economically at industrial scale on commercial growers' fields, herbicides must be registered. Several economical and environmentally safe herbicides for this crop have been identified in extensive field research studies.[11]

CONCLUSION

Agricultural research and extension personnel, USDA/ARS scientists, and agribusiness representatives have identified 21 counties in Arizona, New Mexico, and Texas believed to hold the greatest potential for lesquerella commercialization.[12] Irrigated acreage in these counties totals almost 1.4 million. Lesquerella will not necessarily replace current commodity crops but will be placed in rotation (e.g., a 2-year, 3-crop rotation of lesquerella, grain sorghum, and cotton). Over the long term, as lesquerella markets strengthen, acreage increases, and net returns remain competitive with other crops, lesquerella could replace commodity crops with negative net returns. Wheat, for example, produced on 320,000 acres in 17 of the 21 aforementioned counties, reported the most crop budgets with estimated negative net returns. Abandoned cropland in the Trans Pecos region of west Texas alone accounts for another potential growing base of 180,000 acres in addition to irrigated acres. Soil erosion on lesquerella plantings will not be an issue because the crop provides soil cover during the fall and winter, and stubble can be left on the land after harvest.

Because new industrial products unique to lesquerella will be developed, lesquerella would not be expected to replace castor in the event it is reintroduced as a U.S. crop. Therefore, domestic supplies of HFAs would not depend on a single crop. The two crops could be produced in different seasons—lesquerella would be planted in the fall and harvested in late spring, whereas castor would be planted in the spring and harvested in late fall. There would be rotational opportunities for both crops. Castor has a high water requirement, and would not be produced in arid and semiarid environments where lesquerella is best suited. Multiple production regions would provide a more reliable domestic supply of HFA. This in turn could encourage further research and new product development, thus opening additional markets for both crops.

ARTICLES OF FURTHER INTEREST

Crop Domestication in Prehistoric Eastern North America, p. 314
Genetically Modified Oil Crops, p. 509
Germplasm Acquisition, p. 537
Natural Rubber, p. 778

REFERENCES

1. Roetheli, J.C.; Carlson, K.C.; Kleiman, R.; Thompson, A.E.; Dierig, D.A.; Glaser, L.K.; Blase, M.G.; Goodell, J. *Lesquerella as a Source of Hydroxy Fatty Acids for Industrial Products*; Growing Industrial Material Series, USDA/CSRS, Office of Agricultural Materials, 1991.
2. Thompson, A.E.; Dierig, D.A.; Johnson, E.R. Yield potential of *Lesquerella fendleri* (Gray) Wats., a new desert plant resource for hydroxy fatty acids. J. Arid Environ. **1989**, *16*, 331–336.
3. Erhan, S.Z.; Asadauskas, S. Lubricant basestocks from vegetable oils. Ind. Crops Prod. **2000**, *11*, 277–282.
4. Isbell, T.A.; Edgecomb, M.R.; Lowery, B.A. Physical properties of estolides and their ester derivatives. Ind. Crops Prod. **2001**, *13*, 11–20.
5. Dierig, D.A.; Tomasi, P.M.; Dahlquist, G.H. Registration of WCL-LY2 high oil *Lesquerella fendleri* germplasm. Crop Sci. **2001**, *41*, 604.
6. Hunsaker, D.J.; Alexander, W.L. *Lesquerella* seed production: Water requirement and management. Ind. Crops Prod. **1998**, *8*, 167–182.
7. Nelson, J.M.; Watson, J.E.; Dierig, D.A. Nitrogen fertilization effects on lesquerella production. Ind. Crops Prod. **1999**, *9*, 163–170.
8. Brahim, K.; Ray, D.T.; Dierig, D.A. Growth and yield characteristics of *Lesquerella fendleri* as a function of plant density. Ind. Crops Prod. **1998**, *9*, 63–71.
9. Dierig, D.A.; Shannon, M.C.; Grieve, C.M. Registration of WCL-SL1 salt tolerant *Lesquerella fendleri* germplasm. Crop Sci. **2001**, *41*, 604–605.
10. Grieve, C.M.; Poss, J.A.; Suarez, D.L.; Dierig, D.A. Saline irrigation water composition affects growth, shoot-ion content and selenium uptake of lesquerella. Ind. Crops Prod. **2001**, *13*, 57–65.
11. Foster, M.A. In *Pesticide Registration in New Crops: Herbicides for Lesquerella Production*, Association for the Advancement of Industrial Crops, Annual Meeting Abstracts, 22–25, September 1996: San Antonio, TX, 1996.
12. Van Dyne, D.L. *Comparative Economics of Producing Lesquerella in Various Areas of the United States*; Evans, M., Decker, D., Eds.; Industrial Uses of Agricultural Materials-Situation and Outlook Report, USDA/ERS: Washington, DC, 1997.

Lipid Metabolism

Jean-Claude Kader
Université Pierre et Marie Curie, Paris, France

INTRODUCTION

Lipids can be defined as a group of fatty-acid-derived compounds soluble in nonaqueous solvents. They constitute a broad family of molecules, which is particularly large in higher plants since more than 300 different fatty acids (FAs) have been characterized. Lipids are important constituents of plant cell membranes and are involved in physiological processes linked to the responses of plants to environmental conditions, to defense reactions, and to intracellular signaling pathways.[1] Lipids are also an important form of carbon storage in seeds. Oilseeds are among the most ancient crops, cultivated for thousands of years as sources of products used for food and for industrial products such as cosmetics or lubricants. Both uses have led to an increase in the demand for oilseeds, and world oilseed production (mainly soybean, rapeseed, sunflower, palm, cottonseed, and peanut) has expanded in the last 50 years from 160 to 300 million tons per year.[2] In order to improve the quality of plant oils (seeds and fruits), it is important to increase our knowledge of the metabolism of plant lipids. It is the aim of this article to summarize the lipid biosynthetic pathways and to present the exciting perspectives offered by the metabolic engineering of oils.

STORAGE AND MEMBRANE LIPIDS HAVE DIFFERENT FATTY ACID COMPOSITION

The major lipids of plant cell membranes are phospholipids (mainly phosphatidylcholine, phosphatidylethanolamine, and phosphatidylinositol) and, in the plastid membranes, galactolipids (mainly monogalactosyldiacylglycerol). It is remarkable that the main FAs present in these membrane lipids are also found in triacylglycerols (TAGs), which are the storage lipids. These FAs are predominantly palmitate (16:0) (the first number refers to the number of carbon atoms, the second to the number of double bonds), stearate (18:0), oleate (18:1), linoleate (18:2), and linolenate (18:3). Seeds contain, in addition to these common acyl chains, a very large diversity of FAs referred to as ''unusual.'' These FAs vary by their chain length, from 8 to 14 carbons in *Lauraceae* or from 20 to 24 carbons in *Brassicaceae*. Other oils accumulate FAs with double bonds at positions not usually found in membrane lipids such as petroselinic acid (16:1) or acyl chains containing hydroxy, epoxy, or acetylenic functional groups. The reason for such a large diversity is still unknown, but the fact that these FAs are excluded from membranes and accumulate in TAGs help plants to tolerate high levels of these unusual lipids. These FAs have important industrial applications (lubricants, paints, detergents, cosmetics) due to their unique chemical properties. However, the production of such compounds at a high level is not possible since their sources are plants that are not adapted to classical agricultural practices. This led to the idea to introduce genetic engineering approaches to adapt common crops to industrial needs.[3,4]

AN INTRACELLULAR COOPERATION IS NEEDED FOR LIPID BIOSYNTHESIS

It is remarkable that the biogenesis of plant glycerolipids needs a complex intracellular cooperation between the plastid and the endoplasmic reticulum. Two distinct pathways are operative: the ''prokaryotic'' one, localized within the plastid, and the ''eukaryotic'' pathway, which requires cooperation between the two cell compartments.[5] The elucidation of these metabolic pathways is the result of early biochemical studies and of approaches of molecular biology and molecular genetics, which, in the case of the model plant *Arabidopsis thaliana*, have allowed the characterization of mutants deficient in enzymes of lipid metabolism. In less than 10 years, these investigations resulted in the cloning and characterization of the key enzymes of the lipid pathways. A survey of genes for plant lipid metabolism has recently been published by searching the public databases for sequences coding for 65 polypeptides involved in lipid biosynthesis and degradation.[6]

PLASTIDS PLAY A CENTRAL ROLE IN FATTY ACID SYNTHESIS

In the stroma of plastids, the first steps of fatty acid biosynthesis are carried out (Fig. 1). The stepwise condensation of C_2 units to a growing acyl-chain attached

Encyclopedia of Plant and Crop Science
DOI: 10.1081/E-EPCS 120010400

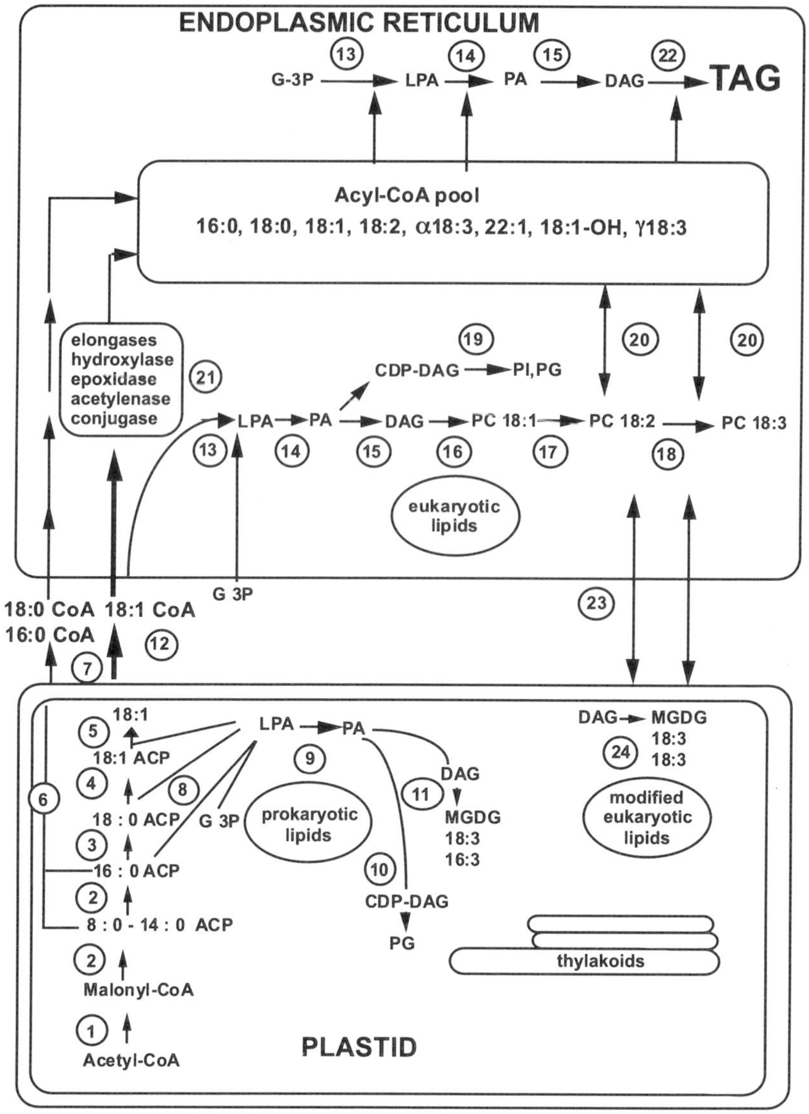

Fig. 1 *A simplified scheme of plant lipid biosynthesis.* An acetyl-CoA carboxylase (1) catalyzes the formation of malonyl-CoA. The elongation process is catalyzed by a fatty acid synthase complex (2). Acyl chains shorter than 16 carbons are formed in some plants. An additional elongation step catalyzed by a β-ketoacyl synthase (3) and a desaturation mediated by stearoyl-ACP desaturase (4) lead to oleoyl-ACP, which can be partly cleaved by a thioesterase (5) to form free fatty acids (FAs). Another thioesterase (6) is more active with shorter acyl-chains. The free FAs are esterified to CoA (7) and exported. However, some acyl chains of ACPs are used within the plastid for the prokaryotic pathway involving a glycerol-3-phosphate acyltransferase (8) and lyso-phosphatidate acyltransferase (9), thus allowing the formation of phosphatidylglycerol (PG) (10) or monogalactosyl-diacylglycerol (MGDG), which is of prokaryotic type (16 carbons at *sn*-2 position) (11). The eukaryotic pathway involves an export of the various acyl-CoAs to the endoplasmic reticulum by an unknown mechanism (12). Acyl-CoAs are inserted into membrane phospholipids, mainly phosphatidyl-choline (PC), through glycerol-3-phosphate acyltransferase (13), lysophosphatidate acyl-transferase (14), phosphatidate phosphatase (15), and CDP-choline : 1,2 diacylglycerol-choline phosphotransferase (16). The FAs of PC are desaturated (17,18). Some of the FAs are used for the synthesis of other phospholipids (19). PC molecules exchange their FAs with those of the acyl-CoA pool by the action of acyl-CoA lyso-PC-acyltransferase (20). The acyl-CoA pool also comprises other FAs modified by various enzymes, depending on the plant species. The FAs allow the synthesis of triacylglycerols (TAGs) through a series of reactions (13–16) including one step unique to TAG formation, the diacylglycerol acyltransferase (22). The final step of the eukaryotic pathway is the reentry of the FAs into the plastid (23) where they form eukaryotic lipids, mainly MGDG esterified by 18 carbons in both positions (24).

to acyl-carrier protein (ACP) leads to the formation of saturated acyl chains with palmitoyl-ACP (16:0-ACP) as a major product. However, shorter acyl chains could be formed depending on the plant species. The group of enzymes involved in fatty acid synthesis is called fatty acid synthase (FAS), which is, in animals and yeast, a multifunctional enzyme complex characterized by large subunits. By contrast, FAS from plants includes at least 12 separate proteins functioning rather like a metabolic pathway. A further step is the elongation of 16:0-ACP to 18:0-ACP by a $\beta-$ketoacyl synthase (the mutant *fab1* of *Arabidopsis* is deficient in this enzyme) followed by a desaturation by a stearoyl-ACP-desaturase (corresponding mutant *fab2*), which introduces a double bond at carbon 9 and forms oleoyl-ACP. This desaturase, soluble in the stroma, is unique among all other acyl-desaturases, which are integral membrane proteins. A key step is then termination, which is catalyzed by thioesterases that hydrolyze the acyl-ACPs to produce free FAs. These FAs are exported from the plastid to the endoplasmic reticulum and participate in the eukaryotic pathway. Some of the acyl-ACPs are used within the plastid and enter into the prokaryotic pathway, allowing the synthesis of plastidial membrane lipids such as phosphatidylglycerol, galactolipids, or sulfolipids.[1,5] The importance of the prokaryotic pathway depends on the plant species and is minor in oilseeds.

THE ENDOPLASMIC RETICULUM IN DIVERSITY OF PLANT LIPIDS

The exported FAs are involved in the eukaryotic pathway and are esterified to CoA. The acyl-CoAs join the acyl-CoA pool, which is used partly for the production of membrane lipids such as phosphatidylcholine and other phospholipids. In addition, the diacylglycerol moiety of phosphatidylcholine partly returns to the plastid by an unknown mechanism and is used for the synthesis of modified eukaryotic plastidial lipids, mainly galactolipids that have 18:3 in both *sn* positions. These galactolipids differ from the prokaryotic species, which have 16:3 at one of the two positions. The other part of the acyl-CoA pool is used to synthesize TAGs, a process highly active in oilseeds. Several enzymes are involved in these pathways, such as membrane-bound desaturases (oleate desaturase (also named FAD2) and linoleate desaturase (FAD3)) and various types of acyltransferases. One step unique to TAG synthesis is the final one catalyzed by a diacylglycerol transferase specific to the third carbon of the glycerol backbone. Depending on the plant species, other specific enzymes may operate, such as elongases, epoxidases, hydroxylases, and acetylenases. This enzyme diversity explains the broad spectrum of TAGs found in oilseeds. TAGs accumulate in subcellular organelles called oil bodies and are hydrolyzed to provide energy for germination and other processes.[7] The intensity of the flux of acyl moieties between the two cell membranes and the channeling of FAs toward either membrane or storage lipids depend on the plant species or the tissue considered.

WHAT ARE THE GENETIC ENGINEERING TARGETS?

The improvement of seed oil quality requires either molecular markers to help breeding or cloned genes to manipulate their expression in planta.[4,7] The two main directions are to modify the level of expression of a gene of interest or to introduce a foreign gene into a common crop to broaden the range of FAs offering industrial applications. Some of the successful experiments were to improve the quality of edible oils by increasing the proportion of monounsaturated FAs. This was achieved, for example, in soybean by underexpressing the oleate-desaturase. Another successful strategy was to increase the amount of saturated FAs, used for margarines and shortenings, by antisense RNA suppression in rapeseed of a gene coding for stearoyl-ACP desaturase. The introduction of a foreign gene was successfully achieved for lauric acid (12:0), which is used for the production of detergents. High laurate oil, commercially sold from 1995, was obtained by expressing a thioesterase isolated from California Bay that terminates the elongation process at a length of 12 carbons. Other FAs interesting for lipochemistry are erucic acid (22:1), a product of an elongase complex in rapeseed; ricinoleic acid (18:1-OH), formed in castor bean by a hydroxylase, and γ-linolenic acid (γ-18:3), synthesized in borage by a specific type of desaturase. These FAs have been produced in model plants by expressing genes corresponding to these enzymes. The latter enzymes belong to the vast family of desaturases (FAD2 type) or are biochemically close to them. A site-directed or random mutagenesis strategy of desaturases offers perspectives for enzyme engineering.[7]

CONCLUSION

It is easy to conclude that the knowledge of plant lipid metabolism, which has greatly progressed in the last decade, offers exciting perspectives for improving oil quality for food and industrial applications and for producing novel compounds that are renewable and environmentally friendly. Additional examples are the synthesis of polyhydroxybutyrate, a biodegradable plastic, by expressing in plants several bacterial genes or the development of oils suitable for use as lubricants or sources of fuels. Future research will be devoted to understanding the role of membrane lipids in the responses of plants to environmental factors or to pathogens. Other promising investigations will be to study the involvement of phosphoinositides in the signal transduction pathways and the participation of fatty acid derivatives (oxylipins, jasmonate) and enzymes (phospholipases, lipoxygenases) in these important processes.

ARTICLES OF FURTHER INTEREST

Biosafety Programs for Genetically Engineered Plants: An Adaptive Approach, p. 160

Functional Genomics and Gene Action Knockouts, p. 476

Genetically Modified Oil Crops, p. 509
Plant Response to Stress: Phosphatidic Acid as a Secondary Messenger, p. 995
Transgenic Crop Plants in the Environment, p. 1248
Transgenic Crops: Regulatory Standards and Procedures of Research and Commercialization, p. 1251

REFERENCES

1. Somerville, C.; Browse, J.; Jaworski, J.G.; Ohlrogge, J.B. Lipids. In *Biochemistry and Molecular Biology of Plants,* 1st Ed.; Buchanan, B., Gruissem, W., Jones, R., Eds.; American Society of Plant Physiologists: Rockville, 2000; 456–527.
2. Murphy, D.J. Production of novel oils in plants. Curr. Opin. Biotechnol. **1999**, *10*, 175–180.
3. Millar, A.A.; Smith, M.A.; Kunst, L. All FAs are not equal: Discrimination in plant membrane lipids. Trends Plant Sci. **2000**, *5*, 95–101.
4. Somerville, C.R.; Bonetta, D. Plants as factories for technical materials. Plant Physiol. **2001**, *125*, 168–171.
5. Miquel, M.; Browse, J. Arabidopsis lipids: A fat chance. Plant Physiol. Biochem. **1998**, *36*, 187–197.
6. Mekhedov, S.; Martinez de Ilatrduya, O.; Ohlrogge, J.B. Toward a functional catalog of the plant genome. A survey of genes for lipid biosynthesis. Plant Physiol. **2000**, *122*, 389–402.
7. Voelker, T.; Kinney, A.J. Variations in the biosynthesis of seed-storage lipids. Annu. Rev. Plant Physiol. Plant Mol. Biol. **2001**, *52*, 335–361.

Male Gametogenesis

M

David Honys
Academy of Sciences of the Czech Republic, Praha, Czech Republic

David Twell
University of Leicester, Leicester, U.K.

INTRODUCTION

The haploid male gametophytes of higher plants play a vital role in plant fertility and crop production through the generation and transport of the male gametes to ensure fertilization and seed set. There has been an evolutionary tendency toward reduction of the male gametophyte and its increasing functional dependence on the sporophyte. This trend is most acute within flowering plants, such that the male gametophyte consists of just two or three cells when shed as pollen grains. Despite its diminutive form, the functional specialization of the male gametophyte is thought to be a key factor in the evolutionary success of flowering plants through mechanisms that promote rigorous selection of superior haploid genotypes and outbreeding.

This article describes the sequential phases of angiosperm pollen development—microsporogenesis and microgametogenesis—emphasizing the vital role of the cell wall interface and sporophytic-gametophytic interactions. This article further describes recent progress in genomewide studies of haploid gene expression, genetic approaches that are being used to identify genes required for key cellular processes, and aspects of pollen biotechnology in crop improvement.

POLLEN DEVELOPMENT: FROM MICROSPOROCYTE TO MATURE POLLEN

Microsporogenesis and microgametogenesis take place inside the anther loculi that are lined by the tapetal cell layer (tapetum). Microsporogenesis is initiated upon meiotic division of the diploid pollen mother cell (microsporocyte) that produces four haploid microspores composing a tetrad[1,2] (Fig. 1). During microgametogenesis, microspores released from the tetrads undergo cell expansion, cell wall synthesis, asymmetric division, and differentiation of the vegetative and generative cells before partial dessication and release from the anther. The tapetal cells play a major role in pollen development through their contribution to microspore release, nutrition, pollen wall synthesis, and pollen coat deposition. Disturbance of tapetal cell functions usually results in reduced pollen fertility or male sterility through a variety of mechanisms, including arrest of microgametogenesis at the microspore stage or altered pollen hydration through modified pollen coat composition.

Microsporogenesis

A unique feature of the walls surrounding the microsporocytes and newly formed microspores within the tetrad is that they consist largely of callose, a β-1-3-glucan. The callose wall is secreted by microsporocytes before meiosis I and separates the microspores within the tetrad following meiosis II (Fig. 1). Microspores begin to synthesize the first elements of the sculptured outer pollen wall layer (exine), starting with primexine that functions as a template for subsequent exine elaboration. When young microspores are still developing the exine within the tetrad, an enzyme complex (callase) is secreted by the tapetal cells, allowing individual microspores to be released from the tetrads. Correct timing of callase secretion is critical because premature or delayed dissolution of the callose wall results in male sterility.[3]

Microgametogenesis

Once released, free microspores increase in size and their multiple small vacuoles enlarge and fuse into a single large vacuole, occupying most of the volume of the cell. In concert, the microspore nucleus migrates to a peripheral position that is required for the subsequent asymmetric division at pollen mitosis I (PMI)[4,5] (Fig. 1). PMI results in two morphologically and functionally distinct cells—a large vegetative cell and a small generative cell. The generative cell subsequently becomes engulfed within a membrane-bound compartment in the cytoplasm of its vegetative sister. This involves dissolution of the hemispherical callose wall separating the vegetative and generative cells, inward migration, and membrane fusion events. The asymmetric division at PMI is a key determinative event in generative cell fate.[6]

Encyclopedia of Plant and Crop Science
DOI: 10.1081/E-EPCS 120012916

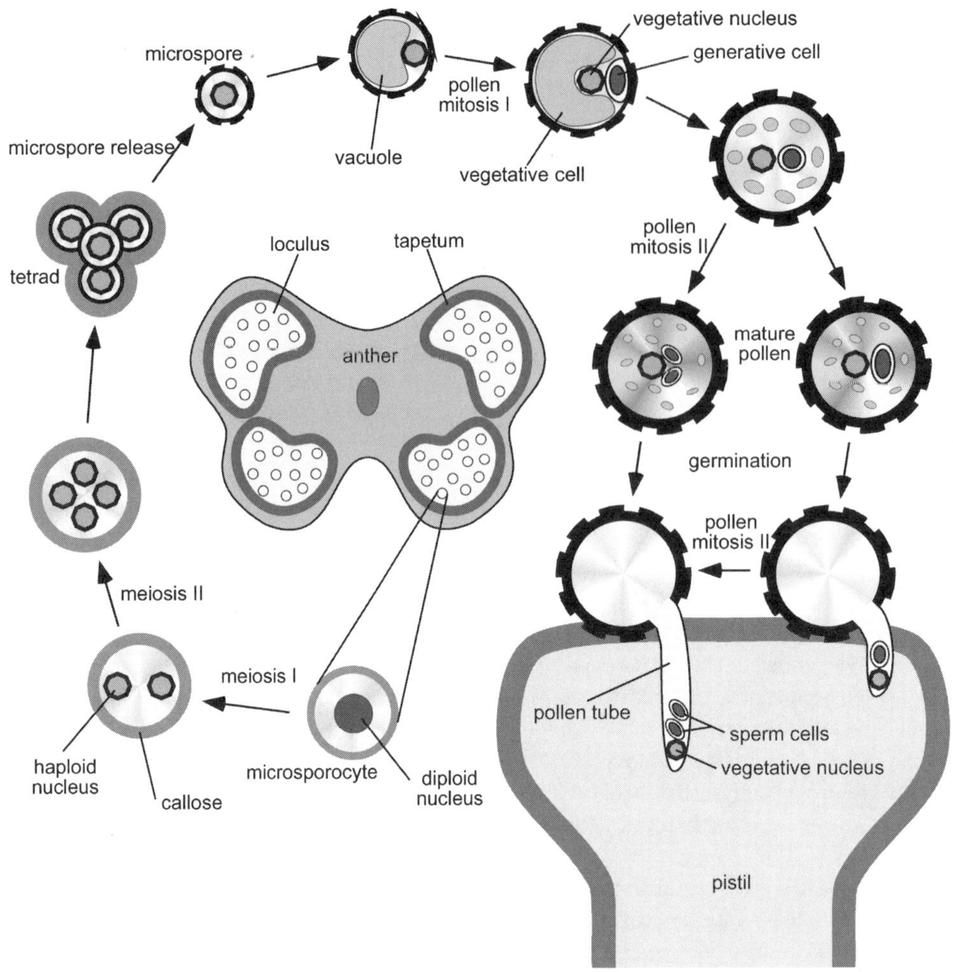

Fig. 1 Schematic diagram illustrating pollen development. (*View this art in color at www.dekker.com.*)

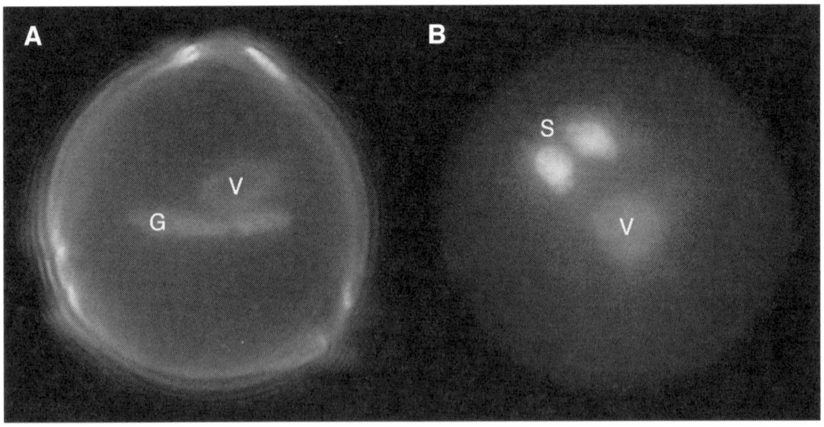

Fig. 2 Bicellular and tricellular pollen. (A) Bicellular tomato and (B) tricellular oilseed rape pollen-stained with the DNA stain DAPI. Nuclear DNA within the vegetative (V) and generative (G) or vegetative (V) and sperm cells (S) are highlighted. (*View this art in color at www.dekker.com.*)

In microspores and immature pollen cultivated in vitro, gametophytic development can be switched to a sporophytic pathway by heat stress and/or starvation treatment, leading to microspore embryogenesis and haploid plant formation.[2] Such techniques are routinely applied to accelerate breeding programs through the rapid generation of double haploid plants and selection among large numbers of homozygous lines.

The generative cell undergoes further mitotic division at pollen mitosis II (PMII) to produce the two sperm cells. In tricellular pollen this division occurs within the anther, whereas in bicellular pollen it occurs within the growing pollen tube. Although the majority of flowering plants produce bicellular pollen, many important food crop plants such as rice, wheat, and maize produce advanced but often short-lived tricellular pollen grains (Fig. 2).[2]

During pollen maturation the vegetative cell accumulates considerable carbohydrate and/or lipid reserves that are transient or are stored in the mature pollen grain. Transient reserves are thought to provide metabolites for energy-demanding developmental events such as asymmetric division and pollen cell wall (intine) synthesis. Osmoprotectants including proline also accumulate in mature pollen grains to protect vital membrane and proteins from damage. In mature pollen grains the extensive stores of lipids and polysaccharides are required to supply the extensive demands for plasma membrane and pollen tube wall synthesis.

During dehydration, the final phase of pollen maturation, pollen grains are finally prepared for release from the anthers. This represents an adaptation to survive exposure to the hostile terrestrial environment. The extent of dehydration varies widely in different species; for example, in poplar the water content is reduced to only 6%, maize loses 50%, whereas cucumber pollen remains fully hydrated. The degree of dehydration and the levels of cytoplasmic reserves positively correlate with pollen fitness and viability. Hydrated pollen is very susceptible to dehydration stress and generally survives only a few hours, whereas fully dehydrated pollen may survive for months or even years under certain conditions.

THE POLLEN WALL—A VITAL INTERFACE

The unique activities and biological role of the pollen grain are reflected in the unique composition of the pollen wall. The pollen wall and its coatings isolate and protect the male gametophyte and its associated gametes and mediate the complex communication with the stigma surface. The pollen wall consists of an inner intine and outer exine layer. Its synthesis begins at the microspore stage, when the pectocellulosic intine and the primexine are formed. The primexine serves as a matrix for subsequent deposition of sporopollenin. Sporopollenin is a highly resistant biopolymer containing fatty acids and phenylpropanoids; its synthesis involves tight cooperation between microspore cytoplasm and tapetal cells. The exine is not evenly distributed over the pollen grain surface, and regions lacking sporopollenin form apertures that are used as sites for pollen tube emergence. The number and size of apertures and exine patterning are under strict sporophytic control.

The formation of pollen coatings is completed at later stages of microgametogenesis. Remnants of degenerating tapetal cells are deposited onto the pollen grain surface creating the pollen coat. The pollen coat is involved in pollen–pistil signalling, self-incompatibility, pollen hydration, adhesivity, color, and odor. The yellow or purple colors of mature pollen grain results from the presence of both carotenoid and phenylpropanoid compounds. These features, as well as the elaborate patterning of sporopollenin, are highly variable among different plant species. In animal-pollinated species, pollen is often decorated with elaborate structures that facilitate vector adhesion, whereas in wind-pollinated species, pollen lacks such sculpturing or may be decorated with air sacs to increase buoyancy (Fig. 3).

HAPLOID GENE EXPRESSION

Development of the male gametophyte is associated with an extensive haploid gene expression program. To date, approximately 150 pollen-specific genes falling into 50 functional classes have been cloned from various species.[4,7] Recent genomewide studies using microarray hybridization technology have comprehensively demonstrated the scale and diversity of haploid gene expression in *Arabidopsis thaliana*.[8] Mature *Arabidopsis* pollen grains express approximately 5000 different mRNA species out of more than 27,000 predicted from the *Arabidopsis* genome. Approximately 40% of these transcripts are predicted to be preferentially or specifically expressed in pollen. Moreover, there is significant overlap between sporophytic and male gametophytic gene expression that reflects the large proportion of genes that are required for basic cellular functions. The most abundant classes of pollen-specific genes are predicted to have functions associated with transcriptional regulation, signal transduction, cytoskeleton organization, and cell wall synthesis. This highlights the importance of these functions for the unique cellular specialization required for pollen differentiation and function.

Interestingly, some of these pollen-specific genes encode proteins representing major allergens, the cause of hayfever and allergic asthma. Recent work has shown that the expression of one class of allergens from ryegrass

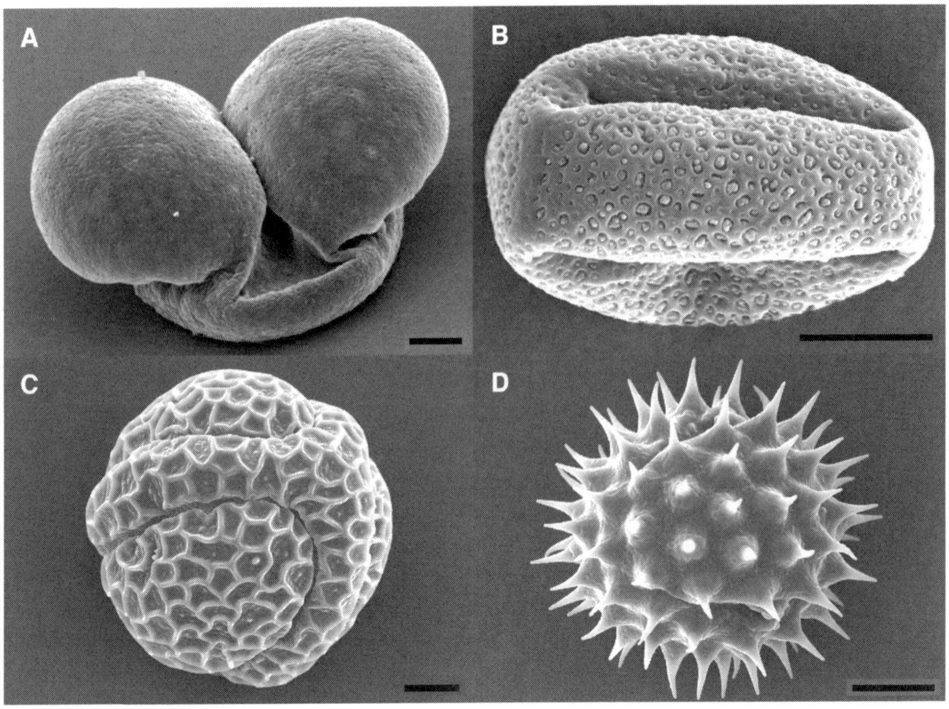

Fig. 3 Pollen morphology and wall patterning. Scanning electron micrographs of (A) pine, (B) papaya, (C) passion flower, and (D) sunflower pollen. Bar length = 10 μm.

pollen can be reduced without significant impact on plant fertility,[9] indicating that the genetic engineering of hypoallergenic cultivars is possible.

Pollen-expressed genes have also been classified according to their temporal regulation. In addition to constitutively expressed genes, groups of early genes active in microspores and late genes acting after PMI have been identified. The late-pollen genes and their regulation have been most intensively studied, resulting in the functional dissection of several pollen-specific transcriptional and translational regulatory elements.[7]

GENETIC APPROACHES: MUTANTS AND DEVELOPMENT

Renewed interest in genetic analysis and the use of the model species *A. thaliana* have had a significant impact on the identification of gametophytic genes controlling pollen development. Successful mutant selection approaches have been devised that involve screening for aberrant pollen morphogenesis or marker-segregation ratio distortion in populations of plants treated with chemical or physical mutagens or by T-DNA or transposon-mediated mutagenesis.[4]

Morphogenesis screens have led to the identification of a number of mutants with novel phenotypes that affect processes throughout microgametogenesis (Table 1). These include mutants that disturb asymmetric cell division at pollen mitosis I,[6,10,11] division of the generative cell division,[4,6] the positioning and structure of the male germ unit,[12] and the repression of pollen germination within the anther.[13] Among these, the *gemini pollen1* (*gem1*) mutation has been tracked to a microtubule-associated protein with strong similarity to the human chTOG and *Xenopus* XMAP215 family of microtubule-associated proteins that stimulate plus-end microtubule growth.[11] The GEM1 protein plays a vital role in microspore polarity and cytokinesis through its involvement in microtubule assembly, and is associated with interphase, spindle, and phragmoplast microtubule arrays.[11] Further exciting discoveries are expected to arise from the identification of the mutant genes responsible for other gametophytic phenotypes described in Table 1.

Screening for gametophytic mutants using marker-segregation ratio distortion has also been used to identify genes involved in microgametogenesis and/or postpollination events including pollen germination, tube growth, and guidance to the ovule. These screens have mostly employed T-DNA or transposon insertion populations that harbor dominant antibiotic or herbicide resistance markers. For example, if a DNA insertion inactivates an essential male gametophytic gene, then the ratio of resistant to sensitive progeny will deviate significantly

Table 1 Gametophytic mutations and genes that affect pollen development in *Arabidopsis thaliana*

Gene	Mutant	Mutant phenotype	Function	Protein identity	Ref.
BOD	*both male and female gametophytic defective* (1,2,3)	Pollen arrested during bicellular development with pleiotropic effects	Pollen maturation after pollen mitosis I	ND	[17]
DUO	*duo pollen* (1,2,3,4,5,6)	Pollen fails to enter or complete generative cell division	Generative cell morphogenesis and cell cycle progression	ND	[4]
EFP	*emotionally fragile pollen*	Pollen shows diffuse callosic staining	Repression of callose synthesis during pollen maturation	ND	[13]
GEM	*gemini pollen* (1,2)	Twin-celled and binucleate pollen formed as a result of abnormal division at pollen mitosis I	Regulation of microspore polarity and cytokinesis through microtubule organization	GEM1: Homologous to chTOGp/XMAP215 family of microtubule-associated proteins	[4,11,12]
GUM	*male germ unit malformed*	Sperm cells separated from the vegetative nucleus	Association of vegetative nucleus and sperm cells	ND	[12]
GWP	*gift wrapped pollen*	Pollen contains internal callosic tube-like structures	Regulation of callose synthesis	ND	[13]
HAM	*halfman*	Pollen degenerates during bicelluar stage	Pollen maturation after pollen mitosis I	~150 kb deletion including 38 predicted genes	[18]
LIP	*limpet pollen*	Generative cell remains attached to pollen wall	Required for GC internalization	ND	[15]
MAD	*male gametophytic defective* (1,2,3)	Pollen arrested during bicellular development	Pollen maturation after pollen mitosis I	ND	[17]
MUD	*male germ unit displaced*	Male germ unit displaced to the cortical cytoplasm in mature pollen	Regulation of nuclear-cytoplasmic organization	ND	[12]
PDP	*polka dot pollen*	Pollen shows localized callose staining	Regulation of callose synthesis	ND	[13]
RTG	*raring-to-go*	Precocious germination of pollen within the anther locule	Regulation of pollen hydration status	ND	[13]
SCP	*sidecar pollen*	Pollen with an extra vegetative cell resulting from early symmetric microspore division	Control of microspore division timing or polarity establishment	ND	[10]
TDT	*T-DNA transmission defective*(6,17,38,40)	Aborted mature pollen	Pollen maturation	ND	[14,16]
TIO	*two-in-one*	Microspores fail to initiate or complete cytokinesis following pollen mitosis I	Regulation of phragmoplast structure and/or function	ND	[4]

Only gametophytic mutants that show phenotypes that are detectable before pollen germination are included. Mutations affecting male meiosis and postpollination events are excluded. In the column labeled 'mutant,' the numbers in parenthesis refer to individual mutants with the same mutant symbol prefix.
ND = not determined.

M

below the expected 3:1 ratio toward 1:1. A number of tagged mutants have now been identified that affect microgametogenesis, pollen germination or pollen tube growth.[14–16] Continued effort in screening and the isolation of tagged genes is expected to lead to a sophisticated genetic map of gametophytic proteins that function during microgametogenesis and postpollination events.

CONCLUSION

The quest to achieve a comprehensive description of the cellular, molecular, and genetic events that control pollen development is not only of fundamental interest, but may find application in the production and genetic improvement of crops. Current knowledge is applied in several areas of crop production, including crop yield optimization and hybrid seed production, and in the production of economic products such as honey and pharmaceuticals. The recent progress in mutational and genome-wide analysis now provides enormous potential to develop new strategies for the molecular dissection and modification of pollen development and functions.

ACKNOWLEDGMENTS

We thank Stefan Hyman at the University of Leicester Electron Microscopy Facility for scanning electron microscopy and the University of Leicester Harold Martin Botanical Gardens for pollen samples. We gratefully acknowledge financial support for our research from the NATO/Royal Society Fellowship Programme and the Grant Agency of the Czech Republic (DH) and the UK Biological and Biotechnology Research Council (DT).

ARTICLES OF FURTHER INTEREST

Anther and Microspore Culture and Production of Haploid Plants, p. 43
Female Gametogenesis, p. 439
Pollen-Stigma Interactions, p. 1035

REFERENCES

1. Shivanna, K.R.; Cresti, M.; Ciampolini, F. Pollen Development and Pollen–Pistill Interaction. In *Pollen Biotechnology for Crop Production and Improvement*; Shivanna, K.R., Sawhney, V.K., Eds.; Cambridge University Press: Cambridge, 1996; 15–39.
2. Touraev, A.; Vicente, O.; Heberle-Bors, E. Initiation of microspore embryogenesis. Trends Plant Sci. **1997**, *2*, 297–302.
3. Goldberg, R.B.; Beals, T.P.; Sanders, P.M. Anther development: Basic principles and practical applications. Plant Cell **1993**, *5*, 1217–1229.
4. Twell, D. *Pollen Developmental Biology. Plant Reproduction*; O'Neill, S.D., Roberts, J.A., Eds.; Annual Plant Reviews, Sheffield Academic Press: Sheffield, 2002; Vol. 6, 86–153.
5. Bedinger, P. The remarkable biology of pollen. Plant Cell **1992**, *4*, 879–887.
6. Twell, D.; Park, S.K.; Lalanne, E. Asymmetric division and cell-fate determination in developing pollen. Trends Plant Sci. **1998**, *3* (8), 305–310.
7. Hamilton, D.A.; Mascarenhas, J.P. Gene Expression During Pollen Development. In *Pollen Biotechnology for Crop Production and Improvement*; Shivanna, K.R., Sawhney, V.K., Eds.; Cambridge University Press: Cambridge, 1996; 40–58.
8. Honys, D.; Twell, D. Comparative analysis of the *Arabidopsis* pollen transcriptome. Plant Physiol. **2003**, *203*, 640–652.
9. Bhalla, P.; Swoboda, I.; Singh, M.B. Antisense-mediated silencing of a gene encoding a major ryegrass pollen allergen. Proc. Natl. Acad. Sci. U. S. A. **1999**, *96*, 11676–11680.
10. Chen, Y.-C.; McCormick, S. *Sidecar pollen*, an *Arabidopsis thaliana* male gametophytic mutant with aberrant cell divisions during pollen development. Development **1996**, *122*, 3243–3253.
11. Twell, D.; Park, S.K.; Hawkins, T.J.; Schubert, D.; Schmidt, R.; Smertenko, A.; Hussey, P.J. MOR1/GEM1 plays an essential role in the plant-specific cytokinetic phragmoplast. Nat. Cell Biol. **2002**, *4*, 711–714.
12. Lalanne, E.; Twell, D. Genetic control of male germ unit organization in *Arabidopsis*. Plant Physiol. **2002**, *129*, 865–875.
13. Johnson, S.A.; McCormick, S. Pollen germinates precociously in the anthers of *raring-to-go*, an *Arabidopsis* gametophytic mutant. Plant Physiol. **2001**, *126*, 685–695.
14. Bonhomme, S.; Horlow, C.; Vezon, D.; de Laissardiere, S.; Guyon, A.; Ferault, M.; Marchand, M.; Bechtold, N.; Pelletier, G. T-DNA mediated disruption of essential gametophytic genes in *Arabidopsis* is unexpectedly rare and cannot be inferred from segregation distortion alone. Mol. Genet. **1998**, *260*, 444–452.
15. Howden, R.; Park, S.K.; Moore, J.M.; Orme, J.; Grosslinklaus, U.; Twell, D. Selection of T-DNA—Tagged male and female gametophytic mutants by segregation distortion in *Arabidopsis*. Genetics **1998**, *149*, 621–631.
16. Procissi, A.; de Laissardiere, S.; Ferault, M.; Vezon, D.; Pelletier, G.; Bonhomme, S. Five gametophytic mutations affecting pollen development and pollen tube growth in *Arabidopsis thaliana*. Genetics **2001**, *158*, 1773–1783.
17. Grini, P.E.; Schnittger, A.; Schwarz, H.; Zimmermann, I.; Schwab, B.; Jurgens, G.; Hulskamp, M. Isolation of ethyl methanesulfonate-induced gametophytic mutants in *Arabidopsis thaliana* by a segregation distortion assay using the multimarker chromosome 1. Genetics **1999**, *151*, 849–863.
18. Oh, S-A.; Park, S-K.; Jang, I.; Howden, R.; Moore, J.M.; Grossniklaus, U.; Twell, D. Halfman, an *Arabidopsis* male gametophytic mutant associated with a 150 kb chromosomal deletion at the site of transposon insertion. Sex Plant Reprod. **2003**, *16*, 99–102.

Management of Bacterial Diseases of Plants: Regulatory Aspects

Jaap D. Janse
Plant Protection Service, Wageningen, The Netherlands

INTRODUCTION

Because most plant diseases caused by bacteria cannot be controlled by chemicals, they pose a serious threat to important agricultural and horticultural crops through yield loss and trade barriers. Use of uninfested planting material, sanitation at production sites, and other preventive measures have often proved to be inadequate to control bacterial diseases and therefore many governments have implemented regulations to prevent the introduction and spread of new or existing pathogens. The role of quarantine regulations in the management of bacterial plant diseases, the organization thereof in different parts of the world, and their success is the subject of this article.

LONG DISTANCE MOVEMENT OF PHYTOPATHOGENIC BACTERIA

For bacteria, the most important mechanisms of global movement are internationally traded plants, seeds, and plant parts. With new free trade agreements between countries in different parts of the world, the risk of introducing exotic pests is increasing. Furthermore, new, difficult to trace pathways are developing. For example, commercial companies produce basic planting material outside their regions under varying disease management conditions, and then introduce the products into their regions for further propagation and sale. This has led to the introduction of *Ralstonia solanacearum* race 3, biovar 2 with *Pelargonium* cuttings from Kenya into Europe and from Guatemala into the United States.

Other possible pathways for the introduction of exotic phytopathogenic bacteria are the use of nonindigenous pollinating insects and biological control agents. For example, it is suspected that *Liberobacter asiaticum* was introduced into the U.S. state of Florida with parasitoids used for control of the Asian citrus psyllid *Diaphorini citri*. Indeed, certain biological control agents, including *R. solanacearum*, used for control of the forest weed *Hedygium gardnerianum* in Hawaii, may attack cultivated hosts. Additionally, genetically modified *Agrobacterium tumefaciens* strains, used for the genetic manipulation of plants, should also be regulated because they may survive in the transformed host.[1]

APPROACHES TO PREVENT INTRODUCTION OF PHYTOPATHOGENIC BACTERIA

Quarantine regulations are mainly aimed at excluding phytopathogenic bacteria from a territory. These may be laws, orders, or decrees that limit the import of plants or plant products and specify the pathogens of interest.[2] More specifically, quarantines facilitate the isolation and inspection of plants and plant products for prohibited organisms. This is only used for small-scale importations of germplasm sent to postentry quarantine stations for breeding or research purposes. The traditional method of controlling the entry of exotic pests, i.e., inspection at the point of import, is not adequate due to latent symptom development and undetectable levels of bacteria on planting materials. If the host plants are not prohibited, then phytosanitary regulations are based on ensuring that the imported plants are symptomless at the point of origin, as certified by the National Plant Protection Organization (NPPO) of the exporting country. An important principle for ensuring that commodities are pathogen-free is the concept of the pest-free area, assigned when a disease has not been observed in a certain country or part of that country. A pest-free area can be created and officially recognized according to international standards, and commodities can be freely exported from it. Similar in concept is the protected zone, that also must be free of a particular bacterium, but must be especially protected from introduction of the bacterium by stricter phytosanitary measures than adjoining areas. Examples of protected zones can be found in Canada and Europe for *Clavibacter michiganensis* subsp. *sepedonicus* and *Erwinia amylovora*, respectively.[3]

If a pest-free area cannot be established or maintained, phytosanitary regulations may allow plants or plant products to originate from a production site that has been free from the particular bacterium for a defined period of time. Quarantine regulations should also include the eradication

Encyclopedia of Plant and Crop Science
DOI: 10.1081/E-EPCS 120010576

or containment of introduced bacterial pathogens. Additionally, they should be used to ensure the production of healthy planting materials by certification schemes and standard production protocols.[4]

Other phytosanitary measures applied to host plants include requirements that the previous generation of the plants or seeds of a commodity be free of the pathogen, and that planting materials be tested or subjected to physical or chemical eradicative treatment.[4] Furthermore, the exchange of bacterial strains from culture collections is subject to greater regulation, mostly at the national level but also through the World Federation for Culture Collections (WFCC). Note that the WFCC also addresses the possibilities that regulations be employed in preventing the use of pathogens in bioterrorism.[5] It seems unlikely, however, that plant pathogenic bacteria will ever be used as bioterror tools due to their slow and unpredictable mode of action.

REGULATIONS AS TRADE BARRIERS

In the Uruguay round of negotiations (1986–1994) of the World Trade Organization, an agreement on Sanitary and Phytosanitary Measures (SPS Agreement) was developed to discourage the use of phytosanitary measures as non-tariff trade barriers. A pest risk analysis (PRA), including statistically sound surveys of pathogens in the areas or countries involved, was established as the basis for developing regulatory measures. The PRA ensures that an organism is not considered a pest to be quarantined without scientific evidence. For example, *P. syringae* pv. *pisi*, causing pea blight, is seedborne, but severe pea blight outbreaks are more dependent on environmental conditions. Therefore, this bacterium has been withdrawn from the quarantine lists of many European countries. However, effort is still needed to remove unimportant pathogens from quarantine lists.

Regulations should be based on thorough knowledge of the incidence of a disease. In many countries, there is insufficient knowledge of endemic pathogens because systematic surveys are rarely done. Radical changes in standard agricultural practices can promote epidemics that are often claimed to be new pathogen introductions. One example of this is bacterial stripe caused by *Xanthomonas translucens* pv. *tranlucens*. At low disease incidence, bacterial stripe symptoms are inconspicuous and easily overlooked. The disease is not important in rain-fed cereals in Syria, but outbreaks occurred when irrigation was intensified, and it was thought that seedborne inoculum was responsible for the outbreaks. The introduction of bacteria on planting material should only be concluded when the bacterium is isolated from the suspected tissues. In the case of bacterial diseases this is difficult, because infested planting material may not display obvious symptoms, and biochemical and molecular tests may yield false positive results. Pathogen isolation, adequate confirmatory tests, and maintenance of the original samples and isolates should be required in all cases.

CREATION OF REGULATIONS

Sovereign states, under the International Plant Protection Convention and the SPS Agreement, have the right and responsibility for preparing phytosanitary regulations to protect themselves against exotic plant pests. In the past, countries prepared their regulations independently, with the consequence that great discrepancies existed between the principles on which the measures were based, and their degree of technical justification. About fifty years ago, the Food and Agricultural Organization (FAO) of the United Nations, the Regional Plant Protection Organizations (RPPOs), and regional economic integration organizations started supporting countries in this task. The Interim Commission on Phytosanitary Measures, established by the IPPC, develops International Standards on Phytosanitary Measures. These standards establish the principles and procedures of phytosanitary measures, ensuring that countries develop their measures consistently and fairly. In addition, most countries belong to RPPOs that coordinate and harmonize the phytosanitary actions of their member countries on a regional basis. With regard to bacterial pathogens, the European Plant Protection Organization (EPPO) advises its members which bacteria should be quarantine pests, dividing them into A1 quarantine pests (those absent from the region and against which all countries are recommended to take phytosanitary action) and A2 quarantine pests (those present in some parts of the region and of concern to only some countries). The EPPO convenes a panel of experts on bacterial diseases that is responsible for developing technical documentation relevant to each of the quarantined bacteria (e.g., geographic distribution, biology, economic impact), pest-specific phytosanitary requirements (measures that should be adopted by countries that wish to export to an EPPO country), phytosanitary procedures (used by the NPPO of the exporting country to ensure absence of the quarantine bacterium), and pathogen-specific diagnostic protocols. The North American Plant Protection Organization (NAPPO) is developing international standards for phytosanitary measures, a list of regulated plant pests including phytopathogenic bacteria (see Tables 1–3), harmonizing pest man-agement programs through the coordination of pest sur-veys and developing criteria for seed potato certification.

Regional economic integration organizations such as the European Union (EU) are formed to allow countries to collaborate for the economic benefit of all, and the organizations take over activities that were once conducted

Table 1 Bacterial pathogens mentioned in the EPPO Quarantine A1 List (2002), Other Regional Plant Protection Organizations, and the European Union (2002)[a,c]

Bacterium	Main hosts	EU	NAPPO[b]	APPPC	CAN	COSAVE	CPPC	IAPSC	NEPPO	PPPO
Citrus variegated chlorosis (probably *Xylella fastidiosa*)	*Citrus*, *Fortunella*, and *Poncirus* spp.	IIA1								
Liberobacter africanium (nonculturable bacterium, vector transmitted, PCR detection)	*Citrus*, *Fortunella*, and *Poncirus* spp. (Citrus greening or Citrus Huanglongbin)	IIA1	X							
Liberobacter asiaticum (nonculturable bacterium, vector transmitted, PCR detection)	*Citrus*, *Fortunella*, and *Poncirus* spp. (Citrus greening or Citrus Huanglongbin)	IIA1	X							
Xanthomonas axonopodis pv. *citri*	*Citrus*, *Fortunella*, and *Poncirus* spp. (Citrus)	IIA1	X		A1	A2	A2	A2		A2
Xanthomonas axonopodis pv. *dieffenbachiae*	*Anthurium* spp., *Dieffenbachia* spp.						A2			
Xanthomonas oryzae pv. *oryzae*	*Oryza* spp. (rice)	IIA1				A1	A2	A2	A1	
Xanthomonas oryzae pv. *oryzicola*	*Oryza* spp. (rice)	IIA1	X			A1		A2	A1	
Xanthomonas translucens pv. *translucens*	*Hordeum vulgare* (barley)									A2
Xylella fastidiosa, vector transmitted. Different forms on grapevine (Pierce's disease), peach (Phony peach), and *Citrus*	*Vitis vinifera* (grapevine), *P. persica* (peach), *Citrus sinensis* (Citrus)	IA1				A2		A1		

[a]Meaning of A1, A2 or IA1, IA2, IIA1, and IIA2 or IIB may slightly differ from the definition by EPPO for A1 and A2; see text.
[b]Regulated pest on draft list in June 2002 (X = present on the list).
[c]EU=European Union; NAPPO=North American Plant Protection Organization; EPPO=European and Mediterranean Plant Protection Organization; APPPC=Asia and Pacific Plant Protection Commission; CAN=Comunidad andina; COSAVE=Comite Regional de Sanidad Vegetal del Cono Sur; CPPC=Caribbean Plant Protection Commission; IAPSC=Interafrican Phytosanitary Council; NEPPO=Near East Plant Protection Organization; PPPO=Pacific Plant Protection Organization.

Table 2 Bacterial pathogens mentioned in the EPPO Quarantine A2 List (2002), other Regional Plant Protection Organizations, and the European Union (2002)[a,e]

Bacterium	Main hosts	EU	NAPPO[b]	APPPC	CAN	COSAVE	CPPC	IAPSC	NEPPO	PPPO
Burkholderia (*Pseudomonas*) *caryophylli*[c]	*Dianthus caryophyllus* (carnation)	IIA2			A1					
Clavibacter michiganensis subsp. *insidiosus*	*Medicago sativa* (alfalfa)	IIA2		A2				A1		
Clavibacter michiganensis subsp. *michiganensis*	*Lycopersicon esculentum* (tomato)	IIA2[d]		A2			A2			
Clavibacter michiganensis subsp. *sepedonicus*	*Solanum tuberosum* (potato)	IA2	A1		A1	A1		A1		
Curtobacterium flaccumfaciens pv. *flaccumfaciens*	*Phaseolus vulgaris, Vigna* spp. (bean)	IIB				A1	A1	A1	A1	
Erwinia amylovora	*Malus* spp. (apple), *Pyrus* spp. (pear), *Crataegus* spp. (hawthorn), *Cotoneaster* spp.	IIA2				A1		A2		
Erwinia chrysanthemi pv. *dianthicola* (*Dianthus* strains)	*Dianthus caryophylus* (carnation)	IIA2[d]								
Pantoea (*Erwinia*) *stewartii* subsp. *stewartii*, also vector transmitted	*Zea mays* (corn)	IIA1						A1		A2
Pseudomonas syringae pv. *persicae*	*Prunus persica* (peach)	IIA2								
Ralstonia (*Pseudomonas*) *solanacearum*	Race 1: many (solanaceous) hosts, race 2: *Musa* spp. and *Heliconia* (banana), race 3: mainly *Solanum tuberosum* (potato), *Lycopersicon esculentum* (tomato)	IA2				Race 1: A2	Race 1 and 2:A2	Race 1: A2		Race 2: A2
Xanthomonas arboricola pv. *corylina*	*Coryllus avellanae* (hazelnut)									
Xanthomonas arboricola pv. *pruni*	*Prunus* spp.	IIA2						A2		
Xanthomonas axonopodis pv. *phaseoli*	*Phaseolus* spp. (bean)	IIA2[d]								
Xanthomonas fragariae	*Fragaria* spp. (strawberry)	IIA2[d]						A1		
Xanthomonas vesicatoria	*Lycopersicon esculentum* (tomato), *Capsicum* spp. (pepper and chilli pepper)	IIA2[d]								
Xylophilus ampelinus (*Xanthomonas ampelina*)	*Vitis vinifera* (grapevine)	IIA2[d]	X			A1		A1		

[a]Meaning of A1, A2 or IA1, IA2, and IIA1, IIA2 or IIB may slightly differ from the definition by EPPO for A1 and A2; see text.
[b]Regulated pest on draft list in June 2002 (X = present on the list).
[c]Older synonyms in brackets.
[d]These bacteria will be transferred from the quarantine list of the EU to a list of regulated nonquarantine pests for which official certification schemes will be developed.
[e]EU = European Union; NAPPO = North American Plant Protection Organization; EPPO = European and Mediterranean Plant Protection Organization; APPPC = Asia and Pacific Plant Protection Commission; CAN = Comunidad andina; COSAVE = Comite Regional de Sanidad Vegetal del Cono Sur; CPPC = Caribbean Plant Protection Commission; IAPSC = Interafrican Phytosanitary Council; NEPPO = Near East Plant Protection Organization; PPPO = Pacific Plant Protection Organization.

Table 3 Other bacteria of quarantine importance in some areas or countries

	Main hosts	EU[a,d]	NAPPO[b]	APPPC	COSAVE	CPPC	IAPSC	NEPPO
Clavibacter michiganensis subsp. *nebraskensis*	*Zea mays* (corn)			A2				
Clavibacter xyli subsp. *xyli*	*Saccharum officinarum* (sugarcane), ratoon stunting disease			A2				
Erwinia carotovora subsp. *atroseptica*	*Solanum tuberosum* (potato)							
Erwinia chrysanthemi pv. *chrysanthemi*	Many hosts					A2		
Erwinia salicis	*Salix* spp. (willow tree)		X		A1			
Erwinia tracheiphila	*Citrullus lanatus* (watermelon), *Cucumis* spp. (cucumber), *Cucurbita* spp. (cucurbits)							
Pseudomonas lignicola[b]	*Ulmus* spp. (elm tree), wood staining		X					
Pseudomonas syringae pv. *pisi*	*Pisum sativum* (pea)						A2	A1
Rathayibacter (Clavibacter) tritici	*Triticum aestivum* (wheat)							
Xanthomonas acernea	*Acer trifidum* (acorn tree)		X					
Xanthomonas albilineans	*Saccharum officinarum* (sugarcane)			A2			A1	
Xanthomonas axonopodis pv. *cajani*	*Cajanus cajan* (pigeon pea)					A1		
Xanthomonas axonopodis pv. *citrumelo*	*Citrumelo* rootstock (*Citrus paradisi* x *Poncirus trifoliata*), *P. trifoliata.*	IIA1			A1			
Xanthomonas axonopodis pv. *manihotis*	*Manihot esculentum* (cassave)		X	A2		A2		A2
Xanthomonas axonopodis pv. *vasculorum*	*Saccharum officinarum* (sugarcane)		X	A2				A2
Xanthomonas campestris pv. *malvacearum*	*Gossypium hirsutum* (cotton)							
Xanthomonas hortorum pv. *carotae*	*Daucus* carota (carrot)					A1		
Xanthomonas populi	*Populus* spp. (poplar tree)		X		A1			
Xanthomonas vasicola (campestris) pv. *holcicola*	*Panicum miliaceum* (millet), *Sorghum* spp. (sorghum), *Zea mays* (corn)						A2	
Wheat yellowing stripe bacterium (*Rickettsia*-like bacterium, described from China)	*Triticum aestivum* (wheat)		X					

[a]Meaning of A1, A2 or IA1, IA2, IIA1, and IIA2 or IIB may slightly differ from the definition by EPPO for A1 and A2; see text.
[b]Regulated pest on draft list in June 2002 (X=present on the list).
[c]Species of uncertain taxonomic position. No reference cultures available.
[d]EU=European Union; NAPPO=North American Plant Protection Organization; EPPO=European and Mediterranean Plant Protection Organization; APPPC=Asia and Pacific Plant Protection Commission; CAN=Comunidad andina; COSAVE=Comite Regional de Sanidad Vegetal del Cono Sur; CPPC=Caribbean Plant Protection Commission; IAPSC=Interafrican Phytosanitary Council; NEPPO=Near East Plant Protection Organization; PPPO=Pacific Plant Protection Organization.

M

by the individual states. The EU has based its lists of quarantine organisms and their specific phytosanitary measures on the phytosanitary requirements of the EPPO.[6] Tables 1–3 show the bacteria that are quarantine pests for the EPPO and the EU, and indicate those that are considered to be quarantine pests for other regional organizations. As internal control measures, the EU has developed special control directives for bacterial brown rot and bacterial ring rot that allow the safe movement of potatoes from parts of the community that have the disease.[7–9] These directives are based on scientific data, and detailed methods for detecting latent infections are provided. Measures are also described for determining pathogen distribution, preventing spread, and suppressing and/or eradicating the disease.

IF QUARANTINE REGULATIONS FAIL

If phytosanitary regulations fail to stop the entry of a pathogen, the success of efforts to prevent long-term establishment depends on early pathogen detection, accurate diagnosis, and rapid implementation of management strategies.[10] Eradication of pathogens is most likely to be successful on small isolated areas, because it is easier to delimit an outbreak and to prevent re-introductions during the eradication campaign. An example is successful eradication of *Xanthomonas axonopodis* pv. *citri* from Thursday Island, Australia.[4] Regulatory eradication actions often fail because of 1) incomplete pathogen eradication; 2) natural reinvasion; and 3) reintroduction through short- or long-distance movement of infected material. In that case, the realistic options are functional eradication or areawide suppression. The problems encountered in eradication campaigns in larger landmasses are illustrated by the history of citrus canker (*Xanthomonas axonopodis* pv. *citri*) in the United States. Despite an intensive eradication program in the 1930s, the disease got a foothold in subsequent years and researchers now question whether the pathogen can ever be eradicated under Florida's climatic conditions.[11]

ACKNOWLEDGMENTS

I thank Ann-Sophie Roy, information officer of EPPO, for helpful information and David McNamara, Assistant-Director of EPPO, for critical review.

ARTICLES OF FURTHER INTEREST

Bacterial Pathogens: Detection and Identification Methods, p. 84

Bacterial Survival and Dissemination in Insects, p. 105

Bacterial Survival and Dissemination in Natural Environments, p. 108

Bacterial Survival and Dissemination in Seeds and Planting Material, p. 111

Bacterial Survival Strategies, p. 115

Crown Gall, p. 379

Fire Blight, p. 443

Management of Bacterial Diseases of Plants: Regulatory Aspects, p. 669

Pierce's Disease and Others Caused by Xylella fastidiosa, p. 928

Plant Diseases Caused by Bacteria, p. 947

Seeds: Pathogen Transmission Through, p. 1142

REFERENCES

1. Leifert, C.; Cassels, A.C.; Doyle, B.M.; Curry, R.F. Quality assurance systems for plant cell and tissue culture: The problem of latent persistence of bacterial pathogens and *Agrobacterium*-based transformation vector systems. Acta Hortic. **2000**, *530*, 87–91.
2. Ebbels, D.L.; King, J.E. *Plant Health. The Scientific Basis for Administrative Control of Plant Diseases and Pests*; Blackwell Scientific Publications: Oxford, UK, 1979.
3. European Communities. *Council Directive 92/76/EEG Protected Zones, Version 6*; October 6, 1992.
4. Janse, J.D. Possibilities in the avoidance and control of bacterial plant diseases when using pathogen tested (certified) or treated planting material. Plant Pathol. **2002**, *51*, 523–536.
5. WFCC, 2002. Bioterrorism. WFCC Newslett. **2002**, *34*, 1–12.
6. European Communities. *Council Directive 2000/29/EG. (earlier version was 77/93/EEG).*
7. European Communities. *Scheme for the Detection and Diagnosis of the Ring Rot Bacterium Corynebacterium sepedonicum in Batches of Potato Tubers. EUR 11288*; Office for Official Publications of the European Communities: Luxembourg, 1987; 1–21.
8. European Communities. Council directive 93/85/EEC of 4 October 1993 on the control of potato ring rot. Luxemb.: Off. J. Eur. Communities **1993**, *L259*, 1–25.
9. European Communities. Council directive 98/57/EC of 20 July 1998 on the control of *Ralstonia solanacearum* (Smith) Yabuuchi et al. (including Annex II, interim testing scheme for the diagnosis, detection and identification of *Ralstonia solanacearum* (Smith) Yabuuchi et al. in potatoes). Publication 97/647/EC. Luxemb.: Off. J. Eur. Communities **1998**, *L235*, 8–39.
10. Hurtt, S.S. The pome fruit tree quarantine program of the plant germplasm quarantine office-fruit laboratory, USDA. Fruit Var. J. **1999**, *53*, 59–63.
11. Schubert, T.S.; Rizvi, S.A.; Sun, X.A.; Gottwald, T.R.; Graham, J.H.; Dixon, W.N. Meeting the challenge of eradicating citrus canker in Florida-again. Plant Dis. **2001**, *85*, 340–356.

Management of Diseases in Seed Crops

M

Lindsey J. du Toit
Washington State University, Mount Vernon, Washington, U.S.A.

INTRODUCTION

The success of modern agriculture is dependent on a vital seed industry that enables the timely production and distribution of high-quality, pathogen-free seed of high-yielding cultivars adapted to specific geographic regions and methods of production. The value of the world seed market is estimated at $40–60 billion, and the demand for seed is expected to keep pace with the world's growing population.[1] The increasingly global nature of the seed industry brings an associated risk of widespread dissemination of seedborne pathogens. As a result, effective disease management in seed crops remains central to the seed industry.

SEED CROP DISEASES

Although management of plant diseases is important for most crops, it is particularly critical for production of high-quality seed. Plant pathogens can reduce the quantity and quality of seed harvested, and many can be seedborne (Fig. 1). A significant proportion of the seed market is associated with worldwide movement of seeds, and seeds are distributed internationally for breeding programs and research purposes. Seeds provide an efficient means of inadvertently disseminating plant pathogens. Numerous examples can be found of plant disease epidemics resulting from the introduction of seedborne pathogens, often culminating in significant economic losses.[2]

Strategies for Managing Seed Crop Diseases

General strategies for disease management in agriculture are pertinent to seed crops, i.e., exclusion of pathogens from regions of seed production, eradication of pathogens from seed crops, protection of seed crops, alleviation of disease pressure using cultural practices, and incorporation of disease resistance into cultivars. However, seed production is a complex process involving meticulous criteria followed rigorously by seed producers.[3] Consequently, disease management programs for seed crops can be more complex than for commercial crops, and require integration of the many tools available. The specific strategies selected are influenced primarily by economic factors, ultimately governing the value and amount of seed produced.

The extremely low tolerance for pathogens in seed crops has resulted in specialized areas of seed production in regions where pathogens are unable to establish or usually remain below threshold levels during seed development.[4,5] In the United States, seed production occurs primarily in the western states where pressure from fungal and bacterial pathogens is reduced by low rainfall and relative humidity. For example, bean seed production occurs in the semiarid regions of Washington and Idaho. Similarly, the mild winters and dry summers of western Washington make this maritime region ideal for production of biennial Brassica seed crops free of black rot (*Xanthomonas campestris* pv. *campestris*) and black leg (*Phoma lingam*). Crucifer seed crops in Denmark are located in coastal areas where winds ventilate the crops, reducing development of black spot (*Alternaria brassicicola*).[2] Furthermore, whenever possible seed crops are isolated from commercial crops for disease control. Lettuce seed produced in the San Joaquin valley of California is isolated from commercial crops in the Salinas valley to prevent infestations of aphids carrying lettuce mosaic virus.[5]

Production of seeds is an energy-intensive process.[3] Consequently, some plant species become increasingly susceptible to certain pathogens at flowering. Development of leaf spot of spinach caused by *Stemphylium botryosum* is greatly exacerbated in the presence of pollen, necessitating initiation of protective fungicide applications prior to pollen shed in spinach seed crops.[6] Although *Fusarium oxysporum* f. sp. *spinaciae* causes damping-off of spinach seedlings, symptoms of Fusarium wilt in the seed crop are not apparent until flowering. Yield losses in the seed crop can be extensive without ≥10-year crop rotations.[7]

The duration of the seed crop season may result in a long window of susceptibility to infection or provide opportunities for infection during periods of stress or injury (e.g., winter injury in biennial carrot seed crops). Overwintered biennial and perennial seed crops may harbor pathogens that can spread to neighboring first-year crops. This green bridge effect led to epidemics of the aphid-vectored beet western yellows luteovirus in the beet seed industry in western Oregon.[8]

DOI: 10.1081/E-EPCS 120019947

Fig. 1 Scape and umbel blight in an onion seed crop, caused by the fungus *Botrytis aclada*, reduces the quantity and quality of seed harvested. (*View this art in color at www.dekker.com.*)

Cultural practices can create less favorable conditions for disease development. Crop inspections and roguing of symptomatic plants or alternative hosts of plant pathogens can reduce disease in a seed crop. Crop rotation is essential for managing many plant diseases. The minimum duration of rotation depends on the longevity of specific pathogens, and may range from a few years to >10 years. Planting seed crops in suppressive soils can assist in production of pathogen-free seed, as demonstrated for pea seed crops free of *F. oxysporum* f. sp. *pisi* in California.[5] Furrow irrigation restricts splash-dispersal of many pathogens compared to overhead irrigation. Increased spacing of plants reduces disease pressure for fungal and bacterial pathogens as a result of increased air circulation. Incorporation of infested crop debris into the soil reduces survival of some pathogens. Fire is used to reduce inoculum levels of certain plant pathogens (e.g., burning of stubble/straw in grass and cereal seed crops).

Foliar applications of fungicides, bactericides, and resistance-inducing chemicals can provide effective management of many plant pathogens in seed crops. The extended season of seed crops, combined with the low tolerance for seedborne pathogens, often necessitates a greater number of pesticide applications than in commercial crops and requires precise timing of applications with periods of increased susceptibility and disease pressure. However, the minor acreage of individual seed crop species impedes attainment of pesticide registrations for seed crops because of the limited returns to pesticide manufacturers. As a result, some states have implemented programs that qualify certain seed crops for non-food, non-feed status to facilitate pesticide registrations.

Fumigation, and biofumigation using cover crops, can eradicate persistent inoculum of some pathogens, but the former may be cost-prohibitive.[2,4,5] Seed treatments provide effective means of eradicating or reducing some pathogens, and can be applied to both the stock and harvested seed. Fungicide seed treatments eradicate seedborne inoculum, inhibit seed-to-seedling transmission of pathogens, or protect emerging seedlings from soilborne or airborne pathogens. Systemic fungicides are valuable for eradication of internal seed infection and protection of subsequent new growth of seedlings. Physical seed treatments (e.g., hot water, steam, or chlorine) can eradicate or reduce inoculum on seeds. The efficacy of seed treatments depends on the degree of internal infection of the seed, the amount of inoculum in a seed lot, specificity of the treatment, and potential phytotoxicity of the treatments.[1,3,4]

Plant breeding has successfully introduced disease resistance into many commercial seed crop cultivars. However, resistance may not be available for some pathogens. In hybrid seed crops, resistance may only be present in one of the parent lines, necessitating disease management practices for the susceptible parent. Furthermore, some pathogens have the ability to overcome single-gene resistance, complicating efforts to manage diseases in seed crops using resistance.

Quality control is fundamental to the seed industry.[2] Seed certification programs encompass field inspections and lab assays so that the history of seed lots can be traced.[4] Emphasis is placed on assays for seedborne pathogens, as many can go undetected during field inspections. To be of value, seed health assays must be specific, sensitive, reliable, cost-effective, and rapid.[5] In 1958, the International Seed Testing Association initiated development of standardized seed health assays.[1] Despite this, most seed testing protocols published in the literature have not been standardized, resulting in a range of methods utilized by different laboratories. This can lead to different results, sometimes with significant ramifications regarding international seed shipments. Techniques for detecting seedborne pathogens include visual inspection, incubation on agar or blotters, staining, pathogenicity tests, serological assays (e.g., enzyme-linked-immunosorbent assays), and nucleic acid assays (e.g., polymerase chain reaction assays).[5,9] Approval of seed lots for shipment may be based on results of seed assays, so rapid turnaround is critical in seed testing facilities. Advances in rapid nucleic acid assays have overcome problems with inhibitory seed-derived compounds, resulting in increasing application of this technology for seed health assays.[9]

CONCLUSION

The new U.S. National Organics Program for certified organic produce states that the ''producer must use organically grown seeds,'' ''may use untreated nonorganic seeds ... when equivalent organic varieties are not commercially available,'' and seeds ''treated with prohibited substances may be used to produce an organic crop when the application of the substance is a requirement of Federal or State phytosanitary regulations.''[10] These standards have increased the demand for organically-produced seed, but have also raised concerns about associated increases in losses from seedborne pathogens because of the limited number of organic options for seed treatment and disease management. Research is needed to investigate alternatives for successful production of pathogen-free organic seed.

The rapid acceleration in worldwide movement of seed can be justification for plant quarantines. However, there is a significant lack of epidemiological research on the economic impacts of specific seedborne pathogens, on which the regulations and quarantines regarding seedborne pathogens need to be based.[1] Consequently, many of a large number of new phytosanitary regulations cannot be justified scientifically and have been interpreted as phytosanitary barriers used only to protect domestic agricultural industries.[1] Epidemiological research is needed to provide a scientific basis on which to implement regulations affecting the global seed industry.

ARTICLES OF FURTHER INTEREST

Bacterial Survival and Dissemination in Seeds and Planting Material, p. 111
Integrated Pest Management, p. 612
Seed Borne Pathogens, p. 1126
Seed Production, p. 1134
Seeds: Pathogen Transmission Through, p. 1142

REFERENCES

1. McGee, D.C. *Plant Pathogens and the Worldwide Movement of Seeds*; APS Press: St. Paul, MN, 1997.
2. Neergaard, P. *Seed Pathology, Volumes I and II*; John Wiley & Sons: New York, 1977.
3. McDonald, M.B.; Copeland, L.O. *Seed Production: Principles and Practices*; Chapman & Hall: New York, 1997.
4. Agarwal, V.K.; Sinclair, J.B. *Principles of Seed Pathology*, 2nd Ed.; Lewis Publishers: New York, 1997.
5. Maude, R.B. *Seedborne Diseases and Their Control: Principles and Practice*; CAB International: Oxon, 1996.
6. du Toit, L.J.; Derie, M.L. Leaf spot of spinach seed crops in Washington state. Phytopathology **2002**, *92* (6), S21.
7. Correll, J.C.; Morelock, T.E.; Black, M.C.; Koike, S.T.; Brandenberger, L.P.; Dainello, F.J. Economically important diseases of spinach. Plant Dis. **1994**, *78* (7), 653–660.
8. Hampton, R.O.; Keller, K.E.; Bagget, J.R. Beet western yellows luteovirus in western Oregon: Pathosystem relationships in a vegetable-sugar beet seed production region. Plant Dis. **1998**, *82* (2), 140–148.
9. Walcott, R.R. Detection of seedborne pathogens. HortTechnology **2003**, *13* (1), 40–47.
10. http://www.ams.usda.gov/nop/NOP/standards.html. (accessed 31 March 2003).

Management of Fungal and Oomycete Diseases: Fruit Crops

Patricia S. McManus
University of Wisconsin, Madison, Wisconsin, U.S.A.

INTRODUCTION

Fruit crops are produced on nearly 50 million hectares worldwide. Fruit plants are susceptible to diseases at every step in their production and distribution, with economic losses occurring at the nursery, in producers' fields, during storage after harvest, and in the marketplace. As with other crops, the majority of diseases of fruit crops are caused by fungi. Fungi primarily cause fruit rots, leaf spots and blights, and stem cankers. However, several fungal genera affect roots of fruit crops, either individually (e.g., *Armillaria*) or as members of a disease complex (e.g., *Fusarium* and *Rhizoctonia*). Oomycetes (water molds) primarily cause root and crown rots, but can also cause fruit rots (e.g., strawberry leather rot caused by *Phytophthora cactorum*) and shoot blights (e.g., grape downy mildew caused by *Plasmopora viticola*).

Consumers have high standards when selecting produce and usually pass over fruit with blemishes or other imperfections. Thus, there is nearly zero tolerance for diseases that directly affect fruit. Aesthetics are less important for fruit that will be processed rather than grown for the fresh market, but growers almost always receive a lower price for fruit destined for processing rather than the fresh market. Since the middle 1990s, the demand for high-quality fruit grown in accordance with organic standards has increased, and organically grown fruit typically is sold at a premium. While growing organic fruit might appear lucrative, managing diseases without synthetic fungicides can be a major challenge depending on the crop and climate where it is produced.

As with other crops, managing diseases of fruit depends on manipulating the host, pathogen(s), and environment to favor plant health and inhibit pathogen growth and spread. Since most fruit plants are long-lived perennials, disease management presents unique challenges. The roles of host resistance, pathogen exclusion, cultural practices, and chemical and biological control in fruit disease management are presented separately. In practice, however, fruit growers integrate multiple methods to manage diseases.

GENETIC RESISTANCE OF THE HOST PLANT

Genetic resistance of the host is often the most effective, least expensive, and safest way to manage diseases of fruit crops. For most fruit crops there are numerous cultivars that vary in susceptibility to important diseases. However, cultivar choice is driven more by processor standards and/or consumer preference than disease resistance. For example, McIntosh is one of the leading apple cultivars for fresh-market sales in the eastern United States, but it is also one of the most susceptible to apple scab. Desirable horticultural traits, such as size-controlling rootstocks on fruit trees, are often higher priorities than disease resistance for growers. The extent to which growers depend on host resistance depends on the on crop, the disease, and how the fruit will be marketed (e.g., fresh versus processed; conventional versus organic). Although some fruit crop species have been genetically transformed to improve disease resistance, transgenic fruit crops currently are not produced commercially.

PATHOGEN EXCLUSION

Pathogen exclusion (i.e., preventing a pathogen from reaching its host) is not a practical means of disease management for most fruit crops. Fruit plantings are usually established from nursery stock rather than seed. Indexing stock for viruses is standard practice, but indexing is not practical for most fungal and oomycete pathogens. Thus, pathogens are often carried on nursery stock and remain with the plant its entire life.

CULTURAL PRACTICES

Cultural practices are especially effective in fruit disease management because they can simultaneously improve plant health, reduce pathogen populations, make the environment unfavorable for pathogens, and in the case of pruning, permit more thorough coverage with chemical or biological pesticides.

Encyclopedia of Plant and Crop Science
DOI: 10.1081/E-EPCS 120019945

Site Selection and Water Management

Annual crop rotation is not an option with perennial fruit plants. Therefore, establishing plantings on suitable sites is essential. Most fruit crops prefer lighter soils with good water drainage. Since oomycete pathogens proliferate in water, good drainage is key in managing diseases caused by oomycetes. Installing drainage tiles is sometimes necessary in heavier soils. At the other extreme, too little water can stress woody plants and exacerbate certain canker and shoot blight diseases. For example, various species of the fungus *Phomopsis* cause shoot dieback or cankers on many different woody fruit crops. Symptoms are usually made worse by drought stress on the host. Trickle irrigation generally is preferred to overhead irrigation, because the latter can contribute to leaf and fruit diseases. However, for crops where frost protection is needed (e.g., cranberry and strawberry in cooler climates), overhead irrigation is used.

Sanitation

Removal and destruction of diseased branches, fruit, and other plant debris reduces pathogen inoculum and thereby decreases disease pressure. For tree and vine crops, sanitation is usually accomplished by manually pruning branches and picking mummified fruit. In the case of apple scab, removing leaves from the orchard floor or shredding them to accelerate their decomposition significantly reduces apple scab inoculum. In perennial strawberry production, leaves are mowed from plants and then either removed from the planting or incorporated into the soil to reduce inoculum of fungal leaf spot pathogens. Cranberry beds are flooded either after harvest or in early spring to remove leaf debris and mummified fruit that harbor fungal fruit rot pathogens.

Air Movement

Fungal pathogens of leaves and fruit depend on free water and/or high relative humidity for germination and growth. Any practice that reduces the length of time that foliage and fruit are wet is likely to hinder disease development. Pruning of woody fruit plants, done primarily for horticultural benefits and sanitation, also improves air circulation and drying of foliage. Planting strawberry and raspberry rows parallel to the direction of prevailing winds enhances drying of foliage and fruit. Controlling weeds also reduces the duration of wetness in the lower parts of fruit plants. This is particularly important for cane and bush crops (e.g., raspberry and blueberry), because lower stems are prone to infection by canker-causing fungi.

CHEMICAL CONTROL

Under some conditions, diseases of fruit crops can be managed by integrating host resistance and cultural practices. However, most fruit have at least one disease caused by a fungal or oomycete pathogen that requires application of a chemical fungicide. Because fungi and oomycetes are very different biologically, different chemicals are used to control them.

Chemical Control of Fungi

Synthetic fungicides commonly used on fruit crops include captan, mancozeb, and chlorothalonil. These are broad-spectrum fungicides that prevent spore germination on the surface of the plant. Nonsynthetic fungicides, such as sulfur and copper compounds, also protect plants from infection and are permitted for use in organic production. Synthetic fungicides are often easier to use and more effective against a range of diseases than sulfur or copper. Nevertheless, sulfur and copper are relatively inexpensive and are highly effective in some situations. For example, under dry conditions powdery mildew may be the only disease that develops, and sulfur is very effective against powdery mildew pathogens of fruit crops.

The sterol demethylation inhibitor (DMI) fungicides (e.g., myclobutanil, fenbuconazole, and tebuconazole) prevent infection but are also taken up by leaves and can kill fungi after infection has occurred. Infection prediction systems are available for some important diseases of apple, grape, and cherry; applying DMI fungicides on a post-infection basis has greatly reduced the amount of fungicide applied to these crops.[1,2] However, infection prediction systems have been developed and validated for relatively few diseases of fruit crops. Also, after many years of use, some fungal pathogens have become less sensitive to the DMI fungicides, and their ability to eradicate fungi after infection may be compromised.

The relatively new strobilurin fungicides (e.g., azoxystrobin, kresoxim methyl, and trifloxystrobin) have a broad spectrum of activity and are relatively nontoxic to applicators, consumers, and the environment. Use of strobilurins on fruit is increasing but with caution, since fungicide resistance is a concern with this class of fungicides.

Chemical Control of Oomycetes

Chemical control of oomycete diseases, especially root and crown rots, is not as effective as good soil drainage. Delivering the chemical to underground plant parts is difficult, and resting spores of oomycetes can withstand chemicals. Moreover, once symptoms of crown rot are detected, the damage is often too great to be mitigated by chemical treatment. However, chemical treatment at the time of planting can assist root establishment. In established plantings, chemicals are applied in the spring and fall when oomycete pathogens are most active. Mefanoxam is applied directly to soil in enough water to reach roots. Depending on the crop and disease, fosetyl-aluminum is applied to roots as a preplant dip, tree trunks as a wound paint, or leaves, where it is taken up and transported to roots.

BIOLOGICAL CONTROL

Biological control of fruit diseases has been researched extensively, but few products are commercially available. The bacterium *Pseudomonas syringae* and the yeast *Candida oleophila* are used as postharvest treatments on citrus and pome fruits to prevent rots caused by *Botrytis cinerea*, *Penicillium* spp., and other fungi. Strains of the fungus *Trichoderma* are used in greenhouses and nurseries to protect roots from both oomycete and fungal pathogens. The fungus *Ampelomyces quisqualis* is a parasite of powdery mildew fungi and is registered on several fruit crops. The bacterium *Bacillus subtilis* suppresses fungal diseases of many fruit crops. Biological control products have not gained widespread acceptance by fruit growers, because they are often less effective against a range of diseases and more expensive than conventional fungicides.

CONCLUSION

Fruit crop disease management is accomplished primarily by integrating host resistance, cultural practices, and chemical control. Concerns regarding the negative effects of fungicides on farm workers, consumers, and the environment have spurred the development of safer fungicides and biological control products. Likewise, nonchemical disease management practices are becoming more important because of the high cost of chemicals, restrictions on their use, and the emergence of fungicide-resistant pathogens. Compendia summarizing diseases of the major fruit crops are published by the American Phytopathological Society.[3]

ARTICLES OF FURTHER INTEREST

Biological Control in the Phyllosphere, p. 130
Biological Control of Oomycetes and Fungal Pathogens, p. 137
Fungal and Oomycete Plant Pathogens: Cell Biology, p. 480
Integrated Pest Management, p. 612
Organic Agriculture As a Form of Sustainable Agriculture, p. 846

REFERENCES

1. Funt, R.C.; Ellis, M.A.; Madden, L.V. Economic analysis of protectant and disease-forecast-based fungicide spray programs for control of apple scab and grape black rot in Ohio. Plant Dis. **1990**, *74* (9), 638–642.
2. Sutton, T.B. Changing options for the control of deciduous fruit tree diseases. Annu. Rev. Phytopathol. **1996**, *34*, 527–547.
3. American Phytopathological Society. http://www.apsnet.org. The site for ordering disease compendia: http://www.shopapspress.org/ (accessed May 2003).

Management of Fungal and Oomycete Diseases: Vegetable Crops

M

Debra Ann Inglis
Washington State University, Mount Vernon, Washington, U.S.A.

INTRODUCTION

The vegetable crop industry is unique and complex. A diversity of plants and production techniques, as well as growing and marketing operations, make the management of vegetable diseases caused by fungi and oomycetes particularly interesting and challenging. The occurrence, spread, and severity of vegetable diseases, and consequently their management, often hinges on the cropping situation in question.

VEGETABLE PRODUCTION

There are hundreds of different types of traditional and speciality vegetables for which edible plant parts (e.g., roots, tubers, bulbs, stems, leaves, flowers, fruits, or seed) are consumed.[1] Vegetables may be grown in home and market gardens, on small truck farms or large farms, or in greenhouses. Production schemes under various climates range from multiple plantings grown sequentially during the same year to annual cropping under rain-fed conditions or irrigation in arid regions. In some cases, crops are hand-planted or direct-seeded; in others, they are hand- or machine-harvested. Marketing schemes for both domestic and export markets affect handling, storage, and packaging practices, and may involve fresh market sales to consumers, retailers or wholesalers, or advanced contracts with processors. All of these factors influence the types of diseases prevalent in specific vegetable crops, as well as the management tools available to specific vegetable producers.

PRINCIPAL PATHOGENS

The number of plant pathogenic fungi and oomycetes affecting various vegetable crops make for hundreds of vegetable diseases. Key diseases are listed in Table 1, although many more comprehensive indices are available.[3–5] Vegetable pathogens can be either seed-, soil- or airborne, initiating diseases that are monocyclic (one disease cycle per growing season) or polycyclic (many disease cycles per growing season). Examples of monocyclic diseases are root rots and wilts, where thick-walled resting structures or spores in the soil or decomposing diseased plant debris provide primary sources of inoculum. New spores usually form only toward the end of a growing season, and disease increase is slow but steady over time. In contrast, many foliar diseases such as mildews, leaf spots, and blights are polycyclic diseases. The primary inoculum is often a sexual spore or fruiting body that can withstand adverse environmental conditions. Numerous asexual spores produced on infected plants provide secondary inoculum. These spores are mostly short-lived, but are typically airborne and spread rapidly by wind and rain to cause multiple infections during the growing season.

APPROACHES IN MANAGEMENT

Generally, plant disease control is directed at excluding a pathogen from a particular host or location, eradicating or destroying a pathogen, or offering protection against a pathogen by manipulating the environment or providing some type of effective barrier.[6] Disease controls directed toward improving the resistance of a host or utilizing microorganisms to reduce a pathogen population are also used.[7] Regulatory, cultural, biological, physical, and chemical measures[3] provide the means for accomplishing plant disease control.

Some examples of the exclusion of vegetable pathogens include: 1) bean seed grown in the arid western states of the United States to avoid anthracnose; 2) cabbage transplants inspected before shipment to production fields to manage black leg; and 3) potato seed tubers certified to limit spread of *Fusarium* and *Verticillium*. The eradication of pathogens is often accomplished through sanitation practices. Hot water treatments eliminate *Septoria* from celery, *Phoma* from cabbage, and *Alternaria* from carrot seed. Crop hygiene (i.e., the use of disease-free planting material and removal of infected plants) mitigates other inoculum sources. Planting healthy seed tubers and destroying volunteer potato plants that harbor late blight are examples of crop hygiene. Sodium hypochlorite, quartenary ammonium compounds, and other disinfec-

Encyclopedia of Plant and Crop Science
DOI: 10.1081/E-EPCS 120019946

Table 1 Examples of fungi and oomycetes that cause diseases on vegetable crops

Groups[a]	Pathogens	Diseases
Kingdom Fungi		
Chytrids	*Synchytrium*	Potato wart
Zygomycetes	*Choanephora Rhizopus*	Storage rots
Ascomycetes	*Erysiphe*	Powdery mildews
	Ascochyta	Leaf and pod blight on pea
	Gibberella	Stalk rot on sweet corn
	Phoma	Black leg on crucifers
	Septoria	Late blight on celery
	Sclerotinia	Stem blights and soft rots
Basidiomycetes	*Puccinia*	Rusts
	Uromyces	
	Urocystis	Smuts
	Ustilago	
Deuteromycetes	*Rhizoctonia*	Seed rots, damping off, root and fruit rots
	Sclerotium	White rot on onion
	Fusarium and *Verticillium*	Vascular wilts
	Alternaria	Leaf spots
	Botrytis Cercospora Cladosporium	
	Colletotrichum	
	Phoma	
	Stemphylium	
	Ulocladium	
Kingdom *Stramenopila*		
Oomycetes (form oospores)	*Aphanomyces*	Damping off and root rots
	Pythium	Damping off, seed, root, and fruit rots
	Phytophthora	Root and fruit rots, and blights
	Bremia Peronospora	Downy mildews
	Albugo	White rusts

[a]From Ref. 2.

tants can be used to eradicate pathogens from seed trays, pots and stakes, greenhouse benches, and storage facilities. Deep plowing and crop rotation allow decomposition of specialized pathogen survival structures. Phytophthora blight and root rot on cucurbits, Fusarium wilt on asparagus, and root rot on pea are all managed in part by rotation with nonhost plants. Steam sterilization in greenhouses, and solarization and fumigation techniques reduce or eradicate fungal and oomycete propagules from soil.

Plant protection can be accomplished by environmental manipulation, cultural practices, and fungicide treatments. Some common ways include selecting healthy seed and transplants; treating, sowing, and transplanting only when soil conditions (temperature, moisture, nutrients, tilth) ensure vigorous plant growth; arranging plant spacing and row orientation to permit ventilation in the canopy (because fungal and oomycete diseases are often favored by high humidity and wet plant surfaces); avoiding overhead irrigation when the crop is vulnerable to infection; avoiding field work when foliage is wet; picking fruit before it overripens; avoiding mechanical damage; removing field heat as quickly as possible at harvest; controlling weeds that harbor plant pathogens or contribute to humid microclimates; and controlling insects that cause wounds.[5] Protecting against fungi and oomycetes by using fungicides is also common in vegetable crops. For some vegetable crops like potatoes and tomatoes, the magnitude of fungicide use is quite high; for others it is quite low.[8] Proper selection, timing, and methods of application are critical to appropriate fungicide use. Seed treatment with fungicides such as captan and thiram minimize the likelihood of damping off caused by *Pythium*, *Rhizoctonia*, and *Fusarium*. Fungicides applied to the foliage such as azoxystrobin, chlorothalonil, copper, fludioxonil, iprodione, mancozeb, sulfur, and thiophanate-methyl) are most likely to succeed if used preventively (before infection) rather than curatively (after infection). Presently, there are few systemic fungicides available in vegetable production that have curative properties. Although mefenoxam is one, diseases caused by oomycetes (such as downy mildew on lettuce and late blight on potato) have developed resistance to mefenoxam.

Use of vegetable cultivars with genetic resistance results in fewer fungicide applications, and is widely employed. Rust-resistant asparagus, *Fusarium*- and *Verticillium*-resistant tomato, angular leaf spot-resistant cucumber, powdery mildew-resistant squash, and downy mildew-resistant lettuce cultivars are among those that are widely available commercially. However, new races of pathogens often pose the threat of overcoming such resistances, and must be monitored carefully.

Biological control agents are becoming more commonly available in vegetable disease management.[7] They provide useful management alternatives, especially for organic production systems. *Trichoderma*, *Gliocladium*, *Streptomyces*, and *Bacillus* spp. are now packaged in commercial products directed mainly at seed- and soilborne pathogens.

Integrated pest management (IPM) programs embrace the principle of integrating multiple crop protection practices to assure minimal pesticide use. By employing genetic, cultural, and biological strategies, and by monitoring pathogen population levels in relation to crop damage, established environmental or pathogen thresholds can serve as guides for appropriately timed chemical applications. IPM programs such as BOTCAST for onion botrytis leaf blight, TOM-CAST for tomato early blight, and BLITE-CAST for potato blight have been used in vegetable disease control.[5]

CONCLUSION

Producing healthy vegetables for a growing population with changing food habits[1] will invite new approaches for managing vegetable diseases in the future. The demand for farm-fresh produce, new alternative specialty vegetables, organic production methods, food that is safe from microbial contaminants and pesticide residues, and food grown under production practices that protect the environment may change the ways in which many vegetable diseases are currently managed. Lack of sufficient farmland to ensure adequate crop rotation schemes, compounded by new strains of pathogens with resistance to commercially desirable cultivars or widely used fungicides, may result in the reemergence of vegetable diseases previously kept adequately in check.

To meet such challenges, the availability of more information-based prediction systems may reduce unnecessary fungicide applications and promote more widespread use of biological controls.[8] Novel integration of management strategies (e.g., combining soil solarization and biological agents, utilizing organic amendments in culture media to improve soil quality and control root diseases in transplant operations, and exploiting plant growth-promoting rhizobacteria to induce systemic disease resistance) may be further exploited.[9] Adapting new fungicides with novel modes of action and compounds that trigger defense mechanisms in plants are on the horizon.[10] Expansion into the areas of biologically intensive IPM, computer and video technology, and biotechnology will likely bring about new vegetable disease control measures that are unavailable today.

ARTICLES OF FURTHER INTEREST

Biological Control of Oomycetes and Fungal Pathogens, p. 137
Breeding for Durable Resistance, p. 179
Fungal and Oomycete Plant Pathogens: Cell Biology, p. 480
Oomycete–Plant Interactions: Current Issues in, p. 843
Seed Borne Pathogens, p. 1126

REFERENCES

1. Swiader, J.M.; Ware, G.W. *Producing Vegetable Crops*, 5th Ed.; Interstate Publishers, Inc.: Danville, IL, 2002.
2. Hanlin, R.T.; Ulloa, M. *Illustrated Dictionary of Mycology*; APS Press: St. Paul, MN, 200.
3. Agrios, G.N. Control of Plant Diseases. In *Plant Pathology*, 3rd Ed.; Academic Press, Inc.: San Diego, CA, 1988; 180–232.
4. http://www.shopapspress.org/, (accessed March 2003).
5. Howard, R.J.; Garland, J.A.; Seaman, W.L. *Diseases and Pests of Vegetable Crops in Canada*; Howard, R.J., Garland, J.A., Seaman, W.L., Eds.; The Canadian Phytopathological Society, Entomological Society of Canada: Canada, 1994.
6. Maloy, O.C. *Plant Disease Control*; John Wiley & Sons, Inc.: New York, 1993.
7. Cook, R.J. Advances in plant health management in the twentieth century. Annu. Rev. Phytopathol. **2000**, *38*, 95–116.
8. Zalom, F.G.; Fry, W.E. Biologically Intensive IPM for Vegetable Crops. In *Food, Crop Pests, and the Environment: The Need and Potential for Biologically Intensive Integrated Pest Management*; Zalom, F.G., Fry, W.E., Eds.; APS Press: Minneapolis, 1992; 108–165.
9. Tuzun, S.; Kloepper, J.W. Potential Application of Plant Growth-Promoting Rhizobacteria to Induce Systemic Disease Resistance. In *Novel Approaches to Integrated Pest Management*; Reuveni, R., Ed.; Lewis Publishers, CRC Press, Inc.: Boca Raton, 1995; 115–127.
10. Ragsdale, N.N.; Sisler, H.D. Social and political implications of managing plant disease with decreased availability of fungicides in the United States. Annu. Rev. Phytopathol. **1994**, *32*, 545–557.

Management of Nematode Diseases: Options

Gregory L. Tylka
Iowa State University, Ames, Iowa, U.S.A.

INTRODUCTION

Plant-parasitic nematodes reduce both the quantity and quality of yield produced by grain, fruit, and vegetable crops and cause discoloration and deformation of plant parts sold for fresh market purposes (Fig. 1). Additionally, many modern, large-scale crop production systems rely on uniform growth and maturation of the crop for efficient mechanical harvesting, and parasitism by nematodes can result in uneven development of plants throughout a field or seed bed, reducing harvest efficiency. Plant-parasitic nematodes in the class Adenophorea also serve as vectors for damaging plant viruses. Consequently, there are direct and indirect yield losses that result from plant-parasitic nematode infestations.

The objective of most nematode management programs is reduction in population densities of the targeted nematode(s), resulting in a direct, concomitant increase in crop yield. Such successful management programs not only increase the short-term economic value of crops being grown, but also maintain the future economic value of the land by maintaining productivity through reduction in pathogen population densities. Successful nematode management also may serve to reduce or eliminate indirect yield losses caused by resistance-breaking pathogen interactions and other disease complexes. For example, tomato cultivars possessing genetic resistance to Fusarium wilt, caused by the fungus *Fusarium oxysporum lycopersici*, will develop the disease if also infected with the root-knot nematode, *Meloidogyne incognita*.[1] However, use of tomato varieties with genetic resistance to both Fusarium wilt and root-knot nematode results in control of both diseases.

There are numerous strategies or tactics that may be used to manage infestations of plant-parasitic nematodes. However, not all strategies are available for every combination of nematode and host crop. Also, most nematode management strategies are not therapeutic; they must be implemented before the crop is planted. Consequently, growers must have advance knowledge of the presence of damaging population densities of nematodes in the fields where crops will be grown. To develop an effective management program for plant-parasitic nematodes, growers must consider what management tactics are available, the economic costs of those tactics in relation to the value of the crop being grown, and whether the management tactics can be implemented without disruption of the production system. There are several categories of nematode management strategies.

EFFECTIVE SCOUTING AND IDENTIFICATION

The first step in a successful nematode management program is accurate and early identification or diagnosis of a problem. However, obvious symptoms of damage may not appear for many months or years after a nematode infestation becomes established due to the obligately parasitic nature of plant-parasitic nematodes (Fig. 2). Extension educators and agribusiness personnel (agronomists, crop scouts, crop advisors, etc.), who routinely communicate and interact with growers, are essential in the development of a nematode management program. These individuals often are the first to notice unusual growth patterns in the crops and usually are knowledgeable about appropriate techniques for scouting for plant-parasitic nematodes, as well as other types of pests. Also, these people often assist growers in determining what specific management tactics are deployed.

Effective scouting for plant-parasitic nematode infestations involves careful observation of plant parts for disease symptoms, but confirmation of a nematode infestation usually requires collection and analysis of plant and/or soil samples. Identification of the genera of plant-parasitic nematodes and enumeration of the number of individuals present is often sufficient for diagnosis and subsequent development of a management program. However, identification of plant-parasitic nematodes to species or subspecific level (race or biotype or pathotype) is necessary in some cases. Accurate knowledge of the nematode or nematodes causing damage is essential in defining what tactics will be effective in a management program. For example, the geographic range of numerous species of the lesion nematode *Pratylenchus* overlap, but the host ranges of the species can vary considerably.[2] Consequently, if a grower wanted to grow a crop that was not a host for the lesion nematode

Encyclopedia of Plant and Crop Science
DOI: 10.1081/E-EPCS 120010394

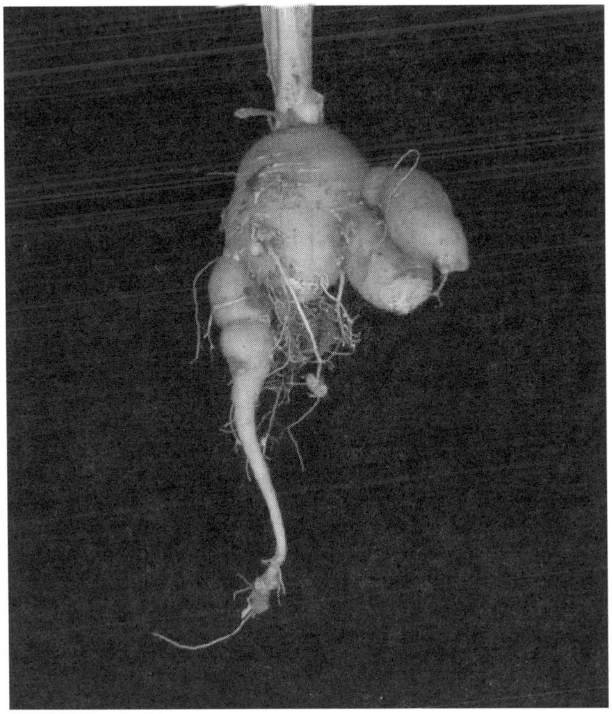

Fig. 1 Deformed carrot root infected with the northern root-knot nematode, *Meloidogyne hapla*. (Photo by G.L. Tylka.) (*View this art in color at www.dekker.com.*)

present in a field as part of a management program, it would be necessary to determine the species of *Pratylenchus* present in order to know what crop species will not serve as a host for the nematode. In addition to host range, other aspects of the biology of the nematode that influence the effectiveness of various management options include the length of the nematode life cycle, the reproductive capability of the nematode females, and the ability of the nematode to survive periods of adverse environmental conditions (i.e., temperature extremes, lack of water, lack of food).

HOST RESISTANCE

For many crops and species of plant-parasitic nematodes, host resistance is an effective and economic management strategy. Nematode resistance is defined as a plant allowing only inefficient reproduction of a nematode species.[3] Because reproduction of the nematode is reduced on resistant plant cultivars, the quantity (and possibly quality) of yield is usually increased, and a reduction in nematode population densities during the course of a growing season usually occurs. Host resistance to plant-parasitic nematodes has been identified and incorporated into many grain, fruit, and fiber crops, and resistance is mostly to sedentary, endoparasitic nematodes, which have a complex feeding relationship with their hosts.[4] Although host plant resistance to nematodes reduces nematode reproduction, the resistance is usually not 100% effective, and a small proportion of the nematode population can successfully reproduce on the resistant plants. Consequently, repeated use of the resistance can result in directional selection of a nematode population that is capable of reproducing on the resistant plant. Nevertheless, host plant resistance is an effective and economical nematode management strategy that warrants serious consideration when available as a management option.

HOST TOLERANCE

In addition to host plant resistance, nematode-tolerant plant cultivars can be a useful management tool. Nematode-tolerant plants are those that are relatively insensitive to parasitism by plant-parasitic nematodes.[3] Tolerant cultivars suffer relatively little yield loss despite being parasitized by nematodes. In addition to providing economically acceptable yields in nematode-infested fields, tolerant plant cultivars can effectively prolong the utility of resistant plant cultivars by not selecting for nematode populations that can reproduce on resistant

Fig. 2 Susceptible soybean cultivars without symptoms despite being grown in field infested with a high population density of the soybean cyst nematode *Heterodera glycines*. (Photo by G.L. Tylka.) (*View this art in color at www.dekker.com.*)

cultivars. Unfortunately, nematode-tolerant plant cultivars are not available for most crops.

NONHOST CROPS

Because plant-parasitic nematodes are obligate parasites, they are unable to complete their life cycles and reproduce in the absence of a host crop. Thus, growing nonhost crops can be an effective management strategy. But the usefulness of this management strategy depends on several factors. Most obviously, the host range of the damaging plant-parasitic nematodes targeted for control will dictate what nonhost crops can be utilized for management purposes. There will be only a few nonhost crop possibilities for nematodes that have a broad host range, such as many species of lesion nematode (*Pratylenchus*), whereas there are several nonhost crop options for nematodes that can reproduce on only a few crop plants, such as the soybean cyst nematode (*Heterodera glycines*). Also, use of nonhost crops will not be an option if the nonhost crops are not of sufficient economic worth in the production systems in which the nematodes need to be controlled. For example, the soybean cyst nematode is a widespread and very damaging parasite of soybeans throughout the entire midwestern United States, and this nematode is unable to reproduce on many crop species, but corn is the only nonhost crop that is grown in alternating years with soybeans throughout most of this region because of agricultural economic constraints. Finally, use of nonhost crops as a nematode management strategy is more effective for plant-parasitic nematodes that cannot survive long periods of time in the absence of a host crop than for those nematodes that have effective long-term survival mechanisms, such as the cyst nematodes (*Globodera* and *Heterodera*).

CHEMICAL CONTROL

Beginning in the middle of the 20th century, pesticides became an effective and cost-efficient means to manage many plant-parasitic nematodes. Many such products, with various modes of action, were developed and marketed beginning in the 1940s. These pesticides generally can be categorized based on their physical properties as either fumigants or contact nematicides.[5] Fumigant nematicides vaporize when applied and permeate quickly through the soil. The effectiveness of fumigants is affected by soil factors such as moisture, temperature, texture, and type. Contact nematicides do not vaporize and, consequently, they diffuse more slowly and are not affected by soil factors as much as fumigants. Some of the more common nematicides include those with the common names aldicarb, carbofuran, 1,3 dichloropropene, dibromochloropropane, ethoprop, ethylene dibromide, fenamiphos, oxamyl, and terbufos.[5,6] Also, broad-spectrum, fumigant biocides used to manage plant-parasitic nematodes include chloropicrin, metam sodium, and methyl bromide.

In the latter part of the 20th century, many nematicides were banned from use in the United States because of environmental concerns. Currently, there are still several nematicidal products registered for use for nematode management, but the legal availability of these products varies greatly depending on the host crop. Nematicides and biocides are an integral part of nematode management programs for high-value fruit and vegetable crops, but very few of these pesticides are still registered for use in the United States for nematode management on low-value row crops and specialty crops grown in relatively small areas. Methyl bromide is widely used for control of insects, weeds, nematodes, and other pathogens in many different crops in the United States, but this biocide is being gradually phased out due to its harmful effects on the Earth's ozone layer. The use of methyl bromide will be completely discontinued in the United States and other developed countries by the year 2005. Research is underway to identify alternatives to methyl bromide for control of these pests and pathogens.

There are several difficulties associated with chemical control of soil-borne, plant-parasitic nematodes.[6] Crop species vary in their sensitivity to the pesticides used for nematode management, and phytotoxicity is more of a concern with the broad-spectrum biocides than with the nematicides. Additionally, nematodes can be distributed a meter or more down into the soil, so incorporation of the pesticides deep into the soil profile is essential to maximize effectiveness, particularly for those products that are not fumigants. No nematicide or biocide will be able to kill all of the individuals in the targeted nematode population and surviving nematodes usually multiply throughout the growing season, so repeated application of these pesticides every time a susceptible crop is grown is usually necessary. Also, individual nematodes that survive nematicide applications may possess some resistance to the chemicals, so selection for nematicide-resistant nematode populations can be a concern. But despite these practical difficulties, limitations, and concerns, nematicides and broad-spectrum biocides can be effective and economical management tactics for plant-parasitic nematodes for some crop production systems, depending on the value of the crop, the crop production practices that are routinely used, the plant-parasitic nematode species that are targeted, and the legal availability of the nematicidal or biocide products.

BIOLOGICAL CONTROL

There are many organisms that live in soil and utilize nematodes as a food source, including bacteria, fungi, mites, and other nematodes. Several natural antagonists, primarily bacteria and fungi, have been investigated for use as biological control agents for plant-parasitic nematodes. Although results of some experiments indicate that there is great potential for use of such natural antagonists for management of plant-parasitic nematodes, few products have been commercialized and are available currently for use in managing nematodes.

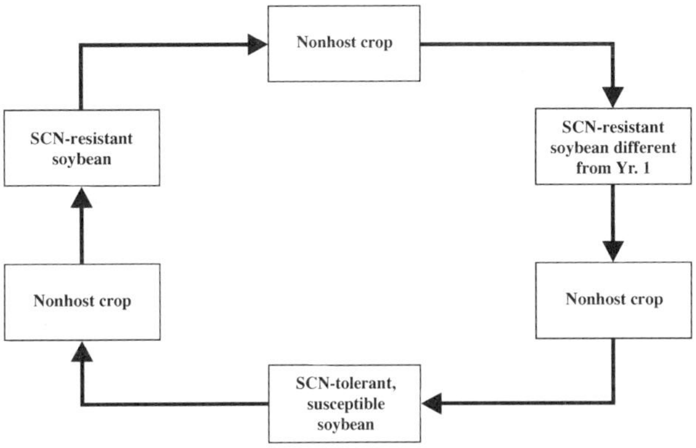

Fig. 4 Crop rotation scheme recommended by the Iowa State University Plant Disease Clinic in conjunction with scouting for early detection of new infestations for integrated management of the soybean cyst nematode (SCN) *Heterodera glycines*.

OTHER TACTICS

There are many other crop production practices that can be employed in attempts to reduce nematode reproduction and/or increase crop yields. As with most every other management strategy, these tactics will not have a beneficial effect for every combination of nematode and host crop. Among the more common practices that have been employed in attempts to achieve management of plant-parasitic nematodes are: 1) planting earlier or later than normal to avoid peak periods of infection of the young crop by the nematode population, 2) growing nematicidal cover crops and trap crops, and 3) leaving infested areas unplanted to reduce nematode population densities directly. Reducing the movement of soil and infected plant tissue, and thus the movement of plant-parasitic nematodes, from field to field by washing off tillage equipment and using only thoroughly cleaned seed (Fig. 3) and propagation materials that are free of nematode infection can also be of great benefit in a nematode management program. But these various tactics vary greatly in their utility based on the biology of the targeted nematode species and crop plant as well as other aspects of the crop production system. Nonetheless, many such tactics can add to the effectiveness of an integrated nematode management program.

Fig. 3 Soybean seed contaminated with soil particles (peds) containing cysts of the soybean cyst nematode *Heterodera glycines*. Planting of such uncleaned seed could introduce the nematode to a previously uninfested field. (Photo by G.L. Tylka.) (*View this art in color at www.dekker.com.*)

CONCLUSION

As is the case with most pests, the most effective management program for plant-parasitic nematodes is one that incorporates several management strategies rather than one that relies on any single tactic. The limitation to integrated pest management (IPM) for plant-parasitic nematodes is the availability of different management tactics, such as host resistance, nonhost crops, and nematicides. Fig. 4 illustrates the recommended program for integrated management of the soybean cyst nematode in Iowa. Coordinated use of all available management strategies serves to reduce the selection pressure on nematode populations to overcome any individual tactic and decreases the risk of failure of any single management tactic.

ARTICLES OF FURTHER INTEREST

Nematode Infestations: Assessment, p. 788
Nematodes and Host Resistance, p. 805

REFERENCES

1. Sidhu, G.; Webster, J.M. Predisposition of tomato to the wilt fungus (*Fusarium oxysporum lycopersici*) by the root-knot nematode (*Meloidogyne incognita*). Nematologica **1977**, *23*, 436–442.
2. *Manual of Agricultural Nematology*; Nickle, W.R., Ed.; Marcel Dekker, Inc.: New York, 1991.
3. Cook, R. Nature and inheritance of nematode resistance in cereals. J. Nematol. **1974**, *6*, 165–174.
4. Young, L.D. Breeding for Nematode Resistance and Tolerance. In *Plant Nematode Interactions*; Barker, K.R., Pederson, G.A., Windham, G.L., Eds.; American Society of Agronomy, Inc., Crop Science Society of America, Inc., Soil Science Society of America, Inc.: Madison, WI, 1998; 187–207.
5. Heald, C.M. Classical Nematode Management Practices. In *Vistas on Nematology*; Veech, J.A., Dickson, D.W., Eds.; Society of Nematologists, Inc.: Hyattsville, MD, 1987; 100–104.
6. Johnson, A.W.; Feldmesser, J. Nematicides—A Historical Review. In *Vistas on Nematology*; Veech, J.A., Dickson, D.W., Eds.; Society of Nematologists, Inc.: Hyattsville, MD, 1987; 448–454.

Marigold Flower: Industrial Applications

James N. BeMiller
Purdue University, West Lafayette, Indiana, U.S.A.

INTRODUCTION

Marigold (*Tagetes erecta*) is grown commercially as a source of lutein, which is used as a poultry feed additive, a yellow food coloring, and a human nutritional supplement. Lutein is absorbed more efficiently from the gastrointestinal tract when the naturally occurring esterified form is saponified. The spent flower meal after lutein ester extraction yields a protein-polysaccharide with citrus oil emulsifying and emulsion-stabilizing properties. This water-soluble gum, called marigold flower polysaccharide (MFP), has been partially characterized.

MARIGOLD FLOWER PRODUCTION AND USE

Marigold (*Tagetes erecta* L., family Asteraceae) is grown not only as an ornamental, cut flower, and landscape plant, but also as a source of lutein for use as an additive to poultry feed, as a yellow food coloring, and as a human nutritional supplement. Marigolds are grown for this purpose in various locations in the Western Hemisphere (primarily in Mexico and Peru) and in Asia, by and for various companies who process flowers into various products. The agronomics of *Tagetes erecta* have been studied rather extensively.

Marigold flower oil, an essential oil obtained by steam distillation, is used for compounding perfumes. The essential oil contains biocidal components (terpenoids) and has been investigated for its bactericidal, fungicidal, larvicidal, insecticidal, and wound healing properties.

PREPARATION OF LUTEIN

Blossoms are collected both by hand and mechanically. Fresh blossoms are either heated or ensiled, then pressed to remove as much water as possible. The resulting cake is dried, pelletized, and subjected to extraction, usually with hexane. The extracted pigment, called marigold oleoresin, is primarily composed of mono- and diesters of lutein (88%).

By far the largest volume use of lutein is as an additive to poultry feed. It is added to intensify the yellow color of egg yolks and broiler skin, especially where white corn is used as feed. However, lutein esters, the primary component of the oleoresin, are not absorbed well, so the pigment is usually saponified. Various degrees of purification are used. For poultry feed, crude saponified extract in the form of an emulsion is usually absorbed on a material such as an earth or the meal remaining after extraction, which has been finely ground to convert it into a dry powder. This product has also been added to fish feed. That used as a food coloring or added to vitamin mixtures must be purified. Lutein is an antioxidant that scavenges singlet oxygen and free radicals. It is claimed to prevent age-related macular degeneration[1] and is accepted as a "generally recognized as safe" substance by the U.S. Food and Drug Administration. Marigold oleoresin has been investigated as a dye for silk.

MARIGOLD FLOWER POLYSACCHARIDE

Water extraction of the meal remaining after commercial removal of the oleoresin from marigold flowers yields a water-soluble gum called marigold flower polysaccharide (MFP).[2] Low-concentration, aqueous solutions of MFP have low viscosity and exhibit emulsifying and emulsion-stabilizing properties similar to those of gum arabic (gum acacia).[2,3] This is important because there is a desire among food processors and other users of gum arabic (gum acacia) to have an alternative to it, because the supply and quality of gum arabic are variable and uncertain. Gum arabic has unique and important properties and applications.[4] Crucial unique properties of gum arabic are its ability to form high-solids, low-viscosity solutions; its emulsifying and protective colloid properties, its ability to withstand relatively high temperatures during spray-drying; and its adhesive and film-forming properties. It is used in food, beverage, confectionary, and related products. Among its unique and important applications are the preparation of spray-dried, free-flowing citrus oil powders (fixed flavors) and bakers' citrus oil emulsions, and as a glaze on confectionary products.

Encyclopedia of Plant and Crop Science
DOI: 10.1081/E-EPCS 120010433

MFP can be extracted from spent meal with warm (50–55°C) water.[2,3,5–7] MFP is a heterogenous material. The major component is a protein-polysaccharide,[2,3] as is the active component of gum arabic.[8,9] The polysaccharide component is a highly branched, acidic arabinoglucogalactan containing D-galactose, D-glucose, L-arabinose, and D-galacturonic acid in approximate mole ratios of 16:6:3:1.[2,6,7] The protein part is composed of at least two hydrophobic polypeptide constituents.[3] The structures of gum arabic and MFP (as protein-polysaccharides) are both unique in the world of industrial and food gums (hydrocolloids) and are key to their properties. In commercial gum arabic, the active component—a protein-polysaccharide—comprises only 1–2% of the total material.

As previously stated, aqueous solutions of MFP, at low concentrations, have low viscosity and have emulsifying and emulsion-stabilizing properties similar to those of gum arabic.[2,6,7] However, while gum arabic forms low-viscosity solutions at concentrations up to 25%, the concentration at which MFP solutions began to exhibit marked increases in viscosity for small increases in concentration is about 4%. Therefore, it cannot produce high-solids, low-viscosity solutions as gum arabic does.

The crude MFP extract is dark. To obtain minimal color, an oxidative pretreatment of the meal prior to gum extraction was developed.[2,3] Bleached MFP also has emulsifying and emulsion-stabilizing powers for limonene equivalent to those of gum arabic at equal concentrations.[2,3] However, the pretreatment increases the viscosity of MFP solutions. Therefore, it is currently not possible to prepare the desired high-solids, low-viscosity solutions of MFP and, hence, the concentrated emulsions. Clearly, this substance warrants further investigation, even though specially modified starch products are acceptable substitutes for gum arabic in many applications.

REFERENCES

1. Fullmer, L.A.; Shao, A. The role of lutein in eye health and nutrition. Cereal Foods World **2001**, *46*, 408–413.
2. BeMiller, J.N.; Gupta, A.K.; Wickramasingha, D.S. Structure and Properties of the Water-Extractable Polysaccharide of Marigold (*Tagetes erecta*) Petals. Frontiers in Carbohydrate Research—1; Millane, R.P., BeMiller, J.N., Chandrasekaran, R., Eds.; Elsevier Applied Science: London, 1989; 1–14.
3. Wickramasingha, D.S. Partial Characterization of the Structure and Properties of the Water Extracted Polysaccharide of Marigold (*Tagetes erecta*) Flower Petals. M.S. Thesis; Purdue University: West Lafayette, IN, 1990; 1–78.
4. Whistler, R.L. Exudate Gums. Industrial Gums; Whistler, R.L., BeMiller, J.N., Eds.; Academic Press: San Diego, 1993; 311–318.
5. Medina Fuentes, A.L. Marigold (*Tagetes erecta*) Flower Polysaccharide: (1) Comparison of Preparations from Different Sources; (2) Viscosity and Emulsion Stabilizing Properties of Sulfated Gum. M.S. Thesis; Purdue University: West Lafayette, IN, 1991; 1–78.
6. Medina, A.L.; BeMiller, J.N. Marigold Flowers as a Source of a Food Gum. New Crops; Janick, J., Simon, J.E., Eds.; John Wiley: New York, 1993; 389–393.
7. Gupta, A.K.; BeMiller, J.N. Unpublished.
8. Connolly, S.; Fenyo, J.-C.; Vandevelde, M.-C. Heterogeneity and homogeneity of an arabinogalactan–protein: *Acacia senegal* gum. Food Hydrocoll. **1987**, *1*, 477–480.
9. Randall, R.C.; Phillips, G.O.; Williams, P.A. The role of the proteinaceous component on the emulsifying properties of gum arabic. Food Hydrocoll. **1988**, *2*, 131–140.

Mass Spectrometry for Identifying Proteins, Protein Modifications, and Protein Interactions in Plants

M

Michael E. Fromm
University of Nebraska, Lincoln, Nebraska, U.S.A.

INTRODUCTION

Proteomics is the large-scale identification and quantitation of the complete set of expressed proteins, defined as the proteome, of the cell. The large number of posttranslational modifications, protein–protein interactions, and subcellular locations make the proteome considerably more complex than the protein-coding capacity of the genome. This article discusses the progress in mass spectrometry in identifying proteins, in protein modifications, and in discovering protein interactions. Not covered here is the progress in other methods.

MASS SPECTROMETRY OF PEPTIDES AND PROTEINS

Mass spectrometry (MS) precisely measures the mass/charge ratio of molecules. Whole protein masses can be measured, but much higher-quality data are obtained by first digesting the proteins to peptides with a site-specific protease such as trypsin. The recent availability of the sequence of most or all of the genome of an organism has revolutionized the interpretation of MS peptide data by allowing the identification of the gene encoding the identified peptide(s).[3,4]

Single-stage MS instruments only measure the peptide's mass/charge ratio. The resulting precise peptide mass measurements are used to calculate possible amino acid compositions, but not the sequence, of the peptide. If the masses of several peptides from the same protein are obtained, the protein can often be identified from informatics analysis of the masses of the predicted protein's digested peptides from the sequenced, annotated genome. Tandem MS-MS instruments separate the peptides in an initial MS stage and then fragment them by collision with a heavy gas, and the mass/charge ratio of the fragments are then determined in a second tandem MS stage. This fragmentation pattern can be used to determine the actual sequence of the peptide. The sequence of a single peptide is often enough to identify the gene from analysis of the genome sequence (Fig. 1).

The ability of MS instruments to detect smaller amounts of peptides and proteins has improved substantially,[5] with sensitivity reported in the attomol to femtomol levels. However, the sensitivity is still less than that of DNA techniques because protein technologies still lack two key aspects of nucleic acid technologies: the ability to be amplified directly, as is done with DNA by PCR, and a simple, sequence-dependent detection method analogous to the hybridization of nucleic acids. The lack of a direct protein amplification technology makes the sensitivity of detection by MS critical. The lack of a protein sequence-dependent detection method makes the relative abundance of the protein of interest critical. The net result is that only 10 to 20% of the cell's proteins can be readily detected in the standard 2-D gel electrophoresis methods. Examination of protein modifications and protein–protein interactions generally requires some additional level of prior purification, either to obtain sufficient amounts of protein for a more detailed MS analysis or to obtain copurification of interacting proteins.

MASS SPECTROMETRY IDENTIFICATION OF PROTEINS

Currently, the standard method of using MS to identify new proteins is to fractionate the proteins by 2-D gel electrophoresis (2DE) to provide sufficient resolution to separate many but not all of the proteins. The protein spots are then visualized by staining with either colloidal Coomassie blue or a MS-friendly silver stain protocol or with some of the newer fluorescent dyes that work well for MS, such as Sypro Ruby.[6] The detection limits of the dyes are generally in the 1–5 ng of protein range, which is sufficient protein for MS analysis. Typically, the 2DE gel patterns of the control and experimental samples are compared by image analysis software to identify individual spots that differ in their expression levels or modification. The protein spots of interest can be picked, preferably by automation, as contamination by skin proteins is a frequent problem. The isolated, protein-containing gel fragments are then chemically modified and

DOI: 10.1081/E-EPCS 120010634

Protein synthesis and modification		
Native Gene or Affinity Tagged Gene →	mRNA →	Proteins & protein complexes
Informatics		
Genome Sequence →	Predicted genes & proteins →	Predicted digested peptides, masses & sequence ↓
Total Proteins		
2-D Gel →	Pick spots & digest peptides →	MS – peptide masses Or MS/MS – peptide sequences: Predicted protein
Affinity-tagged Proteins		
Affinity purify protein complex & 1-D SDS Gel →	Pick bands & digest peptides →	MS/MS – peptide masses & peptide sequences: Predicted proteins in complex

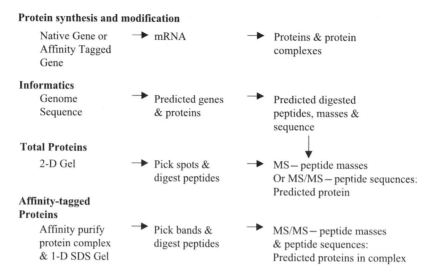

Fig. 1 Mass spectrometry analysis of proteins and affinity-tagged proteins when the genome sequence is known.

digested with site-specific proteases to obtain peptides suitable for MS analysis. Identifying low-abundance proteins still represents considerable separations and sensitivity challenges for current MS instruments, and it is estimated that only 20% of the most abundant proteins are readily detected by these methods. Alternative multistep HPLC fractionation methods that might be more sensitive are emerging.[7] In either case, quantitative comparisons can be done with isotope-coded affinity tagging (ICAT) methods.[8]

MASS SPECTROMETRY IDENTIFICATION OF PROTEIN MODIFICATIONS

Much of the regulation of protein activity occurs at the level of covalent modifications and protein–protein interactions. Covalent modifications alter the mass of the digested peptide, and there are over 100 known modifications, including the more common ones such as glycosylation, phosphorylation, acetylation, and methylation. The challenge with identifying protein modifications is that it is rare to identify all the peptides in a digested protein due to different rates of vaporization and detection, with 20% coverage of the peptides being more typical. Additional problems are the occurrence of mixtures of proteins and peptides in the fractionation methods, making it difficult to analyze peptides present in low amounts in the mixture. Therefore, relatively pure proteins are currently needed for covalent modification studies to assign the modified or low-abundance peptides to the protein of interest.[9] Purifications can be facilitated with either antibody precipitations or by using the high-affinity protein tags described below.

MASS SPECTROMETRY IDENTIFICATION OF PROTEIN INTERACTIONS

The most common separations techniques prior to MS analysis, such as 2DE, destroy protein–protein interactions and are only useful for detecting individual proteins. Additionally, identifying low-abundance proteins still represents considerable separations and sensitivity challenges for current MS instruments. A new approach, which combines affinity-tagged proteins and MS, provides a high level of purification of intact protein complexes. This new approach creates an in-frame fusion of a high-affinity tag to the gene and then transfers the tagged gene back into the cells or the organism. The resulting affinity tag on the fusion protein provides a standard high-affinity, gentle purification method that allows highly purified samples of the tagged protein and any interacting proteins to be obtained as intact complexes. The high degree of purification also allows for processing the large amounts of starting materials necessary to detect low-abundance proteins.

The affinity-tags used have been designed for either single-step or two-step purifications and rely on the copurification of other proteins to be an indication of protein–protein interactions. In the two-step or tandem purifications, a first affinity purification and elution step is followed by a second, different affinity purification and elution step, providing a much higher level of purification than a single-affinity-step purification.[10] For example, the tandem affinity purification or TAP method[11] uses IgG beads to bind to a protein A domain in the affinity tag, and then releases the bound protein by using a site-specific protease to cleave at its specific recognition site in the affinity tag eluting the now slightly shorter fusion protein. The second purification step uses a calmodulin binding protein domain in the remaining portion of the affinity tag to bind to calmodulin beads, in a Ca++-dependent manner, followed by elution by chelation of the Ca++ with EGTA. The two-step purification provides a higher level of purification and thus more reliable data on the proteins likely to be interacting with the tagged protein, as opposed to those that fortuitously copurify. Single and tandem affinity tags have been used in analyzing large numbers of yeast protein interactions[12,13] but are still relatively new in plant applications.[14] In the yeast examples, many of the copurifying yeast proteins found in affinity-tagged protein complexes had been previously identified by genetic or biochemical methods, confirming the validity of the affinity tag method.

It is critical that copurifying proteins are confirmed with additional experiments. The simplest step is to repeat the experiment, as 30% of copurifying proteins are not found again.[12] Further confirmation can be obtained by epitope-tagging the new interacting protein and confirming its interaction with the first tagged protein by precipitation of the in vivo complexes and by verifying that both proteins were precipitated by the presence of the second epitope.[15] If the function of the tagged gene is known, and therefore provides information about what type of pathway the protein is part of, functionally testing the gene encoding the interacting protein is an excellent method to confirm the interaction, using one of the functional genomics technologies described in this book.

SUMMARY

The 2DE approach can be used to identify proteins in any plant, although it is much more efficient when most or all of the genome sequence is known, as this facilitates the identification of the gene encoding the peptide(s) identified. The affinity-tagging methods can be used for any plant into which genes can be stably transformed. Additionally, in some instances transient expression assays can be conducted in cells that have the responses appropriate for the biology being examined, greatly increasing experimental flexibility and speed.[14] For example, *Arabidopsis* suspension culture cells can be infected with *Agrobacterium* in sufficient amounts for proteomic analysis[15,16] and show responses to osmotic and salt stress. We have recently demonstrated that transient expression of several TAP-tagged proteins that are involved in osmotic stress signaling in *Arabidopsis* cells can be purified in amounts sufficient for tandem MS/MS sequencing of the peptides of the isolated and digested proteins.

The post-genome era will herald increasing focus on gene function and biochemical mechanisms. Proteomics tools, particularly mass spectrometry, will play an increasing role in identifying proteins, their modifications, and their protein–protein interactions. Recent results in protein–protein interactions in yeast indicate that an extensive network of interacting proteins exists in cells that we are only now beginning to understand.[17]

REFERENCES

1. Roberts, J.K. Proteomics and a future generation of plant molecular biologists. Plant Mol. Biol. **2002**, *48* (1–2), 143–154.
2. Kersten, B., et al. Large-scale plant proteomics. Plant Mol. Biol. **2002**, *48* (1–2), 133–141.
3. Kuster, B., et al. Mass spectrometry allows direct identification of proteins in large genomes. Proteomics **2001**, *1* (5), 641–650.
4. Choudhary, J.S., et al. Interrogating the human genome using uninterpreted mass spectrometry data. Proteomics **2001**, *1* (5), 651–667.
5. Jonsson, A.P. Mass spectrometry for protein and peptide characterisation. Cell. Mol. Life Sci. **2001**, *58* (7), 868–884.
6. Liu, H.; Lin, D.; Yates, J.R., III. Multidimensional separations for protein/peptide analysis in the post-genomic era. BioTechniques **2002**, *32* (4). pp. 898, 900, 902 passim.
7. Peng, J.; Gygi, S.P. Proteomics: The move to mixtures. J. Mass Spectrom. **2001**, *36* (10), 1083–1091.
8. Turecek, F. Mass spectrometry in coupling with affinity capture release and isotope-coded affinity tags for quantitative protein analysis. J. Mass Spectrom. **2002**, *37* (1), 1–14
9. Zhang, K., et al. Identification of acetylation and methylation sites of histone H3 from chicken erythrocytes by high-accuracy matrix-assisted laser desorption ionization-time-of-flight, matrix-assisted laser desorption ionization-postsource decay, and nanoelectrospray ionization tandem mass spectrometry. Anal. Biochem. **2002**, *306* (2), 259–269.
10. Honey, S., et al. A novel multiple affinity purification tag and its use in identification of proteins associated with a cyclin–CDK complex. Nucleic Acids Res. **2001**, *29* (4), E24.
11. Rigaut, G., et al. A generic protein purification method for protein complex characterization and proteome exploration. Nat Biotechnol. **1999**, *17* (10), 1030–1032.
12. Gavin, A.C., et al. Functional organization of the yeast proteome by systematic analysis of protein complexes. Nature **2002**, *415* (6868), 141–147.
13. Ho, Y., et al. Systematic identification of protein complexes in *Saccharomyces cerevisiae* by mass spectrometry. Nature **2002**, *415* (6868), 180–183.
14. Rivas, S., et al. An approximately 400 kDa membrane-associated complex that contains one molecule of the resistance protein Cf-4. Plant J. **2002**, *29* (6), 783–796.
15. Ferrando, A., et al. Detection of in vivo protein interactions between Snf1-related kinase subunits with intron-tagged epitope-labelling in plants cells. Nucleic Acids Res. **2001**, *29* (17), 3685–3693.
16. Ferrando, A., et al. Intron-tagged epitope: A tool for facile detection and purification of proteins expressed in Agrobacterium-transformed plant cells. Plant J. **2000**, *22* (6), 553–560.
17. von Mering, C., et al. Comparative assessment of large-scale data sets of protein–protein interactions. Nature **2002**, *417* (6887), 399–403

Mechanisms of Infection: Imperfect Fungi

Gyungsoon Park
Jin-Rong Xu
Purdue University, West Lafayette, Indiana, U.S.A.

INTRODUCTION

Imperfect fungi (deuteromycetes) by definition are fungi that do not have known sexual stages. Although their taxonomic status is no longer valid, many of several thousands of known imperfect fungal species are important plant pathogens. In general, infection starts with the adhesion of fungal propagules (usually asexual spores or conidia) to plant surfaces followed by surface recognition and penetration. Some fungi enter plant hosts through natural openings, such as stomata or wounds, while others directly penetrate through plant surfaces and epidermal cells. After penetration, infectious hyphae colonize plant tissues and induce disease symptom development.

ATTACHMENT AND GERMINATION

Attachment

Adhesion of dispersing propagules to plant surfaces is commonly the first step of infection by fungal pathogens, but the composition of the adhesive materials and the development of adhesiveness vary among fungal species. In some fungi, the adhesive materials are preformed and released upon contact with the plant surface or upon hydration of conidia, such as the spore tip mucilage formed by the rice blast fungus *Magnaporthe grisea* and the adhesion knob formed by the nematode trapping fungus *Drechmeria coniospora*. In some fungal pathogens, the extracellular matrix of conidia may contain adhesive materials or enzymes that assist in attachment and preparation of the infection court.[1] In fungi with an active adhesion process, respiration and protein synthesis are required before the conidia or their germ tubes become adhesive. Synthesis of adhesive materials may be stimulated by compounds present at plant surfaces or regulated by specific fungal-plant recognition events.[1]

Adhesion has been recognized as an important step in plant infection, and it is worth noting that fungal adhesion can occur at different stages of infection-related morphogenesis.[2] Various proteins, glycoproteins, and polysaccharides have been identified in different pathogens, but the exact chemical components of adhesive materials have not been determined for most fungi. No fungal adhesive compound has been experimentally proved by directed mutagenesis to mediate plant attachment.

Germination

Conidial germination involves activation, swelling, and germ tube emergence.[3] In a few fungi, such as *Botrytis cinerea* and some *Colletrotrichum* species, conidia can remain dormant or quiescent on plant surfaces until host plants produce certain stimulatory compounds (e.g., ethylene produced during fruit ripening). However, in most fungal pathogens, conidia germinate immediately after attachment and produce germ tubes when moisture, nutrients, temperature, and other environmental conditions are favorable. Conidia of a number of fungi contain endogenous self-inhibitors that prevent conidia from germinating when the spore density exceeds a given threshold. Germination occurs only after these compounds are washed off or diluted, or their inhibitory effects are relieved with nutritional supplements or plant surface molecules (such as cutin monomers). In many fungi, adhesive materials are formed surrounding the germ tubes to mediate adhesion and possibly provide protection against harmful environmental factors.

PENETRATION

Penetration Structures

Fungal pathogens have evolved distinct strategies for penetrating plant leaves, stems, or roots. In some fungi, the undifferentiated germ tube or hyphal tip can directly form a penetration peg to invade plant cells. The penetration peg is a thin, tip-growing cellular protuberance that is generally much narrower than somatic hyphae. However, in many fungi, penetration pegs develop from the thin-walled appressorial pores located at the contact area between the appressorium and the plant surface. Appressoria are infection structures formed on the end of germ tubes or hyphae for adhering to host surface and subsequent penetration. The tight adhesion of appressoria to the plant surface is mediated by a ring of appressorium

Encyclopedia of Plant and Crop Science
DOI: 10.1081/E-EPCS 120019931

mucilage. An appressorium can be a simple or multiple-lobed germ tube tip swelling or a specialized swollen body delimited from the germ tube by a septum. Some fungi form multicellular aggregates that are known as compound appressoria or infection cushions. Multiple penetration pegs can arise from a single compound appressorium. Penetration pegs also can develop from a net-like mycelium without well-differentiated appressoria. Another infection structure common among asexual foliar pathogens is the hyphopodium. Hyphopodia formed on epiphytic somatic hyphae may be terminal, lateral, or intercalary, simple or lobed. Many environmental factors, including temperature, pH, nutrients, plant surface molecules and physical features, are known to affect penetration structure formation. Germ tubes can also recognize surface hydrophobicity, hardness, components of the plant surface, and topographical properties to regulate infection-related morphogenesis.

Mechanisms of Penetration

Both physical and enzymatic forces are used by fungal pathogens to invade plant tissues.[4] In fungi such as *M. grisea*, elevated osmotic pressures within melanized appressoria are used to puncture leaf cuticles physically and plant cell walls.[2] Blocking melanin synthesis by mutation or inhibitors prevents penetration because the melanin layer is essential for lowering the porosity of the appressorial wall. The turgor pressure in melanized appressoria is as high as 80 bars by accumulation of a high concentration of glycerol.[2] Pressure generated by turgor is exerted over the restricted area of the appressorial pore where the penetration peg emerges, resulting in penetration of plant tissue. Other fungi, including *Colletotrichum* species, also develop turgor pressures in appressoria as part of the penetration forces.[4] Forces produced by the cytoskeleton in penetration pegs may also contribute to appressorial penetration, particularly in fungi with low appressorium turgor pressures. A high concentration of cytoskeleton element in penetration peg may also be necessary to compensate for the difference in osmotic pressure between appressorial and host cell protoplasts.

Many fungal pathogens produce degradative enzymes, such as cutinases, laccases, polygalacturonases, and cellulases. For fungi that directly penetrate plant tissue without producing specialized infection structures, it is likely that cell wall–degrading enzymes play important roles in penetration. Some fungi are able to erode the plant cuticle, and localized degradation of plant cell wall materials has been observed along the penetration peg in several fungi. Even in fungi with high appressorium turgor, cell wall–degrading enzymes may be involved in penetration by modifying or softening the plant surface or cell wall. However, determining the importance of individual cell wall–degrading enzymes in plant penetration is complicated by the genetic redundancy and variable regulation of these enzymes.

INFECTIOUS GROWTH

After penetration of host epidermal cells, infectious hyphae are differentiated from the bulbous primary vesicle. Infectious hyphae are often morphologically distinct from somatic hyphae or germ tubes. In some fungi, infectious hyphae may grow only between the cuticle and cell wall or within epidermal cell walls. Some endophytic fungi can grow extensively in plant tissues without damaging host cells and only cause disease symptoms when host plants are under certain stresses or developmental processes. In biotrophic pathogens, an elaborate structure known as the haustorium is formed within the penetrated cell for nutrient absorption. Most asexual fungi, however, do not produce haustoria. In necrotrophic pathogens such as *Alternaria* and *Rhizoctonia* species, plant cells are killed in advance of the invading hyphae, and the infectious hyphae grow between and into dead and dying cells. Some asexual pathogens are hemibiotrophic. During the initial biotrophic stage, the infectious hyphae can grow without killing plant cells. Switching from the biotrophic to necrotrophic phase occurs in the later stages and results in plant cell death. A few genes that are specifically expressed in the biotrophic phase or involved in the transition from biotrophic to necrotrophic phase have been reported in several *Colletotrichum* species.

During growth within the plant, the infectious hyphae must adapt to the plant environment and produce necessary enzymes for nutrient absorption. Infectious hyphae must also use different strategies to overcome constitutive and induced plant defenses, including degradation of preformed or induced antimicrobial compounds (such as pisatin demethylases or saponin-detoxifying enzymes and enzymes for scavenging reactive oxygen species). Some fungal pathogens produce molecules that suppress induced plant defenses.[5] For some pathogens with narrow host ranges, the fungal-plant interaction is governed by specific interactions between fungal avirulence and plant resistance genes. Fungal avirulence genes have been identified from several pathogens. These avirulence genes encode proteins with diverse biological functions and do not share common structural features.

Various fungi produce toxins or elicitors to damage plant cells. While some phytotoxins, such as trichothecenes, interfere with general host cellular functions, some toxins are known as host-specific toxins, such as AAL- and HC-toxins produced by *A. alternata* and HC-toxin produced by *Cochliobolus carbonum*. Host-specific toxins are primary determinants of host range and can elicit

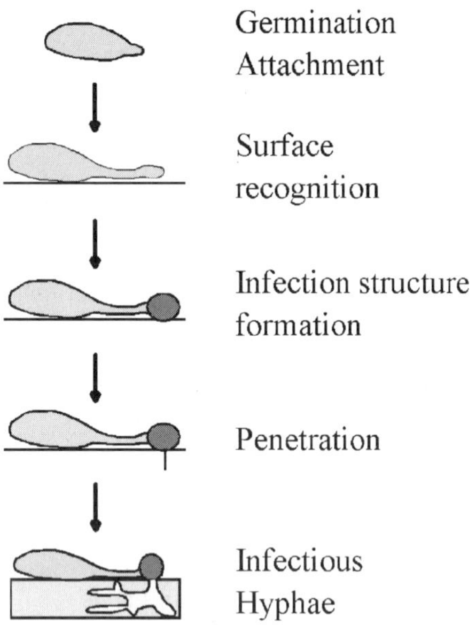

Fig. 1 Major steps of plant infection processes.

disease symptoms with the same specificity as the pathogens. In pathogens causing vascular wilt diseases, such as *Fusarium oxysporum* and *Verticillium* species, the xylem vessels of infected stems and roots may be clogged with hyphae, spores, or polysaccharides produced by the pathogens. Cell wall–degrading enzymes, proteinases, and other enzymes also play important roles during infectious hyphal growth. For example, *B. cinerea* mutants disrupted in the endopolygalacturonase gene BCPG1 are significantly reduced in virulence, and transgenic tomatoes expressing a high level of polygalacturonase inhibitory proteins exhibit increased resistance to *B. cinerea*.[6]

CONCLUSION

In the past few years, both forward and reverse genetics approaches have been used to characterize genes important for plant infection processes and signal transduction pathways regulating infection-related morphogenesis.[7] However, our knowledge about molecular mechanisms involved in the intimate fungal-plant interactions is still very limited, particularly the events after penetration.[8] Recent advances in imaging techniques and genomics studies will be very helpful to further characterize fungal infection structures and efficiently identify genes critical for fungal pathogenesis (Fig. 1).

ARTICLES OF FURTHER INTEREST

Fungal and Oomycete Plant Pathogens: Cell Biology, p. 480
Mechanims of Infection: Oomycetes, p. 697
Mechanims of Infection: Rusts, p. 701

REFERENCES

1. Epstein, L.; Nicholson, R.L. Adhesion of Spores and Hyphae to Plant Surfaces. In *The Mycota V: Plant Relationships A*; Carroll, G.C., Tudzynski, P., Eds.; Springer: Berlin, Germany, 1997; 11–26.
2. Tucker, S.L.; Talbot, N.J. Surface attachment and prepenetration stage development by plant pathogenic fungi. Annu. Rev. Phytopathol. **2001**, *39*, 385–419.
3. Osherov, N.; May, G.S. The molecular mechanisms of conidial germination. FEMS Microbiol. Lett. **2001**, *199*, 153–160.
4. Bastmeyer, M.; Deising, H.B.; Bechinger, C. Force exertion in fungal infection. Annu. Rev. Biopharmacol. Biomed. **2002**, *31*, 321–341.
5. Bouarab, K.; Melton, R.; Peart, J.; Baulcombe, D.; Osbourn, A. A saponin-detoxifying enzyme mediates suppression of plant defenses. Nature **2002**, *418*, 889–892.
6. Powell, A.L.T.; van Kan, J.; ten Have, A.; Visser, J.; Greve, L.C.; Bennett, A.B.; Labavitch, J.M. Transgenic expression of pear PGIP in tomato limits fungal colonization. Mol. Plant-Microb. Interact. **2000**, *13*, 942–950.
7. Kronstad, J.; Maria, A.D.; Funnell, D.; Laidlaw, R.D.; Lee, N.; Mario, M.D.; Ramesh, M. Signaling *via* cAMP in fungi: Interconnections with mitogen activated protein kinase pathways. Arch. Microbiol. **1998**, *170*, 395–404.
8. Kahmann, R.; Basse, C. Fungal gene expression during pathogenesis-related development and host plant colonization. Curr. Opin. Microbiol. **2001**, *4*, 374–380.

Mechanisms of Infection: Oomycetes

Paul R. J. Birch
David E. L. Cooke
Scottish Crop Research Institute, Dundee, U.K.

INTRODUCTION

Oomycetes are a diverse group of organisms that morphologically resemble fungi yet are members of the chromista, more closely related to key organisms in aquatic environments such as brown algae (e.g., kelp), golden-brown algae, and diatoms. The oomycetes include a tremendous range of free-living water molds, as well as saprophytes and pathogens of plants, algae, insects, fish, crustaceans, vertebrate animals, and microbes, including fungi. Plant pathogenic oomycetes cause devastating diseases of crop, ornamental, and native species and are arguably the most important pathogens of dicot plants. This article focuses on the most damaging groups, comprising more than 60 species of *Phytophthora*, several downy mildew genera (the most important being *Peronospora*, *Plasmopora*, and *Bremia*) and more than 100 *Pythium* species.

There is considerable diversity in oomycete infection strategies, from opportunistic or weakly pathogenic necrotrophs with wide host ranges, typified by many soil-borne *Pythium* species, through to the highly specialized, biotrophic, aerially disseminated, foliar *Peronospora* species. The genus *Phytophthora* spans these two extremes, including species colonizing leaf-litter, others that are described as ''root-pruning,'' and a range of highly specialized pathogenic hemibiotrophs. A common feature, however, is the absolute dependence on living plant tissue to complete their life cycle. Such diversity of pathogenic strategies begs the question of whether similar variation is manifest in infection mechanisms.

Oomycetes undergo many developmental stages throughout a successful infection cycle, including formation of sporangia, release of motile zoospores, their encystment and germination to form hyphae and appressoria, production of primary and secondary infection hyphae, haustoria, and finally sporangiophores. These stages facilitate dispersal, host recognition, adhesion, penetration, and colonization, encompassing biotrophic and/or necrotrophic phases of infection, finally returning to dispersal. In Fig. 1 we present a diagrammatic representation of ''typical'' oomycete infection structures and mechanisms and discuss the extent to which variation is apparent, using specific examples selected from key genera.

DISPERSAL AND RECOGNITION: SPORANGIA AND ZOOSPORES

Asexual, multinucleate sporangia develop from simple or compound sporangiophores at the apices of hyphae and are the most rapid means of reproduction (Fig. 1). The sporangia of airborne oomycetes, such as the potato late-blight pathogen *Phytophthora infestans*, are, in general, caducous, i.e., released freely from aerial hyphae and dispersed by wind or insects. In oomycetes predominantly colonizing roots or stem bases, sporangia are not freely released (noncaducous) and zoospores provide the main means of dispersal (Fig. 1). Generally, *Pythium* spp. produce noncaducous sporangia and downy mildews produce caducous sporangia. Within the Phytophthoras there are groups of species that span both extremes.

Sporangia may germinate directly to form an infection hypha or differentiate, through specialized cleavage vesicles, into between 10 and 30 zoospores that are released.[1] The mechanism adopted varies, with zoospore production prevalent in *Pythium* species, through to an inability to form zoospores in *Peronospora* species. All *Phytophthora* species form zoospores but direct germination also occurs. The lack of the aquatic zoospore stage in *Peronospora* may reflect a greater adaptation to a terrestrial habitat, and sporangia thus provide the means of dispersal, recognition, and penetration of the host.[2]

Zoospores are ephemeral, motile, biflagellate cells lacking cell walls and are often the initial agents of contact with the host.[3] Under certain conditions they remain motile for several hours, although the distances traveled rarely exceed a few centimeters; passive movement in mass flow or surface water is a more significant means of dispersal. They can differentiate to form adhesive cysts, involving detachment or resorption of flagella, formation of a cell wall, and secretion of adhesive materials. *Phytophthora cinnamomi* cysts can differentiate into a sporangium that releases a single, secondary zoospore. This process can be repeated several times to increase the distance traveled to colonize a new host.[4]

Zoospores respond to both plant exudates (chemotaxis) and to changes in electric currents (electrotaxis), although the pathological significance is unclear, since they are

Encyclopedia of Plant and Crop Science
DOI: 10.1081/E-EPCS 120019929

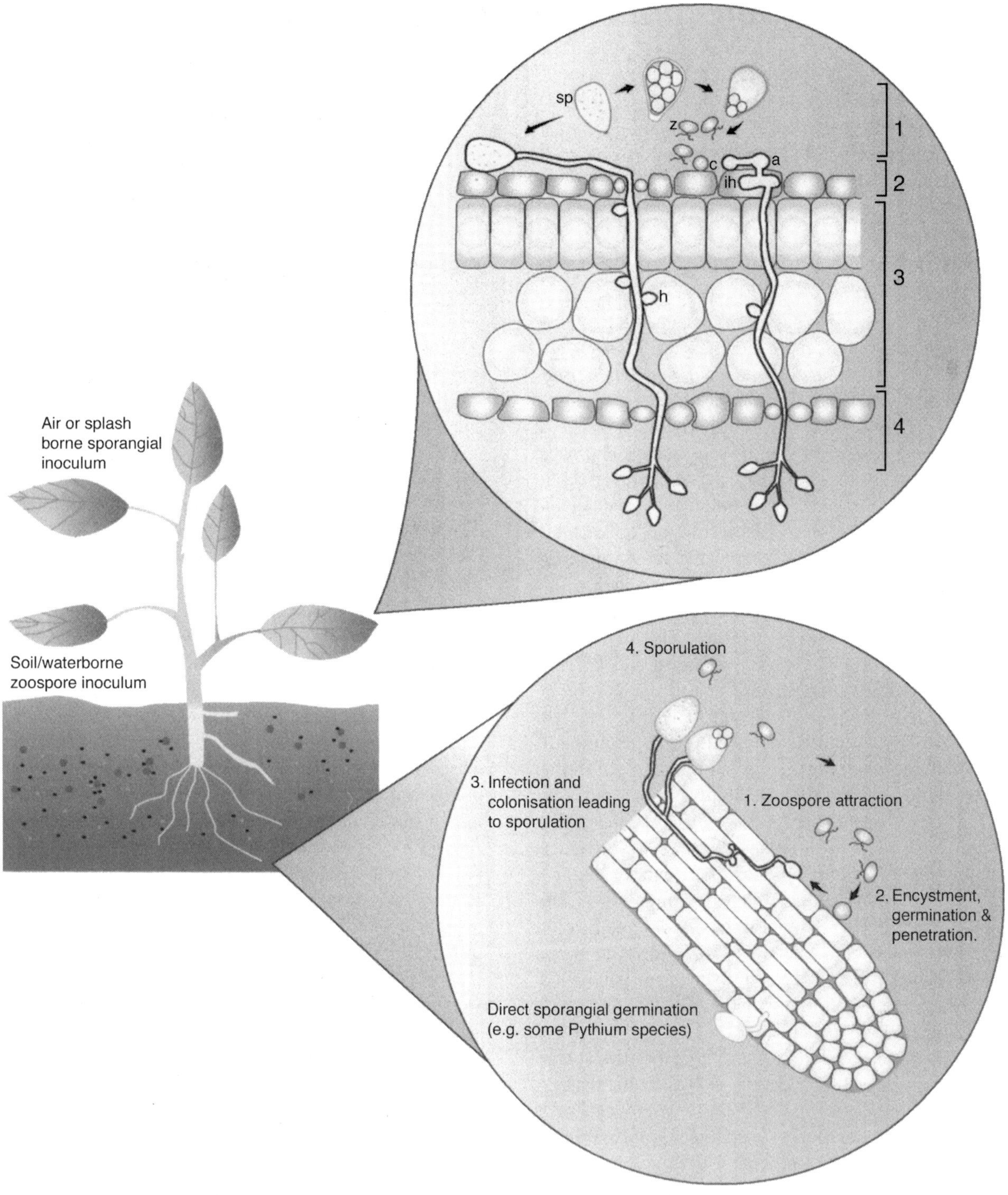

Fig. 1 Typical oomycete infection cycle in host leaf (upper circle) or root (lower circle) tissues. In each circle, **1** represents mechanisms of dispersal of oomycetes, including wind or insect dispersal (upper circle) of sporangia (sp.), and attraction of zoospores (**z**) by electrotaxis or chemitaxis (lower circle); **2** direct germination of sporangia, or encystment (**c**) and germination of zoospores to penetrate the host along anticlinal epidermal cell walls (lower circle), via stomata, or through the periclinal cell wall with the aid of an appressorium (**a**) and infection hypha (**ih**) (upper circle); **3** hyphal colonization of host tissues and the development of intracellular haustoria (**h**); **4** sporulation on the leaf or root surface to complete and reinitiate the infection cycle. (*View this art in color at www.dekker.com.*)

attracted nonspecifically to roots of host and nonhost plants. However, many oomycetes with restricted host ranges respond to specific host-derived chemicals. For example, *Phytophthora sojae* zoospores are attracted to isoflavones in soybean seeds and root exudates.[5] Recently, van West et al.[6] demonstrated that *Pythium aphanidermatum* and *Phytophthora palmivora* differed in their responses to electric currents. *P. palmivora* is anodotactic and attracted to rye grass root tips, whereas *Py. aphanidermatum* is cathodactic and attracted to a region behind rye grass root tips and to wounds, which are cathodic.

CYST GERMINATION, APPRESSORIUM FORMATION, AND HOST PENETRATION

During zoospore encystment and germination, adhesive materials are secreted that are important in preventing dislodgement of cysts and in providing attachment as physical force is exerted during host penetration. The adhesive materials are unknown, although genes encoding mucin-like proteins, termed *Car*, have been postulated to play a role in attachment and are up-regulated in germinating cysts and appressoria.[7] Moreover, a protein called CBEL, identified in *Phytophthora parasitica*, contains a fungal-like cellulose binding domain and is required for attachment to cellulose.[8]

Phytophthora cysts germinate 20 to 30 minutes after encystment, and hyphae directly penetrate host roots either intercellularly along anticlinal cell walls, or intracellularly through the periclinal wall of epidermal cells.[9] Both forms of penetration are also observed in the *Pythium* spp. Aerial plant parts are similarly infected by Phytophthorae and downy mildews but with an additional means of direct entry via stomata.

Both root- and leaf-infecting oomycetes can form appressorium-like structures, hyphal swellings that are manifest prior to penetration of host epidermal cells. In contrast to those formed in fungal plant pathogens such as *Magnaporthe grisea*, oomycete appressorium-like structures lack cross-walls separating them from the spore and germ tube. However, there is little doubt that they are involved in host penetration and thus meet one of the important criteria that define appressoria.[9]

Appressorium-like structures are formed with greater prevalence in oomycetes invading foliar host tissues. Nevertheless, whether in leaf- or root-infecting species, they are more frequently formed when directly penetrating the periclinal wall of epidermal cells, possibly due to the greater demands of overcoming more challenging host cell wall barriers.[9]

Penetration of host cell walls involves not only exertion of physical pressure but also production of enzymes facilitating breakdown of cell wall polymers, including pectin and cellulose. Various extracellular hydrolytic enzymes, including cellulase, β-glucosidases and 1,3-β-glucanases were purified from *P. infestans* culture filtrates,[10] and genes encoding some of these, and also cutinases and polygalacturonases, have been identified in a range of oomycetes.[11]

HOST COLONIZATION AND HAUSTORIUM FORMATION

The infection tube that develops from a *P. infestans* appressorium-like structure penetrates a leaf epidermal cell to form a dilated hyphal structure termed a ''primary hypha'' or ''infection vesicle.'' This ''biotrophic'' structure is surrounded by the host plasma membrane. One or more secondary hyphae develop that traverse the cell and exit, either penetrating a mesophyll cell or entering intercellular spaces. The hyphae ramify through intercellular spaces, where they form a further ''biotrophic'' structure within living plant cells, the haustorium (Fig. 1).[12] Unlike the spherical bodies typical of many true-fungi, those in *Phytophthora* spp., downy mildews, and *Pythium* spp. are fingerlike. As with the infection vesicle, they are surrounded by the host plasma membrane and make no direct contact with host cytoplasm. Their outer cell wall comprises an extrahaustorial matrix, an electron dense substance of unknown function laid down at the initial stage of haustorial penetration of the host cell wall.[9,12] Interestingly, haustoria of *Phytophthora* spp. are reported to lack nuclei,[9,12] whereas those in downy mildews may be uninucleate or multinucleate.[13]

Haustoria are biotrophic, since they are formed within living plant cells and are presumably involved, at least in part, in exchange of nutrients from the host to the pathogen. Subtle penetration of living host cells implies avoidance or suppression of host defences. Doke et al.[14] proposed the production of a glucan by *P. infestans* that suppresses the hypersensitive response, a form of programmed cell death or host cell suicide to which haustoria would be particularly vulnerable. Furthermore, a family of glucanase inhibitor proteins (GIPS) in *P. sojae* inhibit host endo-β-1,3-glucanases.[15] GIP inhibition not only prevents host glucanases from directly attacking pathogen cell walls but also prevents their activity leading to release of elicitor-active oligosaccharides that induce additional host defenses.

In susceptible hosts, oomycetes acquire sufficient nutrients to ramify through plant tissues and sporulate on the root, stem, or leaf surface, releasing sporangia and/or zoospores to complete and reinitiate the cycle of infection.

FUTURE PROSPECTS

The cytology of plant infection by oomycetes has been extensively studied over many years. Infection follows a cycle of clearly defined developmental stages. With each cell type that is formed, countless genes will be up- or down-regulated, not only to drive cellular development but also to facilitate host invasion and counteract host defences. The molecular bases of pathogenicity are poorly understood, although the current era of genomic research promises to accelerate our understanding of molecular processes in two key, model oomycetes, *P. sojae* and *P. infestans*.[5,11]

ARTICLE OF FURTHER INTEREST

Oomycete–Plant Interactions: Current Issues in, p. 843

REFERENCES

1. helios.bto.ed.ac.uk/bto/microbes/zoospores.htm.
2. Lucas, J.A.; Hayter, J.B.R.; Crute, I.R. The Downy Mildews: Host Specificity and Pathogenesis. In *Pathogenesis & Host Specificity in Plant Diseases, Histopathological, Biochemical, Genetic and Molecular Bases*; Singh, U.S., Kohmoto, K., Singh, R.P., Eds.; Pergamon. 1996.
3. Deacon, J.W.; Donaldson, S.P. Molecular recognition and homing responses of zoosporic fungi. Mycol. Res. **1993**, *97*, 1153–1171.
4. Carlile, M.J. Motility, Taxis, and Tropism in *Phytophthora*. In *Phytophthora: Its Biology, Taxonomy, Ecology and Pathology*; Erwin, D.C., Bartnicki-Garcia, S., Tsao, P.H., Eds.; American Phytopathological Society: St Paul, MN, USA, 1983; 95–107.
5. Tyler, B.M. Molecular basis of recognition between *Phytophthora* pathogens and their hosts. Annu. Rev. Phytopathol. **2002**, *40*, 137–167.
6. Van West, P.; Morris, B.M.; Reid, B.; Appiah, A.A.; Osborne, M.C.; Campbell, T.A.; Shepherd, S.J.; Gow, N.A.R. Oomycete plant pathogens use electric fields to target roots. Mol. Plant-Microb. Interact. **2002**, *15*, 790–798.
7. Gornhardt, B.; Rouhara, I.; Schmelzer, E. Cyst germination proteins of the potato pathogen *Phytophthora infestans* share homology with human mucins. Mol. Plant-Microb. Interact. **2000**, *13*, 32–42.
8. Gaulin, E.; Jauneau, A.; Villalba, F.; Rickauer, M.; Esquerré-Tugayé, M.-T.; Bottin, A. The CBEL glycoprotein of *Phytophthora parasitica* var. *nicotianae* is involved in cell wall deposition and adhesion to cellulosic substrates. J. Cell Sci. **2002**, *115*, 4565–4575.
9. Hardham, A.R. The cell biology behind *Phytophthora* pathogenicity. Australas. Plant Pathol. **2001**, *30*, 91–98.
10. Bodenmann, J.; Heiniger, U.; Hohl, H.R. Extracellular enzymes of *Phytophthora infestans*: Endo-cellulase, β-glucosidases and 1,3-β-glucanases. Can. J. Microbiol. **1985**, *31*, 75–82.
11. Kamoun, S. Molecular genetics of pathogenic oomycetes. Eukaryotic Cell **2003**, *2*, 191–199.
12. Coffey, M.D.; Gees, R. The cytology of development. Adv. Plant Pathol. **1991**, *7*, 31–51.
13. Fraymouth, J. Haustoria of the peronosporales. Trans. Brit. Mycol. Soc. **1956**, *39*, 79–107.
14. Doke, N.; Sanchez, L.M.; Yoshioka, H.; Kawakita, K.; Miura, Y.; Park, H.-J. *Molecular Genetics of Host-Specific Toxins in Plant Disease*; Kohmoto, K., Yoder, O.C., Eds.; Kluwer Academic Publishers: Dordrecht, The Netherlands, 1998; 331–341.
15. Rose, J.K.C.; Ham, K.-S.; Darvill, A.G.; Albersheim, P. Molecular cloning and characterisation of glucanase inhibitor proteins: Coevolution of a counterdefence mechanism by plant pathogens. Plant Cell **2002**, *14*, 1329–1345.

Mechanisms of Infection: Rusts

Matthias Hahn
University of Kaiserslautern, Kaiserslautern, Germany

INTRODUCTION

Among the various infection strategies used by plant pathogenic fungi, growth in living host tissue (biotrophy) and formation of haustoria are hallmarks of rust fungi. Rust infection is also unique in the ability of a germ hypha to sense surface topography in order to achieve invasion via stomata. Our understanding of biotrophic parasitism and the crucial role of rust haustoria in the uptake of sugars and amino acids from infected host cells has increased significantly using molecular and cytological methods. This article summarizes our knowledge about host invasion and establishment of rust infection.

BIOLOGY AND INFECTION MECHANISMS OF RUST FUNGI

With more than 5000 described species, rust fungi (Uredinales, Basidiomycetes) form a major group of plant pathogens. Named after their noticeable sori, which are often filled with reddish brown (or yellow, black, etc.) colored spores, rust fungi have destroyed cereal crops since ancient times, and still cause significant damage to a large variety of economically important plants (e.g., legumes, coffee, pines). The most conspicuous property of rust fungi is their highly complex life cycle and their ability to switch hosts. Macrocyclic rust species produce up to five different spore forms, but reduced life cycles also exist. Three types of rust spores are infectious, namely uredo- and aecidiospores (dikaryotic) and basidiospores (monokaryotic, haploid). In heteroecious rusts, which need two different host plants to complete their life cycle, dikaryotic spores infect one species and monokaryotic spores infect another. A well-known example is *Puccinia graminis*, which infects wheat as a dikaryon and barberry as a monokaryon. Infection by uredo- and aecidiospores usually occurs via stomata and involves a series of infection structures (see below). In contrast, basidiospores penetrate epidermal cells directly, giving rise to a mycelium which shows a lower degree of cellular differentiation and which often tends to grow toward vascular host tissue.[1]

SPORE ATTACHMENT, GERMINATION, AND INFECTION STRUCTURE FORMATION

Rust uredospores are spread by air, sometimes over very long distances, to reach their host plants. They have hydrophobic coats, which allow adhesion to aerial hydrophobic plant surfaces. Irreversible attachment, however, requires the release of adhesive gel-like material consisting of carbohydrates and glycoproteins. It accumulates—in the presence of water, by passive hydration and swelling—between spore and host surface. Hydrolytic enzymes such as esterases and cutinases increase the adhesive properties of the secreted material, possibly by ''eroding'' the hydrophobic cuticular layer and converting it to a hydrophilic one.[2] Subsequently, the germ tube emerges and grows over the plant surface, with the tip of the hypha being closely attached to the cuticle.

Localization of stomata as entry points for subsequent invasion is achieved by a unique surface-sensing mechanism of the germinated rust uredospore. Growth direction of the germ tube is aligned perpendicular to repetitive features formed by the ridges or grooves of adjacent epidermal cell walls. This increases the chance that the fungus will hit a stoma. Subsequently, when the germ tube tip reaches a stomatal guard cell, a new type of hypha called an appressorium is formed. Allowing rust germlings to develop on artificial surfaces with ridges or grooves of defined dimensions proved that topographical features provide the key signals for this differentiation process. For instance, a ridge of approximately 0.5 μm height was most effective in the case of the bean rust *Uromyces appendiculatus*, and this height is similar to the height of the guard cell lip of bean leaves.[3] Within a few minutes after topographical recognition, germ tube elongation is stopped and appressorium formation is initiated. There is evidence that ion fluxes across the plasma membrane in the tip region of the germ tube, followed by cytoskeletal rearrangements, are involved in transducing the mechanical signal ''presence of the stoma'' into the response ''appressorium formation and invasion.'' Invasion of host tissue occurs by means of additional infection structures, i.e., penetration hypha, infection hypha, and haustorial mother cell (Fig. 1). In several rust species, these structures are formed without additional signals. The

Encyclopedia of Plant and Crop Science
DOI: 10.1081/E-EPCS 120019934

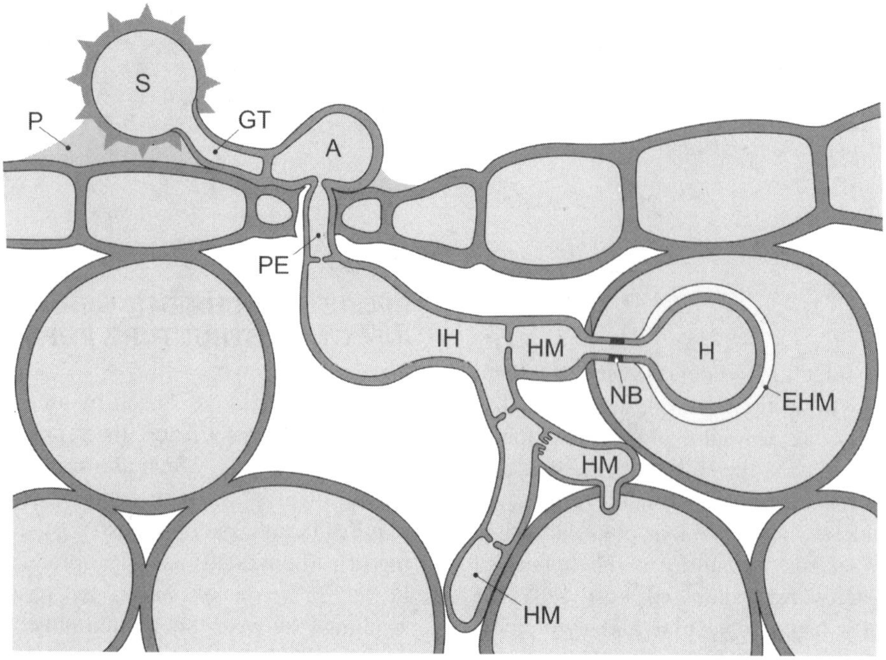

Fig. 1 Infection by rust fungi (*Uromyces* spp.). A germ tube (GT) emerges from an uredospore (S) attached to the leaf surface by an adhesion pad (P). Above the stoma, an appressorium (A) is formed, a penetration hypha (PE) grows into the substomatal chamber, and an infection hypha (IH) extends into the apoplastic space. A haustorial mother cell (HM) attaches to the host cell wall and penetrates into the cell, giving rise to a haustorium (H). Unique features of the haustorium are the neckband (NB) and the interfacial, extrahaustorial matrix (yellow) surrounded by the host-derived extrahaustorial membrane (EHM). After formation of the primary haustorium, the infection hypha forms branches and the biotrophic mycelium proliferates within the leaf tissue. (From Ref. 6.) (*View this art in color at www.dekker.com.*)

formation of infection structures in response to artificial signals has made it possible to study molecular events associated with the early invasion process of rust fungi in the absence of a host plant.

During invasion, rust fungi secrete low amounts of host cell wall degrading enzymes, which is strictly dependent on infection structure formation. For instance, the secretion of several isoforms of cellulases and pectin esterases was not observed until appressorium formation and occurred in a sequential fashion. Another enzyme, pectate lyase, was not produced until haustorium mother cell differentiation. In contrast to the situation in many other fungi, production of these enzymes was not subject to glucose repression, and no substrate induction was observed except for pectate lyase, which was secreted only in the presence of pectate (=polygalacturonate).[4]

Although it has not been proven, it is plausible that the abovementioned enzymes—together with others—aid in the attachment of the invading hyphae to cells walls of the host and in the subsequent penetration of the walls during haustorium formation. The highly controlled secretion of these lytic enzymes could be important in minimizing damage to the host cell that is invaded during haustorium formation.

HAUSTORIA

In contrast to the infection structures mentioned above, haustoria are formed only within living host cells. This makes their experimental analysis difficult, even more so since rusts are obligate biotrophs which cannot be cultivated (with few exceptions) in artificial media. Therefore, it is not possible to apply powerful molecular genetic experimental tools such as genetic transformation and knockout mutagenesis to rust fungi.

Haustoria certainly play a crucial role for the establishment of biotrophic rust infections. They are part of a parasitic mycelium which grows largely intercellularly. While being formed within host cells, they remain surrounded by a modified plasma membrane. By this means, a highly specialized interface is formed. Extensive cytological studies using light and electron microscopy identified the unique structural properties of haustoria that are likely to be correlated with their function. According to our present view, rust haustoria are centrally involved 1) in biotrophic nutrient acquisition; 2) in vitamin biosynthesis; and 3) in maintaining the biotrophic relationship with the host plant. Whereas only little experimental evidence is available about the last

M

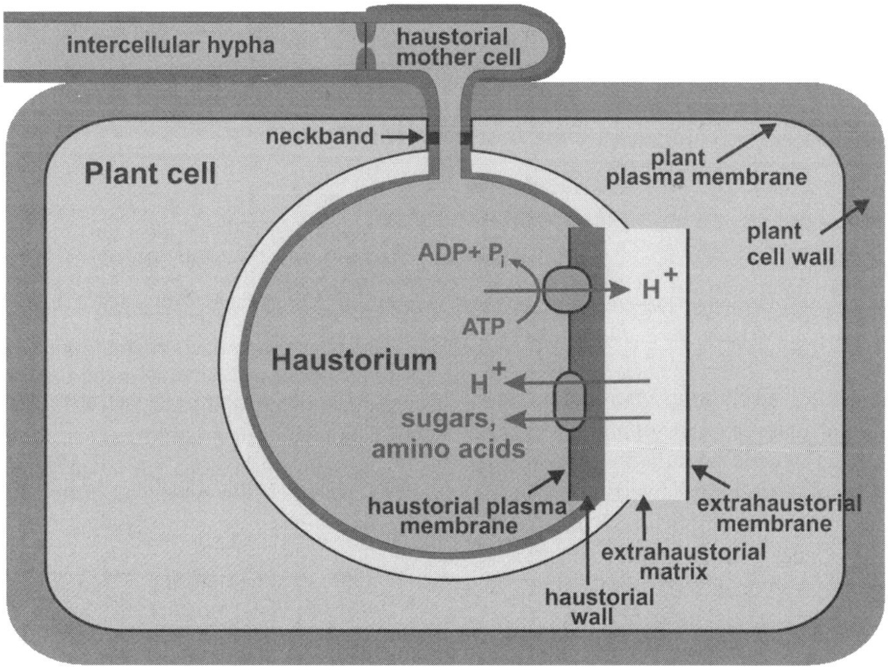

Fig. 2 Model for uptake of nutrients by a rust haustorium from an infected plant cell. The proton-pumping activity of a plasma membrane H^+–ATPase in the haustorial plasma membrane generates a proton gradient and membrane potential. They are used by proton-coupled carrier proteins that transport hexose sugars (glucose, fructose) and amino acids into the haustorium. One hexose transporter and three amino acid transporters are known in the broad bean rust. (*View this art in color at www.dekker.com.*)

aspect, molecular studies have provided significant information about the nutritional and metabolic aspects of haustorium function.

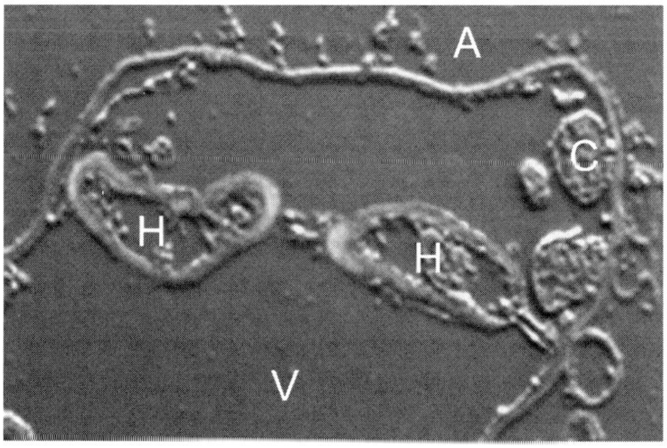

Fig. 3 Immunolocalization of a hexose transporter within the plasma membrane of rust haustoria. A resin-embedded section of a rust-infected broad bean leaf was incubated first with an antibody against a rust hexose transporter, followed by a fluorescence-labeled secondary antibody. Labeling is visible only in the periphery of rust haustoria (H). V, plant vacuole; C, chloroplast; A, apoplast. (From Ref. 6.) (*View this art in color at www.dekker.com.*)

According to the diagram shown in Fig. 2, nutrients released from the infected host cell into the interface (extrahaustorial matrix) diffuse across the interface layer and the haustorial wall to the haustorial membrane, where they are transported by proton symport carriers into the haustorial cytoplasm. The driving force for metabolite uptake is a proton gradient across the haustorial plasma membrane generated by the activity of a plasma membrane H^+–ATPase. This model is supported by the following experimental evidence: Genes encoding amino acid and hexose carriers have been cloned from a rust fungus and were found to be transcribed at high levels in haustoria, but not (or only at low levels) in early infection structures and the intercellular hyphae. With immunolocalization, a proton-coupled hexose carrier was detected exclusively in the haustorial plasma membrane, convincingly showing that this is the major site of sugar uptake into the fungus (Fig. 3).[5] Furthermore, the activity of the rust plasma membrane H^+–ATPase was shown to be highest in a membrane preparation isolated from rust haustoria, so this enzyme could well be responsible for the generation of the membrane potential required for secondary active nutrient uptake.

Evidence for major metabolic activities in haustoria also came from the identification of two highly expressed genes that encode enzymes required for thiamine biosynthesis. One of these enzymes was detected in haustoria in high concentrations, but only at low levels in other parts of the rust mycelium. Thus, haustoria appear to be not only the principal sites of nutrient uptake, but also of the biosynthesis of vitamin B1 and possibly other important metabolites that are not provided by the host plant.[6]

CONCLUSION

The biotrophic lifestyle provides rust fungi with a continuous supply of nutrients and allows them to escape competition with most other herbivorous microorganisms. To achieve this goal, rust fungi need to suppress host cell death, which is usually associated with the most powerful plant defense mechanism—the hypersensitive response. How defense suppression is accomplished remains unknown, but there is evidence that haustoria also play a crucial role in this process.

ARTICLES OF FURTHER INTEREST

Fungal and Oomycete Plant Pathogens: Cell Biology, p. 480
Mechanisms of Infection: Imperfect Fungi, p. 694
Mechanisms of Infection: Oomycetes, p. 697

REFERENCES

1. Hahn, M. The Rust Fungi. Cytology, Physiology and Molecular Biology of Infection. In *Fungal Pathology*; Kronstad, J.W., Ed.; Kluwer Academic Publishers, 2000; 267–306.
2. Deising, H.; Nicholson, R.L.; Haug, M.; Howard, R.J.; Mendgen, K. Adhesion pad formation and the involvement of cutinase and esterases in the attachment of uredospores to the host cuticle. Plant Cell **1992**, *4*, 1101–1111.
3. Dean, R.A. Signal pathways and appressorium morphogenesis. Annu. Rev. Phytopathol. **1997**, *35*, 211–234.
4. Deising, H.; Rauscher, M.; Haug, M.; Heiler, S. Differentiation and cell wall degrading enzymes in the obligately biotrophic rust fungus *Uromyces viciae-fabae*. Can. J. Bot. **1995**, *73*, S624–S631.
5. Voegele, R.T.; Struck, C.; Hahn, M.; Mendgen, K. The role of haustoria in sugar supply during infection of broad bean by the rust fungus *Uromyces fabae*. Proc. Natl. Acad. Sci. U. S. A. **2001**, *98*, 8133–8138.
6. Mendgen, K.; Hahn, M. Plant infection and the establishment of fungal biotrophy. Trends Plant Sci. **2002**, *7*, 352–356.

Medical Molecular Pharming: Therapeutic Recombinant Antibodies, Biopharmaceuticals and Edible Vaccines in Transgenic Plants Engineered via the Chloroplast Genome

M

Henry Daniell
University of Central Florida, Orlando, Florida, U.S.A.

INTRODUCTION

The daily average income of nearly one billion people is less than one U.S. dollar. Globally, about 170 million people are infected with hepatitis C virus, with 3–4 million new infections each year. WHO Department of Communicable Disease Surveillance and Response reports that more than one third of world population is infected with hepatitis B. In Asia, the prevalence of chronic hepatitis B and C is very high (about 110 million infected by HCV and 150 million infected by HBV). A large majority of hepatitis C–infected patients have severe liver cirrhosis and currently there is no vaccine available for this disease. The annual requirement of human, insulinlike growth factor 1 (IGF-1) per cirrhotic patient is 600 mg (1.5–2 mg per day) and the cost of IGF-1 per mg is $30,000. Current annual cost of interferon therapy for viral hepatitis is $26,000 per year. Therefore, agricultural scale production of therapeutic proteins and vaccines is necessary to meet such large demand at a reasonable cost. Recent observation of the potential ability of one acre of transgenic tobacco plants to produce 600 million anthrax vaccines makes this a promising new approach.

CHLOROPLAST-DERIVED THERAPEUTIC PROTEINS

Chloroplast-derived biopharmaceuticals should be inexpensive to produce and store, easy to scale up for mass production, and free of animal or human pathogens.[1,2] Chloroplast genetic engineering is an environmentally friendly approach and offers biological containment of transgenes.[3] Expression of thousands of copies of transgenes per plant cell via chloroplast genomes has yielded the highest level of foreign proteins ever reported in transgenic plants.[1,2] Chloroplast transgene expression is also free of position effect and gene silencing,[1,2] frequently encountered in nuclear transgenic plants that result in low levels of expression. Oral delivery of pharmaceuticals should eliminate the purification steps that usually account for most of production costs. Edible vaccines may be the only practical solution to combat bioterrorism agents. The successful engineering of tomato chromoplasts for high-level transgene expression in fruits,[4] coupled with hyper expression of vaccine antigens[5] and the use of plant-derived, antibiotic-free selectable markers,[6] augurs well for oral delivery of edible vaccines or biopharmaceuticals that are currently beyond the reach of those who need them most. However, because of the concerns of dosage and potential for contamination with the food supply, plant derived pharmaceuticals should be prescribed as a freeze-dried powder (with appropriately identified dosage) and fully contained.

PLANT-DERIVED BIOPHARMACEUTICALS AND HUMAN PROTEINS

Generally, levels of pharmaceutical proteins expressed in nuclear transgenic plants have been less than the levels needed for commercial feasibility, if the protein must be purified. Recombinant hepatitis B surface antigen (HBsAg) expressed via the nuclear genome induced only a low-level serum antibody response in a small human study, probably reflecting the low level of expression (1–5 ng/g fresh weight) in transgenic lettuce. Recent studies increased the level of expression up to 8.35 μg/g fresh tuber (0.000835% fresh weight). Despite such low levels of expression, adequate primary antibody response was observed in animals fed with uncooked potato tubers, but this response was lost when tubers were cooked.[7] Similarly, Norwalk virus capsid protein expressed in potatoes caused oral immunization when fed to mice, but expression levels were too low for large-scale oral administration. Similar to viral proteins, expression of genes encoding several human proteins in transgenic plants has been disappointingly low, e.g., human serum albumin, 0.02%; human protein C, 0.001%; human epidermal growth factor, 0.001% and erythropoietin,

Encyclopedia of Plant and Crop Science
DOI: 10.1081/E-EPCS 120010591

0.0026% of total soluble protein; human interferon-β was 0.000017% of fresh weight.[8–10] Therefore, there is a great need to increase expression levels of human, viral, or bacterial proteins or antigens in order to enable commercial production of pharmacologically important proteins in plants.

CHLOROPLAST TRANSGENIC SYSTEM

Chloroplast genetic engineering was conceived as a novel approach to increase expression levels and overcome problems of nuclear genetic engineering.[11–13] Foreign genes have been integrated into the chloroplast genome of several crop plants, including tobacco, tomato, and potato (up to 10,000 copies of transgenes per cell), resulting in accumulation of recombinant proteins several hundred–fold higher than nuclear transgenic plants (up to 47% of the total soluble protein).[14] Targeted integration of transgenes at specific sites into the chloroplast genomes eliminates the ''position effect'' frequently observed in nuclear transgenic plants resulting from random integration of transgenes.[1,2] In addition, gene silencing has not been observed in transgenic chloroplasts, in spite of extraordinarily high levels of transgene expression, whereas it is a common phenomenon in nuclear transformation.[2] Because of these reasons, expression and accumulation of foreign proteins is uniform in independent chloroplast transgenic lines.[5] It has been shown that multiple genes can be engineered in a single transformation event via the chloroplast genome, regulated by a single promoter; this facilitates coordinated expression of multisubunit proteins or multicomponent vaccines, or engineering new pathways. This was demonstrated by successful expression and assembly of monoclonal antibodies[15] or multigene bacterial operons[14] in transgenic chloroplasts. Yet another advantage is the lack of toxicity of foreign proteins to plant cells when they are compartmentalized within chloroplasts.[5,16] Chloroplast genetic engineering is an environmentally friendly approach, minimizing several environmental concerns, including transgene containment.[3] Most importantly, chloroplasts are able to process eukaryotic proteins, including correct folding and formation of disulfide bridges. Accumulation of large quantities of a fully assembled form of human somatotropin with the correct disulfide bonds provides strong evidence for hyperexpression and assembly of pharmaceutical proteins using this approach.[17] In addition, functional assays showed that chloroplast-synthesized cholera toxin β subunit (CTB) binds to the intestinal membrane GM1-ganglioside receptor, confirming correct folding and disulfide bond formation of the plant-derived CTB pentamers.[5] Such folding and assembly of foreign proteins should eliminate the need for highly expensive in vitro processing of pharmaceutical proteins produced in recombinant organisms. For example, 60% of the total operating cost for the commercial production of human insulin in *Escherichia coli* is associated with in vitro processing (formation of disufide bridges and cleavage of methionine).[18]

Expression of Human Serum Albumin in Transgenic Chloroplasts

Human serum albumin (HSA) accounts for 60% of the total protein in blood serum and it is the most widely used intravenous protein. Currently, HSA is produced primarily by the fractionation of blood serum. Because of unique advantages of the chloroplast genetic engineering approach, HSA was expressed in transgenic chloroplasts.[19,20] Regulation of HSA under the control of a Shine-Dalgarno sequence (SD), 5′ *psbA* region, or the *cry2Aa2* untranslated region (UTR) resulted in different levels of expression in transgenic chloroplasts from seedlings: 0.8, 1.6, and 5.9% of HSA in total protein (tp), respectively. On the other hand, a maximum of 0.02, 0.8, and 7.2% of HSA in tp was observed in transgenic potted plants regulated by SD, *cry2Aa2* UTR, or 5′ *psbA* region, respectively, demonstrating excessive

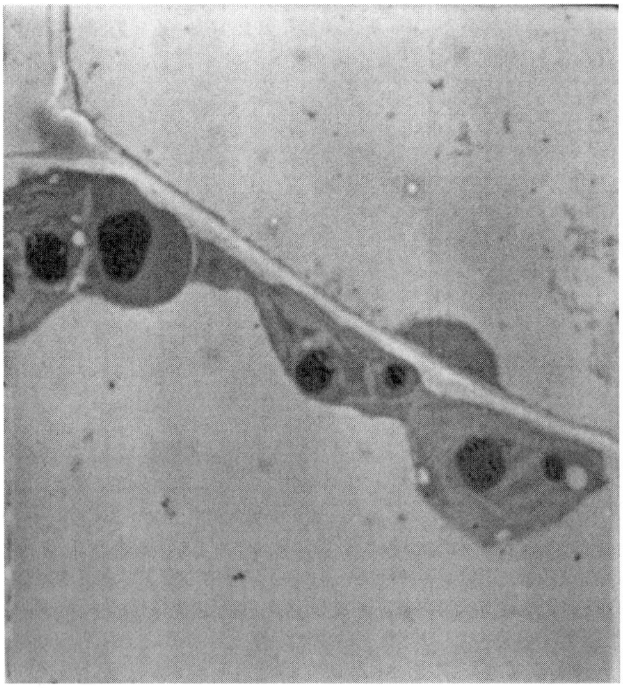

Fig. 1 Electron micrograph of immunogold labeled inclusion bodies of human serum albumin (HSA) in transgenic chloroplasts (magnification × 6300). HSA is the most widely used intravenous protein in human therapies.

proteolytic degradation, unless compensated by enhanced translation. The psbA-HSA expression was subject to developmental and light regulations, with the lowest expression observed in seedlings and maximal expression (11.1% tsp) under continuous illumination. This is the highest expression of HSA so far and five hundred–fold higher than previous reports of HSA expression in leaves of nuclear transgenic plants. Accumulation of HSA was so high that inclusion bodies were observed within transgenic chloroplasts (Fig. 1). Formation of HSA inclusion bodies not only offered protection from proteolytic degradation but also provided a simple method of purification from other cellular proteins by centrifugation. HSA inclusion bodies could be readily solubilized to obtain monomeric form using appropriate reagents. The *cry2Aa2* UTR–mediated expression in seedlings and chromoplasts, although as eficient as 5′ *psbA* region, is independent of light regulation and should therefore facilitate expression of foreign genes in nongreen tissues, thereby enabling oral delivery of pharmaceuticals. The regulatory elements used in this study should serve as a model system for enhancing expression of foreign proteins that are highly susceptible to proteolytic degradation and provide advantages in purification.

Optimization of Codon Composition and Regulatory Elements for Expression of Human Insulinlike Growth Factor 1 in Transgenic Chloroplasts

Human insulinlike growth factor 1 (IGF-1) is a potent multifunctional anabolic hormone produced by the liver. IGF-1 polypeptide is composed of 70 amino acids with a molecular weight of 7.6 kDa and contains three disulfide bonds. IGF-1 is involved in the regulation of cell proliferation and differentiation of a wide variety of cell and tissue types, and plays an important role in tissue renewal and repair. One cirrhotic patient requires 600 mg of IGF-1 per year and the cost of IGF-1 per mg is $30,000.[21] In the past, IGF-1 has been expressed in E. coli but the protein cannot be produced in the mature form, because *E. coli* does not form disulfide bonds in the cytoplasm. In order to increase the expression levels, a synthetic IGF-1 gene with chloroplast optimized codons was made. The integration of the IGF-1 gene into the tobacco chloroplast genome was confirmed using Polymerase chain reaction (PCR) and Southern analyses. The IGF-1 protein was detected in large quantities in transgenic chloroplasts by western blot analysis. The ELISA performed on the transgenic lines showed expression levels up to 32% IGF-1 of the total soluble protein, both in the native and synthetic genes. This is the highest level of pharmaceutical protein reported in transgenic plants.[19,22] Most importantly, these observations show that the chloroplast translation machinery is quite flexible, unlike the bacterial translation machinery that translated only the synthetic chloroplast codon–optimized IGF-1 gene.

Expression and Functionality of Human Interferon in Transgenic Chloroplasts

Interferon alphas (IFNαs) are known inhibitors of viral replication and cell proliferation. In addition, they are potent enhancers of the immune response and have many uses in the clinical treatment. A specific subtype, IFNα2b, was first approved in 1986 for the treatment of hairy cell leukemia by the Food and Drug Administration and has shown efficacy in a growing number of treatments for various viral and malignant diseases. However, the recombinant IFNα2b now on the market is being produced through an *E. coli* expression system and due to necessary in vitro processing and purification, the average cost of treatment is $26,000 per year. Recently, recombinant IFNα2b is used to treat patients suffering from West Nile virus. The 2 week treatment costs $2500 per patient. This drug is usually administered by injection, but the development of severe side effects is quite common. Also, up to 20% of patients produce anti-IFNα antibodies when the IFNα2b aggregates with human serum albumin in the blood. These antibodies are an undesirable response and lessen the effectiveness of the treatment. Evidence links the negative side effects to route of administration and dosage parameters. In fact, oral administration of natural human IFNα has proven to be therapeutically useful in the treatment of various infectious diseases.

Oral delivery of IFNα2b expressed via the chloroplast genome may eliminate some of these problems. Therefore, a recombinant IFNα2b containing a polyhistidine purification tag and a thrombin cleavage site was generated and integrated into the chloroplast genome of petit Havana and into a low nicotine variety of tobacco, LAMD-609.[15,19,23] Chloroplasts correctly process and fold human proteins as well as form the requisite disulfide bonds. In addition, bioencapsulation in plants cells can protect recombinant proteins from degradation in the mammalian gastrointestinal tract, and plant-based production systems are free of human pathogens. Western blots detected monomers and multimers of IFNα2b in both tobacco varieties using interferon alpha monoclonal antibody (MAB). Southern blots confirmed stable, site-specific integration of transgenes into chloroplast genomes and determined homoplasmy or heteroplasmy in the T_0 generation. In the petit Havana transgenic lines, homoplasmy of chloroplast genomes occurs in the first generation, and this corresponds to the highest level of IFNα2b expression. ELISAs were used to quantify up to 18.8% of total soluble protein in petit Havana and up to

12.5% in LAMD-609. IFNα2b functionality was determined by the ability of IFNα2b to protect HeLa cells against the cytopathic effect of encephalomyocarditis virus (EMC) and RT PCR. Chloroplast derived IFNα2b was as active as commercially produced Intron A. The mRNA levels of two genes induced by IFNα2b: 2′-5′ oligoadenylate synthase and STAT-2, were tested by RT-PCR using primers specific for each gene. Chloroplast derived IFNα2b induced expression of both genes similar to commercial IFNα2b. These expression levels and functionality are ideal for purification and further use in oral IFNα2b delivery or preclinical studies.

Expression of *Bacillus anthracis* Protective Antigen in Transgenic Chloroplasts for an Improved Vaccine Against Anthrax

Bacillus anthracis is the causative agent for the anthrax disease, which has become a serious threat due to its potential use in bioterrorism and recent outbreaks among wild-life in the United States. CDC lists *B. anthracis* as a category A agent and estimates the cost of an anthrax attack to exceed $26 billion per 100,000 exposed individuals. Concerns regarding vaccine purity, the current requirement for six injections followed by yearly boosters, and a limited supply of the key protective antigen (PA), underscore the urgent need for an improved vaccine. Therefore, the gene coding for PA (*pag*) was cloned into a chloroplast vector along with the *psb*A regulatory signals to enhance translation. Chloroplast integration of the transgenes was confirmed by PCR and Southern blot analyses. Crude plant extracts contained up to 2500 μg PA/g of fresh leaf tissue. Exceptional stability of full length PA in leaves stored for several months or in crude extracts was observed, even in the absence of protease inhibitors. PA expression was light-regulated, and maximal expression was observed under continuous illumination. Co-expression of the ORF 2 chaperonin from *Bacillus thuringiensis* did not increase PA accumulation or fold it into cuboidal crystals in transgenic chloroplasts. Both trypsin and chymotrypsin proteolytic cleavage sites present in PA were protected in transgenic chloroplasts. Furin or furinlike proteases are absent in chloroplasts because only PA 83 was observed. Both CHAPS and SDS detergents extracted PA with equal efficiency and PA was observed in the soluble fraction. Chloroplast-derived PA efficiently bound to anthrax toxin receptor, underwent proper cleavage, heptamerized, and bound lethal factor, resulting in macrophage lysis; up to 25 μg functional PA per ml crude extract was observed. With anticipated expression level up to 6.24 kg PA per acre (three cuttings), assuming 50% loss of PA during purification from plant extracts and at 5 μg PA per dose (the current vaccine contains 1.75–7 μg/ml PA), 600 million doses of vaccine (free of contaminants) could be produced per acre of transgenic tobacco.[15,24] This opens a new approach to combat bioterrrorism.

Expression of Monoclonals in Transgenic Chloroplasts

Owing to their remarkable specificity and therapeutic nature for defined targets, monoclonal antibodies are emerging as therapeutic drugs at a fast rate. The chloroplast genome was chosen for transformation with antibody genes due to tremendously high levels of foreign protein expression, the ability to fold, process, and assemble foreign proteins with disulfide bridges, and simpler purification and transgene containment via maternal inheritance. To enhance translation, a codon-optimized gene under the control of specific 5′ untranslated regions (UTRs) was used. IgA-G, a humanized, chimeric monoclonal antibody (Guy's 13) has been successfully synthesized and assembled in transgenic tobacco chloroplasts with disulfide bridges.[25] Guy's 13 recognizes the surface antigen *Streptococcus mutans*, the bacteria that causes dental cavities. In this study, integration into the chloroplast genome was confirmed by PCR and southern blot analyses. Western blot analysis revealed the expression of heavy and light chains individually, as well as the fully assembled antibody, thereby suggesting the presence of chaperonins for proper protein folding and enzymes for formation of disulfide bonds within transgenic chloroplasts.

Expression of Anti-Microbial Peptides (AMP) to Combat Drug-Resistant Pathogens

There are several human pathogenic bacteria that are drug resistant or have acquired resistance over a period of time. There is an urgent need to explore alternate ways of combating such bacteria. Magainin and its analogues have been investigated as a broad spectrum topical agent, a systemic antibiotic, a wound-healing stimulant, and an anticancer agent. Magainin's analogue MSI-99, a synthetic lytic peptide, has been recently expressed via the chloroplast genome.[26] This AMP is an amphipathic alpha helix molecule and possesses affinity for negatively charged phospholipids found in the outer membrane of all bacteria. The probability of bacteria adapting to the lytic activity of this peptide is very low. It was observed that the lytic peptide is expressed at high levels, i.e., 21.5% of tsp. A multidrug-resistant gram negative bacteria, *Pseudomonas aeruginosa*, which is an opportunistic pathogen for plants, animals, and humans, was used for in vitro assays to test for the effectiveness of the lytic peptide expressed in the chloroplast. Cell extracts prepared from T1 generation plants resulted in 96% inhibition in growth of this pathogen. This result is highly encouraging in the exploration treatments against drug-resistant bacteria in general and to cystic fibrosis patients in particular because of their high susceptibility to *P. aeruginosa*.

Pharmaceutical companies are exploring the use of lytic peptides as broad-spectrum topical antibiotics and systemic antibiotics. It has been reported that the outer leaflet of melanoma and colon carcinoma cells express three- to seven fold more phosphatidylserine than their noncancerous counter parts. Previous studies have reported that analogues of magainin 2 were effective against hematopoietic, melanoma, sarcoma, and ovarian teratoma lines. Given the preference of this lytic peptide for negatively charged phospholipids, MSI-99 shows potential as an anticancer agent. The minimum inhibitory concentration of MSI-99 was investigated. Based on total inhibition of 1000 *P. syringae* cells, MSI-99 was most effective against *P. syringae*, requiring only 1 μg/1000 bacteria.[26] Because the lytic activity of antimicrobial peptides is concentration dependent, the amount of antimicrobial peptide required to kill bacteria was used to estimate the level of expression in transgenic plants. Based on the minimum inhibitory concentration, it was estimated that transgenic plants expressed MSI-99 at 21.5–43% of the total soluble protein.

Oral Delivery of Recombinant Proteins via Cholera Toxin B Subunit (CTB)

CTB has previously been expressed in nuclear transgenic plants at levels of 0.01 (leaves) to 0.3% (tubers) of the total soluble protein. To increase expression levels, the chloroplast genome was engineered to express the CTB gene.[5] Expression of oligomeric CTB at levels of 4–5% of total soluble plant protein was observed. PCR and Southern blot analyses confirmed stable integration of the CTB gene into the chloroplast genome. Western blot analysis showed that transgenic chloroplast–expressed CTB was antigenically identical to commercially available purified CTB antigen. Also, GM1-ganglioside binding assays confirm that chloroplast synthesized CTB binds to the intestinal membrane receptor of cholera toxin.[5] Transgenic tobacco plants were morphologically indistinguishable from untransformed plants and the introduced gene was found to be inherited stably in the subsequent generation as confirmed by PCR and Southern blot analyses. In addition to establishing unequivocally that chloroplasts are capable of forming disulfide bridges to assemble foreign proteins, the increased production of an efficient transmucosal carrier molecule and delivery system, such as CTB, in plant chloroplasts makes plant-based oral vaccines and fusion proteins with CTB needing oral administration a much more feasible approach. Spontaneously forming CTB pentamers have exhibited intact transcytosis to the external basolateral membrane of intestinal epithelium and have been widely used as oral vaccine vehicles. It is thus feasible to produce CTB fusion proteins for oral delivery.

CONCLUSION

The first pharmaceutical protein expressed in transgenic chloroplasts was a protein-based polymer with varied biomedical applications, including prevention of postsurgical adhesions/scars, use in wound coverings, artificial pericardia, tissue reconstruction, and programmed drug delivery.[27] Since then several biopharmaceutical proteins have been expressed in transgenic chloroplasts. Hyperexpression of several human blood proteins, including human serum albumin, Magainin, interferons, somatotropin and insulinlike growth factors in transgenic chloroplasts for mass production and purification, makes chloroplast genetic engineering an invaluable approach to realize the full potential of plant-derived biopharmaceuticals. The successful engineering of tomato chromoplasts for high-level transgene expression in fruits, coupled with hyperexpression of vaccine antigens (for anthrax, cholera, plague) and the use of plant-derived, antibiotic-free selectable markers, augurs well for the oral delivery of edible vaccines or biopharmaceuticals that are currently beyond the reach of those who need them most.

ARTICLES OF FURTHER INTEREST

Biosafety Applications in the Plant Sciences: Expanding Notions About, p. 146
Biosafety Approaches to Transgenic Crop Plant Gene Flow, p. 150
Biosafety Considerations for Transgenic Insecticidal Plants: Nontarget Herbivores, Detritivores, and Polinators, p. 153
Biosafety Considerations for Transgenic Insecticidal Plants: Nontarget Predators and Parasitoids, p. 156
Biosafety Programs for Genetically Engineered Plants: An Adaptive Approach, p. 160
Gene Flow Between Crops and Their Wild Progenitors, p. 488
Gene Silencing: A Defense Mechanism Against Alien Genetic Information, p. 492
Molecular Farming in Plants: Technology Platforms, p. 753
Tobamoviral Vectors: Developing a Production System for Pharmaceuticals in Transfected Plants, p. 1229

Transgenic Crop Plants in the Environment, p. 1248
Vaccines Produced in Transgenic Plants, p. 1265

REFERENCES

1. Daniell, H.; Khan, M.S.; Allison, L. Milestones in chloroplast genetic engineering: An environmentally friendly era in biotechnology. Trends Plant Sci. **2002**, *7*, 84–91.
2. Daniell, H.; Dhingra, A. Multiple gene engineering: Dawn of an exciting new era in biotechnology. Curr. Opin. Biotechnol. **2002**, *13*, 136–141.
3. Daniell, H. Molecular strategies for gene containment in transgenic crops. Nat. Biotechnol. **2002**, *20*, 581–586.
4. Ruf, S.; Hermann, M.; Berger, U.; Carrer, H.; Bock, R. Stable genetic transformation of tomato plastids and expression of a foreign protein in fruit. Nat. Biotechnol. **2001**, *19*, 870–875.
5. Daniell, H.; Lee, S.B.; Panchal, T.; Wiebe, P.O. Expression and assembly of the native cholera toxin B subunit gene as functional oligomers in transgenic tobacco chloroplasts. J. Mol. Biol. **2001**, *311*, 1001–1009. Also see WO 01/72959.
6. Daniell, H.; Muthukumar, B.; Lee, S.B. Engineering the chloroplast genome without the use of antibiotic selection. Curr. Genet. **2001**, *39*, 109–116. Also see, WO 01/64023.
7. Richter, L.J.; et al. Production of hepatitis B surface antigen in transgenic plants for oral immunization. Nat. Biotechnol. **2001**, *18*, 1167–1171.
8. Daniell, H.; Wycoff, K.; Streatfield, S. Medical molecular farming: Production of antibodies, biopharmaceuticals and edible vaccines in plants. Trends Plant Sci. **2001**, *6*, 219–226.
9. Cramer, C.; Boothe, J.G.; Oishi, K.K. Transgenic plants for therapeutic proteins: Linking upstream and downstream technologies. Curr. Top. Microbiol. Immunol. **1999**, *240*, 95–118.
10. Kusnadi, A.; Nikolov, G.; Howard, J. Production of recombinant proteins in plants: Practical considerations. Biotechnol. Bioeng. **1997**, *56*, 473–484.
11. Daniell, H.; McFadden, B.A. Genetic Engineering of Plant Chloroplasts. U.S. Patents 5,693, 507, December 2, 1997.
12. Daniell, H.; McFadden, B.A. Genetic Engineering of Plant Chloroplasts. U.S. Patents 5,932,479, August 3, 1999.
13. Daniell, H. Universal Chloroplast Integration and Expression Vectors, Transformed Plants and Products Thereof. Australian Patent 748210, September 2000. WO 99/10513.
14. DeCosa, B.; Moar, W.; Lee, S.B.; Miller, M.; Daniell, H. Over-expression of the Bt Cry2Aa2 operon in chloroplasts leads to formation of insecticidal crystals. Nat. Biotechnol. **2001**, *19*, 71–74. Also see WO 01/64024.
15. Daniell, H. Medical Molecular Pharming: Expression of Antibodies, Biopharmaceuticals and Edible Vaccines via the Chloroplast Genome. In *Plant Biotechnology, 2000 and Beyond*; Vasil, I.K., Ed.; Klewer Academic: 2003; 371–376.
16. Lee, S.B.; Byun, M.O.; Daniell, H. Accumulation of trehalose within transgenic chloroplasts confers drought tolerance. Mol. Breed. **2003**, *11*, 1–13. Also see WO 01/64850.
17. Staub, J.M. High yield production of a human therapeutic protein in tobacco chloroplasts. Nat. Biotechnol. **2000**, *18*, 333–338.
18. Petridis, D.; Sapidou, E.; Calandranis, J. Computer aided process analysis and economic evaluation for biosynthetic human insulin production – A case study. Biotechnol. Bioeng. **1995**, *48*, 529–541.
19. Daniell, H. *Pharmaceutical Proteins, Human Therapeutics, Human Serum Albumin, Insulin, Native Cholera Toxin B Subunit on Transgenic Plastids. WO 01/72959*; October 4, 2001.
20. Fernandez-San Millan, A.; Mingeo-Castel, A.; Miller, M.; Daniell, H. A chloroplast transgenic approach to hyper-express and purify Human Serum Albumin, a protein highly susceptible to proteolytic degradation. Plant Biotechnol. J. **2003**, *1*, 71–79.
21. Nilsson, B.; Forsberg, G.; Hartmanis, M. Expression and purification of recombinant insulin like growth factors from *E. coli*. Methods Enzymol. **1991**, *198*, 3–16.
22. Ruiz, G. Optimization of codon composition and regulatory elements for expression of the human IGF-1 in transgenic chloroplasts. M.S. Thesis; University of Central Florida, 2002.
23. Falconer, R. Expression of Interferon alpha 2b in transgenic chloroplasts of a low nicotine tobacco. M.S. Thesis; University of Central Florida, 2000.
24. Watson, J. *Expression of Bacillus anthracis Protective Antigen in Transgenic Chloroplasts for an Improved Vaccine Against Anthrax*; M.S. Thesis, University of Central Florida, 2002.
25. Daniell, H.; Wycoff, K. *Production of antibodies in transgenic plastids. WO 01/64929*; September 7, 2001.
26. DeGray, G.; Rajasekaran, K.; Smith, F.; Sanford, J.; Daniell, H. Expression of an antimicrobial peptide via the chloroplast genome to control phytopathogenic bacteria and fungi. Plant Physiol. **2001**, *127*, 852–862.
27. Guda, C.; Lee, S.B.; Daniell, H. Stable expression of a protein based polymer in tobacco chloroplasts. Plant Cell Rep. **2000**, *19*, 257–262. Also see WO 99/10513.

Meiosis

Wojciech P. Pawlowski
Lisa C. Harper
W. Zacheus Cande
University of California, Berkeley, California, U.S.A.

INTRODUCTION

Meiosis is the specialized pair of cell divisions that alter the genetic content of the nucleus from 2n (diploid) to 1n (haploid) and lead to the production of gametes. Following a premeiotic S phase, homologous chromosomes pair and synapses. This is followed by meiotic recombination and subsequent chiasmata formation between the paired homologous chromosomes. Chromosome pairing and the formation of chiasmata ensure the bipolar attachment of homologous chromosomes to the spindle, which is required for the reductional division of meiosis I that leads to the separation of homologs. This is followed by a modified cell cycle lacking an S phase. A second, equational division leads to the separation of sister chromatids and a halving of the chromosome number per cell. The general progression of meiosis is highly conserved, and is thus similar in yeast, humans, and plants.

PLANTS AS MODEL SYSTEMS FOR STUDYING MEIOSIS

The availability of large numbers of cells undergoing meiosis (meiocytes) in anthers has made plants excellent models for studying meiosis. Maize is one of the few organisms with a large genome where pairing and synapsis is amenable to analysis by a combination of cytological, genetic, and molecular techniques. The features of meiosis in *Arabidopsis*, at least at a cytological level, are conventional and comparable to those of other plants, but it has been the ease of gene discovery in this organism that has led to the adoption of *Arabidopsis* for meiosis research.

CHROMOSOME BEHAVIOR DURING MEIOTIC PROPHASE

The accurate segregation of chromosomes at the first division of meiosis requires that homologous chromosomes pair and recombine with each other to form bivalents (Fig. 1). The pairing and crossing over of homologous chromosomes occurs in the extensive prophase of meiosis. Dramatic changes in chromosome behavior and morphology during meiotic prophase have been used to subdivide the prophase into stages: leptotene, zygotene, pachytene, diplotene, and diakinesis.[1–3] In leptotene, the decondensed clouds of chromatin are organized into long, thin fibers by the assembly of a proteinacious core—the axial element—onto the chromosomes. During zygotene, homologous chromosomes pair and begin to tightly associate, or synapse, along their length when the central element of the synaptonemal complex (SC) is installed between the homologous chromosomes. By pachytene, SC formation is complete and meiotic recombination between homologs is resolved. In diplotene, the synaptonemal complex disassembles and chiasmata, which are responsible for holding the homologous chromosomes together, are visible. Finally, in diakinesis, the chromosomes undergo a final stage of chromosome condensation just prior to metaphase.

TELOMERES CLUSTER ON THE NUCLEAR ENVELOPE AT THE BEGINNING OF MEIOTIC PROPHASE

Before chromosomes synapse during zygotene, the nucleus becomes highly polarized by the formation of the telomere bouquet (Fig. 1). The ends of the chromosomes become tightly clustered together on the inner surface of the nuclear envelope, resulting in a structure resembling a bouquet of flowers. The close relationship between the telomere clustering and chromosome pairing has led to the suggestion that the bouquet may help to facilitate pairing. By coaligning the ends of the chromosomes, homologous regions of the chromosomes are vectorially aligned and within the same region of the nucleus. In addition, the clustering of telomeres on the nuclear envelope may serve to restrict the homology search to a much smaller volume of the nucleus. Consistent with this potential role is the general observation that synapsis is typically initiated near the telomeres. Mutants defective in bouquet formation show delay in the progression of meiosis and severe defects in pairing and synapsis.[4]

Encyclopedia of Plant and Crop Science
DOI: 10.1081/E-EPCS 120005586

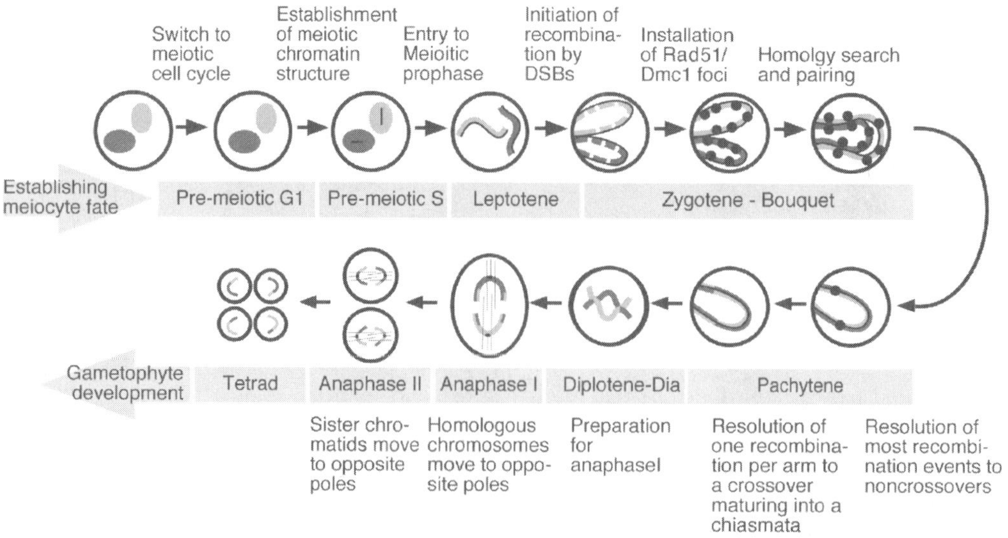

Fig. 1 A diagram showing the key stages and main molecular events in meiosis. Only one homologous chromosome pair is shown, and each homologous chromosome (two sister chromatids) is a different shade of gray. The Rad51/Dmc1 foci are shown as nodules on the extended prophase chromosomes.

CHROMOSOME PAIRING AND SYNAPSIS

In some organisms, such as the hexaploid wheat and *Drosophila*, the pairing of homologous chromosomes occurs prior to meiosis. However, in many other organisms such as maize, oat, humans, and mice, homologous chromosomes are not associated with each other until zygotene. Regardless of when chromosomes pair, a major question in meiosis is, how do the homologous chromosomes identify and associate with each other? The axial cores of the synaptonemal complex do not appear to contain the information for distinguishing homology because they will synapse randomly in a haploid organism undergoing meiosis.[3] The general consensus is that the homology search is DNA-based, which accounts for the correct pairing of chromosome inversions, translocations, and alien chromosomes. In most organisms, including plants, a successful homologous chromosome pairing is linked to the progression of the meiotic recombination pathway, initiated by the action of the topoisomerase Spo11 to generate DNA double-strand breaks (DSBs).[2] Mutants lacking Spo11 or other components of the meiotic recombination pathway that act at subsequent steps show serious defects in homologous chromosome pairing.

In a number of species, the chromosome homology search and synapsis are closely coupled during the zygotene stage. Electron microscopy has revealed the presence of large numbers of small, electron-dense spheres associated with synapsing chromosomes during this stage. These nodules, called early recombination nodules, contain recombination enzymes and are predicted also to function in the chromosome homology search.[2,3]

MEIOTIC RECOMBINATION

Homologous recombination fulfils a number of critical biological functions, including maintaining genetic diversity, repairing DSBs. Stabilizing interactions between homologous chromosomes in the form of chiasmata, promoting proper disjunction, healing a number of types of DNA damage, and removing undesirable mutational load. Because meiosis is a specialized type of cell division and the meiotic recombination machinery is largely derived from the machinery responsible for mitotic recombination and DNA repair in somatic cells, the meiotic and somatic recombination pathways share many components. However, a number of meiotic recombination-specific components also have been identified. Meiotic recombination is initiated by the introduction of DSBs on chromosomes by the topoisomerase Spo11 (Fig. 1). The DSBs are then resected and subsequently repaired by a protein complex containing two homologs of the *E. coli* RecA protein, Rad51 and Dmc1. These two recombination proteins are some of the most studied elements of meiotic recombination and have also been proposed to play roles in the pairing of homologous chromosomes.[5] The Rad51/Dmc1 complex forms distinct foci on chromosomes during the meiotic prophase I, which are thought to mark the positions of DSBs. In maize, there are about 500 Rad51 foci in zygotene when chromosomes are pairing, located mostly on unpaired chromosomes. By early pachytene, when pairing and synapsis are complete, there are only about 10–20 foci, which corresponds well to the number of meiotic crossovers in maize.

METAPHASE I AND II CHROMOSOME SEGREGATION

The reductional separation of homologous chromosomes at the first division of meiosis (MI) requires that homologous chromosomes recombine with each other and form bivalents. Since chiasmata hold the homologs together until the metaphase/anaphase transition during MI, failure to undergo recombination results in the inappropriate attachment of chromosomes to the spindle and their inaccurate segregation, leading to aneuploidy. Sister chromatid cohesion at the centromere and arms proximal to the chiasmata is maintained during the first meiotic division (MI) but is dissolved during the second meiotic division (MII), permitting the separation of sisters. During the MI reductional division, the kinetochores of both sister chromatids face the same spindle pole, and the kinetochores of one homolog face toward one pole while the kinetochores of the other homolog face the other pole. Often, sister kinetochores are fused to form a single structure. Following anaphase of MI, the two sisters end up at the same spindle pole. During the MII equational metaphase, sister kinetochores face the opposite spindle poles, ensuring that when sister chromatid cohesion is released at anaphase, the sister chromatids will move to opposite spindle poles and end up in different daughter cells.[3]

THE IMPORTANCE OF UNDERSTANDING MEIOSIS FOR THE CROP SCIENTIST

Meiosis requires DNA replication and chromosome condensation, homologue pairing and recombination, sister chromatid cohesion and homologue attachment, and spindle assembly and chromosome segregation. Many genes are expected to participate in these events, some of which will be unique to meiosis, whereas others are likely to be involved in other cellular processes. Therefore, understanding gene functions in meiosis will provide new insights into the molecular control of many basic cellular events in plants. Furthermore, knowledge about the genes controlling meiosis also has several potential applications in plant breeding. Such applications would include: 1) developing strategies for acquiring apomixis in maize and other grasses; 2) allowing manipulation of the level of genetic recombination in breeding programs, for example in wide hybrid crosses and for the introgression of quantitative trait loci (QTL) from exotic germplasm; 3) altering genetic incompatibility between species, thus permitting wide hybrid crosses and making it possible to select for desired traits from a broader gene pool; and 4) developing new methods for manipulating transgene integration in random or homologous transformation methods.

CONCLUSION

Meiosis is a highly specialized cell cycle leading to halving of the chromosome number per cell, which is required for the formation of gametes. Homologous chromosomes must pair and recombine, forming bivalents, in order for meiotic chromosome segregation to occur properly.

ARTICLES OF FURTHER INTEREST

Chromosome Structure and Evolution, p. 273
Cytogenetics of Apomixis, p. 383
Female Gametogenesis, p. 439
The Interphase Nucleus, p. 568
Male Gametogenesis, p. 663
Mitosis, p. 734

REFERENCES

1. Zickler, D.; Kleckner, N. The leptotene–zygotene transition of meiosis. Annu. Rev. Genet. **1998**, *32*, 619–697.
2. Zickler, D.; Kleckner, N. Meiotic chromosomes: Integrating structure and function. Annu. Rev. Genet. **1999**, *33*, 603–754.
3. Dawe, R.K. Meiotic chromosome organization and segregation in plants. Annu. Rev. Plant Phys. **1998**, *47*, 371–395.
4. Golubovskaya, I.N.; Harper, L.C.; Pawlowski, W.P.; Schichnes, D.; Cande, W.Z. The *pam1* gene is required for meiotic bouquet formation and efficient homologous synapsis in maize (*Zea mays*, L.). Genetics **2002**, *162* (4), 1979–1993.
5. Franklin, A.E.; McElver, J.; Sunjevaric, I.; Rothstein, R.; Bowen, B.; Cande, W.Z. Three-dimensional microscopy of the Rad51 recombination protein during meiotic prophase. Plant Cell **1999**, *11* (5), 809–824.

Metabolism, Primary: Engineering Pathways

Alberto A. Iglesias
Instituto Tecnológico de Chascomús (IIB-INTECH), Buenos Aires, Argentina

INTRODUCTION

Plants are traditional sources of organic materials used by humans for food and many industrial purposes.[1] The relevance of plants as food producers is clear, since their photosynthetic capacity ultimately represents the basis of life on our planet. Main industrially important compounds obtained from plants are (among others) wood, cotton, cork, and latex. Despite advances in chemical synthesis of organic molecules (i.e., plastics), many plant products are still unique. Thus, production of rubber of a certain quality only utilizes latex as a raw material.

In this scenario, the relevance of manipulating plants to improve their synthetic ability is clear. For years, selective breeding allowed increased productivity of crops, but with limitations. The development of genetic engineering expanded possibilities for improvement, as modification of plant metabolic pathways would allow the manipulation of the quantity and quality of natural products and also the synthesis of novel or heterologous compounds.[1,2]

The feasibility of plant transformation was firstly utilized for the production of genetically modified crops with enhanced resistance to herbicides and pathogens or with longer postharvest life.[3,4] These transgenic plants importantly affected agriculture, because they allowed reduced production costs and increased yields.[4] After these advances, efforts centered on modifying specific plant metabolic routes related to the synthetic proficiency of a particular crop.[3,4]

PLANTS AND METABOLIC ENGINEERING

Plant metabolic engineering is in the period of inital development. The rationale of genetic transformation of organisms requires the understanding of metabolism within the framework of metabolic control analysis, an issue not completely elucidated in plants.[2] Additionally, plant cells are highly compartmentalized, and the whole organism possesses both phototrophic and heterotrophic tissues. These facts determine complex metabolic networks, with partitioning of metabolites (and metabolic routes) intracellularly and between source and sink organs.[5] Despite this complexity, plant metabolic engineering has reached relevant goals and (more important) shows a highly promising future.[1,2] Key issues in the subject are the type of tissue (source or sink) and the intracellular compartment targeted for transformation.

PLANT PRODUCTIVITY IMPROVEMENT

Autotrophy of plants for fixing CO_2 into carbohydrates makes this process a key target for engineering in order to improve productivity. Possibilities include not only improving the carbon fixation process itself (in source tissues, see Fig. 1), but also increasing the carbon demand at sink organs where the product is finally accumulated (Fig. 2). Table 1 details the possibilities and advances made in the manipulation of carbon partitioning and allocation in plants.

Engineering Source Tissues

About 95% of plants (including many crops) are C_3 species that assimilate atmospheric CO_2 via the Calvin cycle or the C_3 photosynthetic pathway. C_3 plants utilize CO_2 with relatively low efficiency, since primary carbon fixation in the C_3 pathway is catalyzed by the chloroplastic enzyme ribulose-1,5-bisphosphate carboxylase/oxygenase (Rubisco), which also utilizes O_2 as a substrate, leading to photorespiration that reduces (by up to 40%) photosynthetic efficiency (Fig. 1).[6]

Engineering of Rubisco to increase its ratio of carboxylase/oxygenase activity seems very unlikely, as predicted from the enzyme's catalytic mechanism. The strategy to increase the carbon photoassimilation efficiency is now following the path traced by natural evolution: the addition of enzymatic steps to increase the CO_2 to O_2 ratio in the chloroplast (Fig. 1). Indeed, C_4 plants evolved a biochemical device that masters photorespiration.[2,6] Briefly, it comprises annexation of a metabolism (C_4 pathway) that pumps atmospheric CO_2 to the chloroplast. Additionally, leaves of C_4 plants have two types of photosynthetic cells (mesophyll and bundle sheath cells) that play a role in the integration of the C_3 cycle and the auxiliary pathway.[2,6]

Encyclopedia of Plant and Crop Science
DOI: 10.1081/E-EPCS 120010399

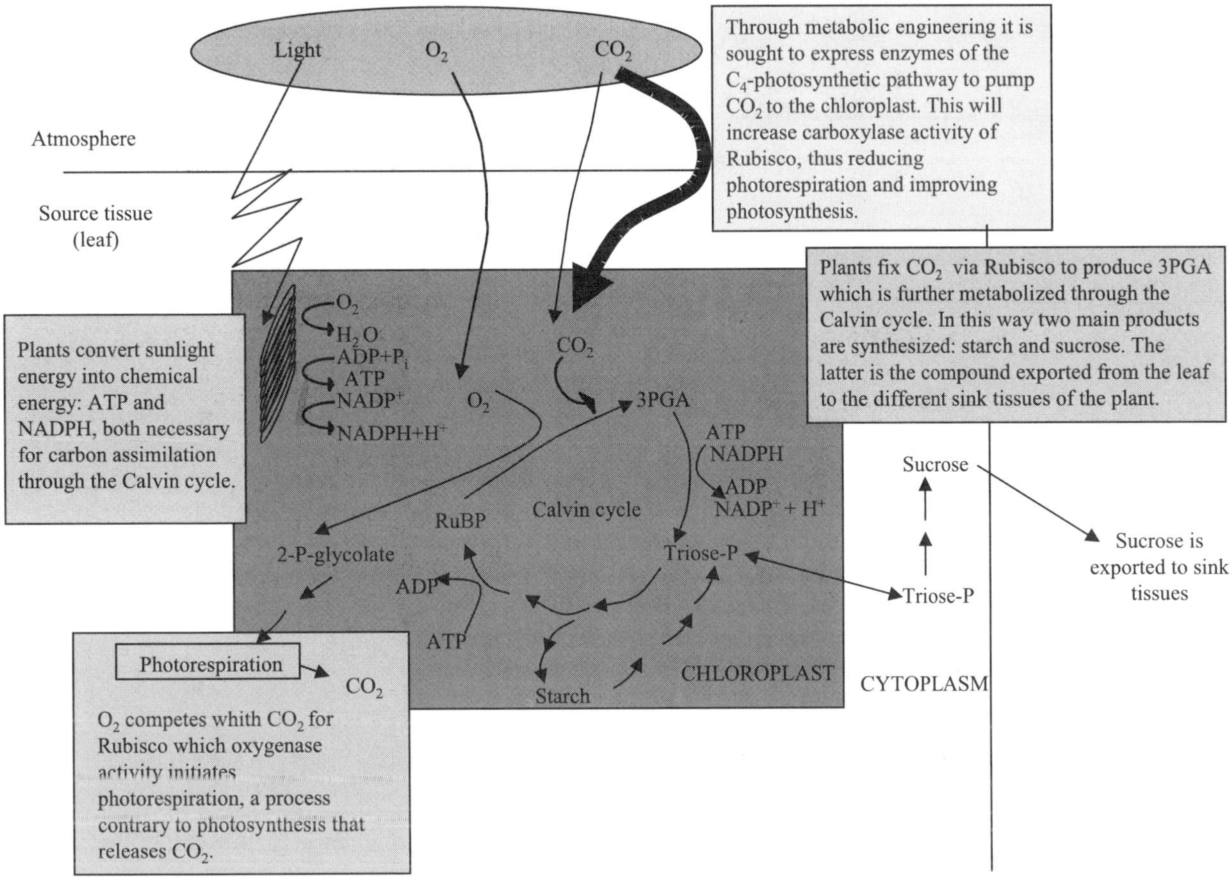

Fig. 1 Carbon photoassimilation in source tissues of plants. Atmospheric CO_2 is converted to 3P-glycerate (3PGA) by the reaction with ribulose-1,5-bisphosphate (RuBP), catalyzed by Rubisco. 3PGA is further metabolized via the Calvin cycle, which consumes ATP and NADPH. Intermediates from the Calvin cycle are utilized for the synthesis of starch and sucrose, the latter being the compound exported to other plant tissues. Strategies to improve carbon assimilation are centered on increasing the CO_2 to O_2 ratio within the chloroplast to reduce photorespiration. (*View this art in color at www.dekker.com.*)

The technology to introduce genes encoding enzymes of the C_4 pathway into C_3 plants was recently developed.[6] Transgenic plants expressing C_4 enzymes at different levels exhibited alterations in carbon metabolism (Table 1). The overall view shows engineering of carbon photoassimilation metabolism at the starting point, with many questions to address in the near future about metabolic fluxes in transformed plants. Anyway, results obtained so far support the possibility of creating efficient CO_2-concentrating mechanisms, via metabolic engineering, to enhance photosynthetic efficiency of crops.[6]

Engineering Sink Tissues

Metabolic engineering in plant sink tissues (Fig. 2), improving production of primary compounds, has successful examples in the manipulation of starch and fatty acid biosynthesis.[1,2]

Engineering Starch Metabolism

Starch, a major product of photosynthesis, is synthesized in plastids of photosynthetic and nonphotosynthesizing cells from glucose-1-phosphate by the consecutive reactions of ADPglucose pyrophosphorylase (producing ADPglucose, the donor of glucosyl residues), starch synthase, and branching enzyme.[5] The pathway is controlled at the level of ADPglucose pyrophosphorylase through allosteric regulation by 3-phosphoglycerate (activator) and Pi (inhibitor).[2,5]

A significant rise in starch accumulation in sink tissues was obtained by increasing ADPglucose pyrophosphorylase activity in plants transformed via the expression of an unregulated mutant of the enzyme from *Escherichia coli*[2,7,8] (Table 1). This is a good example of the relevance of the understanding of enzyme kinetics and regulation for metabolic engineering. Based on the knowledge

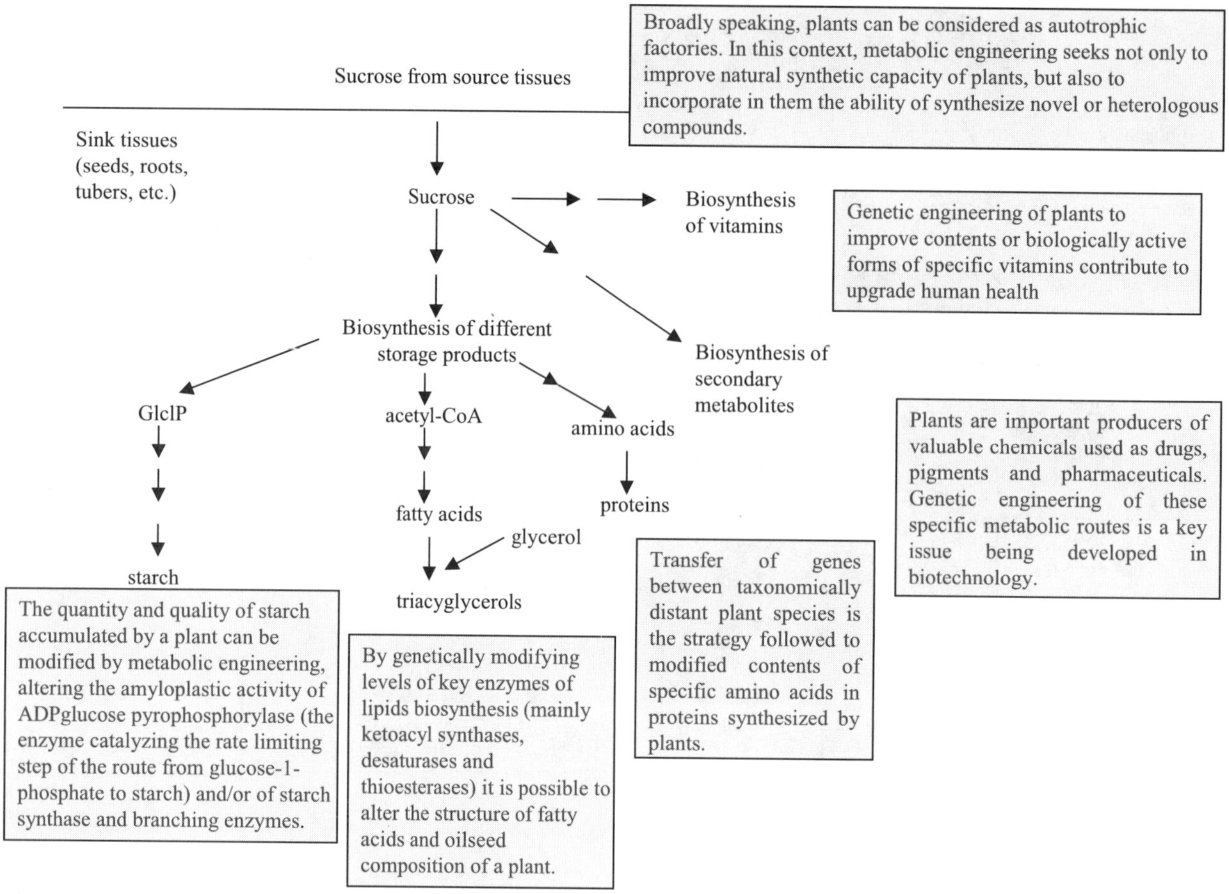

Fig. 2 The pathway of carbon in plant sink tissues and different strategies to be followed by metabolic engineering. Sink tissues synthesize different storage products as well as secondary metabolites and vitamins, representing relevant producers of key natural chemicals. Through metabolic engineering, it is sought to improve the synthetic capacity of plants by modifying the quantity and quality of the compounds they produce. (*View this art in color at www.dekker.com.*)

that this metabolic route is regulated at the level of ADPglucose pyrophosphorylase, its activity was permanently increased by expressing a mutant enzyme (lacking activator requirement for maximal activity) and thus enhancing ADPglucose and starch production.[2,7] Additionally, expression of the foreign gene was induced in storage tissue amyloplasts by using a tuber-specific promoter, as detailed in Table 1.[2,7,8]

Transgenic products obtained by this methodology exhibited substantial improvement in their food quality traits. Thus, potato tubers with more starch (and solids) increased processing efficiency and resistence to cold storage, whereas transformed tomato fruits showed improved flavor.[1,8] Another objective is to engineer starch metabolism to modify the polysaccharide structure. Current strategies propose manipulating starch synthase and branching enzyme levels to alter contents of amylose and amylopectin and the branching degree in the latter polymer.[2,3]

Engineering Lipid Biosynthesis

Plants accumulate oils as triacylglycerols, with properties determined by the type of fatty acids they contain. Fatty acid biosynthesis occurs within plastids, by two-carbon elongation steps (initially derived from acetyl-CoA) catalyzed by a multienzyme complex. Length and desaturation, level of the product is determined by ketoacyl synthases, desaturases, and thioesterases, which render the final fatty acid that is exported to the cytoplasm for triacylglycerol biosynthesis.[8]

Cloning of genes encoding enzymes involved in fatty acid biosynthesis allowed the modification of oilseed

Table 1 Metabolic engineering in plant source and sink tissues to improve biosynthetic pathways

Objective (*O*)/Strategy (*S*)	**Results**
SOURCE TISSUES	
(O) To increase carbon photoassimilation by reducing photorespiration.	— Transgenic tobacco and potato plants, overexpressing phosphoenolpyruvate carboxylase, exhibited better capacity to assimilate CO_2 at high temperatures than nontransformed plants.[6]
(S) Expression of enzymes of the C_4 photosynthetic pathway in C_3 plants, to concentrate CO_2 at the intracellular site where Rubisco is operative. The increase of CO_2 over O_2 will reduce photorespiration.	— Transgenic rice, expressing high levels of phosphoenolpyruvate carboxylase in the leaves, showed no alteration in the rate of CO_2 assimilation, but significant reduction in O_2 inhibition of net carbon fixation.[2,6] — The expression of phosphoenolpyruvate carboxykinase into chloroplasts of leaf cells in rice significantly increased the production of C_4 acids and the synthesis of sucrose from these compounds.[6]
SINK ORGANS	
(O) To improve carbon allocation by increasing the accumulation of reserve products. *(S)* To increase synthesis of ADPglucose in almyloplasts by expressing ADPglucose pyrophosphorylase, the limiting regulatory enzyme in starch biosynthesis. Expression of a mutant enzyme, insensitive to regulation, to ensure constant levels of activity in plastids.	— Potato plants transformed with *glgC16* gene [from *Escherichia coli*, encoding a mutant ADPglucose pyrophosphorylase less dependent on the activator (fructose-1,6-bisphosphate) and relatively insensitive to inhibition by AMP] accumulated up to 30% more starch in tubers, when the foreign gene was introduced under the control of a tuber-specific patatin promotor. Constitutive expression of the gene was detrimental to plant growth, because excess accumulation of polysaccharide in leaves diminished sucrose synthesis, which decreased export of carbon to growing parts of the plant.[2,7,8] — Tomato fruits with higher starch content during the green stage of development and canola seeds with low ratio oil/starch were obtained by transformation of plants with identical strategy as above indicated.[8]
(O) To alter oilseed composition by manipulating length and desaturation degree of fatty acids.	— Seed-directed expression of specific antisense desaturase genes rendered transgenic soybean and canola plants with higher content of oleic acid (18:1) and reduced levels of natural polyunsaturated oils [mainly linoleic acid (18:2)].[4,8]
(S) To manipulate activity levels of ketoacyl synthases, desaturases, and/or thioesterases in plastids.	— *Arabidopsis* and canola plants transformed by seed-specific expression of lauroyl-acyl carrier protein thioesterase accumulated up to 50% laurate. Plants further manipulated to express lysophosphatidic acid acyltransferase from coconut accumulated triacylglycerols with three laurate substituents.[8] — Expression of FatB class thioesterases produced large amounts of C8 and C10 fatty acids in canola. Transformation with oleyl-acyl carrier protein thioesterase increased accumulation of palmitate and stearate.[4,8,9]
(O) To improve plant protein quality. *(S)* Expression of genes coding for proteins with high content of selected amino acids.	— Seed-directed expression (by using phaseolin promoter) of a gene encoding a methionine-rich protein (from Brazil nut) into tobacco and rapeseeds rendered transgenic plants with enhanced contents (up to 30%) of sulfur amino acids.[8]
(O) To improve vitamin content in plant products used for food purposes.	— *Arabidopsis* plants engineered by expression of the gene coding for γ-tocopherol methyltransferase converted γ-tocopherol pools into α-tocopherol, with the respective increase in vitamin E activity.[9]
(S) Expression of genes coding for enzymes of vitamin metabolism to drive its biosynthesis.	— Carotenoid biosynthetic enzymes (two from plants and one from *Escherichia coli*) were simultaneously expressed in rice endosperm. Transgenic rice contained increased levels of provitamin A (β-carotene) visualized by the yellow color of the endosperm and named "golden-rice."[8,9]

M

composition of certain plants by overexpression, antisense, and cosuppression technologies (Table 1).[1,8,9] By this approach, oilseeds more resilient to oxidation, healthier for human consumption, or with industrial applications were obtained. This demonstrates the potentiality of plant metabolic engineering for managing production of renewable and biodegradable compounds.[1,9] The next step envisions the modification of fatty acids to produce specific chemical structures, such as compounds containing epoxy functions or conjugated double bonds[1,2,4]

Improving Protein Quality and Vitamin Content

Plants produce about 65% of the protein utilized for food by humans and livestock. Plant proteins are deficient in certain essential amino acids, mainly lysine, tryptophan, cysteine, and methionine.[8] Table 1 shows strategies developed to correct this deficiency, by transferring genes between taxonomically distant plant species.

Manipulation of vitamin content in plants is an important issue where metabolic engineering may contribute to upgrading human health.[1,9] Table 1 describes increases in vitamin E levels obtained in natural oils. This vitamin is an antioxidant whose presence in the human diet is relevant, especially under stress conditions such as those affecting populations in the developed world.[7] Transgenic rice with high β-carotene content (Table 1) will also contribute to improving human health, mainly in developing countries where deficiency of vitamin A would be alleviated with the use of a major staple containing it.[7]

CONCLUDING REMARKS AND FUTURE PERSPECTIVES

The wide metabolic diversity and the autotrophy exhibited by plants made them key tools in biotechnology. In the framework of metabolic engineering, this relevance is enhanced, as plants can be considered as potential factories for the production of compounds useful for food and industrial processes.[1,2] Examples of the goals already reached and the potentiality of genetic manipulation for improving the quantity and quality of compounds synthesized and accumulated by plants have been detailed and are summarized in Table 1.

Two additional fields need consideration because of their current and future relevance in plant metabolic engineering (Fig. 2). First is production of secondary metabolites, as certain plants synthesize valuable chemicals (including alkaloids, anthocyanins, anthraquinones, polyphenols, steroids, and terpenoids) widely utilized as drugs, pigments, and pharmaceuticals.[1,2,7] Secondly, plants are important biological systems for the production of heterologous compounds; they exhibit advantages for the production of recombinant proteins used for different industries (mainly pharmaceutical), as they synthesize fully folded functional products with low costs while avoiding ethical and safety problems.[2]

Considering that the above analysis represents a relatively novel field, the future of plant metabolic engineering can be visualized as very promising. It is tempting to envision that the complete development of the technology and the better understanding of the biological process will make it possible to widely manipulate amounts and types of carbohydrates, oils, proteins, and many other compounds synthesized by plants.

ACKNOWLEDGMENTS

The author thanks support from CONICET and ANPCyT (PICT'99 No. 1-6074).

ARTICLES OF FURTHER INTEREST

Transgenic Crop Plants in the Environment, p. 1248
Transgenic Crops: Regulatory Standards and Procedures of Research and Commercialization, p. 1251
Vaccines Produced in Transgenic Plants, p. 1265

REFERENCES

1. Somerville, C.R.; Bonetta, D. Plants as factories for technical materials. Plant Physiol. **2001**, *125* (1), 168–171.
2. Aon, M.A.; Cortassa, S.; Iglesias, A.A.; Lloyd, D. *An Introduction to Metabolic and Cellular Engineering*; World Scientific Publishing Co. Pte. Ltd.: New York, 2002; 264 pp.
3. Willmitzer, L. Plant biotechnology: Ouput traits—The second generation of plant biotechnology is gaining momentum. Curr. Opin. Biotechnol. **1999**, *10* (2), 161–162.
4. Ohlrogge, J. Plant metabolic engineering: Are we ready for phase two? Curr. Opin. Plant Biol. **1999**, *2* (2), 121–122.
5. Iglesias, A.A.; Podestá, F.E. Photosynthate Formation and Partitioning in Crop Plants. In *Handbook of Photosynthesis*; Pessarakli, M., Ed.; Marcel Dekker, Inc.: New York, 1996; 681–698.
6. Matsuoka, M.; Furbank, R.T.; Fukayama, H.; Miyao, M. Molecular engineering of C_4 photosynthesis. Annu. Rev. Plant Physiol. Plant Mol. Biol. **2001**, *52*, 297–314.
7. Stark, D.M.; Timmerman, K.P.; Barry, G.F.; Preiss, J.; Kishore, G.M. Regulation of the amount of starch in plant tissues by ADPglucose pyrophosphorylase. Science **1992**, *258* (5080), 287–292.
8. *Engineering Plants for Commercial Products and Applications*; Collins, G.B., Shepherd, R.J., Eds.; Ann. N.Y. Acad. Sci., 1996; Vol. 792, 183.
9. Della Penna, D. Plant metabolic engineering. Plant Physiol. **2001**, *125* (1), 160–163.

M

Metabolism, Secondary: Engineering Pathways of

David Rhodes
Gregory J. Peel
Natalia Dudareva
Purdue University, West Lafayette, Indiana, U.S.A.

INTRODUCTION

The term engineering refers to metabolic engineering, a field pioneered by microbial systems biologists and chemical engineers. Metabolic engineering is defined as "the directed improvement of product formation or cellular properties through the modification of specific biochemical reactions or the introduction of new ones through the use of recombinant DNA technology." This field has a strong foundation in mathematical analysis of metabolic fluxes and control in biochemical systems. Its emphasis is on understanding the function of biochemical reaction networks in vivo, thereby synthesizing information on the interactions of the various system components. Frequently this synthesis is achieved by an iterative cycle of genetic engineering, and by metabolic flux analysis using isotopic labeling. Metabolic modeling is employed to organize and conceptualize the data, facilitating the generation of new hypotheses guiding the next round of engineering. Plant metabolic engineering is in its infancy, but is receiving increased attention, as highlighted in the recent special issue of the journal "Metabolic Engineering," which is devoted to plants.

SECONDARY METABOLISM TERMINOLOGY

Secondary metabolism is a loose term generally referring to pathways that have no essential role in plant growth and development. Secondary metabolism is thus distinguished from primary metabolism, which refers to the network of catabolic and anabolic reactions essential for growth and development.[1] Plant metabolites typically referred to as secondary metabolites include a diverse array of alkaloids, phenylpropanoids, terpenoids, glucosinolates, and cyanogenic glucosides that may play roles in plant defense against pathogens and herbivores.[1–3] Their synthesis is often restricted to distinct plant taxonomic groups. The term natural products is preferred to describe this array of compounds.[1] Many are of great potential value as pharmaceuticals and/or precursors in the synthesis of industrial chemicals.[1,3] Because secondary metabolites are invariably derived from primary metabolites, effective engineering of secondary pathways cannot be readily undertaken without careful consideration of the competition between primary and secondary pathways.[4,5]

EXAMPLES OF METABOLIC ENGINEERING OF PATHWAYS OF PLANT SECONDARY METABOLISM

Recent advances in the metabolic engineering of plant secondary pathways will be illustrated by considering the natural products derived from tryptophan, tyrosine, and phenylalanine, the three aromatic amino acids (Fig. 1).

Alkaloids, Glucosinolates, and Cyanogenic Glucosides

Alkaloids, glucosinolates, and cyanogenic glucosides are strongly implicated in plant herbivore and pathogen defense.[1] Manipulation of fluxes to these compounds is of interest not only in modulating herbivore and pathogen resistance, but also in improving the nutritional quality of plants. Early successes in plant metabolic engineering involved overexpressing tryptophan decarboxylase (TDC) in canola, in order to divert flux away from the bitter indole glucosinolates, thus improving the quality of canola oil and making the product more palatable as an animal feed[1] (Fig. 1). Conversely, antisense downregulation of TDC has been used to eliminate indole alkaloid biosynthesis.[5] Overexpression of TDC in combination with other enzymes of the indole alkaloid pathway (e.g., stricosidine synthase) is being vigorously pursued as a strategy to manipulate flux to important terpenoid indole alkaloid compounds such as vinblastine.[1,3,5] Promising approaches also involve overexpression of transcription factors, resulting in induced levels of a number of genes of the pathway.[3,5] Daunting challenges in engineering metabolic flux to the terpenoid indole alkaloids will involve coordinating the supply of the indole moiety (derived from tryptophan) with the supply of

Encyclopedia of Plant and Crop Science
DOI: 10.1081/E-EPCS 120005946

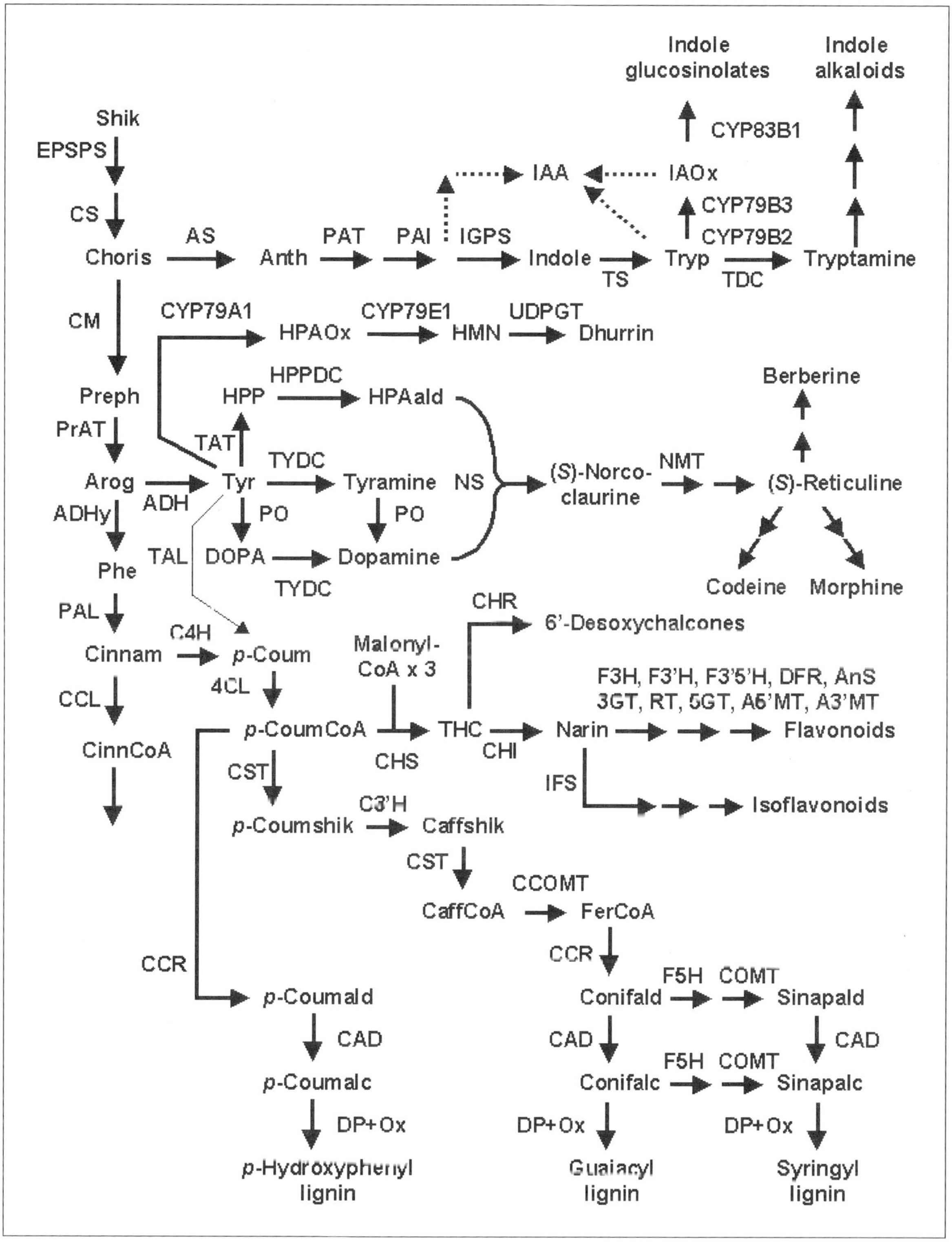

Fig. 1 Plant secondary products derived from the amino acids tryptophan (Tryp), tyrosine (Tyr), and phenylalanine (Phe). Intermediate abbreviations appear on the next page.

the terpenoid precursors (derived from geraniol via the nonmevalonate pathway), and ensuring appropriate intracellular compartmentation and tissue-specific expression of the numerous pathway enzymes.[1,5]

The pathway of synthesis of the indole glucosinolates in *Arabidopsis thaliana* proceeds via the catalytic action of O_2 and NADPH-dependent cytochrome P450s acting on tryptophan and indole-3-acetaldoxime.[6] CYP79B2 and CYP79B3 N-hydroxylate tryptophan, converting it to indole-3-acetaldoxime. CYP83B1 subsequently N-hydroxylates indole-3-acetaldoxime to its N-oxide[6] (Fig. 1). Mutants (*superroot*2 (*sur*2); *runt*1) of *Arabidopsis* are defective in CYP83B1 and have elevated levels of IAA, and decreased levels of indole glucosinolates. In contrast, overexpression of CYP83B1 causes a bushy, short phenotype consistent with reduced levels of IAA, suggesting that the glucosinolate pathway competes with the auxin biosynthesis pathway for the indole-3-acetaldoxime intermediate.[6] These engineering efforts illustrate the difficulty in manipulating fluxes around the tryptophan node of metabolism without detrimental effects on plant growth via perturbations of auxin levels. This is further complicated by the existence of multiple pathways of auxin biosynthesis (both tryptophan-dependent and -independent) which may operate at different stages of plant development, as indicated by the dashed lines at the top of Fig. 1.[6]

The cytochrome P450 catalyzing the conversion of tyrosine to *p*-hydroxyphenylacetaldoxime in the biosynthesis of the cyanogenic glucoside dhurrin in sorghum, has been designated CYP79A1. *p*-hydroxyphenylacetaldoxime is subsequently converted to *p*-hydroxymandelonitrile by a second multifunctional P450 monooxygenase, CYP71E1. The final step in dhurrin synthesis in sorghum is the transformation of the labile cyanohydrin into a stable storage form by O-glucosylation of (*S*)-*p*-hydroxymandelonitrile at the cyanohydrin function, by a UDP-glucose:*p*-hydroxymandelonitrile-O-glucosyltransferase[7] (Fig. 1). The entire pathway for synthesis of dhurrin has been transferred from sorghum to *Arabidopsis thaliana* by expressing CYP79A1, CYP71E1 and UDP-glucose:*p*-hydroxymandelonitrile-O-glucosyltransferase.[7] The accumulation of dhurrin in the transgenic plants confers resistance to the flea beetle.[7] This clearly demonstrates the importance of cyanogenic glucosides in plant defense against insect herbivores, and it illustrates the potential for introduction of multiple genes to install a complete pathway in plants. Interestingly, when only CYP79A1 is expressed in *Arabidopsis*, the transgenic plants produce *p*-hydroxylbenzyl glucosinolates.[7] This is presumably because the intermediate *p*-hydroxyphenylacetaldoxime is recognized as a substrate by native glucosinolate-forming enzymes normally involved in indole and/or alykl glucosinonate biosynthesis.

Considerable progress has been made in identifying the enzymes and genes involved in the production of pharmaceutically important isoquinoline alkaloids derived from tyrosine, including berberine, codeine, and morphine.[1,3] Engineering of this pathway is currently focused on manipulating expression of tyrosine/DOPA decarboxylase (TYDC) and fluxes around the (*S*)-reticuline node.[3]

Fig. 1 Intermediate abbreviations: Anth = anthranilate; Arog = arogenate; CaffCoA = caffeoyl-CoA; Caffshik = caffeoyl shikimate; Choris = chorismate; Cinnam = cinnamate; CinnCoA = cinnamoyl-CoA; *p*-Coum = *p*-coumarate; *p*-CoumCoA = *p*-coumaroyl-CoA; *p*-Coumald = *p*-coumaraldehyde; *p*-Coumalc = *p*-coumaryl alcohol; *p*-Coumshik = *p*-coumaroyl shikimate; Conifald = coniferaldehyde; Conifalc = coniferyl alcohol; DOPA = dihydroxyphenylalanine; FerCoA = feruloyl-CoA; HMN = *p*-hydroxylmandelonitrile; HPAald = *p*-hydroxyphenylacetaldehyde; HPAOx = *p*-hydroxyphenylacetaldoxime; HPP = *p*-hydroxyphenylpyruvate; IAA = indole 3-acetic acid; IAOx = indole 3-acetaldoxime; Narin = naringenin; Preph = prephenate; Shik = shikimate; Sinapald = sinapaldehyde; Sinapalc = sinapyl alcohol; THC = 4,2′,4′6′-tetrahydroxychalcone. Enzyme abbreviations: AnS = anthocyanidin synthase; A3′MT = anthocyanin 3′ methyltransferase; A5′MT = anthocyanin 5′ methyltransferase; AS = anthranilate synthase; ADHy = arogenate dehydratase; ADH = arogenate dehydrogenase; COMT = caffeate/5-hydroxyferulic acid O-methyltransferase; CCOMT = caffeoyl CoA 3-O-methyltransferase; CHI = chalcone isomerase; CHR = chalcone reductase; CHS = chalcone synthase; CM = chorismate mutase; CS = chorismate synthase; CCL = cinnamate:CoA ligase; C4H = cinnamate 4-hydroxylase; CCR = cinnamoyl CoA reductase; CAD = cinnamyl alcohol dehydrogenase; DFR = dihydroflavonol 4-reductase; DP + Ox = dirigent protein + oxidase; EPSPS = 5-*enol*pyruvylshikimate-3-phosphate synthase; F5H = ferulate 5-hydroxylase; F3′H = flavonoid 3′-hydroxylase; F3′5′H = flavonoid 3′,5′-hydroxylase; F3H = flavanone 3-hydroxylase; 3GT = 3-glucosyltransferase; 5GT = 5-glucosyltransferase; HPPDC = *p*-hydroxyphenylpyruvate decarboxylase; CST = hydroxycinnamoyl-CoA: shikimate hydroxycinnamoyltransferase; 4CL = 4-(hydroxy)cinnamoyl:CoA ligase; IGPS = indole-3-glycerol-phosphate synthase; IFS = isoflavone synthase; NS = (*S*)-norcoclaurine synthase; NMT = norcoclaurine 6-O-methyltransferase; PAL = phenylalanine ammonia-lyase; PO = phenol oxidase; PAI = phosphoribosylanthranilate isomerase; PAT = phosphoribosylanthranilate transferase; PrAT = prephenate aminotransferase; RT = rhamnosyltransferase; TDC = tryptophan decarboxylase; TS = tryptophan synthase; TAT = tyrosine aminotransferase; TAL = tyrosine ammonia-lyase; TYDC = tyrosine decarboxylase; UDPGT = UDP-glucose:*p*-hydroxymandelonitrile-O-glucosyltransferase.

Phenylpropanoids

Phenylalanine is the starting point for the synthesis of a large number of phenylpropanoid secondary products, including flavonoids, isoflavonoids, and lignin. Examples of successful engineering of these pathways in plants include the introduction of chalcone reductase (CHR) in acyanic petunia to generate a novel, yellow flower color due to the accumulation of 6′-desoxychalcones[8] (Fig. 1), and the overexpression of transcription factors to globally activate or repress anthocyanin biosynthesis.[3] However, recent attempts to engineer isoflavone synthesis in *Arabidopsis* by introducing the soybean isoflavone synthase (IFS) gene have encountered a problem. It seems that IFS does not compete effectively with flavanone 3-hydroxylase (F3H) for the naringenin substrate, perhaps because chalcone synthase (CHS), chalcone isomerase (CHI), and F3H may exist as a multienzyme complex, causing metabolite channeling.[9]

Lignin biosynthesis is a key branch-point from the phenylpropanoid pathway.[1,2] The lignin biosynthetic pathway has been recently clarified, and is now proposed to involve a 3′-hydroxylase (C3′H) acting on *p*-coumaroyl shikimate or *p*-coumaroyl quinate as substrates[10] (Fig. 1). The revised metabolic scheme for lignin biosynthesis is beginning to explain some of the phenotypes of altered lignin composition of mutants defective in the various metabolic steps, although our understanding of metabolic control in the pathway remains far from complete.[4] A detailed analysis of transgenic plants and mutants suggests that PAL, C4H and C3′H may be especially important in regulating carbon allocation in the pathway, with CCOMT, 4CL, CCR, F5H, and COMT playing subsidiary processing roles.[2]

FUTURE PROSPECTS

A significant number of the genes of plant primary and secondary metabolism have now been cloned, and they are being used in metabolic engineering experiments. Urgently needed are mathematical models that integrate the stoichiometric coefficients of the pathways with the kinetic properties of the enzymes encoded by these genes, and with the regulatory circuits that control both gene expression and enzyme activity. Such models will considerably aid future efforts devoted to rational flux manipulations of these pathways.

ARTICLES OF FURTHER INTEREST

Amino Acid and Protein Metabolism, p. 30
Metabolism, Primary: Engineering Pathways of, p. 714
New Secondary Metabolites: Potential Evolution, p. 818
Phenylpropanoids, p. 868
Phytochemical Diversity of Secondary Metabolites, p. 915
Secondary Metabolites as Phytomedicines, p. 1120

REFERENCES

1. Croteau, R.; Kutchan, T.M.; Lewis, N.G. Natural Products (Secondary Metabolites). In *Biochemistry & Molecular Biology of Plants*; Buchanan, B., Gruissem, W., Jones, R., Eds.; American Society of Plant Physiologists: Rockville, MD, 2000; 1250–1318.
2. Anterola, A.M.; Lewis, N.G. Trends in lignin modification: A comprehensive analysis of the effects of genetic manipulations/mutations on lignification and vascular integrity. Phytochemistry **2002**, *61*, 221–294.
3. Verpoorte, R.; Memelink, J. Engineering secondary metabolite production in plants. Curr. Opin. Biotechnol. **2002**, *13*, 181–187.
4. Stephanopoulos, G.N.; Aristidou, A.A.; Nielsen, J. *Metabolic Engineering Principles and Methodologies*; Academic Press: San Diego, CA, 1998; 1–725.
5. Hanson, A.D.; Shanks, J.V. Plant metabolic engineering—Entering the S curve. Metab. Eng. **2002**, *4*, 1–2. [See also other articles in this special issue.]
6. Celenza, J.L. Metabolism of tyrosine and tryptophan—New genes for old pathways. Curr. Opin. Plant Biol. **2001**, *4*, 234–240.
7. Tattersall, D.B.; Bak, S.; Jones, P.R.; Olsen, C.E.; Nielsen, J.K.; Hansen, M.L.; Hoj, P.B.; Moller, B.L. Resistance to an herbivore through engineered cyanogenic glucoside synthesis. Science **2001**, *293*, 1826–1828.
8. Mol, J.; Cornish, E.; Mason, J.; Koes, R. Novel colored flowers. Curr. Opin. Plant Biol. **1999**, *10*, 198–201.
9. Liu, C.-J.; Blount, J.W.; Steele, C.L.; Dixon, R.A. Bottlenecks for metabolic engineering of isoflavone glycoconjugates in Arabidopsis. Proc. Natl. Acad. Sci. U. S. A. **2002**, *99*, 14578–14583.
10. Humphreys, J.M.; Chapple, C. Rewriting the lignin roadmap. Curr. Opin. Plant Biol. **2002**, *5*, 224–229.

Milkweed: Commercial Applications

Herbert D. Knudsen
Natural Fibers Corporation, Ogallala, Nebraska, U.S.A.

INTRODUCTION

Milkweed, especially the *Asclepias syriaca* and *Asclepias specosia* varieties, has excellent potential for commercialization as a new crop. Milkweed raw materials include pod floss, seed, and stem fiber. This article reviews the progress over the last 15 years in developing milkweed production, products, and markets to create a new agricultural business.

HISTORICAL

Asclepiadaceae is the milkweed plant family native to many countries in the world.[1] The various species range from small shrubs, to vine covers, to plants ten feet tall. Two varieties—*Asclepias syriaca* and *Asclepias specosia*—are particularly of interest because they are native to many areas in the United States, have erect stems four to six feet tall, and bear multiple large milkweed pods per stem.

Small-volume use of milkweed raw materials from native stands is widely reported. North American Indians used bast fibers from milkweed stems in clothing more than 1000 years ago. Flowers, sprouts, pods, and seeds have been used as food and medicine.

Boris Berkman, president of Milkweed Floss Corporation of America, organized the first effort to produce milkweed floss on a large scale.[2] In 1943, the U.S. Government built a milkweed processing plant in Petoskey, Michigan as part of the war effort. This plant produced over two million pounds of milkweed floss for military life jackets in its one year of operation. Pods for the production facility were hand-collected in onion bags and shipped to Michigan. War volunteers throughout the United States collected milkweed pods under the slogan ''Two bags save one life.'' Berkman envisioned great potential for milkweed raw materials in industrial products, but no record of significant milkweed floss production or commercial use of milkweed in products after World War II has been found.

Standard Oil of Ohio conducted extensive research on growing milkweed and extracted the total harvested milkweed biomass with hexane to produce a crude oil substitute. Technically, the process was feasible, but the yield in terms of barrels per acre was too low to make the process economical. Standard Oil also worked with Kimberly-Clark to develop nonwoven products from milkweed floss, but no commercial products were produced.

In 1987, Standard Oil sold its milkweed venture to Natural Fibers Corporation (NFC), a privately held corporation formed to commercialize milkweed as a new crop. In 1990, a consumer products company, Ogallala Down Comforter Company, was formed as a subsidiary of NFC to make and sell goose down/milkweed floss comforters to specialty bedding stores in the United States and to the Four Seasons Hotels throughout the world.

MARKETS

The raw materials from milkweed are floss, seed, pod hulls, and stems, and markets for milkweed as a raw material needed to be created. The options are 1) to sell raw materials to others for their use in products they would develop; or 2) to develop, produce, and sell value-added products with these new raw materials. The downlike qualities of milkweed floss allowed for application in the down industry. Because of the up to 30-fold increase in value of down bedding compared to the milkweed floss content and because of the need to ensure a base load of floss demand, goose down/milkweed floss comforters and pillows as a new consumer product were produced by NFC internally and sold commercially.

Companies, research centers, and organizations have worked to develop milkweed raw material demand. Current projects include 1) working with Monarch butterfly organizations to expand milkweed habitat; 2) pressing milkweed seed to produce oil useful in cosmetics; 3) chopping milkweed floss to a fine powder for body care products; 4) developing milkweed floss nonwoven pads for bedding; 5) researching pulp made from milkweed for paper products; and 6) experimenting with milkweed biomass for nematode control in soil.

PRODUCTION FACILITY

To create a business at startup, a processing system for milkweed pods and a facility for manufacturing consumer products were needed. These issues were more important than cultivating milkweed, because wild collection of milkweed pods could produce enough dry pods to start and run the business for a number of years.

Encyclopedia of Plant and Crop Science
DOI: 10.1081/E-EPCS 120010434

The processing system goal was to develop continuous mechanical separation of milkweed raw materials from dried pods. Boris Berkman's plant in Michigan appeared to be continuous, but all production of milkweed raw materials for Standard Oil was done in a laborious batch process.

For the processor, information about Boris Berkman's cylinder production system was collected, but experiments in similar, but smaller, cylindrical processing systems were not successful. A lower-technology approach of modifying a 1940 John Deere harvester, however, *did* work. See U.S. Patent 4,959,038. Since 1992, an automatic pod feeder was added, the floss hopper was eliminated in favor of direct bagging, and the footprint of the processor was reduced in a move to a new facility. The processor now operates to produce 40 pounds of floss per hour while recovering over 95% of the floss and seed. This processor is more than adequate for current needs, and the unit can be duplicated or modernized as raw material demand increases.

A second mechanical challenge was the blender. The fill developed for comforters and pillows was an intimate blend of milkweed floss and goose down. From bench scale data, a commercial-size blender was invented from wood and blowers. The blender was later upgraded to a cylindrical plastic chamber.

A final manufacturing issue was the production of comforters and pillows. A major issue was developing dependable comforter and pillow shell vendors and down suppliers.

CULTIVATED PRODUCTION

Cultivated production is still in the experimental stage, so milkweed pods are collected by hand from native stands in Nebraska, Wyoming, and Illinois. Wild sources provide time for our agricultural production research to develop proven farming practices. The estimated break-even point between the economics of wild collection and cultivated production is a cultivated yield of 50 pounds of floss per acre.

Cultivated production focused on maintaining milkweed in rows until 1996. However, because milkweed is a perennial plant that spreads through rhizomes, cultivation between rows disrupts the milkweed's natural growing mechanism. With this understanding and the newly available preemergence herbicides that do not affect established milkweed, a change was made to plant broadcast stands of milkweed. Planting milkweed seed in stubble after early July wheat harvest protects the fragile seedlings from strong western Nebraska winds. The goal is to establish enough growth in half of a growing season so that the milkweed plant will produce pods the following year.

Systems for harvest and drying of milkweed pods were developed primarily by Kenneth Von Bargen and David Jones at the University of Nebraska.[3] A row crop harvester for milkweed was modified from a New-Idea Uni-System ear corn harvester. The row crop harvester achieves 80% pod recovery in the broadcast stands of experimental fields.

Pods are generally harvested in September. During harvest, the harvester strips whole, green pods from the milkweed plant. The stalks are left behind and can be harvested by swathing and baling the dried stems. The pods are air dried and processed into raw materials. Raw materials can be stored under warehouse conditions for years.

A field life of 5–10 years is expected. Development of Roundup-ready soybeans is a major step forward in cultivated milkweed production. Before this development, a fallow year was required to eradicate the milkweed field before planting soybeans. Now milkweed can be eradicated while soybeans grow, and a production year is not lost.

Target economics provide the milkweed grower at least the same revenue as corn, with variable farming costs less than with corn. Contract farming would supply milkweed seed and technical advice to the farmer and would guarantee the purchase of the field's output.

CONCLUSION

Increasing the demand for milkweed raw materials is the key challenge in developing the milkweed business. The strategy is to concentrate on higher-value (>$5 per pound of milkweed floss) and lower-volume uses. On the agricultural side, the goal is to develop growing technology to the point where innovative farmers can profitably grow the crop.

ARTICLES OF FURTHER INTEREST

Biological Control of Nematodes, p. 134
Lesquerella Potential for Commercialization, p. 656
Non-wood Plant Fibers: Applications in Pulp and Papermaking, p. 829

REFERENCES

1. Knudsen, H.D.; Zeller, R.D. The Milkweed Business. In *New Crops*; Janick, J., Simon, J.E., Eds.; John Wiley and Sons: New York, 1992; 442–448.
2. Berkman, B. Milkweed—A war strategic material and a potential industrial crop for sub-marginal lands in the United States. Econ. Bot. **1949**, *3* (3), 223–239.
3. Von Bargen, K.; Jones, D.; Zeller, R.; Knudsen, P. Equipment for milkweed floss-fiber recovery. Ind. Crops Prod. **1994**, *2*, 201–210.

Minor Nutrients

Wolfgang Schmidt
Humboldt University, Berlin, Germany

INTRODUCTION

In contrast to major (macro) mineral nutrients, which (according to their function in osmoregulation, electrochemical equilibria, or as constituents of organic molecules) are required in relatively large amounts, micronutrients are needed in relatively small quantities; in some cases in only minute amounts. Micronutrients serve mainly as catalytic components of enzymes and play a structural role that is indispensable for enzymic function. Mineral nutrients that are required by higher plants for specific functions—such as Fe, Mn, Zn, Cu, B, Mo, Cl, and Ni—can be distinguished from those that have beneficial effects but are not needed by plants to complete a full life cycle and can be substituted by other nutrient elements or are needed by only few taxa, such as Na, Si, and Co. Research into micronutrients serves several goals. Knowledge of how micronutrients are acquired and transported helps in ameliorating the effects of reduction in crop yield caused by nutrient deficiencies, and improves our understanding of the distribution of species in natural habitats. In edible plant products there is a need for an adequate supply of those micronutrients essential for human and animal nutrition in order to eliminate the requirement for dietary supplements and postharvest treatments. Thus, increasing the metal content of our food helps to decrease health-threatening human deficiencies. A further aspect of the study of micronutrients concerns the use of plants for extracting metals from soil, for either phytoremediation or phytomining. Cleaning soils contaminated with trace metals by plants is an inexpensive method of soil remediation. Similarly, plants engineered to accumulate metals return an economic profit from waste rock.

Concerning the molecular background of micronutrient uptake and homeostasis, most knowledge comes from investigations into iron, zinc, manganese, and copper. Therefore, and due to space limitation, this article focuses on these nutrients. For a detailed overview of micronutrients, see the excellent textbook by Marschner and the recent review by Reid.

FUNCTION OF MICRONUTRIENTS IN PLANTS

Iron and copper are essential redox components in photosynthetic and respiratory electron transport and an indispensable cofactor in fundamental cellular processes. Due to its ability to pass readily between the divalent and trivalent oxidation states, iron participates in a wide range of electron transfer reactions that are essential for almost all organisms. Besides electron transport, copper is involved in free radical elimination, lignification, pollen formation, and hormone perception and signaling. Iron and copper metabolism appears to be interlinked in various respects. High levels of free Cu^{2+} causes iron deficiency, possibly by competition between Fe^{2+} and Cu^{2+} for binding sites on proteins or complexing compounds. In yeast and *Chlamydomonas*, copper deficiency also results in iron deficiency because copper is required for the synthesis of a multicopper oxidase that oxidizes ferrous iron prior to uptake of iron.[1] The valency of zinc cannot be changed and the biological role of zinc is based on its tendency to form tetrahedral complexes. Zinc functions in the regulation of gene expression and is essential for a large number of enzymes, including alcohol dehydrogenase, carbonic anhydrase, and CuZn-superoxide dismutase. Manganese is required for a variety of essential processes, including light-induced oxygen evolution in phothosynthesis and CO_2 fixation in C4 and Crassulacean acid metabolism (CAM) plants via phosphoenol-pyruvate carboxylase. In addition, manganese is important for the detoxification of free oxygen radicals via Mn-superoxide dismutase. The function of molybdenum is related to valency change. Molybdenum plays an essential role in nitrate reduction and nitrogen fixation via nitrogenase. Chloride is essential for water splitting in PSII. A further function of chloride is associated with stomatal regulation. In species that lack functional chloroplasts for malate synthesis, Cl^- is transported as a counter anion for K^+. Due to the ubiquitous presence of chloride in soils, toxicity is much more frequent than deficiency. Nickel is required in extremely low

Encyclopedia of Plant and Crop Science
DOI: 10.1081/E-EPCS 120010508

amounts and is the most recent plant nutrient to be included in the list of essential elements. The only nickel-containing enzyme is urease. Boron has been implicated in a diverse array of functions and is dealt with in a separate article. Despite the fact that the requirement for boron is highest among the micronutrients on a molar basis, the role of boron is still not completely understood.

UPTAKE AND TRANSPORT OF MICRONUTRIENTS

Several classes of transporters that mediate the uptake of micronutrients from the soil solution and their distribution around the plant have been identified by sequence similarities and functional complementation.[2,3] IRT1, the founding member of the ZRT-IRTlike proteins (ZIP) family, mediates iron uptake in *Arabidopsis*, and orthologs have been subsequently identified in other taxa. AtIRT1 can also transport manganese, zinc, and cobalt but not copper. Interestingly, the substrate specificity of IRT1 is defined by only few residues, offering the possibility of constructing plants with altered selectivity. Depending on the species, iron is either taken up as a divalent cation after reduction of the ferric form that prevails in most soils (strategy I),[4] or as ferric chelate, complexed with plantborne siderophores with high affinity to ferric ions (phytosiderophores, PS, strategy II). Transporters of the natural resistance-associated macrophage protein (Nramp) family participate in iron uptake and iron homeostasis and have potential roles in the distribution of other metal cations, but their physiological functions are not yet clearly defined.[5] Zinc is taken up as a divalent ion and accumulates in high amounts by some hyperaccumulators, such as *Thlaspi caerulescens*. ZIP proteins confer zinc uptake in yeast and appear to represent the major route of entry for zinc into plants. ZIP1 and ZIP3 are expressed in roots in response to zinc deficiency.[6]

Copper enters the cell either complexed by organic ligands or as a free metal ion. Uptake of copper is mediated by a family of transporters named COPT1-COPT5, some members of which are repressed by excess copper.[7] In yeast, Cu^{2+} must be reduced before uptake. Whether such a reduction is also required for copper uptake by plants is unclear. Manganese is taken up mainly as the free Mn^{2+} ion; a transporter specific for manganese has not been identified. Less is known concerning the uptake of those micronutrients that are taken up as anions. Molybdenum is transported across the plasma membrane as the MoO_4^{2-} anion. Chloride is taken up as a monovalent Cl^-, most likely by an active process in symport with protons. No specific transporters involved in the uptake of micronutrient anions have been identified to date. Boron crosses membranes as an uncharged molecule. Besides simple permeation, boron is thought to be taken up by a transporter or via aquaporins.[8]

CELLULAR TRAFFICKING AND HOMEOSTASIS OF MICRONUTRIENTS

The availability of micronutrients in the environment varies, necessitating regulated uptake systems to ensure that adequate amounts of essential ions are required. Knowledge of the mechanisms for maintaining cytoplasmic homeostasis of micronutrients is still fragmentary. The regulation of transporters mediating the uptake of metal cations appears to be complex, including both transcriptional and posttranslational control. For example, in transgenic plants overexpressing the iron transporter gene *IRT1*, the IRT1 product is rapidly degraded and represents a safety module when the iron concentration is increased rapidly.[9] Moreover, IRT1, which is not expressed in zinc deficient plants, can be downregulated by high zinc levels, indicating interaction between signalling cascades of various nutrients. Such cross talks have also been revealed by cDNA microarrays with plants grown in the absence of different essential nutrients, showing that several genes with potential regulatory roles are induced by deficiencies in various nutrients.[10]

To prevent toxicity, metal cations have to be chelated after entering the cells. This is mainly achieved by peptides such as phytochelatin and metallothioneins, organic acids, polyphenolics, and an array of yet unidentified low molecular weight anionic ligands. A specific role in the distribution of heavy metals has been suggested for the amino acid nicotianamine (NA). In some plants, NA concentrations are considerably increased by high external copper and iron, pointing to the role of NA in the detoxification of excess metals. Immunostaining of the NA–Fe complex has revealed that labelling density was highest in the cytoplasm under adequate iron supply, whereas most of the labelling was present in the vacuole when the plants were iron loaded.[11] Accordingly, sequestration of iron and possibly other metals into the vacuole appears to represent a mechanism regulating the cytoplasmic level of metal ions. Besides Nramps, members of the cation diffusion facilitator (CDF) family, yellow stripelike (YSL) transporters, and heavy metal (CPx-type) ATPases have been implicated in the uptake, efflux, and sequestration of metal cations, but (with few exceptions) a specific physiological function of these transporters has not yet been proven.

LONG-DISTANCE TRANSPORT

Little is known about how metals are transported over long distances, either from roots to the shoot after uptake from the soil via the xylem or, during remobilization, from source to sinkin the phloem Generally, metal micronutrients are complexed by ligands to decrease interaction with charged groups during transport. In the xylem, iron is complexed with citrate as the $FeCit_2^{3-}$ ion in a 1:1 ratio. Zinc and manganese are also most likely bound by citrate during xylem transport. Unlike other heavy metals, copper appears to be bound by NA during translocation, a complex with an extremely high stability constant. Other low molecular-weight ligands that have been suggested to function in xylem transport are histidine and phytosiderophores of the mugineic acid family. Coloading of NA and micronutrients into the phloem supports a role for NA in this process, but a function of NA as a ligand for phloem transport has not been confirmed. A protein with high affinity for iron that can also complex copper, zinc, and manganese in vitro has recently been isolated from *Ricinus* phloem exudates, representing the first identified micronutrient ligand mediating phloem-mediated long-distance transport.[12]

CONCLUSION

Although unravelling the intra- and intercellular transport of micronutrients has recently begun at the molecular level, many important questions concerning the mechanisms underlying uptake, trafficking, and sequestration still remain unanswered. Little is known about how the level of micronutrients—either within the cell or in its immediate vicinity—is sensed and how this signal is translated to adjust uptake and transport within the cell. The elucidation of the mechanisms that determine the concentration and distribution of micronutrients in plants opens new avenues to deepen our understanding of both basic and applied aspects of plant mineral nutrition as diverse as the autecology of species and the improvement of human nutrition.

ACKNOWLEDGMENTS

I would like to thank Ernest Kirkby for comments on the manuscript.

ARTICLES OF FURTHER INTEREST

Phytoremediation: Advances Toward a New Cleanup Technology, p. 924
Rhizosphere: An Overview, p. 1084
Rhizosphere: Nutrient Movement and Availability, p. 1094

REFERENCES

1. Herbik, A.; Bölling, C.; Buckhout, T.J. The involvement of a multicopper oxidase in iron uptake by the green algae *Chlamydomonas reinhardtii.* Plant Physiol. **2002**, *130*, 2039–2048.
2. Fox, T.C.; Guerinot, M.L. Molecular biology of cation transport in plants. Ann. Rev. Plant Physiol. Plant Mol. Biol. **1998**, *49*, 669–696.
3. Williams, L.E.; Pittman, J.K.; Hall, J.L. Emerging mechanisms for heavy metal transport in plants. Biochim. Biophys. Acta **2000**, *1465*, 104–126.
4. Marschner, H. *Mineral Nutrition of Higher Plants*, 2nd Ed.; Academic Press: London, 1995.
5. Curie, C.; Briat, J.F. Iron transport and signaling in plants. Ann. Rev. Plant Biol. **2003**, *54*, 183–206.
6. Guerinot, M.L.; Eide, D. Zeroing in on zinc uptake in yeast and plants. Curr. Opin. Plant Biol. **1999**, *2*, 244–249.
7. Sancenón, V.; Puig, S.; Mira, H.; Thiele, D.J.; Peñarrubia, L. Identification of a copper transporter family in *Arabidopsis thaliana.* Plant Mol. Biol. **2003**, *51*, 577–587.
8. Reid, R.R. Mechanisms of micronutrient uptake in plants Aust. J. Plant Physiol. **2001**, *28*, 659–666.
9. Connolly, E.L.; Fett, J.P.; Guerinot, M.L. Expression of the IRT1 metal transporter is controlled by metals at the levels of transcript and protein accumulation. Plant Cell **2002**, *14*, 1347–1357.
10. Wang, Y.H.; Garvin, D.F.; Kochian, L.V. Rapid induction of regulatory and transporter genes in response to phosphorus, potassium, and iron deficiencies in tomato roots. Evidence for cross talk and root/rhizosphere-mediated signals. Plant Physiol. **2002**, *130*, 1361–1370.
11. Pich, A.; Manteuffel, R.; Hillmer, S.; Scholz, G.; Schmidt, W. Fe homeostasis in plant cells: Does nicotianamine play multiple roles in the regulation of cytoplasmic Fe concentration? Planta **2001**, *213*, 967–976.
12. Krüger, C.; Berkowitz, O.; Stephan, U.W.; Hell, R. A metal-binding member of the late embryogenesis abundant protein family transports iron in the phloem of *Ricinus communis* L. J. Biol. Chem. **2002**, *227*, 25062–25069.

Mitochondrial Respiration

Agepati S. Raghavendra
University of Hyderabad, Hyderabad, India

M

INTRODUCTION

Respiratory metabolism occurs in two compartments of plant cells: cytoplasm and mitochondria. Cytoplasm is the site of glycolysis, which produces pyruvate from glucose. The mitochondria carry out further oxidation of pyruvate to CO_2 and H_2O, and they produce adenosine triphosphate (ATP).

Mitochondria are threadlike or rod-shaped organelles, enveloped by an outer and an inner membrane. The space between these two membranes is called intermembrane space. The outer membrane is unselective in permeability, but may be regulatory because of porins (pore-forming proteins). The inner membrane is highly selective and directs metabolite movement through translocators. The inner membrane encloses the granular matrix and has invaginations called cristae. Besides the enzymes of the tricarboxylic acid (TCA) cycle, the matrix contains mtDNA and machinery for protein synthesis and fatty acid metabolism. The components of oxidative electron transport and phosphorylation are located in the inner membrane.

Mitochondrial respiration is most important in terms of contribution, complexity, regulation, and interaction with other organelles. Of about 64 molecules of ATP generated from the oxidation of one molecule of sucrose, 56 molecules of ATP are produced in mitochondria. Besides meeting cellular energy demands, mitochondrial respiration provides the carbon skeletons for primary and secondary metabolism, forms a link in interorganelle interaction, and helps in the homeostasis of redox in cytoplasm.[1,2]

SUBSTRATES AND PRODUCTS

Mitochondrial respiration consists of three steps: the TCA cycle, electron transport, and oxidative phosphorylation.[1,2] The first step is oxidation of substrates like pyruvate or malate through the TCA cycle to generate reduced nicotinamide adenine dinucleotide (NADH) and reduced flavin adenine dinucleotide ($FADH_2$), which are oxidized through electron transport to generate a proton electrochemical gradient across the inner membrane. In the last step, the energy from the proton gradient is used for ATP formation through oxidative phosphorylation.

The substrates for mitochondrial respiration in plant cells range from pyruvate, malate, glutamate, succinate, and glycine, coming from the cytoplasm, chloroplasts, glyoxysomes, or peroxisomes.[2,3] The predominance of substrate depends on the developmental stage of the organ, the type of tissue, and the presence of light. Pyruvate, malate, or succinate is oxidized through the TCA cycle to yield NADH or $FADH_2$, CO_2, and H_2O. Glutamate is converted to 2-oxoglutarate and fed into the TCA cycle. Glycine is oxidized to serine and sent into peroxisomes. NADH and $FADH_2$ generated during the oxidation of substrates are utilized in the electron transport to produce ATP.

The final products of mitochondrial respiration are ATP, CO_2, and H_2O. However, intermediates of the TCA cycle are exported to provide carbon skeletons for nitrate assimilation and other processes.

TCA CYCLE

The major activity of mitochondria is to form citrate, by condensation of acetyl CoA (from pyruvate) and oxaloacetate, and oxidize it through the TCA cycle (Fig. 1). Pyruvate is oxidized (to acetyl CoA) by pyruvate dehydrogenase, a multienzyme complex composed of three components: pyruvate dehydrogenase, dihydrolipoyl transacetylase, and dihydrolipoyl dehydrogenase. Acetyl CoA condenses with oxaloacetate to form citrate, an irreversible reaction catalyzed by citrate synthase. The enzyme aconitase isomerizes citrate to isocitrate, which is oxidized to 2-oxoglutarate by isocitrate dehydrogenase. Another multienzyme complex, 2-oxoglutarate dehydrogenase oxidizes 2-oxoglutarate to succinyl CoA. The hydrolysis of succinyl CoA by succinyl CoA thiokinase yields succinate and ATP. The next step of succinate oxidation to fumarate is catalyzed by succinate dehydrogenase (also known as Complex II), an enzyme bound to the inner membrane. Fumarate is converted to malate by fumarase. Malate is oxidized to oxaloacetate by NAD-malate dehydrogenase to complete the cycle.[1,2]

The complete operation of the TCA cycle does not always occur, because some intermediates are exported to meet the carbon demands of cytoplasm.[2,3] For example, citrate and 2-oxoglutarate are needed for nitrate assimilation. It becomes essential to replenish the intermediates

Encyclopedia of Plant and Crop Science
DOI: 10.1081/E-EPCS 120010404

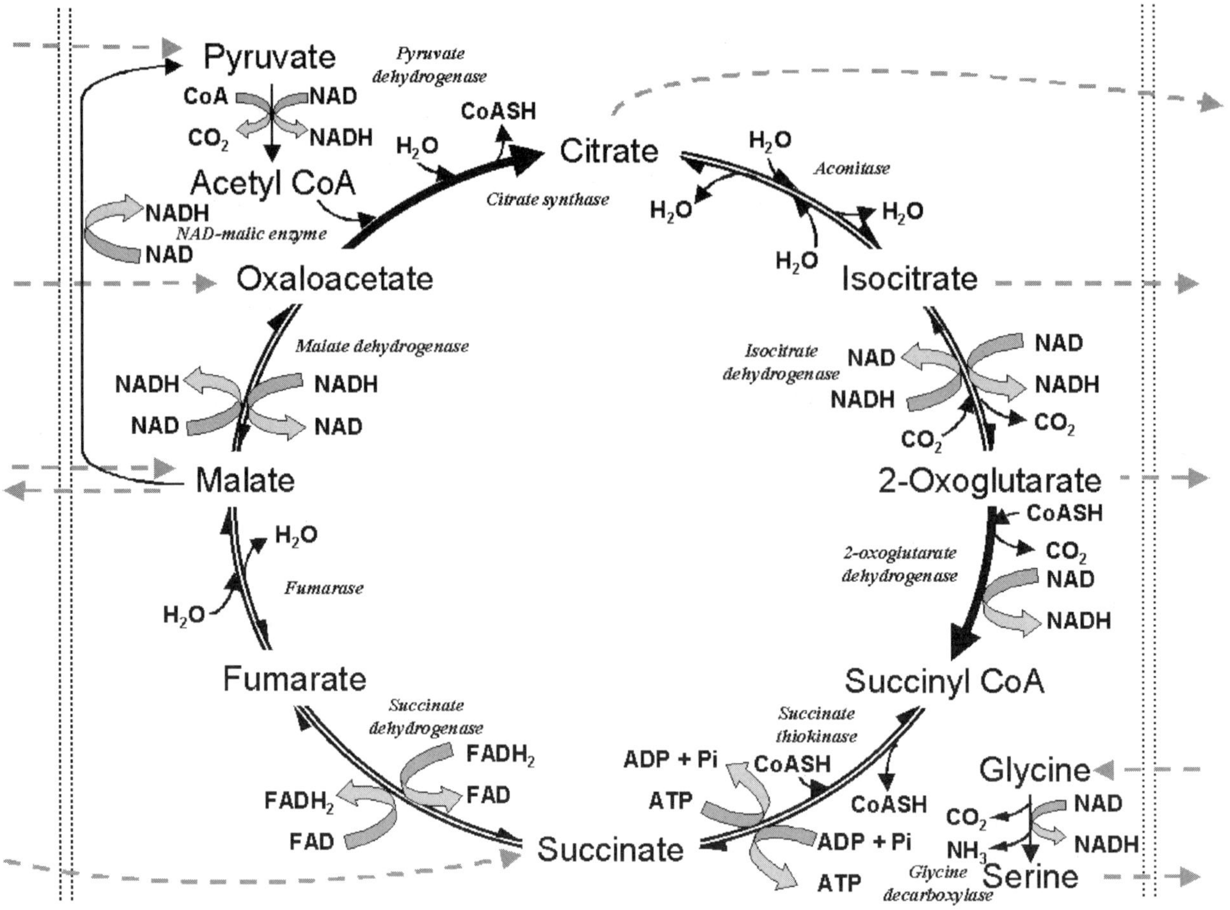

Fig. 1 Overview of the oxidation of pyruvate through the TCA cycle. The dotted lines represent the boundaries of mitochondria; enzymes are indicated by italics; and the inputs and outputs are indicated by gray broken arrows.

of the TCA cycle. Mitochondria, therefore, import malate (formed from oxaloacetate through PEP carboxylase), glutamate (formed from nitrate assimilation and present in high quantities in the cytoplasm), and glycine (formed in photorespiration). Glycine is the major substrate for mitochondrial respiration in leaves under light.

ELECTRON TRANSPORT CHAIN

The oxidative electron transport chain, located in the mitochondrial inner membrane, is formed of complexes, as in chloroplasts.[2,3] The mitochondrial electron transport chain consists of four modules: NADH-dehydrogenase complex (Complex I), the cyt-b/c complex (Complex III), and the terminal cyt-a/a_3 complex (Complex IV). The oxidation of succinate is initiated by the succinate dehydrogenase (Complex II), and the electrons pass through Complexes II and IV (Fig. 2).

Complex I accepts electrons from NADH and passes them on to ubiquinone within the membrane. The ubiquinone is reduced and traverses across the inner membrane to reduce the cyt-b/c complex and to pull protons from the matrix side into the intermembrane space. The NADH dehyrogenase complex is made up of more than 40 subunits. Complex II is the smallest and has 4 subunits. The cyt-b/c complex is made up of 9–11 subunits. The reduced cytochrome c diffuses along the surface of the inner membane to reach and reduce Complex IV (earlier known as cyt-a/a_3, consisting of 7–9 subunits), which is in turn oxidized by O_2. Protons move from the matrix into the intermembrane space during the reduction/oxidation of cyt-a/a_3.

Plant mitochondria can oxidize NAD(P)H on the cytoplasmic side, by an external NAD(P)H dehydrogenase, distinct from Complex I. The external NAD(P)H dehydrogenase is regulated by calcium and the levels of NADH/NADPH.[4]

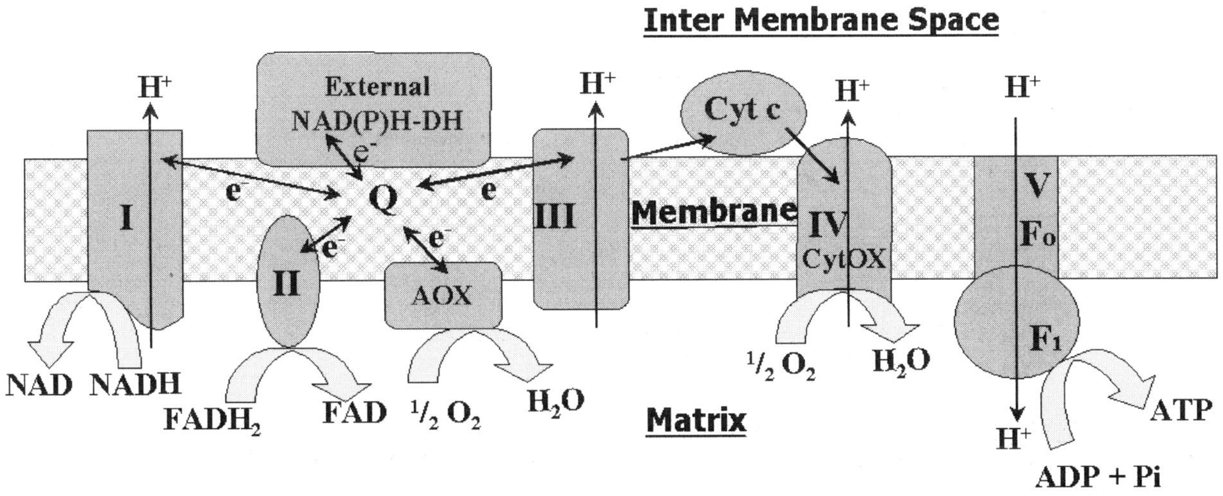

Fig. 2 Electron flow from NADH or FADH through the mitochondrial electron transport chain involving either the cytochrome pathway (Complex IV or CytOX) or the alternative pathway (alternative oxidase, AOX). Plant mitochondria also have an external NAD(P) dehydrogenase.

OXIDATIVE PHOSPHORYLATION

The mitochondrial electron transport creates an electrochemical proton gradient and facilititates ATP formation, as in chloroplasts. The oxidation of NADH/FADH drives protons from the matrix into the intermembrane space, creating a strong proton gradient (the intermembrane space being acidic) and depolarizing the membrane.[2,4] Protons diffuse back into the matrix through the F_0-F_1-ATP synthase complex. The F_0-F_1-ATP synthase of mitochondria is a multienzyme complex with two major components. The F_0 is membrane-embedded and is the proton channel. The F_1 is peripheral, extends as a stalk into the matrix, and is the site of ATP formation. F_1 contains five different subunits in a ratio of $\alpha_3\beta_3\gamma_1\delta_1\varepsilon_1$, while F_0 has three, with a composition of $a_1b_2c_{10-12}$.

The function of F_0-F_1 is quite fascinating. F_1 can synthesize ATP from ADP and Pi without additional energy, but a change in conformation driven by H^+ transport across the complex is needed to release ATP bound to the ATP synthase. The oxidation of each NADH molecule through the electron transport and phosphorylation yields 2.5 molecules of ATP and the oxidation of each of FADH molecule yields 1.5 of ATP.

ALTERNATIVE PATHWAY OF OXIDATIVE ELECTRON TRANSPORT

The mitochondrial respiration in plants is not completely inhibited by compounds such as cyanide, because their electron transport occurs through two different routes: the cyanide-sensitive cytochrome path and a cyanide-insensitive alternative path. The alternative pathway is facilitated by an alternative oxidase (a 32 kDa polypeptide), which has been isolated, cloned, and characterized in detail.[5,6]

Both pathways use Complex I/II and diverge at the site of ubiquinone (Fig. 2). Since these electrons do not use Complex III or IV, the extent of ATP formed from NADH through the alternative pathway is much less than that through the cytochrome pathway. Most of the energy is released as heat. The main function of the alternative electron transport is to oxidize surplus NADH (even if without ATP formation) and to produce heat. Since the alternative oxidase expression is boosted under environmental stress, it also may have other functions.

FUNCTION AND REGULATION OF MITOCHONDRIAL RESPIRATION

Unlike chloroplasts, mitochondria export ATP to the cytoplasm at high rates through an adenylate translocator, complemented by a phosphate translocator. The translocators are specialized proteins of the inner membrane of mitochondria that facilitate the movement of key metabolites. Among the translocators (and the compounds translocated) present on the inner membrane of mitochondria are malate or dicarboxylate (malate/Pi), 2-oxoglutarate (2-oxoglutarate/malate), citrate/tricarboxylate (citrate/dicarboxylate), oxaloacetate (oxaloacetate/malate), pyruvate (pyruvate/OH-), and glutamate (glutamate/aspartate). The presence of a glycine/serine translocator is expected.

Table 1 Metabolic inhibitors that block mitochondrial electron transport, phosphorylation, or other associated processes

Site of action	Inhibitor
Complex I	Rotenone, Amytal A, Piericidin
Complex II	Malonate
Complex III	Antimycin A, Myxothiazol
Complex IV	Cyanide, Azide, Carbon monoxide
Alternative oxidase	Salicylhydroxamic acid, Propyl gallate
F_1 of F_0-F_1-ATP synthase	Oligomycin
F_0 of F_0-F_1-ATP synthase	Dicyclohexyl carbodiimide
Proton gradient	2,4-dinitrophenol, Carbonyl cyanide-m-chloro-phenylhydrazone, Carbonyl cyanide p-trifluormethoxy-phenylhydrazone, Ammonium chloride, Methyl amine
Oxaloacetate translocator	Phthalonate
Malate translocator	Butyl malonate
Adenylate transclocator	Bongkrekic acid, Carboxyatractyloside

The operation of the TCA cycle and oxidative electron transport are highly regulated[1,2,4] by the energy status (relative levels of ATP and ADP) of the cell and the redox status of mitochondria (the levels of NADH) and acetyl–CoA. Increase in ATP promotes the phosphorylation and inactivation of pyruvate dehydrogenase. Similarly, increase in NADH causes feedback inhibition of dehydrogenases. Thus the turnover of the TCA cycle depends on the rate of NADH reoxidation by electron transport and the balance between ATP production and utilization. The availability of substrates (e.g., carbohydrate) is important. The regulation of oxidative electron transport by the availability of ADP and Pi is well known (''respiratory control''). Further, the movement of ATP/ADP/Pi into and out of mitochondria is modulated by the electrochemical proton gradient.

The alternative oxidase facilitates a distribution of electrons through the oxidative electron transport. The extent of flow through cytochrome or alternative pathways is modulated by metabolites such as pyruvate and reducing agents such as dithiothreitol.[5,6] The operation of alternative pathways is regulated by environmental stresses and factors modulating the carbon metabolism. Several metabolic inhibitors are available which block or inhibit different steps of the TCA cycle, translocators, oxidative electron transport, or ATP formation (Table 1).

IMPORTANCE IN PHOTOSYNTHESIS AND INTERORGANELLE INTERACTION

In plants, photorespiration is essential for retrieving half of the carbon diverted through the Rubisco oxygenase activity, with mitochondrial oxidation of glycine as the key. Mitochondria are the site of malate decarboxylation in NAD-malic enzyme type C_4 and CAM plants. Thus, mitochondrial respiration is essential for C_3-, C_4-, and CAM photosyntheses.[2,3]

Mitochondria interact beneficially with chloroplasts and peroxisomes.[7] Chloroplasts supply the substrates for mitochondria, which in turn dissipate the excess reductants generated in chloroplasts. Such cooperation optimizes the photosynthetic carbon assimilation and protects chloroplasts against photoinhibition. Another process, which involves the cooperation of mitochondria with chloroplasts and cytoplasm, is nitrate assimilation.

CONCLUSION

The focus of plant science research has shifted from physiology and biochemistry to the areas of molecular biology and biotechnology. Stupendous progress has occurred in understanding the molecular biology of the function and regulation of mitochondrial respiration. Genomics of plant mitochondria are well known, and the present focus is on proteomics.[8] Genetic manipulation of enzymes, such as citrate synthase, has been done.[9] Mutants (e.g., those deficient in some components of F_0-F_1) have long been used to analyze the mitochondrial function.[10] The mitochodrial transformation is bound to catch up soon with chloroplast transformation.

ACKNOWLEDGMENTS

Supported by a grant (No. SP/SO/A12/98) from Department of Science and Technology, New Delhi.

ARTICLES OF FURTHER INTEREST

Amino Acid and Protein Metabolism, p. 30
Glycolysis, p. 547

REFERENCES

1. Mackenzie, S.; McIntosh, L. Higher plant mitochondria. Plant Cell **1999**, *11*, 571–586.
2. Siedow, J.N.; Day, D.A. Respiration and Photorespiration. In *Biochemistry and Molecular Biology of Plant*; Buchanan, B.B., Gruissem, W., Jones, R.L., Eds.; American Society of Plant Physiologists: Maryland, USA, 2000; 676–728.
3. Bowsher, C.G.; Tobin, A.K. Compartmentation of metabolism within mitochondria and plastids. J. Exp. Bot. **2001**, *52*, 513–527.
4. Affourtit, C.; Krab, K.; Moore, A.L. Control of plant mitochondrial respiration. Biochim. Biophys. Acta **2001**, *1504*, 58–69.
5. Siedow, J.N.; Umbach, A.L. The mitochondrial cyanide-resistant oxidase: Structural conservation amid regulatory diversity. Biochim. Biophys. Acta **2000**, *1549*, 432–439.
6. Affourtit, C.; Albury, M.S.; Crichton, P.G.; Moore, A.L. Exploring the molecular nature of alternative oxidase regulation and catalysis. FEBS Lett. **2002**, *510*, 121–126.
7. Padmasree, K.; Padmavathi, L.; Raghavendra, A.S. Importance of mitochondrial oxidative metabolism in the optimization of photosynthesis and protection against photoinhibition. Crit. Rev. Biochem. Mol. Biol. **2002**, *37*, 71–119.
8. Kruft, V.; Eubel, H.; Werhahn, W.; Braun, H.-P. Proteomic approach to identify novel mitochondrial proteins in *Arabidopsis*. Plant Physiol. **2001**, *127*, 1694–1710.
9. Koyama, H.; Kawamura, A.; Kihara, T.; Takita, E.; Shibata, D. Overexpression of mitochondrial citrate synthase in *Arabidopsis thaliana* improved growth on a phosphorus-limited soil. Plant Cell Physiol. **2000**, *41*, 1030–1037.
10. Sabar, M.; De Paepe, R.; de Kouchkovsky, Y. Complex I impairment, respiratory compensation, and photosynthetic decrease in nuclear and mitochondrial male sterile mutants of *Nicotiana sylvestris*. Plant Physiol. **2000**, *124*, 1239–1250.

Mitosis

Dénes Dudits
Attila Fehér
Hungarian Academy of Sciences, Szeged, Hungary

INTRODUCTION

Duplication of somatic plant cells by mitotic division is a fundamental cellular function that provides a continuous supply of cells to create the plant body during the life cycle. Mitosis ensures the faithful transmission of genetic information by packaging replicated DNA strands into high-order chromatin structures leading to the condensation of mitotic chromosomes that are distributed into the daughter cells. Chromosome organization is an active condensation/decondensation process that depends on histone H3 phosphorylation and the functions of topoisomerases and condensin complexes. Parallel with chromosome condensation, organization of microtubule (MT) arrays, such as the plant-specific preprophase bands (PPB), the phragmoplast, and the mitotic spindle provide the apparatus for chromosome separation and new cell plate formation in cytokinesis. The MT function is dependent on the dynamic structure of tubulins and the large number of MT-associated proteins (MAPs). Cyclin-dependent kinases (CDKs) and various members of plant mitogen-activated protein kinases (MAPKs) act as mitotic regulators in cooperation with phophatases (PP2A). The transition from metaphase to anaphase is controlled by the degradation of cohesin by separase. This cysteine protease can be activated by degradation of an inhibitor protein securin, through the anaphase-promoting complex (APC) as a specific ubiquitin ligase complex. During cytokinesis, the phragmoplast serves as a scaffold for the maturation of the vesicle fusion-generated network binding of the new cell wall.

VISUALIZATION OF STRUCTURAL CHANGES IN THE NUCLEUS AND MICROTUBULES OF MITOTIC PLANT CELLS

The series of micrographs in Fig. 1 highlight the major mitotic events and reveal the coordination between chromosome dynamics and MT organization. The DNA was stained by DAPI and the MTs were visualized by fluorescent labeled antitubulin antibodies.

The interphase nucleus is accompanied by cortical and cytoplasmic MTs. In preprophase (late G_2), chromatin condensation is significantly advanced, as shown by dot formation or organization of chromatid strands. Concurrent with the organization of chromatin, MTs assemble to form the preprophase band. This cytoskeletal structure is a unique characteristic of dividing plant cells. The PPB appears transiently and its position determines the organization site of the phragmoplast and the new cell plate. Condensed chromosomes can be recognized during prophase inside the nuclear envelope surrounded by assembled MTs. The nuclear envelope breaks down during prometaphase, and microtubules attach to the kinetochore at the centromeric region of chromosomes. Simultaneously with the start of mitotic spindle construction, the chromosomes move to the spindle equator. In metaphase cells the chromosomes align at the spindle equator and the barrel-shaped mitotic spindles are formed from the spindle and kinetochore. The MTs follow a bipolar organization with several microtubule-nucleation sites. At anaphase, the replicated sister chromatids separate synchronously and chromosomes move to the poles that subsequently separate. The chromosome movement is dependent on MT depolarization. Between the two spindle poles, the phragmoplast is formed from microtubules, actin, and myosin during late anaphase. Phragmoplast formation in telophase occurs when chromosomes decondense in daughter nuclei and the newly formed nuclear envelope develops. At this time, the mitotic spindle breaks down and the phragmoplast serves as a scaffold for the collection of Golgi-derived vesicles containing materials required for the construction of the cell plate. Cytokinesis in plant cells is completed by the formation of a new wall between daughter cells.

GENE EXPRESSION PROFILES DURING G2-M PROGRESSION

The identification of transcriptionally active genes in mitotic cells provides essential information about the functionally important molecules and their roles in mitosis.

Encyclopedia of Plant and Crop Science
DOI: 10.1081/E-EPCS 120005585

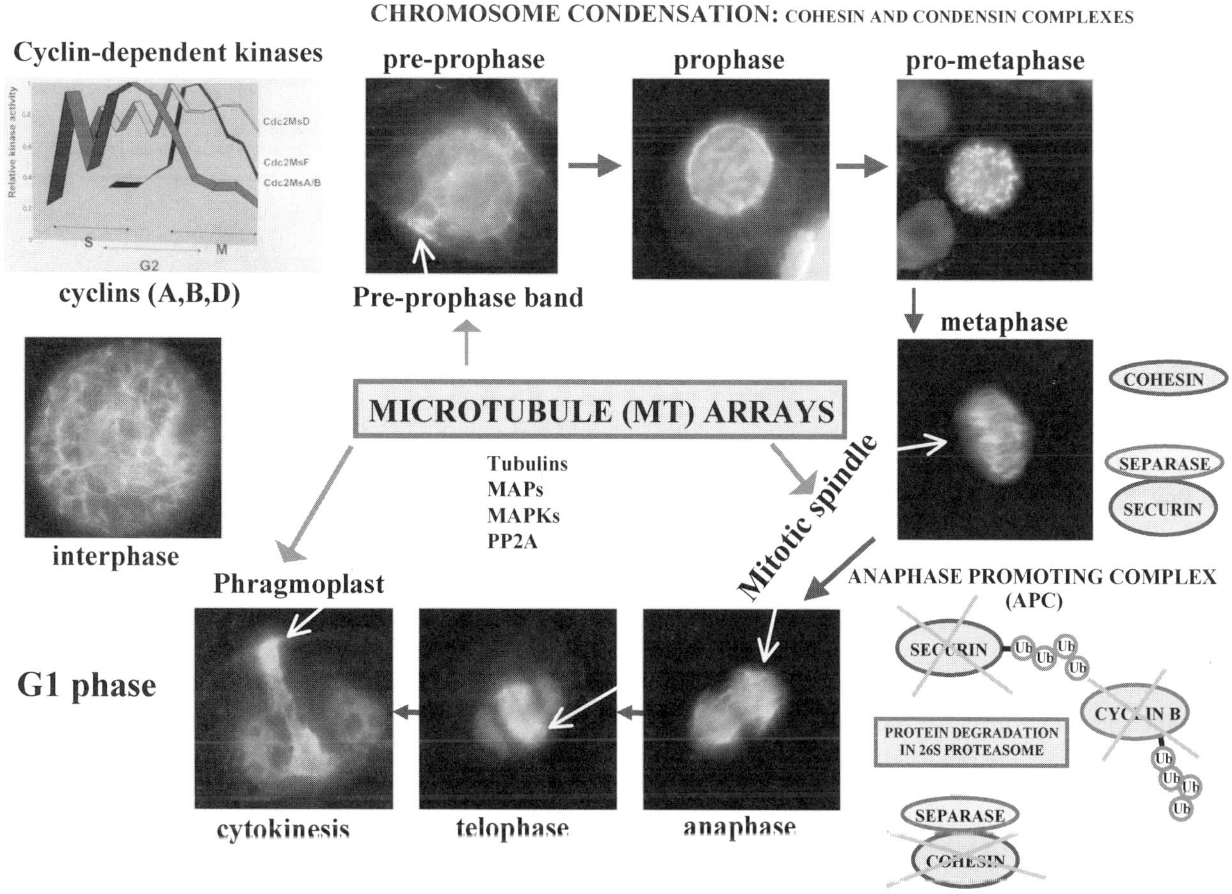

Fig. 1 Components of mitotic events in plant cells. In alfalfa cells the DNA is stained with DAPI (orange) and the microtubules are visualized by fluorescent-labeled antitubulin antibodies (green). MAP: microtubule-associated protein; MAPK: mitogen-activated protein kinase; PP2A: phosphatase. (Photos by Ferhan Ayaydin, Ph.D.) (*View this art in color at www.dekker.com.*)

The recent genomewide transcriptome analysis of synchronized tobacco and *Arabidopsis* cells highlighted a set of genes that are preferentially expressed in G2 and mitotic phases of the cell cycle. In tobacco cells, the cDNA-AFLP-based transcript profiling revealed more than 1300 transcript tags with significant modulation during the cell cycle.[1] The G2 cells exhibited 16% of these tags enriched in transcriptional factors. Mitotic cells provided the largest group of cell cycle tags (53%) representing genes with functions in cell cycle control (4.17%), RNA processing (5.37%), protein synthesis (8.65%), protein phosphorylation (5.67%), proteolysis (5.07%), cytoskeleton (11.34%), cell wall (7.46%), and transport secretion (6.26%). A large portion (33.43%) of tags corresponded to proteins with unknown function. By the use of an Affymetrix Gene Array in synchronized *Arabidopsis* cells, nearly 500 gene signals (12% of total expressed genes) were identified that have a significant cell cycle phase-dependent fluctuation in expression.[2] A high portion (30%) of these genes showed a peak of expression in the G2 phase and mitosis. These data convincingly show that transcriptional control has a critical role in regulation of mitotic functions. Furthermore, the list of genes expressed during mitosis and the prediction of the encoded proteins can provide a comprehensive insight into mitosis at the molecular level. In addition to mitotic B-type cyclins, plant-specific CDKs (in *Arabidopsis*: CDKB1 and CDKB2; in alfalfa: cdc2MsD, cdc2MsF) with unique cyclin-binding motifs (PPTALRE, PPTTLRE) exhibit clear transcript accumulation in G2/M cells. Plants differ from yeasts and animals in this feature. Similarly, the β-tubulin genes from plants are highly induced in G2 and early M-phase, whereas the tubulin-gene expression fluctuates moderately during the yeast cell cycle. The class of kinesin genes showed mitosis-linked expression profiles. The *Arabidopsis* homologue of the CDC20 gene from budding yeast was actively transcribed in mitotic cells. The encoded protein is involved in the activation of the anaphase-promoting complex. The mitotic function of various kinases such as NPK1, MAPKs with a role in phragmoplast and cell plate growth, or the homologue of Aurora kinase involved in chromosome segregation and cytokinesis, can be inferred from the gene expression profile. The regulatory role of protein phosphorylation is also emphasized by the activity of genes encoding various phosphatases (e.g., protein phosphatase 2C). Activity of the ubiquitin-proteasome complexes during mitosis can be seen from expression of genes of the M-phase-specific, protein-degradation system. The hormonal control of mitosis can be related to the high level of expression of histidine kinase gene encoding the cytokinin receptor CRE1. The mitotic peak in expression of auxin-related genes (IAA17; At2g46690) is in accordance with the auxin dependence of this cell cycle phase.

As extended analyses of cell cycle, phase-dependent gene that expression patterns demonstrate, significant number of plant genes are under a tight transcriptional control. The Mitosis-Specific Activator (MSA) *cis* element in promoters of various B-type cyclins and the tobacco kinesin-like protein genes (NACK 1,2) has been identified as responsible motif for the G2/M phase-specific gene expression. The Myb-related transcription factors (NtmybA1 and A2) with three imperfect repeats in the DNA-binding domain can activate while the Ntmyb factor represses transcript accumulation from genes with MSA-promoter element.

MOLECULAR MECHANISMS UNDERLYING SISTER CHROMATID COHESION AND CHROMOSOME CONDENSATION

The two initial phases in chromosome organization include cohesion and condensation—both events based on distinct, yet structurally similar proteins forming the cohesin and condensin complexes. These components of mitotic chromosome dynamics and functions show high evolutionary conservation from yeast to vertebrates, and recent studies on the titan (ttn) mutants of *Arabidopsis* demonstrate the functional significance of plant cohesin and condesin as well.

The duplicated sister chromatids produced by replication of chromosomal DNA are tightly paired, at both the centromere and along the arms throughout the G2 phase and during condensation of chromosomes. The cohesin complex, which physically holds sister chromatids together, involves the SMC1 and SMC3 proteins (structural maintenance of chromosomes) in yeast, Aspergilus, Xenopus, and humans. In *Arabidopsis*, a single orthologue of these genes has been identified and the SMC1 (ttn8) and SMC3 (ttn7) knockouts show a strong titan phenotype with defects in embryo and endosperm development.

In mitotic plant chromosomes, condensation is a highly ordered and active process that starts during early prophase from the centromeric region. The complete remodelling of the chromatin structure results in a high degree of DNA packing into tightly folded chromatin.[3] The condensation process depends on histone modifications (phosphorylation, de-ubiquitination) and the activity of topoisomerases and condensin complexes. In mitotically dividing cells, the process of chromosome condensation has been shown to be coupled with phosphorylation of histone H3. Mitotic CDK complexes and serine/threonine phosphatases can alter the degree of chromosome condensation, as demonstrated by experiments based on the microinjection of kinase complexes or treatment with phosphatase inhibitors (endothall, sodium vanadate, cantharidin).[4] The phenomenon of premature chromosome condensation (PCC) can be induced by fusion of mitotic and interphase plant protoplasts or by exposing cells to phosphatase inhibitors.

The functional role of topoisomerase II during the condensation process is to induce super coils into the DNA helix that allow an increase in the packing density of DNA. Treatment of plant cells with inhibitors of this enzyme aborted the chromosome condensation. The presence of an other topoisomerase (type I) is required for the function of the multisubunit condensin complex that organizes the topology of DNA loops in the chromosomes. The core condensin complex consists of two structural maintenance of chromosome (SMC) subunits—SMC2 and SMC4—that form the 13S condensin complex, and includes three other proteins (Cnd1–Cnd3). The SMC core complex can bind to naked DNA and exhibits ATPase activity that is increased in association with regulatory subcomplex formed by Cnd proteins. In *Arabidopsis*, the TTN3 gene corresponds to a SMC2 condensin homologue expressed in most of the plant organs. Apart from CDKs, other enzymes such as the Aurora B kinase can be required for condensin activity during metaphase. A plant orthologue of this kinase gene has been found in transcript-profiling experiments with cultured tobacco cells.

As shown by electron micrographs in Fig. 2., condensed mitotic plant chromosomes are organized into chromosome arms separated by a centromeric region. Sister chromatids of metaphase chromosomes are joined at the centrome where the centromeric DNA is closely associated with a protein-rich structure called the kinetochore, the attachment site for MTs. At the end of the linear chromosome, telomeres protect the chromosome from shortening. The length of repeated telomeric sequences (TTTAGGG in *Arabidopsis*) may change during development. Telomerases can copy several telomere repeats with the help of complementary RNA molecules.

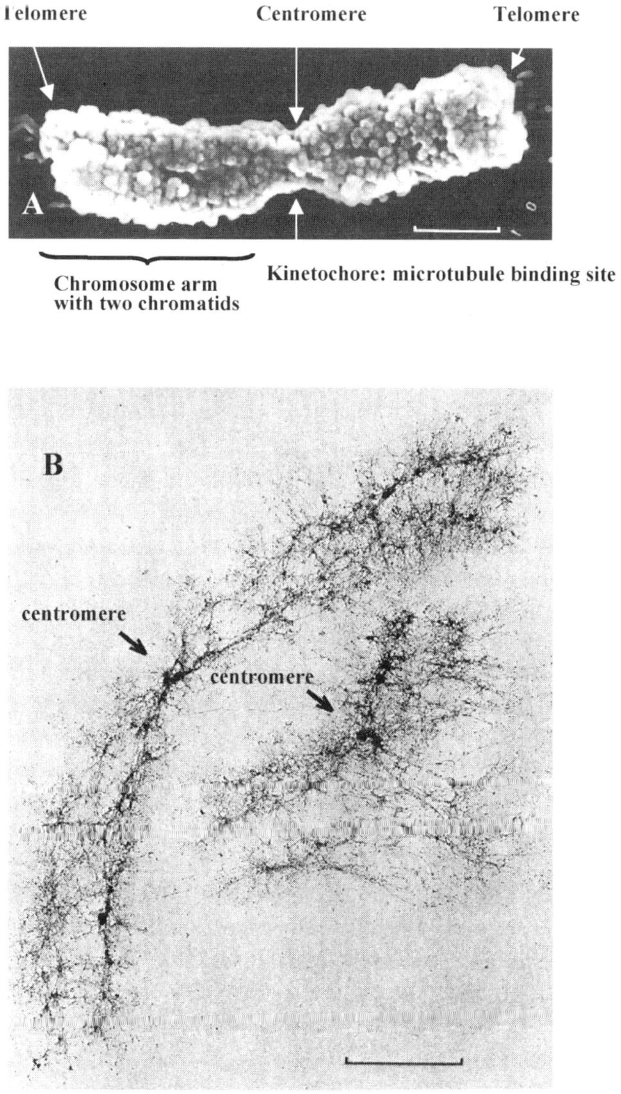

Fig. 2 A) Scanning electron micrograph of isolated wheat chromsome, Bar = 1 μ; B) Electron micrograph of protein-depleted wheat chromosome, Bar = 5 μ. (Figure from Gy Hadlaczky.)

MODES OF MICROTUBULE FUNCTIONS

In coordination with the condensation of chromosomes, the highly dynamic structure of the MTs is organized in dividing plant cells to serve as one of the key functional components for segregation of chromosomes. The heterodimers of α- and β-tubulin subunits constantly assemble, disassemble, and rearrange (polymerize/depolymerize) during the organization of MT arrays such as the preprophase bands, the mitotic spindle and the phragmoplast (Fig. 1). The MT organization and dynamics require the functional contribution of a large number of MT-associated proteins.[5] The ATP-dependent motor proteins move unidirectionally along the surface of MTs and function as transporters or MT organizers. The kinesin-like motor proteins such as KCBP/2-WICHEL can induce the transient and local stabilization of MTs to be rearranged or focused. Mutation in the TANGLED1 (TAN1) gene causes misorientation of MT arrays. This protein is associated with mitotic MT arrays and repeats in cytokinesis.

The central role of different MAPs in phragmoplast organization and cytokinesis is shown by the characteristics of cell division mutants or immunolocalization-studies.[6] The *Arabidopsis* Microtubule Organization (MOR1) protein and its homologue, the tobacco MT-bundling protein (TMBP200), participate in the organization of MT arrays (e.g., by cross-bridging between phragmoplast MTs). Inhibition of a motor protein called TKRP15 interferes with translocation of phragmoplast MTs. One of the phragmoplast-localized, kinesin-related proteins of *Arabidopsis* (AtPAKRP2) has been suggested to act as transporter of Golgi-derived vesicles to the newly forming division plate.

Several essential mitotic events including MT dynamics are under a defined phosphorylation control, based on a series of kinases and phosphatases with regulatory and/or coordinatory function.[7] The pivotal role of CDKs in mitotic plant cells can be concluded from G2/M phase-specific expression of the corresponding genes, identification of mitotic kinase complexes, protein interaction, and immunolocalization studies. In synchronized G2/M alfalfa cells, at least three types of Cdc2-related kinase complexes can be detected with characteristic timing of activity peaks (Fig. 1). The Cdc2MsA/B kinases are highly active in mid-G2 phase, whereas the Cdc2MsD kinase activity peaks before the start of mitosis. The CdcMsF kinase represents a typical mitotic kinase.

The alfalfa Cdc2MsD and F kinases have *Arabidopsis* orthologues (CDKB1 and CDKB2) and exhibit plant-specific, cyclin-binding motives (PPTALRE or PPTTLRE) instead of the PSTAIRE sequence element. The alfalfa Cdc2MsF mitotic kinase was shown to be colocalized with MTs in all mitotic stages.[8] In general, the B-type cyclins are considered regulators of mitotic CDKs; however, a role for A-type and D-type cyclins can be inferred from protein-interaction studies. Gene predictions based on sequence data of the Arabidopsis genome suggest 10 A-type cyclins, 9 B-type cyclins, and 10 D-type cyclins. As far as the number of cyclin gene variants and kinase complexes is concerned, the mitotic cell cycle control machinery in plants shows higher complexity than in animals; however, the specificity and redundancy in cyclin gene functions have not been elucidated so far. The formation of active CDK complexes is dependent on the phosphorylation of selected amino-acid residues of the interacting partners. The CDK-activating kinases (CAKs) and inhibitory phosphorylation by the Wee1 (Mik1-type) kinases can alter the CDK functions that are also controlled by the set of inhibitor proteins. The Kip-related proteins (in *Arabidopsis*, KRP1-KRP7) can differentially regulate the various CDK complexes. The list of kinases involved in the control of mitosis is not restricted to the CDKs. Various members of plant mitogen-activated protein kinase (MAPK) cascades can contribute to the establishment and function of the phragmoplast during cytokinesis. Nucleus-and phragmoplast-localized Protein Kinase 1 (NPK1) activity is required for expansion of both the cell plate and the phragmoplast. Other MAPKs, such as $p43^{Ntf6}$ from tobacco and MMK3 from alfalfa, are activated during cytokinesis. The *Arabidopsis* HINKEL (HIK) protein and its orthologue in tobacco, the Nicotinana protein kinase 1 (NPK1)-activating kinase-like protein 1 (NACK1), can initiate a signaling cascade involved in the cytokinesis.

An increasing number of experimental data support the significance of phosphatases as counteracting enzymes of kinases in regulating mitotic events in plant cells. The formation of active CDK complexes requires the removal of inhibitory phosphate from the Tyr-14 and Thr-15 residues of the CDKs. Cdc25-related phosphatases can activate CDKs in yeast, Drosophila, and in mammalian cells. Attempts to identify plant orthologues of these phosphatases are expected to be successful. However, expression of the yeast Cdc25 enzyme in tobacco transformants caused cytokinin-independent premature mitosis, reduced cell size, and altered lateral root development. In general, cytokinins are needed to form an active CDK complex. The role of serine/threonine phosphatases (e.g., PP2A) in MT organization can be concluded from inhibitor experiments. In endothall-treated alfalfa cells, the formation of PPBs and the phragmoplast is disturbed, and the premature polarization of the mitotic spindle is coupled with the early activation of mitotic Cdc2MsF kinase.[4] Mutation in the *Arabidopsis* TONNEAU2 (TON2)/FASS gene encoding a putative B-regulatory subunit of PP2A phosphatases causes similar abnormalities in mitosis.

METAPHASE-TO-ANAPHASE TRANSITION AND CYTOKINESIS

The release of cohensin-mediated chromosome cohesion at the metaphase plate is a prerequisite for the chromosome segregation in anaphase.[9] The dissolution of cohesion between sister chromatids can be induced through degradation of cohensin by separases belonging to the CD clan of cysteine proteases. The precise timing of separase activation is crucial, because it can occur only after the formation of bipolar attachments between the sister chromatids and the spindle MTs. This tightly controlled regulatory process includes an inhibitory protein called securin that blocks separase activity until the Anaphase-Promoting Complex (APC), a specific ubiquitin-ligase complex, links a polyubiquitin chain to securin and targets it for degradation by the 26S proteasome. The composition of APC has not been identified in plants, but we expect functional conservation with the 10 subunits of mammals and 13 subunits of yeasts. The APC activity also regulates cyclin degradation in mitosis. In plants, both the B-type and A-type cyclins can carry a short peptide motif of nine amino acids called the destruction (D)-box, which serves as a recognition site for APC. Reduction in the availability of cyclins could contribute to the loss of CDK activity at the end of mitosis. The APC function is under complex regulatory control, including an association with activator WD-repeat proteins such as Cdc20p (Fizzy) or Cdh1 (Fizzy-related) proteins that recruit substrates to APC. Components of the mitotic spindle checkpoint such as MAD (mitotic arrest deficient) and BUB (budding uninhibited by benomyl) proteins can inhibit APC activity. The APC is also regulated by cell cycle-specific phosphorylation; both kinases (Polo-like kinase, Plk1) and phosphatases (PP1) can activate APC.

After completion of mitosis, the phragmoplast serves as a scaffold to direct the movement of Golgi-derived vesicles containing cell wall components to form the new cell wall between the daughter nuclei.[10] The phragmoplast complex is composed of two bundles of antiparallel MTs and actin filaments that are colocalized with number of kinesins. The process of phragmoplast expansion by reorganization from a solid cylinder into a ring-shaped structure is dependent on MT depolymerization in the center and reassembly-polymerization of MTs on the outside of phragmoplast. This lateral expansion of the phragmoplast can be arrested by brefelding treatment, which prevents the establishment of a ring-shape and MT disassembly. Caffeine can disrupt cell-plate formation by interfering with the maturation of the fusion-generated vesicle membrane network into a wide tubular network in the phragmoplast. The cell plate grows outward by incorporating vesicles into the edge until it reaches the parental cell walls. The primary cell walls consist of cellulose, cellulose-binding hemicellulose, and pectins. Cellulose is generated at the plasma membrane in the form of paracrystalline microfibrils, whereas hemicellulose and pectins are synthesized within Golgi cisternae. After the formation of a new primary cell wall, the division procedure is complete and the separated daughter cells can either enter into a new division cycle or undergo differentiation.

CONCLUSION

Our present knowledge about the major regulatory mechanisms of mitotic events is increased by the use of improved cytological techniques, based mainly on immunolocalization of key proteins. Detailed characterization of cell division mutants (primarily in *Arabidopsis*), and the cloning of the responsible genes provide essential functional information about control of the mitotic cell cycle. Overexpression of genes encoding proteins with roles in mitosis can be an alternative approach in functional analysis. In the future, detailed characterization of protein complexes and outlining cascades of phosphorylation/dephosphorylation during the perception of developmental or environmental signals will be needed. The DNA damage checkpoint will attract more attention. Discovery of the major controlling elements in endoreduplication is expected to highlight plant-specific elements in mitosis. Activity of meristems as major foci of mitotic events is under the precise control of a developmental program that can be altered by environmental influences.

ACKNOWLEDGMENT

The authors thank Keczánné Zsuzsa Czakó for excellent secretarial work during preparation of the manuscript.

ARTICLES OF FURTHER INTEREST

B Chromosomes, p. 71
Chromosome Manipulation and Crop Improvement, p. 266
Chromosome Rearrangements, p. 270
Chromosome Structure and Evolution, p. 273
Cytogenetics of Apomixis, p. 383
The Interphase Nucleus, p. 568
Meiosis, p. 711
Polyploidy, p. 1038
Sex Chromosomes in Plants, p. 1148
Somaclonal Variation: Origins and Causes, p. 1158

REFERENCES

1. Breyne, P.; Dreesen, R.; Vandepoele, K.; De Veylder, R.; Van Breusegem, F.; Callewaert, L.; Rombauts, S.; Raes, J.; Cannoot, B.; Engler, G.; Inzé, D.; Zabeau, M. Transcriptome analysis during cell division in plants. PNAS **2002**, *99* (23), 14825–14830.
2. Menges, M.; Hennig, L.; Gruissem, W.; Murray, J.A.H. Cell cycle regulated gene expression in *Arabidopsis*. J. Biochem. Chem. **2002**, *277* (44), 41987–42002.
3. Doerner, P. Cell Division Regulation. In *Biochemistry & Molecular Biology of Plants*; Buchanan, B.B., Gruissem, W., Jones, R.L., Eds.; John Wiley & Sons Inc.: Somerset, NJ, 2000; 528–565.
4. Ayaydin, F.; Vissi, E.; Mészáros, T.; Miskolczi, P.; Kovács, I.; Fehér, A.; Dombrádi, V.; Erdödi, F.; Gergely, P.; Dudits, D. Inhibition of serine/threonine-specific protein phosphatases causes premature activation of cdc2MsF kinase at G2/M transition and early mitotic microtubule organization in alfalfa. Plant J. **2000**, *23*, 85–96.
5. Azimzadeh, J.; Traas, J.; Pastuglia, M. Molecular aspects of microtubule dynamics in plants. Curr. Opin. Plant Biol. **2001**, *4*, 513–519.
6. Mayer, P.; Jürgews, G. Microtubule cytoskeleton: A track record. Curr. Opin. Plant Biol. **2002**, *5*, 494–501.
7. Criqui, M.C.; Genschik, P. Mitosis in plants: How far we have come at the molecular level? Curr. Opin. Plant Biol. **2002**, *5*, 487–493.
8. Mészáros, T.; Miskolczi, P.; Ayaydin, F.; Pettkó-Szandtner, A.; Peres, A.; Magyar, Z.; Horváth, G.V.; Bakó, L.; Fehér, A.; Dudits, D. Multiple cyclin-dependent kinase complexes and phosphatases control G2/M progression in alfalfa cells. Plant Mol. Biol. **2000**, *43*, 595–605.
9. Baskin, T. The Cytoskeleton. In *Biochemistry & Molecular Biology of Plants*; Buchanan, B.B., Gruissem, W., Jones, R.L., Eds.; John Wiley & Sons Inc.: Somerset, NJ, 2000; 202–256.
10. Nishihama, R.; Machida, Y. Expansion of the phragmoplast during plant cytokinesis: A MAPK pathway may MAP it out. Curr. Opin. Plant Biol. **2001**, *4*, 507–512.

Molecular Analysis of Chromosome Landmarks

James A. Birchler
Juan M. Vega
Akio Kato
University of Missouri, Columbia, Missouri, U.S.A.

INTRODUCTION

The major landmarks of a plant chromosome include the primary constriction, which represents the centromere; the ends of the chromosomes or telomeres; and blocks of heterochromatin, which, depending on the circumstances, can be rather discrete or diffuse. In many plant species a secondary constriction that represents the nucleolus organizer regions (NORs) is observed in one or more pairs of chromosomes. This secondary constriction encodes the major species of ribosomal RNA. The different landmarks consist of various types of repetitive sequences. This fact allows their molecular analysis to be facilitated by visualizing them using hybridization of probes to the mitotic metaphase or meiotic pachytene chromosomes.

CENTROMERES

The centromere is the structure present at the primary constriction of a chromosome.[1] At this site, the proteins are organized in a complex called the kinetochore, which serves to move the chromosome to the poles at mitosis and meiosis. Moreover, at meiosis I, it serves to hold the two sister chromatids together during anaphase.

Among plant species, early molecular studies were of the B-specific repeat of the B chromosome centromere of maize[2] and of the satellite repeat in Arabidopsis.[3] The B-specific unit is approximately 1.4 kilobases in length and is only found in the centromere of the supernumerary B chromosome.[2] Its location in the centromeric region was confirmed by analysis of B chromosome derivatives that have suffered from misdivision of the centromere. This process results when a centromere is present as a univalent in meiosis. The spindle attaches from both poles and ruptures the centromere transversely instead of separating it lengthwise. A number of misdivision derivatives have been studied and all change the restriction pattern identified by the B-specific repeat probe confirming the location of this unit within the B centromere.[4]

The B centromeric repeat arrays extend over approximately 9 megabases in length. The misdivision derivatives reduce this size to as small as a few hundred kilobase pairs.[5] Although other sequences are also included, the B-repeat cluster is likely to be throughout the centromere, based on the fact that misdivision can fracture this sequence and both fractured centromeres of the broken chromosome can still function.[4]

The centromeric satellite sequence of Arabidopsis is 180 base pairs in length and is repeated many times at the site of the primary constriction of each of the five chromosomes of Arabidopsis. The length of the repeat cluster ranges from 1–2 megabases.[3] The satellite cluster falls within the genetically defined centromere as determined by tetrad analysis.[6]

For the A chromosomes of maize, there are other repeats present at the primary constriction of the respective chromosomes.[7,8] One of the more prominent is the CentC unit[8] (see Fig. 1). It is present at the primary constriction of all of the ten chromosome pairs to varying degrees. Another element, CentA, is present and variable as well. A third related element is CRM.[9] The latter two repeats are retrotransposons that are specific in their location to the centromeric regions.

The centromeres of rice are also composed of a collection of repeats that are quite similar to those of maize.[10] The satellite sequence referred to as CentO is a small repetitive element that contributes to slightly more than 1% of the rice genome. The representation of CentO in different centromeres of the twelve chromosomes is highly variable. Interspersed among the CentO repeats are copies of the CRR retroelement. Thus, a common feature of cereal centromeres is the presence of both conserved retrotransposons and species-specific tandem repetitive sequences. Functional centromeres are associated with a particular version of histone 3.[11] A portion of the amino acid sequence of this histone is highly conserved relative to the normal version present in nucleosomes outside the centromere. However, one end of the molecule has

Encyclopedia of Plant and Crop Science
DOI: 10.1081/E-EPCS 120005588

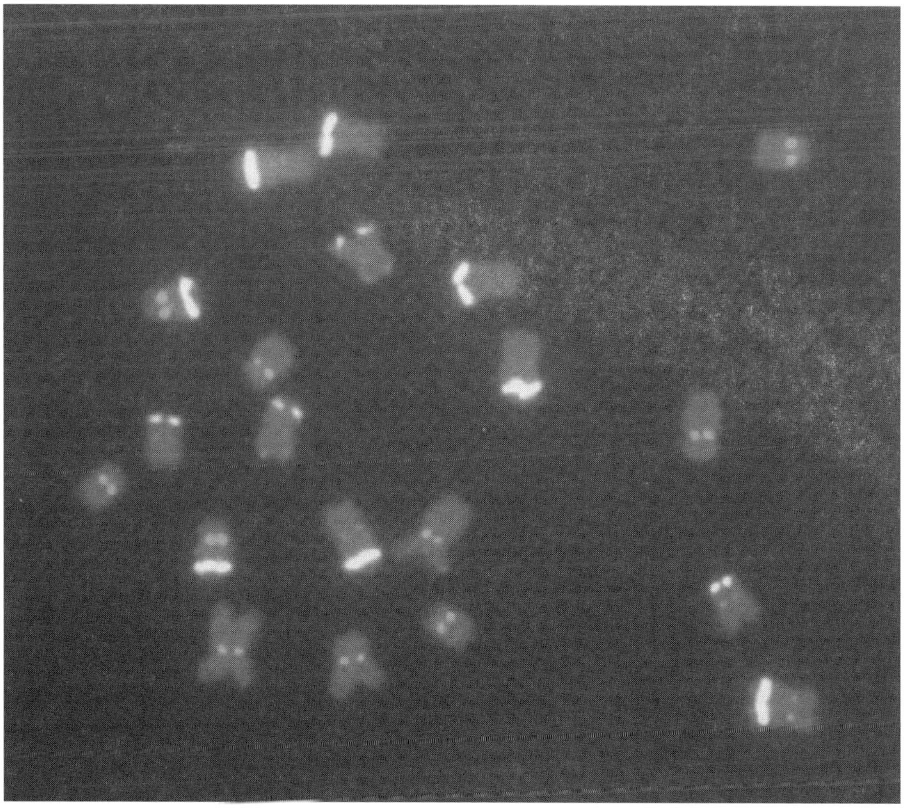

Fig. 1 Fluorescent in situ hybridization of maize root tip metaphase chromosomes illustrating the locations of the centromeres (red) and the knob heterochromatic sites (yellow green). A probe for CentC was used to mark the centromeres. (*View this art in color at www.dekker.com.*)

evolved very rapidly. To date, antibodies have been produced against the centromere histone 3 from Arabidopsis and from maize.[9,11]

NUCLEOLAR ORGANIZER REGIONS (NORS)

The ribosomal RNA genes are clustered by the hundreds or thousands of copies, often spanning millions of basepairs of chromosomal DNA at loci known as nucleolus organizer regions (NORs).[12] The number of NORs in a haploid chromosome set is characteristic of a species.[13,14] As a rule, at least one NOR is present per genome. During metaphase of mitosis, the NOR is usually observed as a secondary constriction on the chromosome. Upon exit from mitosis, transcription of the rRNA genes initiates the formation of a nucleolus. In interspecific hybrids, the NORs of one progenitor species are often inactivated, regardless of whether that species served as the maternal or paternal parent.[15] In this phenomenon, known as nucleolar dominance, silenced rRNA genes can be efficiently derepressed by chemical inhibitors of DNA methylation or histone deacetylation, implicating chromatin modifications in the maintenance of nucleolar dominance.

TELOMERES

The ends of chromosomes are capped by a specific sequence, called the telomere, which appears to stabilize the terminus. Typically broken chromosome ends will fuse with each other to generate chromosomal changes in structure. Alternatively, sister chromatids of a broken chromosome fuse at the ends after replication to initiate a bridge–breakage–fusion cycle. The addition of a telomere at the broken site will prevent such fusions.[16,17] Moreover, the presence of a telomere allows complete replication of the chromosome ends. Telomeres also cluster in meiosis before homologue synapsis.[18]

The first telomere sequence isolated from plants was from Arabidopsis.[19] The sequence is composed of the repeat TTTAGGG in many tandem copies. Most plant species have the same structure.[20] Adjacent to the telomere repeats are telomere-associated sequences that consist of degenerate versions of the repeats. Some few plant species, including members of Alliaceae and Aloe spp. (Asphodelaceae), were found to lack the Arabidopsis-type telomeric repeats.[21]

Mutations in telomerase—the enzyme that synthesizes telomeric DNA—cause telomere shortening.[22] This ultimately limits cell proliferation capacity because uncapped chromosome ends activate DNA damage checkpoints. Arabidopsis mutants that lack telomerase have unstable genomes, but manage to survive up to 10 generations with increasingly shortened telomeres and cytogenetic abnormalities.[22]

HETEROCHROMATIN

Pericentric Heterochromatin

Most plant chromosomes have deeply staining heterochromatin surrounding the centromeres that displays gradual transition to more lightly staining regions called euchromatin. There are species, however—tomato being a prime example—in which the demarcation between centric heterochromatin and the euchromatin is very distinct. Typically there are few genes located in the pericentric regions and recombination values are low. Consequently the majority of genes are placed distally on the chromosome. The centric heterochromatin is likely composed of retrotransposon copies.[23] In maize, some retroelements such as PREM[24] are preferentially distributed around the centromeres.

Knob Heterochromatin

Some plant chromosomes contain blocks of heterochromatin within the chromosome arms that are referred to as knobs.[1] The most thoroughly studied knobs from the molecular point of view are those of maize (see Fig. 1). These chromosomal landmarks were first used as a means to distinguish the members of the maize karyotype in meiosis.[1] Subsequently, they were found to act as neocentromeres and cause preferential segregation of linked markers through the female side.[25]

This neocentromeric activity only occurs during meiosis. The knobs proceed to the poles ahead of the true centromeres. For preferential segregation to occur, heterozygosity is necessary for the presence or absence of a knob or at least for the size of knobs. A recombination event must occur between the true centromere and the knob to generate a heteromorphic dyad that pulls the knobbed chromatid to the outer poles during female meiosis. Because the most basal megaspore gives rise to the female gametophyte, the knob containing chromatid is recovered more often than random in the egg. The knobs act as neocentromeres only in the presence of an abnormal version of chromosome 10, which itself has a novel large knob near the end of the long arm.

Peacock et al.[26] identified a repeat of 180 base pairs in length that is limited to knob heterochromatin. These units are arranged in tandem arrays interrupted by retroelements.[27] Subsequently, another version was found (referred to as TR1) that is the major component of a few knobs.[27] The neocentric activity in the knobs is not associated with the presence of the centromeric version of histone 3. Moreover, their attachment to the spindle is at an angle, in contrast to the perpendicular nature of spindle attachment to a normal centromere.[28]

Constitutive Heterochromatin

Constitutive heterochromatin refers to chromosome regions that remain condensed and transcriptionally inactive during interphase. They are enriched in tandemly repeated sequences of which the heterochromatic state is a heritable chromosomal trait. These regions include the already cited knobs and other heterochromatic blocks that reside in interstitial and terminal positions in many plant chromosomes. They can be visualized by conventional staining methods and a number of chromosomal banding techniques.

In plants, the first systematic analysis of the relationship between different heterochromatic bands and repetitive DNA sequences was done in rye.[29] Several repeats arranged in tandem arrays, with repeating units of a few hundred base pairs, are located within the blocks of telomeric heterochromatin that can be observed on all seven pairs of rye chromosomes by C-banding. Most heterochromatic blocks contain more than one class of repeated elements. The size of the blocks varies among different rye accessions, indicating that the copy number of the DNA sequences may change extensively.

The heterochromatic state is highly stable. However, an interstitial C-band present in rye 5R chromosome is unusually decondensed when this chromosome is added to wheat. In the wheat background, this region appears as a constriction, which coorients with the true centromere and shows neocentric activity at meiosis.[30] This phenomenon resembles the situation observed with maize knobs. However, whereas maize neocentromeres interact with spindle microtubules in a lateral manner, the 5R neocentromere shows end-on contact with microtubules, in a way similar to the binding of true centromeres. Tandem arrays of a 118 base pair monomeric unit are present at the 5R neocentromere.

CONCLUSION

The major landmarks of plant chromosomes are typically composed of repeated units of one type or another. The telomeres are synthesized by an enzyme (telomerase) that adds the appropriate sequence to the ends of the chromosomes. The basis for the maintenance of the centromeric repeats, especially considering their uniformity on nonhomologous chromosomes, is unknown. Also unknown are the exact sequence requirements to specify a centromere. Because the putative *cis*-acting centromeric sequences are not conserved, even in closely related species, it is possible that centromeric identity is not determined by DNA sequence but rather by chromatin assembly during replication. The accumulation of some repeated sequences under these sites might be unrelated to the specification of the centromere location. Future work is needed to reveal the nature of the centromere determination.

ACKNOWLEDGMENTS

The authors thank The National Science Foundation (USA) Plant Genome Initiative for research support. Juan Vega was a recipient of a European Molecular Biology Organization (EMBO) Postdoctoral Fellowship.

ARTICLES OF FURTHER INTEREST

Chromosome Banding, p. 263
Chromosome Rearrangements, p. 270
Chromosome Structure and Evolution, p. 273

REFERENCES

1. McClintock, B. Chromosome morphology in *Zea mays*. Science **1929**, *69*, 629–630.
2. Alfenito, M.R.; Birchler, J.A. Molecular characterization of a maize B-chromosome centric sequence. Genetics **1993**, *135*, 589–597.
3. Round, E.K.; Flowers, S.K.; Richards, E.J. *Arabidopsis thaliana* centromere regions: Genetic map positions and repetitive DNA structure. Genome Res. **1997**, *7*, 1045–1053.
4. Kaszas, E.; Birchler, J. Misdivision analysis of centromere structure in maize. EMBO J. **1996**, *15*, 5246–5255.
5. Kaszas, E.; Birchler, J. Meiotic transmission rates correlate with physical features of rearranged centromeres in maize. Genetics **1998**, *150*, 1683–1692.
6. Copenhaver, G.P.; Nickel, K.; Kuromori, T.; Benito, M.I.; Kaul, S.; Lin, X.; Bevan, M.; Murphy, G.; Harris, B.; Parnell, L.D.; McCombie, W.R.; Martienssen, R.A.; Marra, M.; Preuss, D. Genetic definition and sequence analysis of Arabidopsis centromeres. Science **1999**, *286*, 2468–2474.
7. Jiang, J.M.; Nasuda, S.; Dong, S.F.M.; Scherrer, C.; Woo, S.S.; Wing, R.; Gill, B.; Ward, D.C. A conserved repetitive DNA element located in the centromeres of cereal chromosomes. Proc. Natl. Acad. Sci. U.S.A. **1996**, *93*, 14210–14213.
8. Ananiev, E.; Phillips, R.L.; Rines, H. Chromosome-specific molecular organization of maize (*Zea mays* L.) centromeric regions. Proc. Nat. Acad. Sci. U.S.A. **1998**, *95*, 13073–13078.
9. Zhong, C.X.; Marshall, J.B.; Topp, C.; Mroczek, R.; Kato, A.; Nagaki, K.; Birchler, J.A.; Jiang, J.; Dawe, R.K. Centromeric retroelements and satellites interact with maize kinetochore protein CENH3. Plant Cell **2002**, *14*, 2825–2836.
10. Cheng, Z.; Dong, F.; Langdon, T.; Ouyang, S.; Buell, C.R.; Gu, M.; Blattner, F.R.; Jiang, J. Functional rice centromeres are marked by a satellite repeat and a centromere-specific retrotransposon. Plant Cell **2002**, *14*, 1691–1704.
11. Talbert, P.B.; Masuelli, R.; Tyagi, A.P.; Comai, L.; Henikoff, S. Centromeric localization and adaptive evolution of an Arabidopsis histone H3 variant. Plant Cell **2002**, *14*, 1053–1066.
12. Phillips, R.L.; Kleese, R.A.; Wang, S.S. The nucleolus organizer region of maize (*Zea mays* L.): Chromosomal site of DNA complementary to ribosomal RNA. Chromosoma **1971**, *36*, 79–88.
13. Kenton, A.; Parokonny, A.S.; Gleba, Y.Y.; Bennett, M.D. Characterization of the *Nicotiana tabacum* L. genome by molecular cytogenetics. Mol. Genet. Genet. **1993**, *240*, 159–169.
14. Dubcovsky, J.; Dvorak, J. Ribosomal RNA multigene loci: Nomads of the Triticeace genomes. Genetics **1995**, *140*, 1367–1377.
15. Chen, Z.J.; Pikaard, C.S. Epigenetic silencing of RNA polymerase I transcription: A role for DNA methylation and histone modification in nucleolar dominance. Genes Dev. **1997**, *11*, 2124–2136.
16. Werner, J.E.; Kota, R.S.; Gill, B.S.; Endo, T.R. Distribution of telomeric repeats and their role in the healing of broken chromosome ends in wheat. Genome **1992**, *35*, 844–848.
17. Wang, S.; Lapitan, N.L.V.; Roder, M.; Tsuchiya, T. Characterization of telomeres in *Hordeum vulgare* chromosomes by in situ hybridization. II. Healed broken chromosomes in telotrisomic 4L and acrotrisomic 4L4S lines. Genome **1992**, *35*, 975–980.
18. Bass, H.W.; Marshall, W.F.; Sedat, J.W.; Agard, D.A.; Cande, W.Z. Telomeres cluster de novo before the initiation of synapsis: A three dimensional spatial analysis

M

of telomere positions before and during meiotic prophase. J. Cell. Sci. **1997**, *137*, 5–18.

19. Richards, E.J.; Ausubel, F.M. Isolation of a higher eukaryotic telomere from *Arabidopsis thaliana*. Cell **1988**, *53*, 127–136.
20. Richards, E.J. Plant Telomeres. In *Telomeres*; Blackburn, E.H., Greider, C.W., Eds.; Cold Spring Harbor Laboratory Press: Cold Spring Harbor, 1995; 371–387.
21. Adams, S.P.; Leitch, I.J.; Bennett, M.D.; Leitch, A.R. Aloe L.—A second plant family without (TTTAGGG)n telomeres. Chromosoma **2000**, *109*, 201–205.
22. Riha, K.; McKnight, T.D.; Griffing, L.R.; Shippen, D.E. Living with genome instability: Plant responses to telomere dysfunction. Science **2001**, *291*, 1797–1800.
23. Fuchs, J.; Strehl, S.; Brandes, A.; Schweizer, D.; Schubert, I. Molecular-cytogenetic characterization of the *Vicia faba* genome-heterochromatin differentiation, replication patterns and sequence localization. Chromosome Res. **1998**, *6*, 219–230.
24. Ananiev, E.; Phillips, R.L.; Rines, H. Complex structure of knob DNA on maize chromosome 9: Retrotransposon invasion into heterochromatin. Genetics **1998**, *149*, 2025–2037.
25. Rhoades, M.M.; Vilkomerson, H. On the anaphase movement of chromosomes. Proc. Natl. Acad. Sci. U.S.A. **1942**, *28*, 433–435.
26. Peacock, W.J.; Dennis, E.S.; Rhoades, M.M.; Pryor, A.J. Highly repeated DNA sequence limited to knob heterochromatin in maize. Proc. Natl. Acad. Sci. U.S.A. **1981**, *78*, 4490–4494.
27. Ananiev, E.; Phillips, R.L.; Rines, H. A knob-associated tandem repeat in maize capable of forming fold-back DNA segments: Are chromosome knobs megatransposons? Proc. Nat. Acad. Sci. U.S.A. **1998**, *95*, 10785–10790.
28. Yu, H.G.; Hiatt, E.N.; Chen, A.; Sweeney, M.; Dawe, R.K. Neocentromere-mediated chromosome movement in maize. J. Cell Biol. **1997**, *139*, 831–840.
29. Bedbrook, J.R.; Jones, J.; O'Dell, M.; Thompson, R.D.; Flavell, R.B. A molecular description of telomeric heterochromatin in Secale species. Cell **1980**, *19*, 545–560.
30. Manzanero, S.; Vega, J.M.; Houben, A.; Puertas, M.J. Charaterization of the constriction with neocentric activity of 5RL chromosome in wheat. Chromosoma **2002**, *111*, 228–235.

Molecular Biology Applied to Weed Science

Patrick J. Tranel
Federico Trucco
University of Illinois, Urbana, Illinois, U.S.A.

M

INTRODUCTION

Weed science is a very practical discipline that has the general goal of improving weed management. To achieve this goal, however, weed scientists historically have relied on many basic scientific disciplines, such as plant physiology, biochemistry, chemistry, and genetics. In more recent years, the tools afforded by molecular biology techniques also have been brought to weed science. In fact, among the first and most widely adopted outcomes of biotechnology were weed science products, namely herbicide-resistant crops. Molecular biology tools are also being widely used in weed science to investigate fundamental questions regarding the biology, ecology, and evolution of weeds. The application of molecular biology tools to weed science is described and illustrated with several examples.

RECENT EVENTS IN THE HISTORY OF WEED SCIENCE

Although humans have contended with weeds for millennia, rapid growth of weed science as a discipline did not occur until the middle of the 20th century, after the discovery of the first synthetic herbicides. Throughout much of the latter half of the 1900s, weed science was focused on herbicide discovery and herbicide physiology. Weed scientists were tremendously successful, increasing the efficacy and range of new herbicidal chemistries and thereby simplifying and improving weed management. As a result, soil conservation practices expanded and yields improved. Herbicides also aided basic biological research. For example, much of what we now know about the shikimic acid pathway (which is disrupted by the herbicide glyphosate) and photosynthetic electron transport (disrupted by triazine and other herbicides) was discovered by weed scientists investigating herbicide phytotoxicity.

Herbicide-Resistant Weeds

A repercussion of the wide adoption of herbicides for weed management was the evolution of herbicide-resistant weed populations. This phenomenon continues to be an area of intense weed science research, and is greatly aided by molecular biology. Illustrative of this point are investigations of resistance to herbicides that inhibit the enzyme, acetolactate synthase (ALS).

The ALS enzyme is necessary for the production of certain amino acids, so inhibition of this enzyme leads to plant death. Numerous herbicides, most of which belong to either the sulfonylurea or imidazolinone chemical group, target this enzyme and have been widely used since the early 1980s. Although these herbicides have been very effective, populations of more than 70 weed species have evolved resistance to these herbicides.[1] In most cases in which it has been investigated, resistance was determined to be due to an altered target site. More specifically, mutations in the gene encoding ALS result in the production of herbicide insensitive versions of the ALS enzyme. Multiple mutations have now been identified from resistant weed populations and are being catalogued.[2] Identification of these mutations and the corresponding patterns of resistance to the various ALS-inhibiting herbicides has greatly improved our understanding of how these herbicides interact with their target site.

Herbicide-Resistant Crops

In recent years, application of molecular biology research methods to herbicide physiology and herbicide resistance in weeds has furthered our understanding in these areas. Taking this idea a step further, our understanding of herbicide phytotoxicity at the molecular level has enabled the most recent revolution in weed science: the development and commercialization of herbicide-resistant crops.

Herbicide-resistant crops are perhaps the most consequential outcome of molecular biology applied to weed science. Among these, glyphosate resistance technology stands as the best example. Glyphosate's ability to control a broad spectrum of weeds, its low toxicity to humans and other nontarget organisms, and its limited environmental persistence have made it a good candidate for efforts at engineering herbicide-resistant crops.

Glyphosate targets the enzyme 5-enolpyruvylshikimate-3-phosphate synthase (EPSPS). First attempts to

Encyclopedia of Plant and Crop Science
DOI: 10.1081/E-EPCS 120020254

engineer glyphosate resistance involved overexpression of the gene encoding EPSPS and expression of mutated forms of the gene encoding glyphosate-insensitive variants of the enzyme.[3] Although marginally successful, such attempts did not result in commercially acceptable levels of glyphosate resistance. The screening of microorganisms resulted in identification of a second type of EPSPS enzyme (Class II EPSPS) that was highly resistant to glyphosate. A Class II EPSPS gene was cloned from *Agrobacterium* spp. strain CP4 and expressed in crop plants. Success in using the CP4 EPSPS gene led to the commercialization of glyphosate-resistant soybean in the United States in 1995, followed by canola and cotton in 1997 and corn in 1998.[4] Since 1996, the worldwide adoption of glyphosate-resistant crops has been immense (Fig. 1).

Transgenic approaches also have been used to obtain other commercial forms of herbicide-resistant crops. Additionally, mutagenesis has been used with success in nontransgenic attempts to obtain herbicide resistance. In this approach, resistant variants are selected from a population of the crop of interest. To improve the chance for success, genetic variation in the population is increased by using one of several techniques to induce mutations in the plants.

Regardless of the specific technique used, the development of herbicide-resistant crops has increased options for weed management, and several such crops are now commercially available (Table 1). Although often criticized by environmental groups, such crops—if used wisely—offer great promise for reducing detrimental impacts of weeds and for developing more sustainable weed management systems.

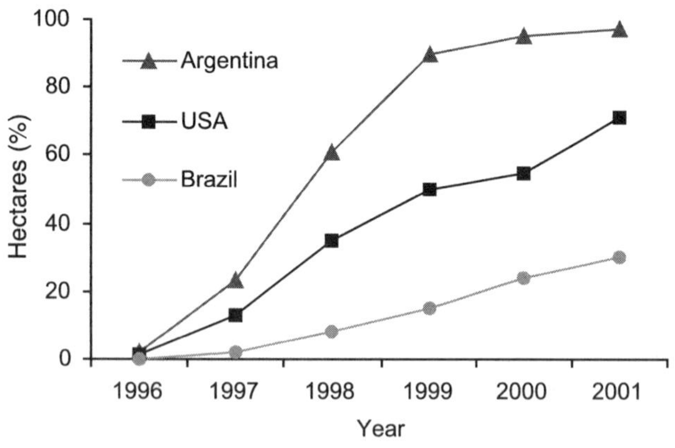

Fig. 1 Glyphosate-resistant soybean hectares as a percentage of total soybean hectares. The United States, Argentina, and Brazil produce 90% of the world's soybean exports. Note the high adoption in Brazil, despite the fact that glyphosate-resistant soybean is not legal in that country. (From http://soystats.com, accessed March 2003.)

Table 1 Partial list of herbicide-resistant crops obtained using molecular biology tools

Herbicide	Resistance mechanism	Crops
Bromoxynil	Herbicide detoxification	Cotton
Glufosinate	Herbicide detoxification	Corn, canola
Glyphosate	Altered herbicide target site	Cotton, corn, canola, soybean
Imidazolinones/ sulfonylureas	Altered herbicide target site	Corn, canola, sunflower, wheat
Sethoxydim	Altered herbicide target site	Corn

MOLECULAR BIOLOGY APPLIED TO THE BIOLOGY, ECOLOGY, AND GENETICS OF WEEDS

Weed scientists have often wondered what the underlying mechanisms are that contribute to the success of weeds. Tools of molecular biology offer novel ways of approaching this question.

Tools to Study Genetic Diversity

It is often hypothesized that large genetic diversity in weed species allows them to survive and adapt to changing agroecosystems. In the past, phenotypic markers were used to measure such diversity. However, because such markers may be influenced by environmental conditions, they do not always reflect genetic diversity. Molecular markers are a significant improvement for the study of genetic diversity.[5]

The first widely used molecular marker system used isozymes. Different versions of a particular enzyme can be separated by electrophoresis, due to different migration rates, and thus can provide a scoreable genetic marker. The discovery of restriction endonucleases (enzymes that cleave DNA) led to development of DNA markers based on restriction fragment length polymorphisms (RFLPs). The RFLP system detects genetic diversity based on the presence or absence of specific nucleotide sequences that are recognized by the restriction endonucleases.

Development of polymerase chain reaction (PCR) technology led to the improvement of DNA-based molecular markers into finer-scale genetic assessment techniques, such as amplified fragment length polymorphism (AFLP) and random amplified polymorphic DNA (RAPD). In general, AFLPs are based on the same principles as RFLPs, but with enhanced resolution provided by PCR amplification. RAPD markers rely on random amplification of DNA by means of short, arbitrary primers. Other variants of these PCR-based techniques have been used to target extremely diverse regions of genomes constituted by tandemly repeated sequences. Genetic diversity can be examined at the finest possible level by comparing nucleotide sequence information (i.e., by direct DNA sequencing).

Application of Genetic Diversity Tools: Studying Gene Flow

Studies that explore the mechanisms by which genetic diversity is acquired gained new force with the use of molecular biology tools. In particular, molecular biology has been invaluable in research examining gene flow among populations of the same species and between different species. Interspecific gene flow has been studied between some crops and their weedy relatives and among weed species themselves.[6] In most cases, gene flow has been studied in the context of herbicide resistance transmission. Hybrid intermediates between herbicide-susceptible and herbicide-resistant populations have been detected using a variety of molecular techniques. Such studies have made it possible to understand the significance of gene flow in the evolution of weed species. Furthermore, such studies are instrumental in predicting the risk that a herbicide-resistant crop may outcross with a weedy relative.

Gene Expression Profiling

Molecular biology can greatly facilitate understanding of how weeds respond to their environment, whether it be growing in competition with crops, surviving herbicide treatment, or initiating flowering after an early frost. In particular, a variety of techniques that can be loosely grouped in a category called gene expression profiling are ideally suited for such studies. The aim of gene expression profiling is to determine what genes are "turned on" or "turned off" in response to a particular treatment or environment. Results from such studies could provide insight at the molecular level as to how a weed responds to a particular treatment.

Modern gene expression profiling techniques (e.g., DNA microarrays) allow monitoring of thousands of genes simultaneously.[7] Such so-called "genomics" analyses have been adopted recently with model organisms—including plants such as Arabidopsis and several crop species—but have seen scant use in weed science research. As this technology becomes more commonplace, it likely will be applied to several aspects of weed science. Basic questions regarding biological phenomena important to weediness, such as induction of flowering, seed dormancy, and control of vegetative reproduction could be readily addressed with genomic techniques.

CONCLUSION

As illustrated by the foregoing examples, molecular biology research techniques provide many opportunities for advancements in weed science. In the future, molecular approaches will continue to provide new tools for managing weeds and will provide new information about the weeds themselves.

ARTICLES OF FURTHER INTEREST

Biological Control of Weeds, p. 141
Gene Flow Between Crops and Their Wild Progenitors, p. 488
Genetic Diversity Among Weeds, p. 496
Herbicide Resistant Weeds, p. 551
Molecular Technologies and Their Role in Maintaining and Utilizing Genetic Resources, p. 757
Parasitic Weeds, p. 864
Seed Banks and Seed Dormancy Among Weeds, p. 1123
Transgenic Crops: Regulatory Standards and Procedures of Research and Commercialization, p. 1251
Weed Management in Less Developed Countries, p. 1295
Weeds in Turfgrass, p. 1299

REFERENCES

1. http://www.weedscience.org/in.asp (accessed March 2003).
2. http://www.weedscience.org/mutations/MutDisplay.aspx (accessed March 2003).
3. Bradshaw, L.D.; Padgette, S.R.; Kimball, S.L.; Wells, B.H. Perspectives on glyphosate resistance. Weed Technol. **1997**, *11* (1), 189–198.
4. http://www.monsanto.com (accessed March 2001).
5. Jasieniuk, M.; Maxwell, B.D. Plant diversity: New insights from molecular biology and genomics technologies. Weed Sci. **2001**, *49* (2), 257–265.
6. Darmency, H. Movement of Resistance Genes Among Plants. In *Molecular Genetics and Evolution of Pesticide Resistance*; Brown, T.M., Ed.; American Chemical Society: Washington, DC, 1996; 209–220.
7. Shu-Hsing, W.; Ramonell, K.; Gollub, J.; Somerville, S. Plant gene expression profiling with DNA microarrays. Plant Physiol. Biochem. **2001**, *39* (11), 917–926.

Molecular Evolution

Michael T. Clegg
University of California, Riverside, California, U.S.A.

INTRODUCTION

The field of molecular evolution combines the tools of chemistry and molecular biology with the theoretical constructs of population genetics and systematics. It borrows concepts and methods from many areas of science—including mathematics, statistics, computer science, systematics, molecular biology, genetics, and chemistry—in addressing questions about the evolution of molecules and the evolution of the organisms that possess and transmit those molecules. Molecular evolution has both static and dynamic dimensions, ranging from the study of phylogeny to the experimental investigation of novel catalytic RNA molecules in test-tube environments. Molecular evolution is a comparative science in which inferences are made from the analysis of molecular data sampled from different organisms or from different gene copies within a genome. At its broadest, the science of molecular evolution can be defined as the study of the processes that cause molecular change and the application of that knowledge to our understanding of the history of life on Earth.

The goal of this article is to provide a brief overview of the current state of knowledge of plant molecular evolution. It will also discuss several exemplar problems relevant to crop evolution. A number of excellent books have been published in recent years that provide a detailed treatment of various aspects of molecular evolution, and these should be consulted by readers who wish to study molecular evolution in greater depth.

PLANT GENOMES AND THEIR EVOLUTIONARY PATTERNS

Plants possess three genomes: a chloroplast genome (denoted cpDNA), a mitochondrial genome (denoted mtDNA), and a nuclear genome (denoted nDNA). Each of these genomes has different origins and different evolutionary dynamics. A few of the most important features of the evolution of these genomes will be described briefly in the following sections.

HOW THE ANCESTORS OF PLANTS LEARNED TO CAPTURE ENERGY THROUGH PHOTOSYNTHESIS

A number of convergent lines of evidence provide overwhelming support for the hypothesis that the cpDNA genome originated from an endosymbiotic association between a cyanobacterialike photosynthetic organism (a prokaryote) and an eukaryotic cell. Thus a prokaryote donated the capacity for photosynthesis to the eukaryotic lineage that ultimately evolved into algae and land plants. As the endosymbiotic association evolved it became an obligate relationship, and many genes were transferred from the prokaryotic genome of the endosymbiont to the genome of the eukaryotic host. This transfer process probably occurred over hundreds of millions of years and may still be ongoing.[3] It represents one of the clearest and most massive cases of horizontal gene transfer in plant evolution. Phylogenetic analysis provides one of the most important lines of evidence for gene transfer, because it reveals that many nuclear-encoded components of photosynthesis are most closely related to cyanobaterial genes with the same (homologous) function.

The cpDNA molecule is conservative in size and in gene content. In land plants and green algae the cpDNA is a reduced molecule that averages about 150,000 base pairs (bp) in size and includes approximately 80 protein coding genes; 30 to 32 tRNA genes; and operons encoding the 16S, 23S, and 4.5S ribosomal RNA genes.[4] This inventory reveals that the chloroplast has its own protein synthesizing machinery. However, many of the functional components of photosynthesis are nuclear-encoded, owing to the prior history of gene transfer to the nuclear genome, and these polypeptides must be imported into the chloroplast for photosynthetic competence. There is considerable evidence of gene transfer from the cpDNA into both nDNA and mtDNA, but evidence for gene transfer into the cpDNA is essentially absent.

The cpDNA molecule was the first genome to be studied in plant molecular biology and evolution because it was relatively easy to purify and it encoded a limited number of genes, thus facilitating molecular cloning and analysis. An important early discovery was a rate of

Encyclopedia of Plant and Crop Science
DOI: 10.1081/E-EPCS 120010471

evolution for proteins encoded on the cpDNA molecule about five times slower than is typical for plant nuclear protein-coding genes. The significance of the conservative (slower) rate of evolution is that it allows the alignment and analysis of gene sequences that span much greater periods of evolutionary time. This discovery led to a major effort to construct land plant phylogenies based on cpDNA genes. The gene initially selected for this effort was *rbc*L. The *rbc*L gene encodes the large subunit of the enzyme ribulose-1, 5-bisphosphate carboxylase that fixes CO_2 in the carbon cycle. *rbc*L was the first cpDNA protein-coding gene to be cloned and sequenced, and thus it was a natural candidate for these early efforts.[5] To date more than 2500 *rbc*L sequences are available from land plants, and these data provide an unprecedented picture of land plant phylogeny. An entire field of plant molecular systematics has grown up around the use of cpDNA (and more recently certain nuclear genes) to infer plant phylogenetic history.

ORIGINS OF THE PLANT MITOCHONDRIAL GENOME

The plant mitochondrial genome also owes its origins to an early endosymbiotic association. In this case the enzymatic machinery for oxidative phosphorylation was donated to the eukaryotic lineage by an α-proteo bacterial (prokaryotic) source prior to the separation of plant and animal lineages.[6] Thus the evolutionary origins of the machinery that permitted the eukaryotic lineage to utilize oxygen and to adapt to an increasing partial pressure of oxygen in the atmosphere were also the result of a prokaryotic association. Many lines of evidence, including phylogenetic analysis, support this conclusion.

The plant mitochondrial genome reveals a strikingly different pattern of evolution when compared to either the chloroplast genome or animal mtDNA. First, the plant mtDNA exhibits a least a tenfold range in total DNA content (from 200,000 bp to 2,000,000 bp). Second, plant mtDNA appears to acquire and integrate DNA fragments from both cpDNA and nDNA sources, and this is thought to account for the large size variations. Third, plant mtDNA shows very high frequencies of genomic rearrangements, in contrast to remarkable stability for cpDNA. Fourth, protein-coding genes on the plant mtDNA are among the slowest-evolving genes characterized to date (roughly ten times slower than nDNA protein-coding genes). Fifth, the frequent loss of certain classes of genes from mtDNA is apparent over flowering plant evolution. And sixth, the mtDNA has repeatedly been invaded by a mobile element (group I intron) that appears to derive from a fungal source, providing clear documentation of horizontal gene transfer in recent evolutionary time.[7] So the general picture is one of great plasticity at the genomic level, but remarkable conservation at the gene level.

An important digression: It is remarkable that two of the major biochemical adaptations—photosynthesis and oxidative phosphorylation—that allowed eukaryotes to become dominant life forms on the terrestrial earth were acquired through endosymbiotic associations that each involved massive horizontal gene transfer to the eukaryotic nuclear genome. This history illustrates an important fact about evolution. Evolution is modular and occurs at several hierarchical levels. Molecular evolution is not simply the process of nucleotide substitution or the accumulation of individual amino acid substitutions in proteins over time. It is an open-ended process where genetic change can and does occur at all levels of organismal complexity. The analytical tools for studying the regular processes of nucleotide change allow us to infer time and to reconstruct history, albeit imperfectly. Inferences based on these tools tell us a rich story about the complexity of biological evolution. Perhaps the most important lesson is the hierarchical nature of genetic change.

EVOLUTION OF THE PLANT NUCLEAR GENOME

The eukaryotic nuclear genome is a mosaic of different histories. We have already seen that prokaryotic genes from both cyanobacterial and α-proteobacterial sources have been incorporated into nDNA. Very early in organismic evolution an association between eubacterial and archeabacterial cells appears to have arisen as an adaptation to oxygen poisoning. The eubacterium was motile and could avoid high oxygen concentrations while the archeabacterium provided an energy source in the form of sulfide. Ultimately the two genomes fused, and the nuclear envelope compartmentalized the integrated genome.[8] So once again we are reminded of the great plasticity of genetic evolution in leading to novel life forms.

We are beginning to acquire a detailed picture of the genetic complexity of the plant nuclear genome as a result of the recent sequencing of the *Arabidopsis* genome.[9] Moreover, a public version of the rice genome is now available. The ability to compare monocot (rice) and dicot (*Arabidopsis*) genomes should tell us much about the elaboration of gene function in the roughly 150 million years of angiosperm evolution. At the time of this writing our most comprehensive knowledge derives from analyses of the *Arabidopsis* genome sequence. A number of important facts have emerged from this project. First, many plant genes are present in multiple copies as a result of duplication over evolutionary time. Some of these gene families have diversified and expanded so that they contain in excess of one hundred gene duplicates, as discussed below. Second, gene-rich and gene-poor regions characterize the chromosomes of *Aribidopsis*, so the distribution of coding information appears to be heterogeneous across plant genomes. Third, transposable elements are a ubiquitous feature of plant genomes.

Evolution of Plant Gene Families

Multiple related genes (gene families) encode many important metabolic and regulatory functions. Gene families originate through gene duplication and subsequent divergence in function or in developmental expression. An example of a gene family that encodes an enzyme important in plant secondary metabolism is chalcone synthase (CHS). The CHS enzyme catalyzes a condensation reaction that yields a 15-carbon three-ring structure that is the first committed step in flavonoid metabolism. The CHS gene family exists in multiple copies in most plant genomes studied to date. For example, in plants of the morning glory genus (*Ipomoea*) there are at least six copies. It is possible to study the evolution of these genes using the computational tools of bioinformatics, and these analyses reveal two distinct subfamilies that duplicated and diverged from one another very early in flowering plant evolution. One subfamily (comprising genes *chs* D and E) performs the chalcone condensation reaction, but the two genes have diverged in tissues-specific patterns of expression.[10] The second subfamily (comprising genes *chs* A, B, C and a pseudogene) appears to have evolved a new catalytic function. This illustrates the fact that duplicate genes can provide a substrate for adaptive divergence by providing new developmental or metabolic capabilities to the plant.

The MYB family of transcriptional activators provides a second example that illustrates the very high levels of redundancy achieved by some plant gene families. MYB proteins are involved in DNA binding by a helix-turn-helix configuration. All known MYB proteins—including those found in plants and animals—have common structural features defined as the R1, R2, and R3 subdomains that are essential for the DNA binding function. The redundancy of MYB genes in plant genomes is much higher than observed in animal genomes. There appears to have been a proliferation of these genes that began early in land plant evolution so that more than 100 duplicate copies of MYB genes are found in the *Arabidopsis* genome. Analyses of the molecular evolution of this gene family reveal at least eight cases where rates of protein evolution accelerated following duplication events, as indicated by excess rates of amino acid change compared to synonymous change, suggesting positive selection at the protein level. Moreover, it is possible to identify specific nucleotide sites where presumably adaptive amino acid substitutions have occurred. It appears that this family of transcriptional activators has proliferated through duplication and evolved through nucleotide substitution to meet a wide array of regulatory needs within plant genomes. The computational methods of molecular evolution provide a powerful approach to the identification of the precise amino acid changes that are adaptive.

APPLICATIONS IN THE STUDY OF PLANT EVOLUTION AND CROP IMROVEMENT

Studies of Molecular Diversity

Application of the coalescent framework to samples of nucleotide sequence data are important in

- Screening loci for evidence of past selection
- Determining effective population sizes
- Evaluating general levels of genetic diversity
- Designing germplasm conservation programs
- Tracing the genetic correlates of crop domestication

A second question of importance concerns the structure of molecular diversity. In particular, it is of considerable practical utility to measure the effectiveness of recombination in randomizing different nucleotide sites on a chromosome as a function of physical or genetic distance. Put a different way, it is important to estimate a quantity known as linkage disequilibrium (LD) that measures the correlation in nucleotide state among pairs of nucleotide sites. This follows because correlations within or among genes may act to retard the operation of selection and thereby slow plant breeding progress. To see this, imagine a deleterious gene that is correlated in transmission with a selectively favored gene. The net effect, obviously is, to retard the increase of the favored gene or to slow the elimination of the deleterious gene. Finally, there is substantial contemporary interest in using single nucleotide polymorphisms (SNPs) as a tool for the genetic mapping of useful traits and for marker-assisted selection. This application requires that the marker SNP be correlated in transmission with the actual gene of interest.

There is a small but growing body of data on nucleotide sequence diversity in plants, most of which derives from the model plant *Arabidopsis*, although data increasingly are becoming available from crop plants such as maize and barley. Both *Arabidopsis* and barley are predominantly self-fertilizing plants, and population genetic theory predicts reduced levels of variation in inbreeding plant species. Moreover, population genetic theory predicts high LD in inbreeding species. Both of these predictions follow from the fact that self-fertilization reduces heterozygosity by a factor of 50% per generation. A population is expected to be composed of a collection of homozygous lines very quickly. Effective recombination occurs only in heterozygotes, so inbreeding greatly reduces the effective rate of recombination—and this should promote higher levels of LD. The effective population size of self-fertilizing populations is half that of comparable random-mating populations, and selection is more effective among homozygous lines. Accordingly, we expect reduced levels of genetic diversity in self-fertilizing species. Current data are equivocal with regard to these two predictions. Levels of LD in *Arabidopsis* are high, but not as high as might be expected relative to, say, human populations.[11] Preliminary evidence from barley also indicates only modest increases in LD relative to maize.[12] Levels of genetic diversity are reduced in inbreeding species, and this generalization is well supported by massive isozyme data analyses.[13] Nevertheless, most self-fertilizing species possess considerable stores of genetic variation and numerous experiments show that self-fertilizing species are able to respond to selection on virtually any phenotypic character.[14]

Plant Relationships and Crop Origins

Molecular data are providing important insights into crop plant domestication. This topic is dealt with at greater length in other sections of this volume and will only be touched on here. One place where markers are especially helpful is unraveling the complex network of relationships associated with hybridization in plant domestication. To cite a single example, the cultivated avocado is a genetically diverse assemblage comprising three botanical varieties. Microsatellite markers provide a means of identifying the genetic origins of various cultivars, because many loci can be screened to provide an informative average picture of the genomic history of particular lineages. Work of this kind shows clearly that many popular cultivars are the result of intervarietal hybridization. In the longer term this kind of work should also provide clear evidence regarding the geographic origin of useful materials, and it should provide a helpful guide for genetic conservation.

CONCLUSION

Molecular evolutionary analyses have provided a rich picture of the evolution of plant genomes. This picture begins with the very early endosymbiotic events that presumably led to the eukaryotic cell and continues to the later acquisition of the machinery for photosynthesis and oxidative phosphorylation. Along the way, gene redundancy is seen to have played a major role in the elaboration of biochemical and developmental novelty. An important lesson is thus gleaned: the recognition that molecular evolution has proceeded at several levels, including the level of protein evolution through nucleotide and amino acid substitution, the level of gene duplication, and the level of whole genome transfer events that have taken place several times in the long course of evolution.

ARTICLES OF FURTHER INTEREST

Agriculture and Biodiversity, p. 1
Arabidopsis Transcription Factors: Genome-Wide Comparative Analysis Among Eukaryotes, p. 51
Bioinformatics, p. 125
Chromosome Rearrangements, p. 270
Chromosome Structure and Evolution, p. 273
Crop Domestication: Fate of Genetic Diversity, p. 333
Evolution of Plant Genome Microstructure, p. 421
Gene Flow Between Crops and Their Wild Progenitors, p. 488
Genome Size, p. 516
Genome Structure and Gene Content in Plant and Protist Mitochondrial DNAs, p. 520
Medical Molecular Pharming: Therapeutic Recombinant Antibodies, Biopharmaceuticals and Edible Vaccines in Transgenic Plants Engineered via the Chloroplast Genome, p. 705
Mitochondrial Respiration, p. 729
Mycorrhizal Evolution, p. 767
New Secondary Metabolites: Potential Evolution, p. 818
Population Genetics, p. 1042

REFERENCES

1. Nei, M.; Kumar, S. *Molecular Evolution and Phylogenetics*; Oxford University Press: Oxford, 2000.
2. Graur, D.; Li, W.-H. *Fundamentals of Molecular Evolution,* 2nd Ed.; Sinauer Associates, Sunderland, Massachusetts, 2000.
3. Palmer, J.D.; Delwiche, C.R. The Origin and Evolution of Plastids and Their Genomes. In *Molecular Systematics of Plants II*; Soltis, D.E., Soltis, P.S., Doyle, J.J., Eds.; Kluwer: Boston, 1998; 375–409.
4. Sugiura, M. The Chloroplast Genome. In *Essays in Biochemistry*; Apps, D.K., Tipton, K.F., Eds.; Portland Press: London, 1995; 49–57.

5. Clegg, M.T. Chloroplast gene sequences and the study of plant evolution. Proc. Natl. Acad. Sci. U. S. A. **1993**, *90*, 363–367.
6. Gray, M.W.; Burger, G.; Lang, B.F. Mitochondrial evolution. Science **1999**, *283*, 1476–1481.
7. Palmer, J.D.; Adams, K.L.Y.; Parkinson, C.L.; Qiu, Y.-L.; Song, K. Dynamic evolution of plant mitochondrial genomes: Mobile genes and introns and highly variable mutation rates. Proc. Natl. Acad. Sci. U. S. A. **2000**, *97*, 6960–6966.
8. Margulis, L.; Dolan, M.F.; Guerrero, R. The chimeric eukaryote: Origin of the nucleus from the karyomastigont in amitochondriate protists. Proc. Natl. Acad. Sci. U. S. A. **2000**, *97*, 6954–6959.
9. Arabidopsis genome initiative. Analysis of the genome sequence of the flowering plant *Arabidopsis thaliana*. Nature **2000**, *408*, 796–815.
10. Durbin, M.L.; McCaig, B.; Clegg, M.T. Molecular evolution of chalcone synthase multigene family in the morning glory genome. Plant Mol. Biol. **2000**, *42*, 79–92.
11. Nordborg, M.; Borevitz, J.O.; Bergelson, J.; Berry, C.C.; Chory, J.; Hagenblad, J.; Kreitman, M.; Maloof, J.N.; Noyes, T.; Oefner, P.J.; Stahl, E.A.; Weigel, D. The extent of linkage disequilibrium in *Arabidopsis thaliana*. Nat. Genet. **2002**, *30*, 190–193.
12. Lin, J.-Z.; Morrell, P.L.; Clegg, M.T. The influence of linkage and inbreeding on patterns of nucleotide sequence diversity at duplicate alcohol dehydrogenase loci in wild barley (*Hordeum vulgare* ssp. *spontaneum*). Genetics **2002**, *in press*.
13. Hamrick, J.L.; Godt, M.J. Allozyme Diversity in Plant Species. In *Plant Population Genetics, Breeding and Germplasm Resources*; Brown, A.H.D., Clegg, M.T., Kahler, A.L., Weir, B.S., Eds.; Sinauer Associates, Sunderland, Massachusetts, 2000; 43–63.
14. Allard, R.W. History of plant population genetics. Annu. Rev. Genet. **1999**, *33*, 1–27.

Molecular Farming in Plants: Technology Platforms

M

Rainer Fischer
Fraunhofer Institute for Molecular Biology and Applied Ecology (IME), Aachen, Germany

Neil J. Emans
Institute for Molecular Biotechnology, Aachen, Germany

Richard M. Twyman
University of York, York, U.K.

Stefan Schillberg
Fraunhofer Institute for Molecular Biology and Applied Ecology (IME), Aachen, Germany

INTRODUCTION

The large-scale production of recombinant proteins in plants is known as molecular farming. Plants have many advantages in terms of cost, practicality, and safety over traditional expression systems and are emerging as a significant force in the commercial sector. Providing adequate yields can be obtained, it is estimated that recombinant proteins can be produced in plants at 2–10% of the cost of microbial fermentation systems and at 0.1% of the cost of mammalian cell cultures/transgenic animals. Plants lack the endotoxins often produced by microbial cultures and, unlike animal cells, do not harbor human pathogens or oncogenic DNA sequences. Posttranslational modification occurs in a similar manner in plant and animal cells with only minor differences in glycan chain structure, which makes plants suitable for the production of complex human glycoproteins. Plants also have a number of unique practical advantages such as the high stability of proteins expressed in seeds and the ability to express pharmaceutical proteins in edible organs for oral administration with minimal processing.

PLANT-BASED EXPRESSION SYSTEMS

Transgenic Plants

In the vast majority of cases, molecular farming has been achieved by stable transformation and the regeneration of transgenic plants.[1] Several crop species have been used as expression hosts, and the best choice for any particular protein must be determined on an empirical basis according to intrinsic properties of the plant (Table 1) and economic factors. This might include the value and intended use of the recombinant protein and the local availability of labor, storage facilities, processing, and distribution infrastructure.[2]

Tobacco has the longest history as a successful expression system for molecular farming and is one of the best candidates for the commercial production of recombinant proteins.[2] The advantages of tobacco include the well-established technology for gene transfer and expression, the high biomass yield (over 100,000 kg per hectare for close-cropped tobacco), and the existence of large-scale infrastructure for processing. However, a major disadvantage of leafy crops is protein instability. The leaf tissue must be frozen or dried for transport or processed at the harvesting site. In contrast, seed-based expression allows long-term storage at ambient temperatures because the proteins accumulate in a stable form.[?] Several different crops have been investigated for seed-based production, including the cereals rice, maize, and wheat. Prodigene Inc. chose to use maize for the first commercial molecular farming venture due to the combination of high biomass yield and the ease of in vitro manipulation; three technical proteins have been marketed successfully.[3] Recombinant proteins have also been produced in fruit and vegetable crops, which are beneficial because the products can be expressed in edible organs for oral administration.[4] Clinical trials have been carried out using vaccine candidates, antibodies, and enzymes expressed in leafy crops, seeds, and vegetables, and several companies have plant-derived pharmaceuticals in the late stages of clinical development.[5]

Transgenic plants are the most cost-effective platform in molecular farming because crops can be established, maintained, harvested, and processed using traditional agricultural practices and existing infrastructure. However, a development phase of 18 months to two years is required to produce the first generation of transformants, and biosafety concerns that transgenes and their

Encyclopedia of Plant and Crop Science
DOI: 10.1081/E-EPCS 120024676

Table 1 Comparison of major transgenic crops used for molecular farming of proteins

Crop	Preferred site of expression	Biomass yield (kg ha^{-1})	Storage	Stability	Other comments
Dicots					
Alfalfa	Leaf	14,400	Dried/frozen	Months	Feed
Canola	Seed	1,440	Ambient	>1 year	Dual purpose, oil body targeting
Pea	Seed	2,400	Ambient	>1 year	Food
Potato	Tuber	36,200	Chilled	Weeks	Food
Soybean	Seed	2,640	Ambient	>1 year	Feed
Tobacco	Leaf	>100,000	Dried/frozen	Months	Not food/feed
Tomato	Fruit	67,500	Chilled	Weeks	Palatable raw, requires glasshouse
Monocots					
Maize	Seed	8,880	Ambient	>1 year	Food
Rice	Seed	6,570	Ambient	>1 year	Food
Wheat	Seed	2,870	Ambient	>1 year	Food

products could spread in the environment are limiting public acceptance.[6] This has prompted research into alternative plant-based technologies for recombinant protein production.

Transplastomic Plants

Chloroplast transformation is a useful alternative to nuclear transformation in molecular farming applications because there are many chloroplasts in the cell, resulting in a high transgene copy number in the transplastomic plants. This, together with the fact that chloroplast transgenes appear not to suffer from position effects or epigenetic transgene silencing, produces unprecedented levels of recombinant protein, in the best cases reaching nearly 50% total soluble protein.[7] Disulfide bonds are formed correctly in chloroplast-derived recombinant proteins, but there is no glycosylation. In biosafety terms, transplastomic plants are advantageous because there are no plastids in the pollen grains of most crops, thus limiting gene flow by pollen dispersal. Chlorogen, Inc. was established in 2003 to commercialize the chloroplast transgenic system.

Transient Expression Systems

Transient expression assays are often used to evaluate the activity of expression constructs or test the functionality of recombinant proteins, but they can also be used as a routine molecular farming platform if moderate amounts of protein can be produced.[8] This is possible using the agroinfiltration method, where recombinant *Agrobacterium tumefaciens* are infiltrated into tobacco leaf tissue and milligrams of protein can be produced within a few weeks.[9] The advantages of agroinfiltration include the minimal set-up costs and the rapid onset of protein expression, but scaling up is neither as economical nor as convenient as is the case with stably transformed plants.

Virus-Infected Plants

In addition to *Agrobacterium*-mediated transformation and direct DNA transfer, plant viruses can also be used as vectors to deliver foreign genes into plants.[10] Although viral gene transfer does not result in stable transformation, the prolific replication and systemic spread of many plant viruses means that the onset of recombinant protein expression is rapid and the total yield is high, two features that are ideal for molecular farming. Thus far, viral vectors have not been widely used for molecular farming, perhaps because there is a limit to the amount of DNA that can be incorporated into such constructs. Tobacco mosaic virus (TMV) and potato virus X (PVX) vectors have both been used for the production of recombinant antibodies in tobacco, including scFv fragments and a full-sized IgG. In the latter case, two different TMV vectors containing the heavy and light immunoglobulin chain genes, respectively, were used to coinfect the plant, and correct assembly of the protein occurred in planta.

Plant Cell Suspension Cultures

The production of recombinant proteins in plant cell suspension cultures is an alternative approach that may be beneficial where defined and sterile production conditions and high-level containment are required, e.g., for the production of pharmaceuticals.[11] Plant cell suspensions are normally derived from callus tissue that has been cultivated on solid medium and then broken up by agitation on rotary shakers or in fermenters. Transformation may be achieved prior to the formation of cultures, in which case the cells already contain the transgene at the callus stage, or the cells can be transformed in culture. The latter procedure is advantageous because it is rapid and avoids the need to regenerate and characterize transgenic plants, so productive cell lines can be generated within a few months. Suspension cell cultures can be maintained in conventional microbial fermentation equipment with only minor technical modifications, and various different culture modes can be applied, including batch, fed-batch, perfusion, and continuous fermentation.[12]

OPTIMIZING PRODUCTION OF ACTIVE RECOMBINANT PROTEINS

Maximizing the Rate of Transcription and Translation

The production of active recombinant proteins in plants can be optimized at each stage of gene expression.[13] Generally, it is useful to use a strong and constitutive promoter to maximize the rate of transcription. The cauliflower mosaic virus 35*S* RNA promoter (CaMV 35S) is widely used in dicots, while the maize ubiquitin-1 promoter (*ubi-1*) is preferred in monocots. However, there are also advantages to the use of tissue-specific and inducible promoters. The use of seed-specific promoters, for example, prevents the accumulation of recombinant proteins in vegetative organs. This limits toxicity effects in the host plant and reduces the likelihood that herbivores and other non-target organisms will be exposed to the protein. Similarly, inducible promoters activated by chemicals such as tetracycline or by mechanical stimuli can be used to initiate protein accumulation prior to or even after harvest.

The rate of protein synthesis can be optimized by eliminating native 5′ and 3′ untranslated regions from the foreign gene and replacing these with elements that increase mRNA stability and enhance translational efficiency, such as the 5′ leader sequence of the tobacco mosaic virus RNA (the omega sequence). It is also useful to change the translational start site of the expression construct to conform to the Kozak consensus for plants. For some transgenes, it may be necessary to modify the coding region to match the codon usage preferences of the expression host.

Protein Targeting and Modification

Subcellular targeting influences protein stability (thus affecting the yield), determines the type of modification that takes place (thus affecting protein structure and activity), and can be exploited to simplify downstream processing (e.g., by including affinity tags or fusions). Targeting to the secretory pathway is beneficial, especially if the protein is normally glycosylated since this modification takes place in the endoplasmic reticulum (ER) and Golgi apparatus. The environment of the ER also promotes correct protein folding and thus increases stability. This has been demonstrated in the case of recombinant antibodies, which accumulate in the secretory pathway at levels up to 100-fold greater than in the cytosol.[14] Targeting to the secretory pathway is achieved by incorporating an N-terminal signal sequence into the expression construct that directs the ribosome to signal receptors on the endoplasmic reticulum. The default destination of proteins targeted to the secretory pathway in plants is the apoplastic space under the cell wall, where they may be retained or secreted depending on their size. Still higher protein levels can be achieved by adding a C-terminal H/KDEL sequence, which causes the recombinant protein to be retrieved from the Golgi and returned to the ER lumen, in the manner of a resident ER protein.[15]

CONCLUSION

Plants have practical and economical advantages over traditional expression systems, making them suitable for the large-scale production of recombinant proteins. Technological issues that remain to be addressed include the optimization of yields, the modification of glycan structures, and improved biosafety. In order to be commercially viable, plants must produce a target protein at a level exceeding 1% total soluble protein. This has been possible for many proteins, but some unstable proteins tend to accumulate at a level of 0.01–0.1%, often reflecting incorrect folding, assembly, or degradation during extraction. Biosafety issues remain at the forefront of current molecular farming research. Such issues include the development of novel ways to prevent transgene spread, to restrict unnecessary exposure of non-target animals and microbes to transgenes and their products, and to prevent transgenic plant material mixing with the food and feed chains.

ARTICLES OF FURTHER INTEREST

Plant-Produced Recombinant Therapeutics, p. 969
Transformation Methods and Impact, p. 1233
Transgenes: Expression and Silencing of, p. 1242

Transgenic Crop Plants in the Environment, p. 1248
Transgenic Crops: Regulatory Standards and Procedures of Research and Commercialization, p. 1251

REFERENCES

1. Giddings, G. Transgenic plants as protein factories. Curr. Opin. Biotechnol. **2001**, *12* (5), 450–454.
2. Stoger, E.; Sack, M.; Perrin, Y.; Vaquero, C.; Torres, E.; Twyman, R.M.; Christou, P.; Fischer, R. Practical considerations for pharmaceutical antibody production in different crop systems. Mol. Breed. **2002**, *9* (3), 149–158.
3. Hood, E.E.; Woodard, S.L.; Horn, M.E. Monoclonal antibody manufacturing in transgenic plants—Myths and realities. Curr. Opin. Biotechnol. **2000**, *13* (6), 630–635.
4. Daniell, H.; Streatfield, S.J.; Wycoff, K. Medical molecular farming: Production of antibodies, biopharmaceuticals and edible vaccines in plants. Trends Plant Sci. **2001**, *6* (5), 219–226.
5. Fischer, R.; Twyman, R.M.; Schillberg, S. Production of antibodies in plants and their use for global health. Vaccine **2003**, *21* (7–8), 820–825.
6. Commandeur, U.; Twyman, R.M.; Fischer, R. The biosafety of molecular farming in plants. AgBiotechNet **2003**, *5*, ABN110. (April).
7. Daniel, H.; Khan, M.S.; Allison, L. Milestones in chloroplast genetic engineering: An environmentally friendly era in biotechnology. Trends Plant Sci. **2002**, *7* (2), 84–91.
8. Fischer, R.; Vaquero-Martin, C.; Sack, M.; Drossard, J.; Emans, N.; Commandeur, U. Towards molecular farming in the future: Transient protein expression in plants. Biotechnol. Appl. Biochem. **1999**, *30* (2), 113–116.
9. Kapila, J.; De Rycke, R.; van Montagu, M.; Angenon, G. An *Agrobacterium*-mediated transient gene expression system for intact leaves. Plant Sci. **1997**, *122* (1), 101–108.
10. Porta, C.; Lomonossoff, G.P. Viruses as vectors for the expression of foreign sequences in plants. Biotechnol. Genet. Eng. Rev. **2002**, *19*, 245–291.
11. Doran, P.M. Foreign protein production in plant tissue cultures. Curr. Opin. Biotechnol. **2002**, *11* (2), 199–204.
12. Fischer, R.; Emans, N.; Schuster, F.; Hellwig, S.; Drossard, J. Towards molecular farming in the future: Using plant-cell-suspension cultures as bioreactors. Biotechnol. Appl. Biochem. **1999**, *30* (2), 109–112.
13. Schillberg, S.; Fischer, R.; Emans, N. Molecular farming of recombinant antibodies in plants. Cell. Mol. Life Sci. **2003**, *60* (3), 433–445.
14. Schillberg, S.; Zimmermann, S.; Voss, A.; Fischer, R. Apoplastic and cytosolic expression of full-size antibodies and antibody fragments in *Nicotiana tabacum*. Transgenic Res. **1999**, *8* (4), 255–263.
15. Conrad, U.; Fiedler, U. Compartment-specific accumulation of recombinant immunoglobulins in plant cells: An essential tool for antibody production and immunomodulation of physiological functions and pathogen activity. Plant Mol. Biol. **1998**, *38* (1–2), 101–109.

Molecular Technologies and Their Role in Maintaining and Utilizing Genetic Resources

M

Amanda J. Garris
Alexandra M. Casa
Stephen Kresovich
Cornell University, Ithaca, New York, U.S.A.

INTRODUCTION

In the broadest sense, the goal of conservation activities is the preservation of diversity at ecosystem, community, and species levels. Conservation of plant genetic resources, with its focus on diversity within crops and their wild relatives, differs in that it is inextricably linked to a mandate for utilization. Perhaps the greatest challenge lies in identification of useful variation not readily assessed at the phenotypic level, due to the complexity of the trait or the masking effects of environment and genetic background. Exploitation of variation in collections has been achieved primarily through phenotypic screens and backcrossing strategies; however, molecular markers may serve to expedite the identification of useful alleles. It is not the scope of this article to provide a detailed description of molecular techniques, but rather to offer some insights on how understanding the structure of diversity allows the generation of data simultaneously useful for both conservation and use, enabling us to dissect gene function and consequently assess the predictive value of diversity for crop improvement.

CLASSES OF MOLECULAR MARKERS

Although morphological and biochemical markers (i.e., isozymes) still play an important role in understanding diversity, their low abundance and unknown selective neutrality have created a need for more specific DNA-based methodologies for characterizing genetic resources. In addition, the growing number of accessions present in germplasm banks has required faster and more cost-effective methodologies for understanding diversity and directing conservation efforts. Among the most widely used classes of molecular markers for germplasm characterization are restriction fragment length polymorphisms (RFLPs), randomly amplified polymorphic DNAs (RAPDs), amplified fragment length polymorphism (AFLPs), simple sequence repeats (SSRs), and more recently DNA sequence polymorphism.[1] The choice of a marker technique will depend on several factors, including what is already known about the species (e.g., mating system and life form), the kinds of questions that are being addressed (e.g., levels of genetic diversity and distribution of variation), and infrastructure (i.e., amount of equipment required, start-up cost, and labor).

Molecular markers have been applied to most crop species in order to characterize the diversity present in nature, in farmers' fields as well as in situ and ex situ collections. These data provide a means to quantify the level of diversity and how it is partitioned within a species, enabling conservation strategies that maximize diversity within collections through reduction of redundancy and identification of unique alleles.[2] Applications include characterization of crop genepools, identification of parentage, and selection of a representative core collection.[3] Much emphasis has been given to the characterization of genetic resources; however, retention of diversity in collections requires effective maintenance strategies. Molecular markers provide a quantitative assessment of the level of diversity and can be used to monitor change over time in collections. In addition, they can provide a basis for setting priorities for additions to collections when used in concert with geographical information system (GIS), which could incorporate agroecological characteristics of collection sites and genetic diversity data to target potentially useful populations.[4]

APPLICATION OF MOLECULAR MARKERS TO UTILIZATION OF GENETIC RESOURCES

Molecular markers, although widespread, have yielded few examples of successful identification of variation that can be exploited in breeding programs. However, population genetics approaches for identifying adaptive variation developed for noncrop species when applied to genetic resources collections may expedite the identification of useful alleles. Generally, these approaches interpret patterns of molecular diversity to evaluate evidence for a causative role in a phenotype, to identify genes for

Encyclopedia of Plant and Crop Science
DOI: 10.1081/E-EPCS 120010459

adaptation, or to predict phenotypic value. The discussion that follows highlights some approaches that use molecular diversity to identify genes of interest that could be utilized in genetic resources collections.

One promising approach is based on association genetics, in which the statistical association of sequence polymorphisms with a phenotype is evaluated. Developed for use in human genetics, association genetics has been successfully applied in maize. For example, a suite of polymorphisms in the *Dwarf8* gene was found to associate with variation in flowering time.[5] The application of statistical analysis to the pattern of DNA polymorphism forged the link between molecular and phenotypic diversity in the case of this quantitative trait.

Another approach is to use patterns of molecular diversity to identify genes that have been targets of selection. Natural selection is expected to reduce diversity at a locus; this reduction in diversity is detectable in comparisons to an outgroup that has not been subject to the same selective pressure. Reduced diversity at SSR loci in maize relative to its wild relative teosinte has been used to identify candidate genes involved in domestication.[6] This approach could be modified to find targets of selection for adaptive traits by comparison of adapted germplasm within a crop. A complementary approach is to identify rapidly evolving genes underlying adaptive variation by their molecular diversity. The approach involves the screening of expressed sequence tags (ESTs) to identify those that have diversified compared to homologues in closely related species. In *Drosophila melanogaster*, this method has been used to identify putative targets of adaptive evolution.[7]

As the structural and active domains of proteins are better understood, it should facilitate *in silico* allele mining in genetic resources collections via analysis of which changes in DNA sequence will effect functional changes in the protein. This is a computational approach akin to reverse genetics, but uses naturally occurring variation in lieu of disruptions to form predictions about gene function. For example, a collection could be screened with primers to amplify the active region in a disease resistance gene to identify as candidates for further study variants that change the protein sequence.

LIMITATIONS

Although the application of molecular markers holds promise for genetic resources conservation and use, a few caveats should be noted. The methods outlined previously require substantial preliminary data on population structure and appropriate sampling. Patterns of diversity can be influenced not only by selection; the influence of population structure, linkage, and drift must be understood in order to correctly interpret results. Although these approaches can identify interesting candidate genes, functional studies will be required to establish causation. A greater limitation is that some differences that affect phenotype are not coded in DNA, including such phenomena as epigenetics and differential splicing of RNA transcripts. In addition, the importance of regulatory elements in plant evolution has been demonstrated, indicating that not only structural but regulatory genes will be important in conservation and breeding.[8]

CONCLUSION

Molecular markers have the potential to build the link between DNA sequence and phenotype, facilitating conservation and use as well as the link between them. Increased access to these technologies, in concert with improved computational methods for analysis, will make molecular techniques both more common and more useful in maintaining and utilizing genetic resources. In the future, important and challenging questions will not be restrained by the lack of appropriate tools, and the allelic diversity in crop collections can be implemented in crop improvement.

ARTICLES OF FURTHER INTEREST

Agriculture and Biodiversity, p. 1
Crop Improvement: Broadening the Genetic Base for, p. 343
Germplasm Collections: Regeneration in Maintenance, p. 541
Germplasm Maintenance, p. 544
Molecular Evolution, p. 748
Population Genetics, p. 1042
Quantitative Trait Location Analyses of the Domestication Syndrome and Domestication Process, p. 1069

REFERENCES

1. Karp, A.; Kresovich, S.; Bhat, K.V.; Ayad, W.G.; Hodgkin, T. *Molecular Tools in Plant Genetic Resources Conservation, A Guide to the Technologies*; IPGRI Technical Bulletin, International Plant Genetic Resources Institute: Rome, Italy, 1997; Vol. 2.

2. Phippen, W.B.; Kresovich, S.; Gonzalez-Candelas, F.; McFerson, J.R. Molecular characterization can quantify and partition variation among genebank holdings, a case study with phenotypically similar accessions of *Brassica oleracea* var *capitata* L (cabbage) 'Golden Acre'. Theor. Appl. Genet. **1997**, *94* (2), 227–234.
3. Brown, A.H.D. Core collections—A practical approach to genetic-resources management. Genome **1989**, *31* (2), 818–824.
4. Jones, P.G.; Beebe, S.E.; Tohme, J.; Galwey, N.W. The use of geographical information systems in biodiversity exploration and conservation. Biodivers. Conserv. **1997**, *6*, 947–958.
5. Thornsberry, J.M.; Goodman, M.M.; Doebley, J.; Kresovich, S.; Nielsen, D.; Buckler, E.S. *Dwarf8* polymorphisms associate with variation in flowering time. Nat. Genet. **2001**, *28* (3), 286–289.
6. Vigouroux, Y.; McMullen, M.; Hittinger, C.T.; Houchins, K.; Schulz, L.; Kresovich, S.; Matsuoka, Y.; Doebley, J. Identifying genes of agronomic importance in maize by screening microsatellites for evidence of selection during domestication. PNAS **2002**, *99* (15), 9650–9655.
7. Swanson, W.J.; Clark, A.G.; Waldrip-Dail, H.M.; Wolfner, M.F.; Aquadro, C.F. Evolutionary EST analysis identifies rapidly evolving male reproductive proteins in Drosophila. PNAS **2001**, *98* (13), 7375–7379.
8. Doebley, J.; Lukens, L. Transcriptional regulators and the evolution of plant form. Plant Cell **1998**, *10* (7), 1075–1082.

M

Mutational Processes

Michael T. Clegg
Peter L. Morrell
University of California, Riverside, California, U.S.A.

INTRODUCTION

Mutation is a permanent heritable change in the DNA of an individual organism. It is the ultimate source of all the biological diversity on earth. Life could not adapt and diversify without the substrate of heritable variation. In agriculture the exploitation of novel heritable variants has been the engine of plant and animal improvement. This article outlines the mechanisms that are responsible for heritable changes in DNA.

NUCLEOTIDE SUBSTITUTION EVENTS

Many different categories of mutation affect DNA molecules. The most regular form of mutational change is when one nucleotide is incorrectly replaced by a different nucleotide during replication (Fig. 1). These nucleotide substitutions are assumed to be stochastically regular in time. (Like radioactive decay, each mutation is assumed to arrive randomly, but with a constant probability in time.) Nucleotide substitution events are also assumed to occur with the same probability at every nucleotide site, independent of the nucleotide state at the mutated site or at neighboring sites. Much of our direct knowledge of the nucleotide substitution processes comes from the study of bacteria or viruses where many generations and very large populations of organisms can be studied in the laboratory. These studies indicate that the DNA replication process has a remarkably high fidelity, with a nucleotide substitution error rate on the order of one error per one billion replications per site (or 10^{-9} errors per site per generation).[1] These studies have also tended to validate the assumptions of constancy and independence previously mentioned. Estimates for eukaryotic organisms like crop plants and animals cannot be obtained from direct observation and must be calculated indirectly. Indirect calculations are based on comparisons between homologous gene sequences separated for some period of evolutionary time. Statistical models of the substitution process must be invoked, together with various assumptions about the absence of natural selection, to arrive at estimates of mutation rates. Nevertheless, indirect estimates from, for example, large numbers of mammalian genes are consistent with those from bacteria, indicating an average nucleotide substitution rate of roughly 4×10^{-9} per site per generation.[1]

INSERTION/DELETION EVENTS

Another major category of mutation is the insertion or deletion of stretches of DNA (indels). Indels are most frequently observed in areas of untranslated DNA (DNA that does not code for a protein). Much of the genome of eukaryotic organisms is made up of untranslated regions, so on a genomic scale, a substantial fraction of mutation events are likely to be indels. Indels arise from a number of causes, including slipped-strand-mispairing in replication, unequal crossover events, or transposable element insertions. In some cases indels have been shown to depend on local nucleotide context, so the assumption of independence over DNA region cannot be invoked.[2] Moreover, the assumption of stochastic regularity in time is also of questionable validity, at least for some forms of transposable element insertions. Finally, the alignment of indels is difficult when sequences have diverged for a substantial length of time, because overlapping events cannot be distinguished. All of these factors severely limit our ability to use mathematical models and to make statistical calculations based on indel data.

Slipped-Strand Mispairing

Despite the fact that slipped-strand mispairing events are context-dependent and are usually inferred indirectly rather than observed, it is desirable to get some sense of the rate at which these kinds of events accumulate. The effects of replication slippage or slipped-strand mispairing generally cannot be observed directly and must be inferred based on the distribution of repetitive nucleotide sequence or base composition surrounding indels (Fig. 2). Estimates from *Drosophila* nucleotide sequence data suggested approximately 0.16 indels per nucleotide substitution.[3,4] Estimates based on chloroplast data from grasses suggested that indels and nucleotide substitutions contributed nearly equal proportions of total mutations; however, the authors point out that the relative contribution of indels

Encyclopedia of Plant and Crop Science
DOI: 10.1081/E-EPCS 120017011

GACACGCCC
GACACGCCG
GGCACCCCC

Fig. 1 An alignment of DNA sequence data. Three nucleotide substitutions are underlined.

appears to diminish among more highly diverged species because of apparent superimposition of indels on previous indels.[5] Slipped-strand mispairing appears to play a major role in indel generation, and may be particularly important in the proliferation of repetitive sequences, such as the short two- and three-base motifs that make up microsatellites.[6]

Unequal Crossover Events

At tandemly repeated genes there is potential for out-of-register recombination, which can result in duplication, deletion, or truncation of a chromosomal region. Such events may be especially common in tandemly repeated arrays of genes such as nuclear ribosomal DNA.

The Interaction Between Mutation and Recombination

Genetic recombination is the process that disassociates mutations along a chromosome, typically through cross over—the breaking and reannealing of homologous portions of parental chromosomes during meiosis. A second form of recombinational exchange is commonly observed between different tandemly repeated gene family members. When such an exchange occurs, a mutant site on one repeated gene copy is transmitted to an adjacent gene copy, causing the repeated copies to become more similar in sequence than would otherwise be the case. This process is called concerted evolution. A related form of recombinational exchange that generates new alleles is observed in microsatellite loci where unequal exchange between tandemly arranged di- or trinucleotide repeats (or higher-order repeats) causes an increase or decrease in repeat number. Finally, unequal exchanges between repeated elements at different chromosomal locations can cause duplications and deficiencies of larger DNA regions.

TRANSPOSON-INDUCED MUTATIONS

Mobile elements (transposons) are a ubiquitous feature of plant and animal genomes. There are two broad categories of mobile elements: class I elements that replicate from an RNA intermediate via reverse transcriptase, and class II or DNA elements that replicate via a cut-and-paste mechanism. Both classes of elements are relatively common in plant and animal genomes. Class II elements tend to be associated with elevated transposition rates, but both categories are clearly implicated as causal agents in many mutations, owing to their ability to insert into or adjacent to genes, and thereby disrupt or alter gene function or gene expression.

One important aspect of transposon-mediated mutation is the ability to alter gene expression patterns. This may occur because the element has inserted into a 5′ UTR (untranslated region in the 5′ region of a gene) and has thereby disrupted transcription factor-binding sequence motifs, or it may occur because sequence motifs within the insertion act as novel sites for transcription factor binding. This latter capacity has been documented in a number of cases in plants and animals, and illustrates the creative acquisition of new gene expression patterns through the transposition of appropriate sequence features around the genome. The fact that transposable elements can move entire sequence motifs into new genic environments and induce novel patterns of expression tells us transposon mutation is modular and acts at a level beyond the individual nucleotide site.

GENE DUPLICATION

One of the important discoveries of the genomics era is the fact that many genes are redundant in plant and animal

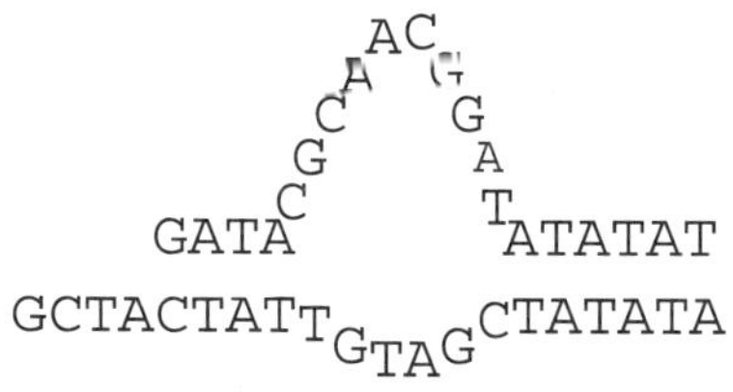

CGATGATACGCAACGGATATATATA
GCTACTATGCGTTGCCTATATATAT

GATAACATCGATATAT
CTATTGTAGCTATATA

Fig. 2 Slipped-strand mispairing between noncontiguous repeats. Excision of the shorter single-stranded loop results in the sequence shown in the second alignment. Excision of the longer loop results in the third alignment. In the second and third alignments, the changed segments are underlined.

genomes. The processes that lead to the duplication of genes range from polyploidy (the doubling of entire genomes owing to the union of unreduced gametes, discussed next), to unequal crossover, to transposon-mediated transposition of genes. There are very few estimates of rates of gene duplication, but one recent estimate suggests a duplication rate of 0.01 per gene per million years.[7]

Transposons are also implicated in the movement of genes or fragments of genes around the genome. This may occur as a consequence of reverse transcription of mRNA molecules or because a gene or gene fragment has been acquired within a mobile element. In either case it represents a kind of higher-level mutation that may provide adaptive flexibility by placing genes into new regions, possibly in association with entirely different expression signals.

POLYPLOIDY

In addition to gene duplication, polyploid formation can potentially induce or accelerate many of the mutational processes outlined in the foregoing discussion. Polyploidy is very common in flowering plants. Crop plants with polyploid origins include bananas, cotton, peanut, and wheat. In addition to extant polyploids, evidence of one or more rounds of ancient polyploid formation have been identified in species such as *Arabidopsis*[8,9] and maize.[10] Some of the ancient polyploid events appear to have occurred in lineages ancestral to the majority of flowering plants.[8,9]

Polyploids are frequently the result of hybridization between well differentiated parental species. Thus, there is a potential for recombination, gene-conversion, and concerted evolution to occur between partially homologous chromosomes from the parental species. Transposons from one parental genome may also invade the genome of the other; with gene duplication, there is the potential for loss of duplicate gene function or even the elimination of coding DNA sequence.[11]

CONCLUSION

Agriculture developed and flourished because humans learned how to exploit and maintain useful hereditary variants in a wide diversity of plant and animal species. Since the rediscovery of Mendelian genetics, a major research goal has been the acquisition of a detailed understanding of the mechanisms that generate hereditary variation. Our knowledge of the mutational process has grown enormously in recent years, revealing that mutation is a complex phenomenon with many mechanisms. One important lesson has been that mutation operates at various levels, ranging from nucleotide changes to transposition to gene duplication to the doubling of entire genomes. This rich variety of biological process has led to the wealth of mutational diversity that we seek to exploit in adapting plants and animals to human needs.

ARTICLES OF FURTHER INTEREST

Bioinformatics, p. 125
Molecular Evolution, p. 748
Polyploidy, p. 1038

REFERENCES

1. Li, W.-H. *Molecular Evolution*; Sinauer Associates, Inc.: Sunderland, MA, 1997; 181.
2. Morton, B.R.; Clegg, M.T. A chloroplast DNA mutational hotspot and gene conversion in a noncoding region near *rbcL* in the grass family (Poaceae). Curr. Genet. **1993**, *24* (4), 357–365.
3. Pritchard, J.K.; Schaeffer, S.W. Polymorphism and divergence at a *Drosophila* pseudogene locus. Genetics **1997**, *147* (1), 199–208.
4. Petrov, D.A.; Hartl, D.L. High rate of DNA loss in the *Drosophila melanogaster* and *Drosophila virilis* species groups. Mol. Biol. Evol. **1998**, *15* (3), 293–302.
5. Golenberg, E.M.; Clegg, M.T.; Durbin, M.L.; Doebley, J.; Ma, D.P. Evolution of a noncoding region of the chloroplast genome. Mol. Phylogenet. Evol. **1993**, *2* (1), 52–64.
6. Levinson, G.; Gutman, G.A. Slipped-strand mispairing: A major mechanism for DNA sequence evolution. Mol. Biol. Evol. **1987**, *4* (3), 203–221.
7. Lynch, M.; Conery, J.S. The evolutionary fate and consequences of duplicate genes. Science **2000**, *290* (5494), 1151–1155.
8. Vision, T.J.; Brown, D.G.; Tanksley, S.D. The origins of genomic duplications in Arabidopsis. Science **2000**, *290* (5499), 2114–2117.
9. Bowers, J.E.; Chapman, B.A.; Rong, J.K.; Paterson, A.H. Unravelling angiosperm genome evolution by phylogenetic analysis of chromosomal duplication events. Nature **2003**, *422* (6930), 433–438.
10. Gaut, B.S.; d' Ennequin; Peek, A.S.; Sawkins, M.C. Maize as a model for the evolution of plant nuclear genomes. Proc. Natl. Acad. Sci. U. S. A. **2000**, *97* (13), 7008–7015.
11. Wendel, J.F. Genome evolution in polyploids. Plant Mol. Biol. **2000**, *42*, 225–249

Mutualisms in Plant–Insect Interactions

W. P. Stephen
Oregon State University, Corvallis, Oregon, U.S.A.

INTRODUCTION

Mutualisms are coevolutionary processes between pairs of organisms that impart joint benefits in the form of increased growth, reproduction, and survival. The relationship between the participants has evolved because the benefits gained are greater than the costs expended.

Mutualisms range from those which are ''facultative,'' in which each member gains a benefit but is not dependent upon the other, to ''obligate,'' in which one or both members is dependent upon the other for survival.

POLLINATION

Pollination involves the transfer of pollen from the anthers of one flower to the stigma of the same or a different flower. Nowhere is mutualism and coevolution more prevalent than in plant–pollinator interactions, and among pollinators insects reign supreme.

Flowers requiring insect pollination provide an abundance of nectar and/or pollen on a platform both visually and aromatically unique and attractive. Coevolution between flowers and insects—especially the bees—proceeded at a rapid pace, resulting in increased specialization in flowers to emphasize their identity and the concomitant specialization of pollinators to excel in their efficient utilization of that specialized resource. Although there are a number of insect groups other than bees that effect pollination (beetles, wasps, moths, flies, etc.) it is the bees which are most adept and abundant.

There are over 16,000 species and subspecies of bees in over 1200 genera worldwide, of which approximately 10% are social and 80% solitary. The social bees—*Apis mellifera*, *Bombus* spp., and the stingless bees—exist as long-lived colonies that have life spans which may endure for years, rather than weeks. Although workers in each colony utilize a wide array of pollen sources, individual pollen gatherers usually confine their activities to collecting pollen from one plant source during any single foraging trip and remaining faithful to that source as long as supplies exist. This transient specificity has probably arisen because of the mutual benefits it conveys to both plant and pollinator: The plant benefits from cross-pollination by bees bearing conspecific pollens, and the bees increase their efficiency in host location and pollen/nectar collection through the associative learning.

Few solitary bee species do not exhibit some degree of specificity to a particular genus or genus-group of plants. Relationships between partners, however, can be judged only by the fidelity shown by a specific pollinator in its pollen collection, for it is this behavior which defines the reproductive continuity of its plant host. Unlike social bees, solitary bees provision their cells with nectar-moistened pollen, and none amass nectar surpluses.

Coevolution between bees and their plant hosts has resulted in varying degrees of pollen collection specificity: *polylecty* (pollen is collected from many diverse plant hosts); *oligolecty* (pollen is limited to groups of related plant hosts); and *monolecty* (pollen is from a single plant species). These terms are relative for intermediate conditions are always found in evolutionary processes, yet they provide guidance.

Adaptations to their floral hosts have occurred in many of the oligolectic bees. In the genus *Diadasia* (Anthophoridae) the hairs of the tibial scopa (areas evolved to transport pollen from floral hosts to nests) have undergone considerable modification to accommodate the various-sized pollen grains of their hosts. Most *Diadasia* that forage on Malvaceae have modestly branched scopal hairs, whereas *D. enevata*, an oligolege on composites, has fine, densely plumose hairs to transport the small pollen grains common to that family. *D. angusticeps* is the only species of this genus found on Onagraceae, and its tibial scopae comprise stiff, nonplumose hairs that carry the cobweb-like pollen masses of the host. Unrelated genera such as *Anthedonia, Diandrena*, and *Anthophora* have species which are oligolectic on *Oenothera*, and all possess sparse, simple, scopal hairs to accommodate the large *Oenothera* pollen with its interconnected viscid threads. *Proteriades* has hooked hairs on its mouthparts (maxillae) used to extract pollen grains from the small, slender corolla tubes of *Cryptantha*. Similarly, the foretarsi of *Calliopsis* (*Verbenapis*) have a series of curled bristles on their forelegs which can be inserted into the flowers of *Verbena* to scrape out the pollen and insert it on the regular pollen-carrying apparatus.[1,2]

A number of obligate plant–bee associations exist. *Duforea versatilis* has been found only on flowers of *Mimulus nanus*, *Anthemurgus passiflorae* on *Passiflora lutea*, and *Diadasia australis* appears to collect both pollen and

Encyclopedia of Plant and Crop Science
DOI: 10.1081/E-EPCS 120010483

nectar from *Opuntia*, rarely visiting other plants—even for nectar.

The foraging period of oligolectic bees is coadapted to the restricted flowering periods of their pollen hosts. Different species of the same bee genus may appear sequentially throughout the summer, utilizing entirely different pollen sources. In northern Utah, *Anthophora pacifica* appears in April and early May on *Astragalus* and stone fruits; *A. bomboides neomexicana* is on *Trifolium*, *Medicago*, and *Vicia* in June; *A. occidentalis* is on *Cirsium* in July; and *A. flexipes* is on late-summer composites in August.

Some species remain dormant for two or more years if the environmental conditions are unfavorable for either plant or insect activity. In 1961 a large nesting population of *Halictus* (*Evylaeus*) *aberrans* failed to appear when their obligate host, *Oenothera latifolia*, failed to germinate. In 1962 *Oenothera* bloomed profusely and the bee was again present. It is not uncommon for a small number of diapausing prepupae of megachilid bees (*Osmia, Megachile*) to emerge during the second rather than the first year in that stage.

Bees with matinal (collect pollen in early morning prior to or at sunrise), crepuscular (collect pollen at dusk), or nocturnal (fly and forage at night) activities have adapted to meet the pollination requirements of their floral hosts. *Xenoglossa* collect pollen from *Cucurbita* at dawn; *Peponapis* work the same flowers in the early sunlight hours; and polyleges such as *Apis* and *Agapostemon* collect pollen until the flowers close. *Halictus* (*Hemihalictus*) *lustrans* is a matinal oligolege on *Pyrrhopappus carolinianus*, the flowers of which close shortly after the sun strikes them in early morning.

Halictus (*Sphecodogastra*) are common crepuscular bees, foraging on *Oenothera* in the early and late evening. The Indo-Maylasian subgenus *Xylocopa* (*Nyctomelitta*) forages much, if not most, of the night on noctural blooming plants.

The most remarkable diversity in flower form (paralleled with the obligate pollinators) is found among the orchids.[3,4] This highly evolved plant group exhibits the most complex and precise adaptations to their pollinator partners of any plant group. Unlike most obligate cross-pollinated plants which rely largely on pollen-collecting females of various bee genera, the pollinating partners of orchids include wasps, beetles, fungus gnats, flies, moths, butterflies, and both males and females of solitary and social bees. The genome flexibility of orchids, especially in the tropical flora, has given rise to the speculation that many of the tight mutualistic associations may be more serendipitous than coevolutionary, or a combination thereof.

The single most complex mutualistic pollination relationship occurs in the fig (*Ficus* spp.) and its pollinator, the wasp, *Blastophaga* spp. (each species of fig has its own species of wasp).[4] There are three types of saclike flower clusters in the fig, each consisting of numerous flowers on a fleshy receptacle. In the spring the first small structures develop, with male flowers located near the entrance and short-styled female flowers near the base. Spring females enter the flower and lay eggs in the ovules that develop into gall-like structures containing either females or wingless males. Males chew out of their galls and enter galls containing females. Upon mating the females leave the gall and become laden with pollen as they pass the male flowers at the fruit entrance. Females then move to the second receptacle type (the true fig), enter, and attempt to—but cannot—oviposit in the long-styled flowers characteristic of this stage. In the process the females pollinate enough of the flowers to yield the mature fruit. A third type of fruit cluster (mother fig), which contains only short-style flowers into which the females oviposit, then develops on the tree. These develop into males and females, which enter diapause and emerge the following spring. In this system the wasp is not only the obligate pollinator, but also a parasite.

A unique obligate relationship exists between plants of the genus *Yucca* and *Tegeticula* moths. Adult yucca moths form the sticky pollen into a small pellet carried in modified mouthparts to the stigma of another yucca flower. Upon deposition of the pellet, the female deposits one to several eggs on the ovary. These hatch and the larvae burrow into the ovary to feed on the developing seeds. The plant tolerates the seed loss in return for pollination services from the adults. The moth consistently limits the number of deposited eggs per ovary and in so doing avoids exploiting its partner and throwing the system into imbalance. Early fruit abortion is a principal cause of yucca moth larval mortality, and it has been suggested that the quality and quantity of pollen provided by the moth are below that necessary for high fruit retention, also helping to promote evolutionary stability in the system.[5]

PROTECTION

A landmark study on the obligate mutualistic relationship between the ant, *Pseudomyrmex ferruginea*, and the bull's-horn acacia, *Acacia cornigera*, was described from Central America.[6] The ant provides protection to the plant and the plant, in turn, offers food and nesting sites. Queens of *Pseudomyrmex* bore holes through the hard covers at the thorn bases, excavate some of the pith, and establish their colonies. The colonies grow, occupying increasing numbers of thorns and often reaching populations of several thousand per tree. The ants provide defense against insect defoliators and clip off the vegetative

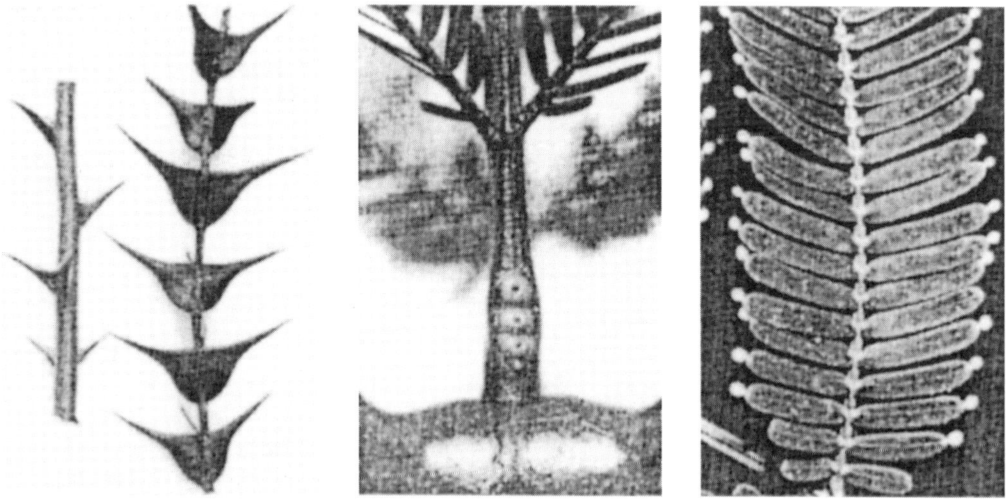

Fig. 1 Thorns of *Acacia hindsii*, which are excavated for ant nests (left); nectaries at leaf base (middle); and leaves of *Acacia collinsii* with protein-rich Beltian bodies at tips of leaflets (right). (From Ref. 6.)

shoots that invade the acacia canopy. The acacia provides food in the form of high-protein Beltian bodies at the tips of some of its pinnate leaves, as well as sugar-rich nectaries at the leaf bases (Fig. 1). Both ant and plant have evolved traits to cement this association; Ants forage 24 hours a day rather than diurnally, and the plant retains its leaves even through the arid portions of the season. Neither ant nor plant can survive without the other in this ant–acacia relationship, although many other facultative ant–acacia associations occur.

Clear evidence of the antiherbivore defense in ant–plant mutualisms is provided in studies on three species of *Macaranga* which have either obligate or facultative relationships with a common ant. Plants, or parts of plants, that were kept ant-free for about one year lost 70–80% of their total leaf area, confirming the requirement of their ant partners for survival.[7]

There is normally a prompt response to the site of herbivore damage by the ant partner. In the obligate *Azteca* ant–*Cecropia* tree relationship there was a fivefold increase in the number of ants on damaged versus undamaged leaves, with activity peaking in 8–12 minutes after damage was inflicted.[8] Rapid recruitment was correlated with the number of ants patrolling the leaves, and the number of ants responding doubled when artificial damage was inflicted on the leaves. Damaged leaves on one tree prompted ant recruitment to leaves in adjacent *Cecropia*, suggesting interplant communication.

There is only one reported example of ants providing protection for an economically important plant. Ants live in close association with the cashew nut tree, *Anacardium occidentale*, which has nectaries on each leaf, the flower stalks, the base of each flower, and on the young nuts. Local ants living in the area ascend the tree to collect nectar, and in the process of foraging they will capture insect herbivore as they are encountered. Commerical crops of cashew nuts in Malaysia and Sri Lanka, solely protected by local ants, have been produced for as long as 20 years without the use of any pesticide.[9]

Most ants with obligate mutualistic associations require in situ nesting sites on their partners. These may be structures such as the greatly enlarged thorns of the bull's horn acacia, or more commonly domatia (or caulinary domatia), expanded hollow stems. It is suggested that ants originally may have colonized later in plant development, when the diameters of the twigs were sufficient to accommodate small colonies. The advantages conferred by the presence of ants resulted in selective pressure for an earlier expression of the domatialike trait.

DISPERSAL

Many ant species carry plant materials, especially seeds, over relatively long distances from the plant source to their nests. Although many of these seeds may ultimately germinate, the habits of these voracious generalist foragers do not fall within the framework of coevolution. However, in South Africa an intimate association has developed between ants and plants in the family Proteaceae that often results in seed dispersal. Many species in this family produce seed with a fleshy, edible appendage termed an elaiosome. Seeds are gathered by the ants and transported to their nests, where the elaiosomes are eaten. The inedible seed is either stored in the underground nest or discarded on the soil surface, where it may ultimately germinate and establish. The chaparral-like habitat of the fynbos is frequently swept by fires, after which only the seed in the underground nests will germinate.

ARTICLES OF FURTHER INTEREST

Breeding Self-Pollinated Crops and Marker-Assisted Selection, p. 202
Coevolution of Insects and Plants, p. 289
Floral Scent, p. 456
In Vitro Pollination and Fertilization, p. 587
Insect–Plant Interactions, p. 609
Pollen–Stigma Interactions, p. 1035

REFERENCES

1. Stephen, W.P.; Bohart, G.E.; Torchio, P.F. *The Biology and External Morphology of Bees with a Synopsis of the Genera of Northwestern America*; Agric. Exper. Sta., Oregon State Univ.: Corvallis, OR, 1969; 140 pp.
2. Michener, C.D. *The Bees of the World*; Johns Hopkins Univ. Press: Baltimore, 2000.
3. van der Pijl, L.; Dodson, C.H. *Orchid Flowers/Their Pollination and Evolution*; Univ. Miami Press: Miami, FL, 1966.
4. Proctor, M.; Yeo, P.; Lack, A. *The Natural History of Pollination*; Timber Press: Portland, OR, 1996.
5. Huth, C.J.; Pellmyr, O. Pollen-mediated selective abortion in yuccas and its consequences for the plant–pollinator mutualism. Ecology **2000**, *81* (4), 1100–1107.
6. Janzen, D.J. Interaction of the bull's-horn acacia (*Acacia cornigera* L.) with an ant inhabitant (*Pseudomyrmex ferruginea* F. Smith) in eastern Mexico. Univ. Kansas Sci. Bull. **1967**, *67*, 315–558.
7. Heil, M.; Fiala, B.; Maschwitz, U.; Linsenmair, K.E. On benefits of indirect defence: Short- and long-term studies on antiherbivore protection via mutualistic ants. Oecologia **2001**, *126*, 395–403.
8. Agrawal, A.A. Leaf damage and associated cues induced aggressive ant recruitment in a neotropical ant-plant. Ecology **1998**, *79* (6), 2100–2112.
9. Rickson, F.R.; Rickson, M.M. The cashewnut, *Anacardium occidentale* (Anacardiaceae), and its perennial association with ants: Extrafloral nectary location and the potential for ant defense. Am. J. Bot. **1998**, *85*, 835–849.

Mycorrhizal Evolution

M

David M. Wilkinson
Liverpool John Moores University, Liverpool, U.K.

INTRODUCTION

The predominately mutualistic relationships between mycorrhizal fungi and vascular plants are very common; indeed, textbooks often cite figures suggesting that 80–90% of all vascular plant species are mycorrhizal. While such figures are approximate and based on extrapolation from a limited number of well studied examples, it is clear that mycorrhizae are widely distributed in modern plants, a fact that hints at a long evolutionary history for the relationship.

FOSSIL EVIDENCE

The fungal fossil record is relatively poor. While some spores are readily preserved, it requires unusual conditions to preserve hyphae (which are usually lacking in morphological features to aid identification even when preserved). However, there is now fossil evidence for arbuscular mycorrhizae contemporaneous with the origin of land plants. Important early fossils come from the Rhynine chert of the Lower Devonian age (approximately 400 million years b.p.), which contains well preserved terrestrial plant fossils. An example is *Aglaophyton major*, an enigmatic plant with features found in both vascular plants and bryophytes. Arbuscules (the sites of exchange of material between plant and fungus) are preserved in the fossilized tissues of this plant.[1]

Recently fossil evidence of fungal hyphae and spores similar to arbuscular mycorrhizae have been found from the mid-Ordovician.[2] These date from between 460 and 455 million years b.p. and were deposited in shallow marine conditions, presumably having washed in from nearby land. The Ordovician apparently occurred before the evolution of vascular plants and the terrestrial vegetation is poorly known, although it appears to have been dominated by plants similar to modern mosses and liverworts. However, the extent of terrestrial vegetation at the time is still controversial.[3] The fossilized fungal remains strongly resemble the modern genus *Glomus*. They suggest that such fungi predate the first vascular plants and may have been either free living or formed mycorrhizal-like relationships with the bryophytes. This second suggestion is supported by the facts that modern arbuscular mycorrhizae are known to form such relationships and that all modern Glomales fungi form mycorrhizae.[2] Other types of mycorrhizae appear to have developed much later,[4] although there is currently a shortage of fossil evidence.

IMPORTANCE FOR THE EARTH SYSTEM

Plants attempting to colonize the land would have faced the twin problems of desiccation and low nutrient substrates. Several authors have suggested that mycorrhizal relationships would have been crucial in surmounting these problems,[4,5] a hypothesis that has acquired the status of textbook orthodoxy. Particular emphasis has been placed on the role of mycorrhizae in acquiring phosphorus.[4]

If the mycorrhizal symbiosis was crucial in the development of widespread terrestrial vegetation, then it indirectly led to major global changes. Terrestrial plants caused the development of soils that retain water and, by maintaining a high surface area of damp mineral grains, greatly enhance chemical weathering. These reactions reduce atmospheric carbon dioxide and thus have important climatic implications.[3,6] As well as contributing to this process indirectly by supporting terrestrial vegetation, mycorrhizae directly contribute to weathering rates by producing organic acids and other chemicals that break down mineral material.[6]

EVOLUTIONARY MECHANISMS

The classic evolutionary problem with any mutualism is the question, why doesn't one of the partners ''cheat''? That is, why doesn't one of them become parasitic, extracting resources from its partner without reciprocating by providing resources in return? There is a well established body of theory that can explain why mutualisms should be evolutionarily stable given certain assumptions such as vertical transmission of symbionts. However, these assumptions are not met by any of the various types of mycorrhizal symbioses. For example mycorrhizal fungi

Encyclopedia of Plant and Crop Science
DOI: 10.1081/E-EPCS 120010473

are always acquired by horizontal transmission (from the environment) rather than by vertical transmission (directly from the parent, e.g., in the seed).[7]

There are a number of mechanisms that could lead to mutualisms based on horizontal transmission being stable;[7] one possibility is retaliation against cheating symbionts (e.g., by expulsion of fungi from the roots). Another is that local dispersal of plants (many seeds are only dispersed small distances) can lead to pseudovertical transmission. Here the plant acquires fungal partners from the soil that are genetically identical to those that infected the parent and so are equivalent to vertical transmission from an evolutionary perspective. It is also possible to view the acquisition of fungal symbionts by a plant as a biological market where the plant forms relationships with fungi that best suit current local soil conditions.[8] The potential importance of this latter mechanism is increased by the recent demonstration of very high levels of genetic diversity in arbuscular mycorrhizae at small spatial scales in some soils;[9] this could mean that the plant has a wide choice of potential fungal partners.

It is becoming increasingly likely that the evolution of mycorrhizae is a dynamic process with repeated gains and losses of the mycorrhizal condition over geological time. This position is supported by a recent molecular phylogeny[10] of ectomycorrhizal fungi. The evolutionary history of this phylum was reconstructed using 161 species (29% were mycorrhizal, the rest free living). This produced a large number of equally parsimonious trees; however, all of these showed repeated transitions between mycorrhizal and free living forms. This analysis strongly suggests that the relationship has arisen many times.

An additional complication in considering mycorrhizal evolution is the presence of genetically distinct nuclei coexisting in the spores of individual arbuscular mycorrhizal fungi.[11] This raises intriguing theoretical questions about what prevents competition between these different genomes. Consider the inheritance of organelles in eucaryotes: These are nearly always inherited from just one parent. The conventional explanation for this is that it removes the potential for conflicts of interest, which could arise if two genetically distinct lines of organelles were found within the same cell. Clearly, arbuscular mycorrhizae don't use this strategy to eliminate the possibility of genetic competition. How they deal with this problem is currently an open question.

MYCORRHIZAL NETWORKS

It has become common for people to suggest that ecologically significant amounts of carbon may be moved from plant to plant through mycorrhizal networks, although the extent to which this happens is still controversial.[12] If such behavior is real this clearly raises a number of interesting evolutionary questions. Why should a fungus give away important resources to the plants and how can it be evolutionarily stable for a plant to give away resources to competitor plants? These questions have received little theoretical attention; however, Wilkinson[13] suggests some tentative answers. It may be to the advantage of the fungi to "invest" surplus carbon in plants that may become a future carbon source for the fungi. It is also possible to see a kin selection advantage in plants passing resources through mycorrhizal networks to nearby plants. Because many seeds only disperse short distances, many of these plants could be related to the donor. However, more complex explanations will be required if the mycorrhizal movement of carbon between different plant species is found to be common.[13]

CONCLUSION

The fossil record suggests that some forms of mycorrhizae have been important since the first terrestrial plants evolved. Indeed, they may have been crucial in this process, which laid the foundation for the complex terrestrial ecosystems we see today and had important implications for weathering rates and through these, for the Earth's climate. One way of viewing the mycorrhizal symbiosis[8] is that it allows plants to adjust their root systems (which are often a complex of root and hyphae) to temporal and spatial changes in the soil. This allows them to adapt to local conditions with much more flexibility than would be possible by relying on their roots alone.

ARTICLES OF FURTHER INTEREST

Mycorrhizal Symbioses, p. 770
Symbioses with Rhizobia and Mycorrhizal Fungi: Microbe/Plant Interactions and Signal Exchange, p. 1213

REFERENCES

1. Remy, W.; Taylor, T.N.; Kerp, H. Four hundred-million-year-old vesicular arbuscular mycorrhizae. Proc. Natl. Acad. Sci. U. S. A. **1994**, *91*, 11841–11843.
2. Redecker, D.; Kodner, R.; Graham, L.E. Glomalean fungi from the Ordovician. Science **2000**, *289*, 1920–1921.

3. Retallack, G.J. Ordovician life on land and early Palaeozoic Global change. Paleontol. Soc. Pap. **2000**, *6*, 21–45.
4. Gryndler, M. The ecological role of mycorrhizal symbiosis and the origin of the land plants. Ces. Mykol. **1992**, *46*, 93–98.
5. Pirozynski, K.A.; Malloch, D.W. The origin of land plants: A matter of mycotrophism. Biosystems **1975**, *6*, 153–164.
6. Lenton, T.M. The role of land plants, phosphorus weathering and fire in the rise and regulation of atmospheric oxygen. Global Change Biol. **2001**, *7*, 613–629.
7. Wilkinson, D.M.; Sherratt, T.N. Horizontally acquired mutualisms, an unsolved problem in ecology? Oikos **2001**, *92*, 377–384.
8. Wilkinson, D.M.; Dickinson, N.M. Metal resistance in trees: The role of mycorrhizae. Oikos **1995**, *72*, 298–300.
9. Vandenkoornhuyse, P.; Leyval, C.; Bonnin, I. High genetic diversity in arbuscular mycorrhizal fungi: Evidence for recombination events. Heredity **2001**, *87*, 243–253.
10. Hibbett, D.S.; Luz-Beatriz, G.; Donoghue, M.J. Evolutionary instability of ectomycorrhizal symbiosis in basidiomycetes. Nature **2000**, *407*, 506–508.
11. Kuhn, G.; Hijri, M.; Sanders, I.R. Evidence for the evolution of multiple genomes in arbuscular mycorrhizal fungi. Nature **2001**, *414*, 745–748.
12. Robinson, D.; Fitter, A. The magnitude and control of carbon transfer between plants linked by a common mycorrhizal network. J. Exp. Biol. **1999**, *50*, 9–13.
13. Wilkinson, D.M. The evolutionary ecology of mycorrhizal networks. Oikos **1998**, *82*, 407–410.

Mycorrhizal Symbioses

Roger T. Koide
Pennsylvania State University, University Park, Pennsylvania, U.S.A.

INTRODUCTION

Mycorrhizas (or mycorrhizae) are symbioses involving plants that become colonized by mycorrhizal fungi. There are several recognized types of mycorrhizas including arbuscular, ecto-, ectendo-, arbutoid, monotropoid, ericoid, and orchid.[1] Each type consists of a unique combination of plant and fungal taxa. As has been well documented for the arbuscular, ecto-, and ericoid mycorrhizas, the symbioses may be mutually beneficial. For these mycorrhizas, the plant's nutrient status may be improved as the fungus absorbs nutrients from the soil and transfers them to the plant. The fungus may also derive a significant amount of carbohydrate from the photosynthetic plant. From a human economic standpoint, the arbuscular and ectomycorrhiza are probably the most important mycorrhizal symbioses because these involve the majority of important food, fiber, and timber plant species, as well as many of the important edible fungi. These symbioses are also very important ecologically because most of the earth's land surface is dominated by vegetation that is largely arbuscular mycorrhizal or ectomycorrhizal.

ARBUSCULAR MYCORRHIZA

Arbuscular mycorrhizas are formed when plants are colonized by members of the fungal phylum Glomeromycota[2], of which there are currently approximately 150 described species. The plants forming arbuscular mycorrhizas include many mosses, ferns, gymnosperms, and angiosperms. The arbuscular mycorrhiza is named for the arbuscule, a highly branched fungal organ usually produced within cortical cells of colonized plant roots. This is the organ across which the fungus absorbs carbohydrate from the plant, and the plant absorbs various nutrients from the fungus including Cu and Zn, possibly N, and especially P. In some cases, P uptake by roots may be downregulated as a consequence of the symbiosis, and the fungus may be the most important organ of phosphate absorption. In P-deficient soils, arbuscular mycorrhizal colonization can significantly improve plant growth and yield.[3] The main reason for this is that phosphate usually occurs in low concentrations in the soil and diffuses slowly. The hyphae of the mycorrhizal fungi extending from colonized roots (Fig. 1) compensate for this by exploring a greater volume of soil than can the roots themselves and by presenting a greater surface area for phosphate uptake. Some, but certainly not all, arbuscular mycorrhizal fungi may produce other structures within the roots of colonized plants including vesicles, which are capable of storing high concentrations of lipid. Thus, some of the arbuscular mycorrhizal fungi are vesicular-arbuscular, or VA.

Common agricultural practices may strongly influence the efficacy of mycorrhizal colonization. For example, vigorous disturbance of the soil such as by tilling can disrupt the fragile fungal mycelium, and this can lead to a significant reduction in phosphorus uptake.[4] Fallow periods can also lead to a reduction in viable mycorrhizal hyphae in the soil. Thus, the planting of mycorrhizal cover crops during normally fallow periods can lead to increased colonization of subsequent crops and to increased yield.[5] Variation in root anatomy and root system architecture result in variation in the extent to which plants benefit from mycorrhizal colonization.[3] Plant species that possess few root hairs, for example, are expected to benefit more from mycorrhizal colonization than do those whose roots are densely covered with long root hairs. Moreover, variation in the way different mycorrhizal fungi explore the soil and thus capture P may lead to variation in their effects on plant growth.[3] In container plant production utilizing soilless media, the effectiveness of arbuscular mycorrhizal fungi may be much lower than in soil. In such production systems fertilizer may supply phosphate in high concentrations and the applied phosphate is often freely available because soilless media typically do not adsorb phosphate as do many natural soils.[6]

ECTOMYCORRHIZA

Unlike the case with arbuscular mycorrhiza, the ability to associate with roots to form an ectomycorrhiza has developed independently in more than one fungal lineage. The several thousands of fungal species that form ectomycorrhizas include members of the Basidiomycota, Ascomycota, and Zygomycota.[7] The plants forming

Encyclopedia of Plant and Crop Science
DOI: 10.1081/E-EPCS 120010509

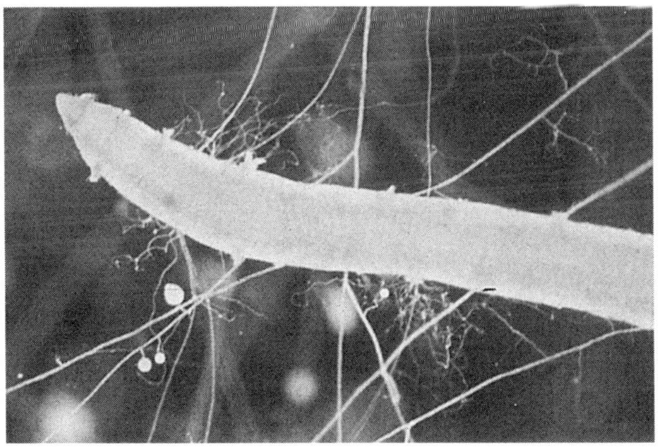

Fig. 1 In vitro carrot/*Glomus intraradices* arbuscular mycorrhiza. (Photograph by Roger Koide.) (*View this art in color at www.dekker.com.*)

ectomycorrhizas are mostly angiosperm and gymnosperm shrubs and trees including members of the economically important and ecologically dominant Pinaceae, Fagaceae, and Myrtaceae. No fungal structures are produced within the root cells, but the fungus grows between the outer cortical and epidermal cells, and may form a dense mantle external to the epidermis. As it does so the root may cease to grow, giving rise to characteristic short mycorrhizal roots (Fig. 2). Hyphae extending from the surface of the mantle into the soil provide surface area for nutrient absorption. Thus, just as in the arbuscular mycorrhiza, this additional surface area can significantly increase the P status of the plant. N absorption may also be enhanced. The fungi also collectively possess a range of hydrolytic abilities that allows them to access some nutrient sources unavailable to uncolonized roots, including the ability to utilize some proteins, amino acids, amino sugars, and organic phosphates.[1] Thus, the ectomycorrhizal fungi may be of particular significance in high latitude or high altitude biomes in which litter decomposition rates are relatively low and thus in which large fractions of the nutrients are bound up in organic forms. Ectomycorrhizal colonization may also be particularly important for developing seedlings that typically have poorly developed root systems. Some fungal species are capable of transporting significant amounts of water to the plant,[8] the lack of which is responsible for a large fraction of seedling mortality.

It is widely recognized that a single tree in a natural ecosystem will be colonized by many ectomycorrhizal fungal species. A single-species plantation may support dozens of ectomycorrhizal fungal species. The causes of this diversity are only now being investigated, but the consequences to ecosystem function are already obvious. Variation among ectomycorrhizal fungi in their physiologies and morphologies give rise to variation in their functions. Whereas some are better at transporting N to the plant, others may be better at transporting P. Still others may be superior at water transport. Thus, nature may select for plants that support a variety of ectomycorrhizal fungal species. Most ectomycorrhizal fungal species produce macroscopic fruiting bodies such as mushrooms, puffballs, and truffles. Many are of great economic importance because of their edibility. These fruiting structures are also of great ecological importance because many serve as major food sources for a number of animal species.

CONCLUSION

Currently the arbuscular mycorrhizal fungi cannot be grown in pure culture. They are cultured on intact plants or on root organ cultures. Consequently, it is very costly to produce arbuscular mycorrhizal fungi for large-scale inoculation such as in field trials. Inoculation may only prove to be economically viable for crops of high value. For other crops, efforts to manipulate field practices to stimulate indigenous mycorrhizal fungi may be more practical. In temperate climates, for example, minimizing fallow periods and reducing soil disturbance are both effective ways to promote colonization by arbuscular mycorrhizal fungi of subsequent crops. In soilless media, such as in the floriculture industry, arbuscular mycorrhizal fungi may not prove to enhance P absorption, but it may influence other economically important traits such as disease resistance.[9]

Nurseries of potentially ectomycorrhizal tree seedlings grown for reforestation or ornamental use may lack the

Fig. 2 Ectomycorrhizas of *Pinus resinosa*. (Photograph by Roger Koide.) (*View this art in color at www.dekker.com.*)

most effective fungi following outplanting. A number of studies have shown that inoculation with appropriate ectomycorrhizal fungi improves survival and growth of tree seedlings.[10] Thus, the view that inoculation of trees with ectomycorrhizal fungi can be helpful and economically viable is widely held. Indeed, there has been some effort to devise methods to artificially inoculate nursery beds and small seedling pots. Often spores or bits of sporocarp material are used as sources of inoculum. In some cases it is possible to grow ectomycorrhizal fungi in pure culture, allowing one to use vegetative mycelium as an inoculum source. Thus far, however, standard inoculum types, and the quantities and qualities of inocula have not been established for many specific tree species. The appropriate species of fungus must also be carefully selected because there is a wide range of specificity among ectomycorrhizal fungi for plant species.[10] Other production methods including rate, form, and placement of fertilizer need to be established to encourage ectomycorrhizal colonization. Moreover, the economic benefit from inoculation needs to be established in more large-scale trials. Until then, inoculation may not become the standard silvicultural practice it probably should be.

ARTICLES OF FURTHER INTEREST

Mycorrhizal Evolution, p. 767
Phosphorus, p. 872
Rhizosphere: An Overview, p. 1084
Rhizosphere: Microbial Populations, p. 1090
Rhizosphere: Nutrient Movement and Availability, p. 1094
Symbiosis with Rhizobia and Mycorrhizal Fungi: Microbe/Plant Interactions and Signal Exchange, p. 1213

REFERENCES

1. Smith, S.E.; Read, D.J. *Mycorrhizal Symbiosis,* Second Ed.; Academic Press: San Diego, 1997; 1–605.
2. Schüssler, A.; Schwarzott, D.; Walker, C. A new fungal phylum, the *Glomeromycota*: Phylogeny and evolution. Mycol. Res. **2001**, *105*, 1413–1421.
3. Koide, R.T. Nutrient supply, nutrient demand and plant response to mycorrhizal infection. New Phytol. **1991**, *117*, 365–386.
4. Evans, D.G.; Miller, M.H. Vesicular-arbuscular mycorrhizas and the soil-disturbance-induced reduction of nutrient absorption in maize: I. Causal relations. New Phytol. **1988**, *110*, 67–74.
5. Boswell, E.P.; Koide, R.T.; Shumway, D.L.; Addy, H.D. Winter wheat cover cropping, VA mycorrhizal fungi and maize growth and yield. Agric. Ecosyst. Environ. **1998**, *67*, 55–65.
6. Biermann, B.; Linderman, R.G. Effect of container plant growth medium and fertilizer phosphorus on establishment and host growth responses to vesicular-arbuscular mycorrhizae. J. Am. Soc. Hortic. Sci. **1983**, *108*, 962–971.
7. Bruns, T.D. Thoughts on the processes that maintain local species diversity of ectomycorrhizal fungi. Plant Soil **1995**, *170*, 63–73.
8. Brownlee, C.; Duddridge, J.A.; Malibari, A.; Reed, D.J. The structure and function of mycelial systems of ectomycorrhizal roots with special reference to their role in assimilate and water transport. Plant Soil **1983**, *71*, 433–443.
9. Hooker, J.E.; Jaizme-Vega, M.; Atkinson, D. Biocontrol of Plant Pathogens Using Arbuscular Mycorrhizal Fungi. In *Impact of Arbuscular Mycorrhizas on Sustainable Agriculture and Natural Ecosystems*; Gianinazzi, S., Schüeppp, H., Eds.; Birkhäuser Verlag: Basel, 1994; 191–200.
10. Grove, T.S.; Le Tacon, F. Mycorrhiza in Plantation Forestry. In *Mycorrhiza Synthesis*; Tommerup, I.C., Ed.; Academic Press: London, 1993; 191–227.

Mycotoxins Produced by Plant Pathogenic Fungi

K. Thirumala-Devi
Nancy Keller
University of Wisconsin, Madison, Wisconsin, U.S.A.

M

INTRODUCTION

Mycotoxins are natural, chemically diverse, fungal products that are defined by their harmful affects on humans and animals. Their existence came into the limelight in 1960 when more than 100,000 turkeys died in the UK as a result of consuming aflatoxin-contaminated peanut meal. Mycotoxins are introduced into the diet through consumption of contaminated produce and are toxic to human beings and livestock by exerting specific effects on a given organ system. Detrimental properties exhibited by mycotoxins include carcinogenesis, mutagenesis, teratogenesis, oestrogenesis and/or immunosuppression. Data suggest that exposure to two or more mycotoxins or an interaction between a mycotoxin and a pathogen exacerbates disease symptoms.

The majority of economically important mycotoxins are produced by *Aspergillus*, *Penicillium*, and *Fusarium* species, although other genera such as *Claviceps* also produce potent toxins. These genera are ubiquitous, and according to the Food and Agriculture Organization of the United Nations (FAO) over 25% of the agricultural commodities worldwide are significantly contaminated by mycotoxins. Although they are a postharvest problem, these fungi also cause significant preharvest contamination. In particular, contamination occurs in countries where climate and poor storage conditions favor the growth of mycotoxin-producing fungi. This article gives an account of a selected group of economically important mycotoxins and refers the reader to recent reviews and the Council for Agricultural Science and Technology (CAST) report published in 2003 for greater detail.

MYCOTOXINS PRODUCED BY *Aspergillus* AND *Penicillium*

Aspergillus and *Penicillium* species are closely related and can produce the same mycotoxins. These molds grow on a wide variety of food- and feedstuffs worldwide, especially cereals and oilseeds (Table 1, Fig. 1).

Aflatoxins

Aflatoxins are a group of polyketides produced by *Aspergillus flavus* and *A. parasiticus* on a variety of crops. The highest levels of contamination have been recorded in peanuts, maize, Brazil nuts, pistachio nuts, cottonseed, and copra. Lower levels are recorded in almonds, pecans, walnuts, raisins, spices, and figs. The four aflatoxins (AF)—AFB1 > AFG1 > AFB2 > AFG2 (ranking based on toxicity)—are the most common, and of these AFB1 and AFG1 are most frequently found in foods. AFM1 (derived from AFB1) contaminates dairy products. Aflatoxins target the liver and immune system, causing hepatocellular carcinoma especially in humans who are infected with the hepatitis B virus.[6] Aflatoxins can also cause several pathological effects on various other organs and tissues.

Sterigmatocystin is the penultimate precursor of AFB1 and is produced most commonly by *A. nidulans* and *A. versicolor*.[7] It contaminates cereal crops and milk products, including cheese.[8] Sterigmatocystin is hepatotoxic and carcinogenic;[9] nevertheless its toxicity is one-tenth that of AFB1.

Ochratoxins

Ochratoxins are a group of dihydroisocoumarins produced by *A. ochraceous* and *Penicillium verruculosum* on a diverse group of crops including barley, oats, rye, maize, wheat, coffee, nuts, olives, and grapes. Ochratoxins bind tightly to serum albumin and are carried in animal tissues and body fluids. Consequently, they can be detected in sausage derived from contaminated meats. Ochratoxin A, the most toxic member of this group of mycotoxins, has been shown to be nephrotoxic, hepatotoxic, teratogenic, carcinogenic, mutagenic, and immmunosuppressive.[10] Of greatest concern for human health is its implicated role in an irreversible and fatal kidney disease referred to as Balkan Endemic Nephropathy. Ochratoxin A has also been shown to contaminate human milk, and thus can cause kidney disorders in breast-fed infants.

Encyclopedia of Plant and Crop Science
DOI: 10.1081/E-EPCS 120019943

Table 1 Selected mycotoxins produced by plant pathogenic fungi, and their biological effects

Fungi	Commodities	Mycotoxins	Biological effects
Aspergillus/Penicillium			
A. flavus	peanuts, maize,	aflatoxins	hepatotoxic
A. parasiticus	cottonseed, tree nuts, milk		carcinogenic
			mutagenic
			teratogenic
A. flavus	maize, wheat, barley	sterigmatocystin	hepatotoxic
A. parasiticus			carcinogenic
A. nidulans			
A. ochraceous	barley, maize, wheat, sorghum,	ochratoxin	nephrotoxic
P. verruculosum	coffee beans, milk		teratogenic
A. flavus	maize, peanuts	cyclopiazonic acid	nephrotoxic
P. cyclopium			cardiovascular lesions
Fusarium			
F. verticillioides	maize, sorghum	fumonisins	neurotoxic
F. proloferatum			hepatotoxic
F. sporotrichioides	barley, maize	trichothecenes	apoptosis
		(T2, DON)	neurotoxic
F. graminearum	maize, cereals	zearalenone	genitotoxic
Claviceps			
C. purpurea	rye	ergot alkaloids	neurotrophic

Cyclopiazonic Acid

Cyclopiazonic acid is an indole tetramic acid produced by several *Penicillium* species, including *P. cyclopium*, *P. aurantiogriseum*, *P. crustosum*, *P. griseofulvum*, and *P. camemberti*, as well as the *Aspergillus* species *A. flavus*, *A. tamari*, and *A. versicolor* on maize and peanuts.[11] Cyclopiazonic acid often co-occurs with aflatoxins and causes necrosis of the liver, spleen, pancreas, kidney, and salivary glands.

MYCOTOXINS PRODUCED BY *Fusarium*

Fusarium species are pathogens on a wide variety of crop plants throughout the world. They cause mainly stem and root rots and frequently infect the grain of cereal crops. Several species of *Fusarium* produce mycotoxins on a range of cereal crops, which include wheat, maize, rice, barley, and oats.

Fumonisins

Fumonisins are pentahydroxyicosanes produced by *F. verticillioides* and *F. proliferatum* on maize and sorghum. Fumonisin B1, the most toxic fumonisin, promotes cancer and causes equine leukoencephalomalacia[12] and porcine pulmonary edema.[13] A high incidence of human esophageal cancer in South Africa and China has been correlated with the presence of fumonisins in foods. Fumonisins exert their toxic effects by altering sphingolipid metabolism. In laboratory tests fumonisins induced cancer of the kidney and liver in rodents.

Trichothecenes

T2 toxin and deoxynivalenol (DON) belong to a large group of structurally related sesquiterpenes known as trichothecenes.

T2 toxin is produced by *F. sporotrichioides* on barley, maize, oats, and wheat. It can severely damage the entire digestive tract and cause rapid death due to internal hemorrhage. It is also implicated in pulmonary hemosiderosis. Damage caused by T-2 toxin is irreversible.

DON is probably the most widely occurring *Fusarium* mycotoxin, contaminating a variety of cereals, especially maize and wheat. It is most frequently produced by *F. graminearum*. The outbreak of emetic syndromes in livestock due to the presence of DON in feeds has resulted in coining the name ''vomitoxin.'' It is implicated in outbreaks of acute human mycotoxicosis in India, China, and rural Japan. However, DON is much less toxic than T2 toxin.

Zearalenone

Zearalenone is a β-resorcyclic lactone produced by *F. graminearum* on maize, wheat, and other cereal grains. It

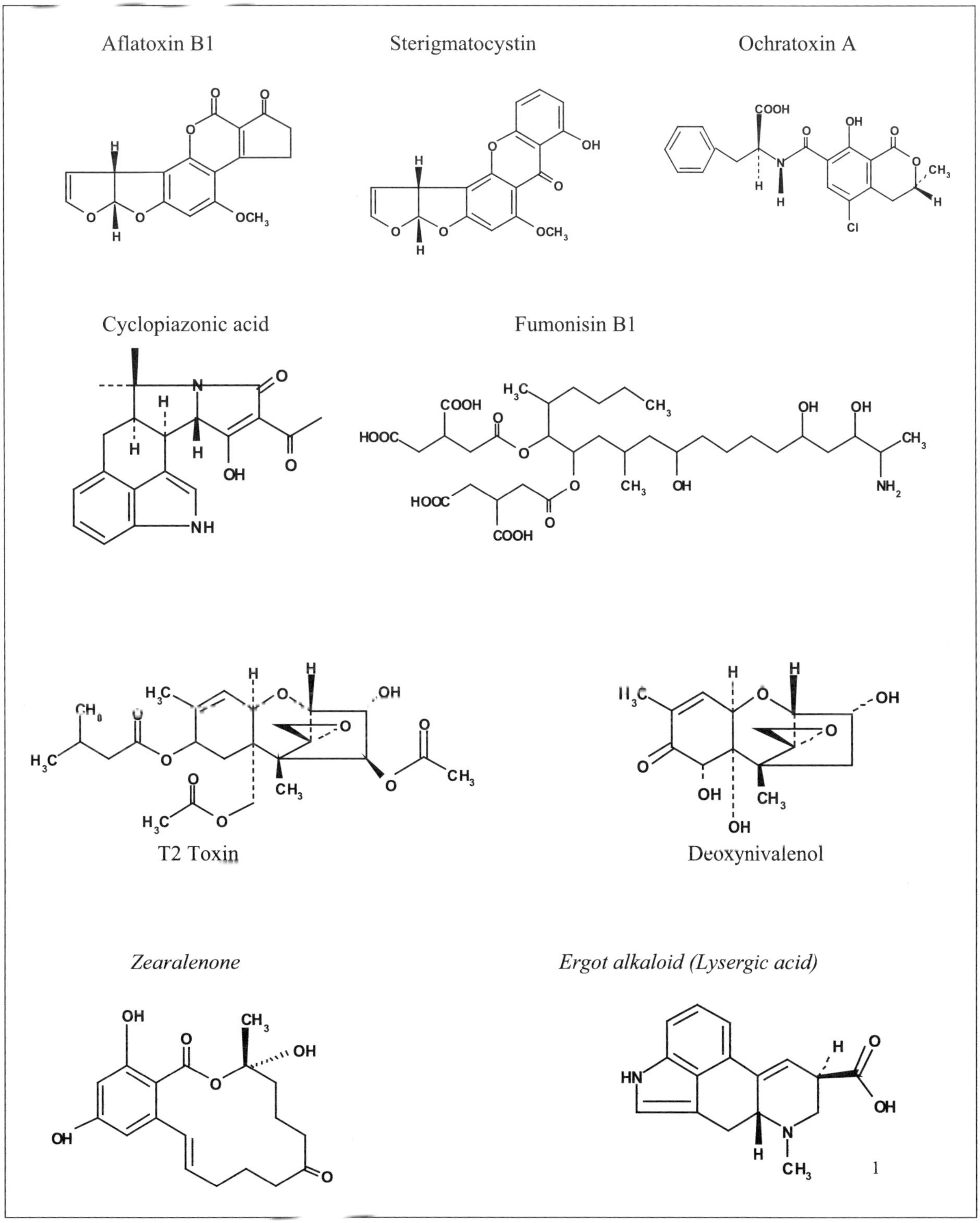

Fig. 1 Chemical structures of selected mycotoxins.

is produced along with DON and has been implicated in the outbreaks of acute human mycotoxicosis. Exposure to zearalenone-contaminated maize elicits estrogenic effects in the mammalian reproductive system and has specifically been associated with hyperoestrogenism in livestock (especially pigs), characterized by vulvar and mammary swelling, infertility, and abortions.

MYCOTOXINS PRODUCED BY *Claviceps purpurea*

Ergot is the common name of the disease caused by *Claviceps purpurea*, a fungus that occurs on rye and other small grains, where it replaces the kernel with a fungal body called a sclerotium.[14] The sclerotium contains ergot alkaloids that when ingested cause toxic reactions, including convulsions, hallucinations, and dry gangrene, that lead to loss of limbs. Epidemics of ergotism have been recorded throughout history. This often fatal condition has since been traced to the presence of alkaloids, derivatives of the four-ring-structure ergoline, in the sclerotium.

GENETIC AND MOLECULAR BIOLOGY OF MYCOTOXIN BIOSYNTHESIS

Aflatoxin and sterigmatocystin biosynthesis is the most thoroughly characterized of any mycotoxin. Molecular genetic analyses of three species of *Aspergillus* have resulted in the cloning of an aflatoxin/sterigmatocystin gene cluster containing all of the enzymatic genes and at least two regulatory genes required for aflatoxin/sterigmatocystin biosynthesis.[15] In general, it appears that genes for fungal secondary metabolism are found in clusters; the trichothecene,[16–18] fumonisin,[19] and ergot alkaloid[20] gene clusters have also been described. Furthermore, additional studies of aflatoxin/sterigmatocystin biosynthesis have shown that mycotoxin formation is genetically linked to asexual spore production through a G protein/cAMP/protein kinase A signaling pathway.[15] Gene cluster expression is also subject to regulation by environmental parameters including nitrogen and carbon source and external pH mediated through the global regulators AreA, CreA, and PacC, respectively.[21]

CONTROL

Regulation of the moisture content of crops is the most effective postharvest management control of mycotoxin formation and is successfully used in developed, but not developing, countries. Preharvest contamination remains a problem worldwide, although some recent success has been documented by using atoxigenic *Aspergillus flavus* strains as biocontrol agents.[22,23] Crop plants which can suppress mycotoxin biosynthesis or resist fungal entry may offer a long-term viable option for reducing mycotoxin contamination. Several research groups are focusing both on traditional plant breeding and on development of engineered plants to reduce mycotoxin contamination.

Once contamination occurs it is not economically feasible to reduce the mycotoxin content of foods. However, potent adsorbents such as activated clays have successfully been used by the feed industry in developed countries to minimize the influence of aflatoxins on livestock.

CONCLUSION

Mycotoxin contamination of foods and feeds has assumed economic importance globally because of its influence on the health of human beings and livestock. This article has given a brief account of the chemical nature of mycotoxins; the deleterious effects caused by economically important mycotoxins; progress in the molecular genetics of mycotoxin biosynthesis; and existing control strategies. Clearly, given the importance of this disease problem, additional controls are needed. Current research focused on identification and characterization of fungal and/or plant genes and gene products important in the production of mycotoxins should yield insight toward the development of additional control strategies.

ACKNOWLEDGMENT

We thank Mr. Justin Gruby and Dr. D. V. R. Reddy for valuable comments.

ARTICLES OF FURTHER INTEREST

Management of Diseases in Seed Crops, p. 675
Mechanisms of Infection: Imperfect Fungi, p. 694
Metabolism, Secondary: Engineering Pathways of, p. 720
New Secondary Metabolites: Potential Evolution, p. 818
Secondary Metabolites as Phytomedicines, p. 1120
Seed Borne Pathogens, p. 1126

REFERENCES

1. Wild, C.P.; Hall, A.J. Hepatitis B virus and liver cancer: Unanswered questions. Cancer Surv. **1999**, *33*, 35–54.
2. Boutrif, E. FAO programmes for prevention, regulation, and control of mycotoxins in food. Nat. Toxins **1995**, *3*, 322–326.
3. Brown, R.L.; Bhatnager, D.; Cleveland, T.E.; Cary, J.W. Recent Advances in Preharvest Prevention of Mycotoxin Contamination. In *Mycotoxins in Agriculture and Food Safety*; Sinha, K.K., Bhatnager, D., Eds.; Dekker: New York, 1998; 351–380.
4. Payne, G.A. Process of Contamination by Aflatoxin-Producing Fungi and Their Impact on Crops. In *Mycotoxins in Agriculture and Food Safety*; Sinha, K.K., Bhatnager, D., Eds.; Dekker: New York, 1998; 278–306.
5. CAST. *Mycotoxins: Risks in Plant, Animal, and Human Systems*; Task Force Report, Council for Agricultural Science and Technology (CAST): Ames, IA, 2003; Vol. 139.
6. Wogan, G.N. Aflatoxins as risk factors for hepatocellular carcinoma in humans. Cancer Res. **1992**, *52* (Suppl.), 2114s–2118s.
7. Cole, R.J.; Cox, E.H. *Handbook of Toxic Fungal Metabolites*; Academic Press: New York, 1981.
8. Jelinek, C.F.; Pohland, A.E.; Wood, G.E. Worldwide occurrence of mycotoxins in foods and feeds—An update. J. Assoc. Off. Anal. Chem. **1989**, *72*, 223–230.
9. Mori, H.; Kawai, K. Genotoxicity in Rodent Hepatocytes and Carcinogencity of Mycotoxins and Related Chemicals. In *Mycotoxins and Phycotoxins 88*; Natori, S., Hashimoto, K., Ueno, Y., Eds.; Elsevier: Amsterdam, 1989; 81–90.
10. Kuiper-Goodman, T.; Scott, P.M. Risk assessment of the mycotoxin ochratoxin A. Biomed. Environ. Sci. **1989**, *2*, 179–248.
11. Huang, X.; Dorner, J.W.; Chu, F.S. Production of aflatoxin and cyclopiazonic acid by various aspergilli: An ELISA analysis. Mycotoxin Res. **1994**, *10*, 101–106.
12. Ross, P.F.; Nelson, P.E.; Richard, J.L.; Osweiler, G.D.; Rice, L.G.; Plattner, R.D.; Wilson, T.M. Production of fumonisins by *Fusarium moniliforme* and *Fusarium proliferatum* isolates associated with equine leukoencephalomalacia and a pulmonary edema syndrome in swine. Appl. Environ. Microbiol. **1990**, *56*, 3225–3226.
13. Harrison, L.R.; Colvin, B.; Greene, J.T.; Newman, L.E.; Cole, J.; Pulmonary edema, R.J. hydrothorax in swine produced by fumonism B1, a toxic metabolite of Fusarium moniliforme. J. Vet. Diagn. Invest. **1990**, *2*, 217–221.
14. Flieger, M.; Wurst, M.; Shelby, R. Ergot alkaloids—Sources, structures and analytical methods. Folia Microbiol. **1997**, *42*, 3–30.
15. Hicks, J.; Shimizu, K.; Keller, N.P. Genetics and Biosynthesis of Aflatoxins and Sterigmatocystin. In *The Mycota; Kempken*; Bennett, Ed.; Spring-Verlag: Berlin, 2002; 55–69.
16. Hohn, T.M.; Desjardins, A.E.; McCormick, S.P.; Proctor, R.H. Biosynthesis of Trichothecenes, Genetic and Molecular Aspects. In *Molecular Approached to Food Safety: Issues Involving Toxic Microorganisms*; Eklund, M., Richard, J.L., Mise, K., Eds.; Alaken: Ft. Collins, 1995; 239–248.
17. Brown, D.W.; McCormick, S.P.; Alexander, N.J.; Proctor, R.H.; Desjardins, A.E. A genetic and biochemical approach to study trichothecene diversity in *Fusarium sporotrichioides* and *Fusarium graminearum*. Fungal Genet. Biol. **2001**, *32*, 121–133.
18. Brown, D.W.; McCormick, S.P.; Alexander, N.J.; Proctor, R.H.; Desjardins, A.E. Inactivation of a cytochrome P-450 is a determinant of trichothecene diversity in Fusarium species. Fungal Genet. Biol. **2002**, *36*, 224–233.
19. Proctor, R.H.; Brown, D.W.; Plattner, R.D.; Desjardins, A.E. Co-expression of 15 contiguous genes delineates a fumonisin biosynthetic gene cluster in *Gibberella moniliformis*. Fungal Genet. Biol. **2003**, *38*, 237–249.
20. Tudzynski, P.; Correia, T.; Keller, U. Biotechnology and genetics of ergot alkaloids. Appl. Environ. Microbiol. **2001**, *57*, 593–605.
21. Calvo, A.M.; Wilson, R.A.; Bok, J.W.; Keller, N.P. Relationship between secondary metabolism and fungal development. Microbiol. Mol. Biol. Rev. **2002**, *66*, 447–459.
22. Brown, R.L.; Cotty, P.J.; Cleveland, T.E. Reduction in aflatoxin content of maize by atoxigenic strains of *Aspergillus flavus*. J. Food Prot. **1991**, *54*, 623–626.
23. Cotty, P.J. Influence of field application of an atoxigenic strain of *Aspergillus flavus* on the populations of *A. flavus* infecting cotton bolls and on aflatoxin content of cottonseed. Phytopathology **1994**, *84*, 1270–1277.

Natural Rubber

Dennis T. Ray
University of Arizona, Tucson, Arizona, U.S.A.

INTRODUCTION

Natural rubber is an essential part of society used in the manufacture of over 40,000 products, more than 400 of which are medical devices.[1] Over 2000 rubber-producing plant species have been described; however, historically only two—the Brazilian rubber tree (*Hevea braziliensis*) and guayule (*Parthenium argentatum*)—have been exploited commercially.[2] The Brazilian rubber tree at present is the only commercial source of this essential natural plant product.[3]

Natural rubber is a renewable resource accounting for approximately 40% of the world's total rubber consumption,[4] the remainder coming from synthetic rubber, a nonrenewable resource derived from petroleum. Synthetic rubber cannot substitute for natural rubber in applications that require high elasticity, resilience, and/or minimum heat buildup (e.g., high-performance tires and medical latex products).[5] Because of these characteristics, the global demand for natural rubber continues to increase,[1] but in the future, because synthetic rubber is a nonrenewable petroleum product, even more natural rubber will be required to replace synthetic rubber's market share.

Having more than one source (biodiversity) of natural rubber is desirable, especially to meet the anticipated need of this essential natural product. At this time guayule has the greatest potential of the over 2000 rubber-producing plants to become an additional commercial source of natural rubber, mainly because of the extensive research and development that have already taken place. Commercialization of guayule would not replace the production of rubber from the Brazilian rubber tree, but would increase total global rubber production.

NEED FOR INCREASE IN BIODIVERSITY OF RUBBER PRODUCTION

Growing guayule as a commercial source of natural rubber would lead biodiversity as well as help stabilize global rubber production. The Brazilian rubber tree is grown exclusively in tropical climates;[5] today commercial production is centered almost exclusively in Asia.[1] Guayule is a xerophytic shrub native to the Chihuahuan desert of north-central Mexico and southwestern Texas;[2] commercial production would be in arid and semiarid environments. The Brazilian rubber tree is grown almost exclusively from clones, which are extremely narrow genetically. This makes the crop particularly susceptible to crop failure due to disease, as in the case of the South American industry's decline due to leaf blight.[3,5] Guayule populations contain a great amount of genetic diversity, making them ideal for selecting plants best suited for cultivation in different areas and environments. Guayule also has a very complicated reproductive system that continually generates genetic diversity among progeny of a single plant.[5,6]

It appears the demand for natural rubber will continue to grow at a rate greater than can be met by new plantings of the Brazilian rubber tree; changing political climates in production areas can also jeopardize the continuity of the natural rubber supply. Growers in developing countries are moving away from plantation farming of the Brazilian rubber tree to higher-value agricultural crops.[3] Guayule, on the other hand, has the potential to add jobs in rural areas with arid and semiarid environments where traditional agriculture is becoming less economical because of water and salt problems. The commercialization of guayule will enhance the sustainability of agriculture in these rural areas, and will benefit more than growers through the concurrent development of local processing facilities and manufacturing plants. Because guayule would be grown in other environments and areas of the world, its production would ensure the continued flow of this essential natural product.

NATURAL RUBBER/LATEX

Natural rubber is an isoprenoid molecule, related to plant compounds essential for plant growth and development.[7] Its desirable qualities are due to its molecular structure and high molecular weight.[5] In the Brazilian rubber tree, natural rubber in the latex form is produced within subcellular rubber particles in a pipelike system of lacticifers, from which the latex can be harvested by tapping.[8] Natural rubber latex in guayule is produced within rubber particles found within the cytoplasm of intact bark parenchyma cells, and does not flow from continuous ducts as in the Brazilian rubber tree.[5] Grinding the stem tissues

Encyclopedia of Plant and Crop Science
DOI: 10.1081/E-EPCS 120010443

and using solvents in order to free the rubber molecules affect extraction of rubber latex in guayule.

Many commercial, medical, transportation, and defense industries are dependent on natural rubber in the production of their products. Although the development of synthetic rubber by the chemical industry after World War II was a great breakthrough, synthetic materials have not been able to achieve the same properties as natural rubber for many high-performance applications.[1]

GUAYULE AS A POTENTIAL RUBBER-PRODUCING CROP

Guayule has been known as a source of natural rubber since pre-Columbian times, its first use dating to the production of rubber balls for indigenous people's games in present-day Mexico.[2] In the early 1900s, guayule was considered an alternative source of natural rubber in the United States due to the high price of rubber imported from the Amazon region.[9] This initial interest resulted in the first of three major efforts to domesticate and commercialize guayule.

This initial attempt started with the harvesting of wild guayule stands in Mexico, and accounted for up to 24% of the total rubber imported to the United States by 1910.[9] At this time, up to 20 extraction plants were either operational or under construction in Mexico, when production came to a halt in 1912 because of the Mexican Revolution.[2] Production then moved across the border to the United States, with efforts centered in Arizona and California. This first effort to commercialize guayule came to a halt in 1929 as a result of the Great Depression.[2]

The second major effort to utilize guayule as a source of natural rubber was the Emergency Rubber Project of World War II. Natural rubber production had moved almost exclusively to large plantations of the Brazilian rubber tree grown in Southeast Asia, and these sources were cut off at the beginning of the war.[2] The Emergency Rubber Project was very successful. It generated the bulk of our knowledge about the basic biology of the guayule plant and developed the germplasm on which current breeding programs are based. The effort ended with the end of the war and the development of synthetic rubber.

Guayule was seriously investigated a third time in the early 1970s, when crude oil prices quadrupled. The fear was that if the oil supply could be manipulated, there might again be a shortage of natural rubber due to either natural disaster or political unrest in Southeast Asia. This led in the United States to the enactment of the Native Latex Commercialization and Economic Development Act of 1978. A tremendous amount of work was again accomplished, resulting in significant yield increases and the refinement of cultural practices to fit modern mechanized agriculture.[2,10] This third effort again showed that guayule could be planted, cultivated, harvested, and processed as a source of natural rubber. However, as the political climate changed, this effort was also terminated, and guayule was again considered a source of natural rubber only in times of emergency.

This appeared to be the end of guayule's commercialization efforts. To be considered a commodity in direct competition with the Brazilian rubber tree, guayule rubber would have to either perform the same functions at lower cost or perform better at the same costs. Although guayule rubber is equivalent in quality to the Brazilian rubber tree latex, it was not competitive economically.[2] However, guayule is once again being considered for commercialization because of the occurrence of latex allergy in the general population.[11]

PRESENT POTENTIAL FOR GUAYULE COMMERCIALIZATION

Allergic reactions are a result of protein contaminants in rubber products, and range from contact dermatitis to anaphylactic shock when susceptible individuals come in contact with these proteins. With over 400 medical devices containing natural rubber, this is potentially a very serious problem. Guayule latex contain's low levels of proteins, none of which elicit an allergic response in subjects who are sensitized to Brazilian rubber tree latex proteins.[12] Therefore, guayule affords a potential new product—hypoallergenic latex—to be used as an alternative in the manufacture of products for individuals with latex allergy. A method to extract natural rubber in latex form from guayule has been developed,[13] and the goal of producing hypoallergenic guayule latex products is moving toward becoming a reality.[14]

CONCLUSION

Natural rubber is an essential part of life in today's society. The demand for natural rubber will continue to increase because of both increased consumption and the eventual need to replace synthetic rubber. At this time the Brazilian rubber tree is the only source of commercial natural rubber. Thus there is a need for additional sources to expand the growing range and worldwide production. At this time guayule is the most promising plant to fill this niche. Guayule makes a natural rubber of at least the quality required for high-end uses, such as high-performance tires and medical products, but it is not directly economically competitive with the Brazilian rubber tree in these markets. However, guayule makes natural rubber latex that is nonreactive to individuals synthesized to proteins found in Brazilian rubber tree latex products. Thus, guayule has the potential to become a commercial crop for the production of hypoallergenic latex products.

ARTICLES OF FURTHER INTEREST

Agriculture and Biodiversity, p. 1
Alternative Feedstocks for Bioprocessing, p. 24
Aromatic Plants for the Flavor and Fragrance Industries, p. 58
Crop Domestication: Fate of Genetic Diversity, p. 333
Isoprenoid Metabolism, p. 625
Lesquerella Potential for Commercialization, p. 656
Marigold Flower: Industrial Applications of, p. 689
Milkweed: Commercial Applications, p. 724
New Industrial Crops in Europe, p. 813
New Secondary Metabolites: Potential Evolution, p. 818
Sustainable Agriculture: Definitions and Goals, p. 1187
Switchgrass as a Bioenergy Crop, p. 1207

REFERENCES

1. Cornish, K. Similarities and differences in rubber biochemistry among plant species. Phytochemistry **2001**, *57*, 1123–1134.
2. Ray, D.T. Guayule: A Source of Natural Rubber. In *New Crops*; Janick, J., Simon, J.E., Eds.; John Wiley and Sons, Inc.: New York, 1993; 338–342.
3. Davis, W. The rubber industry's biological nightmare. Fortune Aug. 4, **1997**, 86–95.
4. Mooibroek, H.; Cornish, K. Alternative sources of natural rubber. Appl. Microbiol. Biotechnol. **2000**, *52*, 355–365.
5. Thompson, A.E.; Ray, D.T. Breeding guayule. Plant Breed. Rev. **1989**, *6*, 93–165.
6. Dierig, D.A.; Ray, D.T.; Coffelt, T.A.; Nakayama, F.S.; Leake, G.S.; Lorenz, G. Heritability of height, width, resin, rubber, and latex in guayule (*Parthenium argentatum*). Ind. Crops Prod. **2001**, *13*, 229–238.
7. Chappell, J. Biochemistry and molecular biology of the isoprenoid biosynthetic pathway in plants. Annu. Rev. Plant Physiol. Plant Mol. Biol. **1995**, *46*, 521–547.
8. d'Auzac, J.; Jacob, J.-L.; Chrestin, H. *Physiology of Rubber Tree Latex*; CRC Press: Boca Raton, FL, 1989.
9. Bonner, J. The History of Rubber. In *Guayule Natural Rubber*; Whitworth, J.W., Whitehead, E.E., Eds.; Office of Arid Lands Studies, Univ. of Arizona: Tucson, AZ, 1991; 1–6.
10. Whitworth, J.W.; Whitehead, E.E. *Guayule Natural Rubber*; Office of Arid Lands Studies, The University of Arizona: Tucson, AZ, 1991.
11. Ownby, D.R.; Ownby, H.E.; McCullough, J.A.; Shafer, A.W. The prevalence of anti-latex IgE antibodies in 1000 volunteer blood donors. J. Allergy Clin. Immunol. **1994**, *93*, 282.
12. Siler, D.J.; Cornish, K. Hypoallergenicity of guayule rubber particle proteins compared to *Hevea* latex proteins. Ind. Crops Prod. **1994**, *2*, 307–313.
13. Cornish, K. Non-Allergenic Natural Rubber Products from *Parthenium Argentatum* (Gray) and Other Non-*Hevea Brasiliensis* Species. U.S. Patent No. 558094, 1996.
14. Cornish, K. Non-Allergenic Natural Rubber Products from *Parthenium Argentatum* (Gray) and Other Non-*Hevea Brasiliensis* species. U.S. Patent No. 5717050, 1998.

Nematode Biology, Morphology, and Physiology

Stephen A. Lewis
Clemson University, Clemson, South Carolina, U.S.A.

David J. Chitwood
USDA-ARS, Beltsville, Maryland, U.S.A.

Edward C. McGawley
Louisiana State University, Baton Rouge, Louisiana, U.S.A.

INTRODUCTION

Nematodes are an extremely successful group of threadlike, worm-shaped animals that are found in many ecological niches, including extreme environments. They may be the most abundant multicellular animals, occuring as parasites of plants and animals and as free-living nematodes feeding on bacteria, fungi, and protozoa. This article presents a description of nematode history and anatomy, including size relationships, life cycle, general external and internal anatomy, survival, and feeding habits.

DISCUSSION

Nematodes are a remarkably successful group of threadlike, worm-shaped invertebrate animals adapted to survival in most ecological niches. Because they require water for locomotion, they occur in marine and freshwater environments, in the film of water between soil particles, and in plant and animal tissues. Nematodes also are found in extreme environments such as desert and polar regions. They may be the most abundant multicellular animals, and the more than 20,000 described species occur as parasites of plants and animals—including humans—and as free-living nematodes feeding on bacteria, fungi, and protozoa. They are inconceivably abundant and can be found in the most unexpected places, forming a recognizable world with all plant and animal life and geographical landmarks visible, even if all the matter on the earth except nematodes were somehow swept away.[1]

A rich history of nematodes derives from early writings on symptoms, signs, and treatment of large human parasitic nematodes, especially *Ascaris*, a large intestinal roundworm. The oldest writings describing these maladies are from the Chinese culture and date back to about 4700 years ago. References to nematodes were also recorded during the dominant civilizations of the Middle East and Mediterranean areas over 2000 years ago. The advent of the microscope in the 17th century resulted in many contributions to the science of nematology. The 18th and early 19th centuries were a time of historic discoveries in nematode anatomy, embryology, taxonomy, and life cycle studies of animal and human nematode diseases, including elucidation of the principles of alteration of generations, intermediate hosts, and vectors. During this time, the study of nematode parasites of livestock and other domestic animals flourished, and the seriousness of the many human diseases caused by nematodes was established.[2]

The study of plant-parasitic nematodes has a relatively recent history dating back only to 1743, when John Turbevill Needham, a clergyman interested in science, observed nematodes emerging from distorted grains of wheat. This seed gall nematode, *Anguina tritici*, had seriously affected wheat yield in the great Middle Eastern civilizations. Over 100 years elapsed until the plant pathologist Julius Kühn described an eelworm disease that caused stunting of teasel, a plant whose spiny fruit were used to raise the nap on wool. Only two years later, a serious disease of sugar beet in Europe was determined to be caused by the nematode later named *Heterodera schactii*. Discovery of another European nematode-induced disease on potato soon followed. Investigations on controlling these and other nematode pathogens dominated nematological research in Europe for about 40 years. The basic principles of disease management tactics such as soil fumigation and crop rotation were established during this period are still applicable today.[3]

The word nematode means ''threadlike'' in Greek and amply describes the appearance of these animals, especially the smaller forms when dispersed in a vial of water. Nematodes are nonsegmented, unlike earthworms for example, and are spindle-shaped, tapering slightly at both ends. Plant-parasitic species vary from about 0.3 μm long in the genus *Paratylenchus* to about 5 mm in the genus *Longidorus*. The longest mammalian parasite occurs in whale placenta and reaches 8 meters in length. Females in

DOI: 10.1081/E-EPCS 120010384

some plant-parasitic genera are kidney-shaped, lemon-shaped, pear-shaped, or otherwise swollen as adults, whereas the males remain vermiform (worm-shaped).

Plant-parasitic nematodes exhibit a simple, direct life cycle that usually includes egg, four juvenile, and adult stages. Nematodes are usually bisexual, existing as distinct males and females. Some nematodes, however, are hermaphroditic, and the female sex organs produce both sperm and eggs that combine during fertilization. In another variation, called parthenogenesis, males (and spermatozoa) are not required for reproduction, although males may be present in the population.

The nematode oral aperture is surrounded by six lips, or fewer if they have become partly combined as in more derived genera. There is typically a single papilla on the inner edge of each lip adjacent to the mouth, and two on the outer side of each lip in some ancestral forms. Certain marine nematodes, which are thought to be ancestral, have the inner row of papillae on the lips surrounding the mouth, an adjacent row of setae, and an outer row of larger setae. Plant-parasitic nematodes have few papillae, no setae, and combined lips that have taken on new forms.

The nematode head skeleton supports the lip region and is composed of radial blades and basal ring, which appear light or dark depending on the degree of hardening. The mouth is the beginning of the alimentary canal and it is followed by the buccal cavity or stoma. In plant-parasitic nematodes the buccal cavity is armed with a hollow, protrusible stylet or mouth spear (http://www.barc.usda.gov/psi/nem/what-nem.htm). The stylet is the most distinctive structure in the head; its shape varies among the genera of plant-parasitic nematodes and can often be used to determine the nematode genus (http://nematode.unl.edu/key/nemakey.htm). The stylet is hollow for part of its length and is used to withdraw nutrients from plant cells. It also can serve as a conduit for digestive gland secretions into the plant cell. The stylet has protractor muscles attached at the base and forward at the anterior wall of the esophagus. Contraction of these muscles thrusts the stylet out of the mouth and into the host plant cell.

The stoma, or mouth cavity, is followed posteriorily by the esophagus, composed of the corpus, the isthmus, and a basal glandular region. The corpus can be further subdivided into the anterior procorpus, followed by the metacorpus, which may or may not contain a distinctive valve. A very narrow (<1 μm) food channel, called the lumen, passes through the esophagus and connects it with the intestine. The metacorpus in plant-parasitic nematodes is often furnished with radial muscles, making it a pump chamber that can pulsate up to several times per second. This pump chamber with its valve withdraws nutrients from cells in the active feeding phase and passes the nutrients through the valve into the intestine. Depending on the nematode, there may be an initial passive-feeding phase that does not require the pumping of the metacorpus to feed the nematode. The esophageal lumen is tri-radiate as it passes through the metacorpus and has one ventral and two subdorsal rays when relaxed. When the radial muscles contract, the lumen becomes more circular and food passes posteriorily. The isthmus that follows the metacorpus is a narrowing of the esophagus before the basal glandular portion of the esophagus. The morphology of the esophagus is one way in which nematodes are identified and is especially important for plant-parasitic genera. The plant virus vectors *Xiphinema, Longidorus, Trichodorus*, and closely related genera have a two-part esophagus, consisting of a narrow anterior portion and an expanded basal glandular portion.

The excretory system of more ancestral nematode forms may contain an H-shaped system of lateral ducts ending in a cuticularized duct and anterior ventromedian pore. Plant-parasitic genera in the class Secernentea (containing most of the plant parasites) may have a duct or a portion of a duct on one side leading to a ventral excretory cell, or the more advanced forms may have only the ventral cell, excretory duct, and pore. Adenophorean nematodes have only the single ventral cell, without collecting tubules (ducts), and a noncuticularized excretory duct leading to the outside pore. There is some debate on whether this system is solely for the function of excretion or possibly also for osmoregulation or secretion of materials other than wastes.[4,5]

The female reproductive system is divided into ovary, oviduct, and uterus. There are one or two ovaries composed of an anterior germinal zone where eggs are produced and an ovum growth zone. If a globular spermatheca for the storage of sperm is present, it is part of the uterus.[6] The tubular reproductive system opens through the ventral vagina and vulva. The tubular male testis joins posteriorly with the digestive system to create a common duct called a cloaca that opens ventromedially through the anus. Males have one or two testes and an external pair of hooklike spicules used to pry open the vulva during copulation. There may be a spicular guide piece called a gubernaculum and another accessory piece called a telamon (capitulum) adjacent to the spicules on the anterior side.

Nerve cells in the nematode are grouped into ganglia that in turn are grouped into a circumesophageal commissure (nerve ring), a collection of coordinated nerve cell bodies that become the primitive nematode ''brain.'' Nerves extend anteriorily toward the mouth and provide chemosensory and tactile information to the nematode. Nerves also extend posteriorily in the dorsal and ventral hypodermal cords and provide chemosensory and tactile sensation. The main nerve is the ganglionated ventral nerve.

Nematodes have an external covering called a cuticle, a nonliving, proteinaceous, multilayered covering secreted by an underlying living, cellular hypodermis ("tissue below the skin"). In order to accommodate growth-necessitated expansion, the passage of a nematode from one juvenile stage to another (or to the adult stage) is accompanied by molting, a process during which the new cuticle is synthesized and the old cuticle is shed. The physiological and biochemical control of nematode molting has not yet been elucidated, unlike the molting process in insects. In many plant parasites, the old cuticle substantially dissolves and is probably absorbed by the hypodermis. The hypodermis invaginates in four sectors and forms cords that contain a limited number of hypodermal cells, separating the bands of muscle fibers into quadrants of two subdorsal and two subventral bands. Nematodes are unique in that the muscle cells send out innervation processes to the nerve cell body, rather than the more typical arrangement of nerve cell axons providing transmission of electrical impulses to the muscles. For most nematodes, muscles of the body are limited to longitudinally oriented fibers.

The ability of nematodes to thrive in a variety of habitats results from their anatomy and physiology. The selective permeability of the nematode cuticle often facilitates the entry of nutrients or the export of wastes, yet impedes the entry of detrimental compounds. Many nematodes can survive extended periods of drought or temperature stress via morphological and biochemical adaptations, such as the accumulation of the sugar trehalose and other specific molecules. The eggs of some plant-parasitic nematodes can survive for years in soil; eggs often hatch when stimulated by exudates from host plants. Identification of these so called hatching factors is an active research area; only one such chemical has been identified to date.

Although some plant-parasitic nematode species climb on and penetrate the aerial parts of plants, most plant-feeding nematodes are root pathogens. Some of these move through plant tissues and feed on different roots, whereas others become established in a permanent feeding site. These sedentary root parasites greatly alter the physiology of host roots. Their esophageal glands secrete proteins and possibly other factors that cause structural and physiological modification of the host cells the nematodes feed upon.

The free-living nematode species *Caenorhabditis elegans* was the first multicellular animal to have its genome completely sequenced. This nematode contains more than 19,000 genes, although their functions are as yet largely uncharacterized. Analyses comparing the sequences of genes from *C. elegans* to those of plant- and animal-parasitic nematodes are yielding insight about the function of genes important in parasite development and pathogenesis (http://elegans.swmed.edu/genome.shtml).

Nematodes possess the same biochemical pathways present in most animals. Unlike their mammalian and plant hosts, nematodes cannot biosynthesize heme or steroids; unlike all other animals, nematodes possess a glyoxylate cycle that enables them to biosynthesize carbohydrates from their storage lipids. Although many nematode parasites of mammals extensively catabolize organic molecules anaerobically and produce short-chain fatty acids as waste products, carbohydrate catabolism in plant-pathogenic species is primarily aerobic, via typical tricarboxylic acid and electron transport pathways.

ARTICLES OF FURTHER INTEREST

Biological of Nematodes, p. 134
Classification and Identification of Nematodes, p. 283
Management of Nematode Diseases: Options, p. 684
Nematode Feeding Strategies, p. 784
Nematode Infestations: Assessment, p. 788
Nematode Population Dynamics, p. 797
Nematode Problems: Most Prevalent, p. 800
Nematodes and Host Resistance, p. 805
Plant Response to Stress: Nematode Infection, p. 1014

REFERENCES

1. Cobb, N.A. Nematodes and Their Relationships. In *Yearbook of the Department of Agriculture for 1914*; Government Printing Office: Washington, 1915; 457–490.
2. Maggenti, A. *General Nematology*; Springer-Verlag: New York, 1981.
3. Thorne, G. *Principles of Nematology*; McGraw-Hill: New York, 1961.
4. Bird, A.F.; Bird, J. Introduction to Functional Organization. In *The Physiology and Biochemistry of Free-living and Plant-Parasitic Nematodes*; Perry, R.N., Wright, D.J., Eds.; CABI Publishing: New York, 1998; 1–24.
5. Wright, D.J. Respiratory Physiology, Nitrogen Excretion and Osmotic and Ionic Regulation. In *The Physiology and Biochemistry of Free-Living and Plant-Parasitic Nematodes*; Perry, R.N., Wright, D.J., Eds.; CABI Publishing: New York, 1998; 103–131.
6. Bird, A.F.; Bird, J. *The Structure of Nematodes*, 2nd Ed.; Academic Press: San Diego, CA, 1991.

Nematode Feeding Strategies

Richard S. Hussey
University of Georgia, Athens, Georgia, U.S.A.

INTRODUCTION

Plant-parasitic nematodes have evolved diverse parasitic strategies and feeding relationships with their host plants to obtain nutrients that are necessary for development and reproduction. Depending on species, these biotrophic parasites feed from the cytoplasm of unmodified living plant cells or have evolved to modify root cells into elaborate, discrete feeding cells. These parasitic nematodes use a hollow, protrusible feeding structure called a stylet to penetrate the wall of a plant cell, inject esophageal gland secretions into the cell, and withdraw nutrients from the cytoplasm of the parasitized cell. Certain species take up cytosol from the parasitized cell directly through a minute perforation created in the plasma membrane at the stylet orifice, while others ingest nutrients through a feeding tube.

NEMATODE FEEDING STRATEGIES

Plant-parasitic nematodes have evolved diverse parasitic strategies and feeding relationships with their host plants to obtain nutrients that are necessary for development and reproduction. Depending on species, these biotrophic parasites feed from the cytoplasm of unmodified living plant cells or have evolved to modify root cells into elaborate, discrete feeding cells.[1] These parasitic nematodes use a hollow, protrusible feeding structure called a stylet to penetrate the wall of a plant cell, inject esophageal gland secretions into the cell, and withdraw nutrients from the cytoplasm of the parasitized cell. Plant-parasitic nematodes can be separated into four general groups according to the evolution of their mode of parasitism: migratory ectoparasites, sedentary ectoparasites, migratory endoparasites, and sedentary endoparasites (Fig. 1). The migratory ectoparasites have the most primitive mode of parasitism, directly feeding from unmodified root cells. In contrast, the more evolutionarily advanced sedentary endoparasites have evolved a very specialized mode of parasitism, dramatically modifying root cells of susceptible hosts into elaborate, unique feeding cells, including modulating complex changes in cell morphology, function, and gene expression. The amount of tissue destruction and the degree of plant response are often related to the type of feeding relationship between the nematode and its host. The migratory ectoparasites remain outside the root and insert their protrusible stylet to feed either on epidermal cells or cells deeper within the root. As a rule, species that possess a short stylet (e.g., *Tylenchorhynchus* spp.) feed on epidermal cells, while species with a long stylet (e.g., *Belonolaimus* spp.) are able to exploit tissues deeper in the root. With the exception of species of a few genera (e.g., *Belonolaimus*) that feed on root tips, nematodes with this type of feeding strategy generally cause little obvious tissue damage. Sedentary ectoparasites feed from a single site or root cell for a prolonged period of time while remaining outside the root. Feeding by *Criconemella xenoplax* causes little tissue damage, but other species, such as *Hemicycliophora arenaria*, induce terminal galls when feeding at root tips.

Another group of nematodes, the endoparasites, invades root tissue with part or all of their body. Although some endoparasites feed as soon as they enter the root, other species feed only after migrating to a preferred feeding site. Migratory endoparasites (e.g., *Pratylenchus* and *Radopholus* spp.) enter roots and feed on cells as they migrate intracellularly through the root tissue. These endoparasites inhabit primarily the cortical tissue of roots. Migratory endoparasites possess a small but robust stylet which is first used to pierce the walls of root cells and then to withdraw food from the cytoplasm. This feeding behavior causes extensive destruction of root tissue along the path of the migrating nematode. Sedentary endoparasites (e.g., *Meloidogyne*, *Globodera*, and *Heterodera* spp.) have evolved very specialized and complex feeding relationships with their host plants, feeding from a single cell or a group of cells for prolonged periods of time. These nematodes invade roots as vermiform second-stage juveniles. Those of the *Meloidogyne* species move within the intercellular space in the cortex and those of the *Globodera* and *Heterodera* species migrate intracellularly within the root cortex. Further nematode development depends upon the elaborate modification of the phenotype of vascular cylinder cells to form specialized feeding cells that become the sole source of nutrients for the parasites.[1,2] The association of feeding sites with the root vascular cylinder supports a more concentrated and

Encyclopedia of Plant and Crop Science
DOI: 10.1081/E-EPCS 120010390

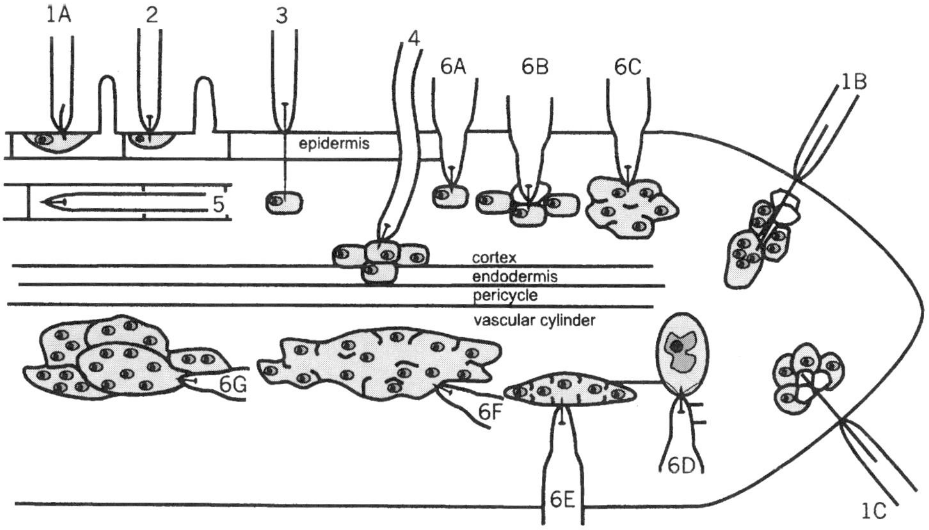

Fig. 1 Schematic representation of feeding sites of selected root-parasitic nematodes. 1: Dorylaimid migratory ectoparasites: 1A, *Trichodorus* spp.; 1B, *Xiphinema index*; 1C, *Longidorus elongatus.* 2–6: Tylenchid nematodes: 2: Migratory ectoparasite: *Tylenchorhynchus dubius*; 3: Sedentary ectoparasites: *Criconemella xenoplax*. 4: Migratory ecto-endoparasites: *Helicotylenchus* spp. 5: Migratory endoparasites: *Pratylenchus* spp. 6: Sedentary endoparasites: 6A, *Trophotylenchulus obscurus*; 6B, *Tylenchulus semipenetrans*; 6C, *Verutus volvingentis*; 6D, *Cryphodera utahensis*; 6E, *Rotylenchulus reniformis*; 6F, *Heterodera* spp.; 6G, *Meloidogyne* spp. (From Ref. 7.)

sustainable supply of nutrients. When feeding commences, the second-stage juvenile grows and becomes saccate and immobile. In this feeding strategy, destruction of root tissue is usually limited to cells around the feeding site and the endoparasitic nematode.

In addition to the protrusible stylet, plant-parasitic nematodes have a well-developed esophagus for feeding on plants. In tylenchid nematodes, which are the most important plant parasite group, the esophagus has a muscular metacorpus containing a triradiate pump chamber

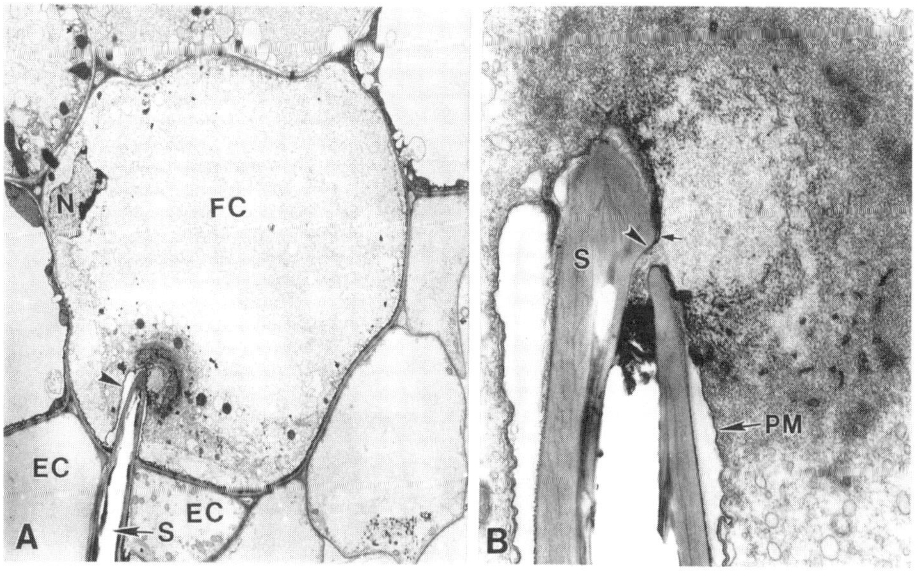

Fig. 2 Fine structure of the *Criconemella xenoplax*–feeding cell relationship. A. Cross-section of a feeding cell (FC) in the cortex of a tomato root. The nematode stylet (S) penetrated between two epidermal cells (EC) and was inserted into the feeding cell without penetrating the plasma membrane (arrowhead). A zone of the cytoplasm of the feeding cell is modified around the stylet tip. A profile of the cell's nucleus (N) is visible. B. Detailed view of the opening created in the plasma membrane (PM), which was invaginated around the stylet (S) tip when the membrane (arrow) became tightly appressed (arrowhead) to the wall of the stylet orifice. (From Ref. 8).

and three large and complex secretory gland cells.[3] The transcriptionally active gland cells, one dorsal and two subventral, are the principal sources of the secretions involved in plant parasitism. During secretion, the gland cells release secretory proteins stored in granules into the lumen of the esophagus to be injected through the stylet into host tissue. Changes in the esophageal gland cells during the parasitic cycle indicate various roles for the gland's secretory proteins during different stages of parasitism. The nature of nematode esophageal gland secretory proteins and their function in parasitism is now beginning to emerge.[4]

Nematode feeding from root cells is a very deliberate process and can be divided into distinct phases that include stylet insertion, injection of esophageal gland cell secretions, ingestion of nutrients, and stylet retraction.[1] The feeding phases of migratory and sedentary parasites are similar, but each phase—and particularly nutrient uptake—is considerably longer for the sedentary than for the migratory feeders. When a potential feeding site is selected by the nematode, the wall of the root cell is penetrated by the stylet, which remains protruded and in contact with the cell cytoplasm during the secretion and ingestion phases. During the secretion phase, secretory proteins synthesized in the esophageal gland cells pass through the stylet into the parasitized cell. After cessation of the secretion activity, rapid maximum dilation of the metacorpal pump chamber creates the suction necessary for the nematode to withdraw nutrients from the cytoplasm of the feeding cell through the stylet lumen. When the metacorpus ceases pumping to terminate the ingestion phase, the stylet is retracted from the plant cell to end the feeding cycle. However, the sedentary ectoparasitic and endoparasitic nematodes establish a prolonged biotrophic feeding association with the elaborate feeding cells they induce.[1,2] During feeding, the sedentary nematode inserts its stylet through the wall of the parasitized cell without piercing the plasma membrane, which becomes invaginated around the stylet tip. Initially in the feeding cycle, esophageal gland cell secretions that may modify the cell are injected through the stylet. Certain species (e.g., *Criconemella xenoplax* (Fig. 2)) take up cytosol of the parasitized cell directly through a minute perforation created in the plasma membrane at the stylet orifice, while others (e.g., root-knot or cyst nematodes (Fig. 3)[8]) ingest nutrients through a feeding tube.[5,9]

Feeding tubes are formed within the cytoplasm of parasitized root cells from stylet secretions injected by sedentary endoparasitic nematodes (*Globodera*, *Meloidogyne*, *Heterodera*, *Rotylenchulus* spp.). These unique structures are used by the nematode to efficiently withdraw nutrients from the feeding cell. Feeding tubes formed by *M. incognita* females are 1 μm wide and up to 110 μm long with a uniform lumen of approximately 450 nm diameter (Fig. 3).[5] The distal ends of feeding tubes are

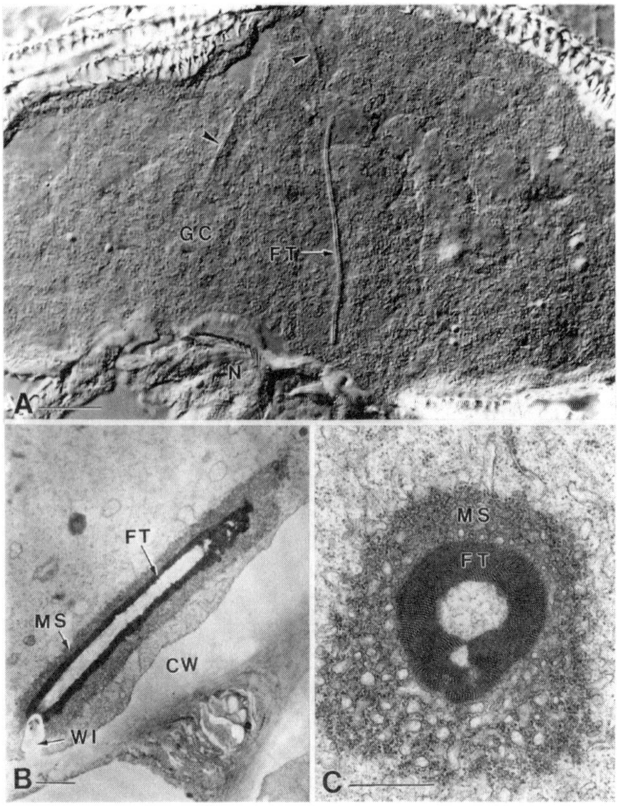

Fig. 3 Feeding tubes formed by *Meloidogyne incognita*. A. Nomarski light micrograph of a cryosection of a giant-cell (GC) from the root of a tomato plant infected by root-knot nematode. One long feeding tube (FT) and sections (arrowheads) of other feeding tubes are visible in the cytoplasm. A section of the head of the adult female nematode (N) that was feeding from the giant-cell is visible. Bar at A = 20 μm. B. Longitudinal section through a feeding tube (FT) with its proximal end attached to a wall ingrowth (WI) where the nematode's stylet penetrated the cell wall (CW) of a giant-cell in a tomato root. The feeding tube is enveloped by a compact membrane system (MS). Bar at B = 1 μm. C. Cross-section of an electron-dense crystalline feeding tube (FT) wall surrounded by a dense membrane system (MS). Bar at C = 0.5 μm. (From Ref. 5.)

closed. The sedentary endoparasitic nematodes feed in cycles from the modified cells and a new feeding tube is formed each time the nematode reinserts its stylet into the parasitized cell to initiate a new feeding cycle. Feeding tubes are formed from dorsal gland cell secretions injected into the cytoplasm of the parasitized cell through a pore in the plasma membrane. Feeding tubes function in facilitating the transport of soluble nutrients from the cytosol of the feeding cells to the stylet orifice. For *M. incognita*, which inserts its stylet only 2 to 3 μm through the wall of the feeding cell without piercing the plasma membrane, the long feeding tube provides the nematode access to more of the cytoplasm of the feeding cell than would be accessible if the nematode used only its stylet to withdraw nutrients. The sealed distal end of the feeding tube and the lack of noticeable wall perforations suggest that the wall of the feeding tube is permeable to soluble assimilates. Microinjection studies, in fact, show that feeding tubes function as a molecular sieve during ingestion of nutrients from the feeding cell by the nematode.[6]

ARTICLES OF FURTHER INTEREST

Classification and Identification of Nematodes, p. 283
Management of Nematode Diseases: Options, p. 684
Nematode Biology, Morphology, and Physiology, p. 781
Nematode Feeding Strategies, p. 784
Nematode Infestations: Assessment, p. 788
Nematode Population Dynamics, p. 797
Nematode Problems: Most Prevalent, p. 800
Nematodes and Host Resistance, p. 805
Plant Response to Stress: Nematode Infection, p. 1014

REFERENCES

1. Hussey, R.S.; Grundler, F.M.W. Nematode Parasitism of Plants. In *Physiology and Biochemistry of Free-Living and Plant Parasitic Nematodes*; Perry, R.N., Wright, D.J., Eds.; CAB International Press: England, 1998; 213–243.
2. Williamson, V.M.; Hussey, R.S. Nematode pathogenesis and resistance in plants. Plant Cell **1996**, *8* (10), 1735–1745.
3. Hussey, R.S. Disease-inducing secretions of plant-parasitic nematodes. Annu. Rev. Phytopathol. **1989**, *27*, 123–141.
4. Davis, E.L.; Hussey, R.S.; Baum, T.J.; Bakker, J.; Schots, A.; Rosso, M.-N.; Abad, P. Nematode parasitism genes. Annu. Rev. Phytopathol. **2000**, *38*, 365–396.
5. Hussey, R.S.; Mims, C.W. Ultrastructure of feeding tubes formed in giant-cells induced in plants by the root-knot nematode *Meloidogyne incognita*. Protoplasma **1991**, *162*, 99–107.
6. Bíckenhoff, A.; Grundler, F.M.W. Studies on the nutrient uptake by the beet cyst nematode *Heterodera schachtii* by in situ microinjection of fluorescent probes into the feeding structures in *Arabidopsis thaliana*. Parasitology **1994**, *109*, 249–254.
7. Wyss, U. Root Parasitic Nematodes: An Overview. In *Cellular and Molecular Aspects of Plant-Nematode Interactions*; Fenoll, C., Grundler, F.M.W., Ohl, S.A., Eds.; Kluwer Academic Publishers: Dordrecht, 1997; 5–22.
8. Hussey, R.S.; Mims, C.W.; Westcott, S.W. Ultrastructure of root cortical cells parasitized by the ring nematode *Criconemella xenoplax*. Protoplasma **1992**, *167*, 55–65.
9. Rebois, R.V. Ultrastructure of a feeding peg and tube associated with *Rotylenchulus reniformis* in cotton. Nematologica **1980**, *26*, 396–405.

Nematode Infestations: Assessment

K. R. Barker
North Carolina State University, Raleigh, North Carolina, U.S.A.

K. Dong
California Department of Food and Agriculture, Sacramento, California, U.S.A.

INTRODUCTION

There is in general an inverse relationship between initial nematode population densities and annual plant growth and crop yield. This relationship results in nematode population assessment being crucial for their management. Assessing nematode populations often involves an initial diagnosis and quantification of diverse kinds and numbers present, after which specific taxa within the nematode community may be monitored. The diversity of nematode species with different morphology, modes of reproduction, survival mechanisms, and feeding habitats increases the challenges of this endeavor.

DISCUSSION

Nematodes that attack higher plants are obligate parasites, and a number of important species within certain genera may vector plant viruses. Ectoparasitic nematodes feed from the outside of plants by inserting stylet into root tissue. Some of these nematodes may become sedentary and may feed on a specific group of cells for an extended period of time, whereas others are browsers and move frequently about the root surfaces as they feed. The latter nematodes are migratory ectoparasites. In contrast, other nematodes invade plant tissues where they feed internally. Nematodes that move throughout cortical or other tissues are called migratory endoparasites, whereas the sedentary endoparasites induce elaborate modifications of certain cells, giving rise to highly specialized feeding sites that are essential for nematodes' life cycles (Fig. 1).

Nematode infestation of soil and infection of plants may be assessed by sampling associated soil and/or plant tissue and extraction of the nematodes from those samples. Infected plant tissue also may be stained to render endoparasitic nematodes readily visible via microscopy. An effective nematode assessment should be accurate, rapid, simple, sensitive, and quantitative at the species/subspecies level. Light microscopic examination of extracted specimens is still the primary means of species identification and quantification. Nematode species identification generally is based on rather specific morphological structures.[1,2] However, differential plant-host responses and biochemical and molecular characteristics are becoming increasingly important for nematode diagnosis; the integration of these techniques with traditional morphological data is ongoing.[3–7]

SIGNS AND SYMPTOMS

The facts that most plant-parasitic nematodes are microscopic, inhabit the soil, and induce relatively nonspecific plant symptoms add to the challenges encountered in assessing their infestation. Assessment of nematode pathogens that attack stem or foliage tissues is less difficult but requires awareness of the potential for this type of disease agent. Thus, associated symptoms and signs of nematode infection are often helpful in initial phases of assessing related maladies. Foliage-infecting nematodes induce specific aboveground symptoms. For example, *Aphelenchoides besseyi* incites the development of pale yellow or white leaf tips on rice, whereas *Ditylenchus dipsaci* causes extensive distortions and swellings of infected tissues on alfalfa and other crops. The key general symptom of nematode infestation is irregular growth patterns within a field and/or a general decline and dieback. The latter often occurs on perennial crops such as fruit trees or ornamentals, e.g., *Radopholus similis* on banana or citrus trees.

Belowground symptoms of nematode infection range from large root galls induced by *Meloidogyne* spp. to root-surface necrosis caused by the migratory endoparasite *Pratylenchus* spp. Surface lesions are also caused by ectoparasites such as *Xiphinema* spp. and *Belonolaimus longicaudatus*. Some of the ectoparasitic nematodes, including *Paratrichodorus minor,* induce characteristic modifications of root size and shape, such as stubby roots. Signs of nematode may include structures such as cysts of *Heterodera* and *Globodera* and egg masses of *Meloidogyne* and *Rotylenchulus*, both of which can be observed on plant roots without a microscope. Thus, an initial assessment might include foliage or root signs and

Encyclopedia of Plant and Crop Science
DOI: 10.1081/E-EPCS 120010388

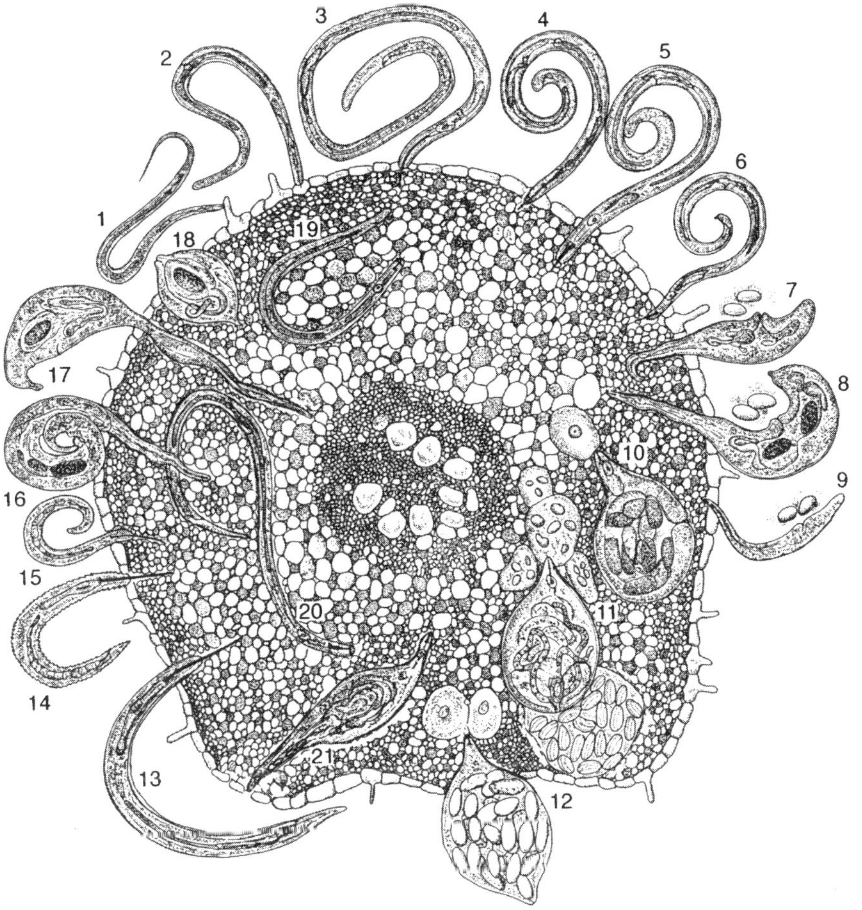

Fig. 1 Diagrammatic presentation of various types of tylenchid nematode feeding on root tissues. 1. *Cephalenchus*; 2. *Tylenchorhynchus*; 3. *Belonolaimus*; 4. *Rotylenchus*; 5. *Hoplolaimus*; 6. *Helicotylenchus*; 7. *Verutus*; 8. *Rotylenchulus*; 9. *Acontylus*; 10. *Meloidodera*; 11. *Meloidogyne*; 12. *Heterodera*; 13. *Hemicycliophora*; 14. *Macroposthonia*; 15. *Paratylenchus*; 16. *Trophotylenchulus*; 17. *Tylenchulus*; 18. *Sphaeronema*; 19. *Pratylenchus*; 20. *Hirschmanniella*; 21. *Nacobbus*. (From Ref. 2.)

symptoms. Sample collection, extraction, and identification of key nematode species, however, should follow this type of preliminary assessment.

SAMPLING

A representative sample is crucial for qualitative and quantitative assessment of nematode infestation, whether soil or plant tissue.[3,8] An excellent sample also is essential for molecular-based as well as traditional nematode-assessment procedures. Generally, a pre-determined number of subsamples of soil is collected with a cylindrical sampling tube and bulked before extraction. The exact number of subunits (cores) is determined for a given area by the purpose of the assessment. For example, soil samples composed of 20 to 30 subunits from 2 hectares (ha) may provide assessments that fall within 50% of the mean of the population levels. However, much greater sample numbers and volume of soil are needed for highly precise assessments.[9] The timing of sample collection is even more critical for securing meaningful data. In temperate and tropical regions, populations of nematodes, especially *Meloidogyne* spp., decline sharply in the absence of a host. Thus, detection failures may be encountered for certain nematodes unless sampling is performed within the root zone at mid to late summer, at harvest, or within a few weeks after harvest. The timing of sampling is less critical for cooler regions where population decline is less severe and with cyst nematodes.

EXTRACTIONS AND BIOASSAYS

The biology and life cycle must be considered in selecting the specific extraction procedures (Fig. 1). Assessments of infestation of ectoparasitic nematodes are limited to soil extractions, whereas most affected soil and/or plant tissue may be assayed for endoparasites. Life stages that occur in soil and/or in plant tissue range from mobile to quiescent forms and include juveniles, adults, cysts, individual eggs, egg masses, and various combinations. Numerous methods for extracting nematodes are available.[1,3,8,10] For most extraction procedures, nematodes are separated from soil by a combination of flotation and sieving, or they are encouraged to emerge from plant tissues by being placed in a wet environment. Specialized extraction procedures are also available for assessing nematode eggs (especially in those highly damaging species, *Meloidogyne, Globodera,* and *Heterodera*[8,10]).

An appropriate host plant can be used in performing a bioassay for assessing low-level infestations, genetic variants, or related host resistance, and for assessing nematicide efficacy for highly aggressive nematodes, including species of *Meloidogyne, Globodera,* and *Heterodera*.[8] Bioassays are invaluable in assessment of nematode races/pathotypes that may attack given sources of host resistance. For an immediate assessment of potential tissue infection by these parasites, a tissue stain such as acid fuchsin-glycerin or cotton blue-glycerin greatly enhances the visibility of endoparasites via microscopy. The gelatinous egg matrix or egg mass produced by *Meloidogyne* spp. may be stained with phloxine B to facilitate the quantification of these structures on plant roots.[8]

NEMATODE IDENTIFICATION

For a detailed treatment on species identification, the reader is referred to the earlier article by J. G. Baldwin that addresses classification and identification of nematodes. Currently, identification of these parasites is based largely on morphology and in some instances on cytology, host response, biochemistry, and molecular diagnostics. As discussed earlier, symptoms and signs on plants may be characteristic of certain nematode species, but identifications should be confirmed by more precise criteria. In addition to nematode morphology, differential host tests are important for assessing infestation of certain nematode species. For many root-knot nematode populations, the North Carolina differential host test has been used widely to differentiate the common species *M. arenaria*, *M. hapla*, *M. incognita*, and *M. javanica*, as well as related host races. This system is especially useful for cropping systems that include peanut, cotton, corn, soybean, and tobacco. In another widely used host-nematode system, 16 host races of *Heterodera glycines,* based on four soybean-genotype differentials, are central to characterizing genetic variants in given fields and facilitating the development of new breeding lines/cultivars with resistance to this pathogen. The subspecies designation system for *H. glycines* was recently revised to include seven host differentials with field populations placed in ''HG Types.''[11]

APPLICATION OF BIOTECHNOLOGY IN NEMATODE ASSESSMENT

After more than a century of taxonomic research on nematodes, only an estimated 10% of all nematode biodiversity has been characterized, primarily morphospecies.[4] Although morphology continues as the primary focus in nematode identification, protein and nucleic acid analyses offer much for the future. These emerging approaches to assessment and characterization of nematode species and populations offer much for advancing our understanding of nematode biodiversity, general ecology, and population assessment for management and regulatory purposes. Molecular methods for nematode identification could be more objective and reliable than traditional systems. Also, molecular probes may be suitable for any life stage present in a sample.

Electrophoretic enzyme phenotypes have proven to be a practical means for identifying species and/or subspecies of *Anguina*, *Meloidogyne, Globodera, Heterodera,* and *Radopholus*.[3,5] The utility of identification of the common *Meloidogyne* species using isozyme phenotypes has been particularly effective in the root-knot nematode research projects. For example, *M. arenaria, M. incognita, M. javanica, M. mayaguensis,* and *M. naasi* can be distinguished via esterase phenotypes. Differential esterase alleles also have proven useful for demonstrating multiple matings among inbred lines of *H. glycines* as well as tracking crosses of certain inbred lines of that species. In addition to the isozyme phenotypes, differences in 2-D protein patterns have proven useful for separating species and infraspecies or variants of *Globodera, Heterodera,* and *Meloidogyne*. Unique proteins are available that could be used to develop serological kits for detection.[7] Monoclonal and/or polyclonal antibodies have been developed for *Bursaphelenchus, Ditylenchus, Globodera, Heterodera,* and *Meloidogyne.* For example, species-specific proteins were used to develop monoclonal antibodies for distinguishing *Globodera rostochiensi* and *G. pallida* in immunoassays.[6] These monoclonal antibodies, in addition to separating the two species when used in an ELISA, have potential for the quantitative determination of potato cyst nematodes in soil samples. This technology has great potential for use in regulatory as well as diagnostic and management programs.

The range of techniques now available for studying DNA has resulted in much progress in characterizing genomic, mitochondrial, and ribosomal DNA of nematodes. In addition to advancing our understanding of nematode phylogenetic relationships, this rapidly expanding information also provides useful tools for nematode diagnostics (Fig. 2). For molecular diagnostic assessments, the objective is to obtain low-cost, user-friendly, reliable molecular data that distinguish nematode species or subspecies levels.[3,5,6,12] The DNA techniques developed and utilized in the diagnostic purposes include amplified fragment length polymorphisms (AFLP); specific oligonucleotide probe; microsatellite and repetitive DNA; polymerase chain reaction (PCR) and multiplexed PCR; randomly amplified polymorphic DNA (RAPD); restriction fragment-length polymorphisms (RFLP); ribosomal DNA (rDNA) intergenic (IGS) and internally transcribed (ITS) spacer region PCR-RFLP and sequence; sequence-characterized amplified region (SCAR); sequence tag site (STS); and variable number tandem repeat (VNTR). In addition to assaying the sedentary endoparasite nematodes *Globodera, Heterodera, Meloidogyne, Nacobbus*, and *Cactodera*, specific DNA diagnostic methods have also been applied to migratory ecto- and endoparasites such as *Anguina, Belonolaimus, Bursaphelenchus, Criconemella, Ditylenchus, Hoplolaimus, Helicotylenchus, Pratylenchus, Radopholus*, and to the virus vector nematodes *Longidorus, Trichodorus*, and *Xiphinema*.[3,5,12]

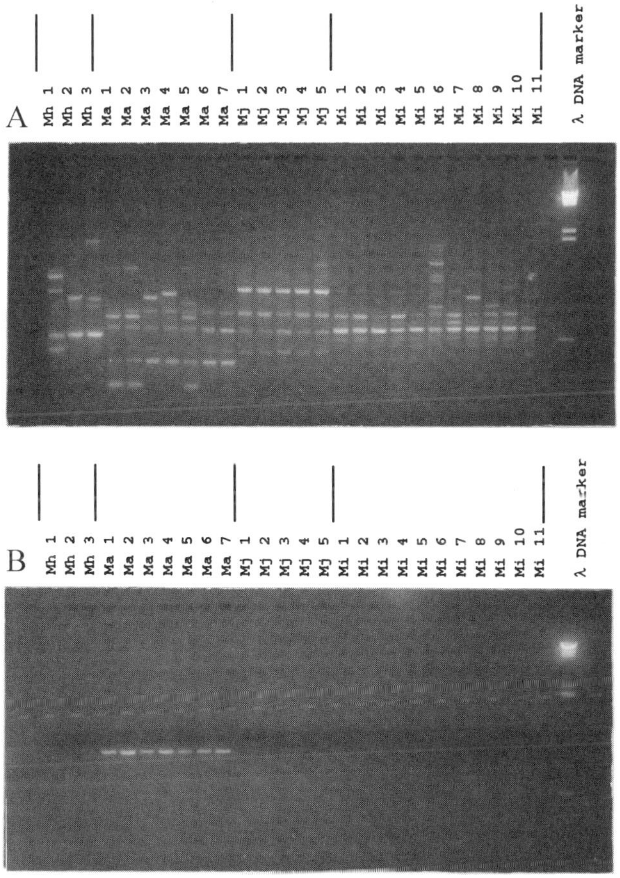

Fig. 2 Electrophoretic patterns from Meloidogyne species. A) RAPD patterns from primer AAAACCGGGC showing variation among nematode species and populations. The 26 tested isolates from left to right were three M. hapla (Mh) (VA, NC, WI); seven M. arenaria (Ma) (Canada race 2, GA race 1, SC race 2-Florence, NC race 2-EM, SC race 2-Govan, SC race 2-83, SC race 2-Rawl); five M. javanica (Mj) (NC, FL, AZ1, AZ2, GA); and 11 M. incognita (MI) (NC race 1, NC, race 2, NC race 3, NC race 4, TN, NC-cotton IA, GA, SC-Cotton, SC-Edisto, NC3-99); B) Specific-specific PCR patterns from M. arenaria primer pair (Ma-f TCGAGGGCATCTAATAAAGG and Ma-r GGGCTGAATATTCAAAGGAA). The same 26 nematode populations tested in Fig. 2-A were utilized in this test. (From Ref. 12.)

CONCLUSION

Based on the ongoing molecular characterization of nematode-plant interactions, the above approaches should prove fruitful for species and population-level assessment of these important crop pathogens.

ARTICLES OF FURTHER INTEREST

Bacterial Pathogens: Detection and Identification Methods, p. 84
Biological Control of Nematodes, p. 134
Classification and Identification of Nematodes, p. 283
Ecology and Agricultural Sciences, p. 401
Integrated Pest Management, p. 612
Management of Nematode Diseases: Options, p. 684
Molecular Technologies and Their Role Maintaining and Utilizing Genetic Resources, p. 757
Nematode Biology, Morphology, and Physiology, p. 781
Nematode Feeding Strategies, p. 784
Nematode Population Dynamics, p. 797
Nematode Problems: Most Prevalent, p. 800
Nematodes and Host Resistance, p. 805
Organic Agriculture As a Form of Sustainable Agriculture, p. 846
Plant Responses to Stress: Nematode Infection, p. 1014

Polyploidy, p. 1038
Rhizosphere: Microbial Populations, p. 1090
Seed Borne Pathogens, p. 1126
Sustainable Agriculture: Ecological Indicators, p. 1191

REFERENCES

1. Shurtleff, M.C.; Averre, C.W., III. *Diagnosing Plant Diseases Caused by Nematodes*; APS Press, The American Phytopathological Society: St. Paul, MN, **2000**; 1–187.
2. Siddiqi, M.R. *Tylenchida Parasites of Plants and Insects,* 2nd Ed.; CABI Publishing: Wallingford, UK, **2000**; 1–833.
3. Barker, K.R.; Davis, E.L. Assessing Plant-Nematode Infestations and Infections. In *Advances in Botanical Research, Incorporating Advances in Plant Pathology, Vol. 23 Pathogen Indexing Technologies*; De Boer, S.H., Andrews, J.H., Tommerup, I.C., Callow, J.A., Eds.; Academic Press: London, **1996**; 103–136.
4. Coomans, A. Nematode systematics: Past, present and future. Nematology **2000**, *2* (1), 3–7.
5. Jones, J.T.; Phillips, M.S.; Armstrong, M.R. Molecular approaches in plant nematology. Fundam. Appl. Nemat. **1997**, *20* (1), 1–14.
6. Robinson, M.P.; Butcher, G.; Curtis, R.H.; Davies, D.G.; Evans, K. Characterization of a 34 kD protein from potato cyst nematodes, using monoclonal antibodies with potential for species diagnosis. Ann. Appl. Biol. **1993**, *123* (2), 337–347.
7. Tastet, C.; Bossis, M.; Renault, L.; Mugníery, D. Protein variation in tropical *Meloidogyne* spp. as shown by two-dimensional electrophoregram computed analysis. Nematology **2000**, *2* (3), 343–353.
8. *An Advanced Treatise on Meloidogyne Vol. II Methodology*; Barker; K.R.; Carter, C.C.; Sasser, J.N.; Eds.; North Carolina State University Graphics: Raleigh, **1985**; 1–223.
9. Schomaker, C.H.; Been, T.H. A model for infestation foci of potato cyst nematodes, *Globodera rostochiensis* and *G. pallida*. Phytopathology **1999**, *89* (7), 583–590.
10. Hooper, D.J. Extraction and Processing of Plant and Soil Nematodes. In *Plant Parasitic Nematodes in Subtropical and Tropical Agriculture*; Luc, M., Sikora, R.A., Bridge, J., Eds.; CAB International: Wallingford, UK, **1990**; 45–68.
11. Niblack, T.L.; Arelli, P.R.; Noel, G.R.; Opperman, C.H.; Orf, J.H.; Schmitt, D.P.; Shannon, J.G.; Tylka, G.L. A revised classification scheme for genetically diverse populations of *Heterodera glycines*. J. Nematol. **2002**, *34*, 279–288.
12. Dong, K.; Dean, R.A.; Fortnum, B.A.; Lewis, S.A. Development of PCR primers to identify species of root-knot nematodes: *Meloidogyne arenaria, M. hapla, M. incognita, and M. javanica*. Nematropica **2001**, *31*, 271–280.

Nematodes: Parasitism Genes

N

Eric L. Davis
North Carolina State University, Raleigh, North Carolina, U.S.A.

INTRODUCTION

Secretions from the stylet (hollow mouth spear) of phytonematodes mediate the process of plant parasitism, including nematode penetration and migration within host plant roots, and dramatic transformations of plant cells into permanent feeding sites for the nutrition of sedentary nematode life stages. The stylet secretions are produced in three elaborate esophageal gland cells that have evolved to enable plant parasitism by tylenchid nematodes. The genes expressed within the gland cells that encode the stylet secretions are termed parasitism genes. The nature and number of parasitism genes in phytonematodes identified to date has been somewhat surprising, and includes significant evidence that phytonematodes may have acquired some of their parasitism genes from other organisms.

PARASITISM GENES

The primary morphological adaptations of nematodes for parasitism of plants include a protrusible stylet (hollow feeding spear) connected to three elaborate esophageal secretory gland cells (Fig. 1). These adaptations allow nematodes to feed on plant cells directly from outside plant roots (as ectoparasites) in soil or to penetrate plant roots (as endoparasites) to modify and feed from internal plant tissues (see R. S. Hussey, this volume). Although the genes that control the formation of the stylet have evolved to promote plant parasitism, it is the genes that encode the secretions synthesized in the esophageal gland cells that control the dynamic interaction of the nematode with its plant host.[1,2] Changes occur within sedentary nematodes during plant parasitism—including atrophy of the two subventral esophageal gland cells shortly after feeding site formation—and the concomitant increase in the size and activity of the single dorsal esophageal gland cell (Fig. 1). Products from the nematode gland cells may be secreted through the stylet (Fig. 2) into plant tissues for relatively simple feeding processes such as the predigestion of plant cell contents by some ectoparasitic nematodes, or for the complex processes of plant tissue penetration and feeding site formation by some sedentary endoparasitic nematodes. The potential range of functions of nematode secretions and their regulated expression during plant parasitism make the genes that encode nematode esophageal gland secretions the primary parasitism genes of plant-parasitic nematodes.[1]

The isolation of nematode stylet secretions to analyze their components has been difficult because the nematodes are microscopic in size and they are obligate parasites. Monoclonal antibodies have been generated that bind to specific secretory proteins within the esophageal gland cells of both cyst and root knot nematodes (Fig. 2), and the antibodies have demonstrated that the proteins can be secreted through the nematode stylet and that the gland cell secretions change during the course of plant parasitism.[1] One gland-specific monoclonal antibody was used to immunoaffinity-purify a protein produced in the subventral esophageal gland cells of both the soybean cyst nematode (*Heterodera glycines*) and the potato cyst nematode (*Globodera rostochiensis*). The amino acid sequence of the purified gland protein was used to design degenerate oligonucleotides to obtain cyst nematode complementary DNA (cDNA) clones from the cyst nematodes that encode the subventral gland protein. Database searches indicated that the cDNA clones encoded β 1,4 endoglucanases (cellulases)—the first endogenous cellulases to be isolated from animals.[3] It was further demonstrated that the cellulases could degrade a carboxymethylcellulose substrate,[3] that the cellulases are secreted into plant tissues (Fig. 3) by cyst nematodes,[4] and that the cellulases are expressed exclusively within the cyst nematode subventral gland cells only during the motile life stages.[5] Interestingly, β-1,4-endoglucanases (and likely other cell wall-modifying enzymes) of plant origin are upregulated in nematode feeding sites when the nematodes become sedentary.[6] The data strongly supported a role for cellulase secretion by cyst nematodes to facilitate their penetration and intracellular migration in plant roots but not for feeding site formation. One striking feature of nematode cellulase genes was their strong similarity to bacterial cellulase genes and relatively weak similarity to cellulases of eukaryotic origin or to any genes identified in the model nematode *Caenorhabditis elegans*.[7] Because the nematode origin of the cellulases was confirmed by both in situ localization studies and analyses of nematode cellulase genomic clones, microbial contamination was ruled out. It has been postulated that one

Encyclopedia of Plant and Crop Science
DOI: 10.1081/E-EPCS 120010392

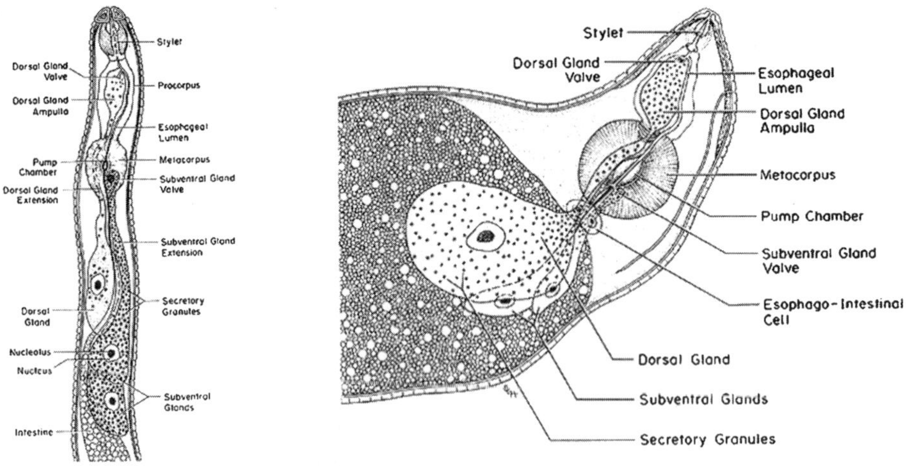

Fig. 1 (Left) Illustration of a plant-parasitic nematode second-stage juvenile showing the stylet used for feeding from plant cells and the three esophageal gland cells used to produce stylet secretions. (Reproduced with permission from Ref. 2.) (Right) Illustration of a plant-parasitic nematode adult female showing the increase in size and secretory granules within the single dorsal esophageal gland cell and decrease in size and secretory granules of the two subventral esophageal gland cells. (Reproduced with permission from R. S. Hussey et al., 1994, Advances in Molecular Plant Nematology, Plenum Press, NY.)

potential origin of the cellulase genes in plant-parasitic nematodes was ancient horizontal gene transfer from prokaryotes.[1,3]

Cloning of several other genes encoding secretions from the esophageal gland cells of plant-parasitic nematodes supports the hypothesis that some nematode parasitism genes—particularly those expressed in the subventral gland cells—may have been derived via horizontal gene transfer. Subtractive hybridization procedures were used to enrich a pool of cDNA clones from the root knot nematode, *Meloidogyne javanica*, for genes expressed in the esophageal gland cell region.[8] One gene that was demonstrated to be exclusively expressed in the esophageal gland cells of *M. javanica* during plant parasitism was homologous to bacterial chorismate mutase.[8] The nematode chorismate mutase gene was able to complement function in a bacterial chorismate mutase-deficient mutant. A cDNA clone derived from the parasitic stages of *Meloidogyne incognita* by RNA

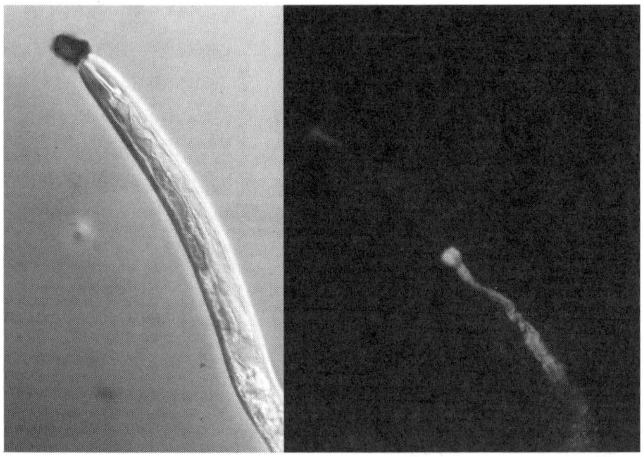

Fig. 2 (Left) Proteinaceous secretions (stained with Coomassie Brilliant Blue) from the stylet of a soybean cyst nematode second-stage juvenile that had esophageal gland activity stimulated by incubation in vitro in the serotonin agonist, 5-methoxy-DMT oxalate; (Right) Binding of a fluorescently labeled monoclonal antibody to secretory granules within the subventral esophageal gland cells of a soybean cyst nematode second-stage juvenile. (*View this art in color at www.dekker.com.*)

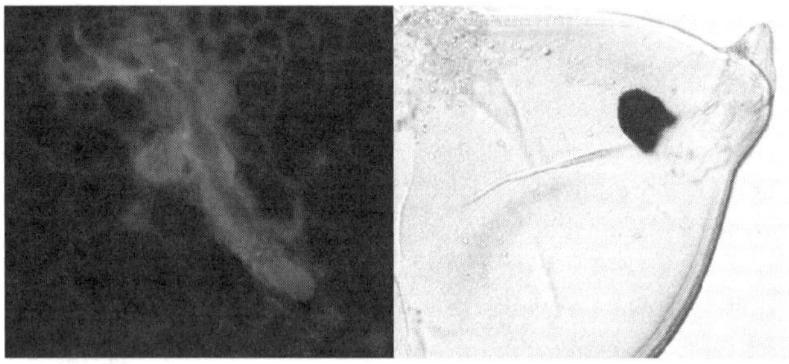

Fig. 3 (Left) Binding of anticellulase sera (green fluorescence) to cellulases secreted in planta by a soybean cyst nematode second-stage juvenile during migration through the soybean root cortex. (Reproduced with permission from Ref. 4.) (Right) In situ hybridization of soybean cyst nematode cDNA probe to transcripts expressed specifically within the dorsal esophageal gland cell of a soybean cyst nematode adult female. (Reproduced with permission from Ref. 12.) (*View this art in color at www.dekker.com.*)

fingerprinting encoded a predicted cellulose-binding protein with homology to bacterial cellulose binding domains, and its expression was localized within the nematode's subventral esophageal gland cells.[9] Genes from cyst nematodes that encode the cell wall–degrading enzyme pectate lyase are also expressed in the subventral gland cells of infective stages, and the nematode pectate lyases have their greatest similarity to pectate lyases of bacteria and fungi.[10]

Although the examples of potential horizontal acquisition of some nematode parasitism genes have been unexpected, the likelihood that some endogenous nematode genes have evolved roles in plant parasitism is great.[1] It is also clear that large-scale isolation and identification of genes expressed in parasitic nematodes is a more efficient and comprehensive way to isolate parasitism genes than to isolate and purify secreted nematode proteins. Expressed sequence tags (ESTs) provide partial sequence of random cDNA clones from a given library, which can be used for homology searches and genome-mapping efforts. cDNA libraries can be generated from whole nematodes of any chosen life stage for direct EST analysis, but the identification of potential parasitism genes (i.e., gland genes) becomes difficult unless the random sequences provide an obvious database homologue (i.e., cellulase, pectinase, etc.). One strategy to narrow the search for parasitism genes is to differentially compare genes expressed in whole nematodes between preparasitic and parasitic life stages—to see what genes are turned on in the nematode during parasitism. This was the basis of the RNA fingerprinting strategy that isolated the cellulose-binding gene expressed in the subventral gland cells of *M. incognita*.[9] It also has been observed that the activity and contents of the esophageal gland cells of *G. rostochiensis* change in preparation for plant parasitism when infective juveniles are incubated in potato root diffusate (PRD). A cDNA-AFLP method was used to compare genes expressed in juveniles that were incubated in water and juveniles that were incubated in PRD, and a number of genes specifically upregulated in *G. rostochiensis* from PRD treatment were isolated and confirmed to be expressed specifically within the nematode's esophageal gland cells.[11]

A tissue-specific way to isolate nematode parasitism genes is to directly target the genes expressed within the nematode esophageal gland cells. The chorismate mutase gene (mentioned earlier) expressed in the gland cells of *M. javanica* was isolated by differentially screening cDNA from sections of nematode esophageal gland regions and (glandless) tail regions that were manually dissected.[8] However, the most direct method has been to exclusively microaspirate the contents of nematode esophageal gland cells and use the gland cell mRNA in RT-PCR to create gland cell cDNA libraries.[12,13] Contents of the esophageal gland cells microaspirated from mixed parasitic stages of *H. glycines* were used to generate a cDNA library that profiled gene expression in the gland cells during the entire parasitic process. Gland cell cDNA clones that contained a secretion signal peptide were selected using a specialized vector in yeast, and 14 unique secretory *H. glycines* gland cell cDNA clones were identified.[12] mRNA in situ hybridization was used to demonstrate that a number of the clones were specifically upregulated within the dorsal gland cell of parasitic stages of *H. glycines* (Fig. 3), but not within preparasitic stages. Another gland-cell cDNA library prepared similarly from mixed parasitic stages of *H. glycines* was used in suppression subtractive hybridization (SSH) against cDNA from the tail region to yield 23 unique clones from the gland cell library.[13] A LD-PCR cDNA library prepared from the same *H. glycines* gland cell cytoplasm was used to obtain full-length clones of the gland cell cDNAs. Sequences of the full-length clones indicated that 10 unique cDNA clones contained a predicted signal peptide, and mRNA in situ hybridizations demonstrated that four of the clones with signal peptides were expressed specifically within the esophageal gland cells of *H. glycines*.[13]

The initial attempts (discussed earlier) to select genes from the *H. glycines* esophageal gland-cell cDNA libraries suggest that the libraries are an enormous source of potential nematode parasitism genes—worthy of extensive EST analysis without prior selection. Bioinformatics development and analysis will be a key component to deduce potential nematode parasitism genes among the many clones, and a scheme to prioritize candidate parasitism genes has been developed. Database homology searches using the Basic Local Alignment Search Tool (BLAST) algorithm may immediately suggest candidate genes with potential function in parasitism. cDNA clones that contain complete 5′-end sequence can be analyzed by the SignalP and PSORT II algorithms for the presence of signal peptides and predicted subcellular (i.e., extracellular) localization, respectively. These analyses provide a manageable number of cDNA clones for high-throughput mRNA in situ hybridization to confirm the expression of secretion genes within the nematode esophageal gland cells.[12,13] For example, one *H. glycines* gland-cell library cDNA clone that had homology to a hypothetical *C. elegans* protein in BLAST analysis, and a second cDNA clone that had homology to a salivary proline-rich glycoprotein of *Rattus norvegicus*, both had predicted extracellular (secreted) localization by PSORT II and were confirmed to be expressed in *H. glycines* esophageal gland cells by in situ hybridization.[13] In other cases, however, sequence of *H. glycines* gland-cell library cDNA clones provided no database homologues (i.e., pioneers) using BLAST analysis, yet the encoded peptides were predicted to be secreted and were expressed exclusively within the dorsal esophageal gland cell of only parasitic stages of *H. glycines*.[12]

CONCLUSION

As bioinformatic analyses are improved and functional assays for candidate nematode parasitism genes are developed in the near future,[1] our understanding of the molecular basis of nematode parasitism of plants will accelerate. Knowledge of the function of nematode parasitism genes will present multiple targets for intervention in the parasitic process and lay the foundation for the development of novel strategies to reduce nematode damage to crops.

ARTICLES OF FURTHER INTEREST

Breeding Plants with Transgenes, p. 193
Functional Genomics and Gene Action Knockouts, p. 476
Management of Nematode Diseases: Options, p. 684
Nematode Biology, Morphology, and Physiology, p. 781
Nematode Feeding Strategies, p. 784
Nematodes and Host Resistance, p. 805
Plant Response to Stress: Nematode Infection, p. 1014
RNA-mediated Silencing, p. 1106
Symbioses with Rhizobia and Mycorrhizal Fungi: Microbe/Plant Interactions and Signal Exchange, p. 1213
Transgenes: Expression and Silencing of, p. 1242

REFERENCES

1. Davis, E.L.; Hussey, R.S.; Baum, T.J.; Bakker, J.; Schots, A.; Rosso, M.N.; Abad, P. Nematode parasitism genes. Annu. Rev. Phytopathol. **2000**, *38*, 341–372.
2. Hussey, R.S. Disease-inducing secretions of plant-parasitic nematodes. Annu. Rev. Phytopathol. **1989**, *27*, 123–141.
3. Smant, G.; Stokkermans, J.; Yan, Y.; de Boer, J.M.; Baum, T.; Wang, X.; Hussey, R.S.; Davis, E.L.; Gommers, F.J.; Henrissat, B.; Helder, J.; Schots, A.; Bakker, J. Endogenous cellulases in animals: Cloning of expressed β-1,4-endoglucanase genes from two species of plant-parasitic cyst nematodes. Proc. Natl. Acad. Sci. U. S. A. **1998**, *95*, 4906–4911.
4. Wang, X.; Meyers, D.M.; Baum, T.J.; Smant, G.; Hussey, R.S.; Davis, E.L. In planta localization of a β-1,4-endoglucanse secreted by *Heterodera glycines*. Mol. Plant-Microb. Interact. **1999**, *12*, 64–67.
5. deBoer, J.M.; Yan, Y.; Wang, X.; Smant, G.; Hussey, R.S.; Davis, E.L.; Baum, T.J. Developmental expression of secretory beta-1,4-endoglucanases in the subventral esophageal glands of *Heterodera glycines*. Mol. Plant-Microb. Interact. **1999**, *12*, 663–669.
6. Goellner, M.; Wang, X.; Davis, E.L. Endo-beta-1,4-glucanase expression in compatible plant–nematode interactions. Plant Cell **2001**, *13*, 2241–2255.
7. *C. elegans* sequencing consortium. Genome sequence of the nematode *C. elegans*: A platform for investigating biology. Science **1998**, *282*, 2012–2018.
8. Lambert, K.N.; Allen, K.D.; Sussex, I.M. Cloning and characterization of an esophageal-gland-specific chorismate mutase from the phytoparasitic nematode *Meloidogyne javanica*. Mol. Plant-Microb. Interact. **1999**, *12*, 328–336.
9. Ding, X.; Shields, J.; Allen, R.; Hussey, R.S. A secretory cellulose-binding protein cDNA cloned from the root-knot nematode (*Meloidogyne incognita*). Mol. Plant-Microb. Interact. **1998**, *11*, 952–959.
10. Popeijus, H.; Overmars, H.; Jones, J.T.; Blok, V.; Goverse, A.; Helder, J.; Schots, A.; Bakker, J.; Smant, G. Degradation of plant cell walls by a nematode. Nature **2000**, *406*, 36–37.
11. Qin, L.; Overmars, H.; Helder, J.; Popeijus, H.; van der Voort, J.; Groenink, W.; van Koert, P.; Schots, A.; Bakker, J.; Smant, G. An efficient cDNA-AFLP-based strategy for the identification of putative pathogenicity factors from the potato cyst nematode *Globodera rostochiensis*. Mol. Plant-Microb. Interact. **2000**, *13*, 830–836.
12. Wang, X.; Allen, R.; Ding, X.; Goellner, M.; Maier, T.; de Boer, J.M.; Baum, T.J.; Hussey, R.S.; Davis, E.L. Signal peptide-selection of cDNA cloned directly from the esophageal gland cells of the soybean cyst nematode, *Heterodera glycines*. Mol. Plant-Microb. Interact. **2001**, *14*, 536–544.
13. Gao, B.; Allen, R.; Maier, T.; Davis, E.L.; Baum, T.J.; Hussey, R.S. Identification of putative parasitism genes expressed in the esophageal gland cells of the soybean cyst nematode, *Heterodera glycines*. Mol. Plant-Microb. Interact. **2001**, *14*, 1247–1254.

Nematode Population Dynamics

Robert McSorley
University of Florida, Gainesville, Florida, U.S.A.

N

INTRODUCTION

Population dynamics focuses on the rise and fall of nematode population numbers over time. Nematode population changes are somewhat predictable, although patterns vary depending on whether crops are annual or perennial. The host status of the crop to plant-parasitic nematodes is critical in determining rates of nematode buildup or decline. Many other biological, physical, and chemical factors affect nematode population dynamics. Knowledge of population dynamics is important in planning crop rotations and sequences with minimal nematode damage. However, much additional biological information may be needed to understand population dynamics in specific sites and situations.

IMPORTANCE OF NEMATODE POPULATION DYNAMICS

Plant-parasitic nematodes cause serious losses on many different crops, and in many instances, thresholds for damage to crops by nematodes have been estimated.[1,2] Above the threshold population density—tolerance limit—the severity of nematode damage increases as population density increases.[3] Properly collected soil samples can reveal the kind and number of nematodes present and indicate whether the threshold for damage has been exceeded. Although nematodes can be moved from place to place in soil or on equipment, they cannot migrate easily on their own. Therefore, the nematode population density in a site is unlikely to be changed much by immigration or emigration, and population growth or decline within the site determines most future trends in nematode population levels.

NEMATODE POPULATION GROWTH OVER TIME

If population density of a plant-parasitic nematode on a favorable host crop is plotted over time, population growth may follow a logistic growth curve (Fig. 1). Individual points typically show deviation from the theoretical logistic curve due to sampling error. Population growth from low initial densities may be quite rapid, approaching exponential growth.[4,5] Ultimately, population growth slows and population density levels off as a carrying capacity—or equilibrium density (E)—is reached.[5] Fluctuations in population growth may be particularly severe as E is first approached, with population density stabilizing over time.

Annual Crops

On an annual crop, the population growth curve (Fig. 1) may correspond with the life cycle of the crop. Once the crop is removed or the roots deteriorate, population levels may fall due to lack of a host or may increase to even greater levels if the next crop planted is a better host and has a greater E value. Over a long period of time, nematode population densities will appear as a sequence of peaks and valleys, increasing to the equilibrium densities appropriate for favorable crops and conditions, and decreasing if poor hosts are grown or during winter or fallow periods.

Perennial Crops

Long-term fluctuations in nematode population levels occur on perennial crops as well. Seasonal growth flushes of active feeder roots provide feeding sources for nematodes and opportunities for population growth. Populations may decline as older roots senesce or as plants approach winter dormancy. Because a root food source may be present most of the time, it is difficult to determine whether nematode population fluctuations are influenced directly by key environmental factors such as temperature and moisture or indirectly by the effect of these factors on root growth.

NEMATODE HOST STATUS

An impression of the host status of a crop cultivar relative to a particular nematode species can be obtained by examining the relationship between the initial population density (P_i) and the final population density (P_f) reached after a constant time interval. The life span of an annual crop is often used as a convenient time interval. If $P_f > P_i$, then the nematode population has increased on that crop,

DOI: 10.1081/E-EPCS 120010386

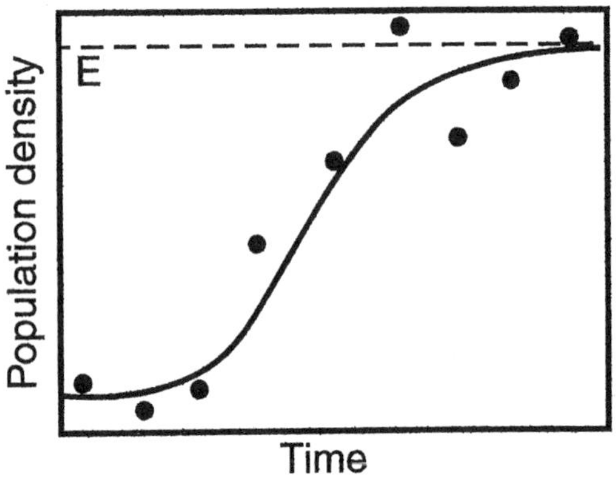

Fig. 1 Typical relationship between nematode population density and time for a series of points, showing smooth logistic growth curve. E=equilibrium density (dashed line).

and the crop could be considered a host of the nematode. A comparison of P_f/P_i for a group of different crop cultivars could be used to provide a comparison of which cultivars were good hosts or poor hosts. However, the relationship between P_f and P_i may be too complex to make such straightforward comparisons.[5] If the natural logarithm of P_f is plotted against the natural logarithm of P_i across a range of different P_i, a curved relationship typically results, leveling off at E (Fig. 2). At a very low P_i, a cultivar may appear to be a good host ($P_f > P_i$), but at a very high P_i, the same cultivar may appear to be a poor host ($P_f < P_i$). The potential for population increase is much greater if P_i is low than if P_i is near E. Therefore, relative comparisons of host status must be made at constant P_i and over a constant time interval. The designation of cultivars as good host, intermediate host, or poor host is relative, and depends on the shape of the relationship between P_f and P_i.[5]

FACTORS AFFECTING POPULATION DYNAMICS

A variety of biological factors affect nematode population dynamics. The genetics, degree of resistance, and resistance mechanisms available are essential features of the crop cultivar that affect its host status to nematodes. The responses of the same nematode species on the same crop cultivar may differ from place to place due to genetic and physiological variation among nematode populations. Recognized biotypes, races, or pathotypes are known in a number of important nematode genera including *Heterodera, Globodera,* and *Meloidogyne*.[6] But even when biotypes are not recognized, isolates (populations from different sources) of the same species may respond differently in their virulence and population growth.

Predators, parasites, and competitors all have the potential to affect nematode population dynamics and change the equilibrium density for a particular situation. When fungal parasites or other nematode antagonists are present, population densities of plant-parasitic nematodes stabilize to lower levels than those present in soils without antagonists or in soils from which antagonists have been removed.[7]

A number of physical and chemical soil properties affect nematode dynamics to varying degrees. These include texture, pH, organic matter, salinity, aeration, bulk density, nutrient status, as well as moisture and temperature.[4] Rainfall and soil type interact to determine soil moisture, which greatly affects nematode activity. Activity and population growth may cease or decline as soil conditions become too dry or too cool. Nematode population growth is directly related to temperature as well as time. Therefore, the use of heat units or degree days often provides a forecast of population growth more accurate than the use of time alone.[4]

POPULATION DYNAMICS AND NEMATODE MANAGEMENT

Many agricultural practices are used to suppress nematode population levels.[2] The use of crop rotation and cover crops for nematode management is based on the disruption of nematode population cycles. By understanding the relative degree of buildup or decline on various crops, an optimum crop sequence can be planned for minimizing nematode impact. The P_f at the end of one crop becomes the P_i for the next crop planted in the site, and so forth. Winter or fallow periods must be included as well,

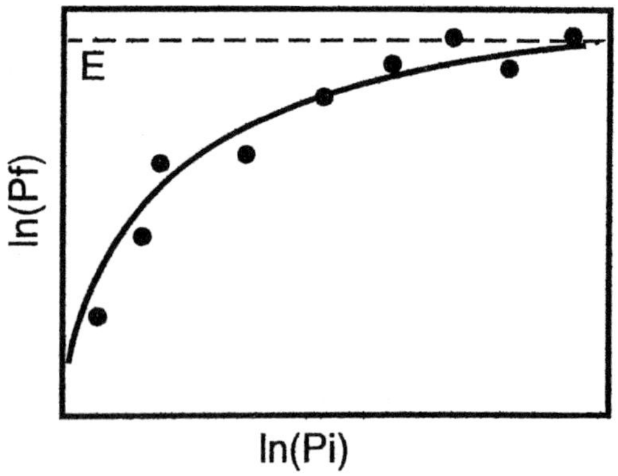

Fig. 2 Typical relationship between the natural logarithm of final nematode population density (P_f) and logarithm of initial population density (P_i) for a series of points, showing smooth curve. E=equilibrium density (dashed line).

because they result in population decline ($P_f < P_i$). Using this approach, simple models of crop sequences and nematode dynamics over time can be constructed.[4] Models forecasting population dynamics are available for a number of plant-nematode systems.[4]

With perennial crops, methods may be directed toward minimizing equilibrium densities of plant-parasitic nematode over time. This may be accomplished by use of resistant rootstocks, if available. The task is more difficult with susceptible perennials, but plant-parasitic nematode numbers in orchards of various fruit crops were suppressed or stabilized at relatively low levels when natural enemies were present.[7] Efforts to conserve and enhance the naturally occurring antagonists that are common in agricultural soils have focused primarily on the use of appropriate organic amendments, tillage practices, and cover crops, but much remains to be learned in this area.[2,7]

CHALLENGES

Measuring changes in nematode population levels and forecasting population growth over time requires sampling to determine P_i and other critical population estimates. The difficulty and variability involved in nematode sampling is well-known, and so estimates of P_i may be associated with substantial error. If P_i values measured with error are used in population dynamics equations and models, the error of the estimate may be magnified over time. Therefore, periodic sampling of nematode population densities in future crops and sequences is useful to be sure that forecast estimates are not drifting far from reality. Alternatively, a more complex and stochastic approach could be used, providing model output as probability distributions or ranges rather than mean population densities.[4]

The greater limitation in applying population dynamics in nematode management is the relative lack of specific information on the dynamics of particular nematode species and isolates on specific crop cultivars. Furthermore, the nematode-host relationship varies with local agricultural practices and environmental conditions. This type of highly specific information is required to better anticipate nematode population changes and design appropriate management strategies.

CONCLUSION

Although deficiencies exist in our current knowledge of nematode population dynamics, new information is continually becoming available on nematode population response and host status on existing crop cultivars, new crop cultivars, and on potential and currently used rotation crops and cover crops. As concern over local variation in nematode isolates and dynamics increases, small on-site tests of host status and nematode buildup on candidate crop cultivars should become more frequent. As chemical nematicides become more limited, demand should increase for alternative practices such as resistant cultivars, rotation crops, cover crops, or allowing land to lie fallow. All of these alternative methods require a detailed understanding of nematode population dynamics to be used effectively.

ARTICLES OF FURTHER INTEREST

Biological Control of Nematodes, p. 134
Classification and Identification of Nematodes, p. 283
Ecology and Agricultural Sciences, p. 401
Nematode Infestations: Assessment, p. 788
Nematode Problems: Most Prevalent, p. 800
Nematodes and Host Resistance, p. 805

REFERENCES

1. *Plant Parasitic Nematodes in Temperate Agriculture*; Evans, K., Trudgill, D.L., Webster, J.M., Eds.; CAB International: Wallingford, UK, 1993.
2. McSorley, R.; Duncan, L.W. Economic Thresholds and Nematode Management. In *Advances in Plant Pathology*; Andrews, J.H., Tommerup, I., Eds.; Academic Press: London, 1995; Vol. 11, 147–171.
3. Seinhorst, J.W. The relation between nematode density and damage to plants. Nematologica **1965**, *11*, 137–154.
4. McSorley, R. Population Dynamics. In *Plant and Nematode Interactions*; Barker, K.R., Pederson, G.A., Windham, G.L., Eds.; American Society of Agronomy, Crop Science Society of America, Soil Science Society of America: Madison, WI, 1998; 109–133.
5. Seinhorst, J.W. Dynamics of populations of plant parasitic nematodes. Annu. Rev. Phytopathol. **1970**, *8*, 131–156.
6. Roberts, P.A.; Matthews, W.C.; Veremis, J.C. Genetic Mechanisms of Host-Plant Resistance to Nematodes. In *Plant and Nematode Interactions*; Barker, K.R., Pederson, G.A., Windham, G.L., Eds.; American Society of Agronomy, Crop Science Society of America, Soil Science Society of America: Madison, WI, 1998; 209–238.
7. Stirling, G.R. *Biological Control of Plant Parasitic Nematodes*; CAB International: Wallingford, UK, 1991.

Nematode Problems: Most Prevalent

Larry Duncan
University of Florida—IFAS, Lake Alfred, Florida, U.S.A.

INTRODUCTION

The phylum Nematoda (Nemata) comprises aquatic roundworms that occupy virtually all habitats on earth. Nematodes are the most numerous metazoans on earth and the phylum is one of the most species-diverse. It has been estimated that only 3% of a half-million species have been described. Nematodes are adapted to numerous life strategies with major trophic groups consisting of primary (plant parasites) and secondary consumers (bacterivores, fungivores, omnivores, predators, and animal parasites). They play a critical role in the fertility of agricultural soils as major contributors to nutrient decomposition in soil food webs, and their ubiquity makes them ideal organisms to study as biological indicators of soil processes. In contrast to the beneficial roles played by nematodes, all plant and animal species have nematode parasites. Parasitism of crop plants by some nematode species is of little economic significance, whereas others are among the most serious agricultural pests.

ECONOMIC IMPORTANCE

Crop losses due to nematodes vary by crop and region. Losses are difficult to estimate because plant-parasitic nematodes are microscopic soilborne organisms and damage frequently goes undetected. Most crop plants in a field are parasitized by several species of nematode, yet spatial patterns and population densities of damaging species are often poorly understood within fields and across regions.[3] Average losses in the major food and forage crops worldwide have been estimated in the range of 11–14%.[4] As for most pests, nematode damage to crops is greater in tropical than in temperate regions due to greater diversity and abundance of nematodes in the tropics. Average estimated losses to nematodes are also higher in less developed countries (13–17%) than in industrial countries (7–11%) due to a lack of management resources. For example, citrus yield loss to nematodes in Florida and California, where resistant rootstocks and nursery certification programs are used, do not exceed 5%,[5] whereas average yield losses of 14% are reported from other regions of the tropics and subtropics.

SELECTED NEMATODE-CROP ASSOCIATIONS

Two classes, Secernentea and Adenophorea, are generally recognized in the Nematoda. The orders Tylenchida and Aphelenchida (Secernentea) contain plant-parasitic nematodes, as does the order Dorylaimida in the Adenophorea (Table 1). The mouthparts of all plant-parasitic nematodes are modified sclerotized structures known as stylets. The stylet functions much as a hypodermic needle, puncturing cells to secrete digestive enzymes and other chemicals into plant cells, and thereafter withdrawing cell contents (Fig. 1A). The generalized life cycle of these parasites is also similar, consisting of the egg, four juvenile stages, and the adult. However, feeding habits of plant-parasitic nematodes differ and are a means to categorize species based on general behavior.

Ectoparasites and Semi-Endoparasites

Genera such as *Criconomella*, *Helicotylenchus*, *Rotylenchus*, *Hoplolaimus*, and *Tylenchorhynchus* are found frequently in the rhizospheres of many types of plants. These nematodes remain in the soil or penetrate only partially into the root cortex as they feed on root epidermal or cortical cells. Frequently they produce no measurable effect on their plant hosts. However, species in these and other ecotoparasitic genera are major pests of some crops. *Mesocriconema curvatum* (part of a group known as ring nematodes) is the most frequently encountered plant-parasitic nematode in Florida citrus orchards where it is of no economic significance. The closely related *Criconomella xenoplax* causes severe root pruning of peach trees, increases susceptibility of trees to bacterial canker by *Pseudomonas syringae*, and results in the disease complex peach-tree short life.[6] More than 1.5 million peach trees were killed by the disease in South Carolina in the decade 1990–99, and the nematode affects orchards throughout the southeastern United States and to a lesser extent in California. *Hoplolaimus* spp. (lance nematodes) parasitize a wide range of plants with little effect; however, *Hoplolaimus galeatus* damages turfgrass, corn, and cotton in the eastern United States, and *Hoplolaimus columbus* is an even more serious pathogen

Encyclopedia of Plant and Crop Science
DOI: 10.1081/E-EPCS 120010383

Table 1 Commonly encountered genera of plant-parasitic nematodes

Tylenchida	
Anguina	Scopoli, 1977
Belonolaimus	Steiner, 1949
Cactodera	Krall and Krall, 1978
Criconema	Hofmänner and Menzel, 1914
Criconomella	De Grisse and Loof, 1965
Ditylenchus	Filipjev, 1936
Dolichodorus	Cobb, 1914
Globodera	Skarbilovich, 1959
Gracilacus	Raski, 1962
Helicotylenchus	Steiner, 1945
Hemicriconemoides	Chitwood and Birchfeld, 1957
Hemicycliophora	de Man, 1921
Heterodera	Schmidt, 1871
Hirschmaniella	Luc and Goodey, 1964
Hoplolaimus	von Daday, 1905
Meloidogyne	Goeldi, 1892
Merlinius	Siddiqi, 1970
Nacobbus	Thorne and Allen, 1944
Pratylenchus	Filipjev, 1936
Punctodera	Mulvey and Stone, 1976
Radopholus	Thorne, 1949
Rotylenchulus	Linford and Oliveira, 1940
Rotylenchus	Filipjev, 1936
Scutellonema	Andrássy, 1958
Tylenchorhynchus	Cobb, 1913
Tylenchulus	Cobb, 1913
Aphelenchida	
Aphelenchoides	Fischer, 1894
Bursaphelenchus	Fuchs, 1937
Dorylaimida	
Longidorus	(Micoletzky, 1922) Thorne and Swanger, 1936
Paralongidorus	Siddiqi, Hooper and Khan, 1963
Paratrichodorus	Siddiqi, 1974
Trichodorus	Cobb, 1913
Xiphinema	Cobb, 1913

of cotton and soybean. *Scutellonema bradys* causes dry rot in stored yam tubers and causes major losses of this food staple in Africa, Asia, the Caribbean, and South America. *Helicotylenchus* spp. are among the most commonly encountered ectoparasitic nematodes worldwide, often in high numbers. There are few reports of significant damage by these nematodes with the exception of *Helicotylenchus multicinctus* on banana. The nematode is widespread throughout tropical America, Africa, the Middle East, and Asia, where it causes stunting, delayed fruit maturation, reduction of yield, plant toppling, and shortening of the productive life of the plant. Stinging nematodes, *Belonolaimus longicaudatus*, are widespread in the southeastern United States and occur also in parts of the U.S. Midwest and California. In sandy soil, the nematode is highly pathogenic to a large number of agronomic and horticultural crops including cotton, potato, sweet potato, corn, turfgrass, strawberry, and citrus. The nematode can break resistance to Fusarium wilt in crops such as cotton. In addition to lesions in the root epidermis and cortex typical of feeding by ectoparasitic nematodes, stinging nematodes frequently feed at the root meristem where they may cause cessation of root growth (stubby root symptoms) and cell hypertrophy, resulting in swelling or small galls at the tips of roots (Fig. 1B).

The economic importance of ectoparasites in the Dorylaimida derives not only from direct damage to the plant root, but also from their ability to vector nepoviruses and tobraviruses responsible for numerous crop diseases (Table 2).[7] Species of *Longidorus* (needle nematodes) also cause direct damage and yield loss in crops such as corn, strawberry, tobacco, sugarbeet, lettuce, and celery. *Xiphinema* spp. (dagger nematodes) are widespread and cause serious damage to citrus, stone and pome fruit trees, forest tree seedlings, strawberry, and numerous other crops. Stubby root nematodes (*Trichodorus* spp. and *Paratrichodorus* spp.) are widespread in many crops including sugarbeet, in which they cause Docking disorder, and potato, where they increase the incidence of corky ringspot (tobacco rattle tobravirus). All dorylaimid plant-parasitic nematodes have feeding habits and cause symptoms similar to those of stinging nematodes. Effective nematode management can be more difficult to

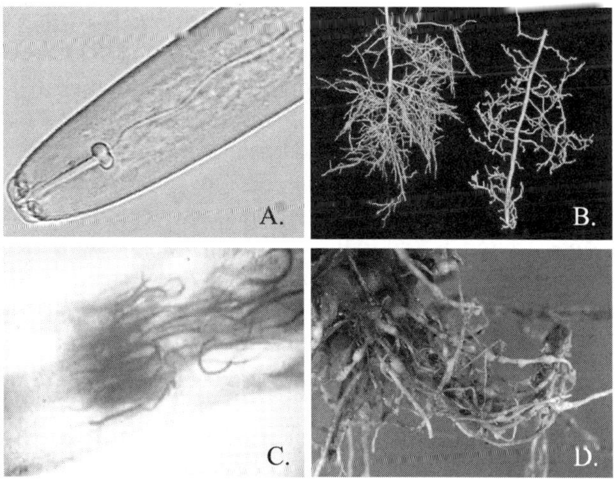

Fig. 1 (A) Anterior end of *Pratylenchus coffeae* showing the stylet used to inject gland secretions into plant cells to withdraw cell contents; (B) Healthy citrus roots (left) and roots with stubby root symptoms (right) caused by feeding of *Belonolaimus longicaudatus*; (C) Migratory endoparasitic nematodes in root cortex; (D) Galls on roots caused by root knot nematodes. (*View this art in color at www.dekker.com.*)

Table 2 Examples of virus diseases of crop plants and their nematode vectors

Vectors of nepoviruses	Virus	Partial list of susceptible crops
Xiphinema americanum	Cherry rasp leaf virus (CRLV)	Apple, peach, plantain, raspberry, cherry
	Tomato ringspot virus (ToRSV)	Raspberry, peach, cherry, almond, apple, blueberry, tobacco
	Peach rosette mosaic virus (PRMV)	Peach, grapevine, blueberry
	Tobacco ringspot virus (TRSV)	Tobacco, soybean, blueberry, cucurbits
Xiphinema diversicaudatum	Strawberry latent ringspot virus (SLRSV)	Strawberry, raspberry, celery, rose
	Arabis mosaic virus (ArMV)	Raspberry, strawberry, cubumber, sugar beet, clover, hop, rose
Xiphinema index	Grapevine fanleaf virus (GFLV)	Grapevine
Longidorus apulus	Artichoke Italian latent virus (AILV)	Artichoke, chicory
Longidorus arthensis	Cherry rosette virus (CRV)	Cherry
Longidorus elongatus	Tomato black ring virus (TBRV)	Artichoke, bean, cabbage, grapevine, lettuce, peach, potato, raspberry, tomato
Longidorus macrosoma	Raspberry ringspot virus (RRSV)	Raspberry, artichoke, red currant, strawberry, gooseberry, grapevine, narcissus
Longidorus martini	Mulberry ringspot virus (MRSV)	Mulberry
Vectors of tobraviruses		
Paratrichodorus spp. and	Pea early browning virus (PEBV)	Bean, clover, pea
Trichodorus spp.	Tobacco rattle tobravirus (TRV)	Potato, gladiolus, pepper, narcissus, hyacinth, tulip

achieve for the dorylaimid species because reducing population numbers below levels that cause direct damage is often insufficient to prevent virus transmission.

Migratory Endoparasites

The most important groups of migratory endoparasitic nematodes are lesion nematodes (*Pratylenchus* spp.), burrowing nematodes (*Radopholus* spp.), and rice root nematodes (*Hirshmaniella* spp.). All life stages of the species in these groups are found within the roots of host plants where the nematodes migrate intercellularly (Fig. 1C). Secondary bacterial and fungal pathogens are important components of the damage resulting from lesions caused by migration of these nematodes. More than half of the world's rice fields have been estimated to be infested by the tropical *Hirshmaniella* spp., with up to 25% yield losses. *Radopholus similis* is another tropical species with more than 250 known plant hosts. With the exception of Sigatoka disease, it is the most important pest of banana worldwide.[8] The nematode is also a serious pest of coconut, tea, and spices such as black pepper, ginger, and turmeric. The citrus race of the nematode causes spreading decline disease of citrus in Florida, where yield reductions >50% are common in the absence of nematode management. The many *Pratylenchus* spp. also often have wide host ranges, with species adapted to both temperate and tropical climates. Species such as *Pratylenchus penetrans*, *Pratylenchus scribneri*, and *Pratylenchus neglectus* can damage a large number of tree fruit, vegetable, and field crops in temperate regions. An important interaction between *Verticillium dahliae* and *P. penetrans*, *P. scribneri*, or *Pratylenchus thornei* results in the potato early dying disease. *Pratylenchus coffeae* and closely related species are major parasites of coffeae, tea, yam, banana, and numerous other major crops throughout the tropics.

Sedentary Endoparasites and Semi-Endoparasites

All of these forms require complex physiological responses by the host plant to initiate and maintain permanent feeding sites.[9] Juvenile nematodes penetrate roots and transform vascular or cortical cells into permanent nutrient transfer cells. Most of the juveniles then develop into sessile adult females that assume a variety of nonvermiform shapes. Frequently, masses of eggs are produced in a gelatinous matrix, usually on the root surface.

The root knot nematodes (*Meloidogyne* spp.) are the most economically important nematodes, worldwide. With over 2000 known host species, relatively few crops are not susceptible to serious damage by one or more species of *Meloidogyne*. Infested plants are easily recognized by galls caused by hyperplasia and hypertrophy of cells adjacent to the feeding cells (Fig. 1D). Widely distributed species such as *Meloidogyne incognita*, *Meloidogyne javanica*, *Meloidogyne hapla*, *Meloidogyne arenaria*, and many others have broad host ranges on weeds and many field, vegetable, tuber, and ornamental crops. Fewer species (i.e., *Meloidogyne coffeicola* on coffee) are more limited in distribution and have few known hosts among crop species. In many crops (i.e., cotton and tomato), the incidence of Fusarium wilt and other soilborne diseases increase dramatically in association with root knot nematodes.

Host ranges of cyst nematodes (*Globodera* spp. and *Heterodera* spp.) tend to be narrow. An important adaptation by these nematodes is an ability of eggs to persist for long periods in the absence of hosts, protected within the body of the female, which forms a resistant cyst. Eggs hatch over a period of several years, and chemical cues from host plants stimulate eclosion of some species. The cyst also provides a very effective means of long-range dispersal by wind, through the alimentary process of birds, and on plant debris. Management of *Globodera rostochiensis* and *Globodera palida* on potato in Europe and elsewhere has been an ongoing research goal for more than half a century. *Heterodera glycines* is well recognized for the remarkable speed with which it spreads to become a limiting factor in soybean growing regions where it has inadvertently been introduced. *Heterodera trifolii* (clover), *Heterodera avaene* (cereal), and *Heterodera goettingiana* (pea) are examples of other important species, as is *Heterodera shachtii*, which has an unusually wide host range (sugarbeet and other Chenopodiaceae as well as many Cruciferae).

Nacobbus aberrans (false root knot nematode) is endemic in the Americas where it has a wide host range and is an important pest of potato in South America, tomato in Central America, and sugarbeet in North America. *Rotylenchulus reniformis* (reniform nematode) is a warm/temperate climate nematode with worldwide distribution. It has a semi-endoparasitic habit; the anterior of the female is embedded deeply in the root while the posterior end and egg mass remain visible on the root surface. It can survive long periods and be transported by wind in a dehydrated condition (anhydrobiosis). It has more that 300 known host plants and is an economic problem on many vegetables, legumes, and cotton.[10] In contrast, the host range of the semi-endoparasite *Tylenchulus semipenetrans* (citrus nematode) is limited to a few species of woody plants. It is ubiquitous in citrus industries throughout the world where it causes the disease slow decline.

Parasites of Aboveground Plant Parts

Ditylenchus dipsaci is a parasite of stems, leaves, or bulbs of onion, garlic, alfalfa, clover, and oat. Stems of infected plants are short and swollen, oat stems tiller prolifically, and infected bulbs of onion and garlic often rot in storage. *Anguina tritici* is known as the seed gall nematode. It transforms wheat and rye ovules into galls containing infective juveniles. Juveniles of *A. tritici* and *D. dipsaci* survive long periods (sometimes many years) between crops in an anhydrobiotic condition.

Aphelenchoides fragariae and *Aphelenchoides besseyi* cause serious losses in strawberry, in which they parasitize leaves and stems causing the disease spring crimp. When combined with the bacterium *Rhodococcus fasciens*, *A. fragariae* causes cauliflower disease on strawberry. *Aphelenchoides ritzemabosi* induces interveinal necrotic lesions in chrysanthemum and other ornamental plants. *Bursaphelenchus xylophilus* causes the devastating pine wilt disease and *Bursaphelenchus cocophilus* causes red ring disease of coconut. Both of these nematodes are transported to new hosts by insect vectors.[11]

MANAGEMENT: CURRENT PROBLEMS AND PROSPECTS

As with all plant pests, some characteristics of nematodes pose special challenges for devising effective management systems.[12] Growers must make management decisions before planting, because there are no remedial options for many nematode problems. However, soil or plant samples must be processed by trained personnel to identify and quantify nematode infestations. As a result, nematodes are monitored less frequently than pests that are readily recognized in the field, and monitoring is beyond the means of farmers in developing countries. The effectiveness of crop rotation to reduce nematode numbers can be compromised by nematodes with wide ranges of crop hosts, or by communities of multiple species with different host ranges, or by nematodes adapted to survive for many years in the absence of a host. Host plant resistance is commercially available for some sedentary endoparasitic nematodes, but not for most other species. Resistance-breaking race development within genera such as *Heterodera* and species shifts in genera such as *Globodera* and *Meloidogyne* commonly occur. A trend toward discontinuation of pesticide-based management tactics due to environmental concerns will likely increase nematode-induced losses in some crops until appropriate alternatives are developed.

Ongoing research is focused on development of new sources of resistance, farming systems that employ cultural practices to reduce nematode numbers and help crops tolerate nematode parasitism, and biochemical/genetic approaches to a variety of management needs.[12] Since 1997, a few genes responsible for host resistance to nematodes have been cloned and sequenced.[13] Genes and gene products involved in nematode parasitism are being discovered that will provide new means to disrupt the nematode life cycle. Numerous nematode species have been characterized using DNA-based methods that will eventually provide growers rapid and simple diagnostic tools to better manage these serious pests.

REFERENCES

1. Poinar, G.O. *The Natural History of Nematodes*; Prentice Hall Inc.: Englewood, NJ, 1983.
2. Ferris, H.; Bongers, T.; de Goede, R.G.M. A framework for soil food web diagnostics: Extension of the nematode faunal analysis concept. Appl. Soil Ecol. **2001**, *18*, 13–29.
3. *Quantitative Studies on the Management of Potato Cyst Nematodes (Globodera spp.) in The Netherlands*; Been, T.H., Schomaker, C.H., Eds.; DLO-Res. Inst. Plant Prot.: Wageningen, The Netherlands, 1998.
4. Sasser, J.N.; Freckman, D.W. A World Perspective on Nematology: The Role of the Society. In *Vistas on Nematology: A Commemoration of the Twenty-Fifth Anniversary of the Society of Nematologists*; Veech, J.A., Dickson, D.W., Eds.; Soc. Nematol, Inc.: Hyattsville, MD, 1987; 7–14.
5. Koenning, S.R.; Overstreet, C.; Noling, J.W.; Donald, P.A.; Becker, J.O.; Fortnum, B.A. Survey of crop losses in response to phytoparasitic nematodes in the United States for 1994. J. Nematol. **1999**, *31* (4S), 587–618.
6. Nyczepir, A.P.; Bertrand, P.F. Preplanting bahia grass or wheat compared for controlling *Mesocriconema xenoplax* and short life in a young peach orchard. Plant Dis. **2000**, *84* (7), 789–793.
7. *Nematode Vectors of Plant Viruses*; Taylor, C.E., Brown, D.J.F., Eds.; CAB Intl.: Wallingford, UK, 1997.
8. Marin, D.H.; Sutton, T.B.; Barker, K.R. Dissemination of bananas in Latin America and the Caribbean and its relationship to the occurrence of *Radopholus similis*. Plant Dis. **1998**, *82* (9), 964–974.
9. Hussey, R.S.; Williamson, V.M. Physiological and Molecular Aspects of Nematode Parasitism. In *Plant Nematode Interactions*; Barker, K.R., Pederson, G.A., Windham, G.L., Eds.; Agronomy, Amer. Soc. Agron, Inc.: Madison, WI, 1998; Vol. 36, 87 108.
10. Robinson, A.F.; Inserra, R.N.; Caswell-Chen, E.P.; Vovlas, N.; Troccoli, A. *Rotylenchulus* species: Identification, distribution, host ranges, and crop plant resistance. Nematropica **1997**, *27* (2), 127–180.
11. Fielding, N.J.; Evans, H.F. The pine wood nematode *Bursaphelenchus xylophilus* (Steiner and Buhrer) Nickle (=*B. lignicolus* Mamiya and Kiyohara): An assessment of the current position. Forestry: J. Soc. Foresters Great Britain **1996**, *69* (1), 35–46.
12. Barker, K.R.; Koenning, S.R. Developing sustainable systems for nematode management. Annu. Rev. Phytopathol. **1998**, *36*, 165–205.
13. Davis, E.L.; Hussey, R.S.; Baum, T.J.; Bakker, J.; Schots, A.; Rosso, M.-N.; Abad, P. Nematode parasitism genes. Annu. Rev. Phytopathol. **2000**, *38*, 365–396.

Nematodes and Host Resistance

Philip A. Roberts
University of California, Riverside, California, U.S.A.

INTRODUCTION

Host resistance to nematodes can be classified in several ways. First, it is important to recognize that parasitic nematodes have finite host ranges. Host plants are susceptible to nematode infection and allow the nematode to feed, develop, and reproduce, using the plant cells and tissues as a substrate. Some nematode species have broad host ranges that include hosts from many diverse plant taxa. Other nematode species, often the more specialized types, have narrow host ranges limited to one or a few plant families. Thus, nematodes do not parasitize all plants. The nonhost plants of a nematode are ones typically immune or resistant in the broadest sense, such that many attributes of the morphology, physiology, and biochemistry of the plant render it unsuitable as a host and prevent the nematode from feeding, developing, and reproducing.

Host resistance is more narrowly defined by its characteristic of being a heritable trait conferred by one or more genes that renders a host plant resistant. That is, the nematode is unable to feed, develop, and reproduce on the resistant plant to some measurable extent compared to a susceptible host plant. The importance of host resistance lies in its ability to protect the plant from full-scale infection, and in preventing the nematode from multiplying its populations. In agriculture and horticulture, resistance is considered a highly valuable character that, through plant breeding, can be transferred into elite crop and horticultural cultivars, varieties, or rootstocks. Significant research effort is invested to identify and quantify nematode-resistance and -tolerance phenotypes in breeding materials, including wild relatives of crops, for use in plant breeding. Genetic and molecular characterization of resistance traits and the matching determinants of pathogenicity and virulence in nematode populations are an important part of the overall advancement of host resistance as a component of genetic improvement of crops and nematode pest management.

TERMS AND DEFINITIONS

Definitions of some important terms used in host resistance to nematodes are given here, and these have been described in detail in several reviews.[1–4] Broader discussions of terminology for use in general plant pathology are also available.[5] Fig. 1 provides a pictorial representation of some common terms. Most plants are immune or nonhost to most nematodes, blocking root invasion, nematode development and reproduction, and plant injury. For example, root-knot nematodes (*Meloidogyne* spp.) avoid roots of Royal Blenheim apricot, rendering the tree immune. ''Resistance'' refers to the ability of a host plant to suppress development or reproduction of the nematode, and it can range from low to moderate (partial or intermediate) resistance, to high resistance. The term ''resistance'' is also used to describe the capacity to suppress the disease, especially root-knot. Partially or moderately resistant plants allow some intermediate levels of nematode reproduction. ''Susceptibility'' is used as the opposite of resistance. A susceptible plant allows normal nematode development to take place, as well as the expression of any associated disease (Fig. 1).

''Tolerance'' and its opposite, ''intolerance,'' are used to describe the ability of the plant to withstand the damage resulting from nematode infection. Tolerant plants grow well despite the presence of heavy infection, whereas intolerant plants are injured and grow less well or even die when infected (Fig. 1). Typically, resistant plants are also tolerant and most susceptible plants are injured to some extent by most nematodes. However, resistance and tolerance are not always linked, being under separate genetic control in some plant-nematode interactions.[4] A useful discussion of concepts of tolerance is given by Wallace.[6]

Resistance defined by mode of inheritance can be monogenic (single gene), oligogenic (a few genes), or polygenic (many genes). These genes may be major genes (large effects) or minor genes (small effects) for phenotypic expression. Other descriptions of resistance follow Vanderplank's[7] classification of vertical resistance (race-specific or qualitative, differentiating intraspecific variants—races, pathotypes, or biotypes—of the pathogen) and horizontal resistance (race-nonspecific or quantitative, effective against all variants of the pathogen). Vertical resistance is usually simply inherited and conforms to a gene-for-gene type of plant-pathogen interaction.[8] Horizontal resistance usually involves several genes with additive effects that express a quantitative

Encyclopedia of Plant and Crop Science
DOI: 10.1081/E-EPCS 120010393

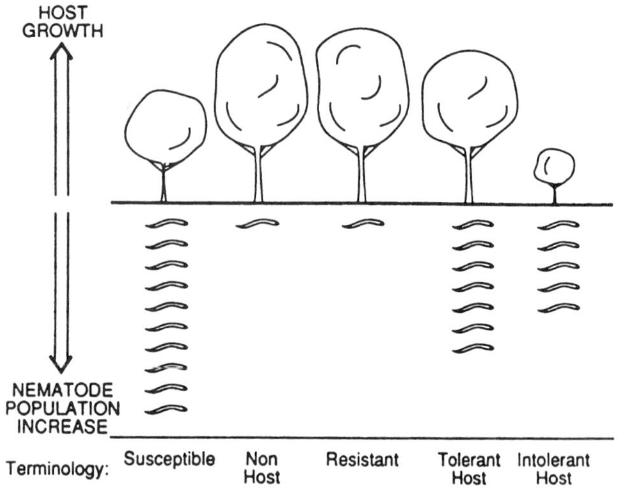

Fig. 1 Diagrammatic representation of terms describing plant growth response to nematodes and nematode reproduction on plants. (From Ref. 18.)

level of resistance. In general, quantitative resistance tends to be less likely to ''break down'' due to selection pressure operating on the nematode parasite population.[9]

Genes for virulence in the nematode match resistance genes in the host plant. Virulent nematodes are able to reproduce, whereas avirulent nematodes are unable to reproduce on a host carrying a specific resistance gene(s).[9] The frequency of virulent individuals in nematode populations will determine the potential for selection of virulence in the presence of resistant host plants. In plant pathology, the genes encoding this trait are typically called avirulence or Avr genes. Nematologists sometimes refer to avirulence genes as genes for parasitism or parasitism genes, but these terms also have broader meanings.[2]

Several terms have been used to categorize the variation within a nematode species based on differential responses to a host or resistance trait. Their use is somewhat confusing because of the indiscriminate use of them for different nematode groups. For example ''race'' or ''host-race'' has been used for categorizing variations within soybean cyst nematode (*Heterodera glycines*); ''pathotype'' has been used for potato cyst nematodes (*Globodera pallida* and *Globodera rostochiensis*) and for cereal cyst nematode (*Heterodera avenae*); and ''biotype'' for variations within the stem and bulb nematode (*Ditylenchus dipsaci*). Triantaphyllou[10] offered the term ''biotype'' as a biological unit consisting of ''a group of genetically closely related individuals sharing a common biological feature or phenotypic trait,'' in this case parasitic ability on given differential hosts. Field populations may consist of individuals of different biotypes, and combinations of biotypes comprising field populations could be designated as races. An individual nematode may be assigned to more than one biotype, depending on the array of genes for avirulence that it possesses in relation to the genetic constitution of the host differentials used to classify the biotypes.[10] Roberts[11] adapted this biotype concept for categorizing variants within species of root-knot nematodes (*Meloidogyne* spp.) defined by reaction to resistance genes in different host plants.

AVAILABILITY OF RESISTANCE

The current availability and use of resistant cultivars and rootstocks for nematode management is well documented.[1,4,9,12–14] Success has been achieved in a significant range of crops for identification of resistance, followed by incorporation into commercially acceptable crop varieties and rootstocks and implementation in management programs. Even so, a large genetic resource of nematode resistance remains to be developed in many crops. Examples of highly successful developments in nematode resistance include the use of the *Mi* gene in tomato and the use of the Nemaguard rootstock for *Prunus* crops (plums, peaches, nectarines, and almonds), both used against root-knot nematodes, and the *H1* gene in potato, used against potato cyst nematodes. These resistance traits are used widely, but they do not protect against all populations and species of the target nematodes. Thus, there continues to be a concerted effort to find additional resistance traits in these crops.[9,14] Technical advances in marker-assisted breeding, resistance-gene cloning, and plant transformation, and in bioengineering novel types of resistance to nematodes will undoubtedly expedite development of nematode resistant crops.[3]

Plant resistance has been developed mainly to the highly specialized parasitic nematodes such as *Globodera*, *Heterodera*, *Meloidogyne*, *Rotylenchulus*, *Tylenchulus*, and *Ditylenchus*, which (except *Ditylenchus*) have a sedentary endoparasitic relationship with their host. The resistance may be effective against nematode species of different genera, against more than one species from the same genus, against a single species, or against certain populations of a species.[13] Resistance to less specialized parasitic groups such as the migratory endoparasitic genera *Aphelenchoides* and *Pratylenchus* has been developed only in a few cases, and also to a few ectoparasitic nematodes, for example, to *Xiphinema* in grapevines.[15] This pattern of resistance reflects the co-evolutionary forces between host and parasite, the more highly specialized relationships having resulted in specific genes for resistance and parasitism.[12] The root-browsing ectoparasitic nematodes, with less specific feeding requirements, apparently have not been a strong selection force for resistance in plant hosts in most interactions, although useful differences in resistance do occur.[16] Recently, the potential for bioengineering novel forms of resistance based on molecular approaches has gained considerable attention and holds much promise, particularly as technologies advance to enable a more streamlined approach, and some progress has been made in developing novel root-knot nematode resistance.[17]

EFFECTS OF RESISTANCE

Resistance has two major attributes that contribute to its success in nematode management programs. These are the tolerance of most resistant plants to nematode infection, which imparts a ''self-protection'' characteristic to enhance growth and yield, and the suppression of nematode multiplication, which results in fewer nematodes in soil that could damage a following susceptible or intolerant crop planted in rotation.

Protection of Yield Potential

The impact of resistance and tolerance traits on crop yield can be quantified, and such knowledge can be used for predictive management approaches based on preplant sampling of nematode population densities in soil. The general relationship of relative yield to initial nematode density, called a damage function, has been defined according to the relative levels of resistance and tolerance and is depicted in Fig. 2[12] For susceptible, intolerant crops, this relationship is linear except at very low or very high population densities. Both the position and the slope of the curve will be governed by the relative tolerance of the particular cultivar. The ideal genotype is one that combines high tolerance with high resistance. The main advantage of this yield protection is that in many cases, the resistant crop can be grown successfully without additional inputs of nematode control. For many crops, resistance and the associated tolerance eases or eliminates the reliance on chemical controls with soil-applied nematicides. In perennial vine, tree fruit, and nut crops, the primary objectives of incorporating nematode resistance into acceptable cultivars and rootstocks are improved yield and longevity, and the majority of forms of resistance in these crops also confer the required tolerance to infection to meet these objectives.[14]

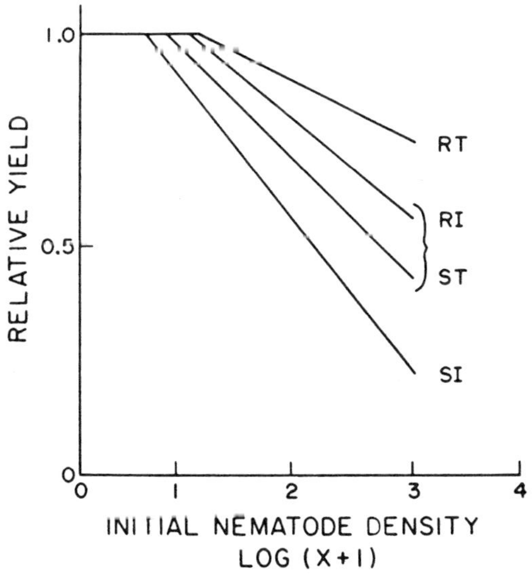

Fig. 2 Hypothetical damage functions (relationship between yield and initial nematode density) for crop cultivars possessing different nematode resistance and tolerance traits. RT = resistant, tolerant; RI = resistant, intolerant; ST = susceptible, tolerant; SI = susceptible, intolerant. (From Ref. 12.)

Suppression of Nematode Populations

In annual cropping systems where from one to several crops per year may be grown on the same ground, nematode problems can be managed by including crops with different levels of resistance. Susceptible crops allow large increases in nematode populations, even from low initial population densities. The large population densities remaining after the susceptible crop often will be highly damaging to a following susceptible crop. Resistance, on the other hand, will suppress nematode multiplication to an extent determined by the level of resistance expressed in the plant. Resistance may be expressed at low, moderate, or high levels, and these expression levels will largely determine the nematode multiplication rate. Therefore, a susceptible crop planted after a highly resistant crop will be protected from nematode infection to a level that reduces or eliminates the need for additional nematode management inputs. These rotation benefits of nematode resistance have been demonstrated in several nematode-plant interactions in annual cropping systems, most notably for root-knot and cyst nematodes and their host crops.[3] In practice, the use of host plant resistance in combination with other control tactics is encouraged, because field research has shown that resistance can break down through selection of virulent nematode populations. An integrated management approach will tend to lower the selection pressure for virulence that may occur when resistance is used alone, thereby promoting the durability of the resistance.[3,9]

ARTICLES OF FURTHER INTEREST

Biological Control of Nematodes, p. 134
Breeding for Durable Resistance, p. 179
Breeding Plants with Transgenes, p. 193

Breeding: Choice of Parents, p. 215
Breeding: Incorporation of Exotic Germplasm, p. 222
Classification and Identification of Nematodes, p. 283
Germplasm: International and National Centers, p. 531
Insect/Host Plant Resistance in Crops, p. 605
Management of Nematode Diseases: Options, p. 684
Nematode Biology, Morphology, and Physiology, p. 781
Nematode Feeding Strategies, p. 784
Nematode Infestations: Assessment, p. 788
Nematode Population Dynamics, p. 797
Nematode Problems: Most Prevalent, p. 800
Nematodes: Ecology, p. 809
Plant Response to Stress: Nematode Infection, p. 1014

REFERENCES

1. Cook, R.; Evans, K. Resistance and Tolerance. In *Principles and Practice of Nematode Control in Crops*; Brown, R.H., Kerry, B.R., Eds.; Academic Press: Sydney, 1987; 179–231.
2. Davis, E.L.; Hussey, R.S.; Baum, T.J.; Bakker, J.; Schots, A.; Rosso, M.N.; Abad, P. Nematode parasitism genes. Annu. Rev. Phytopathol. **2000**, *38*, 365–396.
3. Roberts, P.A. Concepts and Consequences of Resistance. In *Plant Resistance to Parasitic Nematodes*; Starr, J.L., Cook, R., Bridge, J., Eds.; CAB International: Wallingford, UK, 2002; 23–41.
4. Trudgill, D.L. Resistance to and tolerance of plant parasitic nematodes in plants. Annu. Rev. Phytopathol. **1991**, *29*, 167–192.
5. Shaner, G.; Lacy, G.H.; Stromberg, E.L.; Barker, K.R.; Pirone, T.P. Nomenclature and concepts of pathogenicity and virulence. Annu. Rev. Phytopathol. **1992**, *30*, 47–66.
6. Wallace, H.R. A perception of tolerance. Nematologica **1987**, *33*, 419–432.
7. Vanderplank, J.E. *Genetic and Molecular Basis of Plant Pathogenesis*; Springer Verlag: Berlin, 1978; 1–167.
8. Keen, N.T. Gene-for-gene complementarity in plant-pathogen interactions. Annu. Rev. Genet. **1990**, *24*, 447–463.
9. Roberts, P.A.; Matthews, W.C.; Veremis, J.C. Genetic Mechanisms of Host Plant Resistance to Nematodes. In *Plant Nematode Interactions*; Barker, K.R., Pederson, G.A., Windham, G.L., Eds.; American Society of Agronomy: Madison, WI, 1998; 209–238.
10. Triantaphyllou, A.C. Genetics of Nematode Parasitism on Plants. In *Vistas on Nematology*; Veech, J.A., Dickson, D.W., Eds.; Society of Nematologists: Hyattsville, 1987; 354–363.
11. Roberts, P.A. Conceptual and practical aspects of variability in root-knot nematodes related to host plant resistance. Annu. Rev. Phytopathol. **1995**, *33*, 199–221.
12. Roberts, P.A. Plant resistance in nematode pest management. J. Nematol. **1982**, *14*, 24–33.
13. Roberts, P.A. Current status of the availability, development, and use of host plant resistance to nematodes. J. Nematol. **1992**, *24*, 213–227.
14. *Plant Resistance to Parasitic Nematodes*; Starr, J.L., Cook, R., Bridge, J., Eds.; CAB International: Wallingford, UK, 2002; 1–258.
15. Harris, A.R. Resistance of some Vitis rootstocks to Xiphinema index. J. Nematol. **1983**, *15*, 405–409.
16. McKenry, M.V.; Kretsch, J.O.; Anwar, S.A. Interactions of selected rootstocks with ectoparasitic nematodes. Am. J. Enol. Vitic. **2001**, *52*, 304–309.
17. Opperman, C.H.; Conkling, M.A. Bioengineering Resistance to Plant-Parasitic Nematodes. In *Plant Nematode Interactions*; Barker, K.R., Pederson, G.A., Windham, G.L., Eds.; American Society of Agronomy: Madison, WI, 1998; 239–250.
18. McKenry, M.V.; Roberts, P.A. *Phytonematology Study Guide, Publication 4405*; University of California Press: Oakland, 1985; 1–56.

Nemato des: Ecology

N

Howard Ferris
University of California, Davis, California, U.S.A.

INTRODUCTION

Nematodes are among the most successful of multicellular organisms. They inhabit every environment imaginable. Although less diverse in number of species than some other phyla, they are incredibly abundant in number of individuals. An estimated four out of every five multicellular organisms on the planet are nematodes,[1] surely a testament to their success and their ability to adapt physiologically and behaviorally to diverse habitats and niches. Cobb[2] asserted that if all matter other than nematodes were removed, the outlines of geographic features, vegetation, and centers of habitation would be recognizable by the abundance and characteristics of nematode communities. That understanding has grown, and nematode community structure is now recognized as a powerful tool in the biomonitoring of soil history and condition.[3] The ecology of plant and soil nematodes can be considered at several levels of resolution, for example, the physiological and behavioral response of individuals to environmental conditions; the reaction of life history traits to environmental resources, and the resulting consequences to abundance and spatial patterns; and the structural positions and functional roles of the organisms in the ecosystems that they inhabit.

PHYSIOLOGICAL ECOLOGY OF PLANT AND SOIL NEMATODES

Like all nematodes, those in plants and soil respond both physiologically and behaviorally to environmental conditions and cues. As poikilotherms, their metabolic activity is modified by temperature; the efficiency of energy release for cellular and organismal function approaches a maximum when they are exposed to favorable temperature ranges. Nematodes are aquatic organisms. In soil, they inhabit water surrounding soil particles, and their permeable cuticle provides direct contact with their microenvironment. They do not rapidly migrate from stressful conditions, and many species survive dehydration, freezing, or oxygen stress; others are more sensitive. Metabolic activity becomes constrained by limiting factors, including availability of moisture in the soil or plant material in which the nematodes live, or the availability of oxygen to drive metabolic processes. They are also affected physiologically by the osmotic condition and pH of the milieu. Such factors vary considerably as concentrations of the soil solution are affected by rainfall, percolation, and evapotranspiration. The amplitude of tolerance to such varying conditions determines the habitats occupied by different species.[4]

It seems reasonable to assert that population levels or aggregations of nematodes in soil or plant material will be greatest where the integral affect of environmental conditions is optimal for their metabolism. In such locations they should feed more effectively, move most efficiently, reproduce most rapidly, and survive longest. Aggregations of nematodes are also enhanced by kinetic effects (movement). Movement is most rapid in favorable zones and less rapid, or even prevented, in unfavorable zones. Of course, the overriding determinant of population aggregates is the presence of food and the ability of nematodes to detect and access it while potentially constrained by the top-down effects of density-dependent predation by natural enemies.

BEHAVIORAL ECOLOGY OF PLANT AND SOIL NEMATODES

Nematodes respond to environmental (and other organismal) cues through taxes and kineses. Kineses are alterations in rate of activity; taxes are directional responses to stimuli. Chemosensory signals are detected by the amphids, which may be relatively small in environments where signal strength changes slowly (e.g., soil), or elaborate where signal strength is more dilute or changes rapidly (e.g., freshwater or marine environments). Even where amphids are small, as in most soil nematodes, expansions of the neuronal endings and the presence of mucoidal material may enhance their sensitivity. In those nematodes (e.g., *Caenorhabditis elegans*) where the nervous system has been studied intensively, cell bodies of amphidial neurons are aggregated in the nerve ring, which may allow some integration of signal strength from the amphids on either side of the head—or between amphids and the posterior phasmids—and allow generation of a signal for muscular responses.[5,6]

Encyclopedia of Plant and Crop Science
DOI: 10.1081/E-EPCS 120010387

Chemosensory communication between individuals is evident in taxis responses of males to females in sexually reproducing species and in the apparently tactile determination of the location of reproductive structures during mating. Response to host signals is generally strongest in plant-feeding nematodes that have a narrow host range and less obvious in those species with wide host ranges. Nevertheless, coordination of life history events with host availability is very important in species that have evolved sophisticated host–parasite relationships, for example, the root-knot nematodes, *Meloidogyne* spp.[7] In those cases, the food reserves packaged into the egg by the female nematode must be sufficient to drive the embryo through several hundred cell divisions, a molting event from the first to the second juvenile stage, hatch from the egg, detection and migration to a host root, penetration of the root tissues through a complex migration pathway, and induction of a feeding site. All of those energy-intensive activities are necessary before the nematode is able to independently obtain new resources. Clearly, there are many opportunities for disaster, including inability to find a root, a nonhost response by the root, and insufficient residual energy to complete the penetration process.

POPULATION ECOLOGY OF PLANT AND SOIL NEMATODES

Population processes are, of course, regulated by physiological responses to the environment. The dynamics of increase and decrease in nematode populations are the integral of birth and death rates as constrained by environmental conditions, resource availability, and community regulatory pressures. Birth rates are affected by the intrinsic capacity of female nematodes to produce eggs and the duration of the life course over which those eggs are produced. Birth rates may be influenced by availability of males where fertilization is necessary, but not where parthenogenesis occurs.[8] Hermaphroditism occurs in some bacterial-feeding nematodes (e.g., *C. elegans*). In those cases, the number of viable eggs produced is greater when cross- rather than self-fertilization occurs, apparently because the hermaphrodite does not produce and store enough sperm to fertilize all the eggs it is capable of producing.[9]

Most plant and soil nematodes seem to produce eggs for the duration of their adult life course. Induced mutants of *C. elegans* may have extended survival beyond the reproductive period, with possible competitive consequences for food. The occurrence of such variants in nature is unknown and, arguably, there would be strong selection pressure against them, as there does not appear to be any social contribution to the population on the part of elderly nematodes.

The life course of plant and soil nematodes may be as short as a week in opportunistic bacterial-feeding species; is often in the 4–8 week range in plant-feeding species; but may be as long as several years in long-lived, large-bodied plant-feeders, predators, and omnivores inhabiting undisturbed and stable environments.[3] The distribution of death expectancy around the mean life duration is generally not known except in some model systems.[10]

Resource availability in an environment determines its carrying capacity for a nematode population. In closed systems, that carrying capacity may be fixed by the total amount of carbon available. But in dynamic systems, such as those generated by a growing plant that is continually fixing carbon, producing roots, and depositing and leaking materials into the rhizosphere, the carrying capacity changes with time. For a plant-feeding nematode species, there is a delicate balance between the amount of damage or pressure a successful parasite applies to the growth of its host plant and the rate and amount of food that the host can make available to the parasite. At higher starting population levels, the damage to the host may be so great that growth and development of the plant is impaired, total photosynthetic input is compromised, and the nematode population is limited to a lower level than it might have been if initial population pressure were lower.[11,12]

COMMUNITY ECOLOGY OF PLANT AND SOIL NEMATODES

Nematodes in any environment are components of communities of organisms and, depending on their feeding habits and life history attributes, fit into their communities in various functional roles. The essential feature of a community is that the participating organisms are interdependent as food resources or interact in some way in the acquisition of resources.

The community of interacting soil organisms is sometimes known as the soil food web. The primary sources of carbon and energy from soil organisms stem from the photosynthetic activity of plants. The flow of carbon from plant material into the soil food web is influenced by nematodes in various ways. Some nematodes are herbivores that channel carbon and energy into the soil food web by direct herbivory on the plant. Others feed on fungi and bacteria and so constitute important bridges between primary decomposers of organic material and detritus and the balance of the food web. Still others are predators and omnivores consuming a range of higher soil organisms. These exert pressure on the abundance of their prey and potentially regulate or even suppress prey populations. Structure of the community is maintained as the integral of the population ecology of various taxa and the regulatory effects of predation across and within

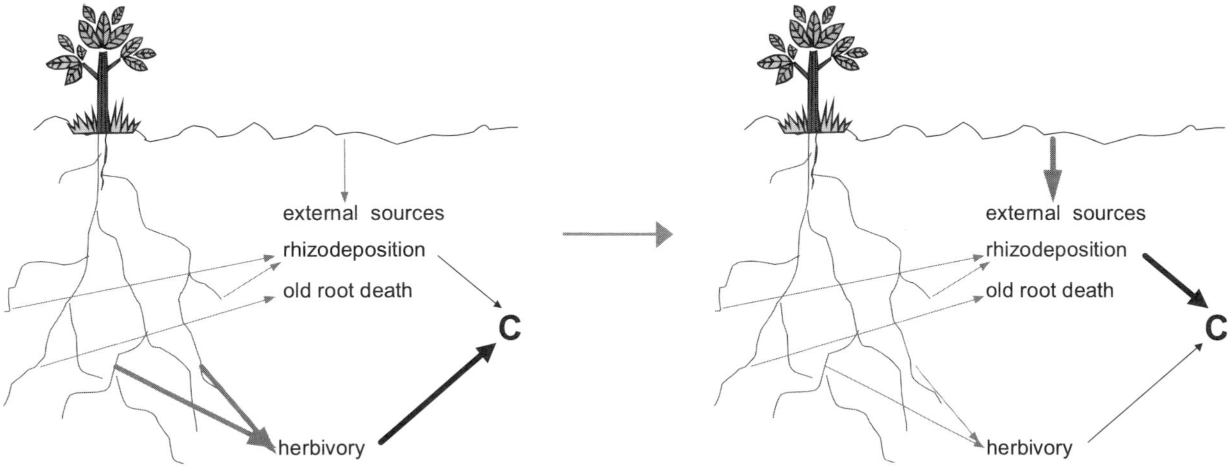

Fig. 1 A management goal in sustainable agricultural production: to maximize the flow of carbon (C) into the soil food web through detritivore rather than herbivore channels.

trophic levels, but ultimately the size and activity of the food web must depend on the amount of carbon entering the soil. A major goal of sustainable production systems that are reliant on the functional roles of soil food webs is to increase carbon flow into the system through detrital channels while minimizing carbon flow resulting from herbivory (Fig. 1).

Besides predation, mechanisms of interaction at the community level include interspecies competition and mutualism. Certainly, nematodes with the same feeding habits compete for resources, but in some situations, competition is mitigated by adaptation of the nematodes to different environmental conditions. Predominance of species in the communities varies with their thermal adaptation, competition is reduced by spatial and seasonal variations in temperature and its diurnal amplitude.[13] One interesting example of mutualism in soil nematodes is exhibited by the so-called entomopathogenic nematodes. These bacterial-feeding nematodes transport mutualistic bacteria into the intestinal tract of insects. There the bacteria are released, which kills the insect and provides an abundance of food for the nematode.[14]

Communities of organisms tend to be centered on sources of food. In the soil, such foci may occur at the interface of organic litter layers and the mineral layers of the soil. They also occur at the metabolically active parts of the root system, the root tips. Because of the multiplicity and dispersion of the various organic matter sources that drive the soil food web (Fig. 1), coupled with the physical heterogeneity of the soil environment, the soil community tends to exist in patches. It might be considered a metacommunity, a series of borderless communities with opportunities for migration of organisms among the patches, depending on their motility and sensory capabilities.

Both resource availability and predation pressure affect soil nematodes, and the interplay between these forces affects the dynamics of the participating populations. Bacterial-feeding nematodes may enter inactive ''dauer'' states, which allows them to conserve resources and, perhaps, renders them less accessible or attractive to predators. Nevertheless, the predators require food, and both they and the prey nematode populations must decline in the absence of resources entering the system.

SUCCESS OF NEMATODE ECOLOGICAL STRATEGIES

Nematodes are found in every habitat, from frozen tundra to hot springs, and from marine sediments to alpine peaks. They obtain sustenance from every form of fixed carbon imaginable, from algae and bacteria to humans. Recent classification systems recognize around 19 orders that constitute the phylum Nematoda. Plant-feeding species occur in three of those orders and are most prevalent in one, the Tylenchida. This relatively small sector of the phylum has developed strategies for feeding on perhaps every species of higher plant and for coexisting with other species on the same plant. Most species feed on root tissues and so avoid the dry conditions of the aboveground habitat, but some species have become behaviorally and physiologically adapted to feeding in stem, leaf, and seed tissues. Enormous diversity exists among root-feeding strategies, from root tip feeders to root hair piercers, and from migratory ectoparasites that withdraw contents of individual cells to sedentary endoparasites that elicit highly specialized feeding-site responses from the host.

ARTICLES OF FURTHER INTEREST

Agriculture and Biodiversity, p. 1
Biological Control of Nematodes, p. 134
Classification and Identification of Nematodes, p. 283
Ecology and Agricultural Sciences, p. 401
Ecology: Functions, Patterns, and Evolution, p. 404
Ecophysiology, p. 410
Nematode Biology, Morphology, and Physiology, p. 781
Nematode Feeding Strategies, p. 784
Nematode Population Dynamics, p. 797
Organic Agriculture As a Form of Sustainable Agriculture, p. 846
Sustainable Agriculture: Ecological Indicators, p. 1191

REFERENCES

1. Platt, H.M. Foreword. In *The Phylogenetic Systematics of Free-living Nematodes*; Lorenzen, S., Ed.; The Ray Society: London, 1994; i–ii, 383p.
2. Cobb, N.A. *Nematodes and Their Relationships*; USDA Yearbook of Agriculture, 1914; 457–490.
3. Bongers, T.; Ferris, H. Nematode community structure as a bioindicator in environmental monitoring. Trends Evol. Ecol. **1999**, *14*, 224–228.
4. Perry, R.N.; Wright, D.J. *The Physiology and Biochemistry of Free-Living and Plant-Parasitic Nematodes*; C.A.B. International: New York, 1998.
5. Wood, W.B. The Nematode. In *Caenorhabditis Elegans*; Cold Spring Harbor Laboratory Press: New York, 1988.
6. Riddle, D.L. *C. Elegans II*; Cold Spring Harbor Laboratory Press: New York, 1997.
7. Sasser, J.N.; Carter, C.C. *An Advanced Treatise on Meloidogyne*; North Carolina State University Graphics, 1985.
8. Ferris, H.; Wilson, L.T. Concepts and Principles of Population Dynamics. In *Vistas on Nematology*; Veech, J.A.Dickson, D.W., Eds.; Society of Nematologists, 1987.
9. Kimble, J.; Ward, S. Germ-Line Development And Fertilization. In *The Nematode Caenorhabditis Elegans*; Wood, W.B., Eds.; Cold Spring Harbor Laboratory Press, 1988.
10. Chen, J.; Carey, J.R.; Ferris, H. Comparative demography of isogenic populations of *Caenorhabditis elegans*. Exp. Gerontol. **2001**, *36*, 431–440.
11. Seinhorst, J.S. Dynamics of populations of plant parasitic nematodes. Annu. Rev. Phytopathol. **1970**, *8*, 131–156.
12. Seinhorst, J.W. The regulation of numbers of cysts and eggs per cyst produced by *Globodera rostochiensis* and *G. pallida* on potato roots at different initial egg densities. Nematologica **1993**, *39*, 104–114.
13. Ferris, H.; Venette, R.C.; Lau, S.S. Dynamics of nematode communities in tomatoes grown in conventional and organic farming systems, and their impact on soil fertility. Appl. Soil Ecol. **1996**, *3*, 161–175.
14. Bedding, R.; Akhurst, R.; Kaya, H. *Nematodes and The Biological Control of Insect Pests*; CSIRO Australia, 1993.

New Industrial Crops in Europe

N

Jan E. G. van Dam
Wolter Elbersen
Agrotechnological Research Institute, Bornsesteeg, Wageningen, The Netherlands

INTRODUCTION

Crop diversification has been one of the promises for European agriculture in recent decades. As a result of surpluses in production of the major food crops, which have arisen from increased productivity in the 1970s and EU pricing policy in the 1980s,[1] alternative crops for nonfood markets have received substantial interest. Overproduction of traditional food crops (sugar beet, grain, and potatoes) in Europe resulted in increased competition and a decline in profitability. Reduction of EU subsidies and the threat of cheap imported products from low-wage countries prompted many farmers to abandon their businesses. Farmers would benefit from additional outlets, especially in novel, industrial nonfood applications. Attention has been given to new markets for existing crops as well as to developing novel crops. Research and development activities have been directed toward valorization of agricultural residues and product development for various ''industrial'' crops to produce specific seed oils, proteins, starches, carbohydrates, or cellulosic fiber.

AGRICULTURAL DIVERSIFICATION AS EU POLICY

In the specific framework RTD programs of the European Commission covering agriculture and fisheries (ECLAIR 1988–1993, 3rd framework 1991–1994 AIR, 4th framework 1994–1998 FAIR, and 5th framework 1998–2002 Quality of Life) a large number of precompetitive projects have been conducted during the last decade to address the various aspects of agricultural crop diversification. Substantial budgets of millions of euros have been spent to enhance the links of agriculture with industry and to encourage collaboration between the institutions from the various member states. Many of the EC-funded projects (with $\geq 50\%$ EU contribution) have been selected to promote industrial participation in innovation to enhance the chances of success for project implementation.[2,3] In addition to the EU framework programs, specific developmental work for agricultural diversification also has been carried out on a national level.

The EU policy for sustainable development promotes the use of renewable resources for production of industrial products that currently are derived from petrochemical or mineral resources. The potential for agricultural production of a wide range of industrial feedstocks has been scrutinized as a source for composites, plastics, and polymers; resins and adhesives; building and insulation materials; energy and fuel feedstocks; paints, coatings, and dyestuffs; soaps, detergents, surfactants, lubricants and waxes; and agro-chemicals, pharmaceuticals, and cosmetics.

AGRO-INDUSTRIAL PRODUCTION CHAINS

A large number of new crops for European farming have been studied for the purpose of agro-industrial production, together with some obsolete crops that have long been neglected. Successful (re)introduction of agricultural raw materials for energy, ''green'' chemicals, and other nonfood uses requires addressing the whole agro-industrial production chain, from primary agricultural production to the perceptions of industrial end-users and consumers.

Generally, research and developmental work is directed at technical bottlenecks in product development. Successful market introduction, however, depends largely on the reliability of supplies and on added value for the consumer. Quality control and product standardization have been considered to overcome problems in the industrial use of renewable raw materials. Crops are a natural product, and intrinsic changes in quality between crops grown under different conditions introduce a certain degree of variability that has to be accounted for.

The production chain for renewable resources can be divided into three main links: agricultural primary production, crop processing, and utilization (see Table 1). Difficulties in the organization of integrated production chains (including storage and transport, marketing and sales) for innovative agro-industrial products have been identified as major constraints in the successful commercial introduction of novel (ecologically enhanced) products to the market. Despite the interdependency of the different links in the chain, the interests of the various

DOI: 10.1081/E-EPCS 120010431

Table 1 Integrated agro-industrial production chains

	Agricultural production
Breeding	Genetics, crossbreeding, and reproduction
Growth	Agronomy: seed density, soil fertility, climate conditions, irrigation, crop protection, weed and pest control, fertilizers
Harvest/storage	Maturity and handling, mechanization
	Postharvest processing and conversion technology
Product extraction	Cleaning and extraction/residue valorization
Product preparation	Conversion technologies (refining, extrusion, steam explosion, etc.), chemical modification/biochemical treatments, etc.
Product processing	Product compilation, finishing, compounding
	Application and use
Utilization	Product performance/marketing and sales
Disposal	Reuse and recycling, incineration/degradation

players may be divergent and counterproductive. The price and performance ratio of the products prevails in terms of competitiveness on the market, rather than in quality or ecological benefits.

Since ecological arguments are not decisive in getting industries or consumers to choose renewable products, legislative measures on the use of less-sustainable products favor the utilization of new crops.

OILSEED CROPS

A number of crops previously established as minor crops for production of fodder, such as rapeseed, have been considered as feedstock for biofuel (biodiesel) production or plant-oil-based surfactants. The increasing area of oil crops in the EU, e.g., sunflower (*Helianthus annuus*) (with a 1993 maximum of 3.3 million ha) and rapeseed (*Brassica oleracea*) (2.4 million ha in 1990 and over 3.0 million in 2000),[4] has also initiated a search for outlets for by-products such as lecithin for cosmetics and fermentation substrates.

The reduction of volatile organic solvents in paints, coatings, and adhesives has enhanced the industrial demand for specific plant oils. Legislation restricts the use of those solvents for professional use in the EU, and alternative products are on the market. However, higher costs and reduced practical value still restrict the more common use of ecological products.

Research has been conducted for the breeding and agronomy of new oilseed crops such as meadow foam (*Limnanthes alba*), crambe (*Crambe abyssinica*), castor (*Ricinus communis*), and *Dimorphotheca pluviales*. The oils and their derived products have been evaluated for the production of pharmaceuticals, cosmetics, oleochemicals, lubricants, coatings, resins, and lacquers. The economic feasibility of primary production and industrial product development has been evaluated, but so far, available varieties have not been able to yield sufficient product to generate larger industrial interest. However, the primary production of crambe has been addressed to enhance its productivity.

Commercial interests in specialty oils containing unusual fatty acids, such as poly-unsaturated fatty acids and hydroxyl- or epoxy fatty acids, has inspired much work on alternative oil crops. Examples are lupin (*Lupinus mutabilis*, *L. albus*), jojoba (*Simmondsia chinensis*), and *Stokesia* sp., investigated for use in cosmetics, pharmaceuticals, and as lubricants; and false flax (*Camelina sativa*), caper spurge (*Euphorbia lathyris*, *E. lagascae*), and marigold (*Calendula officinalis*), studied because of their specialty fatty acids content for medicinal purposes, food, and a variety of nonfood uses as well.

FIBER CROPS

Automotive industries have been the driving force in developing cellulosic fiber production for (thermoplastic) composite car parts. Traditional fiber crops such as flax (*Linum usitatissimum*) and hemp (*Cannabis sativa*), but also crops new to Europe such as kenaf (*Hibiscus cannabinus*) and broom (*Ginestra* sp.), have been studied intensively for nontraditional end uses in compounds and fiber composites with numerous synthetic polymers.[5]

The breeding and agronomy of new hemp varieties has been studied to produce high-yield fiber and low Tetrahydrocannabinol (THC) content. Experimental fields of fiber hemp have been established in many EU countries and commercial production has been introduced at some locations. New fiber production technologies for specific industrial utilization have been studied. Some products have entered the market successfully at a modest scale. Production of high quality fiber hemp for textile use so far has not been commercially successful.

One of the identified issues for sustainable developments and ecological building—apart from the energy conservation aspects—is the use of renewable resources as building materials. Fiberboards, panels, and insulation materials for building applications have been developed based on flax, hemp, miscanthus, or fibers derived from agro-residues (wheat straw, reeds, etc.). The production scale (and related costs) of established building materials is hindering extension of the market share for renewable building products.

The use of plant fibers as geotextiles in civil engineering and as horticultural substrates or biodegradable plant pots is increasingly receiving attention. The competing (synthetic or mineral) market products, however, hold a strong position.

Despite substantial research on the pulping of annual fiber crops and agro-residues, the contribution of nonwood pulp as raw material in European paper and board industries remains marginal. Some flax and hemp fiber is processed into specialty paper.[6] Efforts to upgrade straw pulps for paper and board production have been unsuccessful, and the last straw-pulp mills have been closing in the EU.

Other fiber crops investigated for cultivation in various EU regions are *Hibiscus esculentus*, for making paper and reinforcing fiber, and for its nutritious pods; and domestications of stinging nettle (*Urtica dioica*), with potential as textile fibers.

Profitable EU agricultural production of nonfood crops implies competitiveness with cheaper raw materials produced from other regions. The availability of bulk quantities of fiber products like jute and sisal, and the potential large-scale production in Eastern Europe of flax or hemp is forcing EU agro-industrial production into a specialized niche market. Traditionally, flax fiber production in Western Europe for high-quality linen textiles has been able to cope with competing imported raw materials because of its high quality standard. The concentration of conventional linen promotion on the fashionable textile market in past decades has increased the dependency of the sector on a strongly fluctuating market segment. One way for EU agriculture to compete on the world market in lignocellulosic fibers is to supply high-quality raw materials with added value for the user. Only when the qualitative aspects of each specific end use have been defined in detail can this be achieved.

BIOMASS (ENERGY) CROPS

Promotion of the use of biomass for the production of ''green'' energy is one of the means to achieve a political target: to stop global warming and reduce CO_2 emission levels. Agricultural production of biomass crops for generation of bioenergy has received substantial attention in the EU. Elephant grass (*Miscanthus* sp.) especially has been selected for its potential high productivity and has been intensively studied agronomically to find suitable genotypes for different climatic regions. Breeding, propagation, planting, and harvesting technologies for miscanthus crops have been the subject of study. Other perennial C4 grasses such as sweet and fiber sorghum (*Sorghum* sp.) and giant reed (*Arundo donax*), which may be suitable for biomass production in the southern regions of the EU, have been evaluated. A production system for reed canary grass (*Phalaris arundinaceae*) has been investigated in the more northern regions. More exotic species studied for Europe are switchgrass (*Panicum virgatum*) and bamboo (*Phyllostachys* sp.).

Because of the relatively low costs for fossil fuel and the lack of supportive political policy, dedicated agricultural production of new energy crops still remains questionable. Therefore, dual-crop-use options are being evaluated to increase the total added value. For example, the same crop might provide both energy and paper pulp, or composites, or building materials, or horticultural substrates.

Apart from the fast-growing grasses and tree species such as willow, poplar, and eucalypt, that have been promoted for paper pulp or biofuel production, another high-yielding perennial, cardoon (*Cynara cardunculus*), has been investigated for growing on marginal and set-aside lands in arid areas of southern Europe.

The work on biofuel utilization has been focusing on thermal conversion by burning, gasification, and pyrolysis of biomass. However, the costs of energy production remain a critical component in the development. For the production of liquid fuel, laboratory-scale hydrolysis/fermentation experiments have been conducted to produce ethanol, ABE (acetone, butanol, ethanol), and hydrogen gas from lignocellulosic feedstock.

CARBOHYDRATE CROPS (CEREALS AND TUBERS)

Innovation and industrial product development for established crops such as cereals, sugar beets, and potatoes would offer the advantage of known agricultural production. Possible uses as feedstock for bioethanol fuel production have remained at the R&D stage due to the economics of production and political factors (bioethanol production would become competitive only when crude oil sells at $30–40 per barrel). Other biotechnological methods to valorize agricultural waste products, such as sugar beet pulp or wheat and maize bran, have been investigated in detail, for example, for the production of vanilin.

Much research effort by various groups has been directed into the production and processing of biodegradable plastics based on thermoplastic starches from potato, corn, and other cereals. Biodegradable plastics for packaging and single-use items are being produced on industrial scale, but so far no breakthrough has been accomplished to generate a wider market share.

A production area of 15,000 ha of chicory (*Chicorium intybus*) for inulin production has been established, concentrated in Belgium, France, and the Netherlands. Inulin and fructan for nonfood applications are under investigation.

"GREEN" CHEMICALS AND PHYTOPHARMACEUTICALS

Higher-value-added specialty chemicals that can be derived from plant sources have been investigated for use in textiles, cosmetics, flavors, fragrances, and pharmaceuticals. Some examples:

- Reintroduction of traditional dye plants such as woad (*Isatis tinctoria*), madder (*Rubia tinctorum*), weld (*Reseda luteola*), and goldenrod (*Solidago* sp.) has been attempted by improving the harvesting and extraction technologies. On a modest scale, some production has been realized in Germany.
- Caraway (*Carum carvi*) has been introduced as a crop for the production of carvon, which has been demonstrated to have sprout-supressing properties in potato storing.
- Some research on guayule (*Parthenium argentatum*) as a source for nonallergenic natural rubber production has been conducted.
- Agricultural production of *Stevia rebaudiana* as a promising new crop for southern regions of the EU has been studied, both for its sweetener properties and for nonfood uses.
- Glucomannan from konjac (*Armorphophallus* sp.) has been studied for its use as a texturing agent in food and nonfood applications.
- The tuber and seeds of ahipa (*Pachyrhizus ahipa*) have been evaluated for their specific (glyco)protein content and biocidal action.

WHOLE CROP UTILIZATION

Since it has been concluded that primary production of many crops is not economically feasible to produce only one principal component, valorization of residues from the crop has been considered. Many crops grown for fiber also may yield a useful oil (or the other way around) and produce other biomass residues (stalks, shives) that could tip the overall economic balance.

Bulk quantities of biomass with reduced (or negative) costs are available from many sources. For example, large quantities of verge grasses are being produced annually from roadsides, and many residues from agricultural and horticultural production accumulate without further use. Utilization of these residues for energy production would make a cheaper feedstock, but for a useful fuel preprocessing is a prerequisite.[7] Green tissues are generally richer in water, protein, and minerals, which could cause more difficulties in thermal conversion.

CONCLUSION

Despite the scope for new outlets and prospects for sustainable production of renewable raw materials, the acceptance of new crops at the farm level, in industry, and in the market has been complicated because of the adapted machinery and investments required at the various production levels. Much has been achieved over the last decade in exploration of the potential markets for different new industrial crops, although a real breakthrough still has to come. As a renewable raw material, crops do present strong marketing arguments as "eco-efficient" products for the development of sustainable consumption and production. Diversification of the market for renewable raw materials in the automotive industry, building and construction materials, paper and pulp, bioenergy, geotextiles, etc., are all in the picture if: 1) quality control of the agro-industrial production chain can be organized according to ISO standards; and 2) supplies of specified products to industrial buyers can be guaranteed.

Concerted action by the whole agro-industrial production chain will be necessary to attain the targets of sustainable consumption and ecologically safe production of the whole range of renewable products.

APPENDIX

Oil Crops

Sunflower	*Helianthus annuus*
Rapeseed	*Brassica oleracea*
Meadow foam	*Limnanthes alba*
Crambe	*Crambe abyssinica*
Castor	*Ricinus communis*
	Dimorphotheca pluviales
Lupin	*Lupinus mutabilis*
	L. albus
Jojoba	*Simmondsia chinensis*
	Stokesia sp.
False flax	*Camelina sativa*
Caper spurge	*Euphorbia lathyris*
	E. lagascae
Marigold	*Calendula officinalis*

N

Fiber Crops

Flax	*Linum usitatissimum*
Hemp	*Cannabis sativa*
Kenaf	*Hibiscus cannabinus*
	Hibiscus esculentus
Broom	*Ginestra* sp.
Stinging nettle	*Urtica dioica*

Biomass Crops

Elephant grass	*Miscanthus* sp.
Sorghum	*Sorghum* sp.
Giant reed	*Arundo donax*
Reed canary grass	*Phalaris arundinaceae*
Switchgrass	*Panicum virgatum*
Bamboo	*Phyllostachys* sp.
Cardoon	*Cynara cardunculus*

Other Crops

Chicory	*Chicorium intybus*
Woad	*Isatis tinctoria*
Madder	*Rubia tinctorum*
Weld	*Reseda luteola*
Goldenrod	*Solidago* sp.
Caraway	*Carum carvi*
Guayule	*Parthenium argentatum*
	Stevia rebaudiana
Konjac	*Amorphophallus* sp.
Ahipa	*Pachyrhizus ahipa*

ARTICLES OF FURTHER INTEREST

Marigold Flower: Industrial Applications of, p. 689
Non-wood Plant Fibers: Applications in Pulp and Papermaking, p. 829
Paper and Pulp: Agro-based Resources for, p. 861
Soy Wood Adhesives for Agro-based Composites, p. 1169
Sustainable Agriculture and Food Security, p. 1183
Sustainable Agriculture: Definition and Goals, p. 1187
Sustainable Agriculture: Ecological Indicators, p. 1191
Sustainable Agriculture: Economic Indicators, p. 1195
Sweet Sorghum: Applications in Ethanol Production, p. 1201
Switchgrass as a Bioenergy Crop, p. 1207

REFERENCES

1. Taylor, R.S. Alternative Crops for Europe. In *New Crops for Food and Industry*; Wickens, G.E., Haq, N., Day, P., Eds.; Chapman and Hall Ltd.: London, 1989, Chap. 17.
2. Relevant publications by The European Commission, Directorate-General XII, Science Research Development, summarized at www.nf-2000.org; www.cordis.lu; http://europa.eu.int.
3. Smith, N.O.; Maclean, I.; Miller, F.A.; Carruthers, S.P. Crops for industry and energy in Europe. EUR 17468 **1997**.
4. FAO statistics: http://apps.fao.org.
5. van Dam, J.E.G.; van Vilsteren, G.E.T.; Zomers, F.H.A.; Hamilton, I.T., Shannon, B. Industrial fibre crops, EC DGXII. EUR 16101 EN **1994**.
6. de Groot, B.; van Roekel, G.J.; van Dam, J.E.G. Alkaline Pulping of Fibre Hemp. In *Advances in Hemp Research*; Ranalli, P., Ed.; Haworth Press Inc., 1999; 213–242.
7. van Dam, J.E.G.; Elbersen, H.W.; van Doorn, J. *Cascade use of verge grasses for green chemicals and energy*, In Proc. 1st World Conference on Biomass for Energy and Industry Conference, Seville, Spain, June 5–6, 2000. James & James (Science Publishers) Ltd., p. 1840.

New Secondary Metabolites: Potential Evolution

David R. Gang
University of Arizona, Tucson, Arizona, U.S.A.

INTRODUCTION

When we consider the potential within the plant kingdom to evolve new metabolites, we need to consider two important properties of plants. The first is the diversity of building blocks—of current metabolites—found in plants, and the ability of these building blocks to be put together in new ways to produce new metabolites. The second important property to consider is the biosynthetic machinery within a plant that is responsible for producing metabolites, and the ability of this machinery to be modified so that new metabolite transformations are possible.

PLANT METABOLITE STRUCTURE AND DIVERSITY

Plants produce an amazing array of metabolites with very divergent structures and functions. These metabolites result from the action of several important and diverse biosynthetic pathways. Beyond the primary metabolic pathways—such as glycolysis, gluconeogenesis, and amino acid biosynthesis—the most significant metabolic pathways in plants are the terpenoid pathway (over 30,000 compounds identified), the phenylpropanoid pathway (over 5000 compounds), and the alkaloid pathways (over 12,000 compounds). Over 50,000 such metabolites have been identified in the plant kingdom,[1] and in the whole of nature there may be several hundred thousand or more of such compounds.[2]

Examples of this structural and functional difference are shown in Fig. 1. These molecules (eugenol, curcumin, or indole glucosinolate) can have simple structures, or they can be much more complex, (salvianin, vinblastine, or taxol). Plant secondary metabolites (also called specialized metabolites or natural products) may be the best manifestation of biodiversity within the plant kingdom. The features that most readily distinguish plant species—the colors, odors, medicinal value, and levels of toxicity—are caused by these metabolites. These are just a few of the many important and highly variable plant attributes that directly result from the production of specific metabolites. Such compounds can also be the initial cohort in plant defense arsenals. Many (e.g., eugenol, vinblastine, and taxol) are produced constitutively in the plant to ward off all but the most agile of pests or pathogens. Other metabolites (called phytoalexins) are produced in response to attack, either to kill the potential predator or prevent further damage. Furthermore, many metabolites (e.g., salvianin) are essential for attraction of pollinators and seed dispersers, and thus are vital for sexual reproduction in most angiosperm species. In fact, these latter roles clearly demonstrate that these compounds are not secondary at all, but are requisite parts of a plant's survival strategy.

BIOSYNTHESIS OF PLANT METABOLITES

The majority of plant metabolites are constructed in a manner whereby a basic structural motif is first synthesized and defines a new set or class of compounds. This core molecule is then further modified, often leading to great structural diversity within members of its compound class. This is what happens in the production of the anthocyanins, which are produced by a branch of the flavonoid pathway (Fig. 2). In the flavonoid pathway, a number of structural classes are formed (indicated by parentheses in Fig. 2). These classes serve as core building blocks for production of more elaborate molecules (e.g., salvianin).

The vast majority of chemical reactions involved in the production of plant metabolites are catalyzed by enzymes that are part of specific biosynthetic pathways.[2] For example, a number of steps are known in the pathway that leads to the formation of salvianin in scarlet sage[3] (Fig. 2). All of these appear to be catalyzed by enzymes: 1) polyketide synthases; 2) isomerases; 3) hydroxylases (cytochromes P450); 4) $NADP^+$-dependent dehydrogenases; 5) dehydratases; 6) glycosyltransferases; and 7) acyl-CoA-dependent acyltransferases. Even though the reaction converting the chalcone to the flavanone can occur spontaneously, this reaction is nevertheless catalyzed by an enzyme. Many other enzyme classes and reaction types are involved in the formation of other plant metabolites, but this example clearly shows that many different classes of enzymes are involved in the formation of any given metabolite that accumulates in a given plant. The genes encoding these enzymes are the genetic material in the plant kingdom that can be modified through evolutionary processes to produce new enzymes that in

Encyclopedia of Plant and Crop Science
DOI: 10.1081/E-EPCS 120005952

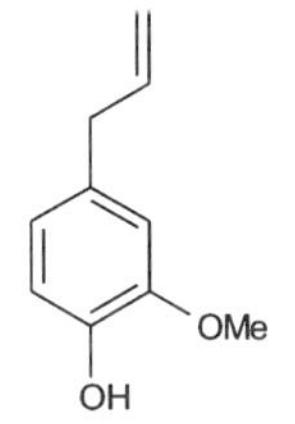

Eugenol
Antimicrobial, Odor and Flavor of Cloves

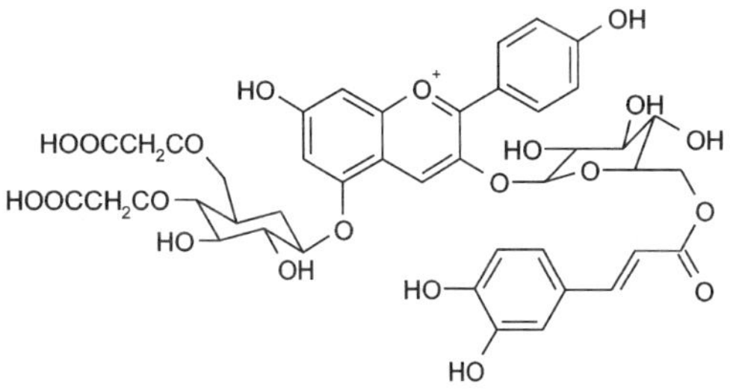

Salvianin
Floral Pigment

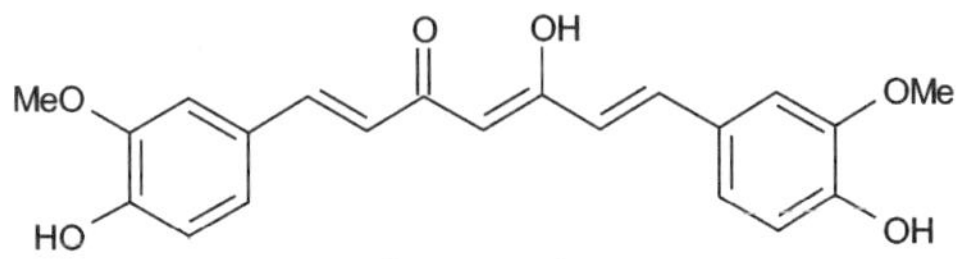

Curcumin
Spice Pigment, Anti-inflammatory

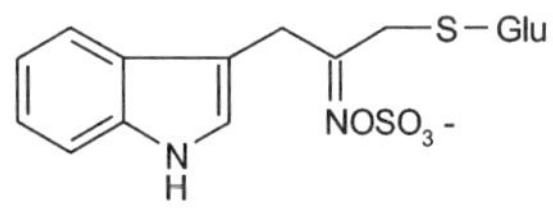

Indole Glucosinolate
Anti-Feedant, Toxin

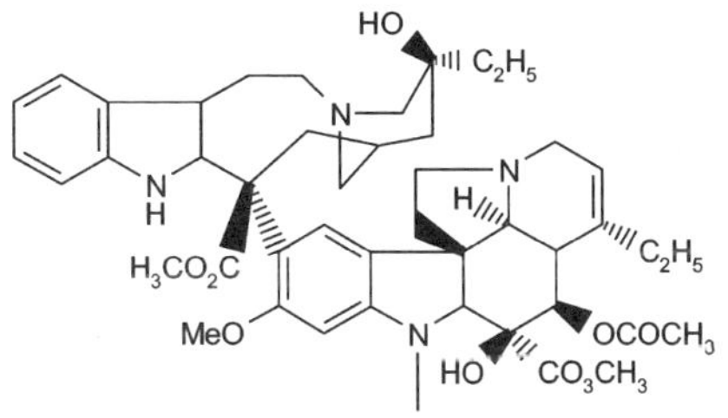

Vinblastine
Cytotoxic, Cancer Drug

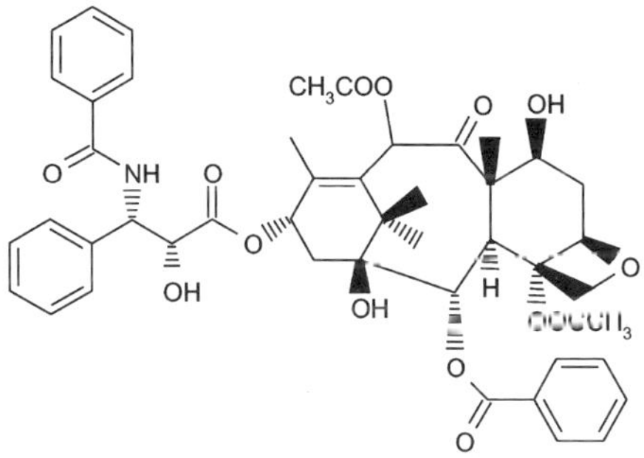

Taxol
Cytotoxic, Cancer Drug

Fig. 1 Examples of the structural and functional diversity in plant secondary metabolites.

turn catalyze the formation of new metabolites.[2,4] The ability to produce new enzymes is the key to the ability of plants to evolve new metabolites.

EVOLUTION OF NEW ENZYMES AND PATHWAYS

Several processes occur in plants that may lead to the formation of new enzymes. These range from polyploidization events and whole genome evolution, to gene duplication and divergence, and allele divergence.

Polyploidization events have been a common phenomenon in plant evolutionary history.[5] These events lead to the merging of whole genomes from two distinct species. When this latter process occurs, large amounts of DNA can be lost or rearranged. This leads to formation of new chimera chromosomes.[5] As can be imagined, this can lead to the duplication of some sets of genes. This sometimes occurs with the concomitant loss of other sets of genes. Only those losses that are not detrimental to the plant's survival will lead to viable progeny. Thus, it is possible that whole groups of biosynthetic enzymes and the resulting pathways and metabolites may be lost, so long as these metabolites are not absolutely required for survival of the plants. In addition to the process of polyploidization, it is possible that segments of chromosomes may be duplicated by other processes. These duplications may be large or small. As a result of all of these processes, the newly duplicated genes become redundant. This sets the stage for the evolution of new functions.

Once duplicate genes are formed, only one copy need retain the original activity for the plant to continue to make the same set of metabolites. The other copy is free to be mutated—to evolve. Many processes may be involved in this evolution. The new gene could simply accumulate random mutations, possibly leading to an altered substrate-binding cavity and new enzyme specificities. The

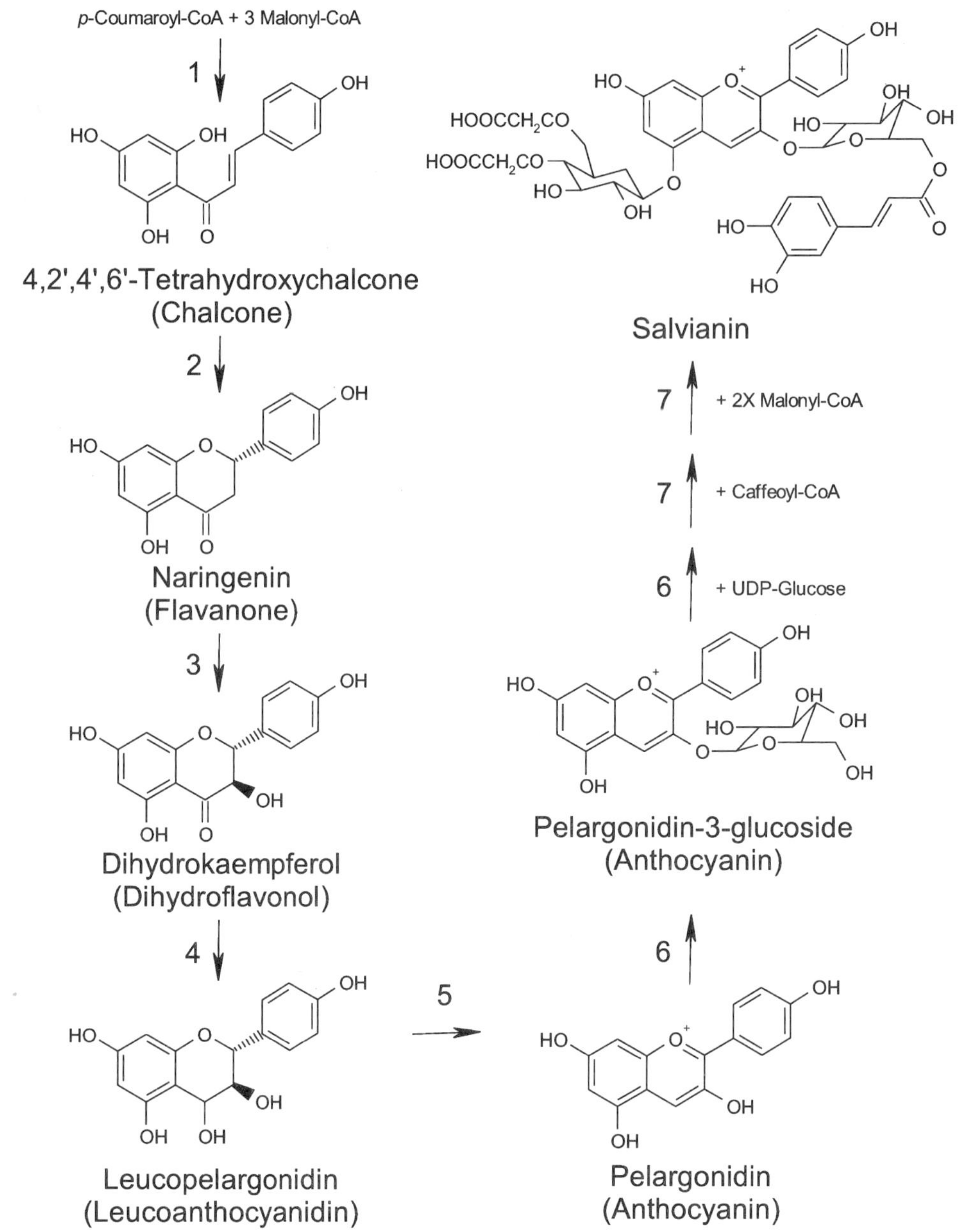

Fig. 2 Biosynthetic pathway to anthocyanin flavonoids such as salvianin, demonstrating the diversity of structural classes and enzymes involved.

results of this process have been observed in the production of a group of phenylpropanoid compounds in sweet basil.[6] In this case, two very similar enzymes (90% identical at the amino acid level) have been found that possess different substrate preferences. The difference in activity is the result of mutation of a single nucleotide in the gene sequence, which led to incorporation of a different amino acid in the active site of the enzyme, and thus to altered enzyme function. In this case, the original enzyme was from a secondary metabolic pathway. Another example of divergence of duplicated genes is the evolution of homospermidine synthase, which is found sporadically in a number of unrelated plant families.[7] This enzyme has very high homology to deoxyhypusine synthase, which is involved in activation of eukaryotic initiation factor 5A (eIF5A). eIF5A is required for proper cell growth. Thus, formation of new enzymatic functions in plant secondary metabolism can be derived directly from enzymes duplicated from primary metabolism.

Other examples of divergence after duplication are found in plant glycosyltransferases,[8] polyketide synthases,[9] cytochrome P450s,[2,4,10] terpene synthases,[2,4] and many other enzyme classes.[2] In fact, for some of these enzyme types, there are dozens to hundreds of individual members in a given plant's genome. For example, there are over 250 different cytochromes P450, over 360 glycosyltransferases, over 60 acyltransferases, and over 50 dehydrogenases in *Arabidopsis* alone.[2] This plant produces only a limited number of secondary metabolites. The function of only a small handful of the genes in these large gene families is known. The diversity in extant enzymes in these classes in the plant kingdom as a whole must be significantly larger.

CONCLUSION

Given the great number of extant metabolites in plants, a great pool of structural building blocks can be modified by future evolution in plants. In addition, a large pool of enzymes in any given plant—many of which are duplicates and redundant—are available for transformation into new enzymes via mutation and other evolutionary processes. Thus, because of the existing complex matrix of metabolites and enzymes in the plant kingdom, there is great potential for the evolution of new enzymes, new pathways, and new metabolites. Advances in genomics, proteomics, and metabolomics allow this area to be investigated in a manner not previously possible.

ARTICLES OF FURTHER INTEREST

Amino Acid and Protein Metabolism, p. 30
Aromatic Plants for the Flavor and Fragrance Industry, p. 58
Coevolution of Insects and Plants, p. 289
Crop Domestication: Fate of Genetic Diversity, p. 333
Crop Improvement: Broadening the Genetic Base for, p. 343
Ecology: Functions, Patterns, and Evolution p. 404
Evolution of Plant Genome Microstructure, p. 421
Floral Scent, p. 456
Genetic Diversity Among Weeds, p. 496
Genetic Resources of Medicinal and Aromatic Plants from Brazil, p. 502
Isoprenoid Metabolism, p. 625
Lipid Metabolism, p. 659
Metabolism, Primary: Engineering Pathways of, p. 714
Metabolism, Secondary: Engineering Pathways of, p. 720
Molecular Evolution, p. 748
Nucleic Acids Metabolism, p. 833
Phenylpropanoids, p. 868
Phytochemical Diversity of Secondary Metabolites, p. 915
Polyploidy, p. 1038
Secondary Metabolites as Phytomedicines, p. 1120

REFERENCES

1. DeLuca, V.; St. Pierre, B. The cell and developmental biology of alkaloid biosynthesis. Trends Plant Sci. **2000**, *5*, 168–173.
2. Pichersky, E.; Gang, D.R. Genetics and biochemistry of secondary metabolites in plants: An evolutionary perspective. Trends Plant Sci. **2000**, *5* (10), 439–445.
3. Suzuki, H.; Nakayama, T.; Yonekura-Sakakibara, K.; Fukui, Y.; Nakamura, N.; Nakao, M.; Tanaka, Y.; Yamaguchi, M.; Kusumi, T.; Nishino, T. Malonyl-CoA : Anthocyanin 5-O-glucoside-6 ‴-O- malonyltransferase from scarlet sage (Salvia splendens) flowers—Enzyme purification, gene cloning, expression, and characterization. J. Biol. Chem. **2001**, *276* (52), 49013–49019.
4. Dixon, R.A. Plant natural products: The molecular genetic basis of biosynthetic diversity. Curr. Opin. Biotechnol. **1999**, *10*, 192–197.
5. Soltis, D.E.; Soltis, P.S. Polyploidy: Recurrent formation and genome evolution. Trends Ecol. Evol. **1999**, *14* (9), 348–352.
6. Gang, D.R.; Lavid, N.; Zubieta, C.; Chen, F.; Beuerle, T.; Lewinsohn, E.; Noel, J.P.; Pichersky, E. Characterization of phenylpropene *O*-methyltransferases from sweet basil: Facile change of substrate specificity and convergent evolution within a plant *O*-methyltransferase family. Plant Cell **2002**, *14* (2), 505–519.
7. Ober, D.; Hartmann, T. Homospermidine synthase, the first pathway-specific enzyme of pyrrolizidine alkaloid biosynthesis, evolved from deoxyhypusine synthase. Proc. Natl. Acad. Sci. U. S. A. **1999**, *96* (26), 14777–14782.
8. Keegstra, K.; Raikel, N. Plant glycosyltransferases. Curr. Opin. Plant Biol. **2001**, *4*, 219–224.
9. Schroeder, J. The Family of Chalcone Synthase-Related Proteins: Functional Diversity and Evolution. In *Evolution of Metabolic Pathways*; Romeo, J.T., Ibrahim, R.K., Varin, L., DeLuca, V., Eds.; Pergamon Press: Amsterdam, 2000; Vol. 34, 55–89.
10. Celenza, J.L. Metabolism of tyrosine and tryptophan—New genes for old pathways. Curr. Opin. Plant Biol. **2001**, *4*, 234–240.

Nitrogen

Peter J. Lea
Lancaster University, Lancaster, U.K.

INTRODUCTION

The availability of water and nitrogen are considered to be the two major factors controlling plant growth and crop yield. In the past, the application of large amounts of nitrogen fertilizer by farmers has been a common practice. However, concern has been raised over the leaching of nitrogen from the soil, which can contaminate drinking water, cause toxic algal blooms, and thus affect human health, commerce, and tourism. There is now a strong move to adjust nitrogen input according to the nitrogen requirement of the plant and corresponding to the target yield.

USING NITROGEN TO OBTAIN HIGH CROP YIELDS

The yield of a high-quality wheat crop grown in Western Europe can reach 10 metric ton hectares (ha^{-1}). In addition, an equivalent amount of straw is produced, with an average 1.5% N in the dry matter, representing a nitrogen (N) requirement of 300 kg ha^{-1}. A grass crop used for animal fodder may produce 20 ton ha^{-1} but will contain 4% N, thus representing a N requirement of 800 kg ha^{-1}.[1] It therefore follows that in order to obtain these high yields, farmers need to apply large amounts of N fertilizers. The global use of N fertilizers has increased dramatically in the last 50 years and has now stabilized in the region of 80 million tons per annum. Nitrogen may be applied to crops in the form of ammonium (NH_4^+) or nitrate (NO_3^-) ions (often both together), urea, or as organic fertilizers derived from farmyard manure. In addition, legume crops are able to fix N_2 directly from the atmosphere through the utilization of symbiotic bacterial species in the root nodules. In well aerated neutral soils, the majority of the N is converted to nitrate. However, in acidic and/or anaerobic soils (e.g., in forests or for rice), ammonium ions are the major form of available N. The use of N fertilizers and the subsequent metabolism of the nitrogen-containing compounds within the plant are covered in two recent books edited by Bacon[2] and Lea and Morot-Gaudry[3] and in a special issue of the *Journal of Experimental Botany*.[4]

Unfortunately, the uptake of N by the crop plant seldom exceeds 60% of that applied. This loss of fertilized N can be due to leaching, erosion, runoff, or by gaseous emissions. The leaching of 20–30 kg N ha^{-1} can increase the nitrate concentration in the groundwater to above the threshold stipulated by the European Union (EU) for drinking water of 50 mg l^{-1}. Nitrogen may be lost to the atmosphere as NH_3 or following denitrification as N_2 or intermediate oxides, e.g., N_2O, NO, and NO_2. In addition, N may be sequestered by soil microorganisms in an organic form, which may then be available for subsequent crops.[5]

BIOMASS AND NITROGEN

The relationship between biomass production and the amount of N applied to the soil has been determined in a large number of field trials for most economically important crop plants. The generalized response curve shown in Fig. 1 clearly demonstrates that when there is a shortage of available N, productivity is very low. However, biomass production increases linearly with the supply of N until the asymptote is reached, at which point additional nitrogen does not increase productivity, and productivity is then limited by genetic potential. If growth is related to the N accumulated by the crop, then the initial slope, which is the true or intrinsic efficiency of nitrogen use by the plant, is steeper (more biomass per unit of N) than if related to the amount of N applied. The difference is a measure of the efficiency with which nitrogen is used, as is the difference between the point at which the plateau for biomass production is reached, when expressed per unit of absorbed and applied N.[6]

An abundant nitrogen supply increases the number of meristems produced by plants, thus encouraging branch and tiller formation and hence growth in most plants. In cereals and grasses the increased production of tillers gives rise to increased biomass and increased numbers of ears in grain-producing plants. As a consequence of the increased N supply, the leaf area index (leaf area/ground area, LAI) increases considerably. In a well watered winter wheat crop receiving 300 kg ha^{-1}N grown in Europe, the LAI can reach 6 by June. However, in a crop receiving only 75 kg ha^{-1} N, the LAI may only reach 2, with values below 1 reported in crops receiving no nitrogen at all.[1] The light energy received by the crop is

Encyclopedia of Plant and Crop Science
DOI: 10.1081/E-EPCS 120010510

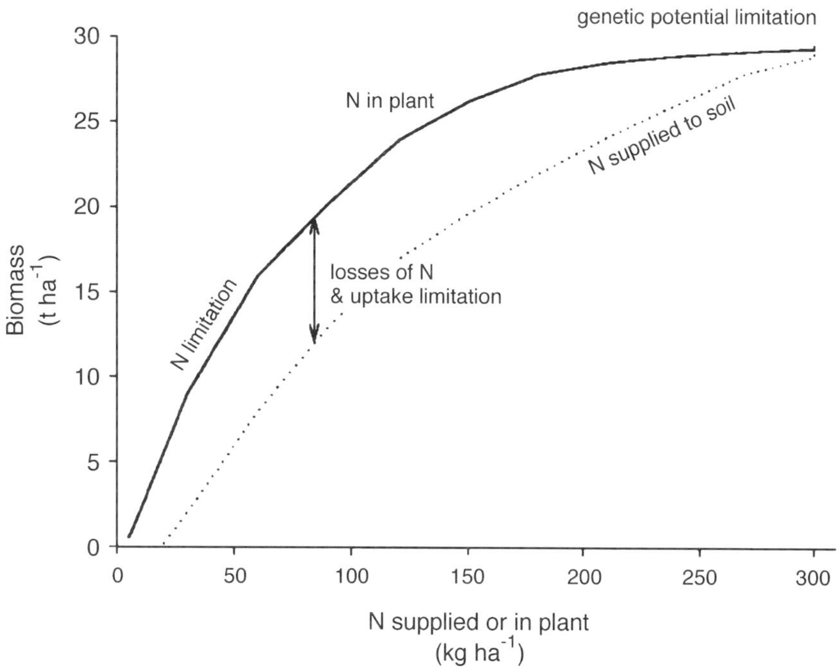

Fig. 1 A generalized response curve relating the production of biomass to the uptake and supply of nitrogen. The differences between the curves is due to losses of N in the soil and limitations of uptake by the plant. (From Ref. 6.)

used by the leaves to assimilate CO_2 in photosynthesis and also to carry out a range of other synthetic reactions. In fact it has been calculated that 20% of the N in the leaf of a C_3 plant is contained in light-harvesting pigments and proteins, 35% of the N is in the enzymes of the Calvin cycle including RuBP carboxylase/oxygenase (which is the major single component), and a further 20% is in other biosynthetic enzymes.[1] Thus the addition of nitrogen not only stimulates the formation of new leaves to intercept all the light available (a LAI of >4 is required), but also raises the amount of enzyme protein available to carry out photosynthesis. Ultimately during leaf senescence, N is mobilized from the enzyme proteins and transported to the seed for incorporation into storage protein.[7]

A range of models of crop growth is currently available based on the concentration of N within a plant. These models utilize the terms $\%N_{max}$, which is the maximum concentration of nitrogen that can be accumulated by the plant, and $\%N_{min}$, which is the minimum concentration of nitrogen below which the plant cannot survive. These values can be obtained from field trials with increasing applications of N fertilizer. The critical value for N concentration $\%N_{crit}$ lies between $\%N_{max}$ and $\%N_{min}$ and corresponds to the concentration of nitrogen, which permits maximal crop growth. Thus crops that are grown in conditions below $\%N_{crit}$ suffer N deficiency and reduced growth, whereas plants that are grown above $\%N_{crit}$ are able to store N. A discussion of the value of the various models available to establish $\%N_{crit}$ has been recently provided by Jeuffroy et al.[8]

VARIABLITY IN NITROGEN UPTAKE

The nitrogen uptake by field crops is highly variable within a single year, between years, between sites, and with different crops even when the N supply is plentiful. This is partly due to the fact that the increase in crop N content with crop mass is not linear. Additional N uptake per unit of additional biomass declines as the crop gets bigger. A plot of $\%N_{crit}$ against biomass provides a gentle downward curve and has been carried out for a range of plant species. C_4 species (e.g., maize and sorghum) were found to have a lower $\%N_{crit}$-biomass curve than C_3 species (e.g., wheat, rape, and pea), presumably due to the lower content of photosynthetic proteins within the leaf. The curves of a wide range of C_3 species, however, exhibit a remarkable similarity.[9]

CONCLUSION

There is a need to adjust the amount and timing of N application to crops to ensure that the maximum growth rate is obtained, while at the same time preventing wastage and hence possible pollution. As earlier, if the actual N content of the crop falls below the critical N content there is a reduction in the growth rate. A small decrease in the uptake of N by wheat could cause a reduction in the grain N content from 11.5% to 10.5%, thus preventing its use for bread making. Thus methods have been developed to establish the N content of the crop, which preclude the use of complex and expensive analyses. Simple colorimetric tests for the concentration of nitrate in leaves are now available, which may be used in conjunction with an analysis of chlorophyll concentration using handheld photometers. Such techniques give a clear result and may be carried out rapidly in the field with a large number of samples.[6]

ARTICLES OF FURTHER INTEREST

Amino Acid and Protein Metabolism, p. 30
C3 Photosynthesis to Crassulacean Acid Metabolism Shift in Mesembryanthenum Crystallinum: A Stress Tolerance Mechanism, p. 241
Photosynthate Partitioning and Transport, p. 897
Rubisco Activase, p. 1117
Symbiotic Nitrogen Fixation, p. 1218
Symbiotic Nitrogen Fixation: Plant Nutrition, p. 1222
Symbiotic Nitrogen Fixation: Special Roles of Micronutrients, p. 1226

REFERENCES

1. Lawlor, D.W.; Lemaire, G.; Gastal, F. Nitrogen, Plant Growth and Crop Yield. In *Plant Nitrogen*; Lea, P.J., Morot-Gaudry, J.-F., Eds.; Springer-Verlag: Berlin, 2001; 343–367.
2. Bacon, P.E. *Nitrogen Fertilization in the Environment*; Marcel Dekker, Inc.: New York, 1995.
3. Lea, P.J.; Morot-Gaudry, J.-F. *Plant Nitrogen*; Springer Verlag: Berlin, 2001.
4. Lea, P.J.; Morot-Gaudry, J.-F.; Hirel, B. Inorganic nitrogen assimilation. J. Exp. Bot. **2002**, *53*, 773–987. special issue.
5. Ter Steege, M.W.; Stulen, I.; Mary, B. Nitrogen in the Environment. In *Plant Nitrogen*; Lea, P.J., Morot-Gaudry, J.-F., Eds.; Springer-Verlag: Berlin, 2001; 379–397.
6. Lawlor, D.W. Carbon and nitrogen assimilation in relation to yield: Mechanisms are the key to understanding production systems. J. Exp. Bot. **2002**, *53*, 773–787.
7. Autran, J.-C.; Halford, N.G.; Shewry, P.R. The Biochemistry and Molecular Biology of Seed Storage Proteins. In *Plant Nitrogen*; Lea, P.J., Morot-Gaudry, J.-F., Eds.; Springer-Verlag: Berlin, 2001; 295–341.
8. Jeuffroy, M.H.; Ney, B.; Oury, A. Integrated physiological and agronomic modelling of N capture and use within the plant. J. Exp. Bot. **2002**, *53*, 809–823.
9. Gastal, F.; Lemaire, G. N uptake and distribution in crops: An agronomical and ecophysiological perspective. J. Exp. Bot. **2002**, *53*, 789–799.

Non-Infectious Seed Disorders

H. K. Manandhar
Nepal Agricultural Research Council, Khumaltar, Lalitpur, Nepal

S. B. Mathur
Danish Government Institute of Seed Pathology for Developing Countries, Copenhagen, Denmark

INTRODUCTION

Noninfectious seed disorders are nonpathogenic and nontransmissible unless genetically controlled. The disorders can be genetic, physiogenic, or caused by injuries. The genetic disorders are usually inherited and transmitted through seeds from one generation to another. The physiogenic disorders are mainly due to environmental conditions, nutrient deficiencies, and aging. In some cases, aging may be genetically controlled and also environmentally influenced. The injuries can be mechanical, chemical, or the result of insect damage.

Seed disorders can be visible or invisible. Visible disorders include cracking or splitting in seed coat; shrunken or wrinkled seeds; discolored seeds; and seeds with holes, abnormal shape, and reduced size. The invisible seed disorders include some genetic disorders, embryoless or embryo-damaged seeds, aging, and some micronutrient deficiencies.

Seed disorders can affect seed germination or produce weak seedlings and plants with symptoms of different kinds. Books such as *Seed Pathology* by P. Neergaard and *Principles of Seed Pathology* by V. K. Agarwal and J. B. Sinclair provide reviews on noninfectious seed disorders. *An Annotated List of Seed-borne Diseases* by M. J. Richardson also provides some relevant information on the topic.

VISIBLE SEED DISORDERS

Seed Coat Cracking or Splitting

This is one of the most common types of seed disorders. It can be either genetic or caused by mechanical injuries or environmental effects. Seed coat cracking due to genetic effects is common in eggplant, pepper, and tomato.[1] Seed coat splitting occurs in flax and is more common in yellow-seeded cultivars.[2] Similar disorder is also found in some cultivars of bean.

Mechanical injuries during harvesting, threshing, processing, and postharvest handling may result in seed coat splitting or cracking of many crops.[1,3] Seeds with low moisture content and thinner seed coat have a higher chance of such damage. Machines with high-cylinder speeds result in damaged seed coats.

Desiccation of seeds during maturation results in cracking of soybean seed coat. Seeds harvested with high moisture and then stacked result in softening or cracking of seed coats.[1]

The major concern over such seed coat abnormality is that it provides entry points for various types of organisms, pathogens, and saprophytes (Fig. 1). Entry of saprophytes is important because such organisms multiply under bad storage conditions (improper ventilation, high temperatures, high humidity). The effect is pronounced when seeds are stored at high seed moisture.

Discolored Seeds

Seed discoloration is often associated with environmental factors and nutritional deficiencies. Cold-damaged canola seeds become brown[1] (see also shrunken seeds, discussed later). Frost-injured peanut seeds are off-white, water soaked or translucent, and off-flavor.[4] Severely damaged seeds can only be used for oil stock. Under humid conditions, radish seeds develop a gray discoloration due to swelling of the subepidermal parenchyma, which results in distortion or cracking of the epidermis.[3] This facilitates invasion by fungi. Seeds that are moist or immature when stored may become discolored due to heat (see also wrinkled and embryo-damaged seeds, discussed later).[3]

Boron deficiency in peanuts results in seed discoloration (see also hollow heart, discussed later). Seed coat browning of broad bean is caused by calcium deficiency.[2] Peanut seeds deficient in calcium often have darkened plumule, and the viability of seeds is directly related to their calcium concentration.[4]

Reduced Seed Size

In general, plants grown in nutrient-deficient conditions produce seeds of reduced size. Soybean seed size is

Encyclopedia of Plant and Crop Science
DOI: 10.1081/E-EPCS 120021914

Fig. 1 Soybean seed showing seed coat cracking (right) that may facilitate the entry of downy mildew fungus, *Peronospora manshurica*, seen as oospore crusts on the surface of both seeds.

reduced when plants are grown in molybdenum-deficient conditions.[5] Such seeds may produce plants that show deficiency symptoms when grown in soils with a marginal supply of this nutrient.

Shrunken and Wrinkled Seeds

Cold-damaged canola seeds become brown and shrunken.[1] Wrinking of soybean seeds may occur due to moderate desiccation during maturation, resulting in lower field emergence than in seeds with sound coat.[3] Self-heating of seeds during storage, particularly when the seed is moist or immature, may lead to wrinkling and discoloration.[3] Severe potassium deficiency tends to result in wrinkled, misshapen soybean seeds.[5]

Hollow Heart

Hollow heart, characterized by cavities or depression on the adaxial face of cotyledons (Fig 2), is a widely known seed disorder of pea.[1,3] The disorder is caused by sudden drying of immature seeds at high temperatures. It is common in wrinkled pea seeds and peas having compound starch grains. Hollow heart symptoms are seen when seeds are soaked in water for 16–18 hours at 20–25°C, testae are separated, and the spilt cotyledons are stored for 24 hours at 20–25°C.[6]

Seeds with hollow heart germinate normally under ideal conditions, but the disorder predisposes seedlings to attack by pathogenic fungi or results in poor emergence.[1,3]

Hollow heart symptoms also occur in peanut seeds, in which case it is caused by boron deficiency. The

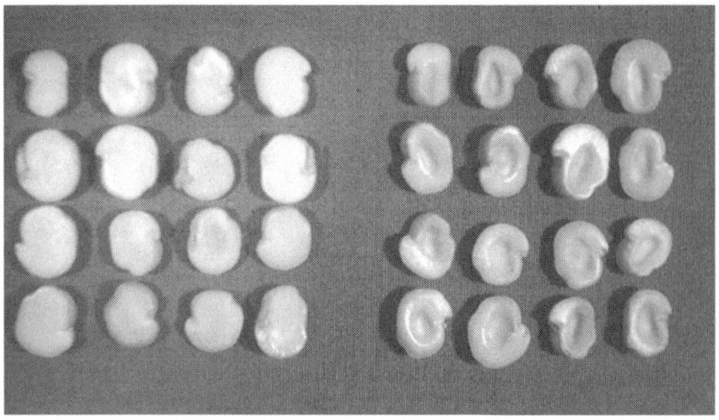

Fig. 2 Hollow heart symptom showing cavitation on the adaxial face of cotyledons of pea (right) and normal seed (left) (From Ref. 14). (*View this art in color at www.dekker.com.*)

symptoms include discoloration and rotting. The inner face of the cotyledon is depressed in the center, which region often turns brown, especially when peanuts are roasted.[4]

Marsh Spot

Marsh spot of pea—characterized by discolored lesions in the center of the adaxial face of the cotyledon—is due to manganese deficiency (Fig. 3). The disorder is common in cultivars having simple and large starch granules in the cotyledons[1] Similar disorder due to manganese deficiency is found in seeds of broad bean, haricot bean, and runner bean.[7] Marsh spot symptoms are also seen by the method described for hollow heart symptoms.[6]

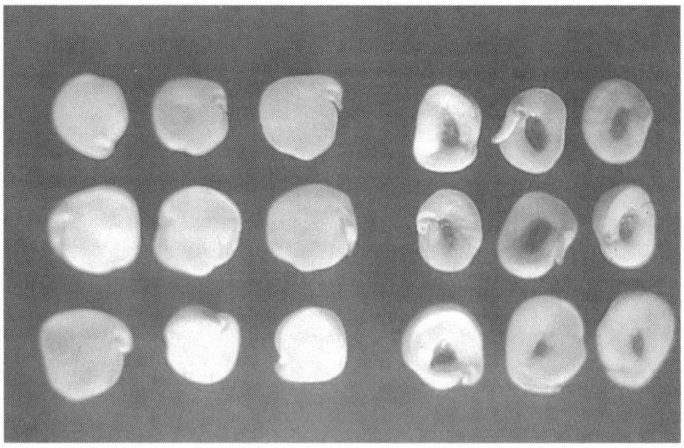

Fig. 3 Marsh spot symptom showing discolored lesions in the center of the adaxial face of cotyledon of pea (right) and normal seed (left) (From Ref. 14). (*View this art in color at www.dekker.com.*)

Fig. 4 Holes in pea seeds made by insects. (*View this art in color at www.dekker.com.*)

Yellow Berry

Yellow berry is the term given to wheat grains that are light-yellow in color, opaque, soft, and starchy. They differ from the normally hard, flinty, translucent grains—clear and dark and reddish-amber in color—described in the U.S. state of Kansas.[8] Yellow berry is associated with nitrogen deficiency and warm temperatures during heading and seed maturation.[9] Such seeds germinate normally, but may result in poor market value. They are also high in starch content, but low in protein. The incidence of yellow berry differs among wheat genotypes.

Seeds with Holes and Pits

This disorder is caused by a wide range of insects feeding on seeds, causing holes and pits and sometimes leaving seeds completely empty (Fig. 4). Examples of such insects are species of *Bruchus*, *Heliothis*, *Lygus*, and *Sitophilus*.[3]

INVISIBLE SEED DISORDERS

Genetic Disorders

These disorders are transmitted through seeds, and include fruit pox, gold fleck, and corky stunt of tomato.[1] Fruit pox appears as incipient lesions on immature fruit, which later rupture and become necrotic before the fruit turns red. Gold fleck appears as small lesions on immature fruit, that turn golden-yellow on ripe fruit. Corky stunt is expressed as shortened internodes, proliferated axillary buds, malformed petioles, roughened lower petiole surfaces, and malformed and corked fruit. See the *Annotated List of Seed-borne Diseases*[10] for other examples of seed-transmitted disorders such as bud failure of almond and stem distortion of wheat.

Embryoless or Embryo-Damaged Seeds

Embryoless or embryo-damaged seeds are often found. The disorder is expressed in the form of either no germination or abnormal seedlings. Embryoless seeds are encountered in carrot, celery, coriander, dill, fennel, parsley, and parsnip seeds when insects such as lygus bugs (various species of *Lygus*) eat embryos but leave the seed coat and endosperm undamaged.[3] Due to some genetic effects, defective fertilization may take place in some crop plants, e.g., barley and wheat, resulting in embryoless seeds.[1]

Embryo damage may be caused by mechanical injuries, and results in abnormal seedlings. For example, mechanical damage of lima bean embryo during threshing results in balhead or snakehead seedlings that may lack a growing point, radicle, or cotyledon.[1,3] Similar damage occurs in peas and clover.[3] Heating of seeds due to metabolic activities during storage may lead to embryo damage and result in the production of abnormal seedlings.[3]

Seed Aging

Old seeds gradually decline in germination capacity and develop plants of reduced vigor and yield. Seeds stored under adverse conditions may also have reduced viability as a result of induced aging. Seeds stored at higher temperatures generally age early and show increased production of lipid peroxidase. This has been demonstrated in seeds of *Phaius tankervilliae*.[11] Blind plant problem (i.e., the lack of normal meristem) in tomato is mostly associated with old seeds.[12] Loss of viability with aging of niger seeds is associated with reduced radicle growth and abnormalities in hypocotyl development.[13] Aged seeds may also exhibit delays in germination, as in the case of meadow steppe grass seed.[14]

Chemical Injuries

Seeds treated with chemicals at excessive doses, stored at higher temperatures after treatment, and stored for a long period after treatment may undergo damage. Such seeds either do not germinate or produce abnormal plants. For example, seed treated with organomercurial fungicides at excessive rates or high moisture either do not germinate or show hypertrophy of the coleoptile and poor root development.[9] Some chemicals may cause phytotoxicity regardless of moisture content of treated seeds, time of sowing after treatment, and temperature, because they are toxic to certain kinds of seeds but beneficial to others. For example, fungicides like carbendazim + thiram and imidacloprid are found to be toxic to pea seeds.[15]

Nutrient Deficiencies

Seeds that look normal may be deficient in nutrients, especially micronutrients. Such seeds may produce abnormal seedlings when sown. Pea seeds deficient in boron develop seedlings with pale stunted shoots without plumule bud.[10] Phaseolus bean seeds produced under molybdenum deficiency conditions give seedlings with scald symptoms.[10]

CONCLUSION

Noninfectious seed disorders can be genetic, physiogenic, or caused by injuries. Although the disorders are nonpathogenic, they may predispose the seeds to invasion by pathogenic and saprophytic organisms. The disorders are responsible for low germination, production of abnormal or diseased plants, and low market value. The types of disorders and their causes suggest that they can be managed by selecting proper genotypes of crop plants; adopting proper measures during harvesting, processing, and storage; and correcting nutrient deficiencies in soil and plants.

ARTICLES OF FURTHER INTEREST

Boron, p. 167
Economic Impact of Insects, p. 407
Integrated Pest Management, p. 612
Minor Nutrients, p. 726
Potassium and Other Macronutrients, p. 1049
Seed Borne Pathogens, p. 1126
Seed Production, p. 1134
Seed Vigor, p. 1139
Seeds: Pathogen Transmission Through, p. 1142

REFERENCES

1. Agarwal, V.K.; Sinclair, J.B. *Principles of Seed Pathology*, 2nd Ed.; CRC Press, Inc.: Boca Raton, 1997.
2. Chishaki, N.; Horiguchi, T. Causes for seed coat browning of broad bean (*Vicia faba* L.) inferred from mineral contents of plants and chemical properties of soils on farmers' fields. Jpn. J. Soil Sci. Plant Nutr. **2000**, *71* (3), 372–377.
3. Neergaard, P. *Seed Pathology*, Revised Edition; The MacMillan Press Ltd.: Houndmills, 1979; Vol. 1 and 2.
4. *Compendium of Peanut Diseases*, 2nd Ed.; Kokalis-Burelle, N., Porter, D.M., Rodríguez-Kábana, Smith, D.H., Subrahmanyam, P., Eds.; APS Press, The American Phytopathological Society: St. Paul, MN, 1997.
5. *Compendium of Soybean Diseases*, 4th Ed.; Hartman, G.L., Sinclair, J.B., Rupe, J.C., Eds.; APS Press, The American Phytopathological Society: St. Paul, MN, 1999.
6. Singh, D. Occurrence and histology of hollow heart and marsh spot in peas. Seed Sci. Technol. **1974**, *2*, 443–456.
7. Wallace, T. *Color Pictures of Mineral Deficiencies in Beans from The Diagnosis of Mineral Deficiencies in Plants by Visual Symptoms*; His Majesty's Office, 1943. (http://www.luminet.net/~wenonah/min-def/beans.htm).
8. Roberts, H.F.; Freeman, G.F. The yellow berry problem in Kansas hard winter wheats. Bull. Exp. Stn. at the Kansas State Agricultural College, Manhattan **1908**, *156*, 1–35.
9. Wiese, M.V. *Compendium of Wheat Diseases*, 2nd Ed.; APS Press, The American Phytopathological Society: St. Paul, MN, 1987.
10. Richardson, M.J. *An Annotated List of Seed-borne Diseases*, 4th Ed.; The International Seed Testing Association: Zürich, Switzerland, 1990.
11. Shan, X.C.; Weatherhead, M.A.; Song, S.Q.; Hodgkiss, I.J. Malondialdehyde content and superoxide dismutase activity in seed of *Phaius tankervillae* (Orchidaceae) during storage. Lindleyana **2000**, *15* (3), 176–183.
12. *Compendium of Tomato Diseases*; Jones, J.B., Jones, J.P., Stall, R.E., Zitter, T.A., Eds.; APS Press, The American Phytopathological Society: St. Paul, MN, 1991.
13. Dhakal, M.R.; Pandey, A.K. Storage potential of niger (*Guizotia abyssinica* Cass.) seeds under ambient conditions. Seed Sci. Technol. **2001**, *29* (1), 205–213.
14. Rice, K.J.; Dyer, A.R. Seed aging, delayed germination and reduced competitive ability in *Bromus tectorum*. Plant Ecol. **2001**, *155* (2), 237–243.
15. Kotlinski, S. Comparison of some seed dressing chemicals combination on germination and weight of pea seedlings as affected by seeds moisture, sowing time and germination temperature. Prog. Plant Prot. **1999**, *39* (2), 905–913.

Non-Wood Plant Fibers: Applications in Pulp and Papermaking

N

Joseph E. Atchison
Atchison Consultants, Inc., Sarasota, Florida, U.S.A.

INTRODUCTION

It is well known that due to a plentiful supply and reasonable costs, the economics of pulp and paper production in North America, and in most other industrial countries, have favored wood as the fibrous raw material, in the past and up to the present time. However, many of the developing countries, as well as a few of the industrial countries, do not have adequate supplies of wood, but they do have large quantities of non-wood plant fibers available. Fortunately, we have found that by choosing the proper blend of non-wood plant fibers, almost all grades of paper and paperboard—ranging from tissue to linerboard, including newsprint—can be produced with as much as 100% fibrous content of these materials. Furthermore, every type of reconstituted panel board—ranging from insulating board to hardboard, including medium-density fiberboard—can likewise be produced, and is being produced, from non-wood plant fibers.

INCREASE IN CAPACITY IN PULPS PRODUCTION FROM NON-WOOD PLANT FIBERS: 1975–1998

As shown in the annual Pulp and Paper Capacities Surveys by the Food and Agriculture Organization of the United Nations for the periods 1975–1980,[1] 1981–1986,[2] 1990–1995,[3] 1993–1998,[4] and 1998–2003,[5] the worldwide capacity for production of non-wood plant fiber pulps for papermaking has increased dramatically since 1975, going from 6.8% of total capacity in 1975 to 11.0% of capacity in 1998.

Table 1 shows this rapid growth in production capacity for non-wood plant fiber pulps. It can also be seen that the average increase in non-wood plant fiber pulping capacity has been greater for many years (up until 1998) than the average increase in wood pulping capacity. For example, during the five-year period from 1988–1993, non-wood pulping capacity increased by 6.0% annually, whereas wood pulping capacity increased by only 2.0% annually. Then during the 1993–1998 period, non-wood pulping capacity increased by 2.7% annually, in contrast to an increase of only 1.6% for wood pulping capacity.

Unfortunately, due to the financial crisis that began in 1997 in the Asian countries and some other developing countries that were major producers of non-wood plant fiber pulp, along with the worldwide recession in the pulp and paper industry (which continues today) and the fact that China has closed several thousand small non-wood plant fiber-based mills that did not have recovery systems, there has been no increase in non-wood plant fiber pulping capacity since 1998. Furthermore, no increased capacity is projected through 2003, as shown in the annual FAO Pulp and Paper Capacities Survey for 1998–2003.[5] However, it is the author's opinion that as soon as the economic situation improves sufficiently in these countries, which are the major producers of non-wood plant fiber pulp, the previous trend upward will resume.

MINIMUM UTILIZATION OF NON-WOOD PLANT FIBERS IN THE UNITED STATES

In contrast to the worldwide use of non-wood plant fiber pulps, the capacity for producing these pulps for papermaking in the United States amounted to only 219,000 metric tons in 2001, whereas wood pulping capacity amounted to 64 million metric tons, as reported in the FAO Pulp and Paper Capacities Survey for 1998–2003.[5] No increase in either non-wood plant fiber pulping capacity or wood pulping capacity in the United States is projected through 2003, due to the continuing worldwide recession in the pulp and paper industry.

WORLDWIDE USE OF NON-WOOD PLANT FIBERS FOR PAPERMAKING PULP

The non-wood plant fibers currently being used worldwide and those which offer potential for future use in

Encyclopedia of Plant and Crop Science
DOI: 10.1081/E-EPCS 120010446

Table 1 World papermaking pulp capacities (1975–1998): Non-wood pulping capacity vs. total and average annual increases

	Total papermaking pulp capacity (millions of metric tons)				Average annual increases (percent)		
Raw materials	**1975**	**1988**	**1993**	**1998**	**75–80**	**88–93**	**93–98**
Total papermaking pulp—all raw materials	136.1	175.7	197.1	214.9	2.2	1.6	1.4
Total papermaking wood pulp	126.8	160.1	176.4	191.3	2.0	2.0	1.6
Total papermaking non-wood plant fiber pulp	9.3	15.6	20.7	23.6	5.2	6.0	2.7
Percentage of non-wood plant fiber pulp to total papermaking pulp	6.8	8.6	10.5	11.0			

(Compiled from Refs. 1–6.)

papermaking pulps include the agricultural residues such as sugar cane bagasse, straw, and corn stover; the naturally growing plants such as reeds, bamboo, and certain grasses; and the fibers grown specifically for their fiber content, such as kenaf, crotalaria, jute, abaca (Manila hemp), sisal, cotton and flax fibers, and cotton linters. In the past, almost every known non-wood plant fiber has been tried for the manufacture of pulp and paper,

Table 2 Leaders in total non-wood plant fiber papermaking pulp production capacity: Percentage of total pulping capacity based on non-wood plant fibers in 1993 and 1998

	1993		1998	
Country	**Non-wood pulping capacity (1000 mt/yr)**	**Percent of total from non-wood plant fibers**	**Non-wood pulping capacity (1000 mt/yr)**	**Percent of total from non-wood plant fibers**
1. China	15,246	86.9%	17,672	84.2%
2. India	1,307	55.5%	2,001	61.3%
3. Pakistan	415	100.0%	491	84.5%
4. Australia	10	0.8%	304	18.9%
5. Venezuela	185	75.2%	255	68.0%
6. U.S.A.	179	0.03%	219	0.03%
7. Colombia	218	45.5%	217	46.4%
8. Mexico	321	29.2%	198	26.4%
9. Turkey	103	16.5%	191	27.4%
10. Italy	120	13.3%	165	23.6%
11. Greece	150	85.7%	160	84.2%
12. Thailand	209	100.0%	148	15.9%
13. Argentina	140	14.6%	140	12.8%
14. Brazil	196	3.1%	136	1,8%
15. Egypt	127	100.0%	127	100.0%
16. South Africa	99	6.4%	115	6.2%
17. Cuba	108	100.0%	108	100.0%
18. Iraq	101	100.0%	101	100.0%
19. France	0	0	100	3.0%
20. Vietnam	86	60.1%	100	40.0%
Total: First 20 Countries	19,320	–	22,948	–
Total: All Countries	20,736	10.6%	23,600	11.0%

(Compiled from Refs. 1–6.)

and a great many of them result in a product having some desirable properties. However, when all of the economic factors and technical problems are taken into consideration, only a very few of the hundreds of thousands of non-wood plant fibers can qualify.

At the present time, on a worldwide basis, straw, sugar cane bagasse, and bamboo are the leading non-wood plant fibers being used. However, many others are being used in countries such as China and India that do not have adequate wood supplies, even for the mass-production grades. These include reeds and grasses of various types. In addition, many other high-cost non-wood plant fibers are being used all over the world, including in the United States. They produce pulps with special properties that are not found in any wood pulps. Such pulps, including flax fibers, cotton fibers, abaca, sisal, and cotton linters, are being used for the production of relatively small quantities of many high-priced specialty papers and paperboards.

WORLDWIDE NON-WOOD PLANT FIBER PULPING CAPACITY BY COUNTRY (1993 AND 1998)

Table 2, shows the total worldwide capacity for production of non-wood plant fiber pulp by the leading producers of such pulps, along with the percentage of their total papermaking pulp capacity devoted to using these raw materials in 1993 and 1998. These figures are based on data from the FAO Pulp and Paper Capacities Survey for 1993–1998,[5] combined with individual reports received by Atchison Consultants in personal correspondence with some of the Pulp and Paper Associations of major non-wood plant fiber pulp-producing countries.[6] As mentioned earlier in this chapter, there has been no increase in non-wood plant fiber pulping since 1998.

Some 42 countries are now producing some papermaking pulp from non-wood plant fibers, with the total capacity approaching 24 million metric tons (Table 2), or 11.0 % of total worldwide papermaking pulp capacity. It can be seen that China is by far the leader, with 84% of its total papermaking pulp capacity devoted to pulping non-wood plant fibers. India is second, with over 64% of its capacity based on pulping non-wood plant fibers. Among the countries shown in Table 2, non-wood plant fiber represents 100% of the pulping raw materials in four of them, and nine of them depend upon these raw materials for more than 50% of their pulping capacity. Table 3 a summary of worldwide papermaking pulp capacities based on using specific non-wood plant fibers, as well as the total of wood-based pulping capacities and the overall percentages of pulping capacities based on using non-wood plant fibers.

CONCLUSION

There appears to be no doubt that non-wood plant fibers will play an increasing role in the world's pulp and paper industry. Certainly, the necessary fiber resources either already exist or can be grown in the wood-poor countries, in order to sustain the pulp and paper requirements in those areas. It has been proved that by selecting the proper

Table 3 Summary of worldwide papermaking pulp capacities using non-wood plant fibers and pulpwood (1983–1998)

	Total papermaking pulping capacities (millions of metric tons)			
Raw materials	**1983**	**1990**	**1993**	**1998**
A. Total wood-based papermaking pulp capacity	151,000	168,600	176,000	191,292
B. Non-wood plant fiber papermaking pulp capacity by raw material				
Straw	6,166	6,787	9,566	10,705
Sugar cane bagasse	2,339	2,739	2,984	3,206
Bamboo	1,545	0.987	1,316	1,474
Miscellaneous non-wood plant fibers	3,302	5,049	6,870	8,174
Subtotal: non-wood plant fiber pulping capacity	13,352	15,562	20,747	23,559
Grand total: papermaking pulping capacity	164,352	184,162	197,736	214,851
Percentage for non-wood plant fiber pulps	8.1	8.5	10.6	11.0

(From Refs. 1–6.)

mixture of these fibers and by the appropriate collecting, handling, storing, and pulping processes, any grade of paper, paperboard, or reconstituted panelboard can be produced from non-wood plant fibers. If circumstances require it, all grades can be produced without the addition of wood pulp. In fact, some grades are already being produced with 100% non-wood plant fiber, especially bagasse pulp. However, for most of the mass production grades it is expected that non-wood plant fiber pulps will be used in blends with at least a small proportion of wood pulp, even in the wood-poor countries.

In regard to the greater use of non-wood plant fiber pulps in North America, the time may be approaching when their greater use will become a reality. With 100 million metric tons of available straw; the potential of some 5 million tons of bagasse, with large concentrations of it available in Louisiana; 150 million metric tons of corn stover; and the exciting possibilities of kenaf, non-wood plant fiber certainly should be considered as supplementary raw materials to be blended with wood pulp.

ARTICLES OF FURTHER INTEREST

New Industrial Crops in Europe, p. 813
Paper and Pulp: Agro-based Resources for, p. 861

REFERENCES

1. FAO *Pulp and Paper Capacities Survey, 1975–1980*; FAO: Rome, Italy.
2. FAO *Pulp and Paper Capacities Survey, 1981–1986*; FAO: Rome, Italy.
3. FAO *Pulp and Paper Capacities Survey, 1990–1995*; FAO: Rome, Italy.
4. FAO *Pulp and Paper Capacities Survey, 1993–1998*; FAO: Rome, Italy.
5. FAO *Pulp and Paper Capacities Survey, 1998–2003*; FAO: Rome, Italy.
6. Personal correspondence with representatives of pulp and paper associations in leading non-wood plant fiber pulp-producing countries, 2000.

Nucleic Acid Metabolism

Hiroshi Ashihara
Ochanomizu University, Tokyo, Japan

N

INTRODUCTION

Nucleic acid metabolism includes biosynthesis of nucleotides (phosphate esters of ribose or deoxyribose in which a purine or pyrimidine base is linked to C1′ of the sugar residue) for RNA and DNA, and degradation of these nucleic acids to simple molecules. The processes of DNA replication and repair, as well as the synthesis of various RNAs related to gene expression, are not included in this review. Compared with microorganisms and mammals,[1–3] only limited research has been carried out on the synthesis and degradation of nucleic acids in higher plants; some reviews of the subject have been published.[4–9]

DE NOVO BIOSYNTHESIS OF RIBONUCLEOTIDES

The pathways in the biosynthesis of nucleotides from their small molecule precursors are traditionally referred to as de novo pathways. In contrast, nucleotide synthetic pathways from preformed purine bases and nucleosides are called salvage pathways.

Pyrimidine Nucleotide Biosynthesis

The de novo pyrimidine biosynthetic pathway, also known as the orotate pathway, is defined as the formation of uridine 5′-monophosphate (UMP) from carbamoyl phosphate. The orotate pathway consists of six reactions, as shown in Fig. 1. The pyrimidine ring is assembled from carbamoyl phosphate and aspartate. The C-2 and N-3 atoms originate from carbamoyl phosphate; N-1, C-4, C-5, and C-6 originate from aspartate. In mammals and many other eukaryotes, the first three enzymes—carbamoylphosphate synthetase, aspartate transcarbamoylase, and dihydroorotase—are present as a multifunctional protein called CAD (after the initial letter of the three constituent enzymes). However, no such complex has been detected in plants, although most of the encoding genes for the individual enzymes have been cloned. Some of the genes contained chloroplast transit sequences. Thus, in higher plants, pyrimidine biosynthesis seems to be operative in chloroplasts.

Although there are two different types of carbamoyl phosphate synthetases that provide substrates for pyrimidine and arginine biosynthesis in most eucaryotes, including mammals, only one is present in higher plants. Therefore, the plant enzyme provides carbamoyl phosphate for both pathways. UMP, an end product of the orotate pathway, is a feedback inhibitor of plant carbamoyl phosphate synthetase, but this inhibition is overcome by ornithine, leading to the utilization of carbamoylphosphate for arginine biosynthesis. The second enzyme—aspartate transcarbamoylase—is also inhibited by high concentrations of UMP. This seems to be feedback control of pyrimidine biosynthesis.

The fourth enzyme—dihydroorotate dehydrogenase—is not well characterized. Subcellular localization studies with heterotrophycally cultured tomato cells suggested that it is located in mitochondria. The fifth and sixth enzymes—orotate phosphoribosyltransferase and orotidine-5′-monophosphate decarboxylase—from plants, reside in a single polypeptide. Recently, a new term, UMP synthase, has been given to this multifunctional protein, which is also observed in animals. UMP produced by the orotate pathway is further phosphorylated by UMP kinase and nucleoside diphosphate kinase to UTP via UDP. CTP is formed from UTP by a one-step reaction catalyzed by CTP synthetase.

Purine Nucleotide Biosynthesis

The de novo purine biosynthetic pathway is defined as the pathway that is responsible for the synthesis of inosine 5′-monophosphate (IMP) from 5-phosphoribosylamine (PRA). PRA is formed from 5-phosphoribosyl-1-pyrophosphate (PRPP), a common phosphoribosyl donor for purine and pyrimidine nucleotide synthesis. The purine ring is assembled from several small molecules. The N-1 atom originates from aspartate; C-2 and C-8 are from activated derivatives of tetrahydrofolate; N-3 and N-9 come from the amide group of the side chain of glutamine; and C-4, C-5, and N-7 are from glycine. Fig. 2 shows the 10 steps of the IMP synthetic pathway from PRPP. Precise details of this pathway in plants are still obscure, but the

Encyclopedia of Plant and Crop Science
DOI: 10.1081/E-EPCS 120010401

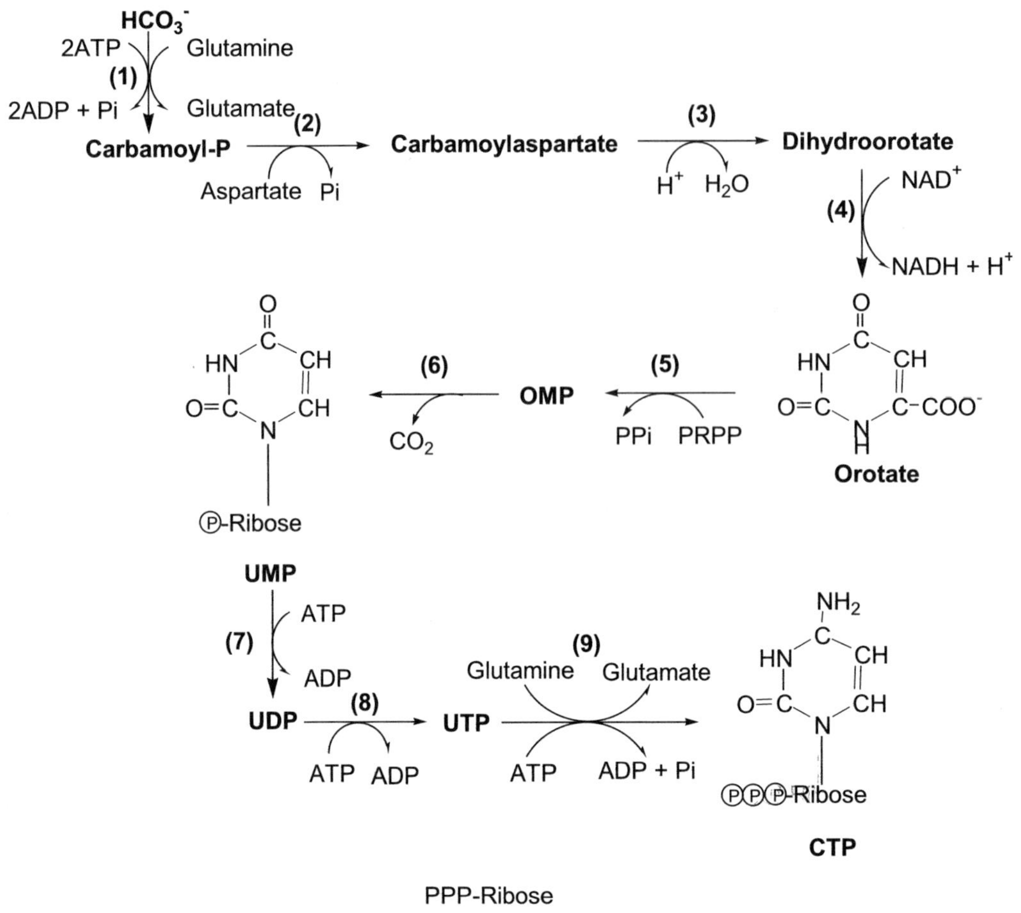

Fig. 1 De novo biosynthetic pathway of pyrimidine biosynthesis in plants. Participating enzymes and genes are as follows: (1) Carbamoyl phosphate synthase (EC 6.3.5.5, *carA, carB* or *pyrAA, pyr AB*); (2) aspartate transcarbamoylase (2.1.3.2, *pyrB*); (3) dihydroorotase (3.5.2.3, *pyrC*); (4) dihydroorotate dehydrogenase (1.3.99.11, *pyrD*); (5) orotate phosphoribosyltransferase (2.4.2.10, *pyrE* or *umps*); (6) orotidine 5′-monophosphate decarboxylase (4.1.1.23, *pyr F* or *umps*); (7) UMP kinase (2.7.4.4, *pmk* or *umpk*); (8) Nucleoside diphosphate kinase (2.7.4.6, *ndk*); (9) CTP synthetase (6.3.4.2, *pyrG*). Large and small subunits of cabamoyl phosphate synthease are coded in car A and car B, respectively. Bifunctional protein consists of orotate phosphoribosyltransferase and orotidine 5′-monophosphate decarboxylase is recently called as UMP synthase. Gene names for plant pyrimidine metabolism have not yet standardized.

same biosynthetic pathway for IMP has been established in animals and microorganisms. In animals, enzymes of the de novo purine nucleotide biosynthetic pathway and those of hydrofolate metabolism are present as a multienzyme complex called a metabolon. However, such a complex has not been observed in higher plants. IMP is further converted to AMP and GMP. These two purine nucleoside 5′-monophosphates are phosphorylated to nucleoside diphosphates and finally to nucleoside triphosphates. Feedback control of the de novo purine biosynthetic pathway is performed in at least three steps: 5-phosphoribosylamine synthase activity is inhibited by IMP, AMP, and GMP. Activities of adenylosuccinate synthetase and IMP dehydrogenase are inhibited by AMP and GMP, respectively.

BIOSYNTHESIS DEOXYRIBONUCLEOTIDES

Synthesis of Deoxyribonucleotides

A single plant ribonucleotide reductase catalyses the reduction of the ribose moiety of the ribonucleotide diphosphates, and deoxyribonucleoside diphosphates (dNDP) are produced. With the exception of dUDP, dNDP are further phosphorylated and the resultants—dCTP, dATP, and dGTP—are utilized as direct precursors for DNA synthesis. For dTTP synthesis in plants and in microorganisms such as *E. coli*, dUDP formed by the ribonucleotide reductase is first converted to dUTP, and then hydrolyzed to dUMP. In contrast, dUMP is formed from dUDP in animals.

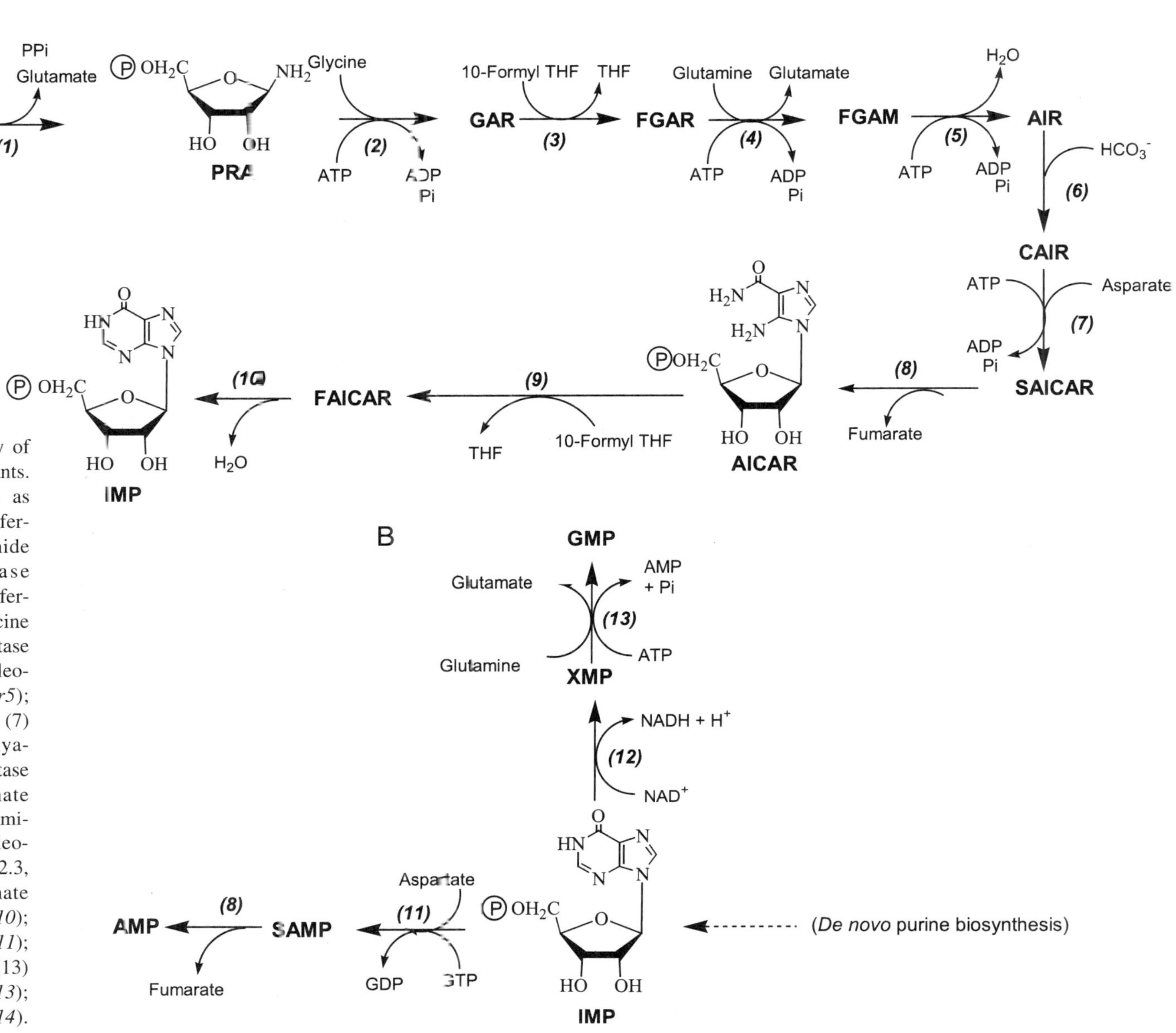

Fig. 2 De novo biosynthetic pathway of purine nucleotide biosynthesis in plants. Participating enzymes and genes are as follows: (1) amido phosphoribosyltransferase (2.4.2.14, *pur1*) (2) glycineamide ribonucleotide (GAR) synthetase (6.3.4.13, *pur2*); (3) GAR formyl transferase (2.1.2.2, *pur3*); (4) formyl glycine amidine ribonucleotide (FGAM) synthetase (6.3.5.3, *pur4*) (5) 5-imidazole ribonucleotide (AIR) synthetase (6.3.3.1, *pur5*); (6) AIR carboxylase (4.1 1.21, *pur6*); (7) 5-aminoimidazole-4-*N*-succinocarboxyamide ribonucleotide (SAICAR) synthetase (6.3.2.6, *pur7*); (8) adenylosuccinate (SAMP) lyase (4.3.2.2, *pur8*); (9) 5-aminoimidazole-4-carboxyamide ribonucleotide (AICAR) formyl transferase (2.1.2.3, *pur9*); (10) inosine 5′-monophosphate (IMP) cyclohydrolase (3.5.4.10, *pur10*); (11) SAMP synthetase (6.3.4.4, *pur11*); (12) SAMP lyase (4.3.2.2, *pur12*), (13) IMP dehydrogenase (1.2.1.205, *pur13*); (14) GMP synthetase (6.3.4.1, *pur14*). Gene names for plant purine metabolism have not yet standardized.

Synthesis of Thymidine Nucleotides

Thymidine nucleotide is synthesized from dUMP by the following reaction catalyzed by thymidylate synthase:

$$\mathrm{dUMP} + \mathrm{N}^5, \mathrm{N}^{10}\text{-methyltetrahydrofolate}$$
$$\rightarrow \mathrm{dTMP} + \text{dihydrofolate}$$

In this reaction, N^5, N^{10}-methyltetrahydrofolate produced by dihydrofolate reductase acts both as donor of the methyl group and as reducing agent. A bifunctional protein that consists of thymidylate synthase and dihydrofolate reductase has been detected in plants.

SALVAGE PATHWAYS

Free nucleosides (nucleobases with sugar attached in a glycosidic linkage) and nucleobases (purines and pyrimidines that are incorporated into nucleic acids and nucleotides) are produced as degradation products of nucleic acids and nucleotides. Pyrimidine and purine nucleotides can be resynthesized using these preformed pyrimidine and purine skeletons by various salvage reactions.

Pyrimidine Salvage

Uracil, one of the pyrimidine bases, is salvaged by uracil phosphoribosyltransferase, but another base—cytosine—is not salvaged by any enzymes. Pyrimidine nucleosides, uridine, cytidine, deoxycytidine, and thymidine are salvaged to their respective nucleotides, UMP, CMP, dCMP, and dTMP. Uridine and cytidine are phosphorylated by a single enzyme, uridine/cytidine kinase, which is present in all plants investigated to date. Deoxycytidine kinase and thymidine kinase may be present in plants, but details of their activity have yet to be reported. Nonspecific nucleoside phosphotransferase activity also participates in the salvage of pyrimidine nucleosides as well as purine nucleosides. Thymidine kinase activity measured in crude plant extracts seems to be due to the activity of nucleoside phosphotransferase and phosphatases.

Purine Salvage

Three purine bases—adenine, guanine, and hypoxanthine—are salvaged to AMP, GMP, and IMP, respectively. Two distinct enzymes—adenine phosphoribosyltransferase and hypoxanthine/guanine phosphoribosyltransferase—participate in these salvage reactions. Purine bases may also be salvaged to nucleotides via nucleosides. Adenosine phosphorylase and inosine-guanosine phosphorylase convert adenine, hypoxanthine, and guanine to respective ribonucleosides using ribose-1-phosphate, but activities of these enzymes are very low in higher plants.

Purine nucleosides, adenosine, guanosine, and inosine are salvaged by kinases and/or nucleoside phosphotransferase. Adenosine kinase is distributed ubiquitously and its activity is high, but inosine-guanosine kinase has only been found in limited plant species. In some plants, inosine and guanosine are mainly salvaged by the nonspecific nucleoside phosphotransferase.

NUCLIC ACID SYNTHESIS

Four ribonucleoside triphosphates—UTP, CTP, ATP, and GTP—are used for the precursor of the synthesis of various RNAs, whereas dTTP, dCTP, dATP, and dGTP are used as the building blocks of DNA. Pool sizes of individual ribonucleotides vary greatly because they are not only the building blocks for RNA but also have other important functions, such as energy transfer. Thus, the ATP pool is always higher than other nucleotide pools, and the CTP pool is very small. The size of deoxyribonucleotide pools is extremely small compared with ribonucleotides; as a consequence, no direct measurements of these nucleotides have yet been carried out with plants.

DEGRADATION OF NUCLEIC ACIDS

DNA and RNA are hydrolyzed by deoxyribonucleases (DNases) and ribonucleases (RNases), respectively. Oligonucleotides formed by these enzymes further hydrolyzed to nucleoside monophosphates by phosphodiesterases with 2′-, 3′-, and 5′-nucleoside monophosphates being produced. Some of these nucleotides are catabolized by nucleotidases and/or phosphatases and nucleosides accumulate. Futhermore, in plants, nucleosides are hydrolyzed exclusively by nucleosidases to nucleobases. Significant amounts of free nucleosides and nucleobases are produced, because the degradation products of nucleic acids seem to be salvaged within the same cells or transported to other organs, e.g., from senescent leaves to young shoots. The remainder will be completely degraded to CO_2 and NH_3.

DEGRADATION OF NUCLEOTIDES

Pyrimidine Catabolism

In plants, pyrimidine nucleotides are degraded to nucleoside and nucleobases. Pyrimidine bases, uracil, and thymine are catabolized by a reductive pathway. There is

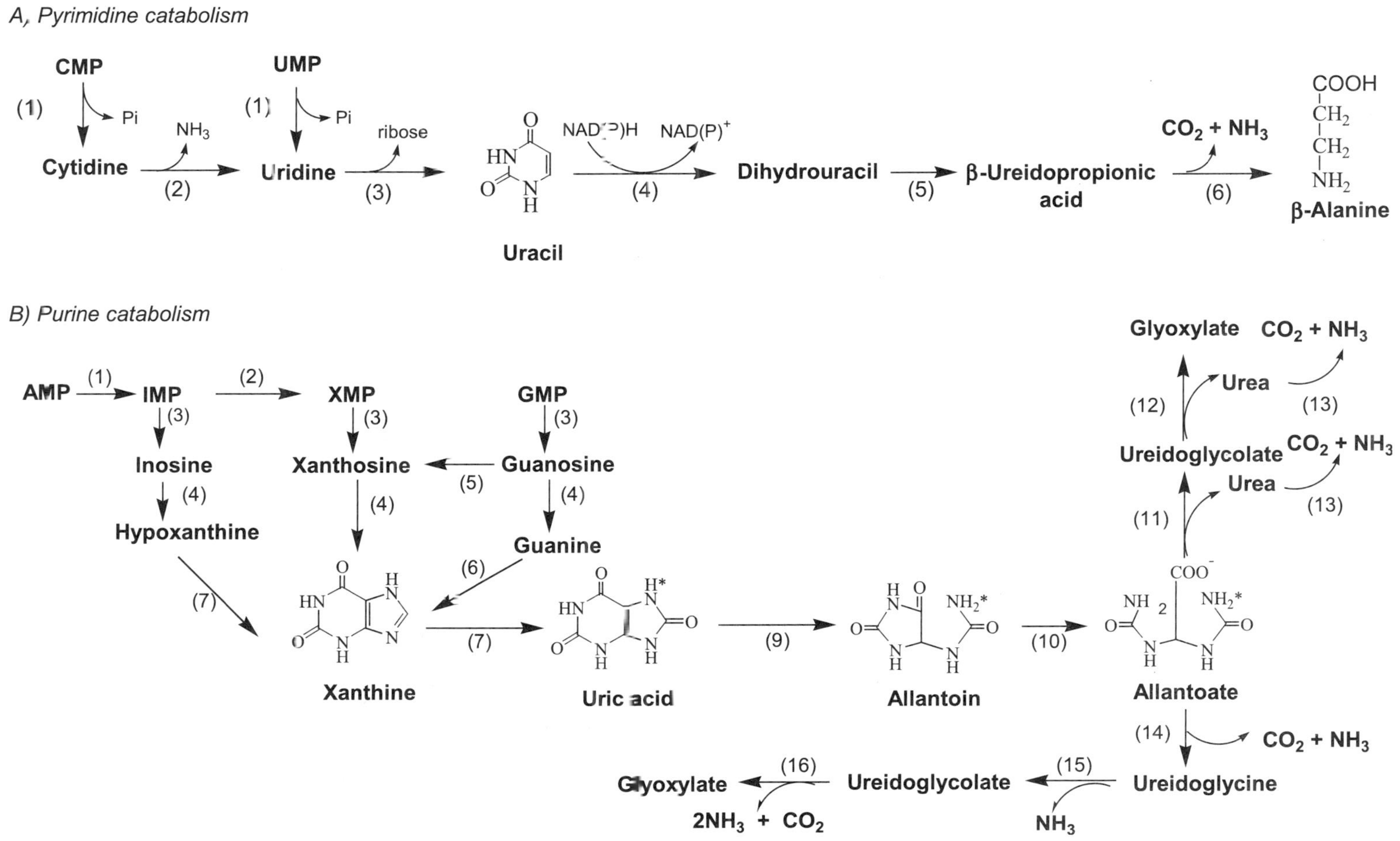

Fig. 3 Catabolic pathways of pyrimidine and purine nucleotides in plants. Pyrimidine and purine catabolism are shown in A and B separately. Participating enzymes are as follows: A: (1) 5′-nucleotidase (3.1.3.5); (2) cytidine deaminase (3.5.4.5); (3) uridine nucleosidase (3.2.2.3); (4) dihydrouracil dehydrogenase (1.3.1.1); (5) dihydropyriminase (3.5.2.2); (6) β-ureidopropionase (3.5.1.6). B: (1) AMP deaminase (3.5.4.6); (2) IMP dehydrogenase (1.1.1.205); (3) 5′-nucleotidase (3.1.3.5); (4) inosine-guanosine nucleosidase (3.2.2.2); (5) guanosine deaminase (3.5.4.3); (6) guanine deaminase (3.5.4.3); (7) xanthine dehydrogenase (1.1.1.204); (8) uricase (1.7.3.3); (9) allantoinase (3.5.2.5); (10) allantoicase (3.5.3.4); (11) ureidoglycolate lyase (4.3.2.3); (12) urease (3.5.1.5); (13) allantoin deaminase (3.5.3.9); (14) ureidoglycine amidohydrolase (no EC number given); (15) ureidoglycolate hydrolase (3.5.3.19). Gene names for most enzymes of nucleotide catabolism in plants have not yet been given.

no catabolic pathway of cytosine in plants; thus, catabolism of CMP must be performed after conversion of cytidine to uridine. Plants have cytidine deaminase, but not cytosine deaminase, so conversion at nucleoside level is essential. Uracil and thymine are catabolized by the same three sequential reactions. The end products of this catabolic pathway are either β-alanine or β-aminoisobutyrate (Fig. 3A). In both cases, CO_2 and NH_3 are produced as by-products. Because β-alanine is a precursor of the pantothenate moiety of coenzyme A, uracil catabolism seems to be important as the biosynthetic pathway of β-alanine in plants.

Purine Catabolism

Plants possess the complete oxidative purine catabolic pathway to CO_2 and NH_3 via uric acid and allantoin. A key starting compound of purine catabolism is xanthine; thus, all purine nucleotides must be converted to xanthine before their purine ring cleavage is initiated. Deamination, dephosphorylation, and glycosidic bond cleavage are included in this process. In contrast to animals, in most plants no adenosine deaminase is found, and AMP deaminase and guanosine deaminase are the predominant deamination enzymes for adenine and guanine nucleotides, respectively. Guanine deaminase is also detected in plants, but its activity is low. GMP reductase, which catalyses the conversion of GMP to IMP, is not present in plants. There are many enzymes for dephosphorylation reaction. Various phosphatases, 3′-nucleotidase, and 5′-nucleotidase appear to produce purine nucleosides. Adenosine nucleosidase and inosine-guanosine nucleosidase seem to participate in glycosidic bond cleavage (Fig. 3B).

Xanthine is converted to uric acid by xanthine dehydrogenase. Uricase catalyzes the formation of allantoin, and allantoic acid is produced by an allantoinase-catalyzed reaction. Some plant organs, such as roots of the tropical legumes and maple trees, accumulate allantoin and/or allantoic acid, which play an important role in the storage and translocation of nitrogen. Different metabolic fates of allantoic acid are proposed in plants. In the classic allantoicase pathway, allantoic acid is degraded to CO_2, NH_3, and glyoxylate via urea and ureidoglycolate. Recently an alternative route, an allantoic acid amidohydrolase pathway, has been proposed in which allantoic acid is initially converted to ureidoglycine, CO_2, and NH_3. The NH_3 is released directly and urea formation is not involved.

REFERENCES

1. Henderson, J.F.; Paterson, A.R.P. *Nucleotide Metabolism—An Introduction*; Academic Press: New York, 1973; 1–304.
2. *Metabolism of Nucleotides, Nucleosides and Nucleobases in Microorganisms*; Munch-Petersen, A., Ed.; Academic Press: London, 1983; 1–322.
3. Neuhard, J.; Nygaard, P. Biosynthesis and Conversions of Nucleotides. In *Escherichia coli and Salmonella typhimurium, Cellular and Molecular Biology*; Neidhardt, F.C., Ed.; American Society for Microbiology: Washington, DC., 1987; Vol. 1, 445–473.
4. Ross, C.W. Biosynthesis of Nucleotides. In *The Biochemistry of Plants*; Stump, P.K., Conn, E.E., Eds.; Academic Press: New York, 1981; Vol. 6, 169–205.
5. Schubert, K.R.; Boland, M.J. The Ureides. In *The Biochemistry of Plants*; Stump, P.K., Conn, E.E., Eds.; Academic Press: San Diego, CA, 1990; Vol. 16, 197–282.
6. Wasternack, C. Metabolism of Pyrimidines and Purines. In *Encyclopedia of Plant Physiology*; Pirson, A., Zimmermann, M.H., Eds.; Springer: Berlin, 1982; Vol. 14B, 263–301.
7. Wagner, K.G.; Backer, A.I. Dynamics of Nucleotides in Plants Studied on a Cellular Basis. In *International Review of Cytology*; Jeon, K.W., Friedlander, M., Eds.; Academic Press: San Diego, CA, 1992; Vol. 134, 1–84.
8. Ashihara, H.; Crozier, A. Biosynthesis and Metabolism of Caffeine and Related Purine Alkaloids in Plants. In *Advances in Botanical Research*; Callow, J.R., Ed.; Academic Press: London, 1999; Vol. 30, 117–205.
9. Moffatt, B.; Ashihara, H. Purine and Pyrimidine Nucleotide Synthesis and Metabolism. In *Arabidopsis Book,* 2nd Ed.; The American Society of Plant Biologists Online: Rockville, MD, 2001, *in press.*

Nutraceuticals and Functional Foods: Market Innovation

N

Nicholas Kalaitzandonakes
James D. Kaufman
Lucy Zakharova
University of Missouri, Columbia, Missouri, U.S.A.

INTRODUCTION

Functional foods are foods that provide health benefits beyond basic nutrition. Functional foods cater to both ill and healthy, and target prevention and treatment of illness, as well as maintenance of one's well-being. Products in the market range from complete portfolios of prepared meals that battle cardiovascular and type II diabetes, to fortified drinks and cereals that provide supplemental minerals, vitamins, and fiber.

Nutraceuticals are isolates that provide concentrated nutrients in the form of pills, tablets, liquids, or powders for direct consumption or for use as ingredients in functional foods. Nutraceuticals include micro- and macronutrient isolates, herbs and botanicals, and isolated reagents (e.g., hormones).

DRIVERS OF INNOVATION

Traditional remedies utilizing plants and animals to cure disease and improve health and well-being have existed for centuries and have been passed down through generations. In the mid-1980s, clinical studies began to formalize the relationship between nutrition and healthfulness, attracting significant attention from both media and consumers. First among these studies was the 1984 National Cancer Institute certification that high-fiber cereals foster cardiovascular health. Since that time, the body of literature connecting nutrition to specific health benefits and disease treatment has increased substantially, and other foods (e.g., calcium, omega-3 fatty acids, whole oats, and soy protein) have gained similar certification by the U.S. Food and Drug Administration (FDA).

Scientific discovery in the development of new products capable of delivering specific health benefits has also advanced rapidly. Cholesterol reduction, cardiovascular disease, and osteoporosis have become the most frequent targets for functional foods and nutraceuticals, followed by child development, moodiness, low energy, high blood pressure, diabetes, gastrointestinal (GI) disorders, menopause, and lactose intolerance.

Consumer interest has bolstered new product development of nutraceuticals and functional foods.[1–3] An aging and wealthier population in Japan, the European Union (E.U.), and the United States has become increasingly interested in the prevention of disease and health maintenance, representing a receptive audience to health claims. Similarly, escalating health care costs and increasing interest in self-medication have shifted attention to health maintenance and disease prevention instead of treatment. Table 1 provides some examples of nutraceuticals and functional foods introduced in recent years.

MARKET SIZE AND GROWTH

In 2001, the global market for functional foods was estimated at over $47.6 billion. Well developed food markets accounted for the majority of the global consumption of functional foods. The U.S. functional foods market alone was valued at $18.25 billion, representing 3.7% of the $495 billion spent for food at home by U.S. consumers. The European Union and Japan followed with $15.4 billion and $11.8 billion in functional food sales, respectively.[4] Consumer spending on functional foods is expected to grow at a fast rate although not evenly across all food categories. In recent years, functional beverages have experienced the fastest growth (Table 2).

Not all functional foods have been successful in the marketplace. Some products have seen diminished consumer demand due to unexpected negative side effects (e.g., the fat substitute Olestra). Others have been quietly removed from supermarket shelves due to lack of sufficient consumer demand (e.g., McNeil's Benecol in the United States; Kellogg's Ensemble; and Cambell Soup Company's Intelligent Cuisine).

The global market for nutraceuticals reached $50.6 billion in 2001. Vitamin/mineral supplements accounted for 40% ($20.6 billion) of those sales, followed by herbs and botanicals ($19.6 billion) and sports/specialty supplements (10.4 billion).[7] Vitamins are especially important in the U.S. market, whereas herbs are relatively more important in the European and Asian markets, which have a long tradition of herbal use and remedies.

Encyclopedia of Plant and Crop Science
DOI: 10.1081/E-EPCS 120010420

Table 1 Nutraceuticals and functional foods

Category	Product name	Product form	Claim
Health foods (Foods with approved health claims)			
	Tofu	Food	Cardiovascular health
	Oatmeal	Food	Cancer prevention
Nutraceuticals (Food ingredients with specific claims of benefit)			
	Glucosamine	Food ingredient	Joint health
	Soluble fiber	Food ingredient	Cancer prevention
	Ginseng	Herb/ingredient	Physical performance
	Soy protein	Food ingredient	Cardiovascular health
Functional foods (Foods or drinks with specific claims of benefit)			
	Yakult Honsha's Yakult	Probiotic active cultures	Improved digestion
	McNeil's Benecol	Margarinelike spread	Lowers LDL cholesterol
	Proctor & Gamble/Olean's Olestra	Fat substitute	Weight loss
	Tropicana's Tropicana calcium	Calcium fortified juice	Skeletal health
	Kellogg's Ensemble	High-fiber cereal	Cholesterol reduction
	Campbell Soup Company's Intelligent Cuisine	Line of prepared meals	Nutritionally balanced
Drug delivery foods (Food as clinical drug delivery mechanism)			
	(Not yet commercialized)	Potato, banana, tomato	Vaccine for hepatitis B, etc.

PRODUCT INNOVATION AND INDUSTRIAL DEVELOPMENT

For continuing growth in the nutraceuticals and functional foods markets, sustained technological innovation will continue to be critical. The scope of innovation will likely range from product functionality, to taste characterization and manufacturing. This will require the pooling of knowledge, skills, and infrastructure from industries that have not worked closely together in the past. For example, nutraceutical startups and biotechnology firms may contribute discoveries in new products and attributes. Food companies may provide product formulation, manufacturing, and distribution capabilities. Pharmaceutical companies may contribute discovery and knowledge in amplification of efficacy, as well as experience in the validation of safety (e.g., clinical trials).

Table 2 U.S. retail sales of functional foods by category ($billion)

Category	1998[a]	2000[b]
Beverages	4.9	8.2
Breads and grains	6.1	4.8
Packaged/prepared foods	1.1	1.6
Dairy	1.7	1.1
Meat, fish, and poultry	0	0
Snack foods	0.8	1.4
Condiments	0.1	0.2
Total	**14.8**	**17.2**

[a]From Ref. 5.
[b]Form Ref. 6.

Major food companies making significant investments in functional foods are already pursuing integration of skills and assets. Many of these companies have moved to acquire diverse skills through acquisitions or strategic alliances. For instance, PepsiCo recently acquired Gatorade and Sobe functional beverage brands; Heinz acquired a share in Hain Foods, a processor of soy and natural foods; and Quaker and Novartis entered into a joint venture to form Altus Food Company, which is dedicated to the development of functional foods.

To be successful, this emergent industry will have to learn how to develop products that deliver functionality in ways that appeal to changing consumer tastes and lifestyles. It will also have to learn how to protect and recoup its investments (e.g., R & D) through patents, brands and trademarks, and other marketing strategies. Finally, this new industry will have to learn how to position functionality claims within an uncertain and fragmented regulatory environment.

INSTITUTIONAL INNOVATION

It will be up to governments to establish relevant institutional environments that create appropriate incentives for research and development in nutraceuticals and functional foods while they protect public health and consumer interests. Meaningful institutional innovation will therefore be necessary in coming years. Current regulation lacks clear standards on health claims (efficacy), product quality, and safety that can impede market development. For instance, the slow market development experienced in herbs/botanicals (a flat 1.2% rate of growth in 2000) has been attributed to the high incidence of low-quality (and occasionally unsafe) products introduced during the boom market of the late 1990s.[7] Herbs contaminated with heavy metals, heart problems attributed to Ephedra, and safety concerns with sports supplements (e.g., Androstenedione, or Andro) attracted significant negative public attention and reduced consumer interest.

Development of a proper regulatory framework has been complicated by the dualistic food and drug nature of nutraceuticals and functional foods. Complexities are likely to increase in the future as active ingredients become more powerful, requiring some foods to be regulated as pharmaceuticals. For instance, pharmaceutical foods (such as edible vaccines) are already under development. These bioengineered foods are being developed to express specific pharmaceutical compounds, providing a convenient drug manufacturing and delivery mechanism. Accordingly, an institutional environment capable of efficiently regulating both foods as drugs and drugs as foods will be necessary.[8]

Currently, there are few regulatory standards, and those that exist fail to transcend national borders. The U.S. system is characterized by a dichotomy of regulatory stringency, wherein some products can be commercialized with minimal regulatory oversight, while others are strictly regulated. Less-regulated products are allowed by the Dietary Supplement Health and Education Act (DSHEA), which shifted the burden of proving an ingredient unsafe from the manufacturer to the FDA. Under DSHEA guidelines, a dietary supplement may include a vitamin, mineral, herb/botanical, amino acid, metabolite, or an extract. Laws allow nutritional support statements such as those relating to classical nutrient-deficiency diseases, structure/function, or well-being, provided they include the following disclaimers: "This statement has not been evaluated by the Food and Drug Administration" or "This product is not intended to diagnose, treat, cure, or prevent any disease."

In order to make health claims linking a food or dietary supplement to a disease or health-related condition, premarket approval by the FDA is required. The FDA approval process is dependent on scientific consensus identifying a specific functional component responsible for physiological action. The approved health claim must include the appropriate dietary context (e.g., low in saturated fat and cholesterol), which can then be used with any similar product, not just those produced by the petitioner.

The main obstacle for the European system is the lack of standard protocols across national borders. For example, France and Germany differ from the United States and many other European countries in that French and German herbal drugs are distributed via pharmacies, allowing for doctor mediation and reimbursement by health insurance.

In contrast to the U.S. and European models, the Japanese nutraceuticals and functional foods markets are highly regulated and supported. Since 1991, the Japanese government has actively promoted the development of functional foods through the Foods for Specialized Health Use (FOSHU) system. Foods in this product category can secure regulatory approval to make specific health claims, and are educationally supported by the program. FOSHU is an umbrella label for all functional foods, and ensures safety and efficacy; however, FOSHU is not mandatory and the manufacturers of some functional products opt to circumvent it, suggesting that subpar products may still be commercialized. This Japanese system has been popularly proposed as a potential template for the development of international standards. The development of such international standards, however, has progressed slowly (until now).

CONCLUSION

It is clear that both scientific discoveries and consumer interest are creating opportunities for an expanding menu of foods and supplements that claim health benefits beyond nutrition. In globalized markets where products are traded across national borders at an increasing pace, a proper regulatory framework for how such claims are brought forward will be increasingly important. Hence, consumers, scientists, entrepreneurs, and regulators will all shape the rate and direction of product innovation in the nascent markets of nutraceuticals and functional foods for many years to come.

ARTICLES OF FURTHER INTEREST

Echinacea: Uses As a Medicine, p. 395
Herbs, Spices, and Condiments, p. 559
Medical Molecular Pharming: Therapeutic Recombinant Antibodies, Biopharmaceuticals and Edible Vaccines

in Transgenic Plants Engineered via the Chloroplast Genome, p. 705

REFERENCES

1. Heasman, M.; Mellentin, J. *The Functional Foods Revolution: Healthy People, Healthy Profits?* Earthscan Publications, Ltd.: London, 2001.
2. *Nutraceuticals: Developing, Claiming, and Marketing Medical Foods*; Defelice, S., Ed.; Marcel Dekker, 1998.
3. Gilbert, L. Marketing functional foods: How to reach your target audience. Agbioforum **2001**, *3* (1). Available on the WWW at: www.agbioforum.org.
4. Sloan, E. The top 10 functional food trends: The next generation. Food Technol. **2002**, *56* (4), 32–57.
5. Nutrition business journal. Nutraceuticals & functional foods IV. Nutr. Bus. J. **2000**, *5* (5), 1–6.
6. Nutrition business journal. Functional foods V. Nutr. Bus. J. **2001**, *6* (10), 1–8.
7. Gruenwald, J.; Herzberg, F. *The Global Nutraceuticals Market*; World Markets Research Centre, 2002.
8. Leland, S. *Agriceuticals: Designing New Food Concepts. Rabobank International*; Food & Agribusiness Research: New York, 2001.

Oomycete-Plant Interactions: Current Issues in

Mark Gijzen
Agriculture and Agri-Food Canada, London, Ontario, Canada

INTRODUCTION

The oomycetes (water molds) are related to the brown algae and diatoms. Oomycetes resemble fungi morphologically, but are now classified separately based on their natural phylogeny. Oomycetes have coenocytic hyphae with cell walls consisting mainly of cellulose and glucans. They are diploid throughout most of their life cycle and reproduce sexually by formation of thick-walled oospores, or asexually by zoospores or sporangia. Heterothallic species require different mating types to produce oospores, whereas homothallic species are self-fertile. Oomycetes include hundreds of plant pathogenic species distributed across several genera, including *Albugo, Aphanomyces, Bremia, Peronospora, Phytophthora, Plasmopara,* and *Pythium*. Plant pathogenic oomycetes may be necrotrophic, biotrophic, or hemibiotrophic species. Many studies on oomycete–plant interactions focus on the *Phytophthora* species, an especially destructive and widespread genus. Genetic and molecular-level research on oomycetes is gaining momentum; new insights into processes that control virulence and pathogenicity are emerging.

OOMYCETES AS PLANT PATHOGENS

Oomycetes display a range of pathogenic strategies in colonizing host tissues. Necrotrophs such as *Pythium* are considered facultative parasites that kill host cells in advance of infection. *Pythium* also include many saprophytic species that simply live off dead tissue and are not true pathogens. Obligate biotrophs such as *Bremia* and *Peronospora* require living host cells and cannot be grown in culture. Many *Phytophthora* spp. are intermediate types (hemibiotrophs) that first establish in living cells as biotrophs but more closely resemble necrotrophs at later stages of infection.[1] An evolutionary sequence has been proposed whereby saprophytes give rise to necrotrophic organisms, which in turn may further evolve into hemibiotrophs, and ultimately, into obligate biotrophs. The oomycetes present compelling evidence in support of this theory because their spectrum of pathogenic strategies is generally concordant with their phylogeny.[2] It is clear that these different colonization strategies result in different mechanisms of infection, but there may remain commonalities among different pathogen types. For example, the subversion of host plant metabolism and the suppression of many chemical and structural defense responses would benefit any pathogen, whether biotroph or necrotroph. In contrast, pathogen-induced host cell death by secretion of toxins or by manipulation of endogenous cell death programs may be predicted to benefit the pathogen in necrotrophic but not biotrophic interactions.

The oomycetes are considered versatile and adaptable plant pathogens for other reasons as well. These organisms attack a wide range of herbaceous and woody plants in many different environments. Host specificity may be broad or restricted to a single plant species. Most *Pythium* species have wide host ranges, whereas the biotrophic species *Bremia* and *Peronospora* are highly specialized with one or few host plants. *Phytophthora* species may have broad or narrow host ranges.

Contemporary studies on plant pathogenic oomycetes address many different species, but a few organisms are attracting a critical mass of researchers and resources. Emerging as model systems for molecular genetic and genomic studies on oomycetes are *Phytophthora infestans* and *Phytophthora sojae*. *P. infestans* is a heterothallic species that causes late blight of potato and tomato. This is predominantly a foliar pathogen, but is also able to infect tubers. Outbreaks of *P. infestans* are spread via airborne sporangia. *P. sojae* is a homothallic oomycete that causes root and stem rot of soybean, as shown in Fig. 1. This soilborne pathogen produces motile zoospores that spread in water and are attracted to soybean roots. Oospores are also abundantly produced in infected tissues, and these propagules are long-lived in soil and may be spread by animals or machinery.

Genetic studies on oomycetes have benefited from advances in molecular biology.[3] It is now possible to distinguish hybrid progeny from self-fertilized individuals based on DNA markers, and to create F_2 progeny sets. This has enabled researchers to perform classical genetic analyses of oomycetes, and has confirmed that many oomycete–plant interactions are governed by complementary avirulence (*Avr*) and resistance (*R*) genes. Genetic analyses of *Bremia lactucae*, *P. infestans*, *P. sojae*, and *Peronospora parasitica* indicate that, although exceptions have been noted, *Avr* genes segregate as single dominant loci and some *Avr* genes occur in clusters or

Encyclopedia of Plant and Crop Science
DOI: 10.1081/E-EPCS 120019939

Fig. 1 A field infestation of *Phytophthora sojae* on soybean plants, photographed near Windsor, Ontario. (*View this art in color at www.dekker.com.*)

as tightly linked loci. Various *Avr* loci comprise targets for positional cloning projects in these species.[4]

OOMYCETE ELICITORS AND TOXINS

Interactions between pathogens and their hosts involve multiple signaling cues that ultimately determine whether the infection is successful. Elicitors and toxins have overlapping activities and each may trigger a similar set of plant defense responses, such as necrosis, phytoalexin accumulation, and cell death. Functionally, elicitors are generally regarded as molecules that aid the host by initiating defense reactions that contain the pathogen, whereas toxins are considered virulence factors produced by the pathogen to disable the host.[5] Among the first elicitors purified and characterized from a plant pathogen was a β-glucan elicitor from *P. sojae*. This is released from the cell walls of *P. sojae* by glucanase enzymes secreted by soybean cells. As a counter-defense, *P. sojae* produces inhibitors of the soybean-secreted glucanase enzymes, to slow cell wall digestion and elicitor release. This example provides evidence for the ''escalating arms race'' hypothesis of plant-pathogen interactions.

Many oomycetes, including *Phytophthora* and *Pythium* species, secrete small cysteine-rich proteins called elicitins. These 10 kD proteins elicit the hypersensitive response in certain plants, most notably in tobacco and other *Nicotiana* species. Larger proteins containing elicitin domains fused to other sequences may also be active. Most *Phytophthora parasitica* strains that are pathogenic on tobacco do not secrete elicitins because these proteins may act as host-specific avirulence determinants, limiting *Phytophthora* infections on tobacco plants. Elicitins have been shown to possess sterol- and lipid-binding activity, and their elicitor activity is also dependent on sterol binding. It is possible that elicitins function as carriers for the acquisition or transport of sterols or lipids. The elicitin protein family is large and diverse, and it is likely that this diversity is also reflected in their functional roles.

Several other *Phytophthora* proteins with elicitor or toxin activity have been described[3] as follows:

- A 42 kD transglutaminase protein secreted by *Phytophthora* species induces phytoalexin accumulation and otherdefense responses in parsley. The elicitor activity of the protein is dependent on the same residues that are necessary for catalytic activity, and this peptide signature has been proposed to constitute a pathogen-associated molecular pattern for *Phytophthora*.
- A 34 kD glycoprotein first purified from *Phytophthora parasitica* induces necrosis and defense gene expression in tobacco. This protein has cellulose-binding activity and plays a role in the adhesion of *Phytophthora* to plant cells.
- A 25 kD protein purified from *Pythium* and *Phytophthora* species is remarkable in its ability to cause plant cell death in a wide range of dicotyledonous plants. It is similar to proteins discovered in unrelated organisms that are also plant pathogens or saprophytes, such as fungi and bacteria.
- A 5.6 kD phytotoxic protein from *Phytophthora cactorum* has been shown to cause necrosis on strawberry and tomato plants.

These examples illustrate that many different factors may contribute to disease symptoms and cell death in oomycete-plant interactions. The proteins may be important factors that enable *Phytophthora* to invade host plants, or alternatively, provide cues for host plant surveillance systems aimed at limiting pathogen spread.

PLANT RESISTANCE TO OOMYCETE PATHOGENS

Because it is a host to several different oomycetes, including *Per. parasitica* and *Phytophthora brassicae*, *Arabidopsis thaliana* provides a powerful model for genetic dissection of mechanisms of plant resistance to oomycete pathogens. Resistance to the downy mildew pathogen has been most intensively studied and several different *RPP* genes, conditioning resistance to *Per. parasitica*, have been isolated from *A. thaliana*.[6] Other plant genes resistant to oomycete pathogens that have been isolated include the lettuce *Dm3* gene, controlling

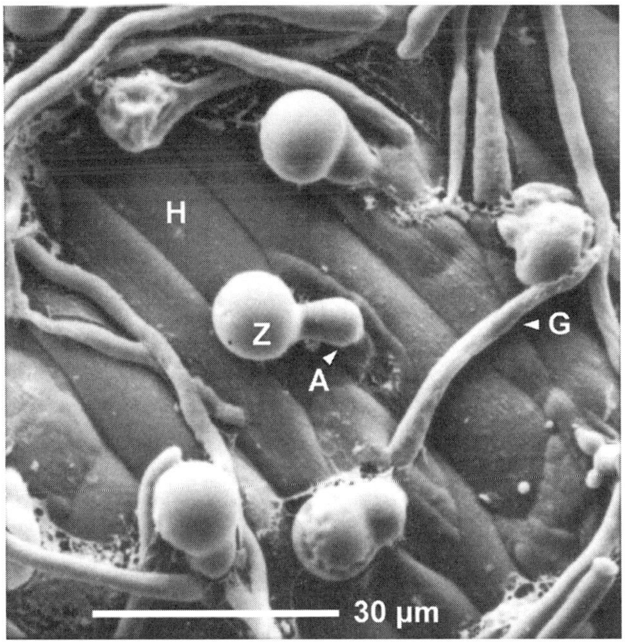

Fig. 2 A scanning electron micrograph showing zoospores of *Phytophthora sojae* on the surface of a soybean hypocotyl, 12 hours after inoculation. The zoospores have formed germ tubes and are penetrating the host tissues. (A: Appresorium like structure; G: Germtube; H: Soybean hypocotyl surface; Z: Encysted zoospore.) (*View this art in color at www.dekker.com.*)

resistance to *Bremia lactucae*, and the potato *R1* gene, for resistance to *P. infestans*. There are common features among these resistance genes, as they all encode proteins that are predicted to occur within the cell, with nucleotide-binding site (NBS) and leucine-rich repeat (LRR) motifs.

The hypersensitive response (HR)—a programmed cell death of plants—has been proposed as a general mechanism conditioning plant resistance to oomycete pathogens. This occurs in the early stages of infection during initial penetration of host cells. Zoospores or sporangia germinate on the surface of host tissues and may form appressoria like structures at penetration points, as shown in Fig. 2. Studies on *P. sojae* and *P. infestans* indicate that when host cells at the initial infection site undergo HR quickly, much more effective containment of pathogen spread results. A lack of host recognition—as evidenced by no HR or a slow HR upon infection—permits the pathogen to grow further and colonize adjacent cells, resulting in disease. Later during the infection process, host cell death occurs extensively but does not limit pathogen spread, especially in the case of hemibiotrophic pathogens such as *P. infestans* and *P. sojae*. At this stage host cell death may even accelerate disease.

CONCLUSION

Oomycetes are destructive plant pathogens that harm crops, ornamental plants, and natural environments. Many have the potential to spread rapidly and cause severe epidemics when presented with suitable conditions. Nineteenth century disease outbreaks caused by oomycetes were important in the development of plant pathology as a scientific discipline and changed the practice of agriculture. Oomycetes continue to cause epidemics and serious problems to this day, most recently exemplified by the emergence of sudden oak death (*Phytophthora ramorum*) in central California. Oomycetes are successful pathogens that present many challenges as experimental organisms in the laboratory, but concerted efforts and advances in technology are leading to a better understanding of their biology. This holds the promise that more effective control measures will be developed to limit the damage caused by oomycetes throughout the world.

ARTICLES OF FURTHER INTEREST

Fungal and Oomycete Plant Pathogens, Cell Biology, p. 480
Mechanisms of Infection: Oomycetes, p. 697

REFERENCES

1. Erwin, D.C.; Ribeiro, O.K. *Phytophthora Diseases Worldwide*; APS Press: St.Paul, USA, 1996.
2. Cook, D.E.L.; Drenth, A.; Duncan, J.M.; Wagels, G.; Brasier, C.M. A molecular phylogeny of *Phytophthora* and related oomycetes. Fungal Genet. Biol. **2000**, *30* (1), 17–32.
3. Kamoun, S. Molecular genetics of pathogenic oomycetes. Eukarot. Cell **2003**, *2* (2), 191–199.
4. Tyler, B. Molecular basis of recognition between *Phytophthora* pathogens and their hosts. Annu. Rev. Phytopathol. **2002**, *40*, 137–167.
5. Nürnberger, T.; Brunner, F. Innate immunity in plants and animals: Emerging parallels between the recognition of general elicitors and pathogen-associated molecular patterns. Curr. Opin. Plant Biol. **2002**, *5* (4), 318–324.
6. Holub, E.B. The arms race is ancient history in *Arabidopsis*, the wildflower. Nat. Rev., Genet. **2001**, *2* (7), 516–527.

Organic Agriculture as a Form of Sustainable Agriculture

John P. Reganold
Washington State University, Pullman, Washington, U.S.A.

INTRODUCTION

During the past decade, organic farming became one of the fastest growing segments of agriculture all over the world.[1,2] Organic agriculture, sometimes called biological or ecological agriculture, combines traditional conservation-minded farming methods with modern farming technologies. It virtually excludes such conventional inputs as synthetic fertilizers, pesticides, and pharmaceuticals. Instead, it uses naturally derived chemicals or products as defined by organic certification programs, and it emphasizes building up the soil with compost additions and animal and green manures, controlling pests naturally, rotating crops, and diversifying crops and livestock.[3] The goal of organic agriculture is to create sustainable agricultural systems. This article discusses organic agriculture and sustainability, organic agricultural practices, the history and extent of organic agriculture, and organic certification.

ORGANIC AGRICULTURE AND SUSTAINABILITY

Organic agriculture addresses many serious problems afflicting world food production: high energy costs, groundwater contamination, soil erosion, loss of productivity, depletion of fossil resources, low farm incomes, and risks to human health and wildlife habitats. However, just because a farm is organic does not mean that it is sustainable. For any farm to be sustainable, whether it be organic or conventional, it must produce adequate high-quality yields, be profitable, protect the environment, conserve resources, and be socially responsible over the long term.[4] So, if an organic farm produces high yields of nutritious food and is environmentally friendly and energy efficient, but is not profitable—then it is not sustainable. Likewise, a conventional farm that meets all the sustainability criteria but pollutes a nearby river with sediment because of soil erosion is not sustainable.

Organic farming systems have been shown to be energy-efficient, environmentally sound, productive, stable, and tending toward long-term sustainability.[5,6] Organic farming systems do not represent a return to the past. They use modern equipment, certified seed, soil and water conservation practices, improved crop varieties, and the latest innovations in feeding and handling livestock. Organic farming systems range from strict closed-cycle systems that go beyond organic certification guidelines by limiting external inputs as much as possible, to more standard systems that simply follow organic certification guidelines. They rely largely on available resources found on or near the farm. The protection of soil and the environment is fundamental to organic farming.

Yields from organic farms are usually somewhat lower than, but sometimes equal to, those from conventional farms, but they are frequently offset by price premiums that lead to equal or greater net returns. Research comparing the agronomic, economic, and ecological performance of organic and conventional farming systems confirms this. In fact, studies[5,7,8] generally have found that organic systems have equal or somewhat lower yields, but less variability in production from year to year; are equally, if not more profitable; cause less erosion and pollution; have better soil quality; are more energy efficient; and rely less on government subsidies than their conventional counterparts. In other words, the organic systems are more sustainable.

ORGANIC AGRICULTURAL PRACTICES

Organic farming systems rely on ecologically based practices. A central component of organic farming systems is the rotation of crops—a planned succession of various crops grown on one field. When crops are rotated, the yields are usually about 10% higher than when they grow in monoculture (growing the same crop on the same field year after year). In most cases, monocultures can be perpetuated only by adding large amounts of fertilizer and pesticide. Rotating crops provides better weed and insect control, less disease buildup, more efficient nutrient cycling, and other benefits. Alternating two crops, such as corn and soybeans, is considered a simple rotation. More complex rotations require three or more crops and often a four- to seven-year (or more) cycle to complete. In growing a more diversified group of crops in rotation, a farmer is less affected by price fluctuations of one or two crops. This may result in more year-to-year financial stability. There are disadvantages, too, however. They include the

Encyclopedia of Plant and Crop Science
DOI: 10.1081/E-EPCS 120012951

need for more equipment to grow a number of different crops, the reduction of acreage planted with government-supported crops, and the need for more time and information to manage more crops.

Organic agriculture not only has a diverse assortment of crops in rotation, but also maintains diversity from mixing species and varieties of crops and from systematically integrating crops, trees, and livestock. When most of North Dakota experienced a severe drought during the 1988 growing season, for example, many monocropping wheat farmers had no grain to harvest. Organic farmers with more diversified systems, however, had sales of their livestock to fall back on or were able to harvest their late-seeded crops or drought-tolerant varieties.

Maintaining healthy soils by regularly adding crop residues, manures, and other organic materials to the soil is another central feature of organic farming. Organic matter improves soil structure, increases its water storage capacity, enhances fertility, and promotes the tilth, or physical condition, of the soil. The better the tilth, the more easily the soil can be tilled or direct-drilled with seed, and the easier it is for seedlings to emerge and for roots to extend downward. Water readily infiltrates soils with good tilth, thereby minimizing surface runoff and soil erosion. Organic materials also feed earthworms and soil microbes.

The main sources of plant nutrients in organic farming systems are green manures, composted animal manures and plant materials, and plant residues. A green manure crop is a grass or legume that is plowed into the soil or surface-mulched at the end of a growing season to enhance soil productivity and tilth. Green manures help to control weeds, insect pests, and soil erosion, while also providing forage for livestock and cover for wildlife.

Organic farmers use disease-resistant crop varieties and biological controls (such as natural predators or parasites that keep pest populations below injurious levels). They also select tillage methods, planting times, crop rotations, and plant-residue management practices to optimize the environment for beneficial insects that control pest species, or to deprive pests of a habitat. Organically certified pesticides, usually used as a last resort to control insects, diseases, and weeds, are applied when pests are most vulnerable or when any beneficial species and natural predators are least likely to be harmed.

HISTORY AND EXTENT OF ORGANIC AGRICULTURE

European organic agriculture emerged in 1924 when Rudolf Steiner held his course on biodynamic agriculture. In the 1930s and 1940s, organic agriculture was developed in Britain by Lady Eve Balfour and Sir Albert Howard, in Switzerland by Hans Mueller, in the United States by J. I. Rodale, and in Japan by Masanobu Fukuoka. Since the beginning of the 1990s, development of organic agriculture in Europe has been supported by government subsidies. In many other countries of the world, organic agriculture was established because of the growing demand for organic products in Europe, the United States, and Japan. Today, organic farming is gaining increasing acceptance by the public at large.

Organic agriculture is practiced in almost all countries of the world, and its share of agricultural land and farms is growing everywhere. In 2001 more than 17 million hectares were managed organically worldwide.[9] The major part of this area is located in Australia (7.7 million hectares), Argentina (2.8 million hectares) and Italy (more than 1 million hectares). Oceania holds 45% of the world's organic land, followed by Europe (25%), and Latin America (22%). In North America more than 1.3 million hectares are managed organically. In most Asian and African countries the area under organic management is still low. Tables 1 and 2 show the top 20 countries in the world with the most organically managed land and with the greatest proportion of total agricultural land in the country in organic production.

ORGANIC CERTIFICATION

Growers are turning to certified organic farming systems as a potential way to lower input costs, decrease reliance on nonrenewable resources, capture high-value markets and premium prices, and boost farm income. Organic certification provides verification that products are indeed produced according to certain standards. For consumers who want to buy organic foods, the organic certification standards ensure that they can be confident in knowing what they are buying. For farmers, these standards create clear guidelines on how to take advantage of the exploding demand for organic products. For the organic products industry, these standards provide an important marketing tool to help boost trade in organic products.

"Certified organic" means that agricultural products have been grown and processed according to the specific standards of various national, state, or private certification organizations. Certifying agents review applications from farmers and processors for certification eligibility, and qualified inspectors conduct annual onsite inspections of their operations. Inspectors talk with producers and observe their production or processing practices to determine if they are in compliance with organic standards.

Organic certification standards detail the methods, practices, and substances that can be used in producing and handling organic crops and livestock, as well as processed products. For example, in December 2000 the

Table 1 Top 20 countries ranked by highest total area (hectares) under organic management, followed by the percentage of total agricultural land in organic production

Country	Hectares of organic farmland	Percent of total agricultural area in organic production
Australia	7,654,924	1.62
Argentina	2,800,000	1.65
Italy	1,040,377	6.76
U.S.A.	900,000	0.22
Brazil	803,180	0.23
Germany	546,023	3.20
U.K.	527,323	3.33
Spain	380,838	1.30
France	371,000	1.31
Canada	340,200	0.46
Austria	271,950	8.64
Sweden	171,682	5.20
Czech Republic	165,699	3.86
Denmark	165,258	6.20
Finland	147,423	6.73
Switzerland	95,000	9.00
Mexico	85,676	0.08
Slovakia	60,000	2.45
Portugal	50,002	1.31
Hungary	47,221	0.77

(From Ref. 9.)

Table 2 Top 20 countries ranked by highest percentage of total agricultural land in organic production, followed by the total area (hectares) under organic management

Country	Percent of total agricultural area in organic production	Hectares of organic farmland
Liechtenstein	17.97	690
Switzerland	9.00	95,000
Austria	8.64	271,950
Italy	6.76	1,040,377
Finland	6.73	147,423
Denmark	6.20	165,258
Sweden	5.20	171,682
Czech Republic	3.86	165,699
Iceland	3.40	3,400
U.K.	3.33	527,323
Germany	3.20	546,023
Slovakia	2.45	60,000
Norway	2.01	20,523
Argentina	1.65	2,800,000
Australia	1.62	7,654,924
Belgium	1.46	20,263
Netherlands	1.42	27,820
Portugal	1.31	50,002
France	1.31	371,000
Spain	1.30	380,838

(From Ref. 9.)

U.S. Department of Agriculture announced final national standards that food labeled "organic" must meet, whether it is grown in the United States or imported from other countries.[10] These standards specifically prohibit the use of genetic engineering methods, ionizing radiation, and sewage sludge for fertilization. They went into labeling effect in October 2002.

On a broader scale, the International Federation of Organic Agriculture Movements (IFOAM) established the IFOAM Accreditation Programme to provide international equivalency of certification bodies worldwide.[2] IFOAM accreditation is based on the international IFOAM standards, which are developed continually by the IFOAM membership, a democratic structure open to all who work in the field of organic agriculture and production. In 2000 the first products with the "IFOAM-accredited" logo came on the market. By early 2002, 17 organizations had been IFOAM-accredited.

ARTICLES OF FURTHER INTEREST

Agriculture and Biodiversity, p. 1
Biological Control of Oomycetes and Fungal Pathogens, p. 137
Biological Control of Weeds, p. 141
Ecology and Agricultural Sciences, p. 401
Integrated Pest Management, p. 612
Social Aspects of Sustainable Agriculture, p. 1155
Sustainable Agriculture and Food Security, p. 1183
Sustainable Agriculture: Definition and Goals of, p. 1187
Sustainable Agriculture: Ecological Indicators of, p. 1191
Sustainable Agriculture: Economic Indicators of, p. 1195
Weed Management in Less Developed Countries, p. 1295

REFERENCES

1. Greene, C. U.S. organic agriculture gaining ground. Agric. Outlook. April **2000**, *3*.
2. www.ifoam.org (accessed August 2002).
3. Lampkin, N. *Organic Farming*; Farming Press: Ipswich, UK, 1990.
4. Reganold, J.P.; Papendick, R.I.; Parr, J.F. Sustainable agriculture. Sci. Am. **1990**, *262* (6), 112–120.
5. Petersen, C.; Drinkwater, L.E.; Wagoner, P. *The Rodale Institute Farming Systems Trial*; Rodale Institute: Kutztown, PA, 1999; 1–40.
6. Mader, P.; Fliessbach, A.; Dubois, D.; Gunst, L.; Fried, P.; Niggli, U. Soil fertility and biodiversity in organic farming. Science **2002**, *296* (5573), 1694–1697.
7. Smolik, J.D.; Dobbs, T.L.; Rickerl, D.H. The relative sustainability of alternative, conventional, and reduced-till farming systems. Am. J. Altern. Agric. **1995**, *10* (1), 25–35.
8. Reganold, J.P.; Glover, J.D.; Andrews, P.K.; Hinman, H.R. Sustainability of three apple production systems. Nature **2001**, *410* (6831), 926–930.
9. Yussefi, M.; Willer, H. *Organic Agriculture Worldwide 2002—Statistics and Future Prospects*; SOEL: Bad Durkheim, 2002.
10. www.ams.usda.gov/nop/ (accessed August 2002).

Osmotic Adjustment and Osmoregulation

Neil C. Turner
CSIRO Plant Industry, Wembley, Australia

INTRODUCTION

Water deficits and lowering the osmotic potential of the external solution have long been known to induce a decrease in the osmotic potential or an increase in the osmotic pressure of higher plant cells.[1–3] This occurs as water is extracted from the cell from the *passive* concentration of the solutes present in the fully turgid cells. Under these circumstances, the osmotic pressure (Π) of a cell is inversely related to the osmotic volume (V):

$$\Pi = \frac{\Pi_{100} V_{100}}{V} \qquad (1)$$

where Π_{100} and V_{100} are the osmotic pressure and osmotic volume, respectively, at full turgor.

RECENT DEVELOPMENTS

More recently it was recognised that higher plants can *accumulate* solutes in response to a water deficit, a process termed osmotic adjustment or osmoregulation. Although previous studies had shown that root turgor was maintained as soil-water deficits increased and seasonal changes in osmotic pressure mirrored seasonal changes in water potential,[3] the possibility of solute redistribution, uptake of solutes from the soil, and changes in elasticity could not be eliminated in these previous studies.[3] The first unequivocal increase in osmotic pressure in response to a soil-water deficit was reported by Jones and Turner,[4,5] who showed that osmotic adjustment in sorghum and sunflower leaves arose from an accumulation of solutes.[5,6]

Osmotic adjustment has now been shown to occur in the leaves of many species[7–10] and has also been observed to occur in roots and fruits.[10–12] Additionally, osmotic adjustment in leaves has been shown to vary among genotypes in a range of species.[13,14] Yield benefits associated with osmotic adjustment[13,14] have led to the suggestion that osmotic adjustment is a desirable trait for turgor maintenance at low water potentials.[15] In cereals, osmotic adjustment appears to be under the control of a single gene or a small number of genes that are simply inherited.[16,17] Studies with rice have identified a single quantitative trait locus (QTL) for osmotic adjustment that appears to be homologous with the single gene for osmotic adjustment in wheat.[16,17] However, to date the only crop in which the gene for osmotic adjustment has been used by breeders is wheat, where a cultivar incorporating the gene for osmotic adjustment has been released in Australia.

BENEFITS OF OSMOTIC ADJUSTMENT

Osmotic adjustment is important because it aids in the maintenance of the turgor pressure of the plant as water deficits develop.[7] As growth is a turgor-dependent process, maintenance of turgor is important in maintaining growth.[3] Osmotic adjustment has been shown to maintain photosynthesis and stomatal conductance, defer leaf rolling and leaf death, and maintain root growth to lower water potentials.[18] In turn, this has been shown to induce root growth into deeper soil layers and increase water uptake.[13]

Osmotic adjustment does not necessarily maintain full turgor. Partial turgor maintenance has been documented in a number of species[7] and leads to zero turgor being reached at lower water potentials than when no osmotic adjustment occurs, but at higher water potentials than in the case of full turgor maintenance (Fig. 1). Also, it should be recognized that water potentials below a threshold may decrease before osmotic adjustment is initiated. Jones and Rawson[19] showed that osmotic adjustment reduced the water potential at which the rate of leaf photosynthesis in sorghum reached low values, but did not prevent the decrease in photosynthesis accompanying a decrease in the leaf water potential. Also, chickpea genotypes have been shown to vary in the degree of osmotic adjustment, but this occurred too late into a drying cycle to influence the rate of leaf photosynthesis.[20]

Additionally, the degree of osmotic adjustment is finite. While osmotic adjustment maintains turgor, there appears to be a point beyond which no further solutes accumulate. Further, because the solutes required for osmotic adjustment may also be required during grain filling, it may be undesirable for the solutes accumulated to remain in the leaf. In chickpea, the degree of osmotic adjustment decreased during late seed filling,[20] presumably so that the solutes could be mobilized to the grain.

Encyclopedia of Plant and Crop Science
DOI: 10.1081/E-EPCS 120010409

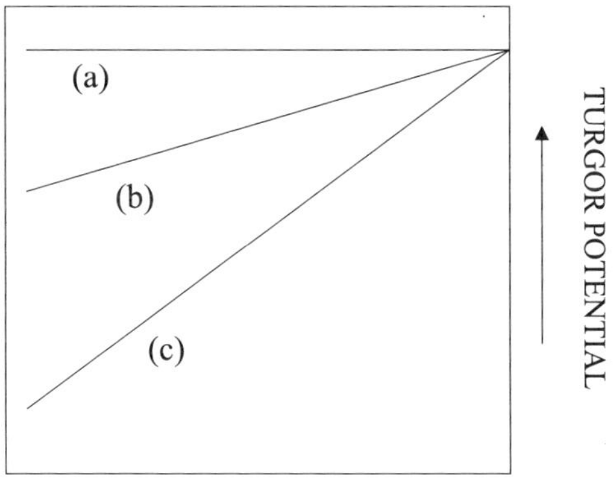

Fig. 1 Schematic representation of the relationship between turgor potential and water potential for (a) full turgor maintenance, (b) partial turgor maintenance, and (c) no turgor maintenance. (Adapted from Ref. 7.)

Indeed, pigeonpea genotypes with high osmotic adjustment toward the end of pod filling had lower yields than those that lost solutes during late pod fill.[21]

Nevertheless, higher yields of wheat, chickpea, and pigeonpea have been associated with greater degrees of osmotic adjustment.[13,14,21,22] In chickpea and pigeonpea, different genotypes were compared, so traits other than osmotic adjustment may have contributed to yield. In wheat, however, comparisons have been made in families selected for differences in osmotic adjustment.[22] In other species, analysis of the benefits to yield of high osmotic adjustment still needs to be undertaken in near-isogenic lines.

SOLUTES INVOLVED IN OSMOTIC ADJUSTMENT

A wide range of solutes have been observed to increase during osmotic adjustment. Jones et al.[6] showed that sucrose, glucose, fructose, potassium, chloride, and amino acids all contributed to osmotic adjustment in sorghum, whereas amino acids, potassium, calcium, magnesium, and nitrate contributed to osmotic adjustment in sunflower. While the amino acid proline has been widely shown to increase when water deficits develop, its contribution to overall osmotic adjustment is small. However, it is considered to be a compatible solute that increases in the cytoplasm to balance organic and inorganic solute changes in the vacuole that would be detrimental if they occurred in the cytoplasm.

It is not clear that the same solutes increase in a particular genotype under a range of environmental conditions. Thus, without further investigation it is not possible to say whether a particular solute can be used as a selection tool in genotypic studies of osmotic adjustment.

MEASUREMENT OF OSMOTIC ADJUSTMENT

In order to measure the extent of osmotic adjustment, the change in osmotic pressure induced by the concentration of solutes resulting from water loss needs to be distinguished from the change induced by solute accumulation. To do this the osmotic pressure at a particular turgor pressure—normally at full turgor—is measured or calculated. To measure or estimate the osmotic pressure at full turgor requires either measurement after rehydration or calculation from measured values of osmotic potential and relative water content (equivalent to the relative osmotic volume in Eq. 1). Because the osmotic pressure at full turgor can change with age and even diurnally, samples from plants subjected to water deficits and those kept adequately watered are taken on the same day, and the degree of osmotic adjustment is considered the difference in osmotic pressure at full turgor between stressed and unstressed plants.

There are errors associated with both of the foregoing methods of measurement. Rehydration may not be complete, or the time required for rehydration may differ between plants that have been stressed to different degrees, so that comparison after the same rehydration time may not be reliable. Also, in some species it is known that solute loss occurs during rehydration, or solutes may be metabolised from osmotically active solutes to osmotically-inactive solutes—or vice versa—leading to errors in the measurement of osmotic pressure. While such errors in osmotic pressure do not occur when the tissue is not rehydrated, similar errors of rehydration can occur in measured values of relative water content.[23] The osmotic pressure is usually measured by vapor pressure osmometry or freezing-point depression osmometry.[23] However, the osmotic pressure at full turgor can be estimated from pressure/volume relations using the pressure chamber technique.[24] Rehydration is necessary and the rehydration errors discussed above are also relevant with this method. Additionally, the osmotic pressure at full turgor is extrapolated from pressure/volume data below zero turgor, and errors can occur in this range, particularly in tissue that is not woody.[24]

Whether the pressure chamber or osmometry is used to measure the osmotic pressure, the procedures are relatively slow, particularly if watered and unwatered plots are required in order to measure the degree of osmotic adjustment. Where a large number of plants need to be compared, measured values of osmotic pressure and either

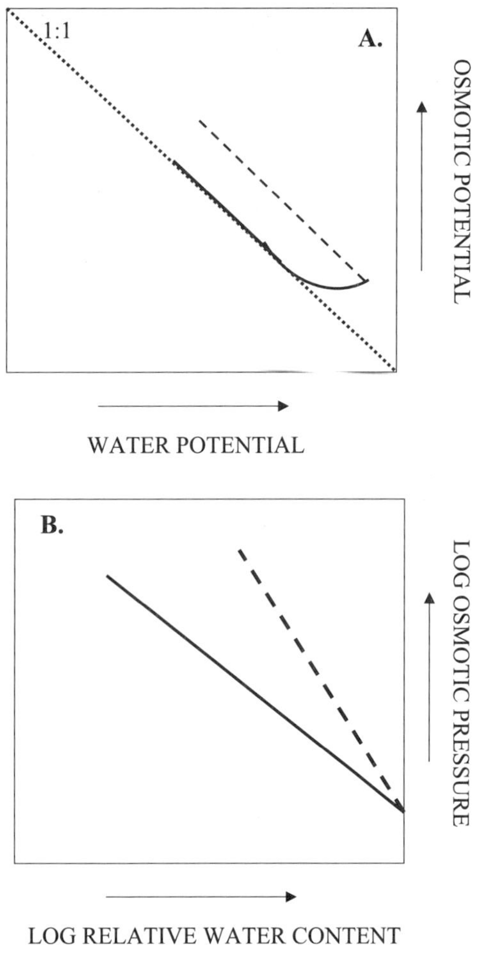

Fig. 2 Schematic representation of the relationship between (A) osmotic pressure and water potential and (B) $\log_n$ osmotic pressure and $\log_n$ relative water content in osmotically adjusting (dashed line) and nonadjusting (solid line) plants. (Adapted from Refs. 27,28.)

relative water content or water potential during a drying cycle have been used to compare the measured values against ones predicted from the concentration of solutes from water loss.[25,26] Fig. 2 shows the changes in osmotic pressure with relative water content and with water potential in nonadjusting and osmotically adjusting plants. Alternatively, at points during the drying cycle leaves may be sampled for osmotic pressure after rehydration. By this method at least 200 plants can be sampled, and it is feasible to screen breeding populations for osmotic adjustment. However, it needs to be recognized that these methods do not account for changes in osmotic pressure induced by age or phenological development, and they are subject to the dehydration errors mentioned above.

Recently, Morgan[29] showed that pollen grains of wheat in the presence of potassium chloride accumulated solutes and became swollen when exposed to stress imposed with polyethylene glycol. Only pollen grains from genotypes that showed osmotic adjustment in their leaves accumulated solutes, so Morgan[29] suggested that the technique could be used as a simple screening technique for osmotic adjustment in breeding programs. The technique needs to be explored in other species that show genetic differences in osmotic adjustment.

CONCLUSION

The maintenance of turgor as water deficits develop through the accumulation of solutes is termed osmotic adjustment or osmoregulation and is widely observed in some higher plant species and genotypes. The maintenance of turgor aids in the maintenance of physiological activity, such as photosynthesis and seed growth, and has been shown to benefit crop yields in water-limited environments. The lack of rapid selection methods has hindered the use of the trait in dryland breeding programs, except in the case of wheat, where a cultivar has been developed with high osmotic adjustment. The identification of a marker for osmotic adjustment in wheat and rice should enable the trait to be used in future marker-assisted breeding programs.

ARTICLES OF FURTHER INTEREST

Drought and Drought Resistance, p. 386

Drought Resistant Ideotypes: Future Technologies and Approaches for the Development of, p. 391

Plant Response to Stress: Absisic Acid Fluxes, p. 973

Plant Response to Stress: Regulation of Plant Gene Expression to Drought, p. 999

Water Deficits: Development, p. 1284

Water Use Efficiency Including Carbon Isotope Discrimination, p. 1288

REFERENCES

1. Walter, H. *Die Hydratur de Pflanze und ihre physiolgisch-ökologische Bedeutung (Untersuchungen über den osmotischen Wert)*; Fischer: Jena, Germany, 1931.
2. Walter, H.; Stadelmann, E. *Desert Biology*; Brown, G.W., Ed.; Academic Press: New York, 1974; Vol. 2, 213–310.
3. Hsiao, T.C.; Acevedo, E.; Fereres, E.; Henderson, D.W. Water stress, growth and osmotic adjustment. Philos. Trans. R. Soc. Lond. Ser. B **1976**, *273*, 479–500.
4. Jones, M.M.; Turner, N.C. Osmotic adjustment in leaves of sorghum in response to water deficits. Plant Physiol. **1978**, *61*, 122–126.

5. Jones, M.M.; Turner, N.C. Osmotic adjustment in expanding and fully expanded leaves of sunflower in response to water deficits. Aust. J. Plant Physiol. **1980**, *7*, 181–192.
6. Jones, M.M.; Osmond, C.B.; Turner, N.C. Accumulation of solutes in leaves of sorghum and sunflower in response to water deficits. Aust. J. Plant Physiol. **1980**, *7*, 193–205.
7. Turner, N.C.; Jones, M.M. Turgor Maintenance by Osmotic Adjustment: A Review and Evaluation. In *Adaptation of Plants to Water and High Temperature Stress*; Turner, N.C., Kramer, P.J., Eds.; John Wiley: New York, 1980; 87–103.
8. Morgan, J.M. Osmoregulation and water stress in higher plants. Annu. Rev. Plant Physiol. **1984**, *35*, 299–319.
9. Subbarao, G.V.; Johansen, C.; Slinkard, A.E.; Rao, R.C.N.; Saxena, N.P.; Chauhan, Y.S. Strategies for improving drought resistance in grain legumes. Crit. Rev. Plant Sci. **1995**, *14*, 469–523.
10. Turner, N.C.; Wright, G.C.; Siddique, K.H.M. Adaptation of grain legumes (pulses) to water-limited environments. Adv. Agron. **2001**, *71*, 193–231.
11. Morgan, J.M. Osmotic adjustment in the spikelets and leaves of wheat. J. Exp. Bot. **1980**, *31*, 655–665.
12. Shackel, K.A.; Turner, N.C. Seed coat cell turgor in chickpea is independent of changes in plant and pod water potential. J. Exp. Bot. **2000**, *51*, 895–900.
13. Morgan, J.M.; Condon, A.G. Water use, grain yield and osmoregulation in wheat. Aust. J. Plant Physiol. **1986**, *13*, 523–532.
14. Morgan, J.M.; Rodriguez-Maribona, B.; Knights, E.J. Adaptation to water-deficit in chickpea breeding lines by osmoregulation: Relationship to grain-yields in the field. Field Crops Res. **1991**, *27*, 61–70.
15. Turner, N.C. Drought Resistance and Adaptation to Water Deficits in Crop Plants. In *Stress Physiology in Crop Plants*; Mussell, H., Staples, R.C., Eds.; John Wiley: New York, 1979; 343–372.
16. Morgan, J.M.; Tan, M.K. Chromosomal location of a wheat osmoregulation gene using RFLP analysis. Aust. J. Plant Physiol. **1996**, *23*, 803–806.
17. Lilley, J.M.; Ludlow, M.M.; McCouch, S.R.; O'Toole, J.C. Locating QTL for osmotic adjustment and dehydration tolerance in rice. J. Exp. Bot. **1996**, *47*, 1427–1436.
18. Turner, N.C. Crop water deficits: A decade of progress. Adv. Agron. **1986**, *39*, 1–51.
19. Jones, M.M.; Rawson, H.M. Influence of the rate of development of leaf water deficits upon photosynthesis, leaf conductance, water use efficiency, and osmotic potential in sorghum. Plant Physiol. **1979**, *45*, 103–111.
20. Leport, L.; Turner, N.C.; French, R.J.; Barr, M.D.; Duda, R.; Davies, S.L.; Tennant, D.; Siddique, K.H.M. Physiological responses of chickpea genotypes to terminal drought in a Mediterranean-type environment. Eur. J. Agron. **1999**, *11*, 279–291.
21. Subbarao, G.V.; Chauhan, Y.S.; Johansen, C. Patterns of osmotic adjustment in pigeonpea—Its importance as a mechanism of drought resistance. Eur. J. Agron. **2000**, *12*, 239–249.
22. Morgan, J.M.; Hare, R.A.; Fletcher, R.J. Genetic variation in osmoregulation in bread and durum wheats and its relationship to grain yield in a range of field environments. Aust. J. Agric. Res. **1986**, *37*, 449–457.
23. Turner, N.C. Techniques and experimental approaches for the measurement of plant water status. Plant Soil **1981**, *58*, 339–366.
24. Turner, N.C. Measurement of plant water status by the pressure chamber technique. Irrig. Sci. **1988**, *9*, 289–308.
25. Morgan, J.M. Differences in osmoregulation between wheat genotypes. Nature **1977**, *270*, 234–235.
26. Takami, S.; Rawson, H.M.; Turner, N.C. Leaf expansion of four sunflower (*Helianthus annuus* L.) cultivars in relation to water deficits. II. Diurnal patterns during stress and recovery. Plant Cell Environ. **1982**, *5*, 279–286.
27. Turner, N.C. The Role of Shoot Characteristics in Drought Resistance of Crop Plants. In *Drought Resistance in Crops with Emphasis on Rice*; International Rice Research Institute: Los Baños, The Philippines, 1982; 115–134.
28. Morgan, J.M. Osmoregulation as a selection criterion for drought tolerance in wheat. Aust. J. Agric. Res. **1983**, *34*, 607–614.
29. Morgan, J.M. Pollen grain expression of a gene controlling differences in osmoregulation in wheat leaves: A simple breeding method. Aust. J. Agric. Res. **1999**, *50*, 953–962.

Oxidative Stress and DNA Modification in Plants

Alex Levine
The Hebrew University of Jerusalem, Jerusalem, Israel

INTRODUCTION

Oxygen constitutes 21% of the atmosphere and is the most abundant chemical in and near the earth's crust. The concentration of oxygen in the atmosphere has risen dramatically since the beginning of life due to the evolution of oxygen during photosynthesis. Because of its chemical properties as an electron acceptor whereby each molecule of dioxygen can accept four electrons producing water, oxygen is fundamentally essential for energy metabolism and respiration in the predominantly aerobic modern biosphere. However, if the gradual reduction process that proceeds by a consecutive addition of electrons is not complete, the partially reduced dioxygen forms highly reactive molecules (Fig. 1). Collectively these molecules are termed *r*eactive *o*xygen *s*pecies (ROS) or *i*ntermediates (ROI).

The continuous flow of electrons in the cell—especially but not exclusively in the energy producing organelles such as mitochondria or chloroplasts—creates a constant leakage of electrons to oxygen, forming superoxide and other ROS derivatives. The leakage of electrons to oxygen is exacerbated during stresses, invoking a complex array of plant responses.

To counteract the potentially dangerous reactions of the oxidants, organisms have evolved elaborate detoxifying systems. Due to the high internal oxygen concentration and the generation of high-energy electrons by photo-activation, plants are particularly prone to damage by the ROS, and have therefore developed effective antioxidant systems.[1]

The major natural as well as anthropogenic environmental factors that cause damage to living organisms arise from pollution of water, soil, and air, and from solar (UV) and ionizing radiations. These conditions cause ROS accumulation and oxidative stress. In addition, ROS are produced during upward or downward shifts in the surrounding temperature, especially when coupled with strong light.

FORMATION OF OXYGEN RADICALS AND ANTIOXIDANT DEFENSES

ROS formation is part of normal cellular metabolism. For example, $O_2^{\cdot -}$ and H_2O_2 are the by-products of enzymatic reactions in nucleotide and sugar metabolism. However, the majority of ROS in nonpathogenic conditions are produced by leakage of electrons from the electron transport chains in mitochondria, chloroplasts, and even in the plasma membrane.[2] Another major source of ROS in plant cells occurs in response to pathogen attack and to mechanical stress, which activate the plasma membrane-localized NADPH oxidase.[3,4] A cell wall peroxidase was also implicated in pathogen-induced generation of ROS.[5] These mechanisms of ROS production are not mutually exclusive.

Plants and other organisms possess elaborate systems to detoxify ROS. However, during environmental stress the rate of ROS production can exceed the rate of their removal.[6] Plants possess ROS-detoxifying enzymes that are also found in animals and other organisms, as well as numerous small-molecule antioxidants (e.g., vitamins C and E) that are products of the secondary metabolism unique to plants. The two systems act in cooperation, whereby the antioxidant molecules react with ROS as a first line of defense and are later regenerated by specific enzymes. Adverse environmental conditions affect the redox metabolism in the cell by interfering with the electron flow in the organelles or with the functioning of the antioxidant systems.

REACTIONS OF OXYGEN RADICALS: DNA DAMAGE AND REPAIR

The initially produced ROS (such as superoxide and hydrogen peroxide) react relatively weakly with the majority of biological molecules. Among their main targets are oxidation of the thiol groups in proteins that can later be reduced back by glutaredoxin or thioredoxin systems. However, $O_2^{\cdot -}$ and H_2O_2 derivatives (such as hydroxyl radicals) react with many biological substrates at a diffusion-limited rate. The $^{\cdot}OH$ damage includes protein oxidation, lipid peroxidation, and formation of DNA adducts and strand breaks. The $^{\cdot}OH$ radicals also cause cross-linking of the DNA to DNA-binding proteins, leading to a replication block.

Although the damage caused to proteins and lipids may be repaired by removal of the impaired molecules and

Encyclopedia of Plant and Crop Science
DOI: 10.1081/E-EPCS 120010641

↓ē ↓ē ↓ē ↓ē

O_2 → $O_2^{\cdot -}$ → H_2O_2 → $\cdot OH$ → H_2O

oxygen superoxide hydrogen peroxide hydroxyl radical water

Fig. 1 Sequential reduction of oxygen to water.

replacement with new ones, such a mechanism is not feasible for DNA because the genetic information must be reproduced precisely. The ˙OH radical reacts with little specificity and can cause mutations in any nucleotide (Fig. 2).

The formation of ˙OH radicals is greatly promoted by Fenton reaction in the presence of heavy metals such as iron, lead, copper, and cadmium. These metals are common soil pollutants. Due to its negative charge, DNA has a high affinity for binding heavy metals, thus promoting hydroxyl radicals' formation in close proximity to the DNA. The most common form of DNA damage is the ˙OH-mediated formation of 8-hydroxyguanosine, which leads to mutagenesis and DNA strand breaks.[7] Especially problematic are the repair of double strand breaks and of mismatches during replication that result in "fixed" mutations. ROS are also involved in DNA strand breaks caused by UV radiation.[8]

In plants, a particularly high rate of mutation occurs during repair of double strand breaks by direct end-joining as part of a nonhomologous recombination.[8] Increased mutation rate is therefore a direct consequence of abiotic/environmental stresses.[9,10] In accord with this, improved antioxidant defenses and decreased oxygen tension were shown to reduce mutability.[11] Interestingly, overexpression of superoxide dismutase is not always beneficial and can even increase DNA damage via accelerated production of H_2O_2, which promotes formation of ˙OH radicals.[12] Thus, a coordinated antioxidant response that includes specific enzymes as well as small molecule antioxidants is essential for genome stability.

GENE SELECTION AND DEVELOPMENT

Evolution depends on genetic variability and subsequent selection of the individuals that are better adapted for the given conditions. In plants, the germ cells differentiate from somatic cells in a process that may take many cycles of mitotic division. Thus, somatic mutations can enter germ cells and be transmitted to the progeny.

Being both sessile and lacking a system that regulates internal temperature, plants are continuously impelled to adapt to changes in their surroundings. Environmental stress results in a reduction of photosynthesis and inhibition of growth, resulting in diminished competitiveness against other members of the same species and against other competitors. Such constantly changing environmental factors constitute a strong selection force. Studies have shown that growth of plants in soils polluted with heavy metals induces phenotypic alterations that stem from stress-dependent mutations.[13] More recently it has been shown that ROS-associated (abiotic) stresses enhance both the rate of recombination and mutagenicity in *Arabidopsis* plants.[14]

SUMMARY

The sessile nature of plants makes them particularly sensitive to the surrounding environment. During evolution plants have developed sophisticated mechanisms to sense changes in environmental conditions and adapt to them. Almost all environmental stresses are associated with the accumulation of reactive oxygen species (ROS), resulting in secondary oxidative stress. Some of the ROS (such as the hydroxyl radical), are highly reactive molecules that can damage cellular components such as lipids, proteins, and DNA. Moreover, the DNA adducts promote strand breaks, augmenting the damage. Whereas the damaged lipids and proteins can be replaced by new molecules, the repair of DNA is prone to introduce mistakes and result in mutations. In eukaryotes the nuclear DNA exists in dynamic association with histones, forming chromatin.

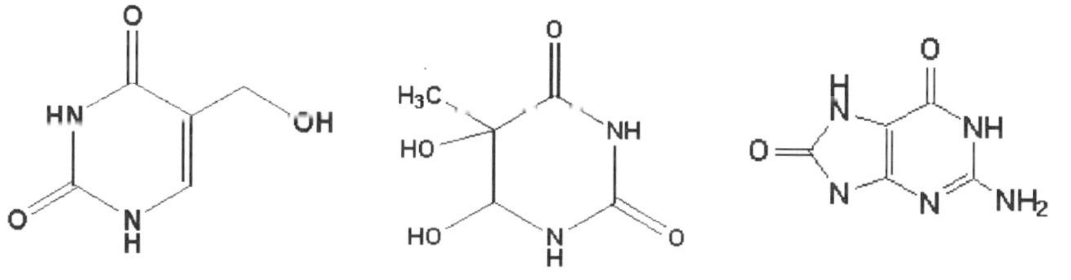

5-Hydroxymethyluracil Thymine glycol 8-Oxoguanine

Fig. 2 Structures of oxidatively damaged DNA bases.

During transcription the active chromatin unfolds, allowing access of the transcription machinery, while the inactive chromatin remains compacted. Thus, genes that are active during stress are specifically exposed to attack by the hydroxyl radicals, leading to a higher mutation rate. Any favorable mutation that will reduce the primary stress will decrease the oxidative stress and stabilize the genome.

CONCLUSION

Changes in environmental conditions invoke an oxidative stress in plant cells. The reactive oxygen species that are generated during stress react with the cellular components. Particularly damaging is the formation of the hydroxyl radical, which reacts with DNA. In addition to causing direct mutations, DNA adducts produce strand breaks resulting in additional matagenicity through nonhomologous recombination.[8,9] The open state of the transcribed genes[15] makes them particularly prone to oxidative damage. Thus, the genes that are expressed during environmental stresses associated with ROS accumulation undergo an increased mutation rate. Improvements in resistance to primary environmental stress by favorable mutations also decrease the oxidative stress and matagenicity, stabilizing plant adaptation to the surrounding environment.

ARTICLES OF FURTHER INTEREST

Arabidopsis thaliana: Characteristics and Annotation of a Model Genome, p. 47
Bacterial Pathogens: Early Interactions with Host Plants, p. 89
Drought and Drought Resistance, p. 386
Genome Rearrangements and Survival of Plant Populations to Changes in Environmental Conditions, p. 513
Molecular Evolution, p. 748
Plant Response to Stress: Mechanisms of Accommodation, p. 987
Plant Response to Stress: Ultraviolet-B Light, p. 1019

REFERENCES

1. Levine, A. Oxidative Stress and Programmed Cell Death. In *Plant Responses to Environmental Stresses: From Phytohormones to Genome Reorganization*; Lerner, H.R., Ed.; Marcel Dekker, Inc.: New York, 1999; 247–261.
2. Moller, I.M. Plant mitochondria and oxidative stress: Electron transport, NADPH turnover, and metabolism of reactive oxygen species. Annu. Rev. Plant Physiol. Plant Mol. Biol. **2001**, *52*, 561–591.
3. Levine, A.; Tenhaken, R.; Dixon, R.; Lamb, C. H_2O_2 from the oxidative burst orchestrates the plant hypersensitive disease resistance response. Cell **1994**, *79*, 583–593.
4. Lamb, C.; Dixon, R.A. The oxidative burst in plant disease resistance. Annu. Rev. Plant Physiol. Plant Mol. Biol. **1997**, *48*, 251–275.
5. Bolwell, G.P.; Butt, V.S.; Davies, D.R.; Zimmerlin, A. The origin of the oxidative burst in plants. Free Radic. Res. **1995**, *23*, 517–532.
6. Noctor, G.; Foyer, C.H. Ascorbate and glutathione: Keeping active oxygen under control. Annu. Rev. Plant Physiol. Plant Mol. Biol. **1998**, *49*, 249–279.
7. Halliwell, B. Oxygen and nitrogen are pro-carcinogens. Damage to DNA by reactive oxygen, chlorine and nitrogen species: Measurement, mechanism and the effects of nutrition. Mutat. Res., Genet. Toxicol. Environ. **1999**, *443*, 37–52.
8. Tuteja, N.; Singh, M.B.; Misra, M.K.; Bhalla, P.L.; Tuteja, R. Molecular mechanisms of DNA damage and repair: Progress in plants. Crit. Rev. Biochem. Mol. Biol. **2001**, *36*, 337–397.
9. Lebel, E.; Masson, J.; Bogucki, A.; Paszkowski, J. Stress-induced intrachromosomal recombination in plant somatic cells. Proc. Natl. Acad. Sci. U. S. A. **1993**, *90*, 422–426.
10. Kovalchuk, I.; Kovalchuk, O.; Hohn, B. Biomonitoring the genotoxicity of environmental factors with transgenic plants. Trends Plant Sci. **2001**, *6*, 306–310.
11. Blanco, M.; Herrera, G.; Urios, A. Increased mutability by oxidative stress in OxyR-deficient *Escherichia coli* and *Salmonella typhimurium* cells—Clonal occurrence of the mutants during growth on nonselective media. Mutat. Res. Lett. **1995**, *346*, 215–220.
12. Kawanishi, S.; Hiraku, Y.; Oikawa, S. Mechanism of guanine-specific DNA damage by oxidative stress and its role in carcinogenesis and aging. Mutat. Res., Rev. Mutat. Res. **2001**, *488*, 65–76.
13. Wurgler, F.E.; Kramers, P.G.N. Environmental—Effects of genotoxins (Ecogenotoxicology). Mutagenesis **1992**, *7*, 321–327.
14. Lucht, J.M.; Mauch-Mani, B.; Steiner, H.Y.; Metraux, J.P.; Ryals, J.; Hohn, B. Pathogen stress increases somatic recombination frequency in Arabidopsis. Nat. Genet. **2002**, *30*, 311–314.
15. Paranjape, S.M.; Kamakaka, R.T.; Kadonaga, J.T. Role of chromatin structure in the regulation of transcription by RNA polymerase II. Annu. Rev. Biochem. **1994**, *63*, 265–297.

Oxygen Production

Charles F. Yocum
University of Michigan, Ann Arbor, Michigan, U.S.A.

O

INTRODUCTION

The ability to oxidize water to oxygen is a unique property of the metabolism of oxygenic photosynthetic organisms. An enzyme complex called photosystem II, which is a component of the thylakoid membranes found in chloroplasts of eukaryotes and in cyanobacteria, catalyzes the reaction. The driving force for oxygen production from water is light absorption by photosynthetic pigments. The absorbed energy causes a special pair of chlorophyll molecules (called P680) to release electrons that are ultimately used to reduce CO_2 to carbohydrate. Electron deficient, oxidized P680 regains electrons by oxidizing manganese atoms that are part of the active site of oxygen production. This site contains four atoms of manganese, as well as one atom each of calcium and chloride. Although the specific details of the mechanism of the oxygen-evolving reaction are still being sought, it is clear that oxidation of the manganese atoms in photosystem II constitutes the key step in oxygen production. Calcium may bind water molecules destined for oxidation by manganese, while chloride is proposed to regulate the oxidation/reduction activity of the manganese atoms.

OXYGENIC PHOTOSYNTHESIS

Oxygenic photosynthesis is the process by which light energy is converted into chemical energy, and oxygen is released as a byproduct.[1] Light-driven electron transfer reactions are localized in the stacked layers of membranes—called thylakoids—of chloroplasts and cyanobacteria. The ultimate electron source is water, which is oxidized to oxygen. This reaction is presented in Eq. 1, which shows that a key step in the process is absorption of light by an enzyme system called photosystem II. The electrons that are withdrawn from water reduce a quinone molecule called plastoquinone, abbreviated as PQ in the equation. Subsequent oxidation of PQH_2 is accomplished by cytochromes and a second light reaction—called photosystem I—that are also found in all organisms that carry out oxygenic photosynthesis.[1]

$$2H_2O + 2PQ_{OXIDIZED} + 4\text{photons}(h\nu) \xrightarrow{\text{Photosystem II}} O_2 + 2PQH_2 \quad (1)$$

PHOTOSYSTEM II AND THE S-STATE CYCLE FOR OXYGEN PRODUCTION

Oxygen production by photosystem II is catalyzed by a linear sequence of reactions.[2] Detection of oxygen produced by illuminating a sample with very short (~10 μs) light flashes gives the result shown graphically in Fig. 1. Upon excitation by three sequential flashes, photosystem II emits a gush of oxygen, and does so again on the seventh and eleventh flashes in a series, although the oscillations become less pronounced with larger numbers of flashes. This oscillating pattern is modeled on a cycle containing intermediate states of the oxygen-producing reaction, called S states.[2] Each state is numbered, S_i ($i = 0$–4), and a flash of light advances the S-state system from S_i to S_{i+1}. The S-state hypothesis is represented schematically by Eq. 2, where $h\nu$ is used to signify the individual flashes of light.

$$2H_2O + S_0 \xrightarrow{h\nu} S_1 \xrightarrow{h\nu} S_2 \xrightarrow{h\nu} S_3 \xrightarrow{h\nu} S_4 \rightarrow S_0 + O_2 + 4H^+ + 4e^- \quad (2)$$

To explain the burst of oxygen on the third flash of light, it has been necessary to postulate that the S_1 state is stable in darkness, and this hypothesis is widely accepted. Higher S states can be deactivated to S_1 by imposing darkness on the system after one or two flashes of light. Typical lifetimes of S_2 and S_3 are about one minute. The gradual loss of the oscillating pattern of oxygen release is due to the intrinsic behavior of the enzyme system, which leads to a random distribution of S states after a large number of light flashes or after exposure of the enzyme to continuous light.

LIGHT-DRIVEN ELECTRON TRANSFER REACTIONS

Advancement of S states requires a series of reactions that are initiated by light. The membrane-associated proteins of photosystem II bind the necessary organic cofactors,

Encyclopedia of Plant and Crop Science
DOI: 10.1081/E-EPCS 120019317

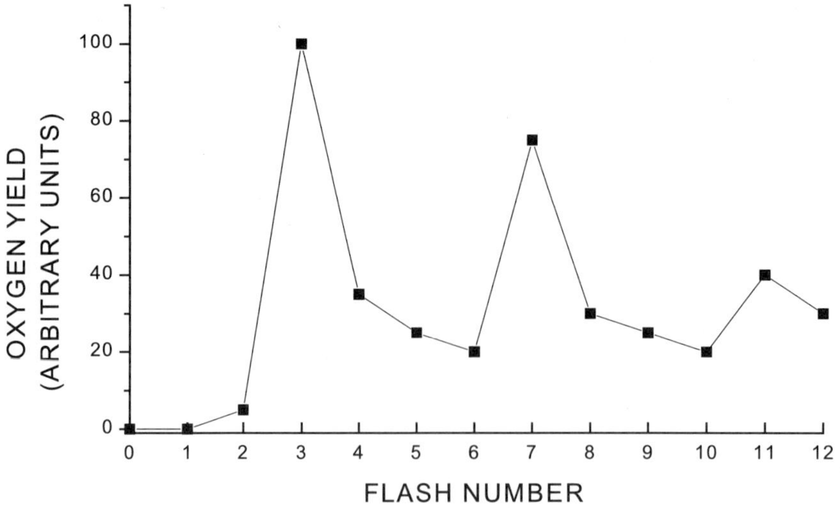

Fig. 1 Oxygen yield from photosystem II as a function of the number of light flashes. The maximum yields of oxygen are obtained on the third, seventh, and eleventh flashes. (See text for details.)

which include tyrosines, chlorophylls, pheophytin (a chlorophyll *a* molecule lacking the central Mg^{2+} atom), and plastoquinone molecules called Q_A and Q_B.[3] Four atoms of manganese and one atom each of calcium and chloride are also bound to the membrane-associated proteins of photosystem II, to form the active site for oxygen production. Additional water soluble proteins assist in stabilizing the binding of these inorganic ions. A 3.8 Å resolution crystal structure of the enzyme is available,[4] which shows the arrangement of the various cofactors that are involved in light-driven electron transfer and oxygen production.

H_2O → [4Mn/Ca^{2+}/Cl^-] → [Tyrosine Z] → [P680] → [Pheophytin] → Q_AQ_B

Reaction 1 (Chlorophyll Excitation) ↓ ◀~~ **h*ν***

H_2O → [4Mn/Ca^{2+}/Cl^-] → [Tyrosine Z] → **[P680]*** → [Pheophytin] → Q_AQ_B

Reaction 2 (Charge Separation) ↓

H_2O → [4Mn/Ca^{2+}/Cl^-] → [Tyrosine Z] → **[P680]$^+$** → **[Pheophytin]$^-$** → Q_AQ_B

Reaction 3 (Stabilization of Charge Separation) ↓

H_2O → [4Mn/Ca^{2+}/Cl^-] → **[Tyrosine Z]$^\bullet$** → [P680] → [Pheophytin] → **Q_A^-**Q_B

Reaction 4 (Manganese Oxidation and Reduction of Q_B) ↓

H_2O → **[4Mn/Ca^{2+}/Cl^-]$^+$** → [Tyrosine Z] → [P680] → [Pheophytin] → Q_A**Q_B^-**

Fig. 2 A schematic representation of the electron transfer reactions in photosystem II that precede oxygen production. **Boldface** type identifies the chemically reactive species at each step in the reaction sequences. (The electron transfer species are identified in text.)

The sequence of electron transfer events that lead to oxygen production is shown in Fig. 2. Light absorption by a special pair of chlorophyll *a* molecules, called P680 after the wavelength of light (680 nm) that it absorbs, triggers a cascade of chemical reactions. The excited chlorophyll molecule created by light absorption in Reaction 1 (**[P680]***) is very reactive, and within a few picoseconds ($\sim 10^{-12}$ s) loses an electron to the acceptor pheophytin molecule, forming **[Pheophytin]**$^-$, as shown in Reaction 2. Oxidized P680 (**[P680]**$^+$) is also very reactive, and removes an electron from a tyrosine residue, called tyrosine Z, forming a radical **[Tyrosine Z]**$^\bullet$ (Reaction 3). Tyrosine Z is one of the amino acids in the sequence of a membrane-associated protein of photosystem II called D1. In a second part of Reaction 3, **[Pheophytin]**$^-$ passes its electron on to the first special quinone molecule, to form $\mathbf{Q_A}^-$. The formation of two stable products is shown in Reaction 4. First, the electron removed from P680 by light absorption reduces a plastoquinone molecule called Q_B to form $\mathbf{Q_B}^-$. Second, an electron is removed from a manganese atom that is part of the group of inorganic ions that make up the active site of the enzyme, forming $\mathbf{[4\ Mn/Ca^{2+}/Cl^-]^+}$. As a result, the energy of the photon absorbed by P680 now resides in two stable species; in a reducing agent, Q_B^-, that forms PQH_2 after accepting another electron and binding two H^+; and in an oxidizing agent, an electron deficient manganese ion. The lifetimes of these species enable them to carry out chemical reactions.

ROLES OF MANGANESE, CALCIUM, AND CHLORIDE IN OXYGEN PRODUCTION

In the dark stable S_1 state, the oxidation states of the four manganese atoms have been determined to be $Mn^{3+}/Mn^{3+}/Mn^{4+}/Mn^{4+}$.[5] Physical techniques for assessing oxidation state changes in these metals indicate that a Mn^{3+} is oxidized to Mn^{4+} on advancement of the S states from S_1 to S_2, and from S_2 to S_3. The oxidation state of S_4 has not been determined, but that of S_0 is currently believed to be $Mn^{2+}/Mn^{3+}/Mn^{4+}/Mn^{4+}$. Removal of either Ca^{2+} or Cl^- from photosystem II inhibits the S-state cycle after formation of S_2. Proposals that calcium may act as a binding site for the water molecules that are oxidized by photosystem II and that Cl^- is present to regulate the oxidation/reduction properties of the manganese atoms[6] may be relevant in explaining why these ions are needed to advance beyond S_2.

The most important question that remains to be answered about photosystem II is how oxidized manganese atoms regain electrons by reacting with water to produce oxygen. One possibility is that an electron deficiency first accumulates among the manganese atoms, after which these atoms regain electrons by oxidizing water to oxygen in a single step. In this mechanism, the manganese atoms would function as a charge accumulator.[7] In the second hypothetical mechanism, water molecules bind to manganese and undergo partial oxidation as the S states advance. In this mechanism, the sequential removal of electrons from water would permit manganese atoms to undergo oxidation, but there would be no accumulation of charge on the system because neutral hydrogen atoms ($H^\bullet$) rather than electrons would be transferred.[8] There is insufficient evidence at present to choose a correct mechanism, and the possibility must remain open that a hybrid system, combining features of both types of hypothetical mechanisms, is responsible for oxygen production.

CONCLUSION

Photosystem II extracts electrons from water to form oxygen by carrying out a highly organized set of reactions. Light absorption by chlorophyll initiates a set of electron transfer reactions. These reactions produce two results. The first of these is formation of reduced quinone molecules that shuttle electrons to other components of the photosynthetic electron transfer apparatus. The second consequence of light absorption is that electrons are withdrawn from a cluster of inorganic ions; loss of four electrons results in oxygen production from two water molecules. The reaction is catalyzed by manganese ions, but requires calcium and chloride as well. The precise mechanism for oxidation of water is unknown, but current models favor either charge accumulation by oxidation of manganese followed by water oxidation, or a sequential oxidation of water by abstraction of hydrogen atoms.

ARTICLES OF FURTHER INTEREST

Chlorophylls, p. 258
Exciton Theory, p. 429
Fluorescence: A Probe of Photosynthetic Apparatus, p. 472
Photosystems: Electron Flow Through, p. 906

REFERENCES

1. Voet, D.; Voet, J. *Biochemistry*, 2nd Ed.; John Wiley and Sons: New York, NY, 1995; Chapter 22.

2. Kok, B.; Forbush, B.; McGloin, M. Cooperation of charges in photosynthetic O_2 evolution. I. A linear four step mechanism. Photochem. Photobiol. **1970**, *11*, 457–475.
3. Diner, B.A.; Babcock, G.T. Structure, Dynamics, and Energy Conversion Efficiency in Photosystem II. In *Oxygenic Photosynthesis: The Light Reaction*; Ort, D.R., Yocum, C.F., Eds.; Kluwer Academic Publishers: Dordrecht, The Netherlands, 1996; 213–247.
4. Zouni, A.; Witt, H.T.; Kern, J.; Fromme, P.; Krauss, N.; Saenger, W.; Orth, P. Crystal structure of photosystem II from *Synechococcus elongatus* at 3.8 angstrom resolution. Nature **2001**, *409*, 739–743.
5. Riggs, P.J.; Mci, R.; Yocum, C.F.; Penner-Hahn, J.E. Reduced derivatives of the manganese cluster in the photosynthetic oxygen-evolving complex. J. Am. Chem. Soc. **1992**, *114*, 10650–10651.
6. Vrettos, J.S.; Limburg, J.; Brudvig, G.W. Mechanism of photosynthetic water oxidation: Combining biophysical studies of photosystem II with inorganic model chemistry. Biochim. Biophys. Acta **2001**, *1503*, 229–245.
7. Nugent, J.H.A.; Rich, A.M.; Evans, M.C.W. Photosynthetic water oxidation: Towards a mechanism. Biochim. Biophys. Acta **2001**, *1503*, 138–146.
8. Tommos, C.; Hoganson, C.W.; Di Valentin, M.; Lydakis-Simantiris, N.; Dorlet, P.; Westphal, K.; Chu, H.A.; McCracken, J.; Babcock, G.T. Manganese and tyrosyl radical function in photosynthetic oxygen evolution. Curr. Opin. Chem. Biol. **1998**, *2*, 244–252.

Paper and Pulp: Agro-based Resources for

George A. White (Retired)
USDA—ARS, Beltsville, Maryland, U.S.A.

INTRODUCTION

Non-wood crops and specialty fiber sources have been utilized for thousands of years in papermaking and other pulp applications. Their use greatly predates that of wood, which now composes about 90% of the worldwide production. As early as the late 1920s in the United States, a mill for producing corrugating medium based on straw began operation. Other mills followed in the Midwest, some of which operated throughout World War II. Unfortunately, they succumbed to high collection, transportation, and storage costs, and to prolonged use of outdated pulping equipment and processes.

Straw, sugarcane bagasse, and bamboo are the leading agricultural fiber sources presently used. China and India, both wood-poor countries, are the main users of straw and bamboo. Several countries use only non-wood fiber resources for papermaking. However, there exists a largely untapped and vast worldwide inventory of these three, plus other crop residues. Additionally, many plant species are suitable for pulping applications. A few, such as kenaf and sunn hemp, have been grown successfully and hence have the potential to add to the overall fiber inventory.

CROP RESIDUES FOR PULPING

Cereal Grain and Flax Straws

By using grain production, the amount of associated straw can be estimated. Data about the 2001 U.S. crop harvest[1] were used to estimate the straw yields for barley, oats, proso millet, rice, rye, and wheat. This computation resulted in an estimated 100 million metric tons (mt) that could be utilized in pulping applications. Most of the straw is either plowed under or used for feed and bedding. On a worldwide basis, the estimate for production of cereal and flax straws is 1.26 billion mt.[2] Straw represents the largest potential supply of crop residues suitable for pulping. The straw from most of the flaxseed oil production in the United States and Canada is processed into high-quality paper products.

Straw from Grass Seed Production

The primary region for grass seed production in the United States is in the tristate area of Oregon, Idaho, and Washington, but is concentrated in the Willamettte Valley of Oregon [Banowetz, G.M. Personal communication, U.S. Department of Agriculture; Agricultural Research Service (USDA-ARS): Corvallis, OR, 2002]. The involved grass species include most of the commonly grown lawn grasses and certain types of forage. The resultant straw from about 202,000 hectares (ha) amounts to about 1.4 million mt based on a yield of 6.7 mt per ha. None of this straw is utilized for pulping purposes. About 85% is baled and exported; the remainder is either burned or chopped and plowed under. Future plans include retooling a facility for pulping grass straw.

Stalks from Corn, Sorghum, and Cotton

These crops are widely grown worldwide and their residues are suitable for many pulping applications.[2] Corn and sorghum stalks and the residue of cotton stalks, lint, and linters appear well suited for bleached pulp and corrugating medium. In 2001 in the United States, 27.8 million ha of corn, 3.3 million ha of sorghum, and 5.6 million ha of cotton were harvested.

Sugarcane Bagasse

Bagasse, the residue from sugar-extracted stalks of sugarcane, has long been recognized for its excellent fiber qualities for various grades of paper and board-type applications. Three Louisiana companies have utilized bagasse: One has produced insulator board since the 1920s, another produces a pressed board, and the third (now out of business) produced writing and printing paper for many years.[a] The case for bagasse is unique in that the use for pulping competes with the use for fuel to power the

[a]Legendre, B.L. Personal communication, Louisiana State University; Sugar Research Station: St. Gabriel, LA, 2002.

Encyclopedia of Plant and Crop Science
DOI: 10.1081/E-EPCS 120010447

sugar mill. The long-range prospect for increasing use of bagasse is promising because of the large worldwide inventory and the availability of superior harvesting, storage, and pulping equipment plus improved processes (Fig. 1). Crop estimates indicate that the 2001 U.S. harvest of 416,510 ha[1] potentially resulted in 31.6 million mt of stalks (fresh weight basis).

STEM/CULM FIBER CROPS

Bamboo

There are many species of bamboo, ranging from the large, very tall, tropical types to the small, shrubby, hardier running types. All are perennial with woody culms (stems). Natural stands or groves, especially of tropical species, occur in many countries and could be exploited for pulping purposes. In more temperate countries, the running species of the genus *Phyllostachys* are fairly hardy and amendable to agricultural practices. Strip and rotational harvests help maintain the long-term productivity of the groves. India and China lead the world in bamboo pulping capacity.

Kenaf and Other Stem Fiber Sources

Kenaf, an old-world fiber crop, was best known as a jute substitute until identified as having good pulping characteristics. American interest in kenaf arose during World War II as imports of jute and allied fibers for maritime uses became jeopardized. In the early 1960s, research emphasis shifted from use for cordage to papermaking. A member of the mallow family, kenaf is a relative of roselle, okra, flowering hibiscus, and cotton. The stalk is composed of outer bast fibers about 2.6 millimeters (mm) in length and an inner thick core of short fibers about 0.6 mm in length.

Fig. 1 Harvesting sugarcane in south Florida. (Photo by USDA-ARS, Beltsville, MD, 2002.) (*View this art in color at www.dekker.com.*)

Fig. 2 Stem fibers of kenaf suitable for newsprint and other pulp products. (Photo by USDA-ARS, Beltsville, MD, 2002.) (*View this art in color at www.dekker.com.*)

Kenaf has been grown successfully in many U.S. states on an experimental and semicommercial or larger scale in about a dozen states. Based on scattered reports, about 4000 and 1500 ha were planted during 2000 and 2001, respectively.[b]

Innovative uses for the bast and core fibers—singularly or in blends—include newsprint (Fig. 2), various papers, absorbents, chicken and small-pet litter, soil amendments,

[b]Columbus, E.P. Personal communication, USDA-ARS: Mississippi State, MS, 2002; Taylor, C. Personal communication, Kenaf Industries of South Texas L.P.: Raymondville, TX, 2002; Sij, J. Personal communication, Texas A&M University: Vernon, TX, 2002.

seeding, and erosion mats, etc. Decorticating systems have been developed to efficiently separate the bast and core fibers. One U.S. company produces papers consisting of about 80% bast and 20% core fibers.[c] The diversity of products manufactured from these fibers should accelerate wider use of kenaf in the future.

The fiber characteristics of roselle and sunn hemp are similar to those of kenaf, but field yields of dry stalks are usually lower. These two crops are apparently immune to root knot nematodes, a kenaf nemesis. Kenaf (a diploid) and roselle (a hexaploid) can be crossed with difficulty. As a legume, sunn hemp production enriches soil through nitrogen fixation. Intense breeding to improve yield and stalk strength in sunn hemp would greatly enhance its potential to become a bona fide fiber crop in the United States. These three fiber sources are photosensitive, thereby requiring short days for floral initiation. Most plantings for seed production are made much later than for fiber. Late planting and lower seeding rates result in short, branched plants that are better suited for machine harvest.

Some other old-world fiber crops that are similar to kenaf in growth habit include hemp, jute, and ramie. Many recall having used jute string or rope. All of these crops are annuals except ramie, a perennial that contains very long fibers. Statistical production and fiber data often lump these four crops with kenaf.

Miscellaneous Fiber Sources

Many plant species are either harvested from natural stands or cultivated for fiber. Most of these are relatively unknown in the Western Hemisphere. The use of these specialty fiber sources, particularly those containing leaf and fruit fibers, is traditionally labor-intensive, thereby tending to keep production low. Sources of these fiber types include leaf-abaca (manila hemp), henequen, maguey and sisal; fruit-coir and kapok; stem-reeds; and whole plants (esparto, papyrus and sabai grass).

CONCLUSION

A great diversity of fibrous plant species with variable fiber characteristics is suitable and potentially available for producing a wide range of pulp-related products. Specific uses have been identified to take advantage of fiber differences. In terms of quantity, the cereal straws, corn and sorghum stalks, and bagasse are predominant.

As the worldwide economic downturn and current recession in the pulp and paper industry eases, greater use of more diverse fibrous raw materials can be anticipated. The most immediate expansion will likely include cereal straws, bagasse, and bamboo. New fiber resources such as kenaf will gradually add to the base supply. Refs. 2 and 3 contain more detailed information about fibrous resources, U.S. and worldwide inventories, and end uses.

REFERENCES

1. Anon. *Crop Production 2001 Summary*, National Agric. Statistics Service Report Cr Pr 2-1 (02); U.S. Department of Agriculture: Washington, DC, 2001; 1–87.
2. Atchison, J.E. *Worldwide Progress in Use of Agricultural Fibers for Manufacture of Pulp, Paper and Paperboard*; Atchison Consultants, Inc.: Sarasota, FL, 2001. Unpublished. Prepared for 5th International Biomass Conference of the Americas (cancelled after tragedy of September 11, 2001).
3. White, G.A.; Cook, C.G. Inventory of Agro-Mass. In *Paper and Composites from Agro-Based Resources*; Rowell, R.M., Young, R.A., Rowell, J.K., Eds.; Lewis Publishers: Boca Raton, 1997; 7–21.

[c]Rymsza, T. Personal communication, KP Products, Inc.; Vision Paper: Albuquerque, NM, 2002.

Parasitic Weeds

James H. Westwood
Virginia Tech, Blacksburg, Virginia, U.S.A.

INTRODUCTION

Parasitic plants acquire all or part of their nutrition through direct connections to a host plant. Parasitic weeds attack economically important crops and severely reduce crop yield and quality. The most destructive parasitic weeds are mistletoes, dodders, witchweeds, and broomrapes. Control of parasitic weeds is especially difficult because the parasites are closely associated with the host plant, even to the point of being largely concealed within the body of the host (e.g., mistletoes) or hidden underground for a significant part of their life cycle (e.g., broomrapes and witchweeds). The general biology and control of these organisms will be considered.

WHAT ARE PARASITIC WEEDS?

Some plant species have evolved the ability to obtain nutrients from other plants rather than rely on their own roots and photosynthetic systems (Fig. 1). Over 1% of plant species are parasitic, and parasitism has evolved multiple times, with 18 botanical families including parasitic members.[1] The multiple evolutionary origins of parasitism account in part for the large diversity among parasite species in growth form, host preference, and reproductive strategy, but they all tend to be highly specialized and may lack leaves, roots, and the ability to photosynthesize. An important characteristic in categorizing parasitic plants is the level of dependency on the host. Some are facultative parasites, and opportunistically parasitize neighboring plants while retaining an ability to live independently. Obligate parasites, in contrast, have an absolute requirement for a host to complete their life cycles. Obligate parasites are further divided into hemiparasites, which are capable of some photosynthesis, and holoparasites, which lack any photosynthetic capacity and must derive all nutrients from the host.

Parasitic plants that use economically important crops as hosts are among the most destructive weeds.[2] Unlike nonparasitic weeds, which merely compete with crops for resources such as light, water, and nutrients, parasitic weeds tap directly into the crop, removing resources and altering host physiology. Parasitic plants connect to their hosts through a specialized structure, called the haustorium, which penetrates the host and forms a physiological bridge between the two plants. The haustorium serves as a conduit for the removal of water, minerals, and photosynthates from the host, thereby draining it of resources it needs to grow and reproduce.[3] In addition to nutrient acquisition, parasite consumption of host water may cause the host to experience drought when environmental water levels are not limiting. The resulting water conservation efforts by the host only serve to suppress its own photosynthesis and arrest growth. Parasites may also disrupt hormone balance in the host, causing deformities and reallocation of resources away from shoots and fruits.

CLASSES OF IMPORTANT PARASITIC WEEDS

Relatively few species of parasitic plants cause most of the problems to agriculture. Three groups that account for the most destructive parasitic weeds are the mistletoes, dodders, and members of the Orobanchaceae, including broomrapes and witchweeds (Table 1). These parasites are quite different from each other, with distinct morphology and parasitic habits.

Mistletoes

The mistletoes are parasites of the shoots of trees and may grow invasively within the tissues of the host as well as externally. After host penetration, a parasite may grow within the host for as long as four years before emerging. Depending on the species of parasite, the shoots may have leaves that are broad (commonly termed leafy mistletoes) or scale-like (dwarf mistletoes) (Fig. 1A–C). Mistletoes are capable of photosynthesis and are generally green in color. Mistletoes may cause a problem in fruit trees and a wide range of deciduous trees, but the greatest economic impact is in coniferous forests parasitized by dwarf mistletoes. This is a major problem for the lumber industry, and several species of dwarf mistletoe decrease timber growth and quality, accelerate incidence of other diseases, and even kill trees. Mistletoes are found worldwide and the value in timber lost on an annual basis in North America alone is estimated to exceed several billion dollars.[4]

Encyclopedia of Plant and Crop Science
DOI: 10.1081/E-EPCS 120020241

Fig. 1 Examples of some weedy parasitic plant species. **A** *Loranthus europaeus* flowers and leaves; **B** *Arceuthobium tsugense* parasitizing *Pinus contorta* ssp. *contorta*; **C** *Phoradendron serotinum* parasitizing *Quercus coccinea*; **D** Flowering *Cuscuta campestris* growing on an ornamental shrub; **E** *Striga asiatica* parasitizing *Zea maize*; and **F** *Orobanche crenata* flowers emerging from roots of *Vicia faba*. (Photo credits: Gerhard Glatzel **A**, Dan Nickrent **B**, **C**, Yaakov Goldwasser **D**, Lytton Musselman **E**, and James Westwood **F**). (*View this art in color at www.dekker.com.*)

Dodders

Dodders are vining parasites that attack the stems and leaves of their hosts (Fig. 1D). These parasites consist of tendril-like stems that wrap around the host and form haustoria at points of contact, with each connection fueling further growth to find additional host shoots. In this way, dense mats of dodder may be formed that simultaneously parasitize many host plants. Although most species of dodder have functional photosynthetic systems, they are generally yellow in color and rely heavily on hosts for resources.[2] Dodders can be very destructive, greatly reducing yields of parasitized crops, and can also be mildly toxic to livestock if consumed with infested forage. Dodders occur throughout the world and affect a wide range of woody and herbaceous dicotyledonous plants.

Broomrapes and Witchweeds

Broomrapes and witchweeds both parasitize plant roots, but while broomrapes are holoparasites that lack photosynthesis and developed leaves, witchweeds have green leaves and the capacity for photosynthesis after emergence from the soil (Fig. 1E–F). The tiny seeds of these parasites lie dormant in the soil until stimulated to germinate by chemical signals exuded from host roots. The haustorium is produced from the parasite radicle as it nears contact with a host root. Once the parasite has established connections to the host, it can grow. Whereas broomrapes develop a bulbous structure adjacent to the host root, witchweed initiates a vegetative shoot that grows out of the soil. The impact of these parasites on their hosts can be dramatic, with estimates ranging from partial to complete yield loss, depending on environmental conditions and level of infestation.[2] Both species flower above-ground and may produce hundreds of thousands of seeds per plant. If measures are not taken to reduce the levels of infestations, seeds of these weeds can build up in the soil to the point that farmers must rotate to non-host crops or abandon the fields completely. These species present the greatest problems in the arid tropics, although their ranges are not restricted to these areas. Witchweeds are centered in Africa and India, while

Table 1 Examples of some of the most economically important parasitic weeds

Family and species	Common name	Host range (crops)
Convolvulaceae		
Cuscuta campestris	Field dodder	Alfalfa; many herbaceous dicot species
Cuscuta reflexa		Citrus, coffee, peach, litchi, and other woody perennials
Cuscuta planiflora	Small-seed dodder	Alfalfa, clover, and other legumes
Loranthaceae	Showy mistletoes	
Loranthus europaeus		Oak
Dendrophthoe falcata		Teak, mango, citrus, eucalyptus, apple, peach, guava, custard apple
Tapinanthus bangwensis		Cocoa, cola
Orobanchaceae		
Alectra vogelii	Alectra	Many, but especially legumes
Striga asiatica	Witchweed	Maize, sorghum, millet, rice, sugarcane
Striga hermonthica	Purple witchweed	Maize, sorghum, millet, rice, sugarcane
Striga gesnerioides	Cowpea witchweed	Cowpea and other legumes
Orobanche aegyptiaca	Egyptian broomrape	Many herbaceous dicots, e.g., from Solanaceae, Fabaceae, Cruciferae
Orobanche crenata	Crenate broomrape	Broad bean and related legumes, carrot, tomato, and related species
Orobanche cernua (syn *O. cumana*)	Nodding broomrape	Solanaceous species, sunflower
Orobanche minor	Clover broomrape	Clover and related legumes, Compositae species
Viscaceae	Leafy and dwarf mistletoes	
Viscum album	European mistletoe	Many deciduous and conifer trees, e.g., apple, poplar, pine, fir, larch
Phoradendron serotinum	American christmas mistletoe	Pear, pecan, walnut, citrus
Arceuthobium americanum	Lodgepole pine dwarf mistletoe	Lodgepole and jack pines, other pines
Arceuthobium tsugense	Hemlock dwarf mistletoe	Western hemlocks, fir, spruce, pine

(Information from Refs. 2, 4, 5, and 7.)

broomrapes occur commonly in regions around the Mediterranean, Middle East, and Eastern Europe.

CONTROL OF PARASITIC WEEDS

Parasitic weeds are extremely difficult to control. Their connection to the host plant makes physical removal nearly impossible. For example, removal of mistletoe shoots or mistletoe-infected branches is common, but does not prevent subsequent regrowth from parasite tissue embedded in the host. Dodder is similarly problematic since the stem is tightly wrapped around and connected to the host, and pulling does not remove all tissue. Witchweed and broomrape are both concealed beneath the ground during the crucial early stages of their life, and are thus hidden from sight and from mechanical control measures.

Chemical control strategies are also challenging because of the close association of host and parasite. With the exception of dodder seedlings, which germinate in the soil and are susceptible to certain pre-emergent herbicides, problems of herbicide delivery and selectivity prevent easy control of parasites.[5] While innovations in herbicide formulation and delivery enable some herbicides to be used effectively, the development of crops with resistance to translocated herbicides such as glyphosate holds promise for parasitic weed control.[6] However, until such crops are widely available, soil fumigation (e.g., methyl bromide) remains the most effective chemical control against root parasites. Soil injection of ethylene gas induces germination of witchweed seeds in the absence of a host (termed ''suicidal'' germination because the parasite cannot survive without a host present) and can be effective where economics justify the high expense.

Cultural practices can reduce parasite populations. Rotations including nonhost crops disrupt the cycle of parasitism. Also, altering planting dates, improving soil fertility, and intercropping have been used to reduce witchweed and broomrape damage. For mistletoes, the selective removal of infected trees and maintenance of distance between trees can reduce spread. Breeding for parasite-resistant crop varieties has received considerable attention and is an ideal solution to the problem, although currently few durably resistant varieties are available for most affected crops. Considering the difficulty in controlling parasitic weeds, preventive practices are important in stopping the spread of parasite seeds to new areas and limiting reproduction where parasites populations are low.

FUTURE DIRECTIONS

Important areas of future research for parasitic weed control include mechanisms of host detection by the parasite, parasite-environment interactions, biological control, and development of resistant host crops. Ironically, it is possible that the close relationship between host and parasite that makes them so difficult to control may lead to new ways to protect crops. This has already been illustrated by the chemical control strategy mentioned above, in which lethal doses of herbicide are delivered to attached parasites via translocation through crops that have been engineered to resist the herbicide.[6] A more direct strategy, requiring no chemical input, is to engineer host crops that are able to prevent parasite attachment and growth at the earliest stages of infection. This may harness the host's own pathogen defense machinery (e.g., phytoalexins or hypersensitive response) or may use entirely novel mechanisms to disrupt parasite growth. Greater understanding of parasitic weeds, including their physiology and host interactions, will facilitate the development of new parasite-resistant crops.

ACKNOWLEDGMENTS

Dan Nickrent and the Parasitic Plant Connection[7] were very helpful in finding photographs of parasitic weeds. Thanks also to Brenda Winkel for critical reading of the manuscript. This chapter is possible because of the support from USDA NRI award 2001-35320-10900 and USDA Hatch Project no. 135657.

ARTICLES OF FURTHER INTEREST

Breeding for Durable Resistance, p. 179
Chemical Weed Control, p. 255
Genetic Diversity Among Weeds, p. 496
Seed Banks and Seed Dormancy Among Weeds, p. 1123
Weed Management in Less Developed Countries, p. 1295

REFERENCES

1. Nickrent, D.L.; Duff, R.J.; Colewell, A.E.; Wolfe, A.D.; Young, N.D.; Steiner, K.E.; de Pamphilis, C.W. Molecular Phylogenetic and Evolutionary Studies of Parasitic Plants. In *Molecular Systematics of Plants II. DNA Sequencing*; Soltis, D.E., Soltis, P.S., Doyle, J.J., Eds.; Kluwer Academic Publishers: Boston, 1998; 211–241.
2. Parker, C.; Riches, C.R. *Parasitic Weeds of the World: Biology and Control*; CAB International: United Kingdom, 1993; 1–235.
3. Graves, J.D. Host-Plant Responses to Parasitism. In *Parasitic Plants*; Press, M.C., Graves, J.D., Eds.; Chapman & Hall: London, 1995; 206–225.
4. Hawksworth, F.G.; Wiens, D. *Dwarf Mistletoes: Biology, Pathology, and Systematics*, U.S. Department of Agriculture Agricultural Handbook 709; Forest Service: Washington, DC, 1996; 123–244. http://www.rms.nau.edu/publications/ah_709/ (accessed April 2003).
5. Foy, C.L.; Jain, R.; Jacobsohn, R. Recent approaches for chemical control of broomrape (*Orobanche* spp.). Rev. Weed Sci. **1989**, *4*, 123–152.
6. Joel, D.M.; Kleifeld, Y.; Losner-Goshen, D.; Gressel, J. Transgenic crops against parasites. Nature **1995**, *374*, 220–221. (16 March).
7. Nickrent, D.L. *Parasitic Plant Connection*; Southern Ilinois University: Carbondale, IL. http://www.science.siu.edu/parasitic-plants/index.html (accessed April 2003).

Phenylpropanoids

Kevin M. Davies
New Zealand Institute for Crop and Food Research Limited, Palmerston North, New Zealand

INTRODUCTION

Plant phenylpropanoids are a major group of secondary metabolites derived from phenylalanine. They have a wide range of functions in plants, both as structural compounds and in interactions with the environment and other organisms. The importance of phenylpropanoids in general biology can be illustrated by a few observations on just one of the phenylpropanoid subgroups, the lignins. Lignins are the second most abundant polymer in plants, after cellulose, and are required for xylem formation and strengthening of cell walls. The evolution of lignin biosynthesis is thought to have been a key step in the colonization of land by plants. Today it is estimated that phenylpropanoids account for over 30% of all organic carbon in the biosphere, and that the majority of this is lignin. Furthermore, the degradation of lignin is a major rate-limiting step in the carbon cycle.

BIOSYNTHESIS AND FUNCTION

The proposed biosynthetic pathway of the major phenylpropanoid groups is represented in Figs. 1 and 2. The initial compounds, hydroxycinnamic acids (HCAs), are formed from phenylalanine by phenylalanine ammonia lyase (PAL), cinnamate 4-hydroxylase (C4H), and 4-coumarate:CoA ligase (4CL). All the other phenylpropanoid types arise from the HCAs. A core section of the phenylpropanoid biosynthetic pathway is common to most plants, including ferns, in particular to the formation of HCAs, flavonoids, lignins, and proanthocyanins (tannins). However, based on these core compounds a huge diversity in structure occurs. For example, about 5000 flavonoid types can arise through glycosylation, hydroxylation and substitution with methyl, sulphur, and aliphatic or aromatic acyl groups. While the functions of the major phenylpropanoid groups are known, the roles of the many variant structures is often unclear. Several excellent in-depth reviews of phenylpropanoid biosynthesis and diversity are available.[1–3]

Lignins and Lignans

Lignins and lignans are polymeric molecules formed by different biosynthetic routes from the same monolignol precursors (Fig. 1). Lignins are high-molecular-weight polymers found in cell walls. Lignans are commonly dimers, although they can form higher oligomers, and are found in abundance in wood resins, in which they may play a plant defence role. Monolignol biosynthesis from the HCAs involves a network of aromatic hydroxylations, NAPH-dependent reductions, CoA ligations, and O-methylations. For lignin biosynthesis, the monolignols are transported from the cytosol to the cell wall, where polymerisation occurs, probably on a proline-rich protein template and catalyzed by peroxidases and laccases. The structures formed can vary much between species, and even within a plant, and the mechanisms that control the process are unclear. However, a dirigent protein was recently characterized that can determine the type of stereospecific coupling linkages that occur between the monolignol precursors.

Flavonoids

The first committed step of flavonoid biosynthesis occurs with the formation of chalcones by chalcone synthase (CHS), when three units of acetate-derived malonyl-CoA combine with the phenylalanine-derived 4-coumaroyl-CoA (Fig. 2). The subsequent steps of flavonoid biosynthesis, resulting in the formation of the anthocyanins, proanthocyanins, flavones, and flavonols, are common to nearly all flowering plants. The notable exception are most species of the Caryophyllales, in which the colored anthocyanins are substituted with betalain pigments. Along with chlorophylls and carotenoids, anthocyanins are major pigments of plants, capable of providing color to all plant tissues. There are three common anthocyanin groups, pelargonodin-derivatives, cyanidin-derivatives and delphinidin-derivates, which usually result in pink-red, red-purple, and mauve-blue colors, respectively. They are typically glycosylated, often further substituted, and localised to the vacuole. However, rare and highly modified anthocyanin forms occur in nature, with over 600 individual anthocyanins reported to date.

Many biosynthetic branches occur in the flavonoid pathway. Of particular note are the action of chalcone reductase (CHR) with CHS forming 6′-deoxychalcones, the aureusidin synthase forming aurones, flavone synthase (FNS) forming flavones, flavanone reductase (FNR)

Encyclopedia of Plant and Crop Science
DOI: 10.1081/E-EPCS 120005948

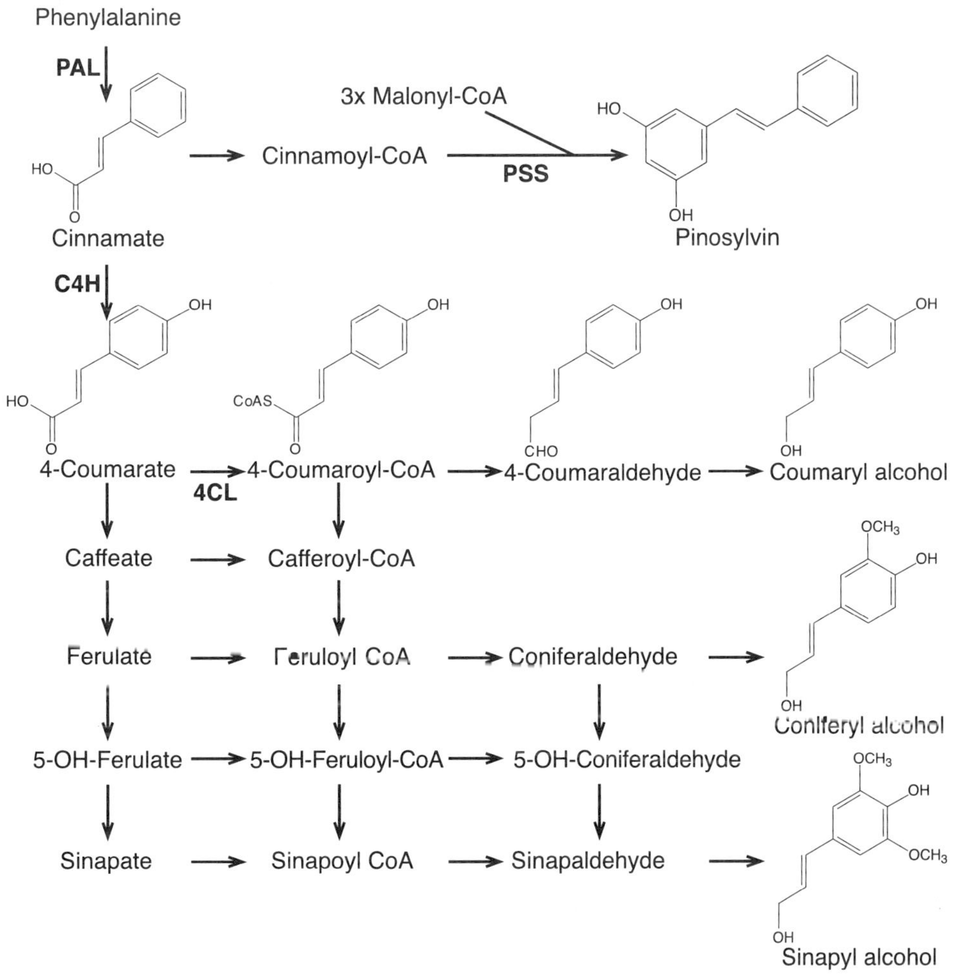

Fig. 1 Diagrammatic representation of a section of the phenylpropanoid pathway leading to hydroxycinnamic acids, monolignols, and pinosylvin. Enzyme abbreviations are as given in the text, except PSS (pinosylvin synthase, a STS type enzyme). The sequence of some steps in monolignol biosynthesis is under debate, and no attempt has been made to represent the alternative pathways.

forming 3-deoxyanthocyanin precursors, flavonol synthase (FLS) forming flavonols, and isoflavone synthase (IFS) forming isoflavone precursors. Many of these enzymes and compounds are taxonomically restricted in their occurrence.

In addition to pigmentation, the functions of flavonoids are diverse. They can protect leaves from oxidative stress and ameliorate the impact of UV-B radiation. They are also key plant defence compounds. The isoflavonoids and flavone C-glycosides are involved in defence against fungal and insect attack, and proanthocyanidins can be herbivore deterrents. The flavonols are essential for fertility in some species, acting by an unknown mechanism during pollen growth in the stigma. Flavonoids are also key signal compounds, most notably to the soil bacterium *Rhizobium*, to trigger nodulation for symbiotic nitrogen fixation. The formation of heartwood in trees is marked by the accumulation of phenylpropanoid rich resins, which are commonly a mix of flavonoids, lignans, and stilbenes. It will be obvious from this list that flavonoids are a key branch of plant secondary metabolism. However, these are probably only some of the functions for these compounds. The action of flavonols in plant fertility came to light only in the early 1990s and, given the great diversity of structural types, many flavonoid functions may yet be determined. For example, there is growing evidence that flavonoids influence auxin transport.

Coumarins, Stilbenes, and Related Compounds

In addition to lignins and flavonoids, a range of other compounds are derived from HCAs. Coumarins are a varied group of compounds, widespread in nature, whose primary function appears to be in plant defense. A well

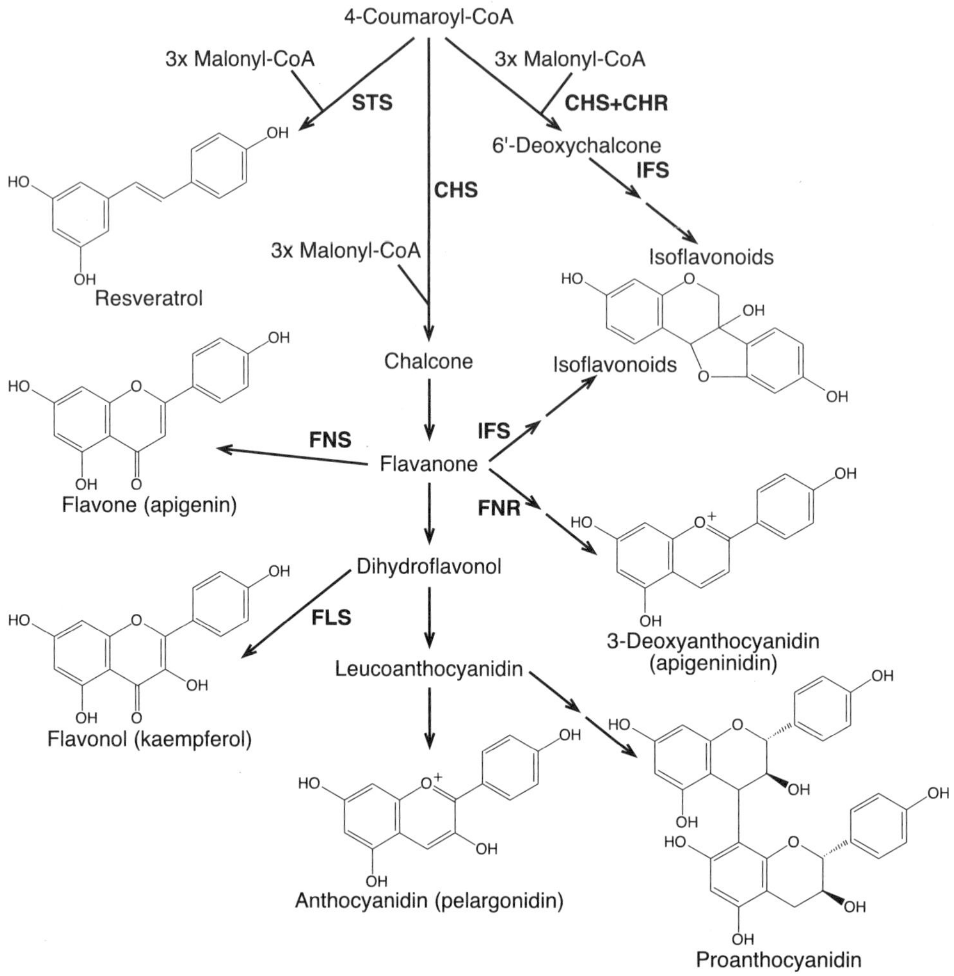

Fig. 2 Diagrammatic representation of the flavonoid and stilbene sections of the phenylpropanoid pathway. Simplifications include showing only 4′-hydroxylated flavonoid types, no indication of the subsequent glycosylation or further modification of compounds, double arrows representing multiple enzyme steps, the absence of some rarer flavonoid types (e.g., aurones), and showing only one isoflavonoid (a pterocarpan phytoalexin). Enzyme abbreviations are as given in the text.

known example of their impact is the photosensitive skin blistering that can result from handling celery (*Apium graveolens*) or giant hogweed (*Heracleum mantegazzianum*). The coumarin biosynthetic pathway(s) have not been elucidated, although it is thought that the initial reactions are *cis/trans*-isomerisation and ring closure of a HCA.

The condensation of malonyl-CoA and HCAs can form compounds other than the flavonoids. A group of CHS-like enzymes use malonyl-CoA and a HCA-CoA to form a range of plant defense compounds.[4] The best characterized are the stilbene synthases (STS) that form compounds such as resveratrol and pinosylvin. Like CHS, they carry out three condensation reactions followed by ring closure, but the nature of the ring closure differs. To date, over 300 different stilbenoid structures have been reported. Malonyl-CoA and HCA-CoA compounds are also the precursors for the arylpyrones and styrylpyrones. These reactions involve only one or two condensations prior to ring closure, but little is known of the biosynthetic enzymes.

The Molecular Biology of Phenylpropanoid Biosynthesis

The colored flavonoids have provided model systems for scientists since the time of Mendel's inheritance studies. More recently, studies on flavonoids have been responsible for some of the most important breakthroughs in plant science, including the cloning of the first plant transcription factor gene, the isolation of one of the first cDNAs for a plant cytochrome P450 enzyme, the first demonstration of antisense RNA technology in a transgenic plant, and the first characterization of a transgene cosuppression mechanism. Studies with the major model systems, *Antirrhinum majus*, *Arabidopsis thaliana*, *Petunia*, and *Zea mays*, have resulted in the isolation of cDNAs or genes for most of the core biosynthetic enzymes of phenylpropanoid biosynthesis, and for several transcription factors involved in the regulation of the pathway.[2,5] Transposon-generated mutations have been key to the successful elucidation of the molecular genetics of phenylpropanoid biosynthesis. A more recent extension of the use of mutants is the linking of metabolic profiling to the genomics systems available for species such as *Arabidopsis* and *Medicago truncatula*.[6] Research is now focusing on aspects such as the regulation of the biosynthetic genes, the subcellular interactions of the biosynthetic enzymes, the biosynthesis of rare phenylpropanoid types, and the control of lignin formation.

The Importance of Phenylpropanoids to Agriculture and Human Health

Phenylpropanoids affect the quality and quantity of the harvested products from most plant crops. Obvious examples include the impacts of lignin and heartwood resins on timber and pulping wood quality, lignin on forage crop digestibility, tannin production for the control of bloat in ruminants, flower and foliage color in ornamental crops, tannin production in wine, and flavor and fragrance compounds in spice crops and teas. Phenylpropanoids are also key plant defense compounds. However, perhaps the area of keenest research interest at present is the influence of phenylpropanoids on human health. Numerous in vitro experiments, animal trials, and epidemiological studies have linked the intake of specific phenylpropanoids with a reduced risk of conditions such as cancer and heart disease. The mechanisms behind such actions are generally unknown, although the antioxidant activity of many phenylpropanoids has often been implicated.

Given the importance of phenylpropanoids in agriculture it is not surprising that the pathway was one of the early targets for genetic modification approaches. Metabolic engineering in transgenic plants has been used to alter the production of nearly all of the different groups of phenylpropanoid compounds.[7,8] Transgenic carnations with modified flower color through the alteration of anthocyanin biosynthesis are on sale in a number of countries, and it is anticipated that these will be joined by some of the transgenic plant lines currently in field trials, such as trees with improved lignin characteristics, forage crops with improved digestibility, and food crops with high levels of health-promoting phenylpropanoids.

ARTICLES OF FURTHER INTEREST

Metabolism, Secondary: Engineering Pathways of, p. 720
Secondary Metabolites As Phytomedicines, p. 1120

REFERENCES

1. Bohm, B.A. *Introduction to Flavonoids*; Harwood Academic Publishers: Amsterdam, 1998.
2. *The Flavonoids: Advances in Research Since 1986*; Harborne, J.B., Ed.; Chapman and Hall: London, 1994.
3. Petersen, M.; Strack, D., Matern, U. Biosynthesis of Phenylpropanoids and Related Compounds. In *Biochemistry of Plant Secondary Metabolism*; Wink, M., Ed.; Sheffield Academic Press: Sheffield, 1999.
4. Schröder, J. The Family of Chalcone Synthase-Related Proteins: Functional Diversity and Evolution. In *Evolution of Metabolic Pathways*; Romeo, J.T., Ibrahim, R., Varin, L., De Luca, V., Eds.; Elsevier Science Ltd.: Oxford, 2000; 55–90.
5. Martin, C.; Jin, H.; Schwinn, K. Mechanisms and Applications of Transcriptional Control of Phenylpropanoid Metabolism. In *Regulation of Phytochemicals by Molecular Techniques*; Romeo, J.T., Saunders, J.A., Matthews, B.F., Eds.; Elsevier Science Ltd.: Oxford, 2001; 155–170.
6. Dixon, R.A.; Steele, C.L. Flavonoids and isoflavonoids—A gold mine for metabolic engineering. Trends Plant Sci. **1999**, *4*, 394–400.
7. Davies, K.M. Plant Colour and Fragrance. In *Metabolic Engineering of Plant Secondary Metabolism*; Verpoorte, R., Alfermann, W., Eds.; Kluwer Academic Publishers: Dordrecht, 2000; 127–164.
8. Boudet, A.M.; Chabannes, M. Gains achieved by molecular approaches in the area of lignification. Pure Appl. Chem. **2001**, *73*, 561–566.

Phosphorus

Carroll P. Vance
USDA/ARS Plant Science Research Unit and University of Minnesota, St. Paul, Minnesota, U.S.A.

INTRODUCTION

Phosphorus (P) is one of 17 essential elements (nutrients) required for plant growth. The P concentration in plants ranges from 0.05 to 0.50% dry weight. It plays a role in a wide array of processes, including energy generation, nucleic acid synthesis, photosynthesis, glycolysis, respiration, membrane synthesis and stability, enzyme activation/inactivation, redox reactions, signaling, carbohydrate metabolism, and nitrogen (N) fixation. The concentration gradient from the soil solution orthophosphate (P_i) to plant cells exceeds 2000-fold, with an average free P_i of 1 μM in the soil solution. This concentration is well below the K_m for plant uptake. Thus, although bound P is quite abundant in many soils, it is largely unavailable for uptake. As such, P is frequently the most limiting element for plant growth and development. Crop yield on 30 to 40% of the world's arable land is limited by P_i availability. Phosphorus is unavailable because it rapidly forms insoluble complexes with cations, particularly aluminum and iron under acid conditions. The acid-weathered soils of the tropics and subtropics are particularly prone to P_i deficiency. Moreover, some 30 to 70% of the P can be bound in insoluble organic forms (such as phytate) due to microbial activity. Application of P-containing fertilizers is the recommended treatment for enhancing soil P_i availability and stimulating crop yields.

THE PHOSPHORUS CONUNDRUM

Application of P fertilizer, however, is problematic for both the intensive and extensive agriculture of the developed and developing worlds, respectively. In intensive agriculture, a grain crop yield of 7 metric tons $\cdot$ Ha^{-1} requires the addition of 90 to 120 kg P $\cdot$ Ha^{-1}.[1–3] But even under adequate P fertilization, only 20% or less of that applied is removed in the first year's growth. This results in P loading of prime agricultural land. Runoff from P-loaded soils is a primary factor in eutrophication and hypoxia of lakes and marine estuaries of the developed world.[1,4,5] Another reason for alarm is that by some estimates, inexpensive rock phosphate reserves could be depleted in as little as 60 to 80 years.[3,6] Phosphorus fertilizer use increased 4- to 5-fold between 1960 and 2000 and is projected to increase further by 20 Tg $\cdot$ $year^{-1}$ by 2030 (Tables 1–3). Several authors have noted that a potential phosphate crisis looms for agriculture in the 21st century.[1,4,6,7] A concern even greater than the overabundant use of P fertilizers by intensive agriculture is the lack of available P fertilizers for extensive agriculture in the tropics and subtropics where the majority of the world's people live. Lack of fertilizer infrastructure, money for purchase, and transportation make P fertilization unattainable for these areas. Sustainable management of P in agriculture requires that plant biologists discover mechanisms that enhance P_i acquisition and exploit these adaptations to make plants more efficient at acquiring P_i, develop P-efficient germplasm, and advance crop management schemes that increase soil P_i availability.

ADAPTATIONS TO LOW P

Because of its ubiquitous importance, plants have evolved two broad strategies for P acquisition in nutrient-limiting environments: 1) those aimed at conservation of use; and 2) those directed toward enhanced acquisition or uptake.[1,8] Processes that conserve the use of P involve decreased growth rate, increased growth per unit of P uptake, remobilization of internal P_i, modifications in carbon metabolism that bypass P-requiring steps, and alternative respiratory pathways.[1,8] By comparison, processes that lead to enhanced uptake include increased production and secretion of phosphatases, exudation of organic acids, greater root growth along with modified root architecture, expanded root surface area by prolific development of root hairs, and enhanced expression of P_i transporters and aquaporins.[1,8–11]

By far the most prevalent evolutionary adaptation by land plants for acquiring P_i (80% of all species) is through mycorrhizal symbioses.

PROTEOID ROOTS OF WHITE LUPIN

Significance

Proteoid roots (also known as cluster roots) are third order lateral roots that resemble bottle brushes in appearance

Encyclopedia of Plant and Crop Science
DOI: 10.1081/E-EPCS 120010511

Table 1 Phosphorus content of plants

Species	Content (% Dry weight)
Maize	0.20
Sunflower	0.28
Lupin	0.51
Barley	0.15
Alfalfa	0.47

(Fig. 1). They develop in response to low P in the Proteaceae, Fabaceae, Casuarinaceae, Betulaceae, Myricaceae, Eleagnaceae, and Moraceae.[11] Along with mycorrhizal association and nitrogen-fixing root nodules, proteoid roots are the third major adaptation by plants for acquisition of scarce nutrients. The well-characterized legume white lupin is an illuminating model for understanding proteoid root formation and plant adaptations to low P habitats.[1,9–11]

White lupin can effectively acquire P_i even though it does not form a mycorrhizal symbiosis.[1,10,11] Instead, its adaptation to P stress is a highly coordinated modification of root development and plant metabolism resulting in proteoid roots that exude copious amounts of organic acids and acid phosphatase.[1,9,10] Proteoid root formation is accompanied by extensive root hair formation that increases root surface area by greater than 100-fold. In addition, P_i uptake within proteoid root zones is much greater than that of normal roots. The developmental and metabolic changes that occur to give rise to proteoid roots are mediated in part by enhanced accumulation of transcripts encoding P_i transporters, exuded acid phosphatase, and enzymes of carbon metabolism.[1,9,10]

Development

Several features distinguish proteoid root development and morphology from that of typical dicot lateral roots. First, lateral roots are randomly initiated from the pericycle of primary roots near the zone of metaxylem differentiation,[1,11] whereas proteoid roots are initiated in waves along secondary roots (Fig. 1). Second, lateral roots are initiated singularly opposite a protoxylem point, in contrast to proteoid roots that are in multiples opposite every protoxylem point within the wave of differentiation. Third, in typical lateral roots, root hair development is highly regulated and occurs from a discrete number of epidermal cells. By comparison, in proteoid roots self-regulation of root hairs has gone awry with superabundant formation. Last, contrasting with the indeterminate growth of lateral roots, proteoid root growth is determinate, ceasing shortly after emergence (Fig. 1). This highly synchronous developmental pattern indicates that proteoid root formation is a finely tuned process. Moreover, because root pericycle cells are arrested in the G2 phase of the cell cycle,[1,11] proteoid root initiation must involve concerted release of multiple pericycle cells from the G2 phase in a wavelike pattern along second order lateral roots.

Table 2 Phosphorus fertilizer application increases crop yield

	Crop yield ($Mt \cdot ha^{-1}$)		
P Application ($Kg \cdot ha^{-1}$)	Wheat	Corn	Alfalfa
0	3.0	4.6	7.4
23	4.4	6.5	8.3
45	4.8	7.5	8.7
90	5.1	9.3	9.2
135	5.5	9.4	11.1

(From Refs. [2] and [3].)

Table 3 Agriculture production and resource use

Item	1960	2000	2030–2040
Food production (Mt)[a]	1.8×10^9	3.5×10^9	5.5×10^9
Population (Billions)	3	6	8 (maybe 10)
Irrigated land (% of arable)	10	18	20
Cultivated land (Hectares)	1.3×10^9	1.5×10^9	1.8×10^9
Water-stressed countries	20	28	52
N fertilizer use (Tg)[b]	10	88	120
P fertilizer use (Tg)	9	40	55–60

(Data from FAO; Refs. [1] and [7].)
[a]Mt = metric tons.
[b]Tg = 10^{12} g or million metric tons.

As might be expected, the internal balance and distribution of the plant growth hormones auxin, ethylene, and cytokinin are thought to play a role in proteoid root formation. Substantial support for the role of auxin in lateral root development is derived from evidence that: 1) exogenous auxin stimulates lateral root formation in most species; 2) auxin transport inhibitors block lateral root formation and this block can be alleviated by exogenous application of auxin; and 3) *Arabidopsis* mutants that overproduce auxin have enhanced lateral root formation, whereas mutants insensitive to auxin have impaired lateral root development. Gilbert et al.[1] have shown that exogenous application of auxin to P-sufficient white lupin mimics proteoid root formation seen under P-deficient conditions. They also showed that auxin transport inhibitors block the formation of proteoid roots on P-deficient

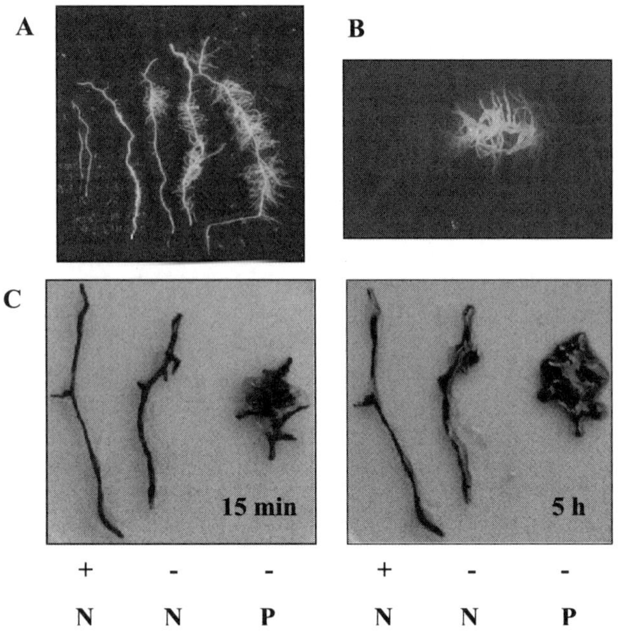

Fig. 1 Proteoid root formation in phosphorus (P)-deficient white lupin and adaptive strategies for acquisition of P. (A) Development of proteoid roots in response to P deficiency. Roots (left to right) are taken from white lupin at 5, 7, 9, 10, 14, and 22 days after initiation of P-stress, respectively. Proteoid roots begin to emerge by day 9 and become more extensive as P-stress progresses. (B) Proteoid roots from P-deficient white lupin acidify the rhizosphere. Proteoid roots from 14-day P-stressed plants are placed on agar plates containing pH indicator. Zone of acidification (yellow color) occurs in the rhizosphere where proteoid roots are in contact with agar. (C) Secretion of acid phosphatase activity from proteoid roots of 14-day P-deficient plants. Normal (N) and proteoid (P) roots from P-sufficient (+) and P-deficient (−) plants are placed in contact with agar media containing acid phosphatase reagent. Photos are taken 15 minutes and 5 hours after roots have been placed in contact with reagents. Red color indicates acid phosphatase activity. Secretion of acid phosphatase from proteoid roots of P-deficient plants is quite evident at 5 hours, whereas normal roots from either P-sufficient or P-deficient plants show little to no secretion of acid phosphatase. (*View this art in color at www.dekker.com.*)

plants. Although the role of ethylene in lateral root formation is less clear, there is convincing evidence that ethylene plays a role in root hair formation and abundance. Consistent with the abundant root hairs found on P-deficient proteoid roots, Gilbert et al.[1] found that ethylene production was increased twofold during the initiation and emergence of proteoid root initials of P-deficient plants compared to P-sufficient plants. Exogenous application of kinetin also inhibits proteoid root formation in P-stressed white lupin. Although we have not measured cytokinin content of P-stressed white lupin, it is revealing that an abundantly expressed EST in white lupin responsive to P-stress is cytokinin oxidase.[9]

Root Exudation of Organic Acids and Acid Phosphatase

Although there is a large reservoir of inorganic and organic P in soil, most of it is unavailable because it is complexed to metals and/or organic residues.[1,10,12] Exudation of organic acids and acid phosphatases from roots is a hallmark adaptation of plants growing in soils with low available P_i.[10] The release of organic acids from roots allows for the displacement of P from Al^{3+}, Fe^{3+}, and Ca^{2+}-phosphates, thus freeing bound P.[1,10] Organically bound P, which can compose as much as 70% of the soil P pool, can be hydrolyzed by acid phosphatases, thus freeing a large pool of unavailable P_i. These root adaptations to insufficient P_i are strikingly displayed by proteoid roots of P-deficient white lupin.

Proteoid roots of P-deficient white lupin synthesize and exude copious amounts of malate and citrate (Fig. 1). In the rhizosphere of white lupin grown in calcareous soil, the citrate concentration can be as high as 50 μmol × g soil^{-1}. Between 10 and 25% of the total carbon fixed during P-stress can be released as malate or citrate. Radiolabeling studies show that nonautotrophic CO_2 fixation in proteoid roots can provide up to 30% of the carbon exuded in citrate and malate. The increased synthesis of organic acids in P-deficient proteoid roots is mediated by highly enhanced activity of malate dehydrogenase (MDH), phosphoenolpyruvate carboxylase (PEPC), and citrate synthase (CS) with an accompanying decrease in aconitase (AC) and respiration rates. Recent molecular studies show that increased PEPC and MDH activities are due in part to enhanced expression of PEPC and MDH genes.

Accompanying the acidification of the rhizosphere of P-deficient white lupin proteoid roots is the excretion of large amounts of acid phosphatase (Fig. 1). Although the release of acid phosphatase and phytase is a common response of plant roots to P_i-stress, only in white lupin proteoid roots have the biochemical and molecular events involved yielded to experimentation. Under P_i-stress, proteoid roots synthesize and secrete a novel isoform of acid phosphatase (SAPase).[1] Characterization of SAPase protein and cDNA from white lupin exudates revealed that the enzyme was synthesized as a 52 kD protein having a 31 amino acid presequence. The presequence targets the protein to outside the cell. Upon exudation, cleavage of the presequence results in a 49 kD processed protein. The formation of a novel SAPase results from increased *SAPase* mRNA and protein synthesis in proteoid roots (Fig. 2). The promoter region of the *SAPase* gene has

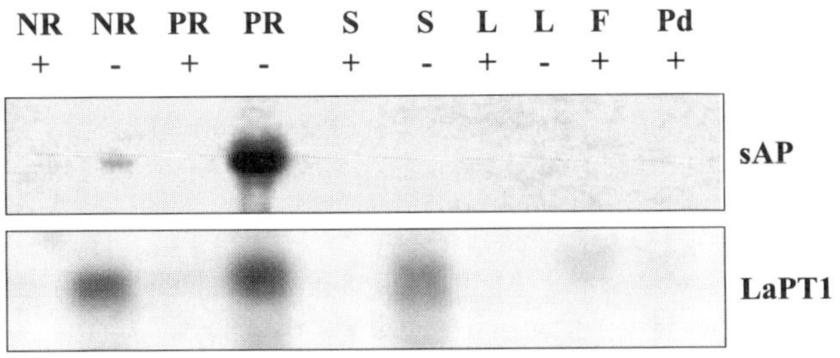

Fig. 2 Expression of mRNA transcripts for secreted acid phosphatase (sAP) and phosphate transporter (LaPT1). The greatest expression of both genes occurred in proteoid roots (PR) of P-deficient (−) plants. There is very little expression of sAP in other tissue from either P-sufficient (+) or P-deficient (−) plants. By comparison, LaPT shows enhanced expression in P-deficient (−) normal roots (N) and stems (S), with little to no expression in P-sufficient (+) plants. Key: NR, normal roots; PR, proteoid roots; S, stems; L, leaves; F, flowers; Pd, pods; +, P-sufficient; −, P-deficient.

cis-acting element(s) that control P-stress-induced gene expression. Similar biochemical and molecular events appear to control P-starvation-induced acid phosphatases in suspension cell cultures of *Brassica nigra*.

Phosphorus Uptake

Inorganic orthophosphate (such as $H_2PO_4^-$ and HPO_4^{2-}) is the form of P_i most generally taken up by plants.[13,14] After diffusion through the soil to the root surface, P_i is rapidly taken up, resulting in a P_i depletion shell of 0.2 to 1.0 mm around the root. Although the soil solution Pi concentration rarely exceeds 2 μM, that in plant cells is much higher, 2 to 20 mM.[12,14] For the plant to surmount this concentration difference, active (energized) transport across the plasmalemma is required. The striking reduction in P_i accumulation of tissues treated with inhibitors reflects this energy requirement for uptake.[12,13] Moreover, kinetic analysis of P_i uptake shows that plants have both a low- and high-affinity uptake system.[10,12,14] The high-affinity system operating at low P_i concentrations has an apparent K_m ranging from 3 to 10 μM. By comparison, the low-affinity system operating at high P_i concentrations has a K_m ranging from 50 to 300 μM. Recently, functional complementation of yeast mutants defective in P_i transport has been used to isolate and characterize P_i transporters from a diverse array of plants.[13,14] Molecular characterization of P_i transporters coupled to the discovery of as many as 16 within the genome of *Arabidopsis* confirms that plants have a multiplicity of P_i transporters functional in specific organs and tissues.

Uptake of P_i by P-deficient proteoid roots is much greater than that of P-sufficient plants.[10] Moreover, the apparent K_m for P_i uptake of proteoid roots is 8.6 μM, compared to a K_m of 30.7 μM for P-sufficient controls. These results suggest that a high-affinity P_i uptake system is induced in proteoid roots of P-deficient plants. We have recently characterized a high-affinity type P_i uptake gene (LaPT1) from proteoid roots of white lupin that shows highly intensified expression in P-deficient plants.[11]

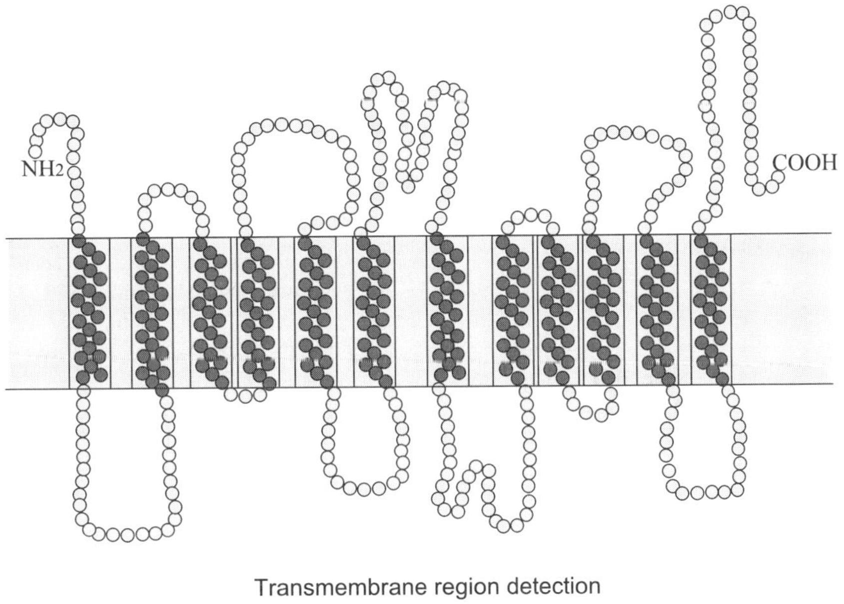

Fig. 3 Deduced structure of white lupin phosphate transporter 1 (*LaPT1*). Note 12 membrane-spanning domains. Both the NH_2 and COOH termini are predicted to be on the outside of the cell. Internal to cell indicated by in; external of cell indicated by out. (*View this art in color at www.dekker.com.*)

Similar to other plant P_i transporters, the white lupin proteoid root P_i transporter has a 1620 base pairs (bp) open reading frame that encodes a protein of 540 amino acids with a Mr of 59 kD. The deduced protein contains 12 transmembrane domains that occurs as two groups (6+6) connected by a hydrophobic domain of 60 amino acid residues (Fig. 3). The deduced amino acid sequence of *LaPT1* is 85% similar to previously reported high-affinity type P_i transporters, but only 75% similar to low-affinity type P_i transporters. Transcripts of *LaPT1* are highly expressed in P_i-deficient proteoid roots, normal roots, and stems with little to no expression in P-sufficient plants (Fig. 2). Thus, the enhanced uptake of P_i displayed by P-deficient proteoid roots can be directly related to increased expression of a high-affinity type P_i transporter.

CONCLUSION

Phosphorus has and will continue to play a major role in crop productivity because of its role in all phases of plant growth. However, the adverse impact that excess P has on water quality, coupled with depletion of cheap sources and lack of availability in the developing world, necessitate that plant scientists unravel the mechanisms leading to improved P acquisition. Proteoid roots of white lupin offer an excellent model for understanding P-acquisition strategies that both improve acquisition and conserve use of the critical nutrient. By discovering how plants control adaptation to low-P environments through changes in metabolism, development, and gene expression, we can begin to identify targets to improve P acquisition through conventional and biotechnological methods.

ARTICLE OF FURTHER INTEREST

Mycorrhizal Symbioses, p. 770

REFERENCES

1. Vance, C.P.; Uhde-Stone, C.; Allan, D.L. Phosphorus acquisition and use: Critical adaptation by plants for acquiring a nonrenewable resource. New Phytol. **2003**, *157*, 423–447.
2. Phosphorus for Agriculture. In *Better Crops with Plant Food*; Armstrong, D.L., Ed.; Potash/Phosphate Institute: Norcross, GA, 1999.
3. Ellington, C.P. Crop Yield Responses to Phosphorus. In *Phosphorus for Agriculture: A Situation Analysis*; Ellington, C.P., Ed.; Potash/Phosphate Institute: Norcross, GA, 1999; 25–41.
4. Runge-Metzger, A. Closing the Cycle: Obstacles to Efficient P Management for Improved Global Security. In *Phosphorus in the Global Environment*; Tiessen, H., Ed.; John Wiley and Sons Ltd.: Chichester, UK, 1995; 27–42.
5. Bumb, B.L.; Baanante, C.A. *The Role of Fertilizer in Sustaining Food Security and Protecting The Environment*, Food, Agriculture and the Environment Discussion Paper 17; International Food Policy Research Institute: Washington, DC, 1996.
6. Council for Agricultural Science and Technology (CAST) *Long Term Viability of U.S. Agriculture*, Report No. 114; CAST: Ames, IA, 1988.
7. Pinstrup-Anderson, P.; Pandy-Lorch, R.; Rosegrant, M.W. *The World Food Situation: Recent Developments, Emerging Issues and Long-Term Prospects*, Vision 2020: Food Policy Report; International Food Policy Research Institute: Washington, DC, 1997.
8. Plaxton, W.C.; Carswell, M.C. Metabolic Aspects of the Phosphate Starvation Response in Plants. In *Plant Responses to Environmental Stress: From Phytohormones to Genome Reorganization*; Lerner, H.R., Ed.; Marcel-Dekker: New York, 1999; 349–372.
9. Uhde-Stone, C.; Zinn, K.E.; Ramirez-Yanez, M.; Li, A.; Vance, C.P.; Allan, D.L. Nylon filter arrays reveal differential gene expression in proteoid roots of white lupin in response to P deficiency. Plant Physiol. **2003**, *131*, 1064–1079.
10. Neumann, G.; Martinoia, E. Cluster roots—An underground adaptation for survival in extreme environments. Trends Plant Sci. **2002**, *7*, 162–167.
11. Skene, K.R. Pattern formation in cluster roots: Some developmental and evolutionary considerations. Ann. Bot. **2000**, *85*, 901–908.
12. Bieleski, R.L. Phosphate pools, phosphate transport, and phosphate availability. Annu. Rev. Plant Physiol. **1973**, *24*, 225–252.
13. Ragothama, K.G. Phosphate acquisition. Annu. Rev. Plant Physiol. Plant Mol. Biol. **1999**, *50*, 665–693.
14. Schactman, D.P.; Reid, R.J.; Ayling, S.M. Phosphorus uptake by plants: From soil to cell. Plant Physiol. **1998**, *116*, 447–453.
15. von Uexküll, H.R.; Mutert, E. Global extent, development and economic impact of acid soils. Plant Soil **1995**, *171*, 1–15.

Photoperiodism and the Regulation of Flowering

P

Isabelle A. Carré
University of Warwick, Coventry, U.K.

INTRODUCTION

Responses to day-length, or photoperiod, enable plants and animals to adapt their physiology in anticipation of seasonal changes in climactic conditions. Many aspects of plant development are regulated by photoperiod, including seed germination, bud dormancy, stem and leaf growth, and formation of bulbs and tubers. The appropriate timing of reproductive activity is particularly important in order to maximize survival of the offspring. Flowering and seed production rely on suitable light and temperature conditions, but are also affected by the degree of competition with other species. In order to minimize such competition, different species have evolved a range of day-length responses that enable them to flower at different times of the year and occupy different ecological niches. For example, short-day plants flower in the spring or autumn when the total duration of light within a day is shorter than a critical photoperiod, whereas long-day plants flower in the summer when the photoperiod is longer.

The discovery of photoperiodism in the early 1920s has had a profound impact on agricultural and horticultural practices. Many commercial crops exhibit day-length responsive flowering, and the systematic testing of photoperiodic requirements can prevent failure of new varieties. The manipulation of day-length using black-outs or supplementary lighting allows flowering to be induced at any time of the year and enables breeders to obtain multiple generations per year. Commercial growers can grow potted ornamental plants out of season and precisely schedule their flowering for commercially important dates. Poinsettia, chrysanthemum, Christmas cactus, and begonias are examples of crops managed in this manner.

Recent studies of *Arabidopsis* (a long-day plant) and rice (a short-day plant) have identified genes that mediate floral responses to day-length. Emerging models for the photoperiodic sensing mechanism are described.

THE TIMING MECHANISM

Perception of photoperiod requires a biological timer measuring the duration of light and (or) darkness. This function is mediated by a 24-hour pacemaker known as the circadian clock. In *Arabidopsis*, the mutation or misexpression of any of the known components of this clock (including LHY, CCA1, TOC1, ELF3, GI, ZTL, LKP2, and FKF1) leads to abnormal floral responses to photoperiod.[1]

Two different mechanisms have been proposed by which a circadian clock might mediate perception of seasonal changes in day-length.[2] The effect of changes in photoperiod may be to bring two different rhythms to a more favorable phase-relationship (internal coincidence). Alternatively, responses may be triggered when light coincides with a sensitive phase of a photoperiodic response rhythm (external coincidence). The photoinducible phase may be determined by the temporal pattern of expression of a regulatory molecule, and rhythmic expression of the floral regulator *CONSTANS* (*CO*) has been proposed to play such a role in *Arabidopsis*.[3] In support of this model, atypical light-dark cycles that altered the

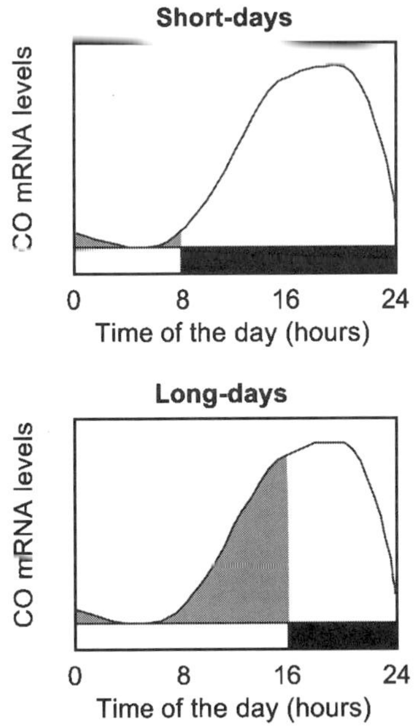

Fig. 1 Rhythmic expression of the *CO* transcript under short and long-day conditions. Similar patterns were observed in *Arabidopsis* and rice.[3,7] White and black boxes under the graphs represent intervals of light and darkness, respectively. Shaded areas under the curve indicate the levels of *CO* coincidence with light.

Encyclopedia of Plant and Crop Science
DOI: 10.1081/E-EPCS 120012965

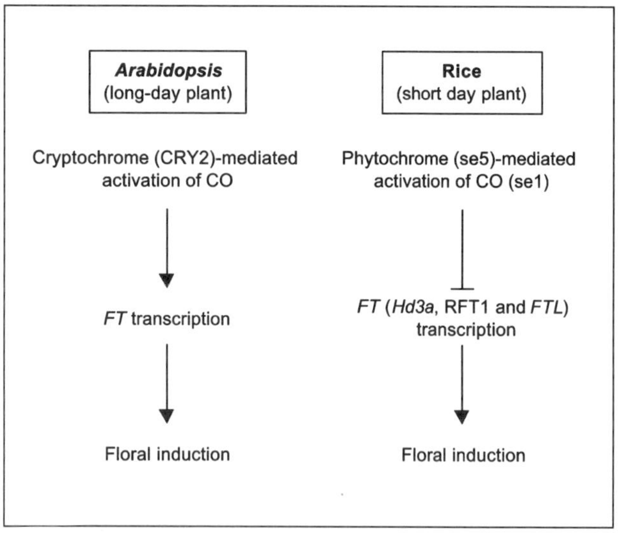

Fig. 2 Comparison of photoperiodic response mechanisms in *Arabidopsis* (a long-day plant) and rice (a short-day plant). Positive interactions are represented by pointed arrows, and negative interactions by blunted arrows. FT functions as a positive regulator of flowering in both species but activation of CO under long-day conditions has opposite effects on *FT* transcription.

phase-relationship between the *CO* expression rhythm and the photoperiod also altered floral responses.[4]

DOWNSTREAM OF THE CLOCK: ROLE OF CONSTANS AND OF ITS TARGET GENES

In *Arabidopsis*, expression of the *CO* transcript is rhythmic.[3] Under 24-hour light-dark cycles, accumulation begins eight hours after dawn, and peak levels are reached after 16 hours. Thus, under noninductive short-day conditions, expression of *CO* is restricted to the dark interval, but the first half of the *CO* mRNA peak coincides with light under inductive long-day conditions (Fig. 1). The CO protein exhibits homology to transcription factors of the zinc finger family.[5] As it is highly unstable, its expression pattern is predicted to resemble that of the mRNA. Coincidence of CO expression with light may lead to its post-translational modification. The light-activated CO protein may become capable of activating transcription of its immediate target genes, *FT* and *SOC1*[3,6] and trigger conversion of vegetative meristems into floral meristems (Fig. 2).

Rice *FT*-like genes (*Hd3A*, *RFT1*, *FTL*) also function as positive effectors of flowering,[7] and the rice counterpart of *CONSTANS* (*SE1*)[8] is expressed rhythmically with a phase similar to that of *CO* in *Arabidopsis* (Fig. 1). However, *SE1* inhibits expression of *FT*-like genes under long days and promotes it under short-day conditions.[7] Thus, the basic mechanism of day-length perception is conserved between rice and *Arabidopsis*, and the change of sign of the flowering response between short- and long-day plants may take place downstream of *CO* at the level of the regulation of *FT* transcription (Fig. 2).

PERCEPTION OF LIGHT SIGNALS

In the external coincidence model, light affects photoperiodic responses by two different mechanisms (Fig. 3). One effect of light is to mediate synchronization (entrainment) of the circadian clock to diurnal changes in

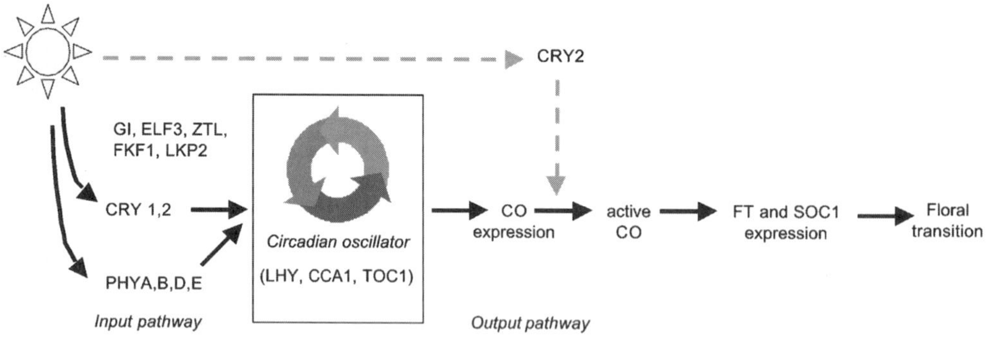

Fig. 3 Genetic interactions underlying photoperiodic time perception in *Arabidopsis* (a long-day plant). In *Arabidopsis*, the circadian oscillator constitutes a transcriptional feedback loop and the *LHY*, *CCA1*, and *TOC1* genes encode some of its key components. This oscillatory feedback loop is entrained to daily changes in environmental conditions via light input pathways from phytochromes and cryptochrome photoreceptors. The *GI*, *ELF3*, *ZTL*, *FKF1*, and *LKP2* genes encode downstream components of these input pathways. The central oscillator drives rhythmic expression of CO, a positive regulator of flowering under long-day conditions. Coincidence of CO expression with light is thought to result in the posttranslational activation of the protein, which becomes capable of promoting expression of floral activators FT and SOC1. Cryptochrome 2 (CRY2) is the main photoreceptor implied in this response (dotted arrows), a function that is distinct from its role in entrainment of the circadian clock (solid arrows). Note that phytochrome (SE5) mediates perception of external coincidence in rice.

Table 1 Types of photoperiodic reponses in plants

Short-day plants	Flower in response to light periods that are shorter than a critical photoperiod.
Long-day plants	Flower in response to light periods that are longer than a critical photoperiod.
Day-neutral plants	Flower at the same stage of development regardless of day-length.
Intermediate day-length plants	Flowering is inhibited by very long and very short photoperiods.
Ambiphotoperiodic plants	Flowering is delayed by intermediate photoperiods.
Dual day-length plants (long-day short-day and short-day long-day plants)	Require sequential exposure to two different inductive photoperiods.
Dark-dominant plants	Require long exposures to uninterrupted darkness. The inductive effect of long nights is inhibited by short exposures to red light. Broadly corresponds to the short-day plant category.
Light-dominant plants	Require long exposures to light. Flowering is promoted by a mixture of red and far-red light. Broadly corresponds to the long-day plant category.
Obligate responses	Where a particular day-length is required for flowering.
Quantitative responses	Where exposure to a favourable day-length accelerates flowering.

Specific examples for each category may be found in Ref. 12.

environmental conditions (solid arrow). Light-dark transitions set the phase of the clock, and thus determine the time of the day when floral responses can be induced. In *Arabidopsis*, this effect of light on the circadian clock is mediated by a broad range of photoreceptors, including phytochromes A, B, D, E, as well as cryptochromes 1 and 2.[9] Another effect of light is to trigger the photoperiodic response when its presence coincides with the photoresponsive phase of the photoperiodic response rhythm (dotted arrow). The blue light photoreceptor cryptochrome 2 (CRY2) plays a major role in the perception of photoperiodic signals in *Arabidopsis*, and plants lacking CRY2 function do not exhibit FT induction under long-days.[6] In contrast, the red/far red photoreceptor phytochrome appears to mediate perception of day-length in rice. As in *cry2* mutants of *Arabidopsis*, loss of phytochrome (SE5) function in rice abolished responses to long days. The *se5* mutant plants flowered as early under inhibitory long-days as wild-type under inductive short-day conditions.[10]

CONCLUSIONS AND PERSPECTIVES

Many aspects of the day-length response remain to be investigated. The mechanism by which light regulates CO activity is not known. There is evidence that diurnal changes in CRY2 protein levels are required for appropriate perception of day-length in *Arabidopsis*,[11] but the mechanism of these changes and their contribution to the photoperiodic response are not understood.

Comparison of patterns of *CO* and *FT* expression in a wider variety of plant species will determine whether rhythmic expression of these proteins may explain the range of day-length responses described in Table 1. The elucidation of photoperiodic timing mechanisms will enable the rational breeding and (or) engineering of crop varieties that are day-length insensitive. Alternatively it may be possible to alter the critical photoperiod for floral induction to allow growth of particular crops or ornamentals under latitudes that differ from their country of origin.

REFERENCES

1. Carré, I.A. Day-length perception and the photoperiodic regulation of flowering in *Arabidopsis*. J. Biol. Rhythms **2001**, *16* (4), 415–423.
2. Pittendrigh, C.S.; Minis, D.H. The entrainment of circadian clocks by light and their role as photoperiodic clocks. Am. Nat. **1964**, *XCVIII*, 261–294.
3. Suarez-Lopez, P.; Wheatley, K.; Robson, F.; Onouchi, H.; Valverde, F.; Coupland, G. *CONSTANS* mediates between the circadian clock and control of flowering in *Arabidopsis*. Nature **2001**, *410* (6832), 1116–1120.
4. Roden, L.C.; Song, H.-R.; Jackson, S.; Morris, K.; Carre, I.A. Floral responses to photoperiod in *Arabidopsis* are correlated with the timing of gene expression relative to dawn and dusk. Proc. Natl. Acad. Sci. U. S. A. **2002**, *99*, 13313–13318.
5. Putterill, J.; Robson, F.; Lee, K.; Simon, R.; Coupland, G. The *CONSTANS* gene of *Arabidopsis* promotes flowering

and encodes a protein showing similarities to zinc finger transcription factors. Cell **1995**, *80* (6), 847–857.
6. Yanovsky, M.J.; Kay, S.A. Molecular basis of seasonal time measurement in Arabidopsis. Nature **2002**, *419*, 308–312.
7. Izawa, T.; Oikawa, T.; Sugiyama, N.; Tanisaka, T.; Yano, M.; Shimamoto, K. Phytochrome mediates the external light signal to repress *FT* orthologs in photoperiodic flowering of rice. Genes Dev. **2002**, *16* (15), 2006–2020.
8. Yano, M.; Katayose, Y.; Motoyuki, A.; Yamanouchi, U.; Monna, L.; Fuse, T.; Baba, T.; Yamamoto, K.; Umehara, Y.; Nagamura, Y.; Sasaki, T. Hd1, a major photoperiod sensitivity quantitative trait locus in rice, is closely related to the *Arabidopsis* flowering time gene *CONSTANS*. Plant Cell **2000**, *12* (12), 2473–2483.
9. Devlin, P.F.; Kay, S.A. Cryptochromes are required for phytochrome signalling to the circadian clock but not for rhythmicity. Plant Cell **2000**, *12* (12), 2499–2509.
10. Izawa, T.; Oikawa, T.; Tokutomi, S.; Okuno, K.; Shimamoto, K. Phytochromes confer the photoperiodic regulation of flowering in rice (a short-day plant). Plant J. **2000**, *22* (5), 391–399.
11. El-Assal, S.E.-D.; Alonso-Blanco, C.; Peeters, A.J.M.; Raz, V.; Koornneef, M. A QTL for flowering time in Arabidopsis reveals a novel allele of CRY2. Nat. Genet. **2001**, *29* (4), 435–440.
12. Thomas, B.; Vince-Prue, D. *Photoperiodism in Plants*; Academic Press: San Diego, U.S.A., 1997.

Photoreceptors and Associated Signaling I: Phytochromes

Karen J. Halliday
University of Bristol, Bristol, U.K.

INTRODUCTION

Plants rely on information from their surrounding environment to synchronize development to daily and seasonal changes. Many of these adaptive changes are mediated in response to alterations in day length and light quality. In higher plants, the need to sense and interpret these environmental cues in a meaningful way has led to the evolution of highly sophisticated light signaling networks. Controlling these networks are families of photoreceptors whose collective action shapes growth and development via a process that is referred to as photomorphogenesis. The phytochrome family of photoreceptors plays a pivotal role in this photosensory network controlling a diverse range of responses, from seed germination, de-etiolation, to cell elongation and flowering. In the model plant *Arabidopsis thaliana* there are five members of the phytochrome family: phyA, phyB, phyC, phyD, and phyE. These photoreceptors, which absorb mainly in the red (600–700 nm) and far-red (700–800 nm) regions of the electromagnetic spectrum, are able to monitor closely and respond to changes in the light environment. In recent years we have made significant inroads into defining roles for individual phytochromes; we have also begun to understand the nature of the signaling networks that they control.

ISOLATION AND CHARACTERIZATION OF PHYTOCHROME PHOTORECEPTOR MUTANTS

Analysis of mutants lacking one or more phytochromes has provided tremendous insights into the contribution of individual phytochromes and the interplay between phytochromes in the regulation of many developmental events. The phytochromes have an important role in de-etiolation and this was exploited to isolate the first phytochrome mutants. When grown in the dark *Arabidopsis* seedlings have elongated hypocotyls and small, folded, unexpanded, pale cotyledons. Exposure to the red, far-red, and blue components of white light triggers series of developmental changes that lead to de-etiolation: inhibition of hypocotyl elongation and opening, expansion, and greening of the cotyledons. Genetic screens designed to select for mutants that were deficient in de-etiolation under specific wavelengths of light led to the identification of the *phyA* and *phyB* null mutants. Several *phyA* mutants were isolated in screens for long hypocotyl mutants under far-red light.[1] Indeed, *phyA* mutants were shown to be completely insensitive to far-red light and they resembled dark-grown seedlings when grown under these conditions. This demonstrated that wild type phyA was important for mediating responses to de-etoliation in FR light.

PhyA has different properties from the other phytochromes. It accumulates to high levels in imbibed seeds, and dark-grown seedlings. Upon exposure to light PfrA is rapidly degraded; thus, phyA is light labile.[2] Careful physiological analysis of mutants null for phyA demonstrated that this phytochrome operates via two modes of action: the Very Low Fluence Response mode (VLFR) and the High Irradiance Response mode (HIR).[3] VLFRs are saturated with very low fluences of light, whilst HIRs require longer periods of far-red irradiation at higher fluence rates. Extensive analysis of *phyA* mutants has shown that phyA is not just important for de-etiolation, but also has an extensive role in controlling development throughout the plant's life. These roles include control of germination, photoperiod-dependent hypocotyl elongation, and the photoperiodic control of flowering.[1]

Similar screens, but under white light, identified mutants that were null for phyB.[1] Seedlings lacking phyB were shown to have a reduced sensitivity to red light and did not de-etiolate fully under those conditions. This indicated that phyB played a major role in controlling seedling de-etiolation under red light. In contrast to phyA, phyB does not degrade as rapidly when seedlings are exposed to light and so is more light stable. Indeed, phyB appears to be the most abundant of the phytochromes in light-grown seedlings.[2] Responses demonstrated to be reversible by far-red light in the wild type were shown to be impaired in the *phyB* mutant.[1] These experiments showed that red/far-red reversibility, the hallmark of phytochrome action, is a characteristic of phyB and other light stable phytochromes. This type of response is referred to as a Low Fluence Response or LFR. It is this property of phytochrome that enables the plant to monitor the proportion of red:far-red ratio accurately, a means of detecting neighboring plants and potential competition.

Encyclopedia of Plant and Crop Science
DOI: 10.1081/E-EPCS 120012960

Such changes in light quality result from selective absorption by green vegetation, which enhances the proportion of far-red wavelengths in the reflected/scattered light. The consequent alteration in red:far-red ratio triggers a striking series of responses that include increased stem and petiole elongation and the acceleration of the transition to flowering. These responses, collectively known as the Shade Avoidance Response, appear to be an important survival strategy under unfavorable shade conditions [1] Mutants deficient in phyB strongly resemble wild type plants that have been exposed to low red:far-red ratio light suggesting a major role for phyB in this important suite of responses (Fig. 1).

The *phyD* mutation was discovered as a naturally occurring mutation in the *Arabidopsis* Wassilewskija accession.[1] The *phyE* mutation was isolated in a genetic screen of mutagenised *phyA phyB* mutants that displayed a constitutively early flowering and elongated internode phenotype.[1] The *phyD* mutant displayed a similar, but less marked seedling phenotype to *phyB* when grown under red light, suggesting a role for phyD in the control of deetiolation. For many responses, when grown under standard laboratory conditions, the *phyD* and *phyE* monogenic mutants were similar to the wild type. However, analysis of mutants lacking phyD or phyE in addition to phyB revealed overlapping roles for these phytochromes with phyB. It has been demonstrated that the relative contributions of individual phytochromes varied according to the response.[1,4,5] Indeed, the hierarchy of photoreceptor action appears to be modified by developmental or environmental changes. For example, under short photoperiods, phyB and phyE appear to have prominent roles in the inhibition of flowering.[4] We have

Fig. 1 The *phyB* mutant phenotype: wild type (left) and *phyB* mutant grown under continuous white light (right). (This figure was kindly donated by Garry C. Whitelam, Leicester University, U.K.) *(View this art in color at www.dekker.com.)*

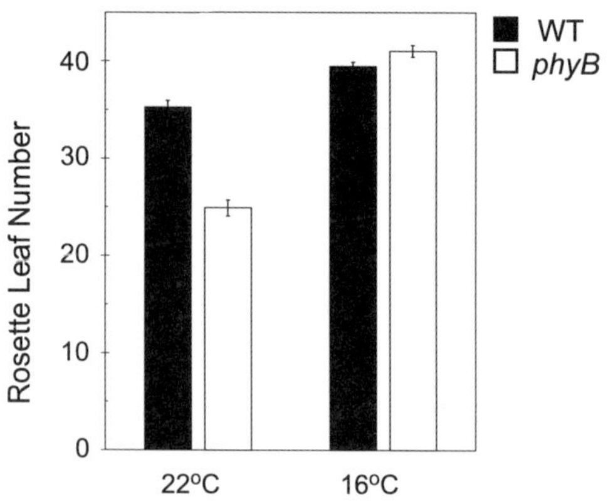

Fig. 2 The *phyB* mutant flowering phenotype is temperature-sensitive: flowering time, measured as rosette leaf number at bolting, in wild type plants and *phyB* mutants grown in 8 hour photoperiods at 22°C or 16°C. Bars represent the SE.

recently shown that the *phyB* mutant–early flowering phenotype is temperature conditional, and that phyB exerts a greater control on flowering at 22°C than at 16°C (Fig. 2). This reduced phyB action observed at cooler temperatures is balanced by a rise in activity of other phytochromes.[4,5] This change in the relative contributions of individual phytochromes appears to be a means via which control of flowering is maintained in the natural environment where temperatures fluctuate. We have yet to determine a role for phyC, but, the recent isolation of *phyC* mutant alleles should enable us to do so.

The picture that is emerging from several elegant genetic studies is one where the phytochromes act not in isolation, but as an integrated network. This network extends beyond the phytochrome family to other photoreceptors: the cryptochromes and phototropins. Indeed, cross-talk between these different photoreceptor pathways is well documented.[6,7] Such an interconnected network has two main functions: It ensures transitional developmental response and it provides a buffer for environmental (e.g., changes in ambient temperature) and genetic variation in the maintenance of the response.[5,8]

PHYTOCHROMES ARE COMPLEX LIGHT-REGULATED CHROMOPROTEINS

Phytochrome was first purified form etiolated oat seedlings by Butler and coworkers over forty years ago.[9] In the intervening period we have discovered much about the properties of this molecule. Phytochromes exist as soluble dimers that comprise two polypeptides of approximately 125 kDa. The phytochromes are unique among the photoreceptors: They are photoreversible switches. Light induces interconversion between a red light-absorbing Pr form and a far-red light–absorbing Pfr form.[9–11] These distinctive spectral properties are conferred by phytochromobilin, a linear tetrapyrrole chromophore that is covalently attached to the phytochrome apoprotein.[9–11] Each phytochrome monomer can be subdivided into an amino-terminal sensor and a C-terminal output domain.[9] It is the sensor domain that houses the chromophore, attached via the bilin lyase domain, also recognized as a GAF domain. Phytochrome GAF domains appear to belong to the bilin-lyase–specific subfamily of these ligand-binding domains. The regulatory or output region comprises two PAS-related domains (PRD) and a histidine-kinase–related domain (HKRD). PAS domains commonly bind small ligands and mediate protein–protein interactions. Similarities in the tertiary structure of PAS and GAF domains suggest there may be an evolutionary link between these two domains.[7,10] The HKRD, which is similar to histidine kinase regions found in bacterial two-component sensor proteins, together with the PRDs, provide a means via which the phytochrome dimer can transduce its signals. Indeed, work from several laboratories has provided strong evidence that support this notion.[7,11]

PHYTOCHROME SIGNAL TRANSDUCTION

In *Arabidopsis* seedlings light induces the radical change in gene expression required for the switch to skotomorphogenic to photomorphogenic development.[1,7] This developmental switch is triggered largely by phytochrome and crytochrome action. Following photoactivation the phytochromes translocate from the cytoplasm to the nucleus where they cluster in nuclear speckles, the putative site of action.[3]

Defining the many phytochrome-controlled pathways is no small task. However, genetic screens and molecular techniques that detect protein–protein interactions have significantly enhanced our understanding of phytochrome signal transduction.[1,7] These studies have identified multiple phytochrome signaling intermediates and revealed that the phytochromes interact directly with several of these components. One such interaction partner is PIF3, a basic helix-loop-helix class of transcription factor, which preferentially binds to phyB in its active form (Pfr).[7] PIF3 also binds to the G-box DNA sequence motif common in the promoters of light-regulated genes. CCA1 and LHY, G-box containing genes and central components of the *Arabidopsis* circadian clock, were shown to be regulated by PIF3.[7] The circadian clock regulates many cellular processes and developmental responses; therefore, PIF3 may represent one mechanism via which light can interact to control these events.

Phytochromes display light-dependent Ser/Thr protein kinase activity.[11] This provides a possible mechanism via which phytochrome signals can be transduced in response to light signals. NDPK2, cry1, cry2, and PKS1 have been shown to interact with phyA and/or phyB. Furthermore, in vitro kinase assays, coupled with red light–mediated in vivo phosphorylation assays (for NDPK2, cry1, and PKS1) indicate that these proteins may be targets for phytochrome kinase activity.[7,11] The precise roles of NDPK2 and PKS1 in phytochrome signaling are not yet known. In contrast, physiological interactions between the cryptochromes and the phytochromes are well documented. It is, therefore, possible that some of these interactions result from modification of cryptochrome action via phytochrome-mediated phosphorylation.

CONCLUSION

Light cues that signal changes in the seasons or the immediate environment strongly influence growth strategy and the timing of development. The phytochromes detect these changes and act as an integrated signaling network to keep development in tune with the environment. Throughout development, it is equally important that responses are maintained under variable conditions. Indeed, the roles for individual photoreceptors and their hierarchy of action vary with developmental stage and environmental conditions.[4–6] This flexibility in the signaling network means that environmental change can be accommodated. Although great strides have been made in recent years we are still at an early stage in our understanding of phytochrome signaling. A major challenge for the future is to understand how the light signaling network is integrated. A multidisciplinary approach, combining molecular genetics, biochemistry, bioinformatics, and mathematical modeling, is required if we are to achieve this goal.

ARTICLES OF FURTHER INTEREST

Circadian Clock in Plants, p. 278
Photoperiodism and the Regulation of Flowering, p. 877
Photoreceptors and Associated Signaling II: Cryptochromes, p. 885
Photoreceptors and Associated Signaling III: Phototropins, p. 889

Photoreceptors and Associated Signaling IV: UV Receptors, p. 893
Phytochrome and Cryptochrome Functions in Crop Plants, p. 920
Shade Avoidance Syndrome and Its Impact on Agriculture, p. 1152

REFERENCES

1. Whitelam, G.C.; Patel, S.; Devlin, P.F. Phytochromes and photomorphogenesis in *Arabidopsis*. Philos. Trans. R. Soc. Lond., B Biol. Sci. **1998**, *353* (1374), 1445–1453.
2. Sharrock, R.A.; Clack, T. Patterns of expression and normalized levels of the five *Arabidopsis* phytochromes. Plant Physiol. **2002**, *130* (1), 442–456.
3. Nagy, F.; Schafer, E. Phytochromes control photomorphogenesis by differentially regulated, interacting signaling pathways in higher plants. Annu. Rev. Plant Biol. **2002**, *53*, 329–355.
4. Halliday, K.J.; Salter, M.G.; Thingnaes, E.; Whitlelam, G.C. The phyB-controlled flowering pathway is temperature sensitive and is mediated by the floral integrator FT. Plant J. **2003**, *33* (5), 875–885.
5. Halliday, K.J.; Whitelam, G.C. Changes in photoperiod or temperature alters the functional relationships between phytochromes and reveals roles for phyD and phyE. Plant Physiol. **2003**, *131* (4), 1913–1920.
6. Casal, J.J. Phytochromes, cryptochromes, phototropin: Photoreceptor interactions in plants. Photochem. Photobiol. **2000**, *71* (1), 1–11.
7. Quail, P.H. Phytochrome photosensory signalling networks. Nat. Rev., Mol. Cell Biol. **2002**, *3* (2), 85–93.
8. Stearns, S.C. Progress on canalization. Proc. Natl. Acad. Sci. U.S.A. **2002**, *99* (16), 10229–10230.
9. Halliday, K.J.; Fankhauser, C. Phytochrome-hormonal signalling networks. New Phytol. Tansley Rev. **2003**, *157*, 449–463.
10. Montgomery, B.L.; Lagarias, J.C. Phytochrome ancestry: Sensors of bilins and light. Trends Plant Sci. **2002**, *7* (8), 357–366.
11. Fankhauser, C. Phytochromes as light-modulated protein kinases. Sem. Cell Dev. Biol. **2000**, *11* (6), 467–473.

Photoreceptors and Associated Signaling II: Cryptochromes

P

Chentao Lin
University of California, Los Angeles, California, U.S.A.

INTRODUCTION

Blue light affects many aspects of plant growth and development. The blue light responses of plants can be roughly divided into two large categories: photomovement responses and photomorphogenetic responses. Photomovement responses, including phototropic curvature, chloroplast relocation, and stomata opening, are mediated by phototropins. Plant photomorphogenetic responses, including inhibition of hypocotyl elongation, stimulation of cotyledon expansion, regulation of flowering time, entrainment of the circadian clock, and regulation of gene expression, are controlled by both phytochromes (in response to red/far-red light) and cryptochromes (in response to blue/UV-A light).

CRYPTOCHROME GENES AND PROTEINS

The term cryptochrome was coined in the late 1970s as a laboratory nickname for blue/UV-A light receptors that mediate plant blue light responses with the specific action spectra of a peak in the UV-A region (approximately 350–400 nm) and a peak with fine structures in the blue region (approximately 400–500 nm). The compound word is composed of -chrome for "pigment" (from the Greek *chroma* meaning color or pigment) and crypto- as in "cryptic" or "cryptogam." This term was chosen because the molecular nature of blue light receptors remained hidden (cryptic) at the time in spite of extensive researches, and because the blue light responses are prevalent in cryptogams (plants without true flowers and seeds, such as ferns, mosses, algae, and fungi).[1]

Cryptochrome now refers to proteins that share sequence similarity to DNA photolyase but lack the photolyase activity. The first cryptochrome gene was isolated from *Arabidopsis thaliana*, which is a small weed widely used in laboratories because of its easy handling and small genome. In 1980, a number of Arabidopsis photomorphogenesis mutants were reported, one of which, called *hy*4, showed long hypocotyls when grown in blue light.[2] The gene corresponding to the *hy*4 mutation was isolated about a decade later, and it was found to encode a protein of 681 amino acids, for which the N-terminal sequence of approximately 500 amino acids was 30% identical to that of *E. coli* DNA photolyase.[3] DNA photolyases catalyze the blue/UV-A light-dependent cleavage of cyclobutane pyrimidine dimers that is a major type of DNA damage caused by short-wavelength (<300 nm) UV light.[4] To investigate whether the *HY*4 gene indeed encoded a blue light receptor, a full-length recombinant *HY*4 gene was expressed in insect cells and the recombinant protein purified.[5] The recombinant *HY*4 gene product is a yellow-colored soluble protein that bound noncovalently to flavin adenine dinucleotide (FAD). Cryptochromes may also contain a pterin as the second chromophore.[4] The facts that the *HY*4 gene product mediates a blue light response, shares sequence similarity to a blue light-dependent enzyme, and contains FAD, which absorbs both blue light and UV-A light, as a prosthetic group, indicated that it is a blue light receptor. The *HY*4 gene was renamed *CRY*1 (for cryptochrome 1) in 1995.[5]

Cryptochromes have been found throughout the plant kingdom, including among the angiosperms, ferns, mosses, and algae, and in animals including fishes, frogs, flies, mice, and humans.[1,4,6] Most plant species studied contain multiple members of the photolyase/cryptochrome gene family. For example, Arabidopsis has two cryptochrome genes, *CRY*1 and *CRY*2, and two photolyase genes; tomato and barley each have at least 3 cryptochrome genes, *CRY*1*a*, *CRY*1*b*, and *CRY*2; ferns and mosses have five and at least two cryptochrome genes, respectively.[1] The amino acid sequences of tomato CRY1 (CRY1a or CRY1b) and CRY2 are more similar to their Arabidopsis counterparts than to each other, indicating that the gene duplication event resulting in *CRY*1 and *CRY*2 occurred more than 100 million years ago at least, before the divergence of Brassicaceae (e.g., Arabidopsis) and Solanaceae (e.g., tomato).

Most plant cryptochromes have two domains, an N-terminal domain called PHR (for photolyase-related) that shares sequence homology with DNA photolyase, and a C-terminal domain called CCT (for cryptochrome C-

Encyclopedia of Plant and Crop Science
DOI: 10.1081/E-EPCS 120012961

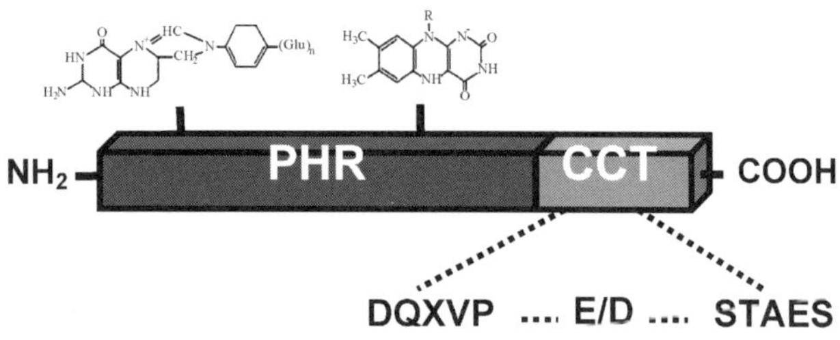

Fig. 1 Diagram depicting the structure of a plant cryptochrome PHR: photolyase-related domain CCT: Cryptochrome C-terminal domain Motifs of the CCT domain are conserved in some plant cryptochromes, and the amino acids of these motifs are shown (X represents any amino acid). (*View this art in color at www.dekker.com.*)

terminus) that is unrelated to photolyase or any known proteins, although many plant cryptochromes contain conserved motifs in this domain (Fig. 1).[1] The PHR domain of cryptochrome is evidently the chromophore-binding domain. The CCT domain is also required for the cryptochrome function. The reaction mechanism and 3-D structure of photolyases are well known,[4] but neither has been extensively studied for cryptochromes. However, given the sequence similarity between cryptochrome and photolyase, at least some aspects of the cryptochrome structure and reaction mechanism may resemble that of a photolyase.

CRYPTOCHROMES MEDIATE VARIOUS PHOTOMORPHOGENETIC RESPONSES

Photomorphogenetic responses mediated by cryptochromes include blue light inhibition of stem elongation (Fig. 2), stimulation of leaf expansion, control of photoperiodic flowering, entrainment of the circadian clock, and regulation of gene expression.[1,6] The functions of cryptochromes are conserved in different plants. It has been demonstrated that cryptochromes regulate hypocotyl inhibition in Arabidopsis and tomato. Different cryptochromes in the same plant can regulate the same light response. For example, in addition to CRY1, Arabidopsis CRY2 also contributes to the blue light inhibition of hypocotyl elongation response; and moss (*Physcomitrella patens*) CRY1a and CRY1b are both required for the blue light induction of side-branching on protonema. Different cryptochromes can also regulate distinct light responses. For instance, Arabidopsis CRY1 plays a more important role in hypocotyl inhibition, whereas Arabidopsis CRY2 is more involved in flowering-time control.

Although phytochromes are known to regulate photoperiodic flowering, it is now clear that cryptochromes also play significant roles in this response. The Arabidopsis laboratory strains Columbia (Col) and Landsberg *erecta* (L*er*), both collected in the Northern Hemisphere, are nonobligate long-day plants that flower earlier in long days than in short days. The *cry*2 mutants isolated from these strains flower later than the wild type in long days but not in short days, so these *cry*2 mutants are late-flowering but more or less day-neutral.[7] Another Arabidopsis ecotype, Cvi, collected from the tropical Cape Verde Islands, flowers earlier than many other Arabidopsis strains, and Cvi plants flower at about the same time in long days or in short days. The major QTL (quantitative trait locus) responsible for the day-neutral early flowering of the Cvi strain was determined to be the *CRY*2 gene.[8] A valine to methionine substitution (V367M) in the CRY2 protein of the Cvi strain was found to be responsible for its day-neutral early-flowering phenotype. A cryptochrome can act as a day-length sensor by changing its relative abundance in response to photoperiods. The abundance of Arabidopsis cry2 protein shows a day-length-dependent diurnal rhythm. In short-day photoperiods, the level of CRY2 is lower in the day but higher in the night. In long-day photoperiods, such a diurnal rhythm of the CRY2 abundance is significantly diminished.[1,8]

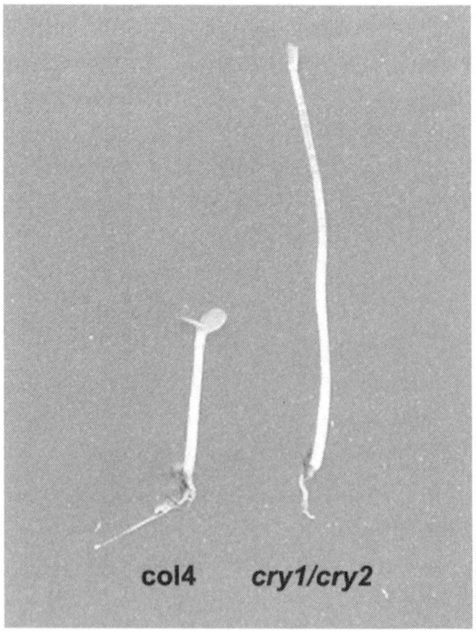

Fig. 2 Arabidopsis cryptochromes mediate blue light inhibition of hypocotyl elongation. Arabidopsis wild-type (col4) and *cry*1*cry*2 double-mutant seedlings grown in continuous blue light for 5 days are shown. (*View this art in color at www.dekker.com.*)

Cryptochrome is a major photoreceptor regulating the circadian clock in both plants and animals. In animals, cryptochromes act redundantly with rhodopsins to regulate the circadian clock.[4] In plants, the entrainment of the circadian clock is controlled by both phytochromes and cryptochromes.[9] Although it has been shown that cryptochrome is a photoreceptor for the entrainment of the circadian clock in the fruit fly, whether cryptochromes also provide light entrainment to the circadian clock in mammals such as humans or mice is still being debated. Plant cryptochromes act mainly as photoreceptors rather than as critical components of the circadian clock; in contrast,[9] mammalian cryptochromes are critical components of the central oscillator.[4]

CRYPTOCHROME SIGNAL TRANSDUCTION

The initial photoreaction of cryptochrome remains unclear. Because electron transport is involved in the photolyase-catalyzed DNA repairing reaction, a redox reaction has been proposed to be likely involved in the cryptochrome photochemistry.[6] It is also not clear what the signal transduction mechanism is underlying cryptochrome regulation of photomorphogenetic responses. It has been proposed that cryptochromes may act to change ion homeostasis in the cell, or that cryptochromes, which are often found in the nucleus, may regulate gene expression to alter developmental processes.[1] And at least two types of biochemical reactions have been demonstrated in the cryptochrome function: protein–protein interactions and a blue light-induced cryptochrome phosphorylation.

Arabidopsis cryptochromes can interact with different proteins, including phytochrome B, COP1 (a putative subunit of E3 ubiquitin ligase complex known to be important for light-regulation of gene expression and hypocotyl inhibition), and ZTL (a PAS-domain-containing protein known to be important for the circadian clock and photoperiodic flowering).[1] The functional significance of cry-COP1 interaction is demonstrated by the constitutive photomorphogenetic phenotype of Arabidopsis transgenic plants overexpressing CCT domain fusion proteins.[10,11] The cry-phyB interaction may provide an explanation of why cry2 function is dependent on phyB.[7,12] Arabidopsis cryptochromes have also been found to undergo a blue light-dependent phosphorylation, and the blue light-induced phosphorylation of Arabidopsis cry2 is important for its function and degradation.[13] It has been proposed that cryptochromes are unphosphorylated and inactive in dark; blue light induces phosphorylation of cryptochromes by an unknown protein kinase; the phosphorylated cryptochrome becomes active in triggering photomorphogenetic responses; and phosphorylation-induced cryptochrome degradation is a mechanism to desensitize the photoreceptor.[13]

CONCLUSION

Cryptochromes are blue/UV-A light receptors regulating various photomorphogenetic responses of plants, but the detailed molecular mechanisms of signal transduction of cryptochromes is not clear at present. Investigation of the interaction among signaling processes of cryptochromes, phytochromes, and phytohormones will also be critical to our understanding of how plants regulate their developmental processes in response to light.

ACKNOWLEDGMENTS

The author thanks Dr. Dror Shalitin and Dr. Todd Mockler for critical readings of the manuscript and for preparing figures. The cryptochrome study in the author's laboratory is supported by NIH, NSF, and USDA.

ARTICLES OF FURTHER INTEREST

Arabidopsis Thaliana Characteristics and Annotation of a Model Genome, p. 47
Circadian Clock in Plants, p. 278
Photoperiodism and the Regulation of Flowering, p. 877
Photoreceptors and Associated Signaling I: Phytochromes, p. 881
Photoreceptors and Associated Signaling III: Phototropins, p. 889
Photoreceptors and Associated Signaling IV: UV Receptors, p. 893
Phytochrome and Cryptochrome Functions in Crop Plants, p. 920

REFERENCES

1. Lin, C. and Shalitin, D. Cryptochrome structure and signal transduction. *Annu. Rev. Plant Biol.* **2003**, *54*, 469–496.
2. Koornneef, M.; Rolff, E.; Spruit, C.J.P. Genetic control of light-inhibited hypocotyl elongation in *Arabidopsis thaliana* (L.). Heynh. Z. Pflanzenphysiol. Bd. **1980**, *100*, 147–160.
3. Ahmad, M.; Cashmore, A.R. HY4 gene of *A. thaliana* encodes a protein with characteristics of a blue-light photoreceptor. Nature **1993**, *366*, 162–166.
4. Sancar, A. Cryptochrome: The second photoactive pigment

in the eye and its role in circadian photoreception. Annu. Rev. Biochem. **2000**, *69*, 31–67.

5. Lin, C.; Robertson, D.E.; Ahmad, M.; Raibekas, A.A.; Jorns, M.S.; Dutton, P.L.; Cashmore, A.R. Association of flavin adenine dinucleotide with the Arabidopsis blue light receptor CRY1. Science **1995**, *269*, 968–970.
6. Cashmore, A.R.; Jarillo, J.A.; Wu, Y.J.; Liu, D. Cryptochromes: Blue light receptors for plants and animals. Science **1999**, *284*, 760–765.
7. Guo, H.; Yang, H.; Mockler, T.C.; Lin, C. Regulation of flowering time by Arabidopsis photoreceptors. Science **1998**, *279*, 1360–1363.
8. El-Din El-Assal, S.; Alonso-Blanco, C.; Peeters, A.J.; Raz, V.; Koornneef, M. A QTL for flowering time in Arabidopsis reveals a novel allele of CRY2. Nat. Genet. **2001**, *29*, 435–440.
9. Somers, D.E.; Devlin, P.F.; Kay, S.A. Phytochromes and cryptochromes in the entrainment of the Arabidopsis circadian clock. Science **1998**, *282*, 1488–1490.
10. Yang, H.-Q.; Wu, Y.-J.; Tang, R.-H.; Liu, D.; Liu, Y.; Cashmore, A.R. The C termini of Arabidopsis cryptochromes mediate a constitutive light response. Cell **2000**, *103*, 815–827.
11. Wang, H.; Ma, L.G.; Li, J.M.; Zhao, H.Y.; Deng, X.W. Direct interaction of Arabidopsis cryptochromes with COP1 in light control development. Science **2001**, *294*, 154–158.
12. Mas, P.; Devlin, P.F.; Panda, S.; Kay, S.A. Functional interaction of phytochrome B and cryptochrome 2. Nature **2000**, *408*, 207–211.
13. Shalitin, D.; Yang, H.; Mockler, T.C.; Maymon, M.; Guo, H.; Whitelam, G.C.; Lin, C. Regulation of Arabidopsis cryptochrome 2 by blue-light-dependent phosphorylation. Nature **2002**, *417*, 763–767.

Photoreceptors and Associated Signaling III: Phototropins

P

Bethany B. Stone
Emmanuel Liscum
University of Missouri, Columbia, Missouri, U.S.A.

INTRODUCTION

Plants have evolved a number of photoreceptor systems that mediate responses to a broad range of wavelengths from ultraviolet to near infrared. Several light-induced processes are specific to the blue (390–500 nm) and ultraviolet-A (320–390 nm) regions of the electromagnetic spectrum, including phototropism, regulation of stomatal aperture, and chloroplast position within mesophyll cells. Phototropism is the bending of a plant organ toward or away from a directional light stimulus. The phototropic response results from increased elongation of cells along one portion of the responding organ relative to cells in an opposing position. Stomata—small pores found along the surfaces of leaves and stems—are created between two guard cells fixed in position relative to each other at opposing ends. Because guard cells can swell or shrink in response to a variety of environmental signals (including light), stomatal pore size can change, allowing control of gas exchange and water release in leaves and stems. Mesophyll chloroplasts have the capacity to move within the cell to either maximize or minimize light capture. Although it has been known for some time that each of these responses is controlled by blue light, molecular studies have only recently identified the responsible blue-light photoreceptors, the phototropins.

ISOLATION AND CHARACTERIZATION OF THE *PHOTOTROPIN 1* (*PHOT1*) MUTANTS

Two genes encoding proteins now classified as phototropins are found in the Arabidopsis genome.[1] The first, *PHOTOTROPIN 1* (*PHOT1*) (formerly called *NPH1*, for *NON-PHOTOTROPIC HYPOCOTYL 1*), was not identified by genome sequencing but by classical genetics in a screen for *Arabidopsis* seedlings that failed to exhibit a hypocotyl phototropic response to low-intensity blue light.[2] Not only are *phot1* mutants impaired in their positive hypocotyl phototropism (bending toward light), but also in their negative root phototropism (bending away from light) in low-light intensities. In high-intensity light, *phot1* mutants remained nonresponsive in the root, but recovered phototropism in the hypocotyl, indicating the presence of a second blue-light photoreceptor acting at higher fluence rates.[3]

ISOLATION AND CHARACTERIZATION OF *PHOT2* MUTANTS

A second likely phototropin gene, *PHOT2* (originally designated *NPL1*, for *NPH1-like1*), was identified by genome sequence analysis by virtue of its 58% DNA sequence identity to *PHOT1*.[4] A T-DNA insertion loss-of-function *phot2* mutant was subsequently shown to have normal phototropic responses under both low and high-light conditions when a *phot1 phot2* double mutant exhibited essentially no phototropism under either light condition.[5] Thus it appears that phot1 is required for phototropism under low light, whereas phot1 and phot2 play redundant roles in high-fluence rates (Fig. 1).[5]

Mutant analyses have also associated phototropins with blue-light–induced chloroplast movements. In low-fluence blue light, mesophyll chloroplasts align along the periclinal walls of cells to maximize light exposure (accumulation response). In contrast, high light causes chloroplasts to relocate to the anticlinal walls in order to minimize light capture and thus avoid photodamage (avoidance response).[6] Kagawa and colleagues[7] identified *phot2* mutants (initially designated as *cav1*, for *chloroplast avoidance 1*) in a screen for plants that failed to exhibit the chloroplast avoidance response. A similar lack of chloroplast avoidance response was found in reverse genetic analyses of *phot2* T-DNA mutants.[5,8] The singular role of phot2 in this high-light response differs from that observed with phototropism, where both phot1 and phot2 are responsible for photoperception in high light (Fig. 1). However, it appears that both phot1 and phot2 can signal the chloroplast accumulation response.[5,7,8] Interestingly, the control of stomatal aperture also requires the activities of phot1 and phot2, but in yet another way. Unlike phototropism (in which the two photoreceptors signal the same response but at two different fluence rates) and unlike chloroplast movement (in which the two photoreceptors trigger two different responses), stomatal opening requires both phot1 and phot2 at all light intensities (Fig. 1).[9]

Encyclopedia of Plant and Crop Science
DOI: 10.1081/E-EPCS 120012962

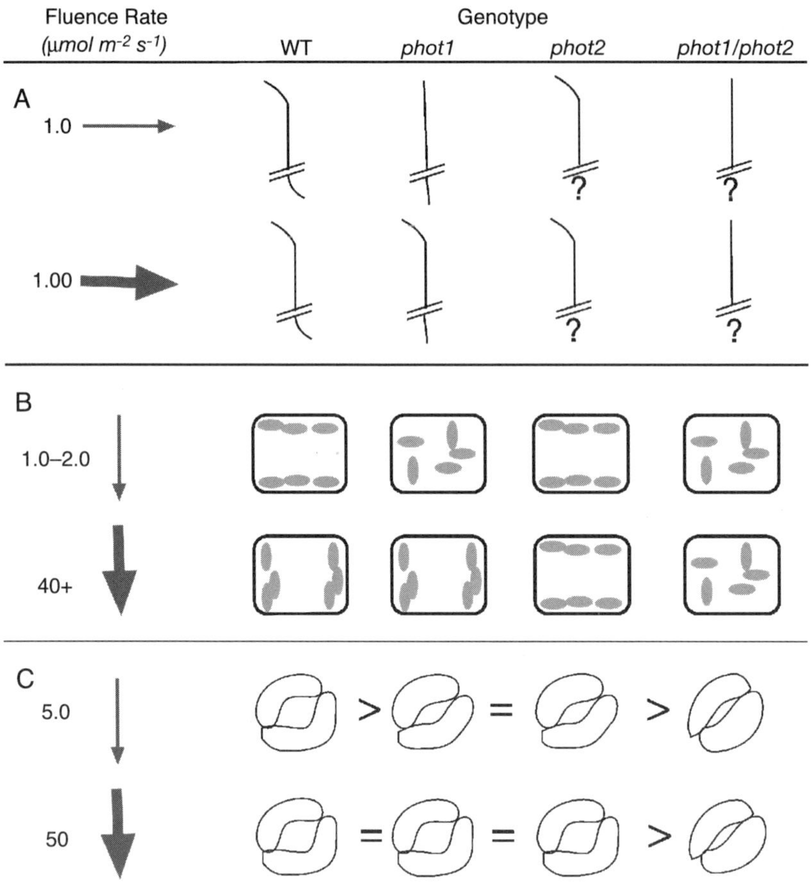

Fig. 1 The effect of fluence rate on wild type, *phot1* mutants, *phot2* mutants, and *phot1/phot2* double mutants with respect to (A) phototropism in the hypocotyl[5] and root;[4] (B) chloroplast movement;[5] and (C) stomatal opening. (*View this art in color at www.dekker.com.*)

PHOTOTROPIN PROTEINS ARE FLAVIN-BINDING LIGHT-ACTIVATED SERINE/THREONINE PROTEIN KINASES

A decade ago, Reymond and associates reported light-induced phosphorylation of a approximately 120-kilodelta(kDa) membrane-associated protein with the phototropic response,[10] and early work with the *phot1* mutants demonstrated that they lacked this light-induced phosphorylation response.[2] When *PHOT1* was cloned by positional cloning it was found to encode a 120 kDa protein with a carboxyl-terminal serine/threonine kinase domain, making it likely that the phosphorylation response was in fact an autophosphorylation response.[11] Christie and colleagues found that phot1 expressed heterologously in insect cells was autophosphorylated in response to blue light, indicating that all components necessary for the light-induced autophosphorylation—photoreceptor, kinase, and substrate—are present in phot1 itself.[12]

The PHOT1 gene also includes two similar amino-terminal domains with 43% sequence identity to each other and similarity to a class of domains found in signaling proteins of various species.[11] These domains are a subset of the PER-ARNT-SIM (PAS) domain superfamily, known to mediate protein-protein interactions and cofactor binding. Because proteins containing this subclass of PAS domains are responsive to light, oxygen, or voltage, these domains have been designated LOV1 and LOV2.[11] In work by Christie and colleagues it has been found that the blue-light sensitivity described above is dependent upon binding of one flavin mononucleotide (FMN) molecule to each LOV domain, providing support for the hypothesis that a dual-chromophoric flavoprotein photoreceptor regulates phototropism in low-fluence blue light.[2,12] Phot2, like phot1, is autophosphorylated in response to blue light and each of its LOV domains binds one FMN molecule.[5]

A number of studies have examined the photochemical and photophysical properties of the LOV domains for phototropin proteins. Both optical and NMR spectroscopy have been used to demonstrate that phot1 LOV domains in solution undergo a dark-reversible photocycle that involves the formation of a C(4a) cysteinyl adduct between the FMN and the phot1 polypeptide.[13,14] The formation of the reversible covalent linkage between phot1 LOV domains and the FMN chromophore has been confirmed by X-ray crystallography at a 2.6Å resolution.[15] The structural changes associated with this light-driven molecular switch presumably lead to activation of the carboxyl-terminal protein kinase domain.[14,15]

PHOTOTROPINS FROM SPECIES OTHER THAN *ARABIDOPSIS*

Phototropins have been identified in several plant species besides *Arabidopsis*, including oat (*Avena sativa*),[16] pea (*Pisum sativum*; accession number U83281), rice (*Oryza sativa*),[16] and maize (*Zea mays*; accession number AF033263). Phototropins are not limited to higher plants, as a phototropin has also been found in the unicellular alga *Chlamydomonas reinhardtii*[13] as well as the nonphotosynthetic bacterium *Bacillus subtillis*.[17] The finding that apparent phototropins (based on similar sequence and predicted structure) exist in diverse taxa from prokaryotes to photosynthetic eukaryotes begs the question: Can we classify these proteins as either phot1 or phot2 and thus predict functions? The simple answer at present is maybe.

The *PHOT* genes in the various taxa can be classified as either *PHOT1* or *PHOT2* based on their sequence similarities to the Arabidopsis sequences.[16] Moreover, in higher plants the phototropins can be classed as phot1 or phot2 based on mRNA expression patterns. For example, *PHOT1* and *PHOT2* also have very different gene expression patterns, with *PHOT2* induced by blue light[7,8,18] and *PHOT1* gene expression reduced by blue light. These expression pattern differences fit nicely with the observed differences in physiological functions of *phot1* and *phot2* mutants in *Arabidopsis* under different light conditions. Kasahara and colleagues have recently suggested that phototropins can also be classified as either phot1 or phot2 based on discrete photochemical properties of the LOV domains in each, such as differences between LOV photoproduct formation and dark recovery. How the biochemical differences between the phototropin proteins relate back to differences in gene sequence has not yet been determined. Time will tell whether these generalizations hold true as more functional studies are done.

SIGNALING DOWNSTREAM OF PHOTOTROPINS

Other nonphototropic hypocotyl (*nph*) mutants were isolated in the same screen that identified *phot1*. One of these, *nph3*, was phenotypically similar to *phot1*, suggesting that NPH3 function occurs early in the phototropism signaling pathway, possibly close to photoperception.[19] Whereas there are no transmembrane domains in the NPH3 protein, it localizes to the plasma membrane,[20] like phot1.[2,21] Motchoulski and Liscum reported a physical interaction between phot1 and NPH3 by both a yeast two-hybrid and in vitro pull-down assays. NPH3 may function as a scaffolding protein to bring the photoperceptor, phot1, in contact with other downstream phototropism signaling components.[20] The *NPH3* homologue, *RPT2* (for ROOT PHOTOTROPISM 2),[3] may perform a similar function with phot2 under high light conditions.[22] It is tempting to speculate that other members of the NPH3/RPT2 family may scaffold early signaling components for other physiological response that the phototropins regulate, such as chloroplast relocalization and stomatal control.

Electrophysiological studies are also providing additional insight for signaling events that occur downstream of phototropin activation. First, Baum and associates reported that high-intensity blue light causes transient increase in cytoplasmic Ca^{2+}within 20 seconds of the ''lights on'' signal.[23] When *phot1* mutants were exposed to blue light, the Ca^{2+}increase did not occur, suggesting that phot1 is the photoreceptor responsible for the increase in cytosolic (Ca^{2+}), and that this transient increase in (Ca^{2+}) may be part of some signaling pathway regulated by phot1. Calcium transients are coincident with increased proton pump activity in guard cells that leads to blue-light–induced stomatal opening.[24] Kinoshita and colleagues have demonstrated that blue light activation of the plasma membrane H^+-ATPase does not occur in *phot1 phot2* double mutants,[9] providing a potential connection between the observed Ca^{2+} transients and responses controlled by phototropins.

CONCLUSION

Therefore, phototropins are involved in the regulation of multicellular (phototropism), cellular (stomatal opening), and even subcellular (chloroplast movement) responses in a semiredundant fashion. Each of these responses uses a unique combination of the two phototropins to respond to low-and high-intensity light. Clearly one of the challenges for the next decade is to identify downstream elements of phototropin-signaling pathways in both plant and nonplant species.

ARTICLES OF FURTHER INTEREST

Photoreceptors and Associated Signaling I: Phytochromes, p. 881
Photoreceptors and Associated Signaling II: Cryptochromes, p. 885

Photoreceptors and Associated Signaling IV: UV Receptors, p. 893
Phytochrome and Cryptochrome Functions in Crop Plants, p. 920
Shade Avoidance Syndrome and Its Impact on Agriculture, p. 1152

REFERENCES

1. Arabidopsis Genome Initiative Analysis of the genome sequence of the flowering plant *Arabidopsis thaliana*. Nature **2000**, *408*, 796–815.
2. Liscum, E.; Briggs, W.R. Mutations in the NPH1 locus of *Arabidopsis* disrupt the perception of phototropic stimuli. Plant Cell **1995**, *7*, 473–485.
3. Sakai, T.; Wada, T.; Ishiguro, S.; Okada, K. RPT2: A signal transducer of the phototropic response in *Arabidopsis*. Plant Cell **2000**, *12*, 225–236.
4. Jarillo, J.A.; Ahmad, M.; Cashmore, A.R. *NPL1* (Accession No AF053941): A second member of the *NPH1* serine/threonine kinase family of *Arabidopsis*. Plant Physiol. **1998**, *117*, 719.
5. Sakai, T.; Kagawa, T.; Kasahara, M.; Swartz, T.E.; Christie, J.M.; Briggs, W.R.; Wada, M.; Okada, K. *Arabidopsis* nph1 and npl1: Blue light receptors that mediate both phototropism and chloroplast relocation. Proc. Natl. Acad. Sci. U. S. A. **2001**, *98* (12), 6969–6974.
6. Briggs, W.R.; Christie, J.M. Phototropins 1 and 2: Versatile plant blue-light receptors. Trends Plant Sci. **2002**, *7* (5), 204–210.
7. Kagawa, T.; Sakai, T.; Suetsugu, N.; Oikawa, K.; Ishiguro, S.; Kato, T.; Tabata, S.; Okada, K.; Wada, M. *Arabidopsis* NPL1: A phototropin homolog controlling the chloroplast high-light avoidance response. Science **2001**, *291*, 2138–2141.
8. Jarillo, J.A.; Gabrys, H.; Capel, J.; Alonso, J.M.; Ecker, J.R.; Cashmore, A.R. Phototropin-related NPL1 controls chloroplast relocation induced by blue light. Nature **2001**, *410*, 952–954.
9. Kinoshita, T.; Doi, M.; Suetsugu, N.; Kagawa, T.; Wada, M.; Shiazaki, K. Phot1 and phot2 mediate blue light regulation of stomatal opening. Nature **2001**, *414*, 656–659.
10. Reymond, P.; Short, T.W.; Briggs, W.R.; Poff, K.L. Light-induced phosphorylation of a membrane protein plays an early role in signal transduction for phototropism in *Arabidopsis thaliana*. Proc. Natl. Acad. Sci. U. S. A. **1992**, *89*, 4718–4721.
11. Huala, E.; Oeller, P.W.; Liscum, E.; Han, I.-S.; Larsen, E.; Briggs, W.R. *Arabidopsis* NPH1: A protein kinase with a putative redox-sensing domain. Science **1997**, *278*, 2120–2123.
12. Christie, J.M.; Reymond, P.; Powell, G.K.; Bernasconi, P.; Raibekas, A.A.; Liscum, E.; Briggs, W.R. *Arabidopsis* NPH1: A flavoprotein with the properties of a photoreceptor for phototropism. Science **1998**, *282*, 1698–1701.
13. Kasahara, M.; Swartz, T.E.; Olney, M.A.; Onodera, A.; Mochizuki, N.; Fukuzawa, H.; Asamizu, E.; Tabata, S.; Kanegae, H.; Takano, M.; Christie, J.M.; Nagatani, A.; Briggs, W.R. Photochemical properties of the flavin monoeotide-binding domains of the phototropins from *Arabidopsis* Rice and *Chlamydomonas reinhardtii*. Plant Physiol. **2002**, *129*, 762–773.
14. Swartz, T.E.; Wenzel, P.J.; Corchnoy, S.B.; Briggs, W.R.; Bogomolni, R.A. Vibration spectroscopy reveals light-induced chromophore and protein structural changes in the LOV2 domains of the plant blue-light receptor phototropin 1. Biochemistry **2002**, *41* (23), 7183–7189.
15. Crosson, S.; Moffat, K. Photoexcited structure of a plant photoreceptor domain reveals a light-driven molecular switch. Plant Cell **2002**, *14*, 1–9.
16. Briggs, W.R.; Beck, C.; Cashmore, A.R.; Christie, J.M.; Hughes, J.; Jarillo, J.A.; Kagawa, T.; Kanegae, H.; Liscum, E.; Nagatani, A.; Okada, K.; Salomon, M.; Rüdiger, W.; Sakai, T.; Takano, M.; Wada, M.; Watson, J.C. The phototropin family of photoreceptors. Plant Cell **2001a**, *13*, 993–997.
17. Losi, A.; Polverini, E.; Quest, B.; Gärtner, W. First evidence for phototropin-related blue-light receptors in prokaryotes. Biophys. J. **2002**, *82*, 2627–2634.
18. Kanegae, H.; Tahir, M.; Savazzini, F.; Yamamoto, K.; Yano, M.; Sasaki, T.; Kanegae, T.; Wada, M.; Takano, M. Rice homologues OsNPH1a and OsNPH1b are differently photoregulated. Plant Cell Physiol. **2000**, *41*, 415–423.
19. Liscum, E.; Briggs, W.R. Mutations of *Arabidopsis* in potential transduction and response components of the phototropic signaling pathway. Plant Physiol. **1996**, *112*, 291–296.
20. Motchoulski, A.; Liscum, E. Arabidopsis NPH3: A NPH1 photoreceptor-interacting protein essential for phototropism. Science **1999**, *286*, 961–964.
21. Sakamoto, K.; Briggs, W.R. Cellular and subcellular localization of phototropin I. Plant Cell **2002**, *14*, 1723–1735.
22. Liscum, E.; Stowe-Evans, E. Phototropism: A "simple" physiological response modulated by multiple interacting photosensory-response pathways. Photochem. Photobiol. **2000**, *72* (3), 273–282.
23. Baum, G.; Long, J.C.; Jenkins, G.J.; Trewavas, A.J. Stimulation of blue light phototropic receptor NPH1 causes a transient increase in cytosolic Ca^{2+}. Proc. Natl. Acad. Sci. U. S. A. **1999**, *96* (23), 13554–13559.
24. Shimazaki, K.; Goh, C.H.; Kinoshita, T. Involvement of intracellular Ca^{2+} in blue light-dependent proton pumping in guard cell protoplasts from *Vicia faba*. Physiol. Plant. **1999**, *105*, 554–561.

Photoreceptors and Associated Signaling IV: UV Receptors

T. H. Attridge
University of East London, Stratford, U.K.

N. M. Cooley
CSIRO (Plant Industry), Victoria, Australia

INTRODUCTION

Interest in the perception of ultraviolet began many years ago with physiological evidence for a putative UV-A/blue photoreceptor. As techniques have improved various receptors can be demonstrated to be operating in this region of the electromagnetic spectrum. In this article we examine what is known of these receptors, their response pathways and the interaction of these pathways.

THE RECEPTORS

It is now known that plants contain at least ten photoreceptors. These are five phytochromes (phy A–E), two cryptochromes, two phototropins, and a photoreceptor with properties akin to both phytochrome and phototropin and dubbed superchrome.

The major developmental photoreceptors, crytochrome and phytochrome are capable of absorbing radiation in the UV part of the spectrum. The UV absorption spectra of Pfr and Pr are very similar and only small changes in Pfr/Ptot can be achieved. No ecological or physiological function has been ascribed to these changes. There is no evidence to date to indicate that the five species of phytochrome differ in their UV absorption although differences have been demonstrated in the visible part of the spectrum.

In recent years it has been revealed that there are two types of cryptochrome, CRY1, which can be stimulated by blue and UV-A radiation, and CRY2, which is stimulated by blue. The presence of this UV-A/blue receptor does not preclude the existence of other specific UV-A receptors that act independently, in association with each other, or in association with known photoreceptors.

Phototropins mediate phototropic responses with blue light, green light, and UV-A radiation. Identification has occurred in a number of species and is reviewed by Briggs et al.[1] Phototropins are proteins that have two LOV domains in the N-terminal region, which binds FNM as a chromophore.

UV-B radiation causes damage to DNA but there is also a good deal of evidence that indicates plants possess receptors to UV-B that cause developmental changes. The UV-B receptor has not yet been isolated and little is known of its signaling pathway. Evidence from current research could be assembled to suggest that a major function of UV receptors is to damp down the perception of the signal by the production of UV-absorbing metabolites. This could be of particular importance in reducing UV-B induced DNA damage. Much of what we know about UV-A and UV-B signaling has come from recent studies of UV-regulated gene expression in *Arabidopsis*. What is known[2–4] of the UV induction of chalcone synthase (CHS) in *Arabidopsis* is shown in Fig. 1. CHS appears to be controlled by a number of different photoreceptors and radiation conditions. Experimental evidence has suggested that CRY2 is regulated in the blue while CRY1 is regulated in the UV-A/blue. Using single and double *phyA* and *phyB* mutants, this group[4] is able to show that phytochrome is a positive regulator of the CRY1 inductive pathway with no preference for phyA or phyB and that there is interaction between CRY1 and phyB. UV-B can also control CHS. When this is so, neither CRY1 or CRY2 appear to be involved, nor is the synergy between blue and UV-A pathways involved with the UV-B response. PhyB is a negative regulator of the UV-B inductive pathway and since phyB acts upstream on the UV-B pathway from the points of synergism with blue and UV-A, it has been proposed that the flux through the UV-B and CRY1 pathways might be controlled by phyB.

UV-A RECEPTION

The study of the putative blue/UV-A has lead to our present understanding of the cryptochrome and these photoreceptors are dealt with elsewhere in this volume. However, any understanding of the integration of the signals received by the plant from the electromagnetic spectrum must take into account that in the presence of UV-A, CRY1 can be stimulated and cause developmental

Encyclopedia of Plant and Crop Science
DOI: 10.1081/E-EPCS 120012963

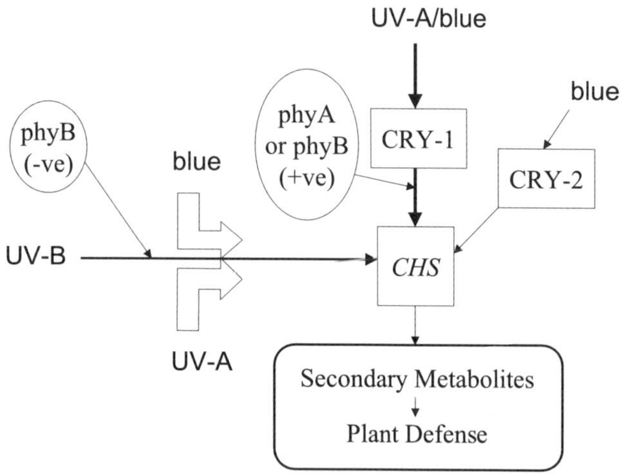

Fig. 1 Summary of the control of CHS gene expression in *Arabidopsis thaliana* showing the interactions of the various pathways and pigments. Nothing is known of the exact points of interaction. (Adapted from Ref. 2 with information from Ref. 4.)

responses. The dual action maxima of CRY1 invites the question ''Do plants have specific UV-A receptors?'' A certain amount of evidence suggests that this may be so.

In an ecological study of a number of meadow species,[5] various growth parameters responded to elevated UV-A under outdoor conditions. These responses were further investigated with action spectra that were produced by growing *Cynosurus cristatus*, a common meadow grass, under polychromatic sources of UV, each with a different $\lambda_{central}$. In the action spectrum shown (Fig. 2), there is a sharp peak on the UV-B/UV-A border and a broad maximum throughout the UV-A. Under outdoor conditions, absorption of blue radiation by cryptochrome would be close to saturation and it is difficult to see, in terms of reciprocity, how the addition of so little energy could cause such significant responses unless UV-A was being received by a specific receptor with its own transduction pathway. Further, in a review of the growth responses of various *Arabidopsis* ecotypes to UV-A and UV-B+A, numerous responses were found to UV-A.[6] One ecotype in particular, Aa-0, was more sensitive to UV-A than UV-B.

UV-B RECEPTORS

UV-B radiation beyond the tolerance of a species will result in damage primarily due to the absorption of UV-B quanta by DNA resulting in dimerization. A secondary cause is the production of reactive oxygen species (ROS), which subsequently cause damage to DNA. However, besides damage, a pivotal role has been suggested for ROS (see below).

Responses to UV-B that are nondamaging or beneficial to the plant may include de-etiolation, flavonoid biosynthesis, leaf and hypocotyl development, regulation of leaf number, and branching frequency.[7] Evidence that these and other plant responses are under the control of a specific UV-B receptor come from a number of action spectra that have an action maximum between 290 and 300 nm.

Fine profile definition of the receptor cannot be expected at these wavelengths since a large number of

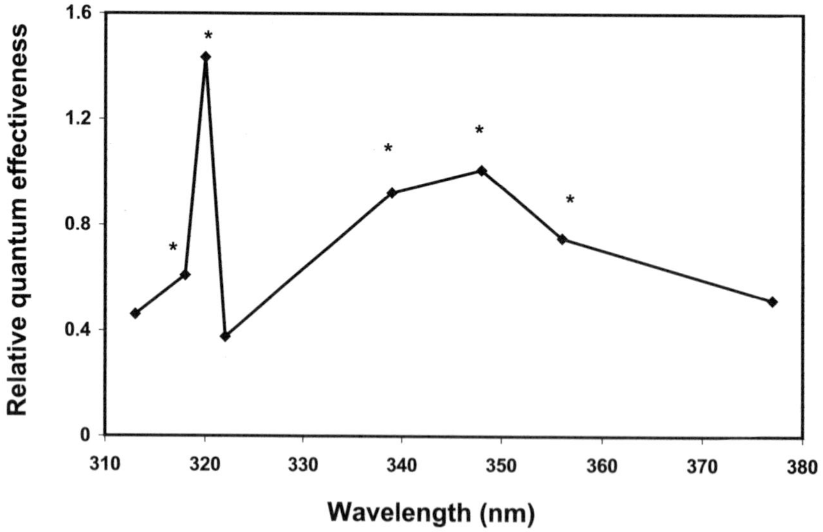

Fig. 2 Polychromatic action spectra for inhibition of leaf area of *Cynosurus cristatus* action spectra showing action maxima in both UV-B and UV-A where *$P<0.05$. (Adapted from Ref. 5.)

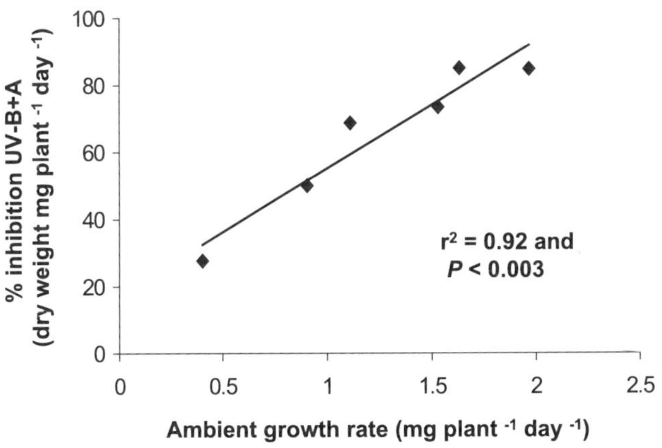

Fig. 3 The % inhibition of dry weight accumulation by supplementary UV-B + A expressed in terms of the accumulation of dry weight by the same ecotypes of *Arabidopsis thaliana* under ambient conditions. (From Ref. 5.)

molecules present in the plant tissues exhibit large changes in absorption and scattering in the UV region and can distort the shape of the action spectra. However, the unassailable argument is that if any of these responses were directly or indirectly linked to DNA damage, then wavelength efficiency would continue to rise with decreasing wavelength rather than form a maximum. Although phytochrome and flavins absorb in the UV there is no coincidence with the 290–300 nm maxima of the action spectra. The absorption of known photoreceptors is poor in the UV spectrum compared to the visible and it is difficult to see how the lesser absorption of these receptors in the UV at lower natural energies could influence their activity via the same induction pathway(s).

A study of responses of various ecotypes of *Arabidopsis* to UV-B[6] demonstrated a straight-line relationship between the growth rate of the control cohort and the level of inhibition induced in the UV-B treatment (see Fig. 3). This relationship indicates that the faster a plant grows the more susceptible it is to UV-B. Higher growth rates infer higher rates of cell division and it has been suggested that there might be a link between UV- B and inhibition of cell division.

Work with the *Arabidopsis* mutants *hy* 42.23N[3] also supports the concept of a specific UV-B receptor in that normal levels of CHS can be generated under UV-B for which there is no other explanation.

TRANSDUCTION

Pharmacological experiments designed to investigate the involvement of calcium in the signal transduction from UV receptors used cell suspensions of *Arabidopsis*, in which CHS was controlled in the same manner as in mature leaves.[3] Using nifedipine, ruthenium red, lanthanum, an ionophore, and the calmodulin antagonist W7 with UV-B and UV-A/blue irradiation, it is argued that calcium at a specific cellular location passes through calcium channels and causes a localised increase in calcium concentration. The calmodulin antagonist W7 strongly inhibits the UV-B induction but not the UV-A/blue induction of CHS trancript. This is in agreement with the genetic evidence that these signals are perceived by separate systems.

As mentioned above, irradiation of plants with UV-B can result in the generation of intracellular ROS (see Ref. 8), which can cause serious damage to membranes. However it has also been suggested that ROS can act as signaling intermediates and are involved in UV-B induced transcript increases in a number of photosynthetic proteins in various plants.[8] ROS may also function as secondary messengers acting up stream in a number of pathways including the synthesis of salicylic acid, jasmonic acid, and ethylene.[8]

SIGNAL DAMPING

CHS is a controling enzyme in the production of flavonoids and phenylpropanoids. If these compounds are produced in the epidermal layers of the plant, they can efficiently absorb UV radiation while also allowing transmission of PAR. The study of *Arabidopsis* mutants deficient in various aspects of flavonoid and lignin synthesis has led to detailed knowledge of the regulation of the phenylpropanoid pathway and its role in plant defense against UV-B.[9] Mutants capable of tolerating high fluences of UV-B have been shown to be able to accumulate high levels of sinapate and flavonoids.[10]

Many of the nondamage plant strategies turned on by UV radiation would include reduction of leaf area, reduction in plant height, branching, and leaf number. These may all be interpreted as strategies that would reduce interception of radiation. Thus stimulation of receptors of UV may well function to produce screens and other developmental changes that would lead to reduced reception of harmful radiation.

ARTICLE OF FURTHER INTEREST

Photoreceptors and Associated Signaling II: Cryptochromes, p. 885

REFERENCES

1. Briggs, W.R.; Cashmore, A.R.; Christie, J.M.; Hughes, J.; Jarolli, J.A.; Kagawa, T.; Kanegae, H.; Liscum, E.; Natatani, A.; Okada, K.; Salonom, M.; Rudiger, W.; Sakai, T.; Takano, M.; Wada, M.; Watson, J.C. The phototropin family of photoreceptors. Plant Cell **2001**, *13*, 993–997.
2. Fuglevand, G.; Jackson, J.A.; Jenkins, G.I. UV-B, UV-A and blue light signal transduction pathways interact synergistically to regulate chalcone synthase gene expression in *Arabidopsis*. Plant Cell **1996**, *8*, 2347–2357.
3. Jenkins, G.I. UV and blue light signal transduction in *Arabidopsis*. Plant Cell Environ. **1997**, *20*, 773–778.
4. Wade, H.K.; Bibikova, T.N.; Valentine, W.J.; Jenkins, G.I. Interactions within a network of phytochrome, cryptochrome and UV-B phototransduction pathways regulate chalcone synthase gene expression in *Arabidopsis* leaf tissue. Plant J. **2001**, *25* (6), 675–685.
5. Cooley, N.M.; Truscott, H.M.F.; Holmes, M.G.; Attridge, T.H. Outdoor ultraviolet polychromatic action spectra for growth responses of *Bellis perennis* and *Cynosurus cristatus*. J. Photochem. Photobiol., B Biol. **2000**, *59*, 64–71.
6. Cooley, N.M.; Higgins, J.T.; Holmes, M.G.; Attridge, T.H. Ecotypic differences in responses of *Arabidopsis thaliana* L. to elevated polychromatic UV-A and UV-B+A radiation in the natural environment: A positive correlation between UV-B+A inhibition and growth rate. J. Photochem. Photobiol., B Biol. **2001**, *60*, 143–150.
7. Holmes, M.G. Non-damaging and Positive Effects of UV-Radiation on Higher Plants. In *Environmental UV Radiation: Impact on Ecosystems and Human Health and Predicitive Models*; Ghetti, F., Vass, I., Eds.; NATO Science Series 2 Mathematics, Physics and Chemistry, Springer-Verlag: Berlin, 2002; Vol. XX, *in press*.
8. A.-H.-Mackerness, S. Plant responses to ultraviolet-B (UV-B:280–320 nm) stress: What are the key regulators? Plant Growth Regul. **2000**, *32*, 27–39.
9. Bharti, A.K.; Khurana, J.P. Mutants of *Arabidopsis* as tools to understand the regulation of phenylpropanoid pathway and UV-B protection mechanism. Photochem. Photobiol. **1997**, *65* (5), 765–776.
10. Bieza, K.; Lois, R. An *Arabidopsis* mutant tolerant to lethal ultraviolet-B levels shows constitutively elevated accumulation of flavonoids and other phenolics. Plant Physiol. **2001**, *126* (3), 1105–1115.

Photosynthate Partitioning and Transport

María Fabiana Drincovich
Rosario National University, Rosario, Argentina

INTRODUCTION

Triose phosphates produced by the Calvin cycle in the chloroplasts may be converted to starch or exported to the cytosol, where they synthesize sucrose to be transported to other parts of the plant. In this way, the pathways for sucrose and starch synthesis are separated in the cytosol and plastids and are fed by different hexose phosphate pools. The activity of the triose phosphate translocator (TPT)—located in the chloroplast inner membrane—is central in communicating the hexose phosphate pools of both compartments by exchanging the three carbon intermediates dihydroxyacetone phosphate (DHAP) and 3-phospho glycerate (3PGA) for inorganic phosphate (Pi).

SUCROSE SYNTHESIS OCCURS IN THE CYTOSOL

Sucrose is a major product of photosynthesis and serves as the long-distance transport compound in most plants. Triose phosphates synthesized in the chloroplasts by the Calvin cycle are transported to the cytosol by the TPT where they produce hexoses and subsequent synthesis of sucrose.

Once in the cytosol, the combined action of triose-phosphate isomerase and fructose 1,6-aldolase converts triose phosphates to fructose 1,6-bisphosphate (F1,6BP). The interconversion of F1,6-BP and fructose 6-phosphate (F6P) involves the regulatory metabolite fructose 2,6-bisphosphate (F2,6BP) and three enzymes: ATP-dependent phospofructokinase (PFK; F6P+ATP $\rightarrow$ F1,6BP+ADP), pyrophosphate-dependent phospofructokinase (PFP; F6P+PPi $\leftrightarrow$ F1,6BP+Pi), and fructose 1,6-bisphosphatase (F1,6BPase; F1,6BP+H_2O $\rightarrow$ F6P+Pi).[1] PFK and F1,6Base are found in both the cytosolic and plastidic compartments in leaves but PFP occurs exclusively in the cytosol. PFP catalyses a reversible reaction and is activated by F2,6BP in the direction of F1,6BP formation, although its direct role in controlling F1,6BP synthesis and degradation in vivo in leaves is still under discussion. PFK is inhibited by PEP and by other metabolites of the last part of the glycolitic pathway. In this way, this pathway is negative-feedback regulated, involving the function of the TPT, which means that if DHAP is supplied by the chloroplast, there is no need to generate F1,6BP through glycolysis (Fig. 1). Cytosolic F1,6BPase regulates the flow of carbon from the triose phosphate to the hexose phosphate pool. This enzyme is strongly inhibited by the metabolite F2,6BP, which is synthesized from F6P by fructose 6-phosphate 2-kinase and converted back to F6P by fructose 2,6-biphosphatase (Fig. 1). Fructose 6-phosphate 2-kinase is activated by Pi and F6P and inhibited by triose phosphates, in contrast to fructose 2,6-biphosphatase, which is inhibited by Pi and F6P. Hence, the concentration of F2,6BP is related to the activity of the TPT, which determines the concentration of triose phosphates and Pi in the cytosol. As the concentration of F2,6BP is controlled by the status of the triose phosphate pool, F1,6BPase is active only when carbon is supplied by the chloroplast (Fig. 1).

After converting triose phosphates to hexose phosphates, the principal route of sucrose synthesis combines the reactions of sucrose-phosphate synthase (SPS, UDP-glucose+fructose 6-P $\leftrightarrow$ sucrose 6-P+UDP) and sucrose-phosphate phosphatase, which has a large negative free energy change (sucrose 6-P $\rightarrow$ sucrose+Pi).[2] The synthesis of UDP-glucose involves the action of the enzyme UDP-glucose pyrophosphorylase (G1P+UTP $\leftrightarrow$ UDP-glucose+PPi). In the absence of a cytosolic pyrophosphatase, the reaction is readily reversible. A second enzyme—sucrose synthase (UDP-glucose+Fructose $\leftrightarrow$ sucrose+UDP)—is capable of catalyzing both sucrose synthesis and degradation. Nevertheless, this enzyme predominates in sucrose-utilizing tissues and is more involved in sucrose degradation. SPS is regulated by both covalent modification and allosteric modulation. Glucose 6-phosphate (G6P) directly activates the enzyme and also inhibits a kinase that phosphorylates and down-regulates the activity of SPS, while Pi inhibits SPS and a phosphatase that up-regulates the activity of SPS (Fig. 1).

STARCH SYNTHESIS OCCURS IN PLASTIDS

When the synthesis of sucrose exceeds the capacity of the leaf to export it, synthesis of starch is used as an overflow mechanism to store carbohydrate. Triose phosphates in the chloroplast are converted to hexose phosphates, which form ADP-glucose to incorporate glucose to starch. The

Encyclopedia of Plant and Crop Science
DOI: 10.1081/E-EPCS 120010402

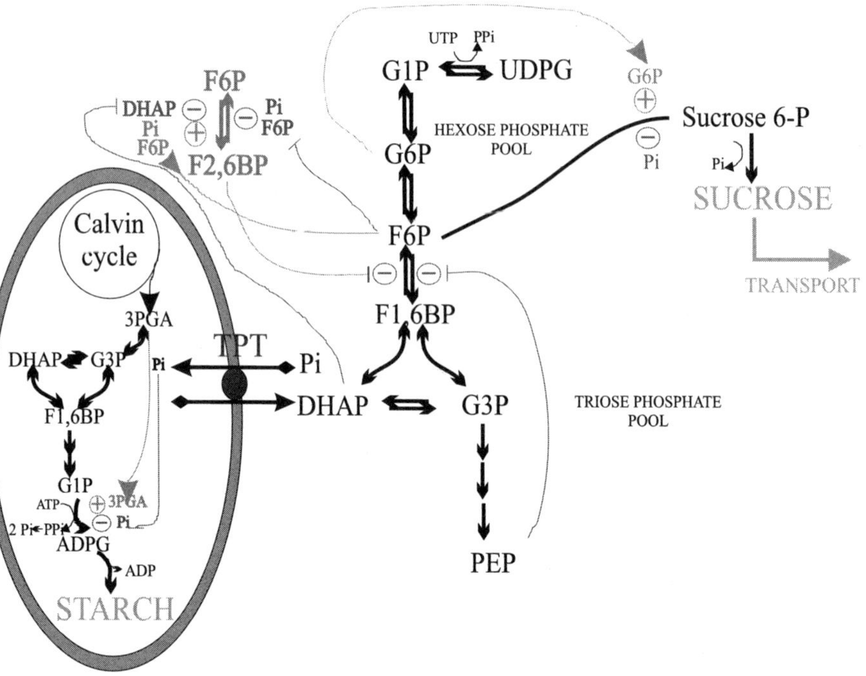

Fig. 1 Regulation of starch and sucrose synthesis. Metabolite levels are constantly sensed in both the cytosolic and chloroplastic compartments and kept in relation to the rate of photosynthesis. 3PGA: 3-phospho glycerate; ADPG: ADP-glucose; DHAP: dihydroxyacetone phosphate; G3P: 3-phospho glyceraldehide; F1,6BP: fructose 1,6-bisphosphate; F6P; fructose 6-phosphate; F2,6BP: fructose 2,6-bisphosphate; G1P: Glucose 1-phosphate; G6P: Glucose 6-phosphate; PEP: phosphoenolpyruvate; Pi: inorganic phosphate; PPi: pyrophosphate; TPT: triose phosphate translocator; UDPG: UDP-glucose. (*View this art in color at www.dekker.com.*)

enzyme that synthesizes ADP-glucose—ADP glucose pyrophosphorylase (Glucose 1-P+ATP$\leftrightarrow$ADP-glucose +PPi)—is the regulatory enzyme in starch biosynthesis.[3] A plastidic-specific pyrophosphatase cleaves pyrophosphate (PPi) driving ADP-glucose synthesis. In turn, starch synthase incorporates glucose from ADP-glucose to the nonreducing end of amylose or amylopectin chain, while a starch-branching enzyme produces $1\rightarrow6$ α linkages. When chloroplastic 3PGA is abundant, ADP glucose pyrophosphorylase is activated. In contrast, when Pi concentration increases, the enzyme is inhibited (Fig. 1). Pi concentration can rise not only when photosynthesis slows, but also when the TPT is active in exchanging triose phosphate from the Calvin cycle by cytosolic Pi.

INTEGRATED CONTROL OF SUCROSE AND STARCH SYNTHESIS INVOLVES TWO CELL COMPARTMENTS

Different processes tightly regulate the balance between sucrose and starch synthesis, which are both related with a constant sensing of metabolite levels in both the chloroplast and the cytosol. The balance between sucrose and starch synthesis varies during the day in response to different environmental conditions that regulate the rate of carbon assimilation and photosynthesis and sucrose transport.

At the beginning of the light period, the Calvin cycle becomes operational and triose phosphates accumulate in the chloroplast. This accumulation triggers the TPT to exchange triose phosphates by Pi, increasing cytosolic concentration of triose phosphates with a consequent decrease in Pi concentration. In turn, F2,6BP concentration decreases, relieving the inhibition of F1,6BPase. This allows the flow of carbon into the hexose phosphate pool (Fig. 1). The increase in G6P and decrease in Pi concentrations activate SPS, allowing the synthesis of sucrose to proceed. Meanwhile, starch synthesis is inhibited by high Pi and low PGA concentration in the chloroplast. Sucrose export is active in this condition (Fig. 2A).

When the increase in photosynthate exceeds the ability of the cell to export sucrose—which may occur at midday at high light conditions—starch begins to be synthesized from excess carbon. Concurrently, sucrose accumulation inhibits its own synthesis, hexose phosphates accumulate, and F6P induces F2,6BP synthesis (Fig. 1). The increase in this metabolite inhibits F1,6BPase, producing an accumulation of triose phosphates in the cytosol that slows the export from the chloroplast. 3PGA accumulates in the chloroplasts, activating ADP-glucose

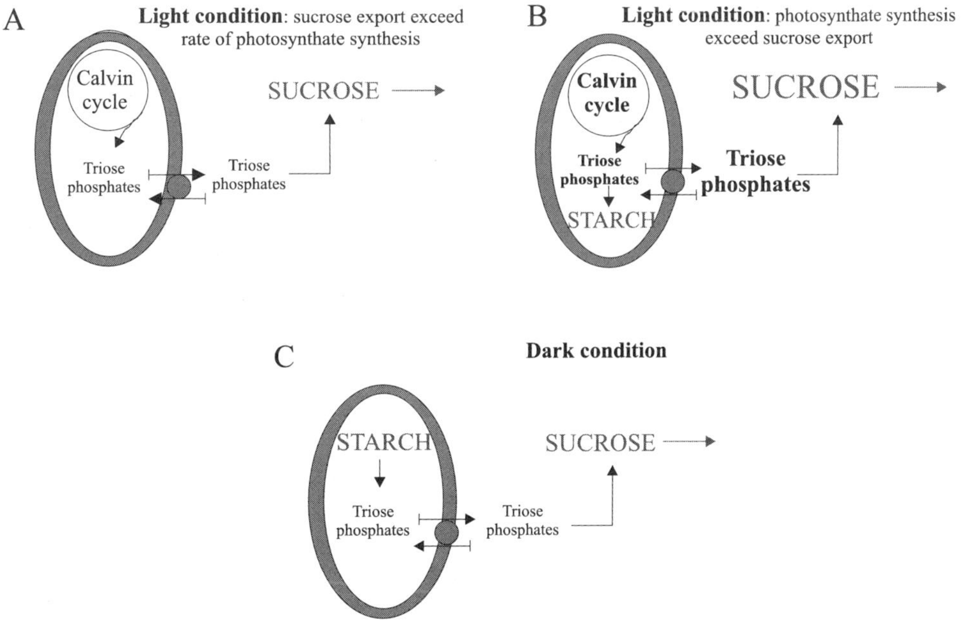

Fig. 2 Three conditions with different balance between starch and sucrose synthesis are shown. A and B: Light condition with different rates of photosynthate synthesis; C: Dark condition in which starch degradation provides triose phosphates for sucrose synthesis. TPT: triose phosphate translocator. (*View this art in color at www.dekker.com.*)

pyrophosphorylase and promoting starch synthesis (Fig. 1). Pi concentration in the cytosol increases as a result of decreased import into the chloroplast, which contributes to the inhibition of SPS and the increase in F2,6BP concentration. Hence, carbon flow is derived from sucrose to starch (Fig. 2B).

When light declines, the rate of photosynthesis slows, and with it the rate of sucrose synthesis. The decrease in 3PGA concentration in the chloroplasts no longer activates ADP-glucose pyrophosphorylase, which, in combination with Pi accumulation, induces the stop of starch biosynthesis. At this time, amylolytic and/or phosphorolytic starch breakdown is triggered. Triose phosphates are transported to the cytosol and F2,6BP decreases the amount of carbon (and hence, sucrose) allowed to flow into the hexose phosphate pool. Sucrose is then transported to other parts of the plant. In this condition, sucrose export exceeds the rate of synthesis (Fig. 2C).

TRANSGENIC PLANTS WITH ALTERED SUCROSE OR STARCH LEVELS

Transgenic plants with altered expression of the TPT have been developed. Potato plants with lower TPT activity presented increased starch levels, indicating a central role of the TPT in the flux of carbon into sucrose and starch.[4] Later, antisense TPT tobacco lines showed that the decrease in TPT levels could be compensated by high turnover of starch and subsequent glucose export from the chloroplast to the cytosol.[5] Recently, transgenic tobacco plants with antisense repression or overexpression of the TPT[6] were used to calculate control coefficients of the TPT on carbon partitioning and photosynthesis.[7] Under high CO_2, tobacco plants with decreased or increased TPT activity presented, respectively, higher or lower carbon incorporation into starch.[6,7] Transgenic plants with altered amounts of F2,6BP have also been developed.[8,9] These plants have directly demonstrated the importance of this regulatory compound in the integration of cytosolic and chloroplastic metabolism during photosynthesis. Elevated F2,6BP levels reduced the rate of sucrose synthesis at the start of the light period, whereas a decreased content of F2,6BP promoted the partitioning of photosynthates into sucrose relative to starch, with a decline in the rate of carbon assimilation. In this way, decisive evidence was provided of the crucial role of F2,6BP in coordinating sucrose synthesis with the rate of carbon fixation and in regulating photosynthate partitioning between sucrose and starch.

APOPLASTIC OR SYMPLASTIC PHLOEM LOADING IN LEAVES DEPENDS ON THE SPECIES

Sucrose synthesized in leaves is transported to the site of consumption and storage via the phloem. Phloem loading in leaves may occur apoplastically or symplastically depending on the species.[10] Anatomical features have been used to categorize plant species based on their mechanism of phloem loading. The symplastic path occurs via the plasmodemata, which connects mesophyll cells and the conducting elements of the phloem. In the apoplastic route, sucrose is first exported into the apoplast and then is taken up by the phloem by an energy-dependent transport system.[10]

ARTICLES OF FURTHER INTEREST

Glycolysis, p. 547
Metabolism, Primary: Engineering Pathways of, p. 714
Starch, p. 1175
Sucrose, p. 1179

REFERENCES

1. Stitt, M. Fructose 2,6-bisphiosphate as a regulatory molecule in plants. Annu. Rev. Plant Physiol. Plant Mol. Biol. **1990**, *41*, 153–185.
2. Huber, S.C.; Huber, J.L. Role and regulation of sucrose-phosphate synthase in plants. Annu. Rev. Plant Physiol. Plant Mol. Biol. **1996**, *47*, 431–444.
3. Martin, C.; Smith, A.M. Starch biosythesis. Plant Cell **1995**, *7*, 971–985.
4. Reismeier, J.W.; Flügge, U.-I.; Schulz, B.; Heineke, D.; Heldt, H.W.; Willmitzer, L.; Frommer, W.B. Antisense repression of the chloroplast triose phosphate translocator affects carbon partitioning in transgenic potato plants. Proc. Natl. Acad. Sci. U. S. A. **1993**, *90*, 6160–6164.
5. Häusler, R.E.; Schlieben, N.H.; Schultz, B.; Flügge, U.-I. Compensation of decreased triose phosphate/phosphate translocator activity by accelerated starch turnover and glucose transport in transgenic tobacco. Planta **1998**, *204*, 366–376.
6. Häusler, R.E.; Schlieben, N.H.; Nicolay, P.; Fischer, K.; Flügge, U.-I. Control of carbon partitioning and photosynthesis by the triose/phosphate translocator in transgenic tobacco plants (*Nicotiana tabacum*). I. Comparative physiological analysis of tobacco plants with antisense repression and overexpression of the triose/phosphate translocator. Planta **2000**, *210*, 371–382.
7. Häusler, R.E.; Schlieben, N.H.; Flügge, U.-I. Control of carbon partitioning and photosynthesis by the triose/phosphate translocator in transgenic tobacco plants (*Nicotiana tabacum*). II. Assessment of control coefficients of the triose/phosphate translocator. Planta **2000**, *210*, 383–390.
8. Scott, O.; Lange, A.J.; Pilkis, S.J.; Kruger, N.J. Carbon metabolism in leaves of transgenic tobacco (*Nicotiana tabacum* L.) containing elevated fructose 2,6-bisphosphate levels. Plant J. **1995**, *7*, 461–469.
9. Scott, P.; Lange, A.J.; Kruger, N.J. Photosynthetic carbon metabolism in leaves of transgenic tobacco (*Nicotiana tabacum* L.) containing decreased amounts of fructose 2,6-bisphosphate. Planta **2000**, *211*, 864–873.
10. Kühn, C.; Barker, L.; Bürkle, L.; Frommer, W.-B. Update on sucrose transport in higher plants. J. Exp. Bot. **1999**, *50*, 935–953.

Photosynthesis and Stress

Barbara Demmig-Adams
Volker Ebbert
William W. Adams III
University of Colorado, Boulder, Colorado, U.S.A.

INTRODUCTION

The impact of stress on photosynthesis can be viewed as the response of photosynthesis to environmental conditions outside the favorable range. For example, although differing greatly in their origin, the stresses of freezing temperatures, low water availability, and high salinity have a common impact in that they all result in low water availability to the plant. In response to this, stomates close to varying degrees and net carbon uptake declines. But what happens behind closed stomates? In some species, the decline of carbon uptake is due merely to a transient closure of stomates—with a full maintenance of internal maximal photosynthetic capacity. In other species, stomatal closure is accompanied by a downregulation of the internal maximal capacity of photosynthesis, including, e.g., the capacity of photosynthetic electron transport. In evergreens, this downregulation of electron transport capacity is typically accompanied by a phenomenon referred to as photoinhibition of photosynthesis.

PHOTOSYNTHESIS BEHIND CLOSED STOMATES

Whereas some species maintain their intrinsic photosynthetic capacity when stomates close under unfavorable environmental conditions, others downregulate it. Those species that maintain electron transport capacity use electrons in processes other than carbon fixation (such as photorespiration or oxygen reduction in the Mehler-peroxidase pathway).[1] This offers the advantage that solar energy can still be used to provide some adenosine triphosphate (ATP) and reducing equivalents—even when there is little or no net carbon gain.

Other species suspend all photosynthetic activity when net carbon uptake ceases and strongly downregulate photosynthetic electron transport capacity. This strategy precludes any utilization of solar energy, but offers the advantage of effectively suppressing the transfer of electrons to potentially toxic superoxide (Fig. 1). This latter strategy necessitates an efficient dissipation of excess absorbed solar energy no longer used in photochemistry as harmless thermal energy (Fig. 2). In many evergreens, the utilization of solar energy in photosynthesis is strongly downregulated under environmental stress, yet the light-harvesting chlorophyll is preserved.[2] This poses a potentially lethal danger in the form of energy transfer from chlorophyll to toxic singlet excited oxygen (Fig. 2).[1,3] Such a transfer of energy from chlorophyll to singlet oxygen can be suppressed effectively by the harmless dissipation of this energy as thermal energy. The syndrome of persistently low utilization of absorbed solar energy in photosynthesis and highly efficient thermal dissipation is a key feature of the photoinhibition of photosynthesis under stress (Fig. 2).[4]

PHOTOINHIBITION OF PHOTOSYNTHESIS

Photoinhibition of photosynthesis can occur in response to exposure to a wide variety of environmental stresses or to sudden increases in growth light intensity. The hallmark of photoinhibition is a combination of two features. First, a decrease in maximal electron transport capacity, resulting from inactivation and subsequent degradation of photosystem II centers. This is accompanied by a second feature—a switch to continuously high levels of harmless dissipation of a potentially lethal excess of light absorbed in the largely preserved light-harvesting antennae of photosystem II (Fig. 2). Whereas these features are generally accepted, current interpretations vary and range from viewing photoinhibition as reflecting damage to photosystem II centers on one hand[5] versus downregulation of photochemistry in response to a low demand for photosynthate on the other.[2,6] Some consequently predict an increase in photosynthetic productivity under stress if photoinhibition could be eliminated. It must be noted that those species exhibiting the strongest photoinhibition in the field are those that possess the highest levels of stress tolerance.[2,4]

PERSISTENT AND FLEXIBLE FORMS OF PHOTOPROTECTION

In evergreens, the switch from utilizing solar energy in photosynthesis to dissipating it as heat can persist for an

Encyclopedia of Plant and Crop Science
DOI: 10.1081/E-EPCS 120019320

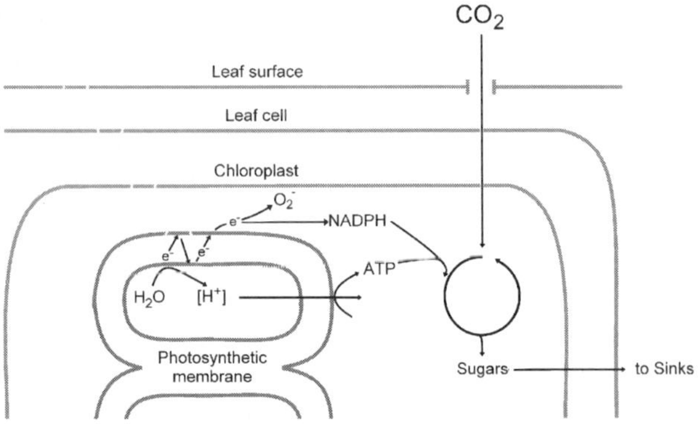

Fig. 1 Schematic depiction of photosynthetic reactions in leaves, including the absorption of solar energy and photosynthetic electron transport, formation of NADP and ATP, and uptake of carbon dioxide and its conversion to sugars. O_2^- is superoxide, formed by transfer of an electron (e^-) to oxygen. (*View this art in color at www.dekker.com.*)

entire unfavorable season. All species, however, utilize a flexible version of this process. All sun-exposed leaves reroute excess absorbed solar energy to thermal dissipation during peak exposure to direct sunlight, and return to efficient harvesting of solar energy for photosynthesis when light levels decline again.[7] The importance of this process is illustrated by the fact that mutants deficient in thermal dissipation have a reduced reproductive fitness.[8] The fraction of absorbed light allocated to thermal dissipation typically increases under a variety of environmental stresses.[7] Thermal dissipation is thus a key photoprotective mechanism in plants and counteracts the formation of toxic reactive oxygen when the utilization of solar energy in photosynthesis declines (Fig. 2).[1,3,9] In addition, all species also possess back-up mechanisms to detoxify reactive oxygen species once formed.[9]

CONTRASTING ACCLIMATION PATTERNS IN SPECIES WITH DIFFERENT GROWTH FORMS AND LIFESPAN

What are the reasons for species-dependent differences in stress response? The answer may lie in the degree to which growth processes and utilization of carbon are affected by environmental extremes in a given species. A key factor regulating maximal photosynthetic capacity is the demand for photosynthate at the whole plant level.[10] As long as photosynthate continues to be consumed in growth processes (i.e., sinks; Fig. 1), a high photosynthetic capacity is maintained. When unfavorable environmental conditions lead to a cessation of growth and carbon utilization, photosynthetic capacity is downregulated. Different species vary in how strongly growth and carbon utilization respond to environmental change.

Freezing Temperatures

In seasonally cold climates, overwintering annual and biennial species exhibit different responses than overwintering long-lived evergreens. (Figs. 3–5). All of these species likely suspend net carbon uptake on subfreezing days. Annuals or biennials, on one hand, maintain intermittent growth as well as maximal photosynthetic capacity (Fig. 3, *Malva neglecta*; Fig. 4), and rapidly resume photosynthesis on milder winter days.[2] Some evergreens, on the other hand, cease growth for the duration of the winter, downregulate their maximal photosynthetic capacity (Fig. 3, Douglas fir; Fig. 5), exhibit strong photoinhibition, and remain photosynthetically inactive even on milder days.[2]

Why don't all species exercise the same flexibility as the annuals and biennials? The acclimation patterns are apparently associated with lifespan. Species with a shorter lifespan tend to invest in maintaining growth and maximal photosynthetic capacities, whereas many species with a longer lifespan simply abandon growth and downregulate maximal photosynthetic capacity during unfavorable seasons. What might be the advantage of the latter strategy? The downregulating evergreens clearly possess a very high level of stress tolerance. Only species employing this latter strategy are found in extreme environments, such as above 3000 m in subalpine forests where soil water remains frozen throughout the winter.

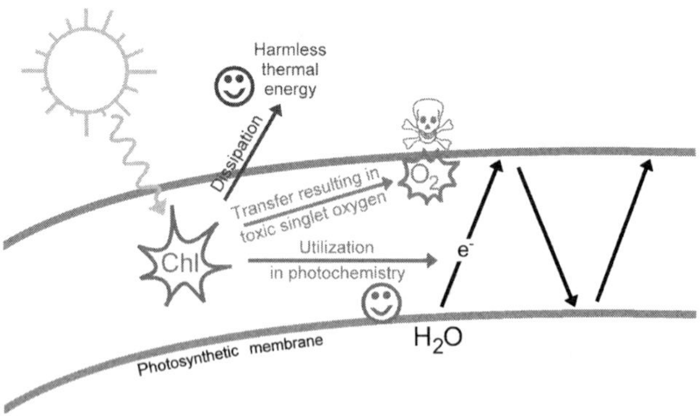

Fig. 2 Schematic depiction of principal fates of solar energy absorbed by light-harvesting chlorophyll. This energy can be utilized in photochemistry, resulting in photosynthetic electron transport. When this utilization via photochemistry is either downregulated or does not occur fast enough to match the rate of light absorption, dangerous transfer of energy to oxygen (leading to the formation of toxic singlet oxygen) is suppressed by harmless dissipation of the excess energy as thermal energy (heat). (*View this art in color at www.dekker.com.*)

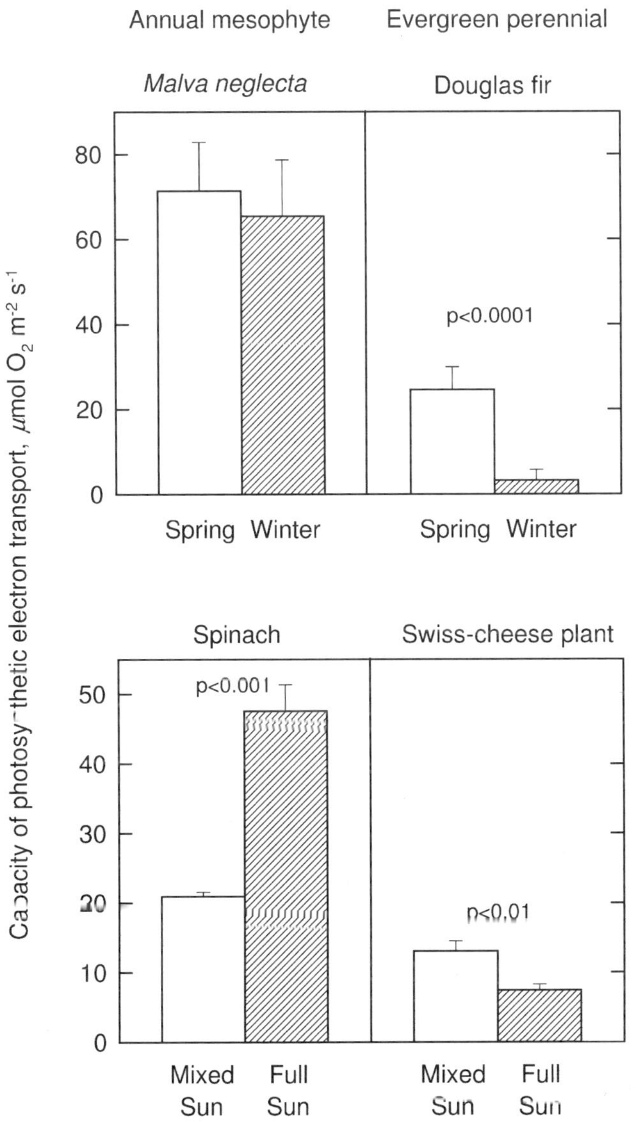

Fig. 3 Response of maximal capacities of photosynthetic electron transport to environmental extremes in different species. The seasonal response of naturally growing populations of the biennial herb *Malva neglecta* (see also Fig. 4) is compared with that of the conifer Douglas fir (*Pseudotsuga menziesii*; see also Fig. 5), both growing at 1700–2000 m near Boulder, Colorado, U.S.A. (data from Ref. 2). The acclimation to different growth light environments in a naturally lit greenhouse of the crop plant spinach is compared with that of the evergreen perennial *Monstera deliciosa* (Swiss-cheese plant; see also Fig. 8; V. Ebbert and D. L. Mellman, unpublished data). Plants were grown with ample water and nutrient supply.

Low Water Availability and Salinity

Contrasting strategies have also been observed under stress conditions other than freezing temperatures, such as low water availability at moderate temperatures. For species adapted to arid (dry) habitats, several principal

Fig. 4 Photograph of *Malva neglecta*, growing during the winter in Boulder, Colorado, U.S.A. (1700 m). (*View this art in color at www.dekker.com.*)

response types have been described. Some species accumulate osmotically active substances to maintain water uptake from the soil (i.e., perform osmotic adjustment), and it is often thought that crop yield under drought may be increased by enhancing solute accumulation.[11] However, osmotic adjustment may be beneficial only in combination with enhanced root development to allow access to water deeper in the soil.[11] An important group of species (phreatophytes) native to arid environments use

Fig. 5 Photograph of Douglas fir, on the south-facing slope of Gregory Canyon west of Boulder, Colorado, U.S.A. (*View this art in color at www.dekker.com.*)

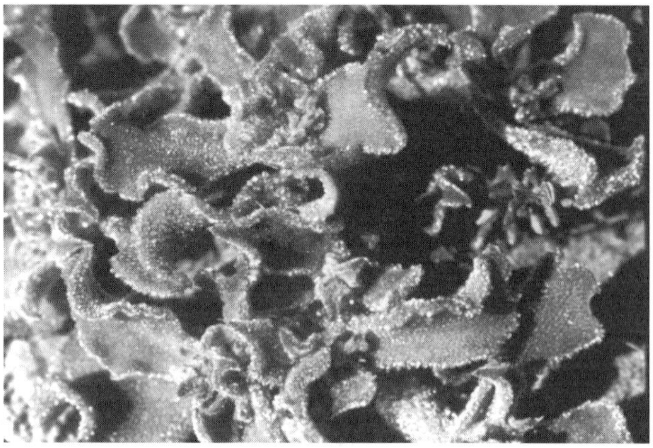

Fig. 6 Photograph of *Mesembryanthemum crystallinum*, growing along the Pacific coast of Baja Norte, Mexico, in late December. (*View this art in color at www.dekker.com.*)

this strategy. But there are also cases where neither osmotic adjustment nor turgor maintenance takes place and strong photosynthetic downregulation and photoinhibition occur, such as in *Nerium oleander* that is native to seasonally dry environments. Salinity is another environmental factor leading to low water availability to the plant. Among species adapted to high levels of salinity (halophytes), different strategies can once again be observed. An example of a species with an extremely flexible strategy is *Mesembryanthemum crystallinum* (Fig. 6) that switches from C3 photosynthesis and rapid growth under more favorable conditions to slowed growth, enhanced succulence, and Crassulacean acid metabolism (CAM) under low water availability/high salinity. Other species, such as mangroves (Fig. 7), instead maintain low growth rates over a wide range of salinities and thermally

Fig. 7 Photograph of the mangrove *Rhizophora stylosa*, growing along the pacific coast of Queensland, Australia. (Photograph by Otto L. Lange.) (*View this art in color at www.dekker.com.*)

Fig. 8 Photograph of 12-week-old spinach plants grown from seed in a sunlit greenhouse and watered with low-nitrate (0.25 mmol/L) (left) versus ample-nitrate (10 mmol/L) (right) nutrient solution. (*View this art in color at www.dekker.com.*)

dissipate a large fraction of the light they absorb when sun-exposed.

Low Nitrogen Availability

Photosynthetic response to low nitrogen availability has been studied extensively in annual species such as spinach.[12] Nitrogen deficiency affects the growing points most severely and leads to stunted overall growth (Fig. 8). This results in severe sink limitation (cf. Fig. 1), and carbon accumulates in the photosynthesizing source leaves. This induces photosynthetic downregulation and favors remobilization of nitrogen from older source leaves to the growing points. Because the downregulation of photosynthesis affects both the capacity of light harvesting and electron transport, and the absorption of light thus decreases along with its utilization, photoinhibition typically does not occur in this species.

Although not studied in mechanistic detail, evergreens adapted to low-nutrient environments tend to exhibit less visible effects and can apparently remain green.

Fig. 9 Photograph of *Monstera deliciosa*, growing in full sun exposure in a greenhouse.(*View this art in color at www.dekker.com.*)

Exposure to High Light Levels

Annuals typically increase photosynthetic capacity with increasing growth light intensity, as long as other factors such as nitrogen availability are not limiting (Fig. 3, spinach). Evergreens tend to show less pronounced increases, even under favorable conditions (Fig. 3, Swiss-cheese plant; Fig. 9), and instead exhibit strong increases in thermal dissipation of excess absorbed light and sometimes strong photoinhibition of photosynthesis.[4] The maximal photosynthetic capacities of the evergreens shown in Fig. 3 are much lower than those of the annuals and biennials. It appears that the life span of leaves is inversely related to maximal photosynthesis rate but positively related to the propensity for photoinhibition.

PHOTOSYNTHETIC PATHWAY AND STRESS RESPONSE OF PHOTOSYNTHESIS

As mentioned earlier, exposure to low water availability triggers a switch to CAM in *M. crystallinum*. Both C4 photosynthesis and CAM are characterized by a higher water-use efficiency and nitrogen-use efficiency than C3 photosynthesis, thus offering advantages when these factors are limiting.[13] In addition, both pathways confer a higher tolerance of high temperatures as a result of minimal photorespiration rates and, in the case of some CAM plants, a higher intrinsic high temperature tolerance.

CONCLUSION

The response of photosynthesis to environmental stress clearly depends on species. Two contrasting examples are long-lived species that endure entire unfavorable seasons in a state of suspended growth and photosynthesis versus short-lived species that continuously maintain growth and photosynthesis but are limited to less extreme environments. Suspension of intrinsic photosynthetic capacity during adverse seasons is dependent on the downregulation of photochemical conversion of solar energy into high-energy electrons and the upregulation of photoprotective thermal dissipation of the absorbed solar energy.

ACKNOWLEDGMENTS

We gratefully acknowledge the financial support of the Andrew W. Mellon Foundation, The National Science Foundation (No. IBN-0235351), and the USDA (CSREES No. 00-35100-9564). We thank David L. Mellman for performing the oxygen evolution analyses on spinach and *Monstera deliciosa* depicted in Fig. 3.

ARTICLES OF FURTHER INTEREST

C_3 Photosynthesis to Crassulacean Acid Metabolism Shift in Mesembryanthemum crystallinum: A Stress Tolerance Mechanism, p. 241
Carotenoids in Photosynthesis, p. 245
Osmotic Adjustment and Osmoregulation, p. 850
Plant Response to Stress: Modifications of the Photosynthetic Apparatus, p. 990
Plant Response to Stress: Source-sink Regulation by Stress, p. 1010

REFERENCES

1. Niyogi, K.K. Safety valves for photosynthesis. Curr. Opin. Plant Biol. **2000**, *3* (6), 455–460.
2. Adams, W.W., III; Rosenstiel, T.N.; Demmig-Adams, B.; Ebbert, V.; Brightwell, A.K. Photosynthesis and photoprotection in overwintering plants. Plant Biol. **2002**, *4*, 545–557.
3. Demmig-Adams, B.; Adams, W.W., III. Antioxidants in photosynthesis and human nutrition. Science **2002**, *298*, 2149–2153.
4. Demmig-Adams, B.; Adams, W.W., III; Ebbert, V.; Logan, B.A. Ecophysiology of the Xanthophyll Cycle. In *The Photochemistry of Carotenoids. Advances in Photosynthesis*; Frank, H.A., Young, H.A., Britton, G., Cogdell, R.J., Eds.; Kluwer Academic Publishers: Dordrecht, 1999; Vol. 8, 245–269.
5. Melis, A. Photosystem-II damage and repair cycle in chloroplasts: What modulates the rate of photodamage in vivo? Trends Plant Sci. **1999**, *4*, 130–135.
6. Adams, W.W., III; Demmig-Adams, B.; Rosenstiel, T.N.; Ebbert, V. Dependence of photosynthesis and energy dissipation activity upon growth form and light environment during the winter. Photosynth. Res. **2001**, *67*, 51–62.
7. Demmig-Adams, B.; Adams, W.W., III. The role of xanthophyll cycle carotenoids in the protection of photosynthesis. Trends Plant Sci. **1996**, *1*, 21–26.
8. Külheim, C.; Ågren, J.; Jansson, S. Rapid regulation of light harvesting and plant fitness in the field. Science **2002**, *297*, 91–93.
9. Niyogi, K.K. Photoprotection revisted: Genetic and molecular approaches. Ann. Rev. Plant Physiol. Plant Mol. Biol. **1999**, *50*, 333–359.
10. Koch, K.E. Carbohydrate-modulated gene expression in plants. Ann. Rev. Plant Physiol. Plant Mol. Biol. **1996**, *47*, 509–540.
11. Serraj, R.; Sinclair, T.R. Osmolyte accumulation: Can it really help increase crop yield under drought conditions? Plant Cell Environ. **2002**, *25* (2), 333–341.
12. Paul, M.J.; Driscoll, S.P. Sugar repression of photosynthesis: The role of carbohydrates in signalling nitrogen deficiency through source:sink imbalance. Plant Cell Environ. **1997**, *20* (1), 110–116.
13. Larcher, W. *Physiological Plant Ecology*, 4th Ed.; Springer-Verlag: Berlin, 2003; 1–513.

Photosystems: Electron Flow Through

Arlene Haffa
Arizona State University, Tempe, Arizona, U.S.A.

INTRODUCTION

The capture of photons and their conversion into an energy form that can be stored and utilized later is the purpose of photosynthesis. Photosynthesis requires electron transfer reactions through photosystems to transduce photon energy to a charge-separated state, the power of which drives the synthesis of organic compounds useful for further chemistry. This transduction requires that the process be fast enough and the involved sites sufficiently spatially separated to prevent charge-separated states from recombining back into charge-neutral states before they are utilized. The segregation of charge is accomplished via a series of oxidation-reduction reactions through several pigment-protein complexes. In plant chloroplasts, electrons are transferred through two pathways: a linear noncyclic series of electron carriers that connect the beginning and the end of the system, and a cyclic pathway that operates around the second half of the system.

ELECTRON TRANSFER REACTIONS

Electron transfer involves an oxidation/reduction reaction (redox for short). One molecule yields an electron and is called the donor (D). A different molecule receives the electron and is called the acceptor (A). D becomes electron deficient, i.e., is oxidized and becomes a cation. A is reduced and becomes an anion. D has a more negative redox potential, i.e., it gives up an electron more readily than A. In photosynthesis, energy for the first reaction, which excites the donor with the most negative redox potential, is provided by photon energy and is called a photochemical reaction. Electrons are then transferred to a series of compounds with increasingly more positive (i.e., more easily reduced) redox potentials. This process is energetically downhill and does not require additional light but, along with the first reaction, is collectively called the light reactions. Good photon absorbers typically have conjugated double bonds, i.e., alternating single and double carbon-carbon bonds. A single photon that has the proper energy to excite an electron can move the electron from a lower to a higher energy orbital. If electron transfer occurs before the excited molecule relaxes to the ground state, then the excited molecule (D*) gives the electron to an acceptor (A). This can be written as:

$$D \xrightarrow{photon} D^* \rightarrow D^+ + A^-$$

D is the primary donor. If conditions are energetically favorable, A can transfer its electron to another nearby molecule. A is referred to as the primary acceptor and the next molecule to receive the electron is a secondary acceptor.

ELECTRON TRANSFER REACTIONS IN PLANTS

Primary electron transfer in photosystems occurs in pigment-protein complexes called reaction centers. The proteins hold the pigments at proper distances and orientations for these reactions and also create electrostatic fields that make the reactions energetically possible. Exitonic energy arrives at the reaction center via an antenna complex where it excites two closely positioned chlorophyll *a* molecules, an arrangement referred to as the special pair (P, or the dimer). Plants have two types of reaction centers called Photosystem I (PSI) and Photosystem II (PSII). The special pairs of these photosystems are called P700 and P680, respectively, based on the wavelength of light necessary to excite them. Both P700 and P680 are on the lumenal side of the thylakoid membranes of higher plants, which is important for creating charge-separated states and proton-motive forces. The entire photosynthetic electron transfer chain also includes a cytochrome b_6f complex, which does not directly require a photon to initiate the electron transfer process. Many types of bacteria are capable of performing photosynthesis with only one photosystem.[1]

NONCYCLIC OR LINEAR ELECTRON TRANSFER

Electron flow through both systems is called noncyclic, or linear, because it is a one-way flow via the three major complexes of the photosynthetic chain: PSII, the cytochrome b_6f complex, and PSI (Fig. 1). Plastoquinone (PQ)

Encyclopedia of Plant and Crop Science
DOI: 10.1081/E-EPCS 120019316

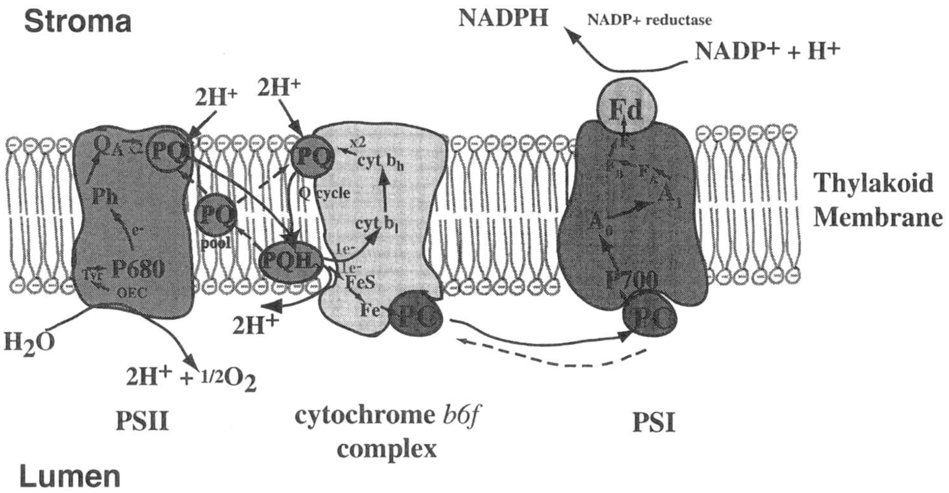

Fig. 1 Linear electron flow from water to $NADP^+$ and the creation of a proton-motive force. Abbreviations are given in the article. (Created using information from Refs. 2–4 and 8–10.)

and plastocyanin (PC) act as shuttles between the first and second and the second and third complexes, respectively.

When PSII donates an electron to the electron transport chain that connects the photosystems, it regains electrons by oxidizing water to produce oxygen. PSI donates electrons to $NADP^+$ to produce NADPH, and regains its electrons from the electron transport carrier plastocyanin.

The first electron transfer within the reaction center of PSII is from P680 to a pheophytin (a chlorophyll molecule lacking the central Mg atom), which occurs in several picoseconds.[5] Electrons are subsequently transferred to quinone acceptors (first Q_A then Q_B) over longer time scales (hundreds of picoseconds and then microseconds) toward the stromal side of the membrane. Q_B is

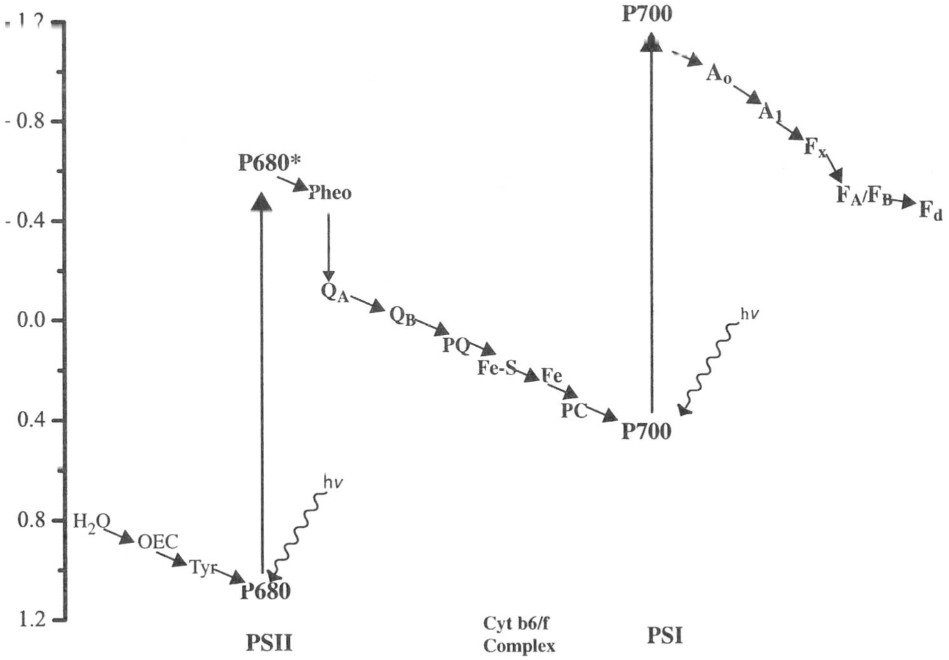

Fig. 2 Z-scheme depicting electron flow in plant photosynthesis from water to $NADP^+$, showing relative energies of pigments along the transfer pathway that make the process thermodynamically favorable. (From Ref. 2.)

not permanently fixed in PSII and is generally called plastoquinone (PQ) when it is not associated with PSII. For simplicity only PQ is used in Fig. 2. Each PQ molecule requires two electrons and two protons to become reduced, which is therefore a two-photon process. Before the second photon can excite P680, the reaction center is brought from an oxidized state back to the ground state by an electron from water via an oxygen-evolving complex (OEC), and a tyrosine residue (Tyr) on one of the protein subunits. After full reduction ($PQ + 2e^- + 2H^+ \rightarrow PQH_2$), PQH_2 leaves PSII, and the binding site is refilled by an unreduced PQ from the plastoquinone pool in the membrane. PQH_2 transfers its electrons to a cytochrome b_6f complex. One electron is transferred to a cytochrome b (b_l for low then b_h for high energy) through the Q cycle and the other is transferred to a Rieske Fe-S center that participates in the linear electron transport. Electron transfer from PQH_2 to the cytochrome complex is accompanied by proton release. The protons, which were taken from the stromal side of the thylakoid membrane, are released on the lumenal side, forming a proton-motive force. The potential energy in the proton-motive force is used to drive conversion of ADP to ATP by ATP synthase. The electron on the Rieske center is transferred from an iron (Fe) atom in the iron-sulfur complex to plastocyanin, located on the lumenal side of the complex. Plastocynanin leaves the complex and docks on PSI where it transfers an electron to P700+, reducing it to P700.

Electron transfer through PSI is similarly initiated by a photon.[2] P700 transfers an electron to A_0 (a chlorophyll *a* molecule), which then transfers it to A_1 (a phylloquinone). A_0 and A_1 designate the primary and secondary electron acceptors. The electron is then passed through bound iron-sulfur centers (F_X to F_A to F_B) to a ferredoxin (Fd), which is soluble. Ferredoxin then transfers the electron to an $NADP^+$ reductase, which completes the electron flow by reducing $NADP^+$ to NADPH.

The Z-scheme (for zigzag) is a description of electron flow in photosynthesis from water to $NADP^+$ (Fig. 2).[3,4] It provides information about the energetics of the pigments in the electron transfer chain.

CYCLIC ELECTRON TRANSFER

Ferredoxin is also capable of performing electron transfer directly back to the electron transfer chain instead of to $NADP^+$. Electrons are then passed to PQ, through the cytochrome b_6f complex to plastocynanin and back to $P700^+$. This process produces a proton gradient (and thus ATP) but not NADPH.

REACTION CENTER STRUCTURE/FUNCTION

The fundamental cores of PSI and PSII consist of symmetrically arranged cofactors imbedded in a dimer of protein subunits. Although there is symmetry in the arrangement of the photosynthetic cofactors, experiments suggest that electron transfer occurs primarily down one side of the reaction center. The symmetry is broken by differences in amino acids in the protein subunits. These differences alter the redox properties of the cofactors and dictate the directionality of the electron transfer reactions.

A detailed understanding of the exact electron transfer mechanisms in PSII has not been elucidated.[5] Evidence for electron transfer occurring down one side of PSII has traditionally been extrapolated from bacterial reaction centers of purple nonsulfur bacteria.[1] In the bacterial reaction center, the sides are called A and B, but the basic overall structure—although simplified and incapable of oxidizing water—possesses homology to PSII. Numerous spectroscopic experiments on the femtosecond timescale have been performed.[6] Although the structure is symmetrical, only the A side is observed as active under normal physiological conditions. One exception is the high energy excitation that results (lifetime <15 ps) in B-side charge separation using the B-side accessory bacteriochlorophyll as the primary donor.[7] Advances in the purification of PSII and experimentation on these pigment/protein complexes leads to the same conclusion in this photosystem (i.e., there is an active and an inactive branch).[8] The spectral properties of PSII differ from those of the bacterial reaction center, however, and the mechanism of electron transfer may be different.[5]

PSI also possesses an approximate symmetry.[2] Experiments on PSI have not led to a clear conclusion as to whether one or both of the electron transfer pathways is used. Data suggest that electron transfer may occur on both sides, but one side seems to be kinetically much faster. Electron transfer on the slower branch cannot be observed without somehow impairing the faster branch.

CONCLUSION

The study of electron flow through photosystems has revealed that plants are capable of utilizing either linear or cyclic electron transfer pathways to convert light into stored energy forms. The two systems allow plants to adapt to a variety of environmental conditions. Primary charge separation reactions occur in symmetric pigment/protein complexes called reaction centers. Further study is needed to determine how the symmetric structure of the reaction center relates to its function.

ARTICLES OF FURTHER INTEREST

ATP and NADPH, p. 68
Chlorophylls, p. 258
Exciton Theory, p. 429
Fluorescence: A Probe of Photosynthetic Apparatus, p. 472

REFERENCES

1. *Anoxygenic Photosynthetic Bacteria*; Blankenship, R.E., Madigan, M.T., Bauer, C.E., Eds.; Kluwer Academic Publishers: Boston, 1995.
2. Diner, B.A.; Rappaport, F. Structure, dynamics, and energetics of the primary photochemistry of photosystem II of oxygenic photosynthesis. Annu. Rev. Plant Biol. **2002**, *53*, 551–580.
3. Chitnis, P.R. Photosystem I: Function and physiology. Annu. Rev. Plant Physiol. Plant Mol. Biol. **2001**, *52*, 593–626.
4. Hill, R.; Bendall, F. Function of the two cytochrome components in chloroplasts: A working hypothesis. Nature **1960**, *186*, 136–137.
5. Blankenship, R.E. *Molecular Mechanisms of Photosynthesis*; Blackwell Science: Ames, 2002.
6. Woodbury, N.; Allen, J.P. Electron Transfer in Purple Nonsulfur Bacteria. In *Anoxygenic Photosynthetic Bacteria*; Blankenship, R.E., Madigan, M.T., Bauer, C.E., Eds.; Kluwer Academic Publishers: Boston, 1995; 527–557.
7. Lin, S.; Katilius, E.; Haffa, A.L.M.; Taguchi, A.K.W.; Woodbury, N.W. Blue light drives B-side electron transfer in bacterial photosynthetic reaction centers. Biochemistry **2001**, *40* (46), 13767–13773.
8. Dekker, J.P.; van Grondelle, R. Primary charge separation in photosystem II—Minireview. Photosynth. Res. **2000**, *63*, 195–208.
9. http://photoscience.la.asu.edu/photosyn/photoweb/default.html (accessed March 2003).
10. http://www.bio.ic.ac.uk/research/barber/barbergroup.html (accessed March 2003).

Physiology of Herbivorous Insects

Cedric Gillott
University of Saskatchewan, Saskatoon, Canada

INTRODUCTION

The long duration of insect–plant coevolution, especially the coevolution of insects and angiosperms since the Cretaceous (150 million years b.p.), has been a major contributor to the diversity of species seen in modern Insecta. Of the approximately 1 million extant insect species identified, more than 80% feed on plant material at some stage in their life. This high proportion, plus the enormous numbers of individuals in populations of a species that occur under appropriate conditions (e.g., up to 10^{10} locusts in a swarm), means that herbivorous insects play a major role in the flow of energy from primary producers to upper-level consumers. All parts of a plant may be used as food: leaves, stems, roots, sap, pollen, nectar, seeds, fruits, and nuts. In many species, but especially Lepidoptera and many Diptera and Hymenoptera, the diet may change from the juvenile to the adult stage. Further, insects may live on or within plants, some species (e.g., gall formers) triggering changes in the structure and physiology of the host for their benefit. To take advantage of the availability of plants as sources of food and shelter, an insect's structure and physiology, especially in respect to feeding and reproduction, have taken on special adaptations.

EXTERNAL FEATURES

Body Form

In most adult insects and juvenile exopterygotes (insects with external wing development, such as grasshoppers, termites, true bugs, and thrips), the division of the body into three major regions (head, thorax, and abdomen) is readily visible (Fig. 1A). However, in many juvenile endopterygotes (insects with a pupal stage, such as ants, bees and wasps, beetles, butterflies and moths, and true flies), the separation of the regions may be indistinct (Fig. 1B, C). Each region comprises a number of segments, each of which primitively possessed a pair of appendages. In modern insects, these appendages may have disappeared; further, new nonsegmental structures may have arisen.

Head

The head is the major sensory center, typically bearing a pair of segmental antennae, which are the principal location of sense organs for touch, taste, and smell, and nonsegmental eyes. Adult insects and juvenile exopterygotes have paired compound eyes, typically having several thousand photosensitive units. Compound eyes are capable of limited form perception, color vision, and in some species, for example the honeybee, perception of polarized light that can be used in navigation and orientation. Simple eyes (up to six pairs) found in juvenile endopterygotes and many adult insects each comprise fewer photosensitive units than compound eyes. In caterpillars and beetle larvae they appear to function like compound eyes. In adults they may measure light intensity, be involved in maintenance of diurnal rhythms, and help to maintain stability during horizontal flight. The mouthparts are derived from ancestral segmental appendages. In herbivorous insects they may be adapted for chewing (as in grasshoppers and beetles), piercing and sucking (aphids), or sucking only (butterflies and moths). In chewing mouthparts the well-developed mandibles typically have both cutting and grinding surfaces (Fig. 2A), while the labrum (upper lip), labium (lower lip), and maxillae taste and manipulate the food. In sucking species, the mandibles and maxillae generally form elongate stylets that pierce tissue. The maxillae are interlocked to form the salivary and food canals, and the labium is developed as a protective tube (Fig. 2B).

Thorax

The thorax is the locomotory center of the insect. Typically, it bears three pairs of legs and, in adults only, the wings (Fig. 1A). Legs are secondarily absent in larvae of Diptera (Fig. 1C) and many Coleoptera and Hymenoptera. Though normally used for walking and running, legs may be modified for other functions, such as jumping (Fig. 1A), sound production, digging (Fig. 3A), and pollen collection (Fig. 3B). The legs of many insects, notably Diptera and Lepidoptera, have many chemosensilla as well allowing these insects to taste the substrate as they walk over it!

Encyclopedia of Plant and Crop Science
DOI: 10.1081/E-EPCS 120010477

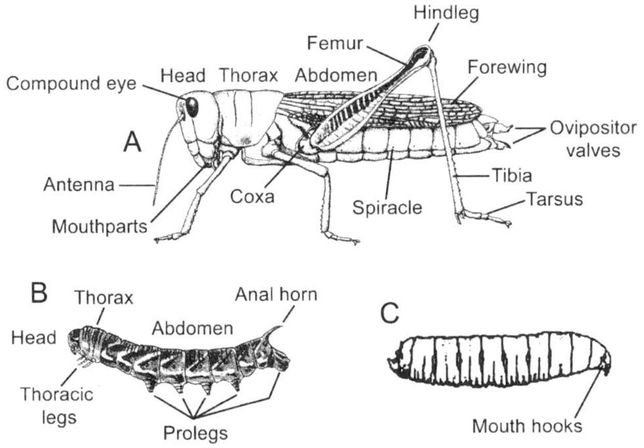

Fig. 1 Insect external features: (A) Grasshopper, (B) hornworm caterpillar, (C) cabbage root maggot. In the grasshopper the three body divisions—head, thorax, and abdomen—are readily seen. Though the caterpillar's head is obvious, the junction between the thorax and abdomen is indistinct; the thorax is recognized by the presence of the three pairs of true legs. In the maggot the only sign of the head are the paired hooks, believed to be remnants of the maxillae. *Note*: The insects are not drawn to scale. (Fig. 1A from Pfadt, R.E. *Field Guide to the Common Western Grasshoppers*; USDA APHIS, Wyoming Agriculture Experiment Station, Bulletin 912, 1988.)

The evolution of wings and flight was a key factor in the success of insects. In the short term, flight facilitates escape from predators as well as dispersal, which is especially useful for species whose food sources and breeding sites are sparsely distributed. In the long term, dispersal leads to colonization of new habitats, geographic isolation, and, ultimately, speciation. Wings, generally two pairs, may be modified for both flight-related and other functions. Reduction to a single pair, or linking of the fore- and hindwings on each side, improves aerodynamic efficiency. In Diptera the greatly reduced hindwings (halteres) function as gyroscopic organs. The hardened forewings (elytra) of beetles provide protection. In crickets and grasshoppers the forewings are used in sound production. In Lepidoptera the wings may be colored and patterned for camouflage or to warn would-be predators that the insect is toxic. Occasionally, a nontoxic species (e.g., the viceroy butterfly) avoids being eaten by having wings that mimic those of a toxic species (e.g., the monarch butterfly).

Abdomen

The abdomen is the metabolic and reproductive center of the insect. The ancestral segmental appendages are lost on most segments, but at the posterior tip they remain as the external genitalia. In males they serve as clasping and intromittent organs, while in females they typically form the ovipositor whose structure reflects the site of egg laying. Thus, in grasshoppers that lay eggs in soil, the short and heavy ovipositor is used to dig a hole; in cicadas it is a hard tubelike structure that places the eggs in plant tissues; and in many Hymenoptera it is variously modified for sawing, piercing, boring, and stinging.

Integument

Covering the entire surface of the insect's body—including the sense organs—as well as lining the foregut, hindgut, tracheal system, and parts of the reproductive system, is an inert, acellular layer called the cuticle. The cuticle may comprise, from inside to outside, the procuticle and a very thin, multilayered epicuticle. Where strength and rigidity are required, as for muscle attachment and general protection, the procuticle may be relatively thick. However, in regions of the body where flexibility or sensitivity is paramount, as at leg joints and over sense organs, only epicuticle is found. The outer layer of the epicuticle is a monolayer of wax molecules which, by reducing water loss from the body, has been another factor contributing to the success of insects as terrestrial organisms. The wax may also be important in preventing entry of microorganisms and in chemical communication where it serves as a pheromone or kairomone (see "Coordinating Systems," following). Color production is also a function of the cuticle. Pigmentary colors result from the occurrence of specific pigments in the cuticle, including carotenoids, pteridines, and ommochromes. Iridescent colors result from the scattering of light due to the laminated nature of the procuticle.

PHYSIOLOGY

Maintenance Systems

Several insect maintenance systems differ markedly from those of vertebrates. For example, the gas-exchange system of insects comprises a tubular air-filled system with segmental openings to the exterior (spiracles) along each side of the body (Figs. 1A, 4). The tubes (tracheae and tracheoles) carry oxygen to and carbon dioxide from the core of tissues, making gas exchange highly efficient. Thus, the blood of insects has no role in gas exchange. Blood in insects does not travel within vessels as in vertebrates. Instead, the simple tubular heart, aided by various undulating membranes, pushes blood sluggishly past the internal organs and through the appendages. Excretion and salt–water balance are achieved via a series of fine Malpighian tubules (Fig. 5A) that filter ions, small organic molecules (including uric acid, the normal form of

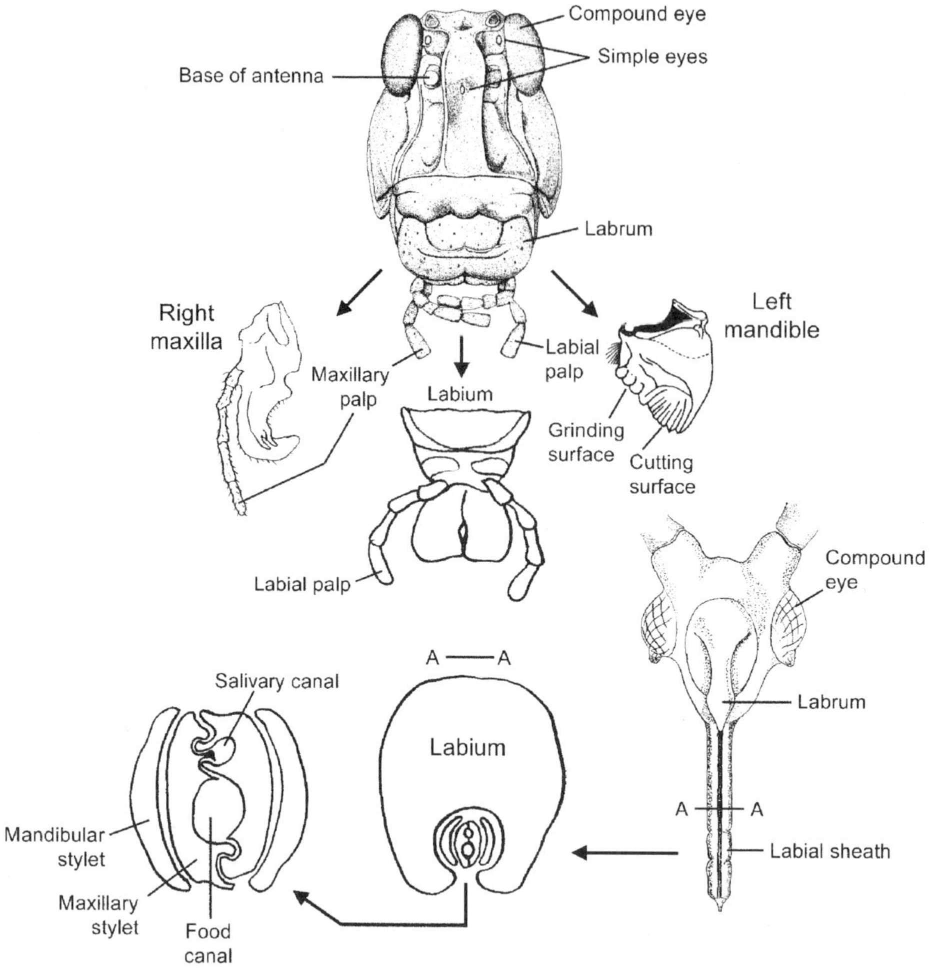

Fig. 2 Head and mouthparts: (A) Grasshopper (cutting and grinding mouthparts); (B) homopteran (piercing and sucking mouthparts). In the grasshopper the mandibles have both shearing and grinding surfaces. The maxillae, labrum (upper lip), and labium (lower lip) are used to test the texture and taste of the food, as well as to manipulate it. In Homoptera the labium serves as a protective sheath for the delicate stylets, which are highly modified mandibles and maxillae. The maxillae interlock to form the salivary and food canals. (Fig. 2A in part redrawn from Pfadt, R.E. *Field Guide to the Common Western Grasshoppers*; USDA APHIS, Wyoming Agriculture Experiment Station, Bulletin 912, 1988. Fig. 2B in part redrawn from Dixon, A.F.G. *Biology of Aphids*; Edward Arnold, London, 1973. Reproduced by permission of Hodder/Arnold Publishers.)

nitrogenous waste), and water from the blood and release these into the hindgut. In the hindgut, resorption of useful materials occurs, and the remaining materials form urine.

Coordinating Systems

Like vertebrates, insects coordinate physiological activity by means of nervous and endocrine systems. The nervous system is relatively simple in both structure and function; much activity is the result of simple reflex pathways, and insects possess very limited ability to learn and memorize. By contrast, insect hormonal systems have roles comparable to those of vertebrates—for example, in reproduction, growth, excretion, and blood–sugar regulation—and appear to be equally complex. Communication and coordination of behavior among insects may include visual, tactile, and auditory stimuli, but chemical signals are usually of greater importance. They include pheromones, allomones, and kairomones, which are chemicals released from the body of one insect to induce a behavioral or physiological response in other insects. Pheromones act intraspecifically and are used by numerous species, but especially Lepidoptera and Coleoptera, to attract mates. Indeed, synthetic sex attractants are widely used in integrated pest management (IPM). Other pheromones regulate sexual maturation and caste differentiation in social insects; mark trails to and from food and indicate its quality and quantity; stimulate aggregation or

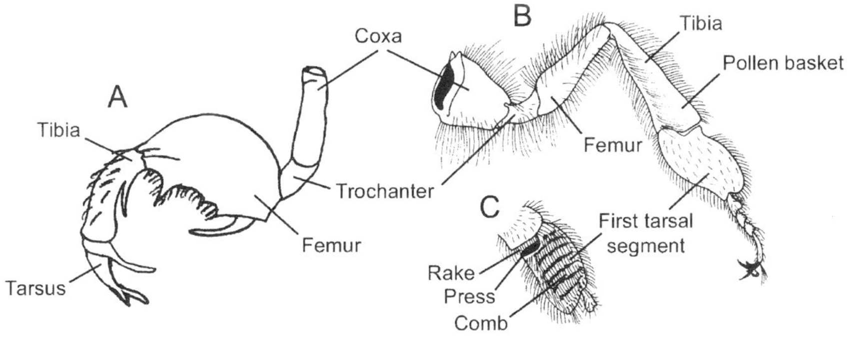

Fig. 3 Insect leg adaptations: (A) Foreleg of cicada (digging and tunneling); (B), (C) hindleg of honey bee (pollen collection). In the cicada the coxa is lengthened to provide increased rotation, the femur is spadelike for moving soil loosened by the tibia and tarsus, which are equipped with spines. On the bee's hindleg, rows of hairs (comb) on the inner side of the first tarsal segment (C) scrape pollen off the abdomen. The rake, a row of hairs at the distal end of the tibia, collects pollen from the comb of the opposite leg and transfers it to the pollen press. The compacted pollen is then moved into the pollen basket (B) where it is stored until the bee returns to the nest.

dispersion; and warn of impending danger. Allomones are used to repel would-be predators, as in the release of formic acid by some ants. Kairomones are chemical signals (including some pheromones) released by one species and detected by another. For example, the aggregation pheromone of bark beetles is used by hymenopteran parasites to locate their hosts.

Food Acquisition and Utilization

The feeding habits of herbivorous insects have special significance for humans from an opposite perspective. Some insects cause enormous damage to crops and stored plant products, whereas others are of massive benefit as pollinators. Generally, olfactory stimuli trigger the initial orientation to a potential food source. Phagostimulants or deterrents on or within the food then determine whether feeding continues. The alimentary canal comprises three regions: foregut, midgut, and hindgut (Fig. 5A). In insects that feed on solid matter, the canal is generally fairly short. By contrast, in most sap feeders, which must both

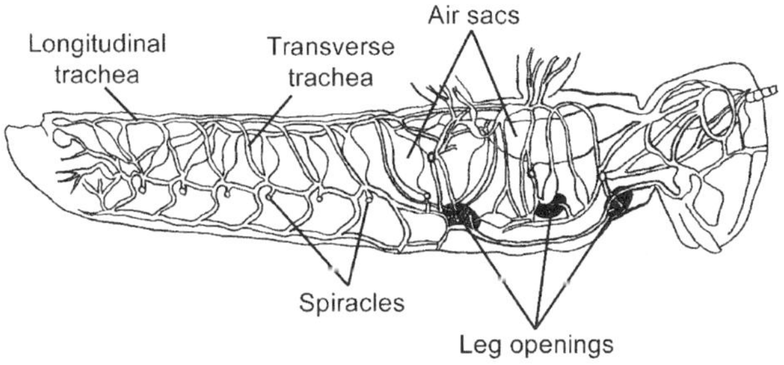

Fig. 4 Grasshopper tracheal system. This is a network of gas-filled tubes that open to the exterior via paired segmental spiracles. The tubes branch and spread as increasingly finer tracheoles to all parts of the body. In larger insects sections of the tubes may be dilated, forming air sacs that increase the volume of air that can be exchanged during ventilation.

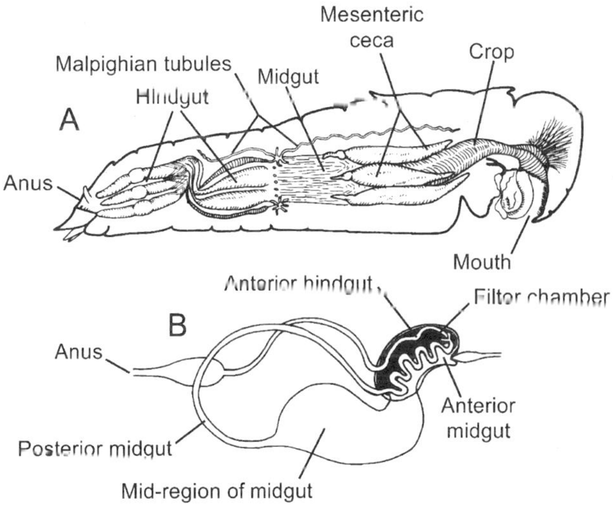

Fig. 5 Insect alimentary canal: (A) Grasshopper, (B) froghopper. The gut of a grasshopper is relatively short and straight. It includes a crop where food is stored and further ground up; the midgut and its diverticula (mesenteric ceca), where digestion and absorption take place; and the hindgut, where feces are produced. By contrast, in the sap-feeding froghopper the gut is very long and coils back on itself, so that the anterior midgut and anterior hindgut are closely adjacent within the filter chamber. This enables excess water in the sap to be rapidly translocated into the hindgut, thereby concentrating the food for digestion and avoiding harmful dilution of the blood. (Fig. 5A redrawn from Hodge, C. The anatomy and histology of the alimentary tract of *Locusta migratoria* L. (Orthoptera: Acrididae). J. Morphol. **1939**, *64*, 375–399. By permission of Wiley-Liss, Inc., a subsidiary of John Wiley & Sons, Inc.)

concentrate the food prior to digestion and prevent the excess water from diluting the blood, the canal is long and coiled back upon itself, and it passes through a filter chamber (Fig. 5B). Here, the anterior midgut and hindgut are closely adjacent, allowing water to pass from the former to the latter and thus to be excreted without entering the body cavity. Generally, the foregut is where food is stored. The midgut, lined with a peritrophic matrix that protects this region from physical damage and entry of microorganisms, is where digestion and absorption occur. The hindgut processes undigested material into feces, though in many herbivores, especially those that feed on woody materials, this region may have special sections that house symbiotic bacteria or protozoa to facilitate digestion. Only a few insects manufacture enzymes capable of splitting cellulose and hemicellulose. The hindgut symbionts degrade such compounds into glucose and organic acids, which are then absorbed through the hindgut wall.

Following their absorption across the gut wall, products of digestion are metabolized in the fat body, a tissue analogous in function to the vertebrate liver. Here, they are converted into appropriate energy-supplying or growth-promoting molecules, or are stored. The fat body is also the principal site for detoxification of insecticides and plant-produced toxins.

Reproduction

Generally, insects have a high reproductive capacity and a short generation time. By facilitating rapid adaptation to changing environmental conditions, including development of insecticide resistance, these factors have also been important in the success of insects. Most insects reproduce sexually and lay eggs. Males locate females using visual, auditory, or chemical cues, especially sex attractants. The latter may be active over distances of several kilometers. Among herbivorous insects, gravid females determine which are suitable host plants by chemical or visual cues. Eggs may be laid in soil, or on or within plant tissues. Embryogenesis, larval development, and metamorphosis are usually rapid, enabling several generations to occur each year. However, in cold climates there may be only one generation annually, with most species overwintering in the egg or pupal stage.

Though most species are bisexual and lay eggs, others—including many pestiferous Homoptera—include parthenogenesis (which produces female-only generations) and sometimes also viviparity in their reproductive repertoire. These unusual strategies lead to massive population increases, hence the insects' ability to exploit a temporary abundance in their food supply and their prominence as major vectors of plant diseases.

ARTICLES OF FURTHER INTEREST

Insect Life History Strategies: Development and Growth, p. 598

Plant Defenses Against Insects: Constitutive and Induced Chemical Defenses, p. 939

Plant Defenses Against Insects: Physical Defenses, p. 944

REFERENCES

1. Chapman, R.F. *The Insects: Structure and Function*, 4th Ed.; Cambridge University Press: New York, 1998, 770 pp.
2. Gillott, C. *Entomology*, 2nd Ed.; Plenum Press: New York, 1995; 798 pp.
3. Gullan, P.J.; Cranston, P.S. *The Insects: An Outline of Entomology*, 2nd Ed.; Blackwell Science: Malden, MA, 2000; 470 pp.
4. *Comprehensive Insect Physiology, Biochemistry and Pharmacology*; Kerkut, G.A., Gilbert, L.I., Eds.; Pergamon Press: Elmsford, NY, 1985; Vols. 1–13.
5. Romoser, W.S.; Stoffolano, J.G., Jr. *The Science of Entomology*, 4th Ed.; McGraw-Hill: Boston, 1998; 605 pp.

Phytochemical Diversity of Secondary Metabolites

Michael Wink
Universität Heidelberg, Heidelberg, Germany

INTRODUCTION

Plants produce a wide variety of secondary metabolites that have evolved as defense compounds against herbivores and microbes. Some secondary metabolites attract pollinating insects or fruit-dispersing animals; others serve as UV protection or as signal compounds. Secondary metabolites can be divided into two main groups—those without nitrogen and those with nitrogen in their structures. Nitrogen-containing compounds include alkaloids, amines, nonprotein amino acids, cyanogenic glycosides, glucosinolates, protease inhibitors, and lectins. Nitrogen-free compounds are various terpenoids (mono-, sesqui-, di-, tri,- and tetraterpenes; saponines; and cardiac glycosides), polyketides (anthraquinones), phenolics (flavonoids, anthocyanins, galloyl and catechol tannins, phenolic acids, lignans, and lignins), and polyacetylenes. The modes of action of these compounds are briefly discussed.

WHAT ARE SECONDARY METABOLITES?

A chemical analysis of low molecular weight substances present in a plant reveals the usual inorganic ions, amino acids, sugars, organic acids, fatty acids, and some other compounds that are needed for primary metabolism or development. In addition, all higher plants produce one or several representatives of the so-called secondary metabolites (SMs), which are also low molecular-weight compounds. In contrast to primary metabolites, SMs are not essential for a plant in terms of primary metabolism; i.e., a plant can usually survive without a particular SMs, especially when grown under protection in a greenhouse.[1–9]

SMs can have ecological functions in that they serve to protect against herbivores, microbes, or competing plants. Some SMs also function as signal compounds to attract pollinating or seed-dispersing animals. SMs usually occur in complex mixtures that also differ among plant organs and developmental stages; e.g., seeds have SMs profiles that often differ from those of leaves and roots. In general, a particular group of plants produces two to three main types of SMs concomitantly, such as phenolics, terpenoids, or alkaloids. Often, a basic structure is present that is varied by several substitutions, such as additional OH-groups, methyl-groups, or double bonds. SMs are dynamic compounds showing turnover, and their biosynthesis can be enhanced upon attack by microbes or herbivores.[1,6–9] Jasmonic and salicylic acid are important signal molecules in this context.[10]

Some SMs occur in phylogenetically related taxa (e.g., glucosinolates in families of the Capparales), whereas others have a broader distribution (often a major radiation in a particular tribe or family and outliers in several others). Considering the whole plant kingdom (including algae, mosses, horsetails, and ferns), an impressive diversity of SMs exists. Fig. 1 gives an estimate of the numbers of known secondary metabolites and illustrates a few representative structures. Three major groups of SMs can be recognized: nitrogen-containing substances, terpenes, and phenolics.

PHYTOCHEMICAL DIVERSITY

Nitrogen-Containing Secondary Metabolites

Over 14,000 nitrogen-containing SMs have been described so far: Alkaloids, amines, nonprotein amino acids, cyanogenic glycosides, and glucosinolates are the main compounds in this group. A few SMs of high molecular weight should also be considered in this context: Some plants accumulate peptides and proteins (protease inhibitors, amylase inhibitors) that play a role in antimicrobial and antiherbivore defense. Their biosynthesis can be induced upon wounding or microbial attack.

Nonprotein amino acids (NPAA)

In addition to the 20 common L-amino acids, more than 400 other amino acids that are not building blocks of proteins exist in plants.[1–9] NPAAs often figure as antinutrients or antimetabolites, i.e., they may interfere with the metabolism of microorganisms or herbivores. Many nonprotein amino acids resemble protein amino acids and can be considered to be their structural analogues. For example, 3-cyanoalanine is an analogue to L-alanine, S-aminoethylcysteine to L-lysine, L-azetidine-2-carboxylic acid to L-proline, and L-canavanine or

Encyclopedia of Plant and Crop Science
DOI: 10.1081/E-EPCS 120005945

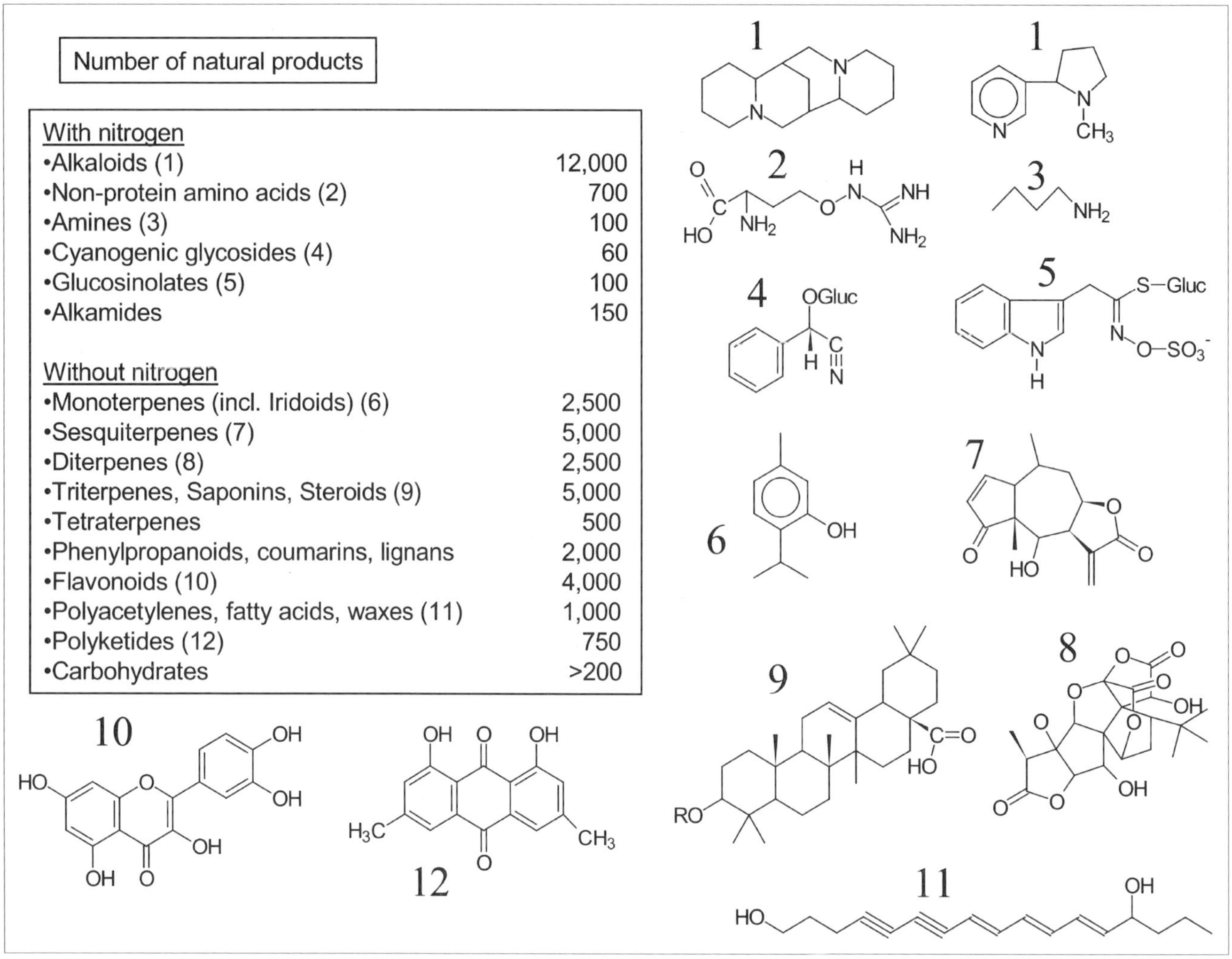

Fig. 1 A number of known secondary metabolites.

L-indospicine to L-arginine. Because NPAAs are often (at least partly) remobilized during germination, they function as N-storage compounds in addition to performing their role as defense chemicals.

Amines

Aliphatic mono- and diamines, polyamines, aromatic amines, and amine conjugates are the main SMs in this group.[1–9] Aliphatic monoamines and diamines are widely distributed SMs in plants, and occur as independent products in their own right or as intermediates in the biosynthesis of alkaloids (as precursors for several groups of alkaloids, such as pyrrolizidine, nicotine, tropane, quinolizidine, *Lycopodium,* and *Sedum* alkaloids). Simple amines serve as attractants in some flowers (e.g., *Arum maculatum*) or as defense compounds in many other plants. Aromatic amines derive from tryptamine and tyramine. Because several tryptamine and tyramine derivatives are structurally identical or similar to those of the neurotransmitters serotonin, dopamine, or noradrenaline they may interfere with the corresponding receptors. Hallucinogenic or stimulating effects have been reported for bufotenine, psilocybin, psilocin, *N,N*-dimethyltryptamine, mescaline, ephedrine, or ergot alkaloids.[11–14]

Cyanogenic glycosides

Cyanogens are derivatives of 2-hydroxynitriles that form glucosides with β-D-glucose.[2–9] More than 60 different cyanogenic glycosides have been described from 2600 plants (mainly Rosaceae, Fabaceae, Gramineae, and Araceae). When plants are wounded by herbivores or other organisms, the cellular compartmentation breaks down and cyanogenic glycosides are hydrolyzed by β-glucosidase and nitrilase to the corresponding aldehyde or ketone and hydroprussic acid (HCN).[15] HCN is highly toxic for animals or microorganisms because it inhibits electron transfer components of the respiratory chain and binds to other enzymes with metal ions. Foods that contain cyanogens, such as manihot (*Cassava esculenta*) and sorghum, have repeatedly caused intoxication and even death in humans and animals.

Glucosinolates

Glucosinolates contain sulfur as an additional element (thioglucosides). When hydrolyzed, glucosinolates liberate D-glucose, sulfate, and an unstable aglycone, which normally converts to isothiocyanate (common name, mustard oil) as the main product.[2–9,15] Isothiocyanates are responsible for the distinctive, pungent flavor and odor of mustards and horseradish. More than 80 different glucosinolates have been found in the Capparales (families Capparidaceae, Brassicaceae, Resedaceae, Moringaceae, and Tropaeolaceae).

Alkaloids

Alkaloids represent a large and diverse class of SMs with a heterocyclic nitrogen that can react as a base.[1–9,11–16] Alkaloids have been detected in about 15% of plants, bacteria, fungi, and even animals. Within the plant kingdom, they occur in primitive groups such as *Lycopodium* or *Equisetum*, in gymnosperms, and most abundantly in angiosperms. Alkaloids often exhibit toxicity toward animals and humans. Their main targets are elements of the neuronal signal transduction system.[13,14] Alkaloids certainly evolved as defense chemicals against herbivores, but also inhibit microorganisms and other plants. In addition, some plants employ them as degradable N-transport and N storage compounds.[5]

Secondary Metabolites Without Nitrogen

Over 20,000 nitrogen-free SMs have been described so far. Terpenoids, phenylpropanoids, flavonoids, polyketides, and polyacetylenes, are the main compounds in this group.[1–9]

Terpenoids

Main terpenoids include monoterpenes (with 10 C atoms), sesquiterpenes (15 C), diterpenes (20 C), triterpenes (30 C), steroids (27 C), tetraterpenes (40 C), and polyterpenes. Although their main building blocks are simple, most terpenes represent complex structures because of various substituents and secondary ring formations.[1–9]

Mono- and sesquiterpenes are often volatile and can be isolated as essential oils. These compounds are usually lipophilic and are stored in specialized oil cells, trichomes, resin channels, or other dead cells.[15] They are especially abundant in Asteraceae, Apiaceae, Lamiaceae, Rutaceae, Lauraceae, Cupressaceae, Pinaceae, Myrtaceae, and Zingiberaceae. An intriguing number of complex ring structures can be found in sesquiterpenes. Some of them are highly reactive compounds, such as ptaquiloside in the fern (bracken) *Pteridium aquilinum* or the sesquiterpene lactone helenaline of *Arnica* species, which can alkylate DNA and proteins. Bracken is known to cause bladder cancer in humans and live-stock. Some sesquiterpenes from gymnosperms mimic the insect juvenile growth hormone (a terpenoid with 17 or 18 C). In general, mono- and sesquiterpenes function as chemical defense compounds against herbivores and microbes, but the volatile terpenes also serve to attract pollinating insects.

Among diterpenes we find extremely complex ring structures, some of which, such as phorbol esters of Euphorbiaceae and Thymelaeaceae, are infamous for their toxicity (they activate protein kinase C and can act as tumor promoters). The plant hormone gibberellic acid belongs to the class of diterpenes, as does forskolin, a potent activator of adenynyl cyclase. Another diterpene, taxol A (paclitaxel, taxol®), can be isolated from several yew species and has been used for almost 10 years with great success in the chemotherapy of various tumors.[7,13,14]

Triterpenes and steroids can occur as aglycons but more often occur as saponins. Whereas free triterpenes and steroids are lipophilic compounds, the saponins are water soluble and amphiphilic molecules that can make biomembranes leaky. They therefore show characteristics of broad antimicrobial and antiherbivore activity. Saponins are widely distributed in the plant kingdom; approximately 70% of all plants produce them. A special case of steroidal saponins is cardiac glycosides that inhibit Na^+,K^+-ATPase and are therefore strong toxins (but are useful in heart medication). Cardiac glycosides can be divided into two classes: cardenolides and bufadienolides.[1–9] Another class of bioactive steroids is the phytoecdysons, which have been isolated from ferns and several higher plants. They mimic the insect moulting hormone ecdysone and are therefore insecticidal.

Carotinoids represent the most important members of tetraterpenes. Carotinoids are the precursors for vitamin A in animals, used to produce retinal and retinoic acid (retinoids bind to nuclear receptors and are local mediators of vertebrate development).

Polyterpenes, consisting of 100 to 10,000 isoprene units, are prominent in latex of Euphorbiaceae and other families; they are used commercially in rubber (from *Hevea brasiliensis*) or gutta-percha.

Phenolic compounds: Phenylpropanoids, coumarins, flavonoids and tannins

Phenylpropanoids can occur as simple compounds (as phenylacrylic acids, aldehydes, or alcohols) that differ by their degree of hydroxylation and methoxylation. Phenylpropanoids can also be conjugated with a second phenylpropanoid, such as rosmarinic acid (an abundant tannin of Lamiaceae) or with amines, such as coumaroylputrescine.[1–9]

Coumarins and furocoumarins are common in certain genera of the Apiaceae, Fabaceae, and Rutaceae. Furocoumarins with a third furane ring form linear psoralen or angular angelicin types. Furocoumarins can intercalate DNA, and upon illumination with UV light can form cross-links with DNA bases and also with proteins.[7] They are therefore mutagenic and possibly carcinogenic. In plants they serve as defense compounds against herbivores and pathogens.

Phenylpropanoids can condense with a polyketide moiety to flavonoids, chalcones, catechins, and anthocyanins. These compounds are characterized by two aromatic rings that carry several phenolic hydroxy, methoxy, and sugar groups. Many of the flavonoids, chalcones, and anthocyanins are widely distributed in higher plants and show colors under visual and UV light and are typical SMs of flowers and fruits. The color of anthocyanins depends on the degree of glycosylation, hydrogen ion concentration, and the presence of certain metals (e.g., aluminium ions) in the vacuole. These SMs function to attract pollinating insects or fruit-dispersing animals. The phenolic hydroxyl groups can interact with proteins to form hydrogen and ionic bonds and can therefore interfere with a multitude of molecular targets.

Isoflavones are common SMs in legumes. They exhibit estrogenic properties and inhibit tyrosine kinases. Therefore, they are often regarded as useful compounds that might play a role in the prevention of certain cancers. Such ingredients are called nutraceuticals (in analogy to phytopharmaceuticals). Other isoflavones, such as rotenone, have potent insecticidal properties and have been used as biorational pesticides for plant protection.

Catechins, which often dimerize or even polymerize to procyanidins and oligomeric procyanidins, form a special class of phenolics. These nonhydrolyzable tannins are characterized by a large number of hydroxyl groups that can interact with proteins to form hydrogen, ionic, and possibly even covalent bonds. These compounds serve as defense chemicals against invading pathogens and herbivores. Another important group of tannins comprises esters of gallic acid with sugars. These hydrolyzable gallotannins are widely distributed in plants, often in bark, leaves, fruits, and galls. Gallotannins contain a large number of phenolic hydroxyl groups, so can form stable protein-tannin complexes and thus interact with a wide variety of protein targets in microbes and animals.[1,5,7]

Polyketides

Important polyketides are anthraquinones, quinones, and flavonoids. SMs with an anthracene skeleton can be present as anthrones, anthraquinones, anthranols, dianthrones, naphthodianthrones, dianthranoles, and corresponding mono- or diglycosides.[1–9] Anthracene derivatives are common in certain genera of the Polygonaceae, Rhamnaceae, Fabaceae, Rubiaceae, Hypericaceae, and Aloes.

Glycosylated monomeric anthrones exhibit a strong laxative activity in vertebrates and can be regarded as potent defense chemicals. They have also been used in medicine as laxatives for several thousand years.

Carbohydrates

Several carbohydrates (such as glucose, galactose, or fructose) form glycosides with SMs and are thus participants in both primary and secondary metabolism.[1–9] Other carbohydrates appear to be allelochemicals in their own right: An example is phytic acid (a myoinositol esterified with up to six phosphate groups) that can complex Ca^{++} and Mg^{++} ions and thus functions as an antinutritive substance. Several di-, tri-, and oligosaccharides—such as stachyose and raffinose, typically found in seeds and roots—produce substantial flatulence in herbivores and thus come closer to typical SMs in being regarded as defense compounds against herbivores.

CONCLUSION

SMs show an impressive structural diversity that can be understood from an ecological and evolutionary perspective. Many SMs have evolved as bioactive compounds that interfere with nucleic acids or proteins (known as evolutionary molecular modelling) and show antimicrobial or insecticidal and pharmacological properties. SMs are therefore of interest in medicine as therapeutics and in agriculture as biorational pesticides. Several SMs have antioxidative, radical scavenging, or estrogenic properties; food and crops containing these nutraceuticals will play an increasingly important role in nutrition.[6,7]

Many crop and food plants originally produced SMs before domestication. During plant breeding, many of the traits leading to SMs have been eliminated because corresponding SMs were toxic, bitter, or unpalatable.[5] Unfortunately, this strategy also removed the natural resistance of many cultivars with the consequence that human-made synthetic plant protectives must replace the natural defense compounds formerly present.

In light of the novel strategies of genetic engineering and food processing, SMs should not be regarded as useless compounds. On the contrary, SMs represent valuable traits that might be used to improve viability, resistance, and the nutritional quality of crop and food plants.

ARTICLES OF FURTHER INTEREST

Amino Acid and Protein Metabolism, p. 30
Aromatic Plants for the Flavor and Fragrance Industry, p. 58
Bacterial Pathogens: Early Interactions with Host Plants, p. 89
Bacterial Products Important for Plant Disease: Toxins and Growth Factors, p. 101
Biological Control of Nematodes, p. 134
Biological Control of Weeds, p. 141
Carotenoids in Photosynthesis, p. 245
Coevolution of Insects and Plants, p. 289
Echinacea: Uses as a Medicine, p. 395
Ecology: Functions, Patterns, and Evolution, p. 404
Genetic Resources of Medicinal and Aromatic Plants from Brazil, p. 502
Insect/Host Plant Resistance in Crops, p. 605
Insect–Plant Interactions, p. 609
Isoprenoid Metabolism, p. 625
Marigold Flower: Industrial Applications of, p. 689
Metabolism, Secondary: Engineering Pathways of, p. 720
Mycotoxins Produced by Plant Pathogenic Fungi, p. 773
New Secondary Metabolites: Potential Evolution, p. 818
Phenylpropanoids, p. 868
Physiology of Herbivorous Insects, p. 910
Secondary Metabolites as Phytomedicines, p. 1120

REFERENCES

1. Harborne, J.B. *Introduction to Ecological Biochemistry*, 4th Ed.; Academic Press: London, 1993.
2. Rosenthal, G.A.; Berenbaum, M.R. *Herbivores: Their Interactions with Secondary Plant Metabolites*; The Chemical Participants. Academic Press: San Diego, 1991; Vol. 1.
3. Rosenthal, G.A.; Berenbaum, M.R. *The Chemical Participants*; Ecological and Evolutionary Processes. Academic Press: San Diego, 1992; Vol. 2.
4. Seigler, D.S. *Plant Secondary Metabolism*; Kluwer: Boston, 1998.
5. Wink, M. Plant breeding: Importance of plant secondary metabolites for protection against pathogens and herbivores. Theor. Appl. Genet. **1988**, *75*, 225–233.
6. Wink, M. *Biochemistry of Plant Secondary Metabolism*; Annual Plant Reviews, Sheffield Academic Press: Sheffield, 1999a; Vol. 2.
7. Wink, M. *Function of Plant Secondary Metabolites and Their Exploitation in Biotechnology*; Annual Plant Reviews, Sheffield Academic Press: Sheffield, 1999b; Vol. 3.
8. Dewick, P.M. *Medicinal Natural Products. A Biosynthetic Approach*; Wiley: New York, 2002.
9. Dey, P.M.; Harborne, J.B. *Plant Biochemistry*; Academic Press: San Diego, 1997.
10. Creelman, R.A.; Mullet, J.E. Biosynthesis and action of jasmonates in plants. Annu. Rev. Plant Physiol. Plant Mol. Biol. **1997**, *48*, 355–381.
11. Mann, J. *Murder, Magic and Medicine*; Oxford University Press: London, 1992.
12. Roberts, M.F.; Wink, M. *Alkaloids-Biochemistry, Ecological Functions and Medical Applications*; Plenum: New York, 1998.
13. Wink, M. Allelochemical Properties and the Raison D'être of Alkaloids. In *The Alkaloids*; Cordell, G., Ed.; Academic press: Orlando, 1993; Vol. 43, 1–118.
14. Wink, M. Interference of Alkaloids with Neuroreceptors and Ion Channels. In *Bioactive Natural Products*; Atta-Ur-Rahman, Ed.; Elsevier: Amsterdam, 2000; Vol. 11, 3–129.
15. Wink, M. Compartmentation of secondary metabolites and xenobiotics in plant vacuoles. Adv. Bot. Res. **1997**, *25*, 141–169.
16. Brown, K.; Trigo, J.R. The Ecological Activity of Alkaloids. In *The Alkaloids*; Cordell, G.A., Ed.; 1995; Vol 47, 227–354.

Phytochrome and Cryptochrome Functions in Crop Plants

Jim Weller
University of Tasmania, Hobart, Tasmania, Australia

INTRODUCTION

Light has profound effects on plant development. Variation in the quality and irradiance of light can influence processes throughout the plant life-cycle, including seed germination, stem elongation, leaf expansion, bud dormancy, synthesis of photosynthetic and photoprotective pigments, and the initiation of flowering. Light is perceived and its developmental effects mediated by two important families of photoreceptor proteins: the phytochromes, which predominantly absorb red (R) and far-red light (FR), and the cryptochromes, which primarily absorb blue light (B). Most recent progress in understanding the roles of these photoreceptors in plant development has come from studies in *Arabidopsis thaliana*, although phytochrome and cryptochrome functions are also being explored by a genetic approach in other model species including tomato, pea, and rice. In most cases, photoreceptor functions in these species have been found to be broadly consistent with their roles in *Arabidopsis*. However, the fact that these species differ in growth habit, physiology, and photoreceptor complement means that they offer a broader perspective on photoreceptor function and how genes controlling photoreceptor function might be useful for modifying specific traits in a diverse range of crop plants.

PHOTORECEPTOR GENE FAMILIES

The phytochrome and cryptochrome photoreceptors are encoded by small gene families in most species so far examined. The full extent of these gene families is known in only a few species, but even so it is clear that they can be somewhat variable in size. For example, different lineages within the phytochrome and cryptochrome families may have undergone independent duplications. The best illustration of this is the existence of phyB-type gene pairs in *Arabidopsis* (phyB and phyD) and tomato (phyB1 and phyB2), where members of each pair are more closely related to each other than to either of the corresponding genes from the other species.[1] There is also evidence for recent duplications within the cryptochrome gene family. Whereas *Arabidopsis* has only two cryptochromes (cry1 and cry2), both tomato and barley have three expressed *CRY* genes, including two *CRY1*-like genes.[2] It is likely that as more species are examined more such instances will be identified. A substantial difference in phytochrome gene complement is also seen in monocots. In contrast to the five found in *Arabidopsis*, rice and sorghum have only three phytochromes, corresponding to phyA, phyB, and phyC lineages, suggesting that monocots may not have undergone the expansion within the phyB lineage that is seen in dicots.

Despite the differences in photoreceptor gene complement, the basic features of specific photoreceptors seem to be generally conserved. For example, where examined, phyB-type phytochromes have generally been found to be relatively stable, whereas phytochrome A is degraded in the light and *PHYA* expression is strongly down-regulated.

STUDYING BASIC PHOTORECEPTOR FUNCTIONS IN CROP PLANTS

Photoreceptor functions in a wide range of crop species have been studied for many years by means of indirect physiological approaches involving irradiation with monochromatic light. However, direct information about photoreceptor functions relies on the use of mutants or transgenic lines with altered photoreceptor levels. Table 1 lists the photoreceptor mutants that have been isolated in *Arabidopsis* and a range of other higher plant species. In general, far less is presently known about cryptochrome function than about phytochrome function, and whereas many phytochrome-deficient mutants are known in crop species, the only known cryptochrome mutant is the *cry1* mutant of tomato.

De-etiolation

As germinating seedlings emerge from the soil, or after dark-grown seedlings are transferred to white light, they display a developmental response called de-etiolation, which includes the co-ordinated inhibition of hypocotyl/

Encyclopedia of Plant and Crop Science
DOI: 10.1081/E-EPCS 120012964

Table 1 Photoreceptor mutants in higher plants

	Phytochrome				Cryptochrome	
Species	**phyA**	**phyB**	**Other phy**	**Chromophore**	**cry1**	**cry2**
Arabidopsis thaliana	*phyA*	*phyB* *phyD*	*phyE*	*hy1* *hy2*	*cry1*	*cry2*
tomato (*Lycopersicon esculentum*)	*phyA*	*phyB1* *phyB2*		*au* *yg2*	*cry1*	
pea (*Pisum sativum*)	*phyA*	*phyB*		*pcd1* *pcd2*		
rice (*Oryza sativa*)	*phyA*			*se5*		
potato (*Solanum tuberosum*)	AS	AS				
Nicotiana plumbaginfolia		*hlg*		*pew1* *pew2*		
sorghum (*Sorghum bicolor*)		*ma3*				
Brassica rapa		*ein*				
cucumber		*lh*				

epicotyl/coleoptile elongation, promotion of leaf expansion, and synthesis of chlorophyll and anthocyanin. The basic features of photoreceptor function appear relatively well conserved between *Arabidopsis* and other species that have been examined to date. Analyses of photoreceptor mutants in tomato and pea have shown that phyA is the only receptor mediating de-etiolation responses to the light in the FR region of the spectrum. PhyA is present at high level in dark-grown seedlings and can mediate de-etiolation responses to low irradiance R and B. As a result it is important during the initial phase of the dark-to-light transition. Under high irradiances, the level and influence of phyA decline rapidly, and phyB and cry1 become the predominant photoreceptors, mediating responses to R and B, respectively.[3,4] A deficiency in either one of these photoreceptors enhances elongation growth and impairs leaf development. This may be quite dramatic (e.g., *lh* in cucumber, *ein* in Brassica) or relatively mild (e.g., *phyB1* in tomato). The difference may have to do with the existence in some species of multiple phyB with overlapping functions.

In contrast, phyA deficiency has little overall effect on seedling morphology in pea, tomato, or rice grown under high irradiance white light, presumably because the effects of phyA are masked by phyB and cry1 under these conditions. However, in the absence of these photoreceptors or under low irradiances of light, a clear phenotype for *phyA* mutants does become evident. The relative contribution of phyA to de-etiolation also differs in different species. In tomato, phyA has a small role in R regardless of which other photoreceptors are present, but in pea, it plays a substantial role in the absence of phyB.[3,4] This may be related to differences in stability of the phyA protein. Also, phyA can interfere with phyB function, an interaction best illustrated in tomato, where *phyA* mutants show a phyB-dependent enhancement of anthocyanin synthesis under R and W.[5]

Shade-Avoidance

Growth of fully de-etiolated plants is also significantly affected by light, in particular by the amount of light and the proportion of FR, in a syndrome termed the shade-avoidance response. In terms of photoreceptor function, the shade-avoidance response actually consists of two distinct photoresponses. PhyB-type phytochromes induce leaf expansion and inhibit stem elongation in response to R. FR activates phyA to induce a similar set of responses, but inactivates phyB phytochromes by converting them to their Pr form. Most phyB mutants thus show a partial shade-avoiding phenotype. Like the seedling response to R, strength of this phenotype varies with species, possibly depending on the number of phyB-like phytochromes present. For example, the *phyA phyB1 phyB2* triple mutant of tomato shows a strong shade-avoidance response, whereas in pea, the *phyA phyB* double mutant is completely unresponsive.[4,5] The activity of phyA may also be an important factor determining the overall strength of the shade-avoidance response in any given species. For example, tomato has a strong shade-avoidance response that is not much influenced by the loss of phyA. In contrast, pea shows a weak shade-avoidance response that is dramatically enhanced in plants lacking phyA.[4,5] This interference of phyA with shade-avoidance has also been examined with a view to practical applications. Studies in several species have shown that elevated levels of phyA can enhance the response to the FR present in canopy shade, thus suppressing the shade-avoidance response and even causing proximity dependent dwarfing.[6]

Flowering and Photoperiodism

The tendency to flower early is often seen as part of the shade-avoidance syndrome. Consistent with the important role for phyB-type phytochromes in shade-avoidance responses, phyB-deficient plants in several species show early flowering. The early-flowering phenotype of phyB-deficient plants is seen under both long and short photoperiod conditions, and in both short-day plants (SDP; e.g., sorghum, rice) and long-day plants (LDP; e.g., *Arabidopsis*, pea). For this reason phyB is not thought to be involved with detection of daylength as such. Instead, the primary function of phyB-type phytochromes may be to detect noncompetitive conditions (high R:FR ratio) and delay flowering. As the FR content increases, these phytochromes are inactivated and plants flower early as a result. One exception is seen in *Nicotiana plumbaginifolia*, in which flowering is promoted by low R:FR, but is delayed in the phyB-deficient *hlg* mutant.[7] In tomato, *phyB* mutations have little effect on flowering, although it is not clear whether this is due to the compensating action of another phytochrome or the possibility that flowering is not responsive to R:FR in this species.

In contrast to phyB, a clear role for phyA has been identified only in LDP. In rice, the only SDP for which *phyA* mutants have been isolated, the loss of phyA does not have any discernable affect on flowering, suggesting that it may not be important for flowering or photoperiodic responses in SDP.[8] In the LDP pea, *phyA* deficient mutants are late flowering in LD, and show pleiotropic characteristics of plants grown in SD.[9] Delayed flowering is also seen in *phyA* mutants of *Arabidopsis* under some long day conditions. PhyA thus appears to have an important role in the promotion of flowering in response to day extensions, and thus acts antagonistically to phyB. Transgenic potato plants with reduced phyA levels failed to inhibit tuberization under long photoperiods,[10] whereas transgenic aspen overexpressing phyA failed to enter dormancy under short photoperiods.[11] Taken together, these results suggest that the level of phyA may determine the critical day length for long-day responses and that phyA acts to promote the transition from growth strategies appropriate to winter (slow vegetative growth, production of vegetative storage organs, dormancy) to those appropriate for summer (rapid vegetative growth and induction of flowering).

Other Responses

There are many other documented responses to light in crop plants that have yet to be investigated from a genetic perspective. Phytochrome has a well known role in germination of many small-seeded species, and there is some limited genetic evidence for the role of phytochromes in tomato seed germination,[12] but pea and rice are large-seeded and show no substantial light effects on germination. There is also evidence from tomato that phytochrome regulates competence for shoot regeneration and specific aspects of fruit ripening.[13,14] Apical dominance is another trait that is clearly modified by photoreceptor action. Activation of the shade-avoidance response results in greater elongation growth and reduced branching or tillering.[15] In contrast, phyA in pea acts to suppress lateral branching, although this can be explained through its effects on the photoperiodic response.[9] In addition, both phyA- and phyB-type phytochromes are likely to influence many other important traits and processes that are responsive to photoperiod, including root development, tuberization, fruit set and development, and partitioning of dry matter between stem and leaf.

CONCLUSION

It is clear that the phytochrome and cryptochrome photoreceptors affect several different responses of applied interest. Modification of photoreceptor activity, either through classical breeding or transgenic manipulation, may offer some prospects for improvement of specific light-responsive traits. A better understanding of light responses may help to modify seedling growth in dense propagation systems, through selection and development of light sources and transparent sheeting materials used in glasshouses and growth. Selection strategies including a consideration of shade-avoidance responses may lead to improved harvest index, increased growth of storage organs, and reduced lodging, features which may in turn allow higher cropping densities and more efficient harvesting. Conversely, in some fiber crops constitutive shade-avoidance may have desirable effects on fiber properties. Modification of photoreceptor activity may help to control photoperiod responsiveness, making it possible to accelerate, delay, or prevent flowering and bolting, and help in the development of crop varieties better suited to a given location or cropping regime. Fruit-specific regulation of photoreceptor activity may provide a means to modify fruit properties. There is no doubt that continuing to expand our knowledge of photoreceptor functions in a broader range of crop plants will offer further useful insights and applications.

ARTICLES OF FURTHER INTEREST

Competition: Responses to Shade by Neighbors, p. 300
Photoreceptors and Associated Signaling I: Phytochromes, p. 881
Photoreceptors and Associated Signaling II: Cryptochromes, p. 885

Shade Avoidance Syndrome and Its Impact on Agriculture, p. 1152

REFERENCES

1. Hauser, B.A.; Cordonnier Pratt, M.M.; Daniel Vedele, F.; Pratt, L.H. The phytochrome gene family in tomato includes a novel subfamily. Plant Mol. Biol. **1995**, *29* (6), 1143–1155.
2. Perrotta, G.; Yahoubyan, G.; Nebuloso, E.; Renzi, L.; Giuliano, G. Tomato and barley contain duplicated copies of cryptochrome 1. Plant Cell Environ. **2001**, *24* (9), 991–997.
3. Weller, J.L.; Perrotta, G.; Schreuder, M.E.L.; van Tuinen, A.; Koornneef, M.; Giuliano, G.; Kendrick, R.E. Genetic dissection of blue-light sensing in tomato using mutants deficient in cryptochrome 1 and phytochromes A, B1 and B2. Plant J. **2001**, *25* (4), 427–440.
4. Weller, J.L.; Beauchamp, N.; Kerckhoffs, L.H.J.; Platten, J.D.; Reid, J.B. Interaction of phytochromes A and B in the control of de-etiolation and flowering in pea. Plant J. **2001**, *26*, 283–294.
5. Weller, J.L.; Schreuder, M.E.L.; Smith, H.; Koornneef, M.; Kendrick, R.E. Physiological interactions of phytochromes A, B1 and B2 in the control of development in tomato. Plant J. **2000**, *24*, 345–356.
6. Robson, P.R.H.; McCormac, A.C.; Irvine, A.S.; Smith, H. Genetic engineering of harvest index in tobacco through overexpression of a phytochrome gene. Nat. Biotechnol. **1996**, *14*, 995–998.
7. Hudson, M.; Robson, P.R.H.; Kraepiel, Y.; Caboche, M.; Smith, H. Nicotiana plumbaginifolia hlg mutants have a mutation in a PHYB-type phytochrome gene: They have elongated hypocotyls in red light, but are not elongated as adult plants. Plant J. **1997**, *12* (5), 1091–1101.
8. Takano, M.; Kanegae, H.; Shinomura, T.; Miyao, A.; Hirochika, H.; Furuya, W. Isolation and characterization of rice phytochrome A mutants. Plant Cell **2001**, *13* (3), 521–534.
9. Weller, J.L.; Murfet, I.C.; Reid, J.B. Pea mutants with reduced sensitivity to far-red light define an important role for phytochrome A in day-length detection. Plant Physiol. **1997**, *114*, 1225–1236.
10. Yanovsky, M.J.; Izaguirre, M.; Wagmaister, J.A.; Gatz, C.; Jackson, S.D.; Thomas, B.; Casal, J.J. Phytochrome A resets the circadian clock and delays tuber formation under long days in potato. Plant J. **2000**, *23*, 223–232.
11. Olsen, J.E.; Junttila, O.; Nilsen, J.; Eriksson, M.E.; Martinussen, I.; Olsson, O.; Sandberg, G.; Moritz, T. Ectopic expression of oat phytochrome A in hybrid aspen changes critical daylength for growth and prevents cold acclimatization. Plant J. **1997**, *12* (6), 1339–1350.
12. Koornneef, M.; Cone, J.W.; Dekens, R.G.; O'Herne-Robers, E.G.; Spruit, C.J.P.; Kendrick, R.E. Photomorphogenic responses of long-hypocotyl mutants of tomato. J. Plant Physiol. **1985**, *120*, 153–165.
13. Bertram, L.; Lercari, B. Phytochrome A and phytochrome B1 control the acquisition of competence for shoot regeneration in tomato hypocotyl. Plant Cell Rep. **2000**, *19* (6), 604–609.
14. Alba, R.; Cordonnier-Pratt, M.M.; Pratt, L.H. Fruit localized phytochromes regulate lycopene accumulation independently of ethylene production in tomato. Plant Physiol. **2000**, *123* (1), 363–370.
15. Foster, K.R.; Miller, F.M.; Childs, K.L.; Morgan, P.W. Genetic regulation of development in *Sorghum bicolor*. VIII. Shoot growth, tillering, gibberellin biosynthesis, and phytochrome levels are differentially affected by dosage of the ma_{3R} allele. Plant Physiol. **1994**, *105*, 941–948.

Phytoremediation: Advances Toward a New Cleanup Technology

Adel Zayed
Paradigm Genetics, Inc., RTP, North Carolina, U.S.A.

INTRODUCTION

Decades of unrestricted use, disposal, and release of industrial-, defense-, and energy production–related chemicals have considerably accelerated the pollution of our environment with radionuclides, toxic metals, and organic pollutants. Environmental exposure to such toxic chemicals is a serious health risk for humans and animals. Unfortunately, existing technologies for the cleanup of contaminated environments are very costly and/or limited in various ways. Cleanup of toxic metal contaminated soils with existing technologies costs up to $1,000,000 per acre without totally removing the pollutant from the environment. In the United States alone, the cost of cleaning up sites contaminated with toxic and radioactive metals is estimated to be $400 billion.

Phytoremediation, the process by which various naturally occurring or genetically modified plants, including trees and grasses, are used to degrade, extract, detoxify, contain, and/or immobilize toxic pollutants from contaminated soil, water, and air, offers an attractive and cost-effective solution for the clean up of contaminated sites. The potential economic benefits of using plants for remediation are impressive. Growing a crop on an acre of land can be accomplished at a cost ranging from two to four orders of magnitude less than the current engineering cost of excavation and reburial. Plants provide a robust, solar-powered system that has little or no maintenance requirements. With their copious root systems, plants can scavenge large areas and volumes of soils, removing the toxic chemical. The rhizosphere soil (soil near plant roots) has microbial populations orders of magnitude greater than bulk soil (nonroot soil). Furthermore, plant-based systems are welcomed by the public due to their superior aesthetics and the societal and environmental benefits that their presence provides. There are several different ways that phytoremediation can be achieved (Table 1).

Research into phytoremediation may offer the remediation solution for at least 30,000 contaminated sites in the United States alone. Phytoremediation is applicable to a number of hazardous waste and other remedial scenarios, including remediation of metals, organics, and radionuclides from soils and water, which together offer a total U.S. market opportunity of up to $10 billion. Phytoremediation is also potentially applicable to all types of water treatments including industrial, agricultural, and municipal wastewaters, landfill leachate, stormwater, and drinking water with a potential market opportunity of up to $40.7 billion (Table 2). Similar markets exist overseas, which are smaller but which offer greater long-term potential for growth.

TRANSGENIC PHYTOREMEDIATION APPROACHES

The ideal plant species for phytoremediation is one that can accumulate, degrade, or detoxify large amounts of the toxic chemical, grow rapidly on contaminated sites and produce a large biomass, tolerate salinity and other toxic conditions, and provide a yield of economic value, e.g., fibers. Metal hyperaccumulator plants occur naturally on metal-rich soils and accumulate metals in their above-ground tissues to concentrations between one and three orders of magnitude higher than surrounding "normal" plants grown at the same site.[1,2] Unfortunately, most hyperaccumulator plants studied to date grow very slowly and accumulate little biomass. For phytoremediation to become a viable technology, dramatic improvements would be required in either hyperaccumulator biomass yield or nonaccumulator metal accumulation. The introduction of novel traits into high biomass or hyperaccumulator plants in a transgenic approach is a promising strategy for the development of effective phytoremediation technologies.

There are several conceivable strategies for the use of genetic engineering to improve phytoremediation. Strategies to improve metal phytoremediation involve the optimization of a number of processes, including trace element mobilization in the soil, uptake into the root, detoxification, and translocation within the plant.[3] Strategies for enhancing phytoremediation of organics are potentially more straightforward and involve the introduction and/or overexpression of genes encoding biodegradative enzymes in transgenic plants to enhance their biodegradative abilities.[4] In the last few years, progress has been made toward this goal and several types of

Encyclopedia of Plant and Crop Science
DOI: 10.1081/E-EPCS 120005584

Table 1 Summary of phytoremediation processes and their definitions

Process	Definition	Contaminated substrate	Target contaminant
Phytoextraction	Use of plants to take up contaminants from soil and water and accumulate them in aboveground plant tissues, which may then be harvested and removed from the site.	Soils Sediments Sludges Wastewater	Toxic trace elements Radionuclides
Phytostabilization	Use of plants to immobilize contaminants chemically and physically at the site, thereby preventing their movement to surrounding areas.	Soils Sediments Sludges	Heavy metals
Phytovolatilization	Use of plants and their associated microbes to remove contaminants, e.g., Se and Hg, from the environment in volatile forms.	Soils Sediments Sludges Industrial wastewater Agricultural wastewater	Metalloids (e.g., Se, Hg) Chlorinated solvents
Phytodetoxification	Use of plants to change the chemical species of the contaminant to a less toxic form.	Soils Sediments Sludges Industrial wastewater Agricultural wastewater	Chromium Organic compounds Explosives Pesticides
Rhizofiltration	Use of plant roots to absorb and adsorb pollutants from water and aqueous streams.	Surface water Industrial wastewater Agricultural wastewater Acid-mine drainage	Water soluble organics Heavy metals

genetic modifications of plants for phytoremediation have been reported. For example, bacterial genes were transferred to plants to enhance their potentials for 1) phytovolatilization (e.g., transfer of *merA* and *merB* genes from bacteria into three different plant species to increase their tolerance to and volatilization of the toxic mercuric ion after its transformation into the much less toxic elemental mercury); 2) phytoextraction (e.g., transfer of *gsh1* and *gsh2* genes from *Escherichia coli* to *Brassica juncea* plants to increase Cd accumulation in shoots); and 3) detoxification (e.g., transfer of bacterial genes for removal of nitroso groups from explosive compounds into plants to detoxify trinitrotoluene, TNT). Another approach was to express mammalian genes in plants, e.g., genes encoding cytochrome P450 2E1, the liver enzyme responsible for activating trichloroethylene (TCE), resulting in an increase in TCE oxidation by two orders of magnitude. A more straightforward approach is to overexpress existing plant genes that control key-limiting processes in order to overcome these limitations and accelerate the phytoremediation process (e.g., overexpression of AtAPS1 in *B. juncea* resulted in twofold increase in selenium accumulation and the overexpression of NtCBP4 in *Nicotiana tabacum* resulted in 2.5-fold increase in Ni tolerance and two-fold increase in Pb accumulation).[5] These examples provide dramatic evidence of the potential of genetically engineered plants for bringing the promise of phytoremediation to fruition.

THE PHYTOREMEDIATION GENOME/PROTEOME

The vast majority of plant genes affecting the remediation of toxic elements and organics have not yet been identified. Identification of these genes is vital to the success of the transgenic phytoremediation approaches. The recent completion of the *Arabidopsis* genome sequencing project has paved the way for the development of high throughput functional genomic approaches to accelerate the determination of gene function.[6] It is estimated that the *Arabidopsis* genome contains 80–90% of all the phytoremediation genes and gene families found in any macrophytes that could be used in phytoremediation applications.[4] The *Arabidopsis* genome is estimated to have 25,500 genes in 11,000 gene families.[4] Approximately

Table 2 Summary of the size of potential U.S. markets for phytoremediation

Market sector	Annual U.S. potential market
Metals from soils	\$1.2–1.4 billion
Metals from groundwater	\$1.2–1.4 billion
Organics from soils	\$2.3–2.6 billion
Organics from groundwater	\$2.3–2.6 billion
Radionuclides (all media)	\$1.5–2.0 billion
Industrial wastewater	\$7.0–10.0 billion
Municipal wastewater	\$18–28 billion
Landfill leachate control	\$0.6–1.2 billion
Agricultural runoff	\$0.4–0.5 billion
Stormwater management	\$0.2–1.0 billion
Treatment of drinking water	\$0.6–1.0 billion
Total	*\$35.3–51.7 billion*

From Ref. 2.

5% of the genome appears to encode membrane transport proteins responsible for the transport of metals and ions across plant plasma and organellar membranes.[7] These proteins are classified in 46 unique families containing approximately 880 members. In addition, several hundred putative transporters have not yet been assigned to families. Only a small number of these transporter proteins have their function fully characterized. The vast majority, however, have no known or presumed function, and genetic and physiological analyses will be needed to determine their functions and their relative contributions to toxic metal uptake and transport in plants.[4,7]

An analysis of the *Arabidopsis* genome using the amino acid sequences of proteins with known roles in remediation suggests that approximately 700 genes encode phytoremediation-related proteins (PRP). This portion of the *Arabidopsis* proteome has been termed the phytoremediation proteome and it consists of approximately 450 enzymes catalyzing redox reactions (e.g., cytochrome P-450s, oxygenases, and dehalogenases), approximately 250 transport proteins, and several metal-binding proteins, e.g., metallothioneins. The availability of the *Arabidopsis* genome and EST sequences and the identification of the phytoremediation proteome provide enormous opportunities to accelerate greatly our effort to clean up environmental pollution.[4]

PHYTOREMEDIATION IN CONSTRUCTED WETLANDS

Constructed wetlands represent a cost-effective phytoremediation solution for the cleanup of large volumes of wastewaters contaminated with low levels of toxic trace elements. They offer an efficient alternative to conventional water treatment systems because they are 1) relatively inexpensive to construct and operate, 2) easy to maintain, 3) effective and reliable wastewater treatment systems, 4) tolerant of fluctuating hydrologic and contaminant loading rates, and 5) providers of green space, wildlife habitat, and recreational and educational areas. In addition, ecosystems dominated by aquatic macrophytes are among the most productive in the world, largely as a result of ample light, water, nutrients, and the presence of plants that have developed morphological and biochemical adaptations enabling them to take advantage of these optimum conditions. Furthermore, these ecosystems are highly rich in microbial activities and therefore have high capacities to decompose organic matter and stabilize toxic trace elements. Constructed wetlands have been used for many years with great success to remove conventional pollutants (e.g., nitrate and phosphorus) from agricultural nutrient-laden runoff, drinking water, and domestic wastewater. Recently, there has been increasing interest in using constructed wetlands for the treatment of industrial wastewaters and acid-mine drainage containing heavy metals and toxic trace elements. A recent study demonstrated the successful use of wetlands in this area where a 36-ha constructed wetland removed 90% of the toxic selenium from 10 million liters/day of selenite-contaminated oil refinery effluent.[8] Constructed wetlands remove metals primarily through immobilization of the sulfide for Cu, Fe, Mn, Zn, and Cd. Sulfides of most metals are very stable under the anoxic waterlogged conditions of the wetlands.[9] In the case of metalloids such as Se, the soluble ions are taken up by plants and volatilized as a less toxic gas. Less is known about the use of constructed wetlands for the removal of toxic organics or pesticides although recent studies indicate that wetlands efficiently remove some chlorinated compounds present at low levels that are difficult to remove by other means.[9] A significant amount of research is currently underway to determine the best wetland plant species, including algae and vascular plants, that can be used to maximize pollutant removal by wetlands.[10]

FUTURE PROSPECTS

New uses of plants are emerging as a result of the recent advances in agricultural biotechnology. Phytoremediation is one of these new uses that has been extensively and effectively tested in the cleanup of real contamination sites and is becoming commercially feasible. The technology is currently being adopted and promoted by several government agencies (e.g., U.S. Environmental Protection Agency), and field demonstration projects are currently underway in many ''superfund'' sites across the country.[3] However, the existing knowledge about the rate and extent of degradation or extraction of pollutants by the phytoremediating crop is still limited, and specific data are needed on more plants, contaminants, and climate conditions. Additional progress in phytoremediation is likely to come from utilizing the recent genomic information to create new ''superplants'' overexpressing genes responsible for the removal, degradation, and/or sequestration of various contaminants. Further optimization of in-field performance of phytoremediation will require improvements in a number of agronomic practices, ranging from traditional crop management techniques to approaches more specific to phytoremediation, such as amendment of soil with chelators.

REFERENCES

1. Salt, D.E.; Smith, R.D.; Raskin, I. Phytoremediation. Annu. Rev. Plant Physiol. Plant Mol. Biol **1998**, *49*, 643–668.
2. Glass, D.J. *The 2000 Phytoremediation Industry*; D. Glass Associates, Inc.: Massachusetts, 1999; 1–92.
3. Salt, D.E.; Blaylock, M.; Kumar, N.P.B.A.; Dushenkov, V.; Ensley, B.D.; Chet, I.; Raskin, I. Phytoremediation: A novel strategy for the removal of toxic metals from the environment using plants. Biotechnol **1995**, *13*, 468–474.
4. Cobbett, C.S.; Meagher, R.B. Arabidopsis and the Genetic Potential for the Phytoremediation of Toxic Elemental and Organic Pollutants. In *The Arabidopsis Book American Society of Plant Biologists*; 2002; 1–32. http://www.aspb.org/publications/arabidopsis/toc.cfm.
5. Krämer, U.; Chardonnens, A.N. The use of transgenic plants in the bioremediation of soils contaminated with trace elements. Appl. Microbiol. Biotechnol. **2001**, *55*, 661–672.
6. Boyes, C.D.; Zayed, A.M.; Ascenzi, R.; McCaskill, A.J.; Hoffman, N.E.; Davis, K.R.; Gorlach, J. Growth stage-based phenotypic analysis of Arabidopsis: A model for high throughput functional genomics in plants. Plant Cell **2001**, *13*, 1499–1510.
7. Maser, P.; Thomine, S.; Schroeder, J.; Ward, J.; Hirschi, K.; Sze, H.; Talke, I.; Amtmann, A.; Maathuis, F.; Sanders, D.; Harper, J.; Tchieu, J.; Gribskov, M.; Persans, M.; Salt, D.; Kim, S.; Guerinot, M. Phylogenetic relationships within cation transporter families of Arabidopsis. Plant Physiol. **2001**, *126*, 1646–1667.
8. Zayed, A.M.; Pilon-Smits, E.; deSouza, M.; Lin, Z.-L.; Terry, N. Remediation of Selenium-Polluted Soils and Waters by Phytovolatilization. In *Phytoremediation of Contaminated Soil and Water*, 1st Ed.; Terry, N., Banuelos, G., Eds.; Lewis Publishers: New York, 2000; 61–83.
9. Horne, A.J. Phytoremediation by Constructed Wetlands. In *Phytoremediation of Contaminated Soil and Water*, 1st Ed.; Terry, N., Banuelos, G., Eds.; Lewis Publishers: New York, 2000; 13–39.
10. Qian, J.-H.; Zayed, A.; Zhu, Y.-L.; Yu, M.; Terry, N. Phytoaccumulation of trace elements by wetland plants: III. Uptake and accumulation of ten trace elements by twelve plant species. J. Environ. Qual. **1999**, *28*, 1448–1455.

Pierce's Disease and Others Caused by *Xylella fastidiosa*

John S. Hartung
USDA—ARS, Beltsville, Maryland, U.S.A.

INTRODUCTION

In the past decade, *Xylella fastidiosa* Wells et al. has become one of the world's most important plant pathogens, causing devastating diseases of grapevine, sweet orange, coffee, almond, peach, plum, several shade trees, and oleander. The Gram-negative bacterium was formally named only in 1987, and is characterized by being nutritionally fastidious and extremely slow-growing in culture. These traits have made the pathogen difficult to study, and have contributed to its previous obscurity. Pierce's disease (PD) of grapevine, (*Vitis vinifera* L.), caused by *X. fastidiosa* has recently reemerged as a major problem for the California grape industry because of the introduction of a new insect vector.

The Spanish introduced the European grapevine into southern California early in the 18th century, where it became the basis for an important industry. In 1884, the "Mission" grapevines in Anaheim suddenly began to die rapidly. The United States Department of Agriculture dispatched Mr. Newton B. Pierce to investigate and report on the disease in 1889. Pierce reported that this previously undescribed vine disease resulted from blockage of the water-carrying vessels in diseased plants. Pierce was unable to culture a pathogen from diseased plants, and he reported the novel observation of growers that grafting contaminated buds onto healthy vines could transmit the disease. He also noted that unlike the susceptible, higher-value European varieties, the North American grape varieties were tolerant or resistant to the disease.

SYMPTOMS

The initial symptom of PD is a leaf scorch that begins at the outer margins of the lower leaves, then progresses inward on the leaves and upward along the vine. Eventually the affected leaves fall, beginning at the bottom of the vine, but the petioles remain attached. Irregular "green islands" remain on the stem where the bark fails to mature properly. Fruit dries on the vine, and in susceptible varieties the disease leads to rapid vine death.[1]

DISEASE EPIDEMIOLOGY AND IMPORTANCE TO VITICULTURE

In the absence of a bacterium or fungus, PD was logically, though incorrectly, attributed to a virus for nearly a century. Early workers established important facts about the disease. First, the "PD virus" was shown to have an extraordinary range of plant hosts in California. This included 36 species belonging to 18 families that were naturally infected, and 75 species from 23 families that could be experimentally infected by insects.[2] However, the majority of hosts were symptomless carriers. The transmission of the pathogen by 20 insect species[3] was taken as further evidence that the pathogen was a virus, since insect transmission is typical of viral plant diseases. Leafhopper insect vectors (family Cicadellidae) feed exclusively in the plant xylem, consistent with Pierce's initial observation that the primary pathology was a dysfunction of the xylem. The extraordinary range of hosts and vectors make PD exceptionally difficult to control.

PD is endemic throughout North America where grapevines are grown in areas without a cold winter. There is evidence that temperatures below 10°C reduce *X. fastidiosa* populations, while freezing temperatures help grapevines recover from infection. Cold winters prevent the pathogen from overwintering in adult insect vectors, since they are killed by cold temperatures. It is believed that PD originated in the southeastern United States, and was introduced into California with native grape species that were being evaluated for resistance to *Phylloxera* root weevils. Similar shipments of North American grape species were sent to Europe in the same era for the same purpose, but PD did not become established there.

The introduction of PD to California in 1884 caused the failure of the grape industry in the Los Angeles basin. The industry in northern California has suffered only intermittent severe PD outbreaks during the past century. Growers in the southeastern United States have never been able to grow *V. vinifera*, because PD has killed every attempted vineyard. The southeastern United States is the home of the glassy-winged sharpshooter, *Homalodisca coagulata* Say, the most effective insect vector for the bacterium. This vector is so effective because it overwinters

Encyclopedia of Plant and Crop Science
DOI: 10.1081/E-EPCS 120010583

as infective adults, allowing it to transmit the bacterium in early spring. This insect also feeds on mature wood, resulting in systemic infections. Introduced into California in the early 1990s, this insect has become established in at least nine counties, where it threatens the $30 billion grape industry.

DETECTION, IDENTIFICATION, AND CONTROL

X. fastidiosa can be detected and identified by serological methods as well as by the polymerase chain reaction. The bacterium can also be cultured in vitro,[4] but this takes several weeks. PD, like other diseases caused by *X. fastidiosa*, is difficult to control. Because the pathogen lives deep inside woody tissue, it is not affected by externally applied pesticides. Pruning of vines at the first sign of infection has provided some control in northern California, but will likely fail in the presence of the glassy-winged sharpshooter. Insecticides may be used to kill vectors, but they interfere with the long-term suppression of the vectors by other insects. PD management relies on planting *X. fastidiosa*–free grapevines, frequent monitoring of the vineyards, and removal of diseased vines before they can serve as inoculum sources. No useful levels of resistance have been found in varieties of wine grapes or in their rootstocks.

OTHER DISEASES CAUSED BY *Xylella fastidiosa*

Strains of *X. fastidiosa* cause important diseases of a wide range of perennial horticultural plants in North America. These include ''phony'' disease of peach, leaf scald of plum, and leaf scorch of almond, oak, maple, and mulberry.[5] Oleander leaf scorch has become a plant disease problem in the southwestern United States recently, killing oleanders that are widely used as horticultural barriers on highways. The outbreak of oleander leaf scorch coincided with the introduction of the glassy-winged sharpshooter.

Perhaps the most economically important diseases caused by *X. fastidiosa* are found in Brazil. Citrus variegated chlorosis (CVC) was first described in sweet orange in São Paulo State in 1987.[6] São Paulo is home to the largest sweet orange industry in the world, and currently 35% of the trees are infected with *X. fastidiosa*. The coffee industry in Brazil is also seriously affected by a disease known as requiema do café, also caused by *X. fastidiosa*.[7] Both diseases are now widespread in Brazil and occur in several other South and Central American countries.

Symptoms of CVC appear as a bright yellow leaf mottle resembling symptoms of zinc deficiency. Localized, brown lesions appear on the upper surface of affected leaves. These lesions extend through the leaf and a sticky exudate is found on the underside. Fruit on affected trees fails to fill to proper size and remains attached to the tree. In contrast with PD, CVC does not cause defoliation or leaf scorch and it is not fatal to the tree. Symptoms of requiema do café include progressive defoliation from the lower part of a branch upward and conspicuously shortened branch internodes. The fruit on affected coffee bushes fails to fill to proper size, but requiema do café does not kill affected bushes.

THE *Xylella fastidiosa* GENOME

The geographical range of *X. fastidiosa* throughout the Americas, and its wide range of economic and asymptomatic hosts, have led researchers to investigate the population structure of *X. fastidiosa*. These studies have revealed four or five clonal lineages associated with the geographical region and economic host from which the strains were first isolated.[8] Thus, although the pathogen can survive in a wide range of plants in nature, there are mechanisms acting to maintain separate clonal lineages associated with economic hosts.

The impact of CVC on the Brazilian sweet orange industry prompted the state of São Paulo to organize a consortium of researchers to study the pathogen. This effort led to the citrus strain of *X. fastidiosa* becoming the first plant pathogenic bacterium to have its genome completely sequenced.[9] Analysis of the sequence data showed that *X. fastidiosa* lacks genes that are characteristic of many animal and plant pathogenic bacteria. These include genes that define and limit host range and those that encode the system required to export molecules involved in pathogenesis.

CONCLUSION

Conventional plant breeding to produce plant varieties resistant to *X. fastidiosa* is not practical for horticultural reasons. However, it may be possible to genetically engineer plants with resistance to *X. fastidiosa* by producing antimicrobial substances in the xylem fluid. In the case of grapevine or sweet orange, the genetic engineering could be done on the rootstock only, leaving the fruiting portion of the plant unaltered. This approach would more likely be accepted by consumers and would also be more efficient, since each grape or orange variety could be propagated on a common rootstock. However, the possibility that the antimicrobial substances could accumulate in the fruit would have to be carefully monitored and addressed during the research and development phase of such a project, to ensure the safety of consumers.

The economically important hosts of *X. fastidiosa* have been introduced into their respective ecosystems. In contrast, species native to these regions are commonly infected by the bacterium, but suffer no apparent ill effects.[3] The pathogen exploits an unusual ecological niche, living only in the water-conducting vessels of plants and in the mouthparts of certain plant sap-feeding insects. These facts, and the observation that *X. fastidiosa* lacks characteristic pathogenicity genes,[9] have led to the suggestion that the bacterium may be an endophyte that has co-evolved with the flora of the Americas. Hence, the diseases caused by *X. fastidiosa* could be viewed as nature's way of defending the ecosystem from invasive plant species.[10] A corollary of this view is that if *X. fastidiosa* is an endophyte that is somehow out of balance with its host, then perhaps the diseases could be managed by biological control, using other antagonistic endophytic organisms.[11]

ARTICLES OF FURTHER INTEREST

Bacterial Pathogens: Detection and Identification Methods, p. 84
Bacterial Pathogens: Early Interactions with Host Plants, p. 89
Bacterial Products Important for Plant Disease: Cell-Surface Components, p. 92
Bacterial Products Important for Plant Disease: Extracellular Enzymes, p. 98
Bacterial Survival and Dissemination in Insects, p. 105
Bacterial Survival and Dissemination in Seeds and Planting Material, p. 111
Bacteria-Plant Host Specificity Determinants, p. 119
Genomic Approaches to Understanding Plant Pathogenic Bacteria, p. 524
Plant Diseases Caused by Bacteria, p. 947
Plant Response to Stress: Regulation of Plant Gene Expression to Drought, p. 999

REFERENCES

1. Pierce, N.B. The California vine disease. USDA Div. Veg. Pathol. Bull. **1892**, *21*. 222 pp.
2. Freitag, J.H. Host range of the Pierce's disease virus of grapes as determined by insect transmission. Phytopathology **1951**, *41*, 920–934.
3. Freitag, J.H.; Frazier, N.W. Natural infectivity of leafhopper vectors of Pierce's disease virus of grape in California. Phytopathology **1954**, *44*, 7–11.
4. Davis, M.J.; Purcell, A.H.; Thomson, S.V. Pierce's disease of grapevines: Isolation of the causal bacterium. Science **1978**, *199*, 75–77.
5. Hopkins, D.L.; Purcell, A.H. *Xylella fastidiosa*: Cause of Pierce's disease of grapevine and other emergent diseases. Plant Dis. **2002**, *86*, 1056–1066.
6. Hartung, J.S.; Beretta, J.; Brlansky, R.H.; Spisso, J.; Lee, R.F. Citrus variegated chlorosis bacterium: Axenic culture, pathogenicity, and serological relationships with other strains of *Xylella fastidiosa*. Phytopathology **1994**, *84*, 591–597.
7. De Lima, J.E.O.; Miranda, V.S.; Hartung, J.S.; Brlansky, R.H.; Coutinho, A.; Roberto, S.R.; Carlos, E.F. Coffee leaf scorch bacterium: Axenic culture, pathogenicity, and comparison with *Xylella fastidiosa* of citrus. Plant Dis. **1998**, *82*, 94–97.
8. Qin, X.; Miranda, V.S.; Machado, M.A.; Lemos, E.G.M.; Hartung, J.S. An evaluation of the genetic diversity of *Xylella fastidiosa* isolated from diseased citrus and coffee in São Paulo, Brazil. Phytopathology **2001**, *91*, 599–605.
9. Simpson, A.J.G., et al. The genome sequence of the plant pathogen *Xylella fastidiosa*. Nature **2000**, *406*, 151–157.
10. Chen, J.C.; Hartung, J.S.; Hopkins, D.L.; Vidaver, A.K. An evolutionary perspective of Pierce's disease of grapevine, citrus variegated chlorosis, and mulberry leafscorch diseases. Curr. Microbiol. **2002**, *45*, 423–428.
11. Araujo, W.; Marcon, J.; Maccheroni, W., Jr.; Elsas, J.D.v.; Vuurde, J.W.L.v.; Azevedo, J.L. Diversity of endophytic bacterial populations and their interaction with *Xylella fastidiosa* in citrus plants. Appl. Environ. Microbiol. **2002**, *68*, 4906–4914.

Plant Cell Culture and Its Applications

James M. Lee
Washington State University, Pullman, Washington, U.S.A.

P

INTRODUCTION

Plant cell cultures, also called cell suspension cultures, can be applied for the production of valuable pharmaceutical products. The large-scale cultivation technique of plant cells is similar to that of microbial suspension cultures. Plant cells can also be used as hosts for the production of mammalian protein products such as interleukins, monoclonal antibodies, and enzymes. In addition to its use as a production system for valuable compounds, plant cell cultures have been successfully employed for the development of mutants, production of polyploids, and genetic engineering of crops.

In the beginning of the 20th century, German botanist Gottlieb Haberlandt envisaged the possibilities and potentials of plant cell culture and made pioneering attempts to isolate and grow plant cells in culture. Since then, spectacular advancements have occurred in this field and plant cell cultures are now used for different industrial applications. It is now also possible to regenerate whole plants from single cells of a large number of plant species.

The use of isolated plant cells to investigate the physiological, biochemical, and molecular aspects of various cellular functions—especially in the absence of the influence of other cells—has been recognized for a long time. For instance, isolated plant cells have been extensively used for studies on photosynthesis, ion transport, secondary metabolite production, and cytodifferentiation. Plant cell cultures have also been exploited for the isolation of mutants, production of polyploids, and genetic engineering. However, from the current research trend, the most significant application of plant cell culture appears to be in the area of commercial production of industrial compounds.

Plant cells can be cultivated as suspension cultures in a large-scale bioreactor for the production of various primary and secondary metabolic products such as pharmaceuticals, food products, and agricultural chemicals. One of the major constraints for plant-cell based production systems is the low productivity of commercially attractive metabolites.

Another potential application of plant cell cultures is in the production of foreign proteins. Plant cells can offer advantages over mammalian cells because the plant cell medium is very simple and inexpensive. This article reviews large-scale cultivation techniques, secondary metabolite production, and foreign protein production from plant cell cultures.

LARGE-SCALE CULTIVATIONS

The first step in creating the suspension culture is to develop a callus culture, which can be initiated by placing a small piece of plant tissue on solid media containing nutrients, salts, vitamins, and growth factors. After several days, an unorganized amorphous mass of cells will develop on the plate. Plant suspension cultures can be initiated by adding well-developed calli into a liquid medium containing ingredients similar to the solid medium, with the exception of agar.

Plant cell cultures are typically grown in shaker flasks in laboratories. Gentle shaking in shaker flasks is a very effective way to suspend the cells, enhance oxygenation through the liquid surface, and aid the mass transfer of nutrients without damaging the structure of delicate plant cells.[1] The typical batch growth of tobacco cells includes approximately one to two days of lag phase, three to five days of exponential growth phase, and a stationary phase. Maximum cell concentration can be as high as 60% (wet cell weight) or 2.4% (dry cell weight).

For a large-scale cultivation, we can use many different types of conventional stirred-tank fermenters, composed mainly of a cylindrical vessel, impellers, an air sparger, and a means of controlling the temperature (Fig. 1a). The growth rates of plant cells in a stirred bioreactor are not as great as those achieved in a shaker flask, due to shear damage caused by agitation. If the agitation speed is reduced to avoid shear damage, the growth rate can also be decreased through inadequate mixing and air dispersion. One way to avoid this problem is to use an impeller with large blades instead of the standard flat-bladed impeller, as suggested by Hooker, et al.[2]

Another type of bioreactor suitable for plant cell cultivation is the airlift bioreactor, composed of a cylindrical column with a draft cylinder as shown in Fig. 1b. The liquid circulation of the airlift fermenter is induced by sparged air that creates a different density in the bubble-rich part of the liquid in the riser than in the denser bubble-depleted part of the liquid in the downcomer. One major

Encyclopedia of Plant and Crop Science
DOI: 10.1081/E-EPCS 120010552

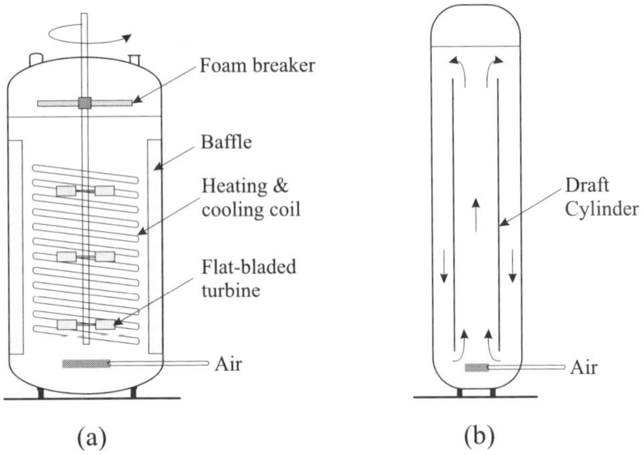

Fig. 1 Typical bioreactors for large-scale plant cell cultures: (a) stirred tank and (b) airlift. (*View this art in color at www.dekker.com.*)

problem with the airlift bioreactor is foaming. Cells tend to rise with the air bubbles, become trapped in the foam, and eventually die due to lack of nutrients. The addition of antifoam agents can reduce this problem, although the agents can also adversely affect cell growth.

THE PRODUCTION OF SECONDARY METABOLITES

Many secondary metabolites can be produced from plant cell cultures.[3] These include alkaloids, flavors, fragrances, organic acids, and steroids (Table 1). Secondary metabolites are seldom produced during the active cell growth phase. They are normally produced during the stationary phase. The product either accumulates within the cells or leaks into the medium.

Some metabolites—such as anthocyanin, anthraquinone, berberine, rosmarinic acid, and shikonin—can be produced with a higher titer than those in the parent plants.[4] For example, shikonin can accumulate up to 14% of dry cell weight in plant cell cultures compared to 1.5% in a whole plant.

However, the major hindrance in the development of commercial processes is the low productivity of secondary metabolites from cell cultures. Therefore, a plant-cell based process can only be justified if it offers an economic advantage over chemical synthesis or traditional extraction processes, or if no other alternative production route exists.[5]

The production levels of secondary metabolites can be improved by adding elicitors to the medium, selecting the best cell line, optimizing the medium formulation, and designing the most appropriate bioreactor. However, efforts to further increase the level of metabolite production are obstructed by lack of basic knowledge of biosynthetic routes and the mechanisms regulating metabolite accumulation, which also prevents the use of modern metabolic or genetic engineering techniques.

THE PRODUCTION OF FOREIGN PROTEINS FROM GENETICALLY MODIFIED PLANT CELLS

Instead of producing natural metabolites, plant cells can be employed as host cells to express mammalian proteins. Several research groups tested the feasibility of producing foreign protein products such as immunoglobulin, monoclonal antibodies, enzymes, interleukins, and vaccines by genetically modifying tobacco cells.[6]

There are several advantages to using plant—rather than mammalian—cells for the production of foreign proteins. Plant cell media are composed mainly of simple sugars and salts and are much less expensive than complex mammalian media. It is therefore much easier and more economical to purify secreted transgenic proteins from plant cell media. Additionally, owing to their rigid exterior walls, plant cells are more resistant than mammalian cells to the shear forces involved in large-scale cell culture. Furthermore, plant cell-derived transgenic proteins are likely to be safer for human use than those derived from mammalian cells, because plant cell contaminants and viruses are not pathogenic to humans.

In general, foreign proteins produced from genetically modified plant cells are correctly folded, glycosylated, and biologically active. The foreign proteins tend to be produced during the exponential growth period of batch cultures and secrete into the medium for separation more easily than do intracellular proteins. The level of intracellular transgenic protein production ranges from 0–0.28 mg/L culture, and that of extracellular protein released into the medium ranges from 0–2 mg/L culture. The percent of extracellular transgenic product relative to the amount of total soluble protein released into the medium ranges from 0–0.84%.[6]

Table 1 Potential product from plant cell cultures

Secondary metabolites	Ajmalacine, anthocyanin, anthraquinones quinoline, berberine, codeine, digitalis, ginsengoside, jasmine, L-dopa, quinine, rosmarinic acid, saponin, serpentine, shikonin, taxol, vanillin, vinblastine, vincristine
Foreign proteins	Enzymes, interleukins, GM-CSF, monoclonal antibodies, vaccines

The productivity of foreign proteins from plant cell cultures is significantly lower than that of animal cell cultures. However, it is still too early to draw any definitive conclusions in this regard because the research efforts in plant host systems have barely begun. Based on our research findings, this low productivity may be due to the protein instability in plant cell media and low expression levels of mammalian genes in plant cell host systems.[7] Nevertheless, when making the above comparisons of transgenic protein production, we must consider that the cost of plant cell media is several orders of magnitude lower than the cost of animal cell media. Furthermore, the cost of the downstream processing of plant cell cultures is also significantly lower than that of animal culture systems. Therefore, plant cell systems may be economically competitive with animal culture systems, even at their current production level.

CONCLUSION

The field of plant cell culture has witnessed remarkable progress in the past few decades. Methods for the establishment of cell cultures and for regeneration of plants from single cells are now available for a large number of plant species. Plant cell cultures have been successfully used for the development of mutants and for molecular plant improvement, and their use as a system for the production of natural plant metabolites or foreign protein products is gaining considerable momentum. One major hindrance for commercial application is their low productivity, which must be overcome before the plant-cell based production system comes an economically viable process.

ARTICLES OF FURTHER INTEREST

Molecular Farming in Plants: Technology Platforms, p. 753
Plant Cell Tissue and Organ Culture: Concepts and Methodologies, p. 934
Plant-Produced Recombinant Therapeutics, p. 969
Protoplast Culture and Regeneration, p. 1065
Secondary Metabolites As Phytomedicines, p. 1120

REFERENCES

1. Lee, J.M.; An, G. Industrial application and genetic Engineering of plant cell culture. Enzyme Microb. Technol. **1986**, *8*, 260–265.
2. Hooker, B.S.; Lee, J.M.; An, G. Cultivation of plant cells in a stirred vessel: Effect of impeller design. Biotechnol. Bioeng. **1990**, *35*, 296–304.
3. Sahai, O.; Knuth, M. Commercializing plant tissue culture processes: Economics, problems and prospects. Biotechnol. Prog. **1985**, *1*, 1–9.
4. Zhong, J.J. Biochemical Engineering of the Production of Plant-Specific Secondary Metabolites by Cell Suspension Cultures. In *Advances in Biochemical Engineering/Biotechnology*; Zhong, J.J., Ed.; Plant Cells, Springer: Berlin, Germany, 2001; Vol. 72, 1–26.
5. Kieran, P.M.; MacLoughlin, P.F.; Malone, D.M. Plant cell suspension cultures: Some engineering considerations. J. Biotechnol. **1997**, *59*, 39–52.
6. James, E.A.; Lee, J.M. The Production of Foreign Protein from Genetically Modified Plant Cells. In *Advances in Biochemical Engineering/Biotechnology*; Zhong, J.J., Ed.; Plant Cells, Springer: Berlin, Germany, 2001; Vol. 72, 127–156.
7. Magnuson, N.S.; Linzmaier, P.M.; Gao, J.W.; Reeves, R.; An, G.; Lee, J.M. Enhanced recovery of a secreted mammalian protein from suspension culture of genetically modified tobacco cells. Protein Expr. Purif. **1996**, *7*, 220–228.
8. Haberlandt, G. Culturversuche mit isolierten Pflanzenzellen. Sitzungsber. Math. Naturwiss. Kl. kais. Akad. Wiss. Wien. **1902**, *111*, 69–92.
9. Gnanam, A.; Kulandaivelu, G. Photosynthetic studies with leaf cell suspensions from higher plants. Plant Physiol. **1969**, *44*, 1451–1456.

Plant Cell Tissue and Organ Culture: Concepts and Methodologies

Darren O. Sage
Horticulture Research International, Wellesbourne, Warwick, U.K.

Ian J. Puddephat
Syngenta, Bracknell, Berkshire, U.K.

INTRODUCTION

In 1904 Hannig isolated immature embryos in vitro from several members of the brassicacae and recovered fertile plants. Since then a range of plant culture techniques has been developed. These techniques are based in part upon the ability of plants to reproduce asexually (vegetative reproduction). Familiar examples would include potato tubers, offsets produced by bulbous plants and cuttings that can be rooted and established as new plants. For many plants, specialized techniques have been developed that allow for the multiplication of isolated cells, tissues, and organs. These techniques share a set of common characteristics; principal among them is that they are conducted using some form of culture vessel. Initially vessels were of glass construction, hence the term in vitro, literally meaning ''in glass.'' Plant culture techniques range from relatively simple systems such as the sowing of orchid seeds in vitro to the regeneration of whole plants from isolated protoplasts, ovules, or microspores (from immature pollen). These diverse methodologies are grouped under the term ''plant cell, tissue, and organ culture,'' although the terms ''in vitro culture of plants'' and ''plant tissue culture'' are also used. These methodologies are defined as the culture of isolated plant parts that include whole plants, seeds, embryos, organs, tissues, cells, protoplasts, and microspores on (or in) a nutrient medium under aseptic conditions. In addition to performing plant tissue culture in vitro, the techniques are further characterized by several other features:

- The environmental conditions are optimized with regard to physical (temperature and light period and quality) as well as nutritional and chemical (plant growth regulators) factors.
- Microorganisms, particularly bacteria and fungi, are excluded, as are other pests such as insects.
- The normal pattern of plant development is usually interrupted; isolated cells, tissues, and organs may enter new patterns of development that lead to regeneration of new organs or result in the formation of embryos.

This article describes various techniques and concepts of plant cell, tissue, and organ culture.

PLANT TISSUE CULTURE TECHNIQUES

In vitro culture of plants has been applied as a practical tool in agriculture and horticulture and as a technique in many studies of plant biology. The basis of these applications is the ability (at least in theory) to clone or proliferate plant cells, tissues, and organs in large numbers. Ultimately, complete new plants can be produced from either preexisting shoot buds and meristems, or following regeneration via shoot organogenesis (morphogenesis) or through embryogenesis. The process of regeneration from cells that would not normally have participated in such events is an indication that plant cells are 'totipotent', retaining a latent capacity to reproduce a whole plant from somatic cells. Totipotency, however, is not a universal property of all plant cells.

The various techniques of plant tissue culture can be classified simply into organized and unorganized cultures with respect to the morphology of the established culture (Table 1). Strictly, unorganized cultures are tissue cultures, although the term is often used more generally to include all culture types. Organ culture generally describes cultures in which some form of organized growth is maintained, as in the continued growth of shoot or root apices. The definitions are not mutually exclusive, and organized cultures can contain mixtures of single cells and unorganized cell clumps, as well as organized structures. Equally, unorganized cells may also be induced to organize and regenerate plant organs through organogenesis or embryogenesis.

Regeneration via organogenesis or embryogenesis may take place either directly from isolated cells, tissues, or

Encyclopedia of Plant and Crop Science
DOI: 10.1081/E-EPCS 120010546

Table 1 Classification and types of plant cell, tissue, and organ cultures

Type of culture	Key features	Selected applications
Unorganized		
Callus (or tissue)	Growth of amorphous cell masses arising from uncoordinated cell divisions from isolated plant parts (e.g., leaf, stem, or root sections) or cultured cells	Cloning of plants through embryogenesis or organogenesis Creation of genetic variation Secondary metabolite production Protoplast production
Cell suspension	Cells (and cell clumps) grown in an agitated liquid medium	As for callus
Protoplast	Culture of isolated cells without cell walls	Somatic hybridization Creation of cybrids and genetic variation
Organized		
Seed	Culture of seeds to produce whole plants	Excluding competition from microorganisms Replacing symbiosis (mycorrhiza)
Shoot tip (or shoot)	Isolated shoot tips or buds that continue shoot growth and multiplication	Induction of multiple shoot formation Production of clonal plants
Node	Cultures of lateral buds on stem tissue that maintain shoot growth	Axillary branching for shoot proliferation Single shoot formation
Shoot meristem	Culture of isolated shoot meristem plus one or two associated leaf primordia	Elimination of pathogens Phytosanitary transport
Embryo	Culture of fertilized or unfertilized zygotic embryos	Preventing embryo abortion Overcoming incompatibility As a source of callus
Root	Proliferation of branching roots in isolation of shoots	Production of secondary metabolites As a source of callus Mycorrhizal production
Anther and microspore	Culture of complete anthers or isolated microspores	Production of haploid plants by androgenesis
Ovule	Culture of isolated ovaries or ovules	Production of haploid plants by gynogenesis Overcoming incompatibility

organs or following an intervening callus/tissue phase. Direct regeneration may occur from isolated tissues or organs that regenerate shoots or roots, or from cells induced to form embryos. Indirect regeneration occurs following the proliferation of unorganized cells established in tissue (callus) or cell suspension culture, and may also occur from semiorganized callus or nodules[1] produced on pieces of tissue or organ isolated in culture.

REQUIREMENTS FOR PLANT CELL TISSUE AND ORGAN CULTURE

Cultures are started from small pieces of plant material usually referred to as explants. Plant cells will only grow in culture when provided with specialized media. Typically, culture media consist of mineral salts for supplying the macro- and micronutrients, and a carbon source (usually sucrose). Vitamins and amino acids in the culture medium may improve growth but are not essential. Growth and development of plant cultures usually depends on the addition of plant growth regulators (PGRs). These compounds, both natural and synthetic, are capable of modifying plant growth and development (morphogenesis) at very low concentrations. Many culture media formulations have been developed for specific techniques or individual species.[2,3] A formulation developed by Murashige and Skoog[4] has been used more widely than others.

Plant material may be cultured in a liquid medium or in a semisolid medium partially solidified with a gelling agent such as agar or gellan gum. Cultures grown in a semisolid medium are kept static, whereas liquid cultures are usually agitated to ensure adequate gaseous exchange. Bioreactors have been developed in conjunction with liquid media to facilitate scaled up operations.[5,6] Promising results have been obtained recently with temporary immersion vessels.

Plant culture media can also support the growth of microorganisms; consequently they are sterilized prior to use, usually by autoclaving. Components of media susceptible to degradation at high temperature are filter-sterilized and added after autoclaving of the media. Manipulations of culture media and plant material are carried out in laminar-flow cabinets that provide sterile working environments. Plant cultures are incubated in controlled-environment growth cabinets.

GENERAL METHODOLOGY USED IN PLANT CULTURE

The processes of plant tissue culture have been divided into five stages (0–IV), as shown in Table 2. The stages are useful for defining the procedural steps, and here we have identified the critical aspects of the biology at each stage for the principal culture techniques illustrated in Fig. 1.

Stage 0: Pretreatment of Motherplant and Explant

Attention must be paid to selection of appropriate starting material. In general, problems with infection of the cultures are greater with material taken from plants grown in the field than with plant material maintained under controlled environments. Growth, morphogenesis, and rates of propagation can be improved by appropriate environmental and chemical pretreatment of stock plants.

Stage I: Establishment of Aseptic Culture

Choice of starting material will depend on the technique used and the plant species being cultured. Material should be disease-free and capable of active growth. Material taken from seedling parts (such as hypocotyls or cotyledons) is usually more responsive than material from mature adult plants. During establishment, the aim is to obtain an aseptic culture of the plant material. Surface sterilization of plant material is achieved through aqueous washing in germicidal agents such as alcohol or hypochlorite. Solutions of sodium and calcium hypochlorite containing between 0.25–1.5% available chlorine are commonly used. Traces of the surface sterilant are removed from plant material by washing several times in sterile distilled water. Explant preparation leads to the induction of wounding responses, some of which can be detrimental to the continued growth of the culture. In these circumstances material may be treated with antioxidants or adsorbents to remove toxic compounds. Antibiotics and fungicides are sometimes used to remove or control contaminants that occur in established cultures. A full account of the procedures used can be found in George.[2]

In addition to achieving aseptic cultures, the establishment stage aims to initiate growth of the culture as well. This may involve the use of stress treatments, including cold, osmotic, and nutrient starvation. Such stress treatments are common for the induction of embryogenesis in isolated cells, particularly for androgenesis and gynogenesis. Plant growth regulators usually play an important part in the induction of organogenesis and embryogenesis, with auxins and cytokinins, often employed to induce them, either singly or in combination.

Table 2 Stages in process of plant cell, tissue, and organ culture

Stage	Activity
0:	Pretreatment
I:	Establishment
II:	Generation of propagules
III:	Preparation for external environment
IV:	Acclimatization in the external environment

Stage II: Generation of Suitable Propagules

This stage concerns the generation of new plant units (propagules) that can give rise to whole plants. These could be axillary or adventitious shoots; branched or adventitious roots; haploid, double haploid, or somatic embryos; or storage or propagative organs (Fig. 1). Propagules may be derived via organogenesis, by shoots, roots, storage or propagative organs produced adventitiously, or by their de novo formation from callus, nodules, meristems, or suspension cultures. Alternatively, propagules may be derived via somatic embryogenesis, androgenesis, or gynogenesis, wherein reprogramming of the cells occurs, to form embryos from somatic cells, microspores, or ovules, respectively. Cytokinins are frequently involved in shoot production and proliferation, whereas growth regulators are often reduced in concentration or omitted for embryogenic development. Some propagules produced during stage II are used in further rounds of multiplication to increase the number of propagules.

Stage III: Preparation for External Environment

In the external environment, the plant material will be required to photosynthesize and survive away from an artificial supply of carbohydrate. Storage and propagative organs may need little if any preparation for this stage;

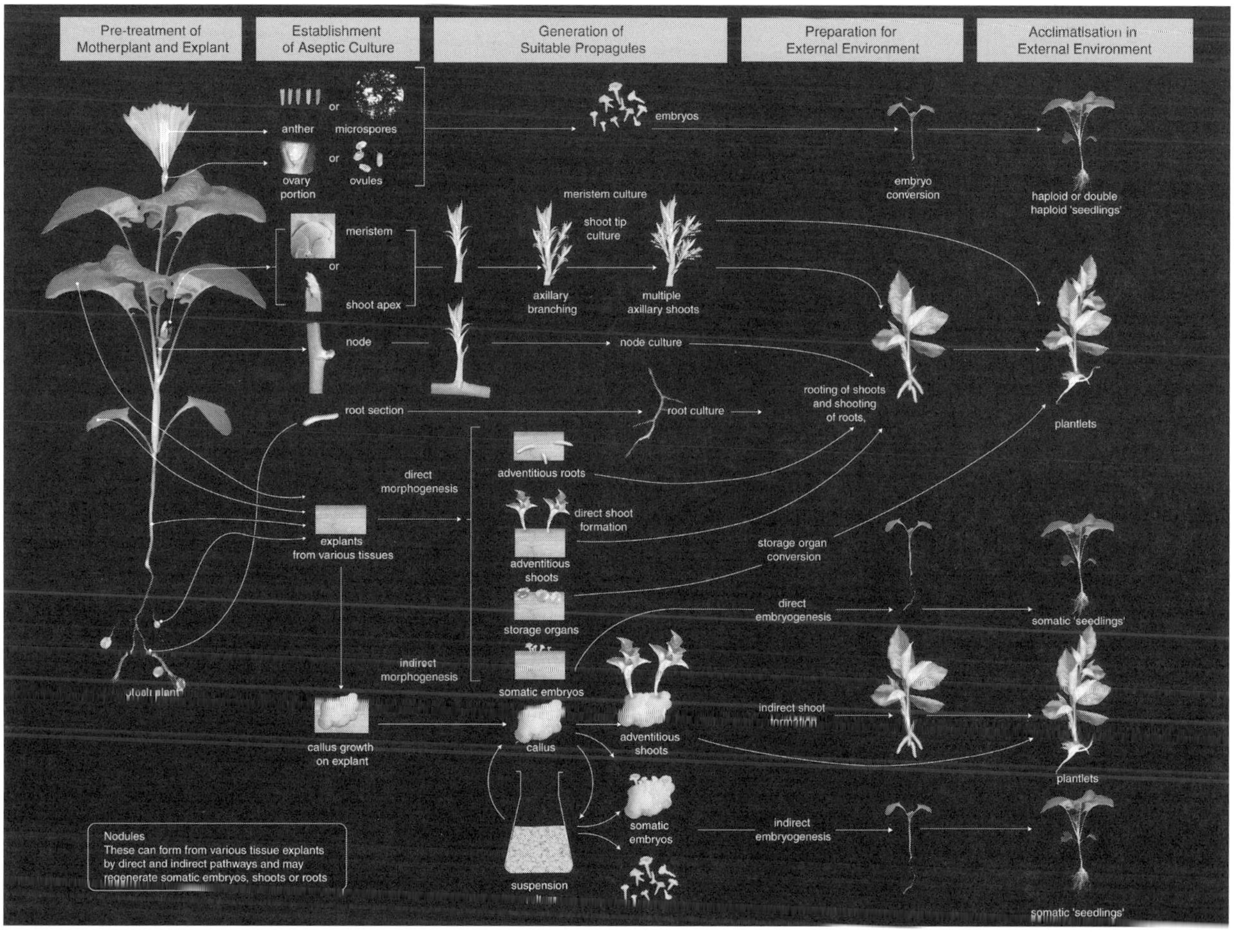

Fig. 1 Plant cell, tissue, and organ culture techniques. (*View this art in color at www.dekker.com.*)

however, shoots must be of adequate size and may be rooted prior to transfer. Some species form adventitious roots on shoots in culture, but are frequently induced by supplying exogenous auxin to the culture media. In root cultures, shoots are regenerated prior to planting out. Culture-derived embryos sometimes spontaneously convert (''germinate'') into plantlets, whereas in other instances they may need to be induced to convert. In some instances, encapsulated propagules are planted out as synthetic seed.[7]

Stage IV: Acclimatization in External Environment

Plants growing in tissue culture have often been exposed to high humidity and low light levels, which may result in poor development and limited photosynthetic capacity on transfer to the external environment. Agar is usually washed from the roots prior to planting and the plants are gradually acclimatized to higher light levels and lower humidity. Some shoots are rooted directly in the weaning stage and not in aseptic culture.

CONCLUSION

A wide range of techniques for the culture of isolated plant cells, tissues, and organs has been developed, as illustrated in Fig. 1. Although they share a set of common features, the underlying biology of the various techniques differs greatly. Various methods are used in the commercial production of a wide range of plant species, including ornamentals, agricultural and forestry species, and natural plant products.[8–11] The techniques are also a core research tool for manipulation of gene function and regeneration of transgenic plants. From their beginnings a century ago, plant cell, tissue, and organ cultures have emerged as important practical tools that have provided fundamental insights into the biology of plants.

ARTICLES OF FURTHER INTEREST

Anther and Microspore Culture and Production of Haploid Plants, p. 43
Commercial Micropropagation, p. 297
In Vitro Chromosome Doubling, p. 572
In Vitro Flowering, p. 576
In Vitro Morphogenesis in Plants—Recent Advances, p. 579
In Vitro Plant Regeneration by Organogenesis, p. 584
In Vitro Pollination and Fertilization, p. 587
In Vitro Production of Triploid Plants, p. 590
In Vitro Tuberization, p. 594
Plant Cell Culture and Its Applications, p. 931
Protoplast Culture and Regeneration, p. 1065
Somaclonal Variation: Origins and Causes, p. 1158
Somatic Embryogenesis in Plants, p. 1165
Transformation Methods and Impact, p. 1233

REFERENCES

1. McCown, B.H.; Zeldin, E.L.; Pinkalla, H.A.; Dedolph, R. Nodule Culture: A Developmental Pathway with High Potential for Regeneration, Automated Micropropagation and Plant Metabolite Production from Woody Plants. In *Genetic Manipulation of Woody Plants*; Hanover, J.W., Keathley, D.E., Eds.; Plenum: New York, 1988; 149–166.
2. George, E.F.; Puttock, D.J.M.; George, H.J. *Plant Tissue Culture Media*; Exergenetics: Westbury, England, 1987; Vol. 1.
3. George, E.F.; Puttock, D.J.M.; George, H.J. *Plant Tissue Culture Media*; Exergenetics: Westbury, England, 1987; Vol. 2.
4. Murashige, T.; Skoog, F. A revised medium for rapid growth and bioassays and tobacco tissue culture. Physiol. Plant. **1962**, *15*, 437–497.
5. George, E.F. *Plant Propagation by Tissue Culture*, 2nd Ed.; The Technology, Exegetics Limited: Edington, England, 1996; Vol. 1, 1–575.
6. George, E.F. *Plant Propagation by Tissue Culture*; The Technology, Exegetics Limited: Edington, England, 1993; Vol. 1.
7. Sharma, S.; Kashyap, S.; Vasudevan, P. Development of clones and somaclones involving tissue culture, mycorrhiza and synthetic seed technology. J. Sci. Ind. Res. **2000**, *59* (7), 531–540.
8. Collin, H.A. Secondary product formation in plant tissue cultures. Plant Growth Regul. **2001**, *34* (1), 119–134.
9. Govil, S.; Gupta, S.C. Commercialization of plant tissue culture in India. Plant Cell, Tissue Organ Cult. **1997**, *51* (1), 65–73.
10. Brown, D.C.W.; Thorpe, T.A. Crop improvement through tissue culture. World J. Microbiol. Biotechnol. **1995**, *11* (4), 409–415.
11. Handley, L.W.; Becwar, M.R.; Chesick, E.E.; Coke, J.E.; Godbey, A.P.; Rutter, M.R. Research and development of commercial tissue culture systems in loblolly pine. Tappi J. **1995**, *78* (5), 169–175.

Plant Defenses Against Insects: Constitutive and Induced Chemical Defenses

Linda L. Walling
University of California, Riverside, California, U.S.A.

INTRODUCTION

Plants have elaborate mechanisms to recognize pathogens and pests and to deploy defense strategies to limit damage. Effective defenses maximize survival from immediate and future challenges and should not significantly compromise plant vitality, longevity, or reproductive success. To achieve this goal, plants utilize two lines of defense to control insects and mites. Both constitutive defenses (constantly expressed) and induced defenses (transiently expressed) channel carbon and nitrogen resources from vegetative and reproductive growth into protective mechanisms. The balance of constitutive and induced responses provides the plant flexibility to cope with a single stress or multiple environmental stresses that occur simultaneously. There are several recent reviews that provide molecular and ecological perspectives on plant defenses to insect and mite feeding.

The constitutive and induced defenses encountered by insects are dependent on their mode and site of feeding. Some insects cause tissue damage by crushing, tearing, rasping, or lacerating plants cells. Other insects use less destructive modes of feeding (piercing and sucking) to consume cellular fluids. It is not surprising that the constitutive and induced defenses encountered by tissue-damaging and piercing/sucking insects are often distinct. While chewing insects release stored chemical defenses, piercing/sucking insects encounter stored chemicals only if they damage cells along the path to their feeding site.

CONSTITUTIVE DEFENSES

Constitutive defenses include both physical barriers and stored chemicals.[1] The cuticle and plant cell wall form the front line of plant defense. Cell wall composition and species-specific compounds that reinforce the wall (lignin, tannins, silicon, suberin) can make it difficult for insects to tear, chew, or penetrate tissue to reach feeding sites, such as the phloem or xylem. The chemical composition and cross-linking of cell wall constituents change after attack, making it a more formidable barrier for subsequent microbial or insect invaders.[2] ''Plant Defenses Against Insects: Physical Defenses'' provides additional details.

Stored chemicals (secondary metabolites) can influence insect growth, development, reproduction, survival, or colonization of plants.[3] The chemical nature and location of these chemicals is dependent on the plant species, organ, or cell-type, and levels can increase after insect feeding. Genes that control secondary metabolite accumulation can be utilized in breeding programs to enhance insect resistance.

Trichomes (leaf hairs) act as physical barriers and chemical storehouses. Trichomes can impale or interfere with the movements of small insects, including insect predators and parasites. Trichomes and secretory glands are also rich sources of chemicals that are insect deterrents and can reduce insect feeding, oviposition, growth, or viability. See ''Plant Defenses Against Insects: Physical Defenses'' for details.

Finally, many defense compounds are stored as inactive glucose-conjugated molecules and are activated only after tissue damage. For example, over 2500 plant species store glucose-conjugated hydrogen cyanide (HCN) precursors and their activating enzymes in separate cells or cellular compartments.[4] Upon insect attack, these molecules mix to produce HCN, a potent toxin. Another example is the glucose conjugates of hydroxamic acid (Hx-Glu), which are correlated with resistance to phloem-feeding and tissue-damaging insects.[5] Hx-Glu levels are highest in young monocot leaves, which need enhanced defenses; young leaves are the preferred site for insect feeding due to their high levels of soluble nitrogen-containing compounds. Older leaves are also protected because Hxs are induced after insect attack.

INDUCED DEFENSES

Insects are active participants in plant-insect interactions. Insects introduce oral secretions into feeding sites. Some compounds (elicitors) in these secretions stimulate defense signaling and volatile production, while other elicitors clearly suppress defense mechanisms. Therefore, induced defenses are complex, often resulting in changes

Encyclopedia of Plant and Crop Science
DOI: 10.1081/E-EPCS 120010489

in plant gene expression or volatile emission that are specific for a insect plant interaction.[6–9]

The duration and intensity of the mechanical and chemical signals generated in each plant-insect interaction influences the nature of the local-and systemic-induced defenses. Plants respond to insect elicitors by modulating levels of plant-produced defense signals, such as salicylic acid (SA; an aspirin derivative) and jasmonic acid (JA) (Fig. 2). SA and JA control opposing defense strategies, which are not mutually exclusive. The magnitude and duration of the increases in JA and SA determine which defense pathways are activated. With tissue-damaging insects, JA-mediated wound responses often predominate; this provides a systemic protection against insects and increased susceptibility to pathogens. With insects that do little tissue damage, SA-mediated or JA-mediated bacterial defenses are activated, providing enhanced protection against pathogens and susceptibility to tissue-damaging herbivores[1,7] See ''Plant Responses to Stress: Role of the Jasmonate Signal Transduction Pathway'' for details.

Activation of Wound Responses

The insects that damage plant tissue induce changes locally (in the damaged tissue) and systemically (in non-damaged tissue) (Fig. 1). The local wound response facilitates healing at the wound site and restoration of cellular homeostasis by increasing cell divisions, cell wall synthesis, and basal metabolism.[2]

Wounding also causes the release of the lipid linolenic acid from plant membranes. Linolenic acid is used to synthesize a variety of biologically active oxylipins, including 12-oxo-phytodienoic acid (12-OPDA) and jasmonic acid (JA) (Fig. 2). These oxylipins induce a set of genes (including the bioactive wound peptide gene prosystemin) that amplify the octadecanoid (oxylipin) pathway in damaged leaves.[1,10] Analyses of mutants that cannot perceive or produce abscisic acid or ethylene have shown that these phytohormones are essential for the wound response. In contrast, as described above, SA antagonizes the wound response (Fig. 2).

While JA, systemin, and 12-OPDA are important in activating wound-response genes in the damaged leaf, the nature of the signal that activates systemic wound responses remains controversial. Current data indicate that systemic signaling must rely on enhanced perception of oxylipins or a novel chemical signal[11] (Fig. 2). Hydrogen peroxide, electrical signals, and hydraulic signals are also possible modulators of systemic wound responses.[10,12]

Many wound-response proteins and secondary metabolites directly interfere with insect feeding or have

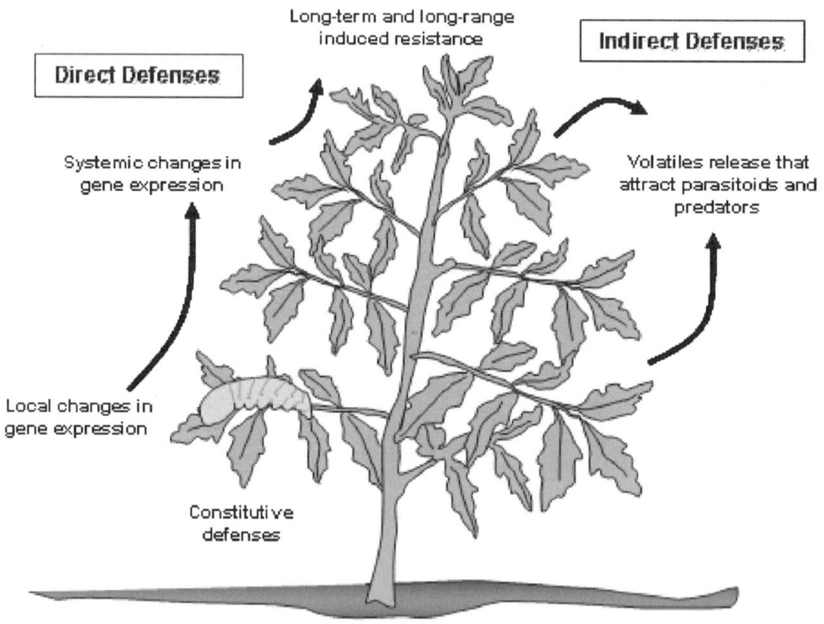

Fig. 1 Constitutive and induced defense responses to insect feeding. Insects encounter constitutive defenses at the site of feeding. Upon feeding, insects provide mechanical signals, introduce chemical signals from their oral secretions at their feeding site, and/or stimulate plants to produce signals that activate defense responses locally and systemically. This often results in a systemic resistance to further attacks. The nature of the systemic signals is unknown at the present time. Genes may be suppressed or activated. Induced proteins and chemicals may have antinutritive or antifeeding effects (direct defense) or may stimulate volatile release, influencing insect densities and attractiveness to natural enemies (indirect defense).

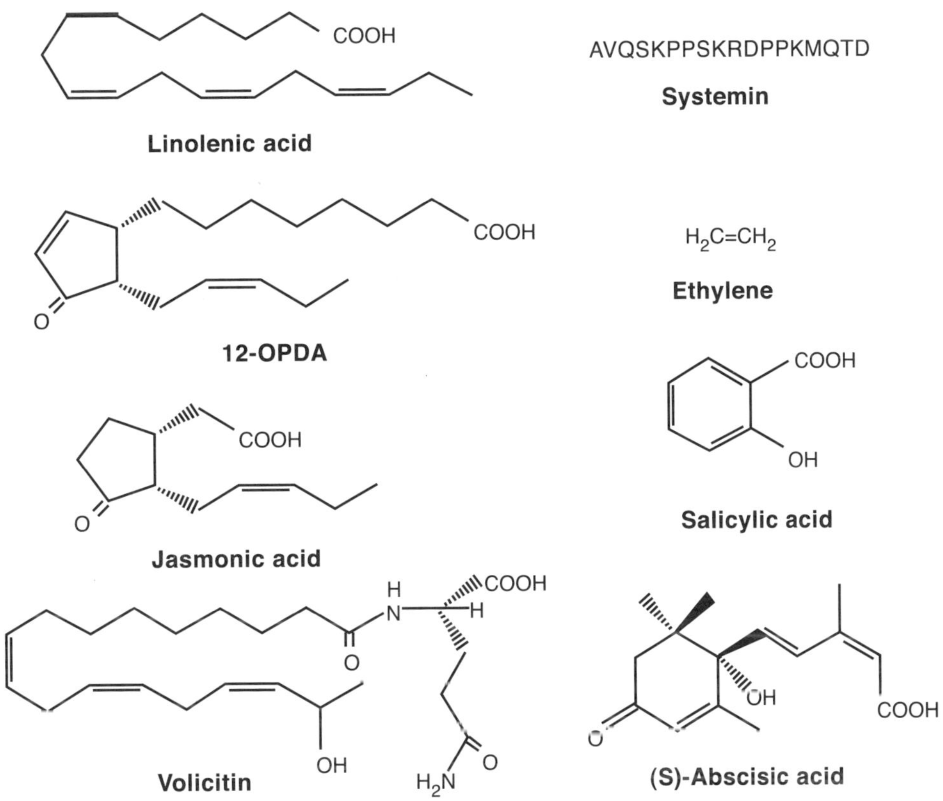

Fig. 2 Molecules important in defense-response signaling in plants. Linolenic acid is released from membranes after insect feeding and is catabolized via hydroperoxide lyase to produce C_6 volatiles (Fig. 3) or the octadecanoid pathway to produce a series of biologically active oxylipins, including 12-OPDA and jasmonic acid. The octadecanoid pathway increases the levels of the wound peptide systemin. Systemin, JA, and 12-OPDA enhance local expression of wound-response genes encoding proteins that have antinutritive or antifeeding effects. Systemin is an 18-amino acid peptide derived from prosystemin that activates the wound response. The amino acid sequence is shown using the single letter amino acid code. Ethylene, abscisic acid, and salicylic acid are known regulators of the octadecanoid pathway. Volicitin is an elicitor of volatile production found in lepidopteran oral secretions. It is formed by modifying the plant-derived linolenic acid.

antinutritive effects.[1,6,10,13] Anti-feeding compounds (alkaloids, C_6 volatiles, phytoalexins) limit plant injury. Anti-nutritive compounds (proteinase inhibitors, polyphenol oxidase, proteases) reduce food quality or digestibility. These compounds restrain insect population expansion by slowing insect development and increasing the time available for attack by predators and parasites. Not surprising, plants that have increased or decreased levels of wound-response proteins (i.e., lipoxygenase, proteinase inhibitors, or polyphenol oxidase) have enhanced resistance or susceptibility to insects, respectively.

Activation of Pathogen-Response Pathways

Piercing/sucking insects use modified mouthparts called stylets to consume large quantities of fluids from plant cells to recover nutrients.[1] The feeding sites may be veins of the phloem (whiteflies, aphids, mealy bugs, pysllids), epidermal and mesophyll cells (thrips), mesophyll parenchyma (scale insects), or xylem (leafhoppers). The amount of damage caused by piercing/sucking insects varies considerably.

The piercing/sucking insects that cause little tissue damage (whiteflies, aphids) activate different signaling pathways than tissue-damaging insects and appear to be perceived as pathogens.[1] Wound responses are not activated or are present transiently. Furthermore, these insects induce pathogenesis-related protein (*PR*) genes, which are usually induced by microbial pathogens. Their roles in insect defense are not understood. *PR* genes are regulated by a variety of defense signals including JA, ethylene, SA, and reactive oxygen species. Recent insect-plant studies indicate that even closely related insect species deliver different chemical or mechanical signals to plants, which are translated into distinct changes in plant gene expression.[1,9]

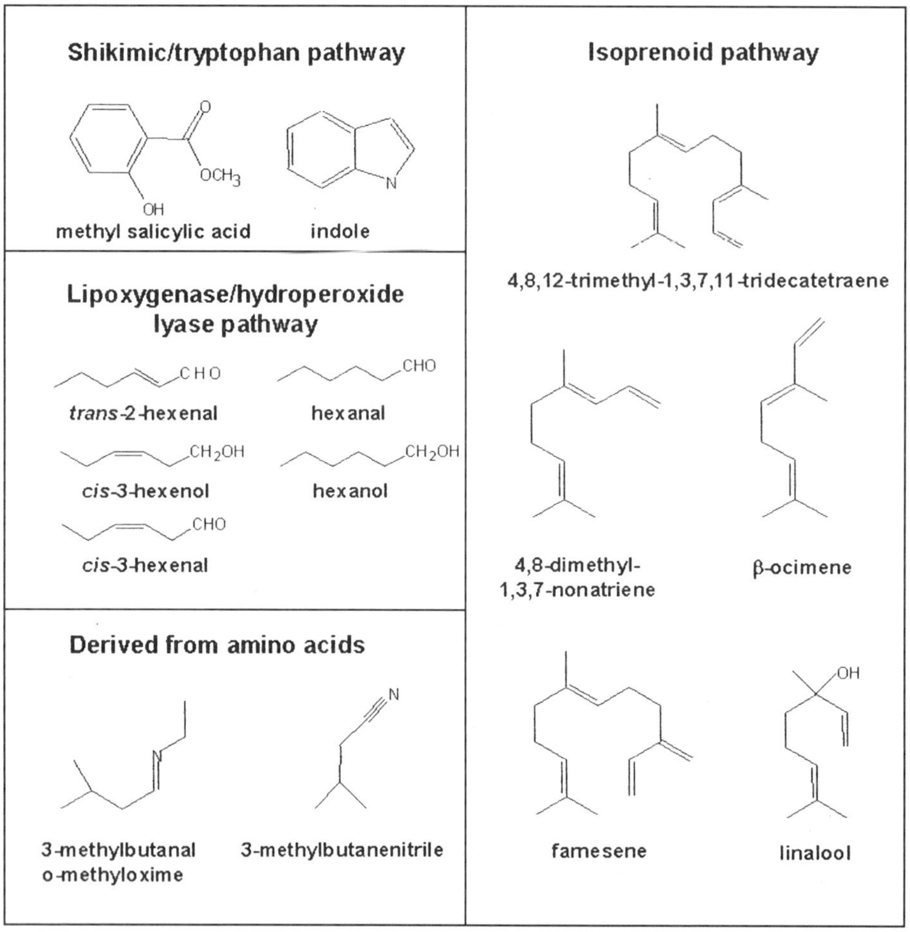

Fig. 3 Classes of volatiles released after herbivore feeding. During herbivore attack, plants synthesize and emit a complex blend of organic volatile compounds. Volatiles are synthesized by the shikimic/tryptophan pathway, or lipoxygenase/hydroperoxide pathway, or isoprenoid pathway, or are derived from amino acids. Representative volatiles and their pathways are illustrated.

Volatile Synthesis and Release

In response to insect and mite attack, wounding or JA treatment, plants release a complex blend of volatiles.[1,8] Six-carbon (C_6) volatiles (the odors of freshly cut grass) are released after wounding and are generated from linolenic acid (Fig. 3). The C_6 volatiles listed in Fig. 3 are known to influence insect reproduction or feeding (for a review, see Ref. 1). Some C_6 volatiles serve as insect attractants. Terpenoids are major constituents of the volatile emissions after insect or mite attack (Fig. 3). In addition, methyl salicylate, indoles, nitriles, and oximes are commonly detected (Fig. 3).

The quantities of each compound in a volatile blend creates important species-specific cues for insects, mites, and their natural enemies. Volatiles dictate insect feeding densities and attract insect parasites and predators. Predators/parasites are able to discriminate between the volatile emissions from plants that are healthy, wounded, or infested with a host or nonhost insect.

Since the volatile blend emitted during each plant-insect interaction is unique, volatile synthesis must be influenced by insect-specific signals. Two elicitors of volatile production are known: volicitin and β-glucosidases. The discovery of volicitin has changed our perceptions of insect elicitors (Fig. 2). Volicitin is made from plant-derived linolenic acid, which is modified in the insect and subsequently reintroduced into the site of feeding via insect oral secretions.[14] This coordinated biochemical initiative of the insect and plant is unique.

CONCLUSION

A broad range of technologies in molecular biology, genetics, and chemical ecology are currently being applied to understand the complexity of plant responses to insect attack. These technologies will certainly be used to investigate the chemical diversity and biochemical origin of the elicitors in insect oral secretions that stimulate or suppress host-plant defenses. Integrated with the rapidly expanding knowledge about plant responses to known defense signals, such as JA, SA, ethylene, and reactive oxygen species, we will soon understand the nature and balance of the signaling pathways induced and suppressed by insects. These data will be useful to understand the cross-talk between insect-and pathogen-defense signaling pathways that ultimately dictate the antinutritive and antifeeding compounds that accumulate after attack. It is anticipated that these findings will identify novel mechanisms that can be used in classical or molecular breeding strategies to enhance plant defenses to insect feeding.

ARTICLES OF FURTHER INTEREST

Economic Impacts of Insects, p. 407
Genetically Engineered Crops with Resistance Against Insects, p. 506
Insect Life History Strategies: Development and Growth, p. 598
Insect/Host Plant Resistance in Crops, p. 605
Insect–Plant Interactions, p. 609
Leaf Cuticle, p. 635
Plant Defenses Against Insects: Physical Defenses, p. 944
Plant Response to Stress: Role of the Jasmonate Signal Transduction Pathway, p. 1006
Trichomes, p. 1254

REFERENCES

1. Walling, L.L. The myriad plant responses to herbivores. J. Plant Growth Regul. **2000**, *19* (2), 195–216.
2. Bostock, R.M.; Stermer, B.A. Perspectives on wound healing in resistance to pathogens. Annu. Rev. Phytopathol. **1989**, *27*, 343–371.
3. Ponda, N.; Khush, G. *Host Plant Resistance to Insects*; CAB International/International Rice Institute: Wallingford, UK, 1995.
4. Vetter, J. Plant cyanogenic glycosides. Toxicon **2000**, *38* (1), 11–36.
5. Gianoli, E.; Papp, M.; Niemeyer, H.M. Costs and benefits of hydroxamic acids-related resistance in winter wheat against the bird cherry-oat aphid, *Rhopalosiphum padi*. L. Ann. Appl. Biol. **1996**, *129* (1), 83–90.
6. Baldwin, I.T.; Halitschke, R.; Kessler, A.; Schittko, U. Merging molecular and ecological approaches in plant-insect interactions. Curr. Opin. Plant Biol. **2001**, *4* (4), 351–358.
7. Felton, G.W.; Korth, K.L. Trade-offs between pathogen and herbivore resistance. Curr. Opin. Plant Biol. **2000**, *3* (4), 309–314.
8. Dicke, M.; Van Poecke, R.M.P.; de Boer, J.G. Inducible indirect defence of plants: From mechanisms to ecological functions. Basic Appl. Ecol. **2003**, *4* (1), 27–42.
9. van de Ven, W.T.G.; LeVesque, C.S.; Perring, T.M.; Walling, L.L. Local and systemic changes in squash gene expression in response to silverleaf whitefly feeding. Plant Cell **2000**, *12* (8), 1409–1423.
10. Ryan, C.A. The systemin signaling pathway: Differential activation of defensive genes. Biochim. Biophys. Acta **2000**, *1477* (1–2), 112–122.
11. Strassner, J.; Schaller, F.; Frick, U.B.; Howe, G.A.; Weiler, E.W.; Amrhein, N.; Macheroux, P.; Schaller, A. Characterization and cDNA-microarray expression analysis of 12-oxophytodienoate reductases reveals differential roles for octadecanoid biosynthesis in the local versus the systemic wound response. Plant J. **2002**, *32* (4), 585–601.
12. Orozco-Cardenas, M.L.; Ryan, C.A. Nitric oxide negatively modulates wound signaling in tomato plants. Plant Physiol. **2002**, *130* (1), 487–493.
13. Karban, R.; Baldwin, I.T. *Induced Responses to Herbivory*; University of Chicago Press: Chicago, 1997.
14. Páre, P.W.; Alborn, H.T.; Tumlinson, J.H. Concerted biosynthesis of an insect elicitor of plant volatiles. Proc. Natl. Acad. Sci. U. S. A. **1998**, *95* (23), 13971–13975.

Plant Defenses Against Insects: Physical Defenses

Sanford D. Eigenbrode
University of Idaho, Moscow, Idaho, U.S.A.

INTRODUCTION

Only nine of the 29 insect orders include plant-eating species or phytophages, suggesting there are unique challenges facing potential herbivores. Four principal challenges can be identified: 1) the exceptional hardness and toughness of many plant tissues; 2) the problem of obtaining adequate attachment to the surfaces of plants in order to feed or oviposit; 3) the nutritional deficiency of plant tissues as a sole source for building animal bodies; and 4) the diverse and abundant secondary chemicals in many plant tissues that appear to serve as plant defenses. The first two of these challenges are physical.

The degree to which the physical properties of plants have been shaped to provide defense against insects is difficult to ascertain. Certainly, stiffened or hardened tissues that support the aerial plant organs and a tough, waterproof integument are requisite for successful exploitation of a terrestrial environment. Nonetheless, these characteristics demonstrably limit insect herbivory and have apparently provoked the evolution of specialized adaptations by phytophagous insects for coping with plant physical characteristics. This article reviews the physical plant characteristics that have been shown to impart defense against insect herbivores.

TOUGHNESS AND HARDNESS OF PLANT TISSUES

Plant tissues are both tough and hard. Toughness is measured in terms of resistance to shearing, whereas hardness is measured in terms of a material's resistance to deformation. Structural polysaccharides in plants (cellulose, hemicelluloses, and pectins), lignin (various polymers with phenylpropane units), and cutin and suberin (polymers saturated and unsaturated long-chain fatty acids) are tough materials that resist shearing. The polysaccharides and lignin contribute to the stiffening of all cell walls, especially in vascular tissues, whereas cutin comprises the protective cuticle covering the primary epidermal cells of fruits and other reproductive structures, leaves, and shoots. Suberin performs a similar function for the epidermis of roots. Some plant tissues, notably within Graminae, are infused with crystals of silica, or other hard minerals.

ADAPTATIONS

The majority of plant-feeding insects do not digest cellulose and other hard or tough materials but must process these tissues to access nutrients in the plant cytosol. Chewing insects must be able to shear these materials to remove pieces of plant material and crush or lacerate the cells to extract nutrients. Adaptations to achieve this include unique forms of mandibular teeth and powerful adductor muscles. The enlarged and strongly sclerotized head capsules of larval lepidoptera, symphyta (sawflies), and phytophagous coleoptera, and attine ants specialized for cutting leaves are evidence for this type of adaptation.

In addition to being tough, structural materials in plants are hard. Hardness causes physical wear to mandibular dentition. Microscopic examination of chewing insects reveals dramatic erosion of mandibular dentition as a result of a few weeks of feeding on hard plant tissues.[1]

Hardness and toughness potentially act synergistically to impede chewing herbivores. The force required to shear a material is considerably greater with a dull rather than a sharp mandible. The tips of the teeth on some phytophagous insects are infused in turn with their own mineral constituents, including zinc and iron, as an apparent response to the challenge of feeding on hardened plant tissues.[2]

Other strategies for dealing with toughened tissues include those of homopterans and heteropterans adapted for extracting phloem or xylem tissues, and mites and thrips, which lacerate plant epidermal cells to extract nutrients. These strategies largely circumvent the problem of cutting or shearing large amounts of cell wall. Some of the earliest insect herbivores—the Paleodictyoptera—extracted nutrients from plants by piercing and sucking, possibly to cope with plant toughness and hardness.

Typically, plant toughness is measured crudely, using a penetrometer that measures the maximum force required to pierce plant tissue. This method has been criticized because it undoubtedly does not measure the physical factors that come into play precisely during mastication of

Encyclopedia of Plant and Crop Science
DOI: 10.1081/E-EPCS 120010490

plant tissue by an insect. Nonetheless, there are good examples in the plant resistance literature in which plant toughness measured this way has been related to susceptibility to chewing injury.[3]

SURFACE STRUCTURES INFLUENCING ATTACHMENT AND HERBIVORY

The epidermis of plants is often covered with hairlike structures known as trichomes. Trichomes may be simple filaments or branching structures, minute and supracellular or large and multicellular, sparsely distributed or forming a dense pubescence on the plant surface. Some are complex structures that terminate in single- or multicellular secretory glands. Trichomes have been shown to be involved in defense against insects.[4] The density, erectness, length, and shape of trichomes can influence their effects on insects. Trichomes may impart defense by forming a barrier that prevents some insects contacting the plant surface. Trichomes tend to be composed of a high proportion of cutin with relatively little cytoplasm. As a result, trichomes contribute to overall plant toughness and toughness of the plant integument, therefore interfering with ingestion. Simple trichomes can also interfere with attachment by feeding or ovipositing insects, and by eggs laid on the leaf surface. Individual insect-plant associations are differently affected by the presence of trichomes. This was assessed by Webster and by Norris and Kogan,[5] who reviewed the results of 59 reports in the host plant resistance literature comparing insect attack on pubescent and hairless or glabrous varieties of crop plants. In 32 of these cases, trichomes provided some kind of resistance against insects, but in 13 cases trichomes conferred increased susceptibility (the remainder of the cases were equivocal). Most of the cases in which trichomes conferred susceptibility involved increased oviposition preference for pubescent plants. Apparently, some herbivorous insects have adapted to cope with plant trichomes. For example, the dense vesture of trichomes on the lower surface of *Quercus ilex* cause the aphid *Myzocallis annulatus* to frequently fall from this surface, whereas the *Q. ilex* specialist *M. schreiberi* walks freely on the plant.[6]

Modified trichomes with hooks or secretory glands can affect insects uniquely. Hooked trichomes can help climbing plants attach to supports. In *Phaseolus vulgaris*, hooked trichomes have also been shown to provide protection against leafhopper nymphs by fatally impaling them.[7] The exudates of glandular trichomes may be sticky and trap small insects. In *Solanum* and *Lycopersicon*, the stickiness is caused by polymerization of phenolics catalyzed by polyphenol oxidases. Substrate and enzyme are compartmentalized in the trichome gland but come into contact and react when insects rupture the glands during feeding. Trichome exudates also often contain topical toxins, repellents, or deterrents that contribute to defense against insects. Although these are beyond the scope of this article, they can work in concert with the physical characteristics of trichomes to enhance defense. For example, in *Solanum berthaultii*, sesquiterpenes produced by one type of glandular trichome cause aphids to move more rapidly, thus making it more likely for them to become trapped in sticky exudates produced by another type of glandular trichome.[8]

The cuticles of terrestrial plants are covered with a layer of long-chain hydrocarbons, related long-chain oxygenated alkyl compounds, and other lipophilic chemicals known as epicuticular waxes. These materials undoubtedly play a primary role in waterproofing the cuticle, but they can also influence insects on plants.[9] Epicuticular waxes sometimes occur as dense vestures of minute crystals or wax blooms that confer a whitish appearance to the plant surface. Prominent wax blooms typically occur on crop plants such as *Brassica* spp., *Sorghum bicolor*, and peas. Such wax blooms can strongly reduce the ability of insects to attach to the plant surface, thereby providing some protection against herbivory. As a result, crops with genetically reduced wax blooms are more susceptible to certain insects.[10]

INDUCIBILITY OF PHYSICAL DEFENSES

Plant defenses against insect herbivores are frequently inducible in response to insect feeding, and physical defenses apparently are no exception. Both trichome density[11] and leaf toughness[12] have been reported to increase following herbivory. As is true for other types of defenses, induced trichome density can be species-specific; for example, feeding by larvae of the cabbage butterfly *Pieris rapae* induced increased densities of trichomes in black mustard, *Brassica nigra*, whereas equivalent injury from flea beetles, *Phyllotreta cruciferae*, had no such effect.[11]

EFFECTS OF PHYSICAL DEFENSES ON INSECT NATURAL ENEMIES

The net defensive effect of most plant characteristics depends not only on their direct effects on insect herbivores, but also on how these characteristics influence the natural enemies of the herbivores. The effect of trichomes and wax blooms on insect herbivores is complicated because these features can also disrupt attachment or impair the mobility of predators and parasitoids. For example, on a range of *Brassica oleracea* genotypes varying in surface waxes, the effectiveness of two predator species was significantly correlated with the attachment forces these predators could generate on the plant surfaces.[13] Wax blooms on some of these genotypes reduced predator attachment forces by as much as two orders of magnitude and severely impaired predator mobility. Apparently as a result of this type of interaction, genetic variants in *Brassica* spp. and in peas with reduced wax bloom are more resistant to certain types of herbivory in the field than are varieties with typical waxy bloom [Rutledge, 2002 #3734]. Both simple and glandular trichomes also have been found to impede natural enemies. Repeatedly, in solanaceous crops and wild species, it has been shown that glandular trichomes producing sticky exudates trap predators and parasitoids, thereby reducing the effectiveness of these natural enemies at suppressing herbivores. Simple trichomes also can alter predator and parasitoid movement patterns and affect their ability to locate prey.

CONCLUSION

Whether physical characteristics of plants have evolved as defenses against insect herbivores is difficult to ascertain. Nonetheless, certain plant physical factors negatively affect herbivores and reduce herbivory. Some of these characteristics have been harnessed by plant breeders as part of their efforts to breed crops with resistance to insects. The possible negative effects of some physical factors such as toughness and trichomes on crop plant quality complicate these breeding efforts. In addition, individual herbivore species respond differently to physical characteristics. Moreover, some physical characteristics have strong effects on the natural enemies of insect herbivores, further complicating the net effects of physical traits on insect herbivory. Thus, as is the case with most efforts to achieve crop protection through breeding, it is necessary to understand the response to physical factors by at least the dominant species in the arthropod community targeted for management.

ARTICLES OF FURTHER INTEREST

Breeding for Durable Resistance, p. 179
Crop Domestication: Role of Unconscious Selection, p. 340
Insect/Host Plant Resistance in Crops, p. 605
Insect–Plant Interactions, p. 609
Leaf Cuticle, p. 635
Leaf Structure, p. 638
Plant Defenses Against Insects: Constitutive and Induced Chemical Defenses, p. 939
Trichomes, p. 1254

REFERENCES

1. Raupp, M.J. Effects of leaf toughness on mandibular wear of the leaf beetle, *Plagiodera versicolora*. Ecol. Entomol. **1985**, *10*, 73–79.
2. Fontaine, A.R.; Olsen, N.; Ring, R.A.; Singla, C.L. Cuticular metal hardening of mouthparts and claws of some forest insects of British Columbia. J. Entomol. Soc. B.C. **1991**, *88*, 45–55.
3. Bergvinson, D.J.; Arnason, J.T.; Hamilton, R.I.; Mihm, J.A.; Jewell, D.C. Determining leaf toughness and its role in maize resistance to the European corn borer (Lepidoptera: Pyralidae). J. Econ. Entomol. **1994**, *87*, 1743–1748.
4. Levin, D.A. The role of trichomes in plant defense. Q. Rev. Biol. **1973**, *48*, 3–15.
5. Norris, D.M.; Kogan, M. Biochemical and Morphological Bases of Resistance. In *Breeding Plants Resistant to Insects*; Maxwell, F., Jennings, P.R., Eds.; John Wiley & Sons: New York, 1980; 23–61.
6. Kennedy, C.E.J. Attachment may be a basis for specialization in oak aphids. Ecol. Entomol. **1986**, *11*, 291–300.
7. Pillemer, E.A.; Tingey, W.M. Hooked trichomes: A physical barrier to a major agricultural pest. Science **1976**, *193*, 482–484.
8. Ave, D.A.; Gregory, P.; Tingey, W.M. Aphid repellent sesquiterpenes in glandular trichomes of *Solanum berthaultii* and *S. tuberosum*. Entomol. Exp. Appl. **1987**, *44*, 131–138.
9. Eigenbrode, S.D.; Espelie, K.E. Effects of plant epicuticular lipids on insect herbivores. Annu. Rev. Entomol. **1995**, *40*, 171–194.
10. Stoner, K.A. Glossy leaf wax and host-plant resistance to insects in *Brassica oleracea* L. under natural infestation. Environ. Entomol. **1990**, *19*, 730–739.
11. Traw, M.B.; Dawson, T.E. Differential induction of trichomes by three herbivores of black mustard. Oecologia **2002**, *131*, 526–532.
12. Robison, D.J.; Raffa, K.F. Characterization of hybrid poplar clones for resistance to the forest tent caterpillar. For. Sci. **1994**, *40*, 686–714.
13. Eigenbrode, S.D.; Kabalo, N.N. Effects of *Brassica oleracea* waxblooms on predation and attachment by *Hippodamia convergens*. Entomol. Exp. Appl. **1999**, *91*, 125–130.
14. Ruthledge, C.E.; Robinson, A.; Eigenbrode, S.D. Effects of a simple morphological mutation on the arthropod community and the impacts of predators on a principal insect herbivore. Oecologia **2003**, *135*, 39–50.

Plant Diseases Caused by Bacteria

Clarence I. Kado
University of California, Davis, California, U.S.A.

INTRODUCTION

In the microbial world, bacteria feed on dead or living substrates to proliferate and survive. Some bacteria have evolved to become highly specialized in their feeding activities. Those that feed on living plants are pathogenic members of the gram-negative bacteria families Rhizobiaceae, Enterobacteriaceae, Pseudomonadaceae, Xanthomonadaceae, Ralstoniaceae, and Burkholderiaceae. Gram-positive pathogens are corneform bacteria in the families Microbacteriaceae and Coyrnebacterineae. They form irregular rods and are non-spore formers. To recognize, infect, and consume plants, pathogenic species are equipped to sense and attach themselves to the host plant, invade and colonize host plant parts, secrete virulence effectors, and move into the intercellular spaces between cells or inside vessel elements. These processes culminate in disease symptoms. This article introduces the major types of disease symptoms caused by phytopathogenic bacteria. References for some specialized texts are provided for readers desiring more detailed information.

FOUR CLASSES OF DISEASE SYMPTOMS

Plant disease symptoms caused by pathogenic bacteria are classified into four general categories: 1) Overgrowths—non–self-limiting growths of plant tissues caused by cell enlargement (hypertrophy) and cell division (hyperplasia), resulting in excessive organ proliferation (fasciations) and/or undifferentiated tissue formation into tumors; 2) Rapid cell death (necrosis)—the rapid killing of cells caused by toxic proteins, effectors (e.g., hormones, proteases, kinases), and organic compounds produced by the invading pathogenic bacteria; 3) Generalized wilt—water stress of plant parts caused by blockage of vessel elements in the host due to exocellular polysaccharides, plant hormones, and other unidentified vessel occlusion factors elaborated by invading phytopathogenic bacteria; and 4) Soft-rot—maceration of plant tissue by invading bacteria that secrete powerful degradative enzymes capable of dissolving pectin (cell cementing material), cellulose, and complex proteins. In addition to these four types of symptoms, some plants become infected but remain symptomless. However, the pathogen isolated from these symptomless plants (which are often wild species,) can cause visible disease symptoms in cultivated species. Finally, there are plant diseases whose symptoms appear only after the crop is harvested and processed.

EXAMPLES OF EACH DISEASE TYPE

Overgrowths

Representative members of plant pathogenic bacteria that cause overgrowths can be found in the Rhizobiaceae, Enterobacteriaceae, Pseudomonadaceae, and Corynebacterineae families. *Agrobacterium tumefaciens*, a member of the Rhizobiaceae, causes a tumor disease called crown gall in a wide range of plant hosts when tested experimentally. In nature, crown gall occurs mainly on woody species such as peaches, plums, apricots, pears, cherries, almonds, walnuts, roses, brambles, kiwi, apples, and grapevines. *Agrobacterium rhizogenes* causes an interesting disease called hairy root, due to the abnormal proliferation of numerous roots from the site of infection. Besides root proliferation, fasciations in the form of leaf and flower proliferation into a witch's broom (Fig. 1) can occur when plants are infected by *Rhodococcus fascians*. *R. fascians* harbors a linear plasmid that encodes a number of virulence effectors, such as isopentenyl adenosine monophosphate, a precursor of the cell growth hormone cytokinin. Olive knot, a tumor disease of olive trees and oleander shrubs, is caused by *Pseudomonas syringae* pv. savastanoi, a member of the Pseudomonadaceae (Fig. 2). *Pantoea herbicola* pv. gypsophilae is a member of the Enterobacteriaceae that causes galls on the ornamental baby's breath plant, *Gypsophila paniculata*. *P. herbicola* pv. gypsophilae produces cytokinin during infection.[1]

Rapid Cell Death

Necrosis-producing bacteria are mainly represented in the Enterobacteriaceae, Pseudomonadaceae, and Xanthomonadaceae. *Erwinia amylovora*, a member of the

Encyclopedia of Plant and Crop Science
DOI: 10.1081/E-EPCS 120010564

Fig. 1 Fasciation disease of carnation caused by *Rhodococcus fascians*.

Enterobacteriaceae and the causal agent of the fire blight disease of Rosaceae plants such as pears and apples, elaborates proteins, toxins, and virulence effectors that rapidly kill cells and assist in advancing the invading bacteria. The pathogen gains access into the host via blossoms and wounds. Necrosis of the blossom petiole followed by drooping of blackened leaves is typical of fire blight (Fig. 3). *Pseudomonas syringae* pv. syringae, a member of the Pseudomonadaceae that causes leaf blights on a wide range of plants and bacterial canker of peaches

(a) (b)

Fig. 2 (a) Olive knot disease on olive tree and (b) oleander knot on oleander, caused by *Pseudomonas syringae* pv. savastanoi. (*View this art in color at www.dekker.com.*)

Fig. 3 Fire blight of pear caused by *Erwinia amylovora*. The scorched appearance of the foliage is typical of the disease.

and other *Prunus* species, produces peptide toxins (e.g., syringomycin), polysaccharides, plant hormones, and at least 11 virulence effector proteins during tissue colonization. The effectors cause cell death by altering the movement of ions across cell membranes, initiating water leakage from within cells (which temporarily fills intercellular spaces, resulting in a symptom called water-soaking), and by altering normal cellular processing mechanisms. *Xanthomonas axonopodis* pv. citri, a member of Xanthomonadaceae, causes citrus canker whose

Fig. 4 Citrus canker of young fruit caused by *Xanthomonas axonopodis* pv. citri. (Photo courtesy of Dr. Dean Gabriel, University of Florida, Gainesville.)

Fig. 5 Early symptoms of angular leaf spot of cotton caused by *Xanthomonas campestris* pv. malvacearum.

Fig. 6 Early leaf spot symptom on the edge of an *Anthirium* leaf caused by *Xanthomonas campestris* pv. dieffenbachiae.

symptoms are typified by raised pustulelike or blisterlike lesions on fruits and leaves. Lesions appear about 7 to 14 days after inoculation followed by the formation of craterlike spots lined with tan-colored tissue. An oily water-soaked margin accompanied by a yellow halo occurs around the spots. On fruit, the blisterlike lesions are pronounced (Fig. 4). Another member of the Xanthomonadaceae is *X. campestris* pv. malvacearum, the causal agent of angular leaf spot of cotton (Fig. 5). The many lesions on the leaf enlarge and coalesce into angular regions, and leaves later drop off prematurely. *X. campestris* pv. dieffenbachiae causes leaf spots on *Anthirium* species (Fig. 6), from where the pathogen prolifically invades the vascular system, in which massive amounts of ooze are produced (Fig. 7).

Generalized Wilt

Vascular wilt-inducing bacteria are as economically important as necrogenic pathogens. *Clavibacter michiganense* subspecies, which are non–spore-forming, gram-positive members of the Microbacteriaceae, cause wilt diseases of specific plants. For example, *C. michiganense* subsp. *michiganensis* causes a wilt disease of tomato plants called bacterial canker. The wilt symptoms are followed by tissue necrosis producing scorched-appearing foliage, cankers (depressions in the stem caused by

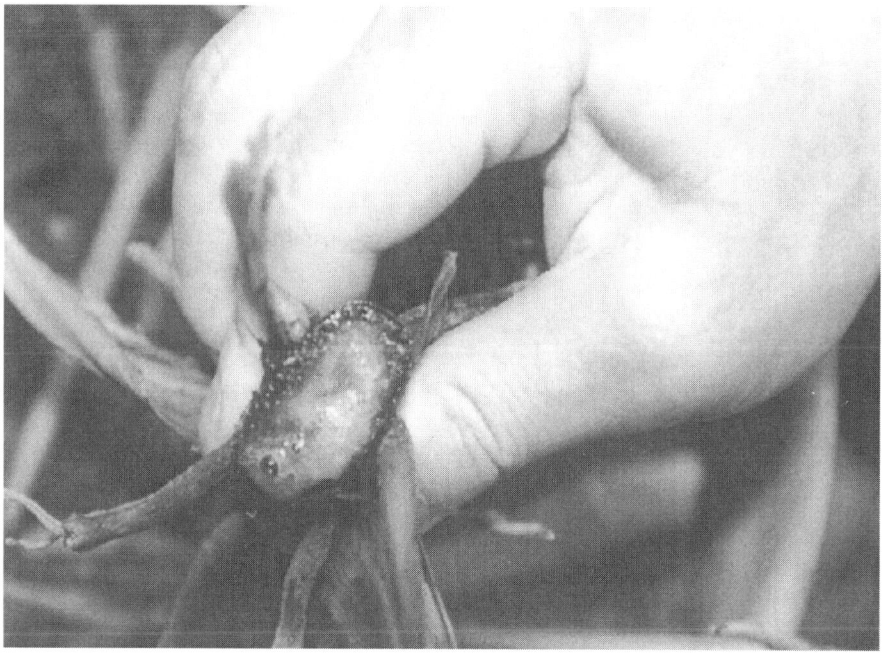

Fig. 7 Ooze from *Anthirium* stem systemically infected by *Xanthomonas campestris* pv. dieffenbachiae.

death of tissue beneath the epidermis), killing of the pith, and spots on the fruit. Another member of this gram-positive group is *C. michiganense* subsp. *insidiosus*, which causes a wilt disease of alfalfa. Infected alfalfa plants are often severely stunted (Fig. 8). *Ralstonia solanacearum*, the sole plant pathogenic member of the family Ralstoniaceae, causes wilt disease of at least 30 plant families. Plants of the Solanaceae family such as potato, tomato, eggplant, and bell pepper are very susceptible to *R. solanacearum* strains classified as Race 1. *R. solanacearum* strains of Race 2 infect banana and cooking banana, causing Moko and Bugtok diseases, respectively (Fig. 9). *Erwinia salicis*, a member of the Enterobacteriaceae, causes wilt of the cricket bat willow tree through vessel occlusion, which is followed by necrosis of the vascular cylinder.

Soft-Rot

Bacteria that cause tissue dissolution resulting in soft rot are mainly members of the Enterobacteriaceae and Pseudomonadaceae. A prototypical example of the Enterobacteriaceae is *Erwinia carotovora* subsp. *carotovora*, a relatively fast growing, rod-shaped bacterium that initiates infection through wounds. Infection progresses as the pathogen elaborates a number of enzymes that break down pectin into consumable oligosaccharides, as well as cell wall material and proteins into utilizable substrates. *E. carotovora* subsp. *carotovora* infects tubers, roots, and corms. The organism also causes the blackleg disease of potato, typified by the rapid collapse of the potato plant in the field (Fig. 10). A member of the Burkholderiaceae, *Burkholderia cepacia* is a complex species that is currently divided into genomovars (genetic species) I, III, and VI. Plant pathogenic strains in genomovar I cause a soft rot of onion bulbs called sour skin disease. Like other soft rot pathogens, *B. cepacia* produces a number of enzymes and virulence effectors that facilitate dissolution of the onion bulb tissues (Fig. 11). Certain strains of *B. cepacia* are used as biocontrol agents, whereas other strains are opportunistic human pathogens that infect individuals with cystic fibrosis.[2]

SYMPTOMLESS DISEASES

Besides the four basic plant disease symptoms, some bacterial diseases do not produce symptoms in the field. For example, the economically important pink disease of pineapple becomes apparent only when the fruit is canned. Pink disease is caused by *Pantoea citrea*, a member of the Enterobacteriaceae that infects the fruit via the florets and produces enzymes that oxidize glucose into ketogluconates, which upon heating results in the

Fig. 8 Alfalfa wilt caused by *Clavibacter michiganense* subsp. *insidiosus*. Stunted plant on the left is infected. Healthy plant is on the right.

formation of a red pigment that stains the canned pineapple fruit, resulting in an unmarketable product[3,4] (Fig. 12).

Xylella fastidiosa, a member of the Xanthomonadaceae, remains latent in a large number of symptom-free wild plants such as California blackberry, Russian thistle, Creek Nettle, Sudan grass, and at least 36 distinct plant species, including wild grape (*Vitis californica* Benth).[5] When experimentally inoculated with *X. fastidiosa*, these plants remain symptomless, yet remain as carriers of the pathogen.[5] Asymptomatic wild plant species apparently tolerate the pathogen, even though their xylem is filled with the bacterium. Hence, these wild plants serve as the reservoir of *X. fastidosa*. In contrast, cultivated plants when infected by *X. fastidiosa* develop severe leaf symptoms such as marginal scorching (Fig. 13) and fruit wilt (raisining) (Fig. 14). *X. fastidiosa* is transmitted from diseased and symptomless plants to crops by the waxy-winged and sharpshooter leaf hoppers. Pierce's disease of grapevines, leaf scorch of almond and plum trees, and variegated chlorosis of citrus are some of the diseases caused by *X. fastidiosa*. *Clavibacter xyli* subsp. *xyli* (now called *Leifsonia xyli* subsp. *xyli* by some researchers) causes ratoon stunting disease of sugarcane. The disease is difficult to diagnose, because modern commercial varieties of sugarcane usually exhibit no external symptoms of infection. Producers may not know they have a problem until yields come in below expectations. Even though older varieties of sugarcane may show reduced vigor (stunting), this symptom is

Fig. 9 Moko disease of a young banana plant caused by *Ralstonia solanacearum* (Race 2). (*View this art in color at www.dekker.com.*)

Fig. 10 Blackleg disease of potato caused by *Erwinia carotovora* subsp. *carotovora*. (*View this art in color at www.dekker.com.*)

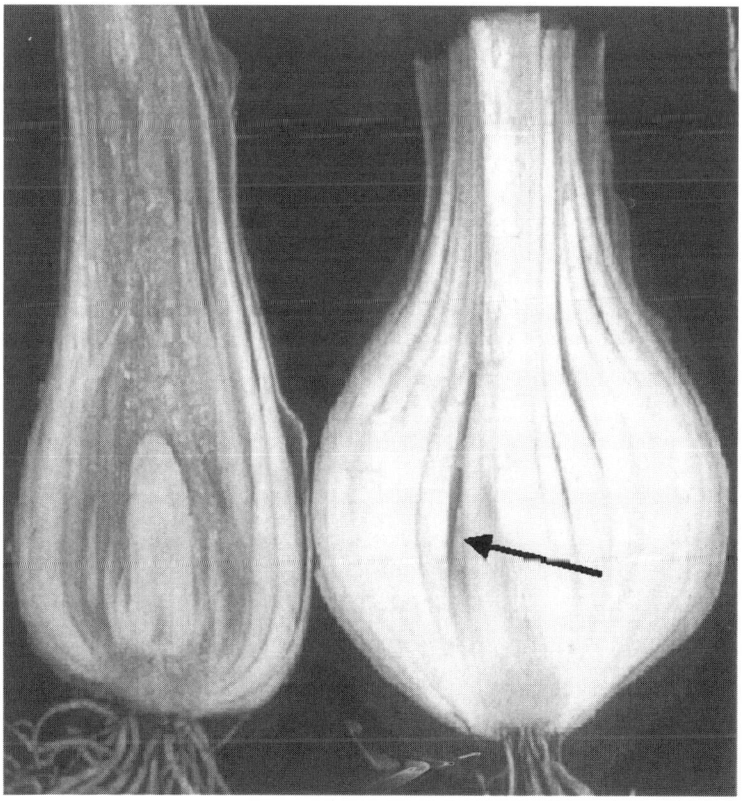

Fig. 11 Sour skin of onion leading to progressive internal rotting caused by *Burkholderia cepacia*. Bulb on right shows initial signs of disease.

Fig. 12 Pink disease of pineapple caused by *Pantoea citrea*. (*View this art in color at www.dekker.com.*)

not diagnostic, because it can also be caused by poor growing conditions.

CONCLUSION

Overgrowths, rapid necrosis, generalized wilt, and soft rot are four distinct symptom classes of plant diseases caused by plant pathogenic bacteria. Each type of symptom reflects the specific type of infecting bacterial species. Overgrowth or tumor-inducing bacteria are generally equipped to elaborate phytohormones. Necrogenic bacteria are known to inject an array of toxic peptides and virulence effectors. Wilt-causing bacteria invade and proliferate in the vascular elements, interrupting fluid transport in the plant. Soft-rot–causing bacteria elaborate a series of degradative enzymes that hydrolyze pectin, cell wall components, and complex proteins. All plant pathogenic bacteria infect their respective plant hosts to create a niche for competitive survival and perpetuation in nature. The host generally provides the food substrates for bacterial survival. In this article, broad examples of diseases caused by plant pathogenic bacteria were provided. Also, examples of asymptomatic diseases are addressed and represent a significant area. Here, native, symptom-free plants are not harmed by the latently infecting bacteria, but modern day, genetically-altered crops are highly susceptible to the same plant pathogenic bacteria that resides in asymptomatic wild species. It appears, therefore, that crop breeding programs have reduced natural resistance to various pathogenic effectors of the plant pathogen.

Fig. 13 Leaf symptom of Pierce's disease of grapevine caused by *Xylella fastidiosa*, a pathogen vectored by the sharpshooter and waxy-winged leafhoppers from diseased and from symptomless wild plant species. (Photo by W. Doug Gubler.)

Fig. 14 Raisining symptom of a grape cluster caused by *Xylella fastidiosa*. (Photo by W. Doug Gubler.)

ARTICLES OF FURTHER INTEREST

Crown Gall, p. 379
Fire Blight, p. 443
Pierce's Disease and Others Caused by Xylella fastidiosa, p. 928

REFERENCES

1. Guo, M.; Manulis, S.; Barash, I.; Lichter, A. The operon for cytokinin biosynthesis of *Erwinia herbicola* pv. gypsophilae contains two promoters and is plant induced. Can. J. Microbiol. **2001**, *47* (12), 1126–1131.
2. Parke, J.L. *Burkholderia cepacia: Friend or Foe?* The Plant Health Instructor 2000. http://www.apsnet.org/education/feature/BurkholderiaCepacia/top.html (accessed October, 2002).
3. Cha, J.-S.; Pujol, C.; Ducusin, A.R.; Macion, E.A.; Hubbard, C.H.; Kado, C.I. Studies on *Pantoea citrea*, the causal agent of pink disease of pineapple. J. Phytopathol. **1997**, *145* (1), 313–319.
4. Pujol, C.J.; Kado, C.I. Genetic and biochemical characterization of the pathway in *Pantoea citrea* leading to pink disease of pineapple. J. Bacteriol. **2000**, *182* (8), 2230–2237.
5. Freitag, J.H. Host range of Pierce's disease virus of grapes as determined by insect transmission. Phytopathology **1951**, *41* (12), 920–934.

Plant Diseases Caused by Subviral Agents

Peter Palukaitis
Scottish Crop Research Institute, Invergowrie, U.K.

INTRODUCTION

Plant diseases also can be caused by other viral-like pathogens referred to collectively as subviral agents. These include replicating small RNAs without genes, known as viroids, and other subviral pathogens known as satellites, which require helper viruses for parts of their life cycle. The latter include satellite viruses, which encode their own coat protein but depend on another virus for their replication and movement, and satellite RNAs and DNAs, which depend on another virus for all of these functions. Satellites can modify the diseases induced by their helper viruses.

VIROIDS AND DISEASES CAUSED BY VIROIDS

Viroids were first described conceptually in the early 1970s, although the diseases that many caused were known for most of the last century.[1,4] Viroids are naked, highly structured RNAs of small size (245–401 nucleotides), which do not encode proteins and thus depend on polymerases encoded by their host plants for their replication. They appear to contain the ability to move throughout their host plants.[4] At present, there are 28 viroid species[2,4] (Table 1). The two families of viroids (Avsunviroidae and Pospiviroidae) show differences in structure, conserved sequences, and ability to undergo autocatalytic cleavage, via ribozymes.[2,4] These physical differences also correlate with their sites of replication; i.e., the viroids in the Avsuniviroidae are replicated by a polymerase in chloroplasts, and at least some viroids in the Pospiviroidae have been shown to be replicated by a nuclear RNA polymerase.[4]

Viroids have been found to be the causal agents of a number of plant diseases, primarily in tropical zones, and to a lesser extent in temperate zones.[4] The major diseases caused by viroids include the spindle tuber disease of potatoes, the sunblotch disease of avocados, various diseases of citrus, cucumber grapevine, hops, pear, and apple, and a number of diseases of ornamental plants[4] (Table 1). However, the most devastating disease caused by viroids is the decline observed in coconut palm in the Philippines. It is estimated that about 40 million trees here have been lost due to this disease, known locally as ''cadang-cadang.''[4] The mechanisms by which viroids induce disease are not known, although it is believed that they alter developmental processes in their host plants, as well as activate disease response genes.[4] In various crop species, the control of the spread of viroids has been accomplished by a combination of removing infected materials, cleaning up breeding lines (through either meristem tip culture coupled with thermotherapy, or by obtaining new sources of disease-free germplasm), and other rigorous phytosanitary measures, especially for orchard plants.[4] This has led to the control of a number of viroid diseases in Australia, Europe, and North America.

SATELLITES AND DISEASES CAUSED BY SATELLITES

In contrast to viroids, which are not encapsidated into particles, satellites depend upon ''helper viruses'' either for their replication as well as encapsidation into the particles of their helper virus (satellite RNAs and satellite DNAs), or for their replication alone, as they encode and are encapsidated by their own coat proteins (satellite viruses).[1,3] Satellites differ from defective RNAs and subgenomic RNAs of plant viruses in not having extensive nucleotide sequence identity with their helper virus.[3,5] There are only four known satellite viruses (Table 2). Except for Panicum mosaic satellite virus, which intensifies the disease elicited by its helper virus satellite viruses, the other satellite viruses do not significantly affect the diseases caused by their respective helper viruses. Thus far, there are six definitive satellite DNAs (Table 2), but there more being characterized.[3] Satellite DNAs all contain singlestranded DNA genomes and all are associated with singlestranded, circular, DNA viruses (geminiviruses). Except for the satellite DNA of tomato leaf curl virus, which contains 682 nucleotides, the other satellite DNAs all contain about 1300 nucleotides.

By contrast, there is considerably more variation in type, size, number, and helper viruses among the satellite RNAs. These are grouped on the basis of whether they encode proteins (large satellite RNAs) or do not do so, and are either linear molecules (small linear satellite RNAs) or form circular molecules (circular satellite RNAs), at least

Encyclopedia of Plant and Crop Science
DOI: 10.1081/E-EPCS 120020064

Table 1 Viroids and their main host plant species

Viroid	Main host
Apple dimple fruit viroid[a]	Apple
Apple scar skin viroid[a]	Apple, pear
Australian grapevine viroid[a]	Grapevine
Avocado sunblotch viroid[b]	Avocado
Chrysanthemum chlorotic mottle viroid[c]	Chrysanthemum
Chrysanthemum stunt viroid[d]	Chrysanthemum
Citrus bent leaf viroid[a]	Grapefruit
Citrus exocortis viroid[d]	Citrus spp.
Cirus viroid III[a]	Citrus
Citrus viroid IV[e]	Citrus
Coconut cadang-cadang viroid[e]	Coconut palm
Coconut tinangaja viroid[e]	Coconut palm
Coleus blumei viroids 1, 2, and 3[f]	*Coleus blumei*
Columnea latent viroid[d]	*Columnea*
Grapevine yellow speckle viroid 1[a]	Grapevine
Grapevine yellow speckle viroid 2[a]	Grapevine
Hop latent viroid[e]	Hop
Hop stunt viroid[g]	Hop, citrus, grapevine, cucumber
Iresine viroid[d]	*Iresine herbstii* (beefsteak plant)
Mexican papita viroid[d]	*Solanum cardiophyllum*
Peach latent mosaic viroid[c]	Peach, plum, apricot, cherry, pear
Pear blister canker viroid[a]	Pear
Potato spindle tuber viroid[d]	Potato
Tomato apical stunt viroid[d]	Tomato
Tomato chlorotic dwarf viroid[d]	Tomato
Tomato planta macho viroid[d]	Tomato

[a]Family Pospiviroidae, Genus Apscaviroid.
[b]Family Avsunviroidae, Genus Avsunviroid.
[c]Family Avsunviroidae, Genus Pelamoviroid.
[d]Family Pospiviroidae, Genus Pospiviroid.
[e]Family Pospiviroidae, Genus Cocadviroid.
[f]Family Pospiviroidae, Genus Coleviroid.
[g]Family Pospiviroidae, Genus Hostuviroid.

at some stage in their replication cycle[3,5] (Table 2). The large satellite RNAs vary in size from about 800–1500 nucleotides. Some encode proteins required for their replication, while others do not. The small linear satellite RNAs contain about 300–900 nucleotides and the circular satellite RNAs are 220–457 nucleotides.[3–5] In a number of cases, members of these various classes of satellite RNA have the ability to alter the diseases induced by their helper viruses. In some cases, the satellite RNAs intensify such disease, while in other cases, they may attenuate the virus-induced disease.[5] In the case of some satellite RNAs, different variants can attenuate symptoms or exacerbate the disease induced by the helper virus, and in one case, the same satellite RNA species can intensify symptoms on one host and attenuate symptoms on another host.[5]

The small, circular, satellite RNAs have sizes and structures similar to that of a number of viroids.[4,6] In common with some of these viroids, these satellite RNAs replicate through a rolling circle mechanism, generating multimeric copies of these RNA molecules. These multimeric copies may or may not be encapsidated by the helper virus.[4] The major satellite RNA molecules present in the cell and inside virus particles are of the monomer species, processed and circularized via ribozymes.

RIBOZYMES: AUTOCATALYTIC RNA

Ribozymes, or RNA enzymes, are RNA sequences that fold into particular structures, allowing them to cleave and/or ligate RNA molecules. A number of these have been found associated with viroids in the family Avsunviroidae, and with the circular satellite RNAs.[1,4–7] The ribozymes associated with subviral agents are classified into two types of autocatalytic structures called hammerhead ribozymes and hairpin ribozymes.[7] These have been studied quite extensively in recent years, in order to understand their mechanisms of cleavage. Various applications for the use of these ribozymes have been suggested, including in molecular medicine.[8,9]

CONCLUSION

Viroids and satellites have provided some valuable tools in our understanding of basic molecular processes, and in furthering our knowledge about pathogenic agents and the roles of RNAs in pathogenicity. More of these agents are being discovered each year, and these pathogens are being recognized as the etiological agents of either new or previously unclassified diseases.[1–4] There is a web site available that contains the nucleotide sequences of all of the known isolates of viroids and satellite RNAs, as well as the secondary structure of these agents. This site also provides information on the presence and structure of any ribozymes associated with such RNAs.[6] Although considerable work has gone into modifying viroids or satellite RNAs for agricultural applications, most of these efforts have not as yet yielded any success. On the other hand, one satellite virus has been modified successfully to express additional RNA sequences in plants, resulting in suppression of the expression of target genes by the RNA silencing pathway.[10] The discovery of ribozymes in several subviral agents has generated considerable interest, although applications of this technology to

Table 2 Satellites of plant viruses

Satellite	Helper virus
Maize white line mosaic satellite virus[a]	Maize white line mosaic virus
Panicum mosaic satellite virus[a]	Panicum mosaic virus
Tobacco mosaic satellite virus[a]	Tobacco mild green mosaic virus
Tobacco necrosis satellite virus[a]	Tobacco mosaic virus
Arabis mosaic virus large satellite RNA[b]	Arabis mosaic virus
Beet ringspot virus satellite RNA[b]	Beet ringspot virus
Bamboo mosaic virus satellite RNA[b]	Bamboo mosaic virus
Chicory yellow mottle virus large satellite RNA[b]	Chicory yellow mottle virus
Grapevine Bulgarian latent virus satellite RNA[b]	Grapevine Bulgarian latent virus
Grapevine fanleaf virus satellite RNA[b]	Grapevine fanleaf virus
Myrobalan latent ringspot virus satellite RNA[b]	Myrobalan latent ringspot virus
Strawberry latent ringspot virus satellite RNA[b]	Strawberry latent ringspot virus
Tomato black ring virus satellite RNA[b]	Tomato black ring virus
Cucumber mosaic virus satellite RNA[c]	Cucumber mosaic virus
Cymbidium ringspot virus satellite RNA[c]	Cymbidium ringspot virus
Groundnut rosette virus satellite RNA[c]	Groundnut rosette virus
Panicum mosaic virus satellite RNA[c]	Panicum mosaic virus
Pea enation mosaic virus satellite RNA[c]	Pea enation mosaic virus
Peanut stunt virus satellite RNA[c]	Peanut stunt virus
Tomato bushy stunt virus satellite RNA[c]	Tomato bushy stunt virus
Turnip crinkle virus satellite RNA[c]	Turnip crinkle virus
Arabis mosaic virus small satellite RNA[d]	Arabis mosaic virus
Cereal yellow dwarf virus-RPV satellite RNA[d]	Cereal yellow dwarf virus
Chicory yellow mottle virus satellite RNA[d]	Chicoy yellow mottle virus
Lucerne transient streak virus satellite RNA[d]	Lucerne transient streak virus
Solanum nodiflorum mottle virus satellite RNA[d]	Solanum nodiflorum mottle virus
Subterranean clover mottle virus satellite RNA[d]	Subterranean clover mottle virus
Tobacco ringspot virus satellite RNA[d]	Tobacco ringspot virus
Velvet tobacco mottle virus satellite RNA[d]	Velvet tobacco mottle virus
Ageratum yellow vein virus satellite DNA β[e]	Ageratum yellow vein virus
Bhendi yellow vein mosaic virus satellite DNA β[e]	Bhendi yellow vein mosaic virus
Cotton leaf curl Multan virus satellite DNA β[e]	Cotton leaf curl Multan virus
Eupatorium yellow vein virus satellite DNA β[e]	Eupatorium yellow vein virus
Honeysuckle yellow vein mosaic virus satellite DNA β[e]	Honeysuckle yellow vein mosaic virus
Tomato yellow leaf curl virus satellite DNA[e]	Tomato yellow leaf curl virus

[a]Satellite viruses.
[b]Large satellite RNAs.
[c]Small linear satellite RNAs.
[d]Circular satellite RNAs.
[e]Satellite DNAs.

either agriculture or medicine have yet to be demonstrated.[8,9] Nevertheless, these various subviral agents may yet yield many dividends that surpass expectations, given their small size.

ACKNOWLEDGMENTS

The author thanks Dr. Ricardo Flores for his sagacious advice. The author was supported by a grant-in-aid from the Scottish Executive Environment and Rural Affairs Department.

ARTICLES OF FURTHER INTEREST

Plant DNA Virus Diseases, p. 960
Virus Assays: Detection and Diagnosis, p. 1273
Virus-Induced Gene Silencing, p. 1276
Virus Movement in Plants, p. 1280

REFERENCES

1. Hull, R. *Matthews' Plant Virology*, 4th Ed.; Academic Press: San Diego, 2002.
2. Flores, R.; Randles, J.W.; Bar-Joseph, M.; Diener, T.O. Subviral Agents: Viroids. In *Virus Taxonomy. Seventh Report of the International Committee on Taxonomy of Viruses*; Van Regenmortel, M.H.V., Fauquet, C.M., Bishop, D.H.L., Carstens, E., Estes, M., Lemon, S., Maniloff, J., Mayo, M.A., McGeoch, D., Pringle, C.R., Wickner, R.B., Eds.; Academic Press: San Diego, 2000; 1009–1024.
3. Mayo, M.A.; Fritsch, C.; Leibowitz, M.J.; Palukaitis, P.; Scholthof, K.-B.G.; Simons, A.E.; Taliansky, M. Subviral Agents: Satellites. In *Virus Taxonomy. Seventh Report of the International Committee on Taxonomy of Viruses*; Van Regenmortel, M.H.V., Fauquet, C.M., Bishop, D.H.L., Carstens, E., Estes, M., Lemon, S., Maniloff, J., Mayo, M.A., McGeoch, D., Pringle, C.R., Wickner, R.B., Eds.; Academic Press: San Diego, 2000; 1025–1032.
4. Hadidi, A.; Flores, R.; Randles, J.W.; Semancik, J.S. *Viroids*; CSIRO Publishing: Collingwood, Australia, 2003; 1–370.
5. Vogt, P.K.; Jackson, A.O. *Satellites and Defective Viral RNAs*; Current Topics in Microbiology and Immunology, Springer: Berlin, 1998; Vol. 239, 1–179.
6. Pelchat, M.; Rocheleau, L.; Perreault, J.; Perreault, J.-P. Subviral RNAs: A database of the smallest known autoreplicable RNA species. Nucleic Acids Res. **2003**, *31* (1), 444–445.
7. Symons, R.H. Plant pathogenic RNAs and RNA catalysis. Nucleic Acids Res. **1997**, *25* (14), 2683–2689.
8. Castanotto, D.; Li, J.R.; Michienzi, A.; Langlois, M.A.; Lee, N.S.; Puymirat, J.; Rossi, J.J. Intracellular ribozyme applications. Biochem. Soc. Trans. **2002**, *30* (6), 1140–1145.
9. Lewin, A.S.; Hauswirth, W.M. Ribozyme gene therapy: Applications for molecular medicine. Trends Mol. Med. **2001**, *7* (5), 221–228.
10. Gossele, V.; Fache, I.; Meulewaeter, F.; Cornelissen, M.; Metzlaff, M. SVISS—A novel transient gene silencing system for gene function discovery and validation in tobacco plants. Plant J. **2002**, *32* (5), 859–866.

Plant DNA Virus Diseases

Robert L. Gilbertson
Maria R. Rojas
University of California, Davis, California, U.S.A.

INTRODUCTION

The majority of plant virus diseases are caused by viruses with single-stranded RNA genomes. Three families of plant viruses have DNA genomes, and some DNA viruses cause economically important diseases, particularly in tropical and subtropical regions. Members of the family Caulimoviridae have a circular double-stranded (ds) DNA genome. The type member, *Cauliflower mosaic virus* (CaMV), is the source of the 35S promoter, which is widely used in plant biotechnology. Banana streak and rice tungro are important diseases caused by caulimoviruses. Members of the family Geminiviridae have a circular single-stranded (ss) DNA genome and distinctive twinned icosahedral virions. The whitefly-transmitted geminiviruses (genus *Begomovirus*) have emerged as one of the most economically important groups of plant viruses. Geminiviruses cause devastating diseases such as African cassava mosaic, bean golden mosaic, beet curly top, cotton leaf curl, maize streak, and tomato yellow leaf curl. Viruses in the family Circoviridae have a multipartite circular ssDNA genome, and members of the genus *Nanovirus* cause plant diseases. Banana bunchy top, the most important viral disease of banana, is caused by *Banana bunchy top virus*. Management of diseases caused by DNA viruses involves an integrated approach involving virus-free propagative material, resistant varieties, synchronized planting dates, insect vector management, sanitation, and host-free periods.

CAULIFLOWER MOSAIC: THE FIRST PLANT DISEASE SHOWN TO BE CAUSED BY A DNA VIRUS

In 1968, Shepherd and colleagues established that CaMV, the causal agent of cauliflower mosaic disease, has a genome composed of dsDNA.[1] Subsequently, a number of diseases were shown to be caused by dsDNA viruses related to CaMV. These viruses were placed in the family Caulimoviridae, which is the only recognized family of plant-infecting dsDNA viruses. The family Caulimoviridae includes two major genera: *Caulimovirus* and *Badnavirus* (Table 1). CaMV is the type species of the genus *Caulimovirus*. The genome of these viruses is composed of a single circular dsDNA (approximately 8.0 kilobase (kb) pair), which is encapsidated in isometric (spherical) particles approximately 50 nm in diameter.[1,2] Caulimoviruses cause mosaic-type diseases (mosaic, mottle, ringspots, and malformation in leaves and stunted plant growth) of dicot crop, ornamental, and weed plants; examples include cauliflower mosaic, carnation-etched ring, dahlia mosaic, and strawberry vein banding.[1,2] Caulimoviruses have narrow host ranges and are spread, plant-to-plant, by aphids or via propagative material, but not through seed.

BADNAVIRUSES: BACILLIFORM dsDNA VIRUSES CAUSING BANANA STREAK AND RICE TUNGRO DISEASES

In the late 1980s, a new type of DNA virus was identified with a circular dsDNA genome of approximately 7.5 kb, and bacilliform (bullet)-shaped virions ($100–300 \times 30$ nm).[3,4] These viruses were placed into the genus *Badnavirus* (ba [bacilliform]-dna [DNA]-virus) in the family Caulimoviridae (Table 1).[4] Badnaviruses cause diseases including banana streak, rice tungro, and cacao swollen shoot (Table 2).[2,4] Banana streak is a mealybug-transmitted disease found in many banana growing regions. In addition to causing losses due to reduced size and production of banana fruit, the dsDNA genome of *Banana streak virus* (BSV) can become integrated into the banana genome, complicating the production of BSV-free propagative material. Rice tungro is a devastating disease of rice in certain areas of Asia, and its impact was particularly severe on high-yielding rice varieties introduced during the green revolution.[5] The disease is caused by a complex of two viruses: *Rice tungro bacilliform virus* (badnavirus) and *Rice tungro spherical virus* (ssRNA virus), and is spread by leafhoppers. The development of rice varieties resistant to the insect vector has been an effective management tool in certain areas (Table 2).[5] Cacao swollen shoot is a mealybug-transmitted disease of considerable importance in West Africa and Sri Lanka, where it reduces yield and quality of cacao by inducing swelling and necrosis of stems and roots.[2]

Encyclopedia of Plant and Crop Science
DOI: 10.1081/E-EPCS 120019988

P

Table 1 Classification and characteristics of DNA viruses of plants

Family	Genus	Genome	Virion shape and size	Insect vector and mode of transmission	Plants infected
Caulimoviridae	*Caulimovirus*	dsDNA monopartite	Spherical 50 nm in diameter	Aphids nonpersistent	Dicots, narrow host range
	Badnavirus	dsDNA monopartite	Bacilliform 130 × 30 nm	Mealybugs semipersistent	Monocots and dicots, individual viruses have narrow host range
	Rice tungro bacilliform virus	dsDNA monopartite	Bacilliform 130 × 30 nm	Leafhoppers semipersistent	Monocots, mostly rice, narrow host range
Geminiviridae	*Mastrevirus*	ssDNA monopartite	Germinate 18 × 30 nm	Leafhoppers persistent, nonpropagative	Monocots, including maize and wheat, narrow host range
	Curtovirus	ssDNA monopartite	Germinate 18 × 30 nm	Beet leafhopper persistent, nonpropagative	Dicots, wide host range
	Begomovirus	ssDNA most bipartite	Germinate 18 × 30 nm	Whiteflies (*Bemisia* spp.) nonpropagative?	Wide range of dicots, individual viruses have narrow host ranges
Circoviridae	*Nanovirus*	ssDNA multipartite	Spherical 18–20 nm in diameter	Aphids persistent, nonpropagative	Monocots, bananas, and *Musa* spp., narrow host range

Table 2 Plant diseases of major economic importance caused by DNA viruses

Disease	Host of economic importance	Geographical distribution	Causal agent	Spread	Management
Rice tungro	Rice	South and Southeast Asia, China	Complex: *Rice tungro bacilliform virus* (caulimovirus) and *Rice tungro spherical virus* (ssRNA virus)	Leafhoppers, propagative material	Vector resistance, planting date, rice-free period
African cassava mosaic	Cassava	Africa	*African cassava mosaic virus* (geminivirus)	Whiteflies, propagative material	Virus-free planting material, sanitation and roguing, resistance
Bean golden mosaic	Common bean	South and Central America, Mexico, SE U.S.A.	*Bean golden mosaic virus* and *Bean golden yellow mosaic virus* (geminivirus)	Whiteflies	Host-free period, time of planting, resistance, vector management
Cotton leaf curl	Cotton	Pakistan, India	Complex. *Cotton leaf curl virus* (geminivirus) and nanovirus-like satellite DNA	Whiteflies	Host-free period, resistance
Tomato yellow leaf curl	Tomato	Asia, Africa, Caribbean basin, SE U.S.A.	Various species of *Tomato yellow leaf curl virus* and *Tomato leaf curl virus* (geminivirus)	Whiteflies, propagative materials (transplants)	Host-free period, time of planting, resistance, vector management
Banana bunchy top	Banana	Australia, Africa, Asia, South Pacific	*Banana bunchy top virus* (nanovirus)	Aphids, propagative material	Quarantine, virus-free planting material, roguing

GEMINIVIRUSES: THE FIRST SINGLE-STRANDED DNA VIRUSES SHOWN TO CAUSE DISEASES IN PLANTS

Viruses in the family Geminiviridae have circular ssDNA genomes (3.0–5.5 kb) encapsidated in small (18×30 nm) twinned icosahedral virions.[3,6,7] The family name comes from the distinctive virion shape and from the latin word ''geminus,'' meaning twin. Plant diseases caused by geminiviruses were recognized long before the nature of the causal agent was determined, and the first recorded observation of a plant disease—an aesthetically pleasing yellow vein symptom described in a Japanese poem in 752 A.D.—may have been caused by a geminivirus.[3,6] Today, the yellow variegation (mosaic) of the ornamental

Fig. 1 Symptoms of diseases caused by various geminiviruses. A. Abutilon mosaic caused by *Abutilon mosaic virus*; B. African cassava mosaic caused by *African cassava mosaic virus*; C. Maize streak caused by *Maize streak virus*; D. Bean golden mosaic caused by *Bean golden yellow mosaic virus*; E. Curly top caused by *Beet mild curly top virus* (previously Worland strain of *Beet curly top virus*); F. Tomato yellow leaf curl caused by *Tomato yellow leaf curl virus*. (*View this art in color at www.dekker.com.*)

flowering maple (*Abutilon* spp.) is due to infection by the geminivirus *Abutilon mosaic virus* (Fig. 1A).

In the late 1970s, geminiviruses were identified and characterized and shown to be the causal agents of many important diseases, including African cassava mosaic, maize streak, and bean golden mosaic (Fig. 1).[2,6,7] Geminiviruses have been classified into four genera based on the insect vector, host range, and genome structure (Table 1),[2,3,6,7] and are now among the best characterized and most economically important plant viruses.

GEMINIVIRUS DISEASES HAVE EMERGED AS MAJOR THREATS TO CROP PRODUCTION

Geminiviruses cause many important diseases (Table 2, Fig. 1).[2,6,7] In southern Africa, India, and islands in the Indian Ocean, maize streak is the most damaging viral disease of maize, causing streaking of leaves, stunted growth, and reduced yields (Fig. 1C).[2,6] The development of moderately resistant varieties has helped reduce losses due to this leafhopper vectored mastrevirus. Curly top disease is caused by three curtovirus species vectored by the beet leafhopper (*Circulifer tenellus*) (Table 1). Disease symptoms include twisted, crumpled, and yellowed leaves; stunted and distorted growth; and vascular discoloration (Fig. 1E).[6] The disease occurs in the western United States, certain Mediterranean countries, and South America. In the early 1900s, curly top nearly destroyed the sugar beet industry in the United States until resistant varieties were developed. The disease still causes losses in tomato and other crops in the U.S. State of California, and an annual spray program is used in an attempt to control the disease by reducing populations of the leafhopper vector.

Diseases caused by whitefly-transmitted geminiviruses (genus *Begomovirus*) (Table 1) have emerged as major constraints on vegetable and field crop production in tropical and subtropical regions throughout the world (Table 2).[6–8] This is due to a worldwide increase in the population and distribution of whiteflies (*Bemisia* spp.) and monoculture of susceptible crops in areas with indigenous weed-infecting geminiviruses. Diseases caused by begomoviruses are characterized by mosaic/mottle, curling, crumpling, and/or yellowing of leaves; stunted and distorted growth; and a reduction in yield quantity and quality (Fig. 1).[2,3,7] African cassava mosaic is a devastating disease in many countries of Africa, causing significant yield losses (e.g., approximately 50% of total production) (Table 2, Fig. 1B).[6] In the late 1990s new, highly pathogenic forms of *African cassava mosaic virus* appeared, arising by genetic recombination. They have caused even greater losses. Bean golden mosaic (Fig. 1D) causes significant losses (as high as 100%) to common bean production in South and Central America and Mexico, and management options remain limited. Some bean varieties possess moderate resistance, but one of the most effective strategies has been a bean-free period of 2–3 months (Table 2).[2] Cotton leaf curl has devastated cotton production in some parts of Pakistan and India, and is caused by a complex of a begomovirus and a nanovirus-like satellite DNA (Table 2).

Tomato yellow leaf curl disease (TYLCD), caused by *Tomato yellow leaf curl virus* (TYLCV), was first recognized in Israel around 1940 and is the most damaging viral disease of tomato. The disease is characterized by stunted and erect growth; small, chlorotic leaves that roll upward; and flower abortion (Fig. 1F). In plants infected early in development, yield loss may reach 100%. In some regions, TYLCD limits the commercial cultivation of tomatoes. TYLCD is now found throughout the Middle East, Southeast Asia, India, many countries of Africa, southern Europe, the Caribbean Basin, and the southeastern United States (Table 2).[2]

In the early 1990s, TYLCV was inadvertently introduced into the Dominican Republic where it destroyed a flourishing tomato processing industry.[8,9] Using the tools of biotechnology, it was established that the virus in the Dominican Republic was identical to TYLCV from the eastern Mediterranean, and that tomato was the primary host.[9] A regional integrated management strategy was implemented that involved 1) a mandatory three-month whitefly host-free period; 2) planting of early maturing hybrid varieties (early season) and TYLCV-resistant varieties (late season); and 3) the selective use of new systemic neonicotinoid insecticides (e.g., imidacloprid). The disease has been effectively managed in the Dominican Republic, and yields are higher than before the introduction of the virus.[9]

BANANA BUNCHY TOP DISEASE: AN ECONOMICALLY IMPORTANT DISEASE CAUSED BY A NANOVIRUS

Banana bunchy top disease (BBTD) is the most important viral disease of banana. First described in the South Pacific in the late 1800s, the disease spread to Australia, Africa, India, the Philippines, and Sri Lanka.[10] BBTD is characterized by the bunchy appearance of the upper portion of the plant, narrow and dwarfed leaves, and dark green streaks on leaves and stems. Infected plants become severely stunted and yield losses can reach 100%. The disease is transmitted by aphids and via propagative materials. Originally thought to be caused by a luteovirus,

BBTD was subsequently shown to be caused by a spherical virus with a multipartite circular ssDNA genome. The virus was named *Banana bunchy top virus* (BBTV). Effective management of BBTD involves the use of quarantines, virus-free propagative material, and a strict regime of sanitation and eradication of diseased plants (Table 2). This approach has been highly effective in Australia, but less effective in other areas.[10] Other diseases caused by viruses similar to BBTV include faba bean necrotic yellows, coconut foliar decay, and subterranean clover stunt.[2,3] These viruses have been placed into the genus *Nanovirus* in the family Circoviridae (Table 1).[3]

CONCLUSION

Although DNA viruses represent a minority of the total viruses that infect plants, they cause many economically important plant diseases, particularly in tropical and subtropical regions. Management of these diseases has been difficult and successful cases have always involved an integrated regional approach that is based on a thorough understanding of viral biology and disease epidemiology.[5,9,10]

ARTICLES OF FURTHER INTEREST

Plant Virus: Structure and Assembly, p. 1029
Plant Viruses: Initiation of Infection, p. 1032
Virus Assays: Detection and Diagnosis, p. 1273
Virus Movement in Plants, p. 1280

REFERENCES

1. Shepherd, R.J.; Lawson, R.H. Caulimoviruses. In *Handbook of Plant Virus Infections and Comparative Diagnosis*; Kurstak, E., Ed.; Elsevier/North Holland Biomedical Press: Amsterdam, 1981; 847–878.
2. Agrios, G.N. *Plant Pathology*, 4th Ed.; Academic Press: San Diego, CA, 1997; 1–635.
3. Hull, R. *Matthews' Plant Virology*, 4th Ed.; Academic Press: San Diego, CA, 2002; 1–1001.
4. Lockhart, B.E.; Olszewski, N.E. Plant Pararetroviruses-Badnaviruses. In *Encyclopedia of Virology*, 2nd Ed.; Granoff, A., Webster, R.G., Eds.; Academic Press: San Diego, 1999; 1296–3000.
5. Koganezawa, H. Present Status of Controlling Rice Tungro Virus. In *Plant Virus Disease Control*; Hadidi, A., Khetarpal, R.K., Koganezawa, H., Eds.; APS Press: St. Paul, MN, 1998; 459–469.
6. Harrison, B.D. Advances in geminivirus research. Annu. Rev. Phytopathol. **1985**, *23*, 55–82.
7. Buck, K.W. Geminiviruses (*Geminiviridae*). In *Encyclopedia of Virology*, 2nd Ed.; Granoff, A., Webster, R.G., Eds.; Academic Press: San Diego, 1999; 597–606.
8. Polston, J.E.; Anderson, P. The emergence of whitefly-transmitted geminiviruses in tomato in the Western Hemisphere. Plant Dis. **1997**, *81*, 1358–1369.
9. Salati, R.; Nahkla, M.K.; Rojas, M.R.; Guzman, P.; Jaquez, J.; Maxwell, D.P.; Gilbertson, R.L. Tomato yellow leaf curl virus in the Dominican Republic: Characterization of an infectious clone, virus monitoring in whiteflies, and identification of reservoir hosts. Phytopathology **2002**, *92*, 487–496.
10. Dale, J.L.; Harding, R.M. Banana Bunchy Top Disease: Current and Future Strategies for Control. In *Plant Virus Disease Control*; Hadidi, A., Khetarpal, R.K., Koganezawa, H., Eds.; APS Press: St. Paul, MN, 1998; 659–669.

Plant–Pathogen Interactions: Evolution

Ian Crute
Institute of Arable Crops Research, Rothamsted, Harpenden, U.K.

INTRODUCTION

All plant pathogens cause disease on a restricted range of plant species but the extent of host specificity differs considerably, presumably as a consequence of different sorts of evolutionary trade-off. The driving force in the evolution of plant–pathogen interactions is conceived as an "arms race" between defense and pathogenesis with the ability to cause diseases of plants almost certainly originating several times during evolution. Two distinct evolutionary routes are likely: 1) acquisition of the capacity to invade living plant tissues by saprotrophic organisms formerly deriving nutrition from dead plant material, and 2) acquisition of pathogenic traits by organisms involved in previously established mutualistic symbioses. Organisms pathogenic on plants may derive their nutrition by either biotrophic or necrotrophic processes. The former involves an intimate association with living cells, and the capacity to stop them dying, while the latter involves the capacity to kill cells, sometimes without direct contact.

DISCUSSION

Co-evolution can be conceived as a process whereby components of non-host resistance are sequentially assembled in evolutionary time and give rise to the degree of host specificity observed in present-day host–pathogen associations. The suggestion is that the diversity of these defenses parallels the evolution of plant species and the pathogens to which they are accessible. An alternative hypothesis is that all plant cells have the capacity to recognize non-self and to actively elaborate a defense response. In this hypothesis, it is envisaged that the defense response of all plants is essentially similar except for subtle variations on the theme; host specificity and the process of co-evolution reside in the recognition process. It is known that micro-evolutionary change over short time periods is mediated by a gene-for-gene relationship involving the recognition of pathogen derived molecules for which within-taxa variation occurs. It is envisaged that recognition events mediating non-host resistance could involve taxon-specific molecules that are invariant and biologically essential to the potential pathogen. Consequently, genotype-specific interactions between host plants and their pathogens mediated by gene-for-gene recognition may simply represent the "tip of an iceberg" in this type of surveillance and response based defense. Recognition specificity may be contributed primarily, but not exclusively, by a leucine-rich repeat domain (LRR) common to the proteins encoded by different classes of plant resistance (R) gene. This domain comprises a variable number of repeats of the motif xxLxLxx (where L is a conserved aliphatic residue, leucine or isoleucine) putatively providing an array of solvent-exposed ligand-binding surfaces. For several R gene families, regions of the LRR have been shown to be under diversifying selection as indicated by a ratio of greater than one for the synonymous to non-synonymous substitutions of the nucleotides encoding residues in the region. The way in which the molecular machinery for signal recognition and transduction is assembled within and between individual plants may be subject to combinatorial variation. Such variation may allow individual plants to recognize and respond effectively to molecules that represent the invariant signature of potential pathogens to which the plant is a non-host. At the same time, the system can provide access to the necessary genetic variation to facilitate a micro-evolutionary response to the tiny sub-set of pathogens of the host species that are detected only by genotypically variant molecular cues on which selection acts to enhance pathogen success.

EVOLUTION AS AN "ARMS RACE"

Plants are intrinsically resistant to the vast majority of agents known to have pathogenic competence. Despite the constant assault on aerial and subterranean plant organs by potential pathogens, most plants are healthy; this healthy state is a consequence of effective defense mechanisms. The driving force in the evolution of plant-pathogen interactions is therefore an "arms-race" between defense and pathogenesis. This "arms-race" commenced before plants colonized land, and certain proteins involved in the mediation of defense against microbial pathogens by both vertebrate and invertebrate animals, as well as plants,

Encyclopedia of Plant and Crop Science
DOI: 10.1081/E-EPCS 120010474

share structural homology[1,2] that may represent conservation of function dating from before divergence of the plant and animal kingdoms.

EVOLUTIONARY ROUTES TO ESTABLISHMENT OF PATHOGENIC INTERACTIONS WITH PLANTS

The ability to cause diseases of plants undoubtedly originated several times during evolution given the phylogenetic diversity of present-day plant pathogens. However, despite this polyphyletic origin there are likely to have been two distinct routes to the establishment of plant pathogenesis: 1) acquisition of the capacity to invade living plant tissues by saprotrophic organisms formerly deriving nutrition from dead plant material, and 2) acquisition of pathogenic traits by organisms involved in previously established mutualistic symbioses. There is evidence for both bacteria and fungi that horizontal transfer of blocks of genes (''pathogenicity islands'') has contributed to the evolution of plant pathogenic capability.

MODES OF NUTRITION AND HOST SPECIFICITY

Two other general observations about the diversity of present-day plant–pathogen interactions are relevant to consideration of their evolutionary origins: mode of nutrition and host specificity. Organisms pathogenic on plants may derive their nutrition by either biotrophic or necrotrophic processes. The former involves an intimate association with living cells and the capacity to stop them dying, while the latter involves the capacity to kill cells, sometimes without direct contact.

All plant pathogens cause disease on a restricted range of plant species but the extent of host specificity differs considerably, presumably as a consequence of different sorts of evolutionary trade-off. For example, there may be trade-off between a pathogen's host range, its capacity to kill plant tissues, its capacity to persist and reproduce on any particular host, and its mode of transmission between hosts. In diverse ecosystems (unlike agriculture) propagules of a pathogen with a wide host range will more readily make contact with a new susceptible host than will a pathogen with a restricted host range. However, pathogen reproductive success will not necessarily always be correlated with wide host range or host mortality. Based on observation of existing host–pathogen associations, it frequently appears to be the case that co-evolution results in an increasing degree of host specific adaptation. For this to be true there must be a fitness deficit for maintaining pathogenic capacity on a wide range of hosts. Investigations of bacterial pathogens provide some evidence for this contention.

Pseudomonas syringae and *Xanthomonas campestris* comprise host-specific biotypes (pathovars). A few genes encoding ''effector proteins'' have been characterized from these bacteria that have pleiotropic effects on virulence and host range.[1] Specific gene-for-gene mediated recognition (discussed later) of the ''effector protein'' by plants carrying a particular ''matching'' resistance gene (R gene) results in an incompatible interaction—the host is classed as resistant and the pathogen is classed as avirulent. For this reason, genes encoding ''effector proteins'' were first recognized by this avirulence trait (and called *avr* genes). However, in addition to being the means by which plants recognize and respond to the presence of the bacterium, ''effector proteins'' also fulfill a virulence function as demonstrated by comparative studies of the growth in planta of bacterial isolates with or without functional *avr* genes.

PRESENT-DAY HOST–PATHOGEN INTERACTIONS: SOME EXAMPLES

A few observations from present-day host–pathogen interactions, placed into a phylogenetic context, serve to illustrate some of the points developed above.

Among the Peronosporales (Oomycetes), there are species of *Pythium* that are considered to be saprotrophs, those that are weak pathogens and those that are destructive, necrotrophic, polyphagous root pathogens. Within the same taxonomic family (Pythiaceae), *Phytophthora cinnamomi* is also a necrotrophic, polyphagous root pathogen (known to attack over 950 plant species) but, while most other species of *Phytophthora* share the necrotrophic mode of nutrition, they exhibit a much more restricted host range. Species in the closely related family Peronosporaceae (the downy mildews) are, by comparison, biotrophic pathogens that exhibit a high degree of host–species specificity with individual pathogen species adapted to distinct host families and subspecific variants being pathogenic on only one or a few closely related plant species within that family.

Within the Ascomycotina, the tribe Helotiales comprises a large number of destructive necrotrophic pathogens but some, such as *Sclerotinia sclerotiorum* and *Botryotinia fuckeliana*, invade a large diversity of host species while others are restricted to particular host families, genera, or individual species. In contrast the tribe Erysiphales (the powdery mildews) comprises a number of exclusively biotrophic genera. While the pathogenesis of most species and sub-specific variants (*formae speciales*) is restricted to one or a few closely related species, others, such as *Erysiphe orontii*, *E. cicheracearum* and, *Oidium lycopersicae* have a wide host range.

HOST SPECIFICITY AND CO-EVOLUTION

There is an assumption that to derive nutrition from living plant tissues, a potential pathogen must be able to evade or render ineffective a multiplicity of constitutive physical and chemical defenses in addition to those that are provoked in active response to attempted colonization. The inaccessibility of most plant species to all but a small number of potential pathogenic agents is referred to as "non-host resistance." Co-evolution can be conceived as a process whereby the components of non-host resistance are sequentially assembled in evolutionary time and give rise to the degree of host specificity observed in present-day host–pathogen associations. The suggestion is that the diversity of these defenses parallels the evolution of plant species and the pathogens to which they are accessible. Such defenses might include: insensitivity to pathogen-derived toxins or hydrolytic enzymes, failure to deliver physical or chemical cues necessary for pathogen development, as well as synthesis of preformed or induced anti-microbial metabolites. In fact, experimental evidence to support these ideas is scarce. In a few specific cases, the importance of plant surface topography as a stimulus for elaboration of infection structures has been clearly demonstrated[3] as has the importance of anti-microbial detoxification mechanisms for successful pathogenesis.[4]

An alternative hypothesis is that all plant cells have the capacity to recognize non-self and to actively elaborate a defense response.[5] In this hypothesis, it is envisaged that the defense response of all plants is essentially similar except for subtle phylogenetic variations on the theme; host specificity and the process of co-evolution reside in the recognition process. Pathogenesis results when an invader fails to deliver a signal that the plant's surveillance apparatus can detect or, alternatively, when the cellular machinery required for the surveillance apparatus to function is destroyed (in the case of pathogens that rapidly kill cells without requiring close contact). It is known that micro-evolutionary change over short time periods is mediated by a gene-for-gene relationship involving the recognition of pathogen derived molecules for which within-taxa variation occurs (discussed later). It is envisaged that recognition events mediating non-host resistance could involve taxon-specific molecules that are invariant and biologically essential to the potential pathogen.[6] Consequently, genotype-specific interactions between host plants and their pathogens mediated by gene-for-gene recognition (discussed in the following) may simply represent the "tip of an iceberg" in this type of surveillance and response based defense. The identification of a plant receptor involved in response to a conserved domain of bacterial flagellin provides evidence for this notion.[7]

THE EVOLUTION OF SPECIFIC RECOGNITION: THE GENE-FOR-GENE RELATIONSHIP

The gene-for-gene relationship is a compelling and unifying concept in plant pathogenesis. Table 2 in Crute et al. provides a summary of the basic properties of a gene-for-gene interaction.[8] Over the last century, many hundreds of so-called plant R genes have been identified by their capacity to mediate pathotype-specific resistance to particular pathogens. Over the last ten years, some tens of these R genes have been isolated and sequenced[1] with details of their evolutionary origins and the way they are organized within plant genomes emerging from associated studies.[9–12] Several different classes of R genes exist in plants[1,12] and access to whole genome sequences now indicates the existence of about 150 "resistance gene analogs" (RGAs) in one genotype of *Arabidopsis thaliana* and about 1700 in rice (of which function has only been attributed to relatively few).[12] R genes with structural homology are now known to mediate resistance to parasitic agents as diverse as fungi, oomycetes, bacteria, viruses, nematodes, and aphids.

Although allelic variation at single R gene loci does occur, RGAs are commonly clustered within complex loci comprising a variable number of paralogs that have arisen by gene duplication.[9,10,12] Out-crossing and recombination, random mutation, interallelic recombination, gene conversion and unequal crossing-over (resulting in gene duplications and deletions) provide mechanisms for the generation of allelic and haplotypic variation on which selection can act. Time-scales over which variation is generated and selected as well as the processes involved in selection are the subject of debate and investigation.[9–12] In addition, selection may favor the observed physical association of genes that are structurally unrelated but contribute to R gene function.[7,9]

CONCLUSION

The protein products of R genes are believed to function as components of a signal-transduction system. A putative receptor system recognizes molecules indicative of pathogen presence and activates the plant's generic defense responses. The receptor and signaling system has numerous components that are being revealed by genetics and cell biology[1] with previously characterized R genes representing one such component. Recognition specificity may be contributed primarily, but not exclusively, by a LRR common to the proteins encoded by different classes of R gene. This domain comprises a variable number of repeats of the motif xxLxLxx putatively providing an array of solvent-exposed ligand-binding surfaces. For several R gene families, regions of the LRR have been shown to be under diversifying selection as indicated by a ratio of greater than one for the synonymous to non-synonymous substitutions of the nucleotides encoding residues in the region (excluding the conserved aliphatic residues).[2,9,10] On the basis of phenotypic specificity, there is no evidence that any one R allele has affinity for more than one ligand. However, there is no intrinsic reason why this would not be the case and the degree of ligand affinity is likely to be another attribute of RGAs on which selection will act.

Astronomical sequence variation is likely to be manifested by RGAs represented within a single plant genome, within the genepool of a particular plant population and among the diversity of a genotypes represented by any particular plant species. Within and between individual plants the way in which the molecular machinery for signal recognition and transduction is assembled may be subject to combinatorial variation. Such variation may allow individual plants to recognize and respond effectively to molecules that represent the invariant signature of potential pathogens (such as bacterial flagellin) to which the plant is a non-host.[7] At the same time, the system can provide access to the necessary genetic variation to facilitate a micro-evolutionary response to the tiny sub-set of pathogens of the host species that are detected only by genotypically variant molecular cues on which selection acts to enhance pathogen success. Experimental evidence will probably be forthcoming to establish whether genotype-specific and non-host resistance are indeed mediated through essentially similar signaling pathways differing only in the number of ligands detected or the extent to which the ligand molecule is variant and dispensable or invariant and indispensable to the potential pathogen.

ARTICLES OF FURTHER INTEREST

Bacteria-Plant Host Specificity Determinants, p. 119
Coevolution of Insects and Plants, p. 289
Insect–Plant Interactions, p. 609
Mechanisms of Infection: Imperfect Fungi, p. 694
Mechanisms of Infection: Oomycetes, p. 697
Mechanisms of Infection: Rusts, p. 701
Molecular Evolution, p. 748
Mutualisms in Plant–Insect Interactions, p. 763
Mycorrhizal Evolution, p. 767
Mycorrhizal Symbioses, p. 770
Oomycete–Plant Interactions: Current Issues in, p. 843
Plant Defenses Against Insects: Constitutive and Induced Chemical Defenses, p. 939
Plant Defenses Against Insects: Physical Defenses, p. 944
Population Genetics of Plant Pathogenic Fungi, p. 1046
Viral Host Genomics, p. 1269

REFERENCES

1. Staskawicz, B.J.; Mudgett, M.B.; Dangl, J.L.; Galan, J.E. Common and contrasting themes of plant and animal disease. Science **2001**, *292*, 2285–2289.
2. Meyers, B.B.; Dicjerman, A.W.; Michelmore, R.W.; Sivaramakrishnan, S.; Sobral, B.W.; Young, N.D. Plant disease resistance genes encode members of an ancient and diverse protein family within the nucleotide-binding superfamily. Plant J. **1999**, *20*, 317–332.
3. Hoch, H.C.; Staples, R.C. Signaling for Infection Structure Formation in Fungi. In *The Fungal Spore and Disease Initiation in Plants and Animals*; Cole, G.T., Hoch, H.C., Eds.; Plenum Press: New York, USA, 1991; 25–46.
4. Morrissey, J.P.; Osbourn, A.E. Fungal resistance to plant antibiotics as a mechanism of pathogenesis. Microbiol. Mol. Biol. Rev. **1999**, *63*, 708–724.
5. Somssich, I.E.; Hahlbrock, K. Pathogen defence in plants—A paradigm of biological complexity. Trends Plant Sci. **1998**, *3*, 86–90.
6. Crute, I.R.; Pink, D.A.C. Genetics and utilization of pathogen resistance in plants. Plant Cell **1996**, *8*, 1747–1755.
7. Gómez-Gómez, L.; Boller, T. A LRR receptor-like kinase involved in recognition of the flagellin elicitor in Arabidopsis. Mol. Cell **2000**, *5*, 1–20.
8. *The Gene-for-Gene Relationship in Plant Parasite Interactions*; Crute, I.R., Holub, E.B., Burdon, J.J., Eds.; CAB International: Wallingford, UK, 1997; 427 pp.
9. Parniske, M.; Hammond-Kosack, K.E.; Golstein, C.; Thomas, C.M.; Jones, D.A.; Harrison, K.; Wulff, B.B.H.; Jones, J.D.G. Novel disease resistance specificities result from sequence exchange between tandomly repeated genes at the Cf-4/9 locus of tomato. Cell **1997**, *91*, 821–832.
10. Michelmore, R.W.; Meyers, B.C. Clusters of resistance genes in plants evolve by divergent selection and a birth-and death process. Genome Res. **1998**, *8*, 1113–1130.
11. Dodds, P.N.; Lawrence, G.J.; Ellis, J.G. Contrasting modes of evolution acting on the complex N locus for rust resistance in flax. Plant J. **2001**, *27*, 439–453.
12. Holub, E.B. The arms race is ancient history in *Arabidopsis*, the wildflower. Nat. Rev. **2001**, *2*, 516–527.

Plant-Produced Recombinant Therapeutics

Lokesh Joshi
Miti M. Shah
C. Robert Flynn
Alyssa Panitch
Arizona Biodesign Institute, Arizona State University, Tempe, Arizona, U.S.A.

INTRODUCTION

The technology of bioengineering plants to produce recombinant therapeutics has become commonplace. Transgenic plants are emerging as suitable alternatives to transgenic animals and bioreactor-based systems for recombinant protein production (e.g., bacterial, yeast, and insect cells). This trend is yielding a recombinant protein expression system that has increased economic viability and a greater capacity for producing lifesaving protein therapeutics and biopolymers. In addition, plant products are free of mammalian viruses, prions, or other adventitious organisms harmful to humans, making them safer production systems. A wide variety of recombinant protein molecules have been successfully produced in plants (Table 1). This section briefly describes some representative proteins and the role of PTMs on these molecules.

IMMUNOGLOBULINS

IgG and IgA are multifunctional glycoprotein immune molecules that bind to antigens, form immune complexes, and activate classical and alternative complement pathways, respectively. This activity leads to the destruction and clearance of the pathogen. Both IgG and IgA are glycosylated, and although glycosylation is not required for antigen binding, it is critical for activation of the complement pathway.[1] Deglycosylated or underglycosylated immunoglobulins (IgG and IgA) are unable to activate the effector mechanism and therefore fail in the clearance of the pathogen and other antigenic moieties. Plants are able to perform expression and assembly of immunoglobulin (Ig) heavy and light chains into functional antibodies.[2] Plants have been used for production of different forms of antibodies that include full-size IgG and IgA, chimeric IgG and IgA, single-chain Fv fragments (ScFv), Fab, and heavy-chain variable domains; the last three types do not require glycosylation. Among the antibodies that have been successfully produced in plants are those against surface antigen of *Streptococcus mutans*, the causative agent of dental caries;[3] herpes simplex virus;[4] carcinoembryonic antigen (CEA), a tumor-associated marker;[5] and single-chain Fv fragments against non-Hodgkins lymphoma.[6] The majority of plant-derived antibodies possessed high-mannose and hybrid-type, α1-3 fucose– and β1-2 xylose–containing structures that lack β1–4 galactose and the terminal sialic acid residues. Expression of mammalian β1-4 Galtransferase in plants resulted in synthesis of galactose-containing antibodies. About 30% of N-glycans of the antibodies were galactosylated.[7] Plant-derived antibodies do not show any alterations in affinity toward antigens or stability in vivo or in vitro.

VACCINES

Plants are a prime candidate for the production of vaccines because of the lower cost of production and the feasibility of producing ''edible vaccines.'' This technology can be transferred to developing nations relatively easily compared to other expression systems. Some of the representative examples are provided here:

Rabies vaccine: The rabies virus glycoprotein G and nucleoprotein N are both very important in designing an appropriate vaccine. G protein is the major antigen responsible for induction of protective immunity, and N protein is responsible for induction of virus-specific T cells. A plant-derived oral vaccine against a fusion protein consisting of glycoprotein G, nucleoprotein N, and alfalfa mosaic virus coat protein was made, which produced significant immune response and rabies virus–specific antibodies in both mice and humans.[8]

Human cytomegalovirus: Human cytomegalovirus (HCMV) is a member of the herpes virus family that is transmitted by blood and body secretions. In immuno compromised individuals, infection of HCMV can lead to damage of the central nervous system and death. Glycoprotein B, a transmembrane envelope protein, was used to produce oral vaccine against the viral infection. The affinity of recombinant glycoprotein B was examined in

Encyclopedia of Plant and Crop Science
DOI: 10.1081/E-EPCS 120024846

Table 1 A brief list of representative biomolecules engineered in plants

Biomolecule	Engineered plant
Human and animal vaccines	Tobacco, potato, corn, lettuce, tomato, alfalfa, *Arabidopsis*
Immunoglobulins	Tobacco, alfalfa, rice, wheat, soybean
Other therapeutic proteins	Tobacco, canola, rice, turnip, alfalfa, *Arabidopsis*
Biomaterial:	
Collagen	Tobacco
Spider silk	Tobacco, potato

See Ref. 18 for a more complete list of plant-derived pharmaceuticals.

vitro by its ability to bind monoclonal antibodies. Plant-derived glycoprotein B showed a high affinity to antibodies.[9]

Gastroenteritis coronavirus: Swine-transmissible gastroenteritis virus (TGEV) is the causative agent of acute diarrhea of newborn piglets. Neutralizing antibodies against this virus is mainly directed toward a surface component, glycoprotein S. Glycoprotein S from TGEV, produced recombinantly in plants, can function as a vaccine against infection when injected intramuscularly in animals. Glycoprotein S has three glycosylation sites, which play an essential role in conformation of this protein. However, no further information on glycosylation of the recombinant glycoprotein S is available.[10]

Measles virus: Measles is a highly contagious viral disease. Severe infection may lead to pneumonia, encephalitis, and death. Hemagglutinin, a surface protein from measles virus, was expressed in plants. For recognition of hemagglutinin by B lymphocyte cells, an appropriate conformation of the protein is essential. Folding, stability, and protease susceptibility of this protein is dependent on four N-glycosylation sites. Plant-derived hemagglutinin was found to be stable and able to induce an immune response in animal model systems upon oral administration. This indicates that plants are able to achieve sufficient glycosylation required for stability and folding of the protein.[11]

OTHER PROTEIN THERAPEUTICS

Follicle-stimulating hormone: Follicle-stimulating hormone (FSH) is a pituitary heterodimeric glycoprotein hormone that requires N-glycosylation for subunit folding, assembly, targeting, and stability. FSH produced in plants possesses terminal mannose residues and is reported to be biologically active.[12]

Lactoferrin: Lactoferrin is a milk protein. In humans, it is known to contain two N-acetyllactosamine-type N-glycans that also contain fucose and sialic acid residues. Lactoferrin belongs to the family of transferrin with iron-binding properties. It has also been found to possess antibacterial, antifungal, antiviral, and antiinflammatory activities. Full-length lactoferrin with antimicrobial properties is produced in plants.[13]

Erythropoeitin: Erythropoeitin (EPO), an N-glycosylated protein that regulates the formation of erythrocytes in mammals, was recombinantly produced in plants. EPO produced in tobacco cells was glycosylated with N-linked oligosaccharides that did not possess terminal sialic acid residues. EPO produced in tobacco exhibited in vitro biological activities by inducing the differentiation and proliferation of erythroid cells. However, it did not show in vivo biological activities.[14] The authors speculated that this was attributable to the different glycosylation of EPO produced in tobacco cells as compared to authentic human EPO and was cleared from circulation by asialo-receptors.

Macrophage Activating Factor: Macrophage activating factor (MAF), also known as vitamin D–binding protein and Gc-globulin, is a multifunctional abundant serum protein. Threonine-420 of domain III of MAF is O-glycosylated. Upon infection with pathogen or tumor cells, activated T and B lymphocyte cell surface exoglycosidases remove sialic acid and galactose residues, leaving a GalNAc attached to the threonine.[15] This GalNAc is essential for MAF to bind to the C-type lectin on the cell surface of mononuclear phagocytes, which leads to their activation. Activated phagocytes remove pathogens and tumor cells from the body. Efforts are underway to express this serum glycoprotein in plants.

Acetylcholinesterase: Acetylcholinesterase (AChE) catalyzes the degradation of acetylcholine, a neurotransmitter. AChE is a potential therapeutic against organophosphate-based chemicals and warfare agents. Sialic acid residues on the N-glycans of AChE are crucial in determining the circulatory clearance rate. Production of AChE in HEK-293 cells led to hyposialylation of the protein, which was responsible for rapid clearance of recombinant AChE from the circulation. Recombinant human AChE was produced in tomato plants by Mor et al.[16] Kinetic studies of the recombinant AChE using inhibitors showed that the plant-produced AChE had an inhibition profile similar to the commercially available AChE purified from human erythrocytes. However, in vivo activity, serum half life, and glycosylation studies of plant-produced AChE have not yet been reported.

BIOMOLECULES

Collagen: Collagen is the major protein in the extracellular matrix and is thus the most prominent animal protein. There are more than 20 different types of collagens described so far. The triple-helical domains of collagens have the repeating amino acid sequence (Gly- X-Y)$_n$, where X and Y are frequently proline and hydroxyproline, respectively. Collagen molecules go through multiple PTMs involving at least eight enzymes and cofactors. Specific proline and lysine residues are hydroxylated, and some of the hydroxylysine residues are further modified by glycosylation with galactose-glucose disaccharide units. Hydroxylation of proline to hydroxyproline residues assists in trimer assembly of collagen polypeptide chains. Collagens and their derivative molecules (gelatin) have multiple applications but are currently only available from animal sources. There is a need for alternative sources to produce large quantities of pure and safe collagens. Human collagen α-1 (C1α1) has been expressed by itself and with the multisubunit recombinant prolyl-4-hydroxylase,[17] in transgenic tobacco, collagen trimers have been produced that are stable up to 37°C. These reports suggest that the production of human collagen is possible in plants. However, further research is required to understand and improve recombinant collagen synthesis and assembly in plants.

CONCLUSION

The discovery of and demand for medically important proteins are growing rapidly. It is becoming increasingly critical to ensure their availability in sufficient quantity for research, therapeutics, and diagnostic uses. Transgenic plants are emerging as suitable systems for the production of functional recombinant human proteins. Plants have the distinct advantages over other expression systems of product safety, economical production, and ease of scale-up. Plants are also able to perform most PTMs that are carried out by mammals, such as glycosylation, phosphorylation, and hydroxylation. However, the PTMs in plants are under-studied and require more attention because it is important to control the addition of and evaluate the structural and functional roles of these modifications on recombinant products for regulatory purposes as well as for desired physiological activities, when used for therapeutic purposes in humans and animals.

ARTICLES OF FURTHER INTEREST

Genetically Modified Oil Crops, p. 509
Human Cholinesterases from Plants for Detoxification, p. 564
Medical Molecular Pharming: Therapeutic Recombinant Antibodies, Biopharmaceuticals, and Edible Vaccines in Transgenic Plants Engineering via the Chloroplast Genome, p. 705
Tobamoviral Vectors: Developing a Production System for Pharmaceuticals in Transfected Plants, p. 1229
Vaccines Produced in Transgenic Plants, p. 1265

REFERENCES

1. Miletic, V.D.; Frank, M.M. Complement-imunoglobulin interaction. Curr. Opin. Immunol. Feb 1995, 7 (1), 41–47.
2. Ma, J. Generation and assembly of secretory antibodies in plants. Science **1995**, *268*, 716–719.
3. Ma, J.; Hikmat, B.; Wykoff, K.; Vine, N.; Chargelegue, D.; Yu, L.; Hein, M.; Lehner, T. Characterization of a recombinant plant monoclonal antibody and preventive immunotherapy in humans. Nat. Med. **1998**, *4*, 601–606.
4. Zeitlin, L. A humanized monoclonal antibody produced in transgenic plants for immunoprotection of the vagina against genital herpes. Nat. Biotechnol. **1998**, *16*, 1361–1364.
5. Stoger, E.; Vaquero, C.; Torres, E.; Sack, M.; Nicholson, L.; Drossard, J.; Williums, S.; Keen, D.; Perrin, Y.; Christou, P.; Fischer, R. Cereal crops as viable production and storage systems for pharmaceutical scFv antibodies. Plant Mol. Biol. **2000**, *42*, 583–590.
6. McCormick, A.A.; Kumagai, M.H.; Hanley, K.; Turpen, T.H.; Hakim, I.; Grill, L.K.; Tuse, D.; Levy, S.; Levy, R. Rapid production of specific vaccines for lymphoma by expression of the tumor-derived single-chain Fv epitopes in tobacco plants. PNAS **1999**, *96*, 703–708.
7. Cabanes-Macheteau, M.; Fitchette.-Lain, A.-C.; Loutelier-Bourhis, C.; Lange, C.; Vine, N.D.; Ma, J.; Lerouge, P.; Faye, L. N-glycosyaltion of a mouse IgG expressed in transgenic tobacco plants. Glycobiology **1999**, *9*, 365–372.
8. Yusibov, V.; Hooper, D.C.; Spitsin, S.V.; Fleysh, N.; Kean, R.B.; Mikheeva, T.; Deka, D.; Karasev, A.; Cox, S.; Ransdall, J.; Koprowski, H. Expression in plants and immunogenicity of plant virus-based experimental rabies vaccine. Vaccine **2002**, *20*, 3155–3164.
9. Tackaberry, E.S.; Dudani, A.K.; Prior, F.; Tocchi, M.; Sardana, R.; Altosaar, I.; Ganz, P.R. Development of biopharmaceuticals in plant expression systems: Cloning,

expression and immunological reactivity of human cytomegalovirus glycoprotein B (UL55) in seeds of transgenic tobacco. Vaccine **1999**, *17*, 3020–3029.

10. Gomez, N.; Carrillo, C.; Salinas, J.; Parra, F.; Borca, M.V.; Escribano, J.M. Expression of immunogenic glycoprotein S polypeptides from transmissible gastroenteritis Coronavirus in transgenic plants. Virology **1998**, *249*, 352–358.
11. Huang, Z.; Dry, I.; Webster, D.; Strungnell, R.; Wesselingh, S. Plant-derived measles virus hemagglutinin protein induces neutralizing antibodies in mice. Vaccine **2001**, *19*, 2163–2171.
12. Dirnberger, D.; Steinkellner, H.; Abdennebi, L.; Remy, J.-J.; Van de Wiel, D. Secretion of biologically active glycoforms of bovine follicle stimulating hormone in plants. Eur. J. Biochem. **2001**, *268*, 4570–4579.
13. Samyn-Petit, B.; Gruber, V.; Flahaut, C.; Wajda-Dubos, J.-P.; Farrer, S.; Pons, A.; Desmaizieres, G.; Slomianny, M.-C.; Theisen, M.; Delannoy, P. N-glycosylation potential of maize: The human lactoferrin used as a model. Glycoconjugate Journal. **2001**, *18* (17), 519–527.
14. Matsumoto, S.; Ikura, K.; Ueada, M.; Sasaki, R. Characterization of a human glycoprotein (erythropoeitin) produced in cultured tobacco cells. Plant Mol. Biol. **1995**, *27*, 1163–1172.
15. Yamamoto, N.; Kumashiro, R. Conversion of vitamin D3 binding protein (group-specific component) to a macrophage activating factor by the stepwise action of beta-galactosidase of B cells and sialidase of T cells. J. Immunol. **Sept. 1, 1993**, *151* (5), 2794–2802.
16. Mor, T.S.; Sternfeld, M.; Soreq, H.; Arntzen, C.J.; Mason, H.S. Expression of recombinant human acetylcholinesterase in transgenic tomato plants. Biotechnol. Bioeng. **2001**, *75*, 259–266.
17. Merle, C.; Perret, S.; Lacour, T.; Jonval, V.; Hudaverdian, S.; Garrone, R.; Ruggiero, F.; Theisen, M. Hydroxylated human homotrimeric collagen I in *Agrobacterium tumefaciens*-mediated transient expression and in transgenic tobacco plant. FEBS Lett. **2002**, *515*, 114–118.
18. Daniell, H.; Streatfield, S.J.; Wykoff, K. Medical molecular farming: Production of antibodies, biopharmaceuticals and edible vaccines in plants. Trends Plant Sci. **2001**, *6*, 219–226.
19. Wamsley, A.M.; Arntzen, C.J. Plant cell factories and mucosal vaccines. Curr. Opin. Biotechnol. **Apr 2003**, *14* (2), 145–150.

Plant Response to Stress: Abscisic Acid Fluxes

Wolfram Hartung
Universität Würzburg, Würzburg, Germany

P

INTRODUCTION

Water relations and stress tolerance of plant cells and tissues of crop plants are significantly improved when abscisic acid (ABA) concentrations are high. This occurs by closing the stomata, stimulation of water uptake into roots, and induction of developmental changes that improve the stress tolerance in plants. Such plants perform particularly well under arid conditions. Stress-tolerant crop varieties often exhibit high ABA concentrations and treatments that intensify long distance ABA signaling reduce water consumption and, in the case of grapevine, improve the quality of the berries. The knowledge of mechanisms that increase ABA concentration in organs under mild and severe stress are also of special interest for agricultural and horticultural scientists.

ABA accumulation in cells, tissues, and organs may be a result of an altered ABA metabolism (increased biosynthesis and/or decreased degradation). The rates of these biochemical processes, however, are rather slow. ABA fluxes that may be influenced under stress can be a much more effective, sensitive, and rapid mechanism to maintain ABA homeostasis and to change rapidly ABA concentrations in cellular compartments, cells tissues, and organs. In this article, ABA fluxes will be discussed at the membrane, tissue, and organ level. Mild (hypoosmotic) stress conditions that are not sufficiently severe to stimulate ABA biosynthesis in leaves and root tips are discussed predominantly.

ABA FLUXES ACROSS BIOMEMBRANES

Roots

ABA fluxes across membranes have been investigated in detail using the efflux compartmental analysis. When barley roots were stressed hypoosmotically with sorbitol ABA, fluxes across the plasmalemma of both cortical and xylem parenchyma cells where reduced, and the flux across the tonoplast was stimulated, resulting in an increased cytosolic ABA concentration in the root cells.[1,2]

ABA fluxes into the surrounding rhizosphere can be particularly severe when the soil solution is alkaline. Under such conditions, ABA leaks from roots into the soil solution, especially when they lack Casparian bands in their hypodermis. An undesirable efflux can be avoided when the ABA concentration in the soil solution of the rhizosphere is sufficiently high. Thus, an equilibrium between internal and external ABA is maintained.[3]

Mesophyll Cells, Epidermal Cells, and Guard Cells

ABA fluxes across plasma membranes and tonoplasts of mesophyll and guard cells have also been investigated using the efflux compartmental analysis. Again, sorbitol was used as an osmoticum to establish hypoosmotic and hyperosmotic stress. ABA flow across the plasma membranes of the guard cells of *Valerianella locusta* is reduced significantly under hypoosmotic stress, whereas the ABA flux across the tonoplast is only slightly affected. This results in a transiently increased apoplastic and a reduced cytosolic ABA concentration. It was concluded[2] that the ABA increase in the apoplast is of special physiological importance in terms of stress, because those structures that may recognise ABA at the guard cells seem to be located on the outer surface of the guard cell plasma membrane.[3] The ABA fluxes across the plasma membrane of barley epidermis cells have also been investigated. They are facilitated by a carrier that becomes active when the apoplastic pH of stressed leaves is increased and excess ABA has to be removed from the leaf apoplast to the surrounding tissues.[4]

It should be pointed out that efflux experiments described here have been performed under equilibrium conditions. In an intact plant, however, delivery of ABA from other organs and changes of ABA metabolism must also be considered. All of these additional processes that have an impact on ABA gradients across membranes have been incorporated into a mathematical model that quantitatively describes ABA flows under the complex fluctuating conditions that occur in the natural habitat.[3]

Stem and Mesocotyl Parenchyma

During its long distance transport in the xylem of bean internodes and maize mesocotyls, ABA can be redistributed to the surrounding tissues. If ABA_{xyl} is low, those

Encyclopedia of Plant and Crop Science
DOI: 10.1081/E-EPCS 120010651

tissues can release ABA into the xylem with rates of approximately 1 pmol $m^{-2}s^{-1}$.[5,6] These tissues play an important role for the ABA homeostasis in the xylem vessels of stems.

ABA FLUXES BETWEEN TISSUES

The ABA flows between root cortex and root stele are of particular physiological interest with regard to stress. Large quantities of root ABA have to be transported laterally from the cortex across the endodermis into the xylem parenchyma and into the xylem vessels. In the case of a symplastic transport, ABA has to pass the plasmalemma of xylem parenchyma cells to reach the xylem vessels. This is a rate limiting step that is much slower than water transport into the xylem. When the water flow is increased by transpiration, ABA_{xyl} is diluted. An apoplastic transport of ABA by solvent drag across the endodermis would compensate or even overcompensate this dilution, provided their Casparian bands are permeable for ABA. Indeed, a significant apoplastic bypass flow of ABA could be detected in maize roots. The intensity of the apoplastic bypass flow of ABA depends on the species, the pH of the root cortical apoplast and the apoplastic ABA concentration.

The other important apoplastic barrier of the roots, the exodermis, prevents ABA loss into the soil solution, especially under nontranspiring conditions. This explains the tolerance of plants with exodermises to alkaline soils of arid habitats compared to those without an exodermis, mainly the Fabaceae, which are unprotected.[5–8]

LONG DISTANCE ABA FLOWS BETWEEN ORGANS

Root-to-shoot flows of ABA are of special interest under conditions of mild stress when water relations of the leaves are not affected. The roots, however, experience stress as the soil is drying. ABA flows between organs have been investigated predominantly with *Ricinus communis* and *Lupinus albus*, which allow harvest of both transport fluids, phloem and xylem sap. Long-distance signaling has been reviewed recently in detail.[5,6,9]

The ABA flow from the root to the shoot in the xylem is increased by moderate salt stress in lupins tenfold, and in castor bean fivefold. This ABA did not originate exclusively from biosynthesis in the roots. A significant portion is recirculated in the phloem from the leaves via roots back to the shoot. This portion of recirculated ABA was increased under salt stress in *Lupinus* and in *Ricinus* two to three fold. The ABA signal in the xylem of *Ricinus* is also increased threefold when nitrate in the nutrient medium is replaced by ammonium. Under phosphate deficiency that often occurs in alkaline soils of extreme arid habitats, the root-to-shoot flow of ABA is increased tenfold, whereas nitrate deficiency has only a weak negligible effect. Potassium deficiency causes a fourfold higher root-to-shoot ABA flow in the xylem.[10]

When nitrate or ammonium is sprayed to the leaves of *Ricinus* and the root medium is kept N-free, ABA biosynthesis of nitrate-sprayed plants is increased, resulting in stimulated ABA transport in the phloem to the roots where most of the ABA is metabolised. In ammonium-sprayed plants, a significant amount of ABA biosynthesis can be observed in the roots, resulting in an increased root-to-shoot flow compared to nitrate-sprayed plants. This resembles strongly the situation of plants where roots were supplied with ammonium. A shoot-to-root signal of ammonium-sprayed plants that causes ABA biosynthesis in the roots and ABA transport in the xylem must be postulated. This signal is unknown.[9]

ABA Flows from the Shoot to the Root

A role of ABA as a shoot-to-root signal has been postulated for *Ricinus* seedlings that increases the ABA flow in the phloem under mild water deficiency by fiftyfold. Intact lupins and castor beans showed a two- to fivefold increase of phloem ABA flow under conditions of salt and ammonium stress but not under conditions of phosphate and nitrate deficiency. Maize plants that were cultivated with only their seminal roots experience water stress in the leaves because water flow was restricted by the constant number of xylem vessels. As a result, not only was ABA increased in the leaves (tenfold), but also surprisingly in the roots. This is most likely due to increased phloem transport because an increase of ABA synthesis in the unstressed, well-hydrated and nutrient-supplied roots seems to be extremely unlikely. The ABA delivered by phloem transport to the roots may stimulate the development and the hydraulic conductivity of the root systems under stress.[7,9]

LONG DISTANCE FLOW OF ABA CONJUGATES

It has been postulated repeatedly that ABA conjugates, predominantly the ABA-glucose ester (ABA-GE), act as hormonal root-to-shoot stress signals. Indeed, an increased xylem transport of ABA-GE has been observed in the xylem sap of drought- and salt-stressed barley, maize, rice, and sunflower plants. Since ABA-GE is extremely hydrophilic, it is transported in the xylem over long distances without any loss by redistribution to the surrounding stem parenchyma. On the other hand, it cannot pass apoplastic barriers in roots and be taken up into mesophyll cells after its arrival if the leaf apoplast is extremely small. Specific esterases in the root and leaf apoplast release free ABA from its conjugate. Afterward, ABA can be redistributed to the surrounding cells. In leaves the activity of these esterases is increased significantly under salt stress which provides a further mechanism to increase the ABA at the site of action in stressed leaves.[7,9,11]

CONCLUSION

Environmental stresses influence flows of the stress hormone ABA on cellular, tissue, and organ levels, which results in altered concentrations of this universal plant stress hormone at the primary site of the target cells and tissues. Flows are altered already under mild stress conditions. Thus, ABA can be increased without affecting the biosynthesis and metabolism of ABA. The root-to-shoot ABA signal can be intensified by special cultivation techniques, such as partial root drying (PRD). In both the greenhouse (tomato[12]) and under field conditions (grapevine[13]), when water was withhold from only a part of the root system while the other was irrigated, ABA concentration in the xylem was increased markedly. Changes of fertilization (ammonium versus nitrate) may also enhance long-distance flows from the roots to the shoots [9,10] The resulting ABA increase in the leaves and the fruit reduces the water consumption without significant negative effects on the yield. Additionally, PRD has a positive effect on those factors that control the fruit quality of tomato and grapevine (sugars, acids, pigments, etc.). The knowledge of the mechanisms that modify ABA fluxes in crop and fruit plants is of practical and commercial interest, especially for agriculture in arid climates.

REFERENCES

1. Behl, R.; Hartung, W. Transport and compartmentation of abscisic acid in roots of *Hordeum distichon* under osmotic stress J. Exp. Bot. **1984**, *35*, 1433–1440.
2. Behl, R.; Hartung, W. Movement and compartmentation of abscisic acid in guard cells of *Valerianella locusta*: Effects of osmotic stress, external H^+-concentration and fusicoccin Planta **1986**, *168*, 360–368.
3. Slovik, S.; Daeter, W.; Hartung, W. Compartmental redistribution and long distance transport of abscisic acid (ABA) in plants as influenced by environmental changes in the rhizosphere. A biomathematical model J. Exp. Bot. **1995**, *46*, 881–894.
4. Dietz, K.J.; Hartung, W. The leaf epidermis, its ecophysiological significance Prog. Bot. **1996**, *57*, 32–53.
5. Hartung, W.; Sauter, W.; Hose, E. Abscisic acid in the xylem: Where does it come from, where does it go to? J. Exp. Bot. **2002**, *53*, 27–32.
6. Sauter, A.; Davies, W.J.; Hartung, W. The long distance abscisic acid signal: The fate of the hormone on its way from root to shoot J. Exp. Bot. **2001**, *52*, 1991–1997.
7. Hose, E.; Sauter, A.; Hartung, W. Abscisic Acid in Roots—Biochemistry and Physiology. In *Plant Roots—The Hidden Half*; Waisel, Y., Eshel, A., Kafkafi, U., Eds.; Marcel Dekker: New York, 2002; 435–448.
8. Hose, E.; Clarkson, D.T.; Steudle, E.; Schreiber, L.; Hartung, W. The exodermis—A variable apoplastic barrier J. Exp. Bot. **2001**, *52*, 2245–2264.
9. Hartung, W.; Peuke, A.D.; Davies, W.J. Abscisic Acid—A Hormonal Long Distance Stress Signal in Plants Under Drought and Salt Stress. In *Handbook of Crop Stress*, 2nd Ed.; Pessarakli, M., Ed.; Marcel Dekker: New York, 1999; 731–747.
10. Peuke, A.D.; Jeschke, W.D.; Hartung, W. Flows of elements, ions and abscisic acid in *Ricinus communis* under potassium limitation J. Exp. Bot. **2002**, *53*, 241–250.
11. Sauter, A.; Dietz, K.J.; Hartung, W. A possible stress physiological role of abscisic acid conjugates in root to shoot signalling Plant Cell Environ. **2002**, *25*, 223–228.
12. Davies, W.J.; Bacon, M.A.; Thompson, D.S.; Sobeih, W.; González Rodriguez, L. Regulation of leaf and fruit growth in plants growing in drying soil: Exploitation of the plants chemical signalling system and hydraulic architecture to increase the efficiency of water use in agriculture J. Exp. Bot. **2000**, *51*, 1617–1626.
13. Stoll, M.; Loveys, B.; Dry, P. Hormonal changes induced by partial root zone drying of irrigated grapevine J. Exp. Bot. **2000**, *51*, 1627–1634.

Plant Response to Stress: Biochemical Adaptations to Phosphate Deficiency

William C. Plaxton
Queen's University, Kingston, Ontario, Canada

INTRODUCTION

Phosphate (Pi) is an essential macronutrient that plays a central role in virtually all metabolic processes in plants, including photosynthesis and respiration. Despite its importance, Pi is one of the least available nutrients in many ecosystems, and is a frequent limiting factor for plant productivity. Although plentiful in the earth's crust, soil Pi often exists in insoluble mineral forms that render it unavailable to plants. Agricultural Pi deficiency is alleviated by the massive application of Pi fertilizers, currently estimated to be about 40 million metric tons per year worldwide. However, the assimilation of Pi fertilizers by crops is quite inefficient, as a large proportion of applied Pi becomes immobile, or may runoff into and thereby pollute nearby surface waters. Moreover, the world's reserves of rock Pi (mined for production of Pi fertilizers) are expected to be depleted within the next century. Thus, studies of the complex mechanisms whereby plants acclimate to nutritional Pi deficiency are of great importance. This could lead to the development of rational strategies for engineering Pi-efficient transgenic crops that would reduce or eliminate our current overreliance on expensive, polluting, and nonrenewable Pi fertilizers. The aim of this article is to provide a brief overview of the fascinating biochemical adaptations of Pi-starved (−Pi) plants.

THE PLANT PHOSPHATE-STARVATION RESPONSE

Plants have evolved the ability to acclimate, within species-dependent limits, to extended periods of Pi deficiency. Pi deprivation elicits a complex array of morphological, physiological, and biochemical adaptations, collectively known as the *Pi-starvation response*. Plant morphological/physiological adaptations that enhance the acquisition of limiting Pi from the soil include increased root growth relative to shoot growth, as well as root colonzization by mycorrhizal fungi.[1,2]

Mycotrophic Versus Nonmycotrophic Plants

Symbiotic mycorrhizal fungi help mycotrophic plants increase Pi-uptake from Pi deficient soils. In return, the host plant supplies sucrose to fuel the energy demands of the mycorrhizal symbiont. Mycorrhizae colonize the roots of most plants, except for the nonmycotrophic minority that include members of the Cruciferae, Chenopodiaceae, and Proteaceae families. It is notable that many nonmycotrophs, such as buckwheat and white lupin, are notorious for their ability to thrive on infertile soils. This reflects the view that relative to mycotrophic plants, the nonmycotrophs appear to have evolved to allow more efficient acclimation to low Pi conditions.[2]

BIOCHEMICAL ADAPTATIONS OF PHOSPHATE-STARVED PLANTS

Adaptation #1: Increased Efficiency of Cellular Phosphate Uptake

Multiple plasmalemma Pi transporters are differentially expressed under varying Pi nutritional regimes.[1] The widely accepted dual Pi uptake model is characterized by constitutive low-affinity and Pi-starvation inducible (PSI) high-affinity Pi transporters that respectively function at high (mM) and low (μM) concentrations of external Pi. High-affinity Pi transporters likely play a crucial role in the acquisition of limiting external Pi by −Pi plants.[1,2] Genome sequencing has indicated that *Arabidopsis* contains nine members of the high-affinity Pi transporter gene family, but only a single low-affinity Pi transporter gene.

Adaptation #2: Induction of Phosphate Scavenging and Recycling Enzymes

2A: Acid phosphatase

Acid phosphatase (APase) induction is a universal symptom of plant Pi stress.[2,3] APases function as intracellular

Encyclopedia of Plant and Crop Science
DOI: 10.1081/E-EPCS 120010648

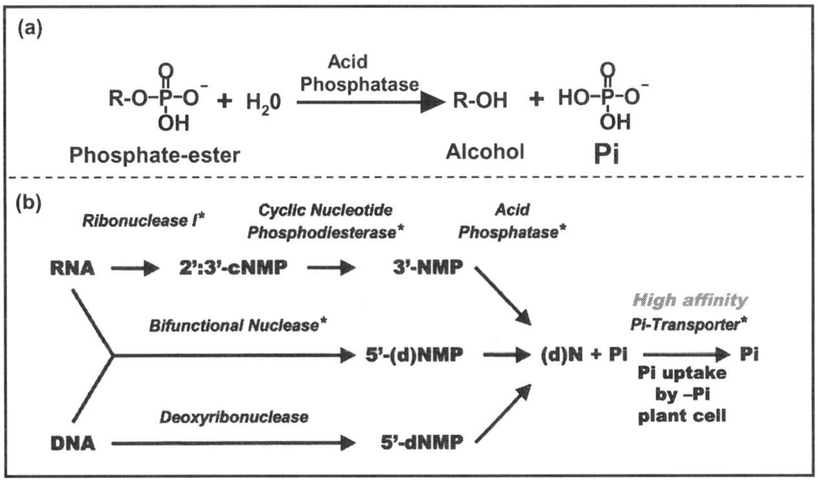

Fig. 1 Phosphate scavenging and recycling enzymes of −Pi plant cells. (a) General reaction catalyzed by acid phosphatase. R denotes an organic molecule. (b) Model of Pi scavenging from extracellular nucleic acids by secretory nucleolytic enzymes. Asterisks denote PSI proteins. (Adapted from Ref. 4; modified figure reproduced with permission of The American Society of Plant Biologists.)

(vacuolar) or extracellular (secreted) Pi salvage systems that catalyze the hydrolysis of Pi from phosphate-monoesters (Fig. 1(a)). The existence of 22 *Arabidopsis* genes that putatively encode different APases indicates that plant APase biochemistry is relatively complex.

2B: Secreted nuclease and phosphodiesterase

Nucleic acids present in decaying organic matter represent an important source of extracellular Pi that may be exploited by −Pi plants. Degradation of extracellular DNA and RNA by PSI-secreted nucleases, phosphodiesterase, and APase liberates Pi from nucleic acids for its subsequent uptake by PSI high-affinity Pi transporters of −Pi plant roots (Fig. 1(b)).[2,4]

2C: Replacement of membrane phospholipids with nonphosphorus galacto- and sulfonyl-lipids

−Pi plants can scavenge and conserve Pi by replacing their membrane phospholipids with amphipathic galacto- and sulfonyl lipids. *Arabidopsis* mutants defective in sulfolipid synthase (the terminal enzyme of sulfonyl lipid synthesis) were recently reported to show impaired growth during Pi deprivation.[5]

2D: Induction of metabolic phosphate recycling enzymes

Several PSI glycolytic bypass enzymes such as PPi-dependent phosphofructokinase (PPi-PFK), phosphoenolpyruvate (PEP) phosphatase, and PEP carboxylase (PEPCase) may facilitate intracellular Pi recycling, because Pi is a by-product of the reactions catalyzed by each of these enzymes (Fig. 2). Their reactions may also facilitate respiration and/or organic acid excretion, while generating free Pi for its reassimilation into the metabolism of the −Pi cells.[2]

2E: Organic acid excretion

Scavenging of Pi from extracellular sources may be aided by the enhanced excretion of organic acids due to PEPCase induction. Roots and suspension cell cultures of −Pi plants have been demonstrated to markedly upregulate PEPCase.[1,2,6–8] PEPCase induction during Pi stress has been correlated with the excretion of significant levels of organic acids such as malate and citrate. This leads to acidification of the rhizosphere, which thereby contributes to the solubilization and assimilation of mineral Pi from the environment.[1,2,7]

Adaptation #3: Induction of Alternative Pathways of Cytosolic Glycolysis

As a consequence of the large decline (up to 50-fold) in cytoplasmic Pi levels that follows severe Pi stress, large reductions in intracellular levels of ATP and related nucleoside phosphates can also occur (Table 1).[2,7,8] This may hinder carbon flux through the enzymes of classical glycolysis that are dependent upon adenylates or Pi as cosubstrates (Fig. 2). Despite depleted intracellular Pi

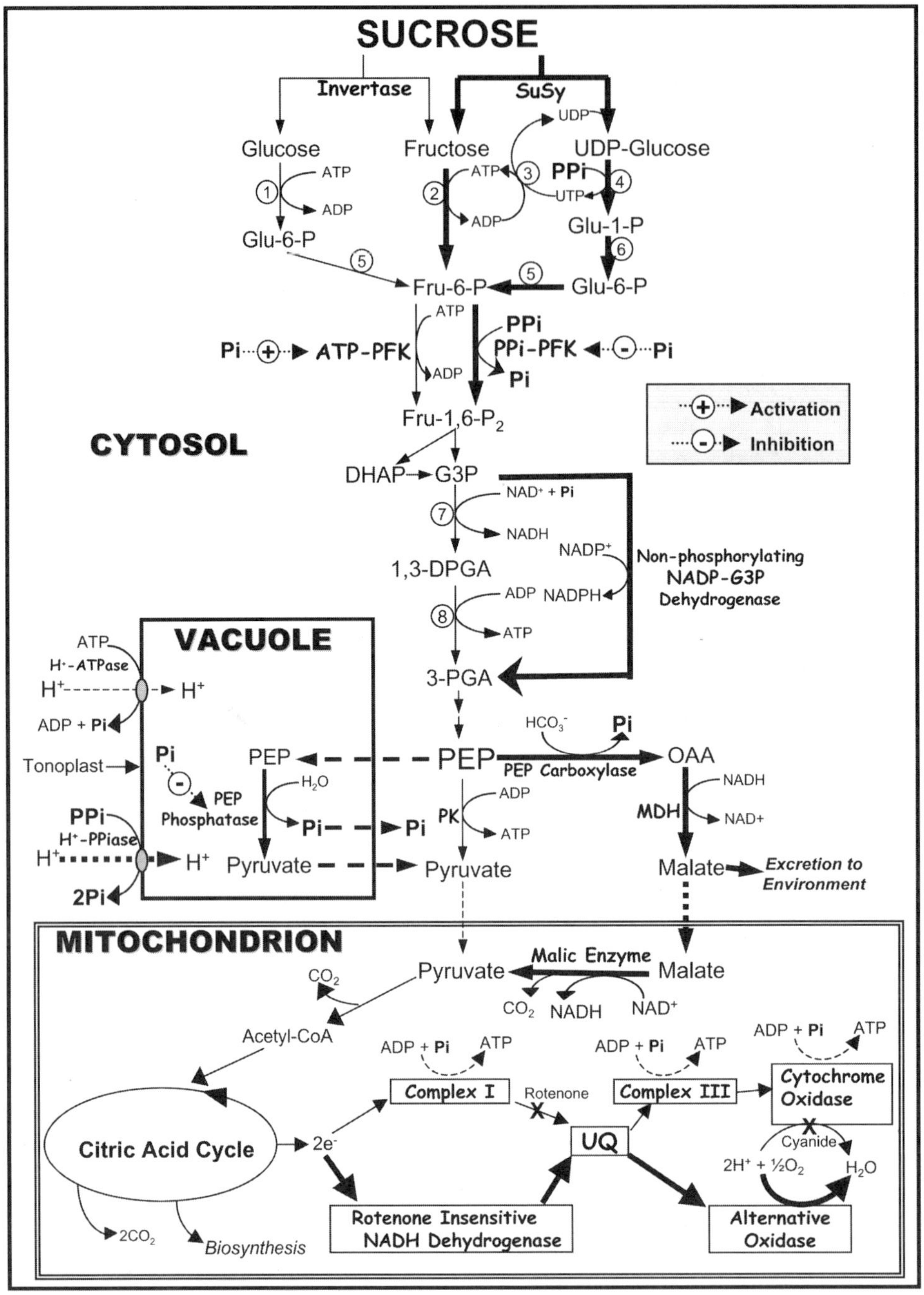

Fig. 2 A model suggesting various adaptive metabolic processes (indicated by bold arrows) that may promote the survival of $-$Pi plants. Alternative pathways of cytosolic glycolysis and mitochondrial electron transport, and tonoplast H^+-pumping facilitate respiration and vacuolar pH maintenance by $-$Pi plant cells because they negate the dependence on adenylates and Pi, the levels of which become markedly depressed during severe Pi starvation. Organic acids produced by PEPCase may also be excreted by roots to increase the availability of mineral-bound Pi (by solubilizing Ca-, Fe- and Al-phosphates). A key component of this model is the critical secondary role played by metabolic Pi recycling systems during Pi deprivation. Enzymes that catalyze the numbered reactions are as follows: 1) hexokinase; 2) fructokinase; 3) nucleoside diphosphate kinase; 4) UDP-glucose pyrophosphorylase; 5) phosphoglucose isomerase; 6) phosphoglucose mutase; 7) NAD-dependent G3P dehydrogenase (phosphorylating); and 8) 3-phosphoglycerate kinase. Abbreviations are as in the text and as follows: DHAP, dihydroxyacetone-phosphate; Fru, fructose; Glu, glucose; MDH, malate dehydrogenase; OAA, oxaloacetate; 3-PGA, 3-P-glycerate; PK, pyruvate kinase; UQ, ubiquinone.

Table 1 Metabolite levels in black mustard suspension cells[a]

Metabolite	nmol.gram fresh wt^{-1}		Change due to Pi-deprivation
	+Pi cells	−Pi cells	
Pi	17,400 ± 1,200	400 ± 60	44-fold decrease
ATP	138 ± 12	36 ± 5	4-fold decrease
ADP	53 ± 9	4.7 ± 0.8	10-fold decrease
PPi	136 ± 12	124 ± 9	Not significant

[a]Metabolite levels in *Brassica nigra* (black mustard) suspension cells cultured for 7 days in the presence and absence of 10 mM Pi (+Pi and −Pi, respectively). All values represent means ± S.E.M. of duplicate determinations performed on three separate cell cultures.
(Data from Ref. 8. Reproduced with Permission of The American Society of Plant Biologists.)

and adenylate pools, −Pi plants must continue to generate energy and carbon skeletons for key metabolic processes. As indicated in Fig. 2, at least six Pi- and adenylate-independent glycolytic bypass enzymes (sucrose synthase (SuSy), UDP-glucose pyrophosphorylase, PPi-PFK, nonphosphorylating NADP-glyceraldehyde-3-phosphate (G3P) dehydrogenase, PEPCase, and PEP phosphatase) have been reported to be plant PSI enzymes.[2,6–8] These enzymes are hypothesized to represent PSI bypasses to the adenylate or Pi-dependent glycolytic enzymes (i.e., invertase/hexokinase, ATP-PFK, phosphorylating NAD-G3P dehydrogenase, and pyruvate kinase), thereby facilitating glycolysis during severe Pi stress. Furthermore, as Pi exerts reciprocal allosteric effects on the activity of ATP-PFK (potent activator) and PPi-PFK (potent inhibitor) (Fig. 2),[7] the large reduction in cytoplasmic Pi in −Pi plants should promote the in vivo activity of PPi-PFK while curtailing that of ATP-PFK.[2,8]

Pyrophosphate helps Pi-starved plants to conserve ATP

PPi is a byproduct of a host of anabolic reactions, including the terminal steps of macromolecule synthesis. In animals, the high energy phosphoanhydride bond of PPi is never utilized because PPi is always hydrolyzed by abundant inorganic pyrophosphatase (PPiase), making macromolecule synthesis thermodynamically favorable.[7] However, the large amounts of PPi produced during biosynthesis may be employed by plants to enhance the energetic efficiency of several cytosolic processes. In contrast to animals, the plant cytosol lacks soluble PPiase and thus contains PPi concentrations of up to about 0.5 mM.[2,7–9] Furthermore, plant cytosolic PPi levels are remarkably insensitive to abiotic stresses such as anoxia or Pi starvation, which elicit significant reductions in cellular ATP pools (Table 1).[2,7–9]

Adaptation #4: Induction of Tonoplast H^+-Pumping Pyrophosphatase

In addition to the SuSy pathway of sucrose conversion to hexose-monophosphates, and PPi-PFK, PPi could be employed as an alternative energy donor for the active transport of protons from the cytosol into the vacuole (Fig. 2). That PPi-powered processes may be a crucial facet of the metabolic adaptations of plants to environmental extremes causing depressed ATP (but not PPi) pools is further indicated by the significant induction of the tonoplast H^+-PPiase by anoxia, or by severe Pi starvation.[7,9] As indicated in Fig. 2, the induction of PPi-dependent cytosolic bypasses (i.e., tonoplast H^+-PPiase, PPi-PFK, and SuSy) may serve −Pi plants by: 1) circumventing ATP-limited reactions; 2) conserving limited cellular pools of ATP; while 3) recycling valuable Pi from PPi.

Adaptation #5: Induction of Alternative Pathways of Respiratory Electron Transport

Respiratory O_2 consumption by plant mitochondria can be mediated by the phosphorylating cytochrome pathway or by nonphosphorylating alternative pathways (Fig. 2). The significant reductions in cellular Pi and ADP pools that follow extended Pi deprivation will impede respiratory electron flow through the cytochrome pathway at the sites of coupled ATP synthesis. However, the presence of nonphosphorylating pathways of electron transport provides a mechanism whereby respiratory flux can be maintained under conditions when the availability of ADP and/or Pi are restrictive. Plants acclimate to Pi stress by increased engagement of the nonphosphorylating (i.e., rotenone- and cyanide-insensitive) alternative pathways of respiratory electron transport[2,10,11] (Fig. 2). Moreover, increased levels of alternative oxidase protein may occur in −Pi plants.[10,11] This allows continued functioning of the citric acid cycle and respiratory electron transport chain with limited ATP production. By preventing severe respiratory restriction, the alternative oxidase has been hypothesized to prevent undesirable redirections in carbon metabolism as well as the excessive generation of harmful reactive O_2 species in the mitochondrion of −Pi plants.[10,11] This has been corroborated by the impaired growth and metabolism of −Pi transgenic tobacco plants that are unable to synthesize a functional alternative oxidase protein.[10·]

CONCLUSION

Studies of plant responses to nutritional Pi deprivation have revealed some remarkably adaptive mechanisms that contribute to the survival of −Pi plants. Although these adaptations are not identical in all plants, certain aspects are conserved in a wide variety of plants from very different environments. The biochemical adaptations of −Pi plants provides an excellent example of how the unique flexibility of plant metabolism and energy transduction helps them cope in a typically stressful environment. These adaptations also provide a useful system for studies of plant signal transduction and gene expression. Future investigations of these pathways should provide further links between the biochemical and molecular control of plant metabolism. A better understanding of the extent to which changes in flux through alternative enzymes and pathways influences plant stress tolerance is of significant practical interest. This knowledge is relevant to the ongoing efforts of agricultural biotechnologists to engineer transgenic crops that have improved resistance to environmental extremes, including Pi starvation.

ACKNOWLEDGMENTS

I am indebted to past and present members of my laboratory who have examined the biochemical adaptations of vascular plants, green algae, and yeast to Pi scarcity. Research and Equipment Grants funding by the Natural Sciences and Engineering Research Council of Canada (NSERC) is also gratefully acknowledged.

ARTICLES OF FURTHER INTEREST

Glycolysis, p. 547
Mitochondrial Respiration, p. 729
Phosphorus, p. 872
Plant Response to Stress: Mechanisms of Accommodation, p. 987
Rhizosphere: An Overview, p. 1084

REFERENCES

1. Raghothama, K.G. Phosphate acquistion. Annu. Rev. Plant Physiol. Plant Mol. Biol. **1999**, *50*, 665–693.
2. Plaxton, W.C.; Carswell, M.C. Metabolic Aspects of the Phosphate Starvation Response in Plants. In *Plant Responses to Environmental Stresses: From Phytohormones to Genome Reorganization*; Lerner, H.R., Ed.; Marcel Dekker, Inc.: New York, NY, 1999; 349–372.
3. Duff, S.M.G.; Sarath, G.; Plaxton, W.C. The role of acid phosphatases in plant phosphorus metabolism. Physiol. Plant. **1994**, *90*, 791–800.
4. Abel, S.; Nčrnberger, T.; Ahnert, V.; Gerd-Joachim, K.; Glund, K. Induction of extracellular nucleotide phosphodiesterase as an accessory of ribonucloeolytic activity during phosphate starvation of cultured tomato cells. Plant Physiol. **2000**, *122*, 543–552.
5. Yu, B.; Xu, C.; Benning, C. Arabidopsis disrupted in *SQD*2 encoding sulfolipid synthase is impaired in phosphate-limited growth. PNAS-USA **2002**, *99*, 5732–5737.
6. Moraes, T.; Plaxton, W.C. Purification and characterization of phospho*enol*pyruvate carboxylase from *Brassica napus* (rapeseed) suspension cell cultures. Implications for phospho*enol*pyruvate carboxylase regulation during phosphate starvation, and the integration of glycolysis with nitrogen assimilation. Eur. J. Biochem. **2000**, *267*, 4465–4476.
7. Plaxton, W.C. The organization and regulation of plant glycolysis. Annu. Rev. Plant Physiol. Plant Mol. Biol. **1996**, *47*, 185–214.
8. Duff, S.M.G.; Moorhead, G.B.G.; Lefebvre, D.D.; Plaxton, W.C. Phosphate starvation inducible 'bypasses' of adenylate and phosphate dependent glycolytic enzymes in *Brassica nigra* suspension cells. Plant Physiol. **1989**, *90*, 1275–1278.
9. Palma, D.A.; Blumwald, E.; Plaxton, W.C. Upregulation of vacuolar H^+—Translocating pyrophosphatase by phosphate starvation of *Brassica napus* (rapeseed) suspension cell cultures. FEBS Letts. **2000**, *486*, 155–158.
10. Parsons, H.L.; Yip, J.Y.H.; Vanlerberghe, G.C. Increased respiratory restriction during phosphate-limited growth in transgenic tobacco lacking alternative oxidase. Plant Physiol. **1999**, *121*, 1309–1320.
11. González-Meler, M.A.; Giles, L.; Thomas, R.B.; Siedow, J.N. Metabolic regulation of leaf respiration and alternative pathway activity in response to phosphate supply. Plant Cell Environ. **2001**, *24*, 205–215.

Plant Response to Stress: Critical Periods in Plant Development

P

G. Nissim Amzallag
Judean Regional Center for Biological Research, Carmel, Israel

INTRODUCTION

Development is not a continuous phenomenon. Discrete phases, also termed phenophases, are each characterized by a specific pattern of organogenesis or by specific sensitivity to environmental factors. Replacement of a phenophase by another one generates an intermediate stage of transition that may be termed a critical period. "Between successive critical levels [periods] the system retains its qualitative properties; it must be characterized here by a low sensibility to external and internal changes in developmental conditions (high resistance)." From this observation, Zhirmunsky and Kuzmin concluded that critical periods, characterized by a strong increase in sensitivity to external factors, must be the privileged time for the emergence of new structures and physiological characteristics.

Critical periods have not been given attention in physiology, partly because they are so evanescent, and partly because they are interpreted as a passive transition in the expression of two successive developmental programs. However, their investigation reveals a quite different reality: Critical periods appear to be crucial for maturation of an emerging phenophase and for its adaptive adjustment to environmental constraints.

PHYSIOLOGICAL ADAPTATION TO STRESS

In many plant species, an increase in tolerance to a stress may be achieved following preexposure to a moderate stress.[1] In *Sorghum bicolor*, for example, a three-week pretreatment with 150 mM NaCl (a sublethal concentration) induces an ability to tolerate a concentration of 300 mM NaCl, which is lethal for non-pretreated plants. Further investigations[1] reveal that increased tolerance is induced by pretreatments starting during a short period of competence, termed a developmental window. During this short period, growth and development are strongly affected in plants exposed to stress, and a large increase in diversity emerges within an initially homogeneous population.[1] These observations suggest that the developmental window is not a simple phase of competence for expression of genes for salt resistance. Self-organizing processes seem inherent to maturation of the adaptive response. In *Sorghum*, the developmental window corresponds to the transition from juvenile to mature vegetative development. Other developmental windows have been identified during transition periods,[1] linking them to critical periods in development.[2,3]

WHAT IS A DEVELOPMENTAL WINDOW?

A direct link is not always observed between whole plant and cellular levels of tolerance to moderate salinity.[4] Similar findings are reported concerning heavy metal toxicity or cold tolerance.[1] In all these cases, the whole plant level of organization remains the determining factor for tolerance to moderate levels of stress. This suggests that the main perturbation following exposure to a sublethal level of stress concerns integrative physiology: between-organs relationships and their controlling factors, the plant growth regulators (PGRs). This is why alleviating effects to moderate levels of stress have been frequently observed following exogenous supply of PGRs.[5] Moreover, physiological adaptation may result from a simple change in hormone metabolism or in cellular sensitivity to the hormone.

Indeed, changes in hormone sensitivity are observed in plant tissues. Some authors consider these changes to be the main mode of hormone action, in contrast with the well accepted dose-response mode.[6] These two antagonistic approaches may be integrated in the same framework when considering that dose-response characterizes phenophases and that changes in cellular sensitivity to PGRs are specific to a developmental window. This assumption has been confirmed by observation of changes in hormone sensitivity during this period and by following emergence of new physiological functions on the basis of these changes.[7,8]

Plants cannot move toward a more appropriate environment. In an ecological perspective, the existence of a developmental window after germination may be justified as a process of adjustment of the seedling to a necessarily unpredictable environment. However, this developmental

Encyclopedia of Plant and Crop Science
DOI: 10.1081/E-EPCS 120010645

window has also been identified as a period of hormone resetting for plants growing in optimal environments.[7,8] In *Sorghum bicolor*, the developmental window identified during early vegetative development corresponds to the transition from seminal to adventitious root systems. As a new organ, the adventitious root system is integrated in the regulation network during aperture of the developmental window already identified for salt-adaptation. For this reason, the developmental window is identified as a critical period of transition between two phenophases. In this context, physiological adaptation to a moderate environmental stress becomes a companion phenomenon of adaptive adjustments to integration of the adventitious roots during normal development. Moreover, the new emerging phenophase is not the simple expression of a preexisting program of development. It is partly elaborated as a function of the initial phenophase and the internal factors disturbed by their emergence. This requirement of an adaptive dimension of development has been suggested from theoretical considerations ("... adaptive changes in the system's structure and the character of its regulations will be required to balance the new developmental conditions.").[9]

CHARACTERIZATION OF A CRITICAL PERIOD

Intuitively, a critical period includes two steps. The first corresponds to dismantlement of the initial hormone regulation network, leading to a transient increase in autonomy of the different parts of the network. The second corresponds to emergence of a new network of regulation through integration of disturbing factors, endogenous as well as exogenous. This dynamic may be followed through measurement of changes in strength of the regulation network. An index, termed connectance, has been defined as the mean strength of linkage between organs generating together the organism. Connectance is measured on a population that is homogeneous as possible exposed to an environment that is as uniform as possible. In such conditions, the r coefficient for linear regression provides information about the degree of linkage in the regulation network. This linkage may be direct or indirect, so that connectance cannot be correctly estimated by analysis of a single couple of parameters. However, when considered together, the r values calculated for all the possible couples inform about the strength of the network.

The r coefficients are not normally distributed, so that connectance cannot be correctly evaluated by averaging their absolute values before transforming them into z values (normally distributed), according to the formula $z = 0.5\ \mathrm{Ln}[(1+r)/(1-r)]$. Connectance is defined as the mean value of all these calculated z coefficients.

Time variations in connectance may be followed after harvesting populations every 2–3 days. By this method, a drop in connectance is especially observed during the beginning of a critical period, and is followed by a gradual increase in emergence of a new phenophase.[10,11] This result does not only confirm initial assumptions, but also provides a tool for investigating factors influencing the expression of a critical period, its intensity (amplitude of drop in connectance), and its duration (time required for stability of the new regulation network). This approach reveals that expression of a critical period varies according to environmental conditions and the genotype considered. Consequently, beyond developmental contingencies related to internal disturbing events, expression of a critical period appears as regulated.

Evidence indicates that brassinosteroids (BRs) are involved in expression of a critical period through their effect on cellular sensitivity to other growth factors. This is why BRs differ so much from other PGRs. They interact with almost all the PGRs already identified, and are involved in stress alleviation.[12] In consequence, BRs should be considered as metahormones of plant development.

CONCLUSION

Development is generally understood to be the expression of a genetic program, so that an adaptive dimension is not required. In such a context, when developmental modifications and an explosion of diversity are transiently observed in an initially homogeneous population, they are automatically related to a genetic or environmental (hypothetic) heterogeneity. In this context, it is not surprising that so few studies about critical periods have been published.

Two redundant levels of regulation (local and global) of development coexist in plants, so that a transient decrease at the global level does not immediately generate an observable modification.[13] This is why changes in regulation network are frequently unobtrusive. The redundancy in regulations hides the transient changes induced in the network, at least as long as the environmental conditions do not affect this latter level of regulation. However, a strong perturbation in many physiological and developmental factors is observed during this period as soon as plants are exposed to suboptimal conditions.

The concept of critical periods is well known in physics and in chemistry, in which nonlinear processes now play a central role. Ignoring such a reality in biology invites contradictory conclusions about the effects of stress and plant response to it. It is unlikely that the concept of critical period will be restricted to interorgan relationships and response to stress at the whole plant level. The concept is probably useful in a series of phenomena, from molecular interactions in the cell[14] to species interactions in a biocoenosis.[15]

ARTICLES OF FURTHER INTEREST

Bacterial Products Important for Plant Disease: Diffusable Metabolites As Regulatory Signals, p. 95
Bacterial Products Important for Plant Disease: Toxins and Growth Factors, p. 101
Breeding Plants and Heterosis, p. 189
C3 Photosynthesis to Crassulacean Acid Metabolism Shift in Mesembryanthenum Crystallinum: A Stress Tolerance Mechanism, p. 241
Chromosome Rearrangements, p. 270
Circadian Clock in Plants, p. 278
Crop Responses to Elevated Carbon Dioxide, p. 346
Crops and Environmental Change, p. 370
Drought and Drought Resistance, p. 386
Drought Resistant Ideotypes: Future Technologies and Approaches for the Development of, p. 391
Ecophysiology, p. 410
Oxidative Stress and DNA Modification in Plants, p. 854
Plant Response to Stress: Abscisic Acid Fluxes, p. 973
Plant Response to Stress: Biochemical Adaptations to Phosphate Deficiency, p. 976
Plant Response to Stress: Critical Periods in Plant Development, p. 981
Plant Response to Stress: Genome Reorganization in Flax, p. 984
Plant Response to Stress: Mechanisms of Accommodation, p. 987
Plant Response to Stress: Regulation of Plant Gene Expression to Drought, p. 999
Plant Response to Stress: Role of Jasmonate Signal Transduction Pathway, p. 1006
Plant Response to Stress: Source-Sink Regulation by Stress, p. 1010

REFERENCES

1. Amzallag, G.N. Plant Evolution: Toward an Adaptive Theory. In *Plant Response to Environmental Stresses: From Phytohormones to Genome Reorganization*; Lerner, H.R., Ed.; Marcel Dekker: New York, 1999; 171–241.
2. Brink, R.A. Phase change in higher plants and somatic cell heredity. Q. Rev. Biol. **1962**, *37*, 1–22.
3. Slafer, G.A.; Rawson, H.M. Sensitivity of wheat phasic development to major environmental factors: A re-examination of some assumptions made by physiologists and modelers. Aust. J. Plant Physiol. **1994**, *21*, 393–426.
4. Munns, R. Physiological processes limiting plant growth in saline soils: Some dogmas and hypotheses. Plant Cell Environ. **1993**, *16*, 15–24.
5. Amzallag, G.N. Tolerance to Salinity in Plants: New Concepts for Old Problems. In *Strategies for Improving Salt Tolerance in Higher Plants*; Jaiwal, P.K., Singh, R.P., Gulati, A., Eds.; Oxford and IBH Publishing: New Delhi, 1997; 1–24.
6. Trewavas, A.J.; Cleland, R.E. Is plant development regulated by changes in the concentration of growth substances or by changes in the sensitivity to growth substances?. Trends Biochem. Sci. **1983**, *8*, 354–357.
7. Amzallag, G.N. Maturation of integrated functions during development. I. Modifications of the regulatory network during transition periods in *Sorghum bicolor*. Plant Cell Environ. **2001**, *24*, 337–345.
8. Amzallag, G.N. Developmental changes in effect of cytokinin and gibberellin on shoot K^+ and Na^+ accumulation in salt-treated *Sorghum* plants. Plant Biol. **2001**, *3*, 319–325.
9. Zhirmunsky, A.V.; Kuzmin, V.I. *Critical Levels in the Development of Natural Systems*; Springer-Verlag: Berlin, 1988.
10. Amzallag, G.N. Adaptive nature of the transition phase in development: The case of *Sorghum bicolor*. Plant Cell Environ. **1999**, *22*, 1035–1042.
11. Amzallag, G.N. Connectance in *Sorghum* development: Beyond the genotype-phenotype duality. Biosystems **2000**, *56*, 1–11.
12. Clouse, S.D.; Sasse, J.M. Brassinosteroids: Essential regulators of plant growth and development. Annu. Rev. Plant Physiol. Plant Mol. Biol. **1998**, *49*, 427–451.
13. Amzallag, G.N. Data analysis in plant physiology: Are we missing the reality? Plant Cell Environ. **2001**, *24*, 881–890.
14. Shirane, K.; Tokimoto, T. Dynamic approach to self-organization or a phase transition. Model for a nonequilibrium phase transition in phosphatidylserine membranes by the binding of Ca2+ and Cu2+. Chem. Phys. Lett. **1986**, *128*, 291–294.
15. Fonseca, C.R.; Leighton John, J. Connectance: A role for community allometry. Oikos **1996**, *77*, 353–358.

Plant Response to Stress: Genome Reorganization in Flax

Christopher A. Cullis
Case Western Reserve University, Cleveland, Ohio, U.S.A.

INTRODUCTION

The plant genome is in a dynamic state, and a range of environmental pressures can stimulate particular alterations. Flax, a bifunctional crop that can be grown for fiber (flax) or oil (linseed), is particularly susceptible to such alterations. Some varieties undergo heritable genomic and phenotypic alterations during a single generation's growth in certain particular destabilizing environments. These genomic changes have been shown to occur in all types of sequences, including highly repetitive tandem arrays, intermediately repetitive dispersed regions, and low-copy number sequences. The genomic changes have been shown to occur during the vegetative growth, resulting in chimeric plants. However, in all the cases that have been characterized to date, the genomic alterations have become homozygous in the plant prior to the reproductive structures being differentiated. Therefore, because flax is essentially a self-fertilizing plant, all the progeny for a particular plant are identical, but could be different from the parent plant. More important, all the progeny from a series of plants grown under the same environment are identical, illustrating that there is reproducibility in response to such environmental pressures. Some of the genomic alterations appear to be adaptive to the conditions under which they are induced, but in every case these possible adaptive changes are part of a large number of genomic regions that are disrupted under any given growth conditions.

Why Study Flax?

The initial observations of flax were made by Durrant when he investigated the previously reported effect of growth conditions on the vigor of the flax seed. A series of flax varieties were grown under a variety of conditions and allowed to self. The seed was collected and grown under a uniform set of conditions. The progeny for the different parental treatments had different phenotypes. For one particular variety, Stormont cirrus, a small number of the treatments (including an imbalance of nutrients or particular temperature regimes) resulted in lines that bred true for the altered phenotypes, irrespective of subsequent growth conditions, although the phenotype resulting from most of the environmental conditions depended on growth conditions in the immediately preceding generation. These stably altered types were termed genotrophs, and the progenitor line termed plastic, Pl. These lines were, for all intents and purposes, different genetic types. The lines differed in many characteristics, including plant height. The plant height characteristic illustrates a number of points relating to the induced changes. As seen in Fig. 1, all the plants derived from a single set of growth conditions are uniform, but different from those grown under a different set of conditions. This height difference is observed irrespective of the growth conditions, and is mainly due to a change in the length of the first few internodes.

WHAT CHANGES OCCUR IN RESPONSE TO NUTRITIONAL AND TEMPERATURE STRESS?

Although much remains to be learned about the induction process in flax, four aspects are established. First, Stormont Cirrus is a predominantly self-fertilizing plant because anther dehiscence and pollination usually occur during flower opening. Second, nearly all of the seeds planted grow under the inducing conditions and can contribute to the next generation.[1] Thus it is extremely unlikely that any form of selection, in the conventional sense, from a heterogeneous population of plants, is the causative agent for the observed change. Third, all of the self-fertilized progeny from all the individuals growing in a specific environment are identical, but different from all the progeny of individuals grown in a different environment. Fourth, the induction of the changes has been repeated with Pl, resulting in the appearance of similar phenotypic, biochemical, and molecular changes.[1–3]

The described phenotypic variations among the genotrophs (height, weight, capsule hair, septa number, and isozyme mobility) are but the tip of the iceberg when compared to the extent of variation seen at the DNA level. With regard to nuclear DNA, the extreme types differ by 15% of their total DNA as determined by Feulgen

Encyclopedia of Plant and Crop Science
DOI: 10.1081/E-EPCS 120010644

Fig. 1 Mature plants of the small (left two pots) and large (right two pots) genotrophs grown under the same conditions demonstrating the uniformity of the plants and the height difference. These plants were ten generations post the initial inducing generation.

staining.[4] This level of variation has been confirmed by the characterization of nuclear DNA by renaturation analysis, the cloning of families of repetitive sequences,[5] the detailed characterization of the families of large (25S+18S) and small (5S) ribosomal RNA genes,[6] RAPD analysis,[7] and representational difference analysis.[8] A notable characteristic of all the DNA comparisons is the specificity of the changes between Pl and the genotrophs. Independent induction events have resulted in identical new structures at many loci, confirming the precise nature of the induced changes. For example, specific polymorphisms have been repeatedly generated in independently induced genotrophs in subsets of the 5SRNA gene family,[6] and specific RAPD polymorphisms have repeatedly appeared.[7] Within the flax genome, therefore, there exists a set of regions that are particularly labile and that can be altered in response to environmental stresses.[3] Such a labile subset that responds during in vitro regeneration in barley has also been described.[9] An important observation is that the availability of these sites is dependent on the physiological status of the plant. Therefore, the array of genomic alterations that occurs in response to any particular set of conditions will depend on the particular stress conditions applied.

WHEN IN DEVELOPMENT DO THE INDUCED CHANGES OCCUR?

Two of the genomic changes that have been characterized in detail are variations in ribosomal RNA gene number and a 5.8 kb insertion sequence named LIS-1. The latter is present in a number of genotrophs and in many other flax and linseed varieties. Both of these sequences have been followed during the growth of plants under inducing conditions, particularly under reduced mineral nutrition, and have shown similar responses. The variation in ribosomal RNA gene number occurred during vegetative growth of Pl under inducing conditions,[10] and was complete before the plants flowered. Similarly, the appearance of LIS-1 was followed in Pl while growing under three different nutrient regimes; leaves were sampled as the plants grew. All three conditions resulted in the appearance of LIS-1 in the DNA extracted from leaves. Under two of these regimes, however, all the plants became homozygous for the insertion, and the inserted site was transmitted to all the progeny, whereas under the third regime, none of the progeny had LIS-1 inserted.

CONTROL OF THE GENERATION OF VARIATION

Within the flax genome there exists a set of regions that are particularly labile and able to be altered in response to environmental stresses.[3] These sites vary frequently, and are therefore difficult to map by conventional genetic crosses, as they can vary in the generation in which they are scored.[11] However, some flax varieties do not respond by destabilizing the genome when grown under these conditions. One of the crosses examined confirms that the loci controlling the response can also segregate, and can therefore be mapped and isolated. An understanding of the control of these genomic changes in flax will guide the search for similar mechanisms in other plants.

ARE THESE RAPID CHANGES ADAPTIVE?

What are the circumstances wherein such reorganization would confer a substantial advantage in an evolutionary context? First, it would need to be an inducible system so that when the organism is surviving well in a given environment there is no continuous generation of variation (although one could question whether any environment is really optimum). Ideally, the variation would only be generated when the organism senses the need. The way in which such a sensory mechanism may work is not clear at present. Second, the regions of the genome that are targeted for restructuring need to be delineated. It appears unlikely that restructuring can occur at random. Random mutagenesis will mainly generate deleterious events rather than useful variation. Therefore, to have an advantageous mechanism, it is important to delineate regions of the genome that will be especially labile. The identification of these regions will be essential in understanding how the genomic reorganizations result in phenotypic variation. Third, the breeding system of the organism is also likely to be important. In the case of an inbreeding species, there will be little variation left when growth occurs in a favorable environment over an extended period. Thus, in such a circumstance, a significant change in the environment is likely to be disastrous, unless a mechanism is present by which either variation can be introduced or maintained. Therefore, it is likely that evolutionary selection will favor the presence of a controlled plasticity in those inbreeding species that have survived the slings and arrows of outrageous environments.

A question arises over the adaptive nature of any of these nuclear changes. In flax, it is clear that not all of the changes can be adaptive, as many of the variants are common to lines that have different phenotypes. The obligatory appearance of LIS-1 after some treatments, however, versus its complete absence after others indicates that it is likely to be of importance in the selection of that specific genomic variation under those conditions. It is therefore either one of the regions that confers adaptive advantage under those conditions, or is closely linked to the advantageous change. Because other heritable epigenetic changes such as imprinting can act at a distance from their chromosomal site, a characterization of the genomic context of this element may prove to be instructive.

CONCLUSION

These results in flax are the first demonstration that a series of genomic locations are labile and specifically responsive to particular environmental conditions. Within the set of variants, a particular combination of changes can be selected that are adaptive in those conditions. Under these circumstances the environment acts as an inducer of variation and then as the subsequent selective agent among the variants generated to genetically alter all or the majority of the population. These results have clear evolutionary implications for any organism in which the germline is not set aside very early in development, as this mechanism can give rise to a coordinated set of variations under particular environmental conditions. The understanding of the control of these genomic changes in flax will be invaluable in the search for and understanding of the same processes in other species.

ARTICLES OF FURTHER INTEREST

Air Pollutants: Responses of Plant Communities, p. 20
Breeding: Genotype-by-Environment Interaction, p. 218
Evolution of Plant Genome Microstructure, p. 421
Genome Rearrangements and Survival of Plant Populations to Changes in Environmental Conditions, p. 513
Genomic Imprinting in Plants, p. 527
Oxidative Stress and DNA Modification in Plants, p. 854
Plant Response to Stress: Mechanisms of Accommodation, p. 987
Somaclonal Variation: Origins and Causes, p. 1158

REFERENCES

1. Durrant, A. The environmental induction of heritable changes in *Linum*. Heredity **1962**, *17*, 27–61.
2. Cullis, C.A. Molecular aspects of the environmental induction of heritable changes in flax. Heredity **1977**, *38*, 129–154.
3. Cullis, C.A. Environmental Stress—A Generator of Adaptive Variation. In *Plant Adaptations to Stress Environments*; Lerner, H.R., Ed.; Marcel Dekker: New York, 1999; 149–160.
4. Evans, G.M.; Durrant, A.; Rees, H. Associated nuclear changes in the induction of flax genotrophs. Nature **1966**, *212*, 697–699.
5. Cullis, C.A.; Cleary, W. Rapidly varying DNA sequences in flax. Can. J. Genet. Cytol. **1986**, *28*, 252–259.
6. Schneeberger, R.; Cullis, C.A. Specific DNA alterations associated with the environmental induction of heritable change in flax. Genetics **1991**, *128*, 619–630.
7. Cullis, C.A.; Song, Y.; Swami, S. RAPD polymorphisms in flax genotrophs. Plant Mol. Biol. **1999**, *41*, 795–800.
8. Oh, T.J.; Cullis, C.A. Labile DNA sequences in flax identified by combined sample representational difference analysis (csRDA). Plant Mol Biol, **2003**, *52*, 527–536.
9. Linacero, R.; Alves, E.F.; Vazquez, A.M. Hot spots of DNA instability revealed through the study of somaclonal variation in rye. Theor. Appl. Genet. **2000**, *100*, 506–511.
10. Cullis, C.A.; Charlton, L.M. The induction of ribosomal DNA changes in flax. Plant Sci. Lett. **1981**, *20*, 213–217.
11. Oh, T.J.; Gorman, M.; Cullis, C.A. A RAPD/RFLP map in flax. Theor. Appl. Genet. **2000**, *101*, 590–593.

Plant Response to Stress: Mechanisms of Accommodation

P

H. R. Lerner
The Hebrew University, Jerusalem, Israel

INTRODUCTION

Whole plants and cultured plant cells have the capacity of increasing their tolerance to stress. This may occur through reversible, preexisting switches, as in phosphate starvation that induces a series of new enzymes. Other examples are the facultative halophytes that modify their carbon-fixing mechanism from C3 to CAM under drought or saline stress. Other examples of overcoming stress are through irreversible changes, such as the effects on flax of imbalanced mineral nutrition or particular temperature regimes (discovered by Durrant in the late 1950s and further studied by Cullis); the discovery by McClintock that stress may cause nonrandom reorganization of the genome and chromosome aberrations in corn; and adaptation of sorghum to NaCl concentrations that are lethal for the unadapted plant. At the cultured plant cell level it has been shown that tobacco cells can be adapted to grow in 500 mM NaCl, and McCoy showed that exposure of cultured alfalfa cells to salt stress enhances dramatically the frequency of both chromosome aberrations and altered isozymes. This article discusses mechanisms of irreversible changes induced by stress.

THE PLANT'S DEVELOPMENTAL TRAJECTORY

The physiological response of a plant is modified as a function of its development. This change as a function of time is the developmental trajectory of the plant, often referred to as phasic development. Moreover, there are limited time periods—developmental windows—during which a plant's response is temporarily modified.[1] The unfolding of internal developmental programs and external signals affect the developmental trajectory, which reflects the genetic makeup of the plant as well as its environmental history. All cells contain two types of information—the genetic information and the epigenetic system that is affected by environmental and developmental cues.[2]

During the unfolding of the developmental trajectory the plant undergoes continuous modulation in its phytohormonal balance as well as in its sensitivity (its response) to phytohormones; in protein, posttranslational modifications such as phosphorylation/dephosphorylation; in histone, modifications such as acetylation, methylation, phosphorylation, and probably DNA methylation. At the same time there is a reorganization of the genome that entails amplification and deletion of DNA (including A+T- and G+C-rich DNA) and DNA inversion.[3,4] All these changes are mechanisms that modulate genome expression during development.

STRESS-RESPONSE MECHANISMS

Somaclonal Variations

Following plant regeneration from cultured cells, protein modifications and aberrations in chromosomes and DNA sequences are seen. These changes have been called somaclonal variations because they arise within clones of somatic cells. Modified phenotypes are sometimes, but not always, heritable.[5] The number of aberrations increases with the length of time cells are maintained in culture and with intensity of stress conditions.[5,6] Reoccurrence of similar types of somaclonal variations in cultured cells, which are sometimes similar to known heritable mutations in whole plants,[5] indicates that there are hot spots for these changes; the phenomenon appears to be nonrandom. Observation of bridges at mitotic and meiotic anaphase suggests that a mechanism called breakage–fusion–bridge cycle (BFB) could be responsible for some of these mutations.[5,7] The appearance of new isozymes and restriction fragment length polymorphism (RFLP) and rapid amplification of polymorphic DNA (RAPD) techniques are sensitive methods that reveal slight modifications in DNA sequences due to somaclonal variations.[5,6] The mechanisms inducing somaclonal variations are unknown; however, several causes have been suggested, including imbalanced nutrient concentrations, phytohormones, leakage of cellular constituents, DNA amplification/deletion, chromosome breakage (including BFB), transposon movement, and 5-methylcytosine deamination.[5]

Changes in Concentration of Phytohormone and Other Molecules Under Stress

Under stress, the concentration of abscisic acid and ethylene increases while the concentration of auxins,

Encyclopedia of Plant and Crop Science
DOI: 10.1081/E-EPCS 120010662

cytokinins, and gibberellins decreases.[8] Other molecules such as polyamines, brassinosteroid, methyl jasmonate/jasmonic acid, and salicylic acid are sometimes produced. Reactive oxygen species (ROS) such as superoxide radical, hydroxyl radical, and hydrogen peroxide are also produced under stress. See the contribution of A. Levine in this volume for the consequence of ROS on genome organization.

Genome Modifications Under Stress

For many plant species the haploid genome size (1C-value), in picograms (pg) or base pairs, has been reported in the literature;[9,10] 1 pg = $965*10^6$ bases. 1C-values for flowering plants range from <0.2 pg in *Arabidopsis thaliana* to 127.4 pg in *Fritillaria assyriaca*.[10] From these values, the genome size varies by more than 600-fold, even though these plants express more or less similar genes, suggesting that the difference in chromatin content is due to noncoding DNA. Reasons for such differences could be changes in ploidy level, repetitive DNA such as tandemly repeated sequences that are grouped in one or a few sites (e.g., rDNA; satellite DNA that can be either A+T- or G+C-rich), or dispersed repeated sequences that are spread throughout the genome (e.g., amplification of transposable elements).

Growth of the inbred flax variety Stormont Cirrus in an inducing environment (such as imbalanced mineral nutrition or specific temperatures) causes stable and genetically altered plants after a single generation.[11] The phenomenon is reproducible. Following similar treatment, identical changes are observed in plant height and weight at maturity and in the number of hairs on the false septa of the seed capsules. This is not due to selection because >90% of the seeds germinate and undergo the same changes. These phenotypic changes are accompanied by alterations in the number of repeats of repetitive DNA of both the tandem rDNA and in families of dispersed repeated sequences. These phenotypic/genotypic modifications are maintained for all generations tested.

The discovery that stress results in chromosome aberrations in the whole corn plant[7] and in cultured alfalfa cells[6] shows that stress modifies genome organization. This is also demonstrated by somaclonal variations[5,6] because, for a plant cell, being cultured constitutes a stress. Adaptation of whole sorghum plants to NaCl[1] is passed on to progenies for all generations tested (Amzallag personal communication). This suggests the inclusion of a genetic effect. Development of cultured cells from regenerated plants obtained from NaCl-adapted tobacco cells[12] yields already adapted cells (Watad personal communication), also suggesting that this adaptation is a genetic effect. The course of accommodation to stress entails modifications in both genome organization and in the process of modulation of genome expression. Another example of genome modification under stress has been shown for cold hardening.[13]

Range of Stress-Response Process

Not all flax or sorghum varieties undergo the phenotype/genotype modifications described here; these changes are specific properties of particular varieties. Although the processes occurring in flax and sorghum seem similar, they are not identical. In flax all the plants submitted to an inducing condition undergo the same modification, whereas in sorghum there is an increase in the variability of parameters within the population (e.g., shoot height, shoot dry weight) following the adaptation treatment,[14] indicating that each individual plant within the population undergoes its own stress–response pathway.

There is a wide range of plants' mechanisms of accommodation to stress—from reversible preexisting switches (e.g., phosphate starvation,[15] facultative halophytes[16]), to irreversible preexisting switches (e.g., flax[11]), to the irreversible elaboration within a population of new patterns of modulations of genome expression (e.g., sorghum[1]).

Nuclear Architecture

Using in situ radioactive DNA hybridization, the organization of parental chromosomes in plant hybrids is shown to occupy distinct, nonrandom, positions. Throughout the cell cycle, parental genomes are maintained in different domains, including during interphase (Ref. 9, Refs. 67,88 in Ref. 9). Moreover, using fluorescent in situ hybridization (FISH) on animal cells, it has been confirmed that the folding of DNA in chromatin is highly specific.[17] It seems probable that nuclear organization may affect or be affected by gene expression.[17]

CONCLUSION

The mechanisms of accommodation of plants to stress include all the mechanisms functioning during the plant's developmental trajectory, including changes in phytohormonal balance and sensitivity to phytohormones; protein posttranslational modification; DNA reorganization through amplification, deletion, and inversion; histone modification; DNA methylation; and movement of transposons. All these changes modulate genome organization and expression. The nucleus is a dynamic, plastic, highly organized organelle that, as described by McClintock, is capable of reorganizing itself to overcome adverse conditions. Although transformation is a very interesting and useful tool in plant research, for agriculture it is more advantageous to allow crop plants to modify their genomes by themselves because they are better equipped than humans to do so.

ARTICLES OF FURTHER INTEREST

C3 Photosynthesis to Crassulacean Acid Metabolism Shift in Mesembryanthenum Crystallinum: A Stress Tolerance Mechanism, p. 241
Chromosome Rearrangements, p. 270
Chromosome Structure and Evolution, p. 273
Evolution of Plant Genome Microstructure, p. 421
Gene Flow Between Crops and Their Wild Progenitors, p. 488
Genomic Imprinting in Plants, p. 527
Genome Rearrangements and Survival of Plant Populations to Changes in Environmental Conditions, p. 513
Genome Size, p. 516
Oxidative Stress and DNA Modification in Plants, p. 854
Plant Response to Stress: Biochemical Adaptations to Phosphate Deficiency, p. 976
Plant Response to Stress: Critical Periods in Plant Development, p. 981
Plant Response to Stress: Genome Reorganization in Flax, p. 984

REFERENCES

1. Amzallag, G.N.; Seligman, H.; Lerner, H.R. A developmental window for salt-adaptation in *Sorghum bicolor*. J. Exp. Bot. **1993**, *44*, 645–652.
2. Trewavas, A.J.; Malhó, R. Signal perception and transduction: The origin of the phenotype. Plant Cell **1997**, *9*, 1181–1195.
3. Arnholdt-Schmitt, B. Physiological aspects of genome variability in tissue culture. II. Growth phase-dependent quantitative variability of repetitive *Bst*NI fragments of primary cultures of *Daucus carota* L. Theor. Appl. Genet. **1995**, *91*, 816–823.
4. Ceccarelli, M.; Giordani, T.; Natali, L.; Cavalini, A.; Cionini, P.G. Genome plasticity during seed germination in *Festuca arundinacea*. Theor. Appl. Genet. **1997**, *94*, 309–315.
5. Olhof, P.M.; Phillips, R.L. Genetic and Epigenetic Instability in Tissue Culture and Regenerated Progenies. In *Plant Responses to Environmental Stresses: From Phytohormones to Genome Reorganization*; Lerner, H.R., Ed.; Marcel Dekker, Inc.: New York, 1999; 111–148.
6. McCoy, T.J. Characterization of alfalfa (*Medicago sativa* L.) plants regenerated from selected NaCl tolerant cell lines. Plant Cell Rep. **1987**, *6*, 417–422.
7. McClintock, B. Mechanisms that rapidly reorganize the genome. Stadler Symp. **1978**, *10*, 25–47.
8. Itai, C. Role of Phytohormones in Plant Responses to Stress. In *Plant Responses to Environmental Stresses: From Phytohormones to Genome Reorganization*; Lerner, H.R., Ed.; Marcel Dekker, Inc.: New York, 1999; 287–301.
9. Dean, C.; Schmidt, R. Plant genomes: A current molecular description. Annu. Rev. Plant Physiol. Plant Mol. Biol. **1995**, *46*, 395–418.
10. Bennett, M.D.; Leitch, I.J. Nuclear DNA amounts in angiosperms. Ann. Bot. **1995**, *76*, 113–176.
11. Cullis, C.A.; Swami, S.; Song, Y. RAPD polymorphism detected among the flax genotrophs. Plant Mol. Biol. **1999**, *41*, 795–800.
12. Watad, A.A.; Lerner, H.R.; Reinhold, L. Stability of salt-resistance character in *Nicotiana* cell lines adapted to grow in high NaCl concentrations. Physiol. Veg. **1985**, *23*, 887–894.
13. Laroche, A.; Geng, X.-M.; Singh, J. Differentiation of freezing tolerance and vernalization responses in Cruciferae exposed to low temperature. Plant Cell Environ. **1992**, *15*, 439–445.
14. Amzallag, G.N.; Seligmann, H.; Lerner, H.R. Induced variability during the process of adaptation in *Sorghum bicolor*. J. Exp. Bot. **1995**, *46*, 1017–1024.
15. Plaxton, W.C.; Carswell, M.C. Metabolic Aspects of the Phosphate Starvation Response in Plants. In *Plant Responses to Environmental Stresses: From Phytohormones to Genome Reorganization*; Lerner, H.R., Ed.; Marcel Dekker, Inc.: New York, 1999; 349–372.
16. Bohnhert, H.; Cushman, J.C. The ice plant cometh: Lessons in abiotic stress tolerance. J. Plant Growth Regul. **2000**, *19*, 334–346.
17. Lemon, B.; Tjian, R. Orchestrated response: A symphony of transcription factors for gene control. Genes Dev. **2000**, *14*, 2551–2569.

Plant Response to Stress: Modifications of the Photosynthetic Apparatus

Eevi Rintamäki
University of Turku, Turku, Finland

INTRODUCTION

Because plants are sessile organisms, their growth and development greatly depend on abiotic and biotic environmental factors. Plants have developed efficient mechanisms for sensing changes in growth conditions and for induction of adaptive processes in order to cope with various kinds of environmental stress. Plants are ultimate harvesters of sunlight, and maintenance of a functional photosynthetic apparatus is fundamental for plant survival under unfavorable conditions. Light is the main environmental regulator in modifying the structure and function of plant photosystems. Other abiotic factors, e.g., extreme temperatures and oxidative stress, also induce changes in the photosynthetic machinery, but their action depends on light. Structural and functional modifications of Photosystem I and Photosystem II under abiotic stress conditions will be discussed in this article, with emphasis on protein phosphorylation, changes in subunit and pigment composition, and induction of protecting compounds.

PHOTOSYSTEM II

Photosystem II (PSII) functions as a light-driven water-plastoquinone-oxidoreductase in plant thylakoid membranes. PSII is composed of more than 25 proteins, including the subunits of the reaction center responsible for primary photochemistry, the oxygen evolving complex, and the polypeptides of chlorophyll a/b antenna.[1] Functional PSII exists as a dimeric complex. In nature, plants experience continual short-term and long-term fluctuations in light intensity. PSII is a main regulated unit in the light reactions, and is thus also a primary target for modifications induced by environmental factors. The best-documented modifications of PSII are associated with photoinhibition of photosynthesis, including both the induction of photoprotective mechanisms and oxidative damage to PSII. Some of the changes are rapidly reversible on a time scale ranging from seconds to hours under transient changes of irradiance, whereas long-term exposure to new light conditions causes stable adjustments of the complex. Protein phosphorylation and changes in pigment composition belong to the reversible modifications occurring in PSII under variable environmental conditions. On the other hand, protein synthesis is required to cope with stress-induced photodamage to PSII, including the replacement of the oxidized protein with the new copy (see below) as well as the stimulation of the expression of protective proteins called ELIPs (early light inducible proteins) and a PSII subunit of 22 kDa (psbs).

Light-induced Protein Phosphorylation in Thylakoid Membranes

Light induces phosphorylation of a number of PSII proteins in thylakoid membranes. Four PSII core proteins—D1 and D2 reaction center proteins, 43-kDA chlorophyll a-binding protein (CP43), and the *psbH* gene product (psbH protein)—undergo reversible phosphorylation in thylakoid membranes.[2] The PSII antenna consists of six pigment-binding proteins designated Lhcb1 to Lhcb6, from which three (Lhcb1, Lhcb2, and Lhcb4) have been shown to become reversibly phosphorylated in light-dark transitions.[2,3] Light regulates this phosphorylation via the redox state of electron carriers between PSII and PSI.[2] Reduction of plastoquinone activates serine/threonine protein kinase(s) that phosphorylate the PSII core phosphoproteins and Lhcb4 protein,[3] whereas both the reduction of plastoquinone and subsequent binding of plastoquinol to the Q_o site of the cytochrome b_6f complex are required to initiate the phosphorylation of Lhcb1 and 2 proteins.[2,3] Moreover, the kinase that phosphorylates the Lhcb1 and 2 proteins is specifically inhibited in the presence of thiol reductants.[3] Significantly, thylakoid membranes contain several membrane-bound protein kinases with specific regulation mechanisms and at least partial substrate specificity for PSII phosphorproteins. Decreasing redox potential in the chloroplast, e.g., by shifting plants into darkness, induces the dephosphorylation of PSII phosphoproteins via protein phosphatases.[2] Both soluble and membrane-bound phosphatases capable of dephosphorylating PSII phosphoproteins have been found in chloroplast. However, these enzymes are still poorly characterized.

Previously, the reversible phosphorylation of PSII proteins was assumed to be coupled to light stress of the chloroplast through prevention of the imbalance between

Encyclopedia of Plant and Crop Science
DOI: 10.1081/E-EPCS 120010657

excitation rates of the photosystems under fluctuations in light intensity, and via protection of PSII from photoinhibition in high light (discussed later). This scheme presumes that an exposure to high light would induce the phosphorylation of PSII proteins. However, screening of the environmental conditions activating the phosphorylation of PSII proteins in vivo has indicated that maximal phosphorylation of the PSII core proteins occurs at light conditions prevailing during growth of plants without any evident symptoms of light stress.[4] This maximal phosphorylation level, corresponding to 70–80% of PSII complexes with phosphorylated PSII core proteins, is maintained under conditions with redox potential equivalent to or higher than observed at growth light conditions.[4] Moreover, the antenna proteins Lhcb1 and 2 are maximally phosphorylated only at low redox potential in the chloroplast, e.g., after a shift of the plant to a light intensity lower than experienced during growth, while high light strongly inhibits phosphorylation of these proteins.[4] Only the phosphorylation of the Lhcb4 protein seems to be specifically stress-stimulated.[3]

What are the physiological implications for the PSII complex modified by protein phosphorylation? A number of functional roles have been suggested for PSII core protein phosphorylation. Chlorophyll fluorescence and low temperature electron paramagnetic resonance (EPR) measurements have demonstrated that phosphorylation of D1, D2, and CP43 proteins does not influence the electron transfer capacity in PSII.[3] Stabilization of a dimeric form of PSII—and direct protection of PSII from photoinhibition—have also been linked to the phosphorylation of PSII core proteins, but discrepancies in the experimental results mean that these hypotheses should still be questioned. A number of experimental data, however, support the hypothesis that phosphorylation of the PSII core proteins regulates the PSII photoinhibition repair cycle by controlling the timing and location of proteolytic degradation of the photodamaged D1 protein in thylakoid membranes.[3] This regulation prevents a collapse and total proteolysis of the subunits of photodamaged PSII complex, especially under the stress conditions that cause deterioration of protein synthesis, e.g., low temperature combined with illumination of plant.

Phosphorylation of Lhcb1 and 2 proteins has been proposed to induce state transition that modulates energy distribution between PSII and photosystem I (PSI) in plant thylakoid membranes.[2] Phosphorylation of these proteins results in partial dissociation of the phosphorylated Lhcb1 and 2 proteins from PSII, and subsequent association of these proteins with PSI.[2] Originally, the state transition was thought to protect PSII from photoinactivation by reducing the antenna size of PSII, and thereby the excitation of PSII. However, this function is not supported by the finding that excessive light intensity strongly inhibits the phosphorylation of these proteins.[3] It has been proposed that Lhcb1 and 2 protein phosphorylation regulates the synthesis of Lhcb proteins during the long-term acclimation of plants to various light intensities.[3] A putative role of this phosphorylation in the relay of signals between chloroplast and nucleus is based on the distinct positive correlation between the phosphorylation level of Lhcb1 and 2 proteins and the expression of *Lhcb* genes.

Phosphorylation of Lhcb4 protein has been attributed to the protection of PSII complex against photoinhibition, especially in chilling-sensitive plants at low temperatures. This protection may be attained by altered distribution of excitation energy between PSII and its antenna after phosphorylation of Lhcb4 protein.

The Xanthophyll Cycle and Photoprotection of Chloroplasts

Besides the photochemistry in thylakoid photosystems, light reactions also generate reactive intermediates and oxygen species that have a potential to damage the photosynthetic machinery, especially in excess light. To prevent their detrimental effect, plants have evolved a number of protection mechanisms—including antioxidant systems and processes for thermal dissipation of light energy.

The extent of the thermal dissipation in PSII can be measured as a quenching of chlorophyll fluorescence in PSII. The energy-dependent component (q_E) of non-photochemical quenching (NPQ) depends on the amount and composition of xanthophyll pigment in membranes, as well as the acidification stage of the thylakoid lumen.[5] The xanthophyll cycle is a universal pigment-conversion mechanism in plants that is involved in the dissipation of excess light energy to heat in thylakoid membranes. In higher plants under light stress, zeaxanthin is formed from violaxanthin via antheraxanthin in the reactions catalyzed by violaxanthin de-epoxidase (VDE), which is activated through acidification of the thylakoid lumen.[6] The cycle is reversible under low light conditions, resulting in the regeneration of violaxanthin via the action of zeaxanthin epoxidase.[6] Xanthophylls bind to chlorophyll a/b binding proteins of both PSII and PSI antennae, but this cyclic conversion of pigments has been proposed to occur in a free form of the molecule, resulting in the reorganization of xanthophyll pigments in Lhcb proteins.[5] Studies with *Arabidopsis* mutants with reduced levels of VDE have shown the involvement of zeaxanthin in thermal dissipation of excess light energy.[6] However, the exact mechanism for this energy dissipation is still obscure. According to the direct model, chlorophyll transfers excitation energy to zeaxanthin, which dissipates it as heat. Alternatively, the production of zeaxanthin may stimulate the proton-dependent structural changes in Lhcb proteins (aggregation, protein–protein interactions) resulting in the formation of dissipation centers in specific antenna complexes.[5]

Recently, a new component involved in dissipation of thermal energy in PSII has been established. q_E was absent in a *psbS*-deletion mutant in Arabidopsis.[7] psbS protein is a component of PSII with homology to chlorophyll a/b binding proteins. However, the chlorophyll-binding domains are not conserved, which excludes the possible role of psbS protein in light harvesting. Transgenic *Arabidopsis* plants overexpressing *psbS* gene had significantly increased q_E and enhanced resistance to PSII photoinhibition.[8] Moreover, low-temperature stress increases the expression of *psbS* gene in illuminated potato leaves. These results strongly support the role of q_E in photoprotection and also the hypothesis that the key function of psbS protein is in the induction of energy-dependent thermal dissipation. The putative function of this protein is in sensing the luminal acidification, which induces a conformational modification of psbS protein and/or of Lhcb proteins, finally resulting in the stabilization of the dissipation centers in PSII.[5]

Photoinhibition of Photosystem II

Despite the photoprotection mechanisms discussed above, irreversible inhibition of photosynthesis occurs in illuminated plants. Both PSII and PSI can be targets of oxygen-dependent photodamage of the reaction center, but under different environmental conditions. Photodamage to PSII can be detected at all physiological light intensities; moreover, the absorption of light in excess of the capacity to utilize it in plants leads to the accumulation of inactive PSII complexes in thylakoid membranes.[3] PSI photoinhibition is induced especially at low temperatures under moderate light intensities.[9]

Irreversible photoinactivation of PSII ultimately results in oxidative damage to the reaction center protein D1 of PSII. To restore the activity of PSII, the photodamaged D1 protein is degraded by proteolysis and substituted by a newly synthesized D1 protein.[3] Strict coordination of the degradation and synthesis of the D1 protein is necessary to prevent loss of other polypeptides in this multisubunit complex, which would lead to total collapse of the complex and thus to more severe damage to the photosynthetic machinery. This coordination is attained via regulation of the turnover of D1 protein by phosphorylation of PSII core proteins.[3] The key feature in this regulation is that the phosphorylated form of the photodamaged D1 protein is not susceptible to proteolytic degradation. Inactivation of PSII takes place in grana lamellae, whereas the co-translational incorporation of the new D1 copy is spatially conceivable only in stroma-exposed thylakoid membranes (Fig. 1). Phosphorylation of PSII core proteins stabilizes the partially disassembled inactive PSII complex upon migration from the grana region to the stroma-exposed

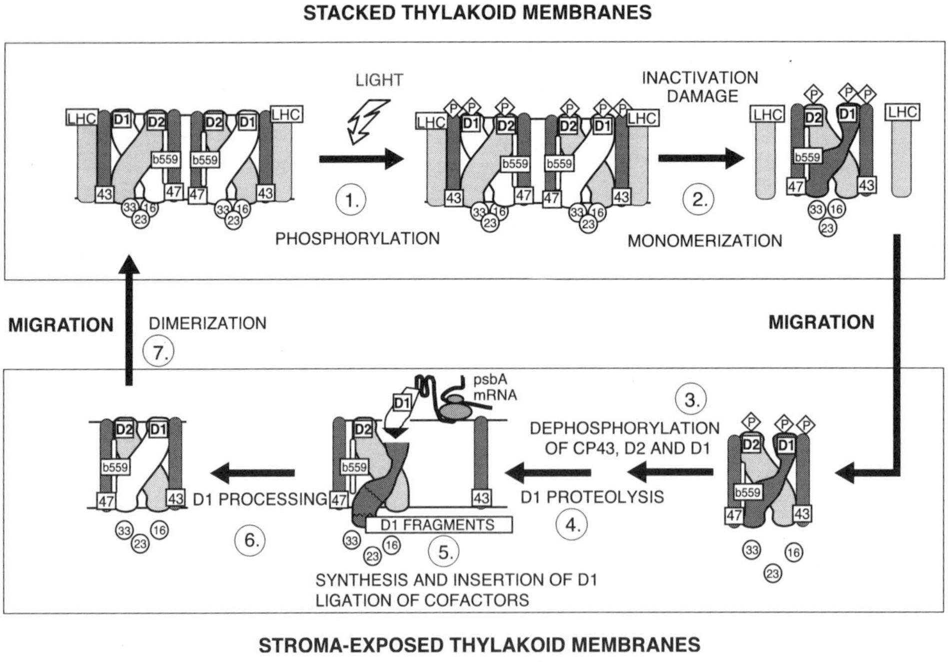

Fig. 1 Phosphorylation of the PSII core proteins and photoinhibition repair cycle of Photosystem II in higher plant thylakoid membranes. (From Ref. 3.)

membranes of thylakoids, in which repair of the complex is induced by dephosphorylation of PSII core proteins, followed by degradation of photodamaged D1 and synthesis of a new copy of protein (Fig. 1).

PHOTOINHIBITION OF PHOTOSYSTEM I

Photosystem I is a light-driven plastocyanin-ferredoxin oxidoreductase involved in the production of NADPH in photosynthetic light reactions. The enzyme complex is composed of 14 subunits, from which the PsaA, PsaB, and PsaC proteins bind the electron carriers of PSI.[1] PsaA-PsaB heterodimer functions also as a core antenna by binding approximately 90 chlorophyll a molecules and 10–15 β-carotenes per reaction center. The chlorophyll a/b-binding proteins of PSI antenna are encoded by four *Lhca* genes designated *Lhca*1 to *Lhca*4.[1]

PSI is not regarded as a rate-limiting component of the light reactions; thus, the dynamic and regulatory aspect of PSI has been investigated less extensively. Low temperature combined with light, however, inhibits PSI activity specifically—both in chilling-sensitive and tolerant plant species.[9] This oxygen-dependent PSI inactivation results in the degradation of the psaA-psaB heterodimer of the PSI reaction center, but a loss of some extrinsic PSI proteins has also been reported.[9] The unique features of PSI photoinhibition are that; 1) bleaching of chlorophyll is observed after return of plants to the growth temperature; 2) degradation of PSI proteins occurs several days after exposure to low temperature; 3) the recovery is much slower than in PSII photoinhibition; and 4) it can ultimately result in a definitive decrease in the amount of PSI complexes in leaves. This more severe photoinhibition of PSI under specific environmental conditions may be due to lack of control over the repair of photodamaged PSI reaction center proteins, compared to the mechanism that has evolved for PSII complexes.

EARLY LIGHT-INDUCED PROTEINS: A STRESS PROTEIN FAMILY IN THYLAKOID MEMBRANES

Originally, the early light-induced proteins (ELIPs) were found upon greening of etiolated seedlings. Expression of these proteins was induced very early after the shift of seedlings to light, although they were not present in mature leaves under ambient growth conditions. However, it was observed later that the expression of these proteins is induced under several different conditions of stress, including light, cold, and salt stress, as well as during nutrient deprivation and desiccation.[10] ELIPs and chlorophyll a/b-binding proteins have sequence similarity, but only binding of chlorophyll a and lutein has been detected in purified proteins.[10] It is generally agreed that these proteins are not involved in energy transfer in photosynthesis. They are localized to the stroma-exposed membranes of thylakoids, probably in close vicinity to PSII complexes, but the direct interaction of ELIPs with the photosystems is still obscure. Furthermore, a number of protective roles of ELIPs under stress conditions has been suggested,[10] although the exact function of these proteins remains to be elucidated. One tempting hypothesis is that ELIPs might bind pigments released from the photosystems during the reorganization of complexes (e.g., photoinhibition, the xanthophyll cycle) under different stress conditions.

CONCLUSION

Light is a key element in modifications of the photosynthetic apparatus that are induced by various abiotic stress factors. The molecular aspects of changes in the structure and function of photosystems are studied mainly by combining light with one of the environmental variables at a time. In nature, stresses can appear contemporaneously, compounding the symptoms and allowing the development of new modifications in plants. In the near future, the establishment of new technologies such as DNA microarray analysis, proteomics, and metabolomics will vastly increase our knowledge of the dynamic aspects of the photosynthetic apparatus, even under the complex environmental conditions that exist in nature.

ARTICLE OF FURTHER INTEREST

Photosystems: Electron Flow Through, p. 906

REFERENCES

1. http://photoscience.la.asu.edu/photosyn/photoweb/default.html (accessed December 2002).
2. Bennett, J. Protein phosphorylation in green plant chloroplasts. Annu. Rev. Plant Physiol. Plant Mol. Biol. **1991**, *42*, 281–311.
3. Rintamäki, E.; Aro, E.-M. Phosphorylation of Photosystem II Proteins. In *Regulation of Photosynthesis*; Aro, E.-M., Andersson, B., Eds.; Advances in Photosynthesis and Respiration, Kluwer Academic Publishers: Dordrecht, 2001; Vol. 11, 395–418.

4. Rintamäki, E.; Salonen, M.; Suoranta, U.-M.; Carlberg, I.; Andersson, B.; Aro, E.-M. Phosphorylation of light-harvesting complex II and photosystem II core proteins shows different irradiance-dependent regulation in vivo. Application of phosphothreonine antibodies to analysis of thylakoid phosphoproteins. J. Biol. Chem. **1997**, *272*, 30476–30482.
5. Bassi, R.; Caffarri, S. Lhc proteins and the regulation of photosynthetic light harvesting function by xanthophylls. Photosynth. Res. **2000**, *64*, 243–256.
6. Eskling, M.; Emanuelsson, A.; Åkerlund, H.-E. Enzymes and Mechanisms for Violaxanthin-Zeaxanthin Conversion. In *Regulation of Photosynthesis*; Aro, E.-M., Andersson, B., Eds.; Advances in Photosynthesis and Respiration, Kluwer Academic Publishers: Dordrecht, 2001; Vol. 11, 433–452.
7. Li, X.-P.; Björkman, O.; Shih, C.; Grossman, A.R.; Rosenquist, M.; Jansson, S.; Niyogi, K.K. A pigment-binding protein essential for regulation of photosynthetic light harvesting. Nature **2000**, *403*, 391–395.
8. Li, X.-P.; Muller-Moule, P.; Gilmore, A.M.; Niyogi, K.K. PsbS-dependent enhancement of feedback de-excitation protects photosystem II from photoinhibition. Proc. Natl. Acad. Sci. U. S. A. **2002**, *99*, 15222–15227.
9. Kudoh, H.; Sonoike, K. Irreversible damage to photosystem I by chilling in the light: Cause of the degradation of chlorophyll after returning to normal growth temperature. Planta **2002**, *215*, 541–548.
10. Adamska, I. The ELIP Family of Stress Proteins in the Thylakoid Membranes of Pro- and Eukaryota. In *Regulation of Photosynthesis*; Aro, E.-M., Andersson, B., Eds.; Advances in Photosynthesis and Respiration, Kluwer Academic Publishers: Dordrecht, 2001; Vol. 11, 487–505.

Plant Response to Stress: Phosphatidic Acid As a Second Messenger

Christa Testerink
Teun Munnik
University of Amsterdam, Amsterdam, The Netherlands

INTRODUCTION

Phosphatidic acid (PA) is emerging as the most important lipid second messenger in plants. It is formed within minutes in response to a wide array of stress conditions, including ethylene, abscisic acid (ABA), cold, Nod factor, pathogen-derived elicitors, wounding, drought, salt, and hypoosmotic stress. PA and the enzymes that form and metabolize this molecule seem to play a key role in plant stress signaling.

FORMATION OF PA

Phospholipids maintain the integrity of biological membranes around cells and their organelles. Relatively new is the awareness that several of them play a crucial role in signal transduction.[1,2] One of the latest additions to this field is the lipid second messenger PA, a minor lipid that occurs in different cellular membranes. When a plant cell is subjected to stress, two different phospholipases can be activated to hydrolyze structural membrane lipids, resulting in the production of PA. Phospholipase D (PLD) hydrolyzes phosphatidylcholine (PC) or phosphatidylethanolamine (PE) to release PA and a free headgroup (choline or ethanolamine), while phospholipase C (PLC) hydrolyzes phosphatidylinositol 4,5-bisphosphate (PIP_2) to generate inositol 1,4,5-trisphosphate (IP_3) and diacylglycerol (DAG), which is immediately phosphorylated by DAG kinase (DGK) to produce PA. The increase is transient because PA is rapidly converted to diacylglycerol pyrophospate (DGPP) by PA kinase (Fig. 1).

Phospholipase D

Several genes encoding PLDs have been cloned from plants. In *Arabidopsis thaliana*, there are twelve members that have been grouped into five classes.[3–5] Individual isoforms have different expression patterns and different biochemical properties. A recent discovery is the association of a PLD with microtubuli, where it could mediate cytoskeletal rearrangements, which are known to occur in response to several stress treatments.[6] When PLD hydrolyzes its substrate, it can use primary alcohols instead of water as an acceptor for the phosphatidyl group (transphosphatidylation). Thus, in the presence of *n*-butanol, phosphatidylbutanol is formed. This characteristic has been used to measure in vivo PLD activity, but also to inhibit PA production and to distinguish PLD-generated PA from PLC-generated PA.[7]

Phospholipase C

In the *Arabidopsis* genome, there are nine PLCs, of which four have been cloned and characterized. The PA they generate together with DGK can be distinguished from that generated by PLD by a differential ^{32}P-labeling protocol and by using specific PLC inhibitors.[7]

THE ROLE OF PHOSPHATIDIC ACID IN STRESS SIGNALING

PA formation has been linked to biotic and abiotic stress treatments, suggesting it is a general stress signal. In support, inhibiting or silencing PLD in plants reduces stress responses, while exogenous application of PA induces stress responses. Some of the best characterised PA responses are discussed here.

ABA and Osmotic Stress

The phytohormone ABA is involved in aging, seed development, germination, and the responses to cold and water stress. It induces the formation of PA in stomatal guard cells and aleurone cells. Osmotic stress, in the form of salt or drought, also increases the level of PA. Both PLC and PLD pathways are activated in response to osmotic stress and ABA. Moreover, in plants in which PLD expression is suppressed, ethylene and ABA-induced senescence, as well as other ABA- and osmotic stress-related responses,

Encyclopedia of Plant and Crop Science
DOI: 10.1081/E-EPCS 120010659

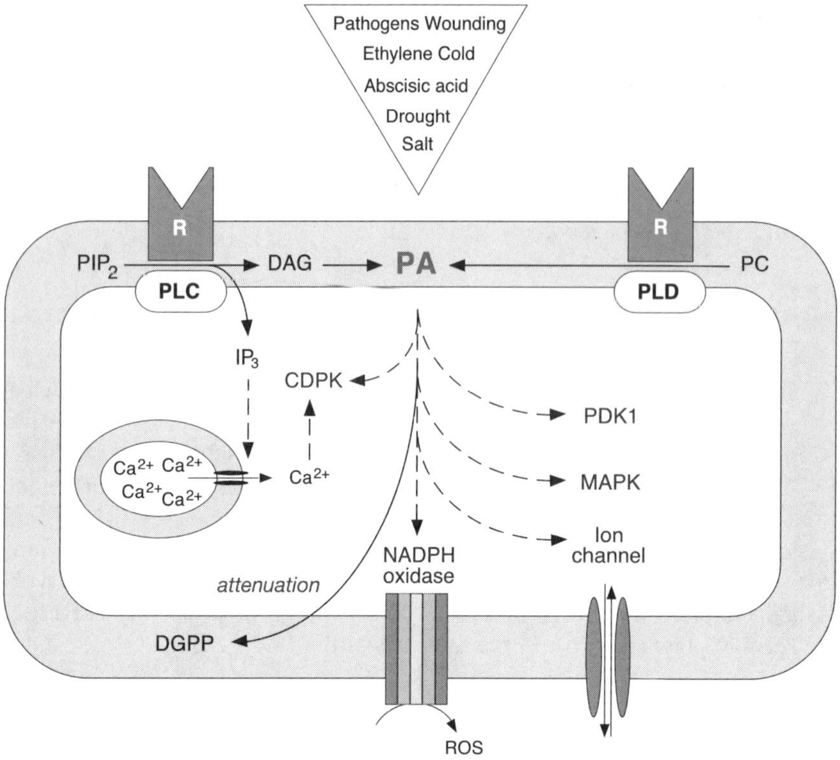

Fig. 1 PA signaling cascades. Different stress treatments activate phospholipid signaling pathways. PLC hydrolyzes PIP_2 into IP_3 and DAG. IP_3 diffuses into the cytosol, where it induces calcium release from intracellular stores. DAG remains in the membrane and is phosphorylated by DGK to PA. PLD hydrolyzes structural phospholipids, thereby generating PA directly. PA formation affects targets such as PDK1, MAPK, CDPK, ion channels, and NADPH oxidase. The PA signal is attenuated by conversion to DGPP. (R = receptor. Solid arrows indicate metabolic conversions; dashed arrows indicate activation (either directly or indirectly) of downstream targets.) (*View this art in color at www.dekker.com.*)

are reduced. Interestingly, exogenously applied PA mimics ABA action in guard cells; it inhibits ion channel activity and induces stomatal closure.[5]

Defense Against Pathogens

A well-studied mechanism of plant defense against pathogens are pathogen-derived elicitors, which activate the plant's defense responses. Several elicitors, such as xylanase, chitotetraose, flagellin, and the *Cladosporium fulvum* race-specific elicitor Avr4, induce PA formation within minutes.[8] Interestingly, Nod factors that are signal molecules of symbiotic *Rhizobium* bacteria also induce the formation of PA in leguminous plants. Wounding a plant triggers PA accumulation, as well, both locally and systemically.[5]

The PA produced in response to pathogen elicitors and Nod factor is mainly due to PLC activation. Blocking PLC enzyme activity inhibits activation of several defense responses, including the formation of reactive oxygen species. Conversely, exogenously applied PA induces an oxidative burst and triggers a MAP kinase pathway, showing that PA is involved in the early response to pathogen attack. PLD is involved in the response to some elicitors and to wounding. When rice is infected by *Xanthomonas oryzae*, PLD is recruited to the site of infection on the plasma membrane. Silencing PLD reduces the wound-induced accumulation of jasmonic acid.[5,8]

PA ACTION AND SPECIFICITY

It is becoming clear that PA plays a major role in cell signaling even though little is known about its working mechanism. Presumably, the formation of PA creates docking sites in the membrane to which specific proteins bind (Fig. 2). Membrane recruitment then leads to activation, either directly by PA or indirectly by other proteins.[8] In addition, PA's shape and negative charge could affect membrane properties, promoting curvature and vesicle formation.

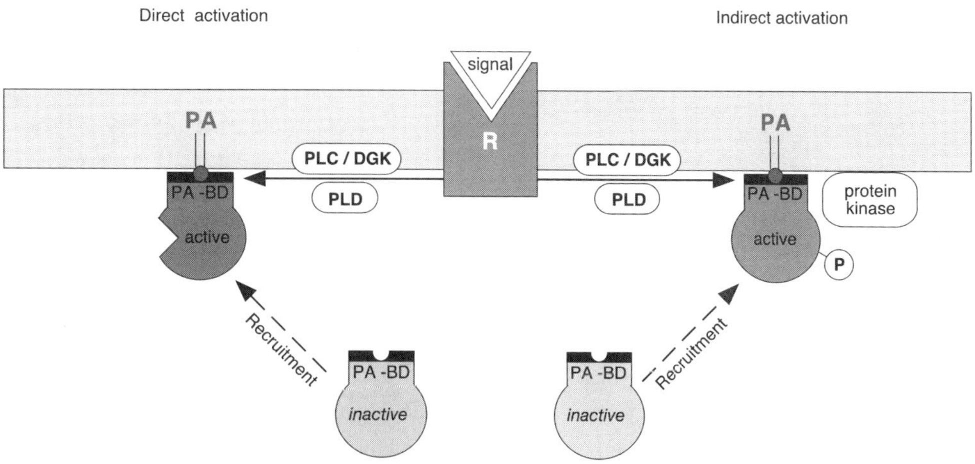

Fig. 2 Possible mechanisms for activation of PA targets. PA is generated locally via PLD and/or PLC/DGK activation. Proteins that are able to bind PA via their PA-binding domain are recruited to the membrane. This can activate them directly by inducing a conformational change, or indirectly, by promoting their activation by other proteins, e.g., a protein kinase. In the latter case, PA acts by increasing the local concentration of target protein(s), bringing it in close vicinity to its activator, thus allowing signaling to occur. (Adapted from Ref. 8, with permission.) (*View this art in color at www.dekker.com.*)

Another major question regards the specificity of the PA signal. Since a variety of environmental conditions induce the formation of PA, one must ask how response specificity is maintained. Of course, first there will be a level of tissue and developmental information that determines the final response to environmental cues, the hard wiring of the cell. And every stress treatment will activate both specific and general signaling pathways. Thus, PA may represent a general, multifunctional stress signal that integrates different pathways, similar to the action of calcium and MAP kinases.[9]

PA Targets

Several mammalian proteins that directly interact with PA have been identified. These so-called PA targets are involved in various cellular processes, such as mitogenic signaling, vesicular trafficing, and the oxidative burst, and include protein kinases and phosphatases, lipid kinases, cAMP-phosphodiesterases, small G-proteins, mTOR, and NADPH oxidase. PA-binding regions in some of them have been identified by deletion studies, but their sequences do not show significant homology to each other.[10]

Few PA targets have been identified in plants. PA affects ion-channel and MAP-kinase activity in vivo, but it is not known whether this is due to direct binding. Phosphoinositide-dependent protein kinase 1 (PDK1) binds PA and a calcium-dependent protein kinase (CDPK) is activated by PA in vitro but clearly more targets must be identified to distinguish a general PA-binding domain and to understand how PA signals stress (Fig. 1).

CONCLUSION AND PERSPECTIVES

PA is a secondary messenger that has all the qualities of a general stress-signaling molecule in plants. It is formed in response to stress conditions and it induces downstream responses typically related to stress, while prevention of PA formation reduces these responses. One of the major challenges will be to find the molecular targets for PA.

Other main goals in this field are to find the upstream regulators of the PA-generating enzymes, and to localize PA in the living cell, using fusion constructs of PA-binding domains with green fluorescent protein. Overexpression of a PA-binding domain can be used to inhibit PA signaling, and knocking out individual PLC and PLD isoforms will establish their contribution to the PA signal under different conditions. These type of approaches will help us to understand what the role of PA is in stress and other responses.

ACKNOWLEDGMENTS

We thank Alan Musgrave and Frank Takken for their help in writing this manuscript.

ARTICLES OF FURTHER INTEREST

Drought and Drought Resistance, p. 386
Lipid Metabolism, p. 659

Osmotic Adjustment and Osmoregulation, p. 850
Oxidative Stress and DNA Modification in Plants, p. 854
Plant Response to Stress: Abscisic Acid Fluxes, p. 973
Plant Response to Stress: Regulation of Plant Gene Expression to Drought, p. 999
Plant Response to Stress: Role of Jasmonate Signal Transduction Pathway, p. 1006
Symbioses with Rhizobia and Mycorrhizal Fungi: Microbe/Plant Interactions and Signals Exchange, p. 1213

REFERENCES

1. Munnik, T.; Irvine, R.F.; Musgrave, A. Phospholipid signalling in plants. Biochim. Biophys. Acta **1998**, *1389*, 222–272.
2. Meijer, H.J.G.; Munnik, T. Phospholipid-based signaling in plants. Annu. Rev. Plant Biol. **2003**, *54*, 256–306.
3. Elias, M.; Potocky, M.; Cvrckova, F.; Zarsky, V.V. Molecular diversity of phospholipase D in angiosperms. BMC Genomics **2002**, *3*, 2.
4. Wang, X. Plant phospholipases. Annu. Rev. Plant Physiol. Plant Mol. Biol. **2001**, *52*, 211–231.
5. Wang, X. Phosholipase D in hormonal and stress signaling. Curr. Opin. Plant Biol. **2002**, *5*, 408–414.
6. Munnik, T.; Musgrave, A. Phospholipid signaling in plants: Holding on to phospholipase D. Sci. STKE **2001**, *111*, PE42.
7. Munnik, T. Phosphatidic acid: An emerging plant lipid second messenger. Trends Plant Sci. **2001**, *6*, 227–233.
8. Laxalt, A.M.; Munnik, T. Phospholipid signaling in plant defence. Curr. Opin. Plant Biol. **2002**, *5*, 332–338.
9. Munnik, T.; Meijer, H.J.G. Osmotic stress activates distinct lipid and MAPK signalling pathways in plants. FEBS Lett. **2001**, *498*, 172–178.
10. Ktistakis, N.T.; Delon, C.; Manifava, M.; Wood, E.; Ganley, I.; Sugass, J.M. Phospholipase D1 and potential targets of its hydrolysis product, phosphatidic acid. Biochem. Soc. Trans. **2003**, *31*, 94–97.

Plant Response to Stress: Regulation of Plant Gene Expression to Drought

Kazuo Shinozaki
RIKEN Tsukuba Institute, Koyadai, Tsukuba, Japan

Kazuko Yamaguchi-Shinozaki
Japan International Research Center for Agricultural Sciences, Ohwashi, Tsukuba, Japan

INTRODUCTION

Plant growth is greatly affected by water deficit. Plants respond and adapt to dehydration in order to survive under drought stress by the induction of various biochemical and physiological responses. Various genes are induced in response to drought at the transcriptional level. Their gene products are thought to function in response to drought and in tolerance to drought. It is important to analyze functions of stress-inducible genes not only to understand higher plants' molecular mechanisms of stress response but also to improve the stress tolerance of transgenic plants. The expression and functions of stress-inducible genes have been studied at the molecular level. Complex mechanisms seem to be involved in gene expression and signal transduction in response to drought stress. This article describes recent progress on *cis*- and *trans*-acting factors involved in drought-inducible gene expression.

EXPRESSION PROFILES OF DROUGHT-INDUCIBLE GENES USING MICORARRAY TECHNOLOGY

Microarray technology has been developed extensively and has become a powerful tool for the global analysis of expression profiles of many genes. This micoroarray technology allows the large-scale comparative analysis of gene expression under various stress conditions. Recently, many drought-inducible genes have been identified in *Arabidopsis*.[1] Microarray analysis has revealed the existence of strong cross talk between drought and high salinity responses, and moderate cross talk between drought and cold stress responses. Various kinds of genes with different functions are involved in stress tolerance and stress response.[1] Among the stress-inducible genes, many transcription factor genes have been identified, suggesting that various transcriptional regulatory machineries function in drought-inducible gene expression.

ABA-DEPENDENT AND ABA-INDEPENDENT GENE EXPRESSION IN RESPONSE TO DROUGHT STRESS

Dehydration triggers the production of abscisic acid (ABA), which in turn, not only causes stomata closure but also induces various genes. As shown in Fig. 1, it is hypothesized that at least four independent signal pathways function in the activation of stress-inducible genes under dehydration conditions: Two are ABA-dependent (pathways I and II) and two are ABA-independent (pathways III and IV).[2] One of the ABA-dependent pathways requires protein biosynthesis (pathway I). ABA-independent regulatory systems are thought to function in the early process of drought-stress signaling before the accumulation of ABA.

ABA-INDEPENDENT GENE EXPRESSION

One of the ABA-independent pathways of drought-stress response overlaps with that of cold-stress response (pathway IV). Promoter analysis of drought-inducible *rd29A* gene reveals that a *cis*-acting element including A/GCCGAC, named the Dehydration Responsive Element (DRE) and C-Repeat (CRT), is essential for regulation of the induction of *rd29A* under drought, low-temperature, and high-salt stress conditions in an ABA-independent pathway.[3] All the DRE/CRT-binding proteins (DREBs and CBFs) contain a conserved DNA-binding motif (AP2/ERF motif).[4,5] These five cDNA clones that encode DRE/CRT-binding proteins are classified into two groups, DREB1/CBF and DREB2. Expression of the *DREB1A* (*CBF3*) gene and its two homologues (*DREB1B* = *CBF1*, *DREB1C* = *CBF2*) is induced by low-temperature stress, whereas expression of the *DREB2A* gene and its single homologue (*DREB2B*) was induced by dehydration. These results indicate that two independent families of DREB proteins, DREB1/CBF and DREB2, function as *trans*-acting factors in two separate signal transduction pathways under low temperature and dehydration conditions,

Encyclopedia of Plant and Crop Science
DOI: 10.1081/E-EPCS 120010660

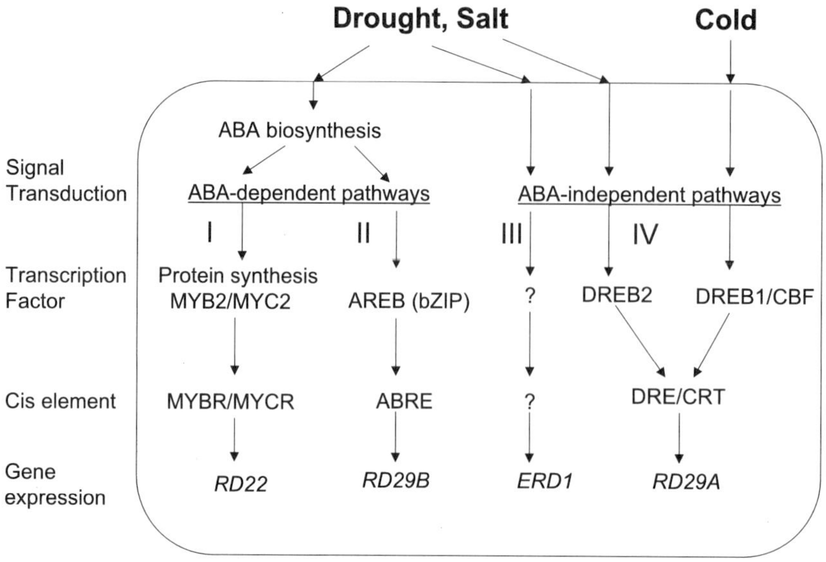

Fig. 1 Signal transduction pathways between initial drought-stress signal and gene expression in *Arabidopsis*. There are at least four signal transduction pathways: Two are ABA-dependent and two are ABA-independent. Stress-inducible genes *RD29A/COR78/LTI78, RD29B/LTI65, RD22*, and *ERD1* are used for the analysis of each signaling process. (From Ref. 2.)

respectively.[5] Overproduction of the DREB1A/CBF3 and CBF1/DREB1B cDNAs in transgenic plants significantly improves stress tolerance to drought and freezing.[5–7] Micorarray analysis identifies many genes with various functions as DREB/CBF target genes.[8,9]

Several drought-inducible genes including *erd1* do not respond to either cold or ABA treatment, which suggests the existence of an additional ABA-independent pathway in dehydration stress response (pathway III). Promoter analysis of these genes will give us more information on pathway III.

ABA-DEPENDENT GENE EXPRESSION

Endogenous ABA increases significantly under drought and high salinity conditions. Many drought-inducible genes are upregulated by exogenous ABA treatment. In one of the ABA-dependent pathways (Fig. 1, pathway II), drought-inducible genes do not require protein biosynthesis for their expression. These dehydration-inducible genes contain potential ABA-responsive elements (ABREs; PyACGTGGC) in their promoter regions. ABRE functions as a *cis*-acting DNA element involved in ABA-regulated gene expression. ABRE was first identified in wheat *Em* and rice *rab* genes, and its DNA-binding protein EmBP1 was shown to encode a bZIP protein.[2] Two ABREs function in the ABA-dependent gene expression of *rd29B*. Two ABRE-binding proteins (AREB1 and AREB2) are involved in ABA-responsive gene expression and contain the bZIP structure.[10] AREB1 and AREB2 genes are induced by drought; their gene products seem to be activated by phosphorylation. AREB1 and AREB2 function downstream from *abi1/abi2* and *era1* in ABA signal transduction pathways. Similar ABRE-binding proteins have been identified and named as ABFs that function in stress tolerance.[11,12]

Biosynthesis of novel protein factors is necessary for the expression of ABA-inducible genes in one of the ABA-dependent pathways (pathway I). The induction of an *Arabidopsis* drought-inducible gene, *rd22*, is mediated by ABA, and requires protein biosynthesis for its ABA-dependent expression. MYC and MYB recognition sequences are essential for the ABA- and drought-responsive expression of *rd22*. Transcription factors—MYB (ATMYB2) and MYC (rd22BP1)—are thought to function in the regulation of slowly stress-inducible genes after the production of ABA-inducible transcription factors (pathway I).[13]

Many drought- and ABA-inducible genes encoding various transcription factors have been reported. These contain conserved DNA-binding motifs, such as zinc finger, homeo box, and WRKY motif. Among them, ATHB6 (containing the homeodomain) functions as a negative regulator downstream of ABI1 protein phosphatase in the ABA signal transduction pathway.[14]

GENETIC ANALYSIS OF DROUGHT-STRESS SIGNAL TRANSDUCTION

Genetic analysis of *Arabidopsis* mutants with the *rd29A* promoter::luciferase transgene reveals complex cross talk among signaling pathways in drought-, salt-, and cold-stress responses, and suggests that stress-signaling pathways, (including ABA-independent and ABA-dependent pathways) are not completely independent.[15] Recently, many genes have been identified using map-based cloning, some of which encode putative transcription factors.[16] Genetic analysis of these mutants is thought to provide more information on drought-stress signal transduction.

CONCLUSION

Many genes are induced by drought stress. Analyses of stress-inducible gene expression have revealed the presence of multiple signal-transduction pathways between the perception of drought-stress signal and gene expression. At least four different transcription systems function in the regulation of dehydration-inducible genes; two are ABA-responsive and two are ABA-independent. This explains the complex stress response observed after the exposure of plants to drought stress. Some genes are rapidly induced by drought stress, whereas others are slowly induced after the accumulation of endogenous ABA. Genetic analysis of Arabidopsis mutants also suggests complex signaling pathways in drought-, salt-, and cold-stress responses. Functional genomics has become a powerful approach to elucidate complex stress-signaling cascades.

ARTICLES OF FURTHER INTEREST

Arabidopsis Transcription Factors: Genome-Wide Comparative Analysis among Eukaryotes, p. 51
Drought and Drought Resistance, p. 386
Plant Response to Stress: Abscisic Acid Fluxes, p. 973
Water Deficits: Development, p. 1284

REFERENCES

1. Seki, M.; Narusaka, M.; Ishida, J.; Nanjo, T.; Fujita, M.; Oono, Y.; Kamiya, A.; Nakajima, M.; Enju, A.; Sakurai, T.; Satou, M.; Akiyama, K.; Taji, T.; Yamaguchi-Shinozaki, K.; Carninci, P.; Kawai, J.; Hayashizaki, Y.; Shinozaki, K. Monitoring the expression profiles of 7000 Arabidopsis genes under drought, cold, and high-salinity stresses using a full-length cDNA microarray. Plant J. **2002**, *31*, 279–292.
2. Shinozaki, K.; Yamaguchi-Shinozaki, K. Molecular responses to dehydration and low temperature: Difference and cross-talk between two stress signaling pathways. Curr. Opin. Plant Biol. **2000**, *3*, 217–223.
3. Yamaguchi-Shinozaki, K.; Shinozaki, K. A novel *cis*-acting element in an Arabidopsis gene is involved in responsiveness to drought, low-temperature, or high-salt stress. Plant Cell **1994**, *6*, 251–264.
4. Stockinger, E.J.; Glimour, S.J.; Thomashow, M.F. *Arabidopsis thaliana* CBF1 encodes an AP2 domain-containing transcription activator that binds to the C-repeat/DRE, a *cis*-acting DNA regulatory element that stimulates transcription in response to low temperature and water deficit. Proc. Natl. Acad. Sci. U.S.A. **1997**, *94*, 1035–1040.
5. Jaglo-Ottosen, K.R.; Gilmour, S.J.; Zarka, D.G.; Schabenberger, O.; Thomashow, M.F. Arabidopsis CBF1 overexpression induces coe genes and enhances freezing tolerance. Science **1998**, *280*, 104–106.
6. Liu, Q.; Sakuma, Y.; Abe, H.; Kasuga, M.; Miura, S.; Yamaguchi-Shinozaki, K.; Shinozaki, K. Two transcription factors, DREB1 and DREB2, with an EREBP/AP2 DNA binding domain, separate two cellular signal transduction pathways in drought- and low temperature-responsive gene expression, respectively, in Arabidopsis. Plant Cell **1998**, *10*, 1406–1491.
7. Kasuga, M.; Liu, Q.; Miura, S.; Yamaguchi-Shinozaki, K.; Shinozaki, K. Improving plant drought, salt, and freezing tolerance by gene transfer of a single stress-inducible transcription factor. Nat. Biotechnol. **1999**, *17*, 287–291.
8. Seki, M.; Narusaka, M.; Abe, H.; Kasuga, M.; Yamaguhci-Shinozaki, K.; Carninci, P.; Hayashizaki, Y.; Shinozaki, K. Monitoring the expression pattern of 1300 Arabidopsis genes under drought and cold stresses by using a full-length cDNA microarray. Plant Cell **2001**, *13*, 61–72.
9. Fowler, S.; Thomashow, M.F. *Arabidopsis* transcriptome profiling indicates that multiple regulatory pathways are activated during cold acclimation in addition to the CBF cold response pathway. Plant Cell **2002**, *14*, 1675–1690.
10. Uno, Y.; Furihata, T.; Abe, H.; Yoshida, R.; Shinozaki, K.; Yamaguchi-Shinozaki, K. Arabidopsis basic leucine zipper transcriptional transcription factors involved in an abscisic acid-dependent signal transduction pathway under drought and high-salinity conditions. Proc. Natl. Acad. Sci. U.S.A. **2000**, *97*, 11632–11637.
11. Choi, H.; Hong, J.H.; Ha, J.; Kang, J.Y.; Kim, S.Y. ABFs, a family of ABA-responsive element binding factors. J. Biol. Chem. **2000**, *275*, 1723–1730.
12. Kang, J.Y.; Choi, H.I.; Im, M.Y.; Kim, S.Y. *Arabidopsis* basic leucine zipper proteins that mediate stress-responsive abscisic acid signaling. Plant Cell **2002**, *14*, 343–357.
13. Abe, H.; Yamaguchi-Shinozaki, K.; Urao, T.; Iwasaki, T.; Hosokawa, D.; Shinozaki, K. Role of Arabidopsis MYC and MYB homologs in drought- and abscisic acid-regulated gene expression. Plant Cell **1997**, *9*, 1859–1868.
14. Himmelbach, A.; Hoffmann, T.; Leube, M.; Hoehener, B.; Grill, E. Homeodomain protein ATHB6 is a target of the protein phosphatase ABI1 and regulates hormone responses in Arabidopsis. EMBO J. **2002**, *21*, 3029–3038.
15. Ishitani, M.; Xiong, L.; Stevenson, B.; Zhu, J.K. Genetic analysis of osmotic and cold stress signal transduction in Arabidopsis: Interactions and convergence of abscisic acid-dependent and abscisic acid-independent pathways. Plant Cell **1997,** (9), 1–16.
16. Xiong, L.; Schumaker, K.S.; Zhu, J.K. Cell signaling during cold, drought, and salt stress. Plant Cell Suppl. **2002**, *14*, S165–S183.

P

Plant Response to Stress: Role of Molecular Chaperones

Barbara Despres
Pierre Goloubinoff
University of Lausanne, Lausanne, Switzerland

INTRODUCTION

Unlike algae, terrestrial plants do not benefit from a well-buffered environment provided by large water bodies. The atmosphere cannot as efficiently attenuate sudden changes in the environment, and land plants—notably desert plants—are exposed to a plethora of combined environmental stresses: heat-shock, water loss, excess light, accumulation of oxygen radicals, osmotic shock, freezing, mechanical stress, etc. Unlike animals, plants have limited motility and cannot seek shelter to escape damage from daily and seasonal variations in the environment. One powerful defense mechanism involves a network of molecular chaperones, which can prevent stress-induced protein aggregation and actively scavenge and refold stable aggregates into functional native proteins.

PROTEIN AGGREGATES AND ACQUIRED STRESS RESISTANCE

Because proteins maintain a delicate equilibrium between structural rigidity and functional flexibility, external stresses may cause them to lose their native conformation and seek alternatively stable inactive structures, called aggregates. Because of their insoluble hydrophobic nature, aggregates may interfere with membrane functions such as photosynthesis and occupy vital cellular space. Some protein aggregates from mammals, the amyloids and prions, cause neural degeneration. Deleterious properties of stress-induced protein aggregates in plants are unknown. Yet the fact that high levels of heat shock proteins (Hsps), which prevent stress-induced aggregation and regulate plant development, are induced in stressed plants strongly suggests that Hsps act to minimize interference with vital cellular processes by protein aggregates.

Most organisms, including plants, have evolved defense mechanisms that can react to various mild doses of stress and build up responses in anticipation of more extreme stresses to come.[1] Although more economical than maintaining various defense mechanisms at all times, acquired resistance is transient by nature. Hence, an abrupt, unannounced stress may not allow enough time to establish proper defense mechanisms, while as time passes, previously induced defense mechanisms may subside and lose their ability to withstand a renewed challenge.

MAJOR MOLECULAR CHAPERONES IN PLANTS AND THEIR ROLES UNDER HEAT STRESS

The effect of heat shock has been extensively studied as a paradigm of other types of environmental stresses. Nearly all organisms, including plants, arrest protein synthesis during mild, nonlethal heat shock, except for a subset of proteins termed heat shock proteins (Hsps), which are produced in large amounts. Induction of heat shock genes in response to heat stress is mediated by transcriptional activator proteins (HSF) that trimerize and bind to cis-regulatory promoter elements (HSEs) conserved in all eukaryotes.[2] Molecular chaperones, many of which are Hsps, belong to several classes of proteins initially classified according to their molecular weight: Hsp110, Hsp100, Hsp90, Hsp70, Hsp60, and small Hsps. Hsp chaperones have also constitutively expressed cognates (Hsc), which are sequence-related and carry housekeeping chaperone functions under nonstressed conditions.

Hsp70

Hsp70 is a highly conserved molecular chaperone present in all living organisms (with the exception of some archeabacteria). In the model plant *Arabidopsis thaliana*, about 14 highly homologous (>45% identity) Hsp70 members are expressed constitutively (Hsc70), or in a stress-dependent manner (Hsp70). The encoded proteins are localized to the cytosol, endoplasmic reticulum (ER), mitochondria, and chloroplasts (Table 1). In addition, there are four members of the Hsp110 family, which is sequence-related to the Hsp70 family. In neuronal cells, Hsp110 prevents stress-induced apoptosis and may regulate the Hsp70/40 chaperone system.

Hsp70 (DnaK in *E. coli*) is composed of a 40-kDa ATPase domain and a 30-kDa protein-binding domain. DnaK collaborates with two co-chaperones, DnaJ (Hsp40)

Encyclopedia of Plant and Crop Science
DOI: 10.1081/E-EPCS 120010643

Table 1 Arabidopsis Hsp70 family

Gene name	TIGR number	Predicted molecular weight (kDa)	Putative localization
DnaK subfamily			
Athsp70-1	At5g02500	71.4	Cytoplasm
Athsp70-2	At5g02490	71.4	Cytoplasm
Athsp70-3	At3g09440	71.1	Cytoplasm
Athsp70-4	At3g12580	71.1	Cytoplasm
Athsp70-5	At1g16030	70.9	Cytoplasm
Athsp70-18	At1g56410	68.3	Cytoplasm
Athsp70-6	At5g49910	76.5	Chloroplast
Athsp70-7	At4g24280	77.0	Chloroplast
Athsp70-8	At2g32120	61.0	Chloroplast
Athsp70-9	At4g37910	71.2	Mitochondria
Athsp70-10	At5g09590	73.0	Mitochondria
Athsp70-11	At5g28540	73.6	ER
Athsp70-12	At5g42020	73.6	ER
Athsp70-13	At1g09080	73.2	ER
Hsp110 subfamily			
Athsp70-14	At1g79920	91.8	Cytoplasm
Athsp70-15	At1g79930	81.8	Cytoplasm
Athsp70-16	At1g11660	85.2	Cytoplasm
Athsp70-17	At4g16660	96.7	ER

(From Ref. 11.)

and GrpE. DnaJ is thought to bind locally to hydrophobic regions in loops of misfolded proteins. DnaJ-bound loops are then transferred to the protein-binding pocket of DnaK, triggering ATP hydrolysis. Release of the bound polypeptide from DnaK requires GrpE, which accelerates nucleotide exchange in DnaK. Reiteration of such local binding/release events likely leads to a gradual disentanglement of the polypeptides from the aggregates and correct refolding of the polypeptide.[3] Hsp70 collaborates with other molecular chaperones and can be considered as the core element of the chaperone network in the cell, toward which are funneled various forms of misfolded proteins for disaggregation (Fig. 1).

The *Arabidopsis* genome expresses a plethora of Hsp40-like proteins, which because of their DnaJ-like domains are suspected to interact with the various members of Hsp70s in the cell and drive specific chaperone reactions. Plant genomes contain several prokaryoticlike GrpE genes, likely in the chloroplasts and mitochondria. Hip (Hsp70 interacting protein) proteins have also been identified in plants and might regulate Hsp70 activity, as in animals, by stabilizing Hsp70 in an ADP-conformation with high affinity for the polypeptidic substrate.[4] In *Arabidopsis* and various other organisms, Hsp70 seems to act as a negative feedback regulator of HSF activity. Indeed, when excess Hsp70 (and other Hsps) is synthetised under heat stress, its interaction with heat shock factors could repress transcription of heat shock genes.[5]

Hsp100

Hsp100 proteins from prokaryotes and eukaryotes are also termed Clp (for *c*aseino*l*ytic *p*rotease) because the first member described was *E.coli* ClpA/P protease, which can hydrolyze casein in vitro. Hsp100/Clp proteins are divided in two major classes: ClpA/B/C/D containing two ATP-binding domains, and ClpM/N/X/Y containing only one ATP-binding domain.[6] Most ClpB proteins are Hsps, which, together with Hsp70, carry the solubilization and refolding of large protein aggregates.[3]

Several Hsp100s have been isolated in different plant species. They are found in the cytosol, in mitochondria, and in chloroplasts. In plants, expression of Hsp100 is developmentally regulated.[7] *Arabidopsis* Hsp100s present four ClpB-like, three ClpA-like, and eight related proteins whose functions cannot be predicted by sequence analysis (Table 2). Plant ClpB-like Hsp101 (At1g74310) plays a crucial role in the acquisition of thermotolerance, but it is dispensable for development and germination in

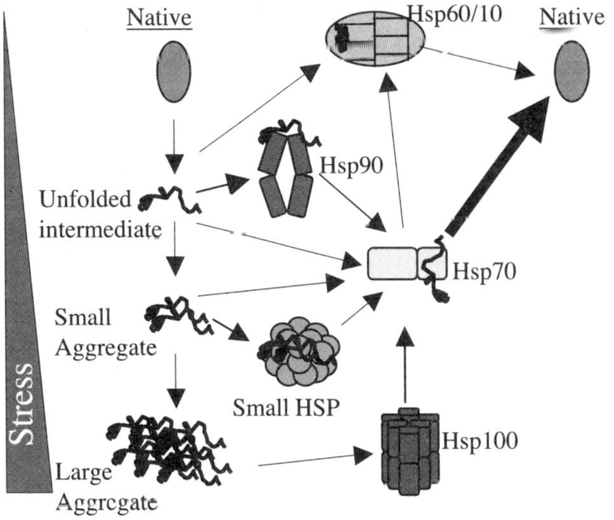

Fig. 1 Functional hierarchy of the chaperone network: Stress unfolds native proteins, which form intermediates that associate into insoluble aggregates. Hsp60, Hsp90, Hsp40, Hsp70, or sHsps can bind small aggregates to prevent further aggregation. Hsp70 is the core of the chaperone network, to which are funneled early intermediates stabilized by Hsp90 or Hsp40, small aggregates stabilized by small Hsps, or large aggregates processed by Hsp100. Following unfolding of misfolded structures by Hsp70, individual polypeptides may refold into the native state. (From Ref. 11.) (*View this art in color at www.dekker.com.*)

Table 2 Arabidopsis Hsp100 and Hsp90 families

Gene name	TIGR number	Predicted molecular weight (kDa)	Putative localization(s)
ClpA related			
C2	At3g48870	93.3	Chloroplast
C1	At5g50920	93.3	Chloroplast
D	At5g51070	94.7	Chloroplast
ClpB related			
Hsp101	At1g74310	101.2	Cytoplasm/nucleus
	At4g14670	92.7	Cytoplasm
	At2g25140	99.5	Mitochondria
	At5g15450	100.5	Chloroplast/mitochondria
Clp related proteins			
	At1g07200	106.3	Cytoplasm
	At2g29970	111.9	Cytoplasm
	At2g40130	51.6	Cytoplasm
	At3g52490	89.9	Microbody/nucleus
	At4g29920	111.8	Nucleus
	At4g30350	98.6	Nucleus
	At5g57130	101	Nucleus
	At5g57710	108.7	Microbody/nucleus
Hsp90			
Athsp90-1	At5g52640	81.1	Cytoplasm/nucleus
Athsp90-2	At5g56030	81.2	Cytoplasm/nucleus
Athsp90-3	At5g56010	81.3	Cytoplasm/nucleus
Athsp90-4	At5g56000	81.4	Cytoplasm/nucleus
Athsp90-5	At2g04030	88.1	Chloroplast
Athsp90-6	At3g07770	89.1	Mitochondria
Athsp90-7	At4g24190	94	ER

(From Ref. 11.)

the absence of stress.[7] It will be interesting to address the protein-refolding and/or proteolytic functions of the many Clp-like genes in *Arabidopsis* and their contribution to various stress resistances.

Small Hsps

In most prokaryotes and eukaryotes, various stresses induce low-molecular-weight (small) 10–30 kDa heat shock proteins (sHsps), which all share a conserved C-terminal domain similar to mammalian α-crystallin. They often form loose, hollowed, oligomeric structures comprising 9 to 32 subunits.

In plants, sHsps are most abundantly expressed under various stresses in the cytosol, endoplasmic reticulum, mitochondria, and chloroplasts. The *Arabidopsis* genome encodes for at least 50 proteins that include a crystallinlike domain. While it is not yet clear which among them are stress-related proteins, about two dozen fall into two groups of cytoplasmic and organellar sHsps, respectively. In the chaperone network, small Hsps complement Hsp100 by presenting aggregates to Hsp70 for disaggregation. sHsps prevent the formation of large aggregates in the first place, and thus ensure optimal interaction with Hsp70.

A remarkable property of the sHsps in vitro, in *Chlamydomonas* chloroplasts and in *Synechocystis*, is their ability to transiently bind membranes during stress and confer membrane stability against thermally or chemically induced hyperfluidity.[8] Thus, sHsps may also provide a mechanism for short-term adaptation of membranes to heat shock.

Hsp60

Hsp60 belongs to the chaperone subfamily named chaperonins. Its prokaryotic homologue, GroEL, mediates polypeptide folding by a well-described cage-sequestering mechanism. Unlike Hsp70 and Hsp100, it cannot act upon large complexes of misfolded protein and cannot solubilize aggregates. Rather, Hsp60 can assist Hsp70-mediated protein disaggregation by accelerating the final refolding steps.[3] GroEL-like Hsp60 can be found in chloroplasts and mitochondria, but not in other compartments of the plant cell, suggesting that it primarily contributes to the stress protection of these organelles.[1]

Hsp90

Hsp90 is a highly conserved ATPase chaperone, very abundant in the cytoplasm of eukaryotic and prokaryotic cells, even under nonstressed conditions. Like many chaperones, Hsp90 can minimize stress-induced protein aggregation and, together with Hsp70 and ATP, assist the correct refolding of stress-induced misfolded proteins. Under nonstress conditions, mammalian Hsp90 acts as the central component of a supercomplex involved in signal transduction via interaction with steroid hormone receptors or protein kinases. Plant cells also contain similar complexes composed of Hsp90, Hsp70, and FKBP-type prolyl-isomerase.[9] Hsp90 controls the morphology and development of complex eukaryotes. In *Arabidopsis* are present four highly homologous cytoplasmic Hsp90 genes (>90%) and one form in each organelle: ER, chloroplasts, and mitochondria (Table 2). Hsp90 is the only chaperone for which there are specific inhibitors (geldanamycin, radicicol) that reveal a central role for Hsp90 in the mechanism of evolution.[10]

CONCLUSION

We describe here a coherent molecular mechanism by which Hsp chaperones can limit damages and actively repair lesions in proteins of plants subjected to rapid and extreme variations of temperature. sHsps may also limit stress damage in membranes. Because chemically modified unrefoldable proteins may occur as a result of other stresses, such as oxidative stress, chaperones may not always act in refolding, but may serve to recognize misfolded proteins that need to be targeted to proteases for degradation.

Chaperones Link Stress with Evolution

The mechanism of evolution implies a balance between accumulation of mutations in a population and their selection by environmental stress. It is thought that protein stabilization by molecular chaperones allows numerous mild mutations to accumulate in unstressed organisms, including plants.[10] During prolonged environmental stresses, the housekeeping and mutant-buffering functions of chaperones are affected by their recruitment for aggregate prevention/repair duties. Thus, environmental stresses may reveal large arrays of new alleles and of corresponding phenotypes, some of which may better fit the new, extreme environmental conditions. Hence, alongside providing efficient short-term protection against environmental stresses, molecular chaperones can also mediate the long-term evolutionary adaptation of plants to ever more extreme and challenging terrestrial climates.

ARTICLES OF FURTHER INTEREST

Plant Response to Stress: Genome Reorganization in Flax, p. 484
Osmotic Adjustment and Osmoregulation, p. 850
Plant Response to Stress: Phosphatidic Acid as a Secondary Messenger, p. 995
Plant Response to Stress: Regulation of Plant Gene Expression to Drought, p. 999

REFERENCES

1. Vierling, E. The roles of heat shock proteins in plants. Annu. Rev. Plant Physiol. Plant Mol. Biol. **1991**, *42*, 579–620.
2. Schöffl, F.; Prändl, R.; Reindl, A. Regulation of the heat-shock response. Plant Physiol. **1998**, *117*, 1135–1141.
3. Ben-Zvi, A.P.; Goloubinoff, P. Mechanisms of disaggregation and refolding of stable protein aggregates by molecular chaperones. J. Struct. Biol. **2001**, *135*, 84–93.
4. Webb, M.A.; Cavaletto, J.M.; Klanrit, P.; Thompson, G.A. Orthologs in *Arabidopsis thaliana* of Hsp70 interacting protein Hip. Cell Stress Chaperones **2001**, *6*, 247–255.
5. Kim, B-H.; Schöffl, F. Interaction between Arabidopsis heat shock transcription factor 1 and 70 kDa heat shock proteins. J. Exp. Bot. **2002**, *53*, 371–375.
6. Schirmer, E.C.; Glover, J.R.; Singer, M.A.; Lindquist, S. HSP100/Clp proteins: A common mechanism explains diverse functions. Trends Biochem. Sci. **1996**, *21*, 289–296.
7. Hong, S.-W.; Vierling, E. Hsp101 is necessary for heat tolerance but dispensable for development and germination in the absence of stress. Plant J. **2001**, *27*, 25–35.
8. Török, Z.; Goloubinoff, P.; Horvath, I.; Tsvetkova, N.M.; Glatz, A.; Balogh, G.; Varvasovszki, V.; Los, D.A.; Vierling, E.; Crowe, J.H.; Vigh, L. *Synechocystis* HSP17 is an amphitropic protein that stabilizes heat-stressed membranes and binds denatured proteins for subsequent chaperone-mediated refolding. Proc. Natl. Acad. Sci. **2001**, *98*, 3098–3103.
9. Miernick, J.A. Protein folding in the plant cell. Plant Physiol. **1999**, *121*, 695–703.
10. Queitsch, C.; Sangster, T.A.; Lindquist, S. Hsp90 as a capacitor of phenotypic variation. Nature **2002**, *417*, 618–624.
11. Despres, B.; Goloubinoff, P.

Plant Response to Stress: Role of the Jasmonate Signal Transduction Pathway

Mirna Atallah
Johan Memelink
Leiden University, Wassenaarseweg, Leiden, The Netherlands

INTRODUCTION

Plants differentially activate distinct defense pathways in response to stress. Depending on the type of stress, plants synthesize the signaling molecules jasmonic acid (JA), salicylic acid (SA), or ethylene, which regulate the defense response.

Jasmonates (JAs) are fatty acid derivatives synthesized via the octadecanoid (ODA) pathway. They play pivotal roles in wound and defense responses, and in anther and pollen development. The defense JA pathway comprises several signal transduction events: the perception of the primary stress stimulus and transduction of the signal locally and systemically; the perception of this signal and induction of JA biosynthesis; the perception of JA and expression of responsive genes; and finally, integration of JA signaling with outputs from other signaling pathways.

STRESS-INDUCED JA BIOSYNTHESIS

How stress signals affect JA biosynthesis is largely unknown. In *Catharanthus roseus* cells, elicitor-induced JA biosynthesis depends on an increase in cytoplasmic Ca^{2+} concentration and protein phosphorylation[2] (Fig. 1). In tobacco, wound-induced JA biosynthesis depends on the mitogen-activated protein kinase WIPK.[1]

More is known about the JA biosynthetic pathway itself.[1] The biosynthesis of JAs, which include the biologically active intermediates in the ODA pathway and derivatives of jasmonic acid, begins in the plastids with phospholipase (PL)-mediated release of α-linolenic acid (LA) from membrane lipids.[1] LA is then converted by lipoxygenase (LOX), allene oxide synthase (AOS), and allene oxide cyclase (AOC) into the intermediate 12-oxo-phytodienoic acid (OPDA). This compound is converted in the peroxisomes into JA by OPDA reductase 3 (OPR3), and by three rounds of β-oxidation. JA can be methylated in the cytoplasm to its volatile derivative methyl-jasmonate (MeJA) by *S*-adenosyl-L-methionine: jasmonic acid carboxyl methyltransferase (JMT) (Fig. 2).

Wounding induces the expression of several JA biosynthesis genes. Therefore, one possible mechanism for stress-induced JA biosynthesis is de novo synthesis of biosynthetic enzymes. In addition, the expression of JA biosynthesis genes is induced by JAs themselves, indicating that JA signaling is amplified by a positive feedback mechanism.

Several mutants in *Arabidopsis thaliana* affected in JA biosynthesis have been isolated.[1] The *fad3-2fad7-2fad8* triple mutant, lacking the fatty acid desaturases necessary to synthesize the JA precursor linolenate, contains negligible amounts of LA and JAs. The *opr3* mutant (also known as *dde1*: *delayed dehiscence1*) lacks the OPDA reductase isoform required for JA biosynthesis, but accumulates OPDA when wounded.

JA SIGNAL TRANSDUCTION

How JAs are perceived by plant cells is unknown. The mechanisms whereby JA signaling triggers gene expression are just starting to be elucidated. A JA- and elicitor-responsive element (JERE) in the promoter of the terpenoid indole alkaloid (TIA) biosynthetic gene *Strictosidine synthase* (*Str*) from *C. roseus* interacts with two transcription factors called Octadecanoid-Responsive *Catharanthus* APΣΓΣLA2/Ethylene Response Factor (AP2/ERF)-domain proteins (ORCAs).[3,4] ORCA2 was isolated by yeast one hybrid screening using the JERE as bait[3] and ORCA3 was isolated by a genetic T-DNA activation tagging approach.[2] Both belong to the AP2/ERF family of transcription factors, which are unique to plants and are characterized by the AP2/ERF DNA-binding domain.

Significantly, ORCA gene expression is rapidly induced by (Me)JA. In addition, evidence suggests that JA activates preexisting ORCA transcription factors by inducing a posttranslational modification, for example, phosphorylation.[4] Activated ORCA proteins may autoregulate ORCA gene expression as well as regulating TIA biosynthetic gene expression. Alternatively, JA-induced ORCA gene expression can occur via a transcriptional cascade, including a yet unidentified transcription-activating factor

Encyclopedia of Plant and Crop Science
DOI: 10.1081/E-EPCS 120010658

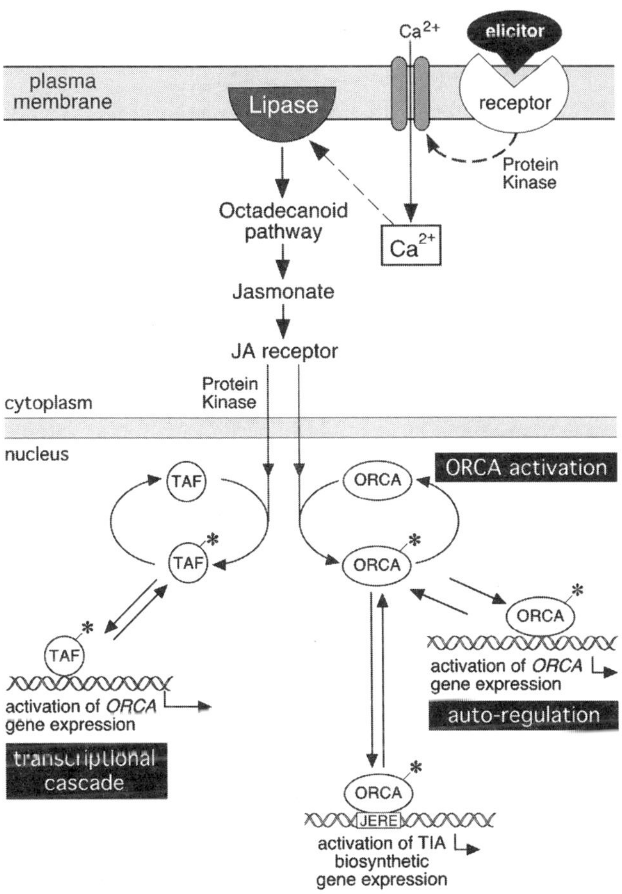

Fig. 1 Model for elicitor signal transduction leading to TIA biosynthetic gene expression in *C. roseus*.

(TAF), which is activated via posttranslational modification (Fig. 1).

In *Arabidopsis*, the AP2/ERF-domain transcription factor ETHYLENE RESPONSE FACTOR 1 (ERF1) was shown to be involved in JA signal transduction as well as in ethylene signaling.[5] Constitutive expression of ERF1 leads to increased expression levels of defense-related genes that are synergistically induced by a combination of ethylene and JA (including *PDF1.2*), and confers resistance to several necrotrophic fungi.[5] Therefore, it appears that *Arabidopsis* also uses a subset of its 124 AP2/ERF-domain transcription factors (including ERF1) to regulate JA-responsive gene expression.

Several JA-insensitive *Arabidopsis* mutants have been found by screening for a reduction in the inhibition of root growth caused by MeJA or by the bacterial toxin coronatine, a structural analogue of JA and OPDA.[1] The *coronatine-insensitive 1* (*coi1*) mutant is affected in a gene encoding a protein with 16 leucine-rich repeats and an F-box motif. The COI1 F-box protein associates with Skp1-like proteins (S) and cullin (C) to form SCF^{COI1} ubiquitin-ligase complexes.[6] F-box proteins are the components of SCF complexes, which recognize substrate proteins and target them for degradation via the ubiquitin-proteasome pathway. Therefore, COI1 seems to recruit one or more repressors of JA responses for degradation.[1,6]

The JA-insensitive mutant *mpk4* was identified by its dwarf phenotype, and is affected in the gene encoding the mitogen-activated protein kinase 4.[7]

JA RESPONSES

A key role for JAs in defense of tomato against insect herbivores and microbial pathogens was proposed in 1992 by Farmer and Ryan, who showed that intermediates and end products of the octadecanoid pathway (but not other closely related lipids) induce proteinase inhibitors that deter insect feeding.[1]

JA is the physiological signal for several wound- and pathogen-induced responses in plants, and it is essential for pollen development in *Arabidopsis*.[1] Exogenously applied (Me)JA results in major reprogramming of gene expression, including defense-related genes that are activated by wounding and pathogen attack. The JA-responsive

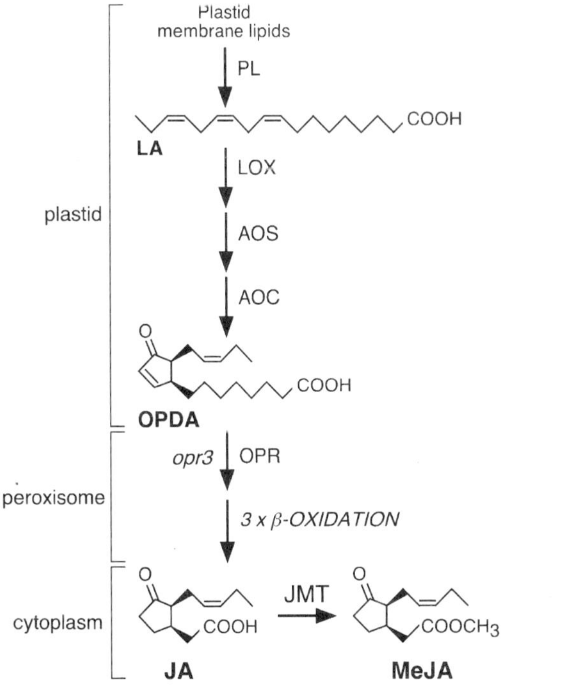

Fig. 2 Schematic representation of the JA biosynthetic pathway. A mutant blocked in a biosynthesis step is in italics.

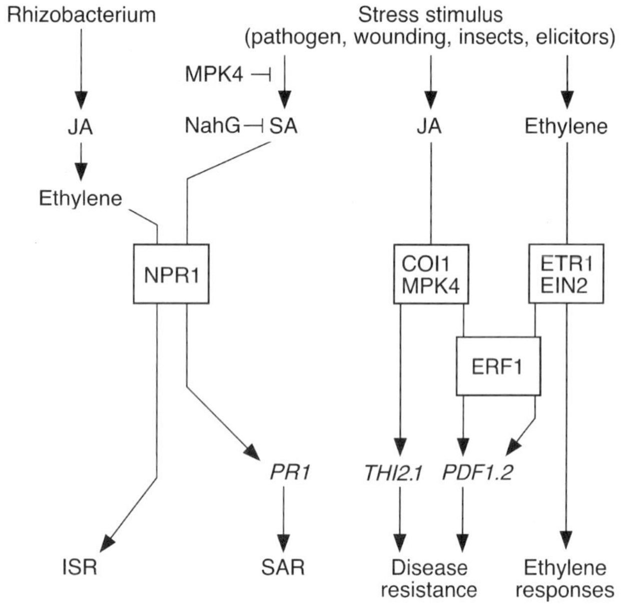

Fig. 3 Model showing signaling in stress responses in *Arabidopsis*.

PDF1.2 and *THI2.1* genes encode antimicrobial plant defensin and thionin proteins, respectively.[1] JAs also induce the expression of biosynthesis genes leading to the accumulation of antimicrobial secondary metabolites, including alkaloids, terpenoids, flavonoids, anthraquinones, and glucosinolates in different plant species.[2,8]

Arabidopsis mutants defective in JA biosynthesis or perception are deficient in certain defense responses and are male-sterile. The *coi1* mutant is defective in its resistance to insects and certain pathogens, and fails to express JA-regulated genes, including *PDF1.2* and *THI2.1*.[1] A single amino acid substitution in COI1, which disrupts SCF^{COI1} complex formation, results in loss of the JA response.[6] The *mpk4* mutant is blocked in the induction of JA-inducible *PDF1.2* and *THI2.1* genes and has reduced fertility[7] (Fig. 3). The *fad* triple mutant shows enhanced sensitivity to the fungus *Pythium irregulare* and the dipteran insect *Bradysia impatiens*.[1] The *opr3* mutant is male-sterile, indicating that pollen development uniquely requires JA and not OPDA. Fertility is restored in the *opr3* mutant and the *fad* triple mutant by application of JA. OPDA plays a major role as a stress signal, because its synthesis in the *opr3* mutant is sufficient to trigger a defense response against *B. impatiens* and the fungus *Alternaria brassicicola*.[9]

JAs play an important role in ISR, a form of induced systemic resistance elicited by nonpathogenic strains of the root-colonizing bacterium *Pseudomonas fluorescens*[10] (Fig. 3).

CROSS TALK BETWEEN DEFENSE SIGNALING PATHWAYS

The JA signaling pathway interacts cooperatively and antagonistically with the ethylene and SA pathways in a variety of responses, leading to fine-tuning of the complex defense response. Together with JA, ethylene plays a crucial role in defense against necrotrophic microbes, in expression of *PDF1.2* and other defense genes, and in ISR (Fig. 3). Mutants affected in ethylene signal transduction (including the ethylene receptor mutant *ethylene-resistant1* (*etr1*) and the *ethylene-insensitive2* (*ein2*) mutant) also have reduced expression of certain JA-responsive genes,[5] are more susceptible to certain microbial pathogens, and cannot mount ISR (Fig. 3). A subset of AP2/ERF-domain transcription factors (including Arabidopsis ERF1)[5] may serve as the platform to integrate the input from the JA and ethylene signaling pathways (Fig. 3).

Systemic acquired resistance (SAR) is a defense response in which, in contrast to ISR, SA is the key regulatory signal. Transgenic *Arabidopsis* NahG plants expressing the bacterial SA-degrading enzyme salicylate hydroxylase cannot mount SAR. SAR provides protection in uninfected plant parts against pathogens and is correlated with the expression of pathogenesis-related (PR) proteins with antimicrobial activity (Fig. 3). The NPR1 (Nonexpressor of PR genes 1) protein has a dual role in systemic resistance mechanisms mediated by either SA (SAR) or JA and ethylene (ISR)[1,10] (Fig. 3). The *mpk4* mutant, blocked in JA signaling, exhibits elevated levels of SA and constitutive SAR.[7]

CONCLUSION

The role of JAs in development, defense responses, and gene expression is currently being delineated through the analysis of additional gain-of-function and loss-of-function *Arabidopsis* mutants,[1,11] and through the analysis of JA-responsive promoters and transcription factors. Future work will focus on the regulation of JA synthesis, the identification of JA receptors, the identification of JA-responsive transcription factors in different plant species and of other signal transduction steps that regulate transcription factor activity, and on the mechanisms of cross talk between different defense signaling pathways.

ARTICLES OF FURTHER INTEREST

Plant Defenses Against Insects: Constitutive and Induced Chemical Defenses, p. 939

Plant Response to Stress: Phosphatidic Acid As a Secondary Messenger, p. 995
Plant Response to Stress: Regulation of Plant Gene Expression to Drought, p. 999

REFERENCES

1. Turner, J.G.; Ellis, C.; Devoto, A. The jasmonate signal pathway. Plant Cell, Suppl. **2002**, *14*, S153–S164.
2. Memelink, J.; Verpoorte, R.; Kijne, J.W. ORCAnization of jasmonate-responsive gene expression in alkaloid metabolism. Trends Plant Sci. **2001**, *6*, 212–219.
3. Menke, F.L.H.; Champion, A.; Kijne, J.W.; Memelink, J. A novel jasmonate- and elicitor-responsive element in the periwinkle secondary metabolite biosynthetic gene *Str* interacts with a jasmonate- and elicitor-inducible AP2-domain transcription factor, ORCA2. EMBO J. **1999**, *18*, 4455–4463.
4. van der Fits, L.; Memelink, J. The jasmonate-inducible AP2/ERF-domain transcription factor ORCA3 activates gene expression via interaction with a jasmonate-responsive promoter element. Plant J. **2001**, *25*, 43–53.
5. Lorenzo, O.; Piqueras, R.; Sanchez-Serrano, J.J.; Solano, R. ETHYLENE RESPONSE FACTOR1 integrates signals from ethylene and jasmonate pathways in plant defense. Plant Cell **2003**, *15*, 165–178.
6. Xu, L.; Liu, F.; Lechner, E.; Genschik, P.; Crosby, W.L.; Ma, H.; Peng, W.; Huang, D.; Xie, D. The SCF^{COI1} ubiquitin-ligase complexes are required for jasmonate response in Arabidopsis. Plant Cell **2002**, *14*, 1919–1935.
7. Petersen, M.; Brodersen, P.; Naested, H.; Andreasson, E.; Lindhart, U.; Johansen, B.; Nielsen, H.B.; Lacy, M.; Austin, M.J.; Parker, J.E.; Sharma, S.B.; Klessig, D.F.; Martienssen, R.; Mattsson, O.; Jensen, A.B.; Mundy, J. Arabidopsis MAP kinase 4 negatively regulates systemic acquired resistance. Cell **2000**, *103*, 1111–1120.
8. Blechert, S.; Brodschelm, W.; Holder, S.; Kammerer, L.; Kutchan, T.M.; Mueller, M.J.; Xia, Z.Q.; Zenk, M.H. The octadecanoid pathway: Signal molecules for the regulation of secondary pathways. Proc. Natl. Acad. Sci. U. S. A. **1995**, *92*, 4099–4105.
9. Stintzi, A.; Weber, H.; Reymond, P.; Browse, J.; Farmer, E.E. Plant defense in the absence of jasmonic acid: The role of cyclopentenones. Proc. Natl. Acad. Sci. U. S. A. **2001**, *98*, 12317–12319.
10. Pieterse, C.M.J.; van Loon, L.C. Salicylic acid-independent plant defense pathways. Trends Plant Sci. **1999**, *4*, 52–58.
11. Berger, S. Jasmonate-related mutants of Arabidopsis as tools for studying stress signaling. Planta **2002**, *214*, 497–504.

Plant Response to Stress: Source-Sink Regulation by Stress

Thomas Roitsch
Universität Würzburg, Würzburg, Germany

INTRODUCTION

Plants are challenged by a variety of biotic and abiotic stress factors. Due to their sessile form of life, higher plants have evolved an enormous metabolic flexibility to cope with adverse stimuli by substantial changes in primary and secondary metabolism. One key target of plant metabolism is the partitioning of assimilates between the autotrophic source tissues and the heterotrophic sink tissues. This results in profound effects on many aspects of growth and development, thereby also affecting crop yield. This article reviews the current knowledge of the underlying regulatory mechanism and discusses possible agricultural and biotechnological implications.

CARBOHYDRATE PARTITIONING IN HIGHER PLANTS

Higher plants are physiological mosaics of source tissues. Their mature leaves, e.g., export carbohydrates to photosynthetically less active or inactive sink tissues such as roots, fruits, and tubers, characterized by net import of sugars. The supply of the transport sugar sucrose is a limiting step for the metabolism and growth of sink tissues. By generating a sucrose gradient to support the unloading of sucrose from phloem, the sucrose-cleaving enzymes invertase and sucrose-synthase are important determinants of sink capacity.[1] Thus they are critical links between photosynthate production in source leaves and growth capacity of sink organs.[2,3] Carbohydrate partitioning between source and sink tissues—competing for a common pool of carbohydrates—is a highly dynamic process that accompanies all stages of the growth and development of higher plants and is influenced by environmental stimuli.

EFFECT OF PATHOGENS ON SOURCE-SINK RELATIONS

Plants may be challenged by a variety of biotic stress factors such as pests and pathogens that result in substantial harvest losses. Fungi and bacteria may act as biotrophic symbionts that utilize carbohydrates synthesized by the host plant. Fungi in particular have evolved sophisticated structures and mechanisms to create an additional strong sink that drains assimilates from the infected plant. In contrast, nectrotrophic pathogens (which may include fungi, bacteria, and viruses) grow on the biomass of the infected plant and ultimately result in the death of the host organism.

Activation of defense responses upon pathogen infection is usually accompanied by a fast induction of sink metabolism, possibly to satisfy the increased demand for carbohydrates as an energy source to activate the cascade of direct defense responses and further mediate physiological adaptations (Fig. 1). The source metabolism is generally inversely regulated.[4] Two mechanisms may account for the observed suppression of photosynthesis in response to pathogens: 1) the direct effect of the pathogen on the expression of photosynthetic genes; and 2) the indirect effect due to induction of sink metabolism mediated via sugar repression of photosynthetic genes (Fig. 1). In addition to the localized source-sink transition, the pathogen infection will also affect whole plant carbon partitioning by creating a new sink that competes with other sinks for the common pool of carbohydrates. During the initial defense response, transient effects on source-sink relations are observed. In addition, biotrophic pathogens will establish persistent changes in assimilate allocations by the permanent establishment of a new sink.[5] The pathogen-induced modulation of the carbohydrate status will also affect the susceptibilty of the plant, which is evident from the phenomenon of high sugar resistance, the finding that various key pathogenesis-related genes are sugar inducible, and the finding that overexpression of a yeast invertase in the plant apoplast results in increased resistance to virus infection.[6] The effects of a systemic viral infection on photosynthesis, carbohydrate accumulation, and assimilate partitioning seem to be mediated by the viral movement proteins.[7]

REGULATION OF SOURCE-SINK RELATIONS BY ABIOTIC STRESS

Plants are exposed to a wide range of different abiotic stimuli including irradiance, ozone, temperature, osmotic

Encyclopedia of Plant and Crop Science
DOI: 10.1081/E-EPCS 120010665

factors, wind, water deficiency and mechanical wounding. Drought and salinity are the two major abiotic stresses that limit plant productivity. Although these two stresses are clearly different in their physical nature, they activate some common reactions in plants.[2] A high root:shoot ratio is an important adaptive response to drought and salinity that, by alleviating stress, permits the plant to recover functional equilibrium.[8] This dry matter redistribution is closely associated with carbohydrate allocation to the roots. Therefore, modulation of the processes involved in carbon metabolism and energy production seems to be a promising approach to the engineering of plants with greater adaptability to water and saline stress.

The photoassimilates produced under salt stress are used to support crucial, mutual processes such as growth and osmotic adjustment. The competition of sink organs for the limited carbon supplies under salinity significantly affects overall plant growth, dry matter distribution, and crop yield. As a consequence, the different growth responses to salinity can be interpreted as resulting from changes in the allocation and partitioning of photoassimilates. In general, salinity causes a reduction in sink enzyme activities, leading to an increase in sucrose in source leaves and a decrease in photosynthesis rate by feedback inhibition. Extracellular invertase activity seems to be particularly involved in salt stress response: The enzyme activity correlates with the redistribution of dry matter and is much higher in the roots of salt-tolerant species. An increase in extracellular invertase activity in transgenic plants improves salt tolerance.[2]

Water deficits in plants lead to physiological modifications such as photosynthesis reduction and osmolyte biosynthesis.[9] Large alterations in source sink relations with source limitations are induced that result in decreased export of assimilates and decreased crop load. It has been shown that an effect of water stress on phloem unloading (via a decline in invertase activity) correlates with male sterility in cereals and blocks pollination and early kernel development in maize.[7]

INTERACTIONS AND SIGNAL INTEGRATION

Under natural conditions plants are simultaneously affected by a variety of both biotic and abiotic stress-related stimuli. Although this fact is neglected when the effect of a single stress is analysed (as is usual), a few studies indicate that naturally occurring multistress situations will differentially affect source-sink relations. For example, it has been shown that elevated levels of CO_2 modulate the effect of powdery mildew infection on carbon partitioning in barley and ameliorate ozone effects on biomass and leaf area.[7]

It is becoming evident that complex regulatory networks operate in plants that link endogenous and exogenous stimuli.[10] The finding that specific subsets of mitogen-activated protein (MAP) kinases are activated in response to different stimuli could provide a molecular mechanism suggesting how diverse signals may be integrated to result in the observed coordinated

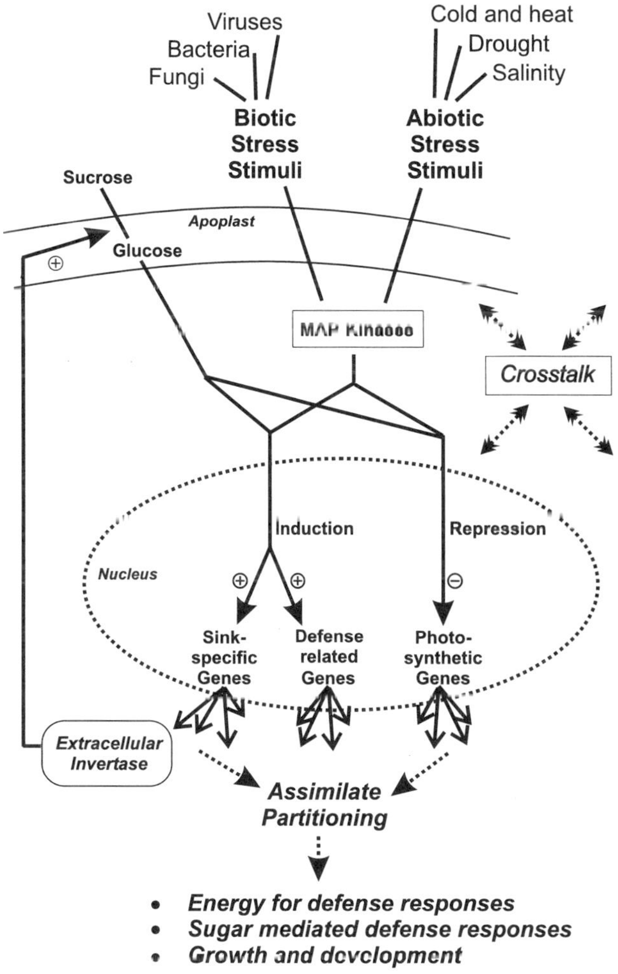

Fig. 1 Model for the inverse regulation of sink metabolism/defense responses and photosynthesis by stress-related stimuli. Abiotic and biotic stress-related stimuli activate different signal transduction pathways that are ultimately integrated to coordinately regulate gene expression. Activation of different subsets of MAP-kinases provides a mechanism to integrate different signals, and results in coordinated and signal-specific gene regulation. Any signal that upregulates extracellular invertase will be amplified and maintained via the sugar-induced expression of this enzyme as outlined in Roitsch et al. (From Ref. 4.)

responses, while simultaneously maintaining the possibility for unique signal-specific downstream effects (Fig. 1; Refs. 11,12).

EXTRACELLULAR INVERTASE: KEY METABOLIC ENZYME AND MODULATOR OF STRESS RESPONSES

Supplying carbohydrates to sink tissues via an apoplastic pathway involves the release of the transport sugar sucrose into the apoplast by a sucrose transporter, cleavage of the disaccharide by an extracellular invertase, and uptake of the hexose monomers by monosaccharide transporters. Experimental data demonstrate not only that the extracellular invertase is crucial in supplying carbohydrates to symplastically isolated sink tissues and actively growing tissues, but that this sucrose-cleaving enzyme also plays a crucial role in mediating stress responses. As outlined by Roitsch et al.,[3] the function of extracellular invertase in stress responses includes the regulation of source-sink transition and the integration of signals that regulate source-sink relations and defense responses. In addition, the sugar inducible expression of extracellular invertase and defense-related genes provides a feed-forward mechanism for maintaining or amplifying the effect of stress-related stimuli: Upregulation of extracellular invertase by any stimulus enhances sink strength and thus results in an elevated sugar concentration. This metabolic signal further induces extracellular invertase for signal amplification and also results in induction of defense-related genes as depicted in Fig. 1.

CONCLUSION

Evidence is accumulating that both biotic and abiotic stress-related stimuli are important exogenous factors that affect source-sink relations (Fig. 1). This interaction modulates plant growth and development and harvest yield, and thus has profound effects on many aspects of the plant life cycle, and also interferes with productivity in agriculture.

Determination of photosynthetic parameters by in vivo fluorescence imaging[13] could be a valuable noninvasive technique to detect effects of pathogen infection on source-sink relations at early stages, for practical applications and to elucidate the underlying regulatory mechanisms in planta.[12] It will be important to further study the effect of the naturally occurring interactions of various signals to determine possible additive, synergistic and compensating effects between stress-related stimuli and endogenous factors on source-sink relations.

Only scattered information is available about the intracellular transduction of stress-related signals and molecular and cellular mechanisms that contribute to the physiological responses. Numerous attempts in many laboratories to modify carbon fluxes by modifying individual enzymatic steps have been essentially unsuccessful.[6] The challenge remains therefore to unravel the underlying sophisticated network of highly flexible regulatory circuits (with complex interactions and crosstalk) for insight into source-sink regulation at the molecular level.[10] This will allow the predictable genetic engineering of resistant plants by manipulating signal transduction pathways to increase the partitioning of fixed carbon into harvestable sinks despite the ubiquitous presence of biotic and abiotic stresses.

ACKNOWLEDGMENTS

Due to space limitations, the author apologizes for not being able to cite the many important original publications that have contributed to rapid progress in the field of plant response to stress. The great effort of all past and current co-workers, stimulating discussions in weekly Laborseminars, and helpful comments on the manuscript by M. Hofmann and S. Berger are gratefully acknowledged.

ARTICLES OF FURTHER INTEREST

Drought and Drought Resistance, p. 386
Metabolism, Primary: Engineering Pathways of, p. 714
Photosynthate Partitioning and Transport, p. 897
Plant Response to Stress: Regulation of Plant Gene Expression to Drought, p. 999
Water Deficits: Development, p. 1284

REFERENCES

1. Farrar, J. Regulation of shoot: Root ratio is mediated by sucrose. Plant Soil **1996**, *185*, 13–19.
2. Balibrea, M.E.; Dell'Amico, J.; Bolarín, M.C.; Pérez-Alfocea, F. Carbon partitioning and sucrose metabolism in tomato plants growing under salinity. Physiol. Plant **2000**, *110*, 503–511.
3. Roitsch, T.; Ehneß, R.; Goetz, M.; Hause, B.; Hofmann, M.; Sinha, A.K. Regulation and function of extracellular

invertase from higher plants in relation to assimilate partitioning, stress responses and sugar signalling. Aust. J. Plant Physiol. **2000**, *27*, 815–825.
4. Ehness, R.; Ecker, M.; Godt, D.E.; Roitsch, T. Glucose and stress independently regulate source/sink relations and defense mechanisms via signal transduction pathways involving protein phosphorylation. Plant Cell **1997**, *9*, 1825–1841.
5. Hall, J.L.; Williams, L.E. Assimilate transport and partitioning in fungal biotrophic interactions. Aust. Plant Physiol. **2000**, *27*, 549–560.
6. Herbers, K.; Sonnewald, U. Altered gene expression brought about inter- and intracellulary formed hexoses and its possible implications for plant-pathogen interactions. J. Plant Res. **1998**, *111*, 323–328.
7. Roitsch, T. Source-sink regulation by sugars and stress. Curr. Opin. Plant Biol. **1999**, *2*, 198–206.
8. Geiger, D.R.; Koch, K.E.; Shieh, W.J. Effect of environmental factors on whole plant assimilate partitioning and associated gene expression. Exp. Bot. **1996**, *47*, 1229–1238.
9. Bohnert, H.J.; Nelson, D.E.; Jense, R.G. Adaptations to environmental stresses. Plant Cell **1995**, *7*, 1099–1111.
10. Knight, H.; Knight, M.R. Abiotic stress signalling pathways: Specificity and cross-talk. Trends Plant Sci. **2001**, *6*, 262–267.
11. Zwerger, K.; Hirt, H. Recent advances in plant MAP kinase signalling. Biol. Chem. **2001**, *382*, 1123–1131.
12. Link, V.L.; Hofmann, M.G.; Sinha, A.K.; Ehness, R.; Strnad, M.; Roitsch, T. Biochemical evidence for the activation of distinct subsets of mitogen activated protein kinases by voltage and defense related stimuli. Plant Physiol. **2002**, *128*, 271–281.
13. Lichtenthaler, H.K.; Miehé, J.A. Technical focus: Fluorescence imaging as a diagnostic tool for plant stress. Trends Plant Sci. **1997**, *2*, 285–323.

Plant Responses to Stress: Nematode Infection

Godelieve Gheysen
Ghent University, Ghent, Belgium

INTRODUCTION

Plant-parasitic nematodes are cosmopolitan pests that have evolved various feeding strategies to obtain nutrients from their host plants. These nematodes possess a stylet capable of penetrating plant cell walls and are all obligate biotrophs, that is, they need to feed on living cytoplasm. Some nematodes do this by just withdrawing the plant cell contents, usually causing cell death, after which they move to another cell. In contrast to this "hit-and-run" strategy, other types of nematodes carefully puncture selected plant cells and transform them into elaborate feeding cells. According to their lifestyle, nematodes are considered migratory or sedentary, and endo- or ectoparasites. Most research to date has been focused on root-parasitic nematodes.[1] Here, we will focus on the sedentary endoparasitic species from the family Heteroderidae: the cyst nematodes (*Globodera* and *Heterodera*) and the root-knot nematodes (*Meloidogyne*). The infective stage is the motile second-stage juvenile that penetrates the plant root, usually close to the tip, and moves to the vascular cylinder. Upon initiation of feeding, the nematode loses motility and becomes sedentary. These are the most damaging plant-parasitic nematodes, and they evoke complicated cellular responses in the infected roots.

SYMPTOMS OF NEMATODE-INFECTED PLANTS

Root-parasitic nematodes are called hidden enemies because the above-ground symptoms are not often visible before major damage has been done. Endoparasitic nematodes either destroy root tissues (migratory species) or distort the vascular tissue (sedentary species). In addition, heavily infected plants usually have a less developed root system than normal. An obvious effect of deficient water transport is wilting of the plants in dry weather conditions. Heavily infected plants are stunted and exhibit symptoms of nutritional deficiencies. Lower infection levels often stay undetected because the resulting symptoms are more subtle: lower yields due to a decreased water transport, a reduction in photosynthesis, and nutrient removal by the nematodes.

Although plant-parasitic nematodes are microscopically small and usually inside the roots, their effects can often be seen on the outside (Fig. 1). Root-knot nematodes are named after the characteristic root knots or galls they induce on susceptible plants. In severe cases, the roots become stunted and completely deformed. Mature cyst nematode females transform into cysts that protect hundreds of eggs against adverse environmental conditions. Depending on the species, these cysts are visible as golden to brown globular or lemon-shaped structures loosely attached to the root. The induction of additional lateral roots is another typical symptom seen after infection with certain nematodes (such as *Heterodera avenae*, *H. schachtii*, *Meloidogyne hapla*), which gives a bearded appearance to taproots.

WOUND AND DEFENSE RESPONSES OF INFECTED PLANTS

Nematode infection always causes wounding of plant tissues, especially during the migration step. In the case of migratory nematodes, this leads to massive necrosis throughout the infection process. Root-knot nematodes circumvent strong plant responses during migration because they move between the cells by separating them at the middle lamella.[2] The juveniles usually move down in the cortex to the root tip, where they turn 180° to migrate up in the vascular cylinder. At this point, the root meristem can be severely damaged if several juveniles pass through the same tip and try to find their way up. Cyst nematodes take a shorter route to the vascular tissue: they go straight through the cells toward the center of the root, causing strong necrosis from their point of entry to the place where they settle down.[2]

Plants are usually able to recognize and react to these parasites by switching on defense responses.[3] When the response is too weak or too late, a successful infection (compatible interaction) will result. A rapid and strong defense response (e.g., due to the presence of a resistance gene) will result in a resistant (incompatible) interaction. An incompatible interaction is often characterized by a hypersensitive reaction (cell death and necrosis), that impedes induction of feeding cells or further migration of the nematode.

Encyclopedia of Plant and Crop Science
DOI: 10.1081/E-EPCS 120010391

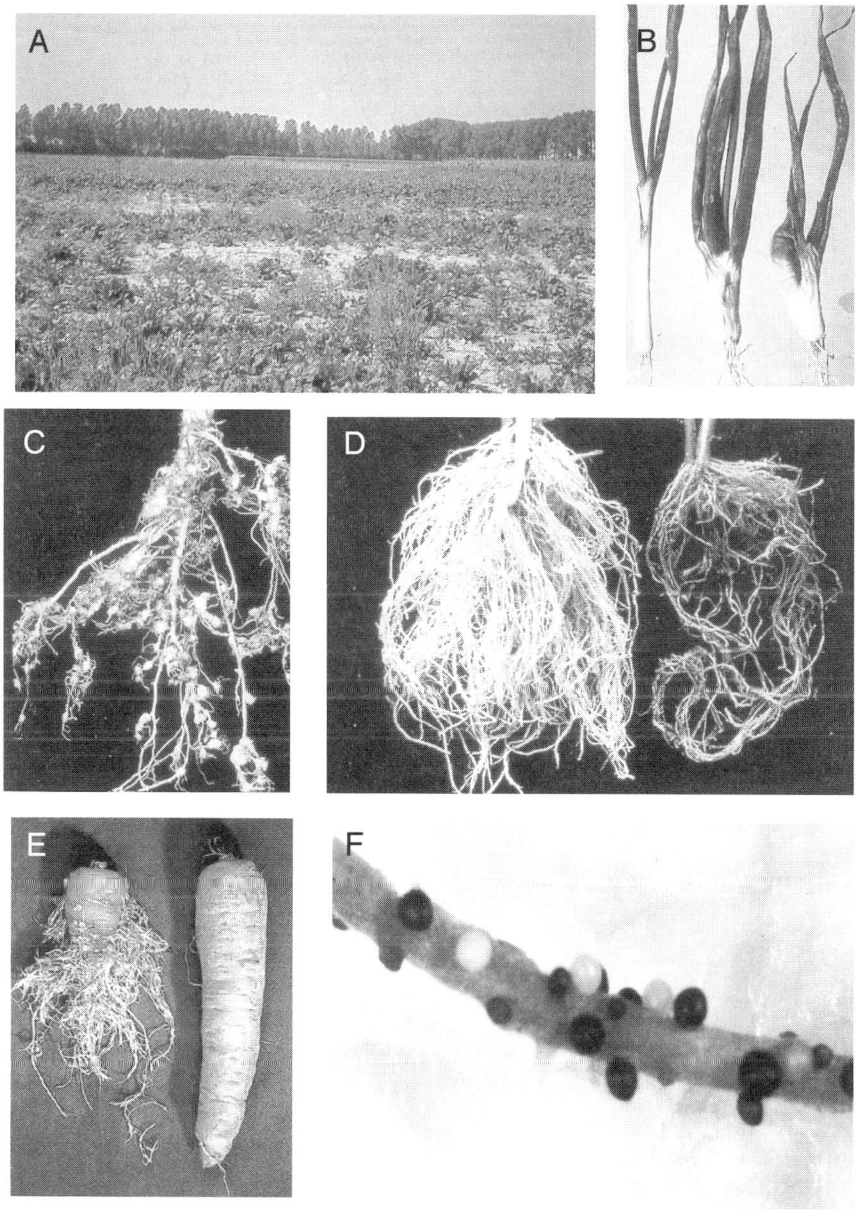

Fig. 1 Symptoms of nematode-infected plants. A. A field of sugarbeets infested with the beet cyst nematode *Heterodera schachtii*. Spots with infection are clearly visible as open spaces because of poor growth of plants. B. Onion plants affected by the stem nematode *Ditylenchus* show thickening (plant at the left is the uninfected control). C. Root galls caused by the root-knot nematode *Meloidogyne*. D. Strong necrosis (browning) in potato roots (right) due to infection by the migratory root-lesion nematode *Pratylenchus penetrans* (healthy plant at left). E. Bearded root effect on carrot caused by northern root-knot nematode (*Meloidogyne hapla*) (left: infected; right: control). F. Microscopic picture of cysts (*G. pallida*) on potato roots. [Photos (A)–(E) courtesy of A. F. Van der Wal (Wageningen).] (*View this art in color at www.dekker.com.*)

Changes in gene expression correlated with wound or defense response have been studied in several plant nematode interactions.[3] For example, the wound-inducible wun1 promoter was rapidly induced by cyst nematodes in potato but only weakly responded to root-knot nematode infection. Induction of pathogenesis-related proteins has been detected in leaves and roots of potato plants infected with different pathotypes of *Globodera pallida*. Analysis of tomato roots 12 hours after infection with root-knot nematodes identified several defense genes to be upregulated. Most of these were induced in the compatible as well as in the incompatible interaction, although with differences in levels and timing.[3] Examples of activated defense genes include peroxidase, chitinase, lipoxygenase, extensin, and proteinase inhibitors. Induced defenses against nematodes are not limited to upregulation of defense proteins; they also include pathways resulting in phytoallexin biosynthesis (such as glyceollin in soybean) and deposition of callose or lignin as a physical barrier.

DEVELOPMENT OF LARGE FEEDING CELLS IN RESPONSE TO SEDENTARY NEMATODES

Nematodes that feed for a prolonged period from the same cell(s) induce cytological modifications to increase

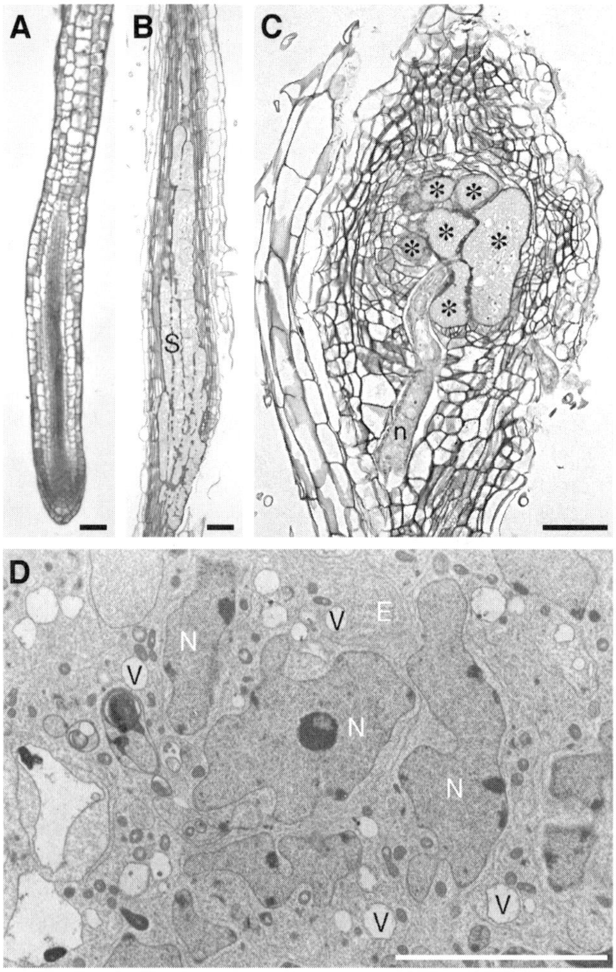

Fig. 2 Cellular alterations induced by sedentary endoparasitic nematodes. A. Section through an uninfected *Arabidopsis thaliana* root. The root tip is shown, as well as the region above the tip where sedentary nematodes induce feeding cells. B. Section through syncytium (S) induced by *Heterodera schachtii* in *A. thaliana*. The large cell in the middle is the syncytium that originated from fusion of neighbouring cells. Pieces of cell wall remnants can be seen as dashed lines inside the syncytium. C. Section through root gall induced by *Meloidogyne incognita* in *A. thaliana*. The nematode (n) is visible in the middle of the gall, feeding from one of the giant cells (*). About six giant cells are seen, as well as the increased number of cell layers that cause the gall. D. Electron microscopy picture of a section through a giant cell, showing typical features: multiple large-lobed nuclei (N), granular cytoplasm, small vacuoles (V), many organelles, and endoplasmic reticulum (E) membranes. (Bars: 50 μm in A–C, 5 μm in D). (Photo courtesy of J. de Almeida-Engler.) (*View this art in color at www.dekker.com.*)

the metabolic and transport capacities of that cell.[1] *Criconemella xenoplax* is a sedentary ectoparasite that feeds for up to eight days from a single cortical cell. The nematode uses its stylet to withdraw nutrients from this cell via a zone of modified cytoplasm. Plasmodesmata between the food cell and surrounding cells are modified in a special way to facilitate solute transport. The ectoparasites *Xiphinema index* and *X. diversicaudatum* feed on root tips and transform them into terminal galls. The galls contain enlarged multinucleate cells that arise from repeated mitosis without cell division. The best studied and most pronounced cellular modifications are induced by the sedentary endoparasitic nematodes (Fig. 2). The successful induction of an elaborate feeding site is essential for the survival of these nematodes, because once they become sedentary they are unable to move to another cell. Probably in response to salivary secretions from the nematode, alterations of the nuclei, the cytoplasm, and the cell walls turn normal root cells into huge multinucleate feeding cells. A syncytium (induced by cyst nematodes) is formed by the breakdown of plant cell walls and subsequent fusion of neighbouring cells. Giant cells (induced by root-knot nematodes) are formed as the result of repeated nuclear divisions without cell division.[4]

NUCLEAR AND CYTOPLASMIC CHANGES IN FEEDING CELLS

Aberrant cell cycles that lead to a higher DNA content are typical for plant tissues that have a nutritional function, such as endosperm and tapetum. This phenomenon also occurs in developing feeding cells, although under a different form in giant cells and syncytia.[5] The first sign of giant cell development is the formation of binucleate cells.[6] Cell plate vesicles initially line up between the two daughter nuclei but then disperse, resulting in the abortion of the new cell plate. Additional mitoses uncoupled from cell division result in multinucleate large cells. The mean number of nuclei per mature giant cell is between 30 and 60 in most studied plant hosts, and the nuclei are also larger than normal. In sharp contrast to giant cells, no mitosis has been seen in syncytia induced by cyst nematodes. The enlargement of nuclei indicates that DNA synthesis (uncoupled from mitosis) is taking place within the syncytial tissue during and after the incorporation of new cells through cell wall dissolution.[7] Both types of feeding cells thus contain many large nuclei, and since the transcriptional and translational activity increases at each doubling of the DNA, these cells are functionally equivalent to hundreds of diploid cells. The resulting high metabolism is reflected in a dense granular cytoplasm with many mitochondria, an increase in rough endoplasmic reticulum, and small vacuoles.

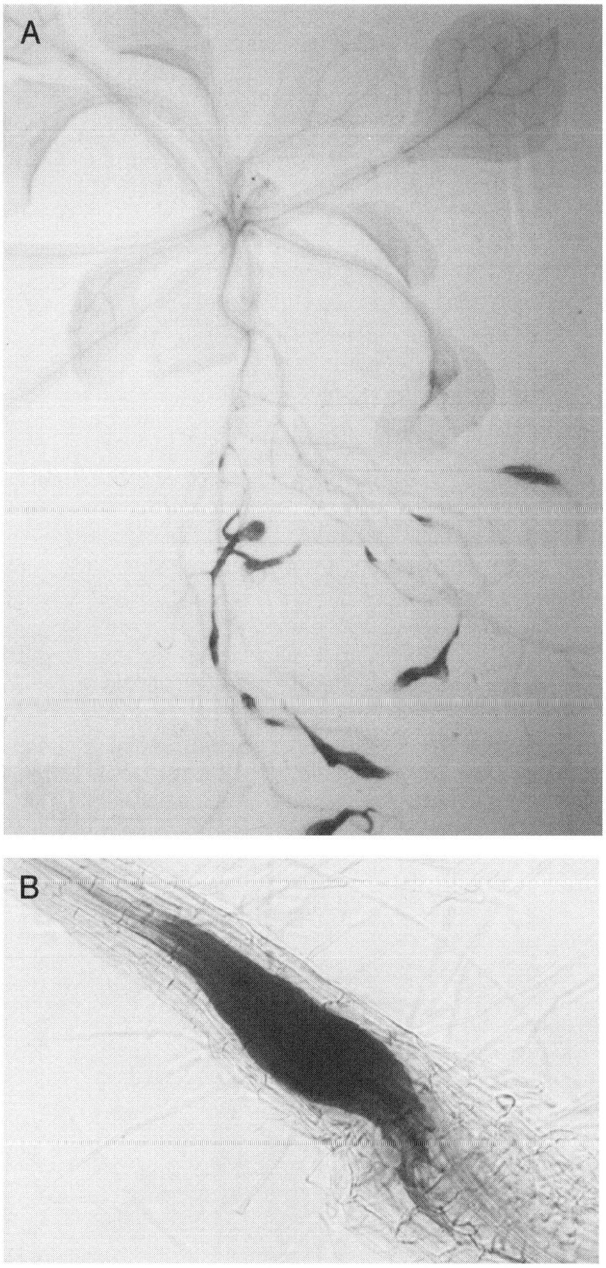

CELL WALL REMODELING IN FEEDING CELLS

When a cyst nematode selects an initial syncytial cell, the plant cell responds by gradually widening some plasmodesmata to neighbouring cells.[2] The protoplasts of adjacent cells then fuse through the developing wall openings. At later stages, cell wall openings are formed de novo and the syncytium continues to grow by integrating neighboring cells. Giant cell and syncytium cell walls adjacent to the xylem increase their thickness by fingerlike cell wall invaginations that are followed by the plasma membrane.[7] These invaginations enlarge the surface for water transport from the xylem into the feeding cell.

GENES EXPRESSED IN FEEDING CELLS

Various strategies have been used to identify plant genes that are upregulated in feeding cells and could thus be important in feeding-cell development or functioning[8] (Fig. 3). Many different plant genes have been found to be activated in feeding cells; the challenge now is to elucidate the role and the importance of these genes in the infection process. Examples include genes for metabolic enzymes, cell wall-modifying enzymes, a water channel protein, cytoskeleton proteins, and genes that are putatively involved in early feeding-cell development, such as cell cycle genes and transcription factors.[9] With the ongoing application of genomics tools to the study of plant–nematode interactions, the number of known plant genes involved in plant–nematode interactions will increase dramatically.

Fig. 3 Gene expression in nematode feeding sites. Promoter trapping is based on the random integration of a promoterless-glucuronidose (*gus*) gene in the DNA of a plant species, often *Arabidopsis thaliana*.[10] When inserted downstream from a plant promoter that is inducible by nematode infection, a higher GUS activity (blue color) can be seen at the infection site. The elegance of this method lies in the ability to directly visualize induced expression in nematode feeding sites at various stages of the interaction, while analyzing the specificity of expression using uninfected parts of the same plant and control plants. A. GUS-assay on *A. thaliana* plant tagged with a promoterless *gus*-construct. The analysis was performed 3 days after inoculation with *Meloidogyne incognita*. Galls turn blue because of local promoter activity. B. Detail of a gall, the nematode has been counterstained with acid fuchsin. (Photo courtesy of Mansour Karimi.) (*View this art in color at www.dekker.com.*)

ACKNOWLEDGMENTS

I thank Rebecca Verbanck for help in preparing the artwork of this manuscript.

ARTICLES OF FURTHER INTEREST

Bacterial Pathogens: Early Interactions with Host Plants, p. 89
Biological Control of Nematodes, p. 134
Breeding for Durable Resistance, p. 179
Classification and Identification of Nematodes, p. 283
Insect–Plant Interactions, p. 609
Management of Nematode Diseases: Options, p. 684
Nematode Biology, Morphology, and Physiology, p. 781
Nematode Feeding Strategies, p. 784
Nematode Infestations: Assessment, p. 788
Nematode Population Dynamics, p. 797
Nematode Problems: Most Prevalent, p. 800
Nematodes and Host Resistance, p. 805
Nematodes: Ecology, p. 809
Plant Defenses Against Insects: Constitutive and Induced Chemical Defenses, p. 939
Root-Feeding Insects, p. 1114

REFERENCES

1. Sijmons, P.C.; Atkinson, H.J.; Wyss, U. Parasitic strategies of root nematodes and associated host cell responses. Annu. Rev. Phytopathol. **1994**, *32*, 235–259.
2. Hussey, R.S.; Grundler, F.M.W. Nematode Parasitism of Plants. In *The Physiology and Biochemistry of Freeliving and Plant-Parasitic Nematodes*; Perry, R.N., Wright, D.J., Eds.; CABI Publishing: Wallingford, 1998; 213–243.
3. Williamson, V.M.; Hussey, R.S. Nematode pathogenesis and resistance in plants. Plant Cell **1996**, *8*, 1735–1745.
4. Huang, C.S.; Maggenti, A.R. Mitotic aberrations and nuclear changes of developing giant cells in Vicia faba caused by root knot nematode, Meloidogyne javanica. Phytopathology **1969**, *59*, 447–455.
5. Gheysen, G.; de Almeida Engler, J.; Van Montagu, M. Cell Cycle Regulation in Nematode Feeding Sites. In *Cellular and Molecular Aspects of Plant–Nematode Interactions*; Fenoll, C., Grundler, F.M.W., Ohl, S.A., Eds.; Developments in Plant Pathology, Kluwer Academic Publishers: Dordrecht, 1997; Vol. 10, 120–132.
6. Jones, M.G.K.; Payne, H.L. Early stages of nematode-induced giant-cell formation in roots of *Impatiens balsamina*. J. Nematol. **1978**, *10*, 70–84.
7. Jones, M.G.K.; Northcote, D.H. Nematode-induced syncytium—A multinucleate transfer cell. J. Cell. Sci. **1972**, *10*, 789–809.
8. Fenoll, C.; Aristizábal, F.A.; Sanz-Alférez, S.; del Campo, F.F. Regulation of Gene Expression in Feeding Sites. In *Cellular and Molecular Aspects of Plant–Nematode Interactions*; Fenoll, C., Grundler, F.M.W., Ohl, S.A., Eds.; Developments in Plant Pathology, Kluwer Academic Publishers: Dordrecht, 1997; Vol. 10, 133–149.
9. Gheysen, G.; Fenoll, F. Gene expression in nematode feeding sites. Annu. Rev. Phytopathol. **2002**, *40*.
10. Barthels, N.; van der Lee, F.M.; Klap, J.; Goddijn, O.J.M.; Karimi, M.; Puzio, P.; Grundler, F.M.W.; Ohl, S.A.; Lindsey, K.; Robertson, L.; Robertson, W.M.; Van Montagu, M.; Gheysen, G.; Sijmons, P.C. Regulatory sequences of Arabidopsis drive reporter gene expression in nematode feeding structures. Plant Cell **1997**, *9*, 2119–2134.

Plant Responses to Stress: Ultraviolet-B Light

P

Brian R. Jordan
Lincoln University, Canterbury, New Zealand

INTRODUCTION

Ultraviolet-B radiation (UV-B: 280–320 nm) is highly energetic and can cause damage to a wide range of cellular components, such as DNA, amino acids, and lipids. Plants, because of their sessile nature, are potentially very susceptible to UV-B exposure, and the increase in UV-B as a result of ozone depletion could have severe consequences. Over the last few decades there has been a substantial amount of research on the effects of UV-B on plants.[1,2] These studies show a large number of responses including changes in growth and development, increase in protective pigment biosynthesis, effects on photosynthesis, DNA damage, etc. These responses are also very variable, both between species and even within varieties of the same species. Most of these studies, however, have been carried out in controlled environment cabinets. Unfortunately, while providing experimental control, this approach does not provide a realistic comparison to the natural environment and consequently over-states the damaging influence of UV-B. Despite this disadvantage, controlled environment cabinets have provided valuable insight into the cellular changes induced by UV-B exposure, e.g., changes in gene expression. Recently more environmentally significant data is being generated on UV-B stress and a complex picture is emerging of multiple impacts at the physiological and ecological level.[2,3] This short review will focus on changes at the molecular level as a result of UV-B–induced stress.[1,4]

MOLECULAR RESPONSE TO UV-B STRESS

Gene Activation

It is now apparent that many of the UV-B induced responses require gene activation. Initial studies showed that genes that were required for protection against UV-B were activated, while those for photosynthesis, primary metabolism, etc, were down-regulated. To change gene expression, UV-B, unlike other wavelengths of light, can be absorbed directly by DNA and nonspecifically inhibit DNA transcription. Alternatively, UV-B could be perceived by a specific photoreceptor or in some nonspecific manner (e.g., increased oxidative stress) that subsequently leads to gene regulation. Although no UV-B photoreceptor has been identified, there is some limited evidence that UV-B perception may operate through such a mechanism.[5] After perception of the UV-B radiation, a signal transduction pathway(s) must be initiated that eventually changes gene activity. Recent studies on UV-B–induced gene activation has unveiled a complex response involving numerous signal transduction pathways.[4,6] These responses are similar in some ways to other stress responses, e.g, to pathogens,[7] but differ in various details such as the chemical components of the transduction pathway. At present, a number of characteristics of UV-B–induced signal transduction pathways have been described, with a variety of different components, which include ROS (reactive oxygen species), jasmonate, salicylate, and ethylene (Fig. 1). ROS is generated through a variety of mechanisms, including NADPH oxidase and peroxidase. Changes in ROS appear to be important in the early stages of signal transduction for many but not all genes.[8] Although these pathways are specific for the activation of a particular gene, ''cross-talk'' between transduction pathways induced by other stress factors must also take place.

Irrespective of the pathway there must eventually be an interaction with the gene involving transcription factors and cis-acting elements within the gene promoter. UV-B–inducible promoter elements and candidate transcription factors have now been identified in plants.[1,4] The cis-acting elements for light responsive *chs* in parsley are called ACE (ACGT-containing elements that recognize common plant regulatory factors) and MRE (Myb recognition elements). Although some differences exist, consistent structural similarities are found in *chs* promoters between species. A number of other UV-B–inducible transcription factors have recently been identified, including a novel 11-bp, GC-rich sequence identified in the pea *sad*A and C genes. The *sad* genes are particularly interesting as they are up regulated rapidly by UV-B and at a relatively low irradiance level. The *sad* genes are also affected by other environmental stresses, and, therefore, elements in the promoter must be a point of convergence

Encyclopedia of Plant and Crop Science
DOI: 10.1081/E-EPCS 120010642

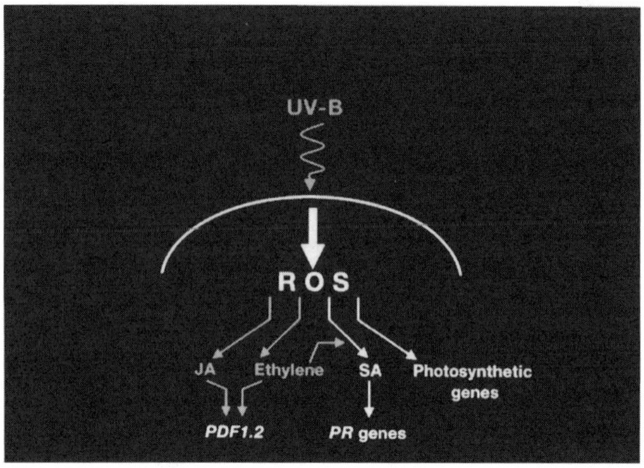

Fig. 1 UV-B–induced signal transduction pathways. (From Ref. 15) (*View this art in color at www.dekker.com.*)

for multiple environmental signals.[9] This is supported by an inversely regulated promoter unit that has been identified and responds positively to UV and negatively to a pathogen-derived elicitor.[10] This promoter unit could function as a convergence for largely distinct signal pathways and may also operate through common plant regulatory factors. These and similar studies are being further advanced by the application of rapid throughput screening techniques such as DNA array analysis[11] and suppression subtractive hybridization[12] that can identify UV-B–specific changes in gene activity.

UV-B Defense Mechanisms

Plants have a wide range of protective mechanisms that they use to defend themselves against UV-B radiation (Table 1). One of the first changes that take place in response to UV-B stress is an enhancement in secondary metabolism, particularly in phenylpropanoid-metabolism. This change produces predominantly water-soluble polyphenolic compounds that are synthesized in the epidermal cells. The biosynthetic enzymes are regulated by transcription factors and these may respond to UV-B levels.[13] In addition, in some desert plants hydrophobic derivatives may also be localized in the cuticle. Studies have shown that these compounds are very effective in absorbing UV-B.[1,4] However, there are substantial differences between species in the degree of UV-B penetration into the underlying mesophyll and palisade tissue. This seems to depend upon a lack of UV-B-absorbing pigments allowing passage through the anticlinal cell walls.[1]

One of the main consequences of UV-B exposure is the generation of free oxygen radicals (e.g., from polyunsaturated fatty acids), leading to a cascade of cellular damage (see the previous section on signal transduction). The plant defense response is to synthesize a number of antioxidants or antioxidant enzymes, such as SOD, catalase, glutathione, ascorbate, flavonoids, or carotenoids. This response is a similar cellular response to that stimulated by many other biotic and abiotic stresses.[7] A more specific change, however, that occurs during UV-B stress is that dihydroxylated flavonoid production is enhanced as they are stronger antioxidants.[4]

A significant difference of UV-B to other stresses, is that UV-B radiation directly impinges on and is absorbed by the DNA of the cell. Many types of DNA lesion are created, but the most common are cyclobutane pyrimidine dimers (CPD) and 6–4 photoproducts.[1] These lesions would have severe consequences for cellular function if not repaired. Most plants have a very effective DNA repair system with light-activated photolyase as the major repair mechanism, dark excision repair enzyme activity also being present. In most studies, DNA damage by exposure to UV-B can be detected, but the repair mechanisms seem to remove the lesions very efficiently.[4]

Overall, the plant defense against UV-B stress depends on protective pigments, antioxidants, and DNA repair mechanisms. Each of these approaches relies on the biosynthesis of enzymes and consequently is dependent on UV-B–induced gene activation. It is apparent that UV-B protection comes at a substantial cost in terms of conversion from primary to secondary metabolism and production of antioxidant systems and repair enzymes. UV-B light appears to stimulate extensive reprogramming of cellular metabolism. This however, is, not as extensive as is the reprogramming induced by pathogens and indicates a clear hierarchy of stress responses with pathogen defense overriding UV-B by selective transcriptional modification of one or more metabolic pathways.[10]

Table 1 UV-B–induced defense mechanisms

- Reflectance by cuticular layer and surface structures.
- Synthesis of UV-B absorbing pigments, primarily in the epidermal layers.
- Changes in gene expression to produce enzymes involved in UV-B defense.
- Production of antioxidant systems to protect against free radical damage.
- DNA repair enzymes activated to remove UV-B–induced DNA lesions.
- Amelioration mechanisms stimulated by other environmental parameters, e.g., high PAR.

INTERACTION BETWEEN UV-B AND OTHER ENVIRONMENTAL FACTORS

The response of plants to UV-B is significantly influenced by the interaction with other environmental factors. There are many examples in the literature[1] and a few examples will be illustrated here. A unique response of plants to UV-B radiation relates to the property of high photosynthetically active radiation (PAR) to ameliorate the damaging impact of UV-B on plants. This was initially discovered at the physiological level, and later molecular studies noted that the down-regulation of gene expression was also reduced by high PAR. This ''protection'' did not involve synthesis of protective pigments, but was related to the function of the photosynthetic apparatus itself, probably associated with electron transport and photophosphorylation.[4] The ability of different wavelengths to modulate the photosynthetic system to sense the external environment and act as a photoreceptor has been documented and can in some instances change chloroplast gene expression.[4] Another important aspect of UV-B–induced changes in gene expression is that it varies depending upon the developmental stage of the tissue. UV-B does not appear to down-regulate genes for photosynthetic proteins in etiolated tissue.[1] Thus, etiolated tissue exposed to light will strongly express genes for chloroplast protein biosynthesis. This expression continues even in the presence of supplementary UV-B radiation. This result is indicative of a strong link between the development of the photosynthetic apparatus and UV-B–induced gene expression. In addition to light quantity, light quality has a strong influence, both at the molecular and the physiological level.

The impact of UV-B stress is also modified by the drought status of the plants. Although some studies show additive effects, most show ameliorating antagonistic interactions. The growth of plants under no-water stress can be severely inhibited by UV-B. However, the same plants exposed to water stress will not be as severely affected. The reduced impact may be due to the drought limiting growth and cell division, consequently limiting the potential of UV-B to damage cellular processes. Other drought stress responses, such as increased levels of osmoprotectants (e.g., proline, GABA, etc.) may also play a role in reducing the impact of UV-B damage. Studies show that the effect will vary on the duration of exposure to both UV-B and drought. In addition, the plant's genetic background will influence the response. For instance, slow growing ecotypes exposed to these stresses are likely to be more tolerant.

Another interactive stress that has not been extensively investigated is that between UV-B and pathogens/herbivores. Studies have indicated that UV-B–induced changes in plant chemistry can have an impact on pathogens or herbivores. Changes may take place in primary or secondary metabolism to alter the chemical constituents. Such changes could have subtle, but substantial impact on the interactions between a number of trophic levels.[2,3]

SUMMARY

Although substantial progress has been made over the last decade to elucidate the molecular mechanisms involved in UV-B–induced stress responses[1,4,6,14] this area still lags behind comparable areas of plant biology. Through greater knowledge at the molecular level the variation in responses at the whole plant level, plant pathogen/herbivore interactions level, and even at the ecological community level will be understood more fully.

ARTICLES OF FURTHER INTEREST

Phenylpropanoids, p. 868
Photoreceptors and Associated Signaling IV: UV Receptors, p. 893
Plant Response to Stress: Modifications of the Photosynthetic Apparatus, p. 990
Plant Response to Stress: Regulation of Plant Gene Expression to Drought, p. 999
Plant Response to Stress: Role of Jasmonate Signal Transduction Pathway, p. 1006

REFERENCES

1. Jordan, B.R. The Effects of Ultraviolet-B Radiation on Plants: A Molecular Perspective. In *Advances in Botanical Research*; Callow, J.A., Ed.; Academic Press Ltd, 1996; Vol. 22, 97–162.
2. Rozema, J.; van de Staaij, J.; Björn, L.O.; Caldwell, M. UV-B as an environmental factor in plant life: Stress and regulation. Tree **1997**, *12*, 22–28.
3. Day, T.A. Multiple trophic levels in UV-B assessments—Completing the ecosystem. New Phytol. **2001**, *152*, 181–186.
4. Jordan, B.R. Molecular response of plant cells to UV-B stress. Funct. Plant Biol. **2002**, *29*, 909–916.
5. Boccalandro, H.E.; Mazza, C.A.; Mazzella, M.A.; Casal, J.J.; Ballare, C.L. Ultraviolet-B radiation enhances a phytochrome-B-mediated photomorphogenic response in *Arabidopsis*. Plant Physiol. **2001**, *126*, 780– 788.

6. Mackerness, S.A.-H. Plant responses to ultraviolet-B (UV-B: 280–320 nm) stress: What are the key regulators? Plant Growth Regul. **2000**, *32*, 27–39.
7. Lamb, C.; Dixon, R.A. The oxidative burst in plant disease resistance. Annu. Rev. Plant Physiol. Plant Mol. Biol. **1997**, *48*, 251–275.
8. Mackerness, S.A.-H.; John, C.F.; Jordan, B.; Thomas, B. Early signaling components in ultraviolet-B responses: Distinct roles for different reactive oxygen species and nitric oxide. FEBS Lett. **2001**, *489*, 237–242.
9. Gittins, J.R.; Schuler, M.A.; Strid, Å. Identification of a novel nuclear factor-binding site in the *Pisum sativum sad* gene promoters. Biochim. Biophys. Acta **2002**, *1574*, 231–244.
10. Logemann, E.; Halhbrock, K. Crosstalk among stress responses in plants: Pathogen defense overrides UV protection through an inversely regulated ACE/ACE type of light-responsive gene promoter unit. Proc. Natl. Acad. Sci. U. S. A. **2002**, *99*, 2428–2432.
11. Brosché, M.; Schule, M.A.; Kalbina, I.; Connor, L.; Strid, Å. Gene regulation by low level UV-B radiation: Identification by DNA array analysis. Photochem. Photobiol. Sci. **2002**, *1*, 656–664.
12. Sävenstrand, H.; Brosché, M.; Strid, Å. Regulation of gene expression by low levels of ultraviolet-B radiation in *Pisum sativum*: Isolation of novel genes by suppression subtractive hybridisation. Plant Cell Physiol. **2002**, *43*, 402–410.
13. Hailing, J.; Cominelli, E.; Bailey, P.; Parr, A.; Mehrtens, F.; Jones, J.; Tonelli, C.; Weisshaar, B.; Martin, C. Transcriptional repression by AtMYB4 controls production of UV-protecting sunscreens in *Arabidopsis*. EMBO J. **2000**, *19*, 6150–6161.
14. Brosché, M.; Strid, Å. Molecular events following perception of ultraviolet-B radiation by plants. Physiol. Plant. **2003**, *117*, 1–10.
15. Mackerness, S.A.H.; Surplus, S.L.; Blake, P.; John, C.F.; Buchanan-Wollaston, V.; Jordan, B.R.; Thomas, B. Ultraviolet-B-induced stress and changes in gene expression in *Arabidopsis thaliana*: Role of signalling pathways controlled by jasmonic acid, ethylene and reactive oxygen species. Plant, Cell and Environment. **1999**, *22*, 1413–1423.

Plant RNA Virus Diseases

Bryce W. Falk
University of California, Davis, California, U.S.A.

Roger Hull
John Innes Centre, Colney, Norwich, U.K.

INTRODUCTION

Plant viruses are widespread and economically important plant pathogens. Virtually all plants that humans grow for food and fiber are affected by at least one virus; depending on the plant species, geographic location, growing season, etc., virus disease(s) can preclude the ability to grow specific crops in certain locations. Hence, significant time and monetary resources are invested in efforts to control plant virus diseases. About 80% of the approximately 1000 currently recognized plant-infecting viruses have ribonucleic acid (RNA) as their genetic material. This contrasts with viruses affecting animals and prokaryotes, in which the most common type of genetic material is deoxyribonucleic acid (DNA). This article discusses RNA plant viruses and diseases, factors affecting disease development, and approaches for controlling these diseases.

VIRUS DISEASES IN PLANTS

The natural host ranges of RNA viruses vary from very wide (e.g., *Cucumber mosaic cucumovirus* (CMV) infects more than 800 species in 85 families) to very narrow, infecting one or a few species (e.g., *Apple stem grooving capillovirus*). Virus infection results in symptoms that range from no or mild symptoms (latent or cryptic infection) to death of the plant. The symptom type is usually characteristic of the virus, the most common being a mottle or mosaic pattern of light and dark green areas on the leaves delimited by the vein structure; in monocotyledonous plants this results in chlorotic or light green stripes. Mosaic in flowers can give color breaking. Other common symptoms include chlorosis of the leaves, vein clearing, and chlorotic or necrotic ringspots. The most common growth abnormality is stunting; viruses can also cause deformation of leaves, swelling of the stem, and even outgrowths on the underside of leaves. All of these result in reduced yield quality or quantity.[1]

At the cellular level virus infection often causes perturbation of chloroplasts (associated with mosaic symptoms) and disruption or proliferation of membrane systems such as the endoplasmic reticulum. Virus gene products often accumulate to give characteristic inclusion bodies (e.g., the pinwheel structures of infections with *Potato virus Y potyvirus* (PVY) and related viruses).

DIVERSITY, COMPOSITION, AND REPLICATION

Most RNA viruses have positive-sense single-stranded genomes ((+)-sense ssRNA) (Table 1); the remainder have either negative-sense single-stranded ((−)-sense ssRNA) or double-stranded RNA (ds-RNA). These viruses are classified into genera based on their genome properties, together with characters such as particle shape and biological vectors. Some of the genera with important characters in common are grouped into families. The (−)-sense RNA and dsRNA viruses are closely related to groups of viruses that infect insects and vertebrates.

The particles of most (+)-sense ssRNA viruses are simple, comprising the RNA genome surrounded by subunits of usually a single species of coat protein. There are two basic particle types. In rod-shaped particles, the coat protein subunits are arranged in a helix with the RNA embedded between them. The particles can be either rigid or flexible with a length and diameter characteristic of the genus or family. Rigid particles (e.g., *Tobacco mosaic virus* (TMV) (300×18 nm)) range from less than 100 nm to a few hundred nm in length and flexuous particles from a few hundred to almost 2 microns (e.g., PVY (750×13 nm) and *Citrus tristeza closterovirus* (CTV, 2000×12 nm)). The structure of isometric spherical particles is based on icosahedral symmetry, with particles usually comprising 180 protein subunits and having a diameter of 25–30 nm (e.g., CMV); the particles of some viruses, e.g., *Tobacco necrosis satellite virus*, are smaller (ca. 18 nm) and comprise 60 subunits. Some viruses, e.g., *Alfalfa mosaic alfamovirus*, have bacilliform particles ($25–60 \times 18$ nm), the structure of which is based on the principles of icosahedral symmetry. The RNA in isometric particles is usually associated with the inner part of the subunits or even with the inner cavity of the particle.

Encyclopedia of Plant and Crop Science
DOI: 10.1081/E-EPCS 120020005

Table 1 Genomes of RNA plant viruses

Genome	No. of species	No. of genera	No. of families
+-sense ss RNA	635	49	7
−-sense ss RNA	100	5	2
Ds RNA	45	6	2

The plant (−)-sense ssRNA viruses have more complex particles made up of protein and lipid and are either spherical (genus *Tospovirus*) or bacilliform (family *Rhabdoviridae*) in shape. The dsRNA viruses also have complex isometric particles comprising two or more layers of different coat protein species.

Many (+)-strand ssRNA viruses, e.g., TMV and PVY, have undivided genomes. However, one unusual feature of other (+)-strand ssRNA viruses is that the genome required for full infection is divided between two and five segments that are packaged into different particles. Thus, the genome of CMV is divided into three segments, each packaged in a separate isometric particle, and that of *Beet necrotic yellow vein benyvirus* (BNYVV) is divided between five rod-shaped particles. The genomes of many (−)-sense ss and ds RNA viruses are divided between several segments, but these are all packaged in the same particle.

RNA viruses replicate by transcribing RNA from RNA. Thus, (+)-strand ssRNA viruses transcribe a (−) strand from the viral template, which in turn is transcribed to give progeny (+) strands. Similarly, (−)-strand ssRNA and dsRNA viruses replicate via a (+)-strand intermediate. Replication is effected by a virus-coded RNA-dependent RNA polymerase (RdRP). The replicase complex comprises RdRP and often a helicase that unwinds double-stranded intermediates and a methyltransferase that caps the 5′ end of the (+) strand.

The genomes of RNA viruses encode the information for virus replication and contain genes for usually five to seven proteins. Those viruses with undivided genomes are faced with the problem of eukaryotic translation systems being able to express only the 5′ open reading frame (ORF). The input RNA of (+)-strand ssRNA viruses acts as an mRNA expressing the genetic information of the virus and giving early virus products such as the RdRP and associated enzymes, the ORFs of which are usually at the 5′ end of the genomic RNA. Downstream ORFs are expressed by processes such as readthrough of stop codons, frameshift from the reading frame of the 5′ ORF, and formation of subgenomic mRNAs during RNA replication. These subgenomic RNAs usually encode late products, e.g., coat protein. The segmentation of the RNAs in divided genome viruses results in most of the ORFs being 5′, thus overcoming the eukaryotic translational constraints.

mRNAs must to be transcribed from the viral genomes for expression of the viruses with (−)-strand and ds RNA genomes. The particles of these viruses contain RdRP, which is used in the early stages of infection to produce mRNAs.

RNA PLANT VIRUS TRANSMISSION

Plant viruses must spread among sedentary plant hosts. As viruses lack self-motility, they are dependent on other means for moving from plant to plant; virus movement is effected in several specific ways. By far, the majority of plant viruses are spread by specific, plant-feeding arthropod vectors, mainly Homopteran insects, in particular, aphids; others include leafhoppers, planthoppers, and whiteflies. Some viruses are transmitted by non-Homopterans such as mites, thrips, and beetles. Whereas dispersal by insect vectors is the most common means, a few plant viruses are soilborne, being transmitted by root-feeding nematodes or unicellular, root-infecting fungi.

The ability of a virus to be vector transmitted is determined by specific interactions between the virus and the corresponding vector. However, even among the aphid-transmitted viruses a number of types of interactions exist and significantly affect disease epidemiology. For example, rapid short-distance virus spread is typical for noncirculative aphid-transmitted viruses, whereas circulative viruses that are borne internally within the vector can sometimes be transmitted for the life of the vector. Thus, for the latter, long-distance spread is quite possible.

CONTROLLING RNA VIRUS INFECTIONS OF PLANTS

Viruses are molecular intracellular obligate parasites. As such, they are dependent on their hosts' normal cellular machinery for aspects of their replication and gene expression. Because virus activities are so intimately associated with those of the host cell, opportunities are not readily available for using chemicals (viricides) to interfere with virus infections without also adversely affecting normal host cell metabolism. The most effective approaches for controlling RNA plant virus infections are to avoid or prevent infections.

Avoiding or eliminating sources of virus inoculum is essential for virus disease control, especially if the virus is contained within propagation materials, including seeds or vegetative cuttings. Most woody species (e.g., grapes, roses, fruit trees) are propagated vegetatively and thus, the simple approach of ensuring that propagation sources are virus-free can provide effective disease control. Many RNA plant viruses such as CTV have been disseminated worldwide by using virus-infected propagation materials. Some viruses are seedborne in crop hosts or in weeds. Both types of host plants can be important for dissemination and subsequent disease development, but if seed-borne virus in the crop host is an imporant source of primary inoculum, then planting seed that is virus-free can provide effective disease control (e.g., lettuce seeds free of *Lettuce mosaic potyvirus* (LMV)). Although eliminating virus inoculum can be very effective, it is not this simple for most plant virus diseases.

Because most plant viruses are transmitted to plants by specific vectors, vector control can also be useful. However, vector control is often expensive and inefficient and may adversely affect non-target organisms. Also, because many RNA plant viruses (particularly those in the genus *Potyvirus*) can be inoculated to plants by their aphid vectors in as little as a few seconds, protective insecticide applications cannot kill aphids before viruses are transmitted, thereby failing to provide disease control.

Host plant resistance (HPR) is a common and often most desirable means for controlling virus diseases in plants. Several important sources of genetic resistance have been identified and are used in crop plants, including beans, tobacco, potatoes, and tomatoes. Although HPR is very desirable and can be very effective (i.e., *N* gene resistance in *Nicotiana* spp. to viruses in the genus *Tobamovirus*), there are limits to its usefulness. For example, *N* gene resistance is specific, and thus effective against most tobamoviruses but not against unrelated viruses. Although the *N* gene can be transferred among related *Nicotiana* spp. by conventional plant breeding, it cannot be similarly transferred to unrelated species lacking effective resistance genes (i.e., cowpeas (*Vigna unguiculata*) for *Sunnhemp mosaic tobamovirus*). In other instances, HPR may be multigenic, and can be very difficult to effectively manipulate.

In recent years genetically engineered resistance has been shown to be a powerful, and potentially very useful, alternative strategy that can complement other approaches to controlling virus diseases in plants. A good source of antivirus resistance genes has proven to be the viruses themselves. By expressing in plants, genes, or gene fragments derived from RNA virus genomes, very high levels of resistance against the donor virus have been obtained. This approach has worked experimentally for many different viruses in several plant species, although only a few genetically engineered virus-resistant plants have been used so far in commercial agriculture. One good example is resistance in papaya (*Carica papaya*) to *Papaya ringspot potyvirus* (PRV) in Hawaii.[2] The use of genetically engineered resistance so far appears very promising for complementing other approaches to virus disease control in plants.

CONCLUSION

Virus infections continue to be a serious constraint on crop production in spite of the wide range of control measures in use. The application of new technologies such as genetic manipulation and genomics should lead to more successful ways of controlling viruses. These will be of particular importance in developing countries where losses due to viruses are relatively great.

The study of the interaction of plant viruses with their hosts has already led to greater understanding of how plants function at the molecular level. With the increasing sophistication of molecular technologies, it is likely this will continue at an ever increasing rate. RNA viruses will play a major role in this, especially in understanding defense systems against foreign RNAs.[3]

ARTICLES OF FURTHER INTEREST

Plant DNA Virus Diseases, p. 960
Plant Virus: Structure and Assembly, p. 1029

REFERENCES

1. Hull, R. *Matthews' Plant Virology*; Academic Press: London, 2002.
2. Gonsalves, D. Control of papaya ringspot virus in papaya: A case study. Ann. Rev. Phytopathol. **1998**, *36*, 415–437.
3. Baulcombe, D. Viral suppression of systemic silencing. Trends Microbiol. **2002**, *10*, 306–308.

Plant Viral Synergisms

John L. Sherwood
University of Georgia, Athens, Georgia, U.S.A.

INTRODUCTION

Synergism is one of the many interactions that may occur between viruses that affects both the expression of disease and the replication of the viruses involved. In medical virology the occurrence of synergism is relatively rare, as the occurrence of two viruses in vertebrate cells is rather uncommon outside of laboratory cell culture. With plant viruses, however, the phenomenon of synergism has been noted since the early days of plant virus research in the 1920s. Synergism was intermittently studied until the tools of molecular biology became available to dissect this interesting virus interaction. Posttranscriptional gene silencing is now thought to modulate the virus replication and accumulation resulting in a synergistic interaction between plant viruses.

INTERACTIONS BETWEEN PLANT VIRUSES

A number of interactions may occur between plant viruses.[2] Cross-protection refers to the protection of a plant by a mild strain of a virus from a related severe strain. When plant viruses are inoculated concurrently, a protection may ensue that results from the reduction of replication rate or final quantity of the challenge virus. Satellite viruses or satellite RNAs require a helper virus to replicate, but generally satellites do not have much sequence homology with the helper virus. The symptomatic outcome is usually a reduced severity of disease, although there are many exceptions involving satellites where in more severe disease symptoms are expressed. The outcome of a synergistic interaction between plant viruses, however, is by definition a more severe expression of disease than either virus alone or the expected additive effect produced by the infection of two viruses. In addition, the increase in symptom severity is usually associated with an increased accumulation of one or both of the coinfecting viruses.

SYNERGISTIC INTERACTIONS ARE NOT LIMITED TO A FEW VIRUSES

The literature is replete with examples of synergism between different plant viruses that include viruses with genomes of RNA or DNA.[3] Some interactions that have been noted are significant in the field, resulting in severe and recurring losses to producers; others are limited to the greenhouse; and some are laboratory oddities that have been produced by inoculation with viruses that normally would not be found in the same host plant. The model system that has been most frequently investigated is the interaction that occurs between *Potato X virus* (PVX), a species in the *Potexvirus* genus, and *Potato Y virus* (PVY), a species in the *Potyvirus* genus, family *Potyviridae* (Fig. 1). PVX and PVY were noted to cause a severe symptom in potato coined "crinkle," "rugose mosaic," and "streak."[4] Initial investigations of this system elucidated that both viruses replicated in the same cells, and an increase in the amount of PVX detected was likely due to an increase in the amount of virus produced.[5] PVX has no allegiance to PVY in producing a synergistic interaction, and coinfection with a number of Potyvirus species (including *Tobacco vein mottling virus* (TVMV), *Tobacco etch virus* (TEV), or *Pepper mottle virus*) all result in a synergistic manifestation of disease.

A CONVERGENCE OF IDEAS LEADS TO INSIGHT INTO THE MECHANISM OF SYNERGISM

Gene Silencing

With the advent and common availability of methodology and reagents for gene cloning and plant transformation, a myriad of transgenic plants was produced with a variety of virus gene constructs, with the goal of overexpression of the gene of interest. In many of the experiments, however, the outcome was the opposite of what was anticipated, and gene expression was suppressed. A number of terms have been used in the literature to describe this phenomenon, including cosuppression, RNA interference (RNAi), posttranscriptional gene silencing (PTGS), sense suppression, quelling, and virus-induced gene silencing (VIGS).[6] Double-stranded RNA (dsRNA) is a key component of the proposed pathway by which RNA silencing occurs, and dsRNA is a key component during replication of RNA plant viruses. RNA silencing has been extensively reviewed[6,7] and will therefore be only briefly treated

Encyclopedia of Plant and Crop Science
DOI: 10.1081/E-EPCS 120019987

Fig. 1 Synergism between *Potato X virus* (PVX), a species in the *Potexvirus* genus, and *Potato Y virus* (PVY), a species in the *Potyvirus* genus, family *Potyviridae*. Leaf at left was inoculated with PVY; leaf at right was inoculated with PVX; middle leaf was inoculated with both PVY and PVX. Photo taken 12 days after inoculation. (Photo by A. F. Ross, courtesy of R. M. Goodman.) (*View this art in color at www.dekker.com.*)

here. RNA silencing is a naturally occurring mechanism of genetic regulation in eukaryotes that is directly targeted at nucleic acid. It can be induced by viruses, transgenes, or the transcripts normally produced in the plant that result in the formation of dsRNA. The dsRNA is acted upon by an enzyme coined DICER that processes the dsRNAs into short (approximately 25-nucleotide) RNAs. The small interfering RNAs (siRNA) derived from the processing of the dsRNA enters into a RNase complex, RISC (RNA interference specificity complex), that further mediates sequence-specific RNA degradation. Associated with gene silencing is a mobile signal that may move throughout the host plant and hence act as the systemic signal in gene silencing. It has also been noted in plants that methylation of the transcribed region is associated with gene silencing, a state that may be recurring in subsequent generations of an affected organism and may be the basis for maintaining the gene silencing.

This system is of great interest because the silencing signal can function in other points in the plant wherein the cascade of events was initiated. The unfolding understanding of the parameters that modulate RNA silencing provides an exciting avenue for the regulation of gene expression, from transgenes or endogenous genes.

Potyvirus and *Potexvirus* Genomes

Virions of the members of the Family *Potyviridae* are a long, flexuous particle whose size ranges from 650–900 nm in length and 11–15 nm in diameter. Most have a single-stranded plus-sense RNA that codes for one polyprotein that is cleaved into a number of proteins. The replication strategy and viral genome are well characterized, as the order of the proteins in the polyprotein and their function have been extensively studied.[2] The 5′ terminus encodes protease-1 (P1, molecular weight 35K), helper component protease (HC-Pro, molecular weight 52K), and protein-3 (P3, molecular weight 50K). P1 is involved in the processing of the polyprotein, as is HC-Pro, which is also involved in transmission of the virus by aphids. Virions of the genus *Potexvirus* have flexuous particles about 470–580 nm long and 13 nm in diameter. The viral RNA is a single-stranded plus-sense RNA. However, unlike members of the Family *Potyviridae*, one protein—the replicase—is produced from the genomic RNA for members of the genus *Potexvirus*; other proteins are produced from subgenomic RNAs that are produced during virus replication.

Synergism

To elucidate the nature of interaction between PVX and members of the *Potyviridae* that result in a synergistic interaction, tobacco was transformed to express a variety of portions of the genome if TVMV or TEV were challenge-inoculated with PVX. Transgenic plants expressing the 5′-proximal region of the TVMV genome that included the P1, HC-Pro, and P3 genes were challenged with PVX, resulting in symptoms typical of the PVX-*Potyviridae* synergism. Typical synergistic symptoms were also observed when transgenic tobacco plants expressing the similar 5′-proximal region of the TEV genome were inoculated with PVX. Hence, the interaction was linked to the 5′-proximal end of the potyvirus genome. Subsequent investigations showed that the P1 and P3 regions were not necessary for the synergistic symptoms to be produced, as expression of HC-Pro alone could facilitate the synergistic interaction.[8] In addition, expression of HC-Pro was found to enhance the accumulation of *Cucumber mosaic virus* and *Tobacco mosaic virus*. These viruses are unrelated to PVX; thus HC-Pro appears to act to enhance virus replication irrespective of the infecting virus.

The enhancement of virus replication in the synergistic interaction appears to result from the suppression of gene silencing. During unaltered virus replication as dsRNA is produced, the natural defense mechanisms of plants likely modulate the accumulation of virus through PTGS as briefly outlined in the foregoing. As the dsRNA are not processed, resulting in the production of the siRNAs necessary for the degradation complex, the resulting accumulation of PVX appears to result in the synergistic interaction and severe symptoms.[9] Grafting experiments indicate that the mobile silencing signal continued to be produced, and thus is not dependent on the production of the small RNAs.

CONCLUSION

Although the possible basis of synergism has been partially unveiled, there remains much to understand about the applied aspects mitigating the effects of synergism in the field. The epidemiological consequences synergism has been modeled, and viruses sharing an infected host may be in a perplexing ecological dilemma.[10] Increased symptom severity can result in reduced availability of the host, whereas increased replication and increase in virus titer may represent greater opportunity for subsequent transmission. In regard to disease management, severe symptom expression is generally accompanied by loss of plant yield and quality. Hence, the basis of synergism must be determined so that it can be effectively managed to address recurring problems.

ARTICLES OF FURTHER INTEREST

Gene Silencing: A Defense Mechanism Against Alien Genetic Information, p. 492
Plant RNA Virus Diseases, p. 1023
Plant Virus: Structure and Assembly, p. 1029
Plant Viruses: Initiation of Infection, p. 1032
RNA-mediated Silencing, p. 1106
Viral Host Genomics, p. 1269
Virus Movement in Plants, p. 1280

REFERENCES

1. Griffiths, P.D. Biological synergism between infectious agents. Rev. Med. Virol. **2000**, *10* (6), 351–353.
2. Hull, R. *Matthew's Plant Virology,* 4th Ed.; Academic Press: New York, 2002.
3. Zang, X.-S.; Holt, J.; Colvin, J. Synergism between plant viruses: A mathematical analysis of the epidemiological implications. Plant Pathol. **2001**, *50* (6), 732–746.
4. Markham, R. Landmarks in plant virology: Genesis of concepts. Annu. Rev. Phytopathol. **1977**, *15*, 17–39.
5. Goodman, R.M.; Ross, A.F. Enhancement of potato virus X synthesis in doubly infected tobacco occurs in doubly infected cells. Virology **1974**, *58* (1), 16–24.
6. Baulcombe, D. RNA silencing. Curr. Biol. **2002**, *12* (3), R82–R84.
7. Vance, V.; Vaucheret, H. RNA silencing in plants—Defense and counterdefense. Science **2001**, *292* (5525), 2277–2280.
8. Pruss, G.; Ge, X.; Shi, X.M.; Carrington, J.C.; Vance, V.B. Plant viral synergism: The potyviral genome encodes a broad-range pathogenicity enhancer that transactivates replication of heterologous viruses. Plant Cell **1997**, *9* (6), 859–868.
9. Mallory, A.C.; Ely, L.; Smith, T.H.; Marathe, R.; Anandalakshmi, R.; Fagard, M.; Vaucheret, H.; Pruss, P.; Bowman, L.; Vance, V.B. HC-Pro suppression of transgene silencing eliminates the small RNAs but not transgene methylation or the mobile signal. Plant Cell **2001**, *13* (3), 571–583.
10. Zhang, X.S.; Holt, J.; Colvin, J. Synergism between plant viruses: A mathematical analysis of epidemilogical implications. Plant Pathol. **2001**, *50* (6), 732–746.

Plant Virus: Structure and Assembly

A. L. N. Rao
University of California, Riverside, California, U.S.A.

P

INTRODUCTION

The structure of coat protein (CP) had been thought to protect its infectious genome from the adverse conditions of the intra- and extracellular milieu. Advances made possible by recombinant DNA technologies have established that the viral CP is multifunctional and, most important, that the architecture of mature virions is critical for the survival of the virus, insect transmission, and disease. Knowledge of virion structure is an important prerequisite for understanding the overall biology of plant viruses.[1] One way to elucidate the final structure of stable virions is to dissect the assembly process through in vitro and in vivo studies.

INTERACTIONS PROMOTING VIRUS ASSEMBLY

The assembly of mature virions, which involves nucleic acid–protein interactions, is an important phase in the virus life cycle.[2] This process has been shown to be obligatory for several plant viruses to move from cell to cell, be transported over long distances within a plant through the phloem, and be acquired by insect vectors for dissemination to new hosts. The following interactions involving protein and RNA influence the assembly process: 1) Protein–protein interactions play a major role in virion assembly, although their contribution relative to RNA–protein interactions varies among different plant virus groups (e.g., *Tymovirus* and *Comovirus* virions are predominantly stabilized by protein–protein interactions and can therefore form empty capsid shells without the cognate genomic RNA); 2) RNA–protein interactions predominate in guiding the assembly for viruses such as *Tobamoviruses*, *Alfamoviruses*, and *Bromoviruses*. In these viruses capsid formation requires RNA and empty virions are therefore not formed in vivo; 3) Sequence-independent RNA–protein interactions are typically theorized as stabilizing encapsidated RNAs. Basic N-terminal arms found in the CPs of several RNA viruses are implicated in the interaction with RNA phosphates, facilitating the assembly of infectious virions;[3] 4) Finally, sequence-dependent RNA–protein interactions are often critical in initiating the viral assembly process. Because viral RNA and CP subunits are localized in the same compartment of the cell as cellular tRNA, mRNAs, or rRNA, these species could potentially be copackaged with viral genomic RNAs. Specific sequence and structure–dependent interactions between RNA and CP are thought to ensure that the majority of assembled virions exclusively contain viral RNA.

PATHWAYS

Tobacco mosaic tobamovirus (TMV) is a helical plant virus in which assembly is initiated by specific interaction between a disk of structural protein subunits and an internal sequence in the RNA genome (Fig. 1A). The study of virus assembly using TMV was pioneered by Frankel-Conrat, who reconstituted TMV particles in vitro from dissociated protein and RNA.

The first experimental evidence that purified RNA and CP of an icosahedral virus (*Cowpea chlorotic mottle virus*, CCMV, a species in the *Bromovirus* genus) can reassemble in vitro to produce infectious particles was provided as early as 1967 by Bancroft and Hiebert.[4] *Bromovirus* virions are predominantly stabilized by RNA–protein interactions; RNA is therefore required for the formation of icosahedral capsids in vivo. RNA–protein interactions are thought to neutralize the negative charge of the RNA and allow condensation of the nucleic acid within the virus particle. The assembly process is independent of the virus replication complex; there is no indication of the involvement of any other virus-encoded proteins. Bromoviruses assemble into icosahedral particles with T=3 quasisymmetry (Fig. 1B). The structure of CCMV has been determined by X-ray crystallography to have 3.3 Å resolution.[5] Based on polymerization kinetics observed by light scattering and gel filtration assays, it has been suggested that capsid assembly in CCMV is nucleated by the formation of a pentamer of dimmers.[6] Recent in vitro assembly studies with *brome mosaic virus* (BMV), the prototype Bromovirus, indicated that assembly of infectious virions requires a highly conserved 3′ coterminal tRNA-like structure (TLS).[7] The transient yet critical involvement of the TLS or tRNAs in BMV assembly suggests a role in the formation of CP dimers that serve as intermediaries in the encapsidation pathway.[6]

Encyclopedia of Plant and Crop Science
DOI: 10.1081/E-EPCS 120019906

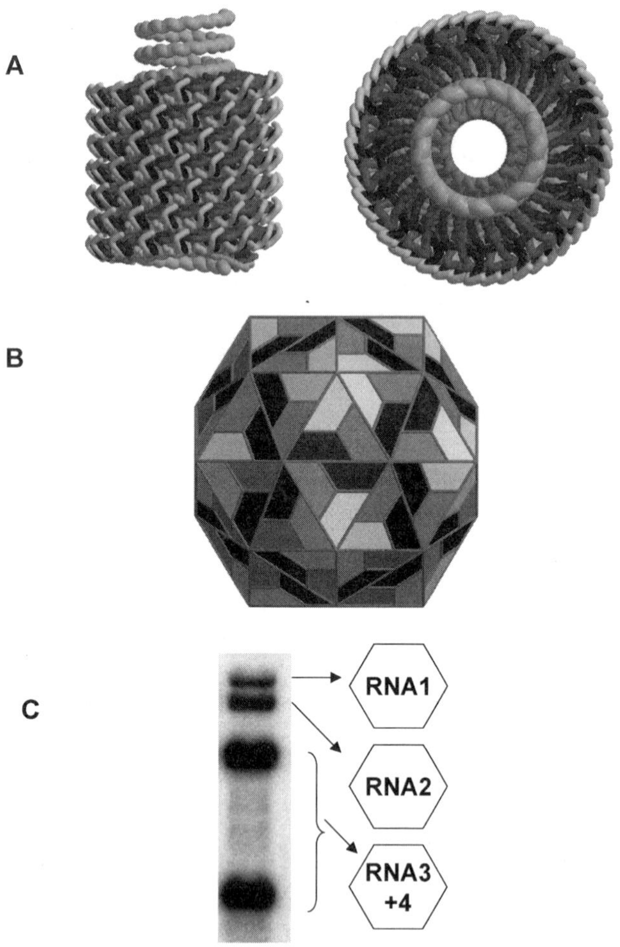

Fig. 1 A. An artistic illustration of TMV based on the structure determined by Pattanayek and Stubbs (23). Shown on the left is a view perpendicular to the helix axis and on the right is a view down the helix axis. The TMV subunits are shown in a tube representation with a color gradient going from black to light gray. The spiral in the middle shown as a thick gray tube corresponds to +ss RNA. B. Schematic representation the T=3 icosahedral lattice of BMV. Each trapezoid corresponds to a subunit. The chemically identical subunits (same gene products) occupy three structurally distinct environments in the T=3 capsids, shown in different colors, light gray, dark gray and black. C. Packing schematic for four BMV RNAs into three morphologically indistinguishable virions.

GENOME PACKAGING

Genome packaging is considered to be a highly specific process. During packaging, viral nucleic acids must be distinguished from other cellular RNA molecules present in the compartment in which assembly takes place. Such discrimination is thought to result from specific recognition of sequences or structures unique to viral nucleic acids, termed origin of assembly sequences (OAS) or packaging signals. These signals are often represented by small stretches of RNA that fold into unique secondary or tertiary structures specifically recognized only by the cognate CP. Although these specific sequences might promote packaging of viral genomic RNAs, their presence does not guarantee packaging, because several other parameters also govern the encapsidation process. For example, the fixed dimensions of icosahedral capsids impose an upper limit on the size of viral nucleic acid that can be accommodated. Consequently, nucleic acids substantially larger than a wild-type genome cannot be packaged, even if appropriate packaging signals are present. In contrast to viruses having one genomic segment (e.g., TMV, TCV, SBMV), those with a genome divided among multiple nucleic acid species have evolved mechanism(s) to balance distribution of the genome segments into either a single virion or among multiple virions. The genomes of plant viruses belonging to the *Bromo* (*icosahedral*), *Cucumo-* (*icosahedral*), *Hordei-* (*rigid rods*), and *Alfamoviruses* (*bacilliform*) genera are divided among three RNA segments. In each of these genera, the genomic and subgenomic RNAs are distributed into separate particles, and their size and the number vary with genera. For example, in bromo- and cucumovirsues, the largest two genomic RNAs are packaged individually into two virions and the third genomic RNA and its subgenomic RNA are hypothesized to copackage into a third virion (Fig. 1C).[8] These virions do not display any physical heterogeneity in either size or appearance. In hordeiviruses, the three genomic RNAs are encapsidated into three distinct-sized rods. Whether the subgenomic is packaged or not is unknown because of its low concentration. In members of the genus *Alfamovirus*, the three genomic RNAs and the subgenomic RNA are packaged individually into four distinctly sized virions. The mechanism(s) involved in maintaining a high degree of precision in distributing the four RNAs into three or four individual capsids is currently obscure.

WHY STUDY VIRUS STRUCTURE?

Structural information on viruses has provided inspiration for many novel approaches in biology, such as development of viruses as epitope carriers for gene therapy and for nanoengineering of materials. Two RNA plant viruses—TMV and *cowpea mosaic virus* (CPMV)—accumulate in very high concentrations in their hosts, and large quantities of highly purified virions can be obtained in less than 48 hours. Because the CP structure of these two viruses is known in detail, they have been used as vectors for the insertion and presentation of foreign peptides on the virion surface.[9,10]

The structure of a given virus must be flexible enough to disassemble in order to release its contents upon entry to a healthy host cell. Structural transitions are therefore thought to occur for many virions due to defined chemical switches resulting in unique gating mechanisms that control the containment and release of the contents. Thus, knowledge gained from understanding the principles of the architecture has helped in using viruses as protein cages for material synthesis and molecular entrapment.[10,11]

Finally, structure has been implicated in playing an important role in controlling many virus–host interactions. In several viruses, experimental evidence has indicated that virion structure assembled from mutated CP subunits has failed to optimally interact with host machinery, resulting in defective local and long-distance movement, and in altered symptom expression in susceptible hosts.

CONCLUSION

Despite many advances, knowledge of the mechanism by which plant viruses package their genomes is still in its infancy. Most of the progress in this direction is limited to viruses containing RNA genomes rather than DNA genomes. Understanding assembly and disassembly mechanisms will likely shed light on the development of novel strategies for virus disease control.

ACKNOWLEDGMENTS

I would like thank Dr. Vijay Reddy of Scripps Institute for providing the illustrations and George Grantham for critical reading of the manuscript.

REFERENCES

1. Hull, R. *Plant Virology*; Academic Press: San Diego, 2002.
2. Baker, T.S.; Olson, N.H.; Fuller, S.D. Adding the three dimension to virus life cycles: Three-dimensional reconstruction of icosahedral viruses from cryo-electron micrographs. Microbiol. Mol. Biol. Rev. **1999**, *63*, 862–922.
3. Fox, J.M.; Johnson, J.E.; Young, M. RNA/protein interactions in icosahedral virus assembly. Semin. Virol. **1994**, *5*, 51–60.
4. Bancroft, J.B.; Hiebert, E. Formation of an infectious nucleoprotein from protein and nucleic acid isolated from a small spherical virus. Virology **1967**, *32*, 354–356.
5. Speir, J.A.; Munshi, S.; Wang, S.; Baker, T.S.; Johnson, J.E. Structure of native and swollen forms of cowpea chlorotic mottle virus determined by X-ray crystallography and cryo-electron microscopy. Structure **1995**, *3*, 63–78.
6. Zlotnick, A.; Aldrich, R.; Johnson, J.M.; Ceres, P.; Young, M. Mechanism of capsid assembly for an icosahedral plant virus. Virology **2000**, *277*, 450–456.
7. Choi, Y.G.; Dreher, T.W.; Rao, A.L.N. tRNA Elements mediate the assembly of an icosahedral RNA virus. PNAS **2002**, *99*, 655–660.
8. Rao, A.L.N. Bromoviruses. In *Encyclopedia of Plant Pathology*; Maloy, O.C., Murry, T.D., Eds.; John Wiley & Sons: Canada, 2001; 155–158.
9. Pogue, G.P.; Lindbo, J.A.; Dawson, W.O.; Turpin, T.H. Tobamovirus Transient Expression Vectors: Tools for Plant Biology and High-Level Expression of Foreign Proteins in Plants. In *Plant Molecular Biology Manual*, 2nd Ed.; Kluwer Academic Publishers: Dordrecht, 1998.
10. Wang, Q.; Lin, T.; Tang, L.; Johnson, J.E.; Finn, M.G. Icosahedral virus particles as addressable nanoscale building blocks. Angew. Chem. Int. Ed. **2002**, *41*, 459–462.
11. Douglas, T.; Young, M. Virus particles as templates for materials synthesis. Adv. Mater. **1999**, *11*, 679–681.

Plant Viruses: Initiation of Infection

James N. Culver
University of Maryland Biotechnology Institute, College Park, Maryland, U.S.A.

INTRODUCTION

Initiation of plant virus infection is a complex process involving the entry and disassembly of the virus particle. Specifically, a virus particle must remain stable within the extracellular environment to protect its genome. However, upon entry into a host cell the particle must destabilize and disassemble in order to initiate the infection process. The mechanisms responsible for shifting particle stability in response to changing environmental conditions have been identified for a number of plant viruses. Furthermore, methods to prevent disassembly have been investigated as a means to develop disease resistance. In this article, the processes that contribute to the initiation of a plant viral infection will be discussed.

VIRUS ENTRY

Unlike animal viruses, which utilize cell surface receptors to gain entry into a cell, plant viruses typically gain access to a new host via nondestructive cellular wounds that penetrate the cell wall and plasmamembrane.[1] Wounding and subsequent infection can occur by mechanical means; for example, through wind-driven rubbing of infected on healthy plants or by damage from contaminated farm implements. Although not of major economic importance, mechanical transmission using some type of abrasive, such as carborundum or celite, has been an important experimental method used in the study of numerous plant viruses. In comparison to mechanical transmission, the majority of economically important plant viruses are vector transmitted. Invertebrates that feed by sucking on plants, for example, aphids and whiteflies, make up the majority of plant virus vectors. However, other invertebrates such as nematodes and beetles as well as fungi—including members in the genera *Olpidium* and *Polymyxa*—also are capable of vectoring plant viruses. Generally, interactions between viruses and their vectors are highly specific, with a vector species transmitting only specific types of viruses.[1] However, all of these vectors provide a direct method of entry for the virus into the cell cytoplasm.

VIRUS DISSASSEMBLY

In general, virus particles are stable macromolecules that function to protect their genomes from degradation. However, once inside a cell, the virus particle must disassemble in order to initiate virus replication and spread. Cellular entry therefore requires the normally stable virus particle to destabilize to permit the virus genome access to the host's molecular machinery. The process of switching from a stable to an unstable particle requires a sophisticated mechanism for sensing changes in the surrounding environment. For many viruses this switching mechanism is thought to involve clusters of negatively charged carboxylate groups that reside at the interfaces between adjacent coat protein subunits (Fig. 1).[2] In the extracellular environment, the repulsive negative charges of the carboxylate groups are neutralized by divalent cations (such as Ca^{++}) or protons. However, upon entry into a cell the lower concentration of cations and the higher pH of the surrounding environment result in the loss of stabilizing cations and protons, allowing the negatively charged carboxylate groups to directly interact. The repulsive electrostatic charges of these groups then function to destabilize the virus particle.

For rod-shaped viruses such as *Tobacco mosaic virus* (TMV), the role of interacting carboxylate residues has been well studied. For each of the approximately 2140 coat protein subunits that make up a TMV particle, there are three known carboxylate groups: one interacting axially, one interacting laterally, and one interacting between the coat protein and the viral RNA (Fig. 1).[3] All of these interacting carboxylate groups have a role in controlling the stability of the TMV particle.[4] Interestingly, destabilization does not happen evenly along the axis of the rod-shaped TMV particle, but occurs predominantly at the 5′ end of viral RNA. The polar nature of virion disassembly is due to instabilities in interactions between the TMV coat protein and the viral RNA. Each TMV coat protein binds three nucleotides of the viral RNA with the strongest binding occurring when a guanine residue is present in the third position.[3] Although there is a preponderance of guanine residues in the third position throughout the viral genome, there are none in the 5′ leader sequence of the viral RNA. This added instability,

Encyclopedia of Plant and Crop Science
DOI: 10.1081/E-EPCS 120019909

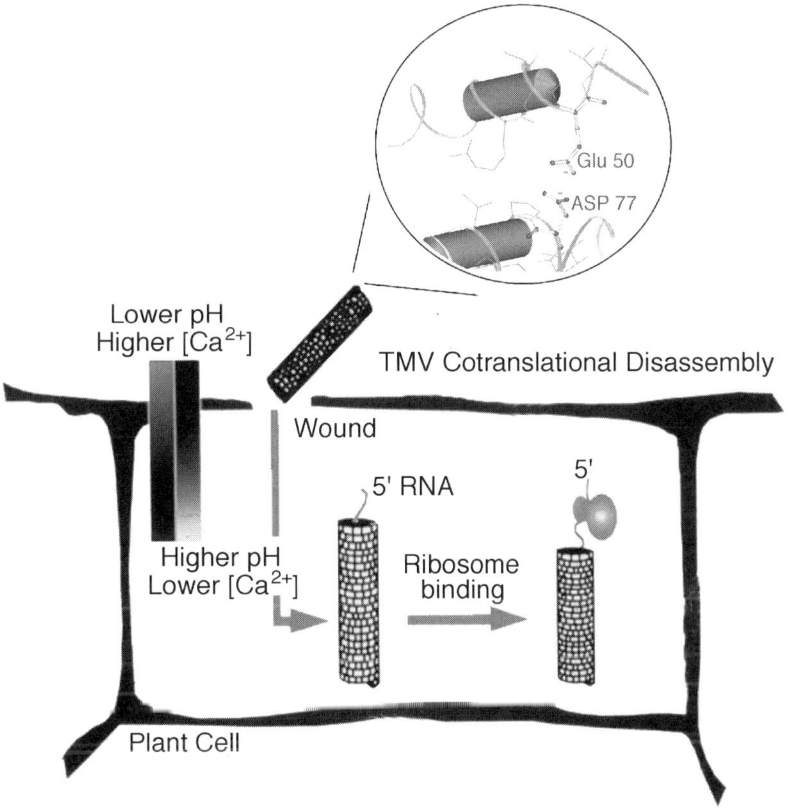

Fig. 1 Model for the cotranslational disassembly of TMV. Expanded circle depicts a carboxylate interaction between aspartate 77 on one coat protein subunit and glutamate 50 on an adjacent coat protein subunit.

along with the repulsive interactions derived from the carboxylate groups, leads to the loss of coat protein subunits from the 5′ end of the viral RNA.

The repulsive interactions of carboxylate groups identified within icosahedral viruses such as *Cowpea chlorotic mottle virus* (CCMV) have been shown to control particle swelling.[5] Particle swelling by as much as 10–15% has been postulated to open channels that allow the release of the viral genome. However, swelling of the virus particle by itself does not always correlate with disassembly because a CCMV mutant deficient in virion swelling readily disassembles in vivo.[6] In addition, some other icosahedral viruses, such as *Cucumber mosaic virus*, do not display particle swelling.[5] Based on these and other findings, it has been proposed that the N termini of CCMV coat proteins, located at the pentameric vertices of the virus particle, undergo a major structural transition from the interior to the exterior of the virus particle.[6] This structural transition likely provides a channel for the release of the viral RNA into the cellular environment. Interestingly, this structural transition is still induced by changes in pH and cation concentration, factors that suggest the involvement of a carboxylate interaction.

TRANSLATION- AND REPLICASE-MEDIATED DISASSEMBLY

Instability mediated by carboxylate interactions does not result in the complete disassembly of either icosahedral or rod-shaped viruses. Instead, the induced instability results in the partial release of the viral genome. For viruses with positive-strand RNA genomes such as TMV and CCMV, this leads to the exposure of ribosome binding sites on the genomic RNA and results in the initiation of protein translation. Ribosome translocation along the length of the genomic RNA results in the removal of the viral RNA from the remaining coat protein/virus particle. The linking of disassembly with translation of the viral genome has been termed cotranslational disassembly, and has the advantage of allowing the viral RNA to remain protected in the cytoplasm until replication is initiated.[7] A similar process, termed replicase-mediated disassembly, also may contribute to the uncoating of the virus particle.[8] For TMV, genome binding and transcription by the viral replicase proteins appear to result in the removal of coat protein subunits from the 3′ nontranslated region of the genomic RNA. Thus, complete disassembly of an RNA viral genome may involve both translation- and replicase-mediated removal of coat protein subunits from the viral RNA.

CROSS PROTECTION: DISRUPTING VIRUS DISASSEMBLY

Cross-protection, defined as the ability of one virus to prevent or delay the infection of a subsequent virus, has been studied as a means to generate novel forms of virus resistance. More recently, plant transformation technologies have significantly expanded our abilities to investigate this phenomenon. In general, cross-protection mechanisms have been linked to either RNA-mediated mechanisms—such as the activation of host-derived gene-silencing—or to protein-mediated mechanisms that disrupt the virus life cycle. The coat protein of TMV has been shown conclusively to play a role in conferring protection in both classical cross-protection and transgenic coat protein-mediated resistance.[9] Specifically, studies have demonstrated that protection can be overcome using infectious uncoated viral RNA as inoculum. Structural studies have also demonstrated that protection is dependent upon the capability of the protecting TMV coat protein to bind the viral RNA and assembly into a virus particle.[10] Mutant coat proteins deficient in assembly cannot confer protection. More important, coat proteins with enhanced abilities to assemble significantly enhance observed levels of protection.[9,10] Together these findings support a recoating protection model in which the presence of the protecting coat protein functions to block the disassembly of the infecting virus particle.

CONCLUSION

The ability of a plant virus to initiate an infection clearly is dependent upon a complex set of mechanisms. These mechanisms also may affect virus movement, disease development, and host resistance.[9] For example, TMV moves systemically within its host's vascular tissues as an assembled virus particle; hence, upon exiting the vascular tissues the virus particle must undergo disassembly in order to initiate a systemic infection. Thus, understanding the processes involved in the initiation of a plant virus infection will enhance our overall knowledge of the entire infection process. Unfortunately, for most viruses our understanding of these processes is minimal. One major limitation is the lack of structural information available for the majority of plant viruses. Thus, there is a need for more information regarding the structure and function of virus particles during the infection process.

ARTICLES OF FURTHER INTEREST

Plant Virus: Structure and Assembly, p. 1029
Virus-Induced Gene Silencing, p. 1276
Virus Movement in Plants, p. 1280

REFERENCES

1. Matthews, R.E.F. *Plant Virology*, 3rd Ed.; Academic Press: San Diego, 1991.
2. Bancroft, J.B. Plant virus structure. Adv. Virus Res. **1970**, *16*, 99–134.
3. Namba, K.; Stubbs, G. Structure of tobacco mosaic virus a 3.6 Å resolution: Implications in assembly. Science **1986**, *231*, 1401–1406.
4. Lu, B.; Stubbs, G.; Culver, J.N. Carboxylate interactions involved in the disassembly of tobacco mosaic tobamovirus. Virology **1996**, *225*, 11–20.
5. Tama, F.; Brooks, C.L., III. The mechanism and pathway of pH induced swelling in cowpea chlorotic mottle virus. J. Mol. Biol. **2002**, *318*, 733–747.
6. Albert, F.G.; Fox, J.M.; Young, M.J. Virion swelling is not required for cotranslational disassembly of cowpea chlorotic mottle virus in vitro. J. Virol. **1997**, *71*, 4296–4299.
7. Shaw, J.G.; Plaskitt, K.A.; Wilson, T.M.A. Evidence that tobacco mosaic virus particles disassemble cotranslationally in vivo. Virology **1986**, *148*, 326–336.
8. Shaw, J.G. Tobacco mosaic virus and the study of early events in virus infection. Philos. Trans. R. Soc. Lond., B **1999**, *354*, 603–611.
9. Culver, J.N. Tobacco mosaic virus assembly and disassembly: Determinants in pathogenicity and resistance. Annu. Rev. Phytopathol. **2002**, *40*, 281–308.
10. Lu, B.; Stubbs, G.; Culver, J.N. Coat protein interactions involved in tobamovirus cross-protection. Virology **1998**, *248*, 188–198.

Pollen-Stigma Interactions

Simon J. Hiscock
University of Bristol, Bristol, U.K.

INTRODUCTION

The pollen-stigma interaction, as defined by Heslop-Harrison, consists of: 1) pollen capture; 2) pollen adhesion; 3) pollen hydration; 4) pollen germination; and 5) pollen tube penetration of the stigma. Successful completion of these events normally results in pollen tube growth to the ovary, leading to fertilization unless incompatibility/incongruity factors preclude. The pollen-stigma interaction is thus a necessary prezygotic courtship between the male gametophyte and the maternal tissues of the sporophyte. Molecules regulating the interaction reside at the surfaces of pollen and stigma (self-incompatibility proteins excluded), but our knowledge of their identity remains largely fragmentary.

THE STIGMA SURFACE

Stigmas can be broadly classified as wet or dry, depending on whether or not they possess a surface secretion.[1,2] Such a classification is useful in the context of this review because the nature of the pollen-stigma interaction depends largely on the microecology of the stigma.[2] For wet stigma species, investigations have focused largely on species from the Solanaceae (notably *Nicotiana*), whereas for dry stigma species, *Brassica* and *Arabidopsis* define the model.

POLLEN-STIGMA INTERACTIONS IN SPECIES WITH WET STIGMAS

Secretions on wet stigmas are primarily lipid-rich (e.g., Solanaceae) or primarily carbohydrate-rich (e.g., Liliaceae) and are clearly important for pollen capture and adhesion. This is because, on making contact with a wet stigma, pollen is quickly trapped and immersed within it. In *Nicotiana*, the absence of a stigmatic secretion results in aberrant pollen development on the stigma, preventing fertilization. Lipids have been shown to be the essential component of the stigmatic secretion necessary for pollen tube penetration because ablated stigmas of female-sterile *Nicotiana* can be restored to full fertility by the application of a range of lipids, particularly *cis*-unsaturated triacylglycerides.[3,4] Lipids within the secretion also appear to direct pollen tube growth toward the stigma by providing an increasing gradient of water potential between the desiccated pollen grain and the turgid cells of the stigma. In vitro simulations of the stigmatic environment using oil-in-water emulsions confirmed that *Nicotiana* pollen tubes grow down gradients of water potential toward an aqueous phase.[4] Whether the carbohydrates in lily-type secretions play a similar role in guidance and penetration of pollen tubes remains to be determined. Wet stigma species usually lack a stigmatic cuticle, so once the pollen tube has germinated and begun to grow toward the stigma, its entry and passage into the transmitting tissue are largely unimpeded. Within the transmitting tissue, chemical cues—notably TTS (a Transmitting Tissue-Specific arabinogalactan glycoprotein)—may guide pollen tubes toward the ovules; physical cues such as cell surfaces may also guide pollen tubes toward ovules.[4,5] It is also within the transmitting tissues of the style that pollen tubes of the self-incompatible (SI) members of the Solanaceae are inhibited by the action of style-specific S-RNases.[5]

POLLEN-STIGMA INTERACTIONS IN SPECIES WITH DRY STIGMAS

For nearly thirty years, *Brassica* has been the model for studies of pollen-stigma interactions on the dry stigma. In *Brassica*, studies of the default state of (compatible) pollen-stigma interactions have been combined with studies of sporophytic self-incompatibility (SSI) because incompatible pollen is recognized and rejected at the stigma surface. More recently, studies of compatibility have shifted to the *Arabidopsis* model.

The stigma of *Brassica* and *Arabidopsis* consists of a dome of epidermal papillae covered by a continuous cuticle that forms a major barrier to pollen tube penetration. Upon the cuticle sits a thin membrane-like layer of protein, the pellicle, which forms the first site of molecular contact between the stigma and alighting pollen grains (Fig. 1). The protein composition of the pellicle has yet to be determined, but cytochemical studies have revealed the presence of esterases and glycoproteins.[1,2]

Encyclopedia of Plant and Crop Science
DOI: 10.1081/E-EPCS 120012919

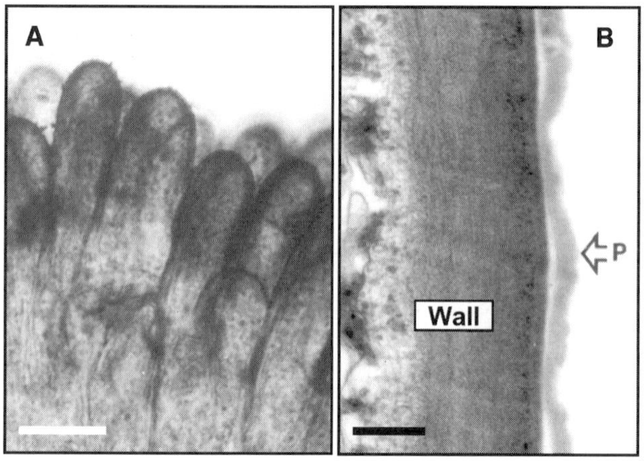

Fig. 1 The dry stigma surface of *Brassica*. A) Stigmatic papillae stained with α-naphthyl actetate and fast blue RR salt to visualize nonspecific esterase activity. Bar = 10 μm. B) Transmission electron micrograph (TEM) of oblique section through cell wall of stigmatic papilla stained with α-naphthyl actetate and hexazatized pararosanalin to visualize esterase activity of the pellicle (P). Bar = 0.2 μm.

In the absence of a sticky stigmatic secretion, adhesion of pollen to the stigma appears to be largely a function of the pollen wall. Structural studies in *Brassica* suggested that the lipidic pollen coating mediates adhesion because immediately after making contact with the stigma, pollen coating is extruded from the exine onto the surface of the stigma where it forms an appresoria-like "foot"[6] (Fig. 2). Nevertheless, pollen adhesion assays using the *Arabidopsis* mutant *cer6*-2, which fails to produce a pollen coating, did not indicate loss of adhesion in the coatless pollen.[7] This observation points to the sporopollenin exine as being responsible for pollen adhesion. Indeed, *Arabidopsis lap* (*l*ess *a*dherent *p*ollen) mutants, compromised in pollen adhesion, show aberrant patterns of exine development.[7] The adhesive function of the exine could be attributable to the exinic outer layer (EOL)—a thin surface layer surrounding the exine, which, like the stigmatic pellicle, has yet to be characterized.[6] Interestingly, structural studies in *Brassica* suggest that upon contact, the EOL and pellicle fuse (S. J. Hiscock unpublished observation).

Two stigmatic members of the *Brassica S* gene family have been implicated in pollen adhesion. Gene knockouts of *SLR1* (*S* Locus Related glycoprotein 1) resulted in reduced adhesion of pollen on transgenic stigmas, and pretreatment of stigmas with antisera to SLR1 or SLG (*S* Locus Glycoprotein) also resulted in reduced pollen adhesion.[8] Importantly, SLR1 and SLG each bind specific members of a group of pollen coat proteins (PCPs)—an interaction that may contribute to the adhesive effects of SLR1 and SLG on pollen.

The main role of the pollen coating appears to be in pollen hydration, because all *Arabidopsis cer* mutants are defective in pollen hydration and consequently do not produce a pollen tube on the stigma. Mutant *cer6*-2 pollen, which has virtually no pollen coating, cannot stimulate water release from the stigma but germinates readily in liquid medium in vitro, and under high humidity in vivo. All *cer* mutants (*cer1*, *cer3*, *cer6*, and *pop1*) are unable to synthesize long-chain lipids[5,7] The fact that *cer* mutants can be rescued by application of triacylglycerides suggests that lipids in the pollen coating play a role in pollen development similar to the lipids in the stigmatic secretions of *Nicotiana*.[3,7] Lipids within the pollen foot may set up a gradient of water potential between the pollen grain and the stigma that acts as a guide to the growing pollen tube, indicating the direction of the stigma. In support of this assertion, pollen tubes of *Brassica* always grow through the pollen foot prior to penetration of the stigma.[6]

Regulated hydration of pollen is clearly essential for normal pollen development on stigmas of *Brassica* and *Arabidopsis*, but pollen coat proteins (in addition to lipids) appear to play a key role in this process. GRP17 is an oleosin-domain protein present in *Arabidopsis* pollen coating that is necessary for normal hydration of the pollen grain. GRP17 belongs to a family of oleosin-domain *Arabidopsis* pollen coat proteins, orthologues of which are also present in the pollen coating of *Brassica*. It has been suggested that this variable group of oleosin proteins and a group of similarly variable pollen coat lipases may form

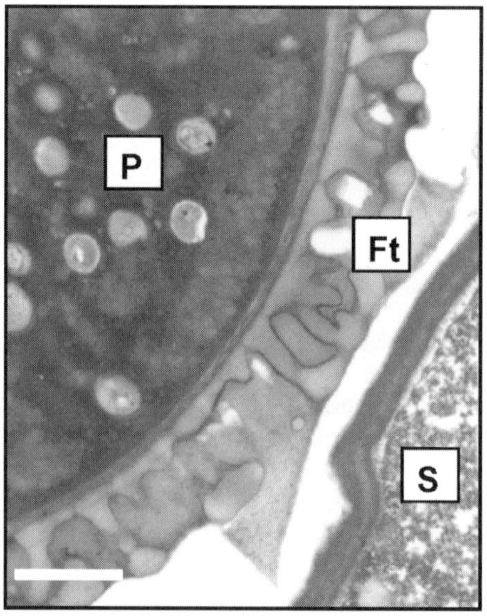

Fig. 2 The *Brassica* pollen foot. TEM of oblique section through pollen-stigma interface 1 hour after compatible pollination. P = pollen grain, Ft = pollen foot, S = stigmatic papilla. Bar = 2 μm.

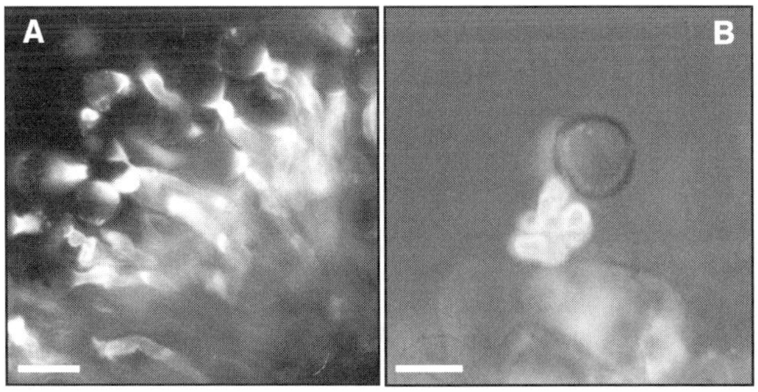

Fig. 3 Pollen tube penetration on the *Brassica* stigma. Fluorescence micrographs of pollen stained with decolorized aniline blue to visualize callose. A) Control pollination showing pollen tubes freely penetrating stigmatic papillae. Bar = 20 μm. B) Stigma treated with serine esterase inhibitor ebelactone B prior to pollination to inhibit cutinase; the convoluted pollen tube is unable to penetrate the stigma. Bar = 20 μm.

species-specific or subspecies-specific pollen recognition cassettes.[5] How these species-specific pollen tags are read by the stigma is a matter for speculation, but an interaction with proteins of the pellicle is appealing.

Unlike species with wet stigmas, pollen tubes of *Brassica* and *Arabidopsis* must breach a cuticle during penetration of the stigma. In *Brassica* an intine-held cutinase is released at the pollen tube tip during penetration. This pollen cutinase is very similar to fungal cutinases, particularly in its sensitivity to serine esterase inhibitors—treating stigmas with DIPF or ebelactone B prior to pollination reduces pollen tube penetration by up to 70%[9] (Fig. 3). Once the cuticle has been breached, the pollen tube grows within the papilla cell wall toward the base, where it emerges and continues to grow intercellularly toward the ovary. Pollen-held enzymes again appear to be involved in these processes, most notably a pectin esterase.[9] In *Arabidopsis*, pollen tubes penetrate directly into the papilla cells, growing through the cytosol and then forcing through the basal wall region and into the transmitting tissues. Similar pollen tube behavior can be seen in *Brassica* when immature stigmas are pollinated.[6]

Self-incompatible *Brassica* pollen can be arrested at any point during the pollen-stigma interaction. Pollen is recognized as self by means of a haplotype-specific protein-protein interaction between SRK, a stigmatic (*S*) receptor kinase, and SCR (*S* cysteine-rich), a small pollen coat protein similar to PCPs.[5] Downstream events following this recognition that directly inhibit pollen development have yet to be characterized. However, the inhibitory machinery of SSI is preserved in self-compatible *Arabidopsis*, even though it lacks an *S* locus, because plants transformed with *SRK* and *SCR* derived from an *S* haplotype of close SI relative *A. lyrata* become SI.[10] This major discovery will now make it possible to dissect the SSI response in the more tractable model *Arabidopsis*.

CONCLUSION

As more is becoming known about the molecular mechanisms of SI, interest is focusing on molecules that mediate the default state of the pollen-stigma interaction—compatibility. The importance of lipids on the stigma in early pollen development is becoming clear, and insight into pollen adhesion and hydration is being gained through studies of *Arabidopsis* male-sterile mutants. Further insight will follow as more subtle screens are devised for *Arabidopsis* mutants compromised at various stages of the pollen-stigma interaction. Key mutants will be those that show penetration defects, possibly resulting from lesions in pollen genes that encode enzymes such as cutinase. Another key challenge will be to identify and functionally characterize the proteins that constitute the pellicle of dry stigmas.

ACKNOWLEDGMENTS

The Biotechnology and Biological Sciences Research Council (BBSRC) is acknowledged for the award of a David Phillips Research Fellowship.

REFERENCES

1. Heslop-Harrison, J.S. Incompatibility and the pollen-stigma interaction. Annu. Rev. Plant Physiol. **1975**, *26*, 403–425.
2. Heslop-Harrison, Y. Control gates and micro-ecology: The pollen-stigma interaction in perspective. Ann. Bot. **2000**, *85* (A), 5–13.
3. Wolters-Arts, M.; Lush, W.M.; Mariani, C. Lipids are required for directional pollen tube growth. Nature **1998**, *392*, 818–821.
4. Lush, W.M.; Spurck, T.; Joosten, R. Pollen tube guidance by the pistil of a solanaceous plant. Ann. Bot. **2000**, *85* (A), 39–47.
5. Franklin-Tong, V.E. The difficult question of sex: The mating game. Curr. Opin. Plant Biol. **2002**, *5*, 14–18.
6. Elleman, C.J.; Franklin-Tong, V.E.; Dickinson, H.G. Pollination in species with dry stigmas: The nature of the early stigmatic response and the pathway taken by pollen tubes. New Phytol. **1992**, *121*, 413–424.
7. Zinkl, G.M.; Preuss, D. Dissecting *Arabidopsis* pollen-stigma interactions reveals novel mechanisms that confer mating specificity. Ann. Bot. **2000**, *85* (A), 15–21.
8. Heizmann, P.; Luu, D.-T.; Dumas, C. The clues to species specificity of pollination among Brassicaceae. Sex. Plant Reprod. **2000**, *13*, 157–161.
9. Hiscock, S.J.; Bown, D.; Gurr, S.J.; Dickinson, H.G. Serine esterases are required for pollen tube penetration of the stigma in *Brassica*. Sex. Plant Reprod. **2002**, *15*, 65–74.
10. Nasrallah, M.E.; Liu, P.; Nasrallah, J.B. Generation of self-incompatible *Arabidopsis thaliana* by transfer of two *S* locus genes from A. *lyrata*. Science **2002**, *297*, 247–249.

Polyploidy

Jan Dvorak
University of California, Davis, California, U.S.A.

INTRODUCTION

Polyploid organisms contain more than two genomes. The gametophytic (haploid) chromosome number is n, and the sporophytic (somatic) chromosome number is $2n$. The basic chromosome number in a phylogenetic lineage is x. Both n and x should be presented to describe ploidy unequivocally. For example, $2n = 4x = 40$ indicates that chromosome number is being described in a sporophyte that is tetraploid and has 40 chromosomes.

TYPES OF POLYPLOIDY

Two basic types of polyploidy are recognized: autopolyploidy (autoploidy) and allopolyploidy (alloploidy). They are defined according to either the origin of a polyploid or its meiotic chromosome behavior. According to the former definition, autoploids are of an intraspecific origin, whereas alloploids are of a hybrid origin. According to the latter definition, autoploids show mutivalents at MI of meiosis and polysomic inheritance, whereas alloploids show bivalents at MI and disomic inheritance. The term segmental alloploidy,[1] proposed for intermediate states between autoploidy and alloploidy, is inappropriate in most cases and should not be used.[2]

The genomic constitution of polyploids is expressed by formulas in which genomes are designated by capital letters according to their relatedness (Table 1). An autotetraploid may have a formula AAAA, reflecting the fact that its genomes originated from within a single species. Its chromosomes are homologous and it likely shows polysomic inheritance. In contrast, an allotetraploid may have a formula AABB, reflecting the fact that its genomes were contributed by two different species, one contributing the A genomes and the other contributing the B genomes. The corresponding chromosomes in the A and B genomes are very likely homoeologous and the species likely shows disomic inheritance.

The sources of genomes are identified by genome analysis of the polyploid.[3] Description of various methods of genome analysis can be found in a compendium on this subject.[4] In its classical form, it consists of analyses of chromosome pairing at MI in hybrids from crosses between a polyploid and its relatives.[3] Sharing of isozyme alleles—chloroplast genome restriction fragments and nucleotide sequences—and restriction fragments of nuclear repeated nucleotide sequences—in situ hybridization of labeled heterologous genomic DNA—can provide additional valuable information on the origin of polyploid species. Once a diploid species (or its closest extant relative) that was the source of a genome in a polyploid is identified, the same letter is assigned to that genome in the polyploid and the diploid (Table 1). Genome evolution that may have occurred since the origin of a polyploid, extinction of a diploid ancestor, or hybridization may result in failure to find a diploid whose genome fully fits a genome of a polyploid. An imperfect fit between genomes is indicated by assigning superscripts or subscripts to genome designations or by other notation in genome formulas.

GENETIC TRANSMISSION

Polyploidy greatly impacts gene segregation. In an autotetraploid, three heterozygous states—simplex (*Aaaa*), duplex (*AAaa*), and triplex (*AAAa*)—are possible in addition to the two homozygous states, nulliplex (*aaaa*) and quadruplex (*AAAA*). Additionally, there is an opportunity for double reduction if a crossover occurs between the locus and centromere. Double reduction creates an opportunity for sister alleles getting into a gamete instead of homologous alleles. The maximum equational segregations of a duplex heterozygote *AAaa* (*A* being completely dominant over *a*) is 19.5 dominant phenotypes:1 recessive phenotype, as compared to the 35 dominant phenotypes:1 recessive phenotype for strictly reductional segregation of the duplex.[5]

In an allotetraploid, a chromosome can pair during meiosis either homogeneticaly, i.e., with its homologue (e.g., chromosome 1A pairs with 1A), or heterogenetically, i.e., with a homoeologue (e.g., chromosome 1A pairs with 1B). A vast majority of alloploids are meiotically diploidized, meaning that only homogenetic pairing occurs. In some polyploids, heterogentic chromosome pairing is genetically suppressed. An example of a heterogenetic pairing suppressor is the wheat *Ph1* locus. If the gene is mutated or removed by aneuploidy, heterogentic

Encyclopedia of Plant and Crop Science
DOI: 10.1081/E-EPCS 120005591

Table 1 Genomes of important polyploid field and tree crops

Botanical name	Crop	Ploidy (2*n*)	Ploidy type	Genome formula (2*n*)	Ancestors
Arachis hypogaea	Peanut	$4x = 40$	Allo.	AABB	A = *A. villosa*, B = *A. ipaensis*
Avena abyssinica	Oats	$4x = 28$	Allo.	AABB	A = *A. strigosa*, B = closely related to A
Avena sativa, *Avena byzantina*, *Avena nuda*	Oats	$6x = 42$	Allo.	AACCDD	A = ?, C is related to *A. ventricosa*, D = ?, AD = *A. marocana*
Beta vulgaris	Sugar beet	$2x = 18$, $3x = 36$*	Auto.		*Cultivars
Brassica napus	Canola and swede	$4x = 38$	Allo. x = paleo $3x$	AACC	A = *B. campestris*, C = *B. oleracea*
Chenopodium spp.	Quinoa	$4x = 36$	?		
Citrullus lanatus	Watermelon	$2x = 22$, $3x = 33$ cultivars	Auto.		
Coffea arabica	Coffee	$4x = 44$	Allo.		
Colocasia esculenta	Taro	$2x = 28$, $3x = 42$	?		
Dioscorea spp.	Yams	$2x$, $4x$, $5x$, $6x$, $7x$, $8x$, $9x$, $10x$, $12x$, $14x = 20$, 40, 50, 60, 70, 80, 90, 100, 120	?		
Eleusine coracana	Finger millet	$4x = 36$	Allo.	AABB	A = *E. indica*, B = *E. floccifolia*
Fragaria ananassa	Strawberry	$8x = 56$	Auto-allo?	AAAABBBB	A is related *F. vesca*
Glycine max	Soybeans	$2x = 40$	Paleo. $4x$		
Gossypium hirsutum, *Gossypium barbadense*	Gotton	$4x = 52$	Allo. x is paleo	AADD	A = *G. arboreum* or *G. herbaceum*, *D* is related to *G. raimondii*
Helianthus annuus	Sunflower	$2x = 34$	x = paleo $4x$?	BB	
Helianthus tuberosus	Jerusalem artichoke	$6x = 102$	Allo.	$A_1A_1A_2A_2BB$	A_1 and A_2 = a perennial *Helianthus*, B = an annual *Helianthus*
Ipomoea batatas	Sweet potato	$6x = 90$	Auto.?	BBBBBB	B = *I. batatas*
Malus pumila	Apple	$2x = 34$, $3x = 51$*	Auto.		*Cultivars
Manihot esculenta	Cassava	$2x = 36$	Paleo?		
Medicago sativa	Alfalfa	$4x = 32$	Auto.		Autoploid of *M. sativa*
Musa sp.	Bananas	$2x = 22$, $3x = 33$	Auto. and Allo.	AAA, AAB, ABB	A = *M. acuminata*, B = *M. balbisiana*
Nicotiana tabacum	Tobacco	$4x = 48$	Allo.	SSTT	S = *N. silvestris*, T = *N. tomentosiformis*
Nicotiana rustica	Tobacco	$4x = 48$	Allo.	PPUU	P = *N. paniculata*, U = *N. undulata*
Prunus avium	Sweet cherries	$2x = 16$, $3x = 24$* $4x = 32$*	Auto. and allo.		*Some cultivars are allotetraplid *P. avium* and *P. cerasus*
Prunus cerasus	Sour cherries	$4x = 36$*	?		
Prunus domestica	Plums	$6x = 48$	Auto.		Autoploid of *P. cerasifera*
Saccharum officinarum	Sugarcane	Either $8x$ or $10x = 80$	?		Unknown
Solanum tuberosum	Potato	$4x = 48$	Auto.	AAAA	A = *S. stenotomum*
Sorghum bicolor	Sorghum	$2x = 20$	Paleo. 4x?		
Trifolium repens	White clover	$4x = 32$	Allo		One genome = *T. nigrecens*, the other may be = *T. occidentale*
Triticum turgidum	Emmer and other tetraploid wheats	$4x = 28$	Allo	AABB	A = *T. urartu*, B is related to *Aegilops speltoides*
Triticum aestivum	Common wheat, spelt, and other hexaploid wheats	$6x = 42$	Allo.	AABBDD	AB = *T. turgidum*, D = *Ae. tauschii*
Triticum timopheevii		$4x = 28$	Allo.	AAGG	G = *Ae. speltoides*
Triticum zhukovskyi		$6x = 42$	Allo.	AAA^mA^mGG	A^m = *T. monococcum*
Zea mays	Maize	$2x = 20$	Paleo. $4x$		

meiotic pairing occurs at MI. Diploidized allotetraploids show disomic inheritance, provided that the locus duplicated in the other genome pair is homozygous recessive. If both loci are heterozygous (*AaAa*) and *A* is completely dominant over *a*, an allotetraploid segregates 15 dominant:1 recessive phenotypes.

INCIDENCE

Polyploidy is widespread in plants but rare in animals.[6] Its incidence may be as high as 70% in angiosperms. It is claimed that as much as 95% of all species of homosporous ferns are polyploid, although this figure is still being debated. The highest chromosome number and highest ploidy among plants was recorded in fern *Ophioglosum pycnostichum* ($2n = 84x = 1260$). In angiosperms, polyploidy is more abundant in monocots than in dicots. For example, as many as 80% of all grass species were estimated to be polyploid. Polyploidy is more abundant in herbaceous angiosperms, particularly those that are perennial, than in woody species. Polyploidy is rare in gymnosperms. For example, only 1.5% of coniferous species are polyploid.

ORIGINS

The principal mechanism by which natural polyploids originate is the union of an unreduced gamete with a reduced gamete or two unreduced gametes.[7] The union of an unreduced diploid gamete with a reduced monoploid gamete of the same species results in an autotriploid. Autotriploids are usually highly male-sterile but partially female-fertile, and may serve as bridges in the origin of autotetraploids.[8] Autotetraploids were estimated to originate with a frequency of about 10^{-5}.[8] The frequency of unreduced gametes is greatly elevated in interspecific hybrids. Therefore, the frequency of the emergence of alloploids is more dependent on the frequency of interspecific hybridization than on the production of unreduced gametes by the parental species. The cross-pollinating mating system enhances the incidence of allopolyploidy. Both auto- and alloloids originating from cross-pollinating species, including those that are self-incompatibile, are usually self-pollinating, which assists nascent polyploids in their propagation and establishment.

EVOLUTIONARY SUCCESS

Polyploids have sometimes been portrayed as genetically depauperized taxa exploiting an available habitat at the expense of long-term evolutionary success. This view is incongruous with the tendency of alloploids to become geographically widespread. Moreover, molecular mapping has demonstrated that some apparently diploid plant lineages are actually paleopolyploid (Table 1), showing that polyploids can establish successful phylogenetic lineages. It has been demonstrated for more than 30 species that polyploids originate recurrently in natural populations.[9] Recurrent origin and interploidy hybridization, followed by introgression, broadens the polyploid gene pools and enhances the potential of polyploids to adapt to changing environment. Nascent alloploids are notorious for their genetic and meiotic instability, caused by recombination and subsequent segregation, DNA rearrangement, and transposition, and epigenetic variation caused by differential methylation These mechanisms increase variation in nascent polyploid populations and contribute to their adaptation and evolutionary success.

INDUCED POLYPLOIDY

Polyploidy can be induced by temperature shocks and numerous other treatments interfering with mitosis. The most effective is treatment with the alkaloid colchicines, which blocks entry of a mitotic cell into the anaphase. Because colchicine is always applied on multicellular tissues, treated plants become polyploid chimeras. Nitrous oxide dissolved in cytosol under pressure also blocks polymerization of microtubules and results in polyploidization. The gas can be applied at the time of zygotic division, resulting in the entire plant being polyploid. Endopolyploid cells, which are frequent in somatic tissues of plants, can also be a source of polyploids regenerating from them. Endopolyploidy is the principal cause of polyploidy of adventitious shoots and plants regenerated from tissue culture. Autotriploids can be produced by crossing diploids with autotetraploids. In some cases, this cross is difficult (triploid block) or is possible in only one direction.

PHENOTYPIC EFFECTS

The primary effect of autoploidy is larger cell size. Therefore, autoploids are often more robust, have thicker and darker leaves and larger stomata and pollen size, and are slower developing and less tolerant of environmental stresses than their diploid counterparts. There is often a specific ploidy level for increase in plant size due to autoploidy. Alloploids tend to be intermediate between parents, although they may resemble one of the parents in specific traits or may show altogether new traits.

POLYPLOID CROPS

Polyploidy was about as frequent in crops as in their noncrop relatives.[10] Polyploid field crops and fruit trees, their genome formulas, and synopses of their genome analysis are listed in Table 1. The majority of polyploid crops are alloploid. Low fertility of triploids has been exploited in the production of seedless triploid varieties of watermelon and bananas. In sugar beet, triploid varieties are more productive than their diploid and autotetraploid counterparts, and have been commercially used to produce monogerm varieties. Some varieties of apples and cherries are triploid.

CONCLUSION

Polyploidy is an important evolutionary strategy in the plant kingdom. Many important crops are polyploid, although the incidence of polyploidy in crops does not appear to be any higher than in their noncrop immediate relatives. Polyploidy greatly modifies inheritance. In autoploid populations, recessive phenotypes appear in much lower frequencies than in diploid populations with similar allele frequencies. A general tendency of natural alloploids is to genetically ''diploidize,'' (i.e., to evolve bivalent meiotic chromosome pairing and disomic inheri- tance). The molecular and genetic mechanisms of this process remain poorly understood. Also poorly understood are gene expression and genome evolution in alloploid species.

ARTICLES OF FURTHER INTEREST

Aneuploid Mapping in Diploids, p. 33
Aneuploid Mapping in Polyploids, p. 37
Genome Size, p. 516
Interspecific Hybridization, p. 619
In Vitro Chromosome Doubling, p. 572
In Vitro Production of Triploid Plants, p. 590

REFERENCES

1. Stebbins, G.L. Types of polyploids: Their classification and significance. Adv. Genet. **1947**, *1*, 403–429.
2. Dubcovsky, J.; Luo, M.-C.; Dvorak, J. Differentiation between homoeologous chromosomes 1A of wheat and $1A^{m}$ of *Triticum monococcum* and its recognition by the wheat *Ph1* locus. Proc. Natl. Acad. Sci. U. S. A. **1995**, *92*, 6645–6649.
3. Kihara, H. *Wheat Studies—Retrospect and Prospects*; Elsevier Scientific Publishing Company: Amsterdam, 1982; 1–308.
4. Jauhar, P.P. *Methods in Genome Analysis in Plants*; CRC Press: Boca Raton, 1996; 1–386.
5. Burnham, C.R. *Discussions in Cytogenetics*; Burgess Publishing Company: Minneapolis, 1962; 1–393.
6. Otto, S.P.; Whitton, J. Polyploid incidence and evolution. Annu. Rev. Genet. **2000**, *34*, 401–437.
7. Harlan, J.R.; DeWet, J.M.J.; On, O. Winge and a prayer: The origins of polyploidy. Bot. Rev. **1975**, *41*, 361–390.
8. Ramsey, J.; Schemske, D.W. Pathways, mechanisms, and rates of polyploid formation in flowering plants. Annu. Rev. Ecol. Syst. **1998**, *29*, 467–501.
9. Soltis, D.E.; Soltis, P.S. Polyploidy: Recurrent formation and genome evolution. Trends Ecol. Evol. **1999**, *14*, 348–352.
10. Hilu, K.W. Polyploidy and the evolution of domesticated plants. Am. J. Bot. **1993**, *80*, 1494–1499.

Population Genetics

Michael T. Clegg
University of California, Riverside, California, U.S.A.

INTRODUCTION

Genes are the fundamental hereditary units transmitted from parent to offspring in every generation. Population genetics is concerned with the statistical rules that govern the transmission of genes within collections of interbreeding individuals over time. Changes in the frequencies of genes within populations are the basis for evolutionary change, as well as the basis for the genetic improvement of plants and animals that provide human sustenance.

Population genetics employs mathematical models to describe trajectories of genetic change under various scenarios. These models are also used to design experiments to dissect the processes of evolution. The basic mathematical models of population genetics were developed long before it was established that DNA is the chemical basis of heredity. The challenge today is to analyze and learn from the rapidly accumulating body of genome sequence data. The basic models of population genetics are highly relevant to this enterprise because they are general, in the sense that change in gene frequency can be substituted with change in individual nucleotide frequency without any alteration in the underlying mathematics. This article provides a brief outline of the theory of population genetics as it applies to changes in the composition of DNA molecules over time. Several excellent books provide a comprehensive introduction to population genetics theory, including works by Hartl and Clark and Hedrick.

EVOLUTIONARY FORCES THAT DRIVE DNA SEQUENCE CHANGE

Five forces determine genetic change.[1,2] These begin with mutation, defined as a permanent heritable change in a gene. A second force is recombination, which is a randomizing force that can also lead to permanent heritable changes in a gene. Genetic random drift, defined as the small fluctuations in gene frequencies caused by the sampling of individuals in reproduction, is third. The fourth force is migration, or the movement of genes among different populations. Fifth is natural selection, in which some individuals are more successful in transmitting their genes owing to superior adaptation to their environment and hence improved survival and reproduction. Natural selection is the force that leads to adaptive genetic change; it essentially sorts among mutations to eliminate those that are deleterious and increase the frequency of those that are advantageous.

MUTATION: THE SOURCE OF GENETIC NOVELTY

Gene or nucleotide sequences are subject to constant change. To begin, mutation takes many forms, each of which has some likelihood of occurring at each DNA replication cycle. From this it is apparent that any DNA molecule is subject to mutational erosion over time; this erosion process is a form of evolutionary change. DNA sequences that code for functionally important molecules are expected to be conserved because natural selection rejects most deleterious mutational changes in functionally important genes. Current genome sequencing efforts suggest that much of the DNA of organisms does not code for proteins or regulatory signals—these regions may be evolving relatively rapidly.

GENETIC RANDOM DRIFT: THE ROLE OF CHANCE IN EVOLUTIONARY CHANGE

Once a new mutation has arisen in a single DNA molecule (to create a new allelic form of a gene or a new allele), it may be transmitted to the next generation or it may be lost. Assuming the new mutation has no effect on fitness, the probability of transmission to the next generation is $1 - [1 - 1/2N]^{2N}$ in diploid organisms, where N is the population size. This result follows from the fact that a new mutation will initially occur in a single individual (frequency = $1/2N$). The pool of gametes is assumed to be much larger than the N reproductive individuals in the population, so sampling with replacement is appropriate. (This expression is the probability of not being lost in the first generation, derived assuming binominal sampling.) In the vast majority of cases a new mutation will be lost through the vagaries of sampling within a generation or

Encyclopedia of Plant and Crop Science
DOI: 10.1081/E-EPCS 120006084

two. A very small proportion will drift to high frequency and be fixed (i.e., become the sole allele in the population or species).

An important result of mathematical genetics indicates that the probability of ultimate fixation of a new neutral mutation is $1/2N$ (the concept of neutrality is discussed later). Because there is a constant flux of new mutations, there will also be a constant rate of fixation of new allelic variants over the long term. In fact, the rate of fixation of new alleles is simply μ, the mutation rate. This result provides a formal justification for the molecular clock hypothesis (see article on molecular evolution), because it indicates that the rate of mutational change should be constant over time if the mutation rate is constant.

A second important theoretical result from population genetics demonstrates that the average time until the fixation of a new mutation is $T \sim 4N_e$, so for large N_e, many mutant alleles are expected to be in transition to fixation at any point in time. (N_e is a population size adjusted to take account of variation in reproductive output among different members of a breeding population.) For example, many plant species must have population sizes in the order of 10^6 to 10^7. The rate of mutation per nucleotide site per generation is approximately $\mu = 5 \times 10^{-9}$. If a typical gene is 10^3 nucleotide sites in length, the total probability of a new mutation in the gene in any single generation is 5×10^{-6}. For a species size ($2N$) of 10^6, five new mutations in this particular gene are expected in the total species per generation. Moreover, the average new mutation that is destined to be fixed will drift in the species of size 10^6 for 4,000,000 generations. The probability of drawing two different alleles of the gene in a population or species can be shown to be approximately $1 - 4N_e\mu/(1 + 4N_e\mu) = 19/21 = 0.90$ for this numerical example, so almost every pair of genes sampled is expected to be different. Thus, high levels of molecular polymorphism are expected in species. Current efforts to define all human single nucleotide polymorphisms (SNPs) verify these theoretical predictions.

THE NEUTRALITY HYPOTHESIS: AN INTERACTION BETWEEN MUTATION AND DRIFT

The theoretical results cited in the preceding discussion are based on the assumption that selection is absent. In the 1970s, Motoo Kimura[3] and associates advanced a theory called the neutrality hypothesis that claims that a large proportion of molecular evolutionary change is neutral to natural selection. This hypothesis was in part motivated by an important calculation by J. B. S. Haldane,[4] who showed that there is an upper bound to the rate of evolutionary change under natural selection. Because selective change requires differential survival and reproduction, the reproductive excess of the species defines this upper bound. Stated differently, some or all of the reproductive excess of a species is consumed by the process of selection. Empirical evidence on rates of molecular change appear to be too high to be consistent with this upper bound, especially for long-lived low-fecundity organisms like the human. As a consequence, it was postulated that a large proportion of molecular change is not driven by selection, but by random drift and mutation.

The test of the neutral theory is to ask how well observed data fit the theoretical calculations outlined earlier. Because the neutral theory provides precise mathematical predictions, the theory is testable; moreover, these tests provide a very useful context for identifying particular genes that are selected. This is because it is known precisely how neutral genes should behave, and departures from neutrality are likely to arise from selection. To date, available data do suggest that many polymorphisms are neutral or only very weakly selected. However, the statistical power to detect departures from neutrality increases with database size, which are growing very rapidly at present.

NATURAL SELECTION: THE ENGINE OF ADAPTIVE CHANGE

Natural selection operates at the level of phenotypic differences, so an allelic change must alter some aspect of phenotype to be perceived by selection. The phenotypic dimensions of organisms, and the ways these may be perceived by various environmental factors, are very complex. These range from subtle changes in rates of flow through a metabolic pathway, to resistance to disease, to the ability to withstand extreme stress environments, to aspects of behavior, and so on. The list is virtually endless. Moreover, each aspect of phenotype may be determined by a number of different genes and by the way these genes interact in development and metabolism. It is similarly difficult to identify the precise facets of an environment that may act as a selective agent. As a consequence, selection is often identified retrospectively. That is, the analytical tools of population genetics may support the inference that selection has affected a genetic locus based on the pattern of change through time. Such an analysis rarely tells the reasons why a particular change is adaptive.

A number of mathematical results that provide useful predictions about the statics and dynamics of selective genetic change are available from population genetic theory. We will mention just one result from this large body of work. The probability of fixation of a new mutant that is favored by selection (assuming the heterozygous genotype is intermediate in its fitness between the two homozygous genotypes) is $\sim 2s$, where s is the selective intensity favoring the new mutant allele. So if $s = 0.02$ (implying a 2% selective advantage), the favored allele will still be lost 96% of the time. This result demonstrates that most new favored mutations will still be lost owing to drift.

The average time to fixation of a new favored mutant is approximately $T \sim (2/s)\text{Ln}(2N_e)$ generations (where Ln denotes the natural logarithmic function). Numerical iterations of this equation, compared to the neutral case, reveal that the time to fixation of selectively favored alleles is much shorter than for a neutral allele. Indeed, fixation of a selected allele can take place in orders of magnitude less time than a neutral allele. This calculation suggests that in contrast to the high levels of neutral polymorphism expected from the neutral theory outlined in the fore going, it may be relatively rare to find a selected gene in transition when one allele is favored over the other allele. The exception to this statement is the case where two or more alleles are favored, either because they lead to a higher fitness among heterozygotes or because their advantage is frequency-dependent (frequency dependent means that s is not constant, but instead depends on allele frequencies). Both of these situations are referred to as balanced polymorphism. Classic examples of a balanced polymorphism are often associated with disease resistance, where the resistance allele causes some other disadvantage or where there is a constant arms race between the disease agent and the target organism favoring new mutations at a resistance locus (a case of frequency dependence). An example of such a balanced polymorphism arises from the multiple major histocompatibility polymorphism (MHC) alleles in human populations.[5] In this case, disease resistance is strongly implicated because the MHC locus is involved in the recognition of potential disease agents.

RECOMBINATION AS BOTH RANDOMIZING AND CREATIVE FORCE

Recombination refers to the rearrangement of genes with respect to one another on chromosomes. Recombination is a randomizing force, but it is also much more. It is a creative force that operates at many levels within plant genomes.[6] Thus, molecular evidence has revealed that recombination is responsible for the generation of many gene novelties. For example, studies of different allelic sequences of a gene often show traces of past intragenic recombination events where new allelic forms have arisen, owing to the exchange of stretches of sequence between two different alleles. This reciprocal exchange can yield a new allele that is different from both parental forms.

THE COALESCENT: LOOKING BACKWARD IN TIME

An important theoretical construct from population genetics is known as the coalescent. Consider a set of DNA sequences that are drawn from a single locus within a species. The sample of sequences must have all descended from a common ancestor. The point where all alleles trace back to the most recent common ancestor is known as the coalescent. It is straightforward to estimate the genealogical history of the sample by using phylogeny estimation tools (discussed in the bioinformatics article). If the nucleotide sequence diversity in the sample is the result of a drift/mutation process at equilibrium, then the arrival times of new mutation events are given by an exponential process.[7] This result provides a useful framework for testing sequence samples to see if they conform to a neutral process.

A particularly important statistic calculated from a sample of sequences is θ, a measure of the nucleotide sequence diversity. A different statistic that also measures nucleotide sequence diversity is π. (Both of these quantities are related to the probability of drawing two different nucleotides at a single site averaged over all sites examined.) If we calculate π and θ as an average over many independent replications of the evolutionary process that led to the observed sample, and if the underlying evolutionary process is the result of a drift/mutation process at equilibrium, then both θ and π should be approximately equal to $4N_e\mu$. Thus, the difference between π and θ should be zero under the drift/mutation assumption. A test statistic introduced by Tajima[8] based on a function of the difference between π and θ allows one to ask whether a data set conforms to the drift/mutation assumption by testing whether this difference is zero. If the difference is nonzero, then other forces such as selection or demographic change must be invoked. Many standard computer programs for sequence analysis implement these calculations (i.e., DnaSP).

Another statistic that features importantly in coalescence theory is T_{MARCA} (time to the most recent common ancestor of the sample). Mathematical calculations show that T_{MARCA} is also approximately equal to $4N_e$ when the drift/mutation assumption is satisfied. Finally, if μ is known it is possible to estimate N_e for the species or population from which the data came. Estimates of N_e from a few important crop plants vary over nearly an order of magnitude from 2×10^5 (wild barley) to 3×10^5 (pearl millet) to 18×10^5 (maize).[9] These data are derived from broad samples of the entire species so the estimates can be viewed as the best estimates of long-term effective population size for each species.

MIGRATION: THE SPATIAL DIMENSION OF GENETIC CHANGE

Every new mutation must have originated in a single individual at a specific location in space. This means that the spread of a mutation through a species is a spatial process, and it is possible to gain some insight into historical migration patterns through the analysis of geographic samples of gene sequences. Moreover, the coalescent framework has been refined to include spatial considerations, so migration rates can be quantified. Efforts to measure migration are still in very early stages, but contemporary data suggest that plants and their genes have been quite mobile over long periods of time.

There is great contemporary concern over the spread of so-called transgenes from agricultural plants into wild or weedy relatives through migration. This question has stimulated much work on the direct measurement of migration in plants. Current results suggest that gene migration rates may be higher than previously believed.[10]

CONCLUSION

Although population genetics is a mature science with a history of nearly 100 years, it is more relevant today than ever. The theoretical models of population genetics provide the conceptual and statistical tools to address important questions ranging from concerns about transgene escape to detection of the influence of selection on specific genes. As the body of genome sequence data expands, these tools will become ever more important in interpreting and mining this new resource.

ARTICLES OF FURTHER INTEREST

Agriculture and Biodiversity, p. 1
Bioinformatics, p. 125
Breeding Self-Pollinated Crops Through Marker-Assisted Selection, p. 202
Breeding: Choice of Parents, p. 215
Breeding: Mating Designs, p. 225
Crop Domestication: Fate of Genetic Diversity, p. 333
Crop Domestication: Role of Unconscious Selection, p. 340
Gene Flow Between Crops and Their Wild Progenitors, p. 488
Gene Silencing: A Defense Mechanism Against Alien Genetic Information, p. 492
Molecular Evolution, p. 748
Mutational Processes, p. 760
Population Genetics of Plant Pathogenic Fungi, p. 1046

REFERENCES

1. Hartl, D.L.; Clark, A.G. *Principles of Population Genetics*, 3rd Ed.; Sinauer Associates Inc.: Sunderland, MA, 1997; 542.
2. Hedrick, P.A. *Genetics of Populations*, 2nd Ed.; Jones and Bartlett: Boston, MA, 2000; 553.
3. Kimura, M. *The Neutral Theory of Molecular Evolution*; Cambridge University Press: Cambridge, England, 1983; 367.
4. Haldane, J.B.S. The cost of natural selection J. Genet. **1957**, *55*, 511–524.
5. Nei, M. DNA Polymorphism and Adaptive Evolution. In *Plant Population Genetics, Breeding and Genetic Resources*; Brown, A.H.D., Clegg, M.T., Kahler, A., Weir, B.W., Eds.; Sinauer Assoc.: Sunderland, MA, 1990; 128–142.
6. Clegg, M.T. The Role of Recombination in Plant Genome Evolution. In *Evolutionary Processes and Theory*; Waser, S.P., Ed.; Kluwer Academic Publishers: The Netherlands, 1999; 33–45.
7. Hudson, R.R. Gene genealogies and the coalescent process Oxf. Surv. Evol. Biol. **1990**, *7*, 1–44.
8. Tajima, F. Statistical method for testing the neutral mutation hypothesis by DNA polymorphism Genetics **1989**, *123*, 585–595.
9. Cummings, M.P.; Clegg, M.T. Nucleotide sequence diversity at the alcohol dehydrogenase I locus in wild barley (*Hordeum vulgare* ssp. *spontaneum*): An evaluation of the background selection hypothesis Proc. Natl. Acad. Sci. U. S. A. **1998**, *95*, 5637–5642.
10. Ellstrand, N. When transgenes wander, should we worry? Plant Physiol. **2001**, *125*, 1543–1545.

Population Genetics of Plant Pathogenic Fungi

Bruce A. McDonald
Swiss Federal Institute of Technology (ETH), Zürich, Switzerland

INTRODUCTION

Pathogens evolve. As a result of pathogen evolution, resistance genes and fungicides can rapidly lose their effectiveness in agroecosystems. To understand the processes that drive pathogen evolution, we must understand pathogen population genetics. The population genetics of all organisms is determined by interactions among five evolutionary factors: mutation, random drift, mating/reproduction system, gene flow, and natural selection. The end result of interactions among these factors is the population genetic structure of a species.

GENETIC DIVERSITY AND POPULATION GENETIC STRUCTURE

Genetic structure (often called population structure) refers to the amount and distribution of genetic diversity within and among populations. Because many pathogens have an asexual reproduction phase, pathogen genetic diversity has two components, gene diversity and genotype diversity. Gene diversity is determined by the number and frequencies of alleles at individual loci within a population. Genotype diversity is determined by the number and frequencies of genetically distinct individuals in a population. Measures of genetic diversity are especially important when considering the evolutionary potential of a pathogen population. According to Fisher's Fundamental Theorem of Natural Selection,[1] the rate of increase in fitness of any organism is proportional to its genetic variance in fitness. In other words, pathogen populations with greater genetic diversity are expected to evolve more quickly than populations with less genetic diversity. Thus we begin to learn about pathogen population genetics and evolutionary potential by analyzing the genetic structure of pathogen populations.

THE FIVE EVOLUTIONARY FORCES

Mutation is the ultimate source of genetic variation, causing changes in the DNA sequence of a gene and creating new alleles. Mutations are rare and would not cause pathogen evolution if they operated in isolation. But when mutation is coupled with selection, mutants can increase in frequency rapidly and cause significant changes in allele frequency. Mutations from avirulence to virulence and from fungicide sensitivity to fungicide resistance are especially important in agriculture.

Population size affects the probability that mutants will be present, and also influences the diversity of genes in populations through a process called random genetic drift. Large populations have more mutants than small populations because mutation rates are relatively constant and usually quite low. Thus, large populations are expected to have greater gene diversity (more alleles) than smaller populations. Genetic drift occurs when a small random subset of a population survives a catastrophic event that causes a severe reduction in population size (a bottleneck) or when a small random subset of a pathogen population colonizes a new host population (founder event). The frequency of mutant alleles in the surviving or founding population can differ significantly from the frequency of the mutant alleles in the original population. Founder events often occur in plant pathology when a disease is introduced into a new area by accident or as a result of a breach of quarantine. Bottlenecks often occur at the end of a growing season when the crop is harvested or during a crop rotation that reduces overseasoning pathogen inoculum.

Gene flow is a process in which particular alleles (genes) or individuals (genotypes) are exchanged among geographically separated populations. For strictly asexual organisms that do not recombine genes with the recipient population, entire genotypes are exchanged among populations (genotype flow). Gene flow substantially increases population size by increasing the size of the genetic neighborhood over which genes or genotypes are exchanged. Gene/genotype flow is the process that moves newly arisen virulent mutant alleles among different field populations. It is clear that the size of the genetic neighborhood is affected by the method of natural dispersal of pathogen propagules. Pathogens with air-dispersed spores are likely to have a larger genetic neighborhood than soilborne pathogens. But humans move many pathogens beyond their natural dispersal limits as a result of intercontinental travel and global commerce.

Encyclopedia of Plant and Crop Science
DOI: 10.1081/E-EPCS 120019937

Reproduction and mating systems affect the way gene diversity is distributed within and among individuals in a population, leading to differences in genotype diversity. Pathogen reproduction can be sexual, asexual, or mixed (having both sexual and asexual components). Mating system is relevant only to the sexual component of reproduction, and can vary from strict inbreeding to obligate outcrossing. Some pathogens (e.g., all *Fusarium oxysporum* formae speciales and most bacteria) reproduce only asexually in agroecosystems. These pathogens exist as a series of discrete clones or clonal lineages with little evidence of recombination among clonal lineages. Populations of these pathogens usually exhibit a low level of genotype diversity. In strictly asexual pathogens, measures of genotype diversity are more meaningful than measures of gene diversity because most of the genetic diversity is distributed among clonal lineages. An advantage of sex is that, with each generation, new combinations of alleles come together through recombination, leading to an increase in genotype diversity that may enable some component of the pathogen population to survive in a variable and potentially hostile environment (e.g., on a resistant host). A disadvantageous result of sex is that fit combinations of alleles (coadapted gene complexes) are broken up through recombination, so it becomes difficult to maintain groups of alleles that offer an advantage in specific environments. Sexual pathogens usually exhibit a high degree of genotype diversity, so measures of gene diversity are needed to compare populations. Pathogens with mixed reproduction systems have significant advantages over strictly asexual or strictly sexual pathogens. During the sexual cycle, new combinations of alleles (genotypes) are produced that can be tested in different environments. During asexual reproduction, combinations of alleles (genotypes) that are most fit are held together through clonal reproduction and may increase to a high frequency. Thus these pathogens can exhibit high levels of both gene and genotype diversity. The spatial and temporal distribution of clonal lineages within and among populations will depend mainly on the dispersal of the asexual propagules. If asexual spores are capable of long-distance dispersal, then the clone with highest fitness can become distributed over a wide area through genotype flow relatively quickly.

Selection is the force that is most easily managed in agroecosystems. The strong directional selection that occurs when a major plant disease resistance gene becomes widely distributed drives the increase in frequency of the virulent pathogen mutant that has the corresponding virulence allele. The many examples of resistance genes that are overcome by new pathogen strains offer abundant evidence that selection is efficient in agricultural ecosystems that are based on monoculture and genetic uniformity. Resistance gene deployment strategies that impose disruptive selection, such as cultivar mixtures, can slow the evolution of pathogen populations and extend the usefulness of resistance genes.[2]

AN EXAMPLE OF GENETIC STRUCTURE

By using selectively neutral genetic markers and hierarchical sampling to determine the genetic structure of pathogen populations, we can begin to understand the evolutionary forces that shaped these populations, and infer the importance of the individual evolutionary factors for each pathogen.[3] Commonly used DNA-based marker systems include RFLPs, microsatellites, AFLPs, and DNA sequences. DNA fingerprints are used to identify clones within pathogen populations. To determine whether a pathogen population has a sexual component of reproduction, it is common to clone-correct the dataset, using only one representative of each clone in the analysis of underlying allele frequencies. For diploid pathogens, the starting point is to test for departures from Hardy Weinberg equilibrium. If a population is at Hardy–Weinberg equilibrium, then there is no deviation between expected and observed genotypic proportions within loci, and associations among loci are random, a state called gametic equilibrium. For haploid pathogens, it is common to test for departures from gametic equilibrium (a state called gametic disequilibrium) in a clone-corrected dataset to determine whether a pathogen population is undergoing sexual recombination. Comparisons between populations are made by comparing allele frequencies to obtain measures of genetic similarity. Population subdivision is measured using F-statistics (diploid pathogens) or Nei's G_{ST} (haploid pathogens).

From a population genetics perspective, the best characterized fungal pathogen at this time is *Mycosphaerella graminicola* (anamorph *Septoria tritici*), which causes the septoria leaf blotch disease on wheat.[4] Over 5000 isolates of *M. graminicola* from field populations around the world have been characterized for diversity in mitochondrial and nuclear genomes using a combination of RFLPs, DNA fingerprints, and DNA sequence loci. Field experiments were conducted to differentiate among the five evolutionary forces and obtain empirical estimates for each factor in an agricultural setting. Through population genetic analysis, we now know that *M. graminicola* has a mixed reproduction system, with sexual recombination occurring mainly between growing seasons but also during the growing season. Asexual spores move between plants by splash-dispersal and sexual ascospores move by wind. As a result, the typical field population is composed of a mosaic of overlapping clones, with few clones distributed over a spatial scale of more than a few meters. On average, each leaf is colonized by a different pathogen genotype, and it is common to find many different genotypes within the same lesion on a leaf. Regular sexual reproduction maintains the two mating types at a 1:1 ratio through frequency-dependent selection, and unlinked loci are at gametic equilibrium. The genetic neighborhood is large as a result of long-distance movement of windblown ascospores, so that populations separated by 1000s of kilometers on the same continent are genetically very similar. Populations on different continents also are very similar, suggesting that the pathogen has been spread globally through infected seed. Approximately 80% of the world's genetic diversity is found within an area 1 m^2 in any field sampled anywhere in the world. Populations are stable over time and a large number of alleles are found at individual RFLP loci, suggesting that effective population sizes are large (at least 70 individuals per square meter). Selection coefficients associated with specific clones competing in field experiments were large, illustrating the potential for rapid evolution in field populations. Taken together, these findings well explain the observed rate of evolution in populations of *M. graminicola.*

CONCLUSION

Knowledge of pathogen population genetics can be applied to the development of resistance breeding strategies and gene deployment that may lead to resistance that is durable over long periods in the field, and can guide disease management strategies.[1] For example, a wilt pathogen such as *Fusarium oxysporum* that has low evolutionary potential as a result of strict asexual reproduction, low gene flow potential, and small population size may best be controlled by using single major resistance genes. A powdery mildew pathogen such as *Blumeria graminis* that has high evolutionary potential as a result of a mixed reproduction system, high genotype flow, and large population size may require quantitative resistance and an intensive resistance gene management strategy (such as cultivar mixtures) to obtain durable resistance.

ARTICLES OF FURTHER INTEREST

Breeding for Durable Resistance, p. 179
Population Genetics, p. 1042

REFERENCES

1. Fisher, R.A. *The Genetic Theory of Natural Selection*, 1st Ed. Clarendon: Oxford, 1930.
2. McDonald, B.A.; Linde, C. Pathogen population genetics, evolutionary potential, and durable resistance. Annu. Rev. Phytopathol. **2002**, *40*, 349–379.
3. McDonald, B.A. The population genetics of fungi: Tools and techniques. Phytopathology **1997**, *87*, 448–453.
4. Linde, C.; Zhan, J.; McDonald, B.A. Population structure of *Mycosphaerella graminicola*: From lesions to continents. Phytopathology **2002**, *92*, 946–955.

Potassium and Other Macronutrients

Patrick H. Brown
University of California, Davis, California, U.S.A.

INTRODUCTION

The mineral nutrients K, Mg, Ca, and S are unique and highly varied with respect to their acquisition from the soil, distribution within the plant, metabolic roles, and deficiency responses. An exhaustive analysis of these processes is beyond the scope of this article, thus only a brief overview is provided. For each element, however, a more detailed description of a single physiological process that is either unique or broadly illustrative of an important principle in plant nutrition is provided.

POTASSIUM

Potassium is the most abundant univalent cation in the cytoplasm of plants. It forms only weak complexes and does not compete strongly with other cations in metabolic reactions. As a consequence of these physical and chemical characteristics, K is uniquely suited to serve as the primary inorganic cellular osmoticum and, through thereversible formation of complexes with soluble and insoluble anions, as a cytoplasmic pH buffer. The critical functions of K in plants include osmoregulation, turgor-related processes (e.g., tropisms, stomatal opening, and cell expansion) and counterbalancing of ions in the movement of sugars into the phloem.

Uptake and Cellular Compartmentation

The importance of K as a cellular osmoticum and pH buffer requires that the uptake and cellular compartmentation of K be tightly regulated and highly responsive to changes in local K availability and cellular K demand. The mechanism of transport of K across membranes is the best studied of all mineral elements and is broadly illustrative of the process of nutrient acquisition and regulation in plants.

In the early 1950s Emmanuel Epstein and colleagues at the University of California, Davis, provided important early insights into nutrient uptake by plants.[1] Two mechanisms were postulated. The first, or high-affinity, system operated at low external concentrations of K, was saturable, and could be described by Michaelis-Menton kinetics. A second mechanism operating at higher concentrations of K was postulated to be present either in the same or in different membranes (i.e., plasma membrane and tonoplast, respectively). With the completion of the genomic sequence from *Arabidopsis* and the advent of powerful molecular and biophysical techniques, it is now clear that transport of K is even more complex than originally envisioned. At least three gene families with a total of 34 individual genes are known to contribute to K uptake and transport. The various K transporters differ in their affinity for K and the specific mechanisms by which they facilitate uptake of K. These can be divided into high-affinity carriers (such as HKT1) and low-affinity channels, of which there are many examples (e.g., the KAT and SAT families). The HKT1 gene functions as a Na-K cotransporter that at high concentrations of Na preferentially transports Na and as such may contribute to salinity stress. The exact mode of action of these transporters remains uncertain, although the large number of K transporters with apparent redundancy likely provides plants with the capacity for closely regulated uptake of K under a variety of environmental conditions.[2]

Functions of Potassium

The majority of functions associated with K occur by virtue of its role as an osmoticum for turgor-driven processes, as a counterion for charge balance associated with H^+ and solute transport, and as a stabilizer of enzymes and proteins. The processes of both cell expansion and stomatal opening involve the active transport of large amounts of K into the vacuole. The subsequent accumulation of water in the cell results in the increase in cell turgor required for stomatal opening and cell expansion. Potassium functions in photosynthesis as the dominant counterion to photosynthetic H^+ flux across the thylakoid membranes of chloroplasts. Potassium facilitates the loading of sucrose into phloem by providing the required osmotic potential in source tissues and through its role as a counterion for sucrose and H^+ transport into phloem sieve tubes.

Encyclopedia of Plant and Crop Science
DOI: 10.1081/E-EPCS 120010513

Deficiencies of Potassium

Potassium is sometimes required in amounts that equal or even exceed the requirement for nitrogen in crop plants that produce large amounts of carbon-rich structures such as grains, fleshy fruits, tubers, and nuts. The characteristic symptoms of K deficiency include marginal and tip burn of mature leaves. This is a consequence of the high mobility of K in the plant that favors K movement from old to young plant parts and of the role of K in water relations of the leaf.

CALCIUM

Plants require substantial amounts of Ca. Among the mineral nutrients, only N and K are required at higher concentrations. The functions of calcium in the plant fall broadly into two categories—structural and signaling. The structural role involves the formation of stable but reversible complexes in the cell wall and membrane. The signaling role is critical in several signal transduction pathways.

Uptake and Cellular Compartmentation

The majority of Ca in the cell is found in the apoplasm where it is either strongly bound in structures or present in exchangeable pools in cell walls or at the external surface of the plasma membrane. Calcium is not freely mobile in the phloem and is delivered to growing cells largely through xylem flow. As a consequence, Ca concentrations increase with tissue age and the supply of Ca may be inadequate to satisfy demand in apical meristems, leaf

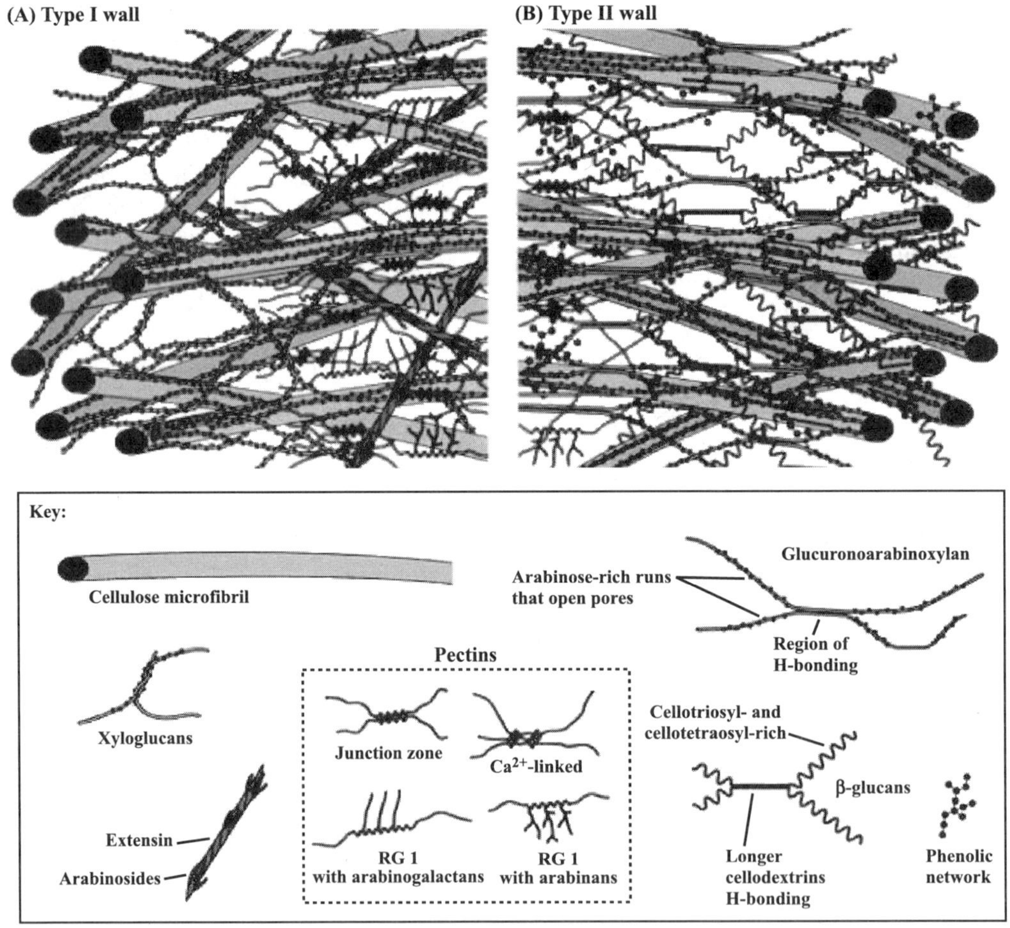

Fig. 1 A three dimensional molecular model of the Type I (non-grass) and Type II (grass) wall shows the molecular interactions between cellulose, xyloglucans, pectins and wall proteins. Pectin in the cell wall provides sites for Ca binding which then cross link with adjacent pectin HGA to form junction zones. The density of these junction zones and the degree of esterification as well as the distribution of borate di-esters (not shown) influences cell wall pore size, structure and function. (*View this art in color at www.dekker.com.*)

tips, and fruits with poor vascular connectivity and rapid growth. In contrast to the highly variable and unregulated accumulation of Ca in the apoplast, cytoplasmic Ca concentrations are extremely low and highly regulated. No fewer than four families of Ca channels, as well as a Ca^{2+}-ATPase, have been found in the plasma membrane, rough ER, tonoplast, and mitochondrial membranes. Two of these channels can be activated by ligands, whereas others are sensitive to voltage changes or activated by mechanical stretching.

Functions of Calcium

In plant cell walls the cellulose-xyloglucan framework is embedded in a pectin matrix. Pectin methylesterase in the cell wall cleaves some of the methyl groups from homogalacturonan (HGA) and provides sites for Ca binding, which then cross-link with adjacent HGA to form junction zones. The density of these junction zones and the degree of esterification as well as the distribution of borate diesters influences cell wall pore size[3] (Fig. 1). In fruit ripening there is a solubilization of Ca that precedes the softening process. Spray application of Ca to fruits is widely practiced to increase the concentration of Ca in tissue and to delay fruit softening.

Calcium plays a critical role in the stabilization of the plasma membrane and is essential for the activity of membrane-bound enzymes and selectivity of ion uptake. In the absence of adequate Ca, membranes become leaky and ultimately disintegrate with a loss of cell

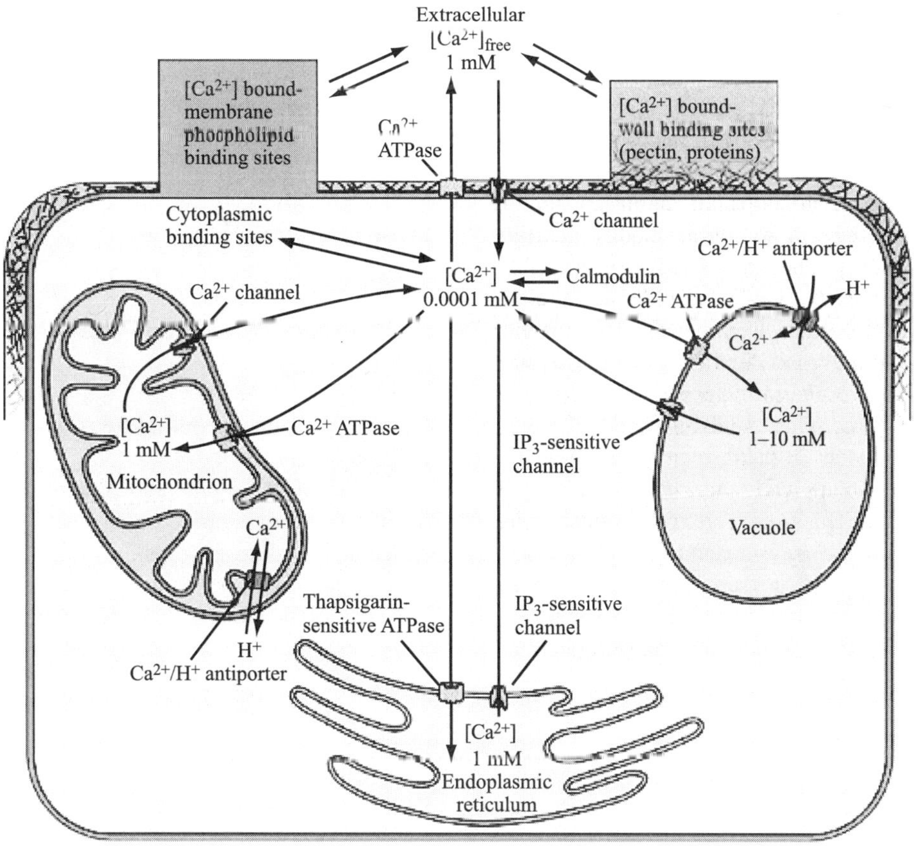

Fig. 2 Interactions of intracellular and extracellular Ca^{2+} in cell signaling. The relationships of Ca^{2+} stores in plant cells are known to be complex, concentrations of Ca^{2+} being high in organelles and in the cell wall and low in the cytoplasm. When the cell is signaled, channels are opened in various organelles or in the plasma membrane, allowing Ca^{2+} to enter the cytoplasm by diffusing down its electrochemical gradient. Ca^{2+} ATPases and perhaps Ca^{2+}/H^+ antiporters return the cytoplasmic concentration to resting value. Where known, subcellular concentrations of Ca^{2+} are indicated (quoted values in the ER vary from 0.1 to 1 mM). The concentration of cytoplasmic binding sites has been measured at about 0.15 to 1 mM. Free cytoplasmic Ca^{2+} is in equilibrium with these binding sites. Increases of cytosolic calcium, $[Ca^{2+}]_i$, activate calmodulin, thereby initiating subsequent downstream events. IP_3, inositol 1,4,5-triphosphate. (*View this art in color at www.dekker.com.*)

compartmentation. A primary consequence of saline stress occurs when Na replaces Ca at membrane-binding sites, resulting in a loss of membrane functionality.

The internal concentration of Ca, $[Ca^{2+}]_i$, is a principle component in signal transduction pathways in plant cells. This role of Ca requires that cytosolic Ca concentrations (100–200 nM) be maintained many orders of magnitude lower than in the cell wall (1–10 mM). Changes in cytosolic Ca occur in response to a variety of environmental and developmental cues that initiate an opening of Ca channels and a rapid transient increase in $[Ca^{2+}]_i$. The increase in cytosolic $[Ca^{2+}]_i$ activates numerous Ca^{2+}-binding proteins (calmodulin and calmodulinlike protein kinases) that bind and activate a variety of target proteins. In this manner, changes in $[Ca^{2+}]_i$ initiate a wide range of responses, including stomatal closure, gravitropism, pollen tube orientation, and many others. The increase in $[Ca^{2+}]_i$ is dissipated by Ca-ATPases located in the tonoplast, ER, and plasma membrane (Fig. 2).

Deficiencies of Calcium

The critical role of Ca in cell wall formation and membrane stability requires that Ca always be present in adequate amounts in all meristematic tissues. Calcium transport within the plant, however, is largely passive, being driven by water movement and local diffusion. Only very limited Ca movement occurs in the phloem. Calcium deficiency often results in a rapid inhibition of growth and subsequent tissue necrosis. In fleshy organs (fruits, tubers), Ca deficiency results in membrane leakiness, reduced cellular cohesion, and enhanced tissue breakdown. This is particularly evident in fast-growing tissues where demand might exceed supply. Leaf tip burn in lettuce, bitter pit in apple, and blossom end rot in tomato are examples of disorders that are induced by Ca deficiency. Correction of these disorders requires that Ca supply be maintained throughout plant growth. Supplemental Ca applications postharvest can significantly increase storage life of fruits and vegetables.

MAGNESIUM

Magnesium is a divalent cation that functions primarily through the formation of ionic complexes with nucleophilic ligands and as a bridging molecule between phosphoryl groups and other molecules or enzymes. Magnesium uptake likely occurs through both channels and ATP-dependent pumps; however, the specific mechanisms of Mg transport are poorly understood. K^+, NH_4^+, Ca^{2+}, and Mn^{2+} each compete with Mg^{2+} for uptake and cause Mg^{2+} deficiency in plants from many environments.

Functions of Magnesium

Magnesium is essential for chlorophyll. Consequently, chlorosis and reduced photosynthesis are primary consequences of Mg deficiency. Depending upon Mg supply, Mg in chlorophyll varies from 6–60% of total cellular Mg. Magnesium also influences photosynthesis through its role as a regulator of Calvin cycle enzymes such as Rubisco and fructose 1,6 bisphosphatase. Magnesium functions as a bridging moiety between enzymes and ATP. Hence, the substrate for ATP-ases, as well as inorganic polyphosphatases, is Mg-ATP rather than free ATP. In roots with sufficient Mg, as much as 90% of cytoplasmic ATP is complexed to Mg. Furthermore, ATP synthesis has an absolute requirement for Mg.

Deficiency of Magnesium

Interveinal chlorosis of the oldest leaves is a characteristic symptom of Mg deficiency. This is a consequence of the role of Mg in chlorophyll synthesis and the high mobility of Mg that results in export of Mg from old tissues to young when Mg becomes limiting. In many Mg-deficient crops, photosynthate accumulates in source leaves as transport to sinks is inhibited.

SULFUR

Sulfur is primarily acquired by plants as the sulfate anion (SO_4^{2-}). The majority of the SO_4^{2-} that enters the plant is then reduced to cysteine and is subsequently converted to methionine and a variety of secondary sulfur-containing molecules. Unlike nitrogen, which is always utilized in the reduced form, SO_4^{2-} may be directly incorporated into a variety of sulfated compounds, including sulfolipids in plant membranes. Animals and humans cannot reduce sulfur, and hence require dietary sources of cysteine and methionine to provide their sulfur needs.

Sulfur Uptake and Transport

Sulfate is relatively abundant in the environment, particularly in industrialized regions where gaseous SO_4^{2-} is present as a pollutant. Sulfate uptake is primarily mediated by a family of proton cotransporters located in various plant membranes. 13 SO_4^{2-} transporter genes have been identified in *Arabidopsis*. Many of these genes respond differently to SO_4^{2-} availability, and the full complex of transporters is likely essential for the normal regulation of sulfur metabolism.

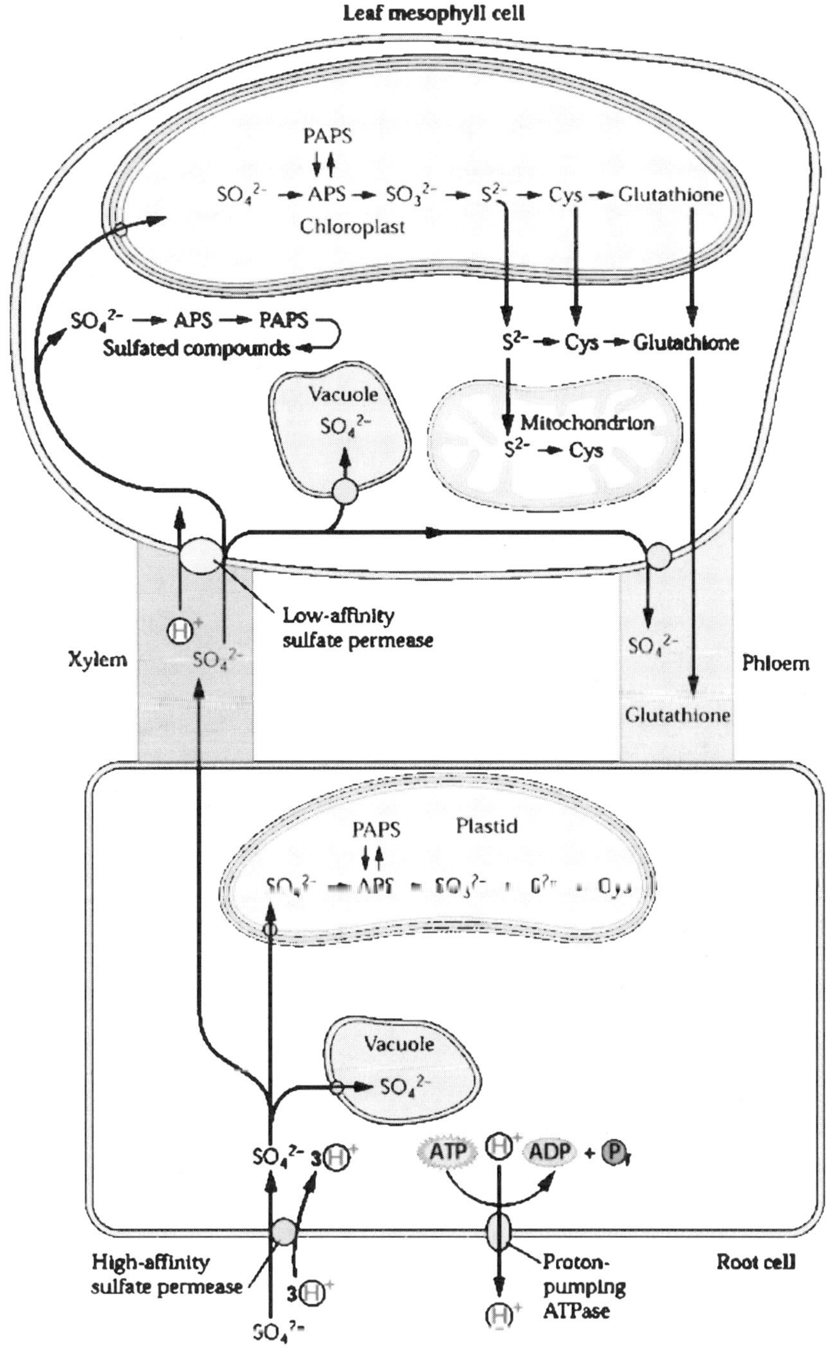

Fig. 3 Overview of sulfur metabolism, reduction, and transport in plants. Like nitrate, sulfate uptake across the plasma membrane is energized by an electrochemical gradient that is maintained by a proton-pumping ATPase. Sulfate is stored in vacuoles. Reduction of sulfate and its assimilation into cysteine take place in plastids of root and leaf cells. APS,5-adenylsulfate; PAPS, 3phoshoadenosine-5′-phosphosulfate. (*View this art in color at www.dekker.com.*)

Functions of Sulfur

Sulfur-containing compounds function in a diverse array of metabolic processes, including as structural components of proteins and membranes; in redox and methylation reactions; and in the detoxification of herbicides, heavy metals, and oxidants. Sulfur-containing coenzymes and vitamins—including the ethylene and polyamine precursor S-adenosyl-L-methionine, coenzyme A, and S-methylmethionine—participate in group transfer and methylation reactions in a variety of metabolic pathways (Fig. 3). The thiol group of cysteines present in proteins can be oxidized, allowing for the formation of disulfide bonds between adjacent cysteine residues. These cross-links are central in determining the tertiary and quaternary structure of proteins. The dithiol–disulfide interchange also occurs in a variety of sulfur-containing molecules (thioredoxin, glutaredoxin, and glutathione) and is critical for the maintenance of cellular redox potential. Glutathione, the major nonprotein thiol in plants, is central for cytoplasmic redox regulation, sulfur storage, heavy metal detoxification, environmental stress tolerance, and as a precursor to phytochelatins. Glutathione detoxifies many toxins and herbicides through the formation of glutathione conjugates that are transported to the vacuole by an ATP-binding cassette (ABC) transporter. Glutathione is also the precursor to phytochelatin ($(\gamma$-Glu-Cys$)_n$ Gly), which plays a critical role in the detoxification of heavy metals such as Cd.

Sulfur Deficiency

Sulfur requirements (0.1 to 0.5%) vary between families (Gramineae < Leguminosae < Cruciferae) and reflect differences in sulfur content of proteins in these families. The most prominent feature of S deficiency is chlorosis, which occurs uniformly throughout the plant. The the sulfur content and concentration of proteins declines in grains during storage and results in important changes in quality (in the case of baking flour) and value as animal feed. Deficiencies of S are most common in acid and leached soils but are becoming more prominent worldwide as levels of pollutant S in the air decline.

ARTICLES OF FURTHER INTEREST

Cropping Systems: Irrigated Continuous Rice Systems of Tropical and Subtropical Asia, p. 349
Minor Nutrients, p. 726
Nitrogen, p. 822
Phosphorus, p. 872
Photosynthate Partitioning and Transport, p. 897
Root Membrane Activities Relevant to Plant-Soil Interactions, p. 1110

REFERENCES

1. Epstein, E.; Rains, D.W.; Elzam, O.E. Resolution of dual mechanisms of potassium absorption by barley roots. Proc. Natl. Acad. Sci. U. S. A. **1963**, *49*, 684–692.
2. Schroeder, J.; Buschmann, P.; Eckelman, B.; Kim, E.; Sussman, M.; Uozumi, N.; Maser, P. Molecular Mechanisms of Potassium and Sodium Transport in Plants. In *Plant Nutrition: Food Security and Sustainability of Agro-ecosystems Through Basic and Applied Research. XIV International Plant Nutrition Colloquium*; Horst, W.J., Schenk, M.K., Burkert, A., Eds.; Kluwer Academic Publishers: Dordrecht, 2001; 10–11.
3. Buchanan, B.B.; Gruissem, W.; Jones, R.J. *Biochemistry and Molecular Biology of Plants*; Amer. Soc. Plant Physiol: Rockville, MD, 2000; 1–1367.

Pre-agricultural Plant Gathering and Management

M. Kat Anderson
Natural Resources Conservation Service, University of California, Davis, California, U.S.A.

INTRODUCTION

During most of the 150,000 year history of humanity, people survived by the hunting and gathering lifestyle. Throughout this extensive period, many cultural groups developed a sophisticated understanding of the inner workings of nature. Hunter-gatherer groups around the world have relied upon hundreds of wild plant species for basketry, ceremonies, foods, furniture, poisons, traps, and weapons for millennia. Their hard-earned, detailed traditional ecological knowledge included information about how wild plants respond to gathering and cultivation. For certain plant species, indigenous people practiced judicious harvesting, adjusting the variables (season, frequency, pattern, scale, intensity, tool) of collecting to influence the growth and abundance of native plant populations. Prior to the invention of agriculture, many areas were also actively managed for increased densities and abundance of wild plants to meet an array of cultural needs. The land management practices that predate agriculture included burning, irrigating, pruning, transplanting, weeding, sowing, and tilling, applied to a multitude of habitat types. This article discusses examples of these practices, along with their potential ecological consequences, and puts forth a view of human-plant interactions as part of a continuum from gathering to domestication.

WILD PLANT MANAGEMENT FOR FOODS

Wild plant gathering for food was traditionally a female responsibility in most hunter-gatherer groups, whereas hunting and fishing were male activities. Many different kinds of plant parts were harvested for food. Small seeds and grains of wildflowers and grasses, and underground swollen stems from herbaceous plants are discussed here (Fig. 1, Tables 1 and 2).[1,3]

Wildflower Seed and Grass Grain Management

Today, cereal grains such as wheat, oats, barley, rice, and corn are the most important of all food crops in the world. One of the oldest forms of sowing in the world is the broadcasting of fruits of wild grasses and the seeds of wildflowers for food. This involves grain and seed selection, saving, and dispersing, often after some kind of ground preparation. It is from this repeated act of broadcasting, together with certain modes of harvesting, that cereal domestication arose.[4] The harvesting strategy of seedbeating coupled with the horticultural techniques of saving and sowing seed, in conjunction with fire management, created ecological effects at the species, population, community, and landscape levels (see Fig. 2A). Seedbeating plants, burning vegetation, and sowing seed tended to select for specific genotypes that hold up well to and even thrive under human harvesting and management regimes.

The Bagundji traditionally repeatedly fired the grasslands in what is now New South Wales, Australia, to increase seed production of Mitchell grass (*Astrelba pectinata*). Patches of grass and brush were burned over to ensure a better crop of edible seeds by the Northern Tonto, Southern Tonto, and White Mountain Apache of the American Southwest. The Southern Paiute of the Great Basin sowed the seeds of Indian rice grass (*Achnatherum hymenoides*). California provides many case examples of tribes managing numerous landscapes to foster the growth and abundance of herbaceous plants for edible seed. The Central Sierra Miwok of the Sierra Nevada cultivated six kinds of seeds with burning, sowing, and harrowing in the Sierra Nevada foothills. Two of these are farewell-to-spring (*Clarkia purpurea* ssp. *viminea*) and mule ears (*Wyethia helenoides*). The Modoc, east Shasta, Achomawi, Northern Maidu, and Nisenan tribes of northern California sowed the seeds of many kinds of wild herbaceous plants.

Bulb, Corm, and Tuber Management

Perennial plants with underground storage organs offer rich sources of starch and other carbohydrates, as well as substantial protein and trace minerals. They were dug with a digging stick by hunter-gatherer groups (see Fig. 2B). Prior to domestication of the onion, potato, cassava, yam, and other underground swollen stems, indigenous people all over the world cultivated subterranean organs of wild plants for foods. This activity was perhaps the oldest form of tillage—one that became the precursory management

Encyclopedia of Plant and Crop Science
DOI: 10.1081/E-EPCS 120017089

regime and provided the ecological foundation for the development of root crop agriculture in many areas. Specific strategies were employed to perpetuate populations of favored plants, such as 1) replanting of cormlets and bulblets; 2) deliberately breaking of the bulb above the root crown, leaving some stem and root tissue; 3) leaving an upper section of the tuber with a vine stem; 4) weeding around favored plants; 5) and burning of areas to decrease plant competition and recycle nutrients. Through human selection, protection, and replanting of offsets, certain geophyte species have probably undergone genetic changes. The excavation of plants with vegetative reproductive parts and the replanting or breaking off of such parts would likely select for specific genotypes that benefit under human disturbance regimes.[5]

Gidjingali women in Arnhemland, Australia, still gather wild *ganguri*, a type of yam (*Dioscorea transversa*), with a digging stick, being careful to leave the vine stem and a small section of tuber behind, replanting it so that it will grow again. The San of South Africa traditionally burned the veldt at the end of the dry season to promote the growth and abundance of edible bulbs and tubers. The Nez Perce of the eastern Columbia Plateau in North America harvested camas (*Camassia quamash*), returning immature bulbs and half of the mature ones to the soil. The preferred digging season was after the seeds set to ensure a future supply. The Western Mono and Sierra Miwok of California harvested the corms of blue dicks (*Dichelostemma capitatum*), purposely breaking off the cormlets and replanting them.

WILD PLANT MANAGEMENT FOR BASKETRY

Making baskets is a worldwide phenomenon. Roots and branches of trees and shrubs, and the leaves and rhizomes of grasses, sedges, and ferns must be pliable, straight, long, and uniform, without insect or disease damage. Yet wild plants often exhibit crooked, brittle growth, with evidence of insects or diseases, making them unsuitable

Fig. 1 Native plants used by the pre-agricultural people of California. (A) Soaproot (*Chlorogalum pomeridianum*)—a very important plant to the majority of California tribes. The bulbs provided glue, fish poison, and food; the old leaf sheaths that clothe the bulb formed the bristles for brushes. The young edible leaves were roasted. The plant is still gathered today. (B) Flowerstalks of deergrass (*Muhlenbergia rigens*)—a perennial bunchgrass—are utilized in the foundation of coiled baskets by numerous California Indian tribes. (C) The now rare and endangered Hall's Wyethia (*Wyethia elata*) that grew prolifically in the montane forests kept open by Indian-set fires were significant for their edible seeds, gathered by the Western Mono of the western Sierra Nevada mountains.

P

Table 1 Cultural uses assigned to native plants by pre-agricultural peoples

Cultural use	Example
Adhesives	The Coast Salish of British Columbia used the pitch of western white pine trees (*Pinus monticola*) to fasten arrowheads onto shafts.
Basketry	The Makah and other tribes of Washington and British Columbia used (and still use) the pliable leaves of a sedge (*Carex obnupta*) in twined basketry.
Clothing	A leafy crown from the vines of *Mascagnia macroptera* was made by the Seri of the Gulf of California in Sonora, Mexico, for protection from the sun and to keep the hair in place.
Cordage	Bast fibers in stems of the perennial dogbane (*Apocynum cannabinum*) were twisted together to make a tough string used for fishing line, nets, tump lines, carrying bags, and bowstrings by tribes of the Pacific Northwest, California, and the Great Basin.
Foods	!Kung Bushmen of the northwest region of the Kalahari Desert relied traditionally on mongongo (mangetti) nuts (*Ricinodendron rautanenii*) for 50% of their vegetable diet. (Mongongo nuts have five times the calories and ten times the protein per cooked unit, compared to cereal crops.)
Games	Boomerangs are cut from bark of river red gum trees (*Eucalyptus camaldulensis*) and are used as toy weapons by aboriginal boys of central Australia.
Medicines	The Attikamek of Quebec used smashed roots of harlequin blueflag (*Iris versicolor*), a perennial wildflower, for a poultice to apply to burns.
Musical instruments	Hollow young branches of blue elderberry shrubs (*Sambucus mexicana*) are carved into clapper sticks and flutes by many California Indian tribes.
Structures	The Thompson of southern British Columbia used leaves and stems of broadleaf cattail (*Typha latifolia*) for mats, wall insulators, and in constructing temporary summer houses.
Tools	Paddles for guiding canoes were made of yellow cedar (*Chamaecyparis nootkatensis*) wood by the Tlingit of Alaska.
Utensils	Spoons were fashioned of bear grass (*Nolina microcarpa*) leaves by the Cibecue, White Mountain, and Tonto Apache of Arizona.
Weapons	The aborigines of Groote Eylandt in northern Australia made spear shafts of yellow hibiscus (*Hibiscus tiliaceus*) for use in hunting wallabies.

Table 2 Pre-agricultural peoples' management techniques

Operation	Description	Example
Burning	Application of fire to particular vegetation areas under specified conditions such as seasonality, fire return interval, and aerial extent to achieve select cultural purposes.	The Australian aborigines of Victoria dug up tubers of murnong (*Microseris scapigera*); gathering areas were historically burned over to increase production.
Irrigating	To supply select land areas with water by means of artificial channels.	The Owens Valley Paiute of the Great Basin artificially watered a host of plants such as wheat grass (*Agropyron* sp.), Great Basin wild rye (*Elymus* sp.), and blue dicks (*Dichelostemma capitatum*) to increase their productivity and abundance.
Pruning	To remove dead and living parts from native plants to enhance growth, form, fruit, or seed production.	The Timbisha Shoshone of the Great Basin pruned honey mesquite (*Prosopis glandulosa*), a very important food resource, keeping areas around the trees clear of undergrowth, dead limbs, and lower branches.
Sowing	The broadcasting of seed collected from native plants to an area (usually recently burned ground).	The Lummi Straits, Nooksack, and Nuuwhaha of the Pacific Northwest harvested the bulbs of camas (*Camassia* spp.) and placed broken stalks bearing ripe seed capsules into holes before they were recovered.
Tilling	The moving of soil to harvest underground perennial plant organs (e.g., roots, rhizomes, corms, bulbs), frequently followed by dividing of organs, leaving individual fragments or smaller clumps in soil.	The Dena'ina of Alaska purposely left fragments of underground swollen stems behind when digging to ensure the growth of new plants.
Weeding	Removing unwanted plants near favored plants.	Stands of sedge (*Carex barbarae*) along lowland creeks and rivers in central and coastal California were intensively weeded to encourage production of long, creeping rhizomes for use as weft or lacing in basketry by the Pomo, Ohlone, and other tribes.

for harvest. Many hunter-gatherer groups traditionally managed plant populations with burning, pruning, coppicing (severe pruning), and/or weeding to create useful growth. Clumps of beargrass (*Xerophyllum tenax*), a plant of widespread use for overlay designs in baskets of the Pacific Northwest, were burned periodically by the Hupa, Karuk, Yurok, Chilula, and other tribes of northern California and Oregon to increase the production of stronger, more supple leaves. Over half the tribes in California relied on the harvesting of the flowering culms of deergrass (*Muhlenbergia rigens*) for the foundation (or stuffing) of their coiled baskets, and these culms are still gathered today. An ancient tradition among many tribes was to burn colonies of this perennial bunchgrass every two to five years to enhance flowering stalk production, clear out detritus, and keep shrubs and trees from encroaching. The Cibecue and White Mountain Apache in eastern Arizona burned willow (*Salix* sp.) and sumac (*Rhus trilobata*) to bring out the young shoots for basket weaving. Burning, pruning, or coppicing shrubs and trees such as maple (*Acer macrophyllum*), redbud (*Cercis occidentalis*), deerbrush (*Ceanothus integerrimus*), hazelnut (*Corylus cornuta* var. *californica*), and willow (*Salix exigua*) were standard practice among tribes throughout California. Willows and other species of shrub are subject to attack by a wide range of leaf-eating, stem-sucking, and wood-boring insects. Fire is also an effective agent to combat many kinds of diseases and insects.[6]

A

B

Fig. 2 Plant gathering and management techniques used by the pre-agricultural people of California. (A) Cecilia Joaquin, a Central Pomo woman, collecting seeds with a seedbeater prior to 1924. Beating seeds of wildflowers and grasses with a shallow basket into a collection basket for food was the first in a series of management steps. Afterward, areas were burned to reduce plant competition and recycle nutrients; some of the seed was saved and broadcast in the burned area. These practices are recorded for tribes throughout western North America. (Photograph by Edward Curtis, courtesy of the National Anthropological Archives, Smithsonian Institution, #75-14715.) (B) A Wintu couple, Rosa Charles and Billy George, digging for yampah (*Perideridia* spp.) in 1931. The digging of many different kinds of bulbs and tubers with a hardwood digging stick, replanting propagules, and burning over areas to increase numbers, densities, and size of subterranean organs of wild plants for food was a common practice throughout the world. These had subtle yet significant effects in shaping the ecology of many ecosystems. (Photograph by J.P. Harrington, courtesy of the Santa Barbara Museum of Natural History.)

VIEWING HUMAN-PLANT INTERACTIONS AS A CONTINUUM

In the repetitive gathering and cultivating of wild plant populations in specific areas over many years, nature was transformed at different scales of biological organization, from genetic to landscape scale. Thus, domestication of plant species grew out of a set of comprehensive land management systems and complex traditional ecological

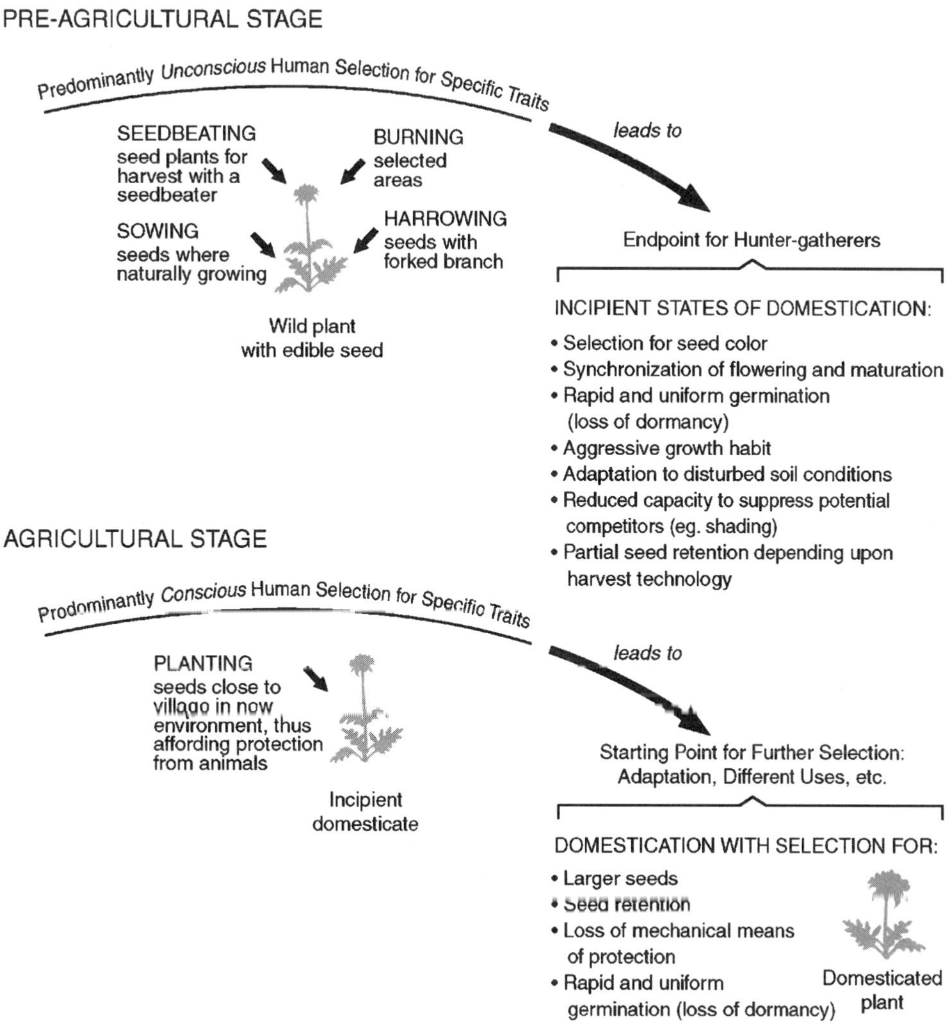

Fig. 3 The continuum of human-induced vegetation change from gathering to domestication can be conceptualized as two major stages on a gradient of ecological change induced by human modification. These stages are not clear-cut, but grade into each other. The diagram does not imply that given enough time, hunter-gatherers would advance from one level of interaction with plants to the next.

knowledge already in place in many different parts of the world for thousands of years prior to the origin of agriculture. Therefore, the distinction between domestic and wild in plant exploitation is not a sudden or marked change, but a gradual transition. Human-plant interactions can be appropriately viewed as a continuum from gathering to tending, to cultivation, to domestication.[7,8]

In areas where plant domestication or the practice of agriculture never occurred, long-term cultivation practices with wild plants would still very likely lead to genetic changes that make plants better suited to the conditions of human-disturbed environments—such as burned over areas—and less adapted to the conditions of natural environments. The pathway of vegetation change that leads to seed domestication, for example, is probably a similar one in different areas. This pathway starts with selection pressures on plants in areas that are harvested repeatedly. Nevertheless, certain types of harvest methods will not automatically lead to selection of the corresponding domestication trait (i.e., lack of seed dispersal). For example, beating wild plants to harvest seeds (Fig. 2A) will select against any mutant that has a stiff rachis. Thus, the selection pressure exerted in this case will not lead to domestication, at least from the standpoint of seed dispersal. Further along the pathway to vegetation change is the actual burning of areas, preparation of soil, sowing of seed, and harrowing of brush. These activities can, over time, favor specific features of the seed or plant, and produce faster or more robust growth, which favors survival and flourishing with reliable production to the point where full domestication is not necessary. The evolutionary modifications are significant enough to lead to incipient or intermediate states of domestication. It is proposed that these intermediate states are evident in certain kinds of native plants labeled as wild. It can be postulated that systems of seed production of hunter-gatherer groups in many parts of the world evolved to a point on the continuum of human-plant interaction that stopped short of the endpoint of full domestication (see Fig. 3). Thus, domestication is but one kind of biologically defined symbiotic relationship, and the manipulation of wild plants for edible seeds is an example of the many other types.[9]

ARTICLES OF FURTHER INTEREST

Agriculture: Why and How Did It Begin?, p. 5
Crop Domestication: Role of Unconscious Selection, p. 340

REFERENCES

1. Denevan, W.M. *Cultivated Landscapes of Native Amazonia and the Andes*; Oxford University Press: Oxford, 2003.
2. Doolittle, W.E. *Cultivated Landscapes of Native North America*; Oxford University Press: Oxford, 2000; 1–574.
3. Anderson, M.K. *Tending the Wild: Indigenous Management of California's Natural Resources and Biodiversity*; University of California Press: Berkeley, *in press*.
4. Doebley, J.F. ''Seeds'' of wild grasses: A major food of Southwestern Indians. Econ. Bot. **1984**, *38* (1), 52–64.
5. Anderson, M.K. From tillage to table: The indigenous cultivation of geophytes for food in California. J. Ethnobiol. **1997**, *17* (2), 149–169.
6. Anderson, M.K. The fire, pruning, and coppice management of temperate ecosystems for basketry material by California Indian tribes. Hum. Ecol. **1999**, *27* (1), 79–113.
7. Harris, D.R. An Evolutionary Continuum of People–Plant Interaction. In *Foraging and Farming: The Evolution of Plant Exploitation*; Harris, D.R., Hillman, G.C., Eds.; Unwin Hyman: Boston, 1989; 11–26.
8. Ford, R.I. The Processes of Plant Food Production in Prehistoric North America. In *Prehistoric Food Production in North America*; Ford, R.I., Ed.; Museum of Anthropology, University of Michigan: Ann Arbor, 1985; 1–18.
9. Rindos, D. *The Origins of Agriculture: An Evolutionary Perspective*; Academic Press, Inc.: New York, 1984; pp. xiv–xv, 99.

Protoplast Applications in Biotechnology

M. R. Davey
P. Anthony
J. B. Power
University of Nottingham, Loughborough, U.K.

K. C. Lowe
University of Nottingham, Nottingham, U.K.

INTRODUCTION

Isolated plant protoplasts have several biotechnological applications, including their fusion to generate novel somatic hybrid and cybrid plants and the production of transgenic plants expressing specific characteristics. Normally, plant breeders rely upon sexual hybridization to combine useful genetic traits from different species or genera, but hybridization may be impeded by natural complex incompatibility barriers. The fusion of protoplasts isolated from somatic cells circumvents such barriers. Unique combinations of nuclear and organellar genomes generate novel germplasm; during protoplast fusion, there is no strict maternal inheritance of organelles, unlike in sexual hybridization.

FUSION OF ISOLATED PROTOPLASTS

Protoplast fusion can be induced chemically or electrically, or by a combination of these techniques, using small- or large-scale procedures.[1] Plasma membranes of protoplasts destabilize during fusion, establishing cytoplasmic continuity between tightly adhering protoplasts. Polyethylene glycol (PEG) is a common fusogen, sometimes in combination with high pH and Ca^{+2}.

Electrical fusion of protoplasts is more reproducible than chemical fusion and often gives greater fusion frequency. Versatile electrofusion instruments can be constructed.[2] Prior to electrofusion, isolated protoplasts are suspended in a medium of low conductivity [e.g., mannitol with $CaCl_2$] to stabilize plasma membranes. Exposure of protoplasts to an alternating current of 0.5–2.0 MHz at 100–400 V cm^{-1} induces protoplasts to align into ''pearl chains'' perpendicular to the electrodes; increasing the field strength induces close membrane-to-membrane contact. A subsequent direct current pulse of 10–200 micro-seconds and 500–2000 V cm^{-1} coalesces adhering plasma membranes at the poles of the protoplasts. Protoplast fusion normally occurs about 10 minutes after electrical treatment.

NOVEL HYBRIDS AND CYBRIDS GENERATED BY PROTOPLAST FUSION

Symmetric nuclear hybrid plants resulting from fusion are generally rare, with most being asymmetric hybrids. Plant breeders often require the introgression of a limited number of chromosomes, parts of chromosomes, or only organelles (chloroplasts and/or mitochondria) from one species into another. Consequently, effort has focused on generating asymmetric nuclear hybrids and cytoplasmic hybrids (cybrids). Cybrids are those with a nuclear genome of a given genus or species (recipient) with plastids from the other partner, or, more rarely, a mixed population of plastids from the recipient and donor. Cybrids harbor a mitochondrial genome partly or totally from the recipient or the donor, or novel mitochondria following recombination of recipient–donor mitochondrial DNA.

Asymmetric hybrid production is stimulated by treatment, before fusion, of protoplasts of one partner with X or gamma irradiation and exposure of the protoplasts of the other parent to a metabolic inhibitor (e.g., iodoacetamide). Irradiation stimulates partial genome transfer, the irradiated and nondividing but metabolically active protoplasts overcoming the inability of inhibitor-treated protoplasts to undergo mitosis. Frequently, somatic hybrid plants resemble the partner used as the source of the inhibitor-treated protoplasts.

GAMETOSOMATIC HYBRIDIZATION

Transfer of genetic information into cultivated plants can be achieved by generating addition or substitution lines via

DOI: 10.1081/E-EPCS 120010549

the sexual production of generating triploids. Infrequently, triploids arise during the fusion of diploid protoplasts, as in *Citrus*. They can also be generated by fusing diploid with haploid protoplasts. The latter can be isolated from haploid plants.[3]

PRODUCTS OF PROTOPLAST FUSION: SELECTION OF SOMATIC HYBRID TISSUES AND PLANTS

Fusion of protoplasts of the same genetic composition generates homokaryons, producing plants with increased ploidy. In contrast, heterokaryons from the fusion of protoplasts of different genetic composition have application in plant improvement, as they contain the nuclei of both parentals, initially in a mixed cytoplasm. The selection of heterokaryon-derived tissues and somatic hybrid plants remains a difficult aspect of somatic hybridization. In some combinations, heterosis (hybrid vigor) results in heterokaryon-derived tissues developing first in culture. Such tissues can be selected manually prior to plant regeneration.

Micromanipulation is useful when parental protoplasts are morphologically distinct. For example, fusion of suspension cell protoplasts with green leaf protoplasts generates heterokaryons initially with colorless plastids in one half of their cytoplasm and chloroplasts in the other half. Fluorochromes and flow cytometry can also be used in heterokaryon identification. Hormone autotrophism has also been exploited for hybrid selection.[4] Complementation systems have been devised to select somatic hybrids. The most simple involves fusion of protoplasts of nonallelic albino mutants to generate hybrid cells that complement to chlorophyll proficiency in the light.

Auxotrophic mutants have been exploited in selection, the nitrate reductase deficiency of the *nia-63* mutant of *Nicotiana tabacum* being complemented by the chlorate-resistant line (*cnx-68*). Protoplasts of the mutants fail to grow on nitrate-supplemented medium. However, somatic hybrid cells undergo complementation and can be selected on nitrate-containing medium. Antibiotic resistant cells are also useful in selection. Dominant antibiotic resistant genetic markers can be introduced by transforming parental cells prior to protoplast isolation.

In addition to antimetabolites being used to promote cybridization, such compounds that inhibit cell development have been exploited in hybrid selection. In the somatic hybridization of *Lactuca sativa* with the wild species *Lactuca virosa*, hybrids were selected by inactivation of *L. sativa* protoplasts with iodoacetamide, combined with the inability of *L. virosa* protoplasts to divide in culture.[5] Transformation to kanamycin resistance has also been combined with metabolic inhibition to select hybrid cells and plants.

CHARACTERIZATION OF SOMATIC HYBRID PLANTS

Characterization of plants generated by protoplast fusion necessitates morphological, cytological, and molecular analyses. Traits characteristic of both parents may be readily apparent in somatic hybrid plants, with leaves, flowers, and pigmentation being intermediate between those of both parents. In other cases, characteristics from one parent may be dominant.

Theoretically, plants regenerated by fusing two diploid somatic cell protoplasts should be tetraploid, which is the case in some combinations. However, plants with complete chromosome complements of both parents are generally rare. Hybrids often possess an asymmetric combination of parental chromosomes, following elimination of parts of genomes during culture.[6] The reasons for chromosome elimination are unclear. Some somatic hybrid plants are fertile; others may be sterile.[7] Cytological analyses of backcross progeny provide evidence of parental chromosomal behavior during mitiosis and meiosis following introduction of somatic hybrids into breeding programs. Flow cytometry can estimate the ploidy of plants and provide a baseline for cytological analyses. DNA fingerprinting[8] permits detailed characterization of nuclear and organellar genomes of parental and somatic hybrid plants.

Organellar events are complex in somatic hybrids. Initially, heterokaryons contain a mixed population of organelles. Subsequently, plastids usually segregate, with those of one partner becoming dominant;[7] in some cases, the plastids of both partners persist in hybrids. Recombination of plastid DNA is rare. Analysis of ribulose biphosphate carboxylase–oxygenase, linked to the chloroplast DNA restriction enzyme profile, confirms the parental origin of chloroplasts in somatic hybrids. Recombination of mitochondrial DNA commonly generates ''new'' organelles, as evidenced by the restriction enzyme digestion patterns.[7] Importantly, DNA recombination in organelles increases genetic diversity arising from protoplast fusion.

Examples exist of the transfer of agronomical traits by protoplast fusion. Considerable effort has focused on the Solanaceae, such a potato and tobacco, together with the Brassicaceae. An excellent example is provided by studies in which protoplasts of *Solanum tuberosum* were fused with those of the wild potato, *Solanum bulbocastanum*,[9] generating hybrids having improved resistance to late blight. Importantly, this resistance was transferred to other breeding lines by back-crossing.

TRANSFORMATION OF ISOLATED PROTOPLASTS BY DNA UPTAKE

Induction of DNA Uptake

The induction of transient pores in the plasma membrane permits uptake of DNA by chemical and/or physical procedures.[10] Several agents induce DNA uptake into protoplasts, including salt solutions with Ca^{2+} at high pH and PEG. Electroporation, using short duration, high-voltage electrical pulses, often with PEG, is also used for DNA uptake into protoplasts. DNA has also been microinjected into protoplasts.[11] PEG and electroporation remain the most successful procedures for protoplast transformation, even though the frequency is low and, at best, only about 1 in 10^4 protoplasts develops into transformed tissues. Protoplasts can be transformed simultaneously with more than one gene, the genes being either on the same or on separate vectors.

Factors Influencing Protoplast Transformation

Parameters have been identified that influence protoplast transformation. The stage in the cell cycle is important, efficiency being higher when protoplasts are in the S or M phases. Consequently, it is beneficial to synchronize cells prior to or immediately following protoplast isolation. Heat shock or irradiation of protoplasts before DNA uptake also stimulates transformation, irradiation probably increasing recombination of genomic DNA with incoming DNA or initiating repair mechanisms that favor DNA integration. Complex integration patterns have been observed following DNA uptake into protoplasts.

MISCELLANEOUS STUDIES WITH PROTOPLASTS

Isolated protoplasts have also featured in physiological, ultrastructural, and genetic studies. Their development into colonies of single cell origin has been exploited to isolate clonal lines and plants, including those for increased secondary product synthesis. Exposure of isolated protoplasts to mutagenic agents or irradiation permits the induction and selection of mutants. Protoplasts take up macromolecules, which has been exploited in studies of endocytosis and virus infection and replication. The osmotic fragility of isolated protoplasts permits their controlled lysis for isolating cell components. In physiological studies, isolated vacuoles have been used to investigate sugar accumulation; protoplasts from barley aleurone cells contain protein storage vacuoles and a lysosome-like organelle, designated the secondary vacuole. Protoplasts are ideal for studying ion transport and regulation of the osmotic balance of cells.

Light-induced proton pumping has been investigated in guard cell protoplasts. Other studies have focussed on cell fusion and metabolism in microgravity, elicitor binding sites, binding of fungal phytotoxins to plasma membranes, and auxin accumulation and metabolism. Protoplasts have also provided unique material for studying cell wall synthesis and the role of microtubules during cell development. Protoplasts from totipotent cells, are useful for assessing the effects of pharmaceuticals, food additives, cosmetics, and agrochemicals on plant cells, whole plants, and their progeny over seed generations.[12]

ARTICLES OF FURTHER INTEREST

Agriculture and Biodiversity, p. 1
Biosafety Approaches to Transgenic Crop Plant Gene Flow, p. 150
Breeding for Durable Resistance, p. 179
Breeding for Nutritional Quality, p. 182
Breeding Hybrids, p. 186
Breeding Plants and Heterosis, p. 189
Breeding Plants with Transgenes, p. 193
Breeding: Incorporation of Exotic Germplasm, p. 222
Chromosome Manipulation and Crop Improvement, p. 266
Commercial Micropropagation, p. 297
Crop Improvement: Broadening the Genetic Base for, p. 343
Fluorescence In Situ Hybridization, p. 468
Gene Flow Between Crops and Their Wild Progenitors, p. 488
In Vitro Chromosome Doubling, p. 572
In Vitro Morphogenesis in Plants—Recent Advances, p. 579
In Vitro Plant Regeneration by Organogenesis, p. 584
In Vitro Pollination and Fertilization, p. 587
In Vitro Production of Triploid Plants, p. 590
Interspecific Hybridization, p. 619
Mitosis, p. 734

REFERENCES

1. Blackhall, N.W.; Davey, M.R.; Power, J.B. Fusion and Selection of Somatic Hybrids. In *Plant Cell Culture, A Practical Approach,* 2nd Ed.; Dixon, R.A., Gonzales, R.A., Eds.; IRL Press at Oxford University Press: New York, 1994; 41–48.
2. Jones, B.; Lynch, P.T.; Handley, G.J.; Malaure, R.S.; Blackhall, N.W.; Hammatt, N.; Power, J.B.; Cocking, E.C.;

Davey, M.R. Equipment for the large-scale electromanipulation of plant protoplasts. BioTechniques **1994**, *6*, 312–321.

3. Davey, M.R.; Blackhall, N.W.; Lowe, K.C.; Power, J.B. Gametosomatic Hybridisation. In *In Vitro Haploid Production in Higher Plants Volume 2: Application*; Mohan Jain, S., Sopory, S.K., Vielleux, R.E., Eds.; Kluwer Academic Publishers: Dordrecht, The Netherlands, 1996; 309–320.
4. Carlson, P.S.; Smith, H.H.; Dearing, R.D. Parasexual somatic hybridization. Proc. Natl. Acad. Sci. U. S. A. **1972**, *69*, 2292–2294.
5. Matsumoto, E. Interspecific somatic hybridization between lettuce (*Lactuca sativa*) and wild-species *L. virosa*. Plant Cell Rep. **1991**, *9*, 531–534.
6. Oberwalder, B.; Schilde-Rentschler, L.; Loffelhardt-Ruoss, B.; Ninnemann, H. Differences between hybrids of *Solanum tuberosum* L. and *Solanum circaeifolium* Bitt. obtained from symmetric and asymmetric fusion experiments. Potato Res. **2000**, *43*, 71–82.
7. Liu, J.H.; Dixelius, C.; Eriksson, I.; Glimelius, K. *Brassica napus* (+) *B. tournefortii*, a somatic hybrid containing traits of agronomic importance for rapeseed breeding. Plant Sci. **1995**, *109*, 75–86.
8. Oberwalder, B.; Ruoss, B.; Schilde-Rentschler, L.; Hemleben, V.; Ninnemann, H. Asymmetric fusion between wild and cultivated species of potato (*Solanum* spp.)-detection of asymmetric hybrids and genome elimination. Theor. Appl. Genet. **1997**, *94*, 1104–1112.
9. Helgeson, J.P.; Pohlman, J.D.; Austin, S.; Haberlach, G.T.; Wielgus, S.M.; Ronis, D.; Zambolim, L.; Tooley, P.; McGrath, J.M.; James, R.V.; Stevenson, W.R. Somatic hybrids between *Solanum bulbocastanum* and potato: A new source of resistance to late blight. Theor. Appl. Genet. **1998**, *96*, 738–742.
10. Davey, M.R.; Rech, E.L.; Mulligan, B.J. Direct DNA transfer to plant cells. Plant Mol. Biol. **1989**, *13*, 273–285.
11. Rakoczy-Trojanowska, M. Alternative methods of plant transformation—A short review. Cell. Mol. Biol. Lett. **2002**, *7*, 849–858.
12. Lowe, K.C.; Davey, M.R.; Power, J.B.; Clothier, R.H. Plants as toxicity screens. Pharm. News **1995**, *2*, 17–22.

Protoplast Culture and Regeneration

P

J. B. Power
M. R. Davey
P. Anthony
University of Nottingham, Loughborough, U.K.

K. C. Lowe
University of Nottingham, Nottingham, U.K.

INTRODUCTION

Plant protoplasts represent the living contents of cells, each bounded by a plasma membrane and enclosed by a cell wall. The plasma membrane is involved in wall synthesis. Consequently, there is usually intimate contact between these two structures. However, when cells are stressed osmotically, their plasma membranes contract away from their surrounding walls. Subsequent removal of the walls enclosing plasmolyzed protoplasts enables the latter to be isolated as spherical, osmotically fragile, ''naked'' cells. The isolation of large populations of protoplasts from a range of plants is now routine. Such isolated protoplasts are ideal for studies of cell development, physiology, and cytogenetics. When cultured in the laboratory, isolated protoplasts undergo wall resynthesis and mitotic division. Protoplast-derived cells may express their totipotency, regenerating into fertile plants under the correct physiological and physical stimuli. This feature, unique to plant cells, is exploited in genetic manipulation through somatic hybridization and cybridization, both involving protoplast fusion, and transformation by direct uptake of foreign DNA.

ISOLATION OF PLANT PROTOPLASTS

The physiological status and age of tissues is crucial for the isolation of viable protoplasts. While leaves of glasshouse-grown plants are a convenient source of protoplasts, seasonal variation in illumination, temperature, and humidity may necessitate the use of environmental cabinets to ensure uniformity of material. Generally, cultured shoots and in vitro grown seedlings are more uniform as source material than pot-grown plants. Haploid pollen tetrads and mature pollen will release protoplasts; cell suspensions are convenient and frequently exploited as a source of protoplasts.[1]

Combinations of commercially available cellulase, hemicellulase and pectinase enzymes are used to release large populations of protoplasts. The cell wall composition of source tissues dictates the enzyme mixture required. Consequently, enzyme concentrations and conditions must be determined empirically for protoplast isolation from specific plant tissues.[2] The time of enzyme digestion, usually at 25–28°C, may be of short duration (e.g., 4–6 hours) or overnight (12–20 hours). Removal of the lower epidermis or dissection of leaves into thin strips facilitates tissue digestion and protoplast release. Preconditioning of donor plants or explants by exposure to reduced illumination or preculture of donor explants on suitable media may increase protoplast yield and viability.

Passage of enzyme-protoplast mixtures through sieves of suitable pore sizes following enzyme incubation removes undigested cells. Gentle centrifugation (e.g., $100 \times g$; 10 min) through a suitable osmoticum [e.g., 13% (w/v) mannitol] generally pellets the protoplasts, leaving fine debris in suspension. Mixing protoplasts with 21% (w/v) sucrose in a salts solution,[3] or with a solution of Percoll or Ficoll, followed by centrifugation, causes protoplasts of many species to float, facilitating their collection.

CULTURE OF ISOLATED PLANT PROTOPLASTS

Nutritional Requirements of Protoplasts: Culture Media

Protoplasts commence wall regeneration within hours of being introduced into culture. However, they require osmotic protection until they have regenerated a new primary wall of sufficient strength to counteract the turgor pressure exerted by the living cytoplasm/vacuoles.

Encyclopedia of Plant and Crop Science
DOI: 10.1081/E-EPCS 120018559

Many media have been reported for protoplast culture. The nutrient-rich KM-type formulations[4] are beneficial for culture of protoplasts at low densities; other media are often based on the well-tested MS[5] and B5[6] formulations. Ammonium ions are detrimental to some protoplasts, particularly those of woody species. Sucrose is the most common carbon source, although glucose may act as both a carbon source and osmotic stabilizer. Maltose stimulates shoot regeneration from protoplast-derived cells, especially in cereals.[7] Most protoplasts require one or more auxins or cytokinins in the culture medium to sustain mitotic division.

Systems for Protoplast Culture

Incubation of isolated protoplasts in liquid medium in Petri dishes or in a shallow liquid layer overlaying semi-solidified medium are simple methods of culture. The inclusion of a filter paper at the interface between the liquid and semisolid phases may stimulate cell colony formation. Isolated protoplasts can also be embedded in semisolidified media. Several gelling agents are available, with agarose often enhancing protoplast plating efficiencies (calculated as the percentage of the protoplasts originally plated that develop into cell colonies), compared with agar. Semisolid medium containing the protoplasts may be dispensed as layers or droplets (the latter about 100 μl in volume) in Petri dishes. Alginate is a useful gelling agent for heat-sensitive protoplasts. Following suspension of the protoplasts, the alginate-culture medium mixture is semisolidified by pouring the warm mixture in which the protoplasts are suspended over an agar layer containing Ca^{2+} or by gently dropping the molten medium into a solution of such ions.

Plating Density and Nurse Cells

The density at which protoplasts are plated in the culture medium is crucial for cell colony formation. Generally, the optimum plating density is 5×10^2–1.0×10^6 ml^{-1}. At greater densities, protoplast-derived cells often fail to undergo sustained mitotic division because of rapid depletion of nutrients from the medium. Protoplasts also fail to grow when plated below a minimum density. Medium previously ''conditioned'' by supporting the culture of actively dividing cells for a limited period, will stimulate the growth of isolated protoplasts. ''Nurse'' cells are often employed to promote division of protoplasts in culture.[7] When using nurse cultures, the isolated protoplasts can be embedded in a semisolid layer, suspended in a thin layer of liquid medium, or spread in a liquid layer on a cellulose nitrate membrane, overlaying the semisolid medium containing the nurse cells.

INNOVATIVE APPROACHES TO PROTOPLAST CULTURE

Chemical Supplements for Culture Media: Surfactants and Antibiotics

Some antibiotics (e.g., cefotaxime) stimulate protoplast division, such compounds being thought to be metabolized to growth regulator–like molecule(s). Supplementation of medium with the nonionic, surfactant *Pluronic*® F-68, increases the plating efficiency of protoplasts in culture. Surfactants may increase the permeability of plasma membranes, stimulating uptake of nutrients from the culture medium into protoplasts and protoplast-derived cells.[8]

Manipulation of Respiratory Gases

Gassing of vessels with oxygen after introducing protoplasts into the culture medium increases the plating efficiency of protoplasts of jute and rice. A further novel approach for regulating the supply of respiratory gases to cultured protoplasts involves the use of inert, chemically stable perfluorocarbon (PFC) liquids.[9] Such compounds dissolve large volumes of respiratory gases and have been exploited in animal systems as oxygenation fluids.[10] PFC liquids, being about twice as dense as water, form a distinct layer beneath aqueous culture media. Consequently, protoplasts and protoplast-derived cells can be cultured at the interface between the lower PFC layer and the overlaying aqueous medium. Experiments revealed that protoplasts in such systems exhibited increased superoxide dismutase activity associated with oxygen detoxification. Other regulators of respiratory gases, notably chemically modified haemoglobin solutions, may also stimulate the growth of protoplasts in culture. Supplementation of culture media with commercial bovine haemoglobin solution (*Erythrogen*™) significantly increased the plating efficiency of rice protoplasts, compared to untreated controls.[11]

Physical Methods to Stimulate Protoplast Growth

Physical parameters have been shown to stimulate protoplast growth, including the use of cellulose nitrate filter membranes (0.2 μm pore size) at the liquid/semisolid medium interface, often in conjunction with nurse cells in the underlying semisolid medium. Insertion of glass rods vertically into semisolid agarose medium stimulated mitotic division of cassava leaf protoplasts in the overlaying liquid medium, the protoplasts probably receiving increased aeration where they aggregate in the liquid menisci around the glass rods. Electrical currents, both low and high voltage, also stimulate mitotic division and cell colony formation from protoplasts of many species.

PLANT REGENERATION FROM PROTOPLAST-DERIVED TISSUES

The induction and sustaining of plant regeneration in protoplast-derived tissues by different pathways of morphogenesis (organogenesis; somatic embryogenesis) is dependent, in part, upon the culture conditions, in particular, the composition of the culture medium, and the inherent totipotency of the donor species. Plant regeneration, via organogenesis, has been reported to occur for more than 70% of those species capable of regenerating plants from protoplast-derived tissues, most notably members of the Compositae, Cruciferae, Leguminosae, and Solanaceae. In contrast, plant regeneration from protoplast-derived tissues via somatic embryogenesis is restricted predominantly to members of the Curcurbitaceae, Gramineae, Leguminosae, Rutaceae, and Umbelliferae.[12] In a limited number of genera, protoplasts may develop directly into somatic embryos through early polar growth of their derived cells, as in *Medicago*, *Asparagus*, and *Persea*.

CONCLUSION

During the last 40 years, considerable progress has been made in regenerating plants from protoplast-derived tissues of an increasing number of genera and species, driven by a need for regeneration as a platform for many aspects of biotechnology. It is interesting to note that it is only recently that protoplast-to-plant systems have been developed for specific genera, an example being provided by *Sorghum*, in which the establishment of such a protoplast-to-plant system required nearly 20 years of on-going research.[13] An extensive literature is directed to plant regeneration from isolated protoplasts.[12] The expression of totipotency from protoplast-derived tissues remains an absolute requirement for the multifaceted applications of somatic cell technologies involving protoplasts, such as somatic hybridization, cybridization, and direct DNA uptake, in plant genetic improvement programs. The relevance of these applications are discussed elsewhere in this volume.

ARTICLES OF FURTHER INTEREST

Anther and Microspore Culture and Production of Haploid Plants, p. 43
Biosafety Approaches to Transgenic Crop Plant Gene Flow, p. 150
Breeding Hybrids, p. 186
Commercial Micropropagation, p. 297
Gene Flow Between Crops and Their Wild Progenitors, p. 488
In Vitro Chromosome Doubling, p. 572
In Vitro Flowering, p. 576
In Vitro Morphogenesis in Plants—Recent Advances, p. 579
In Vitro Plant Regeneration by Organogenesis, p. 584
In Vitro Pollination and Fertilization, p. 587
In Vitro Production of Triploid Plants, p. 590
In Vitro Tuberization, p. 594
Interspecific Hybridization, p. 619
Plant Cell Culture and Its Applications, p. 931
Plant Cell Tissue and Organ Culture: Concepts and Methodologies, p. 934
Regeneration from Guard Cells of Crop and Other Plant Species, p. 1081
Somaclonal Variation: Origins and Causes, p. 1158
Somatic Embryogenesis in Plants, p. 1165

REFERENCES

1. Blackhall, N.W.; Davey, M.R.; Power, J.B. Isolation, Culture and Regeneration of Protoplasts. In *Plant Cell Culture, A Practical Approach,* 2nd Ed.; Dixon, R.A., Gonzales, R.A., Eds.; IRL Press at Oxford University Press: New York, 1994; 27–39.
2. Enzymes for the Isolation of Plant Protoplasts. In *Biotechnology in Agriculture and Forestry. Plant Protoplasts and Genetic Engineering*; Ishii, S., Bajaj, Y.P.S., Eds.; Springer-Verlag: Berlin, 1997; Vol. 8, 23–33.
3. Frearson, E.M.; Power, J.B.; Cocking, E.C. The isolation, culture and regeneration of *Petunia* leaf protoplasts. Dev. Biol. **1973**, *33*, 130–137.
4. Kao, K.N.; Michayluk, M.R. Nutritional requirements for growth of *Vicia hajastana* cells and protoplasts at a very low population density in liquid media. Planta **1975**, *126*, 105–110.
5. Murashige, T.; Skoog, F. A revised medium for rapid

growth and bioassays with tobacco tissue cultures. Physiol. Plant. **1962**, *15*, 473–497.
6. Gamborg, O.L.; Miller, R.A.; Ojima, K. Nutrient requirements of suspension cultures of soybean root cells. Exp. Cell Res. **1968**, *50*, 151–158.
7. Jain, R.K.; Khehra, G.S.; Lee, S-H.; Blackhall, N.W.; Marchant, R.; Davey, M.R.; Power, J.B.; Cocking, E.C.; Gosal, S.S. An improved procedure for plant regeneration from indica and japonica rice protoplasts. Plant Cell Rep. **1995**, *14*, 515–519.
8. Lowe, K.C.; Anthony, P.; Davey, M.R.; Power, J.B. Beneficial effects of Pluronic F-68 and artificial oxygen carriers on the post-thaw recovery of cryopreserved plant cells. Art. Cells, Blood Subst. Immob. Biotech. **2001**, *23*, 221–239.
9. Lowe, K.C.; Davey, M.R.; Power, J.B. Perfluorochemicals: Their applications and benefits to cell culture. TIBTECH. **1998**, *16*, 272–277.
10. Lowe, K.C. Engineering blood: Synthetic substitutes from fluorinated compounds. Tissue Eng. **2003**, *9*, 389–399.
11. Al-Forkan, M.; Anthony, P.; Power, J.B.; Davey, M.R.; Lowe, K.C. Haemoglobin (*Erythrogen*™)—Enhanced microcallus formation from protoplasts of Indica rice (*Oryza sativa* L.). Art. Cells, Blood Subs., Immob. Biotech. **2001**, *29*, 399–404.
12. Xu, Z-H.; Xue, H-W. Plant Regeneration from Cultured Protoplasts. In *Morphogenesis in Plant Tissue Cultures*; Soh, W-Y., Bhojwani, S.S., Eds.; Kluwer Academic Publishers: Dordrecht, The Netherlands, 1999; 37–70.
13. Sairam, R.V.; Seetharama, N.; Devi, P.S.; Verma, A.; Murthy, U.R.; Potrykus, I. Culture and regeneration of mesophyll—Derived protoplasts of sorghum [*Sorghum bicolour* (L.) Moench]. Plant Cell Rep. **1999**, *18*, 972–977.

Quantitative Trait Locus Analyses of the Domestication Syndrome and Domestication Process

Valérie Poncet
IRD, Montpellier, France

Thierry Robert
Aboubakry Sarr
Université Paris-Sud, Orsay, France

Paul Gepts
University of California, Davis, California, U.S.A.

INTRODUCTION

Domestication of many of today's main food crops occurred approximately 10,000 years ago with the beginning of agriculture. It has lead to the selection by farmers of a wide range of morphological and physiological traits that distinguish domesticated crops from their wild ancestors. These characteristics are collectively referred to as the domestication syndrome and include changes in plant architecture (e.g., apical dominance in maize), gigantism in the consumed portion of the plant (e.g., fruit size in tomato and eggplant), and reduced seed dispersal (i.e., nonshattering or nondehiscence, as in the common bean, sunflower, and cereals). Many investigations based on multidisciplinary approaches—such as genetic, archaeological, and phytogeographical analyses—have succeeded in identifying progenitor species and centers of domestication. However, the genetic and molecular bases of morphological evolution in plants under domestication are largely unknown.

Advances in genome mapping, which have resulted in high-density molecular-marker linkage maps in most crops, have provided tools for dissecting the genetic basis underlying complex traits into their individual components, i.e., their quantitative trait locus (QTL). This method relies on the frequent ability to cross the crop and its wild progenitor. It enables the characterization of genetic differences in terms of the number and chromosomal location of the genes as well as quantitative estimates of the kind and amount of genetic effects associated with individual loci. Recent studies have analyzed the genetics of the domestication syndrome of crops belonging to diverse families.

GENETIC BASIS AND ORGANIZATION OF THE DOMESTICATION SYNDROME

Many quantitative genetic analyses have focused on individual traits that are related to domestication but comprehensive analyses of the inheritance of the domestication syndrome[1] as a whole are scarce.

Major vs. Minor Genes

Research has revealed that numerous traits that distinguish crop plants from their wild relatives are often controlled by a relatively small number of loci with effects of unequal magnitude.

Many qualitative traits are controlled by one Mendelian locus such as seed shattering in sorghum[2] and pearl millet.[3] But even the traits that are usually considered as exhibiting quantitative inheritance involve few QTLs of large effect plus others of more modest effect. The genetic study of an F_2 population derived from a cross of maize (*Zea mays* ssp. *mays*) and its wild progenitor, teosinte (*Z. mays* ssp. *parviglumis*),[4] revealed that the domestication traits are controlled primarily by five chromosomal segments (Table 1). One of the major QTLs, *tb1* (*teosinte branched 1*), conditions the dramatic alteration in plant architecture from a multistemmed, branched plant to the single-stemmed plant most people are familiar with. Similarly, in common bean, seed dispersal (pod dehiscence), seed dormancy, and photoperiod sensitivity are all determined by a few loci with effects of large magnitude.[5] In eggplant, most of the dramatic phenotypic differences in fruit weight, shape, color, and plant prickliness that distinguish domesticated eggplant,

Encyclopedia of Plant and Crop Science
DOI: 10.1081/E-EPCS 120017088

Table 1 Genomic regions showing QTL clustering for domestication traits

Crop	Biology	Mapping cross (domesticated × wild forms)	Cluster	Attribute of the corresponding traits
Maize[4]	Outcrossing, $2n = 4x = 20$	F_2: *Zea mays* ssp. mays × *Z. mays* ssp. parviglumis	Chr 1	Shattering (ear disarticulation), growth habit, branching pattern (*tb1*), ear and spikelet architecture
			Chr 2S	Number of rows of cupules
			Chr 3L	Growth habit, ear architecture
			Chr 4S	Glume hardness (*tga1*)
			Chr 5	Ear architecture
Common bean[5]	Self-pollinated, $2n = 2x = 22$	F_2: *Phaseolus vulgaris* cultivated form × *P. v.* wild form	LG D1	Growth habit and phenology
			LG D2	Seed dispersal (pod dehiscence) and dormancy
			LG D7	Pod length and size
Rice[7]	Self-pollinated, $2n = 2x = 24$	F_2: *Oryza sativa* × *O. rufipogon*	Chr 1	Growth habit (tillering & height), shattering, panicle architecture
			Chr 3	Shattering, panicle architecture, earliness
			Chr 6	Shattering, panicle architecture, earliness
			Chr 7	Panicle architecture
			Chr 8	Growth habit (height), earliness, shattering
Pearl millet[3]	Outcrossing, $2n = 2x = 14$	F_2: *Pennisetum glaucum* ssp. glaucum × *P. glaucum* ssp. monodii	LG 6	Shattering, spikelet architecture, spike weight, growth habit
			LG 7	Spikelet architecture, spike size, growth habit and phenology
Sunflower[8]	Outcrossing, $2n = 2x = 34$	F_3: *Helianthus annus* var *macrocarpus* × *H. a.* var *annus*	LG 17	Shattering, apical dominance, achene weight, earliness
			LG 09	Achene size and weight, growth habit, head size
			LG 06	Growth habit, achene size and weight, earliness, head size
Eggplant[6]	Self-compatible, $2n = 2x = 24$	F_2: *Solanum melongena* × *S. linnaenum*		No obvious colocalization

Solanum melongena, from its wild relative, *S. linnaeanum*, could be attributed to six loci with major effects.[6] On this basis, early domestication is most likely to have been a process involving major genes, while subsequent changes may have occurred by the accumulation of minor mutations. Some domestication traits result from the loss of wild-type function and are associated with recessive mutations. However, mutations altering gene regulation are also reported (discussed later).

Clustered Distribution of QTLs

Another interesting feature of the inheritance of domestication traits in crop plants is that the loci for such traits are frequently clustered in a few chromosomal regions (Table 1). This pattern of genetic correlations across traits due to linkage has been documented in some cases of domestications (Table 1). For example, in common bean[5] and maize,[4] QTLs underlying domestication traits are largely restricted to three and five genomic regions, respectively. In pearl millet, two regions of the genome control most of the key morphological differences of the spike and spikelet, including seed shattering.[3] Each of the segments identified has an effect on related but also unrelated traits, suggesting that they could carry either a single mutation with pleiotropic effects or several mutations in linked genes. Linkage among domestication QTL is predicted to evolve under strong selection, especially in allogamous species. Eggplant is a predominantly self-pollinated crop and does not provide strong evidence for the colocalization of domestication syndrome traits.[6]

Table 2 Domestication-related loci with putative conservation across the fabaceae,[9,10] poaceae,[2] and solanaceae[6] families[a]

Family	Center of origin	Crop	Corresponding genomic region										
Fabaceae				Seed weight									
	African	Cowpea (*Vigna unguiculata*)		LG vii									
	African	Mung bean (*Vigna radiata*)		LG 2									
	Chinese	Soybean (*Glycine max*)		LG M									
	Near Eastern	Pea (*Pisum sativum*)		LG III									
Poaceae			Seed dispersal (shattering)		Seed mass							Short-day flowering	
	African	Sorghum (*Sorghum bicolor*)	LGC		LGA	LGC	LGF	LGB		LGE		LGD	
	Mesoamerican	Maize (*Zea mays*)	chr. 5/ chr. 1	chr. 4	chr. 1	chr. 1	chr. 4	chr. 7	chr. 1		chr. 9	chr. 10	chr. 9
	Chinese	Rice (*Oryza sativa*)	chr. 9	chr. 2	chr. 1	chr. 10	chr. 2		chr. 3	chr. 5	chr. 3		chr. 6
Solanaceae			Fruit weight			Fruit shape							
	Southeast Asian	Eggplant (*Solanum melongena*)	LG2 (fw2.1)	LG9 (fw9.1)	LG11 (fw11.1)	LG2 (fl2.[illegible])	LG4 (ovs4.1)	LG7 (fs7.1)					
	Mesoamerican	Tomato (*Solanum Lycopersicon*)	LG2 (fw2.2)	LG9 (fw9.2)	LG11 (fw11.1)	LG2 (ovate)	LG10 (fs10.1)	LG7 (fs7.b)					
	Mesoamerican	Pepper (*Capsicum* spp.)	LG2 (fw2.1)				LG10 (fs10.1)						

[a]LG: linkage groups; ch: chromosome.

INDEPENDENT SELECTION OF ORTHOLOGOUS REGIONS UNDER DOMESTICATION?

Comparative genetic mapping provides insights into the evolution of genome organization within the species investigated. A framework of common markers provides a basis for evaluating the correspondence between the locations of genes that confer common phenotypes. Many investigations have highlighted the preservation of the basic gene order, especially across grass species. Because similar traits have been selected during the domestication of crops belonging to the same family (e.g., Poaceae or Solanaceae), a common set of loci may also have been selected under domestication. The initial report of orthologous QTL noted that a genomic region that had the greatest effect on seed weight in mung bean and cowpea spanned the same restricted fragment length polymorphism (RFLP) markers in the same linkage order in both species. Later works showed that pea and soybean also contained this conserved genomic region (Table 2). Another study[2] described the comparative molecular analysis of QTLs associated with domestication of three crops, *Sorghum*, *Oryza*, and *Zea*, each on a different continent. Correspondence was evaluated among QTLs involved in sensitivity to photoperiod, shattering, and increased seed size. Genes/QTLs were found to correspond far more often than would be expected to occur by chance, suggesting that orthologous genes (i.e., homologous genes that trace back to a common ancestral gene as a result of speciation, so that the history of the genes reflects the history of the species) may be involved in the evolution of these phenotypes. In similar manner, comparison of the genomic locations of the eggplant fruit weight, fruit shape, and color QTL with the positions of similar loci in tomato, potato, and pepper revealed that 40% of the different loci have putative orthologous counterparts in at least one of these other crop species.[6]

Overall, the results suggest that domestication within each family has been driven by mutations in a very limited number of homologous loci that have been conserved throughout the evolution of the different species. Correspondence in location of QTLs in different taxa does not prove identity between the underlying genes, but it does suggest the identity of some of them.

FROM MORPHOLOGY EVOLUTION TO MOLECULAR EVOLUTION

Plant domestication offers a powerful system for studying the genetic and developmental basis of morphological evolution. Cloning the genes that largely control the differences between wild and domesticated plants provides the opportunity to examine the effects of selection on domestication genes at molecular and physiological levels, identify characters that are relevant for future crop improvement, and infer the history of domestication. The discovery of loci such as *tb1*[11] in maize or *fw2.2*[12] in tomato represented cases in which the evolution of better adapted phenotype was largely governed by a single locus, and the genes underlying each locus were subsequently identified and analyzed.

In tomato, a QTL, *fw2.2* changes fruit weight by up to 30% and appears to have been responsible for a key transition during tomato domestication. By applying a map-based approach, the gene responsible for the QTL was identified: *ORFX*.[12] This gene has a sequence suggesting structural similarity to the human oncogene c-H-*ras* p21. Alterations in fruit size, imparted by *fw2.2* alleles, are most likely due to changes in upstream regulatory sequences rather than in the sequence and structure of the encoded protein. The large- and small-fruited alleles differ in the timing of *fw2.2* transcription (heterochronic allelic variation) by approximately 1 week and in total transcript level.[13] Moreover, these differences are associated with concomitant changes in mitotic activity and are sufficient to cause a major change in final fruit mass. The large-fruit allele of *fw2.2* arose in wild populations long before being fixed in most domesticated tomatoes.[14] However, despite the fact that this allele was likely a target of selection during domestication, it has not evolved at distinguishably different rates in domesticated and wild tomatoes.

The most thoroughly analyzed domestication gene is *teosinte branched 1* (*tb1*) in maize. It was shown to correspond to a QTL involved in apical dominance. During

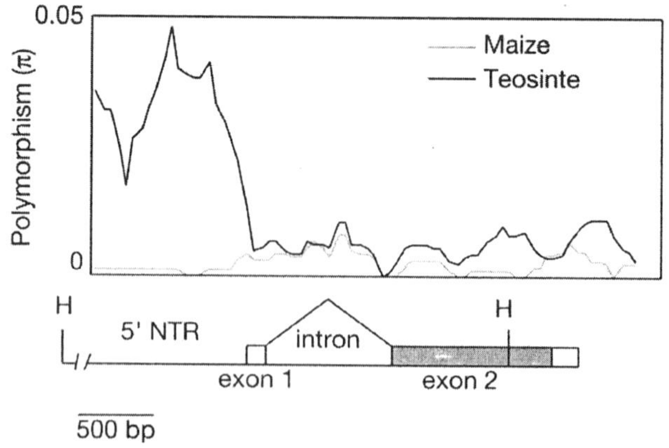

Fig. 1 Predicted structure of *teosinte branched1* (*tb1*) and analysis of polymorphism (π) in maize and teosinte. (From Ref. 11.)

development, *tb1* acts as a repressor of axillary branch growth in those organs in which its RNA messenger accumulates. The difference in mRNA accumulation in between maize and teosinte alleles suggests that the evolutionary switch involved changes in the regulatory regions of *tb1*. Indeed, population–genetic analysis of nucleotid e polymorphism in *tb1* from a diverse sample of maize and teosinte indicates that a strong selective sweep has occurred during domestication in the 5′ NTR region of the gene but not in its coding region (Fig. 1).[11] It was also inferred that the process of domestication could have taken at least several hundred years to bring the maize allele of *tb1* to fixation.

CONCLUSION

The archaeological record indicates that the domestication process of a crop, once initiated, may have been rapid, possibly encompassing only the few hundred years needed to fix relevant alleles of key genes and of their modifiers. A better understanding of the genetic differences between wild plants and domesticated crops adds important facets to the continuing debate on the origin of agriculture and the societies to which it gave rise.[15] The study of crop domestication is also an opportunity for scientists who seek to understand the genetic basis of plant growth and development. The molecular dissection of complex traits through breeding approaches, coupled with the parallel analysis of gene expression, promises to add much to our understanding of the relationships between molecular polymorphism and phenotypic diversity and the genetic basis and evolutionary dynamics of adaptation.

ARTICLES OF FURTHER INTEREST

Agriculture and Biodiversity, p. 1
Crop Domestication in Africa, p. 304
Crop Domestication in China, p. 307
Crop Domestication in Mesoamerica, p. 310
Crop Domestication in Southeast Asia, p. 320
Crop Domestication in Southwest Asia, p. 323
Crop Domestication: Fate of Genetic Diversity, p. 333
Crop Domestication: Role of Unconscious Selection, p. 340
Farmer Selection and Conservation of Crop Varieties, p. 433
Gene Flow Between Crops and Their Wild Progenitors, p. 480
Molecular Evolution, p. 748

REFERENCES

1. Harlan, J.R. *Crops and Man*; 1975. American Society of Agronomy Inc.; Crop Science Society of America Inc.: Madison US.
2. Paterson, A.H.; Lin, R.; Li, Z.; Schertz, K.F.; Doebley, J. Convergent domestication of cereal crops by independent mutations at corresponding genetic loci. Science **1995**, *269*, 1714–1718.
3. Poncet, V.; Lamy, F.; Devos, K.M.; Gale, M.D.; Sarr, A.; Robert, T. Genetic control of domestication traits in pearl millet (*Pennisetum glaucum* L., Poaceae). Theor. Appl. Genet. **2000**, *100*, 147–159.
4. Doebley, J.; Stec, A. Genetic analysis of the morphological differences between maize and teosinte. Genetics **1991**, *129*, 285–295.
5. Koinange, E.M.K.; Singh, S.P.; Gepts, P. Genetic control of the domestication syndrome in common bean. Crop Sci. **1996**, *36*, 1037–1045.
6. Doganlar, S.; Frary, A.; Daunay, M.-C.; Lester, R.N.; Tanksley, S.D. Conservation of gene function in the Solanaceae as revealed by comparative mapping of domestication traits in eggplant. Genetics **2002**, *161*, 1713–1726.
7. Xiong, L.Z.; Liu, K.D.; Dai, X.K.; Xu, C.G.; Zhang, Q.; Zhang, Q.F. Identification of genetic factors controlling domestication-related traits of rice using an F2 population of a cross between *Oryza sativa* and *O. rufipogon*. Theor. Appl. Genet. **1999**, *98*, 243–251.
8. Burke, J.M.; Tang, S.; Knapp, S.J.; Rieseberg, L.H. Genetic analysis of sunflower domestication. Genetics **2002**, *161*, 1257–1267.
9. Timmerman-Vaughan, G.M.; McCallum, J.A.; Frew, T.J.; Weeden, N.F.; Russell, A.C. Linkage mapping of quantitative trait loci controlling seed weight in pea (*Pisum sativum* L.). Theor. Appl. Genet. **1996**, *93*, 431–439.
10. Maughan, P.J.; Saghai, M.; Buss, G.R. Molecular-marker analysis of seed-weight: Genomic locations, gene action, and evidence for orthologous evolution among three legume species. Theor. Appl. Genet. **1996**, *93*, 574–579.
11. Wang, R.L.; Stec, A.; Hey, J.; Lukens, L.; Doebley, J. The limits of selection during maize domestication. Nature **1999**, *398*, 236–239.
12. Frary, A.; Nesbitt, T.C.; Grandillo, S.; van der Knaap, E.; Cong, B.; Liu, J.P.; Meller, J.; Elber, R.; Alpert, K.B.; Tanksley, S.D. fw2.2: A quantitative trait locus key to the evolution of tomato fruit size. Science **2000**, *289*, 85–88.
13. Cong, B.; Liu, J.; Tanksley, S.D. Natural alleles at a tomato fruit size quantitative trait locus differ by heterochronic regulatory mutations. Proc. Natl. Acad. Sci. U. S. A. **2002**, *99*, 13606–13611.
14. Nesbitt, T.C.; Tanksley, S.D. Comparative sequencing in the genus *Lycopersicon*. Implications for the evolution of fruit size in the domestication of cultivated tomatoes. Genetics **2002**, *162*, 365–379.
15. Salamini, F.; Ozkan, H.; Brandolini, A.; Schafer-Pregl, R.; Martin, W. Genetics and geography of wild cereal domestication in the near east. Nat. Rev. Genet. **2002**, *3*, 429–441.

Radiation Hybrid Mapping

Ron J. Okagaki
Ralf G. Kynast
Howard W. Rines
Ronald L. Phillips
University of Minnesota, St. Paul, Minnesota, U.S.A.

INTRODUCTION

Radiation hybrid maps are physical maps of genomes that provide an alternative to traditional genetic maps. These radiation hybrid maps have two important advantages over genetic maps. First, distances on a radiation hybrid map are determined by the frequency of radiation-induced breaks between markers. The distribution of breaks is believed to be random; therefore, distances calculated along a radiation hybrid map are proportional to physical distances. Genetic crossovers (used to determine genetic distances) are not randomly distributed, making it impossible to predict physical distances between markers from genetic maps in maize and many other plant species. Second, any sequence of interest is readily placed on a radiation hybrid map. In contrast, only polymorphic markers can be mapped on a genetic map. Individual markers for genetic mapping must have detectable differences between the parents of the mapping population. A monomorphic marker—one for which there is no apparent difference in the population—cannot be mapped. High-quality maps produced by radiation hybrid mapping have proven their worth in the human genome project, with over 30,000 sequences mapped. It is anticipated that radiation hybrid maps will be similarly useful in plants.

RADIATION HYBRID MAPPING

Radiation hybrid (RH) mapping was developed in the 1970s as a tool for mapping genes in humans. Improved analytical and molecular methods allowed Cox and coworkers to develop RH mapping into an effective tool.[1] This work spurred further developments, and RH maps are now under construction or completed for the human genome and a number of other species.[2,3] Current information about projects may be obtained from the Radiation Hybrid Mapping Page at http://compgen.rutgers.edu/rhmap/. One reason for the popularity of radiation hybrid maps is that these are physical maps. This property makes RH maps particularly useful in assisting the construction of bacterial artificial chromosome (BAC) or yeast artificial chromosome (YAC) contigs. In this role RH maps are contributing to genome sequencing projects. Map information has other uses; with the degree of gene, chromosome, and genome duplication in plants, positional information is vital for identifying candidate sequences.

In the widely used whole genome approach (WG-RH), a donor cell line from the species being mapped is irradiated to kill the cells and fragment the chromosomes. Then donor cells are fused with recipient cells. Some of the chromosome fragments become integrated into the host chromosomes or are retained as a new chromosome consisting of numerous fragments joined together. Each radiation hybrid cell line retains between 10% and 50% of the donor genome, and a group of approximately 100 cell lines forms a mapping population or panel.[3,4]

The likelihood that a break will occur between markers is directly proportional to the physical distance between them at a given radiation dose. This property of radiation hybrid lines underlies the methodologies used to analyze RH mapping data. Plus-minus PCR assays are used to detect the presence of markers in RH lines; polymorphic markers are not required for mapping. These data are analyzed by one of several approaches to produce an RH map. Distances between markers are expressed in centirays (cR). One cR represents a 1% frequency of breakage between two markers at a specific radiation dose. Breaks are assumed to be distributed randomly, and this makes it possible to relate centirays to kilobases of DNA. Examples of methods used to analyze RH mapping data may be found in Barrett[4] and in references contained in Walter and Goodfellow.[2] Computer programs for analysis of RH data may be accessed through the Radiation Hybrid Mapping Page.

Whole plants instead of cell lines may be used to create RH lines.[5] The starting materials here were oat-maize addition lines containing a single maize chromosome in an oat background. RH lines were selected from among the progeny of irradiated oat-maize addition lines (Fig. 1). RH lines produced by this method have different characteristics from WG-RH lines. First, most RH lines

Encyclopedia of Plant and Crop Science
DOI: 10.1081/E-EPCS 120005601

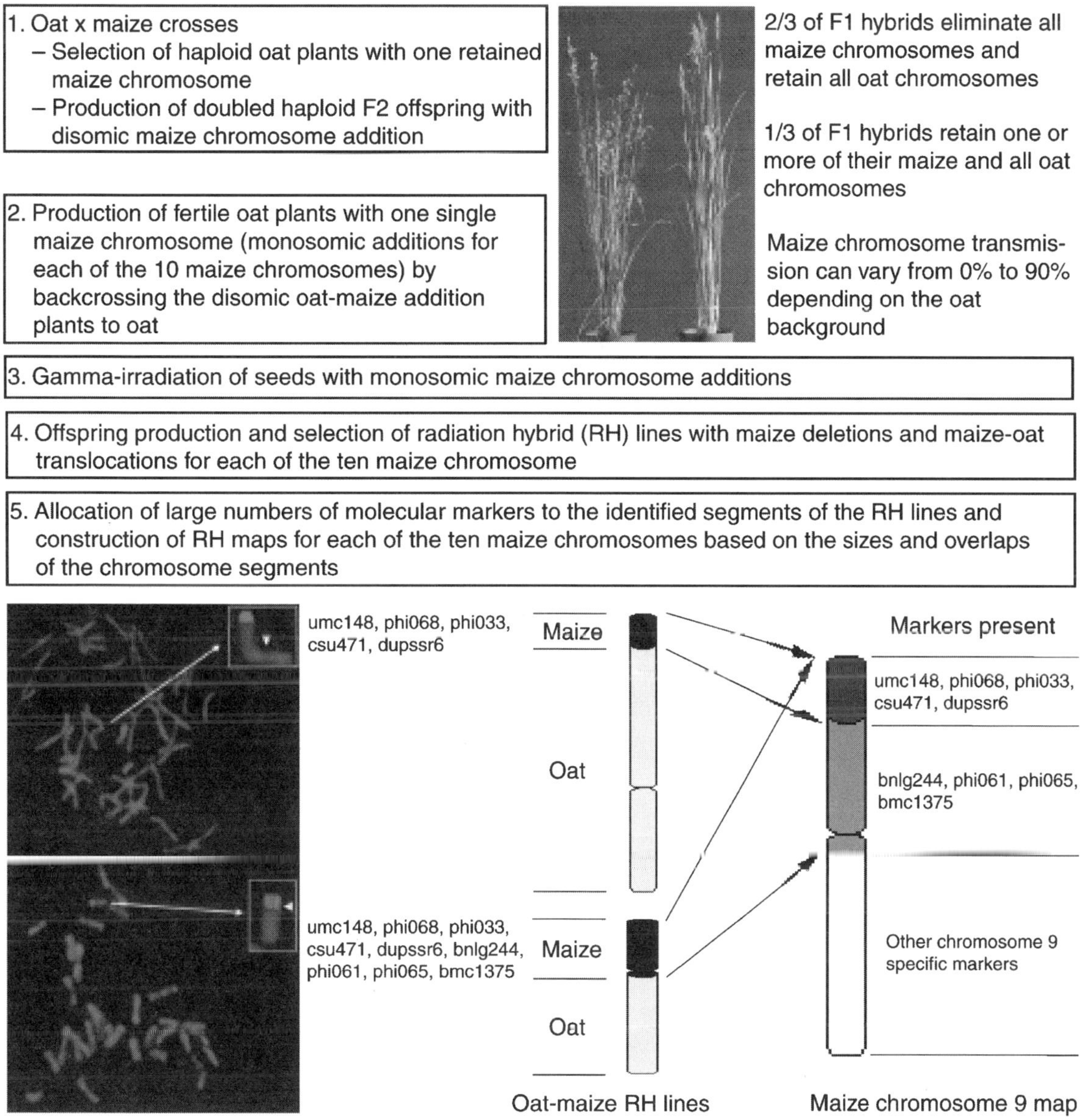

1. Oat x maize crosses
– Selection of haploid oat plants with one retained maize chromosome
– Production of doubled haploid F2 offspring with disomic maize chromosome addition
2/3 of F1 hybrids eliminate all maize chromosomes and retain all oat chromosomes
1/3 of F1 hybrids retain one or more of their maize and all oat chromosomes
2. Production of fertile oat plants with one single maize chromosome (monosomic additions for each of the 10 maize chromosomes) by backcrossing the disomic oat-maize addition plants to oat
Maize chromosome transmission can vary from 0% to 90% depending on the oat background
3. Gamma-irradiation of seeds with monosomic maize chromosome additions
4. Offspring production and selection of radiation hybrid (RH) lines with maize deletions and maize-oat translocations for each of the ten maize chromosome
5. Allocation of large numbers of molecular markers to the identified segments of the RH lines and construction of RH maps for each of the ten maize chromosomes based on the sizes and overlaps of the chromosome segments
umc148, phi068, phi033, csu471, dupssr6
Maize
Oat
umc148, phi068, phi033, csu471, dupssr6, bnlg244, phi061, phi065, bmc1375
Maize
Oat
Markers present
umc148, phi068, phi033, csu471, dupssr6
bnlg244, phi061, phi065, bmc1375
Other chromosome 9 specific markers
Oat-maize RH lines
Maize chromosome 9 map

Fig. 1 Principal steps for development and use of oat-maize RH lines. A set of oat plants with a monosomic addition of one maize chromosome is produced from an oat x maize interspecies hybrid. Monosomic seed are irradiated to induce breaks in the maize chromosome. The maize chromosome fragments are transmitted either through their association with the maize centromere or through translocation to a centric oat chromosome fragment. In the segregating offspring, plant genotypes are screened for breaks by molecular and cytological means. The example shows two different oat-maize translocation plants that define three maize chromosome segments. (*View this art in color at www.dekker.com.*)

produced from addition lines appear to be stable (personal communication), whereas cell culture lines produced by the WG-RH approach are unstable. Second, lines produced from oat-maize addition lines have fewer chromosome breaks than do WG-RH lines. The need to maintain seed viability places limits on the level of chromosome breakage in oat-maize RH lines. A single contiguous fragment may be retained in these RH lines. With a few of these lines, markers may be quickly localized to a chromosome segment. WG-RH lines retain many small fragments. Many WG-RH lines are required to map a marker, but the mapping can be very precise. The pattern of chromosome breaks in RH lines produced from addition lines may limit the practical resolution of these RH maps.

No WG-RH map from a plant has yet been reported. A variation of the WG-RH protocol is being developed in barley, where barley donor protoplasts are irradiated and fused with tobacco recipient protoplasts.[6] Results suggest that RH mapping can become a useful tool in plants.

OTHER PHYSICAL MAPPING APPROACHES

RH mapping represents only one approach to developing physical maps. Alien addition lines, such as the oat-maize addition lines, have been used for locating markers to chromosomes in several species. Aneuploid lines, nullisomic lines, and B–A translocation lines are also used to locate genes to chromosomes or chromosome arms. Examples of these systems are discussed elsewhere in this encyclopedia. Second, systems that fractionate chromosomes have been exploited to place sequences in subchromosome regions. Notable among this work are the wheat deletion stocks where hundreds of lines have been produced with cytologically characterized deletions.[7] Work is ongoing to place molecular markers on these materials. A third approach is molecular cytogenetic maps. These are described in another article. Interest in plant genomics has fostered a desire for better physical maps of plant genomes. There are now several methods to make physical maps. Each approach has its own strengths and weaknesses for resolution, ease of development, and utility in particular species and applications.

STABLE RADIATION HYBRID LINES HAVE MANY USES

The primary reason for producing the oat-maize radiation hybrid lines was to develop a physical mapping system for maize. But mapping is not the only use for these materials. The stability of these materials allows other uses, three descriptions of which follow. First, chromosome pairing has been studied.[8] Using the addition lines, the single pair of maize chromosomes can be specifically labeled and followed through meiosis in the addition lines. This greatly simplifies the task of observing chromosome behavior because only the maize chromosome is visible. With the RH lines it will be possible to study at the effect of removing specific chromosome regions. Specific chromosome landmarks (such as centromeres) can be isolated from other chromosome constituents in RH lines and observed. Second, flow cytometry can physically separate the maize chromosome from the oat chromosomes.[9] It will be possible to make chromosome-specific libraries or libraries enriched for specific chromosome regions. Third, expression of maize genes in a foreign environment can be investigated. Ectopic expression of the *liguleless3* gene in chromosome 3 oat-maize chromosome addition lines produces a characteristic hooked panicle phenotype reminiscent of the *liguleless3* mutant phenotype in maize.[10] Other phenotypes have been associated with particular chromosome addition lines.[11] Expression studies may determine the gene(s) responsible, and radiation hybrid lines may be an effective means to transfer traits from one species to another.

CONCLUSION

The set of oat-maize chromosome addition lines and RH lines is making important contributions to our knowledge of the physical organization of the maize genome. However, it is likely that BAC contig maps and genome sequencing will supplant RH maps and other physical maps of the maize genome for the purpose of mapping. RH mapping and other physical mapping techniques may remain useful in specific situations—for example, creating maps in species with very large genomes or where there is not the economic incentive to invest in contig maps and genome sequencing. The long-term value of RH mapping may therefore derive from other uses envisioned for the oat-maize radiation hybrid lines.

ACKNOWLEDGMENTS

We thank Oscar Riera-Lizarazu and Isabel Vales for sharing unpublished data with us, and The National Science Foundation for funding. Any opinions, findings, and conclusions or recommendations expressed in this material are those of the author(s) and do not necessarily reflect the views of The National Science Foundation.

ARTICLES OF FURTHER INTEREST

Aneuploid Mapping in Diploids, p. 33
Aneuploid Mapping in Polyploids, p. 37
Flow Cytogenetics, p. 460

REFERENCES

1. Cox, D.R.; Burmeister, M.; Price, E.R.; Kim, S.; Myers, R.M. Radiation hybrid mapping: A somatic cell genetic method for constructing high-resolution maps of mammalian chromosomes. Science **1990**, *250*, 245–250.

2. Walter, M.A.; Goodfellow, P.N. Radiation hybrids: Irradiation and fusion gene transfer. Trends Genet. **1993**, *9* (10), 352–356.
3. McCarthy, L.C. Whole genome radiation hybrid mapping. Trends Genet. **1996**, *12* (12), 491–493.
4. Barrett, J.H. Genetic mapping based on radiation hybrid data. Genomics **1992**, *13*, 95–103.
5. Riera-Lizarazu, O.; Vales, M.I.; Ananiev, E.V.; Rines, H.W.; Phillips, R.L. Production and characterization of maize chromosome 9 radiation hybrids derived from an oat-maize addition line. Genetics **2000**, *156* (1), 327–339.
6. Waldrop, J.; Snape, J.; Powell, W.; Machray, G.C. Constructing plant radiation hybrid panels. Plant J. **2002**, *31* (2), 223–228.
7. Endo, T.R.; Gill, B.S. The deletion stocks of common wheat. J. Heredity **1996**, *87*, 295–307.
8. Bass, H.W.; Riera-Lizarazu, O.; Ananiev, E.V.; Bordoli, S.J.; Rines, H.W.; Phillips, R.L.; Sedat, J.W.; Agard, D.A.; Cande, W.Z. Evidence for the coincident initiation of homolog pairing and synapsis during the telomere-clustering (bouquet) stage of meiotic prophase. J. Cell. Sci. **2000**, *113* (6), 1033–1042.
9. Li, L.J.; Arumuganathan, K.; Rines, H.W.; Phillips, R.L.; Riera-Lizarazu, O.; Sandhu, D.; Zhou, Y.; Gill, K.S. Flow cytometric sorting of maize chromosome 9 from an oat-maize chromosome addition line. Theor. Appl. Genet. **2001**, *102* (5), 658–663.
10. Muehlbauer, G.J.; Riera-Lizarazu, O.; Kynast, R.G.; Martin, D.; Phillips, R.L.; Rines, H.W. A maize chromosome 3 addition line of oat exhibits expression of the maize homeobox gene liguleless3 and alteration of cell fates. Genome **2000**, *43* (6), 1055–1064.
11. Kynast, R.G.; Riera-Lizarazu, O.; Vales, M.I.; Okagaki, R.J.; Maquieira, S.B.; Chen, G.; Ananiev, E.V.; Odland, W.E.; Russell, C.D.; Stec, A.O.; Livingston, S.M.; Zaia, H.A.; Rines, H.W.; Phillips, R.L. A complete set of maize individual chromosome additions to the oat genome. Plant Physiol. **2001**, *125* (3), 1216–1227.

R

Reconciling Agriculture with the Conservation of Tropical Forests

Wil de Jong
Center for International Forestry Research, Bogor, Indonesia

INTRODUCTION

The expansion of agriculture is widely held to be one of the causes of tropical deforestation. Simply stated, the progressive expansion of agriculture into areas of tropical forest can only occur when forests are slashed. There is little doubt that in many places an increased need for agricultural land goes hand in hand with expansion into forestland. However, the relationship between agricultural expansion and tropical deforestation is more complex than a simple addition and subtraction formula suggests. Many forms of tropical agriculture are integrated with forestry activities. Forests play an important role in the complex of resource management among many farmers in the tropics. Although such tropical agroforestry is no alternative for strict forest conservation in protected areas, it is a viable alternative—especially from a conservation point of view—to other agricultural development schemes that are often proposed or blindly copied in many tropical countries.

SWIDDEN AGRICULTURE AND FOREST MANAGEMENT

One of the most common types of agriculture in areas of tropical rainforest remains swidden agriculture. Swidden agriculture implies using a single field for intensive cultivation for one or a few years, after which the field is fallowed. During the intensive production cycle farmers grow crops like rice, manioc, and bananas, usually in combination with a few or many other crops. The reason farmers practice swidden—rather than permanent—agriculture is that soils' natural nutrient reserves do not permit permanent annual cropping. In most cases, the main purpose of the fallow period is to allow the restoration of the nutrient stock in the vegetation. This nutrient stock is released again into the soil when, at the end of the fallow period, the vegetation is slashed and burned. This new nutrient boost allows a subsequent round of intensive crop production. An additional benefit of the fallow vegetation is the reduction of obnoxious weeds that put high pressure on farmers' labor reserves and may also reduce crop yields.

Forests do constitute an important role in many swidden fallow agricultural systems. They are where swidden fields are originally developed. The vegetation that grows on swidden fields that are being fallowed will eventually develop into secondary forests. Besides the nutrient restoration, these swidden fallow secondary forests are also important for forestry production. Forestry production is an important component of many swidden agricultural systems. This forestry production may occur in forest remnants that have been spared from conversion to agricultural lands. It may also occur in so-called secondary forest gardens—fields that were once used for agriculture, but that have changed to tree vegetation resulting from the partly spontaneous/partly actively pursued regrowth of trees.[1]

Forest Remnants and Primary Forest Gardens

In the remote corners of the world, where most swidden agriculturists reside, forests supply an important number of products for daily needs. They supply food, like edible leaves, mushrooms, berries, and game. In addition, they also supply construction material to build houses, medicinal plants, and sites where spirits are believed to reside and where graveyards are made. Considering this importance of forests, many swidden farmers may spare important areas of forests when advancing their agricultural fields into the forest. Examples of these forest remnants can be found in many locations. They are reported mostly in countries in Southeast Asia.[2,3] They are also common in many rural areas in Africa, where they are kept as sacred groves. These groves hold important spiritual meanings for the people who protect them.

Most primary forest gardens are safeguarded from conversion to agricultural land for the sake of assuring the supply of a wide range of forest products. Some forest gardens, however, are known for their production of economically important species. For instance, small producers cultivate *Psychotria ipecacuanha*, a medicinal

Encyclopedia of Plant and Crop Science
DOI: 10.1081/E-EPCS 120010504

herb, in tropical forests in several countries in South and Central America.[4] Originally from Brazil, this species is now being produced in small areas in fully grown primary forests in Costa Rica and Nicaragua. In similar fashion, indigenous Dayak farmers in areas of West Kalimantan, who partly rely on rubber production for their monetary income, may plant rubber in their primary forest gardens.[3]

Swidden Fallow Secondary Forests

The most common type of forests cooccurring with swidden agriculture is the swidden fallow secondary forest.[a] These forests regenerate largely through natural processes in woody fallows of swidden agriculture for the purpose of restoring the land for cultivation again. In a typical swidden fallow land use system, swidden fallow secondary forests often constitute the largest part of the area under some kind of vegetation cover. Besides their role in restoration of nutrient stocks, this type of forest is habitat for an important number of species that have some use value for their owners. Because these forests are usually located closer to settlements than are primary forests, they often may be the first place where farmers will turn to collect forest products.

Although swidden fallow secondary forests have often been classified as degraded forests, there is increasing evidence of their importance in fulfilling the functions that primary forests used to fulfill: conservation of biodiversity, regulation of the water flow that affects downstream areas, and carbon sequestration. It can indeed be expected that in many places of the world these functions will increasingly be fulfilled by some kind of secondary forest.[5]

Secondary Forest Gardens

Secondary forest gardens are managed forests that appear on land once used for cropping that has subsequently been designated for forestry production. In addition to spontaneous vegetation, these forest gardens hold a number of species that may be planted or tended after they occur spontaneously. They play an important role in the livelihood of many farmers, precisely because they provide a larger suite of products than natural forests or primary forest gardens. They may occupy significant areas in swidden agricultural villages all over the world.[3,6]

[a]Chokkalingam and de Jong[1] distinguish six types of secondary forests, two of which are related to swidden agriculture: swidden fallow secondary forests and secondary forest gardens.

TRANSITION, INTENSIFICATION, AND THE FOREST LANDSCAPE

It has been a widely held belief that farmers who practice swidden agriculture will progressively expand their activities into natural forests, or that increased population pressure will lead to the progressive decline of forests in regions where swidden farmers have still managed to preserve areas of forest. None of these general belief is correct. The myth of progressive encroachment into natural forests is largely wrong. In most cases, swidden farmers will limit themselves to occupying an adequate area of agricultural land on which to rotate their swidden fields. When populations increase, there follows in many instances a period during which the relative length of the fallow period will decline, leading to wider presence of vegetation that has trees of lower height and less biomass. Many examples can be found, however, in which a land use intensification process eventually occurs.

Van Noordwijk[7] described three possible options for intensification of swidden agriculture. These three options include tree-cash crop production, food crop production, and fodder-pasture production. In cases where tree crop intensification takes place, the outcome is often beneficial for the general presence of forests in the landscape and related biodiversity and environmental functions. A recent study on secondary forests in tropical Asia[8] provides several examples in which tree crop intensification may actually contribute to the presence of forests in a landscape that had been gradually loosing its forest cover. For instance, swidden agriculturists in some villages in West Kalimantan, Indonesia, have gradually been changing their traditional production of subsistence agriculture complemented with commercial forest product extraction to production of rubber in mixed secondary forest gardens. Data from one village in West Kalimantan show that the reforestation of land with these mixed rubber–secondary forest gardens outweighed the conversion of primary forests for agricultural purposes.[3,8] The same study describes examples of how farmers in the Himalayas of northern India gradually are changing their traditional swidden agriculture to mixed agriculture–forestry production, centered around the production of *Alnus nepalensis*.[9] Additonal examples have been reported from several countries in Africa where, as a result of farmers' activities, forests have expanded into savannas.[6]

CONCLUSION

Although a diverse swidden agriculture landscape with its mixture of anthropogenic forests may not be a better alternative to true conservation in protected areas, such a landscape suggests an important potential in other instances. The kind of agriculture–forestry interaction described here has many conservation and environmental advantages over other large-scale estate plantations and other kinds of intensified agricultural production such as oil palm production, pulp-wood plantations, and the like. Where it is necessary to make alternative choices, it is useful to consider the benefits that intensified swidden agriculture production may offer, including intensified forestry production as described here. Second, where efforts at development of swidden agriculture are pursued, it matters to an important degree to what extent traditionally managed forests are being taken into consideration. This will have different impacts on the forest, the biodiversity in the landscape, and the watershed protection that this landscape provides.

ARTICLES OF FURTHER INTEREST

Agriculture and Biodiversity, p. 1
Columbian Exchange: The Role of Analogue Crops in the Adoption and Dissemination of Exotic Cultigens, p. 292
Cropping Systems: Slash-and-Burn Cropping Systems of the Tropics, p. 363
Ecology and Agricultural Sciences, p. 401
Farmer Selection and Conservation of Crop Varieties, p. 433
Pre-agricultural Plant Gathering and Management, p. 1055
Sustainable Agriculture: Ecological Indicators of, p. 1191
Weed Management in Less Developed Countries, p. 1295

REFERENCES

1. Chokkalingam, U.; de Jong, W. Secondary forest: A working definition and typology. Int. For. Rev. **2001**, *3* (1), 19–26.
2. Padoch, C.; Peters, C. Managed Forest Gardens in West Kalimantan. In *Perspectives on Biodiversity: Case Studies in Genetic Resource Conservation and Development*; Potter, C.S., Cohen, J.I., Janczewski, J., Eds.; American Association for the Advancement of Science: Washington, DC, 1993; 167–176.
3. de Jong, W. *Forest Products and Local Forest Management in West Kalimantan, Indonesia: Implications for Conservation and Development*; Tropenbos Kalimantan Series, 2002; Vol. 6, Wageningen, The Netherlands.
4. Ocampo, R. *Informe del estudio de caso de ipecuana*; Center for International Forestry Research: Bogor, Indonesia, 2001.
5. de Jong, W.; van Noordwijk, M.; Sirait, M.; Suyanto; Liswanta, N. Farming secondary forests in Indonesia. J. Trop. For. Sci. **2001**, *13* (4), 705–726.
6. Fairhead, J.; Leach, M. *Reframing Deforestation: Global Analysis and Realities, Studies in West Africa*; Routledge: London, UK, 1998.
7. van Noordwijk, M. Productivity of intensified crop fallow rotations in the Trenbath model. Agrofor. Syst. **1999**, *47*, 223–237.
8. Chokkalingam, U.; de Jong, W.; Smith, J.; Sabogal, C. A conceptual framework for the assessment of tropical secondary forest dynamics and sustainable development potential in Asia. J. Trop. For. Sci. **2001**, *13* (4), 577–600.
9. Ramakrishnan, P.S.; Kushwaha, S.P.S. Secondary forests of the Himalaya with emphasis on the north-eastern hill region of India. J. Trop. For. Sci. **2001**, *13* (4), 727–747.

Regeneration from Guard Cells of Crop and Other Plant Species

Jim M. Dunwell
University of Reading, Reading, U.K.

INTRODUCTION

Guard cells, the components of the stomata within the leaf epidermis, are usually considered to be terminally differentiated and only of interest to plant physiologists concerned with gas exchange within the leaf. However, as will be described, these cells are of much greater general interest for their capability of being induced into division and even regeneration into complete plants.

HISTORY

Examination of the botanical literature from the last hundred years reveals several reports of abnormal development of guard cells from a variety of species. These references have been largely overlooked and only relatively recently[1] has there been any real discussion of their significance. Among the early studies was the first attempt to induce division in mature guard cells in vitro.[2] It was more than sixty years before this specific ambition was achieved, although during that period, evidence of guard cell division was reported in wounded leaves of various magnoliaceous species[3] and in cucumber (*Cucumis sativus*) hypocotyls exposed to particular red light conditions.[4]

DIVISION AND REGENERATION IN VITRO

There are three examples in which complete plants have been produced from guard cells in vitro; these will be considered in sequence below.

Tree Tobacco (*Nicotiana glauca*)

The first report of successful cell division in vitro involved an accidental discovery with cultured guard cell protoplasts of *N. glauca*. These were being grown in culture in an attempt to mimic normal development, but surprisingly some were found to develop into undifferentiated callus. Subsequently, experiments were performed to optimize conditions for culturing such protoplasts and to determine whether they were indeed totipotent.[5] Protoplasts were isolated from adaxial epidermal tissue of leaves of plants grown under fluorescent light (800–900 μmol m^{-2} s^{-1} of photons of photosynthetically active radiation [PAR]). To increase the probability that the guard cells were of uniform osmotic potential at the time of harvest, leaves were collected in darkness, just before the beginning of each light period. Protoplasts were cultured in liquid media similar to those used for culturing mesophyll cell protoplasts of *Nicotiana tabacum* but with modified pH and KCl, $CaCl_2$, sucrose, and glycine concentrations; concentrations of growth regulators were 0.3 mg l^{-1} α-naphthaleneacetic acid (NAA) and 0.075 mg l^{-1} of 6-benzylaminopurine (BAP). Protoplasts were incubated in darkness at 25°C in 8-well microchamber slides at a density of 1.25×10^5 cells ml^{-1}. Cell divisions began within 72–96 hours (h) of initiation of cultures, with an average cell survival of 57% at this time. After 8–10 weeks of culture, cell colonies were transferred to a callus initiation medium containing agar and incubated under continuous white fluorescent light (25 μmol m^{-2} s^{-1} of photons of PAR) for another 8–10 weeks. Green callus tissue was then transferred to a callus growth medium and incubated under the same conditions. After a further 8–10 weeks, the callus was transferred to a shoot differentiation medium and incubated similarly. Two weeks later multiple shoots appeared and were transferred to a root differentiation medium. When roots were sufficiently developed (6–8 weeks), plants were transplanted to soil and grown in a growth chamber. It was concluded that guard cell protoplasts could survive and divide in culture and are totipotent.

In a later study,[6] additional modifications in protocol were reported that allowed directed development either to callus production or to maintain normal guard cell behavior. In particular, it was shown that temperature is an important determinant of survival, growth, and differentiation. As the temperature was increased from 24 to 32°C, the survival of cells cultured for 7 days (d) was increased from approximately 20% to approximately 80%. At all these temperatures, approximately 90% of surviving cells divided to form callus tissue. Cells cultured for 7 d at 34 to 40°C also survived in high percentages (approximately 80%), but in contrast, they retained a morphology similar

Encyclopedia of Plant and Crop Science
DOI: 10.1081/E-EPCS 120010551

to that of guard cells and they did not divide. These observations were explored further in a later study using differential display technology to define the various cell types.[7]

Sugar Beet (*Beta vulgaris*)

The most valuable practical application of guard cell culture is that developed for sugar beet. This procedure was founded initially on the application of advanced microscopical techniques to address a difficult culture problem.[8] Using a computer-controlled microscope system to assist in the positioning and rapid relocation of large numbers of cultured cells, it was found to be possible to identify those specific leaf protoplasts with the capacity to divide within a highly recalcitrant culture in which only a tiny fraction (0.5%) of the total population was able to produce viable microcalli. It was discovered that such regenerable microcalli come only from guard cell protoplasts (60% capacity for cell division), and not from protoplasts derived from mesophyll or other leaf tissue. This important discovery led to the development of an optimized protocol for the efficient and rapid genetic modification of this species.[9,10] This involved a polyethylene glycol-mediated DNA transformation technique applied to protoplast populations enriched specifically for this single totipotent cell type in order to achieve high transformation frequencies. Bialaphos resistance, conferred by the *pat* gene, produced a highly efficient selection system. The majority of plants were obtained within 8 to 9 weeks of selection and were appropriate for plant breeding purposes; all were resistant to glufosinate-ammonium–based herbicides. Detailed genomic characterization revealed verified transgene integration, and progeny analysis showed Mendelian inheritance. This system has since been applied to the large-scale production of a range of transgenic sugar beet.

Cotton (*Gossypium hirsutum*)

The most recent extension of this technology is its application to cotton,[11] a species generally considered recalcitrant to many regeneration methods. Rather than relying on protoplast isolation, this method comprised development of a regeneration system from stomatal guard cells directly on epidermal strips, especially from bracts. The most important factors affecting embryogenic callus initiation in both of the varieties tested (Coker 312 and 315) were the source of the epidermal tissue, including plant age (4–5 months old), the developmental stage of the flower (opening flower stage) from which bracts were obtained, the composition of the culture medium, and light irradiance. The flower developmental stage was critical for callus formation, which was observed only from bracts obtained from opening flowers. In addition, epidermal strips excised from the bract basal region were more responsive in culture than those obtained from the upper region. Improved callus initiation was obtained on epidermal strips, which had their cuticle in contact with the culture medium. Light irradiance was a limiting factor for embryogenic callus formation, which was observed only in calli cultured under the lower light irradiance (15.8 μmol m^{-2} s^{-1}). Somatic embryogenesis was observed on callus cultures subcultured consecutively in a culture medium containing NAA (10.7 μM) and isopentenyladenine (4.9 μM). Histodifferentiation of somatic embryos was improved in a medium containing NAA (8.1 μM), isopentenyladenine (2.5 μM), and abscisic acid (0.19–0.38 μM). Somatic embryo germination and plantlet development were obtained using established protocols with few modifications, and on average, one fully developed plant was obtained from the culture of about 100 epidermal strips in both cultivars.

CONCLUSION

It is likely that future investigations will extend the number of species from which totipotent guard cells can be isolated, either in intact epidermis or as protoplasts.[1,12] In many ways the development of this procedure has mirrored the development of regeneration methods for another unusual and isolated cell type—namely microspores—that were also once considered to be a developmental dead end but were then shown to have a high level of totipotency.

It can also be assumed that methods for the isolation of RNA from guard cells[13] will be applied to this particular culture process in order to investigate the details of this profound switch from the normal differentiated state to one capable of division.

Apart from the value of guard cells as a source of totipotent material for use in transformation experiments (e.g., sugar beet) and perhaps the regeneration of other transgenic plants in the future,[1,12] stomatal formation in *Arabidopsis thaliana* is also emerging as an elegant and powerful model system to study the genetic and molecular control of cell-fate specification and pattern formation in multicellular organisms.[14] Eventually it is hoped that the combination of these emerging analytical methods will help to not only uncover the nature of the molecular signals that restrict the developmental potential of guard cells in situ, but also allow more predictable methods for the induction of division in such cells when they are separated from the plant.

ARTICLES OF FURTHER INTEREST

In Vitro Morphogenesis in Plants—Recent Advances, p. 579
In Vitro Plant Regeneration by Organogenesis, p. 584
Plant Cell Culture and Its Applications, p. 931
Plant Cell Tissue and Organ Culture: Concepts and Methodologies, p. 934
Somatic Embryogenesis in Plants, p. 1165

REFERENCES

1. Hall, R.D. Biotechnological applications for stomatal guard cells. J. Exp. Bot. **1998**, *49*, 369–375.
2. Thielmann, M. Über Kulturversuche mit Spaltöffnungszellen. Arch. Exp. Zellforsch. **1925**, *1*, 66–108.
3. Tucker, S.C. Dedifferentiated guard cells in magnoliaceous leaves. Science **1974**, *185*, 445–447.
4. Kazama, H.; Mineyuki, Y. Alteration of division polarity and preprophase band orientation in stomatogenesis by light. J. Plant Res. **1997**, *110*, 489–493.
5. Sahgal, P.; Martinez, G.V.; Roberts, C.; Tallman, G. Regeneration of plants from cultured guard-cell protoplasts of *Nicotiana glauca* (Graham). Plant Sci. **1994**, *97*, 199–208.
6. Roberts, C.; Sahgal, P.; Merritt, F.; Perlman, B.; Tallman, G. Temperature and abscisic-acid can be used to regulate survival, growth, and differentiation of cultured guard-cell protoplasts of tree tobacco. Plant Physiol. **1995**, *109*, 1411–1420.
7. Taylor, J.E.; Abram, B.; Boorse, G.; Tallman, G. Approaches to evaluating the extent to which guard cell protoplasts of *Nicotiana glauca* (tree tobacco) retain their characteristics when cultured under conditions that affect their survival, growth, and differentiation. J. Exp. Bot. **1998**, *49*, 377–386.
8. Hall, R.D.; Riksen Bruinsma, T.; Weyens, G.; Lefebvre, M.; Dunwell, J.M.; Van Tunen, A.; Krens, F.A. Sugar beet guard cell protoplasts demonstrate a remarkable capacity for cell division enabling applications in stomatal physiology and molecular breeding. J. Exp. Bot. **1997**, *48*, 255–263.
9. Hall, R.D.; Verhoeven, H.A.; Krens, F.A. Computer-assisted identification of protoplasts responsible for rare division events reveals guard-cell totipotency. Plant Physiol. **1995**, *107*, 1379–1386.
10. Hall, R.D.; Riksen Bruinsma, T.; Weyens, G.J.; Rosquin, I.J.; Denys, P.N.; Evans, I.J.; Lathouwers, J.E.; Lefebvre, M.P.; Dunwell, J.M.; van Tunen, A.; Krens, F.A. A high-efficiency technique for the generation of transgenic sugar beets from stomatal guard cells. Nat. Biotechnol. **1996**, *14*, 1133–1138.
11. Nobre, J.; Keith, D.J.; Dunwell, J.M. Morphogenesis and regeneration from stomatal guard cell complexes of cotton (*Gossypium hirsutum* L.). Plant Cell Rep. **2001**, *20*, 8–15.
12. Dunwell, J.M. Future prospects for transgenic crops. Phytochem. Rev. **2002**, *1*, 1–12.
13. Smart, L.B.; Nall, N.M.; Bennett, A.B. Isolation of RNA and protein from guard cells of *Nicotiana glauca*. Plant Mol. Biol. Rep. **1999**, *17*, 371–383.
14. Serna, L.; Torres-Contreras, J.; Fenoll, C. Specification of stomatal fate in *Arabidopsis*: Evidences for cellular interactions. New Phytol. **2002**, *153*, 399–404.

Rhizosphere: An Overview

Roberto Pinton
Università di Udine, Udine, Italy

INTRODUCTION

Plant survival and adaptation to adverse soil conditions are strictly dependent on the capability of roots to interact with biotic and abiotic components of soil. Processes at the basis of the root-soil interaction concern a very limited area surrounding the root tissue. In this particular environment, exchanges of energy, nutrients, and molecular signals take place, rendering the chemistry and biology of this environment different from the bulk soil. In this chapter, an overview of the key processes occurring at the rhizosphere, which can account for the complexity of this environment, will be presented.

COMPLEXITY, DEFINITIONS, AND BOUNDARIES

The rhizosphere is a complex environment where almost every process involves multiple interactions between the soil-root-microbe triumvirate.[1] It is well known that movement of water, nutrients, and microbial dynamics are more convoluted around the roots than in the bulk soil. Changes in pH and redox potential often occur. Furthermore, the rhizosphere generally experiences higher mineral weathering rates than bulk soil and is characterized by variable rates of mineralization of the native organic matter.

The term rhizosphere is generally used to define the field of action or influence of a root. Roots vary enormously in their morphology, longevity, activity, and influence on soil as a result of physiological, environmental, and genetic differences. It can be assumed that the rhizosphere forms around each root as it grows because each root changes the chemical, physical, and biological properties of the soil in its immediate vicinity.

The rhizosphere lacks physically precise delimitation; rather it can be described in terms of the longitudinal and radial gradients that develop along the axis of each root as a result of root growth and metabolism, nutrient and water uptake, rhizodeposition, and subsequent microbial growth.[2] These will be gradients with depletion profiles (i.e., the solute concentrations will be lowest at the root surface), as in the case of most essential plant nutrients, and accumulation gradients (i.e., solute concentrations are highest at the root surface), as in the case of the soluble organic solutes released by the roots.[3]

The volume of the rhizosphere also depends on the rate of exudation and impact utilization of the rhizodeposits by microorganisms.[4] Rhizodeposition includes lysates liberated by autolysis of sloughed cells and tissues, as well as biochemically defined root exudates released passively (diffusates) or actively (secretions) from intact cells. Exudation is not uniformly distributed along the whole root system. Usually apical root zones are characterized by a higher capacity for release of low-molecular-weight exudates, whereas basal parts of the root system generally show a higher microbial activity. This might affect rhizosphere microbial population. The longitudinal gradient of exudation may reflect the gradient in microbial catabolism of root exudates as well as differences in root cell activity. The gradients of microbial population and exudates along the root axis can have important implications for the uptake of nutrients by plants.[5]

SOURCE OF NUTRIENTS (AND TOXIC ELEMENTS)

The rhizosphere also defines the soil volume from which plants take up mineral nutrients. In order to provide adequate supply of each essential nutrient, the plant can adjust the root architectural form and/or increase the uptake capacity.[6,7] These modifications in soil-grown plants may occur at the local rhizosphere scale.

Roots can also modify the pH and redox state of the rhizosphere, thus altering the solubility of some nutrients.[8]

Limitation in nutrient availability, as well as excessive availability of toxic elements, can be overcome by release from the roots of low-molecular-weight compounds, which include carboxylates, amino acids, and phenolic compounds.[9] These processes can be greatly enhanced by adverse soil conditions.[6]

Although genotypical differences have been observed in the capacity to acquire limiting nutrients and release of exudates from the roots, it has to be taken into account that these solutes can undergo microbial degradation, enzymatic breakdown, or sorption by soil colloids. Thus, the effectiveness of root exudates appears to be dependent on

Encyclopedia of Plant and Crop Science
DOI: 10.1081/E-EPCS 120012974

the spatial and temporal characteristics of exudation[5] and possibly on the presence of some protection provided by the spatial arrangement of root and soil surfaces.

ROOT-MICROBES: INTERACTIONS AND MOLECULAR SIGNALS

During their growth in natural conditions, plants continuously interact with soil microorganisms.

Microbial population in the rhizosphere may be one or two orders of magnitude higher than that of the bulk soil. Some of these microbes are deleterious, while others are beneficial to plants.[10]

Root colonization is dependent on the release of root exudates. On the other hand, it has been demonstrated that root exudation can be stimulated by microbial colonization of plant roots and by the presence of microbial metabolites, the effects of microbes being species-specific.

Due to differences in rhizodeposition, not only different plant species but also different parts of the root system of the same plant may have distinctive rhizosphere microfloras,[11] possibly reflecting the utilization profile of root exudates.

Among the beneficial microorganisms, bacteria belonging to *Rhizobium* or *Frankia* genera, as well as mychorrizal fungi, are able to establish a symbiotic relation with their host plant.

It is well known that *Rhizobia*-legume symbiosis involves a molecular cross talk between the plant and the microbe; recently it has been shown that a similar signaling process might be involved in establishing a symbiosis with endomychorrizal fungi.

Other rhizobacteria, generally designated as ''plant growth-promoting rhizobacteria'' (PGPR), stimulate plant growth by producing phytohormones or act indirectly as biocontrol agents through the production of antibiotics and siderophores and even by inducing plant resistance mechanisms.[12]

Saprophytic microorganisms in the rhizosphere are responsible for the decomposition of organic residues and nutrient mineralization/turnover processes.

Many plant-associated bacteria use N-acetylated-homoserine-lactones (AHL) as quorum-sensing signals to control processes that are linked both to host interactions and to survival. Interestingly, exudates from some plants were found to contain AHL signal-mimic molecules. This suggests that molecular signals could also be exchanged between plants and non-pathogenic and non-symbiotic rhizobacteria.

Deciphering the molecular basis of root-microbes interactions is a major challenge for understanding the rhizosphere ecology. This would in turn allow the management of the rhizosphere and possibly offer agronomic applications.

The complexity of the rhizosphere and its unique features are emphasized also by the biochemical events and molecular cross talk between host and parasitic plants: It has been shown that compounds released by host roots may act as molecular signals for parasitic plants.[13]

It is worthwhile to say that caution should be used when transferring data obtained using artificial microcosms to the situation occurring in the soil. In fact, any bacterial species living in a mixed microbial population, such as that of the rhizosphere soil, may encounter not only the molecular signal produced by a cell of the same species but also molecular signals produced by cells of different species. The situation is made more complex by the presence of plant molecular signals and by the fact that the same AHL molecule can be used to regulate the expression of different biological processes in different bacterial species. On the contrary, some bacterial species can produce multiple AHLs, each having different effects on the phenotype. In soil, the situation is even more complex because the molecular signals can interact with surface active soil particles through electrostatic interactions, hydrogen bonds, and van der Waals bonds. In addition, molecular signals with anionic groups can be adsorbed by inorganic soil colloids by a ligand exchange mechanism.

ROOT SENSING OF THE ENVIRONMENT

To cope with the uneven distribution of ions in the surrounding root, plants have evolved mechanisms which include expression of transporter genes in specific root zones or cells and synthesis of enzymes involved in the uptake and assimilation of nutrients. Root development can also be significantly modified in response to variations in the supply of inorganic nutrients in the soil. These plant responses are largely controlled by the internal status of the plant. On the other hand, it has been recently shown that this behavior is also dependent on the capacity to locally sense the changes in the external concentration of the nutrient.[14] These evidences support the existence in plants of sensing mechanisms for environmental signal(s) coming from the rhizosphere, including variations in the availability and distribution of nutrients.

ARTICLES OF FURTHER INTEREST

Root Membrane Activities Relevant to Plant-Soil Interactions, p. 1110

Symbioses with Rhizobia and Mycorrhizal Fungi: Microbe/Plant Interactions and Signal Exchange, p. 1213
Symbiotic Nitrogen Fixation: Plant Nutrition, p. 1222

REFERENCES

1. Lynch, J.M. *The Rhizosphere*; John Wiley: Chichester, 1990.
2. Darrah, P.R.; Roose, T. Modeling the Rhizosphere. In *The Rhizosphere. Biochemistry and Organic Substances at the Soil Plant Interface*; Pinton, R., Varanini, Z., Nannipieri, P., Eds.; Marcel Dekker: New York, 2001; 327–372.
3. Tinker, P.B.; Nye, P.H. *Solute Movement in the Rhizosphere*; Oxford University Press: Oxford, UK, 2000.
4. Badalucco, L.; Kuikman, P.J. Mineralization and Immobilization in the Rhizosphere. In *The Rhizosphere. Biochemistry and Organic Substances at the Soil Plant Interface*; Pinton, R., Varanini, Z., Nannipieri, P., Eds.; Marcel Dekker: New York, 2001; 159–196.
5. Römheld, V. The role of phytosiderophores in acquisition of iron and other micronutrients in graminaceous species: An ecological approach. Plant Soil **1991**, *130*, 127–134.
6. Marschner, H. *Mineral Nutrition of Higher Plants,* 2nd Ed.; Academic Press: London, 1995.
7. Hell, R.; Hillebrand, H. Plant concepts for mineral acquisition and allocation. Curr. Opin. Biotechnol. **2001**, *12*, 161–168.
8. Hinsinger, P. How do plant roots acquire mineral nutrients? Chemical processes involved in the rhizosphere. Adv. Agron. **1998**, *64*, 225–265.
9. Jones, D.L.; Ryan, P.R.; Delhaize, E. Function and mechanism of organic anion exudation from plant roots. Annu. Rev. Plant Physiol. Plant Mol. Biol. **2001**, *52*, 527–560.
10. Brimecombe, M.J.; De Leij, F.A.A.M.; Lynch, J.M. The Effect of Root Exudates on Rhizosphere Microbial Populations. In *The Rhizosphere. Biochemistry and Organic Substances at the Soil Plant Interface*; Pinton, R., Varanini, Z., Nannipieri, P., Eds.; Marcel Dekker: New York, 2001; 95–140.
11. Crowley, D. Function of Siderophores in the Plant Rhizosphere. In *The Rhizosphere. Biochemistry and Organic Substances at the Soil Plant Interface*; Pinton, R., Varanini, Z., Nannipieri, P., Eds.; Marcel Dekker: New York, 2001; 223–261.
12. Persello-Cartieaux, F.; Nussaume, L.; Robaglia, C. Tales from the underground: Molecular plant-rhizobacteria interactions. Plant Cell Environ. **2003**, *26*, 189–199.
13. Yoder, J.I. Parasitic plant responses to host plant signals: A model for subterranean plant-plant interactions. Curr. Opin. Plant Biol. **1999**, *2*, 65–70.
14. Forde, B. Local and long-distance signaling pathways regulating plant responses to nitrate. Annu. Rev. Plant Biol. **2002**, *53*, 203–224.

Rhizosphere: Biochemical Reactions

Paolo Nannipieri
Università degli Studi di Firenze, Florence, Italy

INTRODUCTION

The rhizosphere soil is a site of complex interactions among plant roots, microflora, and microfauna. Therefore, all known biochemical reactions of living cells also occur in the rhizosphere soil. Pathways of primary and secondary metabolisms of microbial, animal, and plant cells also occur in the rhizosphere soil, including microbial degradation of natural and man-made recalcitrant compounds (pesticides, synthetic polymers, etc.), nitrification by autotrophic bacteria, or dissimilatory reactions of nitrate and sulfate. However, the biochemistry of rhizosphere soil shows some distinctive properties such as the presence of: 1) a variety of compounds released from roots (exudates, secreted lysates, chelators, and signal molecules) and from rhizospheric microorganisms (signal molecules, antibiotics, antifungal metabolites, toxins, antibiotics, phytostimulators, chelators, etc.), and 2) active enzymes in living root, microbial and animal cells, free extracellular enzymes and enzymes adsorbed by clays or entrapped by humic molecules. Another peculiar aspect of rhizosphere biochemistry is its metabolism, which is the result of complex microflora-microfauna interactions.

RHIZODEPOSITION AND OTHER PARTICULAR ORGANIC COMPOUNDS IN THE RHIZOSPHERE

Rhizodeposition is the term used to indicate the overall compounds released by plant roots into soil. In addition to compounds of primary and secondary metabolism and polymers (mucilage), plant roots, in response to nutrient stresses, can also release chelators. Thus, grasses produce phytosiderophores under iron deficiency whereas dicotyledonous plants respond by increasing the release of organic acids, acidifying the rhizosphere and secreting reductants to increase iron availability to plants.[1] Under iron deficiency, microorganisms of rhizosphere soil also produce chelators, called siderophores, which have a high affinity for iron; the complex interactions between plants and microorganisms under iron stress have been discussed by Crowley.[1]

Microbes interacting with plants can be classified based on their effects on plants, as pathogenic, saprophytic, and beneficial; the latter are also denominated plant growth-promoting rhizobacteria (PGPR). When microorganisms interact with plants, no matter if this interaction is beneficial or pathogenic, they use the same mechanism.[2] The key event in the response of plants to microorganisms is recognition, which can occur through physical interaction or in response to specific compounds (signal molecules) released by the roots of the specific plant species to be infected. Thus, flavonoids and isoflavonoids are specific signal molecules in the legume-rhizobia symbiosis.[3] Different plant species release different organic compounds that vary in their ability to induce or inhibit effective signalling by different soil bacteria.[3,4]

Other compounds of the rhizosphere soil include antibiotics and hydrogen cyanide produced by plant growth-promoting rhizobacteria for the biological control of plant pathogen; antifungal metabolites are very important for the interactions of bacteria in the rhizosphere.[2] Rhizobacteria can also produce phytostimulators, such as auxin derivatives.[5] *Azospirillum*, a N_2-fixing bacteria, may promote plant growth by releasing phytostimulators rather than by its N_2-fixing capacity.[2] Indeed, cytokinins, auxins, and gibberellines have been found in the supernatants of these bacteria.

In addition, as any microbial environment, the rhizosphere is characterized by the presence of signal molecules, which permit bacteria to communicate with cells of the same species and sometimes with other bacterial species.[2] Intercellular signal compounds include N-acyl homoserine lactones, cyclic dipeptides and quinolones, and also volatile fatty acyl methyl esters.[2,4] Microbial interactions have been studied in laboratory conditions without the presence of plants and soil colloids such as clay minerals and humic molecules. According to Pinton et al.,[4] future research is needed to understand how the presence of plants and adsorbing surface-active colloids affects molecular cross talk among microbial cells.

ENZYMES IN THE RHIZOSPHERE

The enzyme activity is higher in the rhizosphere than in the bulk soil due to the higher number and activity of microorganisms, in response to the greater substrate

DOI: 10.1081/E-EPCS 120012889

availability, and to the release of root-derived enzymes.[6] Thus, acid and alkaline phosphatase activities were higher in the rhizosphere soil than in the bulk soil; these enzyme activities increased by approaching to the rhizoplane, whereas organic P concentration was negatively correlated with the increase in the enzyme activities. The inorganic P was depleted as the result of the root P uptake.[7] The increases in enzyme activities were very high reaching, as the highest values among the different tested crop plants, 911% and 262% for acid and alkaline phosphatase, respectively.

The main problem in measuring an enzyme activity of rhizosphere soil is to distinguish the site of the various enzymes contributing to the measured enzyme activity. Indeed, enzymes can be active inside a plant, microbial, or animal cell or even in cell debris.[8,9] In addition, enzymes not requiring cofactors for their activity, such as hydrolases, can remain active for a short period in the extracellular soil environment unless they are adsorbed by clay minerals or entrapped into humic molecules.[10,11] Enzymes can be located in the soil matrix by combining ultracytochemical tests with electron microscopy. Unfortunately, it is problematic to apply these techniques to soil because of the presence of minerals, which are naturally electron-dense; the other problems of these techniques are reactions of the counterstaining compounds, such as OsO_4, with humic molecules or aspecific reactions with the heavy metal component of the enzyme-specific medium.[8] In spite of these limitations, Forster and collaborators were capable of detecting acid phosphomonoesterase, succinic dehydrogenase, peroxidase, and catalase in microorganisms of rhizosphere soil.[8,12] In addition, acid phosphomonoesterase was also detected in roots, mycorrhizae, and fragments of small (7×20 nm) microbial membranes.[8,12,13]

The enzymes adsorbed by clay minerals or entrapped into the humic matrix can remain active even under unfavorable conditions for microbial activity, because they are more resistant to thermal and pH denaturation than the microbial enzymes.[9,11] A role for these extracellular stabilized enzymes in microbial ecology has been hypothesized; indeed, their presence ensures the extracellular degradation of a series of substrates and decreases the need of microbial cells to synthesize and release extracellular enzymes to carry out these reactions.[10] It is noteworthy to say that soil is an inhospitable environment for extracellular enzymes because nonbiological denaturation, adsorption and inactivation, and degradation by proteases all conspire to end the life of the free extracellular enzyme.

A better interpretation of changes in enzyme activities in the rhizosphere soil requires to distinguish the activity of the enzyme stabilized by soil colloids and that of free extracellular or intracellular enzymes.[11] Unfortunately, the present assays do not allow distinguishing both enzyme activities.

METABOLISM OF RHIZOSPHERE SOIL AS AFFECTED BY MICROFLORA AND MICROFAUNA INTERACTIONS

The plant uptake of nutrients such as N depends on the interactions between microflora and microfauna. Microbial growth in rhizosphere soil is promoted by the release of organic materials by plant roots; since the C/N ratio of the root-derived C is higher than the C/N ratio of microbial biomass, a substantial amount of N is immobilized during microbial growth.[6] This can result in a temporary decrease in the amount of plant-available N. According to Clarholm,[14] this microbial N immobilization is followed by a stimulation of the N mineralization promoted by protozoan grazing, with the release of ammonium, which is taken up by the plant root. The release of ammonium occurs because protozoa have a higher C/N ratio than bacteria. This series of events seems to be confirmed by the increase in protease and deaminase activities observed in the rhizosphere soil.[6]

Plants can take up and assimilate several N sources (ammonium, nitrate, urea, amino acids), but the particular form taken up depends on the plant species; thus, tomato growth is inhibited if the ammonium concentration is too high.[6] Competition between nitrifiers, heterotrophs, and plants occurs in the rhizosphere soil and can be particularly evident in N-limited systems such as forests. In this case, nitrifiers converting ammonium to nitrate may limit the amount of N immobilized by heterotrophs and favor the N uptake by plant species preferring nitrate to ammonium as a N source.

CONCLUSIONS AND FUTURE RESEARCH NEEDS

Rhizosphere metabolism includes pathways of animal, microbial, and plant cells. However, the overall metabolism of the rhizosphere soil is peculiar because it reflects the complex interactions among microflora, plant roots, and microfauna. Another distinctive characteristic of rhizosphere soil is the presence of a variety of compounds released from plant roots, microorganisms, and microfauna affecting these complex interactions. The most well-known molecular cross talk is that between legumes and rhizobia. However, future research is needed to understand exchanges of molecular signals between the various microorganisms and plant roots, among microbial cells of the same and different species, and between microbial cells and microfauna. It is also needed to study the effect of surface-active colloids on the exchange of these molecular signals. Future research should also be directed in setting up enzyme assays, which allow the distinction of the activity due to enzymes stabilized by soil colloids from that of enzymes present in active animal, microbial, and plant cells.

ARTICLE OF FURTHER INTEREST

Symbioses with Rhizobia and Mycorrhizal Fungi: Microbe/Plant Interactions and Signal Exchange, p. 1213

REFERENCES

1. Crowley, D. Function of Siderophores in the Plant Rhizosphere. In *The Rhizosphere. Biochemistry and Organic Substances at the Soil-Plant Interface*; Pinton, R., Varanini, Z., Nannipieri, P., Eds.; Marcel Dekker: New York, 2001; 223–261.
2. Lugtenberg, J.J.; Chin-A-Woeng, T.F.C.; Bloemberg, G.V. Microbes-plant interactions: Principles and mechanisms. Antoine van Leewenhoek **2002**, *81*, 373–383.
3. Werner, D. Organic Signals Between Plants and Microorganisms. In *The Rhizosphere. Biochemistry and Organic Substances at the Soil-Plant Interface*; Pinton, R., Varanini, Z., Nannipieri, P., Eds.; Marcel Dekker: New York, 2001; 197–222.
4. Pinton, R.; Varanini, Z.; Nannipieri, P. The Rhizosphere as a Site of Biochemical Interactions Among Soil Components, Plants, and Microorganisms. In *The Rhizosphere. Biochemistry and Organic Substances at the Soil-Plant Interface. Interface*; Pinton, R., Varanini, Z., Nannipieri, P., Eds.; Marcel Dekker: New York, 2001; 1–17.
5. Brimecombe, M.J.; De Leij, F.A.; Lynch, J.M. The Effect of Root Exudates on Rhizosphere Microbial Populations. In *The Rhizosphere. Biochemistry and Organic Substances at the Soil Plant Interface*; Pinton, R., Varanini, Z., Nannipieri, P., Eds.; Marcel Dekker: New York, 2001; 95–140.
6. Badalucco, L.; Kuikman, P.J. Mineralization and Immobilization in the Rhizosphere. In *The Rhizosphere. Biochemistry and Organic Substances at the Soil-Plant Interface*; Pinton, R., Varanini, Z., Nannipieri, P., Eds.; Marcel Dekker: New York, 2001; 159–196.
7. Tarafdar, J.C.; Jungk, A. Phosphatase activity in the rhizosphere and its relation to the depletion of soil organic phosphorus. Biol. Fertil. Soils **1987**, *3*, 199–202.
8. Ladd, J.N.; Foster, R.C.; Nannipieri, P.; Oades, M.J. Soil Structure and Biological Activity. In *Soil Biochemistry*; Stotzky, G., Bollag, J.-M., Eds.; Marcel Dekker: New York, 1996; Vol. 9, 23–78.
9. Nannipieri, P.; Kandeler, E.; Ruggiero, P. Enzyme Activity and Microbiological Processes and Biochemical Processes in Soil. In *Enzymes in the Environment. Activity, Ecology and Applications*; Burns, R.G., Dick, R.P., Eds.; Marcel Dekker: New York, 2002; 1–33.
10. Burns, R.G. Enzyme activity in soil: Location and possible role in microbial ecology. Soil Biol. Biochem. **1982**, *14*, 423–427.
11. Nannipieri, P.; Fusi, P.; Sequi, P. Humus and Enzyme Activity. In *Humic Substances in Terrestrial Ecosystems*; Piccolo, A., Ed.; Elsevier Applied Science: New York, 1995; 293–328.
12. Foster, R.C. In situ localization of soil organic matter. Quaest. Entomol. **1985**, *21*, 609–633.
13. Gianinazzi-Pearson, V.; Smith, S.E.; Gianinazzi, S.; Smith, F.A. Enzymatic studies on the metabolism of vesiscular-arbuscular mycorrhizas. New Phytol. **1990**, *117*, 61–74.
14. Clarholm, M. The Microbial Loop in Soil. In *Beyond the Biomass*; Ritz, K., Dighton, J., Giller, K.E., Eds.; John Wiley & Sons: New York, 1994; 221–230.

Rhizosphere: Microbial Populations

James M. Lynch
Forest Research, Farnham, Surrey, U.K.

INTRODUCTION

The rhizosphere—the field of action of plant roots—can be viewed as a trinity whereby the biotic and abiotic factors from the soil, plant, and microbial population interact to determine the population structure and composition of the rhizosphere. The microbiota includes bacteria, fungi, actinomycetes, algae, protozoa, nematodes, and some small mites. This diverse assemblage can be influenced by agricultural practices and is influential in determining the sustainability of agricultural systems. The rhizosphere is the powerhouse of soil microbiological activity and offers major opportunities in soil biotechnology. A major research problem is that only about 5% of bacteria are culturable, and the fungi that are isolated often derive from spores that do not represent the negative states of the organism.

DISCUSSION

The bacterial population of the rhizosphere differs from that in the bulk soil, in terms of both population numbers and composition.[1–4] It contains large numbers of Gram-negative nonspore formers with a higher proportion of chromogenic and motile forms, and more ammonifiers, nitrifiers, denitrifiers, and aerobic cellulose decomposers. However, it sustains fewer Gram-positive cocci and organisms capable of utilizing aromatic acids than found in the bulk soil.[5] Short Gram-negative rods respond most to rhizosphere conditions, and invariably make up a greater percentage of the rhizosphere microflora than that in the bulk soil population. The three genera *Pseudomonas*, *Achromobacter*, and *Agrobacterium* are examples of the Gram-negative rods that are highly stimulated within the rhizosphere.

Given that rhizosphere exudations and secretions—components of rhizodeposition—serve to enhance and maintain a higher population than that found in the bulk soil, it must also be assumed that there is a high level of microbial competition within the rhizosphere. This in turn enhances selection pressure to select for those microorganisms most suited to the rhizosphere environment. Selection of bacteria whose growth is enhanced by amino acids is a reflection of the amount of amino acids in the rhizosphere environment in comparison to in the environment of bulk soil. In general terms, bacterial populations within the rhizosphere are considered less fastidious than those found outside.

In contrast to the quantitative rhizosphere effect on bacteria, the nature of the effect for fungi is more qualitative with little effect on fungal numbers but an enhanced stimulation effect for certain genera. The imperfect fungi *Fusarium* spp, which are often pathogenic, are prominent members of the rhizosphere and a common example of this phenomenon, although numerous other genera are represented. There can also be an ecological link between fungal components and the bacterial components of the rhizosphere. The symbiotic mycorrhizal fungi, which can be ectotrophic and endotrophic, are particularly important in effectively increasing the root's absorbing power to promote the uptake of phosphorus and water.[6]

ROOT COLONIZATION

It is generally understood that the organisms associated with seeds (such as the common fungi (*Penicillium* spp, *Aspergillus* spp)) and bacteria contribute very little to the established rhizosphere. The young seedling is first colonized by chance bacteria that first encounter the influence of the developing root, with a more species-stable population establishing as the root matures. The developing root tip and root hairs are commonly free of microorganisms, with bacteria colonizing at sites of lateral root emergence from the main root stem. The rhizosphere effect then increases with the age of the plant, reaching a peak at the height of vegetation and declining as senescence approaches. This is, however, a generalization; specific plant species display peaks at alternative times during their life cycles, with crop species generally exhibiting greater rhizosphere effects than do trees. The most pronounced population increases involve the amino acid–requiring bacteria and other organisms with simple nutritional requirements and rapid growth rates.[7] The dominance of specific bacteria in certain plants is another a phenomenon that has been noted (Fig. 1).

The nature of secretions changes through a root's life cycle; consequently, the microorganisms that it can

Encyclopedia of Plant and Crop Science
DOI: 10.1081/E-EPCS 120012976

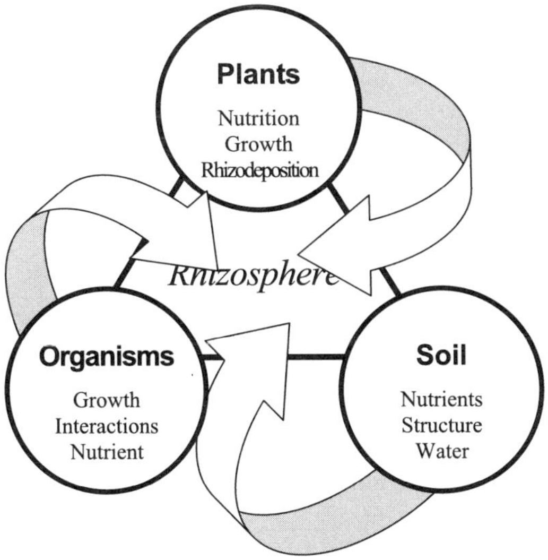

Fig. 1 The rhizosphere trinity: The interacting factors that determine the environmental conditions of the rhizosphere. (Based on Ref. 1.)

support change. It is considered that most root colonization is triggered as the root tip advances through the soil, with the root tip performing as a slowly advancing source of substrate that serves to stimulate the growth and activity of dormant bacteria within the sphere of its influence. The process of root colonization can be conceptually divided into a series of steps: migration toward plant roots, attachment, distribution along the root, and growth and establishment of the microbial population. A number of microbial traits involved in these steps has been postulated, including chemotaxis, agglutinability, the use of flagella, and the ability to utilize complex carbohydrates. The phenomenon of rhizosphere competence describes the ability of microorganisms to associate with roots; the quantative measurement of this is the colonization potential. The challenge is to stimulate organisms beneficial to the plant rather than those that are harmful, such as toxin producers and pathogens. In this respect some rhizosphere-competent strains of *Pseudomonas*[8] and *Trichoderma*[9] have proven particularly beneficial in disease control and growth stimulation.

NUTRIENT CYCLING

The rhizosphere is an environment created and maintained by the plant at cost to the plant. What benefits does the plant receive from this input of energy? The acquisition of nutrients is the preliminary goal of the investment that plants make in their environmental modifications. Interactions with the microbial populations of the rhizosphere enhance the availability and uptake of nutrients that might otherwise have been unavailable to the plant root system. This happens in a number of ways and for a number of nutrients. Probably the most important nutrient in terms of plant growth is nitrogen, the discussion of which will serve here to also exemplify the importance of other nutrients.

Plant roots and the microbial populations that inhabit them are highly integral to the nitrogen cycle. The microbial population of the bulk soil is involved in the conversion of nitrogen through its various forms in the nitrogen cycle, as well as in fixing free nitrogen from the atmosphere (free-living nitrogen fixers). The enhanced populations within the rhizosphere provide the host plant with nitrogen in a readily available form for immediate utilization and subsequent release when the plant dies or is consumed. Far more nitrogen passes through this link in the cycle than does from the bulk soil.

Nutrient cycling in soil is extremely dependent on the supply of energy to the soil biota. Given that the source of this energy is carbon from organic compounds, it can be seen that the nitrogen and carbon cycles are inextricably linked, with carbon as the impetus.

MINERALIZATION AND IMMOBILIZATION

The decomposition of plant material and accompanying mineralization and immobilization of the organic N it contains are key processes in the soil–nitrogen cycle. Once incorporated into the soil matrix, plant residues will bring about a rapid increase in microbial biomass as the material is colonized. In the case of heterotrophic microorganisms using organic material as a source of combined carbon for both respiration and cell synthesis, the progressive decomposition of nitrogen-containing compounds will result in ammonium for use by the organism itself, with the excess released as waste. If insufficient nitrogen is contained within the substrate, then the organisms will draw upon the soil mineral nitrogen, resulting in nitrogen immobilization. The balance between release of nitrogen and immobilization depends on the carbon-to-nitrogen ratio of the material under decomposition, the energy efficiency of the organism, and the carbon-to-nitrogen ratio of the cells being synthesized. The carbon-to-nitrogen ratio of the substrate is the most influential of these factors, with materials varying from 5:1 for animal wastes to 100:1 for cereal straws. The ratio at which nitrogen is neither released nor immobilized has been shown to be approximately 35:1. This highlights the importance of this microbial-mediated process as it supplies nitrogen to the plant in what can be considered its preferred form (which is a less susceptible form than leaching) as well as providing the raw material for the nitrification sequence.

Given that ammonifying bacteria have a higher rhizosphere population than the bulk soil or R:S ratio (in excess of 50:1), it would be expected that there would be greater mineralization of organic matter within the rhizosphere. In the field situation, however, less mineralization can be shown to have occurred in soil under crops than in fallow soil, even allowing for the nitrogen taken up by the plants. This anomaly can be explained by ^{15}N studies that have shown that, although the net amount of mineralization in cropped soil is only half that in fallow soil, the total quantity mineralized is greater under cropping. The difference between the two soil types is due to the rapid assimilation of the mineralized matter by the microbiota within the rhizosphere of the crop.

This illustrates that the availability of nitrogen to the plant is a result of the opposing processes of mineralization and immobilization and on the plant's ability to compete for the available nitrogen. This suggests that the overall microbial effect in the rhizosphere is detrimental as often as it is beneficial. However, experimental comparisons of rhizosphere and nonrhizosphere soils have demonstrated that the rhizosphere effect on the net N mineralization from indigenous soil organic matter is positive in many cases but not in all.[10] The actual microbial N immobilization generated by root-derived carbon seems to be counterbalanced by stimulation of nitrogen mineralization by promoted grazing of protozoa and the consequential release of ammonium by assimilation, as well as a higher microbial turnover of soil organic matter. It is hypothesized that the effect can also be beneficial, as a consequence of the nitrogen form involved (nitrate), and because the continual cycling of the nitrogen compounds via the microbial components prevents their loss from the system through leaching, thus providing a more constant supply to the plant.

Pseudomonas fluorescens is a very common soil bacterium; some strains can increase the mineralization of nitrogen in the pea rhizosphere, but reduce it in wheat.[11] The likely interpretation is governed by the differential effects in the two rhizospheres on the bacterial-feeding nematodes and protozoa.

DINITROGEN FIXATION

The biological fixation of molecular nitrogen is carried out by several free-living bacterial genera, some of which are rhizosphere-associated, such as other *Azotobacter* and *Azospirillum*.[12] In terrestrial systems, however, the largest contribution of combined nitrogen comes from the symbiotic fixation of dinitrogen by rhizobia, with fixation rates often two to three orders of magnitude higher than those exhibited by free-living bulk soil dinitrogen fixers. It is the increase in plant-available nitrate that is the most consistent effect on soil nitrogen levels brought about by both crop and pasture legumes. A combination of conserved soil nitrogen, greater mineralization potential, and the return of fixed nitrogen explain why the benefits of crop legumes to nonlegume crops can be considerable.

The best studied examples of symbiotic nitrogen fixation are the soil bacteria of the genera *Rhizobium*, *Bradyrhizobium*, and *Azorhizobium*, collectively referred to as rhizobia. These organisms have the ability to induce the formation of nodules on the roots of leguminous plants. In these nodules the host plant provides the bacteria with their nutrient requirements and an environment conducive to nitrogen fixation. In return, the bacteria fix dinitrogen in excess of their requirements. The dinitrogen is subsequently released as ammonia and taken up by the plant.

While the rates of N_2 fixation for free-living bacteria are relatively low in comparison to those in a symbiotic or mutualistic relationship, the bacteria that conduct fixation under such conditions are widespread in soils, so do contribute to soil nitrogen inputs. This said, the rate of free-living bacteria such as *Azotobacter* and *Azospirillum* is increased under rhizosphere conditions, due to increased efficiency as a result of more readily available organic compounds exuded from roots. Consequently, it can be seen that the symbiotic fixing of N_2 in the rhizosphere brings substantial benefit to the plant system's ability to acquire and utilize nitrogen.

CONCLUSION

The rhizosphere community is a very complex structure that has great metabolic diversity. Some of this diversity can be beneficial to plants and the environment, although other facets are harmful. The population can now be investigated at the molecular level by determining nucleic acids present. The challenge is to link conventional population biology and metabolic diversity to this genetic capability. The study of the functional genomics of the rhizosphere provides tantalizing prospects for the future.

ARTICLES OF FURTHER INTEREST

Rhizosphere Management: Microbial Manipulation for Biocontrol, p. 1098

R

Symbiotic Nitrogen Fixation, p. 1218
Symbiotic Nitrogen Fixation: Plant Nutrition, p. 1222

REFERENCES

1. *The Rhizosphere*; Lynch, J.M., Ed.; Wiley: Chichester, 1990.
2. Pankhurst, C.E.; Lynch, J.M. The Role of the Soil Biota in Sustainable Agriculture. In *Soil Biota: Management in Sustainable Farming Systems*; Pankhurst, C.E., Doube, B.M., Gupta, V.V.S.R., Grave, P.R., Eds.; CSIRO: East Melbourne, 1994; 3–9.
3. Sylvia, D.M.; Fuhrmann, J.J.; Hartel, P.G.; Zuberer, D.A. *Principles and Applications of Soil Microbiology*; Prentice Hall: New Jersey, 1998.
4. Lynch, J.M. *Soil Biotechnology: Microbial Factors in Crop Productivity*; Bleulewell Scientific: Oxford, 1983.
5. *Russell's Soil Conditions and Plant Growth*, 11th Ed.; Wild, A., Ed.; Longman: Harlow, 1988.
6. Smith, S.E.; Read, D.J. *Mycorrhizal Symbiosis*; Academic Press: San Diego, 1997.
7. Curl, E.A.; Truelove, B. *The Rhizosphere*; Springer-Verlang: Berlin, 1986.
8. O'Sullivan, D.J.; O'Gara, F. Traits of *Pseudomonas* spp involved in the suppression of root pathogens. Microbiol. Rev. **1992**, *56*, 662–676.
9. *Trichoderma and Gliocladium*; Harman, G.E., Kubiceck, C.P., Eds.; Taylor & Francis: London, 1998; Vols. 1 and 2.
10. Badalucco, L.; Kuikman, P. Mineralization and Immobilization in the Rhizosphere. In *The Rhizosphere: Biochemistry and Organic Substances at the Soil–Plant Interface*; Pinton, R., Varanini, Z., Nannipieri, P., Eds.; Marcel Dekker: New York, 2001; 159–196.
11. Brimecombe, M.J.; De Leij, F.A.A.M.; Lynch, J.M. Effect of introduced strains on soil nematode and protozoan populations in the rhizosphere of wheat and pea. Microb. Ecol. **2000**, *38*, 387–397.
12. Basham, Y.; Holguin, G. *Azospirillum*—Plant relationships: Environmental and physiological advances (1990–1996). Can. J. Microbiol. **1997**, *43*, 103–121.

Rhizosphere: Nutrient Movement and Availability

Philippe Hinsinger
INRA–ENSA. M, Montpellier, France

INTRODUCTION

The absorption of nutrients by plant roots is the process by which nutrients enter root cells across the plasma membrane. For soil-grown plants and most nutrients, it is considered that the limiting step in plant nutrition is not the absorption, but rather the processes occurring in the rhizosphere prior to absorption, namely nutrient mobilization and movement. The acquisition of nutrients embraces all these major steps, which depend on physical, chemical, and biological processes and factors.

BIOLOGICAL PROCESSES AND FACTORS INVOLVED IN NUTRIENT ACQUISITION

The acquisition of nutrients by plants combines three strategies. The first strategy is to increase the surface of absorption, which is essential for the acquisition of poorly mobile nutrients such as P and micronutrients.[1,2] The architecture of the root system is one of these means: Its diversity and plasticity among plant species show that this strategy is differently exploited. The actual surface of absorption is not simply the external surface developed by root system: Root hairs are extensions of epidermal cells that can considerably extend the surface of absorption and the actual volume of soil prospected by roots. Their quantitative influence on the acquisition of P has been clearly demonstrated.[2] In addition, nutrients such as Ca and Mg move through the apoplasm of root cortex before actually being absorbed. Thus, the surface developed by the apoplasm should be taken into account,[3] although it does not contribute to enlarge the volume of prospected soil. Another extension of the surface of absorption is provided by the symbiosis with mycorrhizal fungi, which is a widespread feature as more than 90% of plant species can form such symbiosis.[1,4] Mycorrhizal hyphae have much finer diameter than roots and root hairs and can extend several centimeters from the root surface, providing access to nutrients that would otherwise not be spatially available to the root. Their importance for the acquisition of poorly mobile nutrients such as P has been demonstrated.[4] The actual volume of the rhizosphere therefore depends on these complex features of the geometry of roots and mycorrhizae.

The second strategy involved in the acquisition of nutrients relies on the physiological equipment of the absorbing surface,[1] i.e., transporters and ion channels of the membranes of root hairs, epidermal and cortical cells, and those of mycorrhizal hyphae in mycorrhizal plants. Their density and absorption characteristics largely determine the plant's capability to acquire nutrients. For many nutrients that occur at rather low concentrations in the soil solution, uptake systems exhibiting a high affinity and able to function at low concentrations are crucial for the plant's acquisition efficiency.[2,5] The absorption is, however, often considered to be nonlimiting, compared with the physical and chemical processes involved in the mobility of nutrients in the soil. The third strategy of the plant for acquiring nutrients relies indeed on physical and chemical processes that affect the availability of nutrients in the rhizosphere.

PHYSICAL PROCESSES AND FACTORS OF NUTRIENT MOVEMENT

Plant roots are taking up mineral nutrients dissolved in the soil solution. Therefore, in the absence of any movement of nutrients in the soil, the roots can access only a minor amount of nutrient from soil solution in the immediate vicinity of roots. This process, called root interception, is therefore considered to contribute only a minor proportion of the overall acquisition of most nutrients (Table 1). In addition to this process, the uptake of water by roots results in a movement of water from the bulk soil towards the root surface, along a gradient of water potential. As the soil solution contains water and dissolved nutrients, the uptake of water by roots generates a convective movement of solutes, i.e., mass-flow. The corresponding flux of nutrient depends on the flux of water and nutrient concentration in the soil solution.[2,6] For those nutrients that are present at elevated concentrations in the soil solution, which is typically the case of nitrate in many agricultural soils, mass-flow can contribute a major proportion of plant acquisition (Table 1). For Ca and Mg, it can even exceed the actual flux of nutrients absorbed by the plant, thereby resulting in a buildup of their concentration in the rhizosphere (Fig. 1). This may ultimately result in a precipitation of Ca salts around roots, such as $CaCO_3$ in calcareous soils.[7]

Encyclopedia of Plant and Crop Science
DOI: 10.1081/E-EPCS 120012890

Table 1 Estimated contributions of interception, mass-flow, and diffusion to the acquisition of nutrients by a maize crop yielding 9500 kg of grain yield per ha

Nutrient	Acquisition	Interception	Mass-flow	Diffusion
	kg ha^{-1}			
Nitrogen	190	2	150	38
Potassium	195	4	35	156
Phosphorus	40	1	2	37
Calcium	40	60	150	0
Magnesium	45	15	100	0
Sulfur	22	1	65	0

(Adapted from Ref. 6 by permission of John Wiley and Sons, Inc.)

On the contrary, for those nutrients that are present at low concentrations in the soil solution compared with the plant's requirement, such as K and P (Table 1), root absorption results in a decrease in concentration in the rhizosphere (Fig. 1). The resulting concentration gradient in the rhizosphere is the driving force for diffusion. The diffusive flux of nutrient also depends on the diffusion coefficient.[2,6] The actual diffusion of any nutrient in soils is, however, much slower than in water, because of physical and chemical interactions with the soil solid phase:[8] The soil water content, tortuosity, and buffer power (ability of soil solid phase to replenish the soil solution) determine the effective diffusion coefficient in soils.[2,9] Table 2 shows that whereas the diffusion coefficient in water is fairly similar for ions such as NO_3^-, K^+, and $H_2PO_4^-$, their diffusion coefficient in soils is different because of the lower mobility of K^+ ions, and more so of $H_2PO_4^-$ ions compared with NO_3^- ions. Such differences are essentially due to differences in buffer power that arise from chemical interactions with soil constituents. $H_2PO_4^-$ ions are prone to strongly react with soil solid phase, via a whole range of precipitation/dissolution and adsorption/desorption reactions, while NO_3^- ions are not. The buffer power of soils for $H_2PO_4^-$ ions is thus much larger than for NO_3^- ions, and hence their effective diffusion coefficient is much smaller (Table 2). Therefore, the radial distance from which a nutrient can diffuse towards the root surface considerably varies among nutrients and soils (Table 2)—from centimeters for NO_3^- ions and millimeters for K^+ ions to a fraction of a millimeter for $H_2PO_4^-$ ions (Fig. 1). The physical factors that influence the movement of nutrients towards the root surface via both mass-flow and diffusion are soil water content, porosity, tortuosity, viscosity, and temperature. All but the last can be altered by roots (Fig. 2).

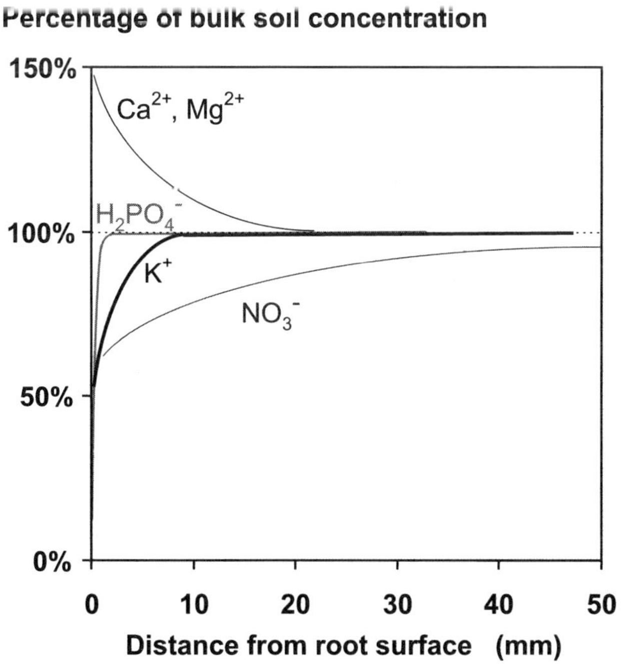

Fig. 1 Typical profiles of nutrient concentrations (expressed in % of the concentration in the bulk soil solution) in the rhizosphere showing an accumulation of Ca and Mg and a more or less steep depletion of N, K, and P. (*View this art in color at www.dekker.com.*)

CHEMICAL PROCESSES AND FACTORS INVOLVED IN NUTRIENT AVAILABILITY

The acquisition of nutrients is much dependent on chemical processes and factors that determine the actual concentration and speciation of nutrients in the soil solution and the soil buffer power. These vary with nutrients, soil types, and properties and as a consequence of root and microbial activities in the rhizosphere.

The uptake activity of the root can result in considerable changes in nutrient concentrations, which can either increase or decrease in the rhizosphere (Fig. 1).

Table 2 Diffusion coefficients of some major nutrient ions in water (D_w), range of values of effective diffusion coefficients in soils (D_e), and corresponding radial distances of diffusion (Δx) within one week ($t = 604{,}800$ s), according to the equation: $\Delta x = \sqrt{(\pi D_e t)}$

Nutrient ion	D_w (m^2 s^{-1})	D_e (m^2 s^{-1})	$\Delta x = \sqrt{(\pi D_e t)}$ (m)
NO_3^-	1.9 10^{-9}	10^{-10}–10^{-11}	4.4–13.8 10^{-3}
K^+	2.0 10^{-9}	10^{-11}–10^{-12}	1.4–4.4 10^{-3}
$H_2PO_4^-$	0.9 10^{-9}	10^{-12}–10^{-15}	0.1–1.4 10^{-3}

(Adapted from Ref. 2 with kind permission from Marcel Dekker, Inc.)

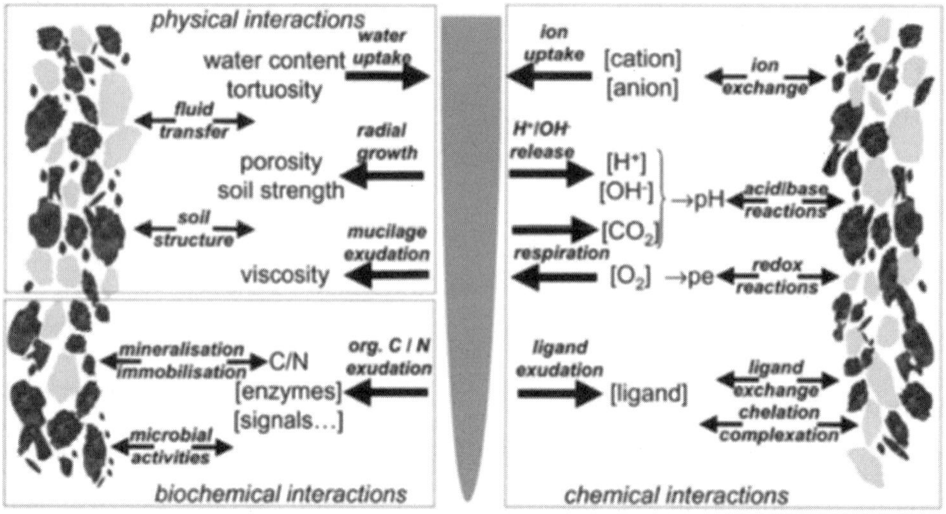

Fig. 2 Schematic diagram of the physical, chemical, and biochemical interactions between roots and soil components that determine the movement and availability of nutrients in the rhizosphere. Note that the microbial activities that are stimulated by root exudation themselves comprehend physical (e.g., aggregation), biochemical (e.g., mineralization), and chemical (e.g., chemical reactions related to microbial respiration or release of acids) processes. Although of equal importance, they have not been drawn for the clarity of the diagram. (*View this art in color at www.dekker.com.*)

This will have a dramatic effect on all the reaction equilibria governing the dynamics of nutrients in soils.[7] It has been estimated that the concentration of K decreases 100– 1000-fold in the rhizosphere, shifting dramatically the equilibria of adsorption/desorption of exchangeable K and even fixation/release of nonexchangeable K.[10] This explains the rapid depletion of exchangeable K and substantial contribution of nonexchangeable K to K availability in the rhizosphere (Fig. 3), which would not occur under bulk soil conditions.[10]

The availability of many nutrients such as P and metal micronutrients (Fe, Mn, Zn, Cu, etc.) is strongly pH-dependent, as protons are implied in their speciation and in many reactions, e.g., precipitation/dissolution at the soil solid–soil solution interface. Roots and rhizosphere microbes can be responsible for considerable changes of pH and, hence, of the availability of nutrients such as P.[5] The origins of root-induced pH decrease are as follows: 1) primarily proton release to counterbalance an excess of cation over anion uptake; 2) release of organic acids; and 3) release of CO_2 via respiration, which leads to a buildup of rhizosphere concentration of H_2CO_3, which dissociates in all but acid soils.[7] The availability of Ca phosphates can be much increased as a consequence of the lower rhizosphere pH found in ammonium-fed plants than in nitrate-fed plants.[5]

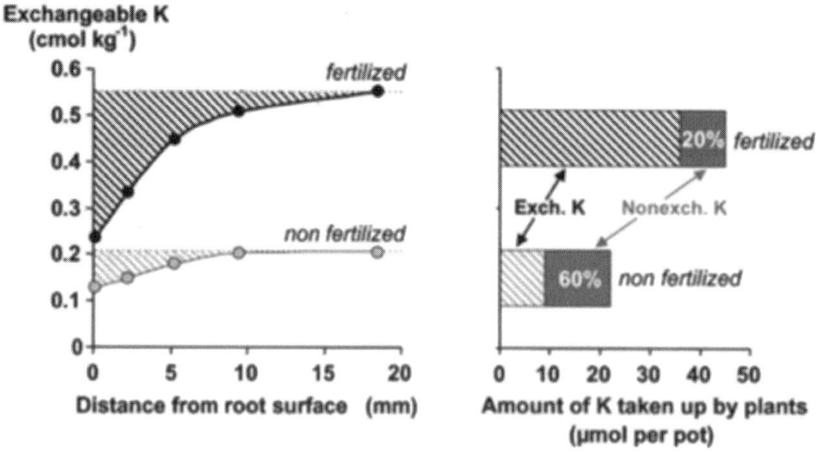

Fig. 3 Depletion of exchangeable K in the rhizosphere of oilseed rape after four days of growth in a fertilized and nonfertilized soil sampled at the long-term fertilizer trial of Gembloux, Belgium. The extension and intensity of the depletion zone were used to calculate the amount of exchangeable K depleted by the plants. Subtracting this value from the actual amount of K taken up (determined by plant analysis) provided an estimate of the contribution of the release of nonexchangeable K within the four days of the pot experiment.

The availability of many metal micronutrients is also governed by redox reactions and complexation processes, which can be considerably affected by the activities of roots and microbes in the rhizosphere.[7] Complexing compounds released by roots and rhizosphere microbes have been shown to potentially increase the availability of metal nutrients.[11] Grasses are capable of releasing Fe-chelating compounds called phytosiderophores.[1,7] This process is enhanced under Fe deficiency and is thus part of an efficient strategy to acquire Fe in soils where its availability is minimal, such as calcareous soils. These root exudates can also strongly chelate other metal micronutrients such as Zn and Cu and thereby increase their availability.[1,7] Carboxylic anions have also been extensively studied for their role in the dynamic of nutrients such as P. Their effect is based on the complexation of metal cations that play a major role in binding phosphate ions in soils, such as Ca, Fe, and Al, but can also rely on ligand exchange reactions that promote the desorption of phosphate ions from soil constituents.[5,12] In many plant species, the rate of release of such exudates may be too low to be of any significance, though. Some species of plants and ectomycorrhizal fungi have been shown to exude large amounts of citrate, malate, or oxalate and significantly increase the availability of soil P.[5,12] Some major nutrients can occur in the soil as organically bound nutrients, such as N, P, and S. These organic molecules will require biochemical processes such as enzymatically controlled hydrolysis to evolve into smaller molecules and ultimately release N, P, and S. Plant roots can exude some enzymes that will catalyze such breakdown processes, as shown for phosphatases that help plants make use of organic P compounds.[1,11] However, soil microorganisms play a major role in these processes, through a variety of enzymatic pathways. Their stimulation by root exudation definitely contributes an enhanced release of organically bound nutrients in the rhizosphere (e.g., phosphatases, proteases, aminases produced by ectomycorrhizal fungi).[4,11] Fig. 2 summarizes the numerous root-mediated, biochemical, and chemical processes and factors that determine the availability of nutrients in the rhizosphere.

CONCLUSION

Measuring the peculiar properties of the rhizosphere is challenging because of its spatial and temporal dynamics. Considerable advances in the identification of the factors and processes that determine the availability of nutrients in the rhizosphere have been made in recent decades, though. Comparing model outputs with actual measurements often shows large discrepancies, which suggest that there is still a need to better quantify the processes involved in nutrient acquisition.[3] Another challenge for future research is to make better use of such knowledge to determine how to manage those rhizosphere processes to improve a plant's acquisition efficiency, growth, and health. Both microbial and plant components are important tools to do so, in order to minimize the inputs of fertilizers and pesticides and increase the sustainability of plant-based productions.

ARTICLES OF FURTHER INTEREST

Minor Nutrients, p. 726
Nitrogen, p. 822
Phosphorus, p. 872
Potassium and Other Macronutrients, p. 1049
Rhizosphere: An Overview, p. 1084
Rhizosphere: Biochemical Reactions, p. 1087
Rhizosphere: Microbial Populations, p. 1090
Rhizosphere Management: Microbial Manipulation for Biocontrol, p. 1098
Root Membrane Activities Relevant to Plant-Soil Interactions, p. 1110
Symbiotic Nitrogen Fixation: Plant Nutrition, p. 1222

REFERENCES

1. Marschner, H. *Mineral Nutrition of Higher Plants*, 2nd Ed.; Academic Press: London, 1995; 889 pp.
2. Jungk, A. Dynamics of Nutrient Movement at the Soil-Root Interface. In *Plant Roots: The Hidden Half*, 3rd Ed.; Marcel Dekker, Inc.: New York, USA, 2002; 587–616.
3. Darrah, P.R. The rhizosphere and plant nutrition: A quantitative approach. Plant Soil **1993**, *155/156*, 1–20.
4. Marschner, H.; Dell, B. Nutrient uptake in mycorrhizal symbiosis. Plant Soil **1994**, *159*, 89–102.
5. Hinsinger, P. Bioavailability of soil inorganic P in the rhizosphere as affected by root-induced chemical changes: A review. Plant Soil **2001**, *237*, 173–195.
6. Barber, S.A. *Soil Nutrient Bioavailability. A Mechanistic Approach*, 2nd Ed.; John Wiley & Sons: New York, USA, 1995; 414 pp.
7. Hinsinger, P. How do plant roots acquire mineral nutrients? Chemical processes involved in the rhizosphere. Adv. Agron. **1998**, *64*, 225–265.
8. Tinker, B.T.; Nye, P.H. *Solute Movement in the Rhizosphere*; Oxford University Press: Oxford, UK, 2000; 444 pp.
9. Nielsen, N.E. Nutrient Diffusion, Bioavailability and Plant Uptake. In *Encyclopedia of Soil Science*; Marcel Dekker Inc.: New York, USA, 2002; 878–884.
10. Hinsinger, P. Potassium. In *Encyclopedia of Soil Science*; Marcel Dekker Inc.: New York, USA, 2002; 1035–1039.
11. Pinton, R.; Varanini, Z.; Nannipieri, P. *The Rhizosphere: Biochemistry and Organic Substances at the Soil-Plant Interface*; Marcel Dekker, Inc.: New York, USA, 2002; 424 pp.
12. Jones, D.L. Organic acids in the rhizosphere—A critical review. Plant Soil **1998**, *205*, 25–44.

Rhizosphere Management: Microbial Manipulation for Biocontrol

Ben J. J. Lugtenberg
Faina D. Kamilova
Leiden University, Institute of Biology, Leiden, The Netherlands

INTRODUCTION

Several microorganisms—such as viruses, bacteria, and fungi—cause diseases on various plants, including economically important ones. Strategies to bring the disease level to economically acceptable levels include the use of chemical pesticides, plant cultivars resistant to a certain pathogen, and natural enemies. In the United States alone, over $600 million is spent annually on agricultural fungicides. The use of chemicals is increasingly discouraged because some are a threat to the environment and human beings. Nongenetically engineered disease-resistant plants are widely used. Drawbacks are that resistance is usually accompanied by a yield decrease (of approximately 3%) and that after a few years, forms of the pathogen take over to which the resistant plant is sensitive. A modern approach—the use of genetically engineered resistant plants—is not accepted in some economically important regions in the world, especially in Europe.

Biological control of plant diseases is becoming increasingly popular, wherein cultivated natural enemies of the pathogenic microbes are applied to seeds, seedlings, or plants in a concentration sufficiently high to compete successfully with the pathogen for several months.

BIOCONTROL AGENTS (BCAs)

Biocontrol agents represent all organisms, (including microbes, nematodes, and insects) that are able to decrease disease levels of crop plants. These diseases can be caused by pathogenic microbes, weed plants, or animals such as nematodes and insects. This article focuses on the biocontrol of plant diseases caused by soilborne pathogens. This control takes place in the rhizosphere (i.e., the root and its immediate surrounding area). The rhizosphere is relatively rich in nutrients and therefore attracts many microbes—pathogens as well as beneficials. Certain insects feed on plants to the extent that plants become seriously diseased or even die. *Bacillus thuringiensis* is a gram-positive soil bacterium able to produce crystalline inclusions during sporulation. These consist of proteins with highly specific insecticidal activity. They can be active against larvae of *Lepidoptera*, dipteran and coleopteran species. The crystals dissolve in the larval midgut where they are processed into smaller toxic proteins that generate pores in the midgut cell membrane. The larva stops feeding and consequently dies.[1] Economically this is the most important application of biocontrol.

Many other plant diseases are caused by fungi, such as *Alternaria*, *Botrytis*, *Fusarium*, *Gaeumannomyces graminis* var. *tritici*, *Pythium*, *Rhizoctonia*, and *Verticillium*. Some pathovars of bacteria such as *Agrobacterium*, *Erwinia*, and *Pseudomonas* can cause crop damage also.

The BCAs used most frequently to control root diseases caused by fungi include the bacteria *Bacillus*, *Pseudomonas*, and *Streptomyces*, and the fungi *Gliocladium* and *Trichoderma* (Table 1). For more information, consult www.nal.usda.gov; www.epa.gov; www.bisolbi.com and www.biconet.com.

PROS AND CONS OF BIOLOGICAL CONTROL

In contrast to various chemical fungicides, BCAs are environmentally friendly. In cases, when they act through the production of an antifungal metabolite (AFM), this molecule is produced locally on the plant surface in small amounts. The molecule is biodegradable, does not reach drinking water, and is not harmful for workers.

In comparison with resistant plants, the use of BCAs has the advantages that: 1) they are inexpensive; 2) the decision to use them can be made relatively late; and 3) they do not cause a yield decrease, but have resulted in unexpected yield increases. Finally: 4) plants resistant to all pathogens do not exist; and 5) public perception of biologicals is positive.

Disadvantages also exist: 1) performance is not always consistent; 2) in comparison with chemicals, the shelf life of biologicals is short; 3) adoption of legislation governing the use of biologicals is an extremely slow process; and 4) the cost of registration is high, taking into account the small market share that biologicals currently represent.

Encyclopedia of Plant and Crop Science
DOI: 10.1081/E-EPCS 120012973

Table 1 Commercially available biocontrol agents

Biocontrol agent	Formulation	Target pathogen	Crop
Bacillus subtilis	Dry powder; water-dispersible granules	*Pythium*; *Phytophtora*; *Fusarium*; *Streptomyces scabies*	Cotton, peanuts, soybeans, vegetables, potato, ornamentals
Burkholderia cepacia	Peat-based dried biomass from solid fermentation	*Rhizoctonia*; *Pythium*; *Fusarium*	Alfalfa, barley, beans, clover, cotton, peas, grain sorghum, vegetable crops, wheat
Pseudomonas fluorescens	Powder	*Erwinia amylovora*	Tomato
Pseudomonas chlororaphis	Powder	*Pythium*; *Rhizoctonia*; *Cylindrocladium*; *Fusarium*	Pea, ornamentals, cucumber
Streptomyces grisioviridis	Powder	*Fusarium*; *Pythium*	Field, ornamental and food crops
Gliocladium virens	Granules; granular fluid	*Pythium*; *Rhizoctonia*	Ornamental and food plants in greenhouses and nurseries
Trichoderma harzianum	Granules or dry powder	*Pythium*; *Rhizoctonia*; *Fusarium*	Tomato, cucumber, cabbage

ISOLATION OF BCAs

To test microbes for their action as BCAs, a system must be developed in which the plant is grown in the presence of the target pathogen. The disease pressure initially used is unnaturally high (60–80% diseased plants) in order to limit the number of plants required to produce a statistically significant result. Later, when the number of candidate BCAs is limited, larger experiments are carried out with lower disease pressure (20–40%).

Because biocontrol experiments are labor-intensive and time-consuming, scientists often make a prescreening or preselection, taking advantage of knowledge of mechanisms. For example, one can prescreen for organisms that inhibit growth of the pathogen (the antagonists) in vitro[2] or one can preselect enhanced colonizers.[3] This approach is biased by limited knowledge of BCAs' mechanisms of action.

APPLICATION OF BCAs

After growth, biocontrol microbes usually are not applied immediately on the plant but are stored. To avoid major losses of viable organisms, they are treated by special methods (often protected by patents) that increase their survival during storage (their shelf life). This process is referred to as formulation. Survival is relatively easy when dealing with dormant forms called spores. When non-sporeformers are to be formulated, successful formulation is much more difficult, although major progress has recently been made. BCAs can be formulated in solid or liquid state and in materials such as alginate, bentonite, carragheenans, peat, or vermiculate. Alternatively, liquid seed-applied inoculants can be used. Ideally formulated, microbes can be used for more than one growth season so that leftovers from one season can be used at the beginning of the next. Formulated BCAs usually contain at least 10^9 viable microbes per gram or ml.[4] One of the most successfully applied microbes is *Bradyrhizobium japonicum*, a biofertilizer of soybean.[5]

Microbes can be applied on the seed or in the furrow before planting. Alternatively, they can be added to the plant during growth. BCAs are also used postharvest; for example, to protect oranges or potatoes from rotting during storage.

To monitor the success of BCAs' application, it is important to monitor plant performance as well. In case of poor performance, data on the viability and activity of the BCA are crucial for understanding whether poor performance is due to the BCA or to other circumstances. Early discovery of the cause of poor performance should allow additional control measures. Because any individual form of disease control can fail, it is advisable to have a combination of methods available.

MECHANISMS USED BY BCAs

Important techniques to study mechanisms of biocontrol are visualization of differentially labeled pathogens (Fig. 1A) and BCAs (Fig. 1B), genetics, and biochemistry. The best studied form of biocontrol is the use of antagonistic microbes. These organisms produce one or more molecules (e.g., AFMs) that inhibit growth of the target pathogen.[6] To protect the whole root system against soilborne pathogens it is important that the BCA deliver the AMF along the whole root system.[7] This can

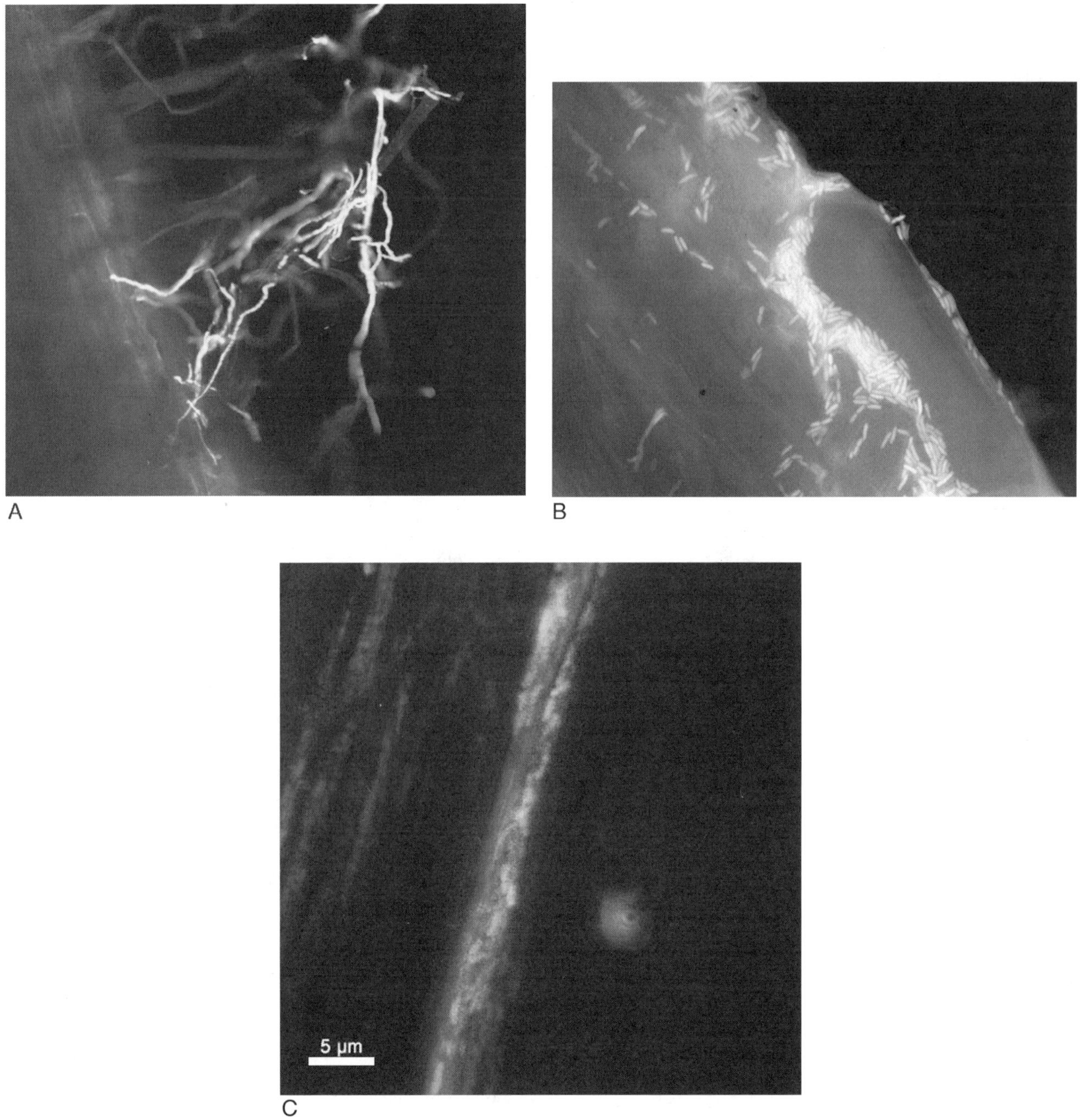

Fig. 1 Visualization of *Pseudomonas fluorescens* biocontrol strain WCS365 and the pathogenic fungus *Fusarium oxysporum* f.sp. *radicis lycopersici* on tomato roots. The microbes were marked with an autofluorescent protein. The root shows red autofluorescence. (A) Initial stage of the contact: Fungal hyphae attach to root hairs. (From Lagopodi et al. Mol. Plant-Microbe Interact. 2002. *15*, 172–179.) (B) Formation of a bacterial biofilm on the root. (From Bloemberg et al. Appl. Environm. Microbiol. 1997. *63*, 4543–4551.) (C) Colonization of fungal hyphae (blue) by biocontrol bacteria (green). (From Lagopodi, A. and Bloemberg, G.V., unpublished). (*View this art in color at www.dekker.com.*)

be done by applying a successful root-colonizing BCA on the seed, which subsequently colonizes the whole growing root. The production of AFMs often requires high cell densities of the BCA.[7] Indeed, bacterial BCAs are often present as microcolonies (Fig. 1B). AFMs are diverse in structure. AFMs produced by several BCAs are 2,4 diacetyl phloroglucinol, phenazine, HCN, and biosurfactants.[6] The latter molecules are thought to be particularly active against zoospores (e.g., from the pathogen *Pythium*).[8]

Some BCAs act through parasitism or predation. For example, because many fungi contain chitin in their cell walls, the secretion of chitinase is considered a trait of some BCAs (e.g., of *Trichoderma*).

Other BCAs colonize the root system very well but fail to produce AFMs or exoenzymes in vitro. Although one cannot exclude that some of these BCAs do produce AFM in the rhizosphere, it is often thought that this class of BCAs acts through competition with the pathogen for niches on the root or for nutrients secreted by the root. These so-called exudates often contain amino acids, sugars, and organic acids, albeit in low concentrations.[7]

BCAs of another class do not produce extracellular toxins, but induce a systemic reaction in the plant (e.g., the presence of certain *Pseudomonas* strains on the root triggers in the plant a faster, more severe response to pathogens present on the leaf). A variety of products of the BCA, including lipopolysaccharide, siderophores, and flagella, have been claimed to be involved in the induction of this form of resistance. It is assumed that bacteria that act through this mechanism, designated Induced Systemic Resistance,[9,10] consist of subclasses that act by different mechanisms.

Using microbes labeled with autofluorescent proteins, confocal laser scanning microscopy studies showed that during biocontrol, hyphae of *Fusarium* become colonized by *Pseudomonas* BCAs (Fig. 1C). The resulting disabling of the pathogen may represent a novel biocontrol mechanism.

CONCLUSION

Mechanisms that allow for the isolation and construction of more robust strains are being developed. Clearly, the reduction of plant diseases by microbial control agents has a bright future.

ARTICLES OF FURTHER INTEREST

Bacterial Attachment to Leaves, p. 75
Bacterial Survival and Dissemination in Natural Environments, p. 108
Bacterial Survival and Dissemination in Seeds and Planting Material, p. 111
Bacteria-Plant Host Specificity Determinants, p. 119
Biological Control of Weeds, p. 141
Fungal and Oomycete Plant Pathogens: Cell Biology, p. 480
Leaf Surface Sugars, p. 642
Management of Bacterial Diseases of Plants: Regulatory Aspects, p. 669
Management of Fungal and Oomycete Diseases: Fruit Crops, p. 678
Management of Fungal and Oomycete Diseases: Vegetable Crops, p. 681
Mycotoxins Produced by Plant Pathogenic Fungi, p. 773
Rhizosphere: An Overview, p. 1084
Rhizosphere: Biochemical Reactions, p. 1087
Rhizosphere: Microbial Populations, p. 1090
Rhizosphere: Nutrient Movement and Availability, p. 1094

REFERENCES

1. Höfte, H.; Whiteley, H.R. Insecticidal crystal proteins of *Bacillus thuringiensis*. Microbiol. Rev. **1989**, *53*, 242–255.
2. Chin-A-Woeng, T.F.C.; Bloemberg, G.V.; van der Bij, A.J.; van der Drift, K.M.G.M.; Schripsema, J.; Kroon, B.; Scheffer, R.J.; Keel, C., Bakker, P.A.H.M.; Tichy, H.-V.; de Bruijn, F.J.; Thomas-Oates, J.E.; Lugtenberg, B.J.J. Biocontrol by phenazine-1-carboxamide-producing *Pseudomonas chlororaphis* PCL1391 of tomato root rot caused by *Fusarium oxysporum* f.sp. *radicis-lycopersici*. Mol. Plant-Microb. Interact. **1998**, *11* (11), 1069–1077.
3. Kuiper, I.; Bloemberg, G.V.; Lugtenberg, B.J.J. Selection of a plant-bacterium pair as a novel tool for rhizostimulation of polycyclic aromatic hydrocarbon-degrading bacteria. Mol. Plant-Microb. Interact. **2001**, *14* (10), 1197–1205.
4. Burges, H.D. *Formulation of Microbial Biopesticides*; Kluwer Academic Publishers: Dordrecht, 1998.
5. Smith, R.S. Inoculant Formulations and Applications to Meet Changing Needs. In *Nitrogen Fixation: Fundamentals and Applications*; Tikhonovich, I.A., Provorov, N.A., Romanov, V.I., Newton, W.A., Eds.; Kluwer Academic Publishers: Dordrecht, 1995; 653–657.
6. Thomashow, L.S.; Weller, D.M. Current Concepts in the Use of Introduced Bacteria for Biological Disease Control: Mechanisms and Antifungal Metabolites. In *Plant-Microbe Interactions*; Stacey, G., Keen, N.T., Eds.; Chapman & Hall, 1996; Vol. 1, 187–235.
7. Lugtenberg, B.J.J.; Dekkers, L.; Bloemberg, G.V. Molecular determinants of rhizosphere colonization by *Pseudomonas*. Annu. Rev. Phytopathol. **2001**, *39*, 461–490.
8. Stanghellini, M., Miller, R. Biosurfactants. Their identity and potential efficacy in the biological control of zoosporic plant pathogens. Plant Dis. **1997**, *81*, 4–12.
9. Van Loon, L.C.; Bakker, P.A.H.M.; Pieterse, C.M.J. Systemic resistance induced by rhizosphere bacteria. Annu. Rev. Phytopathol. **1998**, *36*, 453–483.
10. Audenaert, K.; Pattery, T.; Cornelis, P.; Höfte, M. Introduction of systemic resistance to *Botrytis cinerea* in tomato by *Pseudomonas aeruginosa* 7NSK2: Role of salicylic acid, pyochelin, and pyocyanin. Mol. Plant Microb. Interact. **2002**, *15* (11), 1147–1156.

Rice

Takuji Sasaki
National Institute of Agrobiological Sciences, Tsukuba, Ibaraki, Japan

INTRODUCTION

Rice is one of the most important cereal crops in the world because it is the main staple for about half of the world's population. Rice is also one of the most extensively analyzed plants in terms of its molecular biology, genetics, and nearly complete genome sequence and is thus considered a model plant for the cereal crops. Recent technological advances in genome analysis, such as high-throughput genome sequencing and genomewide analysis of transcripts, have paved the way for the characterization of many of the genes expressed in response to various environmental stresses. These innovations are expected to accelerate the breeding of new rice varieties with high yields and adaptability to unfavorable environmental conditions, thereby ensuring sufficient food to feed a burgeoning population, particularly in Asia and Africa. For this purpose, a thorough understanding of the rice plant's complex biological pathways, such as the genetic mechanisms of reproduction and physiological mechanisms of flowering time, is of utmost importance. In this article, the recent progress in rice molecular genetics is reviewed with emphasis on the development of rice genomics and its applications in the future.

MAPPING THE RICE GENOME CLONE-BY-COLNE

A molecular genetic analysis of rice has been undertaken to dissect the rice genome using DNA markers such as RFLP (Restriction Fragment Length Polymorphism), SSR (Simple Sequence Repeat), AFLP (Amplified Fragment Length Polymorphism), or CAPS (Cleaved Amplified Polymorphic Sequence). As a result, a high-density and precise genetic map with RFLP markers is now available,[1] and soon these markers will be converted to PCR-based markers using the genome sequence information to facilitate wider applications. In addition, a new kind of marker based on SNP (Single Nucleotide Polymorphism) or In/Del (Insertion or Deletion of nucleotide) generated by comparison of nucleotide sequences among rice varieties or rice subspecies will also become available soon using the genome sequence data. DNA markers are extremely useful in determining the precise allocation of target genomic region within the genome, particularly in gene isolation by map-based cloning.[2] Also, DNA markers are indispensable in ordering large insert clones for construction of a sequence-ready physical map of the genome.[3]

Physical maps of the rice genome have been established using genomic DNA fragments cloned in YAC (Yeast Artificial Chromosome), BAC (Bacterial Artificial Chromosome), or PAC (P1-derived Artificial Chromosome). At the Rice Genome Research Program (RGP), a YAC-based physical map covering 80% of the whole rice genome has been constructed by combining the results of RFLP marker and EST (Expressed Sequence Tag) screening.[4] Initially, a skeleton YAC physical map with 63% coverage was constructed using 1439 RFLP markers in the genetic map.[5] Thereafter, the 3' untranslated, unique sequences of rice ESTs were used as PCR primers to screen the YAC library and assign corresponding clones to previously mapped YAC contigs. As a result, YAC clones that carried both mapped and unmapped ESTs were newly allocated on the chromosomes so that the YAC contigs were extended and the genome coverage was increased to 80%.[4] By mid-2002, this YAC-based physical map contained 6591 ESTs, and it is highly informative for map-based cloning. For example, the gene corresponding to rice dwarfism, *d1*, is one of the genes that has been cloned and characterized by this approach.[6]

The rice physical map for genome sequence analysis has been constructed with BAC or PAC clones using methods for identifying overlaps and contiguous clones. In order to assign the clones to specific regions of the chromosome, fingerprinting by restriction enzymes is generally used. The fingerprint data obtained by electrophoresis is then analyzed by an fpc (fingerprinted contigs) program[7] under an empirically adopted condition to make correct assignments of clones and overlaps. In this method, it is necessary to confirm the true overlaps by two parameters involved in fpc, namely, tolerance and cutoff values. These two parameters determine whether the contigs are reliable or not. A rice physical map based on fingerprints has been constructed at Clemson University Genomics Institute.[8]

Clones were first assigned to their chromosomal positions by using the DNA markers described above. The availability of many PCR-based DNA markers

Encyclopedia of Plant and Crop Science
DOI: 10.1081/E-EPCS 120010636

facilitated accurate anchoring of BACs/PACs by PCR prior to fingerprinting. The RGP screened Nipponbare PAC and BAC libraries with EST markers assigned on each of the 12 chromosomes. For example, in the case of chromosome 1, about 900 kinds of PCR primers were used for screening, and the fingerprint data of the anchored BACs/PACs were used for estimation of the overlaps among them. By this procedure, about 80% of chromosome 1 has been covered by BACs/PACs.

The physical map can be further refined using sequence information of the anchored clones. For this purpose, a database of BAC/PAC end sequences was established. CUGI has sequenced both ends of 92,000 BAC clones[9] to help select the next target clone for sequencing among the aligned and mapped clones. This procedure is also useful to fill gaps on the RGP's physical map. Additionally, the accuracy of the physical map is determined by the consistency of the sequences of aligned clones.

Rice genome sequencing is performed by an international consortium, the International Rice Genome Sequencing Project (IRGSP), in which 11 countries contribute to generate a high-quality sequence (99.99%) of the 12 rice chromosomes.[10] The genome sequence data are accessible through a centralized database named INE (*IN*tegrated rice genome *E*xplorer) working by Java applet.[11,12] Users of INE can find a PAC/BAC clone mapped to its chromosomal location by DNA markers and see the result of the annotation of the corresponding PAC/BAC sequence. Annotation of the sequence is performed using an automated annotation system, RiceGAAS (Rice Genome Automated Annotation System),[13,14] which combines the results of both gene prediction and similarity search programs to assign the location of predicted gene and its function. Manual curation is also performed to come up with the most plausible gene model. The annotation map for each sequenced clone can be viewed graphically through INE, and each predicted gene is shown with the complete coding and protein sequence as well as a list of similar genes with information from the nonredundant protein database of NCBI. Two other databases, Oryzabase[15] and Gramene,[16,17] have been developed to show rice genetic resources and morphology and to integrate cereal genome information based on annotated rice genome sequence and associated biological information, respectively.

WHOLE-GENOME SHOTGUN SEQUENCING

In contrast to the clone-by-clone strategy of constructing a physical map to organize the DNA sequence, an alternative inexpensive and rapid strategy has been recently applied to the rice genome.[18,19] This is called a whole-genome shotgun sequencing and involves the sequence analysis of a large number of randomly chosen, small-insert clones to provide enough data for assembling the entire genome. However, because of the high complexity of the rice genome, it is impossible to make the correct assignment of a large number of short contigs without additional information, such as positional information of specific markers. Furthermore, many repetitive sequences in the genome were not assembled, and the gaps between contigs would be difficult to fill with this strategy. Therefore, supplemental data such as BAC-end sequences and a reference to the standard genome sequence by IRGSP are indispensable to raise the quality of sequence data obtained by a whole-genome shotgun method.

GENETIC AND REVERSE-GENETIC METHODS

Functional analysis of rice genes is mainly by genetic and reverse-genetic methods and is greatly aided by genome sequence data. So far, about 500 phenotypes have been assigned to 12 linkage groups, and about 200 of them have been genetically positioned on 12 chromosomes.[20] In addition, many mutants generated by chemical mutagens, such as N-methyl-N-nitrosourea,[21] and by rice endogenous retrotransposon, such as *Tos17*,[22] have been used as resources for gene isolation. Among the agronomically important genes isolated using these strategies are the bacterial blight resistance genes, *Xa1*[23] and *Xa21*,[24] the rice blast resistance genes, *Pib*[25] and *Pi-ta*,[26] the gibberellin-insensitive dwarf gene, *d1*,[6] the constitutively gibberellin-responsible slender rice gene, *slr1*,[27] and the viviparous mutant genes, *Osaba1* and *Ostatc*.[28] In particular, *Tos17* is an advantageous tool because it has a high probability of being inserted into gene-rich regions,[22] and the insertion site sequences are easily identified by analyzing the flanking sequence of the *Tos17* insert with suppression PCR.[29] The disrupted gene sequence can be easily identified by referring to the genome sequence, and the data of resultant phenotype indicate the disrupted gene function. The insertion lines containing *Tos17* are also used for reverse-genetics methods to identify gene function. PCR screening of large populations of *Tos17* insertional mutants by gene-specific primers often find out target mutant. This strategy has been successfully used to identify the function of rice homeobox gene, *OSH15*,[30] and phytochrome A.[31]

The genes isolated as described above have strong phenotypes to facilitate isolating the responsible gene. However, most genetic variation seen in the field doesn't show such disruptive phenotypes. Natural variants show differences in most plant traits including flowering time, clum height, grain weight, and number of seeds. These traits are generally important in agronomy and are called quantitative trait loci (QTL) or polygenic traits. Before the introduction of molecular tools for genetic analysis, it was difficult to isolate genes responsible for these traits. However, a combination of marker-assisted selection and advanced backcrossing methods makes it possible to isolate each component of a target QTL as a single Mendelian factor similar to the isolation of the mutant genes described above.[32] This strategy has been used for the identification of genes identified as a QTL for rice flowering time.[33,34] At least 14 loci were detected as QTLs of flowering time using progenies derived from a cross between a japonica variety, Nipponbare, and an indica variety, Kasalath. Among them, seven loci were identified as photoperiod sensitivity genes, and the structure and function of three of these genes were elucidated by map-based cloning followed by transformation. They are homologues to zinc-finger protein *CONSTANS*,[35] protein kinase *CK2α*,[36] and flowering time *FT*[34] genes found in *Arabidopsis*. However, these homologous genes appear to have opposite functions: Flowering is triggered under short day-length conditions in rice and under long day-length conditions in *Arabidopsis*. This indicates the existence of a clock gene in both plants, yet to be identified.

CONCLUSION

With its small genome size, rice is considered the prototypic cereal genome. Recent findings show that because of the considerable syntenic relationships between all the cereals, the homologues of rice genes can be anticipated to be found at analogous chromosomal positions in other grass species.[37] Recognition of these relationships has led to the exciting prospect that rice can be used to understand much of the genomic arrangement of the economically important cereals. However, a more detailed comparison of genomic sequences and gene function is necessary to clarify the syntenic relationships among cereal crops. Since the birth of the ancestral cereal plant about 60 million years ago, each species has diverged to its present nature by modifying the ancestral genes to fit its lifestyle. Understanding the similarities and differences among cereal genes using rice as a reference plant may be the key in establishing how gene families diverged and how biological pathways have been modified in the course of evolution.

ACKNOWLEDGMENT

The Rice Genome Research Program has been supported by the Ministry of Agriculture, Forestry, and Fisheries.

REFERENCES

1. http://rgp.dna.affrc.go.jp/publicdata/geneticmap2000/index.html.
2. Martin, G.B. Chromosomal landing: A paradigm for map-based gene cloning in plants with large genomes. Trends Genet. **1995**, *11* (2), 63–68.
3. Gojobori, T. The genome sequence and structure of rice chromosome 1. Nature **2002**, *in press*.
4. Sasaki, T. A comprehensive rice transcript map containing 6591 expressed sequence tag sites. Plant Cell **2002**, *14* (3), 525–535.
5. Sasaki, T. A physical map with yeast artificial chromosome (YAC) clones covering 63% of the 12 rice chromosomes. Genome **2001**, *44* (1), 32–37.
6. Yoshimura, A. Rice gibberellin-insensitive dwarf mutant gene Dwarf 1 encodes the alpha-subunit of GTP-binding protein. Proc. Natl. Acad. Sci. U.S.A. **1999**, *96* (18), 10284–10289.
7. http://www.genome.clemson.edu.fpc.
8. Wing, R.A. An integrated physical and genetic map of the rice genome. Genome Res. **2002**, *14* (3), 537–545.
9. ftp://ftp.genome.clemson.edu/pub/rice/stc/OSJNBa.lib.gz.
10. Burr, B. International rice genome sequencing project: The effort to completely sequence the rice genome. Curr. Opin. Plant Biol. **2000**, *3* (2), 138–141.
11. Sasaki, T. Nucleic Acids Res. **2000**, *28* (1), 97–101.
12. http://rgp.dna.affrc.go.jp/giot/INE.html.
13. Higo, K. Rice GAAS: An automated annotation system and database for rice genome sequence. Nucleic Acids Res. **2002**, *30* (1), 98–102.
14. http://RiceGAAS.dna.affrc.go.jp/.
15. http://www.shigen.nig.ac.jp/rice/oryzabase/.
16. Stein, L. Gramene: A resource for comparative grass genomics. Nucleic Acids Res. **2002**, *30* (1), 103–105.
17. http://www.gramene.org/.
18. Yang, H. A draft sequence of the rice genome (Oryza sativa L. ssp.indica). Science **2002**, *296* (5565), 79–92.
19. Briggs, S. A draft sequence of the rice genome (Oryza sativa L. ssp. Japonica). Science **2002**, *296* (5565), 92–100.
20. Kinoshita, T. Report of committee on gene symbolization, nomenclature and linkage groups. Rice Genet. Newslett. **1995**, *12*, 9–153.
21. Omura, T. Induction of mutation by the treatment of fertilized egg cell with N-methyl-N-nitrosourea in rice. J. Fac. Agric., Kyushu Univ. **1979**, *24*, 165–174.
22. Hirochika, H. Contribution of the *Tos17* retrotransposon ton rice functional genomics. Curr. Opin. Plant Biol. **2001**, *4* (2), 118–122.
23. Sasaki, T. Expression of Xa1, a bacterial blight-resistance gene in rice, is induced by bacterial inoculation. Proc. Natl. Acad. Sci. U.S.A. **1998**, *95* (4), 1663–1668.
24. WaRonald, P.C. Xa21D encodes a receptor-loke molecule with a leucine-rich repeat domain that determines race-specific recognition and is subject to adaptive evolution. Plant Cell **1998**, *10* (5), 765–779.
25. Sasaki, T. The Pib gene for rice blast resistance belongs to the nucleotide binding and leucine-rich repeat class of

plant disease resistance genes. Plant J. **1999**, *19* (1), 55–64.
26. Valent, B. tA single amino acid difference distinguishes resistant and susceptible alleles of the rice blast resistance gene Pi-ta. Plant Cell **2000**, *12* (11), 2033–2046.
27. Yamaguchi, J. *Slender* rice, a constitutive gibberellin response mutant, is caused by a null mutation of the *SLR1* gene, an ortholog of the height-regulating gene *GAI/RGA/RHT/D8*. Plant Cell **2001**, *13* (5), 999–1010.
28. Hirochika, H. Screening of the rice viviparous mutants generated by endogenous retrotransposon *Tos17* insertion. Tagging of a zeaxanthin epoxidasc gene and a novel *OsTATC* gene. Plant Physiol. **2001**, *125* (3), 1248–1257.
29. Hirochika, H. Systematic screening of mutants of rice by sequencing retrotransposon-insewrtion sites. Plant Biotechnol. **1998**, *15* (4), 253–256.
30. Matsuoka, M. Loss-of-function mutations in the rice homeobox gene *OSH15* affect the architecture of internodes resulting in dwarf plants. EMBO J. **1999**, *18* (4), 992–1002.
31. Furuya, M. Isolation and characterization of rice phytochrome A mutants. Plant Cell **2001**, *13* (3), 521–534.
32. Sasaki, T. Genetic and molecular dissection of quantitative traits in rice. Plant Mol. Biol. **1997**, *35* (1–2), 145–153.
33. Yano, M. Genetic and molecular dissection of naturally occurring variation. Curr. Opin. Plant Biol. **2001**, *4* (2), 130–135.
34. Sasaki, T. Genetic control of flowering time in rice, a short-day plant. Plant Physiol. **2001**, *127* (4), 1425–1429.
35. Sasaki, T. Plant Cell **2000**, *12* (12), 2473–2484.
36. Yano, M. Hd6, a rice quantitative trait locus involved in photoperiod sensitivity, encodes the alpha subunit of protein kinase CK2. Proc. Natl. Acad. Sci. U.S.A. **2001**, *98* (14), 7922–7927.
37. Gale, M.D. Cereal genome evolution. Grasses, line up and for a circle. Curr. Biol. **1995**, *5* (7), 737–739.

RNA-Mediated Silencing

Karin van Dijk
Heriberto Cerutti
University of Nebraska, Lincoln, Nebraska, U.S.A.

INTRODUCTION

RNA-mediated silencing involves various epigenetic mechanisms—apparently triggered by double-stranded RNA (dsRNA)—that result in suppression of gene expression. These mechanisms can lead to the degradation of homologous RNAs, translational repression, heterochromatin formation, and DNA methylation. Epigenetic phenomena induced by dsRNA have been observed in many eukaryotes and they may have evolved as defense responses against transposable elements and viruses. Recently, it has become apparent that the basic machinery required for these processes is also involved in the control of development. Moreover, links between RNA-mediated silencing and DNA and chromatin modifications are also starting to emerge. For more on this, the reader is referred to several recent reviews. This article discusses the basic processes leading to RNA silencing, the putative triggers, some of the key plant proteins involved in these pathways, and some potential applications of RNA silencing.

RNA-SILENCING BY RNA DEGRADATION

Posttranscriptional gene silencing (PTGS), cosuppression and RNA-mediated virus resistance in plants and algae, RNA interference (RNAi) in animals, and quelling in fungi are all examples of RNA-mediated silencing. Although these phenomena may differ in the details, central to them appears to be a molecular machinery that processes long dsRNA into smaller dsRNAs—called small interfering RNAs (siRNAs)—which are then used as guides to target homologous RNA molecules for degradation. In the late 1980s and throughout the 1990s, RNA-mediated silencing processes were observed in many eukaryotes. However, it was not realized until the late 1990s that these seemingly different processes are mechanistically related.[1–5] In plants, the initial observation came from research in the Jorgensen lab when an attempt was made to enhance the purple color of petunia flowers by overexpressing chalcone synthase.[4,5] Instead, the reverse happened. The expression of both the transgene and the native gene were diminished, and many flowers lost their pigment. This phenomenon was named cosuppression, and was later found to occur at the posttranscriptional level, hence the name posttranscriptional gene silencing. In the fungus *Neurospora crassa*, a similar phenomenon was observed, named quelling. Around the same time, researchers working on the nematode *Caenorhabditis elegans* found that they could successfully interfere with gene expression by injecting antisense RNA. However, to their surprise, injections with sense RNA could also suppress gene expression. We now know that the gene silencing observed in these cases was triggered by the presence of dsRNA, a contaminant produced during the preparation of sense or antisense RNA.[5] This strategy of using dsRNA to silence a target gene is called RNA interference. Researchers have for many years used yet another RNA-mediated silencing process—antisense technology—to interfere with specific gene expression. To achieve this, transgenic organisms are engineered to produce RNA complementary to the target transcript. Although the precise mechanism by which antisense RNA works remains elusive, in at least some cases dsRNA is likely involved as an intermediate.[3]

THE BASIC RNA-MEDIATED SILENCING MACHINERY

The combination of genetic and biochemical evidence from different organisms is starting to reveal the molecular machinery involved in the dsRNA-induced degradation of cellular RNAs. This process can be divided into an initiation and an effector step (Fig. 1).[1,2] In the initiation step, the ATP-dependent dsRNA nuclease DICER (Table 1), initially described in *Drosophila melanogaster*, chops the dsRNA molecule into siRNAs of about 21–23 nucleotides. In the effector step, the siRNAs are transferred to an enzyme complex—the RNA-induced silencing complex (RISC)—which has an endonuclease activity distinct from DICER. RISC is proposed to unwind the double-stranded siRNAs, and use the antisense strand to identify complementary messenger RNAs (mRNAs). Target mRNAs are then cleaved across from the center of the siRNA-mRNA hybrid and further degraded (Fig. 1). In some eukaryotes, an amplification step appears to be required for efficient silencing (see following discussion).

Encyclopedia of Plant and Crop Science
DOI: 10.1081/E-EPCS 120010637

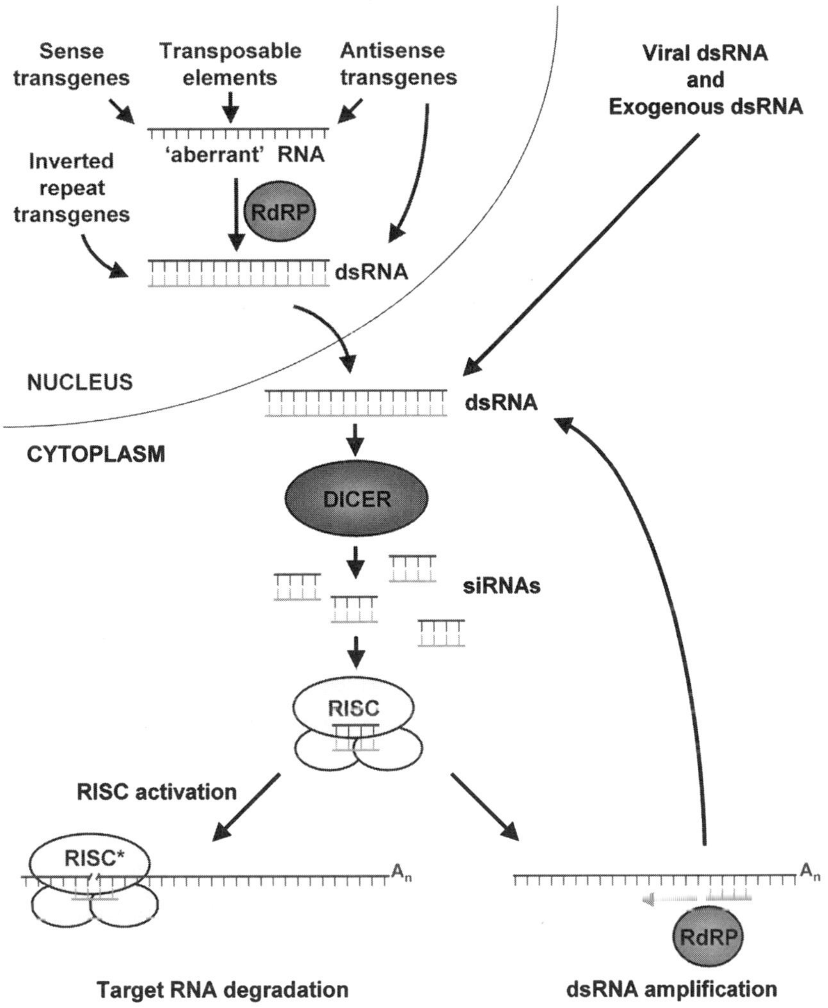

Fig. 1 Model for RNA-mediated silencing resulting in the degradation of target RNAs. Silencing is induced by double-stranded RNA (dsRNA). The dsRNA can be introduced exogenously, produced during viral replication, or transcribed from inverted repeat transgenes. In the case of transposable elements and sense or antisense transgenes, dsRNA can be synthesized by an RNA-directed RNA polymerase (RdRP) that uses aberrant RNA as a template (see article), or it can result from the annealing of single-stranded transcripts to RNA of opposite polarity (not depicted). DICER processes the long dsRNAs into small interfering RNAs (siRNAs), which are transferred to a multiprotein complex, the RNA-induced silencing complex (RISC). Upon unwinding of the siRNAs, the activated RISC (denoted by an asterisk) identifies and effects the degradation of target mRNAs with sequence homology. The dsRNA trigger can also be amplified by an RdRP. (*View this art in color at www.dekker.com.*)

Moreover, there is also spreading of silencing throughout the whole organism. The exact mechanism is unclear, but in plants it involves a sequence-specific signal that travels between cells as well as through the phloem.[6]

TRIGGERS OF RNA-MEDIATED SILENCING

Direct delivery of dsRNA (injected, fed, or the result of transcription from inverted repeat transgenes) can induce silencing in a variety of organisms, including plants. One of the fascinating aspects of RNAi, which is particularly noticeable in *C. elegans*, is that minute amounts of dsRNA can trigger degradation of a vast amount of mRNA. Recent studies suggest that the initial dsRNA trigger needs to be amplified to effectively silence target genes.[7] In this model, primary siRNAs are used as primers by an RNA-directed RNA polymerase (RdRP) to synthesize more dsRNA, using the target mRNA as a template (Fig. 1). DICER then cleaves the dsRNA to produce enough siRNAs to effectively target even abundant RNAs for degradation.

With sense and antisense transgenes, a major question is how dsRNA is produced. One model proposes that an

Table 1 Proteins implicated in RNA-mediated silencing in *Arabidopsis*

Protein	Proposed or known function
Carpel factory (CAF/SIN-1)	RNAse III and RNA helicase activity; homologue of *Drosophila* DICER
AGO1	Protein containing PAZ and PIWI domains; one *Drosophila* homologue is associated with the RISC complex
SDE1/SGS2	RdRP; dsRNA production and/or dsRNA amplification
SGS3 (SDE2)	Coiled-coil protein with unknown function
DDM1	Member of SWI/SNF family of chromatin remodeling proteins; required to maintain DNA methylation
MET1	DNA methyltransferase
SDE3	RNA helicase

RdRP synthesizes dsRNA from aberrant RNA[1,3] (Fig. 1). What makes an RNA molecule aberrant is not known, but it may involve the production of prematurely terminated, nonpolyadenylated or misprocessed RNAs. Because these transcripts are most likely present in the nucleus, the synthesis of dsRNA by an RdRP is proposed to occur in the nucleus.[1] In support of this model, *Arabidopsis* mutants in an RdRP, SDE1/SGS2 (Table 1), are defective in the PTGS of sense transgenes. In contrast, silencing induced by viruses, expressing their own RdRP, is not affected in this mutant background.[3]

MicroRNAs, ANOTHER SMALL RNA SPECIES PRODUCED BY DICER

Additional layers of complexity are emerging as more is discovered about RNA-mediated silencing pathways. For example, DICER can produce another species of small RNAs called microRNAs (miRNAs). MicroRNAs are processed from endogenous precursor RNAs consisting of double-stranded hairpin structures that have regions of imperfect complementarity.[8] MicroRNAs were first identified in *C. elegans* as products of the genes *let-7* and *lin-4*. These miRNAs are involved in regulating developmental timing by binding to the 3′ untranslated regions and preventing translation of their target mRNAs. Thus, in this case, miRNAs induce translational repression rather than mRNA degradation.

Recently, a combined effort has resulted in the isolation of many miRNAs in both animals and plants. The regulatory targets of most of these are unknown, but what has become clear is that the distinction between miRNA and siRNA has become blurred.[1] For instance, one of the *Arabidopsis* miRNAs has perfect complementarity with the transcripts of a group of transcription factors, and it has been implicated in controlling their expression through degradation of the mRNA.[1,8] Intriguingly, many plant miRNAs have a higher level of complementarity with their potential targets than animal miRNAs,[8] and they may function by regulating transcript stability rather than translation. Moreover, in plants, many predicted targets function as developmental regulators, suggesting that miRNAs might play a role in coordinating growth and development.

RNA-MEDIATED SILENCING AND CHROMATIN/DNA MODIFICATIONS

In addition to effects at the posttranscriptional level, it is becoming apparent that RNA can direct DNA methylation and chromatin modifications. In several plant species dsRNA can target the methylation of homologous DNA. This RNA-directed DNA methylation (RdDM) was initially observed in viroid-infected plants, and has now been shown to also occur in plants expressing inverted repeat constructs or infected with replicating viruses containing sequences homologous to genomic DNA.[9] When the RdDM occurs in promoter sequences, the target genes are silenced at the transcriptional level, i.e., no RNA is produced. Promoter-directed dsRNA is also processed into siRNAs, but it is not yet clear whether these are required to induce DNA methylation.[1,9]

In plants, many posttranscriptionally silenced transgenes are also methylated, but in this case, methylation occurs within the coding regions. Interestingly, two *Arabidopsis* mutants, *met1* and *ddm1* (Table 1), initially isolated in a screen for reduced methylation of the genome, are also defective in their ability to post-transcriptionally silence transgenes.[10] The relevance of this methylation is not clear, but it might be involved in the initiation or maintenance of silencing by affecting the production of aberrant RNA. In addition, RNA has been implicated in heterochromatin formation and maintenance in other organisms.[1] For instance, in *Schizosaccharomyces pombe,* mutants defective in components of the RNAi pathway show reactivation of transgenes integrated in the centromeric (heterochromatic) regions. This correlates with the loss of histone H3 methylation, which is commonly associated with a repressive chromatin structure.[1] Thus, RNA-mediated processes may also play a role in chromosomal organization and transcriptional control.

APPLICATIONS OF RNA-MEDIATED SILENCING IN PLANTS

RNA silencing has become an important tool in molecular biology and has the potential to revolutionize functional genomics, because it is an effective way to knock down the expression of any gene.[11] Moreover, since RNA-mediated silencing does not require complete sequence identity between the dsRNA trigger and a target gene, it is even possible to silence the expression of gene families. There are also clear applications of this technology to agriculture. For instance, an RNA-silencing strategy can be used to suppress undesirable traits in plants. As an example, tomatoes that have reduced levels of polygalacturonase (a protein involved in fruit ripening) have been produced by antisense technology.[4] The resulting tomatoes stay firm while ripening on the tomato vine. In the past, the success rate of this technology was somewhat unpredictable. However, a greater understanding of the mechanisms involved in RNA silencing has allowed the design of better constructs. Effective silencing of a particular gene can now be accomplished with inverted repeat transgenes that produce hairpin RNAs (hpRNA)[11] (Fig. 1). One success story, using hpRNA, is the production of cotton seeds with altered fatty acid content due to downregulation of expression of a fatty acid desaturase.[11]

ACKNOWLEDGMENTS

We thank Nandita Sarkar for critical reading of the manuscript. This work was supported by NIH grant GM62915.

ARTICLES OF FURTHER INTEREST

Gene Silencing: A Defense Mechanism Against Alien Genetic Information, p. 492
Transgenes: Expression and Silencing of, p. 1242
Virus-Induced Gene Silencing, p. 1276

REFERENCES

1. Cerutti, H. RNA interference: Traveling in the cell and gaining functions? Trends Genet. **2003**, *19* (1), 39–46.
2. Hannon, G.J. RNA interference. Nature **2002**, *418* (6894), 244–251.
3. Vance, V.; Vaucheret, H. RNA silencing in plants-defense and counterdefense. Science **2001**, *292* (5525), 2277–2280.
4. Baulcombe, D. RNA silencing. Curr. Biol. **2002**, *12* (3), R82–R84.
5. Zamore, P.D. Ancient pathways programmed by small RNAs. Science **2002**, *296* (5571), 1265–1269.
6. Mlotshwa, S.; Voinnet, O.; Mette, M.F.; Matzke, M.; Vaucheret, H.; Ding, S.W.; Pruss, G.; Vance, V.B. RNA silencing and the mobile silencing signal. Plant Cell **2002**, *14* (Suppl), S289–S301.
7. Plasterk, R.H. RNA silencing: The genome's immune system. Science **2002**, *296* (5571), 1263–1265.
8. Jones, L. Revealing micro-RNAs in plants. Trends Plant Sci. **2002**, *7* (11), 473–475.
9. Matzke, M.A.; Matzke, A.J.; Pruss, G.J.; Vance, V.B. RNA-based silencing strategies in plants. Curr. Opin. Genet. Dev. **2001**, *11* (2), 221–227.
10. Vaucheret, H.; Beclin, C.; Fagard, M. Post-transcriptional gene silencing in plants. J. Cell Sci. **2001**, *114* (Pt 17), 3083–3091.
11. Wang, M.B.; Waterhouse, P.M. Application of gene silencing in plants. Curr. Opin. Plant Biol. **2002**, *5* (2), 146–150.

Root Membrane Activities Relevant to Plant-Soil Interactions

Zeno Varanini
University of Udine, Udine, Italy

INTRODUCTION

The plasma membrane of root cells plays a fundamental role in the complex interactions that occur at the soil-root interface, contributing to the transport of solutes into the cell and thus changing the composition of the rhizosphere. Clearly, the biochemical mechanisms regulating the plant-soil interaction must respond and adapt to the conditions both within the root cell and at the rhizosphere; the plant must therefore possess sensing mechanisms that can efficiently modulate transport and enzyme activities at the plasma membrane. There are several classes of membrane-bound proteins, the most important of which in the soil-plant relationship are proton pumps, carriers, channel proteins, and reductases.

PROTON PUMPS

The plasma membrane H^+-ATPase (pmH^+-ATPase) is an electrogenic enzyme, exploiting the energy released upon the hydrolysis of ATP to transport protons from the cytoplasm into the apoplast. This enzyme can catalyze the extrusion of over 10^2 H^+ ions per second: It is also extremely abundant, forming approximately 1% of the overall amount of proteins at the plasma membrane. Moreover, these levels can increase remarkably under certain conditions (discussed later), in particular in the layers of cells at the soil-plant interface (i.e., the epidermal and root hair cells).

The plant pmH^+-ATPase is formed by a single polypeptide approximately 100 kDa in size. Its putative topology suggests that the enzyme possesses ten transmembrane regions and four cytoplasmic domains,[1] the latter including the carboxy-terminal region (Ct), an important domain from a posttranslational viewpoint. The Ct has in fact been observed to have auto-inhibitory properties; its effect can however be cancelled when, once phosphorilized, it becomes the binding site for proteins of the 14-3-3 family.

Plant pmH^+-ATPase is encoded by a multigene family; it is not yet clear if the various isoforms function under specific environmental conditions or are expressed in specific types of cells, tissues, or organs.[1]

The pmH^+-ATPase plays a fundamental role in the soil-plant relationship, supplying energy for nutrient transport and acidifying the rhizosphere in order to facilitate the roots' acquisition of certain mineral nutrients; it is also involved in the plant's response to abiotic stress and signals from the rhizosphere.

Energization of Nutrient Transport

The activity of the pmH^+-ATPase creates a gradient of pH (more acidic outside the cell) and potential (more negative inside the cell) across the plasma membrane, which is then exploited for nutrient transport by means of carriers or channels (Fig. 1). Cations are attracted to the cell by the membrane potential and could simply enter into the cell through protein channels; anions, on the other hand, require carriers (symporters). This type of transport is an active process, and the energy is supplied by the accompanying inflow of protons. The activity of pmH^+-ATPase seems to be modulated according to the plants' nutritional needs. In the case of nitrate, the enhanced uptake brought about by the roots' exposure to this nutrient (induction) is accompanied by higher levels and thus activity of the pmH^+-ATPase, presumably to meet the greater demand for proton-driving force.[2] Recent investigations also reveal that treatment with nitrate induces a differential expression of the various forms of the enzyme, thus suggesting that certain members of the pmH^+-ATPase multigene family may respond specifically to nutrient uptake.[3]

Nutrient Acquisition by Rhizosphere Acidification

By extruding protons into the apoplast, the root pmH^+-ATPase acidifies the rhizosphere, with major consequences: The lower pH stimulates the release of cations from the exchange sites on the cell wall and the soil colloids, which then pass into the soil solution and reach the binding sites of the transporters on the outer surface of the plasma membrane. The level of rhizosphere acidification brought about by the proton pump can be modulated by processes of secondary transport, and in particular by the ratio of cations and anions taken up by

Encyclopedia of Plant and Crop Science
DOI: 10.1081/E-EPCS 120012892

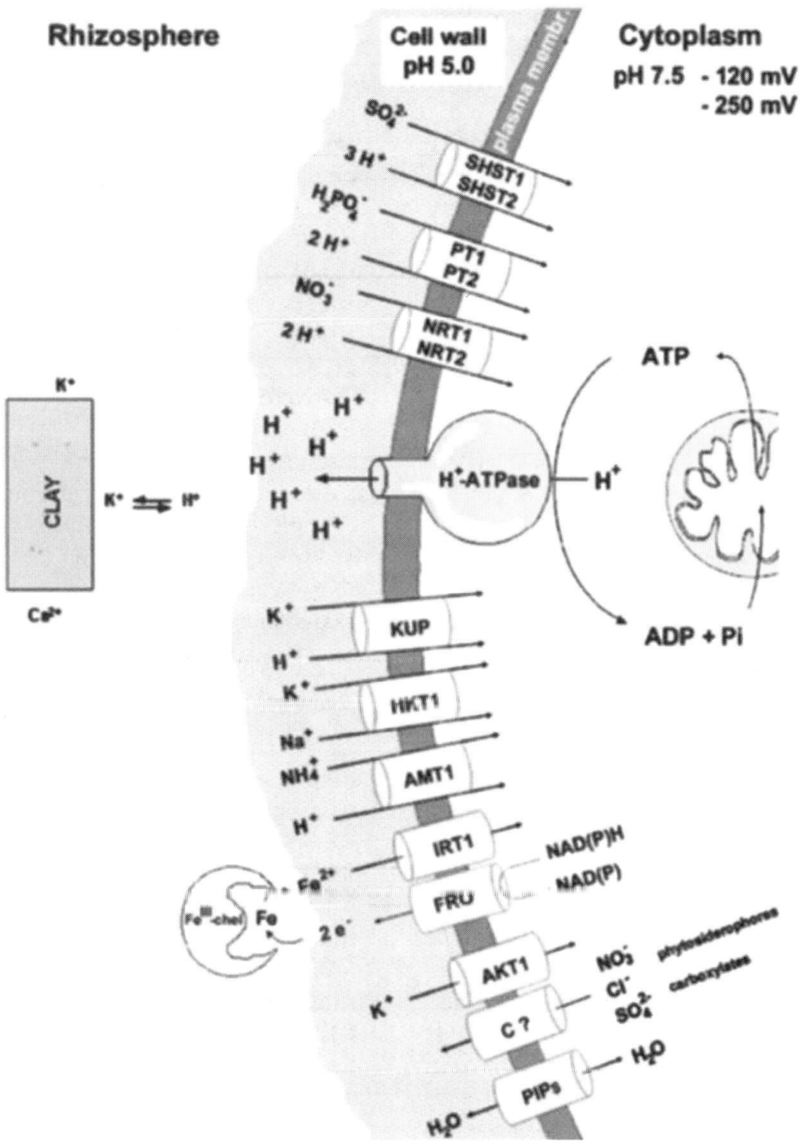

Fig. 1 Schematic representation of the major transport proteins operating at the root plasma membrane. (*View this art in color at www.dekker.com.*)

the root cells. When nitrogen, for example, is taken up as ammonium, rhizosphere acidification increases; the opposite occurs when it is taken up as nitrate.

The acidifying action exerted by the pmH$^+$-ATPase has important consequences on the solubilization of essential nutrients characterized by low availability. A well-investigated case is that of iron. This element is mostly present in the soil as ferric ion (Fe^{3+}), which is extremely insoluble at neutral or alkaline pH values. Dicots and nongraminaceous monocots (Strategy I plants) can, however, respond to Fe-deficiency by acidifying the rhizosphere—the extent of the process depending on the species—and thus solubilizing and rendering available Fe^{3+}. Plant species endowed with a greater ability to acidify can increase root pmH$^+$-ATPase activity under conditions of Fe deficiency.[4] Higher levels of the enzyme are present at the plasma membrane, in particular in the rhizoderma and in the root hairs of the subapical root sections.[5]

Response to Abiotic Stress and Signals from the Rhizosphere

The activity of pmH$^+$-ATPase can also change when anomalous conditions are encountered at the rhizosphere. Salt stress is a common case: Under these conditions, enhanced proton extrusion may help energize the Na^+/H^+

antiporter systems operating at the plasma membrane. Salt stress has been reported to induce pmH^+-ATPase gene expression in specific tissues; in tomato plants, the response was ascribed to a single, specific isoform.[6]

Recent studies moreover suggest that under conditions of osmotic shock, there may also be a posttranslational regulation of the enzyme: The interaction with 14-3-3 proteins would appear to increase the H^+/ATP coupling ratio.[7] Enhanced enzyme activity has also been observed in the roots of plants growing in acidic environments: Root pmH^+-ATPase would therefore also play a fundamental role in preserving cytoplasm pH homeostasis.[8]

Finally, pmH^+-ATPase activity also appears to be modulated by other factors present at the rhizosphere, such as humified organic matter. Low-molecular-weight fractions (<3500) in particular can stimulate the activity of pmH^+-ATPase with positive effects on ion uptake rates and the ability to acquire iron and other metal micronutrients.[9]

CARRIERS

These membrane proteins can move solutes either up or down electrochemical gradients at a rate of 10^2–10^4 molecules per second. There are three different types of carriers: symporters, antiporters, and uniporters. Whereas the latter proteins move substances down a concentration gradient, symporters and antiporters can transport a compound against its concentration gradient by contemporaneously transporting a second substance (usually H^+) down its electric and/or chemical gradient. Symporters in particular play an important role in the root-soil interaction, as their transport activity of oxoanionic and cationic macronutrients helps modify rhizosphere pH.

Carrier proteins (symporters) have a molecular mass of approximately 50–60 kDa. Their structure includes 12 membrane-spanning regions and in some cases, motifs on their cytoplasmic domains suggesting a posttranslational regulation of their activity via phosphorylation and dephosphorylation. Although no evidence has been found that such regulation may occur in vivo, it has been clarified that the expression of symporter-encoding genes is regulated by nutrient availability at the rhizosphere.[10,11] The modulation of gene expression depends on the nutrient: Nitrate, for example, is taken up from the soil through low- and high-affinity transport systems encoded by *NRT1* and *NRT2* genes. Some of these transporters are constitutively expressed, whereas others (*NRT2.1*) are nitrate-inducible and are subjected to negative feedback regulation by the products of nitrate assimilation. On the other hand, the genes encoding high-affinity transporters for phosphate (*PT1*, *PT2*) and sulphate (*SHST1*, *SHST2*) are derepressed when plants are deprived of the nutrient. Potassium and ammonium are also taken up by symporters when present at the rhizosphere at low concentrations (in a μM range). The carriers are encoded by such genes as *HKT1* (Na^+/K^+ symporter), *KUP* (H^+/K^+ symporter), and *AMT1* (H^+/NH_4^+ symporter), which are upregulated under conditions of potassium and ammonium deficiency, respectively. Positively-charged micronutrients are transported into the cell via uniport carriers. An example is *IRT1*, the expression of which is induced in the roots of Fe-deficient plants.[5] The gene encodes a nonspecific Fe uniport carrier that is also capable of transporting other bivalent cations into the cell, including Cd.

CHANNEL PROTEINS

Channels are membrane-spanning proteins that contribute to the diffusion of water and ions down energetically-favorable gradients. Electric or chemical signals control the opening and closing of these proteins; when open, inorganic and organic ions and water molecules pass through them at very high rates (10^6 to 10^7 molecules per second). Channel proteins are involved in several physiological processes: mineral nutrition; cell osmoregulation, by allowing great ion fluxes across the plasma membrane over short periods of time; signalling, by amplifying and propagating electric signals or transporting secondary messengers such as Ca_2^+; and the control of membrane potential. The most important channels for the root-rhizosphere interaction are those allowing the flow of potassium and water across the plasma membrane and the efflux of organic and inorganic anions from the root cells.

Various types of channel proteins for the influx and efflux of potassium have been identified at the root cell plasma membrane,[12] but it has only been possible to ascribe a physiological role to AKT1, a protein that surprisingly displays both high- and low-affinity kinetics for this nutrient. The channel proteins involved in the efflux of anions (carboxylates and phytosiderophores) are key factors in the response to nutrient deficiencies and in the resistance to aluminum toxicity;[13] in the latter case, it has been suggested that aluminum itself may be the signal triggering the opening of the channels.

As regards water transport, channel proteins called aquaporins (encoded by *PIPs*) have been found to be involved in this process. When open, they facilitate the passive movement of water molecules down a water potential gradient. The activity of at least some aquaporins appears to be regulated by processes of phosphorylation and dephosphorylation.

Up to now, 30 genes have been found in *Arabidopsis* for aquaporin homologues. Some of these genes are constitutively expressed, whereas others are known to be temporally and spatially regulated during the plant's development and in response to stress. Aquaporins are very abundant proteins, and appear to be involved not only in cytosol osmoregulation, but also in the bulk flow of water.

REDUCTASES

Various redox activities are present at the plasma membrane that can transport electrons from cytoplasmic donors (e.g., NADH and NADPH) to oxidized acceptors in contact with the outer surface of the membrane. Among these enzymes, there is the turboreductase, induced in the rhizodermal cells of dicots and nongraminaceous monocots under conditions of Fe-deficiency. Its physiological function is to reduce $Fe^{(III)}$-chelates present at the rhizosphere, thus allowing the transport of Fe^{2+} into the cell by the IRT proteins (discussed earlier). Recently, a ferric chelate reductase (FRO2) has been cloned: It belongs to the superfamily of flavocytochromes, a class of proteins that mediate the transport of electrons across the membrane, and appears to be strongly induced in Fe-deficient plants.[5] The roles and electron acceptors of the other reductases that are constitutively expressed in the roots of all plant species and are distributed along the entire root axis are still unknown. Since they produce reactive forms of oxygen, they may be involved in the metabolism of cell wall production and in pathogen control. Other authors, highlighting the ability to reduce nitrate, have instead suggested they may also act as sensors in the regulation of nitrate uptake and assimilation.

CONCLUSION

Biochemical and molecular analyses have clarified many aspects of the structure and regulation of plasma membrane activities modulating the root-soil relationship. However, much remains to be learned, and not only with regard to those systems that have been less investigated (i.e., channel proteins and reductases), but also the sensing mechanisms involved in the perception of the environmental conditions. There are also other objectives in the research on proton pumps and nutrient-transporting carriers: to understand the specific role of each isoform and shed more light on the aspects relative to posttranslational regulation. It would be interesting to assess which are the major systems involved in the root-soil relationship in field conditions. This could be a useful parameter in selecting or obtaining genotypes that are more efficient in their interaction with the soil (e.g. greater ability to recuperate nutrients, greater adaptability to adverse conditions, etc.).

ARTICLES OF FURTHER INTEREST

Minor Nutrients, p. 726
Nitrogen, p. 822
Potassium and Other Macronutrients, p. 1049

REFERENCES

1. Morsomme, P.; Boutry, M. The plant plasma membrane H^+-ATPase: Structure, function and regulation. Biochim. Biophys. Acta **2000**, *1465*, 1–16.
2. Santi, S.; Locci, G.; Pinton, R.; Cesco, S.; Varanini, Z. Plasma membrane H^+-ATPase in maize roots induced for NO_3^- uptake. Plant Physiol. **1995**, *109*, 1277–1283.
3. Santi, S.; Locci, G.; Monte, R.; Pinton, R.; Varanini, Z. Induction of nitrate uptake in maize roots: Expression of a putative high-affinity nitrate transporter and plasma membrane H^+-ATPase isoforms. J. Exp. Bot. **2003**, *54*, 1851–1864.
4. Dell'Orto, M.D.; Santi, S.; De Nisi, P.; Cesco, S.; Varanini, Z.; Zocchi, G.; Pinton, R. Development of Fe-deficiency responses in cucumber (*Cucumis sativus* L.) roots: Involvement of plasma membrane H^+-ATPase activity. J. Exp. Bot. **2000**, *51*, 695–701.
5. Schmidt, W. Iron homeostasis in plant: Sensing and signalling pathways. J. Plant Nutr. **2003**, *26*, 2211–2230.
6. Kalampanayil, B.D.; Wimmers, L.E. Identification and characterization of a salt-stress-induced plasma membrane H^+-ATPase in tomato. Plant Cell Environ. **2001**, *24*, 999–1005.
7. Kerkeb, L.; Venema, K.; Donaire, J.P.; Rodriguez-Rosales, M.P. Enhanced H^+/ATP coupling ratio of H^+-ATPase and increased 14-3-3 protein content in plasma membrane of tomato cells upon osmotic shock. Physiol. Plant. **2002**, *116*, 37–41.
8. Yan, F.; Feuerle, R.; Schäffer, S.; Fortmeier, H.; Schubert, S. Adaptation of active proton pumping and plasmalemma ATPase activity of corn roots to low root medium pH. Plant Physiol. **1998**, *117*, 311–319.
9. Varanini, Z.; Pinton, R. Direct Versus Indirect Effects of Soil Humic Substances on Plant Growth and Nutrition. In *The Rhizosphere: Biochemistry and Organic Substances at the Soil-Plant Interface*; Pinton, R., Varanini, Z., Nannipieri, P., Eds.; Marcel Dekker Inc.: New York, 2001; 141–157.
10. Hirsch, R.E.; Sussman, M.R. Improving nutrient capture from soil by the genetic manipulation of crop plant. Trends Biotech. **1999**, *17*, 356–361.
11. Chrispeels, M.J.; Crawford, N.M.; Schroeder, J.I. Protein for transport of water and mineral nutrients across the membranes of plant cells. Plant Cell **1999**, *11*, 661–675.
12. Véry, A.-A.; Sentenac, H. Cation channels in the *Arabidopsis* plasma membrane. Trends Plant Sci. **2002**, *7*, 168–175.
13. Neumann, G.; Römheld, V. The Release of Root Exudates as Affected by the Plant's Physiological Status. In *The Rhizosphere: Biochemistry and Organic Substances at the Soil-Plant Interface*; Pinton, R., Varanini, Z., Nannipieri, P., Eds.; Marcel Dekker Inc.: New York, 2001; 41–93.

Root-Feeding Insects

Eli Levine
Illinois Natural History Survey, Champaign, Illinois, U.S.A.

INTRODUCTION

Belowground plant parts constitute a major portion (often >50%) of the biomass and primary plant production available to herbivores. Therefore, it is not surprising that many species of insects have taken advantage of this resource. Root feeding is used here to refer to consumption of all belowground plant parts, including roots, rhizomes, and other storage organs. Root-feeding insects can influence plant diversity, plant succession, competitive interactions among plant species, the susceptibility of plants to other herbivores and plant diseases, and crop yield.[1] This article discusses adaptations insects possess to live in the soil environment, examines some of the diverse species of insects that feed on plant roots, and discusses the effects of root feeding on host plants with emphasis on corn rootworms.

ADAPTATIONS TO THE SOIL ENVIRONMENT

Organisms that spend at least some part of their life cycle underground are referred to in this article as edaphic organisms. Edaphic species that live on the soil surface or in the litter are referred to as epedaphic organisms, whereas those that inhabit mineral strata beneath the litter layer are referred to as euedaphic organisms. Distantly related species that have come to share similar soil habitats may have followed parallel or convergent evolutionary paths, and may appear similar in form.[2] A major factor limiting insect occupation of the soil has been the ability of these organisms to move. Most insects are not small enough to travel though interconnected pore spaces and must therefore either tunnel through the soil (by pushing soil particles aside and squeezing through) or excavate it (removing the soil in front and moving it behind them). Burrowing insects are generally round in cross section, but may otherwise vary in shape. Insects that tunnel often have very small or no legs, and may be of two body forms. Wireworms (the larval stage of click beetles; Coleoptera: Elateridae) are often smooth, slender, and hard-bodied; using a serpentine motion, they are able to force their wirelike bodies through the soil. In contrast, corn rootworm larvae (Coleoptera: Chrysomelidae) are soft-bodied and can use peristaltic movements to pass through the soil. Soil insects that move by excavating the soil are often armored and thick-bodied, and may display modified forelegs for digging, such as in mole crickets (Orthoptera: Gryllotalpidae) or nymphal periodical cicadas (Homoptera: Cicadidae). Wings are generally reduced in adult insects that spend much of their lives in the soil, and compound eyes are often reduced or absent. In contrast, tactile sense organs are generally well developed in euedaphic insects, and many of these insects are sensitive to carbon dioxide, which is often used to locate plant roots. In some species, such as the Japanese beetle, *Popillia japonica* (Coleoptera: Scarabaeidae), grubs can move vertically through the soil profile as seasonal soil temperatures change, and thus remain active and able to feed on plant roots.[3]

DIVERSITY OF INSECTS THAT FEED ON PLANT ROOTS

Insects that feed on roots are represented in most of the larger insect orders. The burrower bug, *Cyrtomenus bergi* (Heteroptera: Cydnidae), is a serious pest of cassava, a very important crop in low-income, food-deficit countries. All five nymphal stages and the adult stage of these small oval bugs live in the soil. They damage this root crop by directly feeding on the roots. Fungal pathogens can enter these wounds, reducing both root yield and quality.[4] Some root-feeding scale insects in the genus *Margarodes* (Homoptera: Coccidae) are given the common name of ground pearls, for the pearl-like appearance of the wax cysts of females. Ground pearls often cause injury to vine crops, turf, and sugarcane, and are very difficult to control because the insects are protected by the cysts. Encysted insects are also able to resist extreme temperatures and low moisture levels, and adult females have been able to prolong emergence for several years.[5] Root-feeding insects may also facilitate the transmission of plant pathogens. For example, root feeding by larvae of the striped cucumber beetle, *Acalymma vittatum* (Coleoptera: Chrysomelidae), was found to increase the incidence and severity of *Fusarium* wilt of muskmelon.[6] Other root feeders have intricate life histories with other insect

Encyclopedia of Plant and Crop Science
DOI: 10.1081/E-EPCS 120010487

species. For example, eggs of the corn root aphid, *Aphis maidiradicis* (Homoptera: Aphididae), pass the winter in nests of ants in the genus *Lasius*. In the spring, when the eggs hatch, the ants carry the nymphs to roots of weeds where the aphids feed. Later, the ants transport the aphids to the roots of maize. All the while, the ants feed on honeydew produced by the aphids.[3]

CORN ROOTWORM LARVAL FEEDING ON MAIZE

The western corn rootworm, *Diabrotica virgifera virgifera*, and the northern corn rootworm, *Diabrotica barberi*, are the most serious insect pests of maize in the major maize-producing states of the United States and Canada. Both species have a single generation per year. Adults are present in maize fields from July through frost, and feed on maize foliage, pollen, silks, and developing kernels. Oviposition takes place almost exclusively in maize fields from late-July through mid-September (since the mid-1990s, however, western corn rootworms have also begun to lay eggs outside of maize fields in east-central Illinois and northern Indiana, rendering crop rotation ineffective in the management of this species in those areas). Diapausing eggs spend the fall and winter in the soil until late May or early June when they hatch and the larvae begin to feed on maize roots (larvae can only complete development on maize and a few grassy weeds). Newly eclosed larvae feed primarily on root hairs and the outer tissue of roots, causing little significant injury. As the larvae grow, root tips may be pruned back and larvae may tunnel into larger roots. Growing points at root tips may also be killed; this often leads to fibrous secondary root production. Larval feeding disrupts root system function, reducing the amount of water and nutrients supplied to developing plants, which can in turn reduce grain yield. Larval feeding may also facilitate infection by root and stalk rot fungi, resulting in further damage. Brace roots may also be severely damaged, causing plants to lodge and stalks to gooseneck. Altered leaf orientation can reduce photosynthetic efficiency of the plant; difficulty in mechanically harvesting the lodged maize can result in additional yield losses.[7] In some cases, rootworm injury has slowed plant development, leading to asynchrony between tassels and silks and resulting in a greater percentage of plants with barren ears.[8]

Root-injury ratings have been used to assess larval injury by corn rootworms. Some investigators have estimated that a mean root-injury rating > 2.5 (on a 1-to-6 root-injury rating scale, where 1 = minor injury and 6 = three nodes of roots destroyed) would result in economic loss. Other workers showed that root injury ratings were not consistent predictors of yield. Numerous factors influence the extent of damage and its impact on yield. Number and species of larvae, size of the root system, ability of the maize hybrid to regenerate roots, availability of soil moisture and nutrients, plant population, and weather are all involved in the amount of damage that may occur and the impact this damage may have on yield. If other stress levels are minimal, severe root damage may not result in significant yield losses, particularly if high winds do not cause lodging.[7]

A recent four-year field study using 12 commonly grown maize hybrids and natural infestations of corn rootworms at two locations in Illinois showed that hybrid characteristics and environmental conditions—particularly soil moisture—can strongly affect the level of root injury needed to cause economic damage. During one year of the study when growing conditions were stressful (low soil moisture), economic damage occurred with some hybrids when the average root rating was only 2.5 (on the 1–6 root-injury rating scale). When growing conditions were more favorable, economic damage did not occur with the same hybrids until root-injury ratings were 4.0 or higher.[9] The complex interactive effects that maize genotype, level of rootworm larval injury, and environment have on grain yield suggest that maize hybrids may differ in their ability to tolerate rootworm injury and partition biomass between vegetative and reproductive tissue. That is, injury to the roots can reduce (to varying degrees, depending on hybrid) the allocation of resources toward seed production as more of the plant's resources are allocated to replace injured roots.[10]

CONCLUSION

For corn rootworm management alone, approximately 88% of nonrotated maize hectares (estimated at over a million hectares) in Illinois are treated with a soil insecticide.[9] The introduction of genetically modified crops with activity against root-feeding insects will revolutionize the way these pests are managed. Although these new crops will likely reduce reliance on soil-applied insecticides, effects on non-target organisms and the development of resistant insect populations must be looked out for.

ARTICLES OF FURTHER INTEREST

Breeding Widely Adapted Cultivars: Examples from Maize, p. 211
Economic Impact of Insects, p. 407

Genetically Engineered Crops with Resistance Against Insects, p. 506
Insect–Plant Interactions, p. 609

REFERENCES

1. Hunter, M.D. Out of sight, out of mind: The impacts of root-feeding insects in natural and managed systems. Agric. For. Entomol. **2001**, *3*, 3–9.
2. Villani, M.G.; Allee, L.L.; Diaz, A.; Robbins, P.S. Adaptive strategies of edaphic arthropods. Annu. Rev. Entomol. **1999**, *44*, 233–256.
3. Daly, H.V.; Doyen, J.T.; Purcell, A.H., III. *Introduction to Insect Biology and Diversity,* 2nd Ed.; Oxford University Press: New York, 1998.
4. Bellotti, A.C.; Smith, L.; Lapointe, S.L. Recent advances in cassava pest management. Annu. Rev. Entomol. **1999**, *44*, 343–370.
5. Gullan, P.J.; Kosztarab, M. Adaptations in scale insects. Annu. Rev. Entomol. **1997**, *42*, 23–50.
6. Latin, R.X.; Reed, G.L. Effect of root feeding by striped cucumber beetle larvae on the incidence and severity of *Fusarium* wilt of muskmelon. Phytopathology **1985**, *75* (2), 209–212.
7. Levine, E.; Oloumi-Sadeghi, H. Management of diabroticite rootworms in corn. Annu. Rev. Entomol. **1991**, *36*, 229–255.
8. Spike, B.P.; Tollefson, J.J. Relationship of plant phenology to corn yield loss resulting from western corn rootworm (Coleoptera: Chrysomelidae) larval injury, nitrogen deficiency, and high plant density. J. Econ. Entomol. **1989**, *82* (1), 226–231.
9. Gray, M.E.; Steffey, K.L. Corn rootworm (Coleoptera: Chrysomelidae) larval injury and root compensation of 12 maize hybrids: An assessment of the economic injury index. J. Econ. Entomol. **1998**, *91* (3), 723–740.
10. Urias-Lopez, M.A.; Meinke, L.J. Influence of western corn rootworm (Coleoptera: Chrysomelidae) larval injury on yield of different types of maize. J. Econ. Entomol. **2001**, *94* (1), 106–111.

Rubisco Activase

R

Archie R. Portis
USDA and University of Illinois, Urbana, Illinois, U.S.A.

INTRODUCTION

Rubisco activase is a protein that is required in the higher plants to maintain and regulate the activity of Rubisco, the protein formally known as ribulose 1,5-bisphosphate carboxylase/oxygenase. Rubisco is a very important protein in photosynthesis because it initiates both photosynthetic carbon metabolism via its carboxylase activity and photorespiration via its oxygenase activity.[1] The activity of Rubisco is a major factor limiting photosynthesis and thus the productivity of many of our major crops. The properties of Rubisco activase and the current understanding of the regulation of Rubisco by the activase will be discussed.

WHAT IS KNOWN ABOUT THE STRUCTURE OF THE ACTIVASE?

Rubisco activase is a member of the very large and diverse AAA^+ protein family that contains related sequence domains forming a common structural motif, which binds ATP. AAA^+ proteins use the energy provided by ATP hydrolysis to perform a wide variety of functions.[2] For example, AAA^+ proteins are involved in proteolysis, DNA synthesis, membrane fusion, and microtubule severing and trafficking. Thus they constitute a novel type of molecular chaperone, which acts to manipulate other molecular or macromolecular structures. In performing these functions, large multimeric complexes are often assembled, which may contain other proteins. The AAA^+ proteins in these complexes contain one or more copies of the AAA^+ motif and typically form ring structures consisting of six or seven AAA^+ proteins.

Rubisco activase is highly self-associating and oligomerizaion of the protein is closely linked to its ATP hydrolysis activity.[3] However, the exact structure of activase has not yet been determined and the types of macromolecular structures it can form remain unclear. Based on a wide variety of evidence, a working model for its activity has been proposed (Fig. 1). In this model the activase forms a ring around the Rubisco holoenzyme, which consists of eight large and small subunits that form eight active sites. ATP hydrolysis is proposed to cause conformational changes in the ring, and thereby in Rubisco.

WHY DO PLANTS NEED ACTIVASE AND HOW DOES IT WORK?

The active site of Rubisco is capable of binding a wide variety of sugar phosphates (ligands) as well as its substrate, ribulose 1,5-bisphosphate (RuBP).[1,3] In addition, the amine group of a lysine residue in the active site must be carbamylated by addition of an ''activating'' CO_2 molecule, in order for carboxylation or oxygenation of RuBP to occur. While only minor changes in the conformation of the enzyme occur with carbamylation, much more dramatic and similar changes occur with the binding of certain ligands to both the carbamylated and uncarbamylated forms of the enzyme.[4] The binding of these ligands causes conformational changes that close off the active site so that the ligand is completely surrounded. This results in the ligand being bound very tightly, and its dissociation from the active site is very slow. Catalysis is possible only when RuBP binds to a carbamylated site. Physiologically, two of the more important ligands involved in inhibiting the activity of Rubisco in this manner are RuBP, which also binds to uncarbamylated active sites, and 2-carboxyarabinitol 1-phosphate, which prefers carbamylated sites.[3] Either directly or indirectly, Rubisco activase appears to reverse the conformational changes that sequester these ligands, allowing them to leave the active site more quickly. The activase was named for its ability to rapidly restore activity to Rubisco that has RuBP bound to uncarbamylated sites. Thus, depending on its ATPase activity, the activase can maintain and regulate the activity of Rubisco at a level appropriate for photosynthesis when it is limited by other factors, such as when light intensity is limiting.

Reductions in Rubisco activity can also occur during catalysis.[5] The reaction sequence is quite complex and various ''mistakes'' have been shown to occur.[1] In several cases it has been shown that an aberrant sugar phosphate product remains tightly bound so that in the absence of Rubisco activase activity, catalytic activity decreases to a low level.

Encyclopedia of Plant and Crop Science
DOI: 10.1081/E-EPCS 120019324

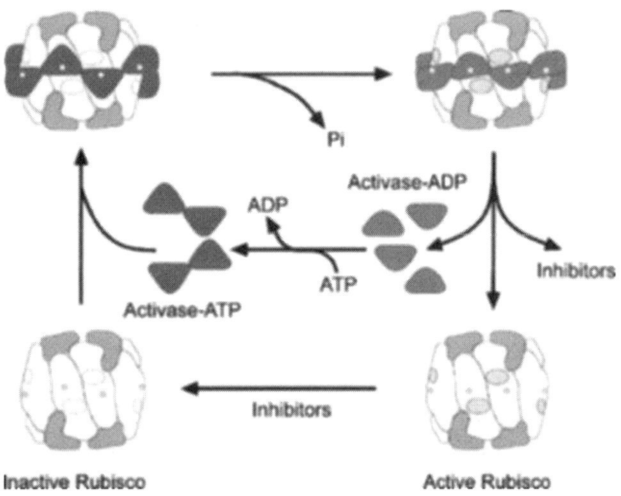

Fig. 1 A proposed model for the mechanism of Rubisco activase. Inhibitors bind (bottom half of the figure) to the open active sites (shaded ellipses) on the large Rubisco subunits (white) and close them off (dotted ellipses). Nucleotide exchange induces small changes in the conformation of the activase (only simulated here by a change in shape and shading, because the actual structure of activase is unknown), enhancing oligomerization and binding to Rubisco with the closed sites (left side), surrounding the holoenzyme in a ring structure. ATP hydrolysis in this supercomplex releases Pi and causes conformational changes in the activase and the ring structure (movement of active units relative to one another), resulting in the opening of the active sites (top half of the figure). The inhibitors can now dissociate more rapidly, and the changes in the activase-activase interactions result in disassembly of the ring and activase release from Rubisco (right side of the figure), restoring Rubisco to an active state. (From Ref. 3.)

The activity of the activase also plays another important role in photosynthesis by promoting a higher carbamylation state for Rubisco (and thus increased catalytic activity) than would otherwise be possible at atmospheric levels of CO_2.[3] Indeed, activase was discovered by characterizing a mutant of *Arabidopsis* that required high CO_2 for growth.[6] Studies of the isolated enzyme showed that less than half-maximal carbamylation is possible with the physiological conditions in the chloroplasts where Rubisco is located. While ligand binding can either increase or decrease carbamylation, the bound ligands also prohibit the needed substrate, RuBP, from binding. A kinetic way out of this apparent conundrum was provided by modeling the effects of activase on ligand binding.[7] The modeling shows that a steady-state level of nearly maximal carbamylation can result from a difference in the ability of activase to promote the release of RuBP from the carbamylated and uncarbamylated forms of the active site.

In addition to playing a critical role in the regulation of Rubisco in response to changes in light intensity and source-sink balance, recent studies indicate that the activase may play an important role in the response of photosynthesis to high-temperature stress.[8] A decrease in the activation state of Rubisco contributes to the reduced rates of photosynthesis at high temperatures. Current evidence indicates that as temperature increases, there is an increased rate of Rubisco inactivation and a reduced activity of the activase, which is quite temperature-labile compared to other photosynthetic enzymes.

HOW IS THE ACTIVASE (AND THUS RUBISCO ACTIVITY) REGULATED?

In most plants surveyed to date, Rubisco activase comprises two isoforms of 41–43 kDa and 45–46 kDa.[1,3] These usually arise from one or more nuclear genes via alternative splicing, and after import and processing of the intermediate forms containing a chloroplast transit peptide. The larger isoform contains an additional domain at the carboxy-terminus, including two cysteine residues. These cysteines provide the ability to regulate the ATPase and Rubisco activation activities via light-induced redox changes mediated by thioredoxin-f, as shown by studies of Rubisco activation in transgenic *Arabidopsis* plants expressing mutant forms of the activase.[9] Oxidation/reduction of these cysteines alters the response of the ATP hydrolysis and Rubisco activation activities to the ATP/ADP ratio in the chloroplast, such that the activase consumes little ATP in the dark when Rubisco activity is not needed. Regulation of the activase via the ATP/ADP ratio in the chloroplast also appears to provide a means to regulate Rubisco activity in response to source-sink changes that alter the demand for the products of photosynthesis. However, several species appear to contain only the smaller isoform, yet the regulation of Rubisco appears to be quite similar to that in the species with both isoforms.[10] Obviously, understanding the regulation of Rubisco in these plants will require more study.

CONCLUSION

Rubisco activase is representative of a rapidly growing class of chaperone-like proteins that are specifically tailored to cover a diverse range of cellular activities requiring alterations of molecular or macromolecular structure. The target of the activase is Rubisco, for which the activase overcomes various problems Rubisco has in carrying out its functions and provides a means to regulate Rubisco activity in response to various environmental factors. Whether or not photosynthesis can be improved by modifying the regulation of Rubisco via the activase is currently being investigated. Meanwhile, there clearly is considerable potential for modifying Rubisco or expressing foreign forms of the enzyme in order to improve crop productivity. In order to achieve this goal, proper interaction between Rubisco and the activase will have to be maintained.

ARTICLES OF FURTHER INTEREST

ATP and NADPH, p. 68
Crop Responses to Elevated Carbon Dioxide, p. 346

REFERENCES

1. Spreitzer, R.J.; Salvucci, M.E. RUBISCO: Structure, regulatory interactions, and possibilities for a better enzyme. Annu. Rev. Plant Biol. **2002**, *53*, 449–485.
2. Ogura, T.; Wilkinson, A.J. AAA$^+$ superfamily ATPases: Common structure—Diverse function. Genes to Cells **2001**, *6*, 575–597.
3. Portis, A.R., Jr. Rubisco activase—Rubisco's catalytic chaperone. Photosynth. Res. **2003**, *75*, 11–27.
4. Duff, A.P.; Andrews, T.J.; Curmi, P.M.G. The transition between the open and closed states of Rubisco is triggered by the inter-phosphate distance of the bound bisphosphate. J. Mol. Biol. **2000**, *298*, 903–916.
5. Edmondson, D.L.; Badger, M.R.; Andrews, T.J. Slow inactivation of ribulosebisphosphate carboxylase during catalysis is caused by accumulation of a slow, tight-binding inhibitor at the catalytic site. Plant Physiol. **1990**, *93*, 1390–1397.
6. Salvucci, M.E.; Portis, A.R., Jr.; Ogren, W.L. A soluble chloroplast protein catalyzes ribulosebisphosphate carboxylase/oxygenase activation in vivo. Photosynth. Res. **1985**, *7*, 193–201.
7. Mate, C.J.; von Caemmerer, S.; Evans, J.R.; Hudson, G.S.; Andrews, T.J. The relationship between CO_2-assimilation rate, Rubisco carbamylation and Rubisco activase content in activase-deficient transgenic tobacco suggests a simple model of activase action. Planta **1996**, *198*, 604–613.
8. Crafts-Brandner, S.J.; Salvucci, M.E. Rubisco activase constrains the photosynthetic potential of leaves at high temperature and CO_2. Proc. Natl. Acad. Sci. U. S. A. **2000**, *97*, 13430–13435.
9. Zhang, N.; Kallis, R.P.; Ewy, R.G.; Portis, A.R., Jr. Light modulation of Rubisco in *Arabidopsis* requires a capacity for redox regulation of the larger Rubisco activase isoform. Proc. Natl. Acad. Sci. U. S. A. **2002**, *99*, 3330–3334.
10. Ruuska, S.A.; Andrews, T.J.; Badger, M.R.; Price, G.D.; von Caemmerer, S. The role of chloroplast electron transport and metabolites in modulating Rubisco activity in tobacco. Insights from transgenic plants with reduced amounts of cytochrome *b/f* complex or glyceraldehyde 3-phosphate dehydrogenase. Plant Physiol. **2000**, *122*, 491–504.

Secondary Metabolites as Phytomedicines

Donald P. Briskin
University of Illinois, Urbana, Illinois, U.S.A.

INTRODUCTION

Throughout history, plants have provided a rich source for the development of human medicines. Given rising demand for phytomedicines, production of medicinal plants as alternative crops could provide important new opportunities in agriculture. This article briefly discusses the fundamental aspects of phytomedicinal chemical production by plant cells and surveys eight medicinal plants that have received considerable attention over the past decade.

HISTORY

From the earliest times of human history to the beginning of the 20th century, plant medicinal products represented a significant component in conventional medicine, but their use declined with the development of modern pharmaceuticals (e.g., aspirin and quinine) containing pure chemical compounds. Many of these modern pharmaceuticals may have been based on active chemicals isolated from plants, and the development of synthetic or semisynthetic derivatives led to drugs with a consistent dose and even higher levels of potency. Nevertheless, the enhanced dose of one or a few active chemicals in these modern pharmaceuticals frequently resulted in problematic side effects, and these modern pure drugs were often expensive.[1]

Over the past decade there has been a strong resurgence in interest in and use of medicinal plants and phytomedicines, especially in North America. Recent surveys of phytomedicinal use by the American public have shown an increase from about 3% of the population in 1991 to over 37% in 1998.[2,3] By 2002, the North American market for plant medicinal products had reached about $4.1 billion/year.[3] No doubt a major factor contributing to this great increase in phytomedicinal use in the United States has been the passage of federal legislation in 1994—the Dietary Supplement Health and Education Act (DSHEA)—that facilitated the production and marketing of phytomedicinal products.[1,2]

PLANT SECONDARY METABOLITES AND PHYTOMEDICINES

The pharmacological actions of plant materials typically result from combinations of secondary metabolites that are present in the plant. The medicinal actions of plants are frequently unique to particular plant species; the combinations of secondary metabolites in a particular plant species or family are often taxonomically distinct.[4,5] This is in contrast to primary metabolic products—such as carbohydrates, lipids, proteins, and nucleic acids—that are common to all plant species, and are involved in the fundamental biochemical processes of building and maintaining plant cells.[5]

Although secondary products can have a variety of functions in plants, it is likely that their ecological roles may have some bearing on potential medicinal effects for humans. For example, secondary products involved in plant defense through cytotoxicity toward microbial pathogens could prove useful as an antimicrobial phytomedicine, if not too toxic to humans. Likewise, secondary products involved in defense against herbivores through neurotoxin activity could have beneficial effects in humans (i.e., as antidepressants, sedatives, muscle relaxants or anesthetics) through their action on the central nervous system. In order to fulfill functions in promoting the ecological survival of plants, molecular structures of secondary products have evolved to interact with molecular targets affecting the cells, tissues, and physiological functions in other competing microrganisms, plants, and animals.[4,5] In this respect, some plant secondary products may exert their action by resembling endogenous metabolites, ligands, hormones, signal transduction molecules, or neurotransmitters, and thus have beneficial medicinal effects on humans due to similarities in their potential target sites.[5]

BENEFITS OF CHEMICAL SYNERGISMS

In contrast to synthetic pharmaceuticals based on single chemicals, many phytomedicinals exert their beneficial effects through the action of several chemical compounds acting additively or synergistically at single or multiple

Encyclopedia of Plant and Crop Science
DOI: 10.1081/E-EPCS 120005947

Table 1 Average sales of specific medicinal plants in the United States

Medicinal plant	% Average of total sales
Ginkgo biloba	20.8
St. John's wort (*Hypericum perforatum*)	17.1
Ginseng (*Panax ginseng* and *P. quinguefolium*)	12.6
Garlic (*Allium sativum*)	11.3
Echinacea sp./Goldenseal (*Hydrastis canadiensis*)	9.7
Saw palmetto (*Serenoa serrulata*)	5.5
Kava (*Piper methysticum*)	2.5
Valerian (*Valeriana officinalis*)	1.4

(Based on data from Ref. 6.)

target sites associated with a physiological process. As pointed out by Tyler,[1] this synergistic or additive pharmacologic effect can promote pharmacological effectiveness without the problematic side effects associated with the predominance of a single xenobiotic compound in the body. In this respect, Kaufman et al.[5] extensively document how synergistic interactions underlie the effectiveness of a number of phytomedicines. This theme of multiple chemicals acting in an additive or synergistic manner likely has its origin in the functional role of secondary products in promoting plant survival.[4,5] For example, in the role of secondary products as defense chemicals, a mixture of chemicals having additive or synergistic effects at multiple target sites would not only ensure effectiveness against a wide range of herbivores or pathogens, but would also decrease the chances of these organisms developing resistance or adaptive responses.[4,5]

MEDICINAL USES AND ACTIVE PHYTOMEDICINAL CHEMICALS OF SEVERAL FREQUENTLY USED MEDICINAL PLANTS

Whereas there are thousands of medicinal plants utilized in Western and non-Western medical approaches, a relatively small number have been the subject of considerable interest in the United States over the past decade.[1,2] As shown in Table 1, eight medicinal plants have tended to dominate the total market for U.S. sales of medicinal plant products since 1998.[2,3,6] In general, the medicinal plants shown in Table 1 have remained within the top 10–15 plants dominating medicinal plant sales, although their percentage of the market has varied from year to year.[6]

Due to the extensive nature of their use in both North America and Europe, these eight medicinal plants have been the subject of considerable study with respect to their

Table 2 Medicinal properties and active phytomedicinal chemicals of eight medicinal plants used in the United States

Plant	Medicinal use	Active phytochemcals	Mode of action
Ginkgo biloba	Improves flow of blood to the brain and extremeties	Gingolides A, B, C Bilobilide	Improves capillary wall elasticity via antioxidant activity
St. John's wort	Antidepressant	Hypericin, Pseudo-Hypericin, Hyperforin, Flavones	Increases serotonin levels via inhibition of reuptake and catabolism
Ginseng	Improves response to stress, increases energy	Ginsenosides	Unknown but may relate to effects involving the adrenal glands
Garlic	Decreases blood cholesterol, lowers blood pressure	Allicin	Decreases cholesterol by inhibiting HMG-CoA reductase, blood pressure decreasing effect unknown
Echinacea	Decreases duration and severity of upper-respiratory infections	Echinacoside, Cichoric Acid, Isobutylamines, Alkylamides	Increases blood levels of several types of white blood cells, increase phagocytic activity of monocytes
Saw palmetto	Reduces symptoms of benign prostatic hyperplasia	Phytosterols present in the berries of this palm tree	Inhibition of 5α-reductase leading to a decrease in dihydrotesterone levels in the prostate
Kava	Treats insomnia and anxiety	Kavapyrones	Binding to GABAa receptors in a similar manner to benzodiazepine drugs (ex. valium)
Valerian	Treats insomnia and anxiety	Valeopotriates (iridoids), Alkaloids?	Stimulation of GABA release and inhibition of GABA reuptake

biochemical characteristics and pharmacological properties. For each of these plants, Table 2 summarizes their possible medicinal actions, active phytomedicinal chemicals that have been identified, and pharmacological mode of action, if known. Detailed information regarding the phytochemistry, pharmacology, and toxicology of these medicinal plants can be found in works by Schultz et al.[7] and Cupp.[8]

CONCLUSION

Although medicinal plants have had long-standing use throughout human history and are of considerable interest as alternatives to synthetic pharmaceuticals, there is a paucity of basic knowledge and research on their physiology and biochemistry. It is clear that few widely used medicinal plants have received the extensive physiological, biochemical, and genetic characterization applied to food crops or model plant systems such as *Arabidopsis*. Although active chemicals have been identified in many medicinal plants, the pathways for their biosynthesis and the genetic and environmental factors regulating their biochemical production are in many cases unclear.

At present, a major concern over the use of phytomedicines regards the maintenance of consistent medicinal quality in botanical medicines.[9] Whereas the focus has tended to be on quality control in herbal manufacturing practices (good manufacturing practices (GMP)), variation in phytomedicinal content due to environmental effects on secondary plant metabolism in plant material can also represent a significant factor in determining the quality of the plant material entering the botanical medicine production process and the efficacy of the resulting product. The use of molecular and biotechnological approaches to medicinal plants would also have wide application and promise, especially with regard to metabolic engineering of phytomedicinal chemical pathways and the in vitro production of phytomedicinals in large-scale tissue culture systems such as bioreactors.[10]

ARTICLES OF FURTHER INTEREST

Echinacea: Uses as a Medicine, p. 395
Herbs, Spices, and Condiments, p. 559
Metabolism, Secondary: Engineering Pathways of, p. 720
New Secondary Metabolites: Potential Evolution of, p. 818

REFERENCES

1. Tyler, V.E. Phytomedicines: Back to the future. J. Nat. Prod. **1999**, *62* (11), 1589–1592.
2. Glaser, V. Billion-dollar market blossoms as botanicals take root. Nat. Biotechnol. **1999**, *17* (1), 17–18.
3. Molyneaux, M. Consumer attitudes predict upward trends for the herbal market place. Herbalgram **2002**, *54*, 64–65.
4. Wink, M. Introduction: Biochemistry, Role and Biotechnology of Secondary Products. In *Biochemistry of Secondary Product Metabolism*; Wink, M., Ed.; CRC Press: Boca Raton, FL, 1999; 1–16.
5. Kaufman, P.B.; Cseke, L.J.; Warber, S.; Duke, J.A.; Brielmann, H.L. *Natural Products from Plants*; CRC Press: Boca Raton, FL, 1999.
6. Blumenthal, M. Herb sales down 3% in mass market retail stores—Sales in natural food stores still growing, but at a lower rate. Herbalgram **2000**, *49*, 68.
7. Schultz, V.; Hänsel, R.; Tyler, V. *Rational Phytotherapy. A Physician's Guide to Herbal Medicine*; Springer-Verlag: Berlin, 1998.
8. Cupp, M.J. *Toxicology and Clinical Pharmacology of Herbal Products*; Humana Press: Totowa, NJ, 2000.
9. Matthews, H.B.; Lucier, G.W.; Fisher, K.D. Medicinal herbs in the United States: Research needs. Environ. Health Perspect. **1999**, *107* (10), 773–778.
10. Walton, N.J.; Alfermann, A.W.; Rhoades, J.C. Production of Secondary Metabolites in Cell and Differentiated Organ Cultures. In *Functions of Plant Secondary Metabolites and Their Exploitation in Biotechnology*; Wink, M., Ed.; CRC Press: Boca Raton, FL, 1999; 311–345.

Seed Banks and Seed Dormancy Among Weeds

S

Lynn Fandrich
Oregon State University, Corvallis, Oregon, U.S.A.

INTRODUCTION

The relationship between seed dormancy and success of a plant as an agricultural weed is significant. Holm et al.[1] estimated that there are about 250 significant weeds in world agriculture. If the top 40 weeds classified as "serious" are selected by ranking them according to the number of countries in which they are considered "serious" weeds, it can be shown, without exception, that each one of the 40 species has dormancy. If weed seed were forever dormant, or simply nondormant, the complexities of weed management would be greatly reduced. However, weed seeds vary widely with respect to degree, duration, and source of dormancy. The existence of large populations of weed seed with varying degrees and states of dormancy is the basis for the annual weed problem. Dormancy allows a weed seed to avoid germination when conditions are not appropriate to complete its life cycle and prolongs population survival in the form of a soil seed bank.

CHARACTERISTICS OF THE SOIL SEED BANK

All of the viable seed present on the surface and in the soil is usually described as the soil seed bank. The number of seeds in the seed bank is determined by the rate of input in the seed rain, minus the losses resulting from disease, predation, and germination. The seed bank consists of new seeds recently shed by a parent plant, as well as older seeds that have persisted in the soil for several years. All of these seeds will vary in depth of burial and state of dormancy. Only a small proportion of seeds, varying between and among species, germinate when conditions are favorable. It is generally assumed that annual species contribute up to 95% of the seeds in the agricultural seed bank and only 2 to 10% of weed seeds emerge as seedlings each year.

Because most weeds are capable of producing a high number of seeds, especially compared to cultivated species, a few uncontrolled plants can rapidly increase the number of seeds in the soil seed bank. Data collected from plants grown in monoculture showed that wild oat (*Avena fatua*) produced 372–623 seeds,[2] common lambsquarters (*Chenopodium album*) produced 75,600–150,400 seeds,[3] and Palmer amaranth (*Amaranthus palmeri*) produced 200,000 to 600,000 seeds **per plant**.[4] It should be noted that biotic (inter- and intraspecific competition) and abiotic (climate) factors affect seed production, and most plants in agricultural situations do not often approach these high seed production values.

Large numbers of seeds are dispersed both horizontally and vertically in the soil. Plowing, chiseling, disking, and other forms of tillage are major ways that seeds are moved and covered with soil. After a number of years of soil disturbance, a relatively stable vertical distribution of seeds is reached in the tillage zone. Early estimates of seed numbers in agricultural soils exceeded 100,000 seeds per hectare in the tillage layer. Samples of soil on well-managed farms in Minnesota averaged 16 million seeds per hectare, and as many as 45 million per hectare have been counted.

PERIODICITY AND INTERMITTENT GERMINATION

Most weed species exhibit annual periodicity in germination and emergence restricted to certain times of the year. This behavior was part of the original evidence for annual cycles in the relief and imposition of seed dormancy. Observations on seed germination and seedling emergence from soil samples placed in the greenhouse revealed that there was a typical periodicity, as well as a period of maximum germination for more than 100 weed species.[5] Exhumed seeds of corn spurry (*Spergula arvensis*) showed clear seasonal changes in dormancy during three successive years of germination tests.[6] Often, seeds are conditionally dormant, i.e., the appropriate conditions for germination are not present; seeds of winter annual species will not germinate in late spring, and spring annual species will not germinate in late fall.

WEED SEEDS AND DEPTH OF BURIAL

Seed germination and emergence of seedlings from various depths below the soil surface depend on the degree of dormancy present and conditions of the immediate

Encyclopedia of Plant and Crop Science
DOI: 10.1081/E-EPCS 120020236

environment. However, dormancy-breaking stimuli are most effectively encountered and germination is rapid when seeds remain at or near the soil surface. There is an overall agreement that the number of seedlings and rate of seedling emergence decrease proportionately with increases in the depths of seed burial. Examination of seeds recovered from 12 cm deep showed that approximately 85% were completely dormant.[7] The results of these investigations also revealed a strong association between seed mass and maximum depth of emergence, with the heaviest seeds emerging from the greatest depths.

PERSISTENCE OF WEED SEEDS IN SOIL

It is quite common for seeds to be dormant when they are fully mature on the parent plant and for dormancy to be lost after they are shed. The term ''after-ripening'' has frequently been used to denote the interaction between environment and seed over time that leads to a loss of dormancy.[8] The length of the after-ripening period varies between and among species and by seed location on the plant. Most weed species show one or two periods of high percent germination in seed produced during any given year, and the percent germination in subsequent years can be quite low. The mean length of the dormancy period in populations drives turnover in the seed bank. Grime[9] makes a distinction between ''transient'' (complete turnover in less than one year) and ''persistent'' seed banks (some seed remains in the bank longer than one year). The term ''aging'' is sometimes used incorrectly to describe after-ripening. Aging reduces seed viability and seedling vigor.

ENVIRONMENTAL FACTORS THAT AFFECT SEED DORMANCY

Temperature

Temperature is the major environmental factor to cause a change in seed dormancy. Temperatures may break dormancy but also send seeds into dormancy when other environmental conditions are limited. Seeds of summer and winter annual weeds may be conditionally dormant due to temperature influences. Experimentally, the effect of chilling is tested through exposure of imbibed seeds to low temperatures, usually around 0 to 1°C, for a substantial period. Seeds of bur buttercup (*Ranunculus testiculatis*) germinated only at very cold incubation temperatures, and maximum germination occurred at 5°C.[10]

Seeds of winter annual species must be exposed to high summer temperatures for several months to germinate at fall temperatures. If lack of moisture prevents seeds of obligate winter annuals from germinating in the fall, low winter temperatures may induce dormancy. Consequently, viable seeds that fail to germinate in the fall cannot germinate in spring because they are dormant. Seeds of the winter annual Japanese brome (*Bromus japonicus*) were forced into dormancy with winter temperatures and would not germinate during the spring and summer, despite favorable germination conditions.[11]

Under natural conditions in the soil, seeds are subjected to fluctuating temperatures. It has been observed that random fluctuations in temperature, in and above a low temperature, are as effective in promoting germination as exposure to constant low temperatures. It has also been found that seeds which remain dormant at constant temperatures may be induced to germinate by a diurnal fluctuation of temperatures. Thompson and Grime[12] investigated the effect of fluctuating temperatures on the germination of 112 species of herbaceous plants. Forty-six of the species examined were found to have their germination stimulated by temperature fluctuations in the light.

Light

Exposure to light breaks dormancy in many weed species, but there may be situations in which light has no effect or even inhibits germination. Baskin and Baskin[13] observed that among 142 species, germination of 107 was promoted by light, 32 were unaffected by the difference in light and dark conditions, and 3 were inhibited by light. Wesson and Wareing[14] concluded that ''under natural conditions in the field, the germination of buried seeds following a disturbance of the soil is completely dependent upon exposure of seed to light.'' Even a short exposure or ''light break'' was sufficient to significantly increase the total number of seedlings emerging from soil.

CONCLUSION

Understanding the processes of dormancy loss and germination is essential to the development of good weed management programs. Currently, most management programs aim to control emerged weed populations and affect the quantity of seed returned to the soil, but it should be recognized that almost all agronomic practices also affect weed seed dormancy and germination through manipulation of the soil environment. Because many agricultural weeds exhibit periodicity of emergence, tillage treatments, planting dates, and other cultural practices can be altered to follow peak germination times and maximize depletion of the seed bank. A natural population of viable weed seeds in the field declined at a rate of approximately 50% per year with frequent cultivation.[15] A review of the utilization of specific agronomic practices to manipulate weed seed dormancy and germination requirements for weed management is available from Dyer[16] and others.[17,18]

ARTICLES OF FURTHER INTEREST

Ecology: Functions, Patterns, and Evolution, p. 404
Seed Dormancy, p. 1130

REFERENCES

1. Holm, L.; Pancho, J.V.; Herberger, J.P.; Plucknett, D.L. *A Geographical Atlas of World Weeds*; John Wiley and Sons: New York, 1979.
2. O'Donnell, C.C.; Adkins, S.W. Wild oat and climate change: The effect of CO_2 concentration, temperature, and water deficit on the growth and development of wild oat in monoculture. Weed Sci. **2001**, *49*, 694–702.
3. Colquhoun, J.; Stoltenberg, D.E.; Binning, L.K.; Boerboom, C.M. Phenology of common lambsquarters growth parameters. Weed Sci. **2001**, *49*, 177–183.
4. Keeley, P.E.; Carter, C.H.; Thullen, R.J. Influence of planting date on growth of Palmer amaranth (*Amaranthus palmeri*). Weed Sci. **1987**, *35*, 199–204.
5. Brenchley, W.E.; Warrington, K. The weed seed population of arable soil. I. Numerical estimation of viable seeds and observations on their natural dormancy. J. Ecol. **1930**, *18*, 235–272.
6. Bouwmeester, H.J.; Karssen, C.M. The effect of environmental conditions on the annual dormancy pattern of seeds of *Spergula arvensis*. Can. J. Bot. **1993**, *71*, 64–73.
7. Roberts, H.A. Studies on the weeds of vegetable crops. II. Effect of six years cropping on the weed seeds in the soil. J. Ecol. **1962**, *50*, 803–813.
8. Simpson, G.M. *Seed Dormancy in Grasses*; Cambridge University Press: Cambridge, England, 1990.
9. Grime, J.P. The role of seed dormancy in vegetation dynamics. Ann. Appl. Biol. **1981**, *98*, 555–558.
10. Young, J.A.; Martens, E.; West, N.E. Germination of bur buttercup seeds. J. Range Manag. **1992**, *45*, 358–362.
11. Baskin, J.M.; Baskin, C.C. Ecology of germination and flowering in the weedy winter annual grass *Bromus japonicus*. J. Range Manag. **1981**, *34*, 369–372.
12. Thompson, K.; Grime, J.P. A comparative study of germination responses to diurnally-fluctuating temperatures. J. Appl. Ecol. **1983**, *20*, 141–156.
13. Baskin, C.C.; Baskin, J.M. Germination ecophysiology of herbaceous plant species in a temperate region. Am. J. Bot. **1988**, *75*, 286–305.
14. Wesson, G.; Wareing, P.F. The role of light in the germination of naturally occurring populations of buried weed seeds. J. Exp. Bot. **1969**, *20*, 402–413.
15. Burnside, O.C.; Wilson, R.G.; Wicks, G.A.; Roeth, F.W.; Moomaw, R.S. Weed seed decline and buildup in soils under various corn management systems across Nebraska. Agron. J. **1986**, *78*, 451–454.
16. Dyer, W.E. Exploiting weed seed dormancy and germination requirements through agronomic practices. Weed Sci. **1995**, *43*, 498–503.
17. Forcella, F. Real-time assessment of seed dormancy and seedling growth for weed management. Seed Sci. Res. **1998**, *8*, 201–209.
18. Mulugeta, D.; Stoltenberg, D.E. Weed and seedbank management with integrated methods as influenced by tillage. Weed Sci. **1997**, *45*, 706–715.

Seedborne Pathogens

S. B. Mathur
Danish Government Institute of Seed Pathology for Developing Countries, Copenhagen, Denmark

H. K. Manandhar
Nepal Agricultural Research Council, Khumaltar, Lalitpur, Nepal

INTRODUCTION

Seedborne pathogens are infectious agents, such as bacteria, fungi, mollicutes, nematodes, viroides, and viruses, that are associated with seeds and which can cause diseases in seeds, seedlings, and plants. These pathogens are capable of causing spots and discoloration in seed coat, change in seed size, loss in seed germination, and reduction in seedling vigor and of producing symptoms often leading to seedling death. Through seeds, such pathogens get transmitted to plants, causing symptoms and ultimately affecting production of seeds in terms of both quality and quantity. Seedborne pathogens can completely or partly replace seeds, such as occurs in ergot, smuts, bunts, and nematode galls.

Some seedborne pathogens can be exclusively seedborne and seed transmitted, while others can be both seedborne and at the same time soilborne or present in collateral hosts. Seedborne pathogens are found in seed lots as contaminations, present on the surface of the seed and/or located in the seed coat, pericarp, endosperm, and embryo. They get transmitted to plants of the next generation systemically, locally, or both locally and systemically.

Many seedborne pathogens are economically important as they are able to cause appreciable losses in crop yield. The losses are enumerated in several large publications that readers should consult because they give comprehensive accounts of different types of seedborne pathogens, mechanisms of seed infection, seed transmission, spread, survival in seeds, and control of diseases caused by such pathogens. Some crop-oriented publications are also of importance. The electronic version of the Crop Protection Compendium (2002) of CAB International provides useful information on a number of aspects of seedborne pathogens that cause diseases in agricultural, horticultural, and industrial crops, medicinal plants, and forest tree species.

IMPORTANCE OF SEEDBORNE PATHOGENS

Seedborne pathogens are important because they affect the planting value of a seed lot and the spread of the disease as seeds get transported from infested areas to noninfested areas. This is also a mechanism by which new races, strains, or pathotypes travel from one region to another region of a country and across international boundaries (Figs. 1–4).

The importance of seedborne pathogens is realized when infected seeds are sown in the field. Depending on seeding rate, even a trace seed infection can bring thousands of infected seeds in a one-hectare area of land, as seen in Table 1. These large numbers of infected seeds, when distributed uniformly in a randomized fashion in the field, increase their role in disease development.

Seedborne pathogens, including some commonly occurring saprophytic organisms, are able to spoil seeds and grains by producing toxins, especially in bad storage conditions. Such spoilage makes the grains unfit for human and animal consumption.

DETECTION OF SEEDBORNE PATHOGENS

Laboratory test methods are available for detecting seedborne pathogens. A recently published book[4] and Seed Health Testing Methods, Annex to Chapter 7 of International Rules for Seed Testing, Edition 2003, published by International Seed Testing Association (ISTA), are essential resources for detecting fungi. Most of the sporulating fungi are detected by incubation methods (Fig. 5). Some fungi, which are located in the embryo, are detected only by the embryo count method.[4]

Bacterial pathogens can be detected by plating of seeds on agar media; liquid assays; growing plants for symptom

Encyclopedia of Plant and Crop Science
DOI: 10.1081/E-EPCS 120021911

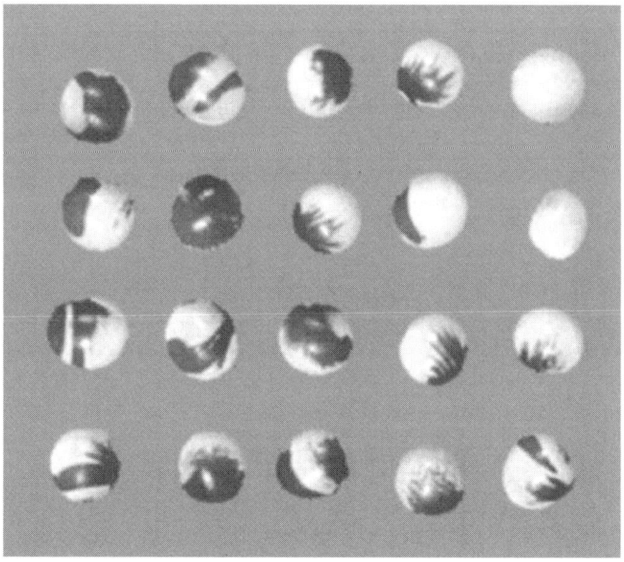

Fig. 1 Mottling of soybean seeds induced by soybean mosaic potyvirus. (*View this art in color at www.dekker.com.*)

development; biochemical, serological, and molecular tests; and using bacteriophages. Viruses are detected by examining dry seeds, germinating seeds, or seedlings raised from seeds, followed by indicator plant test, serology, molecular tests, and electron microscopy.

Most nematodes present in seeds can be extracted using a simple Baermann-funnel technique and by direct inspection of seeds for wheat gall nematode.

CONTROL OF SEEDBORNE PATHOGENS

Pathogens found in seeds can be controlled using physical, chemical, and biological means. In general, it is advisable to grow crops by using, as far as possible, pathogen-free

Fig. 2 Chickpea seeds with reduced size (right) collected from plants infected by *Fusarium oxysporum*, a wilt fungus.[1]

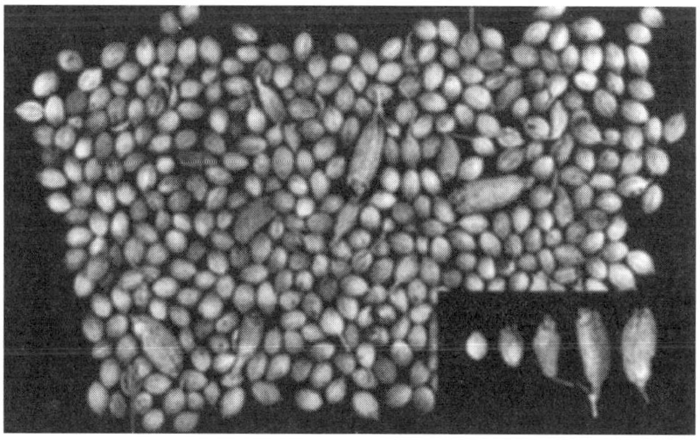

Fig. 3 Hypertrophied seed caused by *Protomyces macrosporus*, mixed with normal seeds of coriander, and hypertrophies of various grades (inset).

seeds, preferably of resistant cultivars, or seeds collected from fields and areas where either specific diseases do not occur or the disease pressure is low.

The choice of control measures depends on the types and nature of pathogens. Those pathogens that are exclusively seedborne are easy to control by appropriate seed treatments. The control, however, is difficult when the pathogens are both seedborne and soilborne or when they are also present in collateral hosts (weeds and hosts other than primary hosts).

As seeds infected by pathogens are often discolored, spotted, and lighter in weight, they can be removed to a great extent by seed-cleaning machines, by soaking in brine solution, by hand (manually), or by winnowing.[5] Clean seeds thus obtained give seedlings of higher vigor and better plant stand (Fig. 6), and the yield from such plants is higher.

Fig. 4 Ears of wheat infected with loose smut fungus, *Ustilago tritici*: (A) a partially smutted ear; (B and C) completely smutted ears.[2] (*View this art in color at www.dekker.com.*)

Table 1 Expected number of infected seeds to enter in a field of one hectare that are capable of producing disease

	Seeding rate					
Percent seed infection	**125 thousand seed/ha**	**250 thousand seed/ha**	**500 thousand seed/ha**	**1 million seed/ha**	**2 million seed/ha**	**5 million seed/ha**
0.1	125	250	500	1,000	2,000	5,000
0.5	625	1,250	2,500	5,000	10,000	25,000
1.0	1,250	2,500	5,000	10,000	20,000	50,000
2.0	2,500	5,000	10,000	20,000	40,000	100,000
5.0	6,225	12,500	25,000	50,000	100,000	250,000
10.0	12,500	25,000	50,000	100,000	200,000	500,000

Usually, the seeding rate is expressed in kg per hectare, but the number of seeds per gram can be calculated in the laboratory. This will help in calculating number of seeds per hectare.
The above table can be expanded. It will help in knowing the number of seeds that will enter a field, but the rate of transmission of seedborne pathogens will depend on the nature of the pathogen, susceptibility of the host, soil conditions, environmental conditions, cultural practices, etc.
(From Ref. 3.)

CONCLUSION

Seedborne pathogens are one of the major constraints in agriculture as they can affect physical appearance and nutritional value of seeds or grains, spoil grains by producing toxins, reduce germination, cause death of seedlings and plants, act as initial inoculum for disease development in the field, and move with seeds from infested areas to noninfested areas. Therefore, seedborne pathogens must receive due consideration in agriculture across the globe. Seeds must be tested before sowing by internationally accepted seed health test methods. Infected seeds must be removed as far as possible from sowing or should be sown only after proper seed treatments.

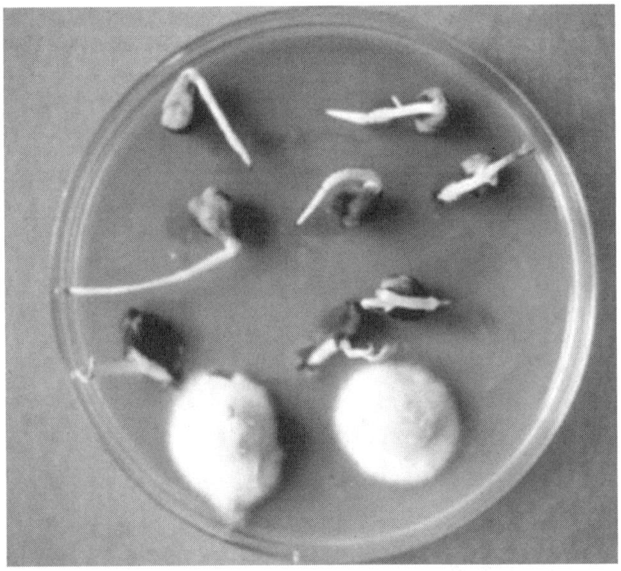

Fig. 5 White cottony growth of *Fusarium oxysporum*, a wilt fungus, usually covering the whole seed of chickpea on Czapek Dox agar (agar test).[1] (*View this art in color at www.dekker.com.*)

ARTICLES OF FUTHER INTEREST

Bacterial Pathogens: Detection and Identification Methods, p. 84
Mycotoxins Produced by Plant Pathogenic Fungi, p. 773
Noninfectious-Seed Disorders, p. 1825
Seed Production, p. 1134

Fig. 6 Rice seedlings grown from differently looking seeds. (*View this art in color at www.dekker.com.*)

Seed Vigor, p. 1139
Seeds: Pathogen Transmission Through, p. 1142

REFERENCES

1. Agarwal, P.C.; Mortensen, C.N.; Mathur, S.B. *Seed-Borne Diseases and Seed Health Testing of Rice*; Technical Bulletin, Danish Government Institute of Seed Pathology for Developing Countries: Copenhagen, 1989; Vol. 3. Phytopathological Papers No. 30; CAB International Mycological Institute: Kew, Surrey.
2. McGee, D.C. *Maize Diseases—A Reference Source for Seed Technologists*; APS Press, The American Phytopathological Society: St. Paul, MN, 1988.
3. Mathur, S.B.; Cunfer, B.M. *Seed-Borne Diseases and Seed Health Testing of Wheat,* 1st Ed.; Jordbrugsforlaget: Frederiksberg, Denmark, 1993.
4. Mathur, S.B.; Kongsdal, O. *Common Laboratory Seed Health Testing Methods for Detecting Fungi,* 1st Ed.; International Seed Testing Association: Bassersdorf, CH-Switzerland, 2003.
5. Manandhar, H.K.; Jørgensen, H.J.L.; Smedegaard-Petersen, V.; Mathur, S.B. Seedborne infection of rice by *Pyricularia oryzae* and its transmission to seedlings. Plant Dis. **1998**, *82* (10), 1093–1099.
6. Agarwal, V.K.; Sinclair, J.B. *Principles of Seed Pathology,* 2nd Ed.; CRC Press, Inc.: Boca Raton, 1997.
7. Maude, R.B. *Seedborne Diseases and Their Control: Principles and Practice*; CAB International: Wallingford, 1996.
8. Neergaard, P. *Seed Pathology,* Revised Ed.; The MacMillan Press Ltd.: Houndmills, 1979; Vols. 1 and 2.
9. Chahal, S.S.; Thakur, R.P.; Mathur, S.B. *Seed-Borne Diseases and Seed Health Testing of Pearl Millet,* 1st Ed.; Kandrups Bogtrykkeri: Frederiksberg, Denmark, 1994.
10. McGee, D.C. *Soybean Diseases—A Reference Source for Seed Technologists*; APS Press, The American Phytopathological Society: St. Paul, MN, 1992.
11. Richardson, M.J. *An Annotated List of Seed-Borne Diseases,* 4th Ed.; The International Seed Testing Association: Zürich, 1990.

S

Seed Dormancy

Alistair J. Murdoch
The University of Reading, Reading, U.K.

INTRODUCTION

Seed dormancy is most easily observed and measured as the absence of germination in an imbibed, viable seed. Its intrinsic nature is, however, less easily defined physiologically. Almost all seeds go through a dormant phase if only to prevent viviparous germination on the mother plant. Many crop species have been selected for low dormancy to ensure rapid and uniform crop establishment. Residual dormancy, however, causes problems for seed germination testing, volunteers in crops, and low germination in the brewing industry.

Prevention of immediate germination through either dormancy or quiescence is a crucial element in the regeneration strategy of many wild plants. The success of such a strategy depends on quantitative and sometimes qualitative differences in dormancy within seed populations.

Stimulation of germination of dormant seeds frequently occurs as a function of fascinating interactions of environmental or chemical stimuli with the annual dormancy cycle of seeds in the soil. These interactions lead to the periodicity of emergence. Quantifying differences *within* seed populations with respect to environmental factors is the key to predictive models for the germination and emergence of dormant seeds.

DEFINING DORMANCY

Dormancy is usually defined in the way it is most often measured and observed—negatively:

> The failure of a viable seed to germinate given moisture, air and a suitable constant temperature for radicle emergence and seedling growth.[1,2]

Attempts to define dormancy positively include "a seed characteristic, the degree of which defines what conditions should be met to make the seed germinate,"[3] but how is the degree of an unspecified characteristic to be measured?

Seed dormancy is sometimes classified by the phenological stage when dormancy is induced or by the putative mechanism.[4] A distinction of particular value is that between primary (innate) and secondary (induced) dormancy. Primary dormancy develops during seed maturation on the mother plant, while secondary dormancy is induced after shedding.

DEFINING QUIESCENCE

Absence of one or more of the three prerequisites for germination of a nondormant seed (moisture, air, and a suitable temperature) usually reduces metabolism.[5] Nongermination of nondormant seeds can, therefore, arguably be called quiescence, again defined negatively:

> The failure of a viable seed to germinate due to shortage of water, poor aeration or an unsuitable temperature for radicle emergence and seedling growth (cf. Ref. 1).

Quiescence is frequently enforced by the environment as in air-dry seeds or in imbibed seeds below the base temperature for the rate of germination of nondormant seed.

Hardseededness, sometimes called physical dormancy,[4] is strictly innate quiescence. Best known in the Leguminosae, hard seeds, which are impermeable to water, occur in several other plant families.[1] Like innate dormancy, impermeability develops during maturation drying on the mother plant and it may increase after shedding in dry environments. Although hard seeds are easily made permeable by mechanical abrasion, some of the greatest longevities of seeds in the soil are achieved by such innately quiescent, hard seeds, which possess no primary dormancy.

MEASURING DORMANCY

Phytochrome is integrally involved in many photomorphological responses in plants, and seed germination is no exception.[5] It has been speculated that the phytochrome molecule in its active, far-red-absorbing form binds to an unidentified receptor protein and that dormancy might be quantified directly by the amount of this protein present in the seed.[3] Current research into the mechanisms and differences in gene expression between dormant and nondormant seeds is likely to revolutionize not only our

Encyclopedia of Plant and Crop Science
DOI: 10.1081/E-EPCS 120021907

understanding of seed dormancy, but perhaps also the way we measure it.

At the time of writing, however, dormancy is usually measured by nongermination of imbibed viable seeds with adequate aeration and at a suitable temperature for seedling growth. If V is percentage viability of a seed lot and G the percentage germination in such conditions, then the percentage dormancy is (V–G). Germination and viability are both quantal responses since an individual seed is either viable or inviable and it either germinates or does not. By inference, the expression (V–G) likewise but incorrectly quantifies dormancy as a quantal response since an individual seed has to be classified as either dormant or nondormant. In reality, it is only the expression of dormancy in the specific testing environment that is a quantal response.[1] Bradford[6] has proposed that dormancy in an individual seed is like

> "a hill that seeds approach from one side (i.e. as the requirements to break dormancy are met) and then progress down the other, gathering momentum (increasing germination rate)."

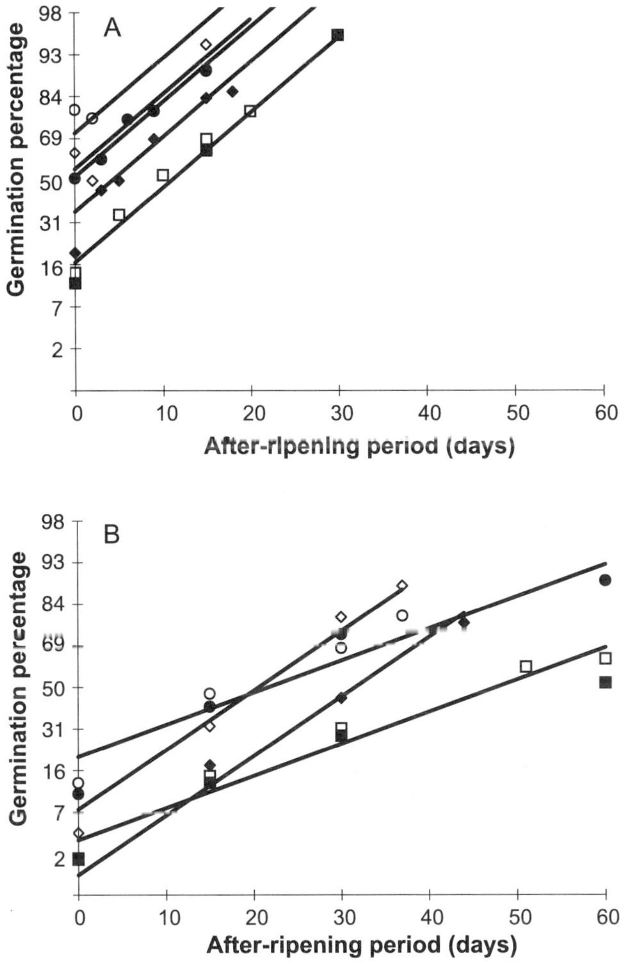

Fig. 1 After-ripening of A) caryopses and B) spikelets of *Cenchrus ciliaris* matured in different thermal, moisture, and fertility environments including 32°/27°C (13 h/11 h) drought stress (□), 27°/20°C drought stress (■), 32°/27°C no water stress (◇), 27°/20°C no water stress (◆), 32°/27°C nutrient (O), 27°/20°C nutrient (●) in Experiment 1. All plants were grown identically at 25°C until 5 d after anthesis on May 31, 1996. They were then transferred to maturation conditions with 13 and 11 h per day at upper and lower temperatures, respectively. Drought stress means a series of cycles in which watering was withheld until the first signs of wilting. Nutrient treatments were watered with a balanced solution containing 100 ppm nitrogen. Seeds were after-ripened as spikelets at 43% r.h., 40°C, and germinated in darkness at 25°C. Germination is plotted on a probability scale showing regression lines fitted by probit analysis. (From Ref. 7; reproduced with permission of CAB International.)

The realism of this view is seen easily when considering the response of seeds as a function of the dose of a dormancy-breaking stimulus. Nondormant seeds in the population should germinate upon imbibition. The dormant seeds vary in dormancy: Some may respond to after-ripening for one week while others may need two (Fig. 1). It is well stated: "We do not yet know how to measure the depth of dormancy in individual seeds but its variation is reflected in the seed-to-seed variation in the expression of dormancy in a seed population."[1]

Seed-to-Seed Variation in Dormancy

Unless 0% or 100% of seeds germinate, the classification of viable seeds into two groups—those that germinate and those that do not—may be thought to imply polymorphism. But polymorphism should be deduced only if there is a discontinuity in dormancy periods or when developmentally or morphologically different seeds vary in the depth of primary dormancy. Polymorphic seeds may be produced on the same or different plants. In the classic case of *Xanthium pensylvanicum*, the two seeds in each capsule are dispersed together. The upper seed is much more dormant and germinates at least 12 months later than the lower one.[2]

There is often, however, a continuum of variation in dormancy even within apparently uniform seed lots (Figs. 1 and 2). Similarly, seed-to-seed variation occurs in the dose of a dormancy-inducing and/or germination-inhibiting treatment required to induce secondary dormancy as in aerobic preconditioning of *Orobanche* seeds (Fig. 2).

This seed-to-seed variation of dormancy responses in seed populations often approximates to the normal frequency distribution. The germination–dose curve is therefore quantified by its mean and variance, and the curves are linearized by transforming the germination or

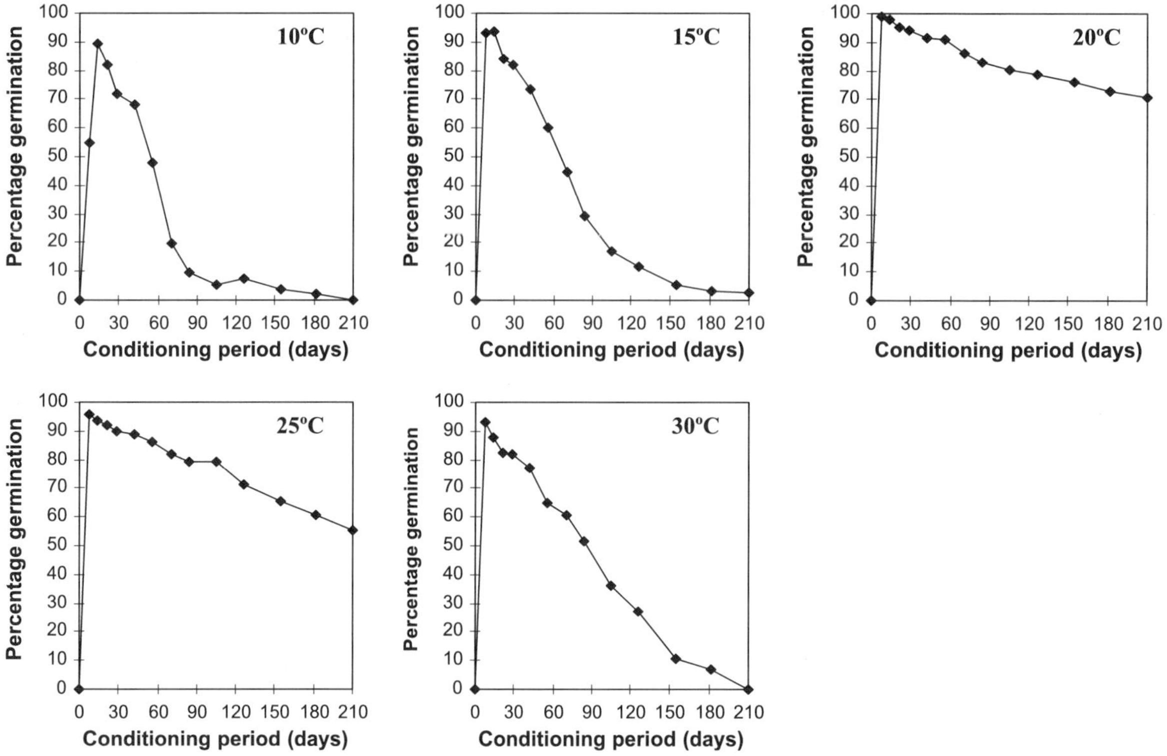

Fig. 2 Germination of *Orobanche aegyptiaca* seeds after conditioning (fully imbibed) for various periods at 10°C, 15°C, 20°C, 25°C, and 30°C. Seeds were surface-sterilized with sodium hypochlorite solution before conditioning for periods shown on the x-axis. Germination tests lasted ten days at 20°C with 3 ppm of GR24. GR24 is an analogue of Strigol—an exudate of cotton roots that stimulates germination of seeds of some parasitic plants. Loss of primary dormancy over the first 15 days (approx.) is followed by induction of secondary dormancy. Primary dormancy is lost most slowly and secondary dormancy induced most rapidly at 10°C. Secondary dormancy is induced most slowly above 20°C, the more rapid declines in germination at 25 and 30°C being due to loss of viability, which was most rapid at 30°C. (Drawn from data published in Ref. 8.)

dormancy percentages to probits (Fig. 1). This (probit) analysis gives a robust measurement of seed dormancy: The intercepts of the probit lines in Fig. 1 give far better estimates of dormancy levels of seeds produced in various maturation environments than the results of a single germination test. The reciprocal of the slope quantifies the standard deviation of the dormancy response to the applied treatment in the seed population, and these two parameters are the key to modeling responses of dormant seeds to environmental variables.[7,8]

PRIMARY DORMANCY

Primary dormancy may simply occur because the embryo is immature so that germination is delayed pending further development as in the linear embryos of the *Ranunculaceae* and *Umbelliferae*.[4] Other seeds are dormant due to a physiological block caused by the seed coat and/or the embryo.

Primary dormancy has two main functions: first, preventing precocious germination,[9] a response linked to the presence of abscisic acid (ABA) during seed maturation; and secondly, assisting the temporal dispersal of seeds by preventing their immediate, synchronous germination. Very few species can therefore reproduce effectively without seed dormancy. One exception is recalcitrant seeds from the tropics. These seeds can neither be dried nor be chilled below 10–15°C without losing viability. Primary dormancy is absent and germination occurs on or before release from the inhibitory effect of the enclosing maternal environment.

Heritability of dormancy is complex because parts of the seeds differ genetically. For example, a diploid embryo may receive nutrients from a triploid endosperm (with two maternal and one paternal sets of chromosomes) and is surrounded by maternal tissues (testa and fruit structures).

Primary dormancy not only varies with genotype but also with maturation environment; in barley (*Hordeum vulgare*) and *Cenchrus ciliaris* (a perennial grass that occurs in arid and semiarid environments), dormancy was greater for seeds produced under water stress, a characteristic which may be of adaptive value in plants growing in arid ecosystems (Fig. 1). The opposite effect of water stress occurs in *Avena fatua.*

Primary dormancy declines both prior to shedding and subsequently. When this loss of dormancy occurs in ''air-dry'' seeds, it is termed after-ripening (Fig. 1). The rate of after-ripening increases with temperature in a predictable manner, the Q_{10} for the relation usually being 2.5–3.8.[1]

SECONDARY DORMANCY

Dormancy may be induced in dormant and in ''quiescent'' nondormant seeds after shedding. Secondary dormancy is induced rapidly by anaerobiosis and more slowly by prolonged moist aerobic treatments in which germination does not occur. Induction of secondary dormancy may be delayed or prevented by intermittent low-intensity laboratory light and nitrate.[1] Rates of induction of secondary dormancy of *Orobanche* seeds were fastest at 10°C and *decreased* to a minimum above about 20°C (Fig. 2).[8] By contrast, rates of induction in imbibed *Rumex crispus* seeds *increased* with an increase in temperature.[1]

Are the mechanisms of primary and secondary dormancies the same? Dry after-ripening and the same chemicals may relieve both primary and secondary dormancy, implying some similarity. It is, however, interesting that while short periods of imbibition are associated with relief of primary dormancy, longer periods lead to secondary dormancy (Fig. 2). The final answer to this question may come when the expression of dormancy genes can be measured in seed lots varying in dormancy.

Annual cycles in which physiologically based dormancy is relieved and secondary induced during the course of a year occur in buried seeds of some annual plants in both temperate and tropical soil environments.[4] Exposure to light, nitrate, and temperature treatments may be needed to relieve residual primary and secondary dormancy at times of low dormancy. Understanding how these physical and chemical factors affect seed populations with different levels of dormancy is crucial if seedling emergence from persistent soil seed banks is to be predicted.[10]

ARTICLES OF FURTHER INTEREST

Cereal Sprouting, p. 250
Endosperm Development, p. 414
Seed Banks and Seed Dormancy Among Weeds, p. 1123
Seed Production, p. 1134
Seed Vigor, p. 1139

REFERENCES

1. Murdoch, A.J.; Ellis, R.H. Dormancy, Viability and Longevity. In *Seeds: The Ecology of Regeneration and Plant Communities*, 2nd Ed.; Fenner, M., Ed.; CAB International: Wallingford, UK, 2000; 183–214.
2. Amen, R.D. A model of seed dormancy. Bot. Rev. **1968**, *34*, 1–31.
3. Vleeshouwers, L.M. Modelling Weed Emergence Patterns. Ph.D. Thesis; Wageningen Agricultural University, The Netherlands, 1997.
4. Baskin, C.C.; Baskin, J.M. *Seeds: Ecology, Biogeography and Evolution of Dormancy and Germination*; Academic Press: San Diego, CA, 1998.
5. Bewley, J.D.; Black, M. *Seeds: Physiology of Development and Germination*; Plenum Press: New York, 1994.
6. Bradford, K.J. Population-Based Models Describing Seed Dormancy Behaviour: Implications for Experimental Design and Interpretation. In *Plant Dormancy: Physiology, Biochemistry and Molecular Biology*; Lang, G.A., Ed.; CABI: Wallingford, UK, 1996; 313–339.
7. Sharif-Zadeh, F.; Murdoch, A.J. The effects of different maturation conditions on seed dormancy and germination of *Cenchrus ciliaris*. Seed Sci. Res. **2000**, *10*, 447–457.
8. Kebreab, E.; Murdoch, A.J. A quantitative model for loss of primary dormancy and induction of secondary dormancy in imbibed seeds of *Orobanche* spp. J. Exp. Bot. **1999**, *50*, 211–219.
9. Bewley, J.D.; Downie, B. Is Failure of Seeds to Germinate During Development a Dormancy-Related Phenomenon? In *Plant Dormancy: Physiology, Biochemistry and Molecular Biology*; Lang, G.A., Ed.; CABI: Wallingford, UK, 1996; 17–27.
10. Murdoch, A.J. Dormancy cycles of weed seeds in soil. Asp. Appl. Biol. **1998**, *51*, 119–126.

Seed Production

Murray J. Hill
New Zealand Seed Technology Institute, Lincoln University, Canterbury, New Zealand

INTRODUCTION

In nature, seeds overcome three major problems for the plant. First, they are the method of multiplication, because a plant is often able to produce very large numbers of seeds. Second, seeds are a survival mechanism for plants because they often survive in the soil, even under adverse conditions, persisting until suitable germination conditions occur. Third, seeds help to disperse plants using agencies such as wind, water, or contact with animals or birds.

In agriculture, the aim of seed production is to multiply seed numbers by establishing a seed crop with sufficient plant density to produce seed to sow new plantings in the following season.

VEGETATIVE AND REPRODUCTIVE DEVELOPMENT

Plants are reproduced in two main ways for use in agriculture. Some are reproduced vegetatively (e.g., potatoes, cassava), although most produce seed by sexual reproduction (e.g., self-fertilization or cross-fertilization). A few crop species reproduce by apomixis, which is asexual but produces seedlike bodies that behave in all respects like normal seed (i.e., reproduction without fertilization)[1] (Fig. 1).

Before seeds can be produced the plant must have the chance to flower. Some plants pass through the vegetative stage to the reproductive phase with no special requirement or stimulus. In other plants, there is a clearly defined transition between the two phases. The early stage (before the plant is receptive to an external flowering stimulus) is the juvenile phase. Attainment of the required physiological phase (''puberty'') is related to plant size (to leaf numbers in particular). At this stage, plants are often stimulated to become reproductive by day length (photoperiod) and/or by exposure to low temperature (vernalization). As a result, there are plants with no required climate stimulus (most annuals) and those that require precise things to happen in terms of low temperature (most perennials) and day length (long-day, short-day, or day-neutral plants).[2] The timing and prediction of flowering in a range of crops is well discussed elsewhere.[2]

Flower morphology, pollen formation and pollination, and fertilization are well described in field crops by Copeland and McDonald[3] and in vegetable seed production by George.[4] A schematic summary of the seed production sequence is shown in Fig. 2.

SEED DEVELOPMENT

In seed crops the sequence of seed development occurs in three stages.[5]

- A growth stage, lasting about 10 days after pollination and characterized by rapid increase in cell division (but not elongation) in cell division, high moisture content, and nonviability (Fig. 3A).
- A food reserve accumulation stage, lasting 10–14 days and involving a great increase in seed dry weight that reaches a maximum at the end of the stage (Fig. 3B). The amount of water in seed changes little but the percentage water content falls rapidly. Seed becomes viable during this stage, which ends with seed becoming physiologically mature.
- A ripening stage, lasting 3–14 days, depending on climate. Seed dehydrates to a moisture content in equilibrium with the environment, resulting in seed that has reached harvest maturity.

SITE SELECTION AND CROP MANAGEMENT

Choice of a seed production site is a basic consideration. It is important because different areas vary in their suitability for seed production. The main requirements are to grow seed crops in areas that have a sufficiently long growing season and suitable day length and temperature to promote strong vegetative growth; strong stimulation of seed head development and flowering; and good climatic conditions during seed development, ripening, and harvesting.

In forage crops, plant breeders have been very successful in producing varieties that grow large amounts of herbage. This is done in many cases by breeding plants that remain vegetative (leaf producing) as long as possible before becoming reproductive (i.e., seed producing).

Encyclopedia of Plant and Crop Science
DOI: 10.1081/E-EPCS 120021910

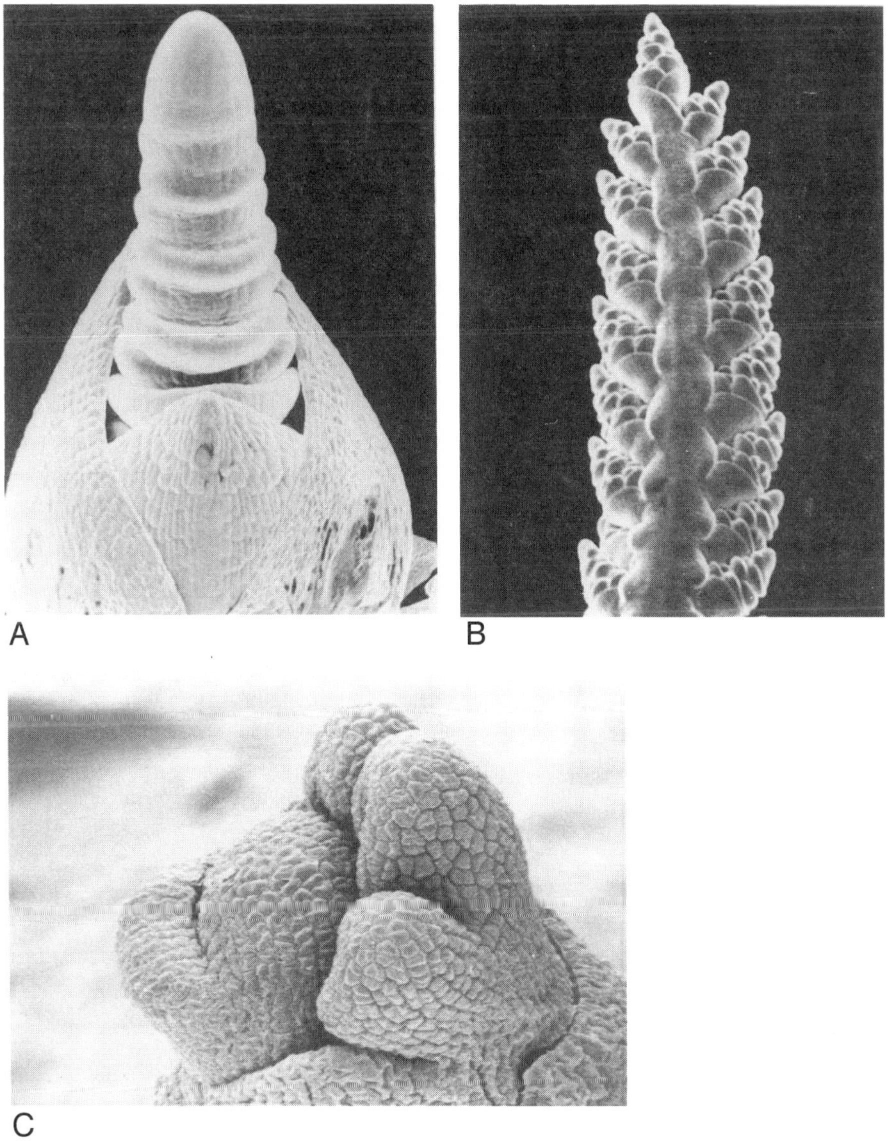

Fig. 1 Apical meristems of ryegrass and white clover. (A) Vegetative growing point of perennial ryegrass (*Lolium perenne* L.) showing leaf primordia ($\times$115). (B) Reproductive growing point of perennial ryegrass (*Lolium perenne* L.) showing spikelet and floret initiation ($\times$75). (C) Vegetative apical bud of white clover (*Trifolium repens* L.) showing expanding leaf initials and central dome ($\times$400).

The objective in specialist seed production systems is to grow large plants, producing flowers and then seeds at as many of the available sites as possible.

Growers who produce seed crops with high yield and quality tend to be those who manage their seed crops to force plants to become more strongly reproductive.[6] Seed management packages are recommended that involve judicious, well timed use of fertilizer (e.g., P and K in legumes and N in grasses); controlled use of water; cutting or grazing to encourage plant branching and provide more places on the plant for flower formation; adjustment of sowing date and rate to take advantage of differences in climate (temperature and day length); specialist herbicides to remove specific weeds; and plant growth regulating chemicals.

CROP ROTATION AND PADDOCK HISTORY

Time intervals between related or similar crops are standard agronomic practice. The many reasons for crop

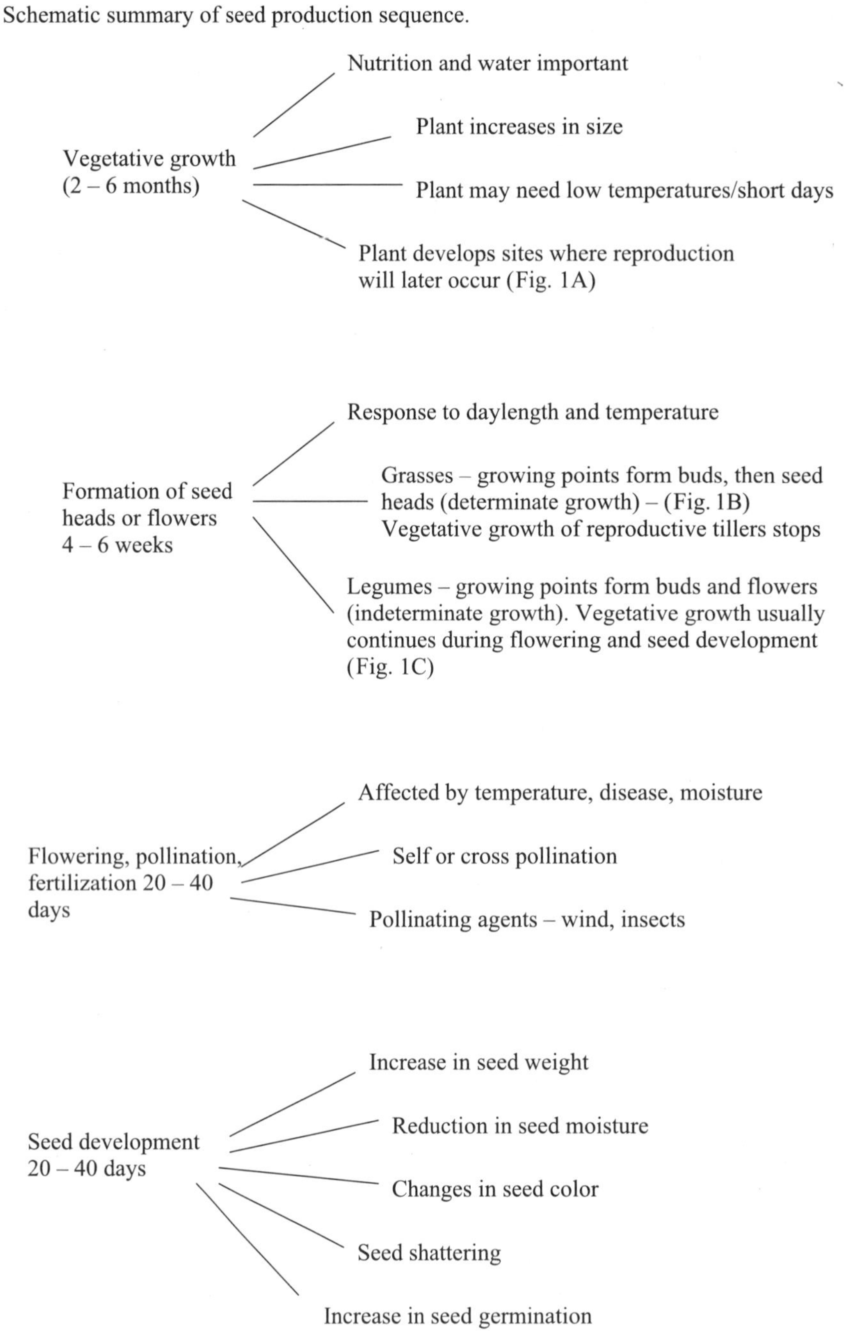

Fig. 2 Schematic summary of seed production sequence.

rotation include plant nutrition; maintenance of soil condition; reduction of disease, pest, and weed seed carryover; and, particularly in the case of seed production, to minimize the risk of plant residues and dormant crop and weed seeds remaining in the soil from previous potentially contaminating crops. Information on previous cropping history is an important consideration in seed production.

Maintaining the genetic purity of the seed crop is also important. This is usually achieved by crop growth regulation and good planning. The major factor is to minimize the possibility of cross-pollination between different crossing-compatible crops. This is done by ensuring that crops likely to cross-pollinate are not flowering at the same time or are planted an acceptably safe distance apart to prevent pollen crossover.

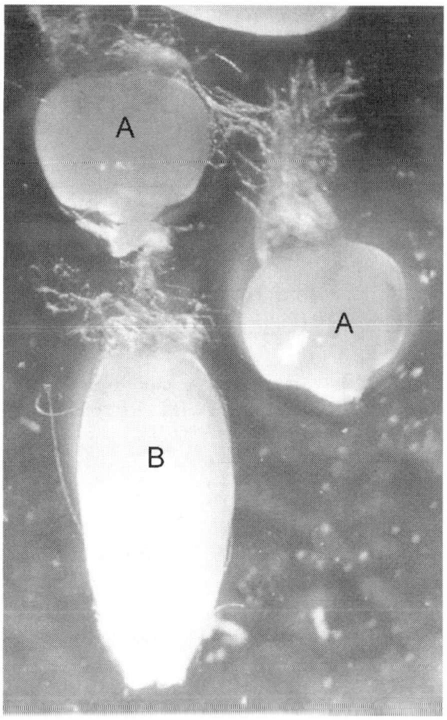

Fig. 3 Seeds of perennial ryegrass (*Lolium perenne*) during the growth stage (A) and food reserve accumulation stage (B) of seed development.

Seed fields, whether they involve an autumn- or spring-sown crop, may be established in rows or in solid stands. The former has the advantage of requiring lower seed sowing and fertilizer rates, and often ensures cleaner weed-free crops and longer stand productivity. Seeding rates lower than those for general crop production are often advocated when crops are sown for specialized seed production. Small seeded grasses such as bentgrass are often sown at 0.25 kg/ha, whereas larger cereals may be sown at 80–100 kg/ha.

PLANT POPULATION

In many seed crops the successful establishment of an adequate or optimum plant density is directly related to high seed yields. With increasing population, the yield of seed per hectare increases to a maximum point or plateau before declining. This parabolic relationship suggests that in seed production, the sowing rate should be sufficient to establish a crop that will produce a seed yield maximum.[6]

Row spacing depends on plant growth habit and the lateral spread of the root system. Taller plants are often grown in wider rows than are shorter plants.

WEEDS

Weeds are objectionable in all crops because they compete for soil, water, and nutrients; smother the crop and cut out light; may delay harvesting; may impede cultivation and harvesting; and may be plant parasites (e.g., dodder (Cuscuta sp)) or hosts for pests and diseases. Their seed may ripen at the same time as the crop and be difficult to remove in seed cleaning (e.g., wild red rice in rice; wild oats in cereals).

Weed control measures include crop rotation, sowing into a clean seedbed, application of presowing or pre-emergence herbicides, inter-row cultivation, rogueing, and sowing clean seed.

PESTS AND DISEASES

The incidence of diseases and pests in a seed crop is affected by climate, their presence in the soil or in the sown seed, and by the presence of alternate hosts growing nearby or in the field. Such control measures as crop rotation, burial of plant debris, seed treatment, foliar sprays (noting risks to pollinating insects if using insecticides), isolation, rogueing, and postharvest hygiene all deserve consideration.

MOISTURE

Moisture stress can reduce seed yield. On the other hand, irrigation can ensure good establishment and vigorous vegetative growth, and ensure that the plant is not stressed for moisture during seed development. Despite this, the induction of some moisture stress immediately flower buds form can be helpful in reducing the protracted flowering period in indeterminate plants (e.g., in legumes such as white clover (Trifolium repens)).

HYBRID SEED PRODUCTION

The discovery of hybrid vigor in field crops such as maize in the early 1900s has probably contributed as much to seed production as any other single factor.[1] A hybrid is produced by crossing inbred lines that have been developed by inbreeding and selection. This results in hybrid seed progeny with greater yield than nonhybrid crops. This occurs due to heterosis or hybrid vigor, due to an accumulation of a large number of dominant factors (genes) that favour growth. Hybrid seed production is now common in many field crops (particularly cereals) and in the production of vegetable seeds.[4]

CONCLUSION

Perhaps the most surprising thing about many of the crops grown for seed is the low yields obtained, compared to the potential yield these crops could produce if all components were maximized.[6]

As an example, the theoretical potential seed yield for alfalfa (Medicago sativa L.) is about 12 tonnes/ha and for Trefoil (Lotus corniculatus L.) about 2500 kg/ha. These can be compared with mean actual seed yields in New Zealand of 670 kg/ha (6% of potential) and 540 kg/ha (22% of potential).

This suggests many farmers still have a long way to go before realizing the production potential of the crops they grow for seed.

ARTICLES OF FURTHER INTEREST

Endosperm Development, p. 414
Floral Induction, p. 452
Flower Development, p. 464
Nitrogen, p. 822
Photoperiodism and the Regulation of Flowering, p. 877
Seed Dormancy, p. 1130

REFERENCES

1. Kelly, A.F. Seed Production of Agricultural Crops. In *Longman Scientific and Technical*; John Wiley & Sons Inc.: New York, 1988; 227 pp.
2. Fairey, D.T.; Griffith, S.M.; Clifford, P.T.P. Pollination, Fertilization and Pollinating Mechanisms in Grasses and Legumes. In *Forage Seed Production. 1. Temperate Species;* Fairey, D.T., Hampton, J.G., Eds.; CAB International: Wallingford, UK, 1997; 153–180. Chapter 7.
3. Copeland, L.O.; Miller, M.B. *Seed Science and Technology,* 3rd Ed.; Chapman & Hall: New York, 1995; 409 pp.
4. George, R.A.T. *Vegetable Seed Production*, 2nd Ed.; CAB Publishing: Wallingford, UK, 1999; 328 pp.
5. Coolbear, P.; Hill, M.J.; WinPe. Maturation of Grass and Legume Seed. In *Forage Seed Production. 1. Temperate Species*; Fairey, D.T., Hampton, J.G., Eds.; CAB International: Wallingford, UK, 1997; 71–104. Chapter 4.
6. Hill, M.J. Temperate Pasture Grass-Seed Crops: Formative Factors. In *Seed Production*; Hebblethwaite, P.D., Ed.; Butterworths: London, 1980; 137–149. Chapter 10.

Seed Vigor

Alison A. Powell
University of Aberdeen, Aberdeen, U.K.

S

INTRODUCTION

The concept of seed vigor arose following observations that, despite having equally high laboratory germination, commercial seed lots of many species show large differences in field emergence, particularly in adverse field conditions. Failure of the germination test to predict differences in field emergence suggested that there is a further physiological aspect to seed quality, not revealed by laboratory germination. This is referred to as seed vigor. Seed lots having high germination but poor emergence are low-vigor seeds; those giving good emergence are high-vigor seeds. Differences in germination rate and storage potential are also indicative of vigor and are included in the definition of vigor accepted by the International Seed Testing Association (ISTA). Seed aging and imbibition damage influence vigor and can be reduced by careful storage and handling of seeds. An understanding of the causes of vigor differences can help to maintain high levels of seed vigor.

DEFINITION OF VIGOR

In 2001 the ISTA formally recognized seed vigor as an important aspect of seed quality when it added vigor as a chapter in the International Rules for Seed Testing.[1] The ISTA definition of seed vigor is as follows:

> "Seed vigor is a sum of those properties that determine the activity and level of performance of seed lots of acceptable germination in a wide range of environments.
>
> Seed vigor is not a single measurable property, but is a concept describing several characteristics associated with the following aspects of seed lot performance.
>
> i. Rate and uniformity of seed germination and seedling growth
> ii. Emergence ability of seeds under unfavorable environmental conditions
> iii. Performance after storage, particularly the retention of the ability to germinate
>
> A vigorous seed lot is one that is potentially able to perform well even under environmental conditions that are not optimal for the species."

The germination of low-vigor seeds is therefore asynchronous and slow and the seeds emerge poorly in unfavorable conditions, in contrast to the rapid uniform germination and good emergence of high-vigor seeds. Low-vigor seeds also have poor storage potential.

CAUSES OF DIFFERENCES IN SEED VIGOR

Aging

Seed deterioration, or aging, is the major cause of differences in seed vigor and can be described as the accumulation of deleterious changes within the seed until the ability to germinate is lost.[2] The aging of a seed lot is described by the seed survival curve (Fig. 1). This illustrates that the germination of a population of seeds initially shows a very slow decline with time during which it is difficult to differentiate between the standard laboratory germination of samples of seeds at different points on the survival curve. However, as the population moves along the survival curve, the seeds are aging, i.e., deleterious changes are accumulating, until, at the end of the slow decline, the proportion of seeds within the population that are incapable of germinating increases and germination falls. At the beginning of the initial slow decline in germination, seeds are described as physiologically young or high-vigor seeds, whereas at the end of this phase they are physiologically old or low-vigor seeds.

Aging may occur before harvest, both during and after postmaturation drying. The degree of preharvest deterioration, frequently described as "weathering," depends on the climatic factors temperature and moisture (humidity/rainfall). Most deterioration occurs after the seeds have dried down to harvest maturity as a result of delayed harvesting.

The greatest incidence of aging arises, however, during seed storage, commonly in commercial storage, but also during brief, temporary periods of storage—for example, after harvesting and before and during processing. The major factors that influence seed aging and hence a decline in vigor and subsequently germination are seed moisture content and temperature. Their effects are described in Harrington's Rules of Thumb,[3] which

Encyclopedia of Plant and Crop Science
DOI: 10.1081/E-EPCS 120021890

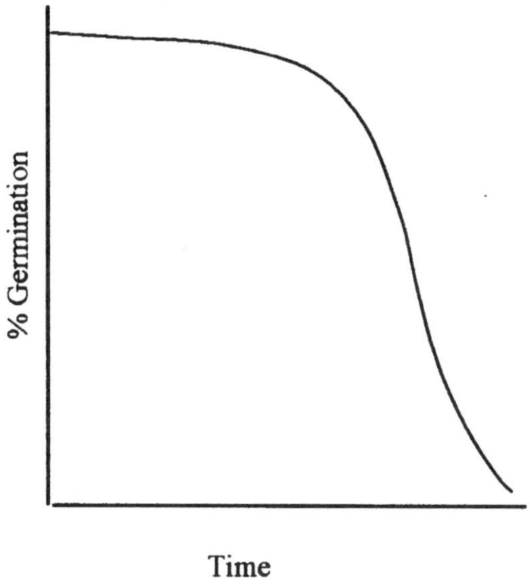

Fig. 1. The seed survival curve.

state that a 1% increase in seed moisture content or a 5°C increase in storage temperature will double the rate of aging, i.e., halve the storage life. These relationships are described more accurately in the viability equation[4] used to predict long-term storage of seeds in gene banks.

Imbibition Damage

Imbibition damage is a major cause of low vigor in grain legumes and arises when water rapidly enters the cotyledons during imbibition leading to physical disruption of membranes, cell death, and high solute leakage from the seeds, particularly at low temperatures.[2] Imbibition damage highlights the role of the testa in protecting the cotyledons from the damaging effect of water uptake.[2] In some species, an intact testa limits the incidence of imbibition damage. High-vigor seed lots have little testa damage, imbibe slowly, show little imbibition damage, and emerge well. Extensive testa damage is associated with rapid water uptake and low vigor.

In other grain legume species (e.g., *Phaseolus* sp., *Cicer ciceris*, *Vigna sesquipedalis*, *Vigna unguicularis*), there is a genotypic component to the susceptibility of seeds to imbibition damage. Cultivars having partially or completely unpigmented testae imbibe more rapidly and show greater levels of imbibition damage compared with cultivars having pigmented testae. As a result, the vigor of unpigmented cultivars is reduced, leading to poorer field emergence than is found in pigmented cultivars.[2]

Interaction of Imbibition Damage and Aging

The two major causes of reduced vigor in grain legumes—aging and imbibition damage—also interact, aged seeds being more susceptible to imbibition damage.[5,6] It has been proposed that the increased susceptibility of aged seeds to imbibition damage arises because membranes that have been weakened by physiological deterioration are more sensitive to physical damage during imbibition.[5,6]

ASSESSMENT OF SEED VIGOR

A range of tests evaluate seed vigor, the theoretical basis of which lies in seed aging. Vigor tests can be divided into three categories: physiological tests, biochemical tests, and tests that apply the whole aging process.

Physiological tests divide into two groups. In the first group, there are those tests that measure the impact of aging on an aspect of germination and early seedling growth. These are essentially a modification of the germination test and reveal the reduced rate and uniformity of germination that is found in aged seeds. Vigor can be assessed as the rate of germination based on the first count from the standard germination test. The slower germination of aged seeds also leads to smaller seedlings, when seedling size is assessed at a particular point in time, as in the seedling growth test. In the second group of tests, such as the cold test for maize and the cool germination test for cotton, stresses that reflect the type of stress that may be encountered in the field are applied to the seed. Low-vigor seeds are revealed by the reduced tolerance of aged seeds to stress.

The second category of vigor tests includes biochemical tests. The Conductivity test uses the increase in solute leakage that occurs from aged seeds to identify seed lots of low vigor. This test is an ISTA-validated test for garden peas[1] and can also be applied to a wide range of grain legumes.[7] The Tetrazolium test indicates the decline in respiratory enzyme activity during aging through the reaction of a colorless solution of 2,3,5-triphenyl tetrazolium chloride with dehydrogenases to produce the red substance, formazan, in living cells.[7] Dead cells do not stain.

The third group of vigor tests evaluates the effect of a period of aging on germination and is based on the manipulation of the rate of seed aging by holding seeds at a raised temperature and high relative humidity (Accelerated Aging test) or moisture content (Controlled Deterioration test). Seeds age rapidly and reductions in germination, normally seen over months or years, occur within days or hours. Both tests predict the field emergence and storage potential of seeds. The Accelerated Aging (AA) test is ISTA-validated for soybean,[1] while

Controlled Deterioration[8] tests the vigor of small seeded vegetable species and is an ISTA suggested test.[7]

MAINTENANCE OF VIGOR AND VIABILITY

Seed moisture content influences the rate of seed aging more than the temperature of storage;[4] therefore, efficient drying of the seed at or after harvest is important to reduce the rate of aging during subsequent storage and prevent a decline in vigor. However, dry seeds have a low water potential and hence high matric potential and readily absorb moisture from the air.[9] Thus, control of the relative humidity of the store is important to maintain low seed moisture content and reduce aging. Alternatively high-cost, low-volume seed may be stored in sealed moisture-proof containers.

Low-vigor seeds show earlier and greater reductions in germination during storage than high-vigor seeds in all storage conditions. Application of the Accelerated Aging or Controlled Deterioration test before storage enables seed producers to identify these low-vigor seed lots with poor storage potential.

Minimizing the incidence of testa damage during harvest and processing can maintain seed vigor by reducing the incidence of imbibition damage. Testa damage is greater when seeds are harvested at low seed moisture content. Modification of harvest timing and handling procedures can reduce testa damage and help maintain seed vigor.

CONCLUSION

Seed vigor has a major effect on successful emergence and crop establishment. Careful harvest and processing of seed offers one approach to maintaining seed vigor. In addition, application of vigor tests during production can identify where any reduction in seed quality has occurred during the production process, whether this is when the seed is still on the plant (weathering), during harvest and processing, or during storage. Thus, guidelines could be developed to minimize any reduction in vigor in subsequent years. In addition, companies may use vigor results in their in-house quality control. This could involve preventing the sale of particularly low-vigor seed or providing guidance regarding the storage potential of seed lots.

ARTICLE OF FURTHER INTEREST

Seed Production, p. 1134

REFERENCES

1. International Seed Testing Association. Rules amendments 2001. Seed Sci. Technol. **2001**, *29* (Suppl. 2), 132 pp.
2. Powell, A.A.; Matthews, S.; Oliveira, M. de A. Seed quality in grain legumes. Adv. Appl. Biol. **1984**, *10*, 217–285.
3. Harrington, J.F. Seed Storage and Longevity. In *Seed Biology, Vol. III*, Kozlowski, T., Ed.; Academic Press: London, 1972; 145–245.
4. Roberts, E.H. Quantifying Seed Deterioration. In *Physiology of Seed Deterioration*; McDonald, M.B., Ed.; Crop Science Society of America: Madison, 1986; 101–123.
5. Powell, A.A. Impaired Membrane Integrity—A Fundamental Cause of Seed Quality Differences in Peas. In *The Pea Crop*; Hebblethwaite, P.D., Heath, M.C., Dawkins, T.C.K., Eds.; Butterworths: London, 1984; 383–395.
6. Asiedu, E.A.; Powell, A.A. Comparisons of the storage potential of cultivars of cowpea (*Vigna unguiculata*) differing in seed coat pigmentation. Seed Sci. Technol. **1986**, *26*, 299–308.
7. *Handbook of Vigour Test Methods*; Hampton, J.G., TeKrony, D., Eds.; ISTA: Zurich, Switzerland, 1995.
8. Matthews, S. Controlled Deterioration: A New Vigour Test for Crop Seeds. In *Seed Production*; Hebblethwaite, P.D., Ed.; Butterworths: London, 1980; 647–660.
9. Roberts, E.H. Storage Environment and the Control of Viability. In *Viability of Seeds*; Roberts, E.H., Ed.; Chapman and Hall: London, 1972; 14–58.

Seeds: Pathogen Transmission Through

Curgonio Cappelli
Dipartimento di Arboricoltura e Protezione delle Piante, Perugia, Italy

INTRODUCTION

Pathogen transmission through seeds occurs when pathogenic microorganisms spread from seeds to seedlings. A developing infected plant can show symptoms within a few days or develop asymptomatically for a variable period of time. In some rare cases, pathogens present in infected plants survive, spreading from seeds to plants and from plants to seeds without causing any external symptom and inoculum production (i.e., *Neotyphodium*, grassess).

Most seed-transmitted pathogens (virus, bacteria, and fungi) in favorable conditions can cause very serious yield losses. For example, *Drechslera maydis* race T (southern leaf blight of corn) in the United States (1970) caused losses of about one billion dollars.

In plant pathology, seed transmission represents a possible mode of primary infection from which dangerous polycyclic microorganisms can start epidemic attacks. The early presence of a small quantity of infected plants (i.e., 0.05% of a crop) is sometimes sufficient for the development of destructive attacks (i.e., viral and bacterial diseases of legumes and vegetables). In some host-pathogen combinations seed transmission occurs every year because seeds represent the only possibility for the organism to survive (i.e., loose smut of wheat). In others this phenomenon occurs sporadically, because pathogens normally survive with other modalities (i.e., soil, other hosts, volunteers and wild plants, etc.). The different types of seed transmission are summarized using different criteria. Some of these diseases (i.e., ergot, bunt, and smut of cereals) were studied some centuries ago by plant pathology pioneers. With the increasing importance of seed pathology and the application of improved methodologies for microorganism detection in seed lots, more diseases were found to be seed transmitted.

MECHANISMS OF PATHOGEN TRANSMISSION AND REPRESENTATIVE EXAMPLES OF IMPORTANT DISEASES

Nine mechanisms of seed pathogen transmission are reported in Table 1.[1] The first six are determined according to the three possible inoculum locations in the seed (intraembrial, extraembryal, and contamination) and the two types of plant infection (local and systemic). The other three are represented by seed contamination with a particular nonparasitic stage of the pathogen in the soil before causing local, systemic, or organospecific infection of the plant.

1. Intraembrial infection causing systemic infection of plants. *Ustilago nuda*, barley (loose smut). When the seed germinates, the fungal mycelium located inside the embryo (mainly in the scutellum) moves to the growing point and then to the young tillers. The mycelium is carried up the plant with its elongation. The first symptoms of the disease appear during spike differentiation, when the teliospores of the fungus replace the seed and are released for the infections of healthy spikes. Embryo infection followed by systemic infection has been observed to be caused by other fungi (*Plasmopara helianthi*, sunflower, Fig. 1), viruses (i.e., Bean common mosaic virus, bean, Soybean mosaic virus, soybean), and bacteria [(i.e., *Xanthomonas axonopodis* pv. *phaseoli*, bean (Fig. 2), *Xanthomonas campestris* pv. *vesicatoria*, pepper (Fig. 3)].
2. Intraembrial infection causing local infection of plants. *Ascochyta pisi*, pea (anthracnose). The mycelium, localized in the embryo, causes lesions on the first true leaves where pycnidia develop under suitable humidity conditions. The spread of conidia by the wind causes the secondary infection of healthy plants.
3. Extraembrial infection causing systemic infection of plants. *Ustilago avenae*, oat (black loose smut). The mycelium, present in the pericarp, develops in the young seedling, causing systemic infection of the plant. During the flowering period, the behavior of the fungus is similar to *U. nuda* (see first mechanism). In *Xanthomonas arboricola* pv. *malvacearum*, cotton, the bacterium, localized mainly in the seed coat, causes systemic infections and the first symptoms on the cotyledons (Fig. 4).
4. Extraembrial infection causing local infection of plants. *Septoria apiicola*, celery (late blight). The pycnidia present in seed coat and pericarp produce conidia that spread to cotyledons and young leaves where local dark lesions may appear after about 10

Encyclopedia of Plant and Crop Science
DOI: 10.1081/E-EPCS 120021915

Table 1 Principal mechanisms of pathogens' seed transmission

Inoculum location	Infection of plant	Example
1—Intraembrial infection	Systemic infection	*Ustilago nuda*—barley
2—Intraembrial infection	Local infection	*Ascochyta pisi*—pea
3—Extraembrial infection	Systemic infection	*Ustilago avenae*—oat
4—Extraembrial infection	Local infection	*Septoria apiicola*—celery
5—Seed contamination	Systemic infection	*Tilletia caries*—wheat
6—Seed contamination	Local infection	*Puccinia carthami*—safflower
7—Seed contamination followed by a period of saprophytism or a dormant stage in the soil	Local infection	*Sclerotinia sclerotiorum*—sunflower
8—Seed contamination followed by a period of saprophytism in the soil	Systemic infection	*Fusarium oxysporum* f.sp. *callistephi*—aster
9—Seed contamination by particular structures of infection of the pathogen followed by nonparasitic phase	Organospecific infection	*Claviceps purpurea*—cereals

days. In favorable conditions, the fungus may produce new pycnidia for the prosecution of its life cycle. A similar mechanism has been recently observed on *Alternaria linicola*, linseed (Fig. 5) and on *Stemphylium herbarum*, red cichory (Fig. 6).

5. Seed contamination causing systemic infection of plants. *Tilletia caries*, wheat (common bunt). Teliospores of *Tilletia caries* are localized on the surface of healthy wheat seeds, where they overwinter. At the time of grain germination, teliospores germinate, producing a dikaryotic hypha that penetrates into the cells of the emerging coleoptiles, reaching the base of the first and second leaves and then the area below the growing point. The infection cycle proceeds asymptomatically, until the production of an ear with kernels that contain the teliospores.
6. Seed contamination causing local infection of plants. *Puccinia carthami*, safflower (rust). Safflower rust is

Fig. 1 Sunflower plants showing typical symptoms of downy mildew (stunt and chlorosis) caused by *Plasmopara helianthi* introduced into the soil by infected seeds of the previous sunflower crop. (Photos courtesy of L. Tosi and A. Zazzerini.) (*View this art in color at www.dekker.com.*)

Fig. 2 Foliar symptoms caused by seedborne inoculum of *Xanthomonas axonopodis* pv. *phaseoli* in bean plants. (Photos courtesy of R. Buonaurio.) (*View this art in color at www.dekker.com.*)

Fig. 3 Pepper cotyledons showing symptoms caused by *Xanthomonas campestris* pv. *vesicatoria*. (Photos courtesy of R. Buonaurio.) (*View this art in color at www.dekker.com.*)

a rare seedborne rust disease transmitted directly from seed to seedling. Teliospores that contaminate the surface of seeds are responsible for the infection of seedling tissues, causing hypertrophy and hyperplasia of hypocotyls (Fig. 7). The inoculum produced above ground spreads to the leaves, where it causes the typical rust symptoms (pustules).

7. Seed contamination followed by a period of saprophytism or by a dormant stage in the soil and then by local infection. *Sclerotinia sclerotiorum*, sunflower (basal stalk rot and head rot). Sclerotia of the fungus, introduced into the soil by the seed, can produce hyphae that can either infect the developing host or generate apothecia, asci, and ascospores. The ascospores resealed in the environment are responsible for local infection of different part of the host, including the head.
8. Seed contamination followed by a period of saprophytism in the soil and then by systemic infection.

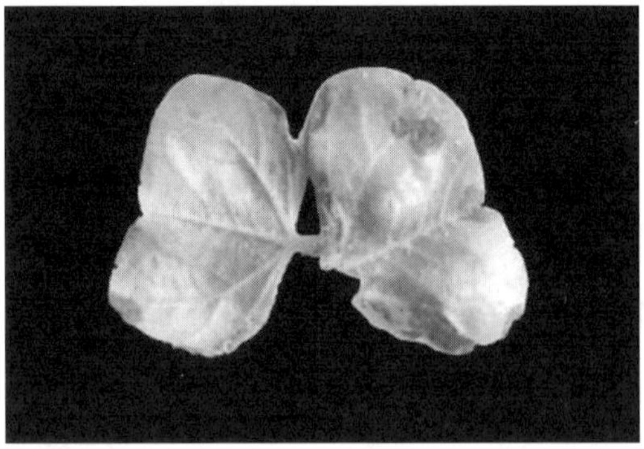

Fig. 4 Cotton cotyledons showing symptoms caused by seedborne *Xanthomonas arboricola* pv. *malvacearum*. (Photo courtesy of R. Buonaurio.) (*View this art in color at www.dekker.com.*)

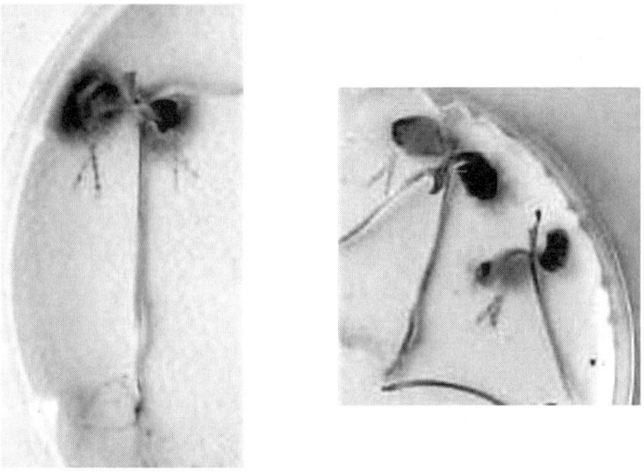

Fig. 5 Linseed cotyledons showing *Alternaria linicola* sporulation. (Seedlings showing no symptoms after the growing on test, were incubated in humid chamber for 48 hrs to detect asymptomatic seedborne infections). (Photos by C. Cappelli.) (*View this art in color at www.dekker.com.*)

Fusarium oxysporum f.sp. *callistephi*, aster (wilt). Conidia and mycelium of the fungus, carried into the soil by the seed, grow saprophytically in the soil, increasing the inoculum level. In favorable conditions, the fungus attacks the roots of young seedlings and develops inside the plant, causing vascular wilt. In *Fusarium oxysporum* f.sp. *gladioli*, saffron (Fig. 8), the same mechanism has been observed.

9. Seed contamination by structures from organospecific seed infection followed by an extramatrical nonparasitic phase and later by direct organospecific seed infection. *Claviceps purpurea*, cereals (ergot). Sclerotia of the fungus, transported into the soil by the

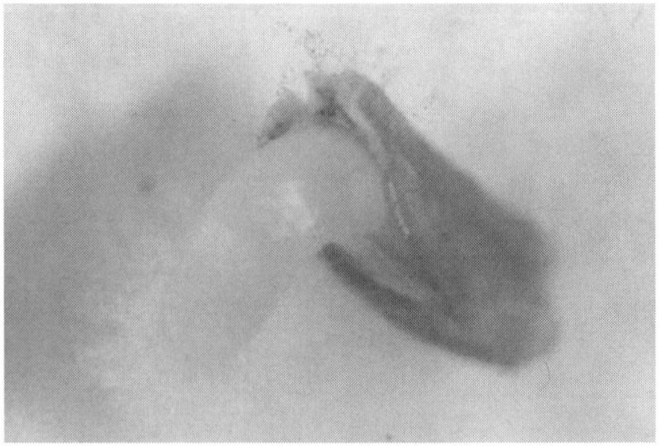

Fig. 6 Red chicory seedling showing *Stemphylium herbarum* conidia on the pericarp, ready for seedling infection. (Photo by C. Cappelli.) (*View this art in color at www.dekker.com.*)

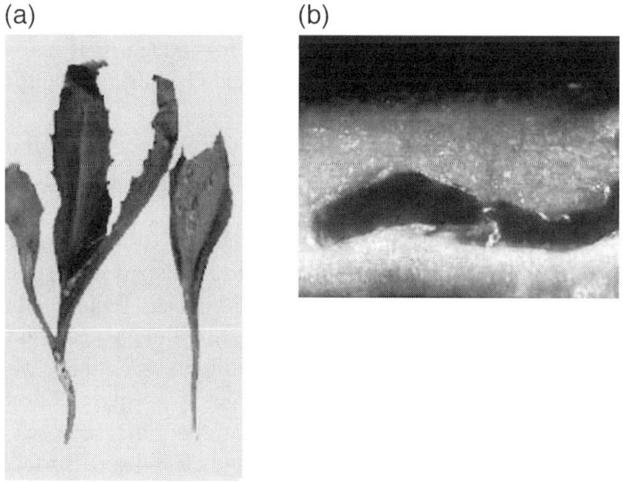

Fig. 7 Seed transmission of *Puccinia carthami* to safflower seedlings. (a) Pycnidia and aecia in the hypocotyls and cotyledons. (b) Detail of pycnidia and aecia on a hypocotyl. (Photos by C. Cappelli and A. Zazzerini.) (*View this art in color at www.dekker.com.*)

seed, remain inactive until the spring when they produce several club-shaped perithecial stromata containing asci and ascospores. Ascospore dispersal occurs during the flowering period of the host when the ovary is susceptible to the fungus, prior to fertilization.

FACTORS AFFECTING THE TRANSMISSION OF SEEDBORNE PATHOGENS

The presence of pathogens in seed lots does not always involve microorganism transmission to the plant in the field and disease development. Research carried out in different host–pathogen combinations showed the importance of some parameters that influence the infection of young seedlings until the development of the first symptoms. A number of factors such as host, pathogen, environmental conditions, and crop management may affect the transmission rates of seedborne pathogens.[2–5]

1. Host Genotype. The development of seedborne infection in the field depends on the genetic constitution of the host cultivar. In plant-resistant genotypes, the activation of defense mechanisms in germinating infected seeds can prevent disease transmission to the plant. Resistance of wheat to loose smut fungus occurs through the reaction of the embryos, which are completely resistant to infection. In other host–pathogen combinations, some cultivars show less susceptibility than others because they require a major quantity of inoculum with respect to the susceptible cultivars (i.e., *Tilletia tritici*, wheat).
2. Quantity of Inoculum. The inoculum level can affect pathogen seed transmission because too little or too much inoculum can result in nontransmission of the disease. For example, in *Tilletia caries*, wheat, infections of seedlings in susceptible cultivars occur when the number of teliospores contaminating the seeds is more than 100 per seed. The quantity of inoculum associated with the seed depends on the different situations occurring during seed differentiation. For example, late flower infections may result in low superficial contamination of seeds only (i.e., *Septoria* spp.). On the other hand, early infections of flowers may cause severe infections of developing seeds, reducing seed vitality: Seeds fail to germinate and do not transmit the disease (e.g., poor quality of cereal seeds infected by *Fusarium* spp.). Between these two extreme situations, cultivar resistance, germination capacity of seeds, and environmental conditions at sowing time determine the amount of inoculum necessary for seed transmission.
3. Inoculum Vitality. Inoculum vitality depends on the longevity of the peculiar structures produced by the pathogens and on the characteristics of each seed lot (Table 2). Mycelia, sclerotia, chlamydospores, oospores, teliospores, virus particles, and bacterial cells that contaminate or infect seeds can maintain their vitality for some years (i.e., 5–10 or more). Often, vitality depends on the quantity of inoculum present in each seed lot and can be reduced by soil microflora (or other factors) when seeds are sown.

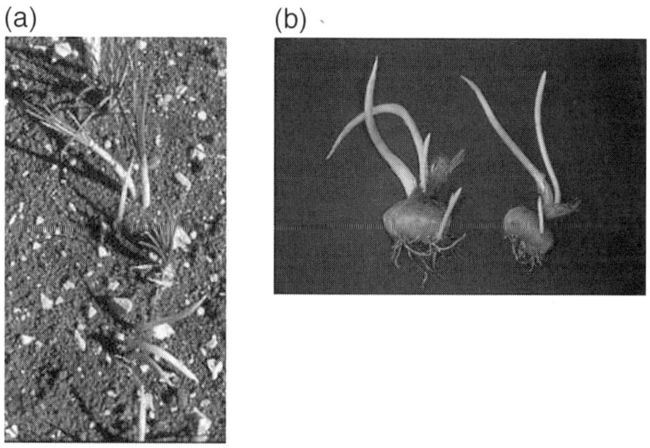

Fig. 8 Abnormal development of saffron shoots caused by *Fusarium oxysporum* f.sp. *gladioli*. (a) Young infected plants in the field. (b) Corms during germination. (Photos by C. Cappelli.) (*View this art in color at www.dekker.com.*)

Table 2 Examples of longevity of some important seedborne pathogens

Pathogen	Host	Viability (years)
Fungi		
Alternaria brassicicola	Cabbage	7
Ascochyta pisi	Pea	7
Botrytis alli	Onion	3.5
Botrytis cinerea	Linseed	3.3
Diaporthe phaseolorum	Soybean	2.5
Drechslera avenae	Oat	10
Drechslera oryzae	Rice	10
Drechslera sorokiniana	Wheat	10
Phoma lingam	Cabbage	1.1
Septoria apiicola	Celery	3
Tilletia caries	Wheat	9
Ustilago nuda	Barley	11
Bacteria		
Pseudomonas savastanoi pv. *phaseolicola*	Bean	3
Xanthomonas campestris pv. *campestris*	Cabbage	3
Xanthomonas arboricola pv. *malvacearum*	Cotton	4.7
Xanthomonas axonopodis pv. *phaseoli*	Bean	15
Pseudomonas syringae pv. *tomato*	Tomato	20
Xanthomonas axonopodis pv. *vesicatoria*	Pepper	10
Viruses		
Alfalfa mosaic virus	Alfalfa	3
Barley stripe mosaic virus	Barley	3
Bean common mosaic virus	Bean	30
Cucumber mosaic virus	Cucumber	1
Soybean mosaic virus	Soybean	2

4. Type of Inoculum. Seed transmission can be influenced by the type of inoculum in seeds because not all the pathogenic structures present in the seed lots can infect the developing seedlings. Safflower seeds by contaminated uredospores of *P. carthami* produce healthy seedlings, whereas teliospores produce the typical symptoms of the disease.
5. Inoculum Location in the Seed. Often, seed transmission occurs only if the inoculum is located in a particular position of the seed. Most viruses, for example, are transmitted to the seedlings only if they are located in the embryo.
6. Soil Microflora. The microflora present in the soil can reduce the transfer of vital inoculum from seeds to seedlings, especially when the contaminating inoculum can be influenced by the presence of an antagonistic microflora in the soil. There are suppressive soils in which some soilborne *F. oxysporum* causing vascular diseases reduce their activity for the presence of antagonistic bacteria present in the rhizosphere.
7. Environmental Factors. Host development is often strongly influenced by environmental conditions. Moisture and temperature are the most important factors and both may affect seed and spore germination, seedling infection, and the first steps of disease development. Common bunt of wheat requires the teliospores to germinate at the same time as the grain, being too short a period of seedling susceptibility. In this case, soil temperatures affect spore germination, and seed transmission rates of the fungi can vary considerably. Teliospores located on the seed surface germinate optimally at 10–15°C, whereas at higher temperatures (20–25°C), germination is reduced. For this reason seeds of the same seed lot used in areas of different climatic conditions produce different percentages of infected plants. Internally localized pathogens normally produce infected plants and there are no significant effects of soil environment on pathogen transmission. Sowing date or geographic crop location do not change the incidence of loose smut of wheat (*Ustilago tritici*) or other similar host–pathogen combinations. In general there are particular geographic and climatic pattern and, temperatures and humidity strongly associated with fungal, viral, and bacterial seed-transmitted diseases. The establishment of infection is favored if, after sowing, the temperatures cause a slow development of seedlings.
8. Cultural Practices. Sowing data, sowing depth, soil type, soil pH, seeding rate, and fertilizers, etc. may affect pathogens seed transmission by having some direct or indirect effects on host, pathogen, and environment. Early sowing and shallow planting of wheat, for example, cause the reduction of bunt transmission, whereas deep sowing may cause the extension of the germination period and consequently seedling susceptibility.

CONCLUSION

Seed lots today move around the world more frequently than in the past. As a consequence, there are more opportunities for the spread of new pathogenic microorganisms in seed crops, as evidenced by the many new seedborne diseases recorded in the last ten years.

According to the practices of "Integrated Pest Management," it will be very useful in the future to increase the use of seed health tests that can forecast the amount of inoculum transmitted to plants in different climatic conditions.

ARTICLES OF FURTHER INTEREST

REFERENCES

1. Neergaard, P. *Seed Pathology*; Vols. I and II, The MacMillan Press Ltd.: London, 1979.
2. Agarwal, V.K.; Sinclair, J.B. *Principles of Seed Pathology 1 and 2*; CRC Press: Boca Raton, FL, 1987.
3. Baker, K.F. Seed Pathology. In *Seed Biology*; Kozlowsky, T.T., Ed.; Academic Press: New York, 1972; Vol. 2, 317–416.
4. Maude, R.B. *Seedborne Diseases and Their Control*; CAB International, UK; 1996.
5. Richardson, M.J. *An Annotated List of Seed-Borne Diseases*, 4th Ed.; International Seed Testing Association: Zürich, Switzerland, 1990.

Sex Chromosomes in Plants

M. Ruiz Rejón
Universidad de Granada, Granada, Spain

INTRODUCTION

Unlike the situation in the animal kingdom, most plants are hermaphrodites. In fact, only about 5% of the 250,000 species of flowering plants are dioecious, having separate females and males. In less than a dozen of these dioecious species, sex is determined by chromosomes such as the X and Y that are so well known in animals. Despite their low number, plant species with sex chromosomes enable investigations of the origin and evolution of these chromosomes, and also advance the genetic improvement of commercial plants.

THE ORIGIN AND EVOLUTION OF SEX CHROMOSOMES

Currently, the most widely accepted hypothesis on the origin of sex chromosomes is that they first arose when a pair of nondifferentiated chromosomes joined together the genes that control male and female sex (i.e., genes that inhibit the development of either female or male sex organs). Second, at some point, differentiation occurred in the zone bearing these genes, which suppressed recombination in this area. Third, suppression of recombination was extended to more areas of the sex chromosomes, followed by a degeneration of the Y chromosome and the appearance of dosage-compensation mechanisms in the X chromosome.[1] In short, this evolutionary process appears to have proceeded in three stages (Fig. 1).

This hypothesis is based on the very old sex chromosomes of animal species, normally in the last stages of this evolutionary process. However, dioecious plants, being at different stages in the process of chromosome differentiation (Table 1), offer a unique opportunity to investigate this process.

The First Stage: Asparagus

In asparagus (*Asparagus officinalis*), males as well as females have 20 chromosomes (Fig. 2A), and sex is controlled by two regulatory genes—the male activator (M) and the female suppressor (F). The males are heterozygous (MF/mf) and the females homozygous (mf/mf). These genes are located in chromosome 5, which has a similar morphology in both males and females.[2]

This similarity indicates that dioecism is of very recent origin in asparagus, and that chromosome 5 can be considered as an example of an incipient sex chromosome.

The Second Stage: White Campion

In the white campion (*Silene latifolia*), females have 22 autosomes (chromosomes with similar morphology in males and females) and two X chromosomes, whereas males have 22 autosomes as well as one X chromosome and one Y chromosome.

The origin of the sex chromosomes in the white campion is recent (some 20 million years ago (M.Y.A.) versus some 100 million years ago in mammals). Thus, although a differentiated sex-chromosome system developed with restriction of recombination in some regions, the process has not advanced to extensive degeneration of the Y and dosage compensation of the X.

The sex chromosomes are the largest chromosomes of the karyotype, with the Y larger than the X. Pairing occurs only during meiosis in one terminal region (Fig. 2B), indicating the suppression of recombination over much of their length.

The Y chromosome became differentiated from the X chromosome in at least three interstitial regions. These regions are thought to contain regulatory genes similar to those mentioned in asparagus. The white campion, however, may have three types of regulatory genes—one gene that suppresses the development of the carpels (female flower parts), and two genes that promote the development of anthers (male flower parts, one early and one late).[3]

At the molecular level, the gene that suppresses the development of the carpels and the gene that promotes the (late) development of anthers are distally located at the ends of the nonrecombinant region of the Y chromosome, whereas these two genes are closely linked in the X.[4] This differential disposition may have come about through chromosome rearrangements such as inversions and/or translocations. These rearrangements have suppressed recombination between the two chromosomes, thereby avoiding breakage of the two allele combinations that determine male or female in the sex-controlling genes.

Encyclopedia of Plant and Crop Science
DOI: 10.1081/E-EPCS 120005599

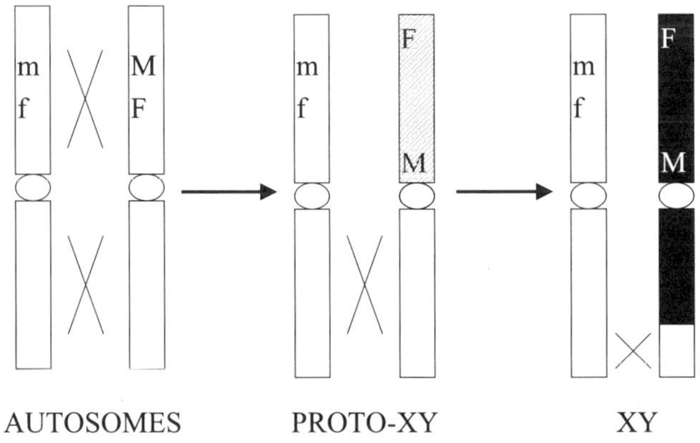

Fig. 1 Hypothesis concerning the origin of sex chromosomes. First stage: Establishment in an autosome pair of two linked genes, one a male activator (M) and the other a female suppressor (F). Second stage: Chromosomal rearrangements can prompt differential disposition of these genes between the proto X and Y chromosomes, suppressing the recombination between them. Third stage: Finally the recombination occurs only in a small region, the Y is degenerated, and some dosage-compensation mechanism appears in the X.

The white campion has no signs of degeneration of the Y chromosome; it is not heterochromatic (i.e., it does not show differential staining with respect to the remaining chromosomes), nor does it possess specific satellite DNA (a kind of repetitive DNA that may lack any function). In addition, although a functional gene has been found in the X chromosome for which its homologue in the Y is degenerated,[5] two other functional genes have been found in the X chromosome for which their copies in the Y are not degenerated.[6,7] The fact that some diplohaploid YY plants (plants obtained by duplication from a haploid male cell with a Y chromosome) are viable and fertile indicates that the process of degeneration of the Y has not been completed for many genes.

With respect to the appearance of dosage-compensation mechanisms in the X chromosome, the data for *S. latifolia* are not conclusive. In females, the two X chromosomes differ in replication time and appear to be differentially methylated (marked with a CH3 group in the cytosine of CG dinucleotides), but there are no signs of facultative heterochromatinization (reversible heterochromatin) in one of the two X chromosomes in the females.

The Third Stage: Sorrel

Sorrel (*Rumex acetosa)* females have 14 chromosomes, two of which (the large pair) are the X chromosomes. The males have 15 chromosomes, including 12 autosomes, one X chromosome, and two Y chromosomes of different sizes (Y1 and Y2). Y1 and Y2 are the largest chromosomes of the karyotype after the X chromosome. This composite chromosome system is believed to have arisen from a simple XX/XY by translocation between the X and an autosome, giving rise to the new X, with the primitive Y being Y1 and the homologue of the autosome being Y2.

In meiosis, the two Y's pair with the end of each arm of the X; thus, there is no meiotic recombination between the two Y's nor with the majority of the X (Fig. 2C). The Y's show clear signs of degeneration, are heterochromatic (Fig. 2D), and are rich in at least two satellite-DNA families with respect to the X and the autosomes, one of these satellites being specific to the Y's.[8,9]

Table 1 Sex chromosomes in representative plant species

Family	Species	Sex chromosomes	Sex determination mechanism
Asparagaceae	*Asparagus offininalis*	5th chromosome pair	Active-Y
Actinidiaceae	*Actinidia deliciosa*	Undifferentiated	Active-Y
Cannabidaceae	*Cannabis sativa*	♀XX/♂XY	Active-Y
	Humulus lupulus	♀XX/♂XY	X/A ratio
	Humulus japonicus	♀XX/♂XY_1Y_2	X/A ratio
Cariaceae	*Carica papaya*	Undifferentiated	Active-Y
Caryophillaceae	*Silene latifolia*	♀XX/♂XY	Active-Y
Chenopodiaceae	*Spinacia oleracea*	Undifferentiated	Active-Y
Cucurbitaceae	*Coccinia indica*	♀XX/♂XY	Active-Y
Dioscoreaceae	*Dioscorea tokoro*	Undifferentiated	Active-Y
Euphorbiaceae	*Mercurialis annua*	Heterocigotic male	Active-Y
Polygonaceae	*Rumex acetosa*	♀XX/♂XY_1Y_2	X/A ratio
	Rumex hastatulus	♀XX/♂XY o ♀XX/♂XY_1Y_2	X/A ratio
Rosaceae	*Fragaria sp.*	Heterogametic female	?
Vitiaceae	*Vitis*	♀XX/♂XY	Active-Y

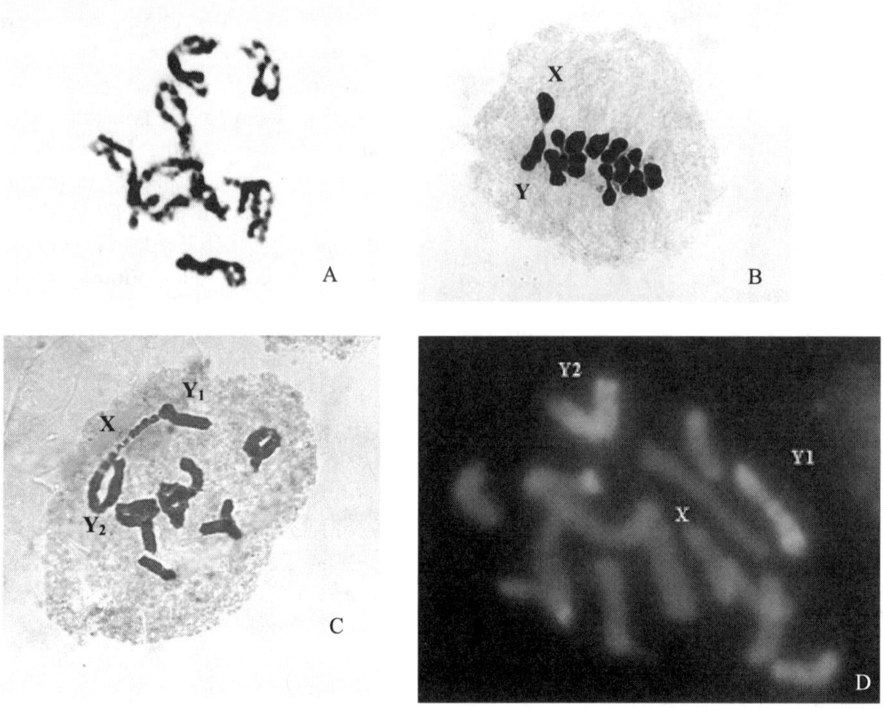

Fig. 2 Sex chromosomes in male cells of some plant species. A: Diplotene of *Asparagus officinalis*. B: Metaphase I of *Silene latifolia*. C: Diplotene of *Rumex acetosa*. D: Mitotic metaphase of *R. acetosa* stained with the blue fluorochrome DAPI. (Note the intense signal of the Y chromosomes indicating its heterochromatic nature.) (*View this art in color at www.dekker.com.*)

The X's show no clear signs of dosage compensation. In females, there is no heteropicnosis (differential staining) of one of the X chromosomes. The genetic-control system in sorrel, however, depends on the ratio of X:autosomes (for this ratio, the females have a value equal to 1 and the males 0.5), in which the compensation, if it occurs, would lead to gene hyperactivity in the X chromosome of the males rather than to the inactivation of an X of the females.

Plants such as *R. acetosa* constitute the third stage in the evolutionary process of the origin of the sex chromosomes characteristic of many animal species. In fact, in the Y chromosomes of sorrel, in addition to the previously mentioned satellite DNA, genes also control male fertility, as is also the case in humans. In this sense, sorrel plants with an X:autosome ratio of 0.5 but without Y's are sterile males.

CONCLUSION

Despite the facts that research on sex chromosomes in plants has lagged behind that in animals and that enigmas persist, recent accelerated investigation using molecular tools is enabling the testing of the general hypothesis for the origin of sex chromosomes. In fact, the different stages proposed by this hypothesis have been found in plants, thus confirming that sex chromosomes originated as a mechanism to avoid breakage of the two alternative combinations that give rise to males or females in sex-determining genes.

The role of the genes themselves in determining the sex of plants and whether they are located in the sex chromosomes or in the autosomes need clarification. For the moment, the genes identified in the sex chromosomes of plants (e.g., the case of the functional genes of *Silene latifolia* discussed above) do not intervene directly in the control of sex. In different dioecious species, genes involved in floral development are not associated with the sex chromosomes nor with sex control.

All these findings are of great interest from scientific and applied standpoints. In scientific terms, the analysis of dioecious plants in which the development of one of the sexes is halted will enable an understanding of the development of hermaphroditic plants in which both sexes appear. From an applied perspective, all these findings are of interest in areas of genetic plant improvement such as early determination of plant sex in crops that benefit from only one sex, in developing single-sex populations, in manipulating the reproductive systems of plants by converting them from dioecious to hermaphroditic and vice versa, and in controlling pollination for the production of hybrid seeds.

ACKNOWLEDGMENTS

Thanks to Drs. M. Garrido-Ramos, R. Lozano, M. Jamilena, and J.L. Santos and to Mr. D. Nesbitt.

ARTICLES OF FURTHER INTEREST

B Chromosomes, p. 71
Chromosome Banding, p. 263
Chromosome Rearrangements, p. 270
Chromosome Structure and Evolution, p. 273
Cytogenetics of Apomixis, p. 383
Female Gametogenesis, p. 439
Flow Cytogenetics, p. 460
Flower Development, p. 463
Fluorescence In Situ Hybridization, p. 468
Male Gametogenesis, p. 663

REFERENCES

1. Charlesworth, B. The evolution of sex chromosomes. Science **1991**, *251*, 1030–1033.
2. Loptien, H. Identification of the sex chromosome pair in asparagus (*Asparagus officinalis L.*). Z. Pflanz. **1979**, *82*, 162–173.
3. Westergaard, M. The mechanisms of sex determination in dioecious flowering plants. Adv. Genet. **1958**, *9*, 217–281.
4. Lebel-Hardenac, K.S.; Hauser, E.; Law, T.F.; Schmid, J.; Grant, S.R. Mapping of sex determination loci on the white campion (*Silene latifolia*) Y chromosome using amplified fragment length polymorphism. Genetics **2002**, *160*, 717–725.
5. Guttman, D.S.; Charlesworth, D. An X-linked gene with a degenerate Y-linked homologue in a dioecious plant. Nature **1998**, *393*, 263–266.
6. Delichère, C.; Veuskens, J.; Hernould, M.; Barbacar, N.; Mouras, A.; Negrutiu, I., et al. SIY1, the first active gene cloned from a plant Y chromosome, encodes a WD-repeat protein. EMBO J. **1999**, *18*, 4169–4179.
7. Atanassov, I.; Delichère, C.; Filatov, D.A.; Charlesworth, D.; Negrutiu, I.; Monéger, F. A putative monofunctional fructose -2-6-bisphophatase gene has functional copies located on the X and Y sex chromosomes in white campion (*Silene latifolia*). Mol. Biol. Evol. **2001**, *18*, 2162–2168.
8. Ruiz Rejón, C.; Jamilena, M.; Garrido Ramos, M.; Parker, J.S.; Ruiz Rejón, M. Cytogenetic and molecular analysis of the multiple sex chromosome system of *Rumex* acetosa. Heredity **1994**, *72*, 209–215.
9. Shibata, F.; Hizume, M.; Kuroki, Y. Differentiation and the polymorphic nature of the Y chromosomes revealed by repetitive sequences in the dioecious plant *Rumex acetosa*. Chromosome Res. **2000**, *8*, 229–236..

Shade Avoidance Syndrome and Its Impact on Agriculture

Garry C. Whitelam
University of Leicester, Leicester, U.K.

INTRODUCTION

The ability to acquire information about their surroundings and use this information to modify behavior or development is a characteristic of all living organisms. Being sessile organisms that are not able to choose their environment, plants need to be especially plastic in their development in order to optimize their growth in response to environmental change. Since plants depend upon light as an energy source for the process of photosynthesis, they are especially sensitive to variations in the light environment. Plants monitor a range of light signals such as intensity, quality, and direction, and use this information to modulate a wide range of developmental and physiological process ranging from seed germination, seedling establishment, shoot architecture, the onset of flowering, and fruit development. Three principal families of signal-transducing photoreceptors have been identified in higher plants: the red/far-red (R/FR) light-absorbing phytochromes and the UV-A/blue light-absorbing cryptochromes and phototropins.[1] In light-grown plants, R/FR light signals acting through the phytochromes enable plants to detect nearby vegetation and to evoke the shade avoidance syndrome of responses.

PROPERTIES OF PHYTOCHROMES

The phytochromes are biliproteins that absorb maximally in the R (600–700 nm) and FR (700–800 nm) regions of the spectrum. The phytochromes exist as a homodimer of two subunits that each consists of a polypeptide (ca. 1200 amino acids) and a linear tetrapyrrole chromophore attached via a thioether linkage to a conserved cysteine residue in the N-terminal globular domain of the protein. Light-induced interconversions between Pr and Pfr involve isomerization of the tetrapyrrole chromophore, accompanied by reversible conformational changes throughout the protein moiety. Phytochromes can exist in either of two relatively stable isoforms: a R light-absorbing form (Pr) with an absorption maximum at about 660 nm, or a FR light-absorbing form (Pfr) with an absorption maximum at about 730 nm. The Pfr form of phytochrome is generally considered to be biologically active, and Pr is considered to be inactive. The absorption spectra of Pr and Pfr show considerable overlap throughout the visible light spectrum, and so under almost all irradiation conditions phytochromes are present in an equilibrium mixture of the two forms that tends to reflect the relative proportions of R and FR wavelengths.

Higher plants contain multiple discrete phytochrome species that make up a family of closely related photoreceptors, the apoproteins of which are encoded by a small family of divergent genes. The size of the phytochrome family varies among different plant species. In dicotyledonous angiosperms, including the model species *Arabidopsis thaliana*, five apophytochrome encoding genes (*PHYA-PHYE*) have been characterized. Apophytochromes A, B, C, and E are evolutionarily divergent proteins, sharing only 46–53% sequence identity, whereas *PHYD* encodes an apoprotein that shares 80% sequence identity with apophytochrome B. Molecular phylogenetic analysis supports the occurrence of four major duplication events in the evolution of phytochrome genes. An initial duplication is believed to have separated *PHYA* and *PHYC* from *PHYB/D/E*. The subsequent separation of *PHYA* from *PHYC* and *PHYB/D* from *PHYE* resulted in three subfamilies: *A/C*, *B/D*, and *E*.[2]

PHYTOCHROMES AND SHADE AVOIDANCE

Phytochromes play a major role in regulating seed germination and early seedling establishment. In light-grown plants, the phytochromes mediate two principal responses: photoperiodic perception, leading to floral induction in some species, and proximity perception, leading to the shade avoidance syndrome. The shade avoidance syndrome, one of the best-studied examples of adaptive phenotypic plasticity in plants, is a suite of developmental responses initiated by the reflected/scattered light signals generated by neighboring vegetation.

The daylight spectrum has more or less equal proportions of R (600–700 nm) and FR (700–800 nm), but the selective absorption of blue and R wavelengths by the chlorophylls causes the radiation that is reflected/scattered by green leaves to be relatively enriched in the FR (Fig.1). This FR-rich light signal (i.e., a reduction in R:FR ratio) is detected by nearby plants as a change in the equilibrium between Pr and Pfr and provides a unique and

Encyclopedia of Plant and Crop Science
DOI: 10.1081/E-EPCS 120012966

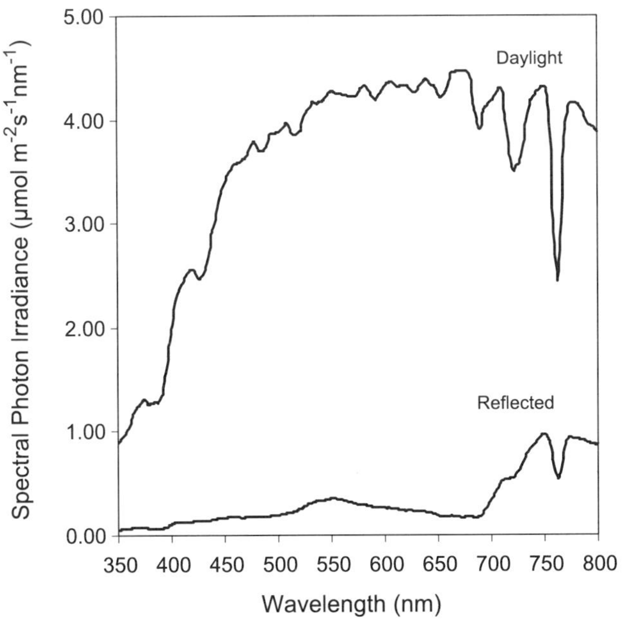

Fig. 1 The spectral photon distributions of daylight and of the radiation reflected from a dense wheat canopy. The R:FR ratio of daylight (600–700 nm:700–800 nm) is 1.10 and the R:FR ratio of the reflected light is 0.28.

unambiguous signal that potential competitors are nearby. The magnitude of the reduction in R:FR ratio is quantitatively related to the proximity and density of the neighboring vegetation. In response to low R:FR ratio signals many plants evoke competitive reactions, including increased elongation of internodes and/or petioles, reduced leaf growth, and increased apical dominance,[3,4] in an attempt to avoid being shaded. In many competitive plant species, such as arable weeds and a number of crops, internodes are especially sensitive to the low R:FR ratio signal. This correlates well with the finding that the reflected low R:FR ratio signal is particularly intense in internodes. The importance of the internode response was demonstrated in field experiments in which individual internodes of *Sinapis alba* or *Datura ferox* were screened for the low R:FR ratio signal by annular cuvettes containing copper sulphate solution, which absorbs strongly in the FR. Such ''blinded'' internodes failed to elongate in response to the proximity signal generated by growing plants at high densities.[5] In some species, prolonged exposure to the low R:FR ratio signals evokes a survival reaction, the acceleration of flowering.

The roles of individual phytochromes in regulating shade avoidance responses have been largely inferred from studies of mutant plants deficient in one or more of the phytochromes. In a number of plant species, deficiency of phytochrome B has been shown to result in plants constitutively displaying phenotypes comparable to those of the shade avoidance syndrome.[6] This indicates a major role for phytochrome B in the perception of the R:FR ratio signal. However, the retention of some sensitivity to low R:FR ratio signals in mutants that are null for phytochrome B indicated the involvement of other phytochromes in shade avoidance. Phytochromes D and E were subsequently shown to play a role in R:FR ratio perception.[7,8]

SHADE AVOIDANCE IN AGRICULTURE

When plants grow in close proximity to one another, the shade avoidance syndrome is invoked. These responses are likely to be of benefit to individual plants growing in mixed communities, by providing a means of competing for light through the reallocation of resources. In an agricultural setting, however, shade avoidance represents an undesirable trait that diverts resources into increased stem and petiole elongation to support competitive growth, and away from agriculturally important activities such as root, leaf, and seed/fruit growth. Despite the strong selection for high yield in monocultures that has characterized the breeding of modern crops, many crop plants have retained the shade avoidance response to nearby vegetation and so show increased stem elongation when grown at increasing densities.

Crop productivity is a complex function of the acquisition and distribution of resources. Shade avoidance responses lead to dramatic alterations in resource allocation that can severely limit harvestable biomass in monocultures. Thus, it can be predicted that by rendering individual plants insensitive to the low R:FR proximity signal, the wasteful allocation of resources into competitive growth should be reduced, with a concomitant increase in harvestable yield. This notion has been tested in an elegant field experiment in which a transgenic approach was used to suppress shade avoidance responses. Tobacco plants were genetically engineered to overproduce phytochrome A, a member of the phytochrome family that normally plays a major role in seedling development in response to FR signals. When overexpressed, phytochrome A antagonizes the action of phytochrome B (and presumably phytochromes D and E), so that the plants do not display increased stem elongation in response to the low R:FR ratio signal. When such plants are grown in the field at different densities, in contrast with wild-type plants, they do not show proximity-induced increases in stem elongation. In fact, they show proximity-conditional dwarfing. Significantly, the transgenic plants display increased harvest index compared with wild-type tobacco plants.[9] It is anticipated that suppression of shade avoidance would have similar benefits in a range of crops.

CONCLUSION

By perceiving the FR-rich light signals reflected from leaves, the phytochromes enable plants to detect the presence of neighboring vegetation and to initiate competitive growth responses: the shade avoidance syndrome. Although beneficial to individuals growing in mixed communities in the competition for light, shade avoidance in many crop plants leads to a wasteful redirection of resources that reduces harvestable biomass. Experimental approaches are being developed that may lead to suppression of shade avoidance in crops, with attendant increases in yield.

ARTICLES OF FURTHER INTEREST

Circadian Clock in Plants, p. 278
Floral Induction, p. 452
Photoperiodism and the Regulation of Flowering, p. 877
Photoreceptors and Associated Signaling I: Phytochromes, p. 881
Photoreceptors and Associated Signaling II: Cryptochromes, p. 885
Photoreceptors and Associated Signaling III: Phototropins, p. 889
Photoreceptors and Associated Signaling IV: UV Receptors, p. 893
Photosynthate Partitioning and Transport, p. 897
Plant Response to Stress: Ultraviolet-B Light, p. 1019

REFERENCES

1. Quail, P.H. Photosensory perception and signalling in plant cells: New paradigms? Curr. Opin. Plant Biol. **2002**, *14*, 180–188.
2. Smith, H. Phytochromes and light signal perception by plants: An emerging synthesis. Nature **2000**, *407*, 585–591.
3. Smith, H. Physiological and ecological function within thc phytochrome family. Annu. Rev. Plant Physiol. Plant Mol. Biol. **1995**, *46*, 289–315.
4. Smith, H.; Whitelam, G.C. The shade avoidance syndrome: Multiple responses mediated by multiple phytochromes. Plant Cell Environ. **1997**, *20*, 840–844.
5. Ballaré, C.L.; Scopel, A.L.; Sánchez, R.A. Far-red radiation reflected from adjacent leaves: An early signal of completion in plant canopies. Science **1990**, *247*, 329–332.
6. Whitelam, G.C.; Devlin, P.F. Roles of different phytochromes in Arabidopsis photomorphogenesis. Plant Cell Environ. **1997**, *20*, 752–758.
7. Devlin, P.F.; Patel, S.R.; Whitelam, G.C. Phytochrome E influences internode elongation and flowering time in *Arabidopsis*. Plant Cell **1998**, *10*, 1479–1487.
8. Devlin, P.F.; Robson, P.R.H.; Patel, S.R.; Goosey, L.; Sharrock, R.A.; Whitelam, G.C. Phytochrome D acts in the shade-avoidance syndrome in *Arabidopsis* by controlling elongation and flowering time. Plant Physiol. **1999**, *119*, 909–915.
9. Robson, P.R.H.; McCormac, A.C.; Irvine, A.S.; Smith, H. Genetic engineering of harvest index in tobacco through overexpression of a phytochrome gene. Nat. Biotechnol. **1996**, *14*, 995–998.

Social Aspects of Sustainable Agriculture

S

Frederick H. Buttel
University of Wisconsin, Madison, Wisconsin, U.S.A.

INTRODUCTION

Agriculture—conventional and sustainable, ''traditional'' and ''modern''—is intrinsically social (or sociotechnical) in nature. The social character of agriculture is particularly apparent, particularly widely recognized, and particularly crucial in the case of sustainable agriculture. Not only must any reasonable definition of sustainable agriculture include both social and economic as well as technological and ecological components, the roots of unsustainability and the pathways to sustainability are fundamentally social processes. This chapter will provide an overview of the most fundamental social dimensions of sustainable agriculture.

THE DEFINITION OF SUSTAINABLE AGRICULTURE

It is widely agreed that sustainable agriculture is intrinsically a joint social and ecological construct. Thus, for example, Ikerd[1] stresses the ''anthropocentric'' as well as ''ecocentric'' nature of agricultural sustainability, noting that the essence of sustainable agriculture is that ''we [in sustainable agriculture] are concerned about sustaining agriculture for the benefit of humans, both now and into the indefinite future.'' Sustainable agriculture is an agriculture that is ''ecologically sound, economically viable, and socially responsible.''[2] Allen similarly emphasizes that a conception of sustainable agriculture that fails to recognize the role of people, social actors, social institutions, and social movements is a limited one.[3]

Sustainable agriculture and organic farming are very closely related but are not exactly coterminous. Organic farming systems tend to be highly sustainable systems according to the ecological and social components of the definition of sustainable agriculture, and are often quite economically viable as well.[4,5] Many in the sustainable agriculture community, however, feel strongly that sustainability improvements in mainstream or conventional agriculture are as important as the expansion of organic farming and organic agro-food systems.

STRUCTURAL CAUSES OF THE LACK OF SUSTAINABILITY

Agricultural sustainability is not only a vision of the long-term goals for agriculture, which imply measuring sticks for or indicators of its achievement.[6] Sustainability also implies a critique of or a set of concerns about ''conventional'' agriculture—that mainstream agriculture has shortcomings in terms of environmental soundness, economic soundness, and social justice—and suggests that there have been trends over time that jeopardize the achievement of these three ends.

The historical U.S. pattern of abundant land, limited and/or relatively expensive rural labor, and a strong tendency toward overproduction and low commodity prices has encouraged a capital-intensive, chemical-intensive monocultural system with a very high degree of enterprise and spatial homogeneity (or specialization). Commodity, trade, and public research policies have historically reinforced these tendencies to monoculture, specialization, and heavy reliance on chemicals, leading to an agriculture that has significant shortcomings on all of the criteria implied in the definition of sustainable agriculture.[7]

SUSTAINABILITY AND THE CORPORATE-INDUSTRIAL AGRICULTURAL ECONOMY

Agricultural sustainability cannot be considered apart from the globalized and highly concentrated corporate economy within which sustainable agriculture, and agriculture as a whole, is enmeshed. Corporate concentration on input provision, food processing, and retailing limit the choices available to sustainable and other farmers. Corporate control over agriculture is increasing through processes such as the disappearance of ''open markets'' for agricultural products; the extension of vertical integration and contractual relationships; and the ever closer alignment of multinational chemical and seed companies on one hand, and food processors, manufacturers, and global trading companies on the other.[8] There is evidence that the increased corporate domination of

Encyclopedia of Plant and Crop Science
DOI: 10.1081/E-EPCS 120012948

agro-food systems has negative implications for achieving agricultural sustainability. Farmer contractees, for example, are typically directed contractually by ''integrators'' to use particular practices which, more often than not, are unsustainable.

It should be stressed, however, that corporations—often small ones, but also large firms—have been major agents in extending the scope of organic food production and organic marketing. There are two major nationwide organic chain grocery stores, and a number of corporations have been very successful in producing organic products such as potatoes, grapes, fresh fruits and vegetables, and so on. The corporate organization of the organic food industry represents a major dilemma for proponents of agricultural sustainability. Highly successful sustainable agriculture ideas will be attractive to private corporations, which will often be successful in diffusing products and practices. At the same time, corporate agricultural sustainability practices may undermine certain aspects of the sustainability agenda (e.g., large-scale monocultural production of organic potatoes in the Northern Intermountain states is associated with a considerable threat of soil erosion and loss of biodiversity).

There is considerable debate in the sustainable agriculture community as to whether the best indicator of the growth of sustainable agriculture is the national market share of organic food—much of which is due to the activities of very large corporations such as Whole Foods—or whether the extent of sustainable agriculture is most accurately measured by the prevalence of community-supported agriculture (CSA), farmers markets, local co-ops, community gardens, and direct marketing of food. Those with the strongest commitments to sustainable agriculture are also most likely to see food system localization and the re-creation of more local ''foodsheds'' as the heart and soul of a genuine and enduring sustainable agriculture.[9]

SUSTAINABILITY AS A SOCIAL MOVEMENT

Agricultural sustainability is as much a social movement as it is a set of technologies and production practices. Social definitions of sustainability—that is, the concerns, views, ideologies, and agendas of movement participants—are as or more important in shaping what sustainability is as ecological or biological indicators such as soil erosion rates, levels of soil organic matter, biodiversity, levels of runoff, and so on. What has put sustainability on the national and global agenda is primarily the organization and activities of three major types of groups: farmer-driven sustainable agriculture organizations and movements, consumer-driven or consumer-oriented sustainable agriculture initiatives (such as many CSA farms and most food co-ops and local food councils), and national and global environmental groups. Environmental groups have arguably done the most to increase the visibility and persuasiveness of the overall notion of sustainability, and of the particular concept of agricultural sustainability, though farmer-oriented groups do not always agree with environmentalists' agendas for agricultural sustainability.

Each of the three main types of sustainable agriculture organizations, however, agrees on the main components of the movement agenda: the need for more public research on sustainable practices, the need for public policy incentives for sustainability, the desirability of family farming, and the imperative to improve the environmental performance of agriculture. Increasingly, this consensus has extended to issues such as opposition to major trade liberalization agreements (the World Trade Organization and NAFTA) and opposition to genetic engineering of crops and foods.[10]

PUBLIC POLICY AND THE FUTURE OF AGRICULTURAL SUSTAINABILITY

Three necessary but insufficient conditions for the continued advance of sustainable agriculture are the development of improved sustainable technology; the presence of a vigorous and dynamic sustainability movement; and the development of a public policy environment that reduces the public policy disincentives to an environmentally sound, socially responsible, and economically viable agriculture. Of these three conditions, the public policy environment of agricultural sustainability is arguably the most important over the long term, even though the macro-public-policy environment is difficult to redirect, and there are areas of public policy disagreement among farmers, consumers, and environmental actors in the sustainable agriculture community. For example, some sustainability advocates, particularly those in the environmental community (and some farm groups in that community), believe that the federal (and global) levels of action are most important or efficacious, whereas others believe that at this time the local (community or regional) arena is where advocates can most easily make a difference. Nonetheless, it is apparent that over the long term sustainable agriculture cannot advance far in a public policy environment that involves the strong disincentives to sustainability that currently prevail. There is a need for more active federal, state, and local regulation of both the on-site and the off-site impacts of agriculture; for ending the commodity-driven pattern of federal agricultural policy that shovels the lion's share of subsidies in the direction of large, monocultural producers of overproduced commodities; and for redirection of the public agricultural research agenda. Some of the most innovative public policy ideas in sustainable agriculture are being developed in Europe. Green payments, taxes on pesticides and fertilizers, and the embracement of a ''multifunctionality'' approach to government agricultural policy are particularly promising policy instruments.[11] Each of these policy instruments would support sustainability and yield other benefits (reduced government outlays, rural development, reduced greenhouse gas emissions). Still, there are no technocratic shortcuts to sustainability. Sustainability is, and must remain, as much a social movement as it is a set of practices and measuring sticks of agro-food system performance.

ARTICLES OF FURTHER INTEREST

Ecology and Agricultural Sciences, p. 401
Sustainable Agriculture and Food Security, p. 1183
Sustainable Agriculture: Definition and Goals, p. 1187
Sustainable Agriculture: Ecological Indicators, p. 1191
Sustainable Agriculture: Economic Indicators, p. 1195

REFERENCES

1. Ikerd, J.E. Assessing the health of ecosystems: A socioeconomic perspective. Paper Presented at the First International Ecosystem Health and Medicine Symposium, Ottawa, Canada, 1994. (http://www.ssu.missouri.edu/faculty/jikerd/papers/Otta-ssp.htm), p. 2 (accessed October 2002).
2. Ikerd, J.E. New farmers for a new century. Paper Presented at the Youth in Agriculture Conference. Ulvik, Norway, February, 2000. (http://www.ssu.missouri.edu/faculty/jikerd/papers/Newfarmer1.htm), p. 3 (accessed October 2002).
3. Allen, P. Connecting the Social and Ecological in Sustainable Agriculture. In *Food for the Future: Conditions and Contradictions of Sustainability*; Allen, P., Ed.; Wiley: New York, 1993; 1–16.
4. *Sustainable Agriculture in Temperate Zones*; Flora, C.B., Francis, C., King, L., Eds.; Wiley-Interscience: New York, 1990.
5. *Interactions Between Agroecosystems and Rural Communities*; Flora, C.B., Ed.; CRC Press: Boca Raton, FL, 2001.
6. *A Life Cycle Approach to Sustainable Agriculture Indicators: Proceedings. Center for Sustainable Systems*; Airstars, G.A., Ed.; University of Michigan: Ann Arbor, MI, 1999.
7. Buttel, F.H. The sociology of agricultural sustainability: Some observations on the future of sustainable agriculture. Agric. Ecosyst. Environ. **1993**, *46*, 175–186.
8. Heffernan, W.D. *Consolidation in the Food and Agriculture System*; National Farmers Union: Denver, CO, 1999. (http://www.nfu.org/images/heffernan_1999.pdf) (accessed October 2002).
9. Kloppenburg, J., Jr.; Hendrickson, J.; Stevenson, G.W. Coming into the foodshed. Agric. Human Values **1996**, *13*, 33–42.
10. Kirschenmann, F. Questioning Biotechnology's Claims and Imagining Alternatives. In *Of Frankenfoods and Golden Rice*; Buttel, F., Goodman, R., Eds.; Wisconsin Academy of Sciences, Arts, and Letters: Madison, WI, 2002; 35–61.
11. Organisation for Economic Cooperation and Development (OECD). *Multidimensionality: A Framework for Analysis*; OECD: Paris, 1998.

Somaclonal Variation: Origins and Causes

Philip Larkin
CSIRO Plant Industry, Canberra, Australia

INTRODUCTION

Somaclonal variation is the genetic variation that is sometimes observed when plants are regenerated from cultured somatic cells, compared to the plant used as a source of the culture. Such variation represents both a problem and an opportunity. For those concerned with micropropagation of valuable elite clones, this variation can result in problematic off-types that diminish the commercial value of propagules. Likewise, genetically enhanced (transgenic) plants are carefully screened to avoid unwanted and unintended somaclonal variation, so that the commercially released clone has only the beneficial effects from the transgene. On the other hand, some somaclonal variants have proved to be of agronomic and commercial advantage and in a limited number of cases have been released as new cultivars. Various types of somaclonal mutation have been described, including point mutations, gene duplication, chromosomal rearrangements, and chromosome number changes. Transposable element movement and changes in DNA methylation have been implicated as mechanisms behind some, but not necessarily all, somaclonal variation. It may be that somatic cells, both in the plant and in culture, naturally accumulate these changes and that the tissue culture environment provides an opportunity for these mutations to be uncovered in tissue culture–derived plants.

BACKGROUND

Recognition that plants regenerated from cell culture could carry mutations began in the 1970s and was initially met with scepticism. The earliest systematic work on somaclonal variation was in sugarcane, tobacco, rice, ryegrass, lettuce and, potato.[1–4] Potato variants were observed after regeneration from cultured protoplasts; this was named protoclonal variation.[5] The more general designation ''somaclonal variation'' was coined when it had become evident that the phenomenon could occur following all forms of culture of somatic cells.[6]

NATURE OF THE GENETIC CHANGES

Table 1 summarizes types and examples of somaclonal variation.[1]

Experiments with multiple regenerated plants from isolated single-leaf protoplasts have demonstrated that somaclonal variation sometimes occurs during the period of culture and sometimes appears to preexist in the leaf. The frequencies of occurrence of these mutations vary widely. The maize ADH (alcohol dehydrogenase) point mutation was found only once among 645 regenerants. One β-amylase mutation in wheat was found among 149 regenerants in which four loci were screened. In some experiments[7] the rate of mutation from somaclonal variation was about tenfold higher than from mutagen ethyl methanesulfonate (EMS) on seed or pollen.

ORIGINS OF THE GENETIC CHANGES

The causes (nature) of the genetic changes can be described more easily than the originating mechanisms. The trigger for somaclonal variation may not be unique to cell cultures but may occur naturally in plant somatic tissues. In other words, cell culture may simply be an effective means to uncover and reveal the genetic changes by giving opportunity for the mutated cells to become mutated plants.

It has been suggested that the causative condition for somaclonal variation is high auxin levels, rapid cell division, or return to a juvenile genomic state, but there is little persuasive evidence. Likewise, the trigger has been described as genomic shock or plasticity, which occurs after the plant has exhausted its ordinary physiological responses to environmental stress.[8] These interpretations of the sometimes massive genetic changes observed with somaclonal variation seem reasonable but do not describe the actual mechanism of unleashing the genetic plasticity.

More recently it has been shown that cell culture can cause the activation of transposable elements.[9] Excision of a transposable element leaves a double-stranded break at the donor site. The cell attempts to repair the breaks

Encyclopedia of Plant and Crop Science
DOI: 10.1081/E-EPCS 120010550

Table 1 Types and examples of somaclonal variation

Type of genetic change	Examples	Comments
Point mutation	Alcohol dehydrogenase (ADH) gene	An altered electrophoretic mobility due to a single amino acid change
Simple genetic changes	Chlorsulfuron resistance in tobacco	These are simply inherited and may be due to one or more base changes
New isozymes	New β-amylase in wheat	A silent member of a multigene family may be activated
Gene amplification	Phosphinothricin resistance in alfalfa	Amplification of the number of copies of glutamine synthase gene
Gene deamplification	Reduction in rDNA spacer region in triticale	An 80% reduction in Nor locus of chromosome 1R
Chromosomal rearrangement	Deletions, telosomics, translocations, inversions	Breakage most often at or near heterochromatic blocks
Chromosome number	Monosomy, trisomy, extreme aneuploidy, gross ploidy changes	Such changes are not simply inherited, i.e., not according to Mendelian ratios

sometimes by strand-to-strand sister chromatid ligation. Subsequent rounds of replication initiate the so-called breakage-fusion-bridge cycle, which ceases only when new telomeres are added to the chromosome ends.

A number of researchers have observed both increases and decreases in C-methylation and A-methylation during cell culture.[3] This phenomenon can result in altered gene expression, although the direct effects are likely to be less stable over time. However, others[8] have described how short-term changes in methylation can result in stable point mutations. "Repeat-induced point mutation" (RIP) was first described in *Neurospora*, and may be a significant underlying cause of somaclonal variation in cultured plant cells; in the plant cell context "methylation-induced mutation" may be a more appropriate name. Newly methylated cytosines can be deaminated to form thymines, resulting in sequence mutation. It is also possible that transient changes in DNA methylation lead to bursts of transposable element movement, with consequent stable genetic changes and chromosome structural alterations, as explained above.

It is therefore difficult to envisage a single originating mechanism that can trigger all the various types of somaclonal variation. The current author therefore favours the view that there are a variety of mechanisms, most or all of which occur naturally in the somatic cells of plants and also in cultured plant cells, perhaps at a higher frequency. Plant regeneration from single somatic cells gives the opportunity to reveal these changes as solid mutant plants. In addition, cell culture itself may be sufficiently stressful to unleash the preprogrammed genomic shock, which is an adaptive mechanism that activates when ordinary physiological responses are insufficient. The occurrence of hot-spots of mutation and recurring menus of alternative alleles, described by some investigators, is consistent with this preprogrammed response.

HOMOZYGOUS MUTATIONS

A number of researchers have observed somaclonal variaton appearing as homozygous mutations (e.g., in rice, wheat, and maize). Homozygosity might be explained in a number of ways: The mutational mechanism is not random and occurs on both homologues; the mutation is rendered homozygous by a type of gene conversion; there are cycles of haploidy (or chromosomal monoploidy) and diploidy such that a mutation on one homologue is duplicated to achieve homozygosity. Some researchers[10] describe a type of somatic meiosis in culture that might explain the ability for mutations to become homozygous. Somaclonal variaton in soybean produces alleles already known in the soybean germplasm, suggesting a type of non-random mutation giving controlled formation of alternative alleles in response to stress.

PROBLEM FOR MICROPROPAGATION AND GENETIC MODIFICATION OF CROPS

Almost invariably micropropagation of horticultural species (and clonally propagated crops such as banana) is intended to produce chosen elite individuals in mass. Somaclonal variation is problematic under these circumstances, where even a low percentage of off-types is unacceptable for commercial use; a high percentage can be very costly, as has been proven in both the banana and oil palm industries.

Likewise, somaclonal variation can be problematic in the genetic modification of crops. A number of studies indicated higher levels of somaclonal variation following transformation procedures than in control populations of regenerants. Here, however, hundreds of individual transgenics are exhaustively tested and only proven elite individuals are chosen to become commercial releases. Somaclonal variation is not a great problem in species where transformation is relatively easy or where the breeding system allows backcrossing to eliminate unwanted changes. The problem is very significant in species such as sugarcane.

EXAMPLES OF SOMACLONAL MUTANTS BEING BENEFICIAL

The literature records instances of somaclonal variation in a large range of plant species, including major food crop species wheat, barley, sugarcane, rice, maize, potato, tomato, soybean, and oilseed rape. The potential improvements reported include enhanced resistance to fungal, bacterial, and viral diseases, improved insect and nematode resistance, enhanced economic yield, improved drought, chilling, salinity, and aluminium tolerance. In a limited number of cases there have been new varieties and cultivars named and released commercially. Examples are listed in Table 2.

Table 2 Examples of commercially available cultivars

Species	Somaclonal trait	Commercial name
Barley	Improved yield and downey mildew resistance	AC Malone
Wheat	Improved yield and agronomy	HeZu 8
Potato	Reduced tuber browning	White Baron
Blackberry	Thornless	Lincoln Logan
Flax	Salt and heat tolerance	Andro
Celery	Fusarium resistance	UC-TC
Celery	Fusarium yellows resistance	MSU-SHK5
Tomato	Fusarium resistance	DNAP-17
Rice	Picularia resistance and improved cooking quality	Dama

SPECIFIC EXPLOITATION OF CHROMOSOMAL CHANGES

Somaclonal variation involving subchromosomal recombination has been exploited to achieve the introgression into wheat of a particular segment of an alien grass chromosome with Barley Yellow Dwarf Virus (BYDV) resistance.[11] In this case it was possible to introduce the whole chromosome by backcrossing the wide hybrid between wheat and *Thinopyrum intermedium*. This alien chromosome neither pairs nor recombines with the wheat chromosomes. However, cells from the monosomic alien addition line (all 42 wheat chromosomes and one from the grass) were cultured, plants regenerated, and progeny families screened for recombination. Five of about 1000 families (0.5%) had a recombination in which a portion of the alien chromosome, carrying the BYDV resistance, had been translocated onto a wheat chromosome. Two new Australian wheat cultivars (cvs. Mackellar and Glover) have just been released with two of these translocations.

CONCLUSION

It was thought for some time that regeneration by somatic embryogenesis might render some protection from somaclonal variation, but this appears not to be the case. Reducing the length of time in culture does seem to reduce the likelihood of gross chromosomal variation but not necessarily the less obvious and fine structural changes. If it is true that cell culture merely enables the recovery of naturally occuring somatic mutations, then it seems unlikely that somaclonal variation can be entirely avoided whenever plants are regenerated from individual somatic cells. If, however, the underlying cause is a stress that triggers changes associated with genomic trauma, then some types of culture may be found to be less traumatic. A recent study in rice indicates that while no form of culture entirely eliminates somaclonal variation, protoplast culture nevertheless results in a much higher incidence. Some vigilence will therefore continue to be required in the selection of transgenic events for commercial release.

ARTICLES OF FURTHER INTEREST

Plant Cell Culture and Its Applications, p. 931
Transformation Methods and Impact, p. 1233

REFERENCES

1. Larkin, P.J. Somaclonal variation: History, method and meaning. Iowa State J. Res. **1987**, *61*, 393–434.
2. Duncan, R.R. Tissue culture-induced variation and crop improvement. Adv. Agron. **1997**, *58*, 201–240.
3. Veilleux, R.E.; Johnson, A.A.T. Somaclonal variation: Molecular analysis, transformation interaction and utilization. Plant Breed. Rev. **1998**, *16*, 229–268.
4. Jain, S.M. Tissue culture-derived variation in crop improvement. Euphytica **2001**, *118*, 153–166.
5. Shepard, J.F. Protoplasts as sources of disease resistance in plants. Annu. Rev. Phytopathol. **1981**, *19*, 145–166.
6. Larkin, P.J.; Scowcroft, W.R. Somaclonal variation—A novel source of variability from cell culture for plant improvement. Theor. Appl. Genet. **1981**, *60*, 197–214.
7. Gavazzi, G.; Tonelli, C.; Todesco, G.; Arreghini, E.; Raffaldi, F.; Vecchio, F.; Barbuzzi, G.; Biasini, M.; Sala, F. Somaclonal variation versus chemical mutagenesis in tomato (*Lycopersicon esculentum* L.). Theor. Appl. Genet. **1987**, *74*, 733–738.
8. Phillips, R.L.; Kaepler, S.M.; Olhoft, P. Genetic instability of plant tissue cultures: Breakdown of normal controls. PNAS **1994**, *91*, 5222–5226.
9. Peschke, V.M.; Phillips, R.L. Genetic implications of somaclonal variation in plants. Adv. Genet. **1992**, *30*, 41–75.
10. Nuti Ronchi, V. Mitosis and Meiosis in cultured plant cells and their relationship to variant cell types arising in culture. Int. Rev. Cytol. **1995**, *158*, 65–140.
11. Banks, P.M.; Larkin, P.J.; Bariana, H.S.; Lagudah, E.S.; Appels, R.; Waterhouse, P.M.; Brettell, R.I.S.; Chen, X.; Xu, H.J.; Xin, Z.Y.; Qian, Y.T.; Zhou, X.M.; Cheng, Z.M.; Zhou, G.H. The use of cell culture for the introgression of BYDV virus resistance from a *Thinopyrum intermedium* to a wheat chromosome. Genome **1995**, *38*, 395–405.

Somatic Cell Genetics of Banana

D. K. Becker
Queensland University of Technology, Brisbane, Queensland, Australia

M. K. Smith
Queensland Department of Primary Industries, Queensland, Australia

INTRODUCTION

Bananas and plantains (*Musa* spp.) are grown throughout the humid tropics and subtropics, where they are of great importance both as subsistence crops and as sources of domestic and international trade. They are monocotyledonous perennial herbs consisting of sympodial rhizomes and large pseudostems composed of tightly clasping leaf sheaths, slightly swollen at the base. Suckers are freely produced. Bracts and flowers are inserted independently on a peduncle and are usually deciduous by abscission, except for functional female ovaries in the basal hands. Basal flowers are generally female with male flowers on distal hands.

Edible, seedless banana, plantain cultivars are derived from intra- and interspecific hybridisation of the two seeded, wild diploid species, *Musa acuminata* (A genome) and *Musa balbisiana* (B genome). The haploid genome of both *M. acuminata* and *M. balbisiana* consist of 11 chromosomes, but the *acuminata* genome has been estimated as being slightly larger, 610 Mbp cf. 560 Mbp for *balbisiana*. Many edible hybrids are parthenocarpic, female sterile and triploid with the relative contribution of each species to the genome being annotated by either A or B. Dessert bananas are usually AAA, plantains AAB and cooking bananas ABB.

Banana breeding programs have largely focused on generating pest and disease resistant cultivars while at the same time retaining acceptable yield and fruit quality. The two major diseases of banana are Sigatoka leaf spots *Mycosphaerella* spp.) and Fusarium wilt (*Fusarium oxysporum* f. sp. *cubense*). Due to infertility, triploidy and long generation time, very few conventionally bred cultivars have reached commercial release. As a result, a great amount of effort has been directed towards improving banana through the manipulation of somatic cells.

SOMACLONAL VARIATION AND MUTATION INDUCTION

Banana has benefited greatly from in vitro-induced mutation. This is largely due to the resources directed towards this approach in view of difficulties associated with breeding triploid cultivars. Genetic variation can arise as a result of the tissue culture process or be intentionally induced by the use of physical and chemical mutagens. All three approaches have been applied to banana.[1,2]

Banana plants regenerated from in vitro culture sometimes exhibit morphological and biochemical variation due to genetic changes, which is of serious concern for clonal micropropagation. These genetic changes, nonetheless, may be exploited as a source of genetic variation for banana improvement. Somaclonal variation in banana usually leads to undesirable characteristics. However, there are several examples of somaclonal variants with advantageous characteristics, some of which have been released commercially. Variants identified include Tai-Chiao No 1 (AAA), TC1-299 (AAA) and Mutiara (AAB) with Fusarium resistance, a variant of SH 3436 (AAAA) with greater resistance to Sigatoka diseases, JD Special (AAA) with larger fruit, and higher yielding variants of Agbagba (AAB).

Of the physical mutagens used in banana and plantain improvement, gamma rays have been most commonly used. Differences in radiosensitivity have been observed among different genotypes and ploidy levels. In one study,[3] the LD_{50} was 20–25 Gy for diploid (AA), 30–35 Gy for triploid (AAA), and 35–40 Gy for the tetraploid (AAAA). Plantain and cooking clones (AAB, ABB) had a radiosensitivity of 25–35 Gy. Among a population of plants that had regenerated from a 60 Gy irradiated explant of ''Grande Naine'' (AAA), an early-flowering plant was identified. The clone was extensively field tested, released as ''Novaria,'' and entered into commercial production in Malaysia in 1993.

Ethyl methanesulphonate (EMS) has been widely used as a chemical mutagen in banana and has resulted, at least at an experimental level, in the generation of clones with increased *Fusarium* tolerance.[4] The concentration required is dependent on exposure time and whether a carrier is used. EMS has been shown to be effective at 8–16 mM with an exposure of 4–5 days and also at 200 mM with a much shorter exposure of 30 min. Dimethylsulphoxide (DMSO) has been used as a carrier agent as

Encyclopedia of Plant and Crop Science
DOI: 10.1081/E-EPCS 120010598

it facilitates greater uptake of EMS. Other mutagens such as sodium azide and diethylsulphate have also been shown to be effective in inducing variation from banana shoot apices.

Chemical methods have also been used to manipulate ploidy in *Musa*, with a view to incorporating the material into conventional breeding programs. By breeding and selecting elite diploid clones, and using them to generate autotetraploids and to cross with diploids, triploid bananas could be resynthesised. Tetraploidy has been readily induced with colchicine in *M. acuminata* and *M. balbisiana* seedlings and also in vitro shoot tips. The in vitro technique has the advantage of rapid multiplication and ease of distribution of new clones for evaluation and breeding.[5] Of importance is the number of in vitro multiplication cycles, usually more than three, following colchicine treatment in order to reduce chimerism. Treatment of shoot cultures with oryzalin has also been effective in producing high frequencies of tetraploids in banana.

SOMATIC EMBRYOGENESIS AND ORGANOGENESIS

Most reports of banana regeneration have been via embryogenic cell suspensions (ECSs). Potential uses of such cultures include micropropagation, as a source of regenerable protoplasts and for genetic transformation. There is also the potential for use as a source of cells for mutagenesis, although, to date, there have been no reports of this application. Banana somatic embryos are thought to be unicellular in origin and therefore offer the potential of generating nonchimeric plants from genetically altered cells.

ECSs have been generated using a range of explants including rhizome tissue, leaf bases, immature zygotic embryos, meristems, and immature male and female flowers. Currently, ECSs are most commonly derived from immature flowers or meristems.[6,7] Both these techniques appear to be applicable to a wide range of genotypes including diploids (AA, AB), triploids (AAA, AAB, ABB) and tetraploids (AAAB). ECSs contain a heterogeneous mix of cell types and their regenerative capacity depends on the proportion of cell types. Single isolated cells and large cell clumps (200 μm to 2 mm) in general do not give rise to somatic embryos, whereas small cell aggregates (50 to 100 μm), consisting of small cells with dense cytoplasms readily form somatic embryos.[8]

The ECSs from immature male flowers entail dissecting out clusters (hands) of immature flowers that are close to the apical meristem of the inflorescence and placing them on induction medium where they remain without subculture until embryogenic complexes appear. Embryogenic complexes are then transferred to liquid medium where a suspension of cells forms. Regeneration involves plating suspension cells on embryo development/maturation medium. Very large numbers of somatic embryos can be generated from these cells with as many as 3.7×10^5 embryos formed per 1 mL Packed Cell Volume (PCV) being reported.[6] When embryos are transferred to germination medium, the germination rate varies depending on their maturity. Leaving embryos on embryo formation/maturation medium for long periods (up to four months) results in a germination rate of over 80%.

To generate ECSs from meristems, highly proliferating clumps of meristems are induced by placing in vitro shoot cultures on multiplication medium that, depending on the cultivar, contains varying concentrations of BAP. Over a series of subcultures on this medium, primordia gradually stop forming leaves and meristems enlarge, resulting in large white meristematic nodules. Following the formation of meristematic clumps, the upper 4–6 mm of these structures (termed ''scalps'') are used as explants for induction of embryogenesis followed by initiation of ECSs in a manner similar to that of flower-derived cultures.[7] Recent work suggests that ''scalps'' are not required, but rather ECSs can be generated directly from longitudinal sections of conventional shoot tip cultures.[9]

In the only reported field trial of plants derived from ECSs, measurements were taken for the growth characteristics of 500 ''Grand Nain'' (AAA) plants regenerated from flower-derived ECS.[10] Overall, there were no significant differences in growth characteristics. The frequency of off-types for ECS-derived plants was nil, while the frequency for those derived from micropropagation was between 1.8% and 3.3% depending on the accession. This study suggests that plants regenerated from ECSs may not be any more prone to somaclonal variation than plants derived from shoot-tip culture.

PROTOPLAST ISOLATION AND CULTURE

Somatic hybridization of bananas is possible and has potential to assist the breeding of edible bananas. In a similar manner to manipulation of ploidy with chemicals, tetraploids could be generated via protoplast fusion and triploids resynthesised by crossing with diploids. Allotetraploids created by the somatic hybridization of two elite diploid clones would be very different from autotetraploids produced by chromosome doubling. As there are now a number of examples of plant regeneration from protoplasts, fusion is now a technique that can potentially contribute to banana-breeding programs. Genotypes for which plants have been regenerated from protoplasts include breeding diploids (AA), dessert bananas (AAA), plantains (AAB), and cooking banana (ABB).[11] All reports of successful plant regeneration have two factors in common. Firstly, protoplasts were derived from ECSs and secondly, feeder cells and/or protoplast culture at high densities was required.

GENETIC TRANSFORMATION

As this article is concerned with genetic manipulation of somatic cells, a brief mention should be made of genetic transformation. Both meristems and embryogenic cells have been targeted for transformation; however, due to greater efficiency and the greatly reduced risk of chimerism, embryogenic cells are the tissue of choice. Following the development of efficient regeneration systems for various cultivars, transformation systems using both microprojectile bombardment[12] and *Agrobacterium*-mediated transformation[9] of embryogenic cultures have been developed. Reports of transformation to date have focussed on the development of the technique itself and the examination of the strength and tissue specificity of various promoter elements. Work in this area has now moved on to conferring useful new traits including disease resistance and altered fruit characteristics. Indeed, potentially disease-resistant transgenic banana are currently undergoing glasshouse and field trials, and delayed fruit ripening using sense suppression of genes involved in ethylene biosynthesis has been reported.

CONCLUSION

The difficulties associated with conventional breeding of banana have been the driving force behind the genetic manipulation of somatic cells. In some cases, such as mutation breeding, new cultivars are already in commercial production. Other techniques, such as ploidy manipulation and protoplast fusion may provide assistance to conventional breeding or be utilised independently to generate new cultivars. Although at times some controversy is associated with genetic transformation, it does provide the potential to circumvent fertility problems and allow the transfer of genes within the banana gene pool or indeed introduce genes from completely unrelated organisms. The possibilities for banana improvement have entered a new and exciting phase based on our understanding of somatic cell genetics.

ARTICLES OF FURTHER INTEREST

Plant Cell Culture and Its Applications, p. 931
Plant Cell Tissue and Organ Culture: Concepts and Methodologies, p. 934
Polyploidy, p. 1038
Protoplast Applications in Biotechnology, p. 1061
Protoplast Culture and Regeneration, p. 1065
Somaclonal Variation: Origins and Causes, p. 1158
Somatic Embryogenesis in Plants, p. 1165

REFERENCES

1. Smith, M.K.; Hamill, S.D.; Becker, D.K.; Dale, J.L. Musacea. In *Biotechnology of Fruit and Nut Crops*; Litz, R.E., Ed.; CAB International: Wallingford, UK. *in press*.
2. Predieri, S. Mutation induction and tissue culture in improving fruits. Plant Cell, Tissue Organ Cult. **2001**, *64*, 185–210.
3. Novak, F.J.; Afza, R.; van Duran, M.; Omar, M.S. Mutation induction by gamma irradiation of in vitro cultured shoot tips of banana and plantain (Musa cvs.). Trop. Agric. (Trinidad) **1990**, *67*, 21–28.
4. Bhagwat, B.; Duncan, E.J. Mutation breeding of banana cv. highgate (*Musa* spp., AAA group) for tolerance to *Fusarium oxysporum* fsp. *cubense* using chemical mutagens. Sci. Hort. **1998**, *73* (1), 11–22.
5. Hamill, S.D.; Smith, M.K.; Dodd, W.A. In vitro induction of banana autotetraploids by cochicine treatment of micropropagated diploids. Aust. J. Bot. **1992**, *40* (6), 887–896.
6. Cote, F.X.; Domergue, R.; Monmarson, S.; Schwendiman, J.; Teisson, C.; Escalant, J.V. Embryogenic cell suspensions from the male flower of *Musa* AAA cv Grand Nain. Physiol. Plant. **1996**, *97* (2), 285–290.
7. Dhed'a, D.H.; Dumortier, F.; Panis, B.; Vuylsteke, D.; De Langhe, E. Plant regeneration in cell suspension cultures of the cooking banana cv. ''Bluggoe'' (*Musa* spp, ABB group). Fruits **1991**, *461* (2), 125–135.
8. Georget, F.; Domergue, R.; Ferriere, N.; Cote, F.X. Morphohistological study of the different constituents of a banana (*Musa* AAA, cv. Grande Naine) embryogenic cell suspension. Plant Cell Rep. **2002**, *19* (8), 748–754.
9. Ganapathi, T.R.; Higgs, N.S.; Balint-Kurti, P.J.; Arntzen, C.J.; May, G.D.; Van Eck, J.M. Agrobacterium-mediated transformation of embryogenic cell suspensions of the banana cultivar Rasthali (AAB). Plant Cell Rep. **2001**, *20* (2), 157–162.
10. Cote, F.X.; Folliot, M.; Domergue, R.; Dubois, C. Field performance of embryogenic cell suspension-derived banana plants (*Musa* AAA, cv. Grande Naine). Euphytica **2000**, *112* (3), 245–251.
11. Assini, A.; Haicour, R.; Wenzel, G.; Cote, F.; Bakry, F.; Forought-Wehr, B.; Bakry, F.; Cote, F.X.; Ducreux, G.; Ambroise, A.; Grapin, A. Influence of donor material and genotype on protoplast regeneration in banana and plantain cultivars *Musa* spp.. Plant Sci. **2002**, *162*, 355–362.
12. Becker, D.K.; Dugdale, B.; Smith, M.K.; Harding, R.M.; Dale, J.L. Genetic transformation of Cavendish banana (*Musa* spp. AAA group) cv. ''Grand Nain'' via microprojectile bombardment. Plant Cell Rep. **2000**, *19* (3), 229–234.

Somatic Embryogenesis in Plants

S

Ray J. Rose
The University of Newcastle, Callaghan, Australia

INTRODUCTION

Somatic embryogenesis refers to the remarkable ability of nonzygotic plant cells (including haploid cells) to develop through characteristic embryological stages into an embryo capable of developing into a mature plant. Somatic embryogenesis is an expression of totipotency and the associated differential gene expression.

Somatic embryos may be produced in nature in certain plant species as a form of apomixis known as adventitious embryony. Somatic embryogenesis in plants usually refers to the induction of somatic embryos in vitro, first demonstrated by both Steward and Reinert in 1958. Research into somatic embryogenesis has intensified as plant regeneration in vitro has come to be widely utilized in transformation and somatic hybridization.

This article emphasizes basic procedures and key variables for inducing somatic embryos, their developmental biology, and their mechanism of induction. Somatic embryogenesis may be direct or indirect. Direct somatic embryogenesis does not require a callus phase to induce somatic embryos from the explant. Indirect somatic embryogenesis has been most extensively studied and is widely used in transformation and somatic hybridization.

PRODUCING SOMATIC EMBRYOS—THE BASIC PROCEDURE

Steward was able to produce embryos from phloem tissue explants from carrot storage roots; Halperin[1] was able to show that auxin is essential for inducing embryos and that subsequent auxin withdrawal or lowering of auxin concentration is required for embryo maturation. Based on Dudits,[2] the basic paradigm is:

Explant + auxin + basal medium
↓
Callus
↓
Embryogenic cells
↓
Auxin reduction or removal
↓
Bipolar embryo
↓
Mature plant

In leguminous plants the first success was with *Medicago sativa* where calli (produced using an auxin + cytokinin) were induced to form embryos with a pulse of high 2,4-D;[2] and in Poaceae, immature embryos cultured in the presence of 2,4-D led to the induction of embryogenic calli from which plants could be regenerated when levels of 2,4-D were greatly reduced.[3,4] There are differences in these examples in the initial explant, the timing of auxin treatments, and the use of special cultivars. The auxin paradigm from carrots has had to evolve; a discussion of the parameters to consider in producing somatic embryos in a given species follows.

PRODUCING SOMATIC EMBRYOS—KEY VARIABLES

Cultivar Selection and Genotype

Within a species, certain cultivars are more amenable to plant regeneration via somatic embryogenesis, reflecting a strong genotypic component in the ability to form somatic embryos. In *M. sativa*, Bingham and co-workers produced Regen S using recurrent selection, and two genes have been invoked in breeding for somatic embryogenesis.[5,6] In *Medicago truncatula*, highly embryogenic plants can be obtained after a single cycle of tissue culture followed by selection.[6] Whether this has an epigenetic or genetic basis is unclear. Plant regeneration via somatic embryogenesis has been achieved in numerous species (Fig. 1)—dicotyledons and monocotyledons, woody and herbaceous species, annuals and perennials.[3,7,8]

Explant Type

The type of explant is a key variable in successful somatic embryogenesis. Immature embryos have been an important source of explants in obtaining somatic embryogenesis in such difficult species as wheat and soybean.[3,4] This is consistent with the idea that immature tissues are less likely to have suffered irreversible somatic genetic change. Explant type is less important in genotypes that are strongly embryogenic (e.g., in *M. sativa* Regen S and *M. truncatula*, 2HA explants from fully expanded leaves

Encyclopedia of Plant and Crop Science
DOI: 10.1081/E-EPCS 120010553

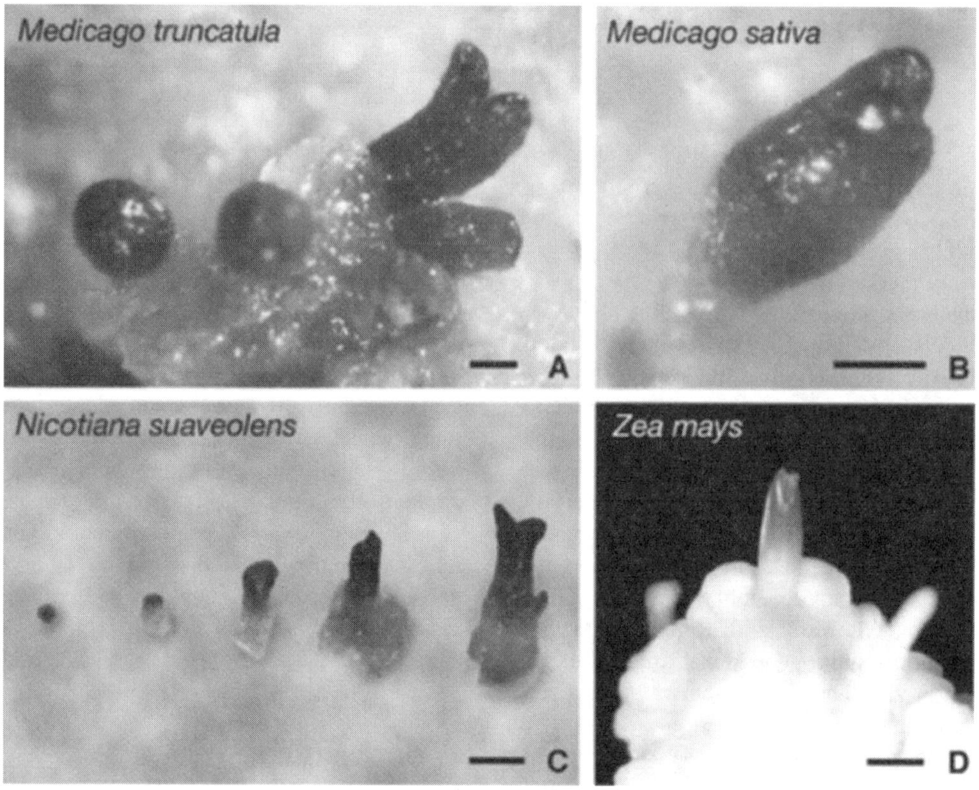

Fig. 1 Somatic embryogenesis in calli of two legumes, (A) the annual *Medicago truncatula* and (B) the perennial *Medicago sativa*; in (C) *Nicotiana suaveolens* and (D) the monocotyledon *Zea mays*. (A), (B), and (C) show examples of globular through (B) late torpedo stage embryos. In (D) an embryonic coleoptile is visible. Bar is 1 mm. (Photographs from author's laboratory by (A) K.E. Nolan, (B) J.T. Fitter, and (C,D) M.R. Thomas.

are fine), whereas in more recalcitrant legumes, immature tissues from embryos are necessary.[3,5,6]

Hormone Requirements

The hormonal regulation of any plant process is physiologically-, developmentally-, and species-dependent so it is not surprising that the auxin paradigm must be qualified. A cytokinin plus auxin may be essential for somatic embryogenesis,[6] or cytokinin may augment the auxin response.[2] In the case of alfalfa, an auxin plus cytokinin induces callus formation, and then a high 2,4-D pulse induces somatic embryos.[2]

Abscisic acid (ABA) also plays a role in somatic embryogenesis. Initially it appeared that the role of ABA was to mimic the in vivo effect where it is part of controlled desiccation during seed formation; it was thought to lead to improvement in embryo quality. ABA studies overall indicate that it stimulates auxin-induced somatic embryogenesis. In some cases, exogenous ABA alone can stimulate direct somatic embryogenesis.[9] This recent work suggests that the role of ABA may reflect its involvement in stress responses, which are known to be capable of inducing somatic embryogenesis.[9]

Nutrition

Nutrition plays an important role in optimizing somatic embryogenesis; a few widely used standardized basal media have been developed.[7] The form of nitrogen is an important variable—reduced nitrogen in addition to nitrate enhances embryo initiation and maturation.[7]

DEVELOPMENTAL BIOLOGY OF SOMATIC EMBRYOGENESIS

Histological studies suggest that the precursors of somatic embryos are frequently provascular cells. In these cells near the vascular tissue, endogenous substances in the phloem may be significant. Epidermal cells may also be sites of somatic embryo precursor cells in direct somatic embryogenesis.[9] In direct somatic embryogenesis, the precursor cells are already embryogenically determined, whereas in indirect somatic embryogenesis, determination has to be induced. In direct somatic embryogenesis a single cell serves as the embryo precursor, whereas in indirect somatic embryogenesis, single cells can be precursors,[10] although this is not unequivocal[8] and must be established by cell tracking studies.

Once somatic embryos are formed by direct or indirect somatic embryogenesis, it is common for secondary cycles of somatic embryogenesis to be initiated, suggesting that embryogenicity has been selected for in some way, similar to the increased embryogenicity of regenerated plants in *Medicago truncatula*.[6] The development of the mature somatic embryo proceeds through stages similar to their zygotic counterparts (Ref. 3, Fig. 1).

THE MECHANISM OF INDUCTION OF SOMATIC EMBRYOGENESIS

An understanding of the induction of somatic embryogenesis requires a number of fundamental effects to be considered, including the consequences of the wounding effect when tissue is excised, hormonal influences, and how the genetic program is reset or redetermined to induce the sequential expression of embryogenic genes (see accompanying flow diagram). Broadly speaking, we can assume that most of the somatic embryo development genes will be similar to in vivo embryogenesis, and that it is the induction of somatic embryogenesis that is special.

Explant (wound stress) + hormone(s)
↓
Redetermination
↓
Early embryo genes

Hormone Action and Stress Influences

Hormones are the key regulators in inducing embryogenic cells. Although progress has been made in understanding basic hormone action, understanding is required in the context of somatic embryogenesis and in relation to genes such as somatic embryo receptor kinase (*SERK*). In *Daucus carota*, it has clearly been illustrated that stress can both induce and stimulate somatic embryogenesis, and in some studies stress has been related to the production of ABA.[9]

Redetermining the Genetic Program

There is clearly a complex of signalling pathways to be unravelled to link the earliest hormone- and stress-induced changes to chromosome remodelling and transcriptional activation. Chromatin remodelling involving changes in methylation may be one of the earliest changes.[3]

Early Embryo Development Genes

An understanding of somatic embryogenesis induction requires that the earliest genes in the embryo pathway be identified. The most substantive work has been carried out on the *SERK* gene, which has been shown to be a marker of embryogenic cells in *Daucus carota*, *Dactylis*, and *Arabidopsis*.[10] An important result has been the demonstration that transformation of the *SERK* gene into *Arabidopsis* can increase somatic embryogenesis.[10] The ligand for this receptor and the substrate for serine-threonine kinase activity is as yet unknown. *SERK* expression initiated in cells destined to become embryos continues until the globular stage, and is also part of normal sexual embryogenesis.

CONCLUSION

Somatic embryogenesis is a characteristic of flowering plants that can be unlocked in most species by defined in vitro culture of a suitable genotype, explant, and appropriate hormone regulatory signals. Although auxin is a key variable, there are repertoires of strategies that require testing in order to obtain a suitable protocol for somatic embryogenesis in a given species. Understanding the mechanism requires more fundamental knowledge of the genetic regulation of cellular dedifferentiation and differentiation.

ARTICLES OF FURTHER INTEREST

Anther and Microspore Culture and the Production of Haploid Plants, p. 43
In Vitro Morphogenesis in Plants—Recent Advances, p. 579
In Vitro Plant Regeneration by Organogenesis, p. 584
Plant Cell Tissue and Organ Culture: Concepts and Methodologies, p. 934
Protoplast Culture and Regeneration, p. 1065
Regeneration from Guard Cells of Crop and Other Plant Species, p. 1081

Somaclonal Variation: Origins and Causes, p. 1158
Transformation Methods and Impact, p. 1233

REFERENCES

1. Halperin, W. Morphogenetic studies with partially synchronized cultures of carrot embryos. Science **1964**, *146*, 408–410.
2. Dudits, D.; Bogre, L.; Gryorgyev, J. Molecular and cellular approaches to the analysis of plant embryo development from somatic cells in vitro. J. Cell Sci. **1991**, *99*, 475–484.
3. Merkle, S.A.; Parrott, W.A.; Flinn, B.S. Morphogenetic Aspects of Somatic Embryogenesis. In *In Vitro Embryogenesis in Plants*; Thorpe, T.A., Ed.; Kluwer Academic Publishers: Dordrecht, 1995; 155–203.
4. Vasil, I.K. Progress in the regeneration and genetic manipulation of cereal crops. Bio/technology **1988**, *6*, 397–402.
5. Reisch, B.; Bingham, E.T. The genetic control of bud formation from callus cultures of diploid alfalfa. Plant Sci. Lett. **1980**, *20*, 71–77.
6. Rose, R.J.; Nolan, K.E.; Bicego, L. The development of the highly regenerable seed line Jemalong 2HA for transformation of *Medicago truncatula*—Implications for regenerability via somatic embryogenesis. J. Plant Physiol. **1999**, *155*, 788–791.
7. Ammirato, P.V. Embryogenesis. In *Handbook of Plant Cell Culture Volume 1, Techniques in Propagation and Breeding*; Evans, D.A., Sharp, W.R., Ammirato, P.V., Yamada, Y., Eds.; MacMillan: New York, 1983; 82–123.
8. Williams, E.G.; Maheswaran, G. Somatic embryogenesis: Factors influencing coordinated behavior of cells as an embryogenic group. Ann. Bot. **1986**, *57*, 443–462.
9. Nishiwaki, M.; Fujino, K.; Koda, Y.; Masuda, K.; Kikuta, Y. Somatic embryogenesis induced by the simple application of abscisic acid. Planta **2000**, *211*, 756–759.
10. Hecht, V.; Vielle-Calzada, J-P.; Hartog, M.V.; Schmidt, E.D.L.; Boutilier, K.; Grossniklaus, U.; de Vries, S.C. The Arabidopsis *SOMATIC EMBRYOGENESIS RECEPTOR KINASE 1* gene is expressed in developing ovules and embryos and enhances embryogenic competence in culture. Plant Physiol. **2001**, *127*, 803–816.

Soy Wood Adhesives for Agro-Based Composites

Monlin Kuo
Deland J. Myers, Sr.
Iowa State University, Ames, Iowa, U.S.A.

INTRODUCTION

Before the advent of synthetic thermosetting adhesives in the 1950s, soybean, blood, and casein glues were the major adhesives for production of plywood and other glued wood products. Uncertainty regarding future supplies of petroleum-derived chemicals and stringent regulations on toxic emissions from building materials bonded with certain synthetic resins has compelled the industry to reevaluate wood adhesives from renewable resources. Rapid progress during the past 10 years in the development of soybean-based wood adhesives will be emphasized. The use of agricultural residues for panel products also will be discussed.

HISTORICAL PERSPECTIVES

Soybean-based adhesives were developed in 1928 for Douglas fir plywood production in the U.S. Pacific Northwest.[1] In the original formulation, partially defatted (extruded and expended) soy meal or flour was mixed with water and dispersed (denatured) with sodium hydroxide. Later formulations also included the use of defatted (solvent-extracted) soy flour and chemicals such as carbon disulfide, hydrated lime, and sodium silicate to impart a better water resistance to the adhesive bonds, and a small amount of preservatives to prevent mold growth.[2] Soy protein, about 50 to 55% in soy flour, is the active bonding agent. Carbohydrates (over 30%) in soy flour provide useful consistency properties for the glue but contribute little adhesion. Soy protein isolates also were used to formulate wood adhesives, but they did not offer enough advantages over soy flour to offset the cost. Annual consumption of soy meal or flour for plywood in North America was about 60 million pounds in 1940 and reached a peak of 100 million pounds in 1950.[2]

Soybean plywood adhesives typically contained 20% soy flour and 10% chemical solids in water and had high pH and viscosity. The plywood's assemblies were either cold- or hot-pressed for up to 9 minutes, but it would take several hours to set the adhesive bonds completely. The plywoods adhesive bond strength was adequate, but the adhesive bond had very poor moisture resistance. Petroleum-based thermosetting adhesives, mainly urea-formaldehyde (UF) and phenol-formaldehyde (PF) resins, because of their low costs, short hot-press time for a high production rate, and water-resistant adhesive bonds, quickly replaced soybean and other protein glues after World War II. In 1999 North America consumed 4.2 billion pounds of UF for interior plywood, particleboard, and fiberboard and 2.99 billion pounds of PF for exterior plywood and oriented strand board (OSB), as well as 195 million pounds of isocyanate resins for OSB and agro-based particleboard.[3] Very little proteinacious wood adhesive is being used today by the wood industry.

CURRENT RESEARCH AND CHALLENGES

Soy-Based Adhesives

Carbohydrates and proteins are the most abundant renewable substances suitable for use as wood adhesives. It is a significant challenge to develop adhesives from these natural substances at reasonable costs to compete with synthetic thermosetting adhesives and to meet stringent performance requirements. The current concept to achieve this goal is to use these natural substances as copolymers with synthetic polymers to reduce the dependency on petroleum-derived chemicals. In this respect, proteins are more suitable than carbohydrates because amino, carboxyl, aliphatic and aromatic hydroxyl, and other functional groups in proteins provide abundant functionality for chemical cross-linking. Research of protein-based adhesives during the past 10 years has been made mainly with soy protein, primarily because of its availability and low cost.

Roland Kreibich (Weyerhaeuser scientist, retired) invented a cold-setting adhesive for finger- jointed lumber containing equal parts of soy protein isolate and phenol-resorcinol-formaldehyde (PRF) resin. In this lumber finger-jointing system, a soy isolate hydrolyzate and a PRF resin are separately applied onto different lumber fingers. Upon joining the two fingers a gel is formed immediately from the cross-linking between soy protein and PRF, and the adhesive bond is cured at the ambient temperature without further application of pressure. This

Encyclopedia of Plant and Crop Science
DOI: 10.1081/E-EPCS 120010448

invention has not been published, but the adhesive is being produced and sold by a chemical company in Oregon. A U.S. patent[4] describes methods of preparing a soybean-based molding compound for rigid biocomposite materials. This soybean-based powder resin was formulated by cross-linking soy flour with 12% methyl diphenyl isocyanate (MDI). The powder resin was mixed with recycled paper fiber in a ratio of 4 parts resin to 6 parts fiber to produce molded composite products. More recently, a PF-cross-linked soy resin composed of 70% defatted soy flour and 30% of a PF cross-linking agent was developed at Iowa State University.[5] This light-colored soy resin can be used as a liquid resin for exterior grade plywood or as a powder resin for molded products, but it is not adequate as a spray resin for OSB and fiberboard. Subsequent research at Iowa State University resulted in an adhesive resin containing 70% soy flour hydrolyzate and 30% PF that could be used as a spray resin for OSB and fiberboard production.[6] A similar PF-cross-linked soy resin composed of 30% soy flour hydrolyzate and 70 % PF has been developed for OSB production.[7] About 3.5% spray-dried animal blood is used as a forming agent in plywood glues for foam extrusion application. Research at the USDA's National Center for Agricultural Utilization Research has shown that soy protein has desirable mixing and foaming properties for plywood glue foam extrusion.[8] The potential to replace blood with soy protein in plywood glue formulation would significantly reduce plywood production cost because soy flour is cheaper than spray-dried animal blood ($0.22/lb vs. $.0.40/lb).

Research on the use of soybean for wood adhesives during the past decade has resulted in the commercialization of the soy isolate/PRF lumber finger jointing system and the MDI-cross-linked soy flour as a compression-molding compound. Development of the PF-cross-linked soy resins and the use of soy flour as a foaming agent in the plywood glues discussed above also are at a stage very close to being commercialized.

Agricultural Fibers

Growing concerns in the United States over the environment have led to changes of forest management practices, resulting in a significant reduction in wood harvest in the midst of a growing demand for wood fiber. The use of agrofibers for pulp and composite materials is commonplace in many parts of the world where there is a lack of forest resources. There is clearly potential in North America to use underutilized agrofibers for industrial purposes, and since 1995 there has been a proliferation of new manufacturing facilities in North America to produce particleboard from wheat straw.[9] The surfaces of crop plant parts are lined with an inert cuticle layer that is very difficult to bond with UF and PF resins, but it can be effectively bonded with more expensive isocyanate resins. However, the straw particleboard industry is experiencing difficulties in the use of isocyanate resins, including high resin cost, lack of tack for mat integrity, and the difficulty of releasing boards from hot presses.

Research at Iowa State University favors the use of agrofibers for the production of fiberboard instead of particleboard. The basic assumption is that the specific cuticle surface area can be very much reduced by refining crop residue (e.g., thermomechanical pulping) into pulp so that the resulting fiber furnish can be bonded with UF, PF, or soybean-based resins to produce fiberboard. It has been shown that cornstalk and switchgrass fibers are inferior to wood fiber for fiberboard, but satisfactory medium-density fiberboard (MDF) and high-density fiberboard (hardboard) bonded with UF, PF, and soybean-based resins could be made by mixing equal parts of either cornstalk or switchgrass fiber with wood fiber.[10]

FUTURE PROSPECTS

The wood products industry in the past 50 years has almost totally depended on petrochemicals for manufacturing wood-based composite products. Since the oil embargo of the mid-1970s, there has been a constant threat of worldwide oil shortage. Therefore, the wood products industry strongly favors research into renewable materials for wood adhesives, and the agricultural industry also is eager to invest in researching such industrial uses to expand their markets. The prices of wood adhesive raw materials fluctuate, but the order of their costs has not changed much in the past 20 years. It ranges from less than $0.10/lb for urea, followed by soy flour, phenol, blood, melamine, and isocyanates (e.g., MDI), to more than $1.5/lb for resorcinol. The current price of soy flour is about $0.15/lb, compared to just under $0.30/lb for petroleum-derived phenol. Soybean clearly has an advantage over phenol, not only in price but also for its stable future supply. Recently developed soybean-based adhesives may not be fully competitive in performance with synthetic adhesives, but they can be implemented if circumstances require.

Agricultural residue has been an important raw material for pulp and composite panel production in China. Agrofibers also are becoming important raw materials for the same purposes in other regions of the world that are poor in forest resources. The agro-based panel industry in North America has just begun. Based on available volumes of agricultural residues, the North American agro-based panel industry could potentially grow to a size about two-thirds as large as the present wood-based panel industry.[9]

S

ARTICLES OF FURTHER INTEREST

Environmental Concerns and Agricultural Policy, p. 418
Non-wood Plant Fibers: Applications in Pulp and Papermaking, p. 829
Paper and Pulp: Agro-based Resources for, p. 861
Sustainable Agriculture: Definition and Goals, p. 1187
Switchgrass As a Bioenergy Crop, p. 1207

REFERENCES

1. Laucks, I.F.; Davidson, G. Vegetable Glue and Method of Making Same. US Pat. No.1,689,732, Oct. 30, 1928.
2. Lambuth, A.L. Soybean Glues. In *Handbook of Adhesives*, 2nd Ed.; Skeist, I., Ed.; Van Nostrand Reinhild: New York, 1977.
3. Johnson, R.S. An Overview of North American Wood Adhesive Resins. In *Wood Adhesives 2000*; Proc. No. 7252, Forest Products Society: Madison, WI, 2001.
4. Riebel, M.J.; Torgusen, P.L.; Roos, K.D.; Anderson, D.E.; Gruber, C. Biocomposite Material and Method of Making. US Pat. No. 5,635,123, June 3, 1997.
5. Kuo, M.L.; Myers, D.L.; Heemstra, H.; Curry, D.G.; Adams, D.O.; Stokke, D.D. Soybean-based adhesive resins and composite products utilizing such adhesives. US Pat. No. 6,306,997, Oct. 23, 2001.
6. Kuo, M.L.; Stokke, D.D. Soybean-Based Adhesive Resins for Composite Products. In *Wood Adhesives 2000*; Proc. No. 7252, 2001, Forest Products Society: Madison, WI, 2001.
7. Hse, C.Y.; Fu, F.; Bryant, B.S. Development of Formaldehyde-Based Wood Adhesives with Co-Reacted Phenol/Soybean Flour. In *Wood Adhesives 2000*; Proceedings. No 7252, Forest Products Society: Madison, WI, 2001.
8. Hojililla-Evangelista, M.P.; Dunn, L.B., Jr. Foaming properties of soybean protein-based plywood adhesives. J. Am. Oil Chem. Soc. **2001**, *78* (6), 567–572.
9. Bowyer, J.L.; Stockmann, V.E. Agricultural residues—An exciting biobased raw material for the global panels industry. For. Prod. J. **2001**, *51* (1), 11–21.
10. Kuo, M.L.; Adams, D.; Myers, D.; Curry, D.; Heemstra, H.; Smith, J.L.; Bian, Y. Properties of wood/agricultural fiberboard bonded with soybean-based adhesives. For. Prod. J. **1998**, *48* (2), 71–75.

Spatial Dimension, The: Geographic Information Systems and Geostatistics

Francis Pierce
Washington State University, Prosser, Washington, U.S.A.

Oliver Schabenberger
SAS Institute Inc., Cary, North Carolina, U.S.A.

Max Crandall
ESRI, Redlands, California, U.S.A.

INTRODUCTION

Crop productivity is regulated by complex interactions among multiple factors that affect plant growth and development. Historically, research has overcome limitations to inherent plant productivity by altering a crop through breeding and genetics, by direct alteration of a crop during its life cycle (bioregulation, pruning), or by altering the growing environment through managed inputs. Scientific advances in these pathways have produced large increases in plant productivity during the 20th century, although there remain considerable differences in crop performance from place to place and from year to year. Variability in crop productivity commonly exists because many of the factors regulating plant performance—soils, weather, pests, and diseases—vary continuously in space and time.

DISCUSSION

Scientists and farmers have made attempts to account for and control the effects of spatio-temporal variation in crop production; however, the tools to effectively deal with this source of variability have lagged behind other advances in plant and crop science. With a suite of technologies and principles composing what is commonly termed precision agriculture,[1] quantifying and managing for spatial variability in crop production is possible at increasingly smaller scales of resolution. The combination of technologies including global position systems (GPS), sensors, controllers, radios, and high-speed computers makes the acquisition, storage, and use of spatial data possible at extraordinary levels of spatial and temporal detail. However, the value of spatial data would be limited without equally impressive tools to manage and analyze these data and subsequently communicate their meaning to users in a variety of formats.

Two tools of great significance in this regard have evolved over the last few decades. These are geographic information systems (GIS), whose roots date back to the 1960s when cartographers first began to adopt computer techniques in mapmaking,[2] and geostatistics, first developed for the mining industry, also in the 1960s.[3] Briefly, a GIS provides a set of tools for collecting, storing, retrieving, transforming, and displaying spatial data for a particular set of purposes,[2] while geostatistics provides a set of tools aimed at understanding and modeling spatial variability.[4] Currently, geostatistics is to some extent offered as an integral part of a GIS but commonly, geostatistical analyses are often performed in an external software application dedicated to geostatistics and the resultant data or functions imported into a GIS.

GEOGRAPHIC INFORMATION SYSTEMS

A GIS allows you to visualize agricultural data. Nearly all agricultural data have some form of spatial component, and a GIS allows you to visualize spatially referenced tabular data in meaningful maps—data that otherwise are very difficult to interpret. The GIS represents agricultural data using geographic coordinates, which serve as the foundation for GIS software. GIS software platforms range from field GIS to desktop GIS, to Internet-enabled GIS, to the popular RDBMS (relational database management system) extended with GIS. These scalable GIS platforms operate independently or integrate together for the enterprise. The common geographical coordinate space of agricultural data is a powerful feature of the GIS that lets users from multiple disciplines manage a project and collaboratively share their work. A farmer, an agronomist, and an agricultural economist can all assess a crop condition and evaluate it using their own professional practices, which are applied through the GIS. Specialized features in the GIS serve to document each sequential

Encyclopedia of Plant and Crop Science
DOI: 10.1081/E-EPCS 120010499

spatial operator and process used. This allows different users to compare and share their spatial analyses and results for optimum decision making.

An agricultural landscape is made up of many interconnected spatial components; a GIS is used to model and analyze this complexity. Georeferenced map layers can include hydrological features, soil characteristics, slope and its derivatives, yield, pests, and anything relevant to a particular investigation. The functionality to extract and highlight certain data features from one layer of information by defining a data feature from a second layer is a powerful analytic tool. Spatial operations are available, like distance buffers that allow the determination of boundaries around a location such as a creek. Some features—an irrigation well, for example—have many associated attributes that cannot be defined as a simple location. Knowledge about a well's depth, the salinity of its water, and the time required for an aquifer to recharge it are some of the data attributes that can be evaluated with a GIS.

A field GIS provides the ability to look at images and maps of fields while actually standing in them. Editing capabilities provide the functionality to map a measurement or observation such as those encountered in crop scouting activities. Users are able to collect in the field whatever information is relevant to them and record it directly using the map to provide the geographic location. The data highlighted and recorded in the field GIS map can be transferred to other GIS platforms, such as an Internet GIS server, or to a desktop GIS.

Desktop GIS software provides tools to carry out more intensive computational procedures such as those made using satellite images or aerial photographs. Images are generally raster based, being made of discrete cells with numeric values. Data layers that are made up of points, lines, and polygons are referred to as vector data. Raster data can be combined with vector data, but require different tools for management and interpretation. Remote sensing/GIS applications include estimating crop vigor, the remote estimation of areas in a field that are under physiological stress, and the documentation of natural resource conditions, such as soil erosion extent.

A GIS is well suited to address significant social and economic issues related to agriculture, although the mechanics of engineering a GIS solution from the farm field to the grocery store shelf can be complex. For example, the geodatabase underlying a GIS can be engineered for the application development of intelligent polygonal objects that can spatially account for transactions in the food supply chain for identity preserve (IP) requirements. This type of GIS implementation is able to record every transaction on every parcel, warehousing metadata and other critical data required for future certification of genetically modified organism (GMO) compliance regulations.

GEOSTATISTICS

Although spatial data are essential to GIS, they are sometimes too expensive or difficult to obtain. Fortunately, techniques are available to model spatial variability based on limited sampling from which estimates of spatial data can be obtained. These so-called geostatistical techniques rely on the concept of spatial continuity (i.e., the notion that two attributes close to each other are more likely to have similar values than are two that are far apart).

The scientific methods used to collect, summarize, analyze, and interpret empirical data that are referenced geographically are assembled under the heading of spatial statistics. The three main components of spatial statistics are distinguished through the data structures as methods applicable to lattice, point pattern, and geostatistical data. A geostatistical data set consists of measurements or classifications of one or more attributes of interest, along with the geographic coordinates at which the observations were obtained throughout a continuous domain.

Statistical methods appeal to a random mechanism in order to separate sources of variation. Consider, for example, the sampling of crop yield on a field at 50 randomly chosen locations. The geostatistical method appeals to a random mechanism that generates the entire (continuous) crop yield surface (a random field). The 50 samples represent an incomplete observation of a randomly generated, continuous surface. It is of interest to reconstruct the surface by predicting the yield values at unsampled locations (i.e., to produce a continuous map of crop yields). Worth noting is that a second realization—in the statistical sense—of the random process would lead to a different crop yield surface.

The prediction task is accomplished by methods of kriging. The term ''kriging'' was coined by Matheron[3] in recognition of mining engineer D.G. Krige. Although kriging is often considered an optimal method of spatial prediction, the classical kriging predictors are best only in the sense of minimizing the mean square prediction error in the class of predictors that are unbiased and linear functions of the observed data.

Computation of the kriging predictor requires: 1) knowledge (simple kriging) or a model of the large-scale trend (constant mean: ordinary kriging; linear mean function: universal kriging), and 2) the ability to evaluate the autocovariance between attributes at spatial locations. In practice, the large-scale trend and the spatial covariance structure are usually subject to estimation. Because estimating the autocovariance or semivariance (one-half the variance of the difference between two attributes) is usually not possible without some form of stationarity assumption (the expected value is constant and does not depend on position), it is important to remove large-scale trends in the data prior to estimating covariances or the semivariogram to prevent these representations of the second-order structure from depicting spurious spatial dependencies. Alternatively, methods that estimate large-scale trend and spatial correlation structure simultaneously can be employed. Likelihood methods belong to this latter class and have recently gained much momentum. Even for stationary processes, the fact that process parameters are estimated is typically not reflected in measures of prediction error. Estimates of the prediction error that consider estimates of semivariogram (covariogram) parameters as fixed quantities tend to overestimate the precision of kriged maps.

Through the years, a large number of kriging variants were developed to cope with the varied situations in which predictions of random attributes are needed. We mention by name only a few: block-kriging, indicator kriging, probability kriging, trans-Gaussian kriging, log-normal kriging, cokriging, and disjunctive kriging. For details on the various kriging methods and their implementation consult, for example, Journel and Huijbregts,[5] Cressie,[6] Chilès and Delfiner,[7] and Schabenberger and Pierce.[8] The basic premise of these variations is the same: to obtain predictions that are best based on the second-order properties of the process under study.

It has been questioned whether the rapid increase in the complexity of scientific questions can be met by current or new geostatistical concepts based on these classical ideas.[9] In particular, as the temporal component in spatial data is becoming increasingly important, new approaches for modeling and analyzing spatio-temporal data structures may be called for that make allowances for nonstationarity in time and/or space. Hierarchical spatial models that incorporate spatial correlation structures implicitly by considering parameters of a spatial model not as fixed quantities but as random variables at different temporal and/or spatial scales offer great promise in this regard.

CONCLUSION

Projected advances in computer processing and storage capacity suggest that in the near future, such massive amounts of data will be collected and stored that very little data will ever see the human eye; rather, data will be automatically processed and relevant information graphically displayed. Thus, spatial data will become increasingly available and the ability of plant and crop scientists to understand the spatio-temporal dimensions of agriculture will become increasingly important in the production and marketing of food. Both GIS and geostatistics will be tools of choice for plant and crop scientists as they strive to manage and understand the spatial nature of agricultural data.

ARTICLES OF FURTHER INTEREST

Crops and Environmental Change, p. 370
Integrated Pest Management, p. 612

REFERENCES

1. NRC. *Precision Agriculture in the 21st Century: Geospatial and Information Technologies in Crop Management*; National Academy Press: Washington, D.C., 1997.
2. Burrough, P.A.; McDonnell, R.A. *Principles of Geographic Information Systems*; Oxford University Press, Inc.: New York, 1998.
3. Matheron, G. Principles of geostatistics. Econ. Geol. **1963**, *58*, 1246–1266.
4. Deutsch, C.V.; Journel, A.G. *GSLIB: Geostatistical Software Library and User's Guide*; Oxford University Press: New York, 1998.
5. Journel, A.G.; Huijbregts, C.J. *Mining Geostatistics*; Academic Press: London, 1978.
6. Cressie, N.A.C. *Statistics for Spatial Data*; John Wiley & Sons: New York, 1993.
7. Chilès, J.P.; Delfiner, P. *Geostatistics. Modeling Spatial Uncertainty*; John Wiley & Sons: New York, 1999.
8. Schabenberger, O.; Pierce, F.J. *Contemporary Statistical Models for the Plant and Soil Sciences*; CRC Press: Boca Raton, FL, 2002.
9. Christakos, G. *Modern Spatiotemporal Geostatistics*; Oxford University Press: New York, 2000.

Starch

Wolfgang Eisenreich
Alfons Gierl
Adelbert Bacher
Technische Universität München, Garching, Germany

INTRODUCTION

The seeds of cereals are an important metabolic sink and an essential source of human and animal nutrition. Information on the biosynthetic pathways of carbohydrates, amino acids, and vitamins in crops provide the basis for the metabolic engineering of plants with improved nutritional profiles. In plants, pathways leading to starch, amino acids, and vitamins have been investigated in some detail at the level of enzymes and their cognate genes, but quantitative aspects of carbon flux including transport processes under in vivo conditions have been investigated less comprehensively.

Although it is clear that isotope incorporation studies can contribute important metabolite flux information, most isotope incorporation studies in the literature are limited to simple concepts that assume metabolism proceeds in a linear and unidirectional fashion. Based on this implicit assumption, the diversion of isotope from a given precursor to a given target compound is accepted as evidence for direct metabolic relatedness between the respective molecular species. In reality, however, metabolism is a complex and nonlinear network, and metabolic flux can occur from any node in the network to virtually any other node. As a consequence, isotope incorporation studies have been repeatedly plagued by remarkable errors of interpretation.

Recently, relatively fail-safe alternatives using general ^{13}C-labeled precursors (such as glucose or acetate) have been developed for the quantitative assessment of carbon fluxes in microorganisms and plants. As an example, this article describes studies on the formation of starch in kernel cultures of maize.

ISOTOPOMER EQUILIBRIA

Terrestrial carbon is an isotopic mixture of 98.9% ^{12}C, 1.1% ^{13}C, and traces of ^{14}C. Consequently, all organic compounds are complex isotopomer mixtures. For any 6-carbon compound (e.g., glucose), the number of nonradioactive carbon isotopomers is $2^6 = 64$. The isotopomer composition of glucose with natural ^{13}C abundance is close to the state of chemical equilibrium; minor deviations caused by isotope selectivity of enzymatic reactions are below the level of sensitivity of the method used in disturbance studies and thus can be disregarded. The approximate abundances of different isotopomers in naturally occurring glucose are summarized in Fig. 1(A). Isotopomers that are labeled with ^{13}C at multiple positions do not occur in significant amounts. Notably, the sum of the concentrations of all naturally occurring isotopomers with two or more ^{13}C atoms in glucose with natural isotope abundance is less than 0.07 mol%.

The quasi-equilibrium isotopomer distribution of biomatter can be perturbed by the introduction of any singly or multiply ^{13}C-labeled metabolite. Cellular metabolism is then conducive to a complex relaxation process in which virtually all chemical reactions occurring in the experimental system are involved. Whereas catabolic processes direct the system to a new equilibrium state (characterized by an increased ^{13}C abundance in the case of closed systems), anabolic processes are conducive to metastable states that can be gleaned from the assimilated biomass (i.e., proteins, polymeric carbohydrates, lipids). It is obvious that enzyme reactions involving the breaking or formation of carbon/carbon bonds play a central role in these relaxation processes.

A detailed analysis of the isotopomer populations formed by the enzyme-catalyzed relaxation processes in the wake of a perturbation (by introduction of a multiply-labeled metabolite) is possible by nuclear magnetic resonance (1) (NMR) analysis[1] and affords an abundance of information that can not be obtained by traditional experimental setups that monitor global isotope enrichment or enrichment at selected positions but fail to document the quantitative composition of the entire isotopomer population.

ANALYSIS OF STARCH BIOSYNTHESIS IN KERNELS OF MAIZE

Using this methodology, the biosynthesis of starch was studied with immature maize kernels that were grown in sterile cultures and then supplied with a mixture of

Encyclopedia of Plant and Crop Science
DOI: 10.1081/E-EPCS 120019331

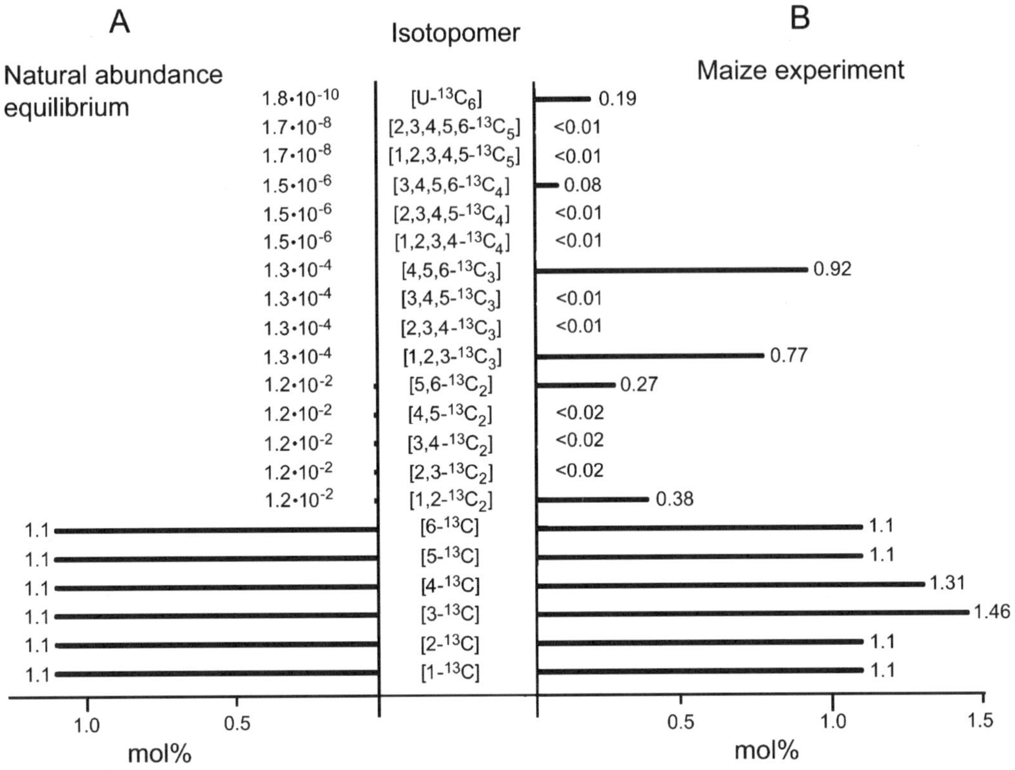

Fig. 1 Molar abundance of ^{13}C isotopomers of glucose. A. In natural abundance material; B. In starch hydrolysate from maize kernels supplied with [U-$^{13}C_6$]glucose and unlabeled glucose at a ratio of 1:40 (w/w).

[U-$^{13}C_6$]glucose and unlabelled glucose at a ratio of 1:40 (w/w).[2] After growth, starch was hydrolyzed and the resulting glucose was isolated and analyzed by quantitative NMR spectroscopy. The detailed description of the NMR analysis has been published,[2] and will not be repeated here.

Of the 57 ^{13}C-isotopomers of glucose that were labeled at more than one position, six occurred with an abundance above 0.08 mol% (Fig. 1(B)), i.e., well above their stochastic occurrence in natural abundance glucose (Fig. 1(A)). Notably, the abundance of [U-$^{13}C_6$]glucose in the sample is nine orders of magnitude above the level of that isotopomer in natural abundance glucose. Two isotopomers carrying single ^{13}C atoms ([3-$^{13}C_1$]- and [4-$^{13}C_1$]glucose) also occurred with increased abundances of 1.46 and 1.31%, respectively, due to the metabolic processes.

The relative paucicity of the [U-$^{13}C_6$]-isotopomer (0.19 mol%) shows that the carbon skeleton of the vast majority of the proffered carbohydrate precursors had been broken and reassembled at least once. Glycolysis followed by glucogenesis results in breaking of the C3/C4 bond of glucose. Passage of glucose through the pentose phosphate pathway during regeneration of a hexose is conducive to breaking the C2/C3 bond (transketolase activity, Fig. 2) and/or the C3/C4 bond (transaldolase activity) of hexoses. Evidence for both processes is immediately obvious from the relatively high abundance of [1,2-$^{13}C_2$]-, [1,2,3-$^{13}C_3$]-, [4,5,6-$^{13}C_3$]-, [3,4,5,6-$^{13}C_4$]-, and [3-$^{13}C_1$]-isotopomers (Fig. 1).

The [5,6-$^{13}C_2$]- and [4-$^{13}C_1$]- isotopomers can be explained by the cooperative action of the pentose phosphate pathway and the glycolysis/glucogenesis pathways (Fig. 3). Thus, the transfer of a [$^{13}C_2$]-fragment from [U-$^{13}C_6$]fructose 6-phosphate to unlabeled erythrose 4-phosphate derived from unlabeled glucose generates [1,2-$^{13}C_2$]pentulose 5-phosphate and, subsequently, [1,2-$^{13}C_6$]hexoses (via transketolase catalysis) (Fig. 2). Glycolysis can produce [2,3-$^{13}C_2$]dihydroxyacetone phosphate from the double-labeled hexose, which yields [2,3-$^{13}C_2$]glyceraldehyde 3-phosphate by the catalytic action of triose phosphate isomerase. Regeneration of a hexose from [2,3-$^{13}C_2$]glyceraldehyde phosphate, either by glucogenesis or via the pentose phosphate pathway could afford the [5,6-$^{13}C_2$]glucose isotopomer observed in our experiment (0.27 mol%) (Fig. 3). The abundance of the [5,6-$^{13}C_2$]-isotopomer is only slightly lower (about 30%) than that of the [1,2-$^{13}C_2$]-isotopomer from which it is proposed to be formed by the sequence of events described earlier. This suggests that the interconnection of the pentose phosphate pathway and the glycolysis/glycogenesis pathways is operating quite efficaciously. More specifically, the data indicate that 87% of the glucose moieties of this

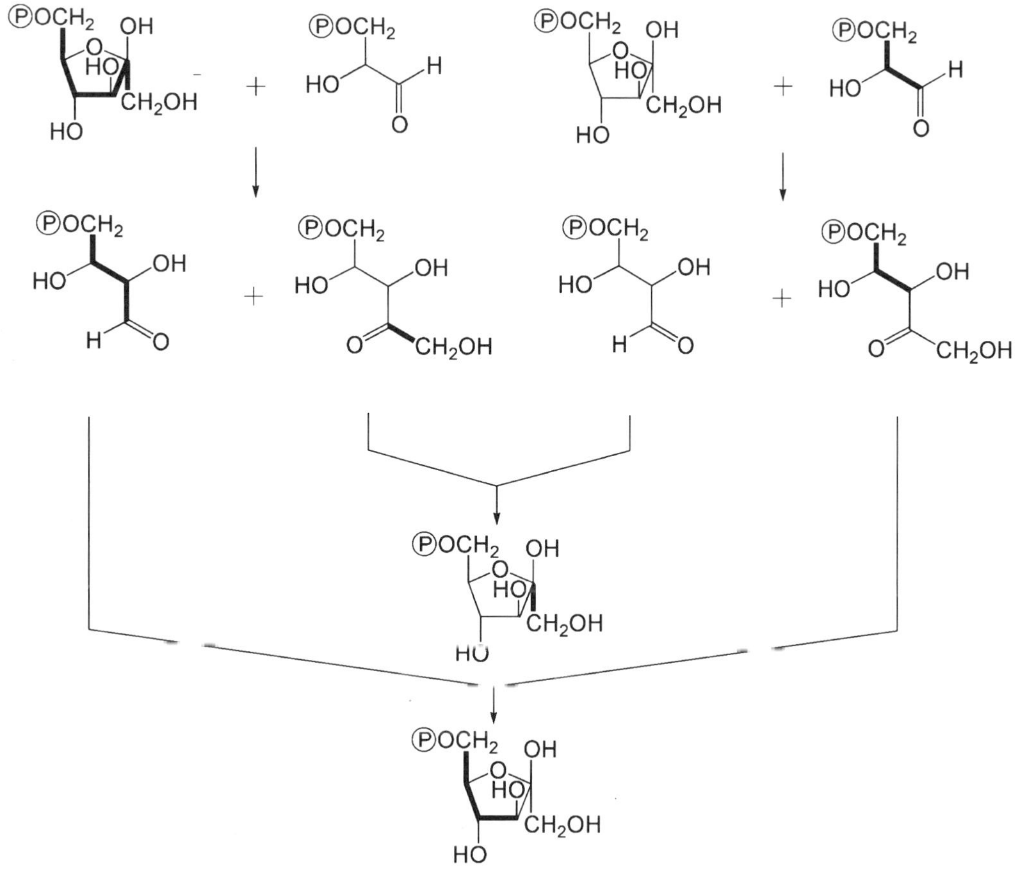

Fig. 2 Formation of [1,2-$^{13}C_2$]- and [3,4,5,6-$^{13}C_4$]hexose phosphate by the catalytic action of transketolase in the pentose phosphate cycle. Bonds in bold type connect ^{13}C atoms in multiply ^{13}C-labeled isotopomers derived from [U-$^{13}C_6$]glucose in the precursor mixture.

precursor are not directly derived from external glucose; the hexose is recycled with high efficiency by the pentose phosphate pathway and glycolysis/glycogenesis, prior to their fixation in starch via the well known intermediate ADP-glucose[3] (Fig. 4).

Plant metabolism is a very complex network, even when only carbohydrate metabolism in a maize endosperm cell is under investigation. Metabolic reactions involving hexoses take place in the cytosol and in the amyloplast of these cells. In fact, with the exception of the nonoxidative branch of the pentose phosphate pathway—which is localized in the maize plastid[4]—both compartments harbor redundant sets of enzymes catalyzing both anabolic and catabolic reactions. The metabolite pools of the cytosol and the amyloplast are efficiently connected by transporters for triose phosphate,

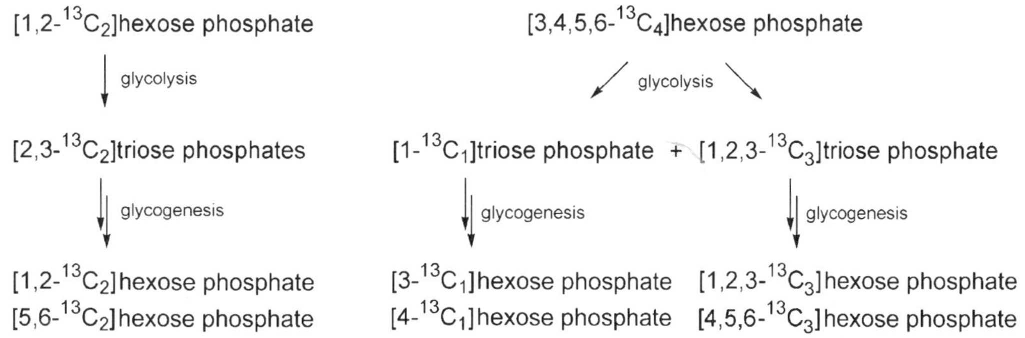

Fig. 3 Pathways for the formation of [1,2,3-$^{13}C_3$]-, [4,5,6-$^{13}C_3$]-, [1,2-$^{13}C_2$]-, [5,6-$^{13}C_2$]-, [3-$^{13}C_1$]-, and [4-$^{13}C_1$]-isotopomers of hexose phosphate in the perturbation experiment with [U-$^{13}C_6$]glucose.

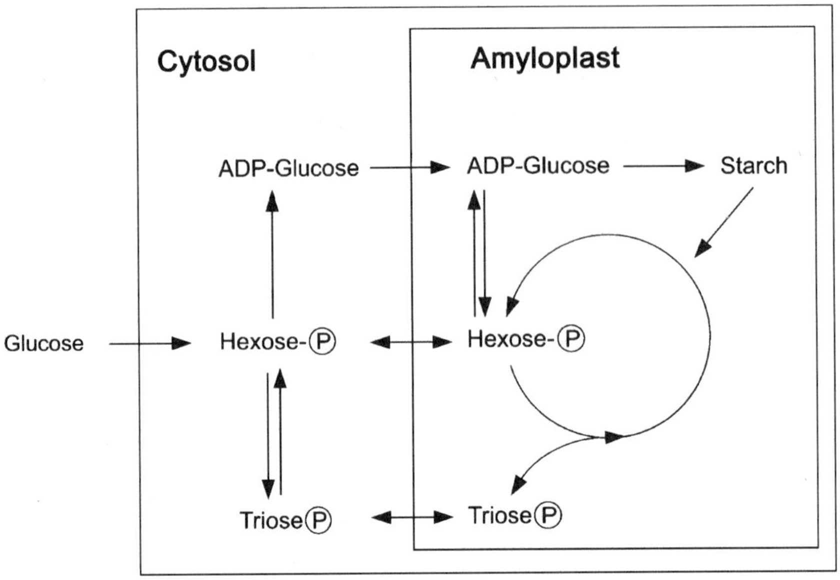

Fig. 4 Metabolic networks involved in starch biosynthesis in growing kernels of maize. The pentose phosphate pathway is shown as a circle in the amyloplast.

hexose phosphate, pentose phosphates, and ADP-glucose that enable metabolite flux and maintain the phosphorus balance in the different compartments of the cell[5] (Fig. 4). In this context, the extensive cycling processes inferred from the glucose labeling pattern describes a flexible metabolic network that serves the physiological needs of the cell.

CONCLUSION

The perturbation of ^{13}C-isotopomer equilibria by general ^{13}C-labeled precursors (e.g., glucose) can serve to analyze carbon fluxes in complex metabolic networks with high resolving power. Many biosynthetic isotopomers of glucose can be assessed quantitatively by NMR analysis and provide detailed information about carbon-carbon connectivities that had been either retained or broken and regenerated during the metabolic conversion of glucose into starch. The technique can now be exploited to investigate and compare glucose flux under different physiological conditions.

ACKNOWLEDGMENTS

This work was supported by the Deutsche Forschungsgemeinschaft, the Fonds der Chemischen Industrie, and the Hans-Fischer-Gesellschaft. The expert help of Angelika Werner and Fritz Wendling with the preparation of the manuscript is gratefully acknowledged.

ARTICLES OF FURTHER INTEREST

Glycolysis, p. 547
Metabolism, Primary: Engineering Pathways of, p. 714
Sucrose, p. 1179

REFERENCES

1. Bacher, A.; Rieder, C.; Eichinger, D.; Fuchs, G.; Arigoni, D.; Eisenreich, W. Elucidation of biosynthetic pathways and metabolic flux patterns via retrobiosynthetic NMR analysis. FEMS Microbiol. Rev. **1999**, *22*, 567–598.
2. Glawischnig, E.; Gierl, A.; Tomas, A.; Bacher, A.; Eisenreich, W. Starch biosynthesis and intermediary metabolism in maize kernels. Quantitative analysis of metabolite flux by nuclear magnetic resonance. Plant Physiol. **2002**, *130*, 1717–1727.
3. Neuhaus, H.E.; Emes, M.J. Nonphotosynthetic metabolism in plastids. Annu. Rev. Plant Mol. Biol. **2000**, *51*, 111–140.
4. Debnam, P.M.; Emes, M.J. Subcellular distribution of enzymes of the oxidative pentose phosphate pathway in root and leaf tissue. J. Exp. Bot. **1999**, *50*, 1653–1661.
5. Tetlow, I.J.; Blissett, K.J.; Emes, M.J. Metabolite pools during starch synthesis and carbohydrate oxidation in amyloplasts isolated from wheat endosperm. Planta **1998**, *204*, 100–108.

Sucrose

Ed Etxeberria
University of Florida, Lake Alfred, Florida, U.S.A.

S

INTRODUCTION

Sucrose plays a unique role in the plant kingdom. Aside from being the primary product of photosynthesis and the main form of carbon transport in plants, sucrose constitutes the most abundant form of soluble storage carbohydrate and also serves as a signaling molecule that triggers essential metabolic events. Furthermore, sucrose plays a key role in plant reproduction and propagation. In nectars, sucrose concentration can determine the type and frequency of visiting pollinators, which may change with the sexual state of the flower, and its presence in fruits serves as an attractant to animals for seed dispersal. A readily available source of energy, sucrose sustains the initial stages of growth after dormant periods in temperate plants. From photosynthetic cells in leaves to heterotrophic root cells, sucrose is found in virtually every living plant cell. There are other soluble saccharides present in plants, and in all cases they are accompanied by high levels of sucrose. In effect, sucrose is the ultimate building block for all other organic compounds in plants and most other carbohydrates in nature, given the position of plants as the cornerstone of the energy food chain.

WHY SUCROSE?

The reason for the ubiquitous position of sucrose in the plant kingdom is not evident. In comparison with trehalose, the other prominent disaccharide found in nature with equivalent functions in insects and fungi, and with raffinose-based saccharides, which are commonly found in various plant species, several conjectures have been made.[1] Based on the process of natural selection to perform equivalent functions, the general premise is that these molecules must share important properties that impart physiological advantages. A common characteristic to all aforementioned saccharides is their nonreducing nature. Nonreducing molecules are less reactive and less susceptible to breakdown by the cellular enzymatic milieu. Their high energy of hydrolysis conserved in their glycosidic linkage makes these molecules more valuable as energy currency and as a readily available carbon source. Two other disaccharides found in living systems, maltose and lactose, have glycosidic linkages with less than half the energy of hydrolysis of sucrose. Finally, both of these molecules have been shown to protect membrane lipids during dehydration and freezing, and to help stabilize organelles and proteins.

FROM PHOTOSYNTHETIC PRODUCT TO THE PHLOEM STREAM

In green cells, glucose 1-phosphate and fructose 6-phosphate are synthesized in the cytosol from triose-phosphates that are produced in the Calvin cycle and exported from the chloroplast. Sucrose is synthesized from UDP-glucose and fructose-6-phosphate in a sequence of two reactions catalyzed by sucrose-phosphate synthase (SPS) and sucrose-phosphate phosphatase (SPP). Both enzymes are localized in the cytosol and appear to form a metabolic unit during synthesis.[2] Regulation of sucrose synthesis is a complex system that involves fine and coarse control.[3] Although some controlling elements have been described, it is likely that other factors yet to be discovered may contribute to the overall regulation and may modify prevailing opinions about sucrose synthesis. In photosynthetic cells, newly synthesized sucrose has two potential fates depending on cellular, physiological, and environmental factors. Sucrose is either stored in the vacuole and/or exported to supply carbon to heterotrophic cells. Temporary storage of excess sucrose in the vacuole usually takes place at times of high photosynthetic activity and limited phloem loading capacity. The mechanisms of sucrose transport into the vacuole of green cells are not fully understood, but the process is believed to be mediated through a passive transport mechanism.[4]

The movement of sucrose from mesophyll cells to the phloem elements can take various routes depending on the plant species, and it involves different cell types (Fig. 1). One route consists of intracellular symplastic movement of sucrose across the plasmodesmata of leaf cells and eventual release into the sieve element/companion cell

Encyclopedia of Plant and Crop Science
DOI: 10.1081/E-EPCS 120019332

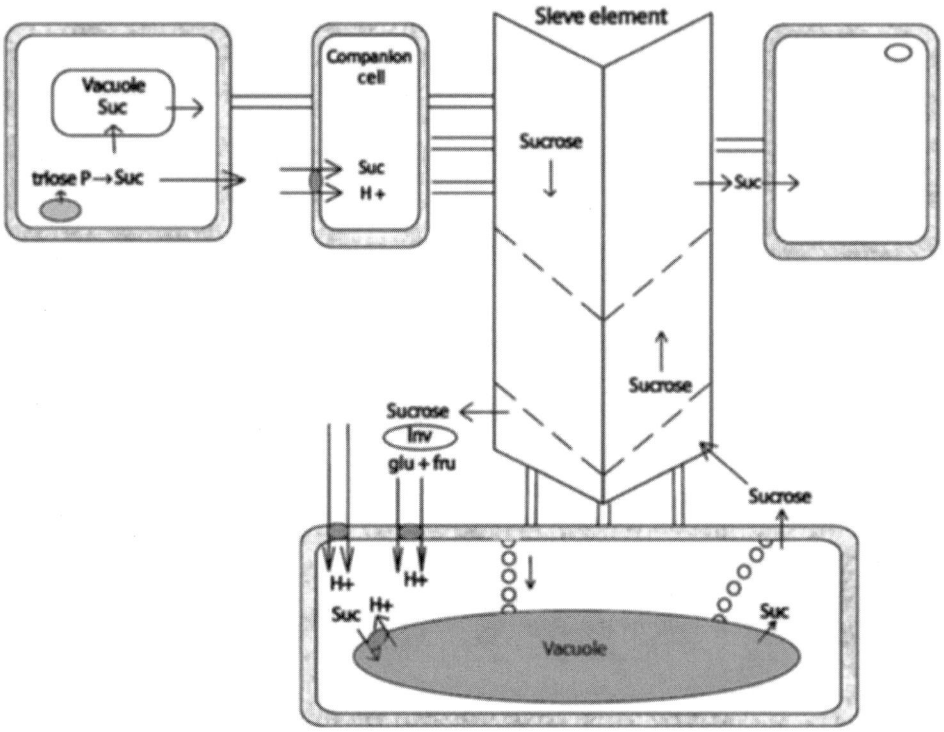

Fig. 1 The cycle of sucrose in a plant, from its synthesis to its storage and utilization. Synthesized in the leaves from photosynthetic products, sucrose is exported to support heterotrophic cells and/or stored temporarily. Along the transport route, the involvement of several carriers is required, either for retrieval of leaked sucrose or as part of the apoplastic route. Once stored, the direction of sucrose transport is reversed to sustain developing plant parts. Loss of sucrose to herbivore consumption can occur at many points along the route. (*View this art in color at www.dekker.com.*)

(SE/CC). In the second route, sucrose is released into the extracellular apoplast at some point, where it diffuses through the cell wall milieu. After reaching the SE/CC complex, sucrose is retrieved by the well characterized plasmalemma-bound sucrose/H^+ symport.[5] Accumulation of sucrose in the SE/CC increases the hydrostatic pressure, which drives mass flow transport to other plant parts via the phloem.

FROM THE PHLOEM TO HETEROTROPHIC CELLS

It is not known whether release of sucrose along the phloem pathway or at the sink end of the phloem route is also proton coupled, occurs by diffusion, or involves transport through the symplast along heterotrophic cells. Sieve element unloading invariably includes an apoplastic component, but its contribution to the overall unloading process depends on many factors and seems to be restricted to specialized circumstances and tissues. Symplastic unloading and transport apparently constitute the principal unloading route. In most cases, there is evidence indicating that sucrose exits the phloem cells and is transported to heterotrophic cells through plasmodesmata connections. Although plasmodesmata connections are present along the entire length of the transport path, efflux seems to occur only at specific regions (Fig. 1).[5]

That sucrose and other photoassimilates are transported into heterotrophic cells through the symplast has been largely inferred from the existence of plasmodesmata connections, the observed transport of large protein or fluorescent probes to the storage cells, and the use of transport inhibitors and transgenic plants. However the presence of plasmodesmata is not a universal characteristic within heterotrophic cells, and transport through the apoplast is undoubtedly required in some instances.[6] The apoplastic route is necessary in cases where there is symplastic discontinuity between two tissues, as is the case for filial and maternal tissue in developing seeds. Once released into the apoplast, sucrose may be hydrolyzed into glucose and fructose by cell wall–bound invertases to maintain a concentration gradient. The situation is quite complicated, given that in some organs (such as potato) both types of postphloem transport take place, depending on developmental stage.

STORAGE OF EXCESS SUCROSE

Whereas some of the sucrose entering the cell is utilized to satisfy immediate metabolic demands, excess sucrose is stored in the vacuole for future needs. The mechanism of sucrose transport into the vacuole of heterotrophic cells depends on its entrance pathway into the cell. Symplastically loaded sucrose needs to traverse only one membrane barrier into the vacuole: the tonoplast, which is believed to possess a sucrose/H^+ antiport. Such an antiport system has been identified at the tonoplast of a few storage cells such as red beet (*Beta vulgaris*) and Japanese artichoke (*Stachys sieboldii*), but this system is conspicuously absent from the high-sucrose-storing cells of sugarcane (*Saccharum officinarum*) and sweet lime (*Citrus limettioides*).

Where sucrose unloading takes the apoplastic route, the plasmalemma offers an additional barrier to accumulation. A sucrose symporter similar to that located at the plasmalemma of SE/CC is presumed to carry sucrose into the cytosol. However, a plasmalemma-bound sucrose symport in storage cells has been inferred from gene expression studies, but its activity has never been demonstrated directly. More recently, an endocytotic system of transport has been proposed to carry sucrose (and other dissolved solutes) from the apoplast to the vacuole of storage cells. Endocytotic vesicles would transport solutes to be stored directly into the vacuole, whereas the plasmalemma-bound sucrose symporter allows the passage of sucrose required by cytosolic activities. In this way, the cytosolic homeostasis is not disrupted by the constant fluctuations of the phloem contents (Fig. 1).

UTILIZATION AND MOBILIZATION OF RESERVE SUCROSE

Metabolic demands for long-term stored vacuolar sucrose occur in vital processes such as resumption of growth in dormant or reproductive tissues, seed germination, and the maintenance of cell viability in stored commodities.[7] Depending on metabolic demand, stored sucrose can be mobilized by storage cells to supply their own physiological requirements and those of remote cells, such as developing shoots, roots, and reproductive organs. Therefore, the mechanisms of sucrose export depend on the ultimate fate of the disaccharide. For internal metabolic use, sucrose can be exported from the vacuole to the cytosol by an ATP-dependent sucrose pump.[7] The exported sucrose can be either released as a disaccharide or catabolized after export by sucrose synthase, which forms a metabolic unit with the ATP-dependent sucrose pump to release UDP-glucose and fructose.

For external transport, a reverse vesicle-mediated system (exocytosis) carries sucrose and other vacuolar substances to the apoplast[8] (Fig. 1). Once released into the apoplast, sucrose is transported to growing points throughout the plant to maintain growth, and in a large number of biennial plants, to sustain the entire second year reproductive activities. In many ways, sucrose secretion as part of flower nectar follows a similar exocytotic route. However, it is believed that some solutes in the nectar originate in other cellular organelles and that a more complex network of vesicle transport is involved.[8]

COMPLETING THE CYCLE

Whether originating from the storage organs or from neighboring exporting photosynthetic leaves, sucrose provides the energy for the development of new leaves until they become fully autotrophic. A series of soluble and wall-bound invertases, in addition to sucrose synthase, channel sucrose to different metabolic pathways. Once the leaf becomes an autotrophic organ, the direction of sucrose flow reverses and export of sucrose renews the cycle. Therefore, as the primary product of photosynthesis, sucrose powers life on earth by virtue of being the basic fuel for life.

ACKNOWLEDGMENTS

This research was supported by the Florida Agricultural Experiment Station, and approved for publication as Journal Series No. R-09528.

ARTICLES OF FURTHER INTEREST

Photosynthate Partitioning and Transport, p. 897
Plant Response to Stress: Source-Sink Regulation by Stress, p. 1010

REFERENCES

1. Pontis, H.G. The Riddle of Sucrose. In *Plant Biochemistry*; Northcote, D.H., Ed.; University Park Press: Baltimore, MD, 1977.
2. Echeverria, E.; Salvucci, M.E.; Gonzalez, P.C.; Paris, G.; Salerno, G.L. Physical and kinetic evidence for an association between sucrose-phosphate synthase and sucrose-phosphate phosphatase. Plant Physiol. **1997**, *115*, 223–227.
3. Huber, S.C.; Huber, J.L. Role and regulation of sucrose-phosphate synthase in higher plants. Annu. Rev. Plant Physiol. Mol. Biol. **1996**, *47*, 431–444.
4. Kaiser, G.; Heber, U. Sucrose transport into vacuoles isolated from barley mesophyll protoplasts. Planta **1984**, *161*, 562–568.
5. Lalonde, S.; Boles, E.; Hellman, H.; Barker, L.; Pattrick, J.W.; Frommer, W.B.; Ward, J.M. The dual function of sugar carriers: Transport and sugar sensing. Plant Cell **1999**, *11*, 707–726.
6. Patrick, J.W. Phloem unloading: Sieve element unloading and post-sieve element transport. Annu. Rev. Plant Physiol. Mol. Biol. **1997**, *48*, 191–222.
7. Echeverria, E.; Gonzalez, P.C. ATP-induced sucrose efflux from red-beet tonoplast vesicles. Planta **2000**, *211*, 77–84.
8. Echeverria, E. Vesicle mediated solute transport between the vacuole and the plasma membrane. Plant Physiol. **2000**, *123*, 1217–1226.

Sustainable Agriculture and Food Security

S

Jules Pretty
University of Essex, Colchester, U.K.

INTRODUCTION

Despite several decades of remarkable agricultural progress, the world still faces a massive food security challenge, with an estimated 790 million people lacking adequate access to food. Most agree that food production will have to increase in the coming years, and that this will have to come from existing farmland. But solving the persistent hunger problem is not simply a matter of developing new agricultural technologies. Most hungry consumers are poor, and so simply do not have the money to buy the food they need. Equally, poor producers cannot afford expensive technologies. They will have to find solutions largely based on locally available natural, social and human resources. The key questions are, therefore, to what extent can agricultural systems become more productive while not causing harm to the environment, and do such system offer any new hope for the hungry?

SUSTAINABLE AGRICULTURE

Something is wrong with our agricultural and food systems. Despite great progress in increasing productivity in the last century, hundreds of millions of people remain hungry and malnourished. More hundreds of millions eat too much, or the wrong sorts of food, and it is making them ill. The health of the environment suffers too, as costly degradation seems to accompany many of the agricultural systems we have evolved in recent years.[1,2] Can nothing be done, or is it time for the expansion of another sort of agriculture, founded more on ecological principles and in harmony with people, their societies and cultures?

In the earliest surviving texts on European farming, agriculture was interpreted as two connected things, *agri* and *cultura*, and food was seen as a vital part of the cultures and communities that produced it. Today, however, our experience with industrial farming dominates, with food now seen simply as a commodity and farming often organized along factory lines. To what extent can we put the culture back into agri-culture without compromising the need to produce enough food? And can we create sustainable systems of farming that are efficient and fair, and founded on a detailed understanding of the benefits of agroecology and people's capacity to cooperate?

As we advance into the early years of the 21st century, we have some critical choices.[3,4] Humans have been farming for some 600 generations, and for most of that time the production and consumption of food has been intimately connected to cultural and social systems. Foods have a special significance and meaning, as do the fields, grasslands, forests, rivers, and seas. Yet in just the last two or three generations, we have developed hugely successful agricultural systems based on industrial principles. In most cases they produce more food per hectare and per worker than ever before, but they look so efficient only if we ignore the harmful side effects—the loss of soils, the damage to biodiversity, the pollution of water, and the harm to human health.

Over these 12,000 years of agriculture, there have been long periods of stability, punctuated by short bursts of rapid change. These changes resulted in fundamental shifts in the way people thought and acted. We are at another such junction. A sustainable agriculture making the best of nature and of people's knowledge and collective capacities has been showing increasingly good promise. But it has been a quiet revolution, because many accord it little credence. It is also silent because those in the vanguard are often the poorest and the marginalized, whose voices are rarely heard in the grand scheme of things. No one can exactly say where this revolution could lead us. Neither do we know whether models of sustainable production would be appropriate for all farmers worldwide, but the principles do apply widely. Once these come to be accepted, then it will be the ingenuity of local people that shapes these new methods of producing food to their own particular circumstances.

We know that most transitions involve trade-offs. A gain in one area is accompanied by a loss elsewhere. A road built to increase access to markets helps remote communities, but it also allows illegal loggers to remove valuable trees more easily. A farm that eschews the use of pesticides benefits biodiversity, but may produce less food. New agroecological methods may mean more labor

Encyclopedia of Plant and Crop Science
DOI: 10.1081/E-EPCS 120012950

is required, putting an additional burden on women. But these trade-offs need not always be serious. If we listen carefully and observe the improvements already being made by communities across the world, we find that it is possible to produce more food while protecting and improving nature. It is possible to have diversity in both human and natural systems without undermining economic efficiency.

RECENT DEVELOPMENTS

In recent research, we examined the extent to which farmers have improved food production in recent years with low-cost, locally available, and environmentally sensitive practices and technologies.[5–7] During 1999–2000 we analyzed by survey 208 projects in 52 developing countries, in which 8.98 million farmers have adopted these practices and technologies on 28.92 million hectares, representing 3% of the 960 million hectares of arable and permanent crops in Africa, Asia, and Latin America.[8,9]

We found improvements in food production occurring through one or more of four mechanisms: 1) intensification of a single component of a farm system; 2) addition of a new productive element to a farm system; 3) better use of water and land, thereby increasing cropping intensity); 4) improvements in per-hectare yields of staples through introduction into farm system of new regenerative elements, locally appropriate crop varieties, and animal breeds. The 89 projects with reliable yield data show an average per-project increase in per-hectare food production of 93% (Fig. 1).

There are several key practices and technologies that have led to these increases: increased water use efficiency, improvements to soil health and fertility, pest control using biodiversity services with minimal or zero-pesticide use, and social organization for collective action.[6,10–13] This research reveals promising advances in the adoption of practices and technologies that are likely to be more sustainable, with substantial benefits for the rural poor. With explicit support through national policy reforms, better markets, and more integrated and cross-disciplinary approaches to science, these improvements in food security could spread to much larger numbers of farmers and rural people in the coming decades.

Social learning is a vital part of the process of adjustment in sustainable agriculture projects.[5,10] The conventional model of understanding technology adoption as a simple matter of diffusion, as if by osmosis, no longer holds. But the alternative is neither simple nor mechanistic. It involves building the capacity of farmers and their communities to learn about the complex ecological and biophysical complexity in their fields and farms, and then to act in different ways. The process of

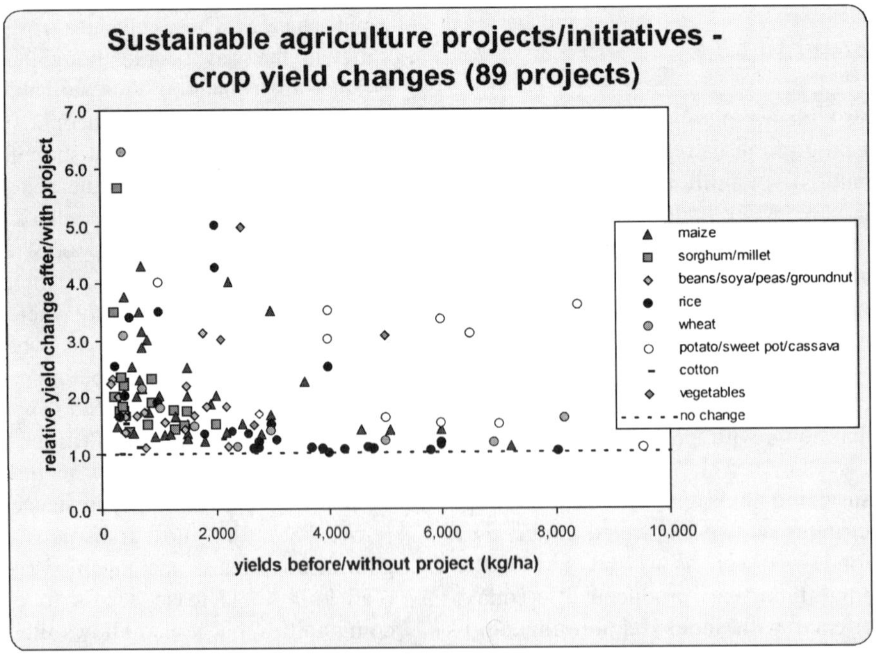

Fig. 1 Sustainable agriculture project/initiatives: relative crop yield changes for 89 projects, where 1.0 represents no change and relative yields above 1.0 represents multiples. (From. Ref. 9.)

learning, if it is socially embedded, provokes wider changes in behavior.

Farmers require timely information on pest–predator relationships, moisture and plants, soil health, and the chemical and physical relationships between plants and animals. These are subject to manipulation—and farmers who understand this, and who are confident about experimentation, are better innovators. The empirical evidence tells us two important things. Social learning leads to greater innovation, and there is increased likelihood that social processes producing these technologies are likely to persist.

CONCLUSION

Several things are now clear with respect to sustainable agriculture:[3]

- The technologies and social processes for local-level agroecological improvements are well-tested and established.
- The social and institutional conditions for spread are less well-known, but have been established in several contexts (in particular social groups at the local level and novel partnerships between external agencies.
- The political conditions for the emergence of supportive policies are the least established, with only a very few examples of real progress.

Most of the sustainable agriculture improvements seen in the past decade have arisen despite existing national policies, which will need major reform. Policies framed to deliver increased food production will have to be changed if they also are to help deliver environmental and social benefits. Food policies framed to help deliver cheap and abundant food, regardless of quality, will have to change as well. And rural development policies and institutions focusing on ''exogenous'' solutions to the economic and social problems of rural communities are ill-suited to the needs of community-based and participatory development.[6,11]

Nevertheless, there has been increasing global recognition of the need for policies to support sustainable agriculture.[3,8] Although almost every country would now say it supports sustainable agriculture, the evidence points toward only patchy reforms. Only three countries—Cuba, Switzerland, and Bhutan—have given explicit national support for sustainable agriculture, putting it at the center of agricultural development policy and integrating policies accordingly. Cuba has a national policy for alternative agriculture, Switzerland has three tiers of support to encourage environmental services from agriculture and rural development, and Bhutan has a national environmental policy coordinated across all sectors.

Some countries, such as in India, Brazil and Sri Lanka, have seen subregional support at the state level for zero-tillage, watershed and soil management, and participatory irrigation management. A much larger number of countries have reformed elements of agricultural policies through new regulations, incentives and/or environmental taxes, and administrative mechanisms, and these are having considerable—though partial—effect. These changes include catchment approaches for soil conservation and bans on selected pesticides, combined with a national program for farmer field schools and integrated pest management (IPM) in rice, support for soybean processing and marketing, and regional integration of agricultural and rural policies. But most countries have not yet explicitly put sustainable agriculture at the center of their policy frameworks.

ARTICLES OF FURTHER INTEREST

Agriculture and Biodiversity, p. 1
Biological Control of Weeds, p. 141
Cropping Systems: Irrigated Continuous Rice Systems of Tropical and Subtropical Asia, p. 349
Cropping Systems: Irrigated Rice and Wheat of the Indo-gangetic Plains, p. 355
Ecology and Agricultural Sciences, p. 401
Herbicides in the Environment: Fate of, p. 554
Organic Agriculture As a Form of Sustainable Agriculture, p. 846
Sustainable Agriculture: Definition and Goals, p. 1187
Sustainable Agriculture: Ecological Indicators, p. 1191
Sustainable Agriculture: Economic Indicators, p. 1195
Symbiotic Nitrogen Fixation, p. 1218
Symbiotic Nitrogen Fixation: Plant Nutrition, p. 1222

REFERENCES

1. Altieri, M. *Agroecology: The Science of Sustainable Agriculture*; Westview Press: Boulder, 1995.
2. Pretty, J.; Brett, C.; Gee, D.; Hine, R.; Mason, C.F.; Morison, J.I.L.; Raven, H.; Rayment, M.; van der Bijl, G. An assessment of the total external costs of UK agriculture. Agri. Syst. **2000**, *65* (2), 113–136.
3. Pretty, J. *Agri-Culture: Reconnecting People, Land and Nature*; Earthscan: London, 2002; 261 pp.
4. McNeely, J.A.; Scherr, S.J. *Common Ground, Common Future. How Ecoagriculture can Help Feed the World and*

Save Wild Biodiversity; IUCN and Future Harvest: Geneva, 2001.

5. *Facilitating Sustainable Agriculture*; Röling, N.G., Wagemakers, M.A.E., Eds.; Cambridge University Press: Cambridge, 1997.
6. *Agroecological Innovations: Increasing Food Production with Participatory Development*; Uphoff, N., Ed.; Earthscan: London, 2002.
7. Ye, X.J.; Wang, Z.Q.; Li, Q.S. The ecological agriculture movement in modern China. Agric. Ecosyst. Environ. **2002**, *92*, 261–281.
8. Pretty, J.; Hine, R. The promising spread of sustainable agriculture in Asia. Nat. Res. Forum **2000**, *24*, 107–126.
9. Pretty, J.; Morison, J.I.L.; Hine, R.E. Reducing food poverty by increasing agricultural sustainability in developing countries. Agric. Ecosyst. Environ. *95* (1), 217–234.
10. Pretty, J.; Ward, H. Social capital and the environment. World Dev. **2001**, *29* (2), 209–227.
11. Scott, J. *Seeing Like a State. How Certain Schemes to Improve the Human Condition Have Failed*; Yale University Press: New Haven, 1998.
12. Stoll, S.; O'Riordan, T. *Protecting the Protected: Managing Biodiversity for Sustainability*; Cambridge University Press: Cambridge, 2002.
13. Pretty, J.; Ball, A.S.; Li, X.; Ravindranath, N.H. The role of sustainable agriculture and renewable resource management in reducing greenhouse gas emissions and increasing sinks in China and India. Trans. Royal Soc. London, A **2002b**, *360*, 1741–1761.

Sustainable Agriculture: Definition and Goals

Glen C. Rains
University of Georgia, Tifton, Georgia, U.S.A.

W. Joe Lewis
Dawn M. Olson
USDA—Agricultural Research Service, Tifton, Georgia, U.S.A.

INTRODUCTION

Agricultural technology has increased farm production to unprecedented levels. However, returns on investments are diminishing and environmental concerns are in conflict with current chemically intensive farm practices. Emerging technologies—such as precision farming and genetically modified crops—can potentially improve farm management through reduced chemical use and better distribution of those chemicals. However, using these farm management strategies to redirect or change the types of chemicals needed continues the reliance on external chemical inputs that are reducing farm profit. To change this trend, management strategies must be redirected to promote sustainable and holistic management that encourages interdependent and diverse properties. With this redirection, use of chemicals can be relegated to a backup role, giving way to a management strategy that enhances and promotes the inherent strengths and processes of the agroecosystem.

NEED FOR CHANGE

Agricultural production in the United States has increased dramatically during the past 75 years. Corn production has nearly quadrupled from 1920 to 1995, while the crop acreage has been reduced by one-third.[1,2] Such impressive crop yields stem primarily from the development and application of agricultural chemicals (pesticides and fertilizers), new and improved agricultural machinery, educational and extension programs, and improved plant cultivars resistant to pests and designed for higher yield.

But these yield gains are showing diminishing returns even with increased inputs (Fig. 1), and have come at the expense of impaired natural resources and diminished soil and water quality.[3] The two farm management practices most devastating to soil and water ecologies are intensive chemical use (pesticides and fertilizers) and tillage practices.[4] High chemical use destroys soil microorganisms that decompose residue and aid in nutrient recycling; kills beneficial insects; and can, with water runoff and leaching, poison water resources. Agriculture is the leading source of wetland water quality impairment.[5] Pesticide use causes destruction of natural enemies and increased pest resistance, thereby resulting in secondary pests, pest resurgence, and a treadmill of pesticide dependency. This dependency on chemical inputs creates an unstable farming enterprise that continues to spiral toward higher and higher chemical inputs to make up for the depletion of inherent resources on the farm. In the United States, crop losses due to pathogens, animal pests, and weeds have increased from 34.9% in 1965 to 42.1% in 1988–1990,[6] despite a 170% increase in pesticide use over roughly the same period (1964–1985).[3] Heavy tillage is an energy-intensive management practice that increases pollution and energy costs and removes soil cover, exposing it to erosion by wind and water. Compounding the issue for farmers is the rising per unit cost of inputs and the recent drop in commodity value. Recent commodity prices have combined with inclement weather patterns to force many farmers out of business. New sustainable management strategies must be developed to reverse the escalating economic and environmental consequences of conventional management practices. As argued by Lewis et al.,[7] these new strategies must go beyond the mere ongoing replacement of therapeutics with alternative newer, more sophisticated intervention materials. Rather, truly satisfactory solutions require a shift to a systems approach to understanding and renewing the inherent strengths and balances of agricultural ecosystems, with therapeutic interventions serving strictly as backups to the natural components. This article defines and discusses the goals of sustainable agriculture as compared to conventional agriculture, and generally outlines the shift in operating philosophy required to achieve the new goals.

SUSTAINABLE VERSUS CONVENTIONAL AGRICULTURE PRACTICES

Sustainable agriculture is a management strategy that seeks to meet current production and profitability needs while protecting the long-term ecological health and

Encyclopedia of Plant and Crop Science
DOI: 10.1081/E-EPCS 120012945

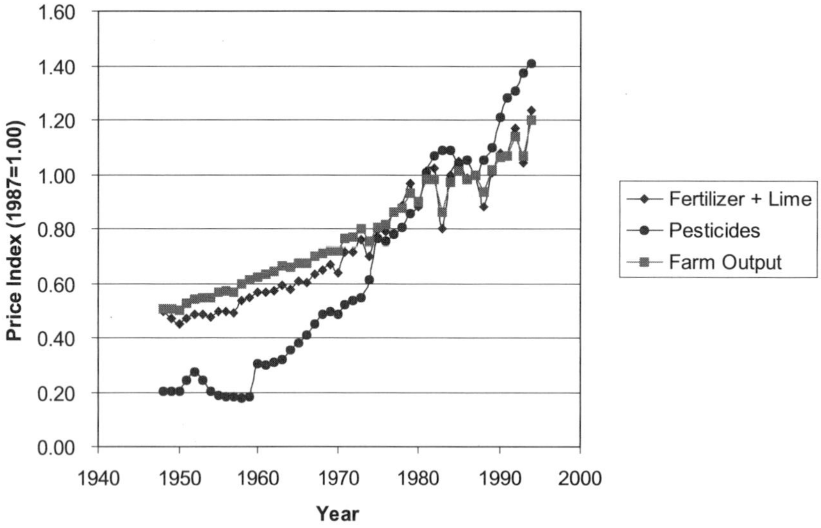

Fig. 1 Price indexes for fertilizers and lime, pesticides, and farm output, 1940–2000. (*View this art in color at www.dekker.com.*)

production capacity of a system through fostering and utilizing inherent and renewable strengths. Sustainable systems are noted for six key attributes: diversity, self-renewal, self-sufficiency, self-regulation, efficiency, and interdependence.[8] Diversity and interdependence develop self-regulating mechanisms such as multitrophic interactions, self-renewing properties through nutrient recycling, and self-sufficiency through reliance on inherent resources. A sustainable system is thereby driven not by chemical and energy-intensive inputs, but by the development of renewable resources and inherent strengths within the agroecosystem. For example, by nurturing soil microbial and earthworm activity and nitrogen-fixing and cover crops, nutrients are replenished within the system, reducing the need for applied fertilizers. This requires an understanding and management of the interactions among vertebrates, invertebrates, plants, water, and soils to build on the inherent renewable strengths and resources of the agroecosystem.

In contrast, current farming practices are generally highly input-dependent and have eroded the inherent strengths of the agroecosystem. Monocultural practices, single-tactic strategies, and reliance on chemical pesticides and fertilizers as well as heavy soil tillage have created a system of farming that requires constant therapeutic inputs. This input-driven farming system is extremely inefficient, costly, and unstable as pest resistance and resurgence require continued heavy pesticide use (Fig. 2, current farming system). Subsequently, inherent strengths of the agroecosytem—such as the self-regulation of pests with beneficial insects—are eroded. As it becomes more difficult to farm for profitability, farmers look to new tools such as better chemical pesticides, precision agriculture, and genetically engineered crops to maintain production. However, these new technologies are still just therapeutic inputs that are not sustainable. Therefore, while efficiency and stability improve initially, the intrinsic ability of agroecosystems to neutralize external interventions causes the new inputs to lose their effectiveness, and instability returns, as the treadmill of input dependency continues (Fig. 2, upper arrow). This is very evident in the reaction of pests to new pesticides. Their efficacy is very high initially, but continually diminishes as pest resistance increases. A redirection in technology development toward sustainable agriculture

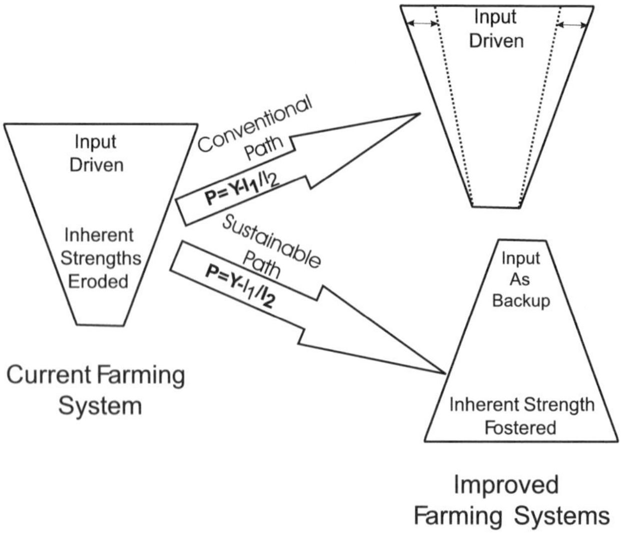

Fig. 2 Conventional versus sustainable paths of farming systems.

will increase farming stability by fostering the natural ecological mechanisms that regulate pests and nutrient recycling. This is accomplished by fostering the inherent strengths of the farming enterprise and reducing therapeutic inputs to a backup role (Fig. 2, lower arrow).

DECIPHERING ISSUES OF PROFITABILITY, YIELD, AND SUSTAINABILITY

As yield increases have begun to plateau, input costs have continued to rise. Consequently, the cost of farming has increased and the price index has begun to flatten (Fig. 1). One factor that is important in determining profitability but that has not been accounted for since the advent of chemical fertilizers and pesticides is the utilization of the built-in, on-farm inherent strengths already described. Farming profit (P) is a function of the commodity yield (Y) and the money spent to produce that commodity (I). Inputs can further be divided into a numerator—therapeutic inputs (I_1)—and a denominator—inputs developed by utilizing the inherent resources on the farm (I_2). I_1 is in the numerator because, as therapeutic input costs increase, profit is reduced. I_2 is placed in the denominator because, as the inherent strengths of the system are utilized, they reduce I_1 and increase the profit margin. The farming profit can then be represented by Eq. 1:

$$P = Y - I_1/I_2 \quad (1)$$

where P = profit (\$); Y = yield (\$); I_1 = input costs for seed, chemical fertilizer, pesticides, equipment, land, and fuel (\$); I_2 = agroecosystem (on-farm) resources and inherent strengths that are normally ignored (\$)

The benefits of inherent strengths and on-farm resources (I_2) have not been utilized in a holistic manner and are practically invisible to the equation as long as therapeutic inputs are the driving force behind agricultural production. The resulting system is inefficient and unsustainable, as inputs continue to eat away at the farming base (Fig. 2, current farming system). New technologies are being developed in an attempt to reduce this dependency and improve the profitability of farming, but are only met with an ever growing dependence on therapeutic and costly inputs (Fig. 2, upper arrow). For example, precision agriculture has helped redistribute chemicals across a field, but has to date not addressed the underlying causes of field variability and has not addressed potentially less intrusive solutions that utilize the inherent strengths of the agroecosystem. The inherent strengths and resources, when properly utilized and fostered in a sustainable manner as defined previously, act to reduce input costs (I_1) as sustainable strategies are put in place. Therapeutic inputs such as chemical pesticides are then reduced to a backup role in the development of inherent strengths and multiple interactions within the agroecosystem as a whole, improving the profitability and sustainability of the agroecosystem (Fig. 2, lower arrow). Eq. 1 becomes

$$P = Y - I_1/I_2 \quad (2)$$

where I_1 (therapeutic) inputs are reduced to a backup role in developing and utilizing inherent strengths (I_2).

It may be difficult to move from a conventional farm management system to a sustainable managed agroecosystem due to cultural biases, economic trends, and technological or knowledge limitations. Economic hardship usually has a negative impact on implementing sustainable practices. Paradoxically, there is a tendency to add more inputs in hopes of increasing production. Unfortunately, profits are typically reduced through increased production costs without a subsequent increase in production. There is also a reluctance to attempt practices that do not yet have a proven track record. Farmers who are successfully managing sustainable agroecosystems are few, making it difficult to demonstrate to farmers real-world, long term benefits. Finally, developing and maintaining a sustainable agroecosystem and making systematic improvements year to year to improve sustainability requires knowledge of the mechanisms of self-regulating systems and tools for monitoring the factors that affect these mechanisms.

Each of these issues must be addressed if there is to be widespread acceptance of such practices. Cultural biases can be reduced by supporting farmers who develop these practices of sustainable and precision management, so they can become mentors for other farmers in the community. As successful practices emanate out, more widespread adoption will become possible. Farmers will also need the support of local community leaders who also practice these ecologically-based principles. Research and educational programs at federal, state, and local levels must be directed to ward developing and understanding agroecosystems so that management practices are well-understood and adopted. Technological tools to monitor and assess the health of agroecosystems can provide databases that provide knowledge of sustainable practices. With increased knowledge, farmers can make informed decisions to manage sustainable agroecosystems more confidently.

CONCLUSION

Current farm production practices are not environmentally friendly and require high levels of chemical and mechanical inputs to maintain and manage. Besides the environmental consequences, the financial burden is high, as chemical inputs are the driving force behind maintaining an acceptable level of production. To reduce the environmental and financial burden of current farm production practices, a fundamental shift in farm management strategies is needed to create a more sustainable farming system that preserves the long-term productivity of existing farmland with less reliance on chemical inputs and more reliance on developing farm resources and inherent strengths. These agroecosystems must be managed in a holistic manner, fostering biodiversity and multiple interactions at several scales. By developing farming systems that are sustainable, with therapeutic inputs reduced to a backup role, farmers can increase profitability and be less susceptible to adverse outside influences.

REFERENCES

1. USDA/NASS. *Field Crops Final Estimates 1992–1997: Statistical Bulletin Number 947*; USDA, 14th & Independence Ave., S.W.: Washington, DC, 1998; 20250.
2. USDA/NASS. *Track Records: United States Crop Production*; USDA, 14th & Independence Ave., S.W.: Washington, DC, 1999; 20250.
3. Edwards, C.A.; Lal, R.; Madden, P.; Miller, R.H.; House, G. *Sustainable Agricultural Systems*; Soil and Water Conservation Society: 7515 Northeast Ankeny Road, Ankeny, IA, 1990; 50021.
4. USEPA. *Water Quality Conditions in the United States, EPA841-F-00-006*; Office of Water: Washington, DC, 2000; 20460.
5. USDA/ERS. *ERS/USDA Briefing Room: Agriculture and Water Quality*; USDA, 14th & Independence Ave., S.W.: Washington, DC, 1996; 20250.
6. Oerke, E.C.; Dehne, H.W.; Schonbeck, F.; Weber, A. *Crop Production and Crop Protection: Estimated Losses in Major Food and Cash Crops*; Elsevier Science: B.V., P.O. Box 521, 1000 AM Amsterdam, The Netherlands, 1994.
7. Lewis, W.J.; van Lenteren, J.C.; Phatak, S.C.; Tumlinson, J.H. *Proceedings of the National Academy of Sciences*; 1997; Vol. 94, 12243–12248.
8. Lewis, W.J.; Jay, M.M. *Ecologically-Based Communities: Putting It All Together at the Local Level*; Kerr Center for Sustainable Agriculture: Poteau, OK, 2000.

Sustainable Agriculture: Ecological Indicators

S

Stephen S. Jones
Washington State University, Pullman, Washington, U.S.A.

INTRODUCTION

Agricultural sustainability is a process rather than an end point. From a purely ecological view, the process leads to a system of food and fiber production that is in such a state of environmental health that it can continue at the given rate of extraction into the measurable future. To evaluate the progress or the status of sustainable systems requires measurable units. Following the precedence of economic models, the term used to describe such units is indicators. To monitor the biological robustness of a system then requires that these units be indicators that reflect on the entire ecology of the system, hence the term ecological indicators.

Ecological indicators go beyond the canary-in-the-coal mine analogy, in that the best indicators do not merely take an immediate assessment of the health of the system but should be able to be monitored over time. They should also be easily measurable, applicable to a wide range of systems, responsive to change, inexpensive, relevant, and easy to interpret. These indicators should also measure ecological health rather than the productivity of a system. Farm productivity, in many cases, may result in the reduced long-term health of a system.

Some very powerful indicators can also be very simple. For example, it is impossible for a system that loses soil more rapidly than it is replaced to be sustainable. As an indicator, then, soil erosion is a very obvious and relevant measurable unit. Other indicators may be far more subtle, however. Shifts in microorganism populations, soil tilth, and above- or below-ground flora and fauna may all have value in assessing the level of sustainability for a particular system. However, each is quite complicated to measure and monitor over time, and no single indicator can monitor an entire system. Thus, the use of indicators based on more holistic approaches is also being considered.

TYPES OF ECOLOGICAL INDICATORS

Indicators can be either biotic or abiotic. Abiotic indicators deal primarily with the general health of the soil relative to sustaining life, while biotic indicators involve the monitoring of various life forms. Holistic approaches to identifying biological indicators can involve such issues as farmer choice of cultural practices and indicators that can be used on a landscape level, as opposed to a single-field level. Holistic approaches are an attempt to move away from reductionist measurements of ecological systems.

Abiotic Indicators

Some abiotic indicators are obvious and require no measurement whatsoever. One such indicator is illustrated in Fig. 1. Here we see massive amounts of soil being lost due to the combined erosive forces of slope and water. This is an excellent example of the stresses, and the results of these stresses, placed on systems. Soil loss at this rate is dramatic, but losses at levels much lower than this can render a practice unsustainable. It is a rare agricultural system that is not losing soil at some rate, and any loss of soil is enough to render agriculture unsustainable.

The loss of soil can be easily seen and measured. By contrast, determining the health of the soil as an indicator is much more complex.[1,2] Changes in pH, organic matter, nitrate leaching, salinization, heavy metals and pesticide accumulation, compaction, and general fertility are just some of the indicators used to determine soil health. Many of these indicators also have confounding effects on the others. For example, as soils become more acid certain metals, such as aluminum, are released.

The interaction of these items with biotic soil factors then complicates the equation even further. As complex as it is, the measure of soil health should be considered the main indicator of sustainability. All nonaquatic agricultural systems rely on this slow-to-build, rapid-to-degrade resource. In fact, nearly all agricultural soils can be viewed as degraded. Indicators then must measure improvements, rather than stability, if a system is to be deemed sustainable.

Biotic Indicators

Species diversity and the relative frequencies of butterflies,[3] spiders,[4] true bugs,[5] and earthworms[6,7] have all been used as indicators of sustainability. Earthworms (Fig. 2) are especially well suited to a role as indicators. More than 3500 species of earthworms exist, they are easy to capture and sort, and they are slow to recolonize. Slow

Encyclopedia of Plant and Crop Science
DOI: 10.1081/E-EPCS 120012947

Fig. 1 Erosion of winter wheat field near Pullman, Washington. Photo by Stephen Jones, Original Data. (*View this art in color at www.dekker.com.*)

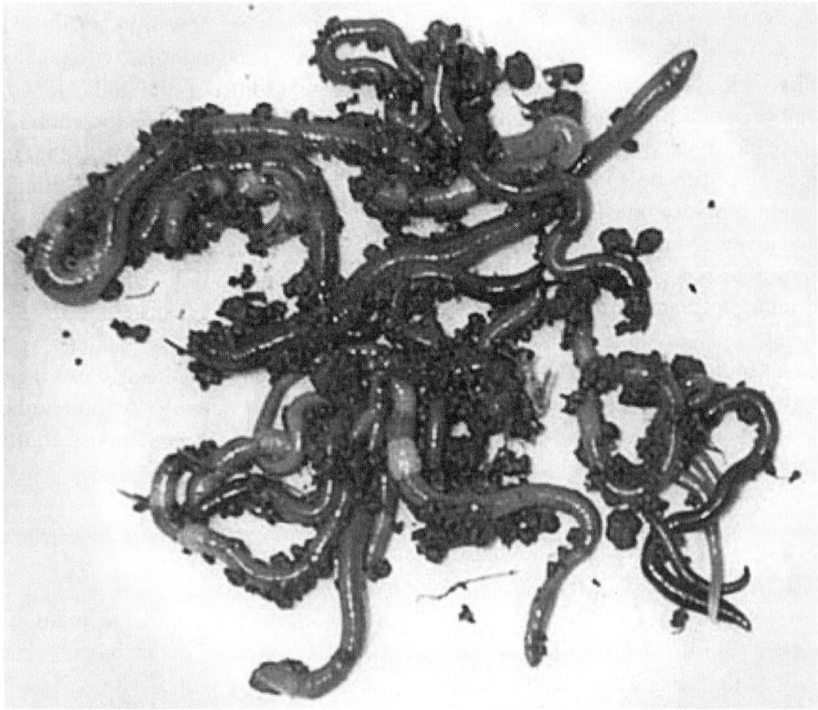

Fig. 2 Many species of earthworms are usually found in individual fields. Each may have different tolerances and susceptibilities to biotic and abiotic pressures. Photo courtesy of Mary Fauci, Washington State University. (*View this art in color at www.dekker.com.*)

recolonization is beneficial in an indicator because the shifts in populations can be more easily studied as they rise and fall over time. Species that recolonize rapidly may cause sampling problems, because rapid shifts may be missed if sampling is infrequent.

As early as the 1600s, earthworms were used as a measure of soil fertility.[7] Today, they are a simple and economic measure of soil life. Individual species vary in their sensitivity to pesticides, tillage, moisture, soil structure, salt, heavy metals, predation, and many other relevant aspects of soil ecology. Paoletti[7] lists 16 unique situations in which earthworms have been used to measure sustainability. The uses range from detecting archeological pesticide residues to comparing soil health at a landscape level in rural and urban interfaces.

Buckerfield et al.,[6] using four species of earthworms in southern Australia, showed a significant inverse relationship between intensity of tillage operation and earthworm abundance. They also recorded a significant correlation between crop productivity and earthworm abundance. They warned, however, that diverse crop management practices and cropping histories over time and space might make simple relationships between earthworms and productivity tenuous. With this in mind, they concluded that on an appropriate scale earthworm communities are useful as indicators of crop sustainability.

Crop and livestock diversity can also be used as indicators of sustainability.[8,9] Ratios of annuals to perennials, permanent pasture to annual crop, and pest plants and animals to useful species may also indicate sustainability.

Aboveground diversity can be highly changeable, however. An individual farmer can go from a total monoculture system to intercropped multilines with perennial borders in a single season. The most important question of crop and animal diversity, therefore, is not if it is here today, but rather, if it will it be available tomorrow. The use and availability of plant and animal varieties bred for long-term sustainability is an important indicator of the long-term health of sustainable systems. Where will farmers get seed and livestock that will be appropriate for sustainable agriculture in the future? Will the biodiversity be available? Will improved varieties be available to the farmers for a reasonable cost?

A MORE HOLISTIC APPROACH TO INDICATORS

Ecological indicators must be combined with economic and other factors when evaluating sustainability. Some also think that they should be viewed over an entire landscape rather than used as a single-field or farm measure. Di Pietro[10] points out that traditional agriculture leads to simplification and that, in general, systems that use simplification as a style are poorly adaptable, and thus not sustainable. Working in the central Pyrenees of France, Di Pietro studied two holistic indicators: 1) the contributions of a field's environmental features to land use choices (i.e., is sloping ground used for pasture or for crops), and 2) the diversity of environmental resources used by farms. The study was performed in two separate valleys. One was organized as a single unit of many farms, and the other as many units of individual farms and fields. Di Pietro showed that decisions at a landscape level are needed for sustainable measures to be effective.

Kemp et al.,[9] however, state that indicators at a farm level should be emphasized, because it is only at this level that environmental management can be applied effectively. This may be true today, but sustainability must eventually get past the single-farm level as the basic unit and move up to landscape levels and beyond. Nonpoint sources of influence, usually detrimental, are all too real in today's agriculture.

CONCLUSION

Ecological indicators can be as simple as soil erosion or as complex as the availability of biologically diverse domestic crops and animals. To be effective, indicators should be used in concert with other relevant measures of sustainability. And as with the components of all models, indicators should be used with the caution that they are only as good as the inferences on which they're based. For example, it is common to find biological indicators of sustainability based on yield and other measures of productivity. Short-term levels of high productivity can be based on unrealistically high amounts of imported inputs. To sustain these levels of production in most cases requires a highly extractive system. The result is anything but sustainable. Once systems have been deemed sustainable, then production would be a valid choice of indicator. Prior to that, though, the choice is ill-timed.

ARTICLES OF FURTHER INTEREST

Rhizosphere: An Overview, p. 1084
Sustainable Agriculture: Economic Indicators, p. 1195

REFERENCES

1. Herrick, J.E. Soil quality: An indicator of sustainable land management? Appl. Soil Ecol. **2000**, *15*, 75–83.

2. Parr, J.F.; Papendick, R.I.; Hornick, S.B.; Meyer, R.E. Soil quality: Attributes and relationship to alternative and sustainable agriculture. Am. J. Altern. Agric. **1992**, *7*, 5–11.
3. Kremen, C. Assessing the indicator properties of species assemblages for natural areas monitoring. Ecol. Appl. **1992**, *2* (2), 203–217.
4. Duelli, P.; Obrist, M.K.; Schmatz, D.R. Biodiversity evaluation in agricultural landscapes: Above-ground insects. Agric. Ecosyst. Environ. **1999**, *74*, 33–64.
5. Fauvel, G. Diversity of heteroptera in agroecosytems: Role of sustainability and bioindication. Agric. Ecosyst. Environ. **1999**, *74*, 275–303.
6. Buckerfield, J.C.; Lee, K.E.; Davoren, C.W.; Hannay, J.N. Earthworms as indicators of sustainable production in dryland cropping in Southern Australia. Soil Biol. Biochem. **1997**, *29*, 547–554.
7. Paoletti, M. The role of earthworms for assessment of sustainability and as bioindicators. Agric. Ecosyst. Environ. **1999**, *74*, 137–155.
8. Cutforth, L.B.; Francis, C.A.; Lynne, G.D.; Mortensen, D.A.; Eskridge, K.M. Factors affecting farmers crop diversity decisions: An integrated approach. Am. J. Altern. Agric. **2001**, *16*, 168–176.
9. Kemp, D.R.; Michalk, D.L.; Charry, A.A. In *The Development of Ecological Performance Indicators for Sustainable Systems*, Proceedings of the 10th Australian Agronomy Conference, Hobart, 2001. http://www.regional.org.au/au/asa/2001/4/c/kemp.htm (Accessed March 2003).
10. Di Pietro, F. Assessing ecologically sustainable agricultural land-use in the Central Pyrénées at the field and landscape level. Agric. Ecosyst. Environ. **2001**, *86*, 93–103.

Sustainable Agriculture: Economic Indicators

David Pimentel
Cornell University, Ithaca, New York, U.S.A.

INTRODUCTION

Food production suffers from overpopulation and environmental degradation at a time when food shortages are critical. The World Health Organization recently reported that more than 3 billion people are malnourished in the world.[1] This is the largest number and proportion of malnourished people ever in history. In the United States and throughout the world, agriculture suffers from many serious environmental problems that reduce the sustainability of crop and livestock production. The environmental resources most notably affected are soil, water, nutrients, energy, and biodiversity—all vital resources for food production.

Degraded agricultural lands require more fertilizers and more irrigation in order to maintain production. This is costly in terms of energy and the economics of agricultural production. In addition, the abandonment of some agricultural technologies (such as crop rotations) has resulted in loss of biodiversity and increased insect pests, plant pathogens, and weeds. These changes in turn require the intensive use of pesticides and other types of pest controls.

In addition to the direct effects of poor environmental resource management on agricultural production are the indirect effects off-site, such as pesticides and nitrogen fertilizer leaching and washing into rivers and groundwater. All of these negative effects reduce the sustainability of agriculture and the natural environment.

The abandonment of productive cropland coupled with the removal of forests is having negative impacts on world forests. Approximately 60% of world forest removal is associated with agricultural spread into forests.[2]

Soil erosion is intensifying, especially in developing countries where poor people depend on biomass energy for cooking and heat.[4] Wood fuel in most developing countries is in short supply, forcing the rural poor to turn to burning crop residues for cooking and heating their homes. The removal of crop residues leaves the soil unprotected from wind and water erosion. Without crop residues, soil erosion rates often increase tenfold. In addition, the crop residues contain nitrogen, phosphorus, and potassium—all essential nutrients for crop production. These vital nutrients are an integral part of the crop residues.

Erosion adversely affects crop productivity by reducing the availability of water, soil nutrients, soil biota, soil organic matter, and soil depth. The reduction in the amount of water available to the crop is considered the most harmful effect of erosion. After water, shortages of soil nutrients are the most important factors in limiting crop production.[3]

Severe soil erosion and associated rapid water runoff problems now seriously diminish the food economy as well as the health of the environment in the United States and world. In total in the United States, soil erosion effects both on the farm and off-site on the environment are estimated to cause more than $45 billion dollars in damages yearly.[3]

SOIL

World and U.S. food supplies depend on the availability of productive soils. Currently, more than 99.7% of world and U.S. food supply comes from the land, whereas less than 0. 3% comes from the oceans and other aquatic ecosystems.[2] Therefore, the dimensions of soil and land destruction in the United States and the world are of increasing concern. At present, the rate of soil erosion on U.S. cropland averages about 13 times greater than soil reformation. In Africa, Asia, and South America, the rate of soil erosion is 30 to 40 times greater than soil reformation. Soil erosion and land degradation result in the loss and abandonment of about 10 million hectares (ha) of cropland each year worldwide.[3]

WATER

All crops require and transpire massive amounts of water. For example, a corn crop that produces about 8000 kg/ha takes up and transpires about 5 million liters of water during the growing season of a little over three months.[5] If the corn crop requires irrigation, close to twice this much water (about 10 million liters) must be applied because not all the water is picked up by the corn plants.

Irrigation is highly energy intensive and costly. Approximately three times as much energy is required for crop production if the crop must be irrigated, compared with requirements for a rain-fed crop. Between $1000 and $1500 is required to irrigate a hectare of land. Approximately 1000 liters of water are required to

DOI: 10.1081/E-EPCS 120012946

Table 1 Energy and economic inputs per hectare for conventional and alternative corn production systems

	Conventional			Ridge planting and rotations		
	Quantity	**10^3 kcal**	**Economics ($)**	**Quantity**	**10^3 kcal**	**Economics ($)**
Labor (hrs)	10	7	50	12	9	60
Machinery (kg)	55	1485	91	45	1215	75
Fuel (litres)	115	1255	38	70	764	23
N (kg)	152	2280	81	27	5591	17
P (kg)	75	450	53	34	214	17
K (kg)	96	240	26	15	38	4
Limestone (kg)	426	134	64	426	134	64
Corn seeds (kg)	21	520	45	21	520	45
Cover crop seeds (kg)	–	–	–	10	120	10
Insecticide (kg)	1.5	150	15	0	0	0
Herbicide (kg)	2	200	20	0	0	0
Electricity (10^3 kcal)	100	100	8	100	100	8
Transport (kg)	322	89	32	140	39	14
TOTAL		6910	532		3712	337
Yield (kg)	7500	26,514		8100	29,160	
Output/Input ratio		3.84			7.86	

(From Ref. 10.)

produce 1 kg of grain. In the United States and world, about 70% of all water used is applied to crops for irrigation.[5]

Producing meat requires more water than crop production. For example, the production of 1 kilogram of beef requires about 43,000 liters of water. (This is not the water that a cow drinks; rather, it is the water required to produce the grain and forage that the cow consumes.)

Water and fossil energy are the two critical resources now limiting world food production.

PESTICIDES

Each year approximately 3 million metric tons of pesticides (insecticide, herbicides, and fungicides) are applied worldwide; approximately 500,000 tons are applied in the United States.[6] Despite the heavy application of pesticides worldwide, pest insects, weeds, and plant pathogens destroy more than 40% of all potential crop production. Losses of crops to pests in the United States are similar to world losses, or about 37% of potential production.[6] It should be pointed out, however, that food losses in each nation are related to cosmetic standards established in each nation. For example, fruits and vegetables sold in markets in India and Guatemala would generally not be salable in markets in the United States.

One concern over the use of pesticides is that less than 0.1% of the pesticide applied actually reaches the target pest. Thus, more than 99.9% disperses widely to contaminate the environment, including water and soil resources.[7]

Although each dollar invested in pesticide control returns about $4 in protected crops, this benefit overlooks the environmental and public health costs of pesticide application. In the United States, pesticide use causes approximately $9 billion in damages to the environment and public heath each year.[6]

Related to the public health problems in the United States are about 110,000 nonfatal human pesticide poisonings each year. In addition, between 10,000 and 12,000 cases of cancer are annually associated with the use of pesticides. Worldwide, about 26 million nonfatal pesticide poisonings occur, resulting in approximately 220,000 deaths each year. During the past 15 years, Sweden has been able to reduce pesticide use by 68%, while at the same time reducing public health problems associated with pesticides by 77%. Several nations (including Norway, Denmark, the Netherlands, and Indonesia) and the Canadian province of Ontario have programs to reduce pesticide use by 50% or more.[6]

FOSSIL ENERGY

The spectacular increases in grain and other crop yields during the past 40 years have been due to fossil energy inputs for fertilizers, irrigation, and pesticides. Thus, the green revolution was in reality due to the use of fossil energy. Grain crops such as rice and wheat were bred to be of short stature so they could produce increased yield in response to higher inputs of nitrogen fertilizer. The old varieties of these crops with large inputs of nitrogen would grow tall and then fall over and rot. Therefore, short varieties were essential to allow large amounts of fertilizers to be applied.

SOUND ENVIRONMENTAL AND ECONOMIC AGRICULTURAL MANAGEMENT

The major difficulties associated with conventional high-input agriculture are 1) high costs of production; 2) serious environmental resource degradation; and 3) instability of crop yields. Numerous agricultural technologies can be implemented to make agriculture environmentally and economically sustainable. These technologies can reduce chemical inputs (including fertilizers and pesticides), reduce soil erosion and rapid water runoff, and make better use of livestock manure.

The economical and environmentally sound agricultural practice of ridge planting and the rotation system are compared with the conventional corn production system in Table 1. Note the high level of inputs in the conventional corn system. The total costs of these inputs average $844 per ha in the United States. The total energy input is 7.0 million kcal/ha. The yields in both systems are the same, 8000 kg/ha.[8]

In the ridge-planting system, employing a cover crop and crop rotation reduced soil erosion from about 20 t/ha/yr to about 1 t/ha/yr. No insecticides or herbicides were employed in this system, demonstrating that corn can be produced without these chemicals.

Environmentally and economically sound corn production required only 4.0 million kcal and cost only about $700 per ha (Table 1). These values were 44% lower for energy and 17% lower for economic inputs than conventional corn production.[9]

ARTICLES OF FURTHER INTEREST

Agriculture and Biodiversity, p. 1
Environmental Concerns and Agricultural Policy, p. 418
Sustainable Agriculture and Food Security, p. 1183
Sustainable Agriculture: Definition and Goals, p. 1187
Sustainable Agriculture: Ecological Indicators, p. 1191

S

REFERENCES

1. WHO. *Malnutrition Worldwide*; 2000. http://www.who.int/nut/malnutrition_worldwide.htm, July 27, 2000.
2. Pimentel, D.; Bailey, O.; Kim, P.; Mullaney, E.; Calabrese, J.; Walman, F.; Nelson, F.; Yao, X. Will the limits of the Earth's resources control human populations? Environ. Dev. Sustain. **1999**, *1*, 19–39.
3. Pimentel, D.; Harvey, C.; Resosudarmo, P.; Sinclair, K.; Kurz, D.; McNair, M.; Crist, S.; Sphritz, L.; Fitton, L.; Saffouri, R.; Blair, R. Environmental and economic costs of soil erosion and conservation benefits. Science **1995**, *267*, 1117–1123.
4. Pimentel, D. The Limitations of Biomass Energy. In *Encyclopedia on Physical Science and Technology*; Academic Press: San Diego, 2001; 159–171.
5. Pimentel, D.; Houser, J.; Preiss, E.; White, O.; Fang, H.; Mesnick, L.; Barsky, T.; Tariche, S.; Schreck, J.; Alpert, S. Water resources: Agriculture, the environment and society. Bioscience **1997**, *47* (2), 97–106.
6. Pimentel, D. *Techniques for Reducing Pesticide Use: Environmental and Economic Benefits*; John Wiley and Sons: Chichester, UK, 1997; 444 pp.
7. Pimentel, D.; Levitan, L. Pesticides: Amounts applied and amounts reaching pests. BioScience **1986**, *36*, 86–91.
8. Pimentel, D.; Doughty, R.; Carothers, C.; Lamberson, S.; Bora, N.; Lee, K. Energy Inputs in Crop Production in Developing and Developed Countries. In *Food Security and Environmental Quality in the Developing World*; Lal, R., Hansen, D., Uphoff, N., Slack, S., Eds.; CRC Press: Boca Raton, 2002; 129–151.
9. Pimentel, D. Environmental and Economic Benefits of Sustainable Agriculture. In *Socio-Economic and Policy Issues for Sustainable Farming Systems*; Paoletti, M.A., Napier, T., Ferro, O., Stinner, B., Stinner, D., Eds.; Cooperativa Amicizia S.r.l: Padova, Italy, 1993; 5–20.
10. Pimentel, D. Environmental and Economic Benefits of Sustainable Agriculture. In *Sustainability in Question: The Search for a Conceptual Framework*; Kohn, J., Gowdy, J., Hinterberger, F., van der Straaten, J., Eds.; Edward Elgar: Northhampon, MA, 1999; 153–170.

Sustainable Agriculture: Philosophical Framework

Paul B. Thompson
Michigan State University, East Lansing, Michigan, U.S.A.

INTRODUCTION

Several different philosophical frameworks are currently being deployed in developing programs and research projects for sustainable agriculture. Although the choice of a framework may depend on the aims and interests of those who develop these projects, it is important to be clear about the various ways in which sustainable agriculture is conceptualized. Successful programs and projects depend upon agreement about the basic meaning of sustainability in agricultural food systems.

APPROACHES TO SUSTAINABLE AGRICULTURE

The adjective ''sustainable'' implies that the activity it modifies can continue, if not indefinitely, then for some foreseeable and generally accepted period of time. Hence ''sustainable agriculture'' should simply indicate ways of producing standard agricultural food and fiber goods that can continue to be practiced for the foreseeable future. But this formulation is both vague and ambiguous. Underlying assumptions, methods, and values for sustainable farming practices and for research undertaken under the rubric of sustainable agriculture have been debated ever since the term came into wide usage in the 1980s.

The primary source of disagreement can be traced to an ambiguity in the basic idea of a sustainable practice that has led to two conflicting fundamental philosophical orientations. To say that an activity can be continued indefinitely is, on the one hand, to predict or estimate the feasibility of undertaking the activity over a future period of time. On the other hand, it may also be understood to mean that the activity should be allowed to continue in the future. This ambiguity has led to two incompatible schools of thought about sustainable agriculture.

In emphasizing prediction and measurement the first approach allows one to draw upon principles and methods developed in the agricultural sciences, as well as for such diverse areas as sustainable development, sustainable structural design, and sustainable urban planning. In this approach articulation of these principles and methods becomes critical to the conceptualization of sustainable agriculture. The alternative view is to understand sustainable agriculture in terms of pluralistic values and social interests that derive coherence from their mutual opposition to a perceived status quo. Advocates of this approach do not typically seek to clarify or unify their understanding of sustainable agriculture in terms of scientific principles. However, they might support the creation of research that would support and further the values and social interests to which they are committed.

SUSTAINABILITY AS A RESEARCH PROBLEM

There is disagreement even among those who have attempted to articulate scientific criteria for sustainability. In one of the first systematic attempts to specify the guiding principles of sustainable agriculture, Gordon Douglass[1] identified three distinct schools of thought: resource sufficiency, or the view that the resources needed to carry out a farming practice are on hand or foreseeably available; ecological, or the view that practices must not disrupt natural biological processes integral to the renewal of organic materials; and social, or the view that farming must be compatible with principles of justice and economic opportunity. Advocates of accounting principles to audit available resources understand sustainable agriculture in terms of resource sufficiency, in much the way that Douglass originally proposed. Other approaches can be categorized as stressing the functional integrity of systems. These can include both ecological and social dimensions of the food system; hence these approaches encompass elements of both ecological and social sustainability.[2]

Resource sufficiency has the advantage of suggesting specific quantitative and auditable measurements. The current stock of available resources can often be measured or estimated, and the rate at which resources such as soil, water, fossil fuel energy, and other inputs are being both consumed and resupplied can be calculated. These basic measures can be made more dynamic with models that estimate fluctuations in climate and the evolution of pathogens, or that predict the rate at which insects become resistant to given control strategies. In principle, it is even possible to incorporate changing economic incentives for producers into such models as the relative costs of various

Encyclopedia of Plant and Crop Science
DOI: 10.1081/E-EPCS 120012949

inputs shift due to increasing scarcity and policy changes. A full integration of such modeling techniques would produce a comparative measure of the relative sustainability of alternative production systems, and an indication of where strategic vulnerabilities may be thought to lie.

However, one weakness of the resource sufficiency approach is that its measure of sustainability is only as good as the data and the accuracy and completeness of the models used to assess the availability and rate of use for critical resources. Furthermore, some of the greatest strategic vulnerabilities of a given farming system may in fact be associated with the most difficult modeling and auditing challenges. What is more, social factors ranging from farm bankruptcy to policy changes and shifting consumer tastes clearly have great influence on producers' ability to continue utilizing any given farming practice over time, yet the underlying processes governing such social changes are so complex that most have never been modeled.

This weakness in the resource sufficiency approach is, in part, why more qualitative conceptualizations that stress the way that various elements of a food and farming system are integrated have proliferated. The theme of functional integrity is that a set of practices or a farming system is sustainable when it is comparatively invulnerable to internal threats. Many of the accounting approaches utilized in measuring the sufficiency of resources reveal threats to the integrity of an agricultural or food system, but stressing functional integrity provides a philosophical framework that points toward a more holistic system and subsystem analysis. Two kinds of qualitative or philosophical value judgment are critical to the formulation of functional integrity. First, one must define system boundaries. Second, one must identify key strategic vulnerabilities.

System boundaries are critical to any conceptualization of sustainability, though boundary judgments are often implicit. The example of climate change illustrates why. Consider a case in which a local farming area suffers from dramatic climate change induced by industrial emissions of greenhouse gases. Drought conditions become commonplace, perhaps, or rising sea levels flood arable lands. Does this imply that the agriculture formerly being practiced there was unsustainable? Most would say not, and the reason is that they have implicitly accepted a boundary condition: Adverse affects on an agricultural system owing to industrial pollution do not count against the sustainability of the system. The industrial emissions certainly threaten the agricultural practice, but this is seen as an external threat—a threat coming from a source outside the system.

Incompatible boundary judgments give rise to disagreements about the sustainability of a given agricultural practice. This may occur especially when biologically trained individuals implicitly assume that all types of social causality are outside the system of interest. This kind of assumption may reflect the classical biologically oriented training of many agricultural researchers, who conduct laboratory research and field trials under conditions in which socioeconomic variables are controlled. Yet for any number of real-world applications, it is clearly the interaction of socioeconomic and ecological factors that will determine whether a system can continue to function under existing conditions.

Sometimes a single social criterion (such as profitability) is taken to determine sustainability irrespective of ecological factors. Here the problem may not be a disagreement about system boundaries, but may reflect implicit linkages between social and ecological system components that are simply being taken for granted. For example, some may assume that increasing scarcity of resources will be reflected in input costs, making profitability an adequate proxy for many ecological system elements. However, this assumption is itself based upon characteristics of policy and market structure that may or may not be reflected in the system of interest. In such instances, it is critical to articulate key vulnerabilities in qualitative terms, and to develop agreement among all participating parties about them.[3]

The need for agreement among those who participate in a sustainable agriculture has led some to argue that sustainability always implies a commitment to democracy. This may simply reflect the assumption that sustainable agriculture is a social movement rather than a researchable topic, but as a component of the feasibility of continuing any agricultural practice, democracy is a situation-specific element of sustainability, at best. In complex systems with many participants, the functional integrity of the system may depend upon including all relevant actors in decision making, and upon conducting decision-making processes in a democratic way. However, if the system of reference is a farmer's field, and the key threats are ecological in nature, democracy may not figure importantly in the operational criteria for sustainability. In general, the specific criteria for functional integrity will vary widely depending upon the system of reference and its boundaries.[3]

SUSTAINABILITY AS A SOCIAL MOVEMENT

An alternative approach has been to see sustainable agriculture purely as a social movement. Deriving in some respects from Douglass's idea of social sustainability, Patricia Allen and Carolyn Sachs[4] have argued that sustainable agriculture should be understood as a banner for interest groups seeking political change. In this view, one should not seek the principles that make agriculture sustainable, but should see the term as a convenient label for politically aligned interests. The interest groups associated with the term often include farm labor, women, consumer groups, small-farm groups, and advocates of animal welfare.

Although this approach to sustainable agriculture is clearly normative, the norms advocated are broad and not necessarily compatible. Sociologists Curtis Beus and Riley Dunlop[5] offer a complex portrait of competing agricultural paradigms. Sustainable agriculture is portrayed as an alternative to the conventional status quo on each of 27 points, summarized in 6 categories: decentralization, independence, community, diversity, restraint, and harmony with nature. To the extent that these points reflect a normative commitment to opposing conventional or industrial agricultural methods, they can be said to reflect a philosophical framework for sustainable agriculture.

CONCLUSION

The choice of a philosophical framework for approaching sustainable agriculture may reflect the specific interests and capabilities of the individuals undertaking a sustainable agriculture project. However, the opportunities for disagreement and misunderstanding are great, and it is always wise to be as explicit in articulating this framework as one possibly can be.

ARTICLES OF FURTHER INTEREST

Social Aspects of Sustainable Agriculture, p. 1155
Sustainable Agriculture: Definition and Goals of, p. 1187

REFERENCES

1. Douglass, G. *Agricultural Sustainability in a Changing World Order*; Westview Press: Boulder, CO, 1984.
2. Thompson, P.B.; Nardone, A. Sustainable livestock production: Methodological and ethical challenges. Livest. Prod. Sci. **1999**, *61*, 111–119.
3. Thompson, P.B. *The Spirit of the Soil: Agriculture and Environmental Ethics*; Routledge Publishing Co.: London, 1995.
4. Allen, P.; Sachs, C.E. The poverty of sustainability: An analysis of current positions. Agric. Human Values **1992**, *94*, 29–35.
5. Beus, C.E.; Dunlap, R.E. Conventional versus alternative agriculture: The paradigmatic roots of the debate. Rural Social. **1990**, *55* (4), 590–616.

Sweet Sorghum: Applications in Ethanol Production

Glen C. Rains
University of Georgia, Tifton, Georgia, U.S.A.

John W. Worley
University of Georgia, Athens, Georgia, U.S.A.

INTRODUCTION

Ethanol demand is currently over 1.8 billion gallons per year and estimated demand is expected to grow at an annual rate of 1.7 percent through the year 2020.[1] Although corn will continue to be the primary feedstock for ethanol, several biomass crops have been investigated for their merits as an ethanol feedstock. Sweet sorghum is an annual crop that has an excellent yield of fermentable sugars, is resistant to drought and adaptable to poor soils, and can thrive in a wide range of climates. It has been found that in certain geographical regions, such as the U.S. Piedmont (a geographical region spanning 170 counties from Virginia to Alabama), sweet sorghum produces more carbohydrates per hectare than corn, and would not directly compete in the food market. However, a system that economically produces and processes sweet sorghum for ethanol production has yet to be developed. This article discusses three systems of sweet sorghum for ethanol production in the U.S. Piedmont, and future developments that could make sweet sorghum an economically viable ethanol feedstock.

CONVERSION OF SWEET SORGHUM TO ETHANOL

The sweet sorghum stalk contains approximately 94% of the total soluble sugars (TSS) in the plant; about 85% of the TSS are located in the stalk pith fraction inside the stalk. Consequently, ethanol production has been examined in terms of both juice expression from whole stalks and pith fraction after separation from the rind and leaf fraction (rind-leaf). Crandell et al.[2] compared juice and sugar expression efficiencies from chopped whole stalks with those from the pith fraction after processing to remove most of the stalk rind and leaves. They found that juice expression as a percentage of whole stalk mass increased from 36 to 45% when the pith was first separated from the rind-leaf fraction. In the same experiment, sugar expression increased from 44 to 61% of whole stalk sugars from the separated material.

POTENTIAL SYSTEMS FOR PROCESSING SWEET SORGHUM FOR ETHANOL

Three systems for sweet sorghum harvesting and processing to produce ethanol have been analyzed.[3] The three systems are referred to as the forage chopper system, pith combine system, and Piedmont system (Fig. 1). The forage chopper system is characterized by harvesting sweet sorghum with a commercially available forage chopper and transporting the chopped stalks to a screw press parked next to or inside a bunk silo. The stalks are fed to the screw press for juice extraction and the residue (presscake) is conveyed into the silo. The ensiled presscake can be utilized for cattle feed, or taken to the fermentation plant for fiber breakdown and conversion to sugars for ethanol.

The pith combine system uses a modified forage chopper that chops the forage and then separates the pith (sugar fraction) from the rind-leaf (fiber fraction) using a straw-walker mechanism.[2] The rind-leaf is left on the field and the pith fraction is transported to the silo. The pith fraction is then fed into the screw press and the presscake conveyed into the bunk silo. By separating the rind-leaf fraction from the pith, screw-press efficiency and capacity are increased. The presscake is then ensiled for cattle feed or for further cellulose breakdown into sugars as in the forage chopper system. The rind-leaf can also be baled for hay and fed to cattle, stored and processed later to break down the cellulose into sugars, or left on the field to increase soil organic matter. The third system is the piedmont system. This system requires the highest capital investment because the harvesting and processing of the sweet sorghum is conducted with a dedicated machine that must be purchased or rented from a contractor who harvests and processes sweet sorghum on a custom basis. The piedmont system operates as follows: A whole-stalk harvester cuts stalks at the base and lays them in continuous windrows around the field. A field loader then loads stalks onto a trailer for transport to a storage facility near a bunk silo. Because the stalks are kept whole and intact, they can be stored 30 to 60 days without substantial loss in fermentables.[4] Stalks are loaded into a

Encyclopedia of Plant and Crop Science
DOI: 10.1081/E-EPCS 120010427

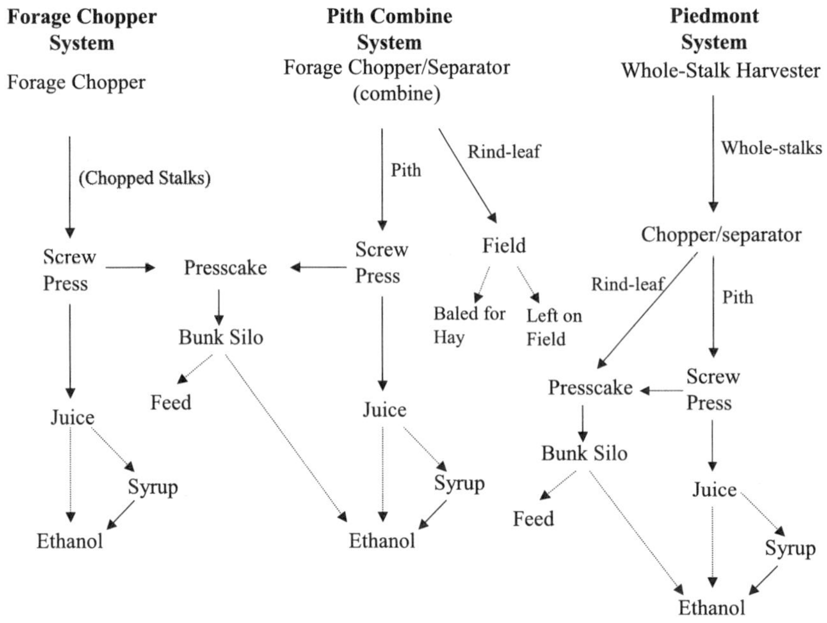

Fig. 1 Three systems for harvesting and processing sweet sorghum for ethanol production.

mobile processor that consists of a feeder, chopper, and pith separator (straw-walker). The pith separator separates the pith from the rind-leaf. The pith is entered into the screw press to express the juice, and the pith presscake is then recombined with the rind-leaf fraction and conveyed into the bunk silo. The presscake and rind-leaf can be handled as cattle feed or as future feedstock for ethanol.

In each system, juice from the screw press can be directly delivered to the fermentation/distillation plant for production of ethanol, or to an evaporation plant where the juice is concentrated into syrup for storage and later carried to the fermentation plant for ethanol production.

ECONOMIC CONSIDERATIONS

Limited data indicate that the expression ratios (mass of juice as a fraction of whole stalk mass) for the screw press using the three systems are 0.35 (forage chopper), 0.45 (pith combine), and 0.55 (piedmont system). An increased expression ratio increases screw press performance and decreases the cost of operation. The piedmont system requires the greatest capital investment and the forage chopper system the least. An important economic consideration is the ability of the piedmont system to store the fermentable sugars in the whole stalk, extending the processing window beyond the harvest season. (Equipment breakdowns do not immediately endanger the processing because stalks can be processed whenever the equipment is ready.) The forage chopper and pith combine systems must either ferment or distill the juice before it begins to sour. A systems model[5] examined the cost of production of each system of sorghum harvesting and processing, and indicates that transportation contributes significantly to the cost of production, especially when moving the by-products for further ethanol production. The model shows that the pith combine system is the cheapest per Mg whole stalk, whereas the piedmont system is the most expensive. All three systems are too expensive to make a profit without government subsidies.

Each system has options for processing the sweet sorghum once it is separated into juice, presscake from the screw press, or rind-leaf fractions (Fig. 1). Juice can be sent to a fermentation/distillation plant for direct conversion to ethanol. In this case, sweet sorghum by-product (materials placed in the bunk silo) can be transported to the plant later, after harvest season peak demand, and be converted to sugars for further ethanol production. A recent study indicates that one or more microorganisms can carry out simultaneous hydrolysis and fermentation of cellulose to ethanol in the same bioreactor.[6] If the by-product is useful as a cattle feed, part of the juice can be sent to an evaporation plant to be concentrated into syrup and stored for processing until peak demand for the fermentation/distillation plant concludes.

Other by-product uses such as source of heat energy, pulp for fiberboard manufacture, hay for animal feed, and organic matter for field amendments are important in determining the economic viability of sweet sorghum for ethanol production. By-product value is difficult to estimate. Based on a model by Worley et al.,[7] the by-product values depend less on the system chosen to produce ethanol than on the intended use of the by-product. The highest value is assigned to by-product silage used as cattle feed on the grower's own farm. Fiber conversion to obtain further ethanol production yields the lowest value. However, this study did not consider some of the newer techniques for bioconversion of presscake that show a yield of 5.1 g ethanol per 100 g of fresh stalks.[6]

CONCLUSION

Sweet sorghum research is in its infancy relative to research on other biomass crops such as corn and sugarcane. As such, it is highly likely that increases in biomass production can be achieved through development of improved genotypes. Studies continue on the use of sweet sorghum as an energy source through conversion to ethanol or electricity.[8,9] Developing more uses for by-products and improved production practices seem to hold promise for improving the economic viability of sweet sorghum.

REFERENCES

1. Energy Information Administration. *Annual Energy Outlook 2002 with Projections to 2020*; DOE/EIA-0383(2002), U.S. Department of Energy: Washington, DC, 2001; 1–255.
2. Crandell, J.H.; Cundiff, J.S.; Worley, J.W.; Vaughan, D.H. In *Methods of Separating Pith from Chopped Sweet Sorghum Stalks*, ASAE International Winter Meeting, New Orleans, LA, Dec. 12–15, 1989; ASAE: St. Joseph, MI. ASAE Paper # 896572.
3. Rains, G.C.; Cundiff, J.S.; Welbaum, G.E. Sweet Sorghum for a Piedmont Ethanol Industry. In *New Crops*; Janick, J., Simon, J.E., Eds.; Wiley: New York, 1993; 394–399.
4. Parrish, D.J.; Gammon, T.C.; Graves, B. Production of fermentables and biomass by six temperate fuelcrops. Energy Agric. **1985**, *4*, 319–330.
5. Worley, J.W.; Cundiff, J.S. System analysis of sweet sorghum harvest for ethanol production in the piedmont. Trans. ASAE **1991**, *34* (2), 539–547.
6. Mamma, D.; Christakopoulos, P.; Koullas, D.; Kekos, D.; Macris, B.J.; Koukios, E. An alternative approach to the bioconversion of sweet sorghum carbohydrates to ethanol. Biomass Bioenergy **1995**, *8* (2), 99–103.
7. Worley, J.W.; Cundiff, J.S.; Vaughan, D.H. Potential economic return from fiber residues produced as by-products of juice expression from sweet sorghum. Bioresour. Technol. **1992**, *41*, 153–159.
8. Türe, S.; Uzun, D.; Türe, I.E. The potential use of sweet sorghum as a non-polluting source of energy. Energy **1997**, *22* (1), 17–19.
9. Nguyen, M.H.; Prince, R.G.H. A simple rule for bioenergy conversion plant size optimization: Bioethnaol from sugar cane and sweet sorghum. Biomass Bioenergy **1996**, *10* (5/6), 361–365.

Sweetgrass and Its Use in African-American Folk Art

Robert J. Dufault
Clemson University, Charleston, South Carolina, U.S.A.

INTRODUCTION

Sweetgrass (*Muhlenbergia filipes*) is the main structural material used in African-coiled basketry. The plant is a native, perennial, warm-season grass found growing sparsely along the ocean in narrow bands between coastal sand dunes from North Carolina to Texas. Much of the native sweetgrass groves have been lost in the region due to oceanfront development. The continuance of this old African-American folk art is jeopardized by the loss of sweetgrass. Since the late 1980s, horticultural research has determined that sweetgrass can be grown successfully as a row crop, but the plant is a short-lived perennial, needing replanting every four years to maintain supply. With knowledge of how to propagate and cultivate sweetgrass, sweetgrass ''gardens'' could be planted in the backyards of individual basketmakers to finally solve the supply problem.

SWEETGRASS BASKETS

The city of Charleston, South Carolina, was founded in 1670 and is bedazzled with lovely old homes and gardens throughout its old and winding streets. However, even older than most homes in Charleston is the ancient tradition of sweetgrass basketmaking. The significance and legacy of producing these historic baskets has been a treasured, beloved birthright held dear by generations of basket makers, the direct descendants of enslaved Africans of plantation days. About 30 years ago, approximately 1200 families were involved in basketmaking, but now only about 300 families are involved in the tradition (personal communication with Mary Jackson, past president of Mt. Pleasant Basketmaking Association). The most insidious factor contributing to the demise of this folk art is the gradual loss of natural sweetgrass habitats in the region. Sweetgrass once was readily available and abundant in the Charleston area, but today sweetgrass is very scarce and difficult to find.

Sweetgrass (*M. filipes*), the main structural material used in African-coiled basketry, is a native, perennial, warm-season grass found growing sparsely along the ocean in narrow bands between coastal sand dunes from North Carolina to Texas. The narrow leaf blades are harvested green and dried before use in sweetgrass baskets. Successive coils of sweetgrass are sewn together in rows, not woven, with strips of leaf material from the palmetto tree (*Sabal palmetto*), the state tree of South Carolina (Fig. 1). The explosion of growth in the Charleston area by new residents and industries has caused a boom in urban and beachfront development to support the demand by the growing population. Urbanization has destroyed much of the natural habitats of sweetgrass, as has development of beachfront communities on the barrier islands off the coast of Charleston. Basketmakers now have to travel hundreds of miles to Georgia and Florida to find adequate supplies. Many basketmakers are old and these long, expensive, arduous trips are impossible and frustrating, which has fueled the loss of basketmakers.

The roots of the sweetgrass basket craft were borne out of slavery. Colonial Charleston was a very prosperous city built on many agricultural industries beginning with rice in the late 1600s and early 1700s, then indigo, and lastly cotton. With Charleston's warm climate, a long growing season, and extensive marshy wetlands, the area was a natural for growing the lucrative rice crop. Rice was a staple crop in Africa, and Africans had cultivated rice for centuries and perfected its production. Slavers identified these people, the ancestors of today's Low Country basketmakers, and abducted them into the slave trade for use on rice plantations in the New World.

The enslaved Africans brought the expertise to produce rice from the planting, harvesting, and processing. African-coiled basketry was critical in the processing of the rice. With flat fanner sweetgrass baskets, the seed was thrown into the air to separate the rice seed from chaff. Slavers were paid premium price for Africans from the West African Rice Kingdoms of the Windward Coast (Senegal to the Ivory Coast) to the mouth of the Congo River (Gabon, Zaire, and Angola).[1] A man or woman who made baskets was worth more than one who did not, with age, strength, and other skills being equal.[2] History has not fully credited the enslaved African for the success of the rice kingdom in the South, but without their knowledge of the crop cultivation and processing, rice would have never flourished.[3] Besides fanner baskets, enslaved Africans sewed all sorts of utilitarian vessels to hold and store agricultural commodities and to use in plantation houses for domestic purposes. The barter of

Encyclopedia of Plant and Crop Science
DOI: 10.1081/E-EPCS 120010442

Fig. 1 A fanner basket (left) and tote baskets (center and right) showing consecutive rows of sweetgrass (light coils) and pine needles (dark coils) woven together with strips of palmetto palm fronds.

baskets among plantations was also a very lucrative trade in colonial days.

Sweetgrass basketry is one of the earliest traditional crafts with a rich documented history from ''carryover'' from African enslavement and colonial plantation days to the present. These African skills remained unchanged and have been passed down from generation to generation for over 300 years from parents to siblings. Though the raw materials are different in American and African baskets, amazingly the construction techniques of sweetgrass baskets are unchanged from their African counterparts to this day. Some American basketmakers have journeyed to Senegal, Africa, and have reported very similar baskets and techniques still produced by today's African basketmaker. Modern basketmakers in the Low Country of South Carolina have introduced a multitude of new basket forms and functions, which are still very useful in the home, but these baskets are now considered ''objects of art'' more than just utilitarian baskets of old.

To basketmakers, sweetgrass basketry symbolizes 1) their heritage and connection to their African ancestry and 2) a remembrance of people who survived centuries of oppression. They have made a reaffirmation of these goals with each new generation with sweetgrass basketry representing the heart and keepsake of their devotion.

Basketmakers have always wondered if they could cultivate sweetgrass inland similarly to vegetables, but individual attempts to transplant the plant were not successful. By 1988, the plight of the basketmakers was common knowledge due to timely newspaper articles. In 1989, I became intrigued with the idea of domesticating sweetgrass and became involved in attempts to learn how to grow it in a cultivated state. After four years of part-time trial and error using many different horticultural approaches and techniques, in 1993 I felt confident to plant a large experimental planting somewhere in the Low Country. The Historic Charleston Foundation offered land at the old McLeod Plantation for our field trials. Established in 1947, the Historic Charleston Foundation is a nonprofit educational organization dedicated to historic preservation. McLeod Plantation, now owned by the Historic Charleston Foundation, is the last intact plantation on James Island and a historic landmark. So, with enthusiasm, members of the Mt. Pleasant Basketmakers Association and I toiled to prepare and plant 2000 sweetgrass plants in an effort to eventually eliminate the shortage. A second acre was planted at McLeod, and in May 1995 two more acres were planted. Many other important benefactors joined us to help support our goal of ending the sweetgrass shortage problem, such as the City of Charleston, the Agricultural Society of South Carolina, the SC Seagrant Consortium, and the Trident Community Foundation. We enlisted the assistance of the National Civilian Community Corps to help prepare and plant the expansion acreage in 1995. By summer 1995, we had four acres of sweetgrass maturing for harvest within two years.

The year of 1995 was a historic year for the basketmakers, as the first acre of sweetgrass at McLeod was finally mature enough for extensive harvesting. In past years, the plants were rather succulent and brittle, which rendered the cultivated grass unsuitable for basketry. However, in 1995, the plants matured and grew into very large clumps. Plant competition caused the leaves to lengthen and become very strong and fibrous, which are critically important characteristics of high-quality sweetgrass for basketry. By midsummer, we finally knew that our grand experiment of cultivating sweetgrass at McLeod was a rousing success. Basketmakers judged the quality of the cultivated sweetgrass excellent for basketry. The downside of this grand experiment was twofold: 1) Sweetgrass in cultivation was found to be a short-lived perennial, and earlier plantings only lived for about four years and then declined by 1998; and 2) maintenance of such large acreage by basketmakers was too difficult for the community to handle. It appeared that large plantings were not the best approach to solve the supply issue.

CONCLUSION

During the years I worked with this crop, I learned many secrets about the plant's growth and development. Although the supply problem still plagues the basketmakers, we now know the horticulture of this plant. I feel that the supply problem can be solved easily by individual basketmakers establishing sweetgrass plantings in their own backyard. A 25′ by 25′ plot would probably sustain the needs of a basketmaker or two. The plant thrives on benign neglect and doesn't require any fertilizer or much water during the growing season. This scheme would provide a free product to supply individuals who would take responsibility and pride in not only producing a beautiful art object but also growing the "basket" from seed.

ARTICLES OF FURTHER INTEREST

Crop Domestication in Prehistoric Eastern North America, p. 314
Ecology and Agricultural Sciences, p. 401
Farmer Selection and Conservation of Crop Varieties, p. 433

REFERENCES

1. Littlefield, D. *Rice and Slaves: Ethnicity and the Slave Trade in Colonial South Carolina*; University of Illinois Press: Chicago, IL, 1991; Baton Rouge, LA.
2. *Charleston Gazette and Advertiser,* Feb. 15 Ed.; South Caroliniana Library: Columbia, SC, 1791.
3. Joyner, C. *Down by the Riverside: A South Carolina Slave Community*; Univ. of Illinois Press: Urbana, IL, 1984.

Switchgrass as a Bioenergy Crop

Samuel B. McLaughlin
Lynn A. Kszos
Oak Ridge National Laboratory, Oak Ridge, Tennessee, U.S.A.

INTRODUCTION

The production of renewable bioenergy from crops grown on American farms to supplement fossil-derived energy sources could provide the United States with many desirable ecological and economic benefits. These include reduced dependency on imported oil and its associated geopolitical and economic risks; reduced damage to sensitive ecosystems associated with the high ecological costs of fossil fuel mining/recovery; reduced emissions of both greenhouse gases and toxic air pollutants associated with combustion of fossil fuels; and increased economic benefits to American farmers who would produce and harvest the renewable energy embodied in crops or crop residues.

Among the many types and sources of renewable energy—which include solar, wind, water, municipal waste, and agricultural residues—perhaps the greatest potential for ecological and economic gains to the agricultural industry is from dedicated energy crops. Dedicated energy crops include short-rotation woody crops as well as annual or perennial herbaceous crops grown specifically for energy industries. These are potentially very attractive to industry because they can be selected to provide desirable feedstock chemical and physical attributes, and can be planted in close proximity to industrial sites to assure adequate supply of feedstock of acceptable quality with reduced transportation costs. Among the potential dedicated energy crops, switchgrass (*Panicum virgatum)* was selected by the U.S. Department of Energy–sponsored Bioenergy Feedstock Development Program (BFDP) in 1991 as a model herbaceous species for further development. Characteristics of switchgrass that reflect its potential to serve as a national source of renewable energy are the focus of this article.

ECOLOGICAL AND AGRONOMIC ATTRIBUTES OF SWITCHGRASS

Switchgrass is a perennial bunch grass native to the United States. It has the highly efficient C4 carbon metabolism typical of warm season grasses and a deep rooting pattern that equips it to thrive in a wide variety of habitats and soils that are unsuitable for many types of row crops.[1] Switchgrass formed an important component of the original tall-grass prairie in the central United States. Much of the original prairie habitat has been converted to cropland, and switchgrass can currently be found in more thinly dispersed and smaller stands in diverse habitats. These include both grasslands and open forests in North America east of the Rocky Mountains. Two main ecotypes occur naturally in this genetically diverse, open-pollinated species: 1) lowland ecotype, that is thicker-stemmed and adapted to warmer, moister sites in the South; and 2) an upland ecotype that has thinner stems and is adapted to somewhat drier soils and cooler climatic conditions. Switchgrass was chosen as a model herbaceous bioenergy crop for research and development based on the relatively low energy and resource requirements associated with high-yield capacity, its ecological value in protecting and improving soil quality and wildlife habitat, and its compatibility with conventional farming equipment and management practices.[2,3]

Switchgrass has been planted as a forage grass in the great plains for over 50 years and is one of the principal grasses planted to restore soil quality in the U.S. Conservation Reserve Program. As a species for bioenergy production, the criteria for successful development shifted toward maximum production of biomass and associated cell wall constituents (such as cellulose that could be converted to fuels such as ethanol, or combusted to produce heat and electricity). In addition, achieving low mineral content in harvested biomass is important in reducing mineral buildup during combustion or conversion processes. Thus, high yields and low mineral nutrient content became relatively more important in selecting, improving, and managing switchgrass as a bioenergy species. This represents a shift in focus from the criteria for traditional forage production for animal consumption, where high mineral nutrient content and low stem:leaf ratios are more desirable attributes.

The past 10 years of BFDP research have focused on strategies for efficiently establishing and managing switchgrass for maximum biomass production; defining and improving breeding techniques to maximize biomass

Encyclopedia of Plant and Crop Science
DOI: 10.1081/E-EPCS 120010429

production potential; and basic studies to better understand the underlying genetics, physiology, and soil–carbon cycles of switchgrass grown as an energy crop. More recently, integrative assessments have evaluated the potential of switchgrass to contribute ecological and economic gains as a component of a national energy strategy.

ESTABLISHING AND MANAGING SWITCHGRASS

One of the principal challenges of producing switchgrass lies in successful establishment of this light-seeded perennial species, which typically attains less than a third of its full production potential during the first year. Because of slow growth in the establishment year, switchgrass typically takes 2–3 years to attain full production capacity. Planting seed of known viability in a firm seedbed, controlling weeds during the initial year, and restricting cutting practices to 1–2 cuts per year are important considerations in maximizing switchgrass production.[4] There are several pre- and post-emergence herbicides available for use on nongrazed grasses, including Plateau®, Paramount®, Roundup®, and atrazine. However, the success of these herbicides and appropriate dosage rates can vary regionally, and latitude should be considered in designing an optimum weed control plan. Higher temperatures in Texas, for example, can increase the toxicity of some herbicides to switchgrass.[5]

Growing within its zone of natural adaptation, switchgrass has very high resistance to foliar diseases, although there has been increasing evidence of leaf diseases on some varieties in larger plantings in the Midwest in recent years. Recently initiated research also suggests that nematodes play some role in inhibiting growth of switchgrass during the establishment year. More research is needed in this area, as important gains in early establishment might be realized from improved understanding of the role of belowground diseases.

Production management research within BFDP has focused on identifying the most productive existing cultivars, their regions of optimum adaptation, and the management practices that optimize production efficiency within the regions. The best commercial varieties determined by 10 years of production in field research plots are the lowland variety Alamo, in the deep South; the lowland varieties Alamo and Kanlow at mid-latitudes; and the upland variety Cave-in-Rock for northern latitudes.[6,7] Yield of the best adapted varieties averages 15–22 Mg ha^{-1} y^{-1} in regional field trials. Yields of currently available varieties in production-scale fields are expected to be 30% lower, around 10–15 Mg ha^{-1} (4.5–7 tons ac^{-1}). Lowland varieties produce highest yields in the deep South, with either a single harvest in late August to mid-September, or two harvests with early July and November target dates. Upland varieties such as Cave-in-Rock may, under the best growing conditions, achieve yields comparable to lowland varieties in the South, but clearly require two harvests annually to attain those yields. In Alabama, row spacings may be increased to as much as 80 cm without yield loss.[6] The combination of flexible harvest frequency and timing is attractive to landowners who are managing multiple crops, whereas the row spacing flexibility improves potential attractiveness of switchgrass stands for game bird cover. From the perspective of resource utilization efficiency, the single-cut system offers higher returns based on reduced energy inputs associated with a single harvest and substantially lower nutrient removal. Nitrogen is the only nutrient that consistently stimulated switchgrass yields, and nitrogen removal has been reduced by approximately 60%, to 30–50 kg N ha^{-1}, with single-harvest systems.[7] Recommended N application rates are 50–100 kg N ha^{-1} y^{-1} in most areas; however, there is some evidence that switchgrass stands may become less dependent on external nitrogen as internal sources in roots, crowns and soil organic matter accumulate.[6,7]

DESIRABLE PRODUCTION CHARACTERISTICS

Production characteristics of primary interest in selecting a bioenergy crop are high yield potential, low yield variability over time in the face of variable environmental conditions, the persistence of the stand over time, and the quality of the feedstock produced. The longest continuous yield plots in the BFDP system were planted to Alamo switchgrass in 1988 and have shown high yields (13 yr avg = 23 Mg ha^{-1}) and an apparent increase in resistence to drought (Fig. 1) over time.[7] Whereas irrigation has not been considered a desirable management option for energy crops, experience in Texas[5] suggests that a single midseason irrigation event may double switchgrass yields in dry years. The potential sharing of irrigation equipment among multiple fields for which only one or two irrigation events are required during the season increases the probability that irrigation would be an economically viable option for switchgrass. The capacity of switchgrass (as well as other warm season grasses) to maintain such high yields even during periods of infrequent rainfall can be attributed to its deep, well developed root system. In the mid-Atlantic states standing root mass has been found to be comparable to annual aboveground production.[6]

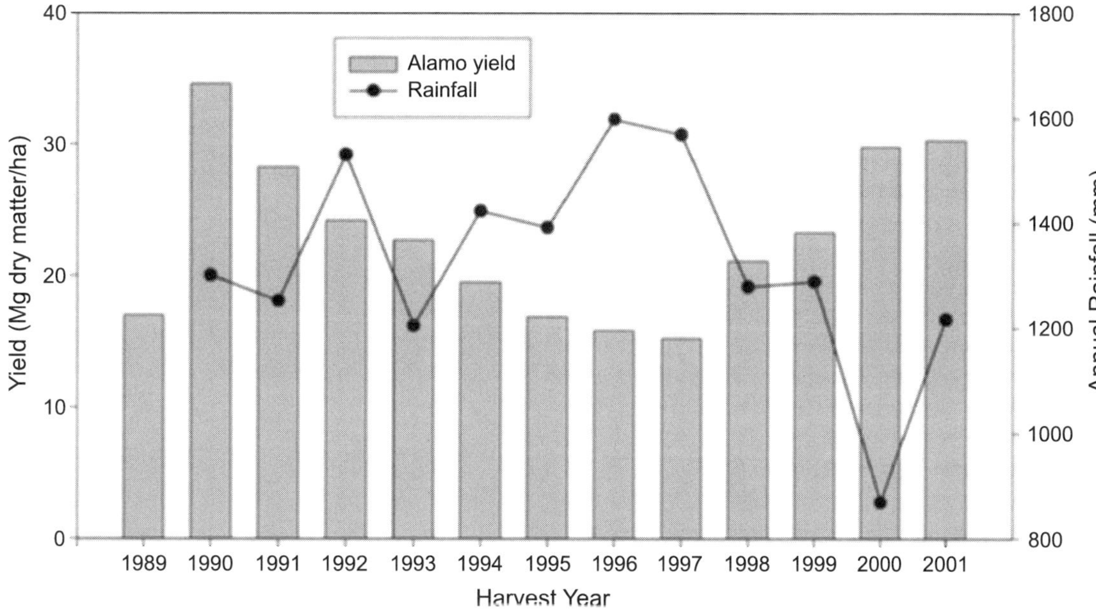

Fig. 1 Long-term yields of Alamo switchgrass from Auburn, Alabama compared to annual rainfall. Note the relative insensitivity of annual yields to the very low rainfall occurring in 2000. This stand was planted in 1988 and was harvested twice each year in early July and September. (Data from Ref. 7.)

High switchgrass root production and turnover are important not only in nutrient and water acquisition from soils but also in increasing soil organic matter (SOM). Studies across diverse soil types in both the mid-Atlantic region and in Texas have documented significant increases in SOM in soils under switchgrass.[5–7] In the mid-Atlantic region, SOM increases ranged from 20% to 125%, and averaged 43% across eight study sites over a 10-year period.[6] The accumulation of SOM with root production and turnover has important implications for improving water and nutrient retention and availability in associated soils. Such increases are also important in the net reduction of carbon dioxide emissions derived from the utilization of switchgrass as a renewable energy

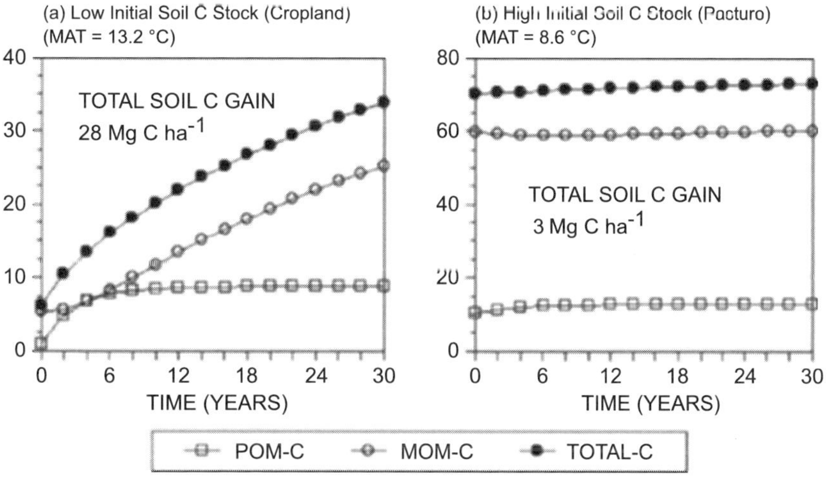

Fig. 2 Through production and turnover of fine roots, switchgrass can contribute significant amounts of soil organic carbon to the farm lands on which it is grown (POM-C = particulate organic matter, MOM-C = mineral-associated organic matter). Modeling studies indicate that these gains are influenced by soil temperature and prior accumulation of soil carbon ((a) cropland; (b) pasture) (MAT = mean annual temperature). (Prediction and analyses from Ref. 8.)

source.[8] An empirically based model of soil carbon accumulation under switchgrass has been developed to estimate relative contributions of both labile and recalcitrant carbon to accumulated soil carbon under switchgrass over time (Fig. 2). The model projects annual accumulation rates that ranged up to 1.4 Mg C ha^{-1} y^{-1} over 10 years on degraded soils in warmer climates and averaged 0.78 Mg C ha^{-1} y^{-1} over 30 years across diverse regions in the U.S. East. Thus, carbon storage in soils under switchgrass can represent a significant contribution to total reduction in carbon emission achieved in utilizing switchgrass for bioenergy.

BREEDING

Breeding research with switchgrass over the past 10 years has contributed significant new knowledge to the basic breeding biology of switchgrass.[9] Switchgrass has been found to typically occur in either tetraploid ($4x = 36n$) or octoploid ($8x = 72n$) multiples of the nine basic chromosome pairs. It is an open-pollinated, self-incompatible species, within which crossing across ecotypes does occur, although crosses between ploidy levels are rare. At present, germplasm derived from accessions from nine clusters that were delineated within a larger population of 115 accessions of switchgrass examined in the BFDP has been submitted to the U.S. Department of Agriculture's National Plant Germplasm System.[9] Early in the breeding effort with switchgrass, fast-track breeding approaches such as Recurrent Restricted Phenotypic Selection based primarily on first-year growth of switchgrass were replaced with a slower but more predictable genotypic selection process based on performance of test plants over at least 2 years, and quantitative evidence of high heritability of desirable agronomic traits from candidate parent plants. In addition, heterosis has now been demonstrated in switchgrass allowing a narrowly restricted breeding base to be used to target specific desirable traits. More than 15 experimental synthetics have been developed, at least three of which are undergoing performance verification for commercial release by seed companies. Expected genetic gains estimated from among half-sib families have been 5–26% above the best commercially available varieties; however, yield gains have also been shown to be strongly influenced by the test environment.[9]

EVALUATING SWITCHGRASS AS A BIOFUEL

To be successful as a biofuel crop, a species like switchgrass must be physically and chemically acceptable as a fuel for industrial use as well as economically competitive in both agricultural and industrial markets. The chemical and physical attributes of switchgrass shown in Table 1 include a relatively high energy content combined with a relatively low ash content.[10] These traits make switchgrass a desirable fuel, both for combustion to produce heat and electricity, and for conversion to ethanol as a transportation fuel. Estimates of the competitive potential of switchgrass as a crop in agricultural markets

Table 1 Chemical and physical properties of switchgrass as a biofuel relative to selected alternate fuels

		Switchgrass	Alternate fuel	
Fuel property	**Units**	**Value**	**Value**	**Fuel type**
Energy content (dry)	$Gj.Mg^{-1}$	18.4	19.6	Wood
			27.4	Coal
Moisture content (harvest)	%	15	45	Poplar
Energy density (harvest)	$Gj.Mg^{-1}$	15.6	10.8	Poplar
Net energy recovery	$Gj.Mg^{-1}$	18	17.3	Poplar
Storage density				
(6′×5′) round bale	$kg.m^{-3}$	133	150	Poplar chips
(4′×5′) round bale	(dry weight)	105		
Chopped		108		
Holocellulose	%	54–67	49–66	Poplar
Ethanol recovery	$L.kg^{-1}$	280	205	Poplar
Combustion ash	%	4.5–5.8	1.6	Poplar
Ash fusion temperature	oC	1016	1350	Poplar
			1287	Coal
Sulfur content	%	0.12	0.03	Wood
			1.8	Coal

(From Ref. 10.)

using an agricultural sector model indicate that at a price of \$44 Mg^{-1}, switchgrass would be more profitable than crops currently grown on approximately 16.9 million ha of American farmland.[8] As reliance on energy from biofuels increases in the future, switchgrass should thus provide an economically attractive alternative cash crop for American farmers when priced at these levels. At present, switchgrass is not being utilized on a large scale for bioenergy production. However, pilot testing of switchgrass combustion characteristics and performance by research teams that have included two commercial power producers has been completed successfully. In addition, a large-scale power production system involving switchgrass produced on 8000 ha by a farmers' cooperative is currently being developed and tested for full-scale operational feasibility in Chariton Valley, Iowa.

The value of a fuel for national energy supply should theoretically include not only its value as a source of energy, but also its impacts on society as a whole. Currently, research in the bioenergy industry is increasingly being directed toward producing not only energy and fuels from biomass, but also a wide variety of industrial chemicals that can add value to the bioenergy cycle. The development of capabilities to transform crops like switchgrass to improve production potential[11] may lead to increased recovery of specific chemicals of industrial interest as well as improved production potential of switchgrass. The true value of bioenergy crops also encompasses ecological and agronomic values that are not traditionally considered in fuel prices. For this reason, life cycle analyses of switchgrass production have been conducted to consider some of these secondary values relative to current fossil fuel scenarios. Included among the effects of production of 16.9 million ha of switchgrass at \$44 Mg^{-1} in the previous example were estimated annual increases of \$6 billion in net farm revenues, \$1.86 billion reduction in government subsidies due to improved farm income, and displacement of 44–150 Tg (1 Tg = 10^{12} g of carbon) emissions to the atmosphere.[9] Benefits of this magnitude substantially reduce the societal costs of renewable biofuels relative to fossil fuels, and should provide economic incentives to expedite the commercialization of switchgrass as a bioenergy crop.

CONCLUSION

Production of dedicated energy crops on American farms represents a potentially important way to decrease America's dependency on imported energy while providing significant economic and ecological benefits to the national economy. Research on management, breeding, and both ecological and economic benefits of a perennial native grass species, switchgrass, as a source of renewable energy suggests that this species could play an important role in the search for greater national energy self-sufficiency. Switchgrass can be used to produce electric power, transportation fuels, and biobased products. Benefits of developing a biobased economy with participation of America's farmers as energy suppliers are substantial and include improved farm income, improved soil productivity and stability, reduced greenhouse gas emissions, and reduced need for government subsidies of crop prices.

ACKNOWLEDGMENTS

This research was sponsored by the Biomass Program of the U.S. Department of Energy. ORNL is managed by UT-Battelle, LLC, for the U.S. Department of Energy under contract DE-AC05-00OR22725.

ARTICLES OF FURTHER INTEREST

Alternative Feedstocks for Bioprocessing, p. 24
Breeding Hybrids, p. 186
Breeding Plants and Heterosis, p. 189
Breeding Plants with Transgenes, p. 193
Breeding: Recurrent Selection and Gain from Selection, p. 232
Molecular Technologies and Their Role Maintaining and Utlizing Genetic Resources, p. 757
Nematodes and Host Resistance, p. 805
New Industrial Crops in Europe, p. 813
Nitrogen, p. 822
Rhizosphere: An Overview, p. 1084
Sustainable Agriculture: Economic Indicators, p. 1195
Transformation Methods and Impact, p. 1233

REFERENCES

1. Moser, L.E.; Vogel, K.P. Switchgrass, Big Bluestem, and Indiangrass. In *Forages Vol. 1, An Introduction to Grassland Agriculture*; Barnes, R.F., Miller, D.A., Nelson, C.J., Eds.; Iowa State Press: Ames, IA, 1995; 409–420.
2. McLaughlin, S.; Bouton, J.; Bransby, D.; Conger, B.; Ocumpaugh, W.; Parrish, D.; Taliaferro, C.; Vogel, K.; Wullschleger, S. Developing Switchgrass as a Bioenergy Crop. In *Perspectives on New Crops and New Uses*; Janick, J., Ed.; ASHS Press: Alexandria, VA, 1999; 282–299. http://www.hort.purdue.edu/newcrop/proceedings1999/v4-282.html (accessed December 2002).
3. Sanderson, M.A.; Reed, R.L.; McLaughlin, S.B.; Wullschleger, S.D.; Conger, B.V.; Parrish, D.J.; Wolf, D.D.;

Taliaferro, C.M.; Hopkins, A.A.; Ocumpaugh, W.R.; Hussey, M.A.; Read, J.C.; Tischler, C.R. Switchgrass as a sustainable bioenergy crop. Bioresour. Technol. **1996**, *56*, 83–93.
4. Wolf, D.D.; Fiske, D.A. *Planting and Managing Switchgrass for Forage, Wildlife, and Conservation*; Virginia Cooperative Extension Publication 418-013; Virginia Polytechnic and State University: Blacksburg, VA, 1995. http://www.ext.vt.edu/pubs/forage/418-013/418-013.html (accessed December 2002).
5. Ocumpaugh, W.; Hussey, M.; Read, J.; Muir, J.; Hons, F.; Evers, G.; Cassida, K.; Venuto, B.; Grichar, J.; Tischer, C. *Evaluation of Switchgrass Cultivars and Cultural Methods for Biomass Production in the South Central U.S.*; Oak Ridge National Laboratory ORNL/SUB-03-19XSY091C/01: Oak Ridge, TN, 2003; 1–137.
6. Parrish, D.J.; Wolf, D.D.; Fike, J.H.; Daniels, W.L. *Switchgrass as a Biofuels Crop for the Upper Southeast: Variety Trials and Cultural Improvements*; Oak Ridge National Laboratory ORNL/SUB-03-19XSY163C/01: Oak Ridge, TN, 2003; 1–139.
7. Bransby, D.I. *Development of Optimal Establishment and Cultural Practices for Switchgrass as an Energy Crop*; Oak Ridge National Laboratory ORNL/SUB-03-19XSY164C/01: Oak Ridge, TN, *in press*.
8. McLaughlin, S.B.; De La Torre Ugarte, D.G.; Garten, C.T., Jr.; Lynd, L.R.; Sanderson, M.A.; Tolbert, V.R.; Wolf, D.D. High-value renewable energy from prairie grasses. Environ. Sci. Technol. **2002**, *36*, 2122–2129.
9. Taliaferro, C.M. *Breeding and Selection of New Switchgrass Varieties for Improved Biomass Production*; Oak Ridge National Laboratory ORNL/SUB-02-19XSY162C/01: Oak Ridge, TN, 2002; 1–69.
10. McLaughlin, S.B.; Samson, R.; Bransby, D.; Weislogel, A. Evaluating Physical, Chemical, and Energetic Properties of Perennial Grasses as Biofuels. In *Bioenergy '96 Partnerships to Develop and Apply Biomass Technologies*, Proceedings of the Seventh National Bioenergy Conference, Vol. I, Nashville, TN, Sept 15–20, 1996; Southeastern Regional Biomass Energy Program: Atlanta, GA, 1996; 1–8.
11. Somleva, M.N.; Tomaszewski, Z.; Conger, B.V. Agrobacterium-mediated genetic transformation of switchgrass. Crop Sci. **2002**, *42*, 2080–2087.

Symbioses with Rhizobia and Mycorrhizal Fungi: Microbe/ Plant Interactions and Signal Exchange

James E. Cooper
Queen's University, Belfast, United Kingdom

INTRODUCTION

By virtue of their ability to capture (fix) atmospheric nitrogen, legumes assume a special significance among agricultural plants. Their productivity is theoretically independent of soil nitrogen status and they provide important grain and forage crops in both temperate and tropical zones. Nitrogen fixation can occur only when these plants are in the symbiotic state and the agents of fixation are soil bacteria, the rhizobia, which invade limited regions of the root cortex via infection threads in root hairs. Successful infections result in the development of root nodules, into which gaseous nitrogen diffuses to be reduced to ammonium by the nitrogenase enzyme of rhizobial bacteroids. An overwhelming majority of land plants, including most legumes, also form root symbioses with fungi, known collectively as mycorrhizae, which improve the supply of phosphate and other nutrients from soil to plant roots and confer increased resistance to plant pathogens. Most mycorrhizal associations can be allocated to one of two broad categories: the ectomycorrhizae, characterised by a fungal sheath and limited intercellular penetration of the root cortex of gymnosperms and woody angiosperms, or the endomycorrhizae, which possess hyphal structures that develop inside cortical cells of woody and herbaceous angiosperms. Arbuscules are examples of such structures; they are highly branched extensions of fungal hyphae that are common, though not ubiquitous, among the endomycorrhizae. Arbuscular mycorrhizal (AM) and rhizobial symbioses are similar in the sense that both are intracellular, but the former, involving the fungal order Glomales, is by far the more ancient of the two systems. In each case the progression to the symbiotic state is governed by reciprocal signal generation and perception, which may be viewed in terms of a ''molecular dialogue'' between the partners. While the early interactions between legumes and rhizobia are now largely understood at the biochemical and molecular genetic levels, details of events leading to the formation of mycorrhizal symbioses have not yet been elucidated.

PRIMARY SIGNALS FROM LEGUMES TO RHIZOBIA

In most examples studied to date, the first identified signals are flavonoids or isoflavonoids in legume root or seed exudates. These compounds are secondary metabolites of the plant phenylpropanoid biosynthetic pathway. (For examples of structural features, see Fig. 1.) Their main role in the initiation of a rhizobial symbiosis is an interaction with the constitutively expressed *nodD* gene product(s) of the microsymbiont to form a protein-phenolic complex: a transcriptional regulator of other rhizobial nodulation (*nod*) genes that are responsible for synthesis of reciprocal signals to the plant root. Different legumes release different compounds that act either as *nod* gene inducers or inhibitors for their specific rhizobia. The first symbiotically active flavonoid to be discovered was a flavone, luteolin, which is a *nod* gene inducer in the *Medicago sativa/Sinorhizobium meliloti* symbiosis.[1] Daidzein, an isoflavone, is an inducer in the *Phaseolus vulgaris/Rhizobium leguminosarum* biovar *phaseoli* symbiosis, whereas kaempferol, a flavonol, is an inhibitor in the *Trifolium/R. leguminosarum* biovar *trifolii* symbiosis. It is the relative proportion of inducing and inhibiting compounds in seed or root exudates that is thought to determine the overall level of *nod* gene induction in compatible rhizobia.

Certain non-flavonoid compounds in legume root exudates also act as *nod* gene inducers: the betaines stachydrine and trigonelline from *Medicago* and the aldonic acids erythronic and tetronic acid from *Lupinus*. However, unlike flavonoids, whose biological activity is expressed at nanomolar concentrations, these compounds elicit *nod* gene induction only at concentrations greater than c. 10 mM. The varying responsiveness to flavonoids of NodD proteins from different rhizobial species and biovars accounts for one of the elements determining the host specificity of these bacteria. For example, *Sinorhizobium meliloti* will form root nodules on alfalfa (*Meddicago*) but not on clovers or soybean. Other responses of rhizobia to legume flavonoids that may be significant for

DOI: 10.1081/E-EPCS 120012977

infection of the host plant include removal of glycosidic residues from flavonoid-sugar conjugates, degradation to yield new flavonoids and other, monocyclic phenolics, positive chemotaxis to some *nod* gene inducers, and increased growth rates in nutrient-limited media supplemented with low micromolar concentrations of inducing or non-inducing compounds.

REVERSE SIGNALS FROM RHIZOBIA TO LEGUME ROOTS—THE CHITOLIPOOLIGOSACCHARIDE NOD FACTORS

The combination of NodD proteins with appropriate plant flavonoids triggers the production of highly specific reverse signal molecules by rhizobia—the chitolipooligosaccharide (CLOS) Nod factors—by means of the transcriptional activation of common and host specific *nod* genes. The common *nodABC* genes are required for the synthesis of a chitinlike template, comprising several *β* 1,4-linked, *N*-acetyl glucosamine residues, which forms the basis of Nod factor structures found in all rhizobia studied to date. NodC is a *β*-glycosyl transferase that forms chitin oligomers and can influence host specificity of the Nod factor to some extent, depending on final chain length (2- to 6-mer). NodB deacetylates the terminal (non-reducing) glucosamine residue while NodA is involved in the transfer of a fatty acid moiety to this position. Fatty acids may be of the type common to bacterial phospholipids (e.g., C18:1), and chain length may vary even in Nod factors produced by a single bacterial species or biovar. In some rhizobia, *nodFEG* genes (which are usually classified as host-specific *nod* genes) are required for the formation of highly unsaturated fatty acids. An example of a typical Nod factor precursor template is shown in Fig. 2.

The major determinants of host specificity lie in the so-called ''decorations'' on the common CLOS template described above. Depending on the producing organism, acetyl, arabinosyl, carbamoyl, fucosyl, glycerol, mannosyl, methyl, or sulphate groups have been reported at various positions. In *S. meliloti*, whose Nod factor structure was the first to be discovered in 1990,[2] an acetyl group is introduced at the 6-C position of the nonreducing end of the CLOS molecule by an *O*-acetyl transferase encoded by the *nodL* gene, while the *nodHPQ* genes control the synthesis of a sulphated residue positioned at the reducing end. Mutations in either *nodQ* or *nodH* result in the production of unsulphated molecules that fail to elicit the responses associated with wild type Nod factor on alfalfa, the normal host legume of *S. meliloti*. An overview of reciprocal flavonoid and Nod factor signaling in the alfalfa/*S. meliloti* symbiosis is presented in Fig. 3.

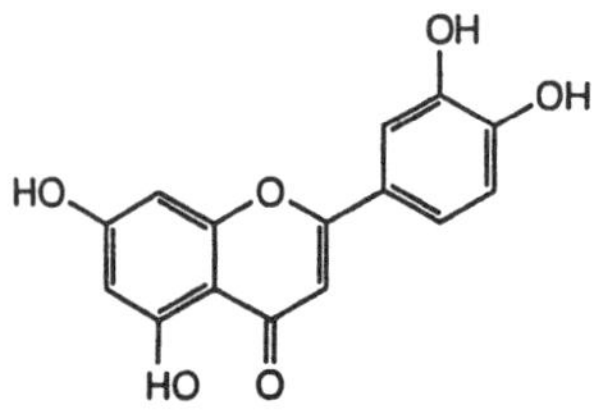

LUTEOLIN
(FLAVONE)

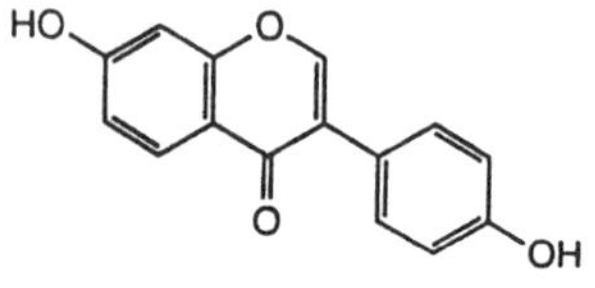

DAIDZEIN
(ISOFLAVONE)

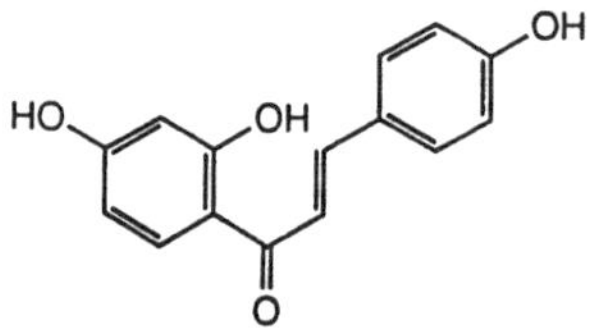

ISOLIQUIRITIGENIN
(CHALCONE)

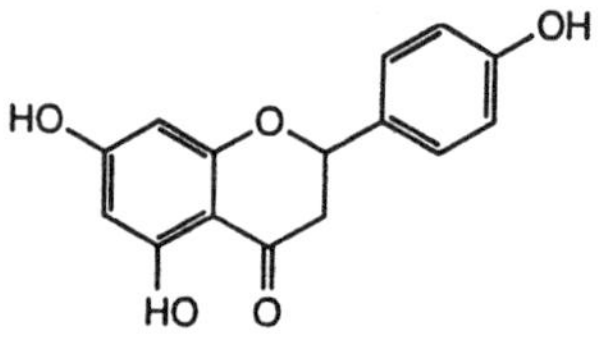

NARINGENIN
(FLAVANONE)

Fig. 1 Examples of compounds, from four flavonoid subgroups, capable of inducing rhizobial nodulation (*nod*) genes at nanomolar concentrations.

LEGUME ROOT RESPONSES TO RHIZOBIAL NOD FACTORS

Nod factors are essential signals for rhizobial entry into legume roots, through infection threads in root hairs, and for the development of root nodules. Rhizobial mutant strains that have lost the ability to produce Nod factor (for example, by deletion of common *nod* genes) cannot nodulate their normal host plant, but this property is restored in the presence of purified Nod factor isolated from the parent strain with its full complement of *nod*

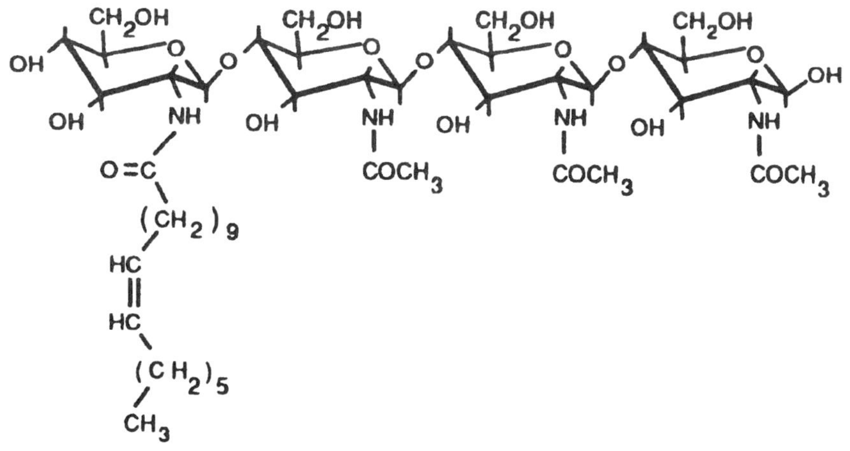

18:1

Fig. 2 A typical Nod factor core template synthesised by enzymes encoded by the common *nodABC* genes of rhizobia.

genes.[3] Application of femtomolar concentrations of purified Nod factor to the roots of a host legume elicits several responses that can be detected by biochemical or microscopical analysis: distortion of root hairs accompanied by membrane depolarization and deformation, calcium ion influx and chloride and potassium ion efflux, preinfection thread formaton in curled root hairs, and localized cortical cell division at the sites of root nodule primordia. Gene expression studies have shown that Nod factors, even in the absence of their producing rhizobia, can induce some of the plant genes (nodulins) that are involved in the preinfection, infection, nodule

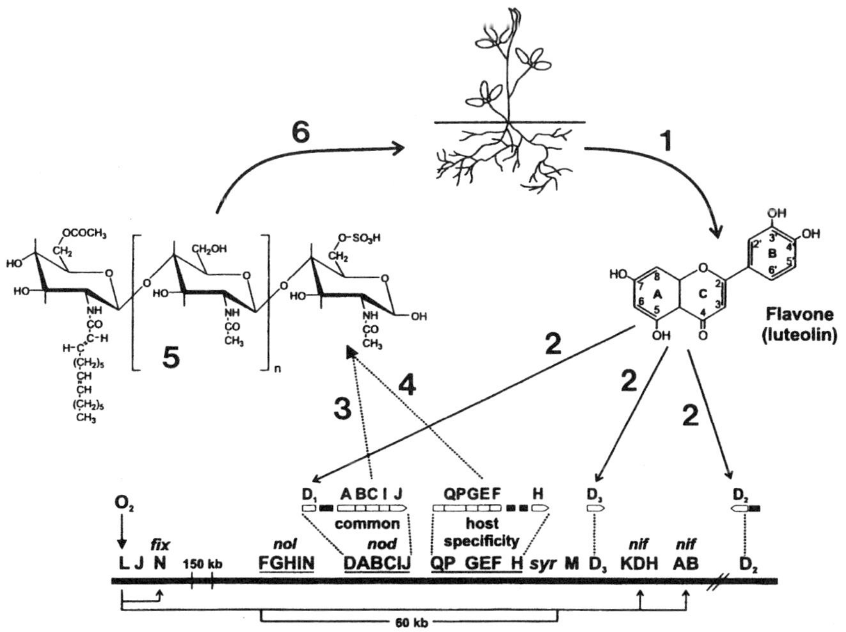

Fig. 3 Overview of reciprocal flavonoid and Nod factor signaling in the alfalfa/*Sinorhizobium meliloti* symbiosis. Flavonoids released from the plant's roots (**1**) are received by the constitutively expressed NodD proteins of the bacterium (**2**). The resulting NodD/flavonoid complex activates transcription of common (**3**) and host-specific (**4**) *nod* genes whose enzymes act collectively to synthesize a chitolipooligosaccharide Nod factor (**5**) that is transmitted back to the plant root (**6**). (From Zance, C.P. Root-bacteria interactions: symbiotic N_2 fixation. In *Plant Roots, the Hidden Half*; Waisel, Y., Eshel, A., Kafkafi, U., Eds.; Marcel Dekker, Inc.: New York, 2002; 839–867.)

development, and nodule function stages of symbiotic interaction. Auxin flow in roots at the earliest stages of nodule formation is perturbed by Nod factors, in conjunction with certain endogenous root flavonoids acting as auxin transport inhibitors.[4] Nod factors also control the number of nodules formed on a root system by inducing an autoregulation signal in the host plant. The nature of signal transduction pathways leading from the perception of Nod factors to symbiosis-related gene activation is currently the subject of intensive research. While a breakthrough was recently achieved with the discovery of a receptor kinase gene in *Lotus*[5] and *Medicago*[6] that is required for early signal transduction subsequent to Nod factor perception, the identity of the direct receptor for the Nod factor itself is yet to be established. Interestingly, this receptor kinase is essential for the formation of both rhizobial and AM symbioses with legumes, and it also shares structural features (leucine-rich-repeats) with animal proteins that function in the innate immune system.

While flavonoids and Nod factors are undoubtedly of prime significance as signal molecules in symbiotic development, they are not the only determinants of early recognition between host plant and bacterium. In addition to the chitolipooligosaccharide Nod factors, rhizobia produce an array of compounds with recognition functions, including extracellular polysaccharides, lipopolysaccharides, K-antigens, and cyclic glucans.[7] In the legume partner carbohydrate-binding proteins on root surfaces, the lectins, appear to mediate host specificity through selective interactions with Nod factors and/or rhizobial extracellular polysaccharides[8] but despite almost 30 years of research on this facet of symbiosis, the details of lectin function remain unresolved.

CROSSTALK IN MYCORRHIZAL SYMBIOSES

It is widely recognised that partially overlapping genetic programmes characterise the development of AM and rhizobial symbioses with terrestrial plants; the latter are thought to have recruited genes from the former during their (more recent) evolution, and the previously cited receptor kinase gene in *Lotus* and *Medicago* is one example. Such shared features notwithstanding, little is known about the mechanisms of reciprocal signaling and perception/response in AM symbioses.[9,10] The obligate biotrophy of AM fungi, which does not allow them to be cultured in the absence of a plant host, and the small quantities of fungal tissue even in highly colonized roots (c. 1% of extracted mRNA is fungal), present obstacles to research into these systems.

AM fungal spore germination, as well as subsequent hyphal growth and branching, are stimulated by root exudates from host plants and CO_2 may act as a volatile signal from the host to direct hyphal growth toward the root surface. Flavonoids are candidates for primary signals from plant root to fungus, mainly on the grounds that they enhance spore germination, but there is no evidence to indicate that they are obligatory compounds or that their mode of action is in any way similar to the one they exhibit in legume-rhizobia symbioses. Contact between fungal hyphae and root epidermal cells is required for the formation of appressoria (fungal structures formed on adhesion to the root surface in advance of host cell penetration) and the plant signal for this stage of infection, though not yet defined, is probably not found in root exudates. The identity of signal molecules from AM fungus to host plant (the equivalent of rhizobial Nod factors) is also unknown, but chitinaceous fragments from fungal cell walls may serve such a purpose. Some pathogen response proteins have been detected in plant roots during the early stages of AM colonization, but the sustained defense response associated with a pathogen attack is either not elicited or is suppressed, as is also the case in legume-rhizobia interactions. Symbiosis-specific gene expression has been detected in both partners and some legume early nodulin genes (e.g., *ENOD2*, *ENOD11*, and *ENOD40* in *Medicago*) are expressed during both rhizobial and AM fungal infections.

As with AM symbioses, the nature of interactive signaling processes in ectomycorrhizae is currently unknown or, at most, ill-defined. Rutin (a flavonoid) and zeatin (a cytokinin) in plant root exudates may stimulate fungal growth in the rhizosphere, while hypaphorine (a fungal alkaloid found in the ectomycorrhizal fungus *Pisolithus*) induces decreased rates and perhaps complete inhibition of root hair elongation in the region of fungal colonization behind the root apex.[11]

Interestingly, the host specificity of mycorrhizal fungi appears to be low, while that of rhizobia, as intimated above, is normally high; six genera from the order Glomales form AM symbioses with 80% of angiosperms, whereas, apart from one known exception (*Parasponia*), rhizobial infections are limited to legumes. Among rhizobial species and biovars, particular specificity is further displayed towards individual or small groups of legume genera. Rhizobia with very broad legume host ranges do exist but they may be the exception.

ARTICLES OF FURTHER INTEREST

Agriculture and Biodiversity, p. 1
Bacterial Pathogens: Early Interactions with Host Plants, p. 89
Bacterial Products Important for Plant Disease: Cell-Surface Components, p. 92

Bacterial Products Important for Plant Disease: Diffusable Metabolites As Regulatory Signals, p. 95
Bacterial Survival and Dissemination in Natural Environments, p. 108
Bacteria-Plant Host Specificity Determinants, p. 119
Mycorrhizal Evolution, p. 767
Mycorrhizal Symbioses, p. 770
Phenylpropanoids, p. 868
Rhizosphere: An Overview, p. 1084
Rhizosphere: Microbial Populations, p. 1090
Symbiotic Nitrogen Fixation, p. 1218
Symbiotic Nitrogen Fixation: Plant Nutrition, p. 1222
Symbiotic Nitrogen Fixation: Special Roles of Micronutrients, p. 1226

REFERENCES

1. Peters, N.K.; Frost, J.W.; Long, S.R. A plant flavone, luteolin, induces expression of *Rhizobium meliloti* nodulation genes. Science **1986**, *233*, 977–980.
2. Lerouge, P.; Roche, P.; Faucher, C.; Maillet, F.; Truchet, G.; Promé, J-C.; Dénarié, J. Symbiotic host-specificity of *Rhizobium meliloti* is determined by a sulphated and acylated glucosamine oligosaccharide signal. Nature **1990**, *344*, 781–784.
3. Relic, B.; Perret, X.; Estradagarcia, M.T.; Kopcinska, J.; Golinowski, W.; Krishnan, H.B.; Pueppke, S.G.; Broughton, W.J. Nod factors of *Rhizobium* are a key to the legume door. Mol. Microbiol. **1994**, *13*, 171–178.
4. Bladergroen, M.R.; Spaink, H.P. Genes and signal molecules involved in the rhizobia-Leguminoseae symbiosis. Curr. Opin. Plant Biol. **1998**, *1*, 353–359.
5. Stracke, S.; Kistner, C.; Yoshida, S.; Mulder, L.; Sato, S.; Kaneko, T.; Tabata, S.; Sandal, N.; Stougaard, J.; Szczyglowski, K.; Parniske, M. A plant receptor-like kinase required for both bacterial and fungal symbiosis. Nature **2002**, *417*, 959–962.
6. Endre, G.; Kereszt, A.; Kevei, Z.; Mihacea, S.; Kaló, P.; Kiss, G.B. A receptor kinase gene regulating symbiotic nodule development. Nature **2002**, *417*, 962–966.
7. Spaink, H.P. Root nodulation and infection factors produced by rhizobial bacteria. Annu. Rev. Microbiol. **2000**, *54*, 257–288.
8. Hirsch, A.M. Role of lectins (and rhizobial exopolysaccharides) in legume nodulation. Curr. Opin. Plant Biol. **1999**, *2*, 320–326.
9. Gadkar, V.; David-Schwartz, R.; Kunik, T.; Kapulnik, Y. Arbuscular mycorrhizal fungal colonization. Factors involved in host recognition. Plant Physiol. **2001**, *127*, 1493–1499.
10. Hirsch, A.M.; Kapulnik, Y. Signal transduction pathways in mycorrhizal associations: Comparisons with the *Rhizobium*-legume symbiosis. Fungal Genet. Biol. **1998**, *23*, 205–212.
11. Martin, F.; Duplessis, S.; Ditengou, F.; Lagrange, H.; Voiblet, C.; Lapeyrie, F. Developmental cross talking in the ectomycorrhizal symbiosis: signals and communication genes. New Phytol. **2001**, *151*, 145–154.

Symbiotic Nitrogen Fixation

Timothy R. McDermott
Montana State University, Bozeman, Montana, U.S.A.

INTRODUCTION

The legume–*Rhizobium* association is an example of mutualism: Both the plant and the bacterium benefit from their interaction. Microbes have a long history of being somewhat selfish and nonfriendly to plants. Indeed, the initiation of this symbiosis (i.e., infection) has many features that are similar to those that initiate pathogenic relationships. When this symbiosis is viewed from the background of the very large number of pathogenic plant-microbe interactions, it is easy to ask questions about how evolution allowed plants to develop divergent relationships with some microbes. In most instances plants appear to go to a lot of trouble to avoid invasion by microbes, whereas in others (i.e., legumes) they will invest significant resources to closely interact with them. This article provides a very general description of the formation and function of the legume–*Rhizobium* symbiosis.

DISTRIBUTION, ECOLOGY, AND IMPORTANCE

Many aspects of the symbiosis are described in great detail in recent comprehensive overviews.[1,2] Indeed, these major contributions are often cited in this article as source material to intentionally guide readers to more intensive treatment of any of the topics briefly touched upon in the following.

The literature documents over 3000 legume–rhizobia root–nodule symbioses (Table 1), ranging from relatively simple plants such as clover to larger organisms such as shrubs and trees.[3] The nodules formed on the roots of legumes are often categorized into two different groups based on nodule shape, nodule meristem activity, and type of fixed nitrogen products transported from the nodule to the shoot (Fig. 1). Continuing research into the ecology of this symbiosis has shown that it can be found in virtually any climate, ranging from the arctic to the tropics, having evolved essentially the same basic phenotype. In each case, these organisms partner to build a specialized structure known as a root nodule. The nodule houses the rhizobia bacteria, serving as a protective environment where an exchange of metabolites occurs. The host legume provides the rhizobia with photosynthetic sugars for carbon and energy; in return the rhizobia utilize some of these energy-yielding substrates to support nitrogen fixation via the enzyme nitrogenase. The ammonia thus generated is released to the plant.

The ecological advantage afforded to the plant is easily imagined. By taking the risk of allowing the bacteria to infect its roots, the plant stands to benefit by acquiring a limiting nutrient from a pool that it does not have to share with other plants. This symbiotically acquired nitrogen (N) also serves as an important N input to the local ecosystem. The value of legumes in this regard has long been recognized in agriculture, where about 80% of the biologically fixed N is obtained from this symbiosis.[4] Legumes have been used in crop rotations to reduce N fertilizer inputs into a variety of cropping systems, where on a worldwide basis legumes account for roughly 30% of the world's dietary protein.[5] This is of greatest proportional importance in Third World countries.

ROOT INFECTION AND NODULE ORGANOGENESIS

The legume–*Rhizobium* symbiosis is species-specific. Some legumes will only nodulate with specific species of rhizobia, and the rhizobia vary, displaying narrow- to broad-host range nodulation capabilities.[6] Nodule formation involves two separate stages—infection and nodule organogenesis. Infection is initiated in the rhizosphere when the bacteria find themselves in close contact with newly emerging root hairs in the rhizosphere. The initial interactions occur as a result of a biochemical dialogue that first involves a class of plant root exudates, known as flavonoids, that act as an inducer of nodulation genes in the rhizobia. The flavonoids bind to key regulatory proteins (e.g., NodD in *Sinorhizobium meliloti*), which then function together to positively regulate the transcription of numerous genes required for infection. Often the nodulation genes are found clustered in a focused location on the chromosome or on large Sym plasmids.[7] These nodulation genes encode proteins that are involved in the synthesis and secretion of lipo-chitgo-oligosacchariade Nod factors, and in protein secretion. Both of these activities are essential for numerous different aspects of infection initiation.[8] Secretion and release of Nod factors represent

Encyclopedia of Plant and Crop Science
DOI: 10.1081/E-EPCS 120005596

Table 1 Examples of *rhizobium*–legume symbioses[a]

Legume	Rhizobia partner
Grain/Food	
Soybean	*Bradyrhizobium japonicum*
	Sinorhizobium fredii
Bean	*Rhizobium etli, Rhizobium tropici*
Lupin	*Mesorhizobium loti*
Pea	*Rhizobium leguminosarum* bv *viciae*
Chickpea	*Mesorhizobium ciceri*
Forage	
Alfalfa	*Sinorhizobium meliloti*
Clover	*Rhizobium leguminosarum* bv *trifolii*
Lotus	*Mesorhizobium loti*
Vetch	*Rhizobium leguminosarum* bv *viciae*
Trees	
Leucaena	*Rhizobium tropici*
Sesbania	*Azorhizobium caulinodans*
Acacia	*Mesorhizobium plurifarium*
Prosopis	*Sinorhizobium kostiense*

[a]This list is not exhaustive, either in terms of legume genera or with respect to nodulation range of some of the rhizobia listed. Some legumes are referred to by their common names, when well-known and available. (For further examples, the reader can refer to www.honeybee.helsinki.fi/users/lindstro/Rhizobium/Names.html).

the first response of the rhizobia in the aforementioned biochemical dialogue. These Nod factors elicit various plant responses, one of the most visible of which is root hair deformation and curling (Fig. 2A), although the curling phenotype apparently requires the presence of live rhizobia.[9] The rhizobia become entrapped within the crook of the curled root hair and in this pocket begin the process of infection thread formation (Fig. 2A).

In addition to root hair curling, the Nod factors also induce plant responses in the root cortex cells. The reestablishment of mitotic activity in highly localized regions of the root cortex marks the initiation of nodule primordia, which is the beginning of organogenesis. In temperate legumes such as pea and alfalfa, nodule primordia become evident in the inner cortex cells, whereas in roots of subtropical legumes like soybean and bean they are apparent in outer cortex cells.[10] Specialized gene transcription also occurs in the primordia cells, leading to the production of host plant nodulins[11] that participate in nodule/nodulation functions such as oxygen binding (hemoglobin)[12] or plant defense-like reactions.[13] Many nodulins have been identified and their expression has been correlated to specific parts of the nodule or to specific times during nodule formation; however, exact functions of many have yet to be determined.[14]

The infection thread (Fig. 1), which is continuous with the root hair cell wall, acts as a pipeline to transfer the rhizobia from the root surface into the root cortex cells that compose the nodule primordia. After the infection thread penetrates a nodule primordia cell, the rhizobia are released from the infection thread in an endocytotic process that involves the encapsulation of the rhizobial cells within host plasma membrane derived from the infected cell.[15] Collectively, the rhizobia bacteria and the host-derived membrane structure are referred to as a symbiosome, to reflect the view that it now behaves somewhat like a plant cell organelle.[16] Initially, a symbiosome typically contains a single (or at most two) rhizobial cell(s), at this stage referred to as bacteroids. It should also be pointed out that the symbiosome membrane serves to maintain separation of the rhizobia bacteria from the host plant, avoiding direct contact between the symbionts. This appears to be a common feature for all plant endosymbionts.[17]

NODULE FUNCTION

Energy Costs

Carbon fixed during photosynthesis (i.e., CO_2) is used as a source of energy and reductant for nitrogen fixation, carbon skeletons for assimilation of fixed nitrogen, and for bacteroid and nodule growth and maintenance. For each gram of N_2 fixed, roughly 5–10 grams of carbon are needed.[18] In terms of biochemical currency, the nitrogenase enzyme requires at least 16 ATP to reduce one mole of N_2 to 2 moles of NH_3. Part of the high cost of biological nitrogen fixation is due to the production of H_2 that results from the reduction of two protons concomitantly with the reduction of N_2.[19] Some rhizobia have hydrogenase enzymes that could recapture some of the evolved H_2 and thus theoretically reduce

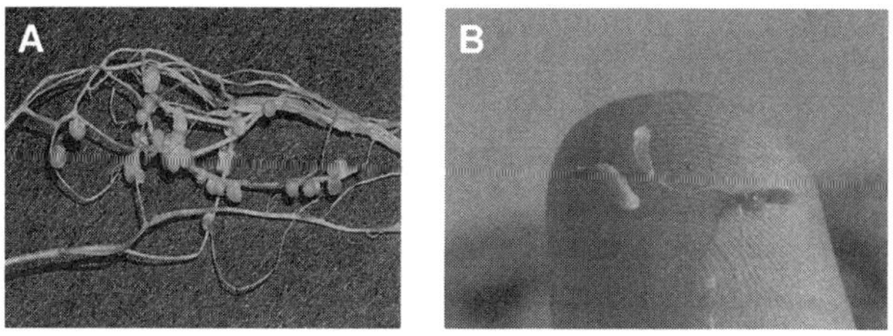

Fig. 1 Examples of legume nodules, demonstrating dominant nodule morphologies. (A) Spherical, determinate meristem, nodules (~2 mm in diameter) on the roots of bean (*Phaseolus vulgaris*) formed by *Rhizobium tropici*. (B) Alfalfa (*Medicago sativa*) nodule (~3 mm long) formed by *Sinorhizobium meliloti*; typical example of a cylindrical, indeterminate meristem-type nodule. (Photo in panel B by Michael L. Kahn.) (*View this art in color at www.dekker.com.*)

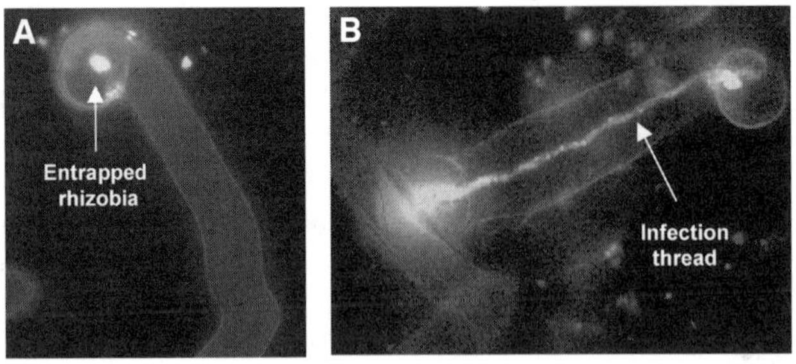

Fig. 2 Fluorescence microscopy photographs of *Sinorhizobium meliloti* infection of alfalfa. Presence of rhizobia is seen as green spots (or bright white in black-and-white image) due to expression of green-fluorescent protein. (A) Root hair curling in response to lipo-chito-oligosaccharides produced by rhizobia. Note the entrapped rhizobia in crook of curled root hair. (B) Infection thread extending from tip to the base of root hair. (Images by Brett J. Pellock and Graham C. Walker.) (*View this art in color at www.dekker.com.*)

total energy costs; however, experimental demonstration of the latter has been variable and inconsistent.[20]

The symbiosome membrane serves a critical role in controlling metabolite traffic between the symbionts. The bacteroids are completely dependent on the symbiosome membrane for supply and flux of carbon substrates, as well as for removal of fixed nitrogen and bacteroid waste metabolites. The symbiosome membrane contains metabolite-specific transporters that are functionally linked with the biochemical activities of the bacteroids. For example, even though glucose and fructose are abundant in the nodule cytosol,[21] the symbiosome membrane selectively transports organic acids such as malate and succinate,[22] which are metabolites that the bacteroids are well prepared to utilize.[20] On the other hand, in most cases the symbiosome membrane transports the aforementioned hexoses relatively poorly. Likely not coincidentally, the biochemical pathways required to catabolize hexoses are found at low levels in bacteroids of most species of rhizobia.[20]

Oxygen Paradox

The role of oxygen in the symbiosis is somewhat of an enigma. The exquisite O_2 sensitivity of the bacterial nitrogenase enzyme is well known and fairly well understood in terms of enzyme function and genetic regulation. However, bacteroids require oxygen as a respiratory electron acceptor for the generation of the large amounts of ATP (i.e., oxidative phosphorylation) required for nitrogen fixation. The problem is solved by control mechanisms that keep oxygen at low levels, yet maintain a steady flow of oxygen to the bacteroids. The outer layers of nodule cells restrict oxygen flux into the nodule by providing a gas diffusion resistance barrier that, when combined with the oxygen consumption of the bacteroids and the plant mitochondria, results in microaerobic conditions.[23] Leghemoglobin is an oxygen-carrying protein in the nodule, composing approximately 20% of the total soluble protein.[24] Its function is similar to that of hemoglobin in blood, serving to transport oxygen throughout the nodule cytoplasm, releasing it as bacteroid and mitochondria respiratory demand requires. Leghemoglobin facilitates increased oxygen flow to the bacteroids and plant mitochondria, without increasing the amount of free oxygen.

Assimilation of Fixed N and Transport to Shoot

Experimental evidence suggests that the ammonium fixed by the bacteroid is released from the bacteroid and exported from the symbiosome,[25] where it is then assimilated by the host into carbon skeletons.[5] However, recent work with soybean suggests that alanine may also serve as a shuttle metabolite that transfers fixed nitrogen from the soybean bacteroid to its host.[26] The ammonia that is released from the symbiosome is immediately incorporated into the amino acids glutamine and glutamate via the host plant enzymes glutamine synthetase (cytoplasm) and glutamate synthase (plastids), respectively.[27,28] In temperate legumes (e.g., pea or alfalfa), the ammonia is transferred to the shoot primarily as the amide asparagine. In tropical legumes (e.g., soybean and bean), the ammonia is incorporated into ureides (allantoin and allantoic acid) and then transported via the xylem to the shoots.[27,28]

ARTICLES OF FURTHER INTEREST

Fixed-Nitrogen Deficiency: Overcoming by Nodulation, p. 448
Mycorrhizal Symbioses, p. 770
Nitrogen, p. 822

Symbiotic Nitrogen Fixation: Plant Nutrition, p. 1222
Symbiotic Nitrogen Fixation: Special Roles of Micronutrients in, p. 1226

REFERENCES

1. *The Rhizobiaceae*; Spaink, H.P., Kondorosi, A., Hooykaas, P.J.J., Eds.; Kluwer Academic Publishers: Dordrecht, 1998.
2. *Prokaryotic Nitrogen Fixation: A Model System for the Analysis of a Biological Process*; Triplett, E.W., Ed.; Horizon Scientific Press: Wymondham, 2000.
3. Allen, O.N.; Allen, E.K. *The Leguminoseae*; University of Wisconsin Press: Madison, 1981.
4. Vance, C.P. Root Bacteria Interactions: Symbiotic Nitrogen Fixation. In *Plant Roots, The Hidden Half*; Waisel, Y., Eschel, A., Kafkafi, U., Eds.; Marcel Dekker, Inc.: New York, 1996; 723–756.
5. Vance, C.P. Legume Symbiotic Nitrogen Fixation: Agronomic Aspects. In *The Rhizobiaceae*; Spaink, H.P., Kondorosi, A., Hooykaas, P.J.J., Eds.; Kluwer Academic Publishers: Dordrecht, 1998; 509–530.
6. Hadri, A.-E.; Spaink, H.P.; Bisseling, T.; Brewin, N.J. Diversity of Root Nodulation and Rhizobial Infection Processes. In *The Rhizobiaceae*; Spaink, H.P., Kondorosi, A., Hooykaas, P.J.J., Eds.; Kluwer Academic Publishers: Dordrecht, 1998; 347–360.
7. Schlaman, H.R.M.; Phillips, D.A.; Kondorosi, E. Genetic Organization and Transcription Regulation of Rhizobial Nodulation Genes. In *The Rhizobiaceae*; Spaink, H.P., Kondorosi, A., Hooykaas, P.J.J., Eds.; Kluwer Academic Publishers: Dordrecht, 1998; 361–386.
8. Downie, J.A. Functions of Rhizobial Nodulation Genes. In *The Rhizobiaceae*; Spaink, H.P., Kondorosi, A., Hooykaas, P.J.J., Eds.; Kluwer Academic Publishers: Dordrecht, 1998; 387–402.
9. Relic, B.; Talmont, F.; Kopcinska, J.; Golinowski, W.; Prome, J.-C.; Broughton, W.J. Biological activity of *Rhizobium* sp. NGR 234 Nod-factor on *Macroptilium atropurpureum*. Mol. Plant-Microb. Interact. **1993**, *6*, 764–774.
10. Kijne, J.W. The *Rhizobium* Infection Process. In *Biological Nitrogen Fixation*; Stacey, G., Burris, R.H., Eds.; Chapman Hall: New York, 1992; 349–398.
11. Vijn, I.; das Neves, L.; Van Kammen, A.; Franssen, H.; Bisseling, T. Nod factors and nodulation in plants. Science **1993**, *260*, 1764–1765.
12. Gualtieri, G.; Bisseling. The evolution of nodulation. Plant Mol. Biol. **2000**, *42*, 181–194.
13. Gamas, P.; deBilly, F.; Truchet, G. Symbiosis-specific expression of two *Medicago truncatula* nodulin genes, MtN1 and MtN13, encoding products homologous to plant defense proteins. Mol. Plant-Microb. Interact. **1998**, *11*, 393–403.
14. Verma, D.P.S. Nodulins: Nodule-Specific Host Gene Products, Their Induction and Function in Root Nodule Symbiosis. In *Prokaryotic Nitrogen Fixation: A Model System for Analysis of a Biological Process*; Triplett, E.W., Ed.; Horizon Scientific Press, 2000; 467–487.
15. Brewin, N.J. Tissue and Cell Invasion by Rhizobium: The Structure and Development of Infection Threads and Symbiosomes. In *The Rhizobiaceae*; Spaink, H.P., Kondorosi, A., Hooykaas, P.J.J., Eds.; Kluwer Academic Publishers: Dordrecht, 1998; 417–429.
16. Roth, L.E.; Jeon, K.; Stacey, G. Homology in Endosymbiotic Systems. The Term ''Symbiosome''. In *Molecular Genetics of Plant-Microbe Interactions*; Palacios, R., Verma, D.P.S., Eds.; Amer. Phytopathol. Soc. Press: St. Paul, 1989; 220–225.
17. Smith, S.E.; Smith, S.A. Structure and function of the interfaces in biotrophic symbioses as they relate to nutrient transport. New Phytol. **1990**, *114*, 1–38.
18. Phillips, D.A. Efficiency of symbiotic nitrogen fixation in legumes. Annu. Rev. Plant Physiol. **1980**, *31*, 29–49.
19. Ruiz-Argueso, T.; Imperial, J.; Palacios, J.M. Uptake Hydrogenases in Root Nodule Bacteria. In *Prokaryotic Nitrogen Fixation: A Model System for Analysis of a Biological Process*; Triplett, E.W., Ed.; Horizon Scientific Press, 2000; 489–507.
20. Kahn, M.L.; McDermott, T.R.; Udvardi, M.K. Carbon and Nitrogen Metabolism in Rhizobia. In *The Rhizobiacae: Molecular Biology of Model Plant-Associated Bacteria*; Spaink, H.P., Kondorosi, A., Hooykaas, P.J.J., Eds.; Kluwer Academic Publishers: Boston, 1998; 461–485.
21. Reibach, P.H.; Streeter, J.G. Metabolism of ^{14}C-labled photosynthate and distribution of enzymes of glucose metabolism in soybean nodules. Plant Physiol. **1983**, *72*, 634–640.
22. Udvardi, M.K.; Day, D.A. Metabolite transport across symbiotic membranes of legume nodules. Annu. Rev. Plant Physiol. Plant Mol. Biol. **1997**, *48*, 493–523.
23. Hunt, S.; Layzell, D.B. Gas exchange of legume nodules and regulation of nitrogenase activity. Annu. Rev. Plant Physiol. Plant Mol. Biol. **1993**, *44*, 483–511.
24. Appleby, C.A. Leghemoglobin and *Rhizobium* respiration. Annu. Rev. Plant Physiol. **1984**, *35*, 443–478.
25. Kaiser, B.N.; Finnegan, P.M.; Tyerman, S.D.; Whitehead, L.F.; Bergersen, F.J.; Day, D.D.; Udvardi, M.K. Characterization of an ammonium transport protein from the peribacteroid membrane of soybean nodules. Science **1998**, *281*, 1202–1206.
26. Waters, J.K.; Emerich, D.W. Transport of Metabolites To and From Symbiosomes and Bacteroids. In *Prokaryotic Nitrogen Fixation: A Model System for Analysis of a Biological Process*; Triplett, E.W., Ed.; Horizon Scientific Press, 2000; 549–558.
27. Atkins, C.; Smith, P. Ureide Synthesis in Legume Nodules. In *Prokaryotic Nitrogen Fixation: A Model System for Analysis of a Biological Process*; Triplett, E.W., Ed.; Horizon Scientific Press, 2000; 559–587.
28. Vance, C.P. Aminde Biosynthesis in Root Nodules of Temperate Legumes. In *Prokaryotic Nitrogen Fixation: A Model System for Analysis of a Biological Process*; Triplett, E.W., Ed.; Horizon Scientific Press, 2000; 589–607.

Symbiotic Nitrogen Fixation: Plant Nutrition

Timothy R. McDermott
Montana State University, Bozeman, Montana, U.S.A.

INTRODUCTION

Symbiotic nitrogen fixation is one of the most fascinating aspects to legumes. This article briefly describes the mutualistic relationship between rhizobia bacteria and host legume plants and the mechanisms that enable plants such as peas and beans to satisfy their nitrogen requirements through this plant-bacteria association. The article explains what is known about the main nutritional requirements for this relationship aside from nitrogen as well as the basic methods for these assessments. Finally, it very briefly reviews the literature on the implications of specific nutrient deficiencies.

THE *RHIZOBIUM*-LEGUME SYMBIOSIS: THE ESSENTIALS

The *Rhizobium*-legume symbiosis forms as a result of a cooperative effort of the rhizobia bacteria and their respective host legume. Beginning with an intricate biochemical dialogue that initiates infection, the bacteria and plant together construct a specialized organ known as the legume root nodule. This symbiosis is viewed to be based on a mutualistic exchange of carbon for nitrogen. The plant utilizes solar energy to fix and reduce CO_2 into carbohydrates, which are subsequently transferred from the leaves to the roots and root nodules. Inside the nodule, these photosynthetic sugars (primarily sucrose) are partially metabolized to organic acids (e.g., malate and succinate), which is the primary form of carbon provided to the bacteroids. The bacteroids utilize these organic acids as an energy source for their nitrogen fixation activity that generates ammonia, which is then released to the plant. The advantage for the legume plant acquiring nitrogen from its symbiosis with the rhizobia bacteria is very clear; nitrogen is most often the limiting nutrient in nonagricultural environments, and thus the plant can gain a competitive advantage in that it does not have to compete for a limited nitrogen pool. For some legumes, the nitrogen acquired from symbiosis can account for almost all of the plant's nitrogen needs. However, many legumes still rely on soil nitrogen (NH_4^+ or NO_3^-) to fill the balance.[1]

ASSESSMENT OF MINERAL NUTRIENT NEEDS FOR THE SYMBIOSIS

Three basic approaches for investigating specific roles of individual nutrients in symbiotic nitrogen fixation are typically used and have been discussed previously[2,3] (Table 1). Examination of nutrient influences might involve assessing interaction effects with inorganic nitrogen: If a particular element corrects an N deficiency, then a reasonable conclusion is that it impacts, directly or indirectly, on the nitrogen fixation process. Another approach is to examine whether correcting a nutrient deficiency improves nitrogen content in the shoot. Finally, effects of a specific nutrient can be viewed from the perspective of its effect on nodule formation and/or levels of nitrogen fixation as directly measured by acetylene reduction, H_2 evolution, or ^{15}N isotope dilution.

There are several phases of nodule formation and several aspects of nodule function where a nutrient deficiency can be manifested, and any or all can place a limit on the functioning of this symbiosis. Nutrient shortfalls that directly influence the soil or rhizosphere populations of the rhizobia can in turn impact infection and total nodulation and, thus, potential nitrogen acquired from symbiosis. Some nutrients have been identified to have specific roles at discrete steps in the infection process and nodule formation; furthermore, bacteroid and nodule function may also have unique nutritional requirements. It is not always easy to sort out effects of nutrient limitation on nitrogen fixation per se versus an impact on general host plant growth and well-being that can have ripple effects on nodule function. A legume plant that is metabolically crippled due to a severe nutrient limitation (e.g., S, P, or K) will not be the optimum symbiotic partner. Such an indirect effect is in contrast to the situation where a critical nodule function such as maintenance of optimum oxygen transport functions by leghemoglobin or nitrogen fixation by nitrogenase are specifically dependent upon iron availability.

NUTRIENT REQUIREMENTS FOR OPTIMAL FUNCTION

Nutrient requirements of this symbiosis include host plant requirements as well as the nutrients required for nodule

Encyclopedia of Plant and Crop Science
DOI: 10.1081/E-EPCS 120010515

Table 1 Proportion of total nitrogen acquired by nitrogen fixation for legumes under field conditions

Legume	Percentage
Soybean	50
Bean	50
Lentil	70
Pea	70
Broad bean	80
Alfalfa	90
Clover	90

(Based on Ref. 2.)

biomass synthesis, including bacteroid nutritional requirements. Further, nodule tissue may have specialized requirements or perhaps nonspecific sink activity that may influence overall functioning and health of the legume host plant. The requirements of several nutrient elements have been investigated, and while exact mechanisms may not always be completely understood, roles for some have been identified. Macronutrients such as Ca, P, S, and K and micronutrients such as Fe, B, Co, Cu, Mo, Ni, Se, and Zn have been recognized to have specific minimum functions in certain stages of nodule formation and/or function.[2] The roles of macronutrient and micronutrient elements in overall plant metabolism are discussed throughout this article, and specific roles of micronutrients in symbiotic nitrogen fixation are discussed by Dalton in the preceding entry. This entry provides a synopsis of Ca, P, S, and K involvement in the *Rhizobium*-legume symbiosis. The role of oxygen is discussed in the preceding entry in this article.

CALCIUM

Early studies showed that Ca deficiency can negatively impact nodule formation on numerous legumes [reviewed in Ref. 2]. Many rhizobia grow well at low Ca concentrations (e.g., 15–30 μM)[2] although growth of *Sinorhizobium meliloti* (alfalfa symbiont) appears especially sensitive to low Ca, and Ca–P interactions further complicate Ca bioavailability.[4] Reduced rhizobia populations might be expected in soils that are deficient in Ca or when the prevailing soil chemistry reduces Ca solubility. Reduced rhizosphere populations could potentially translate into reduced levels of infection. Specific roles for Ca during infection and nodule formation have also been demonstrated. A Ca-dependent adhesin is implicated for attachment of *Rhizobium leguminosarum* biovar viciae to pea root hairs.[5] For clover rhizobia, Ca is also required for *nod* gene expression[6] and in Nod factor-based signaling between plant and bacteria in the rhizosphere.[7] Nod factors are also involved in altering Ca gradients and behavior in the tips of infectible root hair cells, resulting in a Ca spiking phenomenon.[8] In alfalfa, the Nod factor brings about oscillations of free Ca in the cytoplasm and is not observed in an alfalfa nodulation mutant.[8]

PHOSPHORUS

In contrast to the other nutrients discussed here, problems of soil P fertility are widespread. Sanchez and Euhara[9] have estimated that crop production on approximately 33% of the world's arable land is limited by P, with P bioavailability in soil being highly sensitive to the prevailing chemistry. Examples of chronic P fertility problems can be found in the highly weathered soils in the tropics, where acidic pH and high levels of aluminum place significant constraints on P solubility.

Virtually every aspect of the *Rhizobium*-legume association is affected by P bioavailability. In various legumes, P deficiency has been shown to result in reduced nodulation and nitrogen fixation and lower plant N content and dry matter production. The exact mechanism(s) by which P is involved in these responses is not well understood. The uncertainty in part derives from the likelihood that different legumes have varying P requirements for optimum function, as well as the shear ubiquity of cellular processes that involve or require P. It also stems from our lack of exact understanding of how P deficiency is manifested in this symbiosis, i.e., in nodule-specific activities or as general plant-wide metabolic perturbations that have ripple effects on nitrogen fixation.[10]

As discussed above for Ca, P deficiency could potentially influence rhizosphere populations of rhizobia because different rhizobia respond varyingly to low P and have different P requirements for growth.[11] Infection experiments show that nodulation efficiency by *Bradyrhizobium japonicum* (symbiont for soybean) is reduced when cultures are P-stressed prior to inoculation onto soybean roots.[12] Further, P stress negatively influences Nod factor production and excretion by clover rhizobia.[13] Also, extracellular polysaccharide production, which is essential for nodulation by some rhizobia, is also directly controlled by the rhizobial Pho regulon (genetic regulatory system that controls gene expression in response to environmental P levels).[11]

Limiting P will of course diminish overall plant growth. A general consequence of reduced shoot growth could be reduced photosynthesis, which could negatively impact on total photosynthates available for translocation to the nodules. However, there are studies that suggest such a generalized effect is not applicable to all symbioses.[10] Soybean P status is positively correlated with nodule-specific nitrogenase activity (total nitrogen fixation/nodule weight) and with nodule ATP content.[14] A still more specific example of how P limitation affects symbiosis involves O_2 flux into the nodule. Nodules on P-limited soybeans and alfalfa appear to have increased oxygen permeability.[15,16] Part of this change in O_2 flux may be due to decreased nodule size (increased surface-to-volume ratio) that is common with P-limited plants. However, Drevon and Hartwig[16] suggested that this would not account for all of the increase, and other possibilities might include the activation of alternative (i.e., energy wasteful) respiratory pathways that would have the effect of increasing the amount of O_2 consumed. Conversely, however, increased nodule O_2 uptake could perhaps account for experimental observations showing increases in specific nodule activity in clover[17] under low P growth conditions. The differences in nodule-specific nitrogenase activity between soybean and that of clover is but one example of how the different symbioses respond differently to a nutrient limitation.

POTASSIUM

The importance of K in the symbiosis can be inferred at least at a general level because of its roles in water uptake, enzyme activity, and translocation of photosynthates within the plant. A symbiosis-specific function of K has, however, been observed with *Vicia faba* and *Phaseolus vulgaris*, where low levels (0.1 mM) of K in the nutrient solution eliminated nodulation.[18] It is also important to note that in this case and others[19] a supraoptimal K condition was demonstrated in that increasing levels of K increased nodulation without increasing total plant N acquired from symbiotic nitrogen fixation. Other symbiosis-specific functions of K were identified in in vitro studies of *Bradyrhizobium* sp (cowpea) that demonstrated that K is necessary for induction of nitrogenase and heme synthesis,[20] two activities that are of central importance to bacteroid function during symbiosis.

SULFUR

Soil S deficiency can have profound effects on this symbiosis. For *Sinorhizobium* and some species of *Rhizobium*, sulfated functional groups are important host-specificity components of Nod factors that are required for nodulation,[21] and a significant amount of S is also required for synthesis of the nitrogenase protein.[22] Furthermore, experiments with field-grown subterranean clover[23] have shown that both plant dry matter production and symbiotically derived N are very sensitive to severe S deficiency; incremental increases of applied S ranging from 0 to 8 $g \cdot m^{-2}$ soil resulted in tripling of dry matter productivity and nearly a fourfold increase in N derived from symbiosis. Other experiments have shown variable S responses by different legumes. After S addition to S-limited plants, total plant dry matter increases for *Vicia faba*, *Trifolium pratense*, and *Pisum sativum* were roughly twice that of *Medicago sativa*. For each of these legumes, the proportion of nitrogen acquired from symbiosis appeared to change little in response to the added S, i.e., host plant S requirements were more easily demonstrated than any effect on nodule function.[24]

CONCLUSION

Some of the variability in plant and symbiosis response to nutrient deficiency observed in the literature may derive from inherent differences between legume species (and/or cultivars of the same species) for the various nutrients. Indeed, additional in-depth studies with different legumes may ultimately demonstrate that such variability among legumes and symbioses may prohibit convenient nutrient effect modeling. Separating host or host nodule nutritional requirements from those of the bacteroids is not always easy, and indeed understanding source-sink relations in this highly specialized plant organ represents a major challenge to plant biologists. Nutrient flow in and out of the nodule and between bacteroid and host remains poorly understood. It will be important for future studies to dissect the potentially very different needs of bacteroids versus host to better understand the exact role of the various nutrient elements. The roles and specific functions of nutrients will likely vary between symbionts and between symbioses.

ARTICLES OF FURTHER INTEREST

Fixed-Nitrogen Deficiency: Overcoming by Nodulation, p. 448
Mycorrhizal Symbioses, p. 770
Nitrogen, p. 822
Symbiotic Nitrogen Fixation: Special Roles of Micronutrients, p. 1226

REFERENCES

1. Werner, D. *Symbiosis of Plants*; Chapman & Hall: London, 1992; 137.
2. O'Hara, G.W.; Bookerd, N.; Dilworth, M.J. Mineral constraints to nitrogen fixation. Plant Soil **1988**, *108*, 93–110.
3. Chalk, P.M. Integrated effects of mineral nutrition on

legume performance. Soil Biol. Biochem. **2000**, *32*, 577–579.

4. Beck, D.P.; Munns, D.N. Effect of calcium on the phosphorus nutrition of *Rhizobium meliloti*. Soil Sci. Soc. Am. J. **1985**, *49*, 334–337.
5. Smit, G.; Logman, T.J.J.; Boerrigter, M.E.T.I.; Kijne, J.W.; Lugtenberg, B.J.J. Purification and partial characterization of the *Rhizobium leguminosarum* biovar viciae Ca^{2+} -dependent adhesin, which mediates the first step in attachment of cells of the family Rhizobiaceae to plant root hair tips. J. Bacteriol. **1989**, *171*, 4054–4062.
6. Richardson, A.E.; Simpson, R.J.; Djordjevic, M.A.; Rolfe, B.J. Expression of nodulation genes in *Rhizobium leguminosarum* biovar trifolii is affected by low pH and by Ca^{2+} and Al ions. Appl. Environ. Microbiol. **1988**, *54*, 2541–2548.
7. Niebel, A.; Gressent, G.; Bono, J.-J.; Ranjeva, R.; Cullimore, J. Recent advances in the study of Nod factor perception and signal transduction. Biochemie **1999**, *81*, 669–674.
8. Ehrhardt, D.W.; Wais, R.; Long, S.R. Calcium spiking in plant root hairs responding to *Rhizobium* nodulation signals. Cell **1996**, *85*, 673–681.
9. Sanchez, P.A.; Euhara, G. Management Considerations for Acid Soils with High Phosphorus Fixation Capacity. In *Role of Phosphorus in Agriculture*; Khasawneh, F.E., Sample, E.C., Kamprath, E.J., Eds.; American Society of Agronomy: Madison, WS, 1980; 471–514.
10. Vance, C.P.; Graham, P.H.; Allan, D.L. Biological Nitrogen Fixation: Phosphorus—A Critical Future Need? In *Nitrogen Fixation: From Molecules to Crop Productivity*; Pedrosa, F.O., Hungria, M., Yates, M.G., Newton, W.E., Eds.; Kluwer: Dordrecht, The Netherlands, 2000; 509–514.
11. McDermott, T.R. Phosphorus Assimilation and Regulation in the Rhizobia. In *Prokaryotic Nitrogen Fixation: A Model System for the Analysis of a Biological Process*; Triplett, E.W., Ed.; Horizon Scientific Press: Norfolk, England, 1999; 529–548.
12. Mullen, M.D.; Israel, D.W.; Wollum, A.G., II. Effects of *Bradyrhizobium japonicum* and soybean (*Glycine max* (L) Merr.) phosphorus nutrition on nodulation and dinitrogen fixation. Appl. Environ. Microbiol. **1988**, *54*, 2387–2392.
13. McKay, I.A.; Djordjevic, M.A. Production and excretion of nod metabolites by *Rhizobium leguminosarum* bv. *trifolii* are disrupted by the same environmental factors that reduce nodulation in the field. Appl. Environ. Microbiol. **1993**, *59*, 3385–3392.
14. Sa, T.M.; Israel, D.W. Energy status and functioning of phosphorus-deficient soybean nodules. Plant Physiol. **1991**, *97*, 928–935.
15. Ribet, J.; Drevon, J.J. Increase in permeability to oxygen and in oxygen uptake of soybean nodules under limiting phosphorus nutrition. Physiol. Plant. **1995**, *94*, 298–304.
16. Drevon, J.J.; Hartwig, U.A. Phosphorus deficiency increases the argon-induced decline of nodule nitrogenase activity in soybean and alfalfa. Planta **1997**, *201*, 463–469.
17. Almeida, J.P.F.; Hartwig, U.A.; Frehner, M.; Nosberger, J.; Luscher, A. Evidence that P deficiency induces N feedback regulation of symbiotic N_2 fixation in white clover (*Trifolium repens* L.). J. Exp. Bot. **2000**, *51*, 1289–1297.
18. Sangakkara, R.; Hartwig, U.A.; Nosberger, J. Growth and symbiotic nitrogen fixation of *Vicia faba* and *Phaseolus vulgaris* as affected by fertilizer potassium and temperature. J. Sci. Food Agric. **1996**, *70*, 315–320.
19. Chalamet, A.; Audergon, J.M.; Maître, J.P., Domenach, A.M. Study of influence of potassium on the *Trifolium pratense* fixation by $\delta^{15}N$ method. Plant Soil **1987**, *98*, 347–352.
20. Gober, J.W.; Kashket, E.R. K^+ regulates bacteroid-associated functions of *Bradyrhizobium*. Proc. Natl. Acad. Sci. **1987**, *84*, 4650–4654.
21. Perret, X.; Staehelin, C.; Broughton, W.J. Molecular basis of symbiotic promiscuity. Microbiol. Mol. Biol. Rev. **2000**, *64*, 180–201.
22. Howard, J.B.; Rees, D.C. Structure of the Nitrogenase Protein Components. In *Prokaryotic Nitrogen Fixation: A Model System for Analysis of a Biological Process*; Triplett, E.W., Ed.; Horizon Press: Norfolk, 2000; 43–53.
23. Shock, C.C.; Williams, W.A.; Jones, M.B.; Center, D.M.; Phillips, D.A. Nitrogen fixation by subclover associations fertilized with sulfur. Plant Soil **1984**, *81*, 323–332.
24. Scherer, H.W.; Lange, A. N_2 fixation and growth of legumes as affected by sulfur fertilization. Biol. Fertil. Soils **1996**, *23*, 449–453.

Symbiotic Nitrogen Fixation: Special Roles of Micronutrients

David A. Dalton
Reed College, Portland, Oregon, U.S.A.

INTRODUCTION

Nitrogen-fixing plants have special requirements for mineral nutrients beyond those of other plants. The most important nitrogen-fixing plants in terms of agriculture are found in the symbiosis between plants of the legume family (Fabaceae) and bacteria of the genus *Rhizobium* (or related genera such as *Bradyrhizobium* and *Sinorhizobium*, collectively referred to hereafter as rhizobia). Additionally, there are 24 genera of nitrogen-fixing actinorhizal plants that consist of symbioses between various woody plants in 8 families and bacteria of the genus *Frankia*. No crop plants are actinorhizal, but the alder–*Frankia* symbiosis is important in forestry. Other actinorhizal plants have significant ecological roles, particularly in enhancing the nutrient status of soil. The complexity of these symbiotic associations results in trace element requirements that are different from those of the individual organisms that participate in the symbiosis. Nitrogen-fixing plants have a unique requirement for cobalt and selenium and an elevated requirement for molybdenum, iron, and nickel, compared to plants that do not fix nitrogen.

COBALT

Cobalt (Co) is essential for all N_2-fixing plant symbioses. This requirement arises from the needs of the endophyte (rhizobia or *Frankia*) rather than those of the host plant per se. Co plays a major role in symbiotic endophytes as a constituent of vitamin B_{12} (cyanocobalamin). Plants do not contain vitamin B_{12}. Coenzyme B_{12} is essential for three types of biochemical reactions: 1) intramolecular rearrangements (e.g., the conversion of L-methylmalonyl CoA into succinyl CoA), 2) methylations (e.g., the synthesis of methionine by methylation of homocysteine), and 3) reduction of ribonucleotides to deoxyribonucleotides. Co does not participate directly in the process of nitrogen fixation.

The requirement of nitrogen-fixing plants for Co was established in 1961 (Fig. 1)[1] using ultrapure nutrient solutions. Symptoms of Co deficiency match those caused by deficiency of nitrogen (e.g., stunted growth, yellowing of tissues) since the symbiotic bacteria are unable to provide the host plant with nitrogen. Because the amounts of Co required are so small, such deficiencies can be demonstrated only by meticulous exclusion of contaminants. Co deficiencies in the field are rare but have occasionally been observed. In some exceptional cases, dramatic increases in yield can result from addition of supplemental Co to crops.[2]

SELENIUM

The requirements of nitrogen-fixing plants for selenium (Se) are less straightforward than is the case for Co. Se is an essential component of hydrogenase, a bacterial enzyme that reclaims energy lost as hydrogen gas, which is produced as a by-product of nitrogenase activity.[3] Although hydrogenase is clearly beneficial to nitrogen-fixing plants, this enzyme is not an absolute requirement. Some rhizobia lack hydrogenase and the plants containing such rhizobia have a reduced yield.[4] Therefore, the selective use of strains of rhizobia that contain hydrogenase is of practical concern in agriculture. The genes for hydrogenase have recently been introduced into rhizobia that lack them, and such efforts may eventually reach applications in the field.[5]

MOLYBDENUM

The role of molybdenum (Mo) in nitrogen fixation by legumes has been recognized since 1950.[6] Mo is a constituent of the molybdenum–iron cofactor (FeMoCo) that is required for nitrogenase activity (Fig. 2).[7] Mo is also essential for the activity of xanthine dehydrogenase, an enzyme that plays an essential role in the synthesis of ureides (nitrogen transport compounds) in some legumes such as soybean, cowpea, and *Phaseolus* beans. Higher plants and other organisms that utilize nitrates also need Mo for activity of nitrate reductase.

Legumes are particularly susceptible to Mo deficiency (Fig. 3). Symptoms of Mo deficiency have been described

Encyclopedia of Plant and Crop Science
DOI: 10.1081/E-EPCS 120010514

Fig. 1 The effects of cobalt deficiency in soybean plants grown under symbiotic (nitrogen-fixing) conditions. (From Ref. 1.) (*View this art in color at www.dekker.com.*)

from many parts of the world and are especially common on acidic soils. The application of lime induces changes in soil chemistry that may relieve Mo deficiencies. In some other cases, the response to fertilization with small amounts of Mo can be spectacular.

IRON

Nitrogen-fixing plants have an increased requirement for iron (Fe) related to several key proteins involved closely with nitrogen fixation. Fe is a constituent of both of the two components of nitrogenase: dinitrogenase reductase, also called the Fe protein, and dinitrogenase, also called the MoFe protein (Fig. 2). Fe is also a component of ferredoxin, the proximal electron donor to the nitrogenase complex, and of leghemoglobin, a heme (Fe porphyrin) protein that plays a critical role in binding and delivering oxygen for respiration in such a way that nitrogenase is not inactivated. Leghemoglobin may account for up to 50% of the total soluble protein in legume nodules and is responsible for the dark red appearance of nodule interiors that indicates healthy, active nodules. Further indication of the key importance of Fe in nodules is the fact that rhizobia contain siderophores to facilitate the uptake of Fe and that nodules contain phytoferritin to store Fe. This storage is critical; although nodules contain large amounts of Fe, the concentration of free (catalytically active) Fe must be kept low because of its ability to promote the formation of damaging forms of activated oxygen such as superoxide and hydroxyl radicals.

NICKEL

Nickel (Ni) has a dual role in the metabolism of nitrogen-fixing plants. Ni is needed for organisms that utilize urea as a nitrogen source and break down urea in their metabolic processes. The breakdown of urea is catalyzed by urease, a Ni-containing enzyme. Urease is widespread in legume leaves and seeds as well as in bacteroids and free-living rhizobia.[8] The precise metabolic role of urea and urease in regard to nitrogen fixation is not clear, but it is likely that urease is involved in the processing of nitrogenous compounds transported from nodules to leaves. Soybeans and cowpeas grown in purified nutrient solutions in symbiosis with rhizobia develop necrotic leaf tips when the Ni supply is inadequate.[9] This necrosis is due to high concentrations of urea that accumulate due to the lack of urease activity. Soybeans grown with fixed nitrogen also have a Ni requirement, but the magnitude of this requirement appears to be less.

Nitrogen-fixing symbioses also require Ni as a component of hydrogenase—an enzyme that participates in the conservation of energy as discussed earlier with respect to Se. Ni deficiency in soybean plants results in a significant decrease in hydrogenase activity in nodule

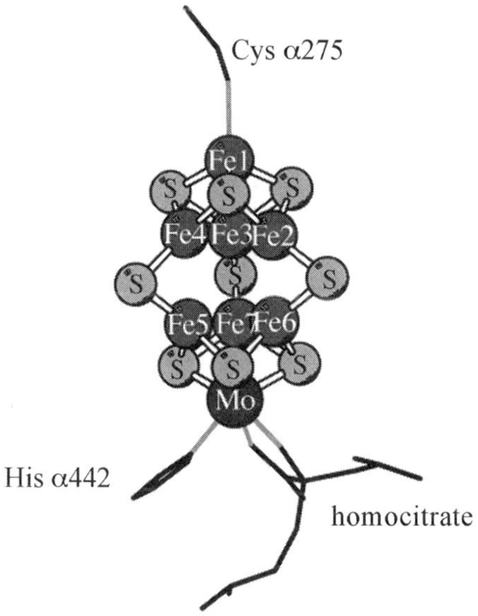

Fig. 2 The structure of the FeMo cofactor in nitrogenase. (Provided by D. C. Rees, CA Inst. of Tech.)

Fig. 3 The effects of molybdenum deficiency in alfalfa plants grown under symbiotic (nitrogen-fixing) conditions. (From Ref. 6.) (*View this art in color at www.dekker.com.*)

bacteroids. The amount of Ni required by plants is extremely low, and there are no reports of Ni deficiency in the field. However, some soils contain inadequate Ni for the optimum activity of urease in the indigenous soil microorganisms. Furthermore, inoculated soybean plants grown on low-nickel soil exhibited increased activities of urease in leaves and hydrogenase in nodule bacteroids when nickel was provided.[8]

OTHER ELEMENTS

Rhizobia also require other micronutrients such as manganese, copper, and zinc,[10] but these elements do not play a direct role in nitrogen fixation. Boron is not required by rhizobia, but is necessary for the formation of legume nodules. Also, it is noteworthy that although some nitrogen-fixing bacteria (e.g., *Azotobacter*) have ''alternative'' nitrogenases that utilize vanadium instead of Mo, there is no evidence for the presence of such alternative nitrogenases in symbiotic bacteria such as rhizobia or *Frankia*.

ARTICLES OF FURTHER INTEREST

Minor Nutrients, p. 726
Nitrogen, p. 822
Symbioses with Rhizobia and Mycorrhizal Fungi: Microbe/Plant Interactions and Signals Exchange, p. 1213
Symbiotic Nitrogen Fixation, p. 1218
Symbiotic Nitrogen Fixation: Plant Nutrition, p. 1222

REFERENCES

1. Ahmed, S.; Evans, H.J. The essentiality of cobalt for soybean plants grown under symbiotic conditions. Proc. Natl. Acad. Sci. U. S. A. **1961**, *47*, 24–36.
2. Gladstones, J.S.; Loneragan, J.F.; Goodchild, N.A. Field responses to cobalt and molybdenum by different legume species, with inferences on the role of cobalt in legume growth. Aust. J. Agric. Resour. **1977**, *28*, 619–628.
3. Boursier, P.; Hanus, F.J.; Papen, H.; Becker, M.M.; Russell, S.A.; Evans, H.J. Selenium increases hydrogenase expression in autotrophically cultured *Bradyrhizobium japonicum* and is a constituent of the purified enzyme. J. Bacteriol. **1988**, *170* (12), 5594–5600.
4. Hanus, F.J.; Albrecht, S.L.; Zablotowicz, R.M.; Emerich, D.W.; Russell, S.A.; Evans, H.J. Yield and N-content of soybean seed as influenced by *Rhizobium japonicum* inoculants possessing the hydrogenase characteristic. Agron. J. **1981**, *73* (2), 368–372.
5. Bascones, E.; Imperial, J.; Ruiz-Argueso, T.; Palacios, J.M. Generation of new hydrogen-recycling Rhizobiaceae strains by introduction of a novel hup minitransposon. Appl. Environ. Microbiol. **2000**, *66* (10), 4292–4299.
6. Evans, H.J.; Purvis, E.R.; Bear, F.E. Molybdenum nutrition of alfalfa. Plant Physiol. **1950**, *25*, 555–566.
7. Rees, D.C.; Howard, J.B. Nitrogenase: Standing at the crossroads. Curr. Opin. Chem. Biol. **2000**, *4*, 559–566.
8. Dalton, D.A.; Russell, S.A.; Evans, H.J. Nickel as a micronutrient element for plants. BioFactors **1988**, *1*, 11–16.
9. Eskew, D.L.; Welch, R.M.; Cary, E.F. Nickel: An essential micronutrient for legumes and possibly all higher plants. Science **1983**, *22*, 621–623.
10. O'Hara, G.W. Nutritional constraints on root nodule bacteria affecting symbiotic nitrogen fixation: A review. Aust. J. Exp. Agric. **2001**, *41* (3), 417–433.

Tobamoviral Vectors: Developing a Production System for Pharmaceuticals in Transfected Plants

T

Jennifer Lee Busto
Monto H. Kumagai
University of Hawaii at Manoa, Honolulu, Hawaii, U.S.A.

INTRODUCTION

Transgenic crops have played a predominant role in the production of pharmaceuticals and other valuable biological molecules. A more efficient strategy has involved inoculating nontransgenic plants with virus-based vectors that carry foreign genes. With the development of infectious cDNA clones, single-stranded RNA plant viruses have become key players in gene function discovery, metabolic engineering, and biomanufacturing. Viral expression vectors provide epigenetic expression of foreign sequences throughout infected plants, leading to gain- or loss-of-function phenotypes due to overexpression or cytoplasmic inhibition of gene expression.

Plant viruses are powerful transfection tools in molecular farming, producing pure, properly folded and glycosylated proteins in planta faster and more economically than other expression systems. They are a highly desirable alternative to transgenic systems that require protracted periods to transform and regenerate whole plants, and that have variation in the expression levels of heterologous proteins. In transgenic systems, once a particular construct is inserted into the plant genome, it may take several crosses to establish a stable line in an elite cultivar. In contrast, plant viral vectors employed in the large-scale production of therapeutic drugs in greenhouse and field-grown crops directly yield high levels of foreign protein, due to the rapid rate of viral replication. In plants transfected with a recombinant tobamovirus, alpha-trichosanthin, a potential anti-AIDS drug, accumulated to approximately 2% of total soluble protein.

Therapeutic compounds stably produced in transfected plants are numerous, and include antiviral drugs such as human interferon-alpha 2 as well as vaccines, proteins, and secondary metabolites. Plant-derived anticancer vaccines have been produced for treatment of human papillomavirus-induced cancer by expressing recombinant E7 fusion oncoproteins in *Nicotiana benthamiana*. The HIV p24 nucleocapsid protein, used as an antigen in the development of HIV vaccines, has been produced in plant protoplasts using tomato bushy stunt virus (TBSV) vector. For viruses that cannot be grown in tissue culture, such as hepatitis C (HCV), tobamoviral vectors are under development to produce a plant-derived vaccine. Recombinant proteins for use in diagnostics have also been expressed in plants. Full-length recombinant monoclonal antibodies (rAbs) directed against a colon cancer antigen and recombinant allergens have been expressed in *N. benthamiana* leaves, using a tobamoviral vector. The binding of IgE from sera of birch pollen- and latex-allergic patients suggests that the plant-produced allergens are properly folded.

TOBACCO MOSAIC VIRUS VECTORS

Several virus groups have been under investigation for design as recombinant plant virus vectors, including geminiviruses, potyviruses, potexviruses, comoviruses, tombusviruses, tobraviruses, alfamoviruses, and hordeiviruses. Members of the tobamovirus group,[1–3] are the most widely studied. The autonomously replicating RNA viral vectors based on the tobacco mosaic virus (TMV) genome have been particularly successful as research and commercial tools.

TMV possesses a positive-sense, single-stranded genome of 6396 nucleotides that encodes replicase enzymes, and movement and coat proteins. Viral genes are expressed through the production of both genomic and subgenomic RNA. Essentially designed as cDNA plasmids, TMV vectors are modified to contain a foreign gene sequence. Originally, vectors were constructed with the gene of interest replacing the capsid protein, until it was recognized that these viral vectors do not move efficiently. Presently, TMV vectors are hybrid versions of several different strains of tobamoviruses[4,5] that include all essential viral genes, a bacterial origin of replication (ori), and an antibiotic resistance marker. Dual heterologous subgenomic promoters from related tobamoviruses have enhanced stability, while an internal ribosome entry site sequence (IRES)[6] has been incorporated into the design to enable expression of multiple proteins.

The transfection process involves mechanically inoculating recombinant in vitro RNA transcripts from viral cDNA clones onto plants. Recombinant virions are

Encyclopedia of Plant and Crop Science
DOI: 10.1081/E-EPCS 120010593

assembled in the plant and move systemically by associating with plasmodesmata and intercellular cytoplasmic channels, producing foreign protein as they travel. One to two weeks after inoculation, recombinant proteins can be isolated from transfected plants. Interstitial fluid containing the desired product can be quickly separated from other cellular proteins by vacuum infiltration and gentle centrifugation. For large agronomic applications, virions can be purified from transfected plants and used for subsequent inoculations using high-pressure sprayers in the field.

Viral Vector Design Construct

For vaccine production, TMV vectors have been developed as coat protein (CP) fusions,[7] with the viral coat providing a flexible framework for the recombinant protein. TMV CP fusions include human immunodeficiency virus type I (HIV-I) peptide, influenza virus hemagglutinin epitope, malaria parasite peptide, and hepatitis C virus peptide.[1] A key technical advance in design has enabled TMV vectors to produce free proteins that are not fused to the coat protein. Instead, genes encoding bioactive compounds are fused to signal peptide sequences that cause the translated protein to be processed via the endoplasmic reticulum and Golgi complex, and to be targeted for cellular secretion. It is recognized that the faithful and efficient expression of such heterologous proteins is influenced by the choice of the signal peptide.

One of the most highly efficient TMV vector constructs fuses the gene of interest to a sequence encoding a rice α-amylase signal peptide adjacent to a 5′ untranslated leader.[3] Viral vectors containing the rice α-amylase signal peptide ORF have been used to express a wide variety of heterologous proteins including mammalian peptides, blood products, glycoproteins, and cytokines. TMV vectors that incorporate this signal peptide have been used successfully to secrete single-chain variable-fragment (Fv) antibodies for the treatment of non-Hodgkin's lymphoma (NHL). Currently in Phase II/III clinical trials, these highly customized patient-specific vaccines act as antitumor agents.[8]

The rice α-amylase 5′ untranslated leader sequence may help to enhance translation of the heterologous protein. The highly expressed viral coat subgenomic RNA has a 5′ cap (m7GpppN) and terminates with a tRNA-like structure instead of a poly (A) tail. The 3′ untranslated region (UTR) has two domains, which contain five RNA pseudoknots. It is possible that interactions between the 34-base pair 5′ leader of α-amylase and the 3′ UTR may cause a synergistic regulation of translation in transfected plants. Significantly, the rice α-amylase signal peptide has been recognized and processed in other transformed organisms, including *Escherichia coli*, *Saccharomyces cerevisiae*, and *Yarrowia lipolytica*, as well as transgenic rice cell suspensions. Subtle differences in the size, source, and sequence of the α-amylase signal peptide can greatly affect the secretion process.

Striving for higher protein yields, researchers are collaborating to develop improvements in vector design for increased production. It is recognized that inclusion of

Fig. 1 Overexpression of phytoene synthase (crtB) in leaves of *Nicotiana benthamiana* produces an orange phenotype. (Photo courtesy of Tony Novelozo—Axiom, Davis.) (*View this art in color at www.dekker.com.*)

foreign genes into TMV vectors reduces efficiencies of replication and movement compared to the wild type. These studies are aimed at improving the ability of these vectors to move and to replicate through gene shuffling of the 30K movement gene.[9] Visible markers for heterologous gene expression in plants are also under development. Tobacco mosaic viral (TMV) vectors have been engineered to overexpress an enzyme involved in carotenoid biosynthesis in *N. benthamiana* and other solaneaceous plants.[4] As the viral genome is translated, the encoded enzyme, phytoene synthase (*psy*), is targeted to the chloroplast, causing an accumulation of phytoene, a colorless compound. However, transfected plants show a characteristic orange phenotype in the leaves and flower sepals, as early as four days post-inoculation (Fig. 1). If fused to a heterologous sequence in a recombinant vector, *psy* may serve as a useful marker for gene expression, particularly in field applications.

Metabolic Engineering

Although plant viruses that are engineered to produce pharmaceutically relevant proteins have proved to be powerful gene expression tools, they are also valuable tools for use in gene discovery and in the metabolic engineering of existing pathways in plants. The biosynthesis of leaf carotenoids in transfected *N. benthamiana* was altered by forced rerouting of the pathway, resulting in the synthesis of capsanthin, a non-native chromoplast-specific xanthophyll. The ectopic expression of capsanthin–capsorubin synthase (Ccs) cDNA caused the plant to develop an orange phenotype and accumulate high levels of capsanthin—up to 36% total carotenoids. By redirecting the existing pathways of plants that produce biologically active compounds, plant virus expression systems can potentially be used to alter or to produce novel enzymes, or cause the accumulation of non-native bioactive compounds.[10]

CONCLUSION

Plant RNA viral vectors have become intensively utilized in several different plant species for large-scale production of high-value therapeutic proteins[11–13] and secondary metabolites.[5] The United States Food and Drug Administration (FDA) has developed a guidance document on plant-derived biologics and has strengthened field-testing controls for permits on those bioengineered traits that are not intended for commodity uses, such as pharmaceuticals, veterinary biologics, or certain industrial products. The human safety of TMV-based expression systems has been documented; plant viruses are not pathogenic to humans. TMV is only transmitted to other plants through mechanical means; proper cleaning of tools and machinery with bleach contains the virus and TMV-based vectors. In addition, FDA demands for recombinant product purity are rigorous.

To date, at least nine separate field trials using viral-based vector systems for the production of biologics have been conducted in three separate states. Concerns regarding the spread of engineered TMV and the persistence of recombinant viruses have been addressed in these studies, and a recent report indicates that recombinant viruses generally delete the foreign gene, have reduced vigor, and are less competitive and pathogenic than the indigenous TMV.[14] Plant viral vectors are also effective tools in metabolic engineering, as well as gene function discovery.[15] Design modifications will lead to improvement of desirable vector traits so that the potential of plant virus vector gene expression systems is fully realized.[2]

ACKNOWLEDGMENTS

We thank Tony Novelozo (Axiom, Davis) for the plant photograph, and Dr. Jon Donson (Ceres, Malibu) and Dr. Guy della-Cioppa for their expert advice and helpful encouragement. We would like to acknowledge David Clary, Damon Harvey, Annie TrucAnh Do, Peter Roberts, and Sherri Wykoff for their expert technical assistance. We thank Larry Grill and other Large Scale Biology colleagues for their support and critical reading of this manuscript. This work was supported in part by the USDA Special Grants Program for Tropical and Subtropical Agriculture, Agreement# 2002-34135-12791.

ARTICLES OF FURTHER INTEREST

Plant RNA Virus Diseases, p. 1023
Vaccines Produced in Transgenic Plants, p. 1265
Virus-Induced Gene Silencing, p. 1276

REFERENCES

1. Nemchinov, L.G.; Liang, T.J.; Rifaat, M.M.; Mazyad, H.M.; Hadidi, A.; Keith, J.M. Development of a plant-derived subunit vaccine candidate against hepatitis C virus. Arch. Virol. **2000**, *145* (12), 2557–2573.
2. Fitzmaurice, W.P.; Holzberg, S.; Lindbo, J.A.; Padgett, H.S.; Palmer, K.E.; Wolfe, G.M.; Pogue, G.P. Epigenetic modification of plants with systemic RNA viruses. Omics **2002**, *6* (2), 137–151.
3. Kumagai, M.H.; Donson, J.; della-Cioppa, G.; Grill, L.K.

Rapid, high-level expression of glycosylated rice alpha-amylase in transfected plants by an RNA viral vector. Gene **2000**, *245* (1), 169–174.

4. Kumagai, M.H.; Donson, J.; della-Cioppa, G.; Harvey, D.; Hanley, K.; Grill, L.K. Cytoplasmic inhibition of carotenoid biosynthesis with virus-derived RNA. Proc. Natl. Acad. Sci. U. S. A. **1995**, *92* (5), 1679–1683.
5. Kumagai, M.H.; Turpen, T.H.; Weinzettl, N.; della-Cioppa, G.; Turpen, A.M.; Donson, J.; Hilf, M.E.; Grantham, G.L.; Dawson, W.O.; Chow, T.P. Rapid, high-level expression of biologically active alpha-trichosanthin in transfected plants by an RNA viral vector. Proc. Natl. Acad. Sci. U. S. A. **1993**, *90* (2), 427–430.
6. Toth, R.L.; Chapman, S.; Carr, F.; Santa Cruz, S. A novel strategy for the expression of foreign genes from plant virus vectors. FEBS Lett. **2001**, *489* (2–3), 215–219.
7. Turpen, T.H.; Reinl, S.J.; Charoenvit, Y.; Hoffman, S.L.; Fallarme, V.; Grill, L.K. Malarial epitopes expressed on the surface of recombinant tobacco mosaic virus. Biotechnology (NY) **1995**, *13* (1), 53–57.
8. McCormick, A.A.; Kumagai, M.H.; Hanley, K.; Turpen, T.H.; Hakim, I.; Grill, L.K.; Tuse, D.; Levy, S.; Levy, R. Rapid production of specific vaccines for lymphoma by expression of the tumor-derived single-chain Fv epitopes in tobacco plants. Proc. Natl. Acad. Sci. U. S. A. **1999**, *96* (2), 703–708.
9. Toth, R.L.; Pogue, G.P.; Chapman, S. Improvement of the movement and host range properties of a plant virus vector through DNA shuffling. Plant J. **2002**, *30* (5), 593–600.
10. Kumagai, M.H.; Keller, Y.; Bouvier, F.; Clary, D.; Camara, B. Functional integration of non-native carotenoids into chloroplasts by viral-derived expression of capsanthin–capsorubin synthase in Nicotiana benthamiana. Plant J. **1998**, *14* (3), 305–315.
11. Franconi, R.; Di Bonito, P.; Dibello, F.; Accardi, L.; Muller, A.; Cirilli, A.; Simeone, P.; Dona, M.G.; Venuti, A.; Giorgi, C. Plant-derived human papillomavirus 16 E7 oncoprotein induces immune response and specific tumor protection. Cancer Res. **2002**, *62* (13), 3654–3658.
12. Zhang, G.; Leung, C.; Murdin, L.; Rovinski, B.; White, K.A. In planta expression of HIV-1 p24 protein using an RNA plant virus-based expression vector. Mol. Biotechnol. **2000**, *14* (2), 99–107.
13. Verch, T.; Yusibov, V.; Koprowski, H. Expression and assembly of a full-length monoclonal antibody in plants using a plant virus vector. J. Immunol. Methods **1998**, *220* (1–2), 69–75.
14. Rabindran, S.; Dawson, W.O. Assessment of recombinants that arise from the use of a TMV-based transient expression vector. Virology **2001**, *284* (2), 182–189.
15. Baulcombe, D.C. Fast forward genetics based on virus-induced gene silencing. Curr. Opin. Plant Biol. **1999**, *2* (2), 109–113.

Transformation Methods and Impact

Karabi Datta
Swapan K. Datta
International Rice Research Institute, Metro Manila, Philippines

INTRODUCTION

Transformation (or genetic engineering) has provided a powerful tool to introduce and modify traits desirable for crop improvement. Transgenes can now be incorporated into any cultivars across the genetic barriers for many crop species. Genetic transformation can be achieved by several methods (protoplast, biolistic, *Agrobacterium tumefaciens*, etc.) and a large number of successful crops have now been developed and commercialized. It is now possible to develop crops engineered with genes for built-in plant protection, durable resistance, and nutrition improvement. In the genomics era, gene discovery will accelerate further crop improvement and precision breeding, using high-throughput genetic transformation.

Plant improvement based on genetics has been occurring for years. In traditional breeding, genetic modifications are made by making crosses between related organisms. This depends mostly on reciprocal crossability and selecting for specific desirable traits. With the use of recombinant DNA technology, it is now possible to remove a piece of DNA containing one or more specific desirable genes from any organism and introduce it into another organism. Transformation is the step in the genetic engineering process that involves the incorporation of genes into genomes by means other than fusion of gametes or somatic cells. The incorporation and expression of foreign genes in plants, which is now possible in at least 35 families, was first described in tobacco. The plant transformation approach is being used to produce plants possessing traits unachievable by conventional plant breeding, especially in cases where the gene pool contains no source of the desired trait. Improved transgenic crops have been developed and field-tested in many countries and will have great impact in the developing countries. Seven transgenic crops—maize, cotton, canola, rapeseed, potato, squash, and papaya—are being produced commercially in 12 countries, including the United States, Argentina, Canada, China, and India. Elite transgenic indica rice has now been evaluated. It showed excellent performance against insect pests and bacterial blight.

Different systems have been developed to integrate foreign DNA into plant cells and to carry out the successive regeneration of transgenic plants, as shown in rice as a model monocot plant. Commonly used methods are based on biological vectors (e.g., *Agrobacterium*, virus), physical techniques (e.g., particle bombardment), and chemical techniques (e.g., PEG-mediated protoplast transformation). Besides the transformation methods, the basic requirements for the production of transgenic plants are: 1) the availability of target tissues to be transformed that are competent for plant regeneration; and 2) a suitable selection system for regenerating transformed plants with reasonable frequency. An efficient system for selecting only a few transformed cells from among a large number of nontransformed cells, and recovery of plants from single transformed cells is essential.

AGROBACTERIUM-MEDIATED TRANSFORMATION

In nature, the gram-negative soil bacterium *Agrobacterium tumefaciens* infects wound sites in dicotyledonous plants, causing the formation of crown gall tumors by a multistep gene transfer procedure. Virulent strains of *A. tumefaciens* and *A. rhizogens* possess the ability to transfer a particular DNA segment (T-DNA) of the extra chromosomal tumor-inducing (Ti) plasmid.[1] The T-DNA is surrounded by 24-base-pair (bp) border repeats and contains two types of genes, oncogenic genes and the gene encoding for the synthesis of opines. The oncogenic genes encode for the enzymes involved in the synthesis of auxins and cytokinins, whose expression leads to tumor formation. Outside the T-DNA, a large number of other genes (*vir* genes) in Ti plasmid are involved in this transfer mechanism. The T-DNA transfer system is determined by *vir* genes, whereas 24-bp direct repeats act as the recognition signals for the transfer.

Binary Vector System for *Agrobacterium*-Mediated Transformation

On the basis of the naturally occurring gene transfer mechanism of crown gall formation, the *Agrobacterium*

Encyclopedia of Plant and Crop Science
DOI: 10.1081/E-EPCS 120010465

Table 1 Achievements in transformation for crop improvement

Achievement	Transgene	Method	Year of development
Transformation demonstrated in tobacco	*gus*	*Agrobacterium*	1983
Transgenic cotton	*nptII*	*Agrobacterium*	1987
Transgenic soybean	*bar*	*Agrobacterium*	1988
Flavr Savr tomato	polygalacturonase	*Agrobacterium*	1988
Transgenic japonica rice	*hph*	Protoplast (Electroporation)	1989
Transgenic indica rice	*hph*	Protoplast (PEG)	1990
Insect-resistant cotton	*Bt*	*Agrobacterium*	1990
Transformation in Brassica	*gus*	*Agrobacterium*	1991
Herbicide-resistant rice	*bar*	Protoplast (PEG)	1992
Herbicide-resistant wheat	*bar*	Biolistic	1992
Transgenic maize	*Bt*	Biolistic	1993
Transgenic barley	*bar, gus*	Biolistic	1994
Insect-resistant sugarcane	*cry1A(b)*	Biolistic	1997
Tissue-specific gene expression in rice	*cry1A(b)/cry1A(c)*	Biolistic/Protoplast	1998
Iron-rich japonica rice	*ferritin*	*Agrobacterium*	1999
Beta-carotene Brassica	*crtB*	*Agrobacterium*	1999
Transgenic rice in field	*Bt, Xa21*	Biolistic	2000
Beta-carotene rich rice	*psy, crt1, lcy*	*Agrobacterium*	2000
Salt-tolerant tomato	*AtNHX1*	*Agrobacterium*	2001
Drought-tolerant rice	*TPSP*	Biolistic	2002
Golden indica rice	*psy, crt1*	Biolistic	2003
Iron-rich indica rice	*ferritin*	Biolistic	2003

system for plant genetic engineering has been designed. This system is based on a disarmed Ti plasmid with genes responsible for crown gall formation being removed, and various disarmed strains of *Agrobacterium* such as LBA4404 and C58C1 being created. Various genes of interest can be placed in a vector and introduced into the disarmed strain. In the binary vector system for T-DNA transfer, a plasmid provides the virulence functions and a small vector carries the artificial T-DNA. The binary vectors can replicate in *Escherichia coli* as well as in *A. tumefaciens*, which allows easy cloning of genes between the T-DNA borders. pBin19, pGA482, and pBI121 are well known vectors used in the *Agro*-transformation system.[2]

A range of improved virulent and binary systems has been developed. The supervirulent wider host-range *Agrobacterium* strain A281 possesses a transformation efficiency higher than that of other strains because of the presence of the pTiB0542 plasmid. The strain EHA101[3] also carries the disarmed version of pTiB0542. In the superbinary system, the *virB*, *virC*, and *virG* genes from the virulence region of pTiB0542 are inserted into a small T-DNA-containing plasmid in addition to the disarmed Ti plasmid with its full set of virulence genes.

Agrobacterium-mediated gene transfer into monocotyledonous plants became possible only recently; efficient methodologies have been reported in rice (Table 1),[4] banana, maize, wheat, and sugarcane.

Agrobacterium-mediated transformation has the advantage that it is relatively simple and can be applied with suitable tissue culture facilities (Fig. 1). This method ensures that a defined region of DNA is precisely transferred to the host genome. It reduces the copy number of the transgene, thus leading to fewer problems of transgene cosuppression and instability. High-molecular-weight DNA can be transferred through this method, but problems such as the transfer of vector sequences beyond the T-DNA border are frequently noticed.

In Planta Transformation

Successful *Agrobacterium*-mediated transformation has recently been reported in *Arabidopsis*, in which the transformation takes place in planta: The transformed tissue is not removed from the plant but left in a natural condition. In the in planta system, panicles at the early flowering stage are usually used as a target tissue. Successful in planta transformation of the legume *Medicago truncatula*

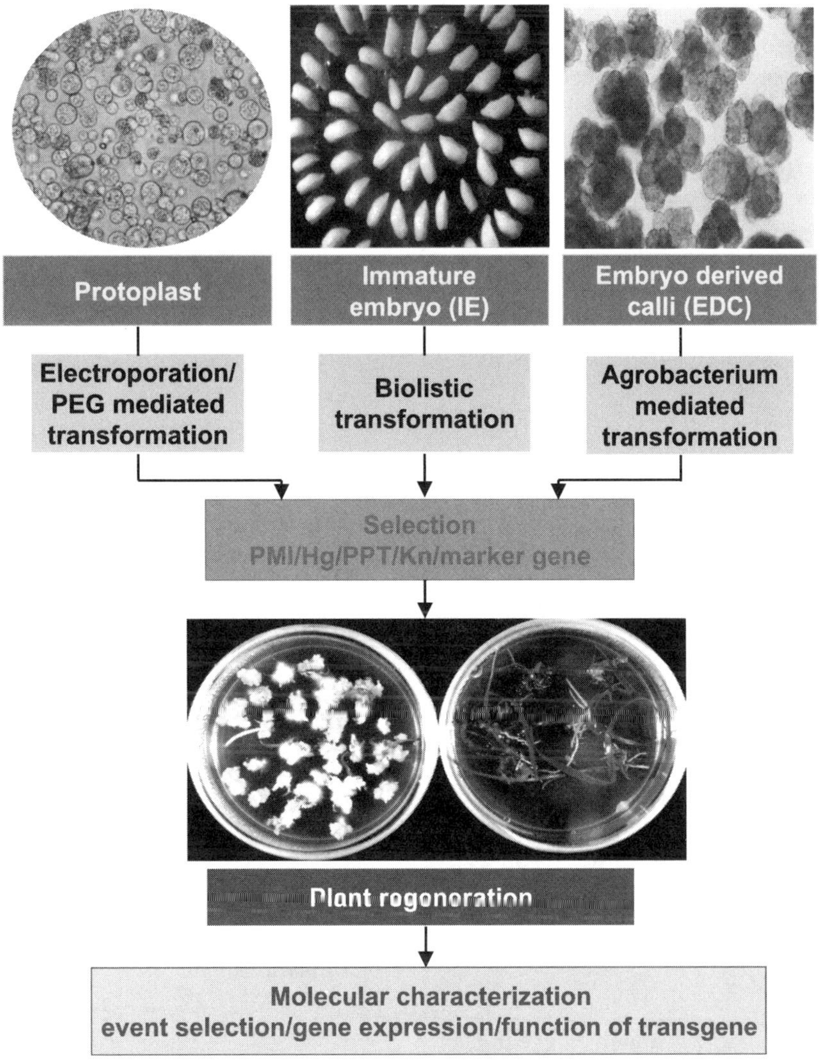

Fig. 1 Development of transgenic plants. (*View this art in color at www.dekker.com.*)

was also reported recently, showing that the method can be adopted in other species.

BIOLISTIC TRANSFORMATION (GENE GUN) SYSTEM

The biolistic (or gene gun) method for transformation involves the direct physical method of delivering nucleic acids into cells. In this system, metallic (gold or tungsten) microparticles are coated with DNA (transgene construct) and fired into target cells, shot with a particle bombardment machine. The particles are accelerated with an electrical discharge or compressed helium gas. Microprojectile bombardment of immature embryos to deliver the DNA into plant cells, and recovery of the whole plants from the transformed cells through selection are becoming almost routine (Fig. 1). Because there is no biological limitation to DNA delivery, genotype-independent transformation may be carried out with this method. Different rice cultivars have been transformed by this system.[5] Bombardment often results in multiple insertions of transgene copies in different loci and extensive rearrangement of the introduced gene, which may in some cases lead to expression instability of the transgene.

MICROINJECTION

In the microinjection method, DNA is injected directly into the nucleus of a cell using an ultrafine needle. The target cells are usually small structures with only a few cells and high regeneration potential (e.g., microspore-derived and zygotic proembryos). This method is not suitable for a plant system but is used routinely to produce transgenic animals.[6]

LIPOSOME-MEDIATED TRANSFER

DNA is encapsulated in liposomes (lipid micels). Liposomes help the DNA to enter through the plasmodesmata or the lipids to impregnate the cell walls, making it easier for DNA to penetrate. This method is not very efficient in transferring DNA to target cells.[6]

WHISKERS METHOD

Tiny microscopic needlelike structures with sharp end fibers of silicon carbide (or whiskers) are used in this method of genetic engineering. Embryogenic tissue culture cells, DNA of interest, and whiskers are suspended and shaken together robustly. The collision between the cells and sharp whiskers creates small holes in the cell wall to deliver the desired gene into the nucleus of the cell. This system has been used successfully for maize.

SELECTION SYSTEM

A suitable selectable marker gene is a prerequisite in distinguishing transformed cells from nontransformed ones. *nptII*, *hph*, and *ppt* are the most effective genes encoding neomycin phosphotransferase, hygromycin phosphotransferase, and phosphinothricin acetyltransferase, respectively.[7] Recently, the nonantibiotic POSITECH™ selection system has been developed using the *phosphomannoseisomerase* (*PMI*) gene available from Syngenta.[8,9]

CONCLUSION

Tremendous improvements in basic transgenic technologies have been made, to the extent that all major crop plants are now transformable and available for improvement with these technologies. Transformation protocols differ from one plant to another and, within species, from one cultivar to another. The optimization of transformation methodologies requires the consideration of several important means, such as the use of suitable explants, a mode of gene delivery, and plant regeneration. Maternal inheritance of foreign genes through chloroplast engineering can be used for those crops in which potential outcrossing is possible. Gene flow or outcrossing is very common in crops such as sunflower and strawberry. This technology is an alternative to overcome concerns regarding nuclear genetic engineering. However, gene flow is a natural phenomenon, and so far has not caused weediness or other negative effect on the environment.

Currently used transformation methods do not allow precise prediction of the position and number of copies of transforming DNA to be integrated into the host-cell genome. The tissue-and development-specific regulation of the gene depends on the choice of the promoter regulating the transcription of the transgene or the transit peptide directing the product to specific organelles. Transgene pyramiding and gene stacking using multiple genes of diverse traits are now possible, and will help in further crop improvement. The *sd1* gene principally involved in the green revolution can now be used to shorten plant height (to avoid lodging) and preserve traditional tall varieties. The main effect of genetic transformation lies in developing an improved product—seeds.[10,11]

ACKNOWLEDGMENTS

Thanks are due to the Rockefeller Foundation (USA), Bundesministerium fuer wirtschaftliche Zusammenarbeit und Entwicklung/Gesselschaft fuer technische Zusammenarbeit (BMZ/GTZ) (Germany), Danida (Denmark), and U.S. Agency for International Development (USAID) for supporting transgenic research; to scientists worldwide for collaborative research; and to Bill Hardy for editorial assistance.

ARTICLES OF FURTHER INTEREST

Protoplast Applications in Biotechnology, p. 1061
Protoplast Culture and Regeneration, p. 1065
Transgenes (GM) Sampling and Detection Methods in Seeds, p. 1238
Transgenes: Expression and Silencing of, p. 1242

REFERENCES

1. Fraley, R.; Schell, J. Plant biotechnology-Editorial overview. Curr. Opin. Biotechnol. **1991**, *2*, 145–146.

2. Datta, K.; Datta, S.K. Plant Transformation. In *Molecular Plant Biology. A Practical Approach*; Gilmartin, P.M., Bowler, C., Eds.; Oxford University Press: United Kingdom, 2002; 13–32.
3. Hood, E.E.; Helmer, G.L.; Fraley, R.T.; Chilton, M.D. The hypervirulence of *Agrobacterium tumefaciens* A281 is encoded in a region of pT1B0542 outside of the T-DNA. J. Bacteriol. **1986**, *168*, 1291–1301.
4. Hiei, Y.; Ohta, S.; Komari, T.; Kumashiro, T. Efficient transformation of rice (*Oryza sativa*) mediated by *Agrobacterium* and sequence analysis of the boundaries of the T-DNA. Plant J. **1994**, *6*, 271–282.
5. Christou, P.; Ford, T.F.; Kofron, M. Production of transgenic rice (*Oryza sativa*) plants from agronomically important Indica and Japonica varieties via electric discharge particle acceleration of exogenous DNA into immature zygotic embryos. Bio/Technology **1991**, *9*, 957–962.
6. Potrykus, I. Gene transfer to plants: Assessment of published approaches and results. Annu. Rev. Plant Physiol. Plant Mol. Biol. **1991**, *42*, 205–225.
7. Datta, S.K. Transgenic Cereals: *Oryza sativa* (Rice). In *Molecular Improvement of Cereal Crops*; Vasil, I.K., Ed.; Kluwer Acad. Pub.: UK, 1999; 149–187.
8. Joersbo, M.; Donaldson, I.; Kreiberg, J.; Peterson, S.G.; Brunstedt, J.; Okkels, F.T. Analysis of mannose selection used for transformation of sugar beet. Mol. Breed. **1998**, *4*, 111–117.
9. Datta, K.; Baisakh, N.; Oliva, N.; Terrizo, L.; Abrigo, E.; Tan, J.; Rai, M.; Rehana, S.; Al-Babli, S.; Beyer, P.; Potrykus, I.; Datta, S.K. Bioengineered 'golden' indica rice cultivars with β-carotene accumulation in the endosperm with hygromycin and mannose selection systems. Plant Biotechnol. J. **2003**, *1*, 81–90.
10. James, C. *Global Review of Commercialized Transgenic Crops*; International Service for Acquisition of Agri-Biotech Applications Publ.: New York, USA, 2002.
11. Tu, J.; Zhang, G.; Datta, K.; Xu, C.; He, Y.; Zhang, Q.; Khush, G.S.; Datta, S.K. Field performance of transgenic elite commercial hybrid rice expressing *Bacillus thuringiensis* δ-endotoxin. Nat. Biotechnol. **2000**, *18*, 1101–1104.

Transgenes (GM) Sampling and Detection Methods in Seeds

Alessandro Pellegrineschi
Applied Biotechnology Center and CRC for Molecular Plant Breeding, CIMMYT, Mexico City, México

INTRODUCTION

Transgenic plants or food crops are genetically modified organisms (GMOs), meaning that genes have been introduced or silenced in their genomes using methods other than sexual crossing. These genetic modifications can provide a variety of useful traits, including herbicide resistance, pest resistance, and enhanced vitamin content.

Genetically modified (GM) crops were grown on an estimated 58.7 million hectares (145 million acres) in 2002 by from 5.5 to 6.0 million farmers in 16 countries. It is interesting to note that more than three-quarters of the farmers who grew GM crops could be considered small-scale and resource-poor. The principal GM crops during this period were soybean (36.5 million hectares), corn (12.4 million hectares), cotton (6.8 million hectares), and canola (3.0 million hectares).[1]

Due to recent debates about safety, some countries are requiring disclosure of GM derivatives in foods, creating the need for reliable detection methods.

METHODS FOR DETECTION OF TRANSGENES

Current methods focus on detecting a molecule—DNA, RNA, or other proteins—associated with or derived from the genetic modification of interest.[2] DNA is the preferred target, because it is very stable and can be purified, and billions of copies of a particular strand can be made in just a few hours using the polymerase chain reaction (PCR), producing quantities large enough for easy detection.[3,4] Multiplication of RNA and other proteins is slower and more complicated. There is also a linear correlation between levels of GM derivatives in a sample and the DNA, if the genetically modified DNA comes from cell nuclei. No such correlation exists in the case of proteins or RNA.

To detect GM DNA using PCR, the double-stranded helix of the target DNA is separated and mixed in a solution with short pieces of synthetic, single-strand DNA known as ''primers.'' The solution contains a cellular enzyme called DNA polymerase, which among other of its natural functions contributes to the repair and replication of DNA. Each primer is ''complementary'' to one or the other end of the single strands of the target DNA. This means that the sequence of nucleotides (the amino acid building blocks of DNA and RNA) in the primer is such that it will bind to the DNA. The first primer matches the start and the coding section of the DNA to be multiplied, while the second primer matches the end and the noncoding portion of the DNA. The portion of the target strand in between is filled in with nucleotides from the polymerase solution, creating two double-stranded DNA molecules that are perfect copies of the original. By raising and lowering the temperature of the solution, the double strands are newly ''unzipped'' and the filling-in process begins again, using the complementary strands as templates. For each temperature cycle the number of copies is doubled, resulting in an exponential multiplication of the original DNA sample. After 20 cycles the copy number is approximately 1 million times higher than at the beginning.

One of the most common techniques to estimate the amount and size of DNA or RNA strands in a sample is gel electrophoresis. Prior to this, the DNA may be ''digested''—that is, put in solution with enzymes known to cut the DNA at sites where there is a specific sequence of nucleotides, resulting in fragments of particular sizes. A more sophisticated technique involves determination of melting point profiles by means of dyes like SYBR Green I, which emits fluorescent light when it intercalates with double-stranded DNA. When the temperature is increased, the DNA strands begin to separate, reducing the fluorescence in readily measurable amounts. The melting point is more characteristic of a specific DNA sequence than of its size, but complete sequencing—determining the exact order of the nucleotides throughout a DNA or RNA fragment—allows for more specific determination of the origin of the molecule. A fourth alternative is to use short, synthetic molecules called ''probes,'' which bind to given sites on a DNA or RNA

Encyclopedia of Plant and Crop Science
DOI: 10.1081/E-EPCS 120021919

strand. If appropriately designed, a probe will be able to discriminate between the correct molecule and almost any other DNA or RNA molecule. Labeling with fluorescence, radioactivity, antibodies, or dyes also facilitates detection of the molecules present.[5]

Protein-based detection of GM derivatives relies on the specific binding between a protein and an antibody, a molecule of the type that protects against bacterial and viral infections in the human body. The antibody recognizes a foreign molecule—the "antigen"—and binds to it. The bound complex is detected in a chemical reaction that causes a particular color to appear. This technique is known as the enzyme linked immunosorbent assay (ELISA) and combines the specificity of antibodies with the sensitivity of simple enzyme assays. ELISA can provide an easy and precise measurement of antigen or antibody concentrations.

One of the most useful of the immunoassays is the two-antibody "sandwich" ELISA. It is a fast and accurate way to determine the antigen concentration in unknown samples. It requires two antibodies that bind to different positions on the same antigen. This can be accomplished using either two monoclonal antibodies—ones that attach to single sites—or a batch of affinity-purified polyclonal antibodies. The antibodies for this technique must be developed using purified samples of the target protein, taken either from the GM derivatives themselves or synthesized, if the exact composition of the protein is known.

RNA based methods rely on specific binding between the RNA molecule and a synthetic RNA or a DNA strand as the primer. The primer nucleotides must again complement those at the starting end of the RNA molecule. Binding results in a double-stranded molecule similar to DNA and, normally, the subsequent conversion of the RNA to a DNA molecule through a process called reverse transcription. Finally the DNA can be multiplied using PCR or transformed into as many as 100 copies of the original RNA molecule and the procedure repeated, using each copy as a template, in a technique called nucleic acid sequence-based amplification (NASBA). The specific primers cannot be developed without knowing the exact composition of the RNA molecule to be detected.[5]

SAMPLING TECHNIQUES

To date, 72 transgenic varieties have been developed in 17 plant species (Table 1). There is insufficient space here to describe one-by-one all sampling methods to detect GM derivatives in seed; in any case, many methods will likely become obsolete soon, with the rapid development of new transgenic varieties and detection methods. This aside, there are two key problems in seed sampling that must be understood: 1) the limit of detection and 2) sampling error.[6,7]

The amount of unreplicated haploid genome (i.e., the 1C value, the amount present in a gamete) in a sample is useful for relating genome copy number to the amount of sample taken. For example, up to 36,697 copies of the haploid *Zea mays* genome (which we will use here as the basis for all examples below) are present in a typical 100 ng DNA analytical sample, given the 1C value of 2.725 picograms. It follows that a single copy of the haploid *Z. mays* genome in a 100 ng DNA sample is present at a level of 0.0027% (wt/wt). Levels of DNA below this threshold cannot be detected reliably in samples of this size.[4]

Sampling error occurs in a perfectly homogeneous preparation, even if a large amount (say, 50 μg) of DNA is extracted from a laboratory sample and simple random sampling procedures are adopted. As the amount of DNA extracted from the sample becomes lower, sampling error becomes proportionally larger. Thus, replicate 100 ng DNA samples containing GMO material at a level of 0.1% (wt/wt) would produce GM DNA estimates no better than approximately 30% of the mean value 95% of the time. This is a poor level of accuracy, even if we ignore other types of error inherent to a real analytical system.

With lower levels of DNA, the problem is even more critical. For a laboratory sample containing DNA at a level of 0.01%, the 100 ng analytical sample would vary between 0.0027% and 0.0191% nearly 95% of the time. These calculations obviously refer to a "best possible" result, as they assume a single sampling step and a perfect analytical system.

When undertaking a dilution series, the assumption of simple random sampling may no longer be valid, as the number of copies available becomes strictly finite. Indeed, the number of copies used to prepare subsequent dilutions heavily influences the sampling error associated with the series. Consequently, the preparation of any dilution series must be undertaken in such a way as to minimize this bias; ideally, dilutions should be made from the primary laboratory sample.

The classical solution to the issue of sampling error is to undertake repetitions and/or use appropriately sized analytical samples. We recommend that in the construction of a dilution series—for example, to determine the "limit of detection" of a method, or for the generation of standard curves—the nominal number of GM-derived copies in the weakest dilution of analytical sample should be set to approximately 20, thus providing good statistical probability that all repetitions contain relevant DNA.

This chapter endeavors to explain the methodologies used for the detection of transgenes, transgene products,

Table 1 List of the common commercial transgenic plants

Name	Scientific name	Number of transgenic varieties	Traits	Developer(s)
Argentine canola	*Brassica napus*	15	High laurate (12:0) and myristate (14:0); high oleic acid and low linolenic acid; herbicide tolerance; male sterility, fertility restoration, pollination control system	Calgene, Inc.; Pioneer Hi-Bred International, Inc.; Monsanto Company; Aventis CropScience
Carnation	*Dianthus caryophyllus*	3	Modified colors; herbicide tolerance	Florigene Pty Ltd.
Chicory	*Chichorium intybus*	1	Male sterility	Bejo Zaden BV
Cotton	*Gossypium hirsutum*	6	Insect resistance; herbicide tolerance	Monsanto Company; DuPont Canada Agricultural Products; Calgene, Inc.
Flax	*Linum usitatissimum*	1	Modification of acetolactate synthase (ALS)	University of Saskatchewan, Crop Dev. Centre
Maize	*Zea mays*	22	Insect tolerance; herbicide tolerance; male sterility; production of the aromatic amino acids	Syngenta Seeds, Inc.; Pioneer Hi-Bred International, Inc.; Dekalb Genetics Corporation; Aventis CropScience; BASF, Inc.; Mycogen; Monsanto Company
Melon	*Cucurbita pepo*	2	Virus tolerance	Asgrow; Seminis Vegetable, Inc.; Upjohn
Papaya	*Carica papaya*	1	Virus tolerance	Cornell University
Polish canola	*Brassica rapa*	1	Herbicide tolerance	Monsanto Company
Potato	*Solanum tuberosum*	4	Insect resistance	Monsanto Company
Rice	*Oryza sativa*	2	Herbicide tolerance	BASF, Inc.; Aventis CropScience
Soybean	*Glycine max*	7	Herbicide tolerance; high oleic acid; low linolenic acid	Aventis CropScience; DuPont Canada Agricultural Products; Monsanto Company
Squash	*Cucurbita pepo*	2	Virus resistance	Asgrow; Seminis Vegetable, Inc.; Upjohn
Tobacco	*Nicotiana tabacum*	1	Herbicide tolerance	Societe National d'Exploitation des Tabacs et Allumettes
Tomato	*Lycopersicon esculentum*	6	Delayed ripening; delayed softening; insect resistance;	DNA Plant Technology Corporation; Agritope, Inc.; Monsanto Company; Zeneca Seeds; Calgene, Inc.
Wheat	*Triticulm aestivum*	1	Herbicide tolerance	Cyanamid Crop Protection

and the problems associated with reliably detecting a GMO in very large samples. Clearly, these methodologies will rapidly improve with the advance of new technologies. For this reason, the author concentrated more on the theory behind the analyses and statistics than the technologies now in use.

REFERENCES

1. James, C. *Global Status of Commercialized Transgenic Crops*; 2002. ISAAA Brief 2002 No. 27.
2. Bertheau, Y.; Diolez, A.; Kobilinsky, A.; Magin, K. Detection methods and performance criteria for genetically modified organisms. J. AOAC Int. **May–Jun 2002**, *85* (3), 801–808.
3. Holst-Jensen, A.; Ronning, S.B.; Lovseth, A.; Berdal, K.G. PCR technology for screening and quantification of genetically modified organisms (GMOs). Anal. Bioanal. Chem. **Apr 2003**, *375* (8), 985–993.
4. Aarts, H.J.; van Rie, J.P.; Kok, E.J. Traceability of genetically modified organisms. Expert Rev. Mol. Diagn. **Jan 2002**, *2* (1), 69–76.
5. Kok, E.J.; Aarts, H.J.; Van Hoef, A.M.; Kuiper, H.A. DNA methods: Critical review of innovative approaches. J. AOAC Int. **May–Jun 2002**, *85* (3), 797–800.
6. Kay, S.; Van den Eede, G. The limits of GMO detection. Nat. Biotechnol. **May 2001**, *19* (5), 405.
7. Ahmed, F.E. Detection of genetically modified organisms in foods. Trends Biotechnol. **May 2002**, *20* (5), 215–223.

T

Transgenes: Expression and Silencing of

George C. Allen
North Carolina State University, Raleigh, North Carolina, U.S.A.

INTRODUCTON

It is hoped that transgenic technology will be a solution for feeding the increasing world population. However, as the use of transgenic crops has increased, evidence has accumulated that the expression of a desired transgene may be unstable. To ensure that plants with unstable transgene expression are kept from reaching market, extensive testing is needed, which requires higher production costs. Even this testing is still not a guarantee that an apparently stable trait will remain stable under all environmental conditions.[1] The purpose of this review is to provide the reader with a greater understanding of the molecular events related to how and why gene silencing occurs. Possible strategies for preventing undesirable gene silencing are also discussed, and examples are given of gene silencing used as a tool to understand gene function.

The discovery of gene silencing in plants laid the groundwork for findings that gene silencing is found in all higher eukaryotes. Historically, unstable transgene expression in plants was ascribed to "chromosomal position effects." According to the chromosomal position-effect model, transgenes integrated into regions of the genome containing uncondensed chromatin were more likely to be expressed. Conversely, transgenes integrated into regions of the genome containing condensed chromatin were less likely to be expressed. While extensive evidence for chromosomal position effects exists, homology-dependent gene silencing is now also known to be a major determinant for transgene expression.

POST-TRANSCRIPTIONAL GENE SILENCING (PTGS)

More than a decade ago a fortuitous observation was made when Jorgensen and his colleagues attempted to "overexpress" chalcone synthase (CHS) in petunia, using a transgene sequence that was identical to the sequence of the natural gene.[2] To achieve overexpression of CHS, a viral promoter known to give high levels of gene expression was used, with the expectation that purple flowers would be produced. Instead, the CHS overexpressing transgenic plants produced flowers that showed various patterns of purple and white. When the CHS messenger RNA (mRNA) levels in the white and purple sections were compared, the mRNA levels from white sectors were drastically reduced for both the CHS transgene and the native petunia CHS gene. The result was verified when van der Krol and colleagues tried to overexpress a petunia flavanoid gene.[3] Both groups concluded that a major requirement for gene silencing was sequence homology between the transgene and the native gene.

Nearly three years later, research on plant virus resistance provided additional clues to the gene silencing puzzle.[4] For several years, it was known that a transgenic plant expressing a viral coat protein for a specific virus had a higher frequency of resistance to the virus. Such observations led to speculations that the viral coat protein prevented the uncoating of the infecting virus and thus interfered with viral replication. Experiments by Dougherty and colleagues resulted in an alternative explanation, in which high levels of viral mRNA expressed from a plant transgene resulted in virus resistance. The transcription rates were similar for both virus-infected and uninfected plants, but no transgene viral mRNA accumulated, indicating that resistance was at the post-transcriptional level.

The model proposed by the Dougherty group suggested that an RNA-dependent RNA polymerase (RdRP) produced an antisense RNA that led to the degradation of both the viral RNA and the transgene "viral" RNA. We now know that the type of gene silencing observed over a decade ago in petunias was post-transcriptional gene silencing (PTGS), which also has been shown to be a natural mechanism plants use for virus resistance.

To circumvent plant defense systems, many viruses have suppressor systems that interfere with PTGS.[5] Thus transgenic plants that constitutively express the viral suppressors of silencing have impaired PTGS, which led to ideas that the suppressors could be used for high-level transgene expression. Unfortunately, plants that express the viral suppressors show developmental abnormalities. Such silencing-suppressed plants may be extremely sensitive to viruses under natural growing conditions, because silencing is a defense against viruses.

Recent work has shown that RdRP uses the mRNA template to produce double-stranded RNA (dsRNA), which is then degraded to small RNAs, leading to the initiation of PTGS.[6] Although the production of the

Encyclopedia of Plant and Crop Science
DOI: 10.1081/E-EPCS 120010462

small RNAs was discovered in plants,[7] the same pathway occurs in animal and fungal systems. Mutant and biochemical analyses are now increasing our understanding of the genes and signals involved in PTGS.

TRANSCRIPTIONAL GENE SILENCING (TGS)

TGS and PTGS typically are distinguished by nuclear run-on assays that compare transcription rates with the steady-state mRNA levels. A gene silenced by TGS has undetectable transcription and undetectable mRNA, whereas a gene silenced by PTGS is transcribed but has greatly reduced levels of mRNA. Additionally, TGS can be inherited, whereas PTGS requires reactivation in every generation.

To understand TGS it is useful to have a basic understanding of chromatin structure. Chromatin consists of proteins that bind and package DNA into successive levels of higher-order structures, which we eventually see as eukaryotic chromosomes. The structure of chromatin can control access of the transcriptional machinery to the DNA. If the chromatin is uncondensed, the transcription apparatus has ready access to promoters. Condensed chromatin, on the other hand, prevents access to a promoter.

Initial work on TGS demonstrated that when a transgene integrates near a region of condensed inactive chromatin, the inactive chromatin could spread into the transgene and prevent transcription. Drosophila studies showed that silencing frequently resulted in variegated expression patterns, known as position-effect variegation. Thus, in typical TGS, the promoter is made inaccessible by the condensed chromatin and cannot be transcribed.

PREVENTION OF TRANSCRIPTIONAL GENE SILENCING

A major limitation in the use of transgenes has been unpredictable variation in expression. To reduce the variation caused by TGS, attempts were made to block the spread of condensed chromatin. DNA elements termed matrix attachment regions (MARs) from the chicken lysozyme gene were used to flank a transgene.[8] The results showed that transfected mammalian cell lines had higher levels of transgene expression and lacked chromosomal position effects when MARs were used.

Work from the Allen and Thompson lab tested whether MARs could prevent silencing in plants using the beta-glucuronidase (GUS) reporter transgene, either flanked by yeast MARs or lacking MARs.[9] Comparison of the GUS transgene copy number and expression levels revealed that the GUS expression levels were much higher in tobacco cell lines transformed with MAR-flanked transgenes. However, even in the cell lines transformed with reporter constructs using the flanking MARs, a drastic reduction in transgene expression was seen when the transgene copy number increased above a threshold of approximately 20 copies. Similar results were found when a tobacco MAR was used.

We now know that silencing is possible, when gene copy numbers are above a threshold even in cell lines with MAR-flanked transgenes, and that this silencing is likely due to PTGS. In contrast to lines containing MAR-flanked transgenes, the control cell lines were subject to silencing by both TGS and PTGS. When cell lines expressing a suppressor of PTGS were retransformed with reporter genes flanked by MARs, silencing was drastically reduced. Presumably the flanking MARs prevented TGS and the viral suppressor prevented PTGS. Thus, MARs appear to prevent TGS but not PTGS.[10]

GENE SILENCING AS A TOOL FOR FUNCTIONAL GENOMICS

With genome sequences rapidly being completed for many organisms, attention is being focused on the area of functional genomics. While microarrays provide information on gene expression patterns and possible gene interactions, understanding how a gene impacts phenotype is limited by the lack of tools for producing functional knockout mutants. PTGS (or RNAi) is now providing scientists with a new tool for knocking out specific genes or groups of genes by using dsRNA to induce gene silencing.[11] The resulting phenotypic knockout can then be examined to determine gene function. By combining the use of high-throughput RNAi with microarray analyses, the effect of knocking out a single gene can be determined for an entire array of expressed genes, allowing scientists to determine the impact the gene has on global gene expression. Thus, gene silencing offers a unique tool for understanding how genes are regulated in complex organisms.

ARTICLES OF FURTHER INTEREST

Arabidopsis Transcription Factors: Genome-Wide Comparative Analysis Among Eukaryotes, p. 51
Breeding Plants with Transgenes, p. 193
Chromosome Manipulation and Crop Improvement, p. 266
Chromosome Rearrangements, p. 270
Chromosome Structure and Evolution, p. 273
Functional Genomics and Gene Action Knockouts, p. 476

Gene Silencing: A Defense Mechanism Against Alien Genetic Information, p. 492
Genetically Modified Oil Crops, p. 509
Genomic Imprinting in Plants, p. 527
Interphase Nucleus, The, p. 568
Meiosis, p. 711
Plant Diseases Caused by Subviral Agents, p. 956
Plant DNA Virus Diseases, p. 960
Plant Viruses: Initiation of Infection, p. 1032
RNA-mediated Silencing, p. 1106
Transformation Methods and Impact, p. 1233
Transgenic Crops: Regulatory Standards and Procedures of Research and Commercialization, p. 1251
Virus-Induced Gene Silencing, p. 1276

REFERENCES

1. Meyer, P.; Linn, F.; Heidmann, I.; Meyer, Z.A.H.; Niedenhof, I.; Saedler, H. Endogenous and environmental factors influence 35S promoter methylation of a maize A1 gene construct in transgenic petunia and its colour phenotype. Mol. Gen. Genet. **1992**, *231*, 345–352.
2. Napoli, C.; Lemieux, C.; Jorgensen, R. Introduction of a chimeric chalcone synthase gene into petunia results in reversible co-suppression of homologous genes in trans. Plant Cell **1990**, *2*, 279–289.
3. van der Krol, A.R.; Mur, L.A.; Beld, M.; Mol, J.N.M.; Stuitje, A.R. Flavonoid genes in petunia: Addition of a limited number of gene copies may lead to a suppression of gene expression. Plant Cell **1990**, *2*, 291–299.
4. Lindbo, J.A.; Silvarosales, L.; Proebsting, W.M.; Dougherty, W.G. Induction of a highly specific antiviral state in transgenic plants—Implications for regulation of gene expression and virus resistance. Plant Cell **1993**, *5*, 1749–1759.
5. Anandalakshmi, R.; Pruss, G.J.; Ge, X.; Marathe, R.; Mallory, A.C.; Smith, T.H.; Vance, V.B. A viral suppressor of gene silencing in plants. Proc. Natl. Acad. Sci. U. S. A. **1998**, *95*, 13079–13084.
6. Matzke, M.; Matzke, A.J.M.; Kooter, J.M. RNA: Guiding gene silencing. Science **2001**, *293*, 1080–1083.
7. Hamilton, A.J.; Baulcombe, D.C. A species of small antisense RNA in posttranscriptional gene silencing in plants. Science **1999**, *286*, 950–952.
8. Stief, A.; Winter, D.M.; Strätling, W.H.; Sippel, A.E. A nuclear DNA attachment element mediates elevated and position-independent gene activity. Nature **1989**, *341*, 343–345.
9. Allen, G.C.; Spiker, S.; Thompson, W.F. Use of matrix attachment regions (MARs) to minimize transgene silencing. Plant Mol. Biol. **2000**, *43*, 361–376.
10. Callaway, A.C.; Allen, G.C. unpublished data.
11. Fraser, A.G.; Kamath, R.S.; Zipperlen, P.; Martinez-Campos, M.; Sohrmann, M.; Ahringer, J. Functional genomic analysis of *C. elegans* chromosome I by systematic RNA interference. Nature **2000**, *408*, 325–330.

Transgenetic Plants: Breeding Programs for Sustainable Insect Resistance

T

Michael B. Cohen
University of Alberta, Edmonton, Alberta, Canada

INTRODUCTION

Genetic engineering of crops with crystal toxin (Cry) genes from *Bacillus thuringiensis* (Bt) has been a revolutionary development in plant breeding for resistance to insect pests.[1] Because Bt toxins are highly effective against several important groups of pests, yet are largely benign to non-target organisms, unprecedented measures have been taken in the United States,[2] Canada,[3] and Australia[4] to delay the evolution of pest adaptation to Bt crops. Resistance management strategies for Bt crops depend on proper deployment of Bt cultivars in farmers' fields and on the development of cultivars with appropriate toxin genes and toxin titers. This article provides an introduction to the principles of resistance management and recommends steps that genetic engineering programs and government regulatory agencies can take to promote the sustainable use of Bt crops.

THE REFUGE/HIGH DOSE STRATEGY

There is widespread agreement that the refuge/high dose strategy (Fig. 1) is the most promising approach for protecting the long-term effectiveness of transgenic insect-resistant crops.[2,5,6] Studies of insect populations have shown that resistance to insecticides, including Bt toxins, is most often attributable to a partially recessive mutation at a single major locus, and that the initial frequency of the resistant allele in unselected populations is low. A high dose in a Bt cultivar is one that is sufficient to make the resistant allele functionally recessive, so that insects heterozygous (RS) at the resistance locus do not survive when feeding on the cultivar. Refuges are non-Bt plants that serve to maintain homozygous susceptible insects (SS) in pest populations. For the refuge/high dose strategy to work, there must be a sufficient number and suitable spatial distribution of SS insects to ensure that the rare homozygous resistant insects (RR) mate with SS insects rather than with each other.

The refuge/high dose strategy has been implemented for Bt corn, potatoes, and cotton in the United States[2] and for Bt corn in Canada.[3] Currently available Bt corn cultivars appear to have a high dose of toxin with respect to the European corn borer *Ostrinia nubilalis*. Farmers who grow Bt corn must plant 20% of their land to non-Bt corn to serve as a refuge. The non-Bt corn fields must be planted at approximately the same time as the Bt fields, be of a similar corn cultivar, and be within approximately 0.5 kilometers (km) of the Bt fields. In Australia, where Bt cotton cultivars do not have a high dose of toxin with respect to the cotton bollworm *Helicoverpa armigera*, larger refuges are required.[4]

The U.S. Environmental Protection Agency established a working definition of a high dose in transgenic plants as "twenty-five times the protein concentration necessary to kill susceptible larvae," and described five procedures for testing whether this criterion has been met.[2] A rough guideline for a high dose that can be used in preliminary assessment of transgenic plants has also been suggested.[7] These authors surveyed the toxin titers of Bt corn, cotton, and potato cultivars that appear to have a high dose based on field performance and noted that these cultivars have titers of at least 0.1–0.2% soluble protein or 1–2 ng/mg fresh weight (see also Table A2 in Ref. 2).

OPTIONS FOR USING MULTIPLE TOXINS

Larvae of Lepidoptera (moths and butterflies) are the target pests for the Bt corn and cotton cultivars that have been released so far and for numerous other Bt crops that are under development. For most lepidopteran pests, there are several Cry toxins that are of sufficient toxicity to achieve a high dose in Bt cultivars. Genetic engineering programs thus face the question of how best to utilize multiple toxin genes. Among the options are sequential release, mosaics, and pyramids. These options have been analyzed using simulation models that incorporate factors such as refuge size, the relative mortality of SS, RS, and RR insects, and the initial frequency of the R allele.[5,6] Two very robust and important conclusions are supported by simulation analyses of most sets of conditions:

- Pyramiding two toxin genes within a single cultivar provides substantially more total years of plant protection than either sequential release of the two genes (i.e., releasing the second gene after the pest

DOI: 10.1081/E-EPCS 120010463

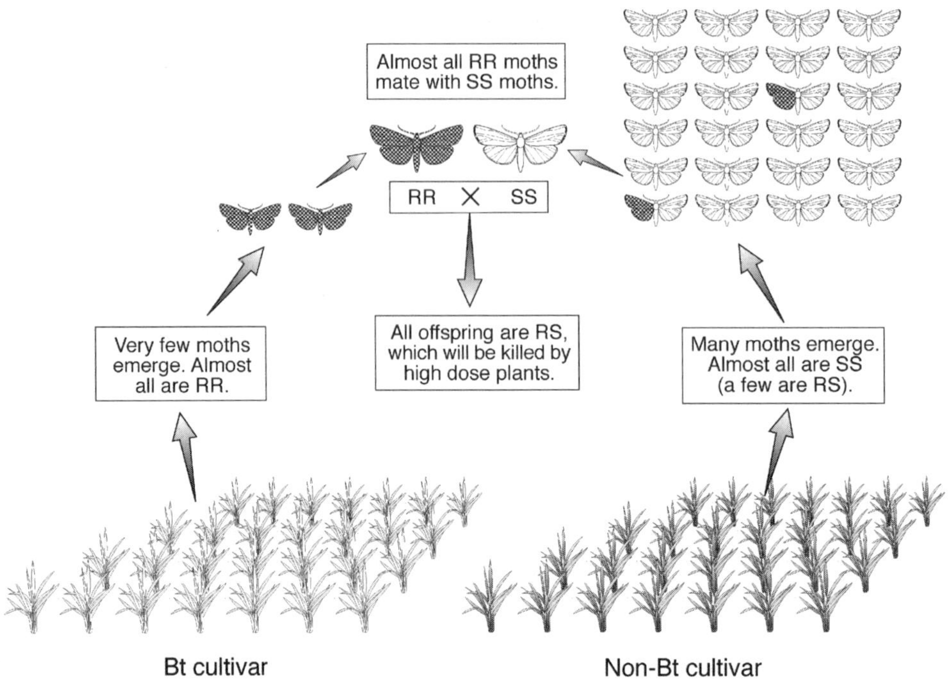

Fig. 1 How the refuge/high dose strategy works to delay the increase in highly resistant insects (RR) in a pest population. The strategy assumes that the dose in the Bt cultivar is sufficient to make the R allele functionally recessive, and that the initial frequency of R is low ($<10^{-3}$). (Modified from Ref. 7.)

population has adapted to the first) or simultaneous use of two cultivars, each with a single gene, as a mosaic.

- Maintaining refuges and using high-dose plants substantially delays the evolution of pest resistance in all three of the above options.

Toxins that are pyramided in a cultivar should be sufficiently different in their properties to make it unlikely that a single insect mutation would result in the loss of effectiveness of both toxins. Cry toxins must bind to particular brush border membrane proteins of the midgut epithelium as one step in their mode of action. The most common basis of insect resistance to Cry toxins are mutations that result in a change in the binding properties of these membrane proteins.[8] Competitive displacement assays of toxin binding to brush border membrane vesicles can be used to identify toxins that do not share binding sites and are therefore suitable for pyramiding. Such assays have generally shown that toxins highly similar in amino acid sequence, such as Cry1Ab and Cry1Ac, compete for the same binding sites, whereas less similar toxins do not.[8]

Broad-spectrum cross-resistance to Bt toxins highly dissimilar in amino acid sequence has been observed only in three lepidopteran species and appears to be caused by resistance mechanisms other than altered binding.[8] However, alleles conferring such cross-resistance are apparently rare, and there are few circumstances under which cross-resistance would result in a pyramid failing in fewer generations than sequential release of two toxins.[6] Cotton cultivars with two pyramided Cry toxins are under development in the United States and Australia.

MAINTAINING REFUGES

Resistance management plans in the United States, Canada, and Australia specify the size and locations of refuges that farmers must maintain. Developing countries will generally need to take a different approach to resistance management, due to smaller farm size and limited extension services. In the case of Bt cotton in China, where the principal pest (*H. armigera*) feeds on many wild and cultivated plant species, it has been argued that non-cotton host plants will provide adequate refuges.[9] Refuges comprised of alternative hosts are not an option for monophagous pests such as the yellow stem borer *Scirpophaga incertulas*, the most important lepidopterous pest of rice in tropical and subtropical Asia. In such cases, innovative approaches to maintain refuges will need to be developed. For example, governments can increase the probability that some farmers will grow non-Bt cultivars if they restrict the number of popular cultivars that are released in Bt form.[7] Farmers who grow non-Bt cultivars would in effect provide refuges for neighboring farms. The non-Bt farmers may in turn benefit from the Bt cultivars grown by their neighbors, because local pest populations may decline substantially in areas where large proportions of fields are planted to Bt cultivars.[10]

Although it is likely that some fields will be planted to non-Bt cultivars in areas where Bt cultivars become available and refuge requirements are not enforced, these fields may be relatively few in number. This highlights another important advantage of cultivars with two pyramided toxins: Such cultivars require smaller refuges. If both toxins are produced at a high dose, then insects able to survive on the cultivar (i.e., an insect homozygous for resistance alleles at two loci) will be extremely rare and thus fewer susceptible insects will be required to reduce the chances of resistant insects mating with each other.

CONCLUSION

The United States, Canada, and Australia have implemented stringent resistance-management strategies to delay the adaptation of pest populations to Bt crops, and these strategies appear to be working. There have been no reports of outbreaks of Bt-resistant pests in these countries since the release of Bt crops in 1996–97. Developing countries may not be able to enforce the use of refuges to the same degree, but can still promote the sustainable use of Bt crops by instituting careful standards for Bt cultivars and limiting the number of popular cultivars that are released in Bt form. For crops in which the target pests are monophagous, it is particularly useful to release Bt cultivars with two pyramided toxin genes—both expressed at a high dose—because such cultivars require smaller refuges.

ARTICLES OF FURTHER INTEREST

Biosafety Science: Overview of Plant Risk Issues, p. 164
Breeding Plants with Transgenes, p. 193
Genetically Engineered Crops with Resistance Against Insects, p. 506
Insect/Host Plant Resistance in Crops, p. 605
Rice, p. 1102
Transgenic Crops: Regulatory Standards and Procedures of Research and Commercialization, p. 1251

REFERENCES

1. Shelton, A.M.; Zhao, J.Z.; Roush, R.T. Economic, ecological, food safety, and social consequences of the deployment of Bt transgenic plants. Annu. Rev. Entomol. **2002**, *47*, 845–881.
2. United States Environmental Protection Agency. *Office of Pesticide Programs, Biopesticides and Pollution Prevention Division. Biopesticides Registration Action Document*; Bacillus thuringiensis (Bt) plant incorporated protectants: Washington, DC, October 15, 2001. http://www.epa.gov/pesticides/biopesticides/pips/bt_brad.htm (accessed June 2003).
3. Canadian Food Inspection Agency. *Plant Health and Production Division, Plant Biosafety Office*; Insect Resistance Management of Bt corn in Canada: Nepean, Ontario, Canada, October 13, 1999. http://www.inspection.gc.ca/english/plaveg/pbo/bt/btcormai1e.shtml (accessed June 2003).
4. Australian Cotton Growers Research Association. *Transgenic and Insect Management Committee. Resistance Management Plan for Ingard® Cotton*; Wee Waa, NSW: Australia, 2001–2002. http://www.cotton.pi.csiro.au/Assets/PDFFiles/IRMS/IRMS0102/Ingard01.pdf (accessed March 2002).
5. Gould, F. Sustainability of transgenic insecticidal cultivars: Integrating pest genetics and ecology. Annu. Rev. Entomol. **1998**, *43*, 701–726.
6. Roush, R.T. Two-toxin strategies for management of insecticidal transgenic crops: Can pyramiding succeed where pesticide mixtures have not? Phil. Trans. Royal Soc. Lond., B **1998**, *353* (1376), 1777–1786.
7. Cohen, M.B.; Gould, F.; Bentur, J.S. Bt rice: Practical steps to sustainable use. Int. Rice Res. Notes **2000**, *25* (2), 4–10.
8. Frutos, R.; Rang, C.; Royer, M. Managing insect resistance to plants producing *Bacillus thuringiensis* toxins. Crit. Rev. Biotechnol. **1999**, *19*, 227–276.
9. Pray, C.; Ma, D.; Huang, J.; Qiao, F. Impact of Bt cotton in China. World Dev. **2001**, *29* (5), 813–825.
10. Andow, D.A.; Hutchinson, W.D. Bt-Corn Resistance Management. In *Now or Never: Serious New Plans to Save a Natural Pest Control*; Mellon, M., Rissler, J., Eds.; Union of Concerned Scientists: Cambridge, MA, 1998; 19–66.

T

Transgenic Crop Plants in the Environment

R. James Cook
Washington State University, Pullman, Washington, U.S.A.

INTRODUCTION

In nearly two centuries of agricultural research, no single technology has offered more benefits and less risk to the environment than the transgenic technologies used to develop disease-, pest-, and herbicide-resistant or -tolerant crop plants. Historically, the negative effects of crop production systems on the environment have been the result of the tillage, pesticide applications, and fertilization used to grow crops, and not the direct effects of crop plants themselves on the environment. The use of a transgenic glyphosate-insensitive *EPSP* gene makes it possible to reduce or eliminate tillage; the use of *Bt* genes for insect resistance in cotton and corn, and coat-protein gene resistance to viruses in squash and papaya make it possible to reduce the use of pesticides for these crops. The use of just eight transgenic varieties of crops based on herbicide tolerance or insect or virus resistance reduced the total use of pesticides in the United States in 2001 by an estimated 20 million kilograms (46 million pounds). Moreover, crop plants genetically protected from pests and diseases are healthy, and therefore less likely than pest-damaged or diseased plants to leave nitrogen unused in the soil to leach into groundwater. In addition, they return more organic matter to the soil, which improves soil structure.

Scientists have rigorously focused on public concerns over the possible negative impact of transgenic crops on the environment, including negative effects on non-target organisms. Thus far, the potential negative environmental effects of so-call input traits (i.e., traits that moderate crop inputs, as for weed, insect, and disease control), remain unconfirmed, inconsequential, or disproved. On the other hand, more research and experience will be needed on possible environmental effects of so-called output traits (i.e., traits that affect end-use products—say, that can convert a food crop into an industrial or medicinal crop). Crops with output traits are not yet in commercial use. Meanwhile, the continued adoption of transgenic plants with their novel input traits is fully consistent with the goals of sustainable growth in agriculture.

AGRONOMICAL USEFULNESS OF TRANSGENES

Transgenic crop plants were grown on 50 million hectares (130 million acres) worldwide in 2001.[1] About two-thirds of the area planted to these crops is in the United States, followed by acreage in Argentina, Canada, China, South Africa, and Australia. More than 99% of this acreage involves the use of just two agronomically useful genes: 1) a glyphosate-insensitive *EPSP* gene used to make varieties of soybean, canola, and cotton insensitive to the herbicide Roundup^R; and 2) variations on a *Bt* gene for production of a protein lethal to certain insects, used to make corn resistant to the European corn borer and corn earworm, and to make cotton resistant to the cotton bollworm, pink bollworm, and tobacco budworm. These genes are from bacteria widely distributed in the environment. The remaining <1% of the area planted to transgenic crop plants grow varieties of squash in the eastern United States and papaya in Hawaii, each transgenic for resistance to one or more of their insect-vectored viruses using the coat-protein gene of that virus.[2]

The traits conferred by these agronomically useful transgenes are commonly referred to as input traits because their use results in a change in the inputs needed to grow the crop. Traits conferred by transgenes intended to improve end use or add value to the harvested product are termed output traits, but are not yet in commercial use.

The rapid adoption of agronomically useful genes is not without precedent. Dwarfing genes *Rht-B1b* and *Rht-D1* for reduced height in wheat and *Sd2* for semidwarf growth habit in rice (introduced by conventional plant breeding) swept into worldwide use in the 1960s. These genes changed the harvest index (ratio of grain/grain + straw) of the plants from about 0.3 to nearly 0.5, thereby sparking what became known as the green revolution. In each of these examples, it is important to understand that, whereas the same or very similar genes are used worldwide, the varieties of wheat, rice, corn, soybean, cotton, and canola with these genes commonly number in the hundreds or thousands. Different varieties are needed for different end uses, and to ensure adaptation of the crop to the local climate, environmental stresses, and diseases. Many if not most of the currently grown transgenic varieties were developed by conventional breeding, wherein the transgene of interest was first introduced by transformation into one line (usually a line easily transformed) and then transferred by standard backcrossing into another line with the desired quality traits, and agronomically adapted to the area where it would be grown. Varieties with transgenes are subjected to the same

Encyclopedia of Plant and Crop Science
DOI: 10.1081/E-EPCS 120010466

rigorous performance evaluations applied to conventional varieties before their release for commercial use.

ENVIRONMENTAL CONCERNS ABOUT TRANSGENIC CROPS

The development of transgenic crops has led to concerns over the possible effects of these crop plants or their new traits on non-target organisms in the environment. Assessments of risk have typically but incorrectly assumed that all factors in the cropping system remain the same except for the replacement of a conventionally bred variety with a transgenic variety. The adoption by farmers of any new variety is invariably accompanied by many modifications in crop management, including changes in date of planting, how or whether the soil is cultivated prior to planting, use of pesticides, and crop rotation. The use of a soybean variety with the glyphosate-insensitive *EPSP* gene results in the use of Roundup^R in place of a mixture of herbicides, and may also result in elimination of preplant and postplant cultivation of the soil. The use of a corn hybrid with the *Bt* gene rather than one without it may also eliminate the need for one or more insecticide treatments and will almost certainly result in greater mass of hardy corn stalks left as field residue, due to reduced European corn borer damage. Consideration of the environmental effects of a crop or gene—whether transgenic or not—must thus be in the context of the cropping system as a whole.

Larvae of the monarch butterfly feed on pollen grains that collect on the leaves of milkweed plants within and outside corn fields. Concern for these insects arose when a laboratory test showed that larvae have higher mortality when fed pollen from a corn hybrid with the *Bt* gene than when fed pollen from a corn hybrid lacking the *Bt* gene.[3] When this risk was examined under field conditions in the context of different cropping systems, management practices such as planting date, weed control, and use of insecticides had large effects on monarch populations, whereas *Bt*-corn pollen from current hybrids had only negligible or no effect on Monarch populations.[4]

Gene transfer because of outcrossing has been raised as an environmental concern specifically for transgenic crops. There are two kinds of concerns: 1) gene transfer by outcrossing from a transgenic crop in one field to a conventional variety of that same crop species in a neighboring field; and 2) gene transfer to a wild or weedy relative of the crop. The transfer by gene flow of input traits between varieties of the same crop raises mainly economic issues, such as when a transgene occurs at some frequency in a crop intended for certification as organic, or when crop plants produced from seed left in the field (volunteers) are resistant to an herbicide intended for use in that field. The transfer by gene flow of an output trait between varieties of the same crop could raise safety issues for human health if the trait expressed in a food crop such as corn or soybean is intended for industrial use only. Geographic isolation or obligatory self-pollination to prevent gene transfer will probably be a prerequisite for production of these kinds of crops.

Gene transfer between related species is a rare event, and fertile hybrids produced by such outcrossing are rarer still. Wheat hybridizes with jointed goat grass, its weedy relative common throughout the wheat-growing Great Plains and Pacific Northwest of the United States, but after growing dwarfed wheat varieties in these regions for 40 years, no evidence exists for natural transfer of a dwarfing gene to jointed goat grass. In the great majority of cases, transfer of a gene agronomically useful in a cropping system will provide little or no survival value to that same plant or its wild relatives in an unmanaged or natural ecosystem. Gene transfer is a moot issue for crops grown where there are no wild or weedy relatives (e.g., corn, soybeans, and cotton in the United States) but can conceivably be an issue for corn grown in Mexico within range of pollen flow to its native progenitor, teosinte, depending on the trait.

The adoption of dwarfed varieties of wheat and rice was driven by the quest for higher yields to feed the growing world population. In addition to greater use of nitrogen fertilizer, which continues to have environmental consequences,[4] the denser crop canopy and lush foliage typical of these cropping systems also led to increased pressures from insects and plant diseases. Where no useful genes for resistance to these pests have been available, or where evolution of the pest into new virulent strains has occurred faster than breeders could develop resistant genotypes of the crop, achieving the high-yield potential of these varieties continues to depend on pesticides.

ECONOMIC INDICATORS FOR TRANSGENIC CROPS

The adoption of crop varieties with the glyphosate-insensitive *EPSP* gene and a *Bt* gene is driven by the quest to increase efficiency through decreased cost of production, although in some cases yields have also increased.[5] These cost-cutting measures are occurring on two fronts: 1) in use of less or no tillage (no-till), except as required to fertilize and plant the crop; and 2) in use of less pesticide or replacement of several pesticide applications with a single, lower-cost, more effective pesticide. Although driven by economics, these changes are also good for the environment. The use by U.S. farmers in 2001 of eight transgenic crop varieties with a *Bt* gene for insect resistance, the *EPSP* gene for Roundup^R tolerance, or coat-protein gene for virus resistance reduced total pesticide use in the United States by an estimated 20 million kilograms.[5]

Of the 75.5 million acres of soybean grown in the United States in 2001, 70% were planted to varieties with the glyphosate-insensitive *EPSP* gene, and therefore relied on Roundup^R as the main or only herbicide for weed control.[6] Nearly 70% of the U.S. cotton crop in 2001 was planted to varieties with the *Bt* gene for resistance to insects. The trend worldwide because of one or two *Bt* genes is to reduce the need for insecticides on cotton from 12–15 applications per season to only two or three. Moreover, stacking two slightly different *Bt* genes in the same plant, as now done with cotton, can greatly delay if not prevent the pest from adapting to this mechanism of plant defense. About one-fourth of the U.S. corn crop is now produced with hybrids with the *Bt* gene. Due to a combination of new genes for pest resistance and integrated pest management, wherein pesticides are applied based on predictive models and decision guides, the global market for pesticides peaked in the 1990s at about $30 billion, is now at about $20 billion annually, and is decreasing at a rate of about 2–3% per year (Ganesh Kishore, personal communication).

CONCLUSION

The negative impact of agriculture on air and water quality results almost exclusively in dust and sediment in water from cultivated land. Sixty percent of the 5.5 million acres of soybean double-cropped with wheat in the United States are now seeded directly (no-till) into wheat stubble.[7] Of the 70 million acres of full-season soybean in 2001, 31% were direct-seeded into undisturbed soil and residue of the previous crop, usually corn. Of the 80 million acres of corn planted in the United States in 2001, 18% was direct-seeded. The European corn borer survives in corn stalks left on the soil surface, and is therefore a greater risk in direct-seed than in plow-based cropping systems. The European corn borer does not survive in Bt corn, however.

No-till cropping systems also result in other changes beneficial to the environment, including greater diversity of soil flora and fauna due to buildup of organic matter; greater water infiltration during intensive rains or snow melt; more or better habitat for birds and other wildlife; and greater sequestration of carbon dioxide, one of the major greenhouse gases identified with global warming.[8] These outcomes are made possible by continual improvement in crops and crop management. The current adoption of transgenic plants with their novel agronomic traits is an important part of this continuum and is fully consistent with the goals of sustainable growth in agriculture.

ARTICLE OF FURTHER INTEREST

Biosafety Programs for Genetically Engineered Plants: An Adaptive Approach, p. 160

REFERENCES

1. James, C. *Global Review of Commercialized Transgenic Crops: 2001*; Briefs, International Service for the Acquisition of Ag-Biotech Applications, 2001; Vol. 24, publications@isaaa.org.
2. Gonsalves, D. Control of papaya ringspot virus in papaya: A case study. Annu. Rev. Phytopathol. **1998**, *36*, 412–437.
3. Losey, J.E.; Raynor, L.S.; Carter, M.E. Transgenic pollen harms monarch larvae. Nature **1999**, *399*, 214.
4. Sears, M.K.; Hellmich, R.L.; Stanley-Horn, D.E.; Oberhauser, K.S.; Pleasants, J.M.; Mattila, H.R.; Siegfried, B.D.; Dively, G.P. Impact of Bt corn pollen on monarch butterfly populations: A risk assessment. Proc. Natl. Acad. Sci. **2001**, *98*, 11937–11942.
5. Gianessi, L.P.; Silvers, C.S.; Sanjula, S.; Carpenter, J.E. *Plant Biotechnology: Current and Potential Impact for Improving Pest Management in U.S. Agriculture—An Analysis of 40 Case Studies*; National Center for Food and Agriculture Policy: Washington, DC, 2002; 698 pp. (www.ncfap.org).
6. Tillman, D.; Fargione, J.; Wolff, B.; D'Antono, C.; Dobson, A.; Howarth, R.; Schindler, D.; Schlesinger, W.H.; Simbeloff, D.; Swackhamer, D. Forcasting agriculturally driven global environmental change. Science **2001**, *292*, 281–284.
7. Conservation Technology Information Center. Purdue University: West Layfayette, IN; http://www.ctic.purdue.edu/CTIC/CTIC.html.
8. Robertson, G.P.; Paul, E.A.; Harwood, R.R. Greenhouse gases in intensive agriculture: Contributions of individual gases to the radiative forcing of the atmosphere. Science **2000**, *289*, 1922–1925.

Transgenic Crops: Regulatory Standards and Procedures of Research and Commercialization

T

Qifa Zhang
Huazhong Agricultural University, Wuhan, P.R. China

INTRODUCTION

The estimated global area of transgenic or genetically modified (GM) crops reached 52.6 million hectares in 2001.[1] The rapid growth of the area planted to GM crops in both developed and developing countries strongly indicates that GM crops are welcome by growers and consumers alike. Utilization of transgenic crops will be a vital alternative to provide the world with adequate food and other agricultural products.

Although there have been widespread controversies regarding many safety aspects of transgenic plants as crops and food, there is an increasingly strong belief that transgenic plants can provide safe crops and food. This is because, in reality, transgenic technique does not pose any higher risk than traditional breeding methods to the safety of the crop products, and also because of the rigorous governmental regulations implemented by the countries that conduct transgenic research and grow GM crops. This article presents a brief overview of the regulatory standards and procedures for the environmental release and commercialization of GM crops.

REGULATORY STANDARDS AND PROCEDURES FOR TESTING AND CULTIVATING GM CROPS

The development of a transgenic crop consists of the following stages: laboratory research, confined field tests, environmental release, and commercialization. Governmental regulations in general address the following concerns in the process of commercialization of GM crops:

- Is the transgenic plant safe for the environment?
- Is the transgenic plant safe for agriculture?
- Is the transgenic plant product safe for use in foods, feeds, or other consumption?

Several strategies have been widely adopted in evaluating the safety of GM crops and foods. One of the strategies is based on the concept of substantial equivalence.[2] This approach acknowledges that the goal of the assessment is not to establish the absolute safety of GM crops and foods, but to evaluate their level of safety relative to their traditional counterparts, where such counterparts exist. Another important strategy commonly used in regulation is that the evaluation be done on a case-by-case basis.

For safeguarding the use of transgenic crops, science- and risk-based regulatory standards have now been established and implemented in many countries, and are being developed in many others. It should be noted that governmental regulation standards are also frequently bound to international agreements. Although there is substantial scientific commonality in the regulatory standards, the procedures followed in one country may be very different from those in another country. The cases of two countries, the United States and China, are given below as examples for demonstrating the common ground and differences in regulation of GM crops.

The United States was the first country to establish functional regulatory machinery for the biosafety of transgenic crops. A coordinated regulatory framework was set up in 1986 that assigned the responsibility to the Environmental Protection Agency (EPA), the U.S. Department of Agriculture (USDA), and the Food and Drug Administration (FDA), based on existing regulatory authorities of the U.S. government. A good starting point for information about regulation in the United States is the Web site ''United States Regulatory Oversight in Biotechnology Responsible Agencies—Overview'' (http://www.aphis.usda.gov/ppq/biotech/usregs.html). The basic premise for regulation of GM crops is that such crops shall not be fundamentally different from unmodified organisms or those produced by conventional methods. The key principle adopted by the regulatory framework is that it is the product, rather than the method of producing the product, that should be regulated.

The regulatory body of the USDA is the Animal and Plant Health Inspection Service (APHIS). This agency regulates the importing, transportation, and field testing of GM crops and determines the likelihood of a transgenic plant having negative agricultural or environmental effects (http://www.aphis.usda.gov/ppg/biotech). For common crops and traits, researchers need only notify the agency of their intention to transport or field test a transgenic plant. The researcher is responsible to ensure

Encyclopedia of Plant and Crop Science
DOI: 10.1081/E-EPCS 120010464

that the gene introduced meets certain technical criteria. For less common crops and for genes or traits that may pose greater risk, the researchers are required to file formal applications for permission to transport or plant the materials. Measures are required in field testing to prevent the spread of the transgene to the environment or into the food supply. Before commercializing a transgenic plant, the developer petitions APHIS for nonregulated status, which requires data on the introduced gene construct, plant biology, likely effects on the ecosystem, and field test reports, as well as data and information for any unfavorable effects. APHIS also has the authority to halt the sale of the GM crop if there is evidence that the GM crop is becoming a pest.

The FDA has authority to determine the safety of foods and food ingredients based on the concept of substantial equivalence, and relevant information can be found at the Web site (http://www.cfsan.fda.gov/~lrd/biotechm.html#label). The FDA adopts a voluntary consultation process with the GM crop developer, and reviews safety and nutritional data. If the introduced gene is from a known allergenic source, the GM food is required to be assessed for allergenicity. The FDA also has the authority to order the GM food's removal from the market, if there is evidence that it is unsafe.

The EPA regulates transgenic plants that contain plant-incorporated protectants, including plants engineered for insect or disease resistance that the agency refers to as plant pesticides (http:/www.epa.gov/pesticides/biopesticides/). This is implemented by granting an experimental use permit for testing, plant propagation registration, and full commercial registration, based on extensive reviews of data for the plant pesticide including its biochemical characteristics, toxicity, and environmental effects. The agency may require a resistance management plan in order to prevent or slow the development of resistance in the target pest. The EPA regulates herbicides, but not herbicide-tolerant plants. In the case of engineering a plant for herbicide tolerance, herbicides must be registered for a new use.

In addition to regulation at the federal level, many states in the United States also apply regulations on GM crops that require additional review and approval at the state level. Moreover, most research institutions have biosafety committees that monitor potentially hazardous transgenic research and ensure compliance with biosafety procedures.

In China, the biosafety of GM crops is regulated by the Ministry of Agriculture (MOA) (http://www.agri.gov.cn/ztzl/2001/0820/0820.htm). The MOA regulates the research, development, field tests, environmental release, and commercialization of GM crops. The MOA also regulates the import of GM crops and their products.

The process from laboratory research to commercialization of GM crops is divided into four stages in the Chinese regulation framework: laboratory research, confined field tests, environmental release, and product demonstration and commercialization. An MOA biosafety committee is responsible for reviewing the applications filed by researchers. In this system, notification of the MOA is required for transgenic research that may pose higher risk. For conducting field tests of a transgenic plant, a formal application must be filed with the MOA, which will be reviewed by the biosafety committee; a permit will subsequently be issued. In filing this application, the researcher is required to provide data about the gene construct, biology of the plant, test site, measures of confinement, and data to be collected. After a two-year field test, the transgenic plant can advance to environmental release when another application is filed with the MOA, submitting all data gained in the field test. Experimentation of environmental release in general will be conducted in another two consecutive years, after which another application should be filed with the MOA to advance the transgenic plant to product demonstration in large scale. At the end of the product demonstration, which usually requires two years to complete, the researcher can apply for a biosafety certificate for commercialization, which may require extensive data and comprehensive reviews. The transgenic plants can be used as parents for breeding purposes only when the biosafety certificate is issued, and the progenies from crosses with the transgenic parent must undergo the product demonstration procedure again and reapply for the biosafety certificate.

Information can also be found at Web sites for regulatory standards and procedures of the European Community (http://www.biosafety.be/Menu/BiosEur.html) and Japan (http://www.s.affrc.go.jp/docs/sentan/).

LABELING OF GM FOODS

Labeling of GM foods is a widely debated issue in many parts of the world. Although different countries' regulatory standards (described in previous sections) address all concerns of consumers and thus provide strong assurance for food safety, there are still diverse opinions about the necessity of labeling GM foods. Again, countries are divided over requirements and procedures for labeling the GM foods.

As described in the previous section, food safety in the United States is regulated by the FDA. The FDA requires that all labeling be truthful, informative and not misleading, and identifies no characteristics of the GM food would justify labeling it as a special class. The FDA requires labeling for GM foods in the same way as for other foods, including information about the composition, nutrition, and allergenicity concerns, but does not require special labeling of GM foods. Additional voluntary labeling is allowed, provided that it is truthful, informative, and not misleading. The FDA has recently developed industry guidance on voluntary labeling.

In Japan, implementation of new regulations on GM food labeling was initiated in April 2001 (http://www.maff.go.jp/soshiki/syokuhin/hinshitu/organic/eng_yuki_top.htm). According to the authority, labeling is not for safety concerns but for consumers' right to choose or right to know. The labeling is implemented according to a positive list system, in which a committee is formed in the Ministry of Agriculture, Forestry, and Fishery (MAFF) to review the positive list of foods that should be labeled. In 2001, 15 soybean products and nine corn products were listed. The regulation requires that the food should be labeled if the DNA/protein of the transgene can be detected, the raw materials are among the top three constituents, and the GM materials are 5% by weight.

In China, labeling of transgenic organisms was enforced in March 2002 (http://www.agri.gov.cn/xxfb/2002/0107/dt0110-2.doc). Labeling is administered by the MOA, and regulation covers the transgenic organisms included in a list that is updated from time to time. The current list includes soybean, corn, rapeseed, cotton, and tomato. It regulates not only GM foods, but also seeds and other products from GMOs that are sold in the marketplace.

In Europe, a series of regulations applies concerning traceability and labeling of GMOs and of GMO food and feed products. (http://www.biosafety.be/Menu/BiosEur4.html). The regulation requires labeling of products consisting of or containing GMOs, foods and food ingredients (including food additives and flavorings produced from GMOs), feed materials, and compound feeding stuffs and feed additives produced from GMOs placed on the market in accordance with European Community legislation. Traceability requires that information about GMO products placed on the market be transmitted from one operator to the other in transaction, and that operators have in place systems and procedures to allow the identification of persons involved in the transactions for a period of five years.

CONCLUSION

Functional regulatory machinery has been established in many countries in the last decade. However, regulatory standards and procedures are still evolving, even in countries with a relatively long history of GM crop cultivation. With rapidly accumulating large-scale adoption and public recognition of the advantages of GM crops, it is believed that goals, standards, and procedures of regulation will gradually become globally harmonized and will promote the utilization and exploitation of the full benefits of GM crops, in order to better meet the demands of the ever-increasing world population.

REFERENCES

1. James, C. Global Review of Commercialized Transgenic Crops: 2001. In *International Service for the Acquisition of Agri-Biotech Application No. 24-2001*; 2001.
2. Food and Agriculture Organization of the United Nations and World Health Organization Safety Aspects of Genetically Modified Foods of Plant Origin. In *Report of a Joint FAO/WHO Expert Consultation on Foods Derived from Biotechnology*; 2000.

Trichomes

Stephen O. Duke
USDA, University, Mississippi, U.S.A.

INTRODUCTION

Most plants are covered with an assortment of structures, varying in morphology from hairlike to globular. Such structures, usually originating from the plant epidermis and projecting outward from the plant, are termed trichomes. They can be found on any plant part, and several different types of trichomes are sometimes found on the same plant surface. At the microscopic level their varied morphologies are truly amazing.

Trichomes may be classified by various criteria; however, they are generally either glandular or nonglandular with regard to function, depending on whether they produce a secretory product. The cotton fiber represents one extreme of a nonglandular trichome; the large, multicellular glandular trichomes found on many species are examples of the other extreme. The economic value of both fibrous trichomes and the contents of glandular trichomes is large. Furthermore, the negative impact of trichomes or trichome products that cause health problems for animals (including humans) is significant.

TRICHOME ANATOMY AND DEVELOPMENT

Plant trichomes have been the subject of at least three entire volumes.[1–3] Several reviews on trichome anatomy and development are available (e.g., Ref. 4). The term trichome has generally been meant to include only epidermal appendages originating from the plant epidermis. Root hairs meet this definition and are beginning to be considered trichomes by some molecular biologists and plant anatomists.[4] Trichome like structures such as the glandular protrusions of several species of *Hypericum* may not originate from epidermal cells, but are considered trichomes by some.

Both nonglandular and glandular trichomes range from unicellular to multicellular (uniseriate, biseriate, or multiseriate) (Fig. 1). Nonglandular trichomes can have many forms, depending on the species, and more than one type can be found on a single species. They can be filamentous, stellate, scalelike, branched, tufted, and combinations of these morphologies. Some are only warty protrusions from epidermal cells. At maturity, the cells of most nonglandular trichomes are dead, with the cell wall and cuticular thickness and composition providing textures ranging from stiff to soft. The cotton fiber is one of the more extreme morphologies of a unicellular, nonglandular trichome. These fibers arise from epidermal cells of ovules, reaching lengths of several centimeters.

Glandular trichomes can be of several types, depending on their morphology and on what they secrete, accumulate, or absorb. The typical multicellular peltate glandular trichome has a specialized secretory cell or cells that secrete products into the space between the cuticle and cell wall of the head cell(s) (Fig. 2).[5] Very early in development of a plant organ, an epidermal cell begins differentiating and dividing into such a trichome.[6] The cells of this structure are initially very similar, but differentiate into stalk and head cells that apparently have different functions. Nevertheless, at maturity each cell type is normally filled with cytoplasm containing very small vacuoles, unlike most plant cell types with a large central vacuole. The apical cells of the head secrete material into the space between the cell wall and cuticle. The cuticle then swells to form a balloonlike structure filled with secretory products over the terminal cells. There is good evidence that many of the compounds secreted by this type of trichome are synthesized only by trichome cells.[7,8] As glandular trichomes age, the cuticle may break, spreading the nonvolatile contents over the plant epidermis. In some species, a new cuticle can form to contain newly synthesized glandular products.

Other specialized types of glandular trichomes are salt glands and nectaries. Salt glands can be of several forms, including bladderlike cells in which salts are compartmentalized in a central vacuole, and epidermal cells that excrete salts into subcuticular spaces or onto the plant surface. Nectaries are glandular trichomes that excrete nectar. They are often associated with flowers, but can also be extrafloral.

The molecular genetics of trichome formation and development is beginning to be understood. More than 20 genes are involved in trichome formation in *Arabidopsis*, and much is known of the genetics of cotton fiber-specific promoters and the genes associated with cotton quality. Some of the genes involved in root hair development are the same or very similar to those involved in trichome development. Genes involved in biosynthesis

Encyclopedia of Plant and Crop Science
DOI: 10.1081/E-EPCS 120010622

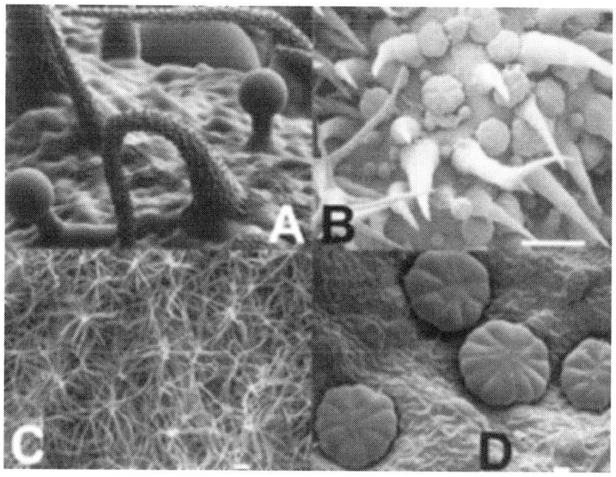

Fig. 1 Nonglandular and glandular trichomes. A. Globular glanded and pointed nonglandular trichomes of *Nepeta racemosa*. B, C, and D. Glandular: Small round and large multicellular (B and D). Nonglandular: Spike shaped (B) and stellate (C) trichomes of *Callicarpa americana*. Bars represent 100 μM. (*C. americana* micrographs from R.N. Paul.) (Photo A from Ref. 2.)

and secretions of glandular trichome-specific compounds are also being identified.[9] Because glandular trichomes appear to be the only location for biosynthesis of many secondary compounds in plants, glandular trichome development and the biosynthetic pathways for these compounds appear to be genetically connected. Knowledge of the molecular genetics and biology of trichomes should eventually provide economic dividends by improving cotton quality and yields, as well as providing the basis for increasing the production of valuable glandular trichome products.

ECONOMIC ASPECTS OF TRICHOME PRODUCTS

Perhaps the most economically important trichome is the cotton fiber. This is the only trichome fiber crop. The contents of glandular trichomes are also often valuable.[4] Table 1 provides a partial list of valuable trichome products. Glandular trichomes can produce many different types of chemical compounds, including terpenoids, alkaloids, and flavonoids. Essential oils (mostly terpenoids) of many plant species are associated primarily with glandular trichomes. Essential oils are important as flavorings and fragrances. Some specific trichome products, such as the antimalarial drug artemisinin, have high value as pharmaceuticals, and others contain the psychoactive compounds of tobacco and marijuana.

Trichomes of a few species contain compounds that are problematic for people and livestock. For example, certain nettles (e.g., *Cnidoscololus texanua* and *Urtica urens*) bear trichomes that cause a painful sting when touched, due to mild neurotoxins (e.g., leukotrienes). Contents of other glandular trichomes are poisonous if ingested in sufficient quantity. Some people are highly allergic to certain glandular trichome compounds such as the sesquiterpene lactones of some *Parthenium* species. These people

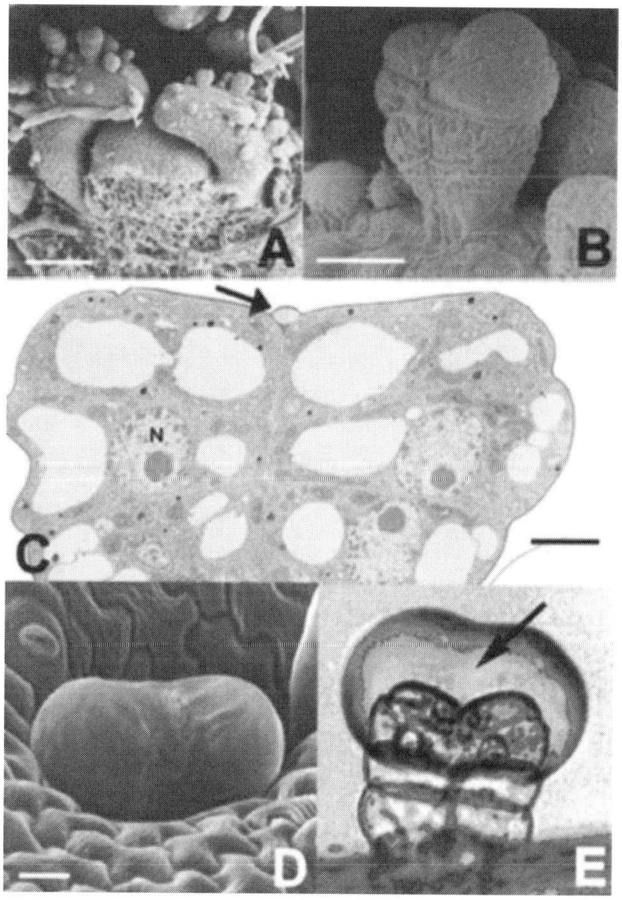

Fig. 2 Development of the peltate, glandular trichome of *Artemisia annua* L. A. Apical meristem with stalk cells being formed; B. Individual, immature trichome; C. Transmission electron micrograph of stalk cell with cuticle beginning to be pushed away from cell wall by excretion of products into the space between the cell wall and cuticle (arrow); D. Scanning electron micrograph of mature trichome with the subcuticular space engorged with secretory product; E. Light micrograph showing the stalk cell and the subcuticular space of the gland (arrow). Scale bars: A=50 μM; B, D=10 μM; C=5 μM. (Photos A–D from Ref. 5. Photo E from Ref. 8.)

Table 1 Contents of glandular trichomes of selected species with economic value

Plant species	Trichome product
Artemisia annua	Antimalarial drug—artemisinin
Basil (*Ocinum* spp.)	Essential oil flavor components
Cannabis sativum	Cannabinoids
Hops (*Humulus lupulus*)	Flavor components
Hypericum peforatum	Hypericin
Marjoram and oregano (*Origanum* spp.)	Essential oil flavor components
Mentha piperita	Mint oil
Nicotiana tabacum	Nicotine and flavor components of tobacco
Cotton (*Gossypium hirsutum*)	Gossypol (a male contraceptive)

develop severe contact dermatitis by direct contact with the plant or by contact with windborne plant material.

TRICHOME FUNCTION

Plant species have evolved trichomes in response to both biotic and abiotic selection pressure. Glandular trichomes represent predominantly chemical strategies for dealing with this pressure, whereas nonglandular trichomes offer physical responses to environmental challenges.

The first contact that plant pathogens, insects, and herbivores have with plants is at the plant epidermis. The chemical contents of glandular trichomes are often highly biologically active against many of the biota from which the plant needs protection. The gland content or its exudate can act as a poison or repellent. It can also attract pollinators or other beneficial insects. Table 2 provides examples of the biological activity of some glandular trichome components. Note that a given compound can have multiple effects, depending on its concentration and the species affected. Glandular trichomes provide a means of concentrating these compounds in the proper place to come in contact with target organisms, while partitioning them from other plant tissues. In this manner, glandular trichomes can protect the plant from autotoxicity. They can also protect it from synthetic phytotoxins. For example, cotton is partially resistant to certain lipophilic herbicides used with this crop, due to partitioning of foliar applications of the herbicide into the lipophilic subcuticular spaces of glandular trichomes. Similarly, lipophilic insecticides can partition into glandular trichomes after foliar application, where they are protected from volatilization and degradation, resulting in increased efficacy.

At sufficient densities, nonglandular trichomes can shield a plant from direct sunlight in severe environments, such as alpine ecosystems. Certain nonglandular trichomes can also reduce water loss by shielding the plant epidermis from wind. However, smaller numbers of trichomes can make the boundary layer of photosynthesizing organs more turbulent as air passes over the epidermis. This can increase transpiration while facilitating photosynthetic gas exchange. The silicon content of some trichomes can be so high as to make the plant unpalatable to insects and other herbivores.

ARTICLES OF FURTHER INTEREST

Aromatic Plants for the Flavor and Fragrance Industry, p. 58
Genetic Resources of Medicinal and Aromatic Plants from Brazil, p. 502
Herbs, Spices, and Condiments, p. 559
Isoprenoid Metabolism, p. 625
Leaf Cuticle, p. 635

Table 2 Glandular trichome products of selected species and their effects on biota

Plant species	Trichome product	Biological effects
Artemisia spp.	Camphene	Antifungal and insecticidal
Artemisia spp.	Camphor	Antimicrobial
Mentha piperita	Menthol	Antifungal and insecticidal repellent
Parthenium hysterophorus	Parthenin	Antifungal
Artemisia spp.	α-pinene	Antifungal, insect repellent and attractant, insecticide
Artemisia spp.	Cineoles	Phytotoxic
Nepeta spp.	Nepetalactone	Insect repellent
Centauria maculosa	Cnicin	Herbivore deterrent
Abutilon spp.	Nectar	Insect attractant

Phytochemical Diversity of Secondary Metabolites, p. 915
Secondary Metabolites as Phytomedicines, p. 1120

REFERENCES

1. Uphof, J.C.T. *Plant Hairs, Encyclopedia of Plant Anatomy, IV, Vol. 5*; Gebrüder Borntraeger: Berlin, 1962; 1–206.
2. *Plant Trichomes—Advances in Botanical Research, Vol. 31*; Hallahan, D.L., Gray, J.C., Eds.; Academic Press: San Diego, 2000; 1–311.
3. *Biology and Chemistry of Plant Trichomes*; Rodriguez, E., Healey, P.L., Mehta, I., Eds.; Plenum: New York, 1984; 1–241.
4. Werker, E. Trichome Diversity and Development. In *Advances in Botanical Research*; Hallahan, D.L., Gray, J.C., Eds.; Academic Press: San Diego, 2000; Vol. 31, 1–135.
5. Duke, S.O.; Paul, R.N. Development and fine structure of the glandular trichomes of *Artemesia annua* L. Int. J. Plant Sci. **1993**, *154*, 107–118.
6. Behnke, H.-D. Plant Trichomes—Structure and Ultrastructure: General Terminology, Taxonomic Applications, and Aspects of Trichome-Bacteria Interaction in Leaf Tips of *Dioscorea*. In *Biology and Chemistry of Plant Trichomes*; Rodriguez, E., Healey, P.L., Mehta, Eds.; Plenum: New York, 1984; 1–21.
7. Duke, S.O.; Canel, C.; Rimando, A.M.; Tellez, M.R.; Duke, M.V.; Paul, R.N. Current and Potential Exploitation of Plant Glandular Trichome Productivity. In *Advances in Botanical Research*; Hallahan, D.L., Gray, J.C., Eds.; Academic Press: San Diego, 2000; Vol. 31, 121–151.
8. Duke, M.V.; Paul, R.N.; ElSohly, H.K.; Sturtz, G.; Duke, S.O. Localization of artemisinen and artemesitene in foliar tissues of glanded and glandless biotypes of *Artemesia annua*. Int. J. Plant Sci. **1993**, *155*, 365–373.
9. Lange, B.M.; Wildung, M.R.; Stauber, E.J.; Sancez, C.; Pouchnik, D.; Croteau, R. Probing essential oil biosynthesis and secretion by functional evaluation of expressed sequence tags from mint glandular trichomes. Proc. Natl. Acad. Sci. U. S. A. **2000**, *97*, 2934–2939.

UV Radiation Effects on Phyllosphere Microbes

Thusitha S. Gunasekera
Michigan State University, East Lansing, Michigan, U.S.A.

INTRODUCTION

The leaf surfaces (phylloplane) of plants are characteristically colonized by epiphytic communities of bacteria, yeasts, and filamentous fungi. Among the different factors that affect microbial survival and growth in the phyllosphere, solar UV radiation (UVR) plays a central role. Colonization of plant surfaces by microorganisms therefore clearly depends on their ability to tolerate UVR stresses or to avoid such stresses by colonizing protected sites. More recently, UVR has received considerable attention because solar UVB wavelengths (290–320 nm) are increasing as a consequence of stratospheric ozone depletion. Current research has shown that variation in UVB radiation significantly affects many organisms and ecosystems including phyllosphere microorganisms.

CELLULAR DAMAGE AND REPAIR MECHANISMS

Only UVA (320–400 nm) and longer wavelengths of UVB (>290 nm) radiation penetrate to the terrestrial environment. UVR is absorbed by vital cell molecules such as nucleic acids and proteins,[1] resulting in various forms of damage including the induction of pyrimidine dimers.[1] UVR-induced lesions distort and deform the DNA helix and interrupt transcription, translation, and replication. However, photodamage by UVR is a wavelength-dependent process. UVA radiation causes indirect damage to DNA, proteins, and lipids through reactive oxygen intermediates. UVB radiation causes both direct and indirect damage because of strong absorption of shorter wavelengths of UVB by DNA. Although the absorption spectra of proteins vary greatly because of variations in content of aromatic amino acids, some proteins are directly vulnerable to UV damage.

Both prokaryotic and eukaryotic organisms have multiple DNA repair pathways. One of the most widely distributed repair mechanisms is photoreactivation (PR), in which pyrimidine dimers are photochemically removed through the mediation of an enzyme, DNA photolyase. Apart from PR, another important and conserved pathway for repair of UV damage to DNA is nucleotide excision repair (NER), which is found in eubacteria, archea, and eukaryotes. In this pathway, UvrABC endonuclease removes a short oligonucleotide encompassing the damaged bases, and the gap is resealed with polymerase and ligase. Additional pathways for UV resistance, such as post-replication, recombinational repair processes, and *rec*A- and *lex*A-mediated inducible SOS stress responses also exist in both prokaryotic and eukaryotic organisms. These processes can quickly repair DNA damage, but they are error prone and so produce mutations.[1]

RESPONSE OF PHYLLOSPHERE MICROORGANISMS TO UV RADIATION

Available information suggests that large inter- and intra-specific variations in UVR sensitivity exist within phyllosphere microbial communities,[2–6] with the UVB component of solar UVR being primarily effective (Fig. 1) in regulating microbial populations. Variation occurs on a range of scales. There is evidence that isolates from different geographic regions with contrasting UVB climate vary in UVB tolerance.[2,3] Large variation was also noticed within the same site or locality. Pathovars of *Pseudomonas syringae* and strains within the pathovar *syringae* have shown large variability to tolerance of UVR.[6] UVB radiation might, therefore, be an important factor influencing the differential survival of phyllosphere organisms and hence the species composition of the phyllosphere community. The effects of UVB radiation on the phyllosphere community colonizing over a commercial tea crop showed that the leaf colonization of *Corynebacterium aquaticum* and *Xanthomonas* spp was influenced by solar radiation and that during certain periods of the year colonization was significantly higher under a UVB-depleted environment than under the natural environment.[2] These results were interpreted as seasonal fluctuation of microorganisms on leaves influenced by UVR. UVR-sensitive species or strains, however, can succeed by colonization on the phyllosphere during the less sunny period of the year.

Existence of marked variation in response suggests that mechanisms have evolved that moderate these potentially damaging effects. UVB levels on the plant surfaces are not uniform; mutual shading of the leaves, depressions, veins, and bases of trichomes provide some

Encyclopedia of Plant and Crop Science
DOI: 10.1081/E-EPCS 120010623

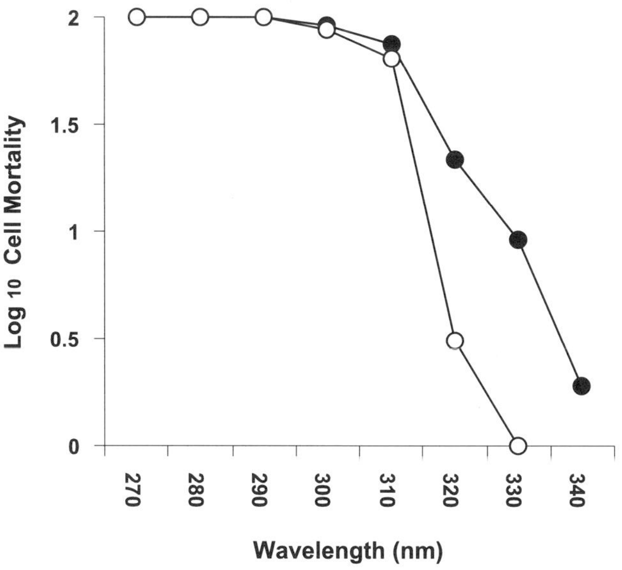

Fig. 1 Effects of different wavebands of UV radiation (10 nm action spectrum) on the survival of colonies of (○) *Sporidiobolus* sp. and (●) *Bullera alba*. Survival is related to that in dark control. Bars indicate standard error of mean of five replicates. (For more detail, see Ref. 3.)

protection against the UVB radiation. Colonizing in protected sites would be useful to avoid such high UV fluxes. The avoidance of UVR by colonization of the abaxial surfaces is thus an important strategy adopted by phyllosphere microorganisms.[5]

Phyllosphere microbes may reduce damage by reducing penetration to target molecules, especially nucleic acids in the nucleus. The most important component of UV protection is the production of UV-absorbing pigments. The putative role of pigments in protecting phyllosphere microorganisms against UVR is supported by the observation that significant numbers of microorganisms isolated from the phyllosphere are pigmented.[5] Phyllosphere yeasts such as *Sporobolomyces* spp., *Rhodotorula* spp., *Sporidiobolus* spp., *Cryptococcus* spp., and the common phyllosphere bacterium *Erwinia herbicola* are capable of producing carotenoids, which protect cells against UVA wavelength–induced reactive oxygen intermediates.

The third strategy for minimizing UVB effects is via the range of repair pathways available to correct primary UV damage. The importance of these pathways for epiphytic growth has been demonstrated for some phyllosphere microorganisms. The common phyllosphere bacterium *P. syringae* contains native plasmids that were shown to confer elevated survival (20- to 30-fold) to UVR.[7] Fig. 2 shows the effect on UVB survival of addition of the native *rulAB*-containing plasmid pPSR1 to *P. syringae* pv. *syringae* FF5 or of an insertional mutation within *rulB* in strain 8B48. The *rulAB* system is also able to promote UV-inducible mutagenesis, suggesting the functional relevance of *rulAB* and its important role in overcoming UVR stress in the phyllosphere.[8] Significantly reduced susceptibility to UV damage in the presence of white light is evidence for PR in the tested phyllosphere microorganisms[3] and in the common phyllosphere bacterium *P. syringae*.[9]

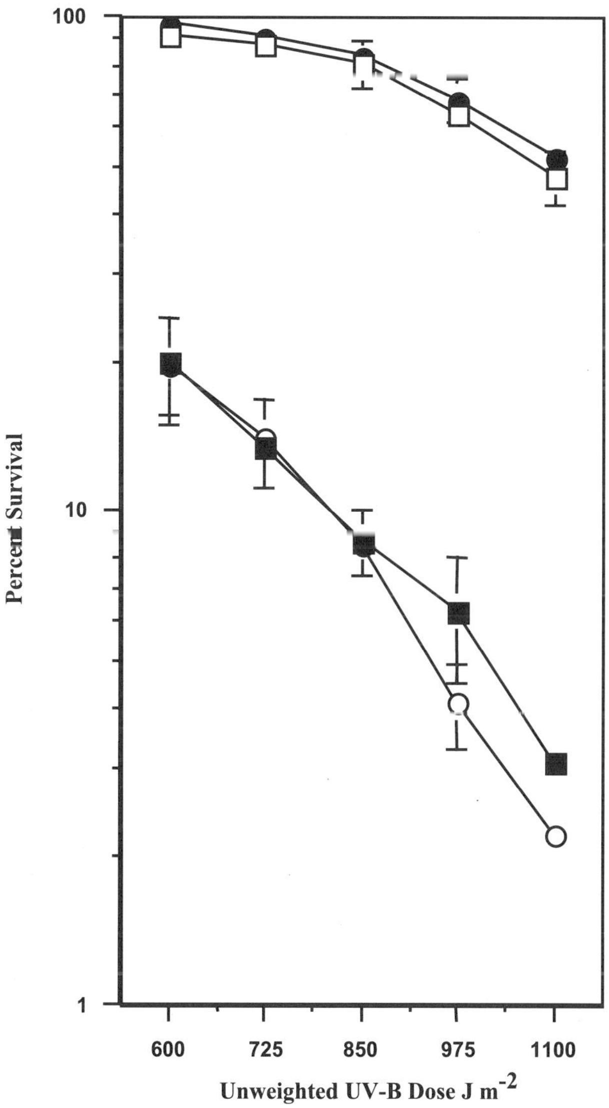

Fig. 2 Response of *Pseudomonas syringae* pv. *syringae* FF5 (○), FF5/pPSR1 (□, *rulAB*+), 8B48 (●, *rulAB*+), and 8B48A (■, *rulB*Km) to UV-B radiation. Cells suspended in saline solution were irradiated as previously described,[6] and survival was calculated as related to a non-irradiated control. Each point represents the mean (± the standard error of the mean) from three replicate experiments.

UVR can also alter the physical and chemical properties of leaf surfaces, with potential indirect effects on colonization of microorganisms on the phyllosphere. These indirect effects are largely unknown. To date, it is not known how UVR affects interactions between phyllosphere microorganisms with specific plant pathogens, but it is likely that such interactions will be complex, depending on host-mediated changes and direct effects on the microbes involved. However, increased understanding of UVR effects on interactions between phyllosphere microbes and pathogens offers the greatest opportunity for exploiting increased understanding of UVR effects on the phyllosphere. Success in controlling plant pathogens, insects, or the ice-nucleation active bacteria depends largely on the establishment of biocontrol agents on plant surfaces. Successful biological control agents should survive and persist in the target sites of the phyllosphere for a long time. UV-resistant strains can possibly be used to increase the efficacy on the phyllosphere. On the other hand, some bacterial pathogens' epiphytic growth is critical for dissemination and their survival away from the host. Therefore, UVR responsible for regulating foliar diseases and more importantly UVR can be used to control UV-sensitive pathogens on plant surfaces.[10]

CONCLUSION

Available information suggests that phyllosphere organisms display inter- and intra-specific variation in response to UVR and increased UVB irradiation above ambient levels, resulting in alterations of the relative abundance and species composition of microorganisms in the phyllosphere. Increase of UVB radiation therefore may change the microbial balance and possible microbial interactions in the phyllosphere or important host-pathogen interactions. Existing information suggests that the ecological succession of phyllosphere microorganisms is in part driven by UVB radiation. Ecological studies also showed that UV tolerance is a prevalent phenotype among phyllosphere microorganisms and that DNA repair or UV-tolerant mechanisms are vital to the phyllosphere microorganisms for survival in this UVB-rich habitat.

ARTICLES OF FURTHER INTEREST

Bacterial Survival and Dissemination in Natural Environments, p. 108
Bacterial Survival and Dissemination in Seeds and Planting Material, p. 111
Bacterial Survival Strategies, p. 115
Biological Control in the Phyllosphere, p. 130
Leaf Cuticle, p. 635
Leaf Structure, p. 638
Leaf Surface Sugars, p. 642
Leaves and Canopies: Physical Environment, p. 646
Leaves and the Effects of Elevated Carbon Dioxide Levels, p. 648
Plant Responses to Stress: Ultraviolet-B Light, p. 1019
UV Radiation Penetration in Plant Canopies, p. 1261

REFERENCES

1. Jagger, J. *Solar UV Actions on Living Cells*; Prager Publishers: New York, USA, 1985.
2. Gunasekera, T.S. *Effects of UV-B (290–320 nm) Radiation on Microorganisms on the Leaf Surface*; Lancaster University, UK, 1996.
3. Gunasekera, T.S.; Paul, N.D.; Ayres, P.G. Responses of phylloplane yeasts to UV-B (290–320) radiation: Inter specific differences in sensitivity. Mycol. Res. **1997**, *101*, 779–785.
4. Jacobs, J.L.; Sundin, G.W. Effect of solar radiation on a phyllosphere bacterial community. Appl. Environ. Microbiol. **2001**, *67*, 5488–5496.
5. Sundin, G.W.; Jacobs, J.L. Ultraviolet radiation (UVR) sensitivity analysis and UVR survival strategies of a bacterial community from the phyllosphere of field-grown peanut (*Arachis hypogeae* L.). Microb. Ecol. **1999**, *38*, 27–38.
6. Sundin, G.W.; Murillo, J. Functional analysis of the *Pseudomonas syringae rulAB* determinant in tolerance to ultraviolet B (290–320 nm) radiation and distribution of *rulAB* among *P. syringae* pathovars. Environ. Microbiol. **1999**, *1*, 75–87.
7. Sundin, G.W.; Kidambi, S.P.; Ullrich, M.; Bender, C.L. Resistance to ultraviolet light in *Pseudomonas syringae*: Sequence and functional analysis of the plasmid-encoded *rulAB* genes. Gene **1996**, *177*, 77–81.
8. Kim, J.J.; Sundin, G.W. Regulation of the *rulAB* mutagenic DNA repair operon of *Pseudomonas syringae* by UV-B (290 to 320 nanometers) radiation and analysis of *rulAB*-mediated mutability in vitro and in planta. J. Bacteriol. **2000**, *182*, 6137–6144.
9. Kim, J.J.; Sundin, G.W. Construction and analysis of photolyase mutants of *Pseudomonas aeruginosa* and *Pseudomonas syringae*: Contribution of photoreactivation, nucleotide excision repair, and mutagenic DNA repair to cell survival and mutability following exposure to UV-B radiation. Appl. Environ. Microbiol. **2001**, *67*, 1405–1411.
10. Gunasekera, T.S.; Paul, N.D.; Ayres, P.G. Effect of UV-B radiation on *Exobasidium vexans* and blister blight disease of tea. Plant Pathol. **1997**, *46*, 179–185.

UV Radiation Penetration in Plant Canopies

Richard H. Grant
Purdue University, West Lafayette, Indiana, U.S.A.

U

INTRODUCTION

Sunlight—or solar radiation—is the ultimate source of all the energy that a plant receives throughout its life. Most would agree that the wavelengths of the solar radiation spectrum utilized by chlorophyll are the most important part of the sunlight reaching plants. There are, however, several other important wavebands of solar radiation that affect plant growth and development—among them, radiation in the ultraviolet (UV) waveband. Solar radiation in the UV band can affect the viability of spores and pollen, the mutation of plant cells, and the chemical composition of plant biomass. Consequently, changes in the intensity of solar radiation in the UV band may have significant impacts on agriculture. We are now experiencing changes in atmospheric ozone that change the intensity of the UV radiation received by plants. How intense that radiation is at the top of the plant canopy, how that radiation is distributed in plant canopies, and the duration of the intense radiation will define how UV radiation will affect crop agriculture.

WHY WORRY ABOUT UV RADIATION IN PLANT CANOPIES?

Solar radiation is largely in wavelengths that the eye can see, but ultraviolet (UV) radiation is in wavelengths shorter than the eye can see. Solar UV radiation reaching the earth's surface is separated into two wavebands: a band from 320 to 400 nm called UV-A (ultraviolet-A), and the band from 280 to 320 nm called UV-B (ultraviolet-B), although some scientific organizations define 315 nm as the division between UV-A and UV-B. Because ozone is the primary absorber of UV-B in the atmosphere, changes in the ozone change the intensity of the UV-B radiation at the earth's surface. Increased UV-B has the potential to decrease the productivity of agricultural crops and viability of spores and pollen in canopies. We will assume here that the plant surface of interest in the canopy could be any phytoelement at any orientation, including stalk, leaf, grain head, silk, or stamen.

ABOVE-CANOPY RADIATION ENVIRONMENT

To understand the penetration of UV radiation into the canopy, we must first consider the nature of UV radiation above the canopy. The atmosphere scatters solar radiation inversely proportional to the wavelength, resulting in more of the direct beam of solar radiation being scattered into the sky hemisphere in the UV than in longer, visible wavelengths. The fraction scattered into the sky, termed diffuse radiation, varies according to atmospheric conditions and solar zenith angle. On a clear day the sky diffuse fraction increases with increasing solar zenith angle, resulting in higher diffuse fractions at the top of plant stands early and late in the day than at midday and at low and mid-latitudes. Generally higher solar zenith angles at higher latitudes result in higher diffuse fractions at high latitudes than at low latitudes for any time of the day.

The slope and aspect of a surface such as a leaf affects the amount of UV-B received above the canopy (Fig. 1). Nonhorizontal leaves at the canopy top receive some diffuse and direct beam radiation reflected off both the canopy phytoelements and underlying soil surface. Irradiance decreases with increasing zenith angle so that leaves (or other phytoelements) perpendicular to the ground and canopy top (erectophile leaves, seed heads, and other vertical phytoelements) have minimal UV-B exposure.[1]

WITHIN-CANOPY RADIATION ENVIRONMENT

The irradiance in the canopy is the cumulative effect of the downward penetration of diffuse and direct beam radiation, the downward phytoelement-scattered radiation, and the upward scattering of radiation penetrating to and scattered from the soil surface and phytoelements below the level of interest in the canopy. The relative importance of each contributing component of the spectral irradiance in the canopy depends on the optical properties of the canopy phytoelements, the canopy density, and structure.[2]

DOI: 10.1081/E-EPCS 120010624

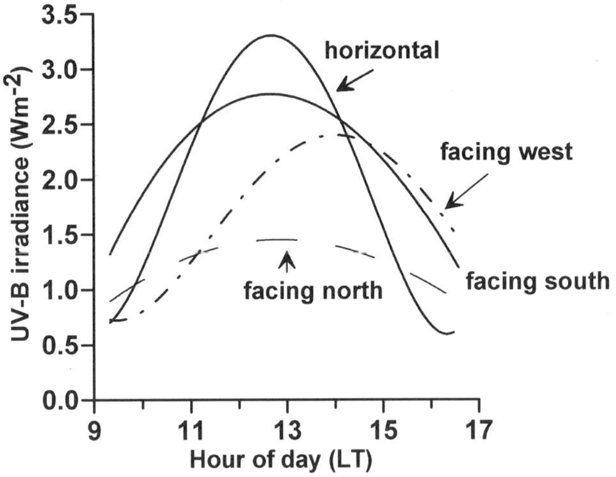

Fig. 1 UV-B irradiance on a clear day above a soybean (*Glycine max* L. (Merr.)) canopy, Northern Hemisphere. Surfaces facing north, west, and south are inclined at 45°, a typical soybean leaf inclination. (Modified from Ref. 1.)

Scattering in the Canopy

The radiation that is scattered from phytoelements and soil within and under the canopy (secondary scattering) depends on the incident radiation and the reflectance and transmittance of the phytoelements and the reflectance of the soil surface. The reflectance of adaxial and abaxial leaf surfaces of maize (*Zea mays*, L.), soybean (*Glycine max* L. (Merr.)), common oats (*Avena sativa* L.), winter wheat (*Triticum aestivum* L.), and sorghum (*Sorghum bicolor* L. (Moench.)), as well as apple (*Malus×domestica* Borkh.) and Callery pear (*Pyrus calleryana* Dcne.) are nearly constant (0.05 to 0.10) throughout the UV-A and UV-B. There is no reported leaf transmittance in the UV for apple, Callery pear, maize, soybean, winter wheat, oats, or sorghum. The reflectance from soils is similar to reflectance from the leaves, with dry light-colored soils having an average UV reflectance of approximately 0.07. Because UV radiation does not penetrate through the plant leaves, the transformation of direct beam radiation to diffuse within the canopy is solely through phytoelement and soil surface reflections. In contrast to the UV-B waveband, leaf transmission and reflection are significant in the photosynthetically active radiation waveband (PAR), resulting in substantial secondary scattering in the canopy.

Downward Penetration in the Canopy

The downward penetration of UV radiation into the canopy varies spatially and temporally due to the distribution of plant phytoelements and the orientation of the elements. The vertical penetration is commonly considered to be the mean or median over time with sunlit or shaded periods, or horizontal space with sunlit and shaded areas.[2,3] If the direct beam radiation cannot penetrate the canopy to a point in the canopy space, then it is considered to be shaded.

The stand density (plants per unit area) influences the view of the sky and the probability of shading in the lower canopy. Low-density canopies typically have large gaps (breaks in the spatial continuity of vegetation) that allow diffuse radiation to penetrate directly to the base of the canopy at all times (in shade) and allow direct beam radiation on occasion (in sunflecks). The size and distribution of gaps depend on the phytoelement orientation, dimension, and clumping. Increasing gap size increases both the diffuse sky radiation received and the probability of the direct beam penetrating the canopy. Shaded areas in the canopy receive only diffuse radiation, and vary from having essentially no irradiance in very dense canopies with no gaps to nearly the same amount of radiation as coming from the sky above the canopy in open canopies (Fig. 2). Sunlit areas (sunflecks) within the canopy have UV-B irradiance varying from the fraction of the above-canopy direct beam UV-B to the global above-canopy UV-B irradiance as the gap increases in size or number (Fig. 2). In sunflecks, the PAR is enriched relative to the UV-B—especially for small gaps with minimal sky view.

For low-density canopies with large gaps, the UV-B penetrates to greater depths than the PAR as a result of the higher fraction of diffuse radiation than in the PAR. In these canopies the UV-B irradiance in shade is closer

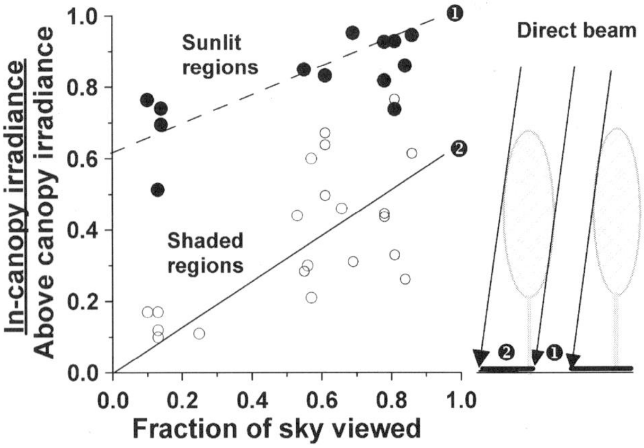

Fig. 2 Effect of sky view on the relative penetration of UV-B radiation in sunlit and shaded regions under tree canopies. Variability is due to the varying solar zenith angle and diffuse radiation fractions. (*View this art in color at www.dekker.com.*)

to that of sunlit regions than for PAR radiation. Consequently, shaded areas are enriched in UV-B relative to the PAR. Such enrichment is evident in young and old orchards. UV-B enrichment is evident in winter wheat, tall fescue (*Festuca arundinacea* Schreb.), and perennial ryegrass (*Lolium perenne* L.) canopies. Sunlit surfaces in an apple orchard[3] and in maize canopies[3,4] had UV-B levels similar to that above the canopy, whereas shaded surfaces had less UV-B than at the top of the canopy—varying with differences in sky view, diffuse fraction, and orientation.

Shading increases with increasing plant stand density and increased cumulative leaf area index (LAI) with depth in the canopy (Fig. 3). The shading in closed canopies (or high-stand densities) is primarily between-plant shading and produces relatively small areas of sunfleck at the stand floor (Fig. 3). In a maize canopy, three distinct radiation regimes were evident for surfaces at a height corresponding to a cumulative LAI of 1; a sunlit regime associated with direct beam penetration through a gap, a deep UV-B shade regime associated with small gaps nearby, and a light UV-B shade regime associated with large gaps nearby.[4] As depth in the maize canopy increased, the light UV-B shade regime disappeared and there was similar (small) penetration of UV-B and PAR. Similar changes in the frequency of gaps and the sunlit and shaded UV-B levels with depth in the canopy were found for winter wheat (Fig. 3). Planophile canopies such as alfalfa (*Medicago sativa* L.), white clover (*Trifolium repens* L.), and sorghum have infrequent gaps and equal or greater penetration in the PAR than UV-B radiation through scattering in the canopy.[5,6]

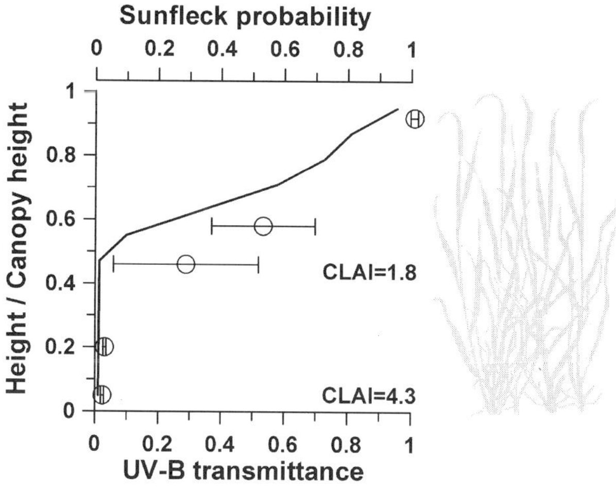

Fig. 3 Mean canopy transmittance of UV (open circles) and the probability of sunfleck (solid line) in a winter wheat canopy. Cumulative leaf area index (CLAI) is indicated. Bars represent the range in UV-B from shaded (low end of bar) to sunlit (high end of bar) regions in the canopy. (*View this art in color at www.dekker.com.*)

Changes in UV Penetration During the Growing Season

Crop canopies begin the season as sparse, low-density canopies. However, the vegetation in many crop canopies fills in (having LAI>1) over the course of the season to leave few gaps between plants at flowering or maturity. Consequently, the relative proportion of UV-B and PAR changes as the crop matures. UV-B penetration of low-density stands is typically greater than PAR because the sunlit regions receive approximately the same amount of PAR and UV-B, whereas the shaded regions receive significantly more sky UV-B than PAR. As the stand fills in, the overall penetration of diffuse sky radiation decreases—increasing the importance of secondary scattering in the canopy that is negligible for the UV-B and significant for the PAR. Consequently, over time the relative penetration of PAR increases compared with the UV-B, as documented for white clover and orchardgrass (*Dactylis glomerata* L.).[6]

CONCLUSION

The penetration of UV radiation into plant canopies is dictated largely by the amount of sky view created by gaps in the canopy. There is little scattering through or off canopy phytoelements within the canopy or off the ground at the canopy base. UV radiation levels in plant canopies are proportional to the size of the gap and view of the sky. Shaded regions near large gaps in the canopy have relatively high levels of UV radiation relative to PAR. Shaded regions near small gaps result in similar levels of UV and PAR. Sunlit regions near large or small gaps typically have greater PAR than UV-B. Increasing stand density corresponds to a decrease in UV-B penetration relative to PAR. Therefore, the highest levels of UV-B occur in the upper portions of dense canopies and in open low-density canopies common to early developmental stages of crops and mature orchards.

ARTICLES OF FURTHER INTEREST

Air Pollutants: Interactions with Elevated Carbon Dioxide, p. 17
Competition: Responses to Shade by Neighbors, p. 300

Ecophysiology, p. 410
Leaf Cuticle, p. 635
Leaf Structure, p. 638
Leaves and Canopies: Physical Environment, p. 646
Trichomes, p. 1254
UV Radiation Effect on Phyllosphere Microbes, p. 1258

REFERENCES

1. Grant, R.H. Ultraviolet irradiance of inclined planes at the top of plant canopies. Agric. For. Meteorol. **1998**, *89*, 281–293.
2. Grant, R.H. The partitioning of biologically-active radiation in plant canopies. Int. J. Biometeorol. **1997**, *40*, 26–40.
3. Gao, W.; Grant, R.H.; Heisler, G.M. A geometric UV-B radiation transfer model applied to agricultural vegetation canopies. Agron. J. **2002**, *94*, 475–482.
4. Grant, R.H. Ultraviolet-B and photosynthetically active radiation environment of inclined leaf surfaces in a maize canopy. Agric. For. Meteorol. **1999**, *95*, 187–201.
5. Grant, R.H.; Jenks, M.; Peters, P.; Ashworth, E. Scattering of ultraviolet and photosynthetically active radiation by sorghum bicolor: Influence of epicuticular wax. Agric. For. Meteorol. **1995**, *75*, 263–281.
6. Deckmyn, G.; Cayenberghs, E.; Ceulemans, R. UVB and PAR in single and mixed canopies grown under different UVB exclusions in the field. Plant Ecol. **2001**, *154*, 125–133.

Vaccines Produced in Transgenic Plants

Hugh S. Mason
Arizona State University, Tempe, Arizona, U.S.A.

INTRODUCTION

It is difficult to underestimate the potential of modern biotechnology for improving human health. We can easily contemplate the development of foods with enhanced nutritional content, new medicines, and vaccines for prevention of infectious diseases. Vaccines are one of the great successes of modern medicine, e.g., worldwide control of poliomyelitis, measles, and smallpox. However, the development and implementation of new vaccines remains prohibitively expensive for economically depressed countries where such measures are needed most. Plant-produced vaccines provide a promising new strategy that combines the innovations in genomics, medicine, and plant biotechnology to create affordable biological products. In the past decade, a growing number of research groups worldwide have studied plant expression and oral delivery of vaccine antigens, some showing very promising potential.

ADVANTAGES AND CHALLENGES OF PLANT VACCINES

For more comprehensive reviews of vaccine antigens produced in plants, please see Refs. 1–4. The development of technology for genetic transformation of plants has provided the opportunity to use agriculture for the production of recombinant proteins. Field production using a stable transgenic line would not require large investments in hardware and culture media, thus making scale-up more economical than fermentation culture. Analysis of the economic potential of different crop plants for production of recombinant proteins[5] provides a basis for comparison and suggests real value in the technology. Moreover, plant systems are much less likely to harbor pathogenic microbes than are mammalian cell or whole animal systems. An attractive possibility with edible plants is the ability to deliver vaccine antigens orally, thus obviating the costly purification process required of injectable vaccines. The delivery of vaccines to mucosal tissues (epithelia that line the gastrointestinal, respiratory, and genitourinary tracts) also has the ability to stimulate secretory IgA at those surfaces, which is rarely observed with injectable vaccines.[6]

The biggest challenge for orally delivered vaccines is the digestive system, whose acidic and proteolytic environment may be too severe for all but the most gut-stable proteins and may require encapsulation of antigens. Furthermore, a robust mucosal vaccine must be readily transported across the epithelium for presentation to the underlying lymphoid tissues. Such transport likely requires special epithelial cell-binding qualities that are not universally present in proteins. Nonetheless, a growing number of candidate vaccine antigens have been shown to be orally active.

PLANT EXPRESSION SYSTEMS

There are two main systems for recombinant protein expression in plants: stable genetic transformation and transient expression. Stable transformation causes integration of foreign DNA into the nuclear or chloroplast chromosomal DNA; these lines can be propagated vegetatively or by seeds and thus can be readily scaled up for protein production. Transient expression uses a plant virus that carries the vaccine gene, which is replicated and expressed during systemic infection of the plant host.[2] Such virus-based expression amplifies gene copy number, often resulting in higher expression than stable transformation. However, the larger foreign genes may be unstable in plant virus systems, which results in deletion of the foreign DNA.

Stable nuclear transformation is often achieved using *Agrobacterium tumefaciens*, which can efficiently transport DNA into plant cells and cause chromosomal integration resulting in Mendelian inheritance. DNA delivery by the ''biolistic'' method (microprojectile bombardment) is frequently used for plant hosts that are recalcitrant to *Agrobacterium*. The biolistic method often causes multiple site integration that may enhance expression, but excessive copies or very high mRNA levels may cause gene silencing.[7]

Encyclopedia of Plant and Crop Science
DOI: 10.1081/E-EPCS 120010379

Stable transformation of the chloroplast genome can yield high levels of recombinant protein[8] due to the high genome copy number. The chloroplast-derived chromoplasts of tomato fruit can also express foreign proteins. Other advantages of chloroplast transformation include maternal inheritance, which limits the potential for transgene escape by dissemination of pollen, and absence of gene silencing. Many foreign genes have been expressed in transplastomic plants, including the vaccine candidate cholera toxin B subunit (CT-B).[8] A limitation for this strategy is that some eukaryotic cellular processing events (e.g., glycosylation) may not be obtained in chloroplasts.

Tobacco and potato are frequently used as convenient systems for vaccine expression because transformation and regeneration of plants are readily achieved. If purification of antigen for injectable delivery is the goal, tobacco is an attractive choice because of its high biomass production. However, the use of edible plant tissues permits oral delivery, wherein one may avoid extensive and costly purification procedures. Raw potato has been used for both published human trials in the United States,[4] but a more palatable system is needed. Tomato is a good alternative and was used to express orally immunogenic respiratory syncytial virus (RSV) fusion (F) protein in fruit. Other fruits, including banana, may be useful if efficient transformation systems can be developed. A study of various economic factors suggests that soybean, alfalfa, and corn are the most efficient systems for recombinant protein expression.[5] A further important consideration is the need to stringently regulate the production of biologicals in plants, in order to prevent the contamination of human food supplies. Thus, the potential for transgene escape by outcrossing must be carefully studied for any potential production system.

ANIMAL AND HUMAN STUDIES

A growing number of vaccine candidates have been expressed in plants and tested in animals, resulting in immune responses after oral delivery.[2,3] In most cases, there was no challenge with infectious pathogens to determine the efficacy of the immunization, but a few reports describe protection from viral or toxin treatment. In a notable recent study,[9] a peptide from a rotavirus enterotoxin fused to cholera toxin B subunit (CT-B) was coexpressed in potato with a peptide from an *E. coli* fimbrial adhesin fused to CT-A, resulting in assembly of functional ganglioside-binding complexes. Pups of mice immunized by ingestion of transgenic potatoes were partially protected from challenge with rotavirus. The transmissible gastroenteritis virus envelope spike (S) protein was produced in transgenic corn and fed to piglets, resulting in 50% of virus-challenged animals free of diarrhea.[3] These studies show potential for edible vaccines as alternatives to injectable vaccines for animals, but they also indicate the need for further optimization of dosage levels or timing to increase efficacy.

Other potential uses for plant-expressed, orally delivered antigens include therapies using autoantigens. For example, treatments for autoimmune disease could be provided by consumption of autoantigens, such as proinsulin fused to CT-B expressed in potato and fed to NOD mice.[10] Reduced inflammation of pancreatic islets in the treated mice suggests that cytotoxic T cells that kill islet cells in NOD mice had been restricted by immune tolerance. Immunization for the treatment of cancer is the subject of promising studies, e.g., single-chain antibodies from B-lymphocyte malignancies expressed in plants with viral vectors, yielding protection from a lethal injection of tumor cells in treated mice.[11]

Only three studies of edible plant vaccines in human subjects have been published,[12–14] each using a different antigen. Two vaccines were directed against agents of gastroenteritis (enterotoxic *E. coli* and Norwalk virus), whose protein antigens are considered the most likely to succeed by oral delivery due to their gut stability and mucosal cell–binding capacity. The other vaccine targeted hepatitis B virus, not usually considered an enteric pathogen.

Cholera and enterotoxigenic *E. coli* (ETEC) cause diarrhea by secretion of cholera toxin (CT) and labile toxin (LT), respectively, which target epithelial cells via their B subunits' ability to bind G_{M1} gangliosides present on cell surfaces. LT-B expressed in transgenic potatoes produced toxin-protective gut antibody responses after ingestion by mice.[4] The LT-B potatoes became the first human test for plant vaccines[12] in which subjects ate 100 g of raw LT-B potato slices containing 750 μg LT-B at each of three weekly doses. The production of toxin-neutralizing serum antibodies in 10 of 11 subjects proved that ingestion of transgenic plant material could be used effectively to deliver a vaccine in humans.

Most cases of viral gastroenteritis are caused by rotavirus and Norwalk-like viruses, which are potential targets for oral subunit vaccines owing to the ability of their capsid proteins to assemble virus-like particles (VLP). These recombinant VLP can mimic the structure and cell-binding of the authentic virus particles, are acid-stable, and stimulate serum responses in humans. NVCP expressed in plants formed VLP that were orally immunogenic in mice,[4] and NVCP potatoes were used in a second clinical trial.[14] The volunteers ate 150 g of raw potato containing up to 750 μg NVCP. Although the average antibody levels were less impressive than in the LT-B potato study, 19 of the 20 subjects showed significant immune responses. The main conclusion of this study was that a recombinant plant-derived protein lacking the ganglioside-binding activity of LT-B can stimulate oral immunization in humans.

The currently used vaccine for hepatitis B is recombinant viral surface antigen (HBsAg) purified from transgenic yeast cultures. Plant-derived HBsAg was described in the first report of a plant vaccine[15] and showed VLP that were immunogenic by potato ingestion in mice.[16] HBsAg expressed in transgenic lettuce and delivered to humans by ingestion of 250 g (containing approximately 1 μg antigen) caused production of serum antibodies at protective levels in two of three volunteers.[13] Since the dose was quite low, it is likely that the HBsAg had assembled into VLP structure, which enhanced sampling by M cells of the gut-associated lymphoid tissue.[6] Another clinical trial used transgenic HBsAg potatoes delivered to volunteers who had been vaccinated earlier with the commercial injectable HBsAg, in order to test the boosting effect of ingested antigen. The results were promising and are now being prepared for publication.[17]

CONCLUSION

Studies during the past 10 years have shown faithful expression of many different vaccine antigens in plants, and many of these have stimulated antibody responses when eaten by humans or animals. A few reports showed partial protection from challenge with infectious pathogens, indicating a strong potential for plant vaccine technology. However, low expression of antigens in transgenic plants limits the dose that can be delivered in crude material; thus, improvements in plant expression technology are needed. Because doses must be uniform and concentrated, the plant stock will be processed to yield a stable product, such as dried powder that can be formulated with adjuvants and perhaps encapsulated for passage through the stomach. Production systems must maintain rigorous containment to prevent contamination of food supplies; thus, pollen-mediated gene flow should be limited (e.g., with male-sterile lines or chloroplast expression), and post-harvest processing facilities must be dedicated for pharmaceutical production.

ACKNOWLEDGMENTS

The author wishes to thank Tsafrir Mor and Heribert Warzecha for research and helpful discussions.

ARTICLES OF FURTHER INTEREST

Biosafety Approaches to Transgenic Crop Plant Gene Flow, p. 150

Medical Molecular Pharming: Therapeutic Recombinant Antibodies, Biopharmaceuticals and Edible Vaccines in Transgenic Plants Engineered via the Chloroplast Genome, p. 705

Transgenic Crop Plants in the Environment, p. 1248

Transgenic Crops: Regulatory Standards and Procedures of Research and Commercialization, p. 1251

REFERENCES

1. Daniell, H.; Streatfield, S.J.; Wycoff, K. Medical molecular farming: Production of antibodies, biopharmaceuticals and edible vaccines in plants. Trends Plant Sci. **2001**, *6* (5), 219–226.
2. Koprowski, H.; Yusibov, V. The green revolution: Plants as heterologous expression vectors. Vaccine **2001**, *19* (17–19), 2735–2741.
3. Streatfield, S.J.; Jilka, J.M.; Hood, E.E.; Turner, D.D.; Bailey, M.R.; Mayor, J.M.; Woodard, S.L.; Beifuss, K.K.; Horn, M.E.; Delaney, D.E.; Tizard, I.R.; Howard, J.A. Plant-based vaccines: Unique advantages. Vaccine **2001**, *19* (17–19), 2742–2748.
4. Mason, H.S.; Warzecha, H.; Mor, T.; Arntzen, C.J. Edible plant vaccines: Applications for prophylactic and therapeutic molecular medicine. Trends Mol. Med. **2002**, *8* (7), 324–329.
5. Kusnadi, A.R.; Nikolov, Z.L.; Howard, J.A. Production of recombinant proteins in transgenic plants: Practical considerations. Biotechnol. Bioeng. **1997**, *56* (5), 473–484.
6. Ogra, P.L.; Faden, H.; Welliver, R.C. Vaccination strategies for mucosal immune responses. Clin. Microbiol. Rev. **2001**, *14* (2), 430–445.
7. Hobbs, S.; Warkentin, T.; CMO, D. Transgene copy number can be positively or negatively associated with transgene expression. Plant Mol. Biol. **1993**, *21*, 17–26.
8. Daniell, H.; Khan, M.; Allison, L. Milestones in chloroplast genetic engineering: An environmentally friendly era in biotechnology. Trends Plant Sci. **2002**, *7* (2), 84–91.
9. Yu, J.; Langridge, W.H. A plant-based multicomponent vaccine protects mice from enteric diseases. Nat. Biotechnol. **2001**, *19* (6), 548–552.
10. Arakawa, T.; Yu, J.; Chong, D.K.; Hough, J.; Engen, P.C.; Langridge, W.H. A plant-based cholera toxin B subunit-insulin fusion protein protects against the development of autoimmune diabetes. Nat. Biotechnol. **1998**, *16* (10), 934–938.
11. McCormick, A.A.; Kumagai, M.H.; Hanley, K.; Turpen, T.H.; Hakim, I.; Grill, L.K.; Tuse, D.; Levy, S.; Levy, R.

Rapid production of specific vaccines for lymphoma by expression of the tumor-derived single-chain Fv epitopes in tobacco plants. Proc. Natl. Acad. Sci. U. S. A. **1999**, *96* (2), 703–708.
12. Tacket, C.O.; Mason, H.S.; Losonsky, G.; Clements, J.D.; Wasserman, S.S.; Levine, M.M.; Arntzen, C.J. Immunogenicity in humans of a recombinant bacterial-antigen delivered in transgenic potato. Nat. Med. **1998**, *4* (5), 607–609.
13. Kapusta, J.; Modelska, A.; Figlerowicz, M.; Pniewski, T.; Letellier, M.; Lisowa, O.; Yusibov, V.; Koprowski, H.; Plucienniczak, A.; Legocki, A.B. A plant-derived edible vaccine against hepatitis B virus [published erratum appears in FASEB J 1999 Dec;13(15):2339]. Faseb J. **1999**, *13* (13), 1796–1799.
14. Tacket, C.O.; Mason, H.S.; Losonsky, G.; Estes, M.K.; Levine, M.M.; Arntzen, C.J. Human immune responses to a novel Norwalk virus vaccine delivered in transgenic potatoes. J. Infect. Dis. **2000**, *182* (1), 302–305.
15. Mason, H.S.; Lam, D.M.K.; Arntzen, C.J. Expression of hepatitis B surface antigen in transgenic plants. Proc. Natl. Acad. Sci. U. S. A. **1992**, *89*, 11745–11749.
16. Kong, Q.; Richter, L.; Yang, Y.F.; Arntzen, C.J.; Mason, H.S.; Thanavala, Y. Oral immunization with hepatitis B surface antigen expressed in transgenic plants. Proc. Natl. Acad. Sci. U. S. A. **2001**, *98* (20), 11539–11544.
17. Thanavala, Y. Personal communication.

Viral Host Genomics

Steven A. Whitham
Iowa State University, Ames, Iowa, U.S.A.

INTRODUCTION

The roles of individual virus-encoded proteins have been established for the key steps of virus life cycles. However, the detailed mechanisms of how viruses cause reduced fitness, quality, and yield are yet to be revealed. A thorough understanding of viral pathogenesis requires that host–virus interactions be understood equally well from the plant side of the interaction. Recent developments in plant genomics promise to provide new insights into the ways that viruses perturb host plants. Broadly defined, genomics includes genome sequencing and subsequent study of genome organization and structure (structural genomics) and gene function (functional genomics). This article discusses the application of functional genomics approaches to study host–virus interactions as well as the exploitation of viruses as useful functional genomics tools to probe plant gene functions.

WHAT IS FUNCTIONAL GENOMICS?

Functional genomics studies use genome-wide, global, and/or high-throughput systems to simultaneously study functions of large numbers of or virtually all genes of an organism. The methodologies rely heavily on computational biology, bioinformatics, and statistics to manage and analyze the enormous data sets that are generated. Viral host genomics encompasses pathogenicity, resistance, and discovery of plant gene functions, and has become possible because of two major factors. First, researchers now have access to the complete genome sequences of many viruses and a few plant species (including rice and *Arabidopsis*), and extensive sequences are available from crop species like maize, wheat, soybean, and tomato. Second, these sequences enable the development and application of new technologies previously not possible to explore the functions of plant genes.

Important technologies for assessing functions of plant genes that have become available recently include DNA microarrays and reverse genetics tools such as insertional mutagenesis (e.g., T-DNA tagging), RNA-induced gene silencing, and virus-induced gene silencing (VIGS).[1] DNA microarrays consist of DNA sequences that are most commonly spotted onto glass slides or synthesized directly onto a silicon or glass substrate.[2] All genes of an organism can be represented on one or a few microarrays, which allows the expression of all genes to be studied in parallel in response to a given condition. Reverse genetics tools, which will be described later in more detail, involve the disruption of genes followed by examination of the effects of loss of gene function on the organism. This suite of technologies allows systematic studies that first associate genes with a plant process and then subsequently determine their roles in that process by observing the mutant phenotype.

FUNCTIONAL GENOMICS APPROACHES TO UNDERSTAND VIRAL PATHOGENICITY

Although viruses are simple entities, interactions with their hosts are relatively complex. Viruses and plants engage in a variety of offensive, defensive, and counter-defensive interactions that ultimately determine the level of colonization.[3] Some specific examples of modifications that plant viruses make in their hosts include gating of plasmodesmata by movement proteins and suppression of RNA interference (RNAi) triggered by replication of viral genomes.[4,5] Viruses also elicit significant changes in patterns of host gene expression—which is expected to influence plant susceptibility and symptom development—and may be a direct result of activities such as gating of plasmodesmata, suppression of RNAi, or interference with other processes.[6]

Recent studies employing genomics approaches have documented changes in the expression of many host genes in response to diverse viruses and viroids. Whitham et al.[7] used Arabidopsis GeneChip® microarrays to assay the expression of about 8300 genes over a five-day infection time course in the susceptible *Arabidopsis* plants. The viruses were ORMV (oilseed rape tobamovirus), TVCV (turnip vein clearing tobamovirus), CMV (cucumber mosaic cucumovirus), PVX (potato virus X), and TuMV (turnip mosaic potyvirus). At least 114 genes were induced more than twofold in response to five viruses at one or more time points over the time course. The results demonstrated that diverse viruses induce the expression of common sets of host genes in inoculated leaves, as has been observed in a pea cotyledon system.[8] Further

Encyclopedia of Plant and Crop Science
DOI: 10.1081/E-EPCS 120019986

analysis of the expression of these 114 genes placed them into seven clusters based on their expression profiles. One interesting cluster was composed of only heat shock genes: HSP70, HSP83 (similar to HSP90), HSP17.6, HSP17.4, and HSP23.6. These heat shock proteins shared the characteristic of being induced by ORMV and TVCV (both tobamoviruses) at one day after inoculation, although they were not induced by the other viruses until later times. This result suggests that tobamoviruses might specifically induce HSP expression early. Another interesting cluster of genes was composed largely of plant defense-related genes. In general, the viruses induced the expression of defense-related genes beginning at 2 days after infection. The differential expression of heat shock and defense-related genes indicates that distinct signaling pathways are modulated in response to viruses in susceptible interactions.

Potato spindle tuber viroid (PSTVd) was shown to elicit a variety of changes in tomato gene expression, some of which occur also in tobacco mosaic virus infection in the same host.[9] Unlike viruses, viroids do not encode proteins. Therefore, the RNA genome of viroids is sufficient to cause symptoms in susceptible plants and elicit changes in host gene expression.

Table 1 lists the functional classes of plant genes induced in response to RNA viruses and viroids in the two studies cited above.[7,9] Direct comparisons of the genes induced in the experiments involving viruses and viroids should be made with caution because the nature of the pathogens, time courses, and experimental methodologies were different. Nevertheless, it is intriguing to point out the induction of stresslike responses observed in both studies. For instance, approximately one-third of the genes induced by PSTVd in tomato and by various viruses in *Arabidopsis* plants are associated with known defense and

Table 1 Functional classification of genes induced by viruses and viroids

	Percentage (%) of genes in each functional class	
Functional class	**RNA viruses (Arabidopsis)**	**PSTVd (tomato)**
Defense/stress	30.7	29.8
Cell wall	2.6	6.4
Chloroplast	0	8.5
Energy	2.6	0
Metabolism	14.9	0
Signaling	14.0	0
Protein destination	7.9	21.3
Transcription	7.9	0
Miscellaneous	0	19.1
Unknown	19.4	14.9

stress responses. These include genes functionally associated with oxidative stress, heat stress, pathogen stress, defense responses, and transcription factors involved in defense and stress responses. From the *Arabidopsis* study, it is clear that these sets of genes are expressed differentially over time, suggesting that a variety of host-signaling pathways is modulated as infections proceed. Furthermore, the ability of diverse viruses and viroids to elicit these changes suggests that a conserved feature of virus–host interactions triggers them. At this time, the mechanisms responsible for these gene expression changes are uncharacterized, and their significance remains to be determined. Reverse genetics approaches will be valuable for addressing these issues.

VIRUS-BASED TECHNOLOGY FOR DISCOVERY OF PLANT GENE FUNCTION

Virus-induced gene silencing (VIGS) is being exploited to study the functions of plant genes in disease resistance pathways, development, and other processes.[1] VIGS is a form of posttranscriptional gene silencing (PTGS) that utilizes viruses engineered to carry sequences from plant genes to disrupt the expression of the cognate plant genes. VIGS is based on the principle that viruses are targeted by PTGS leading to degradation of their RNA. Once PTGS is induced against a specific sequence, it can degrade that sequence from any source, such as a virus in the cytoplasm or a plant mRNA encoded by a nuclear gene. Thus, the engineered VIGS virus will trigger degradation of itself, the foreign plant sequence it carries, and the corresponding mRNA from the endogenous plant gene. These events lead to a loss of function of the plant gene. The plasticity of viral genomes allows them to accom)modate virtually any sequence, enabling most plant genes to be disrupted. VIGS is particularly advantageous for plant species that are not easily transformable (most crops) or not genetically tractable, because of the potential to rapidly generate mutations to facilitate functional analysis of a gene by virtue of a virus infection.

VIGS vectors derived from potato virus X and tobacco rattle virus have been most widely used in dicot species such as *Nicotiana benthamiana*, tobacco, and tomato; many other viruses hold promise for similar applications in other species.[1] Recently, a VIGS vector was developed from barley stripe mosaic virus, demonstrating the utility of this approach in monocots.[10] VIGS is not only useful in the study of individual genes, but is also adaptable to high-throughput genomics-scale experiments, as illustrated in Fig. 1. The first step in the process is to create a library of plant copy DNA (cDNA) clones by extracting mRNA from plant cells and converting mRNA to cDNA with the enzyme reverse transcriptase. These cDNA sequences are then cloned into the VIGS vector to

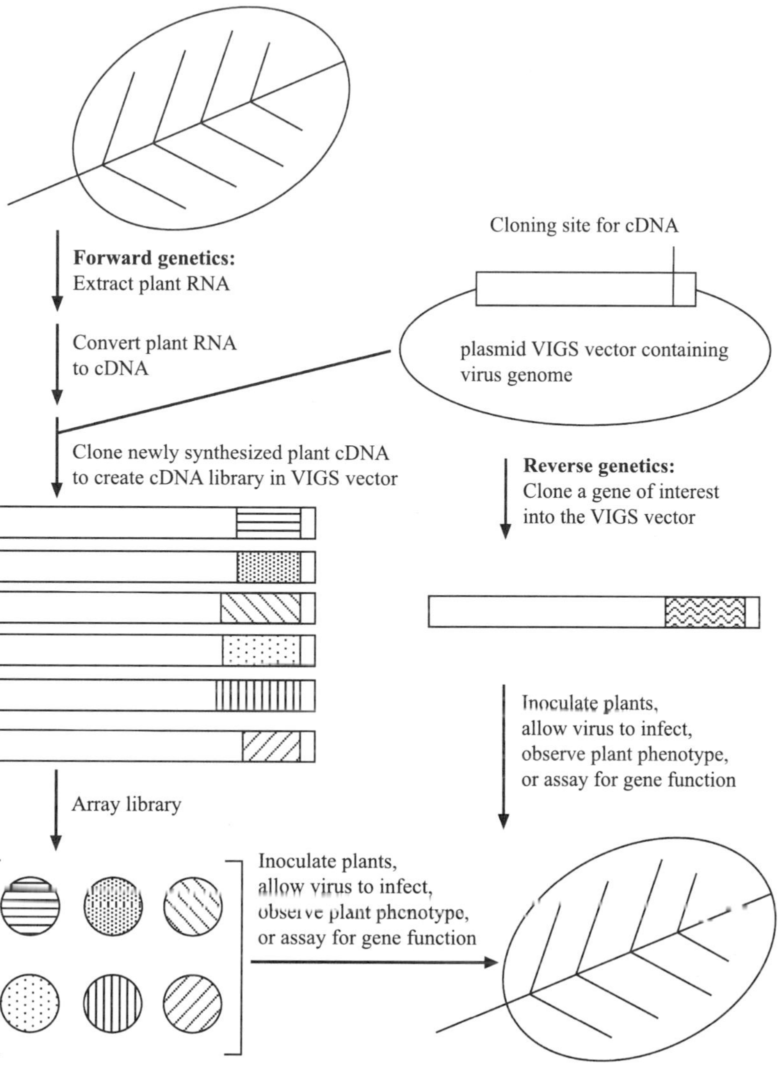

Fig. 1 Use of VIGS for forward and reverse genetic studies to discover plant gene function. See text for description.

create a VIGS library. The libraries can be arrayed so that it is possible to preserve each clone, determine its sequence, and return to it for further analyses. In this example, individual clones are used to inoculate plants to silence the corresponding plant gene and subsequently screen for changes in phenotype. Alternatively, VIGS can be used to systematically study the effects of disrupting selected genes, as illustrated in Fig. 1.

CONCLUSION

Genomics technologies such as DNA microarrays are just beginning to be applied to plant–virus interactions. From initial published report it appears, not surprisingly, that plants are stressed in a variety of ways as viruses invade susceptible hosts. As microarray data are accumulated for both pathogenic and resistant interactions, it will be interesting to compare and contrast the signaling pathways modulated during pathogenesis and resistance. The availability of full genome microarrays will further illuminate ways that viruses perturb the expression of host genes. Reverse genetics tools such as T-DNA insertion lines, RNA-induced gene silencing/RNAi, and VIGS will enable researchers to discover the significance of these gene expression changes and identify genetic components of the signaling pathways involved in regulating the changes. Such integrated studies are expected to reveal mechanisms associated with pathogenesis, onset of disease, and resistance.

On the flip side of the coin, viruses are themselves serving as useful tools for genomics research. Newly developed VIGS vectors include features such as the Gateway™ cloning site, which allows genes and libraries to be shuttled in and out of VIGS vectors. Such systems are expected to enable the production of complete VIGS libraries from which an investigator can order the VIGS clone corresponding to any gene of interest.[1] Comprehensive VIGS libraries will facilitate rapid assessment of a gene's function once it has been associated with a plant trait or response such as viral pathogenesis or resistance.

ACKNOWLEDGMENTS

This work was supported by Hatch Act and State of Iowa Funds, the Iowa State University Plant Sciences Institute, and USDA grant number 2002-35319-12566.

ARTICLES OF FURTHER INTEREST

Functional Genomics and Gene Action Knockouts, p. 476
Gene Silencing: A Defense Mechanism Against Alien Genetic Information, p. 492
Plant RNA Virus Diseases, p. 1023
Virus-Induced Gene Silencing, p. 1276

REFERENCES

1. Waterhouse, P.M.; Helliwell, C.A. Exploring plant genomes by RNA-induced gene silencing. Nat. Rev. Genet. **2003**, *4* (1), 29–38.
2. Schena, M. *Microarray Analysis*; John Wiley & Sons, Inc.: Hoboken, NJ, 2002.
3. Whitham, S.A.; Dinesh-Kumar, S.P. Signalling in Plant-Virus Interactions. In *Plant Signal Transduction*; Scheel, D., Wasternack, C., Eds.; Oxford University Press: Oxford, 2002; 226–249.
4. Lazarowitz, S.G.; Beachy, R.N. Viral movement proteins as probes for intracellular and intercellular trafficking in plants. Plant Cell **1999**, *11* (4), 535–548.
5. Voinnet, O. RNA silencing as a plant immune system against viruses. Trends Genet **2001**, *17* (8), 449–459.
6. Kasschau, K.D.; Xie, Z.; Allen, E.; Llave, C.; Chapman, E.J.; Krizan, K.A.; Carrington, J.C. P1/HC-Pro, a viral suppressor of RNA silencing, interferes with Arabidopsis development and miRNA unction. Dev. Cell **2003**, *4* (2), 205–217.
7. Whitham, S.A.; Quan, S.; Chang, H.S.; Cooper, B.; Estes, B.; Zhu, T.; Wang, X.; Hou, Y.M. Diverse RNA viruses elicit the expression of common sets of genes in susceptible *Arabidopsis thaliana* plants. Plant J. **2003**, *33*, 271–283.
8. Escaler, M.; Aranda, M.A.; Thomas, C.L.; Maule, A.J. Pea embryonic tissues show common responses to the replication of a wide range of viruses. Virology **2000**, *267* (2), 318–325.
9. Itaya, A.; Matsuda, Y.; Gonzales, R.A.; Nelson, R.S.; Ding, B. Potato spindle tuber viroid strains of different pathogenicity induces and suppresses expression of common and unique genes in infected tomato. Mol .Plant Microbe. Interact. **2002**, *15* (10), 990–999.
10. Holzberg, S.; Brosio, P.; Gross, C.; Pogue, G.P. Barley stripe mosaic virus-induced gene silencing in a monocot plant. Plant J. **2002**, *30* (3), 315–327.

Virus Assays: Detection and Diagnosis

Clarissa J. Maroon-Lango
USDA—ARS, Beltsville, Maryland, U.S.A.

INTRODUCTION

Plant viruses and viroids cause significant direct or indirect economic losses. Specific losses attributed to certain viruses have been tabulated and reviewed.[1] Early detection of the causal viruses and accurate diagnosis of disease are important in preventing or minimizing such losses. Current methods for detecting viruses and viroids are classified into serological assays, nucleic acid-based techniques, bioassays, and electron microscopy. Generally, the choice of detection method depends on the causal agent (indicated by the symptoms on the plant), the kind and number of samples, the availability of reagents and/or equipment, the time frame available for testing, and the cost. This article discusses some of the techniques used in detecting viruses and viroids, and their applicability in routine testing of plants.

SEROLOGICAL ASSAYS

Serological techniques for detecting plant viruses involve the use of either polyclonal or monoclonal antibodies raised to detect specific viruses. The serological methods that have been applied in the detection of plant viruses include enzyme-linked immunosorbent assay (ELISA), rapid immunofilter paper assay (RIPA) (also known as lateral flow assay), dot immunoblot assay (DIBA), tissue-blot immunoassay (TBIA), and immunosorbent electron microscopy (ISEM).

ELISA—In direct ELISA, the virus is immobilized by either an uncoated or antibody-coated solid phase (usually a polystyrene wall). The immobilized virus is recognized by an enzyme-labeled antibody, and the reaction is visualized by the addition of an enzyme substrate. In indirect ELISA, the virus is reacted with an unconjugated specific antibody, which is in turn detected by an enzyme-labeled anti-immunoglobulin antibody or antibody fragments. The resulting complex is visualized by the addition of an enzyme substrate.[2] Since its introduction into plant viral diagnosis less than thirty years ago, ELISA has been widely used in routine testing of plants because of its relative sensitivity, cost-effectiveness, and ease of use.

RIPA—An increasingly popular technique for detecting plant viruses is based on RIPA.[3] In the commercially available format of RIPA (known as lateral flow, strip, or stick test), a virus-specific antibody is immobilized on the capture line near the top of a membrane strip. A mixture consisting of the tracer virus-specific antibody conjugated to either dyed latex or colloidal metal is applied to a defined region (conjugate pad) just above the lower end of the strip. When the sample (sap) is applied at the lower end of the strip, the virus moves by capillary action through the conjugate pad forming a labeled antibody-virion complex. The moving virion complex is trapped by the immobilized antibody on the capture line and the capture line develops a positive color reaction. In addition to being a rapid test (results occur within minutes), its design makes it suitable for on-site (greenhouse or field) use. Additionally, the sensitivity of RIPA is similar to that of ELISA. The use of RIPA for detecting viruses is limited, however, by the availability of tests in the commercially available format. To date, only 18 specific virus tests are offered in such a format (e.g., ImmunoStrips, Agdia, Inc; Spot √Check LF™, Adgen, Ltd, UK).

DIBA and TIBA—These virus detection methods involve spotting plant sap (DIBA) or blotting freshly cut tissue (TIBA) on a membrane that is probed directly or indirectly with a virus-specific antibody.[4] The addition of a suitable substrate results in a positive color reaction. Whereas these techniques may be ideal for routine testing because of their suitability for processing a large number of samples on site, they may not be very reliable due to background (false positives) in certain antibody-host combinations.

ISEM—Another serological detection method involves electron microscopy. The virus is visualized by electron microscopy using leaf dips or purified viral preparations on carbon-stabilized, polyvinal formal-coated grids pre-treated with a specific antibody.[5] This technique is suitable for high titer viruses, but is limited by the uneven distribution of the virus in the plant. Furthermore, because ISEM requires sophisticated equipment, its application in commercial virus testing has been very limited.

NUCLEIC ACID–BASED TECHNIQUES

Techniques have been designed to detect viruses and viroids based on their genome. Double-stranded (ds) RNA

Encyclopedia of Plant and Crop Science
DOI: 10.1081/E-EPCS 120019913

analysis, hybridization, and amplification reaction-based assays are some of the nucleic acid–based techniques used in routine virus and viroid testing. The choice of these assays is usually dictated by the need for significantly increased sensitivity and versatility, or the lack of a suitable serological test.

dsRNA Analysis—As a broad-spectrum assay for detecting RNA viruses, dsRNA analysis entails the electrophoresis of dsRNA isolated from infected plants and its visualization on the gel by ethidium bromide staining. The resulting dsRNA pattern suggests the causal viral group;[6] the identity of the virus can be confirmed by cloning and sequencing. Although the assay is not ideal for large volumes of samples, it offers advantages because it does not require the use of antibodies, sequence information, or probes.

Hybridization—In hybridization, the viral nucleic acid is detected by annealing it with a complementary strand of either RNA or DNA, referred to as the probe. The labeled probe allows visualization of the formed hybrid by autoradiography, fluorescence, or enzymatic reaction. Samples may be in the form of membrane-spotted sap or enriched preparations in dot blot hybridization (e.g., viroids) or membrane-squashed tissue in squash blot hybridization (e.g., geminiviruses). A large number of samples can be tested simultaneously by these assays due to simple sample processing. Alternatively, DNA or RNA may be separated on a gel by electrophoresis and transferred to a membrane for Southern or Northern blot hybridization, respectively.

Amplification—Offering significantly enhanced sensitivities are amplification-based tests such as polymerase chain reaction (PCR) and nucleic acid sequence-based amplification (NASBA™). PCR is the amplification of a segment of DNA bound by two regions of known sequence. The process involves numerous (25–40) cycles of denaturation of the target DNA, annealing of primers (oligonucleotides that are complementary to the regions bordering the sequence), and primer extension by Taq polymerase or similar heat-resistant polymerases.[7] In the case of RNA viruses, PCR is preceded by the synthesis of cDNA by reverse transcription (RT). In the isothermal amplification method, NASBA™, cDNA synthesis is carried out using a primer containing a T7 RNA polymerase promoter. The RNA in the cDNA:RNA hybrid is digested, allowing the synthesis of a dsDNA using a second primer. With the resulting dsDNA as template, T7 RNA polymerase transcribes amplicons to form multiple copies of antisense RNA, which are used as templates for more cycles of amplification.[8] Amplicons resulting from both PCR and NASBA™ are routinely electrophoresed on gels and visualized either by ethidium bromide staining or hybridization upon transfer to a membrane. In addition, amplicons may be cloned and sequenced.

Aside from their marked sensitivity, both PCR and NASBA™ can be designed to be strain-, species-, or group-specific tests.[9,10] RT-PCR has been particularly useful in detecting numerous viruses of herbaceous and woody plants, whereas NASBA™ has been used effectively to detect the various strains of PVY.[10] Despite the advantages offered by amplification-based tests, their applicability in routine testing and high throughput analyses is limited by laborious and expensive pre- and post-amplification manipulations. Methods to simplify sample processing (e.g., direct binding or immunocapture of virions) have been evaluated. Detection of RT-PCR-derived amplicons has been simplified by sandwich hybridization in enzyme-linked-oligosorbent-assay (ELOSA) in the case of PVY.[11] ELOSA involves a capture probe covalently linked to the wall of a microtiter well and a 3′-biotinylated-specific detection probe, both of which hybridize to resulting amplicons. The formed hybrid is reacted with a strepavidin-enzyme conjugate that results in a color reaction when given the appropriate substrate. Alternatively, amplicons derived by either PCR or NASBA™ may be specifically detected and quantitated in real time using a probe labeled with fluorescent and quencher dyes at either end (e.g., TaqMan and molecular beacons).[10,12] When the labeled probe binds the amplicon, a sequence-specific fluorescent signal is released in real time with intensity increasing proportionally.

BIOASSAYS AND ELECTRON MICROSCOPY

When serological and nucleic acid-based techniques are unavailable, alternative detection methods include bioassays and electron microscopy. Bioassays entail the grafting or inoculation of certain woody and herbaceous plant species (indicator plants) with cuttings or sap from the infected plant materials. The indicator plants exhibit symptoms characteristic of certain host-virus systems within days or months. In electron microscopy, ultrathin tissue slices or plant sap are examined for viral morphology and cytopathology. Based on these observations, the group to which the causal virus belongs is identified.

CONCLUSION

Increased sensitivity combined with simplified and inexpensive assays should be the goal in developing tests for effective viral detection and diagnosis. These characteristics should extend the applicability of such tests in routine testing of high volumes of samples. Currently, serological lateral flow assays offer the simplest test format, whereas amplification-based tests prove to be the most sensitive. An assay with both of these properties was developed to detect a human pathogen in water. The specific detection assay involved solid phase–based extraction coupled with RT–PCR, and amplicon detection by hybridization in a lateral flow format.[13] The development of a similar assay for detecting plant viruses and viroids will be a practical and effective tool for disease management.

ARTICLES OF FURTHER INTEREST

Bacterial Pathogens: Detection and Identification Methods, p. 84
Plant Diseases Caused by Subviral Agents, p. 956
Plant DNA Virus Diseases, p. 960
Plant RNA Virus Diseases, p. 1023
Seed Borne Pathogens, p. 1126
Transgenes (GM) Sampling and Detection Methods in Seeds, p. 1238

REFERENCES

1. Waterworth, H.E.; Hadidi, A. Economic Losses Due to Plant Viruses. In *Plant Virus Disease Control*; Hadidi, A., Khetarpal, R.K., Koganezawa, H., Eds.; The American Phytopathological Society: Minnesota, 1998; 1–13.
2. Converse, R.H.; Martin, R.R. ELISA Methods for Plant Viruses. In *Serological Methods for Detection and Identification of Viral and Bacterial Plant Pathogens. A Laboratory Manual*; Hampton, R., Ball, E., De Boer, S., Eds.; The American Phytopathological Society: Minnesota, 1990; 179–196.
3. Tsuda, S.; Kameya-Iwaki, M.; Hanada, K.; Kouda, Y.; Hikata, M.; Tomaru, K. A novel detection and identification technique for plant viruses: Rapid immunofilter paper assay (RIPA). Plant Dis. **1992**, *76*, 466–469.
4. Makkouk, K.M.; Hsu, H.T.; Kumari, S.G. Detection of three plant viruses by dot-blot and tissue-blot immunoassays using chemiluminescent and chromogenic substrates. J. Phytopathol. **1993**, *139*, 97–102.
5. Milne, R.G.; Luisoni, E. Rapid immune electron microscopy of virus preparations. Methods Virol. **1977**, *6*, 265–281.
6. Dodds, J.A.; Morris, T.J.; Jordan, R.L. Plant viral double-stranded RNA. Annu. Rev. **1984,** *22,* 151–168.
7. Mullis, K.B.; Faloona, F.A. Specific synthesis of DNA in vitro via a polymerase catalyzed chain reaction. Methods Enzymol. **1987**, *155*, 335–350.
8. Compton, J. Nucleic acid sequence-based amplification. Nature **1991**, *350*, 91–92.
9. Maroon, C.J.M.; Zavriev, S. PCR-based tests for the detection of tobamoviruses and carlaviruses. Acta Hortic. **2002**, *568*, 117–122.
10. Szemes, M.; Klerks, M.M.; van den Heuvel, J.F.J.M.; Schoen, C.D. Development of a multiplex AmpliDet RNA assay for simultaneous detection and typing of potato virus Y isolates. J. Virol. Methods **2002**, *100*, 83–96.
11. Chandelier, A.; Dubois, N.; Baelen, F.; De Leener, F.; Warnon, S.; Remacle, J.; Lepoivre, P. RT-PCR-ELOSA tests on pooled sample units for the detection of virus Y in potato tubers. J. Virol. Methods **2001**, *91*, 99–108.
12. Lie, Y.S.; Petropolous, C.J. Advances in quantitative PCR technology: 5′ nuclease assays. Curr. Opin. Biotechnol. **1998**, *9*, 43–48.
13. Kozwich, D.; Johansen, K.A.; Landau, K.; Roehl, C.A.; Woronoff, S.; Roehl, P.A. Development of a novel, rapid integrated *Crytosporidium parvum* detection assay. Appl. Environ. Microbiol. **2000**, *66*, 2711–2717.

Virus-Induced Gene Silencing

S. P. Dinesh-Kumar
Rajendra Marathe
Radhamani Anandalakshmi
Yale University, New Haven, Connecticut, U.S.A.

INTRODUCTION

The profusion of genome sequences and expressed sequence tags (ESTs) has revealed the existence of a large number of novel genes whose sequences provide few clues as to their specific functions. Therefore, techniques that aid in the quick study of functions of these genes have significant practical value in the post-genomic era. From this perspective, virus-induced gene silencing (VIGS) and RNA interference (RNAi) techniques are promising, as they enable the researcher to link a gene to its function reliably and quickly by silencing its activity.

WHAT IS VIGS?

VIGS was first observed when plants infected with an engineered virus vector carrying host gene sequences showed suppression of the activity of the corresponding endogenous gene.[1] This inhibition of the gene activity was due to a posttranscriptional gene silencing (PTGS) mechanism similar to co-suppression in plants, quelling in fungi, and RNAi in animals. All these related processes involve sequence specific degradation of RNA via highly conserved cellular machinery.[2]

PTGS is currently understood as a form of defense that plants employ to protect themselves from invading foreign nucleic acids such as transposons and viruses, and it is targeted against double-stranded RNA (dsRNA). Most plant viruses make dsRNA intermediates during their replication. The defense machinery of the plant is able to sense the presence of viral dsRNA and cleave it into small RNA (siRNA) by the nuclease activity of enzymes such as DICER.[2] siRNAs help in the priming and amplification of the silencing signal, which then spreads systemically in the whole plant. When a recombinant virus harboring any plant gene sequence is used to infect a host, the dsRNA formed will be homologous to the transcripts from the target gene. Destruction of this dsRNA by PTGS and the formation of the siRNAs will eventually result in the systemic silencing of the target gene's expression. The phenotype of the plant silenced for a particular gene by VIGS mimics its loss-of-function mutant phenotype. Silencing of phytoene desaturase (PDS) gene using a TRV-based VIGS vector is shown in Fig. 1. The photo-bleaching effect seen is comparable to mutant PDS.

VIGS VECTOR

Several plant viruses have been used to develop VIGS vectors. VIGS vectors derived from tobacco mosaic virus (TMV), potato virus X (PVX), tomato golden mosaic virus (TGMV), and tobacco rattle virus (TRV) have shown varying degrees of success to silence genes in *Nicotiana benthamiana*, which is a very good host for most plant viruses. When choosing a virus to develop a VIGS vector, several factors need to be considered. Ideally, the virus should spread uniformly and rapidly throughout the plant, including the meristematic regions of the plant, without producing infection symptoms. Viruses that encode strong silencing suppressors such as HC-Pro in potyviruses or protein 2b in cucumber mosaic virus are undesirable to use as VIGS vectors.

The TRV-based vectors have several advantages over other virus vectors, as TRV-based vectors do not induce chlorotic or necrotic symptoms on plants, which are typical to virus infections. This makes the identification of the VIGS-induced phenotype much easier. In addition, TRV invades every cell of the plant and thereby induces uniform silencing.[3,4] PVX and TMV are unable to reach the meristematic or growing regions of a plant, but TRV and TGMV can induce silencing of genes in the meristems and flowers.[3,5]

TRV is a bipartite positive sense RNA virus. RNA-1 encodes the replicase and the movement protein and thus can multiply and spread in the absence of RNA2 (Fig. 2). The TRV-based VIGS vector has the cDNA clones of RNA1 and RNA2 inserted into a T-DNA expression cassette. The multiple cloning sites (MCS) created in RNA2 allow the cloning of target gene sequences for VIGS.[3,4] When *Agrobacterium tumefaciens* bacterial cultures containing TRV RNA1 and RNA2 constructs are mixed and infiltrated onto the leaves of *N. benthamiana*, viral RNA is

Encyclopedia of Plant and Crop Science
DOI: 10.1081/E-EPCS 120010638

Fig. 1 Silencing of the PDS gene. Infection of recombinant TRV carrying the PDS sequence silences endogenous PDS in *N. benthamiana* and causes inhibition of carotenoid biosynthesis resulting in photo-bleaching phenotype. On the left, the whole plant; on the right, an enlarged single leaf. (*View this art in color at www.dekker.com.*)

synthesized. These RNA transcripts then serve as templates for the further replication of viral RNA by the RNA-dependent RNA polymerase encoded by RNA1. Systemic infection by the recombinant TRV then brings about VIGS of the targeted plant host sequences.

VIGS AS A FUNCTIONAL GENOMICS TOOL

Several techniques are available to study gene function in plants including T-DNA and transposon-based insertion mutagenesis.[6,7] These methods have limitations such as gene target bias and difficulty in disrupting or tagging all genes. In addition, the high degree of gene duplication in plant gene families often results in insertion or deletion mutants of a single gene lacking an observable phenotype. Finally, lethal mutations in the above approaches are almost always lost. Recently, a single-stranded self-complementary (hairpin) RNA was successfully employed to suppress gene function.[8] All of these approaches, however, rely on the generation of transgenic lines, which is time consuming and even challenging in many economically important plants.

VIGS is a more desirable approach to study gene function.[9] It enables a specific gene to be silenced if its sequence is known. Moreover, it is conditional; hence, loss of mutations due to organismal lethality is less likely

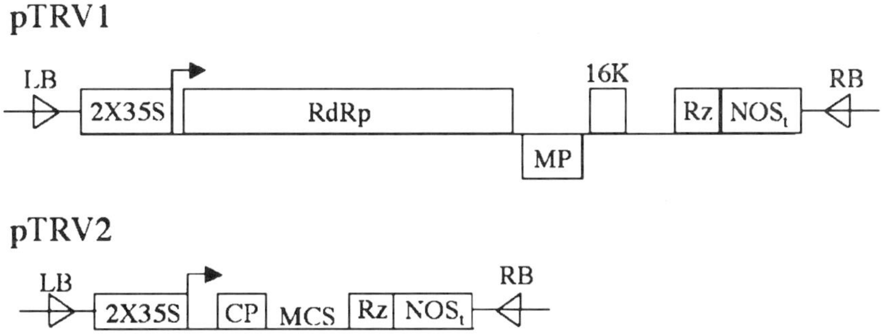

Fig. 2 TRV-based VIGS vectors. TRV cDNA clones were placed in between duplicated CaMV 35S promoter (2X35S) and nopaline synthase terminator (NOS_t) in a T-DNA vector. RdRp, RNA-dependent RNA polymerase; 16K, 16 kDa cysteine-rich protein; MP, movement protein; CP, coat protein; LB and RB, left and right borders of T-DNA; Rz, self-cleaving ribozyme; MCS, multiple cloning sites.

to occur. It is also an excellent system to study gene function in plants that are recalcitrant to Agrobacterium-mediated transformation. VIGS is fast, reliable, and specific. Once a VIGS vector is assembled, any host gene and its close sequence homologues could be efficiently silenced in 2–3 weeks to study its biological function.

VIGS APPLICATION

The VIGS has been successfully used in plants to knock out both endogenous genes and transgenes. VIGS of different components of the signaling pathways in flowering, disease resistance, and cell proliferation has helped in the identification of key factors. For example, suppression of the signaling proteins EDS1, NPR1, Rar1, and Sgt1 and of the COP9 signalosome by VIGS compromised the function of the TMV resistance gene *N* and showed that these proteins are essential for the resistance mechanism to operate.[4,10,11]

HERITABLE VIGS

One of the main disadvantages of the VIGS approach is that the phenotype observed cannot be transmitted to the next generation. Because of this, it is not possible to perform genetic crosses, suppressor or enhancer screens, and other genetic manipulations based on transmissible phenotypes. However, transgenic expression of a replicating PVX (termed PVX amplicon), containing plant exon sequences, consistently induces the silencing of the corresponding endogenous gene in subsequent generations.[12] Therefore, the generation of transgenic lines containing a VIGS vector-based amplicon will provide an invaluable resource for further genetic analyses.

CONCLUSION

Even though VIGS has been used successfully to study gene function in plants, it has not been applied in systematic studies at the whole-genome level. Two impressive studies in nematodes utilized the RNAi approach successfully to investigate loss-of-function phenotypes of all the predicted genes on chromosomes I and III.[13,14] Therefore, in future, we need to develop VIGS vectors that could be used to silence genes *en masse* in fully sequenced genomes, such as those of *A. thaliana* and rice. The VIGS approach, in conjunction with the available collection of insertion mutants, will help us to understand the function and interplay of genes that control plant growth, development, and responses to the environment.

ACKNOWLEDGMENTS

We thank Janet Stewart for editing the manuscript and members of the S. P. D-K lab for comments and critical reading of the manuscript. The National Science Foundation Grant DBI-0211872 supports VIGS work in S.P. D-K's laboratory.

ARTICLES OF FURTHER INTEREST

Functional Genomics and Gene Action Knockouts, p. 476
Gene Silencing: A Defense Mechanism Against Alien Genetic Information, p. 492
Genomic Approaches to Understanding Plant Pathogenic Bacteria, p. 524
Plant RNA Virus Diseases, p. 1023
Plant Virus: Structure and Assembly, p. 1029
Plant Viruses: Initiation of Infection, p. 1032
RNA-Mediated Silencing, p. 1106
Tobamoviral Vectors: Developing a Production System for Pharmaceuticals in Transfected Plants, p. 1229
Transgenes: Expression and Silencing of, p. 1242
Viral Host Genomics, p. 1269
Virus Movement in Plants, p. 1280

REFERENCES

1. Kumagai, M.H.; Donson, J.; della-Cioppa, G.; Harvey, D.; Hanley, K.; Grill, L.K. Cytoplasmic inhibition of carotenoid biosynthesis with virus-derived RNA. Proc. Natl. Acad. Sci. **1995**, *92*, 1679–1683.
2. Waterhouse, P.M.; Wang, M.B.; Lough, T. Gene silencing as an adaptive defence against viruses. Nature **2001**, *411*, 834–842.
3. Ratcliff, F.; Martin-Hernandez, A.M.; Baulcombe, D.C. Technical Advance. Tobacco rattle virus as a vector for analysis of gene function by silencing. Plant J. **2001**, *25*, 237–245.
4. Liu, Y.; Schiff, M.; Marathe, R.; Dinesh-Kumar, S.P. Tobacco Rar1, EDS1 and NPR1/NIM1 like genes are required for N-mediated resistance to tobacco mosaic virus. Plant J. **2002**, *30*, 415–429.
5. Peele, C.; Jordan, C.V.; Muangsan, N.; Turnage, M.; Egelkrout, E.; Eagle, P.; Hanley-Bowdoin, L.; Robertson,

D. Silencing of a meristematic gene using geminivirus-derived vectors. Plant J. **2001**, *27*, 357–366.
6. Azpiroz-Leehan, R.; Feldmann, K.A. T-DNA insertion mutagenesis in *Arabidopsis*: Going back and forth. Trends Genet. **1997**, *13*, 152–156.
7. Martienssen, R.A. Functional genomics: Probing plant gene function and expression with transposons. Proc. Natl. Acad. Sci. **1998**, *95*, 2021–2026.
8. Smith, N.A.; Singh, S.P.; Wang, M.B.; Stoutjesdijk, P.A.; Green, A.G.; Waterhouse, P.M. Total silencing by intron-spliced hairpin RNAs. Nature **2000**, *407*, 319–320.
9. Baulcombe, D.C. Fast forward genetics based on virus-induced gene silencing. Curr. Opin. Plant Biol. **1999**, *2*, 109–113.
10. Liu, Y.; Schiff, M.; Serino, G.; Deng, X.-W.; Dinesh-Kumar, S.P. Role of SCF ubiquitin-ligase and the COP9 Signalosome in the N Gene–mediated resistance response to tobacco mosaic virus. Plant Cell **2002**, *14*, 1483–1496.
11. Peart, J.R.; Cook, G.; Feys, B.J.; Parker, J.E.; Baulcombe, D.C. An EDS1 orthologue is required for N-mediated resistance against tobacco mosaic virus. Plant J. **2002**, *29*, 569–579.
12. Angell, S.M.; Baulcombe, D.C. Technical advance: Potato virus X amplicon-mediated silencing of nuclear genes. Plant J. **1999**, *20*, 357–362.
13. Fraser, A.G.; Kamath, R.S.; Zipperlen, P.; Martinez-Campos, M.; Sohrmann, M.; Ahringer, J. Functional genomic analysis of C. elegans chromosome I by systematic RNA interference. Nature **2000**, *408*, 325–330.
14. Gonczy, P.; Echeverri, G.; Oegema, K.; Coulson, A.; Jones, S.J.; Copley, R.R.; Duperon, J.; Oegema, J.; Brehm, M.; Cassin, E.; Hannak, E.; Kirkham, M.; Pichler, S.; Flohrs, K.; Goessen, A.; Leidel, S.; Alleaume, A.M.; Martin, C.; Ozlu, N.; Bork, P.; Hyman, A.A. Functional genomic analysis of cell division in C. elegans using RNAi of genes on chromosome III. Nature **2000**, *408*, 331–336.

Virus Movement in Plants

Richard S. Nelson
Xin Shun Ding
Shelly A. Carter
Samuel Roberts Noble Foundation, Inc., Ardmore, Oklahoma, U.S.A.

INTRODUCTION

Plant viruses cause the greatest damage to their host and maximize their potential for spread to uninfected hosts through movement from the initially infected cell. Virus systemic movement in plants is a multistep process, consisting of local spread from the initially infected cell (i.e., cell-to-cell movement) and later vascular-mediated spread to all plant tissues (i.e., vascular-dependent movement). Early studies detected virus spread through the appearance of a visual symptom, followed by biological assay to confirm virus presence. Later, the ability to detect virus components in sampled tissue using antibodies and microscopy enhanced the visualization of virus movement at the cellular level. More recently, the availability of confocal microscopes and fluorescent reporter genes expressed from cloned viral sequences has allowed near real-time observation of virus movement in living cells. Combining these resources with molecular and genetic techniques, viral and host factors involved in virus movement have been identified. Results from recent studies support the dictum that the lack of virus accumulation in systemic tissue is due to one of two causes—either the lack of a virus or host factor necessary for virus movement or the lack of a virus or host factor necessary to defeat or support host defenses. With further advances in methods to visualize virus movement in real time, to harvest and analyze the content of specific host cells, and to manipulate host and viral gene expression, scientists will be able to fully understand how viruses move in plants.

METHODS TO ANALYZE VIRUS ACCUMULATION AND MOVEMENT

A still useful method to estimate the rate of virus movement from an inoculated leaf to the remainder of the plant is by detachment of the inoculated leaf at various times postinoculation. Virus exit from the inoculated leaf is assumed by the appearance of visual symptoms in the remainder of the plant. Steam girdling, wherein phloem tissue in the vasculature is killed by heat treatment, is also used to study virus movement in specific vascular tissue (i.e., through phloem versus xylem tissue). For both leaf detachment and steam girdling studies, the presence of virus in the symptomatic tissue can be verified by inoculation of extracts from putatively infected tissue onto hosts known to yield visible lesions after virus infection (i.e., indicator hosts; see Fig. 1A).[1] Indicator hosts are also useful for estimating the rate of cell-to-cell virus spread by observing the expansion of the visible lesion. Virus movement is now determined more precisely through detection of virus-expressed proteins with microscopy.[1,2] Initially, virus movement at the cell level was determined using fixed and wax- or plastic-embedded tissue probed with antibodies specific to a virus protein (e.g., Fig. 1B and C). Newer techniques take advantage of our ability to insert and express reporter genes from cloned virus genomes.[2] The use of genes whose protein products fluoresce (e.g., green fluorescent protein (GFP)) allows the monitoring of virus movement in near real time within living cells (i.e., a signal appearing approximately one hour after synthesis of GFP) (Fig. 2).

CELL-TO-CELL MOVEMENT OF VIRUS

Virus transport between cells requires both viral and host factors. Identification of viral factors necessary for virus cell-to-cell movement has progressed rapidly owing to our ability to alter viral sequences within a cloned full-length genome or an isolated gene followed by infection or transfection of plants to observe the movement of the virus or viral gene product. In addition, our increased understanding of the structure, location, and composition of plasmodesmata (PD)—the intercellular tunnels that connect plant cells to each other—has aided the interpretation of results (Fig. 1D).[3,4] Viral factors necessary for cell-to-cell movement are reviewed in detail by others.[2–4] For most viruses, at least one viral protein that is required for virus cell-to-cell movement has been identified. For an increasing number of viruses, more than one protein is required for this activity. Consideration

Encyclopedia of Plant and Crop Science
DOI: 10.1081/E-EPCS 120019910

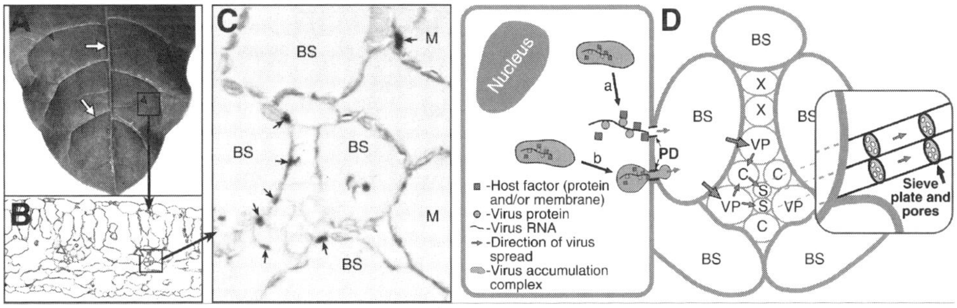

Fig. 1 Virus movement in *Nicotiana tabacum* (tobacco) after inoculation with *Tobacco mosaic virus*. (A) Local symptoms (chlorotic lesions indicated by arrowheads) on the inoculated leaf at 4 days postinoculation; major veins indicated by open arrows. (B) Cross section of a tobacco leaf showing location of minor veins (open arrowheads) relative to nonvascular cells. (C) Magnification of a minor vein showing location of virus movement protein within vascular cells (small arrows pointing to dark bodies). M = mesophyll cell, BS = bundle sheath cell. (D) Diagram indicating potential routes and methods (a and b) of spread by virus into and through phloem sieve elements for its rapid vascular transport. a: transport of virus movement form from replication complex to next cell, b: transport of virus movement form within a replication complex to the next cell. Sieve elements, plates, and pores shown in expanded view at right edge of panel. VP = vascular parenchyma cell, C = companion cell, S = sieve element, X = xylem vessel, PD = plasmodesmata. (*View this art in color at www.dekker.com.*)

is now given to the exciting prospect that cell-to-cell movement of some viruses may be coupled to the existence of virus accumulation complexes at the PD (Fig. 1D).[5]

Only recently have host genes been identified that are necessary for the cell-to-cell spread of viruses.[3,6] One recent example is the identification of a host protein that interacts directly with a viral protein necessary for virus cell-to-cell movement.[7] The host protein also interacts with a host enzyme thought to affect PD size and virus movement through its ability to alter callose deposition. Thus the host protein is proposed as an intermediary between a virus movement protein and a host enzyme considered important for PD-dependent virus movement.

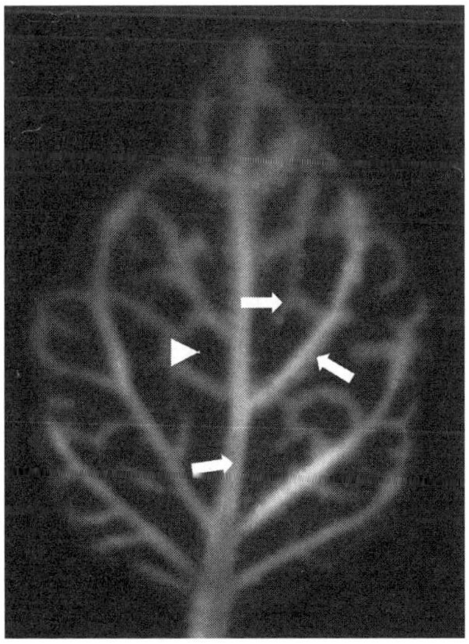

Fig. 2 Detection of *Tobacco mosaic virus* (TMV) accumulation along major veins in a developing, systemically infected tobacco leaf. TMV was modified to express green fluorescent protein whose fluorescence (green areas) was detected using confocal microscopy. Virus was associated with major veins (arrows) but not with minor veins (dark areas, arrowhead). (Photo courtesy of N.H. Cheng (Samuel Roberts Noble Foundation).) (*View this art in color at www.dekker.com.*)

VASCULAR-DEPENDENT ACCUMULATION OF VIRUS

The vasculature in plants is complicated tissue, regulated in its deposition and composed of many cells with unique structures, implying unique functions for each cell (Fig. 1A–D).[1] Within leaves, vascular tissue is often divided into two groups—minor and major veins—based on their structure and function (Fig. 1A and B, Fig. 2). Minor veins are sites of photoassimilate loading in mature leaves, whereas major veins are sites for the transport and release of photoassimilate in, respectively, mature and developing tissues. Using clones of viruses modified to express GFP, it was determined that both minor and major veins in mature leaves function as initial sites for vascular spread from nonvascular cells.[8,9] In addition, exit and accumulation of virus from developing leaves is from major veins only (Fig. 2),[3,8,9] with some exceptions.[10,11]

Virus transport through vascular tissue is either cell-to-cell via PD, or vascular via sieve tube pores, depending on the location of the virus within the tissue (Fig. 1D). Movement between the various cell types (e.g., bundle sheath cells to vascular parenchyma cells, etc.) may require viral and host factors different from those required for movement between nonvascular cells (e.g., mesophyll cells to mesophyll cells). Many viral proteins are necessary for vascular-dependent accumulation.[1] Some of these proteins, however, are also known to defeat a host defense response (discussed later). Further work is needed to determine the relationship between vascular transport and host defense suppression functions for all viral proteins required for the systemic accumulation of viruses in plants. Very few host loci have been identified whose function is required solely for vascular-dependent virus accumulation.[12]

HOST DEFENSIVE AND VIRUS COUNTER-DEFENSIVE MEASURES DURING VIRUS MOVEMENT

The preceding discussions highlight how viral and host factors actively support virus movement throughout a plant. Virus accumulation in systemic tissue also may be affected by the ability of the host to prevent virus accumulation and the ability of the virus to defeat the host defense. Such a host-defense/virus–counter-defense system can yield a vascular-dependent virus accumulation pattern similar to that displayed when factors actively supporting virus movement interact. Some host genes have been identified that prevent vascular-dependent accumulation of virus, but allow its cell-to-cell movement.[13,14] Viral genes that combat the effects of these genes have not been identified. However, the concept of host defensive and virus counter-defensive genes has been more fully developed through the identification of a host enzyme pathway that identifies, targets, and destroys aberrant or overexpressed RNA, including plant virus RNA.[15] RNA silencing is the term for this surveillance system. Recently it was determined that viral proteins, called suppressors, can defeat the surveillance system through multiple ways.[15] Many of the viral proteins that function to defeat RNA silencing were previously shown to support the vascular-dependent accumulation of particular viruses. Thus, in order to understand virus movement it is important to realize that the accumulation of virus in systemic tissue can be influenced by the ability of the virus to interact with host factors necessary for movement and the ability of the virus to avoid host defense factors.

CONCLUSION

Initial studies of virus movement identified strains of viruses with altered movement abilities and loci in plants that controlled movement. In the last 15 years, our understanding of virus movement has increased dramatically due to our ability to clone and alter virus genes and then express these genes alone, as a transgene, or from an infectious virus in plants. Reporter genes such as GFP have aided the analyses of virus spread by allowing near real-time imaging of virus accumulation patterns. Using these technical breakthroughs, researchers have determined rates and routes of virus movement. Systemic accumulation of virus is a balancing act between the expression of viral and host factors that actively support virus movement and those that defend the host or virus from deleterious effects. There will be continued research directed toward identifying viral, and especially host, factors that support or prevent virus accumulation in systemic tissue. However, the next major step in understanding virus movement will be to understand how these factors interact within single cells and at each cellular interphase to control this phenomenon.

ACKNOWLEDGMENTS

We thank Drs. Greg May and Justin Pita for valuable comment on the manuscript, Dr. Ning Hui Cheng (U.S. Department of Agriculture/Agricultural Research Service, Children's Nutrition Research Center, Baylor College of Medicine, Houston, TX 77030) for supplying the image for Fig. 2, Alisha Raney for text preparation, and Cuc Ly for figure preparation.

ARTICLES OF FURTHER INTEREST

Leaf Structure, p. 638
Plant DNA Virus Diseases, p. 960
Plant RNA Virus Diseases, p. 1023
Plant Viral Synergisms, p. 1026
RNA-Mediated Silencing, p. 1106
Viral Host Genomics, p. 1269
Virus-Induced Gene Silencing, p. 1276

REFERENCES

1. Nelson, R.S.; van Bel, A.J.E. The mystery of virus trafficking into, through, and out of vascular tissue. Prog. Bot. **1998**, 476–533.
2. Hull, R. *Matthews' Plant Virology*; Academic Press: San Diego, CA, 2001.
3. Roberts, A.G.; Oparka, K.J. Plasmodesmata and the control of symplastic transport. Plant Cell Environ. **2003**, *26*, 103–124.
4. Haywood, V.; Kragler, F.; Lucas, W.J. Plasmodesmata: Pathways for protein and ribonucleoprotein signaling. Plant Cell **2002**, *14*, 303–325.

5. Szécsi, J.; Ding, X.S.; Lim, C.O.; Bendahmane, M.; Cho, M.J.; Nelson, R.S.; Beachy, R.N. Development of tobacco mosaic virus infection sites in *Nicotiana benthamiana*. Mol. Plant-Microb. Interact. **1999**, *12*, 143–152.
6. Lellis, A.D.; Kasschau, K.D.; Whitham, S.A.; Carrington, J.C. Loss-of susceptibility mutants of *Arabidopsis thaliana* reveal an essential role for eIF(iso)4E during potyvirus infection. Curr. Biol. **2002**, *12*, 1046–1051.
7. Fridborg, I.; Grainger, J.; Page, A.; Coleman, M.; Findlay, K.; Angell, S. TIP, a novel host factor linking callose degradation with the cell-to-cell movement of *Potato virus X*. Mol. Plant-Microb. Interact. **2003**, *16*, 132–140.
8. Cheng, N.H.; Su, C.L.; Carter, S.A.; Nelson, R.S. Vascular invasion routes and systemic accumulation patterns of tobacco mosaic virus in *Nicotiana benthamiana*. Plant J. **2000**, *23*, 349–362.
9. Silva, M.S.; Wellink, J.; Goldbach, R.W.; van Lent, J.W.M. Phloem loading and unloading of *Cowpea mosaic virus* in *Vigna unguiculata*. J. Gen. Virol. **2002**, *83*, 1493–1504.
10. Sudarshana, M.R.; Wang, H.L.; Lucas, W.J.; Gilbertson, R.L. Dynamics of bean dwarf mosaic geminivirus cell-to-cell and long-distance movement in *Phaseolus vulgaris* revealed, using the green fluorescent protein. Mol. Plant-Microb. Interact. **1998**, *11*, 277–291.
11. Rajamäki, M.-L.; Valkonen, J.P.T. Localization of a potyvirus and the viral genome-linked protein in wild potato leaves at an early stage of systemic infection. Mol. Plant-Microb. Interact. **2003**, *16*, 25–34.
12. Lartey, R.T.; Ghoshroy, S.; Citovsky, V. Identification of an *Arabidopsis thaliana* mutation (*vsm*1) that restricts systemic movement of tobamoviruses. Mol. Plant-Microb. Interact. **1998**, *11*, 706–709.
13. Chisholm, S.T.; Parra, M.A.; Anderberg, R.J.; Carrington, J.C. Arabidopsis *RTM*1 and *RTM*2 genes function in phloem to restrict long-distance movement of tobacco etch virus. Plant Physiol. **2001**, *127*, 1667–1675.
14. Ueki, S.; Citovsky, V. The systemic movement of a tobamovirus is inhibited by a cadmium-ion-induced glycine-rich protein. Nat. Cell Biol. **2002**, *4*, 478–485.
15. Mlotshwa, S.; Voinnet, O.; Mette, M.F.; Matzke, M.; Vaucheret, H.; Ding, S.-W.; Pruss, G.; Vance, V.B. RNA silencing and the mobile silencing signal. Plant Cell **2002**. S14, S289–S301.

Water Deficits: Development

Kadambot H. M. Siddique
The University of Western Australia, Crawley, Australia

INTRODUCTION

Despite water being the earth's most abundant compound, water deficit is the single most important factor limiting crop yields worldwide. Of the earth's water supply, 97% is saline and 2.25% is trapped in ice, leaving only 0.75% available in freshwater aquifers, rivers, and lakes. Agriculture is the major consumer of fresh water worldwide (69%). Increasing demands being placed on both food and water resources throughout the world require that agriculture be more efficient in its water use without sacrificing production. To enable more efficient distribution and utilization of water resources, a greater understanding is required of plant–water relations, in particular the way in which water deficits develop and affect plant growth and productivity.

THE SOIL, PLANT, AND ATMOSPHERE CONTINUUM

The plant-water deficit that develops in any particular situation is the result of a complex combination of soil, plant, and atmospheric factors, all of which interact to control the rate of water absorption and loss.[1] In plants, water deficits develop as a consequence of water loss from the leaf as the stomata open to allow the uptake of CO_2 from the atmosphere for photosynthesis. The water lost by transpiration from the leaf mesophyll cells is replaced by water drawn from the soil through the root, stem, and leaf via the xylem. The movement involves both symplastic and apoplastic pathways, i.e., from cell to cell via the symplasm and along the cell walls and xylem vessels, respectively. Both pathways create resistances to flow between the soil and leaf. This pathway of movement is a continuum between the soil, plant, and atmosphere.

It is generally accepted that water moves through the soil–plant–atmosphere continuum along a gradient of decreasing water potential from the soil, through the plant, to the atmosphere. The driving force for the upward movement of water is the negative hydrostatic pressure (tension) that develops in the xylem of transpiring plants. Except for halophytes, the solute content of the xylem sap is usually very low, and thus its solute potential is very close to zero. Water forms a continuous liquid system from the soil up to the evaporating surfaces in the leaf mesophyll. When water evaporates from the leaf cells, the reduction in potential at the evaporating surface (the leaf cell walls) causes movement of water from the xylem to the evaporating surface, which in turn reduces the pressure of water in the xylem and causes the ascent of water. This reduced pressure is transmitted throughout the liquid continuum to the root surfaces, where it reduces the root water potential and causes uptake of water from the soil.

Changing water availability has the greatest effect on the soil–root boundary that forms an interface between the plant and its environment.[2] The difference in water potential between the plant and the soil depends on the evaporative demand, the extent to which the plant can meet that demand, and the water-conducting properties of the soil and plant. It was previously thought that there was no direct relationship between the water potential in the leaf and the water potential in the soil, with the soil–water potential acting only to establish an upper limit of recovery possible by the plant during the night.[1] However, current evidence suggests that plant growth regulators such as abscisic acid (ABA) may play an important role in transferring the message of increasing soil and root water deficits to the shoot.

WATER STATUS OF CROP PLANTS

The water status of a crop plant is usually defined in terms of its water content, water potential, or the components of water potential. The total water potential can be partitioned into components consisting of the osmotic potential, turgor pressure, matric potential, and gravitational potential.[1] In addition to water moving from the soil to the atmosphere, the lower water potential in the leaves and stems results in water moving from the leaf mesophyll cells and the parenchyma cells surrounding the xylem and phloem. This water not only acts as a reservoir of water buffering the plant against diurnal changes in water deficit, but also induces changes in cell volume. Seed growth is the final plant process affected by soil water stress (e.g., in grain legume species the water relations of the embryo are isolated from the water relations of the maternal

Encyclopedia of Plant and Crop Science
DOI: 10.1081/E-EPCS 120010406

plant). There is a close coupling of pod- to plant-water status, but a clear isolation of seed-from pod-water status. This results in relatively constant seed coat turgor despite changes in the water potential of the maternal plant environment. It is thought this homeostasis may be part of a mechanism to ensure continued seed filling and assimilate redistribution, even when low water potentials have reduced the current availability of assimilates from the leaves.[3]

Plants are in a position to extract water from the soil only when their water potential is lower than that present in the soil. Thus water in the plant is seldom in equilibrium with water in the soil. In fact, water deficits occur in the tissues of all transpiring plants as an inevitable consequence of the flow of water along a pathway in which frictional resistances and gravitational potential have to be overcome.[1] At times during the day, all plants (with the exception of those in humid climates) are undergoing some degree of water deficit. The extent of any imbalance between transpiration and water uptake is limited by the storage capacity of the plant: For crop, forage, and pasture species this is usually less than one-third of daily transpiration.

The difference in water potential between the plant and the soil depends on the rate of uptake of water from the soil and the water-conducting properties of the soil and plant. The progressive changes in soil– and plant–water potential as the soil dries out are presented schematically (Fig. 1).

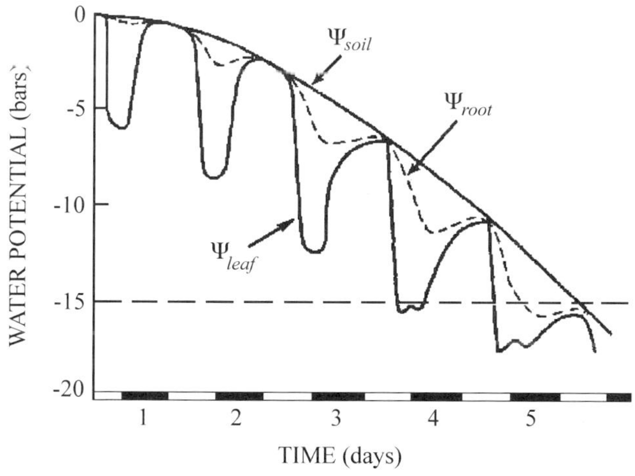

Fig. 1 Schematic representation of changes in leaf–water potential (ψ_{leaf}), root–water potential (ψ_{root}), and soil–water potential (ψ_{soil}) as transpiration proceeds during a drying cycle. The same evaporative conditions are considered to prevail each day; the horizontal dashed line indicates the value of ψ_{leaf} at which wilting occurs. (From Slatyer, R.O. 1967. *Plant–Water Relations*, Academic Press, New York.)

EVAPOTRANSPIRATION

Evaporation from plant communities is frequently termed evapotranspiration, to signify the combined soil-plant nature of the evaporating surfaces. The evaporation of water, whether it be from a free water surface, wet soil, or as transpiration from plants, is an energy-dependent process involving a change in state from the liquid to the vapor phase; the rate is a function of the vapor pressure gradient, the resistance to flow, and the ability of the soil and plant to transport water to the sites of evaporation.[4] The primary source of energy for this process is provided by radiation from the sun. Secondary sources include scattered and reflected radiation from the sky and clouds, as well as sensible heat transferred from the adjacent air, crop, and soil. The transport of water vapor molecules from the wet surfaces to the atmosphere occurs by molecular diffusion and turbulent eddy movement.[5]

MECHANISMS FOR CONTROL OF WATER TRANSPORT

The relative magnitude of water supply and water loss appears to determine the degree to which plants are water-deficient. However, because plants are living organisms, it is important to realize they do not provide solely a pathway for the movement of water from the soil to the atmosphere. Plants possess mechanisms for osmotic regulation that effectively control the flow of water according to their water status.[4]

The additional resistance encountered when evaporation occurs via plant surfaces is of great ecological importance and greatly affects evapotranspiration. In well hydrated plants, resistance is very small. Thus, the evapotranspiration from a soil completely covered by vegetation well supplied with water is very close to the evaporation from an open water surface and is called potential evaporation. Evapotranspiration from non-stressed crops is largely governed by meteorological conditions in the atmosphere external to the soil-plant system rather than by plant and soil factors. Crop canopy photosynthesis, transpiration, or leaf extension growth usually do not show midday depressions caused by high evaporative demand unless the soil water supply is depleted below a threshold of ca. 50% of total extractable soil water. The exact point of the threshold water content is difficult to distinguish and is usually extrapolated from measurements taken when evapotranspiration/maximum evaporation is clearly reduced by drought.[6]

The plant, in contrast to the soil, does exercise control over water loss through its stomatal and cuticular resistances in the water pathway and also through its ability to reduce the radiation load through active and/or passive changes in leaf orientation. Because these controls are located at the leaf surface, the internal water status of the plant is closely coupled to that of the soil and less closely coupled to the atmosphere.[1]

Regulation of stomatal conductance enables the plant to reversibly modify gas exchange with the atmosphere, with a time scale of only a few minutes. This short-term regulation may both optimize the dry matter production and offer some protection from xylem embolism or other detrimental effects of very low leaf water potential.[7] Stomata do not respond to changes in leaf–water potential or relative water content (RWC) until a critical threshold level of these parameters is reached, and then the stomata close over a narrow range of leaf–water potential or RWC. There is now considerable evidence to support the theory that a chemical message—originating in dehydrating roots and conveyed by the water flux towards the stomatal complex—can control stomatal conductance.[8]

Because stomata act as regulators for CO_2 exchange, water deficits sufficient to close stomata must also depress photosynthesis. Photosynthesis declines initially as a result of stomatal closure, but prolonged and severe water stress can lead to depression of chloroplast and enzyme activity and to nonstomatal effects on photosynthesis.[1] In shoot tissues, when water deficits limit the supply of CO_2 for photosynthesis, a number of reversible mechanisms down-regulate the efficiency of excitation energy transfer to Photosystem II.[2] Dark respiration is depressed whenever the water deficit is sufficiently great to close stomata and decrease photosynthesis, but the decrease in dark respiration is less than that of net photosynthesis.[1]

A high cuticular resistance is associated with either thicker cuticles or with cuticles enriched with hydrophobic materials. Although the plant cannot flexibly control this, cuticular resistance has been found to decrease with increasing temperature and atmospheric relative humidity.

MEASUREMENT OF PLANT–WATER RELATIONS

Any study of plant–water relations requires accurate methods for measuring them. Indirect methods that have been used to characterize the leaf water status include visible wilting, color change, leaf rolling, leaf temperature, leaf thickness, stomatal conductance, photosynthetic rate, and leaf permeability. Direct measurements can be undertaken using standard equipment such as the pressure chamber, psychrometers, and infrared gas analyzers. These have now been joined by newer technologies, such as the pressure probe, acoustic emission sensors, portable photosynthesis and fluorescence meters with solid-state circuitry, in addition to sophisticated modeling techniques for extrapolation to larger scales.[2]

Mass spectrometers can be used to trace source water through the measurement of a combination of stable radioactive carbon isotopes suspended in water, and may also have the potential to reveal the quantitative contribution to atmospheric fluxes of transpirational water derived from vegetation. Other approaches such as mass-balance techniques, micrometerological methods, and measurement of liquid flow and vapor fluxes can be used as a basis for scaling up to the level of canopies and stands of vegetation.

EFFECTS OF WATER DEFICITS

Long-term severe water deficits can have a negative effect on crop growth and yield, and may even result in senescence of the plant. It has been suggested that drought resistance of crops may be improved via breeding for specific phenological, morphological, physiological, or biochemical characteristics that improve yields in water-limited environments. These characteristics include early vigor, transpiration efficiency, stomatal control, abscisic acid accumulation, proline accumulation, osmotic adjustment, carbon remobilization, rapid grain growth, changes in hydraulic conductance, membrane stability, and lethal water potential.[9] It is, however, important to note that water deficits do not always lead to detrimental effects on yield. Mild or early water deficits may in fact increase yields or at least have no effect on yield, especially in indeterminate species.

CONCLUSION

The global atmospheric CO_2 was in a steady state at a concentration of about 280 μmol/mol 150 years ago and has been rising steadily since then as a result of human activity. Over the past 20 years, the annual rate of increase has varied, depending on global economics, between 1 and 2 μmol/mol.[10] The rise in atmospheric CO_2 concentration is likely to affect global climate and to cause regional changes in air temperature and humidity, length of growing season, precipitation, and evaporation, all of which will affect plant water relations.[10] At a plant scale we may expect to see substantial but rather variable increases in plant growth and changes in the phenology of growth processes.

Stomatal conductance has long been known to decrease at high levels of CO_2. Empirical data suggest that increases in plant size, particularly in leaf area, may offset any possible advantages to water economy resulting from stomatal closure. There is also evidence to suggest that the growth-enhancing effects of elevated CO_2 may to some extent offset the growth-reducing effects of water stress.[10] Changes in land use and regional climate brought about by anthropogenic incursions mean that crop productivity in more marginal rainfall areas will be of increasing importance in world agriculture. Expansion to these areas requires the ability to identify and breed for characteristics that better enable crop plants to cope with limited water availability.

ARTICLES OF FURTHER INTEREST

Drought and Drought Resistance, p. 386
Osmotic Adjustment and Osmoregulation, p. 850
Water Use Efficiency Including Carbon Isotope Discrimination, p. 1288

REFERENCES

1. Begg, J.E.; Turner, N.C. Crop water deficits. Adv. Agron. **1976**, *28*, 161–217.
2. Karamanos, A.J. The Development of Water Deficits in Plants. In *Water Stress on Plants*; Simpson, G.A., Ed.; Praeger Publishers: New York, 1981; 34–84.
3. Shackel, K.A.; Turner, N.C. Seed coat cell turgor in chickpea is independent of changes in plant and pod water potential. J. Exp. Bot. **2000**, *51* (346), 895–900.
4. Turner, N.C.; Begg, J.E. Plant–water relations and adaptation to stress. Plant Soil **1981**, *58*, 97–131.
5. Gardner, W.R.; Jury, W.A.; Knight, J. Water Uptake by Vegetation. In *Heat and Mass Transfer in the Biosphere. 1. Transfer Processes in Plant Environment*; de Vries, D.A., Afgan, N.H., Eds.; Scripta Book Co.: Washington, 1975; 443–456.
6. Ritchie, J.T. Water dynamics in the soil–plant–atmosphere system. Plant Soil **1981**, *58*, 81–96.
7. Smith, J.A.C.; Griffiths, H. Integrating Plant Water Deficits From Cell to Community. In *Water Deficits: Plant Responses From Cell to Community*; Smith, J.A.C., Griffiths, H., Eds.; BIOS Scientific Publishers Ltd.: Oxford, UK, 1993; 1–4.
8. Tardieu, F.; Davies, W.J. Root–Shoot Communication and Whole-Plant Regulation of Water Flux. In *Water Deficits: Plant Responses from Cell to Community*; Smith, J.A.C., Griffiths, H., Eds.; BIOS Scientific Publishers Ltd.: Oxford, UK, 1993; 147–162.
9. Turner, N.C. Further progress in crop water relations. Adv. Agron. **1997**, *58*, 293–338.
10. Jarvis, P.G. Global Change and Plant Water Relations. In *Water Transport in Plants Under Climatic Stress*; Borghetti, M., Grace, J., Raschi, A., Eds.; Cambridge University Press: Cambridge, UK, 1993; 1–13.

Water Use Efficiency Including Carbon Isotope Discrimination

Anthony G. Condon
CSIRO Plant Industry, Canberra, Australia

INTRODUCTION

Less water available for irrigation and the demand for ever-greater productivity from rainfed and irrigated agriculture are major challenges for the 21st century. Agronomic and engineering improvements can help meet these challenges. There are also several plant attributes that have potential to be exploited for greater water use efficiency of crop production. In this overview, plant attributes and agronomic practices that may contribute to greater water use efficiency of rainfed agriculture will be emphasized, although many are also important for irrigated agriculture. Examples will be highlighted from current research with temperate cereals.

AVENUES FOR GREATER WATER USE EFFICIENCY

The term ''water use efficiency'' has a variety of definitions.[1] In agronomic contexts, most of these definitions describe a ratio of grain yield and some measure of water used or water available for crop growth.[1] Yield formation of rainfed crops grown on a finite supply of water can be described more explicitly as a function of 1) the amount of water used by the crop (ET); 2) the water use efficiency for biomass growth (W); and 3) the harvest index (HI), i.e., how much of the final biomass is partitioned to grain.[2] Closer dissection[3,4] reveals two important avenues for improving crop W: increasing the amount of total water use that is actually transpired by the crop rather than lost as evaporation from the soil surface and improving crop transpiration efficiency (W_T), i.e., biomass production per unit water transpired.

MAXIMIZING TRANSPIRATION

Restricting Evaporation from the Soil Surface

Evaporation from the soil surface can account for between 20 and 70% of growing-season water use, varying with how frequently the soil is rewetted and the rate and extent of canopy closure. So, restricting water lost by evaporation from the soil surface can dramatically improve crop W. A reduction in soil evaporation to promote greater transpiration is most easily achieved through the rapid development of leaf area to shade the soil surface from direct solar radiation.[3,4]

Good stand establishment and vigorous early plant growth will both contribute to rapid leaf area development. Good stand establishment is best achieved by plants with vigorous growth of the primary shoot that extends to the soil surface, i.e., either the coleoptile or the hypocotyl. Desirable traits are primary shoots that reach the soil surface more quickly if seed is sown relatively shallow; shoots that reach the soil surface much more consistently if seed is sown deep, such as when farmers are seeding into receding moisture; and shoots that are better able to penetrate crusted topsoils and retained stubble.[4]

Once plants have emerged, traits important for vigorous early leaf area growth may be associated with a larger embryo size, a greater relative growth rate, or a greater investment in leaf area per unit plant growth. Barley, for example, has a much faster rate of early canopy growth than bread wheat due to a larger embryo size and higher specific leaf area. These desirable traits have been combined in a wheat breeding program in Australia to produce new parental lines with early leaf area growth that is double that of current varieties.[4] Soil shading may be further enhanced by more prostrate leaf display.

Agronomic practices such as retention of residues from previous crops can be effective in shading and insulating the soil surface to restrict soil evaporation. Sowing an adequate density of large, healthy seeds at the optimum depth in a well-prepared seedbed will promote good stand establishment. Vigorous leaf area growth can be promoted and sustained by providing adequate levels of major nutrients such as N and P, removing imbalances of minor nutrients and pH, controlling shoot and root diseases, and minimizing soil compaction and waterlogging.

Capturing and Extracting Subsoil Moisture

Water that drains beyond the depth of rooting is lost for dry matter production by the crop. Practices that minimize

Encyclopedia of Plant and Crop Science
DOI: 10.1081/E-EPCS 120010408

drainage in irrigated and rainfed agriculture are important, not just for greater water use efficiency but also for avoiding environmental degradation. Water extracted by the crop from depth will all be used in transpiration, and so extracting available subsoil moisture is important for increasing W.

Management practices and plant attributes that promote vigorous shoot growth will generally also promote vigorous root growth, encouraging water capture. In many rainfed and irrigated cropping systems, there may be constraints, such as waterlogging, compacted soil layers, nutrient or pH imbalance, and root pathogens, that restrict the depth of rooting or water uptake from depth. Growing "probe" genotypes known to be susceptible to or tolerant of a particular condition can be a useful way of identifying a particular subsoil constraint.

MAXIMIZING TRANSPIRATION EFFICIENCY

At its most basic, W_T may be defined as the ratio of the instantaneous rates of CO_2 assimilation (A) and transpiration (T) at the stomata.[5] These instantaneous rates can be described by relatively simple equations.

$$A = g_c(c_a - c_i) \qquad (1)$$

$$T = g_w(e_i - e_a) \qquad (2)$$

Eqs. 1 and 2 both consist of two components: the stomatal conductance (g) to CO_2 (g_c) or water vapor (g_w) and the concentration gradient of CO_2 ($c_a - c_i$) or water vapor ($e_i - e_a$) between the air and inside the leaf.

Leaf-level W_T can be closely approximated by Eq. 3, which combines Eqs. 1 and 2 and in which the factor 0.6 refers to the relative diffusivities of CO_2 and water vapor in air.[5,6]

$$W_T = 0.6c_a(1 - c_i/c_a)/(e_i - e_a) \qquad (3)$$

Eq. 3 indicates two main avenues for improving W_T. One is to reduce ($e_i - e_a$), the driving gradient for transpiration. The other is to reduce c_i (c_a being assumed constant) to increase the gradient of CO_2 concentration.[1,5]

Changing the Driving Gradient for Transpiration

The most effective way to improve crop W_T is to achieve a larger proportion of crop growth when evaporative demand is low,[1,4] i.e., when e_a is high relative to e_i. Choosing an appropriate crop or variety for the location and sowing the crop at the recommended time are vital management activities. Crop or variety choice can be expanded by breeding to adjust sowing time and crop phenology. Eliminating disease susceptibility may present additional breeding opportunities to adjust sowing time and phenology. Promoting greater early leaf area growth in temperate cereals increases transpiration when evaporative demand is low.

Leaf temperature determines the value of e_i, so traits that reduce leaf temperature will help to reduce e_i. Evaporative demand is usually greatest during the second half of the day. Diurnal changes in stomatal conductance, leaf rolling, and other leaf movements may reduce water loss, although they are also likely to reduce carbon gain.

Changing the Leaf Internal CO_2 Concentration

In theory, large gains in W_T are possible for relatively modest changes in c_i.[5] According to Eq. 3, changing c_i/c_a from 0.7 (a typical value for nonstressed plants of C_3 species) to 0.6 should result in a gain of 33% in W_T, recalling that W_T is a function of $(1 - c_i/c_a)$. Measuring genotypic differences in c_i/c_a is tedious. Thus, any rapid, indirect measure of variation in c_i/c_a has great appeal. Carbon isotope discrimination provides such a measure.[5,6]

Carbon Isotope Discrimination

Carbon isotope discrimination ($\Delta^{13}C$) is a measure of the ratio of the stable isotopes of carbon ($^{13}C/^{12}C$) in plant material relative to the value of the same ratio in the atmosphere.[6] Approximately 1% of atmospheric CO_2 contains ^{13}C. Plants of C_3 species contain fractionally less than this, primarily because of discrimination against ^{13}C during two processes: firstly, during diffusion of CO_2 into the leaf through the stomata and, secondly, during the first key step in CO_2 fixation by C_3 plants, catalyzed by the enzyme Rubisco (ribulose-1,5-bisphosphate carboxylase).[5]

Theory and experimental data indicate that discrimination against ^{13}C is proportional to c_i/c_a (Eq. 4).

$$\Delta^{13}C = 4.4 + 22.6c_i/c_a \qquad (4)$$

Eq. 4 indicates that $\Delta^{13}C$ and c_i/c_a are positively related, whereas Eq. 3 indicates a negative relationship between W_T and c_i/c_a. Thus, $\Delta^{13}C$ and W_T should be negatively related.[5,6] This negative relationship has been confirmed in numerous glasshouse studies with genotypes of several C_3 crop species.[6,7] Genetic variation in $\Delta^{13}C$ is substantial in all C_3 crop species tested. It is likely that $\Delta^{13}C$ is under the control of many genes[7,9] because low $\Delta^{13}C$

(high W_T) may result from low stomatal conductance and/or high photosynthetic capacity. Nonetheless, the heritability of $\Delta^{13}C$ has often been found to be high.

Relationships Between Yield and Carbon Isotope Discrimination

Despite the apparent utility of $\Delta^{13}C$, complications have arisen in scaling up from the leaf and plant to W and yield in field stands.[7–10] There have been few examples where low $\Delta^{13}C$ has been associated with higher yield.[8,10] Examples with cereals have been studies conducted in environments where rainfall is very seasonal, crops were sown after this rainfall, and crop transpiration has dominated total crop water use. For one species, peanut, field studies consistently show greater biomass production to be associated with higher W_T.[11] Positive relationships between yield and $\Delta^{13}C$ have been observed in many instances with cereals and other species, contrary to expectations from the negative relationship between W_T and $\Delta^{13}C$ found in the glasshouse.[7,8,10] Many of these studies have been with temperate cereals in environments where water supply was not a major limitation to growth or in relatively favorable Mediterranean-type environments with substantial rainfall in the pre-anthesis phase.

Several factors could account for the dominance of positive relationships between yield and $\Delta^{13}C$. Low $\Delta^{13}C$ tends to be associated with slower growth rate in the absence of water stress (not the case in peanut). Slow growth rate might be expected if low $\Delta^{13}C$ is associated with low stomatal conductance. Faster crop growth rate associated with high $\Delta^{13}C$ may give a yield advantage if soil water is replenished by rainfall or irrigation or if greater transpiration is compensated by less soil evaporation.[3,9] Low $\Delta^{13}C$ tends to be associated with lower harvest index and later flowering.[8,10] A recent study with wheat has shown that many of these associations may be overcome by using back-crossing to transfer low $\Delta^{13}C$ into high-yielding backgrounds.[10]

CONCLUSION

Maximizing Transpiration

Vigorous early growth to reduce evaporation from the soil surface in favor of transpiration should be targeted in environments where the soil surface is frequently rewetted. Vigorous early growth may risk exhausting soil water reserves too rapidly. Combining earlier flowering with vigorous early growth may help reduce this risk.[4] An alternative strategy may be to combine faster early growth with higher W_T (low $\Delta^{13}C$),[10] but achieving the combination may be difficult in practice.

Exploiting Variation in Transpiration Efficiency

Environment and species combinations should be identified for which changes in crop phenology can be made that maximize growth under conditions of low evaporative demand. Promoting vigorous growth at low evaporative demand through effective crop management and targeted breeding will also prove useful, provided consequent risks are minimized.

Exploiting Variation in Carbon Isotope Discrimination

If $\Delta^{13}C$ is to be used for improving crop W_T, it is likely to have ready application in dry environments where transpiration accounts for a large proportion of crop water use.[8,10] Low $\Delta^{13}C$ is unlikely to be important in environments where a large proportion of water is evaporated from the soil surface unless it can be combined with faster early leaf area growth (see ''Maximizing Transpiration''). More consistent yield advances from selecting for low $\Delta^{13}C$ may be made by identifying those species or genotypes of species in which the association between $\Delta^{13}C$ and crop growth rate is positive, as in peanut. Conversely, it is possible that selection for high $\Delta^{13}C$ may be useful to maximize assimilation rate of C_3 crop species grown in irrigated or moist environments, or if selection for high $\Delta^{13}C$ results in faster canopy growth in short-season environments or where evaporation from the soil surface is substantial.

ARTICLES OF FURTHER INTEREST

Breeding: Genotype-by-Environment Interaction, p. 218
Crop Responses to Elevated Carbon Dioxide, p. 346
Drought and Drought Resistance, p. 386
Drought Resistant Ideotypes: Future Technologies and Approaches for the Development of, p. 391
Leaves and Canopies: Physical Environment, p. 646
Leaves and the Effects of Elevated Carbon Dioxide Levels, p. 648
Water Deficits: Development, p. 1284

REFERENCES

1. Tanner, C.B.; Sinclair, T.R. Efficient Water Use in Crop Production: Research or Re-search? *Limitations to Efficient Water Use in Crop Production*; Taylor, H.M., Jordan, W.R., Sinclair, T.R., Eds.; ASA; CSSA; SSSA: Madison, WI, 1983; 1–27.

2. Passioura, J.B. Grain yield, harvest index and water use of wheat. J. Aust. Inst. Agric. Sci. **1977**, *43*, 117–120.
3. Condon, A.G.; Richards, R.A. Exploiting Genetic Variation in Transpiration Efficiency in Wheat: An Agronomic View. In *Stable Isotopes and Plant Carbon–Water Relations*; Ehleringer, J.R., Hall, A.E., Farquhar, G.D., Eds.; Academic Press: San Diego, CA, 1993; 435–450.
4. Richards, R.A.; Rebetzke, G.J.; Condon, A.G.; van Herwaarden, A.F. Breeding opportunities for increasing the efficiency of water use and crop yield in temperate cereals. Crop Sci. **2002**, *42*, 111–121.
5. Farquhar, G.D.; O'Leary, M.H.; Berry, J.A. On the relationship between carbon isotopic discrimination and the intercellular carbon dioxide concentration in leaves. Aust. J. Plant Physiol. **1982**, *9*, 121–137.
6. Farquhar, G.D.; Richards, R.A. Isotopic composition of plant carbon correlates with water-use efficiency of wheat genotypes. Aust. J. Plant Physiol. **1984**, *11*, 539–552.
7. Hall, A.E.; Richards, R.A.; Condon, A.G.; Wright, G.C.; Farquhar, G.D. Carbon Isotope Discrimination and Plant Breeding. In *Plant Breeding Reviews*; Janick, J., Ed.; Wiley: New York, 1994; Vol. 12, 81–113.
8. Condon, A.G.; Hall, A.E. Adaptation to Diverse Environments: Genotypic Variation in Water-Use Efficiency Within Crop Species. In *Agricultural Ecology*; Jackson, L.E., Ed.; Academic Press: San Diego, CA, 1997; 79–116.
9. Condon, A.G.; Richards, R.A.; Farquhar, G.D. Relationships between carbon isotope discrimination, water use efficiency and transpiration efficiency for dryland wheat. Aust. J. Agric. Res. **1993**, *44*, 1693–1711.
10. Condon, A.G.; Richards, R.A.; Rebetzke, G.J.; Farquhar, G.D. Improving intrinsic water-use efficiency and crop yield. Crop Sci. **2002**, *42*, 122–131.
11. Wright, G.C.; Hubick, K.T.; Farquhar, G.D.; Nageswara Rao, R.C. Genetic and Environmental Variation in Transpiration Efficiency and Its Correlation with Carbon Isotope Discrimination and Specific Leaf Area in Peanut. In *Stable Isotopes and Plant Carbon–Water Relations*; Ehleringer, J.R., Hall, A.E., Farquhar, G.D., Eds.; Academic Press: San Diego, CA, 1993; 245–267.

W

Wax Esters from Transgenic Plants

Kathryn D. Lardizabal
Monsanto, Calgene Campus, Davis, California, U.S.A.

INTRODUCTION

Wax esters are long-chain linear esters of fatty acids and fatty alcohols. They are commonly found as a component of the cuticle that covers plant surfaces where they help provide protection against stresses such as desiccation, wetting, and pathogen attack. A small number of plants produce and store wax esters in the seed instead of the more commonly found triacylglycerols. These seed storage lipids provide energy for use during germination. Of these plants, jojoba (*Simmondsia chinensis*) is the only one that is commercially cultivated; however, the economics involved in the growth and harvesting of this crop have limited the use of jojoba oil as an industrial feedstock. With the advent of biotechnology, the opportunity arose to overcome this limitation and produce wax esters in a temperate crop, such as *Brassica napus*, which is grown on far greater acreage with more favorable economics. The genes involved in the production of wax esters in jojoba were cloned and introduced into *Brassica napus* in order to produce wax esters transgenically.

WAX BIOSYNTHESIS IN JOJOBA

Jojoba (*Simmondsia chinensis*) produces long-chain wax esters exclusively as a seed storage lipid rather than triacylglycerols.[1–3] A cross section of developing jojoba embryos reveals the presence of lipid bodies analogous to those present in all oilseeds.[4] Two proteins are responsible for this biosynthesis.[5,6] A fatty acyl-CoA reductase (FAR) carries out the four-electron reduction of an acyl-CoA to form a primary alcohol. The enzyme requires NADPH as the electron donor and prefers very-long-chain (C20–C24) fatty acyl-CoA substrates. This preference is reflected in the wax composition of jojoba, where greater than 93% of the fatty acids are C20–C24. Though the reaction proceeds through an aldehyde intermediate, free aldehyde is not released in the process. Next, a fatty alcohol acyltransferase (wax synthase, WS) transfers a second acyl-CoA molecule to the primary alcohol to form the wax ester (Fig. 1). Since the FAR prefers very-long-chain substrates, an efficient mechanism for fatty acid elongation system is essential.[7] The endoplasmic reticulum contains such a system, which involves four enzymatic activities; however, the first enzyme, beta-ketoacyl-CoA synthase (KCS), was found to be the rate-limiting and chain-length-determining step.[8]

IDENTIFICATION OF THE GENES RESPONSIBLE FOR WAX BIOSYNTHESIS

The enzyme activities involved in wax production in jojoba were described in the late 1970s; however, the membrane-associated nature of the proteins hindered their identification for another 20 years. Following cell fractionation of developing jojoba embryos, the activities are found in both the membrane and floating wax fractions. Purification of the proteins was achieved by detergent solubilization of the activities and chromatographic separation of the proteins followed by peptide sequencing and cloning of the corresponding genes.

The jojoba FAR was identified as a 1.7 kb gene that encoded a protein with a molecular mass of 56.2 kD and a pI of 8.76.[9] Hydropathy analysis suggested the presence of 1–2 transmembrane domains. Expression of the protein in *E. coli* resulted in the production of a small amount of primary alcohol in the cells. The jojoba FAR was expressed in *Brassica napus* in combination with the KCS from *Lunaria annua*, in order to maximize the substrate available to the enzyme. Low levels of alcohol were detected to about 6%. Further investigation, using nuclear magnetic resonance spectroscopy and GC/mass spectroscopy, showed that a majority of the alcohols were esterified to fatty acids, though the presence of unesterified alcohol indicated the incorporation into wax was incomplete.[9] This confirmed the ability of the jojoba FAR to produce alcohols transgenically in another plant and also identified the presence of an endogenous wax-ester-forming activity in *Brassica napus*.

The jojoba WS was also purified from membrane fractions.[10] This enzyme proved to be more versatile, demonstrating activity toward a wide range of acyl-CoA and alcohol (C8–C24; saturated; monounsaturated and polyunsaturated) substrates in in vitro assays. It was also more difficult to purify due to its inhibition by detergents

Encyclopedia of Plant and Crop Science
DOI: 10.1081/E-EPCS 120010440

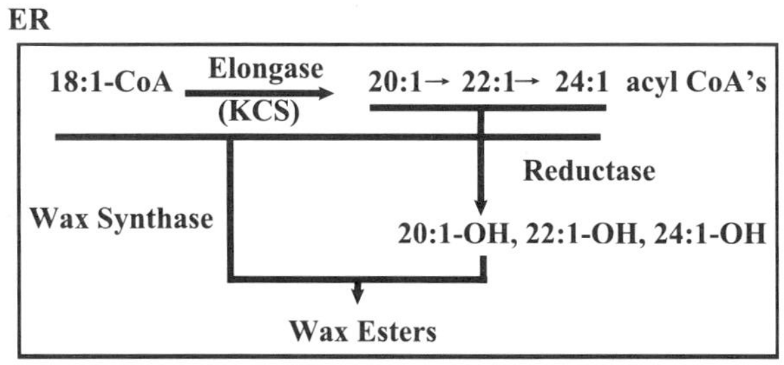

Fig. 1 Wax biosynthetic pathway in jojoba.

and a requirement for phospholipid reconstitution in order to restore enzymatic function. The jojoba WS was identified as a 1.3 kb gene that encoded a protein with a molecular mass of 40.2 kD and a pI of 9.86. Hydropathy analysis suggested the presence of 7–9 transmembrane domains. The jojoba WS was cloned and expressed in the model oilseed plant *Arabidopsis thaliana* in combination with jojoba FAR and *Lunaria* KCS. All three genes were under the control of napin regulatory sequences that directed gene expression to the seed and coordinated the timing of expression to coincide with fatty acid deposition during embryo development. Intact wax levels were determined on seed pools using ^{13}C-NMR. The highest expressing plants stored approximately half of their lipid in the form of wax. Further analysis showed that up to 68% wax was produced in individual seeds.[10]

Table 1 Data from the first (R1) generation of plants transformed with cDNAs encoding wax biosynthetic enzymes

R1 transformant	% Wax
8559-26	13.79
8559-13	9.91
8559-32	9.10
8559-28	7.85
8559-16	5.82
8559-34	5.05
8559-3	4.71
8559-8	4.36
8559-37	3.24
8559-1	3.24
Reston control	0.22

PRODUCTION OF WAX ESTERS IN *Brassica napus*

Identification of the genes involved in wax production provided the tools necessary to produce wax esters transgenically in a commercial crop. The three-gene construct (jojoba WS: jojoba FAR: *Lunaria* KCS) was transformed into *Brassica napus* var. Reston, and a range of wax levels (0.2–14%) was detected in the transgenic R1 seed pools (Table 1). Single seed analysis of the highest

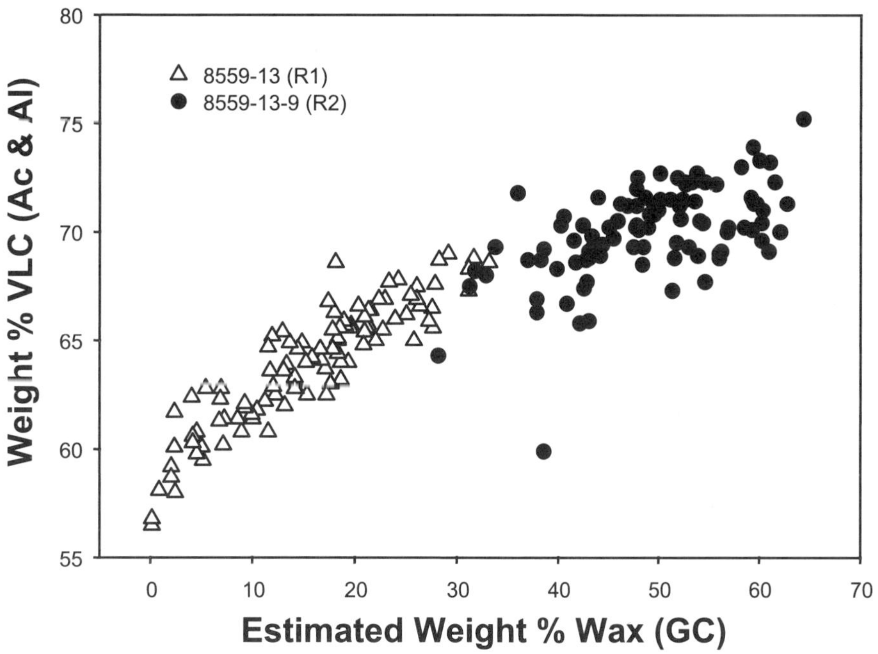

Fig. 2 The weight percent of very-long-chain (VLC) fatty acids and alcohols of individual half-seeds of transgenic *Brassica* determined by GC analysis is plotted against the estimated weight percent wax determined by GC analysis. Data from the first (R1) generation of seed (8559-13) and data from the second (R2) generation of seed (8559-13-9).

wax-containing lines indicated that up to 33% wax ester and 68% very-long-chain-fatty-acids (VLCFA) were made. Selected lines were advanced to the next generation (R2) where wax ester content increased to 64% and VLCFA content increased to 75%. It is important to note that wax and VLCFA levels for a given population represent an average of the single seed content. In R2 seed, wax levels range from 28–64% with an average of 49%, and VLCFA levels range from 64–75% with an average of 70% (Fig. 2).

CONCLUSION

Overexpression of wax biosynthetic genes (WS, FAR, KCS) in the seeds of *Brassica napus* resulted in the storage of approximately 50% of the lipid as wax ester. In jojoba, these genes contribute to the high levels of wax esters (50–60% by weight of the seed) found there. The proteins encoded by these genes clearly are optimized for this production since wax esters represent the entire stored form of lipid in jojoba seeds. Since acyl-CoAs are the substrates for both wax ester biosynthesis and triacylglycerol biosynthesis, the achievement of a 50% diversion ofsubstrate, in developing *Brassica* seeds, into a novel storage product (wax ester) indicates that the wax biosynthetic enzymes compete efficiently for these substrates. The stored wax ester also appears to be stable and nontoxic to *Brassica* embryos since the seeds can be propagated.

The capacity to produce wax esters resides in the genetic makeup of nearly all plants since they are found as a component of the cuticle. In *Brassica*, expression of the jojoba FAR in combination with the *Lunaria* KCS produced a limited amount of wax ester in the absence of jojoba WS. A similar observation was made in *Arabidopsis*. The source of this wax-ester-forming activity was presumed to come from one or more of the seven genes identified in *Arabidopsis* that share homology to jojoba WS, an indication that these genes are present in *Brassica* and, most likely, other plant species.

Many waxes found in nature are used in the lubricant, food, and cosmetic industries. In particular, jojoba oil is used in cosmetics because of its moisturizing ability, despite its high cost. By demonstrating the production of high levels of wax esters in transgenic *Brassica*, the possibility for commercial production of wax esters in a temperate crop has become reality.

ARTICLES OF FURTHER INTEREST

Genetically Modified Oil Crops, p. 509
Lipid Metabolism, p. 659

REFERENCES

1. Post-Beittenmiller, D. Biochemistry and molecular biology of wax production in plants. Ann. Rev. Plant Physiol. Plant Mol. Biol. **1996**, *47*, 405–430.
2. Miwa, T.K. Jojoba oil wax esters and derived fatty acids and alcohols: Gas chromatographic analyses. J. Am. Oil Chem. Soc. **1971**, *48*, 259–264.
3. Kartha, A.R.; Singh, S.P. The in vivo ''quantum'' synthesis of reserve waxes in seeds of *Murraya koenigii*. Chem. Ind. **1969**, 1342–1343.
4. Muller, L.L.; Hensarling, T.P.; Jacks, T.J. Cellular ultrastructure of jojoba seed. J. Am. Oil Chem. Soc. **1975**, *52*, 164–165.
5. Pollard, M.R.; McKeon, T.; Gupta, L.M.; Stumpf, P.K. Studies on biosynthesis of waxes by developing jojoba seed. II. The demonstration of wax biosynthesis by cell-free homogenates. Lipids **1979**, *14*, 651–662.
6. Wu, X.-Y.; Moreau, R.A.; Stumpf, P.K. Studies of biosynthesis of waxes by developing Jojoba seed: III. Biosynthesis of wax esters from acyl-CoA and long chain alcohols. Lipids **1981**, *6*, 897–902.
7. Lassner, M.W.; Lardizabal, K.; Metz, J.G. A jojoba β-ketoacyl-CoA synthase cDNA complements the canola fatty acid elongation mutation in transgenic plants. Plant Cell **1996**, *8*, 281–292.
8. Millar, A.A.; Kunst, L. Very-long-chain fatty acid biosynthesis in controlled through the expression and specificity of the condensing enzyme. Plant J. **1997**, *12*, 121–131.
9. Metz, J.G.; Pollard, M.R.; Anderson, L.; Hayes, T.; Lassner, M.W. Purification of a jojoba embryo fatty acyl-coenzyme A reductase and expression of its cDNA in high erucic acid rapeseed. Plant Physiol. **2000**, *122*, 635–644.
10. Lardizabal, K.; Metz, J.G.; Sakamoto, T.; Hutton, W.C.; Pollard, M.R.; Lassner, M.W. Purification of a jojoba embryo wax synthase, cloning of its cDNA, and production of high levels of wax in seeds of transgenic Arabidopsis. Plant Physiol. **2000**, *122*, 645–655.

Weed Management in Less Developed Countries

Peter R. Hobbs
Robin R. Bellinder
Cornell University, Ithaca, New York, U.S.A.

INTRODUCTION

Weeds cause yield losses in both developed and less developed countries (LDCs). The difference between the two groups of countries lies in the degree of technology available to and utilized by farmers. The primary method of controlling weeds in LDCs continues to be hand weeding. Although farmers utilize crop rotations, intercropping, and mulches to suppress weeds, their wives and children, hired laborers, and the farmers themselves continue to hand-weed crops of significant economic value. With the increasing influence of satellite TV, youth in LDCs are becoming disenchanted with the drudgery of farming and are leaving the villages for jobs in urban centers. The cost of hand weeding is escalating and becoming too expensive. LDC farmers are interested in exploring the use of new tillage and mechanical weeding tools, planting strategies, herbicides, and genetically modified crops.

WEED LOSSES

Crop yield losses due to weeds are difficult to estimate because of interactions with insects, diseases, soils, crops, and the environment. Nevertheless, a 10% yield loss is accepted as a global figure.[1] Pathak et al.[2] calculated that yield losses in transplanted rice range from 10–50%, and from 50–90% in upland rice.

TRADITIONAL WEEDING STRATEGIES

Since the beginning of agriculture, the major weed-control strategy has been hand weeding, and the same is still true today in most LDCs. Farmers employ family members or hired labor using an array of hand tools that have been developed for local conditions, crops, and weeds present. There are differences in weed control strategies between larger commercial farms and smaller farm households in LDCs. In subsistence situations, farmers rely on family labor for weeding and often use weeds for animal fodder or even as human food. Commercial farmers with larger land holdings use more hired labor and herbicides, because with them weed control is easier and less costly. LDC farmers, subsistence and commercial, do understand the need for integrated weed control and use crop rotations, mulches, stale seedbed systems, water, and in some cases herbicides to control weeds.

Although hand weeding is a traditional system, it is also tedious, grueling, and often done by resource-poor labor, especially women and children. In a way, it perpetuates poverty because the employment opportunities it provides hardly compensate for the labor and the long hours in the field (Fig. 1). Crops that receive the highest priority for hand weeding in LDCs are high-value vegetables, paddy rice, and row crops like maize, cotton, and sugarcane. Estimates of time and cost for hand weeding vary depending on weed populations present, when the crop is grown (wet or dry season), and labor availability. Average figures for rice are 30–40 labor-days per hectare, with more for maize and less for wheat. Labor rates also vary with country but are usually below one U.S. dollar per day. It is estimated that one-third to one-half of the labor in rice production is for weed control. With young male labor migrating to cities in search of less tedious and better paying jobs, labor rates for weeding are increasing, making this method of weed control less and less profitable. This same phenomenon happened in developed countries at the time of the Industrial Revolution.

INTEGRATED WEED-CONTROL SYSTEMS

In many LDCs, especially in the tropics, slash-and-burn agriculture is still practiced. As soil fertility declines and weeds increase, farmers abandon land and let it rejuvenate while they move to new, reclaimed land. As agriculture becomes less mobile and farming concentrates on the same piece of land, integrated weed-control systems are needed. In situations where the same crop is grown repeatedly, weeds specific to that system will start to predominate and require control. In the rice–wheat areas of South Asia, where continuous cereal production is practiced, the grassy weed *Phalaris minor* has proliferated and created problems for control, including development of resistance to herbicides.[3,4] Farmers are aware of this problem, and as hand-weeding labor becomes more

Encyclopedia of Plant and Crop Science
DOI: 10.1081/E-EPCS 120020282

Fig. 1 The tedious nature of agriculture in developing countries: (A) Woman threshing, (B) Women transplanting rice, (C) Men handweeding. (*View this art in color at www.dekker.com.*)

costly, many decide to change crops, introducing ones that will better compete with existing weeds. An example of this change is using sugarcane in the rice–wheat rotation in South Asia. Similarly, use of a fodder crop that is cut multiple times can help keep weeds under control. Sieving is another tool for weed control because it effectively separates weed and crop seeds (Fig. 2).

Tillage is another practice designed to control weeds. Using a stale seedbed approach, the first flushes of weeds germinate, and are then plowed down before planting the crop. However, tillage also brings more seed to the surface, enabling germination simultaneously with the crop. There is a movement to introduce new systems that reduce tillage in LDCs[5] which will require innovative integrated weed management strategies. However, it has been shown that when zero-till wheat follows paddy rice, fewer weeds germinate than in conventional tillage.[6] Mechanical weeding also can be introduced to control weeds after germination if crops are line-sown in straight rows. In some LDCs where crop or farm residues are available, farmers use mulch to suppress weeds. However, crop residues are valuable for animal feed, cooking fuel, or thatching and may not be available for weed-control. Selection of competitive varieties is another integrated weed control factor.

Rice is a major crop in LDCs, and there are many ways to grow rice. When it is grown under upland conditions like other cereal crops, weeds are a major problem and cause low yields. Tillage is commonly combined with hand weeding to control weeds in dryland rice. Competitive rice varieties are also favored since they are less affected by weeds. However, most rice in LDCs is grown under lowland conditions, where seedlings are raised in separate beds, uprooted, and then transplanted into soils that have been puddled (plowed when wet). Puddling restricts water percolation so that water can be ponded more easily in the field, and standing water is then used to control weeds. However, new weeds appear under this

Fig. 2 Sieving and mechanical weeding in developing countries: (A) Sieving out weed seed, (B) Rotary weeder in use, (C) Rotary weeders on display. (*View this art in color at www.dekker.com.*)

Fig. 3 Traditional (one-nozzle) spray boom (left) and improved (three-nozzle) spray boom (right). (*View this art in color at www.dekker.com.*)

system and will need to be controlled as weed populations shift. Some farmers plant rice seedlings in rows so that mechanical weeding can be done between rows, using simple, locally manufactured rotary weeders (Fig. 2). Additionally, most rice growers in LDCs will hand-weed the crop at least once, and many do so twice, with 30–40 workdays per hectare commonly needed for weeding. This labor-intensive job is often done by female workers and children.

HERBICIDE USE IN LDCs

Herbicides are also used in LDCs to complement other weed control measures. Use of low-dose herbicides followed by hand weeding can be very effective and less costly than using only hand weeding. Herbicides are used quite extensively in LDCs for transplanted, puddled rice. These are frequently granular formulations that are mixed with soil or sand and spread by hand. Isoproturon (for wheat in South Asia) and butachlor (for rice) are both mixed and applied this way. In many instances herbicide efficacy is variable, and in some cases poor, because of the lack of proper spray equipment and poor knowledge of proper herbicide application methods, with non-uniform distribution as a result. Many farmers apply herbicides the same way they apply insecticides and other plant protection chemicals—with single nozzle booms (Fig. 3). Spray coverage is poor, with many missed plants. What is urgently needed is instruction in how to apply herbicides. Farmers need training in the use of proper nozzle tips, multiple nozzle booms[7] that can be made locally (Fig. 3), and pressure regulators that will improve application uniformity and accuracy. Safety issues and health hazards also need to be part of the training program for farmers and applicators.

GM CROP USE

Herbicide-resistant GM crops (maize, rice, canola, wheat, and sunflower) are not presently grown in LDCs, although interest among farmers is high. Farmers want to reduce the drudgery of farming, including weeding. Governments are wary of the implications that growing GM crops might have for their export markets and so are waiting to see how global markets adapt to these crops. Several LDCs, such as China and India, have invested heavily in research so they are ready with GM crops when acceptance is assured. These crops would be a boon to farmers in LDCs because they would enable farmers to overcome one of the most arduous tasks in farming, that of hand weeding. Direct-seeded rice would be much more feasible and would mean that rice farmers would not have to degrade their lands through puddling, which would contribute to more environmentally sustainable production systems over time.

CONCLUSION

Farmers in LDCs, like farmers in any country, are aware of the losses in yield from weeds. They have traditionally used hand weeding as the major means to control them, but as costs for labor increase and young laborers shun the drudgery of agriculture, alternative integrated systems—including the use of herbicides and herbicide-resistant crops—will be welcomed by LDC farmers, just as they are by commercial farmers in developed countries.

The key to their use and success will be the availability and proper training in their use.

ARTICLES OF FURTHER INTEREST

Biological Control of Weeds, p. 141
Biosafety Applications in the Plant Sciences: Expanding Notions About, p. 146
Chemical Weed Control, p. 255
Cropping Systems: Irrigated Continuous Rice Systems of Tropical Subtropical Asia, p. 349
Cropping Systems: Irrigated Rice and Wheat of the Indo-gangetic Plains, p. 355
Herbicide-Resistant Weeds, p. 551
Herbicides in the Environment: Fate of, p. 554
Molecular Biology Applied to Weed Science, p. 745
Parasitic Weeds, p. 864
Seed Dormancy, p. 1130
Transgenic Crop Plants in the Environment, p. 1248

REFERENCES

1. *Recent Advances in Weed Science. Introductory Chapter*; Fletcher, W.W., Ed.; Commonwealth Agricultural Bureaux. The Gresham Press: Surrey, UK, 1983, 1–3.
2. Pathak, M.D.; Ou, S.H.; de Datta, S.K. *Pesticides and Human Welfare*; Gunn, D.L., Stephens, J.G.R., Eds.; Oxford University Press, 1976.
3. In *Herbicide Resistance Management and Zero-Tillage in Rice-Wheat Cropping System*, Proceedings of an International, Workshop, Hisar, India, March 4–6, 2002; Malik, R.K.; Balyan, R.S.; Yadav, A.; Pahwa, S.K., Eds.; CCSHA Univ.: Hisar, India, 2002a.
4. Vincent, D.; Quirke, D. *Controlling Phalaris minor in the Indian Rice-Wheat Belt*; ACIAR Impact Assessment Series, ACIAR: Canberra, Australia, 2002. 35 pages, Vol. 18. (http://www.aciar.gov.au/publications/db/abstract.asp?pubsID=523).
5. Hobbs, P.R.; Gupta, R.K. Resource Conserving Technologies for Wheat in Rice-Wheat Systems. In *Improving the Productivity and Sustainability of Rice-Wheat Systems: Issues and Impact*; Ladha, J.K., Hill, J., Gupta, R.K., Duxbury, J., Buresh, R.J., Eds.; ASA, Spec. Publ., ASA Madison: WI, USA, 2003; Vol. 65, 149–171. chapter 7.
6. Malik, R.K.; Yadav A.; Singh, S.; Malik, R.S.; Balyan, R.S.; Banga, R.S.; Sardana, P.K.; Jaipal, S.; Sardana, P.K.; Hobs, P.R.; Gill, G.; Singh, S.; Gupta, R.K.; Bellinder, R.R. *Herbicide Resistance Management and Evolution of Zero-Tillage: A Success Story*; Research Bulletin, Haryana Agricultural University: Hisar, India, 2002b.
7. Bellinder, R.R.; Miller, A.; Malik, R.K.; Ranjit, J.D.; Hobbs, P.R.; Brar, L.S.; Singh, G.; Singh, S.; Yadev, A. Improving herbicide application accuracy in South Asia. Weed Technol. **2002**, *16*, 845–850.

Weeds in Turfgrass

John C. Stier
University of Wisconsin, Madison, Wisconsin, U.S.A.

INTRODUCTION

Weeds are the primary pest problem in most turf areas, particularly in lawns. Cultural management including mowing, fertility, and irrigation, along with site and turfgrass species selection, can be used to minimize most weed problems. Herbicides are used for both pre- and postemergent weed control although relatively few active ingredients are available compared to compounds used in conventional agriculture.

IMPORTANCE OF TURF

Turf is a contiguous community of plants, usually grasses, that can withstand routine defoliation (mowing) and traffic. Turf acreage is increasing in the United States, with approximately two-thirds in home lawns. No reliable figures exist for determining the value of turf but the industry is actively growing and annual management inputs are at least $20 to $30 billion.[1] In 1999, turf maintenance costs for Wisconsin's 486,000 hectares (ha) were 1.9 billion.[2] The social and environmental benefits of turf include green space for relaxation and recreation, reduction of stormwater runoff and groundwater recharge, pollutant absorption, dust and erosion control, and oxygen production (Fig. 1).[3]

WEED CONCERNS IN TURF

Unlike agronomic crops, turf is grown for quality, not yield. Thus, weed tolerances are typically low and an aesthetic rather than an economic threshold is used for justifying weed control measures. Weeds are the major pest in lawn areas but on golf courses are less important than diseases. Weeds may pose health hazards directly, as in poison ivy or sandburs when in flower, and indirectly through poor traction on athletic fields or as allergens when in flower.

TYPES OF WEEDS

Most turf weeds are herbaceous, low-growing plants, because mowing prevents high-growing weeds. Turf weeds include both dicots (broadleaves) and monocots (grasses and sedges). Most are perennials, although a few annuals such as annual bluegrass (*Poa annua* L.) and crabgrass (*Digitaria* spp.) are ubiquitous and survive in many well managed turfs. Desirable turfgrasses may be considered weeds when they occur in swards of a different species.

WEED MANAGEMENT

The purpose of a turf dictates weed acceptance levels. A highway median has up to 100% weed tolerance, whereas a golf course putting green has near zero weed tolerance.

Cultural Management

Many weeds can be avoided by using certified, weed-free seed and mulch for establishment. Imported topsoil may contain weed propagules such as quackgrass (*Elytrigia repens* L.) rhizomes and seeds. The establishment period allows many types of weeds to develop and may cause the seeding to fail because weed seeds germinate and establish faster than most turfgrasses. Weed problems are minimized by planting cool season grasses in late summer/early fall and warm season grasses in spring. Weeds often can be avoided for years if a good quality sod is used for establishment.

The key to weed management is to grow a thick, dense turf. Good mowing and fertility practices will reduce up to 90% of potential weed problems, especially if irrigation is supplied. Mowing stimulates hormones that encourage tillering and leaf growth. Each turf species has an optimal mowing height range to achieve vigorous growth and high turf density. Mowers should be set so that no more than one-third of the leaf tissue is removed at any one mowing to prevent scalping. Scalping opens the canopy for weed seed germination or development of new plants from existing vegetative propagules of weeds (e.g., rhizomes and stolons). Nitrogen (N) is almost always the limiting element for turf growth. High-quality turf requires 96–192 kg N per ha, applied incrementally during the growing season to achieve balanced turf growth with high density. This is approximately 1/7 to 1/4 of the N required for maximum growth. Potable water is best for irrigation to

Encyclopedia of Plant and Crop Science
DOI: 10.1081/E-EPCS 120020285

Fig. 1 Weed management in turf is increasing due to urbanization and involves lawns, roadsides, and recreational areas. (*View this art in color at www.dekker.com.*)

avoid salts that may desiccate the turf. On average, turf uses about 2.5 cm water weekly through irrigation, precipitation, or soil moisture withdrawal.

Weed identification is used to select appropriate chemical controls and to diagnose underlying problems that may be corrected to prevent future weed development.[4] Crabgrass infestation is often due to scalping in the spring. Quackgrass, ground ivy (*Glechoma hederacea*), and other weeds invade dry soil whereas many sedges prefer moist soils. Prostrate knotweed (*Polygonum aviculare* L.) indicates soil compaction, red sorrel (*Rumex acetosella* L.) indicates acidic soil, and legumes such as white clover (*Trifolium repens* L.) indicate N-deficient soil.

Biological Controls

Few if any practical biocontrol options for turf weeds exist. A bacterial wilt disease caused by *Xanthomonas campestris* showed promising results in laboratory tests but was relatively ineffective in field trials.[5] Corn gluten meal offers promise as a preemergent herbicide but in current form the effects are ephemeral, timing and moisture are crucial, and high rates may be required.[6]

Chemical Controls

The types and classifications of chemical controls are similar to those used in agronomic crops (Table 1); however, fewer active ingredients are available because manufacturers typically have little financial incentive to support labels for what is essentially a minor use. Homeowners have access to most of the active ingredients available to professionals, although concentrations and formulations may differ. Preemergent herbicides are used to prevent establishment of new weeds (primarily annuals) from seed. Most weed control is performed postemergent, particularly for perennial broadleaf weeds. Relatively few contact herbicides are used because most problem weeds are perennials that require a systemic herbicide. Liquid products often are more effective than granules, although granules are often perceived as being safer for use. Many granular products are formulated as a combination of fertilizer and herbicide for either pre- or postemergent weed control.

A common perception exists that turf requires a large amount of pesticides, especially herbicides, although in fact herbicides often are used in lieu of proper cultural management practices that could reduce the need for herbicides. Golf course purchase of herbicides for 1995/96 was just over $60 million, whereas $130 million was spent on professional lawn care.[7] Data on product purchases by homeowners are unavailable, but likely meet or exceed that used by professional lawn care because only about half of the 22% of home lawns treated with pesticides utilize a professional service.[7] Adoption of integrated pest management practices has led to many of the post-emergent liquid herbicide applications used as spot treatments to individual weeds or patches rather than broadcast applications.

Another common perception is that herbicides are readily leached or lost in runoff from turf surfaces. Data from multiple research projects since the 1980s indicates that relatively little herbicide applied to dense turf

Table 1 Commonly used turf herbicides

Common name	Timing	Target weeds
Benefin	Pre-emergent	Annual grasses
Bensulide	Pre-emergent	Annual grasses
Pendimethalin	Pre-emergent	Annual grasses
Prodiamine	Pre-emergent	Annual grasses
Siduron	Pre-emergent	Annual grasses
Dithiopyr	Pre-(post)-emergent	Annual grasses
Basagran	Post-emergent	Sedges
Clopyralid	Post-emergent	Perennial broadleaves
Dicamba	Post-emergent	Perennial broadleaves
Glufosinate	Post-emergent	Herbaceous annuals
Glyphosate	Post-emergent	Herbaceous annuals and perennials
Halosulfuron	Post-emergent	Sedges
MCPA	Post-emergent	Perennial broadleaves
MCPP	Post-emergent	Perennial broadleaves
2,4-D	Post-emergent	Perennial broadleaves
2,4-DP	Post-emergent	Perennial broadleaves
Triclopyr	Post-emergent	Perennial broadleaves

(<1%) is capable of leaching to groundwater.[8] Runoff losses may be potentially higher, although less than 15% of water soluble herbicides applied to short-cut golf course fairways are likely to run off, even under worst-case scenarios.[8] Higher mowing heights and thatch further reduce leaching and runoff due to higher runoff resistance and the pesticide-binding capacity of organic matter in the turf.[3,9,10] Mowed turfgrasses provide a dense turf cover of 75 million to >20 billion shoots per hectare. Kentucky bluegrass (*Poa pratensis* L.), commonly used for lawns, can have up to 16,000 kg per hectare of root biomass.[3] Soil type, soil saturation, timing of precipitation or irrigation following application, application rate, water solubility, and adsorption to organic matter can substantially affect the amount of herbicide in leachate or runoff.

THE FUTURE

A few lawn care companies have programs to meet the small but growing demand for organic lawn care. No legal definition of organic lawn care exists. Consequently, organic lawn care relies on naturally occurring fertilizers (e.g., composted sewage sludge or manure) and avoids synthetic chemicals. Because increased pesticide restrictions are limiting the development of new synthetic chemicals, the development of natural products for weed control is expected to grow.

CONCLUSION

Turf acreage is expanding as the population grows and turf maintenance has a tremendous economic impact in urbanized states. Weeds are usually the greatest pest problem in turf. Most weed problems can be managed using practices that develop a thick turf sward. Herbicides are used for weed control in all types of turf in either liquid or granular products, sometimes in combination with fertilizer.

ARTICLES OF FURTHER INTEREST

Biological Control of Weeds, p. 141
Chemical Weed Control, p. 255
Integrated Pest Management, p. 612

REFERENCES

1. Watson, J.R.; Kaerwer, H.E.; Martin, D.P. The Turfgrass Industry. In *Turfgrass*; Waddington, D.V., Carrow, R.N., Shearman, R.C., Eds.; Agronomy Society of America, Monograph 32: Madison, WI, 1992; 29–88.
2. Anonymous. *1999 Wisconsin Turfgrass Industry Survey*; Wisconsin Agricultural Statistics Service: Madison, WI, 2001; 1–16.
3. Beard, J.B.; Green, R.L. The role of turfgrasses in environmental protection and their benefits to humans. J. Environ. Qual. **1994**, *23* (3), 452–460.
4. Hall, D.W.; McCarty, L.B.; Murphy, T.R. Weed Taxonomy. In *Turf Weeds and Their Control*; Turgeon, A.J., Ed.; 1994; 1–28. American Society of Agronomy, Inc.; Crop Science Society of Agronomy, Inc.: Madison, WI.
5. Zhou, T.; Neal, J.C. Annual bluegrass (*Poa annua*) control with *Xanthomonas campestris* pv. *poannua* in New York State. Weed Technol. **1995**, *9* (1), 173–177.
6. Gardner, D.S.; Christians, N.E.; Bingaman, B.R. Pendimethalin and corn gluten meal combinations to control turf weeds. Crop Sci. **1997**, *37* (6), 1875–1877.
7. Racke, K.D. Pesticides for Turfgrass Pest Management: Uses and Environmental Issues. In *Fate and Management of Turfgrass Chemicals*; Clark, J.M., Kenna, M.P., Eds.; American Chemical Society: Washington, DC, 2000; 45–64.
8. Smith, A.E.; Bridges, D.C. Movement of certain herbicides following application to simulated golf course greens and fairways. Crop Sci. **1996**, *36* (6), 1439–1445.
9. Linde, D.T.; Watschke, T.L.; Jarrett, A.R.; Borger, J.A. Surface runoff assessment from creeping bentgrass and perennial ryegrass turf. Agron. J. **1995**, *87* (2), 176–182.
10. Carrol, M.J.; Hill, R.L.; Raturi, S.; Herner, A.E.; Pfeil, E. Dicamba Transport in Turfgrass Thatch and Foliage. In *Fate and Management of Turfgrass Chemicals*; Clark, J.M., Kenna, M.P., Eds.; American Chemical Society: Washington, DC, 2000; 228–242.

Yeast Two-Hybrid Technology

Lisa Nodzon
Wen-Yuan Song
University of Florida, Gainesville, Florida, U.S.A.

INTRODUCTION

Over the past decade, protein–protein interactions have drawn broad attention because most cellular proteins function through association with other proteins, and the networks formed by protein complexes control most biological processes. To study protein–protein interactions, a number of technologies have been developed. Among them, the yeast two-hybrid system appears to be a relatively simple, yet powerful tool. This system has recently been used to detect interactions between thousands of protein pairs in the model organism yeast.

MECHANICS OF THE YEAST TWO-HYBRID SYSTEM

The two-hybrid system, performed within the yeast *Saccharomyces cerevisiae*, takes advantage of the independent domains present in many transcription factors, such as the yeast Gal4 protein. The two physically separable, functionally independent domains are a DNA-binding domain and a transcription activation domain.[1] The DNA-binding domain targets the transcription factor to a specific promoter sequence (known as the upstream activation sequence (UAS) in yeast), whereas the activation domain recruits the transcription complex for initiation of transcription. A mutant transcription factor lacking either one of the two domains fails to activate transcription of the downstream gene driven by the promoter. The independence of the two domains was shown when early investigators revealed that a hybrid transcription factor could be reconstituted in yeast cells by fusing the Gal4 DNA-binding and the Gal4 transcription activation domains to two physically interacting proteins, respectively (Fig. 1).[3] To demonstrate activity of the hybrid transcription factor, a reporter gene, such as the bacterial *lacZ* gene, driven by a promoter containing the UAS where the Gal4 DNA-binding domain interacts, was also introduced into the cells. The *lacZ* gene encodes the β-galactosidase enzyme that can hydrolyze the colorless substrate 5-bromo-4-chloro-3-indolyl-β-D-galactoside (known as X-gal) and generate a blue product. Thus, the formation of blue yeast colonies in medium containing X-gal is indicative of reconstitution of a functional transcription factor by the interaction of the two proteins that bring the Gal4 DNA-binding and activation domains into proximity. These studies led to the establishment of the yeast two-hybrid system.[3]

The original yeast two-hybrid idea has been applied to screening of cDNA libraries to identify proteins (commonly called "prey") that interact with a given protein (also called "bait"). The cDNAs are cloned into a Gal4 activation domain vector to generate translational fusions (prey) of the library proteins to the Gal4 activation domain. In addition to *lacZ*, a nutritional selection reporter, such as the yeast *HIS3* gene, is placed under the control of a Gal4-responsive promoter. *HIS3* encodes imidazole glycerol phosphate dehydratase, an enzyme involved in the biosynthesis of the amino acid histidine. Yeast strains with mutations in the *HIS3* gene are unable to survive on medium lacking histidine supplementation. To carry out two-hybrid screening, the prey library is transformed into the *his3*$^-$ mutant that expresses the bait protein and carries an engineered *HIS3* gene driven by a Gal4-responsive promoter.[4] The transformed cells are then selected onto medium lacking histidine. The physical association between the bait and a prey activates transcription of the *HIS3* gene, allowing the transformed cells to grow and form colonies in the absence of histidine. To confirm the interaction identified by the *HIS3* reporter, the selected cells can be subjected to an assay for β-galactosidase activity using the chromogenic substrate X-gal.

USING THE YEAST TWO-HYBRID SYSTEM TO DISSECT PLANT SIGNAL TRANSDUCTION PATHWAYS

Yeast two-hybrid screens have been used to identify proteins involved in plant signal transduction pathways that regulate plant defense and development. The *Arabidopsis* protein RPM1 confers resistance to the bacterial

Encyclopedia of Plant and Crop Science
DOI: 10.1081/E-EPCS 120016292

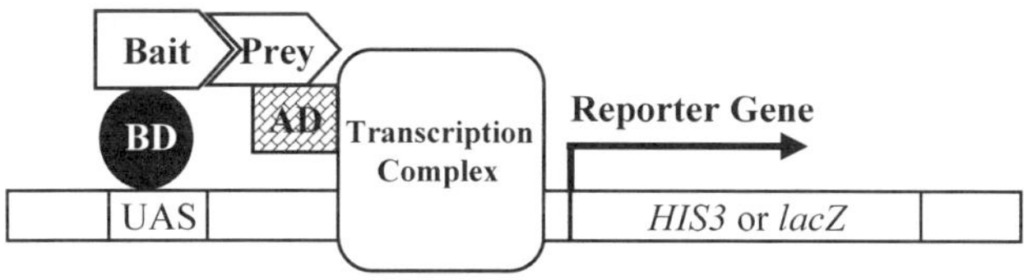

Fig. 1 Diagram of the yeast two-hybrid system. Interaction of the bait and prey proteins brings the DNA-binding domain (BD) and activation domain (AD) into close proximity, thus initiating transcription of the *HIS3* or *lacZ* reporter genes by the transcription complex.

pathogen *Pseudomonas syringae* expressing either of the two effectors AvrB and AvrRpm1.[5] An *Arabidopsis* protein, RIN4 (for RPM1-interacting protein 4), was found from a two-hybrid screen using the bacterial protein AvrB as bait.[6] RIN4 interacts with both AvrB and RPM1. Subsequent genetic analysis confirmed that RIN4 is required for accumulation of RPM1 and for RPM1-mediated disease resistance.[6]

The *Arabidopsis Brassinosteroid-Insensitive 1* (*BRI1*) gene encodes a receptor kinase protein that initiates the brassinosteroid signal transduction pathway.[7] Brassinosteroids are a class of endogenous plant hormones that regulate different physiological processes during the course of plant development.[7] A yeast two-hybrid screen for interactors of the BRI1 kinase domain identified a second receptorlike kinase protein, BAK1 (BRI1 Associated receptor Kinase 1).[8] Further biochemical and genetic studies confirmed that BRI1 interacts with BAK1 in plants and that BRI1 and BAK1 function together in mediating brassinosteroid signaling.[8,9]

DETECTION OF PROTEIN–PROTEIN INTERACTIONS ON A GENOMEWIDE SCALE

The most recent application of the yeast two-hybrid system is the comprehensive study of protein–protein interactions between thousands of protein pairs. The approximately 6000 predicted open reading frames from the yeast *Saccharomyces cerevisiae* were individually expressed as Gal4 DNA-binding and transcription activation domain fusions in yeast strains of opposite mating types.[10,11] To screen for binding between any two yeast proteins, cells expressing the Gal4 DNA-binding fusions were systematically mated with the cells expressing the Gal4 activation fusions. Diploids from the mating were then subjected to an assay for expression of the reporter genes under the control of Gal4-responsive promoters. Using this approach, more than 4000 protein–protein interactions have been discovered.[10,11] These studies are particularly important as a large number of genes have been predicted from genome sequencing projects, ranging from unicellular organisms, such as yeast, to multicellular organisms such as *Arabidopsis* and rice. The identification of protein–protein networks can link distinct biochemical pathways and facilitate the studies of unknown proteins whose binding partners have been characterized.

CONCLUSIONS AND PROSPECTIVES

More than 10 years have passed since the yeast two-hybrid technology was first described. Although some drawbacks, such as false–positive interactions, have been found, the system is still popular for the study of protein–protein interactions. The viability of the two-hybrid system relies on its simplicity in manipulation of libraries and sensitivity in detection of protein interactions. The only prerequisite for examining the binding of two proteins is their predicted open reading frames. Because of these features, the yeast two-hybrid system has evolved from a means to study interactions between small numbers of proteins to a powerful tool that can be used to establish a protein–protein interaction database on a genomewide scale. It is anticipated that such a global approach can be applied to investigate more complex organisms such as *Arabidopsis* and rice whose genomes have been sequenced. Information obtained from yeast two-hybrid analyses and from other sources will undoubtedly contribute to our understanding of cellular networks and to agricultural practices.

REFERENCES

1. Bai, C.; Elledge, S.J. Searching for Interacting Proteins with the Two-Hybrid System I. In *The Yeast Two-Hybrid System*; Bartel, P.L., Fields, S., Eds.; Oxford University Press: New York, NY, 1997; 11–28.
2. Hazbun, T.R.; Fields, S. Networking proteins in yeast. Proc. Natl. Acad. Sci. U. S. A. **2001**, *98*, 4277–4278.
3. Fields, S.; Song, O. A novel genetic system to detect protein–protein interactions. Nature **1989**, *340*, 245–246.
4. Durfee, T.; Becherer, K.; Chen, P.L.; Yeh, S.H.; Yang, Y.; Kilburn, A.E.; Lee, W.H.; Elledge, S.J. The retinoblastoma protein associates with the protein phosphatase type 1 catalytic subunit. Genes Dev. **1993**, *7*, 555–569.
5. Grant, M.R.; Godiard, L.; Straube, E.; Ashfield, T.; Lewald, J.; Sattler, A.; Innes, R.W.; Dangl, J.L. Structure

of the Arabidopsis *RPM1* gene enabling dual specificty disease resistance. Science **1995**, *269*, 843–846.
6. Mackey, D.; Holt, B.F., III; Wiig, A.; Dangl, J.L. RIN4 Interacts with *Pseudomonas syringae* type III effector molecules and is required for RPM1-mediated resistance in Arabidopsis. Cell **2002**, *108*, 743–754.
7. Li, J.; Chory, J. A putative leucine-rich repeat receptor kinase involved in brassinosteroid signal transduction. Cell **1997**, *90*, 929–938.
8. Nam, K.; Li, J. BRI1/BAK1, a receptor kinase pair mediating brassinosteroid signaling. Cell **2002**, *110*, 203–212.
9. Li, J.; Wen, J.; Lease, K.A.; Doke, J.T.; Tax, F.E.; Walker, J.C. BAK1, an Arabidopsis LRR receptor-like protein kinase, interacts with BRI1 and modulates brassinosteroid signaling. Cell **2002**, *110*, 213–222.
10. Uetz, P.; Giot, L.; Cagney, G.; Mansfield, T.A.; Judson, R.S.; Knight, J.R.; Lockshon, D.; Narayan, V.; Srinivasan, M.; Pochart, P.; Qureshi-Emili, A.; Li, Y.; Godwin, B.; Conover, D.; Kalbfleisch, T.; Vijayadamodar, G.; Yang, M.; Johnston, M.; Fields, S.; Rothberg, J.M. A comprehensive analysis of protein–protein interactions in *Saccharomyces cerevisiae*. Nature **2000**, *403*, 623–627.
11. Ito, T.; Chiba, T.; Ozawa, R.; Yoshida, M.; Hattori, M.; Sakaki, Y. A comprehensive two-hybrid analysis to explore the yeast protein interactome. Proc. Natl. Acad. Sci. U. S. A. **2001**, *98*, 4569–4574.

Index

(*Continued from inside front cover*)